HAMILTON & HARDY'S
INDUSTRIAL TOXICOLOGY

HAMILTON & HARDY'S INDUSTRIAL TOXICOLOGY

Sixth Edition

Edited by

RAYMOND D. HARBISON, MARIE M. BOURGEOIS, AND GIFFE T. JOHNSON

WILEY

Library of Congress Cataloging-in-Publication Data

Hamilton & Hardy's industrial toxicology. – Sixth edition / editors,
Raymond D. Harbison, Marie M. Bourgeois, and Giffe T. Johnson.
    p. ; cm.
  Hamilton and Hardy's industrial toxicology
  Industrial toxicology
  Includes bibliographical references and index.
  ISBN 978-0-470-92973-5 (hardback)
  I. Harbison, Raymond D., editor. II. Bourgeois, Marie M., editor. III. Johnson, Giffe T., editor. IV. Title: Hamilton and Hardy's industrial toxicology. V. Title: Industrial toxicology.
  [DNLM: 1. Hazardous Substances–toxicity. 2. Occupational Exposure. 3. Occupational Health. 4. Risk Assessment. WA 465]
  RA1229
  615.9'02–dc23
                                                                                                    2014037175

Printed in the United States of America

10 9 8 7 6 5 4 3 2 1

CONTENTS

CONTRIBUTORS

Alison J. Abritis, MPH, Department of Environmental and Occupational Health, College of Public Health, University of South Florida, Tampa, FL, USA

Mayowa Amosu, Laboratory Manager, Smith Lab Environmental Health Sciences, University of Georgia, Athens, GA, USA

Lisa A. Bailey, PhD, Gradient Corporation, Cambridge, MA, USA

James C. Ball, PhD, Research Scientist, Integrated Risk Information System (IRIS), U.S. Environmental Protection Agency, Washington, DC, USA

Marek Banasik, MD, PhD, Institute of Public Health and Environmental Protection, Warsaw, Poland

Charles Barton, PhD, DABT, Senior Toxicologist & Toxicology Program Manager, Oak Ridge Institute for Science and Education (ORISE), Oak Ridge Associated Universities (ORAU), Oak Ridge, TN, USA

Robert W. Benson, PhD, Toxicologist, Office of Partnerships and Regulatory Assistance, U.S. Environmental Protection Agency, Denver, CO, USA

J. Michael Berg, PhD, Toxicologist, Center for Toxicology and Environmental Health, LLC, Houston, TX, USA

Leslie A. Beyer, MS, DABT, Gradient Corporation, Arlington, MA, USA

Robyn Blain, MS, PhD, Technical Specialist, ICF International, Fairfax, VA, USA

Jacob R. Bourgeois, BS, ChE, Department of Environmental and Occupational Health, College of Public Health, University of South Florida, Tampa, FL, USA

Marie M. Bourgeois, PhD, MPH, Research Assistant Professor, Center for Environmental/Occupational Risk Analysis and Management, Department of Environmental and Occupational Health, College of Public Health, University of South Florida, Tampa, FL, USA

K. Paul Boyev, MD, Associate Professor, Director, Division of Otology-Neurotology, Residency Program Director, Department of Otolaryngology-Head and Neck Surgery, College of Medicine, University of South Florida, Tampa, FL, USA

Michael A. Cardinale, MS, CIH, Environmental Health Officer John F. Kennedy Space Center National Aeronautics and Space Administration NASA-Kennedy Space Center, FL, USA

Vincent Castranova, PhD, National Institute for Occupational Safety and Health, Morgantown, WV, USA

Clara Y. Chan, MSc, DABT, Senior Toxicologist, Veritox, Inc., Redmond, WA, USA

Juhi K. Chandalia, Data Scientist, Zettics, Concord, MA, USA

John Christie, RD, CSSD, CSCS, Performance Nutritionist, Phoenix, AZ, USA

C. Clifford Conaway, PhD, DABT, Department of Pharmacology New, York Medical College, Valhalla, NY, USA

Jayme P. Coyle, MSPH, Department of Environmental and Occupational Health, College of Public Health, University of South Florida, Tampa, FL, USA

Ghazi A. Dannan, PhD, Office of Research and Development, National Center for Environmental Assessment,

U.S. Environmental Protection Agency, Washington, DC, USA

Ushang Desai, MBBS, MPH, Research Assistant, Center of Environmental & Occupational Risk Analysis, Department of Environmental & Occupational Health, College of Public Health, University of South Florida, Tampa, FL, USA

David G. Dodge, MS, DABT, Senior Toxicologist, Gradient, Cambridge, MA, USA

Anna M. Fan, PhD, Chief, Pesticide and Environmental Toxicology Branch, Office of Environmental Health Hazard Assessment, California Environmental Protection Agency, Oakland/Sacramento, CA, USA

Nikolay M. Filipov, PhD, Department of Physiology and Pharmacology, College of Veterinary Medicine, University of Georgia, Athens, GA, USA

M. Rony Francois, MD, MSPH, PhD, Director for Public Health Services, Supervisory Occupational Medicine Physician, Public Health Emergency Officer, Naval Health Clinic, Corpus Christi, TX, USA

Jason M. Fritz, PhD, Toxicologist, Office of Research and Development, National Center for Environmental Assessment/Integrated Risk Information System, U.S. Environmental Protection Agency, Washington, DC, USA

Jason S. Garcia, MPH, CPH, College of Public Health, University of South Florida, Tampa, FL, USA

Phillip T. Goad, PhD, Toxicologist, Center for Toxicology and Environmental Health, LLC, North Little Rock, AR, USA

James Godin, DPT, CSCS, Director of Tactical Performance Rehabilitation, EXOS, Inc., Tampa, FL, USA

Julie E. Goodman, PhD, DABT, Gradient Corporation, Cambridge, MA, USA

Joseph M. Greco, CHP, CLSO, VP, Imaging Physics, Upstate Medical Physics – A Landauer Medical Physics Partner, Victor, NY, USA

Maureen R. Gwinn, PhD, DABT, Associate National Program Director, Community Health, Sustainable and Healthy Communities, National Center for Environmental Assessment, Washington, DC, USA

Kambria K. Haire, MPH, Department of Environmental and Occupational Health, College of Public Health, University of South Florida, Tampa, FL, USA

Kelly W. Hall, MPH, PhD, Associate Director, Toxicology, Global Product Safety and Toxicology, Merck Consumer Care, Memphis, TN, USA

Ali K. Hamade, PhD, DABT, Environmental Public Health Program Manager, Alaska Section of Epidemiology, Division of Public Health, Department of Health and Social Services, Anchorage, AK, USA

Yehia Y. Hammad, DSc, Professor, College of Public Health University of South Florida, Tampa, FL, USA

Raymond D. Harbison, PhD, ATS, Professor, Center for Environmental/Occupational Risk Analysis and Management, Department of Environmental and Occupational Health, College of Public Health, University of South Florida, Tampa, FL, USA

Stephen C. Harbison, PhD, MPH, Southwestern Energy Company, Conway, AR, USA

Jerry F. Hardisty, DVM, Experimental Pathology Laboratories, Inc., Research Triangle Park, NC, USA

Judith W. Henck, PhD, DABT, Research Advisor, Nonclinical Safety Assessment, Eli Lilly and Company, Greenfield, Indiana, USA

Thomas W. Hesterberg, PhD, MBA, Senior Toxicologist, Center for Toxicology and Environmental Health, LLC, Denver, CO, USA

David J. Hewitt, MD, MPH, DABT, Resources for Environmental and Occupational Health (REOH), Missoula, MT, USA

James V. Hillman, MD, FAAP, FACEP, FACMT, Clinical Professor of Pediatrics, University of South Florida College of Medicine, Tampa, FL, USA

R. Timothy Hitchcock, CIH, CLSO, LightRay Consulting, Inc., Cary, NC, USA

Ching-Hung Hsu, PhD, DABT, HEC Pharm Group, China

Kathleen T. Jenkins, MD, MPH, Florida Association of Occupational and Environmental Medicine, Tampa, FL, USA

David R. Johnson, MD, MS, MHA, FACOEM, DABT, Professor Center for Environmental/Occupational Risk Analysis and Management, Department of Environmental and Occupational Health, University of South Florida College of Public Health, Tampa, FL, USA

Giffe T. Johnson, PhD, MPH, Research Assistant Professor, Center for Environmental/Occupational Risk Analysis and Management, Department of Environmental and Occupational Health, College of Public Health, University of South Florida, Tampa, FL, USA

Paul Jonmaire, PhD, Director, Health & Safety/Toxicology, Ecology and Environment, Inc., Lancaster, NY, USA

Sharon S. Kelley, MS, PhD, Chief Executive officer, Associates in Emergency Medical Education, Inc., Tampa, FL, USA

Bruce J. Kelman, PhD, DABT, FATS, European Registered Toxicologist, United Kingdom and EUROTOX Registries, University of Edinburgh, Edinburgh, UK

Laura E. Kerper, PhD, Gradient Corporation, Cambridge, MA, USA

Robert A. Klocke, MD, Emeritus Professor and Chair, Department of Medicine, University at Buffalo, Buffalo, NY, USA

David Y. Lai, PhD, DABT, Senior Toxicologist, Risk Assessment Division, Office of Pollution Prevention and Toxics, U.S. Environmental Protection Agency, Washington, DC, USA

Alex LeBeau, PhD, MPH, Toxicologist/Environmental Scientist, Conestoga-Rovers & Associates, Inc., Dallas, TX, USA

Janice S. Lee, MHS, PhD, National Center for Environmental Assessment, U.S. Environmental Protection Agency, Research Triangle Park, NC, USA

Richard V. Lee, Professor, Department of Medicine, School of Medicine, University of Buffalo, The State University of New York, Buffalo, NY, USA

Thomas A. Lewandowski, PhD, DABT, ERT, Toxicologist, Gradient Corporation, Seattle, WA, USA

Haixia Lin, Division of Genetic and Molecular Toxicology, National Center for Toxicological Research, Jefferson, AR, USA

Christopher M. Long, ScD, DABT, Principal Scientist, Gradient Corporation, Cambridge, MA, USA

April M. Luke, MS, Toxicologist, Office of Research and Development, National Center for Environmental Assessment/Integrated Risk Information System, U.S. Environmental Protection Agency, Washington, DC, USA

Ann Lurati, ARNP, ACNP-BC, DNP, MPH, COHN-S, Occupational Health Nurse Practitioner, University of South Florida, Tampa, FL, USA

Michael H. Lumpkin, PhD, DABT, Senior Toxicologist, Center for Toxicology and Environmental Health, LLC, Little Rock, AR, USA

Jeffrey H. Mandel, MD, MPH, Associate Professor, Division of Environmental Health Sciences, University of Minnesota School of Public Health, Minneapolis, MN, USA

Kathleen MacMahon, DVM, MS, Education and Information Division, National Institute for Occupational Safety and Health, Centers for Disease Control and Prevention Cincinnati, OH, USA

Mugimane G. Manjanatha, Division of Genetic and Molecular Toxicology, National Center for Toxicological Research, Jefferson, AR, USA

Robert R. Maronpot, DVM, MS, MPH, DACVP, DABT, FIATP, Maronpot Consulting, LLC, Raleigh, NC, USA

M.E. (Bette) Meek, Institute of Population Health, University of Ottawa, Ottawa, Canada

Matthew Mifsud, MD, Department of Otolaryngology-Head and Neck Surgery, University of South Florida, Tampa, FL, USA

Steven Morris, MD, RN, Assistant Professor, Basic Sciences, College of Nursing, University of South Florida, Tampa, FL, USA

Carlos A. Muro-Cacho, MD, Professor, Pathology and Cell Biology, University of South Florida; Chief, Anatomic Pathology, James A. Haley Veterans' Hospital, Tampa, FL, USA

Nicole Nation, University of Georgia, Athens, GA, USA.

Daniel A. Newfang, MS, CIH, CSP, Center for Environmental/Occupational Risk Analysis and Management, Department of Environmental and Occupational Health, College of Public Health, University of South Florida, Tampa, FL, USA

Dorian S. Olivera MS, PhD, CLAD, Inhalation Toxicology Program Manager, Technology Solutions Group, Life Sciences Division, Alion Science and Technology Corporation, Durham, NC, USA

Michael K. Peterson, MEM, DABT, Senior Toxicologist, Gradient Corporation, Cambridge, MA, USA

Michael W. Perkins, PhD, Research Physical Scientist, Analytical Toxicology Division, U.S. Army Medical Research, Institute of Chemical Defense, MD, USA

David M. Polanic, JD, College of Law, Michigan State University, East Lansing, MI, USA

Debra J. Price, PhD, MSPH, MEd, Director of Education, Health Services Advisory Group. Inc., Tampa, FL, USA

Joseph A. Puccio, MD, FAAP, Medical Director, Student Health Center, College of Medicine, University of South Florida, Tampa, FL, USA

Erin L. Pulster, MSMS, Senior Chemist and Laboratory Manager, Environmental Laboratory of Forensics, Mote Marine Laboratory, Sarasota, FL, USA

Humairat H. Rahman, MBBS, MPH, Department of Environmental and Occupational Health, University of South Florida College of Public Health, Tampa, FL, USA

Santhini Ramasamy, PhD, MPH, DABT, Senior Toxicologist, Office of Water, Office of Science and Technology, U.S. Environmental Protection Agency, Washington, DC, USA

Lorenz R. Rhomberg, PhD, ATS, Gradient Corporation, Cambridge, MA, USA

Keith D. Salazar, PhD, National Center for Environmental Assessment, U.S. Environmental Protection Agency, Washington, DC, USA

Philip J. Scarpa, Jr, MD, MS, FAsMA, Deputy Chief Medical Officer, John F. Kennedy Space Center, National Aeronautics and Space Administration, NASA-Kennedy Space Center, FL, USA

Alfred M. Sciuto, PhD, Fellow, Academy of the Toxicological Society, Research Physiologist, Medical Toxicology Branch, Analytical Toxicology Division, U.S. Army Medical Research Institute of Chemical Defense, Aberdeen Proving Ground, MD, USA

Davinderjit Singh MBBS, MPH, Toxicology and Risk Assessment, Environmental and Occupational Health College of Public Heath, University of South Florida, Tampa, FL, USA

Mary Alice Smith, PhD, Associate Professor, Environmental Health Science Department, University of Georgia, Athens, GA, USA

Thomas J. Steinbach, DVM, DACVP, DABT, Experimental Pathology Laboratories, Inc., Sterling, VA, USA

Frank D. Stephen, PhD, Assistant Professor of Biology, Department of Math & Natural Sciences, D'Youville College, Buffalo, New York, USA

Daniel L. Sudakin, MD, MPH, FACMT, FACOEM, Associate Professor, Oregon State University, Department of Environmental and Molecular Toxicology, Corvallis, OR, USA

James R. Taffer, MSPH, CIH, Manager, Environmental Health Services, InoMedic Health Applications, Inc. John F. Kennedy Space Center, FL, USA

Thomas Truncale, DO, MPH, Associate Professor of Medicine, Program Director, Occupational Medicine Residency, University of South Florida, James A. Haley Veterans' Hospital, Tampa, FL, USA

Peter A. Valberg, PhD, Principal, Environmental Health, Gradient Corporation, Cambridge, MA, USA

Marco Vinceti, MD, PhD, Dipartimento di Medicina Diagnostica, Clinica e di Sanità Pubblica, Università di Modena e Reggio Emilia, Modena, Italy

Qingli Wang, Center for Drug Evaluation, Chinese Food and Drug Administration (CFDA), Beijing, China

Donald E. Wasserman, MSEE, MBA, Independent Occupational Vibration Consultant, D.E. Wasserman, Inc., Frederick, MD, USA

Joan M. Watkins, DO, MPH, Affiliate Faculty, Occupational and Environmental Medicine Residency, College of Public Health and College of Medicine, University of South Florida, Tampa, FL, USA

Benjamin Wong, BS PhD, Research Physical Scientist, MTB, ATD, United States Army Medical Research Institute of Chemical Defense, Aberdeen Proving Ground, MD, USA

Carol S. Wood, PhD, DABT, Oak Ridge National Laboratory, Oak Ridge, TN, USA

Daniel Woodard, MD, Aerospace Physician Innomedic Health Applications, Kennedy Space Center, FL, USA

Yin-tak Woo, PhD, DABT, Senior Toxicologist, Risk Assessment Division, Office of Chemical Safety Pollution Prevention, U.S. Environmental Protection Agency Washington, DC, USA

ABBREVIATIONS

5-HT	5-Hydroxytryptamine
AAS	Anabolic androgenic steroids
ACGIH	American Conference of Governmental Industrial Hygienists
AIHA	American Industrial Hygiene Association
ARDS	Acute respiratory distress syndrome
ASSE	American Society of Safety Engineers
ATA	Atmospheres absolute pressure
ATP	Adenosine triphosphate
ATSDR	Agency for Toxic Substances Disease Registry
BEI	Biological exposure indices
BZE	Benzoylecgonine
CaO	Calcium oxide
CASRN	Chemical Abstract Service Registry Number
CB	Control banding
CDC	Center for Disease Control and Prevention
CH_2Cl_2	Dichloromethane
CH_3CN	Acetonitrile
CIH	Certified Industrial Hygienist
CN^-	Cyanide anion
CNS	Central nervous system
CO	Carbon monoxide
CO_2	Carbon dioxide
COHb	Carboxyhemoglobin
CONSB	Carbon Monoxide Neuropsychological Screening Battery
CS_2	Carbon disulfide
CSA	Controlled Substances Act
CSP	Certified Safety Professional
CWA	Chemical warfare agent
DEA	Drug Enforcement Agency
DNEL	Derived no effect level
DNS	Delayed neuropsychiatric syndrome
DPE	Delayed postanoxic encephalopathy

ECG	Electrocardiography
EME	Ecgonine methyl ester
EPA	Environmental Protection Agency
EU	European Union
FEV_1	1-Minute forced expiratory volume
FVC	Forced vital capacity
GABA	Gamma-aminobutyric acid
GBL	Gamma-butyrolactone
GC	Gas chromatography
GC–MS	Gas chromatography/mass spectrometry
GHB	Gamma-hydroxybutyric acid
GHS	Globally Harmonized System of Classification and Labelling of Chemicals
HCl	Hydrochloric acid
HCN	Hydrogen cyanide
HDL_c	High-density lipoprotein concentration
HOCl	Hypochlorous acid
H_2S	Hydrogen sulfide
HSE	Health Safety Executive
IDLH	Immediately Dangerous to Life and Health
IHD	Ischemic heart disease
ILO	International Labour Organization
IRIS	Integrated Risk Information System
LC–MS	Liquid chromatography/mass spectrometry
LC–MS/MS	Liquid chromatography/tandem mass spectrometry
LDL_c	Low-density lipoprotein concentration
LOD	Limit of detection
MAPK	Mitogen-activated protein kinase
MGP	Manufactured gas plants
MMF	Midmaximal flow
MRLs	Minimal risk levels
MS	Mass spectrometry
NaCN	Sodium cyanide

NGOs	Non-governmental organizations	$PM_{2.5}$	Particulate matter
NIOSH	National Institute of Occupational Safety and Health	PtD	Prevention through design
		RADS	Reactive airways dysfunction syndrome
NMMAPS	National Morbidity, Mortality, and Air Pollution Study	RBC	Red blood cell
		REL	Recommended exposure limit
NRC	Nuclear Regulatory Commission	ROS	Reactive oxygen species
OEB	Occupational exposure band	SDS	Safety data sheet
OEHS	Occupational and Environmental Health and Safety	SEG	Similar exposure group
		SLs	Screening levels
OEL	Occupational exposure limit	SMR	Standardized mortality ratio
OR	Odds ratio	STEL	Short-term exposure limit
OSHA	Occupational Safety and Health Administration	THC	Delta-9-tetrahydrocannabinoid
		TLV	Threshold limit value
PAH	Polycyclic aromatic hydrocarbon	UK	United Kingdom
PEL	Permissible exposure limit	U.S.	United States
PPE	Personal protective equipment	WHO	World Health Organization

THE HERITAGE OF ALICE HAMILTON, M.D. AND HARRIET HARDY, M.D.

It has been over a century since Alice Hamilton graduated medical school, and almost a century since Harvard appointed her as the first female associate professor. It has only been just under three-quarters of a century since the first female full professor at Harvard, Harriet Hardy, was appointed. Few working people entering the fields of occupational and environmental health and hygiene are old enough to appreciate the incredible strides made in these fields to protect workers, families, flora, and fauna. Even fewer still are old enough to understand the personal and professional battles fought by Dr. Hamilton and Dr. Hardy in just trying to attend college, gain employment in their field, and establish the credibility automatically and generously granted to men with lesser education and experience.

Both women had to limit their choices of educational institutions to those designed or willing to accept women. Both had to withstand accusations of emotionalism and "hysteria" as naysayers sought to discredit them. Dr. Hamilton achieved a notable landmark in being appointed the first woman faculty at Harvard, but her position was without several of the privileges enjoyed by the male faculty, and she retired from Harvard still an associate professor. Dr. Hardy was similarly notable in her appointment as full professor at Harvard, but she suffered from several physical ailments that took a severe toll on her. However, it would be truly short-sighted to view their difficulties and successes only in terms of gender issues. They were scientists, first and foremost, and most of the dissension encountered during their professional lives was primarily due to their challenging the deep-seated socially dismissive attitude toward worker health, the associations between chemicals and illness, and the industries' unwillingness to incur expenses in attending to employee needs.

Certainly, the dimming of memories signifies positive strides in occupational and environmental health goals and gender equality. However, the historical advances in occupational and environmental health are now so much a part of our way of life that we risk losing the appreciation of the sacrifices, and the successes of the people behind these changes. Neither Dr. Hamilton nor Dr. Hardy proffered theories that were gracefully met with acceptance, regardless of the supporting evidence or the number of potential lives at risk. Their strength of character drove them, which in turn propelled the changes they sought.

Industrial Toxicology was first drafted by Dr. Hamilton in 1934. Fifteen years later, she teamed with her colleague Dr. Hardy for the second edition. The subsequent editions maintain their names in the title, not just as courtesy but as a genuine homage to their impressive contributions to the health and welfare of global populations, at times at great expense to their own personal life and health. Perhaps, ultimately, this is the legacy carried forth with these volumes, the reason we need to remember exactly who these women were and what they accomplished—so that we may honor them by our dedication and sacrifice in striving for greater knowledge and safer lives.

IN MEMORY

During the lengthy process of creating this sixth edition, we lost one of our most integral people when, on May 7, 2013, Dr. Richard Vaile Lee passed away. A brilliant physician and educator, Dick Lee was also a valued friend and colleague.

Dr. Lee began his collegiate experience at Yale, where he received both his B.S. and M.D. and gained membership in the Alpha Omega Alpha Honor Medical Society. His internship and residency concentration was internal medicine. Between his residency years, he also squeezed in 2 years of service at the Fort Peck Indian Reservation in Montana for the U.S. Public Health Service, remaining an additional year to assist a colleague in general practice. Upon his return to Yale, he completed his residency and was awarded a fellowship in infectious diseases. While a fellow, he also acted as Director for the medical clinics and emergency room for the Department of Medicine.

In 1976, Dr. Lee left Yale to assume the positions of Professor and Vice Chairman of the Department of Medicine at the State University of New York at Buffalo (UB), and Chief of Medical Services for the Buffalo Veterans Administration Medical Center. In 1979, he took over as Head of the Department of Medicine at Buffalo's Children's Hospital. In 1997, he left Children's Hospital to establish a private practice in obstetrics medicine.

Dr. Lee's practice was not confined to a simple driving radius. He practiced "geographic medicine," bringing much-needed medical care and attention to remote locations. He treated tribal cultures in Northern Kenya and Brazil, and he traveled to Thailand, treating refugees from Laos and Cambodia. He developed an acquaintance with the Dalai Lama through his work with Tibetan refugees and was part of the UB board responsible for Dalai Lama's trip to Buffalo in 2006. He developed the UB's Medical Trek Program, sending students to perform medical care in field expeditions across the world.

The range of scholarship was almost as broad as the countries traveled. In addition to his experience in general practice, infectious diseases, and obstetrics, Dr. Lee also served as Medical Director for Ecology and Environment, Inc., in Lancaster. He taught classes at UB in pediatrics, gynecology, and obstetrics in the School of Medicine and Biomedical Sciences as well as various classes in the Department of Social and Preventive Medicine in the School of Public Health and Health Professions, and in the Department of Anthropology in the College of Arts and Sciences. He also served as a consulting physician to both the Buffalo Zoo and the Bronx Zoo.

His marriage in 1961 to Susan Bradley ultimately produced two sons, Benjamin and Matthew. It also unearthed a family secret. When he announced his impending marriage to his family, Dr. Lee learned that his paternal grandfather was Li Yan Phou, one of a select group of Chinese students to attend school in the United States. Because of strong anti-Asian prejudice (at the time there were laws preventing

Chinese and non-Chinese intermarriages in some states), Dr. Lee's parents had opted to hide the Asian heritage. However, Dr. Lee embraced his Chinese heritage, serving as a trustee for the Yale–China Association and participating in a UB delegation to China to renew an academic affiliation with a Beijing university. He also edited and wrote the foreword for the 2004 edition of his grandfather's book, *When I was a Boy in China*, published under the Anglicized name of Yan Phou Lee.

Dr. Lee produced over 70 publications and garnered several awards. In 2002, he was the Laureate Award winner of the American College of Physicians–American Society of Internal Medicine, and in 2007, he was the winner of the C.G. Barnes Award from International Society of Obstetric Medicine. In the same year, the North American Society of Obstetric Medicine established a lecture in his name for their annual meetings.

So we dedicate this edition to Richard V. Lee, M.D., FACP, FPGS: a scholar, a humanitarian, an educator, a leader, a traveler, a guide, a speaker, a listener, and ultimately an inspiration. By his own words: "a professor." By our words, a great man.

PREFACE

The sixth edition of *Hamilton & Hardy's Industrial Toxicology* has been updated and expanded with new chapters on aspects of regulatory toxicology, toxicity testing, physical hazards, high production volume (HPV) chemicals, and workplace drug use. The format has been modified and now includes information on occupational and environmental sources of exposure, mammalian toxicology, industrial hygiene, medical management, and ecotoxicology where appropriate. The book is organized by substance and includes the latest research on industrial toxicants. The goal was to provide a broad range of professionals with an accessible text. We are extremely grateful to our contributors as they provided integral industry, regulatory, and academic perspectives.

The landscape of industrial toxicology has changed considerably since Dr. Alice Hamilton and Dr. Harriet Hardy published the first edition of the text in 1929. It is estimated that there are now more than 70,000 industrial chemicals in common use. In contrast, regulatory agencies have established fewer than 1000 exposure limits. Determination of an appropriate exposure limit is an exhaustive process; as an illustration, the U.S. Occupational Safety and Health Administration (OSHA) has promulgated fewer than 30 standards (new permissible exposure limits (PELs) for 16 agents, and standards without PELs for 13 carcinogens) since 1970. The European Union's REACH (Regulation on Registration, Evaluation, Authorization and Restriction of Chemicals) framework will require several decades to determine the exposure limits for common chemicals.

The assessment backlog is so critical that OSHA recommends that the employers consider using alternative occupational exposure limits established by the National Institute for Occupational Safety and Health (NIOSH) and American Conference of Industrial Hygienists (ACGIH). Traditional toxicological testing is costly, time consuming, reliant on animal models, and fraught with uncertainty stemming from interspecies extrapolation. Tox21, an interagency project between the National Toxicology Program, the U.S. Environmental Protection Agency, the National Institutes of Health, and the Food and Drug Administration, is a high-throughput testing prototype robot created to address both the knowledge gap and problems of traditional toxicity testing. Tox21 will screen thousands of chemicals using advanced *in vitro* and cell-based assays and non-rodent models such as zebrafish in far less time than previously possible.

The evaluation of risk subsequent to exposure is an essential component of industrial toxicology. It is critical that readers remember that exposure only represents the *opportunity* for contact with a chemical. An exposure does not guarantee that an adverse effect will result. A dose is the amount of the chemical entering the body following an exposure. There are both harmful and safe doses of all chemicals. As Paracelsus said, "All things are poisons, for there is nothing without poisonous qualities . . . it is only the dose which makes a thing poison." Exposure and dose both need to be evaluated for an adequate hazard assessment. This edition of the text includes an assessment of risk where possible.

Hamilton & Hardy's Industrial Toxicology was prepared as a concise reference for academics and professionals alike. There is a wealth of information available on industrial exposures and toxicants. Distillation of these resources into something relevant can be daunting. It is our hope that this text will be that something relevant for our readers.

MARIE M. BOURGEOIS, PH.D.

SECTION I

INTRODUCTION

Section Editor: Raymond D. Harbison

1

THE MODERN APPROACH TO THE DIAGNOSIS OF OCCUPATIONAL DISEASE

RAYMOND D. HARBISON AND JEFFREY H. MANDEL

BACKGROUND

An understanding of the health effects that may occur from occupational exposures is critical in terms of the potential human toll and an industry's success and sustainability. The diagnosis of workplace-induced diseases is necessary if the disease in question is to be prevented. In the context of modern medicine, the diagnosis of an occupational disease is a multidisciplinary process and includes input from professionals in occupational medicine, nursing, industrial hygiene, toxicology, epidemiology, engineering, and others. Though physicians are primarily responsible for making an individual diagnosis, the remaining disciplines are critical parts of the process for establishing the nature and cause of the disease(s). It is the collective group that has become paramount to the understanding and control of occupational disease within modern societies. The diagnosis of an occupational disease may be understood in the context of a public health model, which incorporates the interplay between the agent, the host, and the disease. With this approach, the agent may be physical, chemical, or biological and has the potential to cause harm depending on its characteristics (e.g., corrosive, pathogenic, and carcinogenic), the exposure concentration and duration, and the ability to target organs in exposed individuals. The host is the individual or population exposed to the agent. The disease results from the interaction of these two factors. The host in this model includes healthy individuals as well as susceptible individuals (e.g., genetic predisposition and life stage). With acute or chronic high-level exposures, host susceptibility generally increases, due to a variety of mechanisms, including saturation of detoxification reactions and increased bioactivation. Manufacturing

facilities control exposures through engineering controls, personal protective equipment, chemical substitution, area monitoring, personal monitoring, hazard communication, and employee training and education.

Despite the vast number of professionals involved, the diagnosis of an occupational disease is complicated by important factors. First, many diseases that occur as a result of workplace exposures (e.g., asthma from isocyanates) may also occur from non-workplace exposures to the same compounds or may be caused by other agents (nonspecificity). Second, disease manifestation is often idiopathic and may be attributed to the workplace purely on the basis of the disease postceding employment. Third, the majority of chemicals in commerce either have not been tested in experimental animals or have been tested but lack data on the mode of action and human relevance of adverse effects. The absence of this type of information complicates extrapolations from animal studies to workers. Finally, many tests used in clinical medicine are not specific for identifying exposures to an agent of interest. For example, a blood test that reveals high carboxyhemoglobin (COHb) levels only confirms an internal dose of carbon monoxide. It does not confirm whether the source exposure was to carbon monoxide, methylene chloride, or some other causative agent. Similarly, chest radiograph findings that suggest interstitial lung disease do not confirm exposures to a specific causative agent (e.g., wood dust). Testing of workers is subject to each test's sensitivity, specificity, and positive and negative predictive value. To complicate matters, many tests do not have established "gold standards" to which they may be compared. All of these issues may potentially compromise the accuracy of diagnoses.

Hamilton & Hardy's Industrial Toxicology, Sixth Edition. Edited by Raymond D. Harbison, Marie M. Bourgeois, and Giffe T. Johnson.
© 2015 John Wiley & Sons, Inc. Published 2015 by John Wiley & Sons, Inc.

As alluded to in the above examples, a multilevel assessment is needed in the diagnosis of an occupational disease. The exposure in question must be assessed in terms of what is known about it, whether the worker's complaints are consistent with this exposure, and insights into the actual work environment are the initial necessary perspectives needed. These are followed by a detailed account of the individual's illness, the person's medical history, occupational history, physical findings, laboratory findings, and a review of the epidemiological literature involving this person's exposure–disease relationship. Finally, some type of assessment of causation is necessary to determine whether there is adequate information to suggest that an exposure could produce the disease in question. This assessment typically involves a comprehensive review of the existing epidemiological literature on the topic. All this is theoretically needed before an occupational disease can be considered. Each of these will be considered in greater detail in the subsequent paragraphs.

COMPONENTS OF THE DIAGNOSIS

The History of Illness

The worker's disease history is often the only information available to the team of professionals assigned to determine the etiology of the disease. In some cases, associations with the workplace are based entirely upon this history, hence its importance. Accordingly, the trained interviewer attempts to have as much of this as possible iterated directly in the worker's own words. A description of the worker's symptoms is the basis of this history, but there needs to also be a focus on the occupational aspects of the illness. Identification of the illness occurring temporally with the workplace is helpful since exposure–disease associations may become worse during the workday, the workweek, often with improvement after work, or on weekends away from the job. It is also important to clarify the duration of the exposure and whether coworkers with similar jobs have had health problems. Since some illnesses have long latencies, the examiner must determine the prior work history of an individual. The practitioner may also use other sources of information, including the employer, incident reports, the state or federal Occupational Safety and Health Administration (OSHA), and an industrial hygienist familiar with the workplace controls, practices, and safety data sheets (SDSs). The history should include the following:

1. *Occupational Factors.* Occupational factors must be assessed and understood. These include insights into the agent (exposure) in question as well as the involved work area. Part of this understanding is obtained through an evaluation of processes, engineering controls, personal protective equipment, employee training and compliance, and the identification of potential causative agents (e.g., which materials or chemicals are used?) and opportunities for exposure. The chemicals used must be understood in terms of the potential routes of exposure (e.g., inhalation of volatile organics), toxicokinetics (i.e., absorption, distribution, metabolism, and elimination), and toxicity.

2. *Nonoccupational Factors.* Many diseases that may originate from exposures to a causative agent in the workplace may also occur outside of the workplace. Asthma, for example, may be triggered by occupational or consumer exposure to certain isocyanates, but it may also be triggered by exposure to cat dander. Lung cancer may be related to asbestos exposure, smoking, or both. Few diseases have only an occupational basis. In fact, estimates of 5–10% of all cancers have been attributed to exposures in the occupational setting (Doll, 1984). The vast majority of cancers are not associated with occupational exposures and are thought to be multifactorial in origin (e.g., environmental exposures, lifestyle choices/hobbies, and genetic predisposition). In particular, attention must be given to these factors in the assessment of occupational disease. Especially important are exposures that occur to individuals with preexisting diseases, since these may complicate an illness resulting from an exposure. For example, an individual with underlying chronic bronchitis may have a reduced capacity for pulmonary clearance, which may predispose this individual to an adverse outcome following exposure to specific types of agents (e.g., fibers or insoluble particles).

The Physical and Laboratory Examinations

The physical and laboratory examinations are regarded as means of verifying what is already suspected following the history of illness. A skilled examiner is able to combine these two areas with a high likelihood of an accurate diagnosis, without any additional evaluation. The occupational physical examination, such as the history of illness, needs to focus on the organ where known toxicity occurs. If the worker has breathing difficulties and has exposure to a lung irritant, the examination will need to focus on the respiratory system. Other organ systems are evaluated for the sake of being thorough and to identify additional potential areas of abnormality.

Following the physical examination, the examiner may need to verify the disease suspected with the use of specific tests. In the above case, it would be appropriate to perform a pulmonary function test (e.g., spirometry) to evaluate the worker's inhalation and exhalation during normal breathing to assess air volume and air flow. It may also be helpful to obtain a chest radiograph. There are an infinite array of tests available, each costly and with their own inherent risks. These have to be weighed against the benefit of the information to be obtained. Most of the common laboratory tests have significant benefits, without much risk. Even so, caution must be used in subjecting workers to these tests unnecessarily, as all tests have—as a disadvantage—the possibility of false positive or negative findings. These may result in additional, more risky tests in the case of the former or missed diagnoses in the case of the latter.

Use of Toxicology, Risk Assessment, and Risk Characterization Information

All chemicals, even everyday, seemingly benign substances, can produce an adverse effect at some level and duration of exposure. The adverse effect may be an alteration of normal function or even death. For example, at a certain level and duration of exposure, carbon tetrachloride may cause reversible effects, such as drowsiness and loss of motor function; however, at higher levels of exposure, this chemical can produce irreversible liver damage and respiratory arrest.

Every chemical can produce a spectrum of toxicological effects. The effects vary and all chemicals are toxic at some dose (Table 1.1); that is, all chemicals are capable of altering some function or causing death in some biological organism. Though this may seem to be stating the obvious, it serves to emphasize the basis for risk assessment, which is identifying those circumstances and conditions under which an adverse effect can be produced. As Emil Mrak, chancellor at the University of California at Davis, stated years ago, "There are no harmless substances; there are only harmless ways of using substances." A chemical is toxic and does harm only within prescribed conditions of usage.

Toxicology is the study of the harmful effects of chemicals on the living system. To identify and characterize chemical-induced disease or injury, the practitioner must understand both the chemical reactions and interactions with tissues and cells and the biological mechanisms of cytotoxicity. The vastness and rapid gains in this area have stimulated many new controversies over chemical-induced injury and workplace safety. However, certain principles of toxicology apply to a large number of chemicals, and an understanding of these principles is essential to the development of insightful toxicological judgment.

Risk assessment is the process of determining whether a chemical will produce harm under specified conditions of exposure. Safety, the reciprocal of risk, is the probability that a chemical will not produce harm under specified conditions of exposure. Thus, in determining the risk or safety of a chemical, the critical factor is not necessarily the intrinsic toxicity of the chemical *per se*, but rather the likelihood that the level of exposure to the chemical is sufficient to allow the expression of its intrinsic toxicity.

Risk is determined by evaluating the exposure required to produce toxicity. The evaluation of human risk associated with chemical exposure requires an assessment of the human epidemiological and animal testing data. When available, the following data should be evaluated:

- Breadth and variety of toxic responses reported.
- Species variation or consistency in toxic responses.
- Possible proposed mode(s) of action or mechanism(s) of action.
- Validity of tests performed and their relevance for extrapolation to humans.
- Dosage used in animal tests compared with the expected level of human exposures.
- Outcomes of poisonings and long-term occupational exposures, which may serve as a guide to the expected human consequences and as validation of the human relevance of animal testing data.

Modern science, including medicine, continues to evolve as a result of the accumulation of knowledge and experience. This process of accumulation and evolution of knowledge has resulted in new principles, concepts, and treatments of disease. Discoveries have been made in medicine and toxicology that have changed medical thinking and practice in the past decade.

As a result of the evolution of regulations to protect public health, large numbers of animal tests have been conducted that have produced massive amounts of new information for the practitioner as well as the worker. A safety data sheet used for hazard communication contains the results and

TABLE 1.1   Normal and Lethal Doses of Common Substances

Substance	Normal Dose	Lethal Dose	Safety Factor
Water	1.5 qt.	15 qt.	10
Aspirin (salicylic acid)	2 tablets	90 tablets	45
Beer (ethyl alcohol)	1 beer	33 beers	33
Salt (sodium chloride)	3 tsp	60 tsp	20
Lima beans (cyanide per serving)	1.18 mg	106 mg	90

classification of animal testing data that must be appropriately interpreted and communicated to employees.

Voluntary and enforceable occupational exposure levels are considered the benchmark for determining whether a disease or injury resulted from chemical exposure. These values cannot, however, be used to determine the cause of a disease or ailment. Rather, they provide guidance for protecting workers from harmful levels of exposure. In short, occupational exposure levels protect; they do not predict. Because safety factors and other margins of safety are incorporated into these standards, exceeding the exposure level does not indicate the likelihood of harm. However, when the exposure level is exceeded, harm can occur at some point. Because all chemicals cause harm at some level of exposure, the principles of the dose–response relationship form the basis for workplace protection against chemical-induced injury.

Exposure and Dose An individual's risk of an adverse health effect is determined by evaluating exposure(s) and dose(s). Exposures must be considered based on the likely route(s) of exposure (i.e., ingestion, inhalation, and dermal/ocular). The concept of dose is different from exposure. Exposure is the opportunity to contact and absorb a chemical; this generally means that the individual and the substance are in some physical proximity. Exposure must occur to receive a dose. As used herein, the term dose refers to the actual amount of a substance absorbed by the individual, but it may also be considered in terms of the dose received at the point of contact (e.g., dermal exposure to corrosive agents). Exposure varies according to the source of the chemical, the distance from the source of contamination, and the concentration of the chemical. Even when individuals have the same or similar exposures, the actual dose received will depend on a number of variables (Table 1.2).

Duration of Exposure With some substances, even brief exposures may be harmful; with others, adverse health effects may be manifested only after chronic exposures. Exposures vary between individuals, even those working in the same areas, based on the duration of the exposure. For example, a worker exposed for 3 h in the workplace does not receive the same dose as a worker exposed for 8 h in the workplace. This necessarily varies by a number of factors, including environment. For example, a person who has lived in an area with measurable levels of specific chemicals all his or her life certainly has a different exposure duration from someone who has recently moved to the area.

Type of Contact The opportunity for contact (e.g., inhalation, ingestion, and dermal/ocular) may determine whether the exposure gives rise to adverse health effects, and if so, the nature and severity of these effects. For example, certain chemical compounds may be harmful if inhaled but are relatively harmless if swallowed.

TABLE 1.2   Variables Determining Degree of Exposure[a]

Route of Exposure	Variables
Inhalation	Inhalation absorption coefficient
	Exposure period outdoors
	Vapor concentration outdoors
	Respiratory rate indoors
	Vapor concentration indoors
Dermal	Dermal absorption coefficient
	Vapor concentration outdoors
	Vapor concentration indoors
	Exposure period outdoors
	Exposure period indoors
	Exposed body surface outdoors
	Exposed body surface indoors
Ingestion	Oral absorption coefficient
	Water ingestion rate
	Soil ingestion rate
	Food ingestion rate
	Water concentration indoors and outdoors
	Soil concentration outdoors and indoors
	Food concentrations

[a]Total daily dose, micrograms/day, determined by all routes of exposure.

Level of Exposure The effect of a chemical varies with the amount of the dose. For example, even oxygen can be dangerous to human health in very high concentrations. Potential differences in exposure result from a wide range of different workplace controls and practices as well as from lifestyle choices.

Environmental Exposure Considerable attention has been focused on environmental causes of diseases in recent years, partly because of publicity and partly because of increasing concern over industrial pollution. In the eighteenth century, it was common for children of a mercury miner to develop mercury poisoning and infants in families that used leaded pottery to suffer from central nervous system damage. In Japan during the 1960s, a disease was reported that was caused by industrial methylmercury pollution of fish used for food, a disorder called Minamata disease, after the bay into which the waste was dumped (Matsumoto et al., 1965). Recently, a painful bone disease in rice field workers in Japan, a condition known as Itai-Itai disease, resulted from cadmium pollution by nearby mines that produced zinc and lead (Kobayashi, 1971).

In the United States, a dramatic example of environmentally induced disease involved 60 cases of chronic beryllium poisoning suffered by women, children, and a few men, none of whom had entered a beryllium-using plant (Hardy et al., 1967). Beryllium in nearby community air from factory stacks was one source; clothes brought from the workplace into the home proved to be a more serious source; and it is likely that soil heavily contaminated with beryllium may be an important cause of long-delayed cases of chronic toxicity. Another example of environmentally induced disease is provided by reports of the hazards of asbestos exposure. Wagner et al. (1960) published their finding that correlated 32 cases of malignant mesothelioma with neighborhood asbestos exposure. Native women and children who lived in the South African villages close to plants that refined asbestos from nearby mines were the main victims. Newhouse and Thompson (1965) reported a series of fatal cases of mesothelioma among residents of a dwelling adjacent to a London asbestos factory. As with the beryllium industry experience in the United States, both proximity and contaminated work clothes were etiological factors in these cases. Kiviluoto (1960) found radiographic evidence of pleural calcification in a significant number of inhabitants of a geographically limited area in Finland, with no other evidence of disease. Further study demonstrated that dust from a neighboring asbestos plant was responsible for the radiographic findings.

It cannot be overemphasized that correct diagnosis, rational treatment, and the prognosis of occupational disease rest mainly on the knowledge of workplace and/or incidental exposure. In some cases, assaying blood, urine, or tissue samples may lead to the diagnosis of industrial illness; in other cases, it may simply create confusion, because the findings reflect only exposure or levels also found in individuals who were not industrially exposed. The scientific method must be used to determine the chemical cause of a disease. Failure to use this method may result in incorrect associations and conclusions.

DETERMINING THE CAUSE OF OCCUPATIONAL DISEASE

Causal Inference

The issue of whether a particular chemical exposure causes disease in humans may be approached in different ways. Details of the exposure and the disease must be thoroughly understood. With this information, the existing scientific literature may be used to determine whether an individual's illness is related to a particular exposure. In addition to the consideration of an individual's circumstances, it is also necessary to determine if there is additional general information in the scientific literature to support a chemical–disease relationship. There are several ways of doing this. In some approaches, a more theoretical approach with causal factors described as having multiple forms (sufficient, component) and with causal inference described as a part of the more general process of scientific reasoning is used (Rothman and Greenland, 1998). In other approaches, multiple factors (biologic plausibility, strength of association, exposure response, consistency, specificity, coherence, experimental evidence, analogy, and temporality) have received consideration (U.S. Surgeon General's Report on Smoking, 1964; Hill, 1965).

The multiple-factor approach to causation has limits, since there may be exceptions to nearly all of the factors even though an exposure–disease relationship may exist. For example, lung cancer is strongly related to cigarette smoke, but cigarette smoke is not specific to the association with lung cancer. With this approach, biologic plausibility may include an assessment of results from toxicity (animal) testing. Animal studies often exist in the absence of epidemiological data or may exist with it. In either case, the interpretation of animal data is complicated by the fact that different species process chemical and material exposures differently in terms of their absorption, distribution, metabolism, and elimination. Second, animal studies often involve the administration of high doses to elicit an effect. Once this is established, it becomes challenging to establish an exposure (dose) where no effect is present. This is often a contentious and complicated process with considerable uncertainty involved in the extrapolation of animal toxicity testing results to human beings (Green et al., 2011).

Causal inference in epidemiology relies on hypothesis testing in order for appropriate conclusions to be made between the relationship of an exposure and a disease. Because of a general lack of specific details concerning occupational/environmental epidemiological hypotheses and the subsequent difficulty in performing hypothesis testing, often the "null hypothesis" approach is used in the scientific process. The null hypothesis is used to focus on a negative association between an exposure and a disease (i.e., the exposure is not related to the disease). If an association is observed, the null hypothesis is rejected (subject to control for bias and chance) and an alternative hypothesis is considered (i.e., the exposure is related to the disease) (Rothman and Greenland, 1998). The use of the null hypothesis approach implies that studies are designed and performed where hypothesis testing is possible. In occupational epidemiology, the two designs where this is most feasible are the cohort and case-control designs (Rothman and Greenland, 1998; Elwood, 2007).

As a general rule, a causative relationship is more likely to be present if the statistical relationship is strong and if the relationship occurs in multiple hypothesis testing studies that include multiple populations where sample selection, sample size, bias, and confounding have been adequately assessed. Regardless of the approach to establishing *causation*, the practitioner must answer a series of questions satisfactorily to determine whether an *association* is present. These include the following:

1. Has the patient been exposed to the chemical?
2. Has the exposure resulted in a dose?

3. Is the dose sufficient to cause an effect?
4. Is the effect consistent with the chemical's known effects?
5. Is there objective medical evidence demonstrating a disease or illness?
6. Is the onset of the disease temporally related to the exposure?
7. Is the effect biologically plausible?
8. Have all other confounding or contributing factors been considered or eliminated?

It is apparent from these questions that the practitioner must be knowledgeable about human toxicology in combination with clinical medicine and existing epidemiological studies. The strongest and most appropriate evidence for establishing the cause of human illness comes from human epidemiological studies. In the absence of reliable human data, results from animal studies may be needed with the above caveats.

Epidemiology and Statistical Considerations

In addition to the consideration of association and possibly causation, diseases from chemicals can also be assessed based on the existing epidemiology and statistical relationships in the medical literature. Epidemiology is the study of the distribution and determinants of health-related conditions and events in specified populations along with the application of this information to the control of health problems. It uses a variety of approaches to understand diseases. Some of these approaches are common in the study of occupational groups. Occupational study types and basic statistical terminology are briefly reviewed.

There are two general measures commonly used in epidemiological investigations, incidence and prevalence. Incidence rate refers to the number of people within a specified population who become ill during a period of time (usually 1 year). This number effectively defines the "risk" of that disease within the specified population. Incidence is contrasted with prevalence or the percentage of individuals with disease in a population at a specified point in time. Prevalence is not a good estimate of risk.

There are several types of scientific investigations seen in the medical literature that involve worker populations. The most common types include cohort mortality, case-control, and cross-sectional studies. Case reports also occur in the literature. These typically involve the combination of a unique exposure and a unique disease in a person. Though they may be important in the recognition of a new disease or the exposure–disease relationship, by themselves they are unable to formally test a scientific hypothesis. The case report may occur as a series of individual reports, also known as the case series.

In the cohort mortality study, the investigator determines mortality rates within an exposed population and compares this number with the adjusted mortality rate in a non-exposed population, usually adjusted for age and gender. The ratio of the exposure-specific mortality rate in the exposed population to the standard population (usually adjusted by age, race, and gender) is referred to as the standardized mortality ratio (SMR). If the ratio is greater than 100 (also stated as 1.0), the implication is that there are more deaths in the exposed population than expected. If it is less than 100, the implication is that the death rate is less than expected. If it is equal to 100, it implies that there is no difference between the mortality rates in the exposed and the comparison populations. Cohort studies may be very helpful as they can account for the complete enumeration of all individuals within the group (cohort) along with the length of time they were exposed.

Generally, the larger the number of study participants, the better the statistical capability of determining whether there is a relationship between exposure and disease, if one truly exists. This is referred to as study "power." Power is the ability of a study to detect a true significant association between exposure and outcome. Epidemiologists can be more confident that an association does or does not exist between an exposure and an outcome when the findings are based on large studies rather than small ones.

Another type of common epidemiological investigation within occupational settings is the "case-control" study. In this type of investigation, cases include people who have the disease of interest and the other group (controls) does not have the disease but is ideally similar in all other respects. Data regarding past exposures in both groups are obtained and compared. Exposure status is unknown at the time of defining cases and controls. In this study type, the measure of risk is referred to as an "odds ratio" (OR). The OR is the comparison of odds for cases having been exposed versus the odds of controls having been exposed.

Both SMRs and ORs provide an estimate of relative risk or the risk of disease in an exposed group relative to the risk in an unexposed group. The strongest association between exposure and disease occurs when incremental exposures result in incremental disease (exposure–response). Relative risks must be interpreted within the context of bias, sample size, sample selection, and study design, as these factors may artificially increase or decrease the estimation of risk.

Cross-sectional studies evaluate the presence of diseases and exposures at one point in time. They are usually not useful in the determination of causative relationships, since it is not possible to determine which of these came first, the exposure or the disease.

If risk estimates (SMRs or ORs) are increased (>1.0), the epidemiologist (or the diagnostician) must determine if these represent true or false positive findings. Several things may affect study findings. First, due to the effect of probability on risk estimates, it is possible that findings may be due to chance alone and not to an exposure or factor of interest. One way epidemiologists express this chance is by the *p*-value. The significance level is usually set to 0.05. If a statistical test has a *p*-value less than this, it is considered statistically significant, by convention.

Another way to consider the likelihood of a positive finding (positive risk estimate) is to determine confidence intervals around the estimate. Confidence intervals are a reflection of a study's size and express the range of possibilities for the risk estimate. Typically, confidence intervals are expressed at the 95th percentile, which provides a range in which the "true" relative risk will occur, upon repeated testing, 95% of the time. When the study's sample size is large enough, the confidence interval will be expressed as a narrower interval. This provides assurance to the epidemiologist that findings are statistically more stable. A wide interval suggests uncertainty in the estimate of relative risk. When the interval excludes 1.0, the findings are considered "statistically significant." When the interval includes 1.0, the findings are not statistically significant and may have occurred as a chance finding more than 5% of the time (Green et al., 2011).

Since SMRs and ORs (risk estimates) are calculated without taking bias into consideration, these issues must be assessed before interpretation of results. Bias refers to anything that results in nonrandom error in the design, conduct, or analysis of an investigation. There are dozens of different types of bias. Three common types are selection (how people were selected for participation), information (access to information may differ for different groups under study), and classification (groups in the study may be classified differently with an impact on relative risk). Another important term that can impact the interpretation of epidemiological findings (risk estimates) is confounding. Confounding refers to a factor that is related to both the exposure and the disease of interest, but is unaffected by the exposure. An example is that alcohol has been determined to be related to lung cancer risk. However, since people who drink are more likely to smoke, it is really the smoking that is the true risk for lung cancer. In other words, smoking confounds the relationship between alcohol and lung cancer.

The use of *p*-values and confidence intervals, along with the consideration of the study's size, sample selection, and an assessment of the potential for bias and confounding, are all critical considerations in the determination of whether study findings are likely to be true or false (Green et al., 2011).

REGULATORY INFORMATION

Direct extrapolation of animal data to identify human hazards is common and is done to develop voluntary and enforceable occupational exposure levels. It is a common practice for the incorporation of uncertainty/variability factors when human data are not available for developing these levels. Thus, occupational exposure levels are not bright lines for identifying safe versus harmful levels of exposure, but rather they are levels of exposure that are intended to protect against adverse health effects.

Although it is a current regulatory practice to assume that a chemical that is carcinogenic in one or more animal studies may also be carcinogenic in humans, *this is not equivalent to saying that the chemical is in fact a human carcinogen*. The determination of whether a chemical is actually a human carcinogen is made quite differently from the regulatory procedure of assuming that humans will mimic animal responses. Therefore, although animal data are useful as a surrogate for assessing human health hazards, it is scientifically inappropriate to reach definite conclusions about the cause of human disease solely on the basis of animal studies.

In the absence of adequate human data, both qualitative and quantitative assumptions must be made to estimate human risk from animal studies. Although such assumptions are implicit in any animal-to-human extrapolation, the scientific community is well aware of contradictions for some of these assumptions. This may sound as if the process is somewhat contrary to the intended purpose (i.e., a reasonably accurate assessment of the human risk). These assumptions are, however, still accepted in most cases because, as a matter of public policy, this process protects public health. Some of these public policy assumptions are as follows:

1. When human data are not adequate, adverse effects in experimental animals are regarded as indicative of adverse effects in humans.

2. Results obtained with dose–response models can be extrapolated outside the range of experimental observation to yield estimated upper bounds on low-dose risk.

3. When an appropriate standardized dosage scale is used, observed experimental results can be extrapolated across species.

4. There may be no threshold for some carcinogens, whereas there may be one for others, depending on their mode(s) or mechanism(s) of action, and threshold effects usually apply for other toxicities.

5. When dose rates are not constant, average doses give a reasonable measure of exposure.

6. In the absence of toxicokinetic data, the effective or target tissue dose is assumed to be proportional to the administered dose.

7. Risks from many exposures and from many sources of exposure to the same chemical usually are assumed to be additive.

8. In the absence of evidence to the contrary and regardless of the route of exposure, standardized absorption efficiencies are assumed across species.

9. Results associated with a specific route of exposure are potentially relevant for other routes of exposure.

Many of these assumptions are controversial when applied to evaluation of a specific health risk. Moreover, it is important to consider that the degree of uncertainty for the final risk estimate increases in a multiplicative fashion with the uncertainty of each assumption adopted. Thus, the number of assumptions made in the final risk estimate may lead to an uncertainty so great that the final estimation of risk no longer reflects reality. The consensus of the scientific and regulatory communities is that risk estimates based on animal data represent *worst-case presumptions* rather than *best estimates* of the potential risk. The real risk of cancer is not known and, in many instances, may be zero (US EPA, 1986). For this reason, risk estimates based on animal data are suitable only for regulatory purposes, for setting an upper boundary on the potential risk posed by a chemical, or for ranking the relative risks posed by a number of animal carcinogens (OTA, 1981). The *actual* human risk associated with a particular level of exposure cannot be established with any degree of medical or scientific certainty, without a complete evaluation of the available database, including information on mode(s) of action and human relevance (OTA, 1981; US EPA, 1986). The risk estimates are, however, useful for setting upper limits of exposure that will not result in adverse health effects.

OCCUPATIONAL EXPOSURE LIMITS

Permissible exposure limit (PEL), *recommended exposure limit* (REL), and *threshold limit value* (TLV) are occupational exposure limits developed by OSHA, the U.S. National Institute for Occupational Safety and Health (NIOSH), or the American Conference of Governmental Industrial Hygienists (ACGIH), respectively. These values express the concept that there is a level below which no exposed worker will become ill. Such levels usually refer to exposures during a 40 h workweek over a working lifetime, except in a few instances when exposure restrictions are specified. An extreme position against these limits is taken by some who believe that no amount of evidence can assure that any exposure will be harmless for all workers. This position is not biologically or medically plausible.

PELs, RELs, and TLVs have been published for a large number of chemicals. PELs are the legally enforceable standards, which have been revised in some cases to adopt NIOSH RELs. Individual states also publish such lists of occupational exposure limits, which often defer to OSHA.

In the absence of formally developed values by OSHA, NIOSH, or ACGIH, manufacturers will often develop their own recommended occupational exposure limits.

The evidence that forms the basis for the PELs and TLVs, the safe-dose levels, currently in use is derived from occupational experience and animal testing. The development of a PEL for benzene (C_6H_6) has such a history. Benzene is a widely used solvent and was important in the manufacture of explosives during World War I. Unprotected workers exposed to as much as 1000 parts per million (ppm) or more died of the chemical's narcotic effect. At lower levels of exposure, benzene's unique action on the hematopoietic system caused fatal aplastic anemia. A number of workers who were exposed to benzene escaped these outcomes but at a later date developed acute myelogenous leukemia. As information on this experience with benzene toxicity was collected and publicized, the worker protective level was reduced to 25 ppm; it is now set at 1 ppm.

The experience with benzene illustrates an important point in understanding and characterizing specific chemical hazards. Toluene and xylene, which are structurally related to benzene, do not elicit the same toxic responses except for narcosis at high levels. In general, a toxic effect likely to be produced by an unknown chemical cannot always be predicted from its chemical likeness to a compound of known toxicity.

Though accidental or unanticipated human exposures provide information on toxicity, a systematic study of toxicity may require studies with animals. Whether a material is nearly inert or potentially harmful may be assessed by experimental animal studies. However, the difficulties involved in studying low-level, long-term effects and the variation in response from species to species create additional obstacles to extrapolating from animal exposure to humans.

Some of these problems are illustrated by the animal studies done to assess beryllium toxicity. Despite experience with chronic illness in 800 beryllium workers, some authorities believe that failure to reproduce the disease in animals ruled out beryllium as a cause. Rabbits exposed to beryllium compounds develop osteogenic sarcomas, whereas rats develop pulmonary tumors—a difference that illustrates the difficulty of predicting a response from one species to another.

REFERENCES

Doll, R. (1984) Occupational cancer: problems in interpreting human evidence. *Ann. Occup. Hyg.* 28(3):291–305.

Elwood, M. (2007) Study designs which can demonstrate and test causation. In: *Critical Appraisal of Epidemiological Studies and Clinical Trials*, Chapter 2, Oxford, UK: Oxford University Press, pp. 37–43.

Green, M.D., Freedman, M., and Gordis, L. (2011) *Reference Guide on Scientific Evidence*, 3rd ed., Federal Judicial Center and National Research Council of the National Academies.

Hardy, H.L., Rabe, E.W., and Lorch, S. (1967) United States beryllium case registry (1952–1966): review of its methods and utilization. *Occup. Med.* 9:271–276.

Hill, A.B. (1965) The environment and disease: association or causation? *Proc. R. Soc. Med.* 58:295–300.

Kiviluoto, R. (1960) Pleural calcification as a roentgenologic sign of nonoccupational endemic anthophyllite asbestosis. *Acta Radiol.* 194(Suppl.):1–67.

Kobayashi, J. (1971) Relation between the "Itai-Itai" disease and the pollution of river water by cadmium from a mine. *Proceedings of the International Conference on Water Pollution Research, I-25,* pp. 1–7.

Matsumoto, H., Koya, G., and Takeuchi, T. (1965) Fetal Minamata disease: a neuropathological study of two cases of intrauterine intoxication by a methyl mercury compound. *J. Neuropathol. Exp. Neurol.* 24:563–574.

Newhouse, M.L. and Thompson, H. (1965) Mesothelioma of pleura and peritoneum following exposure to asbestos in the London area. *Br. J. Ind. Med.* 22:261–269.

Office of Technology Assessment (OTA) (1981) *Assessment of Technologies for Determining Cancer Risks from the Environment*, Washington, DC: Congress of the United States, Office of Technology Assessment.

Rothman, K.J. and Greenland, S. (1998) Causation and causal inference. In: *Modern Epidemiology*, 2nd ed., Chapter 2, Philadelphia, PA: Lippincott Williams and Wilkins, pp. 7–28.

United States Environmental Protection Agency (US EPA) (1986) Guidelines for carcinogen risk assessment. *Fed. Reg.* 51:33993–34003.

United States Surgeon General's Advisory Committee on Smoking and Health (1964) *Smoking and Health*, Rockville, MD.

Wagner, J.C., Sleggs, C.A., and Marchand, P. (1960) Diffuse pleural mesothelioma and asbestos exposure in the North Western Cape Province. *Br. J. Ind. Med.* 17:260–271.

2

CONTROLS OF OCCUPATIONAL DISEASES

Daniel A. Newfang, Giffe T. Johnson, and Raymond D. Harbison

BACKGROUND, SCOPE, AND PURPOSE

As presented in Chapter 1, from the early recognition of lead poisonings by Hippocrates (460–370 B.C.) through the Age of Enlightenment (Ramazzini being noted as the "Father of Industrial Hygiene"), the contributions of physicians such as Thackrah, McCready, Hamilton, and Hardy to name a few as well as the many unnamed modern, global contributors have been recognized and used to shift to a more holistic and a proactive approach to risk management and addressing occupational and environmental health issues. Specific methodologies, models, and global trends as well as a greater ability to transfer knowledge using technology, these all contribute to new approaches for the strengthening of assessing and controlling occupational risk. Note that there still are limitations on the time-tested classic approaches (i.e., anticipate, recognize, evaluate, and control) as well as evolving new principles (e.g., hazard assessments, risk assessments, control and exposure banding, generalized harmonization of risk assessment tools, etc.), which still demand the requirement for sound professional judgment. As a final thought, *it needs to be stated* with the start of the twenty-first century well behind us now, risk assessment/exposure strategies are indeed integral parts of almost every traditional and nontraditional profession, craft, trade, or industry in highly developed industrial countries. Who would have ever thought that there would be exposure guidelines for deployed military, humans living and working in low earth orbit as well as beyond our planet? But there are still very traditional occupational health and safety hazards in developing countries with economies in transition. These concerns for human health will most likely increase due to regulatory infancy, the lack of governmental infrastructure, or by the sheer volume of chemicals being introduced into commence with minimal test data. The need for partnerships between the regulators, manufacturing/production leadership, industrial hygienist, safety professional and toxicologist, and occupational medical team is critical in the management to prevent or reduce the onset of occupational risk, diseases, and disorders.

REGULATING BODIES AND PROFESSIONAL ORGANIZATIONS

There are many regulating bodies and mature authoritative advisory organizations throughout the industrialized world sharing information on health and safety programs and management. Additionally there are too many specific states and countries to form a comprehensive list (such as the MAK limits from Germany, WELs from the UK, etc.). Therefore, some of the more prominent entities are listed below. Be sure to check with your local government to identify under whose control and what standards need to be followed.

US EPA: While one may think of the United States Environmental Protection Agency (EPA) as directing environmental regulations, it also establishes health-based risk criteria standards and guidance for the *general public (including sensitive subgroups) during a lifetime of exposure* such as screening levels (SLs), EPA's Integrated Risk Information System (IRIS), National Ambient Air Quality Standards (NAAQS), and Safe Drinking Water Act. The Acute Exposure Guideline Levels (AEGL) exposure index is also developed and published by the EPA.

http://www.epa.gov/reg3hwmd/risk/human/
rb-concentration_table/faq.htm

Hamilton & Hardy's Industrial Toxicology, Sixth Edition. Edited by Raymond D. Harbison, Marie M. Bourgeois, and Giffe T. Johnson.
© 2015 John Wiley & Sons, Inc. Published 2015 by John Wiley & Sons, Inc.

http://www.epa.gov/iris/index.html

http://www.epa.gov/air/criteria.html

http://water.epa.gov/drink/contaminants/index.cfm#List

US ATSDR: The United States Agency for Toxic Substances and Disease Registry (US ATSDR) is a federal public health agency of the U.S. Department of Health and Human Services that serves the public by using the best science, taking responsive public health actions, and providing trusted health information to prevent harmful exposures and diseases related to toxic substances. The agency publishes an Internet web portal that provides comprehensive access to the best science, the latest research, and important information about how toxic substances can affect health (e.g., minimal risk levels, MRLs; ToxGuide, etc.).

http://www.atsdr.cdc.gov

US OSHA: Under the US Occupational Safety and Health Act of 1970 (a.k.a. The OSH Act), employers are responsible for providing a safe and healthful workplace. The United States Occupational Safety and Health Administration's (OSHA) mission is to assure *safe and healthful workplaces* by setting and enforcing standards, and by providing training, outreach, education, and assistance. Employers must comply with all applicable OSHA standards. Employers must also comply with the General Duty Clause of the OSH Act, which requires employers to keep their workplace free of serious recognized hazards. The OSHA publishes permissible exposure limits (PELs) that are legally enforceable workplace exposure limits.

https://www.osha.gov/index.html

US NRC: Although radiation is naturally present in our environment, it can have either beneficial or harmful effects, depending on its use and control. For that reason, the US Congress charged the United States Nuclear Regulatory Commission (US NRC) with protecting people and the environment from unnecessary exposure to radiation as a result of civilian uses of nuclear materials. Toward that end, the US NRC requires nuclear power plants; research reactors; and other medical, industrial, and academic licensees to use and store radioactive materials in a way that eliminates unnecessary exposure and *protects radiation workers and the public.*

http://www.nrc.gov/about-nrc/radiation.html

US CDC: From the food you eat, to the air you breathe, to staying safe wherever you are, the United States Centers for Disease Control and Prevention (US CDC) mission touches all aspects *of daily life.* The US CDC researchers, scientists, doctors, nurses, economists, communicators, educators, technologists, epidemiologists, and many other professionals all contribute their expertise *to improving public health.*

http://www.cdc.gov

EU: The European Union (EU) offers a wide variety of community measures in the field of *safety and health at work* based on Article 153 of the Treaty on the Functioning of the EU (ex: Article 137 TEC). European directives are legally binding and have to be transposed into national laws by the member states. There are numerous directives, guidelines, and standards as well as other practical documents on safety and health from member states or European institutions (e.g., occupational exposure limits (OELs); acceptable operator exposure limits (AOELs); derived no effects levels (DNELs), etc.).

http://osha.europa.eu/

ILO: The International Labour Organization (ILO) is a unique structure of the United Nations that globally gives an equal voice to workers, employers, and governments to ensure that the views of the social partners are closely reflected in labor standards and in shaping policies and programs in the area of *safety and health at work.*

http://www.ilo.org/

http://www.hc-sc.gc.ca/ahc-asc/legislation/index-eng.php

Military, Space, etc.: Most US federal installations will reference either the American Conference of Governmental Industrial Hygienists (ACGIH) guidance or US OSHA regulation, whichever is more stringent. At times, the agency might have a lower exposure limit published within its organization. As of this publishing date, some examples are as follows:

Department of the Army EM 385-1-1 (Sept 2008):

  http://www.usace.army.mil/safetyandoccupational health.aspx

The US Army Center for Health Promotion and Preventative Medicine, Chemical Exposure Guidelines for Deployed Military Personnel (USACHPPM Technical Guide 230, 2013 Revision):

  http://phc.amedd.army.mil/Pages/Library.aspx?Series= PHC+Technical+Guide

NASA-STD-3001, NASA Space Flight Human System Standard, Volume 1:

  http://www.nasa.gov/centers/johnson/slsd/about/ divisions/hefd/standards/

Department of Defense, Human Engineering Design Data Digest (April 2000):

http://www.dtic.mil/cgi-bin/GetTRDoc?
AD=ADA467401

US NIOSH: The United States National Institute for Occupational Safety and Health (US NIOSH) is the U.S. federal agency that conducts research and makes recommendations (e.g., recommended exposure limits (RELs)) to prevent worker injury and illness. The NIOSH is part of the US CDC in the U.S. Department of Health and Human Services.

http://www.cdc.gov/niosh/

ACGIH®: The ACGIH® is a member-based organization that advances occupational and environmental health. Examples of this include annual editions of the threshold limit values (TLV®) and biological exposure indices (BEIs®) work practice guides.

https://www.acgih.org

AIHA: The American Industrial Hygiene Association (AIHA) was founded in 1939; AIHA is a nonprofit organization devoted to achieving and maintaining the highest professional standards for its members. More than half of 10,000 members are certified industrial hygienists (CIHs), and many hold other professional designations. A professional designation of CIH identifies an individual as having met the professional challenge of illustrating competency through education, experience, and examination. The AIHA administers comprehensive education programs that keep occupational and environmental health and safety (OEHS) professionals current in the field of industrial hygiene. The AIHA is one of the largest international associations serving OEHS professionals practicing industrial hygiene and is a resource for those in large corporations, small businesses, and who work independently as consultants. The mission is to create knowledge to protect worker health keeping a vision of eliminating workplace illnesses. Exposure indices such as Workplace Environmental Exposure Levels (WEELs) and Emergency Response Planning Guidelines (EPRGs) are developed and published by the AIHA.

https://www.aiha.org/

ASSE: Founded in 1911, the American Society of Safety Engineers (ASSE) is the world's oldest professional safety society. The ASSE promotes the expertise, leadership, and commitment of its members while providing them with professional development, advocacy, and standards development. It also sets the occupational safety, health, and environmental community's standards for excellence and ethics. A professional designation of certified safety professional (CSP) identifies an individual as having met the professional challenge of illustrating competency through education, experience, and examination.

http://www.asse.org/

ROLE OF OCCUPATIONAL HEALTH PROFESSIONALS

Technological gains have been impressive over the past number of decades, which translate into productivity improvements. With these gains, the sustainability of the world's resources is critical in order to maintain health and wellness for our most valuable resource, the human capital. The roles of the different occupational health professionals in proactively controlling exposures and diseases in the workplace are described below and should be used as best as one can. Obvious to the working health professionals, the need for a safe and healthy work environment requires the control of physical, chemical, biological, psychosocial, illnesses, injury, etc. Not so obvious though are the needs to establish organizational/management systems to assertively and proactively address early recognition and response to work-related health/safety risks critical to preserving the human element and building a sustainable workforce. Hence, a multidisciplinary occupational health and safety team approach is needed.

The goal of a multidisciplinary occupational health and safety team is to design, implement, and evaluate a comprehensive health and safety program, which will maintain and enhance health, improve safety, and increase productivity. Such programs often provide similar results for the families of workers, with resultant financial and other benefits for the corporation. Occupational health and safety professionals include occupational and environmental health nurses, occupational medicine physicians, industrial hygienists, safety professionals, and occupational health psychologists. Other related members of the multidisciplinary team are ergonomists, toxicologists, epidemiologists, human resource specialists, and organizational psychologists (in an attempt to name a few).

Occupational medicine professionals (i.e., physicians, physician assistants, registered nurses including nurse practitioners) include health care professionals qualified to design, manage, supervise, and deliver health care *in occupational settings.* These professionals have usually completed an additional occupational medical training or acquired an on-site experience. Completion of additional training (residency or otherwise) and further practice in occupational medicine enable them to pursue certification in occupational medicine after meeting rigorous qualifying standards and successfully completing an examination in occupational medicine. Many times, these team members will be charged with developing and *managing* the safety and health programs.

Occupational technical team (*professional level*) includes industrial hygienists, safety engineers, professional engineers in the traditional disciplines such as civil, electrical, environmental, chemical, mechanical, etc. Toxicologists, health physicists, ergonomists, health educators, dieticians, dentists, physiologists, psychologists, and professionals from other disciplines are involved directly or indirectly with the prevention of occupational disease. These are key team members who may be employed full time or on call, as and when needed. On the professional level, these team members should have completed a minimum of a bachelor's degree from an accredited university. Advanced degrees; master and doctorate and professional credentials (CIH, CHMM, CSP, PE, DABT, BOHS-COC, etc.) are desirable because these certifications show an accomplished level of education, measurable amount of knowledge, and experience.

Occupational technical team (*technical level*) includes technicians, production, and many other personnel that are full or part time dedicated to *delivering* the elements of the health and safety programs to the operating or manufacturing environment. Typically these roles are staffed with team members that haven't completed formal university training (i.e., bachelor's or advanced degrees), but their importance on the occupational health team is critical to any type of success. These technicians are the heroes of the team and often overlooked. Their role places them directly where the work is being performed and they're typically well received through their relationships with their coworkers performing the work. The knowledge, experience, and execution capabilities these technicians offer to the health and safety program are invaluable.

Research, engineering, and design team includes team members who participate in researching, designing, developing, engineering, and constructing the manufacturing/ production equipment and environment. It is critical that these team members are aware of the potential hazards and risks their processes can produce so that they can eliminate or mitigate them through design. A gold standard approach would be to have an occupational health team member be an active contributing member of the research, engineering, and design team who will enable education opportunities to the other members of the team as well as contribute to focused health and safety input into any designs.

Production leadership team is composed of team members who manage and supervise the production/manufacturing environment. It is *only* by acceptance and approval from these team members that any health or safety control is implemented into the production or manufacturing environment. As with the research, engineering, and design team, a gold standard approach would be to have an occupational health team member as an active contributing member of the production leadership team who will enable education

opportunities as well as contribute to focused health and safety input into any production decisions.

CONTEMPORARY STRATEGIES FOR CONTROLLING EXPOSURES AND DISEASES

Over the past number of centuries, strategies for occupational disease control, managing workplace illnesses and injuries have changed. Historically, the concept of asking the injured person what their occupation was to help identify the "cause of their illness" has now developed into the AIHA's anticipate, recognize, evaluate, and control strategy, which is continuing to mature into newer principles for occupational risk assessments. While some of the more seasoned concepts and principles such as reducing the risk of occupational injury and illness by designing the risk out (e.g., US NIOSHs Prevention through Design, (PtD)) will always be relevant, newer holistic and comprehensive methods are being developed. Some of the more generally accepted strategies are described below.

Current approaches: The AIHA, US OSHA, US EPA, Canadian Centre for Occupational Health and Safety, UK Health and Safety Executive (HSE) as well as other well-established entities discuss many successful strategies for managing occupational exposures via the science of anticipating, recognizing, evaluating, and controlling the workplace conditions. Some of the more germane concepts are as follows:

Regulatory OELs are established by government statute (e.g., US OSHA PELs) and have been established for a limited number of chemicals in commerce. These exposure limits *do consider technical and economic feasibility* thereby have inherent limitations. Many health professionals state the OEL development and setting process/infrastructure is labored, which results in out-of-date standards.

Authoritative OELs (*also referred to as OELs*) are established by research or consensus organizations based on health criteria (e.g., US NIOSH RELs, ACGIH TLVs®, AIHA WEELs®, US EPA Reference Dose (RfD), etc.) and *do not consider economic or technical feasibility*. The developing entity uses professional judgment, elimination mechanisms, and the significance of adverse health outcomes. Additionally, typically the exposure limit is derived based on no-effect-levels or a set ratio of other toxic end points.

Internal OELs are set by private organizations as per their internal policies, guidelines, corporate risk tolerance, and data.

Working OELs are set by the industrial hygienist based upon their professional judgment, knowledge, personal risk tolerance, and experience.

Control banding (*CB*) is a methodology that germinated in 1988 by the US NIOSH, AIHA, ACGIH, ILO, International Occupational Hygiene Association (IOHA), and

World Health Organization (WHO) describing a system that matches a control measure (e.g., ventilation, engineering controls, containment, etc.) to a range or "hazard band" of hazards and exposures (e.g., skin/eye irritation, very toxic, carcinogenic, etc.). The CB system groups chemicals according to similar physical or chemical characteristics, how the chemical will be handled or processed, and what the anticipated exposure is expected to be. This system then determines a set of useful controls that will prevent harm to workers. The methodology uses performance-based criteria for chemicals that do not have established exposure limits but similar hazards to already studied or known chemicals, the idea being to follow the existing control strategy that the "studied or known" chemical is using.

Occupational exposure bands (*OEBs*) is the application of both hazard banding and CB methods in the assessment and management of workplace risks.

Similar exposure groups (*SEGs*) NOTE: This approach was previously referred to as homogeneous exposure groups (HEGs). Similar exposure groups is a risk assessment strategy sponsored by the AIHA that focuses on the concept of similarly exposed workers. It is a method for considering the exposure of a limited number of workers in a group to be representative of everybody in the group. This allows for limited resources to be better used in characterizing the many exposures that are present in the workplace. The approach is intended to identify workers having the same general exposures/exposure profile due to their similarity and frequency of work tasks, same use of materials and processes. Benefits from using the SEG methodology are: with limited resources, it will quickly identify what hazard(s) need to be prioritized and urgently addressed in a complex facility where many hazards exist; and help identify what "control banding" principles for selecting a control technology can be put into place to eliminate or mitigate the exposures. Finally, it is acceptable to provide professional judgment and estimate exposure level, severity of health effects, and the uncertainty associated with the available information.

Comprehensive exposure assessments methods involves a continual process of collecting as much health and safety information as you can, prioritizing controls, and gathering follow-up information to manage the hazards and risks in the workplace.

Qualitative exposure assessment methods can be used when exposure-monitoring data are *not* available or not required, modeling techniques have become accepted. A qualitative exposure assessment may be initially used to assess exposures from an upcoming task. A critical component though is to have the assessment performed by an industrial hygienist who is familiar with the operation or process being evaluated.

Hierarchy of controls is a widely accepted system used to eliminate or minimize workplace exposures. The concept has been a core foundation for decades in many health and

safety organizations and is also drafted into many standards throughout the industrialized world such as ISO 12100-1, ISO 14121/EN 1050, ANSI/ASSE Z244.1, ANSI B11.TR3, ANSI/RIA R15.06, SEMI S2-0200, MIL-STD-882D, ANSI/AIHA Z10, ANSI/PMMI B155.1 . . . just to cite a few. The concept presents hazard controls in the event that the hazard cannot be eliminated altogether. Some methods are more effective than others and typically the solution is to use a combination of one or more of the five approaches. The controls are presented in the order of decreasing effectiveness:

Elimination (eliminating the hazard and removing it is the *most* effective method of control)

Substitution (the second most effective method, replace the hazard with something that is less hazardous to the workers)

Engineering controls a.k.a. safeguarding technology (the third most effective method is to design a control strategy to protect the worker from the hazard such as ventilation, noise dampening, mechanical lifts, or machine guards)

Administrative controls a.k.a. training and procedures (the fourth most effective method is to change the way the work is performed through policy changes, training, warning alarms, signage, etc.)

Personal protective equipment (*PPE*) (the fifth and the *least desirable* method of control is by placing protective equipment directly on the workers' bodies. This is referred to as *the least desirable* because the management of the safety device is solely up to the worker performing the work. There is a high potential for the PPE to not be worn correctly or at all. Examples are hard hats, goggles, earplugs, and protective clothing.)

Prevention through Design is a concept being presented by the US NIOSH that focuses on reducing the risk of occupational injury and illness by integrating decisions affecting safety and health in all stages of the design process. To further elaborate, the concept of PtD can be defined as: Addressing occupational safety and health needs in the design process to prevent or minimize the work-related hazards and risks associated with the construction, manufacture, use, maintenance, and disposal of facilities, materials, and equipment. *This is why it is critical* to have a safety and health team member involved on the research, engineering, and design team as discussed earlier in this chapter in the "Role of Occupational Health Professionals" section.

With the changing global regulatory environment, there is an ever-increasing expectation for proactive risk management, use of more control/hazard banding when OELs are lacking *plus* the fact that only a small number of chemicals are either regulated by a government statute or have a traditional OEL. We must recognize that OELs *are just one* of the components of a larger body of occupational benchmarks and guidance values, which help in the risk assessment process. The future challenges are when a regulatory OEL or authoritative guidance value does not exist or cannot be derived easily, what alternative risk assessment method should be used? New risk management tools are designed around the globalization of risk issues and enhanced abilities for information transfer. There is no doubt that greater expectations for occupational risk assessments will need to be addressed in a systemic manner for a growing list of chemicals being used in the workplace. The future will see developments such as the hierarchy of OELs model, further harmonization of OELs and DNELs at the European Level, US OSHA providing annotated references as a supplement to their health-based PELs, US NIOSH refining and harmonizing their hazard banding techniques, and AIHA and US NIOSH working jointly with OEL experts to categorize OEBs based on the Globally Harmonized System for Classification and Labeling of Chemicals (GHS) within the REACH regulation.

CONCLUSIONS

Updating OELs is a continuing process based on new discoveries and advances in science.

We know:

That the dose makes the poison (Paracelsus)

A hazard is defined as the potential to cause harm posed by some material or situation.

Risk is the probability, or chance, that it actually will harm someone.

And there are still occupational illnesses, injuries, and fatalities happening on a daily basis.

Continued vigilance, learning through scientific advancement, passion, and humility are needed by all involved in the protection of the worker whether terrestrial bound or working in outer space. The health and safety team member's dedication to humanity is admired and appreciated. The future for occupational safety and health risk management is ever changing and will prove to be as challenging as its history has shown it to be.

FURTHER READING

AIHA (2002a) White Paper on Risk Assessment and Risk Management.

AIHA (2002b) White Paper on Permissible Exposure Limits (PELs).

AIHA (2011) *The Occupational Environment: Its Evaluation, Control, and Management*, 3rd ed., AIHA.

Ignacio, J.S. and Bullock, W.H. (2006) *A Strategy for Assessing and Managing Occupational Exposures*, 3rd ed., AIHA.

Lasczc-Davis, C., Maire, A., and Perkins, J. (2014) The hierarchy of OELs, a new organizing principle for occupational risk assessment. *Synergist* 25(3):27–30.

Manuele, F.A. (2005) Risk Assessment & Hierarchies of Control. Professional Safety.

NIOSH (2013) *Prevention Through Design*, CDC. Available at http://www.cdc.gov/niosh/topics/ptd/.

Sullivan, E. and Malik, O. (2007) *Control Banding: Pharmaceutical Caterpillar to Mainstream IH Butterfly*, AIHA.

SECTION II

METALS AND METALLOIDS

SECTION EDITOR: RICHARD V. LEE

3

INTRODUCTION

Richard V. Lee

One of the defining events in the cultural and biological evolution of our species, *Homo sapiens sapiens*, was the refining and use of metals and metalloids. These elements have shaped human history by their malleability, desirability, and toxicity, properties determined by their physical chemical properties and the fashions and passions of their human artisans. Ancient gold jewelry, blue and green copper salts for eye makeup in Egypt, lead-based wine additives and wig powders in the seventeenth and eighteenth centuries, and iron alloys in contemporary steel superstructure construction constitute only a small segment of the history of metallic progress. Although 80 of the 105 elements listed in the periodic table have been labeled as metals, fewer than 30 are documented human toxins.

The oldest toxins recognized by our species were probably derived from foods, from the plants, and flesh our forebears ate. The toxicity of metals was recognized after the working of metals became more than an occasional event. Metals were the first industrial toxins. Gold, copper, lead, silver, tin, and arsenic were the earliest metals systematically extracted and worked into ornaments, tools, and weapons (Table 3.1).

Copper workings have been found in the Middle East dating from 3000 to 4000 B.C. By 2000 B.C., bronze, an alloy of copper and tin, became the dominant useful metal. By 1400 B.C., iron was smelted and worked simultaneously with copper and bronze, and rapidly supplanted them as the preferred material for tools and weapons. About the same time, brass, an alloy of copper and zinc, was created and became a popular material for coins, ornaments, and sculpture.

Early observers of medicine, natural history, and industry described the preparation of metals and metal objects and some of the health consequences of such work. Hippocrates described abdominal colic in a metal worker in the fourth century B.C. Arsenic and mercury products were mentioned by Theophrastus of Erebus (370–287 B.C.) and Pliny the Elder (A.D. 23–79). In *De Medicinia*, Celsus (25 B.C.–A.D. 30) described medicaments containing antimony, lead, mercury, iron, copper, zinc, arsenic, and aluminum.

Many of the metals that cause concern for health are of recent discovery and application (Angerer and Heinrich, 1988; Abrams and Murrer, 1993). Cadmium was identified in 1817. Many of the lanthanides were purified only in the twentieth century. In 1845, Robert Christison's *Treatise on Poisons* devoted 120 pages to arsenic, 73 pages to mercury, 65 pages to lead, 27 pages to copper, 15 pages to antimony, 6 pages to zinc, 4 pages to iron, 2 pages each to tin, bismuth, silver; and a collection of "rarer metals" (osmium, platinum, palladium, iridium, rhodium, molybdenum, manganese, uranium, cobalt, cadmium, nickel, tungsten, titanium, and cerium); and 1 page each to chromium and gold. Contemporary technology is finding novel uses for many metals and metalloids that were thought to be only curiosities. Information about their toxicities will be developed as their applications become more common (Underwood, 1977; Beijer and Jernelov, 1986; Friberg et al., 1986; Seiler and Sigel, 1986; Angerer and Heinrich, 1988; Abrams and Murrer, 1993; Elinder et al., 1994).

The characteristic properties of metals are a shiny crystalline surface and the ability to conduct heat and electricity better than nonmetals. Metals are distinctive because they have freely movable electrons, the so-called conduction electrons. The intrinsic toxicity of metals is determined by the properties of the atom. As a general rule, within the vertical groups of the periodic table, greater toxicity is

Hamilton & Hardy's Industrial Toxicology, Sixth Edition. Edited by Raymond D. Harbison, Marie M. Bourgeois, and Giffe T. Johnson.
© 2015 John Wiley & Sons, Inc. Published 2015 by John Wiley & Sons, Inc.

TABLE 3.1 Ancient Metallurgy

Period	Smelting and Refining	Alloying and Working
Before 4000 B.C.	Occasional use of native gold and copper and of meteoritic iron	Hammering to shape and to harden copper annealed without melting (about 4000 B.C.)
4000–3000 B.C.	Reduction of oxidized ores of copper and lead	Melting and casting of nearly pure copper, followed by creation of copper–arsenic and copper–tin alloys (made from ores)
	Creation of copper–arsenic alloys by smelting of natural mixed ores, followed by creation of bronzes from intentionally mixed copper and tin ores (ca. 3500 B.C.)	Creation of permanent molds of stone and metal
		Invention of lost-wax process for complicated castings
		Use of native silver and gold
		Use of native gold–silver alloys
		Soldering with copper–gold and lead–tin alloys
3000–2000 B.C.	Roasting and smelting of sulfide ores of copper	Most jewelry techniques developed before 2500 B.C. (e.g., inlay, stamping, repoussé, raising, soldering, riveting, granulation, surface coloring)
	Smelting of tin (perhaps earlier), which becomes important trade good	
	Cupellation of lead to extract silver and of gold and silver for purification	Golf leaf
		Wire made by cutting sheet
		Experimental production of sponge iron
2000–1000 B.C.	Bellows used in furnaces by 1800 B.C.	Bronze made from metallic tin and copper (perhaps much earlier), high tin bronze (speculum metal) for mirrors, leaded bronzes for statuary
	Wrought iron becomes important by 1600 B.C. (directly reduced from ore as sponge iron and forged without melting)	
		Brass produced from copper and zinc oxide ores (not important before 2000 B.C.)
		Steel regularly made by carburization in hearth, perhaps hardened by quenching (1200 B.C.)
1000–0 B.C.	Cast iron (China only)	Vast expansion in production of iron, common nonferrous metals, gold and silver
	Cast steel (India only)	
	Gold purified by sulfide process	Iron and steel welded into composite tools and weapons
	Mercury distilled from ores	Stamping of coins (700 B.C.)
	Amalgamation of gold ores	Gilding of bronze and silver with gold amalgam
		Assay by cupellation, cementation, touchstone, and density

Source: From Encyclopaedia Britannica: *Metallurgy and Metals*, 15th ed., vol. 2, 1978.

associated with increasing atomic weight or atomic number (Venugopal and Luckey, 1978). For example, heavy metals tend to form stable, irreversible complexes with biological macromolecules, thus altering their biological characteristics. Toxicity is proportional to the electropositivity and the solubility of metal cations in water or in lipids. The relatively nontoxic properties of the lighter metallic elements (sodium, magnesium, potassium, and calcium) must have determined their ubiquity and importance in the life of virtually all living creatures. The chemical form of the metal determines solubility, absorption, cellular uptake, and excretion (Underwood, 1977; Venugopal and Luckey, 1978; Beijer and

Jernelov, 1986; Friberg et al., 1986; Seiler and Sigel, 1986; Angerer and Heinrich, 1988; Abrams and Murrer, 1993; Elinder et al., 1994; Goyer, 1995). Soluble compounds are more readily absorbed by the mucosa and skin and usually are more rapidly excreted. Insoluble compounds may coat mucosal surfaces and become entrapped in mucus or phagocytosed by polymorphonuclear leukocytes and macrophages. Thus the dose of a metal depends on factors other than just the magnitude or site of the exposure.

The biological processing of metals—absorption, distribution, and excretion—is another determinant of toxicity. The toxic effects of metals are the result of the interactions between free metal ions and a susceptible cellular target. Prolonged contact between target and metallic toxin enhances toxicity. The biological half-life therefore is a major factor in metal toxicity. Lead and cadmium have 20- to 30-year biological half-lives. Insoluble nickel compounds have a much longer half-life than soluble nickel salts, which dictates the differences in their biological effects. The concept of "dose" must be expanded beyond the size of the inoculum to include the duration of the substance in the host. The inoculum of a metal with a prolonged biological half-life may be substantially smaller than the inoculum of a biologically short-lived metal to produce the same toxic effects (Furst and Haro, 1969).

The metallic elements have a wide range of toxic potentials. Direct injury to cellular processes is a consistent effect that can result in necrosis, acute inflammation, and enhanced susceptibility to secondary infection or injury because of loss of the host's defense mechanisms. Indirect injury results from an alteration in molecular configuration, as in the creation of immunogenic haptens and the stimulation of cell-mediated immunity. For example, platinum, cobalt, and chromium can illicit type I anaphylactic/immediate hypersensitivity reactions; gold can illicit type II cytotoxic hypersensitivity; mercury and gold can illicit type III immune-complex hypersensitivity; and nickel, chromium, beryllium, zirconium, titanium, and aluminum can illicit type IV cell-mediated hypersensitivity. Carcinogenesis and mutagenesis are complex, multistep processes in which metal ions act as conspirators with other physical or chemical events (Table 3.2).

Considering their ubiquity in the natural and manmade environments, the truly remarkable feature of the metals and metalloids is how uncommon serious toxicity has become. One of the major success stories of occupational medicine has been the striking reduction in serious occupational metal intoxication in contemporary industry. However, serious environmental consequences from the metal working and industrial revolution of the past two millennia remain. The quantity of environmental lead in the Greenland ice cap has risen dramatically over the past two millennia in synchrony with the growth in the industrial use of lead and lead compounds (Ng and Patterson, 1981).

A major contemporary ecological problem is the existence of large residues of potentially toxic metallic waste at mining, refining, and manufacturing sites, many of which were in operation before enactment or enforcement of industrial hygiene and environmental codes. There is always the danger that old lessons are forgotten as new discoveries and novel applications dominate the interests of research and industry. Even with the explosion in technology and the use of toxicologically unfamiliar metals and metalloids, ancient metals such as lead, mercury, copper, and iron are still produced and used, and the health and environmental safety routines developed for them must continue to be applied. Moreover, the basic principles of industrial hygiene and environmental safety developed for old metals and metallurgy are applicable to the emerging industries using new techniques and novel applications for a variety of metallic elements and should be followed until guidelines for the safe handling of newer metals and processes are established.

TABLE 3.2   Carcinogenic Potential of Some Common Metals[a]

Metal	Evidence for Carcinogenesis	
	Humans	Animals
Arsenic	Sufficient	Limited
Beryllium	Sufficient	Sufficient
Cadmium	Sufficient	Sufficient
Chromium		
Trivalent	Insufficient	Insufficient
Hexavalent	Sufficient	Sufficient
Cisplatin	Insufficient	Sufficient
Lead		
Inorganic	Insufficient	Sufficient
Mercury		
Inorganic	Inadequate	Limited for chloride
Nickel		
Metallic	Insufficient	Sufficient
Compounds	Sufficient	Sufficient
Selenium	Insufficient	Insufficient

Source: From International Agency for Research on Cancer (1987).
[a]As determined by the International Agency for Research on Cancer.

REFERENCES

Abrams, M. and Murrer, B.A. (1993) Metal compounds in therapy and diagnosis. *Science* 261:725–730.

Angerer, J. and Heinrich, R. (1988) Cobalt. In: Seiler, H.G. and Sigel, H., editors. *Handbook on Toxicity of Inorganic Compounds*, New York, NY: Marcel Dekker.

Beijer, K. and Jernelov, A. (1986) Sources, transport and transformation of metals in the environment. In: Friberg, L., Nordberg, G.F., and Vouk, V.B., editors. *Handbook on the Toxicology of Metals. General Aspects*, 2nd ed., vol. 1, Amsterdam: Elsevier.

Elinder, C.-G., et al. (1994) *Biological Monitoring of Metals*, Geneva: World Health Organization.

Friberg, L., Nordberg, G., and Vouk, V.B., editors (1986) *Handbook on the Toxicology of Metals*, 2nd ed., vol. 2, Amsterdam: Elsevier.

Furst, A. and Haro, R.T. (1969) A survey of metal carcinogenesis. *Prog. Exp. Tumor Res.* 12:102–133.

Goyer, R.A. (1995) Nutrition and metal toxicity. *Am. J. Clin. Nutr.* 61(Suppl.):646S–650S.

International Agency for Research on Cancer (1987) *Monograph on the Evaluation of Carcinogenicity: An Update of IARC Monographs*, vols 2–42, Suppl. 7, Lyon: World Health Organization.

Ng, A. and Patterson, C. (1981) Natural concentrations of lead in ancient Arctic and Antarctic ice, *Geochim. Cosmochim. Acta* 45:2109–2121.

Seiler, H.G. and Sigel, H., editors (1986) *Handbook on Toxicity of Inorganic Compounds*, New York, NY: Marcel Dekker.

Underwood, E.J. (1977) *Trace Elements in Human and Animal Nutrition*, 4th ed., New York, NY: Academic Press.

Venugopal, B. and Luckey, T.D. (1978) *Metal Toxicity in Mammals: Physiological and Chemical Basis for Metal Toxicity*, vol. 1, New York, NY: Plenum Press (see also vol. 2, *Chemical Toxicity of Metals and Metalloids*).

4

ALKALI COMPOUNDS

Steve Morris and Ann Lurati

Target organ(s): Respiratory system, skin, eyes, mucous membranes

Occupational exposure limits: See individual compounds

Reference values: See individual compounds

Risk/safety phrases: Caustics, bases, explosives

BACKGROUND AND USES

Alkali compounds comprise a diverse group of corrosive chemicals with the common characteristic of a pH > 7.0. This group of chemicals comes in a variety ranging from gaseous (anhydrous ammonia) to solids (calcium lime). Alkalis include the hydroxides of sodium, potassium and ammonia; sodium carbonate; calcium and barium oxide; calcium chloride; the sulfides of sodium, calcium and arsenic; and alums. Their uses in the workplace are widespread and can easily reach exposure levels requiring interventions, from the production of explosives to the formation of reagents in clinical chemistry tests to the mixtures of common cleaners in the office workplace. Alkali exposures can also reach actionable levels at atypical occupational locales such as ammonia buildup in animal containment houses.

In general, most alkali usage causes mild to moderate skin and eye irritations. However, some of the more caustic alkalis can be considered significant threats to worker health and safety. Inhalation of concentrated anhydrous ammonia mists may be lethal. Alkalis produce injury to the soft tissue through a process called "liquefaction necrosis" that causes loosening of tissue and allows a deeper penetration of the chemical (Palao et al., 2009). Sodium hydroxide (lye) can cause blindness or death after ingestion or extensive skin and eye contact. Lime (calcium oxide, calcium hydroxide, or similar mixtures with magnesium) can cause severe skin burns and eye injuries. Because of their potential toxicity and widespread use in industry, these three are discussed in further detail later in the section.

PHYSICAL AND CHEMICAL PROPERTIES

Alkalis manifest as all forms of matter, from gases to solids. Their definitive common trait is their elevated pH and their corrosivity, the degree of which is determined by their composition and concentration. They are also water soluble and commonly exist in solution. Alkalis can form salts and anhydrides, which maintain the same chemical properties of the original compound. Because of their reactivity, these compounds should be stored in glass or plastic containers rather than metal or rubber ones (Sullivan and Kreiger, 2001).

ENVIRONMENTAL FATE AND BIOACCUMULATION

Few regulatory environmental assessments have been done on these compounds, most likely due to their ability to be neutralized to innocuous compounds over time. For example, while any discharge of lime risks immediate adverse effects to the stability of the environment, time does reduce its toxicity. Calcium oxide will rapidly convert to calcium hydroxide in the presence of moisture, which then reacts with carbon dioxide to reduce pH and precipitate to calcium carbonate. Thus, in time cement and concrete lose their

Hamilton & Hardy's Industrial Toxicology, Sixth Edition. Edited by Raymond D. Harbison, Marie M. Bourgeois, and Giffe T. Johnson.
© 2015 John Wiley & Sons, Inc. Published 2015 by John Wiley & Sons, Inc.

alkalinity and present little corrosivity risk to the surrounding environment. Of course, spills and other unplanned releases can have a devastating effect on the surrounding flora and fauna until the neutralization process is complete.

Ammonia has a distinct impact in environmental studies, but is difficult to assess by degree since it is also a natural by-product of bacterial decomposition. The alkalis can form salts that are common to environmental locales. Therefore, it is the quantity of these agents in the environment that is a factor.

CHEMICAL PATHOLOGY

Alkali injuries are both heat and chemical generated. Heat results from the exothermic chemical reactions. Typically, lipid emulsification and saponification and protein dissolutions result from the oxidation–reductions of available oxygen species (Sullivan and Kreiger, 2001).

MEDICAL MANAGEMENT

Alkalis generally cause more damage than acids when in contact with body tissues, especially when the compound's pH exceeds 11. Unlike most acids, alkali damage to body tissue is a twofold problem. The heat generated from the alkali hydration from body moisture causes heat damage, while the oxidation of the compound causes chemical damage to the surrounding molecules. Additionally, the mechanism of acid burns tends to result in the formation of eschars, which then serve somewhat of a protective function. Alkalis cause liquefactive necrosis, leaving softened tissues, so no protective layer is formed—leaving the tissue subject to both heat and chemical damage (Sullivan and Kreiger, 2001). This liquefaction of tissue also allows deeper penetration of the injurious substance, compounding the determination of the extent of injury and treatment thereof. In fact, alkali injuries can appear deceptively minor in the preliminary stages of contact (Riffat and Cheng, 2009; Maguina and Busse, 2011).

Alkali burn injury detection is further complicated by the varying nature of each compound and thus different ways of exposure can occur. Gaseous ammonia exposure can be through dermal absorption or through inhalation, while ammonium salts can be ingested. Eye exposures and inhalation are perhaps the most dangerous of all exposures due to the fragile systems involved, but this in no way diminishes the potential lethality of ingestion exposures.

In studies of children and ingestion of caustic substances, little correlation was found between the presence of oral lesions and the degree or even presence of esophageal injury (Riffat and Cheng, 2009; Arici et al., 2012). Thus medical follow-up should be sought in any case of possible alkali exposure, especially ingestion, regardless of whether visible injury is present.

Ocular exposures have additional risks when the alkalis are in particulate form, as the embedded particle continues to cause continuing chemical damage to the sensitive cornea until the particle itself is removed. Excessive ocular exposure to alkali usually destroys the corneal epithelial and stromal cells, often severely damaging or killing the corneal endothelial cells. Repair and regeneration of these tissues are extremely slow or impossible. Alkali injury is often followed by degeneration and sloughing of the tissue, ending finally in perforation of the globe, endophthalmitis, and ultimately the loss of the eye (Records, 1979). The importance of ocular alkali burns thus lies not only in the large number of patients seeking care for eye injuries, but also in the total disability that often results and the difficult management problems created. Uniocular injuries generally result in prolonged time off from work, and with binocular injuries, the worker must usually leave the workforce (Pfister and Koski, 1982).

Alkali damage to the esophagus can result in strictures at best and perforation at worst. Although not held in total consensus, induction of vomiting in the case of alkali ingestion is generally discouraged to prevent further damage and potential rupture of the esophagus (Riffat and Cheng, 2009).

Alkalis have not yet been found to have a clear association as carcinogens or teratogens. Cancers have been found in higher incidence with some workers exposed to alkalis, but there is no definitive study that links it to the alkali itself and not to dysfunctions in damaged tissue repair. Some studies have suggested a link between potash mining and lung cancer. Some have speculated a link between the miners and diesel fumes, but Pahwa and McDuffie (2008) conducted a study of 1435 males and found a correlation of potash mining to a higher incidence of lung cancer without regard to diesel exposure. However, the authors noted that they neither measured radon, nor personal levels of potash or diesel exposure.

In the event of an exposure, proper medical management is imperative. Delay in treatment can have significant negative consequences later on. Maguina and Busse (2011) discuss the case study of a 30-year-old male who had recurrent ulcerations and necrotic tissue patches for 2 years after an industrial sodium hydroxide accident resulted in chemical burns to his forearms. The authors stated that immediate irrigation of an affected area until a tissue pH of 7–7.5 is called for; irrigation is of little use when offered after delays. Effective irrigation for exposure to strong alkalis may take several hours before a normal pH is resumed. Thus "adequate knowledge of first responder treatment of alkali burns is of the utmost importance to their initial treatment and final outcome" (Maguina and Busse, 2011).

AMMONIA

Occupational exposure limits:

OSHA: PEL: 50 ppm, 35 mg/m^3 TWA (General Industry and Maritime)

ACGIH: TLV: 25 ppm, 17 mg/m^3 TWA; 35 ppm, 24 mg/m^3 STEL

NIOSH: REL: 25 ppm, 18 mg/m^3 TWA; 35 ppm, 27 mg/m^3 STEL; IDLH: 300 ppm

Reference values: IRIS: RfC $= 1 \times 10^{-1}$ mg/m^3 (UF $= 30$)

BACKGROUND AND USES

Ammonia (CASRN 7664-41-7) is a common chemical of production, holding a third place in world production (Makarovsky et al., 2008). The large-scale use of ammonia is facilitated by large-scale storage tanks containing millions of gallons of anhydrous ammonia with a network of piping for off-loading to train or truck. Industrial exposures come from its use in fertilizers, explosives, intermediary in chemical productions, metallurgy, paper/pulp manufacturing, and refrigeration (Makarovsky et al., 2008).

PHYSICAL AND CHEMICAL PROPERTIES

At atmospheric pressure, ammonia is a colorless gas with a characteristic pungent odor, which is extremely irritating to the eyes and upper respiratory tract. It is approximately half as dense as air and thus rises readily. With pressurization, 13 cubic feet of ammonia vapor can be condensed into 1 cubic foot.

MAMMALIAN TOXICOLOGY

Acute Effects

The irritating effects of high concentrations of ammonia vapor usually prompt exposed individuals to leave the area before significant pulmonary injury occurs (Summer and Haponik, 1981). With an odor typically described as "pungent" and reminiscent of urine, gaseous ammonia is typically detectable by odor at levels as low as 5 parts per million (ppm). The strong odor was thought to be its own safety alarm for the worker; however, later studies ruled this safety factor out as olfactory fatigue occurs readily and thus odor is not a reliable warning system (Makarovsky et al., 2008). Experimental studies show that short-term exposure to 50–100 ppm causes moderate eye, nose, and throat irritation in most people, but acclimation to these effects can occur after 1–2 weeks. Below 50 ppm no significant effects are

noted, even though the odor of ammonia may be detectable (Swotinsky and Chase, 1990).

Acute inhalation injuries have been seen as two distinct clinical presentations. The first, associated with very high exposure levels, is tracheobronchitis with massive swelling of the upper airways and possible laryngospasm (Montague and Macneil, 1980). This is thought to be a protective response to shield the lower airways. Of the patients who can be revived with endotracheal intubations (tracheostomy if necessary), many survive without any severe pulmonary sequelae (Close et al., 1980). People exposed to a lesser concentration of ammonia may remain conscious and continue to breathe in the mist for a longer period. These patients do no manifest acute respiratory distress; rather, they gradually develop obstructive symptoms or pulmonary edema.

Dermal, ocular, or mucous membrane exposure to ammonia often results in significant airway involvement. Superficial injury to the esophagus also has been reported with ingestion of ammonia. These patients typically respond to conservative therapy and do not have permanent injuries (Sugawa and Lucas, 1989).

Mixtures of household ammonia and bleach (sodium hypochlorite) form chloramines, which are released as an unstable, respirable gas. The gas produced is less soluble than ammonia and is not readily absorbed by the mucous membranes of the upper airways. As a result, it reaches the alveoli, where it hydrolyzes to ammonia and hypochlorous acid. Chloramines exert a direct toxic effect that is described as a toxic pneumonitis (Reisz and Gammon, 1986).

Chronic Effects

Although cases of acute exposure to high levels of ammonia are well documented, one study by Rahman, Bratweit, and Moen (2007) suggested that long-term exposure to ammonia may result in obstructive and restrictive respiratory disease. It is not considered carcinogenic or teratogenic (Makarovksy et al., 2008). In one study a group of workers exposed to airborne ammonia at levels generally below 25 ppm and not above 50 ppm were compared with controls. No differences were found in symptoms, sense of smell, or lung function (Holness et al., 1989).

DIAGNOSIS AND TREATMENT

If the victim with an inhalation injury reports a pungent odor, the clinician should consider ammonia as a causative agent. Obtaining ammonia levels is of little value due to a poor correlation with the victim's clinical presentation (Burgess, Pappas, and Robertson, 1997). There are no biomarkers available to identify ammonia exposures (Center for Disease Control, 2004). Chest X-rays appear to offer little help in assessing the immediate status of lung involvement

(Montague and Macneil, 1980). Initial chest X-rays and early blood gas values also correlate poorly in predicting the future clinical course of patients who suffer acute ammonia inhalation. The presence or absence of abnormal chest findings on physical examination is the best prognostic factor (Arwood et al., 1985). There may be a need to establish baseline spirometry with future monitoring (Rahman, Bratveit, and Moen, 2007).

First aid treatment consists of removing the patient from the source of exposure and basic life support measures. If possible, an attempt should be made to remove the ammonia from the eyes and skin by irrigating with copious amounts of water, and affected clothing should be removed. Ammonia causes eye and skin injuries identical to those of other caustic alkalis (O'Kane, 1983). If ammonia is ingested, do not induce vomiting, and consider consulting for an upper and lower endoscopy to assess the injury (Burgess, Pappas, and Robertson, 1997).

SOURCES OF EXPOSURE

Occupational

Exposures to anhydrous ammonia for fertilizer production generally occur during transfer operations (Greenberg et al., 1985). Ammonia used in refrigeration is typically stored under pressure, which means that leaks will disperse ammonia at greater pressures and larger areas than non-pressurized systems. Contact with the leaking liquid at normal atmospheric pressure brings an addition hazard of frostbite, since its temperature then runs about −28 °F.

Ammonia is an additional risk factor in animal management facilities. A by-product of bacterial decomposition, ammonia (as well as other gases) can be produced by decomposing animal wastes. Employee fatalities result from oxygen depletion and toxic gas exposure from cleaning manure pits in animal facilities (Greenberg et al., 1985).

Amshel reported a 28-year-old male that sustained a 45% total body surface area burn due to an anhydrous ammonia explosion (Amshel et al., 2000). It was concluded that the injured worker had not received any formal training regarding the hazardous nature of the anhydrous ammonia, and the company was unprepared to meet the demands for immediate first aid.

REFERENCES

Amshel, C., Fealk, M., Phillips, B., and Caruso, D. (2000) Anhydrous ammonia burns care reports and review of the literature. *Burns* 25(5):493–497.

Arici, A., Ozdemir, D., Oray, N., Buyukdeligoz, M., Tuncok, Y., and Kalkan, S. (2012) Evaluation of caustic and household detergents exposures in an emergency service. *Hum. Exp. Toxicol.* doi: 10.1177/0960327111412803.

Arwood, R., Hammond, J., and Ward, G.G. (1985) Ammonia inhalation. *Trauma* 25:444–446.

Burgess, J., Pappas, G., and Robertson, W. (1997) Hazardous material incidents: the Washington Poison Center Experience and approach to exposure assessment. *J. Occup. Environ. Med.* 39(8):760–768.

Center for Disease Control (2004) *Ammonia. Agency for Toxic Substances and Disease Registry.* Available at http://www.atsdr.cdc.gov/toxprofiles/tp.asp?id=11&tid=2 (accessed 2014).

Close, D.G., Catlin, F.I., and Cohn, A.M. (1980) Acute and chronic effects of ammonia burns of the respiratory tract. *Arch. Otolaryngol.* 106:151–158.

Greenberg, A.E., Trussell, R.R., and Clesceri, L.S., editors. (1985) *Nitrogen (Ammonia) Standard Methods for the Examination of Water and Wastewater*, Washington, DC: American Public Health Association, American Water Works Association, Water Pollution Control Federation, pp. 374–391.

Holness, D.L., Purdham, J.T., and Nethercott, J.R. (1989) Acute and chronic respiratory effects of occupational exposure to ammonia. *Am. Ind. Hyg. Assoc. J.* 50:646–650.

Maguina, P. and Busse, B. (2011) Local tissue flaps definitive treatment for chronic blisters after chemical burns. *J. Burn Care Res.* 32(4):e140.

Makarovsky, I., Markel, G., Dushnitsky, T., and Eisenkraft, A. (2008) Ammonia-when something smells wrong. *Isr. Med. Assoc. J.* 10(7):537–543.

Montague, T.J. and Macneil, A.R. (1980) Mass ammonia inhalation. *Chest* 77:496–498.

O'Kane, G.J. (1983) Inhalation of ammonia vapour. A report on the management of eight patients during the acute stages. *Anaesthesia* 38:1208–1213.

Pahwa, P. and McDuffie, H. (2008) Cancer among potash workers in Saskatchenwan, J. Occup. Env. Med. 50(9), 1035–1041.

Rahman, M., Bratweit, M., and Moen, B. (2007) Exposure to ammonia and acute respiratory effects in a urea fertilizer factory. *Int. J. Occup. Environ. Health* 13(2):153–159.

Reisz, G.R. and Gammon, R.S. (1986) Toxic pneumonitis from mixing house-hold cleaners. *Chest* 89:49–52.

Reisz, G. and Gammon, R. (1989) Toxic pneumonitis from mixing household cleaners. *Chest* 89(1):49–52.

Riffat, F. and Cheng, A. (2009) Pediatric caustic ingestion: 50 consecutive cases and a review of the literature. *Dis. Esophagus* 22(1):89–94.

Sugawa, C. and Lucas, C.F. (1989) Caustic injury of the upper gastrointestinal tract in adults: a clinical and endoscopic study. *Surgery* 106(4):802–806.

Sullivan, J. and Krieger, G. (2001) *Clinical Environmental Health & Toxic Exposure*, 2nd ed., Philadelphia, PA: Lippincott, Williams, & Wilkins, p. 984.

Summer, W. and Haponik, E. (1981) Inhalation of irritant gases. *Clin. Chest Med.* 2:273–285.

Swotinsky, R.B. and Chase, K.H. (1990) Health effects of exposure to ammonia: scant information. *Am. J. Ind. Med.* 17:515–521.

LIME (CALCIUM OXIDE; CALCIUM HYDROXIDE)

Occupational exposure limits: OELs are for the specific constituents:

Calcium Oxide:

OSHA PEL: 5 mg/m^3.

NIOSH REL: 2 mg/m^3;

NIOSH IDLH: 25 mg/m^3

Calcium Hydroxide:

OSHA PEL: 15 mg/m^3 (total); 5 mg/m^3 (resp)

NIOSH REL: 5 mg/m^3

BACKGROUND AND USES

Lime (calcium oxide CASRN 1305-78-8 and calcium hydroxide CASRN 1305-62-0), produced from the heating of mined limestone, is an important industrial product for metal manufacturing, building and construction, and chemical processing. Almost 20 million tons of lime is produced from 29 states and Puerto Rico, with most of it being used for purification processes in steel manufacturing (EPA, 2011).

"Lime" as a general term refers to calcium oxide, calcium hydroxide, or magnesium–calcium oxides or hydroxides (dolomites). Calcium oxide (also known as quicklime, unslaked lime, or burnt lime) is a white to gray hygroscopic crystalline solid. Calcium hydroxide (also known as slaked lime or caustic lime) is a white odorless powder formed from the addition of water to calcium oxide ("slaking"). The hydration process is quite an exothermic process and creates a strongly alkaline product (pH ≈ 12.4).

Workers in the construction industry often sustain dermal injuries from foundation work with cement or concrete, accounting for 1.8% of burns treated in adults (Sherman and Larkin, 2005). The alkalinity of cement comes from calcium oxide, which constitutes 60–70% of cement. Cement acquires its industrial properties when water is added to "slake" the lime, increasing cement strength as the amount of free water declines through progressive hydration and evaporation. The hydration process increases the alkalinity to a pH typically exceeding 12. However, exposure to air allows the calcium hydroxide to react with ambient carbon dioxide, resulting in an inert calcium carbonate and a more neutral pH.

Reports of occupational injuries from lime or cement burns are as varied as the number of occupations themselves. Ng and Koh (2011) presented a case study of contact dermatitis after a worker dispensed calcium oxide in a cloud seeding procedure for a meteorological operation. The case study presented sound arguments for the use of full-face respirators when working in areas with the potential for lime dust. Tindholdt,

Danielsen, and Åbyholm (2005) described a worker who sustained 3% TBSA (total body surface area) burns from placing newly molded flagstones in the rain. The 90 min of work he performed resulted in multiple medical procedures, an 8-day hospital stay, and a 6-week absence from work.

Sherman and Larkin (2005) discussed the case of a cement worker who sustained cement burns from standing in cement in tennis shoes for roughly 4 h, despite his awareness of the potential burn risks. The worker sustained burns on his lower extremities that required skin graphs and 10 days hospitalization. The authors suggested that despite warnings, these types of burns will most likely reoccur in an industry where workers choose to ignore the need for protective equipment. Rigorous education and required adherence to safety policies are perhaps the best prevention for incidents as these.

MAMMALIAN TOXICOLOGY

Acute Effects

Cain et al. (2004) studied 12 human subjects exposed to selected chemical dusts for 20 min and found that the subjects had a "chemesthetic" response to calcium oxide dust at concentrations of 2–5 mg/m^3.

In a series of reported eye injuries, lime was the agent most responsible for burns. The lime in cement reacts avidly to form calcium hydroxide solution, a concentrated solution of which can have a pH of 12.4–12.7 (Benmeir et al., 1993). Lime also differs from the previously discussed agents because, since it is a solid, its particles tend to adhere and continue to damage the area until they are removed.

Cement can damage the skin in several ways. It can act as a primary irritant that causes dermatitis by means of the alkalinity of the material or by the gritty particles, or both, and it also can be a sensitizer that causes a contact allergic dermatitis in workers who become sensitized to the chromium present in many cements (mostly from Europe). Cement can cause burns similar to those caused by other alkalis with two major differences: the cement particles can adhere to the skin, and the injury can go unnoticed for some time (Vickers and Edwards, 1976). The lower legs area is most commonly injured because workmen kneel in wet cement or it splashes inside their work boot. Benmeir et al. (1993) reported on two soccer players who suffered burns to the groin and upper legs when lime powder that was used to line the field penetrated their clothing.

However, the most innocuous of uses can provide a source for injurious exposures to lime. Neubauer et al. (2000) described a rare case of pulmonary injury sustained from a diver's exposure to soda lime (a mixture of calcium hydroxide, sodium hydroxide and potassium hydroxide.) through a failure in his closed circuit rebreather.

Chronic Effects

Cement particles may be a source of occupational asthma as well as other obstructive and restrictive lung diseases (Bauer et al., 2013). It is noted that the use of dust exposure control methods prevents a decrease in lung function (Tungu et al., 2014).

INDUSTRIAL HYGIENE

With the strong alkalinity and hygroscropic nature of lime products, care must be exercised to avoid exposures, especially to moist areas such as eyes and airways. At a minimum, protective clothing and eyewear should be worn and eyewashes and quick drench facilities should be readily available to workers. The National Institute of Occupational Safety and Health (NIOSH) recommendations for respirators should be consulted for those working with calcium oxide.

Spills of solid lime must be handled with care. Unslaked lime, for example, reacts violently with water and so the dry product should be removed as completely as possible before any washing of the area commences.

REFERENCES

Bauer, P., Schiavo, D., Osborn, T., Levin, D., Sauver, J., Hanson, A., Schroeder, D., and Ryu, J. (2013) Influence of interstitial lung disease on outcome in systemic sclerosis. *Chest* 144(2):571–577.

Benmeir, P., et al. (1993) Chemical burn due to contact with soda lime on the playground: a potential hazard for football players. *Burns* 19:358–359.

Cain, W., Jalowayski, A., Kleinman, M., Lee, N., Lee, B., Ahn, B., Magruder, K., Schmidt, R., Hillen, B., Warren, C., and Culver, D. (2004) Sensory and associated reactions to mineral dusts: sodium borate, calcium oxide, and calcium sulfate. *J. Occup. Environ. Hyg.* 1(4):222–236.

Center for Disease Control (2011) *Sodium Hydroxide.* Agency for Toxic Substances & Disease Registry. Available at http://www.atsdr.cdc.gov/substances/toxsubstance.asp?toxid=45 (accessed 2014).

Environmental Protection Agency (EPA) (2011) *Economic Impact Analysis for the Lime Manufacturing MACT Standard.* Available at http://nepis.epa.gov/Exe/ZyNET.exe/P1006JTA.TXT? (accessed 2014).

Neubauer, B., Mutzbauer, T., and Tetzlaff, K. (2000) Exposure to soda-lime dust in closed and semi-closed diving apparatus. *Aviat. Space Environ. Med.* 71(2):1248–1251.

Ng, W. and Koh, D. (2011) Occupational contact dermatitis in manual cloud seeding operations. *Singapore Med. J.* 52(5):e85–87.

Sherman, S. and Larkin, K. (2005) Cement burns. *J. Emerg. Med.* 29(1):97–99.

Tindholdt, T., Danielsen, T., and Åbyholm, F. (2005) Skin burn by contact with flagstones made of cement. *Scand. J. Plast. Reconstr. Surg. Hand. Surg.* 39:373–375.

Vickers, H. and Edwards, D. (1976) Cement burns. *Contact Dermatitis* 2(2):73–78.

SODIUM HYDROXIDE

Occupational exposure limits:

NIOSH REL: (ceiling) $2\,mg/m^3$; IDHL: $10\,mg/m^3$

OSHA PEL: $2\,mg/m^3$ TWA

ACGIH TLV: (ceiling) $2\,mg/m^3$

BACKGROUND AND USES

Sodium hydroxide (NaOH; CASRN 1310-73-2) is commonly available as an aqueous solution known as caustic soda, soda lye, or simple as lye. It has various uses, including neutralization of acid; the manufacture of paper, textiles, plastics, corrosives, dyestuffs, paint, paint remover, and soap; refining of petroleum; electroplating; metal cleaning; laundering; and dish washing. A burgeoning use has been in the illegal manufacture of methamphetamine.

PHYSICAL AND CHEMICAL PROPERTIES

Sodium hydroxide is a white, odorless, nonvolatile alkaline material marketed in solid form as pellets, flakes, clumps, or sticks. Its solubility in water is 111% by weight and a vapor pressure of 0 mmHg (NIOSH, 1994).

It can react with tricholoethylene (TCE) to form flammable dichloroacetylene and with metals to form hydrogen gas (OEHHA, 1993). Its reactivity with metals should be considered in regards to storage units and containers.

TOXICOLOGY

Acute Effects

Sodium hydroxide dusts generally are without odor; therefore, the corrosivity of the dust reacting with moist tissue can be the only preliminary sign of exposure. Concentrations as low as $0.5\,mg/m^3$ can elicit mild to moderate feelings of respiratory irritation (OEHHA, 1993).

A total of 25–50% solutions can cause immediate and overt damage to the skin, while several hours may pass before the damage from exposures to 4% or lower solutions manifests (OEHHA, 1993).

It has been shown that 4% lye solution has an initial anesthetic effect on intact skin and that hours can elapse before a sense of irritation occurs; by comparison, with a 25% solution, the sense of irritation occurs in 3 min. Because of this initial insensitivity, medical attention may not be sought until sometime after the thermal burn occurs and progressive skin destruction has occurred (Harper and Dickson, 1994).

It is recommended that the eyes be flushed for at least 15 min with copious amounts of water (Wax and Young, 2013). Due to the excessive amount of hydrochloric acid that is produced, a patient with a history of gastroesophageal reflux may need treatment for this condition (Messner et al., 1996).

Chronic Effects

Sodium hydroxide has not been determined to be carcinogenic. While some studies have associated esophageal cancer with damage from sodium hydroxide exposure, the general belief is that the cancer arises from mutations in the repairing tissue rather than exposure to the sodium hydroxide itself.

Chemical Pathology

Winemaker et al. (1992) found that irreversible damage to the skin by sodium hydroxide is proportional to the concentration when above 1% and pH when above 11.5.

INDUSTRIAL HYGIENE

The ACGIH and the NIOSH have set the TLV and REL, respectively, with a ceiling of $2\,mg/m^3$. OSHA has set the PEL for sodium hydroxide at $2\,mg/m^3$ for an 8-h, time-weighted average (TWA). Respiratory protective equipment should be worn when working with NaOH mists.

REFERENCES

Harper, R.D. and Dickson, W.A. (1994) Mr Muscle oven cleaner: is he too strong for us? *Burns* 20:336–339.

Messner, A., Browne, J., and Geisinger, K. (1996) Effects of intermittent acid and pepsin exposure on burned esophageal mucosa. *Am. J. Otolaryngol.* 17(10):45–49.

Milner, J.E. (1982) The office treatment of minor chemical skin burns. *Cutis* 29:285–286.

Montague, T.J. and Macneil, A.R. (1980) Mass ammonia inhalation. *Chest* 77:496–498.

National Institute for Occupational Safety and Health (NIOSH) (1994) *Pocket Guide to Chemical Hazards*, NIOSH 94-116, Washington, DC: US Department of Health and Human Sciences.

Office of Environmental Health Hazard Assessment (OEHHA) (1993) *Sodium Hydroxide*. Available at http://oehha.ca.gov/air/acute_rels/pdf/1310932A.pdf (accessed 2014).

Palao, R., Monge, I., Ruiz, J., and Barret, P. (2009) Chemical burns: pathophysiology and treatment. *Burns*, doi: 10.1016/j.burns.2009.07.009.

Pfister, R.R. and Koski, J. (1982) Alkali burns of the eye: pathophysiology and treatment. *South. Med. J.* 75:417–422.

Records, R.E. (1979) Primary care of ocular emergencies. II. Thermal chemical and nontraumatic eye injuries. *Postgrad. Med.* 65(5):157–163.

Swotinsky, R.B. and Chase, K.H. (1990) Health effects of exposure to ammonia: scant information. *Am. J. Ind. Med.* 17:515–521.

Tungu, A., Bratveit, M., Mamuya, S., and Moen, B. (2014) The impact of reduced dust exposure on respiratory health among cement workers: an ecological study. *J. Occup. Environ. Med.* 56(1):101–110.

Wax, P.M. and Young, A. (2013) Caustics. In: Marx, J.A., editor. *Rosen's Emergency Medicine: Concepts and Clinical Practice*, 8th ed., Chapter 153, Philadelphia, PA: Saunders Elsevier.

Winemaker, M. Douglas, L., and Peters, W. (1992) Combination alkali/thermal burns caused by "black liquor" in the pulp and paper industry. *Burns* 18:68–70.

5

ALUMINUM

Giffe T. Johnson

Target organ(s): Dermis, lungs, nervous system

Occupational exposure limits: TWA: 5 mg Al/m^3 from ACGIH (TLV) [United States] inhalation (pyro powders, welding fumes); TWA: 10 mg Al/m^3 from ACGIH (TLV) [United States] inhalation (metal dust)

Reference values: RfD/RfC: N/A (USEPA); MRL: 1 mg Al/kg/day (ATSDR)

Risk/safety phrases: Combustible dust

BACKGROUND AND USES

Occurring naturally in the soil, aluminum makes up 8% of the earth's surface, although higher concentrations may be found in the soil surrounding associated industries. Aluminum is used to make items such as antiperspirants, antacids, cooking utensils, soda cans, aluminum foil, and construction materials.

Transportation accounts for an estimated 40% of aluminum usage in the United States, mostly for automotive applications, with the remaining amounts consumed by packaging (28%), building (13%), consumer durables (7%), electrical (5%), and other applications (7%) (USGS, 2007).

Various aluminum compounds are also found in a diverse number of industrial processes and consumer products. Aluminum chloride, anhydrous form, is used as an acid catalyst to provide a chemical intermediate for other aluminum compounds, in the cracking of petroleum, in the manufacture of rubbers and lubricants, and as an antiperspirant. The hexahydrate form is used as a wood preservative, for disinfecting stables and slaughterhouses, in deodorants and antiperspirants, in cosmetics (such as topical astringents), in refining crude oil, dyeing fabrics, and manufacturing parchment paper (O'Neil et al., 2001). Aluminum chlorohydrate is an ingredient in commercial antiperspirant and deodorant preparations and is also used for water purification and treatment of sewage and plant effluent (Lewis, 2001).

PHYSICAL AND CHEMICAL PROPERTIES

Aluminum (also aluminum metal, aluminum powder, or elemental aluminum) is a silvery-white, odorless metal that is malleable and ductile; its Chemical Abstract Service (CAS) number is 7429-90-5. Aluminum has a vapor pressure of approximately 0 mmHg and is insoluble in water. It is a combustible solid, and as a finely divided dust, it ignites easily and can cause explosions.

MAMMALIAN TOXICOLOGY

Acute Effects

Aluminum may cause death (as observed in laboratory animals) only from exposures that are extremely high compared to what a human might normally encounter. The LD$_{50}$ for aluminum bromide in Sprague-Dawley rats and Swiss Webster mice have been reported as 162 and 164 mg Al/kg, respectively (Llobet et al., 1987). The chloride form has produced LD$_{50}$ values of 370, 222, and 770 mg Al/kg in Sprague-Dawley rats, Swiss Webster mice, and male Dobra Voda mice, respectively (Llobet et al., 1987; Ondreicka et al., 1966). The LD$_{50}$ for aluminum sulfate in male Dobra Voda mice has been reported as 980 mg Al/kg (Ondreicka et al., 1966).

Acute toxicity in humans associated with aluminum is rare, although aluminum may cause local contact dermatitis. Vaccines absorbed to aluminum have been used for many

Hamilton & Hardy's Industrial Toxicology, Sixth Edition. Edited by Raymond D. Harbison, Marie M. Bourgeois, and Giffe T. Johnson.

years for immunization (Böhler-Sommeregger and Linde-mayr, 1986). This procedure is thought to prolong the absorption process and enhance the immunological response. Böhler-Sommeregger and Lindemayr report a case of a 3-year-old boy who received routine triple vaccinations. The mother reported a strong local skin reaction with intense redness after each injection. The development of nodular or granulomatous reactions, or both, after injection with an aluminum adjuvant is not uncommon. However, in this child the effects persisted until age 10. Fawcett et al. (1985) also reported such injection-site granulomas, but those reactions did not persist beyond a few weeks.

Fischer and Rystedt (1982) reported other cases of contact dermatitis. Positive patch tests were found with aluminum chloride and aluminum hydroxide. Cases of contact derma-titis, as well as positive skin reactions, arising from exposure to aluminum filings, have been reported among aircraft workers. Meding et al. (1984) described a similar contact allergy, but they point out that the condition is rare consid-ering the common exposure to aluminum.

Chronic Effects

Neurotoxicity has been linked to exposure to aluminum in both humans and laboratory animals. Aluminum has been docu-mented to accumulate in the central nervous system (brain nuclei and spinal cord) when bound to citrate and administered to laboratory mice (Oteiza et al., 1993). Accumulated tissue concentrations were shown to be dose dependent. Banks and Kastin (1983) also demonstrated that aluminum increases the permeability of the blood–brain barrier to labeled N-Tyr-delta sleep-inducing peptide (DSIP) and beta endorphin. Laboratory rats injected intraperitoneally with aluminum demonstrated this effect. Analysis of variance showed a statistically signifi-cant effect resulting from this treatment. These authors sug-gested that the increase in permeability to the test compounds produced by aluminum may provide a theoretical mechanism by, which changes in brain biochemistry that lead to neuro-toxicity may occur.

Some epidemiological studies have supported the finding of an association between aluminum and neurotoxicity though a causal relationship has not been clearly established. The results of a cross-sectional study undertaken at an aluminum plant in Norway suggested an increased risk of visuospatial organization and a tendency toward decline in psychomotor tempo as a result of long-term, low dose exposure to aluminum (Bast-Pettersen et al., 1994). As well, it has been hypothesized that increased aluminum levels in the brain and the incidence of Alzheimer's disease may be linked. Crapper et al. (1973) were the first to identify increased levels of aluminum in subanatomical regions of the brain in patients who had suffered from Alzheimer's disease. However, this hypothesis has been subsequently refuted by an abundance of publications. It is difficult for

any research group to have access to a sufficient number of patients with Alzheimer's disease and proper controls. Also, a significant amount of the controversy can be attributed to the detection methodologies used to evaluate neurofibrillary tangles and senile plaques, which are the hallmarks of Alzheimer's disease (Savory et al., 1996). Two case-control studies examining occupational exposure to aluminum dust or fumes and the risk of Alzheimer's disease found no significant association (Graves et al., 1998; Salib and Hillier, 1996). A study of former aluminum dust-exposed workers (retired for at least 10 years) found some impairment in some tests of cognitive function; however, the data did not link the exposure to the diagnosis of Alzheimer's disease (Polizzi et al., 2002).

Osteomalacic dialysis osteodystrophy was identified 20 years ago by Ward et al. (1978). It was recognized that many patients suffered from this disease in areas with tap water that had a high concentration of aluminum. Subse-quently, the medical records of patients who had been dialyzed with softened or untreated water were compared with those of patients who had been dialyzed with deionized water. The findings from this comparison supported the original observation. Later, more quantitative studies were undertaken using bone biopsies from subjects known to have been exposed to contaminated drinking water in, which a high level of aluminum was one of the contaminants (Eastwood et al., 1990). These subjects were exposed to aluminum concentrations ranging from 30 to 620 mg/l, although regulatory drinking water standards in the geo-graphical area were established at 0.2 mg/l. The bone biop-sies revealed a period of high gastrointestinal absorption, probably because of the high acidic conditions of the con-taminated drinking water. Bone biopsies in hemodialysis patients from other studies have revealed almost pure osteo-malacia (Gratwick et al., 1978; De Vernejoul et al., 1982). These experiments support some of Ward's original obser-vations of 20 years ago.

In animal models, repeated exposure to high levels of aluminum has been suggested to adversely affect the hemato-logical system of rats and mice. Significant decreases in hemoglobin, hematocrit, and/or erythrocyte osmotic fragility were observed in rats exposed to 420 mg Al/kg/day as alumi-num citrate in drinking water for 15 weeks, mice exposed to 13 mg Al/kg as aluminum citrate administered via gavage 5 days/week for 22 weeks, rats exposed to 230 mg Al/kg/day as aluminum citrate in drinking water for 8 months, and rats exposed via drinking water to 54.7 mg Al/kg/day as aluminum sulfate in a sodium citrate solution for 18 months (Garbossa et al., 1996; Farina et al., 2005; Vittori et al., 1999).

Mechanism of Action(s)

A predominant mechanism for the uptake of aluminum has yet to be described (Exley et al., 1996). The primary route of uptake is believed to be via the small intestine. It serves as a formidable barrier; however, only about 1% of an oral dose of aluminum is absorbed into the body. The absorption of aluminum depends on the intraluminal pH (McKinney, 1993). In addition, certain cofactors increase and limit the absorption of aluminum. Aluminum complexes with oxygen donor ligands, especially phosphates, allowing the formation of compounds such as insoluble aluminum phosphate, which reduces free Al^{3+} and facilitates elimination of aluminum from the body. Citrate, on the other hand, actually is known to increase the bioavailability of aluminum by forming the neutral compound aluminum citrate, which can readily pass through membranes (Oteiza et al., 1993). Aluminum would also be expected to bind with other oxygen donor ligands such as phosphate groups, nucleotides, and polynucleotides. The relationships between aluminum and donor ligands may be especially significant because some of these donors (phosphate and citrate) are known to activate brain hexokinase enzymes (Martin, 1986).

Inhalation of aluminum dust may result in some absorption across lung epithelia and also passage to the gut. Inhalation of the dust also may cause deposition of aluminum in the lungs, and these mucociliary pathways may be the reigning mechanism by, which aluminum becomes systemic (Exley et al., 1996).

The mechanisms for neurotoxic effects have not been elucidated. Strong et al. (1996) have suggested that because aluminum is transported to the brain primarily by means of transferrin, toxicity may occur through the transferrin-mediated receptor system. The aluminum complexation with nucleoside phosphates (adenosine triphosphate [ATP] and adenosine diphosphate [ADP]) and binding proteins has also been examined. The coordination chemistry of Al^{3+} demonstrates a preference for binding to negatively charged oxygen atoms. Because of this characteristic, the phosphate and oxygen atoms of ATP may act as a dominant binder of Al^{3+} in some cells (Martin, 1994). This is of great interest to human biochemistry as we require ATP for a variety of cellular metabolic activities. In addition, many nucleoside phosphate reactions involve Mg^{2+}, and Al^{3+} is an effective surrogate for Mg^{2+} (Nelson, 1996).

Chemical Pathology

No specific target organ or mechanism has been conclusively described for aluminum toxicity (Shi et al., 1993). It seems that aluminum has varying effects on different organs associated with different exposure routes. Studies suggest that aluminum may be associated with many neurodegenerative disorders, including reduced cognitive function, impaired psychomotor function, and dialysis encephalopathy (Bast-Pettersen et al., 1994). Altmann et al. (1989) suggested that reduction of psychomotor function occurs as a result of inhibition of the brain enzyme dihydropteridine reductase.

McDermott et al. (1978) found increased concentrations of aluminum in the brain in seven patients suffering from dialysis encephalopathy. These results suggest a correlation between increased levels of aluminum and untreated or softened dialysis water.

Neurofibrillary degeneration has been found in cats and rabbits similar to that observed in patients with Alzheimer's disease (McDermott et al., 1978). Banks and Kastin (1983) found that when rats were injected with aluminum chloride ($AlCl_3$), the aluminum increased the permeability of the blood–brain barrier to small peptides. However, specific dose-dependent pathways to disease remain unclear.

SOURCES OF EXPOSURE

Occupational

Occupational exposure to aluminum occurs during the refining of the primary metal and industries that use aluminum products, including the aircraft, automotive, and metalwork industries. Exposure to aluminum may also occur from aluminum welding (Nieboer et al., 1995).

In the aluminum refining process, exposure is primarily to aluminum hydroxides and aluminum oxides from the initial extraction and purification process. Aluminum oxide and aluminum fluoride exposure may occur in the potroom, an area that also produces tar-pitch volatiles such as PAHs, which are important confounding exposures when examining the potential health effects from exposure of aluminum (Drablos et al., 1992; IARC, 1984; Nieboer et al., 1995). Most studies of occupational exposure to aluminum have dealt with inhalation of aluminum-containing dust particles, in combination with other dust constituents. Rarely is a worker exposed solely to aluminum-containing dust; exposure to mixtures of aluminum with fine respirable particles or other potentially toxic chemicals is more prevalent.

A great variety of industries employ workers potentially exposed to aluminum and aluminum compounds, including the plumbing, heating, and air conditioning industry; the masonry and stonework industry; electrical workers; machinists; metal and wire production; general medical and surgical hospitals; industrial building construction; and industries that use special dyes, tools, and fixtures (NIOSH, 1991).

Environmental

Aluminum is the most abundant metal in the earth's crust and the third most abundant element making up about 8.8% by weight. It is found in most rocks in the form of aluminosilicate minerals (Lide, 2005; Staley and Haupin, 1992). Aluminum enters the environment naturally through the weathering of rocks and minerals, and through anthropogenic releases from air emissions, waste water effluents, and the solid waste generated from aluminum refinement and production. The aluminum released through natural weathering processes far exceeds the contribution of releases to air, water, and land associated with human activities (Lantzy and MacKenzie, 1979).

The use of alum (aluminum sulfate) as a flocculent in water treatment facilities is a potential source of aluminum in treated water (Letterman and Driscoll, 1988; Miller et al., 1984). However, the concentration of aluminum in surface and groundwater is modulated by the acidity of the water. Higher concentrations of aluminum are typically limited to water with a pH < 5; as a result, aluminum concentrations in most surface water are very low.

Chronic exposure to aluminum is found primarily in patients undergoing dialysis treatment with softened or untreated water. In a study by McDermott et al. (1978), 7 of 19 patients receiving dialysis treatment experienced dialysis-associated encephalopathy. In all seven patients, the aluminum concentration in the brain was significantly higher than in other patients who were uremic but not dialyzed. Abreo et al. (1990) have demonstrated exposure to aluminum by means of inadequately treated domestic water sources containing high concentrations of the metal.

Another source of chronic exposure to aluminum is aluminum-containing phosphate binders. Randall (1983) reported a case of an 11-month-old girl who was not undergoing dialysis but who received aluminum-containing phosphate binders. The child subsequently died of a progressive encephalopathy, and increased levels of aluminum were found in the serum and cerebrospinal fluid. It is now well established that iatrogenic exposure to aluminum, such as through hemodialysis or administration of phosphate binders, may cause encephalopathy, bone disease, and anemia (Flaten et al., 1996).

INDUSTRIAL HYGIENE

Aluminum has been used in the industry for more than 100 years. Electrolytic reduction of aluminum was developed in 1886 by Hall and by Hdroult in France. Aluminum has many uses, either alone or alloyed with other compounds such as copper, magnesium, manganese, silicon, and zinc. These alloys are used in industries such as construction materials, communications, laboratory and diagnostic studies, paper, printing, glass, water purification, and textiles.

Industrial health hazards are largely associated with electrolytic reduction of aluminum oxide, also known as alumina, aluminum trioxide, or bauxite. Health hazards may exist as a result of exposure to condensed pitch volatiles (tars), which are produced during the reduction process by burning the anode in the electrolytic cell (Gibbs, 1985). Gibbs conducted a cohort study on aluminum reduction plant workers between 1950 and 1977. The study found a slight increase in mortality

from respiratory disease among some of the men occupationally exposed to tars during this time period. However, no adjustment was made for lifestyle factors, such as smoking or for other personal traits that might contribute to respiratory disease.

Many early reports describe inhalation of aluminum as a dust to be hazardous to workers. German literature from World War II describes the cases of workers who had inhaled aluminum and who subsequently exhibited X-ray changes much like those seen with silicosis. The workers complained of dyspnea and the development of spontaneous pneumothorax.

These findings are compatible with those of Shaver and Riddell (1947), who described the first identified cases of what has since been called Shaver's disease. This disease is characterized by dyspnea, coughing, substernal pain, weakness, and fatigue. X-ray films show bilateral lacelike shadowing, more frequently in the upper halves of the lungs and with increasing intensity toward the lung roots, and varying degrees of pneumothorax. Outstanding features at autopsy are advanced emphysema and interstitial pulmonary fibrosis.

Currently, the National Institute of Occupational Safety and Health (NIOSH) classifies aluminum compounds as nuisance dusts and sets exposure limits at $10\,mg/m^3$ total and respirable dusts, and $5\,mg/m^3$ respirable dusts. The NIOSH also sets exposure limits for aluminum pyropowders and welding fumes at $5\,mg/m^3$ and for soluble salts and alkyls at $2\,mg/m^3$. Atomic absorption spectrometry can be used to monitor aluminum levels (NIOSH, 1994).

MEDICAL MANAGEMENT

Two important risk groups for aluminum toxicity exist for, which emergency management may be required: those with renal insufficiency and those with excessive and prolonged aluminum exposure, which produce clinical symptoms, including altered mental status, anemia, or osteoporosis.

Treatment of aluminum toxicity includes elimination of aluminum from the diet, total parenteral nutrition, elimination of aluminum from dialysate, medications or antiperspirants, and elimination of any occupational exposure that may be present. An attempt at elimination and chelation of the element from the body's stores may be prescribed depending on the severity of the exposure.

Elimination by chelation is accomplished through the administration of deferoxamine. Patients with serum levels >20 mcg Al/dl may require chelation therapy though symptomatic patients with serum levels >50–60 mcg Al/dl generally indicate that chelation therapy is necessary. Deferoxamine therapy typically ranges from 2.5 to 5 mg/kg/wk (Kan et al., 2010). Chelation therapy is typically conducted in consultation with a nephrologist and a medical toxicologist.

REFERENCES

Abreo, K., Sella, M., and Brown, S. (1990) Aluminum in domestic water. *J. Environ. Health* 53:280–290.

Altmann, P., Dhanesha, U., Hamon, C., Cunningham, J., Blair, J., and Marsh, F. (1989) Disturbance of cerebral function by aluminum in haemodialyxis patients without overt aluminum toxicity. *Lancet* 2:7–11.

Banks, W.A. and Kastin, A.J. (1983) Aluminum increases permeability of the blood-brain barrier to labelled DSIP and gendorphin: possible implications for senile and dialysis dementia. *Lancet* 2:1227–1229.

Bast-Pettersen, R., Drabløs, P.A., Goffeng, L.O., Thomassen, Y., and Torres, C.G. (1994) Neuropsychological deficit among elderly workers in aluminum production. *Am. J. Ind. Med.* 25:649–662.

Böhler-Sommeregger, K. and Lindemayr, H. (1986) Contact sensitivity to aluminum. *Contact Dermatitis* 15:278–281.

Crapper, D.R., Krishnan, S.S., and Dalton, A.J. (1973) Brain aluminum distribution in Alzheimer's disease and experimental neurofibrillary degeneration. *Science* 180:511–513.

De Vernejoul, M.C., Girot, R., Gueris, J., et al. (1982) Calcium phosphate metabolism and bone disease in patients with homozygous thalassemia. *J. Clin. Endocrinol. Metab.* 54:276–281.

Drablos, P.A., Hetland, S., Schmidt, F., et al. (1992) Uptake and excretion of aluminum in workers exposed to aluminum fluoride and aluminum oxide. In: *Proceedings of the Second International Conference on Aluminum and Health, Tampa, FL, February 2–6, 1992*, New York, NY: Aluminum Association, pp. 157–160.

Eastwood, J.B., Levin, G.E., Pazianas, M., Taylor, A.P., Denton, J., and Freemont, A.J. (1990) Aluminum deposition in bone after contamination of drinking water supply. *Lancet* 336:462–464.

Exley, C., Burgess, E., Day, J.P., Jeffery, E.H., Melethil, S., and Yokel, R.A. (1996) Aluminum toxicokinetics. *J. Toxicol. Environ. Health* 48:569–584.

Farina, M., Rotta, L.N., Soares, F.A., Jardim, F., Jacques, R., Souza, D.O., and Rocha, J.B. (2005) Hematological changes in rats chronically exposed to oral aluminum. *Toxicology* 209:29–37.

Fawcett, H.A., McGibbon, D., and Cronin, E. (1985) Persistent vaccination granuloma due to aluminum hypersensitivity. *J. Occup. Environ. Med.* 113:101–102.

Fischer, T. and Rystedt, I. (1982) A case of contact sensitivity to aluminum. *Contact Dermatitis* 8:343.

Flaten, T.P., et al. (1996) Status and future concerns of clinical and environmental aluminum toxicology. *J. Toxicol. Environ. Health* 48:527–541.

Garbossa, G., Gutnisky, A., and Nesse, A. (1996) Depressed erythroid progenitor cell activity in aluminum overloaded mice. *Miner. Electrolyte Metab.* 22:214–218.

Gibbs, G.W. (1985) Mortality of aluminum reduction plant workers, 1950 through 1977. *J. Occup. Med.* 27:761–770.

Gratwick, G.M., Bullough, P.G., Bohne, W.H.O., Markenson, A.L., and Peterson, C.M. (1978) Thalassemic osleoarthropathy. *Ann. Intern. Med.* 88:494–501.

Graves, A.B., Rosner, D., Echeverria, D., Mortimer, J.A., and Larson, E.B. (1998) Occupational exposures to solvents and aluminum and estimated risk of Alzheimer's disease. *Occup. Environ. Med.* 55:627–633.

IARC (1984) *Polynuclear Aromatic Compounds. Part 3. Industrial Exposures in Aluminum Production, Coal Gasification, Coke Production, and Iron and Steel Founding*, vol. 34, Lyon, France: World Health Organization, International Agency for Research on Cancer, pp. 37–64.

Kan, W.C., Chien, C.C., Wu, C.C., Su, S.B., Hwang, J.C., and Wang, H.Y. (2010) Comparison of low-dose deferoxamine versus standard-dose deferoxamine for treatment of aluminium overload among haemodialysis patients. *Nephrol. Dial. Transplant.* 25:1604–8.

Lantzy, R.J. and MacKenzie, F.T. (1979) Atmospheric trace metals: global cycles and assessment of man's impact. *Geochim. Cosmochim. Acta* 43:511–525.

Letterman, R.D. and Driscoll, C.T. (1988) Survey of residual aluminum in filtered water. *J. Am. Water Works Assoc.* 80:154–158.

Lewis, R.J., editor (2001) *Hawley's Condensed Chemical Dictionary*, New York, NY: John Wiley & Sons, Inc., pp. 39–46, 118, 555.

Lide, D.R., editor (2005) *CRC Handbook of Chemistry and Physics*, New York, NY: CRC Press, pp. 4–3 to 4–4, 4–44 to 4–46, 4–79.

Llobet, J.M., Domingo, J.L., Gómez, M., Tomás, J.M., and Corbella, J. (1987) Acute toxicity studies of aluminum compounds: antidotal efficacy of several chelating agents. *Pharmacol. Toxicol.* 60:280–283.

Martin, R.B. (1986) The chemistry of aluminum as related to biology and medicine. *Clin. Chem.* 32:1797–806.

Martin, R.B. (1994) Aluminum: a neurotoxic product of acid rain. *Acc. Chem. Res.* 27:204–210.

McDermott, J.R., Smith, A.I., Ward, M.K., Parkinson, I.S., and Kerr, D.N. (1978) Brain aluminum concentration in dialysis encephalopathy. *Lancet* 1:901–903.

McKinney, J. (1993) Metals bioavailability workshop and disposition kinetics research needs workshop. *Toxicol. Environ. Chem.* 38:1–71.

Meding, B., Aususlxon, A., and Hanson, C. (1984) Patch test reactions to aluminum. *Contact Dermatitis* 10:107.

Miller, R.G., Kopfler, F.C., Kelty, K.C., Stolber, J.A., and Ulmer, N.S. (1984) The occurrence of aluminum in drinking water. *J. Am. Water Works Assoc.* 76:84–91.

Nelson, D.J. (1996) Aluminum complexation with nucleoside di- and triphosphates and implication in nucleoside binding proteins. *Coord. Chem. Rev.* 149:95–1.

Nieboer, E., Gibson, B.L., Oxman, A.D., and Kramer, J.R. (1995) Health effects of aluminum: a critical review with emphasis on aluminum in drinking water. *Environ. Rev.* 3:29–81.

NIOSH (1991) *National Occupational Exposure Survey Matrix*, Cincinnati, OH: Department of Health and Human Services, National Institute for Occupational Safety and Health.

NIOSH (1994) *National Institute of Occupational Safety and Health. Pocket Guide to Chemical Hazards*, NIOSH 94–116,

Washington, DC: US Department of Health and Human Services.

Ondreicka, R., Ginter, E., and Kortus, J. (1966) Chronic toxicity of aluminum in rats and mice and its effects on phosphorus metabolism. *Br. J. Ind. Med.* 23:305–312.

O'Neil, M.J., Smith, A., Heckelman, P.E., et al. (2001) Aluminum and aluminum compounds. In: *The Merck Index. An Encyclopedia of Chemicals, Drugs, and Biologicals*, Whitehouse Station, NJ: Merck & Co., Inc., pp. 59–65.

Oteiza, P.I., Keen, C.L., Han, B., and Golub, M.S. (1993) Aluminum accumulation and neurotoxicity in Swiss-Webster mice after long-term dietary exposure to aluminum and citrate. *Metabolism* 42:1296–1300.

Polizzi, S., Pira, E., Ferrara, M., Bugiani, M., Papaleo, A., Albera, R., and Palmi, S. (2002) Neurotoxic effects of aluminum among foundry workers and Alzheimer's disease. *Neurotoxicology* 23:761–774.

Randall, M.E. (1983) Aluminum toxicity in an infant not on dialysis. *Lancet* 1:1327–1328.

Salib, E. and Hillier, V. (1996) A case-control study of Alzheimer's disease and aluminum occupation. *Br. J. Psychiatry* 168:244–249.

Savory, J., Exley, C., Forbes, W.F., Huang, Y., Joshi, J.G., Kruck, T., McLachlan, D.R., and Wakayama, I. (1996) Can the controversy of the role of aluminum in Alzheimer's disease be resolved? *J. Toxicol. Environ. Health* 48:615–635.

Shaver, C.G. and Riddell, A.R. (1947) Lung changes associated with the manufacture of alumina abrasives *J. Ind. Hyg. Toxicol.* 29:145–157.

Shi, B., Chou, K., and Haug, A. (1993) Aluminum impacts elements of the phosphoinositide signalling pathway in neuroblastoma cells. *Mol. Cell. Biochem.* 12:1109–1118.

Staley, J.T. and Haupin, W. (1992) Aluminum and aluminum alloys. In: Kroschwitz, J.I. and Howe-Grant, M., editors. *Kirk-Othmer Encyclopedia of Chemical Technology. Alkanolamines to Antibiotics (Glycopeptides)*, vol. 2, New York, NY: John Wiley & Sons, Inc., pp. 248–249.

Strong, M.J., Garruto, R.M., Joshi, J.G., Mundy, W.R., and Shafer, T.J. (1996) Can the mechanisms of aluminum neurotoxicity be integrated into a unified scheme? *J. Toxicol. Environ. Health* 48:599–613.

USGS (2007) Aluminum. In: *Mineral Commodity Summaries 2007*, U.S. Geological Survey, pp. 18–19. Available at http://minerals.usgs.gov/minerals/pubs/mcs/2007/mcs2007.pdf.

Vittori, D., Nesse, A., Pérez, G., and Garbossa, G. (1999) Morphologic and functional alterations of erythroid cells induced by long-term ingestion of aluminum. *J. Inorg. Biochem.* 76:113–120.

Ward, M.K., Ellis, H.A., Feest, T.G., Parkinson, I.S., Kerr, D.N. (1978) Osteomalacic dialysis osteodystrophy: evidence for a waterborne aetiological agent, probably aluminum. *Lancet* 1:841–845.

6

ANTIMONY

Marie M. Bourgeois

First aid: Eye: irrigate immediately; skin: water flush promptly; breathing: respiratory support; swallow: medical attention immediately

Target organ(s): Eyes, skin, respiratory system, cardiovascular system

Occupational exposure limits: NIOSH REL: 0.5 mg Sb/m^3 TWA, OSHA PEL: 0.5 mg Sb/m^3 TWA, 1993–1994 ACGIH TLV: 0.5 mg Sb/m^3 TWA

Reference values: Oral RfD of 40,000 mg/kg/day (confidence is low), IDLH: 50 mg Sb/m^3

Risk phrases: R34-51/53-20/22-36/37/38-36/38

Causes burns. Toxic to aquatic organisms, may cause long-term adverse effects in the aquatic environment. Harmful by inhalation and if swallowed. Irritating to eyes, respiratory system, and skin. Irritating to eyes and skin

Safety phrases: S60-61-36/37/39-26

This material and its container must be disposed of as hazardous waste. Avoid release to the environment. Refer to special instructions/safety data sheet. Wear suitable protective clothing, gloves, and eye/face protection. In case of contact with eyes, rinse immediately with plenty of water and seek medical advice.

BACKGROUND AND USES

Antimony (CASRN 7440-36-0, EINECS 231-146-5) is a metal with synonyms, including antimony black; elemental antimony; antimony metal; metallic antimony; stibanylidynestibane; antimony ingot; antimony powder; antimony regulus; stibium and stibium metallicum. It has multiple industrial applications. The trioxide form has primarily been used as a flame retardant in items such as automobile and aircraft upholstery, toys, and clothing (NRCNA, 2008). It also is used in ceramics to achieve hardness and opacity and in glassware to remove bubbles and stabilize color. The trisulfide and pentasulfide forms are used as a pigment to make rubber tires black and as pigment stabilizers in pavement paint. A small amount is also used as a fining agent to remove bubbles in the glass of computer monitors and television picture tubes (Butterman and Carlin, 2004). Antimony trisulfide is an important component of pigments and munitions primers. Metallic antimony is a minor strengthening component in lead and tin alloys used for cable coverings and solder. Unalloyed antimony is used infrequently, principally in doping semiconductor wafers.

Antimony has been used medicinally as an emetic, expectorant (e.g., potassium antimony tartrate and antimony dimercaptosuccinate) and antiparasitic agent in schistosomiasis, trypanosomiasis, and leishmaniasis for centuries (Lizarazo-Jaimes et al., 2014). Meglumine antimoniate and sodium stibogluconate, forms of pentavalent antimony (Sb(v)), have been the drugs of choice for leishmaniasis for more than 60 years (Garcerant et al., 2012). Antimony plays no important role in nutrition (Beliles, 1979; Squires et al., 1993). Intake has been reported as being between 23 and 1250/ag per week (Dickerson, 1994). Beliles (1979) cites an average body burden of 90 mg.

PHYSICAL AND CHEMICAL PROPERTIES

A natural element, antimony (symbol Sb; CASRN 7440-36-0) occurs in valence states of 3, 5, and −3 and has both metallic and nonmetallic properties (Table 6.1). It is commercially

Hamilton & Hardy's Industrial Toxicology, Sixth Edition. Edited by Raymond D. Harbison, Marie M. Bourgeois, and Giffe T. Johnson.

TABLE 6.1 Chemical and Physical Properties of Antimony

Characteristic	Information
Chemical name	Antimony
Synonyms/trade names	Antimony black; elemental antimony; antimony metal; metallic antimony; stibanylidynestibane; antimony ingot; antimony powder; antimony regulus; stibium; and stibium metallicum
Chemical formula	Sb
CASRN	7440-36-0
Molecular weight	121.76 g/mol
Melting point	630 °C
Boiling point	1635 °C
Vapor pressure	0 mmHg (approx)
Density/specific gravity	6.69 g/ml at 25 °C
Flashpoint	Not applicable
Water solubility at 25 °C	Insoluble
Explosion limits	No data available
Octanol/water partition coefficient (log K_{OW})	0.73 (estimated)
Henry's law constant	No data available
Odor threshold	No data available
Conversion factors	1 ppm = 3.082 mg/m^3
	1 mg/m^3 = 0.324 ppm

available as a silver white lustrous solid metal or a dark gray powder (HSDB, 2005; Budavari, 1989). The amount in the earth's crust is <1 parts per million (ppm); its most common ore is stibnite (CASRN 1345-04-6). Antimony trisulfide (symbol Sb2S3; CASRN 1317-86-8), is a chemical form of antimony (Beliles, 1979). In the soils of the conterminous United States, it occurs at a geometrical mean of 0.48 ppm (Shacklett and Boerngen, 1984).

Because antimony is a group VA element, it has many of the same chemical and toxicological properties as arsenic and lead. For example, the toxicity of pentavalent antimony compounds is less than that of trivalent compounds (DeWolff and Edelbroek, 1994). The suggested descending order of toxicity is metalloid antimony (particularly stibine gas), the trisulfide, the pentasulfide, the trioxide, and the pentoxide.

ENVIRONMENTAL FATE AND BIOACCUMULATION

In water, antimony occurs primarily as suspended particulate, although some may occur as products of hydrolysis. It likely accumulates in sediments, and, although some bioconcentration has been observed in some freshwater and marine invertebrates (most likely from methylated forms), antimony does not biomagnify (USEPA, 1985, 1988; Wilson et al., 2010; Feng et al., 2013).

In most soils the trioxide form prevails and persists as a result of precipitation and of the element's lack of reactivity, its stability, and its low vapor pressure (USEPA, 1985; Dragun, 1988; Wilson et al., 2010). Antimony can be found on the 2007 Comprehensive Environmental Response, Compensation, and Liability Act (CERCLA) priority list of hazardous substances (Padilla et al., 2010; Ren et al., 2014).

ECOTOXICOLOGY

Quantitative information on antimony in the environment is not available. Toxicity assessment can be further complicated because antimony and arsenic are frequent environmental co-contaminants (Gebel, 2000). There is limited information on persistence, degradability, bioaccumulative potential, and mobility in soil. Environmental hazard in the event of a spill cannot be ruled out.

MAMMALIAN TOXICOLOGY

Antimony is poorly absorbed from the gastrointestinal tract but more readily absorbed through the respiratory tract. It is distributed in a rather unspecific manner, although trivalent compounds tend to react with red blood cells and the liver, and the pentavalent form tends to remain in plasma. The metabolism of antimony is similar to that of arsenic. Unlike arsenical compounds, antimony does not appear to be methylated *in vivo* (Bailly et al., 1991). Antimony tends to bind to sulfhydryl groups of respiratory enzymes. Antimony is excreted in both bile and urine as a glutathione conjugate (Bailly et al., 1991; NRCNA, 2008). Pentavalent antimony tends to be excreted more rapidly.

Acute Effects

Acute poisoning is rare but can result in death within several hours (Beliles, 1979). The symptomatology is similar to that for the arsenicals: vomiting, colic, diarrhea, and a metallic taste. Acute inhalation can induce rhinitis and pulmonary edema. Potassium antimony tartrate at a dose of 0.53 mg Sb/kg produced vomiting (Dunn, 1928). Inhalation of antimony dust by factory workers produced gastrointestinal irritation, probably the result of antimony dust transported via the mucociliary escalator and swallowed (ATSDR, 1992). No acute dermal effects have been reported.

Chronic Effects

Workers exposed to airborne antimony and antimony compounds can develop dermatitis, eye irritation, obstructive lung changes, and emphysema with increased pneumoconiosis (Cooper et al., 1968; Schroeder et al., 1968; Potkonjak and

Pavlovich, 1983). Transient "antimony spots" also may appear in individuals who are chronically exposed; these spots appear to be the result of internal changes rather than external effects (Sundar and Chakravarty, 2010). The pustular eruptions occur most frequently in areas of sweating. Otherwise, the only skin effect is dermatitis (White et al., 1993). A single study appears to indicate an increase in the number of spontaneous abortions and menstrual cycle disturbances among women workers exposed to antimony aerosols; however, this observation has not been reported elsewhere (Balyaera, 1967). There is limited evidence of carcinogenic, mutagenic, and teratogenic risks of antimony compounds, although additional investigation of pregnancy outcomes in women treated for leishmaniasis with antimony compounds may be indicated (Leonard and Gerber, 1996). Exposure to antimony may induce some cardiac effects, including arrhythmia (Brieger et al., 1954; Huang et al., 1958; Berman, 1988).

The federal Environmental Protection Agency (EPA) has established an oral reference dose (RfD) for antimony (40,000 mg/kg/day) based on a chronic oral bioassay in which 5 ppm of potassium antimony tartrate was administered to rats in their drinking water. The primary study used 100 rats (Schroeder et al., 1970). Confidence in the principal study is considered low because only one species and one dose level were used; a "no observed adverse effects level" (NOAEL) was not documented; and gross pathology and histopathology were not well described. Confidence in the database, and consequently the RfD, is low because no adequate investigations of oral exposure have been done. In 1994, the EPA has established a limit of 145 ppb in surface waters and 0.006 ppm in drinking water. EPA requires reports for environmental discharges of antimony exceeding 5000 pounds. The Agency for Toxic Substances and Disease Registry (ATSDR) has not derived acute, intermediate, or chronic oral minimal risk level for antimony due to insufficient data (NRCNA, 2008).

Existing data suggest that antimony may be an animal carcinogen; however, the data are not sufficient to justify a quantitative estimate of cancer potency at this time. In laboratory rats, inhalation of antimony dust can increase the risk of lung cancer. However, there is no evidence of an increased risk of cancer in animals that eat food or drink water containing antimony. The International Agency for Research on Cancer (IARC) concluded that antimony trioxide is a possible carcinogen in humans (Group 2E) and that the trisulfide form is not classifiable (IARC, 1989).

Mechanism of Action(s)

The primary targets in antimony poisoning are the gastrointestinal and respiratory tracts, and the effects in these regions have been documented in humans and animals. Other targets observed in animals include the cardiovascular system, the blood, and the liver (Goi et al., 2003; Newkirk et al., 2014). Renal toxicity (histological changes) has also been observed

in animals. Only limited data are available which indicate that occupational exposure to antimony may adversely affect the reproductive system of women.

SOURCES OF EXPOSURE

Exposure to antimony in humans can occur through inhalation, ingestion, and eye or skin contact (Majestic et al., 2012; Felix et al., 2013).

Occupational

Prior to automation and improved working conditions, antimony process workers exhibited an increased risk of lung cancer (Schnorr et al., 1995; McCallum, 2005). Excessive exposures may still occur in electrorefining. Smoking enhanced this association (Cooper and Harrison, 2009). Occupational exposures to antimony may occur in battery industries where antimony is alloyed with lead and zinc. In this case, workers are typically exposed to antimony hydride (SbH_3) (Cooper and Harrison, 2009). Also known as stibine, SbH_3 is produced when antimony reacts with an acid in the presence of a reducing agent such as zinc. Workers in semiconductor industries, where antimony is used as a dopant, can be exposed to stibine. Antimony exposure may also occur when antimony replaces lead in soldering (Fowler et al., 1993).

Environmental

Antimony is also a ubiquitous urban air pollutant, which occurs at an average concentration of $0.001/mg/m^3$ (Beliles, 1979; Smichowski, 2008). Antimony commonly is released into the air through combustion of fossil fuels and their by-products and of products containing antimony (Hodzic et al., 2012; Tian et al., 2012; Sun et al., 2014). The resultant form most likely is the trioxide, although the tetroxide and pentoxide forms are possible. Rapidly condensed on suspended particulates, the residence time of antimony in the air is thought to range from 30 to 40 days.

INDUSTRIAL HYGIENE

Both proper industrial hygiene and personal hygiene are important. The approaches used to limit exposure to arsenic apply to antimony as well (Table 6.2). The Occupational Safety and Health Administration (OSHA), the National Institute for Occupational Safety and Health (NIOSH, 2013), and the American Conference of Governmental Industrial Hygienists (ACGIH, 2013) have set the permissible exposure limit (PEL), the recommended exposure limit (REL), and threshold limit value (TLV) for antimony compounds (as Sb) at $0.5 mg/m^3$ for an 8-h workday, 40-h

TABLE 6.2 NIOSH Recommended Respiratory Protection for Workers Exposed to Antimony

Condition/Airborne Concentration of Antimony	Minimum Respiratory Protection
Up to 5 mg/m^3	(APF = 10) Any particulate respirator equipped with an N95, R95, or P95 filter (including N95, R95, and P95 filtering facepieces) except quarter-mask respirators. The following filters may also be used: N99, R99, P99, N100, R100, P100.
	(APF = 10) Any supplied-air respirator
Up to 12.5 mg/m^3	(APF = 25) Any supplied-air respirator operated in a continuous-flow mode
	(APF = 25) Any powered, air-purifying respirator with a high efficiency particulate filter.
Up to 25 mg/m^3	(APF = 50) Any air-purifying, full facepiece respirator with an N100, R100, or P100 filter.
	(APF = 50) Any supplied-air respirator that has a tight-fitting facepiece and is operated in a continuous-flow mode
	(APF = 50) Any powered, air-purifying respirator with a tight-fitting facepiece and a high efficiency particulate filter
	(APF = 50) Any self-contained breathing apparatus with a full facepiece
	(APF = 50) Any supplied-air respirator with a full facepiece
Up to 50 mg/m^3	(APF = 1000) Any supplied-air respirator operated in a pressure-demand or other positive-pressure mode
Emergency or planned entry into unknown concentrations or IDLH conditions	(APF = 10,000) Any self-contained breathing apparatus that has a full facepiece and is operated in a pressure-demand or other positive-pressure mode
	(APF = 10,000) Any supplied-air respirator that has a full facepiece and is operated in a pressure-demand or other positive-pressure mode in combination with an auxiliary self-contained positive-pressure breathing apparatus
Escape	(APF = 50) Any air-purifying, full facepiece respirator with an N100, R100, or P100 filter.
	Any appropriate escape-type, self-contained breathing apparatus

workweek. The ACGIH has listed antimony trioxide production as a suspected source of human carcinogenicity (A2).

MEDICAL MANAGEMENT

Treatment for antimony poisoning follows the same management as that for arsenic intoxication, primarily chelation with British antilewisite (BAL) for extreme exposure, as well as mercaptans such as dimercaptosuccinic acid or dimercaptopropane sulfonic acid (Aaseth, 1983). Stibine intoxication should be treated as arsine exposure.

REFERENCES

Aaseth, J. (1983) Recent advances in therapy of metal poisoning with chelating agents. *Hum. Toxicol.* 2:257–272.

ACGIH (American Conference of Governmental Industrial Hygienists) (2013) *Threshold Limit Values (TLVs) for Chemical Substances and Physical Agents and Biological Exposure Indices (BEIs)*, Cincinnati, OH: American Conference of Governmental Industrial Hygienists.

ATSDR (1992) *Toxicological Profile for Antimony*. Available at http://www.atsdr.cdc.gov/toxfaqs/tf.asp?id=331&tid=58.

Bailly, R., Lauwerys, R., Buchet, J.P., Mahieu, P., and Konings, J. (1991) Experimental and human studies on antimony metabolism: their relevance for the biological monitoring of workers exposed to inorganic antimony. *Br. J. Ind. Med.* 48(2):93–97.

Balyaera, A.P. (1967) The effect of antimony on reproduction. *Gig. Tr. Prof. Zabol.* 11:32–37.

Beliles, R.P. (1979) The lesser metals. In: Oehme, F.W., editor. *Toxicity of Heavy Metals in the Environment*, New York, NY: Marcel Dekker, pp. 547–615.

Berman, J.D. (1988) Chemotherapy for leishmaniasis: biochemical mechanisms, clinical efficacy, and future strategies. *Rev. Infect. Dis.* 10(3):560–586.

Brieger, H., Semisch, C.W., 3rd, Stasney, J., and Piatnek, D.A. (1954) Industrial antimony poisoning. *Ind. Med. Surg.* 23(12):521–523.

Budavari, S. (1989) Antimony. In: Budavari, F S., editor. *The Merck Index: An Encyclopedia of Chemicals, Drugs, and Biologicals*, 11th ed., Rahway, NJ: Merck and Co.

Butterman, W.C. and Carlin, J.F. (2004) *Antimony: Mineral Commodity Profiles*, U.S. Department of the Interior, United States Geological Service.

Cooper, D.A., Pendergrass, E.P., Vorwald, A.J., Mayock, R.L., and Brieger, H. (1968) Pneumoconiosis among workers in an antimony industry. *Am. J. Roentgenol. Radium Ther. Nucl. Med.* 103(3):496–508.

Cooper, R.G. and Harrison, A.P. (2009) The exposure to and health effects of antimony. *Indian J. Occup. Environ. Med.* 13(1):3–10.

DeWolff, F.A. and Edelbroek, P.M. (1994) Neurotoxicity of arsenic and its compounds. In: DeWolff, F.A., editor. *Vinken and Bruyn's Handbook of Clinical Neurology: Intoxications of the Nervous System*, Amsterdam: Elsevier.

Dickerson, O.B. (1994) Antimony, arsenic, and their compounds. In: Dickerson, O.B., Zenz, C., and Horvath, B.P., Jr., editors. *Occupational Medicine*, 3rd ed., St. Louis, MO: Mosby.

Dragun, J. (1988) The soil chemistry of hazardous materials. In: *National Conference on Management of Uncontrolled Hazardous Waste Sites*, Silver Spring, MD: Hazardous Materials Control Research Institute.

Dunn, J.T. (1928) A curious case of antimony poisoning. *Analyst* 531:532–533.

Felix, P.M., Franco, C., Barreiros, M.A., Batista, B., Bernardes, S., Garcia, S.M., Almeida, A.B., Almeida, S.M., Wolterbeek, H.T., and Pinheiro, T. (2013) Biomarkers of exposure to metal dust in exhaled breath condensate: methodology optimization. *Arch. Environ. Occup. Health* 68(2):72–79.

Feng, R., Wei, C., Tu, S., Ding, Y., Wang, R., and Guo, J. (2013) The uptake and detoxification of antimony by plants: a review. *Environ. Exp. Bot.* 96:28–34.

Fowler, B.A., Yamauchi, H., Conner, E.A., and Akkerman, M. (1993) Cancer risks for humans from exposure to the semiconductor metals. *Scand. J. Work Environ. Health* 19(Suppl. 1):101–103.

Garcerant, D., Rubiano, L., Blanco, V., Martinez, J., Baker, N.C., and Craft, N. (2012) Possible links between sickle cell crisis and pentavalent antimony. *Am. J. Trop. Med. Hyg.* 86(6):1057–1061.

Gebel, T. (2000) Confounding variables in the environmental toxicology of arsenic. *Toxicology* 144(1–3):155–162.

Goi, G., Bairati, C., Massaccesi, L., Sarnico, M., Pagani, A., Lombardo, A., and Apostoli, P. (2003) Low levels of occupational exposure to arsenic and antimony: effects on lysosomal glycohydrolase levels in plasma of exposed workers and in lymphocyte cultures. *Am. J. Ind. Med.* 44(4):405–412.

HSDB (2005) *Hazardous Substances Data Bank: Antimony, Elemental. From NLM (National Library of Medicine)*, Bethesda, MD: National Institutes of Health, U.S. Department of Health and Human Services.

Hodzic, A., Wiedinmyer, C., Salcedo, D., and Jimenez, J.L. (2012) Impact of trash burning on air quality in Mexico City. *Environ. Sci. Technol.* 46(9):4950–4957.

Huang, M.H., Chiang, S.C., P'An, J. S., Yu, K.J., Lu, C.W., Hsu, C.Y., and Kao, W.S. (1958) Mechanism and treatment of cardiac arrhythmias in tartar emetic intoxication; with special reference to atropine therapy. *Chin. Med. J.* 76(2):103–115.

IARC (1989) *Some Organic Solvents, Resin Monomers and Related Compounds, Pigments and Occupational Exposures in Paint Manufacture and Painting. IARC Monographs on the Evaluation of Carcinogenic Risks to Humans*, vol. 47, Lyon, France: World Health Organization.

Leonard, A. and Gerber, G.B. (1996) Mutagenicity, carcinogenicity, and teratogenicity of antimony compounds. *Mutat. Res.* 366(1):1–8.

Lizarazo-Jaimes, E.H., Reis, P.G., Bezerra, F.M., Rodrigues, B.L., Monte-Neto, R.L., Melo, M.N., Frezard, F., and Demicheli, C. (2014) Complexes of different nitrogen donor heterocyclic ligands with SbCl and PhSbCl as potential antileishmanial agents against Sb-sensitive and -resistant parasites. *J. Inorg. Biochem.* 132:30–36.

Majestic, B.J., Turner, J.A., and Marcotte, A.R. (2012) Respirable antimony and other trace-elements inside and outside an elementary school in Flagstaff, AZ, USA. *Sci. Total Environ.* 435–436:253–261.

McCallum, R.I. (2005) Occupational exposure to antimony compounds. *J. Environ. Monit.* 7(12):1245–1250.

Newkirk, C.E., Gagnon, Z.E., and Pavel Sizemore, I.E. (2014) Comparative study of hematological responses to platinum group metals, antimony and silver nanoparticles in animal models. *J. Environ. Sci. Health A Tox. Hazard. Subst. Environ. Eng.* 49(3):269–280.

NIOSH (2013) *Pocket Guide to Chemical Hazards*, Washington, DC: National Institute for Occupational Safety and Health.

NRCNA (2008) *Spacecraft Water Exposure Guidelines for Selected Contaminants*, vol. 3, National Academies Press.

Padilla, M. A., Elobeid, M., Ruden, D.M., and Allison, D.B. (2010) An examination of the association of selected toxic metals with total and central obesity indices: NHANES 99-02. *Int. J. Environ. Res. Public Health* 7(9):3332–3347.

Potkonjak, V. and Pavlovich, M. (1983) Antimoniosis: a particular form of pneumoconiosis. I. Etiology, clinical and X-ray findings. *Int. Arch. Occup. Environ. Health* 51(3):199–207.

Ren, J.H., Ma, L.Q., Sun, H.J., Cai, F., and Luo, J. (2014) Antimony uptake, translocation and speciation in rice plants exposed to antimonite and antimonate. *Sci. Total Environ.* 475C:83–89.

Schnorr, T.M., Steenland, K., Thun, M.J., and Rinsky, R.A. (1995) Mortality in a cohort of antimony smelter workers. *Am. J. Ind. Med.* 27(5):759–770.

Schroeder, H.A., Mitchener, M., Balassa, J.J., Kanisawa, M., and Nason, A.P. (1968) Zirconium, niobium, antimony, and fluorine in mice: effects on growth, survival and tissue levels. *J. Nutr.* 95(1):95–101.

Schroeder, H.A., Mitchener, M., and Nason, A.P. (1970) Zirconium, niobium, antimony, vanadium and lead in rats: life term studies. *J. Nutr.* 100(1):59–68.

Shacklett, H.T. and Boerngen, J.G. (1984) *Element Concentration in Soils and Other Surficial Materials of the Conterminous United States*. Alexandria, VA: US Geological Survey.

Smichowski, P. (2008) Antimony in the environment as a global pollutant: a review on analytical methodologies for its determination in atmospheric aerosols. *Talanta* 75(1):2–14.

Squires, K.E., Rosenkaimer, F., Sherwood, J.A., Forni, A.L., Were, J.B., and Murray, H.W. (1993) Immunochemotherapy for visceral leishmaniasis: a controlled pilot trial of antimony versus antimony plus interferon-gamma. *Am. J. Trop. Med. Hyg.* 48(5):666–669.

Sun, F., Yan, Y., Liao, H., Bai, Y., Xing, B., and Wu, F. (2014) Biosorption of antimony(V) by freshwater cyanobacteria *Microcystis* from Lake Taihu, China: effects of pH and competitive ions. *Environ. Sci. Pollut. Res. Int.* 21(9):5836–5848.

Sundar, S. and Chakravarty, J. (2010) Antimony toxicity. *Int. J. Environ. Res. Public Health* 7(12):4267–4277.

Tian, H., Zhao, D., Cheng, K., Lu, L., He, M., and Hao, J. (2012) Anthropogenic atmospheric emissions of antimony and its spatial distribution characteristics in China. *Environ. Sci. Technol.* 46(7):3973–3980.

USEPA (1985) *Health and Environmental Effects Profile for Antimony Oxides*, Cincinnati, OH: Office of Health and Environmental Assessment, Environmental Criteria and Assessment

Office, Office of Research and Development, United States Environmental Protection Agency.

USEPA (1988) *Ambient Aquatic Life Water Quality Criteria for Antimony (III)*, Duluth, MN and Narragansett, RI: Environmental Research Laboratories, Office of Research and Development, United States Environmental Protection Agency.

White, G.P., Jr., Mathias, C.G., and Davin, J.S. (1993) Dermatitis in workers exposed to antimony in a melting process. *J. Occup. Med.* 35(4):392–395.

Wilson, S.C., Lockwood, P.V., Ashley, P.M., and Tighe, M. (2010) The chemistry and behaviour of antimony in the soil environment with comparisons to arsenic: a critical review. *Environ. Pollut.* 158(5):1169–1181.

7

ARSENIC

Janice S. Lee and Santhini Ramasamy

First aid (inorganic and organic arsenic): Eye—irrigate immediately; skin—soap wash immediately; breathing—respiratory support; swallow—medical attention immediately

First aid (arsine): Eye—frostbite; skin—frostbite; breathing—respiratory support

Target organ(s): Liver, kidneys, skin, lungs, lymphatic system (inorganic arsenic); kidneys, liver, thyroid, gastrointestinal tract (organic arsenic)

Occupational exposure limits: OSHA PEL: $0.010 \, mg/m^3$ (inorganic arsenic); OSHA PEL: TWA $0.2 \, mg/m^3$ (arsine); OSHA PEL: TWA $0.5 \, mg/m^3$ (organic arsenic); ACGIH TLV: TWA $0.01 \, mg/m^3$ (inorganic arsenic and inorganic arsenic compounds); ACGIH TLV: TWA $0.02 \, mg/m^3$ (arsine); NIOSH REL: $0.002 \, mg/m^3$ (15 min) (inorganic arsenic and arsine)

Reference values

Inorganic arsenic: EPA MCL: $0.010 \, mg/l$; EPA MCLG: 0; EPA RfD: $0.0003 \, mg/kg/day$; EPA oral cancer slope factor: 1.5 per mg/kg/day; EPA inhalation unit risk: 0.0043 per $\mu g/m^3$; ATSDR acute oral MRL: $0.005 \, mg/kg/day$; ATSDR chronic oral MRL: $0.0003 \, mg/kg/day$; RIVM TDI: $1.0 \, \mu g/kg/day$ for chronic oral exposures and $1.0 \, \mu g/m^3$ for chronic inhalation exposures; Cal EPA acute RELs: $0.2 \, \mu g/m^3$; Cal EPA chronic RELs: $0.015 \, \mu g/m^3$; Cal EPA inhalation unit risk: 3.3 per $\mu g/m^3$; Cal EPA inhalation slope factor: 12.0 per mg/kg/day

Arsine: EPA RfC: $0.05 \, \mu g/m^3$; WHO provisional guidance criteria: $0.05 \, \mu g/m^3$; Cal EPA acute RELs: $0.2 \, \mu g/m^3$; Cal EPA chronic RELs: $0.015 \, \mu g/m^3$

Organoarsenicals: EPA RfD: $0.03 \, mg/kg/day$ for MMA and $0.014 \, mg/kg/day$ for DMA; ATSDR intermediate oral MRL: $0.1 \, mg/kg/day$ for MMA; ATSDR chronic oral MRL: $0.02 \, mg/kg/day$ for DMA

Risk phrases: 23/25-50/53

Toxic by inhalation and if swallowed. Very toxic to aquatic organisms, may cause long-term adverse effects in the aquatic environment

Safety phrases: 20/21-28-45-60-61

When in use, do not eat, drink, or smoke. In case of accident or if you feel unwell, seek medical advice immediately (show the label where possible). This material and/or its container must be disposed of as hazardous waste. Avoid release to the environment. Refer to special instructions/safety data sheets.

BACKGROUND AND USES

Arsenic (As; CAS #7440-38-2) is a brittle solid with a metallic coloring that ranges from silver to gray. It is a naturally occurring element found in the earth's crust, and it cycles rapidly through water, land, air, and living systems. Exposure to it occurs through ingestion, inhalation, and dermal contact.

Arsenic is a metalloid that can exist in the −3, 0, +3, and +5 oxidation states. Arsenic as a free element (0 oxidation state) is rarely encountered in the environment (HSDB, 2005). Compounds that contain arsenic are ranked according to their relative toxicities, depending on the form. Inorganic forms of arsenic, such as arsine gas, are considered more

Hamilton & Hardy's Industrial Toxicology, Sixth Edition. Edited by Raymond D. Harbison, Marie M. Bourgeois, and Giffe T. Johnson.
© 2015 John Wiley & Sons, Inc. Published 2015 by John Wiley & Sons, Inc.

toxic than its organic forms, such as tributylarsine. These different forms of arsenic determine the solubility of the compound and its vapor pressure.

The arsenic metalloid is used for hardening copper and lead alloys (HSDB, 2005). It is also used in glass manufacturing as a decolorizing and refining agent, as a component of electrical devices in the semiconductor industry, and as a catalyst in the production of ethylene oxide. Arsenic compounds are used as a mordant in the textile industry, for preserving hides, as medicinals, pesticides, pigments, and wood preservatives. The production of chromate copper arsenate (CCA), a wood preservative, accounts for approximately 90% of the domestic arsenic consumption (ATSDR, 2007). However, production of this preservative is being phased out. The uses of inorganic arsenical compounds (e.g., lead arsenate) as pesticides were voluntarily cancelled by the industry during late 1980s and early 1990s. A majority of organoarsenicals are used on cotton and turf as herbicides. disodium methanearsenate (DSMA), monosodium methanearsenate (MSMA), and calcium methanearsenate (CAMA) continue to be used as contact herbicides.

Arsenic naturally comprises approximately 3.4 parts per million (ppm) of the earth's crust, where it is the twentieth most abundant element (ATSDR, 2007; The Merck Index, 1989) Arsenic leaches from the natural weathering of soil and rock into water; and low concentrations of arsenic are found in water, food, soil, and air. However, industrial activities such as coal combustion and smelting operations release higher concentrations of arsenic into the environment (Adams et al., 1994). The highest background arsenic levels found in the environment are in soils, with concentrations ranging from 1 to 40 ppm (ATSDR, 2007). Food typically contains total arsenic concentrations of 20–140 parts per billion (ppb), with inorganic arsenic levels being much lower (ATSDR, 2007). The majority of surface and ground waters contain <10 ppb (although levels of 1000–3400 ppb have been reported, especially in areas of the western United States) (ATSDR, 2000). The average arsenic content in drinking water in the United States (U.S.) is 2 ppb with 12% of the water supply from surface water in central portions of the United States and 12% of groundwater sources in western portions of the United States exceeding 20 ppb (ATSDR, 2007). Mean arsenic concentrations in ambient air have generally been found to range from 1 to 2000 ng/m^3 (ATSDR, 2007).

PHYSICAL AND CHEMICAL PROPERTIES

Arsenic (As) is the third element in Group VA of the periodic table. Elemental arsenic can be found in two solid forms: yellow and gray or metallic, with specific gravities of 1.97 and 5.73, respectively (CRC, 1999). Gray arsenic is the ordinary stable form. Arsenic compounds can be categorized as inorganic and organic. Inorganic compounds do not contain an arsenic–carbon bond while organic compounds do. Chemical and physical properties of inorganic arsenic and selected organic arsenic compounds are listed in Table 7.1.

MAMMALIAN TOXICOLOGY

Acute Effects

The effect of acute exposure to arsenic in an industrial setting is local irritation on the skin, mucous membranes, and conjunctivae, wherever the dust is deposited. Tissue necrosis may also occur. Moist surfaces such as the lips, nares, eyelids, pharynx, and scrotum suffer the most. Another manifestation of acute arsenic exposure is the latent appearance of horizontal white bands, called Mee's lines, across the nails as a result of injury to the nail bed. Mee's lines may be visible in the fingernails several weeks to months after acute arsenic poisoning (Rossman, 2007).

Other exposures often are associated with ingestion of contaminated food, usually intentional. Major symptoms of this type of exposure are gastrointestinal inflammation and irritation with possible hemorrhage. Signs and symptoms include abdominal pain, vomiting, watery diarrhea, low blood pressure, and pulse irregularity, all similar to the signs of cholera (Landrigan, 1981). These symptoms can result in stupor, convulsions, coma, and death. Franzblau and Lilis (1989) described two cases of acute arsenic poisoning caused by contaminated well water. In addition to gastrointestinal effects, some neurological and hematopoietic effects were observed.

After acute ingestion, a delayed sensorimotor peripheral neuropathy may occur. Electrophysiological studies in cases that followed a single exposure showed axonal degeneration (Le Quesne and McLeod, 1977). Common symptoms include numbness and tingling of the hands and feet in a stocking-and-glove symmetrical distribution. Motor weakness of the distal extremities may develop and is distinguishable from the effects of lead poisoning by the severe neuralgic pain. Neuropathy may develop whether exposure is inhalation, ingestion, or cutaneous absorption.

Arsine   The principal effect of arsine poisoning is intravascular hemolysis. Varying periods of time elapse after inhalation of the gas before the destruction of red cells results in clinical manifestations. Such intervals have been reported as ranging from 6 to 36 h or longer, depending on the severity and length of exposure. The initial symptoms of headache, malaise, weakness, dizziness, and dyspnea are followed by abdominal pain, nausea, and vomiting (Fowler and Weissberg, 1974). These acute symptoms are followed in 4–6 h by the passage of dark or bloody urine, often the first symptom

TABLE 7.1 Chemical and Physical Properties of Inorganic Arsenic and Selected Organic Arsenic Compounds

Properties	Arsenic	Dimethylarsinic Acid	Methanearsonic Acid	Disodium Methanearsonate	Sodium Methanearsonate
Molecular weight	74.9216	138.00	139.97	183.93	161.95
Color	Silver-gray or tin-white	Colorless	White	White	White
Physical state	Solid	Solid	Solid	Solid	Solid
Melting point	817 °C (triple point)	195 °C	160.5 °C	>355 °C	130–140 °C
Boiling point	603 °C (sublimes)	>200 °C	–	–	–
Density	5.778 g/cm^3	–	–	1.04 g/cm^3	1.55 g/ml
Odor	Odorless	Odorless	–	–	Odorless
Solubility:					
Water	Insoluble	2000 g/l at 25 °C	256 g/l at 20 °C	432 g/l at 25 °C	580 g/l at 20 °C
Organic solvent(s)	–	Soluble in alcohol; insoluble in diethyl ether	Soluble in ethanol	Soluble in methanol; practically insoluble in most organic solvents	Insoluble in most organic solvents
Other	Insoluble in caustic and nonoxidizing acids	Soluble in acetic acid	–	–	–
Partition coefficients:					
Log K_{ow}	–	–	–	<1	−3.10
pK_a	–	1.57	pK_{a1} = 4.1; pK_{a2} = 9.02	pK_{a1} = 4.1; pK_{a2} = 8.94	pK_{a1} = 4.1; pK_{a2} = 9.02
Vapor pressure	7.5×10^{-3} mmHg at 280 °C	–	$<7.5 10^{-8}$ mmHg at 25 °C	10^{-7} mmHg at 25 °C	7.8×10^{-8} mmHg at 25 °C
Flammability limits in air	–	Nonflammable	–	Nonflammable	Nonflammable

Source: ATSDR (2007).

–: No data.

that alarms the victim. Jaundice appears later, after 24–48 h, and the red cell count may fall below 2 million. Jones (1907) described the triad of abdominal pain, hemoglobinuria, and jaundice as characteristic of arsine poisoning. Findings during the physical examination include fever, tachycardia, tachypnea, jaundice, and abdominal tenderness, especially of the enlarged liver. Hemolytic anemia is evidenced by the low hemoglobin values and by the presence of reticulocytosis, red cell fragments, and ghost cells. Toxic granules may be found in the leukocytes. Plasma hemoglobin may be present at concentrations as high as 2 mg/dl. Urinalysis reveals the presence of both hemoglobin and arsenic. Hong et al. (1989) demonstrated similar effects in mice exposed to low concentrations of arsine gas. A decline in the red cell count and hemoglobin level correlated with increasing concentrations.

Arsine intoxication is differentiated from acute poisoning with nitro and amido derivatives of benzene by the absence of methemoglobin formation and cyanosis. A valuable observation that confirms the direct toxic action of arsine on the cardiac muscle has been provided by the reports of lowered T waves or inversion of T waves on the electrocardiogram (Weinberg, 1960). Such changes may persist for 2 months after apparent recovery.

The pulmonary edema that occurs secondary to the effects of arsine on cardiac muscle and the dilation of the heart caused by anemia resulting from hemolysis require heroic measures. Early death after acute arsine poisoning may be the result of myocardial failure accompanied by massive pulmonary edema that precedes the onset of renal failure.

Organoarsenicals MSMA, DSMA, CAMA, and DMA have moderate to low acute toxicity via the oral, dermal, and inhalation routes (LD$_{50}$ ~ 2000 mg/kg and above; LC$_{50}$ > 2 mg/l). They have slight or negligible irritant properties to the eye and skin and are not sensitizers (USEPA, 2006).

Chronic Effects and Cancer

Arsenic has been associated with a number of health hazards, including several cancers, skin lesions, cardiovascular disease, diabetes mellitus, anemia and other hematological changes, neurological symptoms, respiratory changes, and immunotoxicity. Chronic occupational and environmental contact with arsenical compounds has been associated with dermatitis (Holmquist, 1951). Chronic oral arsenic exposure is associated with hyperpigmentation, depigmentation, keratosis, and peripheral vascular disorders (NRC, 1999). One report by Juhlin and Ortonne (1986) links chronic arsenic exposure in a vineyard worker to the darkening of scalp hair from reddish to dark brown. High levels of arsenic were also detected in the vineyard worker's hair. With several compounds and possible exposure routes available for arsenic, the chronic toxicity of this element varies by organ. Cancers of the skin, bladder, liver, and lung can also result from chronic arsenic exposure, in addition to cardiovascular and neurological diseases.

Occupational studies have reported excess risk of lung cancer from inhalation of arsenic (Enterline et al., 1987; Jarup et al., 1989; Lee-Feldstein, 1986; Ott et al., 1974; Taylor et al., 1989). Data from these studies suggest that the risk of lung cancer after inhalation of inorganic arsenic compounds increases with higher concentrations (NRC, 1999). Respiratory disease is also a common hazard in the smelting of metals, in which arsenical compounds are released. Lundgren et al. (1951) studied 1500 Swedish smelter workers exposed to both arsenic trioxide and sulfur dioxide. These researchers described the occurrence of Rönn-skär disease, a worker illness named for the location of the smelter. In contrast to the controls, the exposed men suffered chronic upper respiratory tract disease. A hoarse voice is characteristic of an arsenic worker, and a perforated nasal septum is an occupational sign of the worker who handles arsenic trioxide. After short periods of exposure, the workers notice nasal obstruction, with consequent mouth breathing, and usually some inflammation in the pharynx and larynx, soreness of the tongue, and excessive salivation. Sloughing of the mucous membrane and necrosis of the nasal cartilage may follow. The bony septum is not involved. Inflammation and noncancerous lesions may occur in chronically exposed workers, although cancer mortality is the most studied chronic effect of the respiratory system.

One of the first studies to link arsenic exposure to lung cancer was reported by Lee and Fraumeni in 1969. This study examined 8047 white male smelter workers exposed to arsenic trioxide and found they were three times more likely to develop respiratory cancer than the general population. More recent studies have found similar associations with respiratory cancers in populations of workers in the gold and copper smelting industries (Enterline et al., 1995; Kabir and Bilgi, 1993; Simonato et al., 1994; WHO, 2001). In addition to respiratory cancers, exposure of smelter workers to arsenic has been associated with cancers of the kidney, digestive tract, and lymphatic and hematopoietic systems. Glass workers, hat makers, and pesticide workers exposed to arsenic also have reported excesses of lung and skin cancer (IARC, 1973, 1980).

There are many regions in the United States and abroad that have arsenic in their drinking water at concentrations >100 μg/l. Parts of Maine, Michigan, Minnesota, South Dakota, Oklahoma, and Wisconsin have widespread arsenic concentrations exceeding 10 μg/l (Welch et al., 1999). Populations in Taiwan, Mexico, India, and Chile who consumed

drinking water with high levels of arsenic had high rates of skin cancer (Tseng et al., 1968; Cebrian et al., 1983; Chakaborty and Saha, 1987; Zaldivar, 1974). Elevated rates of liver cancer deaths were found in an area of southwestern Taiwan with high levels of arsenic in the artesian well water supply (Chen et al., 1988; Wu et al., 1989). Elevated rates of lung cancer were found in areas of Argentina with high levels of arsenic in drinking water (Tello, 1986; Biagini et al., 1978). Chen et al. (1988) also found an increase in bladder and kidney cancer with increasing arsenic concentrations. In addition, a few studies have suggested that exposure to arsenic in drinking water is associated with prostate cancer in men and nasal cancer in both men and women (Cantor, 1997).

Neurological effects associated with chronic arsenic exposure may involve a peripheral neuropathy, which may evolve into a sensorimotor distal axonopathy. Mayers (1954) reported a case of peripheral neuropathy in a worker exposed for 20 years to inhalation and skin contact of Paris green. Heyman et al. (1956) reported arsenical neuropathy in seven farm workers exposed occupationally to arsenate sprays or dust. Peripheral and subclinical neuropathies were also noted in arsenic smelter workers by Feldman et al. (1979) and Blom et al. (1985). An assessment of workers chronically exposed to inorganic arsenic found lower motor nerve conduction velocity among the workers compared to the nonexposed control group (Lagerkvist and Zetterlund, 1994). The authors concluded that arsenical neuropathies are dependent on long-term exposure. Recovery from neurological effects from chronic exposure to arsenic may take years.

Vascular effects associated with arsenic have not been seen among occupationally exposed groups. Characteristic vascular effects include peripheral vascular disease (e.g., blackfoot disease), cardiovascular-related mortality, atherosclerosis, QT abnormalities, and elevated blood pressure.

Hematopoietic effects, most often involving a reversible bone marrow suppression with pancytopenia, are noted with chronic exposure. Anemia, leukopenia, eosinophilia, thrombocytopenia, and basophilic stippling also are common hazards in this type of exposure (Ringenberg et al., 1988). Aplastic anemia followed by acute myelogenous leukemia has also been reported (Kveldsberg and Ward, 1972).

The developmental toxicity associated with arsenic has been reviewed, and, although arsenic is teratogenic in some experimental animals, occupational exposure levels and routes of exposure have not been linked to human teratogenic effects (Domingo, 1994). However, drinking water studies in humans have shown an association between arsenic in drinking water and congenital heart defects (Engel and Smith, 1994; Zierler et al., 1988).

Arsine   Chronic exposure to arsine has been reported in workers employed in the cyanide extraction of gold (Bulmer et al., 1940). These workers showed a progressive reduction in erythrocyte levels and hemoglobin values, marked basophilic stippling, and an elevation of urine arsenic. They had exertional dyspnea and the weakness typical of anemia. These chronic effects are similar to those associated with acute exposure to arsenic.

Organoarsenicals   No human studies have been found in the literature relating to organoarsenical poisoning. Animal studies suggest the target organs following oral exposure to MMA (MSMA, DSMA, CAMA) as the gastrointestinal tract, particularly the large intestine, and kidney (USEPA, 2006). The target organs following oral exposure to DMA (cacodylic acid) are believed to be the bladder and thyroid in animals (USEPA, 2006). The IARC classifies MMA and DMA as "possibly carcinogenic to humans" (Group 2B) (Straif et al., 2009).

Mechanism of Action(s)

The metabolism of inorganic arsenic in humans occurs through alternating steps of reduction and oxidative methylation mostly to DMA^V. Less toxic species (i.e., As^V, MMA^V, and DMA^V) can be converted to more toxic species (i.e., As^{III}, MMA^{III}, and DMA^{III}). The trivalent species have been found to be more cytotoxic, genotoxic, and more potent inhibitors of enzyme activity (Thomas et al., 2001). The trivalent methylated metabolites, such as MMA^{III} and DMA^{III}, may play a role in arsenic-mediated health effects along with inorganic arsenic.

Although arsenic has been associated with altering numerous key events (e.g., reactive oxygen species (ROS), changes in genes/proteins, proliferation, or apoptosis), no specific mode of action (MOA) has been identified for any of the noncancer effects. However, the various effects associated with arsenic exposure indicate possible multiple MOAs. The MOA for arsenic carcinogenicity has not been established (NRC, 1999). Possible MOAs for arsenic carcinogenesis include chromosome abnormalities, oxidative stress, altered growth factors, cell proliferation, promotion and/or progression in carcinogenesis, altered DNA repair, p53 gene suppression, altered DNA methylation patterns, and gene amplification (Kitchin, 2001). The MOAs for arsenic may depend on the following biochemical events: (1) binding of trivalent arsenicals to tissue targets such as sulfhydryl groups and/or selenium atoms; (2) formation of superoxide, hydrogen peroxide, and other ROS; and/or (3) hypomethylation of DNA (Kitchin and Conolly, 2010). It is not clear which arsenic species and key events are most important in the target organs because mechanisms of arsenic carcinogenesis in the organs may be quite different.

Chemical Pathology

Arsenic exposure is associated with skin pathology, including hyperpigmentation, hyperkeratosis, and skin cancers (Centeno et al., 2000; Heyman et al., 1956). Liver pathology, not as well described, includes noncirrhotic portal hypertension (Nevens et al., 1990), hepatic enlargement, hepatocellular carcinoma (Centeno et al., 2000), and liver angiosarcoma (Neshiwat et al., 1992). Long-term arsenic exposure has also been found to be associated with peripheral vascular diseases (Yu et al., 2002).

Hyperpigmentation of the skin in patches is a common occurrence with arsenic toxicity. The patch is dark brown with lighter brown spots. Arsenical hyperkeratosis manifests itself as small, corn-like lesions that typically are benign but that may become precancerous and resemble lesions associated with Bowen's disease. Persistence of the skin lesions with hyperpigmentation may result in a basocellular carcinoma or arsenical keratosis even after exposure has been eliminated (Alain et al., 1993).

Three distinct cancers of the skin have been associated with chronic exposure to arsenic, namely, Bowen's disease, basal cell carcinoma, and squamous cell carcinoma. These are all thought to develop secondary to arsenical keratoses (Landrigan, 1981).

Although chronic ingestion of arsenic has been reported to lead to cirrhosis of the liver and to noncirrhotic portal hypertension (Huet et al., 1975; Morris et al., 1974; Cowlishaw et al., 1979), there are a few documented cases of liver cirrhosis that occurred as a result of industrial exposure to arsenic. Jhaveri (1959) reported a case of primary liver carcinoma associated with liver cirrhosis in a man who had handled arsenous oxide and sodium arsenate for more than 20 years in a chemical plant.

Blackfoot disease, an arterial disease with dry gangrene of the lower extremities due to chronic arsenic poisoning, has been endemic in Taiwan. It has been associated with the ingestion of high concentrations of arsenic in drinking water. The pathology of blackfoot disease includes thickening and fibrinoid necrosis of subcutaneous arterioles.

SOURCES OF EXPOSURE

Occupational

Unlike the environmental exposures, where the health concerns are mainly to arsenite/arsenate from oral exposures (e.g., drinking water, food), occupational exposures can occur from many types of arsenic compounds introduced via inhalation and dermal contact. Dermal exposure has been reported from application and handling of pesticides containing arsenicals (e.g., copper chromated arsenic-treated wood). Differences in the exposure routes, metabolism, disposition, and toxicity of these various arsenical compounds in occupational settings (inorganic arsenic oxides, organoarsenicals, arsine, arsenides) must be considered in evaluating the health concerns of the workers (Carter et al., 2003).

U.S. production of arsenic compounds (e.g., arsenic trioxide) ceased in 1985, thus now all arsenic consumed in the United States is imported. Arsenic is imported mainly from China and to some extent from other countries. In 2005, the United States remained the major consumer of arsenic compounds. About 65% of domestic consumption of arsenic trioxide was used for the production of CCA used in wood preservatives. The remainder was used for the production of agricultural chemicals, including herbicides and insecticides (ATSDR, 2007). The use of arsenic compounds as pesticides has been reduced in the past few years due to voluntary cancellation of residential uses of CCA-treated wood (USEPA, 2008). However, copper chromated arsenic-treated wood is still used in industrial and agricultural settings. The United States Environmental Protection Agency (EPA) has also made restrictions on herbicidal uses of organoarsenicals (USEPA, 2006). With the increased use of arsenic in electronic industries, occupational exposures from arsenide (e.g., gallium arsenide) are likely; however, the toxicities of these arsenides are not well studied.

Exposures to arsenical compounds are commonly monitored by measuring total or speciated arsenicals in urine. Speciated arsenicals are preferred in exposure assessment to differentiate background arsenic exposures from food (e.g., arsenobetaine in seafood) to that introduced by occupational exposures. Arsenical exposures are also measured in blood to a limited extent. Because inorganic arsenic binds to sulfhydryl groups, keratin-rich tissues such as skin, hair, and nails, may also serve as indicators of arsenic exposures. However, external contamination may pose a problem in interpretation of the results (ATSDR, 2007).

The World Health Organization (WHO) reported a wide range of airborne arsenic concentrations in various departments of smelter operations (WHO, 2001). The exposure levels may vary depending upon the departments, the operations, and industrial hygiene measures. Farmer and Johnson (1990) measured total and speciated arsenic in urine samples collected from workers in different types of United Kingdom industries. Mean urinary arsenic (total arsenic) concentrations ranged from 4.4 μg/g creatinine for controls to <10 μg/g for those in the electronics industry, 47.9 μg/g for timber treatment workers applying arsenical wood preservatives, 79.4 μg/g for a group of glassworkers using arsenic trioxide, and 245 μg/g for chemical workers engaged in manufacturing and handling inorganic arsenicals. The speciation measurements of arsenic in urine samples for the most exposed groups indicate 1–6% As(V), 11–14% As(III), 14–18% monomethylated metabolite, and 63–70% dimethylarsinic acid. In another study conducted in the United Kingdom, evidence of increased urinary levels of total inorganic arsenic from timber treatment workers exposed to CCA wood preservatives were reported, after adjusting for seafood intake, compared to control subjects (Cocker et al., 2006). The urine and blood arsenic concentrations were also increased in incinerator workers compared to control subjects (Chao and Hwang, 2005).

In addition to urinary arsenic measurements, other urinary biomarkers have been examined in arsenic-exposed workers. These other biomarkers may not be specific to arsenic since other pollutants may influence these effects. For example, increased urinary excretion of porphyrins (uroporphyrin and coproporphyrin) has been reported in 84 smelter workers (Feldman et al., 1979; Telolahy et al., 1993). Also, 8-oxodGuo (e.g., 8-oxo-7,8-dihydro-2″deoxyguanosine), biomarker of oxidative DNA damage, was found elevated in urine samples from 50 smelter workers compared to 40 unexposed workers (Blom et al., 1985; Hu et al., 2006). In this study, Hu et al. (2006) demonstrated a positive correlation between 8-oxodGuo and monomethyl arsenic acid in urine as well as primary methylation index, supporting the existing evidence that accumulation of monomethyl arsenic may play a significant role in arsenic-mediated adverse effects. A positive association between blood arsenic levels and DNA damage in leukocytes and increased frequency of micronuclei in buccal cells was reported in glass workers (Vuyyuri et al., 2006). The measurement of biomarkers of oxidative damage is less commonly used in biomonitoring compared to measuring urinary arsenic and its metabolites in workers.

Environmental

Arsenic occurs naturally in a wide range of minerals in soils. It is the main constituent of more than 200 mineral species, and the most common of the arsenic minerals is arsenopyrite. It has been estimated that about one-third of the atmospheric flux of arsenic is of natural origin. Volcanic action is the most important natural source of arsenic, followed by low temperature volatilization. Mining, smelting of nonferrous metals, and burning of fossil fuels are the major industrial processes that contribute to anthropogenic arsenic contamination of the air, water, and soil. Concentrations of arsenic in air in remote locations range from <1 to 3 ng/m^3, but concentrations in cities may range up to 100 ng/m^3. Arsenic levels in groundwater average about 1–2 μg/l, except in areas with volcanic rock and sulfide mineral deposits where arsenic levels can range up to 3400 μg/l (Ferreccio et al., 2000). Surveys of arsenic concentrations in rivers and lakes indicate that most values are below 10 μg/l (WHO, 2010). Concentrations of arsenic in open ocean water are typically 1–2 μg/l. Drinking water in the United States generally contains an average of 2 μg/l of arsenic. Natural levels of arsenic in soil range from 1 to 40 mg/kg, with a mean of 5 mg/kg, with much higher levels occurring in mining areas, at waste sites, near high geological deposits of arsenic-rich minerals, or from pesticide application (ATSDR, 2007). Sediments in aquatic systems often have higher arsenic concentrations than that of water. Most sediment arsenic concentrations reported for rivers, lakes, and streams in the United States range from 0.1 to 4000 mg/kg, with higher levels occurring in areas of contamination (Franzblau and Lilis, 1989). Arsenic concentrations of <10,000 mg/kg (dry weight) were found in surface sediments near a copper smelter (ATSDR, 2007; WHO, 2001). Arsenic is also found in many foods at concentrations ranging from 20 to 140 µg/kg. Total arsenic concentrations may be substantially higher in certain seafoods. The general impression in the literature is that about 85 to >90% of the arsenic in the edible parts of marine fish and shellfish is organic arsenic (e.g., arsenobetaine, arsenochloline, dimethylarsinic acid) and that approximately 10% is inorganic arsenic (USEPA, 2003).

Environmental Fate and Bioaccumulation

In soil there are many biotic and abiotic processes controlling arsenic's overall fate and environmental impact. Arsenic in soil exists in various oxidation states and chemical species, depending upon the soil pH and oxidation–reduction potential (ATSDR, 2007). It is largely immobile in agricultural soils and tends to remain in upper soil layers (ATSDR, 2007). However, reducing conditions form soluble mobile forms of arsenic and leaching is greater in sandy soil than in clay loam (ATSDR, 2007). Mobility of arsenicals is typically very low to intermediate and sorption is higher in soils with higher percentage of clay or with more iron or aluminum content (USEPA, 2006).

Transport and partitioning of arsenic in water depends upon the chemical form of the arsenic and on interactions with other materials present (ATSDR, 2007). Under normal conditions in water, arsenic is present as soluble inorganic As^{V} because it is thermodynamically more stable in water than As^{III}. Soluble forms may be carried long distances through rivers, but arsenic may also be adsorbed from water onto sediments or soils, especially clays, iron oxides, aluminum hydroxides, manganese compounds, and organic material (USEPA, 1979, 1982b; Welch et al., 1988). Groundwater arsenic concentrations are usually controlled by adsorption rather than by mineral precipitation under oxidizing and mildly reducing conditions (ATSDR, 2007).

High temperature processes, such as coal and oil combustion, smelting operations, and refuse incineration, contribute to most of the anthropogenic arsenic emitted into the atmosphere (Pacyna, 1987). These fine particles, with a mass median diameter of about 1 µm, can reside in the atmosphere for about 7–9 days and be transported thousands of kilometers by wind and air currents until they are returned to earth by wet or dry deposition (Pacyna, 1987). Atmospheric fallout can also be a significant source of arsenic in coastal and inland waters near industrial areas (ATSDR, 2007).

Terrestrial plants may accumulate arsenic by root uptake from the soil or by absorption of airborne arsenic deposited on the leaves (USEPA, 1982a). Pitten et al. (1999) found that when plants were grown on soils high in arsenic, the arsenic level taken up by the plants remained comparably low. Arsenic speciation affects arsenic accumulation in plants. Arsenic uptake increases with increasing arsenic concentration, with organic arsenicals showing higher upward translocation than the inorganic arsenical (Burlo et al., 1999; Carbonell-Barrachina et al., 1999).

Arsenic bioaccumulation depends on the environmental setting, organism type, trophic status within the aquatic food chain, exposure concentrations, and route of uptake (Williams et al., 2006). Bottom-feeding and predatory fish can accumulate arsenic found in water, with arsenic mainly accumulating in the exoskeleton of invertebrates and in the livers of fish (ATSDR, 2007).

Organoarsenicals   Unlike other pesticides that degrade over time, MSMA, DSMA, CAMA, and cacodylic acid (DMA) contain elemental arsenic that does not degrade. The extent and speed of transformation and redistribution of the organic arsenical herbicides in soil is highly variable and depends mostly on the localized environmental conditions. Thus, persistence of applied organic arsenical herbicides can range from days to years, depending on the soil properties and ambient conditions such as soil moisture, temperature, chemical concentration, bacterial population, and amount of organic matter. Environmental fate studies show that organic arsenicals are stable and do not degrade by hydrolysis or by aquatic or soil photolysis. However, soil microorganisms could convert the organoarsenicals to inorganic arsenic and add to the background arsenic levels in the environment (USEPA, 2006).

Potential metabolites of applied organic arsenicals include volatile alkylarsines and inorganic arsenic (as arsenate or arsenite) along with carbon dioxide. Transformation to volatile alkylarsines is the only metabolism route that would directly reduce soil arsenic loading. This transformation has been shown to be possible in certain circumstances but is generally not expected to be a major route of dissipation.

Ecotoxicology

Acute Effects and Chronic Effects   In general, inorganic arsenicals are more toxic than organoarsenicals, and arsenite is more toxic than arsenate. Arsenic compounds cause acute and chronic effects in individuals, populations, and communities at concentrations ranging from a few micrograms to milligrams per liter, depending on species, time of exposure, and end points measured. These effects include lethality, inhibition of growth, photosynthesis and reproduction, and behavioral effects (USEPA, 1985; IPCS, 2001).

The mode of toxicity and mechanism of uptake of arsenate differ greatly from organism to organism (WHO, 2001). This may explain why there are interspecies differences in organism response to arsenate and arsenite. The primary

TABLE 7.2 **Permissible and Recommended Exposure Limits for Arsine and Arsenic Compounds**

Arsenic Compounds	CAS Number	PEL	TLV/BEI	REL/IDLH	ERPG
Arsine	7784-42-1	$0.2\,mg/m^3$ (OSHA, 2014a)	TLV $0.02\,mg/m^3$ (ACGIH, 2010)	REL: $0.002\,mg/m^3$ IDLH: 3 ppm NIOSH, 2007	ERPG(1) N/A; ERPG(2) $2.00\,mg/m^3$ (without serious, adverse effects); ERPG(3) $6.00\,mg/m^3$ (not life threatening) each up to 1 h exposures (AIHA, 2014)
Arsenic and inorganic arsenic compounds	7440-38-2	$0.01\,mg/m^3$ (OSHA, 2014b)	TLV $0.01\,mg/m^3$ BEI: $35\,\mu g$ As/l in urine (ACGIH, 2010)	REL: $0.002\,mg/m^3$ IDLH $5\,mg/m^3$	N/A
Organic arsenic compounds*		$0.5\,mg/m^3$ (OSHA, 2014a)	N/A	N/A	N/A

PEL: permissible exposure limit; TLV: threshold limit value; REL: recommended exposure limit; IDLH: Immediately Dangerous to Life or Health; ERPG: Emergency Response Planning Guidelines; OSHA: Occupational Safety and Health Administration; ACGIH: American Conference of Governmental Industrial Hygienists; AIHA: The American Industrial Hygiene Association; *: Value for organic arsenic is based on inorganic arsenic.

mechanism of arsenite toxicity is considered to result from its binding to protein sulfhydryl groups. Arsenate is known to affect oxidative phosphorylation by competition with phosphate. In environments where phosphate concentrations are high, arsenate toxicity to the biota is generally reduced. As arsenate is a phosphate analog, organisms living in elevated arsenate environments must acquire the nutrient phosphorous yet avoid arsenic toxicity (IPCS, 2001).

The freshwater residue data indicate that arsenic is not bioconcentrated but that lower forms of aquatic life may accumulate higher arsenic residues than fish (USEPA, 1982b). Recently, Culioli (2009) reported that the accumulation of arsenic decreased with increasing trophic level after testing the macroinvertebrates, bryophytes, and fish tissues.

INDUSTRIAL HYGIENE

ACGIH (2010) classifies elemental arsenic and inorganic arsenic compounds as confirmed human carcinogen (A1). NIOSH (2009) classifies elemental arsenic and inorganic arsenic compounds as potential occupational carcinogen. Table 7.2 lists the permissible and recommended exposure limits for arsine and arsenic compounds.

RISK ASSESSMENTS

The regulatory levels as well as the toxicity values determined for inorganic and organoarsenical compounds by various government agencies and organizations are listed below.

U.S. EPA

Inorganic Arsenic The U.S. EPA published the maximum contaminant level (MCL) of 0.01 mg/l (10 ppb) and maximum contaminant level goal (MCLG) of zero for arsenic on January 22, 2001 (USEPA, 2001). The MCL for arsenic was set at the level at which lung and bladder cancer reduction benefits were maximized at a cost that was justified by the benefits.

The reference dose (RfD) for inorganic arsenic for oral exposure is 0.0003 mg/kg/day, and this is based on hyperpigmentation, keratosis, and possible vascular complications in a large number of southwest Taiwanese subjects from blackfoot disease endemic area for arsenic (USEPA, 1993). The EPA concluded that inorganic arsenic is a human carcinogen from oral and inhalation exposures. An oral cancer slope factor for inorganic arsenic of 1.5 per mg/kg/day was derived based on skin cancers observed in a large Taiwanese cohort exposed to arsenic via drinking water (USEPA, 1993). A chronic reference (RfC) concentration for arsenic exposure via inhalation was not provided. The inhalation unit risk was determined as 0.0043 per $\mu g/m^3$ based on the respiratory cancer mortality observed in smelter workers. This cancer risk estimate corresponds to the air arsenic concentration of $0.0002\,\mu g/m^3$ for 1 in 1,000,000 cancer risk.

Arsine The U.S. EPA Integrated Risk Information System (IRIS) assessment indicates the chronic inhalation reference concentration of $0.05\,\mu g/m^3$ based on increased hemolytic effects and spleen changes in animal models (USEPA, 1994).

Organoarsenicals The U.S. EPA Office of Pesticide Programs (USEPA, 2006) derived a chronic RfD of 0.03 mg/kg/day for MMA based on decreased body weights, diarrhea, body weight gains, food consumption, histopathology of gastrointestinal tract and thyroid in rats, and 0.014 mg/kg/day for DMA based on regenerative proliferation of the bladder epithelial tissue in rats. For carcinogenicity effects, EPA determined that there was "no evidence for carcinogenicity" for MMA and that DMA was "not carcinogenic at doses that do not result in enhanced cell proliferation" (USEPA, 2006).

ATSDR

No inhalation risk levels were determined for inorganic or organic arsenical compounds. However, ATSDR (2007) determined oral toxicity values for potential noncancer effects associated with exposure to inorganic and organic arsenical compounds. The acute oral minimal risk level (MRL) for inorganic arsenic was determined as 0.005 mg/kg/day based on edema of the face and gastrointestinal effects (nausea, vomiting, diarrhea) observed in humans exposed to arsenic-contaminated soy sauce in Japan. The chronic oral MRL of 0.0003 mg/kg/day was based on skin lesions (hyperpigmentation and hyperkeratosis) observed in southwest Taiwanese populations exposed to arsenic in drinking water. ATSDR (2007) derived an intermediate oral MRL for MMA of 0.1 mg/kg/day based on diarrhea observed in rats and a chronic oral MRL of 0.01 mg/kg/day based on progressive glomerular nephropathy observed in mice. The chronic oral MRL for DMA was derived as 0.02 mg/kg/day based on vacuolization of the urothelium in the urinary bladder in female mice.

RIVM

Baars (2001) determined the tolerable daily intake (TDI) of 1.0 μg/kg/day for chronic oral exposures and the tolerable concentration in air as 1.0 μg/m^3 for chronic inhalation exposures for inorganic arsenic.

Other Standard Regulations and Guidelines

As a collaborative effort with public and private sectors in handling emergency situations, the U.S. EPA (USEPA, 2010) has derived the acute exposure guideline levels (AEGLs) for arsine and these values are presented in Table 7.3.

WHO

WHO guidance criterion for arsine was derived as 0.05 μg/m^3 based on the hematological effects observed in animal models (WHO, 2002). WHO recommends a provisional guideline for drinking water of 10 ppb because of scientific uncertainties in the assessment (WHO, 2010).

TABLE 7.3   Acute Exposure Guideline Levels (AEGLs) for Arsine

Arsine Exposure Time	AEGL 1 (ppm) (Discomfort)	AEGL 2 (ppm) (Impaired Escape)	AEGL 3 (ppm) (Life Threatening/ Death)
10 min	NR	0.3	0.91
30 min	NR	0.21	0.63
1 h	NR	0.17	0.5
4 h	NR	0.04	0.13
8 h	NR	0.02	0.06

Source: USEPA (2010).

Cal EPA

The Cal EPA has determined the acute inhalation reference exposure level as 0.2 μg/m^3 and the chronic inhalation reference exposure level as 0.015 μg/m^3 for inorganic arsenic compounds, including arsine (CalEPA, 2008). The inhalation unit risk for inorganic arsenic is determined as 0.0033 per μg/m^3. The inhalation slope factor is 12.0 per mg/kg/day based on lung tumors observed in smelter workers (CalEPA, 2009).

REFERENCES

ACGIH (American Conference of Governmental Industrial Hygienists) (2010) *TLVs and BEIs. Threshold Limit Values for Chemical Substances and Physical Agents and Biological Exposure Indices*, Cincinnati, OH: American Conference of Governmental Industrial Hygienists.

Adams, M., Bolger, O.M., and Gunderson, E.L. (1994) Dietary intake and hazards of arsenic. In: Chappell, W., Abernathy, C.O., and Cothem, C.R., editors, *Arsenic: Exposure and Health*, Northwood, UK: Science and Technology Letters, pp. 41–49.

AIHA (American Industrial Hygiene Association) (2014) *ERPG/WEEL Handbook*, Falls Church, VA: American Industrial Hygiene Association.

Alain, G., Tousignant, J., and Rozenfarb, E. (1993) Chronic arsenic toxicity. *Int. J. Dermatol.* 32(12):899–901.

ATSDR (2000) Arsenic. In: *Toxicological Profile (Update)*, Atlanta, GA: Agency for Toxic Substances and Disease Registry, Public Health Service, U.S. Department of Health and Human Services.

ATSDR (2007) Arsenic. In: *Toxicological Profile (Update)*, Atlanta, GA: Agency for Toxic Substances and Disease Registry, Public Health Service, U.S. Department of Health and Human Services.

Baars, A. (2001) *Re-evaluation of human-toxicological maximum permissible risk levels.* RIVM Report 711701025. Bilthoven, The Netherlands: National Institute for Public Health and the Environment.

Biagini, R., Rivero, M., Salvador, M., and Cordoba, S. (1978) Chronic arsenism and lung cancer. *Arch. Argent. Dermatol.* 28:151–158.

Blom, S., Lagerkvist, B. and Linderholm, H. (1985) Arsenic exposure to smelter workers. Clinical and neurophysiological studies. *Scand. J. Work Environ. Health* 11(4):265–269.

Bulmer, F.M.R., Rothwell, H.E., Polack, S.S., and Stewart, D.W. (1940) Chronic arsenic poisoning among workers employed in the cyanide extraction of gold: a report of fourteen cases. *J. Ind. Hyg. Toxicol.* 22:111–124.

Burlo, F., Guijarro, I., Carbonell-Barrachina, A.A., Valero, D., and Martinez-Sanchez, F. (1999) Arsenic species: effects on and accumulation by tomato plants. *J. Agric. Food Chem.* 47(3):1247–1253.

CalEPA (2008) *Inorganic Arsenic Reference Exposure Levels.* Available at http://oehha.ca.gov/air/hot_spots/2008/AppendixD1_final.pdf#page=68 (accessed March 2011).

CalEPA (2009) *Adoption of the Revised Air Toxics Hot Spots Program Technical Support Document for Cancer Potency Factors.* Available at http://www.oehha.ca.gov/air/hot_spots/tsd052909.html (accessed March 2011).

Cantor, K.P. (1997) Drinking water and cancer. *Cancer Causes Control* 8(3):292–308.

Carbonell-Barrachina, A.A., Burlo, F., Valero, D., Lopez, E., Martinez-Romero, D., and Martinez-Sanchez, F. (1999) Arsenic toxicity and accumulation in turnip as affected by arsenic chemical speciation. *J. Agric. Food Chem.* 47(6):2288–2294.

Carter, D.E., Aposhian, H. V., and Gandolfi, A. J. (2003) The metabolism of inorganic arsenic oxides, gallium arsenide, and arsine: a toxicochemical review. *Toxicol. Appl. Pharmacol.* 193(3):309–334.

Cebrian, M.E., Albores, A., Aquilar, M., and Blakely, E. (1983) Chronic arsenic poisoning in the North of Mexico. *Hum. Toxicol.* 2:121–133.

Centeno, J.A., Martinez, L., Ladich, E.R., Page, N.P., Mullick, F.G., Ishak, K.G., Zheng, B., Gibb, H., Thompson, C., and Longfellow, D. (2000) *Arsenic-Induced Lesions,* Washington, DC: Armed Forces Institute of Pathology.

Chakaborty, A.K. and Saha, K.C. (1987) Arsenical dermatosis from tubewell water in West Bengal. *Indian J. Med. Res.* 85:326–334.

Chao, C.L. and Hwang, K.C. (2005) Arsenic burden survey among refuse incinerator workers. *J. Postgrad. Med.* 51(2):98–103.

Chen, C.J., Kuo, T.L., and Wu, M.M. (1988) Arsenic and cancers (letter). *Lancet* 1:414–415.

Cocker, J., Morton, J., Warren, N., Wheeler, J.P., and Garrod, A.N. (2006) Biomonitoring for chromium and arsenic in timber treatment plant workers exposed to CCA wood preservatives. *Ann. Occup. Hyg.* 50(5):517–525.

Cowlishaw, J.L., Pollard, E.J., Cowen, A.E., and Powell, L.W. (1979) Liver disease associated with chronic arsenic ingestion. *Aust. N. Z. J. Med.* 9(3):310–313.

CRC (1999) *Handbook of Chemistry and Physics,* Boca Raton, FL: CRC Press, Inc.

Culioli, J.L. (2009) Trophic transfer of arsenic and antimony in a freshwater ecosystem: a field study. *Aquat. Toxicol.* 94(4):286–293.

Domingo, J.L. (1994) Metal-induced developmental toxicity in mammals: a review. *J. Toxicol. Environ. Health* 42(2):123–141.

Engel, R.R. and Smith, A.H. (1994) Arsenic in drinking water and mortality from vascular disease: an ecologic analysis in 30 counties in the United States. *Arch. Environ. Health* 49(5):418–427.

Enterline, P.E., Day, R., and Marsh, G.M. (1995) Cancers related to exposure to arsenic at a copper smelter. *Occup. Environ. Med.* 52(1):28–32.

Enterline, P.E., Henderson, V.L., and Marsh, G.M. (1987) Exposure to arsenic and respiratory cancer. A reanalysis. *Am. J. Epidemiol.* 125(6):929–938.

Farmer, J.G. and Johnson, L.R. (1990) Assessment of occupational exposure to inorganic arsenic based on urinary concentrations and speciation of arsenic. *Br. J. Ind. Med.* 47(5):342–348.

Feldman, R.G., Niles, C.A., Kelly-Hayes, M., Sax, D.S., Dixon, W.J., Thompson, D.J., et al. (1979) Peripheral neuropathy in arsenic smelter workers. *Neurology* 29(7):939–944.

Ferreccio, C., Gonzalez, C., Milosavjlevic, V., Marshall, G., Sancha, A.M., and Smith, A.H. (2000) Lung cancer and arsenic concentrations in drinking water in Chile. *Epidemiology* 11(6):673–679.

Fowler, B.A. and Weissberg, J.B. (1974) Arsine poisoning. *N. Engl. J. Med.* 291(22):1171–1174.

Franzblau, A. and Lilis, R. (1989) Acute arsenic intoxication from environmental arsenic exposure. *Arch. Environ. Health* 44(6):385–390.

Heyman, A., Pfeiffer, J.B., Jr., Willett, R.W., and Taylor, H.M. (1956) Peripheral neuropathy caused by arsenical intoxication; a study of 41 cases with observations on the effects of BAL (2,3-dimercapto-propanol). *N. Engl. J. Med.* 254(9):401–409.

Holmquist, I. (1951) Occupational arsenical dermatitis: a study among employees at a copper ore smelting work using investigations of skin reactions to contact with arsenic compounds. *Acta Derm. Venereol.* 31:1–214.

Hong, H.L., Fowler, B.A., and Boorman, G.A. (1989) Hematopoietic effects in mice exposed to arsine gas. *Toxicol. Appl. Pharmacol.* 97(1):173–182.

HSDB (2005) *Records for Elemental Arsenic and Arsenic Compounds,* National Library of Medicine.

Hu, C.W., Pan, C.H., Huang, Y.L., Wu, M.T., Chang, L.W., Wang, C.J., et al. (2006) Effects of arsenic exposure among semiconductor workers: a cautionary note on urinary 8-oxo-7,8-dihydro-2′-deoxyguanosine. *Free Radic. Biol. Med.* 40(7):1273–1278.

Huet, P.M., Guillaume, E., Cote, J., Legare, A., Lavoie, P., and Viallet, A. (1975) Noncirrhotic presinusoidal portal hypertension associated with chronic arsenical intoxication. *Gastroenterology* 68(5 Pt 1):1270–1277.

IARC (1973) Arsenic and inorganic arsenic compounds. In: *IARC Monographs on the Evaluation of Carcinogenic Risk of Chemicals to Humans*, vol. 2, Lyon, France: International Agency for Research on Cancer, pp. 48–73.

IARC (1980) Arsenic and inorganic arsenic compounds. In: *IARC Monographs on the Evaluation of Carcinogenic Risk of Chemicals to Humans*, vol. 23, Lyon, France: International Agency for Research on Cancer, pp. 39–141.

IPCS (2001) *Arsenic.* Available at http://www.who.int/ipcs/assessment/public_health/arsenic/en (accessed March 2011).

Jarup, L., Pershagen, G., and Wall, S. (1989) Cumulative arsenic exposure and lung cancer in smelter workers: a dose-response study. *Am. J. Ind. Med.* 15(1):31–41.

Jhaveri, S.S. (1959) A case of cirrhosis and primary carcinoma of the liver in chronic industrial arsenical intoxication. *Br. J. Ind. Med.* 16:248–250.

Jones, N. (1907) Arseniureted hydrogen poisoning: with report of five cases. *JAMA* 48:1099–1105.

Juhlin, L. and Ortonne, J.P. (1986) Red scalp hair turning dark-brown at 50 years of age. *Acta Derm. Venereol.* 66(1): 71–73.

Kabir, H. and Bilgi, C. (1993) Ontario gold miners with lung cancer. Occupational exposure assessment in establishing work-relatedness. *J. Occup. Med.* 35(12):1203–1207.

Kitchin, K.T. (2001) Recent advances in arsenic carcinogenesis: modes of action, animal model systems, and methylated arsenic metabolites. *Toxicol. Appl. Pharmacol.* 172(3): 249–261.

Kitchin, K.T. and Conolly, R. (2010) Arsenic-induced carcinogenesis—oxidative stress as a possible mode of action and future research needs for more biologically based risk assessment. *Chem. Res. Toxicol.* 23(2):327–335.

Kveldsberg, G.R. and Ward, H.P. (1972) Leukemia in arsenic poisoning. *Ann. Intern. Med.* 35:935–937.

Lagerkvist, B.J. and Zetterlund, B. (1994) Assessment of exposure to arsenic among smelter workers: a five-year follow-up. *Am. J. Ind. Med.* 25(4):477–488.

Landrigan, P.J. (1981) Arsenic—state of the art. *Am. J. Ind. Med.* 2(1):5–14.

Lee, A.M. and Fraumeni, J.F. Jr. (1969) Arsenic and respiratory cancer in man: an occupational study. *J. Natl. Cancer Inst.* 42(6):1045–1052.

Lee-Feldstein, A. (1986) Cumulative exposure to arsenic and its relationship to respiratory cancer among copper smelter employees. *J. Occup. Med.* 28(4):296–302.

Le Quesne, P.M. and McLeod, J.G. (1977) Peripheral neuropathy following a single exposure to arsenic. Clinical course in four patients with electrophysiological and histological studies. *J. Neurol. Sci.* 32(3):437–451.

Lundgren, K.D., Richtner, N.G., and Sjostrand, T. (1951) Changes of respiratory tract in workers of Ronnskar smelting workers probably due to arsenic trioxide intoxication. *Nord. Med.* 46:1556–1560.

Mayers, M.R. (1954) Occupational arsenic poisoning; report of a case. *AMA Arch. Ind. Hyg. Occup. Med.* 9(5):384–388.

Morris, J.S., Schmid, M., Newman, S., Scheuer, P.J., and Sherlock, S. (1974) Arsenic and noncirrhotic portal hypertension. *Gastroenterology* 66(1):86–94.

Neshiwat, L.F., Friedland, M.L., Schorr-Lesnick, B., Feldman, S., Glucksman, W.J., and Russo, R.D., Jr. (1992) Hepatic angiosarcoma. *Am. J. Med.* 93(2):219–222.

Nevens, F., Fevery, J., Van Steenbergen, W., Sciot, R., Desmet, V., and De Groote, J. (1990) Arsenic and non-cirrhotic portal hypertension. A report of eight cases. *J. Hepatol.* 11(1):80–85.

NIOSH (2007) DHHS (NIOSH) Publication No. 2005-149. Available at http://www.cdc.gov/niosh/docs/2005-149/pdfs/2005-149.pdf

NIOSH (2009) *NIOSH Pocket Guide to Chemical Hazards*, Washington, DC: Department of Health and Human Services.

NRC (1999) *Arsenic in Drinking Water*, Washington, DC: National Academy Press.

OSHA (2014a). OSHA Publication 29 CFR 1910.1000. Available at http://www.ecfr.gov/cgi-bin/text-idx?SID=0cceb4d61d14745 c59aefca20860eede&node=se29.6.1910_11000&rgn=div8

OSHA (2014b). OSHA Publication 29 CFR 1910.1018. Available at http://www.ecfr.gov/cgi-bin/text-idx?SID=0cceb4d61d14745 c59aefca20860eede&node=se29.6.1910_11018&rgn=div8

Ott, M.G., Holder, B.B., and Gordon, H.L. (1974) Respiratory cancer and occupational exposure to arsenicals. *Arch. Environ. Health* 29(5):250–255.

Pacyna, J. (1987) Atmospheric emissions of arsenic, cadmium, lead and mercury from high temperature processes in power generation and industry. In: Hutchinson, T.C. and Meema, K.M., editors. *Lead, Mercury, Cadmium and Arsenic in the Environment*, New York, NY: John Wiley & Sons, Ltd., pp. 69–87.

Pitten, F.A., Muller, G., Konig, P., Schmidt, D., Thurow, K., and Kramer, A. (1999) Risk assessment of a former military base contaminated with organoarsenic-based warfare agents: uptake of arsenic by terrestrial plants. *Sci. Total Environ.* 226(2–3):237–245.

Ringenberg, Q.S., Doll, D.C., Patterson, W.P., Perry, M.C., and Yarbro, J.W. (1988) Hematologic effects of heavy metal poisoning. *South. Med. J.* 81(9):1132–1139.

Rossman, T. (2007) Arsenic. In: Rom, W. and Markowitz, S., editors. *Environmental and Occupational Medicine*, 4th ed., Hagerstown, MD: Lippincott Williams and Wilkins, pp. 1006–1017.

Simonato, L., Moulin, J.J., Javelaud, B., Ferro, G., Wild, P., Winkelmann, R., and Saracci, R. (1994) A retrospective mortality study of workers exposed to arsenic in a gold mine and refinery in France. *Am. J. Ind. Med.* 25(5):625–633.

Straif, K., Benbrahim-Tallaa, L., Baan, R., Grosse, Y., Secretan, B., El Ghissassi, F., et al. (2009) A review of human carcinogens—part C: metals, arsenic, dusts, and fibres. *Lancet Oncol.* 10(5):453–454.

Taylor, P.R., Qiao, Y.L., Schatzkin, A., Yao, S.X., Lubin, J., Mao, B.L., et al. (1989) Relation of arsenic exposure to lung cancer among tin miners in Yunnan Province, China *Br. J. Ind. Med.* 46(12):881–886.

Tello, E.E. (1986) Hydro-arsenicisms: what is the Argentine chronic hydroarsenicism (HACREA)? *Arch. Argent. Dermatol.* 36:197–216.

Telolahy, P., Javelaud, B., Cluet, J., de Ceaurriz, J., and Boudene, C. (1993) Urinary excretion of porphyrins by smelter workers chronically exposed to arsenic dust. *Toxicol. Lett.* 66(1):89–95.

The Merck Index (1989) *The Merck Index: An Encyclopedia of Chemicals, Drugs, and Biologicals*, 11th ed., Rahway, NJ: Merck & Co.

Thomas, D.J., Styblo, M., and Lin, S. (2001) The cellular metabolism and systemic toxicity of arsenic. *Toxicol. Appl. Pharmacol.* 176:127–144.

Tseng, W.P., Chu, H.M., How, S.W., Fong, J.M., Lin, C.S., and Yeh, S. (1968) Prevalence of skin cancer in an endemic area of chronic arsenicism in Taiwan. *J. Natl. Cancer Inst.* 40:453–463.

USEPA (1979) Introduction and technical background, metals and inorganics, pesticides and PCBs. In: *Water-Related Environmental Fate of 129 Priority Pollutants*, vol. I, Washington, DC: U.S. Environmental Protection Agency.

USEPA (1982a) *Exposure and Risk Assessment for Arsenic*, Washington, D.C.: U.S. Environmental Protection Agency.

USEPA (1982b) *Inductively coupled plasma-atomic emission spectrometric method for trace element analysis of water and wastes- method 200.7*, Cincinnati, OH: U.S. Environmental Protection Agency.

USEPA (1985) *Ambient water quality criteria for arsenic*. 440/5-84-033. Duluth, MN: U.S. Environmental Protection Agency.

USEPA (1993) *Toxicological Review of Inorganic Arsenic*. Available at http://www.epa.gov/iris/subst/0278.htm.

USEPA (1994) *Toxicological Review of Arsine*. Available at http://www.epa.gov/iris/subst/0672.htm.

USEPA (2001) *National Primary Drinking Water Regulations: Arsenic and Clarifications to Compliance and New Source Contaminants Monitoring*, Washington, DC: U.S. Environmental Protection Agency.

USEPA (2003) *Technical summary of information available on the bioaccumulation of arsenic in aquatic organisms*. Available at http://www.epa.gov/waterscience/criteria/arsenic/tech-sum-bioacc.pdf.

USEPA (2006) *Revised Reregistration Eligibility Decision for MSMA, DSMA, CAMA, and Cacodylic Acid*, Washington, DC: U.S. Environmental Protection Agency.

USEPA (2008) *Human Health Risk Assessment and Ecological Risk Assessment for the Reregistration Eligibility. Decision Document for Inorganic Arsenicals and Chromium-Based Wood Preservatives*, Washington, DC: U.S. Environmental Protection Agency.

USEPA (2010) *Acute Exposure Guideline Levels (AEGLs)*. Available at http://www.epa.gov/opptintr/aegl/pubs/results2.htm. (accessed March 2011).

Vuyyuri, S.B., Ishaq, M., Kuppala, D., Grover, P., and Ahuja, Y.R. (2006) Evaluation of micronucleus frequencies and DNA damage in glass workers exposed to arsenic. *Environ. Mol. Mutagen.* 47(7):562–570.

Weinberg, S.L. (1960) The electrocardiogram in acute arsenic poisoning. *Am. Heart J.* 60:971–975.

Welch, A.H., Helsel, D.R., Focazio, M.J., and Watkins, S.A. (1999) Arsenic in ground water supplies of the United States. In: Chappell, W.R., Abernathy, C.O., and Calderon, R.L., editors. *Arsenic Exposure and Health Effects*, New York, NY: Elsevier Science, pp. 9–17.

Welch, A.H., Lico, M.S., and Hughes, J.L. (1988) Arsenic in groundwater of the western United States. *Ground Water* 26(3):333–347.

WHO (2001) Arsenic and arsenic compounds. In: *Environmental Health Criteria 224*. Available at http://www.inchem.org/documents/ehc/ehc/ehc224.htm. (accessed March 2011).

WHO (2002) *Arsine: Human Health Aspects*. Available at http://www.inchem.org/documents/cicads/cicads/cicad47.htm (accessed March 2011).

WHO (2010) Exposure to arsenic: a major public health concern. In: *Public Health and Environment*, Geneva, Switzerland: World Health Organization.

Williams, L., Schoof, R.A., Yager, J.W., and Goodrich-Mahoney, J. (2006) Arsenic bioaccumulation in freshwater fishes. *Hum. Ecol. Risk Assess.* 12(5):904–923.

Wu, M.M., Kuo, T.L., Hwang, Y.H., and Chen, C.J. (1989) Dose-response relation between arsenic concentration in well water and mortality from cancers and vascular diseases. *Am. J. Epidemiol.* 130:1123–1132.

Yu, H.S., Lee, C.H., and Chen, G.S. (2002) Peripheral vascular diseases resulting from chronic arsenical poisoning. *J. Dermatol.* 29(3):123–130.

Zaldivar, R. (1974) Arsenic contamination of drinking water and food-stuffs causing endemic chronic poisoning. *Beitr. Pathol.* 151:384–400.

Zierler, S., Theodore, M., Cohen, A., and Rothman, K.J. (1988) Chemical quality of maternal drinking water and congenital heart disease. *Int. J. Epidemiol.* 17(3):589–594.

8

BARIUM

Marie M. Bourgeois

First aid: Skin: Remove contaminated clothing; wash skin with plenty of water and soap. Eyes: Flush eyes with plenty of water. Inhalation: Fresh air, administer oxygen, if necessary. Ingestion: Rinse mouth; induce vomiting in conscious patients, seek medical attention immediately

Target organ(s): Respiratory system, eye, skin, immune system (allergic reactions), central nervous system, and heart

Occupational exposure limits: OSHA PEL: 0.5 mg/m^3 (soluble compounds), 15 mg/m^3 (barium sulfate dust), TLV: 0.5 mg/m^3 as TWA A4 (not classifiable as a human carcinogen); (ACGIH, 2008). EU OEL: 0.5 mg/m^3 as TWA (EU 2006)

IDHL barium chloride levels of 50 mg/m^3 (NIOSH)

Reference values: 0.56 mg/kg/day with UF of 300

Risk phrases: 11–14/15

Flammable. Reacts violently with water, liberating highly flammable gases

Safety phrases: 7/8-30-43

Keep container tightly closed and dry. Never add water to this product

BACKGROUND AND USES

Barium (symbol Ba; CASRN 7440-39-3) is a dense, silver-white metal that oxidizes rapidly in air, assuming a silver-yellow color, when present in its elemental state. The TSCA inventory, established by the Toxic Substances Control Act of 1976, lists more than 270 compounds containing barium. Barium is naturally found in two minerals, barytes, or barite ("heavy spar") (BaSO$_4$), and witherite (BaCO$_3$). Witherite is no longer mined for commercial purposes. Some barium compounds are soluble in water (barium sulfide, barium chloride, barium hydroxide), but the more common compounds found in nature (barium sulfate and barium carbonate) are less soluble.

Barium is commonly used in the manufacture of cathode ray tubes in the form of barium–aluminum alloy evaporation getters (gas absorbers) (HSDB, 2010). Barium sulfate ore is used primarily in the petroleum industry, in which it is used for drilling muds, which act as lubricants. It also has been used in the manufacture of lithopone, a white pigment used in the paint industry. Barium is used for metal alloys (barium nickel) in spark plugs, as filler for paper, and in textiles, leather, soap, rubber, and linoleum. Compounds such as barium chloride, barium hydroxide, and barium carbonate are used in saltwater-resistant cement, in the manufacture of ceramics and glass, and in pesticides, rodenticides, and dyestuffs. Barium nitrate and barium chlorate produce a green color in pyrotechnics. Barium sulfide phosphoresces following exposure to the light making it ideal for fluorescent lamps. Wool used for stuffing mattresses is treated with barium chloride (BaCl$_2$), which makes the wool whiter, more elastic, and more moth resistant. Barium chloride also has been used in the electronics industry to remove gases from vacuum tubes. Barium compounds (e.g., barium sulfate) are used in the medical industry for radioimaging, because they are radiopaque, extremely insoluble, and essentially nontoxic in humans. Radioactive barium isotopes are useful as bone-scanning agents in the study of bone metabolism.

PHYSICAL AND CHEMICAL PROPERTIES

Barium is a dense alkaline earth metal in Group IIA of the periodic table found at concentrations ranging from 15 to

Hamilton & Hardy's Industrial Toxicology, Sixth Edition. Edited by Raymond D. Harbison, Marie M. Bourgeois, and Giffe T. Johnson.

TABLE 8.1   Chemical and Physical Properties of Barium

Characteristic	Information
Chemical name	Barium
Synonyms/trade names	Bario, baryum, bario [Spanish], baryum [French], barium, soluble compounds, barium and soluble compounds, HSDB 4481, EINECS 231-149-1
Chemical formula	Ba
CASRN	7440-39-3
Molecular weight	137.33 g/mol
Melting point	725 °C
Boiling point	1640 °C
Vapor pressure	2 mmHg at 20 °C
Density/specific gravity	3.5 g/cm^3 at 20 °C
Flashpoint	Not available
Water solubility at 25 °C	Reacts with water; slightly soluble in ethanol
Explosion limits	Not available
Octanol/water partition coefficient (log K_{OW})	Cannot be estimated
Henry's law constant	Not available
Odor threshold	Odorless
Conversion factors	Not available

Source: NIOSH (2010).

3500 parts per million (ppm), dry weight. It is a naturally occurring component of the Earth's crust, particularly in areas rich in sandstone, shale, igneous rock, and coal. Under normal conditions, barium is present in +2 valence state. It does not occur in its free state; it is stable as an inorganic complex such as barite (barium sulfate) and witherite (barium carbonate). Selected chemical and physical properties of barium are presented in Table 8.1.

ENVIRONMENTAL FATE AND BIOACCUMULATION

Barium is released to the environment from natural and anthropogenic sources. Barium is present in the air, water, soil, and food, although background levels are very low (ATSDR, 2007). For example, background barium in ambient air is approximately 0.0015 parts per billion (ppb) or less. Perturbations in pH, redox states, and the availability of oxides of metals such as aluminum, manganese, and silicon may alter environmental partitioning of elemental barium and barium compounds. The presence of sulfate and carbonate elicits similar effects. Barium may enter the environment via wet and dry deposition, leaching, natural weathering, and soil and sediment adsorption. Biomagnification in aquatic and terrestrial food chains is possible.

Research indicates that, although only a limited amount of barium accumulates in most plants, it is actively absorbed from the soil by legumes, grains, forage plants, some fir trees, mushrooms, and legumes such as black walnut, hickory, and Brazil nuts. It does not appear that vegetation can incorporate airborne barium, although dry and wet deposition on foliage has been noted. Barium has been found in animal products such as milk and eggs. Bioconcentration factors have been calculated for some biota (soil to plants 0.4, soil to insects 0.2, and soil to rodent 0.2) but there are significant data gaps. Uptake in marine animals and plants is significantly higher and represents an important clearance mechanism in saltwater.

The primary exposure route in humans is ingestion. Barium has been detected in the blood, urine, feces, and biological tissues of humans; however, there are no data correlating exposure levels with the levels in tissues and fluids. Barium is eliminated in the urine (5%) and the feces (20%) and almost completely eliminated from the bloodstream within 24 h (HSDB, 2010). Barium metabolizes similarly to calcium and has been found to localize to the bone (IPCS, 1991). Barium preferentially deposits at the periosteal surfaces, which are the most active areas of bone growth (HSDB, 2010). The bioavailability of barium and its compounds has been studied extensively in foods, water, and air but there are limited data for soil (McCauley and Washington, 1983). This represents a significant avenue for potential exposures so more information is needed (IPCS, 2001).

ECOTOXICOLOGY

Historically, naturally occurring barium has been found in 99% of surface waters; however, far more barium is released to the groundwater so it is likely that the barium concentrations are even higher there. Most municipal and surface water contain an average of 0.03 ppm barium. Groundwater concentrations in well water may be as high as 10 ppm; this exceeds the Environmental Protection Agency (EPA) limit of 2 ppm. The primary source of barium and barium compounds in groundwater is sedimentary rock leaching and erosion.

The Toxic Release Inventory (TRI) estimated that there were environmental releases of approximately 105,000 metric tons by manufacturing and processing facilities of barium/barium compounds in 2004. This is likely an underestimate given that mandatory reporting of releases is restricted to certain kinds of facilities. These include facilities employing 10 or more full-time employees, facilities that combust coal and/or oil for the generation of electricity for commercial distribution, and businesses performing solvent recovery services for hire (EPA, 2005). Water-soluble forms of barium (hydroxides, nitrates, chlorides) typically do not persist following release unless they combine with other less soluble forms. These insoluble forms (sulfates and carbonates) can persist for a lengthy period. If insoluble forms of barium are released onto land, they will adsorb with particles of soil. Barium concentrations in soil may range from about 15 to 3500 ppm. Barium has been identified in roughly half of the

EPA's proposed National Priority List hazardous waste sites. Barium sulfate and barium carbonate are the most common forms found in the soil and water.

MAMMALIAN TOXICOLOGY

Acute Effects

The soluble barium salts may present an acute hazard, but they have not been prominent in causing occupational illness. Ingestion of barium chloride or carbonate can lead to toxic effects as low as 0.2–0.5 g (IPCS, 1991). Soluble barium compounds such as the oxide, chloride, or carbonate can cause dermal or mucous membrane irritation (Elkins, 1959). It is absorbed via the mucosal tissue in the respiratory system and gastrointestinal tract. The amount absorbed is dependent on factors, including age, solubility of the barium compound, and stomach contents. Stewart and Hummel (1984) reported that a worker who was exposed to molten barium chloride when a furnace exploded had abnormal electrocardiographic tracings for about 9 h, although no other adverse effects were associated with this accident. Barium carbonate is used as a rat poison, and accidental ingestion of this compound has resulted in gastroenteritis, electrocardiographic changes, and muscular paralysis which is possibly from severe hypokalemia (IPCS, 1991; Morton, 1945; Bowen, 2010). Barium compounds profoundly affect all muscle types, especially the heart (IPCS, 1991).

Bertarelli (1931) found that extended contact with wool that had been treated with $BaCl_2$ did not result in inhalation of sufficient quantities of this compound to cause toxic signs or symptoms. Doig (1976) reported a fatal case of a worker exposed for several days in the mid-1920s to dust from barium peroxide crushing operations. The baryta (BaO), BaCO3, and the peroxide presumably were converted to the soluble chloride in the stomach. Death on the third day was preceded by abdominal pain, vomiting, tachycardia, dyspnea, cyanosis, and paralysis of the right arm and leg. Nonoccupational poisoning has resulted from accidental ingestion associated with errors in food preparation. Paralysis has been noted in inhalation of barium carbonate.

Most acute human data are the result of intentional or accidental ingestion of barium chloride and barium carbonate. Intentional ingestion of barium carbonate or barium nitrate, common rodenticides in developing countries, is relatively rare but has been used in suicide attempts (Johnson and VanTassell, 1991; Torka et al., 2009). Barium chloride has multiple industrial applications. Physical effects of ingestion of barium and barium compounds include hypokalemia, tachycardia, hypertension (or hypotension), and muscle weakness. Barium acts as a potassium channel antagonist, decreasing blood potassium levels. This shunts potassium from the extracellular to intracellular compartments. Other symptoms include nausea, vomiting, cramping, and loose stool. Similar symptomology may occur in acute airborne exposures.

Barium sulfate is extremely insoluble, a property that has made it useful as an opaque contrast medium in radiography. Under normal conditions barium sulfate is safe enough for diagnostic X-rays during pregnancy (Han et al., 2011). If barium sulfate or other insoluble compounds enter the bloodstream through a gastrointestinal perforation, there is a chance of toxicity.

Chronic Effects

Baritosis, a benign pneumoconiosis, has been reported in workers who have inhaled barium sulfate in baryta (BaO) mines and also in lithopone workers. This condition presents no respiratory distress, but the chest X-rays show small, sharply circumscribed nodules distributed throughout the lungs. Some reports suggest that baritosis is accompanied by a significant increase in mortality from pneumonia and tuberculosis; however, exposure to fibrogenic, dust-free silica may have occurred in such cases (Chong et al., 2006; Pendergrass and Greening, 1953). Seaton et al. (1986) detected silicosis in barium sulfate miners, which suggests that in some cases other materials associated with barium mining and processing may cause illness.

Doig (1976) studied a group of workers in a barytes crushing factory and found air levels of about 22–92 mg/ m^3. Baritosis was detected radiographically in some cases as early as 18–21 months after initial employment. The chest X-rays were noteworthy for the "intense radiopacity of the discrete opacities," which were profuse and evenly disseminated, and for the absence of any other abnormalities of the lungs. The radiographic opacities slowly disappeared after exposure stopped. Pulmonary function tests were essentially normal even though the X-rays indicated substantial radiopacity. Doig noted that the chest X-rays produced by well-established baritosis are similar to those obtained in cases of stannosis.

Essing et al. (1976) studied workers chronically exposed to barium carbonate dust and found no respiratory symptoms or radiographic findings associated with exposure to this material. In the same study the researchers reported that 3 of the 12 workers had elevated blood pressure and two had abnormal electrocardiographic results. These findings are consistent with those reported for accidental or intentional ingestion of barium compounds and with the results of studies of animals exposed to barium compounds, in which elevated blood pressure and abnormal electrocardiographic findings also were observed.

Overexposure to barium via the inhalation route results in characteristic X-ray findings because of the dense nature of this element. Discrete opacities are seen evenly distributed throughout the chest, and they disappear if exposure is eliminated.

Intoxication associated with barium exposure by other routes is most notably associated with increased blood pressure and electrocardiographic abnormalities (Tsai et al., 2011). One worker who was exposed to very high but unmeasured levels of barium carbonate by the inhalation route suffered a broad spectrum of adverse health effects (Shankle and Keane, 1988). The worker experienced nausea, vomiting, stomach cramps, loss of tendon reflex response, low serum potassium, kidney failure, muscle weakness, and paralysis. These findings are consistent with those associated with accidental or intentional human ingestion of barium compounds and with those seen in oral exposure studies in animals (Hicks et al., 1986; Payden et al., 2011).

Barium stimulates contraction of smooth, cardiac, and striated muscles, which may partly explain the symptoms noted. Barium levels in the body are measured in the blood, urine, feces, and bone. In the worker who was exposed to molten barium chloride from a furnace explosion, plasma barium level was 12.4 mg/l at approximately 2 h postburn. At 4, 8, 12, and 18 h postburn the plasma barium levels were 10.3, 8.2, 8.9, and 4.0 mg/l, respectively. Welders who used welding rods with barium flux for 3 hours had elevated levels of barium at the end of a 3-hour welding session and also the next morning (Dare et al., 1984; Yeo, 1986). Apart from symptomatic treatment for barium overexposure, the most important consideration is monitoring potassium levels and providing potassium supplementation if needed.

Using the EPA's 1986 Guidelines for Carcinogen Risk Assessment, barium has been classified as Group D (not classifiable as to human carcinogenicity). There have been no studies that assess the carcinogenicity of barium following chronic inhalation or dermal exposures. Several long-term studies have examined the potential carcinogenicity of ingested barium using rats and mice (NTP, 1994; Schroeder and Michener, 1975a, 1975b). These studies did not report an increased incidence of neoplastic lesions in either species. Significant data gaps exist for immunotoxicity, genotoxicity, developmental toxicity, and reproductive toxicity.

Chemical Pathology

Barium can mimic calcium and thus affect nerve impulse transmission in the nervous system as well as ventricular automaticity in the heart (IPCS, 1990). Calcium can release a neurotransmitter only after depolarization; barium does not require prior depolarization to release a neurotransmitter from the synapse (IPCS, 1990).

Hypothesized Modes of Action

The mechanism of action underlying barium toxicity has not been elucidated. It is hypothesized that exposure to high doses produces muscle weakness, paralysis, ventricular tachycardia, and alterations in blood pressure. It is thought that these symptoms are the result of barium's action on calcium-activated potassium channels, which increases intracellular potassium. Barium stimulates contractions in muscle tissues, vasoconstriction, peristaltic activity, muscle tremors and fasciculation, and dysrhythmias. These symptoms are more common following exposures to soluble barium compounds.

SOURCES OF EXPOSURE

Occupational

The majority of occupational exposures to barium occur via inhalation in barium sulfate and barium carbonate miners and processors. Recent estimates indicated that more than 10,000 individuals were at risk of occupational exposures to barium and up to 474,000 may have been exposed to barium compounds. When individuals must work around high levels of barium compounds, engineering controls are recommended to limit exposure. Since the respiratory route is the most important occupational route, respirators are recommended if engineering controls are inadequate (Moreton and Jenkins, 1985). The irritating nature of several barium compounds indicates that protective clothing and eye wear are needed for those working with these substances. If the skin becomes contaminated with barium compounds, it should be washed as soon as is practical. Fugitive dust inhalation is a common source of occupational exposures in barium processing and mining facilities.

Environmental

Barium is the fourteenth most abundant element in the earth's crust. As previously stated, barium and barium compounds exist in varying concentrations in all environmental media (Zielinski, 2011). Studies indicate that bioaccumulation is possible. The Third National Health and Nutrition Examination Survey (NHANES) measured barium concentration in the urine of study participants 6 years and older (CDC, 2008; Padilla et al., 2010). The geometric mean for the urinary barium across all ages was 1.44 (1.31–1.58) μg per gram of creatinine. No information has been collected regarding populations living close to hazardous waste sites with high levels of barium. Recent data on barium levels in plants and ambient air, soils, and groundwater, particularly from NPL sites, are necessary to close this data gap.

INDUSTRIAL HYGIENE

The Occupational Safety and Health Administration (OSHA) and the American Conference of Governmental Industrial Hygienists (ACGIH) have set the permissible exposure limit

TABLE 8.2   NIOSH Recommended Respiratory Protection for Workers Exposed to Barium Chloride

Condition/Airborne Concentration of Barium Chloride	Minimum Respiratory Protection
Up to 5 mg/m^3 (10 × PEL)	Any particulate respirator equipped with an N95, R95, or P95 filter (including N95, R95, and P95 filtering facepieces) except quarter-mask respirators
Up to 12.5 mg/m^3 (25 × PEL)	(APF = 25) Any supplied-air respirator operated in a continuous-flow mode
	(APF = 25) Any powered, air-purifying respirator with a high efficiency particulate filter
Up to 25 mg/m^3 (50 × PEL)	(APF = 50) Any air-purifying, full facepiece respirator with an N100, R100, or P100 filter
	(APF = 50). Any supplied-air respirator that has a tight-fitting facepiece and is operated in a continuous-flow mode
	(APF = 50) Any powered, air-purifying respirator with a tight-fitting facepiece and a high efficiency particulate filter
	(APF = 50) Any self-contained breathing apparatus with a full facepiece
	(APF = 50) Any supplied-air respirator with a full facepiece
Up to 50 mg/m^3 (2000 × PEL)	(APF = 2000) Any supplied-air respirator that has a full facepiece and is operated in a pressure-demand or other positive-pressure mode
Emergency or planned entry into unknown concentrations or IDLH conditions	(APF = 10,000) Any self-contained breathing apparatus that has a full facepiece and is operated in a pressure-demand or other positive-pressure mode
	(APF = 10,000) Any supplied-air respirator that has a full facepiece and is operated in a pressure-demand or other positive-pressure mode in combination with an auxiliary self-contained positive-pressure breathing apparatus
Escape	(APF = 50) Any air-purifying, full facepiece respirator with an N100, R100, or P100 filter
	Any escape-type, self-contained breathing apparatus with a suitable service life (number of minutes required to escape the environment)

Source: NIOSH (2010).

(PEL) and the threshold limit value (TLV), respectively, for soluble barium salts, barium nitrate, barium oxide, barium carbonate, and barium chloride at 0.5 mg/m^3 (Table 8.2) (OSHA, 1996). These criteria are based on the fact that excessive exposure to these compounds may result in irritation of the eyes, nose, throat, and respiratory tract. Several of these compounds are strongly alkaline, which contributes to their irritancy. The PEL for insoluble barium sulfate is 10 mg/m^3 total dust or 5 mg/m^3 respirable dust. This is based on the potential for barium sulfate to cause a benign pneumoconiosis associated with very high exposure levels .The greatest potential for exposure to barium compounds exists in the mining and processing of barium-containing ores, in the production of barium compounds from barium sulfate, and in the incorporation of barium compounds into commercial products.

RISK ASSESSMENTS

Recently the EPA updated the oral RfD for barium exposure from 0.7 to 0.6 mg Ba/kg/day. The update was based on long-term animal data in the absence of long-term human data. The EPA determined that human and animals metabolize barium in similar ways; therefore, animal data present an excellent tool for modeling the acute and chronic health effects of exposure in humans. The EPA has evaluated the potential carcinogenicity of barium and classified it as Group D (not classifiable as to human carcinogenicity).

REFERENCES

American Conference of Governmental Industrial Hygienists (ACGIH) (2008) *Documentation of the Threshold Limit Values and Biological Exposure Indices, Cincinnati.* American Conference of Governmental industrial Hygienists.

ATSDR (2007) *Toxicological Profile for Barium*, Atlanta, GA: Agency for Toxic Substances and Disease Registry. Available at http://www.atsdr.cdc.gov/toxprofiles/tp.asp?id=327&tid=57.

Benarelli, E. (1931) Treatment of wool for mattresses with barium chloride from the hygienic aspect. *J. Ind. Hyg. Toxicol.* 12: 6.

Bowen, L.N., et al. (2010) Elementary, my dear Dr. Allen: the case of barium toxicity and Pa Ping. *Neurology* 74(19): 1546–1549.

CDC (2008) Urinary heavy metal. In: *National Health and Nutrition Examination Survey*, Atlanta, GA: Centers for Disease Control.

Chong, S., et al. (2006) Pneumoconiosis: comparison of imaging and pathologic findings. *Radiographics* 25(1):59–77.

Elkins, H.B. (1959) *Chemistry of Industrial Toxicology*, 2nd ed., New York: John Wiley & Sons.

Essing. H.G., et al. (1976) Exclusion of disturbances to health from long years of exposure to barium carbonate in the production of steatite ceramics. *Arbeitsmed. Sozialmed. Praventivmed.* 11: 299–302.

Dare, P.R., et al. (1984) Barium in welding fume. *Ann. Occup. Hyg.* 28(4):445–448.

Doig, A.T. (1976) Baritosis: a benign pneumoconiosis. *Thorax* 31(1):30–39.

EPA (2005) Toxicological review of barium compounds. In: *Integrated Risk Information System*, Washington, DC: United States Environmental Protection Agency.

Han, B.H., et al. (2011) Pregnancy outcome after 1st-trimester inadvertent exposure to barium sulphate as a constrast media for upper gastrointestinal tract radiography. *J. Ostet. Gynaecol.* 31(7):586–588.

Hicks, R., et al. (1986) Cardiotoxic and bronchoconstrictor effects of industrial metal fumes containing barium. *Arch. Toxicol. Suppl.* 9:416–420.

HSDB (2012) *Hazardous Substances Data Bank: Barium, Compounds. From NLM (National Library of Medicine)*, Bethesda, MD: National Institutes of Health, U.S. Department of Health and Human Services.

IPCS (1991) *Barium Health and Safety Guide*, Geneva, Switzerland: World Health Organization. International Programme on Chemical Safety.

IPCS (1999) *Barium, INCHEM. Concise International Chemical Assessment Document 33*, Geneva, Switzerland: World Health Organization.

Johnson, C.H. and VanTassell, V.J. (1991) Acute barium poisoning with respiratory failure and rhabdomyolysis. *Ann. Emerg. Med.* 20(10):1138–1142.

McCauley, P.T. and Washington, I.S. (1983) Barium bioavailability as the chloride, sulfate, or carbonate salt in the rat. *Drug Chem. Toxicol.* 6(2):209–217.

Moreton, J. and Jenkins, N. (1985) Barium in welding fume. *Ann. Occup. Hyg.* 29(3):443–444.

Morton, W. (1945) Poisoning by barium carbonate. *Lancet* 249: 738–739.

NIOSH (2010) *Pocket Guide to Chemical Hazards: Barium Chloride*, Washington, DC: National Institute for Occupational Safety and Health.

NTP (1994) *Toxicology and Carcinogenesis Studies of Barium Chloride Dihydrate in F344/N Rats and B6C3F Mice*, Research Triangle Park, NC: National Toxicology Program, National Institute of Environmental Health Sciences, Department of Health and Human Services.

OSHA (1996) *Barium, Soluble Compounds*, Washington, DC: United States Occupational Safety and Health Administration, United States Department of Labor.

Padilla, M.A., et al. (2010) An examination of the association of selected toxic metals with total and central obesity indices: NHANES 99-02. *Int. J. Environ. Res. Public Health* 7(9): 3332–3347.

Payden, C.A., et al. (2011) Intoxication by large amounts of barium nitrate overcome by early massive supplementation and oral administration of magnesium sulphate. *Hum. Exp. Toxicol.* 30(1):34–37.

Pendergrass, E.P. and Greening, R.R. (1953) Baritosis: report of a case. *AMA Arch. Ind. Hyg. Occup. Med.* 7(1):44–48.

Schroeder, H.A. and Michener, M. (1975a) Life-term effects of mercury, methyl mercury, and nine other trace metals on mice. *J. Nutr.* 105(4):452–458.

Schroeder, H.A. and Michener, M. (1975b) Life-term studies in rats: effects of aluminum, barium, beryllium, and tungsten. *J. Nutr.* 105(4):421–427.

Seaton, A., et al. (1986) Silicosis in barium miners. *Thorax* 41(8):591–595.

Shankle, R. and Keane, J.R. (1988) Acute paralysis from inhaled barium carbonate. *Arch. Neurol.* 45(5):579–580.

Stewart, D.W. and Hummel, R.P. (1984) Acute poisoning by a barium chloride burn. *J. Trauma* 24(8):768–770.

Torka, P.R., et al. (2009) Rodenticide poisoning as a cause of quadriparesis: a rare entity. *Am. J. Emerg. Med.* 27(5):625. e1–625.e3.

Tsai, C.Y., et al. (2011) Acute barium intoxication following accidental inhalation of barium chloride. *Intern. Med. J.* 41(3):293–295.

Wetherhill, S.F., et al. (1981) Acute renal failure associated with barium chloride poisoning. *Ann. Intern. Med.* 95(2):187–188.

Yeo, R.B. (1986) Barium in welding fume. *Ann. Occup. Hyg.* 30(4):515–517.

Zielinski, R.A., et al. (2011) Radionuclides, trace elements, and radium residence in phosphogypsum of Jordan. *Environ. Geochem. Health* 33(2):149–165.

9

BERYLLIUM

Alison J. Abritis and Raymond D. Harbison

Target organ(s): Lung, skin, mucosal membranes

Occupational exposure limits: OSHA PEL: $2\,\mu g/m^3$, TWA. Ceiling: $5\,\mu g/m^3$ STEL (30 min maximum): $25\,\mu g/m^3$; ACGIH: TLV–TWA: $0.05\,\mu g/m^3$; NIOSH TWA–REL: $0.5\,\mu g/m^3$

EPA (air emissions): $10\,g/24\,h$

Reference values: EPA IRIS: Rfd: 2×10^{-3} mg/kg/day

Risk/safety phrases: Inhalation risk, irritant, sensitizer

BACKGROUND AND USES

Beryllium (Be; CASRN 7440-41-7) is the lightest structural metal known to exist at this time. It is stronger than steel by 40%, its dimensional stability, favorable stiffness-to-weight ratio, high thermal conductivity, high melting point, and effectiveness as an electrical insulator make it an ideal industrial element for use in "inertial guidance systems, turbine rotor blades, laser tubes, rocket engine liners, springs, aircraft brakes and landing gear, ball bearings, injection and blow mold tooling, electrical contacts, automotive electronics, X-ray tube windows, spark plugs, electrical components, ceramic applications, gears, aircraft engines, oil and gas industries, welding electrodes, computer electronics, and golf clubs" (Barna et al., 2003).

Golf clubs notwithstanding, the United States Department of Defense considers beryllium as a "strategic and critical material" due to its integral use in industries vital to national security (Boland, 2012). Beryllium is used in control rods for nuclear reactors, in heat shields for space vehicles, for castings of nuclear weapons, and in windows for radiographic treatments. When bombarded with alpha particles it releases electrons. It reduces the energy of the neutrons rather than absorbing neutrons from radioactive material, and is permeable to X-rays.

At least 85% of the world's mined resources of beryllium come from Spor Mountain, Utah (controlled by Materion Brush, Inc., originally Brush Wellman, Inc.) in the United States, with China making up much of the rest and <2% coming from numerous other countries (Boland, 2012). Beryllium is also imported as composites of other industrial commodities; Russia supplies roughly 45%; Kazakhstan, 22%; Japan, 7%; Kenya, 5%; and the remainder from various other countries (Jaskula, 2012).

Three types of beryllium products are in current use: elemental beryllium, beryllium compounds, and beryllium alloys. Elemental beryllium is too reactive to be mined as an isolated ore, and is generally recovered as bertrandite and beryl. Virtually all of the beryllium mined in the United States is of the silicate bertrandite, while beryl (an aluminosilicate) is primarily mined in Russia, Brazil, and China. The green or blue-green color characteristic to emeralds and aquamarines is due to their beryl constituent. Beryllium compounds include beryllium chlorides, fluorides, hydroxides, and oxides. Beryllium hydroxide is used in the production of elemental beryllium, Be oxides, and copper alloys (Gordon and Bowser, 2003). However, beryllium oxides are more versatile and considered as "the most important high purity commercial beryllium chemical" (ATSDR, 2002).

About 75% of the beryllium used in the United States is used for the production of beryllium alloys. These alloys are light and markedly resistant to stress and strain, and most commonly made with copper, aluminum, or nickel. Copper beryllium alloy is commercially the most important and most widely used alloy. The master alloy usually contains 4–4.5%

Hamilton & Hardy's Industrial Toxicology, Sixth Edition. Edited by Raymond D. Harbison, Marie M. Bourgeois, and Giffe T. Johnson.
© 2015 John Wiley & Sons, Inc. Published 2015 by John Wiley & Sons, Inc.

beryllium by weight (ATSDR, 2002) and shows good resistance to corrosion and fatigue, electroconductivity, and machining characteristics. Beryllium copper products are non-sparkling, resistant to metal fatigue, ductile, and highly electrically conductive (Day et al., 2006) making them optimal in the production of tools, manufacture of electrical equipment, and in the construction of aircraft and spacecraft. Beryllium nickel and beryllium–chromium–nickel alloys have been used for applications such as dental crowns, bridges, and prostheses. Most beryllium alloys have a low percentage of beryllium, but "lock-alloy" has 62% Be and 38% aluminum (Bruce et al., 1998).

PHYSICAL AND CHEMICAL PROPERTIES

Elemental beryllium is a hard, brittle, grey-white alkaline metal with a melting point of 2341 °F (1283 °C), and is insoluble in water under normal conditions. With its high charge-to-radius ratio, its behavior mimics that of aluminum more than it does of other elements in its group. Beryllium tends to form stable compounds with smaller ions such as fluoride and oxide, although beryllium fluoride is highly soluble in water while beryllium oxide is virtually insoluble. Beryllium also is likely to form covalent bonds, from which organometallics can be formed.

The solubility of a compound plays a significant role in its potential toxicity and beryllium is no different. While beryllium alloys and beryllium oxide are virtually insoluble in water, beryllium salts (e.g., fluoride, chloride, sulfate, nitrate) can be very soluble. Beryllium is amphoteric, forming positive ions in acids with pH < 5 and negative ions in pH > 8. Beryllium is reactive with oxygen and forms a superficial layer of beryllium oxide when exposed to the atmosphere. It is considered non-volatile, except it may pose an explosive hazard if present as a powder or dust.

MAMMALIAN TOXICOLOGY

Animal studies using a variety of test subjects from rodents to dogs to monkeys have been used to model beryllium toxicity. Wagner et al. (1969) exposed animals of various species to beryllium ores at concentrations of 15 mg/m^3, from which they calculated the actual Be concentration to be 210 μg/m^3 for the bertrandite ore exposure and 620 μg/m^3 for the beryl ore exposure. They acknowledged these levels to be 105 and 310 times greater, respectively, than the existing occupational exposure limit of 2 μg/m^3.

In the Wagner study, squirrel monkeys, two strains of rats, and one strain of golden Syrian hamsters were exposed on a daily basis for 6 h per day for 5 days per week. Not surprisingly considering the elevated exposures, weight loss, and mortality rates for the animals were higher than controls. Monkeys showed similar pathologies in lung tissues for both ores, namely, "aggregates of dust-laden macrophages, lymphocytes, and plasma cells near respiratory bronchioles and small blood vessels" but "no other consistent or significant lesions," which could be related to the inhaled particulates. No such changes were noted in the controls. The hamsters showed different responses to the bertrandite and beryl ore exposures, where the beryl ore exposures caused small areas of atypical cell proliferation akin to bronchiole alveolar cell tumors, and the bertrandite exposures produced "a few" granulomatous lesions. No such responses were observed in the control hamsters. Interestingly, both strains of rats showed multiple forms of lesions, including adenomas, adenocarcinomas, and epidermoid tumors, with all rats showing tumor formation by 17 months. The control rats while not manifesting any "neoplastic or granulomatous pulmonary lesions," did show "moderately chronic bronchitis, which did not involve the respiratory bronchioles." In their discussion, the authors noted that while bertrandite produced granulomatous lesions, the beryl ore exposures did not, although the beryl exposures produced certain lung histopathologies. They concluded that beryl ore exposures should be considered as "an intoxicant and, more seriously, as a pulmonary carcinogen." They did not explore the possibility, drawn from the adverse responses of the control rats, that the exposure chamber design may have its own adverse response capability on test animals.

Grier et al. (1949) tested beryllium's effects on alkaline phosphatase *in vitro* using rat and human serum, and concluded that Be inhibits the production of the enzyme. However, the study by Wagner et al. (1969) noted no change in the phosphatase activity of their beryllium-exposed animals and speculated that *in vivo* results may differ from *in vitro* for this response.

In a summary of animal studies performed at the Inhalation Toxicology Research Institute (ITRI), Finch et al. (1996) noted that both dogs and monkey case studies yielded results similar to those seen in case studies of humans. They found that dogs, however, showed a tendency for resolution in symptoms, which makes exposure modeling to humans unlikely. They also noted that the F344 rat strain used in their studies did not respond to Be exposure in the same way that certain mice strains do, making non-strain-specific murine modeling difficult to apply to humans.

Some human epidemiological data suggest that beryllium may be a factor in some lung cancers, although there are conflicting reports. Studies by Infante et al. (1980), Wagoner et al. (1980), Steenland and Ward (1991), and Ward et al. (1992) reported a higher risk of cancer associated with beryllium exposure. Ward et al. (1992) working under the auspices of the National Institute for Occupational Safety and Health (NIOSH), performed "retrospective cohort mortality study of seven beryllium production facilities" in Pennsylvania and Ohio. Each facility handled all or part of the industrial processes involved in the extraction of beryllium from beryl ore or in the use of beryllium metal, oxide, and copper alloy in the manufacturing of other products. Their study found a "modest excess of lung cancer (SMR = 1.26) in a cohort of 9,225 workers." The standardized mortality ratios (SMRs) for each of the seven plants in the study ranged from 0.82 to 1.69. Since it can be safely assumed that working in a beryllium plant would not be protective from lung cancer (i.e., SMR < 1), other factors must be in play for the determination of mortality ratios. The authors admitted to the limitations in knowledge of occupational history, degree of exposure, and nature of exposure but did include consideration of smoking history and geographic region, which they discounted as a complete explanation for lung cancer risk.

The 1992 NIOSH report was challenged by Levy et al. (2002), who, after reanalyzing the data in consideration of local statistics and adjusted smoking risks, stated that "(o)ur analyses have shown that there is no statistical association between beryllium exposure in these workers and lung cancer when using the most appropriate population cancer rates. There is, at best, an extremely fragile association when the data are corrected for smoking." This analysis was soon countered by Schubauer-Berigan et al. (2008), who stated that average beryllium exposure correlated to a lung cancer risk while cumulative exposure did not. Levy et al. (2009) followed with a Cox proportional hazards single- and multivariant analysis to reevaluate the NIOSH study data without the inclusion of smoking as a confounder and concluded that again the link between beryllium and lung cancer was tenuous. Schubauer-Berigan et al. (2011) examined an expansion of the original data set, which now included data up to 2005, and again found an association between beryllium exposure and lung cancer: "Among these workers, lung cancer SMRs[1] were significantly elevated for workers in the highest cumulative exposure group ($>10,300\,\mu g/m^3$-days) and for workers having an annual maximum DWA[2] of $10\,\mu g/m^3$ or higher." Risks associated with smoking were included, but individual region effects were omitted.

The International Agency for Research on Cancer (IARC), referencing the Ward et al. (1992) and Schubauer-Berigan et al. (2008) analyses among other reports, stated that there "is sufficient evidence in humans for the carcinogenicity of beryllium and beryllium compounds. Beryllium and beryllium compounds cause cancer of the lung. There is sufficient evidence in experimental animals for the carcinogenicity of beryllium and beryllium compounds. Beryllium and beryllium compounds are carcinogenic to humans (Group 1)" (IARC, 2012). Similarly, in their Report of Carcinogens (2011), the National Toxicology Program (NTP) states that beryllium and beryllium compounds "are *known to be human carcinogens*"

based on human studies performed by Steenland and Ward (1991) and Ward et al. (1992) (NTP, 2011).

The EPA, however, has not changed their assessment of the Be's carcinogenicity since 1998 and still rates it as B1: a *probable human carcinogen* (Bruce et al., 1998). They address IARC's categorization of Be as a human carcinogen and counter that "the U.S. EPA, however, considers that the issues of incomplete smoking data and exposure to other potential lung carcinogens are not completely resolvable with the data currently available, and therefore concludes that the evidence of carcinogenicity of beryllium and compounds is limited in humans." Indeed, the epidemiological studies conducted on occupationally exposed workers show a lack of identification and isolation of other carcinogens that may have been present in the industrial settings in which the workers spent their work period. Adhesives, fossil fuels, and even asbestos may be a part of the manufacturing and processing process (Goldberg, 2006), all of which may be potential contributors to carcinogenicity in an exposed worker.

Acute Effects

Since the advent of occupational exposure and community air regulations, the incidence of acute beryllium disease (ABD) has been almost completely eliminated. Acute beryllium disease is a chemical irritant condition from which the patient can recover if the damage is not too extensive. It is usually associated with an isolated high concentration exposure ($>1\,mg\ Be/m^3$) of soluble forms of beryllium (ATSDR, 2002; Wambach and Laul, 2008) such as seen in an accidental release rather than a continuous exposure. The solubility of the compounds allows beryllium clearance from exposed tissues, usually before damage becomes irreparable.

Signs and symptoms of ABD from inhalation exposures are generally similar to a chemical irritant reaction. Coughing, weight loss, shortness of breath, labored breathing, swelling and inflammation of the respiratory tract, and pneumonitis may occur. Dermal exposures usually are associated with the soluble salts and can result in dermatitis, although skin granulomas have occurred from an immune-mediated reaction (ATSDR, 2002). If the patient survives the acute overexposure, the signs and symptoms of acute beryllium disease usually disappear after removal from the exposure source, although sometimes remission can take months for full recovery.

There has been limited speculation that ABD may actually be a preliminary phase in a "continuum of hypersensitivity reactions" leading to chronic beryllium disease (CBD) (Cummings et al., 2009). However, the case studies were limited and involved either repeated exposures or exposures to insoluble forms of beryllium, which may have accounted for the continued low level exposure associated with the development of the chronic form of the disease.

---

[1] Standardized mortality rates.
[2] Daily weighted average.

Chronic Effects

Beryllium exposure is known to result in two forms of chronic effects: beryllium sensitization (BeS) and CBD. Both conditions arise from an immune-mediated response to beryllium exposure, and CBD will not occur without BeS being initiated. It is also speculated to play a role in the development of lung cancer, although there remains some controversy over this finding.

Beryllium sensitization is a condition of a body's immune system's over-response to beryllium exposure. Unlike ABD and CBD, BeS is not dependent upon the length or extent of exposure, but is merely dependent upon the exposure occurrence. As the initial immune response may be minimal, BeS is generally asymptomatic (Day et al., 2006). There is no known dose–response curve for the onset of BeS; neither is there a time frame between exposure and sensitization response (CBAE, 2008). Various studies show a prevalence ranging from 0.47 to 14.9%, varying with the nature of the occupational exposure by profession and product (Kreiss et al., 1997; McCanlies et al., 2003; Day et al., 2006; Taiwo et al., 2010).

Previously, BeS was screened by use of a "beryllium patch test," whereby a patch of soluble beryllium salts was affixed to the test subject's skin and the nature of an inflammatory response was evaluated. However, the test fell out of favor as it was suspected as its own source of induced BeS (Donovan et al., 2007). The current "standard of care" in employee monitoring and screening for BeS and CBD is the Beryllium Lymphocyte Proliferation Test (BeLPT) (Newman et al., 2005).

The Beryllium Lymphocyte Proliferation Test measures the proliferation of tritiated thymidine-marked lymphocytes in three different concentrations of beryllium salts and two different time periods for culturing. The ratio between a test sample and its control is called a stimulation index (SI); six SIs are then produced for each test. Two (or more) SIs of 3.0 or more are considered as an abnormal result, one SI of 3.0 or more is considered borderline, and no SIs meeting or exceeding 3.0 is considered a negative test result (Barna et al., 2003; Middleton et al., 2006). It bears noting that some test protocols may use SIs as low as 2.0 as the "normal limit" (Kreiss et al., 1993; US Department of Energy, 2001). Roughly 10% of workers in beryllium industries are considered to have BeS based on the test results (Cullen, 2005).

While BeLPT use in screening and diagnosis has become common, it is not without its limitations. Although the Department of Energy (DOE) has a procedural protocol that it requires its contracted laboratories to use, it does not require adherence to the protocol by non-DOE contracted laboratories (US Department of Energy, 2001). Variance in test results has been noted between inter- and intra-laboratory reviews and can involve "concentrations of cells, media, or serum; the types of sample wells (i.e., flat-bottom vs. round-bottom); the number of replicate wells; length of incubation; blood sample collection and shipping procedures; the stimulation index threshold that is used to define a positive result; and the statistical methods used to analyze data" (Donovan et al., 2007).

Additional concerns to the use of BeLPT as a routine screening tool has to do with the potential for BeS to develop into CBD. The BeLPT has been found to have only a 40–50% positive predictive value (US Department of Energy, 2001). Those who test positive may not even develop CBD; ranges of associations for BeS developing into CBD run from 14 to 100% (Newman et al., 2005), and such wide range offers no conclusive predictive values. Routinely screening for CBD before symptom development may be without justification for the emotional and financial costs as there is no cure and little success in slowing disease progression (Barna et al., 2003; Cullen, 2005; Borak et al., 2006; Donovan et al., 2007).

Chronic beryllium disease incidence has been estimated to run anywhere from 0.4 to 18% of exposed employees (Day et al., 2006; Mack et al., 2010). Similar to ABD, CBD manifests with respiratory symptoms typical to a reduction of pulmonary function: wheezing, coughing, shortness of breath, fatigue, and weight loss. Other signs of CBD are cough, chest pain, cor pulmonale, followed by cardiac failure, liver and spleen damage, and the development of kidney stones (ATSDR, 2002; Donovan et al., 2007). Histopathological manifestations of non-necrotizing granulomatous pulmonary lesions are characteristic in clinical CBD, but can also present as a "subclinical" CBD, i.e., in the absence of overt symptoms. Skin lesions may also develop.

Chronic beryllium disease is "practically indistinguishable" from sarcoidosis (Barnard et al., 1996). To this respect, BeLPT is considered "highly useful" in eliminating sarcoidosis as a differential diagnosis for CBD (Barna et al., 2003).

It is now "generally accepted" that the variances noted between Be exposures and manifestation of BeS and CBD are due to the genetic makeup of the exposed person and the nature of the exposure (Wambach and Laul, 2008; Van Dyke et al., 2011). Newman et al. (2005) speculated that BeS progresses into CBD in roughly 6–8% of cases per year after the diagnosis.

Although there is no definitive proof that BeS will inevitably lead to CBD, the development of CBD is of the same process as that of BeS. The immunological response of cytokine cascades from the activation of Be-antigen-specific CD4+ T-cells maintains the inflammatory process and subsequent cell damage as Be^{2+} stimulation continues (Wambach and Laul, 2008). The pulmonary granulomas are associated with a massive proliferation of T-cells. One feature of CBD is the continued disease progression even in the absence of continued beryllium exposure. Mack et al. (2010) speculated that the impairment of regulatory T-cells (T_{reg}) may prevent the control of the pro-inflammatory responses. Their study showed a correlation between the increase of T_{reg} cell dysfunction and a corresponding disease severity.

The condition is generally thought to be associated with low soluble compounds that remain in the lung tissues for an extended period of time. It has been associated with both high and low level exposures, but generally is seen with extended low level exposures (Wambach and Laul, 2008).

Beryllium has been classified as a *probable human carcinogen* by the EPA and a *known human carcinogen* by IARC and NTP (Bruce et al., 1998; NTP, 2011; IARC, 2012).

Infante et al. (1980) performed a study and found higher rates of lung cancer (and other respiratory diseases) among deceased white male beryllium workers based on the registration in the Beryllium Case Registry (BCR) at Massachusetts General Hospital. However, there are methodological flaws in their study. First, the population group itself is skewed by definition. Those workers exposed without any illness would not likely be registered. As Mass General is a major hospital, there is a greater likelihood that the more significantly ill would seek services there instead of a smaller general hospital. Second, there is only the word of the patient without corroboration as to the nature of the exposure in regards to concentrations, duration of each exposure, and most importantly, other potential exposures from occupational duties. Finally, they discounted the potential confound of smoking because it "seems unlikely that subjects entered into the BCR with the past diagnosis of acute respiratory illness would have had past smoking habits of sufficient magnitude to account for the excessive lung cancer risk in that subcohort."

Mechanism of Action(s)

Beryllium works as both a chemical irritant and a cell-mediated hypersensitivity phenomenon. The chemical irritant mechanism of beryllium is generally straightforward and is akin to the nature of its group and the compounds it forms. Soluble forms will cause more invasive damage than insoluble forms, and the nature of the oxidative damage is contingent upon the type of compound. Sawyer et al. (2005) demonstrated via *in vitro* studies on mice macrophages that Be exposure initiates the formation of reactive oxygen species (ROS), which induces cell apoptosis.

The mechanism behind the development of the hypersensitivity reaction is not so clear-cut. What is known is that hypersensitivity is the result of the formation of an antigenic beryllium–hapten complex formed with a major histocompatibility class II molecule (MHCII) (Kreiss et al., 1993; Newman, 1993). CD4$^+$ T-cells, in the presence of this complexed molecule, then trigger the production of pro-inflammatory cytokines TNF-α and TNF-γ. It is this cascade of inflammatory responses that is believed to be responsible for the formation of the non-caseating granulomas seen in beryllium-induced dermatitis and chronic disease. Regulatory T-cell dysfunction may also play a role in the perpetuation of the pro-inflammatory cytokine response (Mack et al., 2010).

What remains to be deduced is exactly why this hypersensitivity happens at all. Less than 1 in 6 exposed workers demonstrated any sensitization reaction, and there has been no establishment of a dose–response for the sensitization effect. Sawyer et al. (2005) speculated in their study on Be and ROS that "exogenous oxidative stress" increased the ROS formation in Be-exposed cell lines. They then speculated that the variances in the sensitizations of individuals may be due to "a pulmonary environment already under oxidative stress, for example, in the presence of pulmonary infection, smoking or other environmental stresses, could cause even greater levels of alveolar macrophage apoptosis."

One presiding theory affixes the source of the sensitization to multiple genetic phenotypes in the exposed subject. One gene family of interest is the human leukocyte antigen (HLA) family, supported by the irregular proliferation of leukocytes and other immune response cytokines during exposure to beryllium. Currently, research supports a higher risk for workers with a non-*02-E69 HLA-DPB1 and HLA-DPB1 E69 homozygotic genotype (Van Dyke et al., 2011), although the population prevalence of HLA-DPB1 E69 is high, the "cross-sectional predictive value is relatively low" (McCanlies et al., 2003). Another area of interest may be the interleukin (IL) genes, which may play a part in the formation of the granulomatous lesions and fibrosis characteristic of CBD (McCanlies et al., 2010).

Chemical Pathology

Historically, bertrandite was considered as the source of beryllium toxicity. Beryl ore was once considered "nontoxic" and was even used in a study as a "negative control" (Wagner et al., 1969). Further irony in light of the origins of this text, beryllium poisoning was the reason for the initial meeting between Alice Hamilton and Harriet Hardy (Hardy, 1965).

The primary exposure route associated with adverse effects is through inhalation and dermal absorption, although ingestion is also possible. However, beryllium is typically not noted for adverse effects from ingestions in healthy individuals, as it tends to remain confined to the GI tract until excreted.

The solubility of Be particles is believed to play a significant role with the potential toxicity of the exposure, but the nature of that role appears dual-natured. Small particles ($<10\,\mu m$) made from grinding or machining beryllium compounds are more soluble in acidic environments such as the internal environment of macrophages, to which Wambach and Laul (2008) associated their toxicity to macrophages and thus their longer retention and continued stimulation of the inflammatory processes. Wambach and Laul (2008) also pointed out that small particles are more likely to be deposited in the deepest part of the lung, where oxygen exchange takes place. On the other hand, Taiwo et al. (2010) suggested that more soluble compounds can be attributed to less toxic response, since the compound's dissolution causes less actual contact with the immune system. Accordingly, beryllium oxides calcined at temperatures lower than 500 °C are more associated with the onset of ABD than the development of CBD, which Wambach and Laul (2008) suggested is due to the smaller particles of the low temperature calcined oxide having a greater solubility than beryllium oxides calcined at higher temperatures (>1000 °C).

Solubility is also directly related to the duration of beryllium body load. Beryllium and its compounds are primarily excreted from the body through feces. The remaining is excreted in the urine, which can cause extended body burden until the conversion of insoluble forms into soluble forms is complete (Bruce et al., 1998).

SOURCES OF EXPOSURE

Occupational

Increases and improvements in engineering controls have reduced the nature and degree of employee occupational exposure. However, normal industrial operations coupled with accidental releases do allow for exposures of varying concentrations and types.

Occupational overexposure to beryllium occurs primarily via the inhalation route, although dermal exposure from particulate dust is also a consideration. Occupationally-related ingestion is rare with adequate engineering controls, although it can occur from respiratory clearance mechanisms allowing particulate-embedded mucus to be swallowed. Overexposure can occur during the processing of alloys and ceramics and from the generation of dusts, fumes, and other fine particulates during the manufacture of beryllium-containing materials.

Beryllium sensitization and CBD rates are clearly associated with the nature of the industry and the work process associated with the employee exposure. Newman et al. (2005) found that machinists were most likely to progress from BeS to CBD than other workers within the same industry. Stanton et al. (2006) found that BeS and CBD rates in the employees of beryllium-processing plants exceeded workers handling manufactured products, such as in a beryllium–copper alloy distribution center.

There is also substantial evidence that the form of beryllium exposure is directly related to its ability to induce sensitization; not all beryllium is equivalent and not all beryllium is capable of inducing disease. Deubner et al. (2001) examined 75 workers at a beryllium ore mining and milling facility in Utah. The workers involved in the mining operation were primarily exposed to beryl ore or bertrandite ore, and the milling operation workers were exposed to these ores and beryllium hydroxide. General area, breathing zone, and personal lapel samples were used to estimate the historical beryllium exposure. The

mean general area, breathing zones, and personal lapel samples ranges were 0.3–1.1, 1.1–8.1, and 0.05–6.9 μg/m^3, respectively. No cases of BeS or CBD were found in workers who only worked in the mines, leading the study authors to suggest that the form of beryllium may influence the risk for developing BeS or CBD. Stefaniak et al. (2008) suggested that the difference in particle size between ore dusts and manufacturing dusts may account for a difference in bioavailability, subsequently affecting the alveolar particulate loads.

Aluminum production can be a cause of occupational beryllium exposure as beryllium is found in the bauxite ore mined for aluminum. Taiwo et al. (2010) found a prevalence of 0.47% of BeS in a composite study of nine aluminum-smelting plants located throughout four countries and speculated that the lower sensitization rate may be "related to work practices and the properties of the beryllium found in this work environment."

Biomonitoring of beryllium exposures have been attempted through urine samples, but no correlation between ambient air concentration and urinary levels has yet been identified. Apostoli and Schaller (2001) in a study of metallurgical workers did indicate higher levels of beryllium in "exposed" workers versus controls. They stated, however, that their study had too limited a dataset to make exact correlations "without reservations." To date, there are no biological exposure indices (BEI) for beryllium urinary levels.

Environmental

The average amount of beryllium permeating the earth's crust is roughly 6 mg/kg—calculated from the beryllium ore content in soils as elemental beryllium is too reactive to occur naturally. Natural sources of beryllium in ambient air come from natural rock erosion, volcanic dusts, and combustion ash. Major manmade ambient air sources are manufacturing and processing of ores and from the combustion of coals as they are commonly combustion agents in power plants for the production of electricity. Be concentration in coal averages 2.3 parts per million (ppm) and can be released into the air as fly ash and other coal combustion products (Eskenazy, 2006). Urban and industrial population may experience slightly higher ambient beryllium exposures; Detroit, Michigan averaged ambient air concentrations of 0.02–0.2 ng/m^3 from 1982 to 1992 while the yearly average concentration for the entire United States is below the detection limit of 0.03 ng/m^3 (ATSDR, 2002).

Soil contamination is also a consideration in environmental exposures. Contaminated soils increase the potential for elevated ambient air levels from fine dust, and children are at increased risk as they are more likely to inhale or ingest contaminated soils by their propensity to play in dirt. By 2002 beryllium was identified as a contaminant in at least 535 of the 1,613 current or former National Priorities List (NPL) sites; however, the beryllium form at these sites are "almost entirely" in insoluble forms and thus less likely to be exposure threats (ATSDR, 2002).

One of the primary concerns from environmental intrusion by beryllium is the potential for bioaccumulation. Ishikawa et al. (2004) found "abnormally high" levels of Be in samples of ascidian liver which they speculated may have been due to ascidian feeding on Be-enriched plankton. However, no study has found evidence of bioaccumulation moving up the food chain and ATSDR (2002) stated that Be "does not bioconcentrate in aquatic organisms." Furthermore, a 2008 study of Floridian ospreys that "have frequently been used as indicator species for contamination" showed no significant elevation in trace beryllium. Samples were consistently below 1 μg/g dry feather weight (Lounsbury-Billie et al., 2008).

INDUSTRIAL HYGIENE

The generally accepted time-weighted occupational exposure limit of 2.0 μg/m^3 has been utilized for various federal regulatory agencies since 1949 (Ward et al., 1992). Currently, Occupational Safety and Health Administration (OSHA) still sets the legally enforceable time-weighted average (TWA)–permissible exposure limit (PEL) as 2.0 μg/m^3 with a ceiling of 5.0 μg/m^3 and a 30 min peak level of 25 μg/m^3. However, in light of some studies challenging the protectiveness of these limits, some agencies have modified their recommendations for occupational exposure limits. The NIOSH currently recommends a more conservative limit of 0.5 μg/m^3, and in 2009, American Conference of Governmental Industrial Hygienists (ACGIH) significantly lowered their threshold limit value (TLV) to 0.05 μg/m^3, a 40-fold reduction (ACGIH, 2009).

In their discussion of literature justifying reduction, ACGIH referenced reports by Kelleher et al., (2001); Madl et al., (2007); Schuler et al., (2005); Rosenman et al., (2005); and Stanton et al., (2006). However, Madl et al. (2007) pointed out that "(f)urthermore, no beryllium-sensitized and CBD workers were exposed to a median concentration >2 μg/m^3, but approximately 50% of the beryllium-sensitized and CBD workers were exposed to a 95th percentile concentration greater than 2 μg/m^3 when looking at supplemented data for only the highest exposed year." They also noted that "(t)wenty-six of the 27 beryllium-sensitized and CBD workers had at least a 10% probability of being exposed to beryllium concentrations exceeding 0.2 μg/m^3 during their career, and all cases had a greater than 5% probability of exceeding this level."

Kelleher et al. (2001) noted problems in determining significant differences between cases and control and stated that "(m)achinist history remained the only significant variable," which thus encompasses a potential for particulate size

as a variable. However, Kelleher et al. (2001) used lifetime exposure calculations to compare cases and controls, which might mask any particular exceedance of PELs. In their graphic representation, they noted that "the median cumulative exposure to total beryllium (based on the median exposure estimate) was 2.6 mg/m^3-years in cases compared with 1.2 mg/m^3-years in controls," which implies a preponderance of PEL exceedance.

Borak (2006) in discussing a perceived need for lower regulatory limits admits that some studies that attribute BeS and CBD to exposure levels below 2.0 μg/m^3 may actually have problems in Be contamination creating exposures exceeding 2.0 μg/m^3. Stanton et al. (2006) noted study problems in small sampling sizes and poor historical exposure data. Schuler et al. (2005) also considered the possibilities that exposures may have actually exceeded 2.0 μg/m^3. Despite the limitations in these studies, it is obvious that there is a growing concern as to the protectiveness of the current OSHA standard of 2.0 μg/m^3, and that a revision of the standard to a lower PEL might reasonably be expected in the coming years.

No matter the exposure limit, good employee management and common sense dictate adherence to sound engineering and housekeeping controls. Proper personal protective equipment should be utilized when airborne exposures range near recommended exposure limits, and apparel should be protective against dermal exposures, especially in areas most prone to the presence of particulate matter or fumes. A 4-year program of instituting building design changes, restricted zone access, and required respirator wear in one facility resulted in a fourfold decrease in employee sensitization rates (Bailey et al., 2010). Therefore tracking areas of potential beryllium exposures in the workplace is imperative. Table 9.1 contains a list of the approved sampling protocols for quantification of beryllium in air or dust.

TABLE 9.1 Approved Analytical Protocols for Detecting Beryllium in Air or Dust

NIOSH 7300	Elements by ICP
NIOSH 7301	Elements by ICP (aqua regia ashing)
NIOSH 7303	Elements by ICP (hot block/HCI/HNO$_3$ digestion)
NIOSH 7704	In Air by Field-Portable Fluorometry, 5th ed.
NIOSH 9102	Elements on Wipes
NIOSH 9110	In Surface Wipes by Field-Portable Fluorometry, 5th ed.
NIOSH 7102	Beryllium and Compounds, as Be
OSHA ID125G	Metal and Metalloid Particulates in Workplace Atmospheres (ICP Analysis)
OSHA ID206	ICP Analysis of Metal/Metalloid Particulates from Solder Operations

Source: NIOSH Pocket Guide to Chemical Hazards. Available at http://www.cdc.gov/niosh/npg/npgd0054.html.

MEDICAL MANAGEMENT

The symptoms of acute or chronic beryllium-related disease may mimic other illnesses; for this reason, identification of beryllium exposure is critical in diagnosing the disease. A history of beryllium exposure, combined with weight loss, a nonproductive cough, eye irritation, irritation of the skin and respiratory tract, or pain in the sternum, suggests ABD. Chest X-rays can be useful in identifying pulmonary involvement, but have limited usefulness in diagnosis as the radiographic appearance is non-specific for beryllium disease (Stange et al., 1996). Thin-section computed tomography (TSCS) is a more sensitive technique than normal chest radiography, but it is only practical for confirmation of diagnosis rather than as a screening tool (Maier, 2002).

Chronic beryllium disease may be more difficult to diagnose than ABD because it generally requires months to years after the initial beryllium exposure and may manifest long after such exposure had stopped. Chronic beryllium disease may be precipitated by stress, such as surgery, combat, pregnancy, or another illness. Initially, CBD is characterized by shortness of breath on exertion, weight loss, chest pain, and cough. Reduced pulmonary function and reductions in vital capacity, forced expiratory volume, and maximal mid-expiratory flow are indicators of possible CBD. Abnormal BeLPT test results may be indicators of BeS (the precursor of CBD) and can differentiate between a differential diagnosis of sarcoidosis and CBD.

Beryllium can be measured in the urine or in various tissues. However, as is true with many other metals, beryllium levels may not correlate with the level or duration of exposure. Elevated levels of beryllium do indicate exposure sometime in the past, but the time since exposure cannot be estimated because of beryllium's long half-life.

Currently, the treatment protocol for beryllium disease is cessation of beryllium exposure and immunosuppression/anti-inflammatory treatment with corticosteroids (Sood, 2009). Palliative treatment with supplemental oxygen and respiratory therapy can be helpful, but as of this time there is no curative therapy for BeS or CBD.

REFERENCES

ACGIH (2009) *Beryllium and Compounds*, American Conference of Governmental Industrial Hygienists.

Apostoli, P. and Schaller, K.H. (2001) Urinary beryllium—a suitable tool for assessing occupational and environmental beryllium exposure? *Int. Arch. Occup. Environ. Health* 74:162–166.

ATSDR (2002) *Toxicological Profile for Beryllium*, Atlanta, GA: ATSDR.

Bailey, R.L. et al. (2010) Evaluation of a preventive program to reduce sensitization at a beryllium metal, oxide, and alloy production plant. *J. Occup. Environ. Med.* 52:505–512.

Barna, B.P., Culver, D.A., Yen-Lieberman, B., Dweik, R.A., and Thomassen, M.J. (2003) Clinical application of beryllium lymphocyte proliferation testing. *Clin. Diagn. Lab. Immunol.* 10:990–994.

Barnard, A.E., Torma-Krajewski, J., and Viet, S.M. (1996) Retrospective beryllium exposure assessment at the rocky flats environmental technology site. *Am. Ind. Hyg. Assoc. J.* 57:804–808.

Boland, M.A. (2012) Beryllium—important for national defense. In: *U.S. Geological Survey Fact Sheet 2012–3056*, U.S. Department of the Interior, U.S. Geological Survery. Available at http://pubs.usgs.gov/fs/2012/3056.

Borak, J. (2006) The beryllium occupational exposure limit: historical origin and current inadequacy. *J. Occup. Environ. Med.* 48:109–116.

Borak, J., Woolf, S.H., and Fields, C. A. (2006) Use of beryllium lymphocyte proliferation testing for screening of asymptomatic individuals: an evidence-based assessment. *J. Occup. Environ. Med.* 48:937–947.

Bruce, R.M., Ingerman, L., and Jarabek, A. (1998) *Toxicological Review of Beryllium and Compounds*, Washington, DC: U.S. EPA.

CBAE (2008) *Managing Health Effects of Beryllium Exposure*, The National Academies Press.

Cullen, M.R. (2005) Screening for chronic beryllium disease: one hurdle down, two to go. *Am. J. Respir. Crit. Care Med.* 171:3–4.

Cummings, K.J., Stefaniak, A.B., Virji, M.A., and Kreiss, K. (2009) A reconsideration of acute Beryllium disease. *Environ. Health Perspect.* 117:1250–1256.

Day, G.A., Stefaniak, A.B., Weston, A., and Tinkle, S.S. (2006) Beryllium exposure: dermal and immunological considerations. *Int. Arch. Occup. Environ. Health* 79:161–164.

Deubner, D. et al. (2001) Beryllium sensitization, chronic beryllium disease, and exposures at a beryllium mining and extraction facility. *Appl. Occup. Environ. Hyg.* 16:579–592.

Donovan, E.P., Kolanz, M.E., Galbraith, D.A., Chapman, P.S., and Paustenbach, D.J. (2007) Performance of the beryllium blood lymphocyte proliferation test based on a long-term occupational surveillance program. *Int. Arch. Occup. Environ. Health* 81:165–178.

Eskenazy, G.M. (2006) Geochemistry of beryllium in Bulgarian coals. *Int. J. Coal Geol.* 66:305–315.

Finch, G.L., et al. (1996) Animal models of beryllium-induced lung disease. *Environ. Health Perspect.* 104(Suppl. 5):973–979.

Goldberg, A. (2006) *Beryllium Manufacturing Processes*, United States Department of Energy.

Gordon, T. and Bowser, D. (2003) Beryllium: genotoxicity and carcinogenicity. *Mutat. Res.* 533:99–105.

Grier, R.S., Hood, M.B., and Hoagland, M.B. (1949) Observations on the effects of beryllium on alkaline phosphatase. *J. Biol. Chem.* 180:289–298.

Hardy, H.L. (1965) Beryllium poisoning—lessons in control of man-made disease. *N. Engl. J. Med.* 273:1188–1199.

IARC (2012) Beryllium and beryllium compounds. In: IARC, editor. *Arsenic, Metals, Fibres, and Dusts: A Review of Human Carcinogens*, vol. 100 C, International Agency for Research on Cancer, pp. 95–120.

Infante, P.F., Wagoner, J.K., and Sprince, N.L. (1980) Mortality patterns from lung cancer and nonneoplastic respiratory disease among white males in the beryllium case registry. *Environ. Res.* 21:35–43.

Ishikawa, Y., Kagaya, H., and Saga, K. (2004) Biomagnification of 7Be, 234Th, and 228Ra in marine organisms near the northern Pacific coast of Japan. *J. Environ. Radioact.* 76:103–112.

Jaskula, B.W. (2012) *Beryllium*, USGS.

Kelleher, P.C., et al. (2001) Beryllium particulate exposure and disease relations in a beryllium machining plant. *J. Occup. Environ. Med.* 43:238–249.

Kreiss, K., Mroz, M.M., Zhen, B., Wiedemann, H., and Barna, B. (1997) Risks of beryllium disease related to work processes at a metal, alloy, and oxide production plant. *Occup. Environ. Med.* 54:605–612.

Kreiss, K., Wasserman, S., Mroz, M.M., and Newman, L.S. (1993) Beryllium disease screening in the ceramics industry. Blood lymphocyte test performance and exposure-disease relations. *J. Occup. Med.* 35:267–274.

Levy, P.S., Roth, H.D., and Deubner, D.C. (2009) Exposure to beryllium and occurrence of lung cancer: findings from a cox proportional hazards analysis of data from a retrospective cohort mortality study. *J. Occup. Environ. Med.* 51:480–486.

Levy, P.S., Roth, H.D., Hwang, P.M.T., and Powers, T.E. (2002) Beryllium and lung cancer: a reanalysis of a NIOSH cohort mortality study. *Inhal. Toxicol.* 14:1003–1015.

Lounsbury-Billie, M.J., Rand, G.M., Cai, Y., and Bass, O.L. Jr. (2008) Metal concentrations in osprey (*Pandion haliaetus*) populations in the Florida Bay estuary. *Ecotoxicology* 17:616–622.

Mack, D.G., et al. (2010) Deficient and dysfunctional regulatory T cells in the lungs of chronic beryllium disease subjects. *Am. J. Respir. Crit. Care Med.* 181:1241–1249.

Madl, A.K., Unice, K., Brown, J.L., Kolanz, M.E., and Kent, M.S. (2007) Exposure-response analysis for beryllium sensitization and chronic beryllium disease among workers in a beryllium metal machining plant. *J. Occup. Environ. Hyg.* 4:448–466.

Maier, L.A. (2002) Clinical approach to chronic beryllium disease and other nonpneumoconiotic interstitial lung diseases. *J. Thorac. Imaging* 17:273–284.

McCanlies, E.C., Kreiss, K., Andrew, M., and Weston, A. (2003) HLA-DPB1 and chronic beryllium disease: a HuGE review. *Am. J. Epidemiol.* 157:388–398.

McCanlies, E.C., et al. (2010) Association between IL-1A single nucleotide polymorphisms and chronic beryllium disease and beryllium sensitization. *J. Occup. Environ. Med.* 52:680–684.

Middleton, D.C., Lewin, M.D., Kowalski, P.J., Cox, S.S., and Kleinbaum, D. (2006) The BeLPT: algorithms and implications. *Am. J. Ind. Med.* 49:36–44.

Newman, L.S. (1993) To Be^{2+} or not to Be^{2+}: immunogenetics and occupational exposure. *Science* 262:197–198.

Newman, L.S., Mroz, M.M., Balkissoon, R., and Maier, L.A. (2005) Beryllium sensitization progresses to chronic beryllium disease: a longitudinal study of disease risk. *Am. J. Respir. Crit. Care Med.* 171:54–60.

NTP (2011) Beryllium and beryllium compounds. In: National Toxicology Program, editor. *Report on Carcinogens*, 12th ed., U.S. Department of Health and Human Services, 499 pp.

Rosenman, K. et al. (2005) Chronic beryllium disease and sensitization at a beryllium processing facility. *Environ. Health Perspect.* 113:1366–1372.

Sawyer, R.T., et al. (2005) Beryllium-stimulated reactive oxygen species and macrophage apoptosis. *Free Radic. Biol. Med.* 38:928–937.

Schubauer-Berigan, M.K., Deddens, J.A., Steenland, K., Sanderson, W.T., and Petersen, M. R. (2008) Adjustment for temporal confounders in a reanalysis of a case-control study of beryllium and lung cancer. *Occup. Environ. Med.* 65:379–383.

Schubauer-Berigan, M.K., et al. (2011) Cohort mortality study of workers at seven beryllium processing plants: update and associations with cumulative and maximum exposure. *Occup. Environ. Med.* 68:345–353.

Schuler, C.R., et al. (2005) Process-related risk of beryllium sensitization and disease in a copper-beryllium alloy facility. *Am. J. Ind. Med.* 47:195–205.

Sood, A. (2009) Current treatment of chronic beryllium disease. *J. Occup. Environ. Hyg.* 6:762–765.

Stange, A.W., Hilmas, D.E., and Furman, F.J. (1996) Possible health risks from low level exposure to beryllium. *Toxicology* 111:213–224.

Stanton, M.L., et al. (2006) Sensitization and chronic beryllium disease among workers in copper-beryllium distribution centers. *J. Occup. Environ. Med.* 48:204–211.

Steenland, K. and Ward, E. (1991) Lung cancer incidence among patients with beryllium disease: a cohort mortality study. *J. Natl. Cancer Inst.* 83:1380–1385.

Stefaniak, A.B., et al. (2008) Physicochemical characteristics of aerosol particles generated during the milling of beryllium silicate ores: implications for risk assessment. *J. Toxicol. Environ. Health A* 71:1468–1481.

Taiwo, O.A., et al. (2010) Prevalence of beryllium sensitization among aluminium smelter workers. *Occup. Med.* 60:569–571.

US Department of Energy (2001) *Beryllium Lymphocyte Proliferation Testing*, DOE-SPEC-1142-2001, Washington, DC: US Department of Energy.

Van Dyke, M.V., et al. (2011) Risk of chronic beryllium disease by HLA-DPB1 E69 genotype and beryllium exposure in nuclear workers. *Am. J. Respir. Crit. Care Med.* 183:1680–1688.

Wagner, W.D., Groth, D.H., Holtz, J.L., Madden, G.E., and Stokinger, H.E. (1969) Comparative chronic inhalation toxicity of beryllium ores, bertrandite and beryl, with production of pulmonary tumors by beryl. *Toxicol. Appl. Pharmacol.* 15:10–29.

Wagoner, J.K., Infante, P.F., and Bayliss, D.L. (1980) Beryllium: an etiologic agent in the induction of lung cancer, nonneoplastic respiratory disease, and heart disease among industrially exposed workers. *Environ. Res.* 21:15–34.

Wambach, P.F. and Laul, J. (2008) Beryllium health effects, exposure limits and regulatory requirements. *J. Chem. Health Saf.* 15:5–12.

Ward, E., Okun, A., Ruder, A., Fingerhut, M., and Steenland, K. (1992) A mortality study of workers at seven beryllium processing plants. *Am. J. Ind. Med.* 22:885–904.

10

BISMUTH AND RELATED COMPOUNDS

Paul Jonmaire

First aid: Since there are many bismuth compounds in commercial use there are no special first aid recommendations. General first aid recommendations are often appropriate. For dermal overexposure, remove contaminated clothing and wash with soap and water. For ocular overexposure, flush with water for 15 min.

Target organ(s): Include eye, respiratory tract, kidney, liver, and nervous system

Occupational exposure limits: Bismuth telluride undoped: $15\,mg/m^3$ (OSHA TWA) $10\,mg/m^3$ (ACGIH TWA) $10\,mg/m^3$ total dust (NIOSH TWA); $5\,mg/m^3$ respirable (OSHA TWA), and $5\,mg/m^3$ respirable (NIOSH TWA)

BACKGROUND AND USES

Bismuth is a relatively rare metal found in the earth's crust at about the same level as silver. It is usually associated with copper, lead, tungsten, silver, and gold ores. Bismuth typically is a byproduct of copper and lead refining and smelting (USGS, 2013). About 8000 metric tons are produced worldwide annually. Production of bismuth in the United States ceased in 1997 (USGS, 2014a). In 2013 about 1700 tons of bismuth were imported into the United States, about half of which came from China (USGS, 2014b) In early times it was confused with tin and lead. In 1450, a German monk, Basil Valentine, first made reference to "wismut" or "white lead," which later would be Latinized to *bismentum*. Claude Geoffory showed it to be distinct from lead in 1753 (WebElements Periodic Table of the Elements, 2012) A bronze knife from Machu Picchu with about 20% bismuth suggests that the Incas may have used it for metallurgical purposes (Palmierri, 1993). The most common bismuth minerals are bismite

(Bi_2O_3) and bismuthinite (Bi_2S_3) although it may occur in its metallic state (Bell, 2014). There are more than 100 bismuth-containing compounds on the U.S. EPA TSCA inventory (US EPA, 2013).

Bismuth is a very brittle metal rarely used by itself. The first widespread metallurgical uses for bismuth were in pewter alloys (Bell, 2014). Because of its low melting point, it is a component of "fusible alloys" found in fire sprinkler systems and low melting solders and other alloys of steel and aluminum (Palmierri, 1993). In the early 1990s research began on the evaluation of bismuth as a nontoxic replacement for lead in such uses as ceramic glazes, food processing equipment, free-machining brasses for plumbing applications, lubricating greases, and low toxicity fishing sinkers and shot for waterfowl hunting (USGS, 2014a; Lenntech, 2014). The Safe Drinking Water Act Amendments of 1996 have banned lead from all fixtures, fluxes, pipes, and solders used for installation or repair of facilities providing potable drinking water prompting a movement to replace lead with bismuth (Bell, 2014). In January 2010 both Vermont and California made the use of any product with greater than 0.25% lead unlawful. These regulations are part of a trend to use bismuth containing brass fittings and valves in water systems (AY McDonald, n.d.). As part of their "Design for the Environment" life-cycle assessment program, the Environmental Protection Agency (EPA) has promoted the use of bismuth in several lead-free solders (US EPA, 2005). Recently, new applications for bismuth include the use of bismuth zinc alloys to achieve thinner, more uniform galvanization (USGS, 2014b). Another developing application is the use of bismuth telluride-based compounds in the manufacture of semi-conductor devices. Bismuth telluride is both a mineral and manufactured compound and is used in both an

Hamilton & Hardy's Industrial Toxicology, Sixth Edition. Edited by Raymond D. Harbison, Marie M. Bourgeois, and Giffe T. Johnson.
© 2015 John Wiley & Sons, Inc. Published 2015 by John Wiley & Sons, Inc.

undoped and doped form. ("Doping" is depositing of a thin layer (often selenium) of a second metal on the semiconductor material of interest to modify its conductivity ACGIH, 2001; USGS, 2014b).

Bismuth is also used in cosmetics, flame retardants, and stabilizers in plastics, pigments, and pharmaceuticals. Bismuth trioxide is used in disinfectants, magnets, glass, ceramics, and rubber vulcanization. Bismuth phosphomolybdate is used as a catalyst in the production of acrylonitrile fiber (Digitalfire Hazards Database, 2008; Palmierri, 1993). Some inorganic bismuth compounds are used to treat abdominal problems such as diarrhea, abdominal pain, and ulcers. The most well-known bismuth compound, bismuth subsalicylate, is used to treat diarrhea, heartburn, and upset stomach. Bismuth subsalicylate, the principal ingredient of Pepto-Bismol and Kaopectate is consumed in large quantities around the globe (COT, 2008; East, 2009; NLM HSDB, 2013). Bismuth is used in surgical dressings (subnitrate and carbonate). Organic bismuth preparations such as bismuth tartarate were used more or less successfully until the 1940s to treat syphilis.

PHYSICAL AND CHEMICAL PROPERTIES

Elemental bismuth (Bi; CASRN 7440-69-9) is a soft, heavy, solid, brittle, pink-tinged white metal, insoluble in water. Its atomic number is 83 and atomic weight 209; density is $9.78 \, g/cc^3$. It may be flammable in powder form. Bismuth exists in trivalent and pentavalent states. The trivalent form is more stable and abundant. Like water, it expands upon solidification from the liquid state. It has a relatively low melting point of $271 \, °C$ ($520 \, °F$). It has the lowest thermal conductivity of all metals and is the most diamagnetic of all metals (i.e., repelled by both poles of a magnetic, therefore positioning itself at right angles to the magnetic lines of influence) (Bell, 2014; INCHEM, n.d.; Lenntech, 2014; WebElements Periodic Table of the Elements, 2012).

HEALTH EFFECTS

Bismuth salts are known for their low toxicity (Salvador et al., 2012). Bismuth and its various compounds are considered to be among the least toxic of heavy metals. Many bismuth compounds are water insoluble and minimally absorbed. Because of the typical low toxicity of bismuth and its compounds, pharmacological use is the primary source of information regarding the potential toxicity of bismuth and its compounds. Sano et al. (2005) performed oral administration studies with bismuth on rats. In one study, a single dose of bismuth of up to 2000 mg/kg produced no deaths or unusual findings in any of the tested animal groups. In a second study, rats received up to 1000 mg/kg/day for 28 days. Again there

were no unusual findings in any of the test groups. The authors concluded that the oral LD_{50} of bismuth is greater than 2000 mg/kg and the no observed adverse effect level (NOAEL) was 1000 mg/kg. They also concluded that the adverse toxic effects of bismuth are low compared to lead, a material for which it is often considered a replacement.

Acute Effects

Although bismuth salts were being used for therapeutic purposes for many years, little attention was paid to their potential toxic effects (Beck, 1909). There have been reports as far back as 1786 of effects such as vomiting, twitching, and dizziness associated with ingestion of bismuth salts. Possible impurities such as arsenic or lead were most likely the actual source of these symptoms (Higgins, 1916). The acute toxicity depends upon the nature of the bismuth substance. Acute exposure to bismuth salts may cause contact sensitivity and epigastric discomfort, and substantial ingestion may cause nausea and vomiting. A gray blue color "bismuth gumline" can be associated with both acute and chronic exposures to bismuth compounds. Bismuth subsalicylate also has been linked to ringing in the ears (NLM, 2011). The clinical manifestations of acute bismuth intoxication are similar to those caused by lead and mercury, i.e., neurological abnormalities, especially encephalopathy and renal dysfunction, with nephrotic syndrome progressing to renal failure (INCHEM, n.d.). Numerous reports of eye and skin irritation have been associated with exposure to bismuth and its compound.

Chronic Effects

Use of bismuth compounds in medicine indicates that bismuth and its salts can cause various types of neurotoxicity, kidney and liver damage, usually mild. Symptoms of chronic toxicity include decreased appetite, weakness, rheumatic pain, diarrhea, fever, foul breath, gingivitis, and dermatitis. Chronic excessive ingestion also produces bismuth lines on the gums and can cause stomatitis, excess salivation, osteoarthropathy, and pathological fractures (COT, 2008; NLM HSDB, 2013). Bismuth salts react with sulfides produced by anaerobic bacteria in the gut to produce bismuth sulfides, which turn the stool black. With chronic ingestion bismuth can accumulate until nephrotoxicity and neurotoxicity develop. Neurotoxicity reports associated with chronic oral exposure to bismuth-containing pharmaceuticals a describe a wide variety of symptoms. These include irritability, insomnia, difficulty sleeping, muscle twitching, hallucinations, confusion, myoclonus, ataxia, and progressive dementia (INCHEM, n.d.). With regard to bismuth encephalopathy, Slikkerveer and de Wolff (1992) proposed three distinct phases of toxicity. A prodromal initial stage lasting 2–6 weeks typically manifests itself with memory changes, incoordination, behavioral changes, and psychiatric symptoms. This is followed by fulminant encephalopathy myoclonic jerks, confusion, and dysarthria. The third phase reflects recovery with no lasting effects, which is quite common when the offending bismuth compounds are withdrawn (INCHEM, n.d.).

Mutagenicity Bismuth subsalicylate was tested as part of the National Toxicological Program (NTP) mutagenicity testing program using the standard Salmonella testing protocol. The results did not suggest any mutagenic potential using this bioassay (Zeiger et al., 1987). Several bismuth compounds have been tested for their ability to cause chromosomal aberrations in lymphocytes. Bismuth citrate and bismuth glutathione showed no clastogenic potential; however, exposure to methylated bismuth did produce increased chromosomal aberrations and sister chromatid exchanges. Since bismuth ions are known to be methylated by intestinal flora, this finding is of potential interest (von Recklinghausen et al., 2008).

Carcinogenicity Bismuth and its compounds have not been identified as carcinogenic by the International Agency for Research on Cancer (IARC), the NTP, Occupational Safety and Health Administration (OSHA), the EPA through IRIS, or California's Prop 65. The American Conference of Governmental Industrial Hygienists (ACGIH) has assigned bismuth telluride an A4 carcinogenicity rating, "Not classifiable as a Human Carcinogen," based on a lack of response in chronic studies of dogs, rabbits, and rats. No data are available regarding the carcinogenicity of bismuth or its compounds in humans (National Library of Medicine, 2002).

Reproduction There are no data suggesting that bismuth presents a reproductive hazard. Bismuth levels in fetuses were reported to be the same as the parent following administration of bismuth salts to the mother (INCHEM, n.d.). There are two reports of women exposed to high levels of bismuth having normal pregnancies (NLM HSDB, 2013). In animal studies, bismuth citrate was given to pregnant rats and rabbits at 200 mg/kg and 1200 mg/kg body weight and no developmental anomalies were observed (Secker, 1993).

MODE OF ACTION

What is known about the biological activity bismuth and its compounds has been identified mostly from studies of their pharmacology. Bismuth is not an essential element for either man or animals (Fowler and Vouk, 1986). Almost all absorption of bismuth occurs from the gastrointestinal tract. Gastrointestinal absorption of bismuth compounds is primarily a function of their water solubility (INCHEM, n.d.). Absorbed bismuth is found primarily in the kidney with deposition occurring to a lesser extent in other organs such as liver, brain, lung, and spleen. Elimination occurs primarily via the kidney (absorbed bismuth) and feces (unabsorbed) (Fowler and Vouk, 1986; Slikkerveer and de Wolff, 1992). The pharmacologically active bismuth compounds have been divided into four toxicological groups (Slikkerveer and de Wolff, 1989):

Group 1: Simple organic salts and subsalts that are insoluble in water are minimally absorbed from the gut and that cause virtually no toxicity (e.g., bismuth subcarbonate).

Group 2: Absorbable lipid-soluble organic compounds and complexes that elevate the blood level of bismuth and cause neurotoxicity and hepatotoxicity (e.g., bismuth subgallate).

Group 3: Absorbable water-soluble organic compounds and complexes that elevate the blood level of bismuth and cause nephrotoxicity (e.g., bismuth oxychloride).

Group 4: Water-soluble complexes that hydrolyze in the gastrointestinal tract, with some absorption of bismuth and the hydrolyzed radical, elevating the levels of bismuth and the complex radical (e.g., bismuth subsalicylate).

Most bismuth compounds are considered to be poorly to moderately absorbed following dermal application, inhalation, or ingestion. The mechanism of action of bismuth toxicity is not fully understood. Most cases of toxicity associated with nephrotoxicity are primarily associated with exposure to water soluble bismuth salts (INCHEM, n.d.).

Until the 1940s, bismuth compounds were given parenterally and orally to treat infections such as syphilis and gastrointestinal complaints. Bismuth compounds have bactericidal actions *in vitro* and cytoprotective effects *in vivo*. Bismuth compounds, the subcitrate and subsalicylate, increase mucous and bicarbonate secretion in the gut and accumulate preferentially in ulcers. These compounds have antibacterial effects on *Helicobacter pylori* and help prevent ulcer formation by binding mucous together and inhibiting pepsin activity (Bierer, 1990; East, 2009). Because of problems with acute and chronic bismuth poisoning, medical use of bismuth and bismuth compounds has been restricted in Europe, Canada, and Australia. In the United States doses of bismuth subsalicylate up to 4.8 g/day have been considered safe (Bierer, 1990).

OCCUPATIONAL EXPOSURE

Sources of Exposure

Although occupational exposure may occur in the manufacture of cosmetics, industrial chemicals, metal alloys, and pharmaceuticals, occupational overexposures are rare (Fowler and Vouk, 1986). Bismuth presents a low hazard for usual industrial handling.

Many occupations can present the possibility of over-exposure to bismuth compounds because these compounds are used as coloring agents for ceramics, enamels, and paints; as corrosion inhibitors for water systems; as catalysts in the plastics and petroleum industry; and as flame retardants for resins. Bismuth is used in metallurgy for the manufacture of numerous alloys, some of which are used in welding and production of malleable irons. Bismuth telluride is used as a semi-conductor. Bismuth oxide, hydroxide, oxychloride, trichloride, and nitrate are used in cosmetics. Other salts such as succinate orthoxyquinolate, subnitrate, carbonate, and phosphate also have pharmaceutical applications (Nordberg, 2012).Workers can be exposed to insoluble bismuth material in the mining and processing of ore that contains lead or gold; when working with bismuth or its alloys; when formulating and using lubricants, catalysts, and as a base metal. Although the literature reports some symptoms from occupational overexposure to bismuth, it is uncommon. Overexposure to bismuth compounds in the workplace may cause some short-lived symptoms such as fatigue, joint pain, headaches, backaches, and other nonspecific symptoms (Fowler and Vouk, 1986). On the rare occasion when occupational over-exposure to bismuth or its compounds are linked to human toxicity it is associated with other more toxic compounds. For example, in an article by Bachanek et al. (2000) numerous adverse health effects were reported over time in a worker in a glass factory in Poland. Findings included loss of hair, scars and ulcerations, trigeminal nerve damage, and impalpable lymph nodes. His teeth were extracted over a 13-month period because they showed unusual damage such as brittleness and dark brown color. Heavy metal levels of thallium, lead, cadmium, and bismuth were substantially elevated. Bismuth levels in tooth material were180 times greater than the control group average. Ascribing the patient's medical problems to the various elevated metal levels was difficult although some of the observations could be linked directly to one of the other metals present. For example, hair loss is well known to be linked to thallium overexposure. Yet bismuth is included as one of the metals linked to the worker's illness.

Industrial Hygiene

The permissible exposure limit (PEL) values are established by the OSHA and the threshold limit values (TLV) are set by the ACGIH. The only PELs for bismuth compounds set by OSHA are for bismuth telluride undoped (15 mg/m^3 [total dust and 5 mg/m^3 respirable dust] and selenium doped bismuth [5 mg/m^3]) (OSHA, 2008). This limit is based on the observation that overexposure to these compounds may cause irritation of the eyes, nose, throat, and respiratory tract, as indicated by animal studies (NIOSH/OSHA, 1992). The TLVs set by ACGIH for these compounds are 10 and 5 mg/m^3 undoped and doped, respectively. These criteria were

established in 1973. The criterion set for bismuth telluride undoped is based on the compound being considered an "inert" material lacking any response other than physical irritation in an animal inhalation study. The criterion for "doped" bismuth telluride is based on the results of inhalation studies in several animal species, including dogs, rabbits, and rats, wherein granulomatous lesions were seen in all exposed species at 15 mg/m^3 for 1 year, 5 days per week. Since the lesions were considered mild and reversible, the lower criterion was recommended.

In any industrial operation, excessive exposure to hazardous materials should be avoided through use of appropriate administrative and engineering controls, such as local ventilation. If engineering controls are not feasible, a program requiring respirators and other personal protective equipment is needed. Impervious gloves, clothing, eye protection, and NIOSH-approved respiratory protection should be used to protect workers from overexposure to bismuth and bismuth compounds which may present a health hazard. If protective clothing is to be reused, laundering contaminated clothing should be controlled to avoid additional exposure-related hazards. For operations in which exposure to hazardous materials is possible, good personal hygiene habits are strongly recommended. These include no eating or smoking in the work area; washing hands and face after leaving the work area and before eating, smoking, or using toilet facilities; and removing contaminated clothing and showering at the end of the workday.

RISK ASSESSMENT

Although bismuth has not been the subject of any complete international review by WHO or other key organizations, e.g., US EPA, numerous organizations with a wide range of constituencies have opined on its hazard potential (COT, 2008). According to the Bismuth Institute Information Center, in a world where many elements and compounds present a multitude of hazards, bismuth metal is one of those rare elements that is considered safe because it is non-toxic and non-carcinogenic (Palmierri, 1993). Although the objectivity of a trade association group might be questioned, their conclusion is consistent with others such as Beliles (1979), who concludes that "except for strong acidic salts as bismuth trinitrate or violently reactive compounds such as bismuth tripentafluoride, bismuth compounds do not present a hazard by dermal application, inhalation or injection," the usual routes of occupational exposure. The International Labor Organization concludes that "there have been no reports of occupational (illness) during the production of metallic bismuth and the manufacture of pharmaceuticals, cosmetics and industrial chemicals. Because bismuth and its compounds do not appear to have been responsible for poisoning associated with work, they are regarded as the least toxic of heavy metals currently used in industry" (Nordberg, 2012).

REFERENCES

American Conference of Governmental Industrial Hygienists (ACGIH) (2001) *Bismuth Telluride, Doped and Undoped*, Washington, DC: American Conference of Governmental Industrial Hygienists.

A.Y. McDonald Mfg. Co. (n.d.) *No-Lead Brass Technical Specifications*. Available at http://www.noleadbrass.com/pdf/No%20Lead%20Brochure.pdf.

Bachanek, T., Staroslawska, E., Wolanska, E., and Jarmolinska, K. (2000) Heavy metal poisoning in glass worker characterised by severe dental changes. *Ann. Agric. Environ. Med.* 7(1):1–3.

Beck, E. G. (1909) Toxic effects from bismuth subnitrate with reports of cases to date. *JAMA* 1:14–18, Available at http://jama.jamanetwork.com/article.aspx?articleid=428760.

Bell, T. (2014) *Metal Profile: Bismuth. About.com Metals Guide*. Available at http://metals.about.com/od/properties/a/Metal-Profile-Bismuth.htm.

Beliles, R. (1979) In: Oehem, F., editor. *Toxicity of Heavy Metals in the Environment. Part 2*, New York, NY: Marcel Dekker, pp. 7603–605.

Bierer D.W. (1990) Bismuth subsalicylate: history, chemistry, and safety. *Rev. Infect. Dis.* 12:3–8.

Committte of Toxicity (COT), Food Standards Agency (2008) *The Al-Zn of metal toxicity: a summary of the toxicological information on 12 metals*. Committee on Toxicity in Food, Consumer Products and the Environment, Annex B. Available at http://cot.food.gov.uk/pdfs/tox200829annexb.pdf.

Digitalfire Hazards Database (2008) *Bismuth Trioxide Toxicology*. Available at http://digitalfire.com/4sight/hazards/ceramic_hazard_bismuth_trioxide_toxicology_352.html.

East, J. (2009) *Bismuth: medical uses and side effects. Metals in Medicine and the Environment*. Available at http://faculty.virginia.edu/metals/cases/east2/.html.

Fowler, B.A. and Vouk, V.B. (1986) Bismuth. In: Nordberg, G.F. and Vouk, V.B., editors. *Handbook on the Toxicology of Metals: General Aspects*, 2nd ed., vol. 1, Amsterdam, The Netherlands: Elsevier Science Publishers B.V., pp. 110–126.

Higgins, W.H. (1916) Systemic poisoning with bismuth. *JAMA* V 66 (9):648 650. Available at jama.jamanetwork.com/article.aspx?articleid=436561.

INCHEM Corporation. (n.d.) Bismuth. In: *UKPID Monograph*, Birmingham, UK: National Poisons Information Service.

Lenntech (2014) *Bismuth (Bi)—Chemical Properties, Health and Environmental Effects*. Available at http://www.lenntech.com/periodic/elements/bi.htm.

National Institute for Occupational Safety and Health (NIOSH)/Occupational Safety and Health Administration (OSHA) (1992) *Occupational Health Guideline for Bismuth Telluride, Doped and Undoped*, U.S. National Library of Medicine. Available at http://toxnet.nlm.nih.gov/cgi-bin/sis/search/a?dbs+hsdb:@term+@DOCNO+2078.

National Library of Medicine (2002) *Bismuth, Elemental*. Available at http://toxnet.nlm.nih.gov/cgi-bin/sis/search/a?dbs+hsdb:@term+@DOCNO+2078.

National Library of Medicine (NLM) (2011) *Bismuth subsalicylate*. MedlinePlus. Available at https://www.nlm.nih.gov/medlineplus/druginfo/meds/a607040.html.

National Library of Medicine, NLM HSDB Database (2013) *Bismuth.Human Health Effects*. Available at http://toxnet.nlm.nih.gov/cgi-bin/sis/search/f?.temp/~fw3Yik:1:BASIC.

Nordberg, G. (2012) *Metals: Chemical Properties and Toxicity, Additional Resources. "Bismuth." International Labour Organization Encyclopedia of Occupational Safety and Health*. Available at http://www.ilo.org/oshenc/part-ix/metals-chemical-properties-and-toxicity/item/132-bismuth.

Occupational Safety and Health Administration (OSHA) (2008) Permissible exposure limits. 29 CFR 1910.1000. In: *Bismuth Telluride*, Washington, DC: Occupational Safety and Health Administration.

Palmierri, Y. (1993) *Bismuth: The Amazingly "Green" Environmentally Minded Element*, Belgium: The Bulletin of the Bismuth Institute.

Salvador, J.A., Figueriedo, S.A., Pinto, R.M., and Silvestre. S.M. (2012) Bismuth compounds in medicinal chemistry. *Future Med. Chem.* 11:1495–523. Available at www.ncbi.nlm.nih.gov/pubmed/22857536.

Sano, Y. et al. (2005) Oral toxicity of bismuth in rat: single and 28-day repeated administration studies. *J. Occup. Health* 47(4):293–298. Available at http://www.ncbi.nlm.nih.gov/pubmed/?term=bismuth+tolerable+daily+intake.

Secker, R.C. (1993) Effect of bismuth citrate on pregnant rats and rabbits. *Teratology* 48:38.

Slikkerveer, A. and de Wolff, F.A. (1989) Pharmacokinetics and toxicity of bismuth compounds. *Med. Toxicol. Adverse Drug Exp.* 4(5):303–23.

Slikkerveer, A. and de Wolff, F.A. (1992) Pharmacokinetics and toxicity of bismuth compounds. In: Slikkerveer, A., editor. *Bismuth: Biokinetics, Toxicity and Experimental Therapy of Overdosage*, The Netherlands: Leiden University.

U.S. Environmental Protection Agency (2005) Solders in electronics: a life-cycle assessment summary (U.S. EPA 744-S-05-001). In: Geibig, J.R. and Socolof, M.L., editors. *Solders in Electronics, A Life-cycle Assessment*, U.S. EPA. Available at www.epa.gov/dfe/pubs/solder/lca/lca-summ2.pdf.

U.S. Environmental Protection Agency (2013) *TSCA Chemical Substance Inventory*. Available at http://www.epa.gov/opptintr/existingchemicals/pubs/tscainventory/howto./html.

U.S. Geological Survey (2013) *Revised 2011 Minerals Yearbook*. Available at http://minerals.usgs.gov/minerals/pubs/commodity/bismuth/.

U.S. Geological Survey (2014a) *Mineral Commodities. Bismuth Statistics and Information*.

U.S. Geological Survey (2014b) *Mineral Commodities Summaries, Bismuth*.

von Recklinghausen, U., Hartmann, L.M., Rabieh, S., Hippler, J., Hirner, A.V., Rettenmeir, A.W., and Dopp, E. (2008) Methylated bismuth, but not bismuth citrate or bismuth-glutathione, induces cyto- and genotoxic effects in human cells *in vitro. Chem. Res. Toxicol.* 12(6):1219–28. Available at www.ncbi.nlm.nih.gov/pubmed/18826176.

WebElements Periodic Table of the Elements (2012) *Bismuth, Essential Information*. Available at https://www.webelements.com/bismuth/.

Zeiger, E., Anderson, B., Haworth, S., Lawlor, T., Mortelmans, K., and Speck, W. (1987) Salmonella mutagenicity tests: III. Results from the testing of 225 chemicals. *Environ. Mutagen.* 9(Suppl. 9): 1–109.

11

BORON

Giffe T. Johnson

First aid:

Eye contact: Check for and remove any contact lenses. In case of contact, immediately flush eyes with plenty of water for at least 15 min. Get medical attention if irritation occurs.

Skin Contact: Wash with soap and water. Cover the irritated skin with an emollient. Get medical attention if irritation develops.

Inhalation: If inhaled, remove to fresh air. If not breathing, give artificial respiration. If breathing is difficult, give oxygen. Get medical attention.

Ingestion: Do not induce vomiting unless directed to do so by medical personnel. Never give anything by mouth to an unconscious person. If large quantities of this material are swallowed, call a physician immediately. Loosen tight clothing such as a collar, tie, belt or waistband.

Target organ(s): Skin, eyes, lungs, systemic toxicant

Occupational exposure limits: Boron oxide, total dust (8 h TWA): 15 mg/m^3 (OSHA)

Boron trifluoride (ceiling): 3 mg/m^3 (OSHA)

Reference values: Boron and boron compounds: RfC N/A (EPA); RfD 0.2 mg/kg/day (EPA); Inhalation MRL 0.3 mg/m^3 (ATSDR); Oral MRL 0.2 mg/kg/day (ATSDR)

Risk/safety phrases: Skin irritant; eye irritant; corrosive

BACKGROUND AND USES

Boron occurs naturally as sodium borate ($Na_2B_4O_7$) and calcium borate. Borax ($Na_2B_4O_7 10H_2O$) has been used for many purposes for years. The earth's crust contains approximately 0.0008% boron (Cotton et al., 1999; Jansen, 2003). Background levels of boron in U.S. soils have been reported to average 4.6–26 mg/kg. Boron is an essential nutrient for higher plants. Some studies indicate that it may also be an essential element for humans (Penland, 1994). Borocaptate is used in boron neutron capture therapy for treatment of malignant brain tumors. Approximately 225 compounds containing boron are listed on the Toxic Substances Control Act inventory.

Boron is used in high energy fuels, in nuclear reactors as a shielding material to absorb neutrons, for hardening steel alloys, as an abrasive, for fireproofing, and in the textile and glass industries. Commercial products that use boron include glass, fire retardants, leather finishing, laundry detergents, perborate bleaches, and pesticides. Boron as a trifluoride is used in the dry etching of semiconductors and as a hydride in the defense industry for rocket propellants. Boron compounds are used as catalysts in the production of polymers and in chemical synthesis. They also are used as biological growth control agents in water treatment algaecides, fertilizers, herbicides, and insecticides. Boric acid is used as a surface disinfectant and is a principal ingredient in several pesticides.

PHYSICAL AND CHEMICAL PROPERTIES

Boron (symbol B; CAS #7440-42-8) is a dark brown or black metalloid in the amorphous form. It has a molecular weight of 10.81 g/mole, a boiling point of 2550 °C (4622°F), and a melting point of 2300 °C (4172°F). It is insoluble in water but

Hamilton & Hardy's Industrial Toxicology, Sixth Edition. Edited by Raymond D. Harbison, Marie M. Bourgeois, and Giffe T. Johnson.
© 2015 John Wiley & Sons, Inc. Published 2015 by John Wiley & Sons, Inc.

soluble in nitric and sulfuric acids. It is insoluble in cold water, hot water, diethyl ether, and alcohol. If finely divided, it is soluble in most molten metals such as copper, iron, magnesium, aluminum, and calcium. Borates are relatively soluble in water.

ENVIRONMENTAL FATE AND BIOACCUMULATION

Boron is never found as a free element in nature, and is commonly found attached to oxygen (Cotton et al., 1999). Atmospheric boron may be in the form of particulate matter or aerosols as borides, boron oxides, borates, boranes, organoboron compounds, trihalide boron compounds, or borazines. The half-life of airborne particles is usually on the order of days, depending on the size of the particle and atmospheric conditions (Nriagu, 1979). Many boron minerals are soluble in water, and it is unlikely that mineral equilibria will modulate the presence of boron in water (Rai et al., 1986). Boron has not been found to be significantly removed during conventional waste water treatment (Matthijs et al., 1999; Pahl et al., 2001; Waggott, 1969). Waterborne boron may be adsorbed by soils and sediments and adsorption–desorption reactions are likely to be the principal mechanism that determines the fate of boron in water (Rai et al., 1986). The extent of boron adsorption depends on the pH of the water and the chemical composition of the soil, with the greatest adsorption occurring at pH 7.5–9.0 (Keren and Mezuman, 1981; Keren et al., 1981; Waggott, 1969).

Although data is limited regarding the bioaccumulation of boron and boron-containing compounds, it is not expected to bioconcentrate or bioaccumulate (Tsui and McCart, 1981).

ECOTOXICOLOGY

Acute Effects

Although data are limited regarding the ecotoxicology of boron, evidence suggests that common environmental boron exposures resulting from both natural and anthropogenic sources do not have substantive acute effects on terrestrial and aquatic ecologies (Butterwick et al., 1989).

Chronic Effects

Although data are limited regarding the ecotoxicology of boron, evidence suggests that common environmental boron exposures resulting from both natural and anthropogenic sources do not have substantive chronic effects on terrestrial and aquatic ecologies (Butterwick et al., 1989).

MAMMALIAN TOXICOLOGY

Acute Effects

Boric acid has a relatively low toxicity in humans. In many cases of accidental or suicidal ingestion, the patient may be asymptomatic. If symptoms do occur, they may include abdominal pain, vomiting, diarrhea, headache, lethargy, and lightheadedness.

Workers exposed to borate dust are more likely to report symptoms such as nasal irritation than are nonworkers (Wegman et al., 1994). A decline in lung function was noted in borate workers over a 7-year period, but most of this (83%) was attributed to age and other factors. Decreases in lung function did not correlate with levels of exposure, but the probability of having symptoms did correlate with increasing levels of exposure.

Whorton et al. (1994) conducted an epidemiological study of male reproductive effects in a facility that produced inorganic borate. Using a standardized birth ratio method, they found that the birth rate among facility employees did not differ significantly from that of the general population.

Metaborates are highly alkaline and therefore may cause skin and eye damage. Boron trihalides have been shown to be respiratory irritants in animals and may have similar effects in humans. Borax and boric acid are absorbed from the gastrointestinal tract and damage skin. Borax is also a respiratory irritant. Air concentrations of 1.2–8.5 mg/m^3 have been associated with respiratory irritation. A dry cough, sore throat, and shortness of breath have been reported. More than 1 g of borax or excessive percutaneous absorption may result in severe acute and sometimes fatal intoxication.

The gas diborane (B_2H_6) is a pulmonary irritant, which causes tightness in the chest, shortness of breath, cough, and nausea. Continuous exposure concentrations below, which pulmonary irritation occurs may produce headaches, dizziness, chills, and fatigue.

The vapor from pentaborane (B_5H_9), a liquid, affects the nervous system, causing hyperexcitability and narcosis. The onset of symptoms may be delayed for a couple of days. Exposure to higher concentrations may cause incoordination, tremors, and seizures.

Experiments with rats suggest that the testes are a primary target organ for boric acid toxicity (Ku et al., 1991). High doses of boric acid cause testicular lesions in adult rats. Mechanistic studies suggest a central nervous system-mediated effect rather than a direct hormonal one (Ku and Chapin, 1994).

Because boric acid has a half-life in humans of about 1 day, urine and blood levels do not increase over a work week (Culver et al., 1994; Moseman, 1994). Boron does not accumulate in the soft tissues of animals; however, in cases of boric acid poisoning, brain and liver concentrations may be very high (Moseman, 1994).

Chronic Effects

Boron trifluoride, a gas, is severely irritating to eyes, skin, and lungs. Workers exposed to concentrations of up to 32 ppm (90 mg/m^3) for a decade or longer developed dryness of the nasal mucous membranes and epistaxis. Chronic exposure at low concentrations is not associated with adverse health effects.

Animal studies, including those conducted in rats and dogs exposed to high levels of boric acid or borax in the diet (Weir and Fisher, 1972) and mice exposed to boric acid in the diet (Dieter, 1994; NTP, 1987), have observed systemic effects such as hematological alterations (decreases in hemo-globin in rats and splenic hematopoiesis in mice), desqua-mation of footpad skin and bloody ocular discharge in rats, decreased body weight gain in rats and mice, lung hemor-rhage in mice, and hepatic chronic inflammation and coag-ulative necrosis. In rats, systemic effects occurred from chronic exposure to 81 mg boron/kg/day (NOAEL of 24 mg boron/kg/day); in mice, at 79 mg boron/kg/day. Tes-ticular atrophy was observed in rats exposed to 81 mg boron/kg/day as boric acid or borax (Weir and Fisher, 1972) and in mice exposed to 201 mg boron/kg/day as boric acid (Dieter, 1994; NTP, 1987).

Chemical Pathology

Boron is absorbed across pulmonary tissues into the blood. Workers exposed to borate dusts were found to have higher blood and urine boron concentrations at the end of a work shift compared to the beginning of the shift (Culver et al., 1994). Boron is almost completely absorbed in the gastro-intestinal tract, with up to 95% of the ingested dose being recovered in the urine (Dourson et al., 1998).

Boron is distributed readily to all body tissues. Tissue levels from daily doses achieve steady state with plasma in neurological and reproductive tissues. Bone and adipose tissues have differential affinities for boron with bone having a higher affinity for boron and adipose tissue having a lower affinity for boron than other soft tissues (Ku et al., 1991).

Excretion of boron is accomplished mostly through renal elimination, with minor contributions from saliva, sweat, and feces (Jansen et al., 1984). In workers exposed to borax, the boron level in the urine at the end of a work shift is a more accurate indicator of exposure than is the blood level (Culver et al., 1994).

Mode of Action(s)

Borates may be corrosive and can irritate the skin and mucous membranes. Mechanisms related to potential neurological or reproductive toxicity have not been elucidated.

SOURCES OF EXPOSURE

Occupational

The NIOSH estimates that the number of workers potentially exposed to boron increased from 6500 in the early 1970s to 35,600 in the early 1980s (NOHS, 1989; NOES, 1989). Sittig (1985) reports that NIOSH estimated that the numbers of workers potentially exposed to borax, boron oxide, and boron trifluoride are 2,490,000; 21,000; and 50,000, respectively.

Workers in the industries of manufacture of glass wool, fiberglass and other glass products, cleaning and laundry products, fertilizers, pesticides, and cosmetics may be exposed to boron compounds (Jensen, 2009; Stokinger, 1981). Workers exposed to boron may absorb measurable quantities; end-of-shift boron concentrations in blood and urine of 0.11–0.26 µg/g and 3.16–10.72 µg/mg creatinine, respectively, have been reported in workers at a facility where borax is packaged and shipped (Culver et al., 1994). In occupational settings where higher exposures exist without exposure mitigation, boron exposure can be subs-tantially higher. Workers from Kuandian City, China employed in boron mines or processing plants producing borax or boric acid had total daily boron measurements (in food, fluids, and personal air) averaging 41.2 mg/day for workers in the boron industry, 4.3 mg/day in the surrounding community, and 2.3 mg/day in a control group. Post-shift urine levels of boron measured in the boron industry workers was 499 mg/day, compared to 48.0 mg/day in the control group (Xing et al., 2008).

Large borax mining and refining plants may produce borax dusts in concentrations ranging from 1.1 to 14.6 mg/m^3 for total particulate with a mean boric acid/boron oxide dust concentration of 4.1 mg/m^3 for total particulate (Garabrant et al., 1984, 1985). In air samples from U.S. facilities where borax was packaged and shipped, mean dust concentrations have been observed in the range of 3.3–18 mg/m^3 (Culver et al., 1994).

Environmental

Ingestion of boron from food and water is the most frequent route of environmental human exposure. Boron exposure may also occur from some consumer products, including cosmetics, medicines, and insecticides. Some areas of the western United States have natural boron-rich deposits in the soil may also have higher boron exposure.

Rainey et al. (2002) reported mean daily intakes of boron from food and beverage sources for male and female adults to be 1.28 and 1.00 mg, respectively. Daily dietary boron intakes were estimated at 0.75 mg for infants aged 0–6 months and 0.99 mg for infants aged 7–11 months. Daily boron intakes were estimated as 0.86 mg for children aged

1–3 years, 0.80 mg for children aged 4–8 years, 0.90 and 1.02 mg, respectively, for adolescent males aged 9–13 and 14–18 years, and 0.83 and 0.78 mg, respectively, for adolescent females aged 9–13 and 14–18 years.

Boron is fairly ubiquitous in surface water and groundwater; the average surface water boron concentration in the United States is about 0.1 mg/l (Butterwick et al., 1989; EPA, 1986), but concentrations vary greatly, depending on local geologic and anthropogenic sources of boron (Butterwick et al., 1989). Mean boron concentrations in soil in the United States are about 30 mg/kg, with concentrations ranging up to 300 mg/kg (Eckel and Langley, 1988).

In the United States, Canada, and England, concentrations of boron in tap water have been reported in a range of 0.007–0.2 mg/l (Choi and Chen, 1979; Davies, 1990; Waggott, 1969). Ambient air samples have reported boron concentrations from $<5 \times 10^{-7}$ to 8×10^{-5} mg/m^3, with an average concentration of 2×10^{-5} mg/m^3 (Howe, 1998). It has been estimated that approximately 11,600 tons of boron per year are released into the atmosphere as a component of fly ash produced by coal combustion (Bertine and Goldberg, 1971).

INDUSTRIAL HYGIENE

Borates, or oxygen and inorganic boron-containing compounds, are the most common form of boron found in commercial products. This group includes boric acid, boric oxide, and hydrides of boron (boranes). Elemental boron is rare in the workplace and is not regulated because it is inert and has minimal use. Dust is the most common form of exposure and borate dusts should be kept below regulatory levels through appropriate control measures. The exposure limits and quantity threshold values for various boron compounds are listed in Table 11.1.

REFERENCES

Bertine, K.K. and Goldberg, E.D. (1971). Fossil fuel combustion and the major sedimentary cycle. *Sci.* 173(3993): 233–235.

Butterwick, L., de Oude, N., and Raymond, K. (1989) Safety assessment of boron in aquatic and terrestrial environments. *Ecotoxicol. Environ. Saf.* 17:339–371.

Choi, W.W. and Chen, K.Y. (1979) Evaluation of boron removal by adsorption on solids. *Environ. Sci. Technol.* 13:189–196.

Cotton, F.A., Wilkinson, G., Murillo, C.A., et al., editors (1999) Boron. In: *Advanced Inorganic Chemistry*, 6th ed., New York, NY: John Wiley & Sons, Inc., pp. 131–174.

Culver, B.D., Shen, P.T., Taylor, T.H., Lee-Feldstein, A., Anton-Culver, H., and Strong, P.L. (1994) The relationship of blood- and urine-boron to boron exposure in borax-workers and usefulness of urine-boron as an exposure marker. *Environ. Health Perspect.* 102:133–137.

Davies, K. (1990) Human exposure pathways to selected organochlorines and PCBs in Toronto and Southern Ontario. *Adv. Environ. Sci. Technol.* 23:525–540.

Dieter, M.P. (1994) Toxicity and carcinogenicity studies of boric acid in male and female B6C3F1 mice. *Environ. Health Perspect.* 102(Suppl. 7):93–97.

Dourson, M., Maier, A., Meek, B., Renwick, A., Ohanian, E., and Poirier, K. (1998) Boron tolerable intake: re-evaluation of toxicokinetics for data-derived uncertainty factors. *Biol. Trace Elem. Res.* 66:453–463.

Eckel, W. and Langley, W.D. (1988) A background-based ranking technique for assessment of elemental enrichment in soils at hazardous waste sites. In: *Superfund '88: Proceedings of the 9th National Conference, November 28–30, Washington, DC. Silver Spring, MD: The Hazardous Materials Control Research Institute.* pp. 282–286.

EPA (1986) *Quality Criteria for Water*, EPA440586001. Washington, DC: U.S.Environmental Protection Agency, Office of Water Planning and Standards.

Garabrant, D.H., Bernstein, L., Peters, J.M., and Smith, T.J. (1984) Respiratory and eye irritation from boron oxide and boric acid dusts. *J. Occup. Med.* 26:584–586.

Garabrant, D.H., Bernstein, L., Peters, J.M., Smith, T.J., and Wright, W.E. (1985) Respiratory effects of borax dust. *Br. J. Ind. Med.* 42:831–837.

Howe, P.D. (1998) A review of boron effects in the environment. *Biol. Trace Elem. Res.* 66:153–166.

Jansen, L.H. (2003) Boron, elemental. In: *Kirk-Othmer Encyclopedia of Chemical Technology*, John Wiley & Sons, Inc. Available at http://www.mrw.interscience.wiley.com/emrw/9780471238966/kirk/article/borojans.a01/current/pdf.

TABLE 11.1   Exposure Level Guidance and Regulatory Values

NIOSH REL (mg/m^3)	
Borax (10 h TWA)	5
Boron oxide (10 h TWA)	10
Boron tribromide (ceiling)	10
Boron trifluoride (ceiling)	3
Sodium tetraborate (10 h TWA)	1
Boron oxide	2000
Boron trifluoride	70
Sodium tetraborate	70
OSHA PEL (mg/m^3)	
Boron oxide, dust (8 h TWA)	15
Boron trifluoride (ceiling)	3
Boron tribromide (8 h TWA)	10
Boron trifluoride (ceiling)	3
OSHA Threshold Quantity (lbs)	
Boron trichloride	2500
Boron trifluoride	250

Jansen, J.A., Anderson, J., and Schou, J.S. (1984) Boric acid single dose pharmacokinetics after intravenous administration to man. *Arch. Toxicol.* 55:64–67.

Jensen, A.A. (2009) Risk assessment of boron in glass wool insulation. *Environ. Sci. Pollut. Res. Int.* 16:73–78.

Keren, R., Gast, R.G., and Bar-Yosef, B. (1981) pH-dependent boron adsorption by Na-montmorillonite. *Soil. Sci. Soc. Am. J.* 45:45–48.

Keren, R. and Mezuman, U. (1981) Boron adsorption by clay minerals using a phenomenological equation. *Clays Clay Miner.* 29:198–204.

Ku, W.W. and Chapin, R.E. (1994) Mechanism of the testicular toxicity of boric acid in rats: *in vivo* and *in vitro* studies. *Environ. Health Perspect.* 102:99–105.

Ku, W.W., Chapin, R.E., Moseman, R.F., Brink, R.E., Pierce, K.D., and Adams, K.Y. (1991) Tissue disposition of boron in male Fischer rats. *Toxicol. Appl. Pharmacol.* 111:145–151.

Matthijs, E., Holt, M.S., Kiewiet, A., and Rijs, G.B. (1999) Environmental monitoring for linear alkylbenzene sulfonate, alcohol ethoxylate, alcohol ethoxy sulfate, alcohol sulfate, and soap. *Environ. Toxicol. Chem.* 18:2634–2644.

Moseman, R.F. (1994) Chemical disposition of boron in animals and humans. *Environ. Health Perspect.* 102:113–117.

NOES (1989) *National Occupational Exposure Survey*, Cincinnati, OH: National Institute of Occupational Safety and Health. Available at http://www.cdc.gov/niosh/.

NOHS (1989) *National Occupational Hazard Survey*, Cincinnati, OH: National Institute of Occupational Safety and Health. Available at http://www.cdc.gov/niosh/.

Nriagu, J.O. (1979) Copper in the atmosphere and precipitation. In: Nriagu J.O., editor. *Copper in the Environment. Part I: Ecological Cycling*, New York, NY: John Wiley & Sons, Inc., pp. 43–67.

NTP (1987) *National Toxicology Program—Technical Report Series No. TR324 on the Toxicology and Carcinogenesis Studies of Boric Acid (CAS No. 10043-35-3) in B6C3F1 Mice (Feed Studies)*, NIH Publication No. 88–2580, Research Triangle Park, NC: U.S. Department of Health and Human Services, Public Health Service, National Institutes of Health.

Pahl, M.V., Culver, B.D., Strong, P.L., Murray, F.J., and Vaziri, N.D. (2001) The effect of pregnancy on renal clearance of boron in humans: a study based on normal dietary intake of boron. *Toxicol. Sci.* 60:252–256.

Penland, J.G. (1994) Dietary boron, brain function, and cognitive performance. *Environ. Health Perspect.* 102:65–72.

Rai, D., Zachara, J.M., Schwab, A.P., Schmidt, R.L., Girvin, D.C., and Rogers, J.E. (1986) *Chemical Attenuation rates, Coefficients, and Constants in Leachate Migration: A Critical Review*, Research Project 2198-1, vol. 1, Palo Alto, CA: Electric Power Research Institute.

Rainey, C.J., Nyquist, L.A., Coughlin, J.R., and Downing, G.R. (2002) Dietary boron intake in the United States: CSFII 1994–1996. *J. Food Comp. Anal.* 15:237–250.

Sittig, M. (1985) *Handbook of Toxic and Hazardous Chemicals and Carcinogens*, 2nd ed., Park Ridge, NJ: Noyes Publications, pp. 137–142.

Stokinger, H.E. (1981) The halogens and the nonmetals, boron and silicon. In: Clayton, G.D., and Clayton, F.E., editors. *Patty's Industrial Hygiene and Toxicology: Toxicology*, 3rd ed., vol. 2B, New York, NY: John Wiley & Sons, pp. 2978–3005.

Tsui, P.T. and McCart, P.J. (1981) Chlorinated hydrocarbon residues and heavy metals in several fish species from the Cold Lake area in Alberta, Canada. *Int. J. Environ. Anal. Chem.* 10:277–285.

Waggott, A. (1969) An investigation of the potential problem of increasing boron concentrations in rivers and water courses. *Water Res.* 3:749–765.

Wegman, D.H., Eisen, E.A., Hu, X., Woskie, S.R., Smith, R.G., and Garabrant, D.H. (1994) Acute and chronic respiratory effects of sodium borate particulate exposures. *Environ. Health Perspect.* 102(Suppl. 7):119–128.

Weir, R.J. and Fisher, R.S. (1972). Toxicologic studies on borax and boric acid. *Toxicol. Appl. Pharmacol.* 23:351–364.

Whorton, D., Haas, I., and Trent, L. (1994) Reproductive effects of inorganic borates on male employees: birth rate assessment. *Environ. Health Perspect.* 102(Suppl. 7):129–132.

Xing, X., Wu, G., Wei, F., Liu, P., Wei, H., Wang, C., Xu, J., Xun, L., Jia, J., Kennedy, N., Elashoff, D., and Robbins, W. (2008). Biomarkers of environmental and workplace boron exposure. *J. Occup. Environ. Hyg.* 5:141–147.

12

CADMIUM

Frank D. Stephen

First aid: Report all exposures to medical personnel immediately. Eye: Wash eyes immediately with large amounts of water for 15 min. If irritation persists, seek medical attention. Skin: Wash the exposed area with soap and water for 15 min. Inhalation: Move the person to fresh air.

Target organ(s): Oral: Kidney and bone. Inhalation: Kidney and lung

Occupational exposure limits: OSHA–PEL 5 µg/m^3 (8-h workday). NIOSH–REL 9 mg/m^3 (10 h TWA). ACGIH–TWA 0.01 mg/m^3

Reference values: MRL—short-term inhalation: 0.03 µg Cd/m^3 (<14 day exposure). Chronic inhalation: 0.01 µg Cd/m^3 (>1 year exposure). Chronic oral: 0.1 µg/kg/day (>1 year exposure). RfD: 5×10^{-4} mg/kg/day (water) and 1×10^{-3} (food). RfC: Not established

BACKGROUND AND USES

Cadmium (symbol Cd; CAS #744-43-9) is a naturally occurring element found in combination with zinc in the earth's crust. Most cadmium used in the United States is derived from zinc, copper, or lead extraction. Naturally, approximately 25,000 tons of cadmium are released in to the environment mostly into water sources due to rock weathering and air due to forest fires and volcanoes. Industrial manufacturing by humans is responsible for all other cadmium releases. Since cadmium can be extracted from zinc smelting, its ore is not mined for the metal. Approximately, 14 tons per year are produced worldwide with Canada being the main producer. Other major producers include the United States, Japan, Mexico, and Australia. It's used primarily in nickel cadmium batteries, coating and plating operations, and in pigments, plastics, and other synthetic products and alloys.

Overexposure to cadmium outside of the workplace is rare and mainly occurs through foodstuffs and smoking. Food sources containing cadmium include liver, mushrooms, cocoa powder, and shellfish. Consumption of these foods can increase the cadmium concentration in human bodies. An exposure to significantly higher cadmium levels occurs from smoking tobacco. Initially deposited in the lungs, cadmium is transported throughout the body by blood via the cardiovascular system, where it can enhance effects by potentiating cadmium that is already present from cadmium-rich food. Excessive environmental exposures have also been reported in some locations. Children living near a Mexican smelter were found to have cadmium exposure and body burdens above the acceptable levels set by the United States Environmental Protection Agency (EPA) (Diaz-Barringa et al., 1993). Nonoccupationally exposed Russians living near a metallurgical plant and storage battery factory also had an elevated body burden of cadmium (Bustueva et al., 1994).

Hamilton & Hardy's Industrial Toxicology, Sixth Edition. Edited by Raymond D. Harbison, Marie M. Bourgeois, and Giffe T. Johnson.
© 2015 John Wiley & Sons, Inc. Published 2015 by John Wiley & Sons, Inc.

TABLE 12.1 Physical and Chemical Properties of Cadmium

Molecular weight	112.41
Color	Silver-white
Physical state	Lustrous metal
Flash point	No data
Melting point	321 °C
Boiling point	765 °C
Density at 20 °C	8.65 g/cm^3
Water solubility	Insoluble
Vapor pressure	7.5×10^{-3} mmHg

PHYSICAL AND CHEMICAL PROPERTIES

Calmium is an odorless, silver-white lustrous metal with a bluish tinge, which is ductile and highly malleable with a melting point of 321 °C. The metal is soft enough to cut with a knife and will tarnish in air; as a powder, cadmium is flammable. Burning cadmium results in an odorless yellow-brown cadmium fume (cadmium monoxide or cadmium oxide fume) composed of finely divided particles dispersed in air. Both cadmium and cadmium oxide are insoluble in water and have a vapor pressure of approximately 0 mmHg. Cadmium is insoluble in water but can be solubilized in acid. Cadmium salts (e.g., cadmium sulfate and cadmium chloride) are soluble in water. Important physical and chemical properties are listed in Table 12.1.

ENVIRONMENTAL FATE AND BIOACCUMULATION

Cadmium is found naturally as one of the core components in the earth's crust. Industrial emissions from combustion and leaching from contaminated waste sites are the most likely causes of cadmium release into the environment. Whether released into the air or soil, the cadmium finally gets accumulated mostly in the soil (Morrow, 2001; Wilber et al., 1992). When emitted into the air, it can condense onto respirable particulates and can be transported over distances up to a couple of miles and can persist in the atmosphere for up to 10 days (Steinnes and Friedland, 2006; Wilber et al., 1992; EPA, 1980). Cadmium particles can also settle due to gravity or attach to water droplets that get subsequently deposited terrestrially.

Unlike many other heavy metals, upon release into water sources, cadmium is very mobile. Cadmium in unpolluted fresh water can be found as its hydrated ion $Cd(H_2O)_6^{2+}$ as opposed to polluted water sources where cadmium binds to humic substances leading to its transport and remobilization (EPA, 1979). The persistence of cadmium in water systems has been calculated to range from 4 to 10 years (Wester et al., 1992). Bacteria found in sediments have been found to aid in

the partitioning of cadmium from water to soil sediments (Burke and Pfister, 1988). Disturbance of cadmium adsorbed to clay sediment has been shown to release to water more easily than cadmium associated with carbonate materials (EPA, 1979). Due to its lack of volatility, cadmium does not partition from water back to the atmosphere (EPA, 1979). On land, once deposited in the soil, cadmium can entrain to water or air (EPA, 1985). Acidic conditions can cause cadmium to leach into water (Elinder, 1985), and a reaction with chloride can cause cadmium to poorly adsorb by clay (Roy et al., 1993).

Bioaccumulation through the food chain is of particular concern with regard to the final cadmium deposition in soil and sediments. Plants efficiently derive cadmium from the soil thereby introducing it into the food chain of animals and eventually humans. Acidic soil pH, which occurs due to acid rain, is of increasing concern because this condition makes it easier for plants to take up cadmium (Elinder, 1992). Bioaccumulation in terrestrial and aquatic organisms is well documented. Strikingly, cadmium levels in freshwater and marine animals have been found to be hundred to thousand times higher than that of surrounding water (EPA, 1979).

Bioaccumulation occurs at all levels of the food chain from grass to food crops leading to further accumulation in livestock. Factors influencing uptake by plants include root system, foliar uptake, and translocation rate within the plant as well as how deep cadmium deposition into the soil occurs (Nwosu et al., 1995). Cadmium has been found to accumulate in plant leaves, and when grown in the same soil, leafy vegetables will contain more cadmium than root vegetables or grains (He and Singh, 1994). Passage of cadmium from the soil to plant appears to happen via passive diffusion (Nwosu et al., 1995). Other factors that increase bioavailability and cadmium uptake include acidic as opposed to alkaline soils (Smith, 1994; He and Singh, 1994; Thornton, 1992), sandy soil as opposed to clay (He and Singh, 1994), and non-calcareous as opposed to calcareous (Thornton, 1992).

Although cadmium accumulates through the food chain, biomagnifications among livestock (animal-to-animal consumption) may not be significant primarily because cadmium is poorly absorbed in the intestine (Sprague, 1986) and is predominantly stored in the liver and kidney rather than skeletal muscle (Harrison and Klaverkamp; Sileo and Beyer, 1985; Vos et al., 1990). Rather, human accumulation of cadmium more likely results in consumption of beef and poultry raised in feed crops grown in cadmium-containing soils.

ECOTOXICOLOGY

Cadmium is found naturally in soils, sediments, rocks, and the earth's crust. As such, environmental exposures are always possible. Erosion of cadmium-containing soils into freshwater systems is possible through natural rainfall.

Further, cadmium can be released into the environment from the burning of municipal and industrial wastes, use of fossil fuels, and phosphate-based fertilizer application. Air entrainment of cadmium can result from natural events or human activity. Since cadmium is found in water and soil, it can be emitted into the air through forest fires, volcanoes, and wind-swept erosion of dry soils (EPA, 1985; Morrow, 2001) as well as aerosolizing of sea salt (Shevchenko et al., 2003). The main sources of cadmium introduction to air are from industrial activity such as smelting, coal boilers, municipal and sewage incinerators, and other urban and industrial emissions. Therefore, the risk of acute and chronic effects to ecosystems may exist. However, tolerance appears to play a role in various species exposed to cadmium from natural sources. Also, complicating the field studies is the inability to directly demonstrate that cadmium is the sole source of the effect observed. The following is a brief review of the laboratory studies where direct effects of cadmium were observed.

Acute Effects

Aquatic Species   Laboratory exposures to cadmium produce various acute effects to aquatic organisms. Its toxicity in water is dependent upon a variety of natural factors such as water hardness, salinity and temperature, as well as chelating agents and organic matter. Embryonic and larval stage aquatic animals appear to be more sensitive than adult animals of the same species. Mortality studies found that 7-day exposure to 85.5 µg/l cadmium produced death in the snail *Physa integra* (Spehar et al., 1978). Cadmium LC_{50} have been reported for crayfish (Mirenda, 1986), scallops (Pesch and Stewart, 1980), amphiopds (Robinson et al., 1988), minnows (Pickering and Henderson, 1966), and various forms of amphibian (Westerman, 1977; Slooff and Baerselman, 1980; Canton and Slooff, 1982; Khangarot and Ray, 1987; Woodall et al., 1988).

Terrestrial Organisms   Cadmium toxicity has been reported in several terrestrial species, including nematodes and snails. Cadmium LC_{50}s have been established for several nematode species while other researchers found decreases in fecundity, growth, and reproduction. Snails that were fed cadmium-enriched diets were found to have reduced food consumption, weight loss, shell growth, and reproductive activity. The toxicity to various juvenile avian species was also studied and cadmium LC_{50}s were reported for quail (Hill and Camardese, 1986; Hill et al., 1975), pheasant (Hill et al., 1975), mallard duck (Hill et al., 1975), and chickens (Pritzl et al., 1974).

Chronic Effects

Aquatic Organisms   Laboratory studies have also investigated the potential for cadmium toxicity in animals due to long-term exposure. Snails exposed to cadmium from the embryonic stage to adulthood were found to have delayed hatching times and reduced overall growth (Holcombe et al., 1984). Similar results were reported in minnows (Meteyer et al., 1988). Oysters exposed to cadmium concentrations of at least 5 µg/l for at least 35 weeks demonstrated slower growth than controls (Zaroogian and Morrison, 1981). Exposure to 3.5 µg/l cadmium produced retarded growth and death was observed to brook trout males during spawning (Benoit et al., 1976). Cadmium exposure during development of minnows and medaka produced malformed spinal vertebra (Bengtsson et al., 1975; Hiraoka and Okuda, 1984), and it was shown that calcium levels were decreased in fish with spinal malformations (Muramoto, 1981).

Terrestrial Organisms   The toxicological potential of cadmium has also been studied experimentally in terrestrial species. Generally, in invertebrates there was reduced growth and reproduction observed after cadmium exposure otherwise cadmium had little to no effect. Male avians exposed to cadmium via intramuscular injection (pigeons) or diet (mallard ducks) had reduced testicular development and spermatogenesis while female mallards had reduced egg production (Lofts and Murton, 1967; White and Finley, 1978; White et al., 1978). Kidney damage was reported in various seabirds that had been contaminated in the wild and starlings exposed in the laboratory (Nicholson et al., 1983).

MAMMALIAN TOXICOLOGY

Acute Effects

Human health effects due to cadmium exposure are dependent on the route of exposure. The most common cadmium exposure route is inhalation. Deaths due to severe pulmonary edema and chemical pneumonitis, leading to death due to respiratory failure have been reported for accidental occupational exposure to cadmium (Beton et al. 1966; Patwardhan and Finckh, 1976; Lucas et al., 1980; Seidal et al., 1993). Nonlethal, immediate post-cadmium exposure effects due to acute cadmium inhalation primarily involve the respiratory tract, including, but not limited to, irritation, tight chest, cough, pain in chest on coughing, dyspnea, chills, sweating, and aching pain in back and limbs.

Chronic Effects

Human epidemiological studies have found that long-term cadmium inhalation exposure produces effects on the respiratory and renal systems also leading to cancer. Inhalation does not appear to have significant effects on the cardiovascular, gastrointestinal, hepatic, nervous, immune, or reproductive systems.

Respiratory In humans, inhalation exposure to high levels of cadmium oxide fumes or dust is intensely irritating to the respiratory tissue, but symptoms can be delayed. Occupationally exposed individuals have been reported to show signs of chronic rhinitis and impairment or loss of the sense of smell, progressive pulmonary fibrosis, severe dyspnea and wheezing, chest pain and precordial constriction, persistent cough, weakness and malaise, anorexia, nausea, diarrhea, nocturia, abdominal pain, hemoptysis, and prostration. Interestingly, Edling et al. (1986) did not find that cadmium exposure impaired the respiratory system function, and other studies have shown that cadmium-induced effects to the respiratory system can be partially reversible and even return to normal several years later (Bonnell, 1955; Chan et al., 1988).

Renal Kidney function is the primary target of chronic cadmium overexposure. Initial indications of impaired kidney function are proteinuria and decreased glomerular filtration rate (GFR). Occupational studies have shown that workers overexposed to cadmium had impaired nephron readsorption of calcium, retinol-binding protein, beta 2-microglobulin, and transferrin. Cadmium-induced kidney impairment may continue to progress after exposure has stopped. A prospective study of 23 former cadmium workers reported that the GFR declined at a higher rate than normal (Järup et al., 1993). Further research indicated that glomerular damage correlated with blood cadmium levels (Järup et al., 1995). In this study, the authors found that every worker with cadmium blood levels over 75 nmol/l had glomerular damage compared with 33% of workers with blood concentrations between 50 and 75 nmol/l and only 3.4% of workers with concentrations below 50 nmol/l.

Recent studies have focused on identifying biomarkers of cadmium exposure. Increased urinary cadmium levels have been correlated with urinary human complex-forming glycoprotein (pHC) also called $\alpha1$ microglobulin (Järup et al., 2000; Olsson et al., 2002; Nakadaira and Nishi, 2003; Teeyaakasem et al., 2007).

Kidney stones may form in some groups as a result of disturbed calcium–phosphorus metabolism, specifically phosphaturia, low serum phosphorus, and low tubular reabsorption of calcium. Workers exposed to cadmium have been reported to have a higher prevalence of kidney stones (Järup and Elinder, 1993).

Cancer Epidemiological studies have been undertaken to research the possible effect cadmium has on carcinogenesis. There is little evidence to show that oral or dermal exposures to cadmium can be associated with increased human cancer rates. Several studies have suggested that there is an increase in mortality due to lung cancer in response to cadmium inhalation. The current data are conflicted regarding whether cadmium exposure causes prostate cancer. Based on the available epidemiological studies, the International Agency for Research on Cancer (IARC) lists cadmium as a Group 1 carcinogen: carcinogenic to humans, whereas, the American Conference of Industrial Hygienists (ACGIH) has listed cadmium as an A2 carcinogen: suspected human carcinogen.

Sorahan and Lancashire (1997) reported a significant, exposure-related increase in lung cancer mortality among a cohort of cadmium workers. This increase in cancer incidence was also related to worker cadmium exposure in association with arsenic trioxide. Workers in nickel–cadmium battery facilities in England and Sweden were found to have increased lung cancer mortality (Sorahan, 1987; Järup et al., 1998). Similarly, increased respiratory tract and lung cancer mortality rates were observed amongst employees in cadmium recovery and processing facilities and/or smelter operations (Lemen et al., 1976; Thun et al., 1985; Ades and Kazantzis, 1988; Kazantzis et al., 1988; Stayner et al., 1992). The association between cadmium and lung cancer though is not completely clear. Many of the studies listed failed to reject other confounding factors such as smoking and exposure to other heavy metals.

Studies have also been conducted focusing on the relationship between cadmium exposure and prostate cancer. Early studies reported increases in prostate cancer after occupational exposure to cadmium (Kipling and Waterhouse, 1967; Kjellström et al., 1979; Lemen et al., 1976). However, subsequent investigations failed to find increases in prostate cancer rates among males occupationally exposed to cadmium (Elinder et al., 1985c; Kazantzis et al., 1988; Sorahan, 1987; Sorahan and Esmen, 2004; Thun et al., 1985; Kazantzis et al., 1992).

Skeletal Detailed clinical findings showing skeletal abnormalities in cadmium workers were reported by Gervais and Delpech (1963). Pathological fractures and bone changes were visible on X-rays. In Japan, the itai-itai disease occurred in people who worked in rice fields near a mine that produced zinc, lead, and cadmium (Kobayashi, 1971). Characterized by hypercalciura, itai-itai causes inhibition of proximal tubular reabsorption leading to osteomalacia and painful fractures. Case studies indicate that calcium deficiency, osteoporosis, or osteomalacia can develop in some workers after long-term occupational exposure to high levels of cadmium (Adams et al., 1969; Blainey et al., 1980; Bonnell, 1955; Kazantzis, 1979; Scott et al., 1980). Effects on bone generally arise only after kidney damage has occurred and are likely to be secondary to resulting changes in calcium, phosphorus, and vitamin D metabolism (Blainey et al., 1980).

Chemical Pathology

Once deposited in the tissue, cadmium's main mechanism appears to be related to the original cadmium compound the subject was exposed to as well as the cadmium particle size.

Cadmium oxide (inhalation) and cadmium chloride (oral) are of greatest concern, and the smallest cadmium particles are considered the most dangerous as they seed deeper into the lung bronchial tree. It is generally thought that unbound cadmium ions, and not those bound to metallothionein or other proteins, produce the observed toxic effects (Goyer et al., 1989), including interfering with metal-dependent enzymes. Renal tubular membranes can be damaged by cadmium bound to metallothionein during cellular uptake (Suzuki and Cherian, 1987).

Mode of Action(s)

Toxicity   Cadmium toxicity occurs in a variety of organs and tissues but the primary targets are the kidney, liver, and bone. Kidney deposition results in damage to the cells of the proximal convoluted tubule or interstitial fibrosis (Stowe et al., 1972). The resulting effects include increased urine concentration of low molecular weight proteins (e.g., β2-microglobulin), glucose, enzymes, and amino acids and increase in total urine output. Cadmium appears to interfere with the metabolism of iron, zinc, copper, and selenium (Petering et al., 1979; Jamall and Smith, 1985) and may deplete various enzymes (e.g., glutathione peroxidase and superoxide dismutase). Bone deposition causes decreased osteoblastic activity that results in decreased production of new osteoid and mineralization and osteoclastic bone resorption may be enhanced. Bone mineralization may also be affected indirectly by cadmium-induced blockage of hydroxylation of 25-hydroxy vitamin D to 1,25-dihydroxy vitamin D in the kidney.

Entrance and Metabolism   Cadmium exposure in humans occurs either via oral or inhalation routes, whereas dermal exposure is of little consequence. Workers are typically exposed through inhalation, whereas the general public is more likely to be exposed orally through consumption of foodstuffs such as grains, cereal, and potatoes (Gartrell et al., 1986). Smoking significantly increases cadmium levels in both groups (Hovinga et al., 1993; Qu et al., 1993; Hertz-Picciotto, Hu, 1994). It is estimated that 50% of cadmium from inhalation exposure is absorbed by the lung tissue. A greater percentage of cadmium is absorbed from inhalation rather than the oral route. The amount of cadmium absorption in the lung is dependent on the particle size. Smaller particles are of more danger in that they are delivered to the smallest respiratory bronchioles and alveoli. Subsequent absorption to the bloodstream results in delivery to the rest of the body. Larger particles are trapped in the ciliated mucosal cells of the upper respiratory system, cleared, and eventually enter into the gastrointestinal track. Data regarding cadmium deposition, retention, or absorption in the human lung are not available.

Gastrointestinal track absorption of calcium has been reported to be <10% of the intake (Flanagan et al., 1978;

McLellan et al., 1978; Newton et al., 1984; Rahola et al., 1973; Shaikh and Smith, 1980; Vanderpool and Reeves, 2001). The mechanism by which cadmium is absorbed is not completely understood, but evidence exists showing that divalent metal transporter I protein is involved (Park et al., 2002; Ryu et al., 2004; Kim et al., 2007). Absorbed cadmium is bound and transported via metallothionien, a low molecular weight compound (Cousins, 1979) and also has an affinity for sulfydryl groups on albumin and metallothionien (Nordberg et al., 1985). The metal–protein complexes are deposited in the liver, from which they are slowly released, and subsequently deposited in the kidney. Cadmium does not appear to undergo any metabolic conversion to other molecular species. Once in the kidney, sufficient amounts can cause renal tubular damage. Animal studies have shown a retention halftime of 73 days in liver tissue and a lifetime in the kidney tissue (Shaikh, 1982). Cadmium concentrations reported to cause renal tubular toxicity have been estimated to be over 200 μg/g wet weight in the cortex (Dudley et al., 1985). Lower levels of metallothionien-bound cadmium are considered nontoxic. The exact biochemical mechanism that determines this threshold is unknown (Goyer, 1989).

Excretion   Cadmium is excreted from the body very slowly; its biological half-life in humans is 10–40 years. Urine cadmium concentration in occupationally exposed workers increases proportionally to the total cadmium body burden (Roels et al., 1981). Low level, long-term cadmium exposure results in urine cadmium excretion that reflects the total body burden and typically amounts to 1–2 μg/day (Goyer, 1989). However, when the exposure is more intense, urine cadmium levels may be as high as 50 μg/g of creatinine, and this value is reflective of the recent exposure rather than the body burden.

SOURCES OF EXPOSURE

Occupational

The greatest risk to cadmium exposure by industrial workers includes processes where cadmium-containing materials are heated, including coating and electroplating, smelting operations, nickel–cadmium battery production, and phosphate fertilizer production and application. Another major route of incidental exposure comes from the ingestion of foods and cigarettes introduced to the mouth by contaminated hands and inhalation of dust and fumes.

Environmental

Cadmium is a naturally occurring element in the earth's crust and thus, plants can accumulate large amounts of cadmium in their leaves. Smoking is a major source of environmental cadmium because of the ability of tobacco leaves to accumulate high levels of cadmium. The major source of cadmium is from food intake. Dietary consumption of organ meats and shellfish, leafy vegetables, potatoes and grains, soybeans and sunflower seeds contributes to the total cadmium body burden. Cadmium exposure from drinking water is not a major concern because the EPA mandates control water cadmium concentrations to be <5 µg/l.

INDUSTRIAL HYGIENE

Exposure Limits

ACGIH The ACGIH has established a threshold limit value (TLV) (8-h time-weighted average (TWA)) of 0.01 mg/m^3 for cadmium and 0.002 mg/m^3 for cadmium compounds.

NIOSH The National Institute for Occupational Safety and Health (NIOSH) considers cadmium compounds to be potential occupational carcinogens as defined by the Occupational Safety and Health Administration (OSHA) carcinogen policy (29 CFR 1990). Based on human data, the NIOSH has established an Immediate Threat to Life and Health (IDLH) level 50 mg/m^3 in dust (Barrett and Card, 1947; Bulmer et al., 1938; Reinl, 1961) and 9 mg/m^3 in fumes (Beton et al., 1966) for cadmium compounds.

OSHA The OSHA has established a PEL (8-h TWA) of 5 µg/m^3 for shipyard, construction, and general industry.

RISK ASSESSMENTS

Listings as Hazardous Waste

Cadmium has been classified as a hazardous waste and has been registered in various publications as such including:

Section 112 of the Clean Air Act lists cadmium compounds as one of the 189 chemicals inventoried as a hazardous air pollutant (EPA, 2007).

Title III of the Emergency Planning and Community Right-To-Know Act of 1986 (EPA, 2008c) requires annual reports of cadmium release to environmental release by any manufacture, import, process, or use.

Section 311 of the Clean Water Act requires that any discharge of cadmium and cadmium-containing compounds over a specified threshold level into navigable waters is subject to reporting requirements (EPA, 2008a).

The Resource Conservation and Recovery Act (RCRA) recognizes cadmium as a hazardous waste. Cadmium is also considered a priority persistent, bioaccumulative, and toxic (PBT) chemical under RCRA waste minimization chemical listing (EPA, 1998), thus, requiring groundwater monitoring at municipal solid waste landfills (EPA, 2008b).

IRIS The EPA's Integrated Risk Information System (IRIS) oral and carcinogenicity assessments for cadmium were last revised in 1994 and 1992, respectively. The critical clinical effect for cadmium is significant proteinuria at an oral reference dose (RfD) of 5×10^{-4} mg/kg/day (water) and 1×10^{-3} mg/kg/day (food). Based on the limited human evidence, the EPA has listed cadmium as a class B$_1$, probable human carcinogen, based on an inhalation unit risk of 1.8×10^{-3} per µg/m^3. Carcinogenicity based on human exposures was described by Thun et al. (1985) showing that workplace exposures to cadmium could result in lung, trachea, or bronchus cancer.

WHO The World Health Organization (WHO) has set a drinking water guideline of 3 µg/l as well as an air guideline of 5 ng/m^3 (annual average).

ATSDR The Agency for Toxic Substances and Disease Registry (ATSDR) has established Minimal Risk Levels (MRLs) for chronic and acute, oral and inhalation exposures to cadmium.

Inhalation The acute-duration MRL of 0.03 µg/m^3 is based on a lowest-observed-adverse-effect level (LOAEL) of 0.088 mg/m3 in rats exposed to cadmium oxide for 6 h/day, 5 days/week for 2 weeks (NTP, 1995).

The chronic-duration MRL of 0.01 µg/m^3 was derived using the 95% lower confidence limit of the urinary cadmium level associated with a 10% extra risk of low molecular weight proteinuria.

Oral The intermediate-duration MRL of 0.5 µg/kg/day is based on the skeletal effects in young female rats exposed to 0.05 mg/kg/day cadmium chloride in drinking water for up to 12 months (Brzóska and Moniuszko-Jakoniuk, 2005).

The chronic-duration oral MRL of 0.1 µg/kg/day is based on proteinuria estimated from a meta-analysis of environmental exposure data. Using pharmacokinetic models, the MRL was determined using a calculated intake of 0.33 µg/kg/day using the UCDL10 at age 55 and dividing by an uncertainty factor of 3 for human variability.

IARC The IARC's review of cadmium as a carcinogen was last revised in 2008. Based on the limited human studies, this review listed cadmium as a Group 1 carcinogen: carcinogenic to humans.

ACGIH The ACGIH has listed cadmium as an A2 carcinogen: suspected human carcinogen.

EPA Based on a chronic intake that would result in a kidney concentration of 200 µg/g, the EPA established a RfD of 5×10^{-4} mg/kg/day in water and 1×10^{-3} mg/kg/day in food (IRIS, 2008).

CalEPA The California Environmental Protection Agency (CalEPA) in October 1996 listed cadmium as a carcinogen under Proposition 65. The California Air Resources Board also identified cadmium as a toxic air contaminant.

FDA The United States Food and Drug Administration (FDA) limits cadmium levels in bottled drinking water at 0.005 mg/l (FDA, 2007).

REFERENCES

Adams, R.G., Harrison, J.F., and Scott, P. (1969) The development of cadmium-induced proteinuria, impaired renal function, and osteomalacia in alkaline battery workers. *Q. J. Med.* 152:425–443.

Ades, A.E. and Kazantzis, G. (1988) Lung cancer in a non-ferrous smelter: the role of cadmium. *Br. J. Ind. Med.* 45:435–442.

Barrett, H.M. and Card, B.Y. (1947) Studies on the toxicity of inhaled cadmium. II. The acute lethal dose of cadmium oxide for man. *J. Ind. Hyg. Toxicol.* 29(5):286–293.

Bengtsson, B.E., Carlin, C.H., Larsson, A., and Svanberg, O. (1975) Vertebral damage in minnows, Phoxinus phoxinus L., exposed to cadmium. *Ambio* 4:166–168.

Benoit, D.A., Leonard, E.N., Christensen, G.M., and Fiandt, J.T. (1976) Toxic effects of cadmium on three generations of brook trout (*Salvelinus fontinalis*). *Trans. Am. Fish. Soc.* 105:550–560.

Beton, D.C., Andrews, G.S., Davies, H.J., Howells, L., and Smith G.F. (1966) Acute cadmium fume poisoning: five cases with one death from renal necrosis. *Br. J. Ind. Med.* 23:292–301.

Blainey, J.D., Adams, R.G., Brewer, D.B., and Harvey, T.C. (1980) Cadmium-induced osteomalacia. *Br. J. Ind. Med.* 37:278–284.

Bonnell, J.A. (1955) Emphysema and proteinuria in men casting copper-cadmium alloys. *Br. J. Ind. Med.* 12:181–197.

Brzóska, M.M. and Moniuszko-Jakoniuk, J. (2005) Disorders in bone metabolism of female rats chronically exposed to cadmium. *Toxicol. Appl. Pharmacol.* 202:68–83.

Bulmer, F.M.R., Rothwell, N.F., and Frankish, E.R. (1938) Industrial cadmium poisoning, a report of fifteen cases, including two deaths. *Can. J. Public Health* 29:19.

Burke, B.E. and Pfister, R.M. (1988) *The Removal of Cadmium From Lake Water by Lake Sediment Bacteria.* In: *Proceedings of the Annual Meeting of the American Society for Microbiology*, Miami Beach, Florida, USA, May 8–13, 1988.

Bustueva, K.A., Revich, B.A., and Bezpalko, L.E. (1994) Cadmium in the environment of three Russian cities and in human hair and urine. *Arch. Environ. Health* 49(4):284–288.

Canton, J.H. and Slooff, W. (1982) Toxicity and accumulation studies of cadmium (Cd2+) with freshwater organisms of different trophic levels. *Ecotoxicol. Environ. Saf.* 6:113–128.

Chan, O.Y., Poh, S.C., Lee, H.S., Tan, K.T., and Kwok, S.F. (1988) Respiratory function in cadmium battery workers: a follow- up study. *Ann. Acad. Med. Singapore* 17:283–287.

Cousins, R.J. (1979) Metallothionein synthesis and degradation: relationship to cadmium metabolism. *Environ. Health. Perspect.* 28:131–136.

Diaz-Barriga, F., Santos, M.A., Mejia, J.D.J., Batres, L., Yáñez, L., Carrizales, L., Vera, E., del Razo, L.M., and Cebrián, M.E. (1993) Arsenic and cadmium exposure in children living near a smelter complex in San Luis Potosi. *Mexico. Environ. Res.* 62(2):242–250.

Dudley, R.E., Gammal, L.M., and Klaassen, C.D. (1985) Cadmium-induced hepatic and renal injury in chronically exposed rats: Likely role of hepatic cadmium-metallothionein in nephrotoxicity. *Toxicol. Appl. Pharmacol.* 77:414–426.

Edling, C., Elinder C.G., and Randma, E. (1986) Lung function in workers using cadmium-containing solders. *Br. J Ind. Med.* 43:657–662.

Elinder, C.G., Kjellström, T., Hogstedt, C., Andersson, K., and Spång, G. (1985c) Cancer mortality of cadmium workers. *Br. J. Ind. Med.* 42:651–655.

Elinder, C.G. (1985a) Cadmium: uses, occurrence and intake. In: Friberg, L., Elinder, C.G., and Kjellström, T., et al., editors. *Cadmium and Health: A Toxicological and Epidemiological Appraisal, vol. I. Exposure, Dose, and Metabolism. Effects and Response.* Boca Raton, FL: CRC Press, pp. 23–64.

Elinder, C.G. (1992) Cadmium as an environmental hazard. *IARC. Sci. Publ.* 118:123–132.

EPA (1979) *Water-Related Fate of 129 Priority Pollutants.* Washington, DC: U.S. Environmental Protection Agency, Office of Water Planning and Standards, EPA 440479029a.

EPA (1980) *Atmospheric Cycles of Cadmium and Lead: Emissions, Transport, Transformation and Removal.* U.S. Environmental Protection Agency, pp. 1–6 to 1–17; 2–14 to 2–41.

EPA (1985) *Cadmium Contamination of the Environment: An Assessment of Nationwide Risk.* Washington, DC: U.S. Environmental Protection Agency, Office of Water Regulations and Standards. EPA 440485023.

EPA (1998) *Notice of Availability of Draft RCRA Waste Minimization PBT Chemical List.* U.S. Environmental Protection Agency. Fed Regist 63:60332. Available at http://www.gpoaccess.gov/fr/index.html (accessed May 05, 2008).

EPA (2007) *The Clean Air Act Amendments of 1990 List of Hazardous Air Pollutants.* U.S. Environmental Protection Agency. United States Code.42 USC 7412. Available at http://www.epa.gov/ttn/atw/orig189.html (accessed April 24, 2008).

EPA (2008a) *Determination of Reportable Quantities.* U.S. Environmental Protection Agency. Code of Federal Regulations. 40 CFR 117.3. Available at http://www.epa.gov/lawsregs/search/40cfr.html (accessed April 24, 2008).

EPA (2008b) *Groundwater Monitoring List.* U.S. Environmental Protection Agency. Code of Federal Regulations. 40 CFR 264, Appendix IX. Available at http://www.epa.gov/lawsregs/search/40cfr.html (accessed May 05, 2008).

EPA (2008c) *Toxic Chemical Release Reporting. Chemicals and Chemical Categories to Which This Part Applies.* U.S. Environmental Protection Agency. Code of Federal Regulations. 40 CFR 372.65. Available at http://www.epa.gov/lawsregs/search/40cfr.html. (accessed April 24, 2008).

FDA (2007) *Beverages. Bottled Water. U.S. Food and Drug Administration.* Code of Federal Regulations. 21 CFR 165.110. Available at http://www.accessdata.fda.gov/scripts/cdrh/cfdocs/cfcfr/CFRSearch.cfm (accessed April 24, 2008).

Flanagan, P.R., McLellan, J., Haist, J., Cherian, G., Chamberlain, M.J., and Valberg, L.S. (1978) Increased dietary cadmium absorption in mice and human subjects with iron deficiency. *Gastroenterology* 74:841–846.

Gartrell, M.J., Craun, J.C., Podrebarac, D.S., and Gunderson, E.L. (1986) Pesticides, selected elements, and other chemicals in adult total diet samples, October 1980–March 1982. *J. Assoc. Off. Anal. Chem.* 69:146–161.

Gervais, J. and Delpech, P. (1963) L'intoxication cadmique. *Arch. Mal. Prof.* 24:803.

Goyer, R.A., Miller, C.R., Zhu, S.Y., and Victery, W. (1989) Non-metallothionein-bound cadmium in the pathogenesis of cadmium nephrotoxicity in the rat. *Toxicol. Appl. Pharmacol.* 101:232–244.

Goyer, R.A. (1989) Mechanisms of lead and cadmium nephrotoxicity. *Toxicol. Lett.* 46:153–162.

Harrison, S.E. and Klaverkamp, J.F. (1990) Metal contamination in liver and muscle of northern pike (*Esox lucius*) and white sucker (*Catostomus commersoni*) and in sediments from lakes near the smelter at Flin Flon. *Manitoba. Environ. Toxicol. Chem.* 9:941–956.

He, Q.B. and Singh, B.R. (1994) Crop uptake of cadmium from phosphorus fertilizers. I. Yield and cadmium content. *Water Air Soil Pollut.* 74:251–265.

Hertz-Picciotto, I. and Hu, S.W. (1994) Contribution of cadmium in cigarettes to lung cancer: an evaluation of risk assessment methodologies. *Arch. Environ. Health* 49(4):297–302.

Hill, E.F. and Camardese, M.B. (1986) *Lethal Dietary Toxicities of Environmental Contaminants and Pesticides to Coturnix.* Washington, DC, US Department of Interior, Fish and Wildlife Service (Fish andWildlife Technical Report No. 2).

Hill, E.F., Heath, R.G., Spann J.W., and Williams, J.D. (1975) *Lethal Dietary Toxicities of Environmental Pollutants to Birds.* Washington, DC, US Department of Interior, Fish and Wildlife Service (SpecialScientific Report, Wildlife No. 191).

Hiraoka, Y. and Okuda, H. (1984) Time course study of occurrence of anomalies in medaka's centrum by cadmium or fenitrothion emulsion. *Bull. Environ. Contam. Toxicol.* 32:693–697.

Holcombe, G.W., Phipps, G.L., and Marier, G.W. (1984) Methods for conducting snail (Aplexa hypnorum) embryo through adult exposures: effects of cadmium and reduced pH levels. *Arch. Environ. Contam. Toxicol.* 13:627–634.

Hovinga, M.E., Sowers, M., and Humphrey, H.E. (1993) Environmental exposure and lifestyle predictors of lead, cadmium, PCB, and DDT levels in Great Lakes fish eaters. *Arch. Environ. Health* 48(2):98–104.

IRIS (2008) *Cadmium. Integrated Risk Information System.* Washington, DC: U.S. Environmental Protection Agency. Available at http://www.epa.gov/iris/subst/index.html. (accessed April 24, 2008).

Jamall, I.S. and Smith, J.C. (1985) Effects of cadmium treatment on selenium dependent and selenium independent glutathione peroxidase activities and lipid peroxidation in the kidney and liver of rats maintained on various levels of dietary selenium. *Arch. Toxicol.* 58:102–105.

Järup, L. and Elinder, C.G. (1993) Incidence of renal stones among cadmium exposed battery workers. *Br. J. Ind. Med.* 50:598–602.

Järup, L., Persson, B., Edling, C., and Elinder, C.G. (1993) Renal function impairment in workers previously exposed to cadmium. *Nephron* 64:75–81.

Järup, L., Persson, B., and Elinder, C.G. (1995) Decreased glomerlar filtration rate in solderers exposed to cadmium. *Occup. Environ. Med.* 52:818–822.

Järup, L., Bellander, T., Hogstedt, C., and Spang, G. (1998) Mortality and cancer incidence in Swedish battery workers exposed to cadmium and nickel. *Occup. Environ. Med.* 55:755–759.

Järup, L., Hellstrom, L., Alfven, T., Carlsson, M.D., Grubb, A., Persson, B., Pettersson, C., Spång, G., Schütz, A., and Elinder, C.G. (2000) Low level exposure to cadmium and early kidney damage: The OSCAR study. *Occup. Environ. Med.* 57(10):668–672.

Kazantzis, G., Lam, T.H., and Sullivan, K.R. (1988) Mortality of cadmium-exposed workers. A five-year update. *Scand. J. Work. Environ. Health.* 14:220–223.

Kazantzis, G., Blanks, R.G., and Sullivan, K.R. (1992) Is cadmium a human carcinogen? In: Nordberg, G.F., Herber, R.F.M., and Alessio, L., editors. *Cadmium in the Human Environment: Toxicity and Carcinogenicity.* IARC Scientific Publications No. 118. Lyon, France: International Agency for Research on Cancer, pp. 435–446.

Kazantzis, G. (1979) Renal tubular dysfunction and abnormalities of calcium metabolism in cadmium workers. *Environ. Health. Perspect.* 28:155–159.

Khangarot, B.S. and Ray, P.K. (1987) Sensitivity of toad tadpoles, *Bufo melanostictus* (Schneider), to heavy metals. *Bull. environ. Contam. Toxicol.* 38:523–527.

Kim, D.W., Kim, K.Y., Choi, B.S., Youn, P., Ryu, D.Y., Klaassen, C.D., and Park, J.D. (2007) Regulation of metal transporters by dietary iron, and the relationship between body iron levels and cadmium uptake. *Arch. Toxicol.* 81:327–334.

Kipling, M.D. and Waterhouse, J.A.H. (1967) Cadmium and prostatic carcinoma. *Lancet* 1(7492):730–731.

Kjellström, T., Fribert, L., and Rahnster, B. (1979) Mortality and cancer morbidity among cadmium-exposed workers. *Environ. Health Perspect.* 28:199–204.

Kobayashi, J. (1971) *Relation Between the Itai-Itai Disease and the Pollution of River Water by Cadmium From a Mine.* In: Fifth International Water Pollution Research Conference, San Francisco, July1970. PergamonPress, New York, 1971.

Lemen, R.A., Lee, J.S., Wagoner, J.K., and Blejer, H.P. (1976) Cancer mortality among cadmium production workers. *Ann. N. Y. Acad. Sci.* 271:273–279.

Lofts, B. and Murton, R.K. (1967) The effects of cadmium on the avian testis. *J. Reprod. Fert.* 13:155–164.

Lucas, P.A., Jariwalla, A.G., Jones, J.H., Gough, J., and Vale, P.T. (1980) Fatal cadmium fume inhalation. *Lancet* 2(8187):205.

McLellan, J.S., Flanagan, P.R., Chamberlain, M.J., and Valberg, L.S. (1978) Measurement of dietary cadmium absorption in humans. *J. Toxicol. Environ. Health* 4:131–138.

Meteyer, M.J., Wright, D.A., and Martin, F.D. (1988) Effect of cadmium on early developmental stages of the sheepshead minnow (*Cyprinodon variegatus*). *Environ. Toxicol. Chem.* 7:321–328.

Mirenda, R.J. (1986) Toxicity and accumulation of cadmium in the crayfish, Orconectes virilis (Hagen). *Arch. Environ. Contam. Toxicol.* 15:401–407.

Morrow, H. (2001) Cadmium and cadmium alloys. In: *Kirk-Othmer Encyclopedia of Chemical Technology.* John Wiley & Sons, Inc., pp. 471–507. http://www.mrw.interscience.wiley.com/emrw/9780471238966/kirk/article/cadmcarr.a01/current/pdf?hd=All%2Ccadmium. April 29, 2008.

Muramoto, S. (1981b) Variations of some elements in cadmium-induced malformed fish. *Bull. Environ. Contam. Toxicol.* 27:193–200.

Nakadaira, H. and Nishi, S. (2003) Effects of low-dose cadmium exposure on biological examinations. *Sci. Total. Environ.* 308 (1–3):49–62.

Newton, D., Johnson, P., Lally, A.E., Pentreath, R.J., and Swift D.J. (1984) The uptake by man of cadmium ingested in crab meat. *Hum. Toxicol.* 3:23–28.

Nicholson, J.K. and Osborn, D. (1983) Kidney lesions in pelagic seabirds with high tissue levels of cadmium and mercury. *J. Zool. Lond.* 200:99–118.

Nordberg, G.F., Kjellström, T., and Nordberg, M. (1985) Kinetics and metabolism. In: Friberg, L., Elinder, C.G., and Kjellström, T., et al., editors. *Cadmium and health: A toxicological and Epidemiological Appraisal.* vol. I. *Exposure, Dose, and Metabolism.* Boca Raton, FL: CRC Press, pp. 103–178.

NTP (1995) *NTP Technical Report on Toxicity Studies of Cadmium Oxide (CAS No. 1306-19-0) Administered by Inhalation to F344/ N Rats and B6C3F Mice.* Research Triangle Park, NC: National Toxicology Program. Toxicity Report Series Number 39.

Nwosu, J.U., Harding, A.K., and Linder, G. (1995) Cadmium and lead uptake by edible crops grown in a silt loam soil. *Bull. Environ. Contam. Toxicol.* 54:570–578.

Olsson, I.M., Bensryd, I., Lundh, T., Ottosson, H., Skerfving, S., and Oskarsson, A. (2002) Cadmium in blood and urine–impact of sex, age, dietary intake, iron status, and former smoking–association of renal effects. *Environ. Health Perspect.* 110(12):1185–1190.

Park, J.D., Cherrington, N.J., and Klaassen, C.D. (2002) Intestinal absorption of cadmium is associated with divalent metal transporter 1 in rats. *Toxicol. Sci.* 68(2):288–294.

Patwardhan, J.R. and Finckh, E.S. (1976) Fatal cadmium-fume pneumonitis. *Med. J. Aust.* 1:962–966.

Pesch, G.G. and Stewart, N.E. (1980) Cadmium toxicity to three species of estuarine invertebrates. *Mar. Environ. Res.* 3:145–156.

Petering, H.G., Choudhury, H., and Stemmer, K.L. (1979) Some effects of oral ingestion of cadmium on zinc, copper and iron metabolism. *Environ. Health Perspect.* 28:97–106.

Pickering, Q.H. and Henderson, C. (1966) The acute toxicity of some heavy metals to different species of warm water fishes. *Int. J. Air Water Pollut.* 10:453–463.

Pritzl, M.C., Lie, Y.H., Kienholz, E.W., and Whiteman, C.E. (1974) The effect of dietary cadmium on development of young chickens. *Poult. Sci.* 53:2026–2029.

Rahola, T., Aaran, R.-K., and Miettenen, J.K. (1973) Retention and elimination of 115mCd in man. In: *Health Physics Problems of Internal Contamination.* Budapest: Akademia, pp. 213–218.

Reinl, W. (1961) *Arch. Toxicol.* 19:1952. [From ACGIH [1991]. Cadmium and compounds. In: Documentation of the threshold limit values and biological exposure indices. 6th ed. Cincinnati, OH: American Conference of Governmental Industrial Hygienists, pp. 190–194. Cited by Teisinger et al. [1969] from Documentation of MAC in Czechoslovakia, Prague (1969).].

Robinson, A.M., Laemberson, J.O., Cole, F.A., and Swartz, R.C. (1988) Effects of culture conditions on the sensitivity of a phoxocephalid amphipod, Rhepoxynius abronius, to cadmium in sediment. *Environ. Toxicol. Chem.* 7:953–959.

Roels, H.A., Lauwerys, R.R., Buchet, J.P., Bernard, A., Chettle, D.R., Harvey, T.C., and Al-Haddad, I.K. (1981) In vivo measurement of liver and kidney cadmium in workers exposed to this metal: Its significance with respect to cadmium in blood and urine. *Environ. Res.* 26:217–240.

Roy, W.R., Krapac, I.G., and Steele, J.D. (1993) Soil processes and chemical transport. *J. Environ. Qual.* 22:537–543.

Ryu, D.Y., Lee, S.J., Park, D.W., Choi, B.S., Klaassen, C.D., and Park, J.D. (2004) Dietary iron regulates intestinal cadmium absorption through iron transporters in mice. *Toxicol. Lett.* 152(1):19–25.

Scott, R., Haywood, J.K., Boddy, K., Williams, E.D., Harvey, I., and Paterson, P.J. (1980) Whole body calcium deficit in cadmium-exposed workers with hypercalciuria. *Urology* 15:356–359.

Seidal, K., Jörgensen, N., and Elinder, C. (1993) Fatal cadmium induced pneumonitis. *Scand. J. Work Environ. Health* 19:429–431.

Shaikh, Z.A. and Smith, J. (1980) Metabolism of orally ingested cadmium in humans. *Dev. Toxicol. Environ. Sci.* 8:569–74.

Shaikh, Z.A. (1982) Metallothionein as a storage protein for cadmium: its toxicological implications. *Dev. Toxicol. Environ. Sci.* 9:69–76.

Shevchenko, V., Lisitzin, A., Vinogradova, A., and Stein, R. (2003) Heavy metals in aerosols over the seas of the Russian Arctic. *Sci. Total. Environ.* 306:11–25.

Sileo, L. and Beyer, W.N. (1985) Heavy metals in white-tailed deer living near a zinc smelter in Pennsylvania. *J. Wildlife Diseases* 21:289–296.

Slooff, W. and Baerselman, R. (1980) Comparison of the usefulness of the Mexican axolotl (*Ambystoma mexicanum*) and the clawed toad (*Xenopus laevis*) in toxicological bioassays. *Bull. Environ. Contam. Toxicol.* 24:439–443.

Smith, S.R. (1994) Effect of soil pH on availability to crops of metals in sewage sludge-treated soils. II. Cadmium uptake by crops and implications for human dietary intake. *Environ. Pollut.* 86:5–13.

Sorahan, T. (1987) Mortality from lung cancer among a cohort of nickel cadmium battery workers: 1946–1984. *Br. J. Ind. Med.* 44:803–809.

Sorahan, T. and Esmen, N.A. (2004) Lung cancer mortality in UK nickel-cadmium battery workers, 1947–2000. *Occup. Environ. Med.* 61(2):108–116.

Sorahan, T. and Lancashire, R.J. (1997) Lung cancer mortality in a cohort of workers employed at a cadmium recovery plant in the United States: an analysis with detailed job histories. *Occup. Environ. Med.* 54:194–201.

Spehar, R.L., Leonard, E.N., and Defoe, D.L. (1978b) Chronic effects of cadmium and zinc mixtures on flagfish (*Jordanella floridae*). *Trans. Am. Fish Soc.* 107:354–360.

Sprague, J.B. (1986) *Toxicity and tissue concentrations of lead, zinc, and cadmium for marine molluscs and crustaceans.* Research Triangle Park, NC: International Lead Zinc Research Organization, Inc., pp. 1–74.

Stayner, L., Smith, R., Thun, M., Schnorr, T., and Lemen, R. (1992b) A quantitative assessment of lung cancer risk and occupational cadmium exposure. In: Nordberg, G.F., Herber, R.F.M., and Alessio, L., editors. *Cadmium in the Human Environment: Toxicity and Carcinogenicity.* Lyon, France: International Agency for Research on Cancer, pp. 447–455.

Steinnes, E. and Friedland, A.J. (2006) Metal contamination of natural surface soils from long-range atmospheric transport: existing and missing knowledge. *Environ. Rev.* 14:169–186.

Stowe, H.D., Wilson, M., and Goyer, R.A. (1972) Clinical and morphological effects of oral cadmium toxicity in rabbits. *Arch. Pathol.* 94:389–405.

Suzuki, C.A.M. and Cherian, M.G. (1987) Renal toxicity of cadmium-metallothionein and enzymuria in rats. *J. Pharmacol. Exp. Ther.* 240:314–319.

Teeyakasem, W., Nishijo, M., Honda, R., Satarug, S., Swaddiwud-hipong, W., and Ruangyuttikarn, W. (2007) Monitoring of cadmium toxicity in a Thai population with high-level environmental exposure. *Toxicol. Lett.* 169:185–195.

Thornton, I. (1992) Sources and pathways of cadmium in the environment. In: Nordberg, G.F., Herber, R.F.M., and Alessio, L., editors. *Cadmium in the Human Environment: Toxicity and Carcinogenicity.* IARC Scientific Publications No. 118. Lyon, France: International Agency for Research on Cancer, pp. 149–162.

Thun, M.J., Schnorr, T.M., Smith, A., Halperin, W.E., and Lemen R.A. (1985) Mortality among a cohort of U.S. cadmium production workers—an update. *J. Natl. Cancer. Inst.* 74:325–333.

Vanderpool, R.A. and Reeves, P.G., (2001) Cadmium absorption in women fed processed edible sunflower kernels labeled with a stable isotope of cadmium, 113Cd1. *Environ. Res.* 87:69–80.

Vos, G., Lammers, H., and Kan, C.A. (1990) Cadmium and lead in muscle tissue and organs of broilers, turkeys and spent hens, and in mechanically deboned poultry meat. *Food Addit. Contam.* 7:83–92.

Wester, R.C., Maibach, H.I., Sedik, L., Melendres, J., DiZio, S., and Wade, M. (1992) *In vitro* percutaneous absorption of cadmium from water and soil into human skin. *Fundam. Appl. Toxicol.* 19:1–5.

Westerman, A.G. (1977) *Lethal and Teratogenic Effects of Inorganic Mercury and Cadmium on Embryonic Development of Anurans.* Lexington, University of Kentucky (M.S. Thesis).

White, D.H. and Finley, M.T. (1978) Uptake and retention of dietary cadmium in mallard ducks. *Environ. Res.* 17:53–59.

White, D.H., Finley, M.T., and Ferrell, J.F. (1978) Histopathologic effects of dietary cadmium on kidneys and testes of mallard ducks. *J. Toxicol. Environ. Health* 4:551–558.

Wilber, G.G., Smith, L., and Malanchuk, J.L. (1992) Emissions inventory of heavy metals and hydrophobic organics in the Great Lakes basin. In: Schnoor, J.L., editor. *Fate of Pesticides and Chemicals in the Environment.* John Wiley and Sons, Inc., pp. 27–50.

Woodwall, C., Maclean, N., and Crossley, F. (1988) Responses of trout fry (*Salmo gairdneri*) and *Xenopus laevis* tadpoles to cadmium and zinc. *Comp. Biochem. Physiol.* 89C:93–99.

Zaroogian, G.E. and Morrison, G. (1981) Effect of cadmium body burdens in adult Crassostrea virginica on fecundity and viability of larvae. *Bull. Environ. Contam. Toxicol.* 27:344–348.

13

CHROMIUM

Kathleen Macmahon

Hexavalent Chromium Compounds Only

First aid: Seek medical assistance immediately as effects may be severe or life threatening. Oral exposure: Do not induce vomiting, seek medical assistance. Eye exposure: Remove from exposure, remove contact lenses, flush with water, see ophthalmologist. Inhalation exposure: Remove from exposure, ensure adequate ventilation, oxygen if symptomatic, seek medical assistance. Dermal exposure: Remove from exposure, flush with water, seek medical assistance.

Target organs: Inhalation exposure: Respiratory tract. Dermal exposure: Skin, mucous membranes.

Occupational exposure limits (OELs): Specific Cr(VI) compounds may have distinct OELs.

OSHA: $5 \mu g/m^3$ 8 h time-weighted average (TWA).
NIOSH: $0.2 \mu g/m^3$ 8 h TWA.
ACGIH: $50 \mu g/m^3$ water-soluble 8 h TWA.
$10 \mu g/m^3$ insoluble 8 h TWA.

Reference values:
RfC inhalation, chromic acid mists,
Cr(VI) aerosols: $8 \times 10^{-6} mg/m^3$.
RfC inhalation, Cr(VI) particulates: $1 \times 10^{-4} mg/m^3$.
RfD Cr(VI) soluble salts: $3 \times 10^{-3} mg/kg/day$.
TDI oral: $3 \times 10^{-3} mg/kg/day$.
TDI inhalation: $3 \times 10^{-5} mg/kg/day$.
MRL inhalation, intermediate, Cr(VI) aerosols, mists: $5 \times 10^{-6} mg/m^3$.
MRL inhalation, chronic, Cr(VI) aerosols, mists: $5 \times 10^{-6} mg/m^3$.
MRL inhalation, intermediate, Cr(VI) particulates: $3 \times 10^{-4} mg/m^3$.
MRL oral, intermediate, Cr(VI) compounds: $5 \times 10^{-3} mg/kg/day$.
MRL oral, chronic, Cr(VI) compounds: $9 \times 10^{-4} mg/kg/day$.

Risk phrases: Specific Cr(VI) compounds may have distinct risk phrases.

R49: may cause cancer by inhalation.

R43: may cause sensitization by skin contact.

R50/53: very toxic to aquatic organisms; may cause long-term adverse effects in the aquatic environment.

Safety phrases: Specific Cr(VI) compounds may have distinct safety phrases.

S45: in case of accident or if you feel unwell, seek medical advice immediately (show the label where possible).

S53: avoid exposure—obtain special instructions before use.

S60: this material and its container must be disposed of as hazardous waste.

S61: avoid release to the environment, refer to special instruction/safety data sheet.

BACKGROUND AND USES

Chromium occurs in several different oxidation states from chromium metal to hexavalent chromium (Cr(VI)) compounds. In the environment, chromium exists primarily in the trivalent (Cr(III)) and Cr(VI) oxidation states. In industry,

Hamilton & Hardy's Industrial Toxicology, Sixth Edition. Edited by Raymond D. Harbison, Marie M. Bourgeois, and Giffe T. Johnson.
© 2015 John Wiley & Sons, Inc. Published 2015 by John Wiley & Sons, Inc.

metallic or elemental chromium (Cr(0), Cr(III), and Cr(VI) are most common. Chromium-containing materials include chromite ore, chromium chemicals, ferroalloys, and metal. In the United States most chromite ore is converted into ferrochromium, which is used to produce stainless and heat-resisting steel (USGS, 2011). Trivalent chromium (Cr(III)) is an essential trace element involved in glucose metabolism. Hexavalent chromium compounds are a large group of chemicals with varying industrial uses. Hexavalent chromium-containing materials include chromate paint and primer pigments, graphic art supplies, fungicides, and corrosion inhibitors.

This chapter focuses on Cr(VI) compounds due to their industrial relevance and associated adverse health effects. Hexavalent chromium compounds are occupational carcinogens associated with an increased excess risk of lung cancer. Some industries with occupational exposure to Cr(VI) compounds include electroplating, welding, and chromate painting. Occupational and environmental exposure to Cr(VI) compounds has been extensively reviewed (ATSDR, 2012; EPA, 1998; IARC, 1990, 2012; 71 Fed. Reg. 10099 (2006); NIOSH, 2013; NTP, 2011).

PHYSICAL AND CHEMICAL PROPERTIES

Chromium occurs in oxidation states from Cr^{-2} through Cr^{+6} but exists mainly in the Cr(III) and Cr(VI) states; Cr(III) is the most stable. Hexavalent chromium compounds have varying physical and chemical properties. Most Cr(VI) compounds are solids; chromyl chloride is a liquid. Their properties include corrosion-resistance, durability, and hardness. Sodium dichromate is the most common chromium chemical from which other Cr(VI) compounds are produced. Examples of other Cr(VI) compounds include sodium chromate, potassium dichromate, potassium chromate, ammonium dichromate, and chromic oxide.

The reduction of Cr(VI) to Cr(III) and the oxidation of Cr(III) to Cr(VI) are important sampling considerations when determining Cr(VI) levels in workplace air (Ashley et al., 2003). Factors that affect the reduction of Cr(VI) or oxidation of Cr(III) include the presence of other compounds in the sample (e.g., iron), the ratio of Cr(VI) to Cr(III) concentrations in the sample, and solution pH. The reduction of Cr(VI) is favored in acidic conditions while Cr(VI) is stabilized in basic conditions.

ENVIRONMENTAL FATE AND BIOACCUMULATION

Chromium is distributed to the air, water, and soil from natural and anthropogenic sources. The environmental fate of chromium is dependent on the oxidation state and solubility of the compound and the environmental conditions affecting reduction or oxidation, such as pH. Oxidizing conditions favor the formation of Cr(VI) compounds, particularly at higher temperatures, while reducing conditions favor the formation of Cr(III) compounds. Chemical manufacturing and natural gas, oil, and gas combustion are the primary sources of chromium in the atmosphere. Most of the chromium in air eventually ends up in water or soil. Electroplating, textile manufacturing, cooling water, and leather tanning are major sources of chromium in wastewater discharges to surface waters.

Chromium(III) is the predominant oxidation state of chromium in many soils. Cr(III) binds to soil and has low mobility. A lower soil pH favors the reduction of Cr(VI) to Cr(III). Runoff from soil and industrial processes may transport chromium to surface water. Cr(VI) compounds may leach into groundwater. The pH of the soil and aquatic environment is an important factor in chromium mobility, bioavailability, and toxicity. The chromate form predominates in most natural surface waters that are basic or neutral. The hydrochromate concentration increases in more acidic conditions. Most of the chromium released into surface water will be deposited in the sediment with the potential for accumulation in aquatic life (Velma et al. 2009). Chromium is not expected to biomagnify in the aquatic or terrestrial food chains (ATSDR, 2012).

ECOTOXICOLOGY

Acute Effects

Chromium(VI) is generally more toxic to aquatic life than Cr(III). Crustaceans and other freshwater invertebrates are more sensitive to the effects of Cr(VI) than fish. Of the species tested, the most sensitive freshwater organisms are more sensitive to Cr(VI) than the most sensitive marine organisms. The acute effects of chromium compounds in freshwater fish include cytotoxic, genotoxic, enzymatic, hematologic, and immunologic effects depending on the species and the dose (Velma et al., 2009). These effects are dependent on environmental factors, including pH, water hardness, salinity, and temperature.

Chronic Effects

Chronic effects of chromium in freshwater fish include cytotoxic, genotoxic, growth, and survival effects depending on the species and the dose (Velma et al., 2009). These effects are also dependent on other factors, including the organism age, exposure duration, and environmental conditions. Some fish species, such as rainbow trout, are more sensitive than other fish species to the effects of chromium, particularly Cr(VI).

MAMMALIAN TOXICOLOGY

Acute Effects

Acute inhalation exposure to Cr(VI) is associated with adverse respiratory effects, including nasal and respiratory irritation. Gibb et al. (2000a) reported diagnoses of irritated nasal septum, ulcerated nasal septum, and perforated eardrum in chromate production workers <1 month after employment. Ingestion of high doses of Cr(VI) may affect many organ systems and may be lethal.

Chronic Effects

Cr(VI) is a known human carcinogen by the inhalation route of exposure (NTP, 2011; IARC, 2012; EPA, 1998). Cr(VI) compounds are occupational carcinogens associated with an increased risk of lung cancer (71 Fed. Reg. 10099, 2006; NIOSH, 2013). The available evidence is inconclusive regarding the association of Cr(VI) compounds with nasal and nasal sinus cancers; there is little evidence of an association with stomach or other cancers (IARC, 2012). Nonmalignant chronic respiratory effects include ulcerated or perforated nasal septa and respiratory sensitization.

Chronic dermal effects include skin ulcers and allergic contact dermatitis. Lindberg and Hedenstierna (1983) reported that exposures >2 $\mu g/m^3$ chromic acid were associated with nasal septal ulceration and perforation. Gibb et al. (2000a) reported irritated or ulcerated skin, dermatitis, burn, conjunctivitis, and perforated nasal septum or eardrum in chromate production workers. Other chronic nonmalignant health effects include immunologic, hematologic, reproductive, and developmental effects (ATSDR, 2012).

Beaumont et al. (2008), in a reanalysis of Chinese data, reported an association between Cr(VI)-contaminated drinking water and stomach cancer. Chronic drinking water studies of sodium dichromate dihydrate (Cr(VI)) in rats resulted in an increased incidence of squamous cell epithelium tumors of the oral mucosa or tongue (NTP, 2008). Chronic drinking water studies in mice resulted in an increased incidence of small intestine tumors (NTP, 2008).

Chemical Pathology

Results of laboratory animal studies of the respiratory effects of Cr(VI) compounds vary with the Cr(VI) compound, dose, route of exposure, duration of study, and animal species. Intratracheal instillation of sodium dichromate or calcium chromate in rats resulted in lung tumors, including adenomas, adenocarcinomas, and squamous cell carcinomas (Steinhoff et al., 1986). Inhalation studies of sodium dichromate in rats also resulted in benign and malignant lung tumors (Glaser et al., 1986). Histopathologic respiratory effects included inflammatory, hyperplastic, and fibrotic changes.

Chronic drinking water studies of sodium dichromate dihydrate (Cr(VI)) in rats resulted in an increased incidence of squamous cell carcinoma or papilloma of the oral cavity (NTP, 2008). Similar chronic drinking water studies in mice resulted in an increased incidence of adenoma or carcinoma of the small intestine. Diffuse epithelial hyperplasia was reported in the small intestine of mice. Histiocytic cell infiltration was reported in the duodenum, mesenteric lymph node, and other locations in rats and mice (NTP, 2008).

Mode of Action

Several mechanisms may be involved in Cr(VI) carcinogenesis, including DNA damage, oxidative stress, aneuploidy, and cell transformation (IARC, 2012). Cr(VI) carcinogenicity may result from damage mediated by the bioreactive products of intracellular Cr(VI) reduction, including the Cr(VI) intermediates (Cr(V) and Cr(IV)), and reactive oxygen species (ROS) (O'Brien et al., 2003). These intermediates and ROS may result in DNA damage, including interstrand cross-links, DNA–protein cross-linking, and double-strand breaks. These genotoxic effects are consistent with the mechanistic events associated with carcinogenesis.

SOURCES OF EXPOSURE

Occupational

Workers may be exposed to airborne Cr(VI) when Cr(VI) compounds are manufactured from other forms of chromium (e.g., production of chromates from chromite ore); when using Cr(VI)-containing materials to manufacture other products (e.g., chromate-containing paints, electroplating); or when Cr(VI) is produced as a by-product of a process (e.g., welding). Primary industries with the majority of occupational exposures to airborne Cr(VI) compounds include welding, painting, electroplating, steel mills, iron and steel foundries, wood preserving, paint and coatings production, chromium catalyst production, plastic colorant producers and users, production of chromates and related chemicals from chromite ore, plating mixture production, printing ink producers, chromium metal production, chromate pigment production, and chromated copper arsenate producers (71 Fed. Reg. 10099, 2006).

Welders are the largest group of workers potentially exposed to Cr(VI) compounds (71 Fed. Reg. 10099, 2006). Cr(VI) exposures to welders are dependent on several process factors, including the welding process and shield-gas type, and the Cr content of both the consumable material and the base metal (Keane et al., 2009; Meeker et al., 2010). Other industrial sectors with substantial numbers of workers exposed include painting, electroplating, steel mills, and iron and steel foundries (71 Fed. Reg. 10099, 2006).

Blade et al. (2007) conducted a series of field surveys of workplaces where there was a potential for Cr(VI) exposure. The authors reported that controlling airborne Cr(VI) exposures was moderately difficult in those operations that involved joining or cutting metals with a relatively high chromium content. Airborne Cr(VI) exposures were reported most difficult to control in those operations involving the application of coatings or finishes.

Dermal exposure to Cr(VI)-containing substances may occur with any task where there is a potential for splashing, spilling, or other skin contact. If not adequately protected, workers' skin may be directly exposed to liquid forms of Cr(VI) (e.g., electroplating baths) or solid forms (e.g., Portland cement). A variety of construction materials contain Portland cement, including cement, mortar, stucco, and terrazzo (OSHA, 2008). Dermal exposure may also occur because of contamination of workplace surfaces or equipment.

Environmental

Chromium(VI) compounds are released to the air, water, and soil from natural and industrial sources. Cr(VI) levels are usually low in U.S. ambient air, $<30\,ng/m^3$ (ATSDR, 2012). Chromium in the air is eventually deposited on to the soil and water. Environmental conditions in the soil and water, (e.g., pH), affect the state of chromium, potential exposures, and toxicity. Cr(VI) compounds may be present in surface waters or groundwater. Potential exposures to Cr(VI) may occur from the ingestion of drinking water or the inhalation of water droplets when showering (OEHHA, 2009).

INDUSTRIAL HYGIENE (PPE)

U.S. Occupational Exposure Limits for Cr(VI) Compounds[*]

Agency	OEL	Cr(VI) Compound(s)	TWA, $\mu g/m^3$
NIOSH	REL	All	0.2
OSHA	PEL	All	5
ACGIH	TLV	Water soluble	50
		Insoluble	10

Sources: NIOSH (2005), ACGIH (2011).

[*] Specific Cr(VI) compounds may have distinct OELs.

The control measures for Cr(VI) compounds, consistent with the hierarchy of controls, include: elimination, substitution, engineering controls, administrative controls and appropriate work practices, and the use of protective clothing and equipment (NIOSH, 2013). Employers should provide gloves, chemical protective clothing (CPC), and eye protective equipment for workers with potential skin or eye contact with Cr(VI) compounds. Selection of CPC should be based on permeation properties and other factors, including size, dexterity, cut and tear resistance, and other chemicals being used. Dermal and mucous membrane contact should be prevented by full-body protective clothing, including head, neck, and face protection; coveralls or similar protective body clothing; impermeable gloves with gauntlets; and shoes and apron. When respiratory protection is needed to control workplace exposures, the employer should establish a comprehensive respiratory protection program as described in the OSHA respiratory protection standard (CFR, 2006 29 CFR 1910.134).

RISK ASSESSMENTS

Studies of chromate manufacturing workers in Baltimore, Maryland, and Painesville, Ohio provided exposure-response data for assessing the excess risk of lung cancer associated with occupational exposure to Cr(VI). These cohorts provided the basis for robust quantitative assessments of occupational risk of Cr(VI) exposure due to the quality and quantity of the exposure and worker data, extent of exposure, and years of follow-up.

Park et al. (2004) calculated excess lifetime risk of lung cancer death resulting from occupational exposure to chromium-containing mists and dusts using the data of Gibb et al. (2000b), the Baltimore cohort of chromate chemical production workers. This cohort had extensive historical exposure assessment and smoking information, strong statistical power, and a relative lack of potentially confounding exposures. A linear relative rate model was selected to estimate the exposure response that was the basis for the risk assessment. The excess lifetime risk of lung cancer death was approximately 6 per 1000 workers from a 45-year working lifetime exposure to $1\,\mu g/m^3$ soluble hexavalent chromium.

Crump et al. (2003) calculated estimates of excess lifetime risk of lung cancer death using the data of Luippold et al. (2003), the Painesville cohort of chromate chemical production workers. This cohort had an extensive exposure assessment (Proctor et al., 2003). The estimate of excess lifetime risk of lung cancer death was approximately 2 per 1000 workers from a 45-year working lifetime exposure to $1\,\mu g/m^3$ hexavalent chromium.

The findings and conclusions in this report are those of the author and do not necessarily represent the views of the National Institute for Occupational Safety and Health.

REFERENCES

71 Fed. Reg. 10099 (2006) *Occupational Safety and Health Administration: Occupational exposure to hexavalent chromium; final rule (29 CFR Parts 1910, 1915, 1917, 1918, 1926). Docket No. H-0054A.*

ACGIH (2011) 2011 TLVs and BEIs. Cincinnati, OH: American Conference of Governmental Industrial Hygienists, Publication No. 0111.

Ashley, K., Howe, A.M., Demange, M., and Nygren, O. (2003) Sampling and analysis considerations for the determination of hexavalent chromium in workplace air. *J. Environ. Monit.* 5:707–716.

ATSDR (2012) *Toxicological Profile for Chromium*, Atlanta, GA: U.S. Department of Health and Human Services, Public Health Service, Agency for Toxic Substances and Disease Registry.

Beaumont, J.J., Sedman, R.M., Reynolds, S.D., Sherman, C.D., Li, L.H., Howd, R.A., Sandy, M.S., Zeise, L., and Alexeeff, G.V. (2008) Cancer mortality in a Chinese population exposed to hexavalent chromium in drinking water. *Epidemiology* 19:12–23.

Blade, L.M., Yencken, M.S., Wallace, M.E., Catalano, J.D., Khan, A., Topmiller, J.L., Shulman, S.A., Martinez, A., Crouch, K.G., and Bennett, J.S. (2007) Hexavalent chromium exposures and exposure-control technologies in American enterprise: results of a NIOSH field research study. *J. Occup. Environ. Hyg.* 4(8):596–618.

CFR (2006) *Code of Federal Regulations*, Washington, DC: U.S. Government Printing Office, Office of the Federal Register.

Crump, C., Crump, K., Hack, E., Luippold, R., Mundt, K., Liebig, E., Panko, J., Paustenbach, D., and Proctor, D. (2003) Dose-response and risk assessment of airborne hexavalent chromium and lung cancer mortality. *Risk Anal.* 23(6):1147–1163.

EPA (1998) *Toxicological Review of Hexavalent Chromium (CAS No. 18540-29-9) in Support of Summary Information on the Integrated Risk Information System (IRIS)*, Washington, D.C.: U.S. Environmental Protection Agency. Available at http://www.epa.gov/iris.

Gibb, H.J., Lees, P.S., Pinsky, P.F., and Rooney, B.C. (2000a) Clinical findings of irritation among chromium chemical production workers. *Am. J. Ind. Med.* 38(2):127–131.

Gibb, H.J., Lees, P.S., Pinsky, P.F., and Rooney, B.C. (2000b) Lung cancer among workers in chromium chemical production. *Am. J. Ind. Med.* 38(2):115–126.

Glaser, U., Hochrainer, D., Kloppel, H., and Oldiges, H. (1986) Carcinogenicity of sodium dichromate and chromium (VI/III) oxide aerosols inhaled by male Wistar rats. *Toxicology* 42:219–232.

IARC (1990) *IARC Monographs on the Evaluation of the Carcinogenic Risk of Chemicals to Man: Chromium, Nickel, and Welding*, vol. 49, Lyon, France: World Health Organization, International Agency for Research on Cancer, pp. 49–256.

IARC (2012) *IARC Monographs on the Evaluation of Carcinogenic Risks to Humans: Arsenic, Metals, Fibres, and Dusts*, vol. 100C, Lyon, France: World Health Organization, International Agency for Research on Cancer, pp. 49–256.

Keane, M., Stone, S., Chen, B., Slaven, J., Schwegler-Berry, D., and Antonini, J. (2009) Hexavalent chromium content in stainless steel welding fumes is dependent on the welding process and shield gas type. *J. Environ. Monit.* 11(2):418–424.

Lindberg, E. and Hedenstierna, G. (1983) Chrome plating: symptoms, findings in the upper airways, and effects on lung function. *Arch. Environ. Health* 38(6):367–374.

Luippold, R.S., Mundt, K.A., Austin, R.P., Liebig, E., Panko, J., Crump, C., Crump, K., and Proctor, D. (2003) Lung cancer mortality among chromate production workers. *Occup. Environ. Med.* 60(6):451–457.

Meeker, J.D., Susi, P., and Flynn, M.R. (2010) Hexavalent chromium exposure and control in welding tasks. *J. Occup. Environ. Hyg.* 7(11):607–615.

NIOSH (2005) NIOSH Pocket Guide to Chemical Hazards. Cincinnati, OH: U.S. Department of Health and Human Services, Public Health Service, Centers for Disease Control, National Institute for Occupational Safety and Health, DHHS (NIOSH) Publication No. 2005–151.

NIOSH (2013) *Criteria for a Recommended Standard: Occupational Exposure to Chromium (VI)*, DHHS (NIOSH) Publication No. 2013–128, Cincinnati, OH: U.S. Department of Health and Human Services, Public Health Service, Centers for Disease Control, National Institute for Occupational Safety and Health.

NTP (2008) *Toxicology and Carcinogenesis Studies of Sodium dichromate Dihydrate (CAS No. 7789-12-0) in F344/N Rats and B6C3F1 Mice (Drinking Water Studies)*, TR 546. Research Triangle Park, NC: National Toxicology Program.

NTP (2011) *Report on Carcinogens*, 12th ed., Research Triangle Park, NC: U.S. Department of Health and Human Services, Public Health Service, National Institutes of Health, National Toxicology Program.

O'Brien, T.J., Ceryak, S., and Patierno, S.R. (2003) Complexities of chromium carcinogenesis: role of cellular response, repair, and recovery mechanisms. *Mutat. Res.* 533:3–36.

OEHHA (2009) *Evidence on the Developmental and Reproductive Toxicity of Chromium (Hexavalent Compounds)*, Reproductive and Cancer Hazard Assessment Branch, Office of Environmental Health Hazard Assessment, California Environmental Protection Agency. Available at http://www.oehha.ca.gov/prop65/hazard_ident/pdf_zip/chrome0908.pdf.

OSHA (2008) *Preventing Skin Problems from Working with Portland Cement*, OSHA Publication No. 3351–07, Washington, DC: U.S. Department of Labor, Occupational Safety and Health Administration.

Park, R.M., Bena, J.F., Stayner, L.T., Smith, R.J., Gibb, H.J., and Lees, P.S. (2004) Hexavalent chromium and lung cancer in the chromate industry: a quantitative risk assessment. *Risk Anal.* 24(5):1099–1108.

Proctor, D.M., Panko, J.P., Leibig, E.W., Scott, P.K., Mundt, K.A., Buczynski, M.A., Barnhart, R.J., Harris, M.A., Morgan, R.J., and Paustenbach, D.J. (2003) Workplace airborne hexavalent chromium concentrations for the Painesville, Ohio, chromate production plant (1943–1971). *Appl. Occup. Environ. Hyg.* 18(6):430–449.

Steinhoff, D., Gad, S.C., Hatfield, G.K., and Mohr, U. (1986) Carcinogenicity study with sodium dichromate in rats. *Exp. Pathol.* 30:129–141.

USGS (2011) *Mineral Commodity Summary: Chromium*. Available at http://minerals.usgs.gov/minerals/pubs/commodity/chromium (accessed August 31, 2012).

Velma, V., Vutukuru, S.S., and Tchounwou, P.B. (2009) Ecotoxicology of hexavalent chromium in freshwater fish: a critical review. *Rev. Environ. Health* 24(2):129–145.

14

COBALT

Davinderjit Singh and Raymond D. Harbison

First aid (metal dust and fume): Eye: Irrigate immediately; skin: soap wash immediately; breathing: respiratory support; swallow: medical attention immediately.

Target organ(s): Respiratory tract, skin, heart, blood, thyroid gland, ear, eye

Occupational exposure limits: NIOSH REL: 0.05 mg/m³; OSHA PEL: 0.1 mg/m³; American Conference of Governmental Industrial Hygienists (ACGIH) TLV: Time-weighted average (TWA) 0.02 mg/m³; Cal/OSHA PEL: 0.02 mg/m³

Reference values: No reference dose or reference concentration set by EPA. Agency for Toxic Substances and Disease Registry (ATSDR) MRL: 0.0001 mg/m³ (inhalational), 0.01 mg/kg/day (ingestion); ATSDR MRL: 0.00003 mg/m³ (mice); CalEPA REL: 0.000005 mg/m³ (mice); NIOSH ILDH: 20 mg Co/m³

Risk phrases: 22-42/43

Harmful if swallowed

May cause sensitization by inhalation and skin contact

Safety phrases: 22-24-37

Do not breathe dust. Avoid contact with skin. Wear suitable gloves

BACKGROUND AND USES

Cobalt (Co; CASRN 7440-48-4) is a silvery-gray brittle but hard metal, which is distributed widely in nature, including rocks, soil, plants, and animals. It is a metal with no odor and no vapor pressure at room temperature and is insoluble in water, except for ultrafine cobalt powder, which is water soluble. It is stable in air but chemically reactive with dilute acids and is a nonvolatile metal. Required in small amounts, cobalt is important for the sustenance of life. It is the only metal found in vitamins, most importantly as a part of vitamin B12 and is necessary in the formation of blood. Cobalt is a component of many naturally occurring minerals like sulfides, hydrates, and oxides. Cobalt compounds can occur in various oxidation states (0, +1, +2, +3, and +4) but the most common oxidation states are +2 and +3. Cobalt has one naturally occurring isotope, cobalt-59 and 26 radioactive isotopes. Common ores of cobalt are cobaltite, erythrite [Co_3 ($AsO_4)_2 \cdot 8H_2O$)], chlorathite, and linnealite (Co_3S_4). Cobalt is used extensively in industries, and more than a million workers are exposed to various industrial processes involving cobalt and its compounds. In occupational settings, inhalational and dermal contact are the two most common routes of exposure.

Cobalt and its compounds are used as an important catalyst in various industrial processes. Cobalt catalysts are used in various processes such as a desulfurization catalyst for gas and oil and for polyesters, pottery, plastic, and detergent production. Cobalt is usually recovered as a byproduct of mining and refining of various minerals like copper and iron. It has the unique ability to be ductile as well as a malleable metal. This property allows it to be transformed into both thin wires and thin sheets. Cobalt is used in the production of various household products like paints, hair dyes, glass, ceramics, enamel, pottery, jewelry, and consumer batteries. It is an important component of porcelain, cement, rubber, and chemical industries. It is used in the production of super alloys to retain physical and chemical properties at extreme temperatures. This property is used in the production of drills, cutting tools, and as parts in rocket, gas turbine,

Hamilton & Hardy's Industrial Toxicology, Sixth Edition. Edited by Raymond D. Harbison, Marie M. Bourgeois, and Giffe T. Johnson.
© 2015 John Wiley & Sons, Inc. Published 2015 by John Wiley & Sons, Inc.

and aircraft engines (Harper et al., 2012). Major uses of cobalt worldwide are in the production of super alloys and other alloys (30%), as hard metals (13%), and as catalysts (10%). In the United States, more than 50% of cobalt used is as a super alloy for the production of aircraft gas turbine engines. Cobalt-60 is the radioactive isotope of cobalt and is used in cancer therapy.

PHYSICAL AND CHEMICAL PROPERTIES

Cobalt is a hard magnetic metal, like iron and nickel. It has the properties of both ductility as well as malleability. Due to this property, it can be molded into wires and compressed into sheets. Cobalt occurs in two allotropic forms, hexagonal and cubic. The hexagonal form is more stable than the cubic form at room temperature. The boiling point of cobalt is about $2927\,°C$ ($5301°F$) and the melting point is about $1490\,°C$ ($2723°F$). Cobalt as a metal is mildly reactive with oxygen and will burn only in powder form. It is soluble in dilute acids but is water soluble only in ultrafine powder form (Kyono et al., 1992).

MAMMALIAN TOXICOLOGY

Acute Effects

Animal Acute Toxicity   Cobalt causes both acute and chronic health effects, although chronic effects are more common. Acute effects of cobalt have been studied in experimental animals extensively. The oral LD_{50} ranges from 42.4 to 317 mg/kg body weight for cobalt compounds for laboratory animals (FDRL, 1984a, 1984b, 1984c; Singh and Junnarkar, 1991). Rats exposed to varied concentrations were observed to have cardiac damage, elevated heart rate, degenerative heart lesions, reduction in heart enzyme activity, and lesions of respiratory tract and kidney damage (Morvai et al., 1993; Grice et al., 1969; Domingo et al., 1984; Holly, 1955; Murdock, 1959). Rats administered with fine cobalt particles showed signs of lung abnormalities like focal hypertrophy of lung epithelium, macrophage abnormalities, and type 2 alveolar epithelium proliferations (Kyono et al., 1992). Necrosis and inflammation were noted in rats when doses of cobalt were administered for 16 days. Thymus necrosis and testicular atrophy were also observed (Bucher et al., 1990). Injection of cobalt in rats at doses of 15 mg/kg led to the breakage of red blood cells and oxidation of hemoglobin iron and methemoglobinuria (Horiguchi et al., 2004).

Human Acute Toxicity

Dermal   All forms of cobalt metal, cobalt alloys, and cobalt salts can result in skin manifestations. Cobalt along with nickel and chromium are considered the more common causes of occupational skin diseases such as irritant eczemas and occupational dermatitis (Kanerva et al., 1988; Kiec-Swierczynska and Szymczk, 1995). In a study over an 11-year period in the United Kingdom, cobalt was associated with 4% of all cases of occupational contact dermatitis (Athavale et al., 2007). Allergic contact dermatitis was also observed in hard metal workers exposed to cobalt (Dooms-Goossens et al., 1986; Julander et al., 2009). Urticarial reactions have also been observed when exposed to cobalt and its salts. Combined dermal exposure to cobalt nanoparticles and UV-B can result in the activation and accumulation of skin fibroblasts and thickening of the epidermis layer of the skin (Murray et al., 2009). The mechanism by which cobalt and its compounds cause skin toxicity is not clear, but studies have highlighted the role of reactive oxygen species (Gault et al., 2010).

Cobalt has skin sensitization properties, which have been shown in both human and animal studies. Another study demonstrated that permeation through damaged skin was significantly higher for metals like cobalt (Filon et al., 2009). Oxidation of metallic cobalt into ions has been proposed as the mechanism by which cobalt can permeate through the skin. Cobalt sensitization can also result in chromium cross-sensitization in construction workers exposed to chromium dust (Bock et al., 2003).

Respiratory   Respiratory effects of cobalt and its compounds have been studied extensively, more so in association with other hard metals especially tungsten carbide. A study of male workers exposed to a cobalt dust mixture for at least 2 years reported significantly higher levels ($0.4\,\mu mol/g$) of cobalt in the urine than nonexposed workers ($0.019\,\mu mol/g$); cobalt exposure was reported below 41.5 ppm (Gennart et al., 1993). A study conducted in a tungsten carbide processing industry reported varied respiratory abnormalities (Coates and Watson, 1973). Exposed workers presented with interstitial pulmonary fibrosis, deposition of collagen and elastic tissue in the septal areas, and alteration of type 1 pneumocytes. Hard metal industries have been a common source of cobalt exposure. Hard metal alloy exposure was shown to be associated with abnormal pulmonary function tests and abnormal radiographic findings in workers (Bech et al., 1962). Common symptoms included excessive cough, chest tightness, and shortness of breath.

Early human exposure was ascribed to cobalt when respiratory conditions like interstitial fibrosis and emphysema were noticed in 12 workers exposed to dust containing tungsten and cobalt (Bech, 1974). Workers exposed to 0.005 mg cobalt/m^3 in industries processing tungsten carbide alloy developed asthma, pneumonia, and wheezing. When exposed to 0.007 mg cobalt/m^3 at work, hypersensitivity to cobalt resulted in asthma and skin rashes. An isolated case of fibrosing alveolitis was observed in a hard metal worker employed in a cobalt grinding process (Zanelli et al., 1994). Association between cobalt and occupational asthma has been observed in a retrospective cohort study done in Japan. Workers exposed to cobalt were followed for signs of respiratory disease. 5.6% of workers (18 cases) developed occupational asthma when exposed to air having a concentration ranging from 3 to 1292 µg cobalt/m^3 of air (Kusaka et al., 1986). Hypersensitivity to cobalt has been suggested to be the cause of respiratory signs and symptoms. Sensitization to cobalt is linked with the occurrence of bronchial asthma when exposed to both cobalt as well as hard metal dust (Linna et al., 2004). In addition, studies have shown an association of giant cell interstitial pneumonia with hard metal lung disease (Dai et al., 2009; Sakai et al., 2010). In one study that compared 100 cases of hard metal lung disease, giant cell interstitial pneumonia was observed in 59 of the cases though cobalt exposure was found in only six workers (Naqvi et al., 2008). Individually, one case of giant cell interstitial pneumonitis was observed in a worker exposed to cobalt dust (Sundaram et al., 2001).

In recent years, cobalt has become a major component of nanotube industries. Nanotubes containing cobalt metal are used as a high strength material in the production of electronics and medical equipment (Martin and Kohli, 2003). Nanotubes deposit in the lungs and accumulate because of high biopersistence and cause fibrotic reactions as a result of the inflammatory response (Muller et al., 2005). Studies also found that the presence of cobalt containing nanotubes resulted in the formation of granulomas. Such granulomas were also observed in studies conducted on experimental animals (Warheit et al., 2004). Cobalt exposure is associated with dental technicians who manufacture and handle dental prosthesis. Pneumoconiosis and pulmonary fibrosis were observed in technicians exposed to cobalt dust (Sherson et al., 1990; Selden et al., 1996). Prolonged cough and increased phlegm were also noticed in dental technicians exposed to cobalt compounds (Radi et al., 2002).

Cardiac   Lower left ventricular fraction was observed in workers who were exposed to high levels of cobalt and its compounds (Horowitz et al., 1988; D'Adda et al., 1994). Hypertension along with reversible electrocardiogram (ECG) changes (Alexandersson et al., 1980) and elevated heart rate (Raffn et al., 1988) were observed when workers were exposed to 0.01 mg/m^3 and 0.8 mg/m^3 of cobalt hard metal, respectively. Altered left ventricular diastolic function increased myocardial stiffness but no ECG changes were observed in the study conducted on exposed workers in Finland (Linna et al., 2004).

Chronic Effects

Animal Chronic Effects   Cobalt may induce developmental (stunted growth, lower body weight in new born,

decreased survival) and reproductive toxicity (lower sperm count, decreased epididymis weight) in animals when exposed. Hamsters developed emphysema when exposed to cobalt compounds for prolonged periods of time. When laboratory rats were administered an intramuscular injection of cobalt powder, sarcoma, especially rhabdomyosarcoma, was observed at the injection site after 122 weeks (Heath, 1954, 1956). Similarly, sarcomas were observed after 6 years in rabbits when metallic cobalt was implanted in the femoral artery (Schinz and Uehlinger, 1942).

Human Chronic Effects   Cobalt was shown to be associated with beer drinkers and cardiomyopathy due to its chronic consumption. Beer drinker cardiomyopathy syndrome was observed when cobalt salts were added as a stabilizer. According to reports, it could result in pericardial effusion and congestive heart failure (Barborik and Dusek, 1972; Kennedy et al., 1981; Smith and Carson, 1981) though the review conducted by Seghizzi could not find any evidence of cardiomyopathy associated with cobalt exposure in beer drinkers (Seghizzi et al., 1994). Cobalt is no longer added to beer as a foam stabilizer. Various studies have proposed an association of cobalt and cardiomyopathy in exposed subjects. Cardiomyopathy has been reported in cases of exposure to cobalt and cobalt dust (Kennedy et al., 1981; Jarvis et al., 1992).

Cobalt and its compounds have also been shown to be genotoxic. Cobalt sulfide is associated with breaks in DNA strands in Chinese hamster ovary cells (Costa et al., 1982). DNA base damage was also noticed in human lymphocytes during cobalt compounds exposure in the presence of hydrogen peroxide (Nackerdien et al., 1991; Kawanishi et al., 1994; Lloyd et al., 1997; De Boeck et al., 1998). Cobalt–tungsten combination has been shown to convert osteoblast-like cells to cells with carcinogenic phenotype and damage DNA to impart chromosomal abnormalities (Miller et al., 2001). Studies have also shown the synergistic carcinogenic potential of hard metal dust of tungsten and cobalt carbide (Sheppard et al., 2006). Such synergistic potential has been associated with an increased risk of lung carcinoma (Moulin et al., 1998). Some studies have also highlighted the role of cobalt in substituting essential divalent metal ions, including iron or copper that can alter important cellular functions. No epidemiological study has been able to identify cobalt as a carcinogenic agent though the combination of cobalt with other metals (tungsten) has been shown to have carcinogenic potential (Lasfargues et al., 1994; Moulin et al., 1998; Wild et al., 2000). In the Lasfargues study, the cohort exposed to cobalt had a significantly higher risk of lung cancer (SMR = 2.13). Moulin calculated an odds ratio of 1.93 among workers exposed to tungsten carbide–cobalt. Wild et al. (2000) calculated a higher standardized mortality ratio (SMR) of twofold in workers who were twice as exposed to cobalt compounds. These studies are of uncertain relevance due to the small size of cohorts and confounders like the presence of multiple metals to which the workers were exposed.

Cobalt and cobalt compounds along with soluble cobalt salts are labeled as Group 2B carcinogens (possibly carcinogenic to humans), and cobalt metal with tungsten chloride has been labeled as 2A (probably carcinogenic to humans) by the International Agency for Research on Cancer (IARC, 2006). The European Commission classified cobalt sulfate and cobalt dichloride in the category 1B (presumed to be carcinogenic to humans, classification largely based on animal evidence). The National Institute of Environmental Health Sciences has labeled cobalt sulfate and cobalt–tungsten carbide as "reasonably anticipated to be a human carcinogen." ACGIH has labeled cobalt in the carcinogenic category of A3 (confirmed animal carcinogen with unknown relevance to humans) (ACGIH, 2007).

Mode of Action

The mechanism by which cobalt exposure results in toxic effects is not fully understood. Exposure to cobalt chloride causes a rapid increase of the intracellular levels of hypoxia-inducible factor-1α (Torii et al., 2011). Hypoxic state induced by cobalt is implicated as one of the precursors to lung effects. In another study, exposure to cobalt and its compounds activated phosphatidylinositol-3-kinase pathway and production of reactive oxygen species (Chachami et al., 2004). Studies have suggested the role of reactive oxygen species (Nemery et al., 1994; Sarkar, 1995; Salnikow et al., 2000; Lison et al., 2001). Similarly, exposure of a variety of cell lines to soluble cobalt salts in lungs, neuronal tissue, and liver resulted in an increased reactive oxygen species production (Salnikow et al., 2000; BelAiba et al., 2004). This resulted in an accelerated accumulation of oxidative stress products, release of cytokines, cell damage, and death. In another study, cobalt potentiated the reactivity of neutrophil-derived oxidants resulting in protease and oxidant-mediated tissue injury (Ramafi et al., 2004).

Chemical Pathology

Cobalt exposure is associated with skin pathology, including sensitization, epidermal thickening, and contact dermatitis (Asano et al., 2009). Exposure has also been associated with lung pathology, including interstitial pneumonia and fibrosis (Davison et al., 1983), fibrosing alveolitis (Van Den Eeckhout et al., 1989), and possible lung carcinoma (Wild et al., 2009). Cobalt is soluble in various body fluids, including lung airway lining, gastrointestinal tract, and skin sweat glands and can be absorbed from exposure pathways like inhalation, ingestion, or dermal contact. Absorption of cobalt depends upon the route of exposure with 30% absorption possible via inhalational route, 3–50% via ingestion, and

0.024 and 3.6% via intact and abraded skin, respectively (Stefaniak et al., 2011). Ultrafine cobalt particles have a size of 20 nm; these particles can cause pulmonary inflammation when exposed for prolonged periods of time. (Kyono et al., 1992; Zhang et al., 1998). The effects of cobalt are affected in the presence of other chemicals. Mutual exposure to nickel and cobalt chloride led to a synergistic effect in cell culture and lowered cell viability than when exposed individually (Cross et al., 2001).

SOURCES OF EXPOSURE

Occupational

Occupational exposure to cobalt occurs from diverse industries and occupations such as mining, heavy metal superalloy production, and tungsten carbide processing. In addition, workers are also exposed when cobalt or its compounds are involved in the refining of cobalt and processing of various alloys and its salts (Swennen et al., 1993). Cobalt air levels have been associated with corresponding incidence of occupational asthma and hard metal disease in the hard metal industry. Suspected asthma incidence was highest in powder plant maintenance, in areas of chemical testing, metal pressing, and sintering. Also, cases suspected of hard metal disease were observed in metal pressing, sintering, and grinding. Along with the respiratory route of occupational exposure, dermal exposure is also common. Exposure to cobalt is particularly common in workplaces that involve the use of cutting or grinding tools, refining or processing cobalt metal or ores, and ceramics. Exposure is also prevalent in industries using cobalt to color glass and producing drier for enamel and paint. For most of the cases, except in the processing of cobalt powder, the exposure is not only to cobalt but also to other metals and compounds like tungsten carbide, chromium, iron, etc. A combination of cobalt with these metals may alter the chemical reactivity of cobalt.

Environmental

Cobalt is found throughout the environment in low concentrations in air, food, and water. The amount of cobalt in air is <2 ng/m^3, which is less than the cobalt exposure from food and water. In drinking water, cobalt concentration is <1–2 parts per billion (ppb). An average person consumes about 11 µg of cobalt in the diet in the form of vitamin B$_{12}$. The concentration in food and water may be many times more in areas that are in the vicinity and/or in the areas of mining, smelting, or hard metal industries. Environmental exposure may occur from surgical procedures such as hip implants (Sunderman et al., 1989) and dental prosthesis (Kettelarij et al., 2014). Elevated blood cobalt levels were observed due to metal hip implants in the absence of other etiologies in patients undergoing total hip replacement (Pizon et al., 2013).

INDUSTRIAL HYGIENE

The Occupational Safety and Health Administration (OSHA) has assigned an 8-h permissible exposure limit (PEL) of $0.1\,mg/m^3$, and the National Institute for Occupational Safety and Health (NIOSH) recommends an 8-h TWA of $0.05\,mg/m^3$ (NIOSH, 2007). ACGIH has recommended an 8-h TWA of $0.02\,mg/m^3$. No reference concentration (RfC) or reference dose (RfD) has been associated with cobalt and its compounds though ATSDR has recommended minimal risk levels (MRL) of $0.0001\,mg/m^3$ via inhalational route and $0.01\,mg/kg/day$ via ingestion (ATSDR, 2004).

MEDICAL MANAGEMENT

Diagnosis

The diagnosis of respiratory illness associated with cobalt is by the worker complaints, work history and signs detected from X-ray abnormalities, pulmonary function tests, Bronchoalveolar lavage and lung biopsy, to high resolution tomography scanning (Dunlop et al., 2005; Gotway et al., 2002). Signs may include dyspnea, shortness of breath, and expectoration; physical examination may reveal crackles. Chest X-ray generally presents with a reticulo-nodular pattern predominantly in the lower zones. Pulmonary function tests may show lowered diffusion capacity with restrictive pattern. Bronchial lavage can reveal multinucleate giant cells and elevation in eosinophil count and T-lymphocytes. Lung biopsy may show giant cell interstitial pneumonitis, fibrosing alveolitis, or inflammatory infiltrates. In another study, high resolution tomography scanning indicated large cystic spaces in mid and upper lungs, upper lobe reticulation, and traction bronchiectasis (Gotway et al., 2002).

Biomonitoring can be undertaken to assess exposure and determine the efficacy of health and safety procedure. Concentrations of cobalt and its compounds in blood and urine show a positively linear association with air concentration levels (Ichikawa et al., 1985). Workers exposed to cobalt may have higher IgA levels and lowered IgE levels (Bencko et al., 1986). In one study human leukocyte antigen residue was monitored to distinguish workers exposed to cobalt and those predisposed to develop hard metal disease (Potolicchio et al., 1997).

Treatment

Early diagnosis and treatment is important in the management of cases exposed to cobalt and cobalt compounds. Removal of the worker from the exposure to cobalt and its compounds is an important initial step in the treatment protocol. In a study, cobalt-exposed workers with the symptoms of cough and eczema showed recovery when the exposure was withdrawn. One worker who remained on the job developed lung nodules 2.5 years later. In another study, proper diagnosis of three patients and removal from cobalt exposure resulted in complete recovery for two; the third had some mild lingering effects (Anttila et al., 1986). A fourth patient in the same report was not diagnosed efficiently and developed lingering, severe respiratory effects. Treatment of chronic effects should be symptomatic. Studies have shown the improvement in symptoms in hard metal disease patients with the administration of steroids (Nureki et al., 2013), but no clinical trials have been performed to confirm the effectiveness of such agents.

REFERENCES

ACGIH (2007) *TLVs and BEIs Based on the Documentation of the Threshold Limit Values for Chemical Substances and Physical Agents & Biological Exposure Indices.* Cincinnati, Ohio: ACGIH.

Alexandersson R., and Atterhog. (1983) Comparison of electrocardiograms among wet grinders in Swedish hard metal industry before and after four weeks holiday. Arbetarskyddsverket Stockholm. *Arbete. och. halsa.* 1983:18.

Anttila, S., et al. (1986) Hard metal lung disease: a clinical, histological, ultrastructural and X-ray microanalytical study. *Eur. J. Respir. Dis.* 69(2):83–94.

Asano, Y., et al. (2009) Occupational cobalt induced systemic contact dermatitis. *Eur. J. Dermatol.* 19(2):166–167.

Athavale, P., et al. (2007) Occupational dermatitis related to chromium and cobalt: experience of dermatologists (EPIDERM) and occupational physicians (OPRA) in the U.K. over an 11-year period (1993–2004). *Br. J. Dermatol.* 157(3):518–522.

ATSDR (2004) *Toxicological Profile for Cobalt.* Atlanta, GA: United States Department of Health and Human Services, Public Health Service, Agency for Toxic Substances and Disease Registry.

Barborik, M. and Dusek, J. (1972) Cardiomyopathy accompaning industrial cobalt exposure. *Br. Heart J.* 34(1):113–116.

Bech, A. (1974) Hard metal disease and tool room grinding. *J. Soc. Occup. Med.* 24:11–16.

Bech, A., Kipling, M., and Heather, J. (1962) Hard metal disease. *Br. J. Ind. Med.* 19:239–52.

BclAiba, R.S., et al. (2004) Redox-sensitive regulation of the HIF pathway under non-hypoxic conditions in pulmonary artery smooth muscle cells. *Biol. Chem.* 385(3–4):249–257.

Bencko, V., et al. (1986) Human exposure to nickel and cobalt: biological monitoring and immunobiochemical response. *Environ. Res.* 40(2):399–410.

Bock, M., et al. (2003) Occupational skin disease in the construction industry. *Br. J. Dermatol.* 149(6):1165–1171.

Bucher, J.R., et al. (1990) Inhalation toxicity studies of cobalt sulfate in F344/N rats and B6C3F1 mice. *Fundam. Appl. Toxicol.* 15(2):357–372.

Chachami, G., et al. (2004) Cobalt induces hypoxia-inducible factor-1alpha expression in airway smooth muscle cells by a reactive oxygen species- and PI3K-dependent mechanism. *Am. J. Respir. Cell Mol. Biol.* 31(5):544–551.

Coates, E.O. Jr. and Watson, J.H. (1973) Pathology of the lung in tungsten carbide workers using light and electron microscopy. *J. Occup. Med.* 15(3):280–286.

Costa, M., et al. (1982) Selective phagocytosis of crystalline metal sulfide particles and DNA strand breaks as a mechanism for the induction of cellular transformation. *Cancer Res.* 42(7):2757–2763.

Cross, D.P., et al. (2001) Mixtures of nickel and cobalt chlorides induce synergistic cytotoxic effects: implications for inhalation exposure modeling. *Ann. Occup. Hyg.* 45(5):409–418.

D'Adda, F., et al. (1994) Cardiac function study in hard metal workers. *Sci. Total Environ.* 150(1–3):179–186.

Dai, J.H., et al. (2009) Giant cell interstitial pneumonia associated with hard metals: a case report and review of the literature. *Zhonghua Jie He He Hu Xi Za Zhi* 32(7):493–496.

Davison, A.G., et al. (1983) Interstitial lung disease and asthma in hard-metal workers: bronchoalveolar lavage, ultrastructural, and analytical findings and results of bronchial provocation tests. *Thorax* 38(2):119–128.

De Boeck, M., et al. (1998) Evaluation of the *in vitro* direct and indirect genotoxic effects of cobalt compounds using the alkaline comet assay. Influence of interdonor and interexperimental variability. *Carcinogenesis* 19(11):2021–2029.

Domingo, J.L., et al. (1984) A study of the effects of cobalt administered orally to rats. *Arch. Farmacol. Toxicol.* 10(1):13–20.

Dooms-Goossens, A.E., et al. (1986) Contact dermatitis caused by airborne agents. A review and case reports. *J. Am. Acad. Dermatol.* 15(1):1–10.

Dunlop, P., et al. (2005) Hard metal lung disease: high resolution CT and histologic correlation of the initial findings and demonstration of interval improvement. *J. Thorac. Imaging* 20(4):301–304.

FDRL (1984a) *Acute Oral LD50 Study of Cobalt Sulphate Lot No. S88336/A in Sprague-Dawley Rats*, FDRL Study No. 8005D, Waverly, NY: Food and Drug Research Laboratories, Inc.

FDRL (1984b) *Study of Cobalt(II) Carbonate Tech Gr. CoCO$_3$, Lot #030383 in Sprague-Dawley Rats*, Waverly, NY: Food and Drug Research Laboratories, Inc.

FDRL (1984c) *Acute Oral Toxicity Study of Cobalt Oxide Tricobalt Tetraoxide in Sprague-Dawley Rats*, Waverly, NY: Food and Drug Research Laboratories, Inc.

Filon, F.L., et al. (2009) *In vitro* absorption of metal powders through intact and damaged human skin. *Toxicol. In Vitro* 23(4):574–579.

Gault, N., et al. (2010) Cobalt toxicity: chemical and radiological combined effects on HaCaT keratinocyte cell line. *Toxicol. In Vitro* 24(1):92–98.

Gennart, J.P., et al. (1993) Increased sister chromatid exchanges and tumor markers in workers exposed to elemental chromium-, cobalt- and nickel-containing dusts. *Mutat. Res.* 299(1):55–61.

Gotway, M.B., et al. (2002) Hard metal interstitial lung disease: high-resolution computed tomography appearance. *J. Thorac. Imaging* 17(4):314–318.

Grice, H.C., et al. (1969) Myocardial toxicity of cobalt in the rat. *Ann. N. Y. Acad. Sci.* 156(1):189–194.

Harper, E.M., et al. (2012) Tracking the metal of the goblins: cobalt's cycle of use. *Environ. Sci. Technol.* 46(2):1079–1086.

Heath, J.C. (1954) Cobalt as a carcinogen. *Nature* 173:822–823.

Heath, J.C. (1956) The production of malignant tumors by cobalt in the rat. *Br. J. Cancer* 10:668–673.

Holly, R.G. (1955) Studies on iron and cobalt metabolism. *JAMA* 158:1349–1352.

Horiguchi, H., et al. (2004) Acute exposure to cobalt induces transient methemoglobinuria in rats. *Toxicol. Lett.* 151(3):459–466.

Horowitz, S.F., et al. (1988) Evaluation of right and left ventricular function in hard metal workers. *Br. J. Ind. Med.* 45(11):742–746.

IARC Working Group on the Evaluation of Carcinogenic Risks to Humans, World Health Organization, & International Agency for Research on Cancer. (2006). *Cobalt in Hard Metals and Cobalt Sulfate, Gallium Arsenide, Indium Phosphide and Vanadium Pentoxide: Cobalt in Hard-Metals and Cobalt Sulfate, Gallium Arsenide, Indium Phosphide and Vanadium*. World Health Organization.

Ichikawa, Y., et al. (1985) Biological monitoring of cobalt exposure, based on cobalt concentrations in blood and urine. *Int. Arch. Occup. Environ. Health* 55(4):269–276.

Jarvis, J.Q., et al. (1992) Cobalt cardiomyopathy. A report of two cases from mineral assay laboratories and a review of the literature. *J. Occup. Med.* 34(6):620–626.

Julander, A., et al. (2009) Cobalt-containing alloys and their ability to release cobalt and cause dermatitis. *Contact Dermatitis* 60(3):165–170.

Kanerva, L., et al. (1988) Occupational skin disease in Finland. An analysis of 10 years of statistics from an occupational dermatology clinic. *Int. Arch. Occup. Environ. Health* 60(2):89–94.

Kawanishi, S., et al. (1994) Active oxygen species in DNA damage induced by carcinogenic metal compounds. *Environ. Health Perspect.* 102(Suppl. 3):17–20.

Kennedy, A., et al. (1981) Fatal myocardial disease associated with industrial exposure to cobalt. *Lancet* 1(8217):412–414.

Kettelarij, J.A., et al. (2014) Cobalt, nickel and chromium release from dental tools and alloys. *Contact Dermatitis* 70(1):3–10.

Kiec-Swierczynska, M. and Szymczk, W. (1995) The effect of the working environment on occupational skin disease development in workers processing rockwool. *Int. J. Occup. Med. Environ. Health* 8(1):17–22.

Kusaka, Y., et al. (1986) Respiratory diseases in hard metal workers: an occupational hygiene study in a factory. *Br. J. Ind. Med.* 43(7):474–485.

Kyono, H., et al. (1992) Reversible lung lesions in rats due to short-term exposure to ultrafine cobalt particles. *Ind. Health* 30(3–4):103–118.

Lasfargues, G., et al. (1994) Lung cancer mortality in a French cohort of hard-metal workers. *Am. J. Ind. Med.* 26(5):585–595.

Linna, A., et al. (2004) Altered left ventricular diastolic function, increased myocardial stiffness, but with no ECG changes were observed in the study conducted on exposed workers in Finland. *Occup. Environ. Med.* 61(11):877–885.

Lison, D., et al. (2001) Update on the genotoxicity and carcinogenicity of cobalt compounds. *Occup. Environ. Med.* 58(10):619–625.

Lloyd, D.R., et al. (1997) Generation of putative intrastrand crosslinks and strand breaks in DNA by transition metal ion-mediated oxygen radical attack. *Chem. Res. Toxicol.* 10(4):393–400.

Martin, C.R. and Kohli, P. (2003) The emerging field of nanotube biotechnology. *Nat. Rev. Drug Discov.* 2(1):29–37.

Miller, A.C., et al. (2001) Neoplastic transformation of human osteoblast cells to the tumorigenic phenotype by heavy metal-tungsten alloy particles: induction of genotoxic effects. *Carcinogenesis* 22(1):115–125.

Morvai, V., et al. (1993) The effects of simultaneous alcohol and cobalt chloride administration on the cardiovascular system of rats. *Acta Physiol. Hung.* 81(3):253–261.

Moulin, J.J., et al. (1998) Lung cancer risk in hard-metal workers. *Am. J. Epidemiol.* 148(3):241–248.

Muller, J., et al. (2005) Respiratory toxicity of multi-wall carbon nanotubes. *Toxicol. Appl. Pharmacol.* 207(3):221–231.

Murdock, H. (1959) Studies on the pharmacology of cobalt chloride. *J. Am. Pharm. Assoc.* 48:140–142.

Murray, A., Kisin, E., Leonard, S., Young, S., Kommineni, C., Kagan, V., Castranova, V., and Shvedova, A. (2009a) Oxidative stress and inflammatory response in dermal toxicity of single-walled carbon nanotubes. *Toxicology* 257: (3):161–171.

Nackerdien, Z., et al. (1991) Nickel(II)- and cobalt(II)-dependent damage by hydrogen peroxide to the DNA bases in isolated human chromatin. *Cancer Res.* 51(21):5837–5842.

Naqvi, A. H., et al. (2008) Pathologic spectrum and lung dust burden in giant cell interstitial pneumonia (hard metal disease/cobalt pneumonitis): review of 100 cases. *Arch. Environ. Occup. Health* 63(2):51–70.

Nemery, B., et al. (1994) Cobalt and possible oxidant-mediated toxicity. *Sci. Total Environ.* 150(1–3):57–64.

NIOSH (2007) COBALT and compounds, as Co. NIOSH Manual of Analytical Methods (NMAM), 4TH Edition - 3rd Supplement.

Nureki, S., et al. (2013) Hard metal lung disease successfully treated with inhaled corticosteroids. *Intern. Med.* 52(17):1957–1961.

Pizon, A.F., et al. (2013) Prosthetic hip-associated cobalt toxicity. *J. Med. Toxicol.* 9(4):416–417.

Potolicchio, I., et al. (1997) Susceptibility to hard metal lung disease is strongly associated with the presence of glutamate 69 in HLA-DP beta chain. *Eur. J. Immunol.* 27(10):2741–2743.

Radi, S., et al. (2002) Respiratory morbidity in a population of French dental technicians. *Occup. Environ. Med.* 59(6):398–404.

Raffn, E., et al. (1988) Health effects due to occupational exposure to cobalt blue dye among plate painters in a porcelain factory in Denmark. *Scand. J. Work Environ. Health* 14(6):378–384.

Ramafi, G.J., et al. (2004) Pro-oxidative interactions of cobalt with human neutrophils. *Inhal. Toxicol.* 16(9):649–655.

Sakai, M., et al. (2010) A case of hard metal lung disease in a man who worked as an iron grinder. *Nihon Kokyuki Gakkai Zasshi* 48(4):282–287.

Salnikow, K., et al. (2000) Carcinogenic metals induce hypoxia-inducible factor-stimulated transcription by reactive oxygen species-independent mechanism. *Cancer Res.* 60(13):3375–3378.

Sarkar, B. (1995) Metal replacement in DNA-binding zinc finger proteins and its relevance to mutagenicity and carcinogenicity through free radical generation. *Nutrition* 11(Suppl. 5):646–649.

Schinz, H.R. and Uehlinger, E. (1942) Metals: a new principle of carcinogenesis. *Z. Krebsforsch.* 52:425–437.

Seghizzi, P., et al. (1994) Cobalt myocardiopathy: a critical review of the literature. *Sci. Total Environ.* 150:105–109.

Selden, A., et al. (1996) Three cases of dental technician's pneumoconiosis related to cobalt–chromium–molybdenum dust exposure. *Chest* 109(3):837–842.

Sheppard, P.R., et al. (2006) Elevated tungsten and cobalt in airborne particulates in Fallon, Nevada: possible implications for the childhood leukemia cluster. *Appl. Geochem.* 21:152–165.

Sherson, D., Maltbaek, N., and Heydorn, K. (1990) A dental technician with pulmonary fibrosis: a case of chromium–cobalt alloy pneumoconiosis? *Eur. Respir. J.* 3(10):1227–1229.

Singh, P. and Junnarkar, A. (1991) Behavioral and toxic profile of some essential trace metal salts in mice and rats. *Indian J. Pharmacol.* 23:153–159.

Smith, I.C. and Carson B.L. (1981) *Trace Metals in the Environment*, Ann Arbor, MI: Ann Arbor Science Publishers.

Stefaniak, A.B., et al. (2011) Total-body exposure to metal sensitizers: inhalation, ingestion, and skin contact. *Epidemiology* 22(1):S83–S84.

Sunderman, F.W., Jr. et al. (1989) Cobalt, chromium, and nickel concentrations in body fluids of patients with porous-coated knee or hip prostheses. *J. Orthop. Res.* 7(3):307–315.

Sundaram, P., et al. (2001) Giant cell pneumonitis induced by cobalt. *Indian J. Chest Dis. Allied Sci.* 43(1):47–49.

Swennen, B., et al. (1993) Epidemiological survey of workers exposed to cobalt oxides, cobalt salts, and cobalt metal. *Br. J. Ind. Med.* 50:835–842.

Torii, S., et al. (2011) Inhibitory effect of extracellular histidine on cobalt-induced HIF-1alpha expression. *J. Biochem.* 149(2):171–176.

Van Den Eeckhout, A.V., et al. (1989) Pulmonary pathology due to cobalt and hard metals. *Rev. Mal. Respir.* 6(3):201–207.

Warheit, D.B., et al. (2004) Comparative pulmonary toxicity assessment of single-wall carbon nanotubes in rats. *Toxicol. Sci.* 77(1):117–125.

Wild, P., et al. (2000) Lung cancer mortality in a site producing hard metals. *Occup. Environ. Med.* 57(8).568–573.

Wild, P., et al. (2009) Lung cancer and exposure to metals: the epidemiological evidence. *Methods Mol. Biol.* 472:139–167.

Zanelli, R., et al. (1994) Uncommon evolution of fibrosing alveolitis in a hard metal grinder exposed to cobalt dusts. *Sci. Total Environ.* 150(1–3):225–229.

Zhang, Q., et al. (1998) Differences in the extent of inflammation caused by intratracheal exposure to three ultrafine metals: role of free radicals. *J. Toxicol. Environ. Health A* 53(6):423–438.

15

COPPER

Marie M. Bourgeois

First aid: Inhalation: Remove to fresh air. Provide respiratory support. Ingestion: Induce vomiting immediately as directed by medical personnel. Skin: Immediately flush skin with plenty of soap and water for at least 15 min. Remove contaminated clothing and shoes. Eye contact: Immediately flush eyes with plenty of water for at least 15 min. Get medical attention immediately.

Target organ(s): Eyes, skin, respiratory system, liver, kidneys (increased risk with Wilson's disease)

Occupational exposure limits: NIOSH REL: TWA 1 mg/m^3, OSHA PEL: TWA 1 mg/m^3 and ACGIH TLV: TWA 1 mg/m^3 [REL, PEL, and TLV also apply to other copper compounds (as Cu) except copper fume], TLV 0.2 mg/m^3 (fume), NIOSH IDHL: 100 mg/m^3

Reference values: MRL: 0.01 mg/kg/day (oral)

Risk phrases: 11-36/37

Flammable: Irritating to eyes and respiratory system

Safety phrases: 26-60

In case of contact with eyes, rinse immediately with plenty of water and seek medical advice. This material and/or its container must be disposed of as hazardous waste.

BACKGROUND AND USES

Copper (symbol Cu; CAS #7440-50-8) is distributed widely in nature; it is the twenty-sixth most abundant element in the earth's crust and is an essential element for many life forms. Copper is an abundant reddish, odorless metal that takes on a greenish-blue patina when exposed to the elements. It was the first metal worked by humans, and copper salts were among the first materials regularly used for therapeutic and cosmetic purposes.

Most animals require copper for certain biological processes. A deficiency of copper, as well as an excess, can have adverse health effects. The daily intake of copper in the United States ranges from 2 to 5 mg, almost all of which is excreted in the feces. Shellfish, seeds, nuts, and grains are rich sources of dietary copper. Minute amounts of cupric ion are absorbed and stored, mainly in the liver, blood, and brain. Copper is an essential cofactor in several enzyme systems (Shaligram and Campbell, 2012). Cuproenzymes catalyze important biochemical reactions, including iron absorption and heme biosynthesis (Colotti et al., 2013). Copper deficiency may lead to anemia and neutropenia, and eventually to bone lesions resembling scurvy and to pathological fractures without hemorrhage (Kumar et al., 2005; Halfdanarson et al., 2008). Copper is also found in some intrauterine devices for the prevention of pregnancy (IPCS, 1998; Szymanski et al., 2012).

Extracting copper from ores involves crushing, roasting, and smelting to produce "matte," a mixture of varying proportions of copper and iron sulfides, which is further refined electrolytically to remove the sulfur and iron. Once separated from its ore, copper is used in a variety of products, including electrical machinery and wiring, roof sheeting and water pipes, and chemical and heating applications. Its malleability and luster have made it a valuable metal since humans began working metal. American coins are all copper alloys (ATSDR, 2004). The biocidal properties of copper salts such as copper sulfate and copper naphthenate make it a common ingredient in fungicides and disinfectants (Reigart and Roberts, 1999). Copper also has important industrial properties, including efficient thermal and electrical conductivity and resistance to corrosion. It

Hamilton & Hardy's Industrial Toxicology, Sixth Edition. Edited by Raymond D. Harbison, Marie M. Bourgeois, and Giffe T. Johnson.
© 2015 John Wiley & Sons, Inc. Published 2015 by John Wiley & Sons, Inc.

is used for electroplating because it increases hardness in its alloys, principally brass (with zinc) and bronze (with tin), and in special mixtures with other metals such as nickel, beryllium, and cobalt (HSDB, 2007a and HSDB, 2007b).

PHYSICAL AND CHEMICAL PROPERTIES

Copper is a group IB transition element on the periodic table and exists in four oxidation states: Cu^0 Cu^{1+} (cuprous ion), Cu^{2+} (cupric ion), and Cu^{3+}. In a natural state, copper is likely to be found in a variety of solid salts and compounds, but it can be found in the elemental form as well. Copper compounds generally are blue or green. The common green color of copper on exposure to air is a basic carbonate compound. The more important industrial compounds are copper sulfate and copper chloride; these and others are listed, with their common uses, in Table 15.1.

Copper is insoluble in water but readily dissolves in hot sulfuric and nitric acids. The vapor pressure is negligible at low temperatures, but in an industrial setting, in which very high temperatures are used to smelt copper ore, some potentially harmful copper fumes may be emitted. Although it not combustible in bulk, powdered copper may ignite. Fires and explosions may result from contact with oxidizing agents, strong mineral acids, alkali metals, and halogens (NIOSH, 2010). Selected chemical and physical properties of copper are presented in Table 15.2.

TABLE 15.1   Common Uses of Copper and Copper Compounds

Substance (CAS Number)	Uses
Copper (Cu) (#7440-50-8)	Electrical conductor, electronics, alloys (bronze, brass, bell, and gun metal)
Copper sulfate ($CuSO_4$) (#7758-98-7)	Parasiticide, fungicide, algicide, herbicide, dye mordant, wood preservative, tanning, electroplating, pigment (ink, photography)
Copper cyanide [Cu $(CN)_2$] (#14763-77-0)	Electroplating, antifouling paint, fungicide, insecticide
Cupric acetoarsenite ($Cu_4AS_6C_4H_8O_4$) (#12002-03-8)	Marine pigment, insecticide
Cupric acetate ($CuC_4H_8O_4$) (#142-71-2)	Fungicide pigments, dyes
Copper chloride ($CuCl_2$) (#7447-39-4)	Electroplating, wood preservative, petroleum desulfurizing, catalyst, mordant, pigments
Copper oxide (Cu_2O) (#1317-39-1)	Pigment, antifouling paint, fungicide, catalyst
Copper nitrate [Cu $(NO_3)_2$] (#3251-23-8)	Herbicide, fungicide, wood preservative, colorant, mordant, photocopying, catalyst

TABLE 15.2   Chemical and Physical Properties of Copper

Characteristic	Information
Chemical name	Copper
Synonyms/trade names	Copper metal dusts, copper metal fumes
Chemical formula	Cu
CASRN	7440-50-8
Molecular weight	63.546 g/mol
Melting point	1083 °C (decomposes)
Boiling point	2595 °C (decomposes)
Vapor pressure	0 mmHg at 20 °C
Density/specific gravity	8.94 g/cm^3 at 20 °C
Flashpoint	Not available
Water solubility at 25 °C	Insolubility
Explosion limits	Not available
Octanol/water partition coefficient (log K_{OW})	−0.57
Henry's law constant	0.0245 atm m^3/mol
Odor threshold	Odorless
Conversion factors	Not available

ENVIRONMENTAL FATE AND BIOACCUMULATION

Elemental copper is stable in the environment. There is little danger of volatilization so copper can be expected to adsorb to sediments and suspended solids in water. This copper is primarily Cu^{2+} and is of limited bioavailabilty when bound to organic matter. Other forms of copper are slightly more bioavailable; these can produce adverse effects in the right conditions (e.g., low pH). Water-soluble forms of copper may enter groundwater supplies. No evidence of bioaccumulation has been found, and there is little evidence of biomagnification (ATSDR, 2004).

There is variation in the bioconcentration factors of some species of fish (10–667) and mollusks (up to 30,000) (Perwak et al., 1980; Georgopoulos et al., 2006). The higher value in mollusks is likely due to the fact that they are filter feeders. However, it is thought that long-term exposure in aquatic and terrestrial environments can lead to the release of the constituent copper compounds in more bioavailable forms. These more bioavailable forms have the potential to yield toxic effects under specific chemical conditions. Mobilization is not usually seen without prolonged exposure to a pH value of 3 or less (HSDB, 2007a and HSDB, 2007b). There are no data on the log K_{ow} of copper, so it is difficult to make determinations about soil mobility.

ECOTOXICOLOGY

Environmental copper has anthropogenic and natural origins (sea spray, hazardous waste dumps, wastewater processing, and industry). Industrial processes accounted for approximately 1.4 billion pounds of environmental copper in 2000 alone. Copper and cupric compounds have been found in more than a third of the proposed Environmental Protection Agency(EPA) National Priorities List sites (HSDB, 2007a and HSDB, 2007b). Airborne copper concentrations in rural and urban areas range from 5 to 200 ng/m^3.

Concentrations in fresh and salt water bodies range from 4 to 10 parts per billion (ppb). Copper concentrations in soil vary considerably according to land use, soil composition, pH, depth, and moisture. Copper binds strongly to organic materials in soils. Naturally occurring copper concentrations average 50 ppm with substantially higher concentrations present near agricultural or industrial sites (Davies and Bennett, 1985).

MAMMALIAN TOXICOLOGY

Acute Effects

Although acute toxicity is uncommon, it is more likely to occur with exposure to copper than chronic toxicity (Box 15.1). Dust and fumes, as well as the green carbonate, are irritating to the eyes, skin, and respiratory tract. The National Institute of Occupational Safety and Health (NIOSH) notes that there is little evidence copper dusts and mists (not fume) present a serious occupational risk. Ingestion of copper-contaminated

---

BOX 15.1   TOXICITY OF COPPER

Copper in plasma

> About 7% loosely bound to albumin (cytotoxic)

> About 93% tightly bound to ceruloplasmin (nontoxic)

Copper in tissue

> Most tightly bound to ceruloplasmin (nontoxic) and metallothionein (less toxic)

> Unbound or free (toxic)

The cellular and tissue damage caused by copper is related to free or loosely bound and exchangeable copper ions. Toxicity results when ingested copper salts exceed the binding capacity of ceruloplasmin and metallothionein or when the individual has a genetically determined deficiency in binding protein capacity (e.g., Wilson's disease, Menkes' syndrome).

---

water may result in gastrointestinal upset, most often nausea and vomiting. Irritation of nasal mucous membranes may occur after inhalation of copper salts. Metal fume fever may result from acute exposure to copper fume; however, the incidence is relatively low due to the high temperatures required to volatilize copper. It is characterized by fever, arthralgia, nausea, cough, and lassitude (NIOSH, 2010). Several renal failure patients have developed acute hemolytic anemia subsequent to copper intoxication when the dialysis tubing and filters contained copper (USEPA, 1985).

Animal studies of acute copper toxicity have demonstrated effects similar to those seen in humans (Fieten et al., 2012). The more severe responses to copper overexposure occurred with concentrations greater than what would be expected in the workplace or through nonoccupational exposure. Romeu-Moreno et al. (1994) exposed rats to high inhalation concentrations of $CuSO_4$ and found high concentrations of accumulated copper in the liver and plasma.

Individuals susceptible to copper-induced toxicity include those suffering from a hereditary metabolic disorder called Wilson's disease, in which the pathway for excretion and storage of copper does not function properly and copper accumulates more rapidly (Muller et al., 1998; Wijmenga and Klomp, 2004). Other deficiencies may result in infantile cirrhosis in children from overexposure to copper.

Chronic Effects

Chronic illnesses are not reported with copper overexposure, and if they are present in copper-generating facilities, they are linked to the presence of other harmful metals and, in some cases, to exposure to radon (Wu et al., 1992). There is some indication that increased levels of hepatic copper are associated with cellular injury (Johncilla and Mitchell, 2011). Chen et al. (1993) described a higher than expected number of deaths from lung cancer in copper miners but attributed these deaths to general exposure to dust, not specifically to copper dust. Copper has not been found to have any carcinogenic, teratogenic, or mutagenic properties (HSDB, 2007a and HSDB, 2007b). Ostiguy et al. (1995) did not find any significant respiratory dysfunctions in a population of workers exposed to low levels of metal dust and fumes. Some instances of a greenish discoloration of the skin and hair from prolonged exposure to copper were reported in the past, but this effect was not associated with any noticeable ill effects. There is no evidence of carcinogenicity from chronic copper exposure (Lightfoot et al., 2010).

Gleason (1968) reported general malaise and "head stuffiness" in a worker polishing copper plates. After local exhaust measures were improved, the workers' complaints disappeared. Some reports exist of ulcerated lesions of the nose resulting from inhalation of copper dust. Fine fragments of copper that penetrate the eye during certain industrial operations can cause severe ocular damage. Ingestion of copper dust has been linked to a variety of health problems, including a metallic taste in the mouth, nausea, vomiting, epigastric burning, and diarrhea (Suciu et al., 1981).

Skin eruptions may be accompanied by swelling and itching in workers exposed to "cement copper," which is precipitated from solution by metallic ion (Hostynek and Maibach, 2003). Hypersensitivity reactions to copper in a welder were described by Forstrom et al. (1977); Dhir et al. (1977) described such reactions in furniture polishers using a commercial spirit colored blue with $CuSO_4$ (Dhir et al., 1977; Forstrom et al., 1977).

Pimentel and Menezes (1977) described "vineyard sprayer's lung" in rural workers in Portugal who applied sprays of Bordeaux mixture ($CuSO_4$ and $CaOH$ involving precipitation of $CuOH$ and several basic sulfates of copper) (Pimentel and Menezes, 1977). These workers had interstitial pulmonary lesions consisting of copper-containing histolytic granulomas and nodular fibrohyaline scars, which were regressing, remaining stationary, or progressing to diffuse pulmonary fibrosis and lung cancer. Lesions containing copper also were found in the nasal mucosa, liver, kidneys, spleen, and lymph nodes. Copper-related liver lesions included sarcoid-like granulomas, fibrosis, cirrhosis, and angiosarcomas, along with idiopathic portal hypertension.

Chemical Pathology

No published data are available regarding the health effects of copper on humans exposed orally or by inhalation for chronic durations. Additionally, data from chronic animal studies for oral and inhalation exposures to copper are inadequate. There are chronic animal studies for oral exposures to copper compounds such as copper cyanide.

Mode of Action(s)

Because of copper's low toxicity, few studies of copper-mediated toxicity have been reported, and therefore the mechanism of toxicity is unclear. Most data are from *in vitro* assays or animal studies, neither of which are conclusive in occupational medicine. Copper retained by the liver is bound to metal-binding proteins, specifically metallothionein (-Figure 15.1); this prevents free ions from continually circulating in the body (Boveris et al., 2012). The bound copper is excreted primarily through the bile. Excessive copper may produce liver oxidative damage. The generation of hydroxyl radicals from unbound copper is a possible mechanism for copper toxicity. Sagripanti et al. (1991) found that Cu(II) bound itself to certain sites in the deoxyribonucleic acid and suggested this as a possible mechanism for copper toxicity (Sagripanti et al., 1991). Variations in some tissue pathology as a result of exposure to copper were found *in vitro* to be related to inhibition of the Na+/K(+) pump and calcium homeostasis (Benders et al., 1994).

Increases in urinary and plasma copper levels, detectable by atomic absorption or inductively coupled plasma-atomic emission spectroscopy, can be indicative of occupational overexposure to copper.

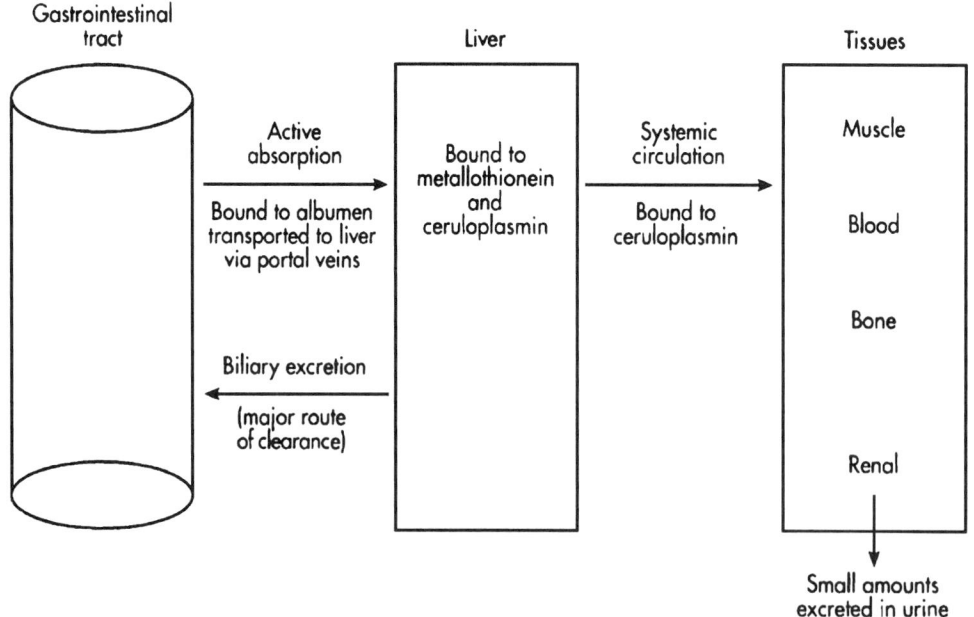

FIGURE 15.1 Mammalian copper.

SOURCES OF EXPOSURE

Exposure to copper in humans can occur through inhalation, ingestion, and eye or skin contact (IPCS, 1998; HSDB, 2007b).

Occupational

Copper exposure is common in occupations, including asphalt makers, electroplaters, wallpaper makers and embossers, solderers, especially miners and smelters of copper ore (ATSDR, 2004). The major industrial route of exposure to copper is inhalation of dust and fumes. Dermal contact and ingestion are less likely to cause adverse effects in an occupational setting (Weir, 1979). The salts of copper generally are considered more toxic than copper dust and fumes, which are relatively nontoxic. The vaporization point of copper is high (about 2350 °C [4262 °F]), and it is rare to encounter these fumes in industry.

Another hazard in copper mining is concurrent exposure to more hazardous metals and compounds. Exposure to arsenic, selenium, lead, and cadmium is associated with copper mining. Industrial hygiene measures are more appropriately tailored to these more hazardous materials than to the copper present. Respiratory protection with face masks and adequate ventilation is recommended in mining operations to protect against exposure to dust and fumes. Protective clothing should be worn, and the skin should be washed after a work shift to reduce dermal contact.

Environmental

Copper is environmentally ubiquitous. The general population may be exposed to copper through the ingestion of drinking water and foods, contact with copper-containing soil, and inhalation of air tainted with copper (HSDB, 2007). Copper concentrations in drinking water range from 20 to 75 ppb and drinking water is the greatest potential exposure source. They may be substantially higher in homes with copper piping/fixtures or lower pH water. Soils, sediments, and dusts commonly contain copper and copper compounds. Many foods, particularly organ meats and mollusks, also contain significant amounts of copper.

INDUSTRIAL HYGIENE

The current Occupational Health and Safety Administration (OSHA) recommended exposure limit(PEL), NIOSH recommended exposure limit(REL), and American Conference of Governmental Industrial Hygienists (ACGIH) threshold limit value(TLV) for copper are $1\,mg/m^3$ as an 8-h time-weighted average (TWA) concentration. This applies to other copper compounds except copper fume. The OSHA PEL and NIOSH REL for copper fume are $0.1\,mg/m^3$ as an 8-h (TWA) concentration, and the ACGIH TLV is $0.2\,mg/m^3$. Recommendations by NIOSH on respiratory protection for workers exposed to copper are listed in Table 15.3.

TABLE 15.3 Recommended Respiratory Protection for Workers

Condition/Airborne Concentration of Copper	Minimum Respiratory Protection
Up to $5\,mg/m^3$	(APF = 5) Any quarter-mask respirator
Up to $10\,mg/m^3$	(APF = 10) Any particulate respirator equipped with an N95, R95, or P95 filter (including N95, R95, and P95 filtering facepieces) except quarter-mask respirators. The following filters may also be used: N99, R99, P99, N100, R100, P100
	(APF = 10) Any supplied-air respirator
Up to $25\,mg/m^3$	(APF = 25) Any supplied-air respirator operated in a continuous-flow mode
	(APF = 25) Any powered, air-purifying respirator with a high efficiency particulate filter
Up to $50\,mg/m^3$	(APF = 50) Any air-purifying, full facepiece respirator with an N100, R100, or P100 filter
	(APF = 50) Any powered, air-purifying respirator with a tight-fitting facepiece and a high efficiency particulate filter
	(APF = 50) Any self-contained breathing apparatus with a full facepiece
	(APF = 50) Any supplied-air respirator with a full facepiece
Up to $100\,mg/m^3$	(APF = 2000) Any supplied-air respirator that has a full facepiece and is operated in a pressure-demand or other positive-pressure mode
Emergency or planned entry into unknown concentrations or IDLH conditions	(APF = 10,000) Any self-contained breathing apparatus that has a full facepiece and is operated in a pressure-demand or other positive-pressure mode
	(APF = 10,000) Any supplied-air respirator that has a full facepiece and is operated in a pressure-demand or other positive-pressure mode in combination with an auxiliary self-contained positive-pressure breathing apparatus
Escape	(APF = 50) Any air-purifying, full facepiece respirator with an N100, R100, or P100 filter
	Any appropriate escape-type, self-contained breathing apparatus

RISK ASSESSMENTS

No chronic reference dose (RfD) or reference concentration (RfC) for copper is currently available on EPA's Integrated Risk Information System (IRIS) or in the Drinking Water Standards and Health Advisories. There are no human carcinogenicity data and inadequate animal data from assays of copper compounds.

REFERENCES

ATSDR (2004) *Toxicological Profile for Copper.* Available at http://www.atsdr.cdc.gov/ToxProfiles/tp.asp?id=206&tid=37.

Benders, A.A., Li, J., Lock, R.A., Bindels, R.J., Bonga, S.E., and Veerkamp, J.H. (1994) Copper toxicity in cultured human skeletal muscle cells: the involvement of Na^+/K^+-ATPase and the $Na^+/Ca2^+$-exchanger. *Pflügers Archiv October (II)* 428(5–6):461–467.

Boveris, A., Musacco-Sebio, R., Ferrarotti, N., Saporito-Magrina, C., Torti, H., Massot, F., and Repetto, M.G. (2012) The acute toxicity of iron and copper: biomolecule oxidation and oxidative damage in rat liver. *J. Inorg. Biochem.* 116:63–69.

Chen, R., Wei, L., and Huang, H. (1993) Mortality from lung cancer among copper miners. *Br. J. Ind. Med.* 50(6):505–509.

Colotti, G., Ilari, A., Boffi, A., and Morea, V. (2013) Metals and metal derivatives in medicine. *Mini Rev. Med. Chem.* 13(2):211–221.

Davies, D.J. and Bennett, B.G. (1985) Exposure of man to environmental copper—an exposure commitment assessment. *Sci. Total Environ.* 46:215–227.

Dhir, G.G., Rao, D.S., and Mehrotra, M.P. (1977) Contact dermatitis caused by copper sulfate used as coloring material in commercial alcohol. *Ann. Allergy* 39(3): 204

Fieten, H., Leegwater, P.A., Watson, A.L., and Rothuizen, J. (2012) Canine models of copper toxicosis for understanding mammalian copper metabolism. *Mamm. Genome* 23(1–2):62–75.

Forstrom, L., Kiistala, R., and Tarvainen, K. (1977) Hypersensitivity to copper verified by test with 0.1% $CuSO_4$. *Contact Dermatitis* 3(5):280–281.

Georgopoulos, P.G., Wang, S.W., Georgopoulos, I.G., Yonone-Lioy, M.J., and Lioy, P.J. (2006) Assessment of human exposure to copper: a case study using the NHEXAS database. *J. Expos. Sci. Environ. Epidemiol.* 16(5):397–409.

Gleason, R.P. (1968) Exposure to copper dust. *Am. Ind. Hyg. Assoc. J.* 29(5):461–462.

Halfdanarson, T.R., Kumar, N., Li, C.Y., Phyliky, R.L., and Hogan, W.J. (2008) Hematological manifestations of copper deficiency: a retrospective review. *Eur. J. Haematol.* 80(6):523–531.

Hostynek, J.J. and Maibach, H.I. (2003) Copper hypersensitivity: dermatologic aspects—an overview. *Rev. Environ. Health* 18(3):153–183.

HSDB (2007a) *Hazardous Substances Data Bank: Copper Compounds.* Available at http://toxnet.nlm.nih.gov.

HSDB (2007b) *Hazardous Substances Data Bank: Copper.* Available at http://toxnet.nlm.nih.gov.

IPCS (1998) *UKPID Monograph Copper.* Available at http://www.inchem.org/documents/ukpids/ukpids/ukpid57.htm.

Johncilla, M. and Mitchell, K.A. (2011) Pathology of the liver in copper overload. *Semin. Liver Dis.* 31(3):239–244.

Kumar, N., Elliott, M.A., Hoyer, J.D., Harper, C.M., Jr., Ahlskog, J.E., and Phyliky, R.L. (2005) "Myelodysplasia," myeloneuropathy, and copper deficiency. *Mayo Clin. Proc.* 80(7): 943–946.

Lightfoot, N.E., Pacey, M.A., and Darling, S. (2010) Gold, nickel and copper mining and processing. *Chronic Dis. Can.* 29(Suppl. 2):101–124.

Muller, T., Muller, W., and Feichtinger, H. (1998) Idiopathic copper toxicosis. *Am. J. Clin. Nutr.* 67(Suppl. 5):1082S–1086S.

NIOSH (2010) *Pocket Guide to Chemical Hazards: Copper Dusts and Fumes*, Washington, DC: US Department of Health and Human Services.

Ostiguy, G., Vaillancourt, C., and Bégin, R. (1995) Respiratory health of workers exposed to metal dusts and foundry fumes in a copper refinery. *Occup. Environ. Med.* 52(3):204–210.

Perwak, J., Byshhe, S., Goyer, M., Nelken, L., Scow, K., Walker, P., Wallace, D., and Delos, C. (1980) *An Exposure and Risk Assessment for Copper*, Washington, DC: US EPA.

Pimentel, J.C. and Menezes, A.P. (1977) Liver disease in vineyard sprayers. *Gastroenterology* 72(2):275–283.

Reigart, J. and Roberts, J. (1999) *Recognition and Management of Pesticide Poisonings*, Washington, DC: EPA, pp. 137–155.

Romeu-Moreno, A., Aguilar, C., Arola, L., and Mas, A. (1994) Respiratory toxicity of copper. *Environ. Health Perspect.* 102(Suppl 3):339–340.

Sagripanti, J.L., Goering, P.L., and Lamanna, A. (1991) Interaction of copper with DNA and antagonism by other metals. *Toxicol. Appl. Pharmacol.* 110(3):477–485.

Shaligram, S. and Campbell, A. (2012) Toxicity of copper salts is dependent on solubility profile and cell type tested. *Toxicol. In Vitro* 27(2):844–851.

Suciu I., Prodan L., Lazar V., Ilea E., Cocîrla A., Olinici L., Paduraru A., Zagreanu O., Lengyel P., Gyõrffi L., and Andru D. (1981) Research on copper poisoning. *Med. Lav.* 72(3):190–197.

Szymanski, P., Fraczek, T., Markowicz, M., and Mikiciuk-Olasik, E. (2012) Development of copper based drugs, radiopharmaceuticals and medical materials. *Biometals* 25(6):1089–1112.

USEPA (1985) *Drinking Water Criteria Document for Copper (Final Draft)*, Washington, DC: US EPA.

Weir, F.W. (1979) Health hazard from occupational exposure to metallic copper and silver dust. *Am. Ind. Hyg. Assoc. J.* 40(3):245–247.

Wijmenga, C. and Klomp, L.W. (2004) Molecular regulation of copper excretion in the liver. *Proc. Nutr. Soc.* 63(1):31–39.

Wu, Z., Hearl, F., Peng, K., McCawley, M., Chen, A., et al. (1992) Current occupational exposures in Chinese iron and copper mines. *Appl. Occup. Environ. Hyg.* 7: 735–743.

16

GALLIUM AND INDIUM

Jayme P. Coyle and Raymond D. Harbison

Target organ(s)

Gallium	Respiratory
Indium[1]	Respiratory, eyes, liver, skin, kidneys, heart, blood

Occupational exposure limits

Gallium

Elemental	No established values
Arsenide[2]	PEL: 0.01 mg/m³ (TWA)
	REL: 0.002 mg/m³ (ceiling)
	TLV: 0.0003 mg/m³ (TWA)

Indium[1,2]

Compounds	PEL: No established limit
	REL: 0.1 mg/m³ (TWA)
	TLV: 0.1 mg/m³ (TWA)
Reference values	No established values

Risk/safety phrases

Gallium (elemental)

Risk phrases	None
Safety phrases	None

Indium (elemental)

Risk phrases	None
Safety phrases	None

[1]NIOSH, 2007. Pocket guide to chemical hazards, NIOSH 2005-149, US Department of Health and Human Services. Washington, DC.

[2]ACGIH, 2009. Threshold limit values (TLVs) for chemical substances and physical agents and Biological Exposure Indices (BEIs). Cincinnati, OH.

BACKGROUND AND USES

Gallium (symbol Ga; CAS# 7440-55-3) and indium (symbol In; CAS# 7440-74-6) are considered together because they are both widely used in the electronics industry, their radio-isotopes are used for diagnostic imaging in medicine, and they are both group III elements (Browning, 1969; Ferm and Carpenter, 1970; Abrams and Murrer, 1993). Both elements typically for compounds with an oxidation state of 3^+. Though neither element is essential for human nutrition, both are widely distributed in low concentrations in the environment (Smith et al., 1978). Gallium is found in concentrations of 0.01–0.7% in bauxite and germanite ores, and also found in zinc-rich ores.

Radioisotopes of gallium are useful for localizing bone and inflammatory lesions and tumor-associated lymphomas (van Amsterdam et al., 1996). Nonradioactive gallium nitrate (CAS# 13494-90-1) is a potent inhibitor of bone resorption that has been used to treat malignancy-associated hypercal-cemia. Radioisotopes of indium have been used to label phagocytes and lymphocytes to localize inflammatory lesions (Dudley and Marrer, 1952; Abrams and Murrer, 1993). Despite early optimism, neither element has found wide use as a treatment for malignancies.

Alloys and compounds containing each element have been used as commercial semiconductors. For example, gallium arsenide (symbol GaAs; CAS# 1303-00-0), indium

Hamilton & Hardy's Industrial Toxicology, Sixth Edition. Edited by Raymond D. Harbison, Marie M. Bourgeois, and Giffe T. Johnson.
© 2015 John Wiley & Sons, Inc. Published 2015 by John Wiley & Sons, Inc.

phosphide (symbol InP; CAS# 22398-80-7), and indium–tin oxide (symbol $In_2O_3 \cdot SnO_2$; CAS# 50926-11-9) are among the most widely used intermetallic semiconductor components (Harrison, 1986; McIntyr and Sherin, 1989). Gallium arsenide is also incorporated into light-emitting diodes and photovoltaic cells, while gallium alloys are used for dental amalgam as a low toxicity replacement for mercury. Indium compounds are gaining increasing use for soldering, flat-panel electronic components, and photovoltaic cells. A substance made by combining indium (24%) and gallium (76%) is liquid at room temperature and is used as a wetting agent for sealing glass to metal or glass to glass when high temperatures cannot be used.

PHYSICAL AND CHEMICAL PROPERTIES

Gallium is a silvery liquid at 29.75 °C (85.55 °F), which boils at 2204 °C (3999.2°F) and has the largest liquid range of any metal. Despite the liquid state, the vapor pressure of elemental gallium is negligible. Indium is a silvery-white, malleable metal that melts at 156 °C (312.8 °F) and boils at 2072 °C (3761.6°F). Oxides of both elements are amphoteric. Both gallium and indium form arsenides, halides, hydroxides, oxides, and phosphides, some of which may be degraded by acids and fire to produce highly toxic gases such as arsine and stibine. Reaction of indium oxide with water produces an insoluble indium hydroxide [$In(OH)_3$], which limits mobilization in solution. Gallium salt solubility increases with increasing ionic strength.

MAMMALIAN TOXICOLOGY

Acute Effects

The oral median lethal dosage (LD_{50}) for gallium arsenide and indium arsenide is >15 g/kg in rodents with an inhalation exposure threshold for acute effects of 152.5 mg/m^3 for gallium arsenide and 139 mg/m^3 for indium arsenide (Wald and Becker, 1986). Indium hydroxide [$In(OH)_3$] and gallium hydroxide [$Ga(OH)_3$] are quickly lethal, teratogenic, and embryocidal in animal models via intravenous administration (Ferm and Carpenter, 1970), despite relative insolubility of the hydroxide salts. Parenterally administered indium chloride has been shown to cause acute tubular necrosis.

Ivanoff et al. (2011) reported the acute symptoms of a human case exposed to aqueous gallium chloride as erythema, nausea, contact dermatitis, and difficulty in breathing, which progressed to cardiomyopathy and neuropathy.

Chronic Effects

Gallium can cause neuromuscular toxicity, including loss of vision and paralysis, as well as constitutional symptoms such as pruritus and anorexia (Browning, 1969). Chronic exposure to gallium oxide (Ga_2O_3) and gallium arsenide (GaAs; ≥0.1 mg/m^3) particles resulted in alveolar proteinosis and pulmonary fibrosis over several weeks or months of exposure in rats with additional incidence of benign respiratory lesions.

Subchronic exposure to indium phosphide (≥1.0 mg/m^3) dusts in rats resulted in pulmonary alveolar proteinosis and fibrosis with an increased incidence of non-neoplasmic hepatic lesions and mononuclear cell leukemia (NTP, 2001).

Epidemiological data suggest increased incidence of pulmonary alveolar proteinosis, interstitial pneumonia, and fibrosis with chronic exposure to indium phosphide dust (Homma et al., 2003; Cummings et al., 2010; Nakano et al., 2009; Tanaka et al., 2010). Progressive dyspnea has been used to signal the onset of pulmonary alveolar proteinosis. Observed lesions are consistent with animal models though a decisive correlation between the exposure and disease is lacking.

Based on the toxicological evidence, the International Agency for Research on Cancer (IARC) has listed gallium arsenide and indium phosphate as Group 1 and Group 2A human carcinogens, respectively. The American Conference of Governmental Industrial (ACGIH) has classified gallium arsenide as A3 (ACGIH, 2009). Neither chemical has been classified as a carcinogen in its elemental form (IARC, 2006a).

Mechanisms of Action

Absorption of gallium and indium compounds from the gastrointestinal tract is generally below 2% of oral intake in humans (Castronovo and Wagner, 1971; Blazka, 1998). Absorbed gallium exhibits a biphasic half-life in blood with an initial half-life phase estimated at 1 h and the second phase ranges from 1 to 5 days (NTP, 2000). Indium is predominantly absorbed across the lung epithelia and, when absorbed, retained within the pulmonary interstitium and sequestered by serum transferrin (Zhang et al., 2004; Nakano et al., 2009; Hoet et al., 2012). Release from biological reservoirs may lead to prolonged secondary exposure.

Indium accumulates in reticuloendothelial and hepatic phagocytes that were found to be less responsive to mitogens and less able to synthesize immunoglobulin (Silberstein et al., 1985). Gottschling et al. (2001) suggests the role of indium in oxidative stress-mediated pulmonary inflammation progressing to neoplastic growth in rats though the exact mechanism for pulmonary indium toxicity has not been completely elucidated. Absorbed gallium arsenide dissociates in the biological fluid whereby gallium becomes rapidly bound and excreted from the body. The anionic partner, i.e., inorganic arsenic, may mobilize after dissociation and induce toxicity.

Animal models have suggested gallium-induced inhibition of δ-aminolevulinic acid dehydratase (ALAD) in the kidney, liver, and erythrocytes (Goering and Rehm, 1990;

Flora et al., 2009). *In vitro* models using GaCl₃ produced cytostatic effects on peripheral blood mononuclear cells (Chang et al., 2003).

Chemical Pathology

Acute gallium poisoning from gallium chloride has been characterized by one case study. This case exhibited erythema, nausea, contact dermatitis, and difficulty in breathing, which eventually progressed to tachycardia and arrhythmia with peripheral neuropathy (leg tremors). Blood tests revealed electrolyte imbalance with decreased serum phosphate levels (Ivanoff et al., 2011).

Chronic exposure to indium compounds has been associated with interstitial pneumonia and pulmonary alveolar proteinosis as assessed by high resolution computed topography (HRCT) scans though definitive causation has not been implied. HRCT scans of patients have revealed typical pathophysiology such as glass opacities, pneumothorax, and intra-/interlobular septal thickening while lung biopsies have confirmed eosinophilic alveolar intrusion and cholesterol clefts. Bronchoalveolar lavage fluid of patients has also confirmed the presence of eosinophilic infiltrate and foamy macrophages (Homma et al., 2003; Cummings et al., 2010; Nakano et al., 2009). The analysis of metal has confirmed the presence of indium within the interstitial tissue.

Workers observed during continued indium exposure bore stable elevated indium plasma levels, which cleared slowly after removal from exposure; levels of the exposed remained higher than the unexposed controls even after several years. Epidemiological studies have associated KL-6 surfactant protein expression and indium exposure, especially to indium–tin oxide, and may act as a biomarker of secondary pulmonary alveolar proteinosis; however, the link (Homma et al., 2005; Hamaguchi et al., 2008; Hoet et al., 2012).

SOURCES OF EXPOSURE

Occupational

Occupational gallium and indium exposure occurs primarily from respirable dusts produced from sawing, grinding, and cleaning of ingots, wafer fabrication, and workroom cleaning operations. Elevated plasma and urine levels were found in an exposed cohort working in a Belgian indium ingot fabrication plant with an ambient air concentration averaging 190 μg/m³ (range: 10–1030 μg/m³) compared to an unexposed cohort. Both plasma and urine levels fell below the limits of detection for the unexposed. Health assessments were not performed to analyze the risk of exposure (Hoet et al., 2012).

In cases of improper handling and storage, exposure by means of smoke and fumes from chemical reactivity or electrical fires may cause human illness. For example, gallium arsenide and indium arsenide may release arsine gas under certain conditions (Harrison, 1986; Wald and Becker, 1986: McIntyr and Sherin, 1989; Scott et al., 1989).

Environmental

Environmental exposure to indium compounds has been measured in ambient air as 43 ng/m³ and in rainwater as 0.59 μg/l. Exposure via the environment is considered negligible (IARC, 2006b).

INDUSTRIAL HYGENE

When working with inorganic compounds and elemental forms of gallium and indium, proper skin and eye/face protection is suggested to avoid direct contact. In grinding and cleaning operations, proper personal respiration equipment should be utilized to avoid inhalation.

No exposure limits have been set for elemental gallium. However, Occupational Safety and Health Administration (OSHA) and ACGIH have set an 8-h time-weighted average (TWA) at 0.01 mg/m³ and 0.0003 mg/m³, respectively, for gallium arsenide, while NIOSH recommends a ceiling exposure value of 0.002 mg/m³.

The ACGIH and NIOSH have set TWA exposure limits for inorganic indium (As In) at 0.1 mg/m³. Currently, permissible exposure limit (PEL) and reference values have not been established for either elemental indium or indium compounds (CDC, 2013).

MEDICAL MANAGEMENT

There is no specific therapy for gallium or indium poisoning though oxalic acid has been suggested for chelating gallium after acute exposures (Ivanoff et al., 2011). Removal to fresh air with respiratory support is suggested after acute inhalation exposure. Ocular exposure should be irrigated immediately for a minimum of 15 min.

In cases of pulmonary alveolar proteinosis from chronic indium and indium compound exposure, whole lung lavage has been shown to attenuate decreased pulmonary function (Trapnell et al., 2003).

REFERENCES

Abrams, M. and Murrer, V.A. (1993) Metal compounds in therapy and diagnosis. *Science* 261:725–730.

American Conference of Governmental Industrial Hygienists (ACGIH) (2009) *Threshold Limit Values (TLVs) for Chemical Substances and Physical Agents and Biological Exposure*

Indices (BEIs), Cincinnati, OH: American Conference of Governmental Industrial Hygienists.

Blazka, M.E. (1998) Indium. In: Zelikoff, J.T. and Thomas, P.T., editors. *Immunotoxicology of Environmental and Occupational Metals*, Philadelphia, PA: Taylor and Francis, pp. 93–109.

Browning, E. (1969) *Toxicity of Industrial Metals*, 2nd ed., New York, NY: Appleton-Century-Crofts.

Castronovo, F.P. and Wagner, H.N. (1971) Factors affecting the toxicity of the element indium. *Br. J. Exp. Pathol.* 52:543–559.

Center for Disease Control and Prevention (CDC) (2013) *Pocket Guide to Chemical Hazards*, Atlanta, GA: Center for Disease Control and Prevention, U.S. Department of Health and Human Services.

Chang, K.-L., Liao, W.-T., Yu, C.-L., Lan, C.-C. E., Chang, L.W., and Yue, H.-S. (2003) Effects of gallium on immune stimulation and apoptosis induction in human peripheral blood mononuclear cells. *Toxicol. Appl. Pharmacol.* 193:209–217.

Cummings, K.J., Donat, W.E., Ettensohn, D.B., Roggli, V.L., Ingram, P., and Kreiss, K. (2010) Pulmonary alveolar proteinosis in workers at an indium processing facility. *Am. J. Respir. Crit. Care Med.* 181(5):458–464.

Dudley, J.C. and Marrer, H.H. (1952). Gallium metabolism: deposition and clearance from bone. *J. Pharmacol. Exp. Ther.* 106:129–134.

Ferm, V.H. and Carpenter, S.J. (1970) Teratogenic and embryopathic effects of indium, gallium, and germanium. *Toxicol. Appl. Pharmacol.* 6:166–170.

Flora, S.J.S., Bhatt, K., and Mehta, A. (2009) Arsenic moiety in gallium arsenide is responsible for neuronal apoptosis and behavioral alterations in rats. *Toxicol. Appl. Pharmacol.* 240:236–244.

Goering, P.L. and Rehm, S. (1990) Inhibition of liver, kidney, and erythrocyte delta-aminolevulinic acid dehydratase (porphobilinogen synthetase) by gallium in the rat. *Environ. Res.* 53:135–151.

Gottschling, B.C., Maronpot, R.R., Hailey, J.R., Peddada, S., Moomaw, C.R., Klaunig, J.E., and Nyska, A. (2001) The role of oxidative stress in indium phosphide-induced lung carcinogenesis in rats. *Toxicol. Sci.* 64:28–40.

Hamaguchi, T., Omae, K., Takebayashi, T., Kikuchi, Y., Yoshioka, N., Nishiwaki, Y., Tanaka, A., Hirata, M., Taguchi, O., and Chonan, T. (2008) Exposure to hardly soluble indium compounds in ITO production and recycling plants is a new risk for interstitial lung damage. *Occup. Environ. Med.* 65:51–55.

Harrison, R.J. (1986) Gallium arsenide. *Occup. Med.* 1:49–58.

Hoet, P., De Graef, E., Swennen, B., Seminck, T., Yakoub, Y., Deumer, G., Haufroid, V., and Lison, D. (2012) Occupational exposure to indium: what does biomonitoring tell us? *Toxicol. Lett.* 213:122–128.

Homma, S., Miyamoto, A., Sakamoto, S., Kishi, K., Motoi, N., and Yoshimura, K. (2005) Pulmonary fibrosis in an individual occupationally exposed to inhaled indium-tin oxide. *Eur. Respir. J.* 25:200–204.

Homma, T., Ueno, T., Sekizawa, K., Tanaka, A., and Hirata, M. (2003) Interstitial pneumonia developed in a worker dealing with particles containing indium-tin oxide. *J. Occup. Health* 45:137–139.

International Agency for Research on Cancer (IARC) (2006a) Gallium arsenide. In: *IARC Monographs on the Evaluation of Carcinogenic Risks to Humans, Cobalt in Hard Metals and Cobalt Sulfate, Gallium Arsenide, Indium Phosphide and Vanadium Pentoxide*, vol. 86, Lyon, France: WHO Press, pp. 163–196.

International Agency for Research on Cancer (IARC) (2006b) Indium phosphide. In: *IARC Monographs on the Evaluation of Carcinogenic Risks to Humans, Cobalt in Hard Metals and Cobalt Sulfate, Gallium Arsenide, Indium Phosphide and Vanadium Pentoxide*, vol. 86, Lyon, France: WHO Press, pp. 197–224.

Ivanoff, C.S., Ivanoff, A.E., and Hottel, T.L. (2011) Gallium poisoning: a rare case report. *Food Chem. Toxicol.* 50:212–215.

McIntyr, A.J. and Sherin, B.J. (1989) Gallium arsenide processing and hazard control. *Solid State Technol.* 32:101–104.

Nakano, M., Omae, K., Tanaka, A., Hirata, M., Michikawa, T., Kikuchi, Y., Yoshioka, N., Nishiwaki, Y., and Chonan, T. (2009) Causal relationship between indium compound inhalation and effects on the lungs. *J. Occup. Health* 51:513–521.

National Institute for Occupational Safety and Health (NIOSH) (2007) *Pocket Guide to Chemical Hazards*, NIOSH 2005-149, Washington, DC: US Department of Health and Human Services.

National Toxicology Program (NTP) (2000) NTP toxicology and carcinogenesis studies of gallium arsenide (CAS No. 1303-00-0) in F344/N rats and B6C3F1 mice (inhalation studies). *Natl. Toxicol. Program Tech. Rep. Ser.* 492:1–306.

National Toxicology Program (NTP) (2001) NTP toxicology and carcinogenesis studies of indium phosphide (CAS No. 22398-80-7) in F344/N rats and B6C3F1 mice (inhalation studies). *Natl. Toxicol. Program Tech. Rep. Ser.* 499:1–340.

Scott, N., Carter, D.E., and Fernando, Q. (1989) Reaction of gallium arsenide with concentrated acids: formation of arsine. *Am. Ind. Hyg. Assoc. J.* 50:379–381.

Silberstein, E.B., Watson, S., Mayfield, G., Kereiakes, J.G., and Bullock, W. (1985) Indium-111 toxicity in the human lymphocyte. *J. Lab. Clin. Med.* 105:608–612.

Smith, I.C., Carson, B.L., and Hoffmeister, F. (1978) *Trace Metals in the Environment: Indium*, vol. 5, Ann Arbor, MI: Ann Arbor Science Publishers.

Tanaka, A., Hirata, M., Kiyohara, Y., Nakano, M., Omae, K., Shiratani, M., and Koga, K. (2010) Review of pulmonary toxicity of indium compounds to animals and humans. *Thin Solid Films* 518:2934–2936.

Trapnell, B.C., Whitsett, J.A., and Nakata, K. (2003) Pulmonary alveolar proteinosis. *N. Engl. J. Med.* 349:2527–2539.

van Amsterdam, J.A., Kluin-Nelemans, J.C., van Eck-Smit, B.L., and Pauwels, E.K. (1996). Role of ^{67}Ga scintigraphy in localization of lymphoma. *Ann. Hematol.* 72:202–207.

Wald, P. and Becker, C. (1986) Toxic gases used in the microelectronics industry. *Occup. Med.* 1:105–117.

Zhang, M., Gumerov, D.R., Kaltashov, I.A., and Mason, A.B. (2004) Indirect detection of protein–metal binding: interaction of serum transferrin with In^{3+} and Bi^{3+}. *J. Am. Soc. Mass Spectrom.* 15:1658–1664.

17

GERMANIUM

Marek Banasik

First aid: Eyes: Irrigate thoroughly with water if discomfort persists, obtain medical attention; skin: wash off thoroughly with soap and water; inhalation: remove from exposure, seek medical advice; ingestion: wash out mouth thoroughly with water—in severe cases, obtain medical attention.

Occupational exposure limit(s): None

Reference value(s): None

Hazard statement(s): None

Precautionary statements(s): None

Risk phrase(s): None

Safety phrase(s): None

BACKGROUND AND USES

Germanium [symbol Ge; Chemical Abstracts Service Registry Number (CASRN): 7440-56-4] is never found in a pure state in the environment. In the refined state, it is a gray–white metalloid. Like its counterparts in group 14 of the periodic table (carbon, silicon, tin, and lead), germanium has both metallic and nonmetallic properties. It has a valence of +2 or +4; the atom usually is quadrivalent.

Germanium is a true semiconductor, is highly transparent to infrared light, and has a high index of refraction. Germanium and its compounds or alloys have found wide use in the optical and electronics industries. In electronic applications, it often is alloyed with antimony, gallium, indium, or arsenic.

Germanium is a low production volume chemical. In the United States (US), the national production volume in 2012, including manufacturing and importing, was 61,890 pounds

(*ca.* 28,073 kg) (EPA, 2012). In the same year, the national production volume for germanium dioxide (GeO_2, CASRN: 1310-53-8), including manufacturing and importing, was 32,536 pounds (*ca.* 14,758 kg) (EPA, 2012). No registration or production volume data are available for germanium or GeO_2 in the European Union (ECHA, 2014).

No formal regulatory risk assessments have been performed on germanium or germanium compounds. The repeated dose toxicology of GeO_2 has primarily been evaluated in exploratory animal studies to evaluate the adverse effects reported in humans, who consumed germanium-containing products for its alleged health protective effects.

PHYSICAL AND CHEMICAL PROPERTIES

Germanium is a grayish-white, lustrous, brittle metalloid. When crystalline, it exists in a diamond-cubic structure. It has a very low vapor pressure and is insoluble in water. A summary of germanium's physicochemical properties is provided in Table 17.1.

ENVIRONMENTAL FATE AND BIOACCUMULATION

Metals are recalcitrant to degradation; therefore, no biodegradation studies have been performed on germanium compounds. Naturally occurring germanium exists in mineral ores; therefore, the levels of free germanium are expected to be low and of low concern for bioaccumulation in aquatic and terrestrial species, due to negligible exposures.

Hamilton & Hardy's Industrial Toxicology, Sixth Edition. Edited by Raymond D. Harbison, Marie M. Bourgeois, and Giffe T. Johnson.
© 2015 John Wiley & Sons, Inc. Published 2015 by John Wiley & Sons, Inc.

TABLE 17.1 Physical and Chemical Properties of Germanium

Characteristic	Information
Name	Germanium
Atomic symbol	Ge
CASRN	7440-56-4
Molecular weight	72.59 g/mol
Physical state	Grayish-white, lustrous, brittle metalloid
Melting point	937.2 °C
Vapor pressure	0 mmHg at 25 °C
Density	5.323 g/cm^3 at 25 °C
Water solubility	Insoluble

Source: HSDB (2002).

ECOTOXICOLOGY

No aquatic toxicology data were identified for germanium or germanium compounds.

MAMMALIAN TOXICOLOGY

Some germanium compounds have been claimed to have bone marrow- and immune-modulating properties (St. Georgiev, 1988). These compounds were incorporated into health food supplements and herbal teas but were subsequently banned by many countries due to numerous cases of adverse effects in humans (reviewed by Tao and Bolger, 1997). In addition to case study findings, several acute and repeated dose toxicity studies have been performed on experimental animals, which generally show the same types of toxicity occurring in humans. A summary of these studies is provided below.

Acute Studies

The acute oral toxicity of GeO_2 is low (reviewed by Tao and Bolger, 1997). In rats and male mice, median lethal dose (LD_{50}) values of 3.7 and 6.3 g/kg body weight (bw) have been reported, respectively. In male and female rats and mice, correspondingly low LD_{50} values ($\geq$6.3 g/kg bw) were reported for bis-carboxyethyl germanium sesquioxide (i.e., germanium-132 or Ge-132).

Lin et al. (1999) evaluated the kinetics and tissue distribution of GeO_2 in male Wistar rats after single and repeated exposures, respectively. In the single dose studies, six rats were administered 100 mg/kg bw by gavage. A maximum serum concentration of 15.5 ± 0.7 µg/ml (mean ± S.E.M.) was reported. The absorption- and elimination-half lives were

0.7 ± 0.1 h and 2.3 ± 0.5 h, respectively. In the repeated dose studies, 15 rats were administered daily doses of GeO_2 at 100 mg/kg bw/day by gavage for 4 weeks. Statistically significant concentrations of GeO_2 were reported in the following tissues, compared to the controls (all tissues <1 µg/g): sciatic nerve (5.5 ± 0.2 µg/g), kidney (4.9 ± 0.2 µg/g), liver (3.5 ± 0.1 µg/g), serum (3.2 ± 0.2 µg/g), heart (3.1 ± 0.1 µg/g), lungs (1.9 ± 0.1 µg/g), gastrocnemius muscle (1.7 ± 0.1 µg/g), and brain (1.5 ± 0.1 µg/g).

As discussed in the following sections, the tissue distribution data reflect some of the target organs identified from repeated dose studies in experimental animals and in human case study reports.

Repeated Dose Studies

Tao and Bolger (1997) reviewed the available repeated dose toxicity studies in animals administered with germanium compounds. As summarized in Table 17.2, the toxicity of inorganic germanium (i.e., GeO_2) was shown to be greater than organic germanium (i.e., Ge-132). The target organs included the kidney, liver, peripheral nerves, and skeletal muscle. Adverse renal effects were reported even after cessation of treatment, up to 24 weeks in one study.

In humans who consumed germanium compounds as a supplement or tonic, the toxicity of germanium reveals similar target organs as reported in experimental animal studies. It should be noted, however, that no reports were identified for systemic toxicity occurring through occupational exposures. As shown in Table 17.3, renal dysfunction and anemia were reported in 100% of patients treated for germanium toxicity. In some cases, renal function was still impaired for 10–31 months after the patients stopped taking germanium products.

SOURCES OF EXPOSURE

Occupational

Elemental germanium may ignite spontaneously and explode when suspended in air as a fine powder (HSDB, 2002). Germanium dioxide dust can form germanic acid, an irritant, if it contacts the eye (HSDB, 2000). Occupational exposures to dusts and fumes may occur during activities that involve separation and purification of the ore concentrate.

Environmental

The major source of germanium exposures in the general population occurs through the consumption of foodstuffs. Daily consumption estimates have been reported between 0.367 and 1.5 mg/day (reviewed by Tao and Bolger, 1997).

TABLE 17.2 Repeated Dose Toxicity Studies in Rats Treated with Germanium Dioxide or Germanium-132

Study Type	Test Substance; Dose	Exposure Regimen	Effect
13-week dietary study	GeO$_2$; calculated intake: 150 mg/kg bw/day	Daily	• ↑ Serum creatinine, BUN[a], and serum phosphate • ↓ Creatinine clearance • ↑ Kidney weight • ↑ GOT[b] and GPT[c] • ↓ Total protein and albumin
4- and 24-week dietary studies	Ge-132; calculated intake: 1 mg/kg bw/day	Daily	• No toxic symptoms or behavior • Slight ↓ in bw in male rats at 6 months • Small ↓ in erythropoiesis and ↑ in cellular metabolism after 28 days • Moderate ↓ in renal efficiency at 6 months
23-week dietary study	GeO$_2$; 0.15% in diet	Daily	• ↓ Muscle weight • ↑ Ragged red muscle fibers • ↑ COX[d]-deficient muscle fibers • No apparent neurological abnormalities reported
24-week dietary study	GeO$_2$; calculated intake: 75 mg/kg bw/day Ge-132; calculated intake: 120 mg/kg bw/day	Daily	GeO$_2$ group: • ↑ BUN and serum phosphate • ↓ Creatinine clearance • ↓ Body weight, anemia, and liver dysfunction • ↑ Degenerated renal distal tubules • Semiquantitative scores of tubular degeneration = 95 ± 9% compared to controls (1 ± 1%) • Renal tubulointerstitial fibrosis was prominent at 16 weeks post-treatment Ge-132: • No toxic or renal histological abnormalities reported • Semiquantitative scores of tubular degeneration = 3 ± 1% compared to controls (1 ± 1%)
32-week dietary study	GeO$_2$; calculated intake: 69 mg/kg bw/day	Daily	Peripheral nerve lesions (e.g., segmented demyelination and remyelination, edema)
40-week dietary study	GeO$_2$; calculated intake: 37.5, 75, or 150 mg/kg bw/day	Daily	LOAEL[e] = 37.5 mg/kg bw/day, based on: ↓ growth, anemia, renal dysfunction, and renal tubular degeneration

Source: Tao and Bolger (1997) and accompanying citations.

[a]Blood urea nitrogen.
[b]Glutamic oxaloacetic transaminase.
[c]Glutamic pyruvic transaminase.
[d]Cytochrome c oxidase.
[e]Lowest-observed-adverse-effect level.

TABLE 17.3 Human Toxicity Data on Germanium

	Male	Female	Total
Number of cases	16	15	31
Age range (years)	5–73	4–58	4–73
Total intake (g)[a]	16–173	15–324	15–324
Cases with ≤70 g of intake	11/14 (79%)	11/13 (85%)	22/27 (81%)
Duration (months)[b]	3–24	2–36	2–36
Effects Death	4/16 (25%)	5/15 (33%)	9/31 (29%)
Gastrointestinal disturbances (nausea, vomiting, no appetite, and diarrhea)	–[c]	–	20/30 (67%)
Anemia	–	–	27/27 (100%)
Renal dysfunction/nephropathy	–	–	31/31 (100%)
Myopathy	–	–	13/21 (62%)
Neuropathy	–	–	13/22 (59%)
Liver dysfunction/necrosis	–	–	7/20 (35%)
Bone marrow (hypoplasia)	–	–	6/8 (75%)

Source: Tao and Bolger (1997).

[a]Unknown amounts ingested in four cases.
[b]Unknown duration in one case.
[c]Gender-specific data not provided.

INDUSTRIAL HYGIENE

The primary routes for potential exposure are through inhalation of dust and contact with skin/eyes. Engineering controls and personal protective equipment are required or recommended. A respirator and polyvinyl alcohol protective gloves are required. If chipping or dust formation is possible, safety goggles or safety glasses with side shields are also required.

REFERENCES

ECHA (2014) *Registered Substances*. Helsinki, Finland: European Chemicals Agency. Available at http://echa.europa.eu/information-on-chemicals/registered-substances. CASRNs: 7440-56-4, 1310-53-8.

EPA (2012) *Chemical Data Access Tool (CDAT)*. Washington, DC: Office of Pollution Prevention and Toxics, US Environmental Protection Agency. Available at http://java.epa.gov/oppt_chemical_search. CASRNs: 7440-56-4, 1310-53-8.

HSDB (2000) *Germanium Dioxide*. Bethesda, MD: Hazardous Substances Data Bank (HSDB), TOXNET®–Toxicology Data Network, Specialized Information Services, US Department of Health and Human Services. Available at http://toxnet.nlm.nih.gov/cgi-bin/sis/htmlgen?HSDB. CASRN: 1310-53-8; HSDB Number: 2119.

HSDB (2002) *Germanium*. Bethesda, MD: Hazardous Substances Data Bank (HSDB), TOXNET®–Toxicology Data Network, Specialized Information Services, US Department of Health and Human Services. Available at http://toxnet.nlm.nih.gov/cgi-bin/sis/htmlgen?HSDB. CASRN: 7440-56-4; HSDB Number: 2118.

Lin, C.-H., Chen, T.-J., Hsieh, Y.-L., Jiang, S.-J., and Chen, S.-S. (1999) Kinetics of germanium dioxide in rats. *Toxicology* 132:147–153.

St. Georgiev, V. (1988) New synthetic immunomodulating agents. *Trends Pharmacol. Sci.* 9:446–451.

Tao, S.-H. and Bolger, P.M. (1997) Hazard assessment of germanium supplements. *Regul. Toxicol. Pharmacol.* 25:211–219.

18

GOLD

Marek Banasik

First aid: Eyes: Dust or powder should be flushed from the eyes with running water for 15 min; skin: skin cuts and abrasions can be treated by standard first aid, skin contamination with dust or powder can be removed with soap and water—if irritation persists, obtain medical assistance; ingestion: obtain medical assistance at once.

Occupational exposure limit(s): None

Reference value(s): None

Hazard statement(s): None

Precautionary statement(s): None

Risk phrase(s): None

Safety phrase(s): None

BACKGROUND AND USES

"There are strange things done 'neath the midnight sun by the men who moil for gold."

So begins *The Cremation of Sam McGee* by British–Canadian poet Robert William Service. Metallic gold [symbol Au; Chemical Abstracts Service Registry Number (CASRN): 7440-57-5] is virtually insoluble, except in aqua regia. Gold exists in three primary forms, elemental, Au (I), and Au (III). As a precious metal, it is resistant to ionization and generally considered biologically benign in its elemental state. Occupational hazards in gold mining and refining arise from mechanical, environmental, and chemical factors and processes involved in the extraction of the precious metal, not from the metal itself. Gold salts, on the other hand, can be irritating and sensitizing and can cause systemic toxicity (Ainsworth et al., 1981; Barelli et al., 1987; Brown and Hill, 1994; Cohen, 1988; Gleichman et al., 1991; Gortenuti et al., 1985; Hall et al., 1987).

Rheumatoid arthritis is the only medical indication for chrysotherapy (i.e., gold therapy), and for some of these long-suffering patients, gold can give impressive, long-lasting relief. Most therapeutic forms are gold salts containing Au (I). Gold has a long biological half-life and can be detected in the blood for as long as 10 months after administration has stopped (Cohen, 1988). Its mode of action is unknown.

In the United States (US) in 2012, the national production volume, including manufacturing and importing, was withheld by the two listed companies. One company classified the manufacturing and importing data, as well as past production volume data, as confidential business information. The same company listed an on-site production volume of 8,228 pounds (*ca.* 3,732 kg) (EPA, 2012). The second company reported a manufacturing production volume of 26,252 pounds (*ca.* 11,908 kg) with a past production volume of 38,884 pounds (*ca.* 17,637 kg); no importing data were reported, and the volume used on-site was listed as 0 pounds (EPA, 2012). In the European Union, gold was not registered (ECHA, 2014).

The potential risks from gold relate primarily to adverse effects that may occur from the various gold compounds that are used therapeutically. A summary of these data is provided below.

PHYSICAL AND CHEMICAL PROPERTIES

A summary of elemental gold's physicochemical properties is provided in Table 18.1.

Hamilton & Hardy's Industrial Toxicology, Sixth Edition. Edited by Raymond D. Harbison, Marie M. Bourgeois, and Giffe T. Johnson.
© 2015 John Wiley & Sons, Inc. Published 2015 by John Wiley & Sons, Inc.

TABLE 18.1   Physical and Chemical Properties of Gold

Characteristic	Information
Name	Gold
Atomic symbol	Au
CASRN	7440-57-5
Molecular weight	196.97 g/mol
Physical state	Soft yellow metal
Melting point	1,064.76 °C
Boiling point	2,700 °C
Density	19.3 g/cm^3
Water solubility	Insoluble

Source: HSDB (2004).

ENVIRONMENTAL FATE AND BIOACCUMULATION

Metals are recalcitrant to degradation; therefore, no biodegradation studies have been performed on gold or its compounds. Naturally occurring gold exists in mineral ores; therefore, the levels of free gold are expected to be low and of low concern for bioaccumulation in aquatic and terrestrial species, due to negligible exposures.

ECOTOXICOLOGY

No aquatic toxicology data were identified for gold or its compounds.

MAMMALIAN TOXICOLOGY

Gold poisoning is most likely to be seen in association with medical uses of gold salts (Jackson et al., 1986; Rollins and Craig, 1991; Stuve and Galle, 1970; Voil et al., 1977). Gold salts are poorly absorbed, and until recently gold therapy for rheumatoid arthritis meant an intramuscular injection of gold sodium thiomalate (CASRN: 12244-57-4) in aqueous solution or gold thioglucose (CASRN: 12192-57-3) in oil. In the past decade an organic gold compound, auranofin, had been developed for oral administration. About 25% of the gold in the compound was absorbed by the gastrointestinal tract (Brent and Parra, 2005).

Episodes of vasodilation with facial flushing, hypotension, nausea, and light-headedness, the so-called nitritoid reaction, occur within minutes after injection of gold sodium thiomalate. Concomitant use of an angiotensin-converting enzyme inhibitor may increase the likelihood of this side effect (Healey and Backes, 1989).

The gold ion is a protoplasmic poison with an affinity for mitochondria. It forms complexes with a variety of proteins and thus can become part of a hapten or an immune complex (Gleichman et al., 1991). Gold salts show a wide range of clinical toxicity such as stomatitis and skin rashes with an element of photosensitivity, glomerulitis with proteinuria and hematuria, bone marrow suppression with thrombocytopenia and neutropenia, inflammatory enterocolitis, and pulmonary infiltrates with dyspnea and restrictive changes on spirometry.

Lymphadenopathy and eosinophilia often accompany gold salt toxicity, suggesting immunological sensitization as a pathogenetic factor. Gold salts interact with other drugs, such as nonsteroidal anti-inflammatory agents, which enhance gold-related renal and pulmonary toxicity (Martin et al., 1981; McFadden et al., 1989).

SOURCES OF EXPOSURE

Occupational

In South Africa, extraordinarily deep gold deposits subject miners to intense heat and pressure changes. In Brazil, liquid mercury is used to separate gold from the earth and sediments. When the gold–mercury amalgam is heated, the mercury vaporizes and is allowed to discharge into the environment, contaminating waterways and posing risks to wildlife and humans. Halogens, especially fluorine, are used in gold refining. Electroplating of jewelry and electronic applications requires gold potassium cyanide salts, which are irritants and systemic toxicants. Because metallic gold is biologically inert, no standards have been recommended by the US Occupational Safety and Health Administration, the US National Institute for Occupational Safety and Health, or the American Conference of Governmental Industrial Hygienists.

Environmental

Environmental exposures to gold are expected to be low. Background levels of $\leq$0.1 µg/ml and <0.0005 µg/ml in whole blood and serum, respectively, have been reported (Brent and Parra, 2005).

INDUSTRIAL HYGIENE

Workplace engineering controls (e.g., local exhaust) are recommended when cutting, grinding, or melting gold or other operations that may generate dusts or fumes. Personal protective equipment include safety glasses and protective gloves. Respiratory protection is not normally required, except in operations where dust or fumes are created.

REFERENCES

Ainsworth, S.K., Swain, R.P., Watabe, N., Brackett, N.C. Jr., Pilia, P., and Hennigar, G.R. (1981) Gold nephropathy: ultrastructural, fluorescent, and energy dispersive x-ray microanalysis study. *Arch. Pathol. Lab. Med.* 105:373–378.

Barelli, A., Calimici, A., and Pala, F. (1987) Gold salts acute poisoning: a case report. *Vet. Hum. Toxicol.* 29:108–110.

Brent, J. and Parra, M. (2005) Gold compounds. In: Brent, J., Wallace, K.L., Burkhart, K.K., Scott, D.P., and Donovan, J.W., editors. *Critical Care Toxicology: Diagnosis and Management of the Critically Poisoned Patient.* Philadelphia, PA, USA: Mosby, Inc. (Elsevier, Inc.), pp. 641–647.

Brown, S.L. and Hill, E.R. (1994) G-CSF in gold-induced aplastic anaemia. *Ann. Rheum. Dis.* 53:213.

Cohen, M.A.H. (1988) Adverse reactions to gold compounds. *Adverse Drug React. Acute Poisoning Rev.* 7:163–178.

ECHA (2014) *Registered Substances.* Helsinki, Finland: European Chemicals Agency. Available at http://echa.europa.eu/information-on-chemicals/registered-substances. CASRN: 7440-57-5.

EPA (2012) *Chemical Data Access Tool (CDAT).* Washington, DC: Office of Pollution Prevention and Toxics, US Environmental Protection Agency. Available at http://java.epa.gov/oppt_chemical_search. CASRN: 7440-57-5.

Gleichman, E., Kubicka-Muranyi, M., Kind, P., Goldermann, R., Goerz, G., Merk, H., and Rau, R. (1991) Insights into the mechanism of gold action provided by immunotoxicology: biooxidation of gold(I) to gold(III) detected from sensitized T-cells. *Rheumatol. Int.* 11:219–220.

Gortenuti, G., Parrinello, A., and Vicentini, D. (1985) Diffuse pulmonary changes caused by gold salt therapy: report of a case. *Diagn. Imaging Clin. Med.* 54:298–303.

Hall, C.L., Fothergill, N.J., Blackwell, M.M., Harrison, P.R., MacKenzie, J.C., and MacIver, A.G. (1987) The natural course of gold nephropathy: long term study of 21 patients. *Br. Med. J.* 295:745–748.

Healey, L.A. and Backes, M.B. (1989) Nitritoid reactions and angiotensin-converting-enzyme inhibitors. *N. Engl. J. Med.* 321:763.

HSDB (2004) *Gold, Elemental.* Bethesda, MD: Hazardous Substances Data Bank (HSDB), TOXNET®–Toxicology Data Network, Specialized Information Services, US Department of Health and Human Services. Available at http://toxnet.nlm.nih.gov/cgi-bin/sis/htmlgen?HSDB. CASRN: 7440-57-5; HSDB Number: 2125.

Jackson, C.W., Haboubi, N.Y., Whorwell, P.J., and Schofield, P.F. (1986) Gold induced enterocolitis. *Gut* 27:452–456.

Martin, D.M., Goldman, J.A., Gilliam, J., and Nasrallah, S.M. (1981) Gold-induced eosinophilic enterocolitis: response to oral cromolyn sodium. *Gastroenterology* 80:1567–1570.

McFadden, R.G., Traher, L.J., and Thompson, J.M. (1989) Gold-naproxen pneumonitis: a toxic drug interaction? *Chest* 96:216–218.

Rollins, S.D. and Craig, J.P. (1991) Gold-associated lymphadenopathy in a patient with rheumatoid arthritis: histologic and scanning electron microscopic features. *Arch. Pathol. Lab. Med.* 115:175–177.

Stuve, J. and Galle, P. (1970) Role of mitochondria in the handling of gold by the kidney: a study by electron microscopy and electron probe microanalysis. *J. Cell Biol.* 44:667–676.

Voil, G.W., Minielly, J.A., and Bistricki, T. (1977) Gold nephropathy: tissue analysis by X-ray fluorescent spectroscopy. *Arch. Pathol. Lab. Med.* 101:635–640.

19

IRON

Steve Morris and Jayme P. Coyle

Target organ(s) iron (elemental)[a]	No established target organs
Occupational exposure limits iron (elemental)	No established exposure limits
Reference values iron (elemental)	No established reference values
Risk/safety phrases iron (elemental)	No established risk/safety phrases

[a] For values for iron compounds, refer to Table 19.1.

BACKGROUND AND USES

Iron (symbol Fe; CASRN 7439-89-6) is the fourth most abundant element in the earth's crust (5%). It is a transition metal and is placed within the d-block of the periodic table and naturally occurs coordinated with other elements. It is mainly extracted from hematite (Fe_2O_3) and limonite ($Fe_2O_3 \cdot 3H_2O$), although other ores, such as magnetite (Fe_3O_4), siderite ($FeCO_3$), and taconite (an iron silicate), are also the key sources. Iron is used principally for structural materials, primarily steel, an iron carbon alloy. It is also used in magnets, dyes, pigments, abrasives, and polishing compounds (e.g., jeweler's rouge).

Apart from being an essential nutrient for erythropoiesis, iron is also important for the proper functioning of myoglobin, heme enzymes, and metalloflavoprotein enzymes (Hardman and Limbird, 1996). Iron oxide (CAS #1309-37-1) is a ubiquitous oxidized form of iron found. Iron chloride (CAS #7705-08-0) and other iron salts are widely used in industrial and chemical synthesis processes. Iron pentacarbonyl (iron carbonyl; CAS #13463-40-6) is used in organic synthesis as an anti-knock fuel additive and as high frequency coils in electronics (O'Neil, 2001).

PHYSICAL AND CHEMICAL PROPERTIES

Iron has a melting point and boiling point of 1538 °C [2800 °F] and 2862 °C [5182 °F], respectively. Iron compounds are generally solid at room temperature.

Iron oxide (Fe_2O_3) is a reddish brown solid that decomposes at 1566 °C and is insoluble in water. Rust is a nonprotective superficial oxidized product, which corrodes and mobilizes over time.

Iron pentacarbonyl ($Fe(CO)_5^{5+}$) is a colorless to dark red, oily water-insoluble liquid with a density of 1.49 g/cm^3 and boiling point of 103 °C. The vapor pressure and flash point of iron carbonyl are 40 mmHg at 30.3 °C (86.5 °F) and −15 °C (5.0 °F), respectively; the latter qualifies it as a class IB flammable liquid. Both air and light can decompose the compound to inorganic iron and gaseous carbon monoxide (NIOSH, 2005).

MAMMALIAN TOXICOLOGY

Acute Effects

Overexposure to therapeutic iron salts by means of ingestion may cause vomiting and intestinal bleeding. In severe cases, acidosis may occur with accompanying damage to the liver, kidneys, and cardiovascular system.

Hamilton & Hardy's Industrial Toxicology, Sixth Edition. Edited by Raymond D. Harbison, Marie M. Bourgeois, and Giffe T. Johnson.
© 2015 John Wiley & Sons, Inc. Published 2015 by John Wiley & Sons, Inc.

Chronic Effects

Chau et al. (1993) studied the mortality of French iron miners. These researchers reported no excess of lung cancer in nonsmokers, moderate smokers, or miners who had worked above or below the ground for <20 years. However, they found a significant excess of lung cancer in moderate smokers who had worked underground for 20–29 years (standardized mortality ratio [SMR] = 349) and in moderate and heavy smokers who had worked underground for over 30 years (SMR = 588 and 478, respectively). The authors attributed the higher risk of lung cancer to diesel engine emissions, dust, and explosions.

Sitas et al. (1989) investigated the mortality patterns in a cohort of iron workers. They found a higher mortality rate from cancer of the trachea, bronchus, and lungs in workers over the age of 65 (1.71 proportional mortality ratio). Workers under the age of 65 had an increased risk of nonmalignant respiratory disease (1.58 proportional mortality ratio). These findings were supported by Adzersen et al. (2003) from a large cohort of German iron foundry workers with an increased standardized mortality rate (151.7) for the same respiratory neoplasms; however, this cohort was not analyzed for either smoking habits or alcohol consumption.

Andjelkovich et al. (1995) studied the mortality of iron foundry workers exposed to formaldehyde. They found an association between lung cancer and exposure to silica but no association between exposure to formaldehyde and any type of mortality from respiratory system disorders. Westberg et al. (2013) studied cancer morbidity from quartz exposure in a Swedish cohort and reported significantly increased incidence of lung and pleural neoplasms (1.58 standardized incidence ratio), particularly with a latency period of 20 years or more. The cohort was incompletely assessed for smoking history, confounding exposures to solvents and respirable dusts, and a discernible dose–response trend could not be determined.

Studies addressing the effects of exposure to organic chemicals, such as isocyanates, in the iron industry are being increasingly researched (Westberg et al., 2005; Liljelind et al., 2010). One study conducted by Andersson et al. (2009) reported an 8-h TWA of 0.98 mg/m^3 respirable dust within a selection of Swedish iron foundries. These studies may offer insight into associations drawn between iron foundry workers and the observed morbidity and mortality figures obtained by the aforementioned studies.

A well-studied series of workers with siderosis showed no increase in morbidity or mortality from respiratory disease (Stanescu et al., 1967; Kleinfeld et al., 1969). Albu and Popescu (1973) found no impairment in pulmonary blood flow in 12 iron oxide pigment workers with radiographic nodular opacities. Morgan (1978) reported a case of pneumoconiosis in a worker exposed for 9 years to extremely dusty atmospheres in a mill that pulverized magnetite with a negligible silica content.

The primary complaints were cough and sputum. Although the chest X-ray showed several rounded opacities throughout both the lung fields, pulmonary function tests were normal. Severe cases of siderosis may be complicated with secondary fibrosis to result in dyspnea.

Mechanisms of Action

Considerable iron oxide dust may be caught in the respiratory tract defenses and expectorated in sputum, a process well known to exposed workers because of the rust color of phlegm. The remaining iron oxide accumulates in the lymphoid tissue along the bronchi, especially at the bifurcation.

Experiments on low level lung exposure to ferric oxide particles in human subjects demonstrated insignificant effects on pulmonary epithelial cells and pulmonary function with iron homeostasis abnormalities. Decreases in transferritin accompanied increases in ferritin and lactoferritin levels, which resolved by day 4 postexposure (Ghio et al., 1998; Lay et al., 2001). When inhaled, iron compounds can accumulate in alveolar macrophages, which may be inactivated if overly burdened.

Unbound iron (Fe^{2+}) in serum is a redox active metal, which readily undergoes Fenton reactions with endogenous oxygen species.

Chemical Pathology

The chest X-ray changes with this benign pneumoconiosis are easily confused with those of silicosis, hemosiderosis, and a number of diseases that produce bilateral densities. Typically, a confirmation of bilateral opacities related to iron particulate or fume exposure does not progress to a significant disability or clinical pathology and remains a rare outcome of exposure.

In a case series of 15 welders, CT scans revealed alveolar septal thickening suggestive of fibrosis. The histopathology confirmed the presence of hemosiderin via Prussian blue-positive lung sections and increased iron-containing alveolar macrophages. Energy dispersive X-ray analysis (EDX) confirmed iron, with an additional few peaks corresponding to low level aluminum and silicon deposition (Buerke et al., 2002).

SOURCES OF EXPOSURE

Occupational

Occupational siderosis is caused by exposure to iron oxide, for example, in paint pigments and jeweler's rouge. Other jobs that can give rise to siderosis are the manufacture of iron oxide and iron shot, the sieving and bagging of powder from emery rock, and grinding with an emery wheel.

TABLE 19.1   Exposure Limits and Threshold Limit Values for Iron Compounds

Iron Compound	CAS Number	PEL (mg/m^3)	REL (mg/m^3)	TLV (mg/m^3)
Iron oxide dust and fumes	1309-37-1	10	5	5
Iron pentacarbonyl	13463-40-6	None	0.23; 0.45[a]	0.23; 0.45*
Iron salts	Varied	None	1	1

Sources: American Conference of Governmental Industrial Hygienists (ACGIH) (2010) and National Institute for Occupational Safety and Health (NIOSH) (2005).

[a]Short-term exposure limit of 15 min.

PEL: permissible exposure limit, set by the Occupational Safety and Health Administration; REL: recommended exposure limit, set by the National Institute for Occupational Safety and Health; TLV: threshold limit value, set by the American Conference of Governmental Industrial Hygienists.

Workers using electric arc or oxyacetylene equipment in welding, cutting, grinding, or polishing are at risk, especially when working in closed spaces such as boilers or tanks or below the decks in ships. The high temperatures of welding operations produce an iron oxide fume that is readily inhaled in the absence of respiratory protection. Because of the heat and physical effort required in this work, it is often impossible for a worker to wear a useful mask except for very short periods.

Workers may be exposed to vapors of iron pentacarbonyl, as it is used in organic synthesis, electronics, and in motor fuels as an additive. Similarly, accidental spills of iron salts in manufacturing and chemical synthesis industry may allow for exposure.

Environmental

Old, unlined lead drinking water distribution pipes release iron compounds, such as hematite and iron hydroxide, over years of use, especially within areas supplying chlorinated water. Water ladened with iron compounds gives the characteristic "red water" (iron may be found in drinking water supplied by old pipes lined with iron (Sarin et al., 2004).

INDUSTRIAL HYGIENE

The exposure limits and threshold limit values for iron compounds are listed in Table 19.1. No established occupational limits have been established for elemental iron.

Iron workers may be at a risk of hearing loss because of their noisy work environment (Burton and Burr, 1993), and impaired hearing may result in poor balance thereby increasing the risk of accidents (Kilburn et al., 1992). For controlling particulates and fume, proper ventilation and, when necessary, respirator use is recommended.

MEDICAL MANAGEMENT

Evacuation from the area is recommended after heavy exposure to iron particulates and fume. Administration of oxygen is recommended if respiratory problems are observed (Currance et al., 2005).

In cases of acute iron salt exposure with associated high serum iron levels, chelation therapy with deferoxamine is recommended. Deferiprone and Exjade are also available as oral chelating agents for iron clearance. Thereafter, titration should be reduced gradually. Avoid chelation therapy for more than 24 hours due to the increased risk of injury.

REFERENCES

Adzersen, K.H., Becker, N., Steindorf, K., and Frentzel-Beyme, R. (2003) Cancer mortality in a cohort of male German iron foundry workers. *Am. J. Ind. Med.* 43:295–305.

Albu, A. and Popescu, H.I. (1973) Lung scanning in occupational siderosis. *Am. Rev. Respir. Dis.* 107:291–294.

American Conference of Governmental Industrial Hygienists (ACGIH) (2010) *Threshold Limit Values (TLVs) for Chemical Substances and Physical Agents and Biological Exposure Indices (BEIs)*, Cincinnati, OH: American Conference of Governmental Industrial Hygienists.

Andersson, L., Bryngelsson, I.L., Ohlson, C.G., Nayström, P., Lilja, B.G., and Westberg, H. (2009) Quartz and dust exposure in Swedish iron foundries. *J. Occup. Environ. Hyg.* 6:9–18.

Andjelkovich, D.A., Janszen, D.B., Brown, M.H., Richardson, R.B., and Miller, F.J. (1995) Mortality of iron foundry workers. IV. Analysis of a subcohort exposed to formaldehyde. *J. Occup. Environ. Med.* 37:826–837.

Buerke, U., Schneider, J., Rösler, J., and Woitowitz, H.J. (2002) Interstitial pulmonary fibrosis after severe exposure to welding fumes. *Am. J. Ind. Med.* 41:259–268.

Burton, N.C. and Burr, G.A. (1993) *Health hazard evaluation report.* HETA 92-157-2304, The General Castings Co., GRA&I 24, 1.

Chau, N., Benamghar, L., Pham, Q.T., Teculescu, D., Rebstock, E., and Mur, J.M. (1993) Mortality of iron miners in Lorraine (France): relations between lung function and respiratory symptoms and subsequent mortality. *Br. J. Ind. Med.* 50:1017–1031.

Currance, P.L., Clements, B., and Bronstein, A.C., editors (2005) *Emergency Care for Hazardous Materials Exposure*, 3rd ed., St. Louis, MO: Elsevier Mosby.

Ghio, A.J., Carter, J.D., Richards, J.H., Brighton, L.E., Lay, J.C., and Devlin, R.B. (1998) Disruption of normal iron homeostasis after bronchial instillation of an iron containing particle. *Am. J. Physiol.* 274:L396–L403.

Hardman, J.G. and Limbird, L.E., editors (1996) *Goodman & Gilman: The Pharmacological Basis of Therapeutics*, 9th ed., New York, NY: McGraw-Hill.

Kilburn, K.H., Warshaw, R.H., and Hanscom, B. (1992) Are hearing loss and balance dysfunction linked in construction iron workers? *Occup. Environ. Med.* 49:138–141.

Kleinfeld, M., Messite, J., Kooyman, O., and Shapiro, J. (1969) Welders' siderosis: a clinical roentgenographic and physiological study. *Arch. Environ. Health* 19:70–73.

Lay, J.C., Zeman, K.L., Ghio, A.J., and Bennett, W.D. (2001) Effects of inhaled iron oxide particles on alveolar epithelial permeability in normal subjects. *Inhal. Toxicol.* 13:1065–1078.

Liljelind, I., Norberg, C., Egelrud, L., Westberg, H., Eriksson, K., and Nylander-French, L.A. (2010) Dermal and inhalation exposure to methylene bisphenyl isocyanate (MDI) in iron foundry workers. *Ann. Occup. Hyg.* 54:31–40.

Morgan, W.K.C. (1978) Magnetite pneumoconiosis. *J. Occup. Environ. Med.* 20:762–763.

National Institute for Occupational Safety and Health (NIOSH) (2005) *Pocket Guide to Chemical Hazards*, NIOSH 2005-149, Washington, DC: U.S Department of Health and Human Services.

O'Neil, M.J., editor (2001) *The Merck Index—An Encyclopedia of Chemicals, Drugs, and Biologicals*, 13th ed., Whitehouse Station, NJ: Merck & Co., Inc.

Sarin, P., Snoeyink, V.L., Bebee, J., Jim, K.K., Beckett, M.A., Kriven, W.M., and Clement, J.A. (2004) Iron release from corroded iron pipes in drinking water distribution systems: effect of dissolved oxygen. *Water Res.* 38: 1259–1269.

Sitas, F., Douglas, A.J., and Webster, E.C. (1989) Respiratory disease mortality patterns among South African iron moulders. *Occup. Environ. Med.* 46:310–315.

Stanescu, D.C., Pilat, L., Gavrilescu, N., Teculescu, D.B., and Cristescu, I. (1967) Aspects of pulmonary mechanics in arc welders' siderosis. *Occup. Environ. Med.* 24:143–147.

Westberg, H., Andersson, L., Bryngelsson, I.-L., Ngo, Y., and Ohlson, C.-G. (2013) Cancer morbidity and quartz exposure in Swedish iron foundries. *Int. Arch. Occup. Environ. Health* 86:499–507.

Westberg, H., Löfstedt, H., Seldén, A., Lilja, B.G., and Nyström, P. (2005) Exposure to low molecular weight isocyanates and formaldehyde in foundries using hot box core binders. *Ann. Occup. Hyg.* 49:719–725.

20

LEAD

Frank D. Stephen

Target organs: Central/peripheral nervous system, cardio-vascular system, kidney, and blood

OSHA PEL: $50 \, \mu g/m^3$ (TWA) (lead); $0.075 \, mg/m^3$ general industry (tetraethyl lead), $0.1 \, mg/m^3$ construction (tetraethyl lead)

ACGIH BEI: $30 \, \mu g/100 \, ml$

ATSDR: $10 \, \mu g/dl$ blood lead action level in children

FDA: $0.005 \, mg/l$ bottled drinking water limit; $0.5–3.0 \, \mu g/ml$ action level (ceramic ware); $0.5–7.0 \, \mu g/ml$ action level (silver-plated hollowware)

BACKGROUND AND USES

Lead has had a multitude of practical uses for over 8000 years and reports of poisoning exist in all ancient civilizations, including Greece, Rome, and China. By the second century in Greece, lead was known to cause colic when swallowed, and lead intoxication also produced paralysis. Piping was made of lead as early as the Roman Empire, and even during Ceasar's reign it was noted that water from earthenware pipes tasted better than that delivered in pipes of lead (Prioreschi, 1998). Lead poisoning outbreaks were common in the Middle Ages due to the fortification of wine with lead-containing additives to improve the taste of poorly vented wines. The Devonshire colic of 1767 was attributed to the storage of cider in earthenware that had been painted with glaze, which contained lead. Colonial America did not escape the ravishes of lead poisoning. Rum distilled in vessels containing lead produced colic and in 1723 prompted the Massachusetts legislature to outlaw the distillation of rum and other hard liquors in lead pipes. French physician Luis Tanquerel des Planches, in 1839, examined 1217 cases of lead poisoning and observed a causative relationship between exposure to lead dust and central nervous system and renal toxicity. Even Beethoven may have been a victim of lead poisoning due to his abusive consumption of wine (Mai, 2006).

Currently, much of the lead used in the United States is found in vehicle batteries. Historically, in addition to pipes and batteries, metallic lead (or lead as an alloy) has been used industrially to produce ammunition, cable covers, weights, radiation shields, and solder. Compounds that include lead have been used in paints, glazes for ceramics, dyes, caulks, glass, and semiconductors, but ironically lead was never used in "lead" pencils. In the United States, tetraethyl- and tetramethyl lead were used as gasoline additives to boost octane ratings and reduce engine "knock" until banned in 1996. Publically, there is a great amount of public awareness with regard to the poisoning of children who have been exposed to old paint, putty, or soils contaminated with lead.

The presence of lead in the human body is considered to be a sign of environmental pollution since lead is not known to have any biological use to humans. Industrial lead intoxication in the United States and elsewhere still occurs. Although acute lead poisoning is almost nonexistent today, low level exposure in some areas of the world is still persistent. Blood lead levels are significantly higher than during preindustrial history, and our understanding of the significance of subclinical blood lead levels is greater. Today, there is no threshold of blood lead levels that is considered safe (Sanborn et al., 2002).

PHYSICAL AND CHEMICAL PROPERTIES

Occurring naturally in the earth's crust, lead (symbol Pb; CAS #7439-92-1) is a heavy bluish-gray metal that is

Hamilton & Hardy's Industrial Toxicology, Sixth Edition. Edited by Raymond D. Harbison, Marie M. Bourgeois, and Giffe T. Johnson.
© 2015 John Wiley & Sons, Inc. Published 2015 by John Wiley & Sons, Inc.

TABLE 20.1 Physical and Chemical Properties of Lead

Molecular weight	207.2
Color	Bluish gray
Physical state	Solid
Flash point	No data
Melting point	327.46 °C
Boiling point	1740 °C
Density at 20 °C	11.34 g/cm^3
Water solubility	Insoluble
Vapor pressure	1.77 mmHg at 1000 °C

lustrous when freshly cut. It is rarely found as a pure metal but rather is complexed with other elements to form lead compounds. Found in ore with copper, zinc, and silver, lead is found in mineral form as galena (PbS), anglesite ($PbSO_4$), and cerussite ($PbCO_3$). It is easily malleable, smelted, and can be added to other metals to form alloys. Resistant to air and water corrosion, it does not mix easily with many solvents but will react with hot acids such as nitric and sulfuric. It has 4 naturally occurring isotopes as well as 17 that have been produced experimentally. Burning with a bluish-white flame, powdered lead displays pyrophoricity and releases toxic fumes when burned. Important physical and chemical properties are listed in Table 20.1.

ENVIRONMENTAL FATE AND BIOACCUMULATION

Lead is found naturally in the earth's crust and can be dispersed throughout the environment as a result of anthropogenic activities, volcanic eruptions, and wind and soil erosion. Anthropogenic activities include mining and smelting operations. Once entrained in the air, lead can be removed by gravitational settling and precipitation. Eventually deposited in the soil, lead adsorbs tightly and rarely leaches to the subsoil or groundwater. In the environment, lead compounds can be converted to other forms of lead compounds. It is constantly cycled through all phases of the environment by such processes as wind spread of dust, stream and river flow, precipitation, and subsequent runoff and weathering.

Air In the air, lead exists naturally as $PbSO_4$ and $PbCO_3$. Lead particles in air can also be emitted from anthropogenic sources such as smelting industries and mining (in the form of inorganic salts such as $PbSO_4$ and PbS) and vehicle/automobiles exhaust (in the form of halides such as $PbBrCl$ and $2PbBrCl \cdot NH_4Cl$) (EPA, 1986).

Water Since lead is found in multiple chemical forms, its chemistry in water can be complex. Preferentially, lead will combine with the anions found in water to form compounds that have low solubility. In freshwater, the divalent form

(Pb^{2+}) of lead is the most stable ionic species; however, $PbCO_3$ forms under alkaline conditions. It can be removed from freshwater through natural settling into sediment or through absorption to polar particulates. In saltwater, lead chloride and carbonate are the predominant species. Lead will form stable complexes with organic matter in water. Species of organic lead (e.g., tetraethyl lead) can degrade from tetraalkyl forms to trialkyl, dialkyl, and finally inorganic lead oxides. Volatilization or photolysis also occurs.

Soil Lead naturally occurs in the soil as a constituent of the earth's crust with additional human contribution coming from lead-containing products (e.g., paint). From natural sources the accumulation of soil lead is dependent on the atmospheric deposition. Once adsorbed to soil particles, lead's soil presence is rather stable but can be affected by chemical exchange with other minerals. Its stability prevents lead from leaching beyond upper levels. It can also form stable organic compounds, which chelate with organic matter.

Atmospheric lead most often enters the soil as lead sulfur compounds (e.g., $PbSO_4$ or PbS). Soil conditions/composition will dictate what types of reactions lead will undergo. Almost all lead released to the atmosphere from human activity will be reactive and after transformation will be adsorbed to soil. The amount of soluble lead in soil is dependent on the pH and the availability of carbonates and phosphates. Organic lead compounds will remain in soil after deposition but can be converted microbially to water-soluble forms, which may in turn leach.

MAMMALIAN TOXICOLOGY

Acute Effects

Symptoms of acute lead poisoning can be found in many different body systems. The encephalopathic effects of central nervous system lead intoxication have been known for hundreds of years. Acute neurological symptoms in adults generally result from inhalation of lead dust or fumes while effects in children occur after ingestion.

Neurobehaviorally, exposed individuals exhibit headaches, hallucinations, poor attention spans, dizziness, irritability, forgetfulness, fatigue, lethargy and general weakness, impotence, decreased libido, paresthesia, and tremors in peripheral muscles due to poor nerve conduction. Similar symptoms have been reported in children exposed to lead containing paint chips, putty, and contaminated soils.

Children absorb lead much faster than adults, thus, the end points of exposure to children are of greater concern. A review conducted in the early 1970s by the National Academy of Sciences (NAS) (NAS, 1972) found that children suffered neurological symptoms such as convulsions, stupor, ataxia, and in some cases even coma and death. Further, children who suffer from lead intoxication were found to have permanent neurological and cognitive impairments. Encephalopathies in children were found to correlate with blood lead levels in excess of 80 μg/dl (Bradley et al., 1956; Bradley and Baumgartner 1958; Rummo et al., 1979; Smith et al., 1983).

Chronic Effects

Almost all human physiological systems have been shown to suffer from negative chronic effects due to long-term lead exposure.

Neurological Lead exposures have been associated with behavioral effects for centuries (Whitfield et al., 1972). Exposure in the workplace produces chronic, long-term negative neuropsychological outcomes such as conflict in interpersonal relationships; poor performance on neurobehavioral tests; disturbances in mood; decreases in memory, IQ test, and cognitive performance; nervousness; ability to cope; reaction time; visual motor performance; and hand dexterity. Kumar et al. (1987) reported that these conditions can then worsen resulting in delirium, convulsions, coma, and death. Traditionally, blood lead levels ranging from 50 to 300 μg/dl were thought to be indicative of acute poisoning (Smith et al., 1938), but blood lead levels as low as 10 μg/dl have been associated with decreased IQ scores and increased aggression (Guidotti and Ragain, 2007).

Neuropsychological effects have been observed in many lead industry workers with blood lead levels between 40 and 80 μg/dl. One difficulty when trying to determine a consensus as to the neurobehavior of adults in response to lead exposure was in the varied methodology employed in these studies. A review of various studies shows that blood lead levels of approximately 60 μg/dl were correlated with decreased reaction time (NIOSH, 1974; Stollery et al., 1989, 1991). High blood and bone lead levels were also positively associated with long-term decreases in cognitive function. In a study of 427 Canadian lead workers, Lindgren et al. (1996) found that visuomotor skills were decreased in those highly exposed. Lead-exposed workers were more likely to get into interpersonal conflicts with fellow coworkers (Parkinson et al., 1986).

A meta-analysis performed by Meyer-Baron and Seeber (2000) found that blood lead concentrations ranging from 30 to 50 μg/dl decreased scores on various behavioral tests and that these decreases were equal to that of people who had aged, without lead exposure, 20 years. Schwartz et al. (2005) followed a cohort of 576 lead workers for 5 years investigating the relationship between blood and tibia lead levels and scores on neurobehavioral tests. They found decreases in lead-exposed individuals at the start of the study and a continued decrease over the course of the study as compared

to the nonexposed control group with respect to decision making and manual dexterity.

Renal With over 40 reviews of occupational studies examined, a relationship between lead exposure and renal function has been well established. Lead nephropathy is an irreversible renal disease that develops over long periods of excessive exposure (Landrigan, 1990). Characterized by a progressive destruction of proximal tubular cells and ensuing interstitial fibrosis, typical lead-induced renal disease results in suppressed glomerular filtration rate, proteinuria/enzymuria, and impaired glucose absorption.

Glomerular filtration rate decreases, as measured by increased serum creatinine concentration, have been used as a barometer of renal dysfunction as related to lead exposure. Payton et al. (1994) reviewed data from 744 subjects and found a decrease in creatinine clearance of 10.4 ml/min (11% from the group mean). Kim et al. (1996) showed an association between blood lead and serum creatinine concentrations in 459 middle-aged men. Increased serum creatinine concentrations were found in subjects with blood lead level concentrations as low as 10 μg/dl (Kim et al., 1996; Muntner et al., 2003). Finally, in a longitudinal study conducted by Tsaih et al. (2004), a 4–8-year follow-up of 448 individuals (mean age 66 years old) found a positive association between blood lead levels and serum creatinine concentrations.

Cardiovascular The available literature is replete with epidemiological studies examining the relationship between blood pressure and body lead burdens. A complete review would be beyond the reach of this chapter. These types of studies pose difficulty for the researcher. The need to control for many confounders and covariables, genetic predisposition, other lifestyle factors and the choice of body lead repository as well as one time blood pressure measurement makes ardent conclusions from the data difficult.

Several meta-analyses were performed, and the results showed that a direct relationship between systolic pressure and increased lead burden exists. This effect is seen more in persons of middle age rather than the young. Pressure increases are about 1 and 0.6 mmHg, for systolic and diastolic, respectively, for every doubling of blood lead concentration (Nawrot et al., 2002; Staessen et al., 1994) A third study found a 1.25 mmHg systolic increase (Schwartz, 1995). However, a study conducted by Staessen using multiple blood pressure measurements did not find this association (Staessen et al., 1996).

Longitudinal studies have examined the relationship of the source of body lead burden and hypertension with slightly differing results. Blood lead levels were found to increase diastolic pressure (Batuman et al., 1983; dos Santos et al., 1994). Weiss et al. (1986, 1988) found that blood lead levels above 30 μg/dl were predictive of an increase in systolic blood pressure in Boston police officers. However, a study of

70 workers at a battery manufacturing plant found no significant correlation between blood lead levels and blood pressure during a 6- to 10-year follow-up period (Navah et al., 1996).

In a study of 590 individuals, Hu et al. (1996) found that average patellar blood lead levels were significantly higher in hypertensive individuals. Similar results were observed in the Nurses Health Study (Korrick et al., 1999). In contrast, Cheng et al. (2001) found a positive association between tibial lead levels and increased blood pressure but no such association was found in blood or patellar lead levels. Finally, Nash et al. (2003) found a positive association between high blood lead levels and hypertension risk in middle-aged women.

A study of 496 workers at an organic lead-manufacturing plant found a positive association between blood and tibial lead levels and systolic blood pressure (Glenn et al., 2003).

Hematological   Anemia may result from decreased hemoglobin content in red blood cells as well as hemolysis (resulting in hematouria). Lead alters the function of three different enzymes, δ-aminolevulinic acid dehydratase (ALAD), δ-aminolevulinic synthetase (ALAS), and ferrochelatase, which interrupt the normal production of heme. The activity of ALAD is also a sensitive indicator of lead exposure. Finally, renal damage may also cause lead-induced anemia due to a decrease in erythropoietin production, which results in poor maturation of erythroid progenitor cells (Osterode et al., 1999).

Musculoskeletal   Lead workers with blood lead levels higher than 40 μg/dl reported muscle soreness, weakness, cramps, and pain (Rosenman et al., 2003; Holness and Nethercott, 1988; Marino et al., 1989; Matte et al., 1989; Pagliuca et al., 1990). Bone is the largest repository of lead in the body accounting for up to 75–95% of the body burden (Barry, 1975). Surprisingly, bone mineral density was increased in a cohort of a highly lead exposed boys in New York City (Campbell et al., 2004) leading to the speculation that lead accelerates bone maturation. Increased blood lead levels are also associated with dental cavities in children (Moss et al., 1999; Campbell et al., 2000; Gemmel et al., 2002). A review of over 10,000 people enrolled in the NHANES III study found that increasing blood lead levels were positively associated with increased periodontal bone loss (Dye et al., 2002).

Gastrointestinal   As described earlier in this chapter, lead ingestion causes colic and abdominal distress. Other gastrointestinal signs and symptoms include decreased appetite (and resultant weight loss), abdominal pain, constipation, diarrhea, nausea, vomiting. Shock may result if there is a loss of water from the gastrointestinal tract.

Immunological   While no conclusive data exist to demonstrate a clinical significance of lead exposure to the immune system, a small bank of research shows that specific end points are affected by lead exposure. Cellular immune parameters such as cellular subpopulations, mitogenic response have all been shown to be altered in lead-exposed workers. Workers exposed to lead had decreased CD4+ T lymphocyte subpopulations (Fischbein et al., 1993; Ündeger et al., 1996; Basaran and Ündeger, 2000) and had decreased response to concanavalin A or phytohemagglutinin-induced mitogenesis (Alomran and Shleamoon, 1988; Fischbein et al., 1993; Mishra et al., 2003).

Reproductive   The reproductive health of men and women as relating to lead exposure for men and women has been evaluated. An inverse relationship between exposure and spermatogenesis exists at blood lead levels exceeding 40 μg/dl. Data also exist demonstrating a relationship between moderately high blood lead levels and spontaneous abortion and preterm delivery. Although associated with adverse reproductive outcomes, there are not data suggesting that lead is a human teratogen (Winder, 1993).

Sallmén et al. (2000) found an inverse relationship between blood lead levels and sperm production in 2111 Finnish lead workers leading to a delay in conception but not decreasing success. Decreased sperm production was also related to those who had been exposed longer occupationally (Gennart et al., 1992). Alterations in sperm number, chromatin quality and physical abnormalities, as well as decreased levels of serum testosterone were found (Alexander et al., 1998). Some of these observed effects might be due to a direct mechanism of action on the testis. Ng et al. (1991) studied 122 lead battery workers in Singapore and found increased serum luteinizing hormone (LH) and follicle-stimulating hormone (FSH) levels—indicative of Leydig and Sertoli cell damage—while no decrease in testosterone level was found.

Nordstrom et al. (1979) reported that women working in a smelter plant in Sweden had a higher incidence of spontaneous abortions than women who became pregnant before being employed or lived further than 10 km away. Further, abortion rates were higher for women who worked in the most contaminated areas of the plant. Borja-Aburto et al. (1999) studied 668 women in Mexico City and found that there was an increased risk of spontaneous abortion related to blood lead levels.

Chemical Pathology

Responsible for a myriad of health effects, lead has no known biochemical or physiological role in the human body. Entrance into the body is dependent on the form of lead. Inorganic forms of lead only enter the body through inhalation and ingestion whereas tetraethyl lead penetrates the skin and can be inhaled. Less than 20% of ingested inorganic lead gets absorbed in adults, but pregnant women and children may absorb as much as 50%. Inhalation results in about 40% of lead being deposited in the lungs and almost all of that lead enters into the bloodstream.

Once having entered the body, lead preferentially partitions to blood, soft tissue, and bone. Deposition time can be measured in days (for blood) to years (for bone). In fact, lead deposited into bone, teeth, nails, and hair is considered safe since that lead is not readily available to other tissues. However, lead can be reintroduced into the blood long after original exposures if released during bone remodeling. While not a concern to adults, this is of particular concern in children whose bones are constantly remodeling during their adolescent years. Lead is also stored very well in liver, kidney, and neural tissue.

Mode of Action(s)

Neurological An enormous number of studies have been published reporting the effects of lead neurotoxicity. Attempting to adequately review all of those studies would be well beyond the intended scope of this chapter and might require the dedication of an entire text to the subject. The following section will discuss the effects of lead on enzyme and neurotransmitter systems.

Enzymatic activation by lead is accomplished by mimicking the action of calcium. Calmodulin is an enzyme that requires a calcium-induced conformational change to become active. Lead mimics calcium by binding to and inappropriately activating calmodulin resulting in altered cAMP second messenger systems and increased protein phophorylation.

Protein kinase C (PKC) is a family of calcium-dependent enzymes that phosphorylates serine and threonine residues in proteins. In neurons involved in spatial learning and memory processing, lead can mimic calcium by binding to and activating PKC-γ. Potentially, the activation of this enzyme during development could lead to a disruption in the formation of the blood–brain barrier.

Similar to lead's ability to activate calmodulin and PKC, lead also has the ability to mimic zinc. Sp1 is a zinc finger transcriptional regulator that when bound to lead will alter the expression of certain genes such as the β-amyloid precursor protein (APP) gene. Interestingly, the lead-induced expression of APP was only observed during neonatal rat brain development while induction was unaltered in older rats.

It is well documented that lead can affect almost all neurotransmitter systems in the brain. Cholinergic, glutaminergic, and dopaminergic systems all play a role in memory formation and each can be affected by lead exposure.

While results from several studies are inconsistent, lead has been found to suppress acetylcholine release thereby decreasing cholinergic function. Acetylcholine release at the neuromuscular junction is dependent on the calcium influx into the synaptic termini. Lead not only blocks action potential-induced calcium influx into the termini but also blocks the binding of acetylcholine to nicotinic receptors.

Lead exposure decreased hippocampal neuron function by blocking glutamate release without affecting subsequent binding to *N*-methyl-D-aspartate (NMDA) receptors (in fact, the number of receptors may be increased in an attempt to compensate for decreased glutamate release). These effects were observed to be greater when exposure occurred during development than after weanling but continued into adulthood even after lead was absent. This suggests a role for lead interfering with normal synaptogenesis.

Dopaminergic systems also have a role in memory and have been shown to be sensitive to lead exposure. Presynaptically, lead appears to decrease dopamine synthesis and release. A compensatory increase on postsynaptic neurons has been observed. Lead may also affect brain reward centers by decreasing dopamine release in the mesolimbic pathway of the frontal cortex and nucleus accumbens.

Hypertension Lead's mechanism of action appears to be dependent on the tissue in which it accumulates. As discussed previously in this chapter, lead may induce hypertension in exposed workers. Vaziri et al. (1997) and Gonick et al. (1997) demonstrated in rats that lead-depleted nitric oxide levels contributed to a decrease in both the peripheral (vasodilation) and central (anti-sympathetic) control of blood pressure regulation. Reactive oxygen species are speculated as being the causative agent in decreasing nitric oxide levels. Lead may also interfere with the ability of nitric oxide to cause vasodilation. Decreased vascular β-adrenergic receptor density and elevated levels of norepinephrine may also contribute to the sympathetic nervous system control of hypertension.

Renal Several studies have indicated that lead preferentially forms inclusion bodies in the nucleus of proximal renal tubule cells as it partitions from the cytoplasmic to nuclear fractions (Choie and Richter, 1972; Goyer et al., 1970a, 1970b). Putnam (1986) has described how the development of intranuclear inclusion bodies in the cells of the proximal renal tubules is used as a marker for lead nephrotoxicity.

Decreased capsular blood flow and glomerular filtration rate have been observed in rats exposed to lead but the mechanisms are not known (Aviv et al., 1980; Khalil-Manesh et al., 1992a, 1992b). Urinary *N*-acetyl-beta-D-glucosoaminidase (NAG), another proposed marker for early lead detection, results from lead-induced oxidative stress possibly related to the depletion of nitric oxide. Finally, mitochondria in the cells of proximal renal tubules accumulate lead resulting in decreased pyruvate/malate- or succinate-mediated respiration (Fowler et al., 1980; Oskarsson and Fowler, 1985).

Hematological As described earlier in this chapter, the inhibition of three enzymes, ALAS, ALAD, and ferrochelatase, leads to decreased heme synthesis. This results in microcytic anemia due to low hemoglobin levels. During the production of heme, lead is able to bind to ALAS and decrease the production of aminolevulinic acid. Further, much like mercury, lead is able to bind sulfhydryl groups on proteins. Lead binds to vicinal sulfhydryls in the active site of ALAD. As zinc is needed for ALAD activity, it only binds to one sulfhydryl. Lead, on the other hand, cross-bridges the sulfhydryls thereby decreasing the production of porphobilinogen. Finally, lead is also able to bind vicinal sulfhydryl groups on the mitochondrial enzyme ferrochelatase resulting in a decrease in the amount of heme produced.

SOURCES OF EXPOSURE

Due to strict regulations, the use of many of the anthropogenic sources of lead has been significantly decreased or eliminated altogether. Lead is an elemental metal and as such cannot degrade; discarded sources of lead leave a footprint indicative of use resulting in higher than expected levels of lead.

Occupational

Occupational exposure to lead may occur during the manufacture, storage, and recycling of batteries; the preparation and use of lead-based ceramic glaze and paint; the addition of lead to gasoline (now banned in the United States); printing operations; the production of capacitors, resistors, and semiconductors; and smelting processes.

Exposure may also occur, during the construction of tank linings, piping, and other equipment used to carry corrosive gases and liquids, when protective clothing is worn to shield against radiographic rays and atomic radiation and in the working of steel and other metals (Budavari et al., 1989).

Construction activities such as residential lead remediation, dismantling of steel structures, and painting/sanding of structures containing lead are also sources of potential lead exposure (Grimsley and Adams-Mount, 1994; Rabin et al., 1994). Other potential occupational exposures to lead include: firearms instructors and students, automobile/radiator repair workers.

Environmental

The primary source of non-soil exposure for many individuals has been human emission into the environment, most notably due to leaded gasoline. As late as 1984 vehicle emissions accounted for over 90% of all environmental anthropogenic lead. The use of lead as an additive in gasoline was banned in the United States as of December 31, 1995.

The general population is exposed to lead in food, soil, ambient air, and drinking water. Due to the presence of lead in soil, children are at a greater risk from this source of exposure as are individuals living in urban areas, near former mining operations, smelting plants, and sites of production and/or disposal. Non-occupationally, people can still be exposed to lead in some paints, ammunition, solder, and pesticides. Other potential sources of lead exposure include environmental tobacco smoke, renovation/remodeling of older homes with lead-based paint and/or putty, consumption of produce from contaminated gardens, recreational use of shooting ranges, pica, and secondary occupational exposure.

INDUSTRIAL HYGIENE

Exposure Limits

ACGIH   The American Conference of Industrial Hygienists (ACGIH) has established a threshold limit value (TLV) of 0.05 mg/m^3 for inorganic lead (ACGIH, 2004).

NIOSH   The National Institute for Occupational Safety and Health (NIOSH) has established a recommended exposure limit (REL) (10-h time-weighted average (TWA) based on a 40-h workweek) for lead of 0.05 mg/m^3 and an Immediately Dangerous to Health or Life (IDHL) value of 100 mg/m^3 (NIOSH, 2005).

OSHA   The Occupational Safety and Health Administration (OSHA) has established a permissible exposure limit (PEL) (8-h TWA) of 50 μg/m^3 (0.006 ppm) for lead (OSHA, 2005d), 0.075 mg/m^3 for the general industry use of tetraethyl lead (OSHA, 2005c), and 0.1 mg/m^3 for construction and all other uses of tetraethyl lead (OSHA, 2005a and b). Supplemental control of lead exposure, provided in the form of respirators, must be instituted when lead exposure levels cannot be maintained at ≤50 μg/m^3.

RISK ASSESSMENTS

Listings as Hazardous Waste

Lead has been classified as a hazardous waste and has been registered in various publications as such including:

Section 112 of The Clean Air Act lists mercury compounds as one of the 188 chemicals inventoried as a hazardous air pollutant.

Section 313 of Superfund Amendments and Reauthorization Act (SARA) designate lead and lead compounds as hazardous substances.

Lead and lead compounds, including tetraethyl lead, are subject to the Emergency Planning and Community Right-to-Know-Act (EPCRA)

Section 307 of The Clean Water Act lists lead as a "priority pollutant."

Lead acetate (U144) and lead phosphate (U145) and tetraethyl lead (P110) are listed as a hazardous waste under Resource Conservation and Recovery Act (RCRA).

IRIS   The Environmental Protection Agency's (EPA) Integrated Risk Information System (IRIS) oral and carcinogenicity assessments for lead were revised in 2005. Based on the human evidence, the EPA has assigned lead a weight-of-evidence carcinogen classification of B2, probable human carcinogen, based on the inadequate information in humans and sufficient data in animals (IRIS, 2005). The EPA has not developed a lead reference concentration (RfC) and decided that since some health effects attributed to lead exposure occur at blood lead levels as low as to be essentially without a threshold the EPA has not developed a reference dose (RfD) for inorganic lead (IRIS, 2005).

ATSDR   The EPA determined that some health effects attributed to lead exposure occur at blood lead levels "as low as to be essentially without a threshold." Thus, the EPA did not develop a RfD or derive a RfC for lead and lead-containing compounds. Likewise, the Agency for Toxic Substances and Disease Registry (ATSDR) did not derive a Minimal Risk Level (MRL) for lead (IRIS, 2005). However, ATSDR has established a blood lead action level in children of 10 µg/dl (ATSDR, 1997).

IARC   The International Agency for Research on Cancer (IARC) review of lead as a carcinogen was last revised in 2004. Based on limited human studies, this review listed organic lead as a Group 3 carcinogen: not classifiable as to their carcinogenicity to humans whereas inorganic lead compounds are listed as Group 2A carcinogens: probably carcinogenic to humans (IARC, 2004).

ACGIH   The ACGIH has listed lead as an A3 carcinogen: confirmed animal carcinogen with unknown relevance to humans (ACGIH, 2004). The biological exposure index (BEI) for blood lead has been established at 30 µg/100 ml.

NTP   The National Toxicology Program (NTP) has listed lead as reasonably anticipated to be a human carcinogen (NTP, 2005).

EPA   The EPA has not established a RfD or RfC for lead. Lead has been listed as a B2 carcinogen: probable human carcinogen (IRIS, 2005). Finally, the EPA has established a safe drinking water standard Maximum Contaminant Level (MCL) of zero and a Maximum Contaminant Level Goal (MCLG) action level of 0.015 mg/l for lead (EPA, 2002).

CalEPA   The California Environmental Protection Agency (CalEPAin October 1992 listed lead and lead-containing compounds as a human carcinogen under Proposition 65.

FDA   The United States Food and Drug Administration (FDA) limits lead levels in bottled drinking water at 0.005 mg/l (FDA, 2004). The FDA has set action levels for lead in ceramic ware (0.5–3.0 µg/ml) and silver-plated hollowware (0.5–7.0 µg/ml) (FDA, 2004).

REFERENCES

ACGIH (2004) Lead. In: *Threshold Limit Values for Chemical Substances and Physical Agents and Biological Exposure Indices*, Cincinnati, OH: American Conference of Governmental Industrial Hygienists.

Alexander, B.H., Checkoway, H., Faustman, E.M., van Netten, C., Muller, C.H., and Ewers, T.G. (1998) Contrasting associations of blood and semen lead concentrations with semen quality among lead smelter workers. *Am. J. Ind. Med.* 34:464–469.

Alomran, A.H. and Shleamoon, M.N. (1988) The influence of chronic lead exposure on lymphocyte proliferative response and immunoglobulin levels in storage battery workers. *J. Biol. Sci. Res.* 19:575–585.

Agency for Toxic Substances and Disease Registry (1997) *Public Health Statement for Lead*, Atlanta, GA: U.S. Department of Health and Human Services, Public Health Service, Agency for Toxic Substances and Disease Registry.

Aviv, A., John, E., Bernstein, J., Goldsmith, D.I., and Spitzer, A. (1980) Lead intoxication during development: its late effect on kidney function and blood pressure. *Kidney Int.* 17:430–437.

Barry, P.S.I. (1975) A comparison of concentrations of lead in human tissue. *Br. J. Ind. Med.* 32:119–139.

Basaran, N. and Ündeger, U. (2000) Effects of lead on immune parameters in occupationally exposed workers. *Am. J. Ind. Med.* 38:349–354.

Batuman, V., Landy, E., Maesaka, J.K., and Wedeen, R.P. (1983) Contribution of lead to hypertension with renal impairment. *N. Engl. J. Med.* 309:17–21.

Borja-Aburto, V.H., Hertz-Picciotto, I., Lopez, M.R., Farias, P., Rios, C., and Blanco, J. (1999) Blood lead levels measured prospectively and risk of spontaneous abortion. *Am. J. Epidemiol.* 150:590–597.

Bradley, J.E. and Baumgartner, R.J. (1958) Subsequent mental development of children with lead encephalopathy, as related to type of treatment. *J. Pediatr.* 53:311–315.

Bradley, J.E., Powell, A.E., Niermann, W., McGrady, K.R., and Kaplan, E. (1956) The incidence of abnormal blood levels of lead in a metropolitan pediatric clinic: with observation on the value of coproporphyrinuria as a screening test. *J. Pediatr.* 49:1–6.

Budavari, S., O'Neil, M.J., Smith, A., et al. editors (1989) *The Merck Index. An Encyclopedia of Chemicals, Drugs, and Biologicals*, 11th ed., Rahway, NJ: Merck & Co., Inc., pp., 851–854.

Campbell, J.R., Moss, M.E., and Raubertas, R.F. (2000) The association between caries and childhood lead exposure. *Environ. Health Perspect.* 108(11):1099–1102.

Campbell, J.R., Rosier, R.N., Novotny, L., and Puzas, J.E. (2004) The association between environmental lead exposure and bone density in children. *Environ. Health Perspect.* 112(11):1200–1203.

Cheng, Y., Schwartz, J., Sparrow, D., Aro, A., Weiss, S.T., and Hu, H. (2001) Bone lead and blood lead levels in relation to baseline blood pressure and the prospective development of hypertension. *Am. J. Epidemiol.* 153(2):164–171.

Choie, D.D. and Richter, G.W. (1972) Lead poisoning: rapid formation of intranuclear inclusions. *Science* 177:1194–1195.

dos Santos, A.C., Colacciopo, S., Dal Bó, C.M., and dos Santos, N.A. (1994) Occupational exposure to lead, kidney function tests, and blood pressure. *Am. J. Ind. Med.* 26(5):635–643.

Dye, B.A., Hirsch, R., and Brody, D.J. (2002) The relationship between blood lead levels and periodontal bone loss in the United States, 1988–1994. *Environ. Health Perspect.* 110(10):997–1002.

EPA (Environmental Protection Agency) (1986) *Air quality criteria for lead, June 1986 and Addendum, September 1986.* Research Triangle Park, NC, EPA 600/8-83-018F.

EPA (2002) *National Primary Drinking Water Regulations*, Washington, DC: U.S. Environmental Protection Agency, EPA816F02013. Available at http://www.epa.gov/safewater/mcl.html (accessed February 15, 2005).

FDA (2004) *Bottled Water*, U.S. Environmental Protection Agency, Code of Federal Regulations, 21 CFR 165.110. Available at http://a257.g.akamaitech.net/7/257/2422/12feb20041500/edocket.access.gpo.gov/cfr_2004/aprqtr/pdf/21cfr165.110.pdf (accessed September 22, 2007).

Fischbein, A., Tsang, P., Luo, J., Roboz, J.P., Jiang, J.D., and Bekesi, J.G. (1993) Phenotypic aberrations of CD3 and CD4 cells and functional impairments of lymphocytes at low-level occupational exposure to lead. *Clin. Immunol. Immunopathol.* 66:163–168.

Fowler, B.A., Kimmel, C.A., Woods, J.S., McConnell, E.E., and Grant, L.D. (1980) Chronic low-level lead toxicity in the rat. III. An integrated assessment of long-term toxicity with special reference to the kidney. *Toxicol. Appl. Pharmacol.* 56:59–77.

Gemmel, A., Tavares, M., Alperin, S., Soncini, J., Daniel, D., Dunn, J., Crawford, S., Braveman, N., Clarkson, T.W., McKinlay, S., and Bellinger, D.C. (2002) Blood lead level and dental caries in school-age children. *Environ. Health Perspect.* 110(10):625–630.

Gennart, J.P., Buchet, J.P., Roels, H., Ghyselen, P., Ceulemans, E., and Lauwerys, R. (1992) Fertility of male workers exposed to cadmium, lead or manganese. *Am. J. Epidemiol.* 135:1208–1219.

Glenn, B.S., Stewart, W.F., Links, J.M., Todd, A.C., and Schwartz, B.S. (2003) The longitudinal association of lead with blood pressure. *Epidemiology* 14:30–36.

Gonick, H.C., Ding, Y., Bondy, S.C., Ni, Z., and Vaziri, N.D. (1997) Lead-induced hypertension: interplay of nitric oxide and reactive oxygen species. *Hypertension* 30:1487–1492.

Goyer, R.A., Leonard, D.L., Moore, J.F., Rhyne, B., and Krigman, M.R. (1970a) Lead dosage and the role of the intranuclear inclusion body. *Arch. Environ. Health* 20:705–711.

Goyer, R.A., May, P., Cates, M.M., and Krigman, M.R. (1970b) Lead and protein content of isolated intranuclear inclusion bodies from kidneys of lead-poisoned rats. *Lab. Invest.* 22(3):245–251.

Grimsley, E.W. and Adams-Mount, L. (1994) Occupational lead intoxication: report of four cases. *South. Med. J.* 87:689–691.

Guidotti, T.L. and Ragain, L. (2007) Protecting children from toxic exposure: three strategies. *Pediatr. Clin. North Am.* 54(2):227–235.

Holness, D.L. and Nethercott, J.R. (1988) Acute lead intoxication in a group of demolition workers. *Appl. Ind. Hyg.* 3:338–341.

Hu, H., Aro, A., Payton, M., Korrick, S., Sparrow, D., Weiss, S.T., and Rotnitzky, A. (1996) The relationship of bone and blood lead to hypertension. The Normative Aging Study. *JAMA* 275:1171–1176.

IARC (2004) *Overall evaluations of carcinogenicity to humans: as evaluated in IARC Monographs volumes 1–82 (at total of 900 agents, mixtures and exposures).* Lyon, France: International Agency for Research on Cancer. Available at http://www-cie.iarc.fr/monoeval/crthall.html (accessed February 15, 2005).

IRIS (2005) Lead. In: *Integrated Risk Information System*, Washington, DC: U.S. Environmental Protection Agency. Available at http://www.epa.gov/iris/ (accessed March 26, 2005).

Khalil-Manesh, F., Gonick, H.C., Cohen, A.H., Alinovi, R., Bergamaschi, E., Mutti, A., and Rosen, V.J. (1992a) Experimental model of lead nephropathy. I. Continuous high-dose lead administration. *Kidney Int.* 41:1192–1203.

Khalil-Manesh, F., Gonick, H.C., Cohen, A., Bergamaschi, E., and Mutti, A. (1992b) Experimental model of lead nephropathy. II. Effect of removal from lead exposure and chelation treatment with dimercaptosuccinic acid (DMSA). *Environ. Res.* 58:35–54.

Kim, R., Rotnitzky, A., Sparrow, D., Weiss, S.T., Wager, C., and Hu, H. (1996) A longitudinal study of low-level lead exposure and impairment of renal function. The Normative Aging Study. *JAMA* 275:1177–1181.

Korrick, S.A., Hunter, D.J., Rotnitzky, A., Hu, H., and Speizer, F.E. (1999) Lead and hypertension in a sample of middle-aged women. *Am. J. Public Health* 89(3):330–335.

Kumar, S., Jain, S., Aggarwal, C.S., and Ahuja, G.K. (1987) Encephalopathy due to inorganic lead exposure in an adult. *Jpn. J. Med.* 26:253–254.

Landrigan, P.J. (1990) Lead in the modern workplace. *Am. J. Public Health* 80(8):907–908.

Lindgren, K.N., Masten, V.L., Ford, D.P., and Bleecker, M.L. (1996) Relation of cumulative exposure to inorganic lead and neuropsychological test performance. *Occup. Environ. Med.* 53(7):472–477.

Mai, F.M. (2006) Beethoven's terminal illness and death. *J. R. Coll. Physicians Edinb.* 36(3):258–263.

Marino, P.E., Franzblau, A., Lilis, R., and Landrigan, P.J. (1989) Acute lead poisoning in construction workers: the failure of current protective standards. *Arch. Environ. Health* 44:140–145.

Matte, T.D., Figueroa, J.P., Burr, G., Flesch, J.P., Keenlyside, R.A., and Baker, E.L. (1989) Lead exposure among lead-acid battery workers in Jamaica. *Am. J. Ind. Med.* 16:167–177.

Meyer-Baron, M. and Seeber, A. (2000) A meta-analysis for neurobehavioural results due to occupational lead exposure with blood lead concentrations <70 microg/100 ml. *Arch. Toxicol.* 73:510–518.

Mishra, K.P., Singh, V.K., Rani, R., Yadav, V.S., Chandran, V., Srivastava, S.P., and Seth, P.K. (2003) Effect of lead exposure on the immune response of some occupationally exposed individuals. *Toxicology* 188:251–259.

Moss, M.E., Lanphear, B.P., and Auinger, P. (1999) Association of dental caries and blood lead levels. *JAMA* 281(24):2294–2298.

Muntner, P., He, J., Vupputuri, S., Coresh, J., and Batuman, V. (2003) Blood lead and chronic kidney disease in the general United States population: results from NHANES III. *Kidney Int.* 63:1044–1050.

NAS (1972) *Lead: Airborne Lead in Perspective*, Washington, DC: National Academy of Sciences, pp. 71–177, 281–313.

Nash, D., Magder, L., Lustberg, M., Sherwin, R.W., Rubin, R.J., Kaufmann, R.B., and Silbergeld, E.K. (2003) Blood lead, blood pressure, and hypertension in perimenopausal and post-menopausal women. *JAMA* 289:1523–1532.

Navah, U., Froom, P., Kristal-Boneh, E., Moschovitch, B., and Ribak, J. (1996) Relationship of blood lead levels to blood pressure in battery workers. *Arch. Environ. Health* 51(4):324–328.

Nawrot, T.S., Thijs, L., Hond, E.M.D., Roels, H.A., and Staessen, J.A. (2002) An epidemiological re-appraisal of the association between blood pressure and blood lead: a meta-analysis. *J. Hum. Hypertens.* 16:123–131.

Ng, T.P., Goh, H.H., Ong, H.Y., Ong, C.N., Chia, K.S., Chia, S.E., and Jeyaratnam, J. (1991) Male endocrine functions in workers with moderate exposure to lead. *Br. J. Ind. Med.* 48:485–491.

NIOSH (1974) Evaluation of behavioral functions in workers exposed to lead. In: Xintaras, C., Johnson, B.L., and de Groot, I., editors. *Behavioral Toxicology: Early Detection of Occupational Hazards*, Cincinnati, OH: U.S. Department of Health, Education and Welfare, National Institute for Occupational Safety and Health, pp. 248–266.

NIOSH (2005) *NIOSH Pocket Guide to Chemical Hazards*, Atlanta, GA: National Institute for Occupational Safety and Health. Available at http://www.cdc.gov/niosh/npg/npgdname.html (accessed February 15, 2005).

Nordstrom, S., Beckman, L., and Nordensen, I. (1979) Occupational and environmental risks in and around a smelter in northern Sweden. V. Spontaneous abortion among female employees and decreased birth weight in their offspring. *Hereditas* 90:291–296.

NTP (2005) *Report on Carcinogens*, 11th ed., Research Triangle Park, NC: U.S. Department of Health and Human Services, National Toxicology Program. Available at http://ntp-server.niehs.nih.gov/ntp/roc/toc11.html (accessed February 15, 2004).

OSHA (2005a) Air contaminants. In: *Occupational Safety and Health Standards for Shipyard Employment*, Washington, DC: Occupational Safety and Health Administration, Code of Federal Regulations. 29 CFR 1915.1000. Available at http://www.osha.gov/comp-links.html (accessed February 15, 2005).

OSHA (2005b) Gases, vapors, fumes, dusts, and mists. In: *Safety and Health Regulations for Construction*, Washington, DC: Occupational Safety and Health Administration, Code of Federal Regulations, 29 CFR 1926.55, Appendix A. Available at http://www.osha.gov/comp-links.html (accessed February 15, 2005).

OSHA (2005c) Limits for air contaminants. In: *Occupational Safety and Health Standards*, Washington, DC: Occupational Safety and Health Administration, Code of Federal Regulations, 29 CFR 1910.1000. Available at http://www.osha.gov/comp-links.html (accessed February 15, 2005).

OSHA (2005d) Toxic and Hazardous Substances. In: *Occupational Safety and Health Standards*, Washington, DC: Occupational Safety and Health Administration, Code of Federal Regulations, 29 CFR 1910.1025. Available at http://www.osha.gov/comp-links.html (accessed March 25, 2005).

Oskarsson, A. and Fowler, B.A. (1985) Effects of lead inclusion bodies on subcellular distribution of lead in rat kidney: the relationship to mitochondrial function. *Exp. Mol. Pathol.* 43:397–408.

Osterode, W., Barnas, U., and Geissler, K. (1999) Dose dependent reduction of erythroid progenitor cells and inappropriate erythropoietin response in exposure to lead: new aspects of anaemia induced by lead. *Occup. Environ. Med.* 56:106–109.

Pagliuca, A., Mufti, G.J., Baldwin, D., Lestas, A.N., Wallis, R.M., and Bellingham, A.J. (1990) Lead-poisoning: clinical, biochemical, and hematological aspects of a recent outbreak. *J. Clin. Pathol.* 43:277–281.

Parkinson, D.K., Ryan, C., Bormet, J., and Connell, M.M. (1986) A psychiatric epidemiologic study of occupational lead exposure. *Am. J. Epidemiol.* 123:261–269.

Payton, M., Hu, H., Sparrow, D., and Weiss, S.T. (1994) Low-level lead exposure and renal function in the normativeaging study, *Am. J. Epidemiol.* 140:821–829.

Prioreschi, P. (1998) *A History of Medicine: Roman Medicine*, vol. 3, Horatius Press, p. 279.

Putnam, R.D. (1986) Review of toxicology in inorganic lead. *Am. Ind. Hyg. Assoc. J.* 47:700–703.

Rabin, R., Brooks, D.R., and Davis, L.K. (1994) Elevated blood lead levels among construction workers in the Massachusetts Occupational Lead Registry. *Am. J. Public Health* 84:1483–1485.

Rosenman, K.D., Sims, A., Luo, Z., and Gardiner, J. (2003) Occurrence of lead-related symptoms below the current Occupational Safety and Health Act allowable blood lead levels. *J. Occup. Environ. Med.* 45(5):546–555.

Rummo, J.H., Routh, D.K., Rummo, N.J., and Brown, J.F. (1979) Behavioral and neurological effects of symptomatic and asymptomatic lead exposure in children. *Arch. Environ. Health* 34:120–125.

Sanborn, M.D., Abelsohn, A., Campbell, M., and Weir, E. (2002) Identifying and managing adverse environmental health effects: 3. Lead exposure. *Can. Med. Assoc. J.* 166:1287–1292.

Sallmén, M., Lindbohm, M.L., Anttila, A., Taskinen, H., and Hemminki, K. (2000) Time to pregnancy among the wives of men occupationally exposed to lead. *Epidemiology* 11:141–147.

Schwartz, J. (1995) Lead, blood pressure, and cardiovascular disease in men. *Arch. Environ. Health* 50:31–37.

Schwartz, B.S., Lee, B.K., Bandeen-Roche, K., Stewart, W., Bolla, K., Links, J., Weaver, V., and Todd, A. (2005) Occupational lead exposure and longitudinal decline in neurobehavioral test scores. *Epidemiology* 16(1):106–113.

Smith, F.L., II, Rathmell, T.K., and Marcil, G.E. (1938) The early diagnosis of acute and latent plumbism. *Am. J. Clin. Pathol.* 8:471–508.

Smith, M., Delves, T., Tansdown, R., Clayton, B., and Graham, P. (1983) The effects of lead exposure on urban children: the Institute of Child Health/Southampton study. *Dev. Med. Child Neurol.* 25(Suppl. 47):1–54.

Staessen, J.A., Lauwerys, R.R., Bulpitt, C.J., Fagard, R., Lijnen, P., Roels, H., Thijs, L., and Amery A. (1994) Is a positive association between lead exposure and blood pressure supported by animal experiments? *Curr. Opin. Nephrol. Hypertens.* 3(3):257–263.

Staessen, J.A., Roels, H., and Fagard, R. (1996) Lead exposure and conventional and ambulatory blood pressure. *JAMA* 275:1563–1570.

Stollery, B.T., Banks, H.A., Broadbent, D.E., and Lee, W.R. (1989) Cognitive functioning in lead workers. *Br. J. Ind. Med.* 46:698–707.

Stollery, B.T., Broadbent, D.E., Banks, H.A., and Lee, W.R. (1991) Short term prospective study of cognitive functioning in lead workers. *Br. J. Ind. Med.* 48:739–749.

Tsaih, S.W., Korrick, S., Schwartz, J., Amarasiriwardena, C., Aro, A., Sparrow, D., and Hu, H. (2004) Lead, diabetes, hypertension, and renal function: the normative aging study. *Environ. Health Perspect.* 112(11):1178–1182.

Ündeger, U., Basaran, N., Canpinar, H., and Kansu, E. (1996) Immune alterations in lead-exposed workers. *Toxicology* 109 (2–3):167–172.

Vaziri, N.D., Ding, Y., Ni, Z., and Gonick, H.C. (1997) Altered nitric oxide metabolism and increased oxygen free radical activity of lead-induced hypertension: effect of lazaroid therapy. *Kidney Int.* 52:1042–1046.

Weiss, S.T., Munoz, A., Stein, A., Sparrow, D., and Speizer, F.E. (1986) The relationship of blood lead to blood pressure in longitudinal study of working men. *Am. J. Epidemiol.* 123:800–808.

Weiss, S.T., Munoz, A., Stein, A., Sparrow, D., and Speizer, F.E. (1988) The relationship of blood lead to systolic blood pressure in a longitudinal study of policemen. *Environ. Health Perspect.* 78:53–56.

Whitfield, C.L., Chien, L.T., and Whitehead, J.D. (1972) Lead encephalopathy in adults. *Am. J. Med.* 52:289–298.

Winder, C. (1993) Lead, reproduction and development. *Neurotoxicology* 14:303–317.

21

LITHIUM

Ushang Desai

Target organ(s)	Inhalation: Respiratory system, eye, and skin
	Ingestion: Nervous system, gastrointestinal system, cardiovascular system, kidney
Occupational exposure limits	OSHA PEL[1]: 0.025 mg/m^3 as lithium hydride (LiH)
	ACGIH TLV[2]: 0.025 mg/m^3 as LiH
	NIOSH REL[3]: 0.025 mg/m^3 as LiH
	NIOSH Immediately Dangerous to Life or Health Concentration (IDLH): 55 mg/m^3 as LiH
Reference values	Lithium (CASRN 7439-93-2)
	Provisional chronic oral RfD: 20 µg/kg/day
Risk/safety phrases	For Lithium Hydride (CASRN 7580-67-8):
	Risk phrases: R 14/15 reacts violently with water liberating extremely flammable gases
	R25 Toxic if swallowed
	R35 Causes severe burns
	Safety phrases: S25 avoid contact with eyes
	S36/37/39 wear suitable protective clothing, gloves, and eye/face protection
	S 45 in case of accident or if you feel unwell, seek medical advice immediately (show the label where possible)
	S7/8 keep container tightly closed and dry
	For lithium nitride (CASRN 26134-62-3)
	1. Risk phrases:
	R14/15 reacts violently with water, liberating extremely flammable gases
	R29 contact with water liberates toxic gas
	R34 causes burns
	2. Safety phrases:
	S16 keep away from sources of ignition—no smoking.
	S22 do not breathe dust.
	S26 in case of contact with eyes, rinse immediately with plenty of water and seek medical advice.
	S27 take off immediately all contaminated clothing.
	S36/37/39 wear suitable protective clothing, gloves, and eye/face protection.
	S45 in case of accident or if you feel unwell, seek medical advice immediately
	(show the label where possible).

[1] OSHA PELs (permissible exposure limits) are 8 h TWA (time-weighted average) concentrations

[2] ACGIH TLVs (time limit threshold values) are 8 h TWA concentrations established by American Conference of Governmental Industrial Hygienists

[3] NIOSH RELs (recommended exposure limits) are 10 h TWA concentrations established by National Institute for Occupational Safety and Health

Hamilton & Hardy's Industrial Toxicology, Sixth Edition. Edited by Raymond D. Harbison, Marie M. Bourgeois, and Giffe T. Johnson.
© 2015 John Wiley & Sons, Inc. Published 2015 by John Wiley & Sons, Inc.

BACKGROUND AND USES

Lithium is the 27th most abundant naturally found element on the earth. It's present in the earth's crust to the extent of about 0.006 wt% (Habashi, 1997). Lithium has several uses but one of the most important uses is for making high energy rechargeable lithium-ion batteries. These rechargeable batteries are much lighter than lead and sulfuric acid batteries and cut use of toxic lead and cadmium. Lithium commonly used in the manufacturing of polyester fibers, as a catalyst, the manufacturing of lubricants and greases, in the production of synthetic rubber, in glass, and aluminum production. Lithium is also used to make alloys with aluminum and magnesium; these alloys are very light, but very strong and used to make armor plates and in aerospace applications. Lithium hydride (LiH, CASRN 7580-67-8) is an important industrial source for hydrogen, as a drying agent, and as a reducing agent in organic synthesis particularly in the form of its derivatives lithium aluminum hydride and lithium borohydride (Buchner et al., 1989). Lithium carbonate (CASRN 554-13-2) is also used as an important pharmaceutical for the treatment of bipolar mood disorders (Amdisen, 1988) and neutropenia caused by chemotherapy and hypersplenism (Herbert et al., 1988).

PHYSICAL AND CHEMICAL PROPERTIES

Lithium is an odorless, soft silver-white solid metal with a melting point of 180.54 °C, boiling point of 1342 °C, and auto ignition temperature of 179 °C (Weast, 1976–1977). Melting point of Li is twice as high as sodium but is considerably lower than the melting temperature of most common metals and is one of the reasons why Li may be used efficiently as a reactor coolant and heat transport medium (Table 21.1). It is an alkali metal similar to magnesium and sodium in chemical and physical properties (Birch, 1988; Arena, 1986). Lithium is stable at room temperature under normal argon or mineral oil storage condition but can react intensively when exposed to water (Messer, 1966; Cowles and Pasternak, 1969). Lithium is the only alkali metal that will react with nitrogen gas to form a lithium nitride (Addison and Davies, 1969). Most common Li compounds formed because of lithium

reactions are lithium hydride (LiH), lithium oxide (Li_2O), lithium nitride (Li_3N), and lithium hydroxide (LiOH). All these compounds are stable but extremely reactive and corrosive.

Lithium hydroxide (LiOH): Corrosive; no metal or refractory material can handle molten LiOH in higher concentrations. Lithium oxide (Li_2O): Extremely reactive with water, carbon dioxide, refractory compounds. Lithium nitride (Li_3N): Highly reactive; no metal or ceramic has been found resistant to molten nitride. Lithium hydride (LiH): Reduces oxides, chlorides, sulfides readily; reacts with metals and ceramic at high temperatures (Cowles and Pasternak, 1969; Weast, 1976–1977; Landolt and Sittig, 1961; Perry, 1973).

MAMMALIAN TOXICOLOGY

Animals were exposed to 5–55 mg LiH/m^3 (4–48 mg Li/m^3) at 50% relative humidity for 4–7 h to assess the toxicology of LiH. All exposure levels of LiH-induced respiratory effects, such as sneezing and coughing. Lithium hydride concentrations of approximately 10 mg/m^3 corroded patches of body fur and skin. Occasionally severe inflammation and irritation of the eyes were observed and in a few animals the external nasal septum was destroyed. This toxicity was because of the alkalinity of the hydrolysis product, lithium hydroxide (Spiegl et al., 1956). Following exposure of approximately 5 mg LiH/m^3 for 5 days (average exposure is 4 h/day), ulceration of nose and forepaws, inflammation of eyes, partial sloughing of mucosal epithelium of trachea, and emphysema was also observed. No histopathological changes were noticed in the lungs after 2–5 months following exposure (Spiegl et al., 1956).

Chmielnicka and Nasiadek's study group found that oral administration of lithium carbonate-induced renal toxicity in rats and the resulting lithium concentration in serum and urine was determined to be dose dependent. They observed higher activity of N-acetyl-β-glucoaminidase (NAG) and a significant increase of copper and protein in the urine of exposed rats compared to the control group. They also found diuretic and proteinuric effects after lithium intoxication. The urinary excretion of lithium affects several factors, including water and sodium balance in the body. The most important in lithium toxicity is dehydration, which will produce sodium and water imbalance in the body (Chmielnicka and Nasiadek, 2003).

Teratogenicity Teratogenicity is the ability of the drugs/chemical substances to cause defects in developing fetus in the mammals. It is a potential side effect of many drugs such as thalidomide. A multicenter study on pregnant women was conducted in the United States and Canada to evaluate teratogenic effects of lithium during the first trimester of the pregnancy. This prospective study on pregnant women

TABLE 21.1 Physical and Chemical Properties of Lithium Compounds

Formula	LiOH	Li_2O	Li_3N	LiH
Molecular weight	23.95	29.88	34.82	7.95
Density (g/cm^3) at 15–20 °C	2.54	2.01	1.38	0.78
Melting point (°C)	471.1	1427	840–850	688
Boiling point (°C)	925	1527		

concluded that lithium is not a major human teratogen; women with major affective disorders who wish to have children may continue lithium during pregnancy, and do not need to terminate the pregnancy provided that level II ultrasound and fetal echocardiography are done on the pregnant women (Jacobson et al., 1992; Kallen and Tandberg, 1983; Kallen, 1988; Zalstein et al., 1990). The outcome of various studies, including cohort, prospective, retrospective and small number of case reports show that lithium is a "weak" teratogen in humans. The main teratogenic effects of lithium are cardiac malformations and babies with increased birth weight. There is a possibility that, in particular, lithium may be associated with the Ebstein anomaly but current evidence cannot definitely affirm or deny this association. Animal studies with lithium using doses comparable to human therapeutic serum levels have not reported any abnormalities. However, higher doses have produced exencephaly, skeletal and craniofacial defects, and abnormalities of blood vessel development in animals (Wright et al., 1970). Experiments with other vertebrates have shown that lithium affects dorsoventral specification and inhibition of vasculogenesis. Both these effects can be prevented by pretreatment with myoinositol indicating that lithium interferes with the phosphatidyl inositol cycle. More recent findings have shown that the effects of lithium on invertebrates may be mediated through inhibition of GSK-3b in the Wnt-GSK-3 pathway (Giles and Bannigan, 2006).

Carcinogenicity, Mutagenicity, and Genotoxicity No cancer studies have been found on possible carcinogenic effects of lithium compounds in the literature (Léonard et al., 1995). Mutagencity is the capability of drugs/chemical substances to cause genetic mutations in sperm, eggs, and other cells. Lithium can affect DNA on several mechanisms such as binding selectively to DNA and competes with Mg^{2+} potentially impairing DNA synthesis and repair (Kuznetsov et al., 1971; Becker and Tyobeka, 1990). However, results on several bacterial strains and isolated systems were inconclusive to support mutagenicity of Li (King et al., 1979; Kanematsu et al., 1980).

Acute Effects

Lithium hydride is respiratory irritant and occupational exposure to workers. A compilation of the signs and symptoms that may occur following lithium exposures is provided in Tables 21.2 and 21.3. Lithium hydride at the high concentrations may also cause pulmonary lesions through caustic action (Birch, 1988).

Dermal exposure to lithium/aluminum alloys dusts cause severe irritant action and long-term exposure to these dusts causes hydrolytic destruction of the epidermis (Bencze et al., 1991). This effect is the result of the formation of strong basic lithium hydroxide in the presence of water. Lithium

TABLE 21.2   Possible Signs and Symptoms at Specific Exposure Levels

Exposure Level (mg/m^3)	Effects
0–0.025	No effect
0.025–0.10	A tickling sensation in the nose, some nasal discharge, tolerated by workers who continuously exposed
0.05	Workers readily adopt, discomfortness for unacclimated individuals
0.10–0.50	Definite nasal irritation with some coughing, not tolerated by workers
>0.1	Eye and nose irritation with some degree of adaption
>0.2	Itching of exposed skin
0.5	Maximal tolerable concentration for brief periods
0.50–1.0	Severe nasal irritation with coughing, in some workers eye irritation
1.0–5.0	Severe irritant effects, skin irritation

Source: Beliles (1994).

carbonate is used for several medical purposes so acute lithium toxicity can develop in patients using lithium for therapeutic purposes.

The lithium as a medicine has very narrow therapeutic index, and the plasma concentration of lithium should not exceed 1.5 mEq/l (10 mg/l) (Goodrich, 1991; Austin et al., 1990; Tilkan, 1976). Lithium toxicity occurred during therapy as attempted suicide, medication errors, and drug interactions (Simon, 1988), and simultaneous medical conditions (Table 21.4), resulting in a higher plasma lithium level (over 2 mEq/l). A patient's lithium level should be checked regularly; particularly patients with a concomitant medical condition or those taking other medications that can interact with lithium.

Chronic Effects

Chronic lithium toxicity occurs in patients receiving chronic lithium therapy (Table 21.5). Chronic lithium toxicity occurs in patients whose lithium dosage has been increased or in

TABLE 21.3   Signs and Symptoms of Lithium Toxicity

Lithium Level (mEq/l)	Signs and Symptoms
2	Nausea, diarrhea, polyuria (nephrogenic diabetes insipidus), confusion and obtundation, muscle fasciculations
2.5	Myoclonus, choreoathetosis, stupor progressing to coma
3	Cardiac arrhythmia, tonic–clonic seizures
4	Cardiovascular collapse (hypotension, cardiac electrochemical dissociation)

TABLE 21.4   Factors that Increase the Risk of Lithium Toxicity

Medical conditions	Acute	Vomiting and diarrhea, leading to dehydration
		Low salt diet
		Low food intake
		Recent increase in lithium dosage
		Intentional overdose
	Chronic	Increased age
		Hypertension
		Diabetes mellitus
		Congestive heart failure
		Chronic renal disease
		Schizophrenia
		Addison's disease
Drugs		Thiazide diuretics
		Nonsteroidal anti-inflammatory drugs (NSAIDs)
		Methyldopa
		Phenytoin
		Carbamazaepine
		Tetracycline
		Angiotensin-converting enzyme inhibitors (ACE inhibitors)

patients whose renal function is impaired, resulting in an increase in serum lithium levels (Ellenhorn et al., 1997). Chronic poisoning developed neurological manifestations such as memory deficits, Parkinson's disease (cogwheel rigidity, tremor), pseudotumor cerebri, and psychosis in the patients (Ellenhorn et al., 1997; Timmer and Sands, 1999; Henry, 1998; Sansone and Ziegler, 1985; Hall et al., 1979). Permanent brain damage has been observed in several patients undergoing long-term lithium therapy (Gosselin et al., 1984).

Chronic lithium therapy has been associated with changes in renal function such as polydipsia/polyuria, nephrogenic diabetes insipidus, transient natriuresis, partial renal tubular acidosis, minimal change disease, and nephrotic syndrome (Ellenhorn and Barceloux, 1988). Chronic use of lithium is also associated with degenerative changes in kidney, which may occur in the glomeruli or in distal convoluted tubules or collecting ducts (Richman et al., 1980; Hestbech et al., 1977).

Mechanisms of Action

Lithium's mechanism of action is not very well understood. Lithium competes and substitutes with other ions such as sodium, potassium, magnesium and calcium, and modifies neurotransmitters actions and also affects the secondary messenger system (cAMP, cGMP) (Ward et al., 1994). At the cellular level, two major lithium transporting proteins are the sodium channel and the sodium-proton exchanger; both these transporters are inhibited by amiloride (Lenox et al., 1998; Holstein-Rathlou, 1990; Greger, 1990). The amiloride-sensitive sodium channel (ENaC) is an important transporter, which is involved in sodium equilibrium in the collecting duct. This channel has the same permeability to lithium and sodium and is a major pathway for lithium accumulation in collecting duct cells (Lenox et al., 1998; Holstein-Rathlou, 1990; Greger, 1990). Another lithium transporter is the Na-K-2Cl co transporter (NKCC2, BSC1) that is found in the apical membrane of the thick ascending limb of loop of Henle and is inhibited by furosemide (Lenox et al., 1998; Holstein-Rathlou, 1990; Greger, 1990). Na/K-ATPase ("the sodium pump") can also play a role in the transport of lithium across the cell membrane (Lenox et al., 1998; Holstein-Rathlou, 1990; Greger, 1990).

Chemical Pathology

There is a correlation between chronic lithium therapy and development of chronic interstitial nephritis in the patients over a period of time (Boton et al., 1987). The biopsy findings in these patients showed tubular atrophy and dilatation, sclerotic glomeruli, cyst formation, and cortical and

TABLE 21.5   Clinical Features of Lithium Poisoning

Target Organ	Acute Poisoning	Chronic Poisoning
Endocrine	None	Hypothyroidism
Gastrointestinal	Nausea, vomiting	Minimal
Cardiac	Prolonged QT interval, ST, and T wave changes	Myocarditis
Hematologic	Leukocystosis	Aplastic anemia
Neurologic	Fine tremor, weakness, apathy, slurred speech, tinnitus, drowsiness, clonus, coma, confusion, seizures, muscular irritability	Memory deficits, Parkisnon's disease, pseudomotor cerebri, psychosis
Renal	Urine concentration defect	Chronic interstitial nephritis, nephorgenic diabetes insipidus, renal failure
Skin	Irritation, itching	Dermatitis, localized edema, ulcers

Source: Hermansen (2001).

medullary fibrosis (Boton et al., 1987; Timmer and Sands, 1999). Renal function of these patients must be followed after the course of lithium therapy because the development of diabetes insipidus and chronic interstitial nephritis is often irreversible, even though lithium is discontinued (Boton et al., 1987; Timmer and Sands, 1999). It is important to detect any changes in renal function early so that lithium can be discontinued, or the dose decreased, before severe loss of kidney function (Boton et al., 1987). While lithium hydride most commonly used in the industry is an alkaline corrosive and may cause liquefaction necrosis. In the gastrointestinal tissue, LiH can develop tissue necrosis followed by granulation and finally stricture formation (Cracovaner, 1964; Hathway and Proctor, 2004).

SOURCES OF EXPOSURE

Occupational

Occupational exposure to lithium may occur during extraction of lithium from its ores, manufacturing of various lithium compounds, and other activities such as welding, brazing, and enameling. Lithium fumes are a significant source of exposure during welding and brazing, particularly from leakage and accidents in the use of lithium hydrides (Beliles, 1994). Occupational exposure to lithium to workers in the pharmaceutical industries is not significant (Léonard et al., 1995). Other possible activities where lithium exposure may occur are extraction of crude oil and natural gas, in the iron and steel industries, in the production of pumps and compressors, in the ship building and reparation industries and in the distribution of electricity (Birgitta and Birgitta, 2002).

Environmental

Lithium is naturally found in the aquatic and terrestrial environment (Bowen, 1979, Wedepohl, 1995; Sposito, 1986; Birch, 1988; Ribas, 1991). Lithium is found mainly in the ionic form in the water; reacts with water, and forms lithium hydroxide and hydrogen. Worldwide mineral water has 0.05–1 mg/l lithium, though higher levels up to 100 mg/l can be found in some natural mineral waters (Schrauzer, 2002). While in terrestrial environments, lithium is found as lithium carbonate ($LiCO_3$), lithium chloride (LiCl), or lithium oxide (Li_2O). Lithium is found in smaller amounts in all soils mainly in the clay fraction, and to a lesser extent in the organic soil fraction in the range from 7 to 200 µg/g (Schrauzer, 2002). The source of lithium in the terrestrial environmental is from sedimentary rocks (Chan et al., 1997). Environmental exposure to lithium can occur through anthropogenic activities such lithium-based grease (lithium hydroxide monohydrate) in vehicles and disposal of lithium batteries in the municipal solid waste (Aral and Vecchio-Sadus, 2008).

Lithium batteries are generally not considered hazardous to the environment, this is supported on the assumption that lithium metal (that reacts extremely with water to produce explosive hydrogen gas) is no longer reactive as the lithium is converted into non-reactive lithium oxide once battery is discharged (Aral and Vecchio-Sadus, 2008). In 1984, the United States Environmental Protection Agency (US EPA) made a regulatory status of spent and/or discarded lithium sulfur dioxide type batteries. These are considered hazardous waste as they contain strips of lithium metal as anode and the nonaqueous sulfur dioxide as an electrolyte (Aral and Vecchio-Sadus, 2008). According to the Australian Capital Territory Environment Protection Regulation, lithium is considered an environmental pollutant with a recommended concentration of ≤2.5 mg/l of lithium entering the waterways (Aral and Vecchio-Sadus, 2008).

INDUSTRIAL HYGENE

Good industrial hygiene practices recommend that engineering controls should be used to reduce occupational exposure to lithium and its compounds to the permissible exposure level. Employees should be provided and required to wear impermeable clothing, gloves, face shields (eight minimum inch), dust, and splash proof goggles, other proper protective clothing to prevent exposure to lithium and its compounds at the work place. Respirators should be used by employees when engineering and work practice control technically not possible and employees required to enter into tanks and closed vessels and in emergency situations. If the use of respirators are mandatory only use respirators approved by the Mine Safety and Health Administration. Preplacement medical examinations are recommended for all employees who will be working with lithium and its compounds. Pre-placement exams should screen for certain medical conditions such as chronic respiratory diseases, skin disorders, and eye disorders which will be exacerbated if they come in contact with lithium and its compounds. Periodic medical evaluations are also recommended for employees working with lithium and its compounds for the above discussed medical conditions. Apart from industrial hygiene, good sanitation at the work place such as washing hands thoroughly before eating, smoking, or using toilet facilities, thoroughly washing body parts that come in contact with lithium and lithium compounds, changing of clothes before leaving from the workplace all reduce occupational exposures.

MEDICAL MANAGEMENT

Acute exposure of patients to lithium and its compounds requires recognition and removal from the source of exposure. In the emergency room, establishing a patient's airway (oropharayngeal or nasopharyngeal airway), looking for respiratory insufficiency, and assisting for ventilations may be required. Monitor for shock, pulmonary edema and seizures, and treat if necessary. For eye contamination with lithium and its compounds, flush eyes immediately with water, if required irrigate eyes with normal saline during treatment. Cover skin burns with dry sterile dressing after decontamination. In severe intoxicated cases (plasma lithium level $\geq$3 mEq/l), hemodialysis is the treatment of choice after fluid rehydration, resuscitation, and intubation (Jaeger et al., 1985; Currance et al., 2005). Look for cardiac arrhythmias, hypotension, and treat if necessary. Treat seizures with diazepam or lorazepam; consider fluid resuscitation and vasopressors for hypovolemia (Currance et al., 2005). In less severe intoxication, discontinue lithium, rehydrate the patients and check serum lithium, sodium, and potassium levels, as well as renal functions as required.

REFERENCES

Addison, C.C. and Davies, B.M. (1969) Reaction of nitrogen with stirred and unstirred liquid lithium. *J. Chem. Soc. A* 1822–1827.

Aral, H. and Vecchio-Sadus, A. (2008) Toxicity of lithium to humans and the environment—a literature review. *Ecotoxciol. Environ. Saf.* 70:349–356.

Arena, J.M. (1986) *Poisoning: Toxicology, Symptoms, Treatments*, Springfield, IL: Charles C. Thomas.

Amdisen, A. (1988) Clinical features and management of lithium poisoning. *Med. Toxicol. Adverse Drug Exp.* 3:18–32.

Austin, L.S., Aran, G.W., and Melvin, J.A. (1990) Toxicity resulting from lithium augmentation of antidepressant treatment in elderly patients. *J. Clin. Psychiatry* 51:344–345.

Becker, R.W. and Tyobeka, E.M. (1990) Lithium enhances proliferation of HL60 promyelocytic leukemia cells. *Leuk. Res.* 14:879–884.

Beliles, R.P. (1994) Lithium, Li. In: Clayton, G.D. and Clayton, F.E., editors. *Patty's Industrial Hygiene and Toxicology*, 4th ed., vol. 2, New York, NY: John Wiley & Sons, Inc., pp. 2087–2097.

Bencze, K., Pelikan, C., and Bahemann-Hoffmeister, A. (1991) Lithium/aluminum alloys. A problem material for biological monitoring. *Sci. Total Environ.* 101:83–90.

Birch, N.J. (1988) Lithium. In: Seiler, H.G., Sigel, H., and Sigel, A., editors. *Handbook on the Toxicity of Inorganic Compounds*, New York, NY: Marcel Dekker, pp. 382–393.

Birgitta, L. and Birgitta, L. (2002) *Lithium and Lithium Compounds*, The Nordic Expert Group for Criteria Documentation of Health Risks from Chemicals, Nordic Council of Ministers.

Boton, R., Gaviria, M., and Batlle, D.C. (1987) Prevalence, pathogenesis, and treatment of renal dysfunction associated with chronic lithium therapy. *Am. J. Kidney Dis.* 10:329–345.

Bowen, H.J.M. (1979) *Environmental Chemistry of the Elements*, New York, NY: Academic Press.

Buchner, W., Schliebs, R., Winter, G., and Buchel, K.H. (1989) Lithium and its compounds. In: *Industrial Inorganic Chemistry*, Weinheim: VCH Verlagsgesellschaft, pp. 215–218.

Chan, L.H., Sturchio, N.C., and Katz, A. (1997) Lithium isotope study of the Yellowstone hydrothermal system. *EOS Trans. Am. Geophys. Union* 78:F802 (Abstract).

Chmielnicka, J. and Nasiadek, M. (2003) The trace elements in response to lithium intoxication in renal failure. *Ecotoxicol. Environ. Saf.* 55:178–183.

Cowles, J. and Pasternak, A.D. (1969) *Lithium Properties Related to Use as a Nuclear Reactor Coolant*, Lawrence Radiation Laboratory, CA.

Cracovaner, A.J. (1964) Stenosis after explosion of lithium hydride. *Arch Otolaryngol.* 80:87–92.

Currance, P.L. Clements, B., and Bronstein, A.C. (2005) *Emergency Care for Hazardous Materials Exposure*, 3rd ed., St. Louis, MO: Elsevier Mosby, pp. 398–390.

Ellenhorn, M.J. and Barceloux D.G. (1988) *Medical Toxicology: Diagnosis and Treatment of Human Poisoning*, New York, NY: Elsevier.

Ellenhorn, M.J., Schonwald, S., Ordog, G., and Wasserberger, J. (1997) Lithium. In: Ellenhorn, M.J., Schonwald, S., Ordog, G., C and Wasserberger, J., editors. *Medical Toxicology: Diagnosis and Treatment of Human Poisoning*, Baltimore, MD: Williams and Wilkins, p. 1579.

Giles, J. and Bannigan, J. (2006) Teratogenic and developmental effects of lithium. *Curr. Pharm. Des.* 12:1531–1541.

Goodrich, P.J. (1991) Schorr-Cain CB: lithium pharmacokinetics. *Psychopharmacol. Bull.* 27:475–491.

Gosselin, R.E., Smith, R.P., and Hodge, H.C. (1984) *Clinical Toxicology of Commercial Products*, 5th ed., Baltimore, MD: Williams and Wilkins.

Greger, R. (1990) Possible sites of lithium transport in the nephron. *Kidney Int. Suppl.* 37:S26–S30.

Habashi, F. (1997) *Handbook of Extractive Metallurgy*, vol. 4, New York, NY: Wiley-VCH.

Hall, H.C., Perl, M., and Pfefferbaum, B. (1979) Lithium therapy and toxicity. *Am. Fam. Pract.* 19:133–139.

Hathway, G. and Proctor, N. (2004) *Chemical Hazards of the Work Place, Proctor & Hughes*, 5th ed., Hoboken, NJ: John Wiley & Sons, Inc., pp. 428.

Henry, G.C. (1998) Lithium. In: Goldfrank, L.R., Flomenbaum, N.E., Lewin, N.A., Weisman, R.S., Howland, M.A., Hoffman, R.S., and Stamford, C.T, editors. *Toxicologic Emergencies*, 6th ed., Stamford, CT: Appleton and Lange, p. 969.

Herbert, V., Hirschman, S., and Jacobson, J. (1988) Lithium for zidovudine induced neutropenia in AIDS. *JAMA* 260:3588.

Hestbech, J., Hansen, H.E., Amdisen, A., and Olsen, S. (1977) Chronic renal lesions following long-term treatment with lithium. *Kidney Int.* 12:205–213.

Hermansen, E. (2001) Lithium, Utox Update, vol. 3, no. 4, Utah Poison Control Center for Health Professionals, Salt Lake City, UT.

Holstein-Rathlou, N.H. (1990) Lithium transport across biological membranes. *Kidney Int. Suppl.* 37:S4–S9.

Jacobson, J., Jones, K., Johnson, K., Ceolin, L., Kaur, P., Sahn, D., Donnenfeld, A.E., Rieder, M., Santelli, R., Smythe, J., Pastuszak, A., Eirnarson, T., and Koren, G. (1992) Prospective multicenter study of pregnancy outcome after lithium exposure during first trimester. *Lancet* 339:530–533.

Jaeger, A., Sander, P., and Kopferschmitt, J. (1985) Toxicokinetics of lithium intoxication treated by hemodialysis. *Clin. Toxicol.* 23:501–517.

Kallen, B. (1988) Comment on teratogen update: lithium. *Teratology* 38:597–598.

Kallen, B. and Tandberg, A. (1983) Lithium and pregnancy. *Acta Psychiatr. Scand.* 68:134–139.

Kanematsu, N., Hara, M., and Kada, T. (1980) Assay and mutagenicity studies on metal compounds. *Mutat. Res.* 77:109–116.

King, M.C., Beikirch, H., Eckhardt, K., Gocke, E., and Wild, D. (1979) Mutagenicity studies with X-ray contrast media, analgesics, antipyretics, antirheumatics and some other pharmaceutical drugs in bacterial *Drosophila* and mammalian test systems. *Mutat. Res.* 66:33–43.

Kuznetsov, I.A., Lukanin, A.S., and Tsurkanov, L.F. (1971) Effect of ions of the alkaline metals on the secondary structure of DNA. IV. Thermal denaturing deoxyribonucleates of alkaline metals in solution with a low ionic strength. *Biofizika* 16:144–145.

Landolt, P.E. and Sittig, M. (1961) Lithium. In: *Rare Metals Handbook*, 2nd ed., New York, NY: Reinhold, pp. 239–252.

Lenox, R.H., McNamara, R.K., Papke, R.L., and Manji, H.K. (1998) Neurobiology of lithium: an update. *J. Clin. Psychiatry* 58:37–47.

Léonard, A., Hantson, P., and Gerber, G.B. (1995) Mutagenicity, carcinogenicity and teratogenicity of lithium compounds. *Mutat. Res.* 339:131–137.

Messer, C.E. (1966) Lithium and its binary compounds, in the alkali metals. In: *International Symposium, Nottingham, July 19–22*, Special Publication No. 22, The Chemical Society, London.

Perry, J.H. (1973) *Chemical Engineer's Handbook*, 4th ed., New York, NY: McGraw-Hill Book Company.

Ribas, B. (1991) Lithium. In: Merian, E., editor. *Metals and Their Compounds in the Environment*, 4th ed., Weinheim: VCH, pp. 1014–1023.

Richman, A.V., Masco, H.L., and Rifkin, S.I. (1980) Minimal change disease and nephrotic syndrome associated with lithium therapy. *Ann. Int. Med.* 92:70–72.

Sansone, M.E.G. and Ziegler, D.K. (1985) A review of neurologic complications. *Clin. Neuropharmacol.* 8:242–248.

Schrauzer, G.N. (2002) Lithium: occurrence, dietary intakes, nutritional essentiality. *J. Am. Coll. Nutr.* 21(1):14–21.

Simon, G. (1988) Combination angiotensin-converting enzyme inhibitor/lithium therapy contraindicated in renal disease. *Am. J. Med.* 56:893–894.

Spiegl, C.J., Scott, J.K., Steinhardt, H., Leach, L.J., and Hodge, H.C. (1956) Acute inhalation toxicity of lithium hydride. *AMA Arch. Ind. Health* 14:468–470.

Sposito, G. (1986) Distribution of potentially hazardous trace metals. In: Sigel, H., editor. *Metal Ions in Biological Systems*, 4th ed., vol. 20, New York, NY: Marcel Dekker, pp. 1–20.

Tilkan, A.G. (1976) The cardiovascular effects of lithium in man: a review of the literature. *Am. J. Med.* 61:665–670.

Timmer, R.T. and Sands, J.M. (1999) Lithium intoxication. *J. Am. Soc. Nephrol.* 10:666–674.

Ward, M.E., Musa, M.N., and Bailey, L. (1994) Clinical pharmacokinetics of lithium. *J. Clin. Pharmacol.* 34:280–285.

Weast, R.C. (1976–1977) *Handbook of Chemistry and Physics*, 57th ed., Cleveland, OH: CRC Press.

Wedepohl, K.H. (1995) The composition of the continental crust. *Geochim. Cosmochim. Acta* 59:1217–1232.

Wright, T.L., Hoffman, L.H., and Davies, J. (1970) Lithium teratogenicity. *Lancet* 1970(II):876.

Zalstein, E., Koren, G., and Einarson, T. (1990) A case control study on the association between first trimester exposure to lithium and Ebstein's anomaly. *Am. J. Cardiol.* 65:817–818.

22

MANGANESE

Alex Lebeau

First aid:

Manganese compounds and fumes

Respiratory: Provide respiratory support

Ingestion: Seek immediate medical attention

Manganese cyclopentadienyl tricarbonyl

Eyes: Irrigate eyes

Skin: Wash with soap and water

Respiratory: Provide respiratory support

Ingestion: Seek immediate medical attention

Manganese tetroxide

Eyes: Irrigate eyes

Skin: Wash with soap and water

Respiratory: Provide respiratory support

Ingestion: Seek immediate medical attention

Methylcyclopentadienyl manganese tricarbonyl (MMT)

Eyes: Irrigate eyes

Skin: Wash with soap and water

Respiratory: Provide respiratory support

Ingestion: Seek immediate medical attention

Main target organ(s): Respiratory system and central nervous system

Occupational exposure limits:

Manganese compounds and fumes:

OSHA PEL: $C = 5 \, mg/m^3$

NIOSH REL:

$TWA = 1 \, mg/m^3$

$ST = 3 \, mg/m^3$

$IDLH = 500 \, mg/m^3$

ACGIH TLV: $TWA = 0.02 \, (mg/m^3)$

Inhalable Manganese: $TWA = 0.1 \, mg/m^3$

Manganese cyclopentadienyl tricarbonyl:

OSHA PEL: $C = 5 \, mg/m^3$

NIOSH REL: $TWA = 0.1 \, mg/m^3$ [skin]

ACGIH TLV: $TWA = 0.1 \, mg/m^3$

Manganese tetroxide:

OSHA PEL: $C = 5 \, mg/m^3$

NIOSH REL: $TWA = 1 \, mg/m^3$ (Appendix D of NIOSH Pocket Guide)

ACGIH TLV: Not established

Methylcyclopentadienyl manganese tricarbonyl:

OSHA PEL: $C = 5 \, mg/m^3$

NIOSH REL: $TWA = 0.2 \, mg/m^3$ [skin]

ACGIH TLV: $TWA = 0.2 \, mg/m^3$

Reference values:

EPA IRIS:

RfD $= 1.4 \times 10^{-1} \, mg/kg\text{-}day$

RfC $= 5.0 \times 10^{-5} \, mg/m^3$

ATSDR:

Chronic inhalation MRL $= 0.3 \, \mu g/m^3$

Risk/safety phrases

Manganese dust or powder: Flammable

Manganese: Reactive

MMT: Moderate irritation to eyes

Hamilton & Hardy's Industrial Toxicology, Sixth Edition. Edited by Raymond D. Harbison, Marie M. Bourgeois, and Giffe T. Johnson.
© 2015 John Wiley & Sons, Inc. Published 2015 by John Wiley & Sons, Inc.

BACKGROUND AND USES

Manganese (CASRN 7439-96-5) is a naturally occurring metal that is widely distributed in environmental media. Pure manganese is not found in nature (Krebs, 2006) but is present in more than 100 ores and minerals, including pyrolusite and rhodochrosite (Krebs, 2006; ATSDR, 2012). Mining for manganese-containing ores occurs in Ukraine, Brazil, Australia, South Africa, Gabon, China, India, and the state of Montana (Krebs, 2006; Lide, 2006). Manganese is primarily used in the metallurgic industry to remove sulfide and oxide impurities from iron, and to create hardened alloys, including carbon, stainless, and high temperature steels (ATSDR, 2012; Krebs, 2006). In addition to iron, manganese has been used to create alloys with aluminum, antimony, and copper (Lide, 2006). Both inorganic and organic manganese have application in a variety of products beyond alloys, including: dry-cell batteries (manganese chloride and dioxide); fertilizer (manganese sulfate); fuel antiknock agent (methylcyclopentadienyl manganese tricarbonyl [MMT]); antiseptic (potassium permanganate); feed additives (manganese sulfate and carbonate); chemical synthesis/catalyst (manganese acetate); dietary supplement (manganese chloride); glass bleacher/pigment (manganese sulfate); pesticides (maneb and mancozeb); fireworks (manganese dioxide); and medical imaging (mangafodipir) (ATSDR, 2012; Toxnet, 2008; Krebs, 2006; Klaassen, 2008; Lide, 2006). In addition to its ubiquitous presence in nature, manganese is considered an essential trace element, found in both plants and animals. Small amounts of manganese are essential in the development of body tissues (bone and connective tissue) (Aschner et al., 2007). Additionally, manganese is essential for fat and carbohydrate metabolism; neurotransmitter synthesis and metabolism (Bowman et al., 2011); and is required for the metalloenzyme manganese–superoxide dismutase, among other enzymes (Davis, 1998).

PHYSICAL AND CHEMICAL PROPERTIES

Manganese is a gray-white, hard, brittle metal that exists in four allotropic forms and has a boiling point of 2061 °C (Lide, 2006). Its brittleness makes it almost impossible to mold it into specific shapes (Krebs, 2006). Manganese must be extracted from the oxide ores that it is colocated with, often requiring reduction with sodium, magnesium, or aluminum, and can also be extracted from the ores via electrolysis (Lide, 2006). Manganese has oxidation states from -3 to $+7$, with the $+2$ form as the most stable (Barceloux, 1999). Manganese in the $+2$ and $+3$ oxidation states are the most predominant forms in biological systems (Michalke et al., 2007).

ENVIRONMENTAL FATE AND BIOACCUMULATION

Manganese is found to comprise approximately 0.1% of the earth's crust, and is found in various concentrations in environmental media (Barceloux, 1999). The average background concentration of manganese in soil ranges from 300 to 600 mg/kg dry weight (WHO, 2004). Emissions from industrial sources, combustion of petroleum fuels, natural erosion of manganese-containing rocks and soils, and volcanic activity contribute to atmospheric and terrestrial concentrations of manganese (ATSDR, 2012). The majority of the manganese particulate matter (PM) in the atmosphere is removed via dry deposition, with transport distance of the PM dependent on the bound particle size (ATSDR, 2012).

Manganese is found to occur naturally in surface water and groundwater at various concentrations. For example, manganese found in bodies of water that were assumed to be relatively free of anthropogenic influences ranged from 10 to >10,000 µg/l, although dissolved natural concentrations were found to rarely exceed 1,000 µg/l and most concentrations were usually <200 µg/l (WHO, 2004). An evaluation of 286 rivers and streams in the United States found a median manganese concentration of 24 µg/l, and seawater concentrations ranged from 0.4 to 10 µg/l (USEPA, 1984). Manganese bound to PM may also contribute to surface water concentrations, depending on the chemical form and solubility of the bound manganese (ATSDR, 2012). Particulate matter deposited in surface water will contribute to sediment deposition of manganese.

Studies have indicated that manganese may bioaccumulate in lower trophic organisms when compared to higher organisms (i.e., plankton, algae, and some fish versus humans) (ATSDR, 2012). Because mollusks have been shown to bioaccumulate manganese, the United states Environmental Protection Agency (USEPA) has established a threshold of 0.1 mg/l for marine water (USEPA, 1984). This threshold has been established to protect individuals who consume marine mollusks. It does not appear that any significant biomagnification occurs from lower organisms to humans.

ECOTOXICOLOGY

There is a paucity of information regarding the aquatic toxicity of colloidal, particulate, and complex manganese; most evaluations have been carried out with inorganic manganese (WHO, 2004). The toxicity of manganese in other forms is assumed to be less toxic than that of the aquo ion form (WHO, 2004). Toxic effects reported below are for fish species. Information for other aquatic organisms exists.

Acute Effects

Acute effects were evaluated in freshwater fish. The 96-h lethal concentration in 50% of the population (LC_{50}) for rainbow trout (*Oncorhynchus mykiss*) was determined to be 4.8 mg/l (form of manganese not reported), and the 96-h LC_{50} for coho salmon (*Oncorhynchus kisutch*) was found to range from 2.4 to 17.4 mg/l (form of manganese not reported) (WHO, 2004). The upper range for the 96-h LC_{50} was found to be 3,350 mg/l for Indian catfish (*Heteropneustes fossilis*) ($MnCl_2$) (WHO, 2004).

Chronic Effects

A 7-day LC_{50} was determined to be 8.2 mg/l for goldfish (*Carassius auratus*), and a 28-day LC_{50} for rainbow trout (*Oncorhynchus mykiss*) was 2.9 mg/l (WHO, 2004).

MAMMALIAN TOXICOLOGY

Acute Effects

Inorganic and organic manganese compounds elicit different toxic effects. Animal and human studies suggest that, overall, exposures to inorganic manganese compounds via all routes result in a very low acute toxicity. Some exceptions do exist. Potassium permanganate, due to its oxidizing potential, has the ability to cause severe corrosion of the skin and mucosa at the point of contact (ATSDR, 2012; Southwood et al., 1987). Potassium permanganate has a high oxidation potential. Overexposures to manganese dust and fumes have been attributed to respiratory inflammation and impairment. Individuals overexposed to manganese may present with cough, bronchitis, pneumonitis, and reductions in the lung function (ATSDR, 2012). Respiratory impairment has been observed among workers handling manganese ores (Rodier, 1955). Studies on manganese dioxide and potassium permanganate-exposed workers have also reported similar respiratory effects (Davies, 1946). A study evaluating potassium permanganate exposures in factory workers observed an 8-year average pneumonia incidence of 26 cases per 1,000. This finding was elevated when compared to the incidence of 0.73 cases per 1,000 in the control group (Davies, 1946; Davies and Harding, 1949).

Pneumonitis and other respiratory impairments resulting from overexposures to dusts are not unique to manganese. Caution is needed when relating these disease outcomes with exposures to manganese dusts. While empirical evidence has indicated that acute overexposures to manganese may result in respiratory impairment, it must be recognized that similar responses have been observed when evaluating other forms of PM. The respiratory impairments observed may not depend on the presence of manganese.

Inhalation of manganese and manganese oxides fumes is reported to cause a condition called metal fume fever (MFF) (OSHA, 1978). Symptoms of this illness include fever, chills, upset stomach, cough, weakness, body aches and headaches, vomiting, and dry throat (OSHA, 1978). These symptoms can appear several hours after exposure to the fumes and typically last no more than a day. This condition is not unique to manganese (reports of MFF are common in the literature following exposures to zinc or cadmium oxide), and there is a paucity of published information relating manganese exposure to MFF.

Organic forms of manganese, MMT and manganese cyclopentadienyl tricarbonyl, have been used as fuel additives in some countries around the world. These compounds are somewhat irritating to the skin and eyes. Additionally, organic forms of manganese can readily pass through the skin. Dermal contact and apparent absorption has been observed to result in nausea, headache, giddiness, and abdominal distress (HSDB, 2001). Animal studies have suggested that the liver and kidney may be affected from exposure to MMT. No study was identified that observed a similar outcome in humans. Data suggest that MMT is only used sparsely in the developed world, including the United States (ATSDR, 2012). Major petroleum refineries in Canada have voluntarily ceased the use of the additive (ICCT, 2004) due to the harm it may cause to the automobile on-board diagnostic equipment (ATSDR, 2012).

Chronic Effects

The deleterious effect most commonly associated with chronic overexposures to inorganic manganese compounds is a disease referred to as manganism, a neurological disorder caused by manganese affecting the central nervous system (CNS). Manganism, an illness similar in presentation to Parkinson's disease (PD), is viewed as a form of Parkinsonism; however, evidence has indicated that manganism is physiologically differentiated from PD. In addition to the duration and concentration of the exposure, the development of neurotoxic effects following chronic overexposures to manganese will depend on individual factors (Michalke and Fernsebner, 2014). Exposure most often occurs by the inhalation route and the disease progresses through several stages, with the onset and timing of symptoms varying in the literature.

Initially, an individual who is overexposed to manganese may develop headache, insomnia, memory loss, muscle cramps and weakness in the lower extremities, bent posture, and irritability (Barceloux, 1999). As the disease progresses further, hypokinesia, hand tremor, loss of balance, postural instability, mask-like face, speech disturbances, a "cock-like" gait, and other gait abnormalities will develop (Barceloux, 1999; Jankovic, 2005). Symptoms may persist even after removing an individual from the exposure. Diminished libido and impotence in male workers is also reported to accompany manganism. An overexposed individual may develop a condition known as manganese madness (*Locura manganica*), which may proceed or accompany the neurological signs. Manganese madness is characterized by irritability, aggression, sporadic laughter, impulses to sing or dance, aimless running, apathy, somnolence, confusion, hallucinations, poor memory, and anxiety (Barceloux, 1999). Individuals with manganese madness appear to be aware of their inappropriate and irregular actions, but are not capable of controlling them.

Chemical Pathology

Evaluating respiratory effects following inhalation of manganese dusts must be done through pulmonary function testing. No biomarker is recognized for evaluating and determining manganese overexposures and any resulting respiratory issues.

Determining biomarkers of exposure or effect is difficult when evaluating the onset of manganism. There is no validated and/or well-recognized biomarker for manganese exposure (Eastman et al., 2013), which will allow the relationship between an exposure and the presence of manganese within the body to be determined. Individual levels of manganese in the blood or urine are not reliable predictors of chronic workplace exposure; however, determining biological levels of manganese may be useful when evaluating a group exposure in a workplace (Jarvisalo et al., 1992). Evaluations of blood and urine in individuals would only aid in confirming the extent of a current exposure (Barceloux, 1999). While manganese is an essential element and is present at low levels in the body, excess is quickly removed from the blood by the liver. The majority of manganese removed by the liver is excreted in bile and eliminated. Further, there are currently no biochemical indicators that would aid in the detection of early manganese neurotoxicity or any biomarker that would indicate long-term overexposures to manganese. Chronically elevated levels of manganese in the blood may indicate that further evaluation of exposures may be warranted. Individuals with potential overexposures to manganese may show a greater signal intensity in the globus pallidus, striatum, and substantia nigra bilaterally when examined with a T1-weighted magnetic resonance imaging (MRI) scan (Michalke and Fernsebner, 2014; Racette, 2013). Decreased excretion of dopamine in manganese-exposed workers has been observed, but there is no clear relationship between central and peripheral dopamine levels and manganese overexposure.

Mode of Action(s)

Neurological damage from overexposures to manganese does occur; however, the exact manner in which manganese exerts its neurotoxicity has not been fully elucidated. The central nervous system is the principal target of manganese toxicity. Inhalation can transfer manganese to target brain tissue via pulmonary uptake and particulate transfer (Farina et al., 2013), although the exact toxicokinetic mechanism is not yet fully known. Neuropathological changes in the brains of those affected by manganism have been found in the globus pallidus, caudate nucleus, the putamen, and, to some extent, the substantia nigra pars reticulate; these areas have been shown to contain the highest manganese concentrations following overexposures (Guilarte, 2013). The degree of which manganese affects some of these regions is still up for debate.

Manganism differentiates itself from PD, although both diseases do affect functional regions of the brain. Parkinson's disease is characterized by the progressive loss of dopamine-containing neurons in the substantia nigra pars compacta, resulting in the structural loss of dopamine terminal markers in the caudate and putamen (Racette, 2013). A proteinaceous biomarker that is found to be associated with dopamine neuron degeneration, the Lewy body, is found in individuals affected by PD (Guilarte, 2013), but they are absent in the majority of manganism cases. As Lewy bodies are not usually found in individuals with manganism, their absence suggest that the disease affects pathways that are post-synaptic to the nigrostral system. This is supported by the fact that some PD patients who were given levodopa as a treatment experienced relief in some symptoms associated with the disease. Studies indicate that, in most cases, individuals affected by manganism and treated with levodopa do not respond to the treatment (Pal et al., 1999), suggesting that the mechanism of the toxicity is different than that of PD. Although the mechanism leading to the destruction of dopaminergic neurons is unknown, research has indicated that the toxicity is exerted through an oxidative stress mechanism (Michalke and Fernsebner, 2014).

SOURCES OF EXPOSURE

Occupational

Occupational exposures to manganese and compounds occur in a limited number of settings. The first identification of manganese-induced Parkinsonism is attributed to Couper in his 1837 description of health effects observed in workers grinding manganese oxide ore (Aschner and Aschner, 1991; Bowman et al., 2011; Guilarte, 2013; Racette, 2013).

Occupational overexposures to manganese will most likely occur by inhalation of fumes or dusts. In the United States, welders, iron- and steel-manufacturing workers, dry-cell battery production employees, ferromanganese workers (WHO, 1999), power plant employees, and operators of coke ovens (USEPA, 2013) have the potential for overexposures from inhalation of manganese-containing dusts and fumes. Employees of ore mining operations also have the chance for overexposures to manganese; however, this exposure is rare in the Unites States as only a small amount of ore is still mined in North America.

Employees working with MMT have the opportunity for an exposure. Due to its use as a fuel additive, mechanics, garage workers, and taxi drivers may also have the potential to inhale particulates containing manganese (ATSDR, 2012). However, indirect exposures to MMT are thought to be low. This is due to the fact that MMT undergoes a rapid photodegredation (Davis, 1998), reducing to manganese in the atmosphere.

Environmental

The ubiquitous nature of manganese in the environment from both naturally occurring and anthropogenic sources suggests that the general population can be exposed to various concentrations. Additionally, because manganese is an essential trace element, human blood will always have a detectable concentration. The average level of manganese in blood is approximately $9\,\mu g/l$ (approximate range of 4–$15\,\mu g/l$) (Barceloux, 1999). Points of exposure to manganese in the general population include ingestion of food and water, inhalation of manganese particulates and fumes, and dermal contact with manganese-containing media. Food is the most significant exposure point for the general population (Frumkin and Solomon, 1997); intake can vary between individuals and will depend on the diet. Food items that contain manganese include nuts, grains, tea, and legumes (Jankovic, 2005). There have been numerous attempts to quantify the average daily intake of manganese. The greatest range reported for daily intake of manganese was 0.7–$10.9\,mg/day$ (WHO, 2004). Although oral exposure is the primary point of exposure for the general population, the absorption of manganese in the gastrointestinal track is low; approximately 2–5% is absorbed while the rest is excreted (Jankovic, 2005).

In addition to manganese-containing foods, consumer products may be an additional exposure source for the general population. Beyond the use of fertilizers, paints, and cosmetics, individuals who smoke tobacco can be exposed to manganese; the concentration in tobacco was found to range from 155 to $400\,\mu g/g$ (Bernhard et al., 2005). Individuals in countries who still use MMT as a fuel additive may be exposed to manganese via automobile exhaust, although exposures are thought to be low due to the assumed limited use of this product.

Residents living near production facilities using manganese may be an additional point of exposure for the general population. A study of residents living in close proximity to a ferromanganese plant had significant increase in respiratory problems (ATSDR, 2012). A follow-up study of those individuals living closest to the plant found that there was a reduction in symptoms after emission controls were implemented at the facility.

INDUSTRIAL HYGIENE

The Occupational Safety and Health Administration's (OSHA) permissible exposure limits (PEL) (NIOSH, 2005), the National Institute for Occupational Safety and Health's (NIOSH) recommended exposure limits (REL) (NIOSH, 2005), and the American Conference of Governmental Industrial Hygienists' (ACGIH) threshold limit values (TLV) (ACGIH, 2013) for manganese and related compounds are summarized below:

Manganese compounds and fumes:

OSHA PEL: Ceiling $(C) = 5\,mg/m^3$
NIOSH REL:
   Time weighted average $(TWA) = 1\,mg/m^3$
   Short term $(ST) = 3\,mg/m^3$
   $IDLH = 500\,mg/m^3$
ACGIH TLV: $TWA = 0.02\,mg/m^3$
Inhalable Manganese: $TWA = 0.1\,mg/m^3$

Manganese cyclopentadienyl tricarbonyl:

OSHA PEL: $C = 5\,mg/m^3$
NIOSH REL: $TWA = 0.1\,mg/m^3$ [skin]
ACGIH TLV: $TWA = 0.1\,mg/m^3$

Manganese tetroxide:

OSHA PEL: $C = 5\,mg/m^3$
NIOSH REL: $TWA = 1\,mg/m^3$ (Appendix D of NIOSH Pocket Guide)
ACGIH TLV: Not established

Methylcyclopentadienyl manganese tricarbonyl:

OSHA PEL: $C = 5\,mg/m^3$
NIOSH REL: $TWA = 0.2\,mg/m^3$ [skin]
ACGIH TLV: $TWA = 0.2\,mg/m^3$

Inhalation is the primary route of exposure for manganese and related compounds in occupational settings. When working with manganese-containing materials, proper ventilation will aid in reducing employee overexposures. For welders, OSHA recommends natural or forced movement of fresh air, or local exhaust ventilation, to remove fumes from the welders breathing zone (OSHA, 2013). If engineering controls cannot reduce the welder's exposure to fumes, respiratory protection and other personal protective equipment (PPE) may be necessary.

RISK ASSESSMENTS

The USEPA Integrated Risk Information System (IRIS) database lists a reference dose (RfD) for manganese of

1.4×10^{-1} mg/kg-day, which is based on human chronic ingestion data. This represents an intake of 10 mg/day; studies have indicated that this is a safe level of intake. As most individuals will consume 2–5 mg manganese per day (USEPA IRIS, 2012), this RfD is conservative in its protection of the human population, including sensitive subgroups. The IRIS database cautions that, when performing a risk assessment, the essential nature of manganese must be taken into consideration.

The inhalation reference concentration (RfC) reported in the EPA IRIS database is 5.0×10^{-5} mg/m^3. This concentration was determined using studies that evaluated the relationship between occupational exposure to manganese dioxide and impairment of the neurobehavioral function. This value was determined using an uncertainty factor of 1,000 and a modifying factor of 1.

The USEPA IRIS system has assigned a "D" classification to manganese with relation to the potential for an exposure to result in the development of cancer in humans: not classifiable as to human carcinogenicity (USEPA IRIS, 2012). IRIS indicates that the basis for this classification stems from the inadequacy of the existing studies that attempt to assess the carcinogenic potential of manganese.

The Agency for Toxic Substances and Disease Registry (ATSDR) lists a chronic inhalation Minimal Risk Level (MRL) concentration of 0.3 μg/m^3 (ATSDR, 2012). The MRL is established as an estimate of a chronic daily exposure of manganese that is likely to be without an appreciable risk of a noncancer adverse health outcome over a specific time period.

REFERENCES

American Conference of Governmental Industrial Hygienists (ACGIH) (2013) *Threshold Limit Values (TLVs) for Chemical Substances and Physical Agents and Biological Exposure Indices (BEIs)*, Cincinnati, OH: American Conference of Governmental Industrial Hygienists.

Aschner, M. and Aschner, J.L. (1991) Manganese neurotoxicity: cellular effects and blood-brain barrier transport. *Neurosci. Biobehav. Rev.* 15(3):333–340.

Aschner, M., Guilarte, T.R., Schneider, J.S., and Zheng, W. (2007) Manganese: recent advances in understanding its transport and neurotoxicity. *Toxicol. Appl. Pharmacol.* 221(2):131–147.

ATSDR (2012) Agency for Toxic Substances and Disease Registry Toxicological Profile for Manganese. Available at http://www.atsdr.cdc.gov/toxprofiles/tp151.pdf.

Barceloux, D.G. (1999) Manganese. *J. Toxicol. Clin. Toxicol.* 37(2):293–307.

Bernhard, D., Rossmann, A., and Wick, G. (2005) Metals in cigarette smoke. *IUBMB Life* 57(12):805–809.

Bowman, A.B., Kwakye, G.F., Herrero Hernandez, E., and Aschner, M. (2011) Role of manganese in neurodegenerative diseases. *J. Trace Elem. Med. Biol.* 25(4):191–203.

Davies, T.A. (1946) Manganese pneumonitis. *Br. J. Ind. Med.* 3:111–135.

Davies, T.A. and Harding, H.E. (1949) Manganese pneumonitis: further clinical and experimental observations. *Br. J. Ind. Med.* 6(2):82–90.

Davis, J.M. (1998) Methylcyclopentadienyl manganese tricarbonyl: health risk uncertainties and research directions. *Environ. Health Perspect.* 106(Suppl. 1):191–201.

Eastman, R.R., Jursa, T.P., Benedetti, C., Lucchini, R.G., and Smith, D.R. (2013) Hair as a biomarker of environmental manganese exposure. *Environ. Sci. Technol.* 47(3):1629–1637.

Farina, M., Avila, D.S., da Rocha, J.B., and Aschner, M. (2013) Metals, oxidative stress and neurodegeneration: a focus on iron, manganese and mercury. *Neurochem. Int.* 62(5):575–594.

Frumkin, H. and Solomon, G. (1997) Manganese in the U.S. gasoline supply. *Am. J. Ind. Med.* 31(1):107–115.

Guilarte, T.R. (2013) Manganese neurotoxicity: new perspectives from behavioral, neuroimaging, and neuropathological studies in humans and non-human primates. *Front. Aging Neurosci.* 5(23):1–10.

HSDB (2001) Hazardous Substances Database: Methylcyclopentadienyl Manganese Tricarbonyl. Available at http://toxnet.nlm.nih.gov/cgi-bin/sis/search/f?./temp/~M8mMC8:1 (accessed February 5, 2014).

ICCT (2004) International Council on Clean Transportation: Status Report Concerning the Use of MMT in Gasoline. Available at http://www.theicct.org/sites/default/files/publications/MMT_ICCT_2004.pdf.

Jankovic, J. (2005) Searching for a relationship between manganese and welding and Parkinson's disease. *Neurology* 64(12):2021–2028.

Jarvisalo, J., Olkinuora, M., Kiilunen, M., Kivisto, H., Ristola, P., Tossavainen, A., and Aitio, A. (1992) Urinary and blood manganese in occupationally nonexposed populations and in manual metal arc welders of mild steel. *Int. Arch. Occup. Environ. Health* 63(7):495–501.

Klaassen, C.D. (2008) *Casarett & Doull's Toxicology: The Basic Science of Poisons*, 7th ed., New York: McGraw-Hill, pp. 955–956.

Krebs, R.E. (2006) *The History and Use of Our Earth's Chemical Elements: A Reference Guide*, 2nd ed., Westport, CT: Greenwood Press.

Lide, D.R. (2006) *CRC Handbook of Chemistry and Physics*, 87th ed., Boca Raton, FL: CRC Press, pp. 4-19–4-31.

Michalke, B. and Fernsebner, K. (2014) New insights into manganese toxicity and speciation. *J. Trace Elem. Med. Biol.* 28(2):106–116.

Michalke, B., Halbach, S., and Nischwitz, V. (2007) Speciation and toxicological relevance of manganese in humans. *J. Environ. Monit.* 9(7):650–656.

National Institute of Occupational Safety and Health (NIOSH) (2005) *Pocket Guide to Chemical Hazards*, Cincinnati, OH: NIOSH Publications.

OSHA (1978) Occupational Health Guideline for Manganese. U.S. Department of Labor. Available at http://www.cdc.gov/niosh/docs/81-123/pdfs/0379.pdf.

OSHA (2013) OSHA Fact Sheet: Controlling Hazardous Fumes and Gases During Welding. Available at https://www.osha.gov/Publications/OSHA_FS-3647_Welding.pdf.

Pal, P.K., Samii, A., and Calne, D.B. (1999) Manganese neurotoxicity: a review of clinical features, imaging and pathology. *Neurotoxicology* 20(2–3):227–238.

Racette, B.A. (2013) Manganism in the 21st century: the Hanninen lecture. *Neurotoxicology*, doi: 10.1016/j.neuro.2013.09.007.

Rodier, J. (1955) Manganese poisoning in Moroccan miners. *Br. J. Ind. Med.* 12(1):21–35.

Southwood, T., Lamb, C.M., and Freeman, J. (1987) Ingestion of potassium permanganate crystals by a three-year-old boy. *Med. J. Aust.* 146(12):639–640.

Toxnet (2008) Manganese Compounds. Available at http://toxnet.nlm.nih.gov/cgi-bin/sis/search/a?dbs+hsdb:@term+@DOCNO+6945 (accessed January 3, 2014).

USEPA (1984) Health Assessment Document for Manganese. Available at http://cfpub.epa.gov/ncea/cfm/recordisplay.cfm?deid=42107#Download.

USEPA (2013) Technology Transfer Network—Air Toxics Web Site: Manganese Compounds. Available at http://www.epa.gov/ttn/atw/hlthef/manganes.html.

USEPA IRIS (2012) United States Environmental Protection Agency Integrated Risk Information System. Available at http://www.epa.gov/iris/ (accessed January 6, 2014).

WHO (1999) Concise International Chemical Assessment Document 12: Manganese and Its Compounds. Available at http://whqlibdoc.who.int/publications/1999/924153012X.pdf.

WHO (2004) Concise International Chemical Assessment Document 63—Manganese and Its Compounds: Environmental Aspects. Available at http://www.who.int/ipcs/publications/cicad/cicad63_rev_1.pdf.

23

MERCURY

Frank D. Stephen

First aid: Eye contact: Immediately flush eyes with plenty of water for at least 15 min, lifting lower and upper eyelids occasionally. Get medical attention immediately; skin contact: Immediately flush skin with plenty of water for at least 15 min and remove contaminated clothing and shoes. Get medical attention immediately. Wash clothing before reuse. Thoroughly clean shoes before reuse. Ingestion: Do *not* induce vomiting. Never give anything by mouth to an unconscious person. Get medical attention immediately. Inhalation: Get medical aid immediately. Remove from exposure and move to fresh air immediately. If breathing is difficult, give oxygen.

Target organ(s): Nervous system and kidney.

Occupational exposure limits: OSHA PEL 0.1 mg/m^3 (8-h workday). NIOSH REL 0.05 mg/m^3 (10-h TWA). ACGIH TLV 0.01 mg/m^3 (alkyl forms of mercury). STEL 0.03 mg/m^3 (inorganic mercury). TWA 0.0025 mg/m^3 (inorganic mercury). TWA 0.1 mg/m^3 (aryl forms of mercury).

Reference values: MRL chronic inhalation: 0.2 µg/m^3. Acute oral: 7 µg/kg/day. Chronic oral: 0.3 µg/kg/day. RfD oral methylmercury: 1×10^{-4}. Phenylmercuric acetate: 0.08 µg/kg/day.

RfC Elemental mercury: 3×10^{-4} mg/m^3.

Risk/safety phrases: *R26-61-48/23-50/53*: Mercury is very toxic if inhaled. May cause harm to the unborn child. Harmful: danger of serious damage to health by prolonged exposure through inhalation. Very toxic to aquatic organisms and may cause long-term adverse effects in the aquatic environment. *S45-53-60-61*: In case of accident or if you feel unwell, seek medical advice immediately (show label where possible). Avoid exposure—obtain special instructions before use. This material and its container must be disposed of as hazardous waste. Avoid release to the environment. Refer to special instructions/safety data sheets.

BACKGROUND AND USES

With all due respect to Lewis Carroll and the World of Alice's Looking-Glass, for almost 300 years the English language has known the colloquial phrase "mad as a hatter," which has been/is used to describe the psychologically unbalanced or insane. The phrase derives from behaviors displayed by hat makers who were exposed daily to trace amounts of mercury that remained in the felt used to produce popular hats of that era. However, recognition that mercury exposure causes poisonings goes back to the time of ancient Rome. Mercurialism has been documented in a diversity of occupations such as miners, hat makers, and in producers of detonators, thermometers, and barometers. Today, the most common risk of mercury exposure for humans comes from eating contaminated flesh from marine mammals and other fish species or through the release of elemental mercury and mercury vapor from dental amalgams.

The dangerous use of mercury and mercury-based compounds can be traced back to the earliest recorded human civilizations. As long ago as 3000 years mercury was used in pharaoh's tombs in Egypt; and to prepare red ink in China (Goldwater, 1972). Mercurialism in Idrian quicksilver miners was described by Paracelsus as early as the 1550s (Hunter, 1974). Similarly, as early as 1713 Bernardino Ramazzini

Hamilton & Hardy's Industrial Toxicology, Sixth Edition. Edited by Raymond D. Harbison, Marie M. Bourgeois, and Giffe T. Johnson.
© 2015 John Wiley & Sons, Inc. Published 2015 by John Wiley & Sons, Inc.

documented that few miners "reach old age, and even when they do not die young their health is so terribly undermined that they pray for death" and that mirror makers "become palsied and asthmatic from handling mercury."

Medicinal use of mercury in treating human diseases is well documented. For example, throughout the Middle Ages mercury-based compounds were incorporated into creams that were used to treat infections. Mercury was once thought to be a treatment for syphilis and calomel (mercurous chloride) has been used in a variety of medicines intended to treat various maladies, including teething, yellow fever, and constipation; as well as in cosmetics and skin lightening creams. Famed signer of the Declaration of Independence, Dr. Benjamin Rush, even used it as a purgative and to induce salivation in patients as a therapy to rid the body of "impurities." Various forms of organic mercury were used as antifungal agents in seed grains. Mercurochrome was used for decades as a topical antiseptic. Most recently, thimerosal (also known as Merthiolate) has been used as a bacteriostatic agent in the preparation of vaccines.

Human avoidance of exposure to mercury is almost impossible. Due to anthropogenic and natural sources mercury is ubiquitously present in the environment as elemental, inorganic, and organic forms. Also known as quicksilver, mercury is an extremely toxic liquid metal (Hg^0) but due to its high vapor pressure is readily released into the surrounding as a vapor. Inorganically, mercury is mined as mercuric sulfide in cinnabar and can be found bound to chlorine and oxygen. Organically, with regard to human exposure, the most important species of mercury is methylmercury. Human health risks have been reported for exposure to all forms of mercury but methylmercury exposure may be of greatest concern.

The use of mercury in the United States is slowly being phased out. Consumption of mercury-contaminated seafood and fish accounts for over 95% of ingested methylmercury in the United States (Mahaffey et al., 2004). Many of the historical uses of mercury have ended and the use of mercury in various instruments (e.g., barometers and thermometers) is ending. Several current uses for mercury include batteries, mercury vapor lamps, dental amalgam, and preservatives. Other folk uses of mercury include religious ceremonies, skin-lightening creams, and herbal medicines. The Agency for Toxic Substances and Disease Registry (ATSDR) concluded that industrial hygiene controls are resulting in a decrease in the amount of and risk from exposure.

PHYSICAL AND CHEMICAL PROPERTIES

Mercury (symbol Hg; Cas # 7439-97-6) is the only metal that is a liquid metal at room temperature. Occurring naturally as

TABLE 23.1   Physical and Chemical Properties of Mercury

Molecular weight	200.59
Color	Silver-white
Physical state	Liquid
Flash point	Not flammable
Melting point	$-38.87\,°C$
Boiling density	$356.72\,°C$
Density at $20\,°C$	$13.534\,g/cm^3$
Water solubility	$0.28\,\mu mol/l$
Vapor pressure	$2 \times 10^{-3}\,mmHg$

mercuric sulfide (in cinnabar ore) because of its low boiling point can be isolated readily from the ore and condensing the resulting vapor to form a metal that is 99.9% pure. Solid mercury is ductile and very malleable and is soft enough to cut with a knife. A poor conductor of heat, mercury can conduct electricity but behaves much like a noble gas. It does not react with most acids but can react with very strong acids and dissolves readily to form amalgams with many metals, including gold, zinc, and aluminum. Important physical and chemical properties are listed in Table 23.1.

ENVIRONMENTAL FATE AND BIOACCUMULATION

Mercury cycles through various environmental phases by exchange from ground to air and back again. Metallic and dimethylmercury, which are volatile, will be released as mercury vapor that can travel long distances before being redeposited. When found in surface waters and soils it will degas into the surrounding air where natural currents and winds spread the materials until they are deposited back on the surface waters and soils. The majority of mercury returned to the soil or water is by wet partition and accounts for almost all of the mercury found in lakes with no other input source. Inert mercury will deposit bound to particulates in aerosols. Once deposited, mercury must adsorb to soil or sediment particulates or be returned to the atmosphere. This cycle continues with a portion of the mercury revolatilizing into the atmosphere in each cycle.

Mercury can exist as either mercurous or mercuric mercury in soil and sediment. Transport is dependent on the form of the compound and the predominant form is mercury in the mercuric state. Bound to particulates in the water table, these compounds will be transported through the water column to the sediments. Very little of the mercury that reaches sediments is returned to the water column (Bryan and Langston, 1992). Conversion of the soil- or sediment-bound mercury to mercury sulfide precedes bioaccumulation into the food streams or bioconversion into more volatile or soluble forms that reenters the atmosphere (EPA, 1984).

Various factors affect the rate of mercury vaporization from and absorption to surface waters, sediments, and soils. In warmer weather, the amount of microbial reduction of mercury increases resulting in more mercury volatilizing from the soil (Lindberg et al., 1991). Once bound to particulate matter, mercury does not desorb readily, therefore, leaching is not likely to occur in soils. Presence of other metals, such as aluminum, is favorable for mercury adsorption. An inverse relationship between acidic soils and mercury absorption exists (Schuster, 1991).

Of great concern with regard to human mercury exposure is the bioaccumulation of methylmercury in the aquatic food chain. Methylmercury, the most common form of mercury, is highly soluble. Once in water columns it can enter and accumulate in the tissue of various sea organisms. Greater biomagnifications have been reported in waters with lower pH (Ponce and Bloom, 1991). Mercury levels will continue to biomagnify through the food chain and be highest in fish at the high end of the food chain, which are then consumed by humans.

ECOTOXICOLOGY

Mercury, in various speciated forms, occurs naturally and is found throughout the environment. As described previously, a continual cycle of deposition/volatilization/reemission occurs in soils and sediments resulting in bioaccumulation and biomagnifications in many terrestrial and aquatic food chains. The presence of mercury in some wetland results in game fish with tissue mercury levels equal to or greater than that seen in waters contaminated by chlor-alkali processing plants. Predatory fish (e.g., shark, tuna, and mackerel) have up to 20 times as much methylmercury as non-predatory fish (e.g., pollock and cod) or shellfish (e.g., scallops) (Mahaffey et al., 2004). Wetlands, waterways adjacent to wetlands, and fresh water lakes with low pH or low alkalinity are extremely sensitive to mercury poisoning. Of great concern in the food chain is methylmercury due to its potential to cause neurotoxic effects in humans.

Acute Effects

Contamination of aquatic systems with high levels of methylmercury is well documented. Over 60 years ago at Minamata Bay, Japan, methylmercury poisoning was observed in a variety of species. Mercury, used as a catalyst by the Chisso Company, was being discharged into the waters of Minamata Bay. In 1956, neurological diseases in over 30 residents around the bay were documented. Fish in the bay area were being killed in such great numbers that they could be picked out of the water by hand. Within several years other terrestrial species, such as crows and cats, also began dying and similar occurrences were reported in nearby areas.

Mercury-based fungicides had also been used for decades as a coating sprayed onto seeds. In Sweden, in the late 1960s, mercury pollution from chlor-alkali plants, the paper industry, and seed coatings were responsible for methylmercury food chain accumulation. As the decade progressed, an increased mortality among seed-eating and predatory birds occurred.

Chronic Effects

A major effect of mercury poisoning on nonmammalian species is reproductive failure. Methylmercury's reproductive effects on various species of fish include decrease in adult spawning success by decreasing hatching success or embryological development and by inhibiting gonadal development in early adult members. Biomagnification of methylmercury in aquatic food systems has led to a parallel increase in methylmercury in avian food systems. Cases of poisonings and reproductive failures (e.g., impaired laying and reduced breeding, hatching, and nesting) were reported in Sweden among Sea eagles, ospreys, and other fish-eating birds. Chronic wasting and emaciation correlated with high tissue mercury levels in Great White Herons in the Florida Everglades, Grey Herons in the Netherlands, and Common loons in Eastern Canada.

MAMMALIAN TOXICOLOGY

Acute Effects

In any chemical form mercury is toxic. Observed effects are dependent on the form of mercury, dosage, and route of exposure. Ingestion of corrosive sublimate, another name for mercuric (II) chloride, produces failure of kidney function and gastric distress and as little as 1 g may be fatal. Upon ingestion other gastrointestinal (GI) manifestations include a burning sensation followed by tightening of the esophagus, "whitening" of the mucous membranes of the mouth and throat, and abdominal pain with bloody vomit may occur. Subsequently, the exposed individual may have a persistent feeling of having to move their bowls and passage of bloody fecal material may result.

Although more likely to produce acute symptoms, inhalation of mercury vapor is not as immediately dangerous as the ingestion pathway. Effects of inhalation can be delayed and exposure manifests itself in nondescript signs and symptoms. Initial signs mimic "metal fume fever" and include coughing and sneezing, fever and chills, dry mouth, difficulty in breathing, and muscle and headache. Lingering effects may include GI distress, stomach pain, nausea and vomiting, and changes in the urinary output, including increased albumin levels.

As observed with fish and avian species, mortality is also found among mammals poisoned with mercury. Exhibition of classical mercury poisoning, including convulsions, weakness, ataxia, lethargy, tremors, and paralysis all occurred in animals that died from methylmercury intoxication. The decline of mink and otter populations in the Southeastern United States has been correlated to soil mercury concentrations.

Chronic Effects

Metallic and Inorganic Mercury   Modern industrial mercury poisonings do not result in the prototypic signs and symptoms of classical mercury poisonings. Workers exposed chronically to low levels of inorganic mercury exhibit inflammation of the mouth, muscle tremors, and psychotic irritability. The severe symptoms associated with other uses of mercury (e.g., GI irritability and pain, nausea, vomiting, bloody stools, and kidney involvement) are rare in industrial poisonings.

Industrial exposures to metallic mercury generally involve mercury vapor or aerosols containing mercury salts. Mercury absorption from the GI tract is poor (Cantor, 1951), thus ingestion of mercury is not considered as an important route of entry for humans. Skin penetration has been calculated to be very slow ($0.01–0.04$ ng $Hg/cm^2/min/ng$ Hg/cm^3 air) and is considered as a minor route of entry. Thus, the most likely and potentially dangerous route of exposure is inhalation of metallic mercury vapor.

Historically, investigators have examined the effects of inhalation of metallic mercury while research into the effects of mercury salt inhalation has not been performed. Death due to respiratory failure has been documented after high level acute exposure to volatile mercury released during heating; however, accurate levels of atmospheric mercury were not obtained for those case studies. Systemic toxicity is rare but specific target organs of mercury toxicity are the kidneys and the central nervous system, however, effects have been seen in the cardiovascular and respiratory systems after very high levels of exposure.

Mercurialism has been observed in a variety of industrial settings and has been scientifically documented for almost 100 years and anecdotally documents for half of a millennia. In the Spanish mines at Almaden, slaves and later convicts, who were used to mine cinnabar for the isolation of mercury, displayed the classic signs of mercury poisoning. More famous, and as described earlier in the chapter, hat makers from France, and later England and the United States, used mercury in the production of felt. An unhealthy occupation at best, hatters suffered from tremors known as "hatters shakes" and studies conducted in the early twentieth century found that American hat makers suffered gum swelling, loss of teeth, malodorous breath, as well as classical "hatter shakes." Studies by the U.S. Public Health Service found that 10% of

all hatters studied suffered various neurological psychosis, as well as the other classical signs and symptoms of mercury poisoning. In 1941, the Connecticut Commissioner of Health conference on mercury poisoning banned the use of mercury in American hat production.

Although the toxicity of mercury and mercury-based chemicals was well known even early twentieth century industry was not been free of the risks to health from mercury use. The manufacture of detonators used in World War I exposed workers to mercury fulminate ($Hg(CNO)_2$) causing dermatitis, ulceration of the face, neck, arms and hands, and genitalia, as well as neurological problems and nephrosis. Paresthesia, erratic eye movements, dexterity and fine motor speed, and verbal memory have been associated with lower body mercury concentrations (as measured by parameters such as blood and hair) than previously thought needed to produce such effects. Due to mercury's ability to volatilize at room temperature workers were easily exposed during the production of several products, including thermometers, barometers, batteries, solder, and chlorine. Exposure to high levels of calomel ($HgCl$) from tracer ammunition and calomel electrodes, as well as low level exposure from its use in teething powders and laxatives, can induce mercury poisoning or acrodynia.

The risk of mercurialism even exists in dental practices and continues to be of concern today. First introduced into dentistry over 150 years ago, dental amalgam is the principal source of mercury vapor exposure to all workers in the dental field, including dentists, hygienists, and technicians. Debate over the safety in preparation and handling of amalgam has raged for over 100 years. Dental amalgam is around half metallic mercury and half other metals, including zinc, silver, and copper. The amalgam is sold, encapsulated, with the liquid mercury separated from the other metals (in powder form) by a plastic divider. During preparation, the divider is broken and the components mixed. Several studies documented cases of mercury poisoning in dental offices. Today, there is still debate and there is no accepted answer as to whether working with amalgam poses any long-term dangers to the dentists, their workers, or their patients. Finally, it is not known how much mercury vapor is given off by the solidified amalgam once set into the teeth and a dentist may never see a case of poisoning in a patient during their practice.

Chronic effects of mercury exposure are not limited to the nervous system. Recently conducted studies have correlated various cardiovascular (increased myocardial infarction and carotid atherosclerosis), endocrine (increased serum T4 levels and T4/T3 ratios), and male (decreased sperm number, motility, and morphology) and female (decreased conception and increased stillbirths) reproductive endpoints, and body mercury concentrations.

Organic Mercury   Similar to metallic mercury the effect of organic mercury is route dependent with target organs, including the central nervous system, liver, and the GI tract. The main forms of organic mercury to which people have been exposed include methylmercury, ethylmercury, and diethylmercury. However, organic mercury compounds are much more toxic than their inorganic counterparts and can produce severe symptoms through skin absorption.

Alkyl mercury compounds (methyl- and ethylmercury) have long been used in agriculture as pesticides and fungicides. Organomercury compounds also can produce a variety of responses when exposure is via the dermal route. Exposed workers exhibited difficulty with speech, ataxia and unsteady gait, slurred speech (dysarthria), and tremors (Hunter et al., 1940). Skin rashes and blisters, burns, itchy and pruritic, papular eruptions, and erythema and urticaria (Morris 1960; Torresani et al., 1993) have all been observed in workers after short-term and long-term exposures.

Consumption of organic mercury-containing food produced widespread poisoning during the mid-twentieth century. Over 900 individuals died and nearly 9000 individuals suffered severe health effects, including paralysis, after eating methylmercury-contaminated fish from Minamata Bay. Epidemics of neurological disease in 1956, 1960, and 1971–1972 were observed in Iraq when grain treated with organic mercury fungicides were ground into flour (Bakir et al., 1973; Al-Mufti et al., 1976). More than 6000 people were affected during the 1971–1972 incident. One of the largest exposures was an outbreak documented by Clarkson in which almost 8000 individual in Iran, Iraq, Pakistan, Ghana, and Guatemala were poisoned when mercury-containing fungicide-coated seed grains intended for planting were eaten (Clarkson, 1977). Workers exposed to organic mercury in pharmaceutical, seed-treating, farming, and pesticides industries also exhibited neurotoxicity due to mercury poisoning (Bidstrup, 1964; Dales, 1972).

The most controversial use of an organic mercury compound involves thimerosal. Used in vaccines for decades as a preservative, concern over thimerosal use peaked in the 1990s as to whether the dose of organic mercury delivered might cause neurological conditions (e.g., autism) in children. Rationale for this concern was based on three major factors, namely, the Minamata Bay poisonings, Iraqi wheat contamination (described above), and an expanded infant vaccine schedule. Several scientific and medical bodies, such as the Institutes of Medicine, U.S. Food and Drug Administration (FDA), and the World Health Organization, conducted extensive investigations and rejected the hypothesis that link between autism and thimerosal use exists. In 2001, vaccine manufacturers eliminated the use of thimerosal as a preservative in products intended for infants. Interestingly, in the decade of thimerosal, removal rates of autism have continued to increase (Doja and Roberts, 2006).

Chemical Pathology

Although occurring in inorganic and organic forms, ultimately it is the mercuric divalent cation, mercuric mercury (Hg^{2+}), that is responsible for the toxic effects of mercury. The divalent ion has a high affinity for sulfhydryl groups and is able to quickly react and move between thiol-containing compounds (Rabenstein et al., 1982; Govindaswamy et al., 1992). In this way, the mercuric ion acts to precipitate protein and to inhibit enzymes-containing sulfhydryl groups. Regardless of the form of mercury to which a person is exposed, ultimately toxicity is dependent on mercury accumulation in sensitive tissues.

Absorption rates of mercury are dependent upon the form of mercury and the route of entrance into the body. Inhaled metallic mercury and mercury vapor rapidly diffuse through the lungs across the alveolar membrane (Berlin et al., 1969; Clarkson, 1989) whereas inorganic mercury compounds are absorbed poorly in the lungs. This difference may be due to the rapid clearing of larger inorganic particles, which are deposited in the upper respiratory system (Friberg and Nordberg, 1973). Inhaled mercury vapor is a monatomic gas released from liquid metallic mercury. Absorbed mercury vapor can also be converted to the toxic divalent cation, and when deposited into tissue can exert the same toxic effects as observed with inorganic mercuric mercury. Due to its property as a vapor, this form of mercury will readily diffuse through tissues and is soluble in lipids. Mercury vapor can dissolve into the bloodstream and pass quickly across the blood–brain barrier.

While the oral absorption rates of metallic and inorganic mercury are low the mechanism for absorption appears to differ. Metallic mercury is poorly absorbed in the GI track and the vapor given off from metallic mercury is of more concern. It is thought that metallic mercury is converted to the divalent form, which is then able to bind sulfhydryl groups and produce its effects. Inorganic mercury absorption in the GI track depends on solubility, molecular dissociation, and intestinal lumen concentration (Piotrowski et al., 1992). Mercurous compounds are not well absorbed due to their low solubility (Friberg and Nordberg, 1973). However, upon entry into the body, dissociation of the molecule occurs releasing a divalent cation and an uncharged mercury atom (which is later converted into the divalent form).

Absorption of mercuric mercury is better understood thanks to a study conducted in Finland in the early 1970s. Volunteers were given known quantities of radioactive mercuric mercury. Two half times were discovered; a 2-day half-time in which unabsorbed mercury was released from the body and a longer 41-day half-time. It was calculated that about 7% of the dose was absorbed (Rahola et al., 1973; Hattula and Rahola, 1975).

The mechanism(s) by which mercury is absorbed by the intestine is not clear. Upon absorption, a diffusible form of mercury readily enters the blood stream while a nondiffusible form remains in the tissues (Halbach and Clarkson, 1978; Magos, 1967). In the first few days post-exposure, body clearance is first accomplished by fecal excretion of unabsorbed mercury. Absorbed mercury can travel to the liver and was shown to be incorporated into bile salts (Harris et al., 2003), and secreted as a conjugate with reduced glutathione (Ballatori and Clarkson, 1985). Although the exact mechanisms by which mercury is dispersed throughout the body are not well understood, once released into the circulatory and digestive systems, mercury can be dispersed throughout the body with storage occurring in the kidneys, liver, spleen, intestinal wall, heart, skeletal muscle, and lungs.

Perhaps the most hazardous form of methylmercury exposure occurs prenatally. Damage included psychomotor retardation and severe mental retardation along with diffuse and widespread damage to the developing brain (Choi, 1991). A prenatal threshold of 10–20 mg Hg/g of hair was calculated; nearly 1/5th that for an adult threshold (Clarkson, 1992). In rodents, the mechanism for this damage appeared to involve a combination of mitotic arrest (Sager et al., 1984) and inhibition of neuronal migration during development, probably due to depolymerization of microtubules (Abe et al., 1975).

Mode of Action(s)

As described earlier in the chapter, divalent mercury has an affinity for the sulfhydryl groups present on proteins. Since the majority of membranous and cellular proteins contain sulfhydryl groups it is not clear whether there is a specific target for mercury binding that helps it get into the cell. Further, this happenstance binding may explain the large potential number of targets affected by mercury. The formation of mercaptides may alter cell membrane permeability (Sahaphong and Trump, 1971). Nonspecific binding to cellular proteins would also explain the wide variety of affected cellular systems such as enzymes, transport processes, and even affecting cells structural proteins (Bulger, 1986). Mercury may also bind to other reactive groups that are not as preferred as sulfhydryl groups, including amine and carboxyl groups, thus, making proteins a prime target. Thus, other potential cellular targets could include any cellular process that is protein dependent such as DNA replication and protein synthesis, immune function, neuronal transmission, and calcium homeostasis.

While the inorganic and organic forms of mercury appear to have a similar mechanism of toxicity the degree of their toxicity appears to be dependent on their accumulation in sensitive tissues. The primary targets of mercury toxicity are the central nervous system and the kidneys and it has been observed that mercury rapidly accumulates in these tissues (Rothstein and Hayes, 1960; Somjen et al., 1973).

In neurons, mercury has been found in the cytoplasm resulting in degradation of cellular organelles. Neuronal protein synthesis was inhibited after rats were exposed to methylmercury (Syversen, 1977). Interperitoneal injections of methylmercuric chloride in rodents produced an increase in oxygen species formation in brain tissue (Ali et al., 1992; LeBel et al., 1990, 1992) leading to the theory that oxidative stress may be a mechanism by which neuronal degeneration occurs (Sarafian and Verity, 1991). Further, neurons may demonstrate high mercury sensitivity due to the lack of endogenous glutathione content and/or redox activity.

Mercury toxicity in the kidney appears to be related to its accumulation in the proximal tubules of the nephron where it binds to proteins resulting in inhibition of protein synthesis. As observed in neurons, oxidative stress is another mechanism by which mercury may exert its effects on cells. Alterations of the mitochondrial membrane result in an increase in hydrogen peroxide formation and increased susceptibility to iron-dependent lipid peroxidation (Lund et al., 1993). Oxidative stress in kidney cells due to alteration in mitochondrial calcium homeostasis has also been postulated. Increases in porphyrinuria have been observed after mercury-induced oxidative damage.

Intracellular mercury may also have direct effects on the function of the cytoskeletal protein, tubulin. Methylmercury binds to tubulin resulting in inhibition of microtubule formation (Vogel et al., 1985), as well as microtubular polymerization inhibition and the disruption of already existing microtubules (Duhr et al., 1993). Mercury binding to tubulin induced microtubule damage (Sager et al., 1982) and subsequently inhibited mitosis (Sager et al., 1983) causing cell injury. Damage to the cerebral cortex, including a decrease in mitotic figures, was also observed in developing rats exposed to methylmercury. Disruption of the developing nervous system by methylmercury has been hypothesized since cell migration is also microtubule dependent.

Another potential mechanism of mercury-induced neurotoxicity is the inhibition of certain nervous system's enzymatic function. Protein kinase C activity was decreased after exposure of rat brain homogenates to either mercuric chloride or methylmercuric chloride (Rajanna et al., 1995). One dose of mercuric chloride given intraperitoneally was sufficient to inhibit, as well as alter the location of cerebral alkaline phosphatase (Albrecht et al., 1994).

DNA synthesis is yet another target of mercury-induced toxicity. In a human *in vitro* model, mercuric ion was found to inhibit elongation of new DNA strands by decreasing DNA polymerase activity and fidelity (Sekowski et al., 1997).

SOURCES OF EXPOSURE

Occupational

Historically, the potential for mercury exposure existed in a large variety of occupations. The highest mercury occupational exposures have been found in chlor-alkali production facilities, cinnabar mining, as well as the manufacture of barometers, thermometers, fluorescent lights, and other industrial instruments. The main route of this occupational exposure is inhaled mercury vapor. At the end of the twentieth century it was estimated that over 150,000 American workers were still being exposed to mercury in the workplace (RTECS, 1998). The use of dental amalgam is still a source of mercury vapor among dentists, hygienists, and technicians. Exposure to mercury vapor can occur in painters using exterior paints containing phenylmercuric acetate. Mercury-based pigments are found in paints and pastels used by commercial artists. The analysis of biological and environmental samples and the synthesis of mercury-containing compounds is a potential source of exposure among chemists. Other occupations with a potential for exposure include individuals handling or analyzing mercury-containing environmental samples, and workers disposing of or recycling mercury-contaminated wastes.

Environmental

Although historically present in a wide variety of commercial and household products, current potential sources of mercury exposure to the general population is limited. It has been estimated that the European and North American daily mercury intake, from all sources, is found to be 6.6 µg (WHO, 1990). Currently, the greatest potential nonoccupational sources of this exposure are consumption of mercury-contaminated seafood and dental amalgam. Subsistence and recreational hunters/anglers consuming mercury-tainted fish and/or marine mammals or other species living near mercury-contaminated waters are at highest risk. Citizens in communities near mercury mining or production sites, public and medical incinerators, or recycling plants risk exposure through airborne contamination. Finally, a risk still exists from other minor use of mercury-containing products such as cosmetics or herbal remedies, religious usage, or unintentional mercury spills.

INDUSTRIAL HYGIENE

Exposure Limits

ACGIH   The American Conference of Industrial Hygienists (ACGIH) has established a threshold limit value (TLV) of $0.01 \, mg/m^3$ for alkyl forms of mercury and a STEL of $0.03 \, mg/m^3$, an 8-h time-weighted average (TWA) of $0.025 \, mg/m^3$ for inorganic forms of mercury, and $0.1 \, mg/m^3$ for aryl forms of mercury.

NIOSH   The National Institute for Occupational Safety and Health (NIOSH) has established a recommended exposure limit (REL) (10-h TWA based on a 40-h work week) of $0.05 \, mg/m^3$ based on the risk of central nervous system damage, eye, skin, and respiratory tract irritation.

OSHA   The Occupational Safety and Health Administration (OSHA) has established a PEL (8-h TWA) of $0.1 \, \mu g/m^3$ for mercury vapor. Engineering controls and work practices must be instituted when workers could be exposed to mercury such that mercury concentrations do not exceed the ceiling value of $0.1 \, mg/m^3$ (OSHA, 1974).

RISK ASSESSMENTS

Listings as Hazardous Waste

Mercury has been classified as a hazardous waste and has been registered in various publications as such, including:

Section 112 of The Clean Air Act lists mercury compounds as one of the 188 chemicals inventoried as a hazardous air pollutant.

Section 313 of Superfund Amendments and Reauthorization Act (SARA) designate mercuric acetate, mercuric chloride, and mercuric oxide as extremely hazardous substances.

Title III of the Emergency Planning and Community Right-To-Know Act of 1986 requires operators of facilities that use or store mercuric acetate, mercuric chloride, and mercuric oxide in or on their facility to establish a program for emergency response and for notifying the public of accidental releases.

Section 307 of The Clean Water Act lists mercury as a "priority pollutant."

Mercury is listed as a hazardous waste under Resource Conservation and Recovery Act (RCRA) and has been assigned Environmental Protection Agency (EPA) Hazardous Waste No. U151 and is banned from land disposal until mercury-containing compounds are treated by retorting or roasting.

IRIS   The EPA's Integrated Risk Information System (IRIS) oral and carcinogenicity assessments for mercury were last revised in 1997. Based on the limited human evidence, the EPA has listed mercury as a class D, not classifiable as to human carcinogenicity based on inadequate data. The reference concentration (RfC) for mercury is $3 \times 10^{-4} \, mg/m^3$ in air and is based on a lowest observed adverse effect level (LOAEL) of 0.025 parts per million (ppm) obtained through human occupational inhalation studies, which found that exposure caused hand tremors and increase in memory disturbances.

WHO   The World Health Organization (WHO, 1984) has set a mercury drinking water guideline of 0.001 mg/l, as well as permissible tolerable weekly uptake of 5 ug/kg.

ATSDR   The ATSDR has established minimal risk levels (MRL) for chronic and acute, oral and inhalation exposures to mercury.

Inhalation The chronic-duration MRL of $0.2\,\mu g/m^3$ was calculated based on chronic-duration industrial inhalation exposure (365 days or more) to metallic mercury vapor for an average of 15.3 years (based on the studies of Fawer et al., 1983).

Oral The acute-duration MRL of $7\,\mu g/kg/day$ is based on a no-observed-adverse effect level (NOAEL) of 0.93 mg mercury/kg for renal effects on rats that were exposed to mercuric chloride for 2 weeks (NTP, 1993).

The intermediate-duration MRL of $2\,\mu g/kg/day$ is based on a NOAEL of 0.23 mg mercury/kg for renal effects in rats (Dieter et al., 1992; NTP, 1993).

The chronic-duration oral MRL of $0.3\,\mu g/kg/day$ is based neurodevelopmental outcomes in a study by Davidson et al. (1998) on *in utero* exposure to methyl mercury from maternal fish ingestion.

IARC The International Agency for Research on Cancer (IARC) review of mercury as a carcinogen was last revised in 1997. Based on limited human studies, this review listed metallic and inorganic mercury as Group 3 carcinogens: not classifiable as to their carcinogenicity to humans whereas methylmercury compounds are listed as Group 2B carcinogens: possibly carcinogenic to humans.

ACGIH The ACGIH has listed mercury as an A4 carcinogen: not classifiable as a human carcinogen (ACGIH, 1996), with a skin notation that indicates the potential for dermal absorption.

EPA The EPA has not established a reference dose (RfD) for metallic mercury. Based on neurological abnormalities in infants an oral RfD of 1×10^{-4} has been established for methylmercury (IRIS, 1997). An oral RfD for mercury has been established from phenylmercuric acetate of 0.08 ug/kg/day (IRIS, 1997). Based on detectable kidney damage in rats the RfD is based on a LOAEL of 0.5 ppm (Fitzhugh et al., 1950). Mercuric chloride and methylmercury have been listed as a C carcinogen, meaning they are possible human carcinogens (IRIS, 1997). Finally, the EPA has established a safe drinking water standard of $2\,\mu g/l$ for mercury (FSTRAC, 1995).

CalEPA The California Environmental Protection Agency in October 1990 listed mercury as a developmental toxicant under Proposition 65.

FDA The U.S. FDA limits mercury levels in bottled drinking water at 0.002 ug/l (FDA, 1995). The FDA has set an action level for methylmercury in foods such as aquatic animals (fish, shellfish) at 1 ppm (FDA, 1995). Further, the FDA regulates the amount of allowable mercury in cosmetics and is generally restricted to eye-area cosmetics and must measure <1 ppm or 0.0001% mercury metal (FDA, 1974).

REFERENCES

Abe, T., Haga, T., and Kurokawa, M. (1975) Blockage of axoplasmic transport and depolymerization of reassembled microtubules by methymercury. *Brain Res.* 86:504–508.

ACGIH (1996) *Threshold Limit Values for Chemical Substances and Physical Agents and Biological Exposure Indices for 1996,* Cincinnati, OH: American Conference of Governmental Industrial Hygienists.

Albrecht, J., Szumanska, G., Gadamski, R., and Gajkowska, B. (1994) Changes of activity and ultrastructural localization of alkaline phosphatase in cerebral cortical microvessels of rat after single intraperitoneal administration of mercuric chloride. *Neurotoxicology* 15(4):897–902.

Ali, S.F., LeBel, C.P., and Bondy, S.C. (1992) Reactive oxygen species formation as a biomarker of methylmercury and trimethyltin neurotoxicity. *Neurotoxicolgy* 13(3):637–648.

Al-Mufti, A.W., Copplestone, J.F., Kazanitzis, G., Mahmoud, R.M., and Majid, M.A. (1976) Epidemiology of organomercury poisoning in Iraq. I. Incidence in a defined area and relationship to the eating of contaminated bread. *Bull. World Health Organ.* 53(Suppl.):23–36.

Bakir, F., Damluji, S.F., Amin-Zaki, L., Murtadha, M., Khalidi, A., al-Rawi, N.Y., Tikriti, S., Dahahir, H.I., Clarkson, T.W., Smith, J.C., and Doherty, R.A. (1973) Methylmercury poisoning in Iraq. *Science* 181:230–241.

Ballatori, N. and Clarkson, T.W. (1985) Biliary secretion of glutathione and of glutathione-metal complexes. *Fundam. Appl. Toxicol.* 5(5):816–831.

Berlin, M., Fazackerly, J., and Nordberg, G. (1969) The uptake of mercury in the brains of mammals exposed to mercury vapor and mercuric salts. *Arch. Environ. Health* 18:719–729.

Bidstrup, P. (1964) *Toxicity of Mercury and Its Compounds,* New York, NY: Elsevier.

Bryan, G.W. and Langston, W.J. (1992) Bioavailability, accumulation and effects of heavy metals in sediments with special reference to United Kingdom estuaries: a review. *Environ. Pollut.* 76(2):89–131.

Bulger, R.E. (1986) Renal damage caused by heavy metals. *Toxicol. Pathol.* 14:58–65.

Cantor, M.O. (1951) Mercury lost in the gastrointestinal tract: report of an unusual case. *J. Am. Med. Assoc.* 146:560–561.

Choi, B. (1991) Effects of methymercury on the developing brain. In: Suzuki, T. and Clarkson, T.W., editors. *Advances in Mercury Toxicology,* New York, NY: Plenum Press.

Clarkson, T.W. (1977) Mercury poisoning. In: Brown, S.S., editor. *Clinical Chemistry Chemical Toxicology of Metals,* New York, NY: Elsevier.

Clarkson, T.W. (1989) Mercury. *J. Am. Coll. Toxicol.* 8(7):1291–1296.

Clarkson, T.W. (1992) Mercury: major issues in environmental health. *Environ. Health Perspect.* 100:31–38.

Dales, L.G. (1972) The neurotoxicity of alkyl mercury compounds. *Am. J. Med.* 53:219–232.

Davidson, P.W., Myers, G.J., Cox, C., Axtell, C., Shamlaye, C., Sloane-Reeves, J., Cernichiari, E., Needham, L., Choi, A.,

Wang, Y., Berlin, M., and Clarkson, T.W. (1998) Effects of prenatal and postnatal methylmercury exposure from fish consumption on neurodevelopment: outcomes at 66 months of age in the Seychelles child development study. *JAMA* 280(8): 701–707.

Dieter, M.P., Boorman, G.A., Jameson, C.W., Eustis, S.L., and Uraih, L.C. (1992) Development of renal toxicity in F344 rats gavaged with mercuric-chloride for 2 weeks, or 2, 4, 6, 15, and 24 months. *J. Toxicol. Environ. Health* 36(4):319–340.

Doja, A. and Roberts, W. (2006) Immunizations and autism: a review of the literature. *Can. J. Neurol. Sci.* 33(4):341–346.

Duhr, E.F., Pendergrass, J.C., Slevin, J.T., and Haley, B.E. (1993) HgEDTA complex inhibits GTP interactions with the e-site of brain beta-tubulin. *Toxicol. Appl. Pharmacol.* 122(2): 273–280.

EPA (1984) Mercury health effects updates. In: *Health Issue Assessment. Final Report*, Washington, DC: U.S. Environmental Protection Agency, Office of Health and Environmental Assessment, Document no. EPA 600/8-84-019F.

FDA (1974) *Food and Drug Administration*. Department of Health and Human Services. Code of Federal Regulations. 21 CFR 700.13.

FDA (1995) *Food and Drug Administration*. Department of Health and Human Services. Code of Federal Regulations. 21 CFR 165.

Fawer, R.F., DeRibaupierre, Y., Guillemin, M., Berode, M., and Lob, M. (1983) Measurement of hand tremor induced by industrial exposure to metallic mercury. *Br. J. Ind. Med.* 40:204–208.

Fitzhugh, O.G., Nelson, A.A., Laug, E.P., and Kunze, F.M. (1950) Chronic oral toxicities of mercuric-phenyl and mercuric salts. *Arch. Ind. Hyg. Occup. Med.* 2:433–442.

Friberg, L. and Nordberg, F. (1973) Inorganic mercury—a toxicological and epidemiological appraisal. In: Miller, M.W. and Clarkson, T.W., editors. *Mercury, Mercurials and Mercaptans*, Springfield, IL: Charles C. Thomas, pp. 5–22.

FSTRAC (1995) *Summary of state and federal drinking water standards and guidelines*. U.S. Environmental Protection Agency. Chemical Communications Subcommittee, Federal State Toxicology and Regulatory Alliance Committee.

Goldwater, L. (1972) *Mercury: A History of Quicksilver*, Baltimore, MD: York Press.

Govindaswamy, N., Moy, J., Millar, M., and Koch, S.A. (1992) A distorted mercury $[Hg(SR)_4]^{2-}$ complex with alkanethiolate ligands: the fictile coordination sphere of monomeric [Hg (SR)$_x$] complexes. *Inorg. Chem.* 26(26):5343.

Halbach, S. and Clarkson, T.W. (1978) Enzymatic oxidation of mercury vapor by erythrocytes. *Biochim. Biophys. Acta* 523:522–531.

Harris, H., Pickering, I., and George, G. (2003) The chemical form of mercury in fish. *Science* 301:1203.

Hattula, T. and Rahola, T. (1975) The distribution and biological half-time of ^{203}Hg in the human body according to a modified whole-body counting technique. *Environ. Physiol. Biochem.* 5(4):252–257.

Hunter, D., Bomford, R.R., and Russell, D.S. (1940) Poisoning by methyl mercury compounds. *Quart. J. Med.* 9:193–213.

IRIS (1997) *Integrated Risk Information System (IRIS). Online*, Cincinnati, OH: U.S. Environmental Protection Agency, Office of Health and Environmental Assessment, Environmental Criteria and Assessment Office.

LeBel, C.P., Ali, S.F., and Bondy, S.C. (1992) Deferoxamine inhibits methyl mercury-induced increases in reactive oxygen species formation in rat brain. *Toxicol. Appl. Pharmacol.* 112(1):161–165.

LeBel, C.P., Ali, S.F., McKee, M., and Bondy, S.C. (1990) Organometal-induced increases in oxygen reactive species: the potential of 2′,7′-dichlorofluorescin diacetate as an index of neurotoxic damage. *Toxicol. Appl. Pharmacol.* 104:17–24.

Lindberg, S.E., Turner, R.R., Meyers, T.P., Taylor, G.E., and Schroeder, W.H. (1991) Atmospheric concentrations and deposition of mercury to a deciduous forest at Walker Branch Watershed, Tennessee, USA. *Water Air Soil Pollut.* 56:577–594.

Lund, B.O., Miller, D.M., and Woods, J.S. (1993) Studies on Hg(II)-induced H_2O_2 formation and oxidative stress *in vivo* and *in vitro* in rat kidney mitochondria. *Biochem. Pharmacol.* 45(10):2017–2024.

Magos, L. (1967) Mercury–blood interaction and mercury uptake by the brain after vapor exposure. *Environ. Res.* 1:323–337.

Mahaffey, K.R., Clickner, R.P., and Bodurow, C.C. (2004) Blood organic mercury and dietary mercury intake: national health and nutrition examination survey, 1999 and 2000. *Environ. Health Perspect.* 112(5):562–570.

Morris, G. (1960) Dermatoses from phenylmercuric salts. *Arch. Environ. Health* 1:53–55.

NTP (1993) *Toxicology and Carcinogenesis Studies of Mercuric Chloride (CAS No. 7487-94-7) in F344/N Rats and B6C3F1 Mice (Gavage Studies)*, NTP TR 408. Research Triangle Park, NC: National Toxicology Program, U.S. Department of Health and Human Services, Public Health Service, National Institutes of Health, NIH Publication No. 91–3139.

OSHA (1974) *Occupational Safety and Health Administration. Code of Federal Regulations. 29 CFR 1910.1000.*

Piotrowski, J.K., Szymanska, J.A., Skrzypinska-Gawrysiak, M., Kotelo, J., and Sporny, S. (1992) Intestinal absorption of inorganic mercury in rat. *Pharmacol. Toxicol.* 70(1):53–55.

Ponce, R.A. and Bloom, N.S. (1991) Effect of pH on the bioaccumulation of low level, dissolved methylmercury by rainbow trout (*Oncorhynchus mykiss*). *Water Air Soil Pollut.* 56:631–640.

Rabenstein, D.L., Isab, A.A., and Reid, R.S. (1982) A proton nuclear magnetic resonance study of the binding of methylmercury in human erythrocytes. *Biochim. Biophys. Acta: Mol. Cell Res.* 720:53–64.

Rahola, T., Hattula, T., Korolainen, A., and Miettinen, J.K. (1973) Elimination of free and protein-bound ionic mercury ^{203}Hg^{2+} in man. *Ann. Clin. Res.* 5:214–219.

Rajanna, B., Chetty, C.S., Rajanna, S., Hall, E., Fail, S., and Yallapragada, P.R. (1995) Modulation of protein kinase c by heavy metals. *Toxicol. Lett.* 81(2–3):197–203.

Rothstein, A. and Hayes, A.L. (1960) The metabolism of mercury in the rat studied by isotope techniques. *J. Pharmacol. Exp. Ther.* 130:166–176.

RTECS (1998) *Registry of Toxic Effects of Chemical Substances (RTECS)*. National Institute for Occupational Safety and Health (NIOSH). Computer Database Online.

Sager, P.R., Aschner, M., and Rodier, P.M. (1984) Persistent, differential alterations in developing cerebellar cortex of male and female mice after methylmercury exposure. *Brain Res.* 314:1–11.

Sager, P.R., Doherty, R.A., and Olmsted, J.B. (1983) Interaction of methylmercury with microtubules in cultured cells and *in vitro*. *Exp. Cell Res.* 146:127–137.

Sager, P.R., Doherty, R.A., and Rodier, P.M. (1982) Effects of methylmercury on developing mouse cerebellar cortex. *Exp. Neurol.* 77:179–193.

Sahaphong, S. and Trump, B.F. (1971) Studies of cellular injury in isolated kidney tubules of the flounder. *Am. J. Pathol.* 63:277–298.

Sarafian, T. and Verity, M.A. (1991) Oxidative mechanisms underlying methyl mercury neurotoxicity. *Int. J. Dev. Neurol.* 9(2):147–153.

Schuster, E. (1991) The behavior of mercury in the soil with special emphasis on complexation and adsorption process—a review of the literature. *Water Air Soil Pollut.* 56:667–680.

Sekowski, J.W., Malkas, L.H., Wei, Y., and Hickey, R.J. (1997) Mercuric ion inhibits the activity and fidelity of the human cell DNA syntheome. *Toxicol. Appl. Pharmacol.* 145:268–276.

Somjen, S.G., Herman, S., Klein, R., Brubaker, P.E., Briner, W.H., Goodrich, J.K., Krigman, M.R., and Haseman, J.K. (1973) The uptake of methylmercury (^{203}Hg) in different tissues related to its neurotoxic effects. *J. Pharmacol. Exp. Ther.* 187:602–611.

Syversen, T. (1977) Effects of methylmercury on *in vivo* protein synthesis in isolated cerebral and cerebellar neurons. *Neuropathol. Appl. Neurobiol.* 3:225–236.

Torresani, C., Caprari, E., and Manara, G.C. (1993) Contact urticaria syndrome due to phenylmercuric acetate. *Contact Dermatitis* 29(5):282–283.

Vogel, D.G., Margolis, R.L., and Mottet, N.K. (1985) The effects of methyl mercury binding to microtubules. *Toxicol. Appl. Pharmacol.* 80:473–486.

WHO (1984) *Guidelines for Drinking Water Quality: Recommendations*, vol. 1, World Health Organization.

WHO (1990) *Methyl Mercury*, vol. 101, Geneva, Switzerland: World Health Organization, International Programme on Chemical Safety.

24

MOLYBDENUM

Paul Jonmaire

First aid: Since there are many molybdenum compounds in commercial use, there are no special first aid recommendations. General first aid recommendations are often appropriate. For dermal overexposure, remove contaminated clothing and flush with plenty of water for 15 min. For ocular overexposure, flush with water for 15 min.

Target organ(s): Eye, respiratory tract, kidney, and liver

Occupational exposure limits: OSHA PEL TWA: 15 mg/m^3; ACGIH TLV TWA: 10 mg/m^3 (dust); 3 mg/m^3 (respirable) [As molybdenum and insoluble compounds]; OSHA PEL TWA: 5 mg/m^3; ACGIH TLV TWA: 0.5 mg/m^3 (respirable) [as soluble compounds]

Reference values: Inhalation RfD: 0.005 mg/kg/day UF 30, MF 1 (EPA [2012]); TDI: 0.009 mg Mo/kg/day

Risk/safety phrases: None

BACKGROUND AND USES

In 1768, the Swedish scientist Carl Wilhelm Steele determined that the mineral molybdenite was a sulfide compound containing an unidentified element by decomposing it in hot nitric acid to yield a white powder (IMOA, 2013b). This powder was found to be molybdenum oxide, the most common molybdenum compound in use today (NTP, 1997). A second scientist, Peter Hjelm, reduced the oxide with carbon obtaining a dark metal compound which he called molybdenum. Late in the nineteenth century, it was shown that molybdenum could replace tungsten as an alloying agent for steel.

The material derives its name from *molybdos*, meaning "lead-like." Originally, the metal was thought to be lead because of its gray color and greasy feel (Kropschot, 2010). After World War I, Bartlett Mountain in Colorado became a major source of molybdenum, with the Climax Molybdenum Company mine providing three-quarters of the world's molybdenum. The Climax Mine stopped supplying molybdenum in 1995, but reopened and again began supplying molybdenum in 2012 as a subsidiary of Freeport-McMoRan (Climax Molybdenum, n.d.). In 2012, approximately 57,000 tons of molybdenum were produced in the United States (USGS). Approximately half or the world's known molybdenum reserves are in the United States (Stiefel and Murray, 2002).

Molybdenum can be obtained from several different minerals, including molybdenite (MoS2), wulfenite ($PbMoO_4$), and powellite ($CaMoWO_4$) (Stiefel and Murray, 2002). Substantial amounts of molybdenum are produced through the processing of molybdenite. Molybdenite is the mineral of greatest commercial importance (Shamberger, 1979).

Additional supplies are obtained in association with the processing of copper, of molybdenum, and tungsten ores. Molybdenum is used primarily to create metal alloys, especially with iron. Moly-steel is known for its hardness and resistance to chemicals. It also is used in the production of ceramics, in pigments, in the electronics industry as corrosion inhibitors and as a lubricant (Browning, 1969; IMOA, 2013a; Stiefel and Murray, 2002). Molybdenum is an essential element for both plants and animals, and is important for the fixation of nitrogen (NTP, 1997; Shamberger, 1979). Soluble molybdates are key elements in some fertilizers (ACGIH, 2008; IMOA, 2013a).

PHYSICAL AND CHEMICAL PROPERTIES

Molybdenum (symbol Mo: CAS #7439-98-7) is a silvery-white metal or gray-black powder of the refractory metal

Hamilton & Hardy's Industrial Toxicology, Sixth Edition. Edited by Raymond D. Harbison, Marie M. Bourgeois, and Giffe T. Johnson.
© 2015 John Wiley & Sons, Inc. Published 2015 by John Wiley & Sons, Inc.

family that is not normally found in its elemental state (Barceloux, 1999). Elemental molybdenum has a molecular weight of 96, a specific gravity of 10.2 g/cc and a melting point of 2610 °C (NIOSH, 2005). It makes up about 0.0015% of the earth's crust (NLM, 2013). Almost 100 compounds of molybdenum are listed in the EPA Toxic Substances Control Act inventory (U.S. EPA, 2013). There are two basic types of molybdenum substances—soluble and insoluble—to be considered from an occupational health perspective. The insoluble type includes elemental molybdenum, molybdenum disulfide, and molybdenum dioxide. Soluble molybdenum compounds include most molybdates, oxides (most notable is molybdenum trioxide), and halides (ACGIH, 2008). Molybdenum functions both as a metal and a metalloid (Eisler, 2000).

MAMMALIAN TOXICOLOGY

Acute Effects

Molybdenum is generally considered to be of low toxicity for animals and plants (CDC, 2013; Stiefel, 2002). Few adverse acute health effects are linked to molybdenum and molybdenum compounds. Molybdenum toxicity is closely related to the amount of copper available in the system, therefore any toxic effects of molybdenum observed must be evaluated in conjunction with copper levels (Vyskocil and Viau, 1999). There have been no reports of accidental occupational deaths due to molybdenum poisoning (Browning, 1969). Insoluble molybdenum substances are considered to be of low toxicity (CDC, 2013). Molybdenum trioxide, a soluble molybdenum compound, is considered to be a skin, eye, and mucous membrane irritant (Honeywell Product Summary, 2012; NIOSH/OSHA, 1978).

Although the literature reports some symptoms from occupational overexposure to molybdenum, they are quite uncommon. Overexposure to molybdenum compounds in the workplace may cause some transient symptoms such as fatigue, joint pain, headaches, backaches, and other nonspecific symptoms (ACGIH, 2008; Friberg, 1975; Lener and Bibr, 1984).

There have been no reports of adverse effects in workers exposed to molybdenum dust in the crushing and milling of ore. Similarly, no adverse health effects have been reported among workers exposed to the fumes of molybdenum oxide in the rolling of red-hot billets of molybdenum steel. (Friberg and Lener, 1986) reported that workers exposed to molybdenum in a copper–molybdenum plant experienced increased blood uric acid levels and gout-like symptoms. Respiratory damage associated with overexposure to molybdenum has been reported as pneumoconiosis (ACGIH, 2008). Workers exposed to molybdenum dust in a molybdenite roasting plant producing molybdenum trioxide and other soluble oxides were found to have a substantially elevated serum ceruloplasmin and a smaller increase in serum uric acid. The estimated daily intake of molybdenum inhaled as dust particles averaged 10.2 mg, and both plasma and urine molybdenum levels were elevated in these workers. The average air concentration of molybdenum in the plant was 9.4 mg/m^3. No long-term adverse health effects were demonstrated by Walravens et al., (1979) in this study. Their major finding in a study of 26 workers was elevated mean serum uric acid levels and serum ceruloplasmin but no gout-like symptoms. A questionnaire of the workers in the study found that joint pain, headache, backache, and skin and hair changes were the most frequently reported complaints.

Chronic Effects

Besides the short-term symptoms reported by Walravens et al. (1979), there are few other consistently reported adverse workplace molybdenum exposure findings. Some reports describe three other effects associated with overexposure to molybdenum which are gout, pneumoconiosis, and the disruption of copper levels in the body. Workers exposed up to 19 mg/m^3 of molybdenum showed signs of pneumoconiosis related to their workplace exposure (ACGIH, 2008).

Selden et al. (2005) describes a case of an electrician who developed gouty symptoms associated with working on the manufacture and maintenance of specialized molybdenum-lined furnaces. Nothing in his history or his family history suggested a predilection to gout. The symptoms subsided after the initial exposure was greatly reduced and returned about a year later when exposures again increased. The authors concluded that the association between gouty symptoms may be circumstantial, but it could also be causal.

Molybdenum and its compounds have not been identified as carcinogenic by the International Agency for Research on Cancer (IARC), the EPA through IRIS, or California's Proposition 65. The American Conference of Governmental Industrial Hygienists (ACGIH) has assigned molybdenum an A3 carcinogen, "Confirmed Animal Carcinogen with unknown relevance to Humans." This rating is based primarily on the NTP (1997) bioassay results. The study discussed below is considered insufficient by ACGIH to assign a higher ranking based on human data. In a 1999 Belgian case-control study investigating lung cancer among workers potentially exposed to 16 carcinogens, including arsenic, nickel, wood dust, asbestos, and polycyclic aromatic hydrocarbons [PAHs], the strongest association for lung cancer after being controlled for smoking cigarettes was having ever been exposed to molybdenum for an extended length of time (Droste et al., 1999). This study suggested that there is some evidence of molybdenum carcinogenicity. Although the study was performed appropriately, it experienced the typical limitations associated with most case-control studies, e.g., lack of documented workplace levels of exposure, self-reporting, and fewer smokers in the control group. Breise (1976) studied molybdenum workers in Colorado and found no evidence that high or low exposures to molybdenum presented mutagenic, teratogenic, or carcinogenic hazards.

Inhalation studies on mice and rats performed as part of the National Toxicology Program using molybdenum trioxide, a compound used in metallurgical alloys, in cosmetics as a pigment, and in contact lens solution, produced equivocal results of carcinogenic activity in male F344/N rats, but no evidence of carcinogenicity in female F344/N rats. In B6C3F mice of both genders there was some evidence of carcinogenic activity. A review, as part of this study, found no indication that molybdenum trioxide is a human carcinogen. Some early studies in Africa, China, and Russia suggested that molybdenum deficiencies could be linked to esophageal cancer (NTP, 1997). Some evidence exists regarding potential mutagenesis of molybdenum compounds. Molybdenum trioxide was studied by Zeiger et al. (1992) in five Salmonella strains with and without metabolic activation, and showed no mutagenicity. Workers exposed to molybdenum, molybdenum trioxide, and molybdenite were reported to have an elevated number of chromosomal aberrations in lymphocyte studies (Babaian et al., 1980).

Mechanism of Action

Molybdenum is an essential nutrient for most forms of life, plant, animal, and microbial. Molybdenum is an essential dietary nutrient and a critical component of several mammalian enzymes, including xanthine oxidase, sulfite oxidase, and aldehyde oxidase (CDC, 2013; Kisker et al., 1997). Soluble molybdenum, usually as a molybdate, is readily absorbed through the intestinal tract into the liver. High levels of copper and sulfur in the diet enhance elimination of molybdenum reportedly by blocking a protein carrier and, therefore, inhibiting reuptake by the kidneys. In mammals, molybdenum can protect against poisoning by metals, such as copper and mercury, and may have anticarcinogenic activity. Excretion of molybdenum occurs primarily through the kidneys, mainly as molybdate (CDC, 2013; Stiefel and Murray, 2002).

The mechanism of action of molybdenum toxicity is not fully understood. Xanthine oxidase is a metalloenzyme that requires molybdenum as a critical cofactor, and is induced by elevated levels of molybdenum. The activity of xanthine oxidase affects the uric acid levels of the urine and blood by converting purines to uric acid. Increased activity may result in symptoms of gout. Overexposure to molybdenum also may upset the copper balance in the body. This occurs as part of a complex interaction involving copper, molybdenum, sulfur compounds, and several enzyme systems (Friberg and Lener, 1986). Higher levels of molybdenum may actually reduce the activity of molybdenum-dependent enzymes as well as induce copper deficiency (Browning, 1969).

Additional investigation of the possible health effects of excessive exposure to molybdenum came from a drinking water study performed by the EPA in two Colorado cities (U.S. EPA, 1979). The urinary levels of molybdenum and copper and the blood levels of ceruloplasmin and uric acid were compared in two groups over a period of 2 years. The lower exposure group drank water with molybdenum levels ranging from 2 µg/l to 50 µg/l; the higher exposure group drank water with a molybdenum level of 200 µg/l or higher. The group with the greater molybdenum exposure was reported to have a higher urinary level of molybdenum. Although the higher daily level of urinary molybdenum was associated with the higher molybdenum intake, no adverse biochemical or systemic effects were noted in either study group. A higher mean serum ceruloplasmin also was associated with the higher molybdenum intake (40.3 mg/dl compared with 30.41 mg/dl for the lower intake group).

Although exposures were not occupational, some of the most informative data describing potential excessive exposure to molybdenum originate from studies in two Armenian villages. Researchers correlated dietary intake of molybdenum with elevated serum uric acid level and gout-like symptoms. The group exposed to higher amounts of molybdenum in plants and soil had greater swelling and inflammation than the control group. The serum uric acid level was much higher in residents from the high exposure area. The average uric acid level was even higher in those who showed symptoms of gout (Kovalskiy et al., 1961). Both serum molybdenum and serum xanthine oxidase were positively correlated with the serum uric acid level. The EPA chronic oral reference dose of 5×10^{-3} mg/kg/day is based upon this 1961 study. A critique of this study performed by EBRC Consulting GmbH (2010) on behalf of the International Molybdenum Association raises questions regarding the validity of the Kovalskiy study for use in identifying a reference dose based on the application of current standards for evaluating the technical performance of the study as well as based on current medical/toxicological knowledge.

SOURCES OF EXPOSURE

Occupational

Many occupations can present the possibility of overexposure to soluble molybdenum compounds because these compounds are used in the production of moly-steel; as electrodes in welding operations; as coloring agents for ceramics, enamels, and paints; as corrosion inhibitors for water systems; as catalysts in the petroleum industry; as flame retardant for resins; and as components of fertilizer (NTP, 1979). Workers can be exposed to insoluble molybdenum material in the mining and processing of ore that contains molybdenum; when working with molybdenum or its alloys; when formulating and using lubricants, catalysts, and the base metal for filaments, rocket nozzles, and ceramics (IMOA, 2013a; Stiefel and Murray, 2002).

INDUSTRIAL HYGIENE

The permissible exposure limit (PEL) values are established by the Occupational Safety and Health Administration (OSHA) and the threshold limit values (TLV) are set by the ACGIH. The PEL for insoluble molybdenum compounds (e.g., molybdenum disulfide), molybdenum dioxide, and metallic molybdenum) is $15\,mg/m^3$ total dust and $5\,mg/m^3$ for respirable dust (OSHA, 2008). These values were promulgated in 1971 as part of the original establishment of the OSHA, and have not changed (Federal Register, 1971). The ACGIH TLV for insoluble molybdenum is $10\,mg/m^3$ total dust and $3\,mg/m^3$ respirable dust. The TLV for insoluble molybdenum compounds is based on ACGIH's intent to bring TLV values for low toxicity, poorly absorbed materials into agreement with each other. The TLV for soluble molybdenum compounds of $0.5\,mg/m^3$ (respirable fraction) is based on the NTP (1979) inhalation study of molybdenum trioxide wherein the observed no effect level was $10\,mg/m^3$ (ACGIH, 2008). Several safety factors were then incorporated for the development of this criterion. The observation that sulfide enhances the toxicity of soluble molybdenum compounds suggests the desirability of exercising special care when handling soluble molybdenum compounds (IMOA, 2013c).

From an occupational health perspective, molybdenum and its compounds present a wide range of potential toxicity from essentially inert (elemental molybdenum) to irritation and potential carcinogenicity (molybdenum trioxide). In any industrial operation, excessive exposure to potentially hazardous materials should be avoided through use of appropriate engineering and administrative controls, such as local ventilation. If engineering controls are not feasible, a program requiring respirators and other personal protective equipment is needed. Impervious gloves, clothing, eye protection, and NIOSH-approved respiratory protection should be used to protect workers from overexposure to molybdenum and its compounds. If protective clothing is to be reused, laundering contaminated clothing should avoid additional exposure-related hazards. Soluble molybdenum compounds are more likely to be irritant hazards, whereas insoluble ones may be associated with pneumoconiosis. Precautions should be based on the hazards associated with the specific molybdenum compound or compounds. For operations in which overexposure to one of these materials is possible, good personal hygiene habits are strongly recommended. These include no eating or smoking in the work area; washing the hands and face after leaving the work area and before eating, smoking, or using toilet facilities; and removing contaminated clothing and showering at the end of the workday (U.S. HHS, 1978).

REFERENCES

ACGIH (American Conference of Governmental Industrial Hygienists) (2008) *Molybdenum and Compounds: Metallic Molybdenum and Insoluble Compounds, as Mo*, Cincinnati, OH: American Conference of Governmental Industrial Hygienists.

Babaian, E.A., Bagramian, S.B., and Pogosian, A.S. (1980) Effect of some chemical hazards involved in molybdenum production on the chromosome apparatus of experimental animals and of man [in Russian]. *Gig. Tr. Prof. Zabol.* 9:33–36.

Barceloux, D.G. (1999) Toxicity of molybdenum towards humans. *J. Toxicol. Clin. Toxicol.* 37:231–237.

Breise, F.W. (1976) In: Chappell, W.R. and Petersen, K.K., editors. *Molybdenum in the Environment*, vol. 1, New York, NY: Marcel Dekker.

Browning, E. (1969) *Toxicity and Industrial Metals*, 2nd ed., New York, NY: Appleton-Century Crofts, pp. 243–248.

CDC (Centers for Disease Control and Prevention) (2013) Biomonitoring Summary—Molybdenum. Available at http://www.cdc.gov/biomonitoring/Molybdehum_BiomonitoringSummary.html (accessed 2013).

Climax Molybdenum (n.d.) Available at http://www.climaxmolybdenum.com/products/molytechlib.asp (accessed 2013).

Droste, J.H., Weyler, J.J., Van Meerbeeck, J.P., Vermeire, P.A., and van Sprundel, M.P. (1999) Occupational risk factors of lung cancer: a hospital based case-control study. *Occup. Environ. Med.* 56:322–327.

EBRC Consulting GmbH (2010) Assessment of Consumer Exposure under REACH for MoCon Substances, Hanover Germany, Prepared for the REACH Molybdenum Consortium.

Eisler, R. (2000) Handbook of chemical risk information: health hazards to humans, plants, and animals. In: *Metalloids, Radiation, Index, "Molybdenum"*, vol. 3, Boca Raton, FL: CRC Press.

Federal Register (1971) 36, 15101.

Friberg, L. (1975) Molybdenum – a toxicological appraisal. U.S. Environmental Protection Agency Report 600/1-75-004.

Friberg, L. and Lener, J. (1986) Molybdenum. In: Friberg, L., Nordberg, G.F., and Vouk, V.B., editors. *Handbook on the Toxicology of Metals*, New York, NY: Elsevier/North Holland Biomedical Press, pp. 446–461.

Honeywell Product Summary (2012) Molybdenum Trioxide: Product Stewardship Summary. Honeywell Performance Materials and Technologies. Available at http://www51.honeywell.com/sm/common/documents/Public_Risk_Summary_GPS0059_Molybdenum_Trioxide_Feb_2012.pdf (accessed 2013).

IMOA (International Molybdenum Association) (2013a) Molybdenum Chemistry & Uses. Overview of Molybdenum Chemistry, vol. 1–4. Available at http://www.imoa.info/moly_uses/moly_chemistry_uses/molybdenum_chemistry_uses.php (accessed 2013).

IMOA (International Molybdenum Association) (2013b) Molybdenum History. Available at http://www.imoa.info/molybdenum/molybdenum_history.php (accessed 2013).

IMOA (International Molybdenum Association) (2013c) Toxicity of Molybdenum Toward Humans. Available at http://www.imoa.info/HSE/environmental_data/human_health/molybdenum_toxicity.php (accessed 2013).

Kisker, C., Schindelin, H., and Rees, D.C. (1997) Molybdenum-cofactor-containing enzymes: structure and mechanism. *Ann. Rev. Biochem.* 66:233–267.

Kovalskiy, V.V., Yarovaya, G.A., and Shmavonyan, D.M. (1961) The change in purine metabolism of humans and animals under the conditions of molybdenum geochemical provinces. *Zh. Obshch. Biol.* 22:179–191.

Kropschot, S.J. (2010) Molybdenum—A Key Component of Metal Alloys. U.S. Geological Survey Fact Sheet 2009–3106. Available at http://pubs.usgs.gov/fs/2009/3106/ (accessed 2013).

Lener, J. and Bibr, B. (1984) Effects of molybdenum on the organism (a review). *J. Hyg. Epidemiol. Microbiol. Immunol.* 28:405–419.

NIOSH/OSHA (National Institute for Occupational Safety and Health) (1978) *Occupational Health Guideline for Molybdenum and Insoluble Molybdenum*, Washington, DC: National Institute for Occupational Safety and Health.

NIOSH (National Institute for Occupational Safety and Health) (2005) *NIOSH Pocket Guide to Chemical Hazards, Molybdenum*, Washington, DC: National Institute for Occupational Safety and Health.

NLM (National Library of Medicine HSDB Database) (2013) Molybdenum. Human Health Effects. Available at http://toxnet.nlm.nih.gov/cgi-bin/sis/search/f?.temp/~fw3Yik:1:BASIC (accessed 2013).

NTP (National Toxicity Program) (1997) *NTP Technical Report (TR-462) on the Toxicology and Carcinogenesis Studies of Molybdenum Trioxide (CAS No. 1313-27-5) in F344 Rats and B6C3F Mice (Inhalation Studies)*. Research Triangle Park, NC: National Toxicity Program.

OSHA (Occupational Safety and Health Administration) (2008) Permissible Exposure Limits. 29 CFR 1910.1000. Molybdenum, Washington, DC: Occupational Safety and Health Administration.

Selden, A.I., Berg, N.P., Soderbergh, A., and Bergstrom, E.O. (2005) Occupational molybdenum exposure and a gouty electrician. *Occup. Med.* 55:145–148.

Shamberger, R. (1979) In: Oehme, F., editor. *Toxicity of Heavy Metals in the Environment*, 2nd ed., New York, NY: Marcel Dekker, pp. 726–729.

Stiefel, E.I. and Murray, H.H. (2002) Molybdenum in the environment. In: Sarkar, B., editor. *Heavy Metals in the Environment*, New York, NY: Marcel Dekker, pp. 503–529.

U.S. HHS (U.S. Department of Health and Human Services) (1978) *Occupational Health Guideline for Molybdenum and Insoluble Molybdenum*, Washington, DC: U.S. Department of Health and Human Services.

U.S. EPA (U.S. Environmental Protection Agency) (1979) Human Health Effects of Molybdenum in Drinking Water, EPA-600A-799-006,Washington, DC: U.S. Environmental Protection Agency.

U.S. EPA (U.S. Environmental Protection Agency) (2013) TSCA Chemical Substance Inventory. Available at http://www.epa.gov/opptintr/existingchemicals/pubs/tscainventory/howto./html, (accessed 2013).

U.S. EPA (U.S. Environmental Protection Agency) (n.d.) Integrated Risk Management System: Molybdenum (CASRN 7439-09-7), Chronic Health Hazard Assessments for NonCarcinogenic Effects. Available at http://www.epa.gov/iris/subst/0425.htm (accessed 2013).

USGS (U.S. Geological Survey) (n.d.) Molybdenum Statistics and Information. Available at http://minerals.usgs.gov/minerals/pubs/commodity/molybdenum/. (accessed 2013).

Vyskocil, A. and Viau, C. (1999) Assessment of molybdenum toxicity in humans. *J. Appl. Toxicol.* 19:185–192.

Walravens, P.A., Moure-Eraso, R., Solomons, C.C., Chappell, W.R., and Bentley, G. (1979) Biochemical abnormalities in workers exposed to molybdenum dust. *Arch. Environ. Health* 34:302–308.

Zeiger, E., Anderson, B., Haworth, S., Lawlor, T., and Mortelmans, K. (1992) Salmonella mutagenicity tests. V. Results from the testing of 311 chemicals. *Environ. Mol. Mutagen.* 19(Suppl. 21): 2–141.

25

NICKEL

ROBYN L. PRUEITT AND JULIE E. GOODMAN

First Aid: *Dermal exposure*: May cause skin sensitization. Dermatitis can become evident upon reexposure. In case of skin contact, immediately flush the skin with plenty of soap and water for at least 15 min. Cover the irritated skin with an emollient. *Inhalation exposure*: Inhalation of mist may cause respiratory irritation. If inhaled, remove to fresh air. If breathing is difficult, give oxygen and call a physician. *Oral exposure*: Can cause gastrointestinal irritation with nausea, vomiting, and diarrhea. If ingested, call a physician immediately. Do not induce vomiting unless directed to do so by medical personnel. *Eye exposure*: May cause eye irritation. In case of eye contact, flush with copious amounts of water for at least 15 min. Ensure adequate flushing by separating the eyelids with fingers. Call a physician.

Target organs: Skin; respiratory system; kidney; liver

Occupational exposure limits: The OSHA permissible exposure limit (PEL) for nickel metal and both water-soluble and insoluble nickel compounds, based on an 8-h time-weighted average (TWA) exposure, is 1.0 mg/m^3. The NIOSH recommended exposure limit for nickel metal and other nickel compounds, based on a TWA exposure of up to 10 h/day, 40 h/week, is 0.015 mg/m^3. The ACGIH threshold limit values (TLVs), based on a TWA inhalable fraction exposure of 8 h/day, 40 h/week, are 1.5 mg/m^3 for nickel metal, 0.2 mg/m^3 for insoluble nickel compounds, and 0.1 mg/m^3 for water-soluble nickel compounds and nickel subsulfide.

Reference values: US EPA reference dose (RfD): 0.02 mg/kg-day (soluble nickel salts). IOM tolerable upper intake level (UL) for adults: 1 mg/day (soluble nickel salts).

Risk/safety phrases: May cause sensitization by skin contact. Irritating to eyes, respiratory system, and skin. Limited evidence of a carcinogenic effect for some nickel compounds but not nickel metal.

BACKGROUND AND USES

Nickel is a metal that is abundant in the environment and widely used in industry. It can combine with other elements to form a large variety of nickel compounds, each with different physical and chemical properties, or with other metals to form alloys. The major source of nickel exposure to the general population is from food, although skin contact with items containing nickel is also common. Inhalation of nickel-containing dusts in occupational settings can result in higher nickel exposures, which are associated with toxicity and carcinogenicity in the respiratory tract. The most common health effect of nickel for the general population is contact dermatitis after dermal exposure, which can be exacerbated by subsequent oral exposure to nickel.

The largest use of nickel is in the manufacture of stainless steel and other corrosion-resistant alloys (Alloway, 1990). Pure nickel metal is used in electroplating, as a chemical catalyst, and in the manufacture of alkaline batteries, coins, jewelry, welding products, magnets, electrical contacts and electrodes, spark plugs, machinery parts, and surgical and dental prostheses (ATSDR, 2005; IARC, 1990). Different forms of nickel are used in electronics, fuel cell electrodes, hydrogenation of fats, colorants in ceramics and glass, catalysts, lithium batteries, electroplating, petroleum products, and dye fixative (Alloway, 1990; ATSDR, 2005; IARC, 1990). Nickel carbonyl is used in the production of high

Hamilton & Hardy's Industrial Toxicology, Sixth Edition. Edited by Raymond D. Harbison, Marie M. Bourgeois, and Giffe T. Johnson.
© 2015 John Wiley & Sons, Inc. Published 2015 by John Wiley & Sons, Inc.

purity nickel powder and for continuous nickel coatings on steel and other metals.

PHYSICAL AND CHEMICAL PROPERTIES

Nickel is a group VIII transition metal and is the 24th most abundant element in the earth's crust (ATSDR, 2005). It can exist in five oxidation or valence states (-1, 0, $+2$, $+3$, and $+4$). The $+2$ valence state is the most common under normal environmental conditions, and many different species of nickel compounds are formed from the Ni^{2+} ion (ATSDR, 2005). Nickel can be extracted from sulfidic nickel ores and lateritic ores. Nickel smelting and refining processes use nickel derived from these ores to produce various chemical forms of nickel. Nickel compounds encountered in nickel production processes are divided into four categories, namely, water soluble, sulfidic, oxidic, and metallic. There is great complexity in the physical and chemical properties of nickel species among these four groups.

Water-soluble nickel refers to highly water-soluble nickel salts, such as nickel chloride hexahydrate ($NiCl_2 \cdot 6\,H_2O$) and nickel sulfate hexahydrate ($NiSO_4 \cdot 6\,H_2O$) (IARC, 1990) but may also describe nickel species that are easily extracted using mild solutions (e.g., ammonium citrate solution). Nickel sulfate hexahydrate is a blue-colored salt that is the main component of the electrolyte solution in electrolytic refining and is a raw material for the production of catalysts. It is also a common source of Ni^{2+} ions used for electroplating. Upon dissolving in water, it forms a solution containing the octahedral $[Ni(H_2O)_6]$ ions (hydrated nickel ions) that are hydrogen bonded to sulfate ions ($SO_4{}^{2-}$).

Sulfidic nickel generally consists of nickel sulfide compounds produced as intermediates in many nickel smelting and refining processes. These compounds include nickel subsulfide (Ni_3S_2), nickel sulfide (NiS), and nickel disulfide (NiS_2). The various nickel sulfide compounds are insoluble in water, although nickel subsulfide is somewhat soluble in biological fluids. Nickel subsulfide exists in two forms, namely, the low temperature green form, α-Ni_3S_2, also known as the mineral heazlewoodite, and the high temperature bronze-yellow form (β-Ni_3S_2) (IARC, 1990). Nickel sulfide occurs in three forms (α, β, and amorphous) as dark green to black crystals or powder.

Oxidic nickel consists of various nickel oxides, nickel hydroxides, nickel silicates, nickel carbonates, and complex nickel oxides such as nickel–copper oxides (NiPERA, 2002). Oxidic nickel compounds are relatively insoluble in water, and their solubility in acids and other properties depend on the method of their preparation. Nickel monoxides include a spectrum of compounds from a high temperature green variety, formed during calcination at temperatures exceeding 900 °C, to the more chemically active, low temperature black nickel oxides formed by calcination at 400 to 800 °C. Nickel hydroxide rarely occurs as a pure compound due to impurities present during industrial production. Nickel carbonates are insoluble in water, but carbon dioxide dissolved in water may dissolve some nickel as bicarbonate (NiPERA, 2002). Complex nickel oxides are often formed as by-products of industrial processes. For example, copper–nickel oxides may be formed during roasting/calcining processes, and spinels (complex nickel oxides such as nickel–chrome–iron oxides and nickel–iron oxides) may form when grinding stainless steel and high nickel alloys (NiPERA, 2002).

Metallic nickel consists of elemental nickel and its alloys. Elemental nickel is a lustrous silvery metal or a gray powder. It is insoluble in water and ammonia but soluble in dilute nitric acid and slightly soluble in hydrochloric acid and sulfuric acid. Powdered nickel is reactive in air, with the extent of reactivity dependent upon its size distribution (ATSDR, 2005). The silvery metal form of elemental nickel is not reactive in air or water at standard temperatures. Nickel alloys contain 3.5–66.5% nickel by weight and can be categorized as nickel–aluminum, nickel–chromium, nickel–chromium–cobalt, iron–nickel–chromium, copper–nickel alloys, and nickel-containing steels (IARC, 1990).

The direct combination of metallic nickel with carbon monoxide forms nickel carbonyl, a volatile liquid with a vapor pressure at 20 °C and a boiling point of 43 °C. Nickel carbonyl is miscible in organic solvents but essentially insoluble in water. Concentrations of nickel carbonyl in ambient air would tend to settle to ground level before being dispersed, because its vapor density is about four times that of air. Once nickel carbonyl is formed, it tends to remain as the metal carbonyl only in the presence of carbon monoxide. In ambient air, nickel carbonyl is relatively unstable and will dissociate to carbon monoxide and nickel metal with a half-life of about 100 s.

ENVIRONMENTAL FATE AND BIOACCUMULATION

Nickel exists in the atmosphere as particulate matter or is adsorbed to other particulate matter. It is removed from the atmosphere by various processes, such as gravitational settling for large particles (>5 μm) and other forms of wet and dry deposition processes for smaller particles (ATSDR, 2005). Nickel particles are present in a broad range of sizes, with an average size distribution measured as a mass median aerodynamic diameter (MMAD) of 1.0 μm in several U.S. cities (Lee et al., 1972).

Nickel is present in soils as water-soluble compounds, including chlorides and nitrates, and insoluble compounds such as nickel oxides and nickel sulfides. Nickel is strongly adsorbed by soil, with stronger adsorption in alkaline soils, which limits nickel's availability and mobility. Water-soluble nickel compounds tend to exhibit greater mobility than insoluble nickel compounds (US EPA, 2007a). Most nickel compounds are not expected to volatilize from soil, with the exception of nickel carbonyl (US EPA, 2007a).

The various forms of nickel in soils are transported into streams and waterways in runoff from natural weathering or from disturbed soil. Nickel may also enter bodies of water through atmospheric deposition. This metal is highly mobile in aquatic environments, and this is controlled largely by the capability of various sorbents to scavenge it from solution (US EPA, 2007b). Nickel is removed from the water column by adsorption onto suspended particles that settle into the sediment. It can also be adsorbed onto dissolved organic matter, limiting the amount of nickel that can be removed by the settling of suspended particles (Martino et al., 2003). Nickel in sediment may be strongly bound or may be remobilized by microbial action under anaerobic conditions (Francis and Dodge, 1990).

Although nickel does bioaccumulate, its bioconcentration factors suggest that nickel is not accumulated in significant amounts by aquatic organisms, and there is evidence indicating that nickel concentrations in organisms decrease with increasing aquatic levels (McGeer et al., 2003). Nickel concentrations also decrease with increasing trophic levels in food webs, indicating minimal transfer of nickel in the food chain (McGeer et al., 2003). Uptake and accumulation of nickel into various plant species is known to occur (Peralta-Videa et al., 2002).

ECOTOXICOLOGY

The effects of metals on ecosystems are related to metal bioavailability and not the total metal concentration. The bioavailability and toxicity of nickel in the aquatic environment are dependent on interactions with alkalinity, hardness, salinity, pH, temperature, and complexing agents such as humic acids. For example, the bioavailability and toxicity of nickel to freshwater organisms are negatively correlated to the hardness of water (US EPA, 1986). Nickel is generally not bioavailable and not likely to become toxic under natural aquatic conditions if it is occluded in minerals, clay, or sand or is strongly sorbed to particulate matter. Because toxicity tests with the same aquatic species performed under different water quality conditions can result in different toxicity end points, the data for freshwater systems must be normalized for differences in water chemistry using bioavailability models (NiPERA, 2012a).

Although the survival of aquatic organisms after acute exposure to various water-soluble nickel compounds has been evaluated in several different species, environmental regulations for nickel ecotoxicity are based on chronic toxicity data. The most sensitive end points for chronic nickel exposure in aquatic organisms are related to growth, development, reproduction, and mortality. For various species of algae in freshwater systems, no observed effects concentrations (NOECs) ranged from 35.9 to 115.3 µg/l nickel (NiPERA, 2012b). Nickel NOECs for several fish species ranged from 99.2 to 324.5 µg/l; for amphibians the NOECs ranged from 293.3 to 1265.9 µg/l (NiPERA, 2012b).

The bioavailability of nickel in natural soils changes over time as bioavailable nickel ions are adsorbed onto natural soil components. Nickel metal and compounds are not considered to be ecotoxic to the terrestrial environment, because they release negligible amounts of bioavailable nickel under naturally occurring soil conditions. In very acidic soils, nickel can cause toxicity to sensitive crops such as oats and barley. Remediation of this toxicity can be achieved by the addition of limestone to increase the soil pH.

MAMMALIAN TOXICOLOGY

In mammals, the health effects associated with exposure to nickel vary greatly depending on the form of nickel and the exposure route. This section describes the acute and chronic health effects of various forms of nickel in humans and experimental animals by the route of exposure.

Acute Effects

Inhalation Respiratory damage is the principal effect of acute inhalation exposure to nickel. Death from respiratory distress syndrome (with histological examination indicating alveolar wall damage and edema in alveolar spaces in the lung as well as tubular necrosis in the kidneys) has been reported after acute exposure to a high concentration of metallic nickel (Rendall et al., 1994). Nickel carbonyl is extensively absorbed after inhalation and is highly toxic. Initial stages of nickel carbonyl poisoning are characterized by headache, chest pain, dizziness, and a metallic taste in the mouth (Morgan, 1992), followed by a second phase of chemical pneumonitis with evidence of cerebral poisoning in severe cases (Shi, 1986). In rat studies, high acute exposures to nickel sulfate resulted in severe hemorrhage of the lungs (Hirano et al., 1994), and acute exposure to nickel chloride reduced lung macrophage activity (Adkins et al., 1979). In 12-day inhalation exposure studies in rats, lung inflammation was observed with the following nickel compounds in increasing order of severity of effects: nickel oxide, nickel subsulfide, nickel sulfate hexahydrate (NTP, 1996a, 1996b, 1996c).

Dermal Absorption of nickel through the skin is very low (1%), even for water-soluble nickel compounds (Tanojo et al., 2001). Contact dermatitis from dermal exposure is the most prevalent effect of nickel in the general population and is more frequently observed in females, particularly younger females, than in males or older individuals (ATSDR,

2005; Uter et al., 2003; Wantke et al., 1996). Nickel contact dermatitis appears to be related to previous nickel exposure rather than increased susceptibility, and the sensitizing source is often prolonged exposure to nickel in consumer products such as jewelry. In animal studies, acute dermal exposure of rats to high concentrations of nickel sulfate resulted in skin effects such as distortion of the epidermis and dermis, hyperkeratinization, vacuolization, hydropic degeneration of the basal layer, and atrophy of the epidermis (Mathur et al., 1977).

Oral   High acute exposure to nickel sulfate by accidental ingestion has resulted in cardiac arrest and death (Daldrup et al., 1983), and lower exposures resulted in gastrointestinal distress symptoms (Sunderman et al., 1988). A single, high oral dose of nickel sulfate can result in a flare-up of dermatitis in nickel-sensitive individuals (ATSDR, 2005). In rats, the following oral lethal dose, 50%, (LD_{50}) values were reported: for nickel metal, >5000 mg/kg; for water-soluble nickel compounds, 362 mg/kg nickel sulfate and 500 mg/kg nickel chloride; and for insoluble nickel compounds, 9990 to >11,000 mg/kg nickel oxide and >11,000 mg/kg nickel subsulfide (Henderson et al., 2012).

Chronic Effects

Inhalation   Occupational exposure to nickel through the inhalation of nickel-containing dust has been associated with toxicity and carcinogenicity in the respiratory tract. Pulmonary fibrosis, based on X-ray findings, was reported in one study to be associated with cumulative exposure to water-soluble nickel compounds or sulfidic nickel (Berge and Skyberg, 2003). A few cases of asthma symptoms also have been reported in workers exposed to water-soluble nickel compounds (Haber et al., 2000). Increased risks of cancers of the lung and nasal sinuses have been reported in some studies of nickel-exposed workers. Because nickel workers were often exposed to different combinations of several forms of nickel and other non-nickel substances and had other dissimilarities (e.g., smoking habits, lifestyle factors), it is not always evident, based on epidemiology data alone, which specific forms of nickel are correlated with increased risk. The occupational epidemiology data indicate clear associations with respiratory cancers for sulfidic nickel as well as oxidic nickel used in the sulfidic ore nickel-producing industry (which includes a nickel–copper oxide) but no consistent associations for nickel oxides that do not contain copper or for metallic nickel (Goodman et al., 2009, 2011; ICNCM, 1990). Although associations between water-soluble nickel exposure and lung cancer risk were noted in several studies of workers who were also exposed to sulfidic and oxidic forms of nickel, it is possible that the increased risks are attributable to these other forms of nickel, are related to water-soluble nickel exposures, or that water-soluble

nickel accentuates risks of other nickel forms (Goodman et al., 2011).

Studies in rodents demonstrate that chronic inflammation in the lungs is the most prominent toxic effect following chronic inhalation exposure to nickel-containing compounds (NTP, 1996a, 1996b, 1996c; Oller et al., 2008). This toxicity can be ranked in increasing order as follows: nickel oxide, nickel subsulfide, nickel metal powder, and nickel sulfate hexahydrate. Atrophy of the olfactory epithelium was also observed in rats with chronic exposure to nickel sulfate hexahydrate and nickel subsulfide (NTP, 1996a, 1996c). Two-year carcinogenicity studies in rodents indicate that nickel subsulfide (NTP, 1996a) and nickel oxide (NTP, 1996b) are carcinogenic in rats, but not mice, and neither water-soluble nickel sulfate hexahydrate (NTP, 1996c) nor nickel metal powder (Oller et al., 2008) are carcinogenic in either species. The tumors with increased incidence rates in the studies of nickel subsulfide and nickel oxide were lung adenomas and carcinomas, but there were no increases in nasal tumor incidence rates in treated animals. Increased incidence rates of benign and malignant pheochromocytomas were observed in male rats exposed to nickel subsulfide, nickel oxide, and nickel metal, but these tumors appeared to be secondary to lung toxicity rather than directly related to nickel exposure (Oller et al., 2008; Ozaki et al., 2002).

Dermal   Dermal sensitivity to nickel remains for many years (Keczkes et al., 1982), but the time interval between exposures can influence the degree of severity. A stronger reaction was observed in nickel-sensitized females after a 1-month period between nickel sulfate exposures compared to a 4-month period, and also with the application of nickel sulfate to an area of the skin with previous allergic contact dermatitis (Hindsén et al., 1997).

Oral   Intermediate-duration studies suggest that long-term oral exposure to nickel can be tolerated by some nickel-sensitive individuals and may even desensitize some individuals. Oral exposure before the dermal sensitizing exposure may also help prevent nickel sensitization in some individuals (van Hoogstraten et al., 1991). Two-year carcinogenicity studies of rodents treated orally with nickel sulfate hexahydrate or nickel acetate did not result in an exposure-related increased incidence of any tumor type (Heim et al., 2007; Schroeder et al., 1964, 1974).

Mode of Action(s)

The bioavailability of nickel ions at target sites is assumed to be the determining factor for the toxicity of nickel-containing compounds. The mechanisms for nickel ion toxicity vary depending on the exposure route and specific effect. With dermal exposure, nickel sensitization usually arises from prolonged contact with nickel or exposure to a large dose of nickel, and the resulting contact dermatitis is an inflammatory reaction mediated by type IV delayed hypersensitivity (ATSDR, 2005). Nickel is bound to albumin in the blood (HSDB, 2007).

The mechanism of adverse respiratory effects following inhalation exposure to nickel has been examined in rabbits exposed to metallic nickel or nickel chloride (Johansson and Camner, 1986; Johansson et al., 1980, 1981, 1983, 1987, 1988, 1989). These studies reported the accumulation of macrophages and granular material (such as phospholipids) in the alveoli and an increase in the volume density of alveolar type II cells with lamellar bodies, indicating a direct effect of the nickel ion on alveolar type II cells. After 6 months of exposure, all of the rabbits had pneumonia, indicating an increased susceptibility to infection, which may be attributable to the decreased function of alveolar macrophages (Johansson et al., 1981).

The mode of action for nickel carcinogenicity is unclear. The bioavailability of nickel ions at the nucleus of respiratory target cells differs among different nickel compounds, which may explain the variation in their carcinogenic effects (Goodman et al., 2009, 2011; Oller, 2002; Oller et al., 2008). This bioavailability has been examined in many *in vivo* and *in vitro* studies (Goodman et al., 2011). Sulfidic forms of nickel are readily taken up by respiratory cells and are predicted to have high nuclear bioavailability. Copper-free nickel oxides are cleared very slowly from the lungs but are not as readily taken up by respiratory cells; therefore, they are predicted to have lower nuclear bioavailability. Water-soluble nickel compounds are predicted to have very low bioavailability in the nucleus because of high toxicity, rapid clearance, and uptake processes that deliver nickel ions to the cytoplasm where they can bind to intracellular ligands. Metallic nickel has relatively high toxicity and is not readily taken up by cells so is predicted to have low nuclear bioavailability. Once in the nucleus of respiratory cells, nickel ions are only weakly genotoxic, but some studies suggest that they are clastogenic and can damage DNA. It is possible that non-genotoxic mechanisms involving indirect interaction of DNA with nuclear-available nickel ions are responsible for nickel-induced carcinogenesis, such as damage to heterochromatin, DNA damage through production of reactive oxygen species, alteration of gene expression through DNA hypermethylation or effects on histones, or inhibition of DNA repair processes.

SOURCES OF EXPOSURE

Nickel releases to the atmosphere occur from natural discharges as well as from anthropogenic activities. Occupational exposures to various nickel compounds occur in nickel production, processing, and using industries, and these industries also release nickel to the environment. The general population is exposed to low levels of nickel in ambient air, water, food, and cigarette smoke, and through contact with nickel-containing items such as coins, stainless steel, jewelry, and orthodontic appliances.

Occupational

Occupational exposure to nickel occurs through the inhalation of nickel-containing dust, fumes, or aerosols or through dermal contact with nickel solutions. Exposure to airborne nickel is highest in operations that involve grinding, welding, and handling powders (ATSDR, 2005). Exposure to airborne nickel in workplaces has been measured by air sampling in more recent years, or estimated by methods such as ranking of likely exposure in different work areas by workplace managers, or by extrapolation of these rankings or measurements from different times, places, or conditions (Goodman et al., 2009). However, determining exposure to specific nickel compounds is difficult. The initial stages in sulfidic ore nickel processing involved mining, milling, and smelting operations that typically involved higher levels of insoluble than water-soluble nickel compounds, and subsequent stages of processing involved a progressively higher ratio of water-soluble to insoluble nickel (Goodman et al., 2009). The electrolytic process involved predominantly water-soluble nickel, while the roasting process involved almost entirely insoluble forms of nickel. Thus, nickel workers in different industries, and those with different duties within the same industry, were often exposed to different combinations of several forms of nickel.

General Population

Most of the nickel exposure to the general population occurs through the diet. The average daily dietary intake of nickel in food ranges from 69–162 μg/day (ATSDR, 2005). Several studies have indicated that the foods with the highest mean nickel levels are nuts, legumes, oatmeal, cocoa, and grains or grain products (Capar and Cunningham, 2000; IARC, 1990; Pennington and Jones, 1987). Although several studies suggest that nickel may be nutritionally important, an essential function for nickel in humans has not been defined clearly (IOM, 2001; Nielsen and Ollerich, 1974; Schnegg and Kirchgessner, 1978; Uthus and Poellot, 1996). Nickel concentrations in drinking water average between 2 and 4.3 μg/l (FDA, 2007; O'Rourke et al., 1999; Thomas et al., 1999), and intake of nickel from drinking water is estimated to be around 4–8 μg/day (ATSDR, 2005). The average nickel concentration in ambient air in the United States is around 2 ng/m^3 (Lippmann et al., 2006; U.S. EPA, 2001). Cigarette smoke is another source of nickel exposure; smoke from five U.S. brands of cigarettes contains a mean concentration of 0.03 μg/g nickel (Torjussen et al., 2003). Exposure to nickel can occur through soil, of which nickel is a natural

TABLE 25.1   Occupational Exposure Limits for Nickel and Other Compounds

Agency	Compound	Parameter	Exposure Guideline	Basis for Guideline	Reference
ACGIH[a]	Elemental nickel	TLV–TWA	1.5 mg/m^3	Dermatitis; pneumoconiosis	ACGIH (2011)
	Soluble inorganic nickel compounds	TLV–TWA	0.1 mg/m^3	Lung damage; nasal cancer	ACGIH (2011)
	Insoluble inorganic nickel compounds	TLV–TWA	0.2 mg/m^3	Lung cancer	ACGIH (2011)
	Nickel subsulfide	TLV–TWA	0.1 mg/m^3	Lung cancer	ACGIH (2011)
	Nickel carbonyl	TLV–TWA	0.12 mg/m^3	Pneumonitis	ACGIH (2011)
OSHA	Nickel metal and insoluble compounds	PEL	1 mg/m^3	Dermatitis; pulmonary effects	OSHA (2009a)
	Soluble nickel compounds	PEL	1 mg/m^3	Dermatitis; cancer; pulmonary effects; respiratory and skin irritation	OSHA (2009b)
	Nickel carbonyl	PEL	0.007 mg/m^3	Acute pulmonary and CNS effects	OSHA (2004)
NIOSH	Nickel metal and other compounds	REL	0.015 mg/m^3	Dermatitis; allergic asthma; pneumonitis	NIOSH (2007)
	Nickel carbonyl	REL	0.007 mg/m^3	CNS, gastric, and pulmonary effects	NIOSH (2007)

Notes: ACGIH: American Conference of Governmental Industrial Hygienists; OSHA: Occupational Safety and Health Administration; NIOSH: National Institute for Occupational Safety and Health; TLV–TWA: Threshold limit value–time-weighted average—average exposure on the basis of an 8 h/day, 40 h/week work schedule; PEL: Permissible exposure limit based on an 8-h TWA exposure; REL: Recommended exposure limit based on a TWA exposure of up to 10 h/day and 40 h/week.
[a]ACGIH TLVs for nickel and nickel compounds are defined as the inhalable aerosol fraction.

constituent, with typical concentrations reported in the range of 4–80 parts per million (ppm). Dermal exposure can also occur through the handling of coins or jewelry. Nickel is present in U.S. coins as a cupronickel alloy containing 25% nickel and 75% copper (United States Treasury, 2010) and is present in gold and silver jewelry to reduce the oxidation of these metals (ATSDR, 2005).

INDUSTRIAL HYGIENE

Occupational limits for inhalation exposures to nickel compounds have been established by the American Conference of Governmental Industrial Hygienists (ACGIH), the Occupational Safety and Health Administration (OSHA), and the National Institute for Occupational Safety and Health (NIOSH). These limits are provided in Table 25.1.

RISK ASSESSMENTS

Several government and international agencies have performed human health risk and carcinogenicity assessments of nickel-containing substances, including the United States Environmental Protection Agency (US EPA), the Agency for Toxic Substances and Disease Registry (ATSDR) of the Centers for Disease Control (CDC), the National Toxicology Program (NTP), the International Agency for Research on Cancer (IARC), and the Danish Environmental Protection Agency (Danish EPA).

U.S. EPA evaluated several nickel-containing substances in its Integrated Risk Information System (IRIS). Soluble nickel salts were the only nickel compounds evaluated for non-cancer effects. U.S. EPA (1996) established an oral RfD of 0.02 mg/kg-day for effects on body and organ weights in a rat chronic oral study (Ambrose et al., 1976). U.S. EPA (1996) noted that many studies demonstrated dermal effects of nickel in sensitive humans after ingestion of nickel, but a dose–response relationship was difficult to establish, so these studies were not adequate to serve as the basis for the quantitative risk assessment. The RfD for soluble nickel salts is believed to be set at a level that would not cause individuals to become sensitized to nickel; however, those who have developed a hypersensitivity already (e.g., from dermal exposure) may not be fully protected. U.S. EPA has not evaluated soluble salts of nickel for effects from inhalation exposure or for potential human carcinogenicity. U.S. EPA (1991a, 1991b) evaluated nickel subsulfide and nickel refinery dust for carcinogenic effects and classified both as human carcinogens based on increased risks of lung and nasal cancer in humans exposed in the workplace and on animal data indicating tumors in animals by more than one route of administration (inhalation and injection). U.S. EPA (1991a) noted that nickel refinery dust is a mixture of many nickel moieties, and it is unclear what the carcinogenic nickel species is in the refinery dust. U.S. EPA (1991c) also evaluated nickel carbonyl for carcinogenicity, classifying it as a probable human carcinogen based on pulmonary carcinomas and malignant tumors at various sites in rats administered with nickel carbonyl by inhalation and intravenously, respectively. U.S. EPA (1991c) stated that the human data are inadequate for assessing carcinogenicity of nickel carbonyl, but it noted that no excess risk of cancer was observed in workers in an area where nickel carbonyl exposure was present (Peto et al., 1984).

Nickel was assessed for risks of non-cancer health effects for the general population by ATSDR. ATSDR (2005) concluded that the respiratory tract is the most sensitive target following inhalation exposures of nickel-containing substances, with nickel sulfate being more toxic to the lungs than nickel subsulfide or nickel oxide. ATSDR (2005) derived a minimal risk level (MRL) of 0.0002 mg Ni/m^3 for intermediate-duration inhalation exposure to nickel based on lung inflammation in a 13-week rat study of nickel sulfate hexahydrate (NTP, 1996c). ATSDR (2005) also derived a chronic duration inhalation MRL of 9×10^{-5} mg Ni/m^3 based on chronic lung inflammation in a 2-year rat study of nickel sulfate hexahydrate (NTP, 1996c). ATSDR (2005) noted that cohort mortality studies did not, in general, find significant increases in risk of dying from nonmalignant respiratory system disease. For effects of nickel from oral exposure, ATSDR (2005) noted that acute oral exposure to water-soluble nickel compounds is associated with allergic dermatitis in previously sensitized individuals and with gastrointestinal upset. Studies in animals indicated potential reproductive and developmental effects with acute and intermediate-duration oral exposures to water-soluble nickel, but reproductive effects were not observed in multigenerational studies of rats and developmental effects often occurred at the same dose levels as maternal toxicity. ATSDR (2005) also noted that chronic toxicity data for ingested nickel were limited at that time, concluding that the available database for effects of oral exposure to nickel is inadequate for deriving an oral MRL for nickel exposure.

The NTP classified nickel compounds as "known to be human carcinogens" based on sufficient evidence of carcinogenicity in epidemiology and mechanistic studies and supported by data from rodent carcinogenicity studies (NTP, 2002). In 2000, NTP classified nickel metal (but not alloys) as "reasonably anticipated to be a human carcinogen" based on sufficient evidence of carcinogenicity from studies in experimental animals, noting that available epidemiology studies of workers exposed to metallic nickel are limited by inadequate exposure information, low exposure levels, small numbers of cases, and short follow-up periods (NTP, 2011).

The IARC classified nickel compounds (including water-soluble and insoluble forms) as "group 1: carcinogenic to humans" based on sufficient evidence in humans for the carcinogenicity of nickel sulfate and the combinations of nickel sulfides and oxides encountered in the nickel refining industry (IARC, 1990). IARC (1990) classified metallic nickel as "group 2B: possibly carcinogenic to humans" based on inadequate evidence in humans but sufficient evidence in animals. The most recent IARC assessment of nickel carcinogenicity confirmed these conclusions (IARC, 2011).

The Danish EPA performed a risk assessment of metallic nickel (Danish EPA, 2008a) and various water-soluble nickel compounds (Danish EPA, 2008b, 2008c, 2008d, 2008e) for the European Union. It concluded that the effects on human health could not be excluded for occupational inhalation exposure to metallic nickel for repeated dose toxicity, carcinogenicity (suspect carcinogen), and development. It further concluded that for these end points, as well as effects of dermal exposures such as irritation and sensitization, there is no need to further limit the risks beyond current risk reduction measures. For exposures to the general population, the Danish EPA concluded that further information was needed to ensure that a threshold was set at an adequate level to prevent new cases of nickel allergy and sufficient to prevent elicitation of symptoms in a significant proportion of nickel-sensitized individuals. Such individuals are potentially at risk from excessive exposure in food and water, and additional risk reduction measures may be needed to limit exposure to nickel released from food contact materials, appliances, and taps for drinking water. For various water-soluble forms of nickel (nickel carbonate, nickel sulfate, nickel dinitrate, and nickel chloride), the Danish EPA concluded that the effects on human health could not be excluded for inhalation exposure of workers to these compounds for the end points of acute and repeated dose toxicity, respiratory sensitization, carcinogenicity, and development. For developmental toxicity and effects of dermal exposures such as irritation and skin sensitization, it was concluded that there was no need to further limit the risks beyond current risk reduction measures. For the general population, it was concluded that individuals with severe nickel sensitization are a particularly sensitive population to oral challenge with nickel and are potentially at risk from excessive exposure in food and water; additional risk reduction measures may be needed to limit this exposure, but there is no concern for those in the general population who are not sensitized to nickel already. The Council of Europe (2002) has issued recommendations for the release of nickel from food contact materials to food and for the release of nickel from electric kettles into water to be as low as reasonably achievable and not more than 0.1 mg/kg to food and 0.05 mg/l to water.

REFERENCES

Adkins, B., Jr., Richards, J.H., and Gardner, D.E. (1979) Enhancement of experimental respiratory infection following nickel inhalation. *Environ. Res.* 20:33–42.

Agency for Toxic Substances and Disease Registry (2005) *Syracuse Research Corp. Toxicological Profile for Nickel (Final)*. Atlanta, GA: ATSDR Division of Toxicology/Toxicology Information Branch.

Alloway, B.J., editor. (1990) *Heavy Metals in Soils*, New York, NY: John Wiley & Sons, Inc.

Ambrose, A.M., et al. (1976) Long term toxicologic assessment of nickel in rats and dogs. *J. Food Sci. Technol.* 13:181–187.

American Conference of Governmental Industrial Hygienists (2011) *2011 TLVs and BEIs: Threshold Limit Values for Chemical Substances and Physical Agents and Biological Exposure Indices*, Cincinnati, OH: ACGIH, Publication No. 0111.

Berge, S.R. and Skyberg, K. (2003) Radiographic evidence of pulmonary fibrosis and possible etiologic factors at a nickel refinery in Norway. *J. Environ. Monit.* 5:681–688.

Capar, S.G. and Cunningham, W.C. (2000) Element and radionuclide concentrations in food: FDA Total Diet Study 1991–1996. *J. AOAC Int.* 83:157–177.

Council of Europe (2002) *Partial Agreement in the Social and Public Health Field, Council of Europe's Policy Statements Concerning Materials and Articles Intended to Come into Contact with Foodstuffs, Policy Statement Concerning Metals and Alloys, Technical Document, Guidelines on Metals and Alloys Used as Food Contact Materials (13.02.2002).* Available at http://www.coe.int/t/e/social_cohesion/soc-sp/public_health/food_contact/TECH%20DOC%20GUIDELINES%20METALS%20AND%20ALLOYS.pdf (accessed September 25, 2012).

Daldrup, T., Haarhoff, K., and Szathmary, S.C. (1983) Toedliche nickel sulfaye-intoxikation [Fatal nickel sulfate poisoning]. *Beitr. Gerichtl. Med.* 41:141–144.

Danish Environmental Protection Agency (2008a) *Nickel (CAS No.: 7440-02-0) (EINECS No.: 231-111-4).Risk Assessment, Final Version*, Copenhagen, Denmark: Chemicals Division. Available at http://www.mst.dk/NR/rdonlyres/23FBC04F-5067-49AA-B8D7-E221374A5803/0/Ni_metal_EU_RAR_HH_March_2008_final_draft.pdf (accessed September 25, 2012).

Danish Environmental Protection Agency (2008b) *Nickel carbonate (CAS No.: 333-67-3) (EINECS No.: 222-068-2). Risk Assessment, Final Version*, Copenhagen, Denmark: Chemicals Division. Available at http://www.mst.dk/NR/rdonlyres/2929A8CB-8A5B-43C9-BB67-506D847E960E/0/Ni_carbonate_EU_RAR_HH_March_2008_finaldraft.pdf (accessed September 25, 2012).

Danish Environmental Protection Agency (2008c) *Nickel chloride (CAS No.: 7718-54-9) (EINECS No.: 231-743-0). Risk Assessment, Final Version*, Copenhagen, Denmark: Chemicals Division, Available at http://www.mst.dk/NR/rdonlyres/BD706A72-29DD-4021-ABAE-BE3D66FD9BAE/0/Ni_chloride_EU_RAR_HH_March_2008_finaldraft.pdf (accessed September 25, 2012).

Danish Environmental Protection Agency (2008d) *Nickel dinitrate (CAS No.: 13138-45-9) (EINECS No.: 236-038-5). Risk Assessment, Final Version*, Copenhagen, Denmark: Chemicals Division. Available at http://www.mst.dk/NR/rdonlyres/21D4A773-AA98-4160-91CA-BB1444238BC9/0/Ni_nitrate_EU_RAR_HH_March_2008_finaldraft.pdf (accessed September 25, 2012).

Danish Environmental Protection Agency (2008e) *Nickel sulphate (CAS No.: 7786-81-4) (EINECS No.: 232-104-9). Risk Assessment, Final Version*, Copenhagen, Denmark: Chemicals Division. Available at http://www.mst.dk/NR/rdonlyres/8DBA0CD9-5845-4C3D-832F-70E12B87F27C/0/Ni_sulfate_EU_RAR_HH_March_2008_finaldraft.pdf (accessed September 25, 2012).

Food and Drug Administration (2007) *Total diet study statistics on element results. Revision 4.1, 1991-3 through 2005-4.* College Park, MD: Center for Food Safety and Applied Nutrition. Available at http://www.fda.gov/downloads/Food/FoodSafety/FoodContaminantsAdulteration/TotalDietStudy/UCM243059.pdf (accessed September 25, 2012).

Francis, A.J. and Dodge, C.J. (1990) Anaerobic microbial remobilization of toxic metals co-precipitated with iron oxide. *Environ. Sci. Technol.* 24:373–378.

Goodman, J.E., et al. (2009) Carcinogenicity assessment of water-soluble nickel compounds. *Crit. Rev. Toxicol.* 39:365–417.

Goodman, J.E., et al. (2011) The nickel ion bioavailability model of the carcinogenic potential of nickel-containing substances in the lung. *Crit. Rev. Toxicol.* 41:142–174.

Haber, L.T., et al. (2000) Hazard identification and dose response of inhaled nickel-soluble salts. *Regul. Toxicol. Pharmacol.* 31: 210–230.

Heim, K.E., et al. (2007) Oral carcinogenicity study with nickel sulfate hexahydrate in Fischer 344 rats. *Toxicol. Appl. Pharmacol.* 224:126–137.

Henderson, R.G., et al. (2012) Acute oral toxicity of nickel compounds. *Regul. Toxicol. Pharmacol.* 62:425–432.

Hindsén, M., Bruze, M., and Christensen, O.B. (1997) The significance of previous allergic contact dermatitis for elicitation of delayed hypersensitivity to nickel. *Contact Dermatitis* 37: 101–106.

Hirano, S., et al. (1994) Pulmonary clearance and inflammatory potency of intratracheally instilled or acutely inhaled nickel sulfate in rats. *Arch. Toxicol.* 68:548–554.

HSDB (2007) *Hazardous Substances Data Bank: Nickel*. Bethesda, MD: NLM (National Library of Medicine), National Institutes of Health, U.S. Department of Health and Human Services.

Institute of Medicine, Standing Committee on the Scientific Evaluation of Dietary Reference Intakes, Food and Nutrition Board (2001) Nickel (excerpt). In: *Dietary Reference Intakes for Vitamin A, Vitamin K, Arsenic, Boron, Chromium, Copper, Iodine, Iron, Manganese, Molybdenum, Nickel, Silicon, Vanadium, and Zinc*, Washington, DC: National Academy Press, pp. 521–529, 668.

International Agency for Research on Cancer (1990) Nickel and nickel compounds. In: *IARC Monographs on the Evaluation of Carcinogenic Risks to Humans: Chromium, Nickel, and Welding*, vol. 49, Lyon, France: IARC Working Group on the Evaluation of Carcinogenic Risks to Humans, pp. 257–445.

International Agency for Research on Cancer (2011) Nickel and nickel compounds. In: *IARC Monographs on the Evaluation of Carcinogenic Risks to Humans: A Review of Human Carcinogens: Arsenic, Metals, Fibres, and Dusts*, vol. 100C, Lyon, France: IARC Working Group on the Evaluation of Carcinogenic Risks to Humans, pp. 169–218.

International Committee on Nickel Carcinogenesis in Man (1990) Report of the International Committee on Nickel Carcinogenesis in Man. *Scand. J. Work Environ. Health* 16:1–82.

Johansson, A. and Camner, P. (1986) Adverse effects of metals on the alveolar part of the lung. *Scan. Electron Microsc. Part 2* 631–637.

Johansson, A., Camner, P., Jarstrand, C., and Wiernik, A. (1980) Morphology and function of alveolar macrophages after long-term nickel exposure. *Environ. Res.* 23:170–180.

Johansson, A., Camner, P., and Robertson, B. (1981) Effects of long-term nickel dust exposure on rabbit alveolar epithelium. *Environ. Res.* 25:391–402.

Johansson, A., et al. (1983) Rabbit lung after inhalation of soluble nickel. II. Effects of lung tissue and phospholipids. *Environ. Res.* 31:399–412.

Johansson, A., et al. (1987) Lysozyme activity in ultrastructurally defined fractions of alveolar macrophages after inhalation exposure to nickel. *Br. J. Ind. Med.* 44:47–52.

Johansson, A., et al. (1988) Alveolar macrophages in rabbits after combined exposure to nickel and trivalent chromium. *Environ. Res.* 46:120–132.

Johansson, A., et al. (1989) Lung lesions after experimental combined exposure to nickel and trivalent chromium. *Environ. Res.* 50:103–119.

Keczkes, K., Basheer, A.M., and Wyatt, E.H. (1982) The persistence of allergic contact sensitivity: a 10-year follow-up in 100 patients. *Br. J. Dermatol.* 107:461–465.

Lee, R.E., et al. (1972) National Air Surveillance cascade impactor network. II. Size distribution measurements of trace metal components. *Environ. Sci. Technol.* 6:1025–1030.

Lippmann, M., et al. (2006) Cardiovascular effects of nickel in ambient air. *Environ. Health Perspect.* 114:1662–1669.

Martino, M., Turner, A., and Millward, G.E. (2003) Influence of organic complexation on the adsorption kinetics of nickel in river waters. *Environ. Sci. Technol.* 37:2383–2388.

Mathur, A.K., et al. (1977) Effect of nickel sulphate on male rats. *Bull. Environ. Contam. Toxicol.* 17:241–248.

McGeer, J.C., et al. (2003) Inverse relationship between bioconcentration factor and exposure concentration for metals: implications for hazard assessment of metals in the aquatic environment. *Environ. Toxicol. Chem.* 22:1017–1037.

Morgan, L. (1992) Problems in the toxicology, diagnosis, and treatment of nickel poisoning. In: Nieboer, E., and Nriagu, J.O., editors. *Nickel and Human Health: Current Perspectives*, New York, NY: John Wiley & Sons, Inc., pp. 261–271.

National Institute for Occupational Safety and Health (2007) *NIOSH Pocket Guide to Chemical Hazards*, Cincinnati, OH: NIOSH. Available at http://www.cdc.gov/niosh/docs/2005-149/pdfs/2005-149.pdf (accessed on September 25, 2012).

National Toxicology Program (1996a) *Toxicology and Carcinogenesis Studies of Nickel Subsulfide (CAS No. 12035-72-2) in F344/N Rats and B6C3F1 Mice (Inhalation Studies)*, Research Triangle Park, NC: National Toxicology Program.

National Toxicology Program (1996b) *Toxicology and Carcinogenesis Studies of Nickel Oxide (CAS No. 1313-99-1) in F344/N Rats and B6C3F1 Mice (Inhalation Studies)*, Research Triangle Park, NC: National Toxicology Program.

National Toxicology Program (1996c) *Toxicology and Carcinogenesis Studies of Nickel Sulfate Hexahydrate (CAS No. 10101-97-0) in F344/N Rats and B6C3F1 Mice (Inhalation Studies)*, Research Triangle Park, NC: National Toxicology Program.

National Toxicology Program (2002) *Tenth Report on Carcinogens*, Research Triangle Park, NC: National Toxicology Program.

National Toxicology Program (2011) *Report on Carcinogens*, 12th ed., Research Triangle Park, NC: National Toxicology Program. Available at http://ntp.niehs.nih.gov/go/roc12 (accessed on September 25, 2012).

Nickel Producers Environmental Research Association (2002) *Physicochemical and Toxicological Properties of "Oxidic" Nickel Compounds*, Durham, NC: NiPERA.

Nickel Producers Environmental Research Association (2012a) *Fact Sheet. European Union Environmental Risk Assessment of Nickel: Incorporation of Bioavailability in the Aquatic Compartment. Fact Sheet #4.* Available at http://www.nipera.org/en/EnvironmentalScience/FS4BioavailabiltyAquaticCompartment.aspx.

Nickel Producers Environmental Research Association (2012b) *Fact Sheet. European Union Environmental Risk Assessment of Nickel: Data Compilation, Selection, and Derivation of PNEC Values for the Freshwater Compartment. Fact Sheet #1.* Available at http://www.nipera.org/EnvironmentalScience/FS1-FreshwaterEffects.aspx.

Nielsen, F.H. and Ollerich, D.A. (1974) Nickel: a new essential trace element. *Fed. Proc.* 33:1767–1772.

Occupational Safety and Health Administration (2004) *Chemical Sampling Information: Nickel Carbonyl.* Available at http://www.osha.gov/dts/chemicalsampling/data/CH_256150.html (accessed on September 25, 2012).

Occupational Safety and Health Administration (2009a) *Nickel, metal and insoluble compounds (as Ni).* Available at http://www.osha.gov/dts/chemicalsampling/data/CH_256200.html (accessed on September 25, 2012).

Occupational Safety and Health Administration (2009b) *Nickel, soluble compounds (as Ni).* Available at http://www.osha.gov/dts/chemicalsampling/data/CH_256300.html (accessed on September 25, 2012).

Oller, A.R. (2002) Respiratory carcinogenicity assessment of soluble nickel compounds. *Environ. Health Perspect.* 110(Suppl. 5):841–844.

Oller, A.R., et al. (2008) Inhalation carcinogenicity study with nickel metal powder in wistar rats. *Toxicol. Appl. Pharmacol.* 233:262–275.

O'Rourke, M.K., et al. (1999) Evaluations of primary metals from NHEXAS Arizona: distributions and preliminary exposures. *J. Expos. Anal. Environ. Epidemiol.* 9:435–445.

Ozaki, K., et al. (2002) Association of adrenal pheochromocytoma and lung pathology in inhalation studies with particulate compounds in the male F344 rat—the National Toxicology Program experience. *Toxicol. Pathol.* 30:263–270.

Pennington, J.A. and Jones, J.W. (1987) Molybdenum, nickel, cobalt, vanadium, and strontium in total diets. *J. Am. Diet. Assoc.* 87:1644–1650.

Peralta-Videa, J.R., et al. (2002) Effect of mixed cadmium, copper, nickel and zinc at different pHs upon alfalfa growth and heavy metal uptake. *Environ. Pollut.* 119:291–301.

Peto, J., et al. (1984) Respiratory cancer mortality of Welsh nickel refinery workers. *IARC Sci. Publ.* 53:37–46.

Rendall, R.E., Phillips, J.I., and Renton, K.A. (1994) Death following exposure to fine particulate nickel from a metal arc process. *Ann. Occup. Hyg.* 38:921–930.

Schnegg, A. and Kirchgessner, M. (1978) Ni deficiency and its effects on metabolism. In: Kirchgessner, M., editor. *Trace Element Metabolism in Man and Animals—3*, Munich, Germany: Technische Universitat, pp. 236–243.

Schroeder, H.A., Balassa, J.J., and Vintin, W.H., Jr. (1964) Chromium, lead, cadmium, nickel and titanium in mice: effect on mortality, tumors and tissue levels. *J. Nutr.* 83:239–250.

Schroeder, H.A., Mitchener, M., and Nason, A.P. (1974) Life-term effects of nickel in rats: survival, tumors, interactions with trace elements and tissue levels. *J. Nutr.* 104:239–243.

Shi, Z. (1986) Acute nickel carbonyl poisoning: a report of 179 cases. *Br. J. Ind. Med.* 43:422–424.

Sunderman, F.W., Jr, et al. (1988) Acute nickel toxicity in electroplating workers who accidentally ingested a solution of nickel sulfate and nickel chloride. *Am. J. Ind. Med.* 14:257–266.

Tanojo, H., et al. (2001) *In vitro* permeation of nickel salts through human stratum corneum. *Acta Derm. Venereol. Suppl.* 212: 19–23.

Thomas, K.W., Pellizzari, E.D., and Berry, M.R. (1999) Population-based dietary intakes and tap water concentrations for selected elements in the EPA Region V National Human Exposure Assessment Survey (NHEXAS). *J. Expos. Anal. Environ. Epidemiol.* 9:402–413.

Torjussen, W., Zachariasen, H., and Andersen, I. (2003) Cigarette smoking and nickel exposure. *J. Environ. Monit.* 5:198–201.

United States Treasury (2010) *Fact sheets: currency & coins. Manufacturing process for U.S. coins.* Available at http://www.treas.gov/education/fact-sheets/currency/manufacturing.shtml (accessed on September 25, 2012).

U.S. EPA (1986) *Ambient Water Quality Criteria for Nickel—1986. EPA-440/5-86-004*, Washington, DC: Office of Water Regulations and Standards.

U.S. EPA (1991a) *IRIS record for nickel refinery dust.* Integrated Risk Information System (IRIS). Available at: http://www.epa.gov/iris/subst/0272.htm (accessed on September 25, 2012).

U.S. EPA (1991b) *IRIS record for nickel subsulfide.* Integrated Risk Information System (IRIS). Available at http://www.epa.gov/iris/subst/0273.htm (accessed on September 25, 2012).

U.S. EPA (1991c) *IRIS record for nickel carbonyl.* Integrated Risk Information System (IRIS). Available at http://www.epa.gov/iris/subst/0274.htm (accessed on September 25, 2012).

U.S. EPA (1996) *IRIS record for nickel, soluble salts.* Integrated Risk Information System (IRIS). Available at http://www.epa.gov/iris/subst/0271.htm (accessed on September 25, 2012).

U.S. EPA (2001) OT 1996 Modeled Ambient Concentration for Nickel Compounds. Available at http://www.epa.gov/ttnatw01/nata/pdf/nicke_conc.pdf (accessed on September 25, 2012).

U.S. EPA (2007a) *Ecological Soil Screening Levels for Nickel (Interim Final). OSWER Directive 9285.7-76*, Washington, DC: Office of Solid Waste and Emergency Response.

U.S. EPA (2007b) *Monitored Natural Attenuation of Inorganic Contaminants in Ground Water: Volume 2. Assessment for Non-Radionuclides Including Arsenic, Cadmium, Chromium, Copper, Lead, Nickel, Nitrate, Perchlorate, and Selenium. EPA 600-R-07-140*, Cincinnati, OH: Office of Research and Development.

Uter, W., et al. (2003) Risk factors for contact allergy to nickel—results of a multifactorial analysis. *Contact Dermatitis* 48: 33–38.

Uthus, E.O. and Poellot, R.A. (1996) Dietary folate affects the response of rats to nickel deprivation. *Biol. Trace Elem. Res.* 52:23–35.

van Hoogstraten, I.M., et al. (1991) Effects of oral exposure to nickel or chromium on cutaneous sensitization. *Curr. Probl. Dermatol.* 20:237–241.

Wantke, F., et al. (1996) Patch test reactions in children, adults and the elderly. A comparative study in patients with suspected allergic contact dermatitis. *Contact Dermatitis* 34:316–319.

26

TANTALUM AND NIOBIUM

Jayme P. Coyle and Raymond D. Harbison

Target organ(s)	
Niobium	None
Tantalum[a]	Respiratory, skin, eyes
Occupational exposure limits	
Niobium	
Elemental	No limits established
Tantalum	
Compounds[a,b]	OSHA PEL: 5 mg/m^3 (TWA)
	OSHA STEL: 10 mg/m^3 (15 min)
	NIOSH REL: 5 mg/m^3 (TWA)
	ACGIH TLV: 5 mg/m^3 (TWA);
	15 mg/m^3 (30 min); 25 mg/m^3 (ceiling)
	IDLH: 2500 mg Ta/m^3
Reference values	None established
Risk/safety phrases	
Niobium (elemental)	
Risk phrases	R17 spontaneously flammable in air
Safety phrases	S33 take precautionary measures against static discharges
Tantalum (elemental)	
Risk phrases	R11 highly flammable
Safety phrases	S43 in case of fire, use metal-fire powder. Never use water

[a] National Institute for Occupational Safety and Health (NIOSH). (2007). *Pocket guide to chemical hazards. DHHS (NIOSH) Publication Number 2005-149*. National Institute for Occupational Safety and Health, Washington, D.C.
[b] American Conference of Governmental Industrial Hygienists (ACGIH). (2011). *Threshold Limit Values (TLVs) for chemical substances and physical agents and Biological Exposure Indices (BEIs)*. American Conference of Governmental Industrial Hygienists, Cincinnati, OH.

BACKGROUND AND USES

Niobium (symbol Nb; CASRN 7440-03-1) and tantalum (symbol Ta; CASRN 7440-25-7) are considered chemically similar as they fall within the same vertical group as vanadium in the periodic table. They are both difficult to separate from one another and, thus, typically found together naturally. Niobium is principally found in columbite ($(Fe,Mn)(Nb,Ta)_2O_6$), columbite–tantalite ($(Fe,Mn)(Ta,Nb)_2O_6$), and pyrochlore ($(Na, Ca)_2Nb_2O_6(OH,F)$), compounds which often consist of both niobium and tantalum. Tantalum is found principally in tantalite ($(Fe, Mn)Ta_2O_6$), as well as in the aforementioned niobium ores.

Niobium is alloyed with steel, aluminum, tungsten, and titanium, among others, which are refractory at high temperatures. High stability in heat makes niobium alloys useful in rocket and jet engine components and as a constituent of nuclear reactor core claddings. Similar to tantalum, niobium alloys have found wide use in joint arthroplasty, especially for high stress applications as composites of niobium and tantalum are both highly biocompatible (Wang et al., 2010).

Given its high weight to strength ratio, a tantalum trabecular metal lattice has been developed to resemble the bone matrix in implants. It is also used as a catalyst for rubber production and in capacitors and rectifiers. Tantalum pentoxide (Ta_2O_5; CASRN 1314-61-0) is utilized within the electronics industry in capacitors.

PHYSICAL AND CHEMICAL PROPERTIES

Tantalum is a blue-gray, ductile metal with melting and boiling points of 2996 °C (5425°F) and 5425 °C (9797°F),

Hamilton & Hardy's Industrial Toxicology, Sixth Edition. Edited by Raymond D. Harbison, Marie M. Bourgeois, and Giffe T. Johnson.

respectively. It is highly resistant to corrosion though corrosion resistance to strong acids and atmospheric oxygen decreases at increasing temperatures. It is incompatible with strong oxidizers and alkalis. Niobium appears as a grey metal with melting and boiling points of 2477 °C (4491 °F) and 4741 °C (8566 °F), respectively. Both elements have negligible vapor pressures, and in pure metallic powder form, both elements are flammable.

Both elements exist primarily in a +5 oxidation state though niobium may assume a +2 state. Under physiological conditions, both elements form insoluble oxides. Similar physiochemistry makes them difficult to separate from their respective ores.

MAMMALIAN TOXICOLOGY

Acute Effects

In the 1960s and 1970s tantalum powder was used for bronchography. This technique has been replaced by other techniques, but no adverse acute or long-term effects were associated with this procedure (Gamsu and Nadel, 1972; Heitzman, 2000; Nadel et al., 1968). An intraperitoneal injection of solubilized tantalum did not elicit an inflammatory response of the rabbit mesothelium but did exhibit slow absorption with preferential accumulation in lymphatic tissue and systemic distribution. Distribution to and deposition within the pulmonary interstitium did not result in pathologic changes (Lawson et al., 1969; Sailer and Stauch, 1973). Animal studies have documented transient bronchial and alveolar inflammation when tantalum oxide (Ta_2O_5; CAS # 1314-61-0) was insufflated by the endotracheal route (Cochran et al., 1950; Gamsu and Nadel, 1972; Schepers, 1955; Smith et al., 1979).

Among some workers who accidentally inhaled radioactive tantalum oxide dust, <10% of the dust remained after 7 days without noticing any adverse effects (Newton, 1977; Sill et al., 1969). The remainder was cleared, as observed by radioactive decay, with a prolonged biological half-life, suggesting that a tiny fraction of inhaled tantalum powder can be retained for years. These results were consistent with experimental animal models (Bianco et al., 1974). No associated health problems have been described.

Niobium pentachloride ($NbCl_5$) (CASRN 10026-12-7) can cause renal damage when ingested or given intraperitoneally to animals (Schroeder and Balassa, 1965). Mild transient hepatocyte hydropic degeneration was observed in mice after an intraperitoneal injection of niobium oxide; assessment for hematological changes were negative (Dsouki et al., 2014). No significant human health problems attributable to niobium metal or niobium alloys have been reported to date.

Chronic Effects

No long-term effects of chronic exposure to either tantalum or niobium in humans have been identified. In one report,

workers exposed to tantalum oxide at concentrations ranging between 0.1 and 7.6 mg/m^3 did not significantly show significant pulmonary injury (Chen et al., 1999). Hamsters exposed to 100 mg/m^3 tantalum oxide presented abnormal pulmonary pathology, including transient interstitial pneumonia, bronchitis, and hyperemia, all of which cleared 1 year after exposure cessation. Only minor hyperplasia was observed as the residual evidence of exposure (Bingham et al., 2001).

Mechanisms of Action

Tantalum demonstrates a triphasic biological half-life. Gastrointestinal absorption of tantalum is generally <1% and clears within days. When absorbed, tantalum binds to plasma proteins with high affinity and can distribute systemically, remaining in the blood for years after periods of high exposure and this accounts for the prolonged half-life (Morrow et al., 1976; Wibowo, 2004). Respiratory clearance is slow in distal bronchioles and alveoli with an effective biological half-life of 2 years or greater in this region. Pulmonary clearance in dogs resulted in translocation to pulmonary lymph nodes where accumulation was noted.

Solid tantalum metal does not significantly elicit an immunological response though tantalum trabecular metal in prostheses has been shown to elicit a PMN-mediated cytokine response; the incidence is rare and only one account has been documented. Rare incidences of allergic contact dermatitis have occurred (Romaguera and Vilaplana, 1995; Schildhauer et al., 2009).

Chemical Pathology

Radiographic imaging of rats administered the radionuclide niobium-95 showed accumulation in the kidneys, bone, and the spleen, while tantalum-182 accumulated predominantly in bone and kidney tissues (Ando and Ando, 1990). Hamsters exposed to tantalum oxide showed interstitial pneumonia with slight, long-term hyperplasia after cessation of exposure (Bingham et al., 2001).

SOURCES OF EXPOSURE

Occupational

Niobium and tantalum are rarely encountered and present few industrial hazards. Workers within the mining and purification of niobium may be exposed to niobium-containing dusts. Production of ferro-niobium alloys can expose workers to aerosols (Lipsztein et al., 2001). However, the extraction and milling of the ores can involve exposure to other toxic substances such as hydrofluoric acid and manganese. Other constituents composed of niobium- or tantalum-containing alloys may be more toxic. Metal fabrication, such as grinding and cutting operations, or use as carbide-based cutting bits are the primary sources of inhalable exposures for both elements (Bochmann et al., 2008).

Environmental

No significant environmental exposures have been described for either compound aside from the improper discard of mining waste.

INDUSTRIAL HYGIENE

For tantalum, the National Institute for Occupational Safety and Health (NIOSH), the Occupational Safety and Health Administration (OSHA), and the American Conference of Governmental Industrial Hygienists (ACGIH) have set the recommended exposure limit (REL), the permissible exposure limit (PEL), and the threshold limit value (TLV), respectively, at $5\,mg/m^3$ (ACGIH, 2011; NIOSH, 2007). Additionally, NIOSH has a 15-min short-term exposure limit (STEL) of $10\,mg/m^3$ and an immediately dangerous to life and health (IDLH) value of $2500\,mg\,Ta/m^3$. Specific exposure limits for niobium have not been established but may be regulated indirectly through the limits established for nuisance dusts ($15\,mg/m^3$).

Generally, general ventilation systems maintain the airborne concentration of niobium and tantalum below regulatory limits. The use of respirators is recommended as protection against particulate (NIOSH, 2007). Sources of ignition should be avoided when working with the metallic powder.

MEDICAL MANAGEMENT

Unlike vanadium, niobium and tantalum are virtually inert in human body tissue and fluids (Cochran et al., 1950; Dales and Kyle, 1958). Therefore, standard hygiene practices should be followed to limit exposure. In case of eye contact, flush with copious amounts of water. No specific advanced medical protocols have been established for either of the substances.

REFERENCES

American Conference of Governmental Industrial Hygienists (ACGIH) (2011) *Threshold Limit Values (TLVs) for Chemical Substances and Physical Agents and Biological Exposure Indices (BEIs)*. Cincinnati, OH: American Conference of Governmental Industrial Hygienists.

Ando, A. and Ando, I. (1990) Biodistribution of [95]Nb and [182]Ta in tumor-bearing animals and mechanisms for accumulation in tumor and liver. *J. Radiat. Res.* 31(1):97–109.

Bianco, A., Gibb, F.R., Kilpper, R.W., Landman, S., and Morrow, P.E. (1974) Studies of tantalum dust in the lungs. *Radiology* 112(3):549–556.

Bingham, E., Cohrssen, B., and Powell, C.H. (2001) *Patty's Toxicology*, 5th ed., vols. 1–9, New York, NY: John Wiley & Sons, Inc.

Bochmann, F., Gabriel, S., Hahn, J.U., Hartig, A., Mittenzwei, V., and Rocker, M. (2008) Hartmetallarbeitsplätze: Exposition und Bewertung. *Gefahrstoffe.-Reinhalt. Luft* 68:7–14.

Chen, Y., Yin, X., Ning, G., Nie, X., Li, Q., and Dong, J. (1999) Effects of tantalum and its oxide on exposed workers. *Chin. J. Prev. Med.* 33(4):234–245.

Cochran, K.W., Doull, J., Mazur, M., and Dubois, K.P. (1950) Acute toxicity of zirconium, columbium, strontium, lanthanum, cesium, tantalum and yttrium. *Arch. Ind. Hyg. Occup. Med.* 1(6):637–650.

Dales, H.C. and Kyle, J. (1958) Late results of using tantalum gauze in the repair of large hernias. *Surgery* 43(2):294–297.

Dsouki, N.A., de Lima, M.P., Corazzini, R., Gascon, T.M., Azzalis, L.A., Junqueira, V.B., Feder, D., and Fonseca, F.L. (2014) Cytotoxic, hematologic and histologic effects of niobium pentoxide in Swiss mice. *J. Mater. Sci. Mater. Med.* doi: 10.1007/s10856-014-5153-0.

Gamsu, G. and Nadel, J.A. (1972) New technique for roentgenographic study of airways and lungs using powdered tantalum. *Cancer* 30(5):1353–1357.

Heitzman, E.R. (2000) Thoracic radiology: the past 50 years. *Radiology* 214(2):309–313.

Lawson, T.L., Margulis, A.R., Nadel, J.A., Rambo, O.N., and Wolfe, W.G. (1969) Intraperitoneal introduction of tantalum powder. A roentgenographic and pathologic study. *Invest. Radiol.* 4(5):293–300.

Lipsztein, J. L., da Cunha, K. M., Azeredo, A. M., Juliao, L., Santos, M., Melo, D. R., and Simoes Filho, F.F. (2001) Exposure of workers in mineral processing industries in Brazil. *J. Environ. Radioact.* 54(1):189–199.

Morrow, P.E., Kilpper, R.W., Beiter, H.E., and Gibb, F.R. (1976) Pulmonary retention and translocation of insufflated tantalum. *Radiology* 121(2):415–421.

Nadel, J.A., Wolfe, W.G., and Graf, P.D. (1968) Powdered tantalum as a medium for bronchography in canine and human lungs. *Invest. Radiol.* 3(4):229–238.

National Institute for Occupational Safety and Health (NIOSH) (2007) *Pocket Guide to Chemical Hazards*, DHHS (NIOSH) Publication Number 2005-149, Washington, DC: National Institute for Occupational Safety and Health.

Newton, D. (1977) Clearance of radioactive tantalum from the human lung after accidental inhalation. *AJR Am. J. Roentgenol.* 129(2):327–328.

Romaguera, C. and Vilaplana, J. (1995) Contact dermatitis from tantalum. *Contact Dermatitis* 32(3):184.

Sailer, R. and Stauch, G. (1973) Beitrag zur indirekten Lymphographie des Mediastinums und des oberen Abdominalraumes beim Kaninchen mit Tantaloxyd- und Tantalkarbidstaub. *Z. Exp. Chir.* 6(5):313–323.

Schepers, G.W. (1955) The biological action of tantalum oxide; studies on experimental pulmonary histopathology. *AMA Arch. Ind. Health* 12(2):121–123.

Schildhauer, T.A., Peter, E., Muhr, G., and Koller, M. (2009) Activation of human leukocytes on tantalum trabecular metal in comparison to commonly used orthopedic metal implant materials. *J. Biomed. Mater. Res. A* 88(2):332–341.

Schroeder, H.A. and Balassa, J.J. (1965) Abnormal trace elements in man: niobium. *J. Chronic Dis.* 18:229–241.

Sill, C.W., Voelz, G.L., Olson, D.G., and Anderson, J.I. (1969) Two studies of acute internal exposure to man involving cerium and tantalum radioisotopes. *Health Phys.* 16(3):325–332.

Smith, P., Stitik, F., Smith, J., Rosenthal, R., and Menkes, H. (1979) Tantalum inhalation and airway responses. *Thorax* 34(4):486–492.

Wang, B.L., Li, L., and Zheng, Y.F. (2010) *In vitro* cytotoxicity and hemocompatibility studies of Ti-Nb, Ti-Nb-Zr and Ti-Nb-Hf biomedical shape memory alloys. *Biomed. Mater.* 5(4):044102.

Wibowo, A.A.E. (2004) Tantalum. In: *Health-Based Reassessment of Administrative Occupational Exposure Limits*, 2000/15OSH/130, The Hague, the Netherlands: Health Council of the Netherlands.

27

PLATINUM GROUP ELEMENTS: PALLADIUM, IRIDIUM, OSMIUM, RHODIUM, AND RUTHENIUM

ALEX LEBEAU

First aid:

Eyes: Irrigate immediately (osmium tetroxide and soluble rhodium compounds)

Dermal: Wash skin with soap (osmium tetroxide). Flush skin with water (ruthenium salts)

Exposure to radioisotopes: Begin decontamination procedures and seek medical attention

Inhalation: Respiratory support (osmium tetroxide, rhodium compounds, metal fumes, and salts)

Ingestion: Seek immediate medical attention (osmium tetroxide, rhodium compounds, metal fumes, and salts)

Target organ(s): Skin, respiratory system, immune system

Occupational exposure limits:

Palladium: No PEL, REL, or TLV established

Iridium: No PEL, REL, or TLV established

Osmium: As osmium tetroxide:

TLV: TWA = 0.0002 ppm; STEL = 0.0006 ppm

REL: TWA = 0.002 mg/m^3; ST = 0.006 mg/m^3

PEL: TWA = 0.002 mg/m^3

IDLH = 1 mg/m^3

Rhodium:

Soluble compounds: PEL = 0.001 mg/m^3; REL = 0.001 mg/m^3; TLV = 0.01 mg/m^3; IDLH = 2 mg/m^3

Metal and insoluble compounds: PEL = 0.1 mg/m^3; REL = 0.1 mg/m^3; TLV = 1 mg/m^3; IDLH = 100 mg/m^3

Ruthenium: No PEL, REL, or TLV established

Reference values: No EPA IRIS reference doses or reference concentrations established. No ASTDR minimal risk levels established.

EPA interim AEGL:

	Osmium Tetroxide, ppm				
	10 min	30 min	60 min	4 h	8 h
AEGL 1	NR	NR	NR	NR	NR
AEGL 2	0.015	0.011	0.0084	0.0033	0.0017
AEGL 3	5.0	5.0	4.0	2.5	2.0

NR: Not recommended due to insufficient data.

Risk/safety phrases:

Osmium tetroxide: Toxic and/or corrosive (nonflammable)

Palladium dust: Flammable

Rhodium powder: Flammable

BACKGROUND AND USES

The platinoid metals (interchangeably referred to as the platinum group elements [PGE] or platinum group metals [PGM]) include palladium, iridium, osmium, rhodium, and ruthenium, and can be found together in nature in various concentrations. Platinum group elements belong to the noble metals group of the periodic table (Dubiella-Jackowska et al., 2009). Platinum group elements are mined in South Africa; North and South

Hamilton & Hardy's Industrial Toxicology, Sixth Edition. Edited by Raymond D. Harbison, Marie M. Bourgeois, and Giffe T. Johnson.
© 2015 John Wiley & Sons, Inc. Published 2015 by John Wiley & Sons, Inc.

America; Siberia (The Ural Mountains); and Sunbury, Ontario (Krebs, 2006; Lide, 2006). They occur naturally in nickel, copper, and iron sulfide seams, and marcasite ore beds, and can also be found in nature as native alloys consisting mainly of platinum (Ravindra et al., 2004; Lide, 2006). Their versatile and unique nature makes PGE attractive to numerous industries. For example, demand for palladium and rhodium has increased in recent years as their use in automobile catalytic converters has grown (Johnson Matthey, 2013). Palladium is used by the jewelry industry, by chemical manufacturers (as a catalyst), by electrical component manufactures, and by the dental restoration industry (Dubiella-Jackowska et al., 2009; Krebs, 2006; Johnson Matthey, 2013). Rhodium is used to make thermocouple elements, electrodes and electric contacts, laboratory crucible, silverware and jewelry, and optical instruments and glass products (including LCD TV screens, mirrors, and other optical devices) (Dubiella-Jackowska et al., 2009; Johnson Matthey, 2013; Krebs, 2006). Ruthenium is used to create alloys with other PGE metals, as a hardening agent, in the creation of electrical contacts, to create jewelry, in the creation of solar cell material (Krebs, 2006), as a catalyst in the chemical industry, and for the creation of computer parts (including hard disk drives and circuit boards) (Johnson Matthey, 2013). Iridium is used to create electrical contacts, crucibles, and to form alloys that are used in situations that require rigid materials (Lide, 2006; Johnson Matthey, 2013). Iridium, palladium, and ruthenium/rhodium radioisotopes are used in the treatment of certain forms of cancer (Michalski et al., 2003; Isager et al., 2006). Osmium is used as a hardening agent for the creation of alloys with other PGE metals (Lide, 2006). It has also been used for phonograph needles and electrical contacts (Lide, 2006). Osmium tetroxide is used as a fixative agent in transmission electron microscopy (Makarovsky et al., 2007) and has been used for fingerprint detection.

PHYSICAL AND CHEMICAL PROPERTIES

Members of the PGE share similar physical and chemical properties with each other, including their ability to resist corrosion, ability to produce a harder material when in alloy with other PGE members, and their ability to serve as catalysts for various reactions in the chemical industry.

Palladium (CASRN 7440-05-3) is steel-white in color, with a boiling point (BP) of 2963 °C and is the least dense of all the PGE (Lide, 2006). Palladium use has increased with its inclusion in automobile catalytic converters. Palladium can be used as a catalyst (it is the most reactive of all the PGE) in cracking petroleum products, in hydrogenation of liquid oils into a solid fat (Krebs, 2006), for the purification of hydrogen gas (palladium can absorb up to 900 times its own volume of hydrogen) (Lide, 2006), and for the formation of hydrogen peroxide and nitric acid (Johnson Matthey, 2013). Palladium can be added to gold to create the alloy most commonly referred to as white gold (Lide, 2006). Its inert state and ability to resist corrosion make it an attractive dental restorative agent.

Iridium (CASRN 7439-88-5) is a hard, brittle, yellowish-white metal that can withstand high temperatures and has a BP of 4428 °C (Lide, 2006). It is the most corrosion-resistant metal known (Lide, 2006). Its resistance to high temperatures makes it attractive for making crucibles and other apparatuses that must withstand extreme temperatures. Iridium may be the densest element discovered (Lide, 2006).

Osmium (CASRN 7440-04-02) is a hard, brittle, bluish-white metal with a BP of 5012 °C (Lide, 2006). It has the distinction of being the element in the PGE, which has the highest BP and lowest vapor pressure (Lide, 2006). Its physical characteristics make it difficult to manipulate; therefore, osmium is primarily used in the creation of alloys with other PGE, notably platinum and palladium.

Rhodium (CASRN 7440-16-6) is a silvery-white metal with a BP of 3695 °C (Krebs, 2006). Similar to osmium, it can be used to create a hardened alloy with platinum and palladium (Lide, 2006). The alloys of rhodium can also be used in high temperature conditions (i.e., thermocouples and crucibles). It also can be used in electroplating glass products due to its reflective properties.

Ruthenium (CASRN 7440-18-8) is a hard, white-colored member of the PGE with a BP of 4150 °C (Lide, 2006). Like osmium, it can be used to create a hardened alloy with platinum or palladium. The addition of a small quantity of ruthenium to titanium makes an alloy with increased corrosion resistance (Lide, 2006). Ruthenium is also a versatile catalyst.

ENVIRONMENTAL FATE AND BIOACCUMULATION

Platinum group elements are found in the environment from both natural and anthropogenic sources. Average PGE concentrations in the lithosphere are estimated for each group member: 0.015 mg/kg for palladium, 0.0001 mg/kg for rhodium, 0.0001 mg/kg for ruthenium, 0.005 mg/kg for osmium, and 0.001 mg/kg for iridium (Ravindra et al., 2004). The increase in demand for the elements has resulted in additional mining activities, which have also contributed to environmental accumulation via waste stream deposits.

The increase in mining for these elements, and their subsequent use in automobile emission control systems, has increased the amount of palladium and rhodium in the environment, both as waste from mining activities and in road side dusts from vehicle exhausts. The rate of emissions from catalytic converters may depend on a variety of factors, including driving speed, age and composition of the catalytic converter, and engine type (Wiseman and

Zereini, 2009; Artelt et al., 1999). Concentrations of PGE in road dust have been found in the pg/m^3 range in both particulate matter (PM) PM_{10} and $PM_{2.5}$ ranges (Wiseman and Zereini, 2009). In high traffic areas, the concentrations of PGE in soils, plants, and dusts have been found to exceed natural background concentrations (Ravindra et al., 2004). Rain and subsequent erosion may wash these dust deposits into local waterways. Some studies suggest that not all PGE road dusts are in metallic form and that they may actually be soluble under certain conditions. Platinum group element's mobility and solubility in soil may depend upon pH, chloride and nitrate content, and the individual PGE (Ek et al., 2004).

Road dust has been studied to evaluate the impact of PGE on the terrestrial biosphere. Of the PGE, palladium has been the most evaluated element (next to platinum) due to its use in vehicle catalytic converters. Some studies have indicated that palladium is the most bioavailable (next to platinum and rhodium) to some aquatic organisms (Ravindra et al., 2004), and PGE have been shown to accumulate in roadside soils and dusts (Dubiella-Jackowska et al., 2009). Platinum group elements may not be in metallic form from exhaust emissions or may be rapidly altered once deposited in the environment (Ravindra et al., 2004). While studies have indicated that anthropogenic PGE can bioaccumulate in aquatic organisms, the deleterious effects associated with the accumulation have not yet been established (Ravindra et al., 2004).

ECOTOXICOLOGY

The recent rise in the use of PGE in vehicle catalytic converters has brought scrutiny towards the deposition of anthropogenic PGE in the environment. Original research into the impact of PGE on the environment suggested that the PM being emitted from vehicle exhaust was inert (Hooda et al., 2008) and would not negatively impact the biota. Recent investigations have suggested that the PM may be more soluble than originally thought and may negatively impact environmental receptors.

Acute Effects

The toxic effects of PGE on ecological receptors have yet to be elucidated. Recent studies have suggested that sensitive receptors can absorb PGE deposited from anthropogenic sources (Ravindra et al., 2004). While uptake of soluble PGE particulates has been observed, the deleterious effects of this uptake, if any, are still being determined.

Chronic Effects

Not enough information has been published to evaluate the chronic effects of anthropogenic PGE on environmental receptors. Studies are ongoing that are attempting to determine the extent and effects of the bioaccumulation of PGE on the ecosystem.

MAMMALIAN TOXICOLOGY

Metallic forms of PGE are biologically inert (Moldovan, 2007); however, PGE salts have the potential to act as sensitizers. Individual PGE salts or radioisotopes of PGE may have specific deleterious effects. Research in recent years has indicated that PGE particles released in automotive exhaust may be more soluble than initially assumed. This may have an impact on the overall toxicity of the PGE.

Acute Effects

While rare in the literature, occurrences of apparent immune-mediated reactions following overexposures to PGE have been reported, including occupational asthma and contact dermatitis (Ravindra et al., 2004).

A case report by Daenen et al. (1999) indicated that an isolated case of occupational asthma developed in a worker following an exposure to palladium fumes from an electroplating bath. Skin prick testing was positive for one palladium salt ($Pd(NH_3)Cl_2$) but not a second ((NH_4)$PdCl_4$), and the exposed worker was skin prick test negative for the corresponding platinum salt (Daenen et al., 1999). Specific bronchial provocation testing resulted in an approximately 35% reduction of forced expiratory volume (FEV_1).

A study on 306 South African refinery workers by Murdoch et al. (1986) found that of the 38 workers who had a positive platinum skin prick test, one worker had a positive test for palladium salts, and six workers had a positive test for rhodium salts. In this study, no independent positive skin prick results were discovered for PGE salts (Murdoch et al., 1986).

A case report detailed an incident involving three individuals that were exposed to a radioactive isotopes of iridium (^{192}Ir) which was the reported cause of individual cases of acute radiation sickness. The source of their exposure was linked to a leaking industrial radiograph machine that was being used in pipeline service and repair (Sevan'kaev et al., 2002).

Osmium tetroxide (OsO_4) is an oxidizer and acutely toxic to those who are overexposed. While osmium, in the metallic form, is considered inert, osmium tetroxide can form upon exposure to air under specific circumstances (spongy metal or powdered osmium metal) (Makarovsky et al., 2007; McLaughlin et al., 1946). Health effects from acute overexposures can lead to irritation of the eyes, respiratory tract, gastrointestinal tract, and skin (McLaughlin et al., 1946). Headaches and dizziness may also present following an acute overexposure (Makarovsky et al., 2007). Irreversible blindness has been observed from overexposure to OsO_4. Its high vapor pressure leads to exposures via inhalation. The symptoms of overexposure may not present immediately and may be delayed for several hours. There have been reports that ruthenium tetroxide has similar health effects from acute overexposures; however, little toxicity information is available in the literature.

Chronic Effects

The chronic effects from exposure to PGE and their salts mainly manifest as an immune response via sensitization. Much like platinum, PGE have the potential to sensitize exposed individuals, with degree of toxicity depending on the exact metal and solubility (Wiseman and Zereini, 2009). Individuals who have nickel allergy also tend to display sensitization effects when exposed to palladium (Kielhorn et al., 2002), including a study that indicated that approximately 90% of the individuals reacting for nickel allergy also displayed palladium sensitivity (Ravindra et al., 2004).

A study by Cristaudo et al. (2005) investigated the hypersensitivity to PGE salts in employees of a catalyst-manufacturing and recycling facility, including platinum, iridium, and rhodium salts. While the majority of those testing skin prick positive were reacting to platinum salts, two workers had positive skin prick reactions to iridium trichloride and rhodium trichloride (Cristaudo et al., 2005).

Chemical Pathology

It is believed that PGE acts in a similar manner to platinum, in which exposure results in sensitization. The World Health Organization (2002) reported that palladium ions target the immune system. Immune responses to platinum have been shown to illicit an immunoglobulin E (IgE) response in those that have a sensitization following an exposure (Calverley et al., 1999; Merget et al., 1999). So far, no clear immune-mediated mechanism has been elucidated for PGE. Detection of PGE sensitivity will need to be conducted with a skin prick test for the specific PGE compound.

Mode of Action(s)

The pathology of PGE is not well studied due to the inert status of their metallic form. However, due to the potential for PGE PM to exist in a soluble form, there has been an increased focus on the mechanism of PGE toxicity. While studies have suggested that PGE can bioaccumulate in certain aquatic organisms, the potential for PGE biomangification up the food chain and the toxicity associated with its detection in body tissue has not not yet been determined.

A target of palladium (and possibly other PGE) is the immune system, as some palladium compounds have been found to be potent sensitizers (WHO, 2002). Overexposures to palladium salts may result in skin and eye irritation (WHO, 2002). Isolated case reports indicate that inhalation of PGE may result in occupational asthma. The immune response following sensitization to PGE has not been fully elucidated.

The effects that PGE will have following an overexposure will depend on the route of exposure and chemical composition of the PGE (i.e., metal versus salt). Uptake of palladium ions can occur from ingestion or dermal contact (WHO, 2002). However, palladium compounds have a low to moderate acute oral toxicity (i.e., 220 to >4,000 mg/kg body weight in rats) and the toxicity will depend on their solubility (Ravindra et al., 2004).

One case report indicated that urticaria and respiratory issues were observed following an exposure to iridium chloride at the subject's place of employment (Bergman et al., 1995). These occasional case reports of immune response following exposures to iridium salts have also indicated that contact dermatitis is possible.

Iridium, palladium, and ruthenium/rhodium radioisotopes are used in brachytherapy to treat certain forms of cancer, including prostate, breast, and ocular (Michalski et al., 2003; Hammer et al., 1994; Isager et al., 2006). The specific PGE radioisotope used will depend on tumor location. The implantation and focused nature of this therapy ensures that the energy from treatment is directed toward the neoplasm.

SOURCES OF EXPOSURE

Occupational

Since the PGE are rare elements, individuals who are occupationally exposed to PGE are limited to only a few select industries. Miners that are involved in the process of extracting these elements from the earth may be exposed during the mining, recycling, and/or refining process (WHO, 2002). Platinum group elements and their compounds (including PGE salts) are used in the chemical industry as catalysts in reactions, resulting in employees in the chemical-manufacturing industry may be exposed to PGE. Dental technicians may be exposed to palladium through the creation of dental restorations. Jewelers may be exposed to palladium during jewelry manufacture.

Employees of electronic, glass, electrochemical, and other similar industries utilizing PGE for technological purposes may also be exposed, either during the creation of these consumer goods or during the assembly process.

The use of ruthenium/rhodium, iridium, and palladium radioisotopes in certain industries may increase the opportunity for exposure to radiation. This could apply to oil and gas pipeline employees using radioisotopes to identify weaknesses in welds of pipes or in a healthcare facility using PGE radioisotopes for brachytherapy in patients.

Osmium tetroxide exposures may take place in pathology laboratories that use the chemical as a staining and/or fixative for electron microscopy. Osmium tetroxide stains and allows the localization of unsaturated lipid bonds and nerves within a tissue (McLaughlin et al., 1946), and allows the microscopist to more easily identify these structures.

Environmental

On a global scale, exposures in the general population are expected to remain low due to the rarity of the elements. While individuals may be exposed via direct contact with palladium-containing jewelry and dental restorations, the metallic form is expected to be biologically inert and not contribute to overall body burden.

Recent studies suggest that the use of PGE in catalytic converters has increased the amount of PGE in deposited road dust. There is also evidence that the PM from vehicle exhaust is in a more soluble form than originally thought. While studies do indicate that concentrations of PGE-containing dust have increased, it still remains unknown what biologically significant contribution this road dust makes to overall human health or how bioavailable the material is for uptake by the exposed flora or fauna.

The use of PGE for the treatment of certain cancers will also be an exposure point for some members of the general population. These implanted therapies have the ability to focus the energy on the neoplasm. This makes the chance of exposure to anyone but the individual undergoing brachytherapy low. Michalski et al. (2003) reported that exposures to family members of patients undergoing brachytherapy with iridium and palladium for prostate cancer were low and were below the U.S. Nuclear Regulatory Commission's recommended threshold of 3.60 millisieverts per year (mSv/year).

INDUSTRIAL HYGIENE

The Occupational Safety and Health Administration's (OSHA) permissible exposure limits (PEL) (NIOSH, 2005), the National Institute for Occupational Safety and Health's (NIOSH) recommended exposure limits (REL) (NIOSH, 2005), and the American Conference of Governmental Industrial Hygienists' (ACGIH) threshold limit values (TLV) (ACGIH, 2012) for PGE are summarized below:

Palladium: No PEL, REL, or TLV established

Iridium: No PEL, REL, or TLV established

Osmium: As osmium tetroxide:

  TLV time-weighted average (TWA) = 0.0002 parts per million (ppm); STEL = 0.0006 ppm

  REL: TWA = 0.002 mg/m^3; ST = 0.006 mg/m^3

  PEL: TWA = 0.002 mg/m^3

    Immediately Dangerous to Life or Health (IDLH) = 1 mg/m^3

Rhodium:

  Soluble compounds: PEL = 0.001 mg/m^3; REL = 0.001 mg/m^3; TLV = 0.01 mg/m^3; IDLH = 2 mg/m^3

  Metal and insoluble compounds: PEL = 0.1 mg/m^3; REL = 0.1 mg/m^3; TLV = 1 mg/m^3; IDLH = 100 mg/m^3

Ruthenium: No PEL, REL, or TLV established

Respiratory protection should be used when working around potential sensitizers, like PGE. Claims of sensitizations from PGE should be investigated and the individual should be removed from exposure. The symptoms of sensitization may persist for several months. Further exposures to PGE following sensitization should be prevented.

RISK ASSESSMENTS

There is a paucity of information relating to the risks from exposures to any of the PGE. No reference dose or reference concentration information is available in the U.S. Environmental Protection Agency's Integrated Risk Information System (EPA IRIS) database for any of the PGE Metals (USEPA, IRIS). The Agency for Toxic Substances and Disease Registry (ATSDR) does not currently have a toxicological profile for any PGE.

An interim Acute Exposure Guideline Level (AEGL) for osmium tetroxide was found in the AEGL database and is listed in the summary section (USEPA AEGL, 2008).

REFERENCES

American Conference of Governmental Industrial Hygienists (ACGIH) (2012) *Threshold Limit Values (TLVs) for Chemical Substances and Physical Agents and Biological Exposure Indices (BEIs)*, Cincinnati, OH: American Conference of Governmental Industrial Hygienists.

Artelt, S., Kock, H., König, H.P., Levsen, K., and Rosner, G. (1999) Engine dynamometer experiments: platinum emissions from differently aged three-way catalytic converters. *Atmos. Environ.* 33(21):3559–3567.

Bergman, A., Svedberg, U., and Nilsson, E. (1995) Contact urticaria with anaphylactic reactions caused by occupational exposure to iridium salt. *Contact Dermatitis* 32(1):14–17.

Calverley, A.E., Rees, D., and Dowdeswell, R.J. (1999) Allergy to complex salts of platinum in refinery workers: prospective evaluations of IgE and Phadiatop status. *Clin. Exp. Allergy* 29(5):703–711.

Cristaudo, A., Sera, F., Severino, V., De Rocco, M., Di Lella, E., and Picardo, M. (2005) Occupational hypersensitivity to metal salts, including platinum, in the secondary industry. *Allergy* 60(2):159–164.

Daenen, M., Rogiers, P., Van de Walle, C., Rochette, F., Demedts, M., and Nemery, B. (1999) Occupational asthma caused by palladium. *Eur. Respir. J.* 13(1):213–216.

Dubiella-Jackowska, A., Polkowska, Z., and Namiennik, J. (2009) Platinum group elements in the environment: emissions and exposure. *Rev. Environ. Contam. Toxicol.* 199:111–135.

Ek, K.H., Morrison, G. M., and Rauch, S. (2004) Environmental routes for platinum group elements to biological materials—a review. *Sci. Total Environ.* 334–335:21–38.

Hammer, J., Seewald, D.H., Track, C., Zoidl, J.P., and Labeck, W. (1994) Breast cancer: primary treatment with external-beam radiation therapy and high-dose-rate iridium implantation. *Radiology* 193(2):573–577.

Hooda, P.S., Miller, A., and Edwards, A.C. (2008) The plant availability of auto-cast platinum group elements. *Environ. Geochem. Health* 30(2):135–139.

Isager, P., Ehlers, N., Urbak, S.F., and Overgaard, J. (2006) Visual outcome, local tumour control, and eye preservation after 106Ru/Rh brachytherapy for choroidal melanoma. *Acta Oncol.* 45(3):285–293.

Johnson Matthey (2013) *Platinum 2013*, Hertfordshine, UK: Johnson Matthey Public Limited Company.

Kielhorn, J., Melber, C., Keller, D., and Mangelsdorf, I. (2002) Palladium—a review of exposure and effects to human health. *Int. J. Hyg. Environ. Health* 205(6):417–432.

Krebs, R.E. (2006) *The History and Use of Our Earth's Chemical Elements: A Reference Guide*, 2nd ed., Westport, Connecticut: Greenwood Press.

Lide, D.R. (2006) *CRC Handbook of Chemistry and Physics*, 87th ed., Boca Raton, FL: CRC Press, pp. 4-19–4-31.

Makarovsky, I., Markel, G., Hoffman, A., Schein, O., Finkelstien, A., Brosh-Nissimov, T., Tashma, Z., Dushnitsky, T., and Eisenkraft, A. (2007) Osmium tetroxide: a new kind of weapon. *Isr. Med. Assoc. J.* 9(10):750–752.

McLaughlin, L.A., Milton, R., and Perry, K.M. (1946) Toxic manifestations of osmium tetroxide. *Br. J. Ind. Med.* 3:183–186.

Merget, R., Schulte, A., Gebler, A., Breitstadt, R., Kulzer, R., Berndt, E.D., Baur, X., and Schultze-Werninghaus, G. (1999) Outcome of occupational asthma due to platinum salts after transferral to low-exposure areas. *Int. Arch. Occup. Environ. Health* 72(1):33–39.

Michalski, J., Mutic, S., Eichling, J., and Ahmed, S.N. (2003) Radiation exposure to family and household members after prostate brachytherapy. *Int. J. Radiat. Oncol. Biol. Phys.* 56(3):764–768.

Moldovan, M. (2007) Origin and fate of platinum group elements in the environment. *Anal. Bioanal. Chem.* 388(3):537–540.

Murdoch, R.D., Pepys, J., and Hughes, E.G. (1986) IgE antibody responses to platinum group metals: a large scale refinery survey. *Br. J. Ind. Med.* 43(1):37–43.

National Institute of Occupational Safety and Health (NIOSH) (2005) *Pocket Guide to Chemical Hazards*, Cincinnati, OH: NIOSH Publications.

Ravindra, K., Bencs, L., and Van Grieken, R. (2004) Platinum group elements in the environment and their health risk. *Sci. Total Environ.* 318(1–3):1–43.

Sevan'kaev, A.V., Lloyd, D.C., Edwards, A.A., Moquet, J.E., Nugis, V.Y., Mikhailova, G.M., Potetnya, O.I., Khvostunov, I.K., Guskova, A.K., Baranov, A.E., and Nadejina, N.M. (2002) Cytogenic investigations of serious overexposures to an industrial gamma radiography source. *Radiat. Prot. Dosimetry* 102(3):201–206.

USEPA AEGL (2008) *United States Environmental Protection Agency Acute Exposure Guideline Levels*. Available at http://www.epa.gov/oppt/aegl/ (accessed January 8, 2014).

USEPA IRIS (n.d.) *United States Environmental Protection Agency Integrated Risk Information System*. Available at http://www.epa.gov/iris/ (accessed January 6, 2014).

WHO (2002) *World Health Organization Environmental Health Criteria 226: Palladium*. Geneva. Available at http://www.who.int/ipcs/publications/ehc/en/ehc226.pdf.

Wiseman, C.L. and Zereini, F. (2009) Airborne particulate matter, platinum group elements and human health: a review of recent evidence. *Sci. Total Environ.* 407(8):2493–2500.

28

PLATINUM

Alex Lebeau

First aid: Eyes: Irrigate

Dermal: Soap wash (platinum); water flush immediately (platinum salts)

Inhalation: Respiratory support

Ingestion: Medical attention

Target organ(s): Respiratory system, eyes, and skin

Occupational exposure limits: Platinum:

REL and TLV TWA = 1 mg/m^3

Platinum salts: PEL, REL, and TLV TWA = 0.002 mg/m^3

IDLH = 4 mg/m^3(as Pt)

Reference values: None established

Risk/safety phrases:

Platinum: Powder may be flammable and/or explosive

Platinum Compounds: Ensure proper ventilation.

BACKGROUND AND USES

Platinum is primarily used in industry as a catalyst, predominately for the production of vehicle catalytic converters (Johnson Matthey, 2012). Platinum is also used in the chemical, electrical, glass, dental, healthcare, pharmaceutical, petroleum, and jewelry fields (Cristaudo et al., 2005; Dubiella-Jackowska et al., 2009; Platinum Applications, 2013). In addition to its catalytic properties, platinum, in combination with cobalt, has powerful magnetic properties (Weast, 1988). Original recognition of occupational illness associated with platinum exposure dates back to 1911 when platinum-treated photographic paper was used in photography studios in Chicago (Karasek and Karasek, 1911; Niezborala and Garnier, 1996).

PHYSICAL AND CHEMICAL PROPERTIES

Platinum (symbol Pt; CASRN 7440-06-4) is a silvery to whitish-gray metal, that is malleable and ductile (Weast, 1988; Krebs, 2006). It is estimated that the average platinum concentration in the earth's lithosphere ranges from 0.001 to 0.005 mg/kg (WHO, 1991). It has a vapor pressure of approximately 0 mmHg (NIOSH, 2005). Platinum, depending on mining location, is found in sperrylite, cooperite, braggite, and copper–nickel sulfide ores, and in alloys with other members of the platinum group elements (PGE; iridium, osmium, palladium, ruthenium, and rhodium) (Dubiella-Jackowska et al., 2009; Weast, 1988; WHO, 1991). Both elemental platinum and platinum salts are used in industry. Elemental platinum is insoluble in water whereas many of the platinum salts are water soluble to varying degrees, depending on their specific chemical properties (WHO, 1991). Platinum metal is inert and resists corrosion (Krebs, 2006). However, platinum halide salts are considered potent allergens that have a high risk of causing sensitization (Linnett, 2005). Platinum bonds with halogens in its +2 and +4 oxidation state (Krebs, 2006). The charge of the salt and the presence and number of halides, especially chloride, in a complex platinum salt may affect the degree of mammalian sensitivity to a specific salt (Cleare et al., 1976; Mapp et al., 1999; Merget et al., 2000; Ravindra et al., 2004; Pepys, 1980).

ENVIRONMENTAL FATE AND BIOACCUMULATION

Platinum is found in the environment naturally and from anthropogenic sources. Beginning in the mid-1970s, platinum

Hamilton & Hardy's Industrial Toxicology, Sixth Edition. Edited by Raymond D. Harbison, Marie M. Bourgeois, and Giffe T. Johnson.
© 2015 John Wiley & Sons, Inc. Published 2015 by John Wiley & Sons, Inc.

was used for catalytic converters in vehicle across the United States and Japan, with Europe utilizing catalytic converters beginning in the mid-1980s (Dubiella-Jackowska et al., 2009; Ravindra et al., 2004). Particulate exhaust from reactions with platinum-containing catalytic converters results in the release of platinum into the environment, often times depositing as road dust (Kümmerer et al., 1999; Ravindra et al., 2004). Particles >10.2 µm have been shown to make up the majority of the total emitted particles in the exhaust from catalytic converters (62–67%), and the particles larger than 10 µm contain the highest concentration of platinum (Artelt et al., 1999; Dubiella-Jackowska et al., 2009). Initial studies on platinum-containing exhaust particulates suggested that the majority of platinum that is released from a catalytic converter is in metallic form. One recent investigation into vehicle exhaust particulate emissions suggested that the platinum found in tunnel road dust was mobile and may be biologically available (Fliegel et al., 2004; Dubiella-Jackowska et al., 2009).

Platinum is used in certain antineoplastics as a treatment for cancer. Platinum can enter hospital wastewater from the urinary excretions of patients receiving treatment with these chemotherapeutics (Kümmerer et al., 1999; Dubiella-Jackowska et al., 2009). However, studies have indicated that hospital effluent is secondary to other environmental sources of platinum, such as those of vehicle catalytic converter emissions (Kümmerer et al., 1999). Industrial use of platinum may also contribute to the presence of anthropogenic platinum compounds in the environment (Dubiella-Jackowska et al., 2009).

Recent investigations into bioavailability suggest that metallic or oxidized platinum may become soluble when deposited as road dust from vehicle exhaust, contaminating soil and water, and ultimately ending up in the food chain (Dubiella-Jackowska et al., 2009). A study investigating the effects of road dust and homogenized catalytic converter material on zebra mussels showed that humic water from a bog lake enhanced the biological availability of particle-bound platinum (Zimmermann et al., 2005). Other similar studies have indicated that bioaccumulation of platinum is possible. Many of these studies have evaluated platinum in combination with other PGE. While it has been shown that anthropogenic platinum is present and can bioaccumulate in the environment, platinum toxicity from this bioaccumulation is not clear (Ravindra et al., 2004).

ECOTOXICOLOGY

Acute Effects

Acute effects on plants or wildlife have not been determined. Recent studies have attempted to evaluate the bioavailability of anthropogenic platinum levels in biota near urbanized areas. One theory is that metallic or oxidized platinum in road dust may transform into a soluble form and enter the food chain (Dubiella-Jackowska et al., 2009). While detectable levels of platinum have been found in plants and animals in urban areas, no deleterious effects have yet been associated with these concentrations.

Chronic Effects

Not enough information has been published to evaluate the chronic effects of anthropogenic platinum on the environment. Ongoing studies are attempting to determine the extent and effects of the bioaccumulation of platinum on the ecosystem.

MAMMALIAN TOXICOLOGY

Acute Effects

Occupational exposure to platinum can occur from inhalation and ingestion. The most common health effects associated with overexposure to platinum salts include asthma, rhinitis, conjunctivitis, urticarial, dermatitis, and nonspecific airway hyperresponsiveness (Brooks et al., 1990; Calverley et al., 1995; Cristaudo et al., 2005; Merget et al., 2000). While the exposure to platinum salts elicits a Type 1 hypersensitivity reaction involving immunoglobulin E (IgE), the sensitization and allergic response to the exposure may not be immediate. The onset of symptom development following exposure may range from a few months to 6 years (Ravindra et al., 2004). This delay in the development of hypersensitivity has also been observed in patients receiving platinum-containing antineoplastics (Makrilia et al., 2010).

Chronic Effects

Chronic effects from exposure are related to the continued sensitivity to platinum salts, even after an individual has been removed from the exposure. One study indicated that asthma symptoms, airway obstruction, elevated serum IgE, positive skin prick tests, and positive cold air challenge persisted for an average of 5 years following removal from the exposure (Brooks et al., 1990). Similar results were observed in other studies, including bronchial hyperresponsiveness and skin sensitization (Merget et al., 1994; Merget et al., 1999).

A historical prospective cohort study by Niezborala and Garnier (1996) examined 77 refinery workers and found that 23 had developed symptoms of platinum salt sensitivity, with the highest incidence within the first 2 years of work. Another study of refinery workers reported that 41% of 78 new workers became sensitized after 2 years of exposure (Calverley et al., 1995).

Healthcare employees working with platinum-containing antineoplastics and patients receiving treatment have the opportunity for chronic exposure. The International Agency for Research on Cancer (IARC) has classified cisplantin as a Class 2A probable human carcinogen (IARC, 1987). Nephrotoxicity, ototoxicity, nausea, and myelotoxicity caused by cisplantin have been observed in both laboratory animals and patients (WHO, 1991). Type 1 hypersensitivity has been reported in patients receiving platinum-containing chemotherapeutics. Hypersensitivity in patients was reported after multiple treatments and the delay in onset of symptoms depended on the type of platinum antineoplastic administered (Makrilia et al., 2010).

Chemical Pathology

Platinum halide salts are potent sensitizers (Pepys, 1980). The common platinum salts used in industry include hexachloroplatinic acid, ammonium hexachloroplatinate (IV), potassium hexachloroplatinate (IV), sodium tertachloroplatinate (II), and potassium tertachloroplatinate (II) (Cristaudo et al., 2005; Dubiella-Jackowska et al., 2009). The immune response from the exposure to platinum salts can elicit a Type 1 hypersensitivity reaction and can increase the total serum IgE. The immune response to platinum salts increased total and platinum-specific IgE in one study; but increased histamine release was not observed (Bolm-Audorff et al., 1992). Sodium tetrachloroplatinate was shown to liberate histamine in another study (Saindelle and Ruff, 1969). Increased total IgE was observed in a study comparing a "high" exposed group to the remainder of the sample groups (Merget et al., 1999).

Mode of Action(s)

The respiratory system, the eyes, and skin are the primary targets following exposure to platinum salts. The degree of allergic response to platinum salts may increase as the number of chlorine atoms bound to the platinum increases (Cleare et al., 1976; Pepys, 1980; Ravindra et al., 2004).

SOURCES OF EXPOSURE

Occupational

The most common occupational exposures to platinum and platinum salts are in refineries and catalyst-production facilities (Cristaudo et al., 2005). Current occupational exposures to platinum include employees of the following: platinum-mining facilities and refineries, catalyst manufacturers and recyclers, jewelers, chemical manufacturers, electronic manufacturers, the pharmaceutical industry, dental offices, and hospitals and healthcare facilities (Calverley et al., 1999;

Cristaudo et al., 2005; Dubiella-Jackowska et al., 2009; Merget et al., 1991; Pethran et al., 2003; Platinum Applications, 2013).

Healthcare personnel may be exposed to platinum through the mixing, handling, administration, or cleanup of the platinum-containing antineoplastics cisplatin, carboplatin, and oxaliplatin, which are used to treat a variety of cancers (NTP, 2011; Pethran et al., 2003). Studies have indicated that exposures in healthcare settings may occur via contact with surfaces and drug vials contaminated with platinum-based antineoplastics (Mason et al., 2005; Naito et al., 2012; Nygren et al., 2002). These studies suggest that drug vial contamination may have occurred during the manufacturing and packaging of the antineoplastics.

Environmental

Patients receiving antineoplastics for the treatment of cancer will be exposed if their therapy involves the administration of platinum-containing chemotherapeutics. Biota may be exposed to platinum from road dust emitted by vehicle catalytic converters and from effluent from hospitals whose patients were administered platinum-containing antineoplastics. It has been estimated that approximately 70% of the platinum dose received by cancer patients is excreted in urine (Dubiella-Jackowska et al., 2009). Industrial sources of platinum may also contribute to environmental exposure. Taking into account the increased concentration of anthropogenic platinum in the environment, the amount found has not been sufficient to elicit an outbreak of allergic reactions in the population (Cristaudo et al., 2005).

INDUSTRIAL HYGIENE

As an important protective measure, prompt removal of sensitized individuals from a platinum salt exposure is highly recommended (Brooks et al., 1990; Niezborala and Garnier, 1996). Skin prick testing with platinum salts can be used to detect sensitization at early stages (Linnett, 2005). Platinum salt exposure and sensitivity have been associated with an increase in total serum IgE levels (Calverley et al., 1999; Merget et al., 1999). Positive skin prick tests have been associated with higher levels of total serum IgE in comparison to workers that were skin prick test negative (Mapp et al., 1999).

Skin prick testing and evaluating an employee's symptoms for platinum salt exposure are better diagnostic tools than evaluating bronchial hyperresponsiveness for determining sensitization for both platinum salt sensitization and for occupational asthma due to exposure (Merget et al., 1995). One study indicated that a positive skin prick result had a 100% positive predictive value for the signs and symptoms of sensitization to platinum salts (Calverley et al., 1995).

Historically, atopic individuals have been excluded from exposure to platinum salts. However, recent studies have indicated that excluding atopic individuals may not necessarily result in a decreased incidence of platinum salt sensitivity and that atopy may not be a viable metric for evaluating a predisposition for platinum salt sensitivity (Merget et al., 2000; Niezborala and Garnier, 1996). Periodic monitoring of employees working with platinum salts is highly recommended as part of a workplace health surveillance program.

The National Institute for Occupational Safety and Health (NIOSH) and the American Conference of Governmental Industrial Hygienists (ACGIH) have set the time-weighted average (TWA) recommended for platinum at $1\,mg/m^3$ (NIOSH, 2005; ACGIH, 2012). The Occupational Safety and Health Administration (OSHA), NIOSH, and the ACGIH have set the permissible exposure limit (PEL), the REL, and the TLV, respectively, for platinum salts at $0.002\,mg/m^3$ (NIOSH, 2005). The Immediately Dangerous to Life and Health (IDLH) concentration for platinum salts is $4\,mg/m^3$ (as Pt) (NIOSH, 2005).

RISK ASSESSMENTS

No information is available for a reference dose or reference concentration relating to the exposure to platinum or platinum salts. New research indicates that anthropogenic platinum in the environment may be bioavailable and may bioaccumulate in the food chain. No study was identified that evaluated risk from environmental exposure.

Cisplatin is classified as *reasonably anticipated to be a human carcinogen* (NTP, 2011). Animal studies have indicated that cisplatin administration has resulted in lung adenomas and leukemia. Human studies have not been conclusive.

Smoking is considered a predictor/risk factor for the development of platinum salt sensitivity. Venables et al. (1989) discovered that smoking was a predictor of a positive skin prick test for platinum salts. Niezborala and Garnier (1996) used smoking status of refinery employees to positively predict a positive skin prick test and the development of symptoms from exposure. Calverley et al. (1995) found an almost eight times increased risk of sensitization from exposure to platinum salts in smokers as compared to nonsmokers. Merget et al. (2000) identified smoking status of an employee as a predictor for skin prick test conversion in highly exposed individuals from a catalyst-production facility (relative risk = 3.9).

REFERENCES

American Conference of Governmental Industrial Hygienists (ACGIH) (2012) *Threshold Limit Values (TLVs) for Chemical Substances and Physical Agents and Biological Exposure Indices (BEIs)*, Cincinnati, OH: American Conference of Governmental Industrial Hygienists.

Artelt, S., Kock, H., König, H.P., Levsen, K., and Rosner, G. (1999) Engine dynamometer experiments: platinum emissions from differently aged three-way catalytic converters. *Atmos. Environ.* 33(21):3559–3567.

Bolm-Audorff, U., Bienfait, H.G., Burkhard, J., Bury, A.H., Merget, R., Pressel, G., and Schultze-Werninghaus, G. (1992) Prevalence of respiratory allergy in a platinum refinery. *Int. Arch. Occup. Environ. Health* 64(4):257–260.

Brooks, S.M., Baker, D.B., Gann, P.H., Jarabek, A.M., Hertzberg, V., Gallagher, J., Biagini, R.E., and Bernstein, I.L. (1990) Cold air challenge and platinum skin reactivity in platinum refinery workers. Bronchial reactivity precedes skin prick response. *Chest* 97(6):1401–1407.

Calverley, A.E., Rees, D., and Dowdeswell, R.J. (1999) Allergy to complex salts of platinum in refinery workers: prospective evaluations of IgE and Phadiatop status. *Clin. Exp. Allergy* 29(5):703–711.

Calverley, A.E., Rees, D., Dowdeswell, R.J., Linnett, P.J., and Kielkowski, D. (1995) Platinum salt sensitivity in refinery workers: incidence and effects of smoking and exposure. *Occup. Environ. Med.* 52(10):661–666.

Cleare, M.J., Hughes, E.G., Jacoby, B., and Pepys, J. (1976) Immediate (type I) allergic responses to platinum compounds. *Clin. Allergy* 6(2):183–195.

Cristaudo, A., Sera, F., Severino, V., De Rocco, M., Di Lella, E., and Picardo, M. (2005) Occupational hypersensitivity to metal salts, including platinum, in the secondary industry. *Allergy* 60(2):159–164.

Dubiella-Jackowska, A., Polkowska, Z., and Namiennik, J. (2009) Platinum group elements in the environment: emissions and exposure. *Rev. Environ. Contam. Toxicol.* 199:111–135.

Fliegel, D., Berner, Z., Eckhardt, D., and Stuben, D. (2004) New data on the mobility of Pt emitted from catalytic converters. *Anal. Bioanal. Chem.* 379(1):131–136.

International Agency for Research on Cancer (IARC) (1987) *Cisplatin*. International Programme on Chemical Safety. Available at http://www.inchem.org/documents/iarc/suppl7/cisplatin.html (accessed January 7, 2013).

Johnson Matthey (2012) *Platinum 2012: Interim Review*, Hertfordshine, UK: Johnson Matthey Public Limited Company.

Karasek, S.R. and Karasek, M. (1911) *The use of platinum paper*. Report of Illinois Commission on Occupational Diseases 97, Cornell University Library Web site. Available at http://stage-csrportal.archive.org/stream/cu31924002339491/cu31924002339491_djvu.txt (accessed January 4, 2013).

Krebs, R.E. (2006) *The History and Use of Our Earth's Chemical Elements: A Reference Guide*, 2nd ed., Westport, CT: Greenwood Press.

Kümmerer, K., Helmers, E., Hubner, P., Mascart, G., Milandri, M., Reinthaler, F., and Zwakenberg, M. (1999) European hospitals as a source for platinum in the environment in comparison with other sources. *Sci. Total Environ.* 225(1–2):155–165.

Linnett, P.J. (2005) Concerns for asthma at pre-placement assessment and health surveillance in platinum refining—a personal approach. *Occup. Med. (Lond.)* 55(8):595–599.

Makrilia, N., Syrigou, E., Kaklamanos, I., Manolopoulos, L., and Saif, M.W. (2010) Hypersensitivity reactions associated with platinum antineoplastic agents: a systematic review. *Met. Based Drugs* 2010:1141–1145.

Mapp, C., Boschetto, P., Miotto, D., De Rosa, E., and Fabbri, L.M. (1999) Mechanisms of occupational asthma. *Ann. Allergy Asthma Immunol.* 83(6 Pt 2):645–664.

Mason, H.J., Blair, S., Sams, C., Jones, K., Garfitt, S.J., Cuschieri, M.J., and Baxter, P.J. (2005) Exposure to antineoplastic drugs in two UK hospital pharmacy units. *Ann. Occup. Hyg.* 49(7): 603–610.

Merget, R., Caspari, C., Kulzer, R., Breitstadt, R., Rueckmann, A., and Schultz-Werninghaus, G. (1995) The sequence of symptoms, sensitization and bronchial hyperresponsiveness in early occupational asthma due to platinum salts. *Int. Arch. Allergy Immunol.* 107(1–3):406–407.

Merget, R., Kulzer, R., Dierkes-Globisch, A., Breitstadt, R., Gebler, A., Kniffka, A., Artelt, S., Koenig, H.P., Alt, F., Vormberg, R., Baur, X., and Schultze-Werninghaus, G. (2000) Exposure-effect relationship of platinum salt allergy in a catalyst production plant: conclusions from a 5-year prospective cohort study. *J. Allergy Clin. Immunol.* 105(2 Pt 1):364–370.

Merget, R., Reineke, M., Rueckmann, A., Bergmann, E.M., and Schultze-Werninghaus, G. (1994) Nonspecific and specific bronchial responsiveness in occupational asthma caused by platinum salts after allergen avoidance. *Am. J. Respir. Crit. Care Med.* 150(4):1146–1149.

Merget, R., Schulte, A., Gebler, A., Breitstadt, R., Kulzer, R., Berndt, E.D., Baur, X., and Schultze-Werninghaus, G. (1999) Outcome of occupational asthma due to platinum salts after transferral to low-exposure areas. *Int. Arch. Occup. Environ. Health* 72(1):33–39.

Merget, R., Schultze-Werninghaus, G., Bode, F., Bergmann, E. M., Zachgo, W., and Meier-Sydow, J. (1991) Quantitative skin prick and bronchial provocation tests with platinum salt. *Br. J. Ind. Med.* 48(12):830–837.

Naito, T., Osawa, T., Suzuki, N., Goto, T., Takada, A., Nakamichi, H., Onuki, Y., Imai, K., Nakanishi, K., and Kawakami, J. (2012) Comparison of contamination levels on the exterior surfaces of vials containing platinum anticancer drugs in Japan. *Biol. Pharm. Bull.* 35(11):2043–2049.

National Institute of Occupational Safety and Health (NIOSH) (2005) *Pocket Guide to Chemical Hazards*, Cincinnati, OH: NIOSH Publications.

National Toxicology Program (NTP) (2011) *Cisplatin: Report on Carcinogens*, 12th ed., pp. 110–111.

Niezborala, M. and Garnier, R. (1996) Allergy to complex platinum salts: a historical prospective cohort study. *Occup. Environ. Med.* 53(4):252–257.

Nygren, O., Gustavsson, B., Strom, L., and Friberg, A. (2002) Cisplatin contamination observed on the outside of drug vials. *Ann. Occup. Hyg.* 46(6):555–557.

Pepys, J. (1980) Occupational asthma: review of present clinical and immunologic status. *J. Allergy Clin. Immunol.* 66(3):179–185.

Pethran, A., Schierl, R., Hauff, K., Grimm, C.H., Boos, K.S., and Nowak, D. (2003) Uptake of antineoplastic agents in pharmacy and hospital personnel. Part I: monitoring of urinary concentrations. *Int. Arch. Occup. Environ. Health* 76(1):5–10.

Platinum Applications (2013) Johnson Matthey: Platinum Today Web site. Available at http://www.platinum.matthey.com/applications/ (accessed January 2, 2013).

Ravindra, K., Bencs, L., and Van Grieken, R. (2004) Platinum group elements in the environment and their health risk. *Sci. Total Environ.* 318(1–3):1–43.

Saindelle, A., and Ruff, F. (1969) Histamine release by sodium chloroplatinate. *Br. J. Pharmacol.* 35(2):313–321.

Venables, K.M., Dally, M.B., Nunn, A.J., Stevens, J.F., Stephens, R., Farrer, N., Hunter, J.V., Stewart, M., Hughes, E.G., and Newman Taylor, A.J. (1989) Smoking and occupational allergy in workers in a platinum refinery. *BMJ* 299(6705):939–942.

Weast, R.C. (1988) *CRC Handbook of Chemistry and Physics*, 69th ed., Boca Raton, FL: CRC Press, p. B-28.

World Health Organization (WHO) (1991) *Platinum. Environmental Health Criteria 125*, International Programme on Chemical Safety Web site. Available at http://www.inchem.org/documents/ehc/ehc/ehc125.htm (accessed on January 4, 2013).

Zimmermann, S., Messerschmidt, J., von Bohlen, A., and Sures, B. (2005) Uptake and bioaccumulation of platinum group metals (Pd, Pt, Rh) from automobile catalytic converter materials by the zebra mussel (*Dreissena polymorpha*). *Environ. Res.* 98(2): 203–209.

29

RARE EARTH METALS

Raymond D. Harbison and David R. Johnson

First aid: Remove from exposure. Give fresh air. Immediately irrigate eyes, irrigate skin with water; wash skin with soap. Provide respiratory support. If ingested by swallowing, seek immediate medical attention.

Target organs: Skin, liver, skeleton

Risk/safety Phrases: Potent contact irritants; anticoagulant effects; GI and respiratory tract absorption; dusts may present and explosion hazard.

Element	CASRN	Atomic Number	Industrial and Commercial Uses
Scandium (Sc) Scandium chloride	10361-84-9	21	High strength aluminum alloys; electron beam tubes; carbon arc rods; electronics; radiopharmaceutical agents; halide lamps; dental lasers
Yttrium (Y)	7440-65-5	39	Capacitors; phosphors; microwave filters; optical glass and ceramics; oxygen sensors; radars; superconductors; color television tubes; magnesium and aluminum alloys; dental lasers; visual displays that give off different colors such as televisions; computer screens; visual displays; fiber optics
Lanthanum (La)	7439-91-0	57	Carbon arc rods; glass; ceramics; phosphor for fluorescent lamps; catalyst for cracking crude petroleum; pigments; accumulators; flint; battery electrodes; camera lenses and telescope lenses; studio lighting and cinema projection; exhaust purification system; water purification
Cerium (Ce)	7440-45-1	58	Optical glasses, polishing abrasives, ceramics, pigments; UV filters; carbon arc rods; alloys with magnesium and iron; phosphors; catalyst for cracking crude petroleum; catalytic converters in cars; ferrocereum flint; oxidizing agent; flatten screen display; exhaust purification system; water purification
Praseodymium (Pr)	7440-10-0	59	Glass colorant, ceramics; pigments; carbon arc rods, alloys; catalyst for cracking crude petroleum; ferrocereum flint; rare earth magnets; lasers; cell phones; portable wireless installation; flatten screen display; MRI and X-ray imaging; hybrid and plug-in electric vehicles; computer disc drive, wireless power tools, integrated starter, wind and hydroelectric power generation
Neodymium (Nd)	7440 00 8	60	Glass colorant; lasers; IR filters; carbon arc rods; catalyst for cracking crude petroleum; high performance magnets such as in loudspeakers, computers; hybrid cars; plug-in electric vehicles; lasers; ceramics; capacitors; flint; cell phones; portable wireless installation; MRI and X-ray imaging; computer disc drive, wireless power tools, integrated starter, wind and hydroelectric power generation
Promethium (Pm)		61	Luminescent and phosphorescent coatings; nuclear batteries; radioactive properties used in luminous paint; other radioactive applications

(continued)

Hamilton & Hardy's Industrial Toxicology, Sixth Edition. Edited by Raymond D. Harbison, Marie M. Bourgeois, and Giffe T. Johnson.
© 2015 John Wiley & Sons, Inc. Published 2015 by John Wiley & Sons, Inc.

(Continued)

Element	CASRN	Atomic Number	Industrial and Commercial Uses
Samarium (Sm) Samarium chloride	10361-82-7	62	Infrared-absorbing glass, lasers; color television phosphors; magnets; microwave filters; catalyst for cracking crude petroleum; alloy with cobalt for magnets; neutron capture; masers; nuclear industry applications
Europium (Eu) Europium chloride	10025-76-0	63	Lasers; phosphor for special X-ray film; mercury vapor lamps; phosphors for fluorescent lamps; NMR relaxation agent; control rods in nuclear reactors, visual displays that give off different colors such as televisions; computer screens; visual displays; cell phones; portable wireless installation; fiber optics; flatten screen display
Gadolinium (Gd)	7440-54-2	64	Reactor control rods; alloys with iron and chromium; glasses; ceramics; crystal scintillators; contrast for magnetic resonance imaging; NMR relaxation agent; X-ray tubes; computer memories; neutron capture; magneto-restrictive alloys; television screens
Terbium (Tb) Terbium chloride	10042-88-3	65	Fluorescent phosphors; lasers; fluorescent lamps; magneto-restrictive alloys; visual displays that give off different colors such as televisions; computer screens; visual displays; cell phones; portable wireless installation; fiber optics; MRI and X-ray imaging; hybrid and plug-in electric vehicles; computer disc drive, wireless power tools, integrated starter, wind and hydroelectric power generation
Dysprosium (Dy)		66	Control rods for nuclear reactors; ceramics; phosphors; magnets; lasers; magneto-restrictive alloys; cell phones; portable wireless installation; MRI and X-ray imaging; hybrid and plug-in electric vehicles; computer disc drive, wireless power tools, integrated starter, wind and hydroelectric power generation
Holmium (Ho) Holmium chloride	10138-62-2	67	Ceramics; lasers; nuclear industry; paramagnetic; research application; calibration for optical spectrophotometers
Erbium (Er) Erbium chloride	10138-41-7	68	Colorant for glass, ceramics; metallurgy; lasers; fiber-optic technology; vanadium steel
Thulium (Tm) Thulium chloride	13537-18-3	69	Radiation source; electron beam tubes; X-ray machines; 16 useful isotopes; lasers; metal-halide lamps
Ytterbium (Yb)	7440-64-4	70	Lasers; metallurgy; radiation source in X-ray/radiation devices; superconductors; chemical reducing agent; nuclear medicine applications; decoy flairs; doping material stainless steel; stress gauges; emits gamma rays; flatten screen display
Lutetium (Lu) Lutctium chloride	10099-66-8	71	Single crystal scintillators; carbon arc rods; positron emission tomography (PET scan detectors); high refractive index glass

Note: Table compiled from Asati et al. (2009), Barducci et al. (2008), Calin and Parasca (2009), Doshi et al. (2000), Gonzalez-Scarano et al. (1987), Hadley et al. (2000), Hartwig et al. (1958), Jenner et al. (2000), Kobayashi et al. (2001), Koch and Hahma (2012), Lowe et al. (2001), Market of Rare-Earth Metals of China (2005), O'Shea et al. (2010), Richter et al. (2007), Røyset and Ryum (2005), Roessl and Proksa (2007), Russel et al. (1987), Schoop et al. (2007), Tebrock and Mackle (1968), Travis et al. (2002), and Wai-Yin Sun et al. (2007).

BACKGROUND AND USES

The rare earth metals (REM) include a group of 17 sequential lanthanide elements beginning with lanthanum (La) and ending with lutetium (Lu), to which scandium (Sc) and yttrium (Y) are added. The REMs incorporate the vertical group IIIB of the periodic table, scandium, yttrium, and lanthanum; and the elements that follow lanthanum in the special horizontal subset of the periodic table because they share some similar physical, chemical, and biological properties (Hirano and Suzuki, 1996). Their unique physical and chemical properties have made them critical to a growing number of modern technologies. In reality these elements are not rare, with the exception of promethium that does not exist in a natural state but is produced by fission, usually from

neodymium. By convention, the elements are designated the REMs, and their oxides, rare earths.

Rare earth elements are present in many minerals, principally monazite and bastnasite (cerium and other lanthanides), gadolinite (yttrium, cerium, chromium, beryllium, and iron), euxenite (vanadium, cerium, europium, niobium, titanium, and uranium), and xenotime (yttrium, thulium, cerium, and other lanthanides). Mineral types include halides, carbonates, oxides, phosphates, and silicates. Although relatively abundant in the earth's crust, they rarely occur in concentrated forms, making them economically challenging to obtain. The metals are extracted from fluorides or oxides by electrolysis, chemical reduction, or precipitation and then purified by ion exchange chromatography. While the United States used to be a world leader in mining and extracting REMs, China now leads in this arena creating a dependency and vulnerability for other countries that need the metals for critical modern applications. New mining opportunities and extraction techniques are being explored and developed.

The value and use of the REMs and their compounds, the rare earths, have been increasing due to their uses in many modern technologies and everyday electronics. The rare earth market has been rising rapidly. From 1964 to 1997, it increased by a factor of 17 and, from 1997 to 2007, it increased by a factor of 20.5 (Naumov, 2008; Savel'eva, 2011). Examples of REM applications include catalytic filter neutralizers of exhaust gases of cars, fiber optics, oxygen sensors, phosphors, superconductors, in lighting, in metallurgy, in glass and ceramic manufacture, in crystallizing synthetic gemstones for lasers and jewelry, and for making long-lasting and special application magnets. The physical properties of the lanthanide series give rise to phosphorescence and luminescence; almost all of the rare earths are used to make television screens, special fluorescent light bulbs, and diagnostic radiographic materials. The rare earths are widely used to impart special colors and properties to glass. For example, didymium, a mixture of praseodymium and neodymium, is used for glass blowers' goggles because it absorbs the intense yellow light of flaming sodium.

Many of the REMs, especially scandium and cerium, produce intense white light when heated or ignited and are used to make the cored carbon arc rods for motion picture projection and other intense lighting equipment. Cerium is pyrophoric and is alloyed with neodymium and lanthanum to make misch metal, which is used in tracer bullets and in lighter flints as an alloy with iron and magnesium. Scandium, yttrium, lanthanum, and cerium are radiopaque, and gadolinium complexes are used extensively as a contrast media in magnetic resonance imaging (MRI).

Other critical applications involve the manufacture of components of cell phones, digital music players, microphones, hard disc drives in computers, hybrid cars and electric vehicles, GPS systems, radar, sonar, and water treatment systems.

Historically, samarium-153 was injected locally to ease the pain caused by skeletal metastases (Turner et al., 1989) and nonradioactive samarium has been used in dental alloys with silver. However, the side effects of the rare metal exposure have limited its therapeutic applications.

PHYSICAL AND CHEMICAL PROPERTIES

The lanthanum series has a similar nuclear structure; each metal has two outer electrons and eight or nine electrons in the next inside electron shell. Each element after lanthanum has an additional electron in the third electron shell from the outside. As the atomic number and weight increase, the primary physical and chemical properties of each metal are conserved. Scandium and yttrium share some properties but do not participate in the unique lanthanide chemistry.

MAMMALIAN TOXICOLOGY

Acute Effects

The rare earths are potent contact irritants and can have significant metabolic effects (Beaser et al., 1942; Chan et al., 1989; Funakoshi et al., 1992; Graca et al., 1957, 1962; Haley et al., 1961, 1965a, 1966; Hirano and Suzuki, 1996; Rim et al., 2013). They are not absorbed from the skin and are slowly absorbed from the gastrointestinal (GI) and respiratory tract. After absorption they concentrate in the liver and the skeleton.

Skin contact produces irritation that progresses to ulceration, delayed healing, and the formation of granulomas. Ocular exposure can cause conjunctivitis, corneal injury, and, ultimately, corneal scarring and opacity. Inhalation of large amounts of rare earth dusts can produce acute irritative bronchitis and pneumonitis. Most REMs are considered to be mild to moderately toxic. Promethium is radioactive, and appropriate exposure precautions should be followed. The free ion form of gadolinium is highly toxic; however, gadolinium used in MRI contrast agents is chelated and considered safe for most persons. Ytterbium compounds are considered highly toxic as they cause irritation to the skin and eye and there is the possibility that some compounds may be teratogenic. Animal studies have shown yttrium to cause lung and liver damage. Yttrium citrate caused pulmonary edema, and while yttrium chloride exposure resulted in liver edema, pleural effusions, and pulmonary hyperemia, and it is thought that exposure to yttrium compounds in humans may cause lung disease (Rim et al., 2013).

Chronic Effects

Repetitive inhalation of large amount of rare earth dusts can cause irritative bronchitis and pneumonitis and can lead to granulomatous disease (Ball, 1966; Rim et al., 2013; Schepers, 1955). There are a number of known case reports of pulmonary fibrosis and pneumoconiosis caused by the rare earths (Rim et al., 2013; Sabbioni et al., 1982; Sulotto et al., 1986; Waring and Watling, 1990). Several of these cases were related to chronic repetitive exposure to the fumes or smoke from rare earth-containing carbon arc lights used in photoengraving, projection, and searchlight operations.

Carcinogenicity The REMs have not been classified as carcinogens. The EPA has found "inadequate information to assess carcinogenic potential" (EPA, 2012).

Mechanism of Action(s) and Chemical Pathology

The REMs all have hepatotoxic and anticoagulant properties (Beaser et al., 1942; Chan et al., 1989; Funakoshi et al., 1992; Graca et al., 1964; Haley et al., 1963, 1964; Kryker and Cress, 1957). Lanthanum, cerium, neodymium, samarium, terbium, dysprosium, erbium, and ytterbium all caused a dose-dependent prolongation of the clotting time of normal human plasma (Beaser et al., 1942). Their presence interferes with activated factor X and thrombin (Hirano and Suzuki, 1996), and they can suppress lymphocyte proliferation and cytokine production (Das et al., 1988; Evans and Ridella, 1985; Palmer et al., 1987; Sedmak and Grossberg, 1981; Yamage and Evans, 1989). Attempts to use rare earth compounds as anticoagulants were unsuccessful because researchers were unable to establish a safe dose and avoid hemorrhage. Similarly, although cerium compounds were historically effective in treating hyperemesis gravidarum and peripheral infection, side effects made their use impractical and dangerous.

Intraperitoneal injection of rare earth chlorides into experimental animals produces peritonitis and hemorrhage ascites (Haley et al., 1961, 1965a, 1965b). With increased doses or absorption, pulmonary edema and liver necrosis occur. The bleeding time and prothrombin time are increased. Recovery is followed by the formation of granulomas of the affected tissues, namely, lung, liver, and peritoneum. The median lethal dose (LD_{50}) for mice, which were given an intraperitoneal injection of chloride salts, ranges from 88 mg/kg for yttrium to 755 mg/kg for scandium (Haley et al., 1961, 1965, 1966). The lanthanide LD_{50} ranges from 315 mg/kg for lutetium to 600 mg/kg for neodymium.

Intravenous injection of neodymium salts (Haley et al., 1961, 1965a, 1966), lanthanum, and cerium in humans resulted in incoagulability of the blood for up to 8 h and unacceptable side effects of rigors, fever, muscle cramps, abdominal pain, hemoglobinuria, and hemoglobinemia.

Chen and Zhu (2008) reported that the long-term intake of low dose rare earth elements may lead to their accumulation in the bone structure, leading to changes in the bone tissue and increased bone marrow micronucleus rate and further to generation of genetic toxicity in bone marrow cells.

SOURCES OF EXPOSURE

Occupational

The growing use of rare earths in a variety of manufacturing processes has increased the potential for industrial or accidental exposure to the rare earths; to mineral contaminants; or to chemicals used in the refining process (Hirano and Suzuki, 1996; Rim et al., 2013; Vennart, 1967). The rarity, in the past, of reports of occupational illness caused by the rare earths has obscured documentation of their toxicity in the medical literature. Nevertheless, potential exposure to REMs and contaminants encountered in production and decommissioning is increasing and occurs in mining, refining processes, manufacturing, transportation, and waste disposal. Some of the risks include exposure to the tailings of rare earth mining; the creation of radioactive dusts and water emissions arising from contaminants during mining processes; dusty environments; poor ventilation; and lack of proper use of protective equipment. While animal studies have shown rare earth exposure to be associated with acute pneumonitis and pulmonary neutrophil infiltration, reports of long-term occupational exposure being associated with pneumoconiosis are limited.

Environmental

Cerium is the 25th most abundant element in the earth's crust. Almost all of the rare earth elements are found in the earth's crust in amounts larger than that of silver, and the four most widespread rare earth elements (yttrium, lanthanum, cerium, and neodymium) are found in amounts greater than that of lead (Naumov, 2008). In nature the REMs are found to be complexed with a variety of minerals and are therefore not a risk for typical exposure via the lungs, skin, or GI tract.

Potential environmental exposures that have been considered involve persons living in close proximity to mining and refining processes; the risk of diesel emissions created by additives; and contamination via radioactive and other chemical wastes of the mining and refining processes.

In some tropical regions, there appears to be an association between a high cerium concentration in vegetables and tubers and an increased prevalence of endomyocardial fibrosis (Valiathan et al., 1989). Postmortem studies have found a high concentration of cerium in the myocardium.

INDUSTRIAL HYGIENE

The REMs in general are considered to have mild to moderate toxicity potential; however, no occupational health standards have been set for the REMs except for yttrium (NIOSH, 1981). Exposure to airborne yttrium has been associated with coughing, shortness of breath, chest pain, and cyanosis. The permissible exposure limit for yttrium was established in 1981 and remains at 1 mg/m^3 as an 8-h time-weighted average. Some common sense guidelines are clearly indicated for occupations involving exposure to the dust and fumes of the rare earths (Vennart, 1967):

1. Skin contact should be avoided.
2. Individuals with skin lesions, contact lenses, and conjunctivitis must protect themselves from contamination of the eyes and skin.

3. Individuals with lung disease and smokers must have respiratory protection.

4. Individuals receiving therapeutic anticoagulation treatment must avoid respiratory and GI exposure.

Rim et al. (2013) recommend control of the level of yttrium, terbium, and lutetium fluorides in workplace air, through maximal admissible concentrations (MACs) for the fluorides of 2.5 mg/m^3 (maximal single concentration) and 0.5 mg/m^3 (average shift concentration), and the level of ytterbium fluoride as moderate fibrogenic dust of 6 mg/m^3.

Promethium has radioactive properties requiring those working with it to wear gloves, foot wear covers, safety glasses, and an outer layer of protective clothing. Further study will be necessary to establish occupational health standards for the REMs.

MEDICAL MANAGEMENT

Medical literature regarding the treatment of acute toxicity from REMs is sparse due to the low number of reported cases of human toxicity. If acute or chronic exposure is suspected, it is important to remove or mitigate human exposure and confirm that the illness is truly due to rare earth exposure. This involves determining that other more common etiologies are not the actual cause of the symptoms. Treatment of acute skin or conjunctival exposure should involve immediate irrigation with water and cleansing the skin with soap and water. If conjunctivitis or corneal injury is apparent, immediate irrigation should be performed in conjunction with ophthalmology consultation. Inhalation exposure should be avoided. If high level or chronic respiratory exposure has occurred or suspected, a CXR should be obtained along with standard treatment for irritative bronchitis and pneumonitis. Chronic exposure has been associated with pneumoconiosis, and bronchoalveolar lavage might be useful in confirming this diagnosis.

If significant absorption is suspected via inhalation or the GI tract, laboratory evaluation should include serum chemistries, liver function test, complete blood count, and coagulation studies (bleeding time, prothrombin time, and clotting time). Significant exposure should be observed in the hospital followed by close outpatient follow-up to assess resolution of any adverse effects and to assure ongoing avoidance of exposure. As with any situation involving exposure toxicity, the local poison control center should be contacted to assist with treatment and for tracking of cases for public health purposes.

REFERENCES

Asati, A., et al. (2009) Oxidase-like activity of polymer-coated cerium oxide nanoparticles. *Angew. Chem. Int. Ed.* 48:2308–2312.

Ball, R.A. (1966) Chronic toxicity of gadolinium oxide for mice following exposure by inhalation. *Arch. Environ. Health* 13:601–610.

Barducci, A., et al. (2008) *The Aerospace Imaging Interferometer Aliseo: Further Improvements of Calibration Methods and Assessment of Interferometer.* The International Archives of the Photogrammetry, Remote Sensing and Spatial Information Sciences. Vol. XXXVII. Part B1. Beijing.

Beaser, S.B., et al. (1942) The anticoagulant effects in rabbits and man of the intravenous injection of salts of the rare earth metals. *J. Clin. Invest.* 21:447–454.

Calin, M.A. and Parasca, S.V. (2009) Light sources for photodynamic inactivation of bacteria. *Lasers Med. Sci.* 24: 453–460.

Chan, K., et al. (1989) Pruritus and paresthesia after IV administration of Gd-DTPA. *AJNR* 10:553.

Chen, Z. and Zhu, X. (2008) Accumulation of rare earth elements in bone and its toxicity and potential hazard to health. *J. Ecol. Rural Environ.* 24:88–91.

Das, T., et al. (1988) Effects of lanthanum in cellular systems. *Biol. Trace Elem. Res.* 18:201–228.

Doshi, N., et al. (2000) Design and evaluation of an LSO PET detector for breast cancer imaging. *Med. Phys.* 27(7): 1535–1543.

Environmental Protection Agency (2012) *Rare Earth Elements: A Review of Production, Processing, Recycling, and Associated Environmental Issues*, EPA/600/R-12/572.

Evans, C.H. and Ridella, J.D. (1985) Inhibition, by lanthanides, of neutral proteinases secreted by human rheumatoid synovium. *Eur. J. Biochem.* 151:29–32.

Funakoshi, T., et al. (1992) Anticoagulant action of rare earth metals. *Biochem. Int.* 28:113–119.

Gonzalez-Scarano, F., et al. (1987) Multiple sclerosis disease activity correlates with gadolinium enhanced magnetic resonance imaging. *Ann. Neurol.* 21:300–306.

Graca, J.G., Garst, E.L., and Lowry, W.E. (1957) Comparative study of stable rare earth compounds (I). *Arch. Ind. Health* 15:9–14.

Graca, J.G., et al. (1962) Comparative toxicity of stable rare earth compounds (II). *Arch. Environ. Health* 5:437–444.

Graca, J.G., et al. (1964) Acute toxicity of injections of chlorides and chelates in dogs. *Arch. Environ. Health* 8:555–564.

Hadley, J., et al. (2000) A laser-powered hydrokinetic system for caries removal and cavity preparation. *J. Am. Dent. Assoc.* 131(6):777–785.

Haley, T.J., et al. (1961) Toxicological and pharmacological effects of gadolinium and samarium chlorides. *Br. J. Pharmacol.* 17:526–532.

Haley, T.J., et al. (1963) Pharmacology and toxicology of terbium, thulium, and ytterbium chlorides. *Toxicol. Appl. Pharmacol.* 5:427–436.

Haley, T.J., et al. (1964) Pharmacology and toxicology of lutetium chloride. *J. Pharmacol. Sci.* 53:1186–1188.

Haley, T.J., et al. (1965a) Pharmacology and toxicology of the rare earth metals. *J. Pharmacol. Sci.* 54:663–670.

Haley, T.J., et al. (1965b) Pharmacology and toxicology of europium chloride. *J. Pharmacol. Sci.* 54:643–645.

Haley, T.J., et al. (1966) Pharmacology and toxicology of dysprosium, holmium, and erbium chlorides. *Toxicol. Appl. Pharmacol.* 8:37–43.

Hartwig, Q.L., et al. (1958) Some toxic effects of yttrium and lanthanum and other rare earths. *AMA Arch. Ind. Health.* 18:505–510.

Hirano, S. and Suzuki, K.T. (1996) Exposure, metabolism, and toxicity of rare earths and related compounds. *Environ. Health Perspect.* 104(Suppl. 1):85–95.

Jenner, A., et al. (2000) Actuation and transduction by giant magnetostrictive alloys. *Mechatronics* 10:457–466.

Kobayashi, H., et al. (2001) Avidin-dendrimer-(1B4M-Gd)$_{254}$: a tumor-targeting therapeutic agent for gadolinium neutron capture therapy of intraperitoneal disseminated tumor which can be monitored by MRI. *Bioconjug. Chem* 12:587–593.

Koch, E. and Hahma, A. (2012) Metal-fluorocarbon pyrolants. XIV: high density-high performance decoy flare compositions based on ytterbium/polytetrafluoroethylene/Viton®. *ZAAC* 638(5): 721–724.

Kryker, G.C. and Cress, E.A. (1957) Acute toxicity of yttrium, lanthanum and other rare earths. *AMA Arch. Ind. Health* 16:475–479.

Lowe, M., et al. (2001) pH-dependent modulation of relaxivity and luminescence in macrocyclic gadolinium and europium complexes based on reversible intramolecular sulfonamide ligation. *J. Am. Chem. Soc.* 123:7601–7609.

Market of Rare-Earth Metals of China (2005) *Market of Rare-Earth Metals of China, Japan, and United States in 2005.* Available at www.metalltorg.ru.

National Institute for Occupational Safety and Health (1981) *Occupational Health Guideline for Yttrium, Occupational Health Guides for Chemical Hazards*, DHHS Pub. No. 81–123, Rockville, MD: National Institute for Occupational Safety and Health.

Naumov, A. (2008) Review of the world market of rare-earth metals. *Russ. J. Non-Ferrous Met.* 49(1):14–22.

O'Shea, F.H., et al. (2010) Short period, high field cryogenic undulator for extreme performance X-ray free electron lasers. *Phys. Rev. Spec. Top. Acc. Beams* 13:070702.

Palmer, R.J., et al. (1987) Cytotoxicity of the rare earth metals cerium, lanthanum, and neodymium *in vitro. Environ. Res.* 43, 142–156.

Richter, A., Heumann, E., and Huber, G. (2007) Power scaling of semiconductor laser pumped praseodymium-lasers. *Opt. Express* 5172(15):8.

Rim, T.K., Koo, K.H., and Park, J.S. (2013) Toxicological evaluations of rare earths and their health impacts to workers: a literature review. *Saf. Health Work* 4(1):12–26.

Roessl, E. and Proksa, R. (2007) K-edge imaging in X-ray computed tomography using multi-bin photon counting detectors. *Phys. Med. Biol.* 52:4679–4696.

Røyset, J. and Ryum, N. (2005) Scandium in aluminium alloys. *Int. Mater. Rev.* 50(1):20.

Russell, E.J., et al. (1987) Multiple cerebral metastases: detectability with Gd-DTPA-enhanced MR imaging. *Radiology* 165:609–617.

Sabbioni, E., et al. (1982) Long-term occupational risk of rare earth pneumoconiosis: a case report as investigated by neutron activation analysis. *Sci. Total Environ.* 26:19–32.

Savel'eva, I.L. (2011) The rare-earth metals industry of Russia: present status, resource conditions of development. *Geogr. Nat. Resour.* 32(1):65–71.

Schepers, G.W.H. (1955) The biological action of rare earths. I. The experimental pulmonary histopathology produced by a blend having a relatively high oxide content. *Arch. Ind. Health* 12:301–305.

Schoop, U., et al. (2007) The use of the erbium, chromium: yttrium-scandium-gallium-garnet laser in endodontic treatment: the results of an *in vitro* study. *J. Am. Dent. Assoc.* 138(7):949–955.

Sedmak, J.J. and Grossberg, S.E. (1981) Interferon stabilization and enhancement by rare earth salts. *J. Gen. Virol.* 52: 195–198.

Sulotto, F., et al. (1986) Rare earth pneumoconiosis: a new case. *Am. J. Ind. Med.* 9:567–575.

Tebrock, H. E. and Mackle, W. (1968) Exposure to europium-activated yttrium orthovanadate: a cathodoluminescent phosphor. *J. Occup. Environ. Med.* 10:692–696.

Travis, J.C., et al. (2002) An international evaluation of holmium oxide solution reference materials for wavelength calibration in molecular absorption spectrophotometry. *Anal. Chem.* 74:3408–3415.

Turner, J.H., et al. (1989) A phase 1 study of samarium-153 ethylenediamine-tetra-methylene phosphonate therapy for disseminated skeletal metastases. *J. Clin. Oncol.* 7:1926–1931.

Valiathan, M.S., et al. (1989) A geochemical basis for endomyocardial fibrosis. *Cardiovasc. Res.* 23:647–648.

Vennart, J. (1967) The usage of radioactive luminous compound and the need for biological monitoring of workers. *Health Phys.* 13:959–964.

Wai-Yin Sun, R., et al. (2007) Some uses of transition metal complexes as anti-cancer and anti-HIV agents. *Dalton Trans.* 4884–4892.

Waring, P.M. and Watling, R. J. (1990) Rare earth deposits in a deceased movie projectionist. A new case of rare earth pneumoconiosis? *Med. J. Aust.* 153:726–730.

Yamage, M. and Evans, C.H. (1989) Suppression of mitogen- and antigen-induced lymphocyte proliferation by lanthanides. *Experientia* 45:1129–1131.

30

SELENIUM AND ITS COMPOUNDS

Anna M. Fan and Marco Vinceti

First Aid: Elemental selenium (Se), Se alloy

Eye: Irrigate immediately

Skin: Soap wash immediately

Breathing: Respiratory support

Swallow: Medical attention immediately

Target organ(s):

Elemental Se, Se alloy: Eyes, skin, respiratory system, liver, kidneys, blood, spleen (OSHA)

Se and compounds: Inhalation—lung as primary; also cardiovascular, hepatic, nervous system, renal involvement. Oral—gastrointestinal, cardiovascular effects; selenosis (dermal and neurological) from chronic exposure.

Occupational exposure limits:

OSHA PEL: TWA $0.2 \, mg/m^3$ (elemental Se and other Se compounds, except Se hexafluoride)

NIOSH REL: TWA $0.2 \, mg/m^3$ (elemental Se and other Se compounds, except Se hexafluoride)

ACGIH TWA $0.2 \, mg/m^3$ (elemental Se).

Reference values (see text for references):

U.S. EPA oral RfD, 5 µg/kg/day

Cal/EPA oral ADD, 5 µg/kg/day

ATSDR chronic oral MRL, 5 µg/kg/day

Dietary reference values (vary for different age groups [six age groups between age 0 and 18], and they include special groups such as pregnant and lactating women):

NRC EAR: Age 0–18, 15–45 µg/day; adults, 45 µg/day; pregnant women, 49 µg/day; lactating women, 59 µg/day

NRC RDA: Age 0–18, 15–55 µg/day; adults, 55 µg/day; pregnant women, 60 µg/day; lactating women, 70 µg/day

NRC UL: Age 0–18, 45–400 µg/day; adults, pregnant women, lactating women, 400 µg/day

NRC AI: 0–6 months, 15 µg/day; 7–12 months, 20 µg/day

Risk/safety phrases: Elemental selenium (Se), Se alloy

Health risk: Potential hazard: Depends on the chemical form and exposure; industrial exposure can be highly toxic, dietary exposure can be beneficial, excess can be toxic (but rare)

Safety: Avoid skin contact. Do not use mouth-to-mouth method if victim ingested or inhaled the substance. Isolate leak or spill area. IDLH: 1 mg Se/m³. The revised IDLH (Immediately Dangerous to Life or Health) for selenium compounds is 1 mg Se/m³ based on acute toxicity data on sodium selenite in animals. It is "an atmosphere that poses an immediate threat to life, would cause irreversible adverse health effects, or would impair an individual's ability to escape from a dangerous atmosphere (OSHA regulation 1910.134(b))." Relevant acute toxicity data for workers are lacking (CDC, 1994; Transport Canada, 2008).

Hamilton & Hardy's Industrial Toxicology, Sixth Edition. Edited by Raymond D. Harbison, Marie M. Bourgeois, and Giffe T. Johnson.
© 2015 John Wiley & Sons, Inc. Published 2015 by John Wiley & Sons, Inc.

BACKGROUND AND USES

Selenium is found ubiquitously in the environment, being released from both natural and anthropogenic sources. Natural sources include the weathering of selenium-containing rocks and soils and volcanic eruptions. The principal anthropogenic source is coal combustion, petroleum mining or refining, fly-ash from coal-burning power plants, and irrigation of high selenium arid farmland soils. It occurs naturally at low concentrations in surface waters and groundwaters of the United States.

Selenium can exist as an inorganic species that includes elemental selenium, selenium salts (selenite, selenate), selenious acid, selenic acid, selenium dioxide, selenium disulfide, selenium hexafluoride, selenium monochloride, selenium monosulfide, selenium oxychloride, and selenium tetrachloride (IPCS, 2001). It can be found in various commercial products, including electronics (photoelectric cells, semiconductors, low voltage rectifiers), ceramics and glass (pigments and dyes), steel (alloy with copper), gun-bluing solutions (selenious acid and copper nitrate and nitric acid), rubber vulcanizing, chemistry (catalyst), photocopy (xerographic properties), and medicine (selenium sulfide for treatment of pityriasis versicolor and palmar plantar mycosis, dandruff and seborrheic dermatitis of the scalp). The gamma-emitting isotope ^{75}Se has been used in diagnostic applications of medicine. Metal industry workers, health service professionals, mechanics, and painters may be exposed to higher levels of selenium than the general population or workers employed in other trades.

Selenium can also exist as an organic species such as methylated selenium compounds, selenoamino acids, selenoproteins, and their derivatives (Pyrzyńska, 2002). The general population is exposed to selenium primarily by dietary intake, mainly as the amino acid selenomethionine in grains, cereals, and forage crops; as selencysteine in meats and dairy products to a lesser extent; and as the inorganic selenate and selenite, which are the main soluble forms in aqueous media. Although selenium dioxide, selenium trioxide, selenious acid, and selenic acid are soluble in water, they have not been detected in drinking water because each becomes the corresponding selenite or selenate once it dissolves in water (Pyrzyńska, 2002; ATSDR, 2003). Selenium is an essential micronutrient for humans and animals, obtained as part of daily dietary intakes and from some dietary supplements. It is also used as a nutritional feed additive for poultry and livestock.

PHYSICAL AND CHEMICAL PROPERTIES

Selenium is a nonmetal (or semimetallic) element (also referred to as a metalloid) with an atomic number of 34 and an atomic mass of 78.96. It belongs to Group 16 (Group VIA) of the periodic table, located between sulfur and tellurium. Six stable isotopes of selenium occur naturally. It exists in several allotropic forms. The stable form at ordinary room temperatures is the gray or hexagonal form; the other two important forms are red (monoclinic) and amorphous selenium. Black amorphous selenium is vitreous and is formed by the rapid cooling of liquid selenium; red amorphous selenium is colloidal and is formed in reduction reactions (Hoffman and King, 1997; CRC, 2000; ATSDR, 2003).

The chemical properties of selenium are similar to sulfur. Selenium combines with metals and many nonmetals directly or in aqueous solution. It does not react directly with hydrogen fluoride or hydrogen chloride but decomposes hydrogen iodide to liberate iodine and yield hydrogen selenide. The selenides resemble sulfides in appearance, composition, and properties. It may form halides by reacting vigorously with fluorine and chlorine, but the reactions with bromine and iodide are not as rapid. It reacts with oxygen to form a number of oxides, the most stable of which is selenium dioxide (Hoffman and King, 1997; ATSDR, 2003).

The important selenium oxidation states are −2, 0, +4, and +6. The commonly occurring selenium species in these oxidation states are: (1) −2: hydrogen selenide, sodium selenide, dimethyl selenide, dimethyl diselenide, selenomethionine, selenocysteine, selenohomocysteine, selenocystine; (2) 0: elemental selenium, selenenic acid, sodium selenate; (3) +4: selenium dioxide, selenious acid, sodium selenite; and (4) +6: selenium trioxide, selenic acid, sodium selenate. Some physical and chemical properties for elemental selenium, sodium selenate, and sodium selenite are shown in Table 30.1. Additional information for these and other chemical forms of selenium can be found in OEHHA (2010; ATSDR, 2003).

ENVIRONMENTAL FATE AND BIOACCUMULATION

The behavior of selenium in the environment is influenced by its oxidation state and the properties of the different chemical compounds (NAS, 1976). The oxidation state of selenium in the environment is dependent on ambient conditions, particularly on pH and microbial activity (Maier et al., 1988).

Selenium, found in nature, can complex with multiple compounds. It can be oxidized to form selenium dioxide, which may be expected at the soil surface. Selenates and selenites are water soluble and can be found in water. Because of their high solubility and low tendency to adsorb onto soil particles, the selenates (especially sodium selenate) are very mobile and are readily taken up by biological systems or leached through the soil. Most selenites are less soluble in water than the corresponding selenates (NAS, 1980). Organic selenium—selenomethionine, selenocysteine, and dimethyl selenides—can be found in plants.

TABLE 30.1   Physical and Chemical Properties of Elemental Selenium, Sodium Selenate, and Sodium Selenite

	Selenium	Sodium Selenate	Sodium Selenite
Synonyms	Elemental selenium, selenium homopolymer, selenium dust, colloidal selenium	Disodium selenate	Disodium selenite, disodium selenium trioxide, selenious acid disodium salt, sodium selenium oxide
CAS number (No.)	7782-49-2	13410-01-0	10102-18-8
Formula	Se	Na_2SeO_4	Na_2SeO_3
Molecular weight	78.96	188.94	172.95
Color	Metallic gray to black; hexagonal crystals	Colorless rhombic crystals	White tetragonal crystals
Physical state	Solid	Solid	Solid
Melting point	144 °C; 221 °C	No data	No data
Boiling point	685 °C	No data	No data
Density (g/cm^3)	4.81 (20 °C)	3.213 (17.4 °C)	No data
Water solubility	Insoluble	84 g/100 ml at 35 °C	Soluble

Selenium sulfides are not very water soluble and are therefore relatively immobile.

Transport and Partitioning. The volatile selenium compounds that partition into the atmosphere include the inorganic compounds, selenium dioxide and hydrogen selenide, and the organic compounds, dimethyl selenide and dimethyl diselenide. Hydrogen selenide is highly reactive in air and is rapidly oxidized to elemental selenium and water (NAS, 1976); other compounds can persist in air. Selenium compounds in the atmosphere can be removed by dry or wet deposition on soils or on surface water where the salts of selenic and selenious acids are the expected forms. Selenic acid (H_2SeO_4) is a strong acid and the soluble selenate salts of this acid are expected to occur in alkaline waters. Selenious acid (H_2SeO_3) is a weak acid, and the diselenite ion predominates in waters between pH 3.5 and 9.

In soils, pH and Eh are determining factors in the transport and partitioning of selenium. Elemental selenium is essentially insoluble and may represent a major inert "sink" for selenium introduced into the environment under anaerobic conditions (NAS, 1977). Heavy metal selenides and selenium sulfides, which are also insoluble, predominate in acidic (low pH) soils and in soils with high amounts of organic matter. They are immobile and will remain in the soil. The selenides of other metals such as copper and cadmium are of low solubility. Sodium and potassium selenites dominate in neutral, well-drained mineral soils. In alkaline (pH > 7.5), well-oxidized soil environments, selenates are the major selenium species.

Transformation and Degradation. Selenium dioxide released into the air from the combustion of fossil fuels is anticipated to be largely reduced to elemental selenium by sulfur dioxide formed during the combustion (NAS, 1977). Hydrogen selenide is unstable in air and is oxidized to elemental selenium and water (NAS, 1976). Hazards from hydrogen selenide are expected to be confined to occupational settings where the gas might build up to hazardous

levels despite oxidative losses. Dimethyl selenide and methyl selenide are volatile organic compounds that can partition into and persist in the atmosphere. In water, in general, the more soluble and mobile forms of selenium (e.g., selenite and selenate) dominate under aerobic (high oxygen concentrations) and alkaline (high pH) conditions.

Data from a simulated laboratory pond showed that bacteria and cyanobacteria have two possible mechanisms for the uptake and transformation of selenate (Bender et al., 1991). The uptake mechanism involves the reduction of selenate to elemental selenium that will be physically held within the biological mat. The microorganisms also caused the transformation of soluble selenium into volatile alkyl selenium compounds.

In soils, elemental selenium and inorganic selenium compounds such as sodium selenite can be methylated by microorganisms and subsequently volatilized into the atmosphere. In general, microorganisms appear to methylate organic selenium compounds more readily than either selenite or selenate (Maier et al., 1988). Demethylation of the trimethylselenonium ion can also occur in soil and microorganisms are evidently required for this reaction (Yamada et al., 1994).

Terrestrial plants take up soluble selenate and selenite and biosynthesize organic selenium compounds, predominantly selenomethionine and, to a lesser extent, selenocysteine. Selenates tend to be taken up by plants from soils more readily than selenites, in part because selenites tend to adsorb more strongly to soils (Dimes et al., 1988; Zhang et al., 1988). These compounds can be released to the soils once the plants die and decay. Water-soluble organic selenium compounds are also probably readily taken up by plants.

Bioaccumulation

Uptake by plants is influenced by factors such as soil type, pH, colloidal content, concentration of organic material, oxidation–reduction potentials in the root–soil environment, and total level of selenium in the soil (Fishbein, 1983). Alkaline and oxidizing conditions favor the plant uptake of soluble forms of selenium and selenate is preferred over selenite (Banuelos and Meek, 1990). In acidic soils (pH 4.5–6.5), and under high moisture conditions, selenium is in the form of selenite bound to colloids as iron hydroxide selenium complexes, which are insoluble and generally not bioavailable to plants (Galgan and Frank, 1995). In basic soils (pH 7.5–8.5), selenium is present as soluble selenate (principally sodium selenate). Although much of the total selenium in soil may be present in other forms, sodium selenate appears to be responsible for most of the naturally occurring accumulation of high levels of selenium by plants (NAS, 1976). The use of lime and plant ash as fertilizers raises soil pH and favors selenate formation.

Selenium in an aquatic environment is bioaccumulated and bioconcentrated by aquatic organisms (Lemly, 1985; Ohlendorf et al., 1986a; Saiki and Lowe, 1987; Maier et al., 1988). Bioconcentration factors (BCFs) of 150–1850 and bioaccumulation factors (BAFs) of 1746–3975 for selenium in freshwater have been reported. Biomagnification is also evidenced by progressively higher concentrations of selenium in organisms at successively higher trophic levels.

ECOTOXICOLOGY

Anthropogenic activities, such as coal combustion, mining, petroleum refining, and irrigation of agricultural lands, have resulted in increased levels of selenium in some aquatic ecosystems. Selenium in the aquatic environment can accumulate in biota. In fish and wildlife, adverse effects from high exposure to selenium have been observed, including reduced reproduction, offspring deformities, mortality, and reduced populations (Lemly, 1998).

In California, accumulation of selenium in agricultural drainage waters led to high levels in evaporation ponds of the Kesterson National Wildlife Refuge in the San Joaquin Valley. More than 50% of that contained in sediments occurred in organic forms (Maier et al., 1988) resulting from the synthesis and bioaccumulation of organic selenium before the plants die and decay. Elevated levels of selenium were measured (dry weight) in algae (average 35 mg/kg), midge larvae (139 mg/kg), dragonfly and damselfly nymphs (average 122 and 175 mg/kg, respectively), and mosquito fish (170 mg/kg) (Ohlendorf et al., 1986b).

Acute Effects

Toxicity of selenium to aquatic organisms can be acute through water column exposure. Excessive selenium has resulted in deformities, reproductive impairment, death, and decreased populations in fish and waterfowl (Lemly, 1993). A dietary concentration of 4 μg/g dry wt. can adversely affect fish reproduction whereas a level below about 0.1–0.5 μg/g dry wt. can cause nutritional deficiencies. Selenium concentration in fish eggs was found to be correlated with the incidence of deformities in larvae and magnitude of reproductive failure. Teratogenic effects included spinal deformities such as lordosis, kyphosis, and scoliosis, and head, mouth, and fin deformities; edema, exophthalmus and cataracts were also observed. Biologic effect thresholds for the health and reproductive success of freshwater and anadromous fish have been suggested for the whole body at 4 μg/g; skeletal muscle at 8 μg/g; ovaries and eggs at 10 μg/g; and liver at 12 μg/g.

Chronic Effects

Selenium toxicity can be chronic via food chain exposure. Often habitats that accumulate selenium best are eutrophic, with shallow slow-moving waters with low flushing rates, such as wetland or reservoirs (Simmons and Wallschlager, 2005). Toxic effects of selenium in the lentic (standing water) environment have been extensively studied in Belews Lake, North Carolina; Hyco Reservoir, North Carline; Martin Reservoir, Texas; Tulare Lake Basin, California; Sweitzer Lake, Colorado; and Kesterson Reservoir in California. In all of these cases, adverse effects resulting in reduced fish and avian populations were observed when waterborne concentrations of selenium exceeded 2–5 μg/l. The acceptable level is believed to be <1 μg/l (Lemly, 1996). In waterfowl, deformities were seen in hair, beak, feathers, and nails (O'Toole and Raisbeck, 1998).

Simmons and Wallschlager (2005) compared the selenium concentrations found in the water and diets of fish in lentic and lotic systems (flowing waters), and the biologic effects observed in fish in these locations. They found that low waterborne exposures in lentic environments often resulted in very high bioaccumulation in tissues, whereas comparable exposures in lotic environments tended to result in lower selenium accumulation. Diet appears to be a more reliable predictor of tissue concentration rather than waterborne exposures, indicating the importance of the transfer of organic selenium through trophic interactions but not bioconcentration from environmental compartments. There is no conclusive evidence that any other dissolved selenium species besides selenite and selenate exist in freshwater. Organic species, such as dimethyselenide, have been reported in natural waters, but they usually constitute a very small fraction of the total selenium and their volatility would suggest that they would escape the water column rapidly.

MAMMALIAN TOXICOLOGY

Acute Effects

In humans and animals, the respiratory tract is the primary site of injury following inhalation exposure to selenium.

Human data are mostly obtained from occupational exposures, and only acute exposure studies were found in animals. Inhalation data are available for elemental selenium, selenium dioxide, selenium oxychloride, hydrogen selenide, and dimethyl selenide but most are on multiple chemical exposures (ATSDR, 2003).

The chemical forms of selenium most likely encountered in air in occupational settings are elemental selenium dusts, selenium dioxide, and hydrogen selenide. Following occupational exposures, in addition to respiratory tract effects, gastrointestinal and cardiovascular effects, and skin and eye irritation also occurred. Acute inhalation of selenium dusts resulted in irritated mucous membranes in the nose and throat and produced coughing, nosebleed, loss of olfaction; high exposures produced dyspnea, bronchial spasm, bronchits, and chemical pneumonia. Selenium dioxide is formed when selenium is heated in air. Selenium forms selenious acid on contact with water, including respiration, causing severe irritation if inhaled. Acute high level exposure to selenium dioxide powder can produce pulmonary edema (following irritation to alveoli), bronchial spasms, asphyxiation symptoms, and bronchitis. Hydrogen selenide and selenium oxychloride are highly toxic. Hydrogen selenide is a gas at room temperature. Selenium oxychloride hydrolyzes to hydrogen chloride, which can then form hydrochloric acid in humid air and in the respiratory tract where it can be very irritating and corrosive.

Workers acutely exposed to selenium dioxide fumes from a fire for, at most, 20 min reported symptoms of shock, including lower blood pressure and elevated pulse rates, which were normalized in 3 h after treatment with oxygen and inhalation of ammonia vapor. Vomiting and nausea were also reported. Stomach pain was reported by workers, in a selenium rectifier plant, exposed to elemental selenium and selenium dioxide. Minimal information was found in the literature on neurological effects, but it included reports of frontal headache by workers exposed to high concentration of selenium of unknown identity for about 2 min.

In experimental animals (rats, rabbits, guinea pigs), acute exposure to selenium fumes or dust ($33 \, \text{mg/m}^3$) resulted in respiratory effects, including hemorrhage and edema of the lungs, interstitial pneumonitis, congestion, and emphysema. Death was also reported. Acute inhalation of guinea pigs to hydrogen selenide ($8 \, \text{mg/m}^3$, 8 h) produced diffuse bronchopneumonia and pneumonitis. Dimethyl selenide at much higher concentrations ($25,958 \, \text{mg/m}^3$) produced minor effects (increased lung and liver weight) in rats after 1 day of exposure. Hepatotoxicity observed as slight liver congestion and mild centrilobular atrophy were reported in rats exposed to elemental selenium dust for 8 h, and in guinea pigs exposed to elemental selenium dust for 4 h/day, for 8 days, both at $33 \, \text{mg/m}^3$.

Dermal exposure to selenium dioxide or selenium oxychloride may produce skin burns (IPCS, 2001). Selenium dioxide is a primary irritant and it causes extremely painful burns on the skin, which always heal without a scar. Although, theoretically, the selenium dioxide powder itself does not burn the skin (and if dropped on the skin, it should be immediately brushed off dry), in practice, in the industrial environment, there is sufficient moisture on the skin from sweating to be taken up by this white solid to form a sticky solution of selenious acid within seconds, or at the most, minutes of coming into contact with the skin. After skin exposure to a 50% solution of selenious acid, signs such as unremitting intense pain, red and swollen skin, blisters develop within several hours, which may be followed by ulcerations. Dermal exposure to selenium oxychloride causes vesicles. Nasal discharge, loss of smell, epistaxis have been described after dermal exposure to selenium dust.

Data on oral exposures to selenium in humans and animals have been compiled for elemental selenium dust, selenium dioxide dissolved in water (selenious acid), sodium selenate, sodium selenite, potassium selenate, and dietary selenium compounds, including selenoamino acids (ATSDR, 2003). Most of the available toxicity information reviewed for oral exposures to selenium compounds comes from domestic or experimental animal studies with exposures to selenite, selenate, selenium sulfides (mixed), and organic forms (selenocystine, selenomethionine). Few acute studies were identified, as the majority of the studies reported were intermediate (weeks to months) and long-term studies.

In animal studies, oral LD_{50}s in the range of 1–12 mg/kg were reported for sodium selenite in rats, mice, guinea pigs, and rabbits (ATSDR, 2003). The minimum oral fatal dose in rats was 6700 mg/kg for elemental selenium. Rats administered selenium sulfide by gavage at the lethal level were reported to have diarrhea and anorexia. Single gavage doses of selenium monosulfide to rats produced death and widespread hepatic necrosis at 75–100 mg/kg.

For acute oral exposures in animals (14 days or shorter), systemic effects have been reported following exposure to selenium (ATSDR, 2003). These include the following: mouse: hematological, hepatic, renal, dermal, immuno/lympho, neurological, death; rat: body weight, death; pig: body weight, neurological; hamster: developmental; guinea pig, rabbit: death.

In humans, acute poisoning with selenium and its compounds is rare, especially fatal poisonings; however, death has occurred following accidental (including errors in manufacturing) or suicidal overdoses. The approximate selenium intake resulting in acute toxicity is reported to be up to 250 mg selenium as a single dose or 27–31 mg as multiple doses (SCF, 2000). Typical chronic selenosis symptoms such as hair and nail changes and loss were noted in some (e.g., U.S. supplements cases) but not all cases.

In 2008, an episode of selenium toxicity caused by a misformulated commercially distributed dietary supplement resulted in the recall by U.S. Food and Drug Administration (FDA) and the manufacturer of 1000 bottles of the supplements that the FDA found to contain selenium at levels up to 200 times greater than the amount stated on the label or up to 40,800 μg per recommended serving (Sutter et al., 2008; U.S. FDA, 2008; MacFarquhar et al., 2010). Ten of the 16 states that received distributions reported cases of selenium toxicity. Symptoms of selenium poisoning can include significant hair loss, muscle cramps, nausea, vomiting, diarrhea, joint pain, fatigue, fingernail changes, and blistering skin. Patients reported that symptoms typically occur within 5–10 days after daily ingestion of these supplements begins. After discontinuing use of the product, the symptoms of selenium toxicity may last for several weeks but improve eventually without treatment for the poisoning. There is no proven antidote or curative treatment for selenium poisoning. A case definition would involve two or more of the following symptoms, such as hair loss, muscle or joint pains, fingernail discoloration or changes, headache, foul breath, weakness, gastrointestinal symptoms (such as nausea, vomiting, diarrhea, or abdominal pain), rash, oliguria/anuria or abnormal renal function tests, jaundice or abnormal liver function tests, anemia or hematological changes, that occurred within 2 weeks of ingesting a dietary supplement manufactured in the United States exclusively for the specific manufacturer. It was noteworthy that there was an array of nonspecific symptoms making the diagnosis difficult as selenium toxicity is rare. For example, 23% of the patients reported fever, a symptom not previously associated with selenium toxicity, and it could not be conclusively determined that selenium toxicity was the cause. Patients often continued to experience symptoms for 90 days for hair and nail changes, and constitutional symptoms such as memory loss, mood swings, fatigue, musculoskeletal complaints, and garlic breath. Selenium is sequestered in different organ tissues, and its slow metabolism accounts for the persistence of symptoms. An earlier report suggested that forced vomiting soon after ingestion of selenate may be helpful as in the case of survival of a patient in New Zealand (Civil and McDonald, 1978).

Chronic Effects

Animals For intermediate oral exposures in animals (15–364 days), systemic effects have been reported following exposure to selenium (ATSDR, 2003). These include the following: rat: hematological, cardiovascular, hepatic, renal, endocrine, metabolic, immune/lympho, developmental, body weight, death; mouse: cardiovascular, hepatic, renal, reproductive, developmental, body weight, death; pig: cardiovascular, musculoskeletal, hepatic, dermal, body weight, neurological, reproductive; rabbit: cardiovascular; monkey: gastrointestinal; cow: dermal; monkey: body weight, neurological, reproductive.

For chronic oral exposures in animals (365 days or longer) (ATSDR, 2003), most of the data on systemic effects have been conducted in mice; effects reported include respiratory, cardiovascular, hepatic, renal, endocrine, and dermal. Hepatic effects and death were reported in rats.

Hepatic The liver has long been recognized as a primary target of selenium toxicity in the pioneering studies carried out in the seleniferous areas of the United States, where intoxicated individuals showed signs such as icteroid discoloration of the skin and a history of jaundice and frank hepatitis (Smith and Westfall, 1936; Smith and Westfall, 1937) and of cirrhosis (Lemley and Merryman, 1941) though the latter observations were less clearly related to selenium intoxication. Subsequent studies carried out in the seleniferous regions of China and the United States have shown biochemical abnormalities of uncertain clinical significance apparently related to liver function, i.e., an increased prothrombin time (Yang et al., 1989a) and serum level of alanine aminotransferase (Longnecker et al., 1991), an observation of interest due to the increased activity of this enzyme observed in selenite-exposed rats (Coudray et al., 1996). The direct correlation between alanine aminotransferase and dietary selenium intake found in the former study carried out in South Dakota and western Wyoming, in particular, was not associated with enzyme levels exceeding the "normal range" (Longnecker et al., 1991). In the study in China, increased prothrombin time was clearly detected in subjects having blood selenium concentrations exceeding 1000 μg/l, corresponding to a daily intake of around 850 μg, but a correlation between prothrombin time and selenium blood levels was not reported by the authors at lower selenium exposures. In a study carried out in adults with prostate cancer supplemented for 12 or more months with 1600 or 3200 μg/day of organic selenium, increased blood total bilirubin and alkaline phosphatase levels in subjects receiving the highest dose were observed, while glutamic pyruvic and glutamic oxaloacetic transaminases were not increased; all blood analytical determinations were within the "normal range" (Reid et al., 2004). In this study, however, there were no comparisons with individuals receiving placebo, and no measurements of prothrombin time or alanine aminotransferase were made. In an investigation carried out in U.S. communities with different levels of exposure to selenium (likely in an inorganic form) through drinking water, no excess prevalence of "liver disease" was found (Valentine et al., 1987). On the other hand, no analysis of liver function and disease in individuals occupationally exposed to selenium has been reported, with the exception of the observation of acute hepatitis with pathologic signs of steatosis, localized liver necrosis, sinusoidal hyperplasia, and elevations of serum transaminases following acute intoxication with selenious acid (Pisati et al., 1988). Overall, epidemiologic and clinical evidence points to the occurrence of hepatotoxicity following selenium overexposure, but available data that would clearly substantiate this association are limited.

Dermatological   The effects on skin and related annexes were considered among the first "markers" of selenium toxicity when the pioneering studies on toxicity of environmental selenium were carried out in U.S. adults (Smith and Westfall, 1936; Smith and Westfall, 1937; Manville, 1939; Lemley, 1940) and in Venezuelan children (Jaffe et al., 1972). Among these effects were dermatitis, skin eruptions and depigmentation, diseased fingernails, and hair loss (Jaffe et al., 1972). Nails of intoxicated subjects, in particular, were atrophic, brittle, and irregular, with frequent occurrence of ridging. Interestingly, Lemley and Merryman (1941) were able to treat the signs and symptoms of selenium intoxication by administering bromobenzene to an affected patient to increase selenium clearance from the body, with immediate relief and recovery of the patient (Lemley and Merryman, 1941). In the seleniferous areas of Venezuela, high intake of selenium through consumption of "Coco de Mono" nuts (*Lecythis ollaria*) has long been shown to be accompanied by gastrointestinal and neurological signs and, in particular, extensive hair loss from several body components (scalp, axillae, chest, abdomen, pubic), loss of eyebrows and eyelashes, and nail streaking and breaking (Kerdel-Vegas, 1966). In studies carried out in the seleniferous areas of Enshi County, Hubei province in central China, a wide spectrum of hair, nail, and skin abnormalities related to selenium exposure was observed (Yang et al., 1983; Yang et al., 1989a; Fordyce, 1995). The hair of selenium-intoxicated individuals was dry and brittle and easily fell from the scalp and other body parts (Yang et al., 1983), making alopecia an extremely common sign in affected individuals (Fordyce, 1995). Nails tended to be brittle, fragile, irregular, and atrophic, with white spots on their surfaces and a clear tendency to drop off. These dermatological abnormalities were ascribed to daily selenium intake ranging from 1500 to 5000 μg/day though nail brittleness was detected in a subsequent study following supplementation with 600 μg/day of selenium (Yang et al., 1989a). Selenium-induced dermatological alterations have also been documented in some seleniferous area of rural Punjab, where consumption on an average of over 600 μg/day of selenium was associated with hair and nail loss and a wide pattern of nail abnormalities such as blackening, breaking, and streaking (Hira et al., 2004). In contrast, no relation between dermatological signs and selenium exposure was detected in seleniferous areas of South Dakota and Wyoming (Longnecker et al., 1991).

The dermatologic toxicity of selenium was actually demonstrated in a U.S. trial, named "SELECT," that was carried out to test possible beneficial effects of selenium on reducing the risk of prostate and other cancers. The trial was terminated before completion due to the lack of efficacy on primary end points and an increased diabetes incidence

among supplemented subjects (Lippman et al., 2009). In that study, dietary supplementation of 200 μg/day selenomethionine in free-living individuals, with an attributable average "background" dietary selenium intake of around 100 μg/day, induced dermatitis and alopecia, with an increased incidence of 17–74% in comparison with placebo-receiving subjects. This study was not only a "controlled" demonstration that long-term administration of selenium induced dermatological effects, but, more importantly, it showed that such effects occurred at doses previously considered to have no adverse effects by some investigators (Yang et al., 1983; Longnecker et al., 1991), i.e., around 300 μg/day (for organic selenium, considered less toxic than the inorganic forms of the element). This dermal effect level of exposure can easily be achieved by subjects living in seleniferous areas, working in an occupationally exposed environment, or self-supplementing with high doses of selenium compounds.

Investigations carried out in workers exposed to high levels of environmental selenium also showed an excess prevalence of dermatological abnormalities such as skin eruptions, redness and burns, generalized allergic body rash, nail pain (Glover, 1970; Rajotte et al., 1996), hair loss, and deformed nails (Srivastava et al., 1995), although not all studies have consistent findings (Holness et al., 1989). Also, acute and subacute selenium intoxications have been characterized by dermatological signs and symptoms such as extensive onycholysis, Mees lines and discoloration in nails, extensive hair loss and discoloration (Schuh and Jappe, 2007; Sutter et al., 2008; Lopez et al., 2010; MacFarquhar et al., 2010; Muller and Desel, 2010; Fitzgerald et al., 1997), and erythema of the hands (Pisati et al., 1988).

Overall, the epidemiologic evidence strongly indicates that selenium exposure in human induces adverse effects on skin and its annexes, hair and nails, and that such signs and symptoms represent early and in some cases rather selective indicators of selenium toxicity. Evidence from the recent large trial carried out in the United States described above also demonstrated that the amount of selenium required to exert such effects is much lower than previously considered.

Caries   The possibility that exposure to environmental selenium may enhance the risk of dental caries has been suggested by early field studies carried out in the seleniferous areas of the United States. These include earlier investigations (Smith and Westfall, 1936) and subsequent investigations specifically carried out to test this association in Oregon and Wyoming (Hadjimarkos, 1965; Hadjimarkos, 1968), plus those in the eastern United States (Ludwig and Bibby, 1969). The results of these studies have confirmed an association between dental caries and environmental selenium in children, possibly indicating a direct effect of the metalloid in increasing susceptibility of children to caries. These observations are corroborated by studies carried out in the seleniferous regions in India (Gauba et al., 1993) but not in

Venezuela (Jaffe et al., 1972) or occupationally exposed individuals (Rajotte et al., 1996), while findings in selenium-intoxicated individuals in China were equivocal (Yang et al., 1983). Overall, the association between selenium and caries is suggestive, but not established, and it needs to be further investigated. It is worth noting that animal studies have provided biological plausibility of an association between selenium overexposure and dental caries (Buttner, 1963; Shearer, 1975).

Neurological The first association between selenium exposure and neurological effects was seen in the early observations in animals in the seleniferous areas of the United States, which documented in livestock a wide spectrum of signs such as wandering, stumbling, blindness, ataxia, disorientation, generalized paralysis, and eventually death due to respiratory failure (Rosenfeld and Beath, 1964; Fan and Kizer, 1990). In humans, chronic selenium exposure has been subsequently linked to a wide spectrum of neurological signs and symptoms, mainly on the basis of studies carried out in selenium-intoxicated individuals living in the seleniferous areas of China (Yang et al., 1983; Fordyce, 1995). Observations in heavily selenium-intoxicated individuals from the seleniferous Enshi region in China showed the occurrence of several neurological signs and symptoms such as peripheral anesthesia, acroparaesthesia, pain to the extremities, tendon hyperreflexia, numbness, convulsions, motor disturbances, and hemiplegia (Yang et al., 1983). These observations, however, were confined only to residents in one heavily contaminated community, with a prevalence of clinically recognized selenosis exceeding 80% (Yang et al., 1983). Unfortunately, Yang et al. (1983) did not mention details of key epidemiologic relevance such as the overall extent of villages and individuals investigated in the survey and the prevalence of signs and symptoms ascribed to selenium exposure in unaffected (control) communities, thus hampering the estimation of the health risks actually attributable to the exposure to environmental selenium. Fordyce (1995), a British geologist who visited the Enshi seleniferous areas in 1995, reported that, on the basis of investigations by local public health officials, the occurrence of various symptoms (paralysis, lack of strength in limbs, and tingling limbs) ranged from 2 to 5% of the 180 individuals sampled from selenium-intoxicated communities in the Enshi area.

The key association between selenium exposure and neurological diseases is related to a rare but extremely severe, neurodegenerative disease involving motor neurons, amyotrophic lateral sclerosis (ALS), also known as Lou Gehrig's disease or motor neuron disease. The first indication of such an association dated back to a 1977 report in JAMA by a physician in private practice and a Harvard neurologist who described the occurrence of a cluster of ALS cases in a sparsely populated South Dakota region, known to be characterized by naturally occurring selenium intoxication in livestock (Kilness and Hochberg, 1977). The etiological significance of this finding was strongly debated for some time (Kurland, 1977; Norris and U, 1978), and limited attention was given to this observation, despite the detection in some studies of higher selenium levels in body tissues of ALS patients (Mitchell et al., 1991; Ince et al., 1994). Interest in the potential relationship was renewed when an excess ALS incidence was detected in an Italian cohort of consumers of drinking water with a selenium content of 8 µg/l (Vinceti et al., 1996), an observation recently confirmed by a case-control study carried out in the same area in a subsequent period (Vinceti et al., 2010).

Besides these epidemiological studies, the plausibility of a selenium–ALS causal relationship found suggestive support in animal studies, particularly those showing that selenium compounds induce selective degeneration of motor neurons in swine (Panter et al., 1996; Nathues et al., 2010; Raber et al., 2010) and possibly cattle (Maag et al., 1960) and horses (Polack et al., 2000). Under this perspective, selenium appears to be the only substance so far identified with a toxicity highly specific for the motor neurons, an observation of strong interest since the human disease ALS is characterized by a selective degeneration of the motor neurons. Furthermore, selenium compounds have been shown to induce *in vitro* selective biochemical abnormalities in neurons such as increased production of inducible nitric oxide synthase, reactive oxygen species (ROS) and 3-nitrotyrosine, as well as type-1 superoxide dismutase translocation into mitochondria; these effects may occur at a concentration as low as 8 µg/l of Se as selenite (Maraldi et al., 2011). Thus, selenium, particularly in its inorganic form, remains under strict scrutiny as a potential risk factor for induction of ALS. Paradoxically (but not unexpectedly due to the extremely controversial "history" of this metalloid in the health fields), selenium has also been proposed in the therapy of this disease as a putative "antioxidant" agent, with null results (Orrell et al., 2008).

Selenium has also been reported to be associated with Parkinson's disease though data are still sparse and unconvincing. In a case-control study, much higher selenium levels in both serum and cerebrospinal fluid were detected in Parkinson's patients compared with controls, irrespective of the clinical response to *L*-dopa therapy (Qureshi et al., 2006). In a cohort study examining mortality patterns in individuals with long-term consumption of drinking water containing an unusually high content of inorganic hexavalent selenium, mortality from Parkinson's disease was strongly increased in men, but not in woman, though the risk estimates were statistically very unstable due to the limited size of the population under study (Vinceti et al., 2000b).

In workers exposed to selenium, the incidence of dizziness and particularly paresthesia was found to be considerably increased compared with controls (Holness et al., 1989), and an association between occupational exposure to selenium and headache was suggested by one study (Srivastava et al., 1995), but not by a subsequent one (Rajotte et al., 1996). In a survey among U.S. consumers of drinking water with selenium content ranging from 194 to 494 µg/l compared with control subjects consuming drinking water with Se content around 2–3 µg/l, Valentine et al. found higher rates (6% vs. 1%) of diseases that were grouped in the general category of "diseases of nerves, paralysis or numbness, etc.," as well as a higher prevalence of depression (18% vs. 11%) and dizziness (22% vs. 10%) (Valentine et al., 1987).

Overall, available epidemiological data indicate that selenium may actually induce neurological abnormalities consisting of different signs and symptoms, with suggestive evidence linking selenium intake to a specific neurological disease, ALS. Since selenium neurotoxicity has been demonstrated in animal and *in vitro* studies (Nehru and Iyer, 1994; Panter et al., 1996; Nehru et al., 1997; Gupta and Porter, 2002; Casteignau et al., 2006; Xiao et al., 2006; Morgan et al., 2010; Maraldi et al., 2011), the possibility of a link between selenium and neurotoxicity in humans appears plausible. Under this perspective, inorganic selenium compounds appear to exert a much stronger neurotoxicity than the organic forms, as shown in experimental studies (Ammar and Couri, 1981; Tsunoda et al., 2000; Nogueira et al., 2003).

Reproductive/Developmental Toxicity The possibility that selenium increases the risk of birth defects in humans has long been suspected, mainly on the basis of observations in livestock in the seleniferous areas and of experimental studies (Willhite, 1993; Usami and Ohno, 1996), but studies in humans on this topic are limited and such evidence has not been substantiated in humans. The first observations on this issue were anecdotal reports of birth defects in a seleniferous area from Colombia (Jaffé and Vélez, 1973) and in female workers occupationally exposed to selenite (Robertson, 1970).

More recently, a few studies have examined, among other outcomes, the teratogenic effects of exposure to selenium. Fan et al. (1988) investigated exposure and health effects in a population potentially exposed to high levels of environmental selenium in the Kesterson National Wildlife Refuge area in northern California (where waterfowl deformities from high selenium exposure were found) and did not observe adverse outcomes, including birth defects. Vinceti et al. (2000) investigated the risk of birth defects, reduced fetal growth, and miscarriage among women exposed and unexposed to drinking water with a selenium content of 8 µg/l (Vinceti et al., 2000a). In that investigation, no evidence of alterations in body weight and length was apparent in 18 newborns to women consuming the high selenium drinking water, neither an excess prevalence of birth defects emerged among 353 newborns in exposed women during a 9-year period, compared with 14,481 control births. A statistically imprecise excess of clubfoot, however, was detected (2 observed cases vs. 0.51 expected), suggesting the need to further investigate this issue, also considering the

previous observation in an occupational environment (Robertson, 1970). The risk of miscarriage among exposed women, which was examined only for 1 year, increased compared to that experienced by unexposed women, though its estimate was statistically very unstable (relative risk 1.73, 95% confidence interval (CI) 0.62–4.80). Overall, the limited study size precluded an accurate estimation of the reproductive effects of such selenium exposure though no major changes in risk of birth defects were evident, and an indication of excess risk of miscarriage emerged. Two case-control studies were carried out in Massachusetts that aimed at identifying the possible association between drinking water chemical composition and adverse pregnancy outcomes (Zierler et al., 1988; Aschengrau et al., 1993). In one study, selenium levels in drinking water were strongly associated with excess risk of integument defects and to a reduced risk of nervous system defects (Aschengrau et al., 1993). In the other investigation, Zierler et al. detected an inverse relation between the drinking water selenium concentrations and risk of congenital heart defects (Zierler et al., 1988).

A possible influence of selenium on semen quality has also been the subject of scientific interest though the observational and intervention studies carried out on this subject yielded conflicting results. Hawkes et al. found that the supplementation with 297 µg/day selenium for 99 days was associated with a reduction in sperm motility of 18%, along with adverse effects of selenium supplementation on thyroid status (decreased serum triiodothyronine and increased thyroid-stimulating hormone levels), two findings of concern also due to the low amounts of additional exposure implicated (Hawkes and Turek, 2001). However, a subsequent study was unable to replicate the occurrence of adverse effects of selenium supplementation on sperm motility, and more generally semen quality, after long-term selenium supplementation (Hawkes et al., 2009). Other experimental and observational human studies have yielded mixed results though no clear pattern of adverse effects emerged (Keskes-Ammar et al., 2003; Akinloye et al., 2005; Wirth et al., 2007; Safarinejad and Safarinejad, 2009).

Overall, the epidemiologic evidence suggesting a relation between selenium and the reproductive health remains elusive and uncertain, and large studies on this issue, focusing among other outcomes on the risk of miscarriage, are clearly needed.

Diabetes The possibility that overexposure to selenium may favor the development of diabetes was first suggested in 1948 following the observation that a chemist acutely intoxicated with hydrogen selenide developed diabetes (Gouffault, 1948), and the subsequent demonstration in rabbits of a strong diabetogenic effect of selenite (Danon et al., 1963). However, this association was overlooked for several years until the report by Stranges and coworkers published in 2007, which examined diabetes risk as a secondary end point of a trial in nonmelanoma skin cancer patients carried out in the United States. During nearly 7 years of follow-up of 1202 patients, 58 subjects supplemented with 200 µg/day of organic selenium and 39 placebo-treated subjects were newly diagnosed with type 2 diabetes, yielding a relative risk of 1.55 (95% CI 1.03–2.33) in the treated group. A direct association between baseline selenium plasma levels and excess diabetes risk associated with selenium supplementation also emerged. Such observations immediately led to a strong concern about the possibility that supplementation with 200 µg/selenium, within a population likely characterized by a dietary intake of around 100 µg/day, could be dangerous (Bleys et al., 2007a, 2007b; Marcason, 2008), contrary to what was previously expected on the basis of some earlier studies and expert committees (Yang et al., 1989a; Yang and Zhou, 1994; NAS, 2000; Norden, 2004). This report was accompanied or followed by large cross-sectional and cohort studies aimed at clarifying the possible relation between selenium and the risk of diabetes and/or related metabolic abnormalities, the results of which supported in all cases such an association (Czernichow et al., 2006; Bleys et al., 2007a, 2007b; Gao et al., 2007; Laclaustra et al., 2009; Lippman et al., 2009; Arnaud et al., 2012; Stranges et al., 2010; Yang et al., 2010), except for one case (Akbaraly et al., 2010). The support of a diabetogenic effect of selenium came, in particular, from an additional experimental investigation carried out in the United States, the *Selenium and Vitamin E Cancer Prevention* (SELECT) trial (Lippman et al., 2009). SELECT was a randomized, placebo-controlled intervention study that aimed at evaluating a possible activity of selenium in preventing cancer and other chronic diseases. It was terminated due to the lack of any activity against prostate cancer or other diseases under investigation, as well as due to an excess risk from diabetes in supplemented individuals (relative risk 1.07, 99% CI 0.94–1.22) (Lippman et al., 2009). Thus, both the selenium trials so far carried out have shown an enhanced risk of diabetes among U.S. residents who received long-term supplementation of a daily dose of 200 µg of organic selenium, providing evidence that such an association was causal. The observational prospective studies also indicated that the amounts of selenium intake enhancing diabetes risk may be considerably lower than those implicated in the above trials that started at around 50 µg/day, thus providing evidence to decrease the allowed highest "safe" dose of dietary selenium (Stranges et al., 2010) and raising a great concern on the toxicity of this metalloid. Interestingly, a cohort study carried out in Rotterdam showed an excess diabetes risk among high consumers of fish, which is an excellent source of dietary selenium (van Woudenbergh et al., 2009). The authors stated that higher selenium intake could have been a cause of such excess diabetes risk, also considering that the adjustment for selenium intake in multivariate analysis decreased the association between fish intake and subsequent diabetes risk. Selenium intake has also been linked to increased concentrations of total and LDL cholesterol in the U.S. adult population (Bleys et al., 2008; Laclaustra et al., 2010), further suggesting a deleterious effect of even low amounts of this metalloid on the metabolic profile in humans.

These observations arising from epidemiologic studies have led to a careful scrutiny of the biological plausibility of such an association, recently reviewed and currently under active investigation (Steinbrenner et al., 2011). First, selenium compounds have been shown to induce several derangements of glucose homeostasis and metabolism (Rasekh et al., 1991; Potmis et al., 1993; Furnsinn et al., 1996) though not all studies have been consistent (Gronbaek et al., 1995; Sheng et al., 2005). Furthermore, studies conducted in plants (Malik et al., 2011) and in fish (Miller et al., 2009) have demonstrated that selenium exposure increases glucose concentrations and production under different experimental conditions. In addition, recent animal studies suggested that overexpression of selenium-dependent glutathione peroxidase represents a potential trigger of insulin resistance, hyperinsulinemia, and obesity, thus providing additional biological plausibility to a relation between selenium exposure and the risk of diabetes (McClung et al., 2004; Wang et al., 2008; Pepper et al., 2011).

Overall, evidence from both epidemiologic and laboratory studies indicates that selenium exposure, even at levels considerably lower than the approximate intake of 300 µg/day, which characterizes overall intake in the two randomized trials carried out in the United States, is associated with derangements of glucose homeostasis and an excess diabetes risk, supporting the conclusion that glucose metabolism may be a major target of selenium toxicity in humans.

Endocrine The possibility that selenium has deleterious effects on the endocrine system, apart from the above-mentioned possible effects on sperm quality and on the increased diabetes risk in particular, has been suggested by some studies originally performed in the seleniferous areas. In a pioneering study carried out in Venezuela, Brätter and Negretti de Brätter observed an inverse association between serum triiodothyronine and two indicators of selenium exposure, dietary intake and serum levels, defined as "significant" at amounts of dietary selenium of about 350 µg/day. More recently, a number of human studies have suggested deleterious effects on thyroid hormone status of both acute (Hofbauer et al., 1997) and chronic selenium exposure (Contempre et al., 1992; Thilly et al., 1992; Ramaekers et al., 1994; Calomme et al., 1995; Duffield et al., 1999). Supplementation of selenium to children with iodine deficiency and to phenylketonuria patients decreased thyroxine levels (Contempre et al., 1992; Thilly et al., 1992; Ramaekers et al., 1994; Calomme et al., 1995), and similar effects were observed in studies carried out in Italy (Olivieri et al., 1995) and New Zealand (Duffield et al., 1999; Thomson et al., 2005). On the contrary, other studies failed to show any association between administration of 100 μg of seleniomethionine for 60 days (Roti et al., 1993) or 90 days (Thomson et al., 2009) and changes of thyroid hormones status in euthyroid individuals. Overall, though conflicting results of the epidemiologic studies preclude definitive conclusions on the relation between selenium and thyroid hormones, it appears that overexposure even to low doses of selenium may adversely affect circulating levels of triiodothyronine and thyroxine in particular, and this might occur only in susceptible populations.

Selenium administration has also been associated with reversible inhibition of growth hormone secretion and irreversible suppression of insulin-like growth factor-1 by Danish investigators on the basis of some animal studies (Thorlacius-Ussing et al., 1987; Thorlacius-Ussing et al., 1988) and following the results of an experimental human study (Thorlacius-Ussing et al., 1989), in which patients with rheumatic disease were randomized to receive 256 μg/day of organic selenium or placebo for 6 months. Such observations raised a concern about the safety of the metalloid, but two subsequent observational studies (Salbe et al., 1993; Maggio et al., 2010) and one additional trial in humans (Meltzer and Haug, 1995) had been unable to confirm an association between organic selenium and growth hormone or insulin-like growth factor-1 inhibition. Such potential relations remain therefore elusive, and additional human studies are clearly needed to elucidate the possible induction of growth retardation by selenium by interfering with growth hormone secretion and activity (Vinceti et al., 2001).

Carcinogenicity The relation between selenium intake/exposure and cancer risk has been the subject of a large body of controversial, promising, and ultimately disappointing literature, with evidence from both epidemiologic and clinical studies ranging from the cancer-enhancing effect to preventive or therapeutic activities of the metalloid (Fan and Kizer, 1990; Clark and Alberts, 1995; Vinceti et al., 2000c; Letavayova et al., 2006; Brozmanova et al., 2010; Dennert et al., 2011). Though the occurrence of controversies on this issue is by no means unexpected due to the usual intriguing and conflicting results of selenium research on most health outcomes, the relation between this metalloid and cancer has generated an enormous body of interest during recent decades, followed by strong disappointment due to the negative results of a large intervention study carried out in the United States, the SELECT trial (Gann, 2009; Lippman et al., 2009).

Presently, the carcinogenicity of selenium compounds is still uncertain, since no substantial reassessment of this issue has been made in most recent years, despite the very large number of human and animal studies published on this topic in the latest two decades (Vinceti et al., 2000c; Lippman et al., 2009). According to major regulatory agencies or scientific bodies, selenium compounds are classifiable as members of

Group 3 by IARC ("Not classifiable as to its carcinogenicity to humans) and, in the form of selenium sulfide and selenium disulfide, as probable human carcinogens by the U.S. Environmental Protection Agency (EPA). Further substantial update of this topic by major regulatory agencies appear to be warranted by recent animal studies that have demonstrated the potential carcinogenicity of selenium compounds (Birt et al., 1989; Woutersen et al., 1999; Chen et al., 2000; Su et al., 2005) and by the substantial advancement of our knowledge of the biological activities of selenium.

The hypothesis of a cancer-preventing effect of selenium on human cancer has been substantially generated by ecologic investigations documenting inverse correlations between average soil selenium content and cancer death rates (Shamberger and Frost, 1969; Clark et al., 1991) and by a randomized trial carried out in the United States (Clark et al., 1996), coupled with laboratory data showing the ability of the metalloid to counteract cancer growth (Jackson and Combs, 2008). The clinical trial, in particular, randomized from 1983 through 1991 a group of 1312 patients in the eastern United States with a history of nonmelanoma skin cancer to receive either 200 μg/day of organic selenium or placebo until the end of the blinded period of treatment February 2, 1996, thus accumulating 9301 person-years of follow-up (Clark et al., 1996; Duffield-Lillico et al., 2002; Duffield-Lillico et al., 2003). This study did not document any beneficial effect of selenium supplementation on the primary end points, i.e., incidence of basal and squamous cell carcinomas of the skin, and it was even eventually associated with an increase in the latter skin cancer type and in overall skin cancer incidence (Duffield-Lillico et al., 2003). However, analysis of secondary end points showed among selenium-supplemented individuals a strong decrease in risk of all cancers and some site-specific cancers such as prostate and colon cancer (Clark et al., 1996; Duffield-Lillico et al., 2002). These findings, along with the new biochemical observations documenting the nutritional importance of selenium and the associated claims of its "antioxidant" properties, generated enormous interest on the metalloid and favored consumption by the general population of selenium-containing supplements, as well as implementation of several observational and experimental epidemiologic studies and laboratory investigations (Vinceti et al., 2000c; Lippman et al., 2009).

The previously mentioned SELECT trial, in particular, followed nearly 35,000 individuals from the United States, Canada, and Puerto Rico to assess if daily supplementation with 200 μg organic selenium in the form of *L*-selenomethionine could decrease the risk of prostate cancer as well as of other cancers. The study, started in July 2001, was unexpectedly discontinued in October 2008 long before its anticipated end date (2013) due to the lack of any effect in preventing cancer, as well as the excess incidence of diabetes in the 8752 supplemented subjects. Its findings have demonstrated that no positive effect in cancer prevention can be ascribed to selenium (Allen and Key, 2009), causing large disappointment in the public and the scientific world (Gann, 2009). Actually the relative risk of prostate cancer in the supplemented individuals was 1.04 (95% CI 0.90–1.18), and slight increases in risk were seen also for colorectal cancer (1.05, 99% CI 0.66–1.67) and lung cancer (1.12, 99% CI 0.73–1.72). In 2014, the SELECT investigators also reported that selenium supplementation increased the risk of high-grade prostate cancer, yielding a relative risk of 1.91 (95% CI 1.20–3.05) for individuals treated with selenium or selenium plus vitamin E supplements compared with placebo (Kristal et al., 2014). Attempts to attribute these null results to the chemical form of selenium or to the selenium status of the supplemented subjects have been subsequently made, but such objections appear to have little basis since the species of selenium used was the same (and in the same amount) as previously used in the trial showing strongly beneficial effects in skin cancer patients (Clark et al., 1996), and also the population was drawn from the same country, the United States.

Moreover, results of the SELECT trial are consistent with the lack of apparent beneficial effects on cancer risk in the Finnish population compared with other Nordic populations, despite the implementation of a nationwide program in Finland aiming at increasing selenium content of locally produced foodstuffs (Vinceti et al., 2000c). In a recent case-control study on vitamin and mineral use and the risk of prostate cancer, an increased risk was seen in long-term (≥10 years) consumers of selenium-containing supplements (relative risk 1.7, 95% CI 0.6–4.3) (Zhang et al., 2009), while no such association was found in two U.S. prospective studies (Lawson et al., 2007; Peters et al., 2008). Finally, a few studies have investigated the cancer risk of an Italian community where drinking water with hexavalent selenium (around 8 µg/l) was distributed (Vinceti et al., 1998; Vinceti et al., 2000b). These investigations have shown an excess risk of site-specific cancers in that community, particularly lymphoid malignancies, melanoma, colorectal cancer, and kidney cancer, raising concerns about the long-term safety of inorganic selenium through drinking water. Finally, the epidemiologic literature has been recently reviewed within a Cochrane Collaboration study group, and the results of this analysis have confirmed that selenium intake does not appear to have effects in cancer prevention (Vinceti et al., 2014).

Overall, the epidemiologic literature indicates that increasing selenium intake over the average daily human consumption in the United States and in Europe (where intake is usually lower) does not lower cancer risk, suggesting that the claim of "chemopreventive" effects of the metalloid were unwarranted, and that unmeasured confounders were likely responsible for the different results of the observational studies (Vinceti et al., 2000c), as well as of the secondary end points noted in the skin cancer patients trial of Clark et al. (1996). On the contrary, a suggestion of potential deleterious effects on cancer risk,

though only through evaluation of biochemical indicators of toxicity, has been provided by a small trial carried out in a Danish county, Funen, by supplementing subjects aged 60–74 with 100–300 µg/day of organic selenium (Ravn-Haren et al., 2008). In that cohort, selenium treatment induced a down-regulation of genes carrying electrophile response elements in the promoter regions, an effect potentially associated with increased cancer risk (Ravn-Haren et al., 2008). Laboratory studies yielded additional support to a potential toxic effect of even very low doses of selenium, particularly in its inorganic forms, due to its genotoxic (Bronzetti et al., 2001; Cemeli et al., 2006) and immunotoxic properties (Nair and Schwartz, 1990; Methenitou et al., 2001), the former effect possibly occurring also in humans (Chandra et al., 1993; Atasever et al., 2006). Finally, the enhanced incidence of skin cancer among selenium-supplemented individuals in the skin cancer trial by Clark et al. (1996) should not be underestimated also considering that this was the primary end point of that investigation, and data on this outcome from the SELECT trial would be of strong importance to evaluate the validity of this finding.

The hypothesis that selenium compounds can counteract cancer growth, acting as pharmacological agents, has also been suggested, on the basis of a number of laboratory studies, though it has not been tested to date in randomized clinical trials. The ability of the metalloid to inhibit cancer cell proliferation through its toxic properties is currently under active investigation (Nilsonne et al., 2009; Olm et al., 2009), as more generally the possibility to use the toxic properties of Se for pharmacological purposes.

In conclusion, the relation between selenium and human cancer remains elusive and unclear despite the enormous number of studies devoted to this issue. Based on the results of the recent SELECT trial, at least one hypothesis has been invalidated, i.e., the possibility that a moderate increase in dietary selenium in free-living individuals in a Western country may reduce cancer risk. On the other hand, open issues in research and public health include the possibility that extreme changes in the intake of the metalloid, i.e., very low and very high exposures, may increase cancer risk, that selenium may influence cancer risk according to its chemical species or only in susceptible or high risk individuals, and that selenium may be able to be used as a therapeutic agent in cancer therapy, calling for further research on these topics (Vinceti et al., 2000c; Letavayova et al., 2006; Brozmanova et al., 2010).

Glaucoma Additional diseases have been associated with excess selenium exposure both in occupational settings and in the general population though frequently on the basis of limited and unconvincing epidemiologic evidence. Among these associations, preliminary data from the first trial with selenium carried out among skin cancer patients (Clark et al., 1996) showed an excess risk of glaucoma among supplemented individuals, particularly in females, thus suggesting that excess selenium may represent a risk factor for glaucoma (Conley et al.,

2004). A study carried out to investigate this issue in a group of hospital-referred patients in Tucson, Arizona, showed a strongly increased relative risk of glaucoma (11.3, *P*-value 0.03) among subjects in the highest tertile of plasma selenium, and a dose–response relation between plasma selenium category and risk, thus confirming the hypothesis tested in the study of a glaucoma-enhancing effect of the metalloid (Bruhn et al., 2009). Surprisingly, subjects in the middle tertile of aqueous humor selenium exhibited a lower risk compared with subjects in the bottom tertile, while risk was not reduced in the highest category. The investigators expressed their concern about the possibility that high selenium may enhance glaucoma risk though they were aware of the study limitations, among which were the study size and the potential for referral bias. The need to further assess this potential end point of selenium toxicity (Bruhn et al., 2009) was also expressed (van Woudenbergh et al., 2009; Sogbesan and Damji, 2010). Biological plausibility supporting a relation between excess selenium exposure and glaucoma has been provided by an *in vitro* study, which showed that even very low doses of selenium decrease the secretion of metalloproteinases in human trabecular meshwork cells, suggesting the capacity of the metalloid to increase intraocular pressure and eventually to cause glaucoma (Conley et al., 2004).

Chemical Pathology

The toxicology of selenium that forms the basis of the chemical's pathology described herein is discussed above and in the derivation of the RfD below. The two principal clinical conditions that characterize selenosis are dermal and neurological effects observed in humans following chronic ingestion of high levels of selenium in food grown in seleniferous soil in China. The dermal manifestations of selenosis include hair loss and nail loss and deformity, while neurological effects include numbness, paralysis, and occasional hemiplegia. Nervous system abnormalities such as hemiplegia, peripheral anesthesia, acroparesthesia, and pain in extremities were observed in some cases. Nail and hair change and loss have also been observed in individuals in the United States, following ingestion of supplements with very high selenium content due to formulation error, and in cases of acute intoxication among residents of a seleniferous area of Venezuela who consumed the high selenium nuts of *L. ollaria*, known as "Coco de Mono."

Hair loss and hoof malformation in pigs, horses, and cattle, and poliomyelomalacia in pigs have been reported to occur following long-term exposure to excessive amounts of the organic selenium compounds found in seleniferous plants. Histologically, swines with selenium-induced neurological signs exhibit bilateral macroscopic lesions of the ventral horn of the spinal cord. Myocardial degeneration has been experimentally produced in cattle, sheep, and swine (as well as in laboratory mammals) by acute and longer-term exposures to inorganic salts of selenium, but it is unclear whether seleniferous grains or forages, or other natural sources of selenium, caused the same cardiomyopathy.

Some effects observed earlier in cattle, sheep, and horses that grazed on plants in areas of high selenium in soil in South Dakota, United States were believed to be associated with selenium. However, these neurological signs and histopathology have not been recorded in laboratory animals. The conditions include "blind staggers" and alkali disease. For blind staggers, the actual involvement of selenium or other compounds is unknown because the effects have not been replicated in experimentally exposed cattles receiving doses of selenium sufficient to induce hoof lesions. Alkali disease is a chronic disease in which the animals become emaciated, stiff, and lame; lose long hair from the mane and the tail; and the hooves become deformed. It is also associated with atrophy of the heart and liver. Congestion and focal necrosis of the liver are more prominent in "blind staggers." No studies were identified regarding definitive hepatic effects in humans after intermediate or long-term exposure to selenium compounds.

In fish, lordosis, kyphosis, scoliosis, and deformities of the head, mouth, and fin were observed. In waterfowl and mammals, abnormalities were observed in keratin-forming cells, which produce hard tissues such as hair, horn, beak, feathers, and nails.

Mode of Action(s)

Selenium in the body can be grouped into three main categories, namely, selenium in proteins, nonprotein selenium species, and selenoamino acids (Lobinski et al., 2000). Selenium functions largely through an association with proteins, known as selenoproteins. Selenoproteins refer only to selenocysteine-containing proteins and do not include selenium-binding proteins or selenomethionine-containing proteins. These proteins are required for health and development, playing a diverse role in various biochemical and physiological processes, including the biosynthesis of coenzyme Q (a component of the mitochondrial electron transport system), regulation of iron flux across membranes, maintenance of the integrity of keratins, and stimulation of antibody synthesis. The molecular basis for the essential role of selenium in higher vertebrates is being a component, in the form of the amino acid residue selenocysteine, of some major redox-regulating enzymes such as the selenium-dependent glutathione peroxidases (GPx) and thioredoxin reductase. The known biological functions include defense against oxidative stress, regulation of thyroid hormone action (selenium-dependent iodothyronine deiodinases), and regulation of the redox status of vitamin C and other molecules.

The possible mechanisms of selenium toxicity include substitution of selenium for sulfur in protein synthesis, inhibition of methylation metabolism resulting in selenide accumulation, or membrane and protein damage from selenium-generated ROS (Spallholz et al., 2004). Substitution of

Se for sulfur may damage sulfhydryl-containing enzymes involved in cellular respiration, affect mitochondrial and microsomal electron transport, and favor apoptotic and free radical formation (Kamble et al., 2009).

The specific biochemical mechanism(s) by which selenium and its compounds exert their acute toxicity is not known. In cases of chronic selenium overexposure, the major clinical manifestations are dermal/integumentary effects such as skin, hair, and nail damage. The mechanism causing these changes is unclear, but it has been postulated to be related to the high selenium concentrations in these tissues as a consequence of the substitution of selenium for sulfur in certain amino acids, including the disulfide bridges that provide tertiary structure and function to proteins. The nails and hair are considered to be routes for excretion of excess selenium (Yang et al., 1989a), and substitution of selenium for sulfur in keratin results in weakened physical protein structure and failure of keratinized tissues such as hair and hoof (Raisbeck, 2000). It might be possible that inactivation of the sulfhydryl enzymes necessary for oxidative reactions in cellular respiration, through effects on the mitochondrial and microsomal electron transport, contributes to selenium's acute toxicity.

Oxidative stress is the key biochemical lesion of selenium intoxication (Raisbeck, 2000; Spallholz et al., 2004). Selenium enzymes and selenoethers that do not readily form a selenide anion are not toxic. Methylation of selenium by both plants and animals serves to detoxify selenium by generating methylated selenides. Alternatively, reduction of selenium to elemental selenium as done by some bacteria and formation of heavy metal selenides such as mercuric selenide (Hg_2Se) results in a noncatalytic nontoxic form of selenium. The catalytic prooxidant attribute of some selenium compounds appears to account for their toxicity when such an activity exceeds plant and animal methylation reactions and antioxidant defenses.

SOURCES OF EXPOSURE

Occupational

Human exposures occur in occupational settings, especially in industries that extract, mine, treat, or process selenium-bearing minerals and in industries that use selenium or selenium compounds in manufacturing (ATSDR, 2003). The selenium compounds that are most likely to be encountered in air in occupational settings are dusts of elemental selenium, hydrogen selenide, and selenium dioxide. Deposits of compounds of selenium could be released off-site in dust or air. Occupational chronic exposure to selenium occurs mainly through the air and in some cases via direct dermal contact.

The National Institute for Occupational Safety and Health (NIOSH) conducted a National Occupation Hazard Survey (NOHS) and estimated that 108,682 workers in 15,127 plants were potentially exposed to selenium in the workplace in 1970 (NIOSH, 1976). NIOSH (1983) conducted a second workplace survey from 1981 to 1983, the National Occupational Exposure Survey (NOES), which indicated that 27,208 workers in 1102 plants were potentially exposed to selenium in the workplace. The majority of these workers were employed in the health services (e.g., nursing), as janitors and cleaners, as machine operators in the metals industry, or in work involving food and kindred products (NIOSH, 1989).

Selenium dioxide, hydrogen selenide, and selenium oxychloride are highly toxic and occupational exposure hazards. Selenium dioxide is formed when selenium is heated in air. It forms selenious acid on contact with water, including perspiration, and can cause severe irritation. Acute inhalation of large quantities of selenium dioxide powder can produce pulmonary edema from local irritant effect on alveoli (Glover, 1970), bronchial spasms, symptoms of asphyxiation, and bronchitis (Kinnigkeit, 1962; Wilson, 1962). Hydrogen selenide is a gas at room temperature with a density much higher than air. Selenium oxychloride is irritating and corrosive to the human respiratory tract because the compound hydrolyzes to hydrogen chloride (HCl), which can then form hydrochloric acid in humid air and in the respiratory tract (Dudley, 1938).

Environmental

Humans are exposed to low levels of selenium daily through food, water, and air, with the majority from dietary intakes, including dietary supplements. Higher exposures may result from consumption of locally grown grains and vegetables, animal products, fish and game in areas where some soils have naturally higher levels of selenium, or working in occupations noted above.

Children are exposed to selenium by the same pathways as adults, with ingestion of food sources as the primary route (Dudley, 1938). Selenium has been identified in pasteurized milk and milk-based infant formulas in the United States and in breast milk of women at different lactation stages. Soil is a potential route of exposure, especially in areas with naturally high selenium content. Young children have a tendency to ingest soil, either through pica or hand-to-mouth activity, and they often play in fields and soils, thus leading to possible dermal exposure and inhalation of dust particles from soil surfaces. The soluble forms such as the inorganic alkali selenites and selenates are more likely to be bioavailable in soils than the relatively insoluble selenides. Children are not likely to be exposed to selenium from their parents' work clothes, skin, hair, tools, or other objects removed from the workplace. Although selenium is contained in some household products such as shampoos and preparations to treat dandruff and eczema, and in some dietary supplements, the potential for overexposure from these products is low unless there is lack of adult supervision or if accidental poisoning occurs.

NUTRITION ESSENTIALITY

The understanding of selenium has been evolving over the last two decades. Selenium was originally known mainly for its high toxicity, then recognized as an essential trace element and studied for its protective role in human health and diseases, but more recently questions have been raised regarding its protective role and possible link to human diseases. Various aspects of toxicity and the studies of the potential role of selenium compounds in the prevention of specific cancers and antitumorigenic effects and its link to human diseases have been described above.

Selenium supplementation in animal nutrition led to the consideration of possible inadequate dietary selenium in human health. The essential functions of selenium for human and animal health are mediated through selenoproteins (discussed above). The Food and Nutrition Board of the Institute of Medicine has established human dietary reference values for selenium intakes for different age groups (NRC, 2000); these are presented above. These values include Estimated Average Requirement (EAR), Recommended Dietary Allowance (RDA), Tolerable Upper Intake Level (UL), and Adequate Intake (AI). The EAR and RDA are the lower bounds of intake compatible with health. The EAR is the intake at which the risk of inadequacy is 50% to an individual. The RDA is the intake at which the risk of inadequacy is 2–3%. The UL is the highest level of daily intake that is likely to pose no risk of adverse health effects (the risks of inadequacy and of excess are both close to zero). The AI is set instead of a RDA if the scientific evidence is not sufficient to calculate an EAR, and is applicable to young infants for whom human breast milk is the recommended sole source of food, based on the nutrient intake of apparently healthy, full-term infants who receive only human milk.

In the United States, most individuals readily meet their nutritional need for selenium. In Asia and Africa, low selenium status has been associated with human conditions such as Keshan disease (cardiomyopathy, a heart disease, responsive to selenium supplementation) and Kashin–Beck disease (osteoarthropathy, a degenerative bone disease). Patients on total parenteral nutrition for prolonged periods can be at a risk of developing selenium deficiency if there is inadequate selenium in the infusion fluids. Selenium deficiency seldom causes overt illness when it occurs in isolation but can lead to biochemical changes that predisposes to illness associated with other stresses (NRC, 2000).

The benefits of selenium nutritional supplements are actively being examined because of potential health issues raised in terms of adverse effects seen in some of the supplement trials or environmental exposures described above, plus the uncertainty about their effectiveness in cancer prevention (Kristal and Moe, 1998).

INDUSTRIAL HYGIENE

Selenium exists in various chemical forms (speciation) and oxidation states that can affect its occurrence, properties, use, and toxicity. Human exposures occur in the workplace, and via food and drinking water, through dermal, inhalation, or ingestion routes involving organic or inorganic forms. Ingestion of the organic form from food, including dietary supplements, is the main source of exposure for the general population. Evaluation of exposure and health effects of selenium has to consider both essentiality and toxicity. The occupational and environmental limits and reference values that are established are associated with the specific route of exposure and the chemical form, where data exist, as shown below.

> Occupational Safety and Health Administration (OSHA) permissible exposure limit (PEL): Time-weighted average (TWA) 0.2 mg/m^3 (elemental Se and other Se compounds, except Se hexafluoride)
>
> NIOSH recommended exposure limit (REL): TWA 0.2 mg/m^3 (elemental Se and other Se compounds, except Se hexafluoride)
>
> American Conference of Industrial Hygienists (ACGIH) TWA 0.2 mg/m^3 (elemental Se).

RISK ASSESSMENTS

Reference Dose (RfD) (U.S. EPA, 2011)

An oral reference dose (RfD) is an estimate (with uncertainty spanning perhaps an order of magnitude) of a daily exposure to the human population (including sensitive subgroups) that is likely to be without an appreciable risk of deleterious effects during a lifetime. An oral RfD of 5 µg/kg/day has been established by the U.S. EPA and presented above. It is based on clinical selenosis in the study of Yang et al. (1989a) at a no-observed effect level, NOEL, of 15 µg/kg/day and an uncertainty factor (UF) of 3. The no-observed-adverse-effect level (NOAEL) of 0.853 mg/day and lowest-observed-adverse-effect level (LOAEL) of 1.261 mg/day were calculated from regression analysis (log $Y = 0.767$ log $X - 2.248$, where $Y =$ blood selenium and $X =$ selenium intake) (Yang et al., 1989b) based upon the correlation ($r = 0.962$) between dietary selenium intake and blood selenium level for data showing incidence of clinical selenosis in adults with an average adult body weight of 55 kg (Yang et al., 1989a).

The study population included approximately 400 individuals living in an area of China with unusually high environmental concentrations of selenium. Three geographical areas with low, medium, and high selenium levels in the soil and food supply were chosen for comparison. The 1989 studies were a follow-up to an earlier study (Yang et al., 1983) conducted in response to endemic selenium intoxication but with a small sample size in the study area. The 1989 study individuals were evaluated for clinical and biochemical signs of selenium intoxication and tissue selenium levels. Selenium levels in soil and approximately 30 typical food types commonly eaten by the exposed population showed a positive correlation with blood and tissue selenium levels. The daily average selenium intakes, based on lifetime exposure, were 70, 195, and 1438 μg for adult males and 62, 198, and 1238 μg for adult females in the low, medium, and high selenium areas, respectively. To estimate a marginal safe level of daily intake, significant correlations demonstrated between selenium concentrations of various tissues were used to estimate the minimal daily intake values, which elicited various alterations in biochemical parameters indicative of possible selenium-induced liver dysfunction (i.e., prolongation of clotting time and serum glutathione titer) and clinical signs of selenosis (i.e., hair or nail loss, morphological changes of the nails, etc.). Persistent clinical signs of selenosis were observed only in 5/349 adults, a potentially sensitive subpopulation. The blood selenium concentration in this group ranged from 1.054 to 1.854 mg/l with a mean of 1.346 mg/l. Clinical signs observed included the characteristic "garlic odor" of excess selenium excretion in the breath and urine, thickened and brittle nails, hair and nail loss, lowered hemoglobin levels, mottled teeth, skin lesions, and CNS abnormalities (peripheral anesthesia, acroparesthesia, and pain in the extremities). Alterations in the measured biochemical parameters occurred at dietary intake levels of 750–850 μg/day, described as a delay in prothrombin time, i.e., increase in blood coagulation time and reduction in blood glutathione concentration. However, these indicators were poorly characterized and are not typically used as an index for clinical selenosis resulting from chronic exposure to selenium (NAS, 1989). Using the blood selenium levels to reflect clinical signs of selenium intoxication, a whole blood selenium concentration of 1.35 mg/corresponding to 1.261 mg of daily selenium intake is the LOAEL for overt signs of selenosis. The NOAEL with no clinical signs of selenosis is 1.0 mg/l, corresponding to 0.853 mg selenium/day.

Minimal Risk Level (MRL) (ATSDR, 2003)

A Minimal Risk level (MRL) represents a daily intake that Agency for Toxic Substances and Disease Registry (ATSDR) considers to be safe for all populations, not a threshold for toxicity. The exact point above the MRL at which effects might occur in sensitive individuals is uncertain. An MRL of 5 μg/kg/day was derived for chronic oral exposure (>365 days) to selenium. This MRL is based on a study (Yang and Zhou, 1994) that examined a group of five individuals recovering from selenosis and who were drawn from a larger population in an area of China where selenosis occurred (Yang et al., 1989a; Yang et al., 1989b). The 1994 study was a follow-up of the two 1989 studies, and thus the data used by ATSDR from the 1989 studies were the same as those by the U.S. EPA for deriving the RfD. Examinations were made on selenium levels in the diet, blood, nails, hair, urine, and milk of residents at three sites with low, medium, and high selenium in soil and food and the incidence of clinical symptoms of selenosis (morphological changes in fingernails) with dietary intake of selenium and selenium levels in blood. Selenium levels in blood were found to correspond to the dietary intake; symptoms of selenosis occurred at or above a selenium intake level of 910 μg/day (0.016 mg/kg/day) (Yang et al., 1989b). In 1992, Yang and Zhou (1994) reexamined five individuals from the high selenium area who were recovering (regrowing nails) from symptoms of selenosis (loss of fingernails and hair). Part of their diet from locally produced corn had been replaced with rice or cereals. The concentration of selenium in the blood of these individuals had lowered from 1346 μg/l (measured in 1986) to 968 μg/l (measured in 1992). Using a regression equation derived from the earlier data (Yang et al., 1989a), the dietary intake of selenium associated with selenosis in these individuals was calculated to be 1270 μg/day, while an intake of 819 μg/day was associated with recovery (Yang and Zhou, 1994). The chronic oral MRL is based on a NOAEL of 819 μg/day (0.015 mg/kg/day) for disappearance of symptoms of selenosis in recovering individuals (Yang and Zhou, 1994), an UF of 3 for human variability, and an average adult body weight of 55 kg (Yang et al., 1989a). An UF of 3 was considered appropriate because the individuals in this study were sensitive individuals drawn from a larger population and because of supporting studies.

Acceptable Daily Dose (ADD) (OEHHA, 2010)

The Office of Environmental Health Hazard Assessment, California Environmental Protection Agency, developed an acceptable daily dose (ADD) of 5 μg/kg/day (OEHHA, 2010) for selenium, which is in agreement to the RfD set by the U.S. EPA and the MRL set by the ATSDR as noted above, also based on the same studies (Yang et al., 1983; Yang et al., 1989a; Yang and Zhou, 1994). Selenosis was reported in approximately 400 adult villagers exposed to excess selenomethionine in locally grown grains and vegetables from seleniferous soils, and to a much lesser extent in the drinking water, in a remote mountainous region of China. A NOAEL of 15 μg/kg/day was derived based on measurements of selenium intakes and whole blood concentrations, a level leading to disappearance of selenosis symptoms in recovering adults (Yang and Zhou, 1994). An UF of 3 was used to account for potentially sensitive subpopulations such as infants, pregnant women and their fetuses, the elderly, the undernourished with respect to proteins and methionine, and patients with liver diseases, resulting in an ADD of 5 μg/kg/day.

REFERENCES

Akbaraly, T.N., Arnaud, J., Rayman, M.P., Hininger-Favier, I., Roussel, A.M., Berr, C., and Fontbonne, A. (2010) Plasma selenium and risk of dysglycemia in an elderly French population: results from the prospective Epidemiology of Vascular Ageing Study. *Nutr. Metab. (Lond.)* 7:21.

Akinloye, O., Arowojolu, A.O., Shittu, O.B., Adejuwon, C.A., and Osotimehin, B. (2005) Selenium status of idiopathic infertile Nigerian males. *Biol. Trace Elem. Res.* 104:9–18.

Allen, N.E. and Key, T.J. (2009) Prostate cancer: neither vitamin E nor selenium prevent prostate cancer. *Nat. Rev. Urol.* 6:187–188.

Ammar, E.M. and Couri, D. (1981) Acute toxicity of sodium selenite and selenomethionine in mice after ICV or IV administration. *Neurotoxicology* 2:383–386.

Arnaud, J., de Lorgeril, M., Akbaraly, T., Salen, P., Arnout, J., Cappuccio, F.P., van Dongen, M.C., Donati, M.B., Krogh, V., Siani, A., and Iacoviello, L. (2012) Gender differences in copper, zinc and selenium status in diabetic-free metabolic syndrome European population—the IMMIDIET study. *Nutr. Metab. Cardiovasc. Dis.* 22:517–524.

Aschengrau, A., Zierler, S., and Cohen, A. (1993) Quality of community drinking water and the occurrence of late adverse pregnancy outcomes. *Arch. Environ. Health* 48:105–113.

Atasever, B., Ertan, N.Z., Erdem-Kuruca, S., and Karakas, Z. (2006) *In vitro* effects of vitamin C and selenium on NK activity of patients with beta-thalassemia major. *Pediatr. Hematol. Oncol.* 23:187–197.

ATSDR (2003) *Toxicological Profile for Selenium*, Atlanta, GA: Agency for Toxic Substances and Disease Registry (ATSDR), U.S. Department of Health and Human Services.

Banuelos, G.S. and Meek, D.W. (1990) Accumulation of selenium in plants grown on selenium-treated soil. *J. Environ. Qual.* 19:772–777.

Bender, J., Gould, J.P., Vatcharapijarn, Y., and Saha, G. (1991) Uptake, transformation and fixation of Sc(VI) by a mixed selenium-tolerant ecosystem. *Water Air Soil Poll.* 59:359–367.

Birt, D.F., Pour, P.M., and Pelling, J.C. (1989) The influence of dietary selenium on colon, pancreas, and skin tumorigenesis. In: Wendel, A., editor. *Selenium in Biology and Medicine*, Berlin: Springer, pp. 297–304.

Bleys, J., Navas-Acien, A., and Guallar, E. (2007a) Selenium and diabetes: more bad news for supplements. *Ann. Intern. Med.* 147:271–272.

Bleys, J., Navas-Acien, A., and Guallar, E. (2007b) Serum selenium and diabetes in U.S. adults. *Diabetes Care* 30:829–834.

Bleys, J., Navas-Acien, A., Stranges, S., Menke, A., Miller, E.R., 3rd, and Guallar, E. (2008) Serum selenium and serum lipids in US adults. *Am. J. Clin. Nutr.* 88:416–423.

Bronzetti, G., Cini, M., Andreoli, E., Caltavuturo, L., Panunzio, M., and Croce, C.D. (2001) Protective effects of vitamins and selenium compounds in yeast. *Mutat. Res.* 496:105–115.

Brozmanova, J., Manikova, D., Vlckova, V., and Chovanec, M. (2010) Selenium: a double-edged sword for defense and offence in cancer. *Arch. Toxicol.* 84:919–938.

Bruhn, R.L., Stamer, W.D., Herrygers, L.A., Levine, J.M., and Noecker, R.J. (2009) Relationship between glaucoma and selenium levels in plasma and aqueous humour. *Br. J. Ophthalmol.* 93:1155–1158.

Buttner, W. (1963) Action of trace elements on metabolism of fluoride. *J. Dent. Res.* 43:453.

Calomme, M., Vanderpas, J., Francois, B., Van Caillie-Bertrand, M., Vanovervelt, N., Van Hoorebeke, C., and Vanden Berghe, D. (1995) Effects of selenium supplementation on thyroid hormone metabolism in phenylketonuria subjects on a phenylalanine restricted diet. *Biol. Trace Elem. Res.* 47: 349–353.

Casteignau, A., Fontan, A., Morillo, A., Oliveros, J.A., and Segales, J. (2006) Clinical, pathological and toxicological findings of a iatrogenic selenium toxicosis case in feeder pigs. *J. Vet. Med. A Physiol. Pathol. Clin. Med.* 53:323–326.

CDC (1994) Selenium compounds (as Se). In: *Documentation for Immediately Dangerous to Life or Health Concentrations (IDLHs)*, Atlanta, GA: National Institute for Occupational Safety and Health (NIOSH), Centers for Disease Control and Prevention (CDC).

Cemeli, E., Marcos, R., and Anderson, D. (2006) Genotoxic and antigenotoxic properties of selenium compounds in the *in vitro* micronucleus assay with human whole blood lymphocytes and TK6 lymphoblastoid cells. *Sci. World J.* 6:1202–1210.

Chandra, R.K., Hambreaus, L., Puri, S., Au, B., and Kutty, K.M. (1993) Immune response of healthy volunteers given supplements of zinc or selenium. *FASEB J.* 7:A23.

Chen, X., Mikhail, S.S., Ding, Y.W., Yang, G.Y., Bondoc, F., and Yang, C.S. (2000) Effects of vitamin E and selenium supplementation on esophageal adenocarcinogenesis in a surgical model with rats. *Carcinogenesis* 21:1531–1536.

Civil, I.D. and McDonald, M.J. (1978) Acute selenium poisoning: case report. *N. Z. Med. J.* 87:354–356.

Clark, L.C. and Alberts, D.S. (1995) Selenium and cancer: risk or protection? *J. Natl. Cancer Inst.* 87:473–475.

Clark, L.C., Combs, G.F.J., Turnbull, B.W., Slate, E.H., Chalker, D.K., Chow, J., Davis, L.S., Glover, R.A., Graham, G.F., Gross, E.G., Krongrad, A., Lesher, J.L. J., Park, H.K., Sanders, B.B.J., Smith, C.L., and Taylor, J.R. (1996) Effects of selenium supplementation for cancer prevention in patients with carcinoma of the skin. A randomized controlled trial. Nutritional Prevention of Cancer Study Group. *JAMA* 276:1957–1963.

Clark, L.C., Cantor, K.P., and Allaway, W.H. (1991) Selenium in forage crops and cancer mortality in U.S. counties. *Arch. Environ. Health* 46:37–42.

Conley, S.M., Bruhn, R.L., Morgan, P.V., and Stamer, W.D. (2004) Selenium's effects on MMP-2 and TIMP-1 secretion by human trabecular meshwork cells. *Invest. Ophthalmol. Vis. Sci.* 45:473–479.

Contempre, B., Duale, N.L., Dumont, J.E., Ngo, B., Diplock, A.T., and Vanderpas, J. (1992) Effect of selenium supplementation on thyroid hormone metabolism in an iodine and selenium deficient population. *Clin. Endocrinol.* 36:579–583.

Coudray, C., Hida, H., Boucher, F., Tirard, V., de Leiris, J., and Favier, A. (1996) Effect of selenium supplementation on biological constants and antioxidant status in rats. *J. Trace Elem. Med. Biol.* 10:12–19.

CRC (2000) In: Lide, D.R., editor. *CRC Handbook of Chemistry and Physics*, 81st ed., New York, NY: CRC Press, pp. 4–27.

Czernichow, S., Couthouis, A., Bertrais, S., Vergnaud, A.C., Dauchet, L., Galan, P., and Hercberg, S. (2006) Antioxidant supplementation does not affect fasting plasma glucose in the Supplementation with Antioxidant Vitamins and Minerals (SU. VI. MAX) study in France: association with dietary intake and plasma concentrations. *Am. J. Clin. Nutr.* 84:395–399.

Danon, G., Paulet, G., and Cormier, M. (1963) Diabetogenic action of sodium selenite. *C. R. Seances Soc. Biol. Fil.* 157:51–54.

Dimes, L., Rendig, V.V., Besgu, G., and Dong, A. (1988) Selenium uptake by subclover, ryegrass, and some *Astragalus* spp. In: Tanji, K.K., Valoppi, L., and Woodring, R.C., editors. *Selenium Contents in Animal and Human Food Crops Grown in California*, Oakland, CA: Cooperative Extension University of California, Division of Agriculture and Natural Resources, pp. 19–24.

Dudley, H.C. (1938) Toxicology of selenium. Toxic and vesicant properties of Se oxychloride. *Public Health Rep.* 53:94–98.

Duffield, A.J., Thomson, C.D., Hill, K.E., and Williams, S. (1999) An estimation of selenium requirements for New Zealanders. *Am. J. Clin. Nutr.* 70:896–903.

Duffield-Lillico, A.J., Reid, M.E., Turnbull, B.W., Combs, G.F. J., Slate, E.H., Fischbach, L.A., Marshall, J.R. and Clark, L.C. (2002) Baseline characteristics and the effect of selenium supplementation on cancer incidence in a randomized clinical trial: a summary report of the Nutritional Prevention of Cancer Trial. *Cancer Epidemiol. Biomarkers Prev.* 11:630–639.

Duffield-Lillico, A.J., Slate, E.H., Reid, M.E., Turnbull, B.W., Wilkins, P.A., Combs, G.F., Jr., Park, H.K., Gross, E.G., Graham, G.F., Stratton, M.S., Marshall, J.R., and Clark, L.C. (2003) Selenium supplementation and secondary prevention of nonmelanoma skin cancer in a randomized trial. *J. Natl. Cancer Inst.* 95:1477–1481.

Fan, A.M., Book, S.A., Neutra, R.R., and Epstein, D.M. (1988) Selenium and human health implications in California's San Joaquin Valley. *J. Toxicol. Environ. Health* 23:539–559.

Fan, A.M. and Kizer, K.W. (1990) Selenium. Nutritional, toxicologic, and clinical aspects. *West. J. Med.* 153:160–167.

Fishbein, L. (1983) Environmental selenium and its significance. *Fundam. Appl. Toxicol.* 3:411–419.

Fitzgerald, E.A., Purcell, S.M., and Goldman, H.M. (1997) Green hair discoloration due to selenium sulfide. *Int. J. Dermatol.* 36:238–239.

Fordyce, F.M. (1995) *Report of Field Visit and Initial Data from Investigations into the Prediction and Remediation of Human Selenium Imbalances in Enshi District, Hubei Province, China, 8–16 November 1995*. Technical Report WC/96/7R, Nottingham, UK: British Geological Survey.

Furnsinn, C., Englisch, R., Ebner, K., Nowotny, P., Vogl, C., and Waldhausl, W. (1996) Insulin-like vs. non-insulin-like stimulation of glucose metabolism by vanadium, tungsten, and selenium compounds in rat muscle. *Life Sci.* 59:1989–2000.

Galgan, V. and Frank, A. (1995) Survey of bioavailable selenium in Sweden with the moose (*Alces alces* L.) as monitoring animal. *Sci. Total Environ.* 172:37–45.

Gann, P.H. (2009) Randomized trials of antioxidant supplementation for cancer prevention: first bias, now chance—next, cause. *JAMA* 301:102–103.

Gao, S., Jin, Y., Hall, K.S., Liang, C., Unverzagt, F.W., Ji, R., Murrell, J.R., Cao, J., Shen, J., Ma, F., Matesan, J., Ying, B., Cheng, Y., Bian, J., Li, P., and Hendrie, H.C. (2007) Selenium level and cognitive function in rural elderly Chinese. *Am. J. Epidemiol.* 165:955–965.

Gauba, K., Tewari, A., and Chawla, H.S. (1993) Role of trace elements Se and Li in drinking water on dental caries experience. *J. Indian. Soc. Pedod. Prev. Dent.* 11:15–19.

Glover, J.R. (1970) Selenium and its industrial toxicology. *IMS Ind. Med. Surg.* 39:50–54.

Gouffault, J. (1948) *Métabolisme des glucides et intoxication par le sélénium*. Thése de Médicine, n. 593, Paris.

Gronbaek, H., Frystyk, J., Orskov, H., and Flyvbjerg, A. (1995) Effect of sodium selenite on growth, insulin-like growth factor-binding proteins and insulin-like growth factor-I in rats. *J. Endocrinol.* 145:105–112.

Gupta, N. and Porter, T.D. (2002) Inhibition of human squalene monooxygenase by selenium compounds. *J. Biochem. Mol. Toxicol.* 16:18–23.

Hadjimarkos, D.M. (1965) Effect of selenium on dental caries. *Arch. Environ. Health* 10:893–899.

Hadjimarkos, D.M. (1968) Effect of trace elements on dental caries. *Adv. Oral. Biol.* 3:253–292.

Hawkes, W.C., Alkan, Z., and Wong, K. (2009) Selenium supplementation does not affect testicular selenium status or semen quality in North American men. *J. Androl.* 30:525–533.

Hawkes, W.C. and Turek, P.J. (2001) Effects of dietary selenium on sperm motility in healthy men. *J. Androl.* 22:764–772.

Hira, C.K., Partal, K., and Dhillon, K.S. (2004) Dietary selenium intake by men and women in high and low selenium areas of Punjab. *Public Health Nutr.* 7:39–43.

Hofbauer, L.C., Spitzweg, C., Magerstadt, R.A., and Heufelder, A.E. (1997) Selenium-induced thyroid dysfunction. *Postgrad. Med. J.* 73:103–104.

Hoffman, J.E. and King, M.G. (1997) Selenium and selenium compounds. In: Kroschwitz, J.I. and Howe-Grant, M.H., editors.

Encyclopedia of Chemical Technology, 4th ed., New York, NY: John Wiley & Sons, Inc., pp. 686–719.

Holness, D.L., Taraschuk, I.G., and Nethercott, J.R. (1989) Health status of copper refinery workers with specific reference to selenium exposure. *Arch. Environ. Health* 44:291–297.

Ince, P.G., Shaw, P.J., Candy, J.M., Mantle, D., Tandon, L., Ehmann, W.D., and Markesbery, W.R. (1994) Iron, selenium and glutathione peroxidase activity are elevated in sporadic motor neuron disease. *Neurosci. Lett.* 182:87–90.

IPCS (2001) *Selenium*. Poison Information Monograph 483, International Programme on Chemical Safety (IPCS), Inter-Organization Programme for the Sound Management of Chemicals (IOMC).

Jackson, M.I. and Combs, G.F., Jr. (2008) Selenium and anticarcinogenesis: underlying mechanisms. *Curr. Opin. Clin. Nutr. Metab. Care* 11:718–726.

Jaffe, W.G., Ruphael, M., Mondragon, M.C., and Cuevas, M.A. (1972) Estudio clinico y bioquimico en ninos escolares de una zona selenifera. *Arch. Latinoam. Nutr.* 22:595–611.

Jaffé, W.G. and Vélez, F.B. (1973) Selenium intake and congenital malformations in humans. *Arch. Latinoam. Nutr.* 23:514–516.

Kamble, P., Mohsin, N., Jha, A., Date, A., Upadhaya, A., Mohammad, E., Khalil, M., Pakkyara, A., and Budruddin, M. (2009) Selenium intoxication with selenite broth resulting in acute renal failure and severe gastritis. *Saudi J. Kidney Dis. Transpl.* 20:106–111.

Kerdel-Vegas, F. (1966) The depilatory and cytotoxic action of "Coco do Mono" (*Lecythis ollaria*) and its relationship to chronic selenosis. *Econ. Bot.* 20:187–195.

Keskes-Ammar, L., Feki-Chakroun, N., Rebai, T., Sahnoun, Z., Ghozzi, H., Hammami, S., Zghal, K., Fki, H., Damak, J., and Bahloul, A. (2003) Sperm oxidative stress and the effect of an oral vitamin E and selenium supplement on semen quality in infertile men. *Arch. Androl.* 49:83–94.

Kilness, A.W. and Hochberg, F.H. (1977) Amyotrophic lateral sclerosis in a high selenium environment. *JAMA* 237: 2843–2844.

Kinnigkeit, G. (1962) Investigation of workers exposed to selenium in a factory producing rectifiers. *Bull. Hyg. (Lond.)* 37:1029–1030 (German).

Kristal, A.R., Darke, A.K., Morris, J.S., Tangen, C.M., Goodman, P.J., Thompson, I.M., Meyskens, F.L. Jr, Goodman, G.E., Minasian, L.M., Parnes, H.L., Lippman, S.M., and Klein, E.A. (2014) Baseline selenium status and effects of selenium and vitamin e supplementation on prostate cancer risk. *J. Natl. Cancer Inst.* 106:djt456.

Kristal, A.R. and Moe, G. (1998) False presumptions and continued surprises: how much do we really know about nutritional supplements and cancer risk? *Cancer Epidemiol. Biomarkers Prev.* 7:849–850; discussion 851–852.

Kurland, L.T. (1977) Amyotrophic lateral sclerosis and selenium. *JAMA* 238:2365–2366.

Laclaustra, M., Navas-Acien, A., Stranges, S., Ordovas, J.M., and Guallar, E. (2009) Serum selenium concentrations and diabetes in U.S. adults: National Health and Nutrition Examination Survey (NHANES) 2003–2004. *Environ. Health Perspect.* 117:1409–1413.

Laclaustra, M., Stranges, S., Navas-Acien, A., Ordovas, J.M., and Guallar, E. (2010) Serum selenium and serum lipids in US adults: National Health and Nutrition Examination Survey (NHANES) 2003–2004. *Atherosclerosis* 210:643–648.

Lawson, K.A., Wright, M.E., Subar, A., Mouw, T., Hollenbeck, A., Schatzkin, A., and Leitzmann, M.F. (2007) Multivitamin use and risk of prostate cancer in the National Institutes of Health-AARP Diet and Health Study. *J. Natl. Cancer Inst.* 99:754–764.

Lemley, R.E. (1940) Selenium poisoning in the human. A preliminary case report. *J. Lancet* 60:528–531.

Lemley, R.E. and Merryman, M.P. (1941) Selenium poisoning in the human. *J. Lancet* 61:435–438.

Lemly, A.D. (1985) Toxicology of selenium in a freshwater reservoir: implications for environmental hazard evaluation and safety. *Ecotoxicol. Environ. Saf.* 10:314–338.

Lemly, A.D. (1993) Teratogenic effects of selenium in natural populations of freshwater fish. *Ecotoxicol. Environ. Saf.* 26:181–204.

Lemly, A.D. (1996) Evaluation of the hazard quotient method for risk assessment of selenium. *Ecotoxicol. Environ. Saf.* 35: 156–162.

Lemly, A.D. (1998) In: Alexander, D.E., editor. *Selenium Assessment in Aquatic Ecosystems: A Guide for Hazard Evaluation and Water Quality*, 1st ed., New York, NY: Springer.

Letavayova, L., Vlckova, V., and Brozmanova, J. (2006) Selenium: from cancer prevention to DNA damage. *Toxicology* 227:1–14.

Lippman, S.M., Klein, E.A., Goodman, P.J., Lucia, M.S., Thompson, I.M., Ford, L.G., Parnes, H.L., Minasian, L.M., Gaziano, J.M., Hartline, J.A., Parsons, J.K., Bearden, J.D., 3rd, Crawford, E.D., Goodman, G.E., Claudio, J., Winquist, E., Cook, E.D., Karp, D.D., Walther, P., Lieber, M.M., Kristal, A.R., Darke, A.K., Arnold, K.B., Ganz, P.A., Santella, R.M., Albanes, D., Taylor, P.R., Probstfield, J.L., Jagpal, T.J., Crowley, J.J., Meyskens, F.L., Jr., Baker, L.H., and Coltman, C.A., Jr. (2009) Effect of selenium and vitamin E on risk of prostate cancer and other cancers: the Selenium and Vitamin E Cancer Prevention Trial (SELECT). *JAMA* 301:39–51.

Lobinski, R., Edmonds, J.S., Suzuki, K.T., and Uden, P.C. (2000) Species-selective determination of selenium compounds in biological materials (technical report). *Pure Appl. Chem.* 72: 447–461.

Longnecker, M.P., Taylor, P.R., Levander, O.A., Howe, M., Veillon, C., McAdam, P.A., Patterson, K.Y., Holden, J.M., Stampfer, M.J., Morris, J.S., et al. (1991) Selenium in diet, blood, and toenails in relation to human health in a seleniferous area. *Am. J. Clin. Nutr.* 53:1288 1294.

Lopez, R.E., Knable, A.L., Jr., and Burruss, J.B. (2010) Ingestion of a dietary supplement resulting in selenium toxicity. *J. Am. Acad. Dermatol.* 63:168–169.

Ludwig, T.G. and Bibby, B.G. (1969) Geographic variations in the prevalence of dental caries in the United States of America. *Caries Res.* 3:32–43.

Maag, D.D., Orsborn, J.S., and Clopton, J.R. (1960) The effect of sodium selenite on cattle. *Am. J. Vet. Res.* 21:1049–1053.

MacFarquhar, J.K., Broussard, D.L., Melstrom, P., Hutchinson, R., Wolkin, A., Martin, C., Burk, R.F., Dunn, J.R., Green, A.L.,

Hammond, R., Schaffner, W., and Jones, T.F. (2010) Acute selenium toxicity associated with a dietary supplement. *Arch. Intern. Med.* 170:256–261.

Maggio, M., Ceda, G.P., Lauretani, F., Bandinelli, S., Dall'Aglio, E., Guralnik, J.M., Paolisso, G., Semba, R.D., Nouvenne, A., Borghi, L., Ceresini, G., Ablondi, F., Benatti, M., and Ferrucci, L. (2010) Association of plasma selenium concentrations with total IGF-1 among older community-dwelling adults: the InCHIANTI study. *Clin. Nutr.* 29:674–647.

Maier, K.J., Foe, C., Ogle, R.S., Williams, M.J., Knight, A.W., Kiffney, P., and Melton, L.A. (1988) The dynamics of selenium in aquatic ecosystems. In: Hemphill, D.D., editor. *Trace Substances in Environmental Health, XXI*, Columbia, MO: University of Missouri, pp. 361–408.

Malik, J.A., Kumar, S., Thakur, P., Sharma, S., Kaur, N., Kaur, R., Pathania, D., Bhandhari, K., Kaushal, N., Singh, K., Srivastava, A., and Nayyar, H. (2011) Promotion of growth in mungbean (*Phaseolus aureus* Roxb.) by selenium is associated with stimulation of carbohydrate metabolism. *Biol. Trace Elem. Res.* 143:530–539.

Manville, I.A. (1939) The selenium problem and its relationship to public health. *Am. J. Public Health Nations Health* 29:709–719.

Maraldi, T., Riccio, M., Zambonin, L., Vinceti, M., De Pol, A., and Hakim, G. (2011) Low levels of selenium compounds are selectively toxic for a human neuron cell line through ROS/RNS increase and apoptotic process activation. *Neurotoxicology* 32:180–187.

Marcason, W. (2008) What is the latest research on the connection between selenium and diabetes? *J. Am. Diet. Assoc.* 108:188.

McClung, J.P., Roneker, C.A., Mu, W., Lisk, D.J., Langlais, P., Liu, F., and Lei, X.G. (2004) Development of insulin resistance and obesity in mice overexpressing cellular glutathione peroxidase. *Proc. Natl. Acad. Sci. USA* 101:8852–8857.

Meltzer, H.M. and Haug, E. (1995) Oral intake of selenium has no effect on the serum concentrations of growth hormone, insulin-like growth factor-1, insulin-like growth factor-binding proteins 1 and 3 in healthy women. *Eur. J. Clin. Chem. Clin. Biochem.* 33:411–415.

Methenitou, G., Maravelias, C., Athanaselis, S., Dona, A., and Koutselinis, A. (2001) Immunomodulative effects of aflatoxins and selenium on human natural killer cells. *Vet. Hum. Toxicol.* 43:232–234.

Miller, L.L., Rasmussen, J.B., Palace, V.P., and Hontela, A. (2009) Physiological stress response in white suckers from agricultural drain waters containing pesticides and selenium. *Ecotoxicol. Environ. Saf.* 72:1249–1256.

Mitchell, J.D., East, B.W., Harris, I.A., and Pentland, B. (1991) Manganese, selenium and other trace elements in spinal cord, liver and bone in motor neurone disease. *Eur. Neurol.* 31:7–11.

Morgan, K.L., Estevez, A.O., Mueller, C.L., Cacho-Valadez, B., Miranda-Vizuete, A., Szewczyk, N.J., and Estevez, M. (2010) The glutaredoxin GLRX-21 functions to prevent selenium-induced oxidative stress in *Caenorhabditis elegans*. *Toxicol. Sci.* 118:530–543.

Muller, D. and Desel, H. (2010) Acute selenium poisoning by paradise nuts (*Lecythis ollaria*). *Hum. Exp. Toxicol.* 29:431–434.

Nair, M.P. and Schwartz, S.A. (1990) Immunoregulation of natural and lymphokine-activated killer cells by selenium. *Immunopharmacology* 19:177–183.

NAS (1976) *Selenium: Medical and Biological Effects of Environmental Pollutants*. Selenium Subcommittee, Washington, DC: National Academy of Sciences (NAS).

NAS (1977) Inorganic solutes. *Drinking Water and Health*. National Research Council Safe Drinking Water Committee, Washington, DC: National Academy Sciences (NAS), pp. 205–488.

NAS (1980) The contribution of drinking water to mineral nutrition in humans. *Drinking Water and Health*, vol. 3. National Research Council Safe Drinking Water Committee, Washington, DC: National Academy of Sciences (NAS), pp. 326–344.

NAS (1989) Trace elements. In: *Recommended Dietary Allowances*, 10th ed., Subcommittee on the Tenth Edition of the Recommended Dietary Allowances, Food and Nutrition Board, Commission on Life Sciences, National Research Council. Washington, DC: National Academy of Sciences (NAS), pp. 217–224.

NAS (2000) *Dietary References Intakes for Vitamin C, Vitamin E, Selenium, and Carotenoids*, Washington, DC: National Academy of Sciences (NAS).

Nathues, H., Boehne, I., Grosse Beilage, T., Gerhauser, I., Hewicker-Trautwein, M., Wolf, P., Kamphues, J., and Grosse Beilage, E. (2010) Peracute selenium toxicosis followed by sudden death in growing and finishing pigs. *Can. Vet. J.* 51:515–518.

Nehru, B., Dua, R., and Iyer, A. (1997) Effect of selenium on lead-induced alterations in rat brain. *Biol. Trace Elem. Res.* 57:251–258.

Nehru, B. and Iyer, A. (1994) Effect of selenium on lead-induced neurotoxicity in different brain regions of adult rats. *J. Environ. Pathol. Toxicol. Oncol.* 13:265–268.

Nilsonne, G., Olm, E., Szulkin, A., Mundt, F., Stein, A., Kocic, B., Rundlof, A.K., Fernandes, A.P., Bjornstedt, M., and Dobra, K. (2009) Phenotype-dependent apoptosis signalling in mesothelioma cells after selenite exposure. *J. Exp. Clin. Cancer Res.* 28:92.

NIOSH (1976) *National Occupational Hazard Survey (1970)*, Atlanta, GA: National Institute for Occupational Safety and Health.

NIOSH (1983) *National Occupational Exposure Survey (NOES)*, Atlanta, GA: National Institute for Occupational Safety and Health.

NIOSH (1989) *National Occupational Exposure Survey*, Atlanta, GA: National Institute for Occupational Safety and Health.

Nogueira, C.W., Meotti, F.C., Curte, E., Pilissao, C., Zeni, G., and Rocha, J.B. (2003) Investigations into the potential neurotoxicity induced by diselenides in mice and rats. *Toxicology* 183:29–37.

Norden (2004) *Nordic Nutrition Recommendations 2004. Integrating Nutrition and Physical Activity*, 4th ed., Copenhagen, Denmark: Nordic Council of Ministers.

Norris, F.H., Jr. and U, K.S. (1978) Amyotrophic lateral sclerosis and low urinary selenium levels. *JAMA* 239:404.

NRC (2000) *Dietary Reference Intakes for Vitamin C, Vitamin E, Selenium, and Carotenoids*. Food and Nutrition Board, Institute of Medicine, National Research Council (NRC), Washington, DC: National Academy Press.

OEHHA (2010) *Public health goal for selenium in drinking water.* Office of Environmental Health Hazard Assessment (OEHHA), California Environmental Protection Agency (Cal EPA).

Ohlendorf, H.M., Hoffman, D.J., Saiki, M.K., and Aldrich, T.W. (1986a) Embryonic mortality and abnormalities of aquatic birds. *Sci. Total Environ.* 52:49–63.

Ohlendorf, H.M., Lowe, R.W., Kelly, P.R., and Harvey, T.E. (1986b) Selenium and heavy metals in San Francisco Bay diving ducks. *J. Wildl. Manage.* 50:64–71.

Olivieri, O., Girelli, D., Azzini, M., Stanzial, A.M., Russo, C., Ferroni, M., and Corrocher, R. (1995) Low selenium status in the elderly influences thyroid hormones. *Clin. Sci.* 89:637–642.

Olm, E., Jonsson-Videsater, K., Ribera-Cortada, I., Fernandes, A.P., Eriksson, L.C., Lehmann, S., Rundlof, A.K., Paul, C., and Bjornstedt, M. (2009) Selenite is a potent cytotoxic agent for human primary AML cells. *Cancer Lett.* 282:116–123.

Orrell, R.W., Lane, R.J., and Ross, M. (2008) A systematic review of antioxidant treatment for amyotrophic lateral sclerosis/motor neuron disease. *Amyotroph. Lateral Scler.* 9:195–211.

O'Toole, D. and Raisbeck, M.F. (1998) Magic numbers, elusive lesions: comparative pathology and toxicology of selenosis in waterfowl and mammalian species. In: Frankenberger, W.T.J. and Engberg, R.A., editors. *Environmental Chemistry of Selenium, Books in Soils, Plants, and the Environment*, New York, NY: Marcel Dekker, Inc., pp. 355–395.

Panter, K.E., Hartley, W.J., James, L.F., Mayland, H.F., Stegelmeier, B.L., and Kechele, P.O. (1996) Comparative toxicity of selenium from seleno-DL-methionine, sodium selenate, and *Astragalus bisulcatus* in pigs. *Fundam. Appl. Toxicol.* 32:217–223.

Pepper, M.P., Vatamaniuk, M.Z., Yan, X., Roneker, C.A., and Lei, X.G. (2011) Impacts of dietary selenium deficiency on metabolic phenotypes of diet-restricted GPX1-overexpressing mice. *Antioxid. Redox Signal.* 14:383–390.

Peters, U., Littman, A.J., Kristal, A.R., Patterson, R.E., Potter, J.D., and White, E. (2008) Vitamin E and selenium supplementation and risk of prostate cancer in the Vitamins and Lifestyle (VITAL) study cohort. *Cancer Causes Control* 19:75–87.

Pisati, G., Baruffini, A., Galli, C., Riboldi, L., and Tomasini, M. (1988) Selenious acid poisoning in galvanization. Description of a clinical case. *Med. Lav.* 79:127–135.

Polack, E.W., King, J.M., Cummings, J.F., Mohammed, H.O., Birch, M., and Cronin, T. (2000) Concentrations of trace minerals in the spinal cord of horses with equine motor neuron disease. *Am. J. Vet. Res.* 61:609–611.

Potmis, R.A., Nonavinakere, V.K., Rasekh, H.R., and Early, J.L. 2nd (1993) Effect of selenium (Se) on plasma ACTH, beta-endorphin, corticosterone and glucose in rat: influence of adrenal enucleation and metyrapone pretreatment. *Toxicology* 79:1–9.

Pyrzyńska, K. (2002) Determination of selenium species in environmental samples. *Microchim. Acta* 140:55–62.

Qureshi, G.A., Qureshi, A.A., Memon, S.A., and Parvez, S.H. (2006) Impact of selenium, iron, copper and zinc in on/off Parkinson's patients on *L*-dopa therapy. *J. Neural Transm. Suppl.* 229–236.

Raber, M., Sydler, T., Wolfisberg, U., Geyer, H., and Burgi, E. (2010) Feed-related selenium poisoning in swine. *Schweiz. Arch. Tierheilkd.* 152:245–252.

Raisbeck, M.F. (2000) Selenosis. *Vet. Clin. North Am. Food Anim. Pract.* 16:465–480.

Rajotte, B.J., P'an, A.Y., Malick, A., and Robin, J.P. (1996) Evaluation of selenium exposure in copper refinery workers. *J. Toxicol. Environ. Health* 48:239–251.

Ramaekers, V.T., Calomme, M., Vanden Berghe, D., and Makropoulos, W. (1994) Selenium deficiency triggering intractable seizures. *Neuropediatrics* 25:217–223.

Rasekh, H.R., Potmis, R.A., Nonavinakere, V.K., Early, J.L., and Iszard, M.B. (1991) Effect of selenium on plasma glucose of rats: role of insulin and glucocorticoids. *Toxicol. Lett.* 58:199–207.

Ravn-Haren, G., Krath, B.N., Overvad, K., Cold, S., Moesgaard, S., Larsen, E.H., and Dragsted, L.O. (2008) Effect of long-term selenium yeast intervention on activity and gene expression of antioxidant and xenobiotic metabolising enzymes in healthy elderly volunteers from the Danish Prevention of Cancer by Intervention by Selenium (PRECISE) pilot study. *Br. J. Nutr.* 99:1190–1198.

Reid, M.E., Stratton, M.S., Lillico, A.J., Fakih, M., Natarajan, R., Clark, L.C., and Marshall, J.R. (2004) A report of high-dose selenium supplementation: response and toxicities. *J. Trace Elem. Med. Biol.* 18:69–74.

Robertson, D.S. (1970) Selenium—a possible teratogen? *Lancet* 1:518–519.

Rosenfeld, I. and Beath, O.A. (1964) *Selenium Geobotany, Biochemistry, Toxicity and Nutrition*, New York, NY: Academic Press.

Roti, E., Minelli, R., Gardini, E., Bianconi, L., Ronchi, A., Gatti, A., and Minoia, C. (1993) Selenium administration does not cause thyroid insufficiency in subjects with mild iodine deficiency and sufficient selenium intake. *J. Endocrinol. Invest.* 16:481–484.

Safarinejad, M.R. and Safarinejad, S. (2009) Efficacy of selenium and/or *N*-acetyl-cysteine for improving semen parameters in infertile men: a double-blind, placebo controlled, randomized study. *J. Urol.* 181:741–751.

Saiki, M.K. and Lowe, T.P. (1987) Selenium in aquatic organisms from subsurface agricultural drainage water, San Joaquin Valley, California. *Arch. Environ. Contam. Toxicol.* 16:657–670.

Salbe, A.D., Hill, C.H., Veillon, C., Howe, M., Longnecker, P.M., Taylor, P.R., and Levander, O.A. (1993) Relationship between serum somatomedin C levels and tissue selenium content among adults living in a seleniferous area. *Nutr. Res.* 13:399–405.

SCF (2000) *Opinion of the Scientific Committee on Food on the tolerable upper intake level of selenium (SCF/CS/NUT/UPPLEV/25 Final).* Scientific Committee on Food (SCF), Unit C3—Management of Scientific Committees II, Directorate C—Scientific Health Opinons, Health and Consumer Protection Directorate—General, European Commission.

Schuh, B. and Jappe, U. (2007) Selenium intoxication: undesirable effect of a fasting cure. *Br. J. Dermatol.* 156:177–178.

Shamberger, R.J. and Frost, D.V. (1969) Possible protective effect of selenium against human cancer. *Can. Med. Assoc. J.* 100:682.

Shearer, T.R. (1975) Developmental and postdevelopmental uptake of dietary organic and inorganic selenium into the molar teeth of rats. *J. Nutr.* 105:338–347.

Sheng, X.Q., Huang, K.X., and Xu, H.B. (2005) Influence of alloxan-induced diabetes and selenite treatment on blood glucose and glutathione levels in mice. *J. Trace Elem. Med. Biol.* 18:261–267.

Simmons, D.B. and Wallschlager, D. (2005) A critical review of the biogeochemistry and ecotoxicology of selenium in lotic and lentic environments. *Environ. Toxicol. Chem.* 24: 1331–1343.

Smith, M.I. and Westfall, B.B. (1936) The selenium problem in relation to public health. *Public Health Rep.* 51:1496–1505.

Smith, M.I. and Westfall, B.B. (1937) Further field studies on the selenium problem in relation to public health. *Public Health Rep.* 52:1375–1384.

Sogbesan, E.A. and Damji, K. (2010) Relationship between glaucoma and selenium levels in plasma and aqueous humor. *Evidence-Based Ophthalmol.* 11:102–103.

Spallholz, J.E., Palace, V.P., and Reid, T.W. (2004) Methioninase and selenomethionine but not Se-methylselenocysteine generate methylselenol and superoxide in an *in vitro* chemiluminescent assay: implications for the nutritional carcinostatic activity of selenoamino acids. *Biochem. Pharmacol.* 67:547–554.

Srivastava, A.K., Gupta, B.N., Bihari, V., and Gaur, J.S. (1995) Generalized hair loss and selenium exposure. *Vet. Hum. Toxicol.* 37:468–469.

Steinbrenner, H., Speckmann, B., Pinto, A., and Sies, H. (2011) High selenium intake and increased diabetes risk: experimental evidence for interplay between selenium and carbohydrate metabolism. *J. Clin. Biochem. Nutr.* 48:40–45.

Stranges, S., Sieri, S., Vinceti, M., Grioni, S., Guallar, E., Laclaustra, M., Muti, P., Berrino, F., and Krogh, V. (2010) A prospective study of dietary selenium intake and risk of type 2 diabetes. *BMC Public Health* 10:564.

Su, Y.P., Tang, J.M., Tang, Y., and Gao, H.Y. (2005) Histological and ultrastructural changes induced by selenium in early experimental gastric carcinogenesis. *World J. Gastroenterol.* 11: 4457–4460.

Sutter, M.E., Thomas, J.D., Brown, J., and Morgan, B. (2008) Selenium toxicity: a case of selenosis caused by a nutritional supplement. *Ann. Intern. Med.* 148:970–971.

Thilly, C.H., Vanderpas, J.B., Bebe, N., Ntambue, K., Contempre, B., Swennen, B., Moreno-Reyes, R., Bourdoux, P., and Delange, F. (1992) Iodine deficiency, other trace elements, and goitrogenic factors in the etiopathogeny of iodine deficiency disorders (IDD). *Biol. Trace Elem. Res.* 32:229–243.

Thomson, C.D., Campbell, J.M., Miller, J., Skeaff, S.A., and Livingstone, V. (2009) Selenium and iodine supplementation: effect on thyroid function of older New Zealanders. *Am. J. Clin. Nutr.* 90:1038–1046.

Thomson, C.D., McLachlan, S.K., Grant, A.M., Paterson, E., and Lillico, A.J. (2005) The effect of selenium on thyroid status in a population with marginal selenium and iodine status. *Br. J. Nutr.* 94:962–968.

Thorlacius-Ussing, O., Flyvbjerg, A., and Esmann, J. (1987) Evidence that selenium induces growth retardation through reduced growth hormone and somatomedin C production. *Endocrinology* 120:659–663.

Thorlacius-Ussing, O., Flyvbjerg, A., and Orskov, H. (1988) Growth in young rats after termination of sodium selenite exposure: studies of growth hormone and somatomedin C. *Toxicology* 48:167–176.

Thorlacius-Ussing, O., Flyvbjerg, A., Tarp, U., Overvad, K., and Ørskov, H. (1989) Selenium intake induces growth retardation through reversible growth hormone and irreversible somatomedin C suppression. In: Wendel, A., editor. *Selenium in Biology and Medicine*, Heidelberg: Springer, pp. 126–129.

Transport Canada (2008) *Emergency Response Guidebook. Guide 152 for Substances—Toxic (Combustible)*, Transport Canada, Transport, Infrastructure and Communities (TIC) Portfolio.

Tsunoda, M., Johnson, V.J., and Sharma, R.P. (2000) Increase in dopamine metabolites in murine striatum after oral exposure to inorganic but not organic form of selenium. *Arch. Environ. Contam. Toxicol.* 39:32–37.

Usami, M. and Ohno, Y. (1996) Teratogenic effects of selenium compounds on cultured postimplantation rat embryos. *Teratog. Carcinog. Mutagen.* 16:27–36.

U.S. EPA (2011) *Selenium and Compounds*, Washington, DC: Integrated Risk Information System (IRIS), U.S. Environmental Protection Agency (U.S. EPA).

U.S. FDA (2008) *FDA finds hazardous levels of selenium in samples of "Total Body Formula" and "Total Body Mega Formula" dietary supplement products linked to adverse reactions.* U.S. Food and Drug Administration (U.S. FDA) News Releases, Washington, D.C.: U.S. Department of Health and Human Services.

Valentine, J.L., Reisbord, L.S, Kang, H.K., and Schluchter, M. (1987) In: Combs, G.F., Levander, O.A., Spallholz, J.E., and Oldfield, J.E., editors. *Effects on Human Health of Exposure to Selenium in Drinking Water*, New York, NY: Van Nostrand Reihold Co., pp. 675–687.

van Woudenbergh, G.J., van Ballegooijen, A.J., Kuijsten, A., Sijbrands, E.J., van Rooij, F.J., Geleijnse, J.M., Hofman, A., Witteman, J.C., and Feskens, E.J. (2009) Eating fish and risk of type 2 diabetes: a population-based, prospective follow-up study. *Diabetes Care* 32:2021–2026.

Vinceti, M., Bonvicini, F., Rothman, K.J., Vescovi, L., and Wang, F. (2010) The relation between amyotrophic lateral sclerosis and inorganic selenium in drinking water: a population-based case-control study. *Environ. Health* 9:77.

Vinceti, M., Cann, C.I., Calzolari, E., Vivoli, R., Garavelli, L., and Bergomi, M. (2000a) Reproductive outcomes in a population exposed long-term to inorganic selenium via drinking water. *Sci. Total Environ.* 250:1–7.

Vinceti, M., Dennert, G., Crespi, M.C., Zwahlen, M., Brinkman, M., Vinceti, M., Zeegers, M.P., Horneber, M., D'Amico, R., and Del Giovane, C. (2014) Selenium for preventing cancer. *Cochrane Database Syst. Rev.*, doi: 10.1002/14651858 .CD005195.pub3.

Vinceti, M., Guidetti, D., Pinotti, M., Rovesti, S., Merlin, M., Vescovi, L., Bergomi, M., and Vivoli, G. (1996) Amyotrophic lateral sclerosis after long-term exposure to drinking water with high selenium content. *Epidemiology* 7:529–532.

Vinceti, M., Nacci, G., Rocchi, E., Cassinadri, T., Vivoli, R., Marchesi, C., and Bergomi, M. (2000b) Mortality in a population with long-term exposure to inorganic selenium via drinking water. *J. Clin. Epidemiol.* 53:1062–1068.

Vinceti, M., Rothman, K.J., Bergomi, M., Borciani, N., Serra, L., and Vivoli, G. (1998) Excess melanoma incidence in a cohort exposed to high levels of environmental selenium. *Cancer Epidemiol. Biomarkers Prev.* 7:853–856.

Vinceti, M., Rovesti, S., Bergomi, M., and Vivoli, G. (2000c) The epidemiology of selenium and human cancer. *Tumori* 86:105–118.

Vinceti, M., Wei, E.T., Malagoli, C., Bergomi, M., and Vivoli, G. (2001) Adverse health effects of selenium in humans. *Rev. Environ. Health* 16:233–251.

Wang, X.D., Vatamaniuk, M.Z., Wang, S.K., Roneker, C.A., Simmons, R.A., and Lei, X.G. (2008) Molecular mechanisms for hyperinsulinaemia induced by overproduction of selenium-dependent glutathione peroxidase-1 in mice. *Diabetologia* 51:1515–1524.

Willhite, C.C. (1993) Selenium teratogenesis. Species-dependent response and influence on reproduction. *Ann. N. Y. Acad. Sci.* 678:169–177.

Wilson, H.M. (1962) Selenium oxide poisoning. *N. C. Med. J.* 23:73–75.

Wirth, J.J., Rossano, M.G., Daly, D.C., Paneth, N., Puscheck, E., Potter, R.C., and Diamond, M.P. (2007) Ambient manganese exposure is negatively associated with human sperm motility and concentration. *Epidemiology* 18:270–273.

Woutersen, R.A., Appel, M.J., and Van Garderen-Hoetmer, A. (1999) Modulation of pancreatic carcinogenesis by antioxidants. *Food Chem. Toxicol.* 37:981–984.

Xiao, R., Qiao, J.T., Zhao, H.F., Liang, J., Yu, H.L., Liu, J., Guo, A.M., and Wang, W. (2006) Sodium selenite induces apoptosis in cultured cortical neurons with special concomitant changes in expression of the apoptosis-related genes. *Neurotoxicology* 27:478–484.

Yamada, H., Hattori, T., Miyamura, T., Yasuda, A., and Yonebayashi, K. (1994) Determination of trimethylselenonium ion and its behavior in soil. *Soil Sci. Plant Nutr.* 40:49–56.

Yang, G.Q., Wang, S.Z., Zhou, R.H., and Sun, S.Z. (1983) Endemic selenium intoxication of humans in China. *Am. J. Clin. Nutr.* 37:872–881.

Yang, G., Yin, S., Zhou, R., Gu, L., Yan, B., Liu, Y., and Liu, Y. (1989a) Studies of safe maximal daily dietary Se-intake in a seleniferous area in China. Part II: relation between Se-intake and the manifestation of clinical signs and certain biochemical alterations in blood and urine. *J. Trace Elem. Electrolytes Health Dis.* 3:123–130.

Yang, G. and Zhou, R. (1994) Further observations on the human maximum safe dietary selenium intake in a seleniferous area of China. *J. Trace Elem. Electrolytes Health Dis.* 8:159–165.

Yang, G., Zhou, R., Yin, S., Gu, L., Yan, B., Liu, Y., and Li, X. (1989b) Studies of safe maximal daily dietary selenium intake in a seleniferous area in China. I. Selenium intake and tissue selenium levels of the inhabitants. *J. Trace Elem. Electrolytes Health Dis.* 3:77–87.

Yang, K.C., Lee, L.T., Lee, Y.S., Huang, H.Y., Chen, C.Y., and Huang, K.C. (2010) Serum selenium concentration is associated with metabolic factors in the elderly: a cross-sectional study. *Nutr. Metab. (Lond.)* 7:38.

Zhang, P., Ganje, T.J., Page, A.L., and Chang, C. (1988) Growth and uptake of selenium by Swiss chard in acid and neutral soils. In: Tanji, K.K., Valoppi, L., and Woodring, R.C., editors. *Selenium Contents in Animal and Human Food Crops Grown in California*, Oakland, CA: Cooperative Extension Unversity of California, Division of Agriculture and Natural Resources, pp. 13–18.

Zhang, Y., Coogan, P., Palmer, J.R., Strom, B.L., and Rosenberg, L. (2009) Vitamin and mineral use and risk of prostate cancer: the case-control surveillance study. *Cancer Causes Control* 20:691–698.

Zierler, S., Theodore, M., Cohen, A., and Rothman, K.J. (1988) Chemical quality of maternal drinking water and congenital heart disease. *Int. J. Epidemiol.* 17:589–594.

31

SILVER

Giffe T. Johnson

Target organ(s): Skin, lungs, eyes, GI tract

Occupational exposure limits: $0.01\,mg/m^3$ OSHA (PEL); $0.01\,mg/m^3$ ACGIH (TLV); EU OEL $0.1\,mg/m^3$; $0.01\,mg/m^3$ NIOSH (REL); $10\,mg/m^3$ NIOSH (IDLH)

Reference values: $0.005\,mg/kg/day$ USEPA (RfD); $0.1\,mg/l$ USEPA (Drinking Water SMCL)

BACKGROUND AND USES

Silver (symbol Ag) is one of the basic elements present in the earth's crust. Silver is rare, but occurs naturally in the environment as a soft, "silver"-colored metal or as a white powdery compound (silver nitrate). Metallic silver and silver alloys are used to make jewelry, eating utensils, electronic equipment, and dental fillings. Nanoparticles of silver have been developed into meshes, bandages, and clothing as an antibacterial.

Silver is used in photographic materials, electric and electronic products, brazing alloys and solders, electroplated and sterling ware, as a catalyst, and in coinage. Silver is alloyed with many other metals to improve strength and hardness and to achieve corrosion resistance. Small amounts of silver are released into the environment during coal gasification, municipal waste incineration, and coal combustion. It is produced in the United States, Mexico, Canada, Peru, and the former Soviet Union. It is often refined from ore containing lead, copper, and gold.

PHYSICAL AND CHEMICAL PROPERTIES

The metal silver (CASRN 7440-22-4) is described as a white, lustrous solid. In its pure form it has the highest thermal and electrical conductivity and lowest contact resistance of all metals. With the exception of gold, silver is the most malleable metal. The melting point of silver is $961.93\,°C$, boiling point is $2212\,°C$, specific gravity is $10.50\,(20\,°C)$, and has valence states consisting of +1 and +2.

MAMMALIAN TOXICOLOGY

Acute Effects

Acute inhalation of colloidal silver aerosols has been reported to damage exposed epithelium in rabbits (Konradova, 1968). Abdominal pain has been reported by workers exposed to silver nitrate and silver oxides from airborne exposure levels of approximately $0.039\,mg\ silver/m^3$ (Rosenman et al., 1979). Death has been observed in rats following ingestion of colloidal silver and inorganic silver compounds at extremely high doses; silver nitrate has an oral LD_{50} of $1173\,mg/kg$ in rats and an LD_{50} of $50\,mg/kg$ in mice (Dequidt et al., 1974; Walker, 1971).

Chronic Effects

The most commonly encountered silver-induced toxicity is argyria, of which two distinct clinical diagnoses have been defined. The first occurs in persons exposed by inhalation or ingestion of silver nitrate, fulminate, or cyanide. This exposure may cause generalized argyria, the systemic distribution and deposition of silver-containing particles in body tissues. Characteristically, this results in a slate gray discoloration of the skin, particularly in sun-exposed areas (Lee and Lee, 1994). The color is the result of silver sulfide as well as metallic silver deposition, and ultimately is produced by

Hamilton & Hardy's Industrial Toxicology, Sixth Edition. Edited by Raymond D. Harbison, Marie M. Bourgeois, and Giffe T. Johnson.
© 2015 John Wiley & Sons, Inc. Published 2015 by John Wiley & Sons, Inc.

photoactivated reduction. Blond people are especially susceptible to this effect. A blue line along the gingival area of the gum may be an early sign of argyria. The second type of argyria is a localized form resulting from dermal exposure. Localized argyria is especially common among jewelry makers who may puncture their skin with a silver wire. The skin discoloration often resembles small dermal nevi (Rongioletti et al., 1992; Sarsfield et al., 1992). Although argyria is a cosmetic effect that does not threaten life or health, it is troublesome because the discoloration is permanent. A specific gray/blue pigmentation of the cornea and conjunctiva resulting from deposition of inert silver is known as argyrosis.

Inhalation of $10\,\mu g/m^3$ of soluble silver for long periods or ingestion of 1–5 g of soluble silver may result in generalized argyria. A study of silver reclamation workers exposed only to insoluble silver revealed no evidence of argyria or any other adverse health effects (Pifer et al., 1989). In another report, a silver reclamation worker with a blood silver level of $49\,\mu g/l$ had no signs of argyrosis (Williams and Gardner, 1995). A second worker at the same facility had a blood silver level of $74\,\mu g/l$ and pronounced argyrosis but no evidence of other health effects.

Dermal exposure to silver fulminate may cause a contact dermatitis sometimes referred to as fulminate itch (White and Rycroft, 1982). Some silver salts can be irritating to the mucous membranes and skin. Silver oxide, which is used in battery manufacturing, glass polishing, and water purification and as a catalyst and germicidal ointment, is a strong oxidizer. Silver nitrate may cause conjunctivitis and blindness. Silver oxide and silver nitrate are both considered irritants. Mucosal irritation, cough, and chest tightness were reported by workers exposed to silver nitrate (Rosenman et al., 1979, 1987). Therefore, both skin contact and inhalation should be controlled. However, the same study failed to find a correlation between silver exposure and nephrotoxic effects. Personal air monitoring conducted 4 months previous to the study determined an 8 h time-weighted average (TWA) concentration range of 0.039–$0.378\,mg$ silver/m^3. There is no research indicating an association between exposure to silver or silver compounds and cancer in humans.

Death has been reported in rats receiving 222.2 mg silver/kg/day as silver nitrate in drinking water beginning at 23 weeks of exposure; the surviving rats showed a decreased weight gain compared to animals receiving only water at the same time point (Matuk and McCulloch, 1981).

Mechanism of Action(s)

Argyria is thought to be the result of a complex biotransformation and deposition of ingested silver. Silver that is ingested orally is oxidized into a salt form that is absorbed by the gastrointestinal tract and distributed by the blood to the dermis. A series of biotransformations may occur in dermal cells, including the photoreduction of thiol–silver complexes and further complexing with selenium which leads to the permanent deposition of silver in this tissue (Liu et al., 2012).

SOURCES OF EXPOSURE

Occupational

Most occupational exposures to silver occur through inhalation of silver-containing dusts or dermal exposure to colloidal silver compounds. Common forms of occupational exposure include elemental silver, silver nitrate, and silver oxide. Humans may be dermally exposed to silver through the use of silver-containing processing solutions for radiographic and photographic materials, dental amalgams, and medicines (e.g., silver sulfadiazine cream and solutions for treating burns). Due to the high electrical conductivity of silver, it is used in many technology-based industries in solders, wiring, and electronic board manufacture that may produce a fume exposure.

Environmental

People are exposed daily to very low levels of silver from food and drinking water, and to some extent, in air—most of which is naturally occurring. Other sources of exposure include the use of silver in medicines and bandages, silver dinnerware and jewelry, nanosilver laced clothing to prevent odor, and hobbies that involve soldering and photography. Exposure from everyday use of silver alloy, such as wearing jewelry or eating with silver-coated flatware, is not expected to result in silver being absorbed into the body. Exposure to fumes, colloidal compounds, or aerosolized forms may, however, result in absorption.

In the natural environment, silver levels of $<0.000001\,mg$ silver/m^3 of air, 0.2–2.0 parts silver per billion parts water (ppb) in surface waters, such as lakes and rivers, and 0.20–0.30 parts silver per million parts soil (ppm) in soils are found from naturally occurring sources (Davidson et al., 1985; Boyle, 1968; Kharkar et al., 1968).

Silver compounds are also found in groundwater, and may be found at increased levels at hazardous waste sites. Sources of drinking water in the United States have been found to contain silver levels of up to 80 ppb. Several sampling surveys have been conducted, indicating that 1/10th–1/3rd of the samples taken from drinking water supplies contain silver at levels >30 ppb (Letkiewicz et al., 1984).

INDUSTRIAL HYGIENE

Overall, the risks of working with silver are low. Workers are most likely to be exposed during silver plating, silver reclamation from photographic materials, and welding. There are minimal risks associated with pure silver; however, silver intensive industries often are hazardous. Many health effects associated with occupational exposure in silver-intensive industries are caused by associated materials, such as cyanide, lead, and cadmium. Silver reclamation workers may be exposed to excessive levels of cyanides (Blanc et al., 1985). Workers in silver soldering operations may be exposed to hazardous levels of cadmium (Vance, 1960). Silver jewelry workers may be exposed to high levels of lead (Flora et al., 1990).

Silver fulminate, an explosive compound, may be formed when silver is treated with nitric acid in the presence of ethyl alcohol. Workers may be exposed to silver during gold casting processes, more so when using an oxyacetylene flame than with electromagnetic induction.

The Occupational Safety and Health Administration (OSHA) and the National Institute for Occupational Safety and Health (NIOSH) have set both the permissible exposure limit and the recommended exposure limit for silver at $0.01 \, mg/m^3$ (NIOSH, 2012). The American Conference of Governmental Industrial Hygienists (ACGIH) has set the threshold limit value (TLV) for the metal silver at $0.1 \, mg/m^3$ and the TLV for soluble silver salts as silver at $0.01 \, mg/m^3$ (ACGIH, 1996). Workers should wear gloves, face shields, and impervious clothing when necessary to prevent contact with powdered metallic silver and soluble silver salts. Respirator use is indicated for airborne exposures that exceed the OSHA PEL (NIOSH, 2012).

MEDICAL MANAGEMENT

In the event of severe exposure to silver or silver compounds, exposure removal is a primary intervention. Symptoms of silver exposure may include the presentation of blue-gray eyes, nasal septum, throat, and skin; irritation or ulceration skin; and gastrointestinal disturbance. First aid for irritant exposure to eyes or skin is to immediately flush affected area with water. After the patient has been removed from the exposure and first aid conducted, medical treatment should be sought for unresolved symptoms. Medical management consists of supportive care for skin and eye irritation and/or acute respiratory impairment.

There is no recognized management for argyria or argyrosis that results from chronic silver exposure. Beyond the physical disfigurement associated with this condition, no further complications exist which require treatment. Removal from known exposures should be emphasized.

REFERENCES

American Conference of Governmental Industrial Hygienists (ACGIH) (1996) *Threshold Limit Values (TLVs) for Chemical Substances and Physical Agents and Biological Exposure Indices (BEIs)*, Cincinnati, OH: American Conference of Governmental Industrial Hygienists.

Blanc P., Hogan M., Mallin K., Hryhorczuk D., Hessl S., and Bernard B. (1985) Cyanide intoxication among silver-reclaiming workers. *JAMA* 253:367–371.

Boyle R.W. (1968) Geochemistry of silver and its deposit notes on geochemical prospecting for the element. Geological Survey of Canada. Ottawa, Canada. *Dep. Energ., Mines Res.* 160:1–96.

Davidson, C.I., Wiersma, G.B., Brown, K.W., Goold, W.D., Mathison, T.P., and Reilly, M.T. (1985) Airborne trace elements in Great Smoky Mountains, Olympic, and Glacier National Parks. *Environ. Sci. Technol.* 19:27–35.

Dequidt, J., Vasseur, P., and Gromez-Potentier, J. (1974) Experimental toxicological of some silver derivatives. *Bulletin de la Societe de Pharmacie de Lille* 1:23–35.

Flora, S.J.S., Singh, S., and Tandon, S.K. (1990) Plumbism among Indian silver jewellery [sic] industry workers. *J. Environ. Sci. Health* 25(2):105–113.

Kharkar, D.P., Turekian, K.K., and Bertine, K.K. (1968) Stream supply of dissolved silver, molybdenum, antimony, selenium, chromium, cobalt, rubidium to the oceans. *Geochim. Cosmochim. Acta* 32:285–298.

Konradova, V. (1968) The ultrastructure of the tracheal epithelium in rabbits following the inhalation of aerosols of colloidal solutions of heavy metals 1. Changes in the ultrastructure of the tracheal epithelium after 2-hour inhalation of colloidal solutions of iron and silver. *Folia Morphol. (Praha)* 16:258–264.

Lee, S.M. and Lee, S.H. (1994) Generalized argyria after habitual use of $AgNO_3$. *J. Dermatol.* 21:50–53.

Letkiewicz, F., Spooner, C., Macaluso, C., and Brown, D. (1984) Occurrence of silver in drinking water, food, and air. In: *Report to US Environmental Protection Agency, Office of Drinking Water, Criteria and Standards Division*. Washington, DC: JRB Associates, McLean, VA.

Liu, J., Wang, Z., Liu, F.D., Kane, A.B., and Hurt, R.H. (2012) Chemical transformations of nanosilver in biological environments. *ACS Nano* 6:9887–9899.

Matuk, Y., Ghosh, M., and McCulloch, C. (1981) Distribution of silver proteins of the albino rat. *Can. J. Ophthalmol.* 16:145–150.

National Institute for Occupational Safety and Health (NIOSH) (2012) *Pocket Guide to Chemical Hazards* NIOSH 94–116, Washington, DC: US Department of Health and Human Services.

Pifer, J.W., Friedlander, B.R., Kintz, R.T., and Stockdale, D.K. (1989) Absence of toxic effects in silver reclamation workers. *Scand. J. Work Environ. Health* 15:210–221.

Rongioletti, F., Robert, E., Buffa, P., Bertagno, R., and Rebora, A. (1992) Blue nevi-like dotted occupational argyria. *J. Am. Acad. Dermatol.* 27:1015–1016.

Rosenman, K.D., Moss, A., and Kon, S. (1979) Argyria: clinical implications of exposure to silver nitrate and silver oxide. *J. Occup. Med.* 21:430–435.

Rosenman, K.D., Seixas, N., and Jacobs, I. (1987) Potential nephrotoxic effects of exposure to silver. *Br. J. Ind. Med.* 44:267–272.

Sarsfield, P., White, J.E., and Theaker, J.M. (1992) Silverworker's finger: an unusual occupational hazard mimicking a melanocytic lesion. *Histopathology* 20:73–75.

Vance, G.H. (1960) Cadmium exposures in silver soldering operations. *Am. Ind. Hyg. Assoc. J.* 21:107–109.

Walker, F. (1971) Experimental argyria: a model for basement membrane studies. *Br. J. Exp. Pathol.* 52:589–593.

White, I.R. and Rycroft, R.J. (1982) Contact dermatitis from silver fulminate-fulminate itch. *Contact Dermatitis* 8:159–163.

Williams, N. and Gardner, I. (1995) Absence of symptoms in silver refiners with raised blood silver levels. *Occup. Med.* 45:205–208.

32

TELLURIUM

Marek Banasik

First aid: CAUTION! First aid personnel must be aware of own risk during rescue! Eyes: make sure to remove any contact lenses from the eyes before rinsing, promptly wash eyes with plenty of water while lifting the eye lids, continue to rinse for at least 15 min—get medical attention if any discomfort continues; skin: remove affected person from source of contamination, remove contaminated clothing, wash the skin immediately with soap and water—get medical attention if any discomfort continues; inhalation: move the exposed person to fresh air at once, rinse nose and mouth with water—get medical attention if any discomfort continues; ingestion: immediately rinse mouth and provide fresh air—get medical attention if any discomfort continues.

Occupational exposure limits: OSHA PEL: 0.1 mg/m^3 time-weighted average (TWA) (except tellurium hexafluoride and bismuth telluride); NIOSH REL: 0.1 mg/m^3 TWA (except tellurium hexafluoride and bismuth telluride); ACGIH TLV: 0.1 mg/m^3 TWA (except hydrogen telluride); see NIOSH (2011)

Reference value(s): None

Hazard statements: H317: may cause an allergic skin reaction; H332: harmful if inhaled; H413: may cause long lasting harmful effects to aquatic life

Precautionary statements: P261: avoid breathing dust/fume/gas/mist/vapors/spray; P280: wear protective gloves/protective clothing/eye protection/face protection; P312: call a poison center/doctor/... if you feel unwell; P333+313: if skin irritation or rash occurs: get medical advice/attention

Risk phrases: R20: harmful by inhalation; R43: may cause sensitization by skin contact; R53: may cause long-term adverse effects in the aquatic environment

Safety phrases: S24: avoid contact with skin; S36/37: wear suitable protective clothing and gloves; S61: avoid release to the environment, refer to special instructions/safety data sheet

BACKGROUND AND USES

Tellurium [symbol Te; Chemical Abstracts Service Registry Number (CASRN): 13494-80-9] is a silvery white metal in group 16 of the periodic table. It shares chemical and clinical properties with selenium (Amdur, 1947, 1958; Schroeder et al., 1967; Shie and Deeds, 1920). Tellurium is a semiconductor and may have multiple electron states (-2, 0, $+2$, $+4$, $+6$). It can react with hydrogen to form hydrogen telluride and with halogens. Tellurite ($+2$) and teleurate ($+4$) compounds are water soluble. Elemental tellurium burns, producing a blue flame and tellurium dioxide (TeO$_2$; CASRN: 7446-07-3).

Tellurium is a common constituent of ores that contain silver, gold, lead, antimony, and bismuth, and it is often present in small amounts in coal. Tellurium is widely used in metallurgy because it improves the properties of copper, tin, lead-based alloys, steel, and cast iron. It is used in rubber manufacturing to increase heat resistance and to retard the aging of rubber hoses and cable coatings. Small amounts are used in the electronics industry for lasers and photoreceptors. Tellurium is not an essential micronutrient; therefore, it is not found in nutritional supplements.

Hamilton & Hardy's Industrial Toxicology, Sixth Edition. Edited by Raymond D. Harbison, Marie M. Bourgeois, and Giffe T. Johnson.
© 2015 John Wiley & Sons, Inc. Published 2015 by John Wiley & Sons, Inc.

In the United States (US), no publicly available production volume data, including manufacturing and importing, are available. Under the US Environmental Protection Agency's 2012 Chemical Data Reporting rule, one US manufacturer was identified, who claimed the production volume as confidential business information (EPA, 2012). In the European Union, the annual production volume of tellurium is between *ca.* 2.2 hundred thousand and *ca.* 2.2 million pounds (100–1,000 tonnes) per annum (ECHA, 2014).

The potential risks from tellurium are primarily limited to occupational exposures because the end uses are not in forms that consumers or the general population will be exposed. Virtually all of the toxicological testing on tellurium has been conducted in accordance with regulatory registration requirements. A summary of these data, along with historical accounts of toxicological effects in humans are provided herein.

PHYSICAL AND CHEMICAL PROPERTIES

Tellurium is a heavy metal, which is processed as a grey powder. It has hexagonal, rhombohedral structure, a low water solubility and high relative density. The particle size ranges from 52.36 to 112.98 µm.

ENVIRONMENTAL FATE AND BIOACCUMULATION

Metals are recalcitrant to degradation; therefore, no biodegradation studies have been performed on tellurium. No aquatic bioaccumulation data exist for tellurium; however, based on its density and low water solubility, it is unlikely to present a concern for bioaccumulation in the water column. No environmental monitoring data are available on the levels of tellurium in sediment or sediment-dwelling organisms. Therefore, it is unclear whether tellurium has the potential to bioaccumulate in this compartment. In humans, tellurium accumulates in the bones. Based on this, it may be assumed that tellurium has the potential to bioaccumulate in vertebrates.

ECOTOXICOLOGY

Elemental tellurium has not been evaluated in standard aquatic or terrestrial toxicity studies. Standard toxicology studies have been performed on TeO_2 and used as a read across for assessing ecotoxicological hazards. A summary of these data is presented in Table 32.2. The following predicted no effect concentrations (PNECs) were derived for tellurium as part of regulatory submissions to the European Union: $PNEC_{aquatic}$ (freshwater) = 5.79 µg/l; $PNEC_{aquatic}$ (marine water) = 0.579 µg/l; $PNEC_{aquatic}$

TABLE 32.1   Chemical and Physical Properties of Tellurium

Characteristic	Information
Name	Tellurium
Synonyms/trade names	Aurum paradoxum, metallum problematum, telloy
CASRN	13494-80-9
Molecular weight	127.60 g/mol
Physical state	Grey powder (at 20 °C and 1,013 hPa)
Melting point	449.8 °C
Relative density	6. 232
Water solubility	20.02 µg/l at 21.2 °C, 183.3 µg/l at 21 °C, and 1.762 mg/l at 21 °C (pH = 8)[a]
Particle size	L_{10} = 52.36 µm; L_{50} = 77.00 µm; L_{90} = 112.98 µm[b]

Source: ECHA (2014); HSDB (2009).

[a]These values were obtained after 7 days with initial loadings of 1, 10, and 100 mg/l, respectively.

[b]$L_{\#}$ represents # % of the particle volume or particle mass with a lower particle diameter.

(intermittent releases) = 57.9 µg/l; and $PNEC_{STP}$ (sewage treatment plant) = 3.2 mg/l (ECHA, 2014).

MAMMALIAN TOXICOLOGY

Animal studies performed in the 1970s or earlier reported that tellurium can cause teratogenic effects in experimental animals and the demyelination of nerves, and it crosses the blood–brain barrier (Agnew and Curry, 1972; De Meio, 1946; Lampert and Garrett, 1971; Uncini et al., 1988). Chronic ingestion of tellurium by rats results in deposition of tellurium in brain tissue, a condition known as "black brain" (Duckett and White, 1974). Inhalation or endotracheal administration of metallic tellurium or TeO_2 dust causes chronic inflammation and blue-gray discoloration of the trachea (Geary et al., 1978). These earlier reports have been evaluated in more recent guideline studies to determine tellurium's potential to cause acute and developmental toxicity in experimental animals. A summary of the acute and developmental studies, performed in accordance with validated testing guidelines, is provided below.

Acute Studies

In a guideline 4 h inhalation study, groups of five male and female Wistar rats were exposed to a single, nose only analytical concentration of tellurium powder at 2.42 g/m^3 (particle size = *ca.* 72% had an aerodynamic diameter between 1.8 and 4.2 µm) (ECHA, 2014). Immediately after exposure, all exposed rats were observed with black heads, half closed eyes, and hunched posture. From postexposure days 7–14, female rats had baldness on the head, back, and posterior. body weight (bw) was decreased in all rats by 1 week postexposure,

TABLE 32.2   Aquatic Toxicity Studies on Tellurium Dioxide

Toxicity to Aquatic Vertebrates			
Test Species	*Test Protocol*	*Endpoint and Exposure Duration*	*Result, mg/l*
Rainbow trout (*Oncorhynchus mykiss*)	OECD[a] 203 (semi-static test)	96 h[b] LC_{50}[c]	>46.4 (M)[d]
Toxicity to Aquatic Plants			
Test Species	*Test Protocol*	*Endpoint and Exposure Duration*	*Result, mg/l*
Algae (*Pseudokirchnerella subcapitata*)	OECD 201 (static)	72 h EC_{50}[e] (growth rate)	>14.65 (W)[h]
		72 h NOEC[f] (growth rate)	4.17 (W)
		72 h EC_{10}[g] (yield)	3.40 (W)
		72 h EC_{50} (yield)	10.25 (W)
		72 h NOEC (yield)	1.42 (W)
Toxicity to Microorganisms			
Test Species	*Test Protocol*	*Endpoint and Exposure Duration*	*Result, mg/l*
Microorganisms from activated sludge obtained from predominantly domestic sewage	OECD 209 (static)	3 h EC_{10} (respiration inhibition)	3.7 (N)[i]
		3 h EC_{50} (respiration inhibition)	320 (N)
		3 h NOEC (respiration inhibition)	1.0 (N)

Source: ECHA (2014).

[a]Organization for Economic Cooperation and Development.
[b]Hour(s).
[c]Median lethal concentration.
[d]Measured value.
[e]Median effect concentration.
[f]No observed effect concentration.
[g]Effect concentration measured as 10% reduction.
[h]Water-accommodated fractions.
[i]Nominal values.

compared to their preexposure bw. During the autopsy, a garlic-like smell was noted. In male rats, the lungs contained brownish spots. In female rats, the lungs appeared pale or contained grey to black stains. In four of the female rats, the thymus was grey. None of the rats died during the exposure or postexposure period. A 4-h LC_{50} >2.42 g/m^3 was reported.

In a guideline acute oral gavage study, five male and five female Wistar rats were administered tellurium in a maize oil vehicle at a single dose of 5,000 mg/kg bw (ECHA, 2014). No mortalities or treatment-related effects on clinical signs or gross pathology were reported after the 14-day posttreatment period. All animals gained weight during the 14-day observation period.

Skin and eye irritation have been assessed using standard guideline *in vitro* methods (ECHA, 2014). Negative results were obtained with each. In support of the *in vitro* skin irritation study, no local skin irritation was observed in mice during the performance of a guideline local lymph node assay, as discussed below. It should be noted, however, that the first-aid measures listed under the European Chemicals Agency tellurium registration state that tellurium skin contact may cause irritation, redness, pain, and discoloration. For eye contact, it states that tellurium may cause redness and pain (ECHA, 2014).

The skin sensitization potential of tellurium powder was evaluated in female CBA mice (four per dose) (ECHA, 2014). Eight animals served as the negative control. Hexyl cinnamic aldehyde was used as a positive control. Tellurium powder was suspended in propylene glycol at 0, 25, 50, or 100% and

applied to the dorsal surface of the ears (25 μl/ear) for 3 consecutive days, followed by 3 consecutive days without treatment. Negative and positive controls were treated in a similar manner with 100% propylene glycol or the positive control, respectively. Proliferation of lymphocytes within the lymph nodes was evaluated by incorporation of ³H-methyl thymidine and measured by β-scintillation counting. A stimulation index ≥3 was used for determining a positive result.

At study termination, the negative controls had normal lymph nodes. In the positive control, a stimulation index value of 13.0 was recorded, which confirmed the validity of the assay. In the treated animals, stimulation index values of 3.8, 3.2, and 3.2 were recorded for the 25, 50, and 100% groups, respectively. These results demonstrated that tellurium has skin sensitization potential.

Repeated Dose Studies

Two guideline developmental toxicity studies have been performed on tellurium in the rat and rabbit (Johnson et al., 1988). In the first study, 32–33 pregnant Crl Sprague-Dawley rats were administered tellurium in the diet at concentrations of 0, 30, 300, 3,000, or 15,000 parts per million (ppm) from gestation days (GDs) 6 through 15. The corresponding daily intake from GDs 6 through 10 was calculated to be 0, 2.2, 19.6, 165.6, or 633.4 mg/kg bw/day. From GDs 11 through 15, the daily intake was 0, 1.9, 18.0, 173.0, or 579.4 mg/kg bw/day. On GD 20, approximately two-thirds of the dams were sacrificed and the fetuses were examined. The remaining dams were allowed to deliver. The pups were observed and sacrificed on postnatal day (PND) 7. The heads of pups sacrificed on PND 7 or that died prior to this were examined.

No treatment-related effects were reported on dam mortality, incidences of pregnancy, mean number of corpora lutea, implantations and resorptions, mean litter size, numbers of live and dead fetuses, or percentage of male fetuses. Maternal bw gain and food consumption statistically significantly decreased in the 300 and 3,000 ppm groups. Weight loss and halved food consumption were statistically significantly decreased in the high dose group.

In the fetuses, no treatment-related effects were observed in the 0, 30, or 300 ppm groups. In the 3,000 ppm group, the incidence of litters with variations was statistically significantly increased. An increased incidence of fetuses with variation (10.6%), malformed fetuses (no details provided), and fetuses with delayed ossification (no details provided) were reported. Non-statistically significant increases were reported for internal hydrocephalus with dilation of the lateral ventricles in 3 litters (14.3%) and 11 fetuses (8.3%). Moderate dilation of the renal pelvis was also reported, but details were not provided. In the 15,000 ppm group, statistically significant decreases in male and female fetal bw and increases in litter variations were reported. Increased incidences of fetuses with variation, malformed fetuses, and fetuses with delayed ossification were reported; no further details were provided. Statistically significant increases in the incidence of internal hydrocephalus with dilation of the lateral ventricles were observed in 17 litters and 67 fetuses. External hydrocephalus was reported for two fetuses; the number of litters was not specified. Moderate dilation of the renal pelvis was reported, but no further details were provided.

In the dams allowed to deliver, no treatment-related effects were reported on duration of gestation, number (percentage) of dams with stillborn, litter size, number of live pups delivered, gross anomalies, external or visceral anomalies, or mean pup weights on PND 7. In the high dose group, pup survival to PND 7 was statistically significantly decreased. A statistically significant increase in the incidence of slight to extreme dilation of the lateral ventricles was reported in pups on PND 7 from the high dose group.

In the second developmental toxicity study, 17 per dose artificially inseminated New Zealand white rabbits were administered tellurium in the diet at concentrations of 0, 17.5, 175, 1,750, or 5,250 ppm from GDs 6 to 18. The average feed intake on GDs 6 through 18 and the average maternal bw on GD 6 were used to calculate intake values of 0, 0.8, 8, 52, and 97 mg/kg bw/day. Dams were sacrificed on GD 29 and the fetuses were examined.

No treatment-related effects were observed on the incidence of abortion, mean numbers of corpora lutea, implantations, resorptions, litter size, or sex ratio (percentage of male fetuses per litter). In the 1,750 and 5,250 ppm groups, statistically

significant decreases in maternal bw and food consumption were reported. Signs of toxicity in these groups included: soft or liquid feces, alopecia, thin appearance, and/or decreased motor activity. In the high dose group, non-statistically significant decreases in mean fetal bw were reported in males (84% of controls) and females (95% of controls). A 46.2% (controls 2%) and 11.8% (controls 6.7%) increase in the incidence of litters with abnormalities and malformed fetuses was reported, respectively. Increased incidences of malformed fetuses and fetuses with delayed ossification were reported; further details were not provided.

The study authors concluded that "[s]ince maternal toxicity was observed at dosages that did not affect the developing conceptus, there were no indications of unique developmental susceptibility upon exposure of pregnant rats or rabbits to tellurium" (Johnson et al., 1988). However, the Health Council of the Netherlands' Subcommittee on the Classification of Reproduction Toxic Substances evaluated the complete developmental toxicity database on tellurium, including the older studies and those performed by Johnson et al. (1988). It recommended that tellurium be classified category 1B (presumed human reproductive toxicant) and be labeled with H360D (may damage the unborn child) (HCN, 2013).

In humans, exposure to tellurium vapor, tellurium dioxide dust, or hydrogen telluride has caused respiratory inflammation, somnolence, and anorexia (Blackadder and Manderson, 1975; Keall et al., 1946; Steinberg et al., 1942). Hydrogen telluride is a hemolytic agent that has the same effects as hydrogen arsenide and hydrogen selenide (Webster, 1946). Fatalities, clinically characterized by cyanosis, vomiting, and stupor progressing to coma, have resulted from accidental parenteral administration of sodium tellurite (Kron et al., 1991). In cases of occupational exposure to the gaseous form of tellurium hexafluoride (CASRN 7783-80-4), patients experienced somnolence, garlic breath odor, and bluish-black patches on the skin of the neck, face, and hands (Blackadder and Manderson, 1975).

Tellurium is excreted in sweat, urine, feces, and exhaled air. In humans, a garlic odor that lasts longer than 200 days has been produced by a dose of about 15 mg of tellurium (De Meio, 1947). The garlic taste and odor are useful indicators of potentially dangerous exposure but are also a social liability. Dr. Harriet Hardy, one of the authors of the previous editions of this book, was asked to consult for the employees of an electronics company that used tellurium because their wives refused to kiss them because of their garlic odor.

SOURCES OF EXPOSURE

Occupational

Tellurium and its various compounds can be absorbed by ingestion, by inhalation, or through the skin (Cerwenka and

Cooper, 1961). Metabolism *in vivo* produces the characteristic "garlic breath" caused by methylation of tellurium to volatile dimethyl telluride (De Meio, 1947). To prevent the "garlic breath" complex, the concentrations of tellurium and its compounds must be kept at or below 0.01 mg/m^3.

Environmental

Background levels in humans are 0.6 μg/ml in urine and 1 μg/ml in serum (ACGIH, 1996; Kron et al., 1991). The calculated total body burden is about 600 mg, 90% of which is found in bone (Kron et al., 1991). Tellurium is a more potent producer of garlic breath than selenium.

INDUSTRIAL HYGIENE

Workplace engineering controls (e.g., ventilation) and personal protective equipment, including suitable respiratory protection, protective gloves, and safety goggles or a face shield, are recommend to reduce respective exposures through inhalation, skin, and eye contact (ECHA, 2014).

REFERENCES

ACGIH (1996) *TLVs and BEIs: Threshold Limit Values for Chemical Substances and Physical Agents and Biological Exposure Indices*. Cincinnati, OH: American Conference of Governmental Industrial Hygienists. ISBNs: 1882417135, 9781882417131.

Agnew, W.F. and Curry, E. (1972) Period of teratogenic vulnerability of rat embryo to induction of hydrocephalus by tellurium. *Experientia* 28:1444–1445.

Amdur, M.L. (1947) Tellurium; accidental exposure and treatment with BAL in oil. *Occup. Med.* 3:386–391.

Amdur, M.L. (1958) Tellurium oxide: an animal study in acute toxicity. *AMA Arch. Ind. Health* 17:665–667.

Blackadder, E.S. and Manderson, W.G. (1975) Occupational absorption of tellurium: a report of two cases. *Brit. J. Ind. Med.* 32:59–61.

Cerwenka, E.A. Jr. and Cooper, W.C. (1961) Toxicology of selenium and tellurium and their compounds. *Arch. Environ. Health* 3:189–200.

De Meio, R.H. (1946) Tellurium; the toxicity of ingested elementary tellurium for rats and rat tissues. *J. Ind. Hyg. Toxicol.* 28:229–232.

De Meio, R.H. (1947) Tellurium effect of ascorbic acid on tellurium breath. *J. Ind. Hyg. Toxicol.* 29:393–395.

Duckett, S. and White, R. (1974) Cerebral lipofucsinosis induced by tellurium: electron dispersive X-ray spectrophotometry analysis. *Brain Res.* 73:205–214.

ECHA (2014) *Registered Substances*. Helsinki, Finland: European Chemicals Agency. Available at http://echa.europa.eu/information-on-chemicals/registered-substances. CASRN: 13494-80-9.

EPA (2012) *Chemical Data Access Tool (CDAT)*. Washington, DC: Office of Pollution Prevention and Toxics, US Environmental Protection Agency. Available at http://java.epa.gov/oppt_chemical_search. CASRN: 13494-80-9.

Geary, D.L. Jr., Myers, R.C., Nachreiner, D.J., and Carpenter, C.P. (1978) Tellurium and tellurium dioxide: single endotracheal injection to rats. *Am. Ind. Hyg. Assoc. J.* 39:100–109.

HCN (2013) *Tellurium-Evaluation of the Effects on Reproduction, Recommendation for Classification [draft]*. The Hague: Subcommittee on the Classification of Reproduction Toxic Substances, Committee of the Health Council of the Netherlands. Available at http://www.gezondheidsraad.nl/sites/default/files/OCR_Tellurium2013.pdf.

HSDB (2009) *Tellurium, elemental*. Bethesda, MD: Hazardous Substances Data Bank (HSDB), TOXNET®–Toxicology Data Network, Specialized Information Services, US Department of Health and Human Services. Available at http://toxnet.nlm.nih.gov/cgi-bin/sis/htmlgen? HSDB. CASRN: 13494-80-9.

Johnson, E.M., Christian, M.S., Hoberman, A.M., DeMarco, C.J., Kilpper, R., and Mermelstein, R. (1988) Developmental toxicology investigation of tellurium. *Fundam. Appl. Toxicol.* 11:691–702.

Keall, J.H.H., Martin, N.H., and Tunbridge, R.E. (1946) A report of three cases of accidental poisoning by sodium tellurite. *Br. J. Ind. Med.* 3:175–176.

Kron, T., Hansen, C., and Werner, E. (1991) Renal excretion of tellurium after peroral administration of tellurium in different forms to healthy human volunteers. *J. Trace Elem. Electrolytes Health Dis.* 5:239–244.

Lampert, P.W. and Garrett, R.S. (1971) Mechanism of demyelination in tellurium neuropathy. Electron microscopic observations. *Lab. Invest.* 25:380–388.

NIOSH (2011) Tellurium. In: *NIOSH Pocket Guide to Chemical Hazards*. Washington, DC: National Institute for Occupational Safety and Health, Centers for Disease Control and Prevention, US Department of Health and Human Services. Available at http://www.cdc.gov/niosh/npg/npgd0587.html.

Schroeder, H.A., Buckman, J., and Balassa, J.J. (1967) Abnormal trace elements in man: tellurium. *J. Chronic Dis.* 20:147–161.

Shie, M.D. and Deeds, F.E. (1920) The importance of tellurium as a health hazard in industry—a preliminary report. *Public Health Rep.* 35:939.

Steinberg, H.H., Massari, S.C., Miner, A.C., and Rink, R. (1942) Industrial exposure to tellurium: atmospheric studies and clinical evaluation. *J. Ind. Hyg. Toxicol.* 24:183–192.

Uncini, A., England, J.D., Rhee, E.K., Duckett, S.W., and Sumner, A.J. (1988) Tellurium-induced demyelination: an electrophysiological and morphological study. *Muscle Nerve* 11:871–879.

Webster, S.H. (1946) Volatile hydrides of toxicological importance. *J. Ind. Hyg. Toxicol.* 28:167–182.

33

THALLIUM

Nikolay M. Filipov

First aid: A. General measures: Immediately remove from exposure source and seek medical attention; antidote (FDA approved): Prussian blue (Radiogardase), given orally as follows: 3 g immediately, followed by 250 mg/kg/day divided into four daily doses (QID); if Prussian blue is unavailable: repeated activated charcoal administration (up to four oral doses (0.5–1.0 g/kg); one dose every 2–3 h). B. Exposure-route specific measures: 1. Ingestion: ensure unobstructed airways; induce emesis (ipecac syrup) within 30–60 min of exposure. 2. Inhalation: ensure unobstructed airways; evaluate respiratory function and pulse. 3. Dermal: decontaminate the patient/victim following standard decontamination procedures.

Target organ(s): CNS, PNS, neuromuscular, eye, cardiac, renal, gastrointestinal, skin

Occupational exposure limits: 8 h shift PEL/TLV: 0.1 mg/m^3 (OSHA/ACGIH); 10 h shift REL/PEL: 0.1 mg/m^3 (NIOSH/OSHA); IDLH: 15 mg/m^3 (NIOSH).

Reference values: Candidate RfD for Tl/Tl salts: 3×10^{-6} mg/kg/day (alopecia), 1×10^{-5} mg/kg/day (alopecia plus hair follicle atrophy); RfC: not available (insufficient information); TDIs: 0.00007 mg/kg/day (for oral Tl), 0.00008–0.00009 mg/kg/day (for oral Tl compounds); TI: 15.4 µg/day; MRL: 0.002 mg/l.

Risk/safety phrases: Very toxic if inhaled of swallowed; danger of cumulative effects; may cause long term effects in the aquatic environment; keep locked up and out of the reach of children; keep away from food, drinks, and feedstuff; avoid contact with skin; in case of accidental exposure seek medical assistance immediately; avoid environmental release.

BACKGROUND AND USES

Thallium (Tl), an element sharing the same family with aluminum (Al), gallium (Ga), indium (In) and boron (B), was discovered by Sir William Crookes in 1861 (Crookes, 1861). Thallium is a rare, but widely dispersed element; it is found in nature in various earth's crust minerals (e.g., lorandite, crookesite, hutchinsonite) from North and South America to Asia (Kazantzis, 2007). Thallium is considered nonessential and its environmental concentrations, similar to Al, Ga, and In, are quite low (Kazantzis, 2007; EPA, 2009). For example, Tl's crustal abundance is approximately 1 mg/kg; in the soil, Tl's concentrations range from 0.1 to 1 mg/kg with higher concentrations occurring in close proximity to metallic ore deposits (Kazantzis, 2007). Thus, Tl concentrations of more than 100 mg/kg in soils from a mining region in China have been reported (Xiao et al., 2004a, 2004b). Detectable levels of Tl are also found in saltwater, freshwater, and air (RIVM, 2005; EPA, 2009). Thallium is taken up by vegetation, with the extent of uptake determined by soil acidity and plant species (Kazantzis, 2007).

All forms of Tl are relatively soluble and capable of causing eco- or mammalian toxicity. In fact, thallium has greater acute mammalian (human) toxicity than, for example, mercury (Hg), cadmium (Cd), or lead (Pb). As a result, Tl has been responsible for many accidental, occupational, deliberate, and therapeutic human intoxications since its discovery in 1861 (Pctcr and Viraraghavan, 2005).

Annual worldwide production of Tl is quite low: approximately 15 tons, but various industrial processes mobilize approximately 2000–5000 tons per year (Kazantzis, 2007; EPA, 2009; Rodriguez-Mercado and Altamirano-Lozano, 2013). Current uses of Tl are numerous and broad ranging.

Hamilton & Hardy's Industrial Toxicology, Sixth Edition. Edited by Raymond D. Harbison, Marie M. Bourgeois, and Giffe T. Johnson.
© 2015 John Wiley & Sons, Inc. Published 2015 by John Wiley & Sons, Inc.

For example, the isotope ^{201}Tl is used for tumor and myocardium imaging (Yildirim et al., 2005). Thallium is also used in medicines and health products, as a pigment and depilatory agent in cosmetics, in specialty alloys and in various electrical and electronic equipment or instrumentation (e.g., semiconductors, fiber optic cables, infrared instruments, laser equipment, ozone meters, and low temperature thermometers). Other uses of Tl include in pyrotechnics (green color), as a catalyst in organic synthesis, and in the production of pesticides and certain phosphate fertilizers (ATSDR, 1992, Kazantzis, 2007; EPA, 2009; Rodriguez-Mercado and Altamirano-Lozano, 2013). Due to favorable physicochemical and toxic properties, Tl salts have been used in war-like preparations (Lawrence and Kirk, 2007) and, either alone or in combination with cocaine or other toxic agents, such as arsenic, for malicious, deliberate intoxications (Rusyniak et al., 2002). Historically, Tl has been used for the treatment of diseases of the scalp, syphilis, tuberculosis, and malaria (Leonard and Gerber, 1997); these treatments have been discontinued due the high toxicity of this metal. Overall, Tl's widespread use and environmental release has led to increased Tl levels in the environment and increased exposure risk to humans and other living organisms (EPA, 2009; Rodriguez-Mercado and Altamirano-Lozano, 2013).

PHYSICAL AND CHEMICAL PROPERTIES

Thallium (Tl), from the Greek word *thallos*, is a bluish-gray, shiny, ductile, malleable, heavy metal with atomic number 81 and atomic weight of 204.38. Its density is 11.85 g/cm^3; other key characteristics of Tl are as follows: melting point of 303.5 °C, boiling point of 1,457 °C, a hexagonal crystalline structure, electron configuration of [Xe] $6s^2\,4f^{14}\,5d^{10}\,6p^1$, and oxidation states of +1 and +3. Thallium belongs to group IIIa of the periodic table with Al, Ga, In, and B. Unlike the other metal cations in this family, which are more stable in oxidation state +3 (Al^{3+}, Ga^{3+}, and In^{3+}), Tl is more stable in oxidation state +1 (thallous, Tl^+) than oxidation state +3 (thallic, Tl^{3+}). Thallium has electronegativity of 2.04, Van der Waals radius of 193 pm, atomic radius of 171 pm, and an ionic radius of 150 pm (Tl^+). There are 26 artificial isotopes of Tl, but in nature 29.52% of Tl exists as [203]Tl and 70.48% exists as [205]Tl. Upon exposure to air, Tl forms brownish black oxide. Thallium is acid-soluble and highly reactive metal; it forms monovalent salts, i.e., Tl acetate, Tl carbonate, Tl sulfate, and trivalent thallic salts, i.e., Tl trichloride and Tl sesquioxide, which are less stable than the thallous salts (Fergusson, 1990; Kazantzis, 2007; EPA, 2009).

ENVIRONMENTAL FATE AND BIOACCUMULATION

Thallium is retained in soil, with large amounts of clay, organic material, and iron and manganese oxides (WHO, 1996). As the soil pH decreases, Tl's uptake by plants increases (WHO, 1996; EPA, 2009) and it could reach up to 45 mg/kg in green kale (Kazantzis, 2007). From acidic soils, Tl can leach into ground and surface waters with the monovalent form being the one predominantly found there (WHO, 1996). Some bioconcentration, but limited biomagnification takes place for Tl. In aquatic environments, thallium's distribution and concentration is poorly defined. However, recent report suggest direct correlation between Tl aqueous concentrations and the proximity of collection sites to metal mines, i.e., water concentrations were 13 and 2640 ng/l in samples from a river with a catchment without no metal mines and from an abandoned mine shaft in England, respectively. Importantly, concentrations of Tl in rivers in the vicinity of mine-related effluents reached about 770 ng/l (Tatsi and Turner, 2014). In lakes, i.e., Lake Ontario, bioavailable concentrations of Tl in the sediments are greater than in the water, with deep-water sediments having the greatest amounts of bioavailable Tl (Borgmann et al., 1998).

In terms of bioaccumulation up the food chain, studies with the bioindicator amphipod crustacean species *Hyalella Azteca* report increased Tl uptake and body burden, but the body concentrations were below the toxic threshold (Borgmann et al., 1998). Regarding plant uptake, Tl^+ is the dominant species, but in extracts of plants cultivated in the presence of tailing sediments about 10% is Tl^{3+}. Some plants are highly tolerant to Tl and bioconcentrate it very effectively; one such example is *Sinapis alba*, which could potentially be used for bioremediation purposes as it bioconcentrates Tl much more effectively than Cd, Zn, or Pb. (Krasnodebska-Ostrega et al., 2012). Another plant, the aquatic macrophyte *Callitriche cophocarpa*, could also be used for phytoremediation of Tl contaminated waters, but in this case the metal selectivity is not directed towards Tl, i.e., this would be a pan-metal phytoremediation approach (Augustynowicz et al., 2014). Clams and mussels also bioconcentrate Tl in aquatic environments with bioconcentration factor (BCF) values of 18.2 and 11.7, respectively, as described by Zitko and Carson (1975). Bioconcentration factors for muscle tissue of juvenile salmon range from 27 to 1430 as described by Zitko et al. (1975). Studies on Tl levels in wildlife are limited, but in wild ducks in Japan contents of Tl in kidney and liver ranged from 0.42 to 119.61 and 0.10 to 33.94 µ/g dry weight, respectively. Importantly, Tl levels were dependent on sampling locations with the highest mean Tl content observed in samples collected from the Ibaraki Prefecture (Mochizuki et al., 2005).

In certain geographical areas, i.e., the Lanmuchang area of southwest Guizhou, China, soil concentrations of Tl are naturally much higher, i.e., 40–124 mg/kg in soils originating from mining areas, 20–28 mg/kg in slope wash materials, and 14–62 mg/kg in alluvial deposits downstream, which are levels substantially higher than the background soil concentrations of <0.2–0.5 mg/kg (Xiao et al., 2004a, 2004b). This is important from a biomagnification point of view as Tl levels in edible plants ranging from green cabbage to rice and corn that are grown in such areas could reach up to 500 mg/kg dry weight (the highest levels are found in cabbage). This would translate to Tl ADI of the local population 50 times higher than the Tl's intake of the background population (Xiao et al., 2004a, 2004b). Ground and surface waters from these "naturally contaminated" areas also contain markedly elevated levels of Tl, further increasing the potential for human overexposure of the local population (Xiao et al., 2004a, 2004b). Preliminary findings in Europe also show Tl uptake by edible vegetables, which is dependent upon the plant type/species and the Tl soil levels (Pavlickova et al., 2006). In the United States and Canada, data on such high levels of thallium in edible plants or animals are lacking, but there may be an increased risk for Tl intoxication by consumption of fish from the Great Lakes (Lin and Nriagu, 1999; Pickard et al., 2001). Due to its interaction with potassium, greater bioaccumulation, or Tl may take place in waters with low K levels (Kazantzis, 2007).

ECOTOXICOLOGY

Acute Effects

Studies examining the acute ecotoxicological effects of Tl are limited. Nevertheless, examination of the aquatic toxicity of Tl using *Daphnia magna* demonstrated that Tl^{3+} is substantially more potent than Tl^+ and, importantly, Tl^{3+} is more toxic to daphids than many other metals, including Cd, Ni, and Cu; Tl^{3+} toxicity was comparable to that of inorganic Hg with 48 h LC_{50} values ranging from 24 to 204 µg/l for Tl chloride and Tl acetate, respectively (Lan and Lin, 2005). In a related study, when the acute toxicity of Tl^+ (as thallium nitrate) was evaluated in two different daphids, *Daphnia magna* and *Ceriodaphnia dubia*, there was 2.5-fold difference in sensitivity, with *C. dubia* being more sensitive; the LC_{50} values, 0.65 and 1.7 mg/l on average for *D. magna* and *C. dubia*, respectively, also confirmed that daphids are more sensitive to Tl^{3+} (Lin et al., 2005). Marine microalgae, i.e., *Ulva lactuca*, are very sensitive to Tl^+ as >10 µg/l concentrations were toxic; toxicity was independent of the water's salinity and it was associated with substantial accumulation of Tl by *U. lactuca* (Turner and Furniss, 2012). To Atlantic salmon, Tl is toxic, with its incipient lethal level estimated to be around 30 µg/l. Importantly, Tl is more toxic to salmon than Zn and is of similar potency with Cd (Zitko et al., 1975). In rainbow trout, the average LC_{50} (96 h exposure) was estimated to be 4.3 mg/l (Pickard et al., 2001).

Chronic Effects

Studies examining the chronic ecotoxicological effects of Tl are extremely limited. A study with the duckweed *Lemna*

minor determined that Tl exposure stuns plant growth, with production of smaller fronds, being the most sensitive plant response to Tl stress. Notably, curtailment of frond expansion was more sensitive than inhibition of frond multiplication; these and other toxic effects of Tl were not permanent at lower (up to EC_{50}) exposure levels (Kwan and Smith, 1988).

There are no true chronic studies with daphids, but a comparison of *C. dubia*'s 48 h LC_{50} of approximately 0.65 mg/l (Lin et al., 2005) with a 7-day LC_{50} in this aquatic organism, which is 0.37 mg/l (Pickard et al., 2001) hints towards the possiblity of lower Tl levels being toxic to aquatic life under chronic exposure circumstances. Overall, chronic ecotoxicity data, especially from areas with dirrerential geochemically and/or anthropogenically driven Tl levels are needed.

MAMMALIAN TOXICOLOGY

Acute Effects

The human oral LD_{50} for Tl is estimated to be between 6 and 40 mg/kg; serious, nonfatal intoxication occurs at lower exposure levels. This estimate is based on cases of suicidal ingestion of Tl salts or homicide attempts (ATSDR, 1992; WHO, 1996). Acute thallium intoxication is characterized with a triad of symptoms: gastrointestinal, polyneuropathy, and alopecia; if left untreated and the exposure level is not lethal, progression through these symptoms will take place (ATSDR, 1992; Hoffman, 2003; Cvjetko et al., 2010). In cases of high deliberate or accidental exposures, contemporary timely therapy may not be successful. For example, despite multiple-dose activated charcoal and Prussian blue (PB) therapy initiated <1 h after initial symptoms, the patient died 3 days post exposure; his last blood and urine Tl concentrations were extremely high: 5369 µg/l and >2000 µg/l, for blood and urine, respectively (Riyaz et al., 2013).

While deliberate intoxications with Tl involve primarily Tl salts; occupational and children exposure to Tl may take place in the form of Tl-containing ceramic superconductor material (CSM) (Mulkey and Oehme, 1998). Comparison of bioavailability of Tl from CSM to that of Tl sulfate suggests that Tl in CSM has significantly lower bioavailability than Tl sulfate implying lower health risk from this source of Tl than the risk from Tl salts such as Tl sulfate (Mulkey and Oehme, 1998).

In neonatal rodents, Tl (Tl^+ acetate) causes long-lasting neuromyopathy manifested with progressive distal axonopathy and muscle fibers myopathic alterations (Barroso-Moguel et al., 1996). Much like in the rodent, signs of polyneuropathy persisted months after acute oral exposure to Tl-contaminated water (about 3124 mg/l) in human patients; damage of sensory nerve endings was still present after all dermatological clinical features have subsided (Lu et al., 2007). In a long-term follow-up of Tl-exposed human patients, using functional imaging with fluorodeoxyglucose positron emission tomography ([18]FDG PET), patients exhibited decreased glucose metabolism in the cingulate gyrus, bilateral frontal, and parietal lobes 2–5 months after exposure. These alterations in brain metabolism were accompanied with deficits in executive function and memory (Liu et al., 2013). Over time, partial, but not complete, recovery was observed (Liu et al., 2013), highlighting the fact that the neurological deficits associated with acute Tl intoxications are of particular concern.

Chronic Effects

Chronic or sub-chronic mammalian toxicity data for Tl are not abundant. Review of data summarized by EPA and ATSDR indicate that (i) long-term exposure to 1.4 mg/kg/day as thallium sulfate for 40–240 days is associated with increased lethality and (ii) chronic inhalation exposure to Tl (unknown amounts) in the workplace is associated with neurological deficits (ATSDR, 1992; EPA, 2009). Chronic animal studies are also scarce. Ninety-day oral gavage exposure of male and female Sprague-Dawley rats to approximately 0, 0.008, 0.04, or 0.20 mg/kg/day Tl was associated with increased incidence of alopecia, which was statistically significant at the mid and high dose for the female rats (EPA, 2009). The biological significance of this finding is not clear. Male reproductive toxicity of Tl was examined in adult male Wistar rats given 10 parts per million (ppm) Tl (as thallium sulfate) in the drinking water for up to 60 days. Sixty, but not 30-day, Tl exposure was associated with reduced sperm motility, increased immature germ cells in the tubular lumen, and with cytoplasmic vacuolation and distension of the smooth endoplasmic reticulum of Sertoli cells (Formigli et al., 1986). These findings indicate that the male reproductive system, Sertoli cells in particular, is susceptible to chronic Tl exposure (Formigli et al., 1986).

There are no reported chronic animal studies or cancer bioassays for Tl; similarly chronic effects of inhalation exposure to Tl in animal models have not been examined (EPA, 2009). There are several studies where sub-chronic or chronic effects of Tl given by injection have been studied. Of note, chronic (up to 26 weeks) s.c. exposure to Tl resulted in liver and kidney pathology in Sprague-Dawley rats; which at the structural level was confined primarily to the mitochondria (EPA, 2009). Two chronic systemic injection studies highlight the fact that the brain is a primary Tl target in both acute and chronic exposure scenarios. Thus, increased brain lipid peroxidation, which at the lower exposure level (0.8 mg/kg) was brain region specific, was observed in Tl exposed rats (Galvan-Arzate et al., 2000); in chronically exposed cats (2.3–4.5 mg/kg Tl[acetate] per week for 4–26 weeks) the main functional disturbances hypotonia and ataxia and the

pathological changes were confined primarily to sensory neurons and were "dying back" type (Kennedy and Cavanagh, 1977).

Due to the scarcity of data and the potential for chronic effects, reliable chronic exposure studies with Tl compounds where relevant exposure routes are employed are clearly needed.

Chemical Pathology

Thallium intoxication targets primarily the nervous system, the digestive system, and the skin (Tromme et al., 1998; Hoffman, 2003; Kazantzis, 2007; Cvjetko et al., 2010; Rusyniak et al., 2010). First clinical signs are observed 1–3 days after exposure and are dominated by nausea and vomiting; if exposure levels are in the LD_{50} range, cardiovascular collapse, coma, and death will ensue within 24 h (Tromme et al., 1998; Cvjetko et al., 2010). Unlike exposure to other metal salts, the gastrointestinal symptoms associated with Tl toxicity are relatively minor and constipation is more characteristic than diarrhea (Hoffman, 2003). Neurological deficits associated with nonlethal Tl overexposure are of greatest concern. When nonlethal exposure takes place, peripheral neuropathy develops, which initially is more severe in the lower extremities. Over time, the neurological symptoms become more wide-spread and may include autonomic neuropathy, various psychiatric symptoms, and seizures. Initial dermatological signs are not specific, i.e., scaling of palms and soles, and acne-like lesions on the face (Tromme et al., 1998), but at higher exposures a characteristic mark, a "thallium band," may appear in the scalp hair about 4 days post exposure (Tromme et al., 1998; Hoffman, 2003; Kazantzis, 2007; Cvjetko et al., 2010; Rusyniak et al., 2010). During the second (third) week after acute exposure, sudden alopecia, which becomes diffuse, develops (Tromme et al., 1998; Hoffman, 2003; Kazantzis, 2007). Another Tl intoxication characteristic, the Mee's lines, takes longer to appear, but are more pronounced than the lines seen in arsenic intoxications. When exposure is terminated and treatment is instituted, Tl tissue levels decline and the hair and nails usually regrow, but the neurological deficits persist and full recovery of neurological function may not take place (Hoffman, 2003; Lu et al., 2007; Liu et al., 2013).

Prussian blue is the treatment of choice for acute Tl intoxications (Hoffman, 2003), but recent studies suggest that DL-penicillamine may be useful addition to PB in combating acute thallotoxicosis (Montes et al., 2011).

Mode of Action(s)

Most thallium salts are water-soluble. Monovalent thallium (Tl[+]) has ionic radius and electrical charge that are similar to potassium (K[+]) and contribute to its toxic mechanism (EPA, 2009). Thallium is absorbed rapidly and distributed widely regardless of exposure route (Kazantzis, 2007). Its elimination rate, which is also exposure route independent, is rather slow; biological half-life is estimated to be 3.3 days (Kazant-zis, 2007). In part due to its slow elimination rate, Tl tends to accumulate in different organs and cellular compartments (ATSDR, 1992). The exact mechanism of Tl toxicity still remains unknown, but interference with potassium homeostasis and glutathione metabolism, as well as oxidative stress, may all play a role in Tl's toxic mechanism (Cvjetko et al., 2010). *In silico,* Tl^{3+} is slowly converted to Tl^+, which is more stable (Harris and Messori, 2002), but the extent of this conversion in a cellular context has not been fully described (Hanzel and Verstraeten, 2006; Hanzel and Verstraeten, 2009; Cvjetko et al., 2010). Thallium (Tl^+ and Tl^{3+}) increases cellular reactive oxygen species (ROS) production (Cvjetko et al., 2010). Lipid peroxidation and interference with intracellular antioxidants are suggested as being key components of Tl's toxic mode of action (Villaverde et al., 2004; Molina and Verstraeten, 2008). Competition of Tl^+ with K^+ has been associated with interference with many cellular processes ranging from modification of important metabolic enzymes, such as pyruvate kinase and fructose-1-6-bisphosphatase, to uncoupling of mitochondrial oxidative phosphorylation (Cvjetko et al., 2010). Thallium also interferes with enzymatic activity of multiple enzymes and interacts with exposed amino-sulfhydryl groups (Mulkey and Oehme, 1993; Cvjetko et al., 2010). *In vivo,* structural degenerative changes in the mitochondria and a "dying back" type of sensory neuron pathology were associated with chronic exposures of laboratory animals to Tl (Kazantzis, 2007). Of note, certain behavioral abnormalities correlate with increased lipid peroxidation in the brain of Tl-exposed rodents (Brown et al., 1985). Information regarding Tl's genotoxicity and mechanism(s) involved is scarce (Rodriguez-Mercado and Altamirano-Lozano, 2013). Scarcity of data and the lack of clearly identifiable mechanism(s) of thallium toxicity in acute and chronic exposure settings need to be addressed with further research.

SOURCES OF EXPOSURE

Occupational

Occupational exposures to Tl may occur through alloy production and machining or through the manufacturing of thallium-containing electronic components, optical lenses, and jewelry (ATSDR, 1992; WHO, 1996). Prior to its elimination, manufacturing of Tl-containing rodenticides was another source of occupational exposure to this metal (ATSDR, 1992; WHO, 1996; Kazantzis, 2007). Air concentration in the workplace of smelting operations is typically $>22 \mu g/m^3$ (WHO, 1996). Inhalation is also an important exposure route for workers engaged in the separation of industrial diamonds, manufacturing of Tl-containing optics, extraction of Tl, or roasting of pyrites (Kazantzis, 2007; EPA,

2009). The two primary routes of exposure to thallium workers that are associated with intoxications are dermal and oral (Ewers, 1988; Kazantzis, 2007).

Environmental

Food in the typical human diet contains $<1 mg/kg$ Tl and human intake is estimated to be $<5 \mu g/day$ (WHO, 1996). Absorption through inhalation in unpolluted environments is fairly low: not $>0.005 \mu g/day$ (WHO, 1996). Detectable, but low levels of Tl are also found in unpolluted saltwater and freshwater (EPA, 2009). However, Tl-polluted atmospheres from emissions from coal-burning power plants and metal smelting operations can contribute to environmental exposures greater than exposures through the diet (Kazantzis, 2007; EPA, 2009). Industrial activity can also lead to substantial increase of Tl levels in soil; i.e., concentrations of more than $100 mg/kg$ in soils from a mining region in China have been reported (Xiao et al., 2004a, 2004b). Some soils contain greater Tl levels from natural sources (Xiao et al., 2012). In either case, soil Tl can be mobilized by edible plants resulting in increased dietary intake of Tl and increased human and ecotoxicological risk for Tl intoxication in such areas (Kazantzis, 2007; EPA, 2009; Xiao et al., 2012). Increased Tl in metal mining effluents and its ineffective waste water removal by conventional means is another source of increased environmental Tl exposures (Peter and Viraraghavan, 2005).

INDUSTRIAL HYGIENE

Exposure Limits:

8 h shift permissible exposure level (PEL): $0.1 mg/m^3$ (OHSA, 2005)

8 h shift threshold limit value (TLV): $0.1 mg/m^3$ (ACGIH, 2010)

10 h shift recommended exposure limit (REL) and PEL: $0.1 mg/m^3$ (NIOSH, 2005; OHSA, 2005)

Immediately Dangerous to Life and Health (IDLH): $15 mg/m^3$ (NIOSH, 2005)

PPEs: NIOSH/OSHA:

$0.5 mg/m^3$: DM^; $1.0 mg/m^3$: DMXSQ^/SA; $2.5 mg/m^3$: SA: CF/PAPRDM^; $5 mg/m^3$: HiEF/SAT: CF/PAPR-THiE/SCBAF/SAF DM^; $15 mg/m^3$: SAF/PD, PP; §: SCBAF:PD, PP/SAF:PD, PP:ASCBA; Escape: HiEF/ SCBAE.

RISK ASSESSMENTS

Thallium is classified as an environmental and industrial pollutant, potentially dangerous to human health and wildlife

(EPA, 2009). From an ecotoxicological perspective, the ecotoxicological serious risk concentrations (SRC$_{ECO}$) values are as follows: (i) fresh water: 6.8 µg/l, ground water: 8.5 µg/l, marine water: 6.5 µg/l (a serious risk addition [SRA$_{ECO}$] value), sediment: 11 mg/kg, and soil: 2 mg/kg (RIVM, 2005). According to EPA and ATSDR, there is inadequate information to assess the carcinogenic potential of Tl/thallium salts (ATSDR, 1992; EPA, 2009). Subsegments of the population, i.e., people with preexisting neurological disease, kidney, and liver damage may be at risk for Tl toxicosis (ATSDR, 1992). Individuals with potassium deficiency may also be at risk (ATSDR, 1992; WHO, 1996, EPA, 2009).

REFERENCES

ACGIH (2010) Thallium and compounds: TLV chemical substances. In: *Threshold Lmit Values and Biological Exposure Indices*, ACGIH, editor. Cincinnati, OH: Americal Conference of Governmentl Industrial Hygienists

ATSDR (1992) *Toxicological Profile for Thallium*, Atlanta, Georgia, USA: Public health service, United States Department of Health and Human Services.

Augustynowicz, J., Tokarz, K., Baran, A., and Plachno, B.J. (2014) Phytoremediation of water polluted by thallium, cadmium, zinc, and lead with the use of macrophyte *Callitriche cophocarpa*. *Arch. Environ. Contam. Toxicol.* doi: 10.1007/s00244-013-9995-0.

Barroso-Moguel, R., Mendez-Armenta, M., Villeda-Hernandez, J., Rios, C., and Galvan-Arzate, S. (1996) Experimental neuromyopathy induced by thallium in rats. *J. Appl. Toxicol.* 16(5):385–389.

Borgmann, U., Cheam, V., Norwood, W.P., and Lechner, J. (1998) Toxicity and bioaccumulation of thallium in *Hyalella azteca,* with comparison to other metals and prediction of environmental impact. *Environ. Pollut.* (Barking, Essex: 1987) 99(1):105–114.

Brown, D.R., Callahan, B.G., Cleaves, M.A., and Schatz, R.A. (1985) Thallium induced changes in behavioral patterns: correlation with altered lipid peroxidation and lysosomal enzyme activity in brain regions of male rats. *Toxicol. Ind. Health* 1(1):81–98.

Crookes, W. (1861) On the existence of a new element, probably of the sulphur group. *Chem. News* 3:193–194.

Cvjetko, P., Cvjetko, I., and Pavlica, M. (2010) Thallium toxicity in humans. *Arh. Hig. Rada Toksikol.* 61(1):111–119.

EPA (2009) *Toxicological Review of Thallium and Compounds*, Washington, DC: U.S. Environmental protection Agency.

Ewers, U. (1988) Environmental exposure to thallium. *Sci. Total Environ.* 71(3):285–292.

Fergusson, J.E. (1990) *The Heavy Elements: Chemistry, Environmental Impact and Health Effects*, Oxford, UK: Pergamon.

Formigli, L., Scelsi, R., Poggi, P., Gregotti, C., Di Nucci, A., Sabbioni, E., Gottardi, L., and Manzo, L. (1986) Thallium-induced testicular toxicity in the rat. *Environ. Res.* 40(2):531–539.

Galvan-Arzate, S., Martinez, A., Medina, E., Santamaria, A., and Rios, C. (2000) Subchronic administration of sublethal doses of thallium to rats: effects on distribution and lipid peroxidation in brain regions. *Toxicol. Lett.* 116(1–2):37–43.

Hanzel, C.E. and Verstraeten, S.V. (2006) Thallium induces hydrogen peroxide generation by impairing mitochondrial function. *Toxicol. Appl. Pharmacol.* 216(3), doi: 485-92, 10.1016/j.taap.2006.07.003

Hanzel, C.E. and Verstraeten, S.V. (2009) Tl(I) and Tl(III) activate both mitochondrial and extrinsic pathways of apoptosis in rat pheochromocytoma (PC12) cells. *Toxicol. Appl. Pharmacol.* 236(1):59–70.

Harris, W.R. and Messori, L. (2002) A comparative study of aluminum(III), gallium(III), indium(III), and thallium(III) binding to human serum transferrin. *Coordin. Chem. Rev.* 228(2):237–262.

Hoffman, R.S. (2003) Thallium toxicity and the role of Prussian blue in therapy. *Toxicol. Rev.* 22(1):29–40.

Kazantzis, G. (2007) Thallium. In: Nordberg, G.F., Fowler, B.A., Nordberg, M., and Friberg, L., editors. *Handbook on the Toxicology of Metals*, New York, NY: Elsevier, pp. 827–837.

Kennedy, P. and Cavanagh, J.B. (1977) Sensory neuropathy produced in the cat with thallous acetate. *Acta Neuropathol.* 39(1):81–88.

Krasnodebska-Ostrega, B., Sadowska, M., and Ostrowska, S. (2012) Thallium speciation in plant tissues-Tl(III) found in *Sinapis alba* L. grown in soil polluted with tailing sediment containing thallium minerals. *Talanta* 93:326–329.

Kwan, K.H. and Smith, S. (1988) The effect of thallium on the growth of *Lemna minor* and plant tissue concentrations in relation to both exposure and toxicity. *Environ. Pollut.* (Barking, Essex: 1987) 52(3):203–219.

Lan, C.H. and Lin, T.S. (2005) Acute toxicity of trivalent thallium compounds to Daphnia magna. *Ecotoxicol. Environ. Saf.* 61(3):432–435.

Lawrence, D.T. and Kirk, M.A. (2007) Chemical terrorism attacks: update on antidotes. *Emerg. Med. Clin. North Am.* 25(2):567–595 (abstract xi).

Leonard, A. and Gerber, G.B. (1997) Mutagenicity, carcinogenicity and teratogenicity of thallium compounds. *Mutat. Res.* 387(1):47–53.

Lin, T.S., Meier, P., and Nriagu, J. (2005) Acute toxicity of thallium to *Daphnia magna* and *Ceriodaphnia dubia*. *Bull. Eviron. Contam. Toxicol.* 75(2):350–355.

Lin, T.S. and Nriagu, J. (1999) Thallium speciation in the Great Lakes. *Environ. Sci. Technol.* 33(19):3394–3397.

Liu, C.H., Lin, K.J., Wang, H.M., Kuo, H.C., Chuang, W.L., Weng, Y.H., Shih, T.S., and Huang, C.C. (2013) Brain fluorodeoxyglucose positron emission tomography ((1)(8)FDG PET) in patients with acute thallium intoxication. *Clin. Toxicol. (Phila)* 51(3):167–173.

Lu, C.I., Huang, C.C., Chang, Y.C., Tsai, Y.T., Kuo, H.C., Chuang, Y.H., and Shih, T.S. (2007) Short-term thallium intoxication:

dermatological findings correlated with thallium concentration. *Arch. Dermatol.* 143(1):93–98.

Mochizuki, M., Mori, M., Akinaga, M., Yugami, K., Oya, C., Hondo, R., and Ueda, F. (2005) Thallium contamination in wild ducks in Japan. *J. Wildl. Dis.* 41(3):664–668.

Molina, L.D.P. and Verstraeten, S.V. (2008) Thallium(III)-mediated changes in membrane physical properties and lipid oxidation affect cardiolipin-cytochrome c interactions. *Biochim. Biophys. Acta* 1778(10):2157–2164.

Montes, S., Perez-Barron, G., Rubio-Osornio, M., Rios, C., Diaz-Ruiz, A., Altagracia-Martinez, M., and Monroy-Noyola, A. (2011) Additive effect of DL-penicillamine plus Prussian blue for the antidotal treatment of thallotoxicosis in rats. *Environ. Toxicol. Pharmacol.* 32(3):349–355.

Mulkey, J.P. and Oehme, F.W. (1993) A review of thallium toxicity. *Vet. Hum. Toxicol.* 35(5):445–453.

Mulkey, J.P. and Oehme, F.W. (1998) A safety study of thallium-containing ceramic superconductor material in rats. *Vet. Hum. Toxicol.* 40(1):11–14.

NIOSH (2005) *Thallium (Soluble Compounds, as Tl)*, Washington, DC: Department of Health and Human Services, Centers for Disease Control and Prevention.

OHSA (2005) *Thallium, Soluble Compounds (as Tl)*, Washington, DC: United States Department of Labor, Occupational Safety, and Health Administration.

Pavlickova, J., Zbiral, J., Smatanova, M., Habarta, P., Houserova, P., and Kuban, V. (2006) Uptake of thallium from naturally-contaminated soils into vegetables. *Food Addit. Contam.* 23(5):484–491.

Peter, A.L. and Viraraghavan, T. (2005) Thallium: a review of public health and environmental concerns. *Environ. Int.* 31(4):493–501.

Pickard, J., Yang, R., Duncan, B., McDevitt, C.A., and Eickhoff, C. (2001) Acute and sublethal toxicity of thallium to aquatic organisms. *Bull. Environ. Contam. Toxicol.* 66(1):94–101.

RIVM (2005) In: van Vlaardingen, P.L.A., Posthumus, R., and Posthuma-Doodeman, C.J.A.M., *Environmental risk limits for nine trace elements* Bilthoven, The Netherlands: National Institute for Public Health and the Environment.

Riyaz, R., Pandalai, S.L., Schwartz, M., and Kazzi, Z.N. (2013) A fatal case of thallium toxicity: challenges in management. *J. Med. Toxicol.* 9(1):75–78.

Rodriguez-Mercado, J.J. and Altamirano-Lozano, M.A. (2013) Genetic toxicology of thallium: a review. *Drug Chem. Toxicol.* 36(3):369–383.

Rusyniak, D.E., Arroyo, A., Acciani, J., Froberg, B., Kao, L., and Furbee, B. (2010) Heavy metal poisoning: management of intoxication and antidotes. *EXS* 100:365–396.

Rusyniak, D.E., Furbee, R.B., and Kirk, M.A. (2002) Thallium and arsenic poisoning in a small midwestern town. *Ann. Emerg. Med.* 39(3):307–311.

Tatsi, K. and Turner, A. (2014) Distributions and concentrations of thallium in surface waters of a region impacted by historical metal mining (Cornwall, UK). *Sci. Total Environ.* 473–474:139–146.

Tromme, I., Van Neste, D., Dobbelaere, F., Bouffioux, B., Courtin, C., Dugernier, T., Pierre, P., and Dupuis, M. (1998) Skin signs in the diagnosis of thallium poisoning. *Br. J. Dermatol.* 138(2):321–325.

Turner, A. and Furniss, O. (2012) An evaluation of the toxicity and bioaccumulation of thallium in the coastal marine environment using the macroalga. *Mar. Pollut. Bull.* 64(12):2720–2724.

Villaverde, M.S., Hanzel, C.E., and Verstraeten, S.V. (2004) *In vitro* interactions of thallium with components of the glutathione-dependent antioxidant defence system. *Free Radic. Res.* 38(9):977–984.

WHO (1996) *Thallium: Environmental Health Criteria*, Geneva, Switzerland: World Health Organization.

Xiao, T., Guha, J., Boyle, D., Liu, C.Q., and Chen, J. (2004a) Environmental concerns related to high thallium levels in soils and thallium uptake by plants in southwest Guizhou. *Sci. Total Environ.* 318(1–3):223–244.

Xiao, T., Guha, J., Boyle, D., Liu, C.Q., Zheng, B., Wilson, G.C., Rouleau, A., and Chen, J. (2004b) Naturally occurring thallium: a hidden geoenvironmental health hazard? *Environ. Int.* 30(4):501–507.

Xiao, T., Yang, F., Li, S., Zheng, B., and Ning, Z. (2012) Thallium pollution in China: a geo-environmental perspective. *Sci. Total Environ.* 421–422:51–58.

Yildirim, M., Ikbal, M., Tos, T., Seven, B., Pirim, I., and Varoglu, E. (2005) Genotoxicity of thallium-201 in patients with angina pectoris undergoing myocardial perfusion study. *Tohoku. J. Exp. Med.* 206(4):299–304.

Zitko, V. and Carson, W.V. (1975) Accumulation of thallium in clams and mussels. *Bull. Environ. Contam. Toxicol.* 14(5):530–533.

Zitko, V., Carson, W.V., and Carson, W.G. (1975) Thallium: occurrence in the environment and toxicity to fish. *Bull. Environ. Contam. Toxicol.* 13(1):23–30.

34

TIN

Raymond D. Harbison and David R. Johnson

First aid: Remove from exposure. Give fresh air. Immediately irrigate eyes, wash the skin with soap. Provide respiratory support. If ingested by swallowing, seek immediate medical attention.

Target organs: Metallic and inorganic tin targets the eyes, skin, respiratory tract. Organic tin compounds target the central nervous system, liver, kidneys, urinary tract, and blood.

Risk/safety phrases: Irritant; inhalation risk; ingestion may be neurotoxic. The solid form is non-combustible; however, the powdered form may ignite.

Exposure limits and threshold limit values for tin compounds:

Tin Compounds	CAS Number	PEL, mg/m^3	REL, mg/m^3	TLV, mg/m^3
Tin	7440-31-5	2	2	2
Inorganic compounds, excluding oxides	7440-31-5	2	2	2
Inorganic compounds, including oxides but excluding SnH$_4$	Varies	–	–	2
Organic compound	Varies	0.1	0.1	0.1; STEL: 0.2
Tin(II) oxide (as Sn)	21651-19-4	15 (maritime shipyards)	2	2
Tin(IV) oxide (as Sn)	18282-10-5	15 (maritime shipyards)	2	2

From National Institute for Occupational Safety and Health (NIOSH, 2010): *Pocket guide to chemical hazards*, NIOSH 94–116, Washington, DC, 2010, US Department of Health and Human Services; American Conference of Governmental Industrial Hygienists (ACGIH, 2003): *Threshold limit values (TLVs) for chemical substances and physical agents and biological exposure index (BEIs)*, Cincinnati, 2003, American Conference of Governmental Industrial Hygienists.

CAS, Chemical Abstract Service; *PEL*, permissible exposure limit, set by the Occupational Safety and Health Administration (OSHA); *REL*, recommended exposure limit, set by the NIOSH;, threshold limit value, set by the ACGIH; *STEL*, short-term exposure limit, set by the ACGIH.

BACKGROUND AND USES

Tin is primarily produced from the ore cassiterite (SnO_2). It has a variety of industrial and domestic uses; its usefulness stems from its pliability and ability to readily form compounds and alloys with other metals, inorganic compounds, and organic compounds. Tin metal is used to line cans for storing and transporting food, beverages, and aerosols. Bronze is an alloy of tin and copper; pewter is an alloy of tin with various amounts of antimony, copper, bismuth, and sometimes lead. Soldering compounds are made with tin alloys. Inorganic tin is used in making toothpaste, perfumes, soaps, coloring agents, food additives, dyes, and in the glass industry (ATSDR, 2005). Artificial organotin compounds are used to make plastics, food packages, as a stabilizer for PVC (pipes), agrochemicals (such as pesticides), paints, wood

Hamilton & Hardy's Industrial Toxicology, Sixth Edition. Edited by Raymond D. Harbison, Marie M. Bourgeois, and Giffe T. Johnson.
© 2015 John Wiley & Sons, Inc. Published 2015 by John Wiley & Sons, Inc.

preservatives, surface disinfectants (including hospital and veterinary disinfectants), bacteriostats, antimicrobials and slimicides, biocides such as rodent repellants, laundry sanitizers, mildewcides, industrial catalysts, scintillation detectors for gamma and X-rays, ballistic additives for solid-rocket engine fuels, and ionophores in liquid membrane ion-selective electrodes (Nath, 2008; Cao et al., 2008; Kotake, 2012). Organotin compounds have been used extensively in boat paints due to their excellent antifouling properties (Okoro et al., 2011; Graceli et al., 2013).

Tributyltin is used as a slime control in paper mills, for disinfection of circulating industrial cooling waters, as antifouling agents, and in the preservation of wood (Antizar-Ladislao, 2008; Nakanishi, 2008). Tributyltin, triphenyltin, and tricyclohexyltin are used in agriculture as fungicides, anti-helminthics, miticides, herbicides, nematocides, ovicides, molluscicides, ascaricides, pesticides, in industry as biocides, and as antifouling agents for large ship bottoms and fishery farm nets (as industrial catalysts). In the past, several organometallic metallocene complexes of tin have been identified as having antitumor activity (Köpf-Maier et al., 1988; Winship, 1988). Tin protoporphyrin has been used to treat acute hepatic porphyria (Dover et al., 1993). Niobium–tin compounds are used to build superconducting magnets (Gaballe, 1993). Zirconium–tin alloys are used for cladding nuclear fuel (Campbell, 2008). Tin is used in advanced Li-ion batteries as the negative electrode (Lucas et al., 2011). Large amounts of organotins are used in the plastics industry, and large amounts of tin are also recycled.

PHYSICAL AND CHEMICAL PROPERTIES

Tin is a ductile, malleable, lustrous solid with a gray to almost silver-white color. It has: atomic number 50; molecular weight 118.7; melting point 449 °F (232 °C); boiling point 4118 °F (2270 °C); and specific gravity 7.28. Tin forms covalent bonds with carbon forming a variety of organometallic or organotin compounds, and the physical and chemical properties differ for various compounds such as Tin II (SnO) and Tin IV (SnO_2), organotins, and inorganic tin compounds. Tin compounds are generally insoluble in water and have vapor pressures of approximately 0 mm Hg. Tin is not easily oxidized and resists corrosion; however, powdered tin has a tendency to oxidize, especially at high humidity.

MAMMALIAN TOXICOLOGY

While tin in its inorganic form is generally considered nontoxic, various organotin compounds have shown significant toxicity in animals and humans.

Acute Effects

Metallic and inorganic tin is of low oral toxicity to humans because of its poor gastrointestinal (GI) absorption and rapid excretion in the feces. About 5% is absorbed from the GI tract, widely distributed throughout the body and then excreted by the kidneys (Winship, 1988). Doses of 100 mg can cause acute GI effects resembling food poisoning. Some reports suggest that the concentration of tin may be more important than the dose in causing acute GI symptoms. Urinary tin levels of 625–1600 µg/l have been associated with illness. There are reports of acute health effects as a result of consumption of highly contaminated fruit juices. As far back as 1890, there were case reports of tin poisoning affecting four individuals who ate cherries from the same large tinned can. Analysis of the juice found it to be strongly acidic due to the presence of malic acid with an estimated "1.9 grains of the oxide of tin or 3.2 grains of the malate of tin per fluid ounce" (1 grain = 65 mg). The authors estimated doses of 4–10 grains of the malate of tin (260–650 mg). The symptoms included nausea, abdominal pain, vomiting, diarrhea, drowsiness, feeble weak irregular pulse, and unconsciousness in one patient. One patient had cramps in the legs and two had a trace of albumin in the urine. Although treatment was unconventional by today's standards, all four patients were practically well by the next morning. The authors concluded that tin salts act as an irritant and cardiac poison. They warned about the dangers of eating fruits preserved in acid juices in tin cans (Luff and Metcalfe, 1890). In 2005 the World Health Organization (WHO) reported that human poisoning from consumption of food and drink contaminated with inorganic tin have resulted in abdominal distension and pain, vomiting, diarrhea, and headache. The symptoms occurred when the food or beverage was found to contain tin at concentrations varying from 250 to 2000 mg/kg. They concluded from available information that the GI irritation was more closely related to the concentration and nature of tin in the product rather than the dose of tin ingested on a body-weight basis. In 2003, Blunden and Wallace suggested that there is little evidence for an association between the consumption of food containing tin at concentrations up to 200 parts per million (ppm) and acute GI effects, noting that further studies needed to be done (Blunden and Wallace, 2003). Food and beverages should not be stored in opened tin-plated cans. The lacquering of tin-plated cans prevents the migration of inorganic tin into food and beverages (WHO, 2005a).

Inorganic tin compounds (excluding oxides) may be skin irritants. Contact with them may cause intense itching and erythematous lesions. People with skin diseases may be particularly susceptible to this irritation. Signs and symptoms typically resolve after exposure has ceased. One case-control study linked occupational exposure to tin with chronic renal failure (Nuyts et al., 1995). Human and animal studies have shown that the ingestion of large amounts of inorganic tin compounds can cause stomach ache, anemia, and liver and kidney problems. Gunay et al (2006) studied the effect of long-term exposure to tin fumes in tinsmiths and concluded that the occupational exposure to heavy tin fumes was associated with left ventricular dysfunction and sclerosis of the aortic valve.

Organotin compounds such as tributyltin, trimethyltin, and triethyltin are more toxic than inorganic tin compounds. The ethyl derivative of organotins is the most toxic, and triethyltin and trimethyltin are known to be neurotoxic (Winship, 1988). As the number of attached alkyl groups increases, the toxicity of organotins compounds also increases (Nath, 2008). Trialkyltin is more toxic than dialkyltin or monoalkyltin. Various organotins are absorbed through the GI tract and skin and may cause irritation of the mucous membranes, eyes and skin, respiratory irritation, GI effects, nausea, vomiting, diarrhea, neurological problems, headache, fatigue, muscular weakness, and paralysis followed by respiratory failure after acute exposure to high amounts (Okoro et al., 2011). Some organotin compounds can cause cerebral edema and produce central nervous system and cardiovascular effects while others may cause hepatic necrosis. Organotins can be highly lipophilic and therefore can easily penetrate the blood–brain barrier ((Nath, 2008; Kotake, 2012). Dermatitis has been reported after exposure to tributyltin phthalate (Hamanaka et al., 1992). Acute burns of skin and eyes have been reported from tributyltin and dibutyltin compounds after brief contact (Lyle, 1958). Lethal cases have been reported due to ingestion of very high amounts of organotins (ATSDR, 2005).

Chronic Effects

Inhalation of tin dust or tin oxides may produce benign pneumoconiosis called *stannosis* (Robertson et al., 1961). Stannosis is *not* associated with symptoms or significant pulmonary impairment; however, there are distinctive chest X-ray findings. The chest X-rays of this condition resemble those of siderosis, showing dense bilateral nodular infiltrates. The opacities are unusually dense because of the high radiopacity of tin. This should facilitate early diagnosis. Blackish or gray macules consisting of dust-laden macrophages are present in the lungs (Morgan and Seaton, 1984). Stannosis has not been associated with morbidity or decreased life expectancy.

Organotin compounds have recently gained research interest due to concerns about their potential for bioaccumulation and biomagnification in food chains coupled with the potential adverse effects on human health and environment. Organotin compounds such as tributyltin are suspected as endocrine disruptors in mammals (Kotake, 2012).

Animal studies have shown that some organotins such as dibutyltins, tributyltins, and triphenyltins can affect the reproductive system (ATSDR, 2005). Other organotins have been associated with neurotoxicity or immunotoxicity in animal studies (Nath, 2008).

Carcinogenicity Tin has not been established as a human carcinogen. Animal studies have not shown associations between exposure to inorganic tin compounds and the development of cancer. Organotin compounds have been associated with cancer in animals in a limited number of studies. A study in rats and another in mice found triphenyltin hydroxide to be associated with the development of cancer after long-term oral administration (ATSDR, 2005). Based on animal studies, Nagano et al. (2011) recently reported an association between the inhalation of indium–tin oxide and the development of benign and malignant lung tumors, preneoplastic lesions of bronchioloalveolar hyperplasia, fibrosis of alveolar walls, and thickening of pleural walls.

Tin Mining and Lung Cancer Extensive research has attempted to explain the high incidence of lung cancer among Chinese tin miners (Wu et al., 1989). Miners and smelter workers over 40 years old with at least 10 years of work experience have annual lung cancer incidence rates >1% (Taylor, 1994). Factors such as smoking, radon, arsenic, and silica exposure may all contribute to these high rates of lung cancer. Qiao et al. (1989) reported that miners in the highest quartile of radon exposure had an odds ratio of 9.5 for lung cancer when controlled for arsenic exposure. In a much larger cohort study, Xuan et al. (1993) reported an increased risk of 0.2% per working month exposed to radon while controlling for arsenic exposure. Pathohistogenetic examination of 100 lung cancer cases also implicated radon rather than ore dust as having a greater role in cancer etiology (Sun et al., 1989). Taylor et al. (1989) reported that tin miners in the highest quartile of arsenic exposure had a relative risk of 22.6 when the variables of radon and smoking were controlled. Silicosis may also be responsible for lung cancer in tin miners (Fu et al., 1994).

One case-control study indicated that the increased fruit and vegetable intake among some tin miners might have a protective effect against the development of lung cancer (Forman et al., 1992)

Mechanism of Action(s)

Animal studies have shown that ingestion of stannous chloride interferes with the body's handling of copper, iron, and zinc; and limited data suggest a possible neurotoxic effect involving calcium channel neuromuscular transmission. Patch testing has shown people to be allergic to tin (WHO, 2005b).

Rabbits treated intraperitoneally with tin were reportedly anemic, but rabbits exposed to tin orally had only transient anemia (Chmielnick et al., 1993).

Some tin compounds, stannic chloride in particular, have been reported to induce chromosomal abnormalities in human peripheral lymphocytes (Ganguly et al., 1992).

Regarding organotins, various mechanisms of action have been described depending on the organotin compound and type of animal being studied. Proposed mechanisms include induction of apotosis, uncoupling of mitochondrial energy transduction and ion exchange, inhibition of membrane functions, interference with ATPase activity, decreases in functional activity of intestinal cytochrome-450, and inhibition of intracellular phospholipid transport affecting the immune systems. Organotins mainly favor fat accumulation and have been associated with endocrine and lipid disruption. They have been reported to affect the heme metabolism, the cardiovascular and central nervous system, and cause a fall in blood pressure due to depression of the vascular smooth muscle (Nath, 2008; Pagliarani et al., 2013; ATSDR, 2005).

Animals dosed with trialkyl tin did not suffer permanent brain damage or demyelination of the spinal cord found with exposure to some other alkyl metals (i.e., mercury). However, postmortem studies confirmed organotin-induced hepatic necrosis and cerebral edema with irreversible cell damage in the area of the nucleus amygdaloideus of humans. These changes correlated with the clinical manifestations observed in organotin-intoxicated workers.

Kotake (2012) reported on the potential mechanisms of organotin toxicity in mammals, including ATP synthase inhibition; disruption of steroid homeostasis; agonistic activities toward nuclear receptors; obesogenic effect and immunotoxicity; neurotoxicity; and GluR2 decrease by endogenous levels of organotins. Nakanishi (2008) describes organotins as endocrine-disrupting chemicals possibly acting as nanomolar agonists for retinoid X receptor and peroxisome proliferator-activated receptors. The critical target molecules for the toxicity of organotin compounds remain unclear and further studies are needed (Kotake, 2012; Pagliarani et al., 2013).

SOURCES OF EXPOSURE

All forms of tin can be found in soil, water, and air in various environmental settings, including areas of natural deposits, near mining operations, manufacturing, or where tin products are being used. However, eating food contaminated with tin is considered the main route of exposure.

Occupational

Reports of oral occupational exposure are lacking. Exposure to tin dust may occur during tin ore extraction, bagging, smelting, and refining operations. As indicated in the preceding discussion regarding cancer, tin miners may be exposed to other potentially toxic substances such as silica, arsenic, radon, thorium, uranium, lead, as well as to bismuth and antimony in the roasting and smelting process. Exposure to these toxic metals may also occur during the preparation and use of tin alloys and solders (Taylor et al., 1989; Dosemeci et al., 1993; WHO, 2005b). Tinsmiths can be exposed to heavy fumes of tin; such exposure should be minimized and workers monitored for heart disease (Gunay et al., 2006). It has been shown that workers involved in electric cable splicing may contaminate their homes with lead and tin (Rinehart and Yanagisawa, 1993).

Workers involved with the use or manufacture of inorganic tin compounds or organotins may be exposed. Potential occupational exposure to organotins is increasing due to the manufacturing of many new organotins with a variety of applications; however, current data on the frequency, concentration, and duration of worker exposure are lacking (ATSDR, 2005).

Environmental

Canned foods make the biggest contribution to the dietary intake of tin. Tin exposure to the general public has been reduced greatly in the past few decades, primarily as a result of improvements in the canning process such as tinning cans with an enamel or lacquered overcoat and crimping lids. Prior to these improvements, tin-canned food could contain tin concentrations of 100 ppm.

Human exposure to organotins can potentially occur via contaminated dietary sources such as seafood, shellfish, and food crops (Nakanishi, 2008). The potential for water contamination with organotins exists due to their use in the manufacture of plastics and PVC pipes.

The daily intake of tin from food is <1 mg. Concentrations of tin in soil range from 5 to 200 ppm. Rural air concentrations are below 10 ng/m^3.

Pewter dinnerware contains >90% tin as an alloy. Stannous fluoride is also used as a toothpaste additive.

Tributyl tin has been reported as one of the most toxic chemicals released into the marine environment by an anthropometric source (Cao et al., 2008), and there are concerns about the increased use of organotins in general resulting in marine contamination via leaching and runoff (Antizar-Ladislao, 2008). Exposure to organotin-contaminated water and sediments induce accumulation of organotins in mollusks and biological effects such as endocrine disruption and imposex (female masculinization). Health effects could potentially result from human consumption of contaminated seafoods (Cao et al., 2008; Nakanishi, 2008; Graceli et al., 2013).

Because of the many increasing applications, organic and inorganic tin has been considered to be the third most important elemental pollutant in the ecosystem (Nath, 2008).

Drinking water is not currently considered a significant source of tin exposure and inhaled environmental air represents very low exposure (WHO, 2005b).

INDUSTRIAL HYGIENE

It is necessary to monitor air concentrations in occupational settings where tin dust exposure is a risk, keeping concentrations within the REL, TLV, and PEL. See the table in the introduction to this chapter. Monitoring methods are available in OSHA and NIOSH publications and websites.

Wet processes can minimize exposures during tin ore extraction; however, dust and fumes may still escape during bagging, smelting, and refining operations. Due to the increasing manufacture and use of alloys and organotins, employers need to be educated about the risk of organotin exposure and implement strategies to minimize exposure to the skin, lungs, and GI tract. Periodic sampling of the worker's environment may be useful in assessing and mitigating risks. Tin mining might involve exposure to radon, thorium, uranium, silica, lead, arsenic, bismuth, and antimony; and industrial hygiene measures should be implemented for these substances as well.

Ultrasensitive analytical techniques have been developed for the measurement of tin in tissues and urine; however, relationships between tin dose and biological indicators have not yet been established (WHO, 2005b). De Azevedo et al. (2013) reported the validation of a procedure for the determination of tin in whole blood and urine utilizing graphite furnace atomic absorption spectrometry.

MEDICAL MANAGEMENT

Because most forms of metallic and inorganic tin are poorly absorbed and acute intoxication is rare, treatment guidelines for acute tin intoxication are sparse or nonexistent in the medical literature. Antidotes are not published. Acute tin intoxication has on rare occasion occurred after ingestion of tin-contaminated food stored in tin cans; most cases have rapidly recovered with removal of exposure. Persons suspected of acute intoxication should be provided close observation and supportive care for symptoms of fever, headache, nausea, vomiting, diarrhea, abdominal cramps, and bloating, which should be short lived. Monitor cardiac function and rhythm. Gastric lavage might be considered in a few specific situations after consulting the local poison control center. A complete workup for other forms of intoxication, including medications and illicit drugs should be included until tin intoxication is proven. A thorough exposure history is imperative to confirm the diagnosis and to identify the source of tin exposure in order to prevent further exposure of the intoxicated individual or others. Contacting the local poison control center may be helpful in the management of the case and is an important reporting responsibility allowing public health authorities to document and track cases. Additionally, the poison control center may be helpful in identifying laboratories that can determine tin levels in urine, blood, and environmental samples.

Although organotins are considered more toxic than inorganic tin compounds and more likely to adversely affect the nervous system and liver, similar medical management principles should be followed as for tin and inorganic tin compounds, including observation, supportive care, confirmation of exposure, confirmation of intoxication, and reporting of the case.

Identifying and eliminating exposure to tin compounds is a must for both acute and chronic intoxication.

REFERENCES

Antizar-Ladislao, B. (2008) Environmental levels, toxicity and human exposure to tributyltin (TBT)-contaminated marine environment: a review. *Environ. Int.* 34:292–308.

American Conference of Governmental Industrial Hygienists (ACGIH) (2003) *Threshold Limit Values (TLVs) for Chemical Substances and Physical Agents and Biological Exposure Indices (BEIs)*, Cincinnati, OH: American Conference of Governmental Industrial Hygienists.

Agency for Toxic Substances and Disease Registry (ATSDR) (2005) *Toxicological Profile for Tin and Tin Compounds*, Atlanta, GA: U.S. Department of Health and Human Services.

Blunden, S. and Wallace, T. (2003) Tin in canned food: a review and understanding of occurrence and effect. *Food Chem. Toxicol.* 41(12):1651–1662.

Campbell, F.C. (2008) Zirconium. *Elem. Metall. Eng. Alloy.* 59.

Cao, D., et al. (2008) Organotin pollution in China: an overview of the current state and potential health risk. *J. Environ. Manage.* 90:516–524.

Chmielnicka, J., et al. (1993) Comparison of tin and lead toxic action on erythropoietic system in blood and bone marrow of rabbits. *Biol. Trace Elem. Res.* 36:73–87.

De Azevedo, S.V., et al. (2013) Direct determination of tin in whole blood and urine by GF AAS. *Clin. Biochem.* 46(1–2):123–127.

Dosemeci, M., et al. (1993) Estimating historical exposure to silica among mine and pottery workers in the People's Republic of China. *Am. J. Ind. Med.* 24:55–66, 1993.

Dover, S.B., et al. (1993) Tin protoporphyrin prolongs the biochemical remission produced by heme arginate in acute hepatic porphyria. *Gastroenterology* 105:500–506.

Forman, M.R., et al. (1992) The effect of dietary intake of fruits and vegetables on the odds ratio of lung cancer among Yunnan tin miners. *Int. J. Epidemiol.* 21:437–441.

Fu, H., et al. (1994) Lung cancer among tin miners in southeast China: silica exposure, silicosis, and cigarette smoking. *Am. J. Ind. Med.* 26:373–381.

Gaballe, T. (1993) Superconductivity: from physics to technology. *Phys. Today* 46(10):52–56.

Ganguly, B.B., Talukdar, G., and Sharma, A. (1992) Cytotoxicity of tin on human peripheral lymphocytes *in vitro*. *Mutat. Res.* 282:61–67.

Graceli, J.B., et al. (2013) Organotins: a review of their reproductive toxicity, biochemistry, and environmental fate. *Reprod. Toxicol.* 36:40–52.

Gunay, N., et al. (2006) Cardiac damage secondary to occupational exposure to tin vapor. *Inhal. Toxicol.* 18(1):53–56.

Hamanaka, S., Hamanaka, Y., and Otsuka, F. (1992) Phthalic acid dermatitis caused by an organostannic compound, tributyl tin phthalate. *Dermatology* 184:210–212.

Köpf-Maier, R., Janiak, C., and Schumann, H. (1988) Antitumor properties of organometallic metallocene complexes of tin and germanium. *J. Cancer Res. Clin. Oncol.* 14:502–506.

Kotake, Y. (2012) Molecular mechanisms of environmental organotin toxicity in mammals. *Biol. Pharm. Bull.* 35(11):1876–1880.

Lucas, I.T., Syzdek, J., and Kostecki, R. (2011). Interfacial processes at single-crystal β-Sn electrodes in organic carbonate electrolytes. *Electrochem. Commun.* 13(11):1271–1275.

Luff, A.P. and Metcalfe, G.H. (1890) Four cases of tin poisoning caused by tinned cherries. *Br. Med. J.* 833–834.

Lyle, W.H. (1958) Lesions of the skin in process workers caused by contact with butyl tin compounds. *Br. J. Ind. Med.* 15:193–196.

Morgan, W.K.C. and Seaton, A. (1984) *Occupational Lung Diseases*, 2nd ed., Philadelphia: W. B. Saunders Company, pp. 480–483.

Nakanishi, T. (2008) Endocrine disruption induced by organotins compounds; organotins function as a powerful agonist for nuclear receptors rather than an aromatase inhibitor. *J. Toxicol. Sci.* 33(3):269–276.

Nagano, K., et al. (2011) Inhalation carcinogenicity and chronic toxicity of indium-tin oxide in rats and mice. *J. Occup. Health* 53:175–187.

Nath, M. (2008) Toxicity and the cardiovascular activity of organotins compounds: a review. *Appl. Organomet. Chem.* 22: 598–612.

National Institute for Occupational Safety and Health (NIOSH) (2010) *Pocket Guide to Chemical Hazards*, NIOSH 94–116, Washington, DC: US Department of Health and Human Services.

Nuyts, G.D., et al. (1995) New occupational risk factors for chronic renal failure. *Lancet* 346:7–11.

Okoro, H.K., et al. (2011) Sources, environmental levels and toxicity of organotin in marine environment—a review. *Asian J. Chem.* 23(2):473–482.

Pagliarani, A., et al. (2013) Toxicity of organotin compounds: shared and unshared biochemical targets and mechanisms in animal cells. *Toxicol. In Vitro* 27:978–990.

Qiao, Y.L., et al. (1989) Relation of radon exposure and tobacco use to lung cancer among tin miners in Yunnan Province, China. *Am. J. Ind. Med.* 16:511–521.

Rinehart, R.D. and Yanagisawa, Y. (1993) Para-occupational exposures to lead and tin carried by electric-cable splicers. *Am. Ind. Hyg. Assoc. J.* 54:593–599.

Robertson, A.J., et al. (1961) Stannosis: benign pneumoconiosis due to tin dioxide. *Lancet* 1:1089–1093.

Sun, S.Q., et al. (1989) Pathohistogenetic approach on the etiology of Yunnan tin miner's lung cancer. *Chin. Med. J.* 102: 347–355.

Taylor, P.R. (1994) *Biologic Specimen Bank for Early Lung Cancer Markers in Chinese Tin Miners*, National Institutes of Health, Crisp Data Base, Washington, D.C.: US Department of Health and Human Services.

Taylor, P.R., et al. (1989) Relation of arsenic exposure to lung cancer among tin miners in Yunnan Province, China. *Br. J. Ind. Med.* 46:881–886.

World Health Organization (WHO) (2005a) The toxicological evaluation of compounds on the agenda: 3.4 inorganic tin. *Tech. Rep. Ser.* 40.

World Health Organization (WHO) (2005b) *Concise International Chemical Assessment Document 65, Tin and Inorganic Tin Compounds*.

Winship, K.A. (1988) Toxicity of tin and its compounds. *Adverse Drug React. Acute Poisoning Rev.* 7(1):19–38.

Wu, K.G., et al. (1989) Smelting, underground mining, smoking, and lung cancer: a case-control study in a tin mine area. *Biomed. Environ. Sci.* 2:98–105.

Xuan, X.Z., et al. (1993) A cohort study in southern China of tin miners exposed to radon and radon decay products. *Health Phys.* 64:120–131.

35

TITANIUM

Ushang Desai and Raymond D. Harbison

Target Organ(s)	Respiratory System
Occupational exposure limits	OSHA PEL[a]: 15 mg/m^3 as titanium dioxide (TiO$_2$, total dust)
	ACGIH TLV[b]: 10 mg/m^3 as TiO$_2$
	NIOSH REL[c]: (Ca) lowest feasible concentration
	NIOSH Immediately Dangerous to Life or Health Concentration(IDLH): 5000 mg/m^3
Reference values	Titanium (CAS # 7440-32-6)
	Oral RfD: 4.0E+00 mg/kg/day
	Inhalation RfD: 8.6E-03 mg/kg/day
	RfC: 3.0E-02 mg Ti/m^3
Risk/safety phrases	For titanium dioxide:
	Risk phrases: R40 limited evidence of a carcinogenic effect
	Safety phrases: S36/37 wear suitable protective clothing and gloves
	For titanium tetrachloride:
	Risk phrases: R-14 reacts violently with water
	R34 causes burns
	2. Safety phrases: S26 in case of contact with eyes, rinse immediately with plenty of water and seek medical advice
	S36/37/39 wear suitable protective clothing, gloves, and eye/face protection
	S45 in case of accident or if you feel unwell, seek medical advice immediately (show the label where possible)
	S7/8 keep container tightly closed and dry

[a] Occupational Safety and Health Administration (OSHA PELs (permissible exposure limits) are 8-h TWA (time-weighted average) concentrations
[b] American Conference of Governmental Industrial Hygienists (ACGIH TLVs (time limit threshold values) are 8-h TWA concentrations established by ACGIH
[c] National Institute for Occupational Safety and Health (NIOSH RELs(recommended exposure limits) are 10-h TWA concentrations established by NIOSH

BACKGROUND AND USES

Titanium (Ti, CASRN 7440-32-6) was first discovered in the eighteenth century and named after the Titan, a god of strength from the Greek mythology. It is the ninth most abundant element in the earth's crust and is known as a space-aged metal. Its enormous strength, low density, and light weight make it the foremost choice for aircraft, space vehicle, and missile programs. Titanium is commonly used in surgical prosthetics and instruments because of its ability to form high

Hamilton & Hardy's Industrial Toxicology, Sixth Edition. Edited by Raymond D. Harbison, Marie M. Bourgeois, and Giffe T. Johnson.
© 2015 John Wiley & Sons, Inc. Published 2015 by John Wiley & Sons, Inc.

strength alloys with other metals. Its versatile properties are used in the manufacturing of jewelry, automobiles, bicycles, firearms, and other merchandise. Its anticorrosive and heat-resistant properties make it useful in marine equipment and wall paints (Gambogi, 1995).

PHYSICAL AND CHEMICAL PROPERTIES

Titanium dioxide (TiO_2, CAS # 13463-67-7) is a white pigment that is water insoluble and very stable. It has a molecular weight 79.9, boiling point (760 mmHg) <3000 °C (<5432 °F), specific gravity of 3.9–4.2, melting point of 1640 °C (2984 °F), and it is not a combustible compound. Titanium tetrachloride ($TiCl_4$, CAS# 7550-45-0) is a colorless liquid with a vapor pressure of 10 mmHg at 20 °C. It has a molecular weight of 189.70, melting point −24 °C, boiling point 136.4 °C, density at 20 °C, and it is nonflammable. It is soluble in both water and alcohol with a penetrating acidic odor.

MAMMALIAN TOXICOLOGY

Acute Effects

Acute inhalation exposure of titanium dioxide (TiO_2) shows pulmonary inflammation with cell damage and an increasing adequately high surface area deposition in rats. Greater toxicity was observed with doses of ultrafine TiO_2 at a given mass dose (Oberdörster et al., 1992; Renwick et al., 2004). Acute exposure of titanium tetrachloride ($TiCl_4$) in rats and dogs resulted in nasal discharge, dyspnea, severe respiratory inflammation, epithelial denudation, discrete inflammatory residue in the lungs, coarsened alveolar septa, and swollen eyelids and death due to pulmonary edema (DuPont, 1980; Kelly, 1980). Acute titanium tetrachloride toxicity causes irritation to the skin, eyes, mucus membranes, and respiratory tract in humans. Acute exposure may result in surface skin burns; marked congestion of mucous membranes of the pharynx, vocal cords, and trachea; and stenosis (constriction) of the larynx, trachea, and upper bronchi in humans. It may also damage the cornea in the eyes.

A worker accidentally exposed to a high concentration of titanium tetrachloride via inhalation later developed endobronchial polyps (Garabrant et al., 1987). Workers exposed to titanium dioxide, showing deposits of dust in the pulmonary interstitium, are associated with cell destruction and slight fibrosis, which also suggested that TiO_2 is a pulmonary irritant (Bermudez et al., 2002, 2004). Acute titanium dioxide toxicity may produce metal fume fever (Otani et al., 2008).

Chronic Effects

Chronic exposures of titanium dioxide (TiO_2) in rodents show dose-related pulmonary responses to fine or ultrafine

TiO_2. The responses in rats include pulmonary inflammation, oxidative stress, tissue damage, fibrosis, and lung cancer (Heinrich et al., 1995; Muhle et al., 1991). Chronic exposure of titanium tetrachloride ($TiCl_4$) in rats resulted in tracheitis and rhinitis. Also, immunological and lymphoreticluar changes in lungs, like foamy lung macrophages with increased $TiCl_4$ dust deposition, were observed (Lee et al., 1986; Driscoll et al., 1991). Chronic exposures to titanium tetrachloride at the workplace produce pleural thickening and decreased pulmonary function in the workers. Chronic inhalation exposure may result in upper respiratory tract irritation, chronic bronchitis, cough, bronchoconstriction, wheezing, chemical pneumonitis, or pulmonary edema in humans (Garabrant et al., 1987).

Mechanisms of Action

Titanium tetrachloride ($TiCl_4$) is extremely corrosive, hydrolyzing upon contact with moisture releasing heat, hydrochloric acid, and orthotitanic acids causing direct tissue damage in the lungs. Studies show that fine oxychloride intermediates generated from $TiCl_4$ hydrolysis can penetrate deep into the lung tissue resulting in direct contact irritation and bronchitis or pneumonia (Mezentseva et al., 1963; Kelly, 1980; Chitkara and McNeela, 1992). The mechanism of toxicity of titanium dioxide (TiO_2) is based on the inflammatory response to TiO_2 particles in the target tissue. Studies have demonstrated a neutrophil-mediated inflammatory response to TiO_2, in addition to increased inflammatory response through particle-laden macrophages (Ferin et al., 1992; Renwick et al., 2004; Grassian et al., 2007a, 2007b).

Chemical Pathology

In vivo studies in animals have demonstrated that exposure to TiO_2 particles can induce inflammatory responses in lung tissue and cytotoxicity in lung cells; the degree of inflammatory response and cytotoxicity elicited depends on the particle size and its surface chemistry (Rehn et al., 2003; Dick et al., 2003; Renwick et al., 2004; Warheit et al., 2005; Monteiller et al., 2007). Exposure to titanium dioxide nanoparticles (<100 nm in diameter) induces genotoxicity and cytotoxicity in cultured human lymphoblastoid (WIL2-NS) cells (Wang et al., 2007).

SOURCES OF EXPOSURE

Occupational

Inhalation and dermal absorption are significant routes of occupational exposure to titanium dioxide (TiO_2). The occupations in which exposure to titanium dioxide (TiO_2) occurs are titanium mining, purification, packaging, and distribution. During the manufacturing of paints, varnishes, lacquers, paper, photographic papers, packaging paper, and cellophane, TiO$_2$ is used as a coating where exposure to TiO$_2$ may occur. Other occupations involved in TiO$_2$ exposure are manufacturing of elastomers for use in tire side walls, footwear, floor mats, gloves, rainwear, wall covering and manufacturing of ceramic and glass for capacitors, electromechanical transducers, welding rod coating, and glass fibers. Titanium dioxide (TiO$_2$) exposure also occurs in the production of cosmetics, food color additives, synthetic diamonds, artificial leather, and printing inks (Petty, 1963; Schwartz et al., 1957). Like titanium dioxide, inhalation and dermal absorption are the significant routes of exposure to TiCl$_4$ (Friberg et al., 1979). Workers who are involved in the process of producing titanium metal from titanium tetrachloride receive the most exposure. Maintenance workers of these facilities are also exposed to a high level of titanium tetrachloride vapor containing its hydrolysis products, titanium dioxide, and hydrochloric acid. (Garabrant et al., 1987) Some of the workers use titanium tetrachloride to check welding machinery. Accidentally, they may be exposed to titanium tetrachloride because of an occupational spill (Ross, 1985).

Environmental

Titanium oxide is the most commonly available form of titanium, which is the ninth most abundant element in the earth. The main source of environmental exposure to titanium dioxide is dermal absorption from sunscreens, paints, enamels, and cosmetics; ingestion of these products may also be a significant route of exposure (IARC, 2010). Populations living nearby titanium mines and industries using titanium dioxide may suffer an increased risk of exposure. However, environmental exposure to titanium tetrachloride is unlikely, as it breaks down rapidly in water (Wilms et al., 1992).

INDUSTRIAL HYGENE

The Occupational Safety and Health Administration set PELs for titanium dioxide and titanium tetrachloride, which are legally enforceable in the industries involving these chemicals. Good industrial hygiene methods like general dilution ventilation and personal protective equipment have been used to keep the environmental concentration below the PELs. Though under certain conditions exposure control may not be technically possible, the workers should use a proper respirator to avoid chemical exposure at that point. During the spillage of TiO$_2$, properly ventilate the area of spill. Thereafter, collect the material and dispose in a secured sanitary landfill; this is the most convenient and safe way of reclamation or disposal. Liquid containing TiO$_2$ should be absorbed in vermiculite, dry sand, earth, or a similar material (Raterman, 1996; Burton, 1997; Hinds, 1999).

MEDICAL MANAGEMENT

Acute exposure of titanium dioxide and titanium tetrachloride requires an immediate removal from the place of exposure and the use of appropriate first aid to the exposed person. The respiratory system is most commonly affected during inhalation exposure. During management, ensure a clear airway and maintain adequate breathing. If the patient has symptoms of bronchospasm, treat symptomatically with bronchodilators or corticosteroids. Endotracheal intubation may be required for a person with life-threatening laryngeal edema. Acute dermal exposures require the removal of the person from the place of exposure, and in case of chemical burn, it should be reviewed by a burn specialist (Bronstein and Currance, 1994).

REFERENCES

Bermudez, E., Mangum, J.B., Asgharian, B., Wong, B.A., Reverdy, E.E., Janszen, D.B., Hext, P.M., Warheit, D.B., and Everitt, J.I. (2002) Long-term pulmonary responses of three laboratory rodent species to subchronic inhalation of pigmentary titanium dioxide particles. *Toxicol. Sci.* 70:86–97.

Bermudez, E., Mangum, J.B., Wong, B.A., Asgharian, B., Hext, P.M., Warheit, D.B., and Everitt, J.I. (2004) Pulmonary responses of mice, rats, and hamsters to subchronic inhalation of ultrafine titanium dioxide particles. *Toxicol. Sci.* 77:347–357.

Burton, D.J. (1997) General methods for the control of airborne hazards. In: DiNardi, S.R., editor. *The Occupational Environment—Its Evaluation and Control*, Fairfax, VA: American Industrial Hygiene Association.

Bronstein, A.C. and Currance, P.L. (1994) *Emergency Care for Hazardous Materials Exposure*, 2nd ed., St. Louis, MO: Mosby Lifeline, pp. 139.

Chitkara, D.K. and McNeela, B.J. (1992) Titanium tetrachloride burns to the eye. *Br. J. Ophthalmol.* 76(6):380–382.

Dick, C.A.J., Brown, D.M., Donaldson, K., and Stone, V. (2003) The role of free radicals in the toxic and inflammatory effects of four different ultrafine particle types. *Inhal. Toxicol.* 15(1):39–52.

Driscoll, K.E., Robert, C., Maurer, J.K., Perkins, L., Perkins, M., and Higgins, J. (1991) Pulmonary response to inhaled silica or titanium dioxide. *Toxicol. Appl. Pharmacol.* 111(2):201–210.

DuPont Company (1980) *Acute inhalation studies with titanium tetrachloride. Haskell. Laboratory Report No. 658–680.*

Ferin, J., Oberdörster, G., and Penney, D.P. (1992) Pulmonary retention of ultra-fine and fine particles in rats. *Am. J. Respir. Cell Mol. Biol.* 6:535–542.

Friberg, L., Nordberg, G.R., and Vouk, V.B. (1979) *Handbook on the Toxicology of Metals*, New York, NY: Elsevier North Holland, pp. 630.

Gambogi, J. (1995) *Titanium and Titanium Dioxide, from Mineral Commodity Summaries*, U.S. Bureau of Mines, January 1995, 180.

Garabrant, D.H., Fine, L.J., Oliver, C., Bernstein, L., and Peters, J.M. (1987) Abnormalities of pulmonary function and pleural disease among titanium metal production workers. *Scand. J. Work Environ. Health* 1347–1351.

Grassian, V.H., Adamcakova-Dodd, A., Pettibone, J.M., O'Shaughnessy, P.T., and Thorne, P.S. (2007a) Inflammatory response of mice to manufactured titanium dioxide nanoparticles: comparison of size effects through different exposure routes. *Nanotoxicology* 1(3):211–226.

Grassian, V.H., O'Shaughnessy, P.T., Adamcakova-Dodd, A., Pettibone, J.M., and Thorne, P.S. (2007b) Inhalation exposure study of nanoparticulate titanium dioxide with a primary particle size of 2 to 5 nm. *Environ. Health Perspect.* 115:397–402.

Heinrich, U., Fuhst, R., Rittinghausen, S., Creutzenberg, O., Bellmann, B., Koch, W., and Levsen, K. (1995) Chronic inhalation exposure of Wistar rats and two different strains of mice to diesel engine exhaust, carbon black, and titanium dioxide. *Inhal. Toxicol.* 4:533–556.

Hinds, W. (1999) *Aerosol Technology: Properties, Behavior, and Measurement of Airborne Particles*, 2nd ed., New York: John Wiley & Sons.

IARC (2010) *IARC Monographs on the Evaluation of Carcinogenic Risks to Humans*, vol. 93: International Agency for Research on Cancer. World Health Organization.

Kelly, D.P. (1980) *Acute inhalation studies with titanium tetrachloride. E.I. du Pont de Nemours and Company, Haskell Laboratory for Toxicology and Industrial Medicine; Haskell Laboratory Report No. 658-80.*

Lee, K.P., Henry, N.W., III, Trochimowicz, H.J., and Reinhardt, C.F. (1986) Pulmonary response to impaired lung clearance in rats following excessive TiO2 dust deposition. *Environ. Res.* 41:144–167.

Mezentseva, N.V., Melnikova, E.A., and Mogilevskaya, O.Y.A. (1963) In: Izraelson, Z.I., editor. *Toxicology of the Rare Metals*, (translated from Russian by the Israel program for scientific translations, Jerusalem, 1967). Gosudarstvennoe Izdatelstvo: Medicinskoi Literatury, Moscova, pp. 35–43.

Monteiller, C., Tran, L., MacNee, W., et al. (2007) The proinflammatory effects of low-toxicity low-solubility particles, nanoparticles and fine particles, on epithelial cells in vitro: the role of surface area. *Occup. Environ. Med.* 64:609–615.

Muhle, H., Bellmann, B., Creutzenberg, O., Dasenbrock, C., Ernst, H., Klipper, R., McKenzie, J.C., Morrow, P., Mohr, U., Takenaka, S., and Mermelstein, R. (1991) Pulmonary response to toner upon chronic inhalation exposure in rats. *Fundam. Appl. Toxicol.* 17(2):280–299.

Oberdörster, G., Ferin, J., Gelein, R., Soderholm, S.C., and Finkelstein, J. (1992) Role of the alveolar macrophage in lung injury: studies with ultrafine particles. *Environ. Health Perspect.* 97:193–199.

Otani, N., Ishimatsu, S., and Mochizuki, T. (2008) Acute group poisoning by titanium dioxide: inhalation exposure may cause metal fume fever. *Am. J. Emerg. Med.* 26:608–611.

Petty, F.A. (1963) *Toxicology, Vol II of Industrial Hygiene and Toxicology*, 2nd ed., New York, NY: Interscience.

Raterman, S.M. (1996) Methods of control. In: Plog, B.A., editor. *Fundamentals of Industrial Hygiene*, Chapter 18, Itasca, IL: National Safety Council.

Rehn, B., Seiler, F., Rehn, S., Bruch, J., and Malter, M. (2003) Investigations on the inflammatory and genotoxic lung effects of two types of titanium dioxide: untreated and surface treated. *Toxicol. Appl. Pharmacol.* 189:84–95.

Renwick, L.C., Brown, D.M., Clouter, A., and Donaldson, K. (2004) Increased inflammation and altered macrophage chemotactic responses caused by two ultrafine particle types. *Occup. Environ. Med.* 61:442–447.

Ross, D.S. (1985) Exposure to titanium tetrachloride. *Occup. Health* 37(11):525.

Schwartz, L., Tulipan, L., and Birmingham, D. (1957) *Occupational Diseases of the Skin*, 3rd ed., Philadelphia: Lea and Febiger.

Warheit, D.B., Brock, W.J., Lee, K.P., Webb, T.R., and Reed, K.L. (2005) Comparative pulmonary toxicity inhalation and instillation studies with different TiO2 particle formulations: impact of NIOSH CIB 63 titanium dioxide 103 surface treatments on particle toxicity. *Toxicol. Sci.* 88(2):514–524.

Wang, J., Zhou, G., Chen, C., Yu, H., Wang, T., Ma, Y., Jia, G., Gao, Y., Li, B., Sun, J., Li, Y., Jiao, F., Zhao, Y., and Chai, Z. (2007) Acute toxicity and biodistribution of different sized titanium dioxide particles in mice after oral administration. *Toxicol. Lett.* 168:176–185.

Wilms, E.B., van Xanten, N.H.W., and Meulenbelt, J. (1992) Smoke producing and inflammable materials. *Govt Reports Announcements & Index (GRA&I)*, Issue 01. NTIS/PB92-104967.

36

TUNGSTEN

Kambria K. Haire and Raymond D. Harbison

Target organ(s): Eyes, skin, lungs, and blood

Occupational exposure limits: Insoluble tungsten (8-h TWA): 5 mg/m³ (OSHA), 5 mg/m³ (NIOSH); 10 mg/m³ STEL (NIOSH)

Risk/safety phrases: Skin irritant; eye irritant; respiratory irritant; gastrointestinal irritant; corrosive

BACKGROUND AND USES

Tungsten (CASRN 7440-33-7), also known as wolfram, occurs as wolframite ($FeWO_4$). It can be found in the earth's crust but not in its pure metal form. It combines with other chemicals and compounds within the rocky earth's crust. It is a transitional hard metal with physicochemical properties and can also be manufactured commercially (Lassner and Schnubert, 1999; Gbaruko and Igwe, 2007; Stefaniak, 2010; Strigul et al., 2010).

Tungsten is most commonly used to increase the hardness of steel. It is available commercially in the form of powder, single crystal, and ultrapure granule grades. It is also used in the manufacturing of alloys, light filaments, and X-ray tubes. A recent use for tungsten is as a lead substitute during the manufacturing of ammunition and sporting good products. Another recent commercial use for tungsten is in the production of wedding bands. It is also used as a catalyst in chemical reactions (Lassner and Schnubert, 1999; Gbaruko and Igwe, 2007; Stefaniak, 2010; Strigul et al., 2010).

There are many chemical configurations for tungsten compounds. The most common compounds are sodium tungstate, tungsten trioxide, and tungsten carbide. Sodium tungstate is commonly used for fire and waterproofing fabrics (Berry and Gailie, 1990; Gbaruko and Igwe, 2007). Tungsten trioxide is used as pigments in ceramics and in the coloring process for textiles and fireproofing fabrics (Berry and Gailie, 1990; Zuyagin, 1999; Gbaruko and Igwe, 2007). Tungsten carbide is used in metal, mining, and petroleum industries. (Cajal, 1982; Gbaruko and Igwe, 2007).

PHYSICAL AND CHEMICAL PROPERTIES

Tungsten is a steel-gray to tin-white metal. It is a heavy metal with a high density and melting point (Table 36.1) (Lassner and Schnubert, 1999; Stefaniak, 2010; Strigul et al., 2010). It is a very hard metal that is stable in dry air at ordinary temperatures. Although very difficult to work with, tungsten can be cut with a hacksaw, forged, or spun. Commercially, tungsten is obtained via the reduction of tungsten oxide with hydrogen or carbon. Due to its hardness, tungsten has the highest boiling point and highest tensile strength of all metals. Although the metal shows resistance to corrosion, it can oxidize in air with an increase in temperature (Robert and Golden, 1980; Gbaruko and Igwe, 2007).

Tungsten is a very hard metal with similar properties and functions to molybdenum and chromium. Unlike molybdenum, tungsten is not an essential trace metal. However, tungsten is a competitive inhibitor of molybdenum in bacteria, plants, and animals (Hainline and Rajagopalan, 1983; Smith, 1991).

At high temperatures, tungsten forms tungsten trioxide from sodium tungstate. Sodium tungstate is an aqueous solution with an alkaline pH (Cajal, 1982; Zuyagin, 1999; Gbaruko and Igwe, 2007). Sodium tungstate is used in the preparation of complex compounds, as a reagent for biological

TABLE 36.1 Physical and Chemical Properties of Tungsten Compounds

Name	Chemical Formula	Molecular Weight/ Molar Mass	Density	Melting Point	Boiling Point
Tungsten	W	183.89 g	19.25 g/cm^3	3422 °C	3422 °C
Sodium tungstate (dihydride)	H$_4$Na$_2$O$_6$W	329.85 g/mol	3.25 g/cm^3	698 °C	–
Tungsten trioxide	WO$_3$	231.84 g/mol	7.16 g/cm^3	1473 °C	1700 °C
Tungsten carbide	WC	195.85 g/mol	15.63 g/cm^3	2870 °C	6000 °C

products, and as a precipitant for alkaloids. It is also used as a catalyst in the oxidation of maleic acid (Berry and Gailie, 1990; Gbaruko and Igwe, 2007).

MAMMALIAN TOXICOLOGY

Acute Effects

As a heavy metal, tungsten is known as an irritant where acute health effects may result in occupational workers. Acute occupational exposures can cause irritation to the skin and eyes. Skin exposures may cause reddening and itching of the skin. Irritation from eye exposures may cause watering and redness of the eye. Inhalational exposures may cause irritation of the lungs through the impairment of the function of the mucus membranes (Gbaruko and Igwe, 2007).

Chronic Effects

Many of the chronic health effects resulting from tungsten exposure impact the lungs and respiratory system. Long-term exposures may impact the cells of the respiratory tract and could lead to lung diseases. Chronic inhalational exposure to tungsten dust may cause hard metal disease or occupational asthma (Stefaniak et al., 2007).

Mechanism of Action(s)

Tungsten is not considered to be an essential nutrient for living organisms; however, some tungsten compounds impact both human and animal health adversely (Leggett, 1997; Tajima, 2001; Domingo, 2002; Lagarde and Leroy, 2002; Strigul et al., 2005). Tungsten exposures may result in the substitution of elements in different biological systems, which may impact specific enzyme processes (Stiefel, 2002; Strigul et al., 2005). Due to its close relation to molybdenum, exposures may impact some of the body's metabolic processes (Friberg et al., 1979; Gossentin et al., 1976; Gbaruko and Igwe, 2007). In the presence of cobalt, tungsten carbide generates high amounts of free radicals (Lison and Lauwerys, 1995; Stefaniak, 2010).

Animal studies conducted in rats exposed to high doses (20 pellets) of heavy metal tungsten alloy (>90%) developed aggressive tumors surrounding the implantation site within 4–5 months. Rats implanted with low dose pellets (four pellets) developed rapidly mestasizing tumors along the respiratory pathways (Nyrén et al., 1995; Witten et al., 2012). Cell culture studies suggest that a combination of tungsten (91%), nickel (6%), and cobalt (3%) alloys in rat muscle cells may result in DNA damage, cell hypoxia, and cell death within 24 h of exposure (Harris et al., 2011; Witten et al., 2012).

Chemical Pathology

Tungsten toxicity depends on the chemical form of the tungsten compounds and the exposure route, with most exposures occurring via the inhalational route. These exposures not only impact the lungs but the other parts of the respiratory system as well. Tungsten is transported through the body by the blood, which can impact the liver, kidneys, spleen, and other organs (Leggett, 1997; Strigul et al., 2005).

SOURCES OF EXPOSURE

Occupational

Although tungsten is considered to be a heavy metal, it is not however considered a heavy metal exposure for industrial and occupational workers. Most occupational exposures to tungsten occur through inhalation of dust or through dermal contact during the processes of production and usage of tungsten and its compounds (Berry and Gailie, 1990; Gbaruko and Igwe, 2007). Industrial or occupational workers who may work with cemented tungsten carbide may be exposed to tungsten as well as its compounds (Kraus et al., 2001; Stefaniak, 2010). The potential for producing adverse health effects following inhalation of fiber-shaped tungsten oxide particles from occupational workers is not understood (Stefaniak, 2010).

Environmental

Exposure to tungsten is possible from the ingestion of products containing tungsten or its compounds for the general public. Beverages such as wine, mineral water, beer, brewed tea, and instant coffee have been found to contribute to the total dietary intake of tungsten (Gbaruko and Igwe, 2007). Currently, there are no environmental regulations in place in the United States addressing tungsten pollution (Strigul et al., 2005). Tungsten may be released into the environmental systems as the result of natural or anthropogenic activities (Bourcier et al., 1980; Ondov et al., 1989; Grimes et al., 1995; Feldman and Cullen, 1997; Peltola and Wikstrom, 2006; Strigul et al., 2010).

If released into the soil, tungsten compounds will have moderate to low mobility, depending upon sorption coefficients. These compounds are not expected to volatize from dry soil surfaces. If released into the water, tungsten compounds may absorb to suspended solids and sediments (Berry and Gailie, 1990; Gbaruko and Igwe, 2007). The production and use of tungsten compounds may result in the release of tungsten into the environment through waste streams.

Industrial emissions may release tungsten in the form of tungsten trioxide into the atmosphere.

Most tungsten compounds exist in the particulate phase in the ambient atmosphere because of their low vapor pressure (Berry and Gailie, 1990; Gbaruko and Igwe, 2007). Metallic tungsten particles may cause adverse environmental effects such as soil acidification (Dermates et al., 2004; Strigul et al., 2005; Ringelberg et al., 2009; Johnson et al., 2009; Strigul et al., 2010).

INDUSTRIAL HYGIENE

Most adverse health effects that result from exposure are due to acute toxic exposures. Acute exposures affect the eyes and skin. The use of personal protective equipment during the handling and use of tungsten and its compounds is considered as safe industrial hygiene practices. Respiratory masks may also be recommended to prevent high levels of exposure when tungsten carbide dust particles combine with cobalt (Stefaniak et al., 2007; Gbaruko and Igwe, 2007).

The National Institute for Occupational Safety and Health (NIOSH) and the American Conference of Governmental Industrial Hygienists (ACGIH) have set the recommended exposure limit (REL) and the threshold limit value (TLV), respectively, for insoluble forms of tungsten at 5 mg/m^3, with a 15-min short-term exposure limit (STEL) of 10 mg/m^3. For soluble forms of tungsten, NIOSH and the ACGIH have set both the REL and the TLV at 1 mg/m^3, with a STEL of 3 mg/m^3.

MEDICAL MANAGEMENT

If tungsten or tungsten compounds come in contact with the eyes, one should immediately wash (irrigate) the eye with water, occasionally lifting the lower and upper lids. If there is contact with the skin, wash the affected skin area with soap and water. If a person inhales large amounts of tungsten or its compounds, move the person to fresh air. If one has

swallowed tungsten or its compounds or suffered a serious injury, seek medical attention immediately.

REFERENCES

Berry, J.P. and Gailie, P. (1990) Subcellular localization of HPA-23 in different rat organs. *Exp. Mol. Pathol.* 53:255–264.

Bourcier, D.R., Hindin, E., and Cook, J.C. (1980) Titanium and tungsten in highway runoff at Pullman, Washington. *Int. J. Environ. Stud.* 15:145–149.

Cajal, S.R. (1982) *Fate and Transport of Tungsten. Fate and Impact of Heavy Metals*, Oxford: Oxford University Press.

Dermates, D., Braida, W., Christodoulatos, C., Strigul, N., Panikov, N., Los, M., and Larson, S. (2004) Solubility, sorption, and soil respiration effects of tungsten and tungsten alloys. *Environ. Forensics* 5:5–13.

Domingo, J.L. (2002) Vanadium and tungsten derivatives as antidiabetic agents: a review of their toxic effects. *Biol. Trace Elem. Res.* 88:97–112.

Feldman, J. and Cullen, W.R. (1997) Occurrence of volatile transition metal compounds in landfill gas: synthesis of molybdenum and tungsten carbonyls in the environment. *Environ. Sci. Techol.* 31:2125–2129.

Friberg, L., Nordber, G.R., and Vouk, V.B. (1979) *Handbook on the Toxicology of Metals*, NY: Elsevier North Holland.

Gbaruko, B.C. and Igwe, J.C. (2007) Tungsten: occurence, chemistry, environmental and health issues. *Global J. Environ. Res.* 1(1):27–32.

Gossentin, R.E., Hodge, H.C., Smith, R.P., and Braddock, J.E. (1976) *Clinical Toxicology of Commercial Products*, 4th ed., Baltimore, MD: Williams and Wilkins.

Grimes, D.J., Ficklin, W.H., Meier, A.L., and McHugh, J.B. (1995) Anomalous gold, antimony, arsenic, and tungsten in ground water and alluvium around disseminated gold deposits along the Getchell Trend, Humboldt County, Nevada. *J. Geochem. Explor.* 52:351–371.

Hainline, B.E. and Rajagopalan, K.V. (1983) Molybdenum in animal and human health. In: Rose, J., editor. *Trace Elements in Health*, London: Butterworths.

Harris, R.M., Williams, T.D., Hodges, N.J., and Waring, R.H. (2011) Reactive oxygen species and oxidative DNA damage mediate the cytotoxicity of tungsten-nickel-cobalt alloys *in vitro*. *Toxicol. Appl. Pharmacol.* 250:19–28.

Johnson, D.R., Inouye, L.S., Bednar, A.J., Clake, J.U., Winfield, L.E., Boyd, R.E., and Ang, C.Y. (2009) Tungsten bioavailability and toxicity in sunflowers (*Helianthus annus*). *J. Land Contam. Reclam.* 17:141–151.

Kraus, T., Schaller, K.H., Zobelein, P., Weber, A., and Angerer, J. (2001) Exposure assessment in the hard metal manufacturing industry with special regard to tungsten and its compounds. *Occup. Environ. Med.* 58:631–634.

Lagarde, F. and Leroy, M. (2002) Metabolism and toxicity of tungsten in humans and animals. In: *Metal Ions in Biological Systems, Molybdenum and Tungsten: Their Role in*

Biological Processes, vol. 39, New York, NY: Marcel Dekker Inc.

Lassner, E. and Schnubert, W. (1999) *Tungsten Properties, Chemistry, Technology of the Element, Alloys, and Chemical Compounds*, Boston: Kluwer Academic/Plenum Publishers.

Leggett, R.W. (1997) A model of the distribution and retention of tungsten in the human body. *Sci. Total Environ.* 206:147–165.

Lison, D. and Lauwerys, R. (1995) The interaction of cobalt metal with different carbides and other mineral particles on mouse peritoneal macrophages. *Toxicol. In Vitro* 9:341–347.

McInturf, S.M., Bekkedal, M.Y., Wilfong, E., Arfsten, D., Chapman, G., and Gunasekar, P.G. (2011) The potential reproductive, neurobehavioral, and systemic effects of soluble sodium tungstate exposure in Sprague-Dawley rats. *Toxicol. Appl. Pharmacol.* 254:133–137.

NIOSH (2010) *NIOSH: Pocket Guide to Chemical Hazards. Tungsten*, Atlanta, GA: National Institute for Occupational Safety and Health, Centers for Disease Control and Prevention. Availavle at http://www.cdc.gov/niosh/npg/npgd0645.html.

Nyrén, O., McLaughlin, J.K., Grindley, G., Ekbom, A., Johnell, O., Fraumeni, J.F. Jr., and Adami, H.O. (1995) Cancer risk after hip replacement with metal implants: a population based cohort-study in Sweden. *J. Natl. Cancer Inst.* 87:28–33.

Ondov, J.M., Choquette, C.E., Zoller, W.H., Gordon, G.E., Bierman, A.H., and Heft, R.E. (1989) Atmospheric behavior of trace elements on particles emitted from a coal-fired power plant. *Atmos. Environ.* 23:2193–2204.

Peltola, P. and Wikstrom, E. (2006) Tyre stud derived tungsten carbide particles in urban street dust. *Boreal Environ. Res.* 11:161–168.

Ringelberg, D.B., Reynolds, C.M., Winfield, L.E., Inouye, L.S., Johnson, D.R., and Bednar, A.J. (2009) Tungsten effects on microbial community structure and activity in a soil. *J. Environ. Qual.* 38:103–110.

Robert, K.J. and Golden, S.M. (1980) Concentrations of tungsten and tungsten compounds in the atmosphere. In: *Atmospheric Exposure to Metals*, 2nd ed., London: Buffer and Tanner Ltd.

Smith, G.R. (1991) Tungsten. In: *Minerals Yearbook, Bureau of Mines, US Department of the Interior*, Washington, DC: US Government Printing Office.

Stefaniak, A.B. (2010) Persistence of tungsten oxide particle/fiber mixtures in artificial human lung fluids. *Part. Fibre Toxicol.* 7(38):1–9.

Stefaniak, A., Day, G., Harvey, C., Leonard, S., Schwegler-Berry, D., Chipera, S., Sahakian, N., and Chisolm, W. (2007) Characteristics of dusts encountered during the production of cemented tungsten carbides. *Ind. Health* 45:793–803.

Stiefel, E.I. (2002) The biogeochemistry of molybdenum and tungsten. In: *Metal Ions in Biological Systems, Molybdenum and Tungsten: Their Role in Biological Processes*, vol. 39, NY: Marcel Dekker Inc.

Strigul, N.S., Koutsospyro, A., Arienti, P., Christodoulatos, C., Dermatas, D., and Braida, W.J. (2005) Effects of tungsten on environmental systems. *Chemosphere* 61:248–258.

Strigul, N., Koutsospyros, A., and Christodoulatos, C. (2010) Tungsten speciation and toxicity: acute toxicity of mono- and poly-tungstates to fish. *Ecotoxicol. Environ. Saf.* 73:164–171.

Tajima, Y. (2001) A review of the biological and biochemical effects of tungsten compounds. *Curr. Top. Biochem. Res.* 4:129–136.

Witten, M., Sheppard, P., and Witten, B. (2012) Tungsten Toxicity. *Chem. Biol. Inter.* 196:87–88.

Zuyagin, B.B. (1999) *Compounds and Reactions of Tungsten. Chemical Properties of Heavy Metals*, New York, NY: Macgre-Hilline.

37

VANADIUM[1]

JAMES C. BALL AND MAUREEN R. GWINN

First aid:

Eyes: Flush eyes with running water for 15 min. If irritation persists, obtain medical assistance.

Skin: Treat skin cuts and abrasions with standard first aid. Skin contamination can be removed with soap and water. If irritation persists, obtain medical assistance.

Ingestion: Obtain medical assistance at once. If conscious, give one to two glasses of milk or water and induce vomiting.

Inhalation: Removal to fresh air is usually sufficient. If breathing difficulty, give oxygen.

Target organ(s): Respiratory system; hematology; developmental effects

Occupational exposure limits: (as vanadium pentoxide):

ACGIH TLVs: TWA $0.05\,mg/m^3$ (respirable dust or fume); STEL not established

OSHA PELs: TWA $0.05\,mg/m^3$ ceiling (fume as vanadium); STEL $0.5\,mg/m^3$ ceiling (respirable dust), $0.1\,mg/m^3$ ceiling (fume)

NIOSH RELs: TWA not established; STEL $0.05\,mg/m^3$ ceiling (15 min, total dust as vanadium)

NIOSH: Immediately Dangerous to Life or Health (IDLH) $35\,mg/m^3$

Reference values:

EPA: RfD $9 \times 10^{-3}\,mg/kg/day$ (vanadium pentoxide)

ATSDR–MRLs: Inhalation: Acute $0.8\,\mu g$ vanadium/m^3, chronic $0.1\,\mu g$ vanadium/m^3; oral: Acute $10\,\mu g$ vanadium/kg/day

Risk phrases:

Harmful by inhalation and if swallowed

Irritating to the respiratory system

Toxic: Danger of serious damage to health by prolonged exposure through inhalation

Possible risk of harm to the unborn child

Toxic to aquatic organisms, may cause long-term adverse effects in the aquatic environment

Possible risk of irreversible effects

Safety phrases:

Wear suitable protective clothing and gloves.

In case of insufficient ventilation, wear suitable respiratory equipment.

In case of accident or if you feel unwell, seek medical advice immediately (show the label where possible).

Avoid release to the environment. Refer to special instructions/safety data sheet.

BACKGROUND AND USES

Vanadium is a naturally occurring element, and is the 22nd most abundant element found in geological formations. The

---

[1] Disclaimer: The findings and conclusions in this report are those of the authors and do not necessarily represent the views of the U.S. Environmental Protection Agency.

Hamilton & Hardy's Industrial Toxicology, Sixth Edition. Edited by Raymond D. Harbison, Marie M. Bourgeois, and Giffe T. Johnson.
© 2015 John Wiley & Sons, Inc. Published 2015 by John Wiley & Sons, Inc.

initial discovery of vanadium occurred in 1801 by a Spanish chemist and mineralogist working as a teacher in Mexico, Andrés Manuel del Río. He described his analyses of lead ores and further chemical tests concluded that he had discovered a new element. He gave samples of his work to Alexander von Humboldt, who after consulting with a French chemist, was convinced that this new element was chromium. Del Río published a retraction of his discovery, but later regretted this action and was convinced that he had indeed discovered what later became known as vanadium. About 30 years later, a Swedish chemist, Nils G. Sefström rediscovered vanadium from iron ore and was given credit for its discovery and the origin of its name (Atomix, 2003; Cintas, 2004). In 1869, the pure metal was isolated by Sir H. E. Roscoe using dry hydrogen gas (oxygen free) to reduce vanadium dichloride (VCl_2) to metallic vanadium (Roscoe, 1869). The first application of vanadium in steel occurred in 1889 under the direction of Professor Arnold of Sheffield Firth College, England. This new vanadium steel had been used by a French luxury car manufacturer and this intrigued Henry Ford after he examined pieces of the car from a crash that occurred during racing. He recognized the potential use of this new, stronger steel to reduce weight in the Model T. Henry Ford learned the process of making vanadium steel, which was subsequently used in Model T cars. The Model T retained the advantage of higher strength steel for at least 5 years before other manufacturers adopted this new steel (Gross and Magazine, 1996).

Currently, a majority of vanadium (>90%) is used in the production of steel, where it is added in small amounts to impart additional strength to steel (Yong and Han, 2010) (Box 37.1). This is due, in part, to the inherent hardness and strength of ferrovanadium (FeV), vanadium metal nitrides (VN), and carbides (e.g., VC or V_3C_4). Much of the steel produced with vanadium as a key component are microalloyed or high strength low alloy steels (Baker, 2009). These kinds of steel also use combinations of other metals such as nickel, tungsten, molybdenum, chromium, manganese, aluminum, and carbon to synthesize unique alloys. Vanadium is present in low concentrations in these steels, typically no more than 0.1%, but these low levels impart significant strength and hardness to the steel. Specific alloys are used in specialized industries. For example, a combination of steel, chromium, molybdenum, and vanadium have been used to make high temperature steels for use in gas turbine engines (Baker, 2009).

Vanadium is also commonly used as a catalyst in the production of sulfuric acid and maleic acid anhydride—the latter compound is used in the synthesis of rubber polymers (Cheng et al., 2009). Vanadium pentoxide is used as a catalyst to oxidize sulfur dioxide (SO_2) to anhydrous sulfur trioxide (SO_3) followed by a reaction with water to generate sulfuric acid (H_2SO_4). Maleic acid anhydride is prepared by the vanadium pentoxide-catalyzed oxidation of benzene. More recently, vanadium pentoxide has been used as a

---

BOX 37.1   APPLICATIONS OF VANADIUM ALLOYS AND CATALYSTS^a

Vanadium Microalloyed Steel
  Structural steel
  Gas turbine engine
  Aerospace applications
  Vehicle components
  High speed tools
  Pipes
Catalytic Uses of Vanadium Compounds
Chemical reduction of oxides of nitrogen (NO_x) in
  diesel engine exhaust to harmless nitrogen gas (N_2).
Oxidation reactions
  Sulfur dioxide to sulfuric acid
  Benzene to maleic anhydride
  Naphthalene to phthalic anhydride
  Butene to phthalic anhydride
  o-Xylene to phthalic anhydride
Synthetic rubber production
Phthalic anhydride
Polyamide

^{a}Weckhuysen and Keller, 2003.

---

selective reducing catalyst to reduce NO and NO_2 emitted from diesel engines to nitrogen gas (N_2) (Matthey, 2010). Other uses of vanadium compounds include the development of new batteries (e.g., vanadium flow and lithium–vanadium batteries), treatment of wastewater for nitrates, and as a colorant in ceramics and enamels.

PHYSICAL AND CHEMICAL PROPERTIES

Vanadium compounds, typically, have a range of oxidation states from +3 to +5 with more exotic compounds having oxidation states from −1 to +2 (IARC, 2006). The solubility of different vanadium compounds in water between 20 and 25 °C differs among different oxidation states (Table 37.1; (Rahman and Skyllas-Kazacos, 1998; WHO, 2001; HSDB, 2008).

The +4 oxidation state (V^{4+}) compound, vanadyl sulfate ($VOSO_4$), is highly soluble in water while vanadium compounds with a +5 oxidation state (V^{5+}) have mixed solubilities. Vanadium pentoxide (V_2O_5) and vanadium trioxide (V_2O_3) can be considered as acid anhydrides and when dissolved in water, these compounds react with water (i.e., decompose) at different rates, depending on the compound, to form mono vanadium compounds. Other V^{5+} compounds such as sodium metavanadate ($NaVO_3$), sodium orthovanadate (Na_3VO_4), and ammonium metavanadate (NH_4VO_3) are highly water soluble with solubilities of 211, 100, and 58 g/l,

TABLE 37.1   Oxidation States and Water Solubility of Various Vanadium Compounds

Vanadium Compound	Formula	CASRN	Oxidation State	Solubility (g/l) at 20–25 °C	Structure
Vanadium	V	7440-62-2	0	Insoluble	
Vanadium pentoxide	V_2O_5	1314-62-1	+5	8, decomposes	
Sodium metavanadate	$NaVO_3$	13718-26-8	+5	211	
Sodium orthovanadate	Na_3VO_4	13721-39-6	+5	100	
Ammonium metavanadate	NH_4VO_3	7803-55-6	+5	5.2 at 15 °C	
Vanadium oxytrichloride	$VOCl_3$	7727-18-6	+5	Decomposes	
Vanadyl sulfate	$VOSO_4$	27774-13-6	+4	535	
Vanadium tetrachloride	VCl_4	7632-51-1	+4	Decomposes	
Vanadyl oxydichloride	$VOCl_2$	10213-09-9	+4	Decomposes	
Vanadium trioxide	V_2O_3	1314-34-7	+3	Slightly soluble, decomposes	

Source: Adapted from Assem and Levy, 2009.

TABLE 37.2   Physical Properties of Metallic Vanadium

Physical Appearance	Solid White, Ductile Metal
Density at 20 °C	6 g/cm³
Melting point	1910 °C
Density at melting point (liquid, 1910 °C)	5.5 g/cm³
Natural isotopes	^{50}V 0.25%
	^{51}V 99.75%
Specific heat capacity	24.89 J/mol
Heat of fusion	21.45 J/mol
Heat of vaporization	459 kJ/mol

respectively. Furthermore, the rate of dissolution (which is distinct from solubility—an equilibrium or thermodynamic parameter) of various vanadium compounds can vary, resulting in different concentrations of specific forms of vanadium depending on how long these compounds have been in water. Additional properties for metallic vanadium are listed in Table 37.2.

ENVIRONMENTAL FATE

Natural sources of vanadium in soil come from minerals (over 50 known minerals containing vanadium; Box 37.2) and from

---

BOX 37.2   REPRESENTATIVE VANADIUM-CONTAINING MINERALS[a]

Mineral	Formula	Location(s) Where Found
Barnesite	$(Na, Ca)_2[V_2O_6V_4O_{10}]\cdot 3H_2O$	Kazakhstan, USA
Carnotite	$K_2(UO_2)_2(V_2O_8)\cdot 1\text{--}3H_2O$	Australia, Kazakhstan, USA
Doloresite	$H_8V_6O_{16}$	Australia, Argentine, USA
Duttonite	$VO(OH)_2$	Argentina, Gabon, USA
Fingerite	$Cu_{11}O_2(VO_4)_6$	El Salvador
Hendersonite	$(Ca, Sr)_{2.6}(V^{5+}, V^{4+})_{12}O_{32}\cdot 12H_2O$	USA
Metadolerite	$CaSrV_2O_6(OH)_2$	USA
Patrónite	VS_4	Peru
Vanadinite	$Pb_5(VO_4)Cl$	Australia, Mexico, Morocco, USA, Namibia, Turkey, USA
Vanalite	$NaAl_9V_{12}O_{42}(OH)_4\cdot 33H_2O$	Kazakhstan

[a]Anthony et al., 2012.

---

the atmospheric deposition of marine aerosols and particulate matter of volcanic origin. Vanadium is not typically mined for directly but is recovered while mining for other minerals. In addition, anthropogenic sources of vanadium also find their way into soil primarily from the deposition of particulate matter from fossil fuel combustion sources (e.g., coal- and oil-fueled power plants). The movement of vanadium in soil is heavily pH dependent since most vanadium minerals are not soluble in neutral water. The pH of water leaching the soil depends primarily on the alkalinity of the soil with more movement of vanadium at higher pH solutions [more alkaline] (NRC, 1974; WHO, 1988; ATSDR, 2012). Acid rain (primarily nitric and sulfuric acids) can contribute to the acidity of leaching fluids by lowering the pH of ground water. Vanadium is found in all natural sources of water. Theoretical estimates of the concentration of vanadium that could be found in sea water, due to the levels found in the earth's crust, suggest that vanadium concentrations should be fairly high, around 60 parts per million (ppm) [60 mg V/l] (NRC, 1974; WHO, 1988). However, concentrations in freshwater and seawater have been measured to be <50 µg V/l and 0.3–3.2 µg V/l, respectively (Holdway and Sprague, 1979; Ünsal, 1982). This discrepancy has been attributed to the binding of vanadium with suspended particulate matter and precipitation of vanadium-enriched silt to the ocean floor (NRC, 1974; WHO, 1988). Many marine organisms bioaccumulate vanadium compared to the surrounding sea water. Sea squirts (*Didemnum* sp.), for example, are known to achieve vanadium levels >10,000× than the concentration of sea water (ATSDR, 2012). Vanadium typically exists in two forms in fresh water depending on the presence of oxidizing or reducing agents in the water. In the presence of reducing agents, vanadium is present as vanadyl compounds (VO^{2+} and $VO(OH)^+$) and in the presence of oxidizing substances vanadium is found in the pentavalent form (V^{+5}) as vanadate compounds such as

$H_2VO_4^-$ and HVO_4^{2-} (Holdway and Sprague, 1979). Vanadium compounds also combines with particulate matter in fresh water with a particular affinity for humic acids (Ünsal, 1982).

The presence of vanadium in the atmosphere is primarily in the form of compounds bound to particulate matter as gaseous species of vanadium do not exist at ambient temperatures. Examples of vanadium compounds bound to particulate matter include vanadium pentoxide (V_2O_5), but most measurements of vanadium bound to particles is measured as elemental vanadium and is not speciated to specific compounds or minerals. Vanadium-enriched particles can dry deposit, as do particles in general, as well as be rained out with particles either acting as condensation nuclei or being absorbed directly into water droplets.

ECOTOXICOLOGY AND BIOACCUMULATION

The toxicity of vanadium to freshwater fish depends on a number of factors, two of which are dominant such as exposure time, (e.g., 96-h exposures) and water hardness, with vanadium generally being more toxic in softer water (defined as $CaCO_3$ equivalents) and longer exposure times. The range of toxicities found for freshwater fish (Table 37.3), depending on these two factors, ranges from 3.2 to 55 mg V/l for 96-h exposures with vanadium being more toxic at longer exposures (e.g., LC_{50} of 48 mg V/l at 24 h compared to 12 mg V/l in the flannelmouth sucker) (Hamilton and Buhl, 1997). In the case of saltwater fish, the salinity of the water is an additional complicating factor. For two saltwater species and one freshwater species tested in 10% salinity (three-spined stickleback, Table 37.3) the LC_{50} toxicities ranged from 0.6 to 18 mg V/l for 96-h exposures. In addition to studies on the lethal effects of vanadium on adult fish, this metal was also assessed for its impact on the growth of juvenile rainbow

TABLE 37.3   Examples of Ecotoxicological Studies of Vanadium

Common Name (Family or *Genus species*)	Comments on Toxicity, Toxicological Endpoint, or Accumulation of Vanadium	Reference
Fish		
Guppies (*Poecilia reticulata*)	LC_{50}: 4.4 mg V/l (7 days)	Perez-Benito, 2006
Three-spined stickleback, (*Gasterosteus aculeatus*); fathead minnow (*Pimephales promelas*); bluegill sunfish (*Lepomis macrochirus*); rainbow trout (*Oncorhynchus mykiss*); brown trout (*Salvelinus fontinalis*); goldfish (*Carassius auratus*); guppy (*Lebistes reticulatus*); American-flag fish (*Jordanella floridae*); striped gourami (*Colisa fasciatus*); mummichog (*Fundulus heteroclitus*); tigerfish (*Terapon jarbua*)	*G. aculeatus*: LC_{50} 3.2 (96 h); 10% salinity, 6–16 mg/V (96 h) *P. promelas*: LC_{50} 4.8 or 30 mg V/l (96 h), depending on water hardness *L. macrochirus*: LC_{50} 6 or 55 mg V/l (96 h), depending on water hardness *O. mykiss*: LC_{50} 2–13 mg V/l (4–20 days); 0.16 mg V/l (28 days) *S. fontinalis*: LC_{50} 2–36 mg V/l (7–720 h) *C. auratus*: LC_{50} 0.37–1.5 mg V/l (6 days) *L. reticulatus*: LC_{50} 2.5–8.1 mg V/l (6 days) *J. floridae*: LC_{50} 11 mg V/l (96 h) *C. fasciatus*: LC_{50} 6.4 mg V/l (96 h) *F. heteroclitus*: LC_{50} 14 and 18 mg V/l at 6% and 22% salinity (96 h) *T. jarbua*: LC_{50}: 1 mg V/l (24 h), 0.97 mg V/l (48 h), 0.8 mg V/l (72 h), 0.62 mg V/l (96 h)	Gravenmier et al., 2005 and references cited
Flannelmouth sucker (*Catostomus latipinnis*)	LC_{50}: 48 mg V/l (24 h); 36 mg V/l (48 h); 19 mg V/l (72 h); 12 mg V/l (96 h)	Hamilton and Buhl, 1997
Catfish (*Clarias batrachus*)	10 mg V/l; lipid peroxidation increased in liver, kidney, and brain	Bishayee and Chatterjee, 1994
Juvenile rainbow trout (*Oncorhynchus mykiss*)	10–900 mg V/Kg (12 weeks); all doses showed reduced growth; toxicity observed at >490 mg V/l	Hilton and Bettger, 1988
Aquatic invertebrates		
Tube-dwelling fan worms (Sabellidae; *Pseudopotamilla occelata, Perkinsiana littoralis*)	Hyper-accumulation of V, >5 mg/g in branchial crowns of *P. occelata*; >10 mg/g were measured in branchial crowns of *P. littoralis*	Fattorini and Regoli, 2012
Mysid shrimp (*Americamysis bahia*)	LC_{50}: 9.4 (7.2–12.6) mg V/l (96 h); the no observed effect concentration was 6.4 mg V/l	Woods et al., 2004
Brine shrimp larvae (*Artemia salina*); Pacific oyster larvae (*Crassostrea gigas*); sea urchin larvae (*Paracentrotus lividus*)	Normally developed larvae and growth of normal larvae; oyster toxic at 50 μg V/l; urchin toxic at 100 μg V/l (2 days); brine shrimp toxic at 250 μg V/l (8 days)	Fichet and Miramand, 1998
Hydroid, brackish water (*Cordylophora caspia*)	Impaired population growth, 2 mg V/l (10 days); >10 mg V/l, reproduction ceased (10 days)	Ringelband and Karbe, 1996
Water plants		
Fresh-water algae (*Scenedesmus acutus*; *Scenedesmus obliquus*)	Growth and photosynthesis are inhibited by vanadium due to competition with phosphorus	Nalewajko et al., 2008
Phytoplankton (cyanobacteria)	0.01–0.84 mg V/l, decreased photosynthesis	Nalewajko et al., 1995
Freshwater vascular plants, (12 species)	Accumulation: 5.8 mg V/kg dry weight (0.30–32 mg/kg dry weight)	Outridge and Noller, 1991
Marine macroalgae brown algae (*Fucus spiralis*); green algae (*Enteromorpha compressa*); red algae (*Polysiphonia urceolata, Goniotrichum alsidii, Nemalion helminthoides*)	400% increased growth of brown algae and 90% increased growth of green algae from 10 μg V/l (53 days)	Fries, 1982
Fresh-water green algae (*Chlorella pyrenoidosa*)	Maximal growth at 0.5 mg V/l; reduced growth at >25 mg V/l; 100 mg V/l is lethal	Meisch et al., 1977

(continued)

TABLE 37.3   *(Continued)*

Common Name (Family or *Genus species*)	Comments on Toxicity, Toxicological Endpoint, or Accumulation of Vanadium	Reference
Terrestrial plants		
Lettuce (*Lactuca sativa*); grasses (*Elymus virginicus*), (*Panicum virgatum*); broad leaf (*Lycopus americanus*), and (*Prunella vulgaris*)	0–265 mg V/g soil; broad-leaf plants more sensitive than grasses or lettuce; increased nutrient levels lowered V toxicity.	Smith et al., 2013
Chinese green mustard (*Brassica campestris* ssp. *chinensis* var. *parachinensis*)	0.43–35 mg V/l (3 weeks) administered to soil, 35 mg V/l reduced leaf and stem weight and plant height; 0–17 mg V/l showed no effects	Vachirapatama and Jirakiattikul, 2008
Soybean (*Glycine max*)	Vanadium added to two soils; >30 mg V/kg in fluvo-aquic soil caused reduced weight of shoots and roots; no effect of 75 mg V/kg in red earth soil on shoots or roots	Wang and Liu, 1999
Lettuce (*Lactuca sativa*)	0.1–1 mg V/l in nutrient solutions, fresh and dry weights inhibited at >0.2 mg V/l	Gil et al., 1995
Soybean (*Glycine max*); bush bean (*Phaseolus vulgaris*)	6 mg V/l in hydroponic solution reduced plant biomass	Kaplan et al., 1990

trout (Hilton and Bettger, 1988). All doses of vanadium (10–900 mg V/kg) inhibited the growth of these fish. An increase in the lipid peroxidation of tissues in the liver, brain, and kidneys was also found for catfish at concentrations that have been shown to be lethal for other freshwater fish (10 mg V/l, Table 37.3 (Bishayee and Chatterjee, 1994).

Vanadium in seawater is present at fairly low concentrations presumably due to the binding of vanadium compounds to silt followed by precipitation to the sea bed (NRC, 1974; WHO, 1988). Some aquatic invertebrates, however, are known to hyper-accumulate vanadium (Table 37.3). For example, tube-dwelling fan worms accumulate >5 mg/g of vanadium in the tissues of branchial crowns when the surrounding sea water concentration of vanadium is about 0.3–3.2 µg V/l (Ünsal, 1982). Studies on adult saltwater invertebrates suggest that adult invertebrates may be less sensitive than larvae invertebrates, although definitive studies comparing life stages of aquatic invertebrates have not been conducted. Mysid shrimps show no observed effects at 6.4 mg V/l with toxic effects beginning to be observed at 7.2 mg V/l. Brackish hydroid showed a reduction in growth at 2 mg V/l with reproduction ceasing at concentrations >10 mg V/l. However, studies on the development of larvae suggest that this life stage may be more sensitive than adult invertebrates. Oyster, sea urchin, and brine shrimp larvae do not develop or grow normally at concentrations of vanadium between 50 and 250 µg V/l (Table 37.3).

At low concentrations (10 µg V/l) marine algae showed mixed results with the largest increase in growth observed with brown algae (*Fucus spiralis*) followed by green algae (*Enteromorpha compressa)* (Table 37.3). No effects were observed for three species of red algae (*Polysiphonia urceolata, Goniotrichum alsidii, and Nemalion helminthoides*). Freshwater green algae (*Chlorella pyrenoidosa)* showed a range of effects from enhanced growth at low concentrations, to reduced growth at intermediate concentrations and lethal effects at high concentrations (Table 37.3). These data suggest that vanadium may be beneficial to algae in low concentrations. In contrast, the effect of vanadium on photosynthesis in phytoplankton, specifically cyanobacteria, was inhibitory even at low concentrations. Freshwater vascular plants (flowering plants, some ferns with a vascular system and well-defined roots) show a wide range of vanadium accumulation (Table 37.3). Two species (*Elodea canadensis* and *Pontederia cordata*) accumulated high levels of vanadium, 18 and 32 mg V/kg dry weight, respectively, compared to the range from the remaining 12 species (0.30–6 mg V/kg).

The effect of vanadium on terrestrial plant growth depends on a number of factors (Table 37.3), including the chemical characteristics of the soil that influence the binding of vanadium making this metal more or less bioavailable. For example, Wang and Liu (1999) showed that soybeans grown in fluvo-aquic soil was toxic to levels of vanadium >30 mg V/kg soil whereas a concentration of 75 mg V/kg in red earth soil was not phytotoxic. Similarly, the level of nutrients in soil showed an inverse toxic effect with vanadium in soils with high nutrient levels showing little, if any, toxic effects of vanadium in contrast to nutrient-poor soils that showed significant phytotoxicity. The conditions of the assay for toxicity also play a complicating factor. For example, studies of plants grown in aqueous solutions (Kaplan et al., 1990; Gil et al., 1995) showed more sensitive effects of vanadium compared to studies using soil to grow plants (Table 36.3). This may be related to the relative bioavailability of the metal with soluble vanadium being essentially 100% bioavailable; all soils will bind, to some extent, vanadium making this metal less bioavailable in soil.

MAMMALIAN TOXICOLOGY

Industrial exposure and illness have most commonly involved vanadium pentoxide; the vast majority of the toxicological literature focuses on vanadium pentoxide, and exposure standards are set accordingly. This partly reflects the importance of vanadium pentoxide in the industrial applications of vanadium. Equally important is the observation that toxicity is proportional to valence: pentavalent vanadium is the most toxic form of the metal. Furthermore, industrial exposures have almost always involved inhalation of dust and fumes containing vanadium or vanadium oxides.

The sections below focus on the major epidemiological and laboratory animal studies available in the published literature on adverse health effects following exposure to all vanadium compounds. The majority of studies focused on the respiratory effects following exposure to vanadium compounds, with the majority examining exposure to vanadium pentoxide. Laboratory animal and mechanistic studies also examined hematological, neurological, and reproductive effects. Limited studies examine other endpoints such as renal, hepatic, and immunological effects following exposure, the results of which were largely negative. Examples of both epidemiological and laboratory animal studies of vanadium compounds are summarized in Table 37.4. For more detailed reviews of the literature, please review EPA, 2011, (IARC, 2006; ATSDR, 2012).

Acute Effects

Zenz et al. (1962) (Zenz and Berg, 1967) have studied industrial vanadium toxicity in the field and in the laboratory and have written the classic summary of acute vanadium toxicity. The clinical picture of acute illness appeared remarkably uniform among these workers. The syndrome consisted of a rapidly developing, mild conjunctivitis, severe pharyngeal irritation, and nonproductive persistent cough, followed by diffuse rales and bronchospasm. After severe exposure, four men complained of skin "itch" and a sensation of "heat" in the face and forearms, but no objective evidence of skin irritation was found. A striking feature among the acute intoxications was the increased severity of symptoms with repeated exposures of lesser time and intensity. Acute inhalation studies have demonstrated respiratory difficulty and irritation of the mucosa in rats (NTP, 2002) and decreased pulmonary function in monkeys (Knecht et al., 1985) following exposure to vanadium pentoxide. Acute inhalation exposure to ammonium metavanadate in rats also resulted in increased inflammatory cells in bronchoalveolar lung fluid (Cohen et al., 1996). Acute oral studies in rats and rabbits did not show any respiratory effects following exposure to vanadium pentoxide (WHO, 2001).

Animal and clinical studies replicate the clinical studies carried out after industrial exposures (Table 36.4). With few exceptions these reports, from 1911 to the present, describe workers with irritation and inflammation of the skin, conjunctivae, and mucosae of the nasopharynx and the respiratory tract (Wyers, 1946; Sjoberg, 1951; Williams, 1952; Sjoberg, 1955; Vintinner et al., 1955; Sjoberg, 1956; Musk and Tees, 1982).

Chronic Effects

Many epidemiological researchers comment on the greenish discoloration of the tongue from coating due to dust containing vanadium. Catarrh, epistaxis, productive cough, occasional hemoptysis, wheezing, dyspnea, and occasionally bronchopneumonia are all part of the respiratory syndrome produced by dust and fumes that contain vanadium. Spirometry consistently revealed decreased forced vital capacity (FVC) and forced expiratory volume (FEV) (Lewis, 1959; Lees, 1980). Continued exposure can result in chronic respiratory changes and symptoms, but cessation of exposure usually allows the respiratory tract to return to normal. However, chronic exposure to high concentrations of dust containing vanadium has been associated with the gradual development of atrophic rhinitis and chronic bronchitis, which only partially remit with cessation of exposure.

Levy et al. (1984) coined the term *boilermakers' bronchitis* to describe the slowly resolving bronchitis that occurred in 74 of 100 workers involved in the conversion of oil-fired to coal-fired boilers. The power plant had used fuel oil, high in vanadium. Workers welding and heating inside the boilers were exposed to concentrations of vanadium pentoxide ranging from 0.05 to 5.3 mg/m^3. The most common symptoms were productive cough, sore throat, dyspnea on exertion, and chest discomfort, and these symptoms were serious enough to cause 70 workers to seek medical attention. Wheezing and diminished FEV were the most common findings. The symptoms lasted from 9 to 82 days, with a median duration of 34 days. None of the workers used adequate respiratory protection.

Kiviluoto reported long-term follow-up on workers making vanadium pentoxide from magnetite (Kiviluoto et al., 1979; Kiviluoto, 1980; Kiviluoto et al., 1981a). These workers were exposed to vanadium pentoxide concentrations ranging from 0.01 to 3.9 mg/m^3 for an average of 11 years. No evidence of cancer or pneumoconiosis was found, although the exposed workers had more complaints of wheezing and cough than nonexposed controls. Vanadium pentoxide inhalation exposure for 2 years was associated with a wide spectrum of nonneoplastic pulmonary lesions in both rats and mice, ranging from hyperplasia to inflammation, fibrosis, and metaplasia. Lesions were detected in the lung, larynx, and nose in both rats and mice exposed to vanadium pentoxide. Bronchial lymph

TABLE 37.4   Examples of Epidemiology and Mammalian Toxicology Studies of Vanadium

Exposure Group	Dose/Exposure Duration/Compound	Observed Effects	Reference
Human			
Controlled human exposure ($n = 9$)	0.1–1 mg/m^3 for 8 h vanadium pentoxide (inhalation)	Coughing, no other signs of irritation; no change in lung function	Zenz and Berg, 1967
Case studies ($n = 36$)	0.6–86.9 mg/m^3 vanadium pentoxide in dust over 3 years (inhalation)	Severe upper respiratory irritation; lower respiratory irritation; skin irritation; fatigue; decrease in hemoglobin levels	Sjoberg, 1951; Sjoberg, 1956
Case studies ($n = 18$)	>0.5 ug/m^3 vanadium pentoxide in air over 2 weeks; lifetime exposure range varied (inhalation)	Upper respiratory irritation; some skin irritation	Zenz et al., 1962
Cross-sectional study ($n = 63$)	0.2–0.5 mg/m^3 vanadium dust as measured on filters over 5 years (inhalation)	Upper respiratory irritation; no hematological changes observed	Kiviluoto et al., 1979; Kiviluoto, 1980; Kiviluoto et al., 1981a
Occupational study ($n = 55$)	0.05–5.3 mg/m^3 vanadium pentoxide in the boiler (inhalation)	Upper respiratory irritation; chest pain; fatigue	Levy et al., 1984
Case-control study ($n = 31$)	0.5 mg vanadium/kg/day 4, 8, 12 weeks vanadyl sulfate (oral)	No effect on blood pressure was observed	Fawcett et al., 1997
Laboratory animal studies			
B6C3F$_1$ mice, M/F ($n =$ up to 50) F344/N rats, M/F ($n =$ up to 50)	0–4 mg/m^3 16 days–2 years vanadium pentoxide (inhalation)	Respiratory effects: chronic inflammation, hyperplasia, adenomas/carcinomas; hematological effects: limited to 3 months only in males.	NTP, 2002
Cynomolgus monkeys, M ($n = 8$)	0.1–1.1 mg/m^3 26 weeks vanadium pentoxide (inhalation)	Impaired pulmonary function	Knecht et al., 1992
CD-1 mice, M ($n = 48$)	2.56 mg/m^3 8 weeks vanadium pentoxide (inhalation)	CNS morphological alterations	Avila-Costa et al., 2004; Avila-Costa ct al., 2005; Avila-Costa et al., 2006
CD-1 mice, M ($n = 20$)	2.56 mg/m^3 6 weeks vanadium pentoxide (inhalation)	Inflammatory, oxidative stress and liver toxicity markers increased	Cano-Gutiérrez et al., 2012
Wistar rats, M ($n = 32$)	0.14 mg vanadium/kg/day 12 weeks ammonium metavanadate (drinking water)	Limited hematological toxicity observed	Dai et al., 1995
Sprague-Dawley rats, F ($n = 14$)	1–75 ug vanadium/g diet through day 21 postpartum sodium metavanadate (oral feed)	Developmental effects (decreased pup weight and survival)	Elfant and Keen, 1987
Wistar rats, M/F ($n = 46$)	0.01–0.05 mg vanadium/kg/day 4–8 week ammonium metavanadate (drinking water)	Hematological effects (decreased erythrocyte and increased reticulocyte levels)	Zaporowska et al., 1993
Wistar rats, M ($n = 12$)	10.7 mg vanadium/kg/day 6 week sodium metavanadate (oral)	Hematological effects (decreased erythrocyte and hemoglobin levels)	Scibior et al., 2006
Swiss mice, F ($n = 17$)	0–60 mg vanadium/kg/day GD 6–15 sodium orthovanadate (oral gavage)	Developmental effects	Sanchez et al., 1991

node changes were detected in mice exposed to vanadium pentoxide. In nonhuman primates and rodents, inhalation or intratracheal instillation of vanadium pentoxide dust or suspensions of vanadium pentoxide, vanadyl sulfate (VOSO4), or sodium metavanadate (NaVO$_3$) rapidly induced an acute inflammatory reaction with a predominance of polymorphonuclear leukocytes (Knecht et al., 1985; Knecht et al., 1992; Pierce et al., 1996; Rondini et al., 2010; Turpin et al., 2010; Schuler et al., 2011; Yu et al., 2011). The inflammation developed more rapidly with the more soluble vanadium salts, NaVO$_3$ and VOSO$_4$, and slowest with the less soluble V$_2$O$_5$. A similar gradient of intensity and rapidity was seen for the induction of proinflammatory cytokines from alveolar macrophages. Changes in pulmonary function tests included reduction in flow rates and vital capacity. Clearance of vanadium from bronchoalveolar fluid was rapid and associated with accumulation in cells and the inflammatory exudate and in lung tissue. No evidence for allergic hypersensitivity has been documented with repetitive experimental exposures.

Limited human studies examined hematological effects following exposure to vanadium pentoxide. No hematological alterations were observed in humans following short-term (Dimond et al., 1963; Zenz and Berg, 1967; Fawcett et al., 1997) or occupational exposure (Sjoberg, 1951; Vintinner et al., 1955; Kiviluoto et al., 1981a) to vanadium pentoxide dusts, ammonium vanadyl tartrate, or vanadyl sulfate through inhalation and oral exposures. Early laboratory animal studies demonstrated a decrease in red blood cells and hemoglobin levels in rats following oral exposure to vanadium pentoxide in feed (Mountain et al., 1953). A series of studies conducted by Zaporowska et al. (1993) examined the hematotoxicity of ammonium metavanadate administered in drinking water to rats for acute or intermediate durations. Results include increases in reticulocyte levels and the percentage of polychromatophilic erythroblasts in the bone marrow in male rats following 2-weeks exposure (Zaporowska and Wasilewski, 1989); a nonsignificant increase in erythrocytes was also observed at this dose level. At exposures up to 4 weeks, decreases in erythrocyte levels and hemoglobin levels and increases in reticulocyte levels were observed (Zaporowska and Wasilewski, 1989; Zaporowska et al., 1993). Similar effects were observed in rats exposed to sodium metavanadate for 6 weeks (Scibior et al., 2006) or 10 weeks (Adachi et al., 2000). Other laboratory animal studies in rats did not observe hematological alterations (Dai et al., 1995; Dai and McNeill, 1994), potentially related to the various doses and animal strains used. Blood pressure changes have also been observed in laboratory animal studies, however, the results are inconsistent. Increased systolic, diastolic, and/or mean blood pressure was observed in rats following exposure to sodium metavanadate or ammonium metavanadate [dose range 0.12–12 mg, up to 210 days] (Boscolo et al., 1994). However, no blood pressure changes were observed in rats following exposure to sodium metavanadate, vanadyl sulfate, or ammonium metavanadate [dose range 10–63 mg, up to 52 weeks] (Dai et al., 2006).

No published studies examined developmental effects in humans following exposure to vanadium. Developmental effects were observed in animal studies following oral exposure to vanadyl sulfate, sodium metavanadate, ammonium metavanadate, or sodium orthovanadate (Paternain et al., 1987; Sanchez et al., 1991; Morgan and El-Tawil, 2003). These effects include delayed ossification, decreased fetal growth, and increase in skeletal and visceral anomalies.

Immunological effects of exposure to vanadium compounds have not been comprehensively studied, but the limited number of epidemiological and laboratory animal study results are summarized below. Case studies of occupational exposure to vanadium pentoxide have reported individuals with dermatitis, positive skin patch reactions, and bronchial reactivity, although this occurrence appears to be uncommon (Sjoberg, 1951; Zenz et al., 1962; Motolese et al., 1993). Increases in the number of inflammatory cells in nasal smears or nasal lavage fluid were observed in exposed workers but no associations were observed in relation to estimates of respirable vanadium dust or vanadium concentrations in nasal fluid (Kiviluoto et al., 1979; Kiviluoto, 1980; Kiviluoto et al., 1981b). The authors did not report increase in eosinophils, suggesting that the inflammation was due to irritation and was not an allergic response. Oral exposure to vanadium pentoxide in drinking water in rats demonstrated a non-statistically significant increase in cell-mediated immune activation (Mravcova et al., 1993). In a study in cynomolgus monkeys, skin sensitivity tests assessed immediate and delayed responses to intradermal injections of vanadium pentoxide–serum albumin conjugate (Knecht et al., 1992). The inhalation exposure in this study also demonstrated an influx of inflammatory cells in the lung. Other laboratory animal inhalation studies have also examined more specific mechanisms related to immune function and demonstrated that exposure to vanadium pentoxide did not impact platelet activation (González-Villalva et al., 2011) but led to decreases in specific immune cells following exposure to vanadium pentoxide suggesting a possible autoimmune effect of exposure (Ustarroz-Cano et al., 2012). These results suggest an impact of vanadium exposure on immune response; however, as there are only a limited number of inhalation studies that examined immune function following exposure to vanadium compounds, increased research is needed to further understand this response.

No human studies examined the impact of exposure to vanadium compounds on neurological function; however, some workers described a variety of neurological systems (e.g., headaches, dizziness) that may or may not be attributed to vanadium exposure (Vintinner et al., 1955; Levy et al., 1984). Laboratory animal studies have demonstrated impaired performance on memory tasks after as little as a 1-h inhalation exposure to vanadium pentoxide (2.56 mg/m^3

vanadium pentoxide) (Avila-Costa et al., 2006). Longer-term studies (up to 8 weeks) have also demonstrated impairment of performance on memory tests and an associated loss of dendritic spines in the hippocampal region of the brain as well as CNS morphological changes in the blood–brain barrier as well as an increase in immunoreactive neurons in the basal ganglia (Avila-Costa et al., 2004; Avila-Costa et al., 2005; Avila-Costa et al., 2006; Colín-Barenque et al., 2008). Only two studies examined neurological function following oral exposure to vanadium (as sodium metavanadate, 8 weeks), with decreased travelling distance observed in one study (1.72 mg vanadium/kg) (Sanchez et al., 1998) but not the other (6.84 mg vanadium/kg) (Sanchez et al., 1999). Rats in both studies demonstrated increases in latency period following stimuli and decreases in the number of avoidance responses to stimuli.

No human epidemiology studies that examined carcinogenesis following exposure to vanadium compounds are available. There is evidence of carcinogenicity in male and female mice exposed to vanadium pentoxide based on observations of alveolar and bronchiolar neoplasms that exceeded historical controls in groups exposed to vanadium pentoxide (NTP, 2002). This study also demonstrated an increase in neoplasms in male rats, although these were not statistically significant. A more recent study has also shown lung tumor promotion in sensitive mouse strains following exposure to vanadium pentoxide (Rondini et al., 2010).

Mode of Action(s)

Currently, evidence is insufficient to establish the mode of action (MOA) for noncancer toxicity or cancer effects following exposure to vanadium compounds. The limited data to inform the mechanisms of various adverse health effects following exposure to vanadium compounds are summarized below.

Chronic Inflammation   Evidence suggests that a potential mechanism of action underlying the formation of pulmonary fibroproliferative lesions may involve inflammation, leading to a regenerative hyperplastic response (i.e., presence of mitogens and observed smooth muscle thickening). Oxidative stress also has been implicated. Oxidative stress induced directly or indirectly by vanadium pentoxide may work in combination with vanadium pentoxide-induced inflammatory responses to activate signaling molecules such as extracellular signal-related kinase 1/2 (ERK1/2) and p38 kinase that lead to induction of growth factors and generation of fibrotic lesions. These potential modes of action are supported by a limited number of mechanistic analyses of inflammation and fibrosis following exposure to vanadium pentoxide.

Genotoxicity   The evidence for mutagenicity in humans is limited (Ivancsits et al., 2002; Kleinsasser et al., 2003; Ehrlich et al., 2008). Few studies have examined genotoxicity in humans in vivo, with equivocal results. Studies have demonstrated a genotoxic effect of vanadium pentoxide on human cells in vitro (Roldán and Altamirano, 1990; Rojas et al., 1996; Ramirez et al., 1997). Based on these studies, vanadium pentoxide-induced mutagenicity could occur at doses higher than those measured in these occupational exposures, could be tissue specific, and could be associated with oxidative stress rather than direct DNA damage. *In vitro* tests in bacterial and yeast systems provide mixed evidence of vanadium pentoxide-induced mutagenicity. In general, classic gene mutation assays were negative, as were tests that assessed sister chromatid exchange (SCE) and other chromosomal aberrations. DNA strand breaks (Rojas et al., 1996; Ivancsits et al., 2002) and micronucleus formation (Zhong et al., 1994) were indicated in some studies in cultured cells but depended on cell type. Fibroblasts appear to be more sensitive to vanadium exposure in vitro than are blood cells. Similarly, experimental data from animal studies is equivocal. NTP (2002) reported that the frequency of micronucleated normochromatic erythrocytes in peripheral blood was not increased in exposed mice compared to control mice. Several studies by Altamirano-Lozano et al. (1993, 1996, 1999), however, have noted DNA damage in specific target tissues in vanadium pentoxide-treated mice, although these studies used intraperitoneal injection as the route of exposure.

SOURCES OF EXPOSURE

Occupational

Occupational exposure to vanadium is predominantly through inhalation, although some oral and dermal exposures are possible. Occupational exposure may occur in workplaces that produce or use vanadium compounds. Vanadium compounds are used as a catalyst in the preparation of vanadium alloys and other compounds as well as in the automobile catalytic converters and in organic synthesis (OSHA, 2007). Vanadium compounds are also used as a component in welding, textile manufacturing, and ceramics production. Occupational exposure may also result from mining and processing of vanadium-containing ores and the cleaning and maintenance of furnaces, boilers, and gas turbines (OSHA, 2007).

Environmental

Vanadium is released into the environment from natural sources such as weathering of geologic formations, volcanoes, and marine aerosols. Vanadium entering the environment from anthropogenic sources is primarily due to emissions from oil and coal power plants and to a lesser extent, from ocean-going ships while in port. However, the primary source of human exposure to vanadium comes from the diet. Vanadium is found, generally at low levels, in most foods (Table 37.5). The concentration of vanadium

TABLE 37.5   Vanadium in Some Foods[a]

Foodstuff	Vanadium Concentration Mean (Range) ng/g Wet Weight
Vegetables	
Garlic, onion, brussel sprouts, cauliflower, peas, leek, lettuce, carrots, cabbage, potato, tomato, radish	1.32 (0.27–3)
Spinach	35
Cereals	
Oats, barley, maize, rice, flour	14 (0.7–40)
Fruits	
Apple, pear, banana, orange, cherry, apricot, peach, hazelnut	0.76 (0.15–3.7)
Meat—beef	
Sirloin, middle rib. steak, liver, lung	7 (0.4–25)
Meat—pork	
Leg, liver, kidney, heart, fat, muscle	3.3 (0.2–8.5)
Meat—chicken	
Breast, liver, kidney, heart	17 (1.7–37)
Meat—fish	
Cod, mackerel, tuna	12 (3.5–28)
Milk and eggs	
Milk	0.17
Egg yolk	2.8 (2–3.6)
Egg white	0.8 (0.3–1.8)
Foodstuff	Vanadium Concentration Mean (Range) ng/g Dry Weight
Spinach, freeze dried	690 (533–840)
Ground parsley	1800
Wild mushrooms, 26 species	50–2000

[a]Adapted from Byrne and Kosta, 1978.

in vegetables and foods varies substantially depending on the moisture content at the time of measurement. This is particularly notable in Table 37.4 for those vegetables measured dry (mushrooms, spinach, and ground parsley) compared to those vegetable measured using wet or freshly ground samples. Lokeshappa et al. (2012) reported much higher levels of vanadium (mean 49 ug V/g; range 43–52 ug V/g) from vegetables (spinach, potato, coriander, ladyfinger, carrots, tulsi, rice, mung bean, and Arhar bean) dried to constant weight. Differences in the vanadium content of vegetables and foods will likely depend on the levels and bioavailability of vanadium in the soil used to grow the food. The amount of vanadium ingested daily from food has been estimated to be about 10–300 ng V/kg/day and is typically in the form of the vanadyl cation (VO^{2+}) or the vanadate anion (HVO_4^{2-}) (ATSDR, 2012).

People can also be exposed to vanadium in drinking water. The primary source of vanadium in ground water comes from leaching or dissolution of vanadium-rich rocks and minerals. Several locations in the world have unusually high concentrations of vanadium in ground water due to the geology of the surrounding area. In Italy, for example, vanadium in ground water is considered high and the health effects of such exposures are of concern. High ($\geq 50\,\mu g$ V/l) and moderate (25–49 μg V/l) levels of vanadium were typically measured in over 8400 ground water samples from California. These samples were collected from sources where the water was in contact with mafic and andesitic rocks and geological formations favored dissolution of vanadium (Wright and Belitz, 2010). The presence of vanadium was detected in 68% of 662 samples taken from wells of the principal aquifers in the United States; the median concentration of vanadium detected in these samples was 1.3 μg V/l (10–90% percentile range, 0.12–20 μg V/l) (DeSimone, 2009). In addition to vanadium from ground and well water, vanadium can also enter drinking water through deposition of vanadium-enriched particles emitted from oil and coal power plants. For example, a study in Mexico suggested that the deposition of vanadium-enriched particulate matter from a thermoelectric plant using a fuel oil with high vanadium content was the origin of increased vanadium in the local groundwater (Mejia et al., 2007). In summary, vanadium concentrations in ground and well water can vary over a wide range (0.12 to >20 μg V/l) depending on the underlying geology and potential for deposition of vanadium-enriched particulate matter (Sepe et al., 2003; DeSimone, 2009). Assuming an average consumption of 2 l of water per day, one can estimate the amount of vanadium ingested from ground or well water to range from nondetectable levels to about 2.6 μg V per day.

People are exposed to vanadium through the air pollution and specifically via small particles ($PM_{2.5}$) generated from the combustion of fossil fuels. Vanadium is only one component of the complex mixture of compounds bound to particles, which makes the contribution of vanadium to adverse human health effects difficult to ascertain. Vanadium is generally recognized as the most abundant metal in petroleum products with some crude oils containing as much as 1500 mg V/kg (Turunen et al., 1995). Heavy fuel oils contain a wide range of vanadium, 18–259 mg V/kg (Shafer et al., 2012; Turunen et al., 1995; Agrawal et al., 2008) while refined petroleum such as commercial gasoline (53 μg V/l) and diesel fuel (88 μg V/l) contain about 1000-fold less vanadium. Vanadium is also found in coal; coal mined from the United States typically contains 22 mg V/kg (Finkelman, 1999). Thus, the major source of atmospheric exposure to vanadium comes from combustion by-products of coal and oil-fueled power plants and off-shore ships in ports. The mean concentration of vanadium in the

TABLE 37.6   Risk Assessments and Exposure Limits for Vanadium Compounds

<table>
<tr><td colspan="2" align="center">**Oral Value**</td></tr>
<tr><td>EPA (1996)</td><td>V_2O_5: Reference dose 9×10^{-3} mg/kg-day based on decreased hair cystine levels in rats (Stokinger et al., 1953).</td></tr>
<tr><td>ATSDR (2012)</td><td>Intermediate MRL: 0.01 mg vanadium/kg/day based on NOAEL for hematological changes following exposure to vanadyl sulfate (Fawcett et al., 1997).</td></tr>
<tr><td colspan="2" align="center">**Inhalation value**</td></tr>
<tr><td>ACGIH (2009)</td><td>Vanadium: A3—confirmed animal carcinogen with unknown relevance to humans. Threshold limit value (TLV)–time-weighted average (TWA) 0.05 mg/m^3, inhalable particulate matter, measured as vanadium
Based on upper and lower respiratory tract irritation.</td></tr>
<tr><td>ATSDR (2012)</td><td>Acute MRL: 0.0008 mg vanadium/m^3 based on LOAEL of incidence of lung inflammation in rats (NTP, 2002).
Chronic MRL: 0.0001 mg vanadium/m^3 based on degeneration of the respiratory epithelium of the epiglottis in rats (NTP, 2002).</td></tr>
<tr><td>NIOSH (2012)</td><td>Vanadium: recommended exposure level (REL) 0.05 mg/m^3, total dust and fume as vanadium</td></tr>
<tr><td>OSHA (2006)</td><td>Vanadium: permissible exposure level (PEL) 0.1 mg/m^3 (fume); 0.4 (respirable dust)</td></tr>
<tr><td>WHO (2000)</td><td>Vanadium: Air quality guideline for vanadium is 1 µg/m^3 based on LOAEL of 20 µg/m^3 from studies on occupationally exposed individuals, using an overall uncertainty factor of 20</td></tr>
<tr><td colspan="2" align="center">**Cancer characterization**</td></tr>
<tr><td>Cal EPA (2005)</td><td>V_2O_5: Added to the Proposition 65 chemical list as "known to the state to cause cancer" (orthorhombic crystalline form)</td></tr>
<tr><td>IARC (v86)
(2006)</td><td>V_2O_5: Possibly carcinogenic to humans (Group 2B) —inadequate evidence in humans for the carcinogenicity of vanadium pentoxide; sufficient evidence in experimental animals for the carcinogenicity of vanadium pentoxide</td></tr>
</table>

atmosphere from cities using oil to generate power is about 620 ng/m^3 compared to 11 ng/m^3 levels in the atmosphere from cities that use a mix of power sources (ATSDR, 2012; Mamane and Pirrone, 1998).

INDUSTRIAL HYGIENE AND RISK ASSESSMENTS

The normal rapid clearance of vanadium means that removal from the exposure source usually is all that is necessary. Chelation therapy can be effective but is not usually required (Gomez et al., 1988). Attention to industrial hygiene and engineering is essential. Warning labels, appropriate ventilation, and properly fitting respiratory protection can prevent problems (Stokinger, 1962). Maintenance and repair workers must be supervised and must wear protective equipment and clothing. A history of the fuels and chemical processes in work sites, even if the site is being used for a different procedure, is an important clue to potential vanadium exposure. The exposure standards are given in Table 37.6.

REFERENCES

ACGIH (American Conference of Governmental Industrial Hygienists) 2009 Documentation of the Threshold Limit Values (TLVs) and Biological Exposure Indices (BEIs) - Vanadium Pentoxide. 2009.

Adachi, A., Asai, K., Koyama, Y., et al. (2000) Subacute vanadium toxicity in rats. *J. Health Sci.* 46(6):503–508.

Agrawal, H., Malloy, Q.G.J., Welch, W.A., Miller, J.W., and Cocker, D.R. III. (2008) In-use gaseous and particulate matter emissions from a modern ocean going container vessel. *Atmos. Environ.* 42:5504–5510. Available at http://dx.doi.org/10.1016/j.atmosenv.2008.02.053.

Altamirano-Lozano, M., Alvarez-Barrera, L., Basurto-Alcántara, F., Valverde, M., and Rojas, E. (1996) Reprotoxic and genotoxic studies of vanadium pentoxide in male mice. *Teratog. Carcinog. Mutagen.* 16:7–17.

Altamirano-Lozano, M., Alvarez-Barrera, L., and Roldán-Reyes, E. (1993) Cytogenetic and teratogenic effects of vanadium pentoxide on mice. *Med. Sci. Res.* 21:711–713.

Altamirano-Lozano, M., Valverde, M., Alvarez-Barrera, L., Molina, B., and Rojas, E. (1999) Genotoxic studies of vanadium pentoxide (V(2)O(5)) in male mice. II. Effects in several mouse tissues. *Teratog. Carcinog. Mutagen.* 19:243–255. Available at http://dx.doi.org/10.1002/(SICI)1520-6866(1999)19:4<243::AID-TCM1>3.0.CO;2-J.

Anthony, J.W., Bideaux, R.A., Bladh, K.W., and Nichols, M.C. (2012) *Handbook of Mineralogy*, Mineralogical Society of America. Available at http://www.handbookofmineralogy.org/.

Assem, F.L. and Levy, L.S. (2009) A review of current toxicological concerns on vanadium pentoxide and other vanadium compounds: gaps in knowledge and directions for future research [Review]. *J. Toxicol. Environ. Health B Crit. Rev.* 12:289–306. Available at http://dx.doi.org/10.1080/10937400903094166.

Atomix, Inc. (2003) *Product data sheet: Vanadium pentoxide (V_2O_5)*. Available at http://www.atomixinc.com/vanadiumpentoxide.htm.

ATSDR (Agency for Toxic Substances and Disease Registry) (2012) *Toxicological Profile for Vanadium (Draft for Public Comment) [ATSDR Tox Profile]*, Atlanta, GA: U.S. Department of Health and Human Services, Public Health Service. Available at http://www.atsdr.cdc.gov/toxprofiles/tp58.pdf.

Avila-Costa, M., Colin-Barenque, L., Montiel-Flores, E., Aley, P., Sanchez, I., Irma, L., Gutierrez, J., Ordones, A., Acevedo-Nava,

S., and Gonzalez-Villalva, A. (2004) Dopaminergic cell death and vanadium inhalation [Abstract]. *Toxicologist* 78:234. Available at http://dx.doi.org/10.1016/j.neulet.2005.01.072.

Avila-Costa, M.R., Fortoul, T.I., Niño-Cabrera, G., Colín-Barenque, L., Bizarro-Nevares, P., Gutiérrez-Valdez, A.L., Ordóñez-Librado, J.L., Rodríguez-Lara, V., Mussali-Galante, P., Díaz-Bech, P., and Anaya-Martínez, V. (2006) Hippocampal cell alterations induced by the inhalation of vanadium pentoxide (V(2)O(5)) promote memory deterioration. *Neurotoxicology* 27:1007–1012, Available at http://dx.doi.org/10.1016/j.neuro.2006.04.001.

Avila-Costa, M.R., Zepeda-Rodriguez, A., Colin-Barenque, L., Pasos, F., Aley, P., Gonzalez-Villalva, A., Mussali-Galante, P., Ordonez-Librado, J., Gutierrez-Valdez, A., Rodriguez-Lara, V., Reyes-Olivera, A., Pinon-Zarate, G.; Rojas-Lemus, M., Delgrado, V., Chavez, B., and Fortoul, T.I. (2005) Blood brain barrier (BBB) disruption after vanadium inhalation [Abstract]. *Toxicologist* 84:124.

Baker, T.N. (2009) Processes, microstructure and properties of vanadium microalloyed steels. *Mater. Sci. Tech.* 25:1083–1107, Available at http://dx.doi.org/10.1179/174328409X453253.

Bishayee, V. and Chatterjee, M. (1994) Increased lipid peroxidation in tissues of the catfish *Clarias batrachus* following vanadium treatment: *in vivo* and *in vitro* evaluation. *J. Inorg. Biochem.* 54:277–284, Available at http://dx.doi.org/10.1016/0162-0134(94)80033-2.

Boscolo, P., Carmignani, M., Volpe, A.R., Felaco, M., Del Rosso, G., Porcelli, G., and Giuliano, G. (1994) Renal toxicity and arterial hypertension in rats chronically exposed to vanadate. *Occup. Environ. Med.* 51:500–503.

Byrne, A.R. and Kosta, L. (1978) Vanadium in foods and in human body fluids and tissues. *Sci. Total Environ.* 10:17–30.

Cano-Gutiérrez, G., Acevedo-Nava, S., Santamaría, A., Altamirano-Lozano, M., Cano-Rodríguez, M.C., and Fortoul, T.I. (2012) Hepatic megalocytosis due to vanadium inhalation: participation of oxidative stress. *Toxicol. Ind. Health* 28:353–360. Available at http://dx.doi.org/10.1177/0748233711412424.

Cal EPA (2005) Proposition 65 Chemical Listed Effective February 11, 2005 as Known to the State of California to Cause Cancer: Vanadium Pentoxide, Sacramento, CA. Available at http://oehha.ca.gov/prop65/prop65_list/021105list.html (accessed October 21, 2014).

Cheng, Q., Lü, Z., and Byrne, H.J. (2009) Synthesis of a maleic anhydride grafted polypropylene-butadiene copolymer and its application in polypropylene/styrene-butadiene-styrene triblock copolymer/organophilic montmorillonite composites as a compatibilizer. *J. Appl. Polymer Sci.* 114:1820–1827. Available at http://dx.doi.org/10.1002/app.30678.

Cintas, P. (2004) The road to chemical names and eponyms: discovery, priority, and credit. *Angew. Chem. Int. Ed. Engl.* 43:5888–5894. Available at http://dx.doi.org/10.1002/anie.200330074.

Cohen, M.D., McManus, T.P., Yang, Z., Qu, Q., Schlesinger, R.B., and Zelikoff, J.T. (1996) Vanadium affects macrophage interferon-gamma-binding and -inducible responses. *Toxicol. Appl. Pharmacol.* 138:110–120. Available at http://dx.doi.org/10.1006/taap.1996.0104.

Colín-Barenque, L., Martínez-Hernández, M.G., Baiza-Gutman, L.A., Avila-Costa, M.R., Ordóñez-Librado, J.L., Bizarro-Nevares, P., Rodriguez-Lara, V., Piñón-Zarate, G., Rojas-Lemus, M., Mussali-Galante, P., and Fortoul, T.I. (2008) Matrix metalloproteinases 2 and 9 in central nervous system and their modification after vanadium inhalation. *J. Appl. Toxicol.* 28:718–723. Available at http://dx.doi.org/10.1002/jat.1326.

Dai, C., Mi, W., Yang, F., Chen, B., Ding, H., and Shan, Y. (2006) A novel inorganic-organic hybrid material: hydrothermal synthesis and properties of [V6O12(CH3O)(4)(phen)(4) center dot 4H(2)O] (phen=1,10-phenanthroline). *J. Coord. Chem.* 59:317–324. Available at http://dx.doi.org/10.1080/00958970500270836.

Dai, S. and McNeill, J.H. (1994) One-year treatment of non-diabetic and streptozotocin-diabetic rats with vanadyl sulphate did not alter blood pressure or haematological indices. *Pharmacol. Toxicol.* 74(2):110–115.

Dai, S., Vera, E., and McNeill, J.H. (1995) Lack of haematological effect of oral vanadium treatment in rats. *Pharmacol. Toxicol.* 76(4):263–268.

DeSimone, L.A. (2009) *Quality of Water from Domestic Wells in Principal Aquifers of the United States, 1991–2004. (SIR 2008–5227)*, Reston, VA: U.S. Department of the Interior, U.S. Geological Survey. Available at http://pubs.usgs.gov/sir/2008/5227/.

Dimond, E.G., Caravaca, J., and Benchimol, A. (1963) Vanadium, excretion, toxicity, lipid effect in man. *Am. J. Clin. Nutr.* 12:49–53.

EPA. (U.S. Environmental Protection Agency). (1996) *Vanadium pentoxide* (CASRN 1314-62-1). Available at http://www.epa.gov/iris/subst/0125.htm (accessed October 21, 2014).

EPA. (U.S. Environmental Protection Agency). (2011) Toxicological Review of Vanadium Pentoxide (External Review Draft), EPA/635/R-11/004A.

Ehrlich, V.A., Nersesyan, A.K., Hoelzl, C., Ferk, F., Bichler, J., Valic, E., Schaffer, A., Schulte-Hermann, R., Fenech, M., Wagner, K.H., and Knasmüller, S. (2008) Inhalative exposure to vanadium pentoxide causes DNA damage in workers: Results of a multiple end point study. *Environ. Health Perspect.* 116:1689–1693. Available at http://dx.doi.org/10.1289/ehp.11438.

Elfant, M. and Keen, C.L. (1987) Sodium vanadate toxicity in adult and developing rats. *Biol. Trace Elem. Res.* 14:193–208.

Fattorini, D. and Regoli, F. (2012) Hyper-accumulation of vanadium in polychaetes. In: Michibata, H., editor. *Vanadium: Biochemical and Molecular Biological Approaches*, Netherlands: Springer, pp. 73–92. Available at http://dx.doi.org/10.1007/978-94-007-0913-3_4.

Fawcett, J.P., Farquhar, S.J., Thou, T., and Shand, B.I. (1997) Oral vanadyl sulphate does not affect blood cells, viscosity or biochemistry in humans. *Pharmacol. Toxicol.* 80:202–206.

Fichet, D. and Miramand, P. (1998) Vanadium toxicity to three marine invertebrates larvae: *Crassostrea gigas, Paracentrotus lividus* and *Artemia salina. Chemosphere* 37:1363–1368. Available at http://dx.doi.org/10.1016/S0045-6535(98)00118-0.

Finkelman, R.B. (1999) Trace elements in coal: environmental and health significance [Review]. *Biol. Trace Elem. Res.*

67:197–204. Available at http://dx.doi.org/10.1007/BF02784420.

Fries, L. (1982) Vanadium an essential element for some marine macroalgae. *Planta* 154:393–396. Available at http://dx.doi.org/10.1007/BF01267804.

Gil, J., Alvarez, C.E., Martinez, M.C., and Pérez, N. (1995) Effect of vanadium on lettuce growth, cationic nutrition, and yield. *J. Environ. Sci. Health, Part A* 30:73–87. Available at http://dx.doi.org/10.1080/10934529509376186.

Gomez, M., Domingo, J.L., Llobet, J.M., et al., (1988) Effectiveness of chelation therapy with time after acute vanadium intoxication. *J. Appl. Toxicol.* 8:439–444.

González-Villalva, A., Piñón-Zárate, G., De La Peña Díaz, A., Flores-García, M., Bizarro-Nevares, P., Rendón-Huerta, E.P., Colín-Barenque, L., and Fortoul, T.I. (2011) The effect of vanadium on platelet function. *Environ. Toxicol. Pharmacol.* 32:447–456. Available at http://dx.doi.org/10.1016/j.etap.2011.08.010.

Gravenmier, J.J., Johnston, D.W., and Arnold, W.R. (2005) Acute toxicity of vanadium to the threespine stickleback, Gasterosteus aculeatus. *Environ. Toxicol.* 20:18–22. Available at http://dx.doi.org/10.1002/tox.20073.

Gross, D. and Magazine, E.o.F. (1996) Henry Ford and the Model T. In: Gross, D.M., editors. *Forbes Greatest Business Stories of All Time*, New York, NY: John Wiley and Sons, Inc., pp. 74–89. Available at http://www.wiley.com/legacy/products/subject/business/forbes/fm.pdf.

Hamilton, S.J., and Buhl, K.J. (1997) Hazard evaluation of inorganics, singly and in mixtures, to flannelmouth sucker Catostomus latipinnis in the San Juan River, New Mexico. *Ecotoxicol. Environ. Saf.* 38:296–308. Available at http://dx.doi.org/10.1006/eesa.1997.1600.

Hilton, J.W. and Bettger, W.J. (1988) Dietary vanadium toxicity in juvenile rainbow trout: a preliminary study. *Aquat. Toxicol.* 12:63–71. Available at http://dx.doi.org/10.1016/0166-445X(88)90020-3.

Holdway, D. and Sprague, J. (1979) Chronic toxicity of vanadium to flagfish. *Water Res.* 13:905–910. Available at http://dx.doi.org/10.1016/0043-1354(79)90226-4.

HSDB (Hazardous Substances Data Bank) (2008) *Vanadium Pentoxide*, Bethesda, MD: National Library of Medicine. Available at http://toxnet.nlm.nih.gov/cgi-bin/sis/search/a?dbs+hsdb:@term+@DOCNO+1024.

IARC (International Agency for Research on Cancer) (2006) *Cobalt in hard metals and cobalt sulfate, gallium arsenide, indium phosphide and vanadium pentoxide [IARC Monograph] (pp. 1–294). (CIS/07/00503).* Lyon, France. Available at http://monographs.iarc.fr/ENG/Monographs/vol86/index.php.

Ivancsits, S., Pilger, A., Diem, E., Schaffer, A., and Rüdiger, H.W. (2002) Vanadate induces DNA strand breaks in cultured human fibroblasts at doses relevant to occupational exposure. *Mutat. Res.* 519:25–35. Available at http://dx.doi.org/10.1016/S1383-5718(02)00138-9.

Matthey, J. (2010) Development of thermally durable copper SCR catalysts. *Global. Emissions Manag.* 2(12):9–11.

Kaplan, D.I., Adriano, D.C., Carlson, C.L., and Sajwan, K.S. (1990) Vanadium: toxicity and accumulation by beans. *Water

Air Soil Pollut.* 49:81–91. Available at http://dx.doi.org/10.1007/BF00279512.

Kiviluoto, M. (1980) Observations on the lungs of vanadium workers. *Br. J. Ind. Med.* 37:363–366.

Kiviluoto, M., Pakarinen, A., and Pyy, L. (1981a) Clinical laboratory results of vanadium-exposed workers. *Arch. Environ. Health* 36:109–113.

Kiviluoto, M., Rasanen, O., Rinne, A., and Rissanen, M. (1979) Effects of vanadium on the upper respiratory tract of workers in a vanadium factory: a macroscopic and microscopic study. *Scand. J. Work Environ. Health* 5:50–58.

Kiviluoto, M., Rasanen, O., Rinne, A., and Rissanen, M. (1981b) Intracellular immunoglobulins in plasma cells of nasal biopsies taken from vanadium-exposed workers: a retrospective case control study by the peroxidase-antiperoxidase (PAP) method. *Anat. Anz.* 149:446–450.

Kleinsasser, N., Dirschedl, P., Staudenmaier, R., Harréus, U., and Wallner, B. (2003) Genotoxic effects of vanadium pentoxide on human peripheral lymphocytes and mucosal cells of the upper aerodigestive tract. *Int. J. Environ. Health Res.* 13:373–379. Available at http://dx.doi.org/10.1080/0960312031000122460.

Knecht, E.A., Moorman, W.J., Clark, J.C., Hull, R.D., Biagini, R.E., Lynch, D.W., Boyle, T.J., and Simon, S.D. (1992) Pulmonary reactivity to vanadium pentoxide following subchronic inhalation exposure in a non-human primate animal model. *J. Appl. Toxicol.* 12:427–434.

Knecht, E.A., Moorman, W.J., Clark, J.C., Lynch, D.W., and Lewis, T.R. (1985) Pulmonary effects of acute vanadium pentoxide inhalation in monkeys. *Am. Rev. Respir. Dis.* 132:1181–1185.

Lees, R.E. (1980) Changes in lung function after exposure to vanadium compounds in fuel oil ash. *Br. J. Ind. Med.* 37:253–256.

Levy, B.S., Hoffman, L., and Gottsegen, S. (1984) Boilermakers bronchitis: respiratory tract irritation associated with vanadium pentoxide exposure during oil-to-coal conversion of a power plant. *J. Occup. Med.* 26:567–570.

Lewis, C.E. (1959) The biological effects of vanadium: II the signs and symptoms of occupational vanadium exposure. *AMA Arch. Ind. Health* 19:497–503.

Lokeshappa, B., Shivpuri, K., Tripath, V., and Dikshit, A.K. (2012) Assessment of toxic metals in agricultural product. *Food Pub. Health* 2(1):24–29.

Mamane, Y. and Pirrone, N. (1998) Vanadium in the atmosphere. In: Nriagu J.O., editor. *Advances in Environmental Science and Technology. Vanadium in the Environment, Part 1: Chemistry and Biochemistry*, vol. 30, New York, NY: John Wiley and Sons, pp. 37–71.

Meisch, H.U., Benzschawel, H., and Bielig, H.J. (1977) The role of vanadium in green plants. II. Vanadium in green algae—two sites of action. *Arch. Microbiol.* 114:67–70.

Mejia, J.A., Rodriguez, R., Armienta, A., Mata, E., and Fiorucci, A. (2007) Aquifer vulnerability zoning, an indicator of atmospheric pollutants input? Vanadium in the Salamanca Aquifer, Mexico. *Water Air Soil Pollut.* 185:95–100. Available at http://dx.doi.org/10.1007/s11270-007-9433-x.

Morgan, A.M. and El-Tawil, O.S. (2003) Effects of ammonium metavanadate on fertility and reproductive performance of adult male and female rats. *Pharmacol. Res.* 47:75–85.

Motolese, A., Truzzi, M., Giannini, A., and Seidenari, S. (1993) Contact dermatitis and contact sensitization among enamellers and decorators in the ceramics industry. *Contact Dermatitis* 28:59–62.

Mountain, J.T., Delker, L.L., and Stokinger, H.E. (1953) Studies in vanadium toxicology; reduction in the cystine content of rat hair. *AMA Arch. Ind. Hyg. Occup. Med.* 8:406–411.

Mravcova, A., Jirova, D., Janci, H., and Lener, J. (1993) Effects of orally administered vanadium on the immune system and bone metabolism in experimental animals. *Sci. Total Environ.* Suppl. Pt 1:663–669.

Musk, A.W. and Tees, J.G. (1982) Asthma caused by occupational exposure to vanadium compounds. *Med. J. Aust.* 1:183–184.

Nalewajko, C., Lee, K., and Jack, T.R. (1995) Effects of vanadium on freshwater phytoplankton photosynthesis. *Water Air Soil Pollut.* 81:93–105. Available at http://dx.doi.org/10.1007/BF00477258.

Nalewajko, C., Lee, K., and Olaveson, M. (2008) Responses of freshwater algae to inhibitory vanadium concentrations: the role of phosphorous. *J. Phycol.* 31:332–343.

NIOSH (2012) *NIOSH Pocket Guide to Chemical Hazard: Vanadium Dust*, Atlanta, GA. Available at http://www.cdc.gov/niosh/npg/npgd0653.html (accessed October 21, 2014).

NRC (National Research Council) (1974) Vanadium in the environment. In: *Vanadium: Committee on Medical and Biological Effects of Environmental Pollutants*, Washington, DC: National Academy of Sciences.

NTP (National Toxicology Program) (2002). Toxicology and Carcinogenesis Studies of Vanadium Pentoxide (CAS No. 1314-62-1) in F344/N Rats and B6C3F1 Mice (Inhalation Studies). Technical Report Series No. 507. NIH Publication No. 02-4441. U.S. Department of Health and Human Services, Public Health Service, National Institutes of Health, Research Triangle Park, NC.

OSHA (Occupational Safety & Health Administration). (2006) *OSHA Occupational Chemical Database: Vanadium, respirable dust & fume*, as V2O5, Washington, DC. Available at https://www.osha.gov/chemicaldata/chemResult.html?recNo=217 (accessed October 21, 2014).

OSHA (Occupational Safety and Health Administration) (2007) *Occupational Safety and Health Guideline for Vanadium Pentoxide Dust*. Washington, DC: Occupational Safety and Health Administration. Available at http://www.osha.gov/SLTC/healthguidelines/vanadiumpentoxidedust/recognition.html.

Outridge, P.M. and Noller, B.N. (1991) Accumulation of toxic trace elements by freshwater vascular plants. *Rev. Environ. Contam. Toxicol.* 121:1–63. Available at http://dx.doi.org/10.1007/978-1-4612-3196-7_1.

Paternain, J.L., Domingo, J.L., Llobet, J.M., and Corbella, J. (1987) Embryotoxic effects of sodium metavanadate administered to rats during organogenesis. *Rev. Esp. Fisiol.* 43:223–227.

Perez-Benito, J.F. (2006) Effects of chromium(VI) and vanadium(V) on the lifespan of fish. *J. Trace Elem. Med. Biol.* 20:161–170. Available at http://dx.doi.org/10.1016/j.jtemb.2006.04.001.

Pierce, L.M., Alessandrini, F., Godleski, J.J., and Paulauskis, J.D. (1996) Vanadium-induced chemokine mRNA expression and pulmonary inflammation. *Toxicol. Appl. Pharmacol.* 138:1–11.

Rahman, F. and Skyllas-Kazacos, M. (1998) Solubility of vanadyl sulfate in concentrated sulfuric acid solutions. *J. Power Sources* 72:105–110. Available at http://dx.doi.org/10.1016/S0378-7753(97)02692-X.

Ramirez, P., Eastmond, D.A., Laclette, J.P., and Ostrosky-Wegman, P. (1997) Disruption of microtubule assembly and spindle formation as a mechanism for the;induction of aneuploid cells by sodium arsenite and vanadium pentoxide. *Mutat. Res.* 386:291–298. Available at http://dx.doi.org/10.1016/S1383-5742(97)00018-5.

Ringelband, U. and Karbe, L. (1996) Effects of vanadium on population growth and Na-K-ATPase activity of the brackish water hydroid Cordylophora caspia. *Bull. Environ. Contam. Toxicol.* 57:118–124.

Rojas, E., Valverde, M., Herrera, L.A., Altamirano-Lozano, M., and Ostrosky-Wegman, P. (1996) Genotoxicity of vanadium pentoxide evaluate by the single cell gel electrophoresis assay in human lymphocytes. *Mutat. Res.* 359:77–84.

Roldán, R.E. and Altamirano, L.M. (1990) Chromosomal aberrations, sister-chromatid exchanges, cell-cycle kinetics and satellite associations in human lymphocyte cultures exposed to vanadium pentoxide. *Mutat. Res.* 245:61–65. Available at http://dx.doi.org/10.1016/0165-7992(90)90001-Z.

Rondini, E.A., Walters, D.M., and Bauer, A.K. (2010) Vanadium pentoxide induces pulmonary inflammation and tumor promotion in a strain-dependent manner. *Part. Fibre Toxicol.* 7:9. Available at http://dx.doi.org/10.1186/1743-8977-7-9.

Roscoe, H.E. (1869) *Researches on Vanadium. Part II.* 159:679–692. Available at http://dx.doi.org/10.1098/rstl.1869.0028.

Sanchez, D.J., Colomina, M.T., and Domingo, J.L. (1998) Effects of vanadium on activity and learning in rats. *Physiol. Behav.* 63(3):345–350.

Sanchez D.J., Colomina M.T., Domingo J.L., et al. (1999) Prevention by sodium 4,5-dihydroxybenzene-1,3-disulfonate (tiron) of vanadium-induced behavioral toxicity in rats. *Biol. Trace Elem. Res.* 69(3):249–259.

Sanchez, D., Ortega, A., Domingo, J.L., and Corbella, J. (1991) Developmental toxicity evaluation of orthovanadate in the mouse. *Biol. Trace Elem. Res.* 30:219–226.

Schuler, D., Chevalier, H.J., Merker, M., Morgenthal, K., Ravanat, J.L., Sagelsdorff, P., Walter, M., Weber, K., and Mcgregor, D. (2011) First steps towards an understanding of a mode of carcinogenic action for vanadium pentoxide. *J. Toxicol. Pathol.* 24:149–162. Available at http://dx.doi.org/10.1293/tox.24.149.

Scibior, A., Zaporowska, H., and Ostrowski, J. (2006) Selected haematological and biochemical parameters of blood in rats after subchronic administration of vanadium and/or magnesium in drinking water. *Arch. Environ. Contam. Toxicol.* 51:287–295. Available at http://dx.doi.org/10.1007/s00244-005-0126-4.

Sepe, A., Ciaralli, L., Ciprotti, M., Giordano, R., Funari, E., and Costantini, S. (2003) Determination of cadmium, chromium, lead and vanadium in six fish species from the Adriatic Sea.

Food Addit. Contam. 20:543–552. Available at http://dx.doi.org/10.1080/0265203031000069797.

Shafer, M.M., Toner, B.M., Overdier, J.T., Schauer, J.J., Fakra, S.C., Hu, S., Herner, J.D., and Ayala, A. (2012) Chemical speciation of vanadium in particulate matter emitted from diesel vehicles and urban atmospheric aerosols. *Environ. Sci. Technol.* 46:189–195. Available at http://dx.doi.org/10.1021/es200463c.

Sjoberg, S.G. (1951) Health hazards in the production and handling of vanadium pentoxide. *AMA Arch. Ind. Hyg. Occup. Med.* 3:631–646.

Sjoberg, S.G. (1955) Vanadium bronchitis from cleaning oil-fired boilers. *AMA Arch. Ind. Health* 11:505–512.

Sjoberg, S.G. (1956) Vanadium dust, chronic bronchitis and possible risk of emphysema: a follow-up investigation of workers at a vanadium factory. *Acta Med. Scand.* 154:381–386.

Smith, P.G., Boutin, C., and Knopper, L. (2013) Vanadium pentoxide phytotoxicity: effects of species selection and nutrient concentration. *Arch. Environ. Contam. Toxicol.* 64:87–96. Available at http://dx.doi.org/10.1007/s00244-012-9806-z.

Stokinger, H.E., Wagner, W.D., Mountain, J.T., Stacksill, F.R., Dobrogorski, O.J., and Keenan, R.G. (1953) Unpublished results. Cincinnati, OH: Division of Occupational Health.

Stokinger, H.E. (1962) Effect of air pollution on animals. In: Stern A., editor. *Air Pollution.* vol. 1, New York, NY: Academic Press, pp. 282–334.

Turpin, E.A., Antao-Menezes, A., Cesta, M.F., Mangum, J.B., Wallace, D.G., Bermudez, E., and Bonner, J.C. (2010) Respiratory syncytial virus infection reduces lung inflammation and fibrosis in mice exposed to vanadium pentoxide. *Respir. Res.* 11:20. Available at http://dx.doi.org/10.1186/1465-9921-11-20.

Turunen, M., Peraniemi, S., Ahlgren, M., and Westerholm, H. (1995) Determination of trace elements in heavy oil samples by graphite furnace and cold vapour atomic absorption spectrometry after acid digestion. *Anal. Chim. Acta* 311:85–91. Available at http://dx.doi.org/10.1016/0003-2670(95)00166-W.

Ünsal, M. (1982). The accumulation and transfer of vanadium within the food chain. *Mar. Pollut. Bull.* 13:139–141. Available at http://dx.doi.org/10.1016/0025-326X(82)90373-3.

Ustarroz-Cano, M., García-Peláez, I., Piñón-Zárate, G., Herrera-Enríquez, M., Soldevila, G., and Fortoul, T.I. (2012) CD11c decrease in mouse thymic dendritic cells after vanadium inhalation. *J. Immunotoxicol.* 9:374–380. Available at http://dx.doi.org/10.3109/1547691X.2012.673181.

Vachirapatama, N. and Jirakiattikul, Y. (2008) Effect of vanadium on growth of Chinese green mustard (Brassica campestris ssp. chinensis var. parachinensis) under substrate culture. *Songklanakarin J. Sci. Technol.* 30:427–431.

Vintinner, F.J., Vallenas, R., Carlin, C.E., Weiss, R., Macher, C., and Ochoa, R. (1955) Study of the health of workers employed in mining and processing of vanadium ore. *AMA Arch. Ind. Health* 12:635–642.

Wang, J.F. and Liu, Z. (1999) Effect of vanadium on the growth of soybean seedlings. *Plant Soil* 216:47–51. Available at http://dx.doi.org/10.1023/A:1004723509113.

Weckhuysen, B.M. and Keller, D.E. (2003) Chemistry, spectroscopy and the role of supported vanadium oxides in heterogeneous catalysis. *Catal. Today* 78:25–46. Available at http://dx.doi.org/10.1016/S0920-5861(02)00323-1.

WHO (World Health Organization) (1988) *Vanadium [WHO EHC]. (Environmental Health Criteria 81).* Geneva, Switzerland: Available at http://www.inchem.org/documents/ehc/ehc/ehc81.htm.

WHO (World Health Organization) (2001) Concise international chemical assessment document 29: Vanadium pentoxide and other inorganic vanadium compounds. (RISKLINE/2001120005). Geneva, Switzerland: World Health Organization, International Programme on Chemical Safety. Available at http://www.inchem.org/documents/cicads/cicads/cicad29.htm.

WHO (2000) Air Quality Guidelines for Europe. WHO Regional Publications, European Series, No. 91, Copenhagen, Denmark. Available at http://www.euro.who.int/__data/assets/pdf_file/0005/74732/E71922.pdf?ua=1 (accessed October 21, 2014).

Williams, N. (1952) Vanadium poisoning from cleaning oil-fired boilers. *Br. J. Ind. Med.* 9:50–55.

Woods, R., Davi, R., and Arnold, W. (2004) Toxicity of vanadium to the estuarine mysid, Americamysis bahia (molenock) (formerly Mysidopsis bahia). *Bull. Environ. Contam. Toxicol.* 73:635–643. Available at http://dx.doi.org/10.1007/s00128-004-0493-y.

Wright, M.T. and Belitz, K. (2010) Factors controlling the regional distribution of vanadium in groundwater. *Ground Water* 48:515–525. Available at http://dx.doi.org/10.1111/j.1745-6584.2009.00666.x.

Wyers, H. (1946) Some toxic effects of vanadium pentoxide. *Br. J. Ind. Med.* 3:177–182.

Yong, G. and Han, D. (2010) Review of the applications of vanadium in steels. In: *Proceedings of International Seminar on Production of High Strength Seismic Grade Rebar Containing Vanadium*, pp. 1–11. Available at http://vanitec.org/wp-content/uploads/2011/09/Review-of-Applications-of-Vanadium-in-Steels1.pdf.

Yu, D., Walters, D.M., Zhu, L., Lee, P.K., and Chen, Y. (2011) Vanadium pentoxide (V(2)O(5)) induced mucin production by airway epithelium. *Am. J. Physiol. Lung Cell Mol. Physiol.* 301:L31–L39. Available at http://dx.doi.org/10.1152/ajplung.00301.2010.

Zaporowska, H. and Wasilewski, W. (1989) Some selected peripheral blood and haemopoietic system indices in Wistar rats with chronic vanadium intoxication. *Comp. Biochem. Physiol. C* 93:175–180.

Zaporowska, H., Wasilewski, W., and Słotwińska, M. (1993) Effect of chronic vanadium administration in drinking water to rats. *Biometals* 6:3–10.

Zenz, C., Bartlett, J.P., and Thiede, W.H. (1962) Acute vanadium pentoxide intoxication. *Arch. Environ. Health* 5:542–546.

Zenz, C. and Berg, B.A. (1967) Human responses to controlled vanadium pentoxide exposure. *Arch. Environ. Health* 14:709–712.

Zhong, B.Z., Gu, Z.W., Wallace, W.E., Whong, W.Z., and Ong, T. (1994) Genotoxicity of vanadium pentoxide in Chinese hamster V79 cells [Abstract]. *Mutat. Res.* 321:35–42.

38

ZINC

Giffe T. Johnson

Target organ(s): Lungs, kidneys, immunological system

Occupational exposure limits:

Zinc Compound		OSHA PEL (mg/m^3)	NIOSH REL (mg/m^3)
Zinc oxide fume		5	5; STEL 10
Zinc oxide dust			
	Respirable	5	5; STEL 15
	Total	15	5; STEL 15
Zinc chloride fume		1	1; STEL 2
Zinc stearate dust			
	Respirable	5	5
	Total	15	10

Reference values: RfD 0.3 mg/kg/day; RfC N/A; intermittent MRL 0.3 mg Zn/kg/day; chronic MRL 0.3 mg Zn/kg/day

Risk/safety phrases: Metal fume fever

BACKGROUND AND USES

Zinc is an essential metal, ubiquitous in modern life, and necessary for human metabolism (NRC, 1989; Sandstead, 1995a). It is widely available as a nutritional supplement, and suggestions that zinc compounds may be effective in treating the common cold have increased the consumption of lozenges and vitamins containing zinc. Zinc compounds are used in many household products besides vitamins (Table 38.1) such as cosmetics, ointments, wood preservatives, solder fluxes, and paint (Goodwin, 1998). Commercially, zinc is sold in a variety of forms, including ingots, lumps, sheets, wire, shot, strips, sticks, granules, and powder (O'Neil et al., 2001).

Zinc is an essential constituent or cofactor for more than 200 metalloenzymes. Humans need zinc throughout their life span, although the amount necessary and the clinical effects of zinc deficiency vary with age, the rate of growth and activity, concomitant medical conditions, and the intake and stores of other metals.

Zinc compounds are commonly found in contemporary daily life. Zinc chloride is used as a soldering flux, a preservative for wood and taxidermy, and a disinfectant and deodorant. Zinc chloride is also used in batteries, dental cement, and petroleum refining. Zinc oxide is used as a topical sun protectant, in cosmetic powders and medicinal ointments, as a pigment and ultraviolet light absorbent for paints and for photocopying processes, as a dental cement and in temporary fillings, and as a fungicide and mildew retardant. Zinc stearate is also used in cosmetic and medicinal preparations. Zinc fluoride and zinc sulfate are used for wood preservation. Zinc chromate, zinc carbonate, and zinc sulfide are used in pigments. Zinc phosphide is used as a rodenticide (Casteel and Bailey, 1986; Cumpston, 1984; Goodwin, 1998; Mack, 1989).

Metallic zinc is used primarily for galvanizing iron and steel to protect them from oxidation and corrosion. Zinc has been used as an alloy with copper and tin. Alloys enhance the dye casting and galvanizing properties of zinc, and various alloys with cadmium, iron, titanium, aluminum, magnesium, and lead, as well as with copper and tin, have industrial and manufacturing uses (Goodwin, 1998).

PHYSICAL AND CHEMICAL PROPERTIES

Zinc (symbol Zn; CASRN 7440-66-6) is a blue-white metal in the same vertical series of the periodic table as cadmium and

Hamilton & Hardy's Industrial Toxicology, Sixth Edition. Edited by Raymond D. Harbison, Marie M. Bourgeois, and Giffe T. Johnson.
© 2015 John Wiley & Sons, Inc. Published 2015 by John Wiley & Sons, Inc.

TABLE 38.1 Commonly Used Zinc Compounds

Zinc Compound	Formula	CAS Number	Uses
Zinc acetate	$ZnC_4H_6O_4$	557-34-6	Wood preservative, glazes, mordant
Zinc carbonate	$ZnCO_3$	3486-35-9	Pigment, feed additive, ceramics and rubber production
Zinc chloride	$ZnCl_2$	7646-85-7	Deodorant; disinfectant; wood preservative; soldering flux; rubber, paper, dye production; petroleum refining; fireproofing; textile treatment; dental cement
Zinc chromate	$ZnCrH_2O_4$	13530-65-9	Pigment (zinc yellow) for paint, linoleum, rubber
Zinc cyanide	ZnC_2N_2	557-21-1	Electroplating, extraction of NH_3 from natural gas
Zinc fluoride	ZnF_2	7783-49-5	Wood preservative, electroplating, fluoridation, phosphors for fluorescent lights
Zinc gluconate	$ZnC_{12}H_{22}O_4$	4468-02-4	Nutritional supplement, cold medications
Zinc oxide	ZnO	1314-13-2	Pigment (zinc white), fungicide, flame retardant, sulfur scavenger, rubber vulcanizing
Zinc phosphide	Zn_3P_2	1314-84-7	Rodenticide
Zinc stearate	$Zn(C_{16}H_{35}O_2)_2$	557-05-1	Cosmetic powders, medicated powders, ointments and tablets, rubber production, waterproofing
Zinc sulfate	$ZnSO_4$	7733-02-0	Bleaching agent (zinc white vitriol), wood preservative, mordant

mercury. It is always divalent. Zinc resists corrosion; at room temperature, in the presence of moisture and carbon dioxide, it forms a superficial layer of hydrated, basic carbonate that protects the metal from further corrosion. The melting point of zinc is 419.5 °C (788 °F); at temperatures above 500 °C (932 °F), it volatilizes to form fumes of zinc oxide.

MAMMALIAN TOXICOLOGY

Acute Effects

Large oral doses of zinc salts (e.g., 200 mg or more) cause symptoms similar to those of staphylococcal food poisoning, such as intense nausea, vomiting, and diarrhea (Murphy, 1970; Potter, 1981). Sources of zinc salts include over-the-counter medications, nutritional supplements, and acidic beverages stored in galvanized containers. Zinc chloride can cause mucosal burns of the esophagus and stomach (Chobanian, 1981; McKinney et al., 1994). There are several case reports of toddlers ingesting substances containing zinc chloride. One child required a partial gastrectomy after surviving ingestion of a zinc chloride/ammonium chloride soldering flux, sustaining a peak plasma level of zinc of 1199 pg/dl (McKinney et al., 1994). The child also suffered hepatic injury and metabolic acidosis. The residual of the esophagitis and gastritis was severe scarring, hence the need for gastrectomy.

Zinc chloride is caustic, and skin contact causes dermatitis and ulcerations. Both zinc chloride and zinc sulfate cause corneal ulcerations and irritative conjunctivitis (Grant, 1993). A splash of concentrated zinc chloride solution (50%) can cause severe eye injury, with residual cataract, glaucoma, and iritis.

Most zinc salts irritate mucous membranes. Inhalation causes upper respiratory and bronchial irritation, with coryza and cough. Inhalation of zinc chloride fumes from smoke bombs used by the military has caused cough and dyspnea, which progressed to fatal respiratory distress syndrome (Evans, 1945; Hjortsø et al., 1988a; Marrs et al., 1988). The acute pulmonary toxicity of zinc chloride fumes also affects animals and is the basis for one of the techniques for eliminating moles and burrowing rodents. Some acute-duration studies of workers exposed to zinc oxide fumes identified metal fume fever as a potentially associated outcome, at exposure levels in excess of 77–600 mg zinc/m^3 (Blanc et al., 1991; Hammond, 1944; Safty et al., 2008; Sturgis et al., 1927). However, there has been concern that other pulmonary toxicants may be present in the workplace and the observed effects may not be wholly or partially related to zinc exposure.

Oliguria, renal failure, and microhematuria have been reported after ingestion or intravenous over administration of zinc chloride and zinc sulfate (Chobanian, 1981; Hjortsø et al., 1988b; McKinney et al., 1994; Potter, 1981). Hepatic injury with jaundice is a result of intravenous zinc administration (Brocks et al., 1977). Chronic ingestion of excess supplemental zinc can produce anemia and leukopenia consequent to induced copper deficiency (Hoffman et al., 1988). Patients with microcytic or hemolytic anemia, or both, and normal iron stores should be suspected of chronically abusing zinc supplements and should be evaluated for plasma zinc and copper levels. Zinc is not teratogenic or carcinogenic (Leonard et al., 1986).

Chronic Effects

Most evidence regarding chronic toxicity involving zinc relates to deficiency or prolonged oral supplementation. Zinc deficiency was clinically characterized in adolescent boys with delayed sexual maturation and growth failure (Prasad, 1996). Multiple nutritional deficiencies also were present, which include pellagra, iron and folate deficiency, and protein–calorie malnutrition. Restoration of normal nutrition, growth, and sexual development required supplementation with zinc, without which other nutritional repletion was incomplete. Recent studies from Southeast Asia have found that undernourished, growth-retarded children responded to zinc supplementation with increased production of insulin-like growth factor I, increased rates of height and weight gain, and fewer episodes of respiratory infection and diarrhea, compared with a placebo-treated control group (Ninh et al., 1996). Zinc repletion also improved wound healing and skin rashes.

Zinc is essential for the biological activity of the lymphopoietic thymic hormone, thymulin; zinc deficiency is associated with reduced biological activity of thymulin and a decrease in the number of CD4+ cells, which can be corrected by zinc supplementation (Mocchegiani et al., 1995; Prasad, 1996). Administration of zinc to patients with acquired immunodeficiency syndrome who also had a zinc deficiency has been shown to increase the number of CD4+ cells and the plasma levels of complexed zinc, as well as active thymulin, and has been shown to reduce the frequency of opportunistic infectious episodes with *Candida* sp. and *Pneumocystix carinii* (Mocchegiani et al., 1995; Prasad, 1996). Studies in the Middle East, Asia, and South America have reported a reduction in infectious diarrhea and respiratory conditions among malnourished children given zinc supplements (Prasad, 1996; Sempértegui et al., 1996).

An association between reduced zinc intake and age-related maculopathy and night blindness has been used to popularize zinc supplements for the elderly (Mares-Perlman et al., 1996). The importance of adequate dietary zinc for older individuals is reinforced by the finding that estrogen replacement therapy in postmenopausal women diminishes zinc excretion and maintains plasma and presumably bone levels of zinc (Herzberg et al., 1996). Some claim that zinc deficiency accelerates the aging process (Fabris and Mocchegiani, 1995).

Large amounts of dietary zinc interfere with the absorption of other essential trace elements such as magnesium, copper, and other metals like mercury and cadmium (NRC, 1989; Sandstead, 1995a; Spencer et al., 1994). Since all of these metals compete for receptor, binding, and transport proteins, an imbalance in the quantities presented to the gastrointestinal mucosa can lead to an unexpected deficiency of copper and magnesium. Copper deficiency is associated with increased low density lipoprotein (LDL) cholesterol and decreased high density lipoprotein cholesterol (Sandstead, 1995b). It is noteworthy that in a study of people using zinc supplements, those with the highest serum levels of zinc had higher levels of total cholesterol, LDL cholesterol, and triglycerides (Hiller et al., 1995). The dangers of regular, excessive zinc supplementation are illustrated by case reports of transient leukopenia and gastrointestinal distress (Forsyth and Davies, 1995; Murphy, 1970).

Occupational exposures to zinc leading to chronic health effects are less prevalent than acute and sub-chronic reporting. Pneumoconiosis in zinc miners has been reported without episodes of metal fume fever or exposure to coal mine dust, though the causative agent in these cases remains uncertain (Joyce et al., 1996). Overall, the consequences of long-term occupational exposure to zinc, in the absence of acute and sub-chronic effects, are not well characterized and have yet to be definitively linked to adverse health outcomes.

Mechanism of Action(s)

Zinc is a catalytic component and a structural constituent of many enzymes, hormones and hormone receptors, neuropeptides, proteins, and polynucleotides. It is not surprising that zinc has multiple biological functions and that there is a mechanistic basis for a variety of clinical manifestations, which can be associated with deficiency or excess. The range of essential activities attributed to zinc is confusing and seemingly hyperbolic. A review of the literature produces a picture of a "supernutrient" or panacea that will act as an antioxidant, biomembrane stabilizer, growth promoter, immune stimulator, sensory restorer, pregnancy protector, osteoporosis marker, and vision protector, though these effects tend to be overstated (Phillips et al., 1996; Simon et al., 1988; Spencer et al., 1994).

About 20–50% of the ingested zinc is absorbed from the gastrointestinal tract. The dietary constituents affect absorption. Zinc absorption is diminished if zinc is consumed with calcium and phosphorus, some vegetable proteins, and phytates, which bind zinc in the gut lumen. Absorption is enhanced when zinc is consumed with animal protein. Zinc forms complexes with metallothionein intracellularly, and absorption is regulated by the availability of binding and transport proteins. In the blood, it is bound to albumin or to α2-macroglobulin.

Tissue concentrations of zinc vary considerably. The liver sequesters about 50% of a tracer dose initially, but the hepatic concentration declines in 5 days to about 25% of the tracer dose. The highest concentration of zinc is found in the prostate gland presumably because of the content of acid phosphatase, one of the many zinc-containing enzymes. Zinc is also concentrated in the pancreas and the retina, but because of the volume of muscle mass, the greatest quantity of zinc is found in the skeletal muscle (Cumpston, 1984).

Zinc is excreted slowly; the biological half-life is more than 300 days. Most of it is excreted via the biliary and gastrointestinal tract. Roughly 15% is excreted in the urine and sweat. In hot climates, as much as 25% can be secreted in sweat alone. Breast milk, the product of a modified sweat gland, contains a substantial concentration of zinc. Urinary excretion by normal adults ranges from 300 to 600 μg/day. Urinary excretion of zinc correlates with zinc consumption in the diet or zinc obtained from other exposures. Excretion can be enhanced by chelation with dimercaprol (BAL), calcium disodium ethylenediamine tetraacetic acid (EDTA), diethylene-triamine pentaacetic acid (DTPA), and N-acetyl-cysteine. Renal tubular reabsorption of zinc is inhibited by diuretics, especially the thiazides (Cumpston, 1984; Domingo et al., 1988; NRC, 1989).

The usual plasma zinc concentration ranges from 85 to 110 µg/dl. Because zinc is sequestered in red blood cells, the zinc concentration in whole blood is as much as five times that in plasma. Zinc deficiency can reduce plasma zinc to 40 µg/dl and urinary excretion may decline to 100 µg/day or less. Laboratory evaluation of zinc metabolism in patients suspected of zinc deficiency or intoxication should include measurement of serum levels and 24-h urinary excretion.

Zinc homeostasis can be altered by clinical disorders such as inflammatory bowel disease, prolonged diarrhea, and malabsorption. Genetically determined metalloprotein disorders can be associated with zinc deficiency (acrodermatitis enteropathica) or zinc overload (juvenile biliary cirrhosis). Chronic disease, such as renal failure, alcoholic cirrhosis, neoplasia, intestinal parasites, and drug therapy, is associated with zinc depletion, as are diuretics, corticosteroids, and some antibiotics. Patients receiving total parenteral nutrition or dialysis, or both, require regular administration of zinc.

Additionally, *in vitro* studies have demonstrated that zinc enhances cytokine production by endotoxin and diminishes monocyte activation by superantigens and staphylococcal enterotoxins (Driessen et al., 1995a, 1995b; Falus and Benes, 1996). These effects are concentration and dose dependent, which in effect mean that extremely large doses of zinc may actually impair response to infectious disease pathogens.

Metal fume fever is believed to be an immunological response to zinc oxide fumes. Bronchoalveolar lavage fluid in zinc-exposed persons who developed metal fume fever demonstrated a correlation between the concentration of airborne zinc and the number of all types of T-cells (helper, inducer, suppressor, and killer) (Blanc et al., 1991). This may be exacerbated by the influx of inflammatory factors such as cytokines TNF, IL-6, and IL-8 from zinc oxide fume inhalation as demonstrated by Kuschner et al., 1997.

SOURCES OF EXPOSURE

Occupational

Most workers are exposed to zinc compounds rather than elemental zinc. Workers exposed to elemental zinc are typically employed in the fabricated metal products industry as millwrights or assemblers. Workers in the primary metal industries that work with fabricated metal products, such as stone, clay, and glass products, or in special trade contractors industries, are typically exposed to a variety of zinc compounds. Other potentially exposed workers include molding and casting machine operators, janitors and cleaners, and machinists within the fabricated metals industry (NIOSH, 1984).

The most common occupational exposure is to zinc oxide fumes, produced during foundry operations in which galvanized steel, brass, or other zinc alloys are melted or produced by torch welding and cutting operations. Metal fume fever has been observed after inhalation of many metallic oxides, but zinc oxide is the most common and a characteristic causal agent when exposures exceed the threshold for toxicity (Bourne et al., 1968; Hamdi, 1969; Marquart, 1989; NRC, 1989).

Environmental

Zinc is widely distributed in nature and the human diet. The average concentration of zinc in the soil and the earth's crust is about 40 mg/kg. The zinc concentration in the air usually is below 1 µg/m^3 but is higher in industrial areas. In water the concentration varies from 1 to 10 µg/l, with higher concentrations around mines, smelters, and industries that use zinc (NRC, 1989). The principal mineral is sphalerite (ZnS), often found with various amounts of cadmium, arsenic, manganese, lead, and copper. The major ore producers are Canada, parts of the Confederated States (formerly the USSR), Mexico, and Peru. The United States produces about 5% of the world's supply of the finished metal.

The principal source of zinc in the general population is food. Zinc is present in seafood, meat, dairy products, whole grains, nuts, and legumes. Garden vegetables take up zinc if it is applied to the soil or is present in high concentrations naturally. The average daily intake of zinc in humans is on the order of 5.2–16.2 mg zinc/day (EPA, 1987; Pennington et al., 1986). Most of this daily dose comes from food, but use of any number of materials containing zinc compounds can increase exposure.

INDUSTRIAL HYGIENE

Health hazards from zinc in the workplace are well known and regulated. Except for accidents or carelessness, health problems caused by zinc are likely to be sporadic and related to domestic use. Proper ventilation is the most effective industrial method to control freshly generated zinc oxide or other metal fumes. Fume hoods and area air filters may help to reduce area exposure levels. A properly fitted cartridge filter respirator or powered, self-contained respirator is effective protection for short-term exposure and required when exposure levels exceed the Occupational Health and Safety Administration (OSHA) permissible exposure levels (PELs) (Fishburn and Zenz, 1969; Safty et al., 2008). To prevent cases of metal fume fever, the National Institute for Occupational Safety and Health (NIOSH) and the OSHA have set the standard 8-h time-weighted average (TWA) for zinc oxide fumes and respirable dust at 5 mg/m^3. The NIOSH additionally recommends a 10 mg/m^3 ceiling limit for zinc oxide fumes (NIOSH, 2007).

Worker education is essential. In any environment where workers are producing zinc oxide fumes or other irritant zinc compounds, protective clothing that covers exposed skin should be used. Eye protection is recommended, particularly for those producing zinc oxide fumes by soldering, welding, or working in the galvanization process.

MEDICAL MANAGEMENT

Acute effects such as abdominal cramps, fever, hematemesis, and hematochezia from large zinc exposures are occasional events. Resolution of symptoms takes several hours or a few days. Removal from exposure and supportive care with fluid and electrolytes usually is all that is needed. This applies to those affected by metal fume fever. Sub-chronic zinc poisoning is treated with supportive management. A severely intoxicated patient may need ventilator support, fluid and electrolyte management, and even administration of corticosteroids, though the benefit of corticosteroids remains controversial. Chelation therapy may be helpful but does not accelerate clinical recovery. As zinc alters absorption of other nutrient metals, supplementation of iron, calcium, magnesium, and copper may be considered to resolve homeostasis (Anthony et al., 1978; Hassaballa et al., 2005; Safty et al., 2008).

REFERENCES

Anthony, J., Zamel, N., and Aberman, A. (1978) Abnormalities in pulmonary function after brief exposure to toxic metal fume. *Can. Med. Assoc. J.* 119:586–588.

Blanc, P., et al. (1991) An experimental human model of metal fume fever. *Ann. Intern. Med.* 114:930–936.

Bourne, H.G., Yee, H.T., and Serferian, S. (1968) The toxicity of rubber additives. Findings from a survey of 140 plants in Ohio. *Arch. Environ. Health* 16:700–706.

Brocks, A., Reid, H., and Glazer, G. (1977) Acute intravenous zinc poisoning. *Br. Med. J.* 1:1390–1391.

Casteel, S.W. and Bailey, E.M. (1986) A review of zinc phosphide poisoning. *Vet. Hum. Toxicol.* 28:151–154.

Chobanian, S.I. (1981) Accidental ingestion of liquid zinc chloride: local and systemic effects. *Ann. Emerg. Med.* 10:91–93.

Cumpston, A.G. (1984) Zinc, alloys and compounds. In: *Encyclopedia of Occupational Safety and Health*, 2nd ed., Geneva: International Labour Office Press.

Domingo, J.L., et al. (1988) Acute zinc intoxication: comparison of the antidotal efficacy of several chelating agents. *Vet. Hum. Toxicol.* 30:224–228.

Driessen, C., et al. (1995a) Divergent effects of zinc on different bacterial pathogenic agents. *J. Infect. Dis.* 171:486–489.

Driessen, C., et al. (1995b) Zinc regulates cytokine induction by superantigens and lipopolysaccharide. *Immunology* 84:272–277.

Environmental Protection Agency (EPA) (1987) *Ambient Water Quality Criteria for Zinc—1987*, Washington, DC: U.S. Environmental Protection Agency, Office of Water Regulations and Standards. EPA440587003.

Evans, E.H. (1945) Casualties following exposure to zinc chloride smoke. *Lancet* 2:368–370.

Fabris, N. and Mocchegiani, E. (1995) Zinc, human diseases and aging. *Aging* 7:77–93.

Falus, A. and Benes, J. Jr. (1996) A trace element preparation containing zinc increases the production of interleukin-6 in human monocytes and glial cells. *Biol. Trace Elem. Res.* 51:293–301.

Fishburn, C. and Zenz, C. (1969) Metal fume fever. *J. Occup. Med.* 11:142–144.

Forsyth, P.D. and Davies, J.M. (1995) Pure white cell aplasia and health food products. *Postgrad. Med. J.* 71:557–558.

Goodwin, F.E. (1998) Zinc compounds. In: Kroschwitz J. and Howe-Grant M., editors. *Kirk-Othmer Encyclopedia of Chemical Technology*, New York, NY: John Wiley & Sons, Inc., pp. 840–853.

Grant, W.M. (1993) *Toxicology of the Eye*, 4th ed., Springfield, IL: Charles C Thomas.

Hamdi, E. (1969) Chronic exposure to zinc of furnace operators in a brass foundry. *Br. J. Ind. Med.* 26:126–134.

Hammond, J.W. (1944) Metal fume fever in crushed stone industry. *J. Ind. Hyg. Toxicol.* 26:117–119.

Hassaballa, H.A., et al. (2005) Metal fume fever presenting as aseptic meningitis with pericarditis, pleuritis and pneumonitis. *Occup. Med. (Lond.)* 55:638–641.

Herzberg, M., et al. (1996) The effect of estrogen replacement therapy on zinc in serum and urine. *Obstet. Gynecol.* 87:1035–1040.

Hiller, R., et al. (1995) Serum zinc and serum lipid profiles in 778 adults. *Ann. Epidemiol.* 5:490–496.

Hjortsø, E., et al. (1988a) ARDS after accidental inhalation of zinc chloride smoke. *Intensive Care Med.* 14(1):17–24.

Hjortsø, E., et al. (1988b) [Zinc chloride poisoning. Observation and treatment]. *Ugeskr. Laeger* 149(36):2381–2384.

Hoffman, H.N., et al. (1988) Zinc-induced copper deficiency. *Gastroenterology* 94:508–512.

Joyce, B.W., et al. (1996) Progressive massive fibrosis in a zinc miner. *J. Ky. Med. Assoc.* 94:144–147.

Kuschner, W.G., et al. (1997) Early pulmonary cytokine responses to zinc oxide fume inhalation. *Environ. Res.* 75:7–11.

Leonard, A., Ferber, G.B., and Leonard, F. (1986) Mutagenicity, carcinogenicity, and teratogenicity of zinc. *Mutat. Res.* 168:343–353.

Mack, R.B. (1989) A hard day's knight. Zinc phosphide poisoning. *N C Med. J.* 50:17–18.

Mares-Perlman, J.A., et al. (1996) Association of zinc and antioxidant nutrients with age-related maculopathy. *Arch. Ophthalmol.* 114:991–997.

Marquart, H. (1989) Lung function of welders of zinc-coated mild steel: cross-sectional analysis and changes over five consecutive work shifts. *Am. J. Ind. Med.* 16:289–296.

Marrs, T.C., et al. (1988) The repeated dose toxicity of a zinc oxide/hexachloroethane smoke. *Arch. Toxicol.* 62:123–132.

McKinney, P.E., Brent, J., and Kulig, K. (1994) Acute zinc chloride ingestion in a child: local and systemic effects. *Ann. Emerg. Med.* 23:1383–1387.

Mocchegiani, E., et al. (1995) Benefit of oral zinc supplementation as an adjunct to zidovudine (AZT) therapy against opportunistic infections in AIDS. *Int. J. Immunopharmacol.* 17:719–727.

Murphy, J.V. (1970) Intoxication following ingestion of elemental zinc. *JAMA* 212:2119–2120.

Ninh, N.X., et al. (1996) Zinc supplementation increases growth and circulating insulin-like growth factor I (IGF-I) in growth-retarded Vietnamese children. *Am. J. Clin. Nutr.* 63:514–519.

National Institute for Occupational Safety and Health (1984) *National Occupational Exposure Survey (1980–83)*, Cincinnati, OH: National Institute for Occupational Safety and Health, Department of Health and Human Services.

National Institute for Occupational Safety and Health (2007) *Pocket Guide to Chemical Hazards. NIOSH No. 2005-149*, Washington, D.C.: US Department of Health and Human Services.

National Research Council (1989) *Recommended Dietary Allowances*, 10th ed., Washington, D.C.: The National Academies Press.

O'Neil, M.J., Smith, A., and Heckelman, P.E. (2001) *Merck Index*, 10th ed., Rahway, NJ: Merck & Co., Inc.

Pennington, J.A., et al. (1986) Mineral content of foods and total diets: the selected minerals in foods survey, 1982 to 1984. *J. Am. Diet Assoc.* 86:876–891.

Phillips, M.J., et al. (1996) Excess zinc associated with severe progressive cholestasis in Cree and Ojibwa-Cree children. *Lancet* 347:866–868.

Potter, J.L. (1981) Acute zinc chloride ingestion in a young child. *Ann. Emerg. Med.* 10:267–269.

Prasad, A.S. (1996) Zinc deficiency in women, infants and children. *J. Am. Coll. Nutr.* 15:113–120.

Safty, A., Maksoud, N., Mahqoub, K., and Helal, S. (2008) Zinc toxicity among galvanization workers in the iron and steel industry. *Ann. N Y Acad. Sci.* 1140:256–262.

Sandstead, H.H. (1995a) Requirements and toxicity of essential trace elements, illustrated by zinc and copper. *Am. J. Clin. Nutr.* 61:621S–624S.

Sandstead, H.H. (1995b) Is zinc deficiency a public health problem? *Nutrition* 11:87–92.

Sempértegui, F., et al. (1996) Effects of short-term zinc supplementation on cellular immunity, respiratory symptoms, and growth of malnourished Ecuadorian children. *Eur. J. Clin. Nutr.* 50:42–46.

Simon, S.R., et al. (1988) Copper deficiency and sideroblastic anemia associated with zinc ingestion. *Am. J. Hematol.* 28:181–183.

Spencer, H., Norris, C., and Williams, D. (1994) Inhibitory effects of zinc on magnesium balance and magnesium absorption in man. *J. Am. Coll. Nutr.* 13:479–484.

Sturgis, C.C., Drinker, P., and Thomson, R.M. (1927) Metal fume fever: I. Clinical observations on the effect of the experimental inhalation of zinc oxide by two apparently normal persons. *J. Ind. Hyg.* 9:88–97.

39

HAFNIUM AND ZIRCONIUM

Kambria K. Haire, Jayme P. Coyle, and Raymond D. Harbison

First Aid: [zirconium and hafnium] Irrigate the eyes immediately for at least 15 min; wash the skin copiously with soap. If inhaled, provide respiratory support. If ingested, seek medical attention immediately; do not induce vomiting.

Target organ(s): [zirconium] Eyes, skin, lungs; [hafnium] eyes, lungs

Occupational exposure limits:

[zirconium (As Zr)] OSHA PEL: 5 mg/m³ TWA; NIOSH REL[a]: 5 mg/m³ (10-h TWA); STEL (15 min) 10 mg/m³ TWA; ACGIH TLV: 5 mg/m³ TWA; STEL 10 mg/m³; IDLH: 25 mg/m³

[hafnium (As Hf)] OSHA PEL: 0.5 mg/m³ TWA; NIOSH REL: 0.5 mg/m³ (10-h TWA); ACGIH TLV: 0.5 mg/m³ (8-h TWA); IDLH: 50 mg/m³

Reference values: None established for either element

Risk/safety phrases:

[zirconium oxide]: R36/37 (irritating to the eyes and respiratory system); S26 (in case of contact with eyes, rinse immediately with plenty of water and seek medical advice); S39 (wear eye/face protection); signal word: flammable

[hafnium (powder)]: R17 (spontaneously flammable in air); S33 (take precautionary measures against static discharges)

BACKGROUND AND USES

Zirconium (symbol Zr; CASRN 7440-67-7) and hafnium (symbol Hf; CASRN 7440-58-6) are commonly found together in hafnon ((Hf,Zr)SiO₄) and zircon (ZrSiO₄; CASRN 14949-68-2); the former typically contains mostly hafnium with approximately 2% zirconium, while the latter typically contains mostly zirconium with 0.5– 5% hafnium. As their respective atomic and ionic radii are similar, their chemical properties are consequently similar. However, their differences, especially regarding neutron absorption cross section and heat refractory properties, require extensive purification and separation prior to end use industrially (Marschner, 2007). Neither of the compounds is biologically essential.

Sand blends containing zirconium are used for sand blasting and foundry moldings. Zirconium compounds are used for reflective surfaces and as a pigment in television tubes. Zirconium and zirconium–aluminum compounds have been used as antiperspirants, antipruritic agents, and abrasives. Incorporation in ceramics and alloys increases thermal protection of furnaces, kilns and crucible mortar, and of jet engine components (Opeka et al., 1999). The nuclear power industry is the leading industry that utilizes hafnium-free zircaloy, which consists of 95% Zr alloyed with nickel (0.05%), chromium (0.1%), iron (0.12%), and tin (1.5%). Zircaloy is used to form nuclear reactor claddings given its low effective neutron absorption cross section (USGS, 2013).

---

[a]Excludes zirconium tetrachloride

Hamilton & Hardy's Industrial Toxicology, Sixth Edition. Edited by Raymond D. Harbison, Marie M. Bourgeois, and Giffe T. Johnson.

Zirconium oxide (ZrO_2) is used as a diamond imitation gemstone and is gaining widespread use in medical industries, especially in dental crowns as an alternative to porcelain and as an alloy in biometallic prosthetics.

Hafnium use is limited due to low abundance. The primary use of hafnium is in the nuclear industry, where it is used in fuel rods to regulate fission given its high neutron absorption cross section. Similar to zirconium, hafnium is alloyed with niobium and carbide to produce high temperature refractory materials for furnaces and jet components as well as for plasma cutters. In addition, hafnium oxide is increasingly being used to augment or replace silicone oxide-based microprocessor chips in certain applications as well as in cathodes and capacitors (Field et al., 2011).

PHYSICAL AND CHEMICAL PROPERTIES

Zirconium is a hard, silvery transition metal with a melting and boiling temperature of $1854\,°C$ ($3369\,°F$) and $4406\,°C$ ($7963\,°F$), respectively. It has a negligible vapor pressure, and is insoluble in water, especially as zirconium oxide. The primary valence state is $4+$. Zirconium and its alloys react violently with strong acids and are incompatible with strong metal alkalis and strong oxidizers but are inert to most weak acids and alkalis. Pure zirconium powder may explode spontaneously in air, while zirconium hydrides react with water to produce a flammable gas. Zirconium tetrachloride is highly corrosive, which hydrolyzes in water to form zirconyl chloride.

Hafnium is a lustrous silvery, ductile metal typically found in the $4+$ oxidation state. In metallic form, the melting and boiling points are $2230\,°C$ ($4046\,°F$) and $4600\,°C$ ($8312\,°F$), respectively, with a negligible vapor pressure. It is poorly soluble in water as a pure metal. It is relatively unreactive to alkalis and strong acids, except hydrofluoric acid and is frequently alloyed with iron, niobium, tantalum, and titanium. Pure hafnium power is pyrophoric and may spontaneously ignite in air, especially under conditions of high moisture. Hafnium powder reacts violently with strong oxidizers and strong acids (Pohanish, 2011). When heated to around $200\,°C$, hafnium metal may react with several period 2 inorganic elements as well as silicone and sulfur. Hafnium oxide (HfO_2) is poorly soluble in water and is a refractory compound, which is highly resistant to corrosion. With halogens, it may react to form tetrahalides.

MAMMALIAN TOXICOLOGY

Both hafnium and zirconium are valued in the industry for their resistance to heat and corrosion, and neither of them is considered particularly toxic to humans. The conventional separation of zirconium and hafnium industrially utilizes sodium isocyanate complexation in methyl isobutyl ketone; exposure to these may potentially be more hazardous than to the elements themselves (Rezaee et al., 2012). Dermal exposure and inhalation may cause local irritation and induce localized granulomas (Bigardi et al., 1992; Leininger et al., 1977; Shelley and Hurley, 1958).

Acute Effects

Rats exposed to zirconium silicate (CAS #14940-68-2) dust in high concentrations for 7 months developed radiographic changes from radiopaque particle accumulation but showed no effects to pneumocytes on histological examination (Zhilova and Kasparoz, 1966). In rabbits, daily short inhalation exposures to a mist containing sodium zirconium lactate produced peribronchial granulomas and alveolar inflammation. In humans, pulmonary and ocular irritation can occur, but these are rare adverse events.

Zirconium-based antiperspirants, typically containing aluminum zirconium tetrachlorohydrex glycerin and sodium zirconium lactate, have been associated with reversible dermal granulomas and hypersensitivity, consisting of papules. Papules disappear after exposure cessation though the reversal of zirconium salt irritation typically takes weeks to months rather than days (Browning, 1969; Epstein et al., 1963; Lisi, 1992; Price and Skilleter, 1986; Sullivan and Krieger, 1999). Similar dermal lesions occurred after the application of zirconium oxide cream for the treatment of poison ivy, which has restricted its use (Williams and Skipworth, 1959). Zirconium(IV) chloride hydrolyzes in water to form a corrosive solution. Upon standing, zirconium(IV) chloride solution liberates hydrochloric acid fumes, which can cause irritation and chemical burns to mucous membranes. However, reports of chemical burns from zirconium (IV) chloride are lacking.

Hafnium dust is an explosive hazard. Like zirconium, hafnium salts can irritate mucosae and skin, although granulomas are not produced. Hafnyl chloride given intravenously was fatal at doses of $10\,mg/kg$ in cats (Haley et al., 1962). *In vitro* exposure of immortalized cells to hafnium oxide nanoparticles does not demonstrate significant cytotoxicity (Field et al., 2011). Reports of acute effects to hafnium exposure apart from dermal and pulmonary irritation are lacking.

Chronic Effects

Chronic particulate or fume exposure typically occurs in industries involving the mining, cutting, and purification of zirconium-containing materials. Chronic exposure has not been associated with significant adverse human health effects though isolated incidences have been documented. Dermal and pulmonary sequelae ranging from mild irritation to focal granulomatous lesions are the most common adverse health outcomes upon exposure to ZrO_2 and zirconium salts

(Bigardi et al., 1992; Leininger et al., 1977; Shelley and Hurley, 1958).

Inhaled zirconium compounds are generally of low toxicity, but occasional reports of severe pulmonary fibrosis exist in those with long-term daily exposure (Bartter et al., 1991; Hadjimichael and Brubaker, 1981; Reed, 1956). Chronic exposure to zirconium compounds in the ceramic and foundry industries has yielded three reports of granulomatous pulmonary lesions with bilateral diffuse radiopacities; any indication of an infective etiological agent was absent. One case had a markedly decreased pulmonary function with bilateral crepitations (Bingham et al., 2001; Liippo et al., 1993; Romeo et al., 1994); one of these workers later died. The cause of death was an acute allergic alveolitis-like hypersensitivity pneumonitis marked by severe bronchioli inflammation and epithelial ulcerations (Liippo et al., 1993). One nonsmoking worker in the nuclear industry chronically exposed to zircaloy dust for 16 years presented with recurrent progressive pneumonia, bilateral lower lobe interstitial infiltrations, and altered alveolar morphology due to the presence of sarcoid granulomatosis. Again, no infective agent was detected. The role of zirconium as an etiological agent in this case cannot be confirmed, as zircaloy contains several constituents (Werfel et al., 1998). Studies on several cohorts did not find associations between zirconium exposure, both particulate and fume, and chronic effects, possibly indicating that adverse effects are rare and influenced by other factors (Mackison et al., 1981; Marcus et al., 1996).

Ceramic fibers that may contain up to 20% ZrO_2 have been classified as B2 (probable human carcinogen) by the U.S. EPA. In one study, chronic exposure to zirconium oxide-containing ceramic fibers produced contact dermatitis. Patch testing ruled out zirconium oxide as the probable allergenic agent, as cobalt and chromium produced profound allergic reactions (Kieć-Świerczyńska and Wojtczak, 2000). However, the pulmonary effects are more consistent with known toxicological data of silicates, which compose a fraction of the ceramic fibers.

Animal studies with rats have documented mild hepatocellular vacuolization with chronic ingestion of hafnyl chloride ($HfOCl_2$; CASRN 13759-17-6), and the mean lethal dose (LD_{50}) of intraperitoneal hafnyl chloride is 112 mg/kg (Haley et al., 1962). These results suggest hepatotoxicity. However, no reports on human chronic effects from exposure to hafnium compounds have been documented to date; thus, animal models may be particularly sensitive to hafnium compounds, especially given the low bioavailability during inhalational and dermal exposure.

Mechanisms of Action

Parenteral injection of zirconium salts results in systemic bioavailability with localized irritation at the site of injection; the absorption fraction accumulates preferentially in the bone. Oral ingestion results in very low bioavailability with the majority excreted in feces (Venugopal and Luckey, 1978). Beryllium sulfate, $BeSO_4$, and zirconium sulfate, $Zr(SO_4)_2$ (CASRN 14644-61-2) act as lymphocyte mitogens in vitro (Shelley and Hurley, 1971), which partly explains the induction of granulomas in vivo by these metals. Zirconium tetrachloride is corrosive to the skin, similar to the action of other strong inorganic acids.

No mechanism of action for hafnium has been elucidated as adverse events are rare.

SOURCES OF EXPOSURE

Occupational

Occupational exposure to zirconium compounds is most likely in foundry workers, ceramic workers, and firefighters. However, the zirconium found in cosmetics and hygienic products is a more common source of health problems than that of occupational setting.

Zirconium dust and shavings ignite easily when exposed to air or moisture, which can result in an explosion. Fires involving zirconium cannot be extinguished by ordinary means; instead, they must be suffocated with a pulverized mineral carbonate such as dolomite. Inhalation and dermal contact are the primary routes of exposure to zirconium.

Both hafnium and zirconium are used to store tritium. Over time, decay and mechanical abrasion leads to radioactive Hf^3H and Zr^3H aerosol and particle formation that can burden the respiratory tract. Clearance was shown to be biphasic with an initial rapid half-life but a long-term secondary half-life (Zhou et al., 2010).

Environmental

Neither of the elements is abundant in the environment to be a significant source of exposure.

INDUSTRIAL HYGIENE

Zirconium oxide powder is often naturally radioactive (Lischinsky et al., 1991). However, its level of radioactivity is so low that the level of radiation at the threshold limit value (TLV) is only about 6% of the Nuclear Regulatory Commission's maximum permissible limit of 0.19 Bq/m^3. Therefore, researchers have concluded that the radioactivity of zirconium oxide powder can safely be ignored for industrial hygiene purposes. Nonradioactive particulates should be controlled similarly to benign dust, using general ventilation, respirators to limit inhalation, and wet cutting to reduce airborne dust.

The American Conference of Governmental Industrial Hygienists (ACGIH), the Occupational Safety and Health

Administration (OSHA), and the National Institute for Occupational Safety and Health (NIOSH) have set the TLV, the permissible exposure limit (PEL), and the recommended exposure limit (REL), respectively at 5 mg/m³ for zirconium compounds. The ACGIH and NIOSH have set an additional 15 min short-term exposure limit (STEL) of 10 mg/m³. An immediately dangerous to life and health (IDLH) limit has been set for zirconium at 50 mg/m³ (ACGIH, 2011; NIOSH, 2007). Zirconium and zirconium compounds are not classified as carcinogens by the ACGIH (class A4) and have not been investigated by the International Agency for Research on Cancer (IARC)

The ACGIH, OSHA, and NIOSH have set the TLV, PEL, and REL, respectively, at 0.5 mg/m³ for hafnium and hafnium compounds (ACGIH, 2011; NIOSH, 2007).

Zirconium has not been classified as a carcinogen by either the ACGIH (group A4) or the IARC (group 3) (IARC, 1999). Hafnium has not been classified as a carcinogen by the ACGIH, and no data are available on hafnium, except that it is more acutely cytotoxic than zirconium (IARC, 1999).

MEDICAL MANAGEMENT

No specific procedures have been devised specifically for zirconium or hafnium; however, dermal and ocular exposures should be rinsed with copious amounts of water. Patients exposed to high air concentrations of dust should be moved to fresh air.

In the case of the nuclear industry worker with recurrent progressive pulmonary pneumonia and sarcoid granulomatosis, a prednisone regime proved palliative with partial restoration of pulmonary function; the worker was able to resume work during treatment (Werfel et al., 1998).

REFERENCES

American Conference of Governmental Industrial Hygienists (ACGIH) (2011) *Threshold Limit Values (TLVs) for Chemical Substances and Physical Agents and Biological Exposure Indices (BEIs)*, Cincinnati, OH: American Conference of Governmental Industrial Hygienists.

Bartter, T., Irwin, R.S., Abraham, J.L., Dascal, A., Nash, G., Himmelstein, J.S., and Jederlinic, P.J. (1991) Zirconium compound-induced pulmonary fibrosis. *Arch. Intern. Med.* 151(6):1197–1201.

Bigardi, A.S., Pigatto, P.D., and Moroni, P. (1992) Occupational skin granulomas. *Clin. Dermatol.* 10(2):219–223.

Bingham, E., Cohrssen, B., and Powell, C.H. (2001) *Patty's Toxicology*, 5th ed., vol. 1–9, New York, NY: John Wiley & Sons.

Browning, E. (1969) *Toxicity of Industrial Metals*, 2nd ed., New York, NY: Appleton-Century-Crofts.

Epstein, W.L., Skahen, J.R., and Krasnobrod, H. (1963) Tile organized epithelioid cell granuloma: differentiation of allergic (zirconium) from colloidal (silica) types. *Am. J. Pathol.* 43:391–405.

Field, J.A., Luna-Velasco, A., Boitano, S.A., Shadman, F., Ratner, B.D., Barnes, C., and Sierra-Alvarez, R. (2011) Cytotoxicity and physicochemical properties of hafnium oxide nanoparticles. *Chemosphere* 84(10):1401–1407.

Hadjimichael, O.C. and Brubaker, R.E. (1981) Evaluation of an occupational respiratory exposure to a zirconium-containing dust. *J. Occup. Med.* 23(8):543–547.

Haley, T.J., Raymond, K., Komesu, N., and Upham, H.C. (1962) The toxicologic and pharmacologic effects of hafnium salts. *Toxicol. Appl. Pharmacol.* 4:238–246.

International Agency for Research on Cancer (IARC) (1999) *IARC Monograph on the Evaluation of the Carcinogenic Risks to Humans: Surgical Implants and Other Foreign Bodies*, vol. 74, Lyon, France: International Agency for Research on Cancer.

Kieć-Świerczyńska, M. and Wojtczak, J. (2000) Occupational ceramic fibres dermatitis in Poland. *Occup. Med. (Lond.)* 50(5):337–342.

Leininger, J.R., Farrell, R.L., and Johnson, G.R. (1977) Acute lung lesions due to zirconium and aluminum compounds in hamsters. *Arch. Pathol. Lab. Med.* 101(10):545–549.

Liippo, K.K., Anttila, S.L., Taikina-Aho, O., Ruokonen, E.L., Toivonen, S.T., and Tuomi, T. (1993) Hypersensitivity pneumonitis and exposure to zirconium silicate in a young ceramic tile worker. *Am. Rev. Respir. Dis.* 148(4 Pt 1):1089–1092.

Lischinsky, J., Vigliani, M.A., and Allard, D.J. (1991) Radioactivity in zirconium oxide powders used in industrial applications. *Health Phys.* 60(6):859–862.

Lisi, D.M. (1992) Availability of zirconium in topical antiperspirants. *Arch. Intern. Med.* 152(2):421–422, 426.

Mackison, F.W., Stricoff, R.S., and Partridge, L.J., editors. (1981) *NIOSH/OSHA—Occupational Health Guidelines for Chemical Hazards*, DHHS(NIOSH) Publication No. 81–123, Washington, DC: Government Printing Office.

Marcus, R.L., Turner, S., and Cherry, N.M. (1996) A study of lung function and chest radiographs in men exposed to zirconium compounds. *Occup. Med. (Lond.)* 46(2):109–113.

Marschner, C. (2007) Hafnium: stepping into the limelight! *Angew. Chem. Int. Ed. Engl.* 46(36):6770–6771.

National Institute for Occupational Safety and Health (NIOSH) (2007) *Pocket Guide to Chemical Hazards*, DHHS (NIOSH) Publication Number 2005-149, Washington, DC: National Institute for Occupational Safety and Health.

Opeka, M.M., Talmy, I.G., Wuchina, E.J., Zaykoski, J.A., and Causey, S.J. (1999) Mechanical, thermal, and oxidation properties of refractory hafnium and zirconium compounds. *J. Eur. Ceram. Soc.* 19(13):2405–2414.

Pohanish, R.P. (2011) *Sittig's Handbook of Toxic and Hazardous Chemicals and Carcinogens*, 6th ed., Norwich, NY: William Andrew.

Price, R.J. and Skilleter, D.N. (1986) Mitogenic effects of beryllium and zirconium salts on mouse splenocytes *in vitro. Toxicol. Lett.* 30(1):89–95.

Reed, C.E. (1956) A study of the effects on the lung of industrial exposure to zirconium dusts. *AMA Arch. Ind. Health* 13(6):578–580.

Rezaee, M., Yamini, Y., and Khanchi, A. (2012) Extraction and separation of zirconium from hafnium using a new solvent microextraction technique. *J. Iran. Chem. Soc.* 9(1):67–74.

Romeo, L., Cazzadori, A., Bontempini, L., and Martini, S. (1994) Interstitial lung granulomas as a possible consequence of exposure to zirconium dust. *Med. Lav.* 85(3):219–222.

Shelley, W.B. and Hurley, H.J. (1958) The allergic origin of zirconium deodorant granulomas. *Br. J. Dermatol.* 70(3):75–101.

Shelley, W.B. and Hurley, H.J. (1971) The immune granuloma: late delayed hypersensitivity to Zr and Be. In: Samter, M., editor. *Immunologic Diseases*, Boston, MA: Little, Brown.

Sullivan, J.B., and Krieger, G.R., editors. (1999) *Clinical Environmental Health and Toxic Exposures*, 2nd ed., Philadelphia, PA: Lippincott Williams and Wilkins.

United States Geological Survey (USGS) (2013) *Mineral commodity summaries, zirconium and hafnium*. Available at http://minerals.usgs.gov/minerals/pubs/mcs/2013/mcs2013.pdf, (accessed January 20, 2014).

Venugopal, B. and Luckey, T.D. (1978) *Metal Toxicity in Mammals: Chemical Toxicity of Metals and Metalloids*, vol. 2, New York, NY: Plenum Press.

Werfel, U., Schneider, J., Rodelsperger, K., Kotter, J., Popp, W., Woitowitz, H.J., and Zieger, G. (1998) Sarcoid granulomatosis after zirconium exposure with multiple organ involvement. *Eur. Respir. J.* 12(3):750.

Williams, R.M. and Skipworth, G.B. (1959) Zirconium granulomas of the glabrous skin following treatment of rhus dermatitis: report of two cases. *Arch. Dermatol.* 80:273–276.

Zhilova, N.A. and Kasparoz, A.A. (1966) Data on the toxicological studies of niobium and zirconium chlorides. *Gig. Sanit.* 31:111–113.

Zhou, Y., Cheng, Y.S., and Wang, Y. (2010) Dissolution rate and biokinetic model of zirconium tritide particles in rat lungs. *Health Phys.* 98(5):672–682.

40

METAL FUME FEVER AND METAL-RELATED LUNG DISEASE

Thomas Truncale

Metal fume fever (MFF) is a self-limiting illness and one of several febrile flu-like respiratory syndromes encountered in the occupational setting. In addition to metal exposure, workers exposed to the dusts of unprocessed cotton, vegetables, contaminated water reservoirs, or exposure to the thermal breakdown products of polytetrafluorethylene (Teflon), develop an acute febrile syndrome (Table 40.1). The shared physiologic mechanism and clinical picture has led to a unifying diagnosis "inhalation fever" which was first proposed in 1978 and gained wider acceptance in 1991 (Rask-Anderson, 1992). Regardless of the cause, inhalation fever is characterized by the onset of symptoms (fever, malaise, myalgias, cough, and chills) within 3–12 h after an exposure. Signs and symptoms are generally short-lived and in most cases resolve spontaneously within 24–48 h (Table 40.2). Because the syndrome results in a blunted response (tachyphylaxis) with continued exposure (Monday morning fever in cotton mill workers), workers with no prior exposure are at particularly high risk (Blanc, 1997).

Metal fume fever is the classic example of inhalation fever. Initially thought to be a malaria type sickness among foundry men or any one exposed to the vapors, which arise from metal, Thackrah was the first to recognize a separate "affection" peculiar to brass workers (Hayhurst, 1911). Over the years, the afflictions came to be known as brass founder's ague, brass chills, smelter shakes, zinc shakes, zinc fume fever, metal malaria, or Monday morning fever among brass foundry workers or welders of galvanized steel (Table 40.3). Metal fume fever has since been recognized as the syndrome of an acute, self-limiting flu-like illness following exposure to metal fumes, primarily zinc oxide but other metal oxides, including manganese, nickel, and chromium have been described (Table 40.4). Welding, brazing, soldering, or thermal cutting of galvanized metal are the main causes of zinc exposure, although exposure can occur in the manufacture of brass (zinc–copper alloy) or industrial processes that use other binary zinc alloys, including aluminum, cobalt, antimony, and nickel. Welders are at particular high risk, as the composition of welding fumes will vary based upon the type of metal being used. Mild steel will produce fumes containing aluminum, magnesium, fluoride, potassium, calcium, manganese, iron, titanium, and trace amounts of cobalt, zinc, and lead. Welding of stainless steel will provide additional exposure to chromium and nickel oxides (Brooks et al., 2007). Galvanized steel, a coating of zinc over steel provides additional protection from rust and corrosion. Because zinc melts and vaporizes at a lower temperature than non-coated steel, the zinc vapor reacts with oxygen and immediately produces zinc oxide as a fine dispersion of dry particles of 1 μm in size, small enough to reach the terminal bronchioles and alveoli when inhaled (Rohrs, 1957; Brown, 1988). Although it is widely accepted that MFF is caused by the exposure to freshly generated zinc oxide fumes, Blanc (Blanc, 1997) in a concise review of inhalation fever references case reports of MFF following heavy exposure to finely ground zinc dust in the absence of a fresh fume exposure.

Despite the general acceptance that an exposure to metal fumes other than zinc oxide causes MFF, there is little scientific evidence to support this conclusion. The association of MFF following exposure to magnesium oxide fumes was first described in the 1920s (Drinker et al., 1927); however, more recent work has failed to validate a systemic

Hamilton & Hardy's Industrial Toxicology, Sixth Edition. Edited by Raymond D. Harbison, Marie M. Bourgeois, and Giffe T. Johnson.

TABLE 40.1 Causes of Inhalation Fever

Disorder	Exposure	Occupation
Metal fume fever	Oxides of metals (most commonly zinc)	Welding (most common), also seen in soldering, melting, casting, grinding, bronzing, and forging
Organic dust toxic syndrome	Organic dusts (moldy hay, grain dusts, wood chips, mulch, silage, compost) contaminated with bacteria, fungi, and endotoxin	Agriculture/farming/environmental
Vegetable and grain dusts		Textile, apparel, and furnishing workers
Mill fever, Gin fever, Monday morning fever, mattress fever	Cotton dust, stained cotton (others include soft hemp (hemp fever), kapok)	Agricultural/yarn-manufacturing, thread mills, fabric manufacturing, textile (upholstery, mattresses)
Heckling fever	Flax	Agricultural/yarn-manufacturing, thread mills, fabric manufacturing, textile (upholstery, mattresses)
Grain fever	Wheat, barley, oat, corn, rice, sorghum, airborne grain dust	Grain workers
Wood trimmer's Disease	Dried contaminated (moldy) wood	Sawmills
Contaminated water reservoirs		
Humidifier fever	*Naegleria gruberi,* Pseudomonas likely other Gram negative species and possibly endotoxin	Printing industry

TABLE 40.2 Characteristics of Inhalation Fever

1. Acute high level exposure
2. Onset of symptoms within 3–12 h (fever, dyspnea, myalgias, malaise, chills)
3. No history of prior sensitization
4. Blunted response with repeated exposure (tachyphylaxis)
5. Pulmonary function is normal
6. Chest X-ray reveals no acute findings
7. Hypoxia is not present
8. Increased pulmonary polymorphonuclear leukocyte response
9. Increased proinflammatory cytokine (IL-6, IL-8, TNF) response
10. ROS, nanoparticles, and TRPA1 activation likely play a role
11. Self-limiting course
12. No chronic sequela expected

Source: Adapted from Blanc PD (1997); Blanc PD (2007).

TABLE 40.3 Metal Fume Fever: Synonyms

Brass founders fever
Brass founders ague
Brass shakes
Monday morning fever
Zinc shakes
Metal shakes
Smelter's shakes
Smelter chills
Glavo
Welders ague
Metal malaria

response, abnormal pulmonary function, or changes in proinflammatory cell or cytokine concentrations in bronchoalveolar lavage (BAL) fluid following inhalation of magnesium oxide particulates (Kuschner et al., 1997). Similar reports documenting the adverse effects of copper fumes are also limited. In a systemic review of the scientific literature, Borak (Borak et al., 2000) concluded that there was insufficient evidence to suggest a causal link between exposure to copper dust or fumes and MFF.

Symptoms

Metal fume fever is characterized by the acute onset (usually within 3–12 h of welding galvanized steel) of respiratory symptoms including cough, wheezing, dyspnea, and chest tightness accompanied by systemic symptoms of a flu-like

TABLE 40.4 Metals Associated With MFF

Zinc (most common)	Magnesium
Copper	Copper zinc alloy (brass)
Aluminum	Silver
Manganese	Antimony
Magnesium	Lead
Nickel	Chromate
Cadmium	Titanium
Selenium	Arsenic
Tin	Thorium
Cobalt	Iron
Arsenic	Vanadium
Beryllium	Boron

TABLE 40.5   Manifestations of Metal Fume Fever

Respiratory
Cough
Shortness of breath
Wheezing
Chest tightness
Systemic symptoms
Fever
Chills
Rigor
Myalgias
Malaise
Dry parched throat
Thirst
Metallic or sweet taste in the mouth
Nausea
Loss of appetite
Headache

illness; fever, malaise, myalgias, chills, metallic taste in the mouth, dry throat, nausea, and headache (Table 40.5). Spontaneous resolution of symptoms occurs within 24–48 h of exposure. Tachyphylaxis develops during the workweek with continued exposure, as patients often complain of symptoms at the start of the week, which improve or completely resolve at the week's end (Krantz and Dorevitch, 2004). New workers, those who have been off for the weekend or for a week's vacation are at highest risk. Once back at work, they develop symptoms upon re-exposure following a 2–3 day period of abstinence.

Diagnosis

On initial presentation, MFF may be difficult to differentiate from other forms of acute febrile respiratory syndromes namely influenza, other viral illnesses, or acute lung injury (ALI) following exposure to cadmium fumes or zinc chloride. The latter are discussed in detail in a following section. It is essential to obtain an accurate occupational history that exhausts all possible exposures to exclude the possibility of acute pneumonitis following exposure to cadmium or zinc chloride. Routine diagnostic tests, including chest X-ray, pulmonary function tests (PFTs), diffusion capacity, and arterial oxygenation in MFF are always normal (Ahsan et al., 2009). In rare instances where the forced vital capacity (FVC) is reduced (Sturgis et al., 1927) or hypoxia and patchy opacification on chest X-ray have been described (Kaye et al., 2002), complete resolution occurs within 24–48 h. In the setting of ALI following exposure to cadmium, nickel carbonyl, and zinc chloride, there is progression of the pulmonary infiltrates and worsening hypoxia in the days following the exposure, where MFF is expected to resolve during that period of time. If abnormal diagnostic studies are present following an exposure to metal fumes, a careful investigation for other causes should be pursued. These conditions are outlined in Table 40.6 and are described below.

Pathogenesis

The pathogenesis of MFF appears to be related to a dose-dependent inflammatory response in the lung followed by the release of various cytokines, which in turn induce the systemic symptoms of fever and flu-like symptoms. Blanc (Blanc et al., 1991) first studied 14 participants who welded galvanized mild steel in an environmental chamber specifically designed to correspond to real work exposures. Varying concentrations of zinc oxide fume were correlated with BAL fluid cell counts at 8 h (early group) and 22 h (late group). They found a dose-dependent increase in the polymorphonuclear leukocyte count in the BAL fluid 22 h following the exposure. First, establishing a dose-dependent pulmonary inflammatory response, Blanc et al. (1993) measured TNF, IL-1, IL-6, and IL-8 in the BAL supernatant of 15 subjects 3, 8, and 22 h following exposure to inhaled zinc oxide fumes. They found a statistically significant dose- and time-dependent increase in TNF in the early follow-up group (3 h) compared to the late (8 and 22 h) group. Additionally, they found a significant exposure response in IL-6 and IL-8, 8 h after the exposure, supporting their earlier hypothesis that the delayed (22 h) increase in the polymorphonuclear count is mediated by the early release of IL-8, a neutrophil chemo-attractant. Subsequent studies measuring TNF, IL-6, and IL-8 using a similar model (Kuschner et al., 1998) and *in vitro* human mononuclear cell lines exposed to zinc oxide (Kuschner et al., 1997) have verified these findings.

Additional mechanisms to explain the inflammatory response seen in MFF appear linked to the formation of oxygen radicals produced by activated neutrophils. Human neutrophils exposed to zinc oxide demonstrate an increased oxidative metabolism causing the release of oxygen radicals within 35 min of exposure (Lindahl et al., 1998). reactive oxygen species (ROS) and the generation of free radicals are also produced from stainless and mild steel welding fumes, and have the potential to cause cellular membrane disruption, protein or DNA damage, and may lead to the generation of ROS or signaling associated changes in airway inflammation. Additionally, combustion derived nanoparticles originating from welding fumes among others, have the potential to cause lung injury through oxidative stress and inflammation (Kim et al., 2010). The possibility that increased oxygen radical formation from activated neutrophils or exposure to welding derived nanoparticles contribute to the pathogenesis of MFF requires further study. Furthermore, the finding of increased glutathione levels, a known antioxidant and free radical scavenger, among welders exposed to zinc may help to explain why MFF is a self-limiting illness (Jin-Chyuan

TABLE 40.6 Metal-Related Lung Disease

Occupational Lung Disease	Cause	PFT's	Chest X-ray	Chronic Sequela
Occupational asthma	Platinum salts, nickel, chromium, tungsten carbide, cobalt, aluminum, palladium, vanadium, zinc	Variable airway obstruction, airway hyper-responsiveness	Normal, may show hyperinflation	Variable: Short duration of exposure correlates to better outcome, longer duration may result in persistent airway responsiveness, obstruction, inflammation, and remodeling
Delayed and immediate anaphylactoid reaction	Chromium, zinc	Normal, exposure induces bronchospasm and airway hyper-responsiveness	Normal	The possibility of re-current symptoms exists upon re-exposure, long-term sequela unknown
Hypersensitivity pneumonitis	MWF contaminated with atypical Mycobacterium	Acute: restriction with decreased DLCO (reversible if removed from the exposure) Chronic: mixed restriction and obstruction with reduced DLCO and hypoxia at rest or with exercise	Acute: normal to mid to upper zone predominance of reticular, nodular or ground glass opacities. Chronic: Reticular nodular with honeycombing and/or emphysema.	Variable: acute presentation is associated with better outcome. Greater intensity and duration of exposure results in accelerated ling function decline
Berylliosis or metal-related granulomatous disease	Beryllium, aluminum, cadmium, hard metal disease (cobalt/tungsten carbide), nickel and mercury(rare)	Restrictive (reduced FEV1, FVC, TLC, FRC, and RV) diminished DLCO	Acute: alveolitis with diffuse infiltrates Chronic: diffuse interstitial opacities with fibrosis and honeycombing	Progressive: workers can have resolution of symptoms if removed from exposure, in long standing exposure restrictive lung function and fibrosis are irreversible
Acute pneumonitis	Aluminum, cadmium, manganese, mercury, zinc chloride, vanadium	Normal to restrictive if patient survives	Diffuse bilateral infiltrates, with or without pulmonary edema	Acute form may be reversible, chronic form will lead to fibrosis, restrictive lung disease, and possibly death
Lung cancer	Aluminum, beryllium, arsenic, cadmium, chromium, nickel	Variable	Lung nodule, mass	Depends of stage, overall poor 5 year survival
Metal fume fever	Oxides of metals (zinc most common)	Normal	Normal	None

et al., 2009). Finally, selective expression of transient receptor potential A1 (TRPA1) in a subpopulation of bronchopulmonary C-fiber afferents following chemical irritant or heavy metal exposure is worth mentioning. Chemical stimulation of the nasal mucosa, glottis, larynx, and lower airways activates TPRA1 and triggers chemosensory nerve endings to release neuropeptides (Substance P and calcitonin gene-related peptide (CGRP)) inducing neurogenic inflammatory dilatation of blood vessels, leaky capillary membranes, and increased secretions. This in turn provokes the cough reflex and can trigger sneezing, tracheal and bronchial constriction, bronchospasm, excessive mucus secretion, and further neurogenic inflammation. TRPA1 can be activated by a variety of chemical irritants, including acrolein, mustard oil, chlorine, aldehydes, ROS, and noxious by-products of smoke (Bessac and Jordt, 2008). Recent studies have shown that heavy metals, zinc, cadmium, and copper stimulate pulmonary sensory neurons through direct activation of TRPA1 (Gu and Ruei-Lung, 2010). Once activated, TRPA1 appears to augment the inflammatory response seen in MFF and may lead to the long-term changes associated with chronic inflammation seen in asthma and COPD (Geppetti et al., 2010). TRPA1 antagonism also offers a promising pharmacologic target for anti-inflammatory and cough medications. Further human study is required, however, to confirm the role of TRPA1 in chemical irritant and heavy metal exposure.

Treatment

Treatment for MFF is supportive, although antipyretics, beta agonists and inhaled or intravenous steroids have been used. No long-term sequela has been found to be associated with this self-limiting illness. The occurrence of MFF suggests poor ventilation or noncompliance with the use of personal protective equipment primarily respirators and warrants a careful engineering and work practice investigation. Lastly, an episode of MFF may identify a worker who may be at risk for developing other welding-related lung diseases, namely asthma (Malo and Cartier, 1987).

METAL-RELATED LUNG DISEASE

Occupational Asthma

Occupational asthma (OA) has been defined by the presence of reversible airflow obstruction and/or airway hyper-responsiveness caused by an exposure to sensitizing agent and/or irritants in the workplace (Chan-Yeung, 1990; Tarlo et al., 2008). Asthma can also be exacerbated in the workplace in those with pre-existing asthma, known as work-aggravated asthma. The reported estimates of all causes of asthma attributed to work ranges from 5 to 29% (Janson et al., 2001). Occupations that appear to be at higher risk include

bakers, laundry workers, shoemakers and repairers, tanners, fell mongers and pelt dressers, and metal plating and coating workers (Karjalainen et al., 2002). In workers exposed to metal aerosols or fumes, OA appears to range from 3 to 10% (Bakerly et al., 2008; El-Zein et al., 2003). While cigarette smoking and atopy do not appear to be important risk factors in most cases of metal-induced OA, smoking, and intensity of exposure definitely play a role in exposure to platinum salts (Calverley et al., 1995). Agents that stimulate an immunologic form of OA are classified into two categories. High molecular weight antigens derived from plant or animal antigens are capable of causing an IgE-mediated response while low molecular weight antigens (LMWA), which are synthetic compounds used in the manufacture of plastics and rubber, act as incomplete antigens (haptens) and combine with human proteins; allergic sensitization, and IgE-mediated OA (Cromwell et al., 1979). Of all exposures to metal dusts, aerosols, or fumes, platinum salts are recognized as the most potent inducers of sensitization and IgE-mediated asthma (Malo, 2005). Exposure to cobalt, nickel, chromium, and palladium (other LMWA) in the form of metal dusts or fumes among welders, electroplaters, metal/hard metal workers, jewelers, or refinery workers are additional causes of metal-induced sensitization and IgE-mediated OA (Chan-Yeung, 1986; Daenen et al., 1999; Kusaka et al., 2001; Leroyer et al., 1998; Malo et al., 1982; Malo and Chan-Yeung, 2009; Novey et al., 1983; Shirakawa et al., 1989). Non-immunologic mechanisms also play a role in the development of metal-induced OA. The demonstration of a 24% reduction in FEV1 following exposure to galvanized heated zinc without evidence of skin sensitization or a specific IgE antibody to zinc in a worker who developed OA suggests an alternative pathway (Malo et al., 1993). At present, the exact immunologic or cellular mechanism to explain how other metals, including zinc, aluminum (potroom asthma), and vanadium are capable of inducing OA is unknown.

IgE-MEDIATED ACUTE AND LATE PHASE REACTIONS

Rarely, exposure to metal fumes has resulted in MFF-like symptoms associated with an immediate and late anaphylactoid reaction. Farrell (Farrell, 1987) describes an employee of a zinc smelting plant who developed generalized pruritus, urticarial lesions, and flu-like symptoms (typical of MFF) after welding galvanized steel. His flu-like symptoms resolved by the next morning, however, his hives and itching worsened. He then developed angioedema of the face, lips, and throat requiring medical treatment with epinephrine and diphenhydramine. On return to work 1 week later where welding of zinc was being done, he immediately developed swelling of the lips, itching, and some swelling in his throat. The presence of an immediate reaction similar to MFF and a late phase reaction (hives and angioedema) suggests an immunologic response distinct from the response seen in MFF.

HYPERSENSITIVITY PNEUMONITIS

Hypersensitivity pneumonitis (HP) or extrinsic allergic alveolitis is an inflammatory lung condition that involves both a type III and type IV hypersensitivity reaction. The reaction is mediated by immune complexes, Th1 T cells and CD8+ predominate lymphocytic inflammation in the lung parenchyma to repeated environmental antigen exposure that results in immunologic sensitization and immune-mediated lung disease (Mohr, 2004; Mason, 2010). Repeated and prolonged exposure may result in interstitial lung disease, which is characterized by an increase in CD4+/CD8+ ratios and a skewing towards Th2 cells as opposed to Th1 (Barrera et al., 2008). In contrast to MFF and acute HP prolonged exposure (10 years or greater in most cases) to an environmental antigen results in interstitial fibrosis on CT scan and restrictive lung disease on PFT.

Most often, all (acute, subacute, and chronic) forms of HP are associated with exposure to moldy or wet hay, grain dust or wood contaminated with fungi. Bacterial contamination of sugar cane, wood dust, and water reservoirs also play a significant role and together they make up the most common causes of HP (Table 40.7). For the years 1990–1999, agricultural production industries (livestock, crops, and farming) were associated with a significantly increased risk for developing HP (DHHS, 2007). Interestingly, HP has been recognized among metalworkers exposed to atypical mycobacterium in contaminated metal working fluids (MWF). Seven cases of HP where identified in three facilities manufacturing automobile parts in Michigan. All of the affected employees had been working in a machining environment for many years before they developed HP. High concentrations of *Mycobacteria immunogenum* were reported from several of the MWF reservoirs (Gupta and Rosenman, 2006). Additionally, *Mycobacteria chelonae* was identified in high numbers in bulk coolant samples and in air samples around colonized machines in a patient who worked in an engine production facility 2 years before developing cough dyspnea and hoarseness. The patient's in-hospital evaluation included

TABLE 40.7   Occupational Related Hypersensitivity Pneumonitis

Etiologic Agents	Disease	Exposure
Microbial agent		
	Bacteria	
Thermophilic actinomycetes	Bagassosis/farmers lung/potato riddlers lung/mushroom workers lung/humidifier lung	Moldy sugarcane, hay, compost, contaminated water
Mycobacterium avium	Hot tub lung	Contaminated water
Mycobacterium immunogenum	Metal working fluid	Contaminated metal working/removing fluids
Mycobacterium chelonae		
Bacillus subtilis	Detergent workers lung	Contaminated detergent
Fungi		
Aspergillus spp	Malt workers lung/tobacco workers lung/compost lung	Moldy malt, barley, Tobacco, compost
Penicilium spp	Cheese workers lung, suberosis	Moldy cheese, moldy cork
Cladosporium spp	Hot tub lung	Contaminated hot tub mists
Alternaria spp	Wood pulp workers disease	Contaminated wood pulp
Cephalosporium spp	Sewer workers lung	Contaminated basement sewage
Trichosporon cutaneum	Japanese summer time HP	Contaminated house dust
	Animal proteins	
Bird proteins	Bird fanciers lung, pigeon breeders lung, Duck fever, lab workers lung, furriers lung	Bird, goose, rat, animal pelt, and other animal proteins
Chemicals		
Isocynates, trimetalic anhydrides, phthalic anhydride, pyrethrum, sodium diazobenzene sulfate (Pauli's reagent)	Chemical workers lung	Isocyanates (in polyurethane paints, adhesives, and foam production), epoxy resins, coatings, paints, Pauli's reagent
Metals		
Zinc fumes	Zinc fume hypersensitivity pneumonitis	Smelter exposed to zinc fumes

a chest X-ray that showed bilateral interstitial infiltrates, a high resolution CT scan consistent with alveolitis and a decreased diffusion capacity on PFTs. A diagnosis of HP was made and the patient's condition improved with corticosteroid treatment and relocation to an office environment at work (Shelton et al., 1999). Of additional interest is a case report of HP in a smelter exposed to zinc fumes while working for 3 years in a nonferrous metal foundry. The patient presented with complaints of acute onset of shortness of breath, fever, cough, and purulent sputum. His chest X-ray showed a mild reticular pattern in the periphery of both lungs. PFTs revealed a moderate obstructive pattern and a significant decrease in diffusion capacity. Bronchoalveolar lavage showed increased lymphocyte counts with a predominance of CD8+ lymphocytes and a decrease in the CD4+/CD8+ ratio resembling HP. After a period of 6 months without contact with zinc fumes, the patient was symptom free and his diffusion capacity returned to normal. The clinical course of this patient, that is; the late emergence of symptoms, similar to HP in metal fluid workers and aluminum workers described below, long-lasting changes in PFT and physiologic findings on BAL are not expected in MFF or ALI from zinc chloride or cadmium exposure. These findings suggest an alternative hypersensitivity reaction (Ameille et al., 1992).

In addition to HP, sarcoid-like granulomatous disease has been identified in workers exposed to aluminum dust. A 50-year-old woman, who worked in a metal reclamation factory for 15 years and was exposed to aluminum, iron, copper, zinc, and nickel dust presented with an intermittent cough, white frothy sputum, and dyspnea. A high resolution CT scan of the chest revealed bilateral ground glass and patchy areas of consolidation with reticular hyperattenuating areas and traction bronchiectasis. Her PFTs showed a restrictive pattern and severe reduction in diffusion capacity. An open lung biopsy showed clusters of well-formed non-necrotizing granulomas with multinucleated giant cells; fibrosis, and honeycombing were absent. Scanning electron microscopy of the granulomas yielded discrete peaks for aluminum. After exclusion of infectious agents and sarcoidosis and based upon her occupational history, the authors conclude there was a distinct relation between the aluminum deposits and pulmonary sarcoid-like granulomas (Cai et al., 2007). The descriptions of sarcoid-like granulomas on histologic examination of the lung following aluminum exposure are not unique to this study and have been described by others (Chen et al., 1978; DeVuyst et al., 1987). Furthermore, chronic beryllium disease (berylliosis or CBD) resulting from long-term exposure to beryllium dust or fumes in the electronic or aerospace industry is also characterized by a cell-mediated or delayed hypersensitivity reaction with granulomatous inflammation. Genetic susceptibility has been established (Fontenot and Maier, 2005; Saltini et al., 1998) through human leukocyte antigen class II marker (HLA-DP Glu69), which has been found to be strongly associated with CBD. Other metals that promote granuloma formation include barium, cobalt, copper, gold, titanium, zirconium, and lanthanides (Newman, 1998).

METAL PNEUMCONIOSIS

The 2002 work-related lung disease (WORLD) surveillance report published by NIOSH (DHHS, 2007) included mortality data focusing on the various occupational-relevant lung diseases, including pneumoconiosis seen in the United States. Reporting on the contributing or cause of death, they identified the majority of cases of dust-associated lung disease (pneumoconiosis) caused by the inhalation of mineral dusts are from asbestos, coal, or silica exposure; however, several metallic dust exposures pose an increased risk of interstitial lung disease, pneumoconiosis, and mortality. The primary pneumoconiotic metallic agents include beryllium and cobalt, although rarely aluminum welding fume-induced pneumoconiosis has been described (Hull and Abraham, 2002). Of all metallic causes of pneumoconiosis, increased risk of mortality is seen in the coal or metal mining industry and in the production of glass and glass products. Occupations at a particular high risk are mining machine operators, welders, and electricians (DHHS, 2007). The diagnosis of pneumoconiosis is suggested when a worker complains of cough, shortness of breath, and activity intolerance with a corresponding history of an occupational exposure. The diagnosis is supported by chest X-ray findings of interstitial fibrosis with diffuse small nodular and reticular patterns and high resolution CT scan revealing bilateral ground glass or consolidation, reticular opacification, traction bronchiectasis, and honeycombing in advanced disease. Pulmonary function tests invariably showed a restrictive pattern with a decreased diffusion capacity (Chong et al., 2006; Gotway et al., 2002). In a worker exposed to hard metal (tungsten carbide particles in a matrix of cobalt metal), the presence of multinucleated giant cells on lung biopsy is pathognomonic for hard metal lung disease (Ohori et al., 1989).

Occupational exposure to other metallic agents may lead to the radiographic appearance of pneumoconiosis without evidence of restrictive lung disease or a diminished diffusion capacity on PFTs or interstitial fibrosis on CT scan of the chest. Potkonjak (Potkonjak and Pavlovich, 1983) described 51 subjects who worked in a smelting plant and were exposed to dust containing predominately antimony oxide. The presence of diffuse, densely distributed punctuate opacities <1 mm, round and irregular in shape characterized the X-ray findings in what was termed "antimoniosis" following exposure to antimony oxide. The clinical manifestations included productive and non-productive cough, upper airway inflammation, wheezing, chest tightness, and shortness of breath. No characteristic pulmonary function abnormalities were identified. Other benign, non-fibrotic causes of pneumoconiosis with distinctive X-ray findings with or without bronchial irritation and no evidence of a ventilatory or diffusion defect on PFTs include tin (stannosis) (Robertson et al., 1961), barium (baritosis) (Doing, 1976), and iron (siderosis) (Nemery, 1990).

There are a few case reports of symptomatic respiratory disease, abnormal pulmonary function with associated interstitial fibrosis in welders exposed to iron (McCormick et al., 2008) and in tinners exposed to tin fumes (Yilmaz et al., 2009). In an attempt to explain how pulmonary fibrosis and restrictive lung disease can develop in those exposed to the benign non-fibrotic causes of pneumoconiosis, it is often suggested that there is cross contamination resulting from non-welding inhalation exposures to mixed dusts (silica, asbestos, coal) termed mixed dust pneumoconiosis (Antonini et al., 2003). A number of studies examining the histology of lung tissue of symptomatic welders have identified a large amount of iron deposits in the fibrotic areas of the lung without evidence of coexisting silicosis or other mixed dusts supporting the possibility for welders exposed to iron to develop interstitial lung disease (Funahashi et al., 1988; Rosler and Woitowitz, 1996).

ACUTE PNEUMONITIS

Acute pneumonitis, acute inhalation pneumonitis, chemical pneumonitis, toxic pneumonitis, chemical or toxic pneumonia, and toxic inhalation syndrome are terms used to describe the development of non-cardiogenic pulmonary edema following an exposure to solvents, welding fumes, grain and fertilizer dusts, smoke, or highly soluble irritants, including chlorine and ammonia. Irrespective of the cause, the physiologic consequence of these exposures results in ALI. Acute lung injury is characterized by the direct toxic effect on lung cells and the indirect acute systemic response; increased permeability that is associated with a collection of clinical, radiographic, and physiologic abnormalities that cannot solely be explained by abnormal heart function or by pulmonary hypertension (Bernard et al., 1994). In its severe form, ALI can progress to adult respiratory distress syndrome (ARDS) where progressive hypoxia, decreased lung compliance, diffuse alveolar damage, and pulmonary fibrosis develop. Table 40.8 describes the recommended criteria for distinguishing ALI and ARDS.

The development of acute chemical pneumonitis has been described following the inhalation of metal fumes containing cadmium and manganese, exposure to mercury vapor or vanadium pentoxide as a byproduct of oil burning furnaces, nickel carbonyl exposure in a waste treatment factory and in military personnel exposed to zinc chloride smoke during combat exercises (Barceloux, 1999; Cooper, 2007; Fernandez et al., 1996; Lilis et al., 1985; Milne et al., 1970; Nemery,

TABLE 40.8 Definition and Characteristics of ALI and ARDS

Onset
Always Acute (ALI and ARDS)
P/F Ratio (measure of oxygenation)
$PaO_2/FiO_2 \geq 200$ mmHg ≤ 300 mm $\Rightarrow$ *Acute lung injury*
$PaO_2/FiO_2 \leq 200$ mmHg $\Rightarrow$ *Adult respiratory distress syndrome*
Chest X-Ray
Bilateral infiltrates without pleural effusions (ALI and ARDS)
Pulmonary Capillary Wedge Pressure
≤ 18 mmHg when measured (ALI and ARDS)
No clinical evidence of left atrial hypertension (ALI and ARDS)

1990; Seet et al., 2005; Zerahn et al., 1999). In most of these cases, patients present acutely with mild symptoms characterized by fever, chills, shortness of breath, and cough similar to those found in MFF. However, their clinical course was complicated by the development of diffuse alveolar infiltrates or non-cardiogenic pulmonary edema on chest X-ray, progressive hypoxia, increasing shortness of breath, and sometimes pneumothorax or fatal hemorrhagic pulmonary edema. In those that survive mercury or zinc chloride exposure, restrictive lung disease and diffusion abnormalities persist despite treatment (Lilis et al., 1985; Zerahn et al., 1999). The pathologic features on post-mortem examination of the lungs in those who do not survive reveal evidence of diffuse alveolar damage, hyaline membranes, and the early stages of intra-alveolar fibrosis consistent with ARDS (Milne et al., 1970; Seet et al., 2005). There are several case reports of diffuse alveolar damage following inhalation of zinc oxide fumes and a description of a severe form of MFF following inhalation of zinc chloride fumes (Barbee and Prince, 1999; Blount, 1990; Bydash et al., 2010; Taniguchi et al., 2003). Although initially patients present with symptoms of MFF, a self-limiting illness, symptoms uniformly progress in the days following the exposure. Hypoxia, radiographic evidence of alveolar injury and pathologic evidence of ALI and ARDS become evident. These findings are not consistent with MFF caused by exposure to zinc oxide and suggest an alternative physiologic mechanism.

METAL-INDUCED LUNG CANCER

Occupational inhalation exposures to several metals are recognized to cause cancer. The International Agency for Research on Cancer (IARC) have classified aluminum, arsenic, beryllium, cadmium, chromium, and nickel as Group 1 known human carcinogens (Straif et al., 2009). These metals along with the occupation and types of cancer they cause are listed in Table 40.9. Cobalt with tungsten carbide is the only

TABLE 40.9   Metal Associated Malignancies

International Agency for Research on Cancer (IARC) Group 1 Metals for Which There is Sufficient Evidence for Cancer in Humans		
Metal Exposure	Occupation	Malignancy
Aluminum compounds	Smelting, mining, construction, manufacturing, explosive and paper industries	Lung, bladder, pancreas, lymphosarcoma/ reticulosarcoma, leukemia
Arsenic and arsenic compounds	Metal smelting, coal production and burning, production and use as wood preservative, glass and semiconductor manufacture	Lung, skin, bladder
Beryllium and beryllium compounds	Mining, electronic, nuclear and aerospace industries	Lung
Cadmium and cadmium compounds	Mining and ore processing, smelting, electroplating, spraying of paints with cadmium pigment, welding, smelting, nickel cadmium battery production	Lung
Chromium VI (hexavalent) compounds	Electroplating, welding, cement production, metallurgical production	Lung, paranasal sinuses
Nickel compounds	Refineries, smelting, welding, electroplating, production of jet engine parts, nickel cadmium battery production	Lung, paranasal sinuses, and nasal cavity
International Agency for Research on Cancer (IARC) Group 2A Metals Probably Carcinogenic in Humans		
Metal Exposure	*Occupation*	*Malignancy*
Cobalt with tungsten carbide*	Aircraft engine production, steel applications, diamond grinding tools, refining and production of alloys in hard metal industry	Current epidemiological evidence suggests an increase lung cancer risk** but falls short of providing convincing evidence for a carcinogenic effect of exposure to cobalt with tungsten carbide in humans

*IARC Monographs 86, 2006.
**Moulin JJ, Wild P, Romazini S, et al. Lung cancer risk in hard-metal workers. Am J Epidmiol 1998;148:241–248.
**Hogstedt D, Alexandersson R. Mortality among hard metalworkers. Arbete Halsa 1990;21:1–26.

metal found in Group 2A: a probable carcinogen in humans (IARC, 2006; Moulin et al., 1998; Hogstedt and Alexandersson, 1990).

CHRONIC OBSTRUCTIVE LUNG DISEASE OR EMPHYSEMA

While cigarette smoking remains the most common cause of airflow limitation, occupational exposure to organic and inorganic dusts, inhaled particulates, and chemicals (vapors, irritants, and fumes) make up an underappreciated risk factor for COPD (Global Initiative for Chronic Obstructive Lung Disease, 2014). The NHANES III survey (Hnizdo et al., 2002) of almost 10,000 US workers aged 30–75 estimated the fraction of COPD attributable to work was 19.2% overall and 31.1% among those who never smoked. After adjusting for age, smoking, body mass index, education, and socio-economic status they found an increased risk for COPD in the following industries: rubber, plastics, and leather manufacturing; utilities; office-building services; textile mill products manufacturing; agriculture, construction, transportation, and trucking; repair services and gas stations. Occupations associated with increased odds ratios for COPD included freight,

stock, and material handlers; records processing and distribution clerks, sales; transportation-related occupations; machine operators; construction trades; and waitresses. Additionally, exposure to metal gases, aluminum production, processing, and welding is also associated with chronic airway obstruction (Meldrum et al., 2005). In a cohort of over 9000 U.S. workers at seven beryllium-processing plants, Schubauer-Berigan (Schubauer-Berigan et al., 2011) found a clear exposure–response association between beryllium exposure and COPD. Emphysema was also reported among survivors from a foundry manufacturing copper–cadmium alloy. Those with the highest exposures recorded a 398 ml drop in FEV1 and a significant reduction in the diffusing capacity (Burge, 1994). As growing evidence suggests that specific industries and occupations are associated with the risk of developing obstructive lung disease, occupational exposure becomes significantly more important as a cause of COPD: independent of exposure to tobacco smoke.

REFERENCES

Ahsan, S.A., Lackovic, M., Katner, A., and Palermo, C. (2009) Metal fume fever: a review of the literature and cases reported to the Louisiana Poison Control Center. *J. La. State Med. Soc.* 161:348–335.

Ameille, J., Brechot, J.M., Brochard, P., et al. (1992) Occupational hypersensitivity pneumonitis in a smelter exposed to zinc fumes. *Chest* 101:862–863.

Antonini, J.M., Lewis, A.B., Roberts, J.R., and Whaley, D.A. (2003) Pulmonary effects of welding fumes: review of the worker and experimental animal studies. *Am. J. Ind. Med.* 43:350–360.

Bakerly, N.D., Moore, V.C., Vellore, A.D., et al. (2008) Fifteen-year trends in occupational asthma: data from the shield surveillance scheme. *Occup. Med.* 58:169–174.

Barbee, J.Y. and Prince, T.S. (1999) Acute respiratory distress syndrome in a welder exposed to metal fumes. *South. Med. J.* 92(5):510–512.

Barceloux, D.G. (1999) Nickel. *Clin. Toxicol.* 37(2):239–258.

Barrera, L., Mendoza, F., Zuniga, J., et al. (2008) Functional diversity of T-cell subpopulations in subacute and chronic hypersensitivity pneumonitis. *Am. J. Respir. Crit. Care Med.* 177(1):44–55.

Bernard, G.R., Artigas, A., Brigham, K.L., et al. (1994) The American-European Consensus Conference on ARDS: Definitions, mechanisms, relevant outcomes, and clinical trial coordination. *Am. J. Respir. Crit. Care Med.* 149:818–824.

Bessac, B.F. and Jordt, S.E. (2008) Breathtaking TRP channels: TRPA1 and PRPV1 in airway chemosensation and reflex control. *Physiology* 23:360–370.

Blanc, P., Wong, H., Berstein, M.S., and Boushey, H.A. (1991) An experimental model of metal fume fever. *Ann. Intern. med.* 114:930–936.

Blanc, P.D., Boushey, H.A., Wong, H., et al. (1993) Cytokines in metal fume fever. *Am. Rev. Respir. Dis.* 147:134–138.

Blanc, P.D. (2007) Inhalation Fever. In: Rom W.N., editor., *Environmental and Occupational Medicine*, 4th ed., Chapter 25.

Blanc, P.D. (1997) Inhalation fevers. *Pulmonary and critical care update*, 12 (Lesson 1), Available at http://www.chestnet.org/educatoin/online/pccu/vol12/lesson01-02.index.php.

Blount, B.W. (1990) Two types of metal fume fever: mild vs. serious. *Mil. Med.* 155(8):372–377.

Borak, J., Cohen, H., and Hethmon, T.A. (2000) Copper exposure and metal fume fever: lack of evidence for a causal relationship. *AIHAJ* 61:832–836.

Brooks, S.M., Truncale, T., and McCluskey, J. (2007) Occupational and Environmental Asthma. In: Rom, W.N., editor., *Environmental and Occupational Medicine*, 4th ed., Chapter 26.

Brown, J.J. (1988) Zinc fume fever. *Br. J. Radiol.* 61:327–329.

Burge, P.S. (1994) Editorial: occupational and chronic obstructive pulmonary disease (COPD). *Eur. Respir. J.* 7:1032–1034.

Bydash, J., Kasmani, R., and Naraharisetty, K. (2010) Metal fume-induced diffuse alveolar damage. *J. Thorac. Imaging* 25:27–29.

Cai, H.R., CAO, M., Meng, F.Q., and Wei, J.Y. (2007) Pulmonary sarcoidlike granulomatosis induced by aluminum dust: report of a case and literature review. *Chin. Med. j.* 120(17):1556–1560.

Calverley, A.E., Rees, D., Dowdeswell, R.J., et al. (1995) Platinum salt sensitivity in refinery workers: incidence and effects of smoking and exposure. *Occup. Environ. Med.* 52(10):661–666.

Chan-Yeung, M. (1986) Occupational asthma. *Clin. Rev. Allergy* 4:25–256.

Chan-Yeung, M. (1990) Occupational asthma. *Chest* 98:148s–161s.

Chen, W.J., Monnat, R.J., Chen, M., and Mottet, N.K. (1978) Aluminum induced pulmonary granulomatosis. *Hum. Pathol.* 9(6):705–711.

Chong, S., Lee, K.S., Chung, M.F., et al. (2006) Pneumoconiosis: comparison of imaging and pathologic findings. *Radiographics* 26:59–77.

Cooper, R.G. (2007) Vanadium pentoxide inhalation. *Indian J. Occup. Environ. Med.* 11(3):97–102.

Cromwell, O., Pepys, J., Parish, E., and Hughes, G. (1979) Specific IgE antibodies to platinum salts in sensitized workers. *Clin. Allergy* 9:109–117.

Daenen, M., Rogiers, P.H., Van de Walle, et al. (1999) Occupational asthma caused by palladium. *Eur. Respir. J.* 13(1):213–216.

DeVuyst, P., Dumortier, P., Schandene, L., et al. (1987) Sarcoidlike lung granulomatosis induced by aluminum dusts. *Am. Rev. Respir. Dis.* 135(2):493–497.

DHHS (2002) *Work Related Lung Disease Surveillance Report*, (NIOSH) publication no. 2003-111, Available at http://www.cdc.gov/niosh/docs/2008-143/pdfs/2008-143.pdf.

Doing, A.T. (1976) Baritosis: a benign pneumoconiosis. *Thorax* 31(1):30–39.

Drinker, P., Thompson, R.M., and Finn, J.L. (1927) Metal fume fever III. The effects of inhaling magnesium oxide fume. *J. Ind. Hyg.* 9:187–192.

El-Zein, M., Malo, J.L., Infante-Rivard, C., and Gautrin, D. (2003) Incidence of probable occupational asthma and changes in airway caliber and responsiveness in apprentice welders. *Eur. Respir. J.* 22(3):513–518.

Farrell, F. (1987) Angioedema and urticaria as acute and late phase reactions to zinc fume exposure with associated metal fume fever-like symptoms. *Am. J. Ind. Med.* 12:331–337.

Fernandez, M.A., Sanz, P., Palomar, M., et al. (1996) Fatal chemical pneumonitis due to cadmium fumes. *Occup. Med.* 46(5): 373–374.

Fontenot, A.P. and Maier, L.A. (2005) Genetic susceptibility and immune-mediated destruction in beryllium induced disease. *Trends Immunol.* 26(10):543–549.

Funahashi, A., Schleuter, D.P., Pintar, K., et al. (1988) Welders' pneumoconiosis: tissue elemental microanalysis by energy dispersive X-ray analysis. *Br. J. Ind. Med.* 45:14–19.

Geppetti, P., Patacchini, R., Nassini, R., and Materazzi, S. (2010) Cough: the emerging role of the TRPA1 channel. *Lung* 188 (Suppl. 1):S63–S68.

Global Initiative for Chronic Obstructive Lung Disease (2014) Retrieved from http:www.goldcopd.org/uploads.users/files/GOLD-Report-2014-Jan23.pdf.

Gotway, M.B., Golden, J.A., Warnock, M., et al. (2002) Hard metal interstitial lung disease: high resolution computed tomography appearance. *J. Thorac. Imaging* 17:314–318.

Gu, Q. and Ruei-Lung, L. (2010) Heavy metals zinc, cadmium, and copper stimulate pulmonary sensory neurons via direct activation of TRPA1. *J. Appl. Physiol.* 108:891–897.

Gupta, A. and Rosenman, K.D. (2006) Hypersensitivity pneumonitis due to metal working fluids: sporadic or under reported? *Am. J. Ind. Med.* 49:423–433.

Hayhurst E. (1911) Report of Commission on Occupational Diseases. Available at https://openlibrary.org/books/OL24152503M/Report_of_Commission_on_occupational_diseases_to_His_Excellency_Governor_Charles_S._Deneen._January_.

Hnizdo, E., Sullivan, P.A., Bang, K.M., and Wagner, G. (2002) Association between chronic obstructive lung disease and employment by industry and occupation in the US population: a study of data from the Third National Health and Nutrition Examination Survey. *Am. J. Epidemiol.* 156:738–746.

Hogstedt, D. and Alexandersson, R. (1990) Mortality among hard metalworkers. *Arbete Halsa* 21:1–26.

Hull, M.J. and Abraham, J.L. (2002) Aluminum welding fume-induced pneumoconiosis. *Hum. Pathol.* 33(8):819–825.

IARC (2006) *Monographs.* 86: 24–30.

Janson, C., Anto, J., Burney, P., et al. (2001) The European community respiratory health survey: what are the main results so far? European community respiratory health survey II. *Eur. Respir. J.* 18(3):598–611.

Jin-Chyuan, J.L., Kuang-Hung, H., and Wu-Shium, S. (2009) Inflammatory responses and oxidative stress from metal fume exposure in automobile welders. *J Occup. Environ. Med.* 51(1):95–103.

Karjalainen, A., Kurppa, K., Martikainen, R., et al. (2002) Exploration of asthma risk by occupation-extended analysis of an incidence study of the Finnish population. *Scan. J. Work Environ. Health* 28(1):49–57.

Kaye, P., Young, H., and O'Sullivan, I. (2002) Metal fume fever; a case report and review of the literature. *Emerg. Med. J.* 19:268–269.

Kim, Y.H., Fazlollahi, F., Kennedy, I.M., et al. (2010) Alveolar epithelial cell injury due to zinc oxide nanoparticle exposure. *Am. J. Respir. Crit. Care Med.* 182:1398–1409.

Krantz, A. and Dorevitch, S. (2004) Metal exposure and common chronic diseases: a guide for the clinician. *Dis. Mon.* 50:215–262.

Kusaka, Y., Kazuhiro, S., Suganuma, N., and Hosoda, Y. (2001) Metal-induced lung disease: lessons from Japan's experience. *J. Occup. Health* 43:1–23.

Kuschner, W.G., D'Alessandro, A., Hambleton, J., and Blanc, P.D. (1998) Tumor necrosis factor-alpha and interleukin-8 release from U937 human mononuclear cells exposed to zinc oxide *In Vitro*: mechanistic implications for metal fume fever. *J. Occup. Environ. Med.* 40(5):454–459.

Kuschner, W.G., Wong, H., D'Alessandra, A., et al. (1997) Human pulmonary responses to experimental inhalation of high concentration fine and ultrafine magnesium oxide particles. *Environ. Health Perspect.* 105:1234–1237.

Kuschner, W.G., D'Alessandro, Wong, H., and Blanc, P.D. (1997) Early pulmonary cytokine responses to zinc oxide fume inhalation. *Environ. Res.* 75:7–11.

Leroyer, C., Dewitte, J.D., Bassanets, A., et al. (1998) Occupational asthma due to chromium. *Respiration* 65(5):403–405.

Lilis, R., Miller A, and Lerman, Y. (1985) Acute mercury poisoning with severe chronic pulmonary manifestations. *Chest* 88(2):306–309.

Lindahl, M., Leanderson, P., and Tagesson, C. (1998) Novel aspect on metal fume fever: zinc stimulates oxygen radical formation in human neutrophils. *Hum. Exp. Toxicol.* 17:105–110.

Malo, J.L., Cartier, A., Doepner, M., et al. (1982) Occupational asthma caused by nickel sulfate. *J. Allergy Clin. Immunol.* 69(1):55–59.

Malo, J.L., Cartier, A., and Dolovich, J. (1993) Occupational asthma due to zinc. *Eur. Respir. J.* 6(3):447–450.

Malo, J.L. and Cartier, A. (1987) Occupational asthma due to fumes of galvanized metal. *Chest* 92(2):375–376.

Malo, J.L. and Chan-Yeung, M. (2009) Agents causing occupational asthma. *J. Allergy Clin. Immunol.* 123(3):545–550.

Malo, J.L. (2005) Editorial: occupational rhinitis and asthma due to metal salts. *Allergy* 60:138–139.

McCormick, L.M., Goddard, M., and Mahadeva, R. (2008) Pulmonary fibrosis secondary to siderosis causing symptomatic respiratory disease: a case report. *J. Med. Case Rep.* 2:257.

Meldrum, M., Rawbone, R., Curran, A.D., and Fishwick, D. (2005) The role of occupation in the development of chronic obstructive pulmonary disease (COPD). *Occup. Environ. Med.* 62(4):212–214.

Milne, J., Christophers, A., and De Silva, P. (1970) Acute mercurial pneumonitis. *Br. J. Ind. Med.* 27:334–338.

Mohr, L.C. (2004) Hypersensitivity Pneumonitis. *Curr. Opin. Pulm. Med.* 10(5):401–411.

Moulin, J.J., Wild, P., Romazini, S., et al. (1998) Lung cancer risk in hard-metal workers. *Am. J. Epidmiol.* 148:241–248.

Nemery, B. (1990) Metal toxicity and the respiratory tract. *Eur. Respir. J.* 3(2):202–219.

Newman, L.S. (1998) Metals that cause sarcoidosis. *Semin. Respir. Infect.* 13(3):212–220.

Novey, H.S., Habib, M., and Wells, I.D. (1983) Asthma and IgE antibodies induced by chromium and nickel salts. *J. Allergy Clin. Immunol.* 72(4):407–412.

Ohori, N.P., Sciurba, F.C., Owens, G.R., et al. (1989) Giant-cell interstitial pneumonia and hard-metal pneumoconiosis. A clinicopathologic study of four cases and review of the literature. *Am. J. Surg. Pathol.* 13(7):581–587.

Potkonjak, V. and Pavlovich, M. (1983) Antimoniosis: a particular form of pneumoconiosis. *Int. Arch. Occup. Environ. Health* 51:199–207.

Rask-Anderson, A. (1992) Inhalation fever: a proposed unifying term for febrile reactions to inhalation of noxious substances. *Br. J. Ind. Med.* 29:296.

Robertson, A.J., Rivers, D., Nagelschmidt, G., and Duncumb, P. (1961) Stannosis: benign pneumoconiosis due to tin dioxide. *Lancet* 1(7186):1089–1093.

Rohrs, L.C. (1957) Metal fume fever from inhaling zinc oxide. *AMA Arch. Intern. Med.* 100. (1):44–49.

Rose, F C.S. and Lara, F A.R. (2010) Hypersensitivity Pneumonitis. In: *Mason: Murray and Nadel's Textbook of Respiratory Medicine*, 5th ed., Chapter 26. Philadelphia: W.B.Saunders.

Rosler, J.A. and Woitowitz, H.J. (1996) *Eur. Respir. J. Suppl.* 9:220S.

Saltini, C., Amicosante, M., Franchi, A., et al. (1998) Immunogentic basis of environmental lung disease: lessons from the berylliosis model. *Eur. Respir. J.* 12(6):1463–1475.

Schubauer-Berigan, M.K., Couch, J.R., Petersen, M.R., et al. (2011) Cohort mortality study of workers at seven beryllium processing plants; update and associations with cumulative and maximum exposure. *Occup. Environ. Med.* 68:345–353.

Seet, R.C.S., Johan, A., Teo, C.E.S., et al. (2005) Inhalational nickel carbonyl poisoning in waste processing workers. *Chest* 128:424–429.

Shelton, B.G., Flanders, D., and Morris, G.K. (1999) Mycobacterium sp. as a possible cause of hypersensitivity pneumonitis in machine workers. *Emerg. Infect. Dis.* 5(2): 270–273.

Shirakawa, T., Kusaka, Y., Fujimura, N., et al. (1989) Occupational asthma from cobalt sensitivity in workers exposed to hard metal dust. *Chest* 95:29–37.

Straif, K., Benbrahim-Tallaa, L., Baan, R., et al. (2009) A Review of human carcinogens-part C: metals, arsenic, dusts, and fibres. *Lancet Oncol.* 10:453–454.

Sturgis, C.C., Drinker, P., and Thompson, R.M. (1927) Metal fume fever. *J. Ind. Hyg.* 9:88–97.

Taniguchi, H., Suzuki, K., Fujisaka, S., et al. (2003) Diffuse alveolar damage after inhalation of zinc oxide. *Nihon Kikyuki Gakki Zasshi* 41(7):447–450.

Tarlo, S.M., Balmes, J., Balkissoon, F B., et al. (2008) Occupational asthma: definitions, epidemiology, causes, and risk factors. *Chest* 134(Suppl. 3):1S.

Yilmaz, A., Gocmen, O.S., and Doruk, A.B. (2009) Is tin fume exposure benign or not? Two case reports. *Tuberk Toraks* 57(4):422–426.

Zerahn, B., Kofoed-Envoldsen, A., Jensen, B.V., et al. (1999) Pulmonary damage after modest exposure to zinc chloride smoke. *Respir. Med.* 93:885–890.

SECTION III

CHEMICAL COMPOUNDS I

SECTION EDITOR: RAYMOND D. HARBISON

41

INTRODUCTION

RAYMOND D. HARBISON

In a fundamental sense, chemicals are essential for human existence. Everything that is not a form of energy, such as sunlight or sound, comprises a chemical or chemicals. Although this is true, it has long been recognized that occupational overexposure to some chemical agents may cause illness and disease. Sections III and IV of this book address the major chemical compounds considered relevant to occupational exposure and disease. Although an effort has been made to cover each chemical or group of chemicals representing an occupational hazard, these sections should not be considered all inclusive. Some chemical compounds are discussed in more specific sections, such as Metals and Metalloids, Pesticides, and Organic High Polymers.

Great advances have been made in understanding the effects of chemical exposures in the recent past. Both industrial experience and animal testing have provided a more complete understanding of the risk of occupational exposure to various chemicals. This improved understanding, in turn, has allowed the development of exposure guidelines and better safety programs for the workplace.

In the field of occupational medicine, attention to chemical agents has shifted from acute overexposure to chronic exposure and illness. Although acute exposure still occurs and must be controlled, it is much less common than during Alice Hamilton's day in the early part of this century. An occupational physician is more likely to be confronted with cancer or a chronic illness that may be work related. As is discussed in the chapter, Diagnosis of Occupational Disease (Chapter 1), these illnesses are much more ambiguous than an acute overexposure. Association with actual occupational exposure is complicated and often inconclusive.

Although our knowledge of the effects of occupational exposure to chemical agents has improved, it is important to remember that thousands of new compounds are produced each year, some of which are destined for widespread industrial use and potential occupational exposure. Thus the workplace must be continually evaluated to identify potential new chemical hazards and to reevaluate existing ones.

While reading the following chapters, a constant reminder is proposed: Exposure is only the opportunity for contact with a chemical; the dose determines the effect or lack of effect. Many factors can affect exposure and, ultimately, the dose of chemical received. As Emil Mrak said, "There are no harmless substances, only harmless ways of using substances."

Hamilton & Hardy's Industrial Toxicology, Sixth Edition. Edited by Raymond D. Harbison, Marie M. Bourgeois, and Giffe T. Johnson.
© 2015 John Wiley & Sons, Inc. Published 2015 by John Wiley & Sons, Inc.

42

CARBON DIOXIDE

Debra J. Price

Target Organ(s)	Respiratory and Cardiovascular System
Occupational exposure limits	OSHA: PEL: 5000 ppm TWA
	NIOSH: REL: 5000 ppm TWA
	STEL: 30,000 ppm TWA
	IDLH: 40,000 ppm
Risk/safety phrases	Dusts of various metals are ignitable and explosive when suspended in carbon dioxide. Forms carbonic acid in water

BACKGROUND AND USES

Carbon dioxide is a colorless, odorless gas at low concentrations; but at higher concentrations, it has a sharp, acidic odor. It is a physiologically important gas produced by the body as a result of cellular metabolism. Carbon dioxide's main mode of action is as an asphyxiate, although it also exerts toxic effects at the cellular level. Low concentrations up to 1% (10,000 parts per million (ppm)) CO2 have little toxicological effects. However, exposure to 2% concentration for several hours can produce headache, increase in blood pressure, and deep respiration, and exposure to 9–10% CO_2 can cause unconsciousness in about 5 min (Patnaik, 1999).

It is widely used in carbonation of beverages, in fire extinguishers, in the manufacture of carbonates, as dry ice for refrigeration, and as a propellant for aerosols. It is a normal constituent of the atmosphere and occurs naturally at a concentration of approximately 0.03% by volume. It is part of the carbon cycle known as photosynthesis and is used by plants, algae, and cyanobacteria which absorb CO_2, light, and water to produce energy for themselves and oxygen as a waster product. It is also produced by the combustion of coal, in the fermentation of liquids, and during exhalation in animals. It is expelled into the atmosphere via springs, wells, and volcanoes. In the atmosphere, it acts as a greenhouse gas and is a source of ocean acidification, dissolving in water to form carbonic acid.

PHYSICAL AND CHEMICAL PROPERTIES

Carbon dioxide (carbonic acid gas, carbonic anhydride, carbonic oxide, dry ice (solid phase); CO_2; CASRN 124-38-9) has a density of 1.527 (air = 1), freezes at −78.5 °C, is soluble in water at 20 °C, is less soluble in alcohol and ether, and is absorbed by alkalis to form carbonates.

MAMMALIAN TOXICOLOGY

Acute Effects

There are no obvious symptoms of toxicity if the CO_2 concentration is below 3%. Carbon dioxide concentrations above 3% for a 30-min exposure can cause dyspnea and an acute headache caused by CO_2-induced vasodilation of meningeal vessels or CO_2-induced acidosis. Hearing and vision impairments may occur within several minutes of 6% CO_2 exposure. With higher concentrations of 15–20% CO_2 in inspired air, respiratory distress is rapid, causing loss of consciousness and spasms of muscle twitching, even if sufficient oxygen is present to prevent hypoxia (Johnston, 1959). A total of 20–30% CO_2 in inspired air causes

Hamilton & Hardy's Industrial Toxicology, Sixth Edition. Edited by Raymond D. Harbison, Marie M. Bourgeois, and Giffe T. Johnson.
© 2015 John Wiley & Sons, Inc. Published 2015 by John Wiley & Sons, Inc.

convulsions within 1–3 min after the onset of exposure, with death occurring after severe acidosis (Marquardt et al., 1999).

Studies of health effects of high levels of CO_2 have made CO_2 poisoning accidents a rare event. However, in earlier years, there were CO_2 poisoning events that occurred. Troisi (1957) reported deaths among workers who entered silos containing green fodder. Another case of CO_2 poisoning was reported by Lillevik and Geddes (1943), where a man was asphyxiated by a 10-min exposure to CO_2 in a grain elevator where 4700 bushels of flaxseed had been stored for 58 days. The man was rescued at once and promptly treated with oxygen inhalation but died within 48 h. The air just over the flaxseed bin had an oxygen content of 1.8% and a CO_2 content of 11.1%; within the bin, the oxygen content was 0.4% and the CO_2 content, 12.6%. Bacteria and saprophytic fungi were believed to be responsible for the abnormally high respiratory activity of the flaxseed. Residues in brewery vats have caused cases of fatal poisoning in the United States and in England; the manufacture of compressed yeast caused six cases of poisoning in a German factory, as reported by Brezina (Hamilton and Hardy, 1949).

Carbon dioxide poisoning also has been reported aboard ships. Dalgaard et al. (1972) reviewed literature dealing with cases of unconsciousness and death in the holds of industrial fishing ships containing trash fish and fish meal. High CO_2 levels (up to 22%) and low O_2 concentrations were blamed although hydrogen sulfide (H_2S) may also have been a factor. In 1958 Williams described a fatal incident aboard a ship with a cargo of onions and crude brown sugar (Williams, 1958). The two dead men had cyanosis of the oral mucosa and nail beds and markedly congested conjunctivae. All those who survived had headache, giddiness, tinnitus, and weakness, signs and symptoms considered consistent with CO_2 poisoning. Another incident involved removal of the cargo from the wrecked liner *Celtic*, which lay for a year at the entrance to Cork Harbor, fastened to the rocks, so that it could not be floated off. While working in one of the holds, 19 men were overcome by poisonous fumes and four died. Since the fumes apparently were generated by rotting apples, it was determined that the gas was carbon dioxide. Similar accidents have been reported when dry ice has been used in large quantities as a refrigerant. An additional hazard of dry ice or carbon dioxide snow is injury to the skin from improper handling.

Medical uses of carbon dioxide have been known to cause CO_2 toxicity as an unusual complication. Carbon dioxide is used during laparoscopic procedures in the abdominal cavity to allow visualization of the peritoneal cavity. A case was reported of a patient with delayed CO_2 elimination that led to signs of toxicity (Rittenmeyer, 1994).

A natural disaster in 1986 emphasized the toxicity of carbon dioxide in high concentrations. A massive release of CO_2 from Lake Nyos, Cameroon, a volcanic crater, killed 1700 people and injured 845. Clinical findings were compatible with an asphyxiant rather than an irritant gas cause. Both survivors and the dead exhibited cutaneous erythema and bullae, which were likely to have occurred secondary to a prolonged coma state resulting from exposure to high levels of CO_2. Survivors also experienced headache, weakness, malaise, limb swelling, and cough (Baxter et al., 1989).

Several studies have been conducted to measure occupational exposure to air pollutants from motor vehicle exhaust, and carbon dioxide has been studied as one of the components of motor vehicle exhaust. In a study of 40 sites visited, Groves and Cain (2000), using personal and background exposure data, found that the principal gaseous components of diesel engine exhaust are CO_2, CO, and NO_x. None of the measurements for CO_2, CO, and NO_x exceeded the UK 8-h time-weighted average (TWA) occupational exposure limits of 5000 ppm CO_2, 50 ppm CO, or 3 ppm NO_x. The measurement of CO_2 may not represent exposure to diesel exhaust as a whole, but it may be an indication of the overall adequacy of control measures. An occupational exposure study by Wheatley and Sadhra (2004) measured nine diesel distribution depots engaged in the supply of drinks to the licensed trade. No smoking was allowed at any of the depots, thereby eliminating confounding sources of pollutants of interest. Personal sampling over 7–10 h measured levels of respirable dust, elemental and organic carbon, PAHs, ultrafine particles, and CO_2 and CO. Current guidance in the UK is that diesel exhaust fumes adverse health effects are unlikely if exposure to CO_2 emissions is less than 1000 ppm. Most levels of CO_2 measured at the nine depots were below 600 ppm. These studies have shown that the role of various diesel exhaust fractions in causing ill health is not yet clear.

Chronic Effects

Chronic poisoning is improbable under most conditions. Prolonged exposure to excessive concentrations is most likely in sealed quarters, such as submarines and spacecraft. Respiratory acclimatization to 3% carbon dioxide has been demonstrated by several investigators (Schaefer, 1963). Schaefer (1961) pointed out that although the health of submarine personnel exposed continually to 3% CO_2 was only slightly affected if the oxygen content of the air was maintained at a normal level, more studies in the U.S. Navy established the need to keep CO_2 concentrations below 1% for conditions of continuous, prolonged exposure. Impairment of performance was noted during prolonged exposure to 3% carbon dioxide even when the oxygen concentration was 21%. Prolonged exposure to 1.5% CO_2 resulted in respiratory acidosis that was not compensated until after the 24th day (Schaefer et al., 1963). Gibbons (1977) reviewed reports that show that actual or potential CO_2 poisoning from the use of carbon dioxide fire extinguishers, and significant loads of dry ice aboard pressurized aircraft could have serious consequences but that small amounts of dry ice in unpressurized planes probably were safe. Similar hazards could be associated with the use of CO_2 fire extinguishers aboard planes. Sevel and Freedman (1967) described cerebral and retinal degeneration in a patient who was in coma for 11 months after asphyxiation in a wellhead chamber during the dismantling of an artesian well in Central London. Earlier, these researchers had described neurological and ocular changes in two patients who survived CO_2 asphyxia in a similar incident (Freedman and Sevel, 1966). Specifically, they reported headache, photophobia, abnormalities of eye movements, constriction of peripheral visual fields, enlargement of blind spots, deficient dark adaptation, and personality changes, largely depression and irritability.

INDUSTRIAL HYGIENE

Carbon dioxide is mainly produced as an unrefined side product of the burning of fossil fuels, in the production of hydrogen by steam reforming, in ammonia synthesis, and fermentation. It is one of the gases found in mines, natural gas wells, silos, the holds of industrial fishing ships, and vats where fermentation has occurred. Miners traditionally have tested the air in shafts with candles that are extinguished when the oxygen level falls to 18% or the carbon dioxide level rises to 8–10%.

It is produced commercially in one of several ways. The combustion of all carbon-containing fuels will yield carbon dioxide, and in most cases, water. In the production of quicklime (CaO), limestone is heated at about 850 °C to produce CaO and CO_2. Iron is reduced with coke in a blast furnace to produce pig iron and CO_2. Yeast metabolizes sugar to produce ethanol and CO_2 in the production of beer, wine, and bioethanol. Finally, all aerobic organisms produce CO_2 in the oxidation of carbohydrates, fatty acids, and proteins.

It is widely used in industry and research laboratories. It is typically stored in cylinders and tanks under pressure but can also be stored at low temperatures in a solid form (dry ice). Dry ice is often used in the wine-making process to cool down bunches of grapes after picking to prevent spontaneous fermentation by wild yeast. Gaseous CO_2 is used in the textile, leather, and paint industries and in the manufacture of drugs. Because it is relatively inert, CO_2 is used in fire extinguishers, as an aerosol propellant, and as a pressure medium in food preservation, in pressurized gas systems, in portable pressure tools, and in purging pipelines and tanks. It is widely used in the production of carbonated soft drinks and soda water. The agricultural industry adds CO_2 to the atmosphere of greenhouses to sustain and increase plant growth.

The quality of indoor air has been a public health issue for over 30 years. Widespread reports of fatigue, headache, and mucous membrane complaints led to a syndrome designated as the sick building syndrome, a phenomenon of sealed buildings and poor ventilation. No single contaminant has been identified as responsible for the syndrome, but a CO_2 level below 0.1% (100 ppm) is the criterion for adequate ventilation (Mendell and Smith, 1990; NIOSH, 1994).

To protect the health of workers, the current tolerance limit is 0.5% of CO_2 for an 8-hr daily exposure. The National Institute for Occupational Safety and Health (NIOSH) has recommended a limit of 1% by volume, determined as a TWA concentration, for up to a 10-h work shift in a 40-h workweek, with a ceiling concentration of 3% by volume for a sampling period not to exceed 10 min (NIOSH, 1994). For higher concentrations of CO_2, respiratory protection is satisfactory only with a self-contained air supply respirator or a hose mask with a blower. The American Conference of Governmental Industrial Hygienists (ACGIH) has set the threshold limit value at 5000 ppm and the short-term exposure limit at 30,000 ppm (ACGIH, 1996).

MEDICAL MANAGEMENT

The treatment of overexposure to CO_2 consists of removing the patient from the area, with care being taken to ensure that the rescuers are not overexposed. Immediate administration of oxygen and cardiopulmonary life support are the basics of emergency care. The best management, however, involves primary prevention, based on proper ventilation and worker education.

REFERENCES

American Conference of Governmental Industrial Hygienists (ACGIH) (1996) *Threshold Limit Values (TLVs) for Chemical Substances and Physical Agents and Biological Exposure Indices (BEIs)*, Cincinnati, OH: American Conference of Governmental Industrial Hygienists.

Baxter, P.J., Kapila, M., and Mfonfu, D. (1989) Lake Nyos disaster, Cameroon, 1986: the medical effects of large scale emission of carbon dioxide. *BMJ* 298:1437–1441.

Dalgaard, J.B., et al. (1972) Fatal poisoning and other health hazards connected with industrial fishing. *Br. J. Ind. Med.* 29:307–316.

Freedman, A. and Sevel, D. (1966) The cerebro-ocular effects of carbon dioxide poisoning. *Arch. Ophthalmol.* 76:59–65.

Gibbons, H.L. (1977) Carbon dioxide hazards in general aviation. *Aviat. Space Environ. Med.* 48:261–263.

Groves, J. and Cain, J. (2000) A survey of exposure to diesel engine exhaust emissions in the workplace. *Ann. Occup. Hyg.* 44(6):435–447.

Hamilton, A. and Hardy, H.L. (1949) *Industrial Toxicology*, 2nd ed., New York, NY: Paul B Hoeber.

Johnston, R.F. (1959) The syndrome of carbon dioxide intoxication: its etiology, diagnosis, and treatment. *Med. Bull (Ann Arbour).* 25:280–292.

Lillevik, H.A. and Geddes, W.F. (1943) Investigation of a death by asphyxiation in a grain elevator bin containing flaxseed. *Cereal Chem.* 20:318–328.

Marquardt, H., et al. (1999) *Toxicology*, San Diego, CA: Academic Press.

Mendell, M.J. and Smith, A.S. (1990) Consistent pattern of elevated symptoms in air-conditioned office buildings: a reanalysis of epidemiologic studies. *Am. J. Public Health* 80:1193–1199.

National Institute for Occupational Safety and Health (NIOSH) (1994) *Pocket Guide to Chemical Hazards*, NIOSH 94–116, Washington, DC: US Department of Health and Human Services.

Patnaik, P. (1999) *A Comprehensive Guide to the Hazardous Properties of Chemical Substances*, 2nd ed., John Wiley & Sons, Inc., p. 371.

Rittenmeyer, H. (1994) Carbon dioxide toxicity related to laparoscopic procedure. *J. Post Anesth. Nurs.* 9:157–161.

Schaefer, K.E. (1961) Concept of triple tolerance limits based on chronic carbon dioxide toxicity studies. *Aerosp. Med.* 32:197–204.

Schaefer, K.E. (1963) Acclimatization to low concentrations of carbon dioxide. *Ind. Med. Surg.* 32:11–13.

Schaefer, K.E., et al. (1963) Respiratory acclimatization to carbon dioxide. *J. Appl. Physiol.* 18:1071–1078.

Sevel, D. and Freedman, A. (1967) Cerebro-retinal degeneration due to carbon dioxide poisoning. *Br. J. Ophthalmol.* 51:475–482.

Triosi, F.M. (1957) Delayed death caused by gassing in a silo containing green forage. *Br. J. Ind. Med.* 14:56–58.

Wheatley, A.D. and Sadhra, S. (2004) Occupational Exposure to Diesel Exhaust Fumes. *Ann. Occup. Hyg.* 48(4):369–376.

Williams, H.I. (1958) Carbon dioxide poisoning: report of eight cases with two deaths. *Br. Med. J.* 2:1012–1014.

43

CARBON MONOXIDE

Erin L. Pulster and James V. Hillman

Target organ(s): Lungs, blood, central nervous system, cardiovascular system

Occupational exposure limits: OSHA PEL: 50 ppm (55 mg/m^3) 8 h TWA, 200 ppm (229 mg/m^3) STEL; ACGHI TLV: 25 ppm (29 mg/m^3) TWA; NIOSH REL: 35 ppm (40 mg/m^3), 200 ppm STEL; NIOSH IDLH: 1200 ppm

Risk/safety phrases: Flammable gas, incompatible and reactive with strong oxidizers, bromine trifluoride, chlorine trifluoride, lithium

BACKGROUND AND USES

Excluding carbon dioxide, carbon monoxide (CO) is considered as one of the most abundant and widely distributed air pollutants in the lower atmosphere (Jaffe, 1970). The formation of CO can be the result of both biogenic and anthropogenic sources. The dominant source of CO in the northern midlatitudes is believed to be from fossil fuel combustion while biomass burning, methane oxidation, biogenic nonmethane hydrocarbons, and other volatile organic carbons are the dominant sources in the tropics (Duncan et al., 2007; Goldstein and Galbally, 2007; Logan et al., 1981). It has been estimated that CO in the atmosphere is produced at a rate of 5×10^{15} g by chemical oxidation, 25 times greater than from combustion sources and has a global average lifetime of approximately 2 months (Drori et al., 2012; Weinstock and Niki, 1972). Between 1988 and 1997, the overall estimated global budget for CO in the troposphere ranged between 305 and 350 Tg depending on the season (Duncan et al., 2007). The seasonal cycle and production of CO by methane

oxidation as well as the removal by carbon monoxide oxidation is associated with hydroxyl radicals (Weinstock and Niki, 1972; Drori et al., 2012). Biomass burning is widely believed to also contribute to the seasonal variations in carbon monoxide levels (Conway et al., 2002–2003). Oxidation of methane is the largest source of CO in the atmosphere and may produce 3 billion metric tons of CO annually in the northern hemisphere alone.

Tremendous amounts of carbon monoxide are released into the atmosphere as a result of human activities. In North America alone, the United States Environmental Protection Agency (USEPA) estimated a total carbon monoxide contribution of 89,170,000 short tons annually, of which 93% is due to anthropogenic sources and the remaining 7% is from biogenic sources (USEPA, 2013). Mobile emission sources account for 86% of the emission source sector (fuel combustion, industrial, fires, solvent, etc.) in the United States (USEPA, 2013) and between 55 and 60% of global man-made emissions of carbon monoxide. Between 1980 and 2012, the national U.S. average of atmospheric carbon monoxide was reduced by 83% (USEPA, 2013). This reduction has been aided by EPA regulations, for example, the installation of catalytic converters in all cars produced since 1975 as well as the 1992 addition of methyl tertiary-butyl ether to gasoline to improve its combustion properties and diminish CO emissions. The CO concentration of automobile exhaust, which once measured as high as 30%, typically is now <1.2%. Over the course of a decade (1988–1997), a slight decrease (2%) was observed in global emissions primarily due to the major North American and European declines being offset by the 51% increase in Asia as a result of rapid economic development (Duncan et al., 2007).

Hamilton & Hardy's Industrial Toxicology, Sixth Edition. Edited by Raymond D. Harbison, Marie M. Bourgeois, and Giffe T. Johnson.
© 2015 John Wiley & Sons, Inc. Published 2015 by John Wiley & Sons, Inc.

The sources of the carbon monoxide budget in the ocean is primarily from photochemical sources (42%), microbial sinks (27%) and microbial and gas exchange (32%) (Zafiriou et al., 2003). Early estimates suggest that a significant oceanic source of CO is generated by kelp, siphonophores (e.g., Portuguese Man of War; jellyfish), and deep-sea coelenterates (Jaffe, 1970; Chapman and Tocher, 1966; Pickwell, 1970). Other biogenic sources include volcanoes, forest and grass fires, marsh gases, electrical storms, terpene oxidation, and the destruction of chlorophyll in autumn. Endogenously produced carbon monoxide is through the normal catabolism of hemoglobin, which in the absence of environmental sources of CO accounts for a blood carboxyhemoglobin (COHb) level of about 0.4–0.7% in a healthy person.

Carbon monoxide emissions can also be produced from stationary sources (space heaters, water heaters, furnaces) and from industrial processes, coal mine explosions, and solid waste disposal procedures. Because most plastics contain carbon, CO is one of the primary gases generated by heating and burning plastics. Appreciable concentrations of CO can result from the pyrolysis of some vinyl plastics. Carbon monoxide plays a necessary part in the production of steel because it is used in the reduction process to take up oxygen. Additional anthropogenic sources of carbon monoxide include the manufacture of synthetic methanol and other organic compounds from CO, industrial and residential fires, charcoal burning, carbide manufacture, distillation of coal or wood, operations near furnaces and ovens, pyrolysis and oxidation of lubricants in air compressors, operation of snow-melting machines, testing of internal combustion engines, and indoor use of propane-powered machines such as forklifts, buffers, portable stoves, ice-resurfacing machines in indoor skating rinks, and internal combustion engines in general.

PHYSICAL AND CHEMICAL PROPERTIES

Carbon monoxide (CO; CASRN 630-08-0) is a nonirritating, odorless, colorless, tasteless gas with no warning properties. It has a vapor pressure >1 atm, and a vapor density of 0.968 (air = 1). Because it is only slightly less dense than air, it mixes readily without stratification. Carbon monoxide has a molecular weight of 28.01, a boiling point of −191.5 °C, and a solubility in water of 27.6 mg/l at 25 °C. Because of its low boiling point, carbon monoxide is shipped as a nonliquified compressed gas. It is also known as carbon oxide, flue gas, and monoxide. Carbon monoxide is a flammable gas and is incompatible or reactive with strong oxidizers, such as bromine trifluoride, chlorine trifluoride, and lithium. Additional chemical and physical properties are presented in Table 43.1.

TABLE 43.1   Carbon Monoxide Chemical and Physical Properties

Characteristic	Properties
Chemical name	Carbon monoxide
Synonyms	Carbon oxide, oxocarbon, oxomethylene, carbonic oxide, flue gas, carbon(II) oxide
CASRN	630-08-0
Molecular formula	CO
Molecular weight	28.01 g/mol
Boiling point	−191.5 °C (−312.7 °F)
Melting point	−205 °C (−337 °F)
Vapor pressure	35 atm (26,600 mmHg)
Density	789 kg/m³ liquid, 1.145 kg/m³ at the rate 25 °C, 1 atm
Ionization potential	14.01 eV
Solubility	27.6 mg/l (25%C) in water

MAMMALIAN TOXICOLOGY

Acute Effects

Toxic effects of CO are primarily caused by its ability to bind to heme, and alter the function and metabolism of proteins (Tian et al., 2013). More specifically, CO interferes with oxygen transport at the cellular level, possibly impairing normal cellular functions (Kealey, 2009; Hardy and Thom, 1994). Before considering the range of symptoms, which develop at various levels of CO, it must be understood that consciousness can be lost with few symptoms if the concentration of CO is high. It is also important to note that high concentrations may be rapidly fatal without producing significant warning symptoms (OSHA, 1978). Symptoms depend on the level and duration of exposure, as well as the overall health status and condition of the victim. Initial symptoms are very nonspecific and may be mistaken for flu-like symptoms, especially in children (Little, 1995). In order of increasing severity, the most commonly reported signs and symptoms are frontal headache, temporal headache, severe headache, roaring in the ears, weakness, dizziness, darkened vision, sleepiness, muscular weakness and incoordination, prolonged reaction time, impaired judgment, collapse, increased pulse and respiration, unconsciousness, involuntary evacuation, muscle contractions, coma, intermittent convulsions, cardiorespiratory depression, and death. As the level of absorbed CO rises, weakness and confusion increases and the victim may become indifferent to the danger and soothed to drowsiness. The majority of CO toxicity victims recover completely.

Acute poisoning has the potential for serious injury to major organs, brain, heart, kidney and muscles, yet it is not believed to accumulate. Tissue accumulation of CO was investigated in rats after acute exposures of CO air concentrations of 1, 0.4, and 0.12% for 4 min, 40 min, and 12 h, respectively (Sokal et al., 1984). After acute intoxication, CO levels did not appear to accumulate in either the heart or skeletal muscles and elimination of CO from the brain was similar to that of blood. Moreover, all the rats exposed for 4 min at 1% and 40 min at 0.4% were unconscious and did not survive higher CO concentrations.

Pulmonary edema is commonly found in victims who die as a result of acute exposure. Aspiration pneumonia, the result of potential vomiting while the victim is unconscious, is also often seen. Ischemia or infarction of brain and cardiac cells can occur, especially if underlying disease is a factor. The effects of CO exposure are the result of the tissue hypoxemia. This can also occur in skeletal and renal tissue producing rhabdomyolysis, compounding potential renal failure. Furthermore, a number of reports have described the delayed neuropsychiatric syndrome (DNS) or the delayed postanoxic encephalopathy (DPE), estimated to occur in 10–30% of CO toxicity victims from acute CO poisoning (Sawa et al., 1981; Choi, 1983; Min, 1986; Hart et al., 1988; Seger and Welch, 1994; Ernst and Zibrak, 1998; Kwon et al., 2004). Neurological complications have been reported to appear between 3 and 240 days after exposure, and 50–75% of the patients have had a full recovery within a year of onset. Symptoms include cognitive and personality changes, incontinence, dementia, psychosis, depression, concentration difficulties, and Parkinsonian movement. The mechanisms are not well understood, although theories include hypoxia, postischemic reperfusion injury, vascular endothelium effects and oxygen radical-mediated brain lipid peroxygenation (Thom, 1993).

Chronic Effects

Well-informed observers and research workers throughout the world have differing opinions as to whether chronic carbon monoxide poisoning exists. Some believe that chronic effects actually are the result of a slow accumulation of daily damage, which results in pathological change. Much more than in acute poisoning, confounding factors such as cigarette smoking, outdoor air pollution, and endogenous production of CO make the study of effects of chronic low level exposure particularly challenging. It is further complicated by the paucity of objective signs that can help the physician make the diagnosis unless an abnormal level of COHb can be detected in the blood.

Symptoms of chronic CO effects that are most commonly reported and observed include headache, irritability, insomnia, and personality disturbances (Table 43.2). Time-interval discrimination, vigilance, tracking, and the ability to drive a vehicle also are reported to be impaired. These manifestations are related to the effects of hypoxia on the nervous system. Lesions of the basal ganglia have been observed in both humans and experimental animals after CO exposure. Additional effects include an increase in the hematocrit, a

TABLE 43.2   Signs and Symptoms of Carbon Monoxide [CO] Exposure

Inhaled CO Concentration		Symptoms
Percent	ppm	Symptoms
0.02	200	Possible mild frontal headache in 2–3 h
0.04	400	Frontal headache and nausea after 1–2 h, occipital headache alter 2–3 h
0.08	800	Headache, dizziness, and nausea in 45 min, with collapse and possibly unconsciousness in 2 h
0.16	1600	Headache, dizziness, and nausea in 20 min; collapse, unconsciousness, and possibly death in 2 h
0.32	3200	Headache and dizziness in 5–10 min; unconsciousness and danger of death in 30 min
0.64	6400	Headache and dizziness in 1–2 min; unconsciousness and danger of death in 10–15 min
1.28	12,800	Immediate effect; unconsciousness and danger of death in 1–3 min

decrease in exercise tolerance, and fetal changes in pregnant women. Exposure to low levels of CO can provoke transient ischemic attacks, strokes, angina, and myocardial infarctions in workers with underlying cardiovascular or cerebrovascular disease. Carbon monoxide also may be atherogenic.

In addition to the effects that may be associated with chronic low level exposure, chronic sequelae can be seen after high level, acute exposure that causes unconsciousness and hypoxia or anoxia. Elimination of CO does not always halt the changes that have been initiated in cells, especially in those of the central nervous system, which may proceed to profound degeneration with late and progressive neuropsychiatric syndromes, motor and sensory involvement, vision loss, cardiac disturbances, or hemorrhage from a sudden rise in blood pressure.

Mechanism of Action(s)

Major endogenous sources of carbon monoxide are produced in small amounts from the natural break down of heme, resulting in normal background levels of COHb (Laing, 2010). Normal CO levels for nonsmokers are typically <2%, while the range for smokers is between 5 and 13% (Prockop and Chichkova, 2007). Endogenously produced CO is believed to have significant protective effects on intracellular signaling processes, including inflammation, proliferation, and apoptosis (Prockop and Chichkova, 2007; Ryter et al., 2006). The few mechanisms most commonly described with the endogenous production of CO include the modulation of mitogen-activated protein kinase (MAPK) activation and the stimulation of calcium-dependent potassium channel activity (Ryter et al., 2006). Because endogenously produced CO has essential physiological functions and is vital to cellular hemostasis, there is increased research investigating the potential therapeutic use of administering low doses of CO (Schmidt et al., 2012).

The main exposure route for exogenous carbon monoxide is inhalation, although dermal absorption and eye contact are plausible. The most widely accepted theory for CO toxicity is due to hypoxia as a result of hypoxemia from COHb formation (Gorman et al., 2003). Carbon monoxide's affinity for hemoglobin is 210–240 times greater than that of oxygen, allowing CO to rapidly bind to hemoglobin thereby displacing oxygen. As the CO exposure increases, the level of CO attached to the blood (COHb) also increases. This interferes with the dissociation of the remaining oxyhemoglobin, thereby reducing the red blood cell's ability to transport oxygen, shifting the oxyhemoglobin dissociation curve to the left. Tissues that consume large amounts of oxygen are particularly susceptible to CO-induced effects. The brain and the heart are the most susceptible to CO toxicity due to their high metabolic rate (Prockop and Chichkova, 2007). Intracellular CO binds with cytochromes of the mitochondria, interrupting normal cellular functions, and ultimately stopping the production of adenosine triphosphate (ATP). Without ATP, aerobic metabolism converts to anaerobic metabolism producing lactic acid, which results in cellular death. If this process is not reversed, it can lead to organ failure and death. Cardiac ischemia, reduction in myocardial contractility, and decreased cardiac output is believed to be caused from CO binding to intracellular myoglobin in cardiac tissues, impairing the oxygen supply to the mitochondria, and consequently the energy source for the heart muscle (Prockop and Chichkova, 2007).

SOURCES OF EXPOSURE

Occupational

Carbon monoxide is primarily the result of incomplete combustion; therefore, exposure is a particular risk for occupations that release CO during such processes. Occupations with the highest risk include acetylene workers, blast furnace workers, boiler room workers, brewery workers, carbon black makers, coke oven workers, customs workers, diesel engine operators, dockworkers, garage mechanics, metal oxide reducers, miners, organic chemical synthesizers, petroleum refinery workers, pulp and paper workers, steelworkers, tollbooth and tunnel attendants, and warehouse workers. However, recent studies suggest welding practices can also generate CO at rates ranging from 386 to 1293 ml/min, depending on equipment configurations (Ojima, 2013).

A study in Washington state, analyzing occupational claims over a 6-year time period, determined that the construction industry (20%), wholesale trade (15%), and agricultural industries (12%) were the major industries responsible for the majority of the incidences (Reeb-Whitaker et al., 2010). Of all the incidents, fuel-powered forklifts and autos/trucks/buses were responsible for 29% and 26% of the cases reported, respectively. Another possible route of exposure to carbon monoxide is metabolism of absorbed methylene chloride (dichloromethane, CH_2Cl_2), a principal ingredient of many paint strippers. Methylene chloride is absorbed through the lungs and converted into CO in the liver after entering circulation (Varon et al., 1999). Barrowcliff and Knell (1979) reported a case of serious cerebral deterioration in a middle-aged man who had used methylene chloride as a solvent for 3 years in a road material laboratory. They suggested the possibility of an immediate risk of anoxia arising from carboxyhemoglobinemia and of a long-term hazard of chronic CO toxicity affecting the brain. Cases of acute myocardial ischemia and infarction related to the use of methylene chloride in poorly ventilated areas have been documented.

Environmental

Carbon monoxide is found in combustion fumes as the result of vehicle emissions, gas appliances, such as stoves, lanterns, and the burning of charcoal or wood. If these fumes are allowed to accumulate in enclosed or semi-enclosed spaces, it can cause sudden illness and even death. Accidental poisonings from these types of sources are preventable with the correct ventilation, installation, and usage. Unintentional, non-fire related poisonings are the leading cause of poisonings in the United States (Iqbal et al., 2012). Accidental exposures tend to be seasonal, with higher incidences occurring during winter months as a result of an increased use of space heaters, furnaces, and alternative energy sources. Exposure to CO results in nearly 440 deaths, more than 2000 hospitalizations, and >20,000 emergency room visits annually (Iqbal et al., 2012). Nonfatal exposures were found to be more common among children and females, whereas hospitalization and fatalities were higher among males and adults over 65 years of age. Hospitalization rates for unintentional, non-fire related, poisonings decreased by approximately 50% between 1993 and 2007. Decreases have been correlated to the increased number of homes with working CO monitors, and the reduction of CO content in vehicle emissions with the introduction of catalytic converters (Iqbal et al., 2012; Hampson, 2005).

Although tobacco smoking contributes negligible amounts of carbon monoxide to the air, cigarette smokers are the most heavily exposed nonindustrial portion of the population, and this exposure obviously affects industrial workers as well. The CO concentration of cigarette smoke is 10,000–40,000 parts per million (ppm). During the smoking process, CO is diluted with air to the extent that alveolar CO concentrations average

400–500 ppm (Klus and Kuhn, 1982). There also have been reports of carbon monoxide exposure in surgical patients given general anesthesia. Lentz (1995) provides supporting evidence of an interaction of halogenated volatile anesthetics and carbon dioxide absorbent in the breathing circuit, especially if an anesthesia machine had been unused for a period of time.

INDUSTRIAL HYGIENE

Carbon monoxide is considered as the oldest of the industrial hazards, because it must have attacked the first people who used heat to break stone for implements or who burned wood with little air to make charcoal. Today, almost any industrial item produced involves the use of fire, oxidation, or combustion at some point during its manufacture. Carbon monoxide poisoning is now the most common of all poisonings in industry and is responsible for more than half of the poisoning fatalities reported each year in the United States. It is a leading agent of lethal inhalation in the United States. The danger of CO lies in its ubiquity and lack of warning properties. In addition, early symptoms can resemble other common disorders, such as the onset of a cold or stomach flu. Carbon monoxide poisoning remains a serious public health problem, and everyone in modern society is at risk (Cobb and Etzel, 1991).

To protect the health of workers, the Occupational Safety and Health Administration (OSHA) has set a permissible exposure limit (PEL) for CO of 50 ppm, or 55 mg/m^3 (8-h time-weighted average [TWA]) and a ceiling limit of 200 ppm, or 229 mg/m^3 (15-min TWA) (OSHA, 1978). Exposure to 1200 ppm or more is considered Immediately Dangerous to Life and Health (IDLH). The National Institute for Occupational Safety and Health (NIOSH) has set the recommended exposure limit (REL) at 35 ppm (40 mg/m^3) with a ceiling limit of 200 ppm (NIOSH, 1994). The American Conference of Governmental Industrial Hygienists (ACGIH) has recommended a threshold limit value (TLV)–TWA of 25 ppm, or 29 mg/m^3 (ACGIH, 1995). Of course, the OSHA-mandated Hazard Communication Standard requires employers to educate workers about any occupational hazard created by CO. The use of carbon monoxide detectors has been increasing, which adds a margin of safety in many situations.

It is a good industrial hygiene to use engineering controls, whenever they are available, to reduce environmental CO concentrations to the lowest level possible. However, when engineering and work practice controls are not technically feasible, respirators may be required to protect the workers' health. If respirators are required, the proper type must be used and a complete respiratory protection program must be instituted that includes regular training, maintenance, inspection, cleaning, and evaluation. Employees should be provided with impervious clothing that would prevent the skin from freezing upon contact with liquid CO or its canisters. Safety goggles should be provided if there is a reasonable possibility of liquid carbon monoxide being splashed into the eyes. Finally, any clothing wet with liquid CO should be removed immediately because of its flammable nature.

MEDICAL MANAGEMENT

Occupational Safety and Health Administration recommends workers who have inhaled large volumes of carbon monoxide be moved to fresh air immediately. If an exposed worker is found unconscious, they should be moved to fresh air immediately and established emergency procedures should be followed.

Victims of a suspected carbon monoxide poisoning should be given 100% oxygen immediately while blood tests to assess the COHb levels are pending. In addition to measuring COHb, complete blood chemistries, cardiac biomarkers, basic chemistries, ethanol and toxicology screens, and a urinalysis are needed (Laing, 2010). Pulse oximetry is not highly recommended, as it can over estimate arterial oxygenation; in contrast, spectrophotometric blood gas analysis is the best method to measure COHb levels (Rosenthal, 2006). Normal CO levels for nonsmokers are typically <2% while the range for smokers is between 5 and 13% (Prockop and Chichkova, 2007). Carboxyhemoglobin levels exceeding 70% are typically associated with cardiopulmonary arrest.

Hyperbaric oxygen is required for pregnant women, regardless of COHb levels and may also be required for victims of severe poisoning. The biological half-life ($T_{1/2}$) of CO varies with minute ventilation, cardiac output, and the concentration of oxygen inhaled. Generally, the $T_{1/2}$ is approximately 4–5 h at room air, 60–90 min at 100% oxygen, and <25 min in hyperbaric oxygen at 2 atm. Unconsciousness or increased respiratory distress would require intubation and mechanical ventilation. Patients receiving oxygen should be consigned to bed rest and monitored closely. In severe poisoning events, serum pH levels should be monitored closely to evaluate for lactic acidosis. High risk patients for cardiovascular complications may require electrocardiography, echocardiography, and additional blood work to assess cardiac health. Patients displaying cognitive dysfunction may require a CAT scan or MRI. In addition, improved emergency department neurological and cognitive dysfunction can be assessed using the carbon monoxide neuropsychological screening battery (CONSB), a tool consisting of six exercises to specifically assess cerebral impairment from CO poisoning (Messier and Myers, 1991). It has been suggested that hyperbaric oxygen treatment may prevent delayed brain injuries (Gorman et al., 2003).

REFERENCES

American Conference of Governmental Industrial Hygienists (ACGIH) (1995) *Threshold Limit Values (TLVs) for Chemical Substances and Physical Agents and Biological Exposure Indices (BEIs)*, Cincinnati, OH: American Conference of Governmental Industrial Hygienists.

Barrowcliff, D.F. and Knell, A.J. (1979) Cerebral damage due to endogenous chronic carbon monoxide poisoning caused by exposure to methylene chloride. *J. Soc. Occup. Med.* 29:12–14.

Chapman, D.J. and Tocher, R.D. (1966) Occurrence and production of carbon monoxide in some brown algae. *Can. J. Bot.* 44:1438–1442.

Choi, I.S. (1983) Delayed neurologic sequelae in carbon monoxide intoxiciation. *Arch. Neurol.* 40:433–435.

Cobb, N. and Etzel, R.A. (1991) Unintentional carbon monoxide-related deaths in the United States, 1979 through 1988. *JAMA* 266:659–663.

Conway, T.J., Andrews, A.E., Bruhwiler, L., Crotwell, A., Dlugokencky, E.J., Hahn, M.P., Hirsch, A.I., Kitzis, D.R., Lang, P.M., Masarie, K.A., Michalak, A.M., Miller, J.B., Novelli, P.C., Peters, W., Tans, P.P., Thoning, K.W., Vaughn, B.H., and Zhao, C. (2002–2003) *Carbon cycle greenhouse gases. Climate Monitoring and Diagnostics Laboratory, Summary Report No. 27*. Available at http://www.esrl.noaa.gov/gmd/publications/annrpt27/.

Drori, R., Dayan, U., Edwards, D.P., Emmons, L.K., and Erlick, C. (2012) Attributing and quantifying carbon monoxide sources affecting the Eastern Mediterranean: a combined satellite, modeling and synoptic analysis study. *Atmos. Chem. Phys.* 12:1067–1082.

Duncan, B.N., Logan, J.A., Bey, I., Megretskaia, I.A., Yantosca, R.M., Novelli, P.C., Jones, N.B., and Rinsland, C.P. (2007) Global budget of CO, 1988–1997: source estimates and validation with a global model. *J. Geophys. Res.* 112:1–29.

Ernst, A. and Zibrak, J.D. (1998) Carbon monoxide poisoning. *N. Engl. J. Med.* 339:1603–1608.

Goldstein, A.H. and Galbally, I.E. (2007) Known and unexplored organic constituents in the Earth's atmosphere. *Environ. Sci. Technol.* 41:1514–1521.

Gorman, D., Drewry, A., Huang, Y.L., and Sames, C. (2003) The clinical toxicology of carbon monoxide. *Toxicology* 187:25–38.

Hampson, N.B. (2005) Trends in the incidence of carbon monoxide poisoning in the United States. *Am. J. Emerg. Med.* 23:838–841.

Hardy, K.R. and Thom, S.R. (1994) Pathophysiology and treatment of carbon monoxide poisoning. *J. Toxicol. Clin. Toxicol.* 32:613–629.

Hart, I.K., Kennedy, P.G.E., Adams, J.H., and Cunningham, N.E. (1988) Neurological manisfestation of carbon monoxide poisoning. *Postgrad. Med. J.* 64:213–216.

Iqbal, S., Clower, J.H., King, M., Bell, J., and Yip, F.Y. (2012) National carbon monoxide poisoning surveillance framework and recent estimates. *Public Health Rep.* 127:486–496.

Jaffe, L.S. (1970) Sources, characteristics and fate of atmospheric carbon monoxide. *Ann. N.Y. Acad. Sci.* 174:76–88.

Kealey, G.P. (2009) Carbon monoxide toxicity. *J. Burn Care Res.* 30:146–147.

Klus, H. and Kuhn, H. (1982) Distribution of various tobacco smoke components in main and sidestream smoke: a review. *Tabakforsch. Int.* 11:229–265.

Kwon, O.Y., Chung, S.P., Ha, Y.R., Yoo, I.S., and Kim, S.W. (2004) Delayed postanoxic encephalopathy after carbon monoxide poisoning. *Emerg. Med. J.* 21:250–251.

Laing, C. (2010) Acute carbon monoxide toxicity: the hidden illness you may miss. *Nursing* 40:38–43.

Lentz, R.E. (1995) Carbon monoxide poisoning during anesthesia poses puzzles. *J. Clin. Monitor.* 11:66–67.

Little, D.N. (1995) Children and environmental toxins. *Prim. Care* 22:69–79.

Logan, J.A., Prather, M.J., Wofsy, S.C., and McElroy, M.B. (1981) Tropospheric chemistry: a global perspective. *J. Geophys. Res.* 86:7210–7254.

Messier, L.D. and Myers, R.A.M. (1991) A neurophyschological screening battery for emergency assessment of carbon-monoxide poisoned patients. *J. Clin. Psychol.* 47:675–684.

Min, S.K. (1986) A brain syndrome associated with delayed neuropsychiatric sequelae following acute carbon monoxide intoxication. *Acta Psychiatr. Scand.* 73:80–86.

National Institute for Occupational Safety and Health (NIOSH) (1994) *Pocket Guide to Chemical Hazards*, NIOSH 94–116, Washington, DC: US Department of Health and Human Services.

Ojima, J. (2013) Generation rate of carbon monoxide from CO2 arc welding. *J. Occup. Health* 55:39–42.

Occupational Safety and Health Administration (OSHA) (1978) *Occupational Health Guideline for Carbon Monoxide*, US Department of Labor. Available at http://www.cdc.gov/niosh/docs/81-123/pdfs/0105.pdf.

Pickwell, G.V. (1970) The physiology of carbon monoxide production by deep-sea coelenterates: causes and consequences. *Ann. N.Y. Acad. Sci.* 174:102–115.

Prockop, L.D. and Chichkova, R.I. (2007) Carbon monoxide intoxication: an updated review. *J. Neurol. Sci.* 262:122–130.

Reeb-Whitaker, C.K., Bonauto, D.K., Whittaker, S.G., and Adams, D. (2010) Occupational carbon monoxide poisoning in Washington State, 2000–2005. *J. Occup. Environ. Hyg.* 7(10):547–556.

Rosenthal, L.D. (2006) Carbon monoxide poisoning. *Am. J. Nurs.* 106:40–47.

Ryter, S.W., Alam, J., and Choi, A.M.K. (2006) Heme oxygenase-1/carbon monoxide: from basic science to therapeutic applications. *Physiol. Rev.* 86:583–650.

Sawa, G.M., Watson, C.P.M., Terbrugge, K., and Chiu, M. (1981) Delayed encephalopathy following carbon monoxide intoxication. *Can. J. Neurol. Sci.* 8:77–79.

Schmidt, R., Ryan, H., and Hoetzel, A. (2012) Carbon monoxide—toxicity of low-dose application. *Curr. Pharm. Biotechnol.* 13:837–850.

Seger, D. and Welch, L. (1994) Carbon monoxide controversies: neuropsychologic testing, mechanism of toxicity, and hyperbaric oxygen. *Ann. Emerg. Med.* 24:242–248.

Sokal, J.A., Majka, J., and Palus, J. (1984) The content of caron monoxide in the tissues of rats intoxicated with carbon monoxide in various conditions of acute exposure. *Arch. Toxicol.* 56:106–108.

Thom, S.R. (1993) Leukocytes in carbon monoxide-mediated brain oxidative injury. *Toxicol. Appl. Pharmacol.* 123:234–247.

Tian, L., Qiu, H., Pun, V.C., Lin, H., Ge, E., Chan, J.C., Louie, P.K., Ho, K., and Yu, I.T.S. (2013) Ambient carbon monoxide associated with reduced risk of hospital admissions for respiratory tract infections. *Am. J. Respir. Crit. Care Med.* 188:1240–1245.

United States Environmental Protection Agency (USEPA) (2013) *2008 National Emissions Inventory: Review, analysis and highlights. Document EPA-454/R-13-005, May 2013, Research Triangle Park, NC.*.

Varon, J., Marik, P.E., Fromm, R.E. Jr., and Gueler, A. (1999) Carbon monoxide poisoning: a review for clinicians. *J. Emerg. Med.* 17:87–93.

Weinstock, B. and Niki, H. (1972) Carbon monoxide balance in nature. *Science* 176:290–292.

Zafiriou, O.C., Andrews, S.S., and Wang, W. (2003) Concordant estimates of oceanic carbon monoxide source and sink processes in the Pacific yield a balanced global "blue-water" CO budget. *Glob. Biogeochem. Cy.* 17:15(1)–15(13).

44

CARBON DISULFIDE

Debra J. Price

Target organ(s): Eyes, nerves, liver, kidney, heart, cardio-vascular system, male reproductive system, female reproductive system

Occupational exposure limits: OSHA: PEL—20 ppm, STEL—10 ppm TWA; NIOSH: REL—1 ppm, STEL—10 ppm TWA, Immediately Dangerous to Life or Health (IDLH)—500 ppm; ACGIH: TWV—10 ppm TWA

Reference values: RfD (oral) 0.1 mg/kg/day; RfC (inhalation) 0.7 mg/m^3

Risk/safety phrases: NIOSH and the ACGIH also assign a "Skin" notation, which indicates that the cutaneous route of exposure, including mucous membranes and eyes, contributes to overall exposure

BACKGROUND AND USES

Carbon disulfide is used as a raw material in the production of such things as rayon, cellophane, semiconductors, and carbon tetrachloride, and to make some pesticides. It is used as an industrial solvent and chemical intermediate to dissolve rubber to produce tires (ATSDR, 1996), as well as in grain fumigation, analytical chemistry research, degreasing, dry cleaning, and oil extraction (Finkel et al., 1983). Natural sources of carbon disulfide include the open ocean, coastal areas of high biological activity, microbial reduction of sulfates in soil, marshlands, and some higher plants where the source of carbon disulfide is the tree roots (Carroll, 1985; Khalil and Rasmussen, 1984).

Previously, carbon disulfide was used as a pesticide, where it was typically mixed with carbon tetrachloride in a 20/80 mixture, respectively. This mixture was used to exterminate insects and rodents from entire boxcars of wheat, corn, rye, and other grains (Peters et al., 1988). Grain fumigators can be acutely intoxicated and may be chronically exposed to carbon disulfide. Therefore, in the late 1980s, all pesticides containing carbon disulfide as an active ingredient were cancelled by the United States Environmental Protection Agency (U.S. EPA) (U.S. EPA, 1999).

PHYSICAL AND CHEMICAL PROPERTIES

Carbon disulfide (carbon bisulfide; CS_2; CASRN 75-15-0), in its pure form is a colorless liquid that evaporates readily at room temperature, with a sweet aromatic odor similar to that of chloroform. In its impure commercial and reagent form, however, carbon disulfide is a yellowish liquid with a foul-smelling odor. It has a vapor pressure of 297 mmHg and solubility in water by weight of 0.3% at 20 °C (68 °F). Once carbon disulfide is in the air it will break down into simpler substances within a few days after release (OEHHA, 2001).

Acute Effects

Although acute exposure is rare, acute effects can range from mild irritation to death and include the following: (1) mild to moderate irritation of skin, eyes, and mucous membranes from liquid or concentrated vapors; (2) headache; (3) garlicky breath, nausea, vomiting, and diarrhea; (4) weak pulse and palpitations; (5) fatigue, weakness in the legs, unsteady gait, and vertigo; (6) hyperesthesia, agitation, mania, hallucinations of sight, hearing, taste, and smell in acute, massive exposures; (7) central nervous system depression with

Hamilton & Hardy's Industrial Toxicology, Sixth Edition. Edited by Raymond D. Harbison, Marie M. Bourgeois, and Giffe T. Johnson.
© 2015 John Wiley & Sons, Inc. Published 2015 by John Wiley & Sons, Inc.

respiratory paralysis; and (8) death may occur during coma or after a convulsion (Gosselin et al., 1984).

Typically the acute effects mentioned above are via inhalation and dermal contact; however, there are a few documented cases of carbon disulfide ingestion. These individuals had spasmodic tremors, Cheyne–Stokes respirations, large pupils, low body temperature, pallor and, finally, coma and death (Foreman, 1886).

Chronic Effects

Neurological Effects   The most widely known and extensively studied aspect of carbon disulfide toxicity is its neurological effects, which were first noted by the French in the mid-1800s (Hamilton, 1937). Auguste Delpech, a physician in Paris, conducted a thorough study of the effects of carbon disulfide exposure between 1856 and 1863. His research involved the study of more than 100 chronically exposed patients and some animal experiments. He described in detail the "carbon disulfide neurosis," a disorder characterized by an early stage of excitation followed by a second stage of depression. Delpech's observations were accurate, but it took almost a century for the details of carbon disulfide poisoning to spread throughout Europe and England and finally to the United States (Davidson and Feinleib, 1972).

It is now accepted that excessive exposure to carbon disulfide causes extensive neurological problems. One study by Aaserud et al. (1990) investigated the neurological symptoms of 24 men formerly and currently employed in the manufacture of viscose rayon. When clinical neurological examinations were performed with the 16 subjects still employed in rayon production, only one was considered normal. Nine had minor neurological deficits. Six had facial palsy, reduced tempo or coordination, asymmetrical reflexes, or a positive Romberg's test. Half of the workers with reflex abnormalities also had distal hyporeflexia.

Peters et al. (1988) investigated the neurological effects of carbon disulfide fumigants in granary workers. They found that 50–80% exhibited a range of symptoms, including cogwheel rigidity, decreased associated movements, distal sensory shading, intention and resting tremulousness, and conduction abnormalities.

Neurological damage caused by chronic carbon disulfide poisoning also is responsible for hearing loss among workers. Experiments have demonstrated that high frequency hearing is most impaired (Zenk, 1970). Some studies have suggested a synergistic or additive ototraumatic interaction between carbon disulfide and noise-induced hearing loss (Morata and Dunn, 1994). Study of brainstem auditory-evoked responses demonstrated that carbon disulfide has an effect on the auditory pathways in the brainstem (Hirata et al., 1992a, 1992b).

Because of the neurotoxicity of carbon disulfide, the exposure can have dramatic effects on a worker's personality. Patients and family members often complain of emotional upset, depression, irritability, sleeplessness, fatigue, and loss of memory. Many of the victims of carbon disulfide poisoning in the 1800s and early 1900s were confined to mental hospitals with diagnoses such as "etiology occupational" or "unknown." Mancuso and Locke (1972) found an association between exposure to carbon disulfide and an increase in the suicide rate.

Cardiovascular Effects   The cardiovascular effects of carbon disulfide exposure were not recognized until long after its neurotoxicity had been established. The current prevalence of lower level exposure has increased awareness of chronic cardiovascular effects.

The first thorough study of ischemic heart disease (IHD) and carbon disulfide exposure was published by Tiller et al. (1968). Investigators found a relative risk of 5.6 for coronary deaths when rayon workers were compared with matched controls. Subsequent research has confirmed this association and clarified the mechanism by which carbon disulfide increases IHD mortality (Balcarova and Halík, 1991). Sweetnam et al. (1987) reported that the association between carbon disulfide exposure and IHD mortality was most pronounced when workers were heavily exposed in the preceding 2 years. A study by Prince et al. (2000) looked at mortality rates among rubber chemical manufacturing workers. They found that IHD mortality among the workers was elevated relative to IHD mortality in the US population, especially in the under 50 years of age category. A significant difference was also found between different age groups of rubber chemical manufacturing workers. The standard mortality rate (SMR) for IHD among all workers in the rubber chemicals department ranged from 1.2 to 1.5, and SMRs in the under 50 age group ranged from 1.9 to 2.5, a significant difference. This evidence suggested that carbon disulfide has a direct, reversible cardiotoxic effect.

A study by Egeland et al. (1992) investigated the low density lipoprotein concentration (LDL_c), and diastolic blood pressure of workers exposed to a carbon disulfide concentration of 0.6–11.8 parts per million (ppm). When the workers were ranked according to level of exposure, researchers found a direct, linear relationship between diastolic blood pressure and the level of exposure to carbon disulfide. A similar relationship was demonstrated for the LDL_c. Carbon disulfide exposure did not correlate with triglyceride, high density lipoprotein concentration (HDL_c), fasting glucose concentration, or systolic blood pressure. These findings are especially significant given the relatively low level of exposure and the potential these factors have for increasing the risk of IHD. Vanhoorne et al. (1992) also found a direct relationship between LDL_c and the level of carbon disulfide exposure, but they reported an inverse association with HDL_c. Krstev et al. (1992) reported an increase in triglyceride levels in highly exposed workers. The difference between these findings could be explained by the fact that workers classified as high exposure in the Krstev study had exposure levels twice than that of the workers in the Egeland report. All of these studies support the conclusion that carbon disulfide has a direct, dose-related, reversible cardiotoxic effect.

Other Effects Although research has failed to find a strong association between carbon disulfide exposure and adverse male reproductive effects (Vanhoorne et al., 1994); the same cannot be said for female exposure to carbon disulfide. In fact, women appear to be more sensitive than men to the neurotoxic effects of carbon disulfide (Gosselin et al., 1984). In one study, women employed in viscose rayon textile jobs had an increased rate of spontaneous abortions, and the wives of men employed in the same textile jobs also had increased rate of spontaneous abortions. No evidence was found, however, that the level of exposure was associated with the risk of spontaneous abortions (Hemminki and Niemi, 1982). In another study of female viscose rayon workers, CAI and BAO (1981) showed placental transfer of carbon disulfide and secretion into mother's milk. The umbilical blood from the female workers, the milk from breast-feeding workers, and the urine from the breast-fed babies all contained carbon disulfide.

Earlier research suggested that the renal toxicity of carbon disulfide is primarily associated with atherosclerosis and its effects on renal function (Browning, 1965). However, Rubin (1990) found that the kidneys of rats and mice were capable of carbon disulfide bioactivation and that this bioactivation was only partly responsible for the total nephrotoxic effect of carbon disulfide.

Effects on eye-sight have been documented prior to other symptoms, with an increase in eye sensitivity to light (Kirk, 1978) and the degeneration of the retina or hemorrhaging of blood vessels (ARD-EHP-11, 2006).

In one study, 350 fiber workers that were exposed to carbon disulfide were examined to determine if there were any alterations to their oral cavity. The group that was exposed for <5 years had significantly lower pH values for both the mucous membrane and the saliva compared to the control group (NIOSH, 1977).

Mechanisms of Action

Even though carbon disulfide is highly lipophilic and can be readily absorbed through the skin, it is not known to bioaccumulate in the body. This may be because most of the carbon disulfide taken into the body is quickly exhaled unchanged. Only about 0.05% of the carbon disulfide absorbed by an individual is excreted in the urine. The remaining absorbed portion not exhaled is metabolized (Bus, 1985; McKenna and DiStefano, 1977).

Carbon disulfide spontaneously reacts with free amine and sulfhydryl groups of amino acids and polypeptides. This reaction produces dithiocarbamates (from amine groups) and trithiocarbonates (from sulfhydryl groups), which may in part account for the neurotoxic effects of carbon disulfide. Both of these metabolites are water soluble and may be excreted efficiently by the kidneys. In rats exposed to carbon disulfide, 90% of carbon disulfide found in the blood was in the form of AL CS_2 in red blood cells (RBCs) (Lam and DiStefano, 1986). The authors also found carbon disulfide bound to fractionated human RBCs in proportion to the amount of hemoglobin.

In rats exposed to carbon disulfide, blood levels increased dramatically and leveled off after 2 h, and then dropped quickly after exposure ceased (McKenna and DiStefano, 1977). Carbon disulfide was concentrated rapidly in fat and organs with high lipid content (e.g., the liver and adrenals). To a lesser extent, it is also collected in the kidneys, brain, muscles, and heart. The availability of free amino acids in these organs may enhance AL CS_2 formation. With the exception of fat and skeletal muscle, the rate of AL CS_2 formation in these organs is greater than the rate of elimination (McKenna and DiStefano, 1977). It is now recognized that neural responses to carbon disulfide are influenced by the mineral content of the animal's diet, where a high mineral content diet offers protection from neurologic effects (ACGIH, 1991). These data suggest that, at least in animals, bioaccumulation may occur with repeated exposure.

Chemical Pathophysiology

Peters et al. (1988) found evidence of central demyelination in cranial magnetic resonance imaging scans. This is consistent with the findings in animal studies that have demonstrated fragmentation and disruption of myelin sheaths in rats exposed to carbon disulfide (Dietzmann and Laass, 1977).

Wilmarth et al. (1993) showed that in rats exposed to carbon disulfide, protein phosphorylation in neurofilaments increased. This finding suggests a physiological route by which carbon disulfide may cause filament disassembly and diminish the transport rate in the axon. Ruijten et al. (1990) demonstrated that exposure to carbon disulfide affected the conduction velocity of slower motor fibers but not the motor nerve conduction velocity or sensory nerve conduction velocity. Graham et al. (1995) found that carbon disulfide causes neurofilament-filled swellings of the distal axon in both the central and peripheral nervous systems. Also, there have been case reports of axonal degeneration and loss of myelinated fibers in patients treated with disulfiram (Ansbacher et al. 1982).

Hoffmann and Klapperstück (1990) reported that carbon disulfide had a depressive effect on intra-cardiac impulse generation, conduction, and contractile force in urethane-anesthetized rats.

Vanhoorne et al. (1993) found that viscose rayon workers exposed to carbon disulfide had significantly lower levels of prolactin. This indicates that carbon disulfide may have an effect on the endocrine system.

In rats, high doses of carbon disulfide have been reported to diminish hepatic cytochrome P-450 (P-450) and the activity of microsomal mixed function oxidase (Kivistö et al., 1995).

Research suggests that carbon disulfide may have a direct, toxic effect on human lymphocytes. Garry et al. (1990) found that exposure to carbon disulfide increased sister chromatid exchanges and chromosomal aberrations *in vitro*. S-9 liver fraction was required to produce these effects suggesting that metabolic activation was responsible for toxicity.

SOURCES OF EXPOSURE

Occupational

The toxicity of carbon disulfide first became apparent in the 1800s, when it was used for "cold" vulcanization of rubber throughout Europe. Cold vulcanization was never in common use in the United States, which partly accounts for the late recognition of the risks associated with carbon disulfide exposure.

Exposure to carbon disulfide is typically occupational via inhalation and dermal contact with the vapor, or dermal contact with the liquid. While workers engaged in any process that uses carbon disulfide have the potential for some degree of exposure, in practice, only workers in the viscose rayon industry are exposed to high concentrations (WHO, 1979). In a 3-year survey, National Institute for Occupational Safety and Health (NIOSH) statistically estimated that 44,441 workers, including 4882 females, were exposed to carbon disulfide in the United States (NOES Survey 1981–1983). The survey excludes trade names that may contain the chemical; therefore, exposure may be considerably higher than the survey results (NIOSH, 1985).

Although many industries use carbon disulfide, currently the largest user, and most common form of occupational exposure to this chemical is in the viscose rayon fiber industry in the technological process. For every kilogram of viscose rayon produced, about 20–30 g of carbon disulfide and 4–6 g of hydrogen sulfide are emitted (NIOSH, 1977). This process is used for more than 80% of the world's supply of rayon (artificial silk) and requires large amounts of carbon disulfide (Davidson and Feinleib, 1972).

The manufacturing of viscose rayon begins with sheets of cellulose pulp. The pulp is steeped in caustic soda and then broken into crumbs, which are aged further in a caustic soda solution. Next, the crumbs are churned with carbon disulfide, becoming a material known as xanthate crumb. The xanthate crumb is dissolved in caustic soda, and a solution is formed that resembles honey in appearance and consistency. After further aging and filtering, this viscose material is pumped into spinning tanks. The solution is forced through nozzles into a sulfuric acid bath, where it coagulates to form filaments. During this step, hydrogen sulfide and carbon disulfide are liberated. Thousands of filaments are hardened simultaneously and guided into a spinning box, where yarn is collected. The yarn continues to release carbon disulfide and hydrogen sulfide gases for hours after removal from the spinning machine. The workers in the spinning rooms are at greatest risk of carbon disulfide exposure. They also constitute the largest group of potentially exposed workers. The churning process is another area of potentially high exposure. A study by Sweetnam et al. (1987) showed that the viscose workers in the spinning rooms have a significantly higher mortality than the least exposed group of workers, and that the excess mortality is largely accounted for by IHD.

Environmental

The general population may also be exposed to carbon disulfide via inhalation of ambient air or pharmaceuticals containing the chemical. Concerning ambient air, one study by Phillips (1992) suggested that urbanites may be constantly exposed to low levels of carbon disulfide. This low level, potentially lifelong exposure could have some of the chronic consequences, but more research is needed to determine whether this is the case.

Concerning pharmaceuticals, the drug disulfiram (Antabuse), which is used to treat alcoholism, and sodium diethyldithiocarbamate (ditiocarb, ditiocarb sodium), which is used in the treatment of patients infected with the human immunodeficiency virus (HIV), both are metabolized to carbon disulfide in the body. The blood level of carbon disulfide in patients treated with disulfiram averaged 9.5 µg/l compared with 0.26 µg/l in controls (Brugnone et al., 1992). The total concentration of carbon disulfide (free and blood bound) averaged 40 µg/l in patients and 0.9 µg/l in controls. For comparison purposes, Campbell et al. (1985) found that workers exposed to an average carbon disulfide level of 10 ppm had acid-labile (AL) blood levels of 332.6 µg/l. Martens et al. (1993) determined that Ditiocarb in aqueous solution at physiological pH and temperature decomposes to carbon disulfide. Chronic exposure to low levels of carbon disulfide can have serious consequences, especially to the nervous system and cardiovascular system. It is uncertain at this time whether the use of these pharmaceuticals poses a risk of carbon disulfide intoxication similar to a low level occupational exposure. However, physicians should be aware of this factor when prescribing Antabuse or ditiocarb for patients who may have been exposed to carbon disulfide.

INDUSTRIAL HYGIENE

To protect the health of workers, the Occupational Safety and Health Administration (OSHA) has set a permissible exposure limit for carbon disulfide of 20 ppm, with a short-term exposure limit (STEL) of 10 ppm for 15 min (time-weighted average [TWA]). The OSHA has also set a 30-min maximum peak at 100 ppm. The standards set by the NIOSH are somewhat stricter, with a recommended exposure limit of 1 ppm and a STEL set at a TWA of 10 ppm for 15 min (NIOSH, 1994). The NIOSH considers a carbon disulfide level of 500 ppm to be immediately dangerous to life and health. The American Conference of Governmental Industrial Hygienists (ACGIH) has set the threshold limit value at a TWA of 10 ppm (ACGIH, 1996).

MEDICAL MANAGEMENT

Diagnosis and Treatment

Diagnosis of carbon disulfide should include a complete history and physical examination in order to detect existing conditions that might place the exposed employee at increased risk, and to establish a baseline for health monitoring. Signs and symptoms for carbon disulfide exposure include upper respiratory burning, cranial nerve palsies, double vision, dizziness, headache, sleeping problems, weakness, exhaustion, anxiety, gastritis, kidney and liver damage, skin and eye irritation, breathing difficulty, increased blood pressure, abdominal pain, nausea, vomiting, diarrhea, neuritis, psychosis, and tremor. Therefore, a complete physical examination should include the central and peripheral nervous systems, kidneys, liver profile, skin examination, eyes, urinalysis, and an electrocardiogram (NIOSH/OSHA, 1981). Evidence of excessive exposure to carbon disulfide combined with the above-mentioned health effects known to be produced by carbon disulfide would be an indication of intoxication.

Chapman et al. (1991) found that individuals exposed to carbon disulfide exhibited markedly different finger tremor amplitude compared with matched controls. The authors believe that this subclinical manifestation may serve as an early indication of carbon disulfide exposure.

There is no specific treatment to enhance the reversibility of the effects of carbon disulfide exposure. In acute situations the patient should be removed from the source of exposure immediately and decontaminated. Special attention should be paid to breathing and circulation. Symptomatic treatment may be given as necessary.

Biomonitoring

For biomonitoring of carbon disulfide, urine, blood, breath, and milk samples can be examined. For urine testing, the iodine-azide test has been available for testing exposure to carbon disulfide for several decades (Vasak et al., 1963). In this test, organosulfur metabolites in the urine, the result of carbon disulfide exposure, are measured by the decolorization of iodine. This method is now considered ineffective because it cannot reliably measure levels of exposure below 15 ppm, levels which are typical of occupational exposures (van Doorn et al., 1981).

Currently, the most effective method for monitoring carbon disulfide exposure is testing the levels of 2-thio thiazolidine-4-carboxylic acid (TTCA) in urine (Campbell et al., 1985; Meuling et al., 1990). This method is extremely sensitive and can detect carbon disulfide exposure as low as 0.5 ppm (Riihimäki et al., 1992). The current biological exposure index is 5 mg TTCA/g creatinine (end of shift) (ACGIH, 1996).

Blood samples also can be used to determine exposure to carbon disulfide accurately, but they are less practical for routine biomonitoring (Campbell et al., 1985).

Breath samples provide an indication that exposure has occurred if measurements are taken shortly after possible exposure (Phillips, 1992). However, because carbon disulfide is eliminated so quickly through the lungs, these measurements are not an accurate reflection of the level of exposure.

Finally, milk samples may be useful as a qualitative indicator of exposure to carbon disulfide (Ryan and Terry, 1997).

REFERENCES

Aaserud, O., et al. (1990) Carbon disulfide exposure and neurotoxic sequelae among viscose rayon workers. *Am. J. Ind. Med.* 18:25–37.

American Conference of Governmental Industrial Hygienists (ACGIH) (1991) *Documentation of the Threshold limit Values (TLVs) and Biological Exposure Indices (BEIs)*, 6th ed, vol. I, II, III, Cincinnati.

American Conference of Governmental Industrial Hygienists (ACGIH) (1996) *Threshold Limit Values (TLVs) for Chemical Substances and Physical Agents and Biological Exposure Indices (BEIs)*, Cincinnati: American Conference of Governmental Industrial Hygienists.

Ansbacher, L.E., et al. (1982) Disulfiram Neurofilamentous Distal Axonopathy. *Neurol.* 32(4):424–428.

ARD-EHP-11 (2006) *Carbon Disulfide: Health Information Summary*, New Hampshire: Department of Environmental Services.

ATSDR (1996) *Toxicological Profile for Carbon Disulfide (Update)*, Atlanta, Georgia: U.S. Department of Health and Human Services, Public Health Service, Agency for Toxic Substances and Disease Registry.

Balcarova, O. and Halík, J. (1991) Ten-year epidemiological study of ischaemic heart disease (IHD) in workers exposed to carbon disulphide. *Sci. Total Environ.* 101:97–99.

Browning, E. (1965) Carbon disulfide. In: Browning, E., editor. *Toxicity and Metabolism of Industrial Solvents*, Amsterdam: Elsevier.

Brugnone, F., et al. (1992) Blood concentration of carbon disulphide in "normal" subjects and in alcoholic subjects treated with disulfiram. *Br. J. Ind. Med.* 49:658–663.

Bus, J.S. (1985) The relationship of carbon disulfide metabolism to development of toxicity. *Neurotoxicology* 6:73–80.

CAI, S.X. and BAO, Y.S. (1981) Placental transfer, secretion into mother milk of carbon disulphide and the effects on maternal function of female viscose rayon workers. *Ind. Health* 19:15–29.

Campbell, L., Jones, A., and Wilson, H., (1985) Evaluation of occupational exposure to carbon disulphide by blood, exhaled air, and urine analysis. *Am. J. Ind. Med.* 8:143–153.

Carroll, M.A. (1985) Measurements of OCS and CS2 in the free troposphere. *J. Geophys. Res.* 90:10483–10486.

Chapman, L.J., et al. (1991) Finger tremor after carbon disulfide-based pesticide exposures. *Arch. Neurol.* 48:866–870.

Davidson, M. and Feinleib, M. (1972) Carbon disulfide poisoning: a review. *Am. Heart J.* 83:100–114.

Dietzmann, K. and Laass, W. (1977) Histochemical studies on the rat brain under conditions of carbon disulfide intoxication. *Exp. Pathol. (Jena.)* 13:320–327.

Egeland, G., et al. (1992) Effects of exposure to carbon disulphide on low density lipoprotein cholesterol concentration and diastolic blood pressure. *Occup. Environ. Med.* 49:287–293.

Finkel, A.J., Hamilton, A., and Hardy, H.L. (1983) *Hamilton and Hardy's Industrial Toxicology*, Littleton, Mass: Yearbook.

Foreman, W. (1886) Notes of a fatal case of poisoning by bisulphide of carbon. *Lancet* 2:118–122.

Garry, V.F., et al. (1990) Preparation for human study of pesticide applications: sister chromatid exchange and chromosome aberrations in cultured human lymphocytes exposed to selected fumigants. *Teratog. Carcinog. Mutagen.* 10:21–29.

Gosselin, R.E., Smith, R.P., and Hodge, H.C. (1984) *Clinical Toxicology of Commercial Products*, 5th ed., Baltimore: Williams and Wilkins.

Graham, D.G., et al. (1995) Pathogenetic studies of hexane and carbon disulfide neurotoxicity. *Crit. Rev. Toxicol.* 25:91–112.

Hamilton, A. (1937) *Occupational Poisoning in the Viscose Rayon Industry, Bulletin 10*, Washington, DC: US Department of Labor.

Hemminki, K. and Niemi, M.L. (1982) Community study of spontaneous abortions: relation to coupation and air pollution by sulfer dioxide, hydrogen sulfide, and carbon disulfide. *Int. Arch. Occup. Environ. Health* 51:55–63.

Hirata, M., et al. (1992a) A cross-sectional study on the brainstem auditory evoked potential among workers exposed to carbon disulfide. *Int. Arch. Occup. Environ. Health* 64:321–324.

Hirata, M., et al. (1992b) Changes in auditory brainstem response in rats chronically exposed to carbon disulfide. *Arch. Toxicol.* 66:334–338.

Hoffmann, P. and Klapperstück, M. (1990) Effects of carbon disulfide on cardiovascular function after acute and subacute exposure of rats. *Biochem. Biophys. Acta* 1:121–128.

Khalil, M.A.K. and Rasmussen, R.A. (1984) Global sources, lifetimes and mass balances of carbonyl sulfide (OCS) and carbon disulfide (CS2) in the earth's atmosphere. *Atmos. Environ.* 18(9):1805–1831.

Kirk-Othmer (1978) *Encyclopedia of Chemical Technology*, 3rd ed., vol. 4 John Wiley and Sons, 1978–1984. 752.

Kivistö, H., et al. (1995) Effect of cytochrome P-450 isozyme induction and glutathione depletion on the metabolism of CS_2 to TTCA in rats. *Arch. Toxicol.* 69:185–190.

Krstev, S., Peruncic, B., and Farkic, B. (1992) The effects of long-term occupational exposure to carbon disulphide on serum lipids. *Eur. J. Drug Metab. Pharmcokinet.* 17:237–240.

Lam, C.W. and DiStefano, V. (1986) Characterization of carbon disulfide binding in blood and to other biological substances. *Toxicol. Appl. Pharmacol.* 86:235–242.

Mancuso, T.F. and Locke, B.Z. (1972) Carbon disulphide as a cause of suicide. Epidemiological study of viscose rayon workers. *J. Occup. Med.* 14:595–606.

Martens, D., Langevin-Bermond, D., and Fleury, M.B. (1993) Ditiocarb: decomposition in aqueous solution and effect of the volatile product on its pharmacological use. *J. Pharm. Sci.* 82:379–383.

McKenna, M.J. and DiStefano, V. (1977) Carbon disulfide. I. The metabolism of inhaled carbon disulfide in the rat. *J. Pharmacol. Exp. Ther.* 202:245–252.

Meuling, W.J.A., Bragt, P.C., and Braun, C.L.J. (1990) Biological monitoring of carbon disulfide. *Am. J. Ind. Med.* 17:247–254.

Morata, T.C. and Dunn, D.E. (1994) Occupational exposure to noise and ototoxic organic solvents. *Arch. Environ. Health* 49:359–365.

National Institute for Occupational Safety and Health (NIOSH) (1977) *Criteria for a Recommended Standard Occupational Exposure to Carbon Disulfide.* Washington, D.C.: US Department of Health, Education, and Welfare.

National Institute for Occupational Safety and Health (NIOSH) (1977) *Criteria Document: Carbon Disulfide, 81.*

National Institute for Occupational Safety and Health (NIOSH) (1985) *National Occupational Health Survey (NOHS).*

National Institute for Occupational Safety and Health (NIOSH) (1994) *Pocket Guide to Chemical Hazards,* NIOSH 94–116, Washington, DC: US Department of Health and Human Services.

NIOSH/OSHA (1981) *Occupational Health Guidelines for Chemical Hazards.* DHHS (NIOSH). Pub No. 81–123. Washington DC: U.S. Government Printing Office.

OEHHA July 5 2001 *Memorandum from Robert A. Howd, PhD, Chief Water Toxicology Unit.* Available at www.oehha.ca.gov/water/pals/carbondisulfide.html.

Peters, H., et al. (1988) Extrapyramidal and other neurologic manifestations associated with carbon disulfide fumigant exposure. *Arch. Neurol.* 45:537–540.

Phillips, M. (1992) Detection of carbon disulfide in breath and air: a possible new risk factor for coronary artery disease. *Int. Arch. Occup. Environ. Health* 64:119–123.

Prince, M.M., et al. (2000) Mortality among rubber chemical manufacturing workers. *Am. J. Ind. Med.* 37:590–598.

Riihimäki, V., et al. (1992) Assessment of exposure to carbon disulfide in viscose production workers from urinary 2-thiothiazolidine-4-carboxylie acid determinations. *Am. J. Ind. Med.* 22:85–97.

Rubin, R.J. (1990) Role of metabolic activation in the renal toxicity of carbon disulfide (CS_2). *Toxicol. Lett.* 53:211–213.

Ruijten, M.W.M.M., et al. (1990) Special nerve functions and colour discrimination in workers with long term low level exposure to carbon disulphide. *Occup. Environ. Med.* 47:589–595.

Ryan, R.P. and Terry, C.E. (1997) *Toxicology Desk Reference*, 4th ed., vol. 1–3, Washington DC: Taylor & Francis.

Sweetnam, P.M., Taylor, S.W., and Elwood, P.C. (1987) Exposure to carbon disulphide and ischaemic heart disuse in a viscose rayon factory. *Occup. Environ. Med.* 44:220–227.

Tiller, J.R., Schilling, R.S.F., and Morris, J.N. (1968) Occupational toxic factors in mortality from coronary heart disease. *Br. Med. J.* 4:407–411.

U.S. EPA, *ESEPA/OPP Pesticide Related Database Queries-OPP's Chemical Ingredients Database (Updated 1999).* Available at www.cdpr.ca.gov/dosc/epa/epamenu.htm.

van Doorn, R.V. et al. (1981) Determination of thio compounds in urine of workers exposed to carbon disulfide. *Arch. Environ. Health* 36:289–297.

Vanhoorne, M., Comhaire, F., and DeBaquer D.D. (1994) Epidemiological study of the effects of carbon disulfide on male sexuality and reproduction. *Arch. Environ. Health* 49:273–278.

Vanhoorne, M., DeBacquer, D., and Barbier, F. (1992) Epidemiological study of gastrointestinal and liver effects of carbon disulfde. *Int. Arch. Occup. Environ. Health* 63:517–523.

Vanhoorne, M., Vermeulen, A., and DeBaquer, D.D. (1993) Epidemiological study of endocrinological effects of carbon disulfide. *Arch. Environ. Health* 48:370–375.

Vasak, V., Vanecek, R., and Kimelova, M. (1963) Assessment of exposure of workers to CS_2 vapours. Part II. Application of iodine-azide reaction to the detection and estimation of CS_2 metabolites in wine. *Prac. Lek.* 15:145–149.

Wilmarth, K.R., Viana, M.E., and Abou-Donia, M.B. (1993) Carbon disulfide inhalation increases Ca^{2+}calmodulin-dependent kinase phosphorylation of cytoskeletal proteins in the rat central nervous system. *Brain Res.* 628:293–300.

World Health Organization (1979) *Environmental Health Criteria 10: Carbon Disulfide,* Geneva.

Zenk, H. (1970) CS_2 effects upon olfactory and auditory functions of employees in the synthetic fiber industry. *Int. Arch. Arbeitsmed.* 27:210–220.

45

HYDROGEN SULFIDE

Stephen C. Harbison and Jacob R. Bourgeois

First aid: Remove from exposure. Give fresh air. Immediately irrigate eyes, wash the skin with soap. Provide respiratory support. If frostbite has occurred, seek medical attention. If ingested by swallowing, seek immediate medical attention.

Target organs: Central nervous system, respiratory system (especially lungs), mucosal membranes, eyes.

Risk/safety phrases: Irritant gas; chemical asphyxiant; flammable. Liquid hydrogen sulfide can cause frostbite.

Exposure limits and threshold limit values for hydrogen sulfide:

	CASRN	PEL[1], ppm	REL, ppm	TLV–TWA, ppm	STEL, ppm	IDLH, ppm
Hydrogen sulfide (general industry)	7783-06-4	20 (ceiling) 50 (peak)	10 (10 min ceiling)	1	5	100
Hydrogen sulfide (construction)	7783-06-4	10 (TWA)				
Hydrogen sulfide (shipyard)	7783-06-4	10 (TWA)				

[1] Exposures must not exceed 20 ppm with the following exception: if no other measurable exposure occurs during the 8 h shift, exposures may exceed 20 ppm, but no more than 50 (peak), for a single time period up to 10 min. *Source*: www.osha.gov/STLC/hydrogensulfide/standards.html.

Chemical Abstract Service (CAS); permissible exposure limit (PEL), set by the Occupational Safety and Health Administration (OSHA); recommended exposure limit (REL), set by the National Institute for Occupational Safety and Health (NIOSH); threshold limit value (TLV), set by the American Conference of Governmental Industrial Hygienists (ACGIH); short-term exposure limit (STEL), set by the ACGIH; IDLH, immediately dangerous to life and health, set by the NIOSH; time-weighted average (TWA).

BACKGROUND AND USES

Hydrogen sulfide (H_2S) is a powerful chemical asphyxiant and irritant gas, which can result in loss of consciousness and sudden death when encountered at high concentrations. Its distinctive rotten egg odor earned it nicknames of swamp gas, sewer gas, and stink damp. Hydrogen sulfide is produced naturally as well as found in industrial settings where it is used as a reactant or is the result of manufacturing or industrial processes. As a natural source, H_2S is one of the principal components in the sulfur cycle and is present in volcanic gases, sulfur springs, undersea vents, swamps, crude petroleum, and natural gas (Lancia et al., 2013). Hydrogen sulfide is also naturally produced from the decomposition of organic sulfur-containing matter and from the reduction of inorganic sulfides (Chen and Wang, 2012). Some scientists have speculated that the extinction of certain species during the planet's evolutionary history may have been due to the natural release of large amounts of built-up H_2S (Wang, 2012). It is recognized as one of the dangers in gypsum, sulfur, and coal mining, since subterranean pools of water collect H_2S formed by the decomposition of iron pyrites.

Hydrogen sulfide is the second most common cause of fatal gas inhalation exposures in the workplace and casualties usually occur in more than one person as would-be rescuers rush to save their coworkers and in their haste neglect to use proper personal protective equipment (PPE) (Guidotti,

Hamilton & Hardy's Industrial Toxicology, Sixth Edition. Edited by Raymond D. Harbison, Marie M. Bourgeois, and Giffe T. Johnson.
© 2015 John Wiley & Sons, Inc. Published 2015 by John Wiley & Sons, Inc.

2010). It is frequently found in the following industrial settings as a by-product of an industrial process: tanneries, waste water treatment facilities, manure and sewage facilities, rayon manufacturing plants, sulfur producers, coke oven plants, Kraft paper mills, iron smelters, food processing plants, tar and asphalt manufacturing plants, petrochemical plants, and oil refineries (Guidotti, 2010). Ramazzini is generally considered the first notable physician to associate its presence with occupational risk through his observation of health changes in sewer workers (Wang, 2012).

PHYSICAL AND CHEMICAL PROPERTIES

Hydrogen sulfide is generally found as a pungent colorless flammable gas, although it is commonly shipped as liquefied compressed gas (NIOSH, 2011). Its characteristic odor of "rotten eggs" cannot be considered indicative of its concentration, as olfactory fatigue occurs quite rapidly, even at relatively low concentration levels (100–150 ppm). It has a vapor density of 1.19, which makes it heavier than air and will cause it to accumulate in low lying areas rather than disperse easily in air and thus can present a hazard for workers in basements, subterranean locations, and confined spaces. It is soluble in water to a maximum of 0.4% by mass at room temperature. The auto ignition temperature is 260 °C and the flammable limits in air, % by volume are 4.5 (lower) and 45.5 (upper). The tendency for belowground accumulation, coupled with its water solubility and flammability has proven to be a significant hazard for those who conduct work in underground tunnels or sewers (Adelson and Sunshine, 1966; Chen and Wang, 2012). Reactivity can occur from contact with strong oxidizers and oxidizing materials and result in fires and explosions. Hydrogen sulfide also attacks many metals resulting in the formation of sulfides and can lead to metal integrity issues.

MAMMALIAN TOXICOLOGY

Hydrogen sulfide is classified as both a chemical asphyxiant and irritant gas. Guidotti (2010) states that H_2S toxicity is one of the most unusual and reliable toxidromes in medical toxicology and remains an important and frequently lethal occupational and environmental hazard.

Asphyxiation is the most frequent cause of death related to occupational exposure to H_2S, where it interferes with cytochrome oxidase and aerobic metabolism (Lancia et al., 2013). At high concentrations, H_2S acts as a chemical asphyxiant as it is a potent inhibitor of cytochrome oxidase. The inhibition interferes with cellular metabolism causing blockage of the proper utilization of oxygen by the cell, ultimately resulting in cytotoxic anoxia despite normal blood flow and oxygen content. At sufficiently high concentrations, these actions may occur before the irritant (i.e., physiological warning) effects of H_2S have time to develop.

Postmortem examinations of fatalities indicate that the primary targets of H_2S poisonings appear to be the central nervous and respiratory systems. Distinctive changes such as a greenish discoloration of the brain, viscera and blood, coupled with pulmonary edema, petechia, and hepatic and visceral congestion are often seen on these postmortem examinations (Adelson and Sunshine, 1966; Arnold et al., 1985; Park et al., 2009).

Acute Effects

Acute H_2S poisoning may present with both irritant and asphyxiant effects. At low concentrations (50 ppm), it can cause local effects (irritative action), acting mainly on the mucous membranes of the airways, lungs, eyes, and digestive organs (Nogue et al., 2011). In tests on small animals, 50 ppm exposures were consistently associated with conjunctival irritation (Evans, 1967). Exposures in excess of 200 ppm are consistently associated with impairment to the respiratory system and mucosal membranes (Dorman et al., 2000; Snyder et al., 1995), although Chen and Wang (2012) noted that respiratory impairment symptoms have been noted in some studies using exposures as low as 10 ppm. Pulmonary edema is another acute effect of H_2S toxicity, because the gas has the ability to travel deep into the respiratory tract and cause alveolar injury especially when exposure is prolonged (Guidotti, 2010).

At higher concentrations, H_2S is a central nervous system depressant, particularly of the respiratory center, and inhibits cytochrome oxidase, preventing mitochondrial oxygen utilization and thereby blocking cellular respiration (Nogue et al., 2011). At sufficiently high concentrations, these actions may occur before the irritant effects of H_2S have time to develop. An example of this is a phenomenon known as "knockdown" which has been reported by oil field workers and described as a sudden, brief loss of consciousness associated with amnesia, followed by immediate full recovery (Doujaiji and Al-Tawfig, 2010). Traumatic injuries from falls have been reported as a result of the sudden loss of consciousness. Dorman et al. (2000) pointed out that the disruption of memory and motor functions, personality changes, and hallucinations that categorize the "knockdown" syndrome were also consistent with anoxia-induced effects of asphyxiants such as carbon monoxide. Significant respiratory tissue and function damage leading to unconsciousness can occur between 250 and 500 ppm, and exposures in excess of 700 are most frequently associated with respiratory arrest and cardiovascular failure and death (Dorman et al., 2000; Snyder et al., 1995). Table 45.1 presents health effects of H_2S at various approximate exposure levels.

Hydrogen sulfide is also transported under high pressure in a liquefied form. Exposure to liquefied H_2S can chill or freeze the skin causing frostbite.

TABLE 45.1 Exposure Concentrations and Health Effects from H$_2$S Exposures

Concentration, ppm	Health Effects
0.01–0.03	Odor threshold (highly variable)
1–5	Moderate offensive odor, may be associated with nausea, tearing of the eyes, or headaches
10	Anaerobic metabolism threshold during exercise
20	Odor very strong; conjunctivitis may occur
20–50	Conjunctivitis and lung irritation. Possible eye damage after several days of exposure; may cause digestive upset and loss of appetite
100	Eye and lung irritation; olfactory paralysis, odor disappears
150–200	Sense of smell paralyzed; severe eye and lung irritation
250–500	Pulmonary edema may occur, especially if prolonged
500	Serious damage to eyes within 30 min; severe lung irritation; "knockdown" and death within 4–8 h; amnesia for period of exposure
1000	Breathing may stop within one or two breaths; immediate collapse

Source: Guidotti (2010).

Chronic Effects

Although the negative effects of acute exposure to high concentrations of H$_2$S have been clearly identified, little information has been confirmed about the deleterious effects of chronic exposure to low level concentrations. Chronic exposures have been shown to affect olfactory senses, and may alter the threshold for odor detection or cause complete anosmia. In either case, the individual may be less able to detect the presence of the H$_2$S; however, recognition of odor does not necessarily correlate with existing concentrations and thus should not be relied upon as a means of ensuring worker safety.

Due to the targeting of the central nervous system and respiratory system in acute exposures, studies have been performed to evaluate potential health effects from chronic low dose exposure. Two studies have attempted to identify effects of low dose exposure to H$_2$S by exposing asthmatic subjects to 2 ppm and healthy exercising individuals to 5 ppm of H$_2$S for 30 min (Jappinen et al., 1990; Bhambani et al., 1994). No respiratory or other adverse health effects were reported in either study. In a 2-year study of a New Zealand city situated among volcanic vents, Bates et al. (2013) found no significant increase in asthma risk among the residents continuously exposed to H$_2$S levels varying between 0 and 764 parts per billion (ppb) in ambient air. The study suggested that low level exposures (i.e., 20 ppb mean) could be associated with a reduction in symptoms of respiratory obstructive diseases (e.g., asthma), potentially due to the induction of smooth muscle effects of H$_2$S. Inserra et al. (2004) performed a neurobehavioral study of residents of a city located in Nebraska who had been routinely exposed during the 1990s to ambient air levels of H$_2$S ranging from 100 to 5000 ppb. Other than the two tests (match to sample and a grip test) which had marginal differences, the remaining 26 tests showed no difference in exposed and nonexposed residents. In a laboratory exposure study, subjects exposed to H$_2$S as low as 0.05 ppm did report more increased anxiety related to ratings of irritation due to odor, but no cognition performance differences were observed in a dose–response association. There were indications of decreased word-learning retention; however, as performance increased with increasing ppm, it is possible that the odor may have been a distraction that faded with olfactory fatigue, not a physiological impediment (Fiedler et al., 2008).

The above studies lend weight to claims that chronic exposure to low to moderate levels of H$_2$S are not associated with significant or permanent injury. In his study of the physiological detoxification process of H$_2$S exposure, Evans (1967) pointed out that the continued respiration of low doses of H$_2$S without adverse effect show that H$_2$S exposures are not cumulative. Adelson and Sunshine (1966) attribute the lack of cumulative damage to the rapid oxidation of H$_2$S in the blood to sulfate and thiosulfate. Bhambani et al., (1994) similarly attributes the body's rapid detoxification of H$_2$S as the protection against cumulative damage.

The USEPA has conducted a human health assessment for H$_2$S (revised 2003) and developed an inhalation reference concentration (2E-3 mg/m^3) with the critical effect being nasal lesions of the olfactory mucosa based on a rat subchronic inhalation study conducted by Brenneman et al. (2000). The USEPA noted that the focus of the human health assessment is on long-term chronic exposures.

Carcinogenicity

Hydrogen sulfide has not been established as a human carcinogen. The USEPA reports that data are inadequate for an assessment of the carcinogenic potential of H$_2$S and that no relevant data could be located from which to develop an assessment.

Mechanism of Action(s)

The primary route of exposure for H$_2$S is inhalation; therefore it will be primarily absorbed through the lungs and broadly distributed in the blood throughout the body. Metabolism of H$_2$S proceeds through three primary pathways which include oxidation, methylation, and reaction with metalloproteins or disulfide-containing proteins. Oxidation in the liver is the major pathway and generally occurs under conditions of low concentrations, causing H$_2$S to be rapidly oxidized into a sulfate or thiosulfate by oxygen bound to hemoglobin in the blood. A small portion of the H_2S may enter the muscles, where it undergoes a similar transformation by oxygen bound to myoglobin. The trace amounts of remaining gas are eliminated by the lungs as dissolved H_2S. This detoxification process occurs rapidly so that at low concentrations, H_2S does not act as a cumulative toxicant. A study by Dorman et al. (2000) showed no evidence of fertility and developmental neurotoxicity effects in Sprague-Dawley rats that were exposed to occupationally relevant exposure concentrations levels (10, 30, or 80 ppm) of H_2S 2 weeks prior to breeding, the 2 week mating period, and the gestation period.

In the event that the detoxification capacity is exceeded, excess H_2S undergoes a third metabolic process involving a protein reaction into toxic metabolites. These metabolites can suppress aerobic metabolism in the various tissues by inhibiting mitochondrial respiration, specifically cytochrome oxidase. In this way, H_2S exerts its systemic toxic effects in a manner similar to cyanide (Beauchamp et al., 1984; Bhambani et al., 1994; Sams et al., 2013).

Ballerino-Regan and Longmire (2010) stated that postmortem blood sulfide levels can be suitably indicative of poisonous exposures if taken within 2 h of exposure, unlike sulfhemoglobin levels, which are not. They reported that the blood levels of thiosulfide recorded in postmortems were found to be 8–77 times the normal level of 0.3 μg/ml, with elevated levels also found in brain and lung tissue samples. Bott and Dodd (2013) also found that thiosulfate has been reported to be a useful indicator for H_2S exposure and that levels in fatal human cases have ranged from 0.025 to 0.143 mmol/l^2. Chaturvedi et al. (2001) noted that blood sulfide levels from 8 to 5.0 μg/ml were associated with H_2S exposure in postmortems.

SOURCES OF EXPOSURE

Hydrogen sulfide is a well-recognized occupational and industrial hazard causing fatalities of workers in a variety of settings. Inhalation is considered the main route of exposure.

Occupational

Perhaps the most significant occupational exposure source of H_2S is within the petroleum and natural gas industry. Crude oil is "saturated" with H_2S while natural gas can contain levels up to 900,000 ppm (Arnold et al., 1985). Doujaiji and Al-Tawfig (2010) stated that H_2S is the primary chemical hazard of natural gas production and found in a retrospective review from the oil and gas industry in Canada (1969–1973) that 221 cases of H_2S exposure were reported and 173 patients were transported to the hospital with 14 of the victims dead on arrival. The presence of high levels of

H_2S offer more than just a direct health risk; H_2S can cause metal stress leading to equipment or pipeline failure, and it can combine with other chemicals to create secondary toxic gases. Sams et al. (2013) states that deaths related to H_2S have traditionally involved accidents from workplace exposure to sewer gas or fatal exposures in commercial manufacturing and that OSHA reported 13 work-related H_2S asphyxiation deaths in 2007.

In any industrial process in which sulfides or sulfates may contact a reducing agent, consideration of the handling and disposal methods of the products or by-products must be carefully examined. For example, a truck driver accidently transferred sodium hydrogen sulfide into a tank with 4% sulfuric acid and ferrous sulfate, causing his immediate collapse and death (Chaturvedi et al., 2001). In another incident, four deaths were attributed to the presences of H_2S generated in the sludge from the combination of a decolorizer and a pH balancer in the waste water (Kage et al., 2004). In the review of cases during two time periods between 1979 and 1983, Arnold et al. (1985) noted that over 75% of H_2S exposures results in 2 days' work loss or less, while 8% resulted in >2 weeks loss, or mortality. Eighty-six percent of the exposures were related to the natural gas and petroleum industry with the remaining exposures, including university laboratories and food processing industries.

In 2011, the USEPA lifted an Administrative Stay of the Emergency Planning and Community Right-to-Know Act (EPCRA) section 313 TRI reporting requirements for H_2S. Therefore facilities that manufacture, process, or otherwise use H_2S in amounts above activity thresholds must report their environment releases and other waste management quantities annually. Hydrogen sulfide is subject to the standard activity thresholds of 25,000 pounds for manufacturing and processing and 10,000 pounds for otherwise use. It falls under the de minimis exemption because it is not considered a carcinogen by OSHA definition and therefore is exempt in a mixture at a concentration lower than 1.0%.

Environmental

Hydrogen sulfide occurs as a natural by-product from the decay of organic material containing sulfur and can be found naturally in the environment. It is found commonly in the later stages of volcanic eruptions. Along with sulfur dioxide and carbon dioxide, it is considered one of the most potentially hazardous volcanic gases because it is heavier than air and has the potential to accumulate in hollows. Its concentrations in ambient air range from 0.1 to 1 μg/m^3 as a result of natural emissions while concentrations above 100 μg/m^3 have been reported near industrial plants (WHO, 2003).

Low levels of H_2S have been found in water due to natural sources and industrial processes; however, drinking water is not currently considered a significant source of H_2S exposure and no drinking water quality standard has been set.

INDUSTRIAL HYGIENE

It is important to monitor air concentrations in occupational settings where H_2S exposure is a risk, keeping concentrations within occupational permissible levels and using appropriate sampling techniques and analytical methods. Standards and guidelines for H_2S in the occupational setting vary between organizations. OSHA, the only organization with enforceable standards, has different PELs depending upon the occupational setting. "General Industry" exposures have a ceiling limit of 20 ppm, with an exception when exposure is sporadic. In such a case where an exposure may only occur once within the work-day, OSHA requires that the exposure not exceed 50 ppm for a single time period up to 10 min, i.e., there is no use of a TWA. For "Construction" and "Maritime" industries, the OSHA PEL is a TWA of 10 ppm. The working level established by ACGIH is a TLV–TWA of 1 ppm and a STEL of 5 ppm, based on its review of upper respiratory irritation and central nervous system involvements. (ACGIH, 2013) The NIOSH REL has a ceiling limit (10 min) of 10 ppm and the concentration IDLH is 100 ppm (NIOSH, 2011).

The pungent odor of H_2S cannot be considered a reliable indicator of the continued presence of the gas as olfactory fatigue occurs rapidly at low levels. Fortunately, most occupational exposures are below an 8 h TWA of 10 ppm (Dorman et al., 2000) and thus decrease the likelihood of olfactory fatigue and paralysis in a routine work environment.

Engineering controls to limit H_2S exposure should be exercised wherever feasible. Prevention of skin and eye contact is recommended through the use of PPE, especially in moist areas where work garments may become damp and eye wash stations should be readily available in the event of an accidental exposure. If engineering controls and work practices are not feasible, respirators may need to be used. In this case, only approved respirators should be permitted and a respiratory protection program should be developed and include training requirements, maintenance, inspection, cleaning, and evaluation. The corrosivity of H_2S, especially in moist environments, should be considered as well in the selection of PPE and respiratory support equipment.

Many of the fatalities due to occupational exposure to H_2S were caused by untrained workers or even first responders succumbing to its effects as they attempted a rescue. Thus, education and training should be conducted for all personnel that work in areas where H_2S levels could exceed allowable levels. General safety awareness should include the characteristics, potential sources, and hazards of H_2S, as well as the proper use of PPE, rescue equipment, detection methods, and occupational permissible levels.

Even in industrial settings not specifically involving H_2S as a product or by-product, H_2S poisoning can occur by the inadvertent mixing of chemicals used for innocuous purposes such as cleaning and site maintenance. Proper storage and training in the handling of these chemicals must also be a priority in the occupational setting.

MEDICAL MANAGEMENT

The diagnosis of H_2S poisoning is primarily dependent upon the awareness of the circumstances of exposure and industry in which the exposure takes place. When the source of injury is unknown, a history of rotten egg smells may provide some insight. No specific physical or laboratory findings are definitive in the diagnostic process making any information of exposure circumstances of paramount importance. The sudden collapse of a worker in an area of potential H_2S production should elicit an immediate assessment for H_2S poisoning (Ballerino-Regan and Longmire, 2010).

Treatment for H_2S exposure is not specific and there is no proven antidote with medical treatment consisting primarily of supporting respiratory and cardiovascular functions. In cases of mild exposure where dermal or conjunctival irritation is the primary complaint, immediate irrigation of any exposed surfaces should be initiated and continued until all burning sensation ceases at which point subsequent medical evaluation should be sought. Concentrated H_2S can cause chemical burns and should be treated in the same manner as a thermal burn. Liquefied H_2S can cause frostbite if it comes into contact with the skin and should be treated by rewarming the affected areas and seeking medical attention. In the case of inhalation exposure, supportive care should be administered and early administration of hyperbaric oxygen, amyl nitrite, and sodium nitrite may be beneficial (Doujaiji and Al-Tawfiq, 2010). According to Nogue et al. (2011), medical treatment should include 100% oxygen and ensuring correct metabolic acidosis with bicarbonate, utilizing CPR if warranted. Evans (1967) also recommends that 100% oxygen be administered immediately, as well as ventilatory support provided if needed as the ability of the body to detoxify H_2S is directly related to the oxygen content of the blood. Due to the corrosivity of H_2S, children may be more vulnerable to respiratory compromise because of the smaller diameter of their airways (ATSDR, 2014). Treatment for hypotension, if present, should involve supportive therapy, such as fluids. Because of the mechanistic similarity to cyanide poisoning, treatment with the cyanide antidote sodium nitrite has been suggested. The rationale is to generate methemoglobin, which competes for H_2S by forming sulfmethemoglobin, freeing the cytochrome oxidase (Chen and Wang, 2012; Evans, 1967). While some anecdotal evidence has been associated with successful treatment, there has been no concrete evidence for the human efficacy of this treatment to date (Burnett et al., 1977; Chen and Wang, 2012; Morii et al., 2010; Stine et al., 1976).

REFERENCES

Agency for Toxic Substances & Disease Registry (ATSDR) (2014) Medical Management Guidelines for Hydrogen Sulfide. Available at http://www.atsdr.cdc.gov/MHMI/mmg114.pdf.

Adelson, L. and Sunshine, I. (1966) Fatal hydrogen sulfide intoxication: report of three cases occurring in a sewer. *Arch. Pathol.* 81:375–380.

American Conference of Governmental Industrial Hygienists (ACGIH) (2013) *Threshold Limit Values (TLVs) for Chemical Substances and Physical Agents and Biological Exposure Indices (BEIs)*, Cincinnati, OH: American Conference of Governmental Industrial Hygienists.

Arnold, I.M., et al. (1985) Health implications of occupational exposure to hydrogen sulfide. *J. Occup. Med.* 27:373–376.

Ballerino-Regan, D. and Longmire, A.W. (2010) Hydrogen sulfide exposure as a cause of sudden occupational death. *Arch. Pathol. Lab. Med.* 134:1105.

Bates, M.N., Garrett, N., Crane, J., and Balmes, J.R. (2013) Associations of ambient hydrogen sulfide exposure with self-reported asthma and asthma symptoms. *Environ. Res.* 122:81–87.

Beauchamp, R.O. et al. (1984) A critical review of the literature on hydrogen sulfide toxicity. *Crit. Rev. Toxicol.* 13:25–97.

Bhambani, Y., et al. (1994) Comparative physiological responses of exercising men and women to 5 ppm hydrogen sulfide exposure. *Am. Ind. Hyg. Assoc. J.* 55:1031–1035.

Bott, E. and Dodd, M. (2013) Suicide by hydrogen sulfide inhalation. *Am. J. Forensic Med. Pathol.* 34(1):23–25.

Brenneman, K.A., James, R.A., Gross, E.A., and Dorman D.C. (2000) Olfactory neuron loss in adult male CD rats following subchronic inhalation exposure to hydrogen sulfide. *Toxicol. Pathol.* 28(2):326–333.

Burnett, W.W., King, E.G., and Grace, M. (1977) Hydrogen sulfide poisoning: review of 5 years' experience. *Can. Med. Assoc. J.* 117:1277–1280.

Chaturvedi, A.K., Smith, D.R., and Canfield, D.V. (2001) A fatality caused by accidental production of hydrogen sulfide. *Forensic Sci. Int.* 123:211–214.

Chen, Y. and Wang, R. (2012) The message in the air: hydrogen sulfide metabolism in chronic respiratory diseases. *Respir. Physiol. Neurobiol.* 184:130–138.

Dorman, D.C., Brenneman, K.A., Struve, M.F., Miller, K.L., James, R.A., Marshall, M.W., and Foster P.M.D. (2000) Fertility and developmental neurotoxicity effects of inhaled hydrogen sulfide in Sprague-Dawley rats. *Neurotox. Teratol.* 22:71–84.

Doujaiji, B. and Al-Tawfiq, J.A. (2010) Hydrogen sulfide exposure in an adult male. *Ann. Saudi Med.* 30(1):76–80.

Evans, C.L. (1967) The toxicity of hydrogen sulfide and other sulfides. *Q. J. Exp. Physiol.* 52:231–248.

Fiedler, N., Kipen, H., Ohman-Strickland, P., Zhang, J., Weisel, C., Laumbach, R., Kelly-McNeil, K., Olejeme, K., and Lioy, P. (2008) Sensory and cognitive effects of acute exposure to hydrogen sulfide. *Environ. Health Perspect.* 116(1):78–85.

Guidotti, T.L. (2010) Hydrogen sulfide: advances in understanding human toxicity. *Int. J. Toxicol.* 29(6):569–581.

Inserra, S.G., Phifer, B.L., Anger, W.K., Lewin, M., Hilsdon, R., and White, M.C. (2004) Neurobehavioral evaluation for a community with chronic exposure to hydrogen sulfide gas. *Environ. Res.* 95:53–61.

Jappinen, P., et al. (1990) Exposure to hydrogen sulfide and respiratory function. *Br. J. Ind. Med.* 47:824–828.

Kage, S., Ikeda, H., Ikeda, N., Tsujita, A., and Kudo, K. (2004) Fatal hydrogen sulfide poisoning at a dye works. *Leg. Med. (Tokyo)* 6:182–186.

Lancia, M., Panata, L., Tondi, V., Carlini, L., Bacci, M., and Rossi, R. (2013) A fatal work-related poisoning by hydrogen sulfide: report on a case. *Am. J. Forensic Med. Pathol.* 34(4):315–317.

Morii, D., Miyagatani, Y., Nakamae, N., Murao, M., and Taniyama, K. (2010) Japanese experience of hydrogen sulfide: the suicide craze in 2008. *J. Occup. Med. Toxicol.* 5:28.

National Institute for Occupational Safety and Health (NIOSH) (2011) *Pocket Guide to Chemical Hazards. NIOSH 94–116*, Washington, D.C.: US Department of Health and Human Services.

Nogue, S., Pou, R., Fernandez, J., and Sanz-Gallen, P. (2011) Fatal hydrogen sulphide poisoning in unconfined spaces. *Occup. Med. (Lond.)* 61(3):212–214.

Park, S.H., Zhang, Y., and Hwang, J. (2009) Discolouration of the brain as the only remarkable autopsy finding in hydrogen sulfide poisoning. *Forensic Sci. Int.* 187:e19–e21.

Sams, R.N., Carver, H.W., Catanese, C., and Gilson, T. (2013) Suicide with hydrogen sulfide. *Am. J. Med. Pathol.* 34:81–82.

Snyder, J.W., Safir, E.F., Summerville, G.P., and Middlebeg, R.A. (1995) Occupational fatality and persistent neurological sequelae after mass exposure to hydrogen sulfide. *Am. J. Emerg. Med.* 13(2):199–203.

Stine, R.J., Slosberg B., and Beachman B.E. (1976) Hydrogen sulfide intoxication: a case report and discussion of treatment. *Ann. Intern. Med.* 85:756–758.

Wang, R. (2012) Physiological implications of hydrogen sulfide: a whiff exploration that blossomed. *Physiol. Rev.* 92:791–896.

WHO (2003) *Hydrogen Sulfide in Drinking-Water*. Background document for development of WHO *Guidelines for Drinking-Water Quality*, Geneva.

46

CYANIDE

Erin L. Pulster and James V. Hillman

Target organ(s): Cardiovascular and respiratory system, thyroid, eyes, skin and blood

Occupational exposure limits: OSHA PEL: 4.7 ppm (5 mg/m^3) 8 h TWA (skin); ACGHI TLV: 4.7 ppm (5 mg/m^3) ceiling (skin); NIOSH REL: 4.7 ppm (5 mg/m^3) STEL (skin) and ceiling (10 min); NIOSH IDLH: 25 mg/m^3.

Risk/safety phrases: Compound dependent, most are incompatible with water, acids, and strong oxidizers.

BACKGROUND AND USES

Cyanide is a rapid acting and potentially fatal chemical, which exists in a variety of chemical complexities and forms (gas, liquid, or solid), all of which contain a CN moiety. While many chemical forms of cyanide exist from both natural and anthropogenic sources, the cyanide anion (CN$^-$) is the primary toxic agent, regardless of origin (Simeonova and Fishbein, 2004).

Natural sources are derived from a variety of bacteria, fungi, algae, and plants (<2000 species). Fruits, vegetables, and seeds contain cyanogenic glycosides which can be released when ingested. Cyanogenic glycosides are present in a number of staple foods consumed in tropical regions, such as cassava (tapioca, manioc) and sorghum (grasses raised for grain). Other foods containing cyanogenic glycosides include sweet potato, lima beans, corn, cabbage, linseed, bitter almonds, bamboo shoots and the pits of wild cherries, apricots, peaches, apples, and pears (Way, 1984; Simeonova and Fishbein, 2004; ATSDR, 2006). Well-described cyanogenic glycosides include amygdalin, linamarin, dhurrin, prunasin, lautaustralin, and taxiphyllin.

Clinical trials tested the efficacy of using a chemically modified form of amygdalin (i.e., laetrile) found primarily in the kernels of apricots, peaches, and almonds as a chemotherapeutic agent; however, it was concluded that it was not an effective cancer treatment (Moertel et al., 1982).

Anthropogenic sources of cyanides consist of the manufacturing and processing of plastic, rubber, paper, textiles, fumigants, insecticides, pesticides, and chemicals for photography processing. Cyanide compounds can be formed from the combustion of synthetic materials (e.g., plastics) and fossil fuels (e.g., vehicle emissions) or the product of solid waste incinerators, biomass burning, fumigation operations, and the production chelating agents, carbon black, coke, or coal carbonization procedures (Simeonova and Fishbein, 2004).

Cyanide compounds have been used in medicine, fishing, pest control, mining, industrial organic chemistry, food (as an additive), and as a poison. For instance, it was used during World War II as a genocidal agent by the Germans. Cyanide salts are used in metallurgical industries for electroplating, metal cleaning, and the extraction of silver and gold. The major end uses of hydrogen cyanide are in the production of adiponitrile, cyanuric chloride, sodium cyanide, chelating agents, and nitrilotriacetic acid.

PHYSICAL AND CHEMICAL PROPERTIES

Cyanides comprise a wide range of compounds, all of which have a CN molecule and exist in either a solid, liquid, or gaseous form. Differing physical characteristics and chemical characteristics of some selected cyanide compounds are shown in Table 46.1. Physical characteristics range from a colorless or

Hamilton & Hardy's Industrial Toxicology, Sixth Edition. Edited by Raymond D. Harbison, Marie M. Bourgeois, and Giffe T. Johnson.

TABLE 46.1 Physical Properties of Selected Cyanide Compounds

Species	Synonyms	CASRN	Molecular Formula	Molecular Weight (Da)	Boiling Point (°C)	Melting Point (°C)	Solubility
Cyanide	Cyanide anion, cyanure, hydrocyanic acid	373-51-3 57-12-5	CN	26.02	26	No data	Soluble at 25 °C (mg/l) in water
Hydrogen cyanide	Prussic acid, hydrocyanic acid, formic anammonide	74-90-8	HCN	27.03	26	−13	Miscible in H_2O at 30 °C
Sodium cyanide	Cyanide of sodium, natriumcyanide	143-33-9	NaCN	49.02	1496	564	48 g/100 ml (10 °C) in H_2O
Potassium Cyanide	Cyanide of potassium, kaliumcyanide	151-50-8	KCN	65.11	1625	634	In water and glycerol
Calcium cyanamide	Calcid, calsyan	156-62-7	$CaCN_2$	80.10	No data	1340	Insoluble
Cyanamide	Dormex, hydrogen cyanamide	420-04-2	NH_2CN	42.04	132–138	43–47	In organic solvents
Cyanogen	Carbon nitrile, dicyanogen, dicyan, oxalic nitrile	460-19-5	NCCN	52.03	−21	−28	In water, alcohol, and ether
Cyanogen chloride	Cyanic chloride, cyanicchlorid, chloride cyanide	506-77-4	ClCN	61.47	14	−6.5	In ethanol and ether

pale blue liquid with a faint bitter almond-like odor (hydrogen cyanide, HCN) to a white solid, powder, or crystalline hygroscopic salt (sodium cyanide, NaCN) to a colorless toxic gas also with an almond-like odor (cyanogens, NCCN). Cyanide compounds are either organic or inorganic. Organic cyanides contain a noncovalent CN functional group and are typically called the cyano group or nitriles. A common nitrile is methyl cyanide, also known as acetonitrile (CH_3CN). Inorganic cyanides have a negatively charged polyatomic cyanide ion (CN^-) and are generally referred to as cyanides. This group includes the cyanide salts (i.e., sodium cyanide, NaCN) which are considered the most toxic form.

MAMMALIAN TOXICOLOGY

Acute Effects

Acute cyanide poisoning is associated with accidental ingestion, homicide or suicide cases, occupational exposures, and

fire smoke (Borron and Baud, 1996; Alarie, 2002). Depending on the type of cyanogen, lethal doses in rats ranged from 57 to >5000 mg/kg, 14 to 2778 mg/kg, and 10 to >5000 mg/kg, for oral, intraperitoneal, or subcutaneous, respectively (Rao et al., 2013, Table 46.2). Inhalation LC_{50} values for rats exposed to hydrogen cyanide were observed at 158 mg/m³ for 60 min exposures and 3778 mg/m³ for 10 s exposures (Ballantyne, 1983). In humans, inhalation of airborne cyanide compounds in excess of 100 ppm can result in death within an hour, whereas concentrations exceeding 300 ppm are generally fatal within minutes (Beasley and Glass, 1998). Similarly, the lethal dose of sodium or potassium cyanide is 200–300 mg within seconds of inhalation. Ingested hydrogen cyanide has a lethal dose of 50 mg with effects beginning within 30 min.

Acute cyanide poisonings primarily affect the central nervous system and result in cardiovascular problems. Inhalation, dermal or dietary (ingestion) exposures to small amounts of cyanide may result in immediate signs and symptoms consisting of dizziness, headache, nausea,

TABLE 46.2 Route-specific LD_{50} (mg/kg) for Various Cyanogens in Rats

	Oral	Intraperitoneal	Subcutaneous
Acetonitrile	>5000	2388 (2053–2778)	>5000
Acrylonitrile	95 (82–111)	95 (82–111)	95 (82–111)
Malononitrile	66 (57–77)	42 (36–49)	48 (41–55)
Propionitrile	84 (72–97)	190 (163–221)	151 (130–175)
Sodium nitroprusside	84 (72–97)	17 (14–19)	12 (10–14)
Succinonitrile	379 (325–440)	836 (718–972)	755 (649–878)

Table modified from Rao et al., 2013.

vomiting, tachypnea, tachycardia, restlessness, and weakness. Exposures to large amounts of cyanide by any exposure route can result in convulsions, loss of consciousness, hypotension, lung injury, respiratory failure, bradycardia, and death. Pulmonary edema and lactic acidosis have been reported after cyanide ingestion (Graham et al., 1977; Beasley and Glass, 1998). Many life-threatening intoxications, including loss of consciousness, have resulted in complete recovery (Peden et al., 1986; Scolnick et al., 1993). Mild cases of exposure in fumigators and factory workers have resulted in symptoms of oxygen deprivation.

Chronic Effects

Toxicological interest primarily focuses on acute cyanide poisoning; however, its most widely distributed toxicological problems are due to chronic toxicity from dietary, industrial, and environmental factors (Way, 1984). Most of the evidence is based on observational studies, with few controlled experiments. The unknown contribution of other factors makes this a complex problem. However, chronic exposure appears to involve the central nervous system and endocrine system.

One of the most widespread pathological conditions associated with cassava consumption is tropical ataxic neuropathy (Osuntokun et al., 1969; Way, 1984; Ernesto et al., 2002). This is a diffuse degenerative neurological disease characterized by irreversible paraparesis that has previously been correlated with concentrations of plasma thiocyanate. Thiocyanate, one of the major detoxification products, acts as a goitrogenic agent by preventing the uptake of iodine. Recent expansion of cassava processing in Nigeria has revealed significantly higher thiocyanate levels in workers compared with non-cassava processors (Adewusi and Akindahunsi, 1994).

Evidence of endocrine toxicity was also apparent in the Nigerian study. Goiter was more prevalent in cassava processors than in other workers. Studies of electroplaters have revealed that workers in this industry, many of whom have enlarged thyroid glands, may be at risk for chronic disease related to cyanide (Way, 1984). A study in rural Mozambique evaluated thyroid function in a population suffering from epidemic spastic paraparesis resulting from the consumption of cassava (Cliff et al., 1986). This study found that subjects had altered thyroid function. However, the authors also noted that if iodine intake is adequate, high levels of dietary cyanide may be tolerated without overt goiter or hypothyroidism.

In addition, a case report of cyanide-induced Parkinson's disease described magnetic resonance imaging changes resulting from either direct cyanide toxicity or from cerebral anoxia, occurring secondary to cyanide intoxication (Rosenberg et al., 1989). Perhaps the most convincing evidence of cyanide CNS toxicity is tobacco amblyopia. Victims have a history of heavy smoking, visual disturbances, and depletion of vitamin B_{12}. The condition is reversed by treatment with a cyanide antagonist, hydroxocobalamin.

El Ghawabi et al. (1975) investigated the effects of chronic cyanide exposure in the electroplating sections of three factories in Egypt that employed 36 workers. A significant correlation was established between urinary excretion of thiocyanates and the concentration of airborne CN (4.2–12.4 ppm) in the workplace. The exposed workers had significantly higher hemoglobin values and lymphocyte counts compared with the unexposed controls, and 78% also had punctate basophilia. Cyanmethemoglobin was only detected in the exposed group, 20 of whom had variable thyroid enlargement, presumably from the thiocyanate. The symptoms noted with greatest frequency were headache, weakness, changes in taste and smell, giddiness, throat irritation, vomiting, and effort dyspnea. Studies with rat hepatocytes suggest that cyanide toxicity depends on adenosine triphosphate (ATP) formation, which maintains mitochondrial membrane potential (Snyder et al., 1993; Nieminen et al., 1994). Fructose was protective against cyanide toxicity through glycolytic ATP formation. Another study by Niknahad et al. (1994) found that all glycogenic and gluconeogenic amino acids and carbohydrates were protective against cyanide toxicity in rat hepatocytes.

Dermatitis is a familiar condition in workers, such as electroplaters, who are chronically exposed to cyanide solutions that tend to be strong irritating agents causing severe itching in sensitive individuals. Cyanide rash is described as usually consisting of pruritic papules and vesicles. Occasionally a blotchy eruption of the face may follow low level exposure to HCN vapors and solutions of cyanide salts. According to Braddock and Tingle (1930), the cyanide rash is more attributed to the caustic action of quicklime rather than cyanides among American gold miners.

Mechanism of Action(s)

The form of cyanide a victim is exposed to does not influence distribution, metabolism, or excretion from the body (Simeonova and Fishbein, 2004). Cyanide is rapidly absorbed through the lungs, skin, and gastrointestinal tract because of its unionized state and low molecular weight. The majority in blood is sequestered in red blood cells and a relatively small amount is transported to target organs via plasma, yet it has not been shown to accumulate in either blood or tissues (ATSDR, 2006; Simeonova and Fishbein, 2004). Symptoms occur within seconds of inhalation and within minutes of ingestion of cyanide salts. Gastrointestinal absorption of nitriles and cyanogens can produce symptoms up to 12 h after ingestion. The nitriles are readily absorbed through skin and through respiratory and gastrointestinal routes.

The central nervous system is particularly vulnerable to cyanide poisoning primarily due to its high dependence on oxidative metabolism and limited anaerobic capacity (Thomas and Brooks, 1970; Way, 1984; Simeonova and Fishbein, 2004; ATSDR, 2006). The principle impairment of cyanide poisoning is the inhibition of cytochrome oxidase, the terminal enzyme of the mitochondrial electron transport chain (Isom and Way, 1974; Way, 1984; Beasley and Glass, 1998). In turn, oxygen available for tissues is impaired resulting in a state of histotoxic anoxia (tissues inability to utilize oxygen). Essentially, cyanide inhibits the enzymes used for basic cellular respiration. Cyanide can also inhibit approximately 40 other enzymes that may be equally, if not more, sensitive to cyanide (Way, 1984; Simeonova and Fishbein, 2004), which include nitrate reductase, myoglobin, horseradish peroxidase, xanthine oxidase, alkaline phosphatase, copper, and iron. Many of the sensitive enzymes are metalloenzymes containing molybdenum or iron. It has been suggested that the inhibiting properties of cyanides may be attributed to its ability to complex with metals (Way, 1984; Hartung, 1994). This binding to metalloenzymes inhibits many additional enzyme reactions, the most critical one being the ferric ion in cytochrome oxidase. This inhibits the final step of oxidative phosphorylation, prevents the production of ATP, and results in anaerobic metabolism. Cellular anoxia results from the inability to use oxygen, and a severe lactic acidosis ensues.

Cyanide can be metabolized through several routes, only two of which are clinically important and also serve as the basis for the treatment of cyanide poisonings. Approximately 80% of cyanide is detoxified through the liver by the mitochondrial enzyme rhodanese, which transfers sulfur from thiosulfate to cyanide, forming the less toxic thiocyanate, a more easily excreted compound through urine. Both cyanide and thiocyanate can also be metabolized by combinations of cyanide and hydroxycobalamin and cyanide and cystein. Thus, the concentration of sulfur containing substrates (i.e., thiosulfate, cysteine, cystine) is the limiting factor in metabolism. Although hydroxocobalamin serves as a minor route of detoxification, it is considered an important antidote. The resulting cyanide metabolites are then normally eliminated through urine while trace amounts may be excreted through the lungs, saliva, and sweat.

SOURCES OF EXPOSURE

Occupational

The principle routes of occupational exposures to cyanides are mainly from inhalation, yet dermal exposures are also probable to a lesser degree. Occupations with the highest risk of cyanide exposures include blacksmithing, metal cleaning, metallurgy, firefighting, pesticide applications, photography or photo engraving and the manufacturing and processing of cassava, plastic, rubber, paper, textiles, fumigants, insecticides, pesticides, steel, pharmaceuticals, chelating agents, the preparation and decomposition of cyanides, the synthesis of acrylonitrile and nitriles (ATSDR, 2006). The National

Occupational Exposure Survey (NOES) has estimated the numbers of workers potentially exposed to the following cyanides: 4005 to hydrogen cyanide; 66,493 to sodium cyanide; 64,244 to potassium cyanide; 3215 to potassium silver cyanide; 3606 to calcium cyanide; 22,339 to copper (I) cyanide; and 1393 to cyanogen chloride (ATSDR, 2006). It has also been suggested that medical and emergency personnel may be exposed to higher levels of cyanides during resuscitation efforts, gastric content removal or postmortem examinations (Andrews et al., 1989; Curry, et al., 1967; Nolte and Dasgupta, 1996). More recently, cyanide has been recognized as a hazardous toxic gas that can be generated in fires. Victims of smoke inhalation have been shown to have toxic levels of both carbon monoxide and hydrogen cyanide (Baud et al., 1991). Firefighters are an occupational group at significant risk of this source of cyanide exposure (Silverman et al., 1988). Exposure to hydrogen cyanide (prussic acid) most often occurs in connection with the fumigation of ships, workshops, and dwellings, but this gas is also encountered in fumigation intended to kill agricultural pests, in chemical laboratories, in blast furnace gas, in the manufacture of illuminating gas, and in the gas from burning nitrocellulose. The presence of HCN in various industrial gases results from an incomplete combustion of nitrogen-containing organic compounds, and its presence are often not suspected until an accident occurs. Many cases of fatal poisoning occurred during World War I because of the extensive use of hydrocyanic acid to destroy vermin, for which it is still used. HCN fumes are formed by combining dilute sulfuric acid and sodium cyanide.

A study of Nigerian cassava processing plants compared levels of urinary and serum thiocyanate (cyanide metabolite) in cassava processing workers versus frequent (once a day) and infrequent (occasionally) consumers of cassava foods (Okafor et al., 2002). Mean urinary thiocyanate concentrations in cassava processors ($154 \pm 25 \, \mu$mol/l) were double those of frequent consumers ($70 \pm 22 \, \mu$mol/l) and almost three times higher than the infrequent consumers ($59 \pm 17 \, \mu$mol/l). Similar levels and trends were observed in the serum. Moreover, cyanide levels in ambient air and natural water sources near the cassava processing facilities ranged from 20 to 46 mg/m^3 and 1.6 to 7.9 mg/l, respectively (Okafor and Maduagwu, 2000; Okafor et al., 2001).

Environmental

Environmental exposures of cyanides include ambient air, water, soil, and food. Cyanide in ambient air is found typically as hydrogen cyanide which is released at the surface and has a mean tropospheric photochemical lifetime of approximately 2.5 years for HCN with concentrations in the trophosphere ranging from 180 to 190 ng/m^3; Mahieu et al., 1995). Enhancements of HCN in the atmosphere have been associated with biomass fires (Rinsland et al., 1998). Hydrogen cyanide from fire smoke is considered the most common cause of acute cyanide poisoning in developed countries (Anseeuw et al., 2013; Alarie, 2002). The concentration of cyanide in unpolluted air averages 0.160–0.166 ppm (ATSDR, 2006). Smoke from U.S. cigarettes also contains cyanide levels ranging from 10 to 400 µg/cigarette for mainstream smoke and 0.006 to 0.27 µg/cigarette for sidestream smoke (Chepiga et al., 2000). Non-filtered cigarettes contain between 400 and 500 µg of HCN per cigarette (Hoffmann, 1997). Similarly, European cigarettes had hydrogen cyanide concentrations in the range of 280–550 µg/cigarette for mainstream smoke (Guerin et al., 1987).

In 1993, 390 of 1300 hazardous waste sites listed on the National Priority List of the Environmental Protection Agency were identified as being contaminated with cyanide (Harper et al., 1993). Many of these locations must be monitored closely to prevent public exposure through drinking water. As these sites undergo remediation, "hazmat" (hazardous material) workers constitute another occupational group at risk of cyanide exposure.

Near a large-scale cassava processing facility in Nigeria, levels of cyanide in ambient air and natural water sources ranged from 20 to 46 mg/m^3 and 1.6 to 7.9 mg/l, respectively (Okafor and Maduagwu, 2000; Okafor et al., 2001). The maximum contaminant level (MCL) for cyanide in drinking water is set at 0.2 mg/l (USEPA, 2013). A survey of 35 water utilities across the United States found cyanogen chloride concentrations in drinking water varies depending on the treatment processes used (Krasner et al., 1989). For instance, median cyanogen chloride concentrations measured in water treated with free chlorine were 0.4 µg/l compared to 2.2 µg/l in water prechlorinated and postammoniated. Cyanide levels up to 100 g/l have been reported for waste water from plating industries (Chen et al., 1994; Grosse, 1986).

Cyanide compounds released to soil are primarily the result of anthropogenic activities such as the disposal of cyanide wastes in landfills, the use of cyanide containing road salts and industrial facilities (Shifrin et al., 1996; Simeonova and Fishbein, 2004). Mean cyanide concentrations at manufactured gas plants (MGP) are below 2000 ppm, although concentrations >20,000 ppm have been reported (Shifrin et al., 1996). However, it is important to note in terms of the potential toxicity of cyanides in soils, there has been no evidence of soil containing free cyanides. Iron-complexed cyanides (i.e., ferrocyanide ion) comprise over 97% of the total cyanides in weathered or unweathered soils at MGP sites (Theis et al., 1994; Shifrin et al., 1996).

Food is another important source of environmental exposures to cyanogenic glycosides. Cyanogenic glycosides are major constituents in variety of staple foods with levels ranging from <1 mg/kg in cereal grains to 7700 µg/g in bamboo shoots (ATSDR, 2006). Cassava is a major source of calories for over 500 million people in tropical regions and total cyanogens can range from 91 to 1515 µg/g in fresh roots

(O'Brien et al., 1992; Okafor et al., 2002). Hydrogen cyanide in soy protein products, soybean hulls, and lima beans have levels measuring 0.07–0.3, 1.24 ± 0.26, and 43 ± 0.26 µg/g, respectively (Honig et al., 1983). Cyanide concentrations have also been reported for apricot pits (89–2170 µg/g), black cherries (78 µg/seed), and canned unpitted fruits (0–4 µg/g) (Lasch and El Shawa, 1981; Swain et al., 1992; Voldrich and Kyzlink, 1992).

INDUSTRIAL HYGIENE

In spite of the highly toxic nature of cyanide, there have been relatively few acute accidental poisonings (Hirsch, 1964). Acute inhalation, ingestion, or dermal exposures of hydrogen cyanide may be rapidly fatal. As such, the Occupational Safety and Health Administration (OSHA) have set a permissible exposure limit (PEL) for HCN of 10 ppm (11 mg/m^3) averaged over an 8 h work shift (OSHA, 1978). The National Institute for Occupational Safety and Health (NIOSH, 1994) and the American Conference of Governmental Industrial Hygienists (ACGIH) recommended an exposure limit (REL) and a threshold limit value (TLV) of 4.7 ppm (5 mg/m^3) for dermal exposure (ACGIH, 1996). Exposures to 50 ppm or greater are considered Immediately Dangerous to Life or Health Limit (IDLH). The current standard guidelines applied to all cyanides (i.e., KCN, NaCN) are 5 mg/m^3 for the OSHA PEL (8 h TWA) and the NIOSH PEL (10 min period).

Prevention remains the best approach to eliminating the potential for cyanide poisoning. However, in the work place engineering controls should be used wherever possible. Ventilating equipment should be used and workers should be provided personal protective equipment, including protective impervious clothing. When liquid hydrogen cyanide is in use, employees should be provided splash proof safety goggles and an emergency eyewash station and shower should be in the immediate vicinity of use. Workers should be informed of the exposure routes and risks (i.e., skin absorption, ingestion, and inhalation) and visible warning signs should be prominent in areas of potential exposure. Although there is no information available concerning the odor threshold for either KCN or NaCN, workers should be trained to recognize the odor of hydrogen cyanide and to evacuate the area immediately. A first-aid antidote kit should be readily available, and workers should be instructed to its use. If exposure to cyanide is possible, p-aminopropiophenone may be taken as a pretreatment antidote (Baskin and Fricke, 1992).

MEDICAL MANAGEMENT

The lack of rapid assays and blood tests to confirm cyanide poisoning has resulted in presumptive diagnosing based on index of suspicion, clinical presentations, and background information (e.g., occupation and mental status) (Borron, 2006). An odor of bitter almonds is typical of cyanide, but up to 40% of the population is unable to detect cyanide by odor. A cherry red color of venous blood (more likely seen in children and, in general, is not a reliable indicator of cyanide toxicity), arising from poor oxygen utilization, unexplained metabolic acidosis, and the absence of cyanosis are clues to the diagnosis. Cyanide is a normal human metabolite which the body can be detoxified and may be produced *in vitro* by blood or *in situ* by organs after death (Dries and Endorf, 2013). Background blood levels in a normal, nonsmoking individual is 0.3 mg/l and increases to 0.5 mg/l in smokers. Normal plasma lactate concentrations range from 0.5 to 2.2 mmol/l, concentrations ≥8 mmol/l suggests the possibility of cyanide poisoning; however, it is not specific and therefore not definitive (Borron, 2006). Fatal cyanide blood levels are typically seen at 3 mg/l, although levels as low as 1 mg/l and as high as 5 mg/l have been reported (Barillo, 2009). Cyanide blood levels of fire survivors or ingestion and inhalation have ranged from 0.02 to 6.5 and 7 to 9 mg/l, respectively (Curry et al., 1967; Barillo et al., 1994; Caravati and Litovitz, 1988; Barillo, 2009). Blood cyanide levels are useful in guiding treatment; however, the majority of cyanide in the bloodstream resides in red blood cells thereby underestimating plasma levels of free cyanide. Nonetheless, laboratory tests confirming an elevated cyanide level are not usually available in a timely manner, and successful treatments must begin early.

Treatment of acute cyanide poisoning involves decontamination, antidotal therapies, and supportive care. Decontamination may include removing the victim from the site of exposure, removing contaminated clothing, rinsing dermal exposures, gastric lavage, and activated charcoal (Borron, 2006). Cyanide antidotes are classified into three groups based on their mechanism of action: sulfur detoxification, methemoglobin formation, and direct combination (Beasley and Glass, 1998). The most important naturally occurring detoxifying mechanism involves the addition of a sulfur atom to form the less toxic thiocyanite ion (Beasley and Glass, 1998). To facilitate this process, various synthetic compounds have been used as additional sources of sulfur (i.e., sodium thiosulfate). Methemoglobin (metHb) formation increases the source of circulating ferric ions to counteract the critical cyanide binding to the ferric ion of cytochrome oxidase. This leads to the formation of cyanmethemoglobin, which has a low toxicity. Cyanide is detoxified by conversion to thiocyanate, and this reaction is enhanced by combined use of nitrite and thiosulfate. Continued or repeated treatment may be required because of liberation of cyanide ions from cyanmethemoglobin and from thiocyanate. Formation of metHb, through the use of nitrites and supplemental thiosulfate, which is excreted via the kidneys, was the primary antidote for acute cyanide poisoning in the United States

(Baskin et al., 1992). Direct combination antidotes previously combined cobalt compounds with cyanide; has a wide therapeutic index and, all factors being equal, would be the preferred cyanide antidote. Although dicobalt edetate is still available in Britain and France to treat cyanide poisoning, significant toxic effects have been observed (Jillman et al., 1974; Beasley and Glass, 1998). It has been suggested that this chelating agent should be reserved for cases of severe poisoning because of its ominous side effects, which include nausea, vomiting, profuse diaphoresis, crushing retrosternal pain, cardiac arrhythmia, facial and palpebral edema, and loss of calcium and magnesium ions (Nagler et al., 1978; Graham et al., 1977). The use of hydroxocobalamin (vitamin B12a) has been used successfully as a direct combiner (Bain and Knowles, 1967; Yacoub et al., 1974; Forsyth et al., 1993). Hydroxocobalamin binds strongly with cyanide to form cyanocobalamin (vitamin B12) and clinical trials suggest it is a safe, effective agent for preventing and treating nitroprusside-induced cyanide toxicity with few toxic effects (Zerbe et al., 1993). Intravenous hydroxocobalamin along with thiosulfate to enhance the antidotal effect is recommended (Evans, 1964; Graham et al., 1977; Beasley and Glass, 1998). However, hydroxocobalamin and thiosulfate need to be given in separated IV lines so as not allow the sodium thiosulfate to bind with hydroxocobalamin thus rendering the hydroxocobalamin inactive.

Hydroxocobalamin is currently used in both the United States and Europe for acute cyanide poisoning; however, Europe's approach is much more aggressive. The amount available in the U.S. (1 mg/ml) limits its usefulness as antidote as 10 l of material would be needed to neutralize a fatal cyanide dose (Barillo, 2009; Dries and Endorf, 2013).

Debate continues over the treatment of concurrent carbon monoxide and cyanide poisoning in victims of smoke inhalation. The carboxyhemoglobin that occurs in these patients reduces the availability of oxygen at the tissue level. The methemoglobinemia induced by the treatment of cyanide poisoning with amyl nitrite and sodium nitrite can worsen tissue hypoxia. In contrast, hydroxycobalamin treatment rapidly neutralizes cyanide without interfering with oxygen use. A study involving mice injected with potassium cyanide and exposed to carbon monoxide, followed by treatment with amyl nitrite, revealed an increase in mortality of 25% compared with the untreated control group (Moore et al., 1987). Breen et al. (1995) suggest that supportive therapy alone should be sufficient in all but those patients who continue to exhibit persistent lactic acidosis. Kirk et al. (1993) reported that seven critically ill victims of smoke inhalation were successfully treated with sodium nitrite. Supportive therapy using the administration of oxygen has also been proven successful. Oxygen itself has a specific antidotal activity by accelerating the reaction of cytochrome oxidase. Hyperbaric oxygen, when available, is advocated in the treatment of acute cyanide toxicity.

TABLE 46.3   Recommended Cyanide Antidote Dosage Regimes for Adults. (Doses of Cyanide Antidotes for Children are Lower and are weight based)

Antidote	Dosage
Sodium thiosulfate	50 ml of 25% solution (12.5 g) IV over 10–30 min
Sodium nitrite	10 ml of 3% solution (300 mg) IV over 5–20 min
4-DMAP	5 ml of 5% solution (250 mg) IV over 1 min
Dicobalt edentate	20 ml of 1.5% solution (300 mg) IV over 1 min
Hydroxocobalamin	12.5 ml of 40% solution (5 g) IV over 15 min in 100 ml of normal saline

Source: Beasley and Glass, 1998.

In summary, there is no single therapy or antidote for all poisoning situations. The choice of an antidote should be influenced by several factors, including (1) the nature of the cyanide compound and exposure scenario; (2) poisoning severity; (3) presence of certain risk factors for antidote toxicity; (4) the number of victims involved, and (5) proximity to medical facilities (Beasley and Glass, 1998). Nitrites or other metHb formers are not recommended for fire victims, because of the risk of concomitant exposure to carbon monoxide, or those with certain enzyme deficiencies which increase susceptibility. Nitrites can be used in children, but care and attention must be given to the dose (0.2 ml/kg of 3% sodium nitrite, not to exceed 10 ml or 300 mg) so as not to cause an excessive and adverse level of methemoglobin. It is also important to consider both the risk of antidote-induced toxicity and risks of withholding antidote treatment during decisions to administer an antidote based on presumptive cyanide poisoning (Beasley and Glass, 1998; Borron, 2006). Borron (2006) outlined dosage regimes (Table 46.3) and the following recommendations for mild, moderate, and severe poisoning. Mild poisonings may only require rest and oxygen treatment; however, if symptoms worsen, amyl nitrite treatment may be needed. In moderate poisonings (unconsciousness, convulsions, or cyanosis) intravenous antidotes may be indicated with sodium thiosulphate as the first choice. Severe poisonings (deep coma, dilate nonreacting pupils, and decrease in cardiorespiratory function) additional intravenous antidotes should be administered. Hydroxocobalamin is highly recommended over methaemoglobin formers and kelocyanor based on considered risks.

REFERENCES

American Conference of Governmental Industrial Hygienists (ACGIH) (1996) *Threshold Limit Values (TLVs) for Chemical Substances and Physical Agents and Biological Exposure Indices (BEIs)*, Cincinnati: American Conference of Governmental Industrial Hygienists.

Adewusi, S.R.A. and Akindahunsi, A.A. (1994) Cassava processing, consumption, and cyanide toxicity. *J. Toxicol. Environ. Health* 43:13–23.

Alarie, Y. (2002) Toxicity of Fire Smoke. *Crit. Rev. Toxicol.* 32(4):259.

Andrews, J.M., et al. (1989) The biohazard potential of cyanide poisoning during postmortem examination. *J. Forensic Sci.* 34(5):1280–1284.

Anseeuw, K., Delvau, N., Burillo-Putze, G., De laco, F., Geldner, G., Holmström, P., Lambert, Y., and Sabbe, M. (2013) Cyanide poisoning by fire smoke inhalation: a European expert concensus. *Eur. J. Emerg. Med.* 20:2–9.

ATSDR (2006) *Toxicological Profile for Cyanide*, Atlanta, GA: Agency for Toxic Substances and Disease Registry.

Bain, J.T.B. and Knowles, E.L. (1967) Successful treatment of cyanide poisoning. *Br. Med. J.* 2:763.

Ballantyne, B. (1983) The influence of exposure route and species on the acute lethal toxicity and tissue concentrations of cyanides. In: Hayes, A.W., Schnell, R.C., and Miya, T.S., editors. *Developments in the Science and Practice Toxicology*, New York, NY: Elsevier Science Publishers.

Barillo, D.J. (2009) Diagnosis and treatment of cyanide toxicity. *J. Burn Care Res.* 148–152.

Barillo, D.J., Goode, R., and Esch, V. (1994) Cyanide poisoning in victims of fire: analysis of 364 cases and review of the literature. *J. Burn Care Rehabil.* 15:46–57.

Baskin, S.I. and Fricke, R.F. (1992) Pharmacology of p-aminopropiophenone in the detoxification of cyanide. *Cardiovasc. Drug Rev.* 10(3):358–375.

Baskin, S.I., Horowitz, A.M., and Neally, E.W. (1992) The antidotal action of sodium nitrite and sodium thiosulfate against cyanide poisoning. *J. Clin. Pharmacol.* 32:368–375.

Baud, F.J., et al. (1991) Elevated blood cyanide concentrations in victims of smoke inhalation. *N. Engl. J. Med.* 325:1761–1766.

Beasley, D.M.G. and Glass, W.I. (1998) Cyanide poisoning: pathophysiology and treatment recommendations. *Occup. Med. (Lond.)* 48:427–431.

Borron, S.W. (2006) Recognition and treatment of acute cyanide poisoning. *J. Emerg. Nurs.* 32:S12–S18.

Borron, S.W. and Baud, F.J. (1996) Acute cyanide poisoning: clinical spectrum, diagnosis, and treatment. *Arh. Hig. Rada Toksikol.* 47(3):307–322.

Braddock, W.H. and Tingle G.R. (1930) So-called cyanide rash in gold mine mill workers. *J. Ind. Hyg.* 12:259–264.

Breen, P.H., et al. (1995) Combined carbon monoxide and cyanide poisoning: a place for treatment? *Anesth. Analg.* 80:671–677.

Caravati, E.M. and Litovitz, T.L. (1988) Pediatric cyanide intoxication and death from an acetonitrile-containing cosmetic. *JAMA* 260:3470–3472.

Chen, S.H., Wu, S.M., Kou, H.S., et al. (1994) Electron-capture gas chromatographic determination of cyanide, iodide, nitrite, sulfide and thiocyanate anions by phase-transfer-catalyzed derivatization with pentafluorobenzyl bromide. *J. Anal. Toxicol.* 18(2):81–85.

Chepiga, T.A., Morton, M.J., Murphy, P.A., et al. (2000) A comparison of the mainstream smoke chemistry and mutagenicity of a representative sample of the U.S. cigarette market with two Kentucky reference cigarettes (K1R4F and K1R5F). *Food Chem. Toxicol.* 38:949–962.

Cliff, J., et al. (1986) Thyroid function in a cassava-eating population affected by epidemic spastic paraparesis. *Acta Endocrinol.* 113:523–528.

Curry, A.S., Price, D.E., and Rutter, E.R. (1967) The production of cyanide in post mortem material. *Acta Pharmacol. Toxicol. (Copenh)* 25:339–344.

Dries, D.J. and Endorf, F.W. (2013) Inhalation injury: epidemiology, pathology, treatment strategies. *Scand. J. Trauma, Resusc. Emerg. Med.* 21:31.

El Ghawabi, S.H., et al. (1975) Chronic cyanide exposure: a clinical, radioisotope, and laboratory study. *Br. J. Ind. Med.* 32:215–219.

Ernesto, M., Cardoso, A.P., Nicala, D., Mirione, E., Massaza, F., Cliff, J., Haque, M.R., and Bradbury, J.H. (2002) Persistent conzo and cyanogen toxicity from cassava in northern Mozambique. *Acta Trop.* 82:357–362.

Evans, C.L. (1964) Cobalt compounds as antidotes for hydrocyanic acid. *Br. J. Pharmacol. Chemother.* 23:455–475.

Forsyth, J.C., el al. (1993) Hydroxocobalamin as a cyanide antidote: safety, efficacy, and pharmacokinetics in heavily smoking normal volunteers. *J. Toxicol. Clin. Toxicol.* 31:277–294.

Graham, D.L., et al. (1977) Acute cyanide poisoning complicated by lactic acidosis and pulmonary edema. *Arch. Intern. Med.* 137:1051–1055.

Grosse, D.W. (1986) Treatment technologies for hazardous wastes part IV. A review of alternative treatment processes for metal-bearing hazardous waste streams. *J. Air Pollut. Control Assoc.* 36:603–614.

Guerin, M.R., Higgins, C.E., and Jenkins, R.A. (1987) Measuring environmental emissions from tobacco combustion: sidestream cigarette smoke literature review. *Atmos. Environ.* 21(2):291–297.

Harper, C., Llados, F., and Pohl, H. (1993) *Toxicological Profile for Cyanide*, GRA&I Issue 24.

Hartung, R. (1994) Cyanides and nitriles. In: Clayton, G.D. and Clayton, F.E., editors. *Patty's Industrial Hygiene and Toxicology*, 4th ed., vol. II, part D, New York, NY: John Wiley & Sons.

Hirsch, F.G. (1964) Cyanide poisoning. *Arch. Environ. Health* 8:622–624.

Hoffmann, D.H.I. (1997) The Changing Cigarette, 1950–1995. *J. Toxicol. Environ. Health* 50(4):307–364.

Honig, D.H., Hockridge, M.E., Gould, R.M., and Rackis, J.J. (1983) Determination of cyanide in soybeans and soybean products. *J. Agric. Food Chem.* 31:272–275.

Isom, G.E. and Way, J.L. (1974) Effect of oxygen on cyanide intoxication. VI. Reactivation of cyanide-inhibited glucose metabolism. *J. Pharmacol. Exp. Ther.* 189:235–243.

Jillman, B., Bardhan, K.D., and Bain, J.T.B. (1974) The use of dicobalt edetate (Kelocyanor) in cyanide poisoning. *Postgrad. Med. J.* 50:171–174.

Kirk, M.A., Gerace, R., and Kulig, K.W. (1993) Cyanide and methemoglobin kinetics in smoke inhalation victims treated with the cyanide antidote kit. *Ann. Emerg. Med.* 22(9):1413–1418.

Krasner, S.W., McGuire, M.J., et al. (1989) The occurrence of disinfection by-products in US drinking water. *J. Am. Water Works Assoc.* 81(8):41–53.

Lasch, E.E. and El Shawa, R. (1981) Multiple cases of cyanide poisoning by apricot kernels in children from Gaza. *Pediatrics* 68:5–7.

Mahieu, E., Rinsland, C.P., Zander, R., Demoulin, P., Delbouille, L., and Roland, G. (1995) Vertical column abundances of HCN deduced from ground-based infrared solar spectra: long-term trend and variability. *J. Atmos. Chem.* 20:299–310.

Moertel, C.G., Fleming, T.R., et al. (1982) A Clinical Trial of Amygdalin (Laetrile) in the Treatment of Human Cancer. *N. Engl. J. Med.* 306(4):201–206.

Moore, S.J., et al. (1987) Antidotal use of methemoglobin-forming cyanide antagonists in concurrent carbon monoxide/cyanide intoxication. *J. Pharmacol. Exp. Ther.* 242(1):70–73.

Nagler, J., Provoost, R.A., and Parizel, G. (1978) Hydrogen cyanide poisoning: treatment with cobalt EDTA. *J. Occup. Med.* 414–416.

Nieminen, A.L., et al. (1994) ATP depletion rather than mitochondrial depolarization mediates hepatocyte killing after metabolic inhibition. *Am. J. Physiol.* 267:67–74.

Niknahad, H., et al. (1994) Prevention of cyanide-induced cytotoxicity by nutrients in isolated rat hepatocytes. *Toxicol. Appl. Pharmacol.* 128:271–279.

National Institute for Occupational Safety and Health (NIOSH) (1994) *Pocket Guide to Chemical Hazards*, NIOSH 94–116, Washington, DC: US Department of Health and Human Services.

Nolte, K.B. and Dasgupta, A. (1996) Prevention of occupational cyanide exposure in autopsy prosectors. *J. Forensic Sci.* 41(1):146–147.

O'Brien, G.M., Mbome, L., et al. (1992) Variations in cyanogen content of cassava during village processing in Cameroon. *Food Chem.* 44(2):131–136.

Occupational Safety and Health Administration (OSHA) (1978) *Occupational Health Guideline for Hydrogen Cyanide*, US Department of Labor. Available at http://www.cdc.gov/niosh/docs/81-123/pdfs/0333.pdf.

Okafor, P.N. and Maduagwu, E.N. (2000) Cyanide contamination of the atmospheric air during large scale "gari" processing and the toxicity of such cyanide equivalent on rat. *Afr. J. Biomed. Res.* 3:19–23.

Okafor, P.N., Okorowko, C.O., Alawem, F.O., and Maduagwu, E.N. (2001) Cyanide contamination of natural water sources during large scale cassava processing. *Afr. J. Biomed. Res.* 4:25–27.

Okafor, P.N., Okorowkwo, C.O., et al. (2002) Occupational and dietary exposures of humans to cyanide poisoning from large-scale cassava processing and ingestion of cassava foods. *Food Chem. Toxicol.* 40(7):1001–1005.

Osuntokun, B.O., Monekosso, G.L., and Wilson, J. (1969) Relationship of a degenerative tropical neuropathy to diet, report of a field study. *Br. Med. J.* 1:547–550.

Peden, N.R., et al. (1986) Industrial exposure to hydrogen cyanide: implications for treatment. *Br. Med. J. (Clin. Res. Ed.)* 293:538.

Rao, P.S., Yadav, S.K., Gujar, N.L., and Bhattacharya, R. (2013) Acute toxicity of some synthetic cyanogens in rats: time-dependent. *Food Chem. Toxicol.* 59:595–609.

Rinsland, C.P., et al., (1998) ATMOS/ATLAS 3 infrared profile measurements of trace gases in the November 1994 tropical and subtropical upper troposphere. *J. Quant. Spectrosc. Radiat. Transfer* 60:891–901.

Rosenberg, N.L., et al. (1989) Cyanide-induced parkinsonism: clinical, MRI, and 6-fluorodopa PET studies. *Neurology* 39:142–144.

Scolnick, B., Hamel, D., and Woolf, A.D. (1993) Successful treatment of life-threatening propionitrile exposure with sodium nitrite/sodium thiosulfate followed by hyperbaric oxygen. *J. Occup. Med.* 35(6):577–580.

Shifrin, N.S., Beck, B.D., et al. (1996) Chemistry, toxicology, and human health risk of cyanide compounds in soils at former manufactured gas plant sites. *Regul. Toxicol. Pharmacol.* 23(2):106–116.

Silverman, S.H., et al. (1988) Cyanide toxicity in burned patients. *J. Trauma* 28:171–176.

Simeonova, F.P. and Fishbein, L. (2004) Hydrogen cyanide and cyanides: human health aspects. *WHO* 1–73.

Snyder, J.W., et al. (1993) ATP-synthase activity is required for fructose to protect cultured hepatocytes from the toxicity of cyanide. *Am. J. Physiol.* 264:709–714.

Swain, E., LI, C.P., and Poulton, J.E. (1992) Development of the potential for cyanogenesis in maturing black cherry (*Prunus serotina Ehrh.*) fruits. *Plant Physiol.* 98(4), 1423–1428.

Theis, T.L., Young, T.C., et al. (1994) Leachate characteristics and composition of cyanide-bearing wastes from manufactured gas plants. *Environ. Sci. Technol.* 28(1):99–106.

Thomas, T.A. and Brooks, N.W. (1970) Accidental cyanide poisoning. *Anaesthesia* 25:110–117.

USEPA (United States Environmental Protection Agency) (2013) *Basic Information about Cyanide in Drinking Water*, Research Triangle, NC: National primary drinking water regulations, Document EPA 816-F-09-004, May, 2009, Available at http://water.epa.gov/drink/contaminants/basicinformation/cyanide.cfm.

Voldrich, M. and Kyzlink, V. (1992) Cyanogenesis in canned stone fruits. *J. Food Sci.* 57(1): 161–162, 189.

Way, J.L. (1984) Cyanide intoxication and its mechanism of antagaonism. *Ann. Rev. Pharmacol. Toxicol.* 24:451–481.

Yacoub, M., et al. (1974) Acute cyanide poisoning: current data on the metabolism of cyanide and treatment with hydroxocobalamin (French). *J. Eur. Toxicol.* 71:22–29.

Zerbe, N.E., Bertil, K.J., and Wagner, J. (1993) Use of vitamin B_{12} in the treatment of nitroprusside-induced cyanide toxicity. *Crit. Care Med.* 21(3):465–467.

47

HALOGENS

Joseph A. Puccio

CHLORINE

Target organ(s): Respiratory system, eyes, skin, and mucous
membranes

Occupational exposure limits:

OSHA: Permissible exposure limit PEL (ceiling): 1 ppm/
3 mg/m^3 (29 CFR 1910.1000)

NIOSH: Recommended exposure limit (REL): 0.5 ppm/
1.45 mg/m^3 for a maximum of 15 min

ACGIH: Threshold limit value (TLV) (8-h TWA) 0.5 ppm/
1.5 mg/m^3 and a short-term exposure limit (STEL) (15 min
TWA) 1 ppm/2.9 mg/m^3)

Reference values:

Reference dose (RfD): 1E-1 mg/kg-day

Reference concentration (RfC): Not available at this time

TDI: 150 µg/kg of body weight

BACKGROUND AND USES

Chlorine (Cl) is not found freely in nature since it is a highly
reactive gas. Chlorine in its gaseous state is produced through
the industrial process of electrolysis of alkali chlorides such
as sodium chloride (Teitelbaum, 2001). It has many industrial
uses particularly in the manufacturing of chlorinated organic
chemicals. It has been widely used as a bleaching agent in the
manufacturing of paper and textiles. It is also used in the
manufacturing of many household items such bleaches,
detergents, antifreeze, plastics, refrigerants and pesticides,

as well as being used for metal extraction, water purification,
and sanitation of sewage wastes.

It reacts explosively or forms explosive compounds
with many common substances such as ether, turpentine,
ammonia, acetylene, and fuel gas. It has been used as a
warfare agent because of its toxic and explosive properties
(Beswick, 1983; Winder, 2001).

PHYSICAL AND CHEMICAL PROPERTIES

Chemical formula: Cl$_2$
Structural formula: Cl-Cl
Molecular weight: 70.906
CASRN 7782-50-5
Physical properties
    Color: Greenish yellow
    State/form: Gas
Description:
    Melting point: −101 °C (−149.8 °F)
    Boiling point: −34.1 °C (−29.3 °F)
    Relative density (specific gravity): 1.467 at 0 °C and
        368.9 kPa (saturated liquefied gas) 0.0032 at 0 °C (gas)
        (water = 1)
    Solubility in water: Slightly soluble (0.73 g/100 g water
        at 20 °C) (reacts)
    Solubility in other liquids: Very soluble in dimethylfor-
        mamide; soluble in benzene, chloroform, carbon
        tetrachloride, tetrachloroethane, chlorobenzene, glacial
        acetic acid (99.84%), sulfuryl chloride, phosphoryl
        chloride, silicon tetrachloride, and metal chlorides
        such as chromyl chloride, titanium tetrachloride, and
        vanadium oxide chloride.

Hamilton & Hardy's Industrial Toxicology, Sixth Edition. Edited by Raymond D. Harbison, Marie M. Bourgeois, and Giffe T. Johnson.
© 2015 John Wiley & Sons, Inc. Published 2015 by John Wiley & Sons, Inc.

MAMMALIAN TOXICOLOGY

Acute Effects

Acute toxicity data have largely come from accidents involving the leakage of chlorine cylinders (Chasis et al., 1947). Tracheobronchitis, pulmonary edema, and pneumonia have been reported with severe overexposure. Exposure may cause burning of the eyes and nose, lacrimation, rhinorrhea, respiratory distress, nausea, and vomiting. All of the immediate signs and symptoms typically subside within 24 h, except for cough, substernal pain, and respiratory distress persisting for up to 2 weeks. High concentrations at approximately 400 parts per million (ppm) can be fatal over a 30 min duration, while a 1000 ppm concentration can cause fatality within minutes (White, 2010).

Beach et al. (1969) studied seven chemical workers exposed to chlorine gas in separate accidents. The usual symptoms of cough, dyspnea, and chest pain started within 10 min of exposure and lasted 2–8 days. Congestion, consolidation, and nodulation were seen on chest X-rays, and pulmonary edema was present in a severe case. Hypoxemia was corrected by oxygen therapy. This treatment had to be continued for 4 days in a one severe case. All recovered completely.

Postmortem changes after fatal massive exposure to chlorine include destruction of the mucous membranes lining the bronchi and bronchioles, focal and confluent areas of edema in the alveoli, patchy superimposed pneumonia, hyaline membrane formation, multiple recent fibrinocellular thromboses of the pulmonary vessels, and ulcerative tracheobronchitis (Dixon and Drew, 1968; Adelson and Kaufman, 1971). Some of the pulmonary damage was attributed to the use of high pressure oxygen in the treatment.

The effects of different levels of chlorine exposure are presented in Table 47.1. In a recent study of chemically related respiratory disease in Michigan, chlorine was one of the most common causative agents (Reilly and Rosenman, 1995).

TABLE 47.1   Effects of Exposure to Chlorine

Concentration, ppm	Effect
0.03–3.5	Range of reported odor thresholds
1	TLV, PEL ceiling
1–3	Slight irritation
3–6	Stinging or burning of eyes, nose, and throat; lacrimation; sneezing, coughing; bloody nose or sputum
5	Severe irritation of eyes, nose, and respiratory tract that becomes intolerable after a few minutes
14–21	Dangerous if exposure lasts 30–60 min
35–50	Lethal in 60–90 min

TLV, threshold limit value; *PEL*, permissible exposure limit.

Construction workers employed at a pulp mill reported eye and throat irritation, cough, congestion, headache, fatigue, and shortness of breath (Courteau et al., 1994). Victims of another acute industrial exposure reported watering eyes, sneezing, cough, production of sputum, retrosternal burning, dyspnea, anxiety, and vomiting (Abhyankar et al., 1989). Formerly healthy subjects with normal lung function had signs of tracheobronchitis that lasted 3–7 days.

Acute exposure to chlorine gas happens quite regularly at homes and in industries. In a review of 97 cases of acute chlorine inhalation reported to a poison control center, 91 patients were adults and 6 were children. The cases were a mixture of home poisonings (74%) and industrial exposures (26%). Chlorine gas was released after mixing sodium hypochlorite bleach and an acidic descaler in 68 cases and during handling of swimming pool tablets in 22 cases. In total 91 patients had symptoms, including cough (71% of symptomatic cases) or cough and dyspnea (26%). Only 26 victims were admitted to the hospital and 15 of these had a preexisting respiratory condition (2 asthma, 3 chronic bronchitis, 2 allergy, 8 smoking). Of these 26 patients, 21 had dyspnea and 5 had expiratory wheeze. X-rays were taken in 9 cases and only 1 was abnormal. Only 2 patients had hypoxemia with an oxygen saturation of 88–90%. Spirometry was performed in 4 cases and 1 patient had an obstructive airway syndrome. Treatment included bronchodilators in 7 patients, corticosteroids in 4 patients, and 15 patients were treated with both. The average stay in the hospital was 24 h. One patient, a 76-year-old female, required mechanical ventilation for 24 h because of an acute respiratory distress. Follow up by telephone 3 months later found 2 patients with symptoms suggestive of reactive airways dysfunction syndrome (RADS) (Guerrero et al., 2003). Female C57BL/6N mice were given 4.8–5.8 mg/kg of body weight/day of hyperchlorinated tap water, which resulted in a reduction of macrophages.

Acute chlorine exposure has resulted at sites of bomb explosions, as chlorine is released following the use of homemade chemicals bombs (CDC, 2003).

Chronic Effects

Chronic exposure to low levels of chlorine does not appear to have any lasting effects. Patil et al. (1970) studied 600 diaphragm cell workers in 25 North American chlorine-manufacturing plants. The exposure level generally was below 1 ppm, and the time-weighted average (TWA) ranged from 0.006 to 1.42 ppm. No statistically significant correlations were found between exposure and signs, symptoms, or abnormal chest X-rays, electrocardiogram results, or pulmonary function tests.

Chronic effects have been reported subsequent to acute overexposure. A group of longshoremen were accidentally exposed when the main valve of a cylinder of liquid chlorine ruptured. Fifty-nine of these workers were studied over a 3-year period after the accident. Eleven heavily exposed individuals showed evidence of alveolocapillary injury and gradually developed increasing airway resistance. Lung function studies of the 59 workers followed showed a decreased vital capacity and diffusing capacity and an increase in the elastic work of breathing (Kowitz et al., 1967).

The effects of chlorine on the mucous membranes were once attributed to the formation of hydrochloric acid and later to the toxic effects of nascent oxygen released by the oxidative action of the gas. This effect of nascent oxygen has since been explained to be the result of a complex series of reactions in which unstable oxidizing agents (e.g., hypochlorous acid and perchloric acid) cause damage by forming stable hydrates of organic chlorine (Chester et al., 1977).

Acute single overexposure to chlorine has been reported to cause RADS, which may persist for several years (Donnelly and FitzGerald, 1990; Malo et al., 1994). This has not been a consistent finding and is concentration and effect dependent.

Pulp mill workers exposed to chlorine or chlorine dioxide in "exposure incidents" had a reduced 1-min forced expiratory volume (FEV_1) and forced vital capacity (FVC) and were more likely to have chest symptoms, such as tightness and wheezing (Enarson et al., 1984; Salisbury et al., 1991). Another study found no significant differences between FEV_1, FVC, the FEV_1/FVC ratio, or midmaximal flow (MMF) among a pulp mill worker cohort and a railroad worker control cohort (Kennedy et al., 1991). However, pulp mill workers exposed to chlorine during "gassing incidents" who were not smokers or were ex-smokers had a significantly lower MMF and FEV_1/FVC ratio.

Wolf et al. (1995) exposed male and female mice and rats to chlorine gas for 6 h a day, 5 days a week. They noted concentration-dependent lesions and respiratory tract irritation. Male mice and female rats were the most susceptible. There was no evidence of carcinogenicity.

Chronic exposure to low levels of chlorine gas can lead the affected individual to develop a dermatitis known as chloracne, tooth enamel corrosion, coughing, severe chest pain, sore throat, hemoptysis, and an increased risk for developing tuberculosis (Genium, 1992).

Mechanism of Action(s)

Chlorine has low solubility in water and penetrates deep into the lungs (Winder, 2001). Tissue damage is caused by the products of its chemical reaction with water: hydrochloric acid (HCl) and hypochlorous acid (HOCl). Hypochlorous acid is unstable and further breaks down to hydrochloric acid and oxygen free radicals:

$$Cl_2 + H_2O \rightarrow HCl + HOCl$$
$$HOCl \rightarrow HCl + O\bullet$$

The acids cause irritation and the oxygen free radicals mediate tissue necrosis (HSDB, 2007; Traub et al., 2002). Hypochlorous acid has been known to penetrate the cell wall that can disrupt the integrity and permeability; it can also react with enzymes through the sulfhydryl groups in cysteine (HSDB, 2007). The severity of the injury depends on the duration of exposure, the concentration, and the water content of the exposed tissues (Winder, 2001). The hydrolysis to hydrochloric and hydrochloric acids increases the solubility of chorine.

It has also been suggested that hypochlorous acid could interact with nitrite (the end auto-oxidation product of nitric acid) to form an intermediate, nitryl chloride (Cl^-NO_2) orchlorine nitrate (Cl^-ONO), which could nitrate, chlorinate, or dimerize aromatic amino acids (Evans, 2005).

Chemical Pathology

The immediate effect of exposure to chlorine gas is acute inflammation of conjunctivae, nose, pharynx, larynx, trachea, and bronchi. In severe cases this can result in effects on the respiratory mucosa with arterial and capillary hyperemia. The plasma exudate can fill the alveoli resulting in pulmonary edema. There is hypersecretion of the seromucoid glands. Mucosal sloughing occurs and exfoliated cells show chromatolysis and multi-nucleation owing to chemical injury. Mucopurulent exudates may cover the areas of sloughing. Chemical pneumonitis may result with the risk of secondary infection. Recovery occurs with epithelial repair and regeneration. This manifests as basal cell hyperplasia in subjects with injury restricted to the surface epithelium. Subjects with erosive tissue damage have fibroblastic and capillary proliferation and repair by fibrosis. Chronic respiratory incapacity following chlorine exposure may be due to fibrosis.

INDUSTRIAL HYGIENE

Chlorine is produced by the electrolysis of sodium chloride (fused or in brine) in diaphragm cells or in mercury cells and by the chemical oxidation of chloride. Production of chlorine via the mercury process may result in exposure to inorganic mercury and associated symptoms (Smith et al., 1970). Environmental monitors for chlorine gas are unnecessary because this compound has excellent warning properties that alert workers to its presence. Ruth (1986) reported a low odor detection value for chlorine of 0.03 mg/m³ (10 ppb). Workplace personal protective equipment is recommended, including respirators, and adequate training is necessary for workers involved in handling chlorine to avoid overexposure and explosion.

The National Institute for Occupational Safety and Health (NIOSH) has set the recommended exposure limit (REL) for chlorine at a ceiling value of 0.5 ppm for a maximum of

15 min (NIOSH, 2005). The Occupational Safety and Health Administration (OSHA) has set a ceiling value for chlorine at 1 ppm (3 mg/m^3). The American Conference of Governmental Industrial Hygienists (ACGIH) has set the threshold limit value (TLV) for chlorine at 0.5 ppm with a 15-min short-term exposure limit (STEL) of 1 ppm (ACGIH, 2006).

MEDICAL MANAGEMENT

Diagnosis and Treatment

Treatment of acute chlorine gas inhalation includes intermittent positive pressure oxygen, use of nebulized bronchodilators, and administration of mild sedatives and a cough medication containing codeine (Kramer, 1967; Noe, 1963). Administration of humidified oxygen is required for most cases of overexposure to a respiratory irritant. Chester et al. (1977) suggested the possible beneficial effects of intramuscular and oral corticosteroid therapy, although these treatments have not proven to be effective in every patient. Arterial blood gas levels are useful in managing patients after chlorine gas exposure (Traub et al., 2002). Intubation of the lungs may be required in some cases in which suctioning to remove fluid is necessary. Anoxia may lead to cardiac and respiratory arrest.

Severe acute exposure to chlorine can result in residual pulmonary dysfunction, particularly RADS, asthma, and low residual volumes in some patients. Sequelae can persist for years, although in most cases pulmonary function tests return to normal within a few months.

Treatment of other effects of chlorine exposure should be supportive and symptomatic.

Dermal and ocular exposure should be immediately treated by flushing the affected surfaces copiously with water. Contact with compressed liquid gas can cause frostbite or burns.

REFERENCES

Abhyankar, A., et al. (1989) Six month follow-up of fourteen victims with short-term exposure to chlorine gas. *J. Soc. Occup. Med.* 39:131–132.

American Conference of Governmental Industrial Hygienists (2006) *Threshold Limit Values (TLVs) for Chemical Substances and Physical Agents and Biological Exposure Indices (BEIs)*, Cincinnati: American Conference of Governmental Industrial Hygienists.

Adelson, L. and Kaufman, J. (1971) Fatal chlorine poisoning: report of two cases with clinicopathologic correlation. *Am. J. Clin. Pathol.* 56:430–442.

Beach, F.X.M., Jones, E.S., and Scarrow, G.D. (1969) Respiratory effects of chlorine gas. *Br. J. Ind. Med.* 26:231–236.

Beswick, F.W. (1983) Chemical agents used in riot control and warfare. *Hum. Toxicol.* 2(2):247–256.

CDC (2003) Homemade chemical bomb events and resulting injuries—selected states, January 1996-March 2003. *MMWR Morb. Mortal Wkly. Rep.* 52(28):662–665.

Chasis, H., et al. (1947) Chlorine accident in Brooklyn. *Occup. Med. (Chic III)* 4:152–176.

Chester, E.H., et al. (1977) Pulmonary injury following exposure to chlorine gas. Possible beneficial effects of steroid treatment. *Chest* 72:247–250.

Courteau, J-P., et al. (1994) Survey of construction workers repeatedly exposed to chlorine over a three to six month period in a pulpmill: I. Exposure and symptomatology. *Occup. Environ. Med.* 51:219–224.

Dixon, W.M. and Drew, D. (1968) Fatal chlorine poisoning. *J. Occup. Med.* 10:249–251.

Donnelly, S.C. and FitzGerald, M.X. (1990) Reactive airways dysfunction syndrome (RADS) due to chlorine gas exposure. *Ir. J. Med. Sci.* 159:275–277.

Enarson, D.A., et al. (1984) Respiratory health at a pulpmill in British Columbia. *Arch. Environ. Health* 39:325–330.

Evans, R.B. (2005) Chlorine: state of the art. *Lung* 183(3): 151–167.

Genium (1992) *Material Safety Data Sheet No. 53*, Schenectady, NY: Genium Publishing Corporation.

Guerrero, J., et al. (2003) Chlorine gas inhalation: a review of 97 cases. *Clin. Toxicol.* 41(4):493.

HSDB (2007) *Hazardous Substances Data Bank: Chlorine. National Library of Medicine.* Available at http://www.toxnet.nlm.nih.gov.

Kennedy, S.M., et al. (1991) Lung health consequences of reported accidental chlorine gas exposures among pulpmill workers.. *Am. Rev. Respir. Dis.* 143:74–79.

Kowitz, T.A., et al. (1967) Effects of chlorine gas upon respiratory function. *Arch. Environ. Health.* 14:545–558.

Malo, J-L., et al. (1994) Bronchial hyperresponsiveness can improve while spirometry plateaus two to three years after repeated exposure to chlorine causing respiratory symptoms. *Am. J. Respir. Crit. Care Med.* 150:1142–1145.

NIOSH (2005) *Pocket Guide to Chemical Hazards*, Cincinnati, Ohio: National Institute for Occupational Safety and Health. NIOSH Publication No 2005-149.

Noe, J.T. (1963) Therapy for chlorine gas inhalation. *Ind. Med. Surg.* 32:411–414.

Patil, L.R.S., et al. (1970) The health of diaphragm cell workers exposed to chlorine. *Am. Ind. Hyg. Assoc. J.* 31:678–686.

Reilly, M.J. and Rosenman, K.D. (1995) Use of hospital discharge data for surveillance of chemical-related respiratory disease. *Arch. Environ. Health* 50:26–30.

Ruth, J.H. (1986) Odor threshold and irritation levels of several chemical substances: a review. *Am. Ind. Hyg. Assoc. J.* 47: A142–A151.

Salisbury, D.A., et al. (1991) First aid reports of acute chlorine gassing among pulp mill workers as predictors of lung health consequences. *Am. J. Ind. Med.* 20:71–81.

Smith, R.G., et al. (1970) Effects of exposure to mercury in the manufacture of chlorine. *Am. Ind. Hyg. Assoc. J.* 31:687–700.

Teitelbaum, D.T. (2001) The halogens. In: Bingham, E., Cohrsson, B., and Powell, C.H., editors., *Patty's Toxicology*, 5th ed., vol. 3, New York, NY: John Wiley & Sons, Inc., pp. 765–775.

Traub, S.J., et al. (2002) Case report and literature review of chlorine gas toxicity. *Vet. Hum. Toxicol.* 44(4):235–239.

White, C.W. (2010) Chlorine gas inhalation. *Proc. Am. Thorac. Soc.* 7(4):257–263.

Winder, C. (2001) The toxicology of chlorine. *Environ. Res.* 85(2):105–114.

Wolf, D.C., et al. (1995) Two-year inhalation exposure of female and male B6C3F1 mice and F344 rats to chlorine gas induces lesions confined to the nose. *Fundam. Appl. Toxicol.* 24:111–131.

HYDROGEN CHLORIDE

Target organ(s): Tissues immediately in contact with the compound: Eyes, skin, and the respiratory system

Occupational exposure limits:

OSHA: PEL (ceiling): 5 ppm or 7 mg/m^3 (29 CFR 1910.1000)

NIOSH: REL: 5 ppm as a ceiling

ACGIH: TLV 5 ppm which should not be exceeded during any part of the working exposure

Reference values:

RfD: Not established

RfC: 0.02 mg/m^3

Risk/safety phrases: Hazardous: Ignitability, corrosivity, reactivity, or toxicity

BACKGROUND AND USES

Hydrochloric acid is used frequently in the production of chlorides. In addition, it is used for refining the ore in the production of tin and tantalum, for pickling and cleaning of metal products; in electroplating, in removing scale from boilers; for the neutralization of basic systems; as a laboratory reagent; as a catalyst and solvent in organic syntheses, in the manufacture of fertilizers and dyes; for hydrolyzing starch and proteins in the preparation of various food products; and in the photographic, textile, and rubber industries. Small amounts of hydrogen chloride are released from waste incinerators (Shy et al., 1995).

PHYSICAL AND CHEMICAL PROPERTIES

The synonyms for an aqueous solution of hydrogen chloride include chlorohydric acid, hydrochloric acid, and muriatic acid.

Chemical properties:

Chemical formula: HCl

Structural formula: H-Cl

Molecular weight is 36.47 g/mol (1,3)

CASRN 7647-01-0

Physical properties:

Color: Colorless to slightly yellow gas

Odor: Pungent, irritating

Water solubility: 67% by weight at 30 °C (86 °F)

Vapor pressure: 40.5 atm at 20 °C (68 °F).

State/form: Nonflammable aqueous solution or gas (1,3,4)

MAMMALIAN TOXICOLOGY

Acute Effects

Exposure to HCl almost always involves the hydrochloric acid aerosols because of the hygroscopic nature of the anhydrous form. These aerosols irritate the mucous membranes of the nose, mouth, and respiratory tract. The concentrations of acute exposure and their likely health effects are listed in Table 47.2. Two workers were exposed to high concentrations of HCl in the degradation products of paint, which were produced by high welding temperatures. These workers complained of fever, dyspnea, fatigue, and respiratory discomfort. The final diagnosis of toxic alveolitis and obstructive lung disease was linked to this exposure (Sjogren et al., 1991). Overheated polyvinyl chloride (PVC) in a fabricating factory is believed to have been the cause of the brief hospitalization of 63 individuals who complained of headache, dizziness, and dyspnea. Hydrogen chloride and inhalation of carbon monoxide were concluded to be the cause of this outbreak of illness (Froneberg et al., 1982).

Dermal exposure to liquid hydrochloric acid can result in severe chemical burns and scarring. Permanent ocular damage may occur if the liquid is splashed into the eyes.

In recent years concern has arisen over increases in the levels of irritant air pollutants and their potential to increase the incidence of asthmatic attacks. The potential effect of HCl was studied in 10 young adult asthmatics (Stevens et al., 1992).

TABLE 47.2   Effects of Exposure to Hydrogen Chloride

Concentration, ppm	Effect
0.067–0.134	Threshold for odor detection and change in respiratory pattern
5	No organic damage
10	Irritation but work undisturbed
10–50	Work difficult but possible
35	Throat irritation after short exposure
50–100	Intolerable; work impossible
1000–2000	Brief exposure dangerous; laryngospasm
13,000–20,000	Lethal after a few minutes

Volunteers were exposed to 0.8–1.8 ppm of HCl during 45-min sessions. Researchers reported no effect on forced expiratory volume, peak flow rate, or total respiratory flow.

Children exposed to the same levels of hydrogen chloride as adults may receive a larger dose because they have greater lung surface area:body weight ratios and increased minute volumes:weight ratios. In addition, they may be exposed to higher levels than adults in the same location because of their short stature and the higher levels of hydrogen chloride found nearer to the ground.

Chronic Effects

Chronic or prolonged exposure to hydrogen chloride gas (above the OSHA PEL) or to mist has been associated with changes in pulmonary function, chronic inflammation of the bronchi, nasal ulceration, and symptoms resembling acute viral infection of the upper respiratory tract as well as inflammation of the skin, discoloration and erosion of dental enamel, and inflammation of the eye membrane. However, these effects were not corroborated in another worker population upon physical examination (Patil et al., 1970).

The reproductive hazards of hydrogen chloride to humans are unknown. Few studies have been directed at the reproductive effects in experimental animals exposed to hydrogen chloride. No data were located pertaining to maternal transfer of hydrogen chloride through the placenta or in breast milk Hydrogen chloride is not included in the Reproductive and Developmental Toxicants, a 1991 report published by the U.S. General Accounting Office (GAO) that lists 30 chemicals of concern because of the widely acknowledged reproductive and developmental consequences.

No information is available on the carcinogenic effects of hydrochloric acid in humans. In one study, no carcinogenic response was observed in rats exposed via inhalation (HSDB, 1993). The Environmental Protection Agency (EPA) has not classified hydrochloric acid with respect to potential carcinogenicity (Bond et al., 1991).

INDUSTRIAL HYGIENE

Hydrogen chloride is a pyrolysis product of PVC and is considered the principal toxic component of PVC smoke (Barrow et al., 1979). Thus it often is encountered by firefighters responding to structure fires and has been implicated as the cause of death in one case (Dyer and Esch, 1976). A positive-pressure, self-contained breathing apparatus (SCBA) is recommended as the respiratory protection in response situations that involve exposure to potentially unsafe levels of hydrogen chloride (Treitman et al., 1980). Chemical-protective clothing is recommended as the appropriate skin protection because hydrogen chloride can cause skin irritation and burns.

MEDICAL MANAGEMENT

There is no antidote for hydrogen chloride poisoning.

The treatment consists of support for respiratory and cardiovascular functions. Victims who are able may assist with their own decontamination. Remove contaminated clothing while flushing exposed skin and hair with water for 3–5 min, wash thoroughly with soap and water. Use caution to avoid hypothermia when decontaminating children or the elderly. Use blankets or warmers when appropriate. Double bag contaminated clothing and personal belongings.

Flush exposed or irritated eyes with tepid plain water or saline for 15 min. Eye irrigation should be carried out simultaneously with other basic care and transport. Remove contact lenses if easily removable without additional trauma to the eye.

Be certain that victims have been decontaminated properly. Victims who have undergone decontamination or who have been exposed only to gas and who have no symptoms of skin or eye irritation pose no serious risk of secondary contamination. In such cases, first responder personnel require no specialized protective gear.

Quickly access for a patent airway. If trauma is suspected, maintain cervical immobilization manually and apply a cervical collar and a backboard when feasible. Ensure adequate respiration and pulse. Administer supplemental oxygen as required and establish intravenous access if necessary. Place on a cardiac monitor.

In cases of respiratory compromise, secure airway and respiration via endotracheal intubation. If not possible, perform cricothyroidotomy if equipped and trained to do so.

Treat patients who have bronchospasm with aerosolized bronchodilators. The use of bronchial-sensitizing agents in situations of multiple chemical exposures may pose additional risks. Consider the health of the myocardium before choosing which type of bronchodilator should be administered. Cardiac-sensitizing agents may be appropriate; however, the use of these agents after exposure to certain chemicals may pose enhanced risk of cardiac arrhythmias (especially in the elderly). Sympathomimetic bronchodilators generally will reverse bronchospasm in patients exposed to hydrogen chloride.

Consider a racemic epinephrine aerosol for children who develop stridor. Administer a dose of 0.25–0.75 ml of 2.25% racemic epinephrine solution in 2.5 cc water, and repeat every 20 min as needed.

Patients who are comatose, hypotensive, or are having seizures or cardiac arrhythmias should be treated according to advanced life support (ALS) protocols.

In cases of ingestion, do not induce emesis. Do not administer activated charcoal or attempt to neutralize stomach contents. Victims who are conscious and are able to swallow should be given 4–8 ounces of water or milk (children's dose is 2–4 ounces).

REFERENCES

Barrow, C.S., Lucia, H., and Alarie, Y.C. (1979) A comparison of the acute inhalation toxicity of hydrogen chloride versus the thermal decomposition products of polyvinyl chloride. *J. Combust. Toxicol.* 6:3–12.

Bond, G.G., et al. (1991) Lung cancer and hydrogen chloride exposure: results from a nested case control study of chemical workers. *J. Occup. Med.* 33:958–961.

Dyer, R.F. and Esch, V.H. (1976) Polyvinyl chloride toxicity in fires: hydrogen chloride toxicity in firefighters. *JAMA* 235:393–397.

Froneberg, B., Johnson, P.L., and Landrigan, P.J. (1982) Respiratory illness caused by overheating of polyvinyl chloride. *Br. J. Ind. Med.* 39:239–243.

HSDB (1993) *Hazardous Substances Data Bank: Hydrogen Chloride*, Bethesda, MD: NLM (National Library of Medicine); National Institutes of Health, U.S. Department of Health and Human Services, Available at http://toxnet.nlm.nih.gov.

Patil, L.R.S., et al. (1970) The health of diaphragm cell workers exposed to chlorine. *Am. Ind. Hyg. Assoc. J.* 31:678–686.

Shy, C.M., et al. (1995) Do waste incinerators induce adverse respiratory effects? An air quality and epidemiological study of six communities. *Environ. Health Perspect.* 103:714–724.

Sjogren, B., et al. (1991) Fever and respiratory symptoms after welding on painted steel. *Scand. J. Work Environ. Health* 17:441–443.

Stevens, B., et al. (1992) Respiratory effects from the inhalation of hydrogen chloride in young adult asthmatics. *J. Occup. Med.* 34:923–929.

Treitman, R.D., Burgess, W.A., and Gold, A. (1980) Air contaminants encountered by firefighters. *Am. Ind. Hyg. Assoc. J.* 41:796–802.

CHLORINE DIOXIDE

Target organ(s): Eyes and the respiratory system

Occupational exposure limits:

OSHA: PEL: 0.1 ppm (0.3 mg/m^3) as an 8-h TWA concentration (29 CFR 1910.1000, 2006)

NIOSH: REL: 0.1 ppm (0.3 mg/m^3 as a TWA for up to a 10-h workday and an hour workweek and a STEL of 0.3 ppm (0.9 mg/m^3)

ACGIH: TLV: (8-h TWA 0.1 ppm mg/m^3) as a TWA for a normal 8-h workday and a 40-h workweek and an STEL of 0.3 ppm (0.83 mg/m^3) for periods not to exceed 15 min. Exposures at the STEL concentration should not be repeated more than four times a day and should be separated by intervals of at least 60 min

Reference values:

RfD: 3×10^{-2} mg/kg-day

RfC: 2×10^{-4} mg/m^3

TDI: 30 µg/kg of body weight

Risk/safety phrases: [list information for classification and labeling]

BACKGROUND AND USES

Chlorine dioxide has been used to bleach cellulose, flour, beeswax, leather, paper pulp, oils, and textiles. It is used as a common compound for water disinfection and in the antimicrobial gel Alcide, and in other bactericides, antiseptics, and deodorizers. It is used in water purification as a taste and odor control agent (HSDB, 2005). Recently, ClO_2 has been shown to inactivate the human immunodeficiency virus (HIV) in medical waste disposal processes (Farr and Walton, 1993).

PHYSICAL AND CHEMICAL PROPERTIES

The synonyms for chlorine dioxide include chlorine oxide, anthium dioxcide, chlorine peroxide, chlorine (IV) oxide, chloroperoxyl, chloryl radical, alcide, doxcide 50. Chloride dioxide decomposes when heated and emits toxic fumes of hydrogen chloride (HSDB, 2005). It is unstable in light but will remain stable in dark if it is pure (HSDB, 2005). However, chlorides can catalyze even in the dark. Chloride dioxide reacts only by oxidation, while it is an efficient disinfectant because it does not vary with pH (HSDB, 2005).

Chemical properties:

Chemical formula: ClO_2

Molecular weight is 67.5 g/mol.

CASRN 10049-04-4

Physical properties:

Color: A yellow or reddish-yellow gas at room temperature. The solid is a yellowish-red crystalline mass. The liquid is reddish-brown in color.

Odor: Unpleasant chlorine-like odor

Boiling point: (at 760 mmHg): 11 °C (51.8 °F)

Specific gravity: 1.64 at 0 °C (32 °F) (liquid)

Vapor density: (air = 1): 2.4

Melting point: −59 °C (−74.2 °F)

Vapor pressure: Greater than 760 mmHg at 20 °C (68 °F)

Solubility: Soluble in water (with decomposition), alkalies, and sulfuric acid

MAMMALIAN TOXICOLOGY

Acute Effects

In general, most effects noted from occupational exposure to ClO_2 are irritation responses to dermal and inhalation exposure. The unpleasant odor at high concentrations can irritate mucous membranes and the upper respiratory tract. Inhalation can produce coughing, wheezing, respiratory distress, and congestion in the lungs. In a group of workers who reported at least one instance of overexposure to chlorine or chlorine dioxide, or both, significant changes were noted in the FEV_1/FVC ratio and chest symptoms increased (Salisbury et al., 1991). No distinction was made between the two gas exposures. Eye irritation may occur and some workers may see halos around lights for a short time after exposure.

Chronic Effects

Ingestion of water with 5 ppm ClO_2 did not have any measurable effect in a group of healthy human volunteers (Lubbers et al., 1984). There has been some concern about chlorine dioxide ingestion and the formation of a thyroid inhibitory substance by means of a redox interaction with iodine and nutritional biochemicals, but this has not been verified (Bercz and Bawa, 1986). ClO_2 is considered not classifiable as to its carcinogenicity in humans.

Workers exposed for 5 years to average chlorine dioxide concentrations below 0.1 ppm but with excursions to higher concentrations had symptoms of eye and throat irritation, nasal discharge, cough, and wheezing; on bronchoscopy, bronchitis was observed in 7 of the 12 workers. Concentrations of 0.25 ppm and less have been reported to worsen mild respiratory ailments. Two adults who ingested 250 ml of a 40 mg/l solution of chlorine dioxide experienced headache, nausea, abdominal discomfort, and lightheadedness within 5 min of ingestion. The symptoms disappeared within another 5 min.

Animal studies have reported the effects of ClO_2 exposure in young laboratory animals. There appears to be some effect on thyroid function and the development of the central nervous system in rats prenatally or postnatally exposed to chlorine dioxide (Orme et al., 1985; Toth et al., 1990).

Mechanism of Action

The metabolism of chlorine dioxide yields a number of different compounds, including chlorite, chlorate, chloramine, and chlorine. These metabolites are of similar low toxicity. There may be some relationship between the endogenous formation of chloroform and ClO_2 exposure, as demonstrated in rats (Suh et al., 1984).

SOURCES OF EXPOSURE

The sources of exposure to chlorine dioxide are as follows:
In the manufacture and transportation of chlorine dioxide.
During its use as a bleaching agent for paper and wood pulp, textiles, cellulose, beeswax, and leather; used to bleach mature flour.
During its use in swimming pool and municipal water treatment for purification and to remove tastes and odors by oxidation; use as a wastewater disinfecting agent; use in bleaching fats and oils; use in the cleaning and de-tanning of leather.

INDUSTRIAL HYGIENE

The reactions used to manufacture ClO_2 include sodium chlorate and sulfuric acid with sulfur dioxide and an alternate mechanism of methanol reacted with $HClO_3$. The primary routes of occupational exposure are inhalation and dermal contact. The general population may be exposed to ClO_2 through residues in ClO_2-treated drinking water. This compound is flammable and combustible and should be treated with caution around flame, heat, and reactive compounds. Protective gear should prevent skin and inhalation exposure. Simon et al. (1987) tested a variety of available respirator cartridges and found all of them to be efficient in removing chlorine dioxide gas. Some situations may require the use of respirators to control exposure. Respirators must be worn if the ambient concentration of chlorine dioxide exceeds prescribed exposure limits. Concentrations of the gas in the workplace can be detected by diffusive samplers followed by nonsuppressed ion chromatography (Bjorkholm et al., 1990; Hekmet et al., 1994; Poovey and Rando, 1995).

The NIOSH, OSHA, and ACGIH have set the allowable exposure limits for ClO_2 at 0.1 ppm (0.3 mg/m^3) with a ceiling value of 0.3 ppm (0.9 mg/m^3) (NIOSH, 1994; ACGIH, 1996). Conveniently, the regulatory guideline for ClO_2 corresponds with the low odor detection limit reported by Ruth (1986). For most workers, odor detection will serve as a warning to possible overexposures.

If chlorine dioxide contacts the skin, workers should immediately wash the affected areas with soap and water. Clothing contaminated with chlorine dioxide should be removed immediately, and provisions should be made for the safe removal of the chemical from the clothing. Persons laundering the clothes should be informed of the hazardous properties of chlorine dioxide, particularly its potential for causing severe irritation.

A worker who handles chlorine dioxide should thoroughly wash hands, forearms, and face with soap and water before eating, using tobacco products, using toilet facilities, applying cosmetics, or taking medication. Workers should not eat, drink, use tobacco products, apply cosmetics, or take medication in areas where chlorine dioxide or a solution containing chlorine dioxide is handled, processed, or stored.

MEDICAL MANAGEMENT

Some populations are more at risk when exposed to ClO_2; individuals with a glucose-6-phosphate dehydrogenase deficiency are sensitive to ClO_2, which can affect the capacity for maintaining significant levels of glutathione (HSDB, 2005).

Remove an incapacitated worker from further exposure and implement appropriate emergency procedures. If the eyes were involved they should be irrigated with water immediately. The skin should be washed with soap and water.

If respiratory irritation occurs, respiratory support should be given including oxygen and aerosolized bronchodilators. In cases of respiratory compromise, secure the airway and respiration via endotracheal intubation. If not possible, perform cricothyroidotomy if equipped and trained to do so.

Rinse out the mouth and drink a few glasses of water or milk immediately, but only if the person is fully conscious. Do not induce vomiting. Seek hospital treatment if more than a minimal amount has been swallowed.

REFERENCES

American Conference of Governmental Industrial Hygienists (1996) *Threshold Limit Values (TLVs) for Chemical Substances and Physical Agents and Biological Exposure Indices (BEIs)*, Cincinnati: American Conference of Governmental Industrial Hygienists.

Bercz, J.P. and Bawa, R. (1986) Iodination of nutrients in the presence of chlorine-based disinfectants used in drinking water treatment. *Toxicol. Lett.* 34:141–147.

Bjorkholm, E., Hultman, A., and Rudling, J. (1990) Evaluation of two diffusive samplers for monitoring chlorine and chlorine dioxide in workplace air. *Appl. Occup. Environ. Hyg.* 5:767–770.

Farr, R.W. and Walton, C. (1993) Inactivation of human immunodeficiency virus by a medical waste disposal process using chlorine dioxide. *Infect. Control Hosp. Epidemiol.* 14:527–529.

Hekmet, M., Smith, R., and Fung, P. (1994) An evaluation of the Occupational Safety and Health Administration method for the "Determination of Chlorine Dioxide in Workplace Atmosphere." *Am. Ind. Hyg. Assoc. J.* 55:1087–1089.

HSDB (2005) *Hazardous Substances Data Bank: Chlorine Dioxide. From NLM (National Library of Medicine); National Institutes of Health*, Bethesda, MD: U.S. Department of Health and Human Services. Available at http://toxnet.nlm.nih.gov.

Lubbers, J.R., et al. (1984) The effects of chronic administration of chlorine dioxide, chlorite and chlorate to normal healthy adult male volunteers. *J. Environ. Pathol. Toxicol. Oncol.* 5:229–238.

National Institute for Occupational Safety and Health (1994) *Pocket Guide to Chemical Hazards*, NIOSH 94–116, Washington, DC: US Department of Health and Human Services.

Orme, J., et al. (1985) Effects of chlorine dioxide on thyroid function in neonatal rats. *J. Toxicol. Environ. Health* 15:315–322.

Poovey, H.G. and Rando, R.J. (1995) Determination of chlorine and chlorine dioxide by nonsuppressed ion chromatography and application to exposure assessment in the paper industry. *J. Liquid Chromatogr.* 18:261–275.

Ruth, J.H. (1986) Odor thresholds and irritation levels of several chemical substances: a review. *Am. Ind. Hyg. Assoc. J.* 47:A142–Al51.

Salisbury, D.A., et al. (1991) First-aid reports of acute chlorine gassing among pulpmill workers as predictors of lung health consequences. *Am. J. Ind. Med.* 20:71–81.

Simon, C.G., Fisher, R.E., and Davison, J.D. (1987) Evaluation of respirator cartridges for effectiveness of chlorine dioxide removal. *Am. Ind. Hyg. Assoc. J.* 48:1–8.

Suh, D.H., et al. (1984) Biochemical interactions of chlorine dioxide and its metabolites in rats. *Arch. Environ. Contam. Toxicol.* 13:163–169.

Toth, G.P., et al. (1990) Effects of chlorine dioxide on the developing rat brain. *J. Toxicol. Environ. Health* 31:29–44.

HYDROGEN FLUORIDE

Target organ(s): Bones, eyes, respiratory system, skin, and teeth

Occupational exposure limits:

OSHA: PEL: 3 ppm; 2 mg/m^3 TWA

NIOSH: REL: 3 ppm; 2.5 mg/m^3 TWA and a ceiling: 15 min of 6 ppm, 5 mg/m^3 (NIOSH, 1994)

ACGIH: TLV: 0.5 ppm TWA with a 2 ppm ceiling

Reference values:

RfD: 8×10^{-2} mg/kg/day

RfC: Not established

Minimal risk level (MRL): 0.02 ppm

Risk/safety phrases: Toxic by inhalation, in contact with skin and if swallowed. Do not breathe dust. Avoid contact with skin and eyes.

BACKGROUND AND USES

Fluoride is used in preventing tooth decay by adding it to drinking water, tooth paste, and mouth washes. The dentist also directly provides fluoride therapy on children's teeth. Fluoride in the form of a pill increases bone mineral density in osteoporosis and is showing promise in rheumatoid arthritis and Crohn's disease. Hydrogen fluoride is manufactured from calcium fluoride; it is mainly used in the production of synthetic cryolite, aluminium fluoride (AlF$_3$), motor gasoline alkylates, and chlorofluorocarbons (IPCS, 2002). Production of chlorofluorocarbons is decreasing due to the restrictions on their effect on the environment (IPCS, 2002). Hydrogen fluoride may also be found in commercial rust removers (IPCS, 2002).

PHYSICAL AND CHEMICAL PROPERTIES

The synonym of hydrogen fluorideis anhydrous hydrofluoric acid: HF-A.

Chemical properties:
  Chemical formula: HF
  Molecular weight is 20.01 g/mol
  CASRN 7664-39-3
Physical properties:
  Color: Colorless gas or fuming liquid
  Odor: Strong, irritating odor
  Boiling point: (at 760 mmHg): 19.4 °C (67 °F)
  Specific gravity: 0.987 at 0 °C (32 °F) (liquid)
  Melting point: −83.3 °C (−118 °F)
  Vapor density: 0.92
  Vapor pressure at 2.5 °C (36.5 °F): 400 mmHg
  Solubility: Very soluble in water, in alcohol, and in most
    organic solvents; slightly soluble in ether
  Evaporation rate: Data not available

MAMMALIAN TOXICOLOGY

Exposure to hydrogen fluoride and its aqueous solution can occur through inhalation, ingestion, and eye or skin contact (Sittig, 1991).

Acute Effects

Exposure to HF or fluorine can produce mucosal, respiratory, and ocular irritation. Acute and chronic conjunctivitis and eczema of the eyelids have been reported. In one case, repeated overexposure to hydrogen fluoride vapor was associated with cartilaginous changes and osteomalacia of the nasal septum and nasal mucosa (Miszke et al., 1984). The authors concluded that repeated irritation of the nasal mucosa by hydrofluoric acid can cause hypertrophic inflammation. In addition, exposure to high concentrations of the vapors of hydrofluoric acid characteristically results in ulcerative tracheobronchitis and hemorrhagic pulmonary edema; this local reaction is equivalent to that caused by gaseous hydrogen chloride (Gosselin et al., 1984).

From accidental, occupational, and volunteer exposures, it is estimated that the lowest lethal concentration for a 5-min human exposure to hydrogen fluoride is in the range of 50–250 ppm. Significant exposures by the dermal or inhalation route may cause hypocalcemia and hypomagnesemia; cardiac arrhythmias may follow. Acute renal failure has also been documented after an ultimately fatal inhalation exposure (Hathaway et al., 1991).

Gaseous F_2 reacts violently with the skin to produce thermal burns. Burns resulting from an exposure to hydrogen fluoride or fluoride may be life threatening and require immediate medical attention. Although actual damage to the skin may be minor, the systemic effects of hydrogen fluoride exposure may quickly create a life-threatening situation. Death typically is the result of respiratory failure or ventricular arrhythmia. An exposure of as little as 2.5% of the body surface to pure hydrogen fluoride has reportedly resulted in death caused due to hypocalcemia (Bertollini, 1992). Tissue injury may increase for days after exposure (Anderson and Anderson, 1988). Exposures to weak solutions of hydrofluoric acid (<20%) may cause considerable pain, erythema, and edema but do not result in long-term morbidity (Upfal and Doyle, 1990). Pain may be delayed for 2–24 h. Although less serious, an exposure to dilute forms of hydrofluoric acid is much more common.

Chronic Effects

Fluorosis is characterized by increased bone density of the spine and pelvis, calcification of ligaments, and hyperostosis. These skeletal changes were first noted in workers exposed to fluoride during cryolite processing as early as the 1930s (Møller and Gudjonsson, 1932). Industrial hygiene practices have since prevented any cases of fluorosis of the crippling severity reported early in this century. However, some skeletal changes may still be noted in workers exposed to fluoride. In one report of a fluorspar processing facility, 22% of the workers had radiological changes indicative of fluorosis (Desai et al., 1983).

Fluorosis of the trabecular bone tissue may be reversible. Grandjean and Thomson (1983) examined 17 cases of skeletal fluorosis 8–15 years after exposure ended. They found that trabecular sclerosis had faded extensively, whereas cortical bone thickening and the calcification of muscle insertions and ligaments remained relatively unchanged.

Grandjean et al. (1992) and Grandjean et al. (1993) detected an excess of lung, larynx, and urinary bladder cancer in a cohort of 522 workers exposed to fluorite during cryolite processing. The higher incidence of respiratory and urinary cancers was especially pronounced in men who were under 35 years of age at first employment. There was no consistent relationship between the length of employment and incidence of cancer, but the highest number of incidences occurred after 10–19 years of employment.

Ingestion of an estimated 1.5 g of hydrofluoric acid produces sudden death; however, repeated ingestion of small amounts of hydrogen fluoride may cause fluoride osteosclerosis (Gosselin et al., 1984).

Mechanism of Action(s)

Hydrogen fluoride can pass easily through the cell membranes by nonionic diffusion. It is absorbed almost completely through the digestive tract (IPCS, 2002). It binds to magnesium and calcium, making them unavailable for various biological functions (Greco et al., 1988; Reinhart, 1992).

Calcium is only weakly buffered in the body by protein-bound and bone stores. Small amounts of fluoride can drastically disrupt metabolism. Fluoride inhibits several enzymes that can diminish tissue respiration and anaerobic glycolysis (HSDB, 2007).

Several investigations demonstrated that fluoride can induce oxidative stress and modulate intracellular redox homeostasis, lipid peroxidation and protein carbonyl content, as well as alter gene expression and cause cellular apoptosis (Barbier et al., 2010).

INDUSTRIAL HYGIENE

Methods that are effective in controlling worker exposures to hydrogen fluoride include process enclosure, local exhaust ventilation, general dilution ventilation, and the use of personal protective equipment.

Medical Surveillance

A study of the Swiss aluminum industry concluded that screening for early detection of fluorosis is more cost effective than allowing employees to progress to a disabling condition (Morabia et al., 1989). This benefit would be realized even if the prevalence of fluorosis in the workforce is <10%.

MEDICAL MANAGEMENT

Hydrofluoric acid burns should be immediately irrigated and the person transported immediately to an emergency care facility, even if he or she is asymptomatic for systemic effects. A subscleral injection of 10% calcium gluconate should be given to manage severe burns (Anderson and Anderson, 1988; Sheridan et al., 1995). Calcium supplementation should be monitored to ensure that hypercalcemia does not result (Greco et al., 1988). Hydrofluoric acid burns are usually treated by traditional means, but severe cases may require surgery and skin grafts (Buckingham, 1988). When treating victims of hydrogen fluoride burns, medical personnel must protect themselves with latex or vinyl gloves and plastic aprons. Oral administration of a high dose of $CaCl_2$ or $MgSO_4$ is an effective method for treating severe oral fluoride poisoning (Kao et al., 2004).

REFERENCES

Anderson, W.J. and Anderson, J.R. (1988) Hydrofluoric acid burns of the hand: mechanism of injury and treatment. *J. Hand Surg.* 13A:52–57.

Barbier, O., Arreola-Mendoza, L., and Del Razo, L.M. (2010) Molecular mechanism of fluoride toxicity. *Chem. Biol. Interact.* 188(2):319–333.

Bertollini, J.C. (1992) Hydrofluoric acid: a review of toxicity. *J. Emerg. Med.* 10:163–168.

Buckingham, F.M. (1988) Surgery: a radical approach to severe hydrofluoric acid burns. A case report. *J. Occup. Med.* 30:873–874.

Desai, V.K., et al. (1983) Clinical radiological observations among workers of fluoride processing industry. *Fluoride* 16:90–100.

Gosselin, R.E., Smith, R.P., and Hodge, H.C. (1984) *Clinical Toxicology of Commercial Products*, 5th ed., Baltimore, MD: Williams & Wilkins.

Grandjean, P. and Thomson, G. (1983) Reversibility of skeletal fluorosis. *Br. J. Ind. Med.* 40:456–461.

Grandjean, P., et al. (1992) Cancer incidence and mortality in workers exposed to fluoride. *J. Natl. Cancer Inst.* 84:1903–1909.

Grandjean, P., Olsen, J.H., and Juel, M.S. (1993) Excess cancer incidence among workers exposed to fluoride. *Scand. J. Work Environ. Health* 19:108–109.

Greco, R.J., et al. (1988) Hydrofluoric acid-induced hypocalcemia. *J. Trauma* 28:1593–1596.

Hathaway, G.J., et al. (1991) *Proctor and Hughes' Chemical Hazards of the Workplace*, 3rd ed., New York, NY: Van Nostrand Reinhold.

HSDB (2007) *Hazardous Waste Data Bank: Hydrogen Fluoride*, From NLM (National Library of Medicine), Bethesda, MD: National Institutes of Health, U.S. Department of Health and Human Services, Available at http://toxnet.nlm.nih.gov.

IPCS (2002) *Environmental Health Criteria 227: Fluorides. International Programme on Chemical Safety*, Geneva, Switzerland: World Health Organization.

Kao, W.F., et al. (2004) A simple, safe and efficient way to treat severe fluoride poisoning—oral calcium or magnesium. *Toxicol, Clin. Toxicol*, 42(1):33–40.

Miszke, A., et al. (1984) Nasal septum and mucosa in fluorosis. *Fluoride* 17:114–118.

Møller, P.F. and Gudjonsson, S.V. (1932) Massive fluorosis of bones and ligaments. *Acta Radiol.* 13:269–294.

Morabia, A., Rey, P., and Bousquet, A. (1989) Screening vs. individual detection of industrial fluorosis: a decision analysis model. *Am. J. Ind. Med.* 15:657–667.

National Institute for Occupational Safety and Health (1994) *Pocket Guide to Chemical Hazards*, NIOSH 94–116, Washington, DC: U.S. Department of Health and Human Services.

Reinhart, R.A. (1992) Magnesium deficiency: recognition and treatment in the emergency medicine setting. *Am. J. Emerg. Med.* 10:78–83.

Sheridan, R.L., et al. (1995) Emergency management of major hydrofluoric acid exposures. *Burns* 21:62–64.

Sittig, M. (1991) *Handbook of Toxic and Hazardous Chemicals*, 3rd ed., Park Ridge, NJ: Noyes Publications.

Upfal, M. and Doyle, C. (1990) Medical management of hydrofluoric acid exposure. *J. Occup. Environ. Med.* 32:726–731.

IODINE

Target organ(s): Cardiovascular system, CNS, eyes, respiratory system, and skin

Occupational exposure limits:

OSHA: PEL: 0.1 ppm; 1 mg/m³ TWA.

NIOSH: REL: 0.1 ppm; 1 mg/m³ TWA (NIOSH, 1994).

ACGIH: TLV: 0.1 ppm TWA with a 1 mg/m³ ceiling.

Reference values:

RfD: Not established

RfC: Not established

MRL: 0.01 mg/kg/day

Risk/safety phrases: Corrosive

BACKGROUND AND USES

Iodine is the largest molecular weight halogen of any industrial significance. Although I_2 itself is not used extensively, salts and other inorganic compounds containing iodine, such as potassium iodide and iodine monochloride, are used in a variety of settings. A large number of I_2 uses are found in the medical setting, where it is used as an antiseptic and biocide in such compounds as USP tincture and Lugol's solution, so named to identify the percent of I_2 in a solution. Other salts are used in photography and explosives and as reaction intermediates and catalysts.

PHYSICAL AND CHEMICAL PROPERTIES

Synonyms include hydrogen fluoride, anhydrous hydrofluoric acid; HF-A

Chemical properties:
   Chemical formula: I_2
   Molecular weight is 253.8 g/mol
   CASRN 7553-56-2
Physical properties:
   Color: Violet solid noncombustible
   Odor: Sharp, characteristic odor
   Boiling point: (at 760 mmHg): 185 °C (365 °F)
   Specific gravity: 4.93 at 0 °C (32 °F)
   Melting point: 133.3 °C (236 °F)
   Vapor density: 0.92
   Vapor pressure at 25 °C (77 °F): 0.3 mmHg
   Solubility: 0.01%.
   Evaporation rate: Data not available

MAMMALIAN TOXICOLOGY

Acute Effects

Acute toxicity typically occurs at dosages of more than 1 g of iodine, according to the Linus Pauling Institute. Acute iodine toxicity may cause symptoms such as diarrhea, nausea and vomiting, as well as a burning sensation in the stomach, throat, and mouth. A weak pulse and coma are also potential complications of acute toxicity.

Dermal exposure occurs primarily in settings where cleaners containing iodine are used. Skin eruptions, fever, and hypersensitivity reactions are commonly reported during overexposure. Occupational contact dermatitis was reported in a butcher who used povidone-iodine as an antibacterial solution for cuts (Lachapelle, 1984). A fisherman exposed to seaweed that contained iodine had extensive lesions on the exposed skin (Van der Willigen et al., 1988). Systemic effects from dermal exposure are rare but have been reported.

Chronic Effects

A case of thyrotoxicosis was observed in a woman who was a beautician and was chronically exposed to cosmetics containing iodine (DelGuerra et al., 1992). A persistent remission was noted after exposure to the cosmetics was discontinued. Iodine monochloride (CASRN 7790-99-0) is harmful to the skin and may cause burns with dark, painful patches.

Inhalation exposure to iodine is irritating to the mucous membranes and may also result in congestion, throat irritation, bronchitis, headache, and other symptoms. The vapor can also irritate the eyes and skin.

An examination of an operating room revealed high concentrations of iodine as a result of scrubbing with surgical cleansing agents containing iodine and of hospital waste and spills of iodine-containing solutions (Morris et al., 1986). The authors suggest that susceptible individuals may experience systemic effects, such as thyrotoxicosis, at estimated intake levels. Thyroid and metabolic disturbances may be possible after an occupational exposure to iodine, but they have not been reported.

Chronic toxicity may develop when intake is >1.1 mg/day. The vast majority of people who ingest excess amounts of iodine remain euthyroid. Occasionally, iodine-deficient individuals who ingest excess amounts of iodine develop hyperthyroidism (Jod-Basedow phenomenon). Excess uptake of iodine by the thyroid may inhibit thyroid hormone synthesis (called Wolff–Chaikoff effect). Thus, iodine toxicity can eventually cause iodide goiter, hypothyroidism, or myxedema. Very large amounts of iodide may cause a distinct taste in the mouth, increased salivation, gastrointestinal irritation, and skin lesions. Patients exposed frequently to large amounts of radiographic contrast dyes or the drug amiodarone need to have their thyroid function monitored (Johnson, 2012).

Mechanism of Action(s)

Ingested iodine not used by the thyroid but is excreted in the urine and other bodily fluids. Inhaled iodine is converted to the less toxic iodide and subsequently excreted. The toxic effects of iodine in high concentrations result from fluid loss in the affected tissues.

INDUSTRIAL HYGIENE

The primary manufacturing procedure for the preparation of iodine is the purification and extraction by acidification with sulfuric acid and precipitation with chlorine. Compounds that contain iodine are considered more hazardous than those that contain chlorine and bromine, but iodine compounds are not used as extensively for industrial applications. The primary route of exposure in most settings is dermal contact, although inhalation of the vapor may occur when the solid is heated or reacted. The use of iodine in some hand cleansers allows for some exposure by both routes when a scrubbing action is required. Protective clothing and gloves should be worn during prolonged exposure. Respirators are recommended for use when airborne concentrations are likely to exceed 1–2 ppm (Ruth, 1986).

MEDICAL MANAGEMENT

In the event of exposure to radioactive iodine, there are several techniques to help protect the patient from too much intake of the radioactive iodine, which can cause the development of thyroid nodules and thyroid cancer. Potassium iodide is a nonradioactive form of iodine, which binds to thyroid receptors to compete with the radioactive iodine. The radioiodine that does not bind to thyroid receptors is eventually cleared from the body in the urine. Potassium iodide isn't a cure-all and is most effective if taken within a day of exposure.

Prussian blue is a type of dye that binds radioactive elements known as cesium and thallium. These radioactive particles are then excreted in the feces. This treatment speeds up the elimination of the radioactive particles and reduces the amount of radiation that cells may absorb.

Diethylenetriamine pentaacetic acid (DTPA) binds to particles of the radioactive elements plutonium, americium, and curium. The radioactive particles pass out of the body in urine, reducing the amount of radiation absorbed.

REFERENCES

DelGuerra, P., et al. (1992) Occupational thyroid disease. *Int. Arch. Occup. Environ. Health* 63:373–375.

Johnson, L.E. (2012) *The Merck Manual for Health Care Professionals Online*: Available at http://www.merckmanuals.com/ professional/nutritional_disorders/mineral_deficiency_and_toxicity/iodine.html.

Lachapelle, J.M. (1984) Occupational allergic contact dermatitis to povidone-iodine. *Contact Dermatitis* 11:189–190.

Morris, D.L., et al. (1986) Inhalation of iodine in the operating theatre. *Occup. Health (Lond.)* 38:314–316.

National Institute for Occupational Safety and Health (1994) *Pocket Guide to Chemical Hazards*, NIOSH 94–116, Washington, D.C.: US Department of Health and Human Services.

Ruth, J.H. (1986) Odor thresholds and irritation levels of several chemical substances: a review. *Am, Ind, Hyg, Assoc, J*, 47:A142–A151.

Van der Willigen, A.H., et al. (1988) Contact allergy to iodine in Japanese Sargassum. *Contact Dermatitis* 18:250–252.

BROMINE

Target organ(s): Cardiovascular system, CNS, kidneys, and respiratory system

Occupational exposure limits:

OSHA: PEL: 0.1 ppm; 0.7 mg/m^3 TWA.

NIOSH: REL: 0.1 ppm TWA, 0.7 mg/m^3; 0.3 ppm, 2.0 mg/m^3 STEL (NIOSH, 1994).

ACGIH: TLV: 0.1 ppm, 0.66 mg/m^3 TWA; 0.2 ppm, 1.3 mg/m^3 STEL (ACGIH, 1996).

Reference values:

RfD: Not established

RfC: Not established

MCL: 0.01 ppm

Risk/safety phrases: Strong oxidizer, corrosive

BACKGROUND AND USES

Bromine compounds are a major component of fire retardants and are also used in gold extraction, drilling fluids (calcium bromide), sanitation, and agriculture (methyl bromide), as laboratory reagents and intermediates, and in ethylene dibromide, an antiknock component of gasoline. The elemental form of this compound is an alternative for chlorine used in swimming pools. A natural element, bromine is found in abundance in the earth's crust and in seawater in a variety of inorganic salts and organic compounds. It is the only halogen that is a liquid at room temperature.

PHYSICAL AND CHEMICAL PROPERTIES

Synonyms include molecular bromine
Chemical properties:

Chemical formula: Br_2

Molecular weight is 159.8 g/mol

CASRN 7726-95-6

Physical properties:

    Color: Dark reddish-brown fuming liquid with suffocating fumes

    Boiling point: (at 760 mmHg): 59.4 °C (139 °F)

    Specific gravity: 3.12 at 0 °C (32 °F)

    Melting point: −7.2 °C (19 °F)

    Vapor pressure: 172 mmHg

    Solubility: 4%.

    Evaporation rate: Data not available

MAMMALIAN TOXICOLOGY

Acute Effects

Bromine is corrosive to human tissue in a liquid state and its vapors irritate the eyes and throat. Bromine vapors are very toxic upon inhalation (Table 47.3). The most important health effects that can be caused by bromine-containing organic contaminants are the malfunctioning of the nervous system and disturbances in genetic materials.

Organic bromines can also cause damage to organs such as liver, kidneys, and lungs. Bromine can cause stomach and gastrointestinal malfunctioning. Some forms of organic bromines, such as ethylene bromine, can even cause cancer.

Acute inhalation exposure to bromine gas can be hazardous, depending on the duration and concentration. An exposure to a lower concentration for a longer period can have the same health effects as an exposure to a higher concentration for a shorter period (Withers and Lees, 1986).

Inflammation of the tongue and palate may be seen as well. Some of the symptoms developed after the individual had a seemingly healthy period of time, and others were reported and treated immediately. More severe respiratory responses to bromine exposure have included case reports of chemical pneumonitis and pneumomediastinum (Kraut and Lilis, 1988; Lossos et al., 1990).

Brief dermal contact with the vapor and liquid form may cause severe chemical burns, manifesting as pustules and vesicles. Prolonged contact with bromine can result in painful, deep ulcers.

TABLE 47.3 Effects of Exposure to Bromine

Concentration, ppm	Effect(s)
0.05–3.5	Odor threshold
0.1	TLV–TWA limit
>1.0	Irritation level
40–60	Toxic pneumonitis and pulmonary edema
1000	Fatal within a few minutes

Chronic Effects

Human studies have not identified bromine as a carcinogenic compound. Some brominated compounds, such as potassium bromate, have been reported to have carcinogenic effects in animal studies (Kurokawa et al., 1990). Inorganic bromines are found in nature. But more bromine has been included in daily life through common use in water and in pesticides. These bromines can damage the nervous system and thyroid gland after prolonged low level exposure. Bromine is potentially capable of extensive damage to the lower respiratory tract. Limited studies have reported diffuse interstitial pulmonary fibrosis, emphysema, and/or airway hyperreactivity secondary to acute exposure to bromine (Kraut and Lilis, 1988; Lossos et al., 1990).

Mechanism of Action(s)

Absorbed bromine can be excreted in the urine or stored in the tissue as bromide. This accounts for the cumulative effect of some bromine exposures. Due to its reactivity, bromine does not persist as an element in living tissue but quickly forms bromide. In this form, it may be deposited in the tissues, displacing other halogens (Sticht and Kaferstein, 1988).

INDUSTRIAL HYGIENE

As with other halogens, bromine can be extracted from brine solutions, first oxidized with chlorine, and then vaporized with air or steam. Many brominated compounds can react violently and form explosive mixtures with a variety of compounds, such as acetylene, ammonia, and organic substances such as rubber and cork. Inhalation and dermal exposures are most common with bromine. Workers should be protected from high concentrations of bromine with proper engineering controls and protective gear. Pool service personnel using elemental bromine should wear gloves and goggles to protect themselves from an accidental splashing exposure. The low odor threshold of this chemical is reported to be 0.392 mg/m^3 (50 ppb), with an irritating concentration of 2 mg/m^3 (0.3 ppm) (Ruth, 1986). This would indicate an inadequate irritation warning of concentrations exceeding the regulatory guidelines.

MEDICAL MANAGEMENT

Acute contact with bromine liquid or vapor requires removal from the source of bromine contamination. An exposure of bromine to the eyes should be irrigated with copious amounts of tepid water for at least 15 min. If irritation, pain, swelling, lacrimation, or photophobia persist, further medical evaluation is recommended. In instances when bromine comes in contact with clothing or the skin, it is necessary to remove the contaminated clothing and thoroughly wash the affected area with copious volumes of water for 20 min. Since the effects may be delayed, close observation for blistering and discoloration of the skin is required for the next 24 h (Sagi et al., 1985).

An exposure to bromine vapors can cause respiratory symptoms and should receive respiratory support in accordance with symptomatology, including maintenance of an adequate airway, oxygen, antibronchospasm therapy, including inhaled beta adrenergic agonist and a short course of corticosteriods and antibiotics if there is evidence of infection. Assisted or supported ventilation with tracheal intubation and positive pressure ventilation may be needed.

REFERENCES

American Conference of Governmental Industrial Hygienists (1996) *Threshold Limit Values (TLVs) for Chemical Substances and Physical Agents and Biological Exposure Indices (BEIs)*, Cincinnati: American Conference of Governmental Industrial Hygienists. Kraut, A. and Lilis, R. (1988) Chemical pneumonitis due to exposure to bromine compounds. *Chest* 94:208–210.

Kraut, A. and Lilis, R. (1988) Chemical pneumonitis due to exposure to bromine compounds. *Chest* 94:208–210.

Kurokawa, Y., et al. (1990) Toxicity and carcinogenicity of potassium bromate: a new renal carcinogen. *Environ. Health Perspect.* 87:309–335.

Lossos, I.S., Abolnik, I., and Breuer, R. (1990) Pneumomediastinum: a complication of exposure to bromine. *Br. J. Ind. Med.* 47:784–790.

National Institute for Occupational Safety and Health (1994) *Pocket Guide to Chemical Hazards*, NIOSH 94–116, Washington, DC: US Department of Health and Human Services.

Ruth, J.H. (1986) Odor thresholds and irritation levels of several chemical substances: a review. *Am. Ind. Hyg. Assoc. J.* 47: A142–A151.

Sagi, et al. (1985) Burns caused by bromine and some of its compounds. *Burns Incl. Therm. Inj.* 11:343–350.

Sticht, G. and Kaferstein, H. (1988). Bromine. In: Seiler, H., Sigel, H., and Sigel, A., editors. *Handbook on Toxicity of Inorganic Compounds*, New York, NY: Marcel Dekker Inc., pp. 143–154.

Withers, R.M.J. and Lees, F.P. (1986) The assessment of major hazards: the lethal toxicity of bromide. *J. Hazard. Mater.* 13:279–299.

48

PHOSGENE

Benjamin Wong, Michael W. Perkins, and Alfred M. Sciuto

First aid:

Eye: Immediately wash (irrigate) the eyes with large amounts of water, occasionally lifting the lower and upper lids. Get medical attention immediately.

Skin: Immediately flush the contaminated skin with water. Remove contaminated clothing and flush the skin with water. Get medical attention promptly.

Inhalation: Move the exposed individual to fresh air at once. Keep the affected person calm and at rest in a half-upright position. Monitor and maintain the airway and blood pressure. If breathing has become difficult administer oxygen; if breathing has stopped, perform artificial respiration. Get medical attention as soon as possible.

Target organ(s): Eyes, skin, respiratory system

Occupational exposure limits:

OSHA PEL: 0.1 ppm (0.4 mg/m^3)

American Conference of Governmental Industrial Hygienists (ACGIH) threshold limit value (TLV): 0.1 ppm (0.4 mg/m^3)

NIOSH REL: 0.1 ppm (0.4 mg/m^3) TWA, 0.2 ppm (0.8 mg/m^3) 15-min ceiling

Reference values:

RfC: $3 \times 10^{-4} - 5 \times 10^{-4}$ mg/m^3

Risk/safety phrases:

R26: Very toxic by inhalation

R34: Causes burns

S1/2: Keep locked up and out of the reach of children

S9: Keep container in a well-ventilated place

S26: In case of contact with eyes, rinse immediately with plenty of water and seek medical advice

S36/37/39: Wear suitable protective clothing, gloves, and eye/face protection

S45: In case of accident or if you feel unwell, seek medical advice immediately (show the label where possible)

BACKGROUND AND USES

Phosgene was first synthesized by John Davy (1790–1868) in 1812 when he exposed a mixture of carbon monoxide and chlorine to sunlight (Davy, 1812). During the nineteenth century, phosgene was slowly integrated into the chemical industry but gained particular notoriety as the most heavily used chemical warfare agent (CWA) of World War I (WWI) (NAS/NRC Board on Environmental Studies and Toxicology, 1984; Tuorinsky and Sciuto, 2008). Although banned from use in warfare by the Geneva Protocol in 1929, phosgene was used again during World War II (WWII). Today, phosgene is an important intermediate in the production of polyurethanes and other chemicals and is commercially prepared by passing chlorine and excess carbon monoxide over activated carbon (Via et al., 2002).

As a CWA in WWI, phosgene was used offensively by both Allied and Axis powers to produce mass casualties. Some reports suggest that phosgene alone was responsible for nearly 80% of all gas casualties, numbering approximately 71,000 in total between 1914 and 1918 (Gilchrist and Matz, 1933; Elsayed and Salem, 2006). The intent of CWAs is not solely to exterminate the opposition, but also to incapacitate the enemy, thus removing soldiers from the battlefield. Phosgene gassing during WWI caused men to

Hamilton & Hardy's Industrial Toxicology, Sixth Edition. Edited by Raymond D. Harbison, Marie M. Bourgeois, and Giffe T. Johnson.
© 2015 John Wiley & Sons, Inc. Published 2015 by John Wiley & Sons, Inc.

lose 311,000 days to hospitalization, an astonishing 852 man-years (Jackson, 1933). After the Geneva Protocol was established in 1929, prohibiting the use of chemical and biological weapons, the only documented use of phosgene as a CWA was by the Imperial Japanese Army against the Chinese during the Second Sino–Japanese War/WWII, with 375 separate occasions occurring during the battle of Wuhan from August to October of 1938 (Tanaka, 1988; Yoshimi and Matsuno, 1997).

Since WWII, phosgene has gained importance in the chemical industry as a reactive chemical intermediate and is used to help supply chlorine groups to reaction products in phosgenation reactions (Babad and Zeiler, 1973). The principal use of phosgene is in the polyurethane industry, which consumes over 85% of the world's phosgene output. The polycarbonate industry accounts for approximately 10% of phosgene used, and the remaining 10% is used in the production of aliphatic diisocyanates, monoisocyanates, chloroformates, agrochemicals, and intermediates for dye-stuffs and pharmaceuticals (NAS/NRC Board on Environmental Studies and Toxicology, 1984). Estimated annual worldwide consumption of phosgene ranges from 5 to 10 million tons (Via et al., 2002; Cotarca and Eckert, 2004; U.S. Environmental Protection Agency, 2005), with U.S. production alone around 1 million tons (Bast and Bress, 2002). Production estimates are difficult to obtain because synthesis of phosgene is almost entirely captive, meaning that more than 99% is used in the manufacture of other chemicals within a plant boundary. In the United States, this is the case for the overwhelming majority of phosgene generation (Cotarca and Eckert, 2004); and only one company sells phosgene in the merchant market.

While phosgene has been a military and terrorist threat for many years due to its low cost and ease of production, it is also an occupational and an environmental hazard. Although phosgene is nearly completely consumed during industrial use, it can be released through failed processes. Phosgene is an environmental pollutant and has been detected worldwide in ambient air samples (Singh, 1976; U.S. Environmental Protection Agency, 2005). It can also be formed by the thermal decomposition of chlorinated hydrocarbons, posing a threat for welders, refrigeration mechanics, and car repairmen (Brown and Birky, 1980). Of recent note in the United States, on January 23, 2010, an operator working at a DuPont chemical plant in the phosgene cylinder storage area was sprayed in the face and upper torso with phosgene when a flexible hose suddenly ruptured. Although exposed to a significant amount of phosgene, he did not exhibit immediate signs of respiratory distress. However, approximately 3 h later, after he had arrived at the hospital, his condition deteriorated and he died the following night (U.S. Chemical Safety and Hazard Investigation Board, 2011). Phosgene gas has been and

TABLE 48.1 Physical Properties of Phosgene

CASRN:	75-44-5
Structural formula:	$COCl_2$
Formula weight:	98.92
Boiling point:	8.2 °C
Liquid density, 0 °C:	1.4187 g/ml
Gas density, 20 °C:	4.39 g/l
Specific gravity, gas, 20 °C:	3.4
Vapor pressure:	1180 mmHg at 20 °C
Solubility:	Slightly soluble in water (5940 mg/l at 25 °C); hydrolyzes to hydrochloric acid and carbon dioxide. Soluble in carbon tetrachloride, chloroform, acetic acid, benzene, and toluene
Color:	Colorless
Odor:	Sweet in low concentrations; sharp, pungent (odor of moldy hay) in higher concentrations
Odor threshold:	0.4–1.5 ppm
Irritation threshold:	3 ppm
Conversion factors:	1 ppm = 4.05 mg/m³ 1 mg/m³ = 0.247 ppm (25 °C, 760 mmHg)

continues to be a potentially serious and occasionally fatal substance (Polednak and Hollis, 1985).

PHYSICAL AND CHEMICAL PROPERTIES

Phosgene, also known as CG, carbon dichloride oxide, carbonic dichloride, carbon oxychloride, carbonyl chloride, carbonyl dichloride, and chloroformyl chloride, is a colorless, volatile, synthetic irritant and asphyxiating gas. In aqueous solution, hydrolysis of phosgene is extremely rapid (Manogue and Pigford, 1960; Schneider and Diller, 1989) producing carbon dioxide and hydrochloric acid ($COCl_2 + H_2O \rightarrow CO_2 + 2HCl$). Some of the physical and chemical properties are listed in Table 48.1 (NAS/NRC Board on Environmental Studies and Toxicology, 1984; National Institute for Occupational Safety and Health, 2005; U.S. Environmental Protection Agency, 2005).

ENVIRONMENTAL FATE AND BIOACCUMULATION

The production and use of phosgene as an intermediate in organic synthesis, especially of isocyanates, polyurethane and polycarbonate resins, carbamates, organic carbonates, and chloroformates may result in its release to the environment through various waste streams (Lewis, 2007). The potential environmental threat of phosgene is compounded by its additional uses in dye and plastics manufacturing, in acid chlorination processes, as a phosgenation reagent, and as a CWA compound (Klaassen, 1995; O'Neil, 2001).

Phosgene is produced in the earth's atmosphere from the degradation of a variety of chlorinated compounds, including tetrachloroethylene, trichloroethylene, chloroform, methyl chloroform, and carbon tetrachloride (Kindler et al., 1995). With a vapor pressure of 1180 mmHg at 20 °C, phosgene will exist solely in the gaseous phase in the atmosphere, where it is slowly degraded by reacting with photochemically produced hydroxyl radicals. The half-life for this reaction in air is estimated to be 44 years (Hazardous Substances Data Bank [Internet], 2008). Phosgene does not absorb at wavelengths above 290 nm and therefore is not expected to be susceptible to direct photolysis by sunlight. If released to soil, phosgene is expected to have very high mobility based upon an estimated soil organic carbon–water partitioning coefficient (Koc) of 2.2. Volatilization from moist soil surfaces is expected to be an important fate process based upon the Henry's Law constant (Hcc) of 17 l·atm/mol and its vapor pressure. If released into water, phosgene is not expected to adsorb on the suspended solids and sediment based upon the estimated Koc; and like for soil, its Hcc indicates that volatilization from water surfaces is expected to be an important fate process. Complete hydrolysis of a 1% solution of phosgene in water occurs within 20 s at 0 °C, suggesting that bioconcentration will not be an important environmental fate process. No biodegradation data are currently available (Hazardous Substances Data Bank [Internet], 2008).

ECOTOXICOLOGY

No information has been reported on either the acute or chronic effects of phosgene on the environment. However, the levels of phosgene currently found in the general environment would not be expected to result in significant effects to aquatic or terrestrial life. Damage to plants and aquatic organisms could result in areas where accidental release of phosgene has occurred, owing its rapid hydrolysis to hydrochloric acid (World Health Organization, 1988).

MAMMALIAN TOXICOLOGY

Acute Effects

In mammals, inhalation or contact (skin or ocular) exposure to phosgene results in a range of symptoms dependent on the dose and time of exposure. An inhaled concentration of phosgene above 3 parts per million (ppm) causes pain and irritation to the mucous membranes of the eyes, nose, throat, and bronchi. Higher concentrations (>200 ppm) of phosgene can cause apnea, bronchoconstriction, inflammation, and thickening of the bronchioles (Banister et al., 1949).

Exposure to high concentrations (>200 ppm) of phosgene results in a reflex, a symptom-free latent, and a terminal phase of injury (Borak and Diller, 2001). The reflex phase results in vagal reflex actions that may cause shallow respiration due to the decrease in the arterial partial pressure, blood pH, vital lung capacity and volume. This phase can last for up to 4 h and is followed by the latent phase. The latent phase initially lacks clinical symptoms and lasts for several hours before the development of early signs of pulmonary edema. In the final terminal or clinical phase, fluid rapidly accumulates in the lungs and gas exchange decreases, resulting in severe pulmonary edema and lobular pneumonia, which peaks at 24 h postexposure (Hamilton et al., 1983). Mortality may occur between 24 and 48 h following phosgene exposure as a result of respiratory and circulatory failure.

Chronic Effects

The chronic effects of phosgene exposure consist of various symptoms that may include emphysema, fibrosis, dyspnea, reduced physical fitness, and alterations in pulmonary function (Diller, 1985; Wyatt and Allister, 1995). An evaluation of six patients who were exposed to an acute dose of phosgene showed that they suffered from cough, tightness of the chest, shortness of breath, and abnormalities in pulmonary function for at least 1 year after the exposure (Galdston et al., 1947a). The author suggested that the abnormalities in pulmonary function were a reflection of the myriad physiological and biochemical changes in the lungs and direct damage to the bronchioles similar to that observed in the early stages of emphysema. Unfortunately, the chronic symptoms resulting from these changes in pulmonary function could not be correlated with the severity of the phosgene intoxication. Another study evaluated five workers who were exposed to multiple occupational doses of phosgene for up to 18 months (Galdston et al., 1947b). These individuals developed symptoms similar to those previously mentioned and included coughs, tightness of the chest, and shortness of breath, all of which gradually improved once the individual was no longer exposed to phosgene.

Victims of phosgene exposure may require months to years to fully recover. It is recommended that all victims of phosgene exposure have yearly medical checkups, including pulmonary function tests (Borak and Diller, 2001). Victims with preexisting lung damage, with pulmonary conditions such as chronic bronchitis, or who are tobacco users may experience more severe and prolonged toxic effects following phosgene exposure (Diller, 1985).

Chemical Pathology

The most prominent symptom of phosgene poisoning is pulmonary edema resulting from increased water permeability of the alveolar capillaries and swelling of lung tissue. It has been reported that the terminal airways, alveolar lining, and alveolar epithelial layer are affected first by the inhalation exposure to phosgene (Gerard, 1948). Damage to the terminal bronchi and alveoli increases the permeability of the blood–air barriers and results in movement of excess fluid and protein into interstitial spaces. Flooding of the interstitial spaces is initially cleared by increased lymphatic drainage, but once fluid volumes exceed the clearance capacity, fluid leaks into the alveolar epithelium, flooding the alveoli. Gas exchange is drastically reduced as a result of both the death of the cells comprising the bronchiolar mucous membranes and migration of leukocytes into the bronchi and alveoli (Mehlman and Fensterheim, 1985). Damage resulting from phosgene exposure is primarily respiratory, but other effects, including cardiovascular, hematological, renal, ocular, and cutaneous toxicities, have also been observed (Borak and Diller, 2001).

Mode of Action(s)

Phosgene is a highly reactive gas that reacts with biological molecules and tissues primarily through acylation and/or hydrolysis of hydrochloric acid. Acylation occurs when amino, hydroxyl, and sulfhydryl groups on biological molecules attack the electrophilic carbon molecule in phosgene (Babad and Zeiler, 1973), and can result in membrane structural changes, protein and lipid denaturation, and irreversible alteration of membrane, enzyme, and cellular functions. The highly reactive nature of phosgene with constituents present in the alveolar lining fluids causes direct damage to the lung surfactant through peroxidation of lipids (Guo et al., 1990; Jugg et al., 1999; Duniho et al., 2002). Phosgene-induced damage leads to the downstream changes in redox enzyme activity and release of arachidonic acid mediators (Sciuto et al., 2003; Sciuto et al., 2005). Exposure to phosgene also depletes and oxidizes lung nucleophiles, particularly glutathione, which has been implicated in alleviating phosgene-induced effects (Sciuto et al., 2003; Sciuto and Hurt, 2004). Phosgene exposure also causes pulmonary cellular glycolysis and oxygen uptake to decrease that results in decreased adenosine triphosphate and cyclic adenosine monophosphate levels (Currie et al., 1987; Schroeder and Gurtner, 1992; Sciuto et al., 1996). These alterations in cellular activity combined with increase in water uptake by epithelial, interstitial, and endothelial cells result in alveolar leakage and pulmonary edema.

Phosgene is hydrolyzed into hydrochloric acid following contact with water and may cause irritation and damage to the eyes, nasopharynx, and respiratory tract. Considering the low water solubility of phosgene, only limited amounts of hydrochloric acid would be produced under normal physiological conditions as a result of phosgene exposure to biological tissues.

SOURCES OF EXPOSURE

Occupational

Occupational exposure to phosgene may occur through inhalation and dermal contact with this compound at workplaces where phosgene is manufactured and/or used. Monitoring data indicate that the general population may be exposed to phosgene via inhalation of ambient air, although the concentrations are low (U.S. Environmental Protection Agency, 2005). Phosgene can be formed by the thermal decomposition of chlorinated hydrocarbons, posing a threat for welders, refrigeration mechanics, and car repairmen (Brown and Birky, 1980). Commonly used industrial degreasers contain chlorinated hydrocarbons such as perchloroethylene, which can form phosgene when heated. Industrial workers, firemen, military personnel are at increased risk for accidental or occupational exposure to phosgene.

Environmental

The only natural source of phosgene in the environment documented to date is from volcanic activity. However, calculations indicate that the preeminent chlorinated species in volcanic gases is HCl and that the concentration of phosgene is roughly 15 orders of magnitude less (Ryan et al., 1996). Nonetheless, phosgene is an environmental pollutant and has been detected in ambient air samples (Singh, 1976; U.S. Environmental Protection Agency, 2005). The source of such pollutants most likely stems from the errant release of phosgene and its precursors during manufacturing processes, as well as the degradation of chlorinated compounds in the atmosphere, as previously described (Kindler et al., 1995). In Poland, the proximity of heavy industrialization to densely populated areas has resulted in the identification of phosgene as one of the most significant threats in the environment in those areas (Krajewski, 1997).

INDUSTRIAL HYGIENE

Recommended personal protective equipment (PPE) for liquid phosgene includes impervious clothing, full facepiece respirator, gloves, face shields (8-inch minimum), splash-proof safety goggles, and the availability of quick-drenching facilities, and eye-wash stations. Table 48.2 shows exposure limits as set forth by several different organizations (National Institute for Occupational Safety and Health, 2005).

TABLE 48.2   Phosgene Exposure Limits

Immediately Dangerous to Life or Health IDLH	2 ppm (8.1 mg/m^3)	
National Institute for Occupational Safety and Health (NIOSH) recommended exposure limits (REL)	Time-weighted average TWA 0.1 ppm (0.4 mg/m^3)	15-m C 0.2 ppm (0.8 mg/m^3)
Occupational Safety and Health Administration (OSHA) permissible exposure limits (PEL)	TWA 0.1 ppm (0.4 mg/m^3)	

RISK ASSESSMENTS

Phosgene is a gas at room temperature, is highly reactive, and hydrolyzes rapidly in water to CO_2 and HCl. As such, oral exposure to phosgene is exceedingly unlikely. The most likely route of phosgene exposure is through inhalation, and the acute effects of inhalational phosgene exposure have been well studied. Exposure results in the development of pulmonary edema and these symptoms increase in severity with both concentration (C) and time (t), as described by Haber's Law. At relatively high concentrations (>200 ppm), irritation of the eyes and alterations in respiratory parameters may occur, and at sufficiently high Ct levels, death may occur as a result of hypoxia or cardiac failure, both believed to be secondary responses resulting from the severe pulmonary edema associated with high levels of inhaled phosgene (U.S. Environmental Protection Agency, 2005). The effects of phosgene inhalation for sub-chronic or chronic durations are less defined, as limited human data and very few animal studies are available. The respiratory tract and the lungs are identified as the primary target organ in all species, and immunotoxicity has been observed in a few animal studies. However, the relevance of these findings to human hazard cannot be addressed at this time. Currently, information to assess the carcinogenic potential of phosgene is inadequate (U.S. Environmental Protection Agency, 2005).

Disclaimer: The views expressed in this article are those of the author(s) and do not reflect the official policy of the Department of Army, Department of Defense, or the U.S. Government.

REFERENCES

Babad, H. and Zeiler, A.G. (1973) Chemistry of phosgene. *Chem. Rev.* 73(1):75–91.

Banister, J., Fegler, G., and Hebb, C. (1949) Initial respiratory responses to the intratracheal inhalation of phosgene or ammonia. *Q. J. Exp. Physiol. Cogn. Med. Sci.* 35(3):233–250, 7 pl.

Bast, C. and Bress, B. (2002) In: Levels, S. o. A. E. G., Toxicology, C.o., and Council, N.R., editors. *Acute Exposure Guideline Levels for Selected Airborne Chemicals: Volume 2*, The National Academies Press, pp. 13–70.

Borak, J. and Diller, W.F. (2001) Phosgene exposure: mechanisms of injury and treatment strategies. *J. Occup. Environ. Med.* 43(2):110–119.

Brown, J.E. and Birky, M.M. (1980) Phosgene in the thermal decomposition products of poly(vinyl chloride): generation, detection and measurement. *J. Anal. Toxicol.* 4(4):166–174.

Cotarca, L. and Eckert, H. (2004) *Phosgenations: A Handbook*, John Wiley & Sons.

Currie, W.D., Hatch, G.E., and Frosolono, M.F. (1987) Changes in lung ATP concentration in the rat after low-level phosgene exposure. *J. Biochem. Toxicol.* 2:105–114.

Davy, J. (1812) On a gaseous compound of carbonic oxide and chlorine. *Phil. Trans. R. Soc. London* 102:144–151.

Diller, W.F. (1985) Late sequelae after phosgene poisoning: a literature review. *Toxicol. Ind. Health* 1(2):129–136.

Duniho, S.M., Martin, J., Forster, J.S., Cascio, M.B., Moran, T.S., Carpin, L.B., and Sciuto, A.M. (2002) Acute changes in lung histopathology and bronchoalveolar lavage parameters in mice exposed to the choking agent gas phosgene. *Toxicol. Pathol.* 30(3):339–349.

Elsayed, N.M. and Salem, H. (2006) Chemical warfare agents and nuclear weapons. In: Salem, H. and Katz, S. A., editors. *Inhalation Toxicology*, 2nd ed., Boca Raton, FL: CRC Press, pp. 521–542.

Galdston, M., Luetscher, J.A. Jr., Longcope, W.T., and Ballich, N.L. (1947a) A study of the residual effects of phosgene poisoning in human subjects. 1. After acute exposure. *J. Clin. Invest.* 26(2):145–168.

Galdston, M., Luetscher, J.A., Longcope, W.T., Ballich, N.L., Kremer, V.L., Filley, G.L., and Hopson, J.L. (1947b) A study of the residual effects of phosgene poisoning in human subjects. 2. after chronic exposure. *J. Clin. Invest.* 26(2):169–181.

Gerard, R. (1948) Recent research on respiratory irritants. In: Andrus, E.C., Bronk, D.W., Carden, G.A. Jr., Keefer, C.S., Lockwood, J.S., Wearn, J.T., and Winternitz, M.C., editors. *Science in World War II: v. II, Advances in Military Medicine*, Boston, MA: Little, Brown and Company.

Gilchrist, H.L. and Matz, P.B. (1933) *The residual effects of warfare gases*. Washington, DC: U.S. Govt. print off.

Guo, Y.L., Kennedy, T.P., Michael, J.R., Sciuto, A.M., Ghio, A.J., Adkinson, N.F., and Gurtner, G.H. (1990) Mechanism of phosgene-induced lung toxicity: role of arachidonate mediators. *J. Appl. Physiol.* 69(5):1615–1622.

Hamilton, A., Hardy, H.L., and Finkel, A.J. (1983) *Hamilton and Hardy's Industrial Toxicology*, 4th ed., Boston: J. Wright.

Hazardous Substances Data Bank [Internet] (2008) *PHOSGENE. In 22 April 2008 ed. National Library of Medicine.*

Jackson, K.E. (1933) Phosgene. *J. Chem. Edu.* 10(10):622.

Jugg, B., Jenner, J., and Rice, P. (1999) The effect of perfluoroisobutene and phosgene on rat lavage fluid surfactant phospholipids. *Hum. Exp. Toxicol.* 18(11):659–668.

Kindler, T.P., Chameides, W.L., Wine, P.H., Cunnold, D.M., Alyea, F.N., and Franklin, J.A. (1995) The fate of atmospheric phosgene and the stratospheric chlorine loadings of its parent compounds: CCl4, C2Cl4, C2HCl3, CH3CCl3, and CHCl3. *J. Geophys. Res.* 100(D1):1235–1251.

Klaassen, C.D. (1995) *Casarett and Doull's Toxicology*, 5th ed., New York, NY: McGraw-Hill.

Krajewski, J.A. (1997) Chemical accidents and catastrophes as a source of the greatest hazard to the environment and human health in Poland. *Med. Pr.* 48(1):93–103. First published on Awarie i katastrofy chemiczne zrodlem nadzwyczajnych zagrozen dla srodowiska i zdrowia ludzi w Polsce.

Lewis, R.J.S. (2007) *Hawley's Condensed Chemical Dictionary*, 5th ed., New York, NY: Wiley-Interscience.

Manogue, W.H. and Pigford, R.L. (1960) The kinetics of the absorption of phosgene into water and aqueous solutions. *Aiche. J.* 6(3):494–500.

Mehlman, M.A. and Fensterheim, R.J. (1985) Phosgene induced edema: diagnosis and therapeutic countermeasures. an international symposium. September 23–24, 1982. *Toxicol. Ind. Health* 1(2):1–160.

NAS/NRC Board on Environmental Studies and Toxicology (1984) *Emergency and Continuous Exposure Limits for Selected Airborne Contaminants*, vol. 2, The National Academies Press.

National Institute for Occupational Safety and Health (2005) *NIOSH Pocket Guide to Chemical Hazards.* Cincinnati, Ohio: U.S. Dept. of Health and Human Services, Public Health Service, Centers for Disease Control and Prevention, National Institute for Occupational Safety and Health.

O'Neil, M.J. (2001) *The Merck Index*, 13th ed., Whitehouse Station, NJ: Merck and Co., Inc.

Polednak, A.P. and Hollis, D.R. (1985) Mortality and causes of death among workers exposed to phosgene in 1943-45. *Toxicol. Ind. Health* 1(2):137–151.

Ryan, T.A., Seddon, E.A., Seddon, K.R., and Ryan, C. (1996) *Phosgene: And Related Carbonyl Halides*, Elsevier Science.

Schneider, W. and Diller, W. (1989) Phosgene. In: *Encyclopedia of Industrial Chemistry*, 5th ed., vol. A, Weinheim, Germany: VCH Verlag, pp. 411–420.

Schroeder, S. and Gurtner, G. (1992) Evidence for a species difference in susceptibility and mechanism of phosgene toxicity between rabbits and dogs. *Am. Rev. Respir. Dis.* 145:A606.

Sciuto, A.M. and Hurt, H.H. (2004) Therapeutic treatments of phosgene-induced lung injury. *Inhal. Toxicol.* 16(8):565–580.

Sciuto, A.M., Cascio, M.B., Moran, T.S., and Forster, J.S. (2003) The fate of antioxidant enzymes in bronchoalveolar lavage fluid over 7 days in mice with acute lung injury. *Inhal. Toxicol.* 15(7):675–685.

Sciuto, A.M., Phillips, C.S., Orzolek, L.D., Hege, A.I., Moran, T.S., and Dillman, J.F. (2005) Genomic analysis of murine pulmonary tissue following carbonyl chloride inhalation. *Chem. Res. Toxicol.* 18(11):1654–1660.

Sciuto, A.M., Strickland, P.T., Kennedy, T.P., Guo, Y.L., and Gurtner, G.H. (1996) Intratracheal administration of DBcAMP attenuates edema formation in phosgene-induced acute lung injury. *J. Appl. Physiol.* 80(1):149–157.

Singh, H.B. (1976) Phosgene in the ambient air. *Nature* 264(5585):428–429.

Tanaka, Y. (1988) Poison gas: the story Japan would like to forget. *B. Atom. Sci.* 44(8):10–19.

Tuorinsky, S.D. and Sciuto, A.M. (2008) Medical aspects of chemical warfare. In: Tuorinsky, S.D. and United States. Dept. of the Army. Office of the Surgeon General., and Borden Institute (U.S.), editors., *Textbooks of Military Medicine*, Washington, DC: Office of the Surgeon General Borden Institute For sale by the Supt. of Docs., U.S. G.P.O., Falls Church, Va, pp. 339–370.

U.S. Chemical Safety and Hazard Investigation Board (2011) *Investigation Report E.I. DuPont de Nemours & Co., Inc. Phosgene Release.*

U.S. Environmental Protection Agency (2005) *TOXICOLOGICAL REVIEW OF PHOSGENE (CAS No. 75-44-5).*

Via, F.A., Soloveichlk, G.L., Kosky, P.G., and Cicha, W.V. (2002) *Method for producing phosgene.* In (Vol. 10056201. JP 02–06307 (Jan, 1990); US 4073806 A (Feb, 1978) Doubovetzky et al.; US 4231959 Λ (Nov, 1980) Obrecht; US 4764308 A (Aug, 1988) Sauer et al.; US 4914070 A (Apr, 1990) Ledoux et al.; US 4978649 A (Dec, 1990) Surovikin et al.; US 5136113 A (Aug, 1992) Rao; US 6022993 A (Feb, 2000) Cicha et al.; US 6054107 A (Apr, 2000) Cicha et al.; US 6054612 A (Apr, 2000) Cicha et al., US.

World Health Organization (1988) *Health and Safety Guide.* In: International Program on Chemical Safety, editor., Geneva: Published by the World Health Organization for the International Programme on Chemical Safety.

Wyatt, J.P. and Allister, C.A. (1995) Occupational phosgene poisoning: a case-report and review. *J. Accid. Emerg. Med.* 12(3):212–213.

Yoshimi, Y. and Matsuno, S. (1997) *Dokugasusen Kankei Shiryô II, Kaisetsu, Jugonen Sensô Gokuhi Shiryoshu.*

49

NITROGEN COMPOUNDS

Kathleen T. Jenkins

First aid: *Inhalation exposure:* Nitrogen is a simple asphyxiant. Remove the victims to fresh air. Immediately eliminate exposure and prevent repeated exposure by removal from exposure. *Dermal/eye exposure:* Nitrogen oxides are irritating to eyes, mucous membranes, lungs, and skin. Nitrogen oxides and acids are skin irritants and corrosive. The hair, skin, and eyes should be irrigated for 20 min and then referred to a physician

Cardiovascular: Nitrogen oxides can cause weak rapid pulse and circulatory collapse. *Hematologic:* High dose exposure can lead to formation of methemoglobinemia and impaired oxygen transport. Methylene blue may be used for treatment

Target organs: Skin, respiratory, cardiovascular, and hematologic

Occupational exposure limits: Nitrogen is a simple asphyxiant and was not assigned a TLV because of the limiting factor of available oxygen

Nitric oxide: OSHA PEL: 25 ppm over 8 h shift, NIOSH IDLH 100 ppm

Nitrogen dioxide: OSHA PEL 5 ppm (9 mg/m^3) ceiling, NIOSH PEL1 ppm (1.8 mg/m^3) STEL, ACGIH TLV 3 ppm (5.6 mg/m^3), STEL, 5 ppm, NIOSH IDLH 50 ppm, Rfd-withdrawn

Nitrous oxide: OSHA, not regulated, NIOSH REL 25 ppm (45 mg/m^3) TWA, ACGIH 50 ppm (90 mg/m^3) TWA

Nitric Acid: OSHA PEL, 2 ppm TWA, NIOSH: REL 2 ppm TWA, NIOSH: IDLH 100 ppm

Risk/safety phrases: Most of the nitrogen oxides are irritants to the eyes, lungs, and skin. Nitrogen oxides are not classified as carcinogenic byBY IARC, EPA, OR DHHS

BACKGROUND AND USES

Nitrogen (CASRN 7727-37-9) is the most abundant element in the atmosphere. It makes up 78% of the atmosphere by volume. It is essential to living systems, including a plant nutrient and utilized by most living systems. Humans and other life forms require the reactive form of nitrogen for a variety of physiological processes. Nitric oxide (NO) is acknowledged as an important endogenous second messenger within several organ systems. It is a ubiquitous signaling molecule produced through metabolism and l-arginine by nitric oxide synthases. Its formation is crucial for the control of blood pressure, blood flow, and other bodily functions (Kovalchin et al., 1997). Nitrogen does not exert a direct toxicological effect and has no toxicological profile. Nitrogen is only safe for humans and animals when mixed with oxygen. Nitrogen is considered a simple asphyxiant and displaces oxygen, which can cause immediate and serious health effects. To improve safety, the commercial use of nitrogen is an inerting agent to keep material free of contaminants, including oxygen. Since oxygen can corrode equipment or cause a fire or explosion hazard; nitrogen is also used to purge air from equipment prior to introducing materials.

Reactive nitrogen includes all forms of nitrogen that are biologically, photochemically, and radioactively active. Compounds of nitrogen that are reactive include the following: nitrous oxide (N_2O), nitrate (NO_3), nitrite (NO_2^-), ammonia (NH_3), and ammonium (NH_4). The nitrogen oxides include elemental nitrogen in five different oxidation states, +1 through +5. Because of its multiple oxygenation states, nitrogen is commonly found in all of the above forms. However, nitric acid (HNO), NO, and NO_2 are the forms of greatest interest to medicine and occupational health. The

Hamilton & Hardy's Industrial Toxicology, Sixth Edition. Edited by Raymond D. Harbison, Marie M. Bourgeois, and Giffe T. Johnson.
© 2015 John Wiley & Sons, Inc. Published 2015 by John Wiley & Sons, Inc.

oxides of nitrogen can best be described as an important group of irritant gases in both industry and the environment. Exposure to these compounds has been reported in a variety of occupational settings and in some cases has been associated with fatal respiratory complications.

Nitric oxide (CASRN 10102-43-9) is a colorless gas with 5% solubility in water and a vapor pressure of 34.2 atm. It is a nonflammable gas that accelerates the burning of combustible materials.

Nitrogen dioxide (CASRN 10102-44-0) may be present in the form of a yellowish-brown liquid or a reddish-brown gas above 21.1 °C (70 °F) with a pungent acrid odor. It reacts with water to form nitric and nitrous acid and has a vapor pressure of 720 mmHg. It is also a noncombustible liquid or gas that accelerates the burning of combustible materials. Nitrogen dioxide is more toxic than nitrogen oxide.

Nitrous oxide (laughing gas) is a colorless gas with a slightly sweet odor and taste. It is transported as cryogenic liquid and used as an anesthetic and foaming agent in whipped cream.

Nitric acid (CASRN 7697-37-2) is commonly used as a strong oxidizing agent. The main use of nitric acid is the production of fertilizers. Other industrial uses include explosives, nylon precursors, and specialty organic compounds. Nitrous acid is also used as an oxidizer in liquid-fuel rockets. In electrochemistry, nitric acid is used as a chemical doping agent for organic semiconductors. It is also used in the purification process for raw carbon nanotubes. During the wood working process, nitric acid is used to artificially age pine and maple. It is also used for etching and pickling metals due to corrosive properties. The food and dairy industry uses nitric acid in combination with other chemicals for cleaning equipment (Safety Bulletin, 2003).

Two of the most toxicologically significant nitrogen oxides are nitrogen oxide and nitrogen dioxide. Nitrogen dioxide (NO_2) is the most abundant and toxic. Nitrogen oxide inhalation is 1/5 as toxic as nitrogen dioxide. Human exposure to nitrogen dioxides varies from indoors to outdoors, from countryside to cities, and the time of day and season. Reactive forms of nitrogen impact the environment through smog, acid rain, and biodiversity loss. Oxides of nitrogen in combination with sunlight may promote the formation of ozone. The formation of ozone may cause coughing, shortness of breath, aggravation of asthma, and other chronic lung diseases.

Nitrogen oxides form naturally during oxidation of nitrogen containing compounds such as coal, diesel, fuel, and silage. It is formed during the burning of fuels in furnaces and internal combustion engines. It is also a component of tobacco smoke. Nitrogen oxides are intermediates in the production of lacquers, dyes, and other chemicals and are important components of photo-oxidant smog. It is used in rocket fuel in nitration production, which is used in nitro-explosives, gun cotton, dynamite, and Trinitrotoluene (TNT).

Nitric acid is used to produce ammonium nitrate in fertilizer. It is also used to dissolve noble metals, etching, and cleaning metals.

Nitric oxide (10–20 ppm) is used to treat pulmonary hypertension (Ichinose et al., 2004).

PHYSICAL AND CHEMICAL PROPERTIES

Nitrogen is a colorless, odorless, and tasteless gas. It is often used under refrigeration as a cryogenic liquid. The boiling point is −195.8 °C and −320 °F. Nitrogen is not combustible. Nitrogen can combine with oxygen at high temperatures to form oxides and may form ammonia in contact with hydrogen at elevated temperatures. Cyanides can form if nitrogen is heated with carbon in presence of alkalies or barium oxide. If nitrogen comes in contact with ozone, nitrogen can oxidize explosively.

The nitrogen oxides include elemental nitrogen in five different oxidation states, +1 through +5. Nitroxyl anion (NO^-) and the anesthetic gas nitrous oxide (N_2O) contain nitrogen in the +1 oxidation state. Nitric oxide (NO) contains nitrogen in the +2 oxidation state. The weak base nitrite (NO_2^-) and its corresponding acid, nitrous acid (HNO_2), the nitrosonium cation (NO^+), dinitrogen trioxide (N_2O_3), and the peroxynitrite anion ($ONOO^-$) all contain nitrogen in the +3 oxidation state. Nitrogen dioxide (NO_2) and its dimer, dinitrogen tetroxide (N_2O_4), are present in equilibrium in the +4 oxidation state. And finally, nitrate (NO_3^-), its corresponding acid, nitric acid (HNO_3), and dinitrogen pentoxide (N2O5) contain nitrogen in the +5 oxidation state. Nitrogen dioxide has a strong harsh odor and is a liquid at room temperature, becoming reddish-brown gas above 70 °F. Nitric oxide is a sharp sweet smelling gas at room temperature. Nitrogen oxide and dioxide are nonflammable gases or liquids; however, they can accelerate combustible materials.

Nitric acid is a highly corrosive strong mineral acid. Most commercially available nitric acid has a concentration of 68%. Fuming nitric acid contains more than 86% of HNO_3. This grade is used in the explosives industry. It is an inorganic acid that is miscible in water and has a vapor pressure of 48 mmHg. It can be a colorless, yellow or red liquid with an acrid, suffocating odor. It is noncombustible but increases the flammability of combustible materials.

ENVIRONMENTAL FATE AND BIOACCUMULATION

Nitrogen oxides are a key component of the formation of ground level ozone or smog. Nitrogen dioxide is a ubiquitous product of combustion, emitted from the exhaust of motor vehicles and emissions from stationary fuel sources such as coal, oil, natural gas, and other industrial sources. In the environment, nitrogen oxides are broken down rapidly in the atmosphere by reacting with substances commonly found in the air. Molecular nitrogen is also modified via discharge of lightening. The electrical discharges of lightening break strong bonds between the nitrogen atoms and cause nitrogen to react with oxygen. The reaction of nitrogen dioxide with chemicals produced by sunlight leads to the formation of nitric acid, which is a major component of acid rain.

ECOTOXICOLOGY

The Environmental Protection Agency (EPA) regulates nitrogen dioxide as a criterion air pollutant. Outdoor sources of nitrogen dioxide include emissions from power plants, oil refineries, and automobile exhaust systems. Indoor sources include emissions from gas stoves and furnaces, kerosene space heaters, and cigarette smoke. Nitrogen dioxide does not exist in water, since it reacts instantaneously with water to form nitric acid and nitrous acids. Nitrogen cycle acceleration affects the environment through eutrophication of terrestrial and aquatic systems and global acidification. Eutrophication is defined as a natural and artificial addition of nutrients to bodies of water. The addition of artificial nitrogen to bodies of water I considered part of water pollution (Gruber and Galloway, 2008). Nitrogen dioxide plays an important role in the formation of photochemical smog, giving it a characteristic brown color. Nitrogen oxides are also major contributors to the formation of acid rain ($NO_X + H_2O \rightarrow HNO_3$). The formation of acid rain, smog, and biodiversity loss occur from reactive forms of nitrogen in the environment.

Acute Effects

The initial symptoms of exposure to nitrogen oxides include cough, burning of the throat and chest, nausea, fatigue, and shortness of breath. Severe exposure can cause drowsiness, hypoxemia, methemoglobinemia causing blue discoloration of the skin and lips, pulmonary edema, mental confusion, unconsciousness, and death. Exposure to liquid nitrogen can cause rapid, severe frostbite to the fingers, hands, and other tissues. The freezing temperature of nitrogen can lead to amputation of fingers despite wearing protective equipment.

MAMMALIAN TOXICOLOGY

The toxicological database for nitric oxide is small, relative to nitrogen dioxide. It is often difficult to obtain pure nitric oxide in air without some contamination with nitrogen dioxide. Most of the experimental animal studies available focused on the therapeutic use of nitric oxide in an animal model of human disease. Lethality studies in dogs, rats, and mice lacked complete concentration-response information,

are confounded by possible NO_2 contamination, or were secondary citations, in which the original source could not be obtained. From these studies, it appears that in the absence of lung injury, the mechanism of toxicity of nitric oxide is methemoglobin formation (IRIS, 2006). There are no apparent species differences in the toxic response to acute inhalation exposure to nitric acid. Nitric acid fumes cause immediate irritation of the respiratory tract, pain, and dyspnea, which are followed by a period of recovery that may last for several weeks. Relapses may occur due to bronchopneumonia and/or pulmonary fibrosis, which may result in death. With chronic exposure, emphysematous lesions were produced in mice after exposure to 10 ppm of nitrogen dioxide for various periods up to 30 weeks (Patty's, 1981).

Acute Effects

Nitrogen oxides are skin irritants and corrosives. Reaction with skin moisture can lead to formation of nitric acids, causing second- and third-degree burns. Nitric acid is a highly corrosive, strong oxidizing acid. Nitric acid may also cause yellowing of the skin and dermal erosion. Nitrogen oxides are respiratory irritants, which cause an immediate response, including cough, fatigue, nausea, choking, headache, abdominal pain, and difficulty breathing. After 3–30 h, pulmonary edema may develop with anxiety, mental confusion, lethargy, and loss of consciousness. Bronchiolitis obliterans (fibrous obstruction of the bronchioles) may occur several weeks later. For exposure to nonlethal concentrations, allergic, or asthmatic individuals appear sensitive to acidic atmospheres. High dose exposure may cause formation of methemoglobin and impaired oxygen transport. Cardiovascular effects include weak, rapid pulse, dilated heart, chest congestion, and circulatory collapse.

Chronic Effects

Victims of inhalation exposure of nitrogen oxides can suffer reactive airway dysfunction (RADS) after a single acute high exposure. Children have an increased risk of respiratory infections. Permanent restrictive and obstructive disease from bronchiolar damage may occur.

Nitrogen oxides are not classified as carcinogenic by IARC, EPA, or DHHS.

Chemical Pathology

Nitrogen dioxide (NO_2) is thought to cause damage to the lungs by three ways. NO_2 is converted into nitric and nitrous oxides in distal airways, which directly damages certain structural and functional lung cells. Nitrogen dioxide initiates free radical generation that causes cell membrane damage; it alters macrophage and immune function by decreasing the resistance to infection. NO_2 is characterized by denudation followed by compensatory proliferation. The persistent injury and remodeling include the development of fibrosis (Persinger et al., 2002) Nitric acid is known to be a highly corrosive inorganic and strong oxidizing acid.

Evidence suggests that nitrogen oxide in the presence of oxygen at physiological pH can produce mutations in *Salmonella* species by the conversion of cytidine to thymidine. These reactions have been proposed as one mechanism by which cigarette smoke causes lung cancer (Wink et al., 1991; Nguyen et al., 1992).

Mode of Action(s)

Nitric oxide is a low molecular weight, lipophilic molecule that has a rapid onset of action and a very short half-life (i.e., seconds). Nitric oxide is an endogenous ubiquitous substance that acts as a biological messenger in the regulation of numerous organ systems. Nitrogen oxide plays central roles in the regulation of vascular smooth muscle tone. Nitric oxide gas is utilized in medicine to selectively dilate the pulmonary vessels by increasing blood flow to areas in the lung and improves oxygen levels in the body (ventilation–perfusion V/Q matching). After NO is inhaled in the lungs and into the blood stream, its effects are quickly deactivated. In recent years, the beneficial effects of inhaled NO to reduce intra-pulmonary shunting and oxygenation have been demonstrated in patients with acute respiratory syndrome and pulmonary hypertension (Ichinose et al., 2004). Nitrogen oxide exposure to the skin and eyes causes inflammation and irritation. Nitrogen oxides are deposited in the lower airways. Sixty percent of the nitrogen dioxide is deposited in the centriacinar region. The amount of exposure increases with exercise. Approximately, 15 times more NO_2 would be delivered to the pulmonary tissue at maximum tidal volume that would occur during heavy exercise. (Bauer, 1986) Nitric oxide has an affinity for heme-bound iron, which is two times higher than that of carbon monoxide. This affinity leads to the formation of methemoglobin and stimulation of guanylate cyclase. Furthermore, nitric oxide reacts with thio-associated iron in enzymes and eventually displaces the iron. This is the possible mechanism for the cytotoxic effects of nitric oxide (WHO, 1997).

SOURCES OF EXPOSURE

Occupational

It has been estimated that more than 1.5 million workers in the United States may be exposed to nitrogen oxides. Nitrogen oxides (NO and NO_2) are formed when nitric acid reacts with reducing agents, as well as by combustion of nitrogen-containing materials. Nitrogen dioxide may be encountered in a wide variety of industrial settings, including: production of nitric acid (by catalytic oxidation of ammonia), manufacture of explosives, in the exhaust of metal cleaning processes, in underground blasting operations, decomposition of agricultural silage, by ice arena resurfacing machines, the exhaust of diesel engines, and burning of gun cotton and cordite. Nitrogen dioxide is also a component of rocket fuels. There were 80 deaths and 50 workers were injured in the industry, between the years 1992 and 2002, from nitrogen asphyxiation hazards. Sixty-two percent occurred in chemical plants, refineries, food processing and storage facilities, and metal and manufacturing sites, including nuclear plants. The majority of workers were working in confined spaces containing nitrogen-enriched air (US Chemical Safety, 2003). Nitric acid is an important industrial acid used in the production of ammonium nitrate fertilizer and explosives. All commonly used explosives, with the exception of gunpowder, are made by the action of nitric acid on cellulose, glycerin, phenol, benzol, toluene, and similar substances. Nitrocellulose is used for lacquers, photographic film, and celluloid. Nitrobenzols and nitrophenols are used in the production of dyes and drugs as well as explosives. Metal etching and photoengraving both require nitric acid, which, in its dilute form, also is used to clean copper and brass.

Nitric oxide has been utilized in the medical field for over 20 years. It is being used worldwide to treat both adult and pediatric patients from pulmonary hypertension and acute respiratory distress syndrome. It is utilized in pulmonary hypertension and congenital heart diseases in newborns. Inhaled gas is delivered via face mask, nasal cannula or through an endotracheal tube (Ichinose et al., 2004).

Environmental

Nitric oxide is formed and combines with oxygen to form nitrogen dioxide. When released to soil, small amounts of nitrogen oxides may evaporate into the air. Most of it is converted into nitric acid and other compounds. The nitrogen dioxide dissolves in water to form nitric and nitrous acids. Nitrogen dioxide compound undergoes direct photolysis in the environment leading to the formation of ozone and smog conditions in the troposphere. Human exposure to nitrogen oxides varies indoors, outdoors, urban, and rural with the time of day and season. Nitrogen dioxide can be mixed with the nitric oxide exposures because nitrogen oxide is oxidize to nitrogen dioxide.

At certain levels, inhaled nitric oxide concentrations can cause vasodilatation in pulmonary circulation without affecting systemic circulation. Relatively high concentrations have been used in clinical applications for brief periods without adverse effects. The lowest effect concentration of NO has not been established. On 01/22/10, EPA strengthened the health-based ambient air standard for NO_2 by setting the standard of 100 parts per billion (ppb). The standard is intended to protect the health of sensitive populations such as people with asthma, children, and elderly.

INDUSTRIAL HYGIENE

Workers should wear personal protective equipment (PPE) when utilizing liquid nitrogen to prevent skin contact due to below freezing temperature. The selection of appropriate PPE is based on the extent of exposure to nitrogen and nitrogen compounds. Safety showers and eye wash stations should be located close to the operations using liquid nitrogen. Engineering controls are the most effective way of reducing nitrogen liquid and gas exposure. The best protection is to enclose operations and or provide local exhaust ventilation at the site of chemical release. In confined spaces where nitrogen is present, sufficient oxygen levels (19%) should be checked prior to entry. Oxygen and nitrogen gases are odorless and not detected by sense of smell. Nitrogen enriched environment, which depletes oxygen can only be detected with special instruments. Asphyxiation can occur if a worker is exposed to a high concentration of nitrogen. Nitric oxide produced in welding arcs and other flame operations, such as glassblowing, is oxidized in air to nitrogen dioxide. It has been reported that the levels of nitrogen dioxide in welding areas can be as high as 7 ppm during flux arc welding. The levels inside the welder's eye protection mask were lower at 2 ppm, illustrating some protection in the breathing zone. Contact between nitric acid and organic matter produces nitrogen oxides. The inhalation effects of nitric acid cannot be separated from those of the nitrogen oxides because they invariably occur together. It is recommended that those working in high concentrations of nitrogen oxide gases wear proper respirators to prevent serious or fatal respiratory injury. Ventilation needs to be adequate in all processes in which nitrogen oxides are formed.

Nitrogen does not have exposure levels listed; it is a simple asphyxiant and was not assigned a TLV, because of the limiting factor of available oxygen (Table 49.1).

HEALTH EFFECTS

Acute overexposure to nitrogen gases may result in death almost immediately. An example of this would be the 1929 fire at the Cleveland Clinic, in which burning nitrocellulose radiographs generated NO, CO, and HCN gases. Ninety-seven workers died immediately from this exposure, and 26 people died within the next month, probably from the effects of nitrogen oxide (Gregory et al., 1969). Individuals who survive acute overexposure often develop delayed symptoms within 48 h, including pulmonary edema, the most common form of occupational injury associated with nitrogen dioxide. In 1975, the presentation of disease

TABLE 49.1   Occupational Exposure Limits for Nitrogen Compounds

Agency	Compound	Exposure Guideline	Parameter
OSHA	Nitrogen dioxide	5 ppm (9 mg/m^3)	PEL
	Nitrogen oxide	25 ppm	PEL
	Nitrous oxide	Not Regulated	
	Nitric acid	2 ppm (5 mg/m^3)	PEL
	Nitric acid	4 ppm (10 mg/m^3)	STEL
NIOSH	Nitrogen dioxide	5 ppm (9 mg/m^3)	PEL
	Nitrogen dioxide	1 ppm (1.8 mg/m^3)	STEL
	Nitrogen dioxide	20 ppm	IDLH
	Nitric oxide	25 ppm (30 mg/m^3)	REL
	Nitric oxide	100 ppm	IDLH
	Nitrous oxide	25 ppm (45 mg/m^3)	REL
	Nitric acid	4 ppm (10 PPM)	STEL
		2 ppm (5 mg/m^3)	REL
ACGIH	Nitrogen dioxide	3 ppm (5.6 mg/m^3)	TLV
	Nitrogen dioxide	1 ppm (1.8 mg/m^3)	REL
	Nitrogen dioxide	5 ppm (9.4 mg/m^3)	STEL
	Nitrogen dioxide	50 ppm	IDLH
	Nitric acid	2 ppm (5.2 mg/m^3)	TLV
ACGIH	Nitric acid	4 ppm (10 mg/m^3)	STEL
	Nitric acid	100 ppm	IDLH
	Nitric oxide	25 ppm (31 mg/m^3)	TLV
	Nitrous oxide	50 ppm (90 mg/m^3)	TLV

ACGIH: American Conference of Governmental Industrial Hygienists
OSHA: Occupational Safety and Health Administration
NIOSH: National Institute for Occupational Safety and Health
TLV-TWA: Threshold limit value, time-weighted average—average exposure on the basis of an 8 h/day, 40 h/week-work schedule
PEL: Permissible exposure limit based on an 8 h-TWA exposure
REL: Recommended exposure limit based on a time-weighted average exposure of up to 10 h/day and 40 h/week
STEL: 15 min TWA exposure that should not be exceeded at any time during the workday even if the 8 h TWA is within TLV–TWA. Exposures above the TLV-TWA up to STEL should not be longer than 15 min and should not occur more than 4 times per day. There should be at least 60 min between successive exposures in this range

occurred when three American astronauts on the Apollo-Soyuz aircraft were inadvertently exposed to nitrogen tetroxide for approximately 4–5 min. They developed pulmonary edema the day after splash down. All of the crew members recovered after 13 days of medical observation.

Several studies have linked air pollution by nitrogen dioxide (NO$_2$) with increased admissions for asthma in children. Exacerbations of asthma in children are often precipitated by upper respiratory infections. It is therefore, possible for NO$_2$ to increases the risk of airway obstruction when children develop upper respiratory infections (Linaker et al., 2000). There is evidence that children who live in households where gas stoves are used, develop more infections before two years of age and have diminished lung function compared with matched controls. Further evaluation is required (Florey et al., 1982). Ambient NO$_2$ and SO$_2$ may be important risk factors for sudden infant death syndrome. One study found increases in both NO_2 and SO2 associated with a 17.72% increase in SIDs incidence in Canada (Dales et al., 2004).

Exposure to liquid nitrogen or liquefied nitrous oxide can cause skin burns or frostbite to exposed areas of the body.

DIAGNOSIS AND TREATMENT

The correct diagnosis is dependent on an accurate detailed occupational and medical history. The most hazardous of the nitrogen oxides are nitric oxides and nitrogen dioxide. Nitrogen oxides (NO_2, N_2O_4, N_2O_3, and N_2O_5) are irritating to the upper respiratory tract and lungs even at low concentrations. Severe toxicity may occur after a brief period of exposure. It is important for the clinician to be aware of the pattern of illness caused by exposure to the nitrogen oxides. Because of the delayed onset of symptoms, people suspected of high dose exposure should be observed for approximately 48 h. The victim could show signs of lung injury such as hypoxemia or rapid respiratory rate. Chest X-rays may not be diagnostically helpful initially, but abnormalities maybe seen after several days or weeks later. Laboratory studies may reveal an elevation in white cell count. The presence of methemoglobinemia is helpful in diagnosing exposure to nitrogen oxides. Treatment involves administration of oxygen, possibly assisted ventilation, hemodynamic monitoring, and use of corticosteroids. Children exposed to the gas are more vulnerable to exposure due to the greater lung surface area: body weight ratios and minute volumes: weight ratios. Since nitrogen dioxide is heavier than air, children may have an increased exposure due to its short statue. The clinical presentation of all the above categories is similar. Since nitrogen dioxide is relatively insoluble in water and moist membrane surfaces, it has little effect on mucous membranes of the eyes and throat. There is no immediate recognition of exposure, and workers or others may remain in the area for several minutes before coughing and chest congestion alert them to leave. Respiratory symptoms are delayed for 3–30 h after exposure, at which time cough, shortness of breath, fever, and elevated white blood cell count may develop. Chest X-rays initially may be negative and over time may reveal pulmonary edema. For this reason, chest X-rays may not be diagnostic in patients evaluated early after initial exposure, and observation may be prudent. Most patients recuperate without sequelae. However, some patients may develop lung disease and die from exposure. Pathologically, the appearance of the lungs in fatal cases reveals hemorrhagic edema and patches of pneumonia. In patients who do not succumb initially, but rather a few weeks after exposure, small, palpable nodules are seen within the lung as well as hemorrhagic areas. Microscopically, there is extensive damage and sloughing of the respiratory epithelium into the bronchioles. Diffuse infiltration of the alveolar walls and numerous macrophages within the alveolar spaces also can be seen, as can numerous areas of bronchiolitis obliterans in various stages of the organization. Bronchiolitis obliterans is characterized by inflammation and fibrosis in and around the small bronchi and bronchioles.

Another clinical manifestation of nitrogen oxide poisonings is development of methemoglobinemia. The nitrogen oxide binds to hemoglobin, causing a sharp leftward shift in the oxygenation dissociation curve resulting in limited delivery of oxygen to tissues. The affinity of nitrogen oxide for hemoglobin renders it incapable of binding oxygen.

Nitric acid spill on the skin should be irrigated with large quantities of water because of the corrosive properties. If nitric acid is accidentally splashed onto the skin or into the eyes, the area should be irrigated immediately and washed with water or saline until all burning sensations cease. Wash the area for 10–15 min to cool surrounding tissue and prevent secondary damage. Contaminated clothing should be removed immediately. Because of its corrosive properties, concentrated nitric acid can injure tissues on direct contact. It can cause a yellow burn to the skin, the result of the formation of denatured colored proteins.

Treat dermal irritation or burns with standard local therapy. Patients developing dermal hypersensitivity may require treatment with systemic or topical steroids or antihistamines.

More important, when nitric acid is splashed in the eye it can cause immediate opacification of the cornea; when severe, this accident may result in blindness. The patient will need to be evaluated by a physician. To prevent such accidents, eye protection and protective clothing should be worn when concentrated nitric acid is used.

If skin is exposed to liquefied nitrogen or nitrous oxide, the skin or area can burn or develop frostbite. The skin should be immersed into warm water to increase the temperature. Medical attention is warranted to evaluate tissue damage.

TARGET ORGANS AND MECHANISM OF ACTION

The severity of the effects of nitrogen oxides depends on the concentration of the agent and the duration of exposure. Because of their relatively low solubility in water, nitrogen oxides have little effect on the oropharyngeal mucosa, which allows the gases to be transmitted to the lower airways. Approximately 80–90% of nitrogen oxides are absorbed by the respiratory epithelium during normal ventilation in healthy subjects. In the lower airways, nitrogen oxides react with water to form nitric acid. The nitrates and nitrites that result from the dissociation of nitric acid can induce direct local tissue inflammation and destruction and also may initiate peroxidation of lung lipids by forming oxygen free radicals. This damage causes noncardiogenic pulmonary edema, which is seen clinically. After inhalation exposure to the lungs, the nitrogen oxides are absorbed into the bloodstream. Most of the absorbed nitrogen dioxide enters the blood as nitrite because of the previous reaction with water, in which nitric acid was formed. The absorbed nitrogen oxide enters the blood unreacted. Both combine with hemoglobin, in the presence of oxygen, and methemoglobinemia results (Gosselin et al., 1984).

BIOLOGICAL MARKERS

Nitrogen oxides have been evaluated as potential biological markers. Because of the turnover of the extracellular matrix of the lungs after exposure to irritant gases, the levels of hydroxyproline and hydroxylysine, collagen breakdown products, in the urine may reflect collagen turnover secondary to lung damage.

A cross sectional study was carried out on two groups of subjects differently exposed to nitrogen dioxide in order to test the urinary hydroxyproline ratio (UHP/mg/24 h/sq m) use as a biomarker of effect after exposure to this pollutant. The UHP was higher in subjects training in polluted areas vs. nonpolluted areas. (Perdelli et al., 2002) Studies in glassworkers exposed to nitrogen oxides have shown increased urinary excretion of nitrate and hydroxyproline compared with controls. (Azari et al., 1994a). However, collagen is found in nearly all organ systems and is not specific for the lung. An increase in the level of lipid peroxidation in the respiratory system, resulting in higher concentrations of pentane in expired air, also was seen (Azari et al., 1994b).

Exhaled nitric oxide is a marker for asthma and utilized as a research tool in evaluation of asthma. It may prove to be a diagnostic tool in the diagnosis of primary ciliary dyskinesia syndromes. The clinical utility of nitric oxide for other lung diseases remains to be established (American Thoracic Society, 2006).

SILO FILLERS' DISEASE (SFD)

Silo fillers' disease (SFD) occurs in workers who enter unopened grain silos. In the United States, SFD is prevalent among farmers during the harvest months of September and October. It is an occupational disease that results from pulmonary exposure of oxides of nitrogen. It is a preventable occupational hazard that can be eliminated by proper work practices. Nitrogen dioxide is formed within hours at toxic and lethal levels after corn and grains are placed in the silos. Nitrogen dioxide gas is heavier than air and is present after the silos are entered or near open hatches. The concentration of nitrogen dioxide in the air partly depends on the nitrate content of the grain. Both carbon dioxide and nitrogen dioxide are heavier than air and accumulate just above the level of the grain, especially within lower uneven surfaces of the grain itself. If the silo has remained unopened, dangerous amounts of gas may be present after a few days to months later. To enter the silo, a worker enters through a chute door and jumps onto the surface of the silage a few feet below. The respiratory rate of the farm worker will increase because of the low levels of oxygen and carbon dioxide causing an elevated amount of nitrogen dioxide to be inhaled into the lungs. If the concentration of gas is high, the worker may be physically unable to get out. Special prevention measures must be taken with forage silos: (1) the tower should be ventilated by means of a blower for at least half an hour before the silo is entered; (2) workers should wear safety harnesses and ropes, and additional workers should be in attendance outside the tower; (3) if the silo must be entered during filling, this should be done after the last load and not the next day, when gases may have evolved. Exposure is usually mild and self-limiting; however, severe exposure can lead to pulmonary edema, bronchiolitis obliterans or asphyxia (Kamanger, 2012).

REFERENCES

American Thoracic Society (2006) American Thoracic Society Workshop Proceedings: exhaled nitric oxide and nitric oxide oxidative metabolism in exhaled breath condensate. *Pro. Am. Thorac. Soc.* 3:131–145.

Azari, M.R., et al. (1994a) Biological and biological effect monitoring of workers exposed to nitrogen oxide. *Hum. Exp. Toxicol.* 13:647–652.

Azari, M.R., et al. (1994b) High breath pentane in workers exposed to nitrogen dioxides. *Respir. Med.* 88:816–817.

Bauer, M.A., et al. (1986) Inhalation of 0.30 ppm nitrogen dioxide potentiates exercise-induced bronchospasm in asthmatics. *Am. Rev. Respir. Dis.* 134(6):1203–1208.

Dales, R., et al. (2004) Air pollution and sudden infant death syndrome. *Pediatrics* 113:6.

Florey, C.D.U. et al. (1982) *Schriftner Ver wasser Boden Lufthg* 53:209–218.

Gosselin, R.E., et al. (1984) *Clinical Toxicology of Commercial Products*, 5th ed., Baltimore, Williams & Wilkins Company.

Gregory, K.C., Malinoski, V.F., and Sharp, C.R. (1969) Cleveland clinic fire survivorship study, 1929–1965. *Arch. Environ. Health* 18:508–512.

Gruber, N. and Galloway, J.N. (2008) With humans having an increasing impact on planet, the interactions between the nitrogen cycle, the important determinant of the Earth System. *Nature* 451:293–296.

Ichinose, F., Roberts, J.D., et al. (2004) Review: cardiovascular drugs, inhaled nitric oxide, a selective pulmonary vasodilator: current uses and therapeutic potential. *Circulation* 109: 3106–3111.

Integrated Risk Information System (IRIS) (2006) *Environmental Protection Agency, Nitric oxide summary, Interim 1:10/ 2006.*

Kamanger, N. *Silo Fillers Disease. Overview, Medscape article, 3012133, January 18, 2012.*

Kovalchin, J.P., Mott,A.R., et al. (1997) Nitric oxide for the evaluation and treatment of pulmonary hypertension in congenital heart disease. *Tex. Heart Inst. J.* 24(4):308–316.

Linaker, C.H., et al. (2000) Personal exposure to nitrogen dioxide and risk of airflow obstruction in asthmatic children with upper respiratory infection. *Thorax* 55(11):930–933.

Nguyen, T., et al. (1992) DNA damage and mutation in human cells exposed to nitric oxide in vitro. *Proc. Natl. Acad. Sci. U. S. A.* 89:3030–3034.

Patty (1981) Patty's industrial hygiene and toxicology: volume 2A, 2B, 2C. In: *Toxicology*, 3rd ed., 1981–1982 . 2236.

Perdelli, F., et al. (2002) Urinary hydroxyproline as a biomarker of effect after the exposure to nitrogen dioxide. *Toxicol. Lett.* 134 (1–3):319–323.

Persinger, R.L., et al. (2002) Molecular mechanisms of nitrogen dioxide induced epithelial injury in the lung. *Moll. Cell. Biochem.* 234–235(1–2).

Safety Bulletin (2003) *US Chemical Safety and Hazard Investigation Board; Hazards of Nitrogen Asphyxiation; 2003-10-B, 2003.*

US Chemical Safety and Hazard Investigation Board (2003) *Hazards of Nitrogen Asphyxiation, 2003-10-B.*

World Health Organization/International Programme on Chemical Safety (1997) *Environmental Health Criteria 188. Nitrogen oxides, 331.* 1–18 270–276.

Wink, D.A., et al., (1991) DNA deaminating ability and genotoxicity of nitric oxide and its progenitors. *Science* 254: 1001–1003.

50

OXYGEN AND OZONE

Robert A. Klocke

OXYGEN

First Aid: Removal from environment with abnormal oxygen level

Target organ(s): Lungs, central nervous system

Occupational exposure limits: OSHA PEL: 19.5% oxygen minimal oxygen percentage; consensus: ACGIH TLV 3.0 ATA maximum pressure breathing 100% oxygen

Reference values: None

Risk/safety phrases: R8: Contact with combustible material may cause fire; S2 keep out of the reach of children; S17 keep away from combustible material

BACKGROUND AND USES

Oxygen is the most abundant element in the outer layer of the earth, largely the result of its combination with hydrogen and silica to form water and sand. It is ubiquitous and is present in the same fractional concentration (0.2094) throughout the earth's atmosphere. However, its partial pressure varies with altitude because of the effect of altitude on barometric pressure. Oxygen is necessary for respiration in all mammals. In medicine, oxygen is the most used pharmaceutical agent and is usually administered under normobaric conditions. However, there is a gradually increasing use of hyperbaric oxygen treatment for a variety of medical illnesses.

Oxygen supports combustion and is frequently used as an oxidizing agent. It is used throughout the industry for processing metal and petroleum products, fabrication of chemicals and pharmaceutical agents, and welding.

PHYSICAL AND CHEMICAL PROPERTIES

Oxygen is a diatomic element (molecular weight 32) normally present in a gaseous form but does exist as a liquid at very low temperatures (boiling point −183 °C) and as a solid (melting point of −218 °C). It is normally used in its gaseous form but liquid oxygen also has some medical and industrial applications. It is relatively insoluble in aqueous solutions with a solubility of 0.0311 and 0.0239 ml/ml at 20 and 37 °C, respectively, at 1.0 atmospheres absolute pressure (ATA) (Battino et al., 1983). As a result, mammalian species and some vertebrates require a protein in blood, such as hemoglobin, to transport adequate quantities of oxygen to support life. Oxygen supports combustion, especially at pressures above 0.21 ATA. The risk increases with increasing oxygen pressure, and avoidance of fire risk is essential at elevated oxygen pressures. Materials that normally do not burn in air may become combustible at elevated oxygen pressures.

ENVIRONMENTAL FATE AND BIOACCUMULATION

Carbon dioxide is the final end product of many biological and industrial processes, thereby depleting atmospheric oxygen and increasing carbon dioxide concentration. Oxygen in the atmosphere is maintained constant through plant photosynthesis in the oxygen–carbon dioxide cycle. Plants use chlorophyll to form carbohydrates through reduction of carbon dioxide and release oxygen into the atmosphere. In addition to the trees and plants that produce oxygen, green algae are also a major natural source of oxygen.

Hamilton & Hardy's Industrial Toxicology, Sixth Edition. Edited by Raymond D. Harbison, Marie M. Bourgeois, and Giffe T. Johnson.
© 2015 John Wiley & Sons, Inc. Published 2015 by John Wiley & Sons, Inc.

Oxygen is used to support combustion of carbon fuels to produce carbon dioxide. As a result of increasing industrial activity, atmospheric carbon dioxide has risen from 300 to 400 parts per million (ppm) in the last century. Incomplete combustion of carbon fuels produces carbon monoxide, a common atmospheric pollutant in urban areas.

ECOTOXICOLOGY

Oxygen concentration in air is essentially constant in the overall environment, and toxic effects have not been observed.

MAMMALIAN TOXICTY

Toxicity caused by oxygen is dependent on its duration of exposure and chemical activity. The latter is commonly estimated by the partial pressure of oxygen in absolute atmospheres. At elevated levels, oxygen principally affects the lungs and central nervous system. Survival data of mammals, especially rodents, when breathing 100% oxygen at 1.0 ATA are limited to several days, but extrapolation of these data to humans has a dubious value (Clark and Lambertsen, 1971). It appears that primates are more resistant to oxygen exposure; monkeys with exposure to pure oxygen have survived as long as 3 weeks. Animal data indicate that the usual cause of mortality is pulmonary edema. There are no comparable studies in humans breathing 100% oxygen at 1.0 ATA for such long periods.

Pulmonary oxygen toxicity is found to be nonexistent in humans at exposures <0.5 ATA (Clark and Lambertsen, 1971), although limited data are available to support a definite threshold. Variability in susceptibility to increased oxygen exposure is substantial in humans. Small changes (4%) in the vital capacity (VC), the volume of gas that can be exhaled after a maximal inhalation, are noted in subjects breathing oxygen at 1.0 ATA for 24 h (Bitterman, 2004). With exposure to oxygen at greater than normal barometric pressure, the time required to produce progressive decrements in VC, as well as symptoms (chest pain, cough, shortness of breath), decreases exponentially (Figure 50.1) (Clark et al., 1999).

Changes in VC following exposures in these experiments returned to normal within 24–30 h. Surprisingly, these changes did not impair pulmonary oxygen exchange, but it must be noted that great care was taken to limit the duration of exposure at higher inspired oxygen levels to prevent serious health threats to the subjects.

In healthy individuals undergoing surgical anesthesia, increased oxygen concentrations (0.6–1.0 ATA) at normal barometric pressure can result in the development of absorption atelectasis, i.e., collapse of pulmonary exchange units

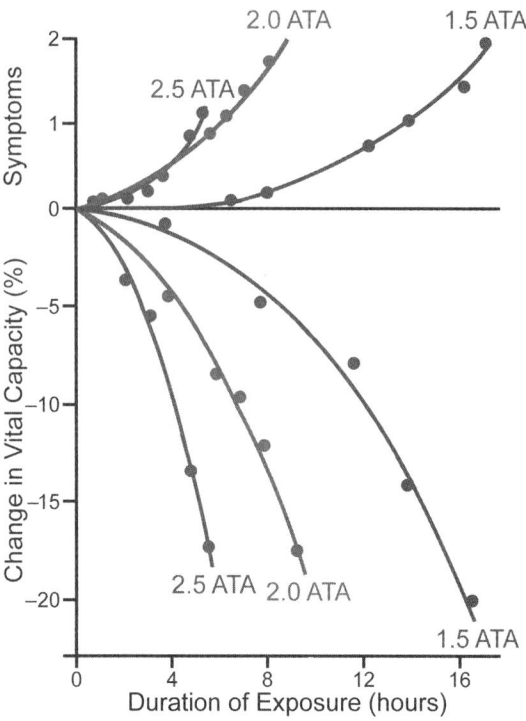

FIGURE 50.1 Mean values of development of symptoms and reductions in vital capacity of young (mean 24.3 years), healthy subjects exposed to 100% oxygen at 1.5 ($n = 9$), 2.0 ($n = 25$) and 2.5 ($n = 8$) ATA for varying periods of time. Curves of responses compared to baseline (0 h) determined while breathing air at 1.0 ATA were fitted empirically. Pulmonary symptoms were self-rated on a 0–3 scale (absent, mild, moderate, severe). Figure redrawn with permission from Clark et al. (1999).

(Edmark et al., 2011). However, this is not a truly toxic effect of oxygen per se. Nitrogen is not absorbed in the lung, and the presence of nitrogen in gas exchange units inhibits the collapse of the units. As oxygen concentration is increased, nitrogen concentration in gas exchange units decreases, and the units are more prone to collapse as oxygen is absorbed into the blood. For this reason, and the likelihood that diseased lungs are more prone to oxidant damage, oxygen is administered at the lowest concentration necessary to maintain acceptable oxygen concentrations in the blood in clinical settings.

Premature infants are more sensitive to increased oxygen pressures than adults. Exposure to increased oxygen levels for prolonged periods is accompanied by a risk of developing a chronic respiratory condition (bronchopulmonary dysplasia). In addition, increased oxygen in the blood of premature infants can cause retinopathy with permanent impairment of vision. This complication is minimized by limiting oxygen administration to the least amount necessary to achieve normal arterial blood oxygenation. The damage to sight can be reduced further by keeping blood oxygenation slightly below normal levels. However, a large controlled study has indicated that this treatment is associated with a small, but significant, increase in neonatal mortality (SUPPORT Study Group, 2010). Current recommendations are to maintain blood oxygenation within the normal range in premature infants.

Oxygen exposure at elevated atmospheric pressures (>2 ATA) causes central nervous system abnormalities ranging from muscular twitching (particularly facial muscles), blurred or tunnel vision, apprehension, dizziness, aphasia to grand mal seizures (Bitterman, 2004). These extreme exposures are limited principally to commercial diving and medical hyperbaric oxygen treatments.

During the World War II era, Donald (1947) investigated the effects of varying oxygen exposures and conditions that precipitated neurological symptoms. With exposure to 3.7 ATA of oxygen in a hyperbaric chamber, 50% of the 36 resting subjects experienced neurological symptoms within 20 min, and all were affected in <100 min. Five of the subjects had grand mal seizures. Anecdotal reports suggest that oxygen-induced convulsions result in no permanent damage, although there have been no long-term, detailed studies of individuals who have suffered this complication. Certainly, convulsions during deep water dives could result in death. In Donald's report (1947), the time required to experience neurological symptoms was reduced by approximately one half if the subjects were submerged in water inside the hyperbaric chamber, simulating wet diving conditions. The mechanism of increased susceptibility to occurrence of neurological symptoms while immersed in water has not been elucidated. Physical labor also reduced the time of exposure prior to the onset of symptoms. Oxygen fractions in inspired gas mixtures are reduced in dives of great depth to prevent neurological and pulmonary toxicity. Since compressed air is used in usual scuba diving, inspired oxygen pressures do not reach a critical level at maximum depths (50–100 ft) associated with recreational dives. This avoids oxygen toxicity but not other pressure-related problems.

In recent years there has been an increasing medical use of hyperbaric oxygenation in the treatment of a wide variety of diseases. The classic uses are treatment of carbon monoxide poisoning and diving-related decompression illness. Short-term (<2.0 h) hyperbaric oxygenation also has been used in the treatment of diabetic ulcers, anaerobic tissue infections, radiation-induced injury, and severe brain trauma. Oxygen exposure is usually limited to a maximum of 3.0 ATA to minimize toxicity, but this does not completely avoid complications (Hampson and Atik, 2003).

Despite extensive studies on the incidence and effects of oxygen toxicity, the actual mechanisms involved remain uncertain. The most common hypothesis postulates production of reactive oxygen species (ROS), including hydrogen peroxide, hydroxyl radicals, and superoxide radicals. These powerful oxidizing agents have important biological functions such as bacterial killing in infectious diseases. Yet, in excess, these agents increase cellular damage and death.

There is a delicate balance between these oxidizing agents and body antioxidant defenses. Oxygen administration can increase the production of ROS and upset the normal balance, leading to toxicity (Auten and Davis, 2009).

Reduced oxygen in inspired air can be fatal. Symptoms and signs of oxygen deprivation are nonspecific and include shortness of breath, increased heart and respiratory rates, fatigue, agitation, cyanosis (bluish discoloration of the lips, finger nails, and skin), and coma. Oxygen uptake in the lungs is dependent on its pressure. Healthy individuals breathing air are exposed to reduced oxygen pressures when flying or residing at an altitude. Federal Aviation Administration regulations limit the exposure in commercial flight to no less than an absolute pressure of 0.75 ATA, equivalent to the reduction in barometric pressure present at an altitude of 8000 ft. Healthy humans tolerate this exposure without any symptoms or difficulty, but individuals with lung disease may require supplemental oxygen during flight. The Occupational Safety and Health Administration (OSHA) has set a minimum of 19.5% oxygen concentration in industrial circumstances. This produces an oxygen pressure in inspired air that is equivalent to an elevation of 2000 ft. This is well above the limit tolerated by healthy humans but is a preventative limit since further fatal reduction can occur rapidly in closed occupational environments.

SOURCES OF EXPOSURE

Normally oxygen at 1.0 ATA does not constitute a risk unless breathed for periods of hours to days, essentially ruling out accidental exposures. At pressures >1.0 ATA, there is an increasing risk as the oxygen partial pressure increases. This limits the risk to hyperbaric chambers (for medicinal or experimental purposes), use of underwater breathing apparatus, or work in caissons.

INDUSTRIAL HYGIENE

There are no National Institute of Occupational Safety and Health (NIOSH) or OSHA permissible exposure limits (PELs) for oxygen. The OSHA limits minimum oxygen concentration to 19.5% in work environments. At lower oxygen concentrations, OSHA mandates the use of a self-contained breathing apparatus. Half-face or full-face respirators should not be used for atmospheres with reduced oxygen concentrations. No personal protective equipment (PPE) are required for oxygen exposure to 1.0 ATA. At higher pressures encountered in commercial diving and work in caissons, it is necessary to reduce the inspired oxygen concentration to avoid the absolute oxygen rising above 3.0 ATA. The U.S. Navy has published data sheets with limits for safe oxygen exposure at increased pressures.

Atmospheres containing increased oxygen concentrations, even at total pressures <1.0 ATA, are hazardous not for inhalational toxicity but as a result of the increased risk of supporting combustion.

RISK ASSESSMENTS

None available.

OZONE

First aid: Removal from contaminated atmosphere

Target organ(s): Eyes, skin, upper respiratory tract, lungs, heart, vasculature

Occupational exposure limits: OSHA PEL 100 ppb TWA; NIOSH PEL 100 ppb ceiling, IDLH 5000 ppb; CAL/ OSHA PEL 100 ppb TWA; ACGIH: TLV 50 ppb heavy work, 80 ppb moderate work, 100 ppb light work, 200 ppb 2-h TWA at all work intensities

Reference values: NAAQS 75 ppb (annual fourth-highest daily maximum 8-h concentration, averaged over 3 years)

Risk/safety phrases: Oxidizer; lung irritation, damage; increases mortality/keep away from skin, eyes; avoid low level chronic inhalation; avoid high level short-term inhalation

BACKGROUND AND USES

Ozone is distributed throughout the earth's atmosphere in both the stratosphere and the troposphere. The vast majority of ozone (~90%) is located in the stratosphere where it has the critical function of absorbing ultraviolet (UV) light, particularly in the UVC bandwidth (<290 nm) (Crutzen et al., 1999). This prevents profound, deleterious effects on both plants and animals. The remainder of the atmospheric ozone is located in the lower troposphere. Excessive concentrations of ozone in the boundary layer surrounding the surface of the earth can cause significant human mortality and morbidity as well as inhibit vegetative growth. However, ozone in the troposphere also has the beneficial effect of oxidizing potentially dangerous organic compounds that arise from both natural and human sources.

Ozone is a potent oxidizing agent. It is increasingly replacing chlorine as an oxidizing agent in industrial processes. It is used in the disinfection and purification of water, sewage, and food products. Its oxidant properties are also used as a bleaching agent in the paper, textile, and chemical industries. Extracting ozone from the atmosphere is difficult,

and in the industry, it is generated by passing an alternating high voltage electric current through a dry stream of air or oxygen. Although ozone is widely used in industrial applications, these potential sources do not contribute significantly to atmospheric contamination because the highly reactant ozone is confined and has a short half-life. Ozone is produced through electrical discharges occurring in lasers printers, copying machines, and similar equipment, but its concentration is low unless the equipment is operated in a confined environment. In general, indoor air contains ozone concentrations that are only 10–40% of the outdoor ozone concentrations (U.S. EPA, 2013), but the indoor concentration varies widely depending on the type of building construction, heating/cooling systems, and the use of open windows for ventilation in a warmer weather when atmospheric ozone levels are usually greater. Ozone generators have been advertised and sold to "purify" indoor air but can produce indoor concentrations well in excess of National Ambient Air Quality Standards (NAAQS) limits. State and federal agencies strongly recommend against their use.

PHYSICAL AND CHEMICAL PROPERTIES

Ozone (O_3) is composed of three oxygen atoms and has a molecular weight of 48. Its boiling point is −114 °C and melting point is −193 °C. Compared to oxygen, ozone is approximately an order of magnitude more soluble (0.2605 ml/ml at 21 °C.) in water at pH 7 and 1.0 ATA (Cremasco and Mochi, 2012). Ozone in air has a relatively short half-life (3 days at 20 °C.). In water at pH 7, its half-life is brief (20 min at 20 °C.). These estimates of longevity are quite variable because ozone is a highly reactive agent. For example, its half-life is highly pH dependent, decreasing with increasing pH. Contaminants in a water solution also decrease its longevity since ozone is highly reactive. In addition, its half-life is inversely related to ambient temperatures and the half-life in solution cannot be measured accurately above 40 °C.

ENVIRONMENTAL FATE AND BIOACCUMULATION

At present, the average overall background concentration of ozone in the troposphere is 20–40 parts per billion (ppb). Background ozone concentrations have doubled with the advent of industrialization in the last 100–150 years, although the rate of increase has slowed modestly in the late twentieth century (Fuhrer, 2009). Ozone concentrations can be substantially greater in locales with fossil fuel energy sources utilized in motor vehicles and industry. Prior to the realization of the deleterious effects of ozone and initiation of environmental controls, ozone levels reached extremely high levels in some areas. In 1955, an ozone level of 680 ppb was recorded in Los Angeles (Bachman, 2007).

Relatively little ozone is created directly from the oxygen in the troposphere. However, substantial ozone can be created in the presence of oxides of nitrogen (NO_x) and volatile organic compounds (VOCs). The principal NO_x involved in ozone generation are nitric oxide (NO) and nitrogen dioxide (NO_2). The atmospheric concentration of NO is much greater than NO_2, and the NO/NO_2 ratio is usually 10:1 or greater (Jenkin and Clemitshaw, 2000). The NO/NO_2 reaction catalyzes the generation of an ozone molecule. When NO_2 is exposed to UV radiation in sunlight (<420 nm), it dissociates into NO and an oxygen atom (Crutzen et al., 1999; Jenkin and Clemitshaw, 2000)

$$NO_2 \xrightarrow{\text{UV}} NO + O \qquad (50.1)$$

The oxygen atom reacts with oxygen in the atmosphere to form ozone

$$O + O_2(+M) \rightarrow O_3(+M) \qquad (50.2)$$

where M is an energy-absorbing substance, usually nitrogen or oxygen. The mean life of an NO_2 molecule in the presence of UV radiation during daylight is estimated to be only a few minutes (Jenkin and Clemitshaw, 2000).

Numerous volatile carbon compounds, such as those present in gasoline, can contribute to ozone generation in the atmosphere. The interaction of NO and VOCs is initiated by a hydroxyl radical attacking a VOC molecule to form an organic peroxy radical (RO_2), where R represents an intermediate in the photochemical oxidation of a VOC (Figure 50.2). Then, the peroxy radical reacts with NO to produce an RO radical and NO_2

$$RO_2 + NO \rightarrow RO + NO_2 \qquad (50.3)$$

In the presence of UV radiation and oxygen, the NO_2 molecule breaks down to form an ozone molecule and NO via Equations (50.1) and (50.2). The RO radical further reacts to form a hydroperoxy radical (HO_2) and carbonyl products, which may decompose or isomerize. The hydroperoxy radical reacts with NO to produce an NO_2 molecule and a hydroxyl radical

$$HO_2 + NO \rightarrow OH + NO_2 \qquad (50.4)$$

This NO_2 degrades to NO in the presence of UV radiation in sunlight and oxygen, producing another molecule of ozone via Equations (50.1) and (50.2). The hydroxyl radical proceeds to initiate the autocatalytic cycle again with another VOC (Jenkin and Hayman, 1999).

Ozone molecules formed with this interaction between VOCs and NO_x can remain free, oxidize particles or chemicals in the atmosphere, or react with NO to regenerate NO_2.

$$NO + O_3 \rightarrow NO_2 + O_2 \qquad (50.5)$$

The actual level of ozone in the atmosphere is a complex function of the atmospheric concentrations of NO_x and VOCs, the type of VOCs, ambient temperature, the amount of UV radiation provided by sunlight, humidity, wind, and the rates of ozone removal during both day and night (Crutzen et al., 1999; Jenkin and Clemitshaw, 2000). An example of the interaction of VOC and NO_x concentrations in ozone generation is shown in Figure 50.3. The isopleths in the figure depict the level of atmospheric ozone produced by different combinations of VOC and NO_x. The ratios of VOC/NO_x are indicated by the dashed lines. In circumstances of low NO_x, there is an insufficient catalytic generation of ozone to maximize its net production even when the VOCs are plentiful. Similarly, low VOC concentrations result in limited production of ozone even when NO_x concentrations are elevated. Maximum ozone production occurs when the ratios of VOCs and NO_x are optimal. The figure serves as an example of the effects of VOC and NO_x on ozone production, but the production also depends on a number of other variables cited above. Sunlight and temperature have dominant effects on ozone levels, and the cited levels are much lower in winter than summer in urban areas with large seasonal changes in these factors. Maximum generation of ozone may occur 10–20 miles downwind of the sites of release of the primary pollutant precursors of ozone.

Outside urban areas, natural sources of ozone contribute to the background level of ozone. Massive electrical

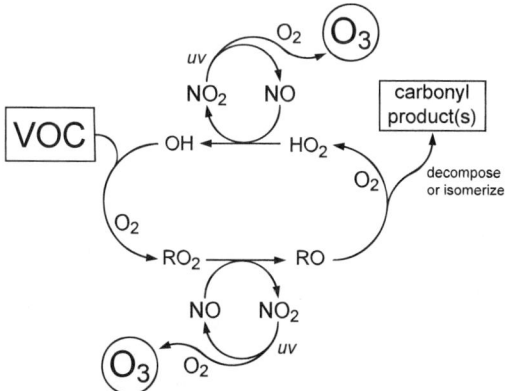

FIGURE 50.2 Role of nitric oxides (NO and NO_2) in the production of ozone (O_3) from a volatile organic compound (VOC). Hydroxyl radical (OH), organic peroxy radical (RO_2), organic oxy radical (RO), and hydroperoxy radical (HO_2) are involved in the cyclical autocatalytic cycle of oxidation. The breakdown of NO_2 requires ultraviolet (UV) radiation and the presence of oxygen (O_2). Figure redrawn from Jenkin and Hayman (1999).

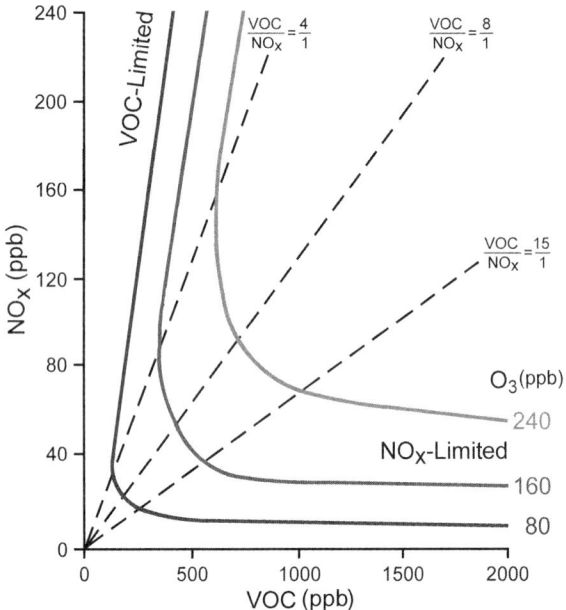

FIGURE 50.3 Example of the relationship between concentrations of nitric oxides (NO_x) and volatile organic compounds (VOC) in determining the levels of ozone (examples of 80, 160, 240 ppb) in the atmosphere under a given set of environmental factors. Ozone production is maximal at the optimal VOC/NO_x ratio and increases relatively little as VOC concentration increases (NO_x-Limited), indicated by the quasi-horizontal lines of production at higher VOC concentrations. Ozone production remains relatively constant (VOC-Limited), indicated by the quasi-vertical lines of production, as increasing NO_x concentration alters the optimal VOC/NO_x ratio. Ozone production and the VOC/NO_x relationship are also affected by a number of other factors, including sunlight, temperature, and humidity. Modified from Jenkin and Clemitshaw (2000). Original diagram from Dodge (1977).

discharges occurring with lightning produce ozone directly; this leads to markedly elevated, but localized, ozone levels. Biomass combustion occurring in wildfires increases atmospheric NO_x and some trees naturally release VOCs that stimulate ozone production. While these sources contribute to ozone levels, the increase noted in the background levels of ozone in the last century is the result of increased human activity. Ozone is widely dispersed in the troposphere by wind and weather patterns and can be transported throughout the earth's atmosphere (WHO Regional Office for Europe, 2008). In spring, the minor contribution of stratospheric ozone to the troposphere may increase with violent storms having a strong vertical component, particularly at high altitudes.

ECOTOXICOLOGY

Even in rural areas, ozone in the atmosphere can affect vegetative growth. Some crops will have reduced yield in regions with ozone levels that are within Environmental Protection Agency (EPA)-approved limits (Morgan et al., 2003). Wheat, corn, and soybean are the major food crops most sensitive to ozone. There is evidence that the yield of these crops in East Asia and Europe has been reduced by as much as 20–25% during the years of severe ozone pollution in the 1990s (Fuhrer, 2009). Other crops, such as rice, are moderately sensitive, but any reduction in rice productivity would be considered serious in view of the importance of this source of nutrition in many countries.

The effect of ozone on plants not only is a factor of the level of exposure, but also depends on the uptake and detoxification of ozone by the particular plant species (Fuhrer, 2009). Ozone is predominantly absorbed by diffusion through the leaf stomata and leads to production of reactive oxygen species (ROS). Stomatal conductance not only varies with different species but is also a complex function of environmental conditions, including solar radiation, temperature, humidity, and carbon dioxide concentration as well as the stage of plant growth. The detoxification of ROS in plants is principally accomplished by reducing agents, e.g., ascorbate. Reactive oxygen species that are not scavenged cause decreased photosynthesis, increased rate of leaf senescence and degradation of proteins and chlorophyll.

Morgan et al. (2003) conducted a meta-analysis of 53 studies of the impact of ozone on soybean growth using the open-top chamber technique. With this methodology, the plants are exposed through the top of the enclosure to prevailing sun and rain. The atmosphere surrounding the plants is maintained constant by continuous ground-level release of a gas with the desired composition. The mean ozone concentration in the experimental group in these studies was 70 ppb and was compared with plants exposed to air free of ozone. At the completion of the growing season, dry plant mass in the ozone-exposed plants had decreased by 38% and seed yield decreased by 24% compared to the control. Other reductions from normal values included photosynthetic rate (20%), leaf chlorophyll concentration (14%), green leaf area (32%) and total leaf area (10%). Interestingly, addition of CO_2 to the gas mixture significantly reduced the damage caused by ozone, most likely the result of a CO_2-induced decrease in stomatal conductance (27%) (Morgan et al., 2003). However, carbon dioxide exposure did not completely eliminate the deleterious effects of ozone.

The influence of ozone on tree growth as opposed to that of ground-level plants presents a more complicated situation. Extrapolating conclusions from juvenile trees grown in exposure chambers does not necessarily correlate with mature tree growth in the field. Matyssek et al. (2008) has criticized this methodology because significant gradients in ozone concentration, sunlight exposure, and humidity occur between the leaf canopy and ground levels in mature trees. These authors argue that conclusions regarding ozone and other air pollutants must be reached by understanding the mechanisms involved in pollutant uptake and detoxification as well as the effect of different species in resistance to pollutant effects. Clearly much more progress has been achieved in the study of ground-level plants compared to trees.

MAMMALIAN TOXICTY

Experimental studies utilizing animal models, principally rodents, have demonstrated that acute, high level exposure to ozone results in mortality through the induction of severe pulmonary edema. Last et al. (2010) have described the pathologic and functional changes in animals exposed to high doses of ozone as well as lesser exposures in humans. In general, reported animal studies are less pertinent to humans, since the levels of exposure in animal work are rarely encountered in current human exposures to ozone. Early experimental studies in humans were conducted with ozone concentrations that no longer occur in developed countries and are present only in a few urban areas in developing countries. Recently, investigations have focused on more pertinent ozone levels and its interaction with other air pollutants, especially particulate matter (PM) less than 2.5 microns in diameter ($PM_{2.5}$).

Ozone has relatively mild irritant effects on the eyes and skin at the current levels of exposures, but these effects are transient without permanent residua. However, ozone produces significant cardiopulmonary morbidity and mortality.

Substantial progress has been achieved in the last decade in defining the effect of ozone exposure on human mortality. Previously published investigations reported time-series studies of mortality, usually performed in a single city or region. The results were variable with both positive and negative findings. Several earlier meta-analyses had suggested an increase in mortality but considerable uncertainty persisted. In 2003 the EPA commissioned three different epidemiological teams to analyze available studies contained in a large database of ozone reports. The teams performed meta-analyses independently using different methodologies (Bell et al., 2005; Ito et al., 2005; Levy et al., 2005). Wherever possible, primary data were analyzed. All three groups reached the conclusion that acute ozone exposure increased mortality within a few days by approximately 0.89% for each 10 ppb increase. Using studies reporting seasonal data, a higher mortality was noted during warmer months when ozone levels are greater. With the limited data available in these primary sources, no significant effect of confounding pollutants could be discerned. One of these three meta-analyses suggested that the calculated mortality rates could partially be due to publication bias that tended to favor studies with positive results (Bell et al., 2005).

A more conclusive approach to investigate the effect of tropospheric ozone is to utilize the data collected from multiple cities in a single uniform study. The National Morbidity, Mortality and Air Pollution Study (NMMAPS) provided 14 years (1987–2000) of information collected in a uniform fashion in 95 large urban areas throughout the United States. Bell et al. (2004) used a distributed-lag model of mortality adjusted for PM, weather, seasons, and long-term trends in each urban area. Individual regional mortalities were combined to calculate an average national mortality that accounted for statistical uncertainty and heterogeneity in each community. The overall national mortality increase was 0.52%/10 ppb increase in a daily average ozone exposure during the previous week. Age (~50% of those dying were ≥75 years) and increased respiratory and cardiovascular mortality were major contributors to the overall mortality. Particulate matter (PM_{10}) did not correlate with ozone levels and inclusion of specific community PM_{10} levels in the model did not affect the calculated ozone mortality. Exclusion of days with temperatures >29 °C. (85 °F.) also did not affect the calculated ozone mortality.

Bell and Dominici (2008) extended their previous study utilizing the data from the NMMAPS study, as well as data from the National Center for Health Statistics, National Climatic Data Center, and EPA, to determine the role of specific community-level characteristics on mortality. This work included data from 40% of the total U.S. population. There was no demonstrable effect of PM, either $PM_{2.5}$ or PM_{10}, on short-term ozone-associated mortality. The highest mortality rates were (1.44 and 0.73%/10 ppb increase in ozone) in the Northeast and Industrial Midwest regions, respectively. The lowest mortalities (essentially no increased risk) were in the Southwest and urban Midwest. A major difficulty in interpreting any ozone-associated effects is the determination of actual exposure, since outdoor ozone concentrations correlate poorly with indoor concentrations. The latter concentration can be much lower than that of outdoors depending upon the extent of air exchange between the two environments (U.S. EPA, 2013).

A study of 48 U.S. cities investigated the timing and factors involved in ozone-associated mortality rates (0.5%/ 10 ppb increase) during the months of May through September (Zanobetti and Schwartz, 2008a). The increase in ozone-associated mortality started in the springtime and peaked in July. Mortality rates decreased in August and the authors calculated a null association in September even though ozone levels were elevated in the latter 2 months. The authors hypothesized that susceptible persons developed biological resistance to ozone, an observation noted in some acute laboratory experiments (Folinsbee et al., 1994). However, this observation also would be compatible with the concept that the susceptible individuals died when first exposed to the elevated ozone levels in the spring and early summer. As noted in earlier studies, the increased mortality was prominent in older individuals (>50 years) in this study.

The influence of PM on ozone-associated mortality has been investigated, but the exact composition of particulate pollutants can vary widely with time and geography. Anderson et al. (2012) studied the effect of the seven major components of $PM_{2.5}$ on ozone-associated mortality in 57 U.S. communities and found no effect of local components of PM, including sulfate. These seven components studied comprise the great majority of small particulate mass in the atmosphere. In addition, levels of particulate pollutants are relatively constant throughout the year while ozone levels and its associated mortality vary substantially throughout the year depending upon the season and ambient temperatures.

The use of advanced statistical analyses and large multicity databases has convincingly described an effect on short-term mortality at ozone levels present in many urban areas. More recent attention has been paid to its role in the morbidity of specific diseases. Patients with asthma, a common and widespread disease, are likely to experience morbidity with pollutant exposure. There are numerous studies relating exposure to automotive air pollution to hospital or emergency room admissions for asthma exacerbation, symptoms such as wheezing and shortness of breath, and the need for acute or chronic asthmatic medications. Despite this large body of evidence that exacerbations of asthma can be precipitated by components of automotive air pollution, especially ozone and small PM, it is still uncertain whether this pollution causes the actual development of asthma rather than only precipitating exacerbations (Ryan and Holguin, 2010). The role of ozone in exacerbations of asthma is still not completely defined due to the lack of many large multicity studies, measurements of only a few of the multiple components of automotive pollution in some studies, the high correlation among some components of atmospheric pollution that obscure individual roles, and the use of retrospective review of medical records to obtain diagnoses. Overall, recent studies are supportive of the concept that ozone itself is a factor in exacerbations of asthma.

Silverman and Ito (2010) analyzed 69,735 non-ICU hospital admissions for asthma to 74 New York City hospitals in the months April through August for the period 1999–2006. They reported an overall increased relative risk of hospital admission (1.06, 95% confidence interval (CI), 1.03–1.10) for an increase in ozone exposure of 22 ppb and an increased relative risk (1.07, 95% CI, 1.05–1.10) for an increased exposure of $PM_{2.5}$ of 12 µg/m^3 in a two-pollutant, 0–1 day lag model. The age effect was greatest in the age group 6–18 years (1.15 and 1.16, respectively, for ozone and $PM_{2.5}$). The correlation (0.59) between ozone and $PM_{2.5}$ increases the uncertainty of which pollutant, or possibly their combination, was responsible for the effect. However, the simultaneous inclusion of both the pollutants in the analytic model argues strongly for both the pollutants having causative roles in exacerbations of asthma, especially in young individuals.

In an analysis of 91,386 emergency room visits for asthma treatment for over 12 years (1993–2004) in 41 hospitals in metropolitan Atlanta, Georgia, Strickland et al. (2010) studied the effects of 11 different ambient air pollutants. In preliminary univariate analysis, 10 of the 11 pollutants had significant positive associations with patient visits. Of these 10 pollutants, ozone had the strongest association with patient visits during the warm weather (relative risk 1.082 for an interquartile change in concentration). Case-crossover and time-series models only reduced significant associations of one and three pollutants, respectively, of the original 10 associations. The association between ozone and emergency room visits for asthma persisted when the data were analyzed in several two pollutant models. In these models carbon monoxide had the strongest association with emergency department visits. Ozone and nitrogen dioxide also exhibited positive associations with visits. The authors concluded that the two-pollutant models support the conclusion that ozone and traffic-related primary pollutants independently contribute to the frequency of exacerbations of asthma.

Investigators have attempted to define the effects of ozone per se in the absence of other pollutants by performing controlled chamber studies in which the same subjects are exposed to filtered air and known concentrations of ozone in air on separate occasions. Since subjects act as their own controls, this approach enables the detection of small decrements in lung function that would not be appreciated in large-scale epidemiological studies investigating the effects of mixed pollutant exposures. Initial studies employed inspired ozone levels as high as 400 ppb for short periods of time and demonstrated substantial decrements in lung function. However, this still did not answer the question of whether exposure to current urban levels of ozone carries a health risk. As urban ozone levels have decreased in most developed countries, investigations have focused on low levels of ozone that span the range currently present in urban atmospheres. A commonly used testing protocol involves a 6.6-h exposure with 50 min of exercise/h plus a 35-min lunch break. The increased ventilation (4–5 times the resting values) obviously leads to a greater overall ozone exposure than experienced by an individual who is not exercising. This serves as an estimate of the potential risk of exposure that could be expected to mimic real-life, outdoor working conditions of a moderate degree.

Using this 6.6-h protocol, Schelegle et al. (2009) studied the responses of 16 healthy females and 15 males to 5 different ozone exposures [mean measured exposure levels: 0 (filtered air), 63, 72, 81, and 88 ppb] on different days in a random order. Exposures were separated by a minimum of 7 days to avoid cumulative effects of the exposures. In this study, exposure was initiated at a level lower than the target level, peaked at 4.6 h, and then decreased to simulate the course of ozone levels during a working day. Spirometry and symptom scores were recorded at the end of each exercise period and 1 and 4 h post exposure. The

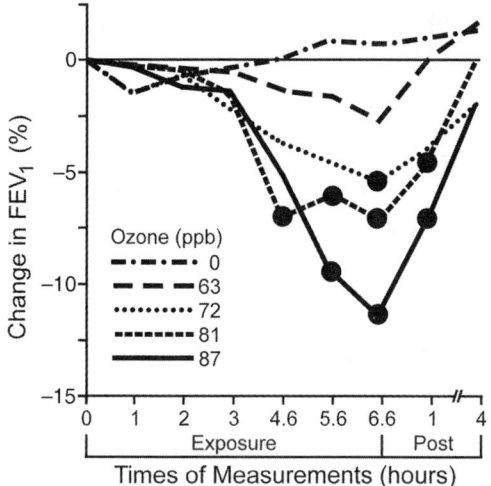

FIGURE 50.4 Mean changes in forced expiratory volume, 1.0 s (FEV$_1$) from baseline pre-exposure values (0 h) with continuing exposure to mean levels of ozone (0, 63, 72, 81, and 87 ppb) during a 6.6 h protocol. Exercise increased minute ventilation in subjects ($n = 31$, mean age 21.4 years) to a mean of 38.7 l/min during 50 min of exercise per hour with a 35 min rest period for food at hour 4. Postexposure values were obtained at rest while breathing air without ozone. The filled circles indicate statistically significant changes ($p < 0.05$) compared with the FEV$_1$ breathing ozone-free inspired air at the same time point. Figure redrawn with permission from Schelegle et al. (2009).

average forced expiratory volume in 1.0 second (FEV$_1$), the most common clinical measurement of airflow obstruction, decreased significantly during the three highest levels of ozone exposure compared to the FEV$_1$ measured during the exposure to filtered air (Figure 50.4). The decrease in the group mean of the FEV$_1$ did not reach statistical significance during exposure to 63 ppb ozone. However, the response to ozone exposure is heterogeneous, and at this level of exposure, 11 of the 31 subjects had decrements in the FEV$_1$ that were significantly greater than the measurement error of the FEV$_1$. Increases in symptom scores of the subjects followed the same pattern as the FEV$_1$ and reached a statistical significance during exposures of 72, 81, and 88 ppb. Reductions in the forced vital capacity (FVC) were less than those of the FEV$_1$ and reached statistical significance only during the exposures to 81 and 88 ppb. Most abnormalities returned to control values by 4 h after the termination of exposures.

A larger study of 59 young (25.0 ± 0.5 SEM years) healthy subjects exposed to a constant level of 60 ppb ozone with the same 6.6-h exercise protocol demonstrated small, but statistically significant, decrements in both FEV$_1$ (-1.7%) and FVC (-1.2%) compared to control experiments breathing ozone-free air (Kim et al., 2011). In addition, the

mean percentage of neutrophils in sputum, an indicator of airway inflammation, was 15% greater than the control percentage 18 h post exposure ($p < 0.002$).

These two studies (Kim et al., 2011 and Schelegle et al., 2009) indicate that exposure to ozone levels that are within the current NAAQS during outdoor exercise can result in decrements of pulmonary function and increases in lung inflammation. While the mean values of functional loss are small, of greater concern is the larger loss of function in a subset of subjects with greater sensitivity to ozone.

Relatively short-term experimental challenges demonstrate loss in function, but this reverts to normal shortly after the termination of the exposure. Long-term effects of modest ozone exposure on lung function and the inflammatory response have not been defined. In fact, repeated exposures to higher levels of ozone (120–200 ppb), over the course of several days, suggest that the initial loss in lung function and the delayed inflammatory response are ameliorated, but not resolved, despite continued exposure (Folinsbee et al., 1994).

The effect of ozone on cardiovascular health is less clear than its influence on respiratory morbidity and mortality. Most epidemiologic data regarding the effect of ozone on the incidence of myocardial infarction, hypertension, emergency department visits, and hospital admission for cardiovascular events do not support an association (U.S. EPA, 2013). Earlier studies investigating a possible association of cardiovascular mortality with ozone exposure were inconsistent, but more recent studies with large populations support a causative association. Jerrett et al. (2009) reviewed data collected from 1.2 million people in an American Cancer Society study. The authors found associations between cardiovascular mortality and exposure to both ozone and PM$_{2.5}$ in single-pollutant analyses, but in a two-pollutant model, only PM$_{2.5}$ had a significant association. Stafoggia et al. (2010) analyzed data from 10 Italian cities, which included both ozone and PM$_{10}$ data. They found increased all-cause and cardiovascular mortalities of 1.5 and 2.3% with an increase of 5 ppb ozone in both single- and two-pollutant models, supporting the role of both the pollutants in total and cardiovascular mortalities. Other investigators (Samoli et al., 2009; Zanobetti and Schwartz, 2008b) also have found significant associations between ozone exposure and cardiovascular mortalities in large studies of multiple European and U.S. cities, respectively. However, neither of these two studies included PM data in the analyses. All of these studies were conducted on data obtained during summer months when ozone levels are the highest. Because of the large effect of PM on cardiovascular mortality and its correlation in many regions with ozone levels, it is difficult to separate completely the effects of both the pollutants on cardiovascular mortalities.

A concern regarding the findings of increased total mortality with ozone exposure is the possibility that the individuals who died were already severely ill and the exposure only hastened inevitable death by a short time. In an analysis of a time-series study, Zanobetti and Schwartz (2008b) investigated this possibility by comparing the number of deaths on a given day to the deaths in the ensuing 3 weeks. If the deaths were premature because of an ozone effect and these individuals were removed from the cohort earlier, then the later number of deaths would be less and the overall mortality for the period would not be increased. In their analysis, they found that the deaths consistently increased over a 3-week window throughout June through August, the time of year when ozone levels are elevated. The authors concluded that the overall mortality increases with ozone exposure. Since their analysis could only include a total period of 3 weeks, they were unable to reach any conclusion regarding longer periods of premature death. Clearly further studies are needed to investigate this possibility. Investigations of the long-term effects of ozone exposure are few in number and difficult to analyze due to the large number of variables involved.

Remaining uncertainties regarding long-term effects, the different levels of indoor and outdoor ozone, and the exact role of ozone in multi-pollutant atmospheres require further study. Air quality in developed countries has improved greatly in the last 50 years. Currently in many areas, average ozone levels are approaching the present background of 40 ppb. Further lowering of ozone levels will become increasingly more difficult and expensive (Frampton, 2011). It is possible that background levels of ozone could have health effects, either independently or in combination with other atmospheric components. Therefore, future research needs to address the most effective means of reducing adverse health effects of ozone. This may not include only an effort to reduce the level of ozone in the troposphere.

SOURCES OF EXPOSURE

Urban atmospheres are the principal sources of exposure to ozone and levels frequently exceed not only the NAAQS of 75 ppb but also the OSHA and NIOSH limits of 100 ppb. Since ozone is a potent oxidizing agent, it is generated as per industrial use and is rarely stored. However, concentrations of ozone generated with an oxygen stream can be as great as 20% weight/total weight. Ozone monitors must be present wherever ozone is generated, transported via conduits, or used, because ozone can leak through corroded valves and pipes. Personal exposures to ozone can be significant during arc welding, but source levels do not approach those occurring with industrial ozone generators. With appropriate ventilation, ozone levels generated in arc welding can be maintained at acceptable levels. Compared to these industrial circumstances, most ozone exposure levels are low. Although ozone is generated by electrostatic printing and

copying procedures, the resulting concentrations are below the current OSHA PEL.

INDUSTRIAL HYGIENE

The OSHA limits for ozone exposure are 100 ppb time-weighted average (TWA). The NIOSH ceiling is 100 ppb. The American Conference of Governmental Industrial Hygienists (ACGIH) has published a sliding scale from 50 to 100 ppb depending on work intensity and a 200 ppb 2-h TWA at all levels of work. For PPE, the Centers for Disease Control and Prevention (CDC)/NIOSH Pocket Guide recommends a chemical cartridge respirator or supplied-air respirator (assigned protection factor (APF) = 10) for ozone levels above acceptable levels up to 1000 ppb; a supplied-air respirator operated in continuous-flow mode or a powered, air-purifying respirator (APF = 25) up to levels of 2500 ppb; a chemical cartridge respirator with full facepiece, an air purifying, full facepiece respirator (gas mask) with chin-style, front or back-mounted canister, a supplied-air respirator with a tight-fitting facepiece operated in a continuous-flow mode, a self-contained breathing apparatus with a full facepiece or a supplied-air respirator with a full facepiece (APF = 50) up to levels of 5000 ppb. For entry into unknown concentrations or Immediately Dangerous to Life or Health (IDLH) conditions, the recommendation is for a self-contained breathing apparatus that has a full facepiece and is operated in a pressure-demand or other positive-pressure mode or a supplied-air respirator that has a full facepiece and is operated in a pressure-demand or other positive-pressure mode in combination with an auxiliary self-contained, positive-pressure breathing apparatus (APF = 10,000).

RISK ASSESSMENTS

The U.S. EPA (2013) published an Integrated Science Assessment for Ozone and Photochemical Oxidants (ISA-Ozone). This document concludes that there is a causal relationship between short-term ozone exposure and respiratory effects. These effects include exacerbations of asthma, respiratory-related hospital admissions, and increased respiratory symptoms. Likely causative effects of short-term ozone exposure included total mortality, cardiovascular effects, and cardiovascular mortality. The conclusion regarding cardiovascular effects was based on toxicological and animal studies. Epidemiologic data were inconsistent. Long-term exposure to ozone was identified as a likely cause of respiratory effects. Elevated ambient ozone exposure was identified as a cause of foliar injury, reduced growth of native plants, and reductions of yield and quality of agricultural crops.

The Regional Health Office for Europe of the World Health Organization (2008) issued a health impact statement that estimated that ozone levels above 35 ppb caused 21,000 premature deaths in 25 European countries in the year 2000. In terms of morbidity, ozone exposures caused an estimated 14,000 annual admissions to hospitals caused by respiratory illness, 54.9 million person-days of minor respiratory distress in adults, and 21.0 million person-days of required respiratory medications in children. In anticipation of the results of current pollution control, the predicted premature mortality in 2020 would decrease only by 5% and hospital admissions would rise by 43%, but the adult and childhood morbidities would decrease by 21 and 38%, respectively.

Berman et al. (2012) calculated the potential effects of adherence to the new NAAQS of 75 ppb that would have resulted if this had been achieved in years 2005–2007 in the United States. The annual premature deaths would have decreased between 1410 and 2480 individuals, cases of respiratory symptoms by 3 million, and days of lost school attendance by 1 million. If a lower ozone exposure limit of 70 ppb had been achieved during 2005–2007, calculated premature deaths would have decreased by an annual average between 2450 and 4130 individuals. It is unlikely that these estimated improvements would be directly applicable to the present time since ozone levels have decreased further since 2005–2007 even though the current NAAQS often is exceeded in areas within the country.

ACKNOWLEDGMENT

The author wishes to acknowledge the editorial and technical assistance of Margaret A. Simons in the preparation of this manuscript.

REFERENCES

Anderson, G.B., Krall, J.R., Peng, R.D., and Bell, M.L. (2012) Is the relation between ozone and mortality confounded by chemical components of particulate matter? Analysis of 7 components in 57 U.S. communities. *Am. J. Epidemiol.* 176:726–732.

Auten, R.L. and Davis, J.M. (2009) Oxygen toxicity and reactive oxygen species: the devil is in the details. *Pediatr. Res.* 66:121–127.

Bachman, J. (2007) Will the circle be unbroken: a history of the U.S. National Ambient Air Quality Standards. *J. Air Waste Manage. Assoc.* 57:652–697.

Battino, R., Rettich, T.R., and Tominaga, T. (1983) The solubility of oxygen and ozone in liquids. *J. Phys. Chem. Ref. Data* 12:163–178.

Bell, M.L. and Dominici, F. (2008) Effect modification by community characteristics on the short-term effects of ozone exposure and mortality in 98 US communities. *Am. J. Epidemiol.* 167: 986–997.

Bell, M.L., McDermott, A., Zeger, S.L., Samet, J.M., and Dominici, F. (2004) Ozone and short-term mortality in 95 US urban communities, 1987–2000. *JAMA* 292:2372–2378.

Bell, M.L., Dominici, F., and Samet, J.M. (2005) A meta-analysis of time-series studies of ozone and mortality with comparison to the National Morbidity, Mortality, and Air Pollution Study. *Epidemiology* 16:436–445.

Berman, J.D., Fann, N., Hollingsworth, J.W., Pinkerton, K.E., Rom, W.N., Szema, A.M., Breysse, P.N., White, R.H., and Curriero, F.C. (2012) Health benefits from large-scale ozone reduction in the United States. *Environ. Health Perspect.* 120:1404–1410.

Bitterman, N. (2004) CNS oxygen toxicity. *Undersea Hyperb. Med.* 31:63–72.

Clark, J.M. and Lambertsen, C.J. (1971) Pulmonary oxygen toxicology: a review. *Pharmacol. Rev.* 23:37–133.

Clark, J.M., Lambertsen, C.J., Gelfand, R., Flores, N.D., Pisarello, J.B., Rossman, M.D., and Elias, J.A. (1999) Effects of prolonged oxygen exposure at 1.5, 2.0, or 2.5 ATA on pulmonary function in men (predictive studies V). *J. Appl. Physiol.* 86:243–259.

Cremasco, M.A. and Mochi, V.T. (2012) Gradient step method to predict the ozone solubility in water. *J. Environ. Sci. Eng. A* 1:256–260.

Crutzen, P.J., Lawrence, M.G., and Poschl, U. (1999) On the background photochemistry of tropospheric ozone. *Tellus* 51 A–B:123–146.

Dodge, M.C. (1977) Combined use of modeling techniques and smog chamber data to derive ozone-precursor relationships. *International Conference on Photochemical Oxidant Pollution and Its Control: Proceedings.* **II B**, EPA/600/3-77-001b. U.S. EPA, Research Triangle, NC, pp. 881–889.

Donald, K.W. (1947) Oxygen poisoning in man. *Br. Med. J.* 1:667–717.

Edmark, L., Auner, U., Enlund, M., Ostberg, E., and Hedenstierna, G. (2011) Oxygen concentration and characteristics of progressive atelectasis formation during anaesthesia. *Acta Anaesthesiol. Scand.* 55:75–81.

Folinsbee, L.J., Horstman, D.H., Kehrl, H.R., Harder, S., Abdul-Salaam, S., and Ives, J.I. (1994) Respiratory responses to repeated prolonged exposure to 0.12 ppm ozone. *Am. J. Respir. Crit. Care Med.* 149:98–105.

Frampton, M.W. (2011) Ozone air pollution: how low can you go? *Am. J. Respir. Crit. Care Med.* 184:150–151.

Fuhrer, J. (2009) Ozone risk for crops and pastures in present and future climates. *Naturwissenschaften* 96:173–194.

Hampson, N. and Atik, D. (2003) Central nervous system oxygen toxicity during routine hyperbaric oxygen therapy. *Undersea Hyperb. Med.* 30:147–153.

Ito, K., De Leon, S.F., and Lippman, M. (2005) Associations between ozone and daily mortality: analysis and meta-analysis. *Epidemiology* 16:446–457.

Jenkin, M.E. and Clemitshaw, K.C. (2000) Ozone and other secondary photochemical pollutants: chemical processes governing their formation in the planetary boundary layer. *Atmos. Environ.* 34:2499–2527.

Jenkin, M.E. and Hayman, G.D. (1999) Photochemical ozone creation potentials for oxygenated volatile organic compounds: sensitivity to variations in kinetic and mechanistic parameters. *Atmos. Environ.* 33:1275–1293.

Jerrett, M., Burnett, R.T., Pope, C.A., Ito, K., Thurston, G., Krewski, D., Shi, Y., Calle, E., and Thun, M. (2009) Long-term ozone exposure and mortality. *N. Engl. J. Med.* 360:1085–1095.

Kim, C.S., Alexis, N.E., Rappold, A.G., Kehrl, H., Hazucha, M.J., Lay, J.C., Schmitt, M.T., Case, M., Devlin, R.B., Peden, D.B., and Diaz-Sanchez, D. (2011) Lung function and inflammatory responses in healthy young adults exposed to 0.06 ppm ozone for 6.6 hours. *Am. J. Respir. Crit. Care Med.* 183:1215–1221.

Last, J.A., Hyde, D.M., and Schelegle, E.S. (2010) Ozone and oxygen toxicity. In: McQueen, C.A., editor. *Comprehensive Toxicology*, 2nd ed., vol. 8, Oxford: Elsevier Ltd., pp. 261–275.

Levy, J.I., Chemerynski, S.M., and Sarnat, J.A. (2005) Ozone exposure and mortality: an empiric Bayes metaregression analysis. *Epidemiology* 16:458–468.

Matyssek, R., Sandermann, H., Wieser, G., Booker, F., Cieslik, S., Musselman, R., and Ernst, D. (2008) The challenge of making ozone risk assessment for forest trees more mechanistic. *Environ. Pollut.* 156:567–582.

Morgan, P.B., Ainsworth, E.A., and Long, S.P. (2003) How does elevated ozone impact soybean? A meta-analysis of photosynthesis, growth and yield. *Plant Cell Environ.* 26:1317–1328.

Ryan, P.H. and Holguin, F. (2010) Traffic pollution as a risk factor for developing asthma: are the issues resolved? *Am. J. Respir. Crit. Care Med.* 181:530–531.

Samoli, E., Zanobetti, A., Schwartz, J., Atkinson, R., LeTertre, A., Schlinder, C., Perez, L., Cadum, E., Pekkanen, J., Paldy, A., Touloumi, G., and Katsouyanni, K. (2009) The temporal pattern of mortality responses to ambient ozone in the APHEA project. *J. Epidemiol. Community Health* 63:960–966.

Schelegle, E.S., Morales, C.A., Walby, W.F., Marion, S., and Allen, R.P. (2009) 6.6-Hour inhalation of ozone concentrations from 60 to 87 parts per billion in healthy humans. *Am. J. Respir. Crit. Care Med.* 180:265–272.

Silverman, R.A. and Ito, K. (2010) Age-related association of fine particles and ozone with severe acute asthma in New York City. *J. Allergy Clin. Immunol.* 125:367–373.

Stafoggia, M., Forastiere, F., Faustini, A., Biggeri, A., Bisanti, L., Cadum, E., Cernigliaro, A., Mallone, S., Pandolfi, P., Serinelli, M., Tessari, R., Vigotti, M.A., Perucci, C.A., and EpiAir Group. (2010) Susceptibility factors to ozone-related mortality: a population-based case-crossover analysis. *Am. J. Respir. Crit. Care Med.* 182:376–384.

Strickland, M.J., Darrow, L.A., Klein, M., Flanders, W.D., Sarnat, J.A., Waller, L.A., Sarnat, S.E., Mulholland J.A., and Tolbert, P.E. (2010) Short-term associations between ambient air pollutants and pediatric asthma emergency department visits. *Am. J. Respir. Crit. Care Med.* 182:307–316.

SUPPORT Study Group of the Eunice Kennedy Shriver NICHD Neonatal Research Network (2010) Target ranges of oxygen saturation in extremely preterm infants. *N. Engl. J. Med.* 362:1959–1969.

U.S. Environmental Protection Agency (2013) *Integrated science assessment for ozone and related photochemical oxidants*. Available at www.epa.gov/ncea/isa/ozone.htm (accessed July 7, 2013).

WHO Regional Office for Europe (2008) *Health risks of ozone from long-range transboundary air pollution*. Available at www .euro.who.int/en/what-we-publish/abstracts/health-risks-of-ozone-from-long-range-transboundary-air-pollution (accessed July 18, 2013).

Zanobetti, A. and Schwartz, J. (2008a) Is there adaptation in the ozone mortality relationship: a multi-city case-crossover analysis. *Environ. Health* 7:22. Available at www.ehjournal .net/content/7/1/22 (accessed February 20, 2013).

Zanobetti, A. and Schwartz, J. (2008b) Mortality displacement in the association of ozone with mortality: an analysis of 48 cities in the United States. *Am. J. Respir. Crit. Care Med.* 177:184–189.

51

PHOSPHORUS COMPOUNDS

M. Rony Francois and Frank Stephen

First aid (white phosphorus): Inhalation: Remove casualty to fresh air and keep at rest. Seek medical advice; Skin: Treat phosphorus burns by immediate immersion under water and call for medical assistance; Eyes: Rinse with plenty of water and seek medical advice; Ingestion: Rinse mouth. Consult a doctor immediately. Do not induce vomiting. Hazards: Danger of circulatory collapse. Treatment: It may be necessary to administer oxygen. After skin contact, mechanically remove any residual attached phosphorus under water; then coat skin with a "Ben Hur" solution. Apply a sterile dressing to the wound. If swallowed, wash out the stomach and administer active charcoal together with a 0.2% solution of copper sulfate

Occupational exposure limits (white phosphorus): 0.1 mg/m^3 (OSHA PEL, NIOSH REL, ACGIH TLV)

Hazard statement (white phosphorus): H250—catches fire spontaneously if exposed to air; H330—fatal if inhaled; H300—fatal if swallowed; H314—causes severe skin burns and eye damage; H400—very toxic to aquatic life

Precautionary statements (white phosphorus): P210—keep away from heat/sparks/open flames . . . /hot surfaces . . . no smoking; P222—do not allow contact with air; P260—do not breath dust/fume/gas/mist/vapors/spray; P264—wash...thoroughly after handling; P270—do not eat, drink, or smoke when using this product; P271—use only outdoors or in a well-ventilated area; P273—avoid releases to the environment; P280—wear protective gloves/protective clothing/eye protection/face protection; P284—wear respiratory protection; P285—in case of inadequate ventilation wear respiratory protection; P302+P334—if on skin, immerse in cool water/wrap in wet bandages; P304+P340—if inhaled, remove victim to fresh air and keep at rest in a position comfortable for breathing; P310—immediately call a poison center or doctor/physician; P320—specific treatment is urgent (see . . . on label); P370+P378—in case of fire, use . . . for extinction; P403+P233—store in a well-ventilated place. Keep container tightly closed; P405—store locked up; P422—store contents under . . . ; P501—dispose of contents/container to . . . ; P301+P310—if swallowed, immediately call a poison center or doctor/physician; P321—specific treatment (see . . . on label); P301+P330+P331—if swallowed, rinse mouth. Do not induce vomiting; P303+P361+P353—if on skin (or hair), remove/take off immediately all contaminated clothing. Rinse skin with water/shower; P305+P351+P338—if in eyes, rinse cautiously with water for several minutes. Remove contact lenses, if present and easy to do. Continue rinsing; P309+P311—if exposed or if you feel unwell, call a poison center or doctor/physician; P391—collect spillage; P363—wash contaminated clothing before reuse

Risk phrases (white phosphorus): R17—spontaneously flammable in air; R26/28—very toxic by inhalation and if swallowed; R35—causes severe burns; R50—very toxic to aquatic organisms

Safety phrases (white phosphorus): S1/2—keep locked up and out of reach of children; S5—keep contents under . . . (appropriate liquid to be specified by the manufacturer); S26—in case of contact with eyes, rinse immediately with plenty of water and seek medical advice; S38—in case of insufficient ventilation, wear suitable respiratory equipment; S45—in case of accident or if you feel unwell, seek medical advice immediately

Hamilton & Hardy's Industrial Toxicology, Sixth Edition. Edited by Raymond D. Harbison, Marie M. Bourgeois, and Giffe T. Johnson.
© 2015 John Wiley & Sons, Inc. Published 2015 by John Wiley & Sons, Inc.

TABLE 51.1 Exposure Limits and Threshold Limit Values for Phosphorus Compounds

Chemical Name and Synonyms	Formula	CASRN	Form	OSHA PEL (mg/m^3)	NIOSH REL (mg/m^3)	ACGIH TLV (mg/m^3)
Phosphorus (yellow), elemental phosphorus, white phosphorus	P$_4$	7723-14-0	Clear, colorless liquid	0.1	0.1	0.1
Phosphoric acid, orthophosphoric acid, white phosphoric acid	H$_3$PO$_4$	7664-38-2	Crystalline solid	1	1.0, 3.0^a	1.0, 3.0^a
Phosphorus oxychloride, phosphorus chloride, phosphorus oxytrichloride, phosphoryl chloride	POCl$_3$	10025-87-3	Clear, colorless to yellow liquid	None	0.6, 3.0^a	0.6
Phosphorus pentachloride, pentachlorophosphorus, phosphoric chloride, phosphorus perchloride	PCl$_5$	10026-13-8	White to pale yellow crystalline solid	1.0	1.0	0.1
Phosphorus pentasulfide, phosphorus persulfide, sulfur phosphide, phosphorus sulfide	P$_2$S$_2$/P$_4$S$_{10}$	1314-80-3	Greenish gray crystalline solid	1.0	1.0, 3.0^a	1.0, 3.0^a
Phosphorus trichloride, phosphorus chloride	PCl$_3$	7719-12-2	Colorless to yellow fuming liquid	3	1.5, 3.0^a	1.5, 3.0^a
Phosphorus anhydride, 1,2-benzenedicarboxylic anhydride, PAN, phthalic acid anhydride	C$_6$H$_4$(CO)$_2$O	85-44-9	White solid or clear colorless liquid	12	6	6
Phosphine, hydrogen phosphide, phosphorated hydrogen, phosphorus hydride, phosphorus trihydride	PH$_3$	7803-51-2	Colorless gas	0.4	0.4, 1.0^a	0.4, 1.4^a

Source: ACGIH (2008) and NIOSH (2005).

^{a}15 min short-term exposure limit.

OSHA, Occupational Safety and Health Administration; NIOSH, National Institute for Occupational Safety and Health; ACGIH, American Conference of Governmental Industrial Hygienists; PEL, permissible exposure limit; REL, recommended exposure limit; TLV threshold limit value.

(show the label where possible); S61—avoid release to the environment, refer to special instructions/safety data sheets.

HISTORY

Essential for life, phosphorus is a nonmetal that, as phosphate (PO$_4$) rocks, is mined for use in producing phosphate-based fertilizers for agriculture. First isolated from urine in the seventeenth century, phosphorus has had common uses in explosives, poisons and nerve agents. Ironically, Because of these uses and the claim that it was the 13th element discovered, it has sometimes been referred to as "the Devil's element" (Emsley, 2002). In its phosphate form, phosphorus is found in ATP, as a constituent of cell membrane phospholipids and part of the sugar-phosphate backbone of DNA and RNA. As a commercial product, phosphate is mainly found in fertilizers where it is used to replace phosphorus removed by growing plants. It is also used in food preservatives, ceramics, cosmetics, animal feed supplements, water treatment (softeners), and metallurgy (steel production). In its organophosphate form,

phosphorus is also used in pesticides, nerve agents, detergents, and matches. The toxicological properties of phosphorus are based upon the form in which it is found. Table 51.1 outlines exposure limits and threshold limit values (TLV) for various phosphorus compounds and definitive examples of toxicity of some compounds are discussed below.

WHITE/YELLOW PHOSPHORUS

Historically, white phosphorus was notorious because its inhalation in the matchmaking industry caused a disfiguring ailment known as *phossy jaw*. Fatal, the ailment is caused when white phosphorus fumes are absorbed through cavities in the teeth. Eventually deposited in bones and causing them to glow a greenish-white color in the dark, white phosphorus exposure can also lead to liver and other organ failure. Victims would also suffer from a foul-smelling discharge from the rotting bone. The use of yellow phosphorus in industry was discontinued early in the twentieth century after the discovery of the safer, but more expensive, red phosphorus.

White phosphorus (CASRN 12185-10-3) is found as a waxy, sometimes translucent solid. However, because it oxidizes quickly and discolors after exposure to light, it is also known as yellow phosphorus. Flammable at temperatures as little as 10 °C above room temperature, white phosphorus is stored underwater. White phosphorus is also pyrophoric and has a melting point of 44.2 °C, a boiling point of 280.5 °C, and a density of 1.823. Used mainly to produce phosphoric acid, white phosphorus has a slight garlic smell. It reacts with oxygen and water to form strong acids (H_3PO_2, H_3PO_3).

White phosphorus can persist in air from minutes to days but may have a longer half-life when aerosolized and coated with a protective layer of oxide (Berkowitz et al. 1981). White phosphorus can exist in water as dissolved phosphorus in amounts ≤ 3 mg/l, in the colloidal state, as large particles of elemental phosphorus at concentrations >3 mg/l, or in the particle-sorbed state (Bullock and Newlands, 1969). Bioconcentration of white phosphorus in aquatic organisms does occur. In soil, white phosphorus will rapidly oxidize as long as oxygen is able to diffuse through (Bohn et al., 1970) and at ambient temperatures, this process can take as little as 2 days (Rodriguez et al., 1972).

NIOSH (1994) and the American Conference of Governmental Industrial Hygienists (ACGIH) (1996) have set the recommended exposure limit (REL) and TLV, respectively, at 0.1 mg/m³. The Environmental Protection Agency (EPA) has derived a chronic oral reference dose (RfD) of 0.00002 mg/kg/day (IRIS, 1993) and listed white phosphorus as a group D carcinogen: not classifiable as a human carcinogen (IRIS, 1995).

PHOSPHORUS PENTACHLORIDE

Found as a white to pale-yellow, crystalline solid, phosphorus pentachloride (PCl_5, CASRN 10026-13-8) is also known as pentachlorophosphorus, phosphoric chloride, and phosphorus perchloride. Noncombustible, phosphorus pentachloride has a very pungent, unpleasant odor with a density of 2.1 g/cm³, a melting point of 148 °C, and a sublimation point of 160.5 °C. The vapors of phosphorus pentachloride decompose in the presence of moisture with the subsequent liberation of hydrochloric and phosphoric acids (Henderson and Haggard, 1943).

Phosphorus pentachloride is used in a wide array of industrial processes. These include the manufacture of chlorophophazenes, penicillin, and cephalosporin antibiotics, industrial catalyst, a mild chlorinating agent, aluminum metallurgy and as a precursor to Mustard gas and O-Mustard. In 2008, NIOSH estimated that approximately 2500 American workers are potentially exposed to phosphorus pentachloride (NIOSH, 2008). OSHA (USDOL, 2014) and NIOSH (2005)

have set the permissible exposure level (PEL) and REL at 1 mg/m³, respectively. ACGIH (2008) established the TLV time-weighted average TWA (TLV–TWA) for phosphorus pentachloride at 0.1 mg/m³. The NIOSH IDLH for phosphorus pentachloride is 70 mg/m³ (NIOSH, 2005). Occupational exposure to phosphorus pentachloride may occur through dermal contact with this compound at workplaces where phosphorus pentachloride is produced or used.

The primary effects of phosphorus pentachloride exposure are eye (conjunctiva) and respiratory passage irritation, deterioration of the mucous membranes of the mouth, throat, and esophagus. Subjects with eye exposures may also experience aversions to light (photophobia) and excessive tear production (lacrimation). Respiratory effects also include rhinitis, cough and dyspnea, mild bronchial spasm, respiratory distress, and pulmonary edema (Sullivan and Krieger, 1992). Bronchitis and bronchopneumonia can also occur. As the exposure concentration to this compound increases, gastrointestinal symptoms and central nervous system effects may develop. Extremely high vapor concentrations may cause tachypnea, tachycardia, arrhythmia, and death as a result of respiratory paralysis.

PHOSPHORUS TRICHLORIDE

Phosphorus trichloride (PCl_3, CASRN 7719-12-2), also called phosphorus chloride, is a clear liquid with a very pungent, hydrochloric acid-like smell. It is a corrosive mostly used in the production of organophosphorus compounds. Listed in schedule 3, it is controlled under the Chemical Weapons Convention. Phosphorus trichloride has a melting point of −93.6 °C, a boiling point of 76.1 °C, and a density of 1.57 g/cm³. It can react with water to form phosphoric acid, hydrochloric acid, and phosphorus oxides.

Phosphorus trichloride is an important intermediate in the production of insecticides, herbicides, and organophosphorus pesticides as well as other chemicals such as phosphoryl chloride, phosphorus pentachloride, thiophosphoryl chloride, and phosphonic acid. It is also used in the production of synthetic surfactants, phosphites, gasoline additives, flame retardants, and silver polish.

OSHA (USDOL, 2014) has set the PEL for phosphorus pentasulfide at 0.5 ppm while NIOSH (2005) and ACGIH (2008) have respectively set the REL and the TLV at 0.2 ppm. ACGIH has set a STEL of 0.5 ppm. The NIOSH IDLH of phosphorus pentachloride is 25 ppm (NIOSH, 2005). An irritant of the eyes, skin, nose, and throat, phosphorus trichloride can also cause eye and skin burns. Other ocular effects include lachrymation and loss of vision. Central nervous system effects include dizziness and headache. Exposure of the respiratory system to phosphorus trichloride can produce cough, sore throat, bronchitis, and even pneumonia. Nausea, vomiting, and abdominal pain are common digestive tract effects.

PHOSPHORUS PENTASULFIDE

Phosphorus pentasulfide (P_2S_5/P_4S_{10}, also known as phosphorus persulfide, CASRN 1314-80-3) is a crystal solid that can be gray, yellow, or green in color with a strong rotten egg smell. It is extremely flammable and can spontaneously ignite because it is very hygroscopic in the presence of moisture. Chemical properties include a melting point of 285 °C (with a vapor pressure of 1 mm), a boiling point of 515 °C, and a density of 2.03 g/cm^3. Phosphorus pentasulfide can be burnt to form phosphorus pentoxide and sulfur dioxide. In water, it will decompose forming H_3PO_4 and H_2S.

Phosphorus pentasulfide is used in the manufacture of lube oil additives, ignition compounds and safety matches, insecticides (e.g., parathion and malathion), and pesticides. It can also be used as a desiccant. OSHA (2014), NIOSH (2005), and ACGIH (2008) have respectively set the PEL, REL, and TLV for phosphorus pentasulfide at 1 mg/m^3. OSHA and ACGIH have set a STEL of 3 mg/m^3. The NIOSH IDLH for phosphorus pentachloride is 250 mg/m^3 (NIOSH, 2005). Potential symptoms of overexposure are irritation of the eyes, skin, and respiratory system. Ocular effects include tearing, aversion to light, conjunctival pain or keratoconjunctivitis, and corneal vesiculation. Prolonged exposure can produce disturbances to the gastrointestinal tract and central nervous system effects such as headache, dizziness, fatigue, convulsions, apnea, and coma.

PHOSPHORUS OXYCHLORIDE

Phosphorus oxychloride (CASRN 10025-87-3, $POCl_3$) is a clear, colorless liquid, which can react with water vapor to form fumes in the air. It is remarkable with regard to its pungent and musky odor. Chemical properties include a melting point of 1.25 °C, a boiling point of 105.8 °C, and a density of 1.645 g/cm^3. It reacts with water to produce products that rapidly corrode steel and most metals.

Phosphorus oxychloride is an important intermediate in the production of triarylphosphate esters (e.g., triphenyl phosphate and tricresyl phosphate), which have been used as flame retardants and plasticizers for PVC. It is acutely toxic to the eyes, throat, and respiratory tract. Phosphorus oxychloride is also used in nuclear reprocessing and the semiconductor industry. OSHA vacated a 1989 PEL of 0.6 mg/m^3 (HSDB, 2008). NIOSH (2005) and ACGIH (2008) have respectively set the REL and TLV for phosphorus oxychloride at 0.6 mg/m^3. NIOSH has also recommended a 15-min STEL of 0.5 mg/m^3. A NIOSH IDLH for phosphorus oxychloride has not yet been determined (NIOSH, 2005).

Acute exposure to phosphorus oxychloride can irritate the eyes and cause conjunctivitis as well as burning of the cornea, skin, and mucus lining of the mouth. Phosphorus oxychloride is also a respiratory irritant that can acutely cause cough, dyspnea, lung edema, and bronchopneumonia. Meanwhile, chronic respiratory exposures cause bronchitis, bronchial pain, and emphysema. Headache, dizziness, abdominal pain, nausea, vomiting, and nephritis have also resulted from exposure.

PHOSPHORUS ANHYDRIDE/PHTHALIC ANHYDRIDE

Phosphorus anhydride (CASRN 85-44-9, $C_6H_4(CO)_2O$) is the anhydride of phthalic acid and is found as white flakes or needles used in the production of plasticizers. It has a mild, acrid odor with a melting point of 130.8 °C, a boiling point of 295 °C, and a density of 1.53 g/cm^3 (as a solid). It reacts with water to form products that rapidly corrode steel and most metals.

In the plastics industry, phosphorus anhydride is important for the production of polyvinyl chloride (PVC) and flexible PVC products such as cables, pipes, and hoses. Other manufacturing uses for phosphorus anhydride include the production of polyester resins, dyes, urethane polyester polyols, rubber flame retarders, insecticides, other synthetic fibers, and leather processing. OSHA (USDOL, 2014) has set a PEL for phosphorus anhydride at 12 mg/m^3 while NIOSH (2005) and ACGIH (2008) have respectively set their REL and TLV at 6 mg/m^3. The NIOSH IDLH for phosphorus anhydride has been set at 60 mg/m^3 (NIOSH, 2005). ACGIH has listed phosphorus anhydride as A4, not classifiable as a human carcinogen (ACGIH, 2008).

The environmental fate of phosphorus anhydride is dependent on the media into which it is mixed. In water, phosphorus anhydride hydrolyzes very rapidly (Towle et al., 1968) and has a half-life of <3 min (Hawkins, 1975). Due to its hydrolyzation in water, it is expected that phosphorus anhydride would also have a short half-life in moist and damp soils. In air, at ambient temperatures, phosphorus anhydride is found as a vapor that has a half-life of approximately 3 weeks (Bidleman, 1988; Meylan and Howard, 1993). Due to its short half-life in water, bioaccumulation of phosphorus anhydride is not likely to occur.

As discussed previously for many of the phosphorus-based compounds, phosphorus acid is an eye, skin, and respiratory irritant. Exposures can produce work-related rhinitis and conjunctivitis (Nielsen et al., 1988). However, workers exposed up to 17 mg/m^3 of phosphorus anhydride did not have any measurable abnormal lung effects (Nielsen et al., 1991). First time exposures irritate the respiratory tract causing sneezing, burning sensations in the nose and throat, cough, and increased mucus secretion. Chronic exposures cause inflammation of the respiratory tract, nose bleeds and ulcers, anosmia, and bronchitis.

Nielsen et al., (1988) also showed an effect on the immune system of exposed workers. Workers classified as having high levels of exposure to phosphorus anhydride had significantly higher levels of phosphorus anhydride-specific IgG than workers with lower exposure levels. There were no significant differences in levels of phosphorus anhydride-specific IgE or IgM. Subjects suffering from occupational asthma had significantly higher levels of phosphorus anhydride-specific IgG.

PHOSPHINE AND PHOSPHIDES

Distinct by its odor of strong garlic or that of rotting fish, phosphine (phosphorus trihydride, CASRN 7803-51-2, PH_3) is atypical of the other compounds described in this chapter. Phosphine can exist as a colorless liquid or as a colorless gas in air. As mentioned above, phosphine has a very distinct odor and has a half-life of about 24 h in air as a gas. It has a melting point of $-133\,°C$, a boiling point of $-87.7\,°C$, and a density of 1.529 g/l. Phosphine is a highly flammable gas that auto ignites at $38\,°C$. Denser than air, phosphine gas will collect in shallow areas and produce phosphorus pentoxide (which, as explained earlier in this chapter, is a respiratory irritant) when burnt. Upon hydrolysis of sodium phosphide, phosphine gas can be released. This reaction can result in fires because it is so exothermic. Phosphine can also be released when calcium phosphide comes in contact with acids or water.

Phosphine is not an actual product with much direct commercial value. Important in the production of semiconductors and plastics, phosphine is a rare source of phosphorus poisoning resulting from the accidental formation of the gas during different industrial applications. Prepared in a variety of ways, phosphine can be generated by reacting white phosphorus with sodium hydroxide. Phosphine can also be produced as a by-product generated during the synthesis of hydriodic acid from iodine and red phosphorus. When red phosphorus reacts with acids or caustics in the presence of a metal, phosphine gas is generated. Phosphine is also formed during the production of methamphetamine (Willers-Russo, 1999; Burgess, 2001).

Occupational phosphine exposure can occur via various forms of metallic phosphides. Aluminum phosphide is used as an insecticide fumigant and use in controlling of *Aspergillus* species, thereby decreasing aflatoxin content of certain grains (Leitao et al., 1987), in animal feeds, tobacco storage, processed foods, and other nonfood products. Aluminum, calcium, magnesium, and zinc phosphides are all used as fumigants in rodenticides. Zinc phosphide is used in the manufacture of photovoltaic cells. Indium phosphide is used in the manufacture of semiconductors and as an intermediate

in flame retardant production. Calcium phosphide is used in autogenous welding and signal flares. Finally, iron phosphide and titanium (III) phosphide are semiconductors used in high power, high frequency applications and in laser and other photo diodes.

OSHA (USDOL, 2014), NIOSH (2005,) and ACGIH (2008) have respectively set their PEL, REL, and TLV for phosphine at 0.3 ppm. The NIOSH IDLH for phosphorus anhydride has been set at 50 ppm (NIOSH, 2005). EPA has classified phosphine as a Group D, not classifiable as to human carcinogenicity (IRIS, 1999) and has recommended an RfD of 0.021 mg/kg/day and RfC of 0.0003 mg/m^3.

Phosphine gas, phosphine derived from reactions using phosphides and phosphides themselves can all cause deleterious health effects in humans. The main route of exposure to phosphine is through inhalation but ocular and dermal effects due to contact and accidental ingestion have also been reported. Acute exposure to phosphine can cause a plethora of effects to the GI tract (e.g., nausea, vomiting, abdominal pain, diarrhea), cardiovascular system (e.g., chest pressure, arrhythmia), respiratory system (e.g., cough with fluorescent green sputum, dyspnea, pulmonary edema), central nervous system (e.g., stupor or syncope, headache, dizziness, fatigue, drowsiness, tremors, convulsions), integumentary system (e.g., chills and frostbite from liquid), numbness and paresthesia in fingers (from touching tablets) and burns. Chronic phosphine exposure is similar to that observed for chronic elemental phosphorus exposure and can produce toothache, phossy jaw, spontaneous fractures of bones, and anemia.

Phosphine gas inhalation can rapidly cause death due to cardiac collapse (Blanke 1970). Exposure to phosphine was shown to be genotoxic in humans. Applicators handling fumigant-containing phosphine were found to have an increase in stable chromosome rearrangements in the applicators (Garry et al., 1989). Exposure to grain that had been fumigated with aluminum phosphide caused the death of a pregnant worker (Garry et al., 1993).

Metal phosphide exposure can also provoke toxic effects in workers. Acute metal phosphide poisoning occurs from ingestion as a result of the formation of phosphine gas upon contact with fluids in the gut. Accidental aluminum phosphide ingestion has been shown to be more than 70% fatal in humans (Anand et al., 2011). Severe inhalation of aluminum phosphide can produce convulsions, acute respiratory distress syndrome (ARDS), heart arrhythmia and failure, and coma. Ntelios et al. (2013) reported a case of leukopenia induced in a worker after exposure to aluminum phosphide.

Phosphine degrades very quickly in all biota and does not bioaccumulate or biomagnify. When found in water, it will move quickly to air where it reacts to form less harmful chemicals. It has been estimated that phosphine in the troposphere has a half-life of <1 day (Fritz et al. 1982). In air, diphosphine (PH_2) can be produced as an intermediate from phosphine via a hydrogen abstraction reaction.

Diphosphine can then react with ozone to produce H_2PO from which hypophosphorus acid (H_2PO2) may be produced. The hypophosphorus acid is ultimately oxidized to phosphorus and phosphoric acid.

Due to its low solubility in water, phosphine will either oxidize or volatize rapidly from water (Lai and Rosenblatt, 1977). It will even oxidize rapidly under anoxic condition and can form diphosphine (P2H4), phosphorus, hypophosphorous acid, phosphorus acid, and phosphoric acid (Kumar et al., 1985). In the presence of oxygen, phosphine will oxidize to form P_2H_4 and finally form phosphoric acid (Garry et al., 1993). Small traces of phosphine can adsorb or chemisorb to soil particles and sediments in water (Berck and Gunther, 1970; Hilton and Robison, 1972). However, volatilization is expected to be the most important loss process for phosphine in water.

REFERENCES

American Conference of Governmental Industrial Hygienists (ACGIH) (1996) *Threshold Limit Values (TLVs) for Chemical Substances and Physical Agents and Biological Exposure Indices (BED),PL Cincinnati, OH*: American Conference of Governmental Industrial Hygienists.

American Conference of Governmental Industrial Hygienists (ACGIH) (2008) *Threshold Limit Values for Chemical Substances and Physical Agents and Biological Exposure Indices*, Cincinnati, OH: American Conference of Governmental Industrial Hygienists, p. 48.

Anand, R., Binukumar, B.K., and Gill, K.D. (2011) Aluminum phosphide poisoning: an unsolved riddle. *J. Appl. Toxicol.* 31(6): 499–505.

Berck, B. and Gunther, F.A. (1970) Rapid determination of sorption affinity of phosphine by fumigation within a gas chromatographic column. *J. Agric. Food Chem.* 18:148–153.

Berkowitz, J.B., Young, G.S., Anderson, R.C., Colella, A.J., Lyman, W.J., Preston, A.L., Steber, W.D., Thomas, R.G., and Vranka, R. (1981) *Research and development for health and environmental hazard assessment, Task Order 5. Occupational and environmental hazards associated with the formulation and use of white phosphorus-felt and red phosphorus-butyl rubber screening smokes*. Cambridge, MA: Arthur D. Little, Inc. DAMD17-79-C-9139. AD-Al 16956.

Bidleman, T.F. (1988) Atmospheric processes. *Environ. Sci. Technol.* 22:361–367.

Blanke, R.V. (1970) Toxicology. In: Tietz, N.W., editor. *Fundamentals of Clinical Chemistry*, Philadelphia, PA: W.B. Saunders Co., pp. 833–889.

Bohn, H.L., Johnson, G.V., and Cliff, J.H. (1970) Detoxification of white phosphorus in soil. *J. Agric. Food Chem.* l8:1172–1173.

Bullock, E. and Newlands, M.J. (1969) Decomposition of phosphorus in water. In: *Proceedings of the Chemical Institute of Canada Conference on Pollution, Halifax, NS, August 24–26*, pp. 23–24.

Burgess, J.L. (2001) Phosphine exposure from a methamphetamine laboratory investigation. *J. Toxicol. Clin. Toxicol.* 39(2): 165–168.

Emsley, J. (2002) *The 13th Element: The Sordid Tale of Murder, Fire, and Phosphorus*, New York, NY: John Wiley & Sons, p. 352.

Fritz, B., Lorenz, K., Steinert, W., and Zellner, R. (1982) Laboratory kinetic investigations of the tropospheric oxidation of selected industrial emissions. In: Versino, B. and Ott, H., editor. *Physico-Chemical Behaviour of Atmospheric Pollutants*. Reidel Publishing Company, pp. 192–202.

Garry, V.F., Good, P.F., and Manivel, J.C., et al. (1993) Investigation of a fatality from nonoccupational aluminum phosphide exposure: measurement of aluminum in tissue and body fluids as a marker of exposure. *J. Lab. Clin. Med.* 122(6):739–747.

Garry, V.F., Griffith, J., and Danzl, T.J., et al. (1989) Human genotoxicity: pesticide applicators and phosphine. *Science* 246:251–255.

Hawkins, M.D. (1975) Hydrolysis of phthalic and 3,6-dimethylphthalic anhydrides. *J. Chem. Soc. Perkin Trans. 2* 75:282–284.

Henderson, Y. and Haggard, H.W. (1943) *Noxious Gases*, 2nd ed., New York, NY: Reinhold Publishing Corporation, p. 134.

Hilton, H.W. and Robison, W.H. (1972) Fate of zinc phosphide and phosphine in the soil–water environment. *J. Agric. Food Chem.* 20:1209–1213.

Hazardous Substances Database (HSDB) (2008) *Hazardous Substances Data Base*, Bethesda, MD: Toxnet, National Library of Medicine, National Toxicology Information Program.

IRIS (1993) *Integrated Risk Information System*, Washington, DC: U.S. Environmental Protection Agency.

IRIS (1995) *Integrated Risk Information System*, Washington, DC: U.S. Environmental Protection Agency.

IRIS (1999) *Integrated Risk Information System (IRIS) on Phosphine*. Washington, DC: National Center for Environmental Assessment, Office of Research and Development.

Kumar, M.D., Somasundar, K., and Rajendran, A. (1985) Stability of phosphorus species in seawater. *Indian J. Mar. Sci.* 14:20–23.

Lai, M.G. and Rosenblatt, D.H. (1977) *Identification of Transformation Products of White Phosphorus in Water*, Silver Spring, MD: Naval Surface Weapons Center, Final Report No. NSWC/WOL/TR-76-103. ADA041068.

Leitao, J., de Saint-Blanquat, G., and Bailly, J.R. (1987) Action of phosphine on production of aflatoxins by various *Aspergillus* strains isolated from foodstuffs. *Appl. Environ. Microbiol.* 53:2328–233l.

Meylan, W.M. and Howard, P.H. (1993) Computer estimation of the atmospheric gas-phase reaction rate of organic compounds with hydroxyl radicals and ozone. *Chemosphere* 26: 2293–2299.

National Institute for Occupational Safety and Health (NIOSH) (1994) *Pocket Guide to Chemical Hazards*, NIOSH 94–116, Washington, DC: US Department of Health and Human Services.

National Institute for Occupational Safety and Health (NIOSH) (2005) *Pocket Guide to Chemical Hazards & Other Databases (CD-ROM)*, Department of Health & Human Services, Centers for Disease Prevention & Control, National Institute for Occupational Safety & Health, DHHS (NIOSH) Publication No. 2005-151.

National Institute for Occupational Safety and Health (NIOSH) (2008) *International Safety Cards. Phosphorus Pentachloride. 10026-13-8*. Available at http//www.cdc.gov/niosh/ipcs/nicstart .html.

Nielsen, J. et al. (1988) Specific serum antibodies against phthalic anhydride in occupationally exposed subjects. *J. Allergy Clin. Imnunol.* 82:126–133.

Nielsen, J. et al. (1991) Serum IgE and lung function in workers exposed to phthalic anhydride. *Int. Arch. Occup. Environ. Health* 63:199–204.

Ntelios, D., Mandros, C., Potolidis, E., and Fanourgiakis, P. (2013) Aluminium phosphide-induced leukopenia. *BMJ Case Rep*. doi: 10.1136/bcr-2013-201229.

Rodriguez, A., Bohn, H.L., and Johnson, G.V. (1972) White phosphorus as a phosphate fertilizer. *Soil Sci. Soc. Am. Proc.* 36:364–366.

Sullivan, J.B., Jr. and Krieger,G.R., editors. (1992) *Hazardous Materials Toxicology—Clinical Principles of Environmental Health*, Baltimore, MD: Williams & Wilkins, p. 938.

Towle, P.H., et al. (1968) *Kirk-Othmer Encyclopedia of Chemical Technology*, C 2nd ed., vol. 15, p. 444.

USDOL (2014) *29 CFR 1910.1000: U.S. National Archives and Records Administration's Electronic Code of Federal Regulations*. Available at http://www.gpoaccess.gov/ecfr (last accessed January 14, 2014).

Willers-Russo, L.J. (1999) Three fatalities involving phosphine gas, produced as a result of methamphetamine manufacturing. *J. Forensic Sci.* 44(3):647–652.

52

SULFUR COMPOUNDS

Frank Stephen and M. Rony Francois

First aid: See text for each substance

Occupational exposure limits: *Sulfur dioxide:* 5.24 mg/m^3 (OSHA PEL, NIOSH REL, ACGIH TLV); 261.76 mg/m^3 (OSHA IDLH); *sulfuric acid:* 1 mg/m^3 (OSHA PEL, NIOSH REL, ACGIH TLV); 15 mg/m^3 (OSHA IDLH); 3 mg/m^3 (ACGIH STEL); *hydrogen sulfide:* 13.91 mg/m^3 (OSHA PEL, NIOSH REL, ACGIH TLV); 139.06 mg/m^3 (NIOSH IDLH); 20.86 (ACGIH STEL)

Reference values: *Hydrogen sulfide:* 2×10^{-3} mg/m^3 (IRIS RfC)

Hazard statements: *Sulfur dioxide:* H280—contains gas under pressure; may explode if heated; H314—causes severe skin burns and eye damage; H331—toxic if inhaled. *Sulfuric acid*: H314–causes severe skin burns and eye damage. *Hydrogen sulfide:* H220—extremely flammable gas; H280—contains gas under pressure; may explode if heated; H330—fatal if inhaled; H400—very toxic to aquatic life

Precautionary statements: *Sulfur dioxide:* P280—wear protective gloves/protective clothing/eye protection/face protection; P304+P340—if inhaled, remove victim to fresh air and give instructions to the victim to rest in a position comfortable for breathing; P305+P351+P338—if in eyes, rinse cautiously with water for several minutes. Remove contact lenses, if present and easy to do. Continue rinsing; P310— immediately call a poison center or doctor/physician; P410+P403—protect from sunlight; store in a well-ventilated place. *Hydrogen sulfide:* P210—keep away from heat/sparks/open flames/. . . /hot surfaces...no smoking; P273—avoid release to the environment; P284—wear respiratory protection; P310—immediately call a poison center or doctor/

physician; P410+P403—protect from sunlight. Store in a well-ventilated place; P501—dispose of contents/container

Risk phrases: *Sulfur dioxide:* R23—toxic by inhalation (>20%); R20—harmful by inhalation (>5%, <20%); R34—causes burns (>5%); R36/37/38—irritating to the eyes, respiratory system, and skin (>0.5%). *Sulfuric acid:* R35—causes severe burns (>15%); R36/38—irritating to eyes and skin (>5%, <15%). *Hydrogen sulfide:* R12—extremely flammable; R26—very toxic by inhalation; R50—very toxic to aquatic organisms

Safety phrases: *Sulfur dioxide:* S9—keep container in a well-ventilated place; S26—in case of contact with eyes, rinse immediately with plenty of water and seek medical advice; S36/37/39—wear suitable protective clothing, gloves, and eye/face protection; S45—in case of accident or if you feel unwell, seek medical advice immediately (show the label where possible). *Sulfuric acid:* S1/2—keep locked up and out of reach of children; S26—in case of contact with eyes, rinse immediately with plenty of water and seek medical advice; S30—never add water to this product; S45—in case of accident or if you feel unwell, seek medical advice immediately (show the label where possible). *Hydrogen sulfide:* S1/2—keep locked up and out of reach of children; S9—keep container in a well-ventilated place; S16—keep away from sources of ignition—no smoking; S36—wear suitable protective clothing; S38—in case of insufficient ventilation, wear suitable respiratory equipment; S45—in case of accident or if you feel unwell, seek medical advice immediately (show label where possible); S61—avoid release to the environment. Refer to special instructions/safety data sheets

Hamilton & Hardy's Industrial Toxicology, Sixth Edition. Edited by Raymond D. Harbison, Marie M. Bourgeois, and Giffe T. Johnson.
© 2015 John Wiley & Sons, Inc. Published 2015 by John Wiley & Sons, Inc.

HISTORY

Anyone familiar with the smell of rotting eggs or well water of poor quality can attest that they are familiar with brimstone. However, the common element to all these three is sulfur. In its raw form, this yellow, nontoxic crystal poses no general threat to humans. Known and used for over 2000 years, it has been used in fumigants, balms, and antiparasitics. Commercially, it is commonly used today in fertilizers, matches, insecticides, and fungicides. It is a nonmetal with a broad range of valence states (from −2 to +6), allow it to form stable compounds with a large number of other elements, except the noble gases. Many of these compounds have very beneficial uses; however, some of these compounds produce deleterious health effects.

SULFUR DIOXIDE

BACKGROUND AND USES

Sulfur dioxide (CASRN 7446-09-5) is an irritant gas commonly found in the environment because of its production during petroleum refining and combustion as well as paper production. It is also known as sulfur oxide, sulfurous acid anhydride, sulfurous anhydride, and sulfurous oxide. It is a nonflammable and colorless gas that is heavier than air at room temperature. It possesses a strong, pungent odor that is a warning sign of its presence. Absorbed quite readily through the upper respiratory tract, sulfur dioxide does not appear to be absorbed dermally. Oral ingestion is possibly due to its presence in some foods. Some of the industrial uses for the approximately 300,000 tons of sulfur dioxide produced each year include oil, metal, and ore refining; refrigeration; waste and water treatment; wood pulp and paper bleaching, fumigation; tanning; metal casting; and the production of sodium sulfate and sulfuric acid. Sulfur dioxide is listed as a U.S. Environmental Protection Agency (EPA) priority air pollutant. The number of American workers potentially exposed occupationally to sulfur dioxide has been estimated at 600,000 (HSDB, 1998).

Naturally emitted from fires and volcanic eruptions, sulfur dioxide is always present in the earth's atmosphere and inhalation is the most common, yet harmless type of exposure. The highest levels of inhalation exposure would be expected to occur near industrial activities that involve fuel combustion. Historically, incidences of general population exposure from industrial activity occurred in London, Meuse Valley, Belgium and Donora, Pennsylvania. In each of these "fog" events, there was an increase in inhalation-induced mortality due to releases of sulfur dioxide, in the presence of other chemicals. Similar episodes occur frequently in winter, especially in densely populated urban areas, possibly because of the higher demand for heat. "Smog" can also occur due to stagnant air masses and atmospheric inversion. During these episodes the sulfur dioxide reacted with the moisture found in fog. This reaction produces sulfurous acid (H_2SO_3), which can irritate the respiratory epithelium and mucous membranes upon inhalation.

PHYSICAL AND CHEMICAL PROPERTIES

Sulfur dioxide (SO_2) is a colorless gas that has a very strong pungent odor at room temperature detectable by most people at levels down to 0.3–1 parts per million (ppm). It is neither flammable nor combustible and has a melting point of −72 °C, a boiling point of −10 °C, and a vapor pressure of 3.2 atm. It is 2.26 times heavier than air. It is 10% soluble in water by weight and easily forms sulfurous acid and sulfuric acid on contact with water and moist surfaces. It is shipped as a liquefied compressed gas (NIOSH, 1994).

ENVIRONMENTAL FATE AND BIOACCUMULATION

Due to the industrial uses of sulfur dioxide, releases are considered to be atmospheric (HSDB, 1998). Once emitted into the air, it remains in the gaseous phase and can move through various biota, including soil, water, grass, and vegetation (WHO, 1979). Oxidation of sulfur dioxide in the atmosphere with water forms sulfuric acid, which can fall as acid rain. Since it has high solubility in water, the oceans are considered to be a sink for sulfur dioxide (Kellogg et al., 1972). However, potential releases of sulfur dioxide from water would be expected to partition to the atmosphere (Kellogg et al., 1972; WHO, 1979). Soil can also be a site of fate for sulfur dioxide; however, the uptake levels would vary dependent on the soil pH and water content (HSDB, 1998). Falling acid rain will also cause increased soil mobility of heavy metals. Heavy metals form soluble sulfates in acidic soils and insoluble sulfate oxides or sulfate hydroxides when the pH of the soil is basic (Grzesiak et al., 1997).

MAMMALIAN TOXICOLOGY

Acute Effects

Due to the ability of sulfur dioxide to dissolve into the moisture on skin, eyes, and mucous membranes, the respiratory tract is the most likely target of effect. After exposure, sulfur dioxide dissolves in water to form sulfurous acid. At levels of 0.6–1.0 ppm, sulfur dioxide irritates the eyes, nose, and throat (Summer and Haponik, 1981). At these relatively low concentrations, the majority of the sulfur dioxide is deposited in the nose and oropharynx. However, at higher exposures, bronchoconstriction can be induced when sulfur dioxide reacts with water in the conduction portions of the lungs to form a bisulfite ion. Symptoms such as sore throat, sneezing, shortness of breath, wheezing, and chest tightness can occur after sulfur dioxide inhalation. Edema and pneumonitis can develop after long-term exposure; and in cases of high acute exposure due to inhalation, sulfur dioxide can lead to reactive airway dysfunction syndrome (RADS). Irritation after sulfur dioxide dermal exposure can cause pain, redness, and blisters. Ocular effects due to sulfur dioxide vapor include corneal burns and conjunctivitis. Finally, sulfur dioxide-induced gastrointestinal effects include nausea, vomiting, and abdominal pain. An industrial accident involving five healthy men exposed to varying concentrations of sulfur dioxide can be used as an example of the range of effects that the gas has on humans. In this reported incident, two men died within 5 min of the exposure with pink, frothy fluid around the mouths. Acute symptoms of the three survivors included chest tightness, dyspnea, and soreness of their eyes, nose, and throat. Rhonchi and wheezing were heard in their lungs, and chest radiographs were normal. On follow-up, one of these individuals developed airway obstruction with associated dyspnea on exertion. Another man with a smoking history developed mild airflow obstruction 4 months after exposure, and the third man who had no smoking history developed no lung abnormalities (Charan et al., 1979). As a rule, individuals who survive an acute exposure recover completely.

Exposure to low concentrations of sulfur dioxide can result in a mild degree of bronchial constriction. This is especially apparent in asthmatic individuals (Linn et al., 1987). Studies of workers chronically exposed to lower levels of sulfur dioxide show little alterations in lung function. It has been proposed that workers who remain in such occupations may adapt to the gas (Ferris et al., 1979).

First Aid

Upon exposure, it is imperative that affected individuals be removed from the site immediately. Contaminated clothing and personal belongings must also be removed and exposed skin, hair, face washed with large amounts of water. Any eye exposure should be treated by 5-min washes with saline or water. Persons displaying any respiratory compromise may require a patent airway secured by endotracheal intubation or tracheotomy. Bronchospasms may be treated with aerosolized bronchodilators. Advanced life-saving protocols may have to be used should an exposed individual become hypotensive, have a seizure, or become comatose. Intravenous saline should be given for any casualty who goes into shock.

Chronic Effects

Very few chronic effects after sulfur dioxide exposure have been noted. As described above the respiratory system is the main target of effect. Altered or lost sense of smell, respiratory infection susceptibility, chronic bronchitis symptoms, and pulmonary function decline can all become chronic after exposure.

Workers exposed to acid mists in steel processing, refinery, and other chemical plants were found to have an increased risk of laryngeal cancer (Ahlborg et al., 1981; Forastiere et al., 1987; Soskolne et al., 1984; Steenland et al., 1988). Some of the highest exposures occur when sulfur dioxide is a by-product, as in the metal smelting industry and in the processing or combustion of high sulfur coal or oil (HSDB, 1998).

MODE OF ACTION(S)

The mechanism by which sulfur dioxide causes its effects on the lungs is still being investigated. Sulfur dioxide-induced reflexive bronchoconstriction during inhalation of sulfur dioxide was demonstrated in humans and in cats (Frank et al., 1962; Nadel et al., 1965). It causes smooth muscle contraction through a parasympathetically vagal-mediated reflex (Roberts et al., 1982). Tachykinin-induced bronchoconstriction was demonstrated using capsaicin to prevent release after exposure to sulfur dioxide (Bannenberg et al., 1994; Wang et al., 1996). Increased airway resistance was shown to result from reduced buffer and H^+ ion absorption capacity of the airway mucus (Holma, 1985). Various studies using atropine indicated a role for the muscarinic (cholinergic) mechanism of action (Snashall and Baldwin, 1982; Nadel et al., 1965; Sheppard, 1988a, 1988b). However, non-cholinergic mechanisms attributed to prostaglandins and leukotrienes have been demonstrated in humans (Field et al., 1996; Lazarus et al., 1997).

As sulfur dioxide is breathed through the nasal passages at low concentrations, it is predominantly deposited in the nose and oropharynx. As it comes in contact with the moist mucous membranes, it is converted, as mentioned previously, to sulfurous acid or sulfuric acid. This can cause local irritation. As this area's capacity to accept the gas is exceeded, it will pass down to the trachea, bronchi, and alveoli. There, the acidic ions will be neutralized by the endogenous production of ammonia by the lungs (Larson et al., 1977). When the buffering ability of the mucous proteins and ammonia is saturated, the acid formed will gain access to the surrounding tissue and initiate adverse reactions.

In studies of cows, intracellular and intercellular edema occurs at pH levels below 6.5, and epithelial cells loosen from each other and the basement membrane at a pH below 6.0 (Holma et al., 1977). In other studies, H^+ ion absorption in the mucus from acid exposure was found to increase mucous viscosity and decrease mucous velocity (Lippmann et al., 1979). In addition to decreased mucous transport, increased mucous viscosity has been demonstrated to correlate with increased airway resistance (Keal, 1974) and reduced pulmonary gas exchange (Pham et al., 1973).

INDUSTRIAL HYGIENE

NIOSH (1994) and the American Conference of Governmental Industrial Hygienists (ACGIH) (ACGIH, 1996) have respectively set the recommended exposure limit (REL) and threshold limit value (TLV) at 2 ppm and the short-term exposure limit (STEL) at 5 ppm. The Occupational Safety and Health Administration (OSHA) has set the permissible exposure limit (PEL) at 5 ppm. Sulfur dioxide gas is considered immediately dangerous to life and health (IDLH set at 100 ppm).

RISK ASSESSMENTS

WHO

The World Health Organization (WHO) has set a 24-h exposure limit for sulfur dioxide inhalation at 100–150 µg/m³.

ATSDR

The Agency for Toxic Substances and Disease Registry (ATSDR) has established a 0.01 ppm minimum risk levels (MRLs) for acute inhalation exposures to sulfur dioxide (ATSDR, 1998).

IARC

The International Agency for Research on Cancer (IARC) has classified sulfur dioxide as a Group 3, not classifiable as to human carcinogenicity.

ACGIH

The ACGIH has listed sulfur dioxide as an A4 carcinogen, not classifiable as a human carcinogen (ACGIH, 1996).

EPA

The EPA has established a 24-h exposure limit National Ambient Air Quality Standards (NAAQS) of 365 µg/m³ for sulfur dioxide.

CalEPA

In July 2011, the California Environmental Protection Agency (CalEPA) listed sulfur dioxide as a chemical known to cause reproductive toxicity for purposes of the Safe Drinking Water and Toxic Enforcement Act of 1986, as listed under Proposition 65.

REFERENCES

Ahlborg, G., Jr., Hogstedt, C., Sundell, L., and Åman, C.-G. (1981) Laryngeal cancer and pickling house vapors. *Scand. J. Work Environ. Health* 7:237–240.

American Conference of Governmental Industrial Hygienists (ACGIH) (1996) *Threshold Limit Values (TLVs) for Chemical Substances and Physiologic Agents and Biological Exposure Indices (BEIs)*, Cincinnati, OH: American Conference of Governmental Industrial Hygienists.

Agency for Toxic Substances and Disease Registry (ATSDR) (1998) *Public Health Statement for Sulfur Dioxide*, Atlanta, GA: U.S. Department of Health and Human Services, Public Health Service, Agency for Toxic Substances and Disease Registry.

Bannenberg, G., Atzori, L., Xue, J., et al. (1994) Sulfur dioxide and sodium metabisulfite induce bronchoconstriction in the isolated perfused and ventilated guinea pig lung via stimulation of capsaicin-sensitive sensory nerves. *Respiration* 6(1):130–137.

Charan, N.B., Myers, C.G., and Lakshminarayan, S. (1979) Pulmonary injuries associated with acute sulfur dioxide inhalation. *Am. Rev. Respir. Dis.* 119:555–560.

Ferris, B.G., Jr., Paleo, S., and Chen, H.Y. (1979) Mortality and morbidity in pulp and paper mill in the United States: a ten year follow-up. *Br. J. Ind. Med.* 36:127–134.

Field, P.I., Simmul, R., Bell, S.C., et al. (1996) Evidence for opioid modulation and generation of prostaglandins in sulphur dioxide (SO_2)-induced bronchoconstriction. *Thorax* 51:159–163.

Forastiere, F., Valesini, S., Salimei, E., Magliola, M.E., and Perucci, C.A. (1987) Respiratory cancer among soap production workers. *Scand. J. Work Environ. Health* 13:258–260.

Frank, N.R., Amdur, M.O., Worcester, J., et al. (1962) Effects of acute controlled exposure to sulfur dioxide on respiratory mechanics in healthy male adults. *J. Appl. Physiol.* 17:252–258.

Grzesiak, P., Schroeder, G., and Hopke, W. (1997) Degradation of the natural environment resulting from the presence of sulphur compounds in the atmosphere. *Pol. J. Environ. Stud.* 6(4):45–48.

Holma, B. (1985) Influence of buffer capacity and pH-dependent rheological properties of respiratory mucus on health effects due to acid pollution. *Sci. Total Environ.* 41:101–123.

Holma, B. Lindegren, M. and Morkholdt, A (1977) pH effects on ciliomotility and morphology of respiratory mucosa. *Arch. Environ. Health* 32:216–226.

HSDB (1998) *Hazardous Substances Data Bank*, Bethesda, MD: National Library of Medicine, National Toxicology Information Program.

Keal, E.E. (1974) Criteria for the testing of mucolytic agents. *Scand. J. Respir. Dis.* 90:49–53.

Kellogg, W.W., Cadel, R.D., and Allen, E.R. (1972) The sulfur cycle: man's contributions are compared to natural sources of sulfur compounds in the atmosphere and oceans. *Science* 175:587–596.

Larson, T.V., et al. (1977) Ammonia in the human airways: neutralization of inspired acid sulfate aerosols. *Science* 197:161–163.

Lazarus, S.C., Wong, H.H., and Watts, M.J., et al. (1997) The leukotriene receptor antagonist zafirlukast inhibits sulfur dioxide-induced bronchoconstriction in patients with asthma. *Am. J. Respir. Crit. Care Med.* 156:1725–1730.

Linn, W.S., et al. (1987) Replicated dose response study of sulfur dioxide effects in the normal, atopic and asthmatic volunteers. *Am. Rev. Respir. Dis.* 136:1127–1134.

Lippmann, M., et al. (1979) Effects of sulfuric acid mists on mucociliary bronchial clearance in healthy nonsmoking humans. In: *Proceedings of the Conference of Aerosols in Science, Medicine, and Technology*, pp. 157–162.

Nadel, J.A., Salem, H., Tamplin, B., et al. (1965) Mechanism of bronchoconstriction during inhalation of sulfur dioxide. *J. Appl. Physiol.* 20:164–167.

National Institute for Occupational Safety and Health (NIOSH) (1994) *Pocket Guide to Chemical Hazards*, Washington, DC: US Department of Health and Human Services.

Pham, Q.T., et al. (1973) Respiratory function and the rheological status of bronchial secretions collected by spontaneous expectoration and after physiotherapy. *Bull. Physiopathol. Respir.* 9:293–314.

Roberts, A.M., Hahn, H.L., and Schultz, J.A. (1982) Afferent vagal C-fibers are responsible for the reflex airway constriction and secretion evoked by pulmonary administration of SO_2 in dogs. *Physiologist* 25:226 (Abstract).

Sheppard, D. (1988a) Mechanisms of airway responses to inhaled sulfur dioxide. In: Loke, J., editor. *Lung Biology in Health and Disease*, vol. 34, New York, NY: Marcel Dekker, pp. 49–65.

Sheppard, D. (1988b) Sulfur dioxide and asthma—a double-edged sword? *J. Allergy Clin. Immunol.* 82:961–964.

Snashall, P.D. and Baldwin, C. (1982) Mechanisms of sulfur dioxide induced bronchoconstriction in normal and asthmatic man. *Thorax* 37:118–123.

Soskolne, C.L., Zeighami, E.A., Hanis, N.M., Kupper, L.L., Herrmann, N., Amsel, J., Mausner, J.S., and Sellman, J.M. (1984) Laryngeal cancer and occupational exposure to sulfuric acid. *Am. J. Epidemiol.* 120:358–369.

Steenland, K., Schnoor, T., Beaumont, J., Halperin, W., and Bloom, T. (1988) Incidence of laryngeal cancer and exposure to acid mists. *Occup. Environ. Med.* 45:766–776.

Summer, W. and Haponik, E. (1981) Inhalation of irritant gases. *Clin. Chest Med.* 2:273–287.

Wang, A.L., Blackford, T.L., and Lee, L.Y. (1996) Vagal bronchopulmonary C-fibers and acute ventilatory response to inhaled irritants. *Respir. Physiol.* 104:231–239.

WHO (1979) *Environmental Health Criteria 8: Sulfur Oxides and Suspended Particulate Matter*, Geneva: World Health Organization.

SULFURIC ACID

BACKGROUND AND USES

One of the most widely used compounds in domestic industry, sulfuric acid (CASRN 7664-93-9) is a colorless to dark brown, odorless liquid. Most commonly, almost 70% of it is used in the fertilizer industry to convert phosphate rock to phosphoric acid (IARC, 1992). Other industrial uses include mineral and ore processing; petroleum refining; production of paper, synthetic rubber, and plastics; wastewater processing, pickling of steel and iron; manufacture of organic sulfonates used in detergents and lubricants; dehydration procedures; and charging of storage batteries (NIOSH, 1974; Budavari, 1989; IARC, 1992). It is also found in common household products such as drain cleaners and other cleaning agents, lead–acid batteries and has been used as a general food additive (HSDB, 1998). The National Institute for Occupational Safety and Health (NIOSH) estimated that approximately 750,000 Americans are occupationally exposed to sulfuric acid (NOES, 1990).

PHYSICAL AND CHEMICAL PROPERTIES

Sulfuric acid (H_2SO_4, CASRN 7664-93-9) is a colorless to dark brown, oily, odorless, and nonflammable liquid. It has a melting point of $10\,°C$, a boiling point of $337\,°C$, and a density of $1.84\,g/cm^3$. Below $10.5\,°C$, the pure compound is found in a solid form. Sulfuric acid is miscible in water.

ENVIRONMENTAL FATE AND BIOACCUMULATION

Upon entering the environment, sulfuric acid enters the natural sulfur cycle. As explained in the section on sulfur dioxide, sulfuric acid is formed when sulfur dioxide mixes with water in the atmosphere. Resident time in the atmosphere can last from 6 (surface air) to 14 days (15 km), and sulfuric acid is the main component of acid rain. Sulfur dioxide is oxidized to sulfur trioxide and then hydrolyzed in the atmosphere, forming sulfuric acid (Treon et al., 1950). Atmospheric sulfur oxides can acidify precipitation to a pH of <5.6 (Chaney et al., 1980). Sulfuric acid can be removed from the atmosphere through wet and dry deposition processes. Acid precipitation is most detrimental to the aquatic ecosystems, marine biota, and some terrestrial species of plants and trees. After falling on the earth and mixing with water, sulfuric acid will dissociate to form a sulfate anion that can combine with other cations. Once in soil, these ions will adsorb to the soil particles or can directly leach into surface water and groundwater. Sulfates will then be taken up by and incorporated into the plant parenchyma. While sulfur is an important constituent found in proteins, food chain bioaccumulation of sulfuric acid is not considered important, and no reports were found.

MAMMALIAN TOXICOLOGY

Acute Effects

Human exposure to sulfuric acid occurs from dermal contact and inhalation with the preliminary targets being the skin, eyes, and respiratory tract. Exposure to sulfuric acid produces many of the same effects described above for sulfur dioxide but can be more severe. Interestingly, these effects occur at lower concentrations of sulfuric acid and shorter exposure times. Data from 905 industrial releases show that the most frequently reported effects were chemical burns, respiratory, and eye irritation (HSEES, 1997).

Ocular burns that resulted in coagulation necrosis, which limits deeper penetration into the cornea, occur after contact with high concentrations of sulfuric acid. However, blindness can still occur. Skin contact with high concentrations of sulfuric acid results in skin burns, the severity of which is directly related to the concentration and quantity of acid, the duration of skin contact, and the extent of skin penetration. Acids are water soluble and able to penetrate the subcutaneous tissue lowering the pH, affecting both the cells and intracellular matrices. Inhalation effects are comparable to those previously described for sulfur dioxide. Studies reporting any effect on the gastrointestinal tract, skin, or musculoskeletal system after inhalation of sulfuric acid have not been found.

First Aid

First aid procedures for exposure to sulfuric acid start with immediate removal of the victim from the source. Initial treatment protocols involve immediate removal of all contaminated clothing followed by continuous washing of exposed skin with water or normal saline for at least 15 min. Exposed persons should also be checked along the hairline and scalp, under the nails, and between the toes for affected areas. After washing to cleanse the area, treatment of minor burns can include application of a topical cream containing antibiotics and a tetanus immunization. If burns are severe, then consultation with a medical specialist may be necessary. After ocular exposure, any debris or particulate matter should be removed from the eye and, if possible, the pH of tears should be checked with litmus paper. Irrigate the eye, preferably with an eyewash, for at least 10 min. The eye pH should be monitored continuously. Ophthalmic antibacterial ointments will be needed after corneal burning. Exposed patients should be referred to an ophthalmologist.

Chronic Effects

Exposures to sulfuric acid mist have recently been observed to increase the risk of laryngeal and upper respiratory tract cancers (IARC, 1992). Workers in the Swedish steel pickling industry showed excess risk for laryngeal cancer (Ahlborg et al., 1981). Pickling operations workers in the U. S. steel industries also showed an elevated risk for laryngeal cancer (Steenland et al., 1988). Finally, it was reported that there were also some nested case-referent studies of workers with increased risk of cancer after exposure to sulfuric acid in a U.S. petrochemical plant (Soskolne et al., 1984). Erosion of incisor and canine teeth may occur after chronic exposure to sulfuric acid mist (Bruggen Cate, 1968).

MODE OF ACTION(S)

Sulfuric acid acts directly on the tissues by changing its pH. It might also act as a direct irritant, resulting in chronic ear, nose, throat irritation, and inflammation. Chronic inflammation may result and cause the release of free radicals, which may be genotoxic and result in an increased risk for some cancers. However, the connection between inflammation and carcinogenesis remains to be explored.

INDUSTRIAL HYGIENE

NIOSH (1997), ACGIH (1996), and OSHA (1998) have respectively set the REL, TLV, and PEL at $1 \, mg/m^3$, and the IDLH level at $15 \, mg/m^3$. A threshold limit value–time-weighted average (TLV–TWA) of $1 \, mg/m^3$ and a STEL of $3 \, mg/m^3$ (ACGIH, 1998) have been recommended by the ACGIH. Engineering controls and good ventilation should be used to minimize contact with the acid and minimize inhalation of acid mists. When this is not possible or when concentrations are high, personal protective equipment should be supplied to all employees. This equipment should include respirators, coveralls, gloves, and goggles (NIOSH, 1994).

RISK ASSESSMENTS

IARC

The IARC considers occupational exposure to strong inorganic mists containing sulfuric acid to be carcinogenic to humans (Group 1, IARC, 1992).

ACGIH

The ACGIH has listed sulfuric acid as an A2 carcinogen, a suspected human carcinogen (ACGIH, 1998).

EPA

The EPA has not classified sulfuric acid for carcinogenic effects but has placed it in the list of hazardous substances under the Clean Federal Water Pollution Control Act (EPA, 1998b) and the list of chemicals in "Toxic Chemicals Subject to Section 3.13 of the Emergency Planning and Community Right-to-Know Act" (EPA, 1998a). An oral reference dose (RfD) and an inhalation reference concentration (RfC) for sulfuric acid have not been derived by the EPA.

CalEPA

The CalEPA in March 2003 listed strong inorganic acid mists containing sulfuric acid as chemicals known to cause cancer for the purpose of the Safe Drinking Water and Toxic Enforcement Act of 1986 as listed under Proposition 65.

REFERENCES

Ahlborg, G., Jr., Hogstedt, C., Sundell, L., and Åman, C.-G. (1981) Laryngeal cancer and pickling house vapors. *Scand. J. Work Environ. Health* 7:237–240.

American Conference of Governmental Industrial Hygienists (ACGIH) (1996) *Threshold Limit values (TLVs) for Chemical Substances and Physical Agents and Biological Exposure Indices (BEIs)*, Cincinnati, OH: American Conference of Governmental Industrial Hygienists.

American Conference of Governmental Industrial Hygienists (ACGIH) (1998) *Threshold Limit Values for Chemical Substances and Physical Agents: Biological Exposure Indices*, Cincinnati, OH: American Conference of Governmental Industrial Hygienists.

Bruggen Cate, H.J. (1968) Dental erosions in industry. *Br. J. Ind. Med.* 25:249–266.

Budavari, S., editor. (1989) *The Merck Index*, 11th ed., Rahway, NJ: Merck & Co., Inc., pp. 8953–8954.

Chaney, S., Blomquist, W., Muller, K., and Goldstein, G. (1980) Biochemical changes in humans upon exposure to sulfuric acid aerosol and exercise. *Arch. Environ. Health* 35:211–216.

EPA (1998a) *Toxic chemical release reporting: community right-to-know. Chemicals and chemical categories to which this part applies. Code of Federal Regulations 40 CFR 372.65*. U.S. Environmental Protection Agency.

EPA (1998b) *Table 116.4B—List of hazardous substances by CAS numbers. Code of Federal Regulations 40 CFR 116.4*. Environmental Protection Agency.

HSDB (1998) *Hazardous Substances Data Bank*, Bethesda, MD: National Library of Medicine, Toxicology Information Program.

HSEES (1997) *Hazardous Substances Emergency Events Surveillance*, Atlanta, GA: U.S. Department of Health and Human Services, Public Health Service, Agency for Toxic Substances and Disease Registry.

IARC (1992) Carcinogenicity of occupational exposures to mists and vapors from strong inorganic acids, including sulfuric acid and hydrochloric acid. *Scand. J. Work Environ. Health* 18:329–330.

National Institute for Occupational Safety and Health (NIOSH) (1974) *Criteria for a Recommended Standard for Occupational Exposure to Sulfuric Acid*, Washington, DC: National Institute for Occupational Safety and Health.

National Institute for Occupational Safety and Health (NIOSH) (1994) *Pocket Guide to Chemical Hazards*, NIOSH 94-116, Washington, DC: US Department of Health and Human Services.

National Institute for Occupational Safety and Health (NIOSH) (1997) *Pocket Guide to Chemical Hazards*, Cincinnati, OH: National Institute for Occupational Safety and Health, pp. 290–291.

NOES (1990) *National Occupational Exposure Survey (1981–1983)*, Cincinnati, OH: U.S. Department of Health and Human Services, National Institute for Occupational Safety and Health.

Occupational Safety and Health Administration (OSHA) (1998) Limits for air contaminants. In: *Occupational Safety and Health Standards, Code of Federal Regulations. 29 CFR 1910.1000*, Washington, DC: Occupational Safety and Health Administration.

Soskolne, C.L., Zeighami, E.A., Hanis, N.M., Kupper, L.L., Herrmann, N., Amsel, J., Mausner, J.S., and Sellman, J.M. (1984) Laryngeal cancer and occupational exposure to sulfuric acid. *Am. J. Epidemiol.* 120:358–369.

Steenland, K., Schnorr, T., Beaumont, J., Halperin, W., and Bloom, T. (1988) Incidence of laryngeal cancer and exposure to acid mists. *Br. J. Ind. Med.* 45:766–776.

Treon, J.F., Dutra, F.R., Cappel, J., Sigmon, H., and Younker, W. (1950) Toxicity of sulfuric acid mist. *Arch. Indust. Hyg. & Occupational Med.* 2:716–734.

HYDROGEN SULFIDE

BACKGROUND AND USES

Hydrogen sulfide (sulfureted hydrogen; symbol H_2S; CASRN 7783-06-4), when found at high concentrations in the air, is an extremely potent chemical asphyxiant that can result in sudden death. Much like the other sulfur compounds described above, it is also a respiratory and skin irritant. The industrial uses for hydrogen sulfide are numerous and can result in its release to the environment through operations involving petroleum (e.g., refining, petrochemical synthesis), production and manufacturing (e.g., coke, sulfur, natural gas, asphalt roofing, rayon, and pulp and paper), vulcanization of rubber-containing sulfur compounds, processing of human and animal waste (e.g., landfilling, sewage pumping and treatment plants, swine containment, and manure handling), processing of animal products (e.g., pelt processing, animal slaughter facilities, leather tanneries), iron smelting, and food processing. It can also be an unwanted by-product formed naturally from the decay of organic sulfur-containing material. Finally, it can be released from natural sulfur deposits found in gypsum, sulfur, and coal mines.

PHYSICAL AND CHEMICAL PROPERTIES

Hydrogen sulfide is a colorless gas that smells like rotten eggs. It is corrosive, flammable, and explosive and has a melting point of $-82\,^\circ C$, a boiling point of $-60\,^\circ C$, and a vapor pressure of 17.6 atm. It is only slightly water soluble (0.4%) at $20\,^\circ C$ (NIOSH, 1994).

ENVIRONMENTAL FATE AND BIOACCUMULATION

Upon release, hydrogen sulfide can partition to the air, surface- and groundwater, and moist soil. Once in the atmosphere, oxygen and ozone can oxidize hydrogen sulfide forming sulfur dioxide (see environmental fate discussed above). Hydrogen sulfide can dissociate in an aqueous solution to form either a bisulfide ion (HS^-) or a sulfide ion (S^{2-}). In soil, hydrogen sulfide enters into the natural sulfur cycle. It does not bioaccumulate or biomagnify in the food chain.

MAMMALIAN TOXICOLOGY

Acute Effects

The nervous system and cardiac tissues are particularly vulnerable to the disruption of oxidative metabolism and death is often the result of respiratory arrest. Hydrogen sulfide also irritates the skin, eyes, mucous membranes, and the respiratory tract. Pulmonary effects may not be apparent for up to 72 h after exposure.

Inhalation of hydrogen sulfide can produce a condition of "olfactory fatigue." In this condition, exposed individuals can no longer detect its presence at concentrations >100 ppm. Thus, they are susceptible to higher exposures more quickly. The inability to detect hydrogen sulfide leads to continued inhalation (dose) and can be followed by major central nervous system (CNS) injury, including, but not limited to, loss of consciousness, respiratory paralysis, and seizures. Coma and death can occur after only a few breaths following exposure to high concentrations. Prior to CNS depression, there may actually be an instance of CNS stimulation manifested as excitation, rapid breathing, and headache.

Hydrogen sulfide exposures can also affect several major organ systems, including the respiratory, cardiovascular, digestive, as well as the skin and eyes. Inhalation of concentrations as low as 50 ppm hydrogen sulfide can rapidly produce irritation of the nose, throat, and lower respiratory tract. Exposed persons may develop shortness of breath and cough, while prolonged exposures may produce bronchial or lung hemorrhage. Bronchitis and fluid accumulation can occur after an exposure to higher concentrations, and this effect may be seen up to 72 h post exposure. Exposures to high doses can produce irregular heartbeat and decreased cardiac output. Nausea and vomiting may also occur after exposure. Skin exposure to very high amounts of hydrogen sulfide can produce itching, redness, and burning. Liquefied hydrogen sulfide gas can cause frostbite. Finally, blurred vision, spasmodic blinking, conjunctivitis, and clouding of the cornea may occur after an exposure to hydrogen sulfide gas.

First Aid

Persons exposed to hydrogen sulfide should be immediately removed from the source. Any contaminated clothing should be removed and double bagged. Exposed skin and hair should be flushed with water for 5 min. Exposed eyes should be irrigated with plain water or saline for at least 5 min. In cases of respiratory compromise, secure the airway and respiration via endotracheal intubation. If not possible, perform cricothyroidotomy, if equipped and trained to do so. Victims who demonstrate bronchospasms should be treated with aerosolized bronchodilators.

Chronic Effects

Contrary to the effects listed above, very little data exist about potential effects of low or chronic exposure. Since hydrogen sulfide does not accumulate in the body, recovery from an initial exposure should not produce any long-term problems. However, a long-term or repeated exposure to hydrogen sulfide has been shown to produce a plethora of effects, including, but not limited to, chronic and possibly intense cough, hypotension, loss of appetite, weight loss, and ataxia. There were no studies found showing carcinogenic, reproductive, teratogenic, or developmental effects in humans.

MODE OF ACTION(S)

At high concentrations, hydrogen sulfide is a potent inhibitor of cytochrome oxidase resulting in an interrupted cellular metabolism when the cells' mitochondria cannot properly utilize oxygen (cytotoxic anoxia) (Sorbo, 1960).

INDUSTRIAL HYGIENE

NIOSH (2006), the ACGIH (2005), and OSHA (2006) have respectively set the REL, TLV, and PEL at 10 ppm. The ACGIH has determined a STEL of 15 ppm and NIOSH has designated an IDLH level of 100 ppm. The OSHA lists the acceptable ceiling limit of hydrogen sulfide at 20 ppm in the workplace.

RISK ASSESSMENTS

WHO

The WHO has set a 24-h exposure limit for hydrogen sulfide inhalation of 0.15 mg/m^3 (WHO, 2000). At levels found in drinking water, WHO does not designate hydrogen sulfide as a human health concern (WHO, 2004).

ATSDR

The ATSDR has derived an acute-duration inhalation MRL of 0.07 ppm and an intermediate-duration inhalation MRL of 0.02 ppm for hydrogen sulfide (ATSDR, 2006).

IARC

The IARC has designated hydrogen sulfide as a Class 3 carcinogen, not classifiable as to its carcinogenicity to humans (IARC, 2006).

ACGIH

Due to the lack of data the ACGIH has not classified hydrogen sulfide as a human carcinogen (ACGIH, 2005).

EPA

The EPA has derived a chronic inhalation RfC of 0.001 ppm based on chronic exposures of rat olfactory mucosa to hydrogen sulfide (Brenneman et al., 2000). A previous RfD of 3×10^{-3} mg/kg/day was withdrawn from the Integrated Risk Information System (IRIS) database based on irreproducibility of effects, and a new RfD has not been derived (IRIS, 2006).

CalEPA

In 2008, the CalEPA established a REL for acute and chronic hydrogen sulfide inhalation at 42 and 10 µg/m^3, respectively.

REFERENCES

IARC (2006) *Agents Reviewed by the IARC Monographs*, vols. 1–88, Lyon, France: International Agency for Research on Cancer. Available at http://monographs.iarc.fr/ENG/Classification/List agentscasnos.pdf (accessed June 13, 2006).

Agency for Toxic Substances and Disease Registry (ATSDR) (1996) *Public Health Statement for Hydrogen Sulfide*, Atlanta, GA: U.S. Department of Health and Human Services, Public Health Service, Agency for Toxic Substances and Disease Registry.

Agency for Toxic Substances and Disease Registry (2006) *Toxicological Profile for Hydrogen Sulfide*, Atlanta, Georgia: U.S. Department of Health & Human Services, Public Health Service, Agency for Toxic Substances and Disease Registry.

American Conference of Governmental Industrial Hygienists (ACGIH) (2005) Hydrogen sulfide. In: *Threshold Limit Values for Chemical Substances and Physical Agents and Biological Exposure Indices*, Cincinnati, OH: American Conference of Governmental Industrial Hygienists, p. 34.

Brenneman, K.A., James, R.A., Gross, E.A., and Dorman, D.C. (2000) Olfactory neuron loss in adult male CD rats following subchronic inhalation exposure to hydrogen sulfide. *Toxicol. Pathol.* 28:326–333.

IRIS (2006) Hydrogen sulfide. In: *Integrated Risk Information System*. Washington, DC: U.S. Environmental Protection Agency. Available at http://www.epa.gov/iris/subst/index.html (accessed June 14, 2006).

National Institute for Occupational Safety and Health (NIOSH) (1994) *Pocket Guide to Chemical Hazards*, NIOSH 94-116, Washington, DC: US Department of Health and Human Services.

NIOSH (2006) Hydrogen sulfide. In: *NIOSH Pocket Guide to Chemical Hazards*, Washington, DC: National Institute for Occupational Safety and Health.

OSHA (2006) Limits for air contaminants. In: *Occupational Safety and Health Standards*, Code of Federal Regulations. 29 CFR 1910.1000, Washington, DC: Occupational Safety and Health Administration.

Sorbo, B. (1960) On mechanisms of sulfide oxidation in biological systems. *Biochim. Biophys. Acta* 38:349 351.

WHO (2000) Hydrogen sulfide. In: *Air Quality Guidelines: Part II. Evaluation of Human Health Risks*, Geneva, Switzerland: World Health Organization.

WHO (2004) *Guidelines for Drinking-Water Quality*, 3rd ed., Geneva, Switzerland: World Health Organization.

SECTION IV

ORGANIC COMPOUNDS

SECTION EDITOR: RAYMOND D. HARBISON

53

ALIPHATIC HYDROCARBONS

Mayowa Amosu, Nicole Nation, and Mary Alice Smith

METHANE

Target organ(s): Skin and eyes, brain, heart, kidney

Occupational exposure limits (OEL): No OEL, simple asphyxiant

Reference values: No reference values reported for this compound

Risk/safety phrases:
Classification: F+
Risk phrases: +R12: Extremely flammable
Safety phrases: +S2: Keep out of the reach of children
                    +S9: Keep container in a well-ventilated place
                    +S16: Keep away from sources of ignition—no smoking
                    +S33: Take precautionary measures against static discharges

BACKGROUND AND USES

Methane gas (marsh gas; CASRN 74-82-8) is produced from a number of natural sources, most notably from anaerobic respiration in marshland areas and in rice production, volcanoes and other natural sources of fire, and from the rudiments of animals, particularly cattle and sheep. Methane is considered an important occupational hazard in farming and mining operations and in sewage treatment facilities because of its asphyxiant and explosive properties. As the principal constituent of natural gas, methane is widely used as a domestic fuel and is also used in chemical reactions due to its reductive properties (Harbison, 1998).

PHYSICAL AND CHEMICAL PROPERTIES

At room temperature, methane is a clear, colorless gas with a weak odor. The liquid form has a high vapor pressure at low temperatures, and will readily evaporate into the atmosphere. Methane maintains a vapor density of 0.554 at $0\,°C$ ($32\,°F$), which means that it is less dense than air and will rise to the upper levels of a contained space. It has a boiling point of $-162\,°C$ and a melting point of $-183\,°C$. If a source of ignition is present while the atmospheric concentrations of methane are between 5 and 15%, an explosion can occur. Methane is incompatible with strong oxidizing agents such as chlorine and bromine.

MAMMALIAN TOXICOLOGY

Acute Effects

Methane has no known toxicity associated with chronic or acute exposures other than asphyxiation. People usually begin to experience symptoms of asphyxiation when oxygen levels in the environment are between 15 and 16%. More severe symptoms of toxicity will begin to develop as oxygen levels decrease to 6–8%. A few symptoms associated with severe methane gas poisoning due to asphyxiation result from respiratory distress and include loss of consciousness, dysrhythmias, myocardial ischemia, pulmonary edema, and seizures. Some clinical cases have reported headache, weakness, myalgia, ataxia, and light-headedness or fainting when confined to a room with a gas heater (Sherman and Harris, 1968).

Methane is one of the few compounds, where most overexposures result in either no adverse health effect or

Hamilton & Hardy's Industrial Toxicology, Sixth Edition. Edited by Raymond D. Harbison, Marie M. Bourgeois, and Giffe T. Johnson.
© 2015 John Wiley & Sons, Inc. Published 2015 by John Wiley & Sons, Inc.

death. Death produced by methane asphyxiation has been reported in all three major industries involving high concentrations of methane release: mining, wastewater treatment, and farming (Manning et al., 1981; Terazawa et al., 1985; CDC, 1989; Perry, 1995). There are no reports of any serious incapacitation or illness resulting from exposure to methane gas. However, an explosion of methane gas has occurred and the death of four workers who were killed in an accidental methane gas explosion in an urban tunnel was described by Nagao et al. (1997).

Liquid methane exposure can result in injury from frostbite, which can cause severe tissue burn and presents as a paling of skin color.

Chronic Effects

Methane has no known toxicity associated with chronic or acute exposures other than asphyxiation.

Mechanism of Action(s)

Asphyxiation from methane inhalation does not result from a direct biological interaction with the gas, but occurs as a simple displacement of oxygen. The metabolism of methane is similar to ethane in that most of the inhaled dose is exhaled unchanged. Absorbed methane is likely converted to its corresponding alcohol and subsequently broken down to carbon dioxide and excreted.

Chemical Pathology

Methane is not considered as a toxic agent and therefore there are no chemical pathologies associated with it.

SOURCES OF EXPOSURE

Occupational

The primary route of exposure to methane is through inhalation. Poor ventilation in a variety of settings is the main cause of an accumulation of hazardous concentrations of methane gas. The measurement of methane production by anaerobic degradation can be a useful indicator of the toxic effects of certain pollutants on the natural flora of bacteria (Vlaardingen and Beelen, 1992).

Environmental

Manure pits and other natural sources of natural gas can release enough methane to cause asphyxiation by displacing adequate concentrations of ambient oxygen. In coal mines, methane is referred to as marsh gas or fire damp and is the principal cause of explosions. Animal confinement areas can also be a hazardous source of methane.

INDUSTRIAL HYGIENE

Personal respirators are generally not required for methane because adequate ventilation is considered as enough protection in most instances, and is needed to avoid explosions.

The liquid form of methane may cause frostbite upon contact, and dermal protection is necessary when the liquid form is present. When liquid methane comes into contact with skin, severe frostbite can result.

Methane is extremely flammable; therefore, ignition sources should be kept away from high concentrations of the gas. Even something as "harmless" as a light bulb breaking when in use has been reported to cause serious explosions (Pettit, 1987). In the presence of light, methane may also react with compounds such as chlorine and bromine to form explosive mixtures. There are no occupational exposure regulations for methane.

While there are no exposure limits associated with concentrations of ambient methane; because of its asphyxiant property in displacing ambient oxygen, it is important to note the limits for oxygen concentration in the work area. Oxygen levels should be no lower than 18% to avoid extreme discomfort or danger. A new method for measuring methane in postmortem samples (Varlet, 2013) may help identify intentional and accidental deaths resulting from methane exposure in the future.

MEDICAL MANAGEMENT

Primary responders should immediately remove the victim to an area of fresh air or administer supplemental oxygen. Generally, individuals recover quickly once oxygen levels are brought back to normal in their environment.

Recently comatose survivors of asphyxia were treated with induced hypothermia with good survival (9 of 14; 65%) and those that survived had good neurological recovery. For asphyxia with methane, this may hold a promising new treatment (Baldursdottir et al., 2010).

In the case of an individual suffering from frostbite, warm water should be used to raise the temperature of the area; hot water should not be used. Digits can be warmed in the armpit and medical attention should be sought immediately.

REFERENCES

Baldursdottir, S., Sigvaldason, K., Karason, S., Valsson, F., and Sigurdsson, G.H. (2010) Induced hypothermia in comatose survivors of asphyxia: a case series of 14 consecutive cases. *Acta Anaesthesiol. Scand.* 54(7):821–826.

Centers for Disease Control (CDC) (1989) Fatalities attributed to methane asphyxia in manure waste pits. *MMWR Morb. Mortal. Wkly Rep.* (Vol. 38, pp. 583–586). Ohio. Michigan.

Harbison, R.D. (1998) *Hamilton & Hardy's Industrial Toxicology,* 5th ed., Mosby, Inc.

Manning, T.J., Ziminski, K., Hyman, A., Figueroa, G., and Lukash, L. (1981) Methane deaths? Was it the cause? *Am. J. Forensic Med. Pathol.* 2(4):333–336.

Nagao, M., Takatori, T., Oono, T., Iwase, H., Iwadate, K., Yamada, Y., and Nakajima, M. (1997) Death due to a methane gas explosion in a tunnel on urban reclaimed land. *Am. J. Forensic Med. Pathol.* 18(2):135–139.

Perry, G.F. (1995) Occupational medicine forum. *J. Occup. Environ. Med.* 37(6):656–660.

Pettit, T. (1987) *Digester explosion kills two*, pp. 27–28. Available at www.cdc.gov/niosh/nioshtic-2/00178569.html.

Sherman, B.H. and Harris E.H. (1968) Gas leak syndrome. *Pediatrics* 42:710–711.

Terazawa, K., Takatori, T., Tomii, S., and Nakano, K. (1985) Methane asphyxia. Coal mine accident investigation of distribution of gas. *Am. J. Forensic Med. Pathol.* 6(3):211–214.

Varlet, V., and Augsburger, M. (2013) Confirmation of natural gas explosion from methane quantification by headspace gas chromatography-mass spectrometry (HS-GC-MS) in postmortem samples: a case report. *Int. J. Legal Med.* 127(2):413–418. doi: 10.1007/s00414-012-0726-2.

Vlaardingen, P.A., and Beelen, P. (1992) Toxic effects of pollutants on methane production in sediments of the river Rhine. *Bull. Environ. Contam. Toxicol.* 49(5):780–786.

ETHANE

Target organ(s): None; not irritating to eyes, nose, or throat

Occupational exposure limits: ACGIH: TWA–TLV: 1000 ppm

Reference values: No reference values reported for this compound

Risk/safety phrases:

Classification: F+; R12
Risk phrases: +R12: Extremely flammable
Safety   +S2: Keep out of the reach of children
  phrases:
                 +S9: Keep container in a well-ventilated place
                 +S16: Keep away from sources of ignition—no smoking
                 +S33: Take precautionary measures against static discharges

BACKGROUND AND USES

Ethane (methylmethane; CASRN 74-84-0) is naturally occurring in mining operations and is a component of natural gas. It is not irritating to mucous membranes, eyes, or skin. Similar to methane, ethane is extremely flammable and is incompatible with chlorine. The production of some compounds, such as vinyl chloride and ethylene, requires ethane as a precursor. The vapor pressure is high at room temperature and has a vapor density of 1.04. Ethane is produced endogenously as an end product of lipid peroxidation (de Ruiter et al., 1980; Sunderman, 1987; Rahimtula et al., 1988).

Inhalation exposures to ethane are most common. Ethane can act as a simple asphyxiant at high concentrations, but has no other reported adverse health effects.

PHYSICAL AND CHEMICAL PROPERTIES

Ethane is a colorless, odorless gas that is generally very stable and chemically inactive. However, in concentrations of 3–12% by volume, it has the potential to be very explosive. It is incompatible with chlorine, dioxygenyl tetrafluoroborate, oxidizing materials, heat, or flame.

MAMMALIAN TOXICOLOGY

Acute Effects

Ethane has no known toxicity associated with acute exposures other than asphyxiation, which can lead to unconsciousness and eventually death. People usually begin to experience symptoms of asphyxiation when oxygen levels in the environment drop between 15 and 16%. More severe symptoms of toxicity will begin to develop as oxygen levels drop to 6–8% (HSDB, 2009). Symptoms of acute asphyxiation due to the displacement of ambient oxygen by ethane include tunnel vision and reduced visual acuity, myocardial infarction due to hypoxia, bronchoconstriction, numbness, and dizziness.

Chronic Effects

Ethane has no known toxicity associated with chronic exposures.

Mechanism of Action(s)

Ethane absorbed by the respiratory tract is primarily excreted unchanged.

Chemical Pathology

Ethane is not considered as a toxic agent and therefore there are no chemical pathologies associated with exposure to ethane.

SOURCES OF EXPOSURE

Occupational

The general public may be exposed to ethane by inhalation of natural gas in crude oil emissions. It is also possible to be exposed through drinking water and food intake, but this exposure is minor in comparison to inhalation. The average daily intake of ethane by one person through breathing ambient air is 183 mg.

There are no reference values for exposures to ethane. However, because of its asphyxiant property in displacing ambient oxygen, it is important to note the limits for oxygen concentration in the working area. Oxygen levels should be no lower than 18% to avoid extreme discomfort and danger.

Environmental

Ethane is a component of crude oil and natural gas and naturally exists in the environment. It is most likely to be released into the environment through waste streams and into the atmosphere in gas form. The combustion of gasoline and the fumes created by waste incinerators might also release ethane into the atmosphere (HSDB, 2009).

INDUSTRIAL HYGIENE

Because ethane only exists as a liquid at below freezing temperatures ($-172\,°C$ to $-88\,°C$) it is important to wear cold-insulating gloves to prevent frostbite or injury while handling the liquid form of ethane. In the case of high vapor concentration, self-contained breathing apparatuses may be worn. In industry a threshold limit value (TLV) at 1000 parts per million (ppm) is used for handling ethane. (Clough, 2005).

MEDICAL MANAGEMENT

In the case of asphyxiation, the victim should immediately be moved to an area with fresh air and medical attention should be sought. Supplemental oxygen can be administered, if necessary. Generally, individuals recover quickly once oxygen levels are brought back to normal. In the event of frostbite, the exposed area should be rinsed in warm water; hot water should not be used. Digits can be warmed in the armpit and medical attention should be sought immediately.

REFERENCES

Clough, S.R. (2005) Ethane. In *Encyclopedia of Toxicology* (2nd ed., pp. 262–263): Elsevier Inc.

de Ruiter, N., Muliawan, H., and Kappus, H. (1980) Ethane production of mouse peritoneal macrophages as indication for lipid peroxidation and the effect of heavy metals. *Toxicology* 17(2):265–258.

HSDB (2009) *Hazardous Substances Data Bank*, Bethesda, MD: National Library of Medicine, National Toxicology Information Program. Available at http://toxnet.nlm.nih.gov/ (accessed April 16, 2013).

Rahimtula, A.D., Bereziat, J.C., Bussacchini-Griot, V., and Bartsch, H. (1988) Lipid peroxidation as a possible cause of ochratoxin A toxicity. *Biochem. Pharmacol.* 37(23):4469–4477.

Sunderman, F.W. (1987) Lipid peroxidation as a mechanism of acute nickel toxicity. *Toxicol. Environ. Chem.* 15(1–2):59–69.

PROPANE

Target organ(s): Eyes, nose, respiratory tract, brain

Occupational exposure limits:	ACGIH TLV–TWA: 1000 ppm 8 h
	National Institute for Occupational Safety and Health (NIOSH) recommended exposure limit (REL)–TWA: 1800 mg/m^3 (1000 ppm) 10 h
	Occupational Safety and Health Administration (OSHA) permissible exposure limit (PEL)–TWA 1800 mg/m^3 (1000 ppm) 8 h

Reference values: None

Risk/safety phrases:	R12 extremely flammable
	S2 keep out of the reach of children
	S9 keep container in a well-ventilated place
	S16 keep away from sources of ignition—no smoking
	S33 take precautionary measures against static discharges

BACKGROUND AND USES

Propane, also called n-propane (CASRN 74-98-6), is a naturally colorless, odorless gas (Harbison, 1998). Because of the potential for explosion, odor is typically added to propane to make its presence more easily detectable. Propane has a wide variety of uses. It is a component of liquid petroleum gas utilized for both commercial and industrial purposes (HSDB, 2009). Propane is often used as a basic material in chemical and organic synthesis. It can be used as a household and industrial fuel, a solvent, a refrigerant, an aerosol propellant, and as a replacement for chlorofluorocarbons (HSDB, 2009). Propane has uses in the agricultural sector for crop drying, weed control, and as a fuel for farm equipment. The most significant negative health effects come from the potential for propane to explode, leading to burns, other injuries, and sometimes death.

PHYSICAL AND CHEMICAL PROPERTIES

Propane is typically found in industrial settings as a gas with a boiling point of $-42\,°C$ ($-43.6\,°F$) and a freezing point of $-189.7\,°C$ ($-309.5\,°F$). Propane has a vapor pressure of 8.4 atm and a vapor density of 1.6. The autoignition point of propane is $432\,°C$ ($842\,°F$), a lower explosive limit of 2.2% and an upper explosive limit of 9.5% (HSDB, 2009). Under normal conditions of storage and use, hazardous decomposition or polymerization products are not produced. However, propane is extremely flammable when exposed to open flames, sparks, static discharge, and oxidizing materials.

MAMMALIAN TOXICOLOGY

Acute Effects

Propane is considered as a simple asphyxiant. This asphyxiant property is related to the direct displacement of oxygen, rather than toxic effects (Harbison, 1998). In confined spaces, high concentrations of propane vapor can accumulate, leading to irritation of the nose and throat, headache, nausea, vomiting, dizziness, and drowsiness. In extreme situations, unconsciousness and asphyxiation can result. However, at concentrations up to 10% or 100,000 ppm, propane has not been shown to cause noticeable irritation to the eyes, nose, or respiratory tract (Snyder, 1987). However, direct skin or mucous contact with liquid propane or with rapidly expanding gas can cause burns or frostbite to the exposed areas.

Chronic Effects

Propane is not considered carcinogenic or mutagenic, nor does it bioaccumulate. There are no known health effects associated with chronic exposure to propane.

Mechanism of Action(s)

As a simple asphyxiant, propane directly displaces oxygen, potentially leading to central nervous system depression, unconsciousness, and death.

Chemical Pathology

No long-term chemical pathologies of industrial exposure to propane have been reported.

SOURCES OF EXPOSURE

Occupational

The most common routes of occupational exposure occur through inhalation and dermal contact in workplaces where propane is present, due to either its production or use. This includes industries utilizing petroleum or natural gas or those using solvents, refrigerants, or aerosol propellants containing propane. Because propane can be used in the agricultural sector, farm workers using propane for crop drying, weed control, or as a fuel for equipment can also be exposed.

Environmental

Due to its use in the petroleum industry, in outdoor gas grills, and as a heating fuel, propane may be released into the environment in multiple ways (HSDB, 2009). Additionally, disposal methods of the natural gas and petroleum industries may release propane into the atmosphere. Propane exists as a gas in the atmosphere. If present in the soil, propane is readily oxidized by soil bacterium into acetone (HSDB, 2009).

INDUSTRIAL HYGIENE

Where exposure might occur, appropriate personal protective equipment should be utilized. This includes goggles or other protective eyewear, chemical protective gloves, and protective clothing to prevent skin freezing in case of dermal contact. Quick-drench facilities and eyewash fountains should be readily available. If working with propane or propane-containing products within a contained area, self-contained breathing apparatuses or other respirator protection is recommended.

MEDICAL MANAGEMENT

If contact with the eyes occurs, contact lenses should be removed immediately. The eyes should be flushed with water for 15 min, and medical attention should be sought immediately.

If contact with the skin occurs, the area should be flushed with water for 15 min while medical attention should be sought. After contaminated clothing is removed, it should be soaked with water to prevent the risk of static discharges and subsequent gas ignition. Contaminated clothing and shoes should be washed thoroughly before reuse. If frostbite occurs, the frozen tissues should be warmed by rinsing in warmed water and medical attention should be sought.

If exposure through inhalation occurs, the person should be moved immediately to fresh air. If the person is not breathing, or if breathing is irregular or respiratory arrest occurs, trained personnel should deliver artificial respiration or oxygen. Medical attention should be sought. If an explosion occurs, seek medical attention.

REFERENCES

Fonseca, C.A., Auerbach, D.S., and Suarez, R.V. (2002) The forensic investigation of propane gas asphyxiation. *Am. J. Forensic Med. Pathol.* 23(2):167–169.

Harbison, R.D. (1998) *Hamilton & Hardy's Industrial Toxicology*, 5th ed., Mosby, Inc.

HSDB (2009) *Hazardous Substances Data Bank.* Retrieved Feb 2013, from National Library of Medicine, National Toxicology Information Program http://toxnet.nlm.nih.gov/

Snyder, R., editor (1987) *Ethyl Browning's Toxicity and Metabolism of Industrial Solvents*, Amsterdam: Elsevier.

BUTANE

Target organ(s): Brain, heart, liver

Occupational exposure limits: NIOSH: REL for 10 h TWA is 800 ppm

Reference values: None

Risk/safety phrases:	R12: Extremely flammable
	S2: Keep out of the reach of children
	S9: Keep container in a well-ventilated place
	S16: Keep away from sources of ignition—no smoking

BACKGROUND AND USES

Butane (CASRN 106-97-8) consists of a chain of four carbons. Like other hydrocarbons, it exhibits narcotic properties. It occurs naturally as a constituent of natural gas, from which it is then refined (HSDB, 2013). It is used as a fuel in disposable lighters, refrigerants, aerosol propellants, and as a gasoline additive. Most reported health effects due to butane result from intentional abuse of butane, primarily from "huffing" by teenagers. There are few reports of industrial health effects. The amounts of butane exposure during intentional abuse far exceed expected occupational exposures.

PHYSICAL AND CHEMICAL PROPERTIES

Butane is a colorless gas with a faint gasoline-like odor. It is a flammable gas, with a flash point of $-69\,°C$ ($-76\,°F$) and an autoignition point of $287\,°C$ ($550\,°F$). Butane has a lower explosive limit of 1.6% and an upper explosive limit of 8.4%. It has a boiling point of $-0.5\,°C$ ($31.1\,°F$). Butane is not considered soluble in water and is an explosion hazard when exposed to a nickel carbonyl/oxygen mixture; it should also be kept away from strong oxidizers.

MAMMALIAN TOXICOLOGY

Acute Effects

Butane is a simple asphyxiant; this asphyxiant property is related to the displacement of oxygen, not a toxic response (Harbison, 1998). Exposure to high concentrations of butane, especially in confined spaces, can result in asphyxiation if the concentration of oxygen is reduced below 16%. Symptoms include those of oxygen deprivation, such as rapid breathing, dizziness, headache, mental confusion, muscular weakness, and numbness of the extremities. If oxygen concentrations fall below 6–8%, unconsciousness, central nervous system damage, and death are possible.

The vapor of butane is not considered to be irritating to the eyes, nose, or throat. However, contact with the liquid form of butane can cause frostbite, freeze burns, and permanent damage to tissues if it comes into contact with the eyes, skin, or mucous membranes.

Several case reports of intentional and unintentional direct butane inhalation have reported myocardial infarction and ventricular fibrillation (Sugie et al., 2004; Girard et al., 2008; De Naeyer et al., 2011; Senthilkumaran et al., 2012). One teenager died after butane abuse from multiple organ failure, which included the central nervous system, the cardiovascular system, the pulmonary system, and the liver (Rieder-Scharinger et al., 2000). Due to the direct nature of these exposures, similar exposures in an occupational setting are unlikely.

Chronic Effects

Because butane is a simple asphyxiant, chronic and carcinogenic effects are not expected. The International Agency for Research on Cancer (IARC), OSHA, and ACGIH do not consider butane to be carcinogenic. There are a few case reports indicating effects after chronic intentional use; one teenage butane abuser developed fatal fulminant hepatic failure after 3 years of butane abuse (McIntyre and Long, 1992). The extent of chronic effects noted in the literature is limited, suggesting that chronic effects are rare.

Mechanism of Action(s)

As a simple asphyxiant, negative effects are related to the displacement of oxygen, not to toxic responses; this lack of oxygen can result in hypoxia or anoxia. Biologically, butane is hydrolyzed by microsomal enzymes to its alcohol, butanol (Tsukamoto et al., 1985; Harbison, 1998). This butanol is exhaled or conjugated and excreted.

Chemical Pathology

Pathologies related to butane are limited. Myocardial infarction and ventricular fibrillation are the most commonly reported pathologies related to acute exposures. Other reported pathologies include fulminant hepatic failure and multiple organ failure after chronic intentional abuse.

SOURCES OF EXPOSURE

Occupational

The most common occupational exposures to butane occur through either inhalation or dermal contact in workplaces, where butane or petroleum- and natural gas-containing butane are used; this could include truck drivers, gas station attendants, and other workers where contact with petroleum and natural gas products are common (HSDB, 2013). One study in France detected butane in a hospital setting (Bessonneau et al., 2013), and a separate study in Poland detected butane in the occupational environment of a painting restorer (Jezewska and Szewczynska, 2012; Bessonneau et al., 2013). All levels detected in these studies were far below regulatory limits.

Environmental

Monitoring data indicate that butane is a widely occurring atmospheric pollutant (HSDB, 2013). The general population may be exposed to butane through the inhalation of ambient air and through contact with liquids containing this compound, such as crude oil, gasoline, and other liquid products.

INDUSTRIAL HYGIENE

Skin barriers and eye protection to prevent exposures to liquid butane are recommended. When working with potentially high concentrations of butane in enclosed spaces, self-contained breathing apparatuses or other respirator protection is recommended. Quick-drench facilities and/or eyewash stations should be located within the immediate work area for emergency use when working with butane.

The vacated 1989 OSHA PEL TWA of 800 ppm is still enforced in some states. National Institute for Occupational Safety and Health has a REL for a 10 h TWA of 800 ppm.

MEDICAL MANAGEMENT

The medical care required will be dependent upon the exposure. For mild to moderate inhalation exposure, removal of the individual away from the source of butane and administration of supplemental oxygen are typically sufficient. For severe inhalation exposure, supplemental oxygen should be administered and immediate medical care should be sought. Treatment will depend on the extent of symptoms, including ventricular fibrillation and central nervous system depression. If dermal exposure occurs, rewarming of the area and topical treatments are recommended. Individuals with severe burns and frostbite should seek professional medical care. For eye exposure, the eyes should be washed with copious amounts of room temperature water for at least 15 min and medical attention should be sought.

REFERENCES

Bessonneau, V., Mosqueron, L., Berrube, A., Mukensturm, G., Buffet-Bataillon, S., Gangneux, J. P., and Thomas, O. (2013). VOC contamination in hospital, from stationary sampling of a large panel of compounds, in view of healthcare workers and patients exposure assessment. *PLoS One* 8(2):e55535.

De Naeyer, A.H., de Kort, S.W., Portegies, M.C., Deraedt, D.J., and Buysse, C.M. (2011) Myocardial infarction in a 16-year old following inhalation of butane gas. *Nederl. Tijdschr. Geneesk.* 155(34):A3443. First published on Myocardinfarct bij een 16-jarige na inhalatie van butaangas.

Girard, F., Le Tacon, S., Maria, M., Pierrard, O., and Monin, P. (2008) Ventricular fibrillation following deodorant spray inhalation. *Ann. Fran. Anesth.* 27(1):83–85. First published on Fibrillation ventriculaire par inhalation de spray deodorant.

Harbison, R.D. (1998) *Hamilton & Hardy's Industrial Toxicology*, 5th ed., Mosby, Inc.

HSDB (2013) *Hazardous Substances Data Bank.* Retrieved March 31, 2013, from National Library of Medicine, National Toxicology Information Program http://toxnet.nlm.nih.gov/.

Jezewska, A. and Szewczynska, M. (2012) Chemical hazards in the workplace environment of painting restorer. *Med. Pr.* 63(5):547–558. First published on Zagrozenia chemiczne w srodowisku pracy konserwatora malarstwa.

McIntyre, A.S. and Long, R.G. (1992) Fatal fulminant hepatic failure in a 'solvent abuser'. *Postgrad. Med. J.* 68(795): 29–30.

Rieder-Scharinger, J., Peer, R., Rabl, W., Hasibeder, W., and Schobersberger, W. (2000) Multiple organ failure following inhalation of butane gas: a case report. *Wien. Klin. Wochenschr.* 112(24):1049–1052. First published on Multiorganversagen nach Butangasinhalation: Ein Fallbericht.

Senthilkumaran, S., Meenakshisundaram, R., Michaels, A.D., Balamurgan, N., and Thirumalaikolundusubramanian, P. (2012) Ventricular fibrillation after exposure to air freshener—death just a breath away. *J. Electrocardiol.* 45(2): 164–166.

Sugie, H., Sasaki, C., Hashimoto, C., Takeshita, H., Nagai, T., Nakamura, S., Furukawa, M., Nishikawa, T., and Kurihara, K. (2004) Three cases of sudden death due to butane or propane gas inhalation: analysis of tissues for gas components. *Forensic Sci. Int.* 143(2–3):211–214.

Tsukamoto, S., Chiba, S., Muto, T., Ishikawa, T., and Shimamura, M. (1985) Study on the metabolism of volatile hydrocarbons in mice—propane, *n*-butane, and *iso*-butane. *J. Toxicol. Sci.* 10(4):323–332.

n-HEXANE

Target organ(s): Lungs, skin

Occupational exposure limits:	OSHA: TWA: 500 ppm
	NIOSH: TWA : 50 ppm (180 mg/m^3)
	ACGIH: TWA: 50 ppm, skin

Reference values: No reference values reported for this compound

Risk/safety phrases:

Classification:	F; R11; Repr. Cat. 3; R62
	Xn; R65: 48/20
	Xi; R38: R67
	N; R51–53
Risk phrases:	+R11: Highly flammable
	+R38: Irritating to the skin
	+R48/20: Harmful: danger of serious damage to health by prolonged exposure through inhalation
	+R62: Possible risk of impaired fertility
	+R65: Harmful: may cause lung damage if swallowed
	+R67: Vapors may cause drowsiness and dizziness
	+R51/53: Toxic to aquatic organisms, may cause long-term adverse effects in the aquatic environment
Safety phrases:	+S2: Keep out of the reach of children
	+S9: Keep container in a well-ventilated place
	+S16: Keep away from sources of ignition—no smoking
	+S29: Do not empty into drains
	+S33: Take precautionary measures against static discharges
	+S36/37: Wear suitable protective clothing and gloves
	+S61: Avoid release to the environment. Refer to special instructions/safety data sheets
	+S62: If swallowed, do not induce vomiting: seek medical advice immediately and show container or label

BACKGROUND AND USES

n-Hexane is used as a solvent in glue, ink, rubber, and adhesives in shoe manufacturing. It is also used in the extraction of a variety of plant oils and in research laboratories. It is commonly found in petroleum products and, as a result of petroleum combustion, *n*-hexane is ubiquitous in ambient air at low concentrations. It may also be released from industrial sources as a result of evaporation.

PHYSICAL AND CHEMICAL PROPERTIES

n-Hexane (CASRN 110-54-3; hexane; hexyl hydride; normal-hexane) is a colorless liquid with a distinct gasoline-like odor. It is the most toxic member of the alkanes. It has a vapor pressure of 150 mmHg and a vapor density of 2.97. It is extremely flammable and easily ignited by heat or sparks. Its vapors are heavier than air; the vapors can accumulate in low areas and may form explosive mixtures with ambient air.

MAMMALIAN TOXICOLOGY

Acute Effects

Exposure to *n*-hexane affects the skin, eyes, and heart, as well as the respiratory and nervous systems. Some symptoms include headache, dizziness, nausea, vomiting, and irritation of the respiratory tract, nose and throat.

The primary route of exposure is inhalation. Symptoms of acute inhalation include numbness, cough, bloody-frothy sputum, tachycardia, and general weakness. Fifty-three workers in a vinyl sandal-manufacturing setting were exposed to *n*-hexane at an average of 118 ppm. They recorded sensory polyneuropathy (Takeuchi, 1993). Acute effects observed in other shoe factory workers included mild weakness and paresthesia of the lower limbs (Governa et al., 1987). Recent studies investigating nerve conduction in workers exposed to *n*-hexane report decreases in sensory nerve conduction in lower limbs when compared to workers not exposed to *n*-hexane (Najafi et al., 2011; Neghab et al., 2013; Neghab et al., 2012). These studies suggest that nerve conduction tests may be predictors of *n*-hexane effects before overt symptoms are seen in workers exposed to chronic low levels of *n*-hexane.

At very high concentrations, *n*-hexane can cause unconsciousness and even death from asphyxiation due to oxygen displacement. Polyneuropathy was observed in Taiwanese workers in a ball-manufacturing factory, where concentrations of *n*-hexane ranged from 89 to 109 ppm as a result of poor ventilation in the workspace (Huang et al., 1991).

Ocular exposure to *n*-hexane can cause corneal injury. Skin irritation and redness may result from dermal contact. Ingestion, though unlikely, can cause nausea and gastrointestinal irritation, as well as CNS depression.

Chronic Effects*

The chronic effects of *n*-hexane are rare in an industrial setting. When they do occur, it is usually in small factories. One example, described by Pezzoli et al. (1995), was of a man who had been complaining of motor impairment and feeling weak. This man had worked in a small, family owned leather-processing factory for 30 years and had been exposed to high levels of *n*-hexane over this period. The chronic nature of this exposure was supported by urinary levels of 2,5-hexanedione (2,5 HD), a metabolite of *n*-hexane, of 785 mg/l. This is 75% higher than the normal population level of 450 mg/l (Pezzoli et al., 1995).

The most frequent and severe chronic effects are seen in cases of solvent abuse. Common products such as glues and correction fluids may be concentrated and inhaled (Ong et al., 1993). This results in high exposure to *n*-hexane and toluene, as well as other solvents. In 1990, volatile substance abuse resulted in the deaths of 118 young people in South Africa (Khedun et al., 1996). In a clinical study of inhalant abuse of *n*-hexane, researchers reported a decrease in nerve conduction, axonal swelling, and demyelination. The patients suffered from muscle weakness and numbness and partial motor conduction block that persisted for 2–4 months after withdrawal from hexane-containing inhalants (Kuwabara et al., 1999).

Mechanism of Action(s)

The oxidation of *n*-hexane forms the toxic metabolite 2,5-hexanedione that can result in both neurotoxicity and testicular toxicity, albeit by different mechanisms. In the brain, 2,5-hexadione disrupts phosphorylation of neurofilaments, which allows neurofilament proteins to accumulate along the axon; this causes it to swell and results in axonopathy (Tshala-Katumbay et al., 2009). The testicular toxicity of 2,4-hexanedione is due to its effect on the Sertoli cells' ability to secrete seminiferous tubule fluid, which ultimately compromises germ cell viability (Boekelheide and Schoenfeld, 2001).

Chemical Pathology

Concentrations greater than 50 g may be lethal (Bingham et al., 2001).

SOURCES OF EXPOSURE

Occupational

The most common occupational exposure to *n*-hexane is by inhalation or dermal contact in industries where it is produced or used. These include automobile repair shops where gasoline tanks are handled, printing factories, manufacturing facilities, and a variety of refineries.

Environmental

n-Hexane is a component of crude oil and natural gas and naturally exists in the environment. Because of its widespread use and production, *n*-hexane is most commonly released into the environment through waste streams and into the atmosphere in vapor form.

INDUSTRIAL HYGIENE

The National Institute for Occupational Safety and Health (NIOSH, 2010) and the ACGIH (2012) have both set the TWA at 50 ppm (180 mg/m^3). The OSHA permissible exposure limit is currently set at a TWA of 500 ppm.

Exposure to hexane is also a concern in the printing industry. Soon after application, the ink sets, and solvents are evaporated into the workroom air. To a large extent, exposure depends on the size and ventilation capabilities of the offset print shops (Wadden et al., 1995).

Health and safety precautions include proper ventilation to maintain exposure limits and proper protective clothing. The hazard of dermal exposure can be reduced through the use of polymeric-containing clothing material such as butyl nomex. This material has been shown to prevent exposure for time periods <21 min. This protection allows short-term contact with *n*-hexane at a reduced risk (Dillon and Obasuyi, 1985).

MEDICAL MANAGEMENT

In the case of acute exposure by inhalation, the affected person should be moved to a well-ventilated area or fresh air and medical attention should be sought to evaluate their respiratory distress. Inhaled corticosteroids (such as albuterol) can be given for severe coughing and wheezing and oxygen administered, if necessary. For skin exposures, contaminated clothing should be immediately removed and affected area should be washed with soap and water. In the case of eye contact, the eyes should be irrigated with plenty of water. Medical personnel should evaluate the treatment.

Recent research has promoted the use of new methods for early detection of *n*-hexane exposure. Mayan (2002) studied factory workers who had occupational exposure to hexane with no apparent health effects. Workspaces were monitored for *n*-hexane and post shift urine samples were collected and analyzed for the presence of 2,5-hexanedione. They found a significant correlation between levels of urinary 2,5-hexanedione and hexane detected in air samples. The authors suggest that urinary 2,5-hexanedione could be used as a biomarker for detection of *n*-hexane exposure and that this method is a more effective approach than air monitoring.

REFERENCES

American Conference of Governmental Industrial Hygienists (ACGIH) (2012) *Threshold Limit Values (TLVs) for Chemical Substances and Physical Agents and Biological Exposure Indices (BEIs)*, Cincinnati, OH: American Conference of Governmental Industrial Hygienists.

Bingham, E., Cohrssen, B., and Powell, C.H. (2001) *Patty's Toxicology*, 5th ed., vol. 4, New York, NY: John Wiley & Sons, Inc., p. 36.

Boekelheide, K. and Schoenfeld, H.A. (2001) Spermatogenesis by Sisyphus: proliferating stem germ cells fail to repopulate the testis after 'irreversible' injury. *Adv. Exp. Med. Biol.* 500:421–428.

Dillon, I.G. and Obasuyi, E. (1985) Permeation of hexane through butyl nomex. *Am. Ind. Hyg. Assoc. J.*, 233.

Governa, M., Calisti, R., Coppa, G., Tagliavento, G., Colombi, A., and Troni, W. (1987) Urinary excretion of 2,5-hexanedione and peripheral polyneuropathies workers exposed to hexane. *J. Toxicol. Environ. Health* 20(3):219–228.

HSDB (2013) *Hazardous Substances Data Bank*, Bethesda, MD: National Library of Medicine, National Toxicology Information Program. Available at http://toxnet.nlm.nih.gov/ (accessed March 8, 2013).

Huang, C.C., Shih, T.S., Cheng, S.Y., Chen, S.S., and Tchen, P.H. (1991) *n*-Hexane polyneuropathy in a ball-manufacturing factory. *J. Occup. Med.* 33(2):139–142.

Huang, J., Kato, K., Shibata, E., Asaeda, N., and Takeuchi, Y. (1993) Nerve-specific marker proteins as indicators of organic solvent neurotoxicity. *Environ. Res.* 63(1):82–87.

Khedun, S.M., Maharaj, B., and Naicker, T. (1996) Hexane cardiotoxicity—an experimental study. *Isr. J. Med. Sci.* 32(2):123–128.

Kuwabara, S., Kai, M.R., Nagase, H., and Hattori, T. (1999) *n*-Hexane neuropathy caused by addictive inhalation: clinical and electrophysiological features. *Eur. Neurol.* 41(3):163–167.

Mayan, O., Teixeira, J.P., Alves, S., and Azevedo, C. (2002) Urinary 2,5 hexanedione as a biomarker of *n*-hexane exposure. *Biomarkers*, 7(4):299–305. doi: 10.1080/13547500210136796.

Najafi, S, Mahmoudabadi, A., Akbarzade, M., Sajadi, S., and Emadi, A. (2011) Evaluation of neuropathy in workers predisposed to inhalation of *n*-hexane. *HBI J.* 9(1):1–5.

National Institute for Occupational Safety and Health (NIOSH) (2010) *Pocket Guide to Chemical Hazards*, NIOSH 2010-168, Washington, DC: US Department of Health and Human Services. Available from: http://www.cdc.gov/niosh/npg/npgd0322.html.

Neghab, M., Soleimani, E., and Khamoushian, K. (2012) Electrophysiological studies of shoemakers exposed to sub-TLV levels of *n*-hexane. *J. Occup. Health* 54:376–382.

Neghab, M., Soleimani, E., and Khamoushian, K. (2013) Subclinical neuropathy in workers occupationally exposed to *n*-hexane. *J. School Public Health Inst. Public Health Res.* 10(4):105–117.

Ong, C., Koh, D., Foo, S., Kok, P., Ong, H., and Aw, T. (1993) Volatile organic solvents in correction fluids: identification and potential hazards. *Bull. Environ. Contam. Toxicol.* 50(6):787–793.

Pezzoli, G., Antonini, A., Barbieri, S., Canesi, M., Perbellini, L., Zecchinelli, A., Mariani, C.B., Bonetti, A., and Leenders, K.L. (1995) *n*-Hexane-induced parkinsonism: pathogenetic hypotheses. *Mov. Disord.* 10(3):279–282.

Takeuchi, Y. (1993) *n*-Hexane polyneuropathy in Japan: a review of *n*-hexane poisoning and its preventive measures. *Environ. Res.* 62(1):76–80.

Tshala-Katumbay, D., Monterroso, V., Kayton, R., Lasarev, M., Sabri, M., and Spencer, P. (2009) Probing mechanisms of axonopathy. Part II: protein targets of 2,5-hexanedione, the neurotoxic metabolite of the aliphatic solvent *n*-hexane. *Toxicol. Sci.* 107(2):482–489.

Wadden, R.A., Scheff, P.A., Franke, J.E., Conroy, L.M., Javor, M., Keil, C.B., and Milz, S.A. (1995) VOC emission rates and emission factors for a sheetfed offset printing shop. *Am. Ind. Hyg. Assoc. J.* 56(4):368–376.

CUTTING OILS

Target organ(s): Skin, respiratory tract

Occupational exposure limits: Oil mist, mineral:	OSHA PEL—5 mg/m^3 TWA
	NIOSH REL—5 mg/m^3 TWA
	NIOSH REL—10 mg/m^3 STEL
	ACGIH TLV—5 mg/m^3 TWA

BACKGROUND AND USES

Cutting oils, also called metal-working fluids, is a term used to refer to a group of compounds that are used to aid in machine operations, specifically when metal pieces are being machined, ground, or milled. There are three main classifications of metal-working fluids: (1) straight oils, also called neat oils, which are insoluble and made up of mineral, animal, marine, vegetable, or synthetic oils; (2) soluble oils, which contain severely refined petroleum oils as well as emulsifiers; and (3) synthetic fluids, which do not contain petroleum oils.

Regardless of the type of fluid, various components, referred to as additives, can be added to the oil to provide or improve specific desirable properties. The hazardous effects of the specific cutting oil are highly dependent on these additives. Common additives include germicides such as formalin, mercurials, or phenolic compounds; emulsifiers such as soaps and sulfonates; corrosion inhibitors such as borates, dichromates, or amines; and extreme pressure compounds of sulfur, chlorine, or phosphorus. The oil–water emulsions tend to become rancid, but they are good coolants, economical, and noncombustible.

PHYSICAL AND CHEMICAL PROPERTIES

Physical and chemical properties are determined by the specific composition of the given cutting oil and thus vary between different oils. Information regarding the physical and chemical properties of particular cutting oil should be obtained from the manufacturer.

MAMMALIAN TOXICOLOGY

Acute Effects

Occupational irritant contact dermatitis is a commonly observed acute effect associated with cutting oil processes (Ramos, 1974; Grattan et al., 1989, Sakakibara et al., 1989; Goh and Gan, 1994). The severity and likelihood of contact dermatitis is dependent upon several factors. Cutting oils have various additives; these additives induce different degrees of inflammatory responses and can influence the absorption of components in the cutting fluid mixture (Baynes et al., 2002, Monteiro-Riviere et al., 2006). Small lacerations of the skin can be caused by sharp metal chips embedded in rags or clothing, increasing the likelihood of irritation or infection following dermal contact with cutting oils. The age of cutting oils may also impact the likelihood and severity of adverse health outcomes after exposure to cutting oils; the composition of the cutting oil, including the concentration of polycyclic aromatic hydrocarbons (PAHs) and the presence of microbes, changes over time (Evans et al., 1989; Apostoli et al., 1993).

Machinists who are prone to acne and seborrhea or are hirsute are relatively susceptible to blockage of hair follicles by insoluble oils and the consequent folliculitis, arising as a result of chemical irritation and mechanical plugging of the follicular canals. Individuals with a previously unresolved skin problem or who have a dry, atrophic skin are vulnerable to soluble fluids, especially clearly alkaline fluids or those containing wetting agents. Onset of the problem usually occurs soon after the first exposure and is marked by acute reactions starting on the dorsal surfaces of the hands and fingers, the extensor surfaces of the forearms and thighs, and the abdomen (i.e., those surfaces in contact with oil or oil-soaked clothing). Comedones and perifollicular papules and pustules ("oil boils") may develop with a possible latent development of melanosis. Secondary infections may occur, but the bacteria in the oil are rarely primary skin pathogens or the single cause of the folliculitis. Clinical manifestations clear rapidly with the termination of exposure and do not resolve if the exposure is continued. Hata et al. (2001) published a case study of a 48-year-old Japanese man who, after exposure to cutting oil on his forearms and dorsa of the hands, developed multiple erythematous lesions on areas that had no direct contact with the cutting oil (Hata et al., 2001). One explanation is that this reaction was mediated through an allergic response following initial contact with the cutting fluid, but this has not been verified. However, there are other reports of contact allergies developing after exposure to cutting fluids (Niklasson et al., 1993).

Exposures to mist sprays of insoluble oils used as coolants, cutting fluids, and lubricants in machine operations have been shown to have some impact on the respiratory tract. Mineral oil droplets <5 pg in diameter may be inhaled and result in fibrotic nodules, paraffinomas, or lipoid pneumonitis (Waldron, 1977). A study of metalworking apprentices showed the development of asthma-like symptoms after exposure to metalworking fluids (Kennedy et al., 1999). Another study showed that 20% of workers exposed to metalworking fluids reported daily or weekly respiratory symptoms, such as bronchitis, shortness of breath, runny nose, and sore throat; these are suggestive of work-related asthma (Rosenman et al., 1997). A number of other studies have investigated hypersensitivity pneumonitis outbreaks in workers exposed to metalworking fluids, especially where mycobacterial contamination occurred (Fox et al., 1999; Shelton et al., 1999; Hodgson et al., 2001; Bracker et al., 2003; O'Brien, 2003; Gupta and Rosenman, 2006).

Chronic Effects

Due to the wide array of cutting oils, the chronic effects are dependent on the individual components of the mixtures. Certain components, or their metabolites or derivatives, are carcinogenic. Specifically, several studies have shown that the concentration of PAHs increases with the age of the cutting oil (Evans et al., 1989; Apostoli et al., 1993). Although neither of these studies found mutagenicity or increased health effects, PAHs are known to be carcinogenic.

The carcinogenicity of cutting oils as a complete mixture has been the subject of a number of studies. These studies typically investigate organs of the aerodigestive tract, including the larynx, esophagus, stomach, pancreas, rectum, and lungs. This pattern of digestive tract cancers in exposed populations has been observed in other epidemiological studies with some respiratory tract cancer links as well (Silverstein et al., 1988; Eisen et al., 1992). A study concentrating on a cohort of automobile-manufacturing workers found an association between cancers of the esophagus, larynx, rectum, skin, brain, liver, prostate, and pancreas and exposure to various metalworking fluids (Eisen et al., 2001). A different study, using data from the same cohort, found an association between cancer of the larynx and metalworking fluid exposure and no association between stomach cancer and the exposure (Zeka et al., 2004).

The carcinogenicity of cutting fluids on other organs has been studied as well. A case-control study involving workers exposed to metalworking fluids showed that dermal exposure was associated with an increased risk for non-seminomatous testicular cancer, particularly among workers with over 5000 h of exposure (Behrens et al., 2012). Another case-control study conducted among Canadian workers found an increased risk of bladder cancer among workers exposed to mineral, cutting, or lubricating oil (Ugnat et al., 2004). A different case-control study in Canada also found an increased risk of renal cell carcinoma after exposure to mineral, cutting, or lubricating oil (Hu et al., 2002).

Dermal lesions can become carcinogenic after chronic exposure to cutting oils. Tsuji et al. (1992) reported multiple keratoses and squamous cell carcinoma on the forearms of a patient, who had worked with cutting oils for 15 years (Tsuji et al., 1992). Excess risk of cancer of the skin and scrotum has also been reported (Jarvholm and Easton, 1990).

The irritant effects and carcinogenicity of cutting oils on the skin have also been demonstrated in animal models. Dermatitides were observed on guinea pigs after repeated exposure to cutting oil, and treatment with emollient creams was not effective for alleviating the response (Goh, 1991). Two different cutting oils were shown to cause skin tumors in mice (Gilman and Vesselinovitch, 1956), and the ability of cutting oil to initiate tumors in mouse skin has been demonstrated (Gupta and Mehrotra, 1989).

Mechanism of Action(s)

Because emulsion and synthetic fluids are potent defatting agents, the skin reaction may include maceration, dryness and "chapping," reddening, and vesiculation. Bacterial growths in the fluid do not appear to be directly injurious to workers, but rancid fluids and products of bacterial action can lead to skin disorders. Individual additives in cutting fluids can be a cause of either primary irritative or hypersensitive dermatitis. Detergents, soaps, and wetting agents defat the skin; and alkaline materials damage the keratin of the upper, protective skin layers. Formalin in germicides is a sensitizer. Additives containing sulfur and chlorine are direct irritants, although chloracne is not associated with cutting fluids. Nickel or chromates derived from materials being cut can be a source of allergic dermatitis. Harsh abrasive soaps and solvents, such as gasoline and kerosene, may contribute to chemical and traumatic dermatitis; these cleaning materials are common in machine shops. Although grime and grease can certainly be removed from the skin with these substances, it is safer to use less hazardous cleansers.

SOURCES OF EXPOSURE

Occupational

Occupational exposures occur among individuals who work with metal-cutting tools, which utilize cutting oils to facilitate machinery performance. The most probable routes of exposure to cutting fluids are dermal contact with the liquid and inhalation of the aerosols.

Environmental

Cutting oils are not naturally occurring and thus have limited environmental exposure. The environmental fate or transport of given cutting oil that has been accidentally or purposefully introduced to the environment will depend heavily on the exact chemical composition of the cutting oil.

INDUSTRIAL HYGIENE

Contact dermatitis can be avoided through the use of appropriate control measures. Workers exposed to cutting fluids should be provided with adequate washing facilities. Machines should be enclosed to reduce the contact through splashing as well as to reduce potential inhalation of aerosolized cutting oil droplets. Sufficient clean clothing and protective garments should be provided. Cutting fluids should be replaced according to manufacturer recommendations to avoid harmful contaminants that increase over time.

MEDICAL MANAGEMENT

The appropriate medical response depends on the route of exposure. For inhalation exposures, the individual should be moved to fresh air, away from the exposure, and help from a medical professional should be sought if breathing difficulties arise. For skin contact, the exposed areas should be washed with soap and water immediately. If the fluid comes into contact with clothing, the clothing should be removed and laundered before the next wear. Medical attention should be sought if irritation or rash appears after contact with skin. If the fluid comes into contact with the eyes, they should be flushed with water for at least 15 min and immediate medical attention should be sought. If the fluid is ingested, medical assistance should be sought before taking any action.

REFERENCES

Apostoli, P., Crippa, M., Fracasso, M.E., Cottica, D., and Alessio, L. (1993) Increases in polycyclic aromatic hydrocarbon content and mutagenicity in a cutting fluid as a consequence of its use. *Int. Arch. Occup. Environ. Health* 64(7):473–477.

Baynes, R.E., Brooks, J.D., Barlow, B.M., and Riviere, J.E. (2002) Physicochemical determinants of linear alkylbenzene sulfonate (LAS) disposition in skin exposed to aqueous cutting fluid mixtures. *Toxicol. Ind. Health* 18(5):237–248.

Behrens, T., Pohlabeln, H., Mester, B., Langner, I., Schmeisser, N., and Ahrens, W. (2012) Exposure to metal-working fluids in the automobile industry and the risk of male germ cell tumours. *Occup. Environ. Med.* 69(3):224–226.

Bracker, A., Storey, E., Yang, C., and Hodgson, M.J. (2003) An outbreak of hypersensitivity pneumonitis at a metalworking plant: a longitudinal assessment of intervention effectiveness. *Appl. Occup. Environ. Hyg.* 18(2):96–108.

Eisen, E.A., Bardin, J., Gore, R., Woskie, S.R., Hallock, M.F., and Monson, R.R. (2001) Exposure-response models based on extended follow-up of a cohort mortality study in the automobile industry. *Scand. J. Work Environ. Health* 27(4):240–249.

Eisen, E.A., Tolbert, P.E., Monson, R.R., and Smith, T.J. (1992) Mortality studies of machining fluid exposure in the automobile industry I: a standardized mortality ratio analysis. *Am. J. Ind. Med.* 22(6):809–824.

Evans, M.J., Hooper, W.B., Ingram, A.J., Pullen, D.L., and Aston, R.H. (1989) The chemical, physical and biological properties of a neat cutting oil during prolonged use in a large manufacturing facility. *Ann. Occup. Hyg.* 33(4):537–553.

Fox, J., Anderson, H., Moen, T., Gruetzmacher, G., Hanrahan, L., and Fink, J. (1999) Metal working fluid-associated hypersensitivity pneumonitis: an outbreak investigation and case-control study. *Am. J. Ind. Med.* 35(1):58–67.

Gilman, J.P. and Vesselinovitch, S.D. (1956) An evaluation of the relative carcinogenicity of two types of cutting oil. *AMA Arch. Ind. Health* 14(4):341–345.

Goh, C.L. (1991) Cutting oil dermatitis on guinea pig skin. II. Emollient creams and cutting oil dermatitis. *Contact Dermatitis* 24(2):81–85.

Goh, C.L. and Gan, S.L. (1994) The incidence of cutting fluid dermatitis among metalworkers in a metal fabrication factory: a prospective study. *Contact Dermatitis* 31(2):111–115.

Grattan, C.E., English, J.S., Foulds, I.S., and Rycroft, R.J. (1989) Cutting fluid dermatitis. *Contact Dermatitis* 20(5):372–376.

Gupta, A. and Rosenman, K.D. (2006) Hypersensitivity pneumonitis due to metal working fluids: sporadic or under reported? *Am. J. Ind. Med.* 49(6):423–433.

Gupta, K.P. and Mehrotra, N.K. (1989) Tumor initiation in mouse skin by cutting oils. *Environ. Res.* 49(2):225–232.

Hata, M., Tokura, Y., and Takigawa, M. (2001) *Erythema multiforme-like eruption associated with contact dermatitis to cutting oil. Eur. J. Dermatol.* 11(3):247–248.

Hodgson, M.J., Bracker, A., Yang, C., Storey, E., Jarvis, B.J., Milton, D., Lummus, Z., Bernstein, D., and Cole, S. (2001) Hypersensitivity pneumonitis in a metal-working environment. *Am. J. Ind. Med.* 39(6):616–628.

Hu, J., Mao, Y., and White, K. (2002) Renal cell carcinoma and occupational exposure to chemicals in Canada. *Occup. Med. (Lond.)* 52(3):157–164.

Jarvholm, B. and Easton, D. (1990) Models for skin tumour risks in workers exposed to mineral oils. *Br. J. Cancer* 62(6): 1039–1041.

Kennedy, S.M., Chan-Yeung, M., Teschke, K., and Karlen, B. (1999) Change in airway responsiveness among apprentices exposed to metalworking fluids. *Am. J. Respir. Crit. Care Med.* 159(1):87–93.

Monteiro-Riviere, N.A., Inman, A.O., Barlow, B.M., and Baynes, R.E. (2006) Dermatotoxicity of cutting fluid mixtures:*in vitro* and *in vivo* studies. *Cutan. Ocul. Toxicol.* 25(4):235–247.

Niklasson, B., Bjorkner, B., and Sundberg, K. (1993) Contact allergy to a fatty acid ester component of cutting fluids. *Contact Dermatitis* 28(5):265–267.

O'Brien, D.M. (2003) Aerosol mapping of a facility with multiple cases of hypersensitivity pneumonitis: demonstration of mist reduction and a possible dose/response relationship. *App. Occup. Environ. Hyg.* 18(11):947–952.

Ramos, H. (1974) Occupational health case report. 5. Cutting oil mists. *J. Occup. Med.* 16(4):273–275.

Rosenman, K.D., Reilly, M.J., and Kalinowski, D. (1997) Work-related asthma and respiratory symptoms among workers exposed to metal-working fluids. *Am. J. Ind. Med.* 32(4):325–331.

Sakakibara, S., Kawabe, Y., and Mizuno, N. (1989) Photoallergic contact dermatitis due to mineral oil. *Contact Dermatitis* 20(4):291–294.

Shelton, B.G., Flanders, W.D., and Morris, G.K. (1999) *Mycobacterium* sp. as a possible cause of hypersensitivity pneumonitis in machine workers. *Emerg. Infect. Dis.* 5(2):270–273.

Silverstein, M., Park, R., Marmor, M., Maizlish, N., and Mirer, F. (1988) Mortality among bearing plant workers exposed to metal-working fluids and abrasives. *J. Occup. Med.* 30(9):706–714.

Tsuji, T., Otake, N., Kobayashi, T., and Miwa, N. (1992) Multiple keratoses and squamous cell carcinoma from cutting oil. *J. Am. Acad. Dermatol.* 27(5 Pt 1):767–768.

Ugnat, A.M., Luo, W., Semenciw, R., Mao, Y., and Canadian Cancer Registries Epidemiology Research Group (2004) Occupational exposure to chemical and petrochemical industries and bladder cancer risk in four western Canadian provinces. *Chronic Dis. Can.* 25(2):7–15.

Waldron, H.A. (1977) Health care of people at work. Exposure to oil mist in industry. *J. Soc. Occup. Med.* 27(2):45–49.

Zeka, A., Eisen, E.A., Kriebel, D., Gore, R., and Wegman, D.H. (2004) Risk of upper aerodigestive tract cancers in a case-cohort study of autoworkers exposed to metalworking fluids. *Occup. Environ. Med.* 61(5):426–431.

LUBRICATING OILS

Risk/safety phrases:	R12: Extremely flammable
	R45: May cause cancer
	R46: May cause heritable genetic damage
	R53: May cause long-term adverse effects in the aquatic environment
	R65: Harmful, may cause lung damage if swallowed
	S45: In case of accident or if you feel unwell seek medical advice immediately (show the label where possible)
	S53: Avoid exposure—obtain special instructions before use

BACKGROUND AND USES

Lubricating oils are based on aliphatic hydrocarbon molecules containing 17 or more carbon atoms. The wide array of lubricating oil mixtures are used to reduce the friction and heat produced when mechanical parts come into contact with one another. Similar to cutting oils, there are two major classifications: (1) mineral, which are made from refined petroleum, and (2) synthetic, which are made from other oil sources. The exact composition of lubricating oils varies, as lubricating oils are complex mixtures that contain small amounts of aromatic and polycyclic substances and a broad variety of dissimilar materials known collectively as additives. Additives serve a variety of functions, including inhibiting corrosion or oxidation, preserving integrity of the oil film, altering the viscosity of the mixture, suppressing bacterial growth within the mixture, or acting as a detergent. The presence of the additives determines the toxicity and possible adverse health outcomes of lubricating oils.

PHYSICAL AND CHEMICAL PROPERTIES

Physical and chemical properties are determined by the specific composition of given lubricating oil and thus vary between different oils. Information regarding the physical and chemical properties of particular lubricating oil should be obtained from the manufacturer.

MAMMALIAN TOXICOLOGY

Acute Effects

Because of the similarity in the two types of mixtures, any adverse health effects seen after exposure to lubricating oils will be similar to those seen after exposure to cutting oils. However, fewer studies have been completed on lubricating oils than on cutting oils, and most of the studies characterize the effects of chronic exposure to lubricating oils. It should be noted that the additives found in lubricating oil mixtures pose the greatest risk for toxicity.

Chronic Effects

There have been a few studies conducted on the chronic effects of lubricating oils. One group found evidence of dermal irritation in mice following chronic application of a lubricating compound (Arfsten et al., 2005). Another study exposed Sprague-Dawley rats to aerosols of base stocks used to create lubricating oils; after exposure, there were a few effects seen in the lungs of the rats, but the limited effects at high doses suggest a relatively low toxicity for these base stock aerosols (Dalbey et al., 1991).

Lubricating oils have also been shown to cause DNA adducts to form in mouse and human skin exposed to these oils (Carmichael et al., 1991). A case-control study conducted among Canadian workers found an increased risk of bladder cancer among workers exposed to mineral, cutting, or lubricating oil (Ugnat et al., 2004). A different case-

control study in Canada also found an increased risk of renal cell carcinoma after exposure to mineral, cutting, or lubricating oil (Hu et al., 2002). However, as with cutting oils, the carcinogenic effects of lubricating oils are likely due to specific compounds found within the mixtures. One study found that after lubricating oils had been treated to significantly reduce PAH levels, the ability of the oil to cause skin tumors in mice was ablated (McKee et al., 1989).

Mechanism of Action(s)

The mechanism of action will depend on the nature of the additive responsible for the adverse health effect. For studies concerning adverse health outcomes following general lubricating oil exposure, mechanisms have not been well characterized.

SOURCES OF EXPOSURE

Occupational

Occupational exposures occur among individuals who work with machines utilizing lubricating oils to enhance machine performance; due to the nature of lubricating oils and modern industry, lubricating oils are likely present in nearly all industrial settings. The most probable routes of exposure to lubricating oils are dermal contact with the liquid and inhalation of the aerosols.

Environmental

Lubricating oils are not naturally occurring and thus have limited environmental exposure. The environmental fate or transport of given lubricating oil that has been accidentally or purposefully introduced to the environment will depend heavily on the exact chemical composition of the lubricating oil.

INDUSTRIAL HYGIENE

The most effective industrial hygiene measures are engineering controls. Adequate ventilation should be ensured, and the fluid should be kept away from ignition sources. Eye and face protection, along with protective clothing such as boots, gloves, and apron or coveralls, should be provided.

MEDICAL MANAGEMENT

Medical action following exposure will depend on the nature of the exposure and the severity of health effects. For inhalation exposures, the individual should be removed from the area, and oxygen should be provided if breathing is difficult; if breathing is not occurring, artificial respiration should be performed. Medical attention should be sought immediately following difficulty or cessation of breathing. For skin exposures, the area should be washed with soap and water. Contaminated clothing should be removed and washed before reuse. Medical attention should be sought immediately. If contact with the eyes, eyes should be flushed immediately with water for at least 15 min. Medical attention should be sought immediately. If swallowed, medical attention should be sought immediately.

REFERENCES

Arfsten, D.P., Johnson, E.W., Thitoff, A.R., Jung, A.E., Still, K.R., Brinkley, W.W., Schaeffer, D.J., Jederberg, W.W., and Bobb, A.J. (2005) Acute and subacute dermal toxicity of break-free CLP: a weapons cleaning and maintenance compound. *J. Appl. Toxicol.* 25(4), 318–327.

Carmichael, P.L., Ni She, M., and Phillips, D.H. (1991) DNA adducts in human and mouse skin maintained in short-term culture and treated with petrol and diesel engine lubricating oils. *Cancer Lett.* 57(3):229–235.

Dalbey, W., Osimitz, T., Kommineni, C., Roy, T., Feuston, M., and Yang, J. (1991) Four-week inhalation exposures of rats to aerosols of three lubricant base oils. *J. Appl. Toxicol.* 11(4):297–302.

Hu, J., Mao, Y., and White, K. (2002) Renal cell carcinoma and occupational exposure to chemicals in Canada. *Occup. Med. (Lond.)* 52(3):157–164.

McKee, R.H., Daughtrey, W.C., Freeman, J.J., Federici, T.M., Phillips, R.D., and Plutnick, R.T. (1989) The dermal carcinogenic potential of unrefined and hydrotreated lubricating oils. *J. Appl. Toxicol.* 9(4):265–270.

Ugnat, A.M., Luo, W., Semenciw, R., Mao, Y., and Canadian Cancer Registries Epidemiology Research Group (2004) Occupational exposure to chemical and petrochemical industries and bladder cancer risk in four western Canadian provinces. *Chronic Dis. Can.* 25(2):7–15.

54

ALCOHOLS AND GLYCOLS

Kelly W. Hall

1-PENTANOL

First aid: See text

Target organ(s): Central nervous system (CNS), lungs, nerves, skin

Occupational exposure limits: Derived no effect level (DNEL, inhalation, systemic/local effects, long term): 73.16 mg/m^3; DNEL, inhalation, systemic/local effects, short term: 292 mg/m^3

Reference values: DNEL, systemic/local effects, long term: 15.4 mg/m^3; DNEL, systemic/local effects, short term: 256.4 mg/m^3; DNEL, oral, systemic effects, long term: 25 mg/kg-bw/day.

Hazard statements: H226—flammable liquid and vapor; H315—causes skin irritation; H319—causes serious eye irritation; H332—harmful if inhaled; H335—may cause respiratory irritation

Risk phrases: R10—flammable; R20—harmful by inhalation; R37/38—irritating to the respiratory system and skin

Safety phrases: S1/2—keep locked up and out of reach of children; S26—in case of contact with eyes, rinse immediately with plenty of water and seek medical advice; S36/37—wear suitable protective clothing and gloves; S46—if swallowed, seek medical advice immediately and show this container or label

BACKGROUND AND USES

Pentanols are irritating, narcotic, and have produced illnesses and fatalities upon ingestion. However, this class of alcohols is of a low order of toxicity in an industrial setting. As the carbon chain lengths of the alcohols increase, the toxicity decreases. Higher chain alcohols are not able to penetrate the skin as readily as smaller molecular weight alcohols and are less likely to be inhaled. There are no current regulations regarding occupational exposures to pentanols.

1-Pentanol (pentyl or amyl alcohol; "fusel oil"; CASRN 71-41-0) is the most common pentanol and exhibits the more prevalent hazards of these compounds. It is a colorless liquid with a pleasant odor.

PHYSICAL AND CHEMICAL PROPERTIES

More soluble in organic solvents than water, 1-pentanol has a vapor pressure of 2.8 mmHg at 20 °C (68 °F) and a vapor density of 3.0. Although reagent grade pentanol can be obtained, a mixture of three isomers is usually present in commercial grade formulations. It is used as an industrial solvent and is also a synthetic flavoring agent. In nature, pentanol is detected in animal waste and in some plant essential oils.

MAMMALIAN TOXICOLOGY

Acute Effects

Acute effects commonly experienced upon exposure to 1-pentanol vapor include irritation of the eyes with lacrimation and corneal disturbances. No significant or permanent damage has been reported as a result of 1-pentanol overexposure. Respiratory tract irritation at high concentrations and some

Hamilton & Hardy's Industrial Toxicology, Sixth Edition. Edited by Raymond D. Harbison, Marie M. Bourgeois, and Giffe T. Johnson.
© 2015 John Wiley & Sons, Inc. Published 2015 by John Wiley & Sons, Inc.

CNS effects, including vertigo, double vision, and headache may occur.

Female rats exposed to 3900 parts per million (ppm) of 1-pentanol vapor for 7 days during gestation did not exhibit any maternal or fetal toxicity (Nelson et al., 1990). Interestingly, the authors intended to test an exposure to a higher concentration of 1-pentanol vapor, but they could not generate high enough concentrations to produce maternal toxicity.

Chronic Effects

Chronic exposure to this chemical has not yielded any significant health hazards except for infrequent mild dermatitis (Fregert et al., 1963).

Mechanism of Action(s)

1-Pentanol can produce irritation and inflammation to the pulmonary, gastrointestinal, and dermal tissue. This chemical may also act as a neurological depressant.

SOURCES OF EXPOSURE

Occupational

1-Pentanol is not considered a significant occupational hazard, except under conditions of high exposure. This chemical is produced in processes in which fractional distillation of mixed alcohols resulting from the chlorination and alkaline hydrolysis of pentane occurs.

Environmental

Environmental exposure to 1-pentanol comes occurs due to the inhalation of ambient air and consumption of drinking water in contaminated areas. Substantial quantities of this chemical are not commonly found in the environment.

INDUSTRIAL HYGIENE

1-Pentanol is not considered a significant occupational hazard, and minimal precautions are required to prevent exposures. Barrier creams and protective clothing should be worn to prevent dermal contact. Any standard organic vapor adsorbent is an appropriate respirator choice. At high concentrations, a self-contained oxygen supply may be necessary.

MEDICAL MANAGEMENT

In case of contact with the eyes, flush with warm running water for at least 20 min, holding the eyelids open during flushing. Take care not to flush contaminated water into unaffected eye or onto skin. In case of skin contact, immediately remove contaminated clothing under running water. Flush exposed area with large amounts of warm running water for at least 20 min. Decontaminate clothing before reuse or discard. For inhalational exposures, remove to fresh air. Give oxygen and get immediate medical attention for any breathing difficulty. If ingested, do not induce vomiting. If the casualty is alert and not convulsing, give 2–4 glasses of water to drink to dilute 1-pentanol. If spontaneous vomiting occurs, have the casualty lean forward to avoid breathing in of emesis. Rinse mouth and administer more water. Get medical attention immediately (ECHA 2013).

REFERENCES

ECHA (2013) *Registered Substances*, Helsinki, Finland: European Chemicals Agency. Available at http://echa.europa.eu/information-on-chemicals/registered-substances (CASRN: 71-41-0).

Fregert, S., et al. (1963) Dermatitis from alcohols. *J. Allergy* 34:404–408.

Nelson, B.K., Brightwell, W.S., and Krieg, E.F., Jr., (1990) Developmental toxicology of industrial alcohols: a summary of 13 alcohols administered by inhalation to rats. *Toxicol. Ind. Health* 6:373–387.

ALLYL ALCOHOL

First aid: See text

Target organ(s): Liver, lungs

Occupational exposure limits: American Conference of Governmental Industrial Hygienists (ACGIH, 1996) threshold limit value (TLV) 0.5 ppm (1.25 mg/m^3), Occupational Safety and Health Administration (OSHA) permissible exposure limit (PEL) 2 ppm (5 mg/m^3), recommended exposure limit (REL) 2 ppm (5 mg/m^3) and 4 ppm, short-term exposure limit (STEL) 9.5 mg/m^3 (skin)

Reference values: Reference dose (RfD) -5×10^{-3} mg/kg/day

Hazard statements: H225—highly flammable liquid and vapor; H301—toxic if swallowed; H311—toxic in contact with skin; H331—toxic in contact with skin; H319—causes serious eye irritation; H335—may cause respiratory irritation; H400—very toxic to aquatic life

Precautionary statements: P210—keep away from heat/sparks/open flames/ . . . /hot surfaces . . . no smoking; P280—wear protective gloves/protective clothing/eye protection/face protection; P303—if on skin (or hair), rinse skin with water/shower and remove immediately all contaminated clothing; P305—if in eyes, rinse cautiously with water for several minutes. Remove contact lenses, if present and easy to do. Continue rinsing; P501—dispose of contents/container to . . . in accordance with local/regional/national/international regulations.

Risk phrases: R23/24/25—toxic by inhalation, in contact with skin, and if swallowed; R36/37/38—irritating to the eyes, respiratory system, and skin; R50—very toxic to aquatic organisms; R10—flammable

Safety phrases: S1/2—keep locked up and out of reach of children; S36/37/39—wear suitable protective clothing, gloves, and eye/face protection; S38—in case of insufficient ventilation, wear suitable respiratory equipment; S45—in case of accident or if you feel unwell, seek medical advice immediately (show the label where possible); S61—avoid release to the environment. Refer to special instructions/safety data sheets

BACKGROUND AND USES

Allyl alcohol ($H_2C=CH-CH_2-OH$; 2-propanol, 1-propen-3-ol; CASRN 107-18-6) is a chemical intermediate with potent irritant properties. Existing as a colorless liquid at room temperature, allyl alcohol has a pungent, mustard-like odor. It is used widely as a chemical intermediate in resins, flavorings, epoxies, and reagent grade herbicide.

PHYSICAL AND CHEMICAL PROPERTIES

Of the olefinic alcohols (those with a double bond), allyl alcohol is considered the most important in the industry. It is miscible with water and organic solvents. At 25 °C (77 °F), allyl alcohol has a vapor pressure of 23.8 mmHg and a vapor density of 2.0. It is not known to occur naturally.

MAMMALIAN TOXICOLOGY

Acute Effects

Inhalation of allyl alcohol vapors produces dose-related effects on the respiratory system and mucous membranes. Signs and symptoms of acute, high concentration exposures may include coughing, dry throat, and irritation of the eyes and nose. Lacrimation, retrobulbar pain, photophobia, and blurring vision may also result from exposures to vapors (Dunlap et al., 1958).

McLaughlin (1952) treated six cases of eye irritation or burns and reported no permanent damage. Resolution of five cases occurred within 48 h. Absorption through the skin leads to deep muscle pain, presumably caused by spasm. Allyl alcohol and many of its esters are hepatic toxicants that cause periportal necrosis and other liver injury in laboratory animals (Kodama and Hine, 1958; Taylor et al., 1964). Occupational exposure to allyl alcohol has not been associated with human liver toxicity.

Although allyl alcohol and its metabolite, acrolein, have been reported to be mutagenic in some assays (Smith et al., 1990), their carcinogenicity has not been established.

Chronic Effects

There have been no signs or symptoms of chronic allyl alcohol exposure reported in the literature.

Mechanism of Action(s)

The metabolism of allyl alcohol to acrolein has been demonstrated in rat liver and may be the cause of hepatotoxicity (Reid, 1972; Patel et al., 1983). The hepatotoxicity of allyl alcohol is apparently dependent on glutathione depletion, followed by lipid peroxidation, which produces free radicals and subsequent liver damage (Pompella et al., 1988; Maellaro et al., 1990). Reports of human organ damage are not available in the literature.

SOURCES OF EXPOSURE

Occupational

Inhalation and dermal exposure where allyl alcohol is produced or used are the main occupation exposures.

Environmental

There are no data on environmental exposures of allyl alcohol as it does not occur naturally.

INDUSTRIAL HYGIENE

Allyl alcohol can be produced from a dehydration and reduction reaction of glycerol. The most likely route of an occupational exposure is inhalation, though dermal contact is also considered an occupational hazard. When allyl alcohol is used in an open system, exhaust ventilation is essential and cleanup of spills requires the use of personal protective equipment. The strong odor of allyl alcohol is a good warning for workers about a potential exposure. Ruth (1986) reported an odor threshold of 1.95 mg/m³ (800 parts per billion (ppb), well below the allowable workplace levels, which are in the order of ppm. The irritation concentration is 12.5 mg/m³ (5 ppm). Workers experiencing eye and nose irritation produced by allyl alcohol would be alerted to possible overexposure and would be able to exit a hazardous environment.

High air concentrations will be highly irritating to the skin, eyes, and respiratory system.

Allyl alcohol can be absorbed through the skin in lethal amounts. Dermal protection should be worn. Butyl rubber, Viton®, and Teflon® have been tested against permeation by allyl alcohol and demonstrated acceptable protection for more than 8 h. Safety glasses, goggles, or face shields should be worn during operations in which allyl alcohol might contact the eyes (Anon, 1996).

MEDICAL MANAGEMENT

Tissue destruction and blindness may result from an exposure to the eye area. The eyes should be flushed gently with large amounts of water. If there is dermal exposure, severe burns, skin corrosion, and absorption of lethal amounts may result. The skin should be washed with soap and water if the skin is intact, or just water if the skin is not intact. For an exposure by ingestion, do not induce vomiting.

REFERENCES

American Conference of Governmental Industrial Hygienists (ACGIH) (1996) *Threshold Limit Values (TLVs) for Chemical Substances and Physical Agents and Biological Exposure Indicies (BEIs)*, Cincinnati, OH: American Conference of Governmental Industrial Hygienists.

Anon (1996) Occupational Safety and Health Guideline for Allyl Alcohol, Washington, DC: US Department of Halth and Human Services.

Dunlap, M.K., et al. (1958) The toxicity of allyl alcohol. I. Acute and chronic toxicity. *AMA Arch. Ind. Health* 18:303–311.

Kodama, J.K. and Hine, C.H. (1958) Pharmacodynamic aspects of allyl alcohol toxicity. *J. Pharmacol. Exp. Ther.* 124:97–107.

Maellaro, E., et al. (1990) Lipid peroxidation and antioxidant systems in the liver injury produced by glutathione depleting agents. *Biochem. Pharmacol.* 39:1513–1521.

McLaughlin, R.S. (1952) Chemical burns of the human cornea. *Am. J. Ophthalmol.* 29:1388–1391.

Patel, J.M., et al. (1983) Comparison of hepatic biotransformation and toxicity of allyl alcohol and [1,1-^{2}H$_2$] allyl alcohol in rats. *Drug Metab. Dipos.* 11:164–166.

Pompella, A., et al. (1988) 4-Hydroxynonenal and other lipid peroxidation products are formed in mouse liver following intoxication with allyl alcohol. *Biochim. Biophys. Acta* 961:293–298.

Reid, W.D. (1972) Mechanism of allyl alcohol-induced hepatic necrosis. *Experientia* 28:1058–1061.

Ruth, J.H. (1986) Odor thresholds and irritation levels of several chemical substances: a review. *Am. Ind. Hyg. Assoc. J.* 47: A142–A151.

Smith, R.A., Cohen, S.M., and Lawson, T.A. (1990) Acrolein mutagenicity in the V79 assay. *Carcinogenesis* 11:497–498.

Taylor, J.M., Jenner, P.M., and Jones, W.I. (1964) A comparison of toxicity of some allyl, propenyl, and propyl compounds in the rat. *Toxicol. Appl. Pharmacol.* 6:378–387.

METHANOL

First aid: See text

Target organ(s): CNS, ocular

Occupational exposure limits: OSHA PEL 200 ppm (260 mg/m^3), NIOSH TWA 200 ppm (260 mg/m^3), 250 ppm (325 mg/m^3 skin)

Reference values: RfD 0.5 mg/kg/day reference concentration (RfC) (inhalation) 298 mg/m^3

Hazard statements: H225—highly flammable liquid and vapor; H331—toxic if inhaled; H311—toxic in contact with skin; H301—toxic if swallowed; H370—causes damage to organs <or state all organs affected, if known> <state route of exposure if it is conclusively proved that no other routes of exposure cause the hazard>. Additional text: Target organs: Optic nerve (nervus opticus), CNS

Precautionary statements: P210—keep away from heat/sparks/open flames/ . . . /hot surfaces . . . no smoking; P233—keep container tightly closed; P240—ground/bond container and receiving equipment; P241—use explosion-proof electrical/ventilating/lighting/ . . . /equipment; P242—use only non-sparking tools; P243—take precautionary measures against static discharge; P260—do not breathe dust/fume/gas/mist/vapors/spray; P261—avoid breathing dust/fume/gas/mist/vapors/spray; P264—wash . . . thoroughly after handling; P270—do not eat, drink, or smoke when using this product; P271—use only outdoors or in a well-ventilated area; P280—wear protective gloves/protective clothing/eye protection/face protection; P301 + P310—if swallowed, immediately call a poison center or doctor/physician; P302 + P352—if on skin, wash with plenty of soap and water; P303 + P361 + P353—if on skin (or hair), remove/take off immediately all contaminated clothing. Rinse skin with water/shower; P304 + P340—if inhaled, remove the victim to fresh air and make him rest in a position comfortable for breathing; P308—if exposed or concerned, +P311—call a poison center or doctor/physician, +P312—call a poison center or doctor/physician if you feel unwell; P321—specific treatment (see . . . on label); P330—rinse mouth; P361—remove/take off immediately all contaminated clothing; P370 +P378—in case of fire, use . . . for extinction;

P403 + P235—store in well-ventilated place. Keep cool; P405—store locked up; P501—dispose of contents/container to . . . hazardous or special waste collection point.

Risk phrases: R11—highly flammable; R23/24/25—toxic by inhalation, in contact with skin, and if swallowed; R39/23/24/25—toxic: danger of very serious irreversible effects through inhalation, in contact with skin, and if swallowed.

Safety phrases: S1/2—keep locked up and out of reach of children; S7—keep container tightly closed; S16—keep away from sources of ignition . . . no smoking; S36/37—wear suitable protective clothing and gloves; S45—in case of accident or if you feel unwell, see medical advice immediately (show label where possible)

BACKGROUND AND USES

Methanol (symbol CH_3OH; methyl alcohol; wood alcohol; CASRN 67-56-1) is used extensively as a solvent for lacquers, paints, varnishes, cements, inks, dyes, plastics, and various industrial coatings. Large quantities are used in the production of formaldehyde and other chemical derivatives such as acetic acid, methyl halides and terephthalate, methyl methacrylate, and methylamines. Methanol is also used as a gasoline additive, as a component of lacquer thinners, in antifreeze preparations of the "nonpermanent" type, and in canned heating preparations of jellied alcohol. It is also used in duplicating fluid, in paint removers, and as a cleaning agent. The potential for methanol as a future alternative for gasoline indicates that the use of this chemical will most likely increase rather than decrease. At room temperature, methanol is a colorless liquid with a pungent odor. It is relatively volatile, with a vapor pressure of 96 mmHg and a vapor density of 1.11. It is miscible with water and soluble with other organic solvents. It is found in nature as a fermentation product of wood and as a constituent of some fruits and vegetables. Some endogenous metabolic reactions, such as a methyltransferase enzyme system, result in methanol formation. Aspartame, an artificial sweetener, may form up to 10% methanol by weight in the gut (Kavet and Nauss, 1990). In the past, methanol was produced from the distillation of wood, leaving a mixture with some additives that made consumption distasteful. Most of the methanol produced in the United States is now prepared synthetically by the reduction of carbon monoxide with hydrogen.

PHYSICAL AND CHEMICAL PROPERTIES

Methanol (CASRN 67-56-1) is a clear, colorless liquid with a characteristic pungent odor (NTP NIEHS web accessed 2/16/2013). Methanol has a molecular weight of 32.04 g/mol and a boiling point of 79 °C. The air odor threshold has been reported as 1500 ppm (approximately 2000 mg/m^3), much higher than the occupational guidelines.

MAMMALIAN TOXICOLOGY

Acute Effects

The most important routes of occupational exposure are inhalation and dermal contact. Accidental or intentional ingestion is also of some concern. Nearly all of the information on methanol toxicity in humans is from acute exposures. Cases of intoxication following oral consumption are the most frequently reported. The minimal lethal dose in humans has not been determined, but it has been suggested that ingestion of about 1 g/kg can cause death if the patient is untreated. Ingestion of methanol causes ataxia, drowsiness, dysarthria, and nystagmus within 30 min, followed by a latent period of 12–24 h before metabolic toxicity becomes apparent. Severe metabolic acidosis occurs after the latent period. Symptoms include nausea, vomiting, and headache. Ocular toxicity generally develops 12–48 h after ingestion. Blurring of the vision, occasionally with changes in color perception and scotomata, and constriction of visual fields also typically develop suddenly with little warning. Poor prognosis includes convulsions, coma, shock, persistent acidosis, bradycardia, and renal failure.

Acute exposures to high concentrations of methanol vapors have been reported to cause primarily headache, dizziness, and nausea (Frederick et al., 1984). Two firefighters exposed to concentrations of vaporized methanol complained only of mild headache, although the exposure concentrations were unknown (Aufderheide et al., 1993). Downie et al. (1992) reported effects in a worker involved in cleaning a methanol tank wearing a respirator but no protective clothing. After a few hours of continuous dermal contact, visual symptoms similar to ingestion exposures were noted. The systemic toxicity of methanol is similar regardless of the route of exposure. The effects of high levels of exposure by inhalation or dermal contact resemble the effects of low level exposure by ingestion.

Chronic Effects

The effects of chronic methanol exposure are not well characterized, as few epidemiological studies have been reported. Information suggests that prolonged exposure to methanol may have similar effects as high concentrations in an acute exposure (Kavet and Nauss, 1990). A series of case reports published by Wood and Buller (1904) and

Zeigler (1921) examined chronic occupational and nonoccupational inhalation, as well as dermal and ingestion exposure to methanol. Visual deterioration, headache, and nausea were commonly reported. The few epidemiological reports by Tyson (1912) and Frederick et al. (1984) revealed similar health effects in chronically exposed worker populations. These investigations revealed no association between the cancer mortality and methanol exposure. One case report linked chronic methanol exposure to multiple sclerosis-like symptoms, and autopsy revealed plaques and lesions of the CNS similar to this disease (Henzi, 1984). There have been no significant findings in animal and human studies of developmental or teratogenic effects in populations exposed to concentrations allowable in the workplace. Lee et al. (1992) have shown in humans that the PEL of 200 ppm does not produce formate accumulation endogenously and therefore does not produce toxicity.

Mechanism of Action(s)

The metabolic pathway and subsequent mechanism of toxicity of methanol in humans have been identified as a series of stepwise oxidation reactions, first to formaldehyde and then to formic acid (Roe, 1943). Methanol oxidation to formaldehyde is catalyzed by alcohol dehydrogenase, a zinc metal enzyme, which oxidizes other alcohols as well (Li and Vallee, 1969). The formic acid formed in humans must be oxidized to carbon dioxide to prevent toxicity. The enzyme responsible for this conversion is hepatic tetrahydrofolate (THF). Tetrahydrofolate is found in lower concentrations in human and nonhuman primates than in rodent species, indicating this deficiency as the cause for the toxicity differences noted between species. Johlin et al. (1987) reported two possible mechanisms for formic acid accumulation in primates, including low THF levels (up to 50% less in primates) and reduced activity of 10-formyltetrahydrofolate dehydrogenase.

Martin-Amat et al. (1978) described the accumulation of formate as the cause of the ocular injury associated with methanol. More recent reports have indicated the Müller cell as the primary target for ocular toxicity in folate-deficient rats (Garner et al., 1995). Similar animal studies have indicated direct retinal damage as the primary effect in acute methanol exposures resulting from formic acid accumulation (Murray et al., 1991).

Animal studies have been conducted using methanol vapor. Inhalation studies with rodents at high concentrations indicate a narcotic effect with methanol similar to other organic solvents but no severe, overt pathological changes systemically or locally in the lung (White et al., 1983; Andrews et al., 1987; Poon et al., 1995; Stanton et al., 1995). Nonhuman primate models have provided more valuable data for identifying methanol hazards. Andrews et al. (1987) exposed

cynomolgus monkeys to methanol vapor at up to 5000 ppm and did not identify specific target organ effects.

Chemical Pathology

Blood and urine samples should be collected for methanol and formic acid determination. Quantitation can be performed by colorimetry or gas chromatography. Normal blood concentration of methanol derived from endogenous production and dietary sources are in the range of 2–30 mg/l. (Baselt, 1982).

SOURCES OF EXPOSURE

Occupational

The most important routes of occupational exposures are inhalation and dermal contact. Accidental ingestion of methanol is also of some concern. The absorption and distribution of methanol by any of these routes occur rapidly.

Environmental

Reese and Kimbrough (1993) indicated a higher incidence of vapor exposure from vehicle-specific emissions as methanol replaces conventional gasoline. Exposures in parking garages, traffic, and at service stations during refueling may present significant inhalation exposures.

INDUSTRIAL HYGIENE

Proper ventilation and protective equipment should be worn when worker exposure is likely to exceed RELs. Frederick et al. (1984) demonstrated the effectiveness of simple exhaust and ventilation engineering controls in decreasing airborne concentrations of methanol near duplicating machines up to 96%. These airborne concentrations of methanol can be monitored using a personal diffusive sampler designed for formaldehyde with an absorbent matrix of water (Kawai et al., 1990).

Unlike many other solvents, concentrations likely encountered in the workplace may not be easily detected by the average worker. Scherberger et al. (1958) found that the average minimum identifiable odor level for methanol was 1500 ppm (roughly 2000 mg/m^3), much higher than the occupational guidelines. Protective gloves and clothing should be worn when a dermal contact is probable. Proper labeling and awareness of ingestion hazards should be sufficient to avoid an accidental methanol intake. This alcohol is designated by the National Institute for Occupational Safety and Health (NIOSH, 1994) as a Class IB flammable compound.

MEDICAL MANAGEMENT

Acute methanol ingestion is likely to be life threatening at high concentrations and should be treated immediately. Methanol has similar CNS depressant effects as ethanol and produces the same euphoria and light-headedness. Because the metabolic acidosis and eye injury resulting from methanol poisoning are related to the metabolites of methanol rather than to the alcohol itself, treatment of methanol intoxication attempts to impair this enzymatic metabolism. Ethanol can compete with methanol for active sites of this enzyme, and in fact the enzyme has a greater affinity for the former than the latter. Consequently, the therapeutic administration of ethanol permits it to be preferentially oxidized and diminishes the production of methanol metabolites. Roe (1943) suggested that sufficient ethanol must be administered to produce a blood level of at least 0.1%. The efficacy of ethanol treatment for methanol intoxication was reported by Bergeron and Cardinal (1982).

In cases of methanol intoxication, the pupils may be dilated and unreactive. Visual loss is associated with hyperemia of the optic disc, and blurring warns of at least some degree of permanent loss of vision. Atrophy of the disc is observed in some 30–60 days after excessive exposure. Pathological changes noted in fatal cases of methanol ingestion were primarily edema, hyperemia, and necrosis of the stomach, intestines, brain, and retinas (Menne, 1938).

Control of acidosis is essential in the treatment of methanol poisoning. The use of intravenous sodium bicarbonate when blood pH drops below 7.2 is recommended and has frequently been lifesaving. The value of hemodialysis or peritoneal dialysis has been emphasized (Cowen, 1964; Keyvan-Larijarni and Tannenberg, 1974). The availability of one or the other form of dialysis and of rapid determinations of blood methanol levels and acid–base balance has dramatically improved the prognosis of methanol intoxication.

Computed tomography of the skull can be helpful in the diagnosis by demonstrating symmetrical areas of low attenuation in the putamens (Aquilonius et al., 1978; McLean et al., 1980).

REFERENCES

Andrews, L.S., et al. (1987) Subchronic inhalation toxicity of methanol. *J. Toxicol. Environ. Health* 20:117–I24.

Aquilonius, S.-M., et al. (1978) Computerised tomography in severe methanol intoxication. *Br. Med. J.* 2:929–930.

Aufderheide, T.P., et al. (1993) Inhalational and percutaneous methanol toxicity in two firefighters. *Ann. Emerg. Med.* 22:1916–1918.

Baselt, R.C. (1982) *Disposition of Toxic Drugs and Chemicals in Man*, 2nd ed., Davis, CA: Biomedical Publications, pp. 492.

Bergeron, R. and Cardinal, J. (1982) Geadah D: prevention of methanol toxicity by ethanol therapy. *N. Engl. J. Med.* 307:1528–1534.

Cowen, D.L. (1964) Extracorporeal dialysis in methanol poisoning. *Ann. Intern. Med.* 61:134–135.

Downie, A., et al. (1992) A case of percutaneous industrial methanol toxicity. *Occup. Med. (Lond.)* 42:47–49.

Frederick, L.J., Schulte, P.A., and Apol, A. (1984) Investigation and control of occupational hazards associated with the use of spirit duplicators. *Am. Ind. Hyg. Assoc. J.* 45:51–55.

Garner, C.D., Lee, E.W., and Louis-Ferdinand, R.T. (1995) Mfiller cell involvement in methanol-induced retinal toxicity. *Toxicol. Appl. Pharmacol.* 130:101–107.

Henzi, H. (1984) Chronic methanol poisoning with the clinical and pathologic–anatomical features of multiple sclerosis. *Med. Hypotheses* 13:63–75.

Johlin, F.C., et al. (1987) Studies on the role of folic acid and folate-dependent enzymes in human methanol poisoning. *Mol. Pharmacol.* 31:557–561.

Kavet, R. and Nauss, K.M. (1990) The toxicity of inhaled methanol vapors. *Crit. Rev. Toxicol.* 21:21–50.

Kawai, T., et al. (1990) Personal diffusive sampler for methanol, a hydrophilic solvent. *Bull. Environ. Contam. Toxicol.* 44:514–520.

Keyvan-Larijarni, H. and Tannenberg, A.M. (1974) Methanol intoxication: comparison of peritoneal dialysis and hemodialysis treatment. *Arch. Intern. Med.* 134: 293–296.

Lee, E.W., et al. (1992) Lack of blood formate accumulation in humans following exposure to methanol vapor at the current permissible exposure limit of 200 ppm. *Am. Ind. Hyg. Assoc. J.* 53:99–104.

Li, T.-K. and Vallee, B.L. (1969) Alcohol dehydrogenase and ethanol metabolism. *Surg. Clin. North Am.* 49:577–582.

Martin-Amat, G., et al. (1978) Methanol poisoning: ocular toxicity produced by formate. *Toxicol. Appl. Pharmacol.* 45:201–208.

McLean, D.R., Jacobs, H., and Mielke, B.W. (1980) Methanol poisoning: a clinical and pathological study. *Ann. Neurol.* 8:161–167.

Menne, F.R. (1938) Acute methyl alcohol poisoning: a report of twenty-two instances, with postmortem examinations. *Arch. Pathol.* 26:77–92.

Murray, T.G., et al. (1991) Methanol poisoning: a rodent model with structural and functional evidence for retinal involvement. *Arch. Ophthalmol.* 109:1012–1016.

National Institute for Occupational Safety and Health (NIOSH) (1994) *Pocket Guide to Chemical Hazards*, NIOSH 94-116, Washington, DC: US Department of Health and Human Services.

Pooh, R., et al. (1995) Short-term inhalation toxicity of methanol, gasoline, and methanol/gasoline in the rat. *Toxicol. Ind. Health* 11:343–361.

Reese, E. and Kimbrough, R.D. (1993) Acute toxicity of gasoline and some additives. *Environ. Health Perspect.* 101(Suppl. 6):115–131.

Roe, O. (1943) Clinical investigations of methyl alcohol poisoning with special reference to the pathogenesis and treatment of amblyopia. *Acta Med. Scand.* 113:558–608.

Scherberger, R.F., et al. (1958) A dynamic apparatus for preparing air–vapor mixtures of known concentrations. *Am. Ind. Hyg. Assoc. J.* 19:494–498.

Stanton, M.E., et al. (1995) Assessment of offspring development and behavior following gestational exposure to inhaled methanol in the rat. *Fundam. Appl. Toxicol.* 28:100–110.

Tyson, H.H. (1912) Amblyopia from inhalation of methyl alcohol. *Arch. Ophthalmol. (NY)* 41:459–471.

White, L.R., et al. (1983) Biochemical and cytological studies of rat lung after inhalation of methanol vapour. *Toxicol. Lett.* 17:1–5.

Wood, C.A. and Buller, F. (1904) Poisoning by wood alcohol: cases of death and blindness from Columbian spirits and other methylated preparations. *JAMA* 43:972–980.

Zeigler, S.L. (1921) The ocular menace of wood alcohol poisoning. *JAMA* 77:1160–1168.

ETHANOL

Target organ(s): CNS, liver

Occupational exposure limits: OSHA PEL: 1000 ppm (1900 mg/m^3), 15 min STEL: 1000 ppm; NIOSH 10-h YWA: 1000 ppm, IDLH: 3300 ppm

Reference values: RfD 0.67 mg/kg-bw/day (proposed)

Hazard statements: H225—highly flammable liquid and vapor; H319—causes serious eye irritation (≥50%)

Precautionary statements: P210—keep away from heat/ sparks/open flames/ . . . /hot surfaces . . . no smoking; P233—keep container tightly closed; P240—ground/bond container and receiving equipment; P241—use explosion-proof electrical/ventilating/lighting/ . . . /equipment; P242—use only non-sparking tools; P243—take precautionary measures against static discharge; P264—wash . . . thoroughly after handling. Additional text: Hands; P280—wear protective gloves/protective clothing/eye protection/face protection; P370 + P378—in case of fire, use . . . for extinction. Additional text: Dry chemical, alcohol foam, all-purpose AFFF, carbon dioxide, or water spray; P305 + P351 + P338—if in eyes, rinse cautiously with water for several minutes. Remove contact lenses, if present and easy to do. Continue rinsing; P337 + P313— if eye irritation persists, get medical advice/attention; P403 + P235—store in well-ventilated place. Keep cool; P405—store locked up; P501—dispose of contents/container to . . . hazardous or special waste collection point

Risk phrase: R11—highly flammable

Safety phrases: S2—keep out of the reach of children; S7— keep container tightly closed; S16—keep away from sources of ignition . . . no smoking

BACKGROUND AND USES

Ethanol (C$_2$H$_5$OH; ethyl alcohol; CASRN 64-17-5) used in industry is synthesized almost entirely by the hydration of ethylene. It is used as a starting material for the synthesis of a wide variety of organic compounds used in industry and is second only to water as an industrial solvent. Beverage alcohol is derived from the fermentation of carbohydrates, and the supply for this purpose is not augmented by synthetic ethanol.

PHYSICAL AND CHEMICAL PROPERTIES

At room temperature, ethanol is a clear, colorless, volatile liquid with an odor similar to that of wine and whiskey. It is volatile, with a vapor pressure of 59.3 mmHg at 25 °C, and is miscible with water and other organic solvents.

MAMMALIAN TOXICOLOGY

Acute Effects

Although exposure to ethanol is common, it is not of great importance as an industrial hazard. Most of the literature on ethanol toxicity is concerned with the consumption of alcoholic beverages. High inhalation exposures to ethanol may cause dizziness, headache, and drowsiness. At high concentrations of ethanol vapor, there may be some irritation of the upper respiratory system, including coughing and dryness of the throat, but no permanent damage. Irritation of the eye in response to ethanol vapors or liquid may occur as well. Contact allergic reactions have been reported, but these were neither occupational exposures nor identified as significantly irritating (Fregert et al., 1969; Phillips et al., 1972; van Ketel and Tan-Lim, 1975).

Lester and Greenberg (1951) computed the absorption via the respiratory tract that would be necessary to cause any continuous increase in blood ethanol levels. They concluded that a workman exposed to 1000 ppm would have to breathe at a rate of 65 l/min. This more than doubles the ventilation rate of 30 l/min associated with hard work; thus, the hazard of systemic effects from airborne ethanol is unlikely. At air concentrations of 5000–10,000 ppm, there may be mild transient coughing or eye irritation. Concentrations of 20,000 ppm are intolerable. Workers sometimes drink ethanol, on hand for manufacturing or other nonbeverage purposes, a practice that adds to the total dose absorbed and increases susceptibility to accidents. Ethanol absorption, regardless of the route of administration, induces hepatic enzymes and has been known to contribute to the toxicities of some chemicals. Coexposures to trichloroethylene, carbon tetrachloride, vinyl chloride, and in workers taking prescription drugs such as disulfiram have been the subject of some discussion of ethanol potentiation (Hills and Venable, 1982; Wilcosky and Simonsen, 1991).

Chronic Effects

Reports of chronic exposures to ethanol in the workplace have not shown an association with any adverse health effect. Specific links to ethanol exposure would be extremely difficult to establish considering the high prevalence of recreational ethanol consumption among the general population. Teta et al. (1992) studied a population of an ethanol production facility. Although small excesses in some cancers were noted, the process of ethanol production and not ethanol itself was examined. An occupational study by Wilcosky and Tyroler (1983) found a modest excess of ischemic heart disease associated with ethanol exposure.

Human epidemiological studies and maternal animal studies have not found any association between the inhalation of ethanol vapors and abnormal fetal development. Chronic ingestion of ethanol in alcoholic beverages by pregnant women has been definitively associated with developmental abnormalities collectively termed fetal alcohol syndrome. However, occupational exposures to ethanol are not of the same magnitude as those required for teratogenicity.

Mechanism of Action(s)

Regardless of the exposure route, ethanol is metabolized primarily in the liver. It is first oxidized to acetaldehyde by the hepatic enzyme alcohol dehydrogenase. The product acetaldehyde is quickly broken down by aldehyde dehydrogenase to acetate. This product is excreted from the body via exhalation and urinary discharge. Genetic variability creates interindividual differences in ethanol metabolism. Certain populations lack adequate aldehyde dehydrogenase activity and may experience a buildup of acetaldehyde. This is especially prevalent in Asian populations and accounts for the low tolerance and reddening or flushing of the face after consumption. Some unchanged ethanol can be detected in the urine and blood, but occupational exposures are likely not to be high enough to significantly change these concentrations from baseline levels

SOURCES OF EXPOSURE

Occupational

The primary routes of an occupational exposure to ethanol are inhalation and dermal contact. Accidental ingestion of ethanol can occur, but it is usually not reported to be fatal. Inhalation exposure to vapors can be minimized with simple engineering controls. Vapor generation can be detected at many levels of ethanol use, from production to its use as a glaze in confections (Wadden et al., 1994). Monitoring of airborne concentrations of ethanol in the workplace is successfully completed with standard protocols (Langvardt and Melcher, 1979; Mueller and Miller, 1979).

INDUSTRIAL HYGIENE

Dermal contact is of concern when exposure time is lengthy and can be minimized with gloves. The glove material must be evaluated because some rubbers, such as neoprene, have longer breakthrough times than others. Ethanol is used as a hand cleanser in some instances, and a short-term exposure is not associated with adverse effects. The strong odor of ethanol is a good warning property. Ruth (1986) reported an odor threshold of $0.342\,mg/m^3$ for ethanol, with an irritation concentration of $9500\,mg/m^3$ (4950 ppm). Although some tolerance for ethanol is expected, reports of inhalation overexposures are infrequent. Because of its flammability, ethanol should be shielded from open flames and conditions where ignition is possible (i.e., reactive chemicals and high temperatures). NIOSH (1994), OSHA, and the ACGIH (1996) have set their respective exposure limits and TLVs for ethanol at 1000 ppm. The ACGIH has also designated ethanol with an A4 classification, not classifiable as to its carcinogenicity.

REFERENCES

American Conference of Governmental Industrial Hygienists (ACGIH) (1996) *Threshold Limit Values (TLVs) for Chemical Substances and Physical Agents and Biological Exposure Indices (BEls)*, Cincinnati, OH: American Conference of Governmental Industrial Hygienists.

Fregert, S., et al. (1969) Alcohol dermatitis. *Acta Derm. Venerol.* 49:493–497.

Hills, B.W. and Venable, H.L. (1982) The interaction of ethyl alcohol and industrial chemicals. *Am. J. Ind. Med.* 3:321–333.

Langvardt, P.W. and Melcher, R.G. (1979) Simultaneous determination of polar and non-polar solvents in air using a two-phase desorption from charcoal. *Am. Ind. Hyg. Assoc. J.* 40:1006–1012.

Lester, D. and Greenberg, L.A. (1951) The inhalation of ethyl alcohol by man. I. Industrial hygiene and medicotegal aspects.

II. Individuals treated with tetraethylthiuram disulfide. *Q. J. Stud. Alcohol* 12:167–178.

Mueller, F.X. and Miller, I.A. (1979) Determination of airborne organic vapor mixtures using charcoal tubes. *Am. Ind. Hyg. Assoc. J.* 40:380–386.

National Institute for Occupational Safety and Health (NIOSH) (1994) *Pocket Guide to Chemical Hazards*, NIOSH 94-116, Washington, DC: US Department of Health and Human Services.

Phillips, L., II, et al. (1972) A comparison of rabbit and human skin response to certain irritants. *Toxicol. Appl. Pharmacol.* 21:369–382.

Ruth, I.H. (1986) Odor thresholds and irritation levels of several chemical substances: a review. *Am. Ind. Hyg. Assoc. J.* 47: A142–A151

Teta, M.I., Perhnan, G.D., and Ott, M.G. (1992). Mortality study of ethanol and isopropanol production workers at twenty facilities. *Scand. J. Work Environ. Health* 18:90–96.

van Ketel, W.G. and Tan-Lim, K.N. (1975). Contact dermatitis from ethanol. *Contact Dermatitis* 1:7–10.

Wadden, R.A., et al. (1994) Ethanol emission factors for glazing during candy production. *Am. Ind. Hyg. Assoc. J.* 55:343–351.

Wilcosky, T.C. and Simonsen, N.R. (1991) Solvent exposure and cardiovascular disease. *Am. J. Ind. Med.* 19:569–586.

Wilcosky, T.C. and Tyroler, H.A. (1983) Mortality from heart disease among workers exposed to solvents. *J. Occup. Environ. Med.* 25:879–885.

PROPANOL

First aid: Inhalation: Keep patient calm, remove to fresh air, seek medical attention. Skin: Wash thoroughly with soap and water. Eyes: Immediately wash affected eyes for at least 15 min under running water with eyelids held open, consult an eye specialist. Ingestion: Rinse mouth immediately and then drink plenty of water, seek medical attention.

Occupational exposure limits: DNEL; DNEL inhalation, systemic effects, long term: 268 mg/m^3; DNEL inhalation, systemic effects, short term: 1723 mg/m^3; DNEL dermal, systemic effects, long term: 136 mg/m^3

Reference values: DNEL inhalation, systemic effects, long term: 80 mg/m^3; DNEL inhalation, systemic effects, short term: 1036 mg/m^3; DNEL dermal, systemic effects, long term: 81 mg/kg-bw/day; DNEL oral, systemic effects, long term: 61 mg/kg-bw/day

Hazard statements: H225—highly flammable liquid and vapor; H318—causes serious eye damage; H336—may cause drowsiness or dizziness

Risk phrases: R11—highly flammable; R41—risk of serious damage to eyes; R67—vapors may cause drowsiness and dizziness

Safety phrases: S7—keep container tightly closed; S16—keep away from sources of ignition . . . no smoking; S24—avoid contact with skin; S26—in case of contact with eyes, rinse immediately with plenty of water and seek medical advice; S39—wear eye/face protection

BACKGROUND AND USES

The propanols are a minor toxicological hazard in industry. Propanol is more toxic to humans than ethanol, probably because of its slower metabolism. The main route of exposure for both alcohols is inhalation and dermal contact, and both are widely used in industry as well as in household products.

PHYSICAL AND CHEMICAL PROPERTIES

Propanol (n-propyl alcohol; CASRN 71-23-8) is a clear, colorless liquid with an odor resembling ethanol. It is primarily used as a specialty solvent in applications such as printing ink and polymer production. It is miscible with water and has a vapor pressure of 15 mmHg at room temperature and a vapor density of 2.1. It occurs naturally in animal wastes and as a product of volcanoes.

MAMMALIAN TOXICOLOGY

Acute Effects

The hazards associated with this compound are low. Propanol produces some effects similar to ethanol on ingestion or inhalation of high concentrations. The concentrations expected in the workplace are unlikely to be high enough to cause any harmful effects. Dermal contact with propanol may have a drying effect on the skin after chronic exposure, but it is not associated with systemic toxicity. There have been no reports of carcinogenicity or teratogenicity among populations exposed to this chemical.

Nelson et al. (1989) reported maternal toxicity and a low incidence of teratogenicity in laboratory animals associated with inhalation of propanol vapors. However, the level required to produce minimal toxicity was 7000 ppm, far above what would be expected in the workplace. The esterification of some fatty acids was observed in rats after administration of propanol (Carlson, 1993). Animal studies have not associated any carcinogenic effects with propanol exposure.

Chronic Effects

There are no studies of chronic effects in humans.

Mechanism of Action(s)

Once absorbed, propanol is oxidized to propionic acid and excreted (Kamil et al., 1953). The intermediate step in this pathway is the conversion of propanol to proponaldehyde by alcohol dehydrogenase.

INDUSTRIAL HYGIENE

Propylene oxide undergoes a hydrogenation reaction over nickel to produce proponaldehyde, which is subsequently reduced to propanol. The odor of this compound can be detected at concentrations as low as 30 ppm (75 mg/m$^{3)}$ (Ruth, 1986), alarming workers of potential overexposures. This compound is flammable and may react with strong oxidizers, such as tertbutoxide to form an explosive mixture. Protective clothing and adequate ventilation are recommended to protect from overexposures.

The NIOSH (1994), OSHA, and ACGIH (1996) have set their respective REL, PEL, and TLV for propanol at 200 ppm (500 mg/m$^{3)}$ with a STEL/ceiling of 250 ppm (625 mg/m^3).

REFERENCES

American Conference of Govermental Industrial Hygienists (ACGIH) (1996) Threshold Limit Values (TLVs) for Chemical Substances and Physical Agents and Biological Exposure Indicies (BEIs), Cincinnati, OH: Americal Conference of Governmental Industrial Hygienists.

Carlson, G.P. (1993) Formation of fatty acid propyl esters in liver, lung and pancreas of rats administered 1-propanol. *Res. Commun. Chem. Pathol. Pharmacol.* 81(1):121–124.

Kamil, I.A., Smith, J.N., and Williams, R.T. (1953) The metabolism of aliphatic alcohols. The glucuronic acid conjugation of acyclic aliphatic alcohols. *Biochem. J.* 53(1):129–136.

National Institute for Occupational Safety and Health (NIOSH) (1994) Pocket Guild to Chemical Hazards, NIOS 94-116, Washington, DC: US Department of Health and Human Services.

Nelson, B.K., Brightwell, W.S., Taylor, B.J., Khan, A., Burg, J.R., Krieg, E.F., Jr., and Massari, V.J. (1989) Behavioral teratology investigation of 1-propanol administered by inhalation to rats. *Neurotoxicol. Teratol.* 11(2):153–159.

Ruth, J.H. (1986) Odor thresholds and irritation levels of several chemical substances: a review. *Am. Ind. Hyg. Assoc.* 47: A142–A151.

ISOPROPANOL

Occupational exposure limits: TLV, PEL, and REL 400 ppm (980 mg/m^3), 25 min STEL 500 ppm (1225 mg/m^3); DNEL; DNEL inhalation, systemic effects, long term: 500 mg/m^3; DNEL dermal, systemic effects, long term: 888 mg/kg-bw/day

Reference values: DNEL inhalation, systemic effects, long term: 89 mg/m^3; DNEL dermal, systemic effects, long term: 319 mg/kg-bw/day; DNEL oral, systemic effects, long term: 26 mg/kg-bw/day

Hazard statements: H225—highly flammable liquid and vapor; H319—causes serious eye irritation; H336—may cause drowsiness or dizziness

Precautionary statements: P102—keep out of the reach of children; P210—keep away from heat/sparks/open flames/ . . . /hot surfaces . . . no smoking; P233—keep container tightly closed; P240—ground/bond container and receiving equipment; P241—use explosive-proof electrical/ventilating/lighting/ . . . /equipment; P242—take precautionary measures against static discharge; P261—avoid breathing dust/fume/gas/mist/vapors/spray; P264—wash . . . thoroughly after handling; P271—use only outdoors or in a well-ventilated area; P280—wear protective gloves/protective clothing/eye protection/face protection; P405—store locked up; P501—dispose of contents/container to . . . ; P303 + P361 + P353—if on skin (or hair), remove/take off immediately all contaminated clothing. Rinse skin with water/shower; P370 + P378—in case of fire, use . . . for extinction; P337 + P313—if eye irritation persists, get medical advice/attention; P305 + P351 + P338—if in eyes, rinse cautiously with water for several minutes. Remove contact lenses, if present and easy to do. Continue rinsing; P304 + P340—if inhaled, remove the victim to fresh air and make him rest in a position comfortable for breathing; P312—call a poison center or doctor/physician if you feel unwell; P403 + P233—store in a well-ventilated place. Keep container tightly closed; P403 + P235—store in a well-ventilated place. Keep cool

Risk phrases: R11—highly flammable; R36—irritating to eyes; R67—vapors may cause drowsiness and dizziness

Safety phrases: S2—keep out of the reach of children; S7—keep container tightly closed; S16—keep away from sources of ignition . . . no smoking; S24/25—avoid contact with skin and eyes; S26—in case of contact with eyes, rinse immediately with plenty of water and seek medical advice

BACKGROUND AND USES

Isopropanol (isopropyl alcohol; CASRN 67-63-0) is a commonly used industrial solvent and the main ingredient in rubbing alcohol. It is a colorless liquid that is miscible with water and has a characteristic odor that is detectable at concentrations as low as $7.84\,\text{mg/m}^3$ (3 ppm) (Ruth, 1986).

PHYSICAL AND CHEMICAL PROPERTIES

At room temperature, this compound has a vapor pressure of 44 mmHg and a vapor density of 2.08. Isopropanol is produced endogenously as a metabolite of other compounds and is also an indirect food additive.

MAMMALIAN TOXICOLOGY

Acute Effects

The health effects associated with isopropanol are primarily the result of irritation. Inhalation overexposures may cause respiratory, eye, and mucous membrane irritation and some CNS depression. Extremely high concentrations may cause dizziness, nausea, hypotension, and hypothermia. A rare instance of allergic sensitivity to isopropanol was encountered in a cosmetic patch testing program (Ludwig and Hausen, 1977). Ito et al. (1979) described a case of polyneuropathy caused by a mixture of organic solvents, the main ingredient being isopropanol. Similar to ethanol, isopropanol stimulates the mixed function oxidase system of the liver and therefore may alter the toxicity of other industrial compounds, such as carbon tetrachloride (Ueng et al., 1983). Concomitant exposures to other chemicals should be evaluated in an environment where the potential for isopropanol exposure exists.

The low toxicity associated with isopropanol has been demonstrated in animals as well. Acute toxicity studies have reported some neurotoxicity and a kidney change but only at extremely high concentrations (Teramoto et al., 1993; Burleigh-Flayer et al., 1994; Gill et al., 1995). Developmental and mutagenic studies have not found any adverse effects associated with isopropanol exposure (Kapp et al., 1993; Tyl et al., 1994).

Chronic Effects

Chronic dermal contact with isopropanol has a drying effect on the skin. No other long-term effects have been found in the literature.

Mechanism of Action

Isopropanol has been found to be quickly metabolized to acetone once it is absorbed (Laham et al., 1980). Acetone is excreted in the urine and exhaled in the breath. Alveolar isopropanol concentrations and acetone levels in the blood and urine have been correlated with workplace concentrations (Brugnone et al., 1983).

INDUSTRIAL HYGIENE

Starting with propylene, production of isopropanol may be accomplished with a strong acid (H_3SO_4), weak acid, or nonacid process. Historically, the strong acid process was favored in the past. More recently, some health effects associated with this process have encouraged the use of other processes. Airborne concentrations of this chemical can be monitored with charcoal desorption tubes and gas chromatography (Langvardt and Melcher, 1979; Mueller and Miller, 1979). When overexposures are likely, protective gear, including an air-purifying respirator with a full face piece and protective clothing, should be worn. This compound is designated a Class IB flammable liquid, indicating caution in areas of open flames and other possible ignition sources. The ACGIH (1996) and NIOSH (1994) have set their FLBA and REL at identical values of 400 ppm ($980\,\text{mg/m}^3$) with a 15-minute STEL of 500 ppm ($1225\,\text{mg/m}^3$). ASHA has set a similar PEL at 400 mg/m^3).

REFERENCES

American Conference of Governmental Industrial Hygienists (ACGIH) (1996) *Threshold Limit Values (TLVs) for Chemical Substances and Physical Agents and Biological Exposure Indices (BEIs)*, Cincinnati, OH: American Conference of Governmental Industrial Hygienists.

Brugnone, F., et al. (1983) Isopropanol exposure: environmental and biological monitoring in a printing works. *Br. J. Ind. Med.* 40:160–168.

Burleigh-Flayer, H.D., et al. (1994) Isopropanol 13-week vapor inhalation study in rats and mice with neurotoxicity evaluation in rats. *Fundam. Appl. Toxicol.* 23:421–428.

Carlson, G.P. (1993) Formation of fatty acid propyl esters in liver, lung and pancreas of rats administered 1-propanol. *Res. Commun. Chem. Pathol. Pharmacol.* 81:121–124.

Gill, M.W., et al. (1995) Isopropanol: acute vapor inhalation neurotoxicity study in rats. *J. Appl. Toxicol.* 15:77–84.

Ito, T., et al. (1979) Polyneuropathy caused by organic solvent, mainly containing isopropyl alcohol. *Neurol. Med.* 10:178–181.

Kapp, R.W., Jr., et al. (1993) *In vitro* and *in vivo* assays of isopropanol for mutagenicity. *Environ. Mol. Mutagen.* 22:93–100.

Laham, S., et al. (1980) Studies on inhalation toxicity of 2-propanol. *Drug Chem. Toxicol.* 3:343–360.

Langvardt, P.W. and Melcher, R.G. (1979) Simultaneous determination of polar and non-polar solvents in air using a two-phase desorption from charcoal. *Am. Ind. Hyg. Assoc. J.* 40:1006–1012.

Ludwig, E. and Hausen, B.M. (1977) Sensitivity to isopropyl alcohol. *Contact Dermatitis* 3:240–244.

Mueller, F.X. and Miller, J.A. (1979) Determination of airborne organic vapor mixtures using charcoal tubes. *Am. Ind. Hyg. Assoc. J.* 40:380–386.

National Institute for Occupational Safety and Health (NIOSH) (1994) *Pocket Guide to Chemical Hazards*, NIOSH 94-116, Washington, DC: US Department of Health and Human Services.

Ruth, J.H. (1986) Odor thresholds and irritation levels of several chemical substances: a review. *Am. Ind. Hyg. Assoc. J.* 47: A142–A151.

Teramoto, K., et al. (1993) Comparison of the neurotoxicity of several chemicals estimated by the peripheral nerve conduction velocity in rats. *Environ. Res.* 62:148–154.

Tyl, R.W., et al. (1994) Developmental toxicity evaluation of isopropanol by gavage in rats and rabbits. *Fundam. Appl. Toxicol.* 22:139–151.

Ueng, T.-H., et al. (1983). Isopropanol enhancement of cytochrome P-450-dependent monooxygenase activity and its effects on carbon tetrachloride intoxication. *Toxicol. Appl. Pharmacol.* 71:204–214.

n-BUTANOL

Target organ(s): Eyes, CNS

Occupational exposure limits: 20 ppm (ACGIH TLV); 100 ppm (OSHA PEL); 50 ppm (OSHA ceiling); 50 ppm (NIOSH ceiling)

Reference Values: RfD 1×10^{-1} mg/kg-bw/day

Hazard statements: H226—flammable liquid and vapor; H302—harmful if swallowed; H335—may cause respiratory irritation; H315—causes skin irritation; H318—causes serious eye damage; H336—may cause drowsiness or dizziness

Precautionary statements: P210—keep away from heat/ sparks/open flames/ . . . /hot surfaces . . . no smoking; P233—keep container tightly closed; P240—ground/bond container and receiving equipment; P241—use explosive-proof electrical/ventilating/lighting/ . . . /equipment; P242—take precautionary measures against static discharge; P280—wear protective gloves/protective clothing/ eye protection/face protection; P303 + P361 + P353—if on skin (or hair), remove/take off immediately all contaminated clothing. Rinse skin with water/shower; P370 + P378—in case of fire, use . . . for extinction; P403 + P235—store in a well-ventilated place. Keep cool; P501—dispose of contents/container to . . .

Risk phrases: R11—highly flammable; R22—harmful if swallowed; R37/38——irritating to the respiratory system and skin; R41—risk of serious damage to eyes; R67—vapors may cause drowsiness and dizziness

Safety phrases: S2—keep out of the reach of children; S7/ 9—keep container tightly closed and in a well-ventilated place; S13—keep away from food, drink, and animal feeding stuff; S26—in case of contact with eyes, rinse immediately with plenty of water and seek medical advice; S37/39—wear suitable gloves and eye/face protection; S46—if swallowed, seek medical advice immediately and show the container or label

BACKGROUND AND USES

Of the butanols, *n*-butanol (*n*-butyl alcohol; CASRN 71-36-3) is the most important in industries and the most extensively studied. *n*-Butanol is a colorless liquid with a strong, mildly alcoholic odor. It is used in chemical derivatives and as a solvent for paints, waxes, brake fluid, and cleaners.

PHYSICAL AND CHEMICAL PROPERTIES

At room temperature, *n*-butanol has a vapor pressure of 6 mmHg and vapor density of 2.6. *n*-Butanol is soluble in water 9% by weight and is a Class IC flammable liquid. Naturally occurring sources of this alcohol include peppermint oil, some teas, and apple aromas.

MAMMALIAN TOXICOLOGY

Acute Effects

Eye irritation, with corneal inflammation, burning sensation, lacrimation, photophobia, and blurred vision, was reported in workers exposed to 200 ppm or more of *n*-butanol (Sterner et al., 1949). This study followed *n*-butanol exposure in a population of workers for 10 years and found no systemic effects with exposures below 100 ppm. Dermatitis of hands and fingers was reported by Tabershaw et al. (1994) but could not be prevented by protective ointments. The apparently specific formation of minute vacuoles in the cornea has been observed in workers coating raincoats with material containing this solvent (Cogan and Grant, 1945). Vacuolar keratopathy was reported to be most significant in the morning immediately upon opening the eyes after waking up. The other butyl alcohols have similar irritative properties.

Animal studies have also demonstrated the low toxicity of *n*-butanol. Wallgren (1960) showed that of the alcohols, the butanols appeared to be the most intoxicating to rats. Of those tested, *n*-butanol was found to be excreted quicker than most but exhibited its intoxicating effects the fastest. Similar to other short-chain aliphatic alcohols, *n*-butanol was not found to be maternally toxic at high concentrations (6000 ppm) and did not produce teratogenicity (Nelson et al., 1989).

Carlson (1994) reported the ability of *n*-butanol to esterify fatty acids in experimental animals.

Chronic Effects

n-Butanol has not been associated with any chronic effects and is not considered classifiable as to its carcinogenicity.

Mechanism of Action(s)

The metabolic pathway for *n*-butanol is similar to other alcohols. First, it is oxidized to its aldehyde via alcohol dehydrogenase. Further oxidation yields butyric acid, which is broken down further and excreted mostly as carbon dioxide and water.

SOURCES OF EXPOSURE

Occupational

In occupations where *n*-butanol is synthesized or used, there is a potential for exposure. The primary routes of exposure to workers are inhalation and dermal contact.

Environmental

No known environmental exposure has been found in the literature.

INDUSTRIAL HYGIENE

n-Butanol may be synthesized from ethanol or acetaldehyde. Inhalation and dermal contact are the primary routes of worker exposure. The odor detection limit for this compound is low, reported at 0.36 mg/m³ (Ruth, 1986). Workers coming into contact with *n*-butanol should be provided with protective clothing, including gloves, goggles, and boots and, if necessary, personal respirators with face masks. NIOSH (1994) and the ACGIH (1996) have set similar ceiling values at 50 ppm (150 mg/m³). OSHA has a less stringent PE for *n*-butanol of 100 ppm (200 mg/m³).

MEDICAL MANAGEMENT

Anyone exposed to vapors should be removed from the area. Symptoms of overexposure usually disappear when the exposure is removed.

REFERENCES

American Conference of Governmental Industrial Hygienists (ACGIH) (1996) *Threshold Limit Values (TLVs) for Chemical Substances and Physical Agents and Biological Exposure Indices (BEIs)*, Cincinnati, OH: American Conference of Governmental Industrial Hygienists.

Carlson, G.P. (1994) *In vitro* esterification of fatty acids by various alcohols in rats and rabbits. *Toxicol. Lett.* 70:57–61.

Cogan, D.G. and Grant, W.M. (1945) An unusual type of keratitis associated with exposure to *n*-butyl alcohol (butanol). *Arch. Ophthamol.* 33:106–109.

National Institute for Occupational Safety and Health (NIOSH) (1994) *Pocket Guide to Chemical Hazards.* NIOSH 94-116, Washington, DC: US Department of Health and Human Services.

Nelson, B.K., et al. (1989) Behavioral teratology investigation of 1-butanol in rats. *Neurotoxicol. Teratol.* 11:313–315.

Ruth, J.H. (1986) Odor threshold and irritation levels of several chemical substances: a review. *Am. Ind. Hyg. Assoc. J.* 47: A142–A151.

Sterner, J.H., et al. (1949) A ten-year study of butyl alcohol exposure. *Am. Ind. Hyg. Assoc. J.* 10:53–59.

Tabershaw, I.R., Fahy, J.P., and Skinner, J.B. (1994) Industrial exposure to butanol. *J. Ind. Hyg. Toxicol.* 26:328–330.

Wallgren, H. (1960) Relative intoxicating effects on tats of ethyl, propyl and butyl alcohols. *Acta Pharmacol. Toxicol. (Copenh.)* 16:217–222.

PROPYLENE GLYCOL

Occupational exposure limits: DNEL; DNEL inhalation, systemic effects, long term: 168 mg/m³; DNEL inhalation, local effects, long term: 10 mg/m³

Reference values: DNEL inhalation, systemic effects, long term: 50 mg/m³; DNEL inhalation, local effects, long term: 10 mg/m³

Hazard statement: None

Precautionary statement: None

Risk phrase: None

Safety phrase: None

BACKGROUND AND USES

Propylene glycol is used for similar applications as other glycols. It is also used in cosmetics and as a heat transfer fluid in solar energy applications (Marshall et al., 1981; Gupta et al., 1987).

PHYSICAL AND CHEMICAL PROPERTIES

Propylene glycol (1,2-dihydropropane; CASRN 57-55-6) is an odorless, colorless, and tasteless liquid. It has a vapor pressure of 0.129 mmHg at 25 °C (77 °F) and is soluble in water and alcohol.

MAMMALIAN TOXICOLOGY

Acute Effects

In one study, rats were exposed to 0.1–2.2 mg/l propylene glycol for 90 days by inhalation (nose only) (Suber et al., 1989). Researchers reported significant differences in some hematological parameters, enzyme activities, and organ weights, but these findings did not reveal a clear dose–response relationship. Another study found that the activity of human natural killer cells was depressed when they were exposed to propylene glycol *in vitro* (Denning and Webster, 1987). Female rats fed dietary levels of propylene glycol up to 2.5 g/kg/day did not have any apparent adverse effects (Gaunt et al., 1972).

Irritant and allergic skin reactions have been associated with propylene glycol exposure (Hannuksela et al., 1975). Propylene glycol may produce effects similar to other glycols at extremely high levels of exposure. In cases of acute exposure, gas chromatography may be used to measure propylene glycol serum and blood levels to confirm a diagnosis and monitor the patients (Houze et al., 1993).

Chronic Effects

No chronic effects of propylene glycol exposure are found in the literature.

Mechanism of Action(s)

Propylene glycol is metabolized by alcohol and aldehyde dehydrogenases to lactaldehyde and then lactic acid (Anonymous, 1994).

INDUSTRIAL HYGIENE

There are no exposure limits or TLVs for propylene glycol from OSHA, NIOSH, or the ACGIH.

REFERENCES

Anonymous (1994) Final report on the safety assessment of propylene glycol and polypropylene glycols. *J. Am. Coll. Toxicol.* 13:437–491.

Denning, D.W. and Webster, A.D.B. (1987) Detrimental effect of propylene glycol on natural killer cell and neutrophil function. *J. Pharm. Pharmacol.* 39:236–238.

Gaunt, I.F., et al. (1972) Long-term toxicity of propylene glycol in rats. *Food Cosmet. Toxicol.* 10:151–162.

Gupta, B.N., et al. (1987) Safety evaluation of a barrier cream. *Contact Dermatitis* 17:10–12.

Hannuksela, M., Pirila, V., and Salo, O.P. (1975) Skin reactions to propylene glycol. *Contact Dermatitis* 1:112–116.

Houze, P., et al. (1993) Simultaneous determination of ethylene glycol, propylene glycol, 1,3-butylene glycol and 2,3-butylene glycol in human serum and urine by wide-bore column gas chromatography. *J. Chromatogr.* 619:251–257.

Marshall, T.C., et al. (1981) Toxicological assessment of heat transfer fluids proposed for use in solar energy applications II. *Toxicol. Appl. Pharmacol.* 58:31–38.

Suber, R.L., et al. (1989) Subchronic nose-only inhalation study of propylene glycol in Sprague-Dawley rats. *Food Chem. Toxicol.* 27:573–583.

ETHYLENE GLYCOL

Target organ(s): CNS, renal

Occupational exposure limits: ACGIH 100 mg/m^3 aerosol; DNEL; DNEL inhalation, local effects, long term: 35 mg/m^3; DNEL dermal, systemic effects, long term: 106 mg/kg-bw/day;

Reference values: DNEL inhalation, local effects, long term: 7 mg/m^3; DNEL dermal, systemic effects, long term: 53 mg/m^3

Hazard statements: H302—harmful if swallowed; H373—may cause damage to organs such as kidneys through prolonged or repeated oral exposure

Risk phrase: R48/22—harmful: danger of serious damage to health by prolonged exposure if swallowed

Safety phrases: S2—keep out of the reach of children; S45—in case of accident or if you feel unwell, seek medical advice immediately (show label where possible)

BACKGROUND AND USES

Ethylene glycol (1,2-ethanediol; glycol alcohol; CASRN 107-21-1) is a colorless, odorless, clear, syrupy liquid. It has a vapor pressure of 0.06 mmHg at room temperature and is miscible in water. The most common and well-known use of ethylene glycol is as an antifreeze in automobiles. It is also widely used in other cooling and heating systems, in hydraulic brake fluids, in electrolytic condensers, as a solvent in paint and plastics, and in many types of inks.

MAMMALIAN TOXICOLOGY

Acute Effects

An accidental or intentional consumption of ethylene glycol is the most common form of an acute overexposure. The first signs and symptoms are associated with CNS depression. They may include ataxia, prostration, cyanosis, stupor, convulsions progressing to coma, and death (Winek et al., 1978). In serious poisonings, CNS depression typically progresses to cardiovascular symptoms and then renal failure over 48 h. The mean lethal dose for an adult has been estimated at about 100 ml (Milles, 1946).

In the first hours after an exposure, up to 1/5th of ethylene glycol may be excreted unchanged. Most is metabolized by alcohol dehydrogenase to glycol aldehyde. Glycol aldehyde is oxidized by aldehyde dehydrogenase to glycolic and glyoxylic acid. Glyoxylic acid is metabolized to oxalic acid. Unmetabolized ethylene glycol is responsible for CNS depression, whereas oxalic acid is the cause of renal toxicity. However, recent studies indicate that metabolites other than oxalic acid contribute to renal toxicity.

The developmental toxicity of ethylene glycol in animal assays differs drastically with the route of exposure. Studies of rats and mice exposed orally have supported the conclusion that ethylene glycol may cause developmental abnormalities (Marr et al., 1992). Neeper-Bradley et al. (1995) established the no observable effect level (NOEL) for developmental effects at 500 mg/kg/day in CD rats and 150 mg/kg/day in CD-1 mice dosed with ethylene glycol via gavage. However, another study of ethylene glycol's developmental toxicity in CD-1 mice concluded that dermal doses of up to 3549 mg/kg did not cause maternal or developmental effects (Tyl et al., 1995c). A third study by Tyl et al. (1995a) investigated the developmental effects of ethylene glycol aerosol in CD-1 mice exposed by nose-only. This study found a NOEL for developmental toxicity (including teratogenicity) of 1000 mg/m^3 and a NOEL for maternal toxicity of 500 mg/m^3. Whole-body exposure of CD-1 mice and CD rats to ethylene glycol aerosol greatly modified the NOEL (Tyl et al., 1995b). The NOEL for maternal toxicity and developmental toxicity in mice was reduced to 150 mg/m^3. The NOEL for rats was 1000 mg/m^3 for maternal toxicity and 150 mg/m^3 for developmental toxicity. The authors suggest that the differences between whole-body and nose-only NOELs could be a result of ingestion during grooming activities. Combined, these studies support the conclusion that a dermal exposure to ethylene glycol is least likely to cause developmental effects, whereas ingestion is most likely to cause adverse effects.

Animal studies indicate that developmental effects are more likely to result from one high dose exposure to ethylene glycol than from several smaller doses (Carney, 1994). The highest peak blood level of ethylene glycol is strongly associated with the degree of the developmental toxicity. This also explains why ingestion, which would create a relatively rapid peak, is potentially more dangerous than a dermal exposure, which is characterized by slower absorption and a lower peak.

Mechanism of Action(s)

The alcohol dehydrogenase pathway has 30–40 times greater affinity for ethanol than for ethylene glycol. This is the basis for the use of ethanol as an antidote for ethylene glycol poisonings. When ethylene glycol metabolism is blocked by ethanol, ethylene glycol is excreted in the urine unchanged.

A recent study investigated the effect of ethanol on ethylene glycol metabolism using Sprague-Dawley rats as an animal model (Kukielka and Cederbaum, 1995). Microsomal preparations from rats given high levels of dietary ethanol had a much greater capacity to metabolize ethylene glycol to formaldehyde and a higher production of hydrogen peroxide. The authors concluded that this effect is most likely the result of ethanol induction of cytochrome P4502E1.

Gas chromatography may be used for determining the serum levels of glycolic acid or ethylene glycol and urine levels of ethylene glycol (Aarstad et al., 1993; Fraser and MacNeil, 1993).

INDUSTRIAL HYGIENE

Occupational injuries associated with ethylene glycol are extremely rare considering its widespread use. Skin contact is the most likely route of worker exposure. Ethylene glycol is poorly absorbed through the skin, and therefore the absorbed dose by this route is low. Inhalation of aerosols may occur, but exposure to ethylene glycol vapor is rare unless it is heated or sprayed as a mist. Ingestion is highly unlikely unless gross negligence is used in handling and labeling this compound. Ethylene glycol is not considered an eye or skin irritant.

MEDICAL MANAGEMENT

Ethanol has been used to block the metabolism of ethylene glycol. Cox et al. (1992) suggest that 1,3-butylene glycol may be a more effective antidote to ethylene glycol intoxication. 1,3-Butylene glycol is more efficient at blocking ethylene glycol metabolism than ethanol, and it is less toxic. Hemodialysis may be used to remove unmetabolized ethylene glycol. Severe acidosis may be treated with sodium bicarbonate or sodium lactate.

REFERENCES

Aarstad, K., et al. (1993) A rapid gas chromatographic method for determination of ethylene glycol in serum and urine. *J. Anal. Toxicol.* 17:218–221.

American Conference of Governmental Industrial Hygienists (ACGIH) (1996) *Threshold Limit Values (TLVs) for Chemical Substances and Physical Agents and Biological Exposure Indices (BEIs)*, Cincinnati, OH: American Conference of Governmental Industrial Hygienists.

Carney, E.W. (1994) An integrated perspective on the developmental toxicity of ethylene glycol. *Reprod. Toxicol.* 8:99–113.

Cox, S.K., Ferslew, K.E., and Boelen, L.J. (1992) The toxicokinetics of 1,3-butylene glycol versus ethanol in the treatment of ethylene glycol poisoning. *Vet. Hum. Toxicol.* 34:36–42.

Fraser, A.D. and MacNeil, W. (1993) Colorimetric and gas chromatographic procedures for glycolic acid in serum: the major toxic metabolite of ethylene glycol. *J. Toxicol. Clin. Toxicol.* 31:397–405.

Kukielka, E. and Cederbaum, A.I. (1995) Increased oxidation of ethylene glycol to formaldehyde by microsomes after ethanol treatment: role of oxygen radicals and cytochrome P-450. *Toxicol. Lett.* 78:9–15.

Marr, M.C., et al. (1992) Developmental stages of the CD (Sprague-Dawley) rat skeleton after maternal exposure to ethylene glycol. *Teratology* 46:169–181.

Milles, G. (1946) Ethylene glycol poisoning with suggestions for its treatment as oxalate poisoning. *Arch. Pathol.* 41:631–638.

National Institute for Occupational Safety and Health (NIOSH) (1994) *Pocket Guide to Chemical Hazards*, NIOSH 94-116, Washington, DC: US Department of Health and Human Services.

Neeper-Bradley, T.L., et al. (1995) Determination of a no-observed-effect level for developmental toxicity of ethylene glycol administered by garage to CD rats and CD-1 mice. *Fundam. Appl. Toxicol.* 27:121–130.

Tyl, R.W., et al. (1995a) Evaluation of the developmental toxicity of ethylene glycol aerosol in CD-I mice by nose-only exposure. *Fundam. Appl. Toxicol.* 27:49–62.

Tyl, R.W., et al. (1995b) Evaluation of the developmental toxicity of ethylene glycol aerosol in the CD rat and CD-1 mouse by whole-body exposure. *Fundam. Appl. Toxicol.* 24:57–75.

Tyl, R.W., et al. (1995c) Assessment of the developmental toxicity of ethylene glycol applied cutaneously to CD-1 mice. *Fundam. Appl. Toxicol.* 27:155–166.

Winek, C.L., Shingleton, D.P., and Shanor, S.P. (1978) Ethylene and diethylene glycol toxicity. *Clin. Toxicol.* 13:297–324.

ETHYLENE CHLOROHYDRIN

Target organ(s): Respiratory system, liver, kidneys, CNS, skin

Occupational exposure limits: 1 ppm (ACGIH TLV, NIOSH REL)

Reference values: None

Hazard statements: H226—flammable liquid and vapor; H290—may be corrosive to metals; H300—fatal if swallowed; H310—fatal in contact with skin; H318—causes serious eye damage; H330—fatal if inhaled; H411—toxic to aquatic life with long-lasting effects

Risk phrases: R10—flammable; R26/27/28—very toxic by inhalation, in contact with skin, and if swallowed; R41—risk of serious damage to eyes; R51/53—toxic to aquatic organisms, may cause long-term adverse effects in the aquatic environment

Safety phrases: S1/2—keep locked up and out of reach of children; S7/9—keep container tightly closed and in a well-ventilated place; S26—in case of contact with eyes, rinse immediately with plenty of water and seek medical advice; S28—after contact with skin, wash immediately with plenty of . . . (to be specified by manufacture); S39—wear eye/face protection; S45—in case of accident or if you feel unwell, seek medical advice immediately (show label where possible); S61—avoid release to the environment. Refer to special instructions/safety data sheets

BACKGROUND AND USES

Ethylene chlorohydrin (CH_2Cl-CH_2-OH; 2-chloroethyl alcohol, CASRN 107-07-3) is used for its special solvent properties. It is a chemical intermediate in the production of ethylene glycol and, in the past, ethylene oxide. This alcohol is used in the manufacturing of insecticides and as a cleaning agent for machines. Indirect exposure can result from release as a by-product in applications involving ethylene oxide and its accumulation in some foodstuffs as a result of sterilization or packaging.

PHYSICAL AND CHEMICAL PROPERTIES

Ethylene chlorohydrin is a colorless liquid with a faint ether-like odor and is miscible with water. At room temperature, it has a vapor pressure of 5 mmHg and a vapor density of 2.78. Among the halogenated alcohols, ethylene chlorohydrin is the most prominent, along with the chloropropanols.

MAMMALIAN TOXICOLOGY

Acute Effects

Dermal contact is the most concerning route of exposure as ethylene chlorohydrin readily penetrates the skin. There is little skin irritation with this compound, decreasing the warning potential of a skin exposure. Inhalation overexposures may produce mucous membrane irritation, but they are usually not sufficient to produce serious systemic poisoning, although there is a potential for these effects. The CNS is especially susceptible, with muscle incoordination and convulsions resulting from exposure (Goldblatt and Chiesman, 1944; Bush et al., 1949). Corneal swelling and irritation may occur upon eye exposure, but no permanent damage has been reported. A primary hazard with ethylene chlorohydrin is its rather small margin of safety. Death may occur as a result of pulmonary edema or edema of the brain. Ingestion of this compound is not likely but has been described as fatal in low doses (Miller et al., 1970).

Chronic Effects

Epidemiological studies of ethylene chlorohydrin exposure have been reported. Greenberg et al. (1990) first reported an increase in pancreatic cancer and leukemia in the chlorohydrin section of an ethylene oxide production facility. Exposure to ethylene chlorohydrin was reported, along with other chlorinated hydrocarbons that occur as by-products or are involved in asynchronous operations. An increase in some cancers, particularly pancreatic cancer and leukemia, was observed in a population of workers in an ethylene chlorohydrin facility (Benson and Teta, 1993). The authors noted that other research in addition to their own has indicated ethylene dichloride as the probable causative agent, along with other chlorinated hydrocarbon by-products.

Animal studies have not shown any associations of carcinogenicity (Allavena et al., 1992) or developmental effects with exposure to ethylene chlorohydrin.

Mechanism of Action(s)

The metabolism of ethylene chlorohydrin has not been well established in humans. Most studies indicate that ethylene chlorohydrin is metabolized to chloroacetaldehyde and chloroacetic acid, which act as inhibitors of the tricarboxylic acid cycle. Orally dosed rats have shown that most of the radiolabeled ethylene chlorohydrin is excreted as cysteine metabolites or other conjugation products of the reactive metabolites (Grunow and Altman, 1982). Alcohol dehydrogenase and glutathione are hypothesized to be important in the metabolism of ethylene cholorohydrin.

INDUSTRIAL HYGIENE

Ethylene chlorohydrin is produced upon addition of hypochlorous acid to ethylene gas. The most probable route of exposure is via inhalation and dermal contact. This is one of the more hazardous alcohols, and caution should be taken during its use. Personal protective gear is necessary when overexposure is likely. Rubber gloves are not recommended because this alcohol can penetrate most glove materials. Adequate ventilation and respiratory protection are recommended for high vapor concentrations, as well as chemical fume hoods. Detection of ethylene chlorohydrin by capillary gas chromatography was shown to be linear (Danielson et al., 1990). Two methods for degrading ethylene chlorohydrin to its less hazardous reduction product, ethanol, are described by Lunn and Sansone (1991). This compound can react with steam and strong caustics to produce corrosive fumes, including phosgene. Engineering controls and informing workers of potentials hazards are essential when using this chemical in an industrial setting. Ethylene chlorohydrin has been given a Class IIIB combustible rating.

Skin exposure is the primary hazard associated with ethylene chlorohydrin. The ACGIH (1996) and NIOSH (1994) have both set the TLV and REL at 1 ppm ($3.3 \, mg/m^3$). The OSHA has regulated this compound less stringently, with a PEL of 5 ppm ($16 \, mg/m^3$). This compound is not classifiable as to its carcinogenicity (ACGIH, 1996).

REFERENCES

Allavena, A., et al. (1992) Evaluation in a battery of *in vivo* assays of four *in vitro* genotoxins proved to be noncarcinogens in rodents. *Teratog. Carcinog. Mutagen.* 12:31–41.

American Conference of Governmental Industrial Hygienists (ACGIH) (1996) *Threshold Limit Values (TLVs) for Chemical Substances and Physical Agents and Biological Exposure Indices (BEIs)*, Cincinnati, OH: American Conference of Governmental Industrial Hygienists.

Benson, L.O. and Teta, M.J. (1993) Mortality due to pancreatic and lymphopoietic cancers in chlorohydrin production workers. *Br. J. Ind. Med.* 50:710–716.

Bush, A.F., Abrams, H.K., and Brown, H.V. (1949) Fatality and illness caused by ethylene chlorohydrin in an agricultural operation. *J. Ind. Hyg. Toxicol.* 31:352–358.

Danielson, J.W., Snell, R.P., and Oxborrow, G.S. (1990) Detection and quantitation of ethylene oxide, 2-chloroethanol, and ethylene glycol with capillary gas chromatography. *J. Chromatogr. Sci.* 28:97–101.

Goldblatt, M.W. and Chiesman, W.E. (1944) Toxic effects of ethylene chlorohydrin (clinical). *Br. J. Ind. Med.* 1: 207–213.

Greenberg, H.L., Ott, M.G., and Shore, R.E. (1990) Men assigned to ethylene oxide production or other ethylene oxide related chemical manufacturing: a mortality study. *Br. J. Ind. Med.* 47:221–230.

Grunow, W. and Altman, H.-J. (1982) Toxicokinetics of chloroethanol in the rat after single oral administration. *Arch. Toxicol.* 49:275–284.

Lunn, G. and Sansone, E.B. (1991) Validated methods for degrading hazardous chemicals: some halogenated compounds. *Am. Ind. Hyg. Assoc. J.* 52:252–257.

Miller, V., Dobbs, R.J., and Jacobs, S.I. (1970) Ethylene chlorohydrin intoxication with fatality. *Arch. Dis. Child.* 45:589–590.

National Institute for Occupational Safety and Health (NIOSH) (1994) *Pocket Guide to Chemical Hazards*, NIOSH 94-1 16, Washington, DC: US Department of Health and Human Services.

DIETHYLENE GLYCOL

Target organ(s): Kidney

Occupational exposure limits: DNEL; DNEL inhalation, local effects, long term: 60 mg/m³; DNEL dermal, systemic effects, long term: 106 mg/kg-bw/day

Reference values: DNEL inhalation, local effects, long term: 12 mg/m³; DNEL dermal, systemic effects, long term: 53 mg/kg-bw/day

Hazard statements: H302—harmful if swallowed; H373—may cause damage to the kidneys through prolonged or repeated oral exposure

Risk phrases: R22—harmful if swallowed

Safety phrases: S2—keep out of the reach of children; S46—if swallowed, seek medical advice immediately and show the container or label.

BACKGROUND AND USES

Diethylene glycol (DEG; CASRN 111-46-6) is a colorless, odorless, syrupy liquid with a sharp sweet taste.

PHYSICAL AND CHEMICAL PROPERTIES

It has a vapor pressure of 1 mmHg at 92 °C (198 °F) and is miscible in water, alcohol, acetone, ether, and ethylene glycol. It is a condensation product of ethylene glycol production. Structurally, DEG is two ethylene glycol moieties joined by an ether linkage.

MAMMALIAN TOXICOLOGY

Acute Effects

Unfortunately, DEG has been the agent of several mass poisoning incidents. The first occurred early this century in the United States and is sometimes referred to as the "Massingale Disaster" (Gelling and Cannon, 1938). Sulfanilamide, an antibacterial "elixir," was prepared with 70% DEG and ingested. Over 100 deaths were attributed to this incident. This incident was largely responsible for the passage of the 1938 Federal Food, Drug and Cosmetic Act (Wax, 1995). More recent mass poisonings have occurred in South Africa, India, Nigeria, and Bangladesh (Okuonghae et al., 1992; Hanif et al., 1995; Wax, 1996). In many of these incidents, DEG was used in pharmaceuticals as an inexpensive substitute for the less toxic propylene glycol.

The most recent incident involved a children's elixir of acetaminophen in Haiti (Scalzo, 1996). A locally made preparation contained imported glycerin contaminated with DEG. Eighty-six children developed a nonspecific febrile illness followed by acute renal failure (Malebranche et al., 1996). Signs and symptoms from DEG intoxication include vomiting, abdominal pain, lethargy, and malaise. Neurological symptoms, including pupil dilation, optic neuritis, and cerebral atrophy, were typically delayed (Scalzo, 1996).

Pregnant rabbits that were treated with up to 1000 mg/kg DEG on days 7 through 19 of gestation showed no statistically significant incidence of fetal abnormalities (Hellwig et al., 1995).

Mechanism of Action(s)

The mechanism of action of DEG is poorly understood. It was once presumed that DEG would first undergo cleavage of the ether bond into two ethylene glycol molecules and then be further metabolized. Although many have concluded that DEG is metabolized similarly to ethylene glycol, clinical findings do not support this conclusion. Patients with confirmed DEG poisoning have not shown oxalate formation or metabolic acidosis characteristic of ethylene glycol poisoning (Scalzo, 1996). Animal studies of DEG metabolism have found only trace amounts of oxalic acid in the urine, suggesting a different primary metabolism. Animal studies have demonstrated that up to 96% of DEG is rapidly excreted in the urine (Mathews et al., 1991; Heilmair et al., 1993). It is primarily excreted unchanged; however, studies on rats indicate that 2-hydroxyethoxyacetic acid and glycolic acid are urinary metabolites (Lenk et al., 1989). 2-Hydroxyethoxyacetic acid is a product of DEG oxidation by alcohol dehydrogenase (Wiener and Richardson, 1989).

Chemical Pathology

Pathological examination of DEG-exposed livers in human necropsy cases revealed central hydropic degeneration with ballooned hepatic cells (Scalzo, 1996). Interestingly, studies of hepatotoxic effects of DEG in rats and dogs do not show hydropic changes in livers.

INDUSTRIAL HYGIENE

Diethylene glycol is poorly absorbed through the skin and is not expected to be an inhalation hazard unless it is heated. There are no occupational exposure limits for DEG from the NIOSH, OSHA, or ACGIH.

REFERENCES

Gelling, E.M.K. and Cannon, P.R. (1938) Pathological effects of elixir of sulfanilamide (diethylene glycol) poisoning. *JAMA* 111:919–926.

Hanif, M., et al. (1995) Fatal renal failure caused by diethylene glycol in paracetamol elixir: the Bangladesh epidemic. *BMJ* 311:88–91.

Heilmair, R., Lenk, W., and Lohr, D. (1993) Toxicokinetics of diethylene glycol (DEG) in the rat. *Arch. Toxicol.* 67:655–666.

Hellwig, J., Klimisch, H.-J., and Jackh, R. (1995) Investigation of the prenatal toxicity of orally administered diethylene glycol in rabbits. *Fundam. Appl. Toxicol.* 28:27–33.

Lenk, W., Lohr, D., and Sonnenbichler, J. (1989) Pharmacokinetics and bio-transformation of diethylene glycol and ethylene glycol in the rat. *Xenobiotica* 19:961–979.

Malebranche, R., et al. (1996) Fatalities associated with ingestion of diethylene glycol-contaminated glycerin used to manufacture acetaminophen syrup—Haiti, November 1995–June 1996. *MMWR Morb. Mortal. Wkly Rep.* 45:649–650.

Mathews, J.M., Parker, M.K., and Matthews, H.B. (1991) Metabolism and disposition of diethylene glycol in rat and dog. *Drug Metab. Dispos.* 19:1066–1070.

Okuonghae, H.O., et al. (1992) Diethylene glycol poisoning in Nigerian children. *Ann. Trop. Paediatr.* 12:235–238.

Scalzo, A.J. (1996) Diethylene glycol toxicity revisited: the 1996 Haitian epidemic. *J. Toxicol. Clin. Toxicol.* 34:513–516.

Wax, P.M. (1995) Elixirs, diluents, and passage of the 1938 Federal Food, Drug and Cosmetic Act. *Ann. Intern. Med.* 122:456–461.

Wax, P.M. (1996) It's happening again—another diethylene glycol mass poisoning. *J. Toxicol. Clin. Toxicol.* 34:517–520.

Wiener, H.L. and Richardson, K.E. (1989) Metabolism of diethylene glycol in male rats. *Biochem. Pharmacol.* 38:539–541.

ALKYL DERIVATIVES OF ETHYLENE GLYCOL

Occupational exposure limits: *Ethylene glycol monomethylether* (EGME): PEL 25 ppm, REL 0.1 ppm, TLV 25 ppm; (DNEL; DNEL inhalation, systemic effects, long term: 3.2 mg/m^3; DNEL inhalation, systemic effects, short term: 10 mg/m^3; DNEL dermal, systemic effects, long term: 0.91 mg/kg-bw/day; *ethylene glycol monoethyl ether* (EGEE): PEL 200 ppm, REL 0.5 ppm, TLV 5 ppm; DNEL inhalation, systemic effects, long term: 0.083 mg/m^3; DNEL dermal, systemic effects, long term: 0.3 mg/kg-bw/day; *ethylene glycol monobutylether* (EGBE): PEL 50 ppm, REL 5 ppm, TLV 25 ppm; DNEL inhalation, systemic effects, long term: 98 mg/m^3; DNEL inhalation, systemic effects, short term: 663 mg/m^3; DNEL inhalation, local effects, short term: 246 mg/m^3; DNEL dermal, systemic effects, long term: 75 mg/kg-bw/day; DNEL dermal, systemic effects, short term: 89 mg/kg-bw/day

Reference values: EGME: DNEL inhalation, systemic effects, long term: 1.6 mg/m^3; DNEL inhalation, systemic effects, short term: 5 mg/m^3; EGEE: none; EGBE: DNEL inhalation, systemic effects, long term: 49 mg/m^3; DNEL inhalation, systemic effects, short term: 426 mg/m^3; DNEL inhalation, local effects, short term: 123 mg/m^3; DNEL dermal, systemic effects, long term: 38 mg/kg-bw/day; DNEL dermal, systemic effects, short term: 44.5 mg/kg-bw/day; DNEL oral, systemic effects, long term: 3.2 mg/kg-bw/day; DNEL oral, systemic effects, short term: 13.4 mg/kg-bw/day

Hazard statements: H226—flammable liquid and vapor (EGME, EGEE); H360—may damage fertility or the unborn child (EGME, EGEE); H331—toxic if inhaled (EGEE); H332—harmful if inhaled (EGME, EGBE); H312—harmful in contact with skin (EGME, EGBE); H302—harmful if swallowed (EGME, EGEE, EGBE); H373—may cause damage to thymus through prolonged or repeated exposures (EGME); H370—causes serious damage to the immune system (EGME); H315—causes skin irritation (EGBE); H319—causes serious eye irritation (EGBE)

Precautionary statements: P201—obtain special instructions before use (EGME, EGEE); P202—do not handle until all safety precautions have been read and understood (EGME); P210—keep away from heat/sparks/open flames/ . . . /hot surfaces . . . no smoking (EGME, EGEE); P233—keep container tightly closed (EGME); P240—ground/bond container and receiving equipment (EGME); P241—use explosion-proof electrical/ventilating/lighting/ . . . /equipment (EGME); P242—use only non-sparking tools (EGME); P243—take precautionary measures against static discharge (EGME); P261—avoid breathing dust/fume/gas/mist/vapors/spray (EGME, EGEE, EGBE); P264—wash . . . thoroughly after handling (EGME, EGBE); P270—do not eat, drink, or smoke when using this product (EGME, EGBE); P271—use only outdoors or in a well-ventilated area (EGME, EGBE); P280—wear protective gloves/protective clothing/eye protection/face protection (EGME, EGBE); P281—use personal protective equipment as required (EGME); P301 + P312—if swallowed, call a poison center or doctor/physician if you feel unwell (EGME,

EGBE); P302 + P352—if on skin, wash with plenty of soap and water (EGME, EGBE); P303 + P361 + P353—if on skin (or hair), remove/take off immediately all contaminated clothing. Rinse skin with water/shower (EGME); P304 + P340—if inhaled, remove victim to fresh air and instruct the victim to rest in a position comfortable for breathing (EGME, EGBE); P305 + P351 + P338—if in eyes, rinse cautiously with water for several minutes. Remove contact lenses, if present and easy to do. Continue rinsing (EGBE); P308 + P313—if exposed or concerned, get medical advice/attention (EGME, EGEE); P312—call a poison center or doctor/physician if you feel unwell (EGME, EGBE); P321—specific treatment (see . . . on label) (EGBE); P322—specific measures (see . . . label) (EGME, EGBE); P330—rinse mouth (EGME, EGBE); P362—take off contaminated clothing and wash before reuse (EGBE); P363—wash contaminated clothing before reuse (EGBE); P403 + P233—store in a well-ventilated place. Keep container tightly closed (EGEE); P501—dispose of contents/container to . . . (EGEE, EGBE).

Risk phrases: R60—may impair fertility (EGME, EGEE); R61—may cause harm to the unborn child (EGME, EGEE); R36/38—irritating to the eyes and skin (EGBE); R20/22—harmful by inhalation and if swallowed (EGEE); R20/21/22—harmful by inhalation, in contact with skin, and if swallowed (EGME, EGBE); R10—flammable (EGME, EGEE).

Safety phrases: S2—keep out of the reach of children (EGBE); S36/37—wear suitable protective clothing and gloves (EGBE); S53—avoid exposure, obtain special instructions before use (EGME, EGEE); S45—in case of accident or if you feel unwell, seek medical advice immediately (show label where possible) (EGME, EGEE); S46—if swallowed, seek medical advice immediately and show the container or label

BACKGROUND AND USES

The three main types of ethylene glycol derivatives are EGME (CASRN 109-86-4), EGEE (CASRN 110-80-5), and EGBE (CASRN 111-76-2). These compounds are important industrial cosolvents used in a variety of applications, including protective coatings, paints, varnishes, domestic and industrial cleaning agents, inks, dyes, some pesticides, and in the semiconductor industry. These glycols are entirely manufactured and are not known to occur naturally.

PHYSICAL AND CHEMICAL PROPERTIES

All three of the ethylene glycol derivatives are liquid at room temperature with mild ether-like odors. The physical properties are similar among these compounds; as a representative example, EGBE has a vapor pressure of 0.6 mmHg, a flash point of 72 °C, a lower flammability limit of 1.1%, and an upper flammability limit of 10.6%.

MAMMALIAN TOXICOLOGY

Acute Effects

The majority of information regarding the toxicity of the monoalkyl glycols is extrapolated from animal data. Recent work has concentrated on evaluating dermal absorption, as it is the primary route of exposure. Historically, alkylated ethylene glycols were first associated with hazardous effects in the shirt collar industry in the 1930s (Greenberg et al., 1938; Parsons and Parsons, 1938). The exposures to EGME resulted in CNS depression and hemolytic anemia. Current regulations have kept workplace exposures sufficiently low to prevent these effects. Zavon (1963) reported cases of workers exposed to EGME exhibiting similar neurological effects, including drowsiness, fatigue, general malaise, and tremors. Eventual recovery was noted in all cases after cessation of the exposure. Routes of exposure in these cases were not characterized. Ohi and Wegman (1978) observed the development of encephalopathy, bone marrow injury, and pancytopenia in two cutaneously exposed workers at a textile printing plant in Massachusetts. The workers additionally developed hematological changes; both suffered bone marrow insult. More recent reports of industrial exposures of EGME have observed similar CNS and hematological effects from dermal and inhalation exposures (Cohen, 1984; Larese et al., 1992). It has also been shown to cause hemopoietic effects in rats after 4 days of oral exposure to 100–500 mg/kg (Grant et al., 1985).

Severe hemoglobinuria has been noted in rats after an exposure to EGBE (Ghanayem et al., 1987a, 1987b). Ingestion of a household product containing EGBE reportedly resulted in hemoglobinuria, metabolic acidosis, hypokalemia, elevated serum creatinine levels, and increased urinary excretion of oxalate crystals (Rambourg-Schepens et al., 1988).

Ethylene glycol monoethyl ether has reportedly caused teratogenic effects in rats exposed by gavage below levels that were maternally toxic (Weir et al., 1987). In the same study, EGBE produced teratogenicity only at levels that were maternally toxic. Teratogenicity has also been reported subsequent to a dermal exposure to EGEE (Hardin et al., 1982).

The testicular toxicity of alkyl derivatives of ethylene glycol has been examined by many researchers. Molsen et al. (1995) reported significant differences in the testicular metabolism of EGME between species and strains. Foster et al. (1983) reported no observed adverse effect level (NOAELs) for testicular effects in rats treated 11 days with EGME and EGEE at 50 and 250 mg/kg, respectively.

Researchers have investigated energy metabolism and hormonal responses as possible mechanisms for male reproductive toxicity (Oudiz and Zenic, 1986; Reader et al., 1991). Dermal exposure to EGME has also reportedly produced a reduction in fertility, reduced testis spermatid count, reduced epididymal sperm count, and increased the sperm count with abnormal morphology in male rats (Feuston et al., 1989).

Chronic Effects

An epidemiological study found no significant health changes in a group of workers exposed to EGME at currently acceptable workplace exposure concentrations (Cook et al., 1982).

The potential reproductive effects of alkylated ethylene glycols have been investigated based on findings of studies of male rat reproduction. Ratcliffe et al. (1989) found a weak association between lowered sperm count and exposure to EGEE in a cross-sectional study. Unfortunately, potentially confounding variables could not be controlled. A similar association with EGEE and EGME exposure and low sperm count in shipyard painters was also reported by Welch et al. (1988). Although both compounds were present, the authors indicated that EGEE exposure was more likely to occur at this facility.

No chronic human carcinogenicity or teratogenicity data were available for these compounds.

Mechanism of Action(s)

Hemoglobinuria in rats correlates closely with the EGBE urinary metabolite butoxyacetic acid (Ghanayem et al., 1987a, 1987b). Ethylene glycol monobutylether requires metabolic activation by alcohol or aldehyde dehydrogenase to exert this effect (Ghanayem, 1989).

Studies of the pharmacokinetics of EGBE have noted significant interspecies differences. Rats have higher predicted peak blood concentrations than humans when receiving the same relative dose (rog/kg) (Corley et al., 1994). In addition, human blood was found to be much less susceptible than rat blood to potential hemolytic effects of EGBE (Ghanayem, 1989; Corley et al., 1994).

The primary urinary metabolite in rats exposed to up to 46 ppm EGEE via inhalation was ethylene glycol monoethyl ether acetate (EGMEA) (Kennedy et al., 1993). Approximately 38% of EGEE was exhaled during or after exposure, and almost half of the retained dose was excreted in the urine.

SOURCES OF EXPOSURE

Occupational

Workers are exposed to these compounds where they are included in solvents for cleaning or manufacturing, as well as the production of these chemicals during solvent manufacture. The most common route of exposure is dermal.

Environmental

Environmental exposure is rare except in cases of sites that have been contaminated by solvents containing these compounds.

INDUSTRIAL HYGIENE

The primary precursor for all alkyl derivatives of ethylene glycol is ethylene oxide. Exposures to these compounds may occur via inhalation, but the low vapor pressures make airborne concentrations low in industrial facilities. Workplace air concentrations of these glycol vapors can be detected through standardized procedures, but the low volatility of these compounds may lead to incomplete assessments of total workplace exposures (Piacitelli et al., 1990). Recent biomonitoring studies indicate that dermal contact contributes to the majority of the absorbed dose (Vincent et al., 1993). One study of percutaneous absorption estimated that it would be possible to absorb enough EGBE by this route to produce acute effects (Johanson et al., 1988). Vapors are easily controlled through engineering and ventilation. During expected high concentration exposures, personal respirators should be worn. Protective clothing, including rubber gloves, should be worn to prevent dermal contact. These controls, in addition to worker education and awareness of these compounds, should provide adequate protection in the workplace

MEDICAL MANAGEMENT

Flush exposed area with large amounts of warm running water for at least 20 min. Decontaminate clothing before reuse or discard. For inhalational exposures, remove to fresh air. Give oxygen and get immediate medical attention for any breathing difficulty.

Measurement of the metabolites present in the urine is the most common method of monitoring worker exposure to these chemicals, as the parent compound is comparatively undetectable. Determination of 2-ethoxyacetic, 2-methoxyacetic acid, and 2-butoxyacetic acid in the urine was found to correlate well with known exposure levels (Smallwood et al., 1988; Groeseneken et al., 1989; Sakai et al., 1994). The latter report involves detection of these acids as pentaflurobenzylesters. The amino acid conjugate of butoxyacetic acid, N-butoxyacetylglutamine, was identified as a viable means of monitoring exposure, correlating well with 2-butoxyacetic acid measurements (Rettenmeier et al., 1993).

REFERENCES

Cohen, R. (1984) Reversible subacute ethylene glycol monomethyl ether toxicity associated with microfilm production: a case report. *Am. J. Ind. Med.* 6:441–446.

Cook, R.R., et al. (1982) A cross-sectional study of ethylene glycol monomethyl ether process employees. *Arch. Environ. Health* 37:346–351.

Corley, R.A., Bormett, G.A., and Ghanayem, B.I. (1994) Physiologically based pharmacokinetics of 2-butoxyethanol and its major metabolite, 2-butoxyacetic acid, in rats and humans. *Toxicol. Appl. Pharmacol.* 129:61–79.

Feuston, M.H., et al. (1989) Reproductive toxicity of 2-methoxyethanol applied dermally to occluded and nonoccluded sites in male rats. *Toxicol. Appl. Pharmacol.* 100:145–161.

Foster, P.M.D., et al. (1983) Testicular toxicity of ethylene glycol monomethyl and monoethyl ethers in the rat. *Toxicol. Appl. Pharmacol.* 69:385–399.

Ghanayem, B.I. (1989) Metabolic and cellular basis of 2-butoxyethanol-induced hemolytic anemia in rats and assessment of human risk *in vitro. Biochem. Pharmacol.* 38:1679–1684.

Ghanayem, B.I., Burka, L.T., and Matthews, H.B. (1987a) Metabolic basis of ethylene glycol monobutyl ether (2-butoxyethanol) toxicity: role of alcohol and aldehyde dehydrogenases. *J. Pharmacol. Exp. Ther.* 242:222–231.

Ghanayem, B.I., et al. (1987b) Metabolism and disposition of ethylene and glycol monobutyl ether in rats. *Drug Metab. Dispos.* 15:478–484.

Grant, D., et al. (1985) Acute toxicity and recovery in the hemopoietic system of rats after treatment with ethylene glycol monomethyl and monobutyl ethers. *Toxicol. Appl. Pharmacol.* 77:187–200.

Greenberg, L., et al. (1938) Health hazards in the manufacture of "fused collars." I. Exposure to ethylene glycol monomethyl ether. *J. Ind. Hyg. Toxicol.* 201:134–147.

Groeseneken, D., et al. (1989) An improved method for the determination in urine of alkoxyacetic acids. *Int. Arch. Occup. Environ. Health* 61:249–254.

Hardin, B.D., et al. (1982) Teratogenicity of 2-ethoxyethanol by dermal application. *Drug Chem. Toxicol.* 5:277–294.

Johanson, G., Boman, A., and Dyndsius, B. (1988) Percutaneous absorption of 2-butoxyethanol in man. *Scand. J. Work Environ. Health* 14:101–109.

Kennedy, C.H., et al. (1993) Effect of dose on the disposition of 2-ethoxyethanol after inhalation by F344/N rats. *Fundam. Appl. Toxicol.* 21:486–491.

Larese, F., Fiorito, A., and De Zotti, R. (1992) The possible haematological effects of glycol monomethyl ether in a frame factory. *Br. J. Ind. Med.* 49:131–133.

Molsen, M.T., et al. (1995) Species differences in testicular and hepatic biotransformation of 2-methoxyethanol. *Toxicology* 96:217–224.

Ohi, G. and Wegman, D.H. (1978) Transcutaneous ethylene glycol monomethyl ether poisoning in the work setting. *J. Occup. Environ. Med.* 20:675–676.

Oudiz, D. and Zenic, H. (1986) *In vivo* and *in vitro* evaluations of spermato-toxicity induced by 2-ethoxyethanol treatment. *Toxicol. Appl. Pharmacol.* 84:576–583.

Parsons, C.E. and Parsons, M.E.M. (1938) Toxic encephalopathy and granulopenia anaemia due to volatile solvents in industry: report of two cases. *J. Ind. Hyg. Toxicol.* 20:124–133.

Piacitelli, G.M., Votaw, D.M., and Krishnan, E.R. (1990) An exposure assessment of industries using ethylene glycol ethers. *Appl. Occup. Environ Hyg.* 5:107–114.

Rambourg-Schepens, M.O., et al. (1988) Severe ethylene glycol butyl ether poisoning: kinetics and metabolic pattern. *Hum. Toxicol.* 7:187–189.

Ratcliffe, J.M., et al. (1989) Semen quality in workers exposed to 2-ethoxyethanol. *Br. J. Ind. Med.* 46:399–406.

Reader, S.C.J., Shingles, C., and Stondard, M.D. (1991) Acute testicular toxicity of 1,3-dinitrobenzene and ethylene glycol monoethyl ether in the rat: evaluation of biochemical effect markers and hormonal responses. *Fundam. Appl. Toxicol.* 16:61–71.

Rettenmeier, A.W., Hennigs, R., and Wodarz, R. (1993) Determination of butoxyacetic acid and *N*-butoxyacetylglutamine in urine of lacquerers exposed to 2-butoxyethanol. *Int. Arch. Occup. Environ. Health* 65:S151–S153.

Sakai, T., et al. (1994) Gas chromatographic determination of butoxyacetic acid after hydrolysis of conjugated metabolites in urine from workers exposed to 2-butoxyethanol. *Int. Arch. Occup. Environ. Health* 66:249–254.

Smallwood, A.W., et al. (1988) Determination of urinary 2-ethoxyacetic acid as an indicator of occupational exposure to 2-ethoxy-ethanol. *Appl. Ind. Hyg.* 3:47–50.

Vincent, R., et al. (1993) Occupational exposure to 2-butoxyethanol for workers using window cleaning agents. *Appl. Occup. Environ. Hyg.* 8:580–586.

Weir, P.J., Lewis, S.C., and Traul, K.A. (1987) A comparison of developmental toxicity evident at term to postnatal growth and survival using ethylene glycol monomethyl ether, ethylene glycol monobutyl ether and ethanol. *Tertog. Carcinog. Mutagen.* 7:55–64.

Welch, L.S., et al. (1988) Effects of exposure to ethylene glycol ethers on shipyard painters: II. Male production. *Am. J. Ind. Med.* 14:509–526.

Zavon, M.R. (1963) Methyl Cellosolve intoxication. *Am. Ind. Hyg. Assoc. J.* 24:36–41.

ALKYL DERIVATIVES OF DIETYHLENE GLYCOL

First aid: See text

Occupational exposure limits: *Diethylene glycol monomethyl ether* (diEGME): DNEL; DNEL inhalation, systemic effects, long term: 50.1 mg/m³; DNEL dermal, systemic effects, long term: 0.53 mg/kg-bw/day; *diethylene glycol monoethyl ether* (diEGEE): DNEL inhalation, systemic effects, long term: 37 mg/m³; DNEL inhalation, local effects, long term: 18 mg/m³; DNEL dermal, systemic effects, long term: 50 mg/kg-bw/day; *diethylene glycol monobutyl ether* (diEGBE): DNEL inhalation, systemic effects, long term: 67.5 mg/m³; DNEL inhalation, local effects, long term: 67.5 mg/m³; DNEL inhalation, local effects, short term: 101.2 mg/m³; DNEL dermal, systemic effects, long term: 20 mg/kg-bw/day

Reference values: diEGME: DNEL inhalation, systemic effects, long term: 25 mg/m³; DNEL dermal, systemic effects, long term: 0.27 mg/kg-bw/day; DNEL oral, systemic effects, long term: 1.5 mg/kg-bw/day; diEGEE: DNEL inhalation, systemic effects, long term: 18.3 mg/m³; DNEL inhalation, local effects, long term: 9 mg/m³; DNEL dermal, systemic effects, long term: 25 mg/kg-bw/day; DNEL oral, systemic effects, long term: 25 mg/kg-bw/day; diEGBE: DNEL inhalation, systemic effects, long term: 34 mg/m³; DNEL inhalation, local effects, long term: 34 mg/m³; DNEL inhalation, local effects, short term: 50.6 mg/m³; DNEL dermal, systemic effects, long term: 10 mg/kg-bw/day; DNEL oral, systemic effects, long term: 1.25 mg/kg-bw/day

Hazard statement: H361—suspected of damaging fertility or to the unborn child (diEGME); H319—causes serious eye irritation (diEGBE)

Precautionary statements: P201—obtain special instructions before use (diEGME); P202—do not handle until all safety precautions have been read and understood (diEGME); P264—wash . . . thoroughly after handling (diEGBE); P280—wear protective gloves/protective clothing/eye protection/face protection (diEGBE); P281—use personal protective equipment as required (diEGME); P305 + P351 + P338—if in eyes, rinse cautiously with water for several minutes; remove contact lenses, if present and easy to do; continue rinsing (diEGBE); P308 + P313—if exposed or concerned, get medical advice/attention (diEGME); P405—store locked up; P501—dispose of contents/container to . . . (diEGME)

Risk phrases: R63—possible risk of harm to the unborn child (diEGME); R36—irritating to the eyes (diEGBE)

Safety phrases: S2—keep out of the reach of children (diEGME, diEGBE); S24—avoid contact with skin (diEGBE); S26—in case of contact with eyes, rinse immediately with plenty of water and seek medical advice (diEGBE); S36/37—wear suitable protective clothing and gloves (diEGME)

BACKGROUND AND USES

The three important alkyl derivatives of DEG are diEGME (CASRN 111-77-3), diEGEE (CASRN 111-90-0), and diEGBE (CASRN 112-34-5). In general, monoalkyls of DEG are less toxic than monoalkyls of ethylene glycol. As a result, they have been substituted for the monoalkyls of ethylene glycol in many industrial processes.

Diethylene glycol monomethyl ether is used in textile dye pastes, in lacquer industrial for thinners and quick drying varnishes; solvent for wood stains, hydraulic brake fluid diluent, and as a coupling agent for miscible organic aqueous systems.

Diethylene glycol monoethyl ether is used as a solvent in lacquer, varnishes, and enamels. It is also used in cosmetic products and dermatological preparations and as a solvent in some medicine products. It enhances the percutaneous absorption through the skin and mucosal barriers. It is used in some drugs to enhance absorption.

Diethylene glycol monobutyl ether is used as a solvent in paints, dyes, inks detergents, and cleaners.

PHYSICAL AND CHEMICAL PROPERTIES

The alkyl derivatives of diethylene glycol are all colorless liquids. The odor of diEGME is mild and pleasant. It has a vapor pressure of 0.2 mmHg at 20 °C (68°F) and is miscible with water, glycerol, alcohol, acetone, and ether. Because of its low vapor pressure, exposure to diEGME via inhalation is unlikely to occur unless it is heated. Diethylene glycol monoethyl ether has an odor that is mild and pleasant. It has a vapor pressure of 0.14 mmHg at 25 °C (77°F) and is soluble in water and alcohol. It has been selected as a replacement for EGME for many uses. Exposure to diEGEE via inhalation is unlikely to occur under normal conditions.

Diethylene glycol monobutyl ether has a faint odor. It has a vapor pressure of 0.02 mmHg at 20 °C (68°F). It is miscible with water and soluble in alcohol, acetone, and ether. Exposure to diEGBE is unlikely to occur via inhalation unless it is heated.

MAMMALIAN TOXICOLOGY

Acute Effects

Rats exposed to up to 216 ppm diEGME vapors for a equivalent of 13 workweeks had no hematological or clinical findings associated with exposure (Miller et al., 1985), nor were there any significant pathological findings.

Rats dosed orally with 2000 mg/kg diEGME for 20 days had significantly reduced absolute and relative liver weights (Kawamoto et al., 1990b). They also had significantly increased microsomal protein and hepatic cytochrome P-450 concentrations. Decreased testes weight (80% of controls) and thymus weight (60% of controls) were also observed after 20 days (Kawamoto et al., 1990a). In another study, rats were dosed with 2000 mg/kg diEGME and the effect on gamma-glutamyl-transpeptidase (GGT) was measured (Kawamoto et al., 1992). Diethylene glycol monomethyl ether exposure caused a 50% increase in brain levels of GGT, but no effect was noted on levels in other organs.

Developmental studies in rats exposed to diEGME orally suggest that it is fetotoxic to rats at levels that are not maternally toxic (Hardin et al., 1986). No toxic or fetotoxic effects were produced by doses up to 720 mg/kg. Another study of the teratogenic potential of diEGME established 50 mg/kg/day as the NOAEL for maternal and developmental toxicity (Scortichini et al., 1986).

In one study, pregnant rats orally exposed to diEGME had offsprings with malformations at levels below those that caused maternal toxicity (Yamano et al., 1993). The NOAEL for fetal and maternal toxicity was reported to be 200 mg/kg.

Various tests of mutagenicity have indicated that diEGBE does not present a genetic or teratogenic risk to humans (Thompson et al., 1984; Nolen et al., 1985). A dose of 2 ml/kg has been shown to produce cyanosis, tachypnea, and slight uremia.

Chronic Effects

Dermal exposure of rats to 1.0 g/kg/day for 3 months resulted in decreased spleen weights and significantly increased serum creatinine kinase and lactate dehydrogenase activity (Hobson et al., 1986).

Mechanism of Action(s)

The mechanism of action of these compounds are poorly characterized but may be related to CYP-P450 enzyme induction and the production of metabolites from enzyme systems not normally associated with ethylene glycols (Bowden et al. 1995).

SOURCES OF EXPOSURE

Occupational

Occupational exposures can potentially occur in settings where DEG-based solvents are used.

Environmental

Environmental exposure could potentially occur from accidental contamination of environmental media with DEG-based solvents.

INDUSTRIAL HYGIENE

No OSHA regulations exist for these compounds. Industrial hygiene guidance for DEG should be observed.

MEDICAL MANAGEMENT

In case of contact with the eyes, flush with warm running water for at least 20 min, holding eyelids open during flushing. Take care not to flush contaminated water into unaffected eye or onto skin. In case of skin contact, immediately remove contaminated clothing under running water. Flush exposed area with large amounts of warm running water for at least 20 min. Decontaminate clothing before reuse or discard. For inhalational exposures, remove to fresh air. Give oxygen and get immediate medical attention for any breathing difficulty. If ingested, do not induce vomiting. If the casualty is alert and not convulsing, give 2–4 glasses of water to drink to dilute. If spontaneous vomiting occurs, have the casualty lean forward to avoid breathing in of emesis. Rinse mouth and administer more water. Get medical attention immediately.

REFERENCES

Bowden, H.C., Wilby, O.K., Botham. C.A., Adam, P.J., and Ross, F.W. (1995) Assessment of the toxic and potential teratogenic effects of four glycol ethers and two derivatives using the hydra regeneration assay and rat whole embryo culture. *Toxicol. In Vitro* 9(5):773–781.

Hardin, B.D., Goad, P.T., and Burg, J.R. (1986) Developmental toxicity of diethylene glycol monomethyl ether (diEGME). *Fundam. Appl. Toxicol.* 6:430–439.

Hobson, D.W., et al. (1986) A subchronic dermal exposure study of diethylene glycol monomethyl ether and ethylene glycol monomethyl ether in the male guinea pig. *Fundam. Appl. Toxicol.* 6:339–348.

Kawamoto, T., et al. (1990a) Acute oral toxicity of ethylene glycol monomethyl ether and diethylene glycol monomethyl ether. *Bull. Environ. Contain. Toxicol.* 44:602–608.

Kawamoto, T., et al. (1990b) Effect of ethylene glycol monomethyl ether and diethylene glycol monomethyl ether on hepatic metabolizing enzymes. *Toxicology* 62:265–274.

Kawamoto, T., et al. (1992) The effect of ethylene glycol mono-methyl ether and diethylene glycol monomethyl ether on hepatic gamma-glutamyl transpeptidase. *Toxicology* 76:49–57.

Miller, R.R., et al. (1985) Diethylene glycol monomethyl ether 13-week vapor inhalation study in rats. *Fumdam. Appl. Toxicol.* 5:1174–1179.

Nolen, G.A., et al. (1985) Fertility and teratogenic studies of diethylene glycol monobutyl ether in rats and rabbits. *Fundam. Appl. Toxicol.* 5:1137–1143.

Scortichini, B.H., et al. (1986) Teratological evaluation of dermally applied diethylene glycol monomethyl ether in rabbits. *Fundam. Appl. Toxicol.* 7:68–75.

Thompson, E.D., et al. (1984) Mutagenicity testing of diethylene glycol monobutyl ether. *Environ. Health Perspect.* 57:105–112.

Yamano, T., et al. (1993) Effects of diethylene glycol monomethyl ether on pregnancy and postnatal development in rats. *Arch. Environ. Contam. Toxicol.* 24:228–235.

55

ALDEHYDES AND KETONES

Jason S. Garcia and Raymond D. Harbison

ALDEHYDES

FORMALDEHYDE

Target organ(s): Eyes, respiratory system

Occupational exposure limits: NIOSH—TWA 0.016 ppm, Ceiling 0.1 ppm; OSHA—TWA 0.75 ppm, STEL 2 ppm; ACGIH—STEL 0.3 ppm

Reference values: RfD—0.2 mg/kg/day; TDI—150 µg/kg; MRL—0.003 ppm

Hazard statements: H301—toxic if swallowed; H311—toxic in contact with skin; H314—causes severe skin burns and eye damage; H317—may cause allergic skin reaction; H331—toxic if inhaled; H351—suspected of causing cancer by inhalation

Risk phrases: R23/24/25—toxic by inhalation, in contact with skin and if swallowed; R34—causes burns; R43—may cause sensitization by skin contact

Safety phrases: S1/2—keep locked up and out of reach of children; S26—in case of contact with eyes, rinse immediately with plenty of water and seek medical advice; S36/37/39—wear suitable protective clothing, gloves, and eye/face protection; S45—in case of accident or if you feel unwell, seek medical advice immediately (show label where possible); S51—use only in well-ventilated areas

BACKGROUND AND USES

Formaldehyde (methyl aldehyde, methylene oxide; CASRN 50-00-0) is a ubiquitous compound found endogenously in the body and environment. It is a colorless, flammable gas with a distinct, pungent odor and is most commonly available in aqueous solutions under the name formalin (37% formaldehyde in water). Formaldehyde has been used as a disinfectant, an embalming agent, and in industry as a precursor in the fabrication of complex compounds. Since scientific research has identified links between formaldehyde and adverse health effects, precautions and protections must be considered during use.

PHYSICAL AND CHEMICAL PROPERTIES

Formaldehyde is colorless gas with a very distinct, pungent odor. It is highly soluble in water and in a variety of organic solvents. It has the potential to react explosively with peroxides and nitrogen oxide. The gas has a molecular weight of 30.0, a boiling point of −6 °F, and a freezing point of −134 °F. The lower explosive limit is 7.0% and the upper explosive limit is 73%. This compound has a vapor pressure of 10 mm Hg at −88 °C and a vapor density of 1.075 as a gas. NIOSH (2010)

Formalin, the aqueous form of formaldehyde, is a colorless liquid with a very distinct, pungent odor. It is incompatible and may react with strong oxidizers, alkalis,

Hamilton & Hardy's Industrial Toxicology, Sixth Edition. Edited by Raymond D. Harbison, Marie M. Bourgeois, and Giffe T. Johnson.

and acids. The liquid has a variable molecular weight, which is dependent on the specific aqueous formulation. Formalin has a boiling point of 214 °F, a vapor pressure of 26 mm Hg at 25 °C, and a specific gravity of 1.08. The flash point of formalin is 185 °F with the lower explosive limit being 7.0% and the upper explosive limit being 73%.

MAMMALIAN TOXICOLOGY

Acute Effects

Due to high water solubility, acute exposure to formaldehyde results primarily in irritant effects of the upper respiratory tract, including the eyes, nose, and throat. These acute effects are avoidable, as the odor threshold is fairly low for this compound (1–2 ppm). Irritant symptoms and slight pulmonary changes were reported in clinical anatomy students and found to be reversible once exposure to formaldehyde (i.e., laboratory experience) was terminated (Kribel et al., 1993). It has been estimated through human nasal computational fluid dynamic models that there would be around 85.3% uptake of 1 ppm formaldehyde (Schroeter et al., 2014). While formaldehyde uptake occurs with little resistance, it does not appear to severely impact the respiratory system once subjects are removed from exposure. Pulmonary function has been the subject of a number of studies, and no significant decreases in lung function were observed (Schacter et al., 1986; Witek, 1987). Even in cases of asthmatic individuals and where exposure to formaldehyde is in resin form, there were no adverse effects on lung function, such as peak expiratory flow (PEF) rate (Imbus and Tochilin, 1988; Harving et al., 1990). Although no systemic effects are normally associated with formaldehyde exposure, four cases of membranous nephropathy occurred months after severe acute effects were observed (Breysse et al., 1994). This association has not been confirmed in other cases or through biochemical analysis, therefore these case reports remain unique.

The most numerous observed health effects during acute and subacute animal testing were related to the upper respiratory tract. Using computational fluid models similar to those described for humans, rats, and monkeys were also considered. Nasal uptake of 1 ppm formaldehyde was determined to be 99.4% for rats and 86.5% for monkeys (Schroeter et al., 2014). Rats exposed to 15 ppm formaldehyde for 3 weeks had decreased mucociliary function in local nasal regions (Morgan et al., 1986a, 1986b). Exposure to 3–4 ppm formaldehyde in one study was associated with rhinitis, degeneration, hyperplasia, and squamous metaplasia of the nasal epithelium (Cassee and Feron, 1994). Prolonged exposure in Rhesus monkeys to 6 ppm formaldehyde in a study by Monticello et al. (1989) resulted in lesions of the transitional and respiratory epithelia of the nasal passages and upper airways. Airway hyperreactivity, although not observed in human studies, was induced by low levels of formaldehyde in guinea pigs after 8-h inhalation exposures (Swiecichowski et al., 1993). Time-dependent results of these studies implicate the duration of exposure as a possible factor for evaluating formaldehyde inhalation hazards. In most cases, elimination of exposure to formaldehyde led to a reduction in symptoms. Exposure to formaldehyde may alter warning effects produced by other chemicals and compounds. Chang and Barrow (1984) first demonstrated in rats the ability of formaldehyde to induce cross-tolerance to the irritant effects of chlorine gas. This was later confirmed also for the aldehydes acrolein and acetaldehyde for formaldehyde-pretreated rats (Babiuk et al., 1985). Cross-tolerance due to formaldehyde exposure may weaken the natural warning signs that help guard workers from hazardous concentrations.

Chronic Effects

The signs, symptoms, and etiology of documented chronic effects of formaldehyde exposure are varied. There is debate whether formaldehyde is the sole cause of observed health effects or whether it is the result of several chemical exposures occurring at the same time. Depending on route and duration of exposure, workers have reported health effects ranging from dermal eczema to reproductive hazards and neurological disorders. Formaldehyde can act as both an irritant and an allergen. Cases of allergic contact dermatitis have been reported among formaldehyde-exposed employees in many industries, such as printing (Reid and Rycroft, 1993; Koch, 1995). In one study, for 10 out of 11 patients with formaldehyde-induced eczema, about 2/3 reported healing or improvement of the affected areas (Flyvholm and Menne, 1992). No systemic effects in humans have been reported after dermal exposures.

Respiratory effects are more commonly reported, as inhalation is a more prominent occupational exposure route. An *in vitro* study observing tracheal ciliary activity and aldehyde exposure determined formaldehyde to be one of the most potent inhibitors of ciliary activity (Dalhamn and Rosengren, 1971). Chronic exposure to lower levels of formaldehyde may produce irritant effects similar to acute exposure and impaired respiratory function. Numerous studies have reported formaldehyde exposure in various industries with decreased pulmonary function and disease. A study of funeral service workers by Holness and Nethercott (1989) reported more frequent cases of chronic bronchitis, dyspnea, and general irritation among these workers than unexposed controls.

It should be recognized that a water-soluble, low molecular weight compound such as formaldehyde would not typically penetrate the lower respiratory system. However, when individuals are exposed simultaneously to larger molecules, such as wood dust, formaldehyde may adhere to these particles deposit deep in the respiratory tract, causing effects not typically seen with formaldehyde exposure alone.

Hence, studies involving products of woodworking industries, particularly particle board and oriented strand board, which are held together with formaldehyde resins, indicate respiratory hazards associated with these activities. Self-reported cases of asthma and coughing and abnormal tests of lung function were identified and dose–response relationships with formaldehyde exposure were suggested (Alexandersson and Hedenstierna, 1989; Herbert et al., 1994). These studies also examined pulmonary signs, such as forced expiratory volume per 1 s and forced vital capacity, and found decreases. Nasal obstruction and mucosal swelling were observed, and mucociliary clearance and olfactory function were significantly reduced in a population of workers with long term exposure to formaldehyde and wood dust (Holström and Wilhelmsson, 1988). Malaka and Kodama (1990) concluded that chronic exposure to formaldehyde in plywood workers induces signs and symptoms similar to those seen in chronic obstructive pulmonary disease. Similar effects are seen in other industries in which formaldehyde is used, such as in application of acid-curing paints and lacquers (Alexandersson and Hedenstierna, 1988).

Unique cases of other respiratory effects are also seen in the literature. A case of lymphocytic alveolitis was reported by Schauble and Rich (1994) in a crematorium worker. The authors concluded that exposure to formaldehyde fumes upon cremation of preserved bodies is a possible explanation for their findings. Some cases of occupational asthma due to formaldehyde gas and dust exposure have also been reported (Lemière et al., 1995). While respiratory effects are seen, no permanent cases of lung impairment or increased severity of symptoms have been documented as a result of occupational formaldehyde exposure. Tolerance to the effects of formaldehyde is not observed among chronically exposed workers.

Systemic effects after inhalation exposure have been reported. Baj et al. (1994) reported formaldehyde and other solvent exposure produced changes in the immune and hematopoietic systems in 22 office workers. Adverse neurobehavioral responses to formaldehyde exposure have been reported, although infrequently. Disturbances in equilibrium, mood, and sleep patterns, as well as memory loss, headache, and in a few cases, occurrence of seizures, have been reported (Kilburn et al., 1985; Kilburn, 1994). Reproductive function in males and congenital birth defects are not associated with formaldehyde exposure, although an increase in spontaneous abortions was reported significant in a study by Taskinen et al. (1994).

The carcinogenicity of formaldehyde is a regularly debated topic. Numerous studies have been unable to establish a definite link between specific cancers and occupational exposures. A vast amount of literature for human and animal data currently exists. Earlier studies demonstrated, particularly in rats, a significant increase in nasal carcinomas after inhalation exposure to formaldehyde (Swenberg et al., 1980; Albert et al., 1982; Kerns et al., 1983). Further investigations

identified the nasoturbinate and nasal septum as primary targets for tumor initiation (Morgan et al., 1986a, 1986b). An article published in 2013 confirmed a possible link between formaldehyde exposure and autophagy in male rats. The study determined a dose-dependent relationship between formaldehyde exposure and abnormal testicular morphology and mass (Han et al., 2013). These animal studies prompted evaluation of similar effects in humans. There have been some case reports of sentinel nasal cancer in humans who were occupationally exposed to formaldehyde for long periods, but these isolated cases cannot be used to definitively establish the carcinogenicity of this compound (Halperin et al., 1983; Holstrom and Lund, 1991). Conflicting studies in the past for this type of cancer have reported significant excesses of sinonasal cancers (Olsen et al., 1984; Hayes et al., 1986) and no associations between these sites and formaldehyde exposure (Vaughn et al., 1986; Roush et al., 1987). A recent study by Hansen and Olsen (1995) noted a relationship with formaldehyde exposure and sinonasal cancers in the absence of a wood dust confounding exposure. A more recent case control study of nasal and paranasal carcinomas found no conclusive evidence to implicate formaldehyde other than as a possible potentiator of the effects of wood dust exposure (Luce et al., 1993). Reviews of epidemiological literature by Blair et al. (1990a, 1990b) and Partanen (1993) calculate some risk associated with sinonasal and nasopharyngeal cancer and formaldehyde exposure, although unequivocal proof has yet to be published.

A study was conducted by the National Cancer Institute to evaluate formaldehyde exposure in workers employed at 10 formaldehyde-producing facilities. The findings did not identify any significant associations with any cancer at the levels of exposure estimated to occur (Blair et al., 1986). This study, the largest of its kind for assessment of formaldehyde exposure, has been the subject of an ongoing debate on the interpretation of the data collected. National Cancer Institute researchers conducted a follow-up study in 2013 and explained that when combined with data from previous investigations, the overall results suggest a link between formaldehyde exposure and nasopharyngeal cancer (Beane Freeman et al., 2013). Sterling and Weinkam (1988) first criticized the original work as possibly masking an association of lung cancer through the design of the study. Reanalysis of the data with particular attention to the "healthy worker effect" and short- and long-term employees shows an increased risk of lung cancer with cumulative exposure to formaldehyde, although some calculation errors were present. Comments from both sides of the issue rebut the opposing position and data interpretation, leading to conflicting opinions among the professional community (Blair and Stewart, 1989; Sterling and Weinkam, 1989). A similar study published in the same year found no association between formaldehyde exposure and cancers (Vaughn et al., 1986). Another study a few years later reported no significant association with respiratory cancers, but there was a possible relationship between formaldehyde exposure and cancers of the buccal cavity and connective tissue in garment workers (Staynor et al., 1988).

Attributing chronic health hazards to occupational formaldehyde exposure is difficult to determine in humans because of coexposures to other compounds such as phenol and particulates such as wood dust. Also, it is important to note that professional exposures (e.g., embalmers) and industrial exposures have shown different mortality patterns. Additional analyses by Blair et al. (1990a, 1990b) are in agreement with the original paper, finding no specific link of formaldehyde to lung cancer. The authors concluded that, although formaldehyde cannot be ruled out, it is more probable that other exposures, such as to phenol and wood dust, are responsible for the excess lung cancer observed in exposed populations. Sterling and Weinkam (1994) reanalyzed the data to show an increased risk of lung cancer with increased formaldehyde exposure. The differences of these two approaches illustrate the difficult task of assigning carcinogenic risk to a compound as common as formaldehyde. An independent analysis of this additional data was performed by Marsh et al. (1992), taking into account coexposures to a variety of substances. In the presence of the coexposures, significant increases in lung cancer were observed, but these were not evident in the analyses that did not include the coexposures. Similar studies showed an increase of lung or nasopharyngeal cancers among occupationally exposed workers and could not conclude, with any certainty, that formaldehyde was the exclusive causative agent for the disease (Bertazzi et al., 1986; Marsh et al., 1994). Andjelkovich et al. (1994) studied mortality of iron foundry workers. An increase in lung cancers was identified, but researchers could not find an association with formaldehyde. Overall, the literature does not seem to support a link between occupational formaldehyde exposure and cancers of the respiratory system. In addition, formaldehyde has not been directly associated with other cancer risks. In a study by Gerin et al. (1989), other cancer sites were examined, including stomach, colon, rectum, liver, and many other systemic sites and no correlation was found with formaldehyde exposure.

This inconclusive and contradictory information regarding this compound has led the ACGIH (2013) to give formaldehyde an A2 designation (probable human carcinogen). The International Agency for Research on Cancer (IARC [1995]) has a similar rating for formaldehyde as probably carcinogenic in humans, Group 2A.

Mechanism of Action(s)

Formaldehyde is an intermediate in a number of metabolic interactions, particularly those involving amino acids and the demethylation of N-, O-, and S-methyl compounds. In a study of mice, the liver was identified as having the highest content of formaldehyde after exposure, with the kidney being second (Johansson and Tjalve, 1978). The fate of formaldehyde can vary, depending on the molecule it comes into contact with after absorption. If it is not quickly metabolized, there is an opportunity for formaldehyde to react with nucleophilic molecules in the cell. Almost all cells are able to metabolize formaldehyde via formaldehyde dehydrogenase and aldehyde dehydrogenase (ADH). The presence of these enzymes in erythrocytes (Uotila and Koivusalo, 1987) may account for the relatively few systemic effects associated with formaldehyde. After inhalation exposure, these enzymes may oxidize the formaldehyde to formate or carbon dioxide and water. This is the major pathway for formaldehyde metabolism. This reaction also involves a glutathione–hemiacetal intermediate (hydroxymethylglutathione), an NAD+ cofactor, and the presence of S-formylglutathione hydrolase, which frees the glutathione molecule from the intermediate. Formate can also be directly incorporated into macromolecules by a tetrahydrofolate-dependent pathway via the one carbon pool. Peroxisomal catalase is capable of oxidizing formaldehyde to formate in the presence of hydrogen peroxide. Production of H_2O_2 is the rate-limiting step in this reaction, which does not account for a great deal of formaldehyde metabolism.

Chemical Pathology

The potential of formaldehyde-induced damage is greatest prior to metabolism. The high reactivity of the compound makes it quite toxic, and is more likely to be hazardous at high exposure concentrations. A review of formaldehyde reactivity by Auerbach et al. (1977) indicates that reactions with amino groups in biological molecules are the ones of most concern. These amino groups form methylol adducts, which can react with a second amino group, producing a molecule of water in the process. A methylene bridge is formed, which may be responsible for deoxyribonucleic acid (DNA) crosslinks, single strand breaks, and mutagenic activity. Induction of nasal carcinomas in rats has been linked to mutation of the p53 tumor suppressor gene (Recio et al., 1992; Wolf et al., 1995). This mechanism of action has been identified in human cancers as well, but not specifically with formaldehyde exposure (Nigro et al., 1989). Since no association of formaldehyde-induced cancer has been noted in humans, the mechanism of action for human carcinogenicity is highly speculative. The biological plausibility of formaldehyde-associated cancers has not been explored in the epidemiologic literature. Data on cell proliferation in humans exposed to formaldehyde have not been reported, although cell proliferation has been observed in the nasal tissue of exposed rats and mice (Chang et al., 1983).

Immunoglobulin (Ig) E or G antibodies have not been identified as being present in some conditions (i.e., occupational asthma and contact dermatitis) associated with formaldehyde exposure (Kramps et al., 1989; Grammer et al., 1990; Liddn et al., 1993). This indicates that the signs and symptoms associated with these conditions may not be immunologically mediated but that they occur through an irritation mechanism. The occurrence of membranous nephropathy is hypothesized to be induced by formaldehyde in genetically susceptible persons with a common amino acid sequence, but clear evidence for this mechanism is unavailable (Breysse et al., 1994).

SOURCES OF EXPOSURE

Occupational

Formaldehyde has a variety of uses in the occupational setting. It is a constituent of many formaldehyde-based resins which may be used in particle board, furniture, plastics, rubbers, and other textiles. Formaldehyde is also used as an intermediate in the synthesis of other chemicals and in dyes, animal feeds, and a wide range of commercial products, such as perfumes and vitamins. The funeral, veterinary, and medical industries use formaldehyde as a disinfectant and preservative of biological tissues. Studies have even shown concentrations of formaldehyde as a pyrolysis product in grill kitchens (Vainiotalo and Matveinen, 1993).

Environmental

Environmentally, this compound is found in combustible engine exhaust and cigarette smoke and as a natural product of photooxidation. An emerging environmental source of formaldehyde exposure is through the use of electronic cigarettes. Electronic cigarettes have become mainstream as an alternative to tobacco cigarettes. Researchers are investigating whether or not exposure to formaldehyde is possible from this route. While formaldehyde was detected in one chamber study, it was below established exposure limits. In addition, researchers were unable to determine if the formaldehyde was the result of endogenous production by the participant or that of the electronic cigarette itself (Burstyn, 2014). In order to make a more concrete determination of a link between formaldehyde exposure and electronic cigarettes, more investigation is needed.

INDUSTRIAL HYGIENE

Humans are exposed to formaldehyde every day since it is part of our environment and bodies. Occupational exposure to excessive amounts of formaldehyde can occur. Formaldehyde is mainly encountered in the workplace as a gas, making inhalation the primary route of exposure. Some compounds,

such as carbamide formaldehyde resins and bromonitropropanediol, are known to release gaseous formaldehyde during decomposition. Inhalation exposure to dusts-containing formaldehyde is also an occupational hazard. In settings where the formalin solution is used, dermal contact with the liquid becomes a concern as well. Production of industrial formaldehyde is accomplished by the catalytic oxidation of methanol, through a variety of techniques. The Occupational Safety and Health Administration (OSHA) requires a permissible exposure limit (PEL) of 0.75 ppm with a short-term exposure limit (STEL) of 2 ppm. A concentration of 0.5 ppm calculated as an 8 h time-weighted average (TWA) has been established by OSHA as the action limit in which increased monitoring should be initiated to prevent unnecessary exposure. The National Institute for Occupational Safety and Health (NIOSH) recommends an exposure limit (REL) of 0.016 ppm and a 15-min ceiling of 0.1 ppm. The American Conference of Governmental Industrial Hygienists (ACGIH) recommends a 0.3 ppm threshold limit value (TLV).

Adequate ventilation and personal respirators are recommended for mitigating inhalation exposures in the occupational environment. Protective clothing, such as aprons, face shields, and gloves, are recommended to prevent dermal contact of formalin. Since each workplace can vary, different control measures will be needed to prevent excessive formaldehyde exposure. Workers in the funeral industry can be exposed to both the gas and liquid form of the chemical. Recommendations for embalmers include all forms of protective equipment. A survey of local exhaust ventilation systems in funeral parlors demonstrated some formaldehyde concentrations exceeding recommended guidelines (Korczynski, 1994). In a gross pathology setting, protective devices, especially respirators, may hinder vision and perspective and therefore may not be a reasonable solution (Akbar-Khanzadeh et al., 1994). Elimination of formaldehyde by substitution with glutaraldehyde or non-aldehyde-containing Kryofix may interfere with embalming techniques. New engineering controls for ventilation, such as a slot hood design with down-draft ventilation or the presence of motors, which cause a down flow of formaldehyde vapors into carbon filtration systems, have been examined and shown to be more effective than air circulation and more cost efficient as well (Gressel and Hughes, 1992; Coleman, 1995). Recent research has focused on hair stylists' exposure to formaldehyde. During a study that simulated use, a popular hair straightening product containing methylene glycol, a hydrated form of formaldehyde, was found to increase formaldehyde levels in the breathing zone above the OSHA 15 min TWA, the NIOSH ceiling limit, and the ACGIH TLV (Stewart et al., 2013). Continued exposure at these levels may require the institution of engineering controls, administrative controls, and/or personal protective equipment.

Protective gloves are necessary to prevent formalin from reaching the surface of the skin. Before selecting protective gloves, it is important to ensure that their construction material can with stand contact with formalin. Schwope et al. (1988) tested a variety of glove materials for permeability to formaldehyde and reported that neoprene-based gloves are more resistant than natural latex, polyvinyl chloride, and polyethylene gloves without compromising tactility or dexterity.

Air monitoring in the workplace for formaldehyde can be accomplished by the recommended NIOSH 3500 pump and impinger method. Alternative tests may be available, such as a passive bubbler technique that has been shown to meet OSHA +25% accuracy requirements (Kollman, 1994). Personal air monitors may be more effective for measuring actual worker hazards. A study in by Nobles et al. (1993) reported that only one of four monitors tested for short-term and TWA sampling was consistent with OSHA performance criteria and the newly revised ACGIH 0.3 ppm ceiling. Chemosorption sampling of formaldehyde in air was found to be reliable for monitoring occupational settings without being compromised by interference from other aldehydes and related compounds (Andersson et al., 1981). Current technology has led to the development of portable new sensors capable of detecting formaldehyde without interferences from other chemical compounds. A handheld formaldehyde detection device has been developed to analyze indoor environments for contamination. The limit of detection for the device was 0.04 ppm during a sampling time of 3 min with no interference from other aldehydes and/or volatile organic compounds (Kawamura et al., 2005). Cataluminescence-based gas sensors have also been developed for online monitoring of formaldehyde in air with satisfactory stability. No interferences from ammonia, ethanol, benzene, carbon monoxide, and sulfur dioxide were found (Zhou et al., 2006).

MEDICAL MANAGEMENT

For accidental formaldehyde exposures, removal from exposure is the primary treatment. Countering measures should be taken if symptoms and signs emerge.

Species differences in the mechanisms of formaldehyde-induced adverse health effects and in exposure estimates make the risk assessment process for carcinogenicity a daunting task. Animal studies in rodents and nonhuman primates yield conflicting results with respect to observed adverse health effects. Inconsistencies of this nature make the extrapolation of animal data to human responses unreliable. For carcinogenicity, the rates of DNA binding and cell proliferation are both possible predictors of cancer and are important parameters to consider for estimating risk (Heck et al., 1990). Animal studies found lower cancer risks associated with primate exposure when compared with rodent exposure.

An additional uncertainty in formaldehyde risk assessments revolves around the quantification of exposure in epidemiologic evaluations. It is important for researchers to consider varying exposure estimates in the occupational environment, where they can be relatively low. Low dose chronic health hazards must continue to be investigated. Very low dose cancer risks are not readily measurable. Evans et al. (1994) described a probabilistic approach that takes into account the uncertainty of carcinogenic action for the compound, as well as the low dose confounder. The authors find that their calculated occupational risk estimates are somewhat lower than the Environmental Protection Agency's estimates for excess cancer risks associated with formaldehyde exposure in the environment. Blair and Stewart (1992) examined exposure misclassification in population-based formaldehyde studies and concluded that analyses should be conducted using several exposure measures, while including data such as duration by intensity to avoid exposure misclassification. Whether positive or negative, these methods would enhance the findings of these investigations.

Endogenous formation of formaldehyde makes it difficult to quantify external exposures to the chemical because of ever-present metabolites. Visual blood and tissue concentrations of formaldehyde are often masked due to rapid biotransformation in the human body. Shifts in formic acid concentrations in the urine of occupationally exposed workers did not reliably monitor formaldehyde exposure (Gottschling et al., 1984). Animal studies of urinary metabolites identified formate and a cysteine adduct upon injection of radiolabeled formaldehyde, but these findings have not been replicated by human data (Hemminki, 1984). More recent studies of formaldehyde biomonitoring have shown an increase in DNA–protein crosslink (DPC) frequencies in the white blood cells of workers exposed to formaldehyde (Shaham et al., 1996). The authors also reported a linear correlation between chronic exposure and the amount of DPC observed. Low level exposure in a study of mortuary students was associated with epithelial cytogenic changes in the mouth and blood lymphocytes, indicating these changes may be useful as markers of biologically effective doses of formaldehyde (Saruda et al., 1993).

REFERENCES

Akbar-Khanzadeh, F., et al. (1994) Formaldehyde exposure, acute pulmonary response, and exposure control options in a gross anatomy laboratory. *Am. J. Ind. Med.* 26:61–75.

Albert, R.E., et al. (1982) Gaseous formaldehyde and hydrogen chloride induction of nasal cancer in the rat. *J. Natl. Cancer Inst.* 68:591–603.

Alexandersson, R. and Hedenstierna, G. (1988) Respiratory hazards associated with exposure to formaldehyde and solvents in acid-curing paints. *Arch. Environ. Health* 43:222–227.

Alexandersson, R. and Hedenstierna, G. (1989) Pulmonary function in wood workers exposed to formaldehyde: a prospective study. *Arch. Environ. Health* 44:5–11.

American Conference of Governmental Industrial Hygienists (ACGIH) (2013) *Threshold Limit Values (TLVs) for Chemical Substances and Physical Agents and Biological Exposure Indices (BEIs)*, Cincinnati, OH: American Conference of Governmental Industrial Hygienists.

Andersson, K., et al. (1981) Chemosorption sampling and analysis of formaldehyde in air. *Scand. J. Work Environ. Health* 7:282–289.

Andjelkovich, D.A., et al. (1994) Mortality of iron foundry workers. *J. Occup. Med.* 36:1301–1309.

Auerbach, C., Moutschen Dahnen, M., and Moutschen, J. (1977) Genetic and cytogenetical effects of formaldehyde and related compounds. *Mutat. Res.* 39:317–361.

Babiuk, C., Steinhagen, W.H., and Barrow, C.S. (1985) Sensory irritation response to inhaled aldehydes after formaldehyde pretreatment. *Toxicol. Appl. Pharmacol.* 79:143–149.

Baj, Z., et al. (1994) The effects of chronic exposure to formaldehyde, phenol and organic chlorohydrocarbons on peripheral blood cells and the immune system in humans. *J. Invest. Allergol. Clin. Immunol.* 4:186–191.

Beane Freeman, L.E., et al. (2013) Mortality from solid tumors among workers in formaldehyde industries: an update of the NCI cohort. *Am. J. Ind. Med.* 56:1015–1026.

Bertazzi, P.A., et al. (1986) Exposure to formaldehyde and cancer mortality in a cohort of workers producing resins. *Stand. J. Work Environ. Health* 12:461–468.

Blair, A. and Stewart, P.A. (1989) Formaldehyde revisited: comments on the reanalysis of the National Cancer Institute study of workers exposed to formaldehyde. *J. Occup. Med.* 31:881–890.

Blair, A. and Stewart, P.A. (1992) Do quantitative exposure assessments improve risk estimates in occupational studies of cancer? *Am. J. Int. Med.* 21:53–63.

Blair, A., et al. (1986) Mortality among industrial workers exposed to formaldehyde. *J. Natl. Cancer Inst.* 76:1071–1084.

Blair, A., Stewart, P.A., and Hoover, R.N. (1990a) Mortality from lung cancer among workers employed in formaldehyde industries. *Am. J. Ind. Med.* 17:683–699.

Blair, A., et al. (1990b) Epidemiologic evidence in the relationship between formaldehyde exposure and cancer. *Scand. J. Work Environ. Health* 16:381–393.

Breysse, P., et al. (1994) Membranous nephropathy and formaldehyde exposure. *Ann. Intern. Med.* 120:396–397.

Burstyn, I. (2014) Peering through the mist: systematic review of what the chemistry of contaminants in electronic cigarettes tells us about health risks. *BMC Public Health* 14:18.

Cassee, F.R. and Feron, V.J. (1994) Biochemical and histopathological changes in nasal epithelium of rats after 3-day intermittent exposure to formaldehyde and ozone alone or in combination. *Toxicol. Lett.* 72:257–268.

Chang, J.C.F. and Barrow, C.S. (1984) Sensory irritation tolerance in F-344 rats exposed to chlorine or formaldehyde gas. *Toxicol. Appl. Pharmacol.* 76:319–327.

Chang, J.C.F., et al. (1983) Nasal cavity deposition, histopathology, and cell proliferation after single or repeated formaldehyde exposures in B6C3F1 mice and F-344 rats. *Toxicol. Appl. Pharmacol.* 68:161–176.

Coleman, R. (1995) Reducing the levels of formaldehyde exposure in gross anatomy laboratories. *Anat. Rec.* 243:531–533.

Dalhamn, T. and Rosengren, A. (1971) Effect of different aldehydes on tracheal mucosa. *Arch. Otolaryngol.* 93:496–500.

Evans, J.S., et al. (1994) A distributional approach to characterizing low-dose cancer risk. *Risk Anal.* 14:25–34.

Flyvholm, M-A. and Menne, T. (1992) Allergic contact dermatitis from formaldehyde: a case study focusing on sources of formaldehyde exposure. *Contact Dermatitis* 27:27–36.

Gerin, M., et al. (1989) Cancer risks due to occupational exposure to formaldehyde: results of a multi-site case-control study in Montreal. *Int. J. Cancer* 44:53–58.

Gottschling, L.M., Beaulieu, H.J., and Melvin, W.W. (1984) Monitoring of formic acid in urine of humans exposed to low levels of formaldehyde. *Am. Ind. Hyg. Assoc. J.* 45:19–23.

Grammer L.C., et al. (1990) Clinical and immunologic evaluation of 37 workers exposed to gaseous formaldehyde. *J. Allergy Clin. Immunol.* 86:177–181.

Gressel, M.G. and Hughes, R.T. (1992) Effective local exhaust ventilation for controlling formaldehyde exposures during embalming. *Appl. Occup. Environ. Hyg.* 7:840–845.

Halperin, W.E., et al. (1983) Nasal cancer in a worker exposed to formaldehyde. *JAMA* 249:510–512.

Han, S.P., et al. (2013) Formaldehyde exposure induces autophagy in testicular tissues of adult male rats. *Environ. Toxicol.*, C doi: 10.1002/tox.21910.

Hansen, J. and Olsen, J.H. (1995) Formaldehyde and cancer morbidity among male employees in Denmark. *Cancer Causes Controls* 6:354–360.

Harving, H., et al. (1990) Pulmonary function and bronchial reactivity in asthmatics during low-level formaldehyde exposure. *Lung* 168:15–21.

Hayes, R.B., et al. (1986) Cancer of the nasal cavity and paranasal sinuses and formaldehyde exposure. *Int. J. Cancer* 37:487–492.

Heck, H.d'A., Casanova, M., and Starr, T.B. (1990) Formaldehyde toxicity—new understanding. *Crit. Rev. Toxicol.* 20:397–426.

Hemminki, K. (1984) Urinary excretion products of formaldehyde in the rat. *Chem. Biol. Interact.* 48:243–248.

Herbert, F.A., et al. (1994) Respiratory consequences of exposure to wood dust and formaldehyde of workers manufacturing oriented strand board. *Arch. Environ. Health* 4t1:465–470.

Holness, D.L. and Nethercott, J.R. (1989) Health status of funeral scrvice workers exposed to formaldehyde. *Arch. Environ. Health* 44:222–228.

Holstrom, M. and Lund, V.J. (1991) Malignant melanomas of the nasal cavity after occupational exposure to formaldehyde. *Br. J. Ind. Med.* 48:9–11.

Holström, M. and Wilhelmsson, B. (1988) Respiratory symptoms and pathophysiological effects of occupational exposure to formaldehyde and wood dust. *Stand. J. Work Environ. Health* 14:306–311.

IARC (1995) Wood dust and formaldehyde. *IARC Monogr. Eval. Carcinog. Risks Hum.* 62:336–349.

Imbus, H.R. and Tochilin, S.J. (1988) Acute effect upon pulmonary function of low level exposure to phenol–formaldehyde-resin-coated wood. *Am. Ind. Hyg. Assoc. J.* 49:434–437.

Johansson, E.B. and Tjalve, H. (1978) The distribution of [^{14}C] dimethylnitrosamine in mice: autographic studies in mice with inhibited and noninhibited dimethylnitrosamine metabolism and a comparison with the distribution of [^{14}C] formaldehyde. *Toxcol. Appl. Pharmacol.* 45:565–575.

Kawamura, K., et al. (2005) Development of a novel hand-held formaldehyde gas sensor for the rapid detection of sick building syndrome. *Sens. Actuators B* 105:495–501.

Kerns, W.D., et al. (1983) Carcinogenicity of formaldehyde in rats and mice after long-term inhalation exposure. *Cancer Res.* 43:4382–4392.

Kilburn, K.H. (1994) Neurobehavioral impairment and seizures from formaldehyde. *Arch. Environ. Health* 49:37–44.

Kilburn, K.H., Seidman, B.C., and Warshaw, R. (1985) Neuro-behavioral and respiratory symptoms of formaldehyde and xylene exposure in histology technicians. *Arch. Environ. Health* 40:229–233.

Koch, P. (1995) Occupational contact dermatitis from colophony and formaldehyde in banknote paper. *Contact Dermatitis* 32:371–372.

Kollman, J.R. (1994) Field evaluation of a diffusive sampler for monitoring formaldehyde in air. *Appl. Occup. Environ. Hyg.* 9(4):262–266.

Korczynski, R.E. (1994) Formaldehyde exposure in the funeral industry. *Appl. Occup. Environ. Hyg.* 9:575–579.

Kramps, J.A., et al. (1989) Measurement of specific IgE antibodies in individuals exposed to formaldehyde. *Clin. Exp. Allergy* 19:509–514.

Kribel, D., Sama, S.R., and Cocanour, B. (1993) Reversible pulmonary responses to formaldehyde. *Am. Rev. Respir. Dis.* 148:1509–1515.

Lemière, C., et al. (1995) Occupational asthma due to formaldehyde resin dust with and without reaction to formaldehyde gas. *Eur. Respir. J.* 8:861–865.

Liddn, S., et al. (1993) Absence of specific IgE antibodies in allergic contact sensitivity to formaldehyde. *Allergy* 48:525–529.

Luce, D., et al. (1993) Sinonasal cancer and occupational exposure to formaldehyde and other substances. *Int. J. Cancer* 53:224–231.

Malaka, T. and Kodama, A.M. (1990) Respiratory health of plywood workers occupationally exposed to formaldehyde. *Arch. Environ. Health* 45:288–294.

Marsh, G.M., Stone, R.A., and Henderson, V.L. (1992) Lung cancer mortality among industrial workers exposed to formaldehyde: a poison regression analysis of the National Cancer Institute study. *Am. Ind. Hyg. Assoc.* 53:681–69l.

Marsh, G.M., et al. (1994) Mortality patterns among chemical plant workers exposed to formaldehyde and other substances. *J. Natl. Cancer Inst.* 86:384–386.

Monticello, T.M., et al. (1989) Effects of formaldehyde as on the respiratory tract of rhesus monkeys. *Am. J. Pathol.* 134:515–527.

Morgan, K.T., Patterson, D.L., and Gross, E.A. (1986a) Responses of the nasal mucociliary apparatus of F-344 rats to formaldehyde gas. *Toxicol. Appl. Pharmacol.* 82:1–13.

Morgan, K.T., et al. (1986b) More precise localization of nasal tumors associated with chronic exposure of F-344 rats to formaldehyde gas. *Toxicol. Appl. Pharmacol.* 82:264–271.

National Institute for Occupational Safety and Health (N1OSH) (2010) *Pocket Guide to Chemical Hazards*, NIOSH 94-116, Washington, DC: US Department of Health and Human Services.

Nigro, J., et al. (1989) Mutations in the p53 gene occur in diverse human tumour types. *Nature* 342:705–708.

Nobles, J.S., Strang, C.R., and Michael, P.R. (1993) A comparison of active and passive sampling devices for full-shift and short-term monitoring of formaldehyde. *Am. Ind. Hyg. Assoc. J.* 54:723–732.

Olsen, J.H., et al. (1984) Occupational formaldehyde exposure and increased nasal cancer risk in man. *Int. J. Cancer* 34:639–644.

Partanen, T. (1993) Formaldehyde exposure and respiratory cancer: a meta-analysis of the epidemiologic evidence. *Scand. J. Work Environ. Health* 19:8–15.

Recio, L., et al. (1992) p53 mutations in formaldehyde-induced nasal squamous cell carcinomas in rats. *Cancer Res.* 52:6113–6116.

Reid, CM. and Rycroft, R.J. (1993) Allergic contact dermatitis from multiple source of MCI/MI biocide and formaldehyde in a printer. *Contact Dermatitis* 28:252–253.

Roush, G.C., et al. (1987) Nasopharyngcal cancer, sinonasal cancer, and occupations related to formaldehyde: a case control study. *J. Natl. Cancer Inst.* 79:1221–1224.

Saruda, A., et al. (1993) Cytogenic effects of formaldehyde exposure in students of mortuary science. *Cancer Epidemiol Biomarkers Prev.* 2:453–460.

Schacter, E.N., et al. (1986) A study of respiratory effects from exposure to 2 ppm formaldehyde in healthy subjects. *Arch. Environ. Health* 41:229–239.

Schauble, T.L. and Rich, E.A. (1994) Lymphocytic alveolitis in a crematorium worker. *Chest* 105:617–619.

Schroeter, J.D., et al. (2014) Effects of endogenous formaldehyde in nasal tissues on inhaled formaldehyde dosimetry predictions in the rat, monkey, and human nasal passages. *Toxicol. Sci.* 1–49.

Schwope, A.D., et al. (1988) Gloves for protection from aqueous formaldehyde: permeation resistance and human factors analysis. *Appl. Ind. Hyg.* 3:167–176.

Shaham, J., et al. (1996) DNA–protein crosslinks, a biomarker of exposure to formaldehyde: *in vitro* and *in vivo* studies. *Carcinogenesis* 17:121–125.

Staynor, L.T., et al. (1988) A retrospective cohort mortality study of workers exposed to formaldehyde in the garment industry. *Am. J. Ind. Med.* 13:667–681.

Sterling, T.D. and Weinkam, J.J. (1988) Reanalysis of lung cancer mortality in a National Cancer Institute study on mortality among industrial workers exposed to formaldehyde. *J. Occup. Med.* 30:895–901.

Sterling, T.D. and Weinkam, J.J. (1989) Formaldehyde revisited: reanalysis of lung cancer mortality in a National Cancer Institute study of "Mortality among industrial workers exposed to formaldehyde": additional comments. *J. Occup. Med.* 31:881–884.

Sterling, T.D. and Weinkam, J.J. (1994) Mortality from respiratory cancer (including lung cancer) among workers employed in formaldehyde industries. *Am. J. Ind. Med.* 25:593–602.

Stewart, M., et al. (2013) Case study—formaldehyde exposure during simulated use of a hair straightening product. *J. Occup. Environ. Hyg.* 10:D104–D110.

Swenberg, J.A., et al. (1980) Induction of squamous cell carcinomas of the rat nasal cavity by inhalation exposure to formaldehyde vapor. *Cancer Res.* 40:3398–3402.

Swiecichowski, A.L., et al. (1993) Formaldehyde-induced airway hyper-reactivity *in vivo* and *ex vivo* in guinea pigs. *Environ. Res.* 61:185–199.

Taskinen, H., et al. (1994) Laboratory work and pregnancy outcome. *J. Occup. Med.* 36:311–319.

Uotila, L. and Koivusalo, M. (1987) Multiple forms of formaldehyde dehydrogenase from human red blood cells. *Hum. Hered.* 37:102–6.

Vainiotalo, S. and Matveinen, K. (1993) Cooking fumes as a hygienic problem in the food and catering industries. *Am. Ind. Hyg. Assoc. J.* 54:376–382.

Vaughn, T.L., et al. (1986) Formaldehyde and cancers of the pharynx, sinus and nasal cavity. I. Occupational exposures. *Int. J. Cancer* 38:677–683.

Witek, T.J., Jr. (1987) An evaluation of respiratory effects following exposure to 2.0 ppm formaldehyde in asthmatics: lung function, symptoms, and airway reactivity. *Arch. Environ. Health* 42:230–237.

Wolf, D.C., et al. (1995) Immunohistochemical localization of p53, PCNA, and TGF-e proteins in formaldehyde-induced rat nasal squamous cell carcinomas. *Toxicol. Appl. Pharmacol.* 132:27–35.

Zhou, K., et al. (2006) On-line monitoring of formaldehyde in air by cataluminescence-based gas sensor. *Sens. Actuators B* 119:392–397.

ACETALDEHYDE

Target organ(s): Eyes, skin, respiratory system, kidneys, central nervous system

Reproductive system

Occupational exposure limits: NIOSH—not established; OSHA—TWA 200 ppm

ACGIH—STEL ceiling 25 ppm

Reference values: RfC—0.009 mg/m^3; TDI—0.1 µg/kg; MRL—1.0 ppm

Hazard statements: H224—extremely flammable liquid and vapor; H319—causes serious eye irritation; H335—may cause respiratory irritation; H351—suspected of causing cancer

Precautionary statements: P210—keep away from heat/sparks/open flames/ . . . /hot surfaces/ . . . no smoking; P281—use personal protective equipment as required; P305+P351+P338—if in eyes, rinse cautiously with water for several minutes; remove contact lenses, if present and easy to do; continue rinsing; P337+P313—if eye irritation persists, get medical advice/attention

Risk phrases: R12—extremely flammable; R36/37—irritating to eyes and respiratory system; R40—limited evidence of a carcinogenic effect

Safety phrases: S2—keep out of the reach of children; S16—keep away from sources of ignition . . . no smoking; S33—take precautionary measures against static discharges; S36/37—wear suitable protective clothing and gloves

BACKGROUND AND USES

Acetaldehyde (acetic aldehyde, ethanol ethyl aldehyde; CASRN 75-07-0) is a low molecular weight aldehyde used as a solvent and chemical intermediate in the manufacturing of compounds. It is also used in the concoction of resins, explosives, fragrances, and flavorings. While widely used in industry, acetaldehyde is regularly found in nature. Higher order plants metabolize acetaldehyde to form an intermediate which can be found in humans after ethanol consumption. It can also be detected in a number of fruits, vegetables, and spices.

PHYSICAL AND CHEMICAL PROPERTIES

Acetaldehyde exists as a colorless liquid or gas with distinct, fruity odor. It is miscible with water and many organic solvents. The chemical has a molecular weight of 44.1 g/mol, a boiling point of 69 °F, and a freezing point of −190 °F. The vapor pressure is 902 mm Hg at 25 °C and the specific gravity is 0.79. The lower explosive limit is 4.0% and the upper explosive limit is 60%. Acetaldehyde may be incompatible with strong oxidizers, acids, and bases. It can also react with hydrogen cyanide and hydrogen sulfide. If acetaldehyde is exposed to air, it has the potential to form peroxides that are explosive.

MAMMALIAN TOXICOLOGY

Acute Effects

Low concentrations of acetaldehyde elicit irritant effects which are primarily a nuisance. Vapor exposure to concentrations of 50 ppm has been shown to cause eye irritation (Sim and Pattle, 1957). Higher concentrations can result in transient conjunctivitis and mild upper respiratory effects (Brabec, 1994). Similar to other aldehydes, high concentrations of acetaldehyde vapor can have narcotic effects. Segel and Mason noted that acetaldehyde caused depressant effects on cardiac mitochondria which may lead to abnormal cardiac biochemistry and function (Segel and Mason, 1979). Dermal contact with the liquid state may result in erythema and burns if the exposure is of a sufficient duration (Pinnas and Meinke, 1992). Similar to formaldehyde, sensitization, and subsequent dermatitis are potential outcomes for certain individuals. Human volunteers who inhaled acetaldehyde in concentrations likely to be encountered in the workplace (1 ppm) were found to retain an average of 60% of the inhaled dose (Egle, 1970).

Chronic Effects

Chronic effects associated with occupational exposure to acetaldehyde have not been reported. One study has shown that inhaled acetaldehyde was responsible for bronchoconstriction in asthmatic patients (Myou et al., 1995). This is achieved through a histamine-releasing action. Chronic exposure to acetaldehyde, a component of cigarette smoke, was shown to be damaging to cilia and lung cells (Rylander, 1973). Chronic alcohol consumption is believed to lead complications, including cancer of the oral cavity, larynx, liver, and breast due to increases in acetaldehyde formation (Voulgaridou et al., 2011). No convincing studies have demonstrated a clear teratogenic effect similar to fetal alcohol syndrome associated with acetaldehyde exposure in the workplace. Continued epidemiologic investigation is necessary to determine if chronic diseases are possible with excessive acetaldehyde exposure.

Animal exposures to acetaldehyde have resulted in some cancers. Feron et al. (1982) reported increased carcinomas of the tracheal epithelium, nasal passages, and larynx upon inhalation exposure in hamsters. The chronic effects that are elicited by acetaldehyde may be increased when mitochondrial aldehyde dehydrogenase deficiencies exist. A 2010 study involving mitochondrial aldehyde dehydrogenase deficient mice identified a potential link between acetaldehyde exposure and increased risk of nasal epithelium and lung cancers (Oyama et al., 2010). Similar results in other animal species have led the IARC (1985) to assign acetaldehyde to Group 2B, sufficient data in animals that acetaldehyde is a carcinogen, but insufficient data in humans. The ACGIH (2013) classifies acetaldehyde in Group A3: confirmed animal carcinogen with unknown relevance to humans.

Mechanism of Action(s)

Acetaldehyde is formed in the liver after the consumption and biotransformation of ethanol. Since most acetaldehyde activity is concentrated in the liver, it is recognized as a hepatotoxin. However, animal studies have shown that metabolism of acetaldehyde can occur in the nasal mucosa and brain mitochondria of rats (Casanova-Schmitz et al., 1984; Quintanilla and Tampier, 1995). Aldehyde dehydrogenase, the enzyme responsible for oxidizing acetaldehyde, is present in many different mammalian tissues and varies widely among individuals.

Effects on dorsal root ganglion neurons similar to those seen with ethanol exposure have been observed in chick embryos (Bradley et al., 1995). The carcinogenicity of acetaldehyde is possibly linked to its ability to produce cross linking in DNA and sister chromatid exchanges after repeated injury (Lambert and He, 1988). It has been shown that acetaldehyde reacts, albeit reversibly, with amino groups of nucleic acid bases (Hemminki and Suni, 1984). The stimulatory effects on the cardiovascular system in animals have been linked to a catecholamine-like response, an increase in heart rate and oxygen consumption (Gailis and Verdy, 1971). Varying concentrations of acetaldehyde can be detected in blood samples.

Chemical Pathology

Acetaldehyde enters the body when fumes of the liquid are inhaled. It can also enter the body if food or drink has been contaminated with the chemical. Several factors determine how acetaldehyde will affect the body: amount of the chemical in the body, the duration of exposure, and the frequency of exposure. Once in the body acetaldehyde is broken down by the enzyme aldehyde dehydrogenase. Acetaldehyde metabolism results in the formation of acetate. If the concentration of acetaldehyde in the liver is too great, individuals may develop symptoms of acute acetaldehyde poisoning, including cardiac arrhythmia, nausea, and anxiety. Physiologically based pharmacokinetic models have been developed that consider acetaldehyde concentration–time data which may help to understand the chemical's pathology (Umulis et al., 2005). The National Institutes of Health explain that highly reactive oxygen species are created by acetaldehyde metabolism and this can lead to the damage of proteins and DNA (2007). In addition, the highly reactive oxygen species have the ability to react with other substances in the body to potentially form harmful compounds.

SOURCES OF EXPOSURE

Occupational

Acetaldehyde is used in industry in the formulation of a variety of chemicals. Any individuals that are involved in the packaging, processing, and/or distribution of the chemical may be at risk for possible exposure. Acetaldehyde is produced by the Hoechst-Wacker process, in which 1.

ethylene is oxidized by aqueous palladium chloride to acetaldehyde, 2. the reduced palladium is reoxidized by cupric chloride to palladium chloride, and 3. the reduced cupric chloride created is reoxidized by oxygen (NIOSH, 1991). Most recently researchers have been interested in worker exposure to acetaldehyde and structurally similar compounds in the production of different food flavoring substances. The chemical has been identified in several microwave popcorn production facilities with concentrations being below the established OSHA PEL (OSHA, 2010). While the workers were not exposed to levels above established limits, consistent low level exposure could potentially lead to adverse health effects.

Environmental

Acetaldehyde is a byproduct found in internal combustion exhaust. Individuals living in highly populated areas may be at risk for exposure as a result of air pollution. Acetaldehyde is formed by photochemical reactions resulting in smog in urban areas. Studies have examined the presence of acetaldehyde in both the indoor and outdoor environments. One study found that the mean personal exposure level of acetaldehyde was 7.9 ppb among 15 research subjects (Jurvelin et al., 2001). Acetaldehyde can also be produced by the burning of wood, tobacco, and the incineration of waste. It may also be produced via the respiration of higher order plants. It was determined that plants subjected to sulfur dioxide produced acetaldehyde when stressed (Kimmerer and Kozlowski, 1982). Acetaldehyde may also be found in perfumes and fragrances as well as baked goods and alcoholic beverages.

INDUSTRIAL HYGIENE

Acetaldehyde is commercially produced through an oxidation reaction of ethylene. Industrial use of acetaldehyde is typically confined to a closed system due to the chemical's explosive nature. The use of a closed system helps to prevent occupational exposures. The most likely route of exposure is inhalation of acetaldehyde fumes. Protective clothing such as aprons and eyewear should be worn to prevent dermal contact from accidental spills. Low concentrations of acetaldehyde vapor produce irritant effects alerting employees to potential exposure before serious health effects occur. The human carcinogenicity of acetaldehyde has not been determined; however, animal carcinogenicity is confirmed. Therefore NIOSH (1995) classifies it as a possible occupational carcinogen and recommends the lowest feasible exposure. OSHA has a less stringent PEL of 200 ppm (360 mg/m^3). The ACGIH (2013) has a stricter ceiling value of 25 ppm, noting acetaldehyde as a definite animal carcinogen with unknown relevance to humans.

MEDICAL MANAGEMENT

Workers that have the potential to be exposed to acetaldehyde should be entered into a medical surveillance program. Medical surveillance can prevent workers from encountering occupational injury and disease by early detection, diagnosis, and treatment. NIOSH explains that employees should be educated by employers about the potential hazards in the work environment and they should be placed in jobs that do not compromise safety and health (NIOSH, 1992). If a worker is using acetaldehyde and experiences any type of adverse health effects, they should be evaluated immediately and preventive measures instituted. Preventive measures may include industrial hygiene monitoring, engineering controls, administrative controls, and personal protective equipment. It should be recognized that worker evaluation should be completed before, during, and after termination from the job site. This will ensure that biological monitoring data above baseline will be investigated immediately. Acetaldehyde can be detected in exhaled air, urine, and blood.

Individuals exposed to acetaldehyde should take various first-aid measures. If skin is exposed to the chemical, it should be rinsed thoroughly with copious amounts of water. The area may then be cleansed with a detergent and water. Any clothing or other apparel that has been contaminated should be removed immediately. Special processes to remove contaminated clothing should be considered. Individuals that regularly use acetaldehyde in the work environment should institute a rigorous personal hygiene regimen. This regimen should include regular washing of the upper extremities and face before eating, drinking, or smoking. No eating, drinking, or smoking should take place in environments where acetaldehyde is present.

REFERENCES

American Conference of Governmental Industrial Hygienists (ACGIH) (2013) *Threshold Limit Values (TLVs) for Chemical Substances and Physical Agents and Biological Exposure Indices (BEIs)*, Cincinnati, OH: American Conference of Governmental Industrial Hygienists.

Brabec, M.J. (1994) Aldehydes and acetals. In: Clayton, G.D. and Clayton, F.E., editors. *Patty's Industrial Hygiene and Toxicology*, vol. 2, Part A, New York, NY: John Wiley & Sons, Inc.

Bradley, D.M., et al. (1995) *In vitro* comparison of the effects of ethanol and acetaldehyde on dorsal root ganglion neurons. *Alcohol. Clin. Exp. Res.* 19:1345–1350.

Casanova-Schmitz, M., David, R.M., and Heck, H.d'A. (1984) Oxidation of formaldehyde and acetaldehyde by NAD$^+$-dependent dehydrogenases in rat nasal mucosal homogenates. *Biochem. Pharmacol.* 33:1137–1142.

Egle, J.L. (1970) Retention of inhaled acetaldehyde in man. *J. Pharmacol. Exp. Ther.* 174:14–19.

Feron, V J., Kruysse, A., and Woutersen, R.A. (1982) Respiratory tract tumors in hamsters exposed to acetaldehyde vapour alone or simultaneously to benzo(a)pyrene or diethylnitrosamine. *Eur. J. Cancer Clin. Oncol.* 18:13–30.

Gailis, L. and Verdy, M. (1971) The effect of ethanol and acetaldehyde on the metabolism and vascular resistance of the perfused heart. *Can. J. Biochem.* 49:227–233.

Hemminki, K. and Suni, R. (1984) Sites of reaction of glutaraldehyde and acetaldehyde with nucleosides. *Arch. Toxicol.* 55:186–190.

IARC (1985) Allyl compounds, aldehydes, epoxides and peroxides. *IARC Monogr. Eval. Carcinog. Risks Hum.* 36:101–131.

Jurvelin, J., et al. (2001) Personal exposure levels and microenvironmental concentrations of formaldehyde and acetaldehyde in the Helsinki Metropolitan area, Finland. *J. Air Waste Manage. Assoc.* 51:17–24.

Kimmerer, T.W. and Kozlowski, T.T. (1982) Ethylene, ethane, acetaldehyde, and ethanol production by plants under stress. *Plant Physiol.* 69:840–847.

Lambert, B. and He, S.M. (1988) DNA and chromosome damage induced by acetaldehyde in human lymphocytes *in vitro. Ann. N. Y. Acad. Sci.* 534:369–376.

Myou, S., et al. (1995) Repeated inhalation challenge with exogenous and endogenous histamine released by acetaldehyde inhalation in asthmatic patients. *Am. J. Respir. Crit. Care Med.* 152:456–460.

National Institute for Occupational Safety and Health (NIOSH) (1991) *Carcinogenicity of Acetaldehyde and Malonaldehyde, and Mutagenicity of Related Low-Molecular-Weight Aldehydes*, NIOSH 91-112, Washington, DC: US Department of Health and Human Services.

National Institute for Occupational Safety and Health (NIOSH) (1992) *Occupational Safety and Health Guideline for Acetaldehyde*, NIOSH 1-8, Washington, DC: US Department of Health and Human Services.

National Institute for Occupational Safety amd Health (NIOSH) (1995) *Pocket Guide to Chemical Hazards*, NIOSH 94-116, Washington, DC: US Department of Health and Human Services.

Occupational Safety & Health Administration (OSHA) (2010) *Occupational Exposure to Flavoring Substances: Health Effects and Hazard Control*, SHIB 10-14-2010, Washington, DC: US Department of Labor.

Oyama, T., et al. (2010) Effects of acetaldehyde inhalation in mitochondrial aldehyde dehydrogenase deficient mice (Aldh2−/−). *Front. Biosci. (Elite Ed.)* 1:1344–1354.

Pinnas, J.L. and Meinke, G.C. (1992) Other aldehydes. In: Sullivan, J.B., Jr. and Krieger, G.R., editors. *Hazardous Materials Toxicology: Clinical Principles of Environmental Health*, Philadelphia, PA: Williams & Wilkins.

Quintanilla, M.E. and Tampier, L. (1995) Acetaldehyde metabolism by brain mitochondria from UChA and UChB rats. *Alcohol* 12:519–524.

Rylander, R. (1973) Toxicity of cigarette smoke components: free lung cell response in acute exposures. *Am. Rev. Respir. Dis.* 108:1279–1282.

Segel, LD. and Mason, D.T. (1979) Acute effects of acetaldehyde and ethanol on rat heart mitochondria. *Res. Commun. Chem. Pathol. Pharmacol.* 25:461–474.

Sim, V.M. and Pattie, R.E. (1957) Effect of possible smog irritations on human subjects. *JAMA* 165:1908–1913.

Umulis, D.M., et al. (2005) A physiologically based model for ethanol and acetaldehyde metabolism in human beings. *Alcohol* 35:3–12.

Voulgaridou, G.P., et al. (2011) DNA damage induced by endogenous aldehydes: current state of knowledge. *Mutat. Res.* 711:13–27.

ACROLEIN

Target organ(s): Eyes, skin, respiratory system, heart

Occupational exposure limits: NIOSH—TWA 0.1 ppm, STEL 0.3 ppm; OSHA—TWA 0.1 ppm; ACGIH—STEL Cicling 0.1 ppm

Reference values: RfD—0.0005 mg/kg/bw/day; RfC—0.00002 mg/m^3; MRL—0.003 ppm

Hazard statements: H225—highly flammable liquid and vapor; H300—fatal if swallowed; H311—toxic in contact with skin; H314—causes severe skin burns and eye damage; H330—fatal if inhaled; H400—very toxic to aquatic life

Precautionary statements: P273—avoid release to the environment; P280—wear protective gloves/protective clothing/eye protection/face protection; P303+P361+P353—if on skin (or hair), remove/take off immediately all contaminated clothing; rinse skin with water/shower; P304+P340—if inhaled, remove victim to fresh air and keep at rest in a position comfortable for breathing; P305+P351+P338—if in eyes, rinse cautiously with water for several minutes; remove contact lenses, if present and easy to do; continue rinsing; P403+P233—store in a well-ventilated place; keep container tightly closed

Risk phrases: R1—highly flammable; R24/25—toxic in contact with skin and if swallowed; R26—very toxic by inhalation; R34—causes burns; R50—very toxic to aquatic organisms

Safety phrases: S23.3—do not breath gas/vapor/aerosol; S26—in case of contact with eyes, rinse immediately with plenty of water and seek medical advice; S28.1—after contact with eyes, rinse immediately with plenty of water and seek medical advice; S36/37/39—wear suitable protective clothing, gloves, and eye/face protection; S45—in case of accident or if you feel unwell, seek medical advice immediately (show label where possible); S61—avoid release to the environment; refer to special instructions/safety data sheets

BACKGROUND AND USES

Acrolein (acraldehyde, acrylaldehyde, acrylic aldehyde, allyl aldehyde, propenal, 2-propenal; CASRN 107-02-8) is another important aldehyde. It is used as a chemical intermediate in many industrial synthetic processes, including in the production of acrylic acid and allyl alcohol. Acrolein may also be used as a biocide and in the production of perfumes and colloidal metals. Residue from industrial emissions and the burning of wood and other organic substrates will contain traces of acrolein. Acrolein is also a constituent of diesel exhaust and photochemical smog. It is also formed during the pyrolysis of cotton and polyethylene and found in cigarette smoke. Much of the human toxicity data is obtained from the formation of acrolein *in vivo* as a metabolite of the anticancer drug, cyclophosphamide.

PHYSICAL AND CHEMICAL PROPERTIES

Acrolein can exist as a colorless or yellow liquid with a penetrating, displeasing odor. The chemical has a molecular weight of 56.1, a boiling point of 127 °F, and a freezing point of −126 °F. The vapor pressure is 274 mm Hg at 25 °C and the specific gravity is 0.84. The lower explosive limit is 2.8% and the upper explosive limit is 31%. Acrolein may be incompatible with oxidizers, acids, alkalis, ammonia, and amines. Acrolein has the ability to polymerize unless inhibited with hydroquinone. Shock-sensitive peroxides may be formed over time NIOSH (2010).

MAMMALIAN TOXICOLOGY

Acute Effects

Lacrimation and severe respiratory irritation can result from acute exposure to acrolein vapor. Pulmonary edema can rapidly occur in air levels of 20 ppm. A concentration of 150 ppm is considered lethal in humans exposed for 10 min or longer. Respiratory effects produced by acrolein can persist for long periods. Ciliary action in the upper airways is significantly reduced due to this highly reactive aldehyde. The result is tissue destruction in all parts of the lungs. Acrolein is a likely contributor to smoke inhalation-induced injury and death. Asphyxiation in animals and humans can occur with high concentrations of acrolein. Skin irritation, painful dermatitis, burns, and epidermal necrosis can occur with dermal contact of high concentrations of acrolein. A pesticide worker exposed to acrolein was diagnosed with thermal skin sensitivity 3 weeks after exposure (CDC, 2013). Several other workers suffered from acute acrolein-related illness with symptoms ranging from eye irritation, headache, and dyspnea.

Acute exposures in animals have resulted in a variety of effects depending on exposure route and concentration. Acrolein was found to have a potent sensory irritant effect in mice after a single exposure (Kane and Alarie, 1977). Spontaneous inhalation of acrolein vapor caused immediate inhibitory effects on breathing rats, bradycardia, and reduced ventilation (Lee et al., 1992). It is a common constituent of smoke, and studies of acrolein-containing smoke in animals produced pulmonary edema and airway damage in both high and low concentration groups (Hales et al., 1988). Excessive exposure to this aldehyde can also impair resistance to bacterial infections in test animals. Astry and Jakab (1983) demonstrated susceptibility to certain bacterial strains that are normally inactivated shortly after contact. Increases in bacterial infections and pneumonia in acrolein-exposed mice were seen. High concentrations of acrolein in tobacco smoke make this of some concern to humans. In addition to sensory irritation, other observations from animal studies include inhibition of protein synthesis and enzyme activity, elevated liver alkaline phosphatase, weight loss, and death. A high concentration of acrolein in the bladder lumen of rats caused acute cytotoxicity and epithelial lesions (Sakata et al., 1989).

Chronic Effects

Similar to acute effects, the chronic effects of acrolein vary as well. Sensitization in some individuals is possible when exposed to acrolein. Since acrolein has such a disagreeable odor, low odor threshold, and strong irritating effects, workers can be alerted to exposure preventing the development of chronic effects. One of the few chronic studies available involves patients treated with cyclophosphamide. Cytotoxicity in the bladder of each patient was noted, although acrolein was not investigated as the causal agent at the time of the study (Beyer-Boon et al., 1978). This type and concentration of acrolein exposure are not likely to occur in the workplace.

Most knowledge about the chronic effects and toxicity of this chemical was generated from animal studies. Subchronic exposures of rats to levels much higher than those likely to be found in the workplace resulted in decreased lung function and suggestive airway obstruction (Costa et al., 1986). In a study by Leach et al., (1987), the adverse effects resulting from excessive exposure were likely to be confined to local nasal tissue pathology, such as exfoliation and erosion of the respiratory epithelium. No human data identifying acrolein-induced adverse developmental toxicity are available, although a study using rabbits completed by Parent et al. (1993) found no adverse developmental effects after ingestion of acrolein.

Animal studies have shown acrolein to be carcinogenic. Acrolein was shown to have initiating activity for bladder carcinogenesis in male rats, but the study was unable to demonstrate promotion or complete carcinogenic action for acrolein (Cohen et al., 1992). This carcinogenic activity is likely a result of the 2,3-epoxy metabolite. Human studies have not explored this link in detail, but it is important to note that cigarette smoking and cyclophosphamide therapy have both been implicated as causes of bladder carcinoma (Seltzer et al., 1978; Auerbach and Garfinkel, 1989). IARC (1987) classification of this compound is a Group 3, inadequate data in humans and animals to classify acrolein as carcinogenic.

Mechanism of Action(s)

The metabolism of acrolein is not well characterized. The first step of acrolein metabolism appears to be a conversion of acrolein to acrylic acid by liver enzymes (Patel et al., 1980). Draminski et al. (1983) proposed the next step as a methylation of acrylic acid to methyl acrylate and then conjugation with glutathione to form its final metabolite, S-carboxymethylmercapturic acid, which was found in the urine of rats exposed to acrolein in the same study. This metabolism is from oral dosing data, which may differ from dermal and inhalation routes. The reaction of acrolein with thiol groups *in vitro* is reported, and this is a detoxification mechanism important for protecting cells and tissues from the toxicity of acrolein.

Many researchers have suspected acrolein as a carcinogenic agent. The carcinogenicity of acrolein is likely related to formation of a 2,3-epoxy metabolite. A study by Wilson et al. (1991) demonstrated DNA adducts produced in human fibroblasts. Wang et al. (2012) found that acrolein not only damages DNA but can also change DNA repair mechanisms resulting in the degradation of the changed repair mechanisms (Wang et al., 2012). The disruption of normal cellular processes by acrolein appears to result in disease manifestation. Acrolein is a highly reactive aldehyde when compared with other aldehydes and their toxicity.

Chemical Pathology

Acrolein is extremely reactive and binds to macromolecules to form thiol ethers. The binding of acrolein to messenger compounds may cause direct cytotoxic effects and interrupts cell signaling (Beauchamp et al., 1985). Disruption of cell signaling pathways can disrupt the normal cellular activity in the human body leading to the generation of different diseases. It is believed that acrolein may have the ability to induce Alzheimer's disease-like manifestations. Huang et al. (2013) found that acrolein caused the inactivation of astrocytes, upregulation of BACE-1, and downregulation of ADAM-10, resulting in Alzheimer's disease-like pathologies both *in vitro* and *in vivo*.

SOURCES OF EXPOSURE

Occupational

Acrolein can be found in industry as a chemical intermediate in the production of acrylic acid and esters. Acrolein itself is prepared commercially by propene oxidation. Acrolein is also used industrially in the manufacture of colloidal metal, plastics, and perfumes. It is also used to control the growth of plants, algae, and other microorganisms making those involved in the plant and wildlife management at risk for exposure. Since acrolein is often a component of smoke, research has focused on firefighter's exposure to the chemical. Materna et al. (1992) found that as a result of wildfires, firefighters may be exposed to harmful components of smoke at or above established occupational exposure limits. Those working in oil or coal fired power plants may also be exposed to acrolein via smoke. Acrolein may be encountered in the food manufacturing industry as a food starch modifier and in the paperboard for the packaging of food products (OSHA, 2013). Individuals in the military or other first responding agencies could be at risk in the event that acrolein is used as a chemical warfare or riot control agent.

Environmental

As previously mentioned, acrolein is a component of smoke. When trees and plants are burned, acrolein exposure can occur if smoke is inhaled. Fireplaces that are indoors may concentrate smoke inside leading to higher exposures. Proper ventilation should be provided. Smoke from oil or coal fired power plants and the burning of tobacco products can contain acrolein. Acrolein levels were found to be elevated in homes that had lower air exchange rate and that housed at least one tobacco smoker (Gilbert et al., 2005). Stevens and Maier (2008) explain that the smoking of tobacco products can equal or exceed the total human exposure to acrolein from other sources. Traces of the chemical can also be found in exhaust from internal combustion engines. Acrolein levels measured at a toll booth during rush hour traffic exceeded established levels in the state of California (Destaillats et al., 2002). Instituted air pollution control measures may have the ability to reduce the potential for acrolein exposure.

INDUSTRIAL HYGIENE

A primary route of exposure is inhalation of acrolein vapor during its manufacture and industrial use. Trace amounts of acrolein, about 0.16 ppm, can normally be detected by a worker because of its strong odor. It is also irritating to the eyes and throat at levels slightly above the odor threshold. Bromination of sampled ambient air followed by gas chromatography analysis is an effective method for monitoring acrolein with sensitivity for measurement of amounts as low as 5 ppb (Nishikawa et al., 1986). OSHA recommends air sampling with a sorbent tube containing 2-(hydroxymethyl) piperidine on XAD-2 adsorbent using toluene to extract acrolein (OSHA, 1992). Recommendations for protection of workers in contact with acrolein are respirators as needed and prevention of skin and eye contact (NIOSH, 2010). The allowable workplace air standards for acrolein are the same for the TLV of the ACGIH (2013), the PEL of OSHA, and the REL of NIOSH (2010), at 0.1 ppm (0.25 mg/m$^{3)}$ TWA and a skin designation of 0.3 ppm (0.8 mg/m^3).

MEDICAL MANAGEMENT

There are no reliable methods for biomonitoring exposure to acrolein; however, some promising techniques have been reported. The inhibition of glutathione-s-transferase in human erythrocytes by acrolein and other chemicals has been used to characterize acrolein exposure (Ansari et al., 1987). The measurement of mercapturic acid metabolites in the urine has also been suggested for biomonitoring, but no reliable quantification of metabolites has been reported as adequate for monitoring industrial exposures. Workers that experience respiratory symptoms should consider investigative chest X-rays (NLM, 2013). Individuals with excessive exposure should be administered blood gases to treat hypoxia. Treatment of exposed persons typically consists of respiratory and cardiovascular support.

REFERENCES

American Conference of Governmental Industrial Hygienists (ACGIH) (2013) *Threshold Limit Values (TLVs) for Chemical Substances and Physical Agents and Biological Exposure Indices (BEIs)*, Cincinnati, OH: American Conference of Governmental Industrial Hygienists.

Ansari, G.A.S., et al. (1987) Human erythrocyle glutathione S-transferase: a possible marker of chemical exposure. *Toxicol. Lett.* 37:57–62.

Astry, C.L. and Jakab, G.J. (1983) The effects of acrolein exposure on pulmonary antibacterial defenses. *Toxicol. Appl. Pharmacol.* 67:49–54.

Auerbach, O. and Garfinkel, L. (1989) Histologic changes in the urinary bladder in relation to cigarette smoking and use of artificial sweeteners. *Cancer* 64:983–987.

Beauchamp, R.O., Jr., Andjelkovich, D.A., Kligerman, A.D., Morgan, K.T., and Heck, H.D. (1985). A critical review of the literature on acrolein toxicity. *Crit. Rev. Toxicol.* 14(4): 309–380.

Beyer-Boon, M.E., De Voogt, H.J., and Schaberg, A. (1978) The effects of cyclophosphamide treatment on the epithelium and stroma of the urinary bladder. *Eur. J. Cancer* 14:1029–1035.

Centers for Disease Control and Preventation (CDC) (2013) Notes from the field: acute pesticide-related illness resulting from occupational exposure to acrolein—Washington and California, 1993–2009. *MMWR Morb. Mortal. Wkly Rep.* 62:313–314.

Cohen, S.M., et al. (1992) Acrolein initiates rat urinary bladder carcinogenesis. *Cancer Res.* 52:3577–3581.

Costa, D.L., et al. (1986) Altered lung function and structure in the rat after subchronic exposure to acrolein. *Am. Rev. Respir. Dis.* 133:286–291.

Destaillats, H., et al. (2002) Ambient air measurement of acrolein and other carbonyls at the Oakland–San Francisco Bay Bridge toll plaza. *Environ. Sci. Technol.* 36:2227–2235.

Draminski, W., Eder, E., and Henschler, D. (1983) A new pathway of acrolein metabolism in rats. *Arch. Toxicol.* 52:243–247.

Gilbert, N.L., et al. (2005) Levels and determinants of formaldehyde, acetaldehyde, and acrolein in residential indoor air in Prince Edward Island, Canada. *Environ. Res.* 99:11–7.

Hales, C.A., et al. (1988) Synthetic smoke with acrolein but not HCl produces pulmonary edema. *J. Appl. Physiol.* 64:1121–1133.

Huang, Y.J., et al. (2013) Acrolein induces Alzheimer's disease-like pathologies *in vitro* and *in vivo*. *Toxicol. Lett.* 217:184–91.

IARC (1987) Overall evaluations of carcinogenicity: an updating of 1ARC monographs volumes 1–42. *IARC Monogr. Eval. Carcinog. Risks Hum.* (Suppl. 7):78–89.

Kane, I.E. and Alarie, Y. (1977) Sensory irritation to formaldehyde and acrotein during single and repeated exposures in mice. *Am. Ind. Hyg. Assoc. J.* 38:509–522.

Leach, C.L., et al. (1987) The pathologic and immunologic effects of inhaled acrolein in rats. *Toxicol. Lett.* 39:189–198.

Lee, B.P., et al. (1992) Acute effects of acrolein on breathing: role of vagal bronchopulmonary afferents. *J. Appl. Physiol.* 72: 1050–1056.

Materna, B.L., et al. (1992) Occupational exposures in California wildland fire fighting. *Am. Ind. Hyg. Assoc. J.* 53:69–76.

National Institute for Occupational Safety and Health (NIOSH) (2010) *Pocket Guide to Chemical Hazards*, NIOSH 94-116, Washington, DC: US Department of Health and Human Services.

Nishikawa, H., Hayakawa, T., and Sakai, T. (1986) Determination of micro amounts of acrolein in air by gas chromatography. *J. Chromatogr.* 370:327–332.

National Library of Medicine (NLM) (2013) *Acrolein*, Bethesda, MD: United States National Institutes of Health.

Occupational Safety & Health Administration (OSHA) (1992) *Occupational Safety and Health Guideline for Acrolein*, OSHA 1-8, Washington, DC: US Department of Labor.

Occupational Safety & Health Administration (OSHA) (2013) *Acrolein and/or Formaldehyde*, OSHA 52, Washington, DC: US Department of Labor.

Parent, R.A., et al. (1993) Developmental toxicity of acrolein in New Zealand white rabbits. *Fundam. Appl. Toxicol.* 20:248–256.

Patel, J.M., Wood, J.C., and Leibman, K.C. (1980) The biotransformation of allyl alcohol and acrolein in rat liver and lung preparations. *Drug Metab. Dispos.* 8:305–308.

Sakata, T., et al. (1989) Rat urinary bladder epithelial lesions induced by acrolein. *J. Environ. Pathol. Toxicol. Oncol.* 9:159–169.

Seltzer, S.E., Bernazzi, R.B., and Kearney, G.P. (1978) Cyclophosphamide and carcinoma of the bladder. *Urology* 11:352–356.

Stevens, J.F. and Maier, C.S. (2008) Acrolein—sources, metabolism, and biomolecular interactions relevant to human health and disease. *Mol. Nutr. Food Res.* 52:7–25.

Wang, H.T., et al. (2012) Effect of carcinogenic acrolein on DNA repair and mutagenic susceptibility. *J. Biol. Chem.* 287: 12379–12386.

Wilson, V.L., et al. (1991) Detection of acrolein and crotonaldehyde DNA adducts in cultured human cells and canine peripheral blood lymphocytes by ^{32}P-postlabeling and nucleotide chromatography. *Carcinogenesis* 12:1483–1490.

CINNAMALDEHYDE

Target organ(s): Eyes, skin, respiratory system

Occupational exposure limits: NIOSH, OSHA, ACGIH—not established; derived no effect level (DNEL); DNEL inhalation, systemic effects, long term: 2.2 mg/m^3; DNEL dermal, systemic effects, long term: 2.51 mg/kg/bw/day

Reference values: RfD, RfC, MRL—not established; DNEL inhalation, systemic effects, long term: 0.54 mg/m^3; DNEL dermal, systemic effects, long term: 0.625 mg/kg/bw/day; DNEL oral, systemic effects, long term: 2.5 mg/kg/bw/day

Hazard statements: H312—harmful in contact with skin; H315—causes skin irritation; H317—may cause an allergic skin reaction; H319—causes serious eye irritation

BACKGROUND AND USES

Cinnamaldehyde (cinnamic aldehyde, benzylideneactaldehyde, phenylacrolein, 3-phenylpropenal, 3-phenyl-2-propenal; CASRN 104-55-2) is a saturated aldehyde with an aromatic ring that is a yellowish, oily liquid at room temperature. It is used as a flavoring and aromatic agent in a variety of food products, perfumes, and household products. It has also seen use as a rubber reinforcing agent. A burning taste that produces the odor and flavor of the spice may be found with this aromatic aldehyde. Cinnamaldehyde has also been used as an attractant for insect control, in the preparation of corrosion inhibitors, and as a coating for metals. Although extensively used in industry, it is also a natural constituent of cinnamon leaves and bark, some essential oils, and other plant products.

PHYSICAL AND CHEMICAL PROPERTIES

Cinnamaldehyde exists as yellowish to greenish-yellow oily liquid with a cinnamon odor. The chemical has a molecular weight of 132.15, a boiling point of 487.4 °F, and a melting point of 19.4 °F. The vapor pressure is 0.0289 mm Hg at 25 °C and the specific gravity is 1.049. It is normally insoluble in water and many organic solvents but is miscible with alcohol and other flavoring oils. Exposure to air will result in thickening and oxidation.

MAMMALIAN TOXICOLOGY

Acute Effects

Metabolism of ingested cinnamaldehyde is believed to involve oxidation of the aldehyde to cinnamic acid via ADH. As a result, hippuric and benzoic acids are formed and excreted in urine. As with many other chemicals, this metabolism is thought to occur mainly in the liver. Rat liver glutathione levels were significantly depressed within 30 min after intraperitoneal administration of cinnamaldehyde (Boyland and Chasseud, 1970). Two urinary metabolites containing sulfur groups, 3-S-(N-acetylcysteinyl)-3-phenylpropyl alcohol and 3-S-(N-acetylcysteinyl)-3-phenylpropionic acid have been identified in a rat model (Delbressine et al., 1980). Nonimmunological contact urticaria is likely to involve a nonallergic histamine-liberating response. Animal models have not clearly demonstrated the mechanism involved, but similar effects of dermal edema and erythema are seen following exposure in these models (Lahti and Maibach, 1984). Research conducted by Subash et al. (2007) indicated that cinnamaldehyde had the ability to reduce hypoglycemic and hypolipidemic effects in diabetic rats.

Acute effects of cinnamaldehyde are noticeable at high concentrations. Inflammation and erosion of the gastrointestinal mucosa may occur after ingestion of undiluted essential oils containing cinnamaldehyde. Systemic reactions are not expected from exposure. Mucous membranes may become irritated due to exposure to high concentrations of inhaled cinnamaldehyde. Contact dermatitis and superficial chemical burns are possible with prolonged dermal contact. Delayed contact dermatitis may also be seen. Cinnamaldehyde-induced sensitization has been reported in a noninvasive mouse ear swelling assay, designed for evaluating fragrances and mixtures (Thorne et al., 1991). Nonimmunological contact urticaria, characterized by redness and wheals, may also be an outcome of acute dermal exposure to cinnamaldehyde (Mathias et al., 1980; Seite-Bellezza et al., 1994). Cinnamaldehyde is a strong skin irritant and sensitizer and may cause urticaria. Cinnamaldehyde was found to arouse spontaneous, immediate pain and induce heat and mechanical hyperalgesia, cold hypoalgesia, and neurogenic axon reflex erythema in human participants (Namer et al. 2005).

Chronic Effects

Chronic effects associated with cinnamaldehyde exposure may result from repeated exposures and subsequent sensitization of atopic individuals. Contact dermatitis has been associated with a number of different occupations, including bakers, chemists, and workers in manufacturing plants (Fisher, 1975; Nethercott et al., 1983). Powered cinnamon contains only 1% of the aldehyde and is not an irritant when ordinarily used. Coal miners provided with a perfumed lotion at their worksite were reported to have a high number of individuals sensitized to cinnamaldehyde (Goodfield and Saihan, 1988). Unfortunately for the affected individuals, reoccurrence of signs and symptoms may develop away from the workplace in response to reexposure to cinnamaldehyde in foods, cosmetics, or other products containing the aldehyde.

Mutagenicity tests have indicated that cinnamaldehyde may actually be an antimutagen, reducing spontaneous or induced mutations (Ohta, 1993). Cinnamaldehyde was found to have a chemoprotective and anti-inflammatory effect on human endothelial cells (Liao et al., 2008). Cinnamaldehyde was also found to have antitumor activity mediated by the inhibition of famesyl transferase (Moon et al., 2006). However, in high doses administered to rats, cinnamaldehyde did induce some genetic alterations at the chromosomal level (Mereto et al., 1994), suggesting a dose-dependent mutagenic effect. *In vivo* developmental toxicity testing in mice has found no overt health effects among dams or offspring after midterm exposure to cinnamaldehyde (Hardin et al., 1987).

Mechanism of Action(s)

Allergic contact dermatitis may be the result of acute and chronic exposure to cinnamaldehyde. Allergic contact dermatitis occurs when the T-lymphocyte immune response reacts to a chemical allergen that contacts the skin (Smith Pease et al., 2003). The allergen binds to a carrier protein and is processed by antigen-presenting cells. The antigen-presenting cells then present the allergen to the T-lymphocyte. This stimulates the production of memory and effector T-lymphocytes. Later contact with the allergen at a sufficient dose will activate an immune response and result in allergic contact dermatitis symptoms. Protecting the skin from cinnamaldehyde will prevent initial and subsequent exposure of allergens, preventing the development of allergic contact dermatitis.

Cinnamaldehyde is used in the preparation of various fragrances. Their use introduces the chemical into the air where it has the potential to be inhaled. Inhalation of cinnamaldehyde can produce irritant effects in the upper airways, resulting in cough. Faruqi and Morice (2008) determined that cinnamaldehyde was a specific agonist of the TRPA-1 receptor and induced cough due to chemesthesis in the airways. Another study found that children's toys that were perfumed with fragrances had the ability to emit fragrance chemicals increasing risk for exposure via inhalation (Masuck et al., 2011). Proper ventilation will help reduce indoor concentrations to acceptable levels. In extreme cases, respiratory protection or cessation of fragrances or perfume may be necessary to eliminate exposure.

Chemical Pathology

Cinnamaldehyde exposure results in effects on the skin from dermal contact and in the respiratory tract from inhalation. Sells et al. (2007) found pigmentation of the olfactory epithelium, specifically in the basal cytoplasm, following exposure to cinnamaldehyde via the noninhalation route. While the skin and the respiratory tract are affected by cinnamaldehyde exposure, additional research studies indicate that other sites may be adversely affected by exposure. Rats provided a dose of 73.5 mg/kg/day showed histological changes in kidney tissue trailed by increased renal, serum, and urinary enzymes (Gowder and Devaraj, 2008). An additional study found that rats exposed to cinnamaldehyde had abnormal liver function leading to a prooxidant state which could increase the probability of cancer development (Raveendran et al., 1993). The kidney and the liver may be other cinnamaldehyde target organs but additional research is needed to understand the complete chemical pathology.

SOURCES OF EXPOSURE

Occupational

Occupational exposures to cinnamaldehyde may occur in the industrial environment as a result of chemical preparation, processing, and distribution. The chemical is easily manufactured as a condensation product of benzaldehyde and acetaldehyde in an acidic or basic solvent. The oxidation of cinnamylic alcohol and condensation of styrene with 2 chloroallylbenzene will also result in the creation of this aldehyde. The United States Environmental Protection Agency (USEPA) issued a report in which they believed that mixers, loaders, and applicators of cinnamaldehyde do not pose an unreasonable hazard when using the biochemical fungicide (USEPA, 2000). However, an individual with acute itching on his right hand was patch tested with various concentrations to cinnamaldehyde and developed immediate reactions to the chemical (Cocchiara et al., 2005). Inhalation of cinnamaldehyde is also possible. Once inhaled, the aldehyde irritates the upper respiratory tract and can induce coughing. Birrell et al. (2009) found that cinnamaldehyde evoked coughing in guinea pig and human volunteers.

Environmental

Cinnamaldehyde can be extracted from natural sources. It is used in many different applications that can be found in an individual's environment. Its use as a flavoring agent is an indication of its low toxicity at levels found in food items. This flavoring agent is used in mouthwashes, dentifrices, throat lozenges, candy, soft drinks, chewing gum, and many other items. This aldehyde is also extensively used in deodorizers, detergents, soaps, and cosmetics. It is also used in the formulation of certain biocides. Cinnamaldehyde appears to have the ability to reduce both Gram-negative and Gram-positive bacterial population, although the mechanism is not completely understood (Gill and Holley, 2004). Inhalation and dermal contact with the aldehyde are possible in different environmental situations. The burning of substrates that contain the aldehyde could cause symptoms, including respiratory irritation and cough. Exposure to cinnamaldehyde and similar compounds may have a relationship with cardiovascular disease (Haberzettl et al., 2011). Increasing ventilation in the indoor environment may help prevent the buildup high concentrations of cinnamaldehyde and other aldehydes.

INDUSTRIAL HYGIENE

A dermal exposure is more frequent and more likely to be a health hazard industrially. Inhalation exposures are not likely to be a source of serious health effects. The strong odor emitted by the chemical serves as a sufficient warning sign of hazardous concentrations. Currently, there are no published guidelines for this aldehyde because concentrations in some items are as high as 5000 ppm (Collins and Mitchell, 1975). Dermal protection is recommended for those individuals who come into contact with high concentrations or who have developed sensitivity to the chemical. Dermal protection may include face protection, gloves, and aprons. Engineering controls and respiratory personal protective equipment may be necessary to further reduce the risk of exposure.

MEDICAL MANAGEMENT

Individuals exposed to cinnamaldehyde who develop symptoms of irritation should be immediately removed from the contaminated environment. The chemical has a strong odor and will alert individuals when concentrations become high making overexposure unlikely. Individuals that are exposed to high concentrations should seek medical assistance if symptoms persist or are troublesome. Individuals that

experience respiratory irritation may require the use of respiratory protection. Those that are exposed to the chemical dermally should be removed from the environment and may require a topical agent to control itching. Covering exposed skin will restrict dermal contact and exposure. Employees should remove contaminated clothing and should cleanse hands and arms thoroughly to remove any chemical that may have contacted the skin. Good personal hygiene measures will prevent unnecessary exposure.

REFERENCES

Birrell, M.A., et al. (2009) TRPA1 agonists evoke coughing in guinea pig and human volunteers. *Am. J. Respir. Crit. Care Med.* 180:1042–1047.

Boyland, E. and Chasseud, L.F. (1970) The effect of some carbonyl compounds on rat liver glutathione levels. *Biochem. Pharmacol.* 19:1526–1528.

Cocchiara, J., et al. (2005) Fragrance material review on cinnamaldehyde. *Food Chem. Toxicol.* 43:867–923.

Collins, F.W. and Mitchell, J.C. (1975) Aroma chemicals. *Contact Dermatitis* 1:43–47.

Delbressine, L.P.C., et al. (1980) Identification of two sulfur containing urinary metabolites of cinnamic aldehyde in the rat. *Br. J. Pharmacol.* 68: 165P.

Faruqi, S. and Morice, A.H. (2008) Cinnamaldehyde: a novel tussive agent in man. *Thorax* 63:A4–A73.

Fisher, A.A. (1975) Dermatitis due Io cinnamon and cinnamic aldehyde. *Cutis* 16:383–388.

Gill, A.O. and Holley, R.A. (2004) Mechanisms of bacterial action of cinnamaldehyde against *Listeria monocytogenes* and of Eugenol against *L. monocytogenes* and *Lactobacillus sakei*. *Appl. Environ. Microbiol.* 70:5750–5755.

Goodfield, M.J.D. and Saihan, E.M. (1988) Fragrance sensitivity in coal miners. *Contact Dermatitis* 18:81–83.

Gowder, J.T. and Devaraj, H. (2008) Food flavor cinnamaldehyde-induced biochemical and histological changes in the kidney of male albine wistar rat. *Environ. Toxicol. Phamacol.* 26:68–74.

Haberzettl, D.J., et al. (2011) Chapter 13 In: *Environmental Aldehydes*, London: Royal Society of Chemistry.

Hardin, B.D., et al. (1987) Evaluation of 60 chemicals in a preliminary developmental toxicity test. *Teratogen. Carcinogen. Mutagen.* 7:29–48.

Lahti, A. and Maibach, H.I. (1984) An animal model for nonimmunologic contact urticaria. *Toxicol. Appl. Pharmacol.* 76:219–224.

Liao, B.C., et al. (2008) Cinnamaldehyde inhibits the tumor necrosis factor-alpha-induced expression of cell adhesion molecules in endothelial cells by suppressing NF-kappaB activation: effects upon IkappaB and Nrf2. *Toxicol. Appl. Pharmacol.* 229:161–171.

Masuck, I., et al. (2011) Inhalation exposure of children to fragrances present in scented toys. *Indoor Air* 21:501–511.

Mathias, C.G.T., Chappler, R.R., and Maibach, H.I. (1980) Contact urticaria from cinnamic aldehyde. *Arch. Dermatol.* 116:74–76.

Mereto, E., et al. (1994) Cinnamaldehyde-induced micronuclei in rodent liver. *Mutat. Res.* 322:1–8.

Moon, E.Y., et al. (2006) Delayed occurrence of H-ras12V-induced hepatocellular carcinoma with long-term treatment with cinnamaldehydes. *Eur. J. Pharmacol.* 530:270–275.

Namer, B., et al. (2005) TRPA1 and TRPM8 activation in humans: effects of cinnamaldehyde and menthol. *Neuroreport* 16:955–959.

Nethercott, J.R., et al. (1983) Contact dermatitis due to cinnamic aldehyde induced in a deodorant manufacturing process. *Contact Dermatitis* 9:241–242.

Ohta, T. (1993) Modification of genotoxicity by naturally occurring flavorings and their derivatives. *Crit. Rev. Toxicol.* 23:127–146.

Raveendran, M., et al. (1993) Induction of prooxidant state by the food flavor cinnamaldehyde in rat liver. *J. Nutr. Biochem.* 4:181–183.

Seite-Bellezza, D. el Sayed, F. and Bazex, J. (1994) Contact urticaria from cinnamic aldehyde and benzaldehyde in a confectioner. *Contact Dermatitis* 31:272–273.

Sells, D.M., et al. (2007) Respiratory tract lesions in noninhalation studies. *Toxicol. Pathol.* 35:170–7.

Smith Pease, C.K., et al. (2003) Contact allergy: the role of skin chemistry and metabolism. *Clin. Exp. Dermatol.* 28:177–83.

Subash, B. et al. (2007) Cinnamaldehyde—a potential antidiabetic agent. *Phytomedicine* 14:15–22.

Thorne, P.S., et al. (1991) The noninvasive mouse ear swelling assay. *Fundam. Appl. Toxicol.* 17:807–820.

United States Environmental Protection Agency (USEPA) (2000) *Cinnamaldehyde (040506) Fact Sheet*, Washington, DC: United States Environmental Protection Agency.

CROTONALDEHYDE

Target organ(s): Eyes, skin, respiratory system

Occupational exposure limits: NIOSH—TWA 2 ppm; OSHA—TWA 2 ppm

ACGIH—STEL ceiling 0.3 ppm

Reference values: RfD/RfC—not established; MRL—0.39 ppm

Risk phrases: R11—highly flammable; R24/25—toxic if contact with skin and if swallowed; R37/38—irritating to respiratory system and skin; R41—risk of serious damage to eyes; R48/22—harmful: danger of serious damage to health by prolonged exposure if swallowed; R50—very toxic to aquatic organisms; R68—possible risk of irreversible effects

Safety phrases: S1/2—keep locked up and out of the reach of children; S26—in case of contact with eyes, rinse immediately with plenty of water and seek medical advice; S28—after contact with skin, wash immediately with plenty of . . . (to be specified by manufacturer; R36/37/39—wear suitable protective clothing, gloves, and eye/face protection; S45—in case of accident or if you feel unwell, seek medical advice immediately (show label where possible); S61—avoid release to the environment; refer to special instructions/safety data sheets

BACKGROUND AND USES

Crotonaldehyde (2-butenal, β-methyl acrolein, propylene aldehyde; CASRN 123-73-9) is similar in structure to acrolein, as both are α,β-unsaturated aldehydes. This structural similarity leads to similar sensitizing and irritating properties of the two compounds. Crotonaldehyde is used industrially in the preparation of other chemicals (chiefly sorbic acid), flavoring agents, and can form endogenously and in the environment.

PHYSICAL AND CHEMICAL PROPERTIES

Crotonaldehyde can exist as a water-white liquid with an asphyxiating odor. The chemical can turn pale yellow when it contacts air. The chemical has a molecular weight of 70.1, a boiling point of 219 °F, and a freezing point of −101 °F. The vapor pressure is 30 mm Hg at 25 °C and the specific gravity is 0.87. The lower explosive limit is 2.1% and the upper explosive limit is 15.5%. Crotonaldehyde may be incompatible with caustics, ammonia, strong oxidizers, nitric acid, and amines. It also has the ability to polymerize at high temperatures.

MAMMALIAN TOXICOLOGY

Acute Effects

Several studies have investigated the mammalian toxicology of crotonaldehyde. The chemical has been implicated in the development of vascular-related disease through lipid and protein oxidation (Lee and Park, 2013). It is believed that crotonaldehyde possesses mutagenic and carcinogenic properties by directly reacting with DNA. Hamster ovary cells exposed to crotonaldehyde formed DNA adducts at a level of 75 µmol/mol deoxyguanosine suggesting potential tumorigenic potential (Foiles et al., 1990). Fernandes et al. (2005) found that mutation yields of DNA adducts were 5–6%, suggesting that adducts could possible contribute to genotoxicity and carcinogenicity. Jha and Kumar (2006) investigated whether exposure to crotonaldehyde caused an increase in the number of abnormal sperm in mice. They found a statistically significant increase in the percentage of abnormal sperm. Based on available scientific research, crotonaldehyde is suspected to be a cancer causing agent and those individuals that may be potentially exposed to it should use proper protective measures.

Acute exposure to small quantities of crotonaldehyde vapor can result in eye and respiratory irritation. Vapors have been known to cause lacrimation. Higher exposure concentrations result in pulmonary edema and more serious respiratory effects. Skin irritation and chemical burns are possible with dermal contact from this aldehyde. Fatal doses can be delivered through all exposure routes. Acute animal exposures to high concentrations have shown a rapid decrease in respiratory rate with little or no recovery (Steinhagen and Barrow, 1984). A male worker accidentally exposed to a high concentration mixture of ethenone and crotonaldehyde presented with decreased vesicular sounds, opacities in the lung, and acute respiratory distress syndrome (Huang et al., 2013). Crotonaldehyde was found to induce apoptosis of endothelial cells resulting in vascular injury (Ryu et al., 2013). Acute effects resulting from crotonaldehyde can be prevented by identification and control of sources of exposure.

Chronic Effects

Chronic exposures in humans are not well documented; animal studies are a large source of current information on chronic crotonaldehyde toxicity. The most notable is the 1986 study by Chung, Tanaka, and Hecht, which demonstrated the induction of liver tumors in male rats. The potential carcinogenicity of this compound is of some concern to humans because of its endogenous formation. Mutagenic assays in Salmonella bacterial strains have shown positive results (Neudecker et al., 1981, 1989); however, the correlation between human carcinogenicity and these tests has not been demonstrated for many chemicals. l, N^2-propanodeoxyguanosine lesions following crotonaldehyde exposure in human and rodent liver DNA have been demonstrated, and these adducts may be related to the carcinogenic effect of crotonaldehyde (Nath and Chung, 1994). This aldehyde has been shown to be tumorigenic in rodents. Clinical cases of sensitization have been observed. Chronic obstructive pulmonary disease (COPD) can result due to chronic exposure to crotonaldehyde by causing lung inflammatory responses by different mechanisms (Yang et al., 2013). Traces of crotonaldehyde that can be found in cigarette smoke play a critical role in inflammatory cytokine production and neutrophilic airway inflammation (van der Toorn et al., 2013). Adequate ventilation to prevent inhalation and personal protective equipment to prevent dermal contact in the occupational workplace and environment can control crotonaldehyde exposure, limiting the potential for chronic disease development. In light of the available evidence, the ACGIH (2013) has placed crotonaldehyde in Group A3: confirmed animal carcinogen with unknown relevance to humans.

Mechanism of Action(s)

The major concern regarding crotonaldehyde exposure is its ability to react with different body systems and with DNA. Boyland and Chasseud (1970) demonstrated that rats injected with crotonaldehyde exhibited subsequent decreased glutathione levels. Crotonaldehyde reacts rapidly with thiol groups. Measurements of urine metabolites in rats have identified 3-hydroxy-l-methyl-propylmercapturic acid as the major metabolite following exposure to crotonaldehyde (Gray and Barnsley, 1971). Wang et al. (1988) found that crotonaldehyde is formed in the liver by metabolic α-hydroxylation of N-nitrosopyrrolidine suggesting that crotonaldehyde is involved in N-nitrosopyrrolidine and macromolecule interactions. Some research indicates that crotonaldehyde can be formed from crotyl alcohol, an unsaturated precursor. Fontaine et al. (2002) demonstrated that crotonaldehyde was formed when crotyl alcohol was oxidized by equine liver alcohol dehydrogenase. This suggests that crotonaldehyde may be encountered in the environment through normal metabolism and biotransformation of other substances in mammals.

Chemical Pathology

The specific chemical pathology of crotonaldehyde is not well understood. What is known regarding the chemical is that it is highly reactive, creates adducts with DNA, and inhibits enzyme activity. Cytochrome P450 and aldehyde dehydrogenase activity is disrupted by crotonaldehyde and human polynuclear leukocytes treated with crotonaldehyde form a dose-related reduction in sulfhydryl groups with subsequent inhibition of superoxide production (Witz et al., 1987). Irritation caused by the chemical is typically immediate with the development of pulmonary edema being delayed. Some research suggests that crotonaldehyde disrupts neural transmission. A rise in oxidative stress and crotonaldehyde formation in glial cells may play a role in the mechanism of action for Alzheimer's disease (Kawaguchi-Niida et al., 2006). A study conducted in 2007 found enhanced crotonaldehyde formation in motor neurons and reactive glia of the spinal cord suggesting a possible relationship between crotonaldehyde and amyotrophic lateral sclerosis development (Shibata et al., 2007).

SOURCES OF EXPOSURE

Occupational

Crotonaldehyde exposure can occur in a variety of occupational settings. Crotonaldehyde is used in industry as an intermediate in the manufacture of butyl alcohol, in polymer technology, as a corrosion inhibitor, and as a stabilizer (OSHA, 1978). In order to protect workers from exposure, processes should be enclosed with the proper ventilation with proper personal protective equipment utilized. Since the chemical is found in smoke from burning wood, fire fighters, forestry workers, and power plant workers are at risk for exposure to crotonaldehyde. Exhaust from internal combustion engines can contain traces of the chemical. Individuals that encounter this exhaust, including truck drivers, heavy equipment operators, and road maintenance workers are at risk to exposure. Individuals that encounter crotonaldehyde are likely to be exposed by way of inhalation or through dermal contact. Proper protective measures should be taken depending on the route of exposure presented. If a worker presents with symptoms indicating exposure, they should be removed from the work environment immediately.

Environmental

Crotonaldehyde, similar to other aldehydes, are emitted by the combustion of petroleum products, wood, and tobacco. Individuals in the population can be exposed to the chemical by inhaling environmental tobacco smoke, petroleum product exhaust, and smoke from wood combustion. Traces of crotonaldehyde found in environmental tobacco smoke can adversely affect mucociliary function, decrease smoke particle removal from the lungs, and serve as a risk factor for the development of asthma in children (Goodwin, 2007). Crotonaldehyde was successfully identified in automobile exhaust gas using gas chromatography with electron-capture detection (Nikshikawa et al., 1987). Environmental exposure to crotonaldehyde is likely to be inhalation of chemical traces from the aforementioned sources. Individuals may experience eye and respiratory irritation if levels are sufficient. If crotonaldehyde levels build up in the indoor environment, increasing the ventilation of the indoor space will help decrease the chemical to acceptable levels. Removal from the contaminated environment may also help improve symptoms.

INDUSTRIAL HYGIENE

Many different chemicals are created using crotonaldehyde as a chemical intermediate. In the occupational environment, exposures to crotonaldehyde are most likely to occur through inhalation, with dermal contact being secondary. Lacrimation and upper respiratory irritation are expected at concentrations slightly above threshold values. The extremely pungent odor and strong irritant properties of crotonaldehyde provide warning signs that make the likelihood of excessive exposure remote. Crotonaldehyde vapors can be detected in air using a solid sorbent technique (Mann and Gold, 1986). Personal respirators and protective clothing to prevent inhalation and dermal exposure are recommended to prevent industrial exposure. Personal air monitoring techniques have been developed using OSHA and the ACGIH exposure limits (Wu and Que Hee, 1995). NIOSH (2010) and OSHA allowable levels of exposure for this compound are identical at 2 ppm (6 mg/m^3). The ACGIH (2013) has established a STEL ceiling of 0.3 ppm. For industrial hygiene monitoring of crotonaldehyde levels, OSHA uses two glass fiber filters coated with 2,4-dinitrophenylhydrazine and phosphoric acid, in open face mode, with a spacer (OSHA, 2006).

MEDICAL MANAGEMENT

There are currently no blood tests that can identify crotonaldehyde exposure but respiratory function tests and chest X-rays may identify damage to the lungs (ATDSR, 2002). Urinary tests may identify crotonaldehyde exposure if the metabolite 3-hydroxy-1-methylpropylmercapturic acid is present (Scherer et al., 2006). Individuals who were exposed to crotonaldehyde, with inhalation or dermal exposure symptoms should be removed from the contaminated environment. Irritation of the eyes, skin, and respiratory system can result but is unlikely since the chemical has a pungent odor. The strong odor will likely alert individuals when concentrations are above normal levels making overexposure unlikely. Medical assistance should be sought if individuals are exposed to high concentrations and their symptoms do not improve. Respiratory protection may be necessary if individuals experience irritation of the upper airways. Covering exposed skin will restrict dermal contact and exposure. As with other aldehydes, individuals should discard contaminated clothing and should cleanse exposed body parts to remove any chemical that may have contacted the skin. Secondary contamination to other individuals due to off gassing of crotonaldehyde is possible from contaminated clothing. Good personal hygiene practices are recommended.

REFERENCES

Agency for Toxic Substances & Disease Registry (ATDSR) (2002) *Crotonaldehyde*, ATDSR 1-18, Atlanta, GA: Centers for Disease Control and Prevention.

American Conference of Governmental Industrial Hygienists (ACGIH) (2013) *Threshold Limit Values (TLVs) for Chemical Substances and Physical Agents and Biological Exposure Indices (BEIs)*, Cincinnati, OH: American Conference of Governmental Industrial Hygienists

Boyland, E. and Chasseud, LF. (1970) The effect of some carbonyl compounds on rat liver glutathione levels. *Biochem. Pharmacol.* 19:1526–1528.

Chung, F.-L., Tanaka, T., and Hecht, S.S. (1986) Induction of liver tumors in F344 rats by crotonaldehyde. *Cancer Res.* 46:1285–1289.

Fernandes, P.H., et al. (2005) Mammalian cell mutagenesis of the DNA adducts of vinyl chloride and crotonaldehyde. *Environ. Mol. Mutagen.* 45:455–459.

Foiles, P.G., et al. (1990) Formation of cyclic deoxyguanosine adducts in Chinese hamster ovary cells by acrolein and crotonaldehyde. *Carcinogenesis* 11:2059–2061.

Fontaine, F.R., et al. (2002) Oxidative bioactivation of crotyl alcohol to the toxic endogenous aldehyde crotonaldehyde: association of protein carbonylation with toxicity in mouse hepatocytes. *Chem. Res. Toxicol.* 15:1051–1058.

Goodwin, R. (2007) Environmental tobacco smoke and the epidemic of asthma in children: the role of cigarette use. *Ann. Allergy Asthma Immunol.* 98:447–454.

Gray, J.M. and Barnsley, E.A. (1971) The metabolism of crotyl phosphate, crotyl alcohol and crotonaldehyde. *Xenobiotica* 1:55–567.

Huang, J.F., et al. (2013) Acute respiratory distress syndrome due to exposure to high-concentration mixture of ethenone and crotonaldehyde. *Toxicol. Ind. Health* 1–3.

Jha, A.M. and Kumar, M. (2006) *In vivo* evaluation of induction of abnormal sperm morphology in mice by an unsaturated aldehyde crotonaldehyde. *Mutat. Res.* 603:159–163.

Kawaguchi-Niida, M., et al. (2006) Crotonaldehyde accumulates in glial cells of Alzheimer's disease brain. *Acta Neuropathol.* 111:422–429.

Lee, S.E. and Park, Y.S. (2013) The role of antioxidant enzymes in adaptive responses to environmental toxicants in vascular disease. *Mol. Cell. Toxicol.* 9:95–101.

Mann, J.H. and Gold, A. (1986) A solid sorbent for crotonaldehyde in air. *Am. Ind. Hyg. Assoc.* 47:832–834.

Nath, R.G. and Chung, F.-L. (1994) Delection of exocyclic 1,N2-propanodeoxyguanosine adducts as common DNA lesions in rodents and humans. *Proc. Natl. Acad. Sci. USA* 91:7491–7495.

National Institute for Occupational Safety and Health (NIOSH) (2010) *Pocket Guide to Chemical Hazards*, NIOSH 94-116, Washington, DC: US Department of Health and Human Services.

Neudecker, T., et al. (1981) Crotonaldehyde is mutagenic in a modified *Salmonella typhimurium* mutagenicity testing system. *Mutat. Res.* 91:27–31.

Neudecker, T., et al. (1989) Crotonaldehyde is mutagenic in *Salmonella typhimurium* TA100. *Environ. Mol. Mutagen.* 14:146–148.

Nikshikawa, H., et al. (1987) Determination of acrolein and crotonaldehyde in automobile exhaust gas by gas chromatography with electron-capture detection. *Analyst* 112:859–862.

Occupational Safety & Health Administration (OSHA) (1978) *Occupational Safety and Health Guideline for Crotonaldehyde*, OSHA 1-6, Washington, DC: US Department of Labor.

Occupational Safety & Health Administration (OSHA) (2006) *Crotonaldehyde*, Washington, DC: US Department of Labor.

Ryu, D.S., et al. (2013) Crotonaldehyde induces heat shock protein 72 expression that mediates anti-apoptotic effects in human endothelial cells. *Toxicol. Lett.* 223:116–123.

Scherer, G., et al. (2006) Influence of smoking charcoal filter tipped cigarettes on various biomarkers of exposure. *Inhal. Toxicol.* 18:821–829.

Shibata, N., et al. (2007) Protein-bound crotonaldehyde accumulates in the spinal cord of superoxide dismutase-1 mutation-associated familial amyotrophic lateral sclerosis and its transgenic model. *Neuropathology* 27:49–61.

Steinhagen, W.H. and Barrow, C.S. (1984) Sensory irritation structure-activity study of inhaled aldehydes in B6C3F1 and Swiss-Webster mice. *Toxicol. Appl. Pharmacol.* 72:495–503.

van der Toorn, M., et al. (2013) Critical role of aldehydes in cigarette smoke-induced acute airway inflammation. *Respir. Res.* 14:45.

Wang, M., et al. (1988) Identification of crotonaldehyde as a hepatic microsomal metabolite formed by α-hydroxylation of the carcinogen *N*-nitrosopyrrolidine. *Chem. Res. Toxicol.* 1:28–31.

Witz, G., Lawrie, N.J., Amoruso, M.A., and Goldstein, B.D. (1987) Inhibition by reactive aldehydes of superoxide anion radical production from stimulated polymorphonuclear leukocytes and pulmonary alveolar macrophages. Effects on cellular sulfhydryl groups and NADPH oxidase activity. *Biochem. Pharmacol.* 36(5):721–726.

Wu, L.-J. and Que Hee, S.S. (1995) A solid sorbent personal air sampling method for aldehydes. *Am. Ind. Hyg. Assoc. J.* 56:362–367.

Yang, B.C., et al. (2013) Crotonaldehyde-exposed macrophages induce IL-8 release from airway epithelial cells through NF-κB and AP-1 pathways. *Toxicol. Lett.* 219:26–34.

FURFURAL

Target organ(s): Eyes, skin, respiratory system

Occupational exposure limits: NIOSH—not established; OSHA—TWA 5 ppm

ACGIH—TWA 2 ppm

Reference values: RfD—0.003 mg/kg/day; RfC—0.05 mg/m^3; MRL—not established

Hazard statements: H226—flammable liquid and vapor; H301—toxic if swallowed; H330—fatal if inhaled; H315—causes skin irritation; H335—may cause respiratory irritation; H319—causes serious eye irritation; H351—suspected of causing cancer

Precautionary statements: P210—keep away from heat/sparks/open flames/ . . . /hot surfaces . . . no smoking; P301+P310—if swallowed, immediately call a poison center or doctor/physician; P403+P233—store in a well-ventilated place; keep container tightly closed; P305+P351+P338—if in eyes, rinse cautiously with water for several minutes; remove contact lenses, if present and easy to do; continue rinsing; P308+P313—if exposed or concerned, get medical advice/attention; P281—use personal protective equipment as required

Risk phrases: R21—harmful in contact with skin; R23/25—toxic by inhalation and if swallowed; R36/37/38—irritating to eyes, respiratory system and skin; R40—limited evidence of a carcinogenic effect

Safety phrases: S1/2—keep locked up and out of reach of children; S26—in case of contact with eyes, rinse immediately with plenty of water and seek medical advice; S36/37—wear suitable protective clothing and gloves; S45—in case of accident or if you feel unwell, seek medical advice immediately (show label where possible)

BACKGROUND AND USES

The presence of furfural (fural, 2-furancarboxyaldehyde, furfuraldehyde, 2-furanaldehyde; CASRN 98-01-1) can be detected in numerous applications, including lubricating oils, resins, pesticides, fuels, and food products. It is also used as a chemical intermediate and reagent in such reactions as rubber vulcanization. Chemically, furfural can also be classified as a furan because of its dienic ether group. Furfural is a natural constituent of a variety of essential oils and can be produced as a by-product of carbohydrate pyrolysis.

PHYSICAL AND CHEMICAL PROPERTIES

Furfural is a colorless to amber color liquid with an odor similar to almonds and a taste similar to caramel. The chemical darkens when it contacts light or air. The chemical has a molecular weight of 96.1, a boiling point of 323 °F, and a freezing point of −34 °F. The vapor pressure is 2.21 mm Hg at 25 °C and the specific gravity is 1.16. The lower explosive limit is 2.1% and the upper explosive limit is 19.3%. Furfural may be incompatible with strong acids, oxidizers, and strong alkalis. Furfural has the ability to polymerize when it contacts strong acids or strong alkalis.

MAMMALIAN TOXICOLOGY

Acute Effects

Due to the low volatility of furfural, most exposures are dermal in nature; therefore the primary health hazards are associated with this route of exposure. Contact dermatitis and eczema may occur, as well as other sensitization reactions. Irritation of the skin, eyes, and mucous membranes may result with acute inhalation exposure. Respiratory effects associated with furfural at typical occupational exposure levels are mild irritation and slight difficulties in breathing. Removal from the contaminated environment will result in abatement of most, if not all, signs and symptoms. Acute exposure in animal studies has shown a dose–response relationship with signs of toxicity, as well as high LC_{50} values (Terrill et al., 1989; Mishra et al., 1991). Signs observed in these animals included discoloration of fur, increased secretory responses, lacrimation, and some

respiratory difficulty. Mishra et al. (1991) further examined the biochemical changes in rat tissue after furfural exposure and found mucous membrane irritation and hepatocyte necrosis. Furfural has also been found to cause histopathological changes in rats at low concentrations with increasing incidence and severity at higher concentrations (Arts et al., 2004). A supporting research study found that rats that inhaled furfural above 126 ppm experienced dose-dependent mortality (Malik et al., 2012).

Chronic Effects

Chronic exposure to furfural in humans has not been the subject of any significant scientific investigation. Furfural's presence in foodstuffs and multiple other sources make determining exposure source difficult. Trace amounts of furfural may be detected in combustion smoke. Naeher et al. (2007) explains that continued inhalation of woodsmoke by firefighters led to chronic pulmonary dysfunction. While furfural is not the sole component in woodsmoke, it possibly plays a role in disease development. Much of the information in this area involves animal and *in vitro* studies. Carcinogenicity studies in rodent assays have not yet shown furfural alone to be a potent carcinogen. However, reports by Feron (1972) and Miyakawa et al. (1991) have shown initiation effects with furfural and known carcinogenic compounds such as benzo(a)pyrene and 12-O-tetradecanoylphorbol-13-acetate in different models. Extrapolation of these results to human risk assessment would be questionable. *In vitro* studies have shown associations with furfural exposure and mutagenic effects on plasmid DNA; however, these mutations were not significantly increased beyond the body's natural repair capabilities to counteract these effects (Khan et al., 1995). A 2-year gavage study found some evidence of carcinogenic potential in male F344/N and B6C3F1 mice for hepatocellular adenoma and bile duct dysplasia (Irwin 1990). The ACGIH (2013) ranks furfural as a Group A3 carcinogen: clear evidence in animals with unknown human relevance.

Mechanism of Action(s)

Furfural metabolism is not well established in humans, and few animal studies have been reported. No systemic effects are known to occur in humans. Nomeir et al. (1992) studied furfural disposition and metabolism in rats. After oral administration of the compound, metabolites of furfural were mainly found in urine, up to 88%. The initial step appears to be oxidation to furoic acid and subsequent excretion unchanged or conjugated with compounds such as glycine. Further scientific investigation is needed to get a better understanding of furfural toxicodynamics.

Chemical Pathology

The pathology of furfural is not well established in the literature. There is some evidence that tissue distribution primarily occurs in the liver and kidneys. These and similar results have focused interest on liver and kidney hazards among workers exposed to furfural. Although no known associations have been discovered, as stated previously, compromised individuals should be cautious. Monitoring of urinary excretion of furfural metabolites, such as furoic acid and 2-furoylglycine, has been reported, although not quantitatively (Sedivec and Flek, 1978; Laham and Potvin, 1989). Liver cells exposed to furfural developed cirrhotic changes after 90 days of exposure indicating a potential site of carcinogenesis (Shimizu and Kanisawa, 1986).

SOURCES OF EXPOSURE

Occupational

Similar to other aldehydes discussed, furfural is a component of smoke. Burning of wood, petroleum, and petroleum products may contain traces of the chemical. Recent scientific investigation has focused on furfural levels in electrocautery smoke. A 2004 study found that electrosurgical smoke had furfural concentrations at 24 ppm, 12 times higher than the occupational exposure limit (Hollman et al., 2004). A supporting study also detected electrosurgical smoke with furfural levels at 24 ppm (Lin et al., 2010). Surgeons and medical staff may be at risk for exposure from this route. Ventilation may be necessary in this situation to decrease concentrations back to allowable levels. Furfural can also be found in industry during the production of oils, pesticides, fuels, and foods. Several individuals working at popcorn popping company developed occupational asthma with furfural, present in environmental air samples, implicated as a potential causative agent (Sahakian et al., 2008). This situation indicates the need for exhaust ventilation and personal protective equipment to protect employees from exposure.

Environmental

Since furfural is used in a variety of products, it can be released into the environment through a variety of waste streams. Like other aldehydes presented, it is a component of smoke from wood burning, tobacco burning, and internal combustion engines. Indoor air quality may be impacted if products are burned inside. A 2002 study found that the burning of incense significantly increased concentrations of aldehydes in indoor air (Ho and Yu, 2002). Increasing air exchange rates and ventilation in indoor environments will prevent furfural and other volatile organic compounds from reaching high concentrations. In some situations, furfural can be off gassed from substrates. Furfural, a product of paper degradation, was found at higher concentrations in library repositories when compared to air-filtered library repositories (Fenech et al., 2010). Clothes exposed to tobacco smoke were found to off gas a significant amount of furfural into the indoor environment (Chien et al., 2011). Irritation of the eyes and respiratory tract may occur at elevated contaminant levels. Removal from the environment is likely to improve symptoms.

INDUSTRIAL HYGIENE

A reaction of agricultural waste residues that contain pentosans with sulfuric acid is capable of producing furfural industrially. Low volatility of furfural vapors makes occupational inhalation exposure rare. Adverse health effects related to furfural are most likely to come from the dermal route. Ingestion of toxic concentrations of furfural is rare. Barrier protection is recommended to protect workers from the irritant properties of furfural. Concentrations from 2 to 14 ppm have caused eye and throat irritation and headaches. Workers who have a history of contact dermatitis, liver maladies, or kidney maladies should be given special consideration. Although liver and kidney effects have not been demonstrated in humans, these organs have been identified as targets in some animal studies, and caution should be taken. Eye, hand, and face personal protective equipment should be worn in case of possible contact with furfural. Respiratory protection equipped with organic vapor cartridges, are recommended for those workers who may be involved in brief work where high concentrations of furfural vapors are present. The use of personal protective equipment coupled with proper exhaust ventilation can keep health hazards to a minimum. The detection of furfural in the workplace can be achieved by adsorption onto activated carbon and desorption with carbon disulfide and 2-propanol followed by GC analysis (Sidhu, 1982). The ACGIH has set TLV guidelines for exposure at 2 ppm ($7.9\,mg/m^3$). OSHA has a higher PEL of 5 ppm ($20\,mg/m^3$), whereas NIOSH has no established REL for this compound. The immediate danger to life and health level for furfural is 250 ppm.

MEDICAL MANAGEMENT

Diagnostic testing may be necessary to identify frequent or high level exposure to furfural. Urine samples that contain furoic acid may indicate exposure to the chemical (Sedivec and Flek, 1978). In addition, liver function tests above baseline values may be a strong indicator. Chest X-rays may identify damage to the lungs from acute exposure. Patch testing conducted by an allergist may be necessary to determine skin allergies or sensitization from furfural (NJDHSS, 2000). Individuals that experience symptoms resulting from exposure should be removed from the contaminated environment immediately. If symptoms persist after removal from the environment, exposed individuals should seek prompt medical treatment. Engineering controls are the best way to control exposure by providing exhaust ventilation and/or by enclosing or isolating the source. Because engineering controls are often quite costly and require adequate amounts of planning, the use of personal protective equipment may be necessary. Clothing, including long shirts and aprons, should be worn to protect the skin from chemical contact and face protection should be worn to prevent the chemical from contacting the face. Respiratory protection may necessary in areas where furfural concentrations are high.

REFERENCES

American Conference of Governmental Industrial Hygienists (ACGIH) (2013) *Threshold Limit Values (TLVs) for Chemical Substances and Physical Agents and Biological Exposure Indices (BEIs)*, Cincinnati, OH: American Conference of Governmental Industrial Hygienists.

Arts, J.H., et al. (2004) Subacute (28-day) toxicity of furfural in Fischer 344 rats: a comparison of the oral and inhalation route. *Food Chem. Toxicol.* 42:1389–1399.

Chien, Y.C., et al. (2011) Volatile organics off-gassed among tobacco-exposed clothing fabrics. *J. Hazard. Mater.* 193:139–148.

Fenech, A., et al. (2010) Volatile aldehydes in libraries and archives. *Atmos. Environ.* 44:2067–2073.

Feron, V.J. (1972) Respiratory tract tumors in hamsters after intratracheal instillations of benzo(*a*)pyrene alone and with furfural. *Cancer Res.* 32:28–36.

Ho, S.S. and Yu, J.Z. (2002) Concentrations of formaldehyde and other carbonyls in environments affected by incense burning. *J. Environ. Monit.* 4:728–733.

Hollman, R., et al. (2004) Smoke in the operating theater: an unregarded source of danger. *Plast. Reconstr. Surg.* 114:458–463.

Irwin, R. (1990) NTP toxicology and carcinogenesis studies of furfural (CAS No. 98-01-1) in F344/N rats and B6C3F1 mice (gavage studies). *Natl. Toxicol. Program Tech. Rep. Ser.* 382: 1–201.

Khan, Q.A., Shamsi, F.A., and Hadi, S.M. (1995) Mutagenicity of furfural in plasmid DNA. *Cancer Lett.* 89:95–99.

Laham, S. and Potvin, M. (1989) Metabolism of furfural in the Sprague-Dawley rat. *Toxicol. Environ. Chem.* 24:35–47.

Lin, Y.W., et al. (2010) A novel inspection protocol to detect volatile compounds in breast surgery electrocautery smoke. *J. Formos. Med. Assoc.* 109:511–516.

Malik, T., et al. (2012) Impact of furfural and kerosene co-exposure through inhalation in lungs of rats. *Biochem. Pharm.* 1:1–4.

Mishra, A., et al. (1991) Pathological and biochemical alterations induced by inhalation of furfural vapor in rat lung. *Bull. Environ. Contam. Toxicol.* 47:668–674.

Miyakawa, Y., et al. (1991) Initiating activity of eight pyrolysates of carbohydrates in a two-stage mouse skin tumorigenesis model. *Carcinogenesis* 12:1169–1173.

Naeher, L.P., et al. (2007) Woodsmoke health effects: a review. *Inhal. Toxicol.* 19:67–106.

New Jersey Department of Health and Senior Services (NJDHSS) (2000) *Hazardous Substance Fact Sheet—Furfural*, NJDHSS, Trenton, NJ.

Nomeir, A.A., et al. (1992) Comparative metabolism and disposition of furfural and furfuryl alcohol in rats. *Drug Metab. Dispos.* 20:198–204.

Sahakian, N., et al. (2008) Asthma arising in flavoring-exposed food production workers. *Int. J. Occup. Med. Environ. Health* 21:173–177.

Sedivec, V. and Flek, J. (1978) Biologic monitoring of persons exposed to furfural vapors. *Int. Arch. Occup. Environ. Health* 42:41–49.

Shimizu, A. and Kanisawa, M. (1986) Experimental studies on hepatic cirrhosis and hepatocarcinogenesis. I. Production of hepatic cirrhosis by furfural administration. *Acta Pathol. Jpn.* 36:1027–1038.

Sidhu, K.S. (1982) A gas chromatographic procedure for the determination of environmental furfural. *Bull. Environ. Contam. Toxicol.* 28:250–255.

Terrill, J.B., et al. (1989) Acute toxicity of furan, 2-methyl furan, furfuryl alcohol, and furfural in the rat. *Am. Ind. Hyg. Assoc. J.* 50:A359–A361.

GLUTARALDEHYDE

Target organ(s): Eyes, skin, respiratory system

Occupational exposure limits: NIOSH—ceiling 0.2 ppm; OSHA—ceiling 0.2 ppm; ACGIH—STEL ceiling 0.05 ppm

Reference values: RfD—0.16 mg/kg/day; RfC/MRL—not established

Hazard statements: H290—may be corrosive to metals; H400—very toxic to aquatic life; H411—toxic to aquatic life with long lasting effects; H301—toxic if swallowed; H331—toxic if inhaled; H314—causes severe skin burns and eye damage; H317—may cause an allergic skin reaction, H334—may cause allergy or asthma symptoms or breathing difficulties if inhaled

Risk phrases: R23/25—toxic by inhalation and if swallowed; R34—causes burns; R42/43—may cause sensitization by inhalation and skin contact; R50—very toxic to aquatic organisms

Safety phrases: S1/2—keep locked up and out of reach of children; S26—in case of contact with eyes, rinse immediately with plenty of water and seek medical advice; S36/37/39—in case of accident or if you feel unwell, seek medical advice immediately (show label where possible); S61—avoid release to the environment; refer to special instructions/safety data sheets

BACKGROUND AND USES

Glutaraldehyde (Symbol GTA; glutaric dialdehyde, 1,5-pentanedial, glutaral; CASRN 111-30-8) is commonly used in the medical industry in cold sterilization and in the X-ray development process. It can also be encountered in the leather industry as a tanning ingredient and in mortuary workers. There are no reports which indicate that glutaraldehyde is a naturally occurring compound. Cidex and Acusol, 2% buffered solutions, use this aldehyde as an active ingredient. Sodium bicarbonate is required to activate the solution, which then has a shelf life of 1–2 weeks. Despite health hazards involved with its use, glutaraldehyde is one of the most effective biocides used. It is particularly effective against bacteria and viruses, including the human immuno-deficiency virus.

PHYSICAL AND CHEMICAL PROPERTIES

Glutaraldehyde is a colorless liquid with a strong, pungent odor. It is miscible in water and organic solvents. The chemical has a molecular weight of 100.1, a boiling point of 212 °F, and a freezing point of 7 °F. The vapor pressure is 0.6 mm Hg at 30 °C and the specific gravity is 1.10. Glutaraldehyde may be incompatible with strong oxidizers and strong bases. It should be noted that alkaline solutions containing glutaraldehyde may react with alcohol, ketones, amines, hydrazines, and proteins.

MAMMALIAN TOXICOLOGY

Acute Effects

A multitude of signs and symptoms may result from acute exposure to glutaraldehyde depending on the exposure route. It is known to irritate the eyes, skin, and upper respiratory tract. Bronchial and laryngeal mucous membrane irritation can result from the inhalation of vapors. A study of the histopathology of rats found a concentration-related acute inflammation and necrosis of the nasal mucosa, larynx, trachea, and bronchi (Ballantyne and Myers, 2001). Vapors of this aldehyde may cause eye irritation, which subsides after exposure is terminated. However, direct contact of glutaraldehyde with the cornea has reportedly been the cause of keratopathy with development of a white plaque that resolved after time (Dailey et al., 1993). Dermal contact with glutaraldehyde solutions used in cold sterilization may

cause a mild skin irritation, but studies report on systemic effects are likely to occur due to low penetration of the skin (Frantz et al., 1993). A study of workers involved in glutaraldehyde sterilization reported general work-related symptoms such as headache and nausea (Norback, 1988). Symptoms of tachycardia and palpitations in workers, which resolve upon discontinued exposure to the chemical, have been reported, although these effects have not been confirmed by others (Connaughton, 1993).

Glutaraldehyde has long been suspected of being a sensitizer with prolonged exposure producing localized edema and other symptoms suggestive of an allergic response. Hilton et al. (1998) conducted a study using the local lymph node assay and demonstrated that glutaraldehyde had a greater potential to cause skin sensitization when compared to formaldehyde. Skin symptoms were 3.6 times more likely to be reported by glutaraldehyde exposed nurses (Waters et al., 2003). Animal studies report that nasal and pulmonary changes occur and persist at exposure levels lower than recommended guidelines (Zissu et al., 1994). Chronic bronchitis and nasal symptoms in humans were noted to be significantly higher at peak concentrations of glutaraldehyde (Takigawa and Endo, 2006). Numerous cases of colitis following endoscopy have been reported (Rosen, 1994; West et al., 1995). These outbreaks have been linked to residual glutaraldehyde on equipment that has not been thoroughly rinsed.

Chronic Effects

Chronic dermal exposure to glutaraldehyde has resulted in many cases of allergic contact dermatitis, particularly among health care workers involved in sterilization of medical equipment (Bardazzi et al., 1986; Nethercott et al., 1988). A study conducted in 2000 found that health care workers were eight times more likely to be allergic to glutaraldehyde when compared to individuals that do not work in the health care field (Shaffer and Belsito, 2000). The hands and arms are the most commonly affected areas. Some cases may take months to resolve and may reemerge with contact outside of the workplace because glutaraldehyde is found in lower concentrations in fabric softeners and hairsprays.

Occupational asthma related to glutaraldehyde exposure has also been reported among health care workers (Chan-Yeung et al., 1993; Gannon et al., 1995). Decreases in the peak expiratory flow (PEF) rate and bronchial challenges have indicated glutaraldehyde as the primary cause of the occupational asthma. There have been some reports of radiography technicians with this illness as well (Cullinan et al., 1992). Other reports include general respiratory irritation and in one case recurrent nosebleeds that subsided after exposure to glutaraldehyde was removed (Wiggins et al., 1989). Recent research indicates that the incidence of occupational asthma resulting from glutaraldehyde has declined indicating that control measures are properly reducing exposure in the work environment (Bakerly et al., 2008).

Glutaraldehyde has not been the subject of human studies evaluating carcinogenicity and teratogenicity. Carcinogenic studies of glutaraldehyde in rodents have reported lesions in the anterior portion of the nose, but no precancerous lesions similar to those seen in rats exposed to formaldehyde have been reported (Gross et al., 1994). A battery of mutagenicity assays *in vitro* revealed no significant increases in genotoxic effects (Slesinski et al., 1983). Gastric intubation of pregnant rats with glutaraldehyde produced no teratogenic effects among the fetuses (Ema et al., 1992). A 2-year study of rats and mice exposed to varying concentrations of glutaraldehyde produced lesions in the nose, none of which were neoplastic (van Birgelen et al., 2000). The ACGIH (2013) declare glutaraldehyde a Group A4 carcinogen: not classifiable as a human carcinogen.

Mechanism of Action(s)

The mechanism of glutaraldehyde toxicity has not been the subject of many studies, as the health hazards associated with occupational exposure to glutaraldehyde are only of recent interest. A possible explanation for the acute effects involving mild irritation is nonspecific corrosive tissue damage. Glutaraldehyde-induced dermatitis and occupational asthma were reported months and, in some cases, years after primary exposure, demonstrating the lengthy latent period required to cause these effects. The biocidal activity of glutaraldehyde is a product of its ability to crosslink membrane proteins, altering the membrane function of the cell (Munton and Russell, 1973). Glutaraldehyde-induced cross-linking of DNA proteins is the proposed mechanism for the possible mutagenicity of this aldehyde. Studies have shown similar mutagenic potential of glutaraldehyde at concentrations less than formaldehyde (St. Clair et al., 1991). However, no evidence has verified the carcinogenicity of glutaraldehyde. Similar intestinal effects as those in human postendoscopy cases were observed in animals treated with glutaraldehyde via the colon. The diarrhea and other health effects were linked to severe mucosal damage and colonic lesions believed to be effects similarly produced in humans (Durante et al., 1992). Studies of glutaraldehyde metabolism have reported expired CO_2 as a major byproduct.

Chemical Pathology

The specific pathology of glutaraldehyde is not well understood. Rodent tissue studies have shown high metabolism of glutaraldehyde in the kidney and liver. Glutaraldehyde is not as good a substrate for the dehydrogenase enzymes compared to other aliphatic aldehydes. Metabolic biotransformation of glutaraldehyde likely involves oxidation to glutaric acid (Beauchamp et al., 1992). Other urinary metabolites have been identified but not characterized. Takigawa and Endo (2006) explain that the major route of metabolism of glutaraldehyde is through oxidation by the kidney and liver resulting in several different byproducts. It is possible that pathological manifestations result by the chemical concentrating in these two organs. Additional research is needed to support this thought. As previously mentioned, glutaraldehyde's ability to crosslink DNA is believed to be the potential mechanism for cancer development. Several research studies have been conducted to examine the carcinogenic pathology of glutaraldehyde but up to this point there is no consensus.

SOURCES OF EXPOSURE

Occupational

Glutaraldehyde is commonly found in the healthcare industry as a disinfecting agent. National Institute for Occupational Safety and Health explains that the chemical is used to sterilize equipment that cannot be heat sterilized, including dialysis instruments, surgical instruments, bronchoscopes, and endoscopes (NIOSH, 2012). Physicians, nurses, medical assistants, and patients are at risk for inhalation or dermal exposure to the chemical. Eight hospital staff members working in an endoscopy unit were exposed to 2% glutaraldehyde and reported lacrimation, rhinitis, dermatitis, respiratory irritation, nausea, and headache (Jachuck et al., 1989). Glutaraldehyde has various uses and can also be found in industries, including tanning, petroleum, water treatment, paper manufacturing, and cosmetic manufacturing. Most worker exposure will be through dermal contact with the chemical; however, inhalation of the chemical is possible. Protective equipment, including gloves, aprons, and face protection may be necessary to prevent dermal contact. In occupations where inhalation may be of concern, proper ventilation or respiratory protection should be considered.

Environmental

Glutaraldehyde is used in a variety of industries and has the potential to be released in the environment from multiple sources. Glutaraldehyde has been found to be toxic to warm and cold water aquatic organisms and is slightly more toxic to freshwater fish when compared to salt water fish (Leung, 2001). Substituting glutaraldehyde with other chemicals with similar properties could help reduce impact on surrounding and aquatic environments. Patients undergoing medical procedures may be at risk for exposure from sources in the health care environment. Medical tools not properly cleansed may have residue that may illicit adverse health reactions in patients. Abdominal pain, diarrhea, and rectal bleeding were symptoms of eight patients receiving either an endoscopy or colonoscopy from a scope-containing glutaraldehyde residue (Stein et al., 2001). Health care professionals should ensure that medical tools are properly sanitized and rinsed prior to patient use. Closed loop systems and engineering controls may be necessary to prevent environmental sources of glutaraldehyde exposure.

INDUSTRIAL HYGIENE

Inhalation and dermal contact are the main exposure routes for this chemical. Manual washing of medical equipment in glutaraldehyde-containing solutions is the most common hazard for all exposures. Mechanical washing and processes in X-ray development are less hazardous for dermal contact, but glutaraldehyde vapors are still released in measurable amounts. Personal respirators and monitors as well as adequate ventilation in and around the work area will provide adequate respiratory protection against chemical vapors. Protective clothing, aprons, and face shields are recommended to protect against dermal contact with glutaraldehyde. Gloves are necessary, but attention should be given to glove materials. Studies have reported glutaraldehyde penetration of glove materials, including latex and nonlatex rubbers (Lehman et al., 1994). Neoprene or butyl rubber gloves are protective. In order to combat this problem, it is recommended that workers frequently replace and/or double glove. Faulty equipment used for sterilization can also be a source of occupational exposure (Lynch et al., 1994).

Workplace standards for glutaraldehyde vary slightly. The ACGIH (2013) has a short-term exposure limit ceiling of 0.05 ppm while OSHA and NIOSH (2010) have short-term exposure limit ceiling values of 0.2 ppm (0.82 mg/m^3). A study reported by Leinster et al. (1993) assessed typical exposure levels in a hospital setting and found that occupational exposure is usually within guidelines. Approximately 35,000 employees are occupationally exposed to glutaraldehyde. Formaldehyde is the most commonly used aldehyde followed by glutaraldehyde. A 2% glutaraldehyde solution used for sterilization can produce airborne concentrations that exceed workplace exposure guidelines.

MEDICAL MANAGEMENT

The hazards associated with glutaraldehyde are well known. It is important to understand that glutaraldehyde is a possible cause of occupational illness. Reports of glutaraldehyde-induced dermatitis and occupational asthma have increased the awareness of its occupational hazards. Occurrence of diseases, such as AIDS, has caused more frequent use of glutaraldehyde as a sterilizing agent in health care settings, thereby increasing risks of exposure (Fowler, 1989). In order to assess occupational exposure, medical surveillance programs should be instituted that include physical examination and pulmonary function testing. For a surveillance program to be successful, it should monitor an employee prior to employment, during employment, and at the termination. Individuals exposed to high concentrations of glutaraldehyde vapors should be removed from the environment. If symptoms persist after removal, medical treatment should be sought. Individuals exposed dermally to glutaraldehyde should remove contaminated clothing and cleanse affected body parts thoroughly. Contaminated clothing should be discarded in an appropriate manner.

REFERENCES

American Conference of Governmental Industrial Hygienists (ACGIH) (2013) *Threshold Limit Values (TLVs) for Chemical Substances and Physical Agents and Biological Exposure Indices (BEIs)*, Cincinnati, OH: American Conference of Governmental Industrial Hygienists.

Bakerly, N.D., et al. (2008) Fifteen-year trends in occupational asthma: data from the Shield surveillance scheme. *Occup. Med. (Lond.)* 58:169–174.

Ballantyne, B. and Myers, R.C. (2001) The acute toxicity and primary irritancy of glutaraldehyde solutions. *Vet. Hum. Toxicol.* 43:193–202.

Bardazzi, F., et al. (1986) Glutaraldehyde dermatitis in nurses. *Contact Dermatitis* 14:319–320.

Beauchamp, R.O., et al. (1992) A critical review of the toxicology of glutaraldehyde. *Crit. Rev. Toxicol.* 22:143–174.

Chan-Yeung, M., et al. (1993) Occupational asthma in a technologist exposed to glutaraldehyde. *J. Allergy Clin. Immmol.* 91:974–978.

Connaughton, P. (1993) Occupational exposure to glutaraldehyde associated with tachycardia and palpitations. *Med. J. Aust.* 159:567–572.

Cullinan, P., et al. (1992) Occupational asthma in radiographers. *Lancet* 640:1477–1479.

Dailey, J.R., Parnes, R.E., and Aminlari, A. (1993) Glutaraldehyde keratopathy. *Am. J. Ophthalnol.* 115:256–258.

Durante, L., et al. (1992) Investigation of an outbreak of bloody diarrhea: association with endoscope cleaning solution and demonstration of lesions in an animal model. *Am. J. Med.* 92:476–480.

Ema, M., Itami, T., and Kawasaki, H. (1992) Teratological assessment of glutaraldehyde in rats by gastric intubation. *Toxicol. Lett.* 63:147–153.

Fowler, J.F., Jr. (1989) Allergic contact dermatitis from glutaraldehyde exposure. *J. Occup. Med.* 31:852–853.

Frantz, S.W., et al. (1993) Glutaraldehyde: species comparisons of *in vitro* skin penetration. *J. Toxicol. Cut. Ocular Toxicol.* 12:349–361.

Gannon, P.F.G., et al. (1995) Occupational asthma due to glutaraldehyde and formaldehyde in endoscopy and X-ray departments. *Thorax* 50:156–159.

Gross, E.A., et al. (1994) Histopathology and cell replication responses in the respiratory tract of rats and mice exposed by inhalation to glutaraldehyde for up to 13 weeks. *Fundam. Appl. Toxicol.* 23:348–362.

Hilton, J. et al. (1998) Estimation of relative skin sensitizing potency using the local lymph node assay: a comparison of formaldehyde with glutaraldehyde. *Am. J. Contact Dermatitis* 9:29–33.

Jachuck, S.J., et al. (1989) Occupational hazard in hospital staff exposed to 2 per cent glutaraldehyde in an endoscopy unit. *J. Soc. Occup. Med.* 39:69–71.

Lehman, P.A., Franz, T.J., and Guin, J.D. (1994) Penetration of glutaraldehyde through glove material: tactylon versus natural latex rubber. *Contact Dermatitis* 30:176–177.

Leinster, P., Baum, J.M., and Baxter, P.J. (1993) An assessment of exposure to glutaraldehyde in hospitals: typical exposure levels and recommended control measures. *Br. J. Ind. Med.* 50: 107–111.

Leung, H.W. (2001) Exotoxicology of glutaraldehyde: review of environmental fate and effects studies. *Ecotoxicol. Environ. Saf.* 49:26–39.

Lynch, D.A.F., et al. (1994) Patient and staff exposure to glutaraldehyde from Keymed auto-disinfector endoscope washing machine. *Endoscopy* 26:359–361.

Munton, T.J. and Russell, A.D. (1973) Effects of glutaraldehyde on cell viability. Iriphenyl tetrazotium reduction, oxygen uptake and 13-galaclosidase activity in *Escherichia coli. J. Appl. Microbiol.* 26:508–511.

National Institute for Occupational Safety and Health (NIOSH) (2010) *Pocket Guide to Chemical Hazards*, NIOSH 94-116, Washington, DC: US Department of Health and Human Services.

National Institute for Occupational Safety and Health (NIOSH) (2012) *Glutaraldehyde*, Washington, DC: US Department of Health and Human Services.

Nethercott, J.R., Holness, D.L., and Page, E. (1988) Occupational contact dermatitis due to glutaraldehyde in health care workers. *Contact Dermatitis* 18:193–196.

Norback, D. (1988) Skin and respiratory symptoms from exposure to alkaline glutaraldehyde in medical services. *Scand. J. Work Environ. Health* 14:366–371.

Rosen, P. (1994) Endoscope-induced colitis: description, probable cause by glutaraldehyde, and prevention. *Gastrointest. Endosc.* 40:547–553.

Shaffer, M.P. and Belsito, D.V. (2000) Allergic contact dermatitis from glutaraldehyde in health-care workers. *Contact Dermatitis* 43:150–156.

Slesinski, R.S., et al. (1983) Mutagenicity evaluation of glutaraldehyde in a battery of *in vitro* bacterial and mammalian test systems. *Food Chem. Toxicol.* 21:621–629.

St. Clair, M.B.G., et al. (1991) Evaluation of the genotoxic potential of glutaraldehyde. *Environ. Mol. Mutagen.* 18:113–119.

Stein, B.L., et al. (2001) Glutaraldehyde-induced colitis. *Can. J. Surg.* 44:113–116.

Takigawa, T. and Endo, Y. (2006) Effects of glutaraldehyde exposure on human health. *J. Occup. Health* 48:75–87.

van Birgelen, A.P., et al. (2000) Effects of glutaraldehyde in a 2-year inhalation study in rats and mice. *Toxicol. Sci.* 55:195–205.

Waters, A., et al. (2003) Symptoms and lung function in health care personnel exposed to glutaraldehyde. *Am. J. Ind. Med.* 43: 196–203.

West, A.B., et al. (1995) Glutaraldehyde colitis following endoscopy: clinical and pathological features and investigation of an outbreak. *Gastroenterology* 108:1250–1255.

Wiggins, P., McCurdy, S.A., and Zeidenberg, W. (1989) Epistaxis due to glutaraldehyde exposure. *J. Occup. Med.* 31: 854–856.

Zissu, D., Gagnaire, F., and Bonnet, P. (1994) Nasal and pulmonary toxicity of glutaraldehyde in mice. *Toxicol. Lett.* 71:53–62.

KETONES

ACETONE

Target organ(s): Eyes, skin, respiratory system, central nervous system

Occupational exposure limits: NIOSH—TWA 250 ppm; OSHA—TWA 1000 ppm

ACGIH—TWA 500 ppm, STEL 750 ppm

Reference Values: RfD—0.9 mg/kg/day; RfC—not established; MRL—inhalation 13–26 ppm, Oral 2 mg/kg/day

Hazard statements: H336—may cause drowsiness or dizziness; H319—causes serious eye irritation; H225—highly flammable liquid and vapor

Precautionary statements: P403+P233—store in a well-ventilated place; keep container tightly closed; P405—store locked up; P305+P351+P338—if in eyes, rinse cautiously with water for several minutes; remove contact lenses, if present and easy to do; continue rinsing; P243—take precautionary measures against static discharge; P210—keep away from heat/sparks/open flames/ . . . /hot surfaces . . . no smoking

Risk phrases: R11—highly flammable; R36—irritating to eyes; R66—repeated exposure may cause skin dryness or cracking; R67—vapors may cause drowsiness and dizziness

Safety phrases: S2—keep out of the reach of children; S9—keep container in a well-ventilated place; S16—keep away from sources of ignition . . . no smoking; S26—in case of contact with eyes, rinse immediately with plenty of water and seek medical advice

BACKGROUND AND USES

Acetone (dimethyl ketone, ketone propane, 2-propanone; CASRN 67-64-1) is the most prominent and simplest ketone in terms of volume, giving it wide industrial importance. It is used as an industrial solvent and as a chemical intermediate in many processes involving oils, rubbers, and dyes. Nonindustrial uses of acetone are omnipresent and it is most notably used in cosmetics such as nail polish remover.

PHYSICAL AND CHEMICAL PROPERTIES

Acetone is a colorless liquid with a fragrant odor, which has been characterized as fruity or mint like. It is miscible with several other solvents. The chemical has a molecular weight of 58.1, a boiling point of 133 °F, and a freezing point of −140 °F. The vapor pressure is 231 mm Hg at 25 °C and the specific gravity is 0.79. The lower explosive limit is 2.5% and the upper explosive limit is 12.8%. Acetone may be incompatible with oxidizers and acids NIOSH (2010).

MAMMALIAN TOXICOLOGY

Acute Effects

Acetone is commonly seen as the least toxic organic solvent used in industry therefore few hazards are associated with occupational exposure. Eye and mucous membrane irritation, nausea, and headache are likely to be caused by exposure to acetone vapor at higher concentrations. Irritation of the trachea, nose, throat, and lungs are likely to be associated with inhalation exposures. Repeated exposures at high concentrations may cause central nervous system (CNS) depression. A study of workers exposed to high concentrations of acetone (average 1000 ppm) over an 8-h work day reported only an increase in a self-reported "annoyance" category of well-being (Kiesswetter et al., 1994). A study in rats found that acetone caused selective damage to olfactory neuroepithelium over a time of exposure (Buron et al., 2009). Skin absorption is not likely to cause systemic effects. Acetone is similar to ethanol in its anesthetic properties and as an inducer of mixed function oxidase enzyme systems. Animal studies have shown that coadministration of acetone with hepatotoxicants has an additive adverse health effect and increased metabolism of cytochrome P450 substrates (Miller and Yang, 1984; Kobusch et al., 1989; Iba et al., 1993). Bruckner and Peterson (1981) evaluated inhalation reactions to acetone in rodents and demonstrated CNS depression and narcosis, as well as a significant 9-h lag time before full recovery, although the levels examined were extremely high (Bruckner and Peterson, 1981). Also known as a defatting agent, acetone may cause skin defatting and dermatitis.

Chronic Effects

Studies of chronic exposures to acetone have been documented. A mortality study of occupationally exposed workers showed no excess mortality risk when compared with the general population (Ott et al., 1983). At concentrations far exceeding those expected to occur occupationally, some developmental effects were observed in mice (Mast et al., 1989). However, in a study of solvent exposure and pregnancy outcome, acetone exposure was not associated with an increased rate of miscarriage or fetal malformations (Axelsson et al., 1984). Epoxy resin carcinogenicity of the skin was studied in mice using acetone as a solvent. A control group of mice dermally exposed to acetone alone exhibited no excess incidence of skin tumors or decreased fitness (Zakova et al., 1985). Long-term animal exposure studies found damage to the kidney and liver, as well as birth defects (Niziaeva 1982; NTP, 1991).

Mechanism of Action(s)

There are several different ways that acetone is metabolized. Some acetone is excreted in the urine or exhaled unchanged. The lactate, methylglyoxal, and propanediol pathways are the three known routes of acetone metabolism in humans. Acetone metabolism occurs primarily in the liver and begins with the oxidation of acetone to acetol catalyzed by acetone monooxygenase (Casazza et al., 1984). This is the rate-limiting step for all three pathways. Acetol monooxygenase catalyzes the oxidation of acetol to methylglyoxyl, which is converted to glucose directly or by lactate. Acetol may also be converted to 1,2-propanediol, which is converted to glucose or lactate or breaks down to acetate and formate. The first two pathways are more common at low concentrations because the propanediol pathway is used after the first two become saturated. High blood levels of acetone may result in the metabolism of the compound to isopropyl alcohol, but this was found to occur in mostly diabetic individuals who were not known to be exposed to acetone occupationally (Lewis et al., 1984). Recent research has focused on the development of portable acetone detectors for the identification of diabetes. Righettoni and Tricoli (2011) summarized the strengths and limitations of the device in their review article.

Chemical Pathology

Acetone is typically present in the human body as a result of the breakdown of adipose tissue. It is one of the important normal processes that make energy for bodily function. The health effects of acetone have been the topic of scientific investigation. Rabbit eyes exposed to acetone had changes consistent with mild irritation and corneal injury limited to the epithelium and superficial stroma (Maurer et al., 2001). In addition to inducing irritation in tissue, acetone is often detected in the breath of diabetic patients. In severe cases of ketoacidosis, acetone is often elevated as a result of fatty acid breakdown for energy (Keltanen et al., 2013). One study found that acetone gas emanating from diabetic rat tails was significantly higher (1134 ppb) when compared to a control group (124 ppb) (Yamai et al., 2012). Animals provided large quantities of acetone to drink developed bone marrow hypoplasia, degeneration of kidneys, and abnormal sized livers (NTP, 1988; NTP, 1991).

SOURCES OF EXPOSURE

Occupational

Acetone is a widely used solvent in industry so occupational exposure can easily occur. It is also used in the application of lacquers and varnishes. A study of 157 Romanian coin and medal factory workers found negative effects on human performance and evidence of neurotoxicity when acetone concentrations over an 8-h work shift ranged from 988 to 2114 mg/m^3 (Mitran et al., 1997). The values reported were above occupational exposure limits in the country. Acetone is commonly used in nail salons to remove nail polish during manicures and pedicures. One study examining acetone concentrations in nail salons found acetone was present in 96% of 70 environmental samples (Gjolstad et al., 2006). In order to prevent exposure to acetone in the work environment, personal respirators and ventilation should be used if inhalation is of concern. If dermal contact is of concern, protective clothing, including aprons, face shields, safety glasses, and gloves are recommended.

Environmental

Volatile organic compounds are known to contaminate indoor environments. In certain environmental conditions, VOCs can produce gas vapors, leading to occupant exposure. Acetone belongs to this class of chemicals. Air fresheners used in homes oxidize 2-methylpropanol to acetone with emission rates ranging from 530 mg/h to 660 mg/h and an acetone concentration of 700 µg/m^3 (Geiss et al., 2014). Increasing the amount of air exchanges or ventilation in the indoor environment can reduce the buildup of acetone vapors. In addition, selecting building materials with low VOCs can also reduce contaminant buildup. Trace amount of acetone can be found in automobile exhaust, tobacco smoke, and landfill material. Acetone-containing waste found in landfills may be a major source of environmental contamination. Soil and water can become contaminated when acetone is leached from landfill waste products. When acetone enters soil or water, it can either evaporate or be removed by microbes (Ramachandriya et al., 2011). Reducing waste or substituting other solvents in place of acetone may reduce the amount that enters the outside environment.

INDUSTRIAL HYGIENE

Exposures to acetone occur mostly via inhalation and dermal contact. This results primarily in an irritant response. Acetone is known to have low toxicity in humans. No known lethalities are associated with occupational exposure to acetone. The odor threshold, 0.2–1.5 ppm, is significantly lower than any exposure guidelines set for acetone. Therefore detecting odor of acetone does not point to a hazard. Since acetone has irritant properties, casual contact with concentrations high enough to be considered as a health hazard would generally not be tolerated by workers.

Workers exposed to high concentrations of acetone should wear recommended protective clothing and personal respirators. Materials used as a barrier for acetone should be evaluated carefully for breakthrough time. Acetone was shown to pass through neoprene rubber independent of the thickness of the material evaluated (Berardinelli and Meyer, 1988). Monitoring for acetone vapor concentrations in the atmosphere can be achieved. Some polymers of sorbent material were not reliable for acetone in normal solid sorbent assays (Levin and Carleborg, 1987). Adsorption of acetone by an activated charcoal medium followed by gas chromatography analysis produced positive reliable results (Whitehead et al., 1984).

The PEL set by OSHA for acetone is 1000 ppm (2400 mg/m^3), with NIOSH having a lower recommendation (REL) of 250 ppm (590 mg/m^3). The ACGIH has set guidelines (TLV/STEL) for acetone exposures at 500 ppm/750 ppm (1780 mg/m^3, 2380 mg/m^3). Acetone is not designated as a suspected carcinogen.

MEDICAL MANAGEMENT

Studies of biomonitoring have identified detectable amounts of unchanged acetone in the urine of workers occupationally and experimentally exposed via inhalation (Pezzagno et al., 1986). These concentrations paralleled rises in environmental concentrations and the amount absorbed. An analysis of urine, alveolar air, and blood concentrations of acetone further demonstrated urine as the most reliable indicator for monitoring exposure (Fujino et al., 1992). The ubiquitous nature of acetone in the environment and body makes acetone an ever-present constituent of urine in those non-occupationally exposed (Ashley et al., 1994). These normal amounts of acetone in the urine make the relationship between urinary and environmental concentrations curvilinear as opposed to linear (Kawai et al., 1992). Increases in urinary acetone will likely not be detectable until atmospheric concentrations reach a level capable of changing endogenous levels and subsequently leading to increases in urinary acetone levels.

REFERENCES

American Conference of Governmental Industrial Hygienists (ACGIH) (2013) *Threshold Limit Values (TLVs) for Chemical Substances and Physical Agents and Biological Exposure Indices (BEIs)*, Cincinnati, OH: American Conference of Governmental Industrial Hygienists.

Ashley, D.L., et al. (1994) Blood concentrations of volatile organic compounds in a nonoccupationally exposed US population and in groups with suspected exposure. *Clin. Chem.* 40: 1401–1404.

Axelsson, G., Ltitz, C., and Rylander, R. (1984) Exposure to solvents and outcome of pregnancy in university laboratory' employees. *Br. J. Ind. Med.* 41:305–312.

Berardinelli, S.R. and Meyer, E.S. (1988) Chemical protective clothing breakthrough time: comparison of several test systems. *Am. Ind. Hyg. Assoc. J.* 49:89–94.

Bruckner, J.V. and Peterson, R.G. (1981) Evaluation of toluene and acetone inhalant abuse. *Toxicol. Appl. Pharmacol.* 61:27–38.

Buron, G., et al. (2009) Inhalation exposure to acetone induces selective damage on olfactory neuroepithelium in mice. *Neurotoxicology* 30:114–20.

Casazza, J.R., Felver, M.E., and Veech, R.L. The metabolism of acetone in the rat. *J. Biol. Chem.* 259:231–236. 1984.

Fujino, A., et al. (1992) Biological monitoring of workers exposed to acetone in acetate fibre plants. *Br. J. Ind. Med.* 49:654–657.

Geiss, O., et al. (2014) Catalytic air freshening diffusers based on isopropyl alcohol—a major source of acetone indoors. *Aerosol Air Qual.* 14:177–184.

Gjolstad, M., et al. (2006) Occupational exposure to airborne solvents during nail sculpturing. *J. Environ. Monit.* 8: 537–542.

Iba, M.M., et al. (1993) Synergistic induction of rat microsomal CYP1A1 and CYP1A2 by acetone in combination with pyridine. *Cancer Lett.* 74:69–74.

Kawai, T., et al. (1992) Curvilinear relation between acetone in breathing zone air and acetone in urine among workers exposed to acetone vapor. *Toxicol. Lett.* 62:85–91.

Keltanen, T., et al. (2013) Assessment of Traub formula and ketone bodies in cause of death investigations. *Int. J. Legal Med.* 127:1131–1137.

Kiesswetter, E., et al. (1994) Acute exposure to acetone in a factory and ratings of well-being. *Neurotoxicology* 15:597–602.

Kobusch, A.B., Plaa, G.L., and Du Souich, P. (1989) Effects of acetone and methyl *n*-butyl ketone on hepatic mixed-function oxidase. *Biochem. Pharmacol.* 38:3461–3467.

Levin, J.-O. and Carleborg, L. (1987) Evaluation of solid sorbents for sampling ketones in work-room air. *Ann. Occup. Hyg.* 31:31–38.

Lewis, G.D., et al. (1984) Metabolism of acetone to isopropyl alcohol in rats and humans. *J. Forensic Sci.* 29:541–549.

Mast, T.J., et al. (1989) Developmental toxicity study of acetone in mice and rats. *Teratology* 39:468–476.

Maurer, J.K., et al. (2001) Pathology of ocular irritation with acetone, cyclohexanol, parafluoroaniline, and formaldehyde in the rabbit low-volume eye test. *Toxicol. Pathol.* 29:187–99.

Miller, K.W. and Yang, C.S. (1984) Studies on the mechanisms of induction of *N*-nitrosodimethylamine demethylase by fasting, acetone, and ethanol. *Arch. Biochem. Biophys.* 229:483–491.

Mitran, E., et al. (1997) Neurotoxicity associated with occupational exposure to acetone, methyl ethyl ketone, and cyclohexanone. *Environ. Res.* 73:181–8.

National Institute for Occupational Safety and Health (NIOSH) (2010) *Pocket Guide to Chemical Hazards*, NIOSH 94-116, Washington, DC: U.S. Department of Health and Human Services.

Niziaeva, I. V. (1982). Hygienic evaluation of acetone. *Gig. Tr. Prof. Zabol.* (6):24–28.

NTP 1988. *Inhalation developmental toxicology studies: teratology study of acetone in mice and rats*. National Toxicology Program Report No. PNL-6768. NTIS DE89-00567 1. Research Triangle Park, NC: U.S. Department of Health and Human Services, Public Health Service, National Institute of Health.

NTP (1991) *Toxicity studies of acetone in F344/N rats and B6C3F1 mice (drinking water studies)*. National Toxicology Program Technical Report No. 3. NIH Publication No. 91-3122. Research Triangle Park, NC: U.S. Department of Health and Human Services, Public Health Service, National Institute of Health.

Ott, M.G., et al. (1983) Health evaluation of employees occupationally exposed to methylene chloride. *Scand. J. Work Environ. Health* 9(Suppl. 1):8–16.

Pezzagno, G., et al. (1986) Urinary elimination of acetone in experimental and occupational exposure. *Scand. J. Work Environ. Health* 12:601–608.

Ramachandriya, K. D., Wilkins, M. R., Delorme, M. J., Zhu, X., Kundiyana, D. K., Atiyeh, H. K., et al. (2011) Reduction of acetone to isopropanol using producer gas fermenting microbes. *Biotechnol. Bioeng.* 108:2330–2338.

Righettoni, M. and Tricoli, A. (2011) Toward portable breath acetone analysis for diabetes detection. *J. Breath Res.* 5:1–8.

Whitehead, L.W., et al. (1984) Solvent vapor exposures in booth spray painting and spray gluing, and associated operations. *Am. Ind. Hyg. Assoc. J.* 45:767–772.

Yamai, K., et al. (2012) Acetone concentration in gas emanating from tails of diabetic rats. *Anal. Sci.* 28:511–514.

Zakova, N., et al. (1985) Evaluation of skin carcinogenicity of technical 2,2-bis-(*p*-glycidyloxyphenyl)-propane in CFI mice. *Food Chem. Toxicol.* 23:1081–1089.

CYCLOHEXANONE

Target organ(s): Eyes, skin, respiratory system, central nervous system, liver, kidneys

Occupational exposure limits: NIOSH—TWA 25 ppm; OSHA—TWA 50 ppm; ACGIH—TWA 20 ppm, STEL 50 ppm

Reference values: RfD—5 mg/kg/bw/day; RfC/MRL—not established

Hazard statements: H226—flammable liquid and vapor; H302—harmful if swallowed; H312—harmful in contact with skin; H332—harmful if inhaled; H315—causes skin irritation; H318—causes serious eye damage

Risk phrases: R10—flammable; R20—harmful by inhalation ($\geq 25\%$)

Safety phrases: S2—keep out of reach of children; S25—avoid contact with eyes

BACKGROUND AND USES

Cyclohexanone (symbol CH; anone, cyclohexyl ketone, pimelic ketone; CASRN 108-94-1) is a primarily used in industry, up to 96%, as a chemical intermediate in the production of nylons 6 and 66. Oxidation or conversion of cyclohexanone yields adipic acid and caprolactam, two of the immediate precursors to the respective nylons. Cyclohexanone can also be used as a solvent in a variety of products, including paints, lacquers, and resins. It has not been found to occur in natural processes.

PHYSICAL AND CHEMICAL PROPERTIES

Cyclohexanone is a colorless liquid with a water-white to pale yellow liquid with a peppermint or acetone-like odor. It is miscible with several other solvents. The chemical has a molecular weight of 98.2, a boiling point of 312 °F, and a freezing point of −49 °F. The vapor pressure is 5 mm Hg at 26.4 °C and the specific gravity is 0.95. The lower exposure limit is 1.1% and the upper exposure limit is 9.4%. Cyclohexanone may be incompatible with oxidizers and nitric acid.

MAMMALIAN TOXICOLOGY

Acute Effects

Exposure to cyclohexanone is known to cause acute effects that are primarily irritation. The most likely areas to be affected by high concentrations of the chemical are the eyes and respiratory system. It has the potential to act as a central nervous system depressant, although its effect is weaker than other ketones. Dizziness and unconsciousness can also result. Mice exposed to extremely high concentrations of cyclohexanone vapors were found to have lung edema and congestion (Gupta et al., 1979). These

concentrations are unlikely to occur occupationally. Neonatal rats were acutely exposed to cyclohexanone intravenously with no adverse health effects reported (Greener et al., 1987). Cutaneous or subcutaneous exposure of cyclohexanone over a period of several days resulted in cataracts in guinea pigs (NIOSH, 1988).

Chronic Effects

There are few available reports on the chronic toxicity of cyclohexanone in humans. Bruze et al. (1988) reported a handful of cases of sensitization contact allergies to a cyclohexanone resin in humans and a follow-up investigation with guinea pigs. Although positive results were found, the resin contained not only cyclohexanone, but other chemicals as well. No associations have been found between cyclohexanone exposure and teratogenicity or carcinogenicity. The ACGIH (2013) has given cyclohexanone an A3 rating, designating this compound is a confirmed animal carcinogen with unknown relevance to humans.

Animal studies involving cyclohexanone exposure are present in the literature. Subchronic intravenous administration of cyclohexanone in rats showed no adverse effects at relatively low doses similar to occupational exposure limits (Greener et al., 1982). A study of rats and mice by Lijinsky and Kovatch (1986) found only a weak carcinogenic effect at extremely high dose levels. There has been some indication that a cumulative toxic effect in mice may occur with cyclohexanone (Gupta et al., 1979). Mitran et al. (1997) conducted a study that showed there may be a relationship between cyclohexanone exposure and neurotoxicity.

Mechanism of Action(s)

The metabolism of cyclohexanone has been characterized in humans and animals. Major metabolites found after exposures to cyclohexanone are cyclohexanol, 1,2- and 1,4-cyclohexanediol, and their glucuronide conjugates (Martis et al., 1980; Sakata et al., 1989; Mraz et al., 1994). These can be detected in breath or in urine by gas chromatography with flame ionization detection. The biological half-life of cyclohexanone is short, leading to a fairly rapid clearance of this compound. Very little, if any, of the parent compound is detectable in plasma or urine.

Chemical Pathology

Research has examined the chemical pathology of cyclohexanone. It has been suspected to cause neurotoxicity in mammals. A study conducted by Koeferl et al. (1981) found that beagle dogs exposed to cyclohexanone developed signs of toxicity, including lacrimation, urination, defecation, stupor, ataxia, and convulsive movements. The chemical is also known to cause skin irritation after contact. It leads to dermatitis be defatting the skin after repeated contact (NLM, 2005). Itching, redness, and inflammation will be present at the site of exposure. Ingestion is not a typical route of exposure for cyclohexanone; however, it is possible. An individual that ingested the chemical developed an altered mental state, shock, acidosis, hepatitis, a renal dysfunction (Zuckerman et al., 1998). The route of exposure appears to determine what body systems are impacted.

SOURCES OF EXPOSURE

Occupational

Cyclohexanone is used in many different industrial processes that may lead to exposure. The chemical is used as to coat fabrics and plastics, in the cleaning of leathers and textiles, and as a solvent in different chemical formulations (NIOSH, 1988). Occupational exposure can occur by inhalation and skin or eye contact. There are two major methods used to control cyclohexanone concentrations in the work environment. Local exhaust ventilation is one way to control exposure. Contaminated air is captured and exhausted to the outside atmosphere. Personal protective equipment is the other way to control exposure. Personal protective equipment, including aprons, gloves, and goggles, prevent the chemical from contacting exposed body parts. Exposed workers should be removed from the contaminated environment and provide medical care. Skin and eyes should be flushed with large amounts of water if they are contacted by cyclohexanone. Personal respiratory protection may be necessary for work in areas where cyclohexanone concentrations are high.

Environmental

Cyclohexanone exposure can occur during the manufacture, processing, distribution, and disposal of the chemical. The use of a closed system can prevent the release of the chemical into the environment. Cyclohexanone can be emitted into the atmosphere during such industrial processes. Those living in surrounding areas could be at risk of inhaling contaminated air or coming into contact with contaminated water. Since cyclohexanone can be used as a solvent, the chemical can evaporate readily into the air. Environmental contamination of the soil and water is possible if cyclohexanone is not disposed off properly. The chemical can volatilize from both soil and water leading to potential inhalation. The National Library of Medicine estimates that the half-life of cyclohexanone in a model river and model lake is 4.1 and 33 days, respectively (NLM, 2005). Industrial companies utilizing the chemical should follow proper procedures to prevent the release of the chemical into the environment.

INDUSTRIAL HYGIENE

Occupational exposure to cyclohexanone is most likely to occur through inhalation, although dermal contact may also have adverse health effects. With its strong odor, cyclohexanone can be detected well below harmful concentrations. The odor threshold for this compound is $0.480\,mg/m^3$ (Ruth, 1986). Air-supplied respirators and area ventilation may be necessary for workers exposed to cyclohexanone. Skin protection should be used when contact is anticipated. Airborne concentrations of cyclohexanone can be monitored using charcoal tubes and carbon disulfide solvent. Contaminants are extracted with simple adsorption techniques followed by gas chromatography analysis.

The OSHA PEL for atmospheric concentrations of cyclohexanone is 50 ppm ($200\,mg/m^3$), the NIOSH (2010) REL is 25 ppm ($100\,mg/m^3$), and the ACGIH (2013) TLV is 20 ppm ($80\,mg/m^3$). The NIOSH and the ACGIH recommendations also include a skin designation.

MEDICAL MANAGEMENT

NIOSH recommends several practices to protect workers from occupational exposure: medical surveillance, preplacement medical evaluation, periodic screening, biological monitoring, and follow-up medical practices at job transfer or termination of employment (NIOSH, 1988). The development of contact and/or allergic dermatitis should be considered a sentinel health event and should warrant investigation into the cause. If cyclohexanone contacts the eyes or the skin, copious amounts of water should be used to flush the chemical from the affected area. Exposed workers should seek prompt medical attention. The metabolites of cyclohexanone can be used as biomarkers for occupational exposure. Ong et al. (1991) found a significant correlation between urinary cyclohexanol and environmental concentrations ($r = 0.77$). A lesser association was found with exhaled breath concentrations of cyclohexanone. Chemical protective clothing should be used to prevent contact with the skin. Face shields or safety glasses should be used to prevent the chemical from splashing into the eyes.

REFERENCES

American Conference of Governmental Industrial Hygienists (ACGIH) (2013) *Threshold Limit Values (TLVs) for Chemical Substances and Physical Agents and Biological Exposure Indices (BEIs)*, Cincinnati, OH: American Conference of Governmental Industrial Hygienists.

Bruze, M., et al. (1988) Contact allergy to a cyclohexanone resin in humans and guinea pigs. *Contact Dermatitis* 18:46–49.

Greener, Y., Martis, L., and Indacochea-Redmond, N. (1982) Assessment of the toxicity of cyclohexanone administered intravenously to Wistar and Gunn rats. *J. Toxicol. Environ. Health* 10:385–396.

Greener, Y., et al. (1987) Assessmenl of the safety of chemicals administered intravenously in the neonatal rat. *Teratology* 35:187–194.

Gupta, P.K., et al. (1979) Toxicological aspects of cyclohexanone. *Toxicol. Appl. Pharmacol.* 49:525–533.

Koeferl, M.T., et al. (1981) Influence of concentration and rate of intravenous administration on the toxicity of cyclohexanone in beagle dogs. *Toxicol. Appl. Pharmacol.* 59:215–29.

Lijinsky, W. and Kovatch, R.M. (1986) Chronic toxicity study of cyclohexanone in rats and mice. *J. Natl. Cancer Inst.* 77:941–949.

Martis, L., et al. (1980) Disposition kinetics of cyclohexanone in beagle dogs. *Toxicol. Appl. Pharmacol.* 55:545–553.

Mitran, E., et al. (1997) Neurotoxicity associated with occupational exposure to acetone, methyl ethyl ketone, and cyclohexanone. *Environ. Res.* 73:181–188.

Mraz, J., et al. (1994) Markers of exposure to cyclohexanone, cyclohexane, and cyclohexanol: 1,2- and 1,4-cyclohexanediol. *Clin. Chem.* 40:1466–1468.

National Institute for Occupational Safety and Health (NIOSH) (1988) *Occupational Safety and Health Guideline for Cyclohexanone*, NIOSH 1-6, Washington, DC: U.S. Department of Health and Human Services.

National Institute for Occupational Safety and Health (NIOSH) (2010) *Pocket Guide to Chemical Hazards*, NIOSH 94-116, Washington, DC: U.S. Department of Health and Human Services.

National Library of Medicine (NLM) (2005) *Cyclohexanone*, NLM 1-1, Bethesda, MD: National Institutes of Health.

Ong, C.N., et al. (1991) Monitoring of exposure to cyclohexanone through the analysis of breath and urine. *Scand. J. Work Environ. Health* 17:430–435.

Ruth, J.H., (1986) Odor thresholds and irritation levels of several chemical substances: a review. *Am. Ind. Hyg. Assoc. J.* 47: A142–A151.

Sakata, M., Kikuchi, J., and Haga, M. (1989) Disposition of acetone, methyl ethyl ketone and cyclohexanone in acute poisoning. *J. Toxicol. Clin. Toxicol.* 27:67–77.

Zuckerman, G.B., et al. (1998) Rhabdomyolysis following oral ingestion of the hydrocarbon cyclohexanone. *J. Environ. Pathol. Toxicol. Oncol* 17:11–5.

METHYL *n*-BUTYL KETONE

Target organ(s): Eyes, skin, respiratory system, central nervous system, peripheral nervous system

Occupational exposure limits: NIOSH—TWA 1 ppm; OSHA—TWA 100 ppm

ACGIH—TWA 5 ppm, STEL 10 ppm

Reference Values: RfD/MRL—not established; RfC— 3 mg/m³

Risk phrases: R10—flammable; R48/23—toxic: danger of serious damage to health by prolonged exposure through inhalation; R62—possible risk of impaired fertility; R67—vapors may cause drowsiness and dizziness

Safety phrases: S1/2—keep locked up and out of the reach of children; S36/37—wear suitable protective clothing and gloves; S45—in case of accident or if you feel unwell, seek medical advice immediately (show the label where possible)

BACKGROUND AND USES

Methyl *n*-butyl ketone (symbol MBK; 2-hexanone, butyl methyl ketone, methyl butyl ketone; CASRN 591-78-6), like most monoketones, is used extensively as an industrial solvent for lacquers, resins, and oils in industries such as plastics, coatings, and leather manufacturing. Methyl *n*-butyl ketone is found in nature in low concentrations and in foodstuffs, such as blue cheese and nectarines.

PHYSICAL AND CHEMICAL PROPERTIES

Methyl *n*-butyl ketone is a colorless liquid with an odor similar to acetone, giving it a low odor threshold. It is slightly soluble in water at 2% by weight and is more soluble in other organic solvents. The chemical has a molecular weight of 100.2, a boiling point of 262 °F, and a freezing point of −71 °F. The vapor pressure is 11.6 mm Hg at 25 °C and the specific gravity is 0.81. The lower flammable limit is 1.2%, and the upper flammable limit is 8%. Methyl *n*-butyl ketone may be incompatible with strong oxidizers.

MAMMALIAN TOXICOLOGY

Acute Effects

Methyl *n*-butyl ketone can be a potent lacrimator at high concentrations and can act as an irritant to the mucous membranes, skin, and eyes. High concentrations of the chemical will be intolerable to the average worker. Central nervous system depression may result from repeated exposures to high concentrations. DiVincenzo et al. (1978) demonstrated methyl *n*-butyl ketone is readily absorbed in humans through the skin (between 75 and 92%), the lungs, and the gastrointestinal tract. Methyl *n*-butyl ketone may also result in coughing, wheezing, headache, dizziness, lightheadedness, and passing out (NJDOH, 2009). Individuals suffering from acute exposure to the chemical should be removed from the area of exposure immediately.

Chronic Effects

Chronic occupational exposure to MBK may be a hazardous to worker health. Unlike other ketones of low toxicity, methyl *n*-butyl ketone has been found to be a potent neurotoxin. Sentinel cases of peripheral neuropathy resulting from solvent exposure dating back to the early 1970s. This prompted epidemiological investigations in populations exposed to, among other solvents, methyl *n*-butyl ketone. Reports by Billmaier et al. (1974) and Allen et al. (1975) were among the first to identify the relationship between methyl *n*-butyl ketone and the polyneuropathies seen among workers in plastic coating factories. Populations of workers were screened with a variety of neurological tests and rated according to the severity of the disease. Muscle weakness, sensory and motor dysfunction, and distal electromyographic abnormalities were the most common health effects noted in investigations. Reflex loss was minimal. No relevant human or animal data existed to suggest a causal association with methyl *n*-butyl ketone. Methyl *n*-butyl ketone along with other solvents has been associated with motor neuron disease in the leather industry and in association with spray painting (Mallov, 1976; Hawkes et al., 1989). The association between exposure to methyl *n*-butyl ketone and the development of neurological abnormalities was discovered in subsequent animal studies.

Mendell et al. (1974) reported data from chickens, rats, and cats exposed to 200, 600, and 600 ppm methyl *n*-butyl ketone, respectively. Clinical weakness was found in all species with cats experiencing some slowing of nerve conduction. Inhalation studies in a rat population showed clinical weakness in the hindlimbs and pathological examination revealed a degeneration of nerve fibers (Spenser et al., 1975). Another study exposed monkeys and rats to 1000 ppm and 100 ppm for up to 10 months 5 days a week for 6 h (Johnson et al., 1977). A progressive decrease in motor conduction velocity developed in monkeys from the high-dose group. Similar but less pronounced effects were found lower exposure groups. Some behavioral effects and weight loss were also noted, although no effects were observed in other neurological tests such as visual effects, electroencephalogram readings, or absolute refractory time. No investigations concerning the carcinogenicity or teratogenicity of this compound have been documented.

Mechanism of Action(s)

The identification of neurotoxic properties has made the study of methyl *n*-butyl ketone metabolism a topic of interest to researchers. The main metabolites of methyl *n*-butyl ketone identified in animal systems are 5-hydroxy-2-hexanone, 2-hexanol, and 2,5-hexanedione, a diketone that is also a metabolite of *n*-hexane (DiVincenzo et al., 1976; Granvil et al., 1994). 2-Hexanol is created through a reduction reaction with the oxidative pathway first producing 5-hydroxy-2-hexanone, followed by a further oxidation to 2,5-hexanedione. Both methyl *n*-butyl ketone and *n*-hexane cause similar neuropathies, and this common 2,5-hexanedione metabolite became the focus of methyl *n*-butyl ketone toxicity studies.

A study of rodents by Couri et al. (1978) demonstrated that oxidation of methyl *n*-butyl ketone to 2,5-hexanedione occurred in the microsomal fraction of the liver and that the reduction of 2-hexanol takes place in the cytosol. It has been shown in mice that, although 2,5-hexanedione appears not to be metabolized further, 2-hexanol may be oxidized back to methyl *n*-butyl ketone or to 2,5-hexane-dione (Granvil et al., 1994). Formation of 2,5-hexanedione is initially slower than 2-hexanol, but the total concentration of 2,5-hexanedione produced may be greater. Krasavage et al. (1978) found a significant relationship between relative neurotoxicity and serum levels of the 2,5-hexanedione metabolite in rodents.

Chemical Pathology

Methyl *n*-butyl ketone research has determined that there is an association between chemical exposure and the development of neuropathies. A prominent feature of these neuropathies is a portion of the axons in the nervous system undergo a "dying back" phenomenon. Distal to proximal dying back axonopathy results when the metabolite 2,5-hexanedione inhibits rapid axon protein transport and glutaraldehyde-3-phosphate dehydrogenase. Neurofilament accumulation, axonal enlargement, and retraction of the myelin sheath can result when the neuropathy is mainly associated with long distal myelinated nerves. A distal motor neuropathy may be formed in the upper and lower extremities. Weakness and loss of sensation of touch and temperatures are insidious in onset and typically symmetrical. Symptoms may progress even after exposure ceases. Confusion, incoordination, and changes in color vision may also be observed when the central nervous system is affected. Reclamation of both central nervous system and peripheral losses may be slow and take several years.

Examination of nerve biopsies from workers with peripheral neuropathy thought to be the result of methyl *n*-butyl ketone exposure revealed accumulation of neurofilaments and enlarged axons (Davenport et al., 1976). Chronic intoxication of cats with methyl *n*-butyl ketone also showed axon swelling with neurofilaments and changes in myelin sheaths and overt fiber breakdown (Spenser and Schaumberg, 1976). Axonal transport is blocked by these accumulations. Dorsal neural tubes were found to be malformed in chick embryos exposed to 2-5-hexanedione, the metabolite of methyl *n*-butyl ketone, indicating a direct effect on nervous system development (Cheng et al., 2012). *n*-Hexane neuropathies are known to have similar pathological findings.

SOURCES OF EXPOSURE

Occupational

Methyl *n*-butyl ketone is widely used as a solvent in industry. It can be found in paints, varnishes, lacquers, and waxes. There are two primary techniques used for methyl *n*-butyl ketone synthesis. The first is a 75% yield reaction of acetyl chloride with butyl magnesium chloride. The second reaction involves a catalyzed reaction of acetic acid and ethylene followed by a distillation reaction to purify the product. Synthesis and industrial application can lead to occupational exposures of methyl *n*-butyl ketone. Occupational exposure to methyl *n*-butyl ketone is likely to occur primarily through vapor inhalation. Dermal exposure is considered as a secondary route and can occur if the chemical is splashed on the skin and not removed. Unintentional ingestion of this compound is not likely. The probability of acute overexposure is remote since the chemical has a strong odor. The strong odor helps to indicate hazardous concentrations of methyl *n*-butyl ketone.

Environmental

Similar to other solvents, methyl *n*-butyl ketone can be released into the environment by a variety of different waste streams. Because of its volatility, vapors are easily released into the atmosphere. The half-life of methyl *n*-butyl ketone in air is estimated to be 2 days (NCBI, 2014). Individuals using the chemical should do so in an area with adequate ventilation. The use of personal respiratory protection may also be necessary. Methyl *n*-butyl ketone volatizes from both soil and water as well. The volatilization of the chemical from the soil is directly dependent on its vapor pressure and the half-life in a model river and a model lake is 7 h and 164 h, respectively (NCBI, 2014). The chemical is not believed to bioaccumulate in aquatic organisms. Those using methyl *n*-butyl ketone should ensure that the chemical is not spilled. If it is spilled, the chemical should be cleaned up immediately to prevent it from negatively impacting the surrounding environment.

INDUSTRIAL HYGIENE

Engineering controls for methyl *n*-butyl ketone should include adequate ventilation to limit inhalation exposures. Proper ventilation will also prevent exposure of the compound to excessive heat and/or sparks, as this compound is flammable. Workers who come into contact with methyl *n*-butyl ketone should be provided with adequate protection, including protective clothing and a supplied-air respirator with a full face piece. The NIOSH (2010) guideline (REL) of 1 ppm (4 mg/m^3) for airborne concentrations of methyl *n*-butyl ketone is significantly lower than the OSHA PEL of 100 ppm (400 mg/m^3). The ACGIH (2013) has set the TLV for methyl *n*-butyl ketone at 5 ppm (mg/m^3) with a STEL of 10 ppm (41 mg/m^3).

MEDICAL MANAGEMENT

Workers diagnosed with peripheral neuropathy in these cases were observed to have a similar pattern of distal motor and sensory disorder with insignificant reflex loss (Allen et al., 1975). Muscle weakness and atrophy, decreased nerve conduction velocities, and loss of body weight are classic signs and symptoms indicating methyl *n*-butyl ketone exposure. For suspected cases of peripheral neuropathy, abnormal electromyogram, and nerve biopsies are useful diagnostic tools. There is no direct treatment for this condition, although cessation of exposure to methyl *n*-butyl ketone is required. A majority of animal and human subjects experienced total recovery from methyl *n*-butyl ketone-induced peripheral neuropathy after a period of discontinued exposure.

Few studies have developed reliable tests to measure worker exposure to methyl *n*-butyl ketone even with its positive association with neuropathies. Plasma and tissue concentrations of methyl *n*-butyl ketone in rats were found to correlate with exposure concentrations in a dose-dependent manner (Duguay and Plaa, 1995). Detection of significant levels of the metabolite 2,5-hexanedione were found to relate to exposure route, particularly via inhalation. Urinary concentrations of 2,5-hexanedione are more readily measurable than unchanged methyl *n*-butyl ketone (DiVincenzo et al., 1978).

The presence of 2,5-hexanedione in the metabolic pathway of *n*-hexane confounds the detection of this compound to indicate exposure to only methyl *n*-butyl ketone. In addition, the extremely low levels of exposure (1–5 ppm) that occur in the workplace have not been shown to correlate with current biomonitoring techniques (Bos et al., 1991). Occupational exposure to methyl *n*-butyl ketone should be carefully monitored because of its direct effect on the central and peripheral nervous system. Methyl *n*-butyl ketone also interacts with other industrial chemicals and may affect the biotransformation of these same materials. Other ketones, such as methyl ethyl ketone, have been shown to potentiate the effects of the neuropathy (Abdel-Rahman et al., 1976).

REFERENCES

Abdel-Rahman, M.S., Hetland, L.B., and Couri, D. (1976) Toxicity and metabolism of methyl *n*-butyl ketone. *Am. Ind. Hyg. Assoc. J.* 37:95–102.

Allen, N., et al. (1975) Toxic polyneuropathy due to methyl *n*-butyl ketone. *Arch. Neurol.* 32:209–218.

American Conference of Governmental Industrial Hygienists (ACGIH) (2013) *Threshold Limit Values (TLVs) for Chemical*

Substances and Physical Agents and Biological Exposure Indices (BEIs), Cincinnati, OH: American Conference of Governmental Industrial Hygienists.

Billmaier, D., et al. (1974) Peripheral neuropathy in a coated fabrics plant. *J. Occup. Med.* 16:665–671.

Bos, P.M.J., Mik, G.D.M., and Bragt, P.C. (1991) Critical review of the toxicity of methyl *n*-butyl ketone: risk from occupational exposure. *Am. J. Ind. Med.* 20:175–194.

Cheng, X., et al. (2012) Exposure to 2,5-hexandione can induce neural malformations in chick embryos. *Neurotoxicology* 33:1239–1247.

Couri, D., Abdel-Rahman, M.S., and Hetland, L.B. (1978) Biotransformation of *n*-hexane and methyl *n*-butyl ketone in guinea pigs and mice. *Am. Ind. Hyg. Assoc. J.* 39:295–300.

Davenport, J.G., Farrell, D.F., and Sumi, S.M. (1976) "Giant axonal neuropathy" caused by industrial chemicals: neurofilamentous axonal masses in man. *Neurology* 26:919–923.

DiVincenzo, G.D., Kaplan, C.J., and Dedinas, J. (1976) Characterization of the metabolites of methyl *n*-butyl ketone, methyl *iso*-butyl ketone, and methyl ethyl ketone in guinea pig serum and their clearance. *Toxicol. Appl. Pharmacol.* 36:511–522.

DiVincenzo, G.D., et al. (1978) Studies on the respiratory uptake and excrelion and the skin absorption of methyl *n*-butyl ketone in humans and dogs. *Toxicol. Appl. Pharmacol.* 44:593–604.

Duguay, A.B. and Plaa, G.L. (1995) Tissue concentrations of methyl isobutyl ketone, methyl *n*-butyl ketone and their metabolites after oral and inhalation exposures. *Toxicol. Lett.* 75:51–58.

Granvil, C.P., Sharkawi, M., and Plaa, G.L. (1994) Metabolic fate of methyl *n*-butyl ketone, methyl isobutyl ketone and their metabolites in mice. *Toxicol. Pharmacol.* 70:263–267.

Hawkes, C.H., Cavanaugh, J.B., and Fox, A.J. (1989) Motoneuron disease: a disorder secondary to solvent exposure? *Lancet* 1:73–76.

Johnson, B.L., et al. (1977) Effects of methyl *n*-butyl ketone on behavior and the nervous system. *Am. Ind. Hyg. Assoc. J.* 38:567–579.

Krasavage, W.J., O'Donoghue, J.L., and Terhaar, C.J. (1978) The relative neurotoxicity of methyl *n*-butyl ketone and its metabolites. *Toxicol. Appl. Pharmacol.* 45:251–260.

Mallov, J.S. (1976) MBK neuropathy among spray painters. *JAMA* 235:1455–1457.

Mendell, J.R., et al. (1974) Toxic polyneuropathy produced by methyl *n*-butyl ketone. *Science* 185:787–789.

National Center for Biotechnology Information (NCBI) (2014) *Methyl* n-*Butyl Ketone*, CID 11583, Bethesda, MD: United States National Library of Medicine.

National Institute for Occupational Safety and Health (NIOSH) (2010) *Pocket Guide to Chemical Hazards*, NIOSH 94-116, Washington, DC: US Department of Health and Human Services.

New Jersey Department of Health (NJDOH) (2009) *Hazardous Substance Fact Sheet—Methyl* n-*Butyl Ketone*, NJDOH 1-6, Trenton, NJ: New Jersey Department of Health.

Spenser, P.S. and Schaumberg, H.H. (1976) Feline nervous system response to chronic intoxication with commercial grades of methyl *n*-butyl ketone, methyl isobutyl ketone and methyl ethyl ketone. *Toxicol. Appl. Pharmacol.* 37:301–311.

Spenser, P.S., et al. (1975) Nervous system degeneration produced by the industrial solvent methyl *n*-butyl ketone. *Arch. Neurol.* 132:219–222.

METHYL ETHYL KETONE

Target organ(s): Eyes, skin, respiratory system, central nervous system

Occupational exposure limits: NIOSH—TWA 200 ppm, ST 300 ppm; OSHA—TWA 200 ppm; ACGIH—TWA 200 ppm, STEL 300 ppm

Reference values: RfD—0.6 mg/kg/day; RfC—5 mg/m^3; MRL—not established

Hazard statements: H225—highly flammable liquid and vapor; H319—causes serious eye irritation; H336—may cause drowsiness or dizziness

Precautionary statements: P102—keep out of reach of children; P210—keep away from heat/sparks/open flames/ . . . /hot surfaces . . . no smoking; P233—keep container tightly closed; P240—ground/bond container and receiving equipment; P241—use explosive-proof electrical/ventilating/lighting/ . . . /equipment; P242—use only non-sparking tools; P243—take precautionary measures against static discharge; P261—avoid breathing dust/fume/gas/mist/vapors/spray; P264—wash . . . thoroughly after handling; P271—use only outdoors or in a well-ventilated area; P280—wear protective gloves/protective clothing/eye protection/face protection; P405—store locked up; P501—dispose of contents/container to . . . ; P303+P361+P353—if on skin (or hair), remove/take off immediately all contaminated clothing; rinse skin with water/shower; P304+P340—if inhaled, remove victim to fresh air and keep at rest in a position comfortable for breathing; P305+P351+P338—if in eyes, rinse cautiously with water for several minutes; remove contact lenses, if present and easy to do; continue rinsing; P312—call a poison center or doctor/physician if you feel unwell; P337+P313—if eye irritation persists, get medical advice/attention; P370+P378—in case of fire; use . . . to extinguish; P403+P233—store in a well-ventilated place; keep container tightly closed; P403+P235—store in a well-ventilated place; keep cool

Risk phrases: R11—highly flammable; R36—irritating to eyes; R66—repeated exposure may cause skin dryness or cracking; R67—vapors may cause drowsiness and dizziness

Safety phrases: S2—keep out of the reach of children; S9—keep container in a well-ventilated place; S16—keep away from sources of ignition . . . no smoking

BACKGROUND AND USES

Methyl ethyl ketone (symbol MEK; 2-butanone, ethyl methyl ketone, methyl acetone; CASRN 78-93-3) is an organic solvent of relatively low toxicity, which is found in many applications. It is used in industrial and commercial products as a solvent for adhesives, paints, and cleaning agents and as a de-waxing solvent. A natural component of some foods, methyl ethyl ketone can be released into the environment by volcanoes and forest fires.

PHYSICAL AND CHEMICAL PROPERTIES

Methyl ethyl ketone is a colorless liquid with an odor that has been described as moderately sharp, fragrant, peppermint, or acetone like. It soluble in water up to 28% by weight and is miscible with many other organic solvents. The chemical has a molecular weight of 72.1, a boiling point of 175 °F, and a freezing point of −123 °F. The vapor pressure is 90.6 mm Hg at 25 °C and the specific gravity is 0.81. The lower explosive limit is 1.4% and the upper explosive limit is 11.4%. Methyl ethyl ketone may be incompatible with strong oxidizers, amines, ammonia, inorganic acids, caustics, isocyanates, and pyridines. When used industrially, methyl ethyl ketone must be handled with caution, as it is a Class lB flammable liquid NIOSH (2010).

MAMMALIAN TOXICOLOGY

Acute Effects

The acute effects of methyl ethyl ketone are similar to acetone. Dermal contact to methyl ethyl ketone can have irritant effects, such as contact urticaria (Varigos and Nurse, 1986). If methyl ethyl ketone comes in contact with the eyes, corneal damage and irritation can occur. When concentrations rise above 200 ppm, eye, nose, and throat irritation occurs. Continual exposure to elevated concentrations may result in central nervous system depression. Most health hazards associated with methyl ethyl ketone occur with inhalation exposure. Implicating methyl ethyl ketone exposure in occupational illness is confounded by the coexposures to other solvents and ketones, including acetone, toluene, and n-hexane. Decrease or change in neurologic function has been noted in most reports. Cases of polyneuropathy in separate shoe factories were suspected to be associated with methyl ethyl ketone, but the author could not conclude whether methyl ethyl ketone was the cause of the disease or acted as a potentiator for acetone, toluene, or n-hexane coexposures (Dyro, 1978; Vallat et al., 1981). A more recent study of volunteers working in a factory found no significant relationship to short duration methyl ethyl ketone exposures

of 200 ppm and neurological effects measured by specific motor tasks (Dick et al., 1988, 1989). A synergistic effect of methyl ethyl ketone was described by Altenkirch et al. (1977) in toxic polyneuropathies associated with individuals sniffing glue thinner as a substitute for illegal drugs. n-Hexane and methyl butyl ketone were reported to be the compounds mainly responsible for the neurological damage. An individual reportedly ingested a high dose of methyl ethyl ketone and was found unconscious with some respiratory and circulatory distress and a high degree of lactic acidosis (Kopelman and Kalfayan, 1983). Sodium bicarbonate was issued as treatment and the patient recovered completely. No residual effects were observed. A 2-year prospective study found that individuals working in a screen printing business had significantly impaired motor functions on dexterity tests, decreased visual memory, and impaired mood following acute exposure to a variety of chemicals, including methyl ethyl ketone (White et al., 1995).

Chronic Effects

Chronic illness associated with methyl ethyl ketone has not been well documented. A retrospective cohort mortality study by Wen (1985) found a decrease in overall cancer deaths and no significant site-specific cancer increase associated with methyl ethyl ketone exposure. A study of rats exposed for up to 7 weeks to extremely high concentrations of methyl ethyl ketone vapor did not show any neurology-related effects, although the animals did die of bronchopneumonia (Altenkirch et al., 1979). This study also showed a similar type of coexposure potentiation with n-hexane-induced neuropathies. Another study found that individuals chronically exposed to chemicals, including methyl ethyl ketone exhibited significantly poor performance on visual memory tasks and mood (White et al., 1995). The implication of methyl ethyl ketone as the causative agent of neurological decline was confounded by coexposures to other solvents during the study. Inhalation studies of pregnant rats did not reveal any significant fetotoxic or embryotoxic effects at exposure levels causing maternal toxicity (Deacon et al., 1981; Stoltenburg-Didinger et al., 1990). Shoe workers exposed to multiple organic substances, including methyl ethyl ketone experienced cytogenetic damage to buccal cells (Burgaz et al., 2002). A potential association may exist between cellular damage and methyl ethyl ketone exposure; however, this could not be determined by the noted study due to coexposures with other solvents.

Mechanism of Action(s)

The metabolism and distribution of methyl ethyl ketone has been studied in both humans and animals. The analysis of excreted urine and exhaled breath may yield small amounts of unchanged methyl ethyl ketone. It is also oxidized by enzymes to 3-hydroxy-2-butanone and 2,3-butanediol and reduced to 2-butanol. Relatively little of these metabolites are detected in the urine or in expired air (Liira et al., 1990). According to a study by Liira et al. (1988), human volunteers exposed to 200 ppm of methyl ethyl ketone for 4 h excreted 3% of absorbed methyl ethyl ketone unchanged in exhaled breath and 2% excreted in the urine. Brown et al. (1987) examined the pharmacokinetics of methyl ethyl ketone in exposed volunteers and found a significant correlation between blood and breath levels. Assuming first order kinetics, the biological half-life for methyl ethyl ketone can be calculated to be approximately 49 min, which is much lower than that of acetone.

Chemical Pathology

Methyl ethyl ketone is known to cause irritation of eyes, skin, and respiratory system. When individuals were exposed to methyl ethyl ketone (200 ppm) along with methyl isobutyl ketone (100 ppm) for 4 h, odor sensations and irritant effects were reported (Dick et al., 1992). The polyneuropathies associated with methyl ethyl ketone are possibly a result of a potentiative effect of methyl ethyl ketone on the toxicity of other compounds. A study by Schmidt et al. (1984) showed more neurological changes in rats exposed to n-hexane and MEK than those exposed to n-hexane alone. Occupational exposure to methyl ethyl ketone is almost always concurrent with other solvents and compounds making it extremely difficult to determine which specific chemical may be responsible for pathological manifestations. The USEPA defines methyl ethyl ketone as a Group D (USEPA, 2000). This means that it is not classifiable as a human carcinogen based on human and animal data.

SOURCES OF EXPOSURE

Occupational

Methyl ethyl ketone is similar to other ketones in that it is widely used as an industrial solvent. It can be found in paints, paint removers, and adhesives. Methyl ethyl ketone is usually produced from a hydrogenation or selective oxidation reaction of sec-butyl alcohol. Major exposure routes include vapor inhalation and dermal contact. Ventilation and personal respiratory protection with organic cartridges are recommended to prevent inhalation exposure. Dermal contact can be prevented by using aprons, gloves, and goggles. Any skin that is exposed to methyl ethyl ketone should be cleansed immediately to limit the amount of the chemical that can be absorbed. Ingestion of methyl ethyl ketone is highly unlikely. The odor threshold for methyl ethyl ketone is low enough to alert workers of workplace concentrations far below those expected to be hazardous. Methyl ethyl ketone

does not typically remain in the body as it is metabolized and excreted in exhaled breath and urine.

Environmental

Much like other organic solvents, methyl ethyl ketone evaporates readily when exposed to the atmosphere. Individuals using the solvent should ensure that there is adequate ventilation to prevent chemical concentrations from building up. If ventilation is not available, personal respiratory protection may be necessary. Organic vapor cartridges must be attached to respirators to prevent inhalation exposure. Individuals using methyl ethyl ketone should avoid spilling the chemical as it could lead to negative environmental impact. The chemical can leach into groundwater and surrounding soil. Methyl ethyl ketone readily dissolves when mixed with water but can evaporate from both water and soil. Microorganisms present in both water and soil can breakdown the chemical and it is not likely to bioaccumulate in plants or animals (USEPA, 1994). Methyl ethyl ketone has the potential to contribute to the development of smog in urban and industrial areas when it reacts with organic substance in the atmosphere.

INDUSTRIAL HYGIENE

Methyl ethyl ketone is widely used in various occupational settings to exposure can easily occur. Workers likely to be exposed to high concentrations of methyl ethyl ketone vapors are advised to wear personal respirators with the appropriate filtering system. Proper ventilation may also be necessary to remove contaminated air from the work environment. Protective clothing, including aprons and gloves, can prevent dermal contact with the chemical. Irritation and injury may result if the chemical comes in contact with the eyes, therefore goggles should be worn. In a study by Vahdat et al. (1995) solid adsorbent tubes were found to be efficient and accurate for detecting atmospheric concentrations of methyl ethyl ketone. OSHA methods 16, 84, and 1004 as well as NIOSH method 2500 are validated sampling procedures for methyl ethyl ketone. Workplace allowable atmospheric concentrations of methyl ethyl ketone are similar for all guidelines. The ACGIH, NIOSH, and OSHA have all set their respective guideline values (TLV, REL, and PEL) at 200 ppm (590 mg/m^3), with a STEL for NIOSH and the ACGIH of 300 ppm (885 mg/m^3).

MEDICAL MANAGEMENT

Methyl ethyl ketone exposure can be monitored by examining contaminant levels in urine. Although the concentrations may be low, they may indicate that exposure has taken place. Ghittori et al. (1987) found a linear relationship between exposure TWA concentrations and urinary concentrations in subjects exposed to a variety of solvents, including methyl ethyl ketone. Ong et al. (1991) examined three different methods for monitoring exposure to methyl ethyl ketone and found that urine was the most reliable exposure indicator when compared to blood and exhaled breath. Another study found that urine was a better indicator of methyl ethyl ketone exposure when compared to blood (Yoshikawa et al., 1995). It can also be detected in expired air by gas chromatography (Brown et al., 1987). Individuals exposed to methyl ethyl ketone may experience irritation of the eyes, skin, and nose, as well as headache dizziness, and vomiting. They should be removed from the contaminated environment and provided medical attention immediately.

REFERENCES

Altenkirch, H., et al. (1977) Toxic neuropathies after sniffing a glue thinner. *J. Neurol.* 214:137–152.

Altenkirch, H., et al. (1979) Experimental data on the neurotoxicity of methyl-ethyl-ketone (MEK). *Experientia* 35:503–504.

American Conference of Governmental Industrial Hygienists (ACGIH) (2013) *Threshold Limit Values (TLVs) for Chemical Substances and Physical Agents and Biological Exposure Indices (BEIs)*, Cincinnati, OH: American Conference of Governmental Industrial Hygienists.

Brown, W.D., et al. (1987) Body burden profiles of single and mixed solvent exposures. *J. Occup. Med.* 29:877–883.

Burgaz, S., et al. (2002) Cytogenic analysis of buccal cells from shoe-workers and pathology and anatomy laboratory workers exposed to *n*-hexane, toluene, methyl ethyl ketone and formaldehyde. *Biomarkers* 7:151–161.

Deacon, M.M., et al. (1981) Embryo- and fetotoxicity of inhaled methyl ethyl ketone in rats. *Toxicol. Appl. Pharmacol.* 159:620–622.

Dick, R.B., et al. (1988) Effects of short duration exposures to acetone and methyl ethyl ketone. *Toxicol. Lett.* 43:31–49.

Dick, R.B., et al. (1989) Neurobehavioral effects of cohort duration exposures to acetone and methyl ethyl ketone. *Br. J. Int. Med.* 46:111–121.

Dick, R.B., et al. (1992) Neurobehavioral effects from acute exposures to methyl isobutyl ketone and methyl ethyl ketone. *Fundam. Appl. Toxicol.* 19:453–473.

Dyro, F.M. (1978) Methyl ethyl ketone polyneuropathy in shoe factory workers. *Clin. Toxicol.* 13:371–376.

Ghittori, S., et al. (1987) The urinary concentration of solvents as a biological indicator of exposure: proposal for the biological equivalent exposure limit for nine solvents. *Am. Ind. Hyg. Assoc. J.* 48:786–790.

Kopelman, P.G. and Kalfayan, P.Y. (1983) Severe metabolic acidosis after ingestion of butanone. *Br. Med. J. (Clin. Res. Ed.)* 286:21–22.

Liira, J., et al. (1988) Kinetics of methyl ethyl ketone in man: absorption, distribution, and elimination in inhalation exposure. *Int. Arch. Occup. Environ. Health* 60:195–200.

Liira, J., et al. (1990) Dose-dependent kinetics of inhaled methyethylketone in human. *Toxicol. Lett.* 50:195–201.

National Institute for Occupational Safety and Health (NIOSH) (2010) *Pocket Guide to Chemical Hazards*, NIOSH 94-116, Washington, DC: US Department of Health and Human Services.

Ong, C.N., et al. (1991) Biological monitoring of occupational exposure to methyl ethyl ketone. *Int. Arch. Occup. Environ. Health* 63:319–324.

Schmidt, R., et al. (1984) Ultrastructural alteration of intrapulmonary nerves after exposure to organic solvents. *Respiration* 46:362–369.

Stoltenburg-Didinger, G. Altenkirch, H., and Wagner, M. (1990) Neurotoxicity of organic solvent mixtures: embryotoxicity and fetotoxicity. *Neurotoxicol. Teratol.* 12:585–589.

United States Environmental Protection Agency (USEPA) (1994) *Chemicals in the Environment: Methyl Ethyl Ketone*, EPA 749-F-94-015, Washington, DC: United States Environmental Protection Agency.

United States Environmental Protection Agency (USEPA) (2000) *Methyl Ethyl Ketone (2-Butanone)*, EPA 78-93-3, Washington, DC: United States Environmental Protection Agency.

Vahdat, N., et al. (1995) Adsorption capacity and thermal desorption efficiency of selected adsorbents. *Am. Ind. Hyg. Assoc. J.* 56:32–38.

Vallat, J.M., et al. (1981) *n*-Hexane- and methylethylketone-induced polyneuropathy: abnormal accumulation of glycogen in unmyelinated axons: report of a case. *Acta Neuropathol. (Berl.)* 55:275–279.

Varigos, G.A. and Nurse, D.S. (1986) Contact urticaria from methyl ethyl ketone. *Contact Dermatitis* 15:259–260.

Wen, C.P. (1985) Long-term mortality study of oil refinery workers. IV. Exposure to the lubricating–dewaxing process. *J. Natl. Cancer Inst.* 74:11–18.

White, R.F., et al. (1995) Neurobehavioral effects of acute and chronic mixed-solvent exposure in the screen printing industry. *Am. J. Ind. Med.* 28:221–231.

Yoshikawa, M., et al. (1995) Biological monitoring of occupational exposure to methyl ethyl ketone in Japanese workers. *Arch. Environ. Contam. Toxicol.* 29:135–139.

METHYL ISOBUTYL KETONE

Target organ(s): Eye, skin, respiratory system, central nervous system, liver, kidneys

Occupational exposure limits: NIOSH—TWA 50 ppm, ST 75 ppm; OSHA—TWA 100 ppm; ACGIH—TWA 20 ppm, STEL 75 ppm

Reference values: RfD/MRL—not established; RfC—3 mg/m^3

Hazard statements: H225—highly flammable liquid and vapor; H332—harmful if inhaled; H319—causes serious eye irritation; H335—may cause respiratory irritation.

Precautionary statements: P210—keep away from heat/sparks/open flames/ . . . /hot surfaces . . . no smoking; P261—avoid breathing dust/fume/gas/mist/vapors/spray; P280—wear protective gloves/protective clothing/eye protection/face protection; P303+P361+P353—if on skin (or hair), remove/take off immediately all contaminated clothing; rinse with water/shower; P312—call a poison center or doctor/physician if you feel unwell; P403+P233—store in a well-ventilated place; keep container tightly closed.

Risk phrases: R11—highly flammable; R20—harmful by inhalation; R37—irritating to respiratory system; R66—repeated exposure may cause skin dryness or cracking.

Safety phrases: S2—keep out of the reach of children; S9—keep container in a well-ventilated place; S16—keep away from sources of ignition . . . no smoking; S29—do not empty into drains.

BACKGROUND AND USES

Methyl isobutyl ketone (symbol MIBK; hexone, isobutyl methyl ketone, 4-methyl-2-pentanone; CASRN 108-10-1) is an organic solvent similar in structure and use to methyl butyl ketone. In addition to its use as a solvent for paints, lacquers, and varnishes, methyl isobutyl ketone is used in extraction processes and as a denaturant for rubbing alcohol. Methyl isobutyl ketone is also used as a synthetic flavoring in some varieties of rum, candy, and cheese. Unlike methyl butyl ketone, methyl isobutyl ketone has not been found to occur naturally.

PHYSICAL AND CHEMICAL PROPERTIES

Methyl isobutyl ketone is a colorless liquid with a pleasant odor. The odor threshold can be as low as 0.10 ppm. It is 2% soluble in water by weight and with several other organic solvents. The chemical has a molecular weight of 100.2, a boiling point of 242 °F, and a freezing point of −120 °F. The vapor pressure is 19.9 mm Hg at 25 °C and the specific gravity is 0.80. The lower explosive limit is 1.2% and the upper explosive limit is 8.0% at 200 °F. Methyl isobutyl ketone may be incompatible with strong oxidizers and potassium tert-butoxide.

MAMMALIAN TOXICOLOGY

Acute Effects

Serious adverse health effects have not been associated with acute exposure to methyl isobutyl ketone. Respiratory tract and eye irritation have been known to result at high concentrations. Unnecessary vapor inhalation has been associated with anesthetic effects. These exposure endpoints were demonstrated in a study by Iregren et al. (1993) in 12 volunteers. After 2-h exposures to low and high concentrations of methyl isobutyl ketone, there was no effect on heart rate or performance of a reaction time task. However, the volunteers did display central nervous system depression and airway irritation. One case report described a sensorimotor neuropathy after a brief, high intensity exposure to methyl ethyl ketone and methyl isobutyl ketone. A study of open-air VOCs from a plastic recycling plant described a potential association with mucocutaneous and respiratory symptoms in individuals living nearby (Yorifuji et al., 2012). A study in primates of response time and accuracy of task performance showed a general decrease in both as a result of methyl isobutyl ketone exposure (Geller et al., 1979).

As noted previously, the distinct odor of methyl isobutyl ketone can be detected at low concentrations. Workers detecting the chemical can remove themselves from an area before concentrations become dangerous. Olfactory perception threshold and adaptation were measured in subjects exposed to methyl isobutyl ketone for 7 h (Gagnon et al., 1994). The study reports a temporary loss of olfactory perception and adaptation to methyl isobutyl ketone, which may hinder workers' ability to detect methyl isobutyl ketone vapors at higher concentrations. A study conducted by Dalton et al. (2000) determined that the best predictor of methyl isobutyl ketone irritation at high concentrations is related to odor rather than threshold for sensory irritation (Dalton et al., 2000). Olfactory fatigue is often common when an individual is exposed to a chemical for an extended period of time. Individuals experiencing acute methyl isobutyl ketone exposure should remove themselves from the contaminated area and seek fresh air.

Chronic Effects

Methyl isobutyl ketone does not produce the neuropathies seen after chronic methyl butyl ketone exposures even though it is a structural isomer of methyl butyl ketone. Clinically, the worker appeared to have a similar ailment, but the etiology of the effect is not certain (Aubuchon et al., 1979). Multiple solvent exposures make it extremely difficult to determine if a specific solvent is responsible for negative health effects or if the solvents act in a synergistic manner to produce adverse health effects. Johnson (2004) conducted a study that described neuropathological changes in distal portions of tibial and ulnar nerves in rats exposed to high concentrations (1500 ppm) of methyl isobutyl ketone for 5 months. With proper controls in place, it is unlikely for exposure to occur at concentrations as high.

Some animal studies of methyl isobutyl ketone toxicity are available. Spenser and Schaumberg (1976) found no nervous system damage in a study comparing methyl butyl ketone, methyl ethyl ketone, and methyl isobutyl ketone-associated effects. There have been some reports of methyl isobutyl ketone exposure acting as a potentiator of other solvent-induced neuropathies, such as n-hexane, and cholestasis-mediated hepatonecrogenesis caused by taurolithocholic acid or manganese–bilirubin concentrations (Lapadula et al., 1991; Duguay and Plaa, 1993). A study of paint factory workers found that those exposed to a variety of organic solvents were 63.04% more likely to develop neuropsychological symptoms when compared to controls (El Hamid Hassan et al., 2013). According to Johnson (2004), several mammalian mutagenicity tests for methyl isobutyl ketone were negative and the chemical did not induce increases in embryotoxicity or fetal malformations. The same study indicated that kidney and liver changes were seen following exposure to the chemical. Additional research is needed to further determine chronic effects of methyl isobutyl ketone. Multiple exposures in the workplace are common, and care should be taken to prevent these potential health hazards.

Mechanism of Action(s)

Methyl isobutyl ketone metabolism has been characterized in some studies and the metabolites identified in animal studies. The two main metabolites of methyl isobutyl ketone metabolism are 4-methyl-2-pentanol (MPOL) and 4-hydroxy-4-methyl-2-pentanone (HMP) (DiVincenzo et al., 1976; Granvil et al., 1994). The carbonyl group of methyl isobutyl ketone is reduced to form MPOL, whereas the HMP metabolite is produced via an oxidation reaction. These biotransformations occur via alcohol dehydrogenase and P450-dependent monooxygenases (Vezina et al., 1990). Methyl isobutyl ketone induces cytochrome P450 expression and thereby increases the hepatotoxicity observed with carbon tetrachloride and chloroform. It has been suggested that methyl isobutyl ketone can cause adverse effects on the kidneys and the testes in males. The chemical was found to increase protein droplets, accumulation of α2u-globulin, and renal cell proliferation, indicating that exposure to methyl isobutyl ketone may be associated with the development of nephropathy (Borghoff et al., 2009). Rats exposed to daily doses of methyl isobutyl ketone, 300 or 600 mg/kg/day, on their tails for 4 months developed changes in the testes and a reduction of spermatocytes, spermatids, and spermatozoa (Johnson, 2004).

Chemical Pathology

Tissue concentrations of methyl isobutyl ketone have been shown to increase in a dose-dependent manner; however, the observed concentrations of each metabolite depend on the route of exposure (Duguay and Plaa, 1995). MPOL is found in higher concentrations in rats after inhalation exposure in relation to oral exposure. The transport and distribution of methyl isobutyl ketone in the body, which occur rapidly, may be dependent on the proteins of the red blood cells, such as hemoglobin (Lam et al., 1990). Methyl isobutyl ketone may play a role in the development of neurological impairment. One study explains that monoketones increase lipid fluidity in synaptic membranes which can disrupt the function of receptors and enzymes (USEPA, 2003). The loss of receptor and enzyme function may possibly be related to neuropathological findings. There has been consideration that methyl isobutyl ketone may lead to liver and renal impairment but a definitive association between exposure and pathological findings has not been established.

SOURCES OF EXPOSURE

Occupational

Methyl isobutyl ketone is used as a solvent in a variety of applications, including in paint, resins, varnishes, and lacquers. It has also been used as a synthetic flavoring in some food applications. Methyl isobutyl ketone is produced industrially via a hydrogenation reaction of mesityl oxide over nickel at a high temperature. It is also an oxidation product of methyl isobutyl carbinol. The major routes of exposure in the occupational setting are through inhalation and dermal contact. Ventilation and personal respiratory protection is recommended to prevent worker exposure to the chemical. Dermal contact can occur if the chemical reaches an individual's skin. Protective clothing, including aprons, gloves, and goggles, can prevent dermal exposure to methyl isobutyl ketone. Ingestion of methyl isobutyl ketone is highly unlikely. The chemical has a noticeable odor capable of alerting workers before concentrations rise to unacceptable levels. Workers exposed to the chemical should be removed from the contaminated work area and provided medical attention.

Environmental

Methyl isobutyl ketone is another solvent that may be released into the environment during manufacturing processes. It is also found in exhaust gas produced by internal combustion engines. Environmental inhalation of methyl isobutyl ketone can occur from these sources. Individuals using methyl isobutyl ketone should ensure that there is adequate ventilation to prevent vapor concentrations from building up. Personal respiratory protection with vapor cartridges may also be used to prevent exposure. Landfills and waterways may contain traces of methyl isobutyl ketone as a result of waste dumping. The chemical can leach into both the soil and water thereby negatively impacting the surrounding environment. Once in the soil or water, methyl isobutyl ketone can volatize into the air. It has been estimated that the half-life of the chemical in a model river and a model lake are 9 h and 6 days, respectively (NCBI, 2014). Attempts should be made to avoid spilling the chemical and after use it should be disposed of in a proper manner.

INDUSTRIAL HYGIENE

The main routes of exposure to methyl isobutyl ketone are inhalation and dermal. The odor threshold of this compound is low enough to alert workers of possible overexposure. The sharp odor produced by methyl isobutyl ketone has been characterized to have an odor detection threshold of 0.410 mg/m^3 (1 ppm) and is an irritant at 410.0 mg/m^3 (Ruth, 1986). Proper ventilation and personal respirators equipped with proper vapor cartridges are recommended to control exposure. Like methyl butyl ketone, methyl isobutyl ketone is also flammable. All attempts should be made to ensure that the chemical does not come into contact with excessive heat, flames, or sparks. Explosive mixtures in air can be formed from vapors. Adequate clothing protection, barrier creams, and proper gloves can easily prevent dermal contact with methyl isobutyl ketone. The PEL set by OSHA for methyl isobutyl ketone air levels in the workplace is 100 ppm (410 mg/m^3), which is the concentration identified as the threshold for irritant effects. The NIOSH (2010) recommended exposure limit is set at 50 ppm with a short-term exposure limit of 75 ppm. The ACGIH (2013) TLV has been set at 20 ppm (82 mg/m^3), with a STEL of 75 ppm (300 mg/m^3).

MEDICAL MANAGEMENT

Monitoring workers for plasma concentrations of methyl isobutyl ketone or its metabolites can be an effective way to evaluate worker exposure. Metabolites in the urine have not yet been characterized for this compound. It may be possible to detect methyl isobutyl ketone exposure by analyzing the urine for the unchanged chemical rather than its metabolites. A linear relationship has described for methyl isobutyl ketone in urine with environmental exposure. A hypothetical example would be that if an exposure occurred at 50 ppm, the mathematical equation predicts that the methyl isobutyl ketone concentration in urine should be 1.7 mg/l (Imbriani and Ghittori, 2005). If methyl isobutyl ketone contacts the skin, contaminated clothing should be removed and the skin should be rinsed thoroughly with soap and water to remove the chemical. If the chemical comes in contact with the eyes, they should be rinsed thoroughly with copious amounts of water. Medical attention should be sought if eye contact occurs. If methyl isobutyl ketone is inhaled, fresh air and medical attention is recommended. Proper controls in the workplace can prevent exposure to methyl isobutyl ketone.

REFERENCES

American Conference of Governmental Industrial Hygienists (ACGIH) (2013) *Threshold Limit Values (TLVs) for Chemical Substances and Physical Agents and Biological Exposure Indices (BEIs)*, Cincinnati, OH: American Conference of Governmental Industrial Hygienists.

AuBuchon, J., Ian Robins, H., and Viseskul, C. (1979) Peripheral neuropathy after exposure to methyl-isobutyl ketone in spray paint. *Lancet* 2:363–364.

Borghoff, S.J., et al. (2009) Methyl isobutyl ketone (MIBK) induction of α2u-globulin nephropathy in male, but not female rats. *Toxicology* 258:131–8.

Dalton, P.H., et al. (2000) Evaluation of odor and sensory irritation thresholds for methyl isobutyl ketone in humans. *AIHAJ* 61:340–50.

DiVincenzo, G.D., Kaplan, C.J., and Dedinas, J. (1976) Characterization of the metabolites of methyl *n*-butyl ketone, methyl *iso*-butyl ketone, and methyl ethyl ketone in guinea pig serum and their clearance. *Toxicol. Appl. Pharmacol.* 36:511–522.

Duguay, A.B. and Plaa, G.L. (1993) Plasma concentrations in methyl isobutyl ketone-potentiated experimental cholestasis after inhalation or oral administration. *Fundam. Appl. Toxicol.* 21:222–227.

Duguay, A.B. and Plaa, G.L. (1995) Tissue concentrations of methyl isobutyl ketone, methyl *n*-butyl ketone and their metabolites after oral and inhalation exposure. *Toxicol. Lett.* 5:51–58.

El Hamid Hassan, A.A., et al. (2013) Health hazards of solvents exposure among workers in paint industry. *OJSST* 3:87–95.

Gagnon, P., Mergler, D., and Lapare, S. (1994) Olfactory adaptation, threshold shift and recovery at low levels of exposure to methyl isobutyl ketone (MIBK). *Neurotoxicology* 15:637–642.

Geller, I., et al. (1979) Effects of acetone, methyl ethyl ketone and methyl isobutyl ketone on a match-to-sample task in the baboon. *Pharmacol. Biochem. Behav.* 1:401–406.

Granvil, C.P., Sharkawi, M., and Plaa, G.L. (1994) Metabolic fate of methyl *n*-butyl ketone, methyl isobutyl ketone and their metabolites in mice. *Toxicol. Lett.* 70:263–267.

Imbriani, M. and Ghittori, S. (2005) Gases and organic solvents in urine as biomarkers of occupational exposure: a review. *Int. Arch. Occup. Environ. Health* 78:1–19.

Iregren, A., Tesarz, M., and Wigaeus-Hjelm, E. (1993) Human experimental MIBK exposure: effects on heart rate, performance, and symptoms. *Environ. Res.* 63:101–108.

Johnson, W., Jr. (2004) Safety assessment of MIBK (methyl isobutyl ketone). *Int. J. Toxicol.* 23(Suppl. 1):29–57.

Lam, C.W., et al. (1990) Mechanism of transport and distribution of organic solvents in the blood. *Toxicol. Appl. Pharmacol.* 104: 117–129.

Lapadula, D.M., et al. (1991) Induction of cytochrome P450 isozymes by simultaneous inhalation exposure of hens to *n*-hexane and methyl *iso*-butyl ketone (MIBK). *Biochem. Pharmacol.* 41:877–883.

National Center for Biotechnology Information (NCBI) (2014) *Methyl Isobutyl Ketone*, Bethesda, MD: United States National Library of Medicine.

National Institute for Occupational Safety and Health (NIOSH) (2010) *Pocket Guide to Chemical Hazards*, NIOSH 94-116, Washington, DC: U.S. Department of Health and Human Services.

Ruth, J.H. (1986) Odor thresholds and irritation levels of several chemical substances: a review. *Am. Ind. Hyg. Assoc.* 47: A142–A151.

Spenser, P.S. and Schaumberg, H.H. (1976) Feline nervous system response to chronic intoxication with commercial grades of methyl *n*-butyl ketone, methyl isobutyl ketone and methyl ethyl ketone. *Toxicol. Appl. Pharmacol.* 37:301–311.

United States Environmental Protection Agency (USEPA) (2003) *Toxicological Review of Methyl Isobutyl Ketone*, EPA/635/R-03/002, Washington, DC: United States Environmental Protection Agency.

Vezina, M., et al. (1990) Potentiation of chloroform-induced hepatotoxicity by methyl isobutyl ketone and two metabolites. *Can. J. Physiol. Pharmacol.* 68:1055–1061.

Yorifuji, T., et al. (2012) Does open-air exposure to volatile organic compounds near a plastic recycling factory cause health effects? *J. Occup. Health* 54:79–87.

OTHER KETONES

ISOPHORONE

Target organ(s): Eyes, skin, respiratory system, central nervous system, liver

Kidneys

Occupational exposure limits: NIOSH—TWA 4 ppm; OSHA—TWA 25 ppm

ACGIH—STEL ceiling 5 ppm

Reference values: RfD—0.2 mg/kg/day; RfC—0.012 mg/m^3; MRL—0.2–3 mg/kg/day

Hazard statements: H351—suspected of causing cancer; H302—harmful if swallowed; H312—harmful in contact with skin; H319—causes serious eye irritation; H335—may cause respiratory irritation

Precautionary statements: P202—do not handle until all safety precautions have been read and understood; P271—use only outdoors or in a well-ventilated area; P281—use personal protective equipment as required; P308+P313—if exposed or concerned, get medical advice/attention; P403+P233—store in a well-ventilated place; keep container tightly closed

Risk phrases: R21/22—harmful in contact with skin and if swallowed; R36/37—irritating to eyes and respiratory system; R40—limited evidence of carcinogenic effect

Safety phrases: S13—keep away from food, drink, and animal feeding stuffs; S23—do not breath gas/fumes/vapors/spray; R36/37/39—wear suitable protective clothing, gloves, and eye/face protection; S46—if swallowed, seek medical advice immediately and show the container or label

Isophorone (CASRN 78-59-1), an alpha, beta-unsaturated ketone, is also an industrial solvent used in a variety of different applications. Most of the isophorone used in industry is used in vinyl coatings and inks. It is an insoluble, colorless to white liquid and has an odor similar to peppermint. Its characteristic odor is detectable at concentrations as low as 1.00 mg/m^3 (0.2 ppm) (Ruth, 1986). Although not highly volatile, with a vapor pressure of 0.438 mm Hg at 25 °C, isophorone is primarily an irritant to the eyes and respiratory system at low concentrations. It may also depress the central nervous system at higher concentrations. Air concentrations of 10 ppm are tolerated for 15 min, but concentrations of 25 ppm produce irritation of the upper airways and eyes. The threshold for eye and nose irritation falls between 35 and 55 ppm. Fatigue and malaise have been associated with air levels of 5–8 ppm. Inhaltion and dermal contact are the primary routes of worker exposure to isophorone. Adequate protection may include exhaust ventilation and respirators. The OSHA PEL for isophorone is 25 ppm (140 mg/m^3) and the NIOSH (2010) REL is 4 ppm (23 mg/m^3). The ACGIH (2013) has set a STEL for isophorone at 5 ppm (28 mg/m^3) with a designation of A3, animal carcinogen.

Chronic effects of exposure in humans have not been reported. The liver, lungs, and kidneys are considered the target organs expected to be affected by isophorone exposure. Animal studies indicate a possible sex-linked carcinogenic effect in male rats, although the mechanism of action is not genotoxic (Bucher et al., 1986). Renal tubular cell adenomas and adenocarcinomas were produced in animal studies. The mechanisms by which these carcinogenic effects are produced cannot be extrapolated to humans. Glutathione levels in the reproductive system were also shown to decrease in male rats exposed to isophorone (Gandy et al., 1990).

BENZOPHENONE

Target organ(s): Eyes, skin, liver

Occupational exposure limits: NIOSH, OSHA, ACGIH—not Established; derived no effect level (DNEL); DNEL inhalation, systemic effects, long term: 0.7 mg/m^3; DNEL dermal, systemic effects, long term: 0.1 mg/kg/bw/day

Reference Values: RfD, RfC, MRL—Not Established; DNEL inhalation, systemic effects, long term: 0.17 mg/m^3; DNEL dermal, systemic effects, long term: 0.05 mg/

kg/bw/day; DNEL oral, systemic effects, long term: 0.05 mg/kg-bw/day

Hazard statements: H411—toxic to aquatic life with long lasting effects: H373—may cause damage to organs through prolonged or repeated exposures

Precautionary statements: P308+P313—if exposed or concerned, get medical advice/attention; P281—use personal protective equipment as required; P260—do not breath dust/fume/gas/mist/vapors/spray; P273—avoid release to the environment; P391—collect spillage; P501—dispose of contents/container to . . .

Risk phrases: R48/22—harmful: danger of serious damage to health by prolonged exposure if swallowed; R51/53—toxic to aquatic organisms, may cause long-term adverse effects in the aquatic environment

Safety phrases: S45—in case of accident or if you feel unwell, seek medical advice immediately (show label where possible); S36—wear suitable protective clothing; S23—do not breath gas/fumes/vapor/spray; S61—avoid release to the environment; refer to special instructions/safety data sheets

A derivative of benzene, benzophenone (CASRN 119-61-9) is found mostly in the cosmetics industry. It is used as a constituent in perfumes and sunscreens. It has a unique and characteristic rose-like odor. While found in products that require dermal application, some derivatives of this compound found in sunscreens have been the cause of sensitivity responses to certain sunscreens (English et al., 1987). Human and animal studies related to benzophenone have not been routinely investigated. Dutta et al. (1993) reported liver toxicity associated with benzophenone exposure in guinea pigs. Alterations occurred that were similar to a chronic hepatitis, with areas of necrosis and regeneration. A study conducted in 2007 found an increased incidence in hepatocellular adenoma in both male and female mice (Rhodes et al., 2007).

REFERENCES

American Conference of Governmental Industrial Hygienists (ACGIH) (2013) *Threshold Limit Values (TLVs) for Chemical Substances and Physical Agents and Biological Exposure Indices (BEIs)*, Cincinnati, OH: American Conference of Governmental Industrial Hygienists.

Bucher, J.R., Huff, J., and Kluwe, W.M. (1986) Toxicology and carcinogenesis studies of isophorone in F344 rats and B6C3F1 mice. *Toxicology* 39:207–219.

Dutta, K., Das, M., and Rahman, T. (1993) Toxicological impacts of benzophenone on the liver of guinea pig (*Caivia porcellus*). *Bull. Environ. Contam. Toxicol.* 50:282–285.

English, J.S.C., White, I.R., and Cronin, E. (1987) Sensitivity to sunscreens. *Contact Dermatitis* 17:159–162.

Gandy, J. et al. (1990) Effects of selected chemicals on the glutathione status in the male reproductive system of rats. *J. Toxicol. Environ. Health* 29:45–57.

National Institute for Occupational Safety and Health (NIOSH) (2010) *Pocket Guide to Chemical Hazards*, NIOSH 94-116, Washington, DC: US Department of Health and Human Services.

Rhodes, M.C., et al. (2007) Carcinogenesis studies of benzophenone in rats and mice. *Food Chem. Toxicol.* 45:843–51.

Ruth, J.H. (1986) Odor thresholds and irritation levels of several chemical substances: a review. *Am. Ind. Hyg. Assoc.* 47: A142–A151.

56

ETHERS AND EPOXIDES

Erin L. Pulster, Jacob R. Bourgeois, and Raymond D. Harbison

ETHYL ETHER

Target organ(s): Eyes, skin, respiratory, central nervous system

Occupational exposure limits: OSHA PEL: 400 ppm (1200 mg/m^3) TWA; ACGHI TLV: 400 ppm (1210 mg/m^3) TWA, 500 ppm (1520 mg/m^3) STEL; NIOSH IDLH: 1900 ppm [LEL]

Reference values: RfD: 2E-1 mg/kg/day, NOAEL: 500 mg/kg/day; LOAEL: 2000 mg/kg/day (USEPA)

Risk/safety phrases: Highly flammable and volatile. Can form explosive peroxides if exposed to air and light for long periods. Incompatible with strong oxidizers, halogens, sulfur, and sulfur compounds

BACKGROUND AND USES

Ethyl ether dates as far back as the thirteenth century when it was first discovered by mixing ethanol and sulfuric acid, historically known as oil of vitriol. During the 1800s, it was first used widely as an anesthetic until the mid-1950s when it was eventually replaced by less flammable anesthetics. Although ethyl ether has a wide range of uses in the chemical industry, it is more commonly used as a laboratory solvent because of its limited solubility in water and its inertness. It is widely used for reactions or extractions in the chemical, fragrance, and pharmaceutical industries. It is commonly used as a solvent for dissolving waxes, resins, gums, fats, oils, dyes, raw rubber, smokeless powder, and perfumes. In addition, it has also been used as an ignition enhancer in petroleum gasoline and diesel engines due to its relatively high cetane number (>125) and low autoignition temperature (160 °C) (Miller Jothi et al., 2008; Bailey et al., 1997). Other applications of ethyl ether include as agricultural insecticides and fumigants, pharmaceutical solvents, paint removal and varnish, medicinal wound sealants, and the production of cellulose plastics.

PHYSICAL AND CHEMICAL PROPERTIES

Ethyl ether, more commonly known as diethyl ether (diethyl oxide, ether, ethyl oxide, solvent ether; CASRN 60-29-7) is a highly volatile and flammable colorless liquid (a gas above 94 °F) with a pungent, sweetish odor. It is extremely flammable with an autoignition temperature of 320 °F. The vapor pressure is 440 mmHg and has limited solubility in water (69 g/l at 68 °F). Additional chemical and physical properties are presented in Table 56.1.

MAMMALIAN TOXICOLOGY

Acute Effects

Typical effects from an acute ethyl ether inhalation exposure include respiratory and eye irritation, vomiting, dizziness, bradycardia, low body temperature, irregular respiration, salivation, drowsiness, muscle relaxation, paleness, acute excitement, loss of consciousness, and even death (Clayton and Clayton, 1982). Effects vary depending on the exposure concentration and duration. Mörch et al. (1956) determined that the concentrations of ethyl ether necessary for surgical anesthesia and respiratory arrest in mice were 6 and 18%. Induction as a human anesthetic is between 10 and 15%

Hamilton & Hardy's Industrial Toxicology, Sixth Edition. Edited by Raymond D. Harbison, Marie M. Bourgeois, and Giffe T. Johnson.
© 2015 John Wiley & Sons, Inc. Published 2015 by John Wiley & Sons, Inc.

TABLE 56.1   Ethyl Ether Chemical and Physical Properties

Characteristic	Properties
Chemical name	Ethyl ether
Synonyms	Diethyl ether; ethyl oxide; diethyl oxide; ether; 3-oxapentane; ethoxyethane; sulfuric ether; 1,1'-oxybisethane; anesthetic ether
CASRN	60-29-7
Molecular formula	$C_4H_{10}O$
Molecular weight	74.12 Da
Boiling point	94 °F (35 °C)
Freezing point	−177 °F (−116 °C)
Melting point	−177 °F (−116 °C)
Flash point	−49 °F (−45 °C)
Autoignition temperature	356 °F (180 °C)
Vapor pressure	440 mmHg at 68 °F (20 °C)
Specific gravity	0.71 at 68 °F (20 °C)
Density	713 kg/m^3
Ionization potential	9.53 eV
Solubility	Limited solubility in water; soluble in sulfuric acid, alcohol, acetone, benzene, and chloroform
Cetane number	>125

(100,000–150,000 parts per million (ppm); Patty et al., 1982). Surgical anesthetic maintenance is 5% (50,000 ppm). Maintenance levels above 10% could lead to fatalities (Flury and Klimmer, 1943). There have been very few reports of ethyl ether anesthesia as a contributing factor to death; however, in Philadelphia between 1957 and 1959, 645 reported deaths were associated with anesthesia (Campbell et al., 1961). Four of those cases were due to excessive ethyl ether and an additional six cases were a combination of ethyl ether and another anesthetic. Acute lethal concentrations (LC_{50}) in animals ranged from 6500 to 73,000 ppm for mice and rats, respectively (Schwetz and Becker, 1970).

Chronic Effects

Typical effects from chronic ethyl ether exposures are similar to those of acute exposures, including drowsiness, dizziness, excitation, headache, exhaustion, anorexia, and psychic disturbances. Chronic dermal exposure can result in dermatitis and dermal burns. Nasal irritation and dizziness are found to occur at concentrations of 200 ppm and over 2000 ppm, respectively. An increase in blood levels of liver enzymes and a decrease in liver weight relative to the body weight were noted in rats exposed to 2000 ppm of ethyl ether chronically over 30 weeks (Clayton and Clayton, 1982).

Mechanism of Action(s)

The use of ethyl ether as an anesthetic has a very slow induction and recovery period; however, an advantage of ethyl ether as an anesthetic agent is its low hepato- and nephrotoxicity. Recovery side effects include headache, nausea, and vomiting. There are many theories on the mechanism of action of general anesthetics, such as ethyl ether, ranging from disrupting lipid membranes to highly selective and specific interactions upon central nervous system proteins (Weir, 2006).

Although ethyl ether has a direct myocardial depressant effect, during clinical anesthesia it increases the heart rate and cardiac output during the induction and emergence phases (Boniface et al., 1954). This effect observed in dogs concluded that the increase in heart rate can be explained by the sympathoadrenal activity. Krishna et al. (1975) observed a dose-dependent increase in the frequency of contractions in isolated rat atrial preparations. Furthermore, they were able to determine that these effects were not mediated by catecholamine release nor via direct stimulation of beta-adrenergic receptors or blocking of cholinergic receptors.

Inhaled ethyl ether is instantaneously absorbed into the bloodstream, from where it is rapidly passed into the brain (Patty et al., 1982). Historically it was believed that ether was not altered once it is absorbed by the body but later discoveries indicated that 8–10% of the absorbed ether is metabolized to ethanol and acetaldehyde (Haggard, 1924; Green and Cohen, 1971). The majority of inhaled ethyl ether is primarily eliminated via expiration (~87%) with nominal excretion via the urine (Haggard, 1924). Prolonged exposure to ether was shown to increase cytochrome P450 enzyme activity in rats (Liu et al., 1993), which are the enzymes also believed to be involved in the metabolism of ethyl ether in the human hepatic system (Chengelis and Neal, 1980; Kharasch and Thummel, 1993).

Chemical Pathology

Potential target organs for ethyl ether exposure include the eyes, skin, respiratory tract, and central nervous system, yet no specific chemical pathologies have been described.

SOURCES OF EXPOSURE

Occupational

Occupations that may result in worker exposures to ethyl ether are mainly from industries involved in the manufacturing or transportation of ethyl ether, in laboratories or industries where it is used as a solvent or an intermediate, and in hospitals where it is used as an anesthetic. Air concentrations starting at 200 ppm triggered nasal irritation and levels of 300 ppm were found to be unacceptable among industry workers (Nelson et al., 1943). Surgical nurses and doctors complained of exhaustion, headache, irritability, appetite loss, and the inability to concentrate with prolonged exposures to ethyl ether (NIOSH, 1993). Results from a national study among over 49,000 exposed operating room personnel and over 23,000 unexposed individuals revealed that exposed female operating room personnel were subject to increased risks of spontaneous abortion, congenital abnormalities in their children, cancer, and hepatic and renal disease (American Society of Anesthesiologists, 1974).

Environmental

Although there are no published reports on environmental exposures to ethyl ether, the general public could be exposed to air concentrations of ethyl ether from its use as an additive in motor fuels, perfumes, and denatured alcohol.

INDUSTRIAL HYGIENE

The current permissible exposure limit (PEL) issued by Occupational Safety and Health (OSHA) for ethyl ether is 400 ppm (1200 mg/m^3) as the time-weighted average (TWA) for an 8-h workday. A recommended exposure limit (REL) has not yet been established by the National Institute for Occupational Safety and Health (NIOSH). Similarly to OSHA limits, the American Conference of Governmental Industrial Hygienists (ACGIH) issued a threshold limit value (TLV) of 400 ppm (1210 mg/m^3) as the TWA for an 8-h workday and a 40-h workweek. In addition, the ACGIH assigned a short-term exposure limit (STEL) of 500 ppm (1520 mg/m^3) for periods not exceeding 15 min and these concentrations should not be repeated more than four times a day and should be separated by intervals of at least 60 min (OSHA; ACGIH, 1994).

The OSHA recommends effective methods in controlling occupational exposures to ethyl ether, which include the implementation of process enclosures, local exhaust ventilation, general dilution ventilation, and personal protective equipment. Workers that may be subject to prolonged or repeated exposure to liquid ethyl ether should be required and provided with impervious clothing, gloves, and face shields or splash-proof safety goggles to prevent skin contact. If ethyl ether leaks or spills occur, remove all ignition sources, ventilate the spill or leak area, and follow cleanup procedures. Small quantities can be absorbed using spill pads or paper towels and placed in the hood to allow evaporation or placed in closed containers for later disposal. Large quantities can be collected and dissolved in an alcohol that has a greater molecular weight than butyl alcohol and atomized in a suitable combustion chamber.

MEDICAL MANAGEMENT

The OSHA recommends workers who handle or have had skin contact with ethyl ether should wash their hands, forearms, and affected areas thoroughly with soap and water. If irritation persists after washing, medical attention may be needed. Any clothing that becomes wet with ethyl ether should be removed immediately and placed in closed containers for storage until it can be discarded or cleaned properly. If the ethyl ether soaks through the clothing, the affected areas should be immediately flushed and washed thoroughly. Contaminated clothing should not be reworn until the ethyl ether is removed. The laundry personnel should be informed of the potential hazards of ethyl ether. If a person has inhaled large volumes of ethyl ether, the person should be moved to fresh air immediately. If an exposed worker is found unconscious, they should be moved to fresh air immediately and established emergency procedures should be followed. If a person has swallowed ethyl ether, prompt them to vomit and seek medical attention immediately.

REFERENCES

American Conference of Governmental Industrial Hygienists (ACGIH) (1994) *Threshold Limit Values for Chemical Substances and Physical Agents and Biological Exposure Indices*, Cincinnati, OH: American Conference of Governmental Industrial Hygienists.

American Society of Anesthesiologists (1974) Occupational disease among operating room personnel: a national study. Report of an Ad Hoc Committee on the Effect of Trace Anesthetics on the Health of Operating Room Personnel, American Society of Anesthesiologists. *Anesthesiology* 41:321–340.

Bailey, B., Eberhardt, J., Goguen, S., and Erwin J. (1997) *Diethyl ether (DEE) as a renewable diesel fuel*. United States Department of Energy Document 972978. Available at http://www.afdec.energy.gov/pdfs/dee.pdf.

Boniface, K.J., Brown, J.M., and Kronen, P.S. (1954) The influence of some inhalation anesthetic agents on the contractile force of the heart. *J. Pharmacol. Exp. Ther.* 113:64–71.

Campbell, J.E., Weiss, W.A., and Rieders, F. (1961) Evaluation of deaths associated with anesthesia: correlation of clinical, toxicological and pathological findings. *Anesth. Analg.* 40:54–68.

Chengelis, C.P. and Neal, R.A. (1980) Microsomal metabolism of diethyl ether. *Biochem. Pharmacol.* 29:247–248.

Clayton, G. and Clayton, F. (1982) *Patty's Industrial Hygiene and Toxicology*, New York, NY: John Wiley & Sons, Inc.

Flury, F. and Klimmer, O. (1943) In: Lehmann, K.B. and Flury, F., editors. *Toxicology and Hygiene of Industrial Solvents*, Baltimore, MD: Williams and Wilkins.

Green, K. and Cohen, E.N. (1971) On the metabolism of ^{14}C-diethyl ether in the mouse. *Biochem. Pharmacol.* 20:393–399.

Haggard, H. (1924) The absorption, distribution, and elimination of ethyl ether. The amount of ether absorbed in relation to the body concentration inhaled and its fate in the body. *J. Biol. Chem.* 59:737–751.

Kharasch, E.D. and Thummel, K.E. (1993) Identification of cytochrome P450 2E1 as the predominant enzyme catalyzing human liver microsomal defluorination of sevoflurane, isoflurane and methoxyflurane. *Anesthesiology* 79:795–807.

Krishna, G., Trueblood, S., and Paradise, R.R. (1975) The mechanism of the positive chronotropic action of diethyl ether on rat atria. *Anesthesiology* 42:312–318.

Liu, P.T., Ioannides, C., Shavila, J., et al. (1993) Effects of ether anaesthesia and fasting on various cytochromes P450 of rat liver and kidney. *Biochem. Pharmacol.* 45:871–877.

Miller Jothi, N.K., Nagarajan, G., and Renganarayanan, S. (2008) LPG fueled diesel engine using diethyl ether with exhaust gas recirculation. *Int. J. Therm. Sci.* 47:450–457.

Mörch, E.T., Aycrigg, J.B., and Berger, M.S. (1956) The anesthetic effects of ethyl vinyl ether, divinyl ether, and diethyl ether on mice. *J. Pharmacol. Exp. Ther.* 117:184–189.

National Institute for Occupational Safety and Health (NIOSH) (1993) *NEG and NIOSH Basis for an Occupational Health Standard: Ethyl Ether.* DHHS (NIOSH) Publication No. 93-103.

Nelson, K.W., Egwe, J.F., Ross, M., Woodman, L.E., and Silverman, L. (1943) Sensory response to certain industrial solvent vapors. *J. Ind. Hyg. Toxicol.* 25:282–285.

Patty, F.A., Clayton, G.D., Clayton, F.E., and Battigelli, M.C. (1982) Ethers. In: *Patty's Industrial Hygiene and Toxicology*, 3rd ed., vol. 2, New York, NY: John Wiley & Sons, Inc.

Schwetz, B.A. and Becker, B.A. (1970) Embryotoxicity and fetal malformations of rats and mice due to maternally administered ether. *Toxicol. Appl. Pharmacol.* 17:275.

Weir, C.J. (2006) The molecular mechanisms of general anaesthesia: dissecting the GABA_A receptor. *Contin. Educ. Anaesth. Crit. Care Pain* 6:49–53.

HALOETHER

Target organ(s): Eyes, skin, and respiratory tract.

BCME occupational exposure limits: ATSDR MRL: 0.0014 mg/m^3 (0.0003 ppm) TWA

Reference values: Human carcinogens

Risk/safety phrases: CMME and BCME are potent carcinogens. BCME is very volatile and may cause fire in presence of excessive heat. Hydrogen chloride and formaldehyde can react to form BCME. BCME hydrolyzes in water forming hydrogen chloride and formaldehyde. CMME is incompatible and will react with surface moisture forming hydrogen chloride. CMME can also contain 1–8% BCME. CMME can also decompose into hydrogen chloride and formaldehyde, which under certain temperatures and humidity can form BCME

BACKGROUND AND USES

Haloethers are compounds that contain an ether moiety and halogen atoms attached to the aryl or alkyl groups. Chloroethers are considered the most commercially significant haloethers and can be divided into alpha- and non-alpha chloroethers. One of the only commercially significant α-haloethers is chloromethyl methyl ether (CMME), which is primarily used as an alkylating agent and solvent for the manufacturing of water repellents, ion-exchange resins, and industrial polymers. It is also of toxicological importance. Bis-chloromethyl ether (BCME) is another α-haloether that is also known for its toxicity. Bis-chloromethyl ether was previously used to make polymers, resins, and textiles. However, it is only used in small quantities in fully enclosed systems to make other chemicals with highly restricted use and used commercially due to its carcinogenicity. A β-chloroether of commercial significance is bis(2-chloroethyl) ether (BCE), which has been used in the textile, petroleum, and paint and varnish industries. It is used as a dewaxing agent, wetting and cleaning agent, and a penetrant. It also serves as a good solvent and extraction solvent for tars, fats, waxes, oils, grease, and resins. In addition, BCE is a by-product of the process for making ethylene oxide. The focus of the discussion in this section will be on the two α-haloethers with the most toxicological relevance, CMME and BCME.

PHYSICAL AND CHEMICAL PROPERTIES

Haloethers have a wide range of physical and chemical properties; however, most are miscible in oils and soluble in organic solvents, such as benzene, carbon tetrachloride, and acetone. Chloromethyl methyl ether) is a clear, colorless liquid that is highly flammable and has a strong noxious and irritating odor. Bis-chloromethyl ether is also a clear, colorless liquid at room temperature and has a strong unpleasant and suffocating odor. Additional chemical and physical properties for both are presented in Table 56.2.

MAMMALIAN TOXICOLOGY

Acute Effects

Acute effects noted for both CMME and BCME have been dermal, mucous membrane, and respiratory tract irritation. Additional short-term effects reported from CMME exposure have been eye irritation, pulmonary edema, and pneumonia in humans. Acute inhalation studies using mice found BCME to be nonirritating to the upper respiratory tract at 10.6 ppm, yet concentrations of 40 ppm of CMME caused slight irritation (Leong et al., 1971). The LC$_{50}$ for the inhalation of CMME in rats and hamsters was 55 and 65 ppm, respectively, after 14 days (Drew et al., 1975). The 14-day LC$_{50}$ for BCME in

TABLE 56.2 Chemical and Physical Properties for Chloromethyl Methyl Ether and Bis-chloromethyl Ether

Characteristic	Chloromethyl Methyl Ether Properties	Bis-chloromethyl Ether Properties
Synonyms	CMME, MOM-chloride, chloro (methoxy)methane, methoxymethyl chloride, methyl chloromethyl ether	BCME, 1,1'-dichlorodimethyl ether, bis (2-chloromethyl) ether
CASRN	107-30-2	542-88-1
Molecular formula	C_2H_5ClO	$C_2H_4CL_2O$
Molecular weight	80.5 Da	115 Da
Boiling point	138 °F (61 °C)	223 °F (106 °C)
Freezing point	−154 °F (−104 °C)	No data available
Melting point	−154 °F (−104 °C)	−43 °F (−42 °C)
Flash point	60 °F (15 °C)	100 °F (38 °C)
Vapor pressure	192 mmHg at 70 °F (21 °C)	33 mmHg at 72 °F (22 °C)
Specific gravity	1.06	1.32
Density	1.06	3.97 (vapor), 1.33 (liquid)
Solubility	Miscible in oils and soluble in organic solvents	

both species (7 ppm) was almost an order of magnitude lower than the LC_{50} for CMME (Drew et al., 1975).

Chronic Effects

Chronic inhalation from CMME exposure may result in bronchitis. Chronic cough, impaired respiratory function, and bronchitis have all been reported from long-term inhalation of BCME. Increased incidences of respiratory cancers have been reported in experimentally exposed animals and in occupationally exposed workers to both of these α-haloethers. Rats chronically exposed to 1 ppm of CMME throughout their lifetime were found to have almost a doubling of bronchial hyperplasia compared to controls, yet mortality and weight gain were not significant (Laskin et al., 1975). In a lifetime observation study, mice and rats chronically exposed to 100 parts per billion (ppb) of BCME vapors for 6 months produced oncogenic responses (Leong et al., 1981). Over 86% of the rats exposed had nasal tumors, whereas pulmonary adenomas were more significant in mice exposed to the same concentrations. No carcinogenic responses were observed in either species at concentrations of 1 or 10 ppb of BCME over the 6-month exposure period. Mice chronically exposed to 1 ppm BCME and 2 ppm of CMME resulted in an increased incidence of pulmonary adenomas (Laskin et al., 1971). Based on animal studies, the Agency for Toxic Substances and Disease Registry (ATSDR) derived a minimal risk level (MRL) of 0.0003 ppm for human chronic inhalation exposure to BCME (ATSDR, 1989).

Mechanism of Action(s)

Specific modes of action for BCME and CMME have not been very well elucidated. Water can rapidly break down CMME into methanol, hydrochloric acid, and formaldehyde and BCME into formaldehyde and hydrochloric acid (Van Duuren, 1980). From the reviewed occupational exposures

and animals studies, BCME and CMME-containing BCME are confirmed potent carcinogens, primarily resulting in lung and nasal cancers. Oat cell (small cell) carcinomas account for a high proportion of the respiratory tumors reported. It has been suggested that BCME may be radiomimetic because of its property similarities with nitrogen mustard, which also had a predominant association with oat cell carcinomas and bronchogenetic cancers (Lemen et al., 1976).

Chloromethyl methyl ether hydrolyzes completely and irreversibly in water to form HCl, methanol, and formaldehyde (Van Duuren, 1980). The frequency of chromosomal aberrations was increased in peripheral lymphocytes of workers exposed to levels of BCME exceeding 0.4 mg (Sram et al., 1985). Chloromethyl methyl ether was also shown to inhibit DNA synthesis in human lymphocytes after being treated for 4 h (Perocco et al., 1983). It was also found to be weakly mutagenic in *Drosophila melanogaster* larvae (Filippova et al. 1967) and mutagenic in *Salmonella typhimurium* TA98 (in the absence of metabolic activation) (Norpoth et al. 1980), *Escherichia coli*, and *S. typhimurium* (Mukai and Hawryluk, 1973).

Chemical Pathology

Target organs of both BCME and CMME are the eyes, skin, and respiratory tract; however, no data are available on their chemical pathologies. Nevertheless, studies have typically exhibited a marked dose response and a histologic specificity for small cell tumors (Steenland et al., 1994; Ward, 1995).

SOURCES OF EXPOSURE

Occupational

Exposures to BCME and CMME are primarily through inhalation and dermal contact. An increased incidence of lung cancer was observed in a group of 125 chemical manufacturing plant workers exposed to CMME (Figueroa et al., 1973). A 17-year follow-up of the cohort revealed a small epidemic of respiratory cancers, which included 14 cases of lung cancer and 2 cases of laryngeal cancer among 91 men (Weiss, 1982). Ninety-eight lung cancers were also observed (24.1 expected; SMR = 4.1) in a study group of over 3000 workers exposed to BCME worldwide (Steenland et al., 1994; Ward, 1995).

Environmental

Although it may be expected that CMME and BCME may be released into the environment via manufacturing and production processes, no data are available on their ambient levels of either in soil, water, or air.

INDUSTRIAL HYGIENE

The OSHA and the NIOSH have not set occupational limits for either CMME or BCME. However, any industry that may use CMME, BCME, or solid or liquid mixtures containing at least 0.1% by weight or volume of either compound are required by both OSHA and NIOSH to implement stringent controls (NIOSH, 1988a,b). A MRL for chronic inhalation of BCME in humans is 0.0003 ppm as the TWA for a normal 8-h workday and a 40-h workweek (ATSDR, 1989, 1999; NIOSH, 1988a,b). The ACGIH has designated CMME as a suspected human carcinogen without having sufficient evidence to assign a TLV.

The OSHA requires workers be provided and use full-body chemical protective clothing (CPC), including smocks, coveralls, or long-sleeved shirts and long pants. In addition, workers should be provided and required to wear respirators and remove CPCs prior to exiting regulated areas. Any products, residues, or container liners contaminated with BCME are considered acute hazardous waste under the Resource Conservation and Recovery Act (RCRA) and must be disposed of by transport to a RCRA waste storage and disposal facility where the preferred method of disposal is incineration. Areas that use either BCME or CMME should be restricted and controlled with only authorized personnel allowed to enter usage areas. Regulated areas should be posted with signs indicating that a potential human carcinogen is present and only authorized personnel with proper CPC are allowed access (NIOSH, 1988a,b). Indoor regulated areas should also operate under negative pressure. Other control methods include process enclosures, restricted access areas, local exhaust ventilation, personal protective equipment, good housekeeping and personal hygiene, and substitiution of a less toxic substance where possible (NIOSH, 1988a,b).

MEDICAL MANAGEMENT

Eyes and skin that become exposed to CMME or BCME should be flushed for at least 15 min and washed thoroughly followed by immediate medical attention. Any industries where employees may be potentially exposed to hazardous chemicals are recommended to have a medical surveillance program. If workers are exposed to levels exceeding regulatory levels, this may be a requirement. Medical surveillance programs include baseline measurements and routine monitoring of air concentrations to determine worker TWA exposure during a workshift. Ceiling levels during peak expected airborne concentrations should also be measured and compared to regulatory levels. The medical surveillance program should also include a preplacement medical evaluation of workers to determine baseline health status. Periodic medical screening and biological monitoring (e.g., blood tests, X-rays, urinalysis) during employment as well as an exit or termination evaluation should also be performed regularly to identify any changes in health status.

REFERENCES

Agency for Toxic Substances and Disease Registry (ATSDR) (1989) *Toxicological Profile for Bis(chloromethyl) Ether.*

Agency for Toxic Substances and Disease Registry (ATSDR) (1999) Bis(chloromethyl) Ether. Available at http://www .atsdr.cdc.gov/toxfaqs/tf.asp?id=918&tid=188.

Drew, R., Laskin, S., Kuschner, M., and Nelson, N. (1975) Inhalation carcinogenicity of alpha halo ethers. I. The acute inhalation toxicity of chloromethyl methyl ether and bis(chloromethyl) ether. *Arch. Environ. Health* 30:61–69.

Figueroa, W.G., Raszkowski, R., and Weiss, W. (1973) Lung cancer in chloromethyl methyl ether workers. *N. Engl. J. Med.* 288:1096–1097.

Filippova, L.M., Pan'shin, O.A., and Kostyankovskii, R.G. (1967) Chemical mutagens. IV. Mutagenic activity of germinal systems. *Genetika* 3:134–148.

Laskin, S., Drew, R.T., Cappiello, M.S., Kuschner, M., and Nelson, N. (1975) Inhalation carcinogenicity of alpha halo ethers. II. Chronic inhalation studies with chloromethyl methyl ether. *Arch. Environ. Health* 30:70–72.

Laskin, S., Kuschner, M., Drew, R.T., Cappiello, V.P., and Nelson, N. (1971) Tumors of the respiratory tract induced by inhalation of bis(chloromethyl) ether. *Arch. Environ. Health* 23:135–136.

Lemen, R.A., Johnson, W.M., Wagoner, J.K., Archer, V.E., and Saccomanno, G. (1976) Cytologic observations and cancer incidence following exposure to BCME. *Ann. N. Y. Acad. Sci.* 271:71–80.

Leong, B.K.J., Kociba, R.I., and Jersey, G.C. (1981) A lifetime study of rats and mice exposed to vapors of bis(chloromethyl ether). *Toxicol. Appl. Pharmacol.* 58:269–281.

Leong, B.K.J., Macfarland, H.N., and Reese, W.H. (1971) Induction of lung adenomas by chronic inhalation of bis(chloromethyl) ether. *Arch. Environ. Health* 22:663–666.

Mukai, F.H. and Hawryluk, I. (1973) The mutagenicity of some halo-ethers and haloketones. *Mutat. Res.* 21:228 (abstract).

National Institute for Occupational Safety and Health (NIOSH) (1988a) *Occupational and Safety Health Guideline for Chloromethyl Methyl Ether, Potential Human Carcinogen.* Available at http://www.cdc.gov/niosh/docs/81-123/pdfs/0129.pdf.

National Institute for Occupational Safety and Health (NIOSH) (1988b) *Occupational and Safety Health Guideline for Bis-chloromethyl Ether, Potential Human Carcinogen.* Available at http://www.cdc.gov/niosh/docs/81-123/pdfs/0128.pdf.

Norpoth, K.H., Reisch, A., and Heinecke, A. (1980) Biostatistics of Ames test data. In: Norpoth, K.H. and Garner, R.C., editors. *Short Term Test Systems for Detecting Carcinogens*, Berlin: Springer, pp. 312–322.

Perocco, P., Bolognesi, S., and Alberghini, W. (1983) Toxic activity of seventeen industrial solvents and halogenated compounds on human lymphocytes cultured *in vitro. Toxicol. Lett.* 16:69–75.

Sram, R.J., Landa, K., Hola, N., and Roznickova, I. (1985) The use of cytogenetic analysis of peripheral lymphocytes as a method for checking the level of MAC in Czechoslovakia. *Mutat. Res.* 147:322.

Steenland, K., Loomis, D., Shy, C., and Simonsen, N. (1994) Occupational causes of lung cancer. In: Schenker, M., Balmes, J., and Farber, P., editors. *Occupational and Environmental Respiratory Disease*, St. Louis: Mosby.

Van Duuren, B.L. (1980) Prediction of carcinogenicity based on structure, chemical reactivity and possible metabolic pathways. *J. Environ. Pathol. Toxicol.* 3:11–34.

Ward, E. (1995) Overview of preventable industrial causes of occupational cancer. *Environ. Health Perspect.* 103:197–203.

Weiss, W. (1982) Epidemic curve of respiratory cancer due to chloromethyl ethers. *J. Natl. Cancer Inst.* 69:1265–1270.

ETHYLENE OXIDE

Target organ(s): Eyes, skin, respiratory system, liver, central nervous system, blood, kidneys, reproductive system

Occupational exposure limits: OSHA PEL: 1 ppm TWA, 5 ppm STEL; ACGHI TLV: 1 ppm (1.8 mg/m^3) TWA; NIOSH REL: 0.1 ppm, 5 ppm STEL; NIOSH IDLH: 800 ppm; EEGL: 20 ppm for 1 h; 1 ppm for 24 h

Reference values: MRL: 0.09 ppm in air (ATSDR, 1990)

Risk/safety phrases: Classified as a Group 1 human carcinogen (IARC, 2012). Very toxic with carcinogenic, mutagenic, teratogenic, and reproductive hazards. Extremely flammable and potentially explosive in the presence of alkali metal hydroxides and highly active catalytic surfaces

BACKGROUND AND USES

Ethylene oxide was first reported in the mid-1800s by a French chemist, Charles-Adolphe Wurtz, who first synthesized it by treating 2-chloroethanol with potassium hydroxide. In 1931, another French chemist, Theodore Lefort, developed a method of direct oxidation of ethylene in the presence of silver catalyst, of which all industrial production uses this process (ATSDR, 1990). Ethylene oxide is mainly used as a chemical intermediate in the manufacturing process of textiles, detergents, polyurethane foam, automotive antifreeze, solvents, medicinals, adhesives, cosmetics, perfumes, pharmaceuticals, lubricants, plasticizers, and paint thinners. Ethylene oxides in the form of ethylene glycols (i.e., automotive antifreeze) account for the majority of the global consumption (75%). According to the Report on Carcinogens (2011), the U.S. production has increased from 4 billion pounds in 1973 to 7.6 billion pounds in 2002.

PHYSICAL AND CHEMICAL PROPERTIES

Ethylene oxide (CASRN 75-21-8), also called oxirane, is an extremely flammable and explosive, colorless gas (25 °C) with a sweet ether-like odor, more distinctive when air concentrations are in excess of 500 ppm. It is miscible in water and most organic solvents, such as ethanol, diethyl ether, acetone, benzene, alcohol, and carbon tetrachloride. Additional chemical and physical properties are presented in Table 56.3.

TABLE 56.3   Ethylene Oxide Chemical and Physical Properties

Characteristic	Properties
Chemical name	Ethylene oxide
Synonyms	Oxirane, epoxyethane, dimethylene oxide, oxacyclopropane, ethene oxide, ethyleneoxide, 1,2-epoxyethane
CASRN	75-21-8
Molecular formula	C_2H_4O
Molecular weight	44.05 Da
Boiling point	51 °F (11 °C)
Freezing point	−171 °F (−113 °C)
Melting point	−168 °F (−111 °C)
Flash point	−4 °F (−20 °C)
Autoignition temperature	804 °F (429 °C)
Vapor pressure	1.46 atm
Specific gravity	0.82 (liquid at 50 °F)
Density	0.882 g/ml
Solubility	Miscible in water

MAMMALIAN TOXICOLOGY

Acute Effects

Acute effects from short-term inhalation of ethylene oxide include nausea, headache, weakness, vomiting, drowsiness, incoordination, and irritation of the eyes and respiratory tract (NIOSH, 1988). Short-term skin exposures can result in blisters, edema, burns, frostbite, and severe dermatitis. Acute lethal concentrations (LC_{50}) of ethylene oxide in animals ranged from 800 to 4000 ppm in rats (Carpenter et al., 1949; Deichmann and Gerarde, 1969). Animal toxicity data concluded that concentrations of 2000 ppm for 60 min, 4000 ppm for 30 min, and 8000 ppm for 10 min are considered to cause injury or death (Clayton and Clayton, 1981). Nasal irritation was reported after 10 s of exposure to 12,500 ppm (Walker and Greeson, 1932).

Chronic Effects

Chronic exposures of ethylene oxide to the skin can cause dermatitis and can even result in burns and blisters. In severe cases, symptoms may include muscle weakness, loss of feeling or prickly sensation in the hands, feet, arms or legs, clumsiness, and paralysis. Hematological abnormalities were observed in 47 hospital workers chronically exposed to EtO levels ranging from <0.01 to 0.06 ppm (Shaham et al., 2000).

The IARC Monograph Volume 97 (IARC, 2008) concluded that there was *limited evidence* in humans for the carcinogenicity of ethylene oxide and was subsequently listed as a Group 2B probable human carcinogen. This conclusion has since changed, and it is now listed as a Group 1 human carcinogen (IARC, 2012). This decision was largely based on a combination of more recent studies with *limited evidence* in humans and *sufficient evidence* from animal experimental studies. These studies produced new data for an association of ethylene oxide with lymphatic and hematopoietic cancers and the genotoxicity of ethylene oxide (IARC, 2012).

Mechanism of Action(s)

Ethylene oxide is typically absorbed through the lungs and into the blood system. Once it is absorbed, it is either biotransformed or eliminated via respiration or excretion. it is rapidly cleared from tissues within 2–3 days. There are two pathways involved in the metabolism of ethylene oxide, hydrolysis to ethylene glycol and conjugation with glutathione. Evidence presented by Hengstler et al. (1994) strongly suggests that a metabolite of ethylene glycol (glycolaldehyde) leads to DNA–protein crosslinks and DNA strand breaks in human blood cells. Several studies using rats and mice have found that elimination predominantly takes place in the hepatic cytosol and is catalyzed by glutathione

S-transferase (GST) (Li et al., 2011; Brown et al., 1996). Although this is also considered the predominant elimination pathway in humans, it is also expected that epoxide hydrolase (EH)-catalyzed hydrolysis accounts for at least 1/4th of the overall ethylene oxide metabolism (Li et al., 2011).

Chemical Pathology

Potential target organs of ethylene oxide include the eyes, respiratory system, central nervous system, reproductive system, blood, and other organs (e.g., liver, kidneys). Most of the cohort studies conducted have not been able to illustrate a significant difference in cancer diagnoses among ethylene oxide-exposed workers in comparison to the general public. In contrast, the increase of genotoxic and mutagenic evidence has contributed to the recent change in classification to a Group 1 human carcinogen (IARC, 2012).

Possible mechanisms of neurotoxicity include the direct exposure of the peripheral nerves through cutaneous absorption and central involvement through inhalation and vascular dissemination (Brashear et al., 1996). Genotoxicity studies of ethylene oxide were reviewed in the IARC Monograph Volume 97 (2008) and revealed an elevated number of chromosomal aberrations and an increase in sister chromatid exchange in the blood of exposed workers. In addition, ethylene oxide is a direct alkylating agent, directly reacting with DNA and proteins and is believed to initiate genotoxicity and cancer (Swenberg et al., 2000). Several studies have reported DNA damage, DNA strand breaks, sister chromatid exchange, cell mutations, and gene mutations in both animal and human studies. It also induced specific locus mutation, micronucleus formation, unscheduled DNA synthesis, and dominant lethal mutations (IARC, 2008).

SOURCES OF EXPOSURE

Occupational

The majority of the occupational exposures are found in the chemical plants where ethylene oxide is produced and used as an intermediate in the manufacturing process of other products (i.e., detergents, polyurethane foam, automotive antifreeze, solvents, adhesives, etc.). The greatest potential for occupational exposures is considered to be during the loading or unloading of transport tanks, product sampling procedures, and equipment maintenance and repair (CDCP, 1981). However, NIOSH (1989) estimated that approximately 270,000 people were exposed to ethylene oxide in the United States, primarily in hospitals and during commercial sterilization.

A cohort study of over 18,000 exposed workers showed little evidence of excess cancer mortality, with the exception of a significant excess of bone cancer (six deaths) compared to the general population (Steenland et al., 2004). This study has provided some evidence for the classification of ethylene oxide as a human carcinogen (IARC, 2012). In addition, this study found positive exposure–response trends for lymphoid tumors among males only and some evidence of a positive exposure–response for breast cancer mortality. Using exposure–response analysis, Stayner et al. (1993) reanalyzed Steenland's previous results (Steenland et al., 1991) and concluded that there was an increase in mortality from lymphatic and hematopoietic cancer in the groups with the highest cumulative ethylene oxide exposure (>8500 ppm/day). In another study evaluating hospital workers, cognitive impairment and personality dysfunction were observed more frequently in hospital workers chronically exposed to EtO, compared to a control group (Klees et al., 1990). Neurotoxicity and central nervous system toxicity were found among 12 nurses and operating technicians (Brashear et al., 1996). Levels measured in their surgical gowns 18 days post sterilization were 298 ppm; however, peak levels during exposure were not known. In addition, EtO was found to be associated, although a weak association, with breast cancer in a cohort study of over 7500 women occupationally exposed for at least 1 year for an average of 10.7 years (Steenland et al., 2003).

Environmental

The majority of ethylene oxide is released into the atmosphere with an estimated 23 tons from industries (WHO, 2003). California reported ethylene oxide levels in air ranging from <0.001 to 0.96 mg/m^3 (Havlicek et al., 1992). Ethylene oxide is not only a synthesized product but it is also metabolically produced from ethylene by microorganisms internally and by the combustion of organic compounds. Furthermore, it is a component of cigarette smoke and is estimated to be present at levels of 7 µg/cigarette (IARC, 2004). In case of smoking, ethene via ethylene oxide (EtO) could be an initiator of approximately 15% of the smoking-related cancer cases (Törnqvist et al., 1986). Moreover, secondhand smoke potentially exposes almost everyone to ethene and ethylene oxide. A preliminary study evaluating fruit store workers exposed to 1 ppm ethene in the ambient air during work hours resulted in the same adduct level in hemoglobin as 0.3 ppm of ethylene oxide (Törnqvist et al., 1989). Hemoglobin adducts are currently used as effective biomarkers of exposure to potential carcinogens and other occupationally and environmentally relevant compounds.

INDUSTRIAL HYGIENE

The current PEL issued by the OSHA for ethylene epoxide is 1 ppm as the TWA for an 8-h workday. A REL established by the NIOSH is set at 0.1 ppm and a STEL of 5 ppm per day.

Similar to the OSHA limits, the ACGIH issued a TLV of 1 ppm as the TWA.

Engineering controls should be instituted to prevent and reduce employee exposures to ethylene oxide. Process enclosures, local exhaust ventilation, and personal protective equipment should be used during the synthesis and handling of ethylene oxide as well as when used as a sterilizing agent. In addition to engineering controls, personal protective clothing and equipment should be provided and used. Respirators may be needed depending on the airborne concentration of ethylene oxide. In the event of a leak or spill, all ignition sources should be removed immediately. Small quantities can be absorbed using paper towels and placed in a fume hood or appropriate container. For large quantities, vermiculite or dry sand may be needed.

MEDICAL MANAGEMENT

Eyes and skin that become exposed to ethylene oxide should be flushed and washed thoroughly followed by immediate medical attention. Any industry where employees may be potentially exposed to hazardous chemicals are recommended to have a medical surveillance program. If workers are exposed to levels exceeding regulatory levels, this may be a requirement. Medical surveillance programs include baseline measurements and routine monitoring of air concentrations to determine worker TWA exposure during a workshift. Ceiling levels during peak expected airborne concentrations should also be measured and compared to the regulatory levels. The medical surveillance program should also include a preplacement medical evaluation of workers to determine their baseline health status. Periodic medical screening and biological monitoring (e.g., blood tests, X-rays, urinalysis) during employment as well as an exit or termination evaluation should also be performed regularly to identify any changes in the health status.

REFERENCES

Agency for Toxic Substances and Disease Registry (ATSDR) (1990) *Toxicological Profile for Ethylene Oxide*, Atlanta, GA: U.S. Public Health Service, U.S. Department of Health and Human Services.

Brashear, A., Unverzagt, F.W., Farber, M.O., Bonnin, J.M., Garcia, J.G. N., and Grober, E. (1996) Ethylene oxide neurotoxicity: a cluster of 12 nurses with peripheral and central nervous system toxicity. *Neurology* 46:992–998.

Brown, C.D., Wong, B.A., and Fennell, T.R. (1996) *In vivo* and *in vitro* kinetics of ethylene oxide metabolism in rats and mice. *Toxicol. Appl. Pharmacol.* 136:8–19.

Carpenter, C.P., Smyth, H.F., Jr., and Pozzani, U.C. (1949) The assay of acute vapor toxicity and the grading and interpretation of results on 96 chemical compounds. *J. Ind. Hyg. Toxicol.* 31:343–346.

Center for Disease Control and Prevention (CDCP) (1981) *Ethylene Oxide (EtO): Evidence of Carcinogenicity.* DHHS (NIOSH) Publication No. 81-130.

Clayton, G.D. and Clayton, F.E. (1981) *Patty's Industrial Hygiene and Toxicology,* 3rd ed., New York, NY: John Wiley & Sons, Inc., pp. 2166–2186.

Deichmann, W.B. and Gerarde, H.W. (1969) Ethylene oxide (epoxyethane; oxirane; dimethylene oxide). In: *Toxicology of Drugs and Chemicals,* New York, NY: Academic Press, Inc., pp. 258–259.

Havlicek, C.S., Hilpert, L.R., Dal, G., and Perotti, D. (1992) *Assessment of Ethylene Oxide Concentrations and Emissions from Sterilization and Fumigation Processes.* Final Report. Sacramento, CA: Research Division, California Air Resources Board.

Hengstler, J. G., Fuchs, J., Gebhard, S., and Oesch, F. (1994) Glycolaldehyde causes DNA–protein cross-links: a new aspect of ethylene-oxide genotoxity. *Mutat. Res.* 304:229–234.

IARC (2004) Tobacco smoke and involuntary smoking. *IARC Monogr. Eval. Carcinog. Risks Hum.* 83:1–1438.

IARC (2008) 1,3-Butadiene, ethylene oxide and vinyl halides (vinyl fluoride, vinyl chloride and vinyl bromide). *IARC Monogr. Eval. Carcinog. Risks Hum.* 97:3–471.

IARC (2012) A review of human carcinogens: chemical agents and related occupations. *IARC Monogr. Eval. Carcinog. Risks Hum.* 100F:379–400.

Klees, J.E., Lash, A., Bowler, R.M., Shore, M., and Becker, C.E. (1990) Neuropsychological impairment in a cohort of hospital workers chronically exposed to ethylene-oxide. *J. Toxicol. Clin. Toxicol.* 28:21–28.

Li, Q., Csanády, G.A., Kessler, W., Klein, D., Pankratz, H., Pütz, C., Richter, N., and Filser, J.G. (2011) Kinetics of ethylene and ethylene oxide in subcellular fractions of lungs and livers of male B6C3F1 mice and male Fischer 344 rats and of human livers. *Toxicol. Sci.* 123:384–398.

National Institute for Occupational Safety and Health (NIOSH) (1988) *Occupational Safety and Health Guideline for Ethylene Oxide. Potential Human Carcinogen.*

National Institute for Occupational Safety and Health (NIOSH) (1989) *National Occupational Exposure Survey: Sampling Methodology.* Publication NIOSH 89-102, Cincinnati, OH: DHHS.

Report on Carcinogens (ROC), 12th ed. (2011) *Ethylene Oxide,* pp. 188–191.

Shaham, J., Levi, Z., Grvich, R. Shain, R., and Ribak, J. (2000) Hematological changes in hospital workers due to chronic exposure to low levels of ethylene oxide. *J. Occup. Environ. Med.* 42:843–850.

Stayner, L., Steenland, K., Greife, A., Hornung, R., Hayes, R.B., Nowlin, S., Morawetz, J., Ringenburg, V., Elliot, L., and Halperin, W. (1993) Exposure-response analysis of cancer mortality in a cohort of workers exposed to ethylene oxide. *Am. J. Epidemiol.* 138:787–798.

Steenland, K., Stayner, L., and Deddens, J. (2004) Mortality analyses in a cohort of 18,232 ethylene oxide exposed workers: Follow up extended from 1987 to 1998. *Occup. Environ. Med.* 61:2–7.

Steenland, K., Stayner, L., Griefe, A., Halperin, W., Hayes, R., Hornung, R., and Nowlin, S. (1991) Mortality among workers exposed to ethylene oxide. *N. Engl. J. Med.* 324:1402–1407.

Steenland, K., Whelan, E., Deddens, J., Stayner, L., and Ward, E. (2003) Ethylene oxide and breast cancer incidence in a cohort study of 7576 women (United States). *Cancer Causes Control* 14:531–539.

Swenberg, J.A., Ham, A., Koc, H., Morinello, E., Ranasinghe, A., Tretyakova, N., Upton, P.B., and Wu, K.Y. (2000) DNA adducts: effects of low exposure to ethylene oxide, vinyl chloride and butadiene. *Mutat. Res.* 464:77–86.

Törnqvist, M. Ã., Almberg, J.G., Bergmark, E.N., Nilsson, S. and Osterman-Golkar, S.M. (1989) Ethylene oxide doses in ethene-exposed fruit store workers. *Scand. J. Work Environ. Health* 15:436–438.

Törnqvist, M., Osterman-Golkar, S., Kautiainen, A., Jensen, S., Farmer, P.B., and Ehrenberg, L. (1986) Tissue doses of ethylene oxide in cigarette smokers determined from adduct levels in hemoglobin. *Carcinogenesis* 7:1519–1521.

Walker, W.J.G. and Greeson, C.E. (1932) The toxicity of ethylene oxide. *J. Hyg. (Lond.)* 32:409–416.

World Health Organization (WHO) (2003) *Ethylene Oxide.* Concise International Chemical Assessment Document 54, Geneva: World Health Organization. Available at http://www.inchem, org/documents/cicads/cidads/cicad54.htm.

DIOXANE

Target organ(s): Eyes, skin, respiratory system, central nervous system, liver, kidneys

Occupational exposure limits: OSHA PEL: 100 ppm ($360\,mg/m^3$) TWA; ACGIH TLV: 25 ppm ($90\,mg/m^3$) TWA; NIOSH REL: 1 ppm ($3.6\,mg/m^3$); NIOSH IDLH: 500 ppm

Reference values: RfD: 0.03 mg/kg/day; NOAEL: 9.6 mg/kg/day; LOAEL: 94 mg/kg/day

Risk/safety phrases: Classified as a Group 2B probable human carcinogen. Hydroscopic and will produce peroxides in the presence of moisture. Highly flammable

BACKGROUND AND USES

Dioxane is produced commercially by either 1. dehydration of ethylene glycol; 2. treatment of bis(2-chloroethyl) ether with alkali; 3. by dimerization of ethylene oxide; 4. or by heating and distilling diethylene glycol with dehydration catalysts (HSDB, 1995). It is primarily used as solvent in chemical processing, as well as a laboratory reagent, as a chemical intermediate, as a polymerization catalyst, and as an extraction medium. It has been used in several industries

TABLE 56.4   Dioxane Chemical and Physical Properties

Characteristic	Properties
Chemical name	Dioxane
Synonyms	Diethylene dioxide; diethylene ether, 1,4-dioxacyclohexane; diethylene oxide; dioxyethylene ether; glycol ethylene ether; p-dioxane; 1,4-dioxane
CASRN	123-91-1
Molecular formula	$C_4H_8O_2$
Molecular weight	88.1
Boiling point	214 °F (101 °C)
Freezing point	53 °F (11.8 °C)
Melting point	53 °F (11.8 °C)
Flash point	54 °F (12 °C)
Vapor pressure	29 mmHg at 68 °F (20 °C)
Specific gravity	1.03
Density	1.03 g/ml
Ionization potential	9.13 eV
Solubility	Miscible in water, acetone, benzene, ether alcohols, and many common solvents

such as plastic, rubber, agriculture, and consumer. Commercial products that may contain dioxane include paint strippers, dyes, greases, varnishes, deodorants, shampoos, and cosmetics. Historically 90% of dioxane was used as a stabilizer for 1,1,1-trichloroethane (ATSDR, 2007). As of 2006, the estimated U.S. production plus imports of dioxane are estimated to be approximately 10 million pounds.

PHYSICAL AND CHEMICAL PROPERTIES

Dioxane (CASRN 123-91-1) is a cyclic ether that exists as a clear, colorless, flammable liquid at ambient temperatures (solid below 12 °C). It is a dimer of ethylene oxide and has a mild, pleasant ether-like odor. It is miscible in water, alcohols, and many other common solvents, such as acetone, ether, and benzene. Additional chemical and physical properties are presented in Table 56.4.

MAMMALIAN TOXICOLOGY

Acute Effects

Acute exposures to dioxane are similar to those of other ethers, which include eye and respiratory irritation, headache, dizziness, drowsiness, nausea, loss of appetite, stomach pain, renal and hepatic damage, and difficulty in breathing. Additional symptoms reported have also included coma and death. The inhalation LC_{50} of dioxane in rats was 46 and 65 mg/l for a 4-h and 2-h exposure period, respectively (Pilipiuk et al., 1977). This same study did not report any

induced skin damage as a result of a single and repeated dermal exposure; however, it was rapidly absorbed into the blood that led to acute poisoning and irritation of the eye.

Chronic Effects

Chronic low dose exposures to dioxane have resulted in liver and kidney damage and chronic dermal exposures have resulted in irritation of the eyes and respiratory system. Torkelson et al. (1974) exposed rats to 111 ppm of dioxane vapors for 2 years and did not find any evidence of carcinogenicity. An additional 2-year inhalation study found that nasal squamous cell carcinomas, hepatocellular adenomas, and peritoneal mesotheliomas statistically increased in rats chronically exposed to 1250 ppm (Kasai et al., 2009). Several long-term (>1 year) studies on exposing animals to dioxane in drinking water have reported cancer in several organs, including liver, kidney, gall bladder, and nasal cavity (DeRosa et al., 1996). Dioxane concentrations ranged from 444 to 1599 mg/kg/day.

Mechanism of Action(s)

An inhalation study in four male volunteers exposed to 50 ppm of dioxane determined that the majority (99.3%) of dioxane is eliminated by metabolism to β-hydroxyethoxy-acetic acid (HEAA) with the remaining 0.7% being excreted through the urine (Young et al., 1977). Further studies suggest that the metabolism of dioxane is mediated by cytochrome P450 (Woo et al., 1978). The concentrations of HEAA were found to be 118% higher than the concentration of dioxane, suggesting rapid and extensive metabolism with a calculated metabolic clearance rate of 75 m/min. This same study concluded that repeated daily exposures to 50 ppm of dioxane would not cause adverse effects because accumulated concentrations would never exceed those attained at 50 ppm or less. β-Hydroxyethoxyacetic acid also accounted for >99% of the total urinary excretion of inhaled dioxane in rats (Young et al., 1978). Conversely, when dioxane is intravenously injected in rats, the metabolic clearance decreased indicating metabolic saturation at high doses (1000 mg/kg). Saturation was found to occur at doses >10 mg/kg/bw resulting in accumulation of 1,4-dioxane (HSDB, 1995).

Chemical Pathology

The chemical pathology has not been described in depth for dioxane. Radiolabeled dioxane in rats was found to be widely distributed relatively uniformly among different tissues, yet the extent of covalent bonding was significantly higher in the liver, spleen, and colon (Woo et al., 1977). In addition, an increase in the hepatocyte labeling index was noted with exposures of 1% (v/v) dioxane in drinking water, suggesting a potential role of cell proliferation in the induction of hepatocellular carcinoma. Increased activity of microsomal enzymes also appears to be induced from repeated exposures of 1000 mg/kg/day in rats and 2000 mg/kg/day in mice (Dietz et al., 1982; Pawar and Mungikar, 1976).

SOURCES OF EXPOSURE

Occupational

Occupational exposures to dioxane may occur through inhalation or dermal contact. Occupations involving exposure include industries involved in the manufacturing and transportation, laboratories, and industries that use dioxane as a solvent, stabilizer, or as a paint and varnish stripper. Reports of occupational exposures are limited. However, eye and respiratory irritations were reported for occupational exposures to dioxane with concentrations ranging from 200 ppm for 15 min to 5500 ppm for 1 min (Silverman et al., 1946).

Environmental

Environmental exposures to dioxane may occur via the inhalation of contaminated air or ingestion of contaminated water. Dioxane can be released into the environment from the manufacturing process through the effluent or into the atmosphere. A study analyzing the effluents from nine different industries in Korea found levels ranging from 0.03 to 2.83 mg/l (Gurung et al., 2012). Average dioxane levels for outdoor air concentrations from across the United States during the 1980s were 0.107 parts per billion by volume (ppbv), which included urban, rural, suburban, and remote areas (Shah and Singh, 1988). This same study found average U.S. indoor concentrations at 1.029 ppbv for both workplace and residential environments. In Colorado, groundwater contaminated near a landfill site showed levels ranging from non-detectable to 79 μg/l (EPA, 1988). Cleanup levels vary from state to state ranging from 3 to 85 μg/l in drinking water or groundwater (EPA, 2009). The Environmental Protection Agency (EPA) regions 3 and 6 have calculated a screening level of 6.1 μg/l for dioxane in tap water.

INDUSTRIAL HYGIENE

The current PEL issued by the OSHA for dioxane is 100 ppm (360 mg/m^3) as the TWA for an 8-h workday with a *skin* notation, indicating that the cutaneous route of exposure contributes to the overall exposure. The NIOSH considers dioxane as a potential occupational carcinogen and has established an exposure limit (REL) of 1 ppm (3.6 mg/m^3) as a 30-min ceiling. The ACGIH issued a TLV of 25 ppm as

the TWA for a normal 8-h workday and a 40-h week and also includes a *skin* notation.

Engineering controls should be instituted where possible to prevent and reduce employee exposures to dioxane. Suggested engineering controls to be used during the synthesis and manufacturing of commercial products include process enclosures, local exhaust ventilation, and personal protective equipment. As with the other ethers, ignition sources should be removed immediately if there is a leak or spill. Small quantities can be absorbed using paper towels and placed in a fume hood or appropriate container. For large quantities, vermiculite or dry sand may be needed.

MEDICAL MANAGEMENT

Eyes and skin that become exposed to dioxane should be flushed with plenty of water and skin washed thoroughly with soap and water. Ingestion may require gastric lavage or use of activated charcoal. Severe dermal burns or blistering may need topical treatment. Dioxane-contaminated clothing should be removed and properly stored until the clothing can be properly laundered or disposed. Medical surveillance programs are recommended, which include worker medical screening, medical baseline and routine evaluations, and biological monitoring. Workplace measurements and monitoring should also be conducted to determine airborne concentrations.

REFERENCES

ATSDR (2007) *Draft Toxicological Profile for 1,4-Dioxane.* Agency for Toxic Substances and Disease Registry.

DeRosa, C.T., Wilbur, S., Holler, J., Richter, P., and Stevens, Y. (1996) Health evaluation of 1,4-dioxane. *Toxicol. Ind. Health* 12:1–43.

Dietz, F.K., Stott, W.T., and Ramsey, J.C. (1982) Nonlinear pharmacokinetics and their impact on toxicology: illustrated with dioxane. *Drug Metab. Rev.* 13:963–981.

Environmental Protection Agency (EPA) (1988) *1,4-Dioxane in Shallow Groundwater Lowry Landfill Superfund Site.* Colorado Department of Public Health and Environment. Available at https://www.colorado.gov/pacific/sites/default/files/HM_LowryLandfill-Lowry-Dixoane-Fact-Sheet.pdf.

Environmental Protection Agency (EPA) (2009) *Emerging Contaminants 1,4-Dioxane Fact Sheet.* Office of Solid Waste and Emergency Response (5106P), United States Environmental Protection Agency (EPA 505-F-09-006).

Gurung, A., Kim, S.-H., Joo, J.H., Jang, M., and Oh, S.-E. (2012) Assessing toxicities of industrial effluents and 1,4-dioxane using sulphur-oxidising bacteria in a batch test. *Water Environ. J.* 26:224–234.

Hazardous Substance Data Base (1995) *1,4-Dioxane.* Hazardous Substance Data Bank.

Kasai, T., Kano, H., Umeda, Y., Sasaki, T. and Ikawa, N. (2009) Two-year inhalation study of carcinogenicity and chronic toxicity of 1,4-dioxane in male rats. *Inhal. Toxicol.* 21:889–897.

Pawar, S.S. and Mungikar, A.M. (1976) Dioxane-induced changes in mouse liver microsomal mixed-function oxidase system. *Bull. Environ. Contam. Toxicol.* 15:762–767.

Pilipiuk, Z.I., Gorban, G.M., Solomin, G.I., and Gorshunova, A.I. (1977) Toxicology of 1,4-dioxane. *Kosm. Biol. Aviakosm. Med.* 11:53–57.

Shah, J.J. and Singh, H.B. (1988) Distribution of volatile organic chemicals in outdoor and indoor air. *Environ. Sci. Technol.* 22:1381–1388.

Silverman, L., Schulte, H.F., and First, M.W. (1946) Further studies on sensory response to certain industrial solvent vapors. *J. Ind. Hyg. Toxicol.* 28:262–266.

Torkelson, T.R., Leong, B.K.J., Richter, W.A., and Gehring, P.J. (1974) 1,4-Dioxane. II. Results of a 2-year inhalation study in rats. *Toxicol. Appl. Pharmacol.* 30:287–298.

Woo, Y.T., Argus, M.F., and Arcos, J.C. (1977) Metabolism *in vivo* of dioxane: effect of inducers and inhibitors of hepatic mixed-function oxidases. *Biochem. Pharmacol.* 25:1539–1542.

Woo, Y.T., Argus, M.F., and Arcos, J.C. (1978) Effect of mixed-function oxidase modifiers on metabolism and toxicity of the oncogene dioxane. *Cancer Res.* 38:1621–1625.

Young, J.D., Braun, W.H., and Gehring, P.J. (1978) The dose-dependent fate of 1,4-dioxane in rats. *J. Environ. Pathol. Toxicol.* 2:263–282.

Young, J.D., Braun, W.H., Rampy, L.W., Chenoweth, M.B., and Blau, G.E. (1977) Pharmacokinetics of 1,4-dioxane in humans. *J. Toxicol. Environ. Health* 3:507–520.

57

ESTERS

Raymond D. Harbison and C. Clifford Conaway

Esters are compounds that result from the reaction of an alcohol or phenol with acids or derivatives of acids. Almost all the aliphatic esters are prepared by the reaction between carboxylic acids and alcohols. The subclasses of esters presented in this chapter include the aliphatic esters, halogenated acid esters, alkyl esters of sulfuric acid, important phosphate esters, nitrate esters, and esters of phthalic acid. Some esters, such as aliphatics esters, may present human health hazards (USEPA, 2007), if exposure occurs via the respiratory or dermal routes.

Many organic esters are used extensively in the plastic industry, either as resins or as plasticizers, and as solvents for lacquers. The group of esters, including succinate, adipate, azelate, sebacate, citrate, and phthalate is generally used as plasticizers. Esters of inorganic acids, such as esters of sulfuric acid and phosphate esters, may have prominent corrosive or pharmacological properties. See Table 57.1 for some industrial uses of esters.

ALIPHATIC ESTERS

METHYL FORMATE

Occupational exposure limits: OSHA PEL 250 mg/m^3; NIOSH recommended exposure limit (REL) 250 mg/m^3; derived no effect level (DNEL); DNEL inhalation, systemic effects, long term: 120 mg/m^3; DNEL inhalation, systemic effects, short term: 120 mg/m^3; DNEL

inhalation, local effects, long term: 120 mg/m^3; DNEL inhalation, local effects, short term: 120 mg/m^3; DNEL dermal, systemic effects, long term: 17 mg/kg/bw/day; DNEL dermal, systemic effects, short term: 17 mg/kg/bw/day

Reference values: DNEL inhalation, systemic effects, long term: 20 mg/m^3; DNEL inhalation, systemic effects, short term: 20 mg/m^3; DNEL inhalation, local effects, long term: 20 mg/m^3; DNEL inhalation, local effects, short term: 20 mg/m^3; DNEL dermal, systemic effects, long term: 2.83 mg/kg/bw/day; DNEL dermal, systemic effects, short term: 2.83 mg/kg/bw/day; DNEL oral, systemic effects, long term: 2.83 mg/kg/bw/day; DNEL oral, systemic effects, short term: 2.83 mg/kg/bw/day

Hazard statements: H224—extremely flammable liquid and vapor; H332—harmful if inhaled; H302—harmful if swallowed; H319—causes serious eye irritation; H335—may cause respiratory irritation

Precautionary statements: P210—keep away from heat/sparks/open flames/ . . . /hot surfaces . . . no smoking; P233—keep container tightly closed; P240—ground/bond container and receiving equipment; P241—use explosion-proof electrical/ventilating/lighting/ . . . /equipment; P242— use only non-sparking tools; P243—take precautionary measures against static discharge; P280—wear protective gloves/protective clothing/eye protection/face protection; P303 + P361 + P353—if in contact with skin (or hair), remove/take off immediately all contaminated clothing; rinse skin with water/shower; P370 + P378—in case of fire, use . . . for extinction; P403 + P235—store in well-ventilated place; keep cool; P501—dispose of contents/container to . . .

Hamilton & Hardy's Industrial Toxicology, Sixth Edition. Edited by Raymond D. Harbison, Marie M. Bourgeois, and Giffe T. Johnson.
© 2015 John Wiley & Sons, Inc. Published 2015 by John Wiley & Sons, Inc.

TABLE 57.1 Some Industrial and Commercial Uses of Esters

Areas of Use	Material Involved
Automotive and aircraft industry	Diphosphates, methyl acrylate phenolates, phosphates, succinates
Flavor and fragrance in food and cosmetics	Methyl and amyl salicylate, methyl anthranilate, benzyl and terpinyl acetate, ethyl butyrate
Lacquer and general solvents	Mainly ethyl and butyl acetates
Pharmaceuticals	*p*-Aminobenzoates, aspirin, methyl, and propyl
	p-hydroxybenzoates, diethyl malonates
Plastics and resins	Cellulose and vinyl acetate, phthalates, and polyesters
Plasticizers	Aliphatic carboxy esters, phthalates, and phosphates
	(Bisesi, 1994)

Risk phrases: R12—extremely flammable; R20/22—harmful by inhalation and if swallowed; R36/37—irritating to eyes and respiratory system

Safety phrases: S2—keep out of the reach of children; S9—keep container in a well-ventilated place; S16—keep away from sources of ignition . . . no smoking; S24—avoid contact with skin; S26—in case of contact with eyes, rinse immediately with plenty of water and seek medical advice; S33—take precautionary measures against static discharges

BACKGROUND AND USES

Methyl formate has a rather pleasant odor. It is an aromatic compound found in apples (Neubeller and Buchloh, 1986), and was identified as a volatile constituent in brewed, roasted, and dried coffee (Lovell et al., 1980); it has also been detected in the volatiles of chicken, beef, and pork flavor (Shahidi et al., 1986). Methyl formate is used primarily to manufacture formamide, dimethylformamide, and formic acid. It is also used as a solvent for quick-drying finishes such as lacquers and in organic synthesis. Other uses include use as a larvicide for tobacco and food crops, in the manufacture of certain pharmaceuticals, and as a blowing agent for foam insulation. It was formerly used as a refrigerant for house appliances.

PHYSICAL AND CHEMICAL PROPERTIES

Methyl formate (formic acid methyl ester; methyl methanoate; $HCOOCH_3$, CASRN 107-31-3, molecular weight 60.05) is a colorless, flammable liquid, with a melting point of $-100\,°C$ and a boiling point (bp) of $31.5\,°C$ (O'Neil, 2001; NIOSH, 2010). The vapor density is 2.07 (air = 1). Its solubility in water is $230\,g/l$ at $25\,°C$ (Riddick et al., 1985), but it reacts slowly with water to form formic acid and methyl alcohol (DOT, 1984). It is soluble in ether, chloroform, and is miscible with ethanol (Lide, 2000).

MAMMALIAN TOXICOLOGY

Acute Effects

Methyl formate is known as an irritant to mucous membranes and skin. May cause pneumonitis, but no known human cases have been reported (if exposure is respiratory). Potential symptoms of acute human overexposure to methyl formate are eye and nose irritation, chest tightness, dyspnea, visual disturbance, and central nervous system (CNS) depression (O'Neil, 2001). In experimental animals, high concentrations of inhaled methyl formate resulted in ocular and upper respiratory tract irritation, dyspnea, convulsions, coma, and death. The oral LD_{50} in rats was 1500 mg/kg bw while the acute rat dermal LD_{50} was >4000 mg/kg bw for 98% pure methyl formate (ECB, 2000). In animal studies, methyl formate caused narcosis and death at airborne concentrations of 5000 parts per million (ppm) to 10,000 ppm in 1–3 h. Inhalation of 1500–50,000 ppm by guinea pigs for 2–3 h caused ocular and upper respiratory tract irritation, retching, incoordination/CNS depression, and death; a concentration of 50,000 ppm was fatal within 20–30 min. Pulmonary irritation of lungs was observed after guinea pigs inhaled nonfatal concentrations, which resolved 4–6 days after exposure (ACGIH, 2001).

Young, healthy subjects exposed to 100 ppm (the American Conference of Governmental Industrial Hygienists' [ACGIH] threshold limit value [TLV]) for 8 h reported increased subjective feelings of fatigue without significant measureable impairment of neurobehavioral performance (Sethre et al., 2000). Methyl formate was attributed to cause irritation of eyes and tearing in 10 of 15 women in a factory in which boiling methyl formate and other esters were employed (Grant, 1986). Occupational effects in workers exposed to methyl formate and other solvents (30% methyl formate, in addition to unspecified amounts of ethyl formate and ethyl and methyl acetate) consisted of visual disturbances, CNS depression, irritation of mucous membranes, and dyspnea (ACGIH, 2001).

Chronic Effects

No records for chronic exposure of methyl formate to laboratory animals were found. Methyl formate (98.4% pure) was negative in the Ames test using *Salmonella typhimurium* strains TA 1535, TA 100, TA 1537, and TA 98 at concentrations of 20–5000 µg/plate with and without metabolic activation (ECB, 2000). It is a known neurotoxicant, and in animal inhalation studies it has been demonstrated to cause CNS solvent syndrome, a disease of the nervous system, which causes confusion, headache, dizziness, and/or stupor/coma. However, no known cases of CNS solvent syndrome in humans have been reported. Methyl formate may cause hemolytic anemia or methemoglobinemia (NCBI, 2014). A study of human volunteers experienced increased fatigue without any other neurobehavioral effects when exposed to 100 ppm for 8 h. There is no available data to indicate the carcinogenic effects, and is considered a pregnancy risk group C (NLM, 2011).

Mechanism of Action

In a mouse inhalation assay, sensory irritation of the upper airways and deep lung to 202–1168 ppm methyl formate was assessed. Sensory irritation, which developed slowly over the 30-min exposure period, was the only parameter observed. The possible role of esterase activity in the production of formic acid and methanol was evaluated by treatment of mice with the esterase inhibitor tri-*o*-cresol phosphate (TOCP). Tri-*o*-cresol phosphate pretreatment reduced the irritation response of methyl formate, suggesting that carboxylesterase-mediated hydrolysis plays a role in the irritation effect of methyl formate. The authors suggested that potential generation of formaldehyde was another source of irritation (Larsen and Nielsen, 2012). Cleavage of methyl formate produces methanol and formic acid, the latter is a critical metabolite responsible for metabolic acidosis.

Chemical Pathology

Application of 1 oz (28.35 g) of an insecticide ointment containing methyl formate to the scalp of a 19-month-old infant suffering from *Pediculosis capitis*, and subsequent covering with a rubber bathing cap, resulted in death 20–30 min later. Both methyl formate and methanol were found by micro-boiling point analysis of a 600 g extract of brain tissue. Further confirmation of the presence of formate was obtained by the detection of mercurous formate crystals (Snyder, 1992).

SOURCES OF EXPOSURE

Occupational

Occupational exposure may occur through inhalation and dermal contact at workplaces where methyl formate is produced or used. Workers may be exposed via, respiratory, dermal, oral, and/or eye contact (NIOSH, 2010). National Institute for Occupational Safety and Health (NIOSH) (1983) has estimated that 7738 workers, consisting of 1402 females, are potentially exposed to methyl formate in the United States.

Environmental

Methyl formate may enter the environment via release through various industrial waste streams. For example,

methyl formate has been released in waste water from urea–formaldehyde resin-manufacturing plants and in vent effluents during commercial production of methanol (Lovell et al., 1980). Methyl formate has been identified as a constituent of cigarette smoke and gasoline engine exhaust (Verschueren, 2001); the amount of methyl formate detected in exhaust gases from the combustion of various hydrocarbon fuels ranged from <0.1 to 0.7 ppm (Seizinger and Dimitriades, 1972). Its use as fumigant and larvicide results in direct release to the environment. Volatilization is probably an important fate process. The half-life of methyl formate in the atmosphere has been estimated to be about 71 days (Meylan and Howard, 1993). The half-life of hydrolysis of methyl formate in water has been estimated to be 5 and 12 days at pH values of 7 and 8, respectively (Mill et al., 1987). The potential for bioconcentration in aquatic organisms is estimated to be low (Franke et al., 1994). No data are available for ecotoxicity.

INDUSTRIAL HYGIENE

The permissible Occupational Safety and Health Administration (OSHA) exposure limit (permissible exposure limit [PEL]) for methyl formate in an 8-h time-weighted average (TWA) exposure is 100 ppm (250 mg/m^3) (NIOSH, 2010). The 8-h TLV is 100 ppm, and the short-term exposure limit (STEL) for methyl formate is 150 ppm (ACGIH, 2008). The NIOSH REL (10-h TWA) is also 100 ppm (NIOSH, 2003). Methyl formate is a very dangerous fire hazard. The National Fire Protection Association (NFPA) hazard classification for methyl formate is Health: 2 (hazardous to health, but areas may be entered freely with self-contained breathing apparatus); Flammability: 4 (very flammable gas or very volatile flammable liquid.), making it a fire hazard; and Reactivity: 0 (material normally stable even under fire exposure conditions, not reactive with water; normal fire-fighting procedures may be used (NFPA, 2002). Immediately dangerous to life or health is 4500 ppm (NIOSH, 2003). Methyl formate is considered hazardous as per the Hazard Communication Standard 29 CFR 1910 1200.

MEDICAL MANAGEMENT

Persons exposed to high levels of methyl formate should be removed as soon as possible to fresh air. Immediate removal of contact lenses to prevent further eye irritation and contaminated clothing, and washing of the skin followed by inhalation of a mist of a 5% bicarbonate solution is recommended. If pulmonary edema develops, medical advice should be sought as quickly as possible. Oxygen should be administered if the patient exhibits signs of respiratory failure (Snyder, 1992). Seek medical treatment immediately. Monitor for shock and treat if necessary. Anticipate seizures, and minimize all stimuli. For eye contamination, flush eyes immediately with water, then irrigate each eye continuously with normal saline if possible; do not use eye ointment. For ingestion, rinse mouth and administer 5 ml/kg up to 200 ml of water for dilution, if the patient can swallow, has a strong gag reflex, and does not drool. Alternatively, administer a 12.5% slurry of activated charcoal (25–100 ml) to minimize absorption from the alimentary tract (Bronstein and Currance, 1994). Do not induce vomiting (NCBI, 2014; NIOSH, 2010).

REFERENCES

American Conference of Governmental Industrial Hygienists (2001) *Documentation of Threshold Limit Values for Chemical Substances and Physical Agents and Biological Exposure. Indices for 2001*, Cincinnati, OH: American Conference of Governmental Industrial Hygienists, p. 11.

American Conference of Governmental Industrial Hygienists (ACGIH) (2008) *Threshold Limit Values (TLVs) for Chemical Substances and Physical Agents and Biological Exposure Indices (BEIs)*, Cincinnati, OH: American Conference of Governmental Industrial Hygienists.

Bisesi, M.S. (1994) Esters. In: Clayton, G.D. and Clayton, F. E., editors., *Patty's Industrial Hygiene and Toxicology*, 4th ed., New York, NY: John Wiley & Sons, Inc.

Bronstein, A.C. and Curranace, P.L. (1994) *Emergency Care for Hazardous Materials Exposure*, 2nd ed., St. Louis, MO: Mosby Lifeline, pp. 203–204.

Department of Transportation (DOT) (1984) *CHRIS-Hazardous Chemical Data*, vol. II, Washington, DC: U.S. Government Printing Office.

European Chemicals Bureau, IUCLID Data set (107-31-3), (2000) CD-ROM edition. It is listed in the References as Chemicals Bureau IUCLID Data set (2000) *Methyl formate (107-31-3)*.

Franke, C., Studinger, G., Berger, G., Böhling, S., Bruckmann, U., Cohors-Fresenborg, D., and Jöhncke, U. (1994) The assessment of bioaccumulation. *Chemosphere* 29:1501–1514.

Grant, W.M. (1986) *Toxicology of the Eye*, 3rd ed., Springfield, IL: Charles C. Thomas Publisher,

Larsen, S.T. and Nielsen, G.D. (2012) Acute airway irritation of methyl formate in mice. *Arch. Toxicol.* 86:285–292.

Lide, D.R., editor. (2000) *CRC Handbook of Chemistry and Physics*, 81st ed., Boca Raton, FL: CRC Press LLC.

Lovell, R.J., Key, J.A., Standifer, R.L., Kalcevic, V., and Lawson, J.F. (1980) *Organic chemical manufacturing selected processes*. Vol 9. Document number *EPA/450/3-80/028D*.

Meylan, W.M. and Howard, P.H. (1993) Computer estimation of the atmospheric gas-phase reaction rate of organic compounds with hydroxyl radicals and ozone. *Chemosphere* 26:2293–2299.

Mill, T., et al. (1987) *Environmental Fate and Exposure Studies Development of a PC-SAR for Hydrolysis: Esters, Alkyl Halides and Epoxides*. (EPA Contract no. 68—02-4254). Menlo Park, CA: SRI International.

National Center for Biotechnology Information (NCBI). *Methyl Formate*. Available at http://pubchem.ncbi.nlm.nih.gov/summary/summary.cgi?cid=7865#x351 (accessed 2014).

Neubeller, J. and Buchloh, G. (1986) Vergleichende Untersuchungen der Apfelsorte Jonagold und ihrer Eltern Jonathan und Golden Delicious. *Mitt. Klosterneuburg* 36:34–46.

National Fire Protection Association (NFPA) (2002) *Fire Protection Guide to Hazardous Materials*, 13th ed., Quincy, MA: National Fire Protection Association.

National Institute for Occupational Safety and Health (NIOSH). *National Occupational Exposure Survey (NOES) 1981–1983 Dataset*. Available at http://www.cdc.gov/noes (accessed 2014).

National Institute for Occupational Safety and Health (NIOSH) (2010) *Pocket Guide to Chemical Hazards*, DHHS (NIOSH) Publication Number 2010-168c, Washington, DC: National Institute for Occupational Safety and Health.

National Library of Medicine (NIH) (2011) *Methyl formate*. Available at http://toxnet.nlm.nih.gov/cgi-bin/sis/search/a?dbs+hsdb:@term+@DOCNO+232 (accessed 2014).

NIOSH Pocket Guide to Chemical Hazards (2003) National Institute for Occupational Safety and Health Publication No. 2004-103.

O'Neil, M.J., editor. (2001) *The Merck Index—An Encyclopedia of Chemicals, Drugs, and Biologicals*, 13th ed., Whitehouse Station, NJ: Merck and Co., Inc.

Riddick, J.A., Bunger, W.B., and Sakano, T.K. (1985) *Techniques of Chemistry*, 4th ed., vol. 2, New York, NY: Wiley-Interscience.

Seizinger, D.E. and Dimitriades, B. (1972) Oxygenates in exhaust from simple hydrocarbon fuels. *J. Air Pollution Control Assoc.* 22(1):47–51.

Sethre, T., Läubli, T., Berode, M., Hangartner, M., and Krueger, H. (2000) Experimental exposure to methyl formate and its neurobehavioral effects. *Int. Arch. Occup. Environ. Health* 73:401–409.

Shahidi, F., Rubin, L.J., and D'Souza, L.A. (1986) Meat flavor volatiles: a review of the composition, techniques of analysis and sensory evaluation. *Crit. Rev. Food Sci. Nutr.* 24:141–243.

Snyder, R., editor. (1992) Alcohols and esters. In: *Ethel Browning's Toxicity and Metabolism of Industrial Solvents*, 2nd ed., vol. 3, New York, NY: Elsevier.

United States Environmental Protection Agency (USEPA) (2007) *Reregistration eligibility decision: aliphatic esters. Prevention, pesticides, and toxic substances*. Available at http://www.epa.gov/oppsrrd1/REDs/aliphatic_esters.pdf (accessed 2014).

Verschueren, K. (2001) *Handbook of Environmental Data on Organic Chemicals*, 4th ed., New York, NY: John Wiley & Sons Inc.

ETHYL FORMATE

Occupational exposure limits: 300 mg/m^3 (OSHA PEL, NIOSH REL, ACGIH TLV)

Reference values: None

Risk phrases: R11—highly flammable; R20/22—harmful by inhalation and if swallowed; R36/37—irritating to the eyes and respiratory system

Safety phrases: S2—keep out of the reach of children; S9—keep container in a well-ventilated place; S16—keep away from sources of ignition . . . no smoking; S24—avoid contact with skin; S26—in case of contact with eyes, rinse immediately with plenty of water and seek medical advice; S33—take precautionary measures against static discharges

BACKGROUND AND USES

Ethyl formate (Ethyl ester, ethyl methanoate; $HCOOCH_2CH_3$; CASRN 109-94-4) is a volatile liquid with a fruity odor and slightly bitter taste (NIOSH, 1994). Ethyl formate is naturally found in some fruits, such as in apples and pears, in Florida orange juice, and in varieties of honey (Furia and Bellanca, 1975). Major industrial uses are in the lacquer industry, and as a solvent for cellulose acetate in the manufacture of artificial silk, safety glass, and celluloid for shoes, as well as in organic synthesis. It is also used as a solvent for various oils and greases (NLM, 2013). It is used as a flavoring agent in lemonade and in artificial rum and arrack, as a substitute for acetone, and in the synthesis of synthetic sex hormones (Budavari, 1996). It is also used as a larvicide or fungicide in agricultural fumigants and sprays for food crops, cereals, dried fruit, and tobacco (Bisesi, 1994).

PHYSICAL AND CHEMICAL PROPERTIES

Ethyl formate, molecular weight 74.09, is a colorless liquid. The bp is 54.5 °C at 760 mmHg, and melting point is −80.5 °C; its density is 0.9168 at 20 °C (Lide, 1990). The ester is slightly soluble in water (9 parts/100 at 18 °C) with gradual decomposition into formic acid and ethanol; it is miscible in ethanol, ether, and acetone (HSDB, 2013), as well as in benzene. The vapor density is 2.55 (air = 1) (Bisesi, 1994), and the vapor pressure is 245 mmHg at 25 °C (Daubert and Danner, 1989).

MAMMALIAN TOXICOLOGY

Acute Effects

It is considered moderately toxic at a lethal dose to humans at 0.5–5 mg/kg between 1 ounce and 1 pint for a 70 kg person (NCBI, 2014). Overexposure may cause irritation of the eyes, nose, throat, and skin (Budavari, 1996; Snyder, 1992). At 330 ppm in air, it causes slight eye irritation that subsides only after several hours (Grant, 1986); and strong persistent nasal irritation, probably due to its hydrolysis to formic acid and ethanol (ACGIH, 2008; Snyder, 1992). In persons with impaired pulmonary function, especially those with obstructive airway disease, breathing ethyl formate might cause exacerbation of symptoms due to its irritant properties (NIOSH, 1981). The acute hazards of ethyl formate are regarded as considerably less than those of methyl formate. Nevertheless, exposure of mice, cats, and dogs to 10,000 ppm for 20 min produced eye irritation and dyspnea, and a 4-h exposure of 10,000 ppm to dogs was fatal after development of pulmonary edema (Bisesi, 1994). Concentrations of 1500 ppm are regarded as immediately dangerous to human life and health (NIOSH, 1997).

Chronic Effects

Potential effects of chronic exposure to ethyl formate have not been adequately reported. When tested by application to the skin of mice for a period of 10 weeks, no tumors were observed (Roe and Salaman, 1955). In another study using A/He mice, total doses of 2.4 or 12 g/kg ethyl formate injected intraperitoneally three times per week for 8 weeks resulted in no primary lung tumors (Snyder, 1992). Ethyl formate is not listed as a human carcinogen by International Agency for Research on Cancer (IARC), and no data were found regarding its genotoxicity. No data are available as per the mutagenic or teratogenic effects of ethyl formate (NCBI, 2014).

Mechanism of Action

Adverse effects of ethyl formate on the eyes and respiratory tract are probably due to rapid hydrolysis of the ester on contact with water, with subsequent formation of ethanol and formic acid. The irritating effects of ethyl formate appear to be less potent than that of methyl formate. The effects appear to be somewhat residual, rather than transient, as irritation of eyes and nose remained for 4 h after exposure to 10,500 ppm ethyl formate (Snyder, 1992). The anesthetic potency of ethyl formate is less than that of chlorinated hydrocarbons and ethyl ether, but it is more active than ethanol, acetone, and aliphatic hydrocarbons (ACGIH, 2008).

Chemical Pathology

The adverse effects of ethyl formate on the eyes, nasal passages, respiratory tract, skin, and other organs is more than likely due to its hydrolysis to formic acid, causing irritation and tissue damage.

SOURCES OF EXPOSURE

Occupational

It has been estimated that 8149 workers are exposed to ethyl formate in the United States (NOES, 1983). Occupational exposure is typically via inhalation and dermal contact.

Adequate personal protective clothing for skin protection and sufficient eye protection to prevent eye contact is required (NIOSH, 1997). Screening of employees for medical conditions that might place them at increased risk from ethyl formate, such as chronic respiratory disease, skin diseases, and liver and kidney disease is recommended (NIOSH, 1981).

Environmental

Release of ethyl formate to the environment will often occur from its transport, disposal, and use as a synthetic flavoring agent, as a fungicide and larvicide for tobacco, cereals, dried fruit and other crops, and manufacturing of other products. Ethyl formate is expected to degrade rapidly by chemical hydrolysis in soil and water. It will also readily volatilize, hence relatively high concentrations are released to the atmosphere, where it is attacked by photochemically produced hydroxyl radicals, with an estimated half-life of 11 days (Atkinson, 1989).

INDUSTRIAL HYGIENE

The PEL for ethyl formate is 100 ppm (300 mg/m^3) (NIOSH, 2010). The 8-h TLV–TWA is 100 ppm; excursions in worker exposure levels may not exceed three times the TLV for more than 30 min during a work day (ACGIH, 2008). The NIOSH REL for a 10-h time-weighted exposure is also 100 ppm (300 mg/m^3) (NIOSH, 2010). The NFPA hazard classification for ethyl formate is Health: 2 (hazardous to health, but areas may be entered freely with self-contained breathing apparatus). Flammability: 3 (easily ignited under fire exposure conditions; water may be ineffective in controlling or extinguishing fires). Reactivity: 0 (material normally stable even under fire exposure conditions, and do not react with water; normal fire-fighting procedures may be used) (NFPA, 1997).

MEDICAL MANAGEMENT

Target organs are eyes, respiratory, and CNS (NIH, 2011). Following inhalation exposure, the patient should be moved to fresh air. Monitor for respiratory distress, and if necessary, administer oxygen and assist with ventilation. Treat bronchospasm with an inhaled beta-2 agonist. Supportive care may be necessary for CNS depression. For eye exposure, irrigate exposed eyes with copious amounts of water at room temperature for 15 min or more. For dermal exposure, remove decontaminated clothing; treat dermal irritation or burns with standard topical therapy. For ingestion, do not induce emesis. Significant esophageal or gastrointestinal irritation and possibly burns may occur following ingestion of ethyl formate. Administration of 4–8 oz of water or milk (children: not to exceed 4 oz) is advised in such instances. Administration of slurry of a 12.5% activated charcoal is also recommended (NLM, 2013).

REFERENCES

American Conference of Governmental Industrial Hygienists (ACGIH) (2008) *Threshold Limit Values (TLVs) for Chemical Substances and Physical Agents and Biological Exposure Indices (BEIs)*, Cincinnati, OH: American Conference of Governmental Industrial Hygienists.

Atkinson, R. (1989) Kinetics and mechanisms of the gas phase reactions of the hydroxyl radical with organic compounds. Number 1. *Journal of Physical and Chemical Reference Data Monographs and Supplements.*

Bisesi, M.S. (1994) Esters. In: Clayton, G.D. and Clayton, F.E., editors., *Patty's Industrial Hygiene and Toxicology*, 4th ed., New York, NY: John Wiley & Sons, Inc.

Budavari, S., editor. (1996) *The Merck Index—An Encyclopedia of Chemicals, Drugs, and Biologicals*, 12th ed., Whitehouse Station, NJ: Merck and Co. Inc.

Daubert, T.E. and Danner, R.P. (1989) *Physical and Thermodyamic Properties of Chemicals Data Compilation*, Washington, DC: Taylor and Francis.

Furia, T.E. and Bellanca, N., editors. (1975) *Fenaroli's Handbook of Flavor Ingredients*, vol. 2, Cleveland, OH: The Chemical Rubber Co.

Grant, W.M. (1986) *Toxicology of the Eye*, 3rd ed., Springfield, IL: Charles C. Thomas Publisher.

Lide, D.R., editor. (1990) *CRC Handbook of Chemistry and Physics*, 71st ed., Boca Raton, FL: CRC Press LLC.

National Center for Biotechnology Information (NCBI) (2014) *Methyl Formate.* Available at http://pubchem.ncbi.nlm.nih.gov/summary/summary.cgi?cid=7865#x351 (accessed 2014).

National Fire Protection Association (NFPA) (1997) *Fire Protection Guide to Hazardous Materials*, 12th ed, Quincy, MA: National Fire Protection Association.

National Institutes of Health (NIH) (2011) Ethyl Formate. Available at http://toxnet.nlm.nih.gov/cgi-bin/sis/search/a?dbs+hsdb:@term+@DOCNO+232 (accessed 2014).

National Institute for Occupational Safety and Health (NIOSH) (1981) *Occupational Health Guidelines for Chemical Hazards*, DHHS (NIOSH) Publication Number 81–123. Washington, DC: National Institute for Occupational Safety and Health.

National Institute for Occupational Safety and Health (NIOSH) (2010) *Pocket Guide to Chemical Hazards*, DHHS (NIOSH) Publication Number 2010-168c, Washington, DC: National Institute for Occupational Safety and Health.

National Institute for Occupational Safety and Health (NIOSH). *National Occupational Exposure Survey (NOES) 1981–1983 Dataset.* Available at http://www.cdc.gov/noes (accessed 2014).

NIOSH (1994) *NIOSH Pocket Guide to Chemical Hazards, Publication No. 94-116.* Washington, DC.

NIOSH (1997) *NIOSH Pocket Guide to Chemical Hazards. DHHS Publication No. 97–140*, Washington, DC: U. S. Government Printing Office, p. 140.

NLM (2013) Hazardous Substances Data Bank of the National Library of Medicine's TOXNET system: Ethyl Formate, CAS Registry number 109-94-4.

NOES (1983) National Occupational Exposure Survey (NOES 1981-1983), NIOSH.

Roe, F.J.C. and Salaman, M.H. (1955) Further studies on incomplete carcinogenesis in the mouse. *Br. J. Cancer* 9:177–203.

Snyder, R., editor. (1992) Alcohols and esters. In: *Ethel Browning's Toxicity and Metabolism of Industrial Solvents*, 2nd ed., vol. 3, New York, NY: Elsevier.

METHYL ACETATE

Occupational exposure limits: ACGIH TLV 610 mg/m^3; ACGIH STEL 757 mg/m^3; DNEL inhalation, systemic effects, long term: 610 mg/m^3; DNEL inhalation, local effects, long term: 305 mg/m^3; DNEL dermal, systemic effects, long term: 88 mg/kg/bw/day

Reference values: DNEL inhalation, systemic effects, long term: 131 mg/m^3; DNEL inhalation, local effects, long term: 152 mg/m^3; DNEL dermal, systemic effects, long term: 44 mg/kg/bw/day; DNEL oral, systemic effects, long term: 44 mg/kg/bw/day

Hazard statements: H225—highly flammable liquid and vapor; H319—causes serious eye irritation; H336—may cause drowsiness or dizziness

Precautionary statements: P210—keep away from heat/ sparks/open flames/ . . . /hot surfaces . . . no smoking; P243—take precautionary measures against static discharge; P280—wear protective gloves/protective clothing/eye protection/face protection; P271—use only outdoors or in a well-ventilated area; P370 + P378—in case of fire: use . . . for extinction; additional text water spray, fire extinguishing powder, foam, or carbon dioxide; P304 + P340—if inhaled, remove victim to fresh air and keep at rest in a position comfortable for breathing; P312—call a poison center or doctor/physician if you feel unwell; P305 + P351 + P338—if in eyes, rinse cautiously with water for several minutes; remove contact lenses, if present and easy to do; continue rinsing; P337 + P313—if eye irritation persists, get medical advice/attention; P403 + P233—store in a well-ventilated place; keep container tightly closed

BACKGROUND AND USES

Methyl acetate (methyl ethanoate; CH_3COOCH_3; CASRN 79-20-9) occurs naturally in low concentrations in mint, fungus, grapes, banana, coffee (Furia and Bellanca, 1975) and is a volatile constituent of nectarines (Takeoka et al., 1988). It is also present in some distilled alcoholic beverages (Shimoda et al., 1993). It is produced industrially via the carbonylation of methanol as a byproduct of acetic acid production or by esterification of acetic acid with methanol in the presence of strong acid such as sulfuric acid. Methyl acetate is mainly used as an effective solvent for the production of nitrocellulose, cellulose acetate, resins, greases, synthetic leathers, perfumes, and in organic synthesis (Budavari, 1996); it is used as a solvent for glues, paints, and nail polish removers.

PHYSICAL AND CHEMICAL PROPERTIES

Methyl acetate is a colorless, volatile liquid, with a molecular weight of 74.08 (Budavari, 1996) and a fragrant, fruity odor and fleeting fruity taste (Furia, 1980). The water solubility of methyl acetate is 243.5 g/l at 20 °C (Stephan and Stephan, 1963). The vapor pressure is 216 mmHg at 25 °C (77 °F) (Budavari, 1996); the melting point is −98 °C and bp is 56.8 °C (Lide, 1995). The specific gravity is 0.9342 at 4 °C (Lide, 1995) and vapor density is 2.8 (air = 1) (ILO, 1983). Methyl acetate is soluble in benzene, acetone, and chloroform (Weast, 1988), and is miscible with alcohol and ether (Budavari, 1996).

MAMMALIAN TOXICOLOGY

Acute Effects

This ester causes irritation to the eyes, skin, nose, and throat; excess exposure causes headache, vertigo, drowsiness, palpitation, and even unconsciousness. Even so, the use of nail polish remover containing 50% w/w methyl acetate is regarded as safe by the Danish EPA (SCCNFP, 2003). Chest tightness, dyspnea, and blurred vision may occur after excess industrial exposure. Metabolic acidosis is a possible symptom of excess exposure, as methanol, which yields formic acid, is a major metabolite (Minns et al., 2013). Employees with chronic respiratory, skin, liver, or kidney disease may be at increased risk for symptoms (NIOSH, 1981). Cases of slight poisoning under industrial conditions have been reported, and one case of blindness had occurred, most likely the result of its biotransformation to methanol. No cases of lethal poisoning have occurred (ILO, 1983).

Chronic Effects

Repeated skin contact with liquid methyl acetate may cause dryness, cracking, and skin irritation. Methyl acetate is not listed in the most recent summary of industrial carcinogens published by the IARC (2012). Methyl acetate was not mutagenic in a robust Ames test (*Salmonella typhimurium*) (WHO, 1998). No additional information was found regarding its potential mutagenic or genotoxic effects. No information was found regarding potential effects of methyl acetate on fetal development in animal studies. Hydrolysis products, such as methanol, may cause permanent damage to the retina (NLM, 2013).

Mechanism of Action

The primary action of methyl acetate in industry is its use as an effective solvent with a rather pleasant smell. In the presence of strong acid, it is converted to methanol and acetic acid, with subsequent toxicity. It reacts slowly with water to form acetic acid and methanol (NIOSH, 2010).

Chemical Pathology

In mammals, methyl acetate is metabolized to methanol and acetate. Methanol is then rapidly converted to formaldehyde by the action of alcohol dehydrogenase. In humans and other primates, formaldehyde is converted to formic acid, with a half-life of about 1.5 min, but formic acid is converted to carbon dioxide much more slowly, with half-life of approximately 20 h. For that reason formic acid builds up in tissues, and causes severe metabolic acidosis, which can lead to retinopathy similar to that of methanol poisoning. Other common clinical manifestations of methanol poisoning include CNS depression, weakness, headache, and vomiting (Fox and Boyes, 2008; SCCNFP, 2003).

SOURCES OF EXPOSURE

In addition to industrial exposure, methyl acetate is permitted as a minimal food additive, as a synthetic flavoring substance, and contaminant adhesives. Natural concentrations of methyl acetate in foods and alcoholic beverages are considered to be negligible. Contaminated drinking water is also a potential source of exposure (USEPA, 1975).

Occupational

National Institute for Occupational Safety and Health has estimated that 17,851 workers (4038 of these are female) are exposed to methyl acetate in the United States, 92% of which are exposed during the use of trade name compounds which contain methyl acetate (NOES, 1983). Most probable routes of occupational exposure are by inhalation and dermal contact. No cases of irritation or systemic injury have been reported for exposures at or below 200 ppm (ACGIH, 1991).

Environmental

The production of methyl acetate and its use as a solvent, synthetic flavoring additive, component of artificial leather, as well as use in organic synthesis may result in its release to the environment. Methyl acetate is expected to have very high mobility in soil, and volatilization from soil surfaces is expected to be rapid (Buttery et al., 1969). If released into water, methyl acetate is expected to volatilize to the atmosphere rather rapidly (Lyman et al., 1982). Hydrolysis of methyl acetate is not expected to be significant except under highly basic conditions, that is, >pH 9 (Drossman et al., 1987). Laboratory experiments indicate that airborne methyl acetate is rapidly degraded by light at 250 nm (half-life ~24 h) or 360 nm (half-life <6 h) (Güsten et al., 1984).

INDUSTRIAL HYGIENE

The ACGIH TLV–TWA is 200 ppm (610 mg/m^3), with a STEL of 250 ppm (757 mg/m^3). Respiratory depression and metabolic acidosis are anticipated as early signs of excess exposure (ACGIH, 2001). Methyl acetate is a highly flammable liquid, and may form dangerous explosive mixtures with air (NFPA, 1997). The NFPA hazard classification for methyl acetate is Health: 1, Flammability: 3, and Reactivity: 0 (NFPA, 1997). The immediate dangerous to life or health level is above 3100 ppm (NIOSH, 2010; NLM, 2013).

MEDICAL MANAGEMENT

In general, gastrointestinal decontamination after oral ingestion is not very useful because methanol is rapidly absorbed and binds poorly to activated charcoal. Very shortly after large ingestions, insertion of a nasogastric tube to aspirate stomach contents may be helpful. Do not use emetics. Treat patients with either ethanol or fomepizole (4-methylpyrazole) to prevent its conversion to formate. This is indicated if plasma methanol concentrations are >20 mg/dL or there is a strong suspicion of methanol poisoning. Ethanol is given to maintain serum ethanol concentration of 100–150 ml/dL. Fomepizole is administered as a 15 mg/kg loading dose, followed by four bolus doses of 10 mg/kg every 12 h. Using fomepizole over ethanol is suggested, as fomepizole does not cause CNS depression or hypoglycemia and may decrease the need for dialysis (NIH, 2003). For inhalation exposure, move patient to fresh air, and monitor for respiratory distress. If cough or difficulty in breathing develops, evaluate for respiratory tract irritation, bronchitis, or pneumonitis. Administer oxygen and assist in ventilation as required. For eye exposure: decontaminate eyes with copious amounts of water at room temperature for at least 15 min. For dermal exposure, remove contaminated clothing and wash exposed area thoroughly with soap and water. If irritation or pain persists, consult a physician.

REFERENCES

American Conference of Governmental Industrial Hygienists (ACGIH) (1991) *Threshold Limit Values (TLVs) for Chemical Substances and Physical Agents and Biological Exposure Indices (BEIs)*, Cincinnati, OH: American Conference of Governmental Industrial Hygienists.

American Conference of Governmental Industrial Hygienists (ACGIH) (2001) *Threshold Limit Values (TLVs) for Chemical Substances and Physical Agents and Biological Exposure Indices (BEIs)*, Cincinnati, OH: American Conference of Governmental Industrial Hygienists.

Budavari, S., editor. (1996) *The Merck Index—An Encyclopedia of Chemicals, Drugs, and Biologicals*, 12th ed., Whitehouse Station, NJ: Merck and Co. Inc.

Buttery, R.G., Ling, L., and Guadagni, D.G. (1969) Food volatiles. Volatilities of aldehydes, ketones, and esters in dilute water solution. *J. Agric. Food Chem.* 17:385–389.

Drossman, H., Johnson, H., and Mill, T. (1987) Structure activity relationships for environmental processes 1. Hydrolysis of esters and carbamates. *Chemosphere* 17:1509–1530.

Environmental Protection Agency (USEPA) (1975) *Preliminary Assessment of Suspected Carcinogens in Drinking Water*. Interim Report to Congress.

Fox, D.A. and Boyes, W.K. (2008) Toxic responses of the ocular and visual system. In: Klaassen, C.D., editor. *Casarett & Doull's Toxicology: The Basic Science of Poisons*, 7th ed., New York, NY: McGraw-Hill Professional.

Furia, T.E., editor. (1980) *CRC Handbook of Food Additives*, 2nd ed., vol. 2, Boca Raton, FL: CRC Press, Inc.

Furia, T.E. and Bellanca, N., editor. (1975) *Fenaroli's Handbook of Flavor Ingredients*, vol. 2, Cleveland, OH: The Chemical Rubber Co.

Güsten, H., Klasinc, L., and Marić, D. (1984) Prediction of the abiotic degradability of organic compounds in the troposphere. *J. Atmos. Chem.* 2:83–93.

International Agency for Research on Cancer (IARC) (2012) *A Review and Update on Occupational Carcinogens*, vol. 100, Lyon, France: International Agency for Research on Cancer.

International Labour Office (ILO) (1983) *Encyclopedia of Occupational Health and Safety*, vol. 1 and 2, Geneva, Switzerland: International Labour Office.

Lide, D.R., editor. (1995–1996) *CRC Handbook of Chemistry and Physics*, 76th ed., Boca Raton, FL: CRC Press Inc., pp. 3–8.

Lyman, W.J., et al. (1990) *Handbook of Chemical Property Estimation Methods*, Washington, DC: American Chemical Society, pp. 15-1–15-29.

Minns, A., McIlvoy, A., Clark, A., Clark, R.F., and Cantrell, F. (2013) Examining the risk of methanol poisoning from methyl acetate-containing products. *Am. J. Emerg. Med.* 31:964–966.

National Fire Protection Association (NFPA) (1997) *Fire Protection Guide to Hazardous Materials*, 12th ed., Quincy, MA: National Fire Protection Association.

National Institutes of Health (NIH) (2003) *Methyl Acetate*. Available at http://toxnet.nlm.nih.gov/cgi-bin/sis/search/a?dbs+hsdb:@term+@DOCNO+95 (accesed 2014).

National Institute for Occupational Safety and Health (NIOSH) (1981) *Occupational Health Guidelines for Chemical Hazards*, DHHS (NIOSH) Publication Number 81–123, Washington, D.C.: National Institute for Occupational Safety and Health.

National Institute for Occupational Safety and Health (NIOSH) (2010) *Pocket Guide to Chemical Hazards*, DHHS (NIOSH) Publication Number 2010-168c, Washington, D.C.: National Institute for Occupational Safety and Health.

National Institute for Occupational Safety and Health (NIOSH) (2010) *Pocket Guide to Chemical Hazards*, DHHS (NIOSH) Publication Number 2010-168c, Washington, D.C.: National Institute for Occupational Safety and Health.

National Library of Medicine (NLM) (2013) *Hazardous substance database: Methyl acetate*. National Library of Medicine. Available at http://toxnet.nln.nih.gov (accessed, 2014).

National Library of Medicine (NLM) (2013) *Methyl Acetate*. Available at http://hazmap.nlm.nih.gov/category-details?table=copytblagents&id=546 (accessed 2014).

NOES (1983) National Occupational Exposure Survey (NOES), NIOSH.

Scientific Committee on Cosmetic Products and Non-food Products intended for Consumers-Denmark (SCCNFP) (2003) *Evaluation and Opinion on: Methyl Acetate*.

Shimoda, M., Shibamoto, T., and Noble, A.C. (1993) Evaluation of headspace volatiles of Cabernet Sauvignon wines sampled by an on-column method. *J. Agric. Food Chem.* 41:1664–1668.

Stephan, M. and Stephan, T., editors. (1963) *Solubilities of Inorganic and Organic Compounds in Binary Systems*, New York, NY: John Wiley & Sons.

Takeoka, G.R., Flath, R.A., Guentert, M., and Jennings, W. (1988) Nectarine volatiles: vacuum steam distillation versus headspace sampling. *J. Agric. Food Chem.* 36:553–60.

World Health Organization (WHO) (1998) *Safety Evaluation of Certain Food Additives, WHO Food Additive Series 40*, Geneva, Switzerland: International Program of Chemical Safety, World Health Organization.

Weast, R.C., editor. (1988) *CRC Handbook of Chemistry and Physics*, 69th ed., Boca Raton, FL: CRC Press, Inc.

ETHYL ACETATE

Occupational exposure limits: 1400 mg/m^3 (OSHA PEL, NIOSH REL, ACGIH TLV); DNEL inhalation, systemic effects, long term: 734 mg/m^3; DNEL inhalation, systemic effects, short term: 1468 mg/m^3; DNEL inhalation, local effects, long term: 734 mg/m^3; DNEL inhalation, local effects, short term: 1468 mg/m^3; DNEL dermal, systemic effects, long term: 63 mg/kg/bw/day

Reference values: DNEL inhalation, systemic effects, long term: 367 mg/m^3; DNEL inhalation, systemic effects, short term: 734 mg/m^3; DNEL inhalation, local effects, long term: 367 mg/m^3; DNEL inhalation, local effects, short term: 734 mg/m^3; DNEL dermal, systemic effects, long term: 37 mg/kg/bw/day; DNEL oral, systemic effects, long term: 4.5 mg/kg/bw/day

Hazard statements: H225—highly flammable liquid and vapor; H319—causes serious eye irritation; H336—may cause drowsiness or dizziness

Precautionary statements: P210—keep away from heat/ sparks/open flames/ . . . /hot surfaces . . . no smoking; P233—keep container tightly closed; P240—ground/bond container and receiving equipment; P241—use explosion-proof electrical/ventilating/lighting/equipment; P242—use only non-sparking tools; P243—take precautionary measures against static discharge; P280—wear protective gloves/protective clothing/eye protection/face protection; P303 + P361 + P353—if in contact with skin (or hair): remove/take off immediately all contaminated clothing; rinse skin with water/shower; P370 + P378—in case of fire: use . . . for extinction; additional text water spray, foam, or carbon dioxide; P403 + P235—store in a well-ventilated place; keep cool; P501—dispose of contents/ container to . . . ; P264—wash . . . thoroughly after handling; P305 + P351 + P338—if in eyes, rinse cautiously with water for several minutes; remove contact lenses, if present and easy to do; continue rinsing; P337 + P313—if eye irritation persists: get medical advice/attention; P261—avoid breathing dust/fume/gas/mist/vapors/spray; P271—use only outdoors or in a well-ventilated area; P304 + P340—if inhaled, remove victim to fresh air and keep at rest in a position comfortable for breathing; P312—call a poison center or doctor/physician if you feel unwell; P403 + P233—store in a well-ventilated place; keep container tightly closed; P405—store locked up

Risk phrases: R11—highly flammable; R36—irritating to eyes; R66—repeated exposures may cause skin dryness or cracking; R67—vapors may cause drowsiness and dizziness

Safety phrases: S2—keep out of the reach of children; S16— keep away from sources of ignition . . . no smoking; S26—in case of contact with eyes, rinse immediately with plenty of water and seek medical advice; S33—take precautionary measures against static discharges

BACKGROUND AND USES

Ethyl acetate (ethyl ethanoate; vinegar naphtha; $CH_3COOC_2H_5$; CASRN 141-78-6) is a colorless, clear liquid with an ether-like fruity odor and a wine-like burning taste. Commercial uses include as an artificial fruit essence, in perfumes, nail polish, and other manicure products. It is also used as a solvent for varnishes and lacquers, in the manufacture of smokeless powder, photographic films and plates, and in artificial silk (Budavari, 1996). Ethyl acetate is found at 25–50 ppm in American-brewed lager beer (Reed, 1983). It is also a component of fresh grapefruit juice (Cadwallader and Xu, 1994). It has been identified as a volatile component of fried bacon, cheeses, kiwi fruit, wines, various cultivars of oranges, and of at least two apple species (NLM, 2013; Mattheis et al., 1991; Moshonas and Shaw, 1994).

PHYSICAL AND CHEMICAL PROPERTIES

At 25 °C, its solubility in water is 64 g/l. It is soluble in alcohol, ether, acetone, and benzene (Weast, 1988) and is miscible with chloroform. The bp for ethyl acetate is 77 °C; the specific gravity of ethyl acetate is 0.902 at 4 °C. Its vapor density is 3.04 (air = 1) (Budavari, 1996).

Acute Effects

Ethyl acetate is regarded as one of the least toxic of the volatile organic solvents, and its irritant effect is less strong than that of propyl acetate or butyl acetate (ILO, 1983). However, it reportedly causes mild eye, nasal, and throat irritation at 400 ppm (ACGIH, 2008), and some workers have complained of a strong odor at 200 ppm. A patch test with 10% ethyl acetate in petrolatum produced no irritation in a 48-h exposure in humans (Opdyke, 1979). However, prolonged exposure may cause skin irritation (NIOSH, 1981). Ocular exposure causes temporary dulling of the cornea, and, on repeated exposure, can cause conjunctival irritation and corneal clouding (Lewis, 1996). Symptoms of acute vapor intoxication include emotional lability, e.g., talkativeness, exhilaration, and boastfulness. In the event that acute exposure produces such effects, these symptoms may progress to flushing of face, impaired coordination, slurred speech, nausea, vomiting, and drowsiness. Ingestion can cause throat irritation and a narcotic effect. Acute exposure can produce more serious effects, such as nausea, vomiting, pulmonary changes, incontinence, shock, hypotension, hypothermia, and coma. Death may occur as a result of respiratory or circulatory failure (Gosselin et al., 1984).

Chronic Effects

No chronic systemic effects have been reported in humans (NIOSH, 1981). It is not listed as a carcinogen by IARC or EPA. Ethyl acetate was negative for mutagenicity in the Ames test (*Salmonella typhimurium*) assays (NTP, n.d.). No reproductive studies of ethyl acetate in animals or humans have been found.

Mechanism of Action

Ethyl acetate causes defatting of the skin, which may be a reason for skin irritation seen after prolonged exposure.

Metabolism of ethyl acetate produces acetic acid and ethanol, which may be partially exhaled. Ethanol is otherwise converted to acetaldehyde and then to acetic acid by aldehyde dehydrogenase. The biological half-life of ethyl acetate was found to be 5–10 min in mammals (The Chemical Society, 1977).

Chemical Pathology

Ethyl acetate is mildly toxic by ingestion, but high concentrations can cause congestion of the liver and kidneys. Chronic poisoning has been described to produce anemia, leukocytosis (transient increase in white blood cell count), cloudy swelling, and fatty degeneration of the viscera in animals (Lewis, 1996).

SOURCES OF EXPOSURE

Occupational

It has been estimated that 375,906 workers (87,691 of these are female) are exposed to ethyl acetate in the United States. Occupational exposure to ethyl acetate may occur primarily through inhalation and dermal contact where ethyl acetate is produced or used, or where adhesives, thinners, degreasers, paints, or inks containing ethyl acetate are used (NOES, 1983). Workers in auto paint shops are routinely exposed. Employees with chronic respiratory, skin, liver, or kidney disease may be at increased risk from the effects of ethyl acetate (NIOSH, 1981).

Environmental

Production and use of ethyl acetate as a flavor additive, solvent, manufacturing constituent or other uses may result in its release to the environment through various waste streams (Budavari, 1996). If released in air, ethyl acetate will exist solely as a vapor in the ambient atmosphere where it will be degraded; its estimated half-life is 9.4 days. If released in soil, ethyl acetate is expected to have high mobility, and to be volatilized rather rapidly from dry soil. If released into water, ethyl acetate is not expected to adsorb on the suspended solids and sediment. The bioconcentration factor (BCF) is estimated to be 3.2, which is low (NLM, 2013). Ethyl acetate is a natural product of fermentation, and natural sources of ethyl acetate include animal waste, plant volatiles, and microbes (Graedel, 1978).

INDUSTRIAL HYGIENE

The PEL, REL, and TLV for ethyl acetate are each 400 ppm (1400 mg/m^3) (ACGIH. 2008 U. S. Department of Labor, 2013). It is considered immediately dangerous to life and health (IDLH) only when the lower exposure limit of 2000 ppm is reached (NIOSH, 1997). The NFPA hazard classification for ethyl acetate is Health: 1, Flammability: 3, and Reactivity: 0 (NFPA, 1997).

MEDICAL MANAGEMENT

Vapors have a more potent intoxicating effect at higher temperatures (Bisesi, 1994). Ethyl acetate is metabolized to acetic acid and ethanol. Because of its low toxicity and high rate of metabolism, people exposed to ethyl acetate rarely require treatment. However, the patient should be monitored for signs of CNS depression. If eyes are exposed, they should be flushed thoroughly with water. Exposed skin should be washed thoroughly with soap and water.

REFERENCES

American Conference of Governmental Industrial Hygienists (ACGIH) (2008) *Threshold Limit Values (TLVs) for Chemical Substances and Physical Agents and Biological Exposure Indices (BEIs)*, Cincinnati, OH: American Conference of Governmental Industrial Hygienists.

Bisesi, M.S. (1994) Esters. In: Clayton, G.D. and Clayton, F.E., editors., *Patty's Industrial Hygiene and Toxicology*, 4th ed., New York, NY: John Wiley & Sons, Inc.

Budavari, S., editor. (1996) *The Merck Index—An Encyclopedia of Chemicals, Drugs, and Biologicals*, 12th ed., Whitehouse Station, NJ: Merck and Co. Inc.

Cadwallader, K.R. and Xu, Y. (1994) Analysis of volatile components in fresh grapefruit juice by purge and trap/gas chromatography. *J. Agric. Food Chem.* 42:782–784.

Gosselin, R.E., Smith, R.P., and Hodge, H.C. (1984) *Clinical Toxicology of Commercial Products*, 5th ed., Baltimore, MD: Williams and Wilkins.

Graedel, T.E. (1978) *Chemical Compounds in the Atmosphere*, New York, NY: Academic Press.

International Labour Office (ILO) (1983) *Encyclopedia of Occupational Health and Safety*, vol. 1 and 2, Geneva, Switzerland: International Labour Office.

Lewis, R.J. (1996) *Sax's Dangerous Properties of Industrial Materials*, 9th ed., vol. 1–3, New York, NY: John Van Nostrand Reinhold.

Mattheis, J.P., et al. (1991) Changes in headspace volatiles during physiological development of Bisbee Delicious apple fruit. *J. Agric. Food Chem.* 39:1902–1906.

Moshonas, M.G. and Shaw, P.E. (1994) Quantitative determination of 46 volatile constituents in fresh, unpasteurized orange juices using dynamic headspace gas chromatography. *J. Agric. Food Chem.* 42:1525–1528.

National Fire Protection Association (NFPA) (1997) *Fire Protection Guide to Hazardous Materials*, 12th ed., Quincy, MA: National Fire Protection Association.

National Institute for Occupational Safety and Health (NIOSH) (1981) *Occupational Health Guidelines for Chemical Hazards*, DHHS (NIOSH) Publication Number 81–123, Washington, DC: National Institute for Occupational Safety and Health.

National Institute for Occupational Safety and Health (NIOSH) (1997) *Pocket Guide to Chemical Hazards*, DHHS (NIOSH) Publication Number 97–140, Washington, DC: National Institute for Occupational Safety and Health.

National Institute for Occupational Safety and Health (NIOSH) (2014) *National Occupational Exposure Survey (NOES) 1981–1983 Dataset*. Available at http://www.cdc.gov/noes (accessed 2014).

National Library of Medicine (NLM) (2013) *Hazardous Substance Database: Ethyl Acetate*, National Library of Medicine. Available at http://toxnet.nln.nih.gov (accessed, 2014).

National Toxicology Program (NTP). *Fiscal year 1987 annual plan*. National Toxicology Program. *NTP Publication Number 87–00170*. Washington, DC: National Toxicology Program.

National Occupational Exposure Survey (NOES) l981-1983, NIOSH. Available at http://www.cdc.gov/noes (accessed 2014).

NOES (1983) National Occupational Exposure Survey (NOES) 1981-1983, NIOSH.

Opdyke, D.L.J., editor. (1979) *Monographs on Fragrance Raw Materials*, New York: Pergamon Press, p. 341.

Reed, G., editor. (1983) *Kirk-Othmer Encyclopedia of Chemical Technology*, 3rd ed., New York, NY: Wiley Interscience.

The Chemical Society (1977) *Foreign Compound Metabolism in Mammals. A Review of the Literature Published During 1974–1975*, vol. 4, London: The Chemical Society.

U.S. Department of Labor (2013) *Ethyl acetate*. Availability at https://www.osha.gov/dts/chemicalsampling/data/CH_239500.html (accessed 2014).

Weast, R.C., editor. (1988) *Handbook of Chemistry and Physics*, 69th ed., Boca Raton, FL: CRC Press, Inc.

n-, *sec*-, AND *tert*-BUTYL ACETATE

Occupational exposure limits: *n*-butyl acetate (*n*-BA): 713 mg/m^3 (OSHA PEL, NIOSH REL, ACGIH TLV); DNEL inhalation, systemic/local effects, long term: 480 mg/m^3; DNEL inhalation, systemic/local effects, short term: 960 mg/m^3; *sec*-butyl acetate (*sec*-BA): 950 mg/m^3 (OSHA PEL, NIOSH REL, ACGIH TLV); *tert*-butyl acetate (*tert*-BA): 950 mg/m^3 (OSHA PEL, NIOSH REL, ACGIH TLV); DNEL inhalation, systemic effects, long term: 159 mg/m^3; DNEL inhalation, systemic effects, short term: 714 mg/m^3; DNEL dermal, systemic effects, long term: 22.5 mg/kg/bw/day

References values: *n*-BA: DNEL inhalation, systemic/local effects, long term: 102.34 mg/m^3; DNEL inhalation, systemic/local effects, short term: 859.7 mg/m^3; *tert*-BA: DNEL inhalation, systemic effects, long term: 47.3 mg/m^3; DNEL inhalation, systemic effects, short term: 710 mg/m^3; DNEL dermal, systemic effects, long term: 13.5 mg/kg/bw/day; DNEL oral, systemic effects, long term: 13.5 mg/kg/bw/day

Hazard statements: H225—highly flammable liquid and vapor (*tert*-BA); H226—flammable liquid and vapor (*n*-BA); H336—may cause drowsiness or dizziness (*n*-BA); additional text—vapor is heavier than air and can travel considerable distance to a source ignition and flashback (*n*-BA); vapors may form explosive mixture with air (*n*-BA); components of the product may be absorbed into the body by inhalation (*n*-BA)

P210—keep away from heat/sparks/open flames/ . . . /hot surfaces . . . no smoking (*n*-BA; *tert*-BA); P233—keep container tightly closed (*n*-BA, *tert*-BA); P235—keep cool (*n*-BA); P243—take precautionary measures against static discharge (*tert*-BA); P261—avoid breathing dust/fume/gas/mist/vapors/spray (*n*-BA); P280—wear protective gloves/protective clothing/eye protection/face protection (*n*-BA, *tert*-BA); P303 + P361 + P353—if in contact with skin (or hair), remove/take off immediately all contaminated clothing; rinse skin with water/shower (*n*-BA); P304 + P340—if inhaled, remove victim to fresh air and keep at rest in a position comfortable for breathing (*n*-BA); P312—call a poison center or doctor/physician if you feel unwell (*n*-BA); P370 + P378—in case of fire, use . . . for extinction; additional text—dry chemicals, CO$_2$, water spray, or alcohol resistant foam (*tert*-BA); P501—dispose of contents/container to . . . (*tert*-BA); additional text—in accordance with local/regional/national/international regulations (*tert*-BA)

Risk phrases: R10—flammable (*n*-BA); R11—highly flammable (*sec*-BA, *tert*-BA); R66—repeated exposure may cause skin dryness or cracking (*n*-BA, *sec*-BA, *tert*-BA); R67—vapors may cause drowsincss and dizziness (*n*-BA)

Safety phrases: S2—keep out of the reach of children (*sec*-BA, *tert*-BA); S16—keep away from sources of ignition . . . no smoking (*sec*-BA, *tert*-BA); S23—do not breath gas/fumes/vapor/spray (appropriate wording to be specified by the manufacturer) (*tert*-BA); S25—avoid contact with eyes (*n*-BA, *sec*-BA); S29—do not empty into drains (*sec*-BA, *tert*-BA); S33—take precautionary measures against static discharges (*sec*-BA, tert-BA)

BACKGROUND AND USES

Butyl acetate (butyl ethanoate, $C_6H_{12}O_2$, molecular weight 116.16) is a clear, flammable ester of acetic acid that occurs in *n*-, *sec*-, and *tert*- forms (INCHEM, 2005). Butyl acetate isomers have a fruity, banana-like odor (Furia, 1980). Isomers of butyl acetate are found in apples (Nicholas, 1973) and other fruits (Bisesi, 1994), as a well as in a number of food products, such as cheese, coffee, beer, roasted nuts, vinegar (Maarse and Visscher, 1989). Butyl acetate is manufactured via esterification of the respective alcohol with acetic acid or acetic anhydride (Bisesi, 1994). *N*-butyl acetate is used as a solvent for nitrocellulose-based lacquers, inks, and adhesives. Other uses include manufacture of artificial leathers, photographic film, safety glass, and plastics (Budavari, 1996). Isomers of butyl acetate are also used as flavoring agents, in manicure products, and as larvicides (Bisesi, 1994). The *tert*-isomer has been used as a gasoline additive (Budavari, 1996). It may be used as a synthetic fruit flavoring in candy, ice cream, cheeses, and baked goods (Dikshith, 2013).

PHYSICAL AND CHEMICAL PROPERTIES

Butyl Acetate Isomers	CAS Number	Vapor Pressure kPa at 20 °C	Solubility in water, g/L at 20 °C, %
n-CH₃COO (CH₂)₃CH₃	123-86-4	1.2	7.0
sec-CH₃COOCH (CH₃)CH₂CH₃	105-46-4	1.33	0 8.0
tert-CH₃COOC (CH₃)₃ (INCHEM, 2005)	540-88-5	6.3 at 25 °C	Practically insoluble

At 20 °C, the density of the *n*-butyl isomer is 0.8825 g/cm^3, and the density of the *sec*-isomer is 0.8758 g/cm^3 (Bisesi, 1994). The *n*-butyl isomer is soluble in most hydrocarbons and acetone, and it is miscible with ethanol, ethyl ether, and chloroform (Haynes, 2010). It dissolves many plastics and resins (NIOSH, 1981). The bp for the *n*-butyl isomer is 126.1 °C (Haynes, 2010). The vapor pressure of the *n*-butyl isomer is 11.5 mmHg at 25 °C, and its vapor density is 4.0 (air = 1) (Verschueren, 2001).

MAMMALIAN TOXICOLOGY

Acute Effects

Butyl acetate isomers have a low toxicity to humans (Iregren et al., 1993); the odor threshold is 7–20 ppm. Nevertheless, *n*-butyl acetate has narcotic properties estimated to be 1.7 times that of ethyl acetate. Brief exposure to 3300 ppm caused marked irritation to the eyes and nose while mild irritation was reported at concentrations of 1400 ppm for 20 min or to 700 ppm for 4 h (NLM, 2013). Inhalation of high concentrations of *n*-butyl acetate can produce headache and drowsiness, eye inflammation and irritation, but it has not been reported to produce systemic ocular toxicity. Exposure can also produce upper respiratory tract irritation and dryness of the skin. Skin contact in high concentrations can also cause a narcotic effect, but 4% *n*-butyl acetate in petrolatum was not irritating or a sensitizer on the skin of human volunteers (unpublished). In the rare case of an acute exposure at high concentrations, symptoms may include headache, nausea, vomiting, diarrhea, and gastrointestinal hemorrhage. Ataxia, confusion, delirium, coma, and disturbances in cardiac rhythm and cardiac failure may also occur after exposure to high air-borne concentrations of butyl acetate. Liver edema and fatty liver, as well as glycosuria and kidney damage have been observed after acute exposure in animal studies. Prolonged high inhalation exposures can cause death to animals, typically occurring from respiratory failure (NLM, 2013).

Chronic Effects

n-Butyl acetate is not regarded as a human teratogen on the basis of studies in rats and rabbits (Hackett et al., 1983). Yang et al. (2007) reported a no-observed-adverse effect level for *tert*-butyl acetate of 400 mg/kg/day for embryo–fetal development in rats. Reproductive toxicity for the *sec*-butyl isomer was found. The isomer *n*-butyl acetate was not found to be mutagenic in the Ames test (McGregor et al., 2005) and it did not induce chromosomal aberrations in Chinese hamster fibroblasts (Ishidate et al., 1984). Negative results were obtained for *n*-butyl acetate in a test for mitotic aneuploidy in *Saccharomyces cerevisiae* strain D61.M (Zimmermann et al., 1985). None of the isomers of butyl acetate (123-86-4, 105-46-4, or 540-88-5) are listed as human carcinogens by IARC (2012).

Mechanism of Action

In the body, *n*-butyl acetate and *sec*-butyl acetate undergo rather rapid hydrolysis to acetic acid and their respective alcohols. The acetic acid is then oxidized via the citric acid cycle to carbon dioxide and water. The major metabolites of butyl acetate excreted by mammals are (RS)-hydroxy(4-hydroxy-3-methoxy-phenyl)acetic acid or vanillylmandelic acid (Bisesi, 1994). In rats exposed to *tert*-butyl acetate by inhalation, the half-life of blood levels was about 15 min (Groth and Freundt, 1994).

Chemical Pathology

Butyl acetates are CNS depressants, and are irritating to the nose, throat, and eyes. Headache, nausea, and other nonspecific symptoms such as fatigue may occur. No specific mechanistic information regarding the adverse effects of butyl acetate was found, however. Dermatitis may occur in sensitive individuals (NLM, 2013).

SOURCES OF EXPOSURE

Occupational

It has been estimated that 822,099 workers (143,606 of these are female) were potentially exposed to *n*-butyl acetate in the United States during the period 1981–1983 (NOES, 1983). Inhalation and dermal contact are the most important routes of exposure in the occupational setting. Spray painting and gluing may offer the highest exposure, but paint manufacturing, silk screen printing, installation of parquet floors, and the application of artificial finger nails by manicurists also offer substantial exposure to butyl acetate (NLM, 2013).

Environmental

The production and use of butyl acetate in the manufacturing of lacquer, artificial leather, photographic films, plastics and safety glass may result in its release to the environment through various waste streams (O'Neil, 2006). The vapor phase of *n*-butyl acetate is degraded by photochemically produced hydroxyl radicals, and the half-life for this reaction is estimated to be 3.8 days. If released to soil, *n*-butyl acetate is expected to have very high mobility, as volatilization is an important fate of *n*-butyl acetate at the soil surface. Volatilization from water surfaces is also expected to be an important fate process. The BCF is estimated to be 7 in fish, suggesting that the potential for bioconcentration in aquatic organisms is expected to be low (HSDB, 2013). In one study, the incidence of *n*-butyl acetate in finished waters using groundwater supplies in the United States was <5% (Dyksen and Hess, 1982).

INDUSTRIAL HYGIENE

The isomer *n*-butyl acetate has an odor threshold of 7–20 ppm. The PELs, RELs, and TLVs for *sec*- and *tert*-isomers are 200 ppm (950 mg/m^3). However, *n*-butyl acetate has a slightly lower PEL, REL, and TLV of 150 ppm (713 mg/m^3) (ACGIH, 2010). All isomers are considered IDLH when they reach their lower explosive limit at 1500–1700 ppm. The NFPA hazard classification for butyl acetate is Health: 2, Flammability: 3, and Instability: 0 (NFPA, 2010). Workplace air borne concentrations that are immediately dangerous to life or health are above 1700 ppm (USDOL, 2013).

MEDICAL MANAGEMENT

Treatment for these esters is seldom needed because of their low toxicity, but patients should be monitored for CNS depression. If ingested, butyl acetate may be diluted with 4–8 ounces of milk or water. In cases of inhalation, the patient should be moved to fresh air. Oxygen should be administered as required. If eyes are exposed, they should be flushed thoroughly with water at room temperature. Dermal exposures should be treated with soap and water. Butyl acetate does not cause significant burns.

REFERENCES

American Conference of Governmental Industrial Hygienists (ACGIH) (2010) *Threshold Limit Values (TLVs) for Chemical Substances and Physical Agents and Biological Exposure Indices (BEIs)*, Cincinnati, OH: American Conference of Governmental Industrial Hygienists.

Bisesi, M.S. (1994) Esters. In: Clayton, G.D. and Clayton, F.E., editors., *Patty's Industrial Hygiene and Toxicology*, 4th ed., New York, NY: John Wiley & Sons, Inc.

Budavari, S., editor. (1996) *The Merck Index—An Encyclopedia of Chemicals, Drugs, and Biologicals*, 12th ed., Whitehouse Station, NJ: Merck and Co. Inc.

Dikshith, T. (2013) *Hazardous Chemicals: Safety Management and Global Regulations*, Boca Raton, FL: CRC Press, Inc.

Dyksen, J.E. and Hess, A.F. III, (1982) Alternatives for controlling organics in groundwater supplies. *J. Amer. Water Works Assoc.* 74:394–403.

Furia, T.E., editor. (1980) *CRC Handbook of Food Additives*, 2nd ed., vol. 2, Boca Raton, FL: CRC Press, Inc.

Groth, G. and Freundt, K.J. (1994) Inhaled *tert*-butyl acetate and its metabolite *tert*-butyl alcohol accumulate in the blood during exposure. *Hum. Exp. Toxicol.* 13:478–480.

Hackett, P.L., Brown, M.G., and Buschbom, R.I. (1983) *Teratogenic Study of Ethylene and Propylene Oxide and Butyl Acetate, Report Number PB83-258038*. Springfield, VA: Department of Commerce, National Technical Information Service.

Haynes, W.M., editor. (2010) *CRC Handbook of Chemistry and Physics*, 91st ed., Boca Raton, FL: CRC Press Inc.

IARC (2012) Monographs on the Evaluation of Industrial and Consumer Products, Food, and Drinking Water, Vol. 101. Lyon, France: International Agency for Research on Cancer, World Health Organization.

INCHEM (2005) Butyl Acetates: Concise International Chemical Assessment Document 64, United Nations Environment Programme (sic) and International Labour (sic) Organization, pp. 1-38. Geneva: World Health Organizaton (draft report, 2005).

Iregren, A., Löf, A., Toomingas, A., and Wang, Z. (1993) Irritation effects from experimental exposure to *n*-butyl acetate. *Am. J. Ind. Med.* 24:727–742.

Ishidate, M.J., Sofuni, T., Yoshikawa, K., Hayashi, M., Nohmi, T., Sawada, M., and Matsuoka, A. (1984) Primary mutagenicity screening of food additives currently used in Japan. *Food Chem. Toxicol.* 22:623–636.

Maarse, H. and Visscher, C.A., editors. (1989) *Volatile Compounds in Food: Qualitative and Quantitative Data*. 6th ed., vol. 1, TNO-CIVO Food Analysis Institute.

McGregor, D.B., Cruzan, G., Callander, R.D., May, K., and Banton, M. (2005) The mutagenicity testing of tertiary-butyl alcohol, tertiary-butyl acetate[TM] and methyl tertiary-butyl ether in *Salmonella typhimurium*. *Mutat. Res.* 565:181–189.

National Fire Protection Association (NFPA) (2010) *Fire Protection Guide to Hazardous Materials*, 14th ed., Quincy, MA: National Fire Protection Association.

Nicholas, H.J. (1973) *Phytochemistry*, vol. 2, New York, NY: Van Nostrand Reinhold.

National Institute for Occupational Safety and Health (NIOSH) (1981) *Occupational Health Guidelines for Chemical Hazards*, DHHS (NIOSH) Publication Number 81–123, Washington, D.C.: National Institute for Occupational Safety and Health.

National Institute for Occupational Safety and Health (NIOSH). *National Occupational Exposure Survey 1981-1983 (NOES), Dataset*. NIOSH. Available at http://www.cdc.gov/noes (accessed 2014).

National Library of Medicine (NLM) (2013) *Hazardous substance database:* toxnet-nlm.nih.gov (butyl acetate-CAS 123-86-4, CAS 105-46-4, and CAS 540-88-5) (Accessed 2014).

National Occupational Exposure Survey 1981-1983 (NOES), NIOSH. Available at http://www.cdc.gov/noes/.

O'Neil, M.J., editor. (2006) *The Merck Index—An Encyclopedia of Chemicals, Drugs, and Biologicals*, 14th ed., Whitehouse Station, NJ: Merck and Co., Inc.

United States Department of Labor (USDOL) (2013) *Ethyl Acetate*. Occupational Safety and Health Administration. Available at https://pubchem.ncbi.nlm.nih.gov/summary/summary.cgi?cid=31272#x332 (accessed 2014).

Verschueren, K. (2001) *Handbook of Environmental Data on Organic Chemicals*, 4th ed., vol. 1 and 2, New York, NY: John Wiley & Sons.

Yang, Y.S., Ahn, T.H., Lee, J.C., Moon, C.J., Kim, S.H., Park, S.C., Chung, Y.H., Kim, H.Y., and Kim, J.C. (2007) Effects of tert–butyl acetate on maternal toxicity and embryo-fetal development in Sprague-Dawley rats. *Birth Defects Res. B: Dev. Reprod. Toxicol.* 80:374–382.

Zimmermann, F.K., Mayer, V.W., Scheel, I., and Resnick, M.A. (1985) Acetone, methyl ethyl ketone, ethyl acetate, acetonitrile and other polar aprotic solvents are strong inducers of aneuploidy in *Scaccharomyces cerevisiae. Mutat. Res.* 149:339–351.

OTHER ALIPHATIC ESTERS

The American Chemical Council and its member companies are presently addressing the toxicological hazards associated with 45 high production volume aliphatic esters, including monoesters, diesters, glycol esters, esters of monoacids and sorbitan, and esters of monoacids and trihydroxy or polyhydroxy alcohols.

HALOGENATED ACID ESTERS

A small number of halogenated acid esters are potent lacrimating and vesicating agents and have the potential capacity for producing pulmonary edema. Compounds in this category are important organic intermediates and include ethyl chloroformate, ethyl chloroacetate (CASRN 105-39-5), and related bromo- and iodo- compounds. These materials have been used as chemical warfare and riot control agents. There is no evidence that they produce chronic diseases, but fatalities have occurred from acute pulmonary edema. The highly hazardous character of these compounds appears to be related to the high reactivity of the halogen atoms.

ETHYL CHLOROFORMATE

Occupational exposure limits: None

Reference values: None

Hazard statements: H225—highly flammable liquid and vapor; H314—causes severe skin burns and eye damage; H301—toxic if swallowed; H330—fatal if inhaled

Precautionary statements: P210—keep away from heat/sparks/open flames/ . . . /hot surfaces . . . no smoking; P233—keep the container tightly closed; P240—ground/bond container and receiving equipment; P241—use explosion-proof electrical/ventilating/lighting/ . . . /equipment; P242—use only non-sparking tools; P243—take precautionary measures against static discharge; P280—wear protective gloves/protective clothing/eye protection/face protection; P284—wear respiratory protection; P271—use only outdoors or in a well-ventilated area; P260—do not breathe dust/fume/gas/mist/vapors/spray; P264—wash . . . thoroughly after handling; P270—do not eat, drink, or smoke when using this product; P310—immediately call a poison center or doctor/physician; P305 + P351 + P338—if in contact with eyes, rinse cautiously with water for several minutes; remove contact lenses, if present and easy to do; continue rinsing; P304 + P340—if inhaled, remove victim to fresh air and keep at rest in a position comfortable for breathing; P363—wash contaminated clothing before reuse; P321—specific treatment (see . . . on the label); P303 + P361 + P353—if in contact with skin (or hair), remove/take off immediately all contaminated clothing; rinse skin with water/shower; P370 + P378—in case of fire, use . . . for extinction; P403 + P235—store in a well-ventilated place; keep cool; P405—store locked up; P501—dispose of contents/container to . . .

Risk phrases: R11—highly flammable; R22—harmful if swallowed; R26—very toxic by inhalation; R34—causes burns

Safety phrases: S1/2—keep locked up and out of reach of children; S9—keep container in a well-ventilated place; S16—keep away from sources of ignition—no smoking;

S26—in case of contact with eyes, rinse immediately with plenty of water and seek medical advice; S28—after contact with skin, wash immediately with plenty of . . . (to be specified by the manufacturer); S33—take precautionary measures against static discharges; S36/37/39—wear suitable protective clothing, gloves and eye/face protection; S45—in case of accident or if you feel unwell, seek medical advice immediately (show the label where possible)

BACKGROUND AND USES

Ethyl chloroformate (chloroformic acid ethyl ester; molecular weight 108.52, $ClCOOC_2H_5$ (CASRN 541-41-3) is used as a solvent in the photographic industry, and as a chemical intermediate in the production of various carbamates, and used in synthesis of dyes, drugs, veterinary medicines, herbicides, and insecticides. It is also used in the production of flotation agents for ores, as a stabilizer for PVC, and in the production of modified penicillins and heterocyclic compounds (Gerhartz, 1985).

PHYSICAL AND CHEMICAL PROPERTIES

Ethyl chloroformate is a colorless liquid that is corrosive and flammable. It is prepared from phosgene and ethanol. It has a sharp odor, like hydrochloric acid, and it decomposes in water. It is miscible with alcohol, benzene, chloroform, and ether. The bp is 95 °C, and the melting point is −80.6 °C (ICSC, 1025); the specific gravity is 1.1403 at 4 °C (Budavari, 1996). The vapor density is 3.7 (air = 1) (NFPA, 1997).

MAMMALIAN TOXICOLOGY

Acute Effects

Vapors of ethyl chloroformate are very strong irritants to the eyes, mucous membranes, and skin (Budavari, 1996). Irritation of the upper respiratory tract, with cough, dyspnea, pulmonary edema, lacrimation, nausea, vomiting, abdominal pain, weakness, and shock may occur. The toxic properties resemble those of the methyl homolog in the degree of corrosiveness, irritation, and lacrimatory qualities (Bisesi, 1994). Onset of symptoms may be delayed, and pulmonary edema of late onset has been reported. Inhalation exposure to 190 ppm for 10 min has caused human fatalities, and exposure to 10 ppm has caused excessive lacrimation (NLM, 2003). Fatal occupational exposures have occurred. The oral LD_{50} for the rat is <50 mg/kg and the intraperitoneal LD_{LO} for the mouse is 5 mg/kg (Sandmeyer and Kirwin, 1981). It reacts with water to produce heat, carbon dioxide, and hydrogen chloride gas (NLM, 2003)

Chronic Effects

No data were found on potential effects on pregnancy and fetal development. Ethyl chloroformate is not listed as a carcinogen by IARC (2012).

Mechanism of Action

Ethyl chloroformate is highly irritant and corrosive by ingestion, inhalation, or eye or skin contact. It is one of the agents used in biochemical carbobenzoxy group removal, without scission of the peptide bond. Analogously, a similar reaction may be responsible for the hepatic and renal effects seen upon exposure (Bisesi, 1994).

SOURCES OF EXPOSURE

Occupational

It has been statistically estimated that 2,247 workers in the USA are potentially exposed in the photographic industry and during its use as a chemical intermediate (NOES, 1983).

Environmental

The use of ethyl chloroformate in the photographic industry could result in release to the environment through evaporation or in various waste streams. In the ambient atmosphere, the estimated half-life of ethyl chloroformate is about 11 days (Meylan and Howard, 1993). Ethyl chloroformate hydrolyzes readily in water; the half-life is 31.5 min at 25 °C (Queen, 1967). Bioconcentration and adsorption on the sediment are not expected to be important fate processes, due to its rapid rate of hydrolysis. Hydrolysis will probably also dominate its fate in moist soils. This chemical has a relatively high vapor pressure of 22.4 mmHg at 25 °C (Daubert and Danner, 1991), which suggests that evaporation from dry soil surface will occur.

INDUSTRIAL HYGIENE

The ACGIH TLV and PEL or MAK values have not been established (ICSC, 1025). Reaction with water or moist air will release hydrogen chloride or other toxic, corrosive, or flammable gases. The NFPA hazard classification for ethyl chloroformate is Health: 4, Flammability: 3, Reactivity: 1; do not use water; use dry chemical, foam, carbon dioxide, or water spray (NFPA, 1997).

MEDICAL MANAGEMENT

If no respiratory compromise is evident, administer no more than 4–8 oz of water or milk to neutralize ingested acid. Vomiting should not be induced, but removal of acid from stomach via gastric suction may be considered. Move patient to fresh air; monitor for respiratory distress, and evaluate possible respiratory tract irritation. Administer oxygen and assist ventilation as required. Irrigate exposed eyes with copious amounts of water at room temperature for at least 15 min. No specific protocol supportive care, respiratory care-may treat bronchospasm with beta 2 agonists and/or corticosteroids (NLM, 2003).

REFERENCES

Bisesi, M.S. (1994) Esters. In: Clayton, G.D. and Clayton, F.E., editors. *Patty's Industrial Hygiene and Toxicology*, 4th ed., New York, NY: John Wiley & Sons, Inc.

Budavari, S., editor. (1996) *The Merck Index—An Encyclopedia of Chemicals, Drugs, and Biologicals*, 12th ed., Whitehouse Station, NJ: Merck and Co. Inc.

Daubert, T.E. and Danner, R.P. (1991) *Physical and Thermodynamic Properties of Pure Chemicals: Data Compilation*, New York, NY: Hemisphere Publishing Corp.

Gerhartz, W., editor. (1985–2012) *Ullmann's Encyclopedia of Industrial Chemistry*, vol. A6, New York, NY: VHC Publishers.

ICSC (1025) International Chemical Safety Cards: Ethyl chloroformate, ICSC 1025. http://www.miss.gov.jm/eoshd/data/nioshdbs/ipcsneng/neng1025.htm.

IARC (2012) Monographs on the Evaluation of Industrial and Consumer Products, Food, and Drinking Water, Vol. 101. Lyon, France: International Agency for Research on Cancer (IARC), World Health Organization.

Meylan, W.M. and Howard, P.H. (1993) Computer estimation of the atmospheric gas-phase reaction rate of organic compounds with hydroxyl radicals and ozone. *Chemosphere* 26:2293–2299.

National Fire Protection Association (NFPA) (1997) *Fire Protection Guide to Hazardous Materials*, 12th ed., Quincy, MA: National Fire Protection Association.

National Institutes of Health (NLM 2003) *Ethyl Chloroformate, HSDB*. Available at http://toxnet.nlm.nih.gov/cgi-bin/sis/search (accessed 2014).

Queen, A. (1967) Kinetics of the hydrolysis of acyl chlorides in pure water. *Can. J. Chem.* 45:1619–1629.

Sandmeyer, E.E. and Kirwin, C.J. Jr. (1981) Esters. In: Clayton, G.D. and Clayton, F.E., editors. *Patty's Industrial Hygiene and Toxicology*, 3rd ed., New York, NY: John Wiley & Sons, Inc.,

ETHYL CHLOROACETATE

Occupational exposure limits: None

Reference values: None

Hazard statements: H226—flammable liquid and vapor; H301—toxic if swallowed; H310—fatal in contact with skin; H331—toxic if inhaled; H315—causes skin irritation; H318—causes serious eye damage; H317—may cause an allergic skin reaction; H400—very toxic to aquatic life

Precautionary statements: P210—keep away from heat/sparks/open flames/ . . . /hot surfaces . . . no smoking; P273—avoid release to the environment; P280—wear protective gloves/protective clothing/eye protection/face protection; P261— avoid breathing dust/fume/gas/mist/vapors/spray P305 + P351 + P338—if in contact with eyes, rinse cautiously with water for several minutes; remove contact lenses, if present and easy to do; continue rinsing; P309 + P311—if exposed or if you feel unwell, call a poison center or doctor/physician; P302 + P352—if in contact with skin, wash with plenty of soap and water

Risk phrases: R23/24/25—toxic by inhalation, when in contact with skin, and if swallowed; R50—very toxic to aquatic organisms

Safety phrases: S7/9—keep container tightly closed and in a well-ventilated place; S45—in case of accident or if you feel unwell, seek medical advice immediately (show the label where possible); S61—avoid release to the environment; refer to special instructions/safety data sheets

BACKGROUND AND USES

Ethyl chloroacetate (chloroacetic acid ethyl ester, molecular weight 122.55, $ClCH_2COOC_2H_5$, CASRN 105-39-5) is a highly flammable clear liquid with a pungent odor (Bisesi, 1994). Ethyl chloroacetate is utilized as a solvent and in organic synthesis; it is also used in vat dyestuffs and as a military poison (Lewis, 1997). In addition, it is a chemical intermediate in the synthesis of sodium fluoroacetate (rodenticide) and benzolin (herbicide) (SRI, 1999).

PHYSICAL AND CHEMICAL PROPERTIES

Ethyl chloroacetate has a vapor pressure of 10 mmHg at 38 °C (Lewis, 1997). It is insoluble in water, but miscible with alcohol, ether, and acetone (Lide, 1998); it is soluble in benzene (Lewis, 1997). The boiling point for ethyl chloroacetate is 143 °C while the melting point is −26 °C; its density is 1.1498 g/ml at 25 °C (Budavari, 1996). The flash point is slightly below 61 °F (Budavari, 1996). The vapor density is 4.3 (air = 1) (Bisesi, 1994). Ethyl chloroacetate readily decomposes in hot water and alkalis (Lewis, 1997).

MAMMALIAN TOXICOLOGY

Acute Effects

This chemical is easily vaporized, and it is a severe eye irritant and potent lacrimator. It is very harmful if inhaled or swallowed; it may cause severe eye, skin, and respiratory tract irritation, with possible burns (Sax, 1979). Exposure to vapor causes immediate ocular burning, blepharospasm (involuntary blinking), lacrimation, rhinorrhea, coughing, sneezing, and pain, but usually no permanent tissue damage. These symptoms may be followed by chest tightness and coughing, sore throat, burning of tongue and mouth, salivation, and vomiting. Symptoms usually subside within 15–30 min, but chemical burns with keratitis, loss of corneal epithelium, and permanently reduced cornea sensation have been reported. One case of contact dermatitis has also been reported (Braun and van der Walle, 1987). Burning sensations of the skin, followed by erythema and bullous dermatitis may occur in 12 h to 3 days following exposure to tear gas (mace) containing ethyl chloroacetate and *o*-chlorobenzylidene malononitrile (CS). Pulmonary edema may occur 12–24 h postexposure. Erythema, vesicle eruptions, and denuded areas with weeping, tender erythematous base may develop 1 day after exposure to mace containing 1-chloroacetophenone (CN). Nausea is common, and vomiting may occur, especially if ethyl chloroacetate is swallowed. Tachycardia and mild hypertension may occur as a result of fear and pain. Agitation and syncope attributed to panic after exposure have also been reported. The acute rat oral LD_{50} for ethyl chloroacetate is 180 mg/kg and the acute dermal LD_{50} in rats is 161 mg/kg bw; the 4-h LC_{50} for rats is 3.3 ml/m^3 (Berufsgenossenschaft der Chemischen Industrie, 1992).

Chronic Effects

Chronic effects of exposure to ethyl chloroacetate have not been reported, as acute exposure is most commonly encountered. There is no evidence that ethyl chloroacetate represents a carcinogenicity risk. The literature does not provide evidence for tumorigenicity or genotoxicity using various in vivo and in vitro test systems and this supports the general conclusion that ethyl chloroacetate is not likely carcinogenic utilizing various laboratory testing methods. Adverse effects in developing animal fetuses were not found in the literature. Based on animal findings, there is no evidence that ethyl chloroacetate represents a carcinogenic risk (NLM, 2003)

Mechanism of Action

Ethyl chloroacetate is irritating and corrosive to all tissues; it can readily cause chemical burns.

Chemical Pathology

Reports of investigations regarding the chemical pathology of ethyl chloroacetate were not found; however, by hydrolysis to form ethanol and chloroacetic acid (NLM, 2003)

SOURCES OF EXPOSURE

Occupational

National Institute for Occupational Safety and Health has statistically estimated that 919 workers (306 of these are female) are potentially exposed to ethyl chloroacetate in the United States (NOES, 1983). Occupational exposure may occur through inhalation and dermal contact in workplaces where ethyl chloroacetate is produced or used.

Environmental

The production and use of ethyl chloroacetate may result in its release to the environment through various waste streams. If released in air, a vapor pressure of 4.87 mmHg at 25 °C indicates that it will exist as a vapor in the ambient atmosphere. Vapor phase ethyl chloroacetate will be degraded in the atmosphere by reaction with photochemically produced hydroxyl radicals; the half-life for this reaction is estimated to be 13 days. If released to soil, ethyl chloroacetate is expected to have high mobility. Ethyl chloroacetate is subjected to hydrolysis, especially under acid or alkaline conditions. If released into water, it is not expected to adsorb rapidly on suspended solids and sediment, and volatilization from the surface of water is expected to be an important fate process. An estimated BCF of 3 for ethyl chloroacetate suggests that the potential for concentration in aquatic organisms is low (Bennett et al., 1984).

INDUSTRIAL HYGIENE

No published PEL or ACGIH exposure standards for ethyl chloroacetate were found. Ethyl chloroacetate is a flammable liquid. The NFPA hazard classification is Health: 3, Flammability: 2, Reactivity: 0 (NFPA, 1997). Sufficient personal protection should be worn when there is a possibility of exposure.

MEDICAL MANAGEMENT

After eye contact, immediately flush with plenty of water for at least 15 min and get medical aid immediately. In the instance of skin contact, flush skin with copious amounts of water for at least 15 min. Remove contaminated clothing and shoes, and get medical attention immediately. If swallowed, do not induce vomiting unless directed to do so by medical personnel. For inhalation exposure, remove to fresh air; give artificial respiration if not breathing. If breathing is difficult, give oxygen. Although permanent injury or death is not common, medical attention may be required for several days after acute exposure.

REFERENCES

Bennett, S.R., Bane, J.M., Benford, P.J., and Pyatt, R.L. (1984) *Environmental hazards of chemical agent simulants.* CRDC-TR-84055, Aberdeen Proving Ground, MD.

Berufsgenossenschaft der Chemischen Industrie (1992) *Chloressigsäureethylester: Toxikologicishe Bewertung*, vol. 190, Heidelberg, Germany: Berufsgenossenschaft der Chemischen Industrie.

Bisesi, M.S. (1994) Esters. In: Clayton, G.D. and Clayton, F.E., editors. *Patty's Industrial Hygiene and Toxicology*, 4th ed., New York, NY: John Wiley & Sons, Inc.

Braun, C.L.J. and van der Walle, H.B. (1987) The ethylester of monochloroacetic acid. *Contact Dermatitis* 16:114–115.

Budavari, S., editor. (1996) *The Merck Index—An Encyclopedia of Chemicals, Drugs, and Biologicals*, 12th ed., Whitehouse Station, NJ: Merck and Co. Inc.

Lewis, R.J. Sr., editor. (1997) *Hawley's Condensed Chemical Dictionary*, 13th ed., New York, NY: John Wiley & Sons, Inc.

Lide, D.R., editor. (1998) *CRC Handbook of Chemistry and Physics*, 79th ed. Boca Raton, FL: CRC Press Inc., pp. 3–6.

National Fire Protection Association (NFPA) (1997) *Fire Protection Guide to Hazardous Materials*, 12th ed., Quincy, MA: National Fire Protection Association.

National Occupational Exposure Survey (NOES) 1981-1983. Available at http://www.cdc.gov/noes (accessed 2014).

National Library of Medicine (NLM) (2003) *Ethyl chloroacetate.* Available at http://toxnet.nlm.nih.gov/cgi-bin/sis/search/f?./temp/~crXbnX:1 (accessed 2014).

Sax, N. (1979) *Dangerous Properties of Industrial Materials*, 5th ed., New York, NY: Van Nostrand Rheinhold.

SRI Consulting (1999) *Directory of Chemical Producers—United States*, Menlo Park, CA: SRI Consulting.

ETHYL 2-BROMOACETATE

Occupational exposure limits: None

Reference values: None

Hazard statements: None

Precautionary statements: None

Risk phrases: R26/27/28—very toxic by inhalation, when in contact with skin, and if swallowed

Safety phrases: S1/2—keep locked up and out of the reach of children; S7/9—keep container tightly closed and in a well-ventilated place; S26—in case of contact with eyes, rinse immediately with plenty of water and seek medical advice; S45—in case of accident or if you feel unwell, seek medical advice immediately (show the label where possible)

BACKGROUND AND USES

Ethyl 2-bromoacetate (2-bromoacetic acid, ethyl ester; $BrCH_2COOC_2H_5$, CASRN 105-36-2) is a flammable liquid that is easily vaporized (Lide, 1999). It has been employed as a tear gas (Grant, 1986). It is also utilized as a chemical intermediate or reactant in the synthesis of an analgesic (Rogers and May, 1974), other pharmaceuticals, and beta-keto acids. Ethyl 2-bromoacetate is also used as a warning agent for other poisonous, odorless gases (SRI Consulting, 1999). It is manufactured by esterification of bromoacetic acid with ethyl alcohol in the presence of sulfuric acid (Lewis, 1997).

PHYSICAL AND CHEMICAL PROPERTIES

Ethyl 2-bromoacetate, molecular weight 167, is colorless, and has a pungent odor. The specific gravity is 1.5032 at 20 °C (Lide, 1999), and the bp is 158 °C (316 °F). The melting point is −38 °C. It is insoluble in water, but is soluble in acetone and benzene (Lide, 1999; Lewis, 1997). Ethyl 2-bromoacetate is miscible with ethanol, ethyl ether, and with other oxygenated and aromatic solvents (Ashford, 1994; Lide, 1999). The vapor density is 5.8 (air = 1) (Lewis, 1997). It is partially decomposed by water (Lewis, 1997).

MAMMALIAN TOXICOLOGY

Acute Effects

Occupational exposure by inhalation or dermal contact may occur in workplaces where ethyl 2-bromoacetate is produced or utilized. Ethyl 2-bromoacetate is a potent irritant to eyes, skin, and all mucous membranes (Lewis, 1996); it is a more potent lacrimator than the chloroacetic acid ethyl ester. A concentration of 8 ppm in air is unbearable for more than a minute (Grant, 1986). High concentrations of vapor have caused keratitis, corneal edema with wrinkling of Descemet's membrane, loss of corneal epithelium, and/or reduced corneal sensation (HSDB, 2013). In most patients, their eyes return to normal. Nevertheless, direct eye contact with the liquid chemical has caused permanent corneal scarring and opacification (Grant, 1986). It is also toxic by inhalation, skin absorption, and ingestion. Upon inhalation, symptoms of chest tightness and coughing, sore throat, shortness of breath, burning of tongue and mouth, salivation, and vomiting may occur. Symptoms usually subside rapidly within 15–30 min of cessation of exposure in most individuals. Dermal contact causes skin irritation and possible bullous dermatitis on all significantly exposed body surfaces. Ethyl bromoacetate, like other esters, reacts with acids to form alcohols and acids (NLM, 2013).

Chronic Effects

Ethyl 2-bromoacetate is not listed as a carcinogen by the IARC (2012). One incomplete report suggests mutagenicity in Ames test or strain TA104 and in *E. coli* strain WP2*uvrA*/pKM101, both with S9 activation. No information on possible developmental toxicity or teratogenicity was found for animal models or humans. Based on animal findings, there is no evidence that ethyl 2-bromoacetate represents a carcinogenic risk (NLM, 2013).

Mechanism of Action

The biochemical mode of action of ethyl 2-bromoacetate has not be elucidated.

Chemical Pathology

Lesions of skin and eyes caused by acute exposure to ethyl 2-bromoacetate are described in Section 3.1.

SOURCES OF EXPOSURE

Occupational

National Institute for Occupational Safety and Health has statistically estimated that 851 workers (187 of these are female) are potentially exposed to ethyl 2-bromoacetate in the United States via inhalation or dermal contact at workplaces where this compound is produced or utilized (NLM 2013). Reaction with water will release heat, and toxic, corrosive or flammable gases.

Environmental

The production and use of ethyl 2-bromoacetate may result in its release to the environment via various waste streams. If released to air, an estimated vapor pressure of 3.37 mmHg at 25 °C indicates that this chemical will exist as a vapor in the ambient atmosphere. Vapor phase ethyl 2-bromoacetate will be degraded photochemically in air with a half-life estimated to be of 14 days (HSDB, 2013). If released into soil, it is expected to have very high mobility. Volatilization from moist soil surfaces and dry soil is expected to be an important fate process. If released into water, ethyl-2-bromoacetate is not expected to absorb by suspended solids and sediment in the water. An estimated BCF of 1 indicates that bioconcentration in aquatic organisms will be quite low.

INDUSTRIAL HYGIENE

No occupational exposure standards exist for ethyl 2-bromoacetate. Since it is highly toxic by ingestion, inhalation, and skin absorption, every effort should be made to provide sufficient worker protection from potential exposure, especially via skin and eyes. Contact lenses should not be worn by individuals potentially exposed to ethyl 2-bromoacetate. Ethyl 2-bromoacetate is regarded as highly flammable, easily ignited by heat, sparks, or flames; vapors of ethyl 2-bromoacetate form potentially explosive mixtures with air. The NFPA hazard classification is Health: 3, Flammability: 2, Instability: 1, Special: none (NFPA, 1997).

MEDICAL MANAGEMENT

For inhalation exposure, move patient to fresh air. Monitor for respiratory distress; if cough or difficulty in breathing develops, evaluate for respiratory tract irritation, bronchitis, or pneumonitis. Laryngospasm may require endotracheal intubation and ventilation. Administer oxygen and assist ventilation as required. Treat bronchospasm with an inhaled beta-2 agonist and oral or parenteral corticosteroids. For exposure to skin, meticulously wash all exposed areas with copious amounts of soap and water. The patient's clothing may contain residual amounts of active chemical. Remove clothing and store in a sealed polyethylene bag to prevent degassing; wash all clothing with cold water. For eye exposure, irrigate exposed eyes with copious amounts of water at room temperature for 15 min or more. If irritation, pain, swelling, lacrimation, or photophobia persist, patient should be referred to a health care facility (HSDB, 2013).

REFERENCES

Ashford, R.D. (1994) Ashford's Dictionary of Industrial Chemicals. London, England: Wavelength Publications, Ltd, p. 390.

Grant, W.M. (1986) *Toxicology of the Eye*, 3rd ed., Springfield, IL: Charles C. Thomas Publisher.

HSDB (Hazardous Substances Data Bank) (2013) Data base of the National Library of Medicine's TOXNET system (http:toxnet.nlm.nih.gov) for ethyl 2-bromo acetate.

Lewis, R.J. (1996) *Sax's Dangerous Properties of Industrial Materials*, 9th ed., vol. 1–3, New York, NY: John Van Nostrand Reinhold.

Lewis, R.J. Sr., editor. (1997) *Hawley's Condensed Chemical Dictionary*, 13th ed., New York, NY: John Wiley & Sons, Inc.

Lide, D.R., editor. (1999) *CRC Handbook of Chemistry and Physics*, 79th ed., Boca Raton, FL: CRC Press LLC.

IARC (2012) *Monographs on the Evaluation of Industrial and Consumer Products, Food, and Drinking Water*, vol. 101, Lyon, France: International Agency for Research on Cancer (IARC), World Health Organization.

National Fire Protection Association (NFPA) (1997) *Fire Protection Guide to Hazardous Materials*, 12th ed., Quincy, MA: National Fire Protection Association.

National Library of Medicine (NLM) (2013) *Hazardous substance data base: Ethyl 2-bromoacetate*. National Library of Medicine. Available at http://tpxnet.nlm.nih.gov (accessed 2014).

Rogers, M.E. and May, E.L. (1974) Improved synthesis and additional pharmacology of the potent analgetic (-)-5-*m*-hydroxyphenyl-2-methylmorphan. *J. Med. Chem.* 17:1328–1330.

SRI Consulting (1999) *Directory of Chemical Producers—United States*, Menlo Park, CA: SRI Consulting.

ALKYL ESTERS OF SULFURIC ACID

DIMETHYL SULFATE

Occupational exposure limits: 1 ppm (OSHA PEL); 0.1 ppm (NIOSH REL); 0.1 ppm (ACGIH TLV with skin notation)

Reference values: None

Hazard statements: H301—toxic if swallowed; H314—causes severe skin burns and eye damage; H317—may cause an allergic skin reaction; H330—fatal if inhaled; H341—suspected of causing genetic defects; H350—may cause cancer

Precautionary statements: P202—do not handle until all safety precautions have been read and understood; P281—use personal protective equipment as required; P303 + P361 + P353—if on skin (or hair), remove/take off immediately all contaminated clothing; rinse skin with water/shower; P304 + P340—if inhaled, remove victim to fresh air and keep at rest in a position comfortable for breathing; P305 + P351 + P338—if in contact with eyes, rinse cautiously with water for several minutes; remove contact lenses, if present and easy to do; continue rinsing; P310—immediately call a poison center or doctor/physician

Risk phrases: R25—toxic if swallowed; R26—very toxic by inhalation; R34—causes burns; R43—may cause sensitisation by skin contact; R45—may cause cancer; R68—possible risk of irreversible effects

Safety phrases: S45—in case of accident or if you feel unwell, seek medical advice immediately (show the label where possible); S53—avoid exposure—obtain special instructions before use

BACKGROUND AND USES

Dimethyl sulfate (DMS; $(CH_3O)_2SO_2$ sulfuric acid dimethyl ester; CASRN 77-78-1) is prepared by distillation of an oleum/methanol mixture; technical production using dimethyl ether and SO_3 has also been reported (NLM, 2013). It is currently used as a methylating agent in the manufacture of organic chemicals, including surfactants,

pesticides, water treatment chemicals, dyes, flavors, pharmaceuticals, rubber chemicals, and photographic films. It was formerly used as a chemical weapon (NTP, 2011).

PHYSICAL AND CHEMICAL PROPERTIES

Dimethyl sulfate has a molecular weight of 126.13 (O'Neil, 2006). It is essentially odorless (Langford, 2004). The specific gravity of this colorless, corrosive, oily liquid is 1.3322 g/cm^3 (O'Neil, 2006); its vapor density is 4.35 (air = 1), and its vapor pressure is 0.54 mmHg at 20 °C. The bp is about 188 °C (O'Neil, 2006). Dimethyl sulfate is soluble in ether, dioxane, acetone, benzene, and other aromatic hydrocarbons. It is sparingly soluble in carbon disulfide and aliphatic hydrocarbons, and only slightly soluble in water (28 g/l at 18 °C) (O'Neil, 2006).

MAMMALIAN TOXICOLOGY

Acute Effects

Dimethyl sulfate is extremely hazardous. Major target organs are the eyes, respiratory system, skin, and CNS. Because of its deficient warning properties, unnoticed exposure to lethal quantities can occur, and appearance of symptoms may be delayed by 4–12 h. Symptoms of industrial poisoning uniformly found after each exposure began with headache, photophobia, and difficulty with breathing and swallowing. Other effects may also include painful coughing, erythema of eyes, lacrimation, blepharospasms (involuntary eye blinking), and chemosis. Corneal ulceration and severe inflammation of the eyes and eyelids with photophobia generally resolve satisfactorily although irreversible loss of vision has been reported. Even burns resulting from direct contact with the skin are delayed in their onset, in spite of immediate, thorough neutralization and irrigation of the exposed area (Littler and McConnell, 1955). Irritation of the mucous membranes of the mouth and respiratory tract may be severe, with pulmonary edema; there may also be hoarseness and oropharyngeal edema, which persists for several weeks. Systemic effects can include convulsions, delirium, coma, analgesia, pyrexia, pulmonary edema, delayed renal and hepatic failure, and cardiac damage. Edema of the tongue, lips, larynx, and lung may also occur later. Delayed effects may consist of respiratory, liver, and kidney damage; severe inflammation, and necrosis of the eyes, mouth, and respiratory tract; severe/fatal pulmonary damage; prostration; convulsions; delirium; paralysis; and coma (Dreisbach, 1987). Other systemic effects such as convulsions; delirium; coma; and renal, hepatic, and cardiac failure have been reported. Clinically, the usual manifestations include skin burns, corneal irritation with photophobia, lacrimation and blurring of vision, oropharyngeal edema with suffocation and hoarseness of the voice, nasal septal necrosis, cough, and dyspnea (Ellenhorn et al., 1997). Major causes of mortality in dimethyl sulfate intoxication include respiratory failure following mucosal inflammation, edema of major airways, and noncardiogenic pulmonary edema.

Chronic Effects

Chronic symptoms include dysphonia, aphonia, dysphagia, chest pain, cyanosis, diarrhea, icterus, albuminuria, and hematuria. Symptoms of hepatic disease may include malaise, lethargy, nausea, vomiting, pruritus, anorexia, hepatic enlargement and tenderness. Other signs of hepatic disease may include upper right quadrant pain, anorexia, scleral icterus, hepatic enlargement and tenderness, dark urine, and light-colored, loose stools (Ellenhorn and Barceloux, 1988). Dimethyl sulfate is a confirmed animal carcinogen (NTP, 2011), and it is a suspected as a human carcinogen (A2) (NLM, 2013). Experimental inhalation studies confirm that DMS vapors at concentrations <8 ppm cause regional DNA methylation; under nose-only conditions N7-methylguanine and N3-methyladenine was found in the nasal mucosa of exposed rats (Mathison et al., 1995). Dimethyl sulfate was also carcinogenic to animals (sarcomas) at the injection site following subcutaneous injection. Furthermore, dimethyl sulfate administered to pregnant rats by intravenous injection caused tumors of the nervous system in the offspring (IARC, 1974). There is no data available for chronic, long-term effects of dimethyl sulfate in humans (USEPA, 2013).

Mechanism of Action

Dimethyl sulfate is a methylating agent that converts phenols, amines, and thiols to the corresponding methyl derivatives (IARC, 1999). Methylation of DNA in animal models is with little doubt a mechanism for its carcinogenic potency. Exposure to 0.2–20 mg/m^3 dimethyl sulfate has caused chromosomal abnormalities in lymphocytes of exposed workers. Nevertheless, there is insufficient clinical or epidemiological evidence to indicate that dimethyl sulfate is a human carcinogen at this time (WHO, 1985).

Besides methylation of DNA, dimethyl sulfate is more than likely an active methyl donor with a number of endogenous nucleophiles, such as oxygen, carbon, sulfur, phosphorus, as well as with some metals.

Chemical Pathology

Severe eye and skin burns are relatively common.

SOURCES OF EXPOSURE

Occupational

Occupational exposure to dimethyl sulfate may occur through inhalation and dermal contact at workplaces where it is produced or utilized. National Institute for Occupational Safety and Health has estimated that 10,481 workers (2455 of these are female) are potentially exposed to dimethyl sulfate in the United States (NOES, 1983). Dimethyl sulfate is highly toxic to humans, but the immediate symptoms of exposure may be non-existent or very mild making it particularly hazardous, especially to the respiratory tract. Some reports of exposure to fumes for only a few minutes have ultimately been fatal (NLM, 2003).

Environmental

The use of dimethyl sulfate as a methylating agent, stabilizer, and chemical intermediate may result in its release to the environment through various waste streams. If released to air, a vapor pressure of 0.667 mmHg at 25 °C indicates that dimethyl sulfate will exist solely as a vapor in the atmosphere, where it will be rapidly degraded by reaction with atmospheric water, or by the action of photochemically produced hydroxyl radicals with a half-life of 82 days. If released to soil, dimethyl sulfate is expected to be hydrolyzed in moist soils or to volatilize from dry soil. If released directly into water, dimethyl sulfate is anticipated to hydrolyze with a half-life of 1.15 h at pH 7, yielding methanol and sulfuric acid (NLM, 2013). Because of its rapid hydrolysis, binding to suspended solids and sediments in water or concentration in biota are not expected to be important fate processes.

INDUSTRIAL HYGIENE

Human exposures are possible during the manufacture, use, handling, or waste management of dimethyl sulfate. The OSHA PEL for DMS is presently 1 ppm (5 mg/m^3), but the NIOSH REL is 0.1 ppm (1990) based upon its potential carcinogenicity. The ACGIH TLV has also been set at 0.1 ppm, with a skin notation (ACGIH, 2008). Dimethyl sulfate is classified as A2 (suspected human carcinogen) (NTP, 2011), and it is a confirmed animal carcinogen that can alkylate DNA (IARC, 1999; ACGIH, 2008). As a suspected carcinogen, dimethyl sulfate should be handled with extreme caution.

MEDICAL MANAGEMENT

After oral exposure, immediately dilute with 4–8 ounces (120–240 ml) of water or milk (not to exceed 4 ounces in a child). Emesis is not recommended due to corrosive nature of dimethyl sulfate. Ipecac is contraindicated. Gastric suction may be considered after recent ingestion, or administration of a 12.5% charcoal slurry. If cough or difficulty in breathing develops, evaluate for respiratory tract irritation and move to fresh air. Remove contaminated clothing and wash exposed skin area thoroughly with soap and water. For skin burns, administer humidified oxygen, and remove from exposure. Determine pulse and obtain a chest X-ray. Administer inhaled beta-2 adrenergic agonists in patients with broncho-spasm. Irrigate exposed eyes with copious amounts of water at room temperature for at least 15 min (NLM, 2013). Once exposed worker is deemed fit to return to work, management should strongly consider reassignment to prevent further exposure to DMS (Agabiklooei et al., 2010).

REFERENCES

ACGIH (2008) Dimethyl sulfate. In: *AGCIH Guide to Occupational Exposure Indices. American Conference of Governmental Industrial Hygienists*, Cincinnati, OH: ACGIH.

Agabiklooei, A., Xamani, N., Siva, H., and Rezaei, N. (2010) Inhalation exposure to dimethyl sulfate vapor followed by reactive airway dysfunction syndrome. *Indian J. Occup. Environ. Med.* 14:104–106.

Dreisbach, R.H. (1987) *Handbook of Poisoning*, 12th ed., Norwalk, CT: Appleton and Lange.

Ellenhorn, M.J. and Barceloux, D.G. (1988) *Medical Toxicology: Diagnosis and Treatment of Human Poisoning*, 1st ed., New York, NY: Elsevier Science Publishing Co., Inc.

Ellenhorn, M.J., Schonwald, S., Ordog, G., and Wasserberger, J. (1997) *Ellenhorn's Medical Toxicology: Diagnosis and Treatment of Human Poisoning*, 2nd ed., Baltimore, MD: Williams and Wilkins.

Hazardous Substances Data Bank (NLM) (2013) Data base of the National Library of Medicine's TOXNET system for dimethyl sulfate, RN 77-78-1 (http:toxnet.nlm.nih.gov).

IARC (1974) Dimethyl sulfate. In: *Some Aromatic Amines, Hydrazine, and Related Substances,* N-*Nitroso Compounds, and Miscellaneous Alkylating Agents. IARC Monographs on the Evaluation of Carcinogenic Risk of Chemicals to Humans*, vol. 4. Lyon, France: International Agency for Research on Cancer, pp. 271–276.

IARC (1999) Dimethyl sulfate. In: *Re-evaluation of Some Organic Chemicals, Hydrazine, and Hydrogen Peroxide. IARC Monographs on the Evaluation of Carcinogenic Risk of Chemicals to Humans*, vol. 71, Lyon France: International Agency for Research on Cancer, pp. 575–588.

Langford, R.E. (2004) *Introduction to Weapons of Mass Destruction*, Hoboken, NJ: Wiley-Interscience.

Littler, T.R. and McConnell, R.B. (1955) Dimethyl sulphate poisoning. *Br. J. Ind. Med.* 12:54–56.

Mathison, B.H., Taylor, M.L., and Bogdanffy, M.S. (1995) Dimethyl sulfate uptake and methylation of DNA in rat respiratory tissues following acute inhalation. *Fundam. Appl. Toxicol.* 28:255–263.

NOES (1983) National Occupational Exposure Survey 1981-1983 Dataset. Available at http://www.cdc.gov/noes (accessed 2014).

National Library of Medicine (NLM) (2003) *Dimethyl sulfate*. Available at http://toxnet.nlm.nih.gov (accessed 2014).

National Toxicology Program (NTP) (2011) *Report on Carcinogens*, 12th ed., Washington, D.C.: National Toxicology Program, US Department of health and human services.

NOES (1983) National Occupational Exposure Survey 1981-83, National Institute for Occupational Safety and Health (NIOSH).

O'Neil, M.J., editor. (2006) *The Merck Index—An Encyclopedia of Chemicals, Drugs, and Biologicals*, 14th ed., Whitehouse Station, NJ: Merck and Co., Inc.

United States Environmental Protection Agency (USEPA) (2013) *Technology transfer network air toxic: Dimethyl sulfate*. Available at http://www.epa.gov/ttn/atw/hlthef/di-sulfa.html (accessed 2014).

World Health Organization (WHO) (1985) *Environmental Health Criteria 38: Dimethyl Sulfate*, Geneva, Switzerland: World Health Organization.

DIETHYL SULFATE

Occupational exposure limits: None

Reference values: None

Hazard statements: H350—may cause cancer; H340—may cause genetic defects; H332—harmful if inhaled; H311—toxic when in contact with skin; H302—harmful if swallowed; H314—causes severe skin burns and eye damage

Precautionary statements: P308 + P313—if exposed or concerned, get medical advice/attention; P310—immediately call a poison center or doctor/physician; P301 + P330 + P331—if swallowed, rinse mouth; do not induce vomiting; P303 + P361 + P353—if in contact with skin (or hair), remove/take off immediately all contaminated clothing; rinse skin with water/shower; P304 + P340—if inhaled, remove victim to fresh air and keep at rest in a position comfortable for breathing; P270—do not eat, drink, or smoke when using this product

Risk phrases: R45—may cause cancer; R46—may cause heritable genetic damage; R20/21/22—harmful by inhalation, when in contact with skin, and if swallowed; R34—causes burns

Safety phrases: S53—avoid exposure—obtain special instructions before use; S45—in case of accident or if you feel unwell, seek medical advice immediately (show the label where possible)

BACKGROUND AND USES

Diethyl sulfate (DES; sulfuric acid diethyl ester; CASRN 64-67-5 ($C_2H_5)_2SO_4$) is produced from ethanol and sulfuric acid, from ethylene and sulfuric acid, or from diethyl ether and fuming sulfuric acid (Budavari, 1996). It is used in the synthesis of ethyl derivatives of phenols, amines, and thiols, and as an accelerator in the sulfation of ethylene, and in some sulfonation processes. It is also used to manufacture dyes, pigments, carbonless paper, and textiles. Smaller quantities are used in household products, cosmetics, agricultural chemicals, pharmaceuticals, and laboratory reagents (NTP, 2011).

PHYSICAL AND CHEMICAL PROPERTIES

Diethyl sulfate is a colorless, oily liquid with a faint peppermint-like odor, which darkens with age (Budavari, 1996). The molecular weight for diethyl sulfate is 154.2, and its density is 1.17 g/cm^3 at 25 °C. The vapor pressure of DES is 1 mmHg at 47 °C, and its vapor density is 5.31 (air = 1). It is miscible with alcohol and ether (O'Neil, 2006). The bp is 210 °C, with some decomposition. The water solubility of DES is 7.0 g/l at 20 °C. At higher temperatures, DES rapidly decomposes into monoethyl sulfate and alcohol (NTP, 2011).

MAMMALIAN TOXICOLOGY

Acute Effects

Diethyl sulfate is highly toxic by inhalation, skin or eye contact, or ingestion. For rats, the lowest oral LD$_{50}$ found was 350 mg/kg, and the lowest toxic dose producing death in the rat by inhalation was 250 ppm for a 4-h exposure (Druckery et al., 1970).

Diethyl sulfate is extremely irritating to the skin, eyes, and mucous membranes. Toxicity is most common following inhalation of vapors or dermal exposure, and may be delayed to several hours. Effects of overexposure can include extreme eye pain, photophobia, lacrimation, blurred vision, corneal edema, and angioneurotic edema, followed by pharyngolaryngeal inflammation, dysphonia, aphonia, dysphagia, productive cough, and cyanosis. Delayed pulmonary edema may also occur. Prolonged inhalation of low concentrations produces nausea and vomiting (IARC, 1992). A human exposure of 97 ppm for 10 min may be fatal (Deichman and Gerade, 1969). Contact with the liquid can cause irritation with severe ocular and dermal chemical burns (NLM, 2013).

Chronic Effects

Diethyl sulfate has been shown to cause cancer in subcutaneous and oral administration studies in BD rats (Druckery et al., 1970). The World Health Organization classified vapors from DES as a Group 2A—probable human carcinogen (IARC, 1992). Ehling and Neuhauser-Klaus (1988) described mutations and chromosomal aberrations induced by DES in male mice. Other authors have also reported DNA damage, DNA repair, and mitotic induction in human cells, experimental animals, and bacterial studies; chromosome aberrations were also induced in humans and animals (NLM, 2013). In one reproductive toxicity study, a single dose of DES (85 mg/kg bw) was administered to three pregnant BD rats on day 15 of gestation. Malignant tumors of the nervous system were seen in 2/30 of the offspring on days 285 and 541. The IARC has classified diethyl sulfate as a Group 2A, probable human carcinogen (IARC, 1992).

Mechanism of Action

Diethyl sulfate is an alkylating agent which induces DNA single-strand breaks, inhibits DNA repair, causes mutations, and attacks other chemical sites. Dominant lethal mutations, specific locus mutations, and aneuploid cells were reported in various animal studies (NLM, 2013).

Chemical Pathology

Diethyl sulfate is a mutagen. Ethylation of DNA at the N^7 position of guanine is a major site for DNA damage *in vitro* (IARC, 1992).

SOURCES OF EXPOSURE

Occupational

National Institute for Occupational Safety and Health has estimated that 2261 workers (164 of these are female) are potentially exposed to diethyl sulfate in the United States (NOES, 1983). Occupational exposure may occur during the production of DES or during its use as an ethylating agent, as an accelerator in the sulfation of ethylene, and in other sulfonation reactions (NLM, 2013).

Environmental

The production and use of diethyl sulfate as an ethylating agent may result in its release to the environment through various waste streams. Diethyl sulfate is listed as a hazardous air pollutant, and is suspected to cause serious health problems. Nevertheless, it is expected to exist as a vapor, where it will be degraded by photochemically produced hydroxyl radicals with a half-life of about 9 days. In addition, it is expected to hydrolyze rapidly in the atmosphere in the presence of water vapor; the half-life for this reaction is <1 day (NLM, 2013). Based on the hydrolysis of DES in aqueous environments, it is not expected to volatilize from moist soil surfaces, and adsorption to soil is not expected to be an important process (Robertson and Sugamori, 1966). If released into water, DES is expected to hydrolyze rapidly with a half-life of 1.7 h at pH 7; the rate of hydrolysis increases under more acidic or more basic conditions (Weisenberger and Mayer, 1987). Volatilization from water surfaces, adsorption on suspended solids and sediments, biodegradation, and bioconcentration are not expected to be important because of the rapid rate of hydrolysis of DES.

INDUSTRIAL HYGIENE

No OSHA PEL or ACGIH TLV values have been established. Although there is inadequate evidence that DES is a human carcinogen, it has been labeled by the EPA as a probable carcinogen in humans (2A). There is sufficient evidence for the carcinogenicity of diethyl sulfate in experimental animals. Adequate personal protection measures must be utilized to minimize exposure at all times. Diethyl sulfate is combustible when exposed to heat or flame (Lewis, 1996). The NFPA hazard classification is Health: 3, Flammability: 1, and Instability: 1 (NFPA, 2002). Use dry chemical, foam, carbon dioxide, or water spray in extinguishing fires that are associated with diethyl sulfate. Runoff from fire control may be toxic. The Environmental Protection Agency has not established a reference concentration or reference dose.

MEDICAL MANAGEMENT

Treatment for DES poisoning is very similar to that for DMS (NLM, 2013); see Section 5. Maintain ventilation and oxygenation, and evaluate with frequent arterial blood gas or pulse oximetry monitoring. Decontaminate eyes with copious amounts of water. Remove clothing and wash exposed skin thoroughly with soap and water. For oral ingestion, immediately dilute with 4–8 oz of water or milk. Do not induce emesis due to the corrosive nature of DES and the potential for seizures (NLM, 2013).

REFERENCES

Budavari, S., editor. (1996) *The Merck Index—An Encyclopedia of Chemicals, Drugs, and Biologicals*, 12th ed., Whitehouse Station, NJ: Merck and Co. Inc.

Deichman, W.B. and Gerade, H.W. (1969) *Toxicology of Drugs and Chemicals*, New York, NY: Academic Press.

Druckery H., Kruse, H., Preussmann, S., Ivankovic, S., and Landschutz, C. (1970) Carcinogenic alkylating substances: Alkyl halides, sulfates,-sulfonates, and ring strained heterocyclic compounds. *Zeit. Krebsforsch.* 74:241–273.

Ehling, U.H. and Neuhauser-Klaus, A. (1988) Induction of specific gene locus and dominant lethal mutations in male mice by diethyl sulfate (DES). *Mutat. Res.* 199:191–198.

International Agency for Research on Cancer (IARC) (1992) *Occupational Exposures to Mists and Vapours from Strong Inorganic Acids and Other Industrial Chemicals*, vol. 54, Lyon, France: International Agency for Research on Cancer.

Lewis, R.J. (1996) *Sax's Dangerous Properties of Industrial Materials*, 9th ed., vol 1–3, New York, NY: John Van Nostrand Reinhold.

National Fire Protection Association (NFPA) (2002) *Fire Protection Guide to Hazardous Materials*, 13th ed., Quincy, MA: National Fire Protection Association.

National Library of Medicine (NLM) (2013) *Diethyl sulfate*. Available at http://toxnet.nlm.nih.gov (accessed 2014).

National Toxicology Program (NTP) (2011) *Report on Carcinogens*, 12th ed., Washington, D.C.: National Toxicology Program, US Department of Health and Human Services.

NOES (1983) National Occupational Exposure Survey 1981-1983 Dataset. National Institute for Occupational Health (NIOSH). Available at http://www.cdc.gov/noes.

O'Neil, M.J., editor. (2006) *The Merck Index—An Encyclopedia of Chemicals, Drugs, and Biologicals*, 14th ed., Whitehouse Station, NJ: Merck and Co., Inc.

Robertson, R.E. and Sugamori, S.E. (1966) The hydrolysis of dimethylsulfate and diethylsulfate in water. *Can. J. Chem.* 44:1728–1730.

Weisenberger, K. and Mayer, D. (1987) *Ullmann's Encyclopedia of Industrial Chemistry*, vol. A8, New York, NY: VCH Publishers, pp. 493–504.

PHOSPHATE ESTERS

Specific phosphate esters constitute a major category of organic insecticides, known as organophosphates (Costa, 2013).

These phosphate and phosphite esters, lacking significant insecticidal properties, are widely used as plasticizers and gasoline additives to control preignition. Included in this group are TOCP, triphenyl phosphate, butylated triphenyl phosphate, tri-2-ethylhexyl phosphate, and 2-ethylhexyl diphenyl phosphate. As plasticizers, these compounds offer an additional flame retardant property and are widely used in vinyl and cellulose formulations. Nonparalytic isomers have been used as plasticizers in the manufacture of vinyl plastics, as solvents in cellulosic molding compositions, as additives to extreme pressure lubricants, as inflammable fluids in hydraulic systems, and also as a lead scavenger in gasoline (Budavari, 1996).

The majority of reported cases of phosphate ester poisonings have been associated with ingestion of esters in contaminated food or drink, not occupational exposures. The most frequent occupational exposures to phosphate esters occur during manufacture, packaging, shipping, and storage, not at the point of product use. Increased absorption of phosphate esters has been noted with increased workplace temperatures. Phosphate esters tend to accumulate in fatty tissues, and hence have a long biological half-life (LeBel and Williams, 1986). Bisesi (1994) provides a more thorough discussion of the phosphate esters and their hazards.

REFERENCES

Bisesi, M.S. (1994) Esters. In: Clayton, D.G. and Clayton, F.E., editors. *Patty's Industrial Hygiene and Toxicology*, 4th ed., vol. II, Part D. New York, NY: John Wiley & Sons, Inc.

Budavari, S., editor (1996) *The Merck Index—An Encyclopedia of Chemicals, Drugs, and Biologicals*, 12th ed., Whitehouse Station, NJ: Merck and Co. Inc.

Costa, L.G. (2013) Toxic Effects of Pesticides, Chapter 22. In: Klaassen, C.D., editor. *Casarett and Doull's Toxicology, The Basic Science of Poisons*, 8th edn, New York: McGraw-Hill, pp. 933–980.

LeBel, G. L. and Williams, D.T. (1986) Levels of triaryl phosphates in human adipose tissue from eastern Ontario. *Bull. Environ. Contam. Toxicol.* 37: 41–46.

TRI-*o*-CRESYL PHOSPHATE

Occupational exposure limits: 0.1 mg/m³ (OSHA PEL, NIOSH REL); 0.1 mg/m³ (ACGIH TLV with skin notation)

Reference values: None

Hazard statements: None

Precautionary statements: None

Risk phrases: R39/23/24/25—toxic: danger of very serious irreversible effects through inhalation, in contact with skin and if swallowed; R51/53—toxic to aquatic organisms, may cause long-term adverse effects in the aquatic environment

Safety Phrases: S1/2—keep locked up and out of the reach of children; S20/21—when the product is in use do not eat, drink, or smoke; S28—after contact with skin, wash immediately with plenty of . . . (to be specified by the manufacturer); S45—in case of accident or if you feel unwell, seek medical advice immediately (show the label where possible); S61—avoid release to the environment; refer to special instructions/safety data sheets

BACKGROUND AND USES

Tri-*o*-cresyl phosphate (TOCP, tri-*o*-tolyl phosphate, CASRN 78-30-8), is by far the most toxic of the three isomers of the tri-tolyl phosphates (ILO, 1983). Historically, flaccid paralysis was reported because of the ingestion of TOCP-contaminated Jamaican ginger alcoholic extract ("Jake") liquor during the prohibition era, causing 10,000–15,000 cases, from which recovery was slow or which resulted in permanent disability (Kidd and Langworthy, 1933). Similar cases of paralytic disease have occurred after accidental ingestion of this ester in contaminated cooking oil, accidental contamination of drinking water, and in adulterated illegal alcoholic beverages. Workers engaged in manufacturing and other occupational uses of TOCP have also been affected (NLM, 2013).

Tri-*o*-cresyl phosphate has been used as a plasticizer in lacquers and varnishes, gasoline additive, fire retardant, extreme pressure additive, and to sterilize certain surgical instruments (Bisesi, 1994).

PHYSICAL AND CHEMICAL PROPERTIES

Tri-*o*-cresyl phosphate is a colorless to pale-yellow, practically odorless liquid, molecular weight 368.37. Its density is 1.1955 at 20 °C, and it is a solid below the 25.6 °C melting point. It is sparingly soluble in water, but is slightly soluble in ethanol and very soluble in ethyl ether (Budavari, 1996). The vapor pressure of TOCP is 0.00002 mm Hg at room temperature, but is 10 mm Hg at 265 °C (Bisesi, 1994). The relative vapor density of TOCP is 12.7 (air = 1) (Kuney and Mullican, 1995). It decomposes slightly upon boiling (bp 410 °C) (Budavari, 1996).

MAMMALIAN TOXICOLOGY

Acute Effects

Symptoms of acute human poisoning include sudden diarrhea, nausea or vomiting, dizziness, tremors of tongue and eyelids, sweating, lacrimation or salivation, respiratory distress, convulsions, and coma that may occur during or shortly after ingestion; these symptoms are seldom prolonged for more than 48 h. Fatalities are quite rare, and occur principally in those who have ingested large quantities over a short period of time. This is followed by an asymptomatic latent period, or with a feverish feeling in spite of normal body temperature, and brief episodes of diarrhea, conjunctivitis, rhinitis, pharyngitis, and laryngitis accompanied by dysphagia (NLM, 2013). Inhalation and skin contact are typical industrial routes of exposure (Bisesi, 1994). Vapors may irritate the eyes, but only at high temperatures.

Chronic Effects

Typically, after a period of 10–40 days, an unusual polyneuritis with flaccid paralysis abruptly appears caused by destruction of anterior horn cells of corticospinal tracts and is accompanied by axonal degeneration. Loss of pain, temperature, and vibration sense may accompany the dysfunction in the motor system. Paralysis, characterized by numbness or tingling in hands and feet, with cramp-like pains that may occur in calves, is followed by weakness, sensory loss of pain, temperature, and vibration sensations. Optic neuritis, with headache, vertigo, loss of appetite, and paresthesia has also been reported (Grant, 1974). The minimum paralytic dose is not known, but has been estimated to be 10–30 mg/kg for an adult human (Bisesi, 1994). Other species, e.g., hens, rats, rabbits, cats, guinea pigs, and cattle, have demonstrated similar neuromuscular symptoms (NLM, 2013). There has been no evidence that TOCP causes effects on pregnancy or fetal development in rats (FMC, 1978). Tri-*o*-cresyl phosphate is not classifiable as a human carcinogen (ACGIH, 2008). It was not mutagenic in the Ames test (Haworth et al., 1983). There are no other genetic studies available for TOCP (EPA, 2013).

Mechanism of Action

A decrease in blood cholinesterase, identical to that caused by organophosphate insecticides, occurs only rarely (NLM, 2013). Tri-*o*-cresyl phosphate has also been reported to cause cholesterol lipidosis in adrenocortical and ovarian interstitial cells (Latendresse et al., 1994). This may affect cholesterol and steroid hormone metabolism (Latendresse et al., 1995). Of the three isomers of tri-cresyl phosphate, TOCP is by far the most toxic, and it alone gives rise to polyneuritis (ILO, 1983).

Chemical Pathology

Six human cases were characterized by the involvement of anterior horn cells; fatty degeneration of white substance cord; tigrolysis/disappearance of chromophil substance of nerve cells with displacement of nucleolus to periphery; and demyelination with marked fragmentation and fatty degeneration of the myelin sheath (Sandmeyer and Kirwin, 1981). In hens, TOCP-induced delayed neurotoxicity and increased calcium ion and calmodulin-dependent kinase II phosphorylation of cytoskeletal proteins in the brain, spinal cord, and sciatic nerve (Abou-Donia et al., 1988). The effects on 2',3'-cyclic-nucleotide-3'-phosphohydrolase and calmodulin kinase II phosphorylation activity in the spinal cord and sciatic nerve were strongly associated with organophosphate-induced delayed neurotoxicity in chickens (Lapadula et al., 1991). Evidence in rats showed that histopathologic testicular damage may also occur after exposure to TOCP; effects on Leydig cells and Sertoli cells of the testis were seen (Chapin et al., 1990, 1991).

SOURCES OF EXPOSURE

Occupational

National Institute for Occupational Safety and Health has estimated that 20 workers (20 of these are female) are potentially exposed to TOCP in the United States (NOES, 1983). Intoxication is a rare occurrence, since production of TOCP is completely enclosed (ILO, 1983). The most frequent occupational exposures to phosphate esters occur during manufacture, packaging, shipping, and storage, not at the point of product use (NLM, 2013). Persons with cardiorenal or nervous system disorders should be specially protected from industrial exposure to TOCP (ILO, 1983).

Environmental

Industrial usage of TOCP may result in its release to the environment through various waste streams. If released into the atmosphere, it will exist in both the vapor and particulate phases (USEPA, 2013). Vapor phase TOCP is degraded in the atmosphere by reaction with photochemically produced hydroxyl radicals with an estimated half-life of about 1.2 days. Air-borne particulate TOCP may be physically removed by both wet and dry deposition. Significant volatilization from dry soil, moist soil, or the surface of streams or lakes is not expected to occur, however, based on the very low vapor pressure. Tri-*o*-cresyl phosphate is expected to biodegrade in both soil and water; the half-life for TOCP in water is about 12 days at 14 °C. Tri-*o*-cresyl phosphate in water is expected to absorb strongly to suspended solids and sediments. Its estimated BCF is very high (3.6×10^4), but available data indicate that bioconcentration may not be an important process.

The general population may be exposed to TOCP via ingestion of contaminated drinking water (NOES, 1983). Several instances of low level TOCP contamination of drinking water or surface water have been reported (NLM, 2013).

INDUSTRIAL HYGIENE

National Institute for Occupational Safety and Health and OSHA have set the REL and PEL for TOCP at 0.1 mg/m³; this standard was vacated in 1989, but is still enforced in some states (NIOSH, 1997). The 8-h TWA is also 0.1 mg/m³, with a skin notation (ACGIH, 2008). Tri-*o*-cresyl phosphate is considered IDLH at levels of 40 mg/m³ and above (NIOSH, 1997). Individuals in workplace environments where exposure to TOCP is possible should wear safety glasses, gas mask, long-sleeve coveralls with tight collars and cuffs, gloves, and boots (International Technical Information Institute, 1988). The NFPA Hazard Classification for TOCP is Health: 2, Flammability: 1, Reactivity: 0, (NFPA, 1997).

MEDICAL MANAGEMENT

For oral exposure, do not induce emesis. If life-threatening amounts of TOCP have been ingested, administration of slurry of 12.5% activated charcoal followed by gastric lavage is suggested. Control any seizures. In the event of skin contact with tri-cresyl phosphates, contaminated clothing should be rapidly removed, and affected body areas copiously irrigated with water. Move patient to fresh air and monitor for respiratory distress. If cough or difficulty in breathing develops, evaluate for respiratory tract irritation. Further treatment is symptomatic and supportive (NLM, 2013). For those that develop significant paralysis, about 3/4th of all cases recover to the point of not requiring further treatment after 1–2 years; only about 5% remain incapacitated (Gosselin et al., 1984). If signs and symptoms of cholinesterase poisoning persist, hospitalization needs to be considered (NLM, 2013).

REFERENCES

Abou-Donia, M.B., Lapadula, D.M., and Suwita, E. (1988) Cytoskeletal proteins as targets for organophosphorus compound and aliphatic hexacarbon-induced neurotoxocity. *Toxicology* 49:469–477.

American Conference of Governmental Industrial Hygienists (ACGIH) (2008) *Threshold Limit Values (TLVs) for Chemical Substances and Physical Agents and Biological Exposure Indices (BEIs)*, Cincinnati, OH: American Conference of Governmental Industrial Hygienists.

Bisesi, M.S. (1994) Esters. In: Clayton, G.D. and Clayton, F.E., editors., *Patty's Industrial Hygiene and Toxicology*, 4th ed., New York, NY: John Wiley & Sons, Inc.

Budavari, S., editor. (1996) *The Merck Index—An Encyclopedia of Chemicals, Drugs, and Biologicals*, 12th ed., Whitehouse Station, NJ: Merck and Co. Inc.

Chapin, R.E., Phelps, J.L., Burka, L.T., Abou-Donia, M.B., and Heindel, J.J. (1991) The effects of Tri-*o*-cresyl phosphate and metabolites on rat Sertoli cell function in primary culture. *Toxicol. Appl. Pharmacol.* 108:194–204.

Chapin, R.E., Phelps, J.L., Somkuti, S.G., Heindel, J.J., and Burka, L.T. (1990) The interaction of Sertoli and Leydig cells in the testicular toxicity of tri-*o* cresyl phosphate. *Toxicol. Appl. Pharmacol.* 104:483–495.

FMC Corporation (1978) *Letter and Attached Appendix to USEPA Regarding Potential Teratogenicity of TOCP, EPA Doc. No. 40–7842476, Fiche No. OTS0519257*. Washington, DC: U.S. Environmental Protection Agency.

Gosselin, R.E., Smith, R.P., and Hodge, H.C. (1984) *Clinical Toxicology of Commercial Products*, 5th ed., Baltimore, MD: Williams and Wilkins.

Grant, W.M. (1974) *Toxicology of the Eye*, 2nd ed., Springfield, IL: Charles C. Thomas Publisher.

Haworth, S., Lawlor, T., Mortelmans, K., Speck, W., and Zeiger, E. (1983) Salmonella mutagenicity test results for 250 chemicals. *Environ. Mutagen.* 5(Suppl. 1):1–142.

International Labour Office (ILO) (1983) *Encyclopedia of Occupational Health and Safety*, vol. 1 and 2, Geneva, Switzerland: International Labour Office.

International Technical Information Institute (1988) *Toxic and Hazardous Chemicals Safety Manual*, Tokyo, Japan: International Technical Information Institute.

Kidd, J.G. and Langworthy, O.R. (1933) *Bull. Johns Hopkins Hosp.* 52:39.

Kuney, J.H. and Mullican, J.M., editors. (1995) *Chemcyclopedia*, Washington, DC: American Chemical Society.

Lapadula, E.S., Lapadula, D.M., and Abou-Donia, M.B. (1991) Persistent alterations of calmodulin kinase II activity in chickens after an oral dose of tri-*o*-cresyl phosphate. *Biochem. Pharmacol.* 42:171–180.

Latendresse, J.R., Brooks, C.L., and Capen, C.C. (1994) Pathologic effects of butylated triphenyl phosphate-based hydraulic fluid and tricresyl phosphate on the adrenal gland, ovary, and testis in the Fischer-344 rat. *Toxicol. Pathol.* 22:341–352.

Latendresse, J.R., Brooks, C.L., and Capen, C.C. (1995) Toxic effects of butylated triphenyl phosphate-based hydraulic fluid and tricresyl phosphate in female F344 rats. *Vet. Pathol.* 32:394–402.

LeBel, G.L. and Williams, D.T. (1986) Levels of triaryl/alkyl phosphates in human adipose tissue from Eastern Ontario. *Bull. Environ. Contam. Toxicol.* 37:41–46.

National Fire Protection Association (NFPA) (1997) *Fire Protection Guide to Hazardous Materials*, 12th ed., Quincy, MA: National Fire Protection Association.

National Institute for Occupational Safety and Health (NIOSH) (1997) *Pocket Guide to Chemical Hazards*, DHHS (NIOSH) Publication Number 97–140, Washington, DC: National Institute for Occupational Safety and Health.

NOES (1983) (National Occupational Exposure Survey 1981-83 Dataset). Available at http://www.cdc.gov/noes (accessed 2014).

National Library of Medicine (NLM) (2013) *Hazardous substance database: Tri-o-cresyl phosphate*. National Library of Medicine. Available at http://toxnet.nln.nih.gov (accessed, 2014).

National Library of Medicine (NLM) (2013) *Glyceryl trinitrate*. Available at http://toxnet.nlm.nih.gov (accessed 2014).

Sandmeyer, E.E. and Kirwin, C.J. Jr. (1981) Esters. In: Clayton, G.D. and Clayton, F.E., editors. *Patty's Industrial Hygiene and Toxicology*, 3rd ed., New York, NY: John Wiley & Sons, Inc.

United States Environmental Protection Agency (USEPA) (2013) *Technology transfer network air toxic: Tri-o-cresyl phosphate*. Available at http://toxnet.nlm.nih.gov/cgi-bin/sis/search/a?dbs+hsdb:@term+@DOCNO+4084 (accessed 2014).

TRIPHENYL PHOSPHATE

Occupational exposure limits: $3\,\text{mg/m}^3$ (OSHA PEL, ACGIH TLV); $0.1\,\text{mg/m}^3$ (NIOSH REL); DNEL inhalation, systemic effects, long term: $0.55\,\text{mg/m}^3$; DNEL dermal, systemic effects, long term: $5.55\,\text{mg/kg/bw/day}$

Reference values: DNEL inhalation, systemic effects, long term: $0.14\,\text{mg/m}^3$; DNEL dermal, systemic effects, long term: $2.77\,\text{mg/kg/bw/day}$; DNEL oral, systemic effects, long term: $0.04\,\text{mg/kg/bw/day}$

Hazard statements: H400—very toxic to the aquatic life; H411—toxic to the aquatic life with long-lasting effects

Precautionary statements: None

Risk phrases: R50/53—very toxic to the aquatic organisms, may cause long-term adverse effects in the aquatic environment

Safety phrases: S61—avoid release to the environment; refer to special instructions/safety data sheets

BACKGROUND AND USES

Triphenyl phosphate (TPP, CASRN 115-86-6, $C_{18}H_{15}O_4P$) is prepared by reacting phosphorus pentoxide and phenol (Budavari, 2001), or by reacting phosphorus oxychloride and phenol (Snyder, 1990). One primary use is as a flame retardant in phenolic- and phenylene oxide-based resins for the manufacture of electrical and automobile components, for auto upholstery, and as a nonflammable plasticizer in cellulose acetate for photographic films. It has also been used to impregnate roofing paper. Triphenyl phosphate occurs as a plasticizer in various lacquers and varnishes (O'Neil, 2006), and as a component of lubricating oil and hydraulic fluids (ACGIH, 2012).

PHYSICAL AND CHEMICAL PROPERTIES

The molecular weight for triphenyl phosphate is 326.28; it is a colorless, crystalline powder with a slight, phenol-like odor (O'Neil, 2006). Its density is 1.2055 g/cm^3 at 50 °C and its melting point is 50.5 °C (Haynes, 2010). Triphenyl phosphate is practically insoluble in water at 1.9×10^{-7} mg/l at 24 °C (Yalkowsky et al., 2010). It is very soluble in carbon tetrachloride (Haynes, 2010) and is soluble in most lacquers, solvents, thinners, and oils, as well as in alcohol, benzene, ether, chloroform, and acetone (Lewis, 1996). The vapor density is 1.19 (air = 1.0) (Bingham et al., 2001). The boiling point is 245 °C (O'Neil, 2006). It begins to decompose at about 600 °C in inert gas, and in a large excess of air, complete combustion to carbon dioxide occurs in the range 800–900 °C (Lhomme et al., 1984). Hydrolysis of TPP occurs very slowly in acidic or neutral solutions, but occurs rapidly in alkaline solutions (Barnard et al., 1966)

MAMMALIAN TOXICOLOGY

Acute Effects

In spite of rather widespread human exposure to triphenyl phosphate, there are minimal reports of human health effects. Triphenyl phosphate does not possess a strong potential for irritation of the skin, and the irritation potential on the mucous membranes of the eyes is very low (EHC111, 1991). Exposure to high concentrations of TPP has been shown to inhibit chlolinesterase *in vitro* and *in vivo*, but TPP is not considered a potent anticholinesterase agent (Bingham et al., 2001). The oral LD_{50} for rats is 3500 mg/kg (Hierholzer et al., 1957).

Chronic Effects

Animal experiments with 99.99% pure triphenyl phosphate did not demonstrate neurotoxicity (Wills et al., 1979). While a statistically significant reduction in red blood cell cholinesterase has been reported in some workers in a TPP-manufacturing plant exposed to 3.5 mg/m³ TPP for an average of 7.4 years, there was no evidence of neurological disease (Sutton et al., 1960). Triphenyl phosphate is not classifiable as a human carcinogen (ACGIH, 2012). Triphenyl phosphate was not active in *Salmonella* (Ames test) mutagenicity studies with and without metabolic activation (Zeiger et al., 1987). Earlier results for mutagenicity in the Ames test and in *Saccharomyces cerevisiae* (strain D4) were also negative (Monsanto, 1979). In a developmental/reproductive study in Sprague-Dawley rats, in which daily intake of TPP from 4 weeks post weaning through mating and gestation was up to 690 mg/kg bw TPP, no toxic effects were seen in the dams or offspring; TPP was thus not a teratogen at the dose levels tested (Welsh et al., 1987).

Mechanism of Action

Statistically significant depression of human erythrocyte cholinesterase was associated with chronic exposure of 32 workers to TPP, but depression of plasma cholinesterase or effects on neurological parameters were not affected (Sutton et al., 1960). In a separate study, in which 39 workers were exposed to 30% TPP and 70% different isopropyl TPPs, statistically significant depressions in serum IgM and borderline significant depressions in erythrocyte cholinesterase and depressions in red blood cell cholinesterase were reported in some workers, but plasma cholinesterase activity and other parameters were not significantly affected (Emmett et al, 1985) .

Chemical Pathology

Minimal human chemical pathology has been reported for TPP.

SOURCES OF EXPOSURE

Occupational

National Institute for Occupational Safety and Health has estimated that 91,754 workers (22,800 were female) are potentially exposed to triphenyl phosphate in the United States (NOES, 1983). Exposure is primarily via inhalation and dermal contact in workplaces where it is produced or utilized, and also by inadvertent ingestion of contaminated food and drinking water. Workers may be exposed to low levels of triphenyl phosphate by inhalation of vapors during injection molding, by dermal contact, or by ingestion of food contaminated by diffusion from packaging (Daft, 1982; Kalman, 1986). The toxicity of TPP is low. With the exception of dermatitis, no reported cases of human poisoning from triphenyl phosphate exposure have been reported. Nevertheless, persons with preexisting neuromuscular disorders may be at increased risk of neurotoxic effects (NIOSH, 1981).

Environmental

Combustion of plastics and volatilization from plastics is a potential source of atmospheric contamination. Entry of high amounts of TPP into the aquatic environment is thought to occur principally via hydraulic fluid leakage. With the exception of relatively high concentrations of TPP in sediments collected near heavily industrialized areas, environmental concentrations of TPP near sites of production or industrial use are generally <100–400 ng/l. The degradation pathway for TPP is reported to involve stepwise enzymatic hydrolysis to orthophosphate and phenolic moieties (EHC111, 1991). Fish, algae, and phytoplankton are subject to toxicity in the aquatic environment. Triphenyl phosphate is regarded as a severe marine pollutant, with a TWA of $3 \, \text{mg/m}^3$ of water (USDOL, 2012).

INDUSTRIAL HYGIENE

National Institute for Occupational Safety and Health has set the REL for triphenyl phosphate at $0.1 \, \text{mg/m}^3$ ($13.35 \, \text{mg/m}^3 = 1$ ppm). The OSHA PEL is $3 \, \text{mg/m}^3$ (NIOSH, 1994). The TWA–TLV is also $3 \, \text{mg/m}^3$ (ACGIH, 2008) in order to minimize possible eye and skin irritation and allergic contact dermatitis (Camarasa and Serra-Baldrich, 1992). The OSHA 8-h PEL for triphenyl phosphate is $3 \, \text{mg/m}^3$ (29CFR 1910.100). The 8 h TWA and TLV is also $3 \, \text{mg/m}^3$ (ACGIH, 2012), with a recommended excursion limit of 3× the TLV for 30 min. Under no circumstances can exposure exceed 5× the TLV–TWA (ACGIH, 2012). The IDLH value is $1000 \, \text{mg/m}^3$. The NFPA hazard classification is Health: 1, Flammability: 1, and Instability: 0; it is regarded as noncombustible (flash point: 428 °F) (NFPA, 2010). To fight a fire of burning triphenyl phosphate, use carbon dioxide or dry chemical. Toxic gases and vapors such as phosphoric acid fumes and carbon monoxide may be released in such a fire.

MEDICAL MANAGEMENT

No cases of human systemic poisoning from TPP exposure have been reported. Skin sensitization has been reported rarely. Triphenyl phosphate is not regarded as an irritant, and is poorly absorbed through intact skin (Carnarasa and Serra-Baldrich, 1992).

REFERENCES

American Conference of Governmental Industrial Hygienists (ACGIH 2008) *Threshold Limit Values (TLVs) for Chemical Substances and Physical Agents and Biological Exposure Indices (BEIs)*, Cincinnati, OH: American Conference of Governmental Industrial Hygienists.

Barnard, P.W.C., Bunton, C.A., Kellerman, D., Mhala, M.M., Silver, B., Vernon, C.A., and Welch, V.A. (1966) Reactions of organic phosphates. Part VI. The hydrolysis of aryl phosphates. *J. Chem. Soc. B* 227–235.

Bingham, E., Cohrssen, B., and Powell, C.H. (2001) Esters. In: Clayton, G.D. and Clayton, F.E., editors. *Patty's Industrial Hygiene and Toxicology*, 5th ed., New York, NY: John Wiley & Sons, Inc.

Budavari, S., editor. (2001) *The Merck Index—An Encyclopedia of Chemicals, Drugs, and Biologicals*, 13th ed., Whitehouse Station, NJ: Merck and Co. Inc.

Carnarsa, J.G. and Serra-Baldrich, E. (1992) Allergic contact dermatitis from triphenyl phosphate. *Contact Dermatitis* 26:264–265.

Daft, J.L. (1982) Identification of aryl/alkyl phosphate residues in foods. *Bull. Environ. Contam. Toxcol.* 29:221–227.

ECH 111 (1991) International Programme (sic) on Chemical Safety (IPCS) Environmental Health Criteria 111, Triphenyl phosphate, World Health Organization, Geneva.

Emmett, E.A., Lewis, P.G., Tanaka, F., Bleecker, M., Fox, R., Darlington, A.C., Synkowski, D.R., Dannenberg, A.M., Taylor W.J., and Levine, M.S. (1985) Industrial exposure to organophosphorus compounds: studies of a group of workers with a decrease in esterase-staining monocytes. *J. Occup. Environ. Med.* 27:905–914.

Haynes, W.M., editor. (2010) *CRC Handbook of Chemistry and Physics*, 91st ed., Boca Raton, FL: CRC Press Inc.

Hierholzer, K., Noetzel, H., and Schmidt, L. (1957) Comparative toxicological study on triphenyl phosphate and tricresyl phosphate. *Arzneimittelforschung* 7:585–588.

Kalman, D.A. (1986) Survey analysis of volatile organics released from plastics under thermal stress. *Am. Ind. Hyg. Assoc. J.*, 47:270–275.

Lewis, R.J. (1996) *Sax's Dangerous Properties of Industrial Materials*, 9th ed., vol. 1–3, New York, NY: John Van Nostrand Reinhold.

Lhomme, V., Bruneau, C., Soyer, N., and Brault, A. (1984) Thermal behavior of some organic phosphates. *Ind. Eng. Chem. Prod. Res. Dev.* 23:98–102.

Monsanto (1979) *Summary of mutagenicity, neurotoxicity study, teratology study, long-term feeding study, and 90-day inhalation study.* Submitted to US EPA (*EPA-OTS document No. 40–7942057*). Monsanto Industrial Chemicals Co., St. Louis, MO.

National Fire Protection Association (NFPA) (2010) *Fire Protection Guide to Hazardous Materials*, 14th ed., Quincy, MA: National Fire Protection Association.

National Institute for Occupational Safety and Health (NIOSH) (1981) *Occupational Health Guidelines for Chemical Hazards*, DHHS (NIOSH) Publication Number 81–123, Washington, DC: National Institute for Occupational Safety and Health.

National Institute for Occupational Safety and Health (NIOSH) (1994) *Pocket Guide to Chemical Hazards*, DHHS (NIOSH) Publication Number 94–116, Washington, DC: National Institute for Occupational Safety and Health.

National Occupational Exposure Survey. (1981–1983) Dataset. National Institute for Occupational Safety and Health, (NIOSH). Available at http://www.cdc.gov/noes (Accessed 2014).

O'Neil, M.J., editor. (2006) *The Merck Index—An Encyclopedia of Chemicals, Drugs, and Biologicals*, 14th ed., Whitehouse Station, NJ: Merck and Co., Inc.

Snyder, R., editor. (1990) Nitrogen and phosphorus solvents. In: *Ethel Browning's Toxicity and Metabolism of Industrial Solvents*, 2nd ed., vol. 3, Amsterdam, NY: Elsevier.

Sutton, W.L., Terhaar, C.J., Miller, F.A., Scherberger, R.F., Riley, E.C., Roudabush, R.L., and Fassett, D.W. (1960) Studies on the industrial hygiene and toxicology of triphenyl phosphate. *Arch. Environ. Health* 1:33–46.

United States Department of Labor (USDOL) (2012) *Triphenyl phosphate. Occupational Safety and Health Administration.* Available at https://www.osha.gov/dts/chemicalsampling/data/CH_274400.html (accessed 2014).

Welsh, J.J., Collins, T.F., Whitby, K.E., Black, T.N., and Arnold, A. (1987) Teratogenic potential of triphenyl phosphate in Sprague-Dawley (Spartan) rats. *Toxicol. Ind. Health* 3:357–369.

Wills, J.H., Barron, K., Groblewski, G.E., Benitz, K.F., and Johnson, M.K. (1979) Does triphenyl phosphate produce delayed neurotoxic effects? *Toxicol. Lett.* 4:21–24.

Yalkowsky, S.H., He, Y., and Jain, P. (2010) *Handbook of Aqueous Solubility Data*, 2nd ed., Boca Raton, FL: CRC Press LLC.

Zeiger, E., Anderson, B., Haworth, S., Lawlor, T., Mortelmans, K., and Speck, W. (1987) Salmonella mutagenicity tests: III. Results from the testing of 255 chemicals. *Environ. Mutagen.* 9(Suppl. 9):1–109.

NITRATE ESTERS

GLYCERYL TRINITRATE

Occupational exposure limits: 0.2 ppm (OSHA PEL); 0.01 ppm (NIOSH REL); 0.05 ppm (ACGIH TLV); DNEL dermal, systemic effects, long term: 0.5 mg/kg/bw/day; DNEL dermal, systemic effects, short term: 2.5 mg/kg/bw/day

Reference values: DNEL oral, systemic effects, long term: 0.5 mg/kg/bw/day

Hazard statements: H200—unstable explosives; H300—fatal if swallowed; H310—fatal when in contact with skin; H330—fatal if inhaled; H373—may cause damage to the circulatory system through prolonged or repeated inhalation exposure; H411—toxic to aquatic life with long-lasting effects

Precautionary statements: P201—obtain special instructions before use; P202—do not handle until all safety precautions have been read and understood; P281—use personal protective equipment as required; P372—explosion risk in case of fire; P373—do not fight fire when fire reaches explosives; P380—evacuate area; P501—dispose of contents/container to . . .

Risk phrases: R3—extreme risk of explosion by shock, friction, fire, or other sources of ignition; R26/27/28—very toxic by inhalation, when in contact with skin, and if swallowed; R33—dangerous and of cumulative effects; R51/53—toxic to the aquatic organisms, may cause long-term adverse effects in the aquatic environment

Safety phrases: S33—take precautionary measures against static discharges; S35—this material and its container must be disposed of in a safe way; S36/37—wear suitable protective clothing and gloves; S45—in case of accident or if you feel unwell, seek medical advice immediately (show the label where possible); S61—avoid release to the environment; refer to special instructions/safety data sheets

BACKGROUND AND USES

Glyceryl trinitrate (GTN; nitroglycerin; CASRN 55-63-0) is a representative compound from the organic nitrate ester family. Introduced in 1879, it has been used as a therapeutic agent in the treatment of angina pectoris. In the later part of the nineteenth century, the explosive property of nitroglycerin was discovered, and it then became the active component of dynamite. Glyceryl trinitrate, ethylene glycol dinitrate, and other nitrate esters are commonly used in military and mining explosives. In preparing dynamite, these nitrate esters are absorbed in a dope of oxidizing salts and various inert fillers.

PHYSICAL AND CHEMICAL PROPERTIES

Glyceryl trinitrate is a colorless to pale-yellow, viscous liquid or solid (below 13.3 °C [56 °F]). It begins to decompose at 50–60 °C (122–140 °F). Both GTN and ethylene glycol trinitrate are volatile and can pass directly through the skin (Hogstedt and Stah, 1980). Because of their volatility, explosive makers, dynamite packagers, miners, and workers handling cordite (a smokeless powder composed of GTN, guncotton, mineral jelly, and acetone) can be exposed to these organic nitrate esters.

MAMMALIAN TOXICOLOGY

Acute Effects

The headache characteristic of exposure to GTN and ethylene glycol trinitrate typically begins as a feeling of warmth or fullness in the head, and develops into a throbbing sensation that progresses from the forehead to the back of the neck. The throbbing headache has been attributed to dilation of intracerebral blood vessels and can be minimized or prevented by pretreatment with an oral vasopressor similar to amphetamine sulfate. The occurrence of headache associated with vasodilator therapy is the most common side effect of the nitrates, appearing in patients after treatment, but tolerance is developed with continued treatment. Development of cross tolerance is a characteristic produced from prolonged use of nitrate esters. It has been demonstrated that the effect of sublingual GTN on blood pressure and heart rate is diminished, during treatment with pentaerythrityl trinitrate and isosorbide dinitrate (Thadani et al., 1980). Increased tolerance to organic nitrate esters is usually attributed to increased biodegradation or elimination of active components and biotransformation by vascular smooth muscle and other target tissues. This controversial topic does not appear to be an important clinical problem for patients being treated properly with GTN (Abrams, 1983).

Many reports have documented characteristic withdrawal symptoms in munitions workers handling nitrate esters. These consist of typical severe headaches in exposed individuals who are absent from the nitrate-laden environment for 2–3 days. To avoid the headaches, employees rubbed GTN into their skin or wore impregnated headbands on weekends or holidays. In addition, more serious sequelae to nitrate withdrawal have been reported. It is well known that GTN-dependent individuals may experience symptoms of angina pectoris leading to cardiovascular collapse, usually within 48–72 h after withdrawal from exposure (Hogstedt and Andersson, 1979; Hogstedt, 1980). Some experts have recommended that nitrates should not be withdrawn abruptly (Abrams, 1980). The adverse effects that do exist are mainly attributed to the vasodilatory effects. Such side effects include headaches, postural hypotension, impaired coronary and cerebral perfusion, dizziness, syncope, flushing, palpitation, and, less frequently, nausea, vomiting, or abdominal distress. The interval between exposure and the onset of symptoms and the duration of the nitrate effect vary with individual compounds. Glyceryl trinitrate typically produces a prompt response of brief duration, whereas other compounds are slower in producing effects.

Chronic Effects

Contact dermatitis is the long-recognized chronic effect produced by dermal exposure to GTN. Ischemic heart disease has been noted to be linked to GTN exposure in munitions workers although no chronic effect was shown for cardiovascular disease risk (Stayner et al., 1992). Glyceryl trinitrate is not classified as a human carcinogen (USEPA, 2013).

Mechanism of Action

The principal biological properties of these esters are their ability to oxidize heme to the ferric state to produce methemoglobinemia and relaxation of various smooth muscles. Most nitrates produce the Heinz bodies commonly associated with methemoglobinemia. Red blood cells containing Heinz bodies have relatively short life spans and appear to be preferentially sequestered by the spleen. There is variability in the ability of nitrate esters to produce methemoglobin. Ethylene glycol dinitrate is more potent than GTN. The basic hemodynamic effect of the nitrates is vasodilation of both venous and arterial vessels in different degrees. This relaxant effect of organic nitrate esters on vascular smooth muscles is responsible for their clinical therapeutic use in the treatment of stable angina pectoris, coronary spasm, silent myocardial ischemia, myocardial infarction, and congestive heart failure.

Nitrate esters have also been used to treat esophageal spasm and spasms in hepatic ducts and the urinary tract and to lower the intraocular pressure in glaucoma (Ahlner, 1991). The two major potential complications associated with the vasodilating nitrate esters are systemic hypotension and intensification of hypoxemia, which can be fatal (Rubin, 1983).

Contact dermatitis has been reported to be an irritant and allergenic. Kanerva et al. (1991) demonstrated an allergic mechanism on patch testing to GTN, ethylene glycol dinitrate, and dinitrotoluene. In most cases, however, the mechanism appeared to be that of an irritation (de la Fuente et al., 1994).

Chemical Pathology

Unable to determine carcinogenetic or mutagenic effects in humans, however, glyceryl trinitrate has been associated with mutagenic effects in the species of *Salmonella typhimurium*. It is speculated that glyceryl trinitrate releases nitric oxide via metabolic reduction (Maragos et al., 1993).

SOURCES OF EXPOSURE

Occupational

Exposures are due to use of nitroglycerin or in the production of explosives.

Environmental

Environmental exposures are due to the use of glyceryl trinitrate in the form of propellants, explosives, and pharmaceuticals. Degraded in the environment by photochemically produced hydroxyl radicals with a half-life of 15 days in the air and biodegradable in water is 13 days. Half-life in soil may be up to 37–96 days under alkaline conditions (USEPA, 2013).

INDUSTRIAL HYGIENE

Hygienic protection from nitrate esters should focus on eliminating skin contact. Clean gloves that are resistant to nitrate esters are necessary, and the hygienic routines should be stringent. Glove changes as often as hourly may be required for some work assignments. Effective measures to prevent skin absorption include easy access to washrooms for cleanup after spills or before breaks, meals, or toilet use. Although skin contact is more common, some industrial applications expose workers to nitroglycerin via inhalation. These workers should wear respirators with a recommended prefilter to an organic vapor cartridge to protect them from adsorbed aerosol particles (Cohen, 1993). The PEL for GTN and ethylene glycol dinitrate is 0.2 ppm, but NIOSH (1994) recommends a TWA of 0.01 ppm for either of these compounds. The ACGIH has set a ceiling TLV of 0.05 ppm (1995).

MEDICAL MANAGEMENT

Exposures occur via dermal, oral, or respiratory routes. If inhalation exposure occurs, remove victim from area to fresh air. Provide respiratory support as necessary. If contact with eyes, wash with water or normal saline. Victims that experience an oral exposure may consider activated charcoal; deliver as a slurry 24 ml water/30 g charcoal. Call regional poison center for exact dosage. Gastric lavage may be considered for life-threatening situations. Protect airway when appropriate. Dermal exposures are to be treated by removing contaminated clothing and wash exposed areas with soap and water (USEPA, 2013). Due to the vasodilation effects of glyceryl trinitrate, there can be a drop of blood pressure, discontinue use of medication, put patient flat and seek medical care immediately. There may be a need to consider using a vasoconstrictor like methoxamine or phenylephrine. If methemoglobinemia is present, consider using intravenous methylene blue (NLM, 2015).

REFERENCES

Abrams, J. (1980) Nitrate tolerance and dependence. *Am. Heart J.* 99:113–123.

Abrams, J. (1983) Nitroglycerin and long-acting nitrates in clinical practice. *Am. J. Med.* 74:85–94.

American Conference of Governmental Industrial Hygienists (ACGIH, 1995) *Threshold Limit Values (TLVs) for Chemical Substances and Physical Agents and Biological Exposure Indices (BEIs).* Cincinnati, OH: American Conference of Governmental Industrial Hygienists.

Ahlner, J. (1991) Organic nitrate esters. Clinical use and mechanisms of actions. *Pharmacol. Rev.* 43:351–423.

Cohen, H.J. (1993) Determining the service lives of organic-vapor respirator cartridges for nitroglycerin under workplace conditions. *Am. Ind. Hyg. Assoc. J.* 54:432–439.

de la Fuente, P.R., Armentia, M.A., and Diez, P.J. (1994) Contact dermatitis from nitroglycerin. *Ann. Allergy* 72:344–346.

Hogstedt, C. (1980) *Dynamite: Occupational Exposure and Health Effects,* Linkoping University Medical Dissertation.

Hogstedt, C. and Andersson, K. (1979) A cohort study on the mortality among dynamite workers. *J. Occup. Med.* 21:553–556.

Hogstedt, C. and Stah, L. R. (1980) Skin absorption and protective gloves in dynamite work. *Am. Ind. Hyg. Assoc. J.* 41:367–372.

Kanerva, L., Laine, R., Jolanki, R., Tarvainen, K., Estlander, T., and Helander, I. (1991) Occupational allergic contact dermatitis caused by nitroglycerin. *Contact Dermatitis* 24:356–362.

Maragos, C., Andrews, A., Keefer, L., and Elespuru, R. (1993) Mutagenicity of glyceryl trinitrate in *Salmonella typhimurium.* *Mutat. Res.* 298:187–195.

Table of the Most Common Phthalates

Name	Abbreviation	Structural Formula	Molecular Weight	CAS No.
Dimethyl phthalate	DMP	$C_6H_4(COOCH_3)_2$	194.18	131-11-3
Diethyl phthalate	DEP	$C_6H_4(COOC_2H_5)_2$	222.24	84-66-2
Diallyl phthalate	DAP	$C_6H_4(COOCH_2CH=CH_2)_2$	246.26	131-17-9
Di-n-propyl phthalate	DPP	$C_6H_4[COO(CH_2)_2CH_3]_2$	250.29	131-16-8
Di-n-butyl phthalate	DBP	$C_6H_4[COO(CH_2)_3CH_3]_2$	278.34	84-74-2
Diisobutyl phthalate	DIBP	$C_6H_4[COOCH_2CH(CH_3)_2]_2$	278.34	84-69-5
Butyl cyclohexyl phthalate	BCP	$CH_3(CH_2)_3OOCC_6H_4COOC_6H_{11}$	304.38	84-64-0

(Continued)

Table of the Most Common Phthalates

Name	Abbreviation	Structural Formula	Molecular Weight	CAS No.
Di-n-pentyl phthalate	DNPP	$C_6H_4[COO(CH_2)_4CH_3]_2$	306.40	131-18-0
Dicyclohexyl phthalate	DCP	$C_6H_4[COOC_6H_{11}]_2$	330.42	84-61-7
Butyl benzyl phthalate	BBP	$CH_3(CH_2)_3OOCC_6H_4COOCH_2C_6H_5$	312.36	85-68-7
Di-n-hexyl phthalate	DNHP	$C_6H_4[COO(CH_2)_5CH_3]_2$	334.45	84-75-3
Diisohexyl phthalate	DIHxP	$C_6H_4[COO(CH_2)_3CH(CH_3)_2]_2$	334.45	146-50-9
Diisoheptyl phthalate	DIHpP	$C_6H_4[COO(CH_2)_4CH(CH_3)_2]_2$	362.50	41451-28-9
Butyl decyl phthalate	BDP	$CH_3(CH_2)_3OOCC_6H_4COO(CH_2)_9CH_3$	362.50	89-19-0
Di(2-ethylhexyl) phthalate	DEHP, DOP	$C_6H_4[COOCH_2CH(C_2H_5)(CH_2)_3CH_3]_2$	390.56	117-81-7
Di(n-octyl) phthalate	DNOP	$C_6H_4[COO(CH_2)_7CH_3]_2$	390.56	117-84-0
Diisooctyl phthalate	DIOP	$C_6H_4[COO(CH_2)_5CH(CH_3)_2]_2$	390.56	27554-26-3
n-Octyl n-decyl phthalate	ODP	$CH_3(CH_2)_7OOCC_6H_4COO(CH_2)_9CH_3$	390.56	119-07-3
Diisononyl phthalate	DINP	$C_6H_4[COO(CH_2)_6CH(CH_3)_2]_2$	418.61	28553-12-0
Di(2-propyl heptyl) phthalate	DPHP	$C_6H_4[COOCH_2CH(CH_2CH_2CH_3)(CH_2)_4CH_3]_2$	418.61	53306-54-0
Diisodecyl phthalate	DIDP	$C_6H_4[COO(CH_2)_7CH(CH_3)_2]_2$		26761-40-0
Diundecyl phthalate	DUP	$C_6H_4[COO(CH_2)_{10}CH_3]_2$	446.66	3648-20-2
Diisoundecyl phthalate	DIUP	$C_6H_4[COO(CH_2)_8CH(CH_3)_2]_2$	474.72	85507-79-5
Ditridecyl phthalate	DTDP	$C_6H_4[COO(CH_2)_{12}CH_3]_2$	474.72	119-06-2
Diisotridecyl phthalate	DIUP	$C_6H_4[COO(CH_2)_{10}CH(CH_3)_2]_2$	530.82	68515-47-9

National Institute for Occupational Safety and Health (NIOSH) (1994) *Pocket Guide to Chemical Hazards*, DHHS (NIOSH) Publication Number 94–116, Washington, D.C.: National Institute for Occupational Safety and Health.

Toxnet, National Library of Medicine, Nitroglycerin CASRN 56-63-0 (NLM, 2015) Available at: http://toxnet.nlm.nih.gov/cgi-bin/sis/search/a?dbs+hsdb:@term+@DOCNO+30.

Rubin, L. (1983) Cardiovascular effects of vasodilator therapy for pulmonary arterial hypertension. *Clin. Chest Med.* 4:309–319.

Stayner, L.T., Dannenberg, A.L., Thun, M., Reeve, G., Bloom, T.F., Boeniger, M., and Halperin, W. (1992) Cardiovascular mortality among munitions workers exposed to nitroglycerin and dinitrotoluene. *Scand. J. Work Environ. Health* 18:34–43.

Thadani, U., Manyari, D., Parker, J.O., and Fung, H.L. (1980) Tolerance to the circulatory effects of oral isosorbide dinitrate: its rate of development and cross tolerance to glyceryl trinitrate. *Circulation* 61:526–535.

PHTHALATES

DIETHYL PHTHALATE

Occupational exposure limits: 5 mg/m³ (NIOSH REL); DNEL inhalation, systemic/local effects, long term: 10.56 mg/m³; DNEL inhalation, systemic/local effects, short term: 52.8 mg/m³; DNEL dermal, systemic effects, long term: 1.5 mg/kg/bw/day; DNEL dermal, systemic effects, short term: 7.5 mg/kg/bw/day

Reference values: 0.8 mg/kg/bw/day (integrated risk information system [IRIS] reference dose [RfD]); DNEL inhalation, systemic/local effects, long term: 2.6 mg/m³; DNEL inhalation, systemic/local effects, short term: 13 mg/m³; DNEL oral, systemic effects, long term: 0.75 mg/kg/bw/day

Hazard statements: None

Precautionary statements: None

Risk phrases: None

Safety phrases: None

BACKGROUND AND USES

Phthalic acid esters (also known as phthalate) are used in enormous quantities, mainly as plasticizers in PVC. Workers in the PVC-processing industry are exposed to phthalates through inhalation in varying concentrations, depending on different job categories. Phthalates are found in brominated flame retardants. Patients in the intensive care units and neonates may be exposed to high levels of phthalates due to exposure to medical devices such as catheters and nasogastric tubes (Talsness et al., 2009). The general population may be exposed to phthalates, for example, bisphenol A, by food packaging and food containers possibly due to leaching effects when heated (Talsness et al., 2009). Phthalates are used in the production of perfumes, nail varnish, hairsprays, and other cosmetics (Bang et al., 2011).

Diethyl phthalate (CASRN 84-66-2) is a colorless oily liquid with a slight aromatic odor. It has been measured both in indoor and outdoor air (NLM, 2014; Schields and Weschler, 1987). Another survey of 39 public water wells identified diethyl phthalate at a maximum concentration of 4.6 μg/l (USEPA, 2010). National Institute for Occupational Safety and Health has specified 5 mg/m^3 as the TWA for diethyl phthalate. Bisphenol A has a "safe dosing" level of 50 μg/kg (NLM, 2011). It causes irritation of the eyes, nose, and skin; nausea; and dizziness. In rats, may cause spasticity in the lungs (Bang et al., 2011). Hepatitis has been attributed to chronic exposure to phthalates (Neilson et al., 1985). Phthalates do bioaccumulate in humans but are metabolized (Bang et al., 2011). Generally, phthalates are rapidly metabolized via oxidation and their metabolites are excreted in the urine and feces (Bang et al., 2011).

Currently, phthalates, such as dioxins and bisphenol A, are labeled as endocrine-disrupting chemicals that interfere with endogenous hormones and may act as anti-androgens (Craig et al., 2011). Some phthalates, such as polybrominated diphenyl ethers (PBDE) and tetrabromo benzoic acid (TBBA), have been suggested to alter thyroid hormone activities (Talsness et al., 2009). Phthalates are being investigated for their possible role in obesity, diabetes, hormonal cancers, such as breast cancer. Bisphenol A is currently being investigated for its role in diabetes, type 2 (Talsness et al., 2009). Bisphenol A may stimulate insulin secretion as well as decrease adiponectin, which is a critical regulatory cytokine. Children may be at increased risk of the effects of phthalates (Talsness et al., 2009). There are ongoing studies investigating the role of bisphenol A and its effect on socio–sexual behavior (Dessi-Fulgheri et al., 2002).

REFERENCES

Bang, D., Lee, I., and Lee, B. (2011) Toxicological characterization of phthalic acid. *Toxicol. Res.* 27:191–203.

Craig, Z., Want, W., and Flaws, J. (2011) Endocrine-disrupting chemicals in ovarian functions: effects on steroidogenesis, metabolism and nuclear receptor signaling. *Reproduction* 142:633–649.

Dessi-Fulgheri, F., Porrini, S., and Farabollini, F. (2002) Effects of perinatal exposure to bisphenol A on play behavior of female and male rats. *Environ. Health Perspect.* 110(Suppl. 3):403–407.

Neilson, J., Ackesson, B., and Skerfving, S. (1985) Phthalate esters exposure-air levels and health of workers processing polyvinyl chloride. *Am. Ind. Hyg. Assoc. J.* 46:643–647.

National Library of Medicine (NLM) (2011) *Phthalates*. Available at http://toxtown.nlm.nih.gov/text_version/chemicals.php?id=24 (accessed 2014).

National Library of Medicine (NLM, 2014) diethyl phthalate CASRN: 84-66-2. Available at: http://toxnet.nlm.nih.gov/cgi-bin/sis/search/a?dbs+hsdb:@term+@DOCNO+926.

Schields, H.C. and Weschler, C.J. (1987) Analysis of ambient concentrations of organic vapors with a passive sampler. *J. Air Pollut. Control Assoc.* 37:1039–1045.

Talsness, C., Andrade, A., Kuriyama, S., Taylor, J., and vom Saal, F. (2009) Components of plastic: experimental studies in animals and relevance for human health. *Phil. Trans. R. Soc. Lond. B Biol. Sci.* 364:2079–2096.

United States Environmental Protection Agency (USEPA) (2010) *Phthalics*. Available at http://www.epa.gov/ttn/atw/hlthef/phthalic.html (access 2014).

MISCELLANEOUS ESTERS

In spring 1992, an epidemic outbreak of papular and follicular rashes caused by a new line of cosmetics occurred throughout Switzerland. Data obtained from epidemiological and clinical studies have suggested that vitamin E linoleate (a mixture of tocopheryl esters, mainly tocopheryl linoleate), used in the cosmetics and other dermatological preparations, was the cause of papular and follicular contact dermatitis (Perrenoud et al., 1994). In most cases, itchy patches were found to be symmetrically distributed and located on the trunk and extremities. Sometimes, secondary extension of these patches to the face had also been observed with pronounced erythema and edema in the ears. Oxidized vitamin E derivatives may act *in vivo* as haptens or as irritants, possibly with synergistic effects.

In addition, other esters of varying chemical classes are also responsible for contact allergic dermatitis. This group includes corticosteroids (Hisa et al., 1993; Lepoittevin et al., 1995), nicotinic acid esters (Audicana et al., 1990), alkyl gallates as antioxidants (Hausen and Beyer, 1992), sodium lauryl sulfate as a surfactant (Rhein et al., 1990), ester of polyethylene glycol 600, and C21 dicarboxylic acid as a component of cutting fluid (Niklasson et al., 1993) and rosin (colophony) esters (Gäfvert et al., 1994).

REFERENCES

Audicana, M., Schmidt, R., and Fernandez de Cortes, L. (1990) Allergic contact dermatitis from nicotinic acid esters. *Contact Dermatitis*, 22:60–61.

Gäfvert, E., Shao, L.P., Karlberg, A.T., Nilsson, U., and Nilsson, J.L. (1994) Allergenicity of rosin (colophony) esters. (II). Glyceryl monoabietate identified as contact allergen. *Contact Dermatitis* 31:11–17.

Hausen, B.M. and Beyer, W. (1992) The sensitizing capacity of the antioxidants propyl, octyl, and dodecyl gallate and some related gallic acid esters. *Contact Dermatitis* 26:253–258.

Hisa, T., Katoh, J., Yoshioka, K., Taniguchi, S., Mochida, K., Nishimuka, T., Kanetomo, H., Kono, T., and Hamada, T. (1993)

Contact allergies to topical corticosteroids. *Contact Dermatitis* 28:174–179.

Lepoittevin, J.P., Drieghe, J., and Dooms-Goossens, A. (1995) Studies in patients with corticosteroid contact allergy: understanding cross-reactivity among different steroids. *Arch. Dermatol.* 131:31–37.

Niklasson, B., Björkner, B., and Sundberg, K. (1993) Contact allergy to a fatty acid ester component of cutting fluids. *Contact Dermatitis* 28:265–267.

Perrenoud, D., Homberger, H.P., Auderset, P.C., Emmenegger, R., Frenk, E., Saurat, J.H., and Hauser, C. (1994) An epidemic outbreak of papular and follicular contact dermatitis to tocopheryl linoleate in cosmetics. *Dermatology* 189:225–233.

Rhein, L.D., Simion, F.A., Hill, R.L., Cagan, R.H., Mattai, J.A.I.R. A.J.H., and Maibach, H.I. (1990) Human cutaneous response to a mixed surfactant system: role of solution phenomena in controlling surfactant irritation. *Dermatology* 180:18–23.

58

CHLORINATED HYDROCARBONS

Michael H. Lumpkin

Chlorinated hydrocarbons are hydrocarbons containing one or more chlorine atoms within their chemical structure. Chlorine is one of the most common halogens found in halogenated hydrocarbons used in industrial settings.

While some chlorinated hydrocarbons are found in nature (i.e., combustion reactions), most are manufactured for industrial use in material synthesis, as chemical intermediates, or are produced as secondary products of industrial chemistry. Many chlorinated hydrocarbons are long-lived in the environment, may accumulate in various organisms, and may cause toxicity at sufficiently high exposures, including central nervous system (CNS) narcosis, reproductive effects, and cancer.

ALLYL CHLORIDE

First aid: Oral: Avoid emesis, administer charcoal slurry. Inhalation: Move the person to fresh air, evaluate for respiratory tract irritation, bronchitis, or pneumonitis if cough or difficulty in breathing develops. Treat bronchospasm with β2 agonists and corticosteroids. Eye: Irrigate with water for 15 min. Dermal: Remove the contaminated clothing and wash the skin with soap and water.

Target organ(s): Eyes, skin, respiratory system, liver, kidneys

Occupational exposure limits: OSHA PEL = 1 ppm (3 mg/m^3) 8-h TWA; ACGIH TLV = 1 ppm 8-h TWA; NIOSH REL = 1 ppm 10-h TWA, IDLH = 250 ppm

Reference values: EPA RfC: 0.0003 ppm (0.001 mg/m^3)

Risk/Safety phrases: 11-26-50; (1/2-)16-29-33-45-61 Note D

BACKGROUND AND USES

Allyl chloride (3-chloropropene; 1-chloro-2-propene; CASRN 107-05-1) is a chemical intermediate used in the synthesis of allyl compounds found in varnish, resins, polymers, pesticides, and pharmaceuticals (O'Neil, 2001).

PHYSICAL AND CHEMICAL PROPERTIES

Allyl chloride (molecular weight = 76.53 g/mol) is a colorless, brown, yellow, or purple liquid with a pungent, unpleasant odor. It has a vapor pressure of 368 mmHg at 25 °C and is miscible in alcohol, ether, chloroform, and has a water solubility of 3370 mg/l at 25 °C. It may dissolve some plastics, coatings, and rubber (O'Neil, 2001).

ENVIRONMENTAL FATE AND BIOACCUMULATION

Allyl chloride released to the environment via manufacturing waste streams exists in the atmosphere as a vapor and is degraded by photochemical hydroxyl radicals. It has high mobility in soil and evaporates from dry (due to vapor pressure) or moist (due to Henry's law constant of 1.1×10^{-2} atm-m^3/mol) soil. Allyl chloride volatilizes significantly from surface water and is not likely to bind to sediments of suspended solids. Bioaccumulation in aquatic

Hamilton & Hardy's Industrial Toxicology, Sixth Edition. Edited by Raymond D. Harbison, Marie M. Bourgeois, and Giffe T. Johnson.
© 2015 John Wiley & Sons, Inc. Published 2015 by John Wiley & Sons, Inc.

organism is low, with a bioconcentration factor (BCF) of <0.14–<1.3 (NLM, 2013).

ECOTOXICOLOGY

Acute Effects

In aquatic organisms, allyl chloride resulted in the following LC_{50} values: 59.3 mg/l/24 h in *Lepomis macrochirus* (Bluegill), 0.34 mg/l/48 h in *Xenopus laevis* (African clawed frog), and 250 mg/l/24 h in *Daphnia magna* (Waterflea).

Chronic Effects

No chronic toxicity data are available for allyl chloride.

MAMMALIAN TOXICOLOGY

Acute Effects

No data for the acute effects of allyl chloride were available.

Chronic Effects

A study of more than 1000 workers exposed to allyl chloride and epichlorohydrin found no increase in mortality for malignant neoplasms, circulatory diseases, or heart disease (Olsen et al., 1994). In one study of occupational exposure to allyl chloride in China, workers reported weakness, numbness, and paresthesia in the extremities (He et al., 1985).

Electroneuromyographic measurements at one facility with high exposure (2.6–6650 mg/m^3) showed a neuropathy prevalence of 53%. At another facility with lower exposure (0.2–25 mg/m^3), 48% of workers were classified as having mild neuropathy.

Chemical Pathology

Distal peripheral nerve degeneration was reported in rabbits injected with 100 mg/kg three times a week for 7.5–11.5 weeks (Fengsheg et al., 1980).

Mode of Action(s)

The mode of action for allyl chloride is unknown.

SOURCES OF EXPOSURE

Occupational

Exposure to allyl chloride may occur from inhalation or to a much lesser extent from dermal absorption.

Environmental

Inhalation of allyl chloride may result from fugitive emissions from chemical manufacturing sites utilizing allyl chloride in product synthesis.

INDUSTRIAL HYGIENE

Neoprene or rubber gloves, gas-tight goggles or full-face shield, and complete skin coverings should be used when handling allyl chloride. Respirator should be used for ambient air levels of 1 parts per million (ppm) for up to 30 min; supplied-air respirators should be used for longer or higher exposure concentrations. The Occupational Safety and Health Administration (OSHA) has set the permissible exposure limit (PEL) for allyl chloride at 1 ppm (3 mg/m^3) as an 8-h time-weighted average (TWA). The National Institute for Occupational Safety and Health (NIOSH) set a recommended exposure limit (REL) of 1 ppm as a 10-h TWA and an Immediately Dangerous to Life and Health (IDLH) level of 250 ppm. The American Conference of Governmental Industrial Hygienists (ACGIH) set a threshold limit value (TLV) of 1 ppm as an 8-h TWA.

RISK ASSESSMENTS

Environmental Protection Agency (EPA) Reference concentration (RfC): 0.001 mg/m^3, C—possible human carcinogen.

International Agency for Research on Cancer (IARC) Group 3: not classifiable as to its carcinogenicity to humans.

REFERENCES

Fengsheg, H., et al. (1980) Toxic polyneuropathy due to chronic allyl chloride intoxication: a clinical and experimental study. *Chin. Med. J.* 93:177–182.

He, F.S., Lu, B.Q., Zhang, S.L., Dong, S.W., Yu, A., and Wang, B.Y. (1985) Chronic allyl chloride poisoning. An epidemiology, clinical, toxicological, and neuropathological study. *G. Ital. Med. Lav.* 7(1):5–15.

NLM (National Library of Medicine) (2013) *Allyl Chloride. Hazardous Substances Data Bank*, Available at http://toxnet.nlm.nih.gov/cgi-bin/sis/search/f?./temp/~EuKWDd:1.

Olsen, G.W., Lacy, S.E., Chamberlin, S.R., Albert, D.L., Arceneaux, T.G., Bullard, L.F., Stafford, B.A., and Boswell, J.M. (1994) Retrospective cohort mortality study of workers with potential exposure to epichlorohydrin and allyl chloride. *Am. J. Ind. Med.* 25(2):205–218.

O'Neil, M.J., editors. (2001) *The Merck Index—An Encyclopedia of Chemicals, Drugs, and Biologicals*, 13th ed., Whitehouse Station, NJ: Merck and Co., Inc.

CARBON TETRACHLORIDE

First aid: Rinse eyes with water. Clean the skin with soap and water. If swallowed, administer gastric lavage and oxygen and artificial respiration (if necessary).

Target organ(s): CNS, eyes, lungs, liver, kidneys, skin

Occupational exposure limits: OSHA: 10 ppm (PEL); NIOSH: 2 ppm (REL), 300 ppm (IDLH); ACGIH: 5 ppm (TLV)

Reference values: ATSDR: Intermediate and chronic inhalation MRLs: 0.03 ppm

EPA: RfC: 0.1 mg/m^3, RfD: 0.004 mg/kg/day

Risk/safety phrases: 23/24/25-40-48/23-52/53-59; (1/2-) 23-36/37-45-59-61

BACKGROUND AND USES

Carbon tetrachloride (tetrachloromethane; CASRN 56-23-5) is used in organic chemical synthesis, in metal recovery, and catalyst regeneration. Historically, it was used in refrigeration and semiconductor production and as a solvent for processing fats, rubber, and various oils. It was formerly widely used as a fumigant, metal degreaser, dry-cleaning solvent, and in fire extinguishers but has been replaced by other organochlorine solvents of lower toxicity for most industrial uses. Because of its ozone-depleting properties, carbon tetrachloride was scheduled for phase out as per the Montreal Protocol; its industrial use has been diminishing since the 1980s.

PHYSICAL AND CHEMICAL PROPERTIES

Carbon tetrachloride (molecular weight = 153.82 g/mol) is a clear, colorless liquid with a distinctive, sweet ether-like odor. It has a high vapor pressure of 91 mmHg at 20 °C. It has a water solubility of 1160 mg/l and is miscible with various organic solvents. It is mildly reactive with lead and copper and can be reduced to chloroform in the presence of zinc and an acid.

ENVIRONMENTAL FATE AND BIOACCUMULATION

In the atmosphere, carbon tetrachloride exists solely as a gas. It is persistent in the atmosphere and degraded by photochemically produced hydroxyl radicals. It can be degraded by high energy irradiation in the troposphere, where it may last for 30–50 years and has been implicated in the breakdown and depletion of ozone. Carbon tetrachloride is mobile in soil and is cleared either by volatilization or by aerobic or methanogenic microbial degradation. It volatilizes quickly from surface waters, is not expected to bind to sediments or suspended particulates, and is not expected to bioaccumulate (BCF = 3.2–7.4) (NLM, 2013).

ECOTOXICOLOGY

Acute Effects

No mortality was seen in water fleas (*Daphnia magna*) exposed to 7.7 mg/l carbon tetrachloride for 2 days, but LC$_{50}$ values of 350–770 mg/l were reported. In bullfrog (*Rana catesbeiana*) embryos exposed for 4 days, no mortality was seen at 0.026 mg/l, but 100% mortality occurred at 66 mg/l. In 28-day exposures of fathead minnows, 0 and 80% reduction in hatch rates were seen at 18 and 73 g/ml, respectively.

Chronic Effects

No chronic-duration studies were identified.

MAMMALIAN TOXICOLOGY

Acute Effects

In humans, acute carbon tetrachloride exposure results in liver, kidney, gastrointestinal (GI), and CNS depressant effects. Acute CNS effects include dizziness, blurred vision, headache, giddiness, and fatigue Stewart et al. (1963). Death occurred in one alcoholic male following a 15-min inhalation of 250 ppm (Stevens and Forster, 1953; Norwood et al., 1950). Gastrointestinal effects were also seen at 250 ppm, including nausea, vomiting, and abdominal cramps (Norwood et al., 1950). Cases of polyneuritis have been associated with acute dermal exposure (Farrell and Senseman, 1944). No CNS effects were observed in humans exposed to 50 ppm for up to 3 h (Stewart et al., 1961). Single oral ingestions of 90–180 mg/kg resulted in nausea, fatty degeneration of the liver, and swelling of renal tubules (Docherty and Nicholls, 1923; Ruprah et al., 1985). Death was reported in cases involving ingestion of as little as 40 mg/kg (Lamson et al., 1928).

In rats, an LC$_{50}$ of 9500 ppm for 8 h was identified (Svirbely et al., 1947). Dogs exhibited CNS depression after 2–10 h inhalation of 15,000 ppm (von Oettingen et al., 1949). Liver changes in rats occurred after a 7-h exposure to 100 ppm (Adams et al., 1952). Oral doses as low as 5 mg/kg in rats results in slight liver vacuolation (Smialowicz et al., 1991). Litter resorption in rats was seen after 25 mg/kg/day ingestion for 9 days (Narotsky et al., 1997).

Chronic Effects

In mice and rats, chronic lifetime inhalation of 25 ppm carbon tetrachloride resulted in frank liver and kidney damage. Liver and kidney tumors also developed in mice and rats chronically inhaling 25 ppm carbon tetrachloride (Nagano et al., 1998). Chronic oral ingestion of 47 mg/kg/day in rats resulted in fatty liver degeneration and cirrhosis and development of liver tumors (ATSDR, 2005).

Chemical Pathology

Gastrointestinal effects in humans arise from GI irritation. The effects of carbon tetrachloride on the liver include cirrhosis, fibrosis, fatty degeneration, and centrilobular necrosis. Liver tumors in rodents include hepatocellular adenoma and carcinoma. In the kidney, oliguria, nephrosis, ketone bodies, protein casts, occult blood, and urobilinogen occurred. Adrenal pheochromocytomas were reported in mouse kidney.

Mode of Action(s)

CNS effects from carbon tetrachloride exposure arise from generalized CNS depression. Liver and kidney toxicity result from activation of the parent chemical by liver (primarily) and kidney (secondarily) cytochrome P-450-dependent monooxygenase to highly reactive free radicals (Azri et al., 1992). These reactive species may form adducts with DNA or other cellular macromolecules and/or initiate oxidative insult to membrane lipids, leading to a cascade of glutathione depletion, release of phagosomal contents, necrosis and cell death, cytokine recruitment (Holden, 2000), and cellular regeneration and proliferation. Non-gentoxic loss of proliferative control at high carbon tetrachloride exposure may lead to exacerbation of genetic mutations and resultant tumor initiation and progression in liver and kidneys.

SOURCES OF EXPOSURE

Occupational

Workers in chemical manufacturing plants using carbon tetrachloride may be exposed to vapors or liquids from accidental spill or contact with process waste streams. Workers performing remediation tasks at hazardous waste sites may be exposed to carbon tetrachloride vapors.

Environmental

With the phase out of carbon tetrachloride in refrigerant systems, aerosol propellants, dry cleaning facilities, and metal cleaning products, environmental exposures are likely to result from contact with older remaining carbon tetrachloride-containing products.

INDUSTRIAL HYGIENE

Rubber or polyvinyl gloves, goggles, and skin coverings should be used when handling carbon tetrachloride. The NIOSH recommends that a self-contained breathing apparatus be used if carbon tetrachloride concentration is above the REL of 2 ppm. OSHA has set a PEL at 10 ppm. NIOSH set an IDLH level of 300 ppm. ACGIH set a TLV of 5 ppm and a short-term exposure limit (STEL) of 10 ppm.

RISK ASSESSMENTS

Agency for Toxic Substances and Disease Registry (ATSDR): Intermediate and chronic inhalation minimal risk levels (MRLs): 0.03 ppm

EPA: RfC: 0.1 mg/m^3, RfD: 0.004 mg/kg/day, Likely to be carcinogenic to humans

IARC: Group 2B (not classifiable as to carcinogenicity in humans).

REFERENCES

Agency for Toxic Substances DIsease Registry (ATSDR) (2005) *Toxicological Profile for Carbon Tetrachloride*, Department of Health and Human Services.

Adams, E.M., Spencer, H.C., Rowe, V.K., et al. (1952) Vapor toxicity of carbon tetrachloride determined by experiments on laboratory animals. *Arch. Ind. Hyg. Occup. Med.* 6:50–66.

Azri, S., et al. (1992) Further examinations of the selective toxicity of CC14 in rat liver slices. *Toxicol. Appl. Pharmacol.* 112:81–86.

Docherty, J.F. and Nicholls, L. (1923) Report of three autopsies following carbon tetrachloride treatment. *Br. Med. J.* 2:753.

Farrell, C.L. and Senseman, L.A. (1944) Carbon tetrachloride polyneuritis—a case report. *R. I. Med. J.* 27:334–336.

Holden, P.R., James, N.H., Brooks, A.N., et al. (2000) Identification of a possible association between tetrachloride-induced hepatotoxicity and interleukin-8 expression. *J. Biochem. Mol. Toxicol.* 14:283–290.

Lamson, P.D., Minot, A.S., and Robbins, B.H. (1928) The prevention and treatment of carbon tetrachloride intoxication. *J. Am. Med. Assoc.* 90:345–346.

Nagano, K., Nishizawa, T., Yamamoto, S., et al. (1998) Inhalation carcinogenesis studies of six halogenated hydrocarbons in rats and mice. In: Chiyotani, K., Hosoda, Y., and Aizawa, Y., editors. *Advances in the Prevention of Occupational Respiratory Diseases*, Elsevier Science B.V., pp. 741–746.

Narotsky, M.G., Brownie, C.F., Kavlock, R.J., et al. (1997) Critical period of carbon tetrachloride-induced pregnancy loss in Fischer-344 rats, with insights into the detection of resorption sites by ammonium sulfide staining. *Teratology* 56(4):252–261.

NLM (National Library of Medicine) (2013) *Allyl Chloride. Hazardous Substances Data Bank*, Available at http://toxnet.nlm.nih.gov/cgi-bin/sis/search/f?./temp/~EuKWDd:1.

Norwood, W.D., Fuqua, P.A., and Scudder, B.C. (1950) Carbon tetrachloride poisoning. *Arch. Ind. Hyg. Occup. Med.* 1:90–100.

Ruprah, M., Mant, T.G.K., and Flanagan, R.J. (1985) Acute carbon tetrachloride poisoning in 19 patients: implications for diagnosis and treatment. *Lancet I.* 1027–1029.

Smialowicz, R.J., Simmons, J.E., Luebke, R.W., et al. (1991) Immunotoxicologic assessment of subacute exposure of rats to carbon tetrachloride with comparison to hepatotoxicity and nephrotoxicity. *Fundam. Appl. Toxicol.* 17:186–196.

Stevens, H and Forster, F.M. (1953) Effect of carbon tetrachloride on the nervous system. *Arch. Neural. Psychiatry* 70:635–649.

Stewart, R.D., Gay, H.H., Erley, D.S., et al. (1961) Human exposure to carbon tetrachloride vapor. *J. Occup. Expos.* 3:586–590.

Stewart, R.D., et al. (1963) Acute carbon tetrachloride intoxication. *JAMA* 183:994–997.

Svirbely, J.L., Highman, B., Alford, W.C., et al. (1947) The toxicity and narcotic action of monochloromonobromomethane with special reference to inorganic and volatile bromide in blood, urine, and brain. *J. Ind. Hyg.* 29:382–389.

von Oettingen, W.F., Powell, C.C., Sharpless, N.E., et al. (1949) *Relation between the toxic action of chlorinated methanes and their chemical and physicochemical properties. National Inst Health Bull No. 191.*

CHLOROFORM

First aid: Oral: Start gastric lavage and administer activated charcoal if ingestion was recent. Inhalation: Move the person to fresh air, evaluate for respiratory tract irritation, bronchitis, or pneumonitis if cough or difficulty breathing develops. Treat bronchospasm with β2 agonists and corticosteroids. Eye: Irrigate with water for 15 min. Dermal: Remove the contaminated clothing and wash the skin with soap and water.

Target organ(s): Liver, kidneys, heart, eyes, skin, CNS

Reference values: ATSDR acute, intermediate, and chronic inhalation MRLs = 0.1, 0.05, and 0.02 ppm; EPA RfD = 0.01 mg/kg

Risk/safety phrases: 22-38-40-48/20/22; (2-)36/37

BACKGROUND AND USES

Chloroform (trichloromethane, CAS# 67-66-3) is used primarily in the production of hydrochlorofluorocarbon-22 (HCFC-22) and secondarily in the production of fluoropolymers like polytetrafluoroethylene (PTFE). It has been used historically as an organic solvent, insecticidal fumigant, and as an anesthetic.

PHYSICAL AND CHEMICAL PROPERTIES

At ambient temperature and pressure, chloroform (molecular weight = 119.38 g/mol) exists as a clear, colorless liquid with a non-irritant ethereal odor and a sweet taste. It is very volatile, with a vapor pressure of 197 mmHg at 25 °C. It may dissolve some plastics and rubber. It is highly water soluble and is miscible in volatile organic compounds (VOCs) and alcohols.

ENVIRONMENTAL FATE AND BIOACCUMULATION

Chloroform may find its way into the environment as part of the waste stream for hydrochlorofluorocarbon (HCFC) and fluoropolymer production, as well as a drinking water disinfection by-product. It exists exclusively as a vapor in the atmosphere and can be photochemically (but not photolytically) degraded with a half-life of about 150 days. It is highly mobile in soil and evaporates from both soil and surface water. Chloroform binds to sediment and suspended solids. It is not expected to bioaccumulate (BCF = 2.9–10.4) (NLM, 2013).

ECOTOXICOLOGY

Acute Effects

Chloroform is lethal in Xenopus frog embryos at 36 mg/l. The LC_{50} for zebrafish was >100 mg/l. The LC_{50} in Japanese Medaka was 215 mg/l. An EC_{50} of 602 mg/l for reduced reproduction was reported for *D. magna* (EPA, 2013).

Chronic Effects

Not chronic ecotoxicological effects were reported for chloroform.

MAMMALIAN TOXICOLOGY

Acute Effects

Occupational exposure to chloroform resulted in symptoms of general lassitude, dry mouth and thirst, gastrointestinal discomfort, and urinary frequency (Challen et al., 1958; Phoon et al., 1983). When used as a general anesthesia, inhalation of 40,000 ppm resulted in respiratory failure and death (Whitaker and Jones, 1965). Nausea and vomiting also occurred during occupational exposures or following anesthesia. Cardiac arrhythmias and hypotension have been reported after ingestion of chloroform or when it was used as an anesthetic (Schroeder, 1965; Smith et al., 1973). Pulmonary obstruction has been reported in cases of very high inhalation or oral exposures (Schroeder, 1965). Mice exposed to 10 ppm or more for 7 days developed liver toxicity (Larson et al., 1994c). Rodent studies have also indicated renal toxicity following acute exposures.

Chronic Effects

Vomiting and chloroform-induced hepatitis was seen in workers exposed to 14 ppm in the workplace for 6 months (Phoon et al., 1983). Fatty liver, hepatitis, hepatomegaly, jaundice, and increased serum enzymes indicative of liver toxicity occurred in chronically exposed workers (Bornski et al., 1967). Lung toxicity has been reported in rodents exposed to ≥50 ppm.

Chronic renal toxicity has been reported in rodents, particularly mice. Both mice and rats exhibited developmental effects in offspring following inhalation exposures (Bornski et al., 1967). Both liver and kidney cancer were seen in rodents during chronic oral drinking water studies (IARC, 1987).

Chemical Pathology

Chloroform exposure in rodents leads to hepatocellular vacuolation, fatty liver, and centrilobular necrosis (Larson et al., 1994c; Culliford and Hewitt, 1957; Kylin et al., 1963). High inhalation levels may lead to interstitial pneumonitis in rodent lungs, resulting in pulmonary surfactant disruption. Rodent kidneys exhibited nephrosis, renal necrosis and regeneration, fatty degeneration, chronic inflammation, and formation of hyaline casts.

Mode of Action(s)

Chloroform depletes hepatic glutathione (Docks and Krishna, 1976). Both liver and kidney toxicity appears to be mediated by oxidative metabolism *in situ* to phosgene, resulting in cytotoxicity and cellular proliferation (Hook and Smith, 1985). Administration of chloroform in a corn oil vehicle via gavage reportedly enhanced cell proliferation, whereas administration of chloroform via drinking water inhibited cell proliferation (Pereira, 1994). Reports have also suggested that the corn oil vehicle may interact with chloroform to increase toxicity (Jorgenson et al., 1985). Gastrointestinal effects seen at low exposure levels are likely due to neurological effects rather than direct mucosal irritation.

SOURCES OF EXPOSURE

Occupational

Occupational exposure may occur from dermal contact or inhalation at sites of HCFC or fluoropolymer manufacturing.

Environmental

The highest expected environmental exposure is from ingestion of water disinfected by chlorination. Other exposure routes include inhalation of ambient air and dermal contact with this compound or other products containing chloroform (Sheperd et al., 1996).

INDUSTRIAL HYGIENE

Gloves, goggles, and skin coverings should be used when handling chloroform. Appropriate cartridge respirators should be used if exposure concentration is >2 ppm. OSHA has set the PEL (ceiling limit) for chloroform at 50 ppm (240 mg/m^3). The NIOSH set a 60-min STEL level of 2 ppm and states that chloroform may be a potential occupational carcinogen. ACGIH set a TLV at 10 ppm as an 8-h TWA.

RISK ASSESSMENTS

ATSDR: Acute, intermediate, and chronic inhalation MRLs: 0.1, 0.05, and 0.02 ppm
  EPA: RfD: 0.01 mg/kg, likely to be a human carcinogen
  IARC: Group 2B, possibly carcinogenic to humans.

REFERENCES

Bornski, H., Sobolewska, A., and Strakowski, A. (1967) Toxic damage of the liver by chloroform in chemical industry workers. *Int. Arch. Arbeitsmed.* 24:127–134.

Challen, P.J.R., Hickish, D.E., and Bedord, J. (1958) Chronic chloroform intoxication. *Br. J. Ind. Med.* 15:243–249.

Culliford, D. and Hewitt, H.B. (1957) The influence of sex hormone status on the susceptibility of mice to chloroform-induced necrosis of the renal tubules. *J. Endocrinol.* 14(4):381–393.

Docks, E.L. and Krishna, G. (1976) The role of glutathione in chloroform-induced hepatotoxicity. *Exp. Mol. Pathol.* 24:13–22.

Environmental Protection Agency (EPA) (2013) *ECOTOX Database.* Available at http://cfpub.epa.gov/ecotox/quick_query.cfm.

Hook, J.B. and Smith, J.H. (1985) Biochemical mechanisms of nephrotoxicity. *Transplant. Proc.* 17:41–50.

IARC (1987) Chloroform. *IARC monographs on the evaluation of the carcinogenic risk to humans, volume 20, Lyon, France, International Agency for Research on Cancer.*

Jorgenson, T.A., et al. (1985) Carcinogenicity of chloroform in drinking water to male Osborne-Mendel rats and female B6C3F$_L$ mice. *Fundam. Appl. Toxicol.* 5:760–769.

Kylin, B., Reichard, H., Suemegi, I., and Yllner, S. (1963) Hepatotoxicity of inhaled trichloroethylene, tetrachloroethylene and chloroform. Single exposure. *Acta. Pharmacol. Toxicol. (Copenh)* 20:16–26.

Larson, J.L., Wolf, D.C., Morgan, K.T., et al. (1994c) The toxicity of l-week exposures to inhaled chloroform in female B6C3F1 mice and male F-344 rats. *Fund. Appl. Toxicol.* 22:431–446.

NLM (National Library of Medicine) (2013) Allyl Chloride. Hazardous Substances Data Bank, Available at http://toxnet.nlm.nih .gov/cgi-bin/sis/search/f?./temp/~EuKWDd:1.

Pereira, M.A. (1994) Route of administration determines whether chloroform enhances or inhibits cell proliferation in the liver of B6C3F1 mice. *Fundam. Appl. Toxicol.* 23:87–92.

Phoon, W.H., et al. (1983) Toxic jaundice from occupational exposure to chloroform. *Med. J. Malaysia* 38:31–34.

Schroeder, H.G. (1965) Acute and delayed chloroform poisoning. *Br. J. Anaesth.* 37:972–975.

Sheperd, J.L., Corsi, R.L., and Kemp, J. (1996) Chloroform in indoor air and waste-water: the rote of residential washing machines. *J. Air Waste Manage. Assoc.* 46:631–642.

Smith, A.A., Volpitto, P.P., Gramling, Z.W., et al. (1973) Chloroform, halothane, and regional anesthesia: a comparative study. *Anesth. Analg.* 52:1–11.

Whitaker, A.M. and Jones, C.S. (1965) Report of 1500 chloroform anesthetics administered with a precision vaporizer. *Anesth. Analg.* 44:60–65.

EPICHLOROHYDRIN

First aid: Oral: Start gastric lavage and administer activated charcoal if ingestion was recent. Inhalation: Move the person to fresh air, evaluate for respiratory tract irritation, bronchitis, or pneumonitis if cough or difficulty in breathing develops. Treat bronchospasm with β2 agonists and corticosteroids. Eye: Irrigate with water for 15 min. Dermal: Remove the contaminated clothing and wash the skin with soap and water.

Target organ(s): Eyes, skin, respiratory system, kidneys, liver, reproductive system

Occupational exposure limits: OSHA PEL: 5 ppm 8 h TWA; NIOSH: lowest feasible concentration; ACGIH TLV: 0.5 ppm 8-h TWA with skin notation

Reference values: EPA RfC—0.001 mg/m^3

Risk/safety phrases: (2-)36/37; 53-45 Note E

BACKGROUND AND USES

Epichlorohydrin (CASRN 106-89-8) is mainly used in the production of epoxy resins. It was also used as a solvent for paints, varnishes, lacquers, cellulose esters and ethers, and gums. Epichlorohydrin was historically used as an insecticide fumigant.

PHYSICAL AND CHEMICAL PROPERTIES

Epichlorohydrin (molecular weight = 92.53 g/mol) is a colorless liquid with a sweet or garlic-like pungent odor. It is a volatile compound with a vapor pressure of 16.4 mmHg at 25 °C. It is soluble in water (6.6×10^4 mg/l at 25 °C) and miscible with most organic solvents.

ENVIRONMENTAL FATE AND BIOACCUMULATION

Epichlorohydrin may be found in the environment as a result of release in waste streams of resin, paint and lacquer, cellulose ester/ether, and gum production. The high vapor pressure of epichlorohydrin maintains it as a gas in the atmosphere, where it is degraded by photochemically produced hydroxyl radicals (half-life = 36 days). Epichlorohydrin is highly mobile in soil, evaporates quickly from dry soil, and is extensively hydrolyzed in moist soil and water. It is not expected to be adsorbed to sediments and suspended particles and is not expected to bioaccumulate in aquatic organisms (NLM, 2013).

ECOTOXICOLOGY

Acute Effects

D. magna exhibited a 2-h LC_{50} of 29 mg/l epichlorohydrin. The LC_{50} for zebrafish was 30.5 mg/l, while no mortality was seen in rainbow trout at 19 mg/l (EPA, 2013).

Chronic Effects

No chronic ecotoxicological data were available for epichlorohydrin.

MAMMALIAN TOXICOLOGY

Acute Effects

Epichlorohydrin has been reported to cause irritation and allergic contact dermatitis (van Joost et al., 1990; Jolanki et al., 1994). The inhalation LC_{50} in rats is 360 ppm for 6 h (Laskin et al., 1980). Inhalation of ≥ 25 ppm epichlorohydrin by male rats resulted in reduced reproduction (Slott et al., 1990; John et al., 1983).

Oral exposures in animals also caused reproductive and developmental effects. Developmental effects were reported in mice and rats receiving ≥ 120 mg/kg/day but were only seen in conjunction with frank maternal toxicity (Marks et al., 1982). Rats exhibited impaired sperm metabolism following 12 mg/kg/day, with total reproductive impairment seen at

50 mg/kg/day (Wester et al., 1985). Shin et al. (2010) reported increased sperm histopathology and reduced reproduction in male rats receiving 30 mg/kg/day.

Chronic Effects

A case control study of epoxy resin line workers indicated that workers involved in the production of anthraquinone dyes had an increased association with the central nervous system neoplasms (CNSNs) (Barbone et al., 1994). Cases of CNSN were histologically similar and found in workers with longer duration exposures. Barbone et al. (1992) studied the relationship between lung cancer and epichlorohydrin production, noting an increased, but not statistically significant risk in lung cancer. A study of workers with potential epichlorohydrin exposure in one facility showed no increase in malignant neoplasms, lung cancer, circulatory diseases, or heart disease (Olsen et al., 1994). Further, Tsai et al. (1990) did not report any significant difference in heart disease or malignant neoplasm mortality in a cohort of workers from the two facilities.

Inhalation studies have demonstrated that epichlorohydrin is an animal carcinogen (Laskin et al., 1980). In that study, short-term exposure to high concentrations of epichlorohydrin was a more potent inducer of neoplastic lesions than chronic, long-term exposures to 10 ppm or less. An inhalation study of epichlorohydrin in rats and rabbits reported no teratogenic effects (John et al., 1983). Chronic oral exposure in animals resulted in the development of stomach tumors. Lifetime exposure of 2–10 mg/kg/day in rats resulted in stomach tumors (Wester et al., 1985).

Chemical Pathology

Rats inhaling 100 ppm epichlorohydrin exhibited reduced sperm velocity (Slott et al., 1990). Gonadal histopathology was seen in rats exposed orally to ≥ 10 mg/kg/day (Shin et al., 2010).

In rats, forestomach squamous carcinomas resulted from lifetime exposure (Konishi et al., 1980; Wester et al., 1985). Dose-related increases in incidences of forestomach hyperplasia and hyperkeratosis were reported in rats (Daniel et al., 1996). Girolamo et al. (2006) suggested that epichlorohydrin causes tissue proliferation at the point of contact in chick embryos.

SOURCES OF EXPOSURE

Occupational

Workers involved in the production of epoxy resins, paints, laquers, varnish, and cellulose gums may be exposed to epichlorohydrin via inhalation or dermal contact.

Environmental

The general population may be exposed to epichlorohydrin via contact with epichlorohydrin-containing products.

INDUSTRIAL HYGIENE

Gloves (preferably butyl rubber), goggles/face shield, and skin coverings should be used when handling epichlorohydrin. Respirators should be used if exposure concentration is >0.5 ppm. OSHA has set the PEL for epichlorohydrin at 5 ppm as an 8-h TWA. NIOSH did not set a REL but recommends that occupational exposures be limited to the lowest feasible concentration. ACGIH set a TLV for epichlorohydrin at 0.5 ppm with a skin notation.

RISK ASSESSMENTS

EPA: RfC: 0.001 mg/m^3, B2 probable human carcinogen
IARC: Group 3, not classifiable as to human carcinogenicity

REFERENCES

Barbone, F., et al. (1992) A case-control study of lung cancer at a dye and resin manufacturing plant. *Am. J. Ind. Med.* 22:835–849.

Barbone, F., et al. (1994) Exposure to epichlorohydrin and central nervous system neoplasms at a resin and dye manufacturing plant. *Arch. Environ. Health* 49:355–358.

Daniel, F.B., Robinson, M., Olson, G.R., and Page, N.P. (1996) Toxicity studies of epichlorohydrin in Sprague-Dawley rats. *Drug Chem. Toxicol.* 19(1–2):41–58.

Environmental Protection Agency (EPA) (2013) *ECOTOX Database*, Available at http://cfpub.epa.gov/ecotox/quick_query .cfm.

Girolamo, F., Elia, G., Errede, M., Virgintino, D., Cantatore, S., Lorusso, L., Roncali, L., Bertossi, M., and Ambrosi, L. (2006) *In vivo* assessment of epichlorohydrin effects: the chorioallantoic membrane model. *Med. Sci. Monit.* 12(1):BR21–BR27.

John, J.A., Gushow, T.S., Ayres, J.A., Hanley, T.R. Jr., Quast, J.F., and Rao, K.S. (1983) Teratologic evaluation of inhaled epichlorohydrin and allyl chloride in rats and rabbits. *Fundam. Appl. Toxicol.* 3(5):437–442.

Jolanki, R., Kanerva, L., Estlander, T., and Tarvainen, K. (1994) Epoxy dermatitis. *Occup. Med.* 9(1):97–112.

Konishi, Y., Kawabata, A., Denda, A., Ikeda, T., Katada, H., Maruyama, H., and Higashiguchi, R. (1980) Forestomach tumors induced by orally administered epichlorohydrin in male Wistar rats. *Gan* 71(6):922–923.

Laskin, S., Sellakumar, A.R., Kuschner, M., Nelson, N., La, M.S., Rusch, G.M., Katz, G.V., Dulak, N.C., and Albert,

R.E. (1980) Inhalation carcinogenicity of epichlorohydrin in noninbred Sprague-Dawley rats. *J. Natl. Cancer Inst.* 65(4):751–757.

Marks, T.A., Gerling, F.S., and Staples, R.E. (1982) Teratogenic evaluation of epichlorohydrin in the mouse and rat and glycidol in the mouse. *J. Toxicol. Environ. Health* 9(1):87–96.

NLM (National Library of Medicine) (2013) *Allyl Chloride. Hazardous Substances Data Bank*, Available at http://toxnet .nlm.nih.gov/cgi-bin/sis/search/f?./temp/~EuKWDd:1.

Olsen, G.W., Lacy, S.E., Chamberlin, S.R., Albert, D.L., Arceneaux, T.G., Bullard, L.F., Stafford, B.A., and Boswell, J.M. (1994) Retrospective cohort mortality study of workers with potential exposure to epichlorohydrin and allyl chloride. *Am. J. Ind. Med.* 25(2):205–218.

Shin, I.S., Lim, J.H., Kim, S.H., Kim, K.H., Park, N.H., Bae, C.S., Kang, S.S., Moon, C., Kim, S.H., Jun, W., and Kim, J.C. (2010) Induction of oxidative stress in the epididymis of rats after subchronic exposure to epichlorohydrin. *Bull. Environ. Contam. Toxicol.* 84(6):667–671.

Slott, V.L., Suarez, J.D., Simmons, J.E., and Perreault, S.D. (1990) Acute inhalation exposure to epichlorohydrin transiently decreases rat sperm velocity. *Fundam. Appl. Toxicol.* 15(3):597–606.

Tsai, S.P., et al. (1990) Morbidity prevalence study of workers with potential exposure to epichlorohydrin. *Br. J. Ind. Med.* 47:392–399.

van Joost, T., Roesyanto, I.D., and Satyawan, I. (1990) Occupational sensitization to epichlorohydrin (ECH) and bisphenol-A during the manufacture of epoxy resin. *Contact Dermatitis* 22(2):125–126.

Wester, P.W., Van der Heijden, C.A., Bisschop, A., and Van Esch, G.J. (1985) Carcinogenicity study with epichlorohydrin (CEP) by gavage in rats. *Toxicology* 36(4):325–339.

ETHYLENE DICHLORIDE

First aid: Oral: Dilute with 4–8 oz of water or milk; administer activated charcoal if ingestion was recent. Inhalation: Take the person out to fresh air, evaluate for respiratory tract irritation, bronchitis, or pneumonitis if cough or difficulty in breathing develops. Treat bronchospasm with β2 agonists and corticosteroids. *Eye*: Irrigate with water for 15 min. *Dermal*: Remove the contaminated clothing and wash the skin with soap and water.

Target organ(s): Eyes, skin, kidneys, liver, CNS, cardiovascular system

Occupational exposure limits: OSHA PEL: 50 ppm 8-h TWA; NIOSH: 1 ppm 10-h TWA; ACGIH TLV: 10 ppm 8-h TWA.

Risk/safety phrases: 45-11-22-36/37/38; 53-45 Note E

BACKGROUND AND USES

Ethylene dichloride (1,2-dichloroethane, CASRN 107-06-2) is used in the synthesis of vinyl chloride and other organic compounds. It is also used as a solvent for nail lacquers, as a degreaser and metal cleaner, in adhesives, in pharmaceutical production, and in tobacco and spice extraction. Ethylene dichloride was historically used as an insecticide fumigant.

PHYSICAL AND CHEMICAL PROPERTIES

Ethylene dichloride (molecular weight = 98.96 g/mol) is a clear and colorless, oily liquid with a sweet odor and taste. It is a volatile compound with a vapor pressure of 78.9 mmHg at 25 °C. It is relatively insoluble in water (8.6×10^3 mg/l at 25 °C) but soluble in various organic solvents and is miscible with alcohol, chloroform, and ether (NLM, 2013).

ENVIRONMENTAL FATE AND BIOACCUMULATION

Ethylene dichloride may be found in the environment as a result of the release in industrial waste streams. It exists in the atmosphere as a gas due to high vapor pressure, where it is degraded by photochemically produced hydroxyl radicals (half-life = 63 days). Ethylene dichloride is highly mobile in the soil and evaporates quickly from dry soil and surface waters. It is not expected to be adsorbed to sediments and suspended particles and is not expected to bioaccumulate in aquatic organisms (BCF = 2).

ECOTOXICOLOGY

Acute Effects

The 2-h LC_{50} for D. magna was 220 mg/l. Similarly, the 4-h LC_{50} for zebrafish was 254 mg/l. The LC_{50} for Japanese medaka was 840 mg/l (EPA, 2013).

Chronic Effects

Reduced reproduction was observed in D. Magna after a 28-day exposure to 21 mg/l (EPA, 2013).

MAMMALIAN TOXICOLOGY

Acute Effects

Acute intoxication with ethylene dichloride may cause headache, dizziness, nausea, vomiting, weakness, trembling, and epigastric cramps (Siegel, 1947; Liu et al., 2010). In severe cases, collapse, coma, and death may follow as a result of respiratory and circulatory failure. Hepatic and renal injuries have been noted in fatal cases, including fatty liver injury.

Acute inhalation of ≥200 ppm ethylene dichloride in rats resulted in CNS depression. Pregnant rats and rabbits exposed to ≤300 ppm produced no embryotoxic or teratogenic effects (Schlachter et al., 1979). Resistance to bacterial infection was reported in mice exposed to a single 5 ppm exposure (Sherwood et al., 1987).

Oral gavage doses of 136 mg/kg ethylene dichloride in rats resulted in pulmonary injury (Salovsky et al., 2002).

Chronic Effects

Hazardous material workers exposed to high levels of ethylene dichloride exhibited possible CNS effects in vision, memory, and behavior (Bowler et al., 2003). Maltoni et al. (1980) exposed rats and mice to ethylene dichloride via inhalation in a long-term carcinogenicity study and reported no exposure-related increase in mortality or tumors. However, Nagano et al. (2006) reported significant increased incidences of benign and malignant lung, mammary gland, and liver tumors in rats and mice inhaling ≥10 ppm for life.

Oral drinking water exposures in mice of 8000 ppm for 13 weeks resulted in significant renal cell regeneration (Morgan, 1991).

Chemical Pathology

Edema, congestion, and inflammatory changes were seen in rats receiving 136 mg/kg/day (Salovsky et al., 2002). DNA single-strand breaks were seen in rats following oral, but not inhalation, exposures (Storer et al., 1984). Olfactory epithelial degeneration and necrosis were seen in rats inhaling ≥200 ppm for 4 h (Hotchkiss et al., 2010). Adenomas and adenocarcinomas developed in the liver, mammary glands, and bronchoalveolar regions of rats and mice inhaling ≥10 ppm over a lifetime (Nagano et al., 2006).

Mode of Action(s)

General CNS depression is the likely mode of action for CNS effects seen at high exposures, although indicators of oxidative stress have been associated with changes in glutamate and GABA expression (Wang et al., 2012). The appearance of fatty liver injury, following acute high level ethylene dichloride exposure, was a result of inhibited lipoglycoprotein secretion by hepatocytes (Cottalasso et al., 1994). The exact mechanism of action of ethylene dichloride carcinogenicity is unclear. Cheng et al. (2000) suggest genotoxic effects in workers as evidenced by increased sister chromatid exchanges. CYP- and GST-mediated chromosomal aberrations, DNA adducts, and gene point mutations were reported in rodents (Gwinn et al., 2011). Markers of lipid peroxidation were seen in rat lungs following oral gavage (Salovsky et al., 2002).

SOURCES OF EXPOSURE

Occupational

Workers involved in the production of epoxy resins, paints, laquers, varnish, and cellulose gums may be exposed to ethylene dichloride via inhalation or dermal contact.

Environmental

The general population may be exposed to ethylene dichloride via food and drinking water and from contact with ethylene dichloride-containing products.

INDUSTRIAL HYGIENE

Gloves, eye protection, and skin coverings should be used when handling ethylene dichloride. Respirators should be used if exposure concentration is >1 ppm. OSHA has set the PEL for ethylene dichloride at 50 ppm as an 8-h TWA. NIOSH set a REL of 1 ppm as a 10-h TWA. ACGIH set a TLV for ethylene dichloride at 10 ppm as an 8-h TWA.

RISK ASSESSMENTS

EPA: B2, probable human carcinogen

REFERENCES

Bowler, R.M., Gysens, S., and Hartney, C. (2003) Neuropsychological effects of ethylene dichloride exposure. *Neurotoxicology* 24 (4–5):553–562.

Cheng, T.J., Chou, P.Y., Huang, M.L., Du, C.L., Wong, R.H., and Chen, P.C. (2000) Increased lymphocyte sister chromatid exchange frequency in workers with exposure to low level of ethylene dichloride. *Mutat. Res.* 470(2):109–114.

Cottalasso, D., Barisione, G., Fontana, L., Domenicotti, C., Pronzato, M.A., and Nanni, G. (1994). Impairment of lipoglycoprotein metabolism in rat liver cells induced by 1,2-dichloroethane. *Occup. Environ. Med.* 51(4):281–285.

Environmental Protection Agency (EPA) (2013) *ECOTOX Database*, Available at http://cfpub.epa.gov/ecotox/quick_query.cfm.

Gwinn, M.R., Johns, D.O., Bateson, T.F., and Guyton, K.Z. (2011) A review of the genotoxicity of 1,2-dichloroethane (EDC). *Mutat. Res.* 727(1–2):42–53.

Hotchkiss, J.A., Andrus, A.K., Johnson, K.A., Krieger, S.M., Woolhiser, M.R., and Maurissen, J.P. (2010) Acute toxicologic and neurotoxic effects of inhaled 1,2-dichloroethane in adult Fischer 344 rats. *Food Chem. Toxicol.* 48(2): 470–481.

Liu, J.R., Fang, S., Ding, M.P., Chen, Z.C., Zhou, J.J., Sun, F., Jiang, B., and Huang, J. (2010) Toxic encephalopathy caused by occupational exposure to 1,2-Dichloroethane. *J. Neurol. Sci.* 292(1–2):111–113.

Maltoni, C., Valgimigli, L., and Scarnato, C. (1980) Long-term carcinogenic bioassays on ethylene dichloride administered by inhalation by rats and mice. In: Ames, B.N., Infante, P., and Ritz, R. editors. *Ethylene Dichloride: A Potential Health Risk?* Cold Spring Harbor, NY: Cold Spring Harbor Lab, pp. 3–33, (Banbury Report No. 5).

Morgan, D. (1991) NTP technical report on the toxicity studies of 1,2-Dichloroethane (Ethylene Dichloride) in F344/N Rats, Sprague Dawley Rats, Osborne-Mendel Rats, and B6C3F1 Mice (Drinking Water and Gavage Studies) (CAS No. 107-06-2). *Toxic. Rep. Ser.* 4:1–54.

Nagano, K., Umeda, Y., Senoh, H., Gotoh, K., Arito, H., Yamamoto, S., and Matsushima, T. (2006) Carcinogenicity and chronic toxicity in rats and mice exposed by inhalation to 1,2-dichloroethane for two years. *J. Occup. Health* 48(6):424–436.

National Library of Medicine (NLM) (2013) *Allyl Chloride. Hazardous Substances Data Bank*, Available at http://toxnet.nlm.nih.gov/cgi-bin/sis/search/f?./temp/~EuKWDd:1.

Salovsky, P., Shopova, V., Dancheva, V., Yordanov, Y., and Marinov, E. (2002) Early pneumotoxic effects after oral administration of 1,2-dichloroethane. *J. Occup. Environ. Med.* 44(5):475–480.

Schlachter, M.M., et al. (1979) *The Effects of Inhaled Ethylene Dichloride on Embryonal and Fetal Development in Rats and Rabbits*, Midland, Mich: Dow Chemical USA.

Sherwood, R.L., O'Shea, W., Thomas, P.T., Ratajczak, H.V., Aranyi, C., and Graham, J.A. (1987) Effects of inhalation of ethylene dichloride on pulmonary defenses of mice and rats. *Toxicol. Appl. Pharmacol.* 91(3):491–496.

Siegel, M. (1947) Ethylene dichloride. *JAMA* 133:577–585.

Storer, R.D., Jackson, N.M., and Conolly, R.B. (1984) *In vivo* genotoxicity and acute hepatotoxicity of 1,2-dichloroethane in mice: comparison of oral, intraperitoneal, and inhalation routes of exposure. *Cancer Res.* 44(10):4267–4271.

Wang, G., Qi, Y., Gao, L., Li, G., Lv, X., and Jin, Y.P. (2012) Effects of subacute exposure to 1,2-dichloroethane on mouse behavior and the related mechanisms. *Hum. Exp. Toxicol.* [epub ahead of print].

METHYL CHLORIDE

First aid: Move the person to fresh air. If seizure occurs, administer benzodiazepine IV, followed by phenobarbital or propofol if seizures reoccur. *Eye*: Irrigate with water for 15 min.

Target organ(s): CNS, liver, kidneys, reproductive system

Occupational exposure limits: OSHA PEL: 100 ppm 8-h TWA; NIOSH: 1 ppm 10-h TWA; ACGIH TLV: 50 ppm 8-h TWA, 100 ppm STEL.

Reference values: EPA: RfC: 9×10^{-2} mg/m^3

Risk/safety phrases: Not available

BACKGROUND AND USES

Methyl chloride (chloromethane CASRN 74-87-3) is used in the synthesis and solvent extraction of various organic compounds. It is also used in the production of plastic and polystyrene foams and processing of timber wood products, and historically, it was used as an insecticide, anesthetic, and aerosol propellant.

PHYSICAL AND CHEMICAL PROPERTIES

Methyl chloride (molecular weight = 50.49 g/mol) is a colorless gas possessing a sweet odor and taste. It is a highly volatile compound with a vapor pressure of 4300 mmHg at 25 °C. It reacts with plastics, rubber, aluminum, zinc, and magnesium. It is relatively insoluble in water (5.3×10^3 mg/l at 25 °C) but soluble in acetone and ethanol and miscible with chloroform, acetone, and ethyl ether.

ENVIRONMENTAL FATE AND BIOACCUMULATION

Methyl chloride found in the environment may result due to the release from chemical synthesis waste streams. It exists in the atmosphere solely as a gas due to its high vapor pressure, where it is slowly degraded by photochemically produced hydroxyl radicals (half-life = 310 days). Methyl chloride is highly mobile in soil and tends to evaporate quickly from dry soil and surface waters. It may be biodegraded under anaerobic conditions. It is not expected to be adsorbed to sediments and suspended particles and is not expected to bioaccumulate in aquatic organisms (BCF = 3) (NLM, 2013).

ECOTOXICOLOGY

Acute Effects

The 4-h LC$_{50}$ for methyl chloride in bluegill fish was 550 mg/l (EPA, 2013).

Chronic Effects

No chronic ecotoxicological data were identified for methyl chloride.

MAMMALIAN TOXICOLOGY

Acute Effects

Methyl chloride has been shown to affect the CNS, liver, kidney, and male reproductive tissues. Inhalation of methyl chloride in humans has resulted in headache, giddiness, drowsiness, ataxia, convulsion, and coma (Repko, 1981).

Inhalation of methyl chloride in laboratory rodents has resulted in decreased reproduction and specific developmental effects. Mouse dams inhaling 500 ppm during gestation gave birth to pups exhibiting structural heart defects. This effect was not seen at 100 ppm (Wolkowski-Tyl et al., 1983a, 1983b).

Chronic Effects

Chronic intoxication, primarily with neurological symptoms, has been reported in methyl chloride and foam plastic workers (Scharnweber et al., 1974). Cases of long-standing neuropsychiatric alterations have also been described with depression, irritability, change of personality, insomnia, and disturbances of vision.

Inhalation of ≥450 ppm methyl chloride in male mice for 10 weeks resulted in a reduced number of litters at mating (Hamm et al., 1985; Chellman et al., 1986, 1987; Working et al., 1985a,b). Heart malformations were seen in mouse pups born to dams inhaling 500 ppm methyl chloride (Wolkowski-Tyl et al., 1983a). In mice, continuous inhalation of ≥100 ppm methyl chloride for ≥11 days resulted in brain lesions (Landry et al., 1985; Jiang et al., 1985).

Chemical Pathology

Heart malformations in mouse pups born to dams inhaling 500 ppm methyl chloride included reduction/absence of the atrioventricular valve, papillary muscle, and chordae tendinea in the bucuspid or tricuspid valve (Wolkowski-Tyl et al., 1983a, 1983b).

In mice, continuous inhalation of 100 ppm methyl chloride for 11 days resulted in cerebral granular cell layer degeneration (Landry et al., 1985). This effect was not seen if the exposure was intermittent rather than continuous. Further, focal and diffuse malacia was observed in these brain tissues following inhalation of 1500 ppm for 2 weeks in rats (Jiang et al., 1985).

Mode of Action(s)

Increased preimplantation fetal losses in rats are likely caused by failure to fertilize, resulting from testicular or epidydimal epithelium cytotoxicity, rather than sperm cell genotoxicity (Working et al., 1985a, 1985b; Working and Bus, 1986), although high inhalation doses may result in dominant lethal genotoxicity mediated by epididymal inflammation (Chellman et al., 1987). Some testicular and epidydimal lesions reported in rats inhaling methyl chloride may result from centrally disrupted testosterone levels, rather than impaired Leydig cell or gonadotrope function (Chapin et al., 1984). *In vitro* assays on bacterial and human cells suggested that methyl chloride is a weak direct acting mutagen (Fostel et al., 1985).

SOURCES OF EXPOSURE

Occupational

Workers involved in the chemical production involving methyl chloride may be exposed primarily via inhalation, with some possible dermal contact.

Environmental

The general population may be exposed to methyl chloride via drinking water and inhalation of ambient air containing methyl chloride.

INDUSTRIAL HYGIENE

Eye protection and skin covering should be used to protect from freezing temperatures associated with liquid methyl chloride releases. Respirators should be used for exposures of up to 5000 ppm. OSHA has set the PEL for methyl chloride at 100 ppm as an 8-h TWA. NIOSH set a REL of 1 ppm as a 10-h TWA. ACGIH set a TLV for methyl chloride at 50 ppm as an 8-h TWA.

RISK ASSESSMENTS

EPA: RfC: 9×10^{-2} mg/m^3, D, not classifiable as to human carcinogenicity

IARC: Group 3, not classifiable as to its human carcinogenicity.

REFERENCES

Chapin, R.E., White, R.D., Morgan, K.T., and Bus, J.S. (1984) Studies of lesions induced in the testis and epididymis of F-344 rats by inhaled methyl chloride. *Toxicol. Appl. Pharmacol.* 76(2):328–343.

Chellman, G.J., Bus, J.S., and Working, P.K. (1986) Role of epididymal inflammation in the induction of dominant lethal mutations in Fischer 344 rat sperm by methyl chloride. *Proc. Natl. Acad. Sci. U.S.A.* 83(21):8087–8091.

Chellman, G.J., Hurtt, M.E., Bus, J.S., and Working, P.K. (1987) Role of testicular versus epididymal toxicity in the induction of cytotoxic damage in Fischer-344 rat sperm by methyl chloride. *Reprod. Toxicol.* 1(1):25–35.

Environmental Protection Agency (EPA) (2013) *ECOTOX Database*. Available at http://cfpub.epa.gov/ecotox/quick_query .cfm.

Environmental Protection Agency (EPA) (2013) *Integrated Risk Information System (IRIS). Online. National Center for Environmental Assessment.*

Fostel, J., Allen, P.F., Bermudez, E., Kligerman, A.D., Wilmer, J.L., and Skopek, T.R. (1985) Assessment of the genotoxic effects of methyl chloride in human lymphoblasts. *Mutat. Res.* 155(1–2): 75–81.

Hamm, T.E. Jr., Raynor, T.H., Phelps, M.C., Auman, C.D., Adams, W.T., Proctor, J.E., and Wolkowski-Tyl, R. (1985) Reproduction in Fischer-344 rats exposed to methyl chloride by inhalation for two generations. *Fundam. Appl. Toxicol.* 5(3):568–577.

Jiang, X.Z., White, R., and Morgan, K.T. (1985) An ultrastructural study of lesions induced in the cerebellum of mice by inhalation exposure to methyl chloride. *Neurotoxicology* 6(1):93–103.

Landry, T.D., Quast, J.F., Gushow, T.S., and Mattsson, J.L. (1985) Neurotoxicity of methyl chloride in continuously versus intermittently exposed female C57BL/6 mice. *Fundam. Appl. Toxicol.* 5(1):87–98.

National Library of Medicine (NLM) (2013) *Allyl Chloride. Hazardous Substances Data Bank*, Available at http://toxnet .nlm.nih.gov/cgi-bin/sis/search/f?./temp/~EuKWDd:1.

Repko, J.D. (1981) Neurotoxicity of methyl chloride. *Neurobehav. Toxicol. Teratol.* 3(4):425–429.

Scharnweber, H.C., Spears, G.N., and Cowles, S.R. (1974) Case reports. Chronic methyl chloride intoxication in six industrial workers. *J. Occup. Med.* 16(2):112–113.

Wolkowski-Tyl, R., Lawton, A.D., Phelps, M., and Hamm, T.E. Jr. (1983a) Evaluation of heart malformations in B6C3F1 mouse fetuses induced by in utero exposure to methyl chloride. *Teratology* 27(2):197–206.

Wolkowski-Tyl, R., Phelps, M., and Davis, J.K. (1983b) Structural teratogenicity evaluation of methyl chloride in rats and mice after inhalation exposure. *Teratology* 27(2):181–195.

Working, P.K. and Bus, J.S. (1986) Failure of fertilization as a cause of preimplantation loss induced by methyl chloride in Fischer 344 rats. *Toxicol. Appl. Pharmacol.* 86(1):124–130.

Working, P.K., Bus, J.S., and Hamm, T.E. Jr. (1985a) Reproductive effects of inhaled methyl chloride in the male Fischer 344 rat. II. Spermatogonial toxicity and sperm quality. *Toxicol. Appl. Pharmacol.* 77(1):144–157.

Working, P.K., Bus, J.S., and Hamm, T.E. Jr. (1985b) Reproductive effects of inhaled methyl chloride in the male Fischer 344 rat. I. Mating performance and dominant lethal assay. *Toxicol. Appl. Pharmacol.* 77(1):133–143.

METHYL CHLOROFORM

First aid: *Oral*: Do not induce vomiting; use gastric lavage only if the ingested amount was large and consumed within the past 30 min. *Inhalation*: Move the person to fresh air. *Dermal*: Wash the exposed skin with soap and water. *Eye*: Irrigate with water for 15 min.

Target organ(s): Eyes, skin, CNS, cardiovascular system, liver

Occupational exposure limits: OSHA PEL: 350 ppm 8-h TWA; NIOSH: 350 ppm 15-min ceiling; ACGIH TLV: 350 ppm 8-h TWA

Reference values: None identified.

Risk/safety phrases: 20–59; (2-)24/25-59-61 Note F

BACKGROUND AND USES

Methyl chloroform (1,1,1-trichloroethane, CASRN 71-55-6) was used historically in degreasing processes and metal cleaning and as a solvent. It was also found in adhesives and household cleaners. It is used in the chemical synthesis as a chemical intermediate in vinylidene chloride productions. It was used as a food and grain fumigant and propellant for pressurized aerosol spray products. Classified in the Montreal Protocol of 1987 as an ozone-depleting chemical, it was to be phased out of production and use by treaty members by 2010.

PHYSICAL AND CHEMICAL PROPERTIES

Methyl chloroform (molecular weight = 133.42 g/mol) is a colorless liquid having a sweet odor. It is a volatile compound with a vapor pressure of 124 mmHg at 25 °C. It is relatively insoluble in water (1.3×10^3 mg/l at 25 °C) but soluble in many common organic solvents.

ENVIRONMENTAL FATE AND BIOACCUMULATION

Methyl chloroform found in the environment may result from the release in electronics manufacturing and metal cleaning waste streams, as well as from the disposal of household products, in which it is an ingredient. It exists in the atmosphere solely as a gas due to its high vapor pressure, where it is slowly degraded by photochemically produced hydroxyl radicals (half-life = 4.7 years). Because of its persistence in the atmosphere, it tends to diffuse into the stratosphere, where it may contribute to the catalytic degradation of ozone. Methyl chloroform is highly mobile in soil and evaporates quickly from dry soil and surface waters. It may get biodegraded slowly in the soil. It is not expected to be adsorbed to sediments and suspended particles and is not expected to bioaccumulate in aquatic organisms (BCF = 3) (NLM, 2013).

ECOTOXICOLOGY

Acute Effects

The 2- and 17-h LC_{50} for methyl chloroform was 11.2 and 5.4 mg/l, respectively. The 2-h LC_{50} in zebrafish and Japanese medaka were 79 and 1 g/l, respectively (EPA, 2013).

Chronic Effects

No chronic ecotoxicological data were identified for methyl chloroform.

MAMMALIAN TOXICOLOGY

Acute Effects

An exposure to high air concentrations of methyl chloroform lead to narcosis with CNS depression and even fatal respiratory depression (Stewart, 1971). Experimental (Reinhardt et al., 1971; Aviado and Belej, 1974) and clinical observations suggest that abusive recreational exposure levels may increase the risk of cardiac arrhythmias. No adverse cardiovascular or neurological effects were seen in volunteers inhaling 200 ppm methyl chloroform while exercising (Laine et al., 1996). Slight sedation was indicated by electroencephalography (EEG) in volunteers inhaling 200 ppm (Muttray et al., 2000). Human data do not indicate carcinogenicity of methyl chloroform in occupational or environmental exposures.

Inhalation of ≥1000 ppm methyl chloroform for 30 min in mice resulted in reduced locomotor activity, muscle tone, and operant behavior (Bowen and Balster, 1996; Warren et al., 1998). Delayed neurological development occurred in pups of mice and rats exposed to ≥6000 ppm during gestation (Jones et al., 1996; Coleman et al., 1999). In dogs, increased pulmonary arterial pressure and cardiac workload

was reported following inhalation exposure (Aoki et al., 1997).

Chronic Effects

Kelafant et al. (1994) reported deficits in memory, balance, and rhythm in a small group of workers exposed to methyl chloroform. Cases of liver disease and sensory neuropathy were reported in individual adults (House et al., 1994; Cohen and Frank, 1994). Tolerance or sensitization to neurological effects was shown in mice inhaling high doses of methyl chloroform (Bowen and Balster, 2006). Although one research group reported significant increases in tumor development in mice and rats when tumor types were grouped for statistical analysis (Ohnishi et al., 2013), studies performed by the NTP were not suggestive of mutagenic or carcinogenic potential in rodents (NTP, 2008).

Chemical Pathology

Postmortem examination following methyl chloroform-induced worker fatalities showed pulmonary, hepatic, and renal congestion and edema (Stahl et al., 1969; Hatfield and Maykoski, 1970; Winek et al., 1997).

Mode of Action(s)

Acute methyl chloroform inhalation depresses CNS function similar to other VOCs. Okuda et al. (2001) suggest that peripheral neuropathy may be mediated by the inactivation of Ca^{2+} channels in the neuron.

SOURCES OF EXPOSURE

Occupational

Workers involved in chemical production involving methyl chloroform may be exposed primarily via inhalation with some possible dermal contact. Tay et al. (1995) showed that end-of-shift blood and alveolar air levels of methyl chloroform, but not urine levels, correlated well with measured exposures.

Environmental

The general population may be exposed to methyl chloroform via drinking water and inhalation of ambient air containing methyl chloroform.

INDUSTRIAL HYGIENE

Eye protection and skin coverings should be used to protect from freezing temperatures associated with liquid methyl chloroform releases. Air-purifying respirators should be used for exposures of up to 700 ppm. OSHA has set the PEL for methyl chloroform at 350 ppm as an 8-h TWA. NIOSH set a REL of 350 ppm as a 15-min ceiling value. ACGIH set a TLV for methyl chloroform at 350 ppm as an 8-h TWA.

RISK ASSESSMENTS

No risk assessments for methyl chloroform were identified.

REFERENCES

Aoki, N., Soma, K., Katagiri, H., Aizawa, Y., Kadowaki, T., and Ohwada, T. (1997) The pulmonary hemodynamic effects of 1,1,1-trichloroethane inhalation. *Ind. Health* 35(4):451–455.

Aviado, D.M. and Belej, M.A. (1974) Toxicity of aerosol propellants on the respiratory and circulatory systems. I. Cardiac arrhythmia in the mouse. *Toxicology* 2:31–42.

Bowen, S.E. and Balster, R.L. (1996) Effects of inhaled 1,1,1-trichloroethane on locomotor activity in mice. *Neurotoxicol. Teratol.* 18(1):77–81.

Bowen, S.E. and Balster, R.L. (1996) Effects of inhaled 1,1,1-trichloroethane on locomotor activity in mice. *Neurotox. Teratol.* 18:77–81.

Cohen, C. and Frank, A.L. (1994) Liver disease following occupational exposure to 1,1,1-trichloroethane: a case report. *Am. J. Ind. Med.* 26(2):237–241.

Coleman, C.N., Mason, T., Hooker, E.P., and Robinson, S.E. (1999) Developmental effects of intermittent prenatal exposure to 1,1,1-trichloroethane in the rat. *Neurotoxicol. Teratol.* 21(6):699–708.

Environmental Protection Agency (EPA) (2013) *ECOTOX Database*. Available at http://cfpub.epa.gov/ecotox/quick_query.cfm.

Environmental Protection Agency (EPA) (2013) *Integrated Risk Information System (IRIS). Online. National Center for Environmental Assessment.*

Hatfield, T.R. and Maykoski, R.T. (1970) A fatal methyl chloroform (trichloroethane) poisoning. *Arch. Environ. Health* 20(2):279–281.

House, R.A., Liss, G.M., and Wills, M.C. (1994) Peripheral sensory neuropathy associated with 1,1,1-trichloroethane. *Arch. Environ. Health* 49(3):196–199.

Jones, H.E., Kunko, P.M., Robinson, S.E., and Balster, R.L. (1996) Developmental consequences of intermittent and continuous prenatal exposure to 1,1,1-trichloroethane in mice. *Pharmacol. Biochem. Behav.* 55(4):635–646.

Kelafant, G.A., Berg, R.A., and Schleenbaker, R. (1994) Toxic encephalopathy due to 1,1,1-trichloroethane exposure. *Am. J. Ind. Med.* 25(3):439–446.

Laine, A., Seppalainen, A.M., Savolainen, K., and Riihimaki, V. (1996) Acute effects of 1,1,1-trichloroethane inhalation on the human central nervous system. *Int. Arch. Occup. Environ. Health* 69(1):53–61.

Muttray, A., Kurten, R., Jung, D., Schicketanz, K.H., Mayer-Popken, O., and Konietzko, J. (2000) Acute effects of 200 ppm 1,1,1-trichloroethane on the human EEG. *Eur. J. Med. Res.* 5(9):375–384.

National Library of Medicine (NLM) (2013) *Allyl Chloride. Hazardous Substances Data Bank*, Available at http://toxnet.nlm.nih.gov/cgi-bin/sis/search/f?./temp/~EuKWDd:1.

National Toxicology Program (NTP) (2008) Final report on the safety assessment of Trichloroethane. *Int. J. Toxicol.* 27(Suppl 4):107–138.

Ohnishi, M., Umeda, Y., Katagiri, T., Kasai, T., Ikawa, N., Nishizawa, T., and Fukushima, S. (2013) Inhalation carcinogenicity of 1,1,1-trichloroethane in rats and mice. *Inhal. Toxicol.* 25(5):298–306.

Okuda, M., Kunitsugu, I., Kobayakawa, S., et al. (2001) Inhibitory effect of 1,1,1-trichloroethane on calcium channels of neurons. *J. Toxicol. Sci.* 26:169–176.

Reinhardt, C.F., et al. (1971) Cardiac arrhythmias and aem',ol "sniffing,". *Atz'h Environ. Health* 22:265–279.

Stahl, C.J., Fatteh, A.X., and Dominguez, A.M. (1969) Trichloroethane poisoning—observations on the pathology and toxicology in six fatal cases. *J. Forensic Sci.* 14:393–397.

Stewart, R.D. (1971) Methyl chloroform intaxication: diagnosis and treatment. *JAMA* 215:1789–1792.

Tay, P., Pinnagoda, J., Sam, C.T., Ho, S.F., Tan, K.T., and Ong, C.N. (1995) Environmental and biological monitoring of occupational exposure to 1,1,1-trichloroethane. *Occup. Med. (Lond.)* 45(3):147–150.

Warren, D.A., Reigle, T.G., Muralidhara, S., and Dallas, C.E. (1998) Schedule-controlled operant behavior of rats during 1,1,1-trichloroethane inhalation: relationship to blood and brain solvent concentrations. *Neurotoxicol. Teratol.* 20(2):143–153.

Winek, C.L., Wahba, W.W., Huston, R., and Rozin, L. (1997) Fatal inhalation of 1,1,1-trichloroethane. *Forensic Sci. Int.* 87(2):161–165.

METHYLENE CHLORIDE

First aid: *Oral*: Do not administer activated charcoal. *Inhalation*: Move the person to fresh air; monitor for CNS symptoms and carboxyhemoglobin levels; administer 100% oxygen for possible CO poisoning. *Dermal*: Remove the contaminated clothing and wash the exposed skin with soap and water. *Eye*: Irrigate with water for 15 min.

Target organ(s): Eyes, skin, cardiovascular system, CNS

Occupational exposure limits: OSHA PEL: 25 ppm 8-h TWA; NIOSH: lowest concentration feasible; ACGIH TLV: 50 ppm 8-h TWA.

Reference values: ATSDR: Acute, intermediate, and chronic inhalation MRLs: 0.6, 0.3, and 0.3 ppm; Acute and chronic oral MRL: 0.2 and 0.06 mg/kg/day

EPA: RfC: 0.6 mg/m^3; RfD: 6 × 10^{-3} mg/kg/day

Risk/safety phrases: 40; (2-)23-24/25-36/37

BACKGROUND AND USES

Methylene chloride (dichloromethane, CASRN 75-09-2) is used in degreasing and metal cleaning and paint stripping and removal. It is also used in caffeine extraction of coffee and the manufacture of polycarbonate plastics. It has been used historically as a pesticide fumigant and refrigerant.

PHYSICAL AND CHEMICAL PROPERTIES

Methylene chloride (molecular weight = 84.93 g/mol) is a colorless liquid having a sweet odor. It is a volatile compound with a vapor pressure of 124 mmHg at 25 °C. It is soluble in water (1.3 × 10^4 mg/l at 25 °C) but insoluble in carbon tetrachloride. It is miscible with ethanol, dimethylformamide, and ethyl ether.

ENVIRONMENTAL FATE AND BIOACCUMULATION

Methylene chloride found in the environment may be emitted from release sites involving degreasing and metal cleaning operations. It exists in the atmosphere as a vapor, where it is degraded by photochemically produced hydroxyl radicals (half-life = 119 days). Methylene chloride is highly mobile in soil and evaporates quickly from dry soil and surface waters although some biodegradation in soil and water is possible. It is not expected to be adsorbed to sediments and suspended particles and is not expected to bioaccumulate in aquatic organisms (BCF = 2) (NLM, 2013).

ECOTOXICOLOGY

Acute Effects

The 2-h EC$_{50}$ (immobility) and 1-h LC$_{50}$ for D. magna was 1.3 and 2.5 g/l, respectively. The 4-h LC$_{50}$s in zebrafish and bluegill were 254 and 220 mg/l (EPA, 2013), respectively.

Chronic Effects

No chronic ecotoxicological data were identified for methylene chloride.

MAMMALIAN TOXICOLOGY

Acute Effects

Acute inhalation effects in humans include eye and throat irritation, nasal discharge, cough, fatigue, and light-headedness (Shusterman et al., 1990). Deaths have been reported from methylene chloride exposure at small furniture stripping facilities (Novak and Hain, 1990). Exposure of volunteers to 500 ppm produced decreased manual performance and attention lapses (Winneke and Fodor, 1976). Sensory and psychomotor function may be impaired at ≥300 ppm. Effects of methylene chloride on the nervous system are reversible.

CNS depression occurred in guinea pigs inhaling 5000 ppm, with sleep cycle disturbances being reported at 1000 ppm (Winneke, 1981). Mice exhibited reduced learning capabilities following inhalation exposures (Alexeeff and Kilgore, 1983). No reproductive, embryotoxicity or fetotoxicity was reported in rats exposed to methylene chloride (Schwetz et al., 1975; Hardin and Manson, 1980; Nitschke et al., 1988a).

Chronic Effects

Epidemiological studies have provided no statistically significant association between chronic methylene chloride exposure and cancer, death, or chronic illness in humans. These data include cohorts of photography workers (Hearne et al., 1990), airline mechanics (Lash et al., 1991), cellulose fiber workers (Lanes et al., 1993), synthetic fiber workers (Soden, 1993), and U.S. Air Force employees (Radican et al., 2008).

In inhalation studies, increases in incidences of fatty lung and mammary gland tumors were reported in rats and liver tumors in mice and rats exposed to ≥500 ppm for 2 years (Nitschke et al., 1988b; Mennear et al., 1988).

Liver effects were seen in drinking water studies involving rodents. Serota et al. (1986) reported fatty liver changes in mice following lifetime exposure at ≥50 mg/kg/day. Malignant mammary gland tumors occurred in mice at 500 mg/kg/day (Maltoni et al., 1988).

Chemical Pathology

Increased carboxyhemoglobin levels occur as a result of metabolism to carbon monoxide. Exposure of volunteers to 1000 ppm methylene chloride for 2 h resulted in higher carboxyhemoglobin levels than permitted in the workplace from exposure to carbon monoxide (Stewart et al., 1972).

Mode of Action(s)

Acute cardiovascular and CNS effects following high inhalation exposure to methylene chloride arise due to solvent-like CNS depression and oxidative metabolism of carbon monoxide, resulting in carboxyhemoglobin. The liver is the primary site for methylene chloride metabolism although it may also be metabolized by the kidneys and lung. Methylene chloride is metabolized via the cytochrome P-450 monooxygenase pathway to carbon monoxide and via the glutathione s-transferase pathway (GSTT1) (Kubic and Anders, 1975; El-Masri et al., 1999) to formaldehyde and a glutathione conjugate. Thus, oxidative metabolism is a detoxification pathway (Jonsson and Johanson, 2001). The methylene chloride–glutathione conjugate (S-chloromethylglutathione) is known to induce single-strand breaks in DNA in some assays (Graves et al., 1994), while DNA adducts were reported in human lung cell lines having an absence of GSH metabolism (Evans and Caldwell, 2010).

SOURCES OF EXPOSURE

Occupational

Workers involved in metal cleaning and degreasing or chemical production involving methylene chloride may be exposed via inhalation and dermal contact.

Environmental

The general population may be exposed to methylene chloride via drinking water and inhalation or dermal contact with paint strippers containing methylene chloride.

INDUSTRIAL HYGIENE

Eye protection, gloves (not PVC or rubber), and skin coverings should be used to protect from freezing temperatures associated with liquid methylene chloride releases. Avoid close-circuit air-rebreathing respirators that employ carbon dioxide absorption media, as neurotoxic compounds may be formed. OSHA has set the PEL for methylene chloride at 25 ppm as an 8-h TWA. NIOSH did not set a REL but recommends exposure to lowest concentration feasible. ACGIH set a TLV for methylene chloride at 50 ppm as an 8-h TWA.

RISK ASSESSMENTS

ATSDR: Acute, intermediate, and chronic inhalation MRLs: 0.6, 0.3, and 0.3 ppm, respectively; Acute and chronic oral MRL: 0.2 and 0.06 mg/kg/day, respectively. EPA: RfC: 0.6 mg/m^3; RfD: 6×10^{-3} mg/kg/day; likely to be carcinogenic to humans. IARC: group 2B: Possibly carcinogenic to humans.

REFERENCES

Alexeeff, G.V. and Kilgore, W.W. (1983) Learning impairment in mice following acute exposure to dichloromethane and carbon tetrachloride. *J. Toxicol. Environ. Health* 11(4–6):569–581.

El-Masri, H.A., Bell, D.A., and Portier, C.J. (1999) Effects of glutathione transferase theta polymorphism on the risk estimates of dichloromethane to humans. *Toxicol. Appl. Pharmacol.* 158(3):221–230.

Environmental Protection Agency (EPA) (2013) *ECOTOX Database*. Available at http://cfpub.epa.gov/ecotox/quick_query .cfm.

Environmental Protection Agency (EPA) (2013) *Integrated Risk Information System (IRIS). Online. National Center for Environmental Assessment.*

Evans, M.V. and Caldwell, J.C. (2010) Evaluation of two different metabolic hypotheses for dichloromethane toxicity using physiologically based pharmacokinetic modeling for in vivo inhalation gas uptake data exposure in female B6C3F1 mice. *Toxicol. Appl. Pharmacol.* 244(3):280–290.

Graves, R.J., Coutts, C., Eyton-Jones, H., and Green, T. (1994) Relationship between hepatic DNA damage and methylene chloride-induced hepatocarcinogenicity in B6C3F1 mice. *Carcinogenesis* 15(5):991–996.

Hardin, B.D. and Manson, J.M. (1980) Absence of dichloromethane teratogenicity with inhalation exposure in rats. *Toxicol. Appl. Pharmacol.* 52(1):22–28.

Hearne, F.T., Pifer, J.W., and Grose, F. (1990) Absence of adverse mortality effects in workers exposed to methylene chloride: an update. *J. Occup. Med.* 32(3):234–240.

Jonsson, F. and Johanson, G. (2001) A Bayesian analysis of the influence of GSTT1 polymorphism on the cancer risk estimate for dichloromethane. *Toxicol. Appl. Pharmacol.* 174(2):99–112.

Kubic, V.L. and Anders, M.W. (1975) Metabolism of dihalomethanes to carbon monoxide. II. In vitro studies. *Drug Metab. Dispos.* 3(2):104–112.

Lanes, S.F., Rothman, K.J., Dreyer, N.A., and Soden, K.J. (1993) Mortality update of cellulose fiber production workers. *Scand. J. Work Environ. Health* 19(6):426–428.

Lash, A.A., Becker, C.E., So, Y., and Shore, M. (1991) Neurotoxic effects of methylene chloride: are they long lasting in humans? *Br. J. Ind. Med.* 48(6):418–426.

Maltoni, C., Cotti, G., and Perino, G. (1988) Long-term carcinogenicity bioassays on methylene chloride administered by ingestion to Sprague-Dawley rats and Swiss mice and by inhalation to Sprague-Dawley rats. *Ann. N.Y. Acad. Sci.* 534:352–366.

Mennear, J.H., McConnell, E.E., Huff, J.E., Renne, R.A., and Giddens, E. (1988) Inhalation toxicology and carcinogenesis studies of methylene chloride (dichloromethane) in F344/N rats and B6C3F1 mice. *Ann. N.Y. Acad. Sci.* 534:343–351.

Nitschke, K.D., Eisenbrandt, D.L., Lomax, L.G., and Rao, K.S. (1988a) Methylene chloride: two-generation inhalation reproductive study in rats. *Fundam. Appl. Toxicol.* 11(1):60–67.

Nitschke, K.D., Burek, J.D., Bell, T.J., Kociba, R.J., Rampy, L.W., and McKenna, M.J. (1988b) Methylene chloride: a 2-year inhalation toxicity and oncogenicity study in rats. *Fundam. Appl. Toxicol.* 11(1):48–59.

National Library of Medicine (NLM) (2013) *Allyl Chloride. Hazardous Substances Data Bank*, Available at http://toxnet.nlm.nih.gov/cgi-bin/sis/search/f?./temp/~EuKWDd:1.

Novak, J.J. and Hain, J.R. (1990) Furniture stripping vapor inhalation fatalities: two case studies. *Appl. Occup. Environ. Hyg.* 5:843–847.

Radican, L., Blair, A., Stewart, P., and Wartenberg, D. (2008) Mortality of aircraft maintenance workers exposed to trichloroethylene and other hydrocarbons and chemicals: Extended follow-up. *J. Occup. Environ. Med.* 50:1306–1313.

Schwetz, B.A., Leong, K.J., and Gehring, P.J. (1975) The effect of maternally inhaled trichloroethylene, perchloroethylene, methyl chloroform, and methylene chloride on embryonal and fetal development in mice and rats. *Toxicol. Appl. Pharmacol.* 32(1):84–96.

Serota, D.G., Thakur, A.K., Ulland, B.M., Kirschman, J.C., Brown, N.M., Coots, R.H., and Morgareidge, K. (1986) A two-year drinking-water study of dichloromethane in rodents. I. Rats. *Food Chem. Toxicol.* 24(9):951–958.

Shusterman, D., Quinlan, P., Lowengart, R., and Cone, J. (1990) Methylene chloride intoxication in a furniture refinisher. A comparison of exposure estimates utilizing workplace air sampling and blood carboxyhemoglobin measurements. *J. Occup. Med.* 32(5):451–454.

Soden, K.J. (1993) An evaluation of chronic methylene chloride exposure. *J. Occup. Med.* 35(3):282–286.

Stewart, R.D., Fisher, T.N., Hosko, M.J., Peterson, J.E., Baretta, E.D., and Dodd, H.C. (1972) Experimental human exposure to methylene chloride. *Arch. Environ. Health* 25(5):342–348.

Winneke, G. (1981) The neurotoxicity of dichloromethane. *Neurobehav. Toxicol. Teratol.* 3(4):391–395.

Winneke, G. and Fodor, G.G. (1976) Dichloromethane produces narcotic effect. *Occup. Health Saf.* 45(2):34–35, 49.

PHOSGENE

First aid: *Inhalation*: Move the person to fresh air. Check for bronchitis, pneumonitis, or severe respiratory tract irritation if cough or breathing difficulty develops or persists. Experimental medical interventions include intratracheal bolus dose of N-acetylcystiene and aminophylline (Sciuto and Hurt, 2004). *Dermal*: Remove the contaminated clothing and wash the exposed skin with soap and water. *Eye*: Irrigate with water for 15 min.

Target organ(s): Eyes, skin, respiratory system

Occupational exposure limits: OSHA PEL: 0.1 ppm 8-h TWA; NIOSH: 0.1 ppm 10-h TWA; ACGIH TLV: 0.1 ppm 8-h TWA

Reference values: EPA: 10 min AEGL 1. NR, AEGL 2. 0.6 ppm, AEGL 3 3.6 ppm;

RfC: 3×10^{-4} mg/m^3

Risk/safety phrases: 26–34; (1/2-)9-26-36/37/39-45

BACKGROUND AND USES

Phosgene (dichloromethane, CASRN 75-44-5) is used as an intermediate in the chemical synthesis of polycarbonates, aniline dyes, isocyanates, carbamate pesticides, perfumes, and pharmaceuticals. It is used in some metal ore separations and has been used historically as a pesticide fumigant and chemical warfare agent.

PHYSICAL AND CHEMICAL PROPERTIES

Phosgene (molecular weight = 98.92 g/mol) is a colorless gas at ambient temperature with an odor of musty hay at very low concentrations. It is a highly volatile compound with a vapor pressure of 1420 mmHg at 25 °C. It is slightly soluble in water but soluble in most hydrocarbon solvents.

ENVIRONMENTAL FATE AND BIOACCUMULATION

Phosgene found in the environment may result due to the release at sites of chemical synthesis of a variety of products. It is also an atmospheric degradation product of a number of chlorinated hydrocarbons. It exists in the atmosphere as a gas, where it is degraded over a half-life of 44 days. Phosgene is highly mobile in soil and evaporates quickly from dry soil and surface waters. It is not expected to be adsorbed to sediments and suspended particles and is not expected to bioaccumulate due to its very high rate of hydrolysis; a 1% phosgene solution is completely hydrolyzed in 20 s at 0 °C (NLM, 2013).

ECOTOXICOLOGY

Acute Effects

No acute ecotoxicological data were identified for phosgene (EPA, 2013a).

Chronic Effects

No chronic ecotoxicological data were identified for phosgene.

MAMMALIAN TOXICOLOGY

Acute Effects

Inhalation of high concentrations (>3 ppm) of phosgene results in shallow respiration, with decreases in respiratory tidal volume and vital capacity occurring (an initial protective phase). This reflexive reaction tends to diminish if exposure ceases. In the second phase, clinical signs and symptoms are typically absent for several hours. The terminal phase involves the development of pulmonary edema (Schneider and Diller, 1989; Diller, 1985). Edema severity increases as gas exchange decreases and fluid rises up the respiratory tract. Cardiac failure mediated by pulmonary congestion may occur at sufficiently high exposures.

Chronic Effects

Occupational epidemiology studies of uranium processing workers chronically exposed to low levels of phosgene did not provide an indication of chronic pulmonary toxicity (Polednak, 1980; Polednak and Hollis, 1985).

Chemical Pathology

Acute histopathology includes edematous swelling of the lung tissues, with infiltration of blood plasma into the pulmonary interstitium and alveoli (Duniho et al., 2002). Alveolar glutathione levels are depleted (Sciuto, 1998). Subsequent damage to the alveolar type I cells and necrosis results (Schneider and Diller, 1989; Diller, 1985; Brown et al., 2002). Rapid respiration may result in the appearance of a frothy fluid.

Mode of Action(s)

Phosgene gas is highly reactive, causing oxidative-like tissue damage. Acrylation of macromolecules and hydrolysis of phosgene to HCl are the main modes of action for acute tissue damage. Chronic exposure may lead to the onset of pulmonary fibrosis. Development of interstitial and intra-alveolar fibrosis may be accompanied by increased lung hydroxyproline content and prolyl hydroxylase and galactosyl-hydroxy-lysyl glucotransferase activity (Kodavanti et al. (1997); Hatch et al., 2001; Borak and Diller, 2001).

SOURCES OF EXPOSURE

Occupational

Chemical workers producing phosgene or using it for the synthesis of other compounds may be exposed via inhalation and dermal contact. Combat personnel may be exposed via inhalation of phosgene dispersed as a chemical warfare agent.

Environmental

The general population is unlikely to be exposed to phosgene, except in the event of deployment of phosgene munitions against the public.

INDUSTRIAL HYGIENE

Complete eye and skin protection is required, as phosgene may cause severe irritation and/or burns of the exposed skin. Phosgene-specific respirator canisters or supplied-air or self-contained air respirators must be used. Specialized safety training is required to work in areas of phosgene exposure. OSHA has set the PEL for phosgene at 0.1 ppm as an 8-h TWA. NIOSH set a REL of 0.1 ppm as a 10-h TWA, and set an IDLH of 2 ppm. ACGIH set a TLV for phosgene at 0.1 ppm as an 8-h TWA.

RISK ASSESSMENTS

EPA: 10 min AEGLs 1: NR, AEGL 2: 0.6 ppm, AEGL 3: 3.6 ppm, RfC: 3×10^{-4} mg/m^3, Inadequate information to assess carcinogenicity.

REFERENCES

Borak, J. and Diller, W.F. (2001) Phosgene exposure of mechanisms of injury and treatment strategies. *J. Occup. Environ. Med.* 43:110–119.

Brown, R.F., Jugg, B.J., Harban, F.M., Ashley, Z., Kenward, C.E., Platt, J., Hill, A., Rice, P., and Watkins, P.E. (2002) Pathophysiological responses following phosgene exposure in the anaesthetized pig. *J. Appl. Toxicol.* 22(4):263–269.

Diller, W.F. (1985) Pathogenesis of phosgene poisoning. *Toxicol. Ind. Health* 1(2):7–15.

Duniho, S.M., Martin, J., Forster, J.S., Cascio, M.B., Moran, T.S., Carpin, L.B., and Sciuto, A.M. (2002) Acute changes in lung histopathology and bronchoalveolar lavage parameters in mice exposed to the choking agent gas phosgene. *Toxicol. Pathol.* 30(3):339–349.

Environmental Protection Agency (EPA) (2013a) *ECOTOX Database.* Available at http://cfpub.epa.gov/ecotox/quick_query .cfm.

Hatch, G.E., Kodavanti, U., Crissman, K., et al. (2001) An "injury-time integral" model for extrapolating from acute to chronic effects of phosgene. *Toxicol. Ind. Health* 17:285–293.

Kodavanti, U.P., Costa, D.L., Giri, S.N., et al. (1997) Pulmonary structural and extracellular matrix alterations in Fischer 344 rats following subchronic phosgene exposure. *Fundam. Appl. Toxicol.* 37:54–63.

National Library of Medicine (NLM) (2013) *Phosgene. Hazardous Substances Data Bank,* Available at http://toxnet .nlm.nih.gov/cgi-bin/sis/search/f?./temp/~EuKWDd:1.

Polednak, A.P. (1980) Mortality among men occupationally exposed to phosgene in 1943–1945. *Environ. Res.* 22:357–367.

Polednak, A.P. and Hollis, D.R. (1985) Mortality and causes of death among workers exposed to phosgene in 1943–1945. *Toxicol. Ind. Health* 1(2):137–151.

Schneider, W. and Diller, W. (1989) Phosgene. In: *Encyclopedia of Industrial Chemistry*, 5th ed., vol. A, 19: Weinheim, Germany: VCH Verlag, 411–420.

Sciuto, A.M. (1998) Assessment of early acute lung injury in rodents exposed to phosgene. *Arch. Toxicol.* 72(5):283–288.

Sciuto, A.M. and Hurt, H.H. (2004) Therapeutic treatments of phosgene-induced lung injury. *Inhal. Toxicol.* 16:565–580.

TETRACHLOROETHYLENE

First aid: *Oral*: Activated charcoal slurry may be given; gastric lavage should be administered if a large amount was ingested within the past hour. Treat CNS and respiratory depression as needed. *Inhalation*: Move the person to fresh air. Administer ventilator assistance if CNS and respiratory depression are present. *Dermal*: Remove the contaminated clothing and wash the exposed skin with soap and water. *Eye*: Irrigate with water for 15 min.

Target organ(s): Eyes, skin, respiratory system, liver, kidneys, CNS

Occupational exposure limits: OSHA PEL: 100 ppm 8-h TWA; NIOSH: lowest feasible concentration; ACGIH TLV: 25 ppm 8-h TWA, 100 ppm STEL

Reference values: ATSDR Acute and chronic inhalation MRLs: 0.2 ppm and 0.04 ppm; Acute oral MRL: 0.05 mg/kg/day

EPA: RfC: 0.04 mg/m^3, RfD: 0.006 mg/kg/day

Risk/safety phrases: 40–51/53; (2-)23-36/37-61

BACKGROUND AND USES

Tetrachloroethylene (1,1,2,2-tetrachloroethylene, CASRN 127-18-4) is used primarily for the commercial dry cleaning of garments, textile preparation, and as feed stock for chlorofluorocarbon production. It is also used in metal cleaning and degreasing and as an insulator in electrical transformers. It has been used historically as a pesticide fumigant and as an anthelmintic treatment option.

PHYSICAL AND CHEMICAL PROPERTIES

Tetrachloroethylene (molecular weight = 165.83 g/mol) is a colorless liquid with a sweet odor. It is a volatile compound with a vapor pressure of 18.5 mmHg at 25 °C. It is relatively insoluble in water (2.1×10^2 mg/l) but is miscible with ethyl ether, ethanol, hexane, chloroform, and benzene.

ENVIRONMENTAL FATE AND BIOACCUMULATION

Tetrachloroethylene may be found in the environment due to the accidental release at dry cleaning sites and waste streams from chemical facilities producing chlorofluorocarbons. It exists in the atmosphere as a gas, where it is degraded over a half-life of 96 days. Tetrachloroethylene is moderately mobile in soil and evaporates quickly from dry soil and surface waters. Very slow biodegradation to trichloroethylene is possible in soil. It is not expected to be adsorbed to sediments and suspended particles. Bioaccumulation in fish may be high (BCF = 26-115) (NLM, 2013).

ECOTOXICOLOGY

Acute Effects

The 2-h EC$_{50}$ (immobility) and LC$_{50}$ for tetrachloroethylene in D. magna was 7.5 and 18 mg/l, respectively. The 2-h LC$_{50}$ in zebrafish was 9.3 mg/l (EPA, 2013).

Chronic Effects

The lowest concentration producing death in flagfish after 28 days was 3.7 mg/l (EPA, 2013).

MAMMALIAN TOXICOLOGY

Acute Effects

Both human and animal data indicate CNS depression as the primary acute effect of high tetrachloroethylene exposure. Changes in visual evoked potentials (Altmann et al., 1990), EEG changes (Hake and Stewart, 1977), and reduced visuospatial memory and cognition errors (Echeverria et al., 1995) were reported in humans. Rats exhibited changes in flash-invoked potential after inhaling 800 ppm (Mattsson et al., 1998).

Liver injuries in dry cleaners were indicated by increased serum levels of a liver enzyme as well as sonographic indicators and were associated with TWA inhalation exposures to 12–16 ppm tetrachloroethylene (Brodkin et al., 1995; Gennari et al., 1992). Mice inhaling 9 ppm or ingesting 100 mg/kg/day for 30 days exhibited liver toxicity more so than rats (Odum et al., 1988; Kjellstrand et al., 1984; NTP, 1986).

Studies of drycleaner workers exposed to tetrachloroethylene were suggestive of increased incidences of spontaneous abortions but are equivocal (Lindbohm et al., 1990; Olsen et al., 1990). Rats inhaling 300 ppm gave birth to pups having significantly reduced birth weights (Tinston, 1994).

Chronic Effects

Chronic exposure in humans may result in vision effects, changes in color vision, and deficits in spatial memory.

Occupational inhalation exposure to 8–15 ppm (TWA) has been associated with kidney toxicity and end-stage renal disease, indicated by increased urinary protein levels (Verplanke et al., 1999; Calvert et al., 2011). Renal damage in rats has been reported at an inhalation exposure of 50 ppm (NTP, 1986).

Tetrachloroethylene-induced bladder cancer (Lynge et al., 2006; Blair et al., 2003), multiple myeloma (Radican et al., 2008), and non-Hodgkin lymphoma (Radican et al., 2008; Calvert et al., 2011) have been suggested in epidemiology studies. Rats inhaling ≥200 ppm tetrachloroethylene exhibited increased incidence of mononuclear cell leukemia, while mice inhaling ≥100 ppm developed increased incidences of liver cancer (NTP, 1986).

Chemical Pathology

Proximal tubule damage in human kidneys was indicated by elevated urinary levels of lysozyme and β-glucuronidase (Franchini et al., 1983). Rats exhibited enlarged renal cell nuclei, cast formation, and glomerular nephrosis (NTP, 1986).

Liver pathology seen in laboratory studies of mice inhaling tetrachloroethylene include liver enlargement, centrilobular necrosis, cytoplasmic vacuolation, inflammatory cell infiltrates, cell pigmentation, hyperplasia, and regenerative foci (Odum et al., 1988; Kjellstrand et al., 1984; NTP, 1986).

Mode of Action(s)

Renal tubule toxicity may be caused by metabolism of tetrachloroethylene to the GSH conjugates TCVC and TCVCS, as indicated by multiple animal studies.

The mode(s) of action for liver and kidney cancer seen in rodents is/are not known but may include mutagenicity, PPARα-activation, α2-globulin nephropathy (in male rats), cytotoxicity from GSH-derived conjugates (Ripp et al., 1999; Dreessen et al., 2003; Lash and Parker, 2001).

SOURCES OF EXPOSURE

Occupational

Workers in the dry-cleaning industry may be exposed to tetrachloroethylene via inhalation and dermal contact.

Environmental

The general population may be exposed to tetrachloroethylene in ambient air or via ingestion of contaminated drinking water. Closet storage of garments dry-cleaned with tetrachloroethylene has resulted in air levels of up to 30 times that of ambient air levels. Some adolescents sniffing typewriter correction fluids for abusive narcosis may be exposed to tetrachloroethylene.

INDUSTRIAL HYGIENE

Use gloves, goggles, and skin-covering clothing when in contact with tetrachloroethylene. OSHA has set the PEL for tetrachloroethylene at 100 ppm as an 8-h TWA. NIOSH recommends exposure to lowest feasible concentrations. ACGIH set a TLV for tetrachloroethylene at 25 ppm as an 8-h TWA.

RISK ASSESSMENTS

ATSDR: Acute and chronic inhalation MRL: 0.2 and 0.04 ppm; Acute oral MRL: 0.05 mg/kg/day

EPA: RfC: $0.04\ mg/m^3$, RfD: 0.006 mg/kg/day, Likely to be carcinogenic to humans.

IARC: Group 2A: Probably carcinogenic to humans.

REFERENCES

Altmann, L., Böttger, A., and Wiegand, H. (1990) Neurophysiological and psychophysical measurements reveal effects of acute low-level organic solvent exposure in humans. *Int. Arch. Occup. Environ. Health* 62:493–499.

Blair, A., Petralia, S.A., and Stewart, P.A. (2003) Extended mortality follow-up of a cohort of dry cleaners. *Ann. Epidemiol.* 13:50–56.

Brodkin, C.A., Daniell, W., Checkoway, H., Echeverria, D., Johnson, J., Wang, K., Sohaey, R., Green, D., Redlich, C., and Gretch, D. (1995) Hepatic ultrasonic changes in workers exposed to perchloroethylene. *Occup. Environ. Med.* 52:679–685.

Calvert, G.M., Ruder, A.M., and Petersen, M.R. (2011) Mortality and end-stage renal disease incidence among dry cleaning workers. *Occup. Environ. Med.* 68:709–716.

Dreessen, B., Westphal, G., Bünger, J., Hallier, E., and Müller, M. (2003) Mutagenicity of the glutathione and cysteine S-conjugates of the haloalkenes 1,1,2-trichloro-3,3,3-trifluoro-1-propene and trichlorofluoroethene in the Ames test in comparison with the tetrachloroethene-analogues. *Mutat. Res.* 539:157–166.

Echeverria, D., White, R.F., and Sampaio, C. (1995) A behavioral evaluation of PCE exposure in patients and dry cleaners: A possible relationship between clinical and preclinical effects. *J. Occup. Environ. Med.* 37:667–680.

Environmental Protection Agency (EPA) (2013) *Integrated Risk Information System (IRIS). Online. National Center for Environmental Assessment.*

Environmental Protection Agency (EPA) (2013) *ECOTOX Database*. Available at http://cfpub.epa.gov/ecotox/quick_query .cfm.

Franchini, I., Cavatorta, A., Falzoi, M., Lucertini, S., and Mutti, A. (1983) Early indicators of renal damage in workers exposed to organic solvents. *Int. Arch. Occup. Environ. Health* 52:1–9.

Gennari, P., Naldi, M., Motta, R., Nucci, M.C., Giacomini, C., Violante, F.S., and Raffi, G.B. (1992) gamma-Glutamyltransferase isoenzyme pattern in workers exposed to tetrachloroethylene. *Am. J. Ind. Med.* 21:661–671.

Hake, C.L. and Stewart, R.D. (1977) Human exposure to tetrachloroethylene: inhalation and skin contact. *Environ. Health Perspect.* 21:231–238.

Kjellstrand, P., Holmquist, B., Kanje, M., Alm, P., Romare, S., Jonsson, I., Mansson, L., and Bjerkemo, M. (1984) Perchloroethylene: Effects on body and organ weights and plasma butyrylcholinesterase activity in mice. *Acta. Pharmacol. Toxicol.* 54:414–424.

Lash, L.H. and Parker, J.C. (2001) Hepatic and renal toxicities associated with perchloroethylene. *Pharmacol. Rev.* 53:177–208.

Lindbohm, M.L., Taskinen, H., Sallmen, M., and Hemminki, K. (1990) Spontaneous abortions among women exposed to organic solvents. *Am. J. Ind. Med.* 17:449–463.

Lynge, E., Andersen, A., Rylander, L., Tinnerberg, H., Lindbohm, M.L., Pukkala, E., Romundstad, P., Jensen, P., Clausen, L.B., and Johansen, K. (2006) Cancer in persons working in dry cleaning in the Nordic countries. *Environ. Health Perspect.* 114(2):213–219.

Mattsson, J., Albee, R.R., Yano, B.L., Bradley, G.J., and Spencer, P.J. (1998) Neurotoxicologic examination of rats exposed to 1,1,2,2-tetrachloroethylene (perchloroethylene) vapor for 13 weeks. *Neurotoxicol. Teratol.* 20:83–98.

National Library of Medicine (NLM) (2013) *Hazardous Substances Data Bank*, Available at http://toxnet.nlm.nih.gov/cgi-bin/sis/search/f?./temp/~EuKWDd:1.

NTP (National Toxicology Program) (1986) *Toxicology and carcinogenesis studies of tetrachloroethylene (perchloroethylene) (CAS no. 127-18-4) in F344/N rats and B6C3F1 mice (inhalation studies). (NTP TR 311). Research Triangle Park, NC: U.S. Department of Health and Human Services, National Toxicology Program.*

Odum, J., Green, T., Foster, J.R., and Hext, P.M. (1988) The role of trichloroacetic acid and peroxisome proliferation in the differences in carcinogenicity of perchloroethylene in the mouse and rat. *Toxicol. Appl. Pharmacol.* 92:103–112.

Olsen, J., Hemminki, K., Ahlborg, G., Bjerkedal, T., Kyyronen, P., Taskinen, H., Lindbohm, M.L., Heinonen, O.P., Brandt, L., Kolstad, H., Halvorsen, B.A., and Egenaes, J. (1990) Low birthweight, congenital malformations, and spontaneous abortions among dry-cleaning workers in Scandinavia. *Scand. J. Work Environ. Health* 16:163–168.

Radican, L., Blair, A., Stewart, P., and Wartenberg, D. (2008) Mortality of aircraft maintenance workers exposed to trichloroethylene and other hydrocarbons and chemicals: Extended follow-up. *J. Occup. Environ. Med.* 50:1306–1319.

Ripp, S.L., Itagaki, K., Philpot, R.M., and Elfarra, A.A. (1999) Methionine S-oxidation in human and rabbit liver microsomes: evidence for a high-affinity methionine S-oxidase activity that is distinct from flavin-containing monooxygenase 3. *Arch. Biochem. Biophys.* 367:322–332.

Tinston, D.J. (1994) *Perchloroethylene: A multigeneration inhalation study in the rat. (CTL/P/4097, 86950000190). Cheshire, UK: Zeneca Central Toxicology Laboratory.*

Verplanke, A.J., Leummens, M.H., and Herber, R.F. (1999) Occupational exposure to tetrachloroethene and its effects on the kidneys. *J. Occup. Environ. Med.* 41:11–16.

TRICHLOROETHYLENE

First aid: *Oral*: Activated charcoal slurry may be given; gastric lavage should be administered if a large amount was ingested within the past hour. Treat CNS and respiratory depression as needed. *Inhalation*: Move the person to fresh air. Administer ventilator assistance if CNS and respiratory depression are present. *Dermal*: Remove the contaminated clothing and wash the exposed skin with soap and water. *Eye*: Irrigate with water for 15 min.

Target organ(s): Eyes, skin, respiratory system, heart, liver, kidneys, CNS

Occupational exposure limits: OSHA PEL: 100 ppm 8-h TWA; NIOSH: lowest feasible concentration, or 2 ppm 60-min ceiling value; ACGIH TLV: 10 ppm 8-h TWA, 25 ppm STEL

Reference values: ATSDR: Acute and intermediate inhalation MRLs: 2 ppm and 0.1 ppm; Acute oral MRL: 0.2 mg/kg/day. EPA: RfC: 1.9×10^{-4} mg/m^3, RfD: 5.1×10^{-4} mg/kg/day

Risk/safety phrases: 40–52/53; (2-)23-36/37-61

BACKGROUND AND USES

Trichloroethylene (CASRN 127-18-4) is used in metal cleaning in the electronics industry and as a chemical intermediate in the production of larger halogenated hydrocarbons. Historically, it was used extensively for the commercial dry cleaning of garments (replaced by tetrachloroethylene), in pesticide formulations, and as an anesthetic.

PHYSICAL AND CHEMICAL PROPERTIES

Trichloroethylene (molecular weight = 131.39 g/mol) is a colorless liquid with a sweet odor. It is a volatile compound with a vapor pressure of 69 mmHg at 25 °C. It is slightly soluble in water $(1.3 \times 10^3 \, \text{mg/l})$ but is miscible with ethyl ether, ethanol, carbon tetrachloride, and chloroform. Open flames and heated surfaces, as well as the presence of an alkaline environment, may cause the pyrolysis of this compound to phosgene.

ENVIRONMENTAL FATE AND BIOACCUMULATION

Trichloroethylene found in the environment may result from historical releases, from dry cleaning and metal cleaning sites, and waste streams from electronic component production. It exists in the atmosphere as a gas, where it is degraded by hydroxyl and nitrate radicals over a half-life of 7 and 114 days, respectively. Trichloroethylene is mobile in soil and evaporates quickly from dry soil and surface waters. Aerobic biodegradation to trichloroethylene in soil may be fairly rapid, while anaerobic degradation is slow. Trichloroethylene is adsorbed to sediments and suspended particles. Bioaccumulation in aquatic organisms may be moderate (BCF = 4-39) (NLM, 2013).

ECOTOXICOLOGY

Acute Effects

No acute ecotoxicological data were identified for trichloroethylene (EPA, 2013a).

Chronic Effects

No chronic ecotoxicological data were identified for trichloroethylene (EPA, 2013b).

MAMMALIAN TOXICOLOGY

Acute Effects

Trichloroethylene vapors are mildly irritating to the mucous membranes in some individuals. Trichloroethylene can act on the skin as a primary irritant and defatting agent, which may aggravate existing eczematous lesions. Narcotic effects produced by acute exposure to trichloroethylene include lack of coordination, dizziness, confusion, drowsiness, and eventual loss of consciousness. The peripheral and central nervous systems may be involved in acute narcosis. Although symptoms may rapidly disappear with the cessation of exposure, persistent headache or hallucinations may remain, likely caused by contaminants (Smith, 1966). Excitatory or euphoric stage narcosis may lead to repeated intentional inhalation of the vapor (Harenko, 1967). Feldman and Mayer (1968) have demonstrated a slowing of conduction time in

peripheral nerves, which is reversible over the course of months. Historical trichloroethylene use as an anesthetic, in high level occupational exposures and suicide attempts involving oral ingestion, has resulted in cardiac arrhythmias leading to death (Smith 1966; Gutch et al., 1965; Pembleton, 1974). Acute trichloroethylene (TCE) poisoning in humans resulted in increased N-acetly-β-D-glucosaminidase (NAG) levels, a nonspecific biomarker of renal damage (Carrieri et al., 2007). High level inhalation exposures ($\geq$1000 ppm) of rodents resulted in significantly increased instances of renal tubule damage (Chakrabarti and Tuchweber, 1988).

Chronic Effects

Chronic exposure may lead to changes in vestibular (as seen in acute exposure) and trigeminal nerve function. Trichloroethylene workers exposed to >100 ppm trichloroethylene exhibited well-defined dose responses for loss in trigeminal nerve function (Barret et al., 1987). In a case report of high level of trichloroethylene exposure (Feldman et al., 1988), facial nerve latency and muscle function returned to normal after 120 weeks but not full facial sensation and pupillary response. However, the cause of these effects could not be attributed to trichloroethylene with certainty. Loss of auditory function was reported in children and workers (Rasmussen et al., 1993) exposed to trichloroethylene via oral or inhalation routes, respectively. However, these data were limited by self-reporting of effects, a paucity of objective measurements, and uncertainties in exposure. Rats exposed to high levels (>2000 ppm) of trichloroethylene have an increased auditory threshold and mid to high frequency loss in brain auditory-evoked responses (Rybak, 1992; Crofton and Zhao, 1997).

Liver effects in metal degreasers exposed to TCE include elevation of serum enzymes indicative of impaired liver function. The dose response for these effects in workers is uncertain due to uncertainties and inconsistencies in exposure estimation as well as confounders such as alcohol consumption. Data for liver cancer in humans are extremely limited to indicate a causal association. Liver toxicity reported in rodents includes liver weight increases, hepatocyte hypertrophy, and proliferation of peroxisomes. Cytotoxicity and regenerative proliferation do not appear to be the significant effects of TCE exposure in rodents. Liver tumors were seen in TCE-cxposed mice more so than rats, possibly because of the higher metabolism of TCE in mice.

Proximal tubule damage was reported in workers exposed to high levels of trichloroethylene alone or in conjunction with other compounds. Increased urinary levels of several nonspecific biomarkers of kidney function were reported, including α1- and β2-microglobulin, total protein, and NAG (Jacob et al., 2007). Mice and rats exhibited significantly increased instances of nephropathy, cytomegaly, and karyomegaly following chronic oral trichloroethylene exposures, while inhalation bioassays produced these effects in male rats only. Epidemiological studies of various designs and locations have reported significant associations between occupational trichloroethylene exposure and renal cancer, particularly in cohorts estimated to have the highest exposure. Renal tubule tumors developed in male rats but not in mice or hamsters.

Occupational exposure has been associated with auto-immune mediated schleroderma, with increased levels of inflammatory cytokines observed similar to those seen in laboratory animal studies of immunotoxicological endpoints.

High TCE exposure in rodents resulted in decreased sperm quality and CNS, ocular, kidney/liver, skeletal, cardiac, and respiratory developmental effects. Human data for TCE-induced birth defects are too limited to show a causal link.

Mode of Action(s)

It is unclear whether the proposed modes of action for carcinogenicity observed in laboratory animals are relevant to human exposures to TCE. In several cases, metabolite pathways are qualitatively similar between humans and animals (although different quantitatively), resulting in the formation of TCE metabolites, which may be involved in toxic insult if produced in sufficient concentrations. The proposed modes of action for liver cancer include peroxisome proliferation-receptor α activation, increased hepatic glycogen stores, oxidative stress, and epigenetic changes (EPA, 2011). However, data are inadequate to indicate which mode of action is responsible for animal hepatocarcinogenicity or if they are relevant to humans.

The mode of action for kidney cancer seen in rats and suggested in TCE workers may be due to sequential *in situ* renal metabolism of TCE to dichlorovinylcysteine (DCVC), downstream glucuronidation of DCVC, and transformation to reactive species that are mutagenic in *in vitro* and *in vivo* animal studies. It is unclear if the mode of action is dominated by mutagenic alterations of DNA of renal tubules, cytotoxic cell proliferation, or a combination of both and if the putative carcinogenic metabolite in rat is also present in human kidneys at sufficiently high levels.

The mode of action for cancer is not well understood. Studies of genotoxic effects in humans, animals, and *in vitro* systems were inconclusive. Chromosomal abnormality reported in studies of trichloroethylene workers was confounded by issues of unbalanced study design, simultaneous exposure to other organic solvents, and tobacco smoke (Rasmussen et al., 1988; Konietzko et al., 1978; Nagaya et al., 1989). Likewise, the dose response for DNA strand breaks in mice and rats was either very different or was not observed. These data suggest that the carcinogenic mechanism for trichloroethylene in rodents may not be genotoxic.

SOURCES OF EXPOSURE

Occupational

Workers in the halogenated hydrocarbon production and electronics manufacturing industries may be exposed to trichloroethylene via inhalation and dermal contact.

Environmental

The general population may be exposed to trichloroethylene via inhalation in ambient air or via ingestion of contaminated drinking water. Exposure via indoor air may also occur via vapor intrusion from migrating groundwater contaminant plumes.

INDUSTRIAL HYGIENE

Use gloves, goggles, and skin-covering clothing when in contact with trichloroethylene. OSHA has set the PEL for trichloroethylene at 25 ppm as an 8-h TWA. NIOSH set 2 ppm as a 60-min ceiling value. ACGIH set a TLV for trichloroethylene at 25 ppm as an 8-h TWA.

RISK ASSESSMENTS

ATSDR: Acute and intermediate inhalation MRLs: 2 ppm and 0.1 ppm; Acute oral MRL: 0.2 mg/kg/day
EPA: RfC: 1.9×10^{-4} mg/m^3, RfD: 5.1×10^{-4} mg/kg/day, carcinogenic to humans
IARC: Group 1: Carcinogenic to humans.

REFERENCES

Barret, M.D., Garrel, S., Dane1, V., et al. (1987) Chronic trichloroethylene intoxication: a new approach by trigeminal-evoked potentials. *Arch. Environ. Health* 42:297–302.

Carrieri, M., Magosso, D., Piccoli, P., Zanetti, E., Trevisan, A., and Bartolucci, G. (2007) Acute, nonfatal intoxication with trichloroethylene. *Arch. Toxicol.* 81:529–532.

Chakrabarti, S. and Tuchweber, B. (1988) Studies of acute nephrotoxic potential of trichloroethylene in Fischer 344 rats. *J. Toxicol. Environ. Health* 23:147–158.

Crofton, K. and Zhao, X. (1997) The ototoxicity of trichloroethylene: extrapolation and relevance of high-concentration, short-duration animal exposure data. *Fundam. Appl. Toxicol.* 38:101–106.

Environmental Protection Agency (EPA) (2011) Toxicological Review of Trichloroethylene (CAS No. 79-01-6). Washington, DC.

Environmental Protection Agency (EPA) (2013a) *ECOTOX Database*. Available at http://cfpub.epa.gov/ecotox/quick_query.cfm.

Environmental Protection Agency (EPA) (2013b) *Integrated Risk Information System (IRIS). Online. National Center for Environmental Assessment.*

Feldman, R.G. and Mayer, R.F. (1968) Studies of trichloroethylene intoxication in man. *Neurol.* 18:309.

Feldman, R., Chirico-Post, J., and Proctor, S. (1988) Blink reflex latency after exposure to trichloroethylene in well water. *Arch. Environ. Health* 43:143–148.

Gutch, C.F., Tomhave, W.G., and Stevens, S.C. (1965) Acute renal failure due to inhalation of trichloroethylene. *Ann. Intern. Med.* 63:128–134.

Harenko, A. (1967) Two peculiar instances of psychotic disturbance in trichloroethylene poisoning. *Acta. Neurol. Scand.* 43 (Suppl 31): 139.

Jacob, S., Hery, M., Protois, J., Rossert, J., and Stengel, B. (2007) New insight into solvent-related end-stage renal disease: Occupations, products and types of solvents at risk. *Occup. Environ. Med.* 64:843–848.

Konietzko, H., Haberlaundt, W., Heilbronner, H., et al. (1978) Chromosome studies on trichloroethylene workers. *Arch. Toxicol.* 40:201–206.

Nagaya, T., Ishikawa, N., and Hata, H. (1989) Urinary total protein and "beta"-2-microglobulin in workers exposed to trichloroethylene. *Environ. Res.* 50:86–92.

National Library of Medicine (NLM) (2013) Allyl Chloride. Hazardous Substances Data Bank, Available at http://toxnet.nlm.nih .gov/cgi-bin/sis/search/f?./temp/~EuKWDd:1.

Pembleton, W.E. (1974) Trichloroethylene anesthesia re-evaluated. *Anesth. Analg.* 53:730–733.

Rasmussen, K., Sabroe, S., Wohlert, M., et al. (1988) A genotoxic study of metal workers exposed to trichloroethylene. Sperm parameters and chromosome aberrations in lymphocytes. *Int. Arch. Occup. Environ. Health* 60:419–423.

Rasmussen, K., Arlien-Søborg, P., and Sabroe, S. (1993) Clinical neurological findings among metal degreasers exposed to chlorinated solvents. *Acta Neurol. Scand.* 87:200–204.

Rybak, L.P. (1992) Hearing: the effects of chemicals. *Otolaryngol. Head Neck Surg.* 106(6):677–686.

Smith, G.F. (1966) Trichloroethylene: a review. *Br. J. Ind. Med.* 23:249–262.

59

OTHER HALOGENATED HYDROCARBONS

Michael H. Lumpkin

Some brominated, chlorinated, and fluorinated hydrocarbons have been used historically in industry, but have since been banned for production and use in the United States and other nations. The Montreal Protocol of 1987 was enacted to phase out ozone-depleting chemicals such as chlorofluorocarbons and hydroflurocarbons. Manufacturing and use bans enacted by the U.S. federal government in the 1970s have removed a substantial volume of brominated insecticides and halogenated biphenyls from commerce. Nonetheless, legacy uses of these compounds (i.e., polychlorinated biphenyls in electrical equipment), as well as their long-lived persistence in the environment, maintain ongoing concern for their exposure to workers today. This chapter includes a subset of chemicals that have been banned but are still of concern, or have ongoing scientific and regulatory decisions to be made regarding their risk to workers and to the general public.

BROMOBENZENE

First aid: *Inhalation*: Remove to fresh air. Take deep breaths. Check for wheezing, coughing, shortness of breath, or burning in the mouth, throat, or chest. Call physician if these symptoms persist. *Dermal*: Remove contaminated clothing while washing exposed skin with soap and water. *Eye*: Irrigate with water or normal saline for 15–20 min

Target organ(s): Eyes, CNS, liver, kidney

Occupational exposure limits: OSHA PEL: None; NIOSH: None; ACGIH TLV: None

Reference values: EPA: subchronic RfC—0.06 mg/m³, subchronic RfD—0.008 mg/kg/day

Risk/safety phrases: 10-38-51/53; (2-)61

BACKGROUND AND USES

Bromobenzene (phenyl bromide, CASRN 108-86-1) is used as a reagent in chemical synthesis and has been used as an additive to motor oils. It has also been used as a high density solvent for chemical recrystallization (NLM, 2013).

PHYSICAL AND CHEMICAL PROPERTIES

Bromobenzene (molecular weight = 157.02 g/mol) is a colorless liquid with a pungent, aromatic odor. It is a volatile organic compound with a vapor pressure of 4.18 mm Hg at 25 °C. It is insoluble in water, but soluble in most organic solvents.

ENVIRONMENTAL FATE AND BIOACCUMULATION

Bromobenzene found in the environment may result from industrial releases. It will exist in the atmosphere in the vapor phase, where it is degraded over a half-life of 21 days. Bromobenzene is moderately mobile in soil; and will evaporate quickly from dry soil and surface waters. It is expected to be absorbed to sediments and suspended particles, and is not expected to bioaccumulate (BCF = 8.8–190) (NLM, 2013).

ECOTOXICOLOGY

Acute Effects

The EC_{50} for immobility in *Daphnia magna* is 1600 µg/l. The LC_{50} for *C. dubia* is 5800 µg/l. The 4-h LC_{50} for fathead minnow is 5600 µg/l (EPA, 2013).

Hamilton & Hardy's Industrial Toxicology, Sixth Edition. Edited by Raymond D. Harbison, Marie M. Bourgeois, and Giffe T. Johnson.
© 2015 John Wiley & Sons, Inc. Published 2015 by John Wiley & Sons, Inc.

Chronic Effects

There are no data for chronic ecotoxicological effects of bromobenzene.

MAMMALIAN TOXICOLOGY

Acute Effects

No acute health effects data for bromobenzene in humans are available. No acute inhalation toxicity data are available in animals. In mice, oral administration of ≥ 300 mg/kg/day bromobenzene for 5 days resulted in liver pathology indicated by centrilobular inflammation and hepatocytic necrosis (NTP, 1985a). In rats ingesting 600 mg/kg/day for 13 weeks, were reported in daily ataxia, malaise, and ocular discharge (NTP, 1985a).

Chronic Effects

No chronic health effects data for bromobenzene in humans are available. Chronic inhalation of up to 300 ppm bromobenzene in rats for 13 weeks was resulted in renal tubule regeneration, but without associated necrosis. No liver pathology was reported (NTP, 1985a). Mice inhaling 300 ppm for 13 weeks exhibited liver damage, including hepatocytic necrosis and mineralization (NTP, 1985b). In rats ingesting 600 mg/kg/day for 13 weeks, resulted in emaciation, as well as liver and kidney cellular damage (NTP, 1985c). Decreased renal glutathione (GSH) levels and increased blood urea nitrogen were observed in rats following bromobenzene administration (Monks and Lau, 1988), suggesting adverse effects to the kidney.

Chemical Pathology

Liver and kidney effects seen in rodents after high inhalation or oral subchronic exposures to bromobenzene were marked by histological effects, including renal tubular degeneration and hepatocyte necrosis (NTP, 1985a–d).

Mode of Action(s)

Bromobenzene is oxidized by cytochrome P450 (CYP) monooxygenases to a reactive 3,4-oxide and subsequently to 4-bromophenol and 4-bromoquinone. One or more of these metabolites may be the putative toxicant binding to macromolecules resulting in cytotoxicity and necrosis in hepatocytes. A GSH conjugate of the quinone may be transported to the kidney, where it is converted to cysteine derivative, undergoing redox cycling, and causing kidney damage through the generation of reactive oxygen species.

SOURCES OF EXPOSURE

Occupational

Workers using bromobenzene in specific chemical synthesis reactions may receive inhalation or dermal exposures.

Environmental

The general population is unlikely to be exposed to bromobenzene in the environment.

INDUSTRIAL HYGIENE

Bromobenzene is highly flammable; heavier than air vapors may travel along the ground to an ignition source. Goggles, rubber gloves, and a protective apron should be used when working with bromobenzene.

RISK ASSESSMENTS

EPA: subchronic RfC—0.06 mg/m^3, subchronic RfD— 0.008 mg/kg/day, inadequate information to assess the carcinogenic potential.

REFERENCES

Environmental Protection Agency (EPA) (2013) *ECOTOX Database.* Available at http://cfpub.epa.gov/ecotox/quick_query.cfm.

Monks, T.J. and Lau, S.S. (1988) The contribution of bromobenzene to our current understanding of chemically-induced toxicities. *Life Sci.* 42:1259-1269.

NLM (National Library of Medicine). (2013) *Allyl Chloride. Hazardous Substances Data Bank.* Available at http://toxnet.nlm.nih.gov/cgi-bin/sis/search/f?./temp/~EuKWDd:1.

NTP (National Toxicology Program). (1985a) *Subchronic Gavage Study of Bromobenzene in Rats,* NC: National Institutes of Health, National Toxicology Program, Research Triangle Park.

NTP. (1985b) *Subchronic Gavage Study of Bromobenzene in Mice,* NC: National Institutes of Health, National Toxicology Program, Research Triangle Park.

NTP. (1985c) *Subchronic Inhalation Study of Bromobenzene in Rats,* NC: National Institutes of Health, National Toxicology Program, Research Triangle Park.

NTP (1985d) *Subchronic Inhalation Study of Bromobenzene in Mice,* NC: National Institutes of Health, National Toxicology Program, Research Triangle Park.

ETHYLENE DIBROMIDE

First Aid: *Inhalation*: Remove to fresh air. Call physician if cough or breathing difficulty develops or persists. *Oral*: Do not induce vomiting; but administer an activated charcoal slurry. *Dermal*: Remove contaminated clothing and wash exposed skin with soap and water. *Eye*: Irrigate with water for 15 min.

Target organ(s): Eyes, skin, respiratory system, liver, kidneys, reproductive system

Occupational exposure limits: OSHA PEL: 20 ppm 8-h TWA; NIOSH: 0.45 ppm 10-h TWA

Reference values: EPA: RfC— 0.009 mg/m^3, RfD— 0.009 mg/kg/day

Risk/safety phrases: 45-23/24/25-36/37/38-51/53; 53-45-61 Note E

BACKGROUND AND USES

Ethylene dibromide (1,2-bromomethane, CASRN 106-93-4) is used as an antiknock additive for leaded fuels, a solvent, and a catalyst for Grignard reagent preparations, and as an intermediate in the production of dyes and pharmaceutical agents and as a pesticide fumigant. It was historically used as a pesticide fumigant for food crops (NLM, 2013).

PHYSICAL AND CHEMICAL PROPERTIES

Ethylene dibromide (molecular weight = 187.86 g/mol) is a colorless liquid at ambient temperature with a sweet chloroform odor. It is a volatile compound with a vapor pressure of 11.2 mm Hg at 25 °C. It is moderately soluble in water, but highly soluble in solvents such as acetone, ethanol, ethyl ether, and benzene.

ENVIRONMENTAL FATE AND BIOACCUMULATION

Ethylene dibromide found in the environment may result from waste stream releases from aviation fuel blending, chemical synthesis, and dye and pharmaceutical manufacturing operations. It will exist in the atmosphere as a vapor, where it is degraded over a half-life of 64 days. Ethylene dibromide is highly mobile in soil, and will evaporate quickly from dry soil and surface waters. It is not expected to be absorbed to sediments and suspended particles, and is not expected to bioaccumulate in aquatic organisms (BCF = <1–14.9) (NLM, 2013).

ECOTOXICOLOGY

Acute Effects

The EC$_{50}$ for behavioral effects in *D.magna* is 116 mg/l. The 4-h LC$_{50}$ for Japanese medaka is 32.1 mg/l. The 4-h LC$_{50}$ for fathead minnow is 5600 µg/l (EPA, 2013).

Chronic Effects

The 28-day no effect concentration (NOAEL) for growth of Japanese medaka is 5.8 mg/l.

MAMMALIAN TOXICOLOGY

Acute Effects

Two workers died cleaning a tank previously filled with ethylene bromide, exhibiting metabolic acidosis, CNS depression, liver damage, and renal failure (Letz et al., 1984). A case of oral lethality was reported in adult female ingesting ethylene dibromide-containing capsules. Reported effects included agitation, abdominal pain, nausea, vomiting, and diarrhea, followed by death at 54 h post-ingestion (Olmstead, 1960).

Chronic Effects

Some epidemiologists suggest that ethylene dibromide adversely affects the male reproductive system. Abnormal sperm morphology, and decreased sperm count, motility, ejaculate volume, and overall reproductive function were reported in workers having a long-term inhalation exposure to ethylene dibromide (Ratcliffe et al., 1987; Schrader et al., 1988; Wong et al., 1979). However, these studies were confounded by possible exposure to other reproductive toxicants.

In animals, ethylene dibromide exposure via inhalation or oral exposure routes resulted in both reproductive and developmental effects (Short et al., 1978). Decreased testicular weight and serum testosterone, testicular atrophy, and reduced reproductive performance were reported in male rats inhaling 89 ppm, while testicular atrophy was observed in male mice inhaling 32 ppm. Oral doses of 4 mg/kg/day ethylene dibromide resulted in decreased sperm motility, altered sperm morphology, and decreased sperm number in seminiferous tubules of bulls (Amir and Ben-David, 1973). Developmental effects in both species at levels that also produced frank parental toxicity (38 ppm) included reduced fetal weights, increased resorptions, and decreased fetus viability.

Inhalation studies in rodents indicate nasal respiratory tissue toxicity as the most sensitive endpoint, with noncancerous lesions also occurring in the liver, kidney, testis, and eyes (NTP, 1982). Oral studies in rodents resulted in forestomach lesions, decreased weight gain, high mortality, and testicular atrophy (NCI, 1978).

No epidemiological studies provided strong evidence for occupational carcinogenicity due to ethylene dibromide exposure, but rodents exhibited tumor development at multiple sites following chronic exposure. Increased nasal tumors and splenic hemangiosarcoma occurred in rats, and lung tumors in mice, inhaling ≥10 ppm for a lifetime. Rats ingesting 40 mg/kg/day over a lifetime developed squamous cell carcinoma of the forestomach, as well as hemangiosarcoma in liver, kidney, spleen, and pancreas when ingesting 80 mg/kg/day (NCI, 1978).

Chemical Pathology

Proliferative lesions in nasal epithelium, including focal epithelial hyperplasia of the nares, and diffuse respiratory hyperplasia were reported in mice inhaling ethylene dibromide for 13 weeks (Stinson et al., 1981; Nitschke et al., 1981). Megalocytic bronchiolar cells in rodents were also reported following inhalation exposures (NTP, 1982). Liver and kidney lesions were reported in mice/rats following inhalation exposures for a lifetime (NTP, 1982). Testicular pathology following chronic inhalation or oral exposure in rodents (NCI, 1978; NTP, 1982).

Mode of Action(s)

Cytotoxic lesions observed in rodent liver, kidney, nasal, and forestomach tissues were likely mediated by cytochrome P-450-dependent metabolism of ethylene dibromide to 2-bromoacetaldehyde, which, in turn, was responsible for lipid peroxidation, GSH depletion, protein binding, and, in the kidney, generation of glutathione-based metabolic toxicants (Khan et al., 1993; Masini et al., 1986; Novotna and Duverger-Van Bogaert, 1994). The genotoxic mode of action for ethylene dibromide appears to be mediated by GSH-mediated metabolites, as evidenced by observation of S-[2-(N$_7$-guanyl)ethyl]glutathione as the prominent DNA adduct (Cmarik et al., 1992).

SOURCES OF EXPOSURE

Occupational

Workers in the chemical and petrochemical industries involving use of ethylene dibromide may experience inhalation or dermal exposures.

Environmental

The general population is not likely to be exposed to ethylene dibromide unless contact is made with waste streams released from industries utilizing it.

INDUSTRIAL HYGIENE

Protective clothing should be worn when working with ethylene dibromide, to include gloves, bib-type aprons, boots, and overshoes. Gloves should not be made of neoprene rubber or many types of plastic. A half-mask respirator should be donned if ethylene dibromide concentrations are ≥10 ppm.

RISK ASSESSMENTS

EPA: RfC—0.009 mg/m^3, RfD—0.009 mg/kg/day, likely to be carcinogenic to humans.

REFERENCES

Amir, D. and Ben-David, E. (1973) The pattern of structural changes induced in bull spermatozoa by oral or injected ethylene dibromide (EDB). *Ann. Biol. Anim. Biochem. Biophys.* 13:165–170.

Cmarik, J.L., Humphreys, W.G., Bruner, K.L., et al. (1992) Mutation spectrum and sequence alkylation selectivity resulting from modification of bacteriophage M13mp18 DNA with S-(2-chloroethyl)glutathione. Evidence for a role of S-(2-N$_7$-guanyl)ethyl) glutathione as a mutagenic lesion formed from ethylene dibromide. *J. Biol. Chem.* 267:6672–6679.

Environmental Protection Agency (EPA) (2013) *ECOTOX Database.* Available at http://cfpub.epa.gov/ecotox/quick_query.cfm.

Khan, S., Sood, C., and O'Brien, P.J. (1993) Molecular mechanisms of dibromoalkane cytotoxicity in isolated rat hepatocytes. *Biochem. Pharmacol.* 45:439–447.

Letz, G.A., Pond, S.M., Osterloh, J.D., et al. (1984) Two fatalities after acute occupational exposure to ethylene dibromide. *JAMA* 252:2428–2431.

Masini, A., Botti, B., Ceccarelli, D., et al. (1986) Induction of calcium efflux from isolated rat-liver mitochondria by 1,2-dibromoethane. *Biochem. Biophys. Acta* 852:19–24.

NCI (National Cancer Institute) (1978) *Bioassay of 1,2-dibromoethane for Possible Carcinogenicity,* Bethesda, MD: National Institutes of Health, U.S. Department of Health, Education, and Welfare, DHEW Publication No. (NIH) 78-1336. Available from: National Technical Information Service, Springfield, VA; PB-288428.8.

Nitschke, K.D., Kociba, R.J., Keyes, D.G., et al. (1981) A thirteen week repeated inhalation study of ethylene dibromide in rats. *Fundam. Appl. Toxicol.* 1:437–442.

NLM (National Library of Medicine) (2013) *Allyl Chloride. Hazardous Substances Data Bank.* Available at http://toxnet.nlm.nih.gov/cgi-bin/sis/search/f?./temp/~EuKWDd:1.

Novotna, B. and Duverger-Van Bogaert, M. (1994) Role of kidney S9 in the mutagenic properties of 1,2-dibromoethane. *Toxicol. Lett.* 74:255–263.

NTP (National Toxicology Program) (1982) *Carcinogenesis bioassay of 1,2-dibromoethane (CAS no. 106-93-4) in F344 rats and B6C3F1 mice (inhalation study). National Institutes of Health, U.S. Department of Health and Human Services, Research Triangle Park, NC; NTP-80-28, NTP 210, NIH Publication No. 82-1766.*

Olmstead, E.V. (1960) Pathological changes in ethylene dibromide poisoning. *AMA Arch. Ind. Health* 21:525–529.

Ratcliffe, J.M., Schrader, S.M., Steenland, K., et al. (1987) Semen quality in papaya workers with long term exposure to ethylene dibromide. *Br. J. Ind. Med.* 44:317–326.

Schrader, S.M., Turner, T.W., and Ratcliffe, J.M. (1988) The effects of ethylene dibromide on semen quality: a comparison of short-term and chronic exposure. *Reprod. Toxicol.* 2:191–198.

Short, R.D., Minor, J.L., Winston, J.M., et al. (1978) Inhalation of ethylene dibromide during gestation by rats and mice. *Toxicol. Appl. Pharmacol.* 46:173–182.

Stinson, S.F., Reznik, G., and Ward, J.M. (1981) Characteristics of proliferative lesions in the nasal epithelium of mice following chronic inhalation of 1,2-dibromoethane. *Cancer Lett.* 12:121–129.

Wong, O., Utidjian, H.M.D., and Karten, V.S. (1979) Retrospective evaluation of reproductive performance of workers exposed to ethylene dibromide (EDB). *J. Occup. Med.* 21:98–102.

METHYL BROMIDE

First aid: *Inhalation*: Remove to fresh air. Check for bronchitis, pneumonitis, or severe respiratory tract irritation if cough or breathing difficulty develops or persists. *Dermal*: Remove contaminated clothing and wash exposed skin with soap and water. *Eye*: Irrigate with water for 15 min.

Target organ(s): Eyes, skin, respiratory system, central nervous system

Occupational exposure limits: OSHA PEL: 20 ppm 8-h TWA with skin notation; NIOSH: Minimize exposure to lowest possible level due to possible carcinogenicity; ACGIH TLV: 1 ppm 8-h TWA with skin notation.

Reference values: EPA: RfC—0.005 mg/m^3, RfD—0.014 mg/kg/day.

ATSDR: Acute, intermediate, and chronic inhalation MRLs—0.05, 0.05, and 0.005 ppm, intermediate oral MRL—0.003 mg/kg/day

Risk/safety phrases: 23–36/37/38-50/53-59; (1/2-)15-27-36/37/39-38-45-59-61

BACKGROUND AND USES

Methyl bromide (bromomethane, CASRN 74-83-9) is used in limited applications as a fumigant pesticide. Historically, it was used as a fumigant for disinfection of products, soil, and crops stored in mills, warehouses, ships, and freight cars. In nature, it is found in the ocean in concentrations of 1–2 ng/l, formed possibly by algae or kelp. As of 2013, the amount of methyl bromide used in the United States allowed by critical use exemptions was 2.1% of 1991 levels (EPA, 2013a).

PHYSICAL AND CHEMICAL PROPERTIES

Methyl bromide (molecular weight = 94.94 g/mol) is a colorless gas (easily liquefied) with a burning taste and a sweet chloroform-like odor at high concentrations. It has a solubility in water by weight of 2% and a vapor pressure of 1.9 atm. Relative vapor density with reference to air is 3.36. It is a highly volatile compound with a vapor pressure of 1620 mm Hg at 25 °C. It is soluble in water, ethanol, ethers, ketones, and other halogenated hydrocarbon solvents.

ENVIRONMENTAL FATE AND BIOACCUMULATION

Methyl bromide is unlikely to be found in the environment, as its production and use have been banned in the United States (with the exception of critical use exemptions) and other countries. It will exist in the atmosphere as a gas, where it is degraded over a half-life of 1.2 years. Methyl bromide is extremely mobile in soil, and will evaporate quickly from dry soil and surface waters. It is not expected to be absorbed to sediments and suspended particles, and is not expected to bioaccumulate in aquatic organisms (BCF = 3) (NLM, 2013).

ECOTOXICOLOGY

Acute Effects

The 2-h EC$_{50}$ for behavioral effects in *D. magna* is 1.7 mg/l. The 2-h LC$_{50}$ for *D. magna* is 2.2 mg/l. No mortality or behavioral effects were seen in Japanese medaka exposed to 320 µg/l for 91 h (EPA, 2013b).

Chronic Effects

No chronic ecotoxicological data for methyl bromide were identified.

MAMMALIAN TOXICOLOGY

Acute Effects

High exposure to methyl bromide via inhalation may result in neurological and respiratory effects, with skin lesions induced by dermal contact. Headaches, eye irritation, as well as persistent throat irritation and cough lasting up to 3 weeks (Rathus and Landy, 1961; Herzstein and Cullen, 1990). High exposures may result in ataxia, tremors, seizures, and death (Behrens and Dukes, 1986). Pulmonary edema and focal hemorrhagic lesions in the lung, progressing to hypoxia, cyanosis, and respiratory failure, have been reported (Greenberg, 1971; Marraccini et al., 1983; Behrens and Dukes, 1986). Oral exposure may result in irritation of the gastrointestinal tract, such as nausea and vomiting.

Skin contact has been reported to cause subepithelial blisters (Alexeeff and Kilgore, 1983; Zwaveling et al., 1987). A diffuse infiltration of neutrophils was also reported (Yamamoto et al., 2000). No systemic effects were reported.

F-344 rats exposed to 150 ppm methyl bromide for 6 h/day had depleted GSH and GST levels in the brain (Davenport et al., 1992). Pretreatment with glutathione S-transferase (GST) inhibitor partially prevented the depletion of GSH and completely blocked the depletion of GST.

Chronic Effects

Fumigators chronically exposed to methyl bromide had a higher prevalence of muscle aches and fatigue tested lower than controls in 23 of 27 neurological tests than unexposed controls (Anger et al., 1986), indicating neurological effects. A worker is a vegetable house fumigated with methyl bromide presented with unstable gait, vertigo, and paresthesia of both feet (Suwanlaong and Phanthumchinda, 2008).

In several 1-year or lifetime animal bioassays, rats exhibited increased incidences of nonneoplastic nasal legions (90 ppm), but did not indicate carcinogenicity (Mitsumori et al., 1990; Reuzel et al., 1991). Neurological effects were seen in mice inhaling 100 ppm over a lifetime, including tremors, hypoactivity, abnormal posture, and limb paralysis (NTP, 1990).

Rats orally treated with 10 mg/kg/day methyl bromide exhibited hyperplastic changes in forestomach epithelium, progressing to severe hyperplasia at 50 mg/kg/day (Danse et al., 1984). Dogs ingesting 178 mg/kg/day for 1 year displayed excessive salivation, lethargy, and periodic diarrhea (Rosenblum et al. 1960).

Chemical Pathology

Goergens et al. (1994) reported increased SCE rates in lymphocytes of fumigators exposed to methyl bromide. DNA adducts have been identified in rats exposed to methyl bromide via inhalation or gavage (Gansewendt et al., 1991).

Inhalation of 30 ppm methyl bromide in rats for 29 months led to mild hyperplasia of the nasal, esophageal, and heart (Reuzel et al., 1991).

Mode of Action(s)

Methyl chloride exerts its acute CNS effects via a solvent depression mode of action.

SOURCES OF EXPOSURE

Occupational

Due to ban on general production and uses, occupational exposures would be limited to pesticide applicators using methyl bromide.

Environmental

Due to ban on general production and uses, the general population is unlikely to be exposed to methyl bromide. Any exposure would likely arise from contact with limited waste stream releases from methyl bromide production.

INDUSTRIAL HYGIENE

The most common route of exposure to methyl bromide is inhalation. Workers who fumigate homes or fields may be exposed to high levels if proper safety precautions are not followed (Fuortes, 1992). Fumigation workers are at risk of accidental exposure even if they wear personal protection devices because methyl bromide may penetrate rubber, tape, and other materials. In soil and greenhouse fumigation, workers not involved in the fumigation may be overexposed when reentering the area if proper precautions are not taken (Van Den Oever et al., 1982; Herzstein and Cullen, 1990; Bishop, 1992; Hustinx et al., 1993). Overexposures typically result from lack of management controls, inadequate training of personnel or inadequate protective equipment. In response to this problem, the EPA allows only licensed professional fumigation workers to buy or use methyl bromide.

RISK ASSESSMENTS

EPA: RfC—0.005 mg/m^3, RfD—0.014 mg/kg/day, not classifiable as to human carcinogenicity

ATSDR: Acute, intermediate, and chronic inhalation MRLs—0.05, 0.05, and 0.005 ppm, intermediate oral MRL—0.003 mg/kg/day

IARC: Group 3—not classifiable as to its carcinogenicity to humans

REFERENCES

Alexeeff, G.V. and Kilgore, W.W. (1983) Methyl bromide. *Residue Rev.* 88:101–153.

Anger, W.K., Moody, L., and Burg, J. (1986) Neurobehavioral evaluation of soil and structural fumigators using methyl bromide and sulfuryl fluoride. *Neurotoxicology* 7:137–156.

Behrens, R.H. and Dukes, D.C.D. (1986) Fatal methyl bromide poisoning. *Br. J. Ind. Med.* 43:561–562.

Bishop, C.M. (1992) A case of methyl bromide poisoning. *Occup. Med.* 42:107–109.

Danse, L.H.J.C., van Velsen, F.L., and van der Heijden, C.A. (1984) Methylbromide: carcinogenic effects in the rat forestomach. *Toxicol. Appl. Pharmacol.* 72:262–271.

Davenport, C.J., et al. (1992) Effect of methyl bromide on regional brain glutathione, glutathione S-transferase, monoamines and amino acids in F344 rats. *Toxicol. Appl. Pharmacol.* 112:120–127.

Environmental Protection Agency (EPA) (2013a) *2013 critical use exemption nominations from the phasout of methyl bromide.* Available at http://www.epa.gov/ozone/mbr/2013_nomination.html.

Environmental Protection Agency (EPA) (2013b) *ECOTOX Database.* Available at http://cfpub.epa.gov/ecotox/quick_query.cfm.

Fuortes, L.J. (1992) A case of fatal methyl bromide poisoning. *Vet. Hum. Toxicol.* 34:240-241.

Gansewendt, B., et al. (1991) Formation of DNA adducts in F-344 rats after oral administration or inhalation of (14C) methyl bromide. *Food. Chem. Toxicol.* 29:557–563.

Goergens, H.W., et al. (1994) Macromolecular adducts in the use of methyl bromide as fumigant. *Toxicol. Lett.* 72:199–203.

Greenberg, J.O. (1971) The neurological effects of methyl bromide poisoning. *IMS Ind. Med. Surg.* 40:27–29.

Herzstein, J., and Cullen, M.R. (1990) Methyl bromide intoxication in four field workers during removal of soil fumigation sheets. *Am. J. Ind. Med.* 17:321–326.

Hustinx, W.N., et al. (1993) Systemic effects of inhalational methyl bromide poisoning: a study of nine cases occupationally exposed to inadvertent spread during fumigation. *Br. J. Ind. Med.* 50:155–159.

Marraccini, J.V., et al. (1983) Death and injury caused by methyl bromide, an insecticide fumigant. *J. Forensic Sci.* 28:601–607.

Mitsumori, K., et al. (1990) Two-year oral chronic toxicity and carcinogenicity study in rats of diets fumigated with methyl bromide. *Food. Chem. Toxicol.* 28:109–119.

NLM (National Library of Medicine) (2013) *Allyl Chloride. Hazardous Substances Data Bank.* Available at http://toxnet.nlm.nih.gov/cgi-bin/sis/search/f?./temp/~EuKWDd:1.

NTP (National Toxicology Program) (1990) *Toxicology and carcinogenesis studies of methyl bromide (CAS No. 74-83-9) in B6C3F1 mice (inhalation studies). NTP TR 385, NIH Publication No. 91-2840. Peer Review Draft.*

Rathus, E.M. and Landy, P.J. (1961) Methyl bromide poisoning. *Br. J. Ind. Med.* 18:53–57.

Reuzel, P.G.J., Dreef-van der Meulen, H.C., Hollanders, V.M.H., Kuper, C.F., Feron, V.J., and van der Heijden, C.A. (1991) Chronic inhalation toxicity and carcinogenicity study of methyl bromide in Wistar rats. *Food. Chem. Toxicol.* 29(1):31–39.

Rosenblum, I., Stein, A.A., and Eisinger, G. (1960) Chronic ingestion by dogs of methyl bromide fumigated food. *Arch. Environ. Health* 1:316–323.

Suwanlaong, K. and Phanthumchinda, K. (2008) Neurological manifestation of methyl bromide intoxication. *J. Med. Assoc. Thai.* 91(3):421–426.

Van Den Oever, R., Roosels, D., and Lahaye, D. (1982) Actual hazard of methyl bromide fumigation in soil disinfection. *Br. J. Ind. Med.* 39:140–144.

Yamamoto, O., Hori, H., Tanaka, I., Asahi, M., and Koga, M. (2000) Experimental exposure of rat skin to methyl bromide: a toxicokinetic and histopathological study. *Arch. Toxicol.* 73(12):641–648.

Zwaveling, J.H., et al. (1987) Exposure of the skin to methyl bromide: a study of six cases occupationally exposed to high concentrations during fumigation. *Hum. Toxicol.* 6:491–495.

POLYCHLORINATED BIPHENYLS

First aid: *Inhalation*: Remove to fresh air. Check for bronchitis, pneumonitis, or severe respiratory tract irritation if cough or breathing difficulty develops or persists. *Oral*: Do not induce vomiting. The efficacy of charcoal slurry administration is unknown. Monitor patient for increased serum liver enzymes and development of chloracne. *Dermal*: Remove contaminated clothing and wash exposed skin multiple times with soap and water. *Eye*: Irrigate with water for 15 min.

Target organ(s): Skin, eyes, liver, reproductive system

Occupational exposure limits: OSHA PEL: $0.5\,mg/m^3$ 8-h TWA with skin notation; NIOSH: $0.001\,mg/m^3$ 10-h TWA.

Reference values (Arochlor 1254): EPA: RfD—20 ng/kg/day

ATSDR: Intermediate and chronic oral MRLs—30 and 20 ng/kg/day

Risk/safety phrases: 33-50/53; (2-)35-60-61 Note C

BACKGROUND AND USES

Polychlorinated biphenyls (PCBs) are a class of chemicals produced in large scale in the United States from 1929 until their ban in 1979. There are 209 individual PCBs, or congeners. PCBs are mixtures of biphenyl congeners having varying degrees of chlorine substituents. The more common mixtures produced historically were Arochlor 1260, 1254,

1242, and 1016 (CASRNs 11096-82-5, 11097-69-1, 53469-21-9, and 12674-11-2), trade names of the Monsanto chemical products whose numerical nomenclature convention denoted the number of carbon atoms by the first two digits, and the percentage of chlorine (by mass) by the last two digits. Polychlorinated biphenyls were used historically in the manufacture of plasticizers, fire retardant fabric treatments, adhesives, paints, and electrical equipment. Since their ban in 1979, PCBs have been utilized for "enclosed uses," such as insulating oil in electrical transformers and capacitors and in hydraulic fluids.

PHYSICAL AND CHEMICAL PROPERTIES

Polychlorinated biphenyls (molecular weight = 291.98–360.86 g/mol) vary in physical appearance from oily liquids to rigid noncrystalline resins, and are generally non-odorous. They are semi-volatile compounds. They are extremely insoluble in water, but soluble in organic solvents. Their low flammability and dielectric properties made them desirable for use in electrical equipment.

ENVIRONMENTAL FATE AND BIOACCUMULATION

Polychlorinated biphenyls are highly persistent compounds, and may be found in the environment as a result of leakage from older electrical equipment, migration from older building materials into surrounding construction substrates, and as a result of historical spills or releases. Mono-, di-, and trichlorinated biphenyls biodegrade rapidly, tetrachlorinated species degrade more slowly, and higher chlorinated species are quite resistant to environmental degradation. They can exist in the atmosphere in the vapor or particulate phase, depending on the level of volatility. Vapor-phase PCBs are photolytically degraded over a half-life of 3.3–490 days. Polychlorinated biphenyls are not mobile in soil, but vapor-phase components will evaporate from dry soil and surface waters. They absorbed extensively to sediments and suspended particles, and are expected to bioaccumulate in aquatic organisms, especially fish (log BCF = 4.4–6.2) (NLM, 2013).

ECOTOXICOLOGY

The ecotoxicological database for PCBs is extensive. Representative values for Arochlor 1254 are provided here.

Acute Effects

Decreased growth was reported in *X. laevis* frogs exposed to 1 mg/l, but not 100 µg/l, Arochlor 1254 for 3 days. *D. magna* exposed for 14 days had an LC_{50} of 24 µg/l. The 2-day LC_{50} for striped bass was 0.5–1.0 mg/l (EPA, 2013).

Chronic Effects

Atlantic croaker exposed to 1.0 mg/l for 30 days displayed reduced organ weights and protein content, relative to body weight. Atlantic char exposed to 100 mg/l for 330 days exhibited significantly increased growth rate (EPA, 2013).

MAMMALIAN TOXICOLOGY

Acute Effects

Acute effects in humans have been reported for individual cases and large-scale incidents. High oral and dermal exposures may result in the development of acne (termed "chloracne") and rashes (Rice and Cohen, 1996). A mass poisoning in Japan in 1968, known as Yusho incident, involved ingestion of PCB- and dibenzofuran-contaminated rice oil and resulted in over 14,000 people reporting skin and eye lesions (conjunctival pigmentation, swollen eyelids), reduced immune function, and changes in menstrual cycles (Chang et al., 1981; Rogan, 1989). Another mass poisoning in 1977 occurred in Taiwan, known as the Yu-Cheng incident, also involved ingestion of contaminated rice oil. Reported effects seen in this cohort include those of the Yusho incident, as well as an increased risk of goiter and increased urinary porphyrin excretion (Gladen et al., 1988; Guo et al., 1999).

Chronic Effects

Effects from long-term exposure of humans to PCBs include possible changes in neurological development, endocrine and immune function and cancer. Cohorts of children born to women who consumed PCB-containing fish from the Great Lakes, New York state, North Carolina, and Europe report an association between PCB body burdens and memory deficits, learning, and IQ scores (Fein et al., 1984; Lonky et al., 1996; Rogan et al., 1986; Koopman-Esseboom et al., 1994; Winneke et al., 1998; Govarts et al., 2012; Forns et al., 2012). Epidemiology studies in electrical capacitor manufacturing workers are suggestive of increased risk for liver, intestinal, and skin cancer (Brown and Jones, 1981; Gustavsson and Hogstedt, 1997; Bertazzi et al., 1987).

Studies in animals indicate that monkeys are more sensitive to PCB-induced immunological, dermatological, and neurological toxicity than other species. Liver damage was indicated by elevated hepatic serum enzymes, fatty liver, fibrosis, and necrosis in rats given ≥7 mg/kg/day (Bruckner et al., 1973) and in monkeys given 0.1–0.2 mg Aroclor 1254 or 1248/kg/day (Tryphonas et al., 1986). Vitamin A storage in the liver was also impaired in rats and rabbits (Bank et al., 1989). Rat fetuses exhibited significantly reduced plasma thyroid hormone levels following maternal ingestion of 2.5 mg/kg/day (Collins and Capen, 1980). Changes in auditory threshold occurred in rats ingesting 10 mg/kg/day (Lilienthal et al., 2011; Gu et al., 2009). Rodent and monkeys exhibited reduced immune function as indicated by decreased antibody responses to SRBC, increased suscepti-bility to bacterial infections, decreased response to mitogens, altered lymphocyte T-cells, and histopathological changes in the thymus, spleen, and lymph nodes (Talcott et al., 1985; Allen and Barsotti, 1976; Tryphonas et al., 1986, 1991).

Reproductive effects reported in rodents and monkeys include impaired conception, reduced implantation, and increased incidences of resorptions and stillbirths (Steinberg et al., 2008). These effects occurred at doses as low as 0.02 mg/kg/day in nonhuman primates (Sager and Girard, 1994; Gellert and Wilson, 1979; Arnold et al., 1990). Decreased birth weight and reduced weight gain was reported in monkeys (Arnold et al., 1993).

Liver cancer occurred in rats ingesting ≥2.5 mg/kg/day Aroclor 1260 for 21 months (Kimbrough et al., 1975; NCI, 1978).

Chemical Pathology

Chloracne reported in occupational exposures presents as inflammatory folliculitis and keratin plug formation in the pilosebaceous orifices (Rice and Cohen, 1996).

Mode of Action(s)

Coplanar PCBs and mono-ortho-PCBs may mediate toxic responses via binding to aryl hydrocarbon receptor (AhR), a gene transcription factor that, if activated abnormally, may disrupt normal cellular control function. The ranking of PCBs using toxic equivalency factors (TEF) is based on their relative ability to activate AhR, with dioxin (ATSDR, 1998) assigned a factor of 1.0.

Di-ortho-substituted noncoplanar PCBs may interfere with calcium-dependent intracellular signal transduction, leading to neurotoxicity (Kodavanti and Tilson, 1997). Ortho-PCBs may disrupt thyroid hormone transport by bind-ing to transthyretin (Cheek et al., 1999).

Some PCB congeners bind to the 17β-estradiol receptor, affecting the regulation of hormone expression and control (Jansen et al., 1993).

SOURCES OF EXPOSURE

Occupational

Occupational exposures to PCBs may occur in workers handling older PCB-containing devices, such as electrical capacitors and transformers, PCB-containing construction materials, or soil/sediment/water at waste sites at which PCBs were disposed.

Environmental

The general population may be exposed to PBCs via contact with or proximity to the failure and leakage of PCB-contain-ing devices, contact with PCB-containing construction mate-rials, and dermal or oral ingestion of PCB-contaminated soil, sediment, or water.

INDUSTRIAL HYGIENE

When working with PCBs, protective clothing should include long-sleeved overalls, boots, overshoes, bib-type aprons, gloves, and safety glasses with side shields. Respirators should be used in the proximity of PCB vapors and when installing, removing, or repairing PCB-containing devices or containers.

RISK ASSESSMENTS

(Arochlor 1254):
  EPA: RfD—20 ng/kg/day
  ATSDR: Intermediate and chronic oral MRLs—30 and 20 ng/kg/day

REFERENCES

Allen, J.R. and Barsotti, B.A. (1976) The effects of transplacental and mammary movement of PCBs on infant Rhesus monkeys. *Toxicology* 6:331–340.

Arnold, D.L., Mes, J., Bryce, F., et al. (1990) A pilot study on the effects of Aroclor 1254 ingestion by Rhesus and Cynomolgus monkeys as a model for human ingestion of PCBs. *Food Chem. Toxicol.* 28:847–857.

Arnold, D.L., Bryce, F., Stapley, R., et al. (1993) Toxicological consequences of Aroclor 1254 ingestion by female Rhesus (*macaca mulatta*) monkeys. Part 1A. Prebreeding phase: Clini-cal health findings. *Food Chem. Toxicol.* 31(11):799–810.

ATSDR (1998) *Toxicological profile for chlorinated dibenzo-*p-*dioxins*, Atlanta, GA: U.S Department of Health and Human Services, Public Health Service, Agency for Toxic Substances and Disease Registry.

Bank, P.A., Cullum, M.E., Jensen, R.K., et al. (1989) Effect of hexachlorobiphenyl on vitamin A homeostasis in the rat. *Biochim. Biophys. Acta* 990:306–314.

Bertazzi, P.A., Riboldi, L., Pesatori, A., et al. (1987) Cancer mortality of capacitor manufacturing workers. *Am. J. Ind. Med.* 11:165–176.

Brown, D.P. and Jones, M. (1981) Mortality and industrial hygiene study of workers exposed to polychlorinated biphenyls. *Arch. Environ. Health.* 36:120–129.

Bruckner, J.V., Khanna, K.L., and Cornish, H.H. (1973) Biological responses of the rat to polychlorinated biphenyls. *Toxicol. Appl. Pharmacol.* 24:434–448.

Chang, K.J., Hsieh, K.H., Lee, T.P., et al. (1981) Immunologic evaluation of patients with polychlorinated biphenyl poisoning: Determination of lymphocyte subpopulations. *Toxicol. Appl. Pharmacol.* 61:58–63.

Cheek, A.O., Kow, K., Chen, J., et al. (1999) Potential mechanisms of thyroid disruption in humans: Interaction of organochlorine compounds with thyroid receptor, transthyretin, and thyroid-binding globulin. *Environ. Health Perspect.* 107(4):273–278.

Collins, W.T. and Capen, C.C. (1980) Fine structural lesions and hormonal alterations in thyroid glands of perinatal rats exposed *in utero* and by the milk to polychlorinated biphenyls. *Am. J. Pathol.* 99:125–142.

Environmental Protection Agency (EPA) (2013) *ECOTOX Database.* Available at http://cfpub.epa.gov/ecotox/quick_query.cfm.

Fein, G.G., Jacobson, J.L., Jacobson, S.W., et al. (1984) *Intrauterine Exposure of Humans to PCBs: Newborn Effects.* Duluth, MN: U.S. Environmental Protection Agency, EPA 600/53-84-060. NTIS PB84-188-887.

Forns, J., Torrent, M., Garcia-Esteban, R., Grellier, J., Gascon, M., Julvez, J., Guxens, M., Grimalt, J.O., and Sunyer, J. (2012) Prenatal exposure to polychlorinated biphenyls and child neuropsychological development in 4-year-olds: an analysis per congener and specific cognitive domain. *Sci. Total Environ.* 432:338–343.

Gellert, R.J. and Wilson, C. (1979) Reproductive function in rats exposed prenatally to pesticides and polychlorinated biphenyls (PCB). *Environ. Res.* 18:437–443.

Gladen, B.C., Rogan, W.J., Hardy, P., et al. (1988) Development after exposure to polychlorinated biphenyls and dichlorodiphenyl dichloroethene transplacentally and through human milk. *J. Pediatr.* 113:991–995.

Govarts, E., Nieuwenhuijsen, M., Schoeters, G., Ballester, F., Bloemen, K., de, B.M., Chevrier, C., Eggesbo, M., Guxens, M., Kramer, U., Legler, J., Martinez, D., Palkovicova, L., Patelarou, E., Ranft, U., Rautio, A., Petersen, M.S., Slama, R., Stigum, H., Toft, G., Trnovec, T., Vandentorren, S., Weihe, P., Kuperus, N.W., Wilhelm, M., Wittsiepe, J., and Bonde, J.P. (2012) Birth weight and prenatal exposure to polychlorinated biphenyls (PCBs) and dichlorodiphenyldichloroethylene (DDE): a meta-analysis within 12 European Birth Cohorts. *Environ. Health Perspect.* 120(2):162–170.

Gu, J.Y., Qian, C.H., Tang, W., Wu, X.H., Xu, K.F., Scherbaum, W.A., Schott, M., and Liu, C. (2009) Polychlorinated biphenyls affect thyroid function and induce autoimmunity in Sprague-Dawley rats. *Horm. Metab Res.* 41(6):471–474.

Guo, Y.L., Yu, M.L., Hsu, C.C., et al. (1999) Chloracne, goiter, arthritis, and anemia after polychlorinated biphenyl poisoning: 14-year follow-up of the Taiwan Yucheng cohort. *Environ. Health Perspect.* 107(9):715–719.

Gustavsson, P. and Hogstedt, C. (1997) A cohort study of Swedish capacitor manufacturing workers exposed to polychlorinated biphenyls (PCBs). *Am. J. Ind. Med.* 32(3):234–239.

Jansen, H.T., Cooke, P.S., Porcelli, J., et al. (1993) Estrogenic and antiestrogenic actions of PCBs in the female rat: in vitro and in vivo studies. *Reprod. Toxicol.* 7(3):237–248.

Kimbrough, R.D., Squire, R.A., Linder, R.E., et al. (1975) Induction of liver tumors in Sherman strain female rats by polychlorinated biphenyl Aroclor 1260. *J. Natl. Cancer Inst.* 55:1453–1459.

Kodavanti, P.R. and Tilson, H.A. (1997) Structure-activity relationships of potentially neurotoxic PCB congeners in the rat. *Neurotoxicology* 18(2):425–441.

Koopman-Esseboom, C., Morse, D.C., Weisglas-Kuperus, N., et al. (1994) Effects of dioxins and polychlorinated biphenyls on thyroid hormone status of pregnant women and their infants. *Pediatr. Res.* 36(4):468–473.

Lilienthal, H., Heikkinen, P., Andersson, P.L., van der Ven, L.T., and Viluksela, M. (2011) Auditory effects of developmental exposure to purity-controlled polychlorinated biphenyls (PCB52 and PCB180) in rats. *Toxicol. Sci.* 122(1), 100–111.

Lonky, E., Reihman, J., Darvill, T., et al. (1996) Neonatal behavioral assessment scale performance in humans influenced by maternal consumption of environmentally contaminated Lake Ontario fish. *J. Great. Lakes Res.* 22(2):198–212.

National Cancer Institute (NCI) (1978) Bioassay of Aroclor 1254 for possible carcinogenicity. Carcinogenesis Tech. Rep. Ser. No. 38.

NLM (National Library of Medicine) (2013) *Allyl Chloride. Hazardous Substances Data Bank.* Available at http://toxnet.nlm.nih.gov/cgi-bin/sis/search/f?./temp/~EuKWDd:1.

Rice, R.H. and Cohen, D.E. (1996) Toxic responses of the skin. In: Klaassen, C.D. editor. *Cassarett and Doull's Toxicology: The Basic Science of Poisons*, New York, NY: McGraw-Hill, pp. 529–546.

Rogan, W.J. (1989) Yu-Cheng. In: Kimbrough, R.D. and Jensen, A.A. editors. *Halogenated Biphenyls, Terphenyls, Naphthalenes, Dibenzodioxins and Related Products*, 2nd ed., Amsterdam, The Netherlands: Elsevier Science Publishers, pp. 401–415.

Rogan, W.J., Gladen, B.C., McKinney J.D., et al. (1986) Neonatal effects of transplacental exposure to PCBs and DDE. *J. Pediatr.* 109:335–341.

Sager, D.B. and Girard, D.M. (1994) Long-term effects on reproductive parameters in female rats after translactational exposure to PCBs. *Environ. Res.* 66(1):52–76.

Steinberg, R.M., Walker, D.M., Juenger, T.E., Woller, M.J., and Gore, A.C. (2008) Effects of perinatal polychlorinated biphenyls on adult female rat reproduction: development, reproductive physiology, and second generational effects. *Biol. Reprod.* 78(6):1091–1101.

Talcott, P.A., Koller, L.D., and Exon, J.H. (1985) The effect of lead and polychlorinated biphenyl exposure on rat natural killer cell toxicity. *Int. J. Immunopharmacol.* 7:255–261.

Tryphonas, L., Charbonneau, S., Tryphonas, H., et al. (1986) Comparative aspects of Aroclor 1254® toxicity in adult Cynomolgus and Rhesus monkeys: A pilot study. *Arch. Environ. Contam. Toxicol.* 15:159–169.

Tryphonas, H., Luster, M.I., White, K.L., Jr., et al. (1991) Effects of PCB (Aroclor® 1254) on non-specific immune parameters in Rhesus (*macaca mulatta*) monkeys. *Int. J. Immmunopharmacol.* 13:639–648.

Winneke, G., Bucholski, A., Heinzow, B., et al. (1998) Developmental neurotoxicity of polychlorinated biphenyls (PCBs): Cognitive and psychomotor functions in 7-month old children. *Toxicol. Lett.* 103:423-428428.

PERFLUOROOCTANOIC ACID

First aid: *Inhalation*: Remove to fresh air. Check for bronchitis, pneumonitis, or severe respiratory tract irritation if cough or breathing difficulty develops or persists. *Dermal*: Remove contaminated clothing and wash exposed skin with soap and water. *Eye*: Irrigate with water for 15 min

Target organ(s): Liver, pancreas, testis, endocrine system (in animals)

Occupational exposure limits: None

Reference values: EPA: Provisional Health Advisory— 0.4 µg/l drinking water

Risk/safety phrases: None

BACKGROUND AND USES

Perfluorooctanoic acid (PFOA) (CASRN 335-67-1) is fluorinated surfactant used, primarily as its ammonium salt (APFO), as an aid in the chemical synthesis of fluoropolymers and fluoroelastomers. As such, it may be found in nonstick cookware and utensils, stain-repellant fabric treatments, and water-proofing treatments for garments. Although an effort is underway by the U.S. EPA to reduce use of and replace perfluoroalkyls with other substances, PFOA is still used in United States industry.

PHYSICAL AND CHEMICAL PROPERTIES

Perfluorooctanoic acid (molecular weight = 414.09 g/mol) is a white to off-white powder. It is very soluble in water.

ENVIRONMENTAL FATE AND BIOACCUMULATION

Perfluorooctanoic acid found in the environment may result from waste stream releases from manufacture of cosmetics, lubricants, paints, polishes, adhesives, fabric treatments, and fire-fighting compounds. It can partition to the vapor phase in the atmosphere, where it is degraded atmospherically with a half-life of 31 days. It is very resistant to hydrolysis, and immobile in soil. It will not likely evaporate from soil (depending on soil pH) or surface waters. It is not expected to be absorbed to sediments and suspended particles. Perfluorooctanoic acid is not expected to bioaccumulate in aquatic organisms (BCF = 3.1–9.4) (NLM, 2013).

ECOTOXICOLOGY

Acute Effects

The 2-h EC_{50} for decreasing mobility of *D. magna* was 477 mg/l. The 4-day LC_{50} for zebrafish was >500 mg/l (EPA, 2013a).

Chronic Effects

An 80–90% decrease in survival of *D. magna* was seen following a 21-day exposure to 12.5–50 mg/l. Decreased survival was seen in Japanese medaka exposed to 0.1 mg/l for 28 days (EPA, 2013a).

MAMMALIAN TOXICOLOGY

Acute Effects

No acute adverse effects have been reported in humans as a result of PFOA exposure.

Hepatocytic necrosis was exhibited in rats inhaling 8 mg/m^3 PFOA for 2 weeks (Kennedy et al., 1986). No teratogenic effects were reported in pups born to rats inhaling up to 25 mg/m^3 (Staples et al., 1984).

Liver effects have been reported in rats administered oral PFOA doses. Rats given 50 mg/kg/day for 21 days exhibited enlarged livers and hepatocyte necrosis (Son et al., 2008).

Rats ingesting ≥30 mg/kg/day for 2 weeks exhibited reduced serum testosterone and increased estradiol levels. The same oral exposures for 2 years resulted in Leydig cell adenomas (Biegel et al., 2001). Pups born to mice given 1–20 mg/kg/day experienced reduced litter weights, live births, and postnatal survival, with delayed ossification of several bones. Male mice born to treated dams matured more quickly than females or control males (Lau et al., 2006; Wolf et al., 2007; Yahia et al., 2010; Macon et al., 2011). Female mice exposed via lactation to PFOA from dams given 5 mg/kg/day had delayed mammary gland development (White et al., 2009).

Immunological effects were reported in mice given ≥50 ppm PFOA in drinking water for 21 days, including changes in T-lymphocyte phenotypes and cytokine expression (Son et al., 2009).

The rat oral LD_{50} was 540 mg/kg (Griffith and Long, 1980); the inhalation LC_{50} was 980 mg/m^3 (Kennedy et al., 1986).

Dermal doses of 0.5 g in rabbits for 24 h resulted in mild irritation, while doses of 2000 mg/kg for 10 days in rats resulted in contact site dermal necrosis as well as hepatic necrosis (Kennedy, 1985).

Chronic Effects

Epidemiological data in occupational and geographic population cohorts have reported associations of PFOA levels in the blood with inconsistent changes in lipid endpoints and renal disease, but convincing evidence of disease or mortality causation has not been shown (Gilliland and Mandel, 1993; Leonard, 2006; Steenland et al., 2010; Steenland and Woskie, 2012; Savitz et al., 2012; Okada et al., 2012).

Multigenerational oral exposures of 5 mg/kg/day in mice resulted in delayed mammary gland development across three generations (White et al., 2011).

Rats administered 0.64 mg/kg/day for 13 weeks had increased liver weight and reversible hepatocyte hypertrophy (Perkins et al., 2004). Cancers of the liver, pancreas, and testis have been reported in rats ingesting APFO in the diet (Biegel et al., 2001; Butenhoff et al., 2012).

Chemical Pathology

Hepatocytic cytomegaly and liquefaction necrosis was seen in rats orally administered ≥50 mg/kg/day for 21 days (Son et al., 2008). Splenic hypertrophy of white, but not red, pulp was reported in mice administered 250 ppm PFOA in drinking water for 21 days (Son et al., 2009).

Mode of Action(s)

The mode of action for liver toxicity and liver, pancreas, and testis cancer in rodents appears to be mediated via PPAR-α expression resulting in changes in lipid metabolism (Klaunig et al., 2012).

Lipid and glucose homeostatic regulation in fetal mice appear to be perturbed by PFOA via a PPAR-α expression (Abbott et al., 2012). However, mammary gland development in mice by PFOA is mediated by increased steroid hormone production in ovaries and increased levels of mammary gland growth factors, independent of PPAR-α expression (Zhao et al., 2010).

SOURCES OF EXPOSURE

Occupational

Occupational exposure to PFOA may occur via inhalation and dermal contact at manufacturing facilities where it is produced or used.

Environmental

Blood monitoring data indicate that the general population has been exposed to PFOA, and may be further exposed due to inhalation of ambient air, dietary and drinking water ingestion, and dermal contact with consumer products that contain PFOA.

INDUSTRIAL HYGIENE

A full-face particle respirator (type N100 or type P3) should be used as a backup to engineering controls. Gloves, a face shield, and safety glasses or goggles should be donned when working with PFOA.

RISK ASSESSMENTS

A risk assessment of PFOA performed by ATSDR (2009) did not provide derivations of toxicity/risk values for PFOA. The EPA is currently performing a risk assessment of PFOA (EPA, 2013b). A risk assessment has been published in the peer-reviewed literature, including a serum PFOA-based margin of exposure (MOE) analysis, identifying an MOE of 1600 for liver weight increases and 8900 for Leydig cell adenomas (Butenhoff et al., 2004).

REFERENCES

Abbott, B.D., Wood, C.R., Watkins, A.M., Tatum-Gibbs, K., Das, K.P., and Lau, C. (2012) Effects of perfluorooctanoic acid (PFOA) on expression of peroxisome proliferator-activated receptors (PPAR) and nuclear receptor-regulated genes in fetal and postnatal CD-1 mouse tissues. *Reprod. Toxicol.* 33(4):491–505.

Agency for Toxic Substances Disease Registry (ATSDR) (2009) *Toxicological Profile for Perfluoroalkyls*, U.S. Department of Health and Human Services.

Biegel, L.B., Hurtt, M.E., Frame, S.R., et al. (2001) Mechanisms of extrahepatic tumor induction by peroxisome proliferators in male CD rats. *Toxicol. Sci.* 60(1):44–55.

Butenhoff, J.L., Gaylor, D.W., Moore, J.A., Olsen, G.W., Rodricks, J., Mandel, J.H., and Zobel, L.R. (2004) Characterization of risk for general population exposure to perfluorooctanoate. *Regul. Toxicol. Pharmacol.* 298(1–3):1–13.

Butenhoff, J.L., Kennedy, G.L. Jr., Chang, S.C., and Olsen, G.W. (2012) Chronic dietary toxicity and carcinogenicity study with ammonium perfluorooctanoate in Sprague-Dawley rats. *Toxicology* 298(1–3):1–13.

Environmental Protection Agency (EPA) (2013a) *ECOTOX Database*. Available at http://cfpub.epa.gov/ecotox/quick_query.cfm.

Environmental Protection Agency (EPA) (2013b) *Website for Perfluorooctanoic acid and fluorinated telomers*. Available at http://www.epa.gov/oppt/pfoa/pubs/pfoarisk.html.

Gilliland, F.D. and Mandel, J.S. (1993) Mortality among employees of a perfluorooctanoic acid production plant. *J. Occup. Med.* 35(9):950–954.

Griffith, F.D. and Long, J.E. (1980) Animal toxicity studies with ammonium perfluorooctanoate. *Am. Ind. Hyg. Assoc. J.* 41(8):576–583.

Kennedy, G.L., Jr. (1985) Dermal toxicity of ammonium perfluorooctanoate. *Toxicol. Appl. Pharmacol.* 81(2):348–355.

Kennedy, G.L. Jr., Hall, G.T., Brittelli, M.R., Barnes, J.R., and Chen, H.C. (1986) Inhalation toxicity of ammonium perfluorooctanoate. *Food Chem. Toxicol.* 24(12):1325–1329.

Klaunig, J.E., Hocevar, B.A., and Kamendulis, L.M. (2012) Mode of Action analysis of perfluorooctanoic acid (PFOA) tumorigenicity and Human Relevance. *Reprod. Toxicol.* 33(4):410–418.

Lau, C., Thibodeaux, J.R., Hanson, R.G., Narotsky, M.G., Rogers, J.M., Lindstrom, A.B., and Strynar, M.J. (2006) Effects of perfluorooctanoic acid exposure during pregnancy in the mouse. *Toxicol. Sci.* 90(2):510–518.

Leonard, R.C. (2006) *Ammonium Perfluorooctanoate: Phase II. Retrospective Cohort Mortality Analyses Related to a Serum Biomarker of Exposure in a Polymer Production Plant,* Wilmington, DE: E.I. du pont de Nemours and Company.

Macon, M.B., Villanueva, L.R., Tatum-Gibbs, K., Zehr, R.D., Strynar, M.J., Stanko, J.P., White, S.S., Helfant, L., and Fenton, S.E. (2011) Prenatal perfluorooctanoic acid exposure in CD-1 mice: low-dose developmental effects and internal dosimetry. *Toxicol. Sci.* 122(1):134–145.

NLM (National Library of Medicine) (2013) *Allyl Chloride. Hazardous Substances Data Bank.* Available at http://toxnet.nlm.nih.gov/cgi-bin/sis/search/f?./temp/~EuKWDd:1.

Okada, E., Sasaki, S., Saijo, Y., Washino, N., Miyashita, C., Kobayashi, S., Konishi, K., Ito, Y.M., Ito, R., Nakata, A., Iwasaki, Y., Saito, K., Nakazawa, H., and Kishi, R. (2012) Prenatal exposure to perfluorinated chemicals and relationship with allergies and infectious diseases in infants. *Environ. Res.* 112:118–125.

Perkins, R.G., Butenhoff, J.L., Kennedy, G.L. Jr., and Palazzolo, M.J. (2004) 13-week dietary toxicity study of ammonium perfluorooctanoate (APFO) in male rats. *Drug Chem. Toxicol.* 27(4):361–378.

Savitz, D.A., Stein, C.R., Bartell, S.M., Elston, B., Gong, J., Shin, H.M., and Wellenius, G.A. (2012) Perfluorooctanoic acid exposure and pregnancy outcome in a highly exposed community. *Epidemiology* 23(3):386–392.

Son, H.Y., Kim, S.H., Shin, H.I., Bae, H.I., and Yang, J.H. (2008) Perfluorooctanoic acid-induced hepatic toxicity following 21-day oral exposure in mice. *Arch. Toxicol.* 82(4):239–246.

Son, H.Y., Lee, S., Tak, E.N., Cho, H.S., Shin, H.I., Kim, S.H., and Yang, J.H. (2009) Perfluorooctanoic acid alters T lymphocyte phenotypes and cytokine expression in mice. *Environ. Toxicol.* 24(6):580–588.

Staples, R.E., Burgess, B.A., and Kerns, W.D. (1984) The embryofetal toxicity and teratogenic potential of ammonium perfluorooctanoate (APFO) in the rat. *Fundam. Appl. Toxicol.* 4(3 Pt 1):429–440.

Steenland, K. and Woskie, S. (2012) Cohort mortality study of workers exposed to perfluorooctanoic acid. *Am. J. Epidemiol.* 176(10):909–917.

Steenland, K., Fletcher, T., and Savitz, D.A. (2010) Epidemiologic evidence on the health effects of perfluorooctanoic acid (PFOA). *Environ. Health Perspect.* 118(8):1100–1108.

White, S.S., Kato, K., Jia, L.T., Basden, B.J., Calafat, A.M., Hines, E.P., Stanko, J.P., Wolf, C.J., Abbott, B.D., and Fenton, S.E. (2009) Effects of perfluorooctanoic acid on mouse mammary gland development and differentiation resulting from cross-foster and restricted gestational exposures. *Reprod. Toxicol.* 27(3–4):289–298.

White, S.S., Stanko, J.P., Kato, K., Calafat, A.M., Hines, E.P., and Fenton, S.E. (2011) Gestational and chronic low-dose PFOA exposures and mammary gland growth and differentiation in three generations of CD-1 mice. *Environ. Health Perspect.* 119(8):1070–1076.

Wolf, C.J., Fenton, S.E., Schmid, J.E., Calafat, A.M., Kuklenyik, Z., Bryant, X.A., Thibodeaux, J., Das, K.P., White, S.S., Lau, C.S., and Abbott, B.D. (2007) Developmental toxicity of perfluorooctanoic acid in the CD-1 mouse after cross-foster and restricted gestational exposures. *Toxicol. Sci.* 95(2):462–473.

Yahia, D., El-Nasser, M.A., bedel-Latif, M., Tsukuba, C., Yoshida, M., Sato, I., and Tsuda, S. (2010) Effects of perfluorooctanoic acid (PFOA) exposure to pregnant mice on reproduction. *J. Toxicol. Sci.* 35(4):527–533.

Zhao, Y., Tan, Y.S., Haslam, S.Z., and Yang, C. (2010) Perfluorooctanoic acid effects on steroid hormone and growth factor levels mediate stimulation of peripubertal mammary gland development in C57BL/6 mice. *Toxicol. Sci.* 115(1):214–224.

60

AROMATIC HYDROCARBONS

Raymond D. Harbison, Amora Mayo-Perez, David R. Johnson, and Marie M. Bourgeois

BENZENE

Target Organ(s): Eyes, skin, respiratory system, blood, CNS, bone marrow.

Occupational exposure limits: NIOSH REL: Ca; TWA: 0.1 ppm; ST: 1 ppm. OSHA PEL: TWA: 1.0 ppm; ST: 5 ppm.

Risk/safety phrases: Leukemia.

First aid: Eye: Immediate irrigation; skin: soap wash immediately; breath: respiratory support; swallow: seek medical attention immediately.

Conversion: 1 ppm = 3.19 mg/m^3.

BACKGROUND AND USES

Benzene is an industrial solvent, which can also be found in cigarette smoke, petroleum, and as a consequence of biomass combustion. Exposure at high levels is generally from occupational sources such as shoe production and petrochemical factories. Airborne benzene is the most common route of exposure. Even at low doses of airborne benzene (<1 parts per million (ppm)), urban populations—especially those working around car exhaust fumes such as taxi drivers, bus drivers, traffic policemen, and gasoline station attendants—show higher levels of benzene metabolites in the urine such as S-phenylmercapturic acid (SPMA) and *trans, trans*-muconic acid (ttMA) (Arnold et al., 2013; Fustinoni et al., 2010; Rappaport et al., 2013). Although inhalation exposure is the most common, dermal exposure is also prevalent. As a result of inhalation exposure, lungs are the major site of benzene metabolism (Arnold et al., 2013).

PHYSICAL AND CHEMICAL PROPERTIES

Benzene (benzol, phenyl hydride, CASRN 71-43-2) is a colorless to light-yellow liquid with a sweet, aromatic odor. Its solubility in water is 0.07% and it has a vapor pressure of 75 mmHg. It is highly flammable and soluble in water at 1.8 g/l. It has a specific gravity of 0.8787 at 15 °C and a vapor pressure of 95.2 mmHg. It has a flashpoint of −11 °C, with a melting point and boiling point of 5.5 °C and 80.1 °C, respectively (IPCS, 2004).

In the environment, benzene partitions to air and photodegrades in approximately 1 week. It has been known to persist for 10–20 days in soil and groundwater under aerobic conditions. Under anaerobic conditions, it may persist for many years. It can be metabolized by the avocado fruit or grapes to carbon dioxide. It is also found to occur naturally in some foods. It can undergo rapid volatilization near the soil surface and in water. Biodegradation occurs in aerobic water conditions but not under anaerobic. It will not be expected to significantly adsorb to the sediment, bioconcentrate in aquatic organisms, or hydrolyze. Varying experiments show that the half-life of benzene is from 1 week to 17 days due to photodegradation. Atmospheric benzene exists predominantly in the vapor phase. Gas phase benzene will not be subject to direct photolysis, but it will react with photochemically produced hydroxyl radicals with a half-life of 13.4 days calculated using an experimental rate constant for the reaction. The reaction time in polluted atmospheres that contain nitrogen oxides or sulfur dioxide is accelerated with the half-life being reported as 4–6 h. Products of photooxidation include phenol, nitrophenols, nitrobenzene, formic acid, and peroxyacetyl nitrate. Benzene is fairly soluble in

Hamilton & Hardy's Industrial Toxicology, Sixth Edition. Edited by Raymond D. Harbison, Marie M. Bourgeois, and Giffe T. Johnson.
© 2015 John Wiley & Sons, Inc. Published 2015 by John Wiley & Sons, Inc.

water and gets removed from the atmosphere during rains. The primary routes of exposure are inhalation of contaminated air, especially in areas with high traffic, and in the vicinity of gasoline service stations and consumption of contaminated drinking water (USEPA, 2009).

MAMMALIAN TOXICOLOGY

Acute Effects

At high concentrations, benzene has a narcotic effect on the central nervous system (CNS). Inhalation of 7500 ppm for 30 min to 1 h may cause loss of consciousness and death from respiratory failure. Levels of 3000 ppm may cause euphoria and excitation followed by fatigue, dizziness, and nausea. Acute effects include light-headedness, headache, excitement, unsteady gait, euphoria, confusion, nausea, vertigo, and drowsiness. If exposure continues, respiratory paralysis and death, sometimes with convulsions, may occur. Inhalation is the most common route of exposure, but dermal exposure can cause edema, burning, and blistering.

In one incident, 15 shipyard workers were exposed to concentrations of benzene over 653 ppm. These workers reported mucous membrane irritation, peculiar taste, dyspnea, dizziness or light-headedness, nausea, headache, cough, drowsiness, fatigue, and skin irritation. One year later, 40% had abnormally high levels of large granular lymphocytes in their hematological profiles (Midzenski et al., 1992).

Chronic Effects

The chronic effects of benzene exposure depend on the level of exposure, and the symptoms are typically vague. Enlarged liver and spleen, loss of appetite, diarrhea, pallor, pain in bones, fever, and hemorrhage have all been reported. Chronic benzene intoxication creates a disturbance in the hematopoietic system, which may affect every cell line (Nwosu et al., 2004). Chronic exposure to benzene has a depressive effect on the bone marrow (Galbraith et al., 2010; Schnatter et al., 2005). Some studies attribute increased risk of acute myelogenous leukemia (AML), aplastic anemia, and myelodysplasia to benzene exposure (Aksoy et al., 1976, 1985; Crump, 1994; Galbraith et al., 2010; Infante et al., 1977; Infante, 2006; Rinsky, 1989; Rinsky et al., 1981, 2002). Some have concluded that benzene increases the incidence of all leukemias. However, the association between benzene exposure and chronic myelogenous leukemia and multiple myeloma is not strong. Additionally, there is a long estimated latency period between benzene exposure and the development of a malignancy (Finkelstein, 2000; Galbraith et al., 2010; Huff et al., 1989; Schnatter et al., 2012; Sharpe, 1993). Benzene-induced leukemia has been associated with a latency period up to 20 years, with the strongest correlation within the first

10 years (Finkelstein, 2000). A cohort of rubber hydrochloride (Pliofilm) workers exposed to benzene without concomitant exposure to other possible carcinogens has served as the basis for estimations of the potency of benzene to produce cancer (particularly AML) in humans. Among the 1868 members of the cohort with benzene exposures beginning in the 1930s and ending in 1965, there have been 481 total deaths as of December 31, 1987, of which 72 were caused by cancer. Among the cancer deaths, 14 were caused by leukemia. For the leukemia deaths, benzene exposure began during or before 1950 and, exposures were considerably higher than after 1950. This information led Paxton et al. (1994) to conclude:

> The newly gathered information continues to be consistent with a threshold model for leukemogenesis by benzene. The leukemia deaths in the entire cohort occurred exclusively among individuals who commenced their work in Pliofilm production in 1950 or earlier. This suggests that the initial occupational environments at the plants in both Akron and St. Mary may have differed from those to which workers starting their Pliofilm employment later may have been exposed.

This conclusion was based on an analysis of standardized mortality ratios (SMRs). Paxton et al. (1994) analyzed SMR data by cumulative benzene exposure (represented as ppm-years benzene exposure). Among the three exposure characterizations that have been developed for the Pliofilm cohort, SMRs increase as the dose increases. However, there was a distinct relationship between cumulative dose and SMR that Paxton et al. (1994) described:

> Note, however, that for none of the three sets of exposure estimates is there a statistically significant increase in the SMRs for cumulative exposures less than 50 ppm-years. This is consistent with the hypothesis that exposure in excess of some threshold value, here suggested to be greater than 50 ppm-years, is necessary for leukemogenesis.

Schnatter et al. (1996) have also concluded that the risk of leukemogenesis is shown only above a critical concentration of benzene exposure of 50–60 ppm using a median exposure estimate. In 2012, Schnatter and his colleagues continued the work on benzene exposure.

The Pliofilm cohort has been reevaluated periodically. Rinsky et al. (2002) reported that the relative risk of leukemia from chronic exposure diminished over time. Since 6 of the 15 workers' deaths due to leukemia were >30 years after exposure, the likelihood of the occupational exposure playing a role in the disease is low. This is due to the latency period estimated for benzene that is much less than 30 years (Galbraith et al., 2010). This was further supported by the mortality studies of California petrochemical workers and French utility workers who showed no significant increase in mortality among these workers from leukemia (Satin et al., 2002). Although Guenel et al. (2002)

found an increased risk of leukemia among workers whose first exposure was before 1964, it was not significant.

Further studies of benzene occupational exposures conclude that leukemia incidence is not statistically different from the general population as seen in the studies performed by Wong (2002), Lewis et al. (2003), and Gun et al. (2006). These studies found that total leukemia incidence and total leukemia mortality are not significantly elevated; additionally two factors such as the duration of employment and magnitude of exposure do not seem to be related to leukemia incidence (Richardson, 2008). Glass et al. (2006) did find that there was a significant increase in the risk of acute nonlymphatic leukemia in workers with exposures >8ppm-years. Lin et al. (2007) also performed a cohort mortality study that included more than 10,621 employees from 1948 to 2003 and found that there was no significant increase in leukemia or leukemia subtype among these workers.

Hematotoxicity   Inhalation exposure targets the hematopoietic system, which can lead to the depression of multiple lineage blood cells known as pancytopenia. Most studies test the effects of higher exposure to the hematopoietic system with outcomes on complete blood counts (CBC) of the participants. Three major blood components are usually monitored, namely, erythrocytes, leukocytes, and platelets; however, the lymphocytes, red blood cells, and hemocrit are also monitored. Rothman et al. (1995) found a decrease in the lymphocytes at a median concentration of 31 ppm with an 8-h time-weighted average (TWA) and those workers exposed to 8 ppm concentration at an 8-h TWA.

The hematologic parameters of this study (total white blood cells [WBC], absolute lymphocyte count, platelets, red blood cells, and hematocrit) were decreased among workers exposed to 31 ppm concentration of benzene compared to the controls. However, workers exposed to lower levels (i.e., 8 ppm) had a higher red blood cell mean corpuscular volume (MCV) than the controls.

Physiologically based pharmacokinetic (PBPK) modeling can be used to determine key toxicological end points. There are some limitations as this model does not accurately determine the fate of benzene metabolites nor repeated exposures; this model also cannot predict urinary benzene—it is best with blood benzene concentrations. Although in several species this model can simulate acute inhaled benzene (Arnold et al., 2013). High exposures to benzene over many years consistently show an increased risk of AML; however, a consistent causal relationship has not been established (Galbraith et al., 2010). Benzene is observed to have as a dose-dependent effect on biological processes (McHale et al., 2010, 2012).

Genotoxicity   One hundred and twenty-five workers exposed to various levels (24.7, 7.2, 0.8, and 0.3 ppm) of benzene were monitored to determine the gene expression profiles of these individuals over the controls. In the study by McHale et al. (2010), various genes were altered compared to controls. These genes represented impacts to the cellular pathways such as T-cell receptor signaling, oxidative phosphorylation, AML, apoptosis, B-cell receptor signaling pathway, and Toll-like receptor signaling. Not surprisingly, genes in the AML pathway are strongly associated with multiple exposure levels of benzene (Glatt et. al., 1989; Nwosu et al., 2004). Additionally, benzene has been shown to have adverse effects on oxidative stress and mitochondria (McHale et al., 2010; Whysner et al., 2004).

Mechanisms of Action

Typically *in vivo* experiments are used to elucidate the possible mechanisms of toxicity of various compounds in humans. Unfortunately, no animal model closely approximates benzene's toxicity in humans. Tremendous differences exist in benzene toxicity and metabolism between experimental animals. Mice have a much higher rate of benzene metabolism and higher levels of biomarkers for metabolic pathways leading to toxic metabolites than rats (Schlosser et al., 1993). Benzene was pancarcinogenic to mice exposed orally and by inhalation in one study (Huff et al., 1989). Benzene in drinking water has also been studied in transgenic mice, and it suppressed the proliferation of progenitor cells due to altered gene expression (Galbraith et al., 2010; Paustenbach et al., 1993). However, cancers developed by mice in long-term studies did not remotely resemble cancers that benzene is suspected of causing in humans (Smith et al., 2007; Snyder, 2007; Henderson et al., 1991). There are also significant differences in pharmacokinetics between various animal species and humans (Nakajima et. al., 1993). For instance, Sato and Nakajima (1979) found that combined exposures to toluene and benzene slowed the rate of metabolism of each in rats. Similar experiments with humans exposed to levels close to the threshold limit value (TLV) found no such antagonistic relationship. Differences in benzene metabolism between male and female rats may be attributed to differences in cytochrome P450 2E1 (CYP2E1) activity (Kenyon et al., 1996; Tamie et al., 1993). One commonality between experimental animals studied and humans is that benzene pathways are saturated at a low exposure level and yield high affinity toxic metabolites (Henderson et al., 1991).

A brief overview of benzene metabolism as described by Snyder and Hedli (1996) illustrates the biotransformation of benzene by the liver enzyme CYP2E1 that adds an oxygen to yield benzene oxide, which spontaneously rearranges to benzene oxepin and finally to phenol (Steinmaus et al., 2008; Weed, 2010) Benzene oxide may be transformed to a metabolite found in urine as a biomarker, ttMA (Snyder et. al., 1993; Snyder, 2012). The candidate mechanisms for benzene hematotoxicity include the generation of reactive oxygen species, direct chromosomal damage, inhibition of topoisomerase II, immune dysfunction, DNA methylation, and the accumulation of toxic metabolites in the liver (Galbraith et al., 2010; McHale et al., 2010; Steinmaus et al., 2008). Numerous *in vitro* studies have investigated the effect of benzene and its metabolites on human cells. HL60 cells, a leukemia cell line, have attracted special attention. Zhang et al. (1993) demonstrated that HL60 cells exposed to the benzene metabolite 1,2,4-benzenetriol *in vitro* had micronuclei and oxidative damage that was potentiated by copper ions. Kolachana et al. (1993) found that the HL60 cells showed a higher level of oxidative deoxyribonucleic acid (DNA) damage when exposed to phenol, hydroquinone, and 1,2,4-benzenetriol. Increased oxidative damage was also observed in mice *in vivo*.

Tough and Brown (1965) reported that benzene may have been responsible for chromosomal aberrations found among factory workers. They found chromosomal aberrations in peripheral lymphocytes similar to those produced by radiotherapy, although there was no evidence of exposure to radiation. Yardley-Jones et al. (1990) also found increased chromosomal aberrations in benzene workers. Benzene has also been shown to induce sister chromatid exchanges and chromosomal aberrations in bone marrow cells of mice (Tice et al., 1980). Rothman et al. (1995) report that benzene produces gene-duplicating mutations but not gene-inactivating mutations at the glycophorin A locus in humans exposed to high benzene levels. This evidence supports the theory that chromosomal damage and mitotic recombination are important in the etiology of benzene-induced leukemia.

Seaton et al. (1994) investigated the effect of low level benzene exposures on human liver microsome metabolism. They demonstrated that there is a great variation between individuals in levels of CYP2E1 and that this affects the metabolites produced by the benzene metabolism. Phenol was usually the predominant metabolite, but hydroquinone was the primary metabolite in some individuals. Metabolism by experimental animal liver microsomes fell within the range of human liver microsome metabolism in this study. The differences between various species and strains may be partially explained by the differences in distribution of cytochrome P-450 isozymes (Nakajima et al., 1993). Mutagenicity assays demonstrated that only the *trans*-1,2 dihydrodiol metabolite of benzene exhibited mutagenic activity (Glatt et al, 1989). This activity was not dependent on the activation by the S9 liver fraction. Several other assays indicate that benzene has a mutagenic potential (Vainio et al., 1985). Benzene can induce aneuploidy and chromosomal aberrations in mice (Nakajima et. al., 1993; Schlosser et. al., 1993).

Although benzene is thought to require bioactivation to cause toxicity, none of its metabolites alone create a similar toxic effect (Medinsky et al., 1994). Combinations of these metabolites are needed to duplicate benzene toxicity. Benzene was not found to be directly embryotoxic to rat fetuses (Chapman et al., 1994).

However, phenol, a primary benzene metabolite, was found to be embryotoxic (Chapman et. al., 1994).

Benzene is not only metabolized in the liver but also in the lungs during the common route of exposure, inhalation. Cytochrome P450 2E1 is an important enzyme that metabolizes benzene. Although other enzymes such as CYP2F1 are also involved in metabolism, CYP2E1 has the highest activity toward benzene (McHale et al., 2012; Sheets et al., 2004). Benzene is oxidized by CYP2E1 to an intermediate molecule, benzene oxide. A series of reactions occurs with intermediate metabolites such as phenol, hydroquinone, catechol, trihydroxybenzene, and -benzoquinones (BQs) (Arnold et al., 2013; McHale et al., 2012). The elimination of benzene and its metabolites varies depending on the route of exposure. However, all metabolites are excreted in the urine no matter the route of exposure. Inhaled benzene has now been proposed to be metabolized in the lung by CYP2F1 and CYP2A13 at low doses (Klaunig, 2010; McHale et al., 2012).

McHale et al. (2012) has proposed a mode of action (MOA) approach to determine the risk assessment. Although it has its limitations, there are five key events associated with benzene toxicity: (a) metabolism of benzene to a benzene oxide metabolite; (b) interaction of benzene metabolites with target cells; (c) formation of mutated target cells; (d) selected clonal proliferation of target cells; and (e) formation of the neoplasm (Lamm et al., 2009, 2005).

Chemical Pathology

Benzene carcinogenicity is thought to be due to 1,4-benzoquinone (BQ), a metabolite of benzene (McHale et al., 2012). NAD(P)H:quinine oxidoreductase 1 (NQO1) has been shown to protect mice against benzene-induced myelodysplasia and can thus prevent benzene hematotoxicity (McHale et al., 2012).

SOURCES OF EXPOSURE

Occupational

Occupational exposures have been examined extensively in case-control cohort studies around the world. Since 1950, workplace benzene concentrations have been reduced significantly in the United States and other Western societies (Galbraith et al., 2010). As a result, the workers exposed to benzene and benzene-related illnesses have consistently been reduced. However, in developing countries where occupational and environmental controls are more lax, health-related illness due to benzene remains a measurable risk (Galbraith et al., 2010; Ruchirawat et al., 2010). Low level exposures have been evaluated as well, which also result in a lower incidence of adverse effects (Galbraith et al., 2010; Raabe and Wong, 1996; Williams et al., 2011).

Environmental

There is mounting evidence that most people are exposed to a sufficiently small dose each day, which will not increase the risk of disease (Johnson et al., 2009; Galbraith et al., 2010). Kim et al. (2006) has shown that benzene metabolism is different in those exposed to low doses versus workers consistently exposed to higher concentrations. Peak exposure periods are considered between 15 and 30 min per day (Djurhuus et al., 2012). These exposures generally are as low as 10 ppm. However, these peak exposures are not appreciably different from 8 ppm-h (Irons, 2000; Knutsen et al., 2013; Rappaport et al., 2009). Environmental exposure from non-occupational sources are typically below $50 \mu g/m^3$; most of these daily exposures are from car exhaust and first- and secondhand smoke (Capleton and Levy, 2005; Liu et al., 2009).

INDUSTRIAL HYGIENE

In the late 1800s and early 1900s, acute exposures to benzene in the United States and Europe were not unusual (Sharp, 1993). Exposures were commonplace in leather making, shoemaking, and rubber manufacture. Workers exposed to benzene inside factories are generally at greater risk than petrochemical workers who typically work outside with better ventilation. Those temporarily exposed to benzene in confined spaces, such as those who degas an oil tanker, are at especially a high risk (Midzenski et al., 1992).

The American Conference of Governmental Industrial Hygienists (ACGIH) (ACGIH, 2009) intends to adopt a benzene-skin TLV of 0.5 ppm ($1.6 mg/m^3$). The ACGIH also plans to change the carcinogen rank of benzene from A2 to A1 (suspected to confirmed human carcinogen). The National Institute for Occupational Safety and Health (NIOSH) classifies benzene as a human carcinogen. The recommended exposure limit (REL) is 0.1 ppm, with a short-term exposure limit (STEL) of 1 ppm for 15 min ($1 ppm = 3.19 mg/m^3$). The Occupational Safety and Health Administration (OSHA) permissible exposure limit (PEL) is 1 ppm with a STEL of 5 ppm. The notable exception is for some employees of the petrochemical industry for which the standards are a PEL of 10 ppm with a ceiling of 25 ppm and a 50 ppm STEL for a maximum of 10 min (ICS, 1993). A benzene level of 500 ppm is considered Immediately Dangerous to Life and Health (IDLH) (ATSDR, 2008; NIOSH, 2007).

Biomonitoring The current OSHA standards have reduced benzene exposure to industrial employees. For routine biomonitoring, the ACGIH recommends measurement of end of shift total urinary phenols or benzene in exhaled air. Urinary ttMA is a major metabolite of benzene, and thus is used as a biomarker (Tunsaringkarn et al., 2013). Blood concentrations of hemoglobin, hemocrit, and eosinophils are all decreased as ttMA urine levels increase (Tunsaringkarn et al., 2013). A level of 20 mg phenol/g creatinine in urine can be expected in a worker consistently exposed to the TLV (ACGIH, 2009). The biological exposure index (BEI) for benzene in exhaled air (measured before next shift) of workers exposed to the TLV should be as follows: mid-exhaled = 0.08 ppm, end exhaled = 0.12 ppm (ACGIH, 2009). A previous controlled experiment indicates that this BEI may not accurately predict the TLV exposure and therefore may not protect the workers (Thomas et al., 1996).

Although biomonitoring through urine and exhaled air may be the most common methods, they are not the most accurate, especially for low levels of exposure. Biomonitoring of urinary phenol may be confounded by some over-the-counter drugs such as Pepto-Bismol® and Chloraseptic® lozenges. These products contain zinc phenyl sulfonate and sodium phenolate, respectively, both of which have been shown to increase urinary phenol levels and interfere with biomonitoring (Fishbeck et al., 1975). Dietary factors may also have a significant impact on urinary phenol (Fishbeck et. al., 1975; Ling and Hänninen, 1992). Another factor confounding the biomonitoring of benzene is an exposure to complex mixtures of chemicals, especially other solvents (Fustinoni et. al., 2010; Ong and Lee, 1994). Studies have shown that an exposure to toluene and xylene significantly complicate the estimation of benzene exposure (Ikeda, 1995).

High performance liquid chromatography (HPLC) determination of ttMA has been shown to be a more accurate indicator of benzene exposure at low concentrations than current standard methods. Direct measurement of benzene in the urine and blood also closely reflects low levels of exposure (Ong and Lee, 1994).

Risk assessment for benzene is complex because of the mechanism of its toxicity (Yardley-Jones et al., 1991). This is complicated further by the low incidence of leukemias, which are suspected to result from chronic benzene exposure (Glass et. al., 2004; Glass et. al. 2005). As mentioned, animal studies are of limited usefulness when trying to elucidate the actual toxicity of benzene in humans.

The lack of a satisfactory animal model and the availability of industrial exposure data have provoked an unprecedented number of risk assessments based on the epidemiological data. The scope of epidemiological studies and risk assessments is too extensive to be discussed here in their entirety. There have been several epidemiological studies of benzene and leukemia risk, although the epidemiological relationship between benzene and leukemia is tenuous (Aksoy et. al., 1985; Infante et al., 1977; Ott et al., 1978; Rinsky et al., 1981; Wong, 1987; Alexander and Wagner, 2010; Galbraith et al., 2010; National Cancer Institute, 2005). The scientific consensus is that Rinsky's data from the Ohio Pliofilm cohort are the most reliable, even though they have some shortcomings. The data from Rinsky et al. (1987) have been reanalyzed for risk assessment many times by the authors and others (Austin et al.,

1988; Crump, 1994; Utterback and Rinsky, 1995). Some claim that the Rinsky data seriously underestimate the levels of exposure at the Pliofilm plants studied, thereby overestimating the risk from low level exposure. Dermal exposures were not taken into account for the Rinsky data; however, many of these claims of higher exposure are based on observations by Pliofilm employees that are of questionable scientific relevance. Depending on the assumptions of the model, the estimated risk from benzene exposure may increase or decrease by an order of magnitude (Bird et al., 2010; Brett et al., 1989; Kane and Newton, 2010; Khalade et al., 2010).

A cohort study from China appears to add new data that may reduce the limitations of the Pliofilm cohort studies (Yin et al., 1994). Perhaps this study may serve as a basis for more risk assessments with greater certainty on the effects of benzene exposure. The problem of risk assessment for benzene will almost certainly continue to provoke numerous epidemiological studies and risk analysis.

MEDICAL MANAGEMENT

A biomarker for benzene is urinary benzene for low exposure testing. The test for urinary benzene has been shown to be relevant at exposures as low as $10\,\mu g/m^3$ (Fustinoni et al., 2012; Lovreglio et al., 2010). Diagnosis of chronic benzene poisoning should include blood cell analysis because of the vague nature of the associated symptomatology. Benzene can adversely alter the hematopoietic system producing aplastic anemia, leukopenia, and thrombocytopenia (Khan, 2007).

Treatment of benzene exposure is primarily symptomatic. If benzene is inhaled, patients should be given oxygen and evaluated for cough, respiratory irritation, bronchitis, and pneumonia. Blood transfusions and corticosteroids may also be given in cases of severe anemia.

In cases of ingestion, induced vomiting is not advised. Gastric lavage may be performed shortly after ingestion.

REFERENCES

American Conference of Governmental Industrial Hygienists (ACGIH) (2009) *Threshold Limit Values (TLVs) for Chemical Substances and Physical Agents Biological Exposure Indices (BEIs)*, Cincinnati, OH: American Conference of Governmental Industrial Hygienists.

Aksoy, M. (1985) Malignancies due to occupational exposure to benzene. *Am. J. Ind. Med.* 7(5–6):395–402.

Aksoy, M., Erdem, S., and Dincol, G. (1976) Types of leukemia in chronic benzene poisoning. A study in thirty-four patients. *Acta Haematol.* 55(2):65–72.

Alexander, D.D. and Wagner, M.E. (2010) Benzene exposure and non-Hodgkin lymphoma: a meta-analysis of epidemiologic studies. *J. Occup. Environ. Med.* 52(2):169–189.

Arnold, S.M., Angerer, J., Boogaard, P.J., Hughes, M.F., O'Lone, R.B., Robison, S.H., and Robert Schnatter, A. (2013) The use of biomonitoring data in exposure and human health risk assessment: benzene case study. *Crit. Rev. Toxicol.* 43(2):119–153.

Austin, H., Delzell, E., and Cole, P. (1988) Benzene and leukemia. A review of the literature and a risk assessment. *Am. J. Epidemiol.* 127(3):419–439.

Bird, M.G., Grein, H., Kaden, D.A., Rice, J.M., and Snyder, R. (2010) Benzene 2009-Health effects and mechanisms of bone marrow toxicity: implications for t-AML and the mode of action framework. Proceedings of a meeting. September 7–12, 2009. Munich, Germany. *Chem. Biol. Interact.* 184 (1–2):1–312.

Brett, S.M., Rodricks, J.V., and Chinchilli, V.M. (1989) Review and update of leukemia risk potentially associated with occupational exposure to benzene. *Environ. Health Perspect.* 82:267–281.

Capleton, A.C. and Levy, L.S. (2005) An overview of occupational benzene exposures and occupational exposure limits in Europe and North America. *Chem. Biol. Interact.* 153–154:43–53.

Chapman, D.E., Namkung, M.J., and Juchau, M.R. (1994) Benzene and benzene metabolites as embryotoxic agents: effects on cultured rat embryos. *Toxicol. Appl. Pharmacol.* 128(1):129–137.

Chen, J.D., Wang, J.D., Jang, J.P., and Chen, Y.Y. (1991) Exposure to mixtures of solvents among paint workers and biochemical alterations of liver function. *Br. Jour. Ind. Med.* 48:696–701.

Crump, K.S. (1994) Risk of benzene-induced leukemia: a sensitivity analysis of the pliofilm cohort with additional follow-up and new exposure estimates. *J. Toxicol. Environ. Health* 42(2):219–242.

Djurhuus, R., Nossum, V., Ovrebo, S., and Skaug, V. (2012) Proposal on limits for chemical exposure in saturation divers' working atmosphere: the case of benzene. *Crit. Rev. Toxicol.* 42(3):211–229.

Finkelstein, M.M. (2000) Leukemia after exposure to benzene: temporal trends and implications for standards. *Am. J. Ind. Med.* 38(1):1–7.

Fishbeck, W., Langner, R., and Kociba, R. (1975) Elevated urinary phenol levels not related to benzene exposure. *Am. Ind. Hyg. Assoc. J.* 36(11):820–824.

Fustinoni, S., Campo, L., Mercadante, R., and Manini, P. (2010) Methodological issues in the biological monitoring of urinary benzene and S-phenylmercapturic acid at low exposure levels. *J. Chromatogr. B Analyt. Technol. Biomed. Life Sci.* 878(27):2534–2540.

Fustinoni, S., Campo, L., Satta, G., Campagna, M., Ibba, A., Tocco, M., Atzeri, S., Avataneo, G., Flore, C., Meloni, M., Bertazzi, P., and Cocco, P. (2012) Environmental and lifestyle factors affect benzene uptake biomonitoring of residents near a petrochemical plant. *Environ. Int.* 39:2–7.

Galbraith, D., Gross, S.A., and Paustenbach, D. (2010) Benzene and human health: a historical review and appraisal of associations with various diseases. *Crit. Rev. Toxicol.* 40(S2):1–46.

Glass, D.C., Sim, M.R., Fritschi, L., Gray, C.N., Jolley, D.J., and Gibbons, C. (2004) Leukemia risk and relevant benzene exposure period-Re: follow-up time on risk estimates, Am J Ind Med 42:481–489, 2002. *Am. J. Ind. Med.* 45: (2):222–222, author reply 224–5, 10.1002/ajim.10327.

Glass, D.C., Gray, C.N., Jolley, D.J., Gibbons, C., and Sim, M.R. (2005) Health Watch exposure estimates: do they underestimate benzene exposure? *Chem. Biol. Interact.* 153–154:23–32.

Glass, D.C., Gray, C.N., Jolley, D.J., Gibbons, C., and Sim, M.R. (2006) The health watch case-control study of leukemia and benzene: the story so far. *Ann. N Y Acad. Sci.* 1076:80–9.

Glatt, H., Padykula, R., Berchtold, G.A., Ludewig, G., Platt, K.L., Klein, J., and Oesch, F. (1989) Multiple activation pathways of benzene leading to products with varying genotoxic characteristics. *Environ. Health Perspect.* 82:81–89.

Guenel, P., Imbernon, E., Chevalier, A., Crinquand-Calastreng, A., and Goldberg, M. (2002) Leukemia in relation to occupational exposures to benzene and other agents: a case-control study nested in a cohort of gas and electric utility workers. *Am. J. Ind. Med.* 42(2):87–97.

Gun, R.T., Pratt, N., Ryan, P., and Roder, D. (2006) Update of mortality and cancer incidence in the Australian petroleum industry cohort. *Occup. Environ. Med.* 63(7):476–481.

Henderson, R.F., Sabourin, P.J., Medinsky, M., Birnbaum, L., and Lucier, G.L. (1991) Benzene dosimetry in experimental animals: relevance for risk assessment. *Prog. Clin. Biol. Res.* 374:93–105.

Huff, J., Haseman, J.K., DeMarini, D.M., Eustis, S., Maronpot, R.R., Peters, A.C., Persing, R.L., Chrisp, C.E., and Jacobs, A.C. (1989) Multiple-site carcinogenicity of benzene in Fischer 344 rats and B6C3F1 mice. *Environ. Health Perspect.* 82:125–163.

Ikeda, M. (1995) Exposure to complex mixtures: implications for biological monitoring. *Toxicol. Lett.* 77(1):85–91.

Infante, P.F. (2006) Benzene exposure and multiple myeloma: a detailed meta-analysis of benzene cohort studies. *Ann. N. Y. Acad. Sci.* 1076:90–109.

Infante, P., Wagoner, J., Rinsky, R., and Young, R. (1977) Leukaemia in benzene workers. *Lancet* 310(8028):76–78.

IPCS (1993) *Benzene. Poisons Information Monograph 63. International Programme on Chemical Safety*, Geneva: World Health Organization.

IPCS (2004) Benzene. Geneva, World Health Organization, International Programme on Chemical Safety (International Chemical Safety Card 0015).

Irons, R.D. (2000) Molecular models of benzene leukemogenesis. *J. Toxicol. Environ. Health A* 61(5–6):391–397.

Johnson, G.T., Harbison, S.C., McCluskey, J.D., and Harbison, R.D. (2009) Characterization of cancer risk from airborne benzene exposure. *Regul. Toxicol. Pharmacol.* 55:361–366.

Kane, E.V. and Newton, R. (2010) Benzene and the risk of non-Hodgkin lymphoma: a review and meta-analysis of the literature. *Cancer Epidemiol.* 34(1):7–12.

Kenyon, E.M., Kraichely, R.E., Hudson, K.T., and Medinsky, M.A. (1996) Differences in rates of benzene metabolism correlate with observed genotoxicity. *Toxicol. Appl. Pharmacol.* 136(1):49–56.

Khalade, A., Jaakkola, M.S., Pukkala, E., and Jaakkola, J.J. (2010) Exposure to benzene at work and the risk of leukemia: a systematic review and meta-analysis. *Environ. Health* 9:31.

Khan, H.A. (2007) Short review: benzene's toxicity: a consolidated short review of human and animal studies. *Hum. Exp. Toxicol.* 26(9):677–685.

Kim, S., Vermeulen, R., Waidyanatha, S., Johnson, B.A., Lan, Q., Smith, M.T., Zhang, L., Li, G., Shen, M., Yin, S., Rothman, N., and Rappaport, S.M. (2006) Modeling human metabolism of benzene following occupational and environmental exposures. *Cancer Epidemiol. Biomarkers Prev.* 15(11):2246–2252.

Klaunig, J.E. (2010) Proposed mode of action of benzene-induced leukemia: interpreting available data and identifying critical data gaps for risk assessment. *Chem. Biol. Interact.* 184(1): 279–285.

Knutsen, J.S., Kerger, B.D., Finley, B., and Paustenbach, D.J. (2013) A calibrated human PBPK model for benzene inhalation with urinary bladder and bone marrow compartments. *Risk Anal.* 33(7):1237–1251.

Kolachana, P., Subrahmanyam, V.V., Meyer, K.B., Zhang, L., and Smith, M.T. (1993) Benzene and its phenolic metabolites produce oxidative DNA damage in HL60 cells *in vitro* and in the bone marrow *in vivo*. *Cancer Res.* 53(5):1023–1026.

Lamm, S.H., Engel, A., and Byrd, D.M. (2005) Non-Hodgkin lymphoma and benzene exposure: a systematic literature review. *Chem. Biol. Interact.* 153–154:231–237.

Lamm, S.H., Engel, A., Joshi, K.P., Byrd, D.M. 3rd., and Chen, R. (2009) Chronic myelogenous leukemia and benzene exposure: a systematic review and meta-analysis of the case-control literature. *Chem. Biol. Interact.* 182(2–3):93–97.

Lewis, C.L., Moser, M.A., Dale, D.E. Jr., Hang, W., Hassell, C., King, F.L., and Majidi, V. (2003) Time-gated pulsed glow discharge: real-time chemical speciation at the elemental, structural, and molecular level for gas chromatography time-of-flight mass spectrometry. *Anal. Chem.* 75(9):1983–1996.

Lin, Y.S., Vermeulen, R., Tsai, C.H., Waidyanatha, S., Lan, Q., Rothman, N., Smith, M.T., Zhang, L., Shen, M., Li, G., Yin, S., Kim, S., and Rappaport, S.M. (2007) Albumin adducts of electrophilic benzene metabolites in benzene-exposed and control workers. *Environ. Health Perspect.* 115(1):28–34.

Ling, W.H. and Hänninen, O. (1992) Shifting from a conventional diet to an uncooked vegan diet reversibly alters fecal hydrolytic activities in humans. *J. Nutr.* 122(4):924–930.

Liu, H., Liang, Y., Bowes, S., Xu, H., Zhou, Y., Armstrong, T.W., Wong, O., Schnatter, A.R., Fang, J., Wang, L., Nie, L., Fu, H., and Irons, R. (2009) Benzene exposure in industries using or manufacturing paint in China—a literature review, 1956–2005. *J.Occup. Environ. Hyg.* 6(11):659–670.

Lovreglio, P., Barbieri, A., Carrieri, M., Sabatini, L., Fracasso, M.E., Doria, D., Drago, I., Basso, A., D'Errico, M.N., Bartolucci, G.B., Violante, F.S., and Soleo, L. (2010) Validity of new biomarkers of internal dose for use in the biological monitoring of occupational and environmental exposure to low concentrations of benzene and toluene. *Int. Arch. Occup. Environ. Health* 83(3):341–356.

McHale, C.M., Zhang, L., Hubbard, A. E., and Smith, M.T. (2010) Toxicogenomic profiling of chemically exposed humans in risk assessment. *Mutat. Res.* 705(3):172–183.

McHale, C.M., Zhang, L., and Smith, M.T. (2012) Current understanding of the mechanism of benzene-induced leukemia in humans: implications for risk assessment. *Carcinogenesis* 33(2):240–252.

Medinsky, M.A., Schlosser, P.M., and Bond, J.A. (1994) Critical issues in benzene toxicity and metabolism: the effect of interactions with other organic chemicals on risk assessment. *Environ. Health Perspect.* 102(Suppl. 9):119–124.

Midzenski, M.A., McDiarmid, M.A., Rothman, N., and Kolodner, K. (1992) Acute high dose exposure to benzene in shipyard workers. *Am. J. Ind. Med.* 22(4):553–565.

Nakajima, T., Wang, R.S., Elovaara, E., Park, S.S., Gelboin, H.V., and Vainio, H. (1993) Cytochrome P450-related differences between rats and mice in the metabolism of benzene, toluene and trichloroethylene in liver microsomes. *Biochem. Pharmacol.* 45(5):1079–1085.

National Cancer Institute (2005) *Occupational Exposure to Benzene and Risk of Leukemia and Lymphoma*, U.S. Public Health Service; U.S. Department of Health and Human Services.

National Institute for Occupational Safety and Health (NIOSH) (2007) *Pocket Guide to Chemical Hazards*, NIOSH 94-116, Washington, DC: US Department of Health and Human Services.

Nwosu, V.C., Kissling, G.E., Trempus, C.S., Honeycutt, H., and French, J.E. (2004) Exposure of Tg.AC transgenic mice to benzene suppresses hematopoietic progenitor cells and alters gene expression in critical signaling pathways. *Toxicol. Appl. Pharmacol.* 196(1):37–46.

Ong, C.-N. and Lee, B.-L. (1994) Determination of benzene and its metabolites: application in biological monitoring of environmental and occupational exposure to benzene. *J. Chromatogr. B Biomed. Appl.* 660(1):1–22.

Ott, M.G., Townsend, J.C., Fishbeck, W.A., and Langner, R.A. (1978) Mortality among individuals occupationally exposed to benzene. *Arch. Environ. Health* 33(1):3–10.

Paustenbach, D.J., Bass, R.D., and Price, P. (1993) Benzene toxicity and risk assessment, 1972–1992: implications for future regulation. *Environ. Health Perspect.* 101(Suppl. 6):177–200.

Paxton, M.B., Chinchilli, V.M., Brett, S.M., and Rodricks, J. V. (1994) Leukemia risk associated with benzene exposure in the pliofilm cohort: I. Mortality update and exposure distribution. *Risk Anal.* 14(2):147–154.

Raabe, G.K. and Wong, O. (1996) Leukemia mortality by cell type in petroleum workers with potential exposure to benzene. *Environ. Health Perspect.* 104(Suppl. 6):1381–1392.

Rappaport, S.M., Kim, S., Lan, Q., Vermeulen, R., Waidyanatha, S., Zhang, L., Li, G., Yin, S., Hayes, R.B., and Rothman, N. (2009) Evidence that humans metabolize benzene via two pathways. *Environ. Health Perspect.* 117(6):946–952.

Rappaport, S.M., Kim, S., Thomas, R., Johnson, B.A., Bois, F.Y., and Kupper, L.L. (2013) Low-dose metabolism of benzene in humans: science and obfuscation. *Carcinogenesis* 34(1):2–9.

Richardson, D.B. (2008) Temporal variation in the association between benzene and leukemia mortality. *Environ. Health Perspect.* 116(3):370–374.

Rinsky, R.A., Smith, A.B., Hornung, R., Filloon, T.G., Young, R.J., Okun, A.H., and Landrigan, P.J. (1987) Benzene and leukemia. An epidemiologic risk assessment. *N. Engl. J. Med.* 316(17):1044–1050.

Rinsky, R.A. (1989) Benzene and leukemia: an epidemiologic risk assessment. *Environ. Health Perspect.* 82:189–191.

Rinsky, R.A., Young, R.J., and Smith, A.B. (1981) Leukemia in benzene workers. *Am. J. Ind. Med.* 2(3):217–245.

Rinsky, R., Hornung, R., Silver, S., and Tseng, C. (2002) Benzene exposure and hematopoietic mortality: a long-term epidemiologic risk assessment. *Am. J. Ind. Med.* 42(6):474–480.

Rothman, N., Haas, R., Hayes, R., Li, G., Wiemels, J., Campleman, S., Quintana, P., Xi, L., Dosemeci, M., and Titenko-Holland, N. (1995) Benzene induces gene-duplicating but not gene-inactivating mutations at the glycophorin A locus in exposed humans. *Proc. Natl. Acad. Sci. U. S. A.* 92(9):4069–4073.

Ruchirawat, M., Navasumrit, P., and Settachan, D. (2010) Exposure to benzene in various susceptible populations: co-exposures to 1,3-butadiene and PAHs and implications for carcinogenic risk. *Chem. Biol. Interact.* 184(1–2):67–76.

Satin, K.P., Bailey, W.J., Newton, K.L., Ross, A.Y., and Wong, O. (2002) Updated epidemiological study of workers at two California petroleum refineries, 1950–95. *Occup. Environ. Med.* 59(4):248–256.

Sato, A. and Nakajima, T. (1979) Dose-dependent metabolic interaction between benzene and toluene *in vivo* and *in vitro*. *Toxicol. Appl. Pharmacol.* 48(2):249–256.

Schlosser, P.M., Bond, J.A., and Medinsky, M.A. (1993) Benzene and phenol metabolism by mouse and rat liver microsomes. *Carcinogenesis* 14(12):2477–2486.

Schnatter, A.R., Nicolich, M.J., and Bird, M.G. (1996) Determination of leukemogenic benzene exposure concentrations: refined analyses of the Pliofilm cohort. *Risk Anal.* 16(6):833–840.

Schnatter, A.R., Rosamilia, K., and Wojcik, N.C. (2005) Review of the literature on benzene exposure and leukemia subtypes. *Chem. Biol. Interact.* 153–154:9–21.

Schnatter, A.R., Glass, D.C., Tang, G., Irons, R.D., and Rushton, L. (2012) Myelodysplastic syndrome and benzene exposure among petroleum workers: an international pooled analysis. *J. Natl. Cancer Inst.* 104(22):1724–1737.

Seaton, M.J., Schlosser, P.M., Bond, J.A., and Medinsky, M.A. (1994) Benzene metabolism by human liver microsomes in relation to cytochrome P450 2E1 activity. *Carcinogenesis* 15(9):1799–1806.

Sharpe, W.D. (1993) Benzene, artificial leather and aplastic anemia: Newark, 1916–1928. *Bull. N. Y. Acad. Med.* 69(1):47–60.

Sheets, P.L., Yost, G.S., and Carlson, G.P. (2004) Benzene metabolism in human lung cell lines BEAS-2B and A549 and cells overexpressing CYP2F1. *J. Biochem. Mol. Toxicol.* 18(2):92–99.

Smith, M.T., Jones, R.M., and Smith, A.H. (2007) Benzene exposure and risk of non-Hodgkin lymphoma. *Cancer Epidemiol. Biomarkers Prev.* 16(3):385–391.

Snyder, R. (2007) Benzene's toxicity: a consolidated short review of human and animal studies by HA Khan. *Hum. Exp. Toxicol.* 26(9):687–696.

Snyder, R. (2012) Leukemia and benzene. *Int. J. Environ. Res. Public Health* 9(8):2875–2893.

Snyder, R. and Hedli, C.C. (1996) An overview of benzene metabolism. *Environ. Health Perspect.* 104 Suppl 6:1165–1171.

Snyder, R., Witz, G., and Goldstein, B.D. (1993) The toxicology of benzene. *Environ. Health Perspect.* 100:293–306.

Steinmaus, C., Smith, A.H., Jones, R.M., and Smith, M.T. (2008) Meta-analysis of benzene exposure and non-Hodgkin lymphoma: biases could mask an important association. *Occup. Environ. Med.* 65(6):371–378.

Tamie, N., Rui-Sheng, W., Elovaara, E., Park, S.S., Gelboin, H.V., and Vainio, H. (1993) Cytochrome P450-related differences between rats and mice in the metabolism of benzene, toluene and trichloroethylene in liver microsomes. *Biochem. Pharmacol.* 45(5):1079–1085.

Thomas, R.S., Bigelow, P.L., Keefe, T.J., and Yang, R.S. (1996) Variability in biological exposure indices using physiologically based pharmacokinetic modeling and Monte Carlo simulation. *Am. Ind. Hyg. Assoc. J.* 57(1):23–32.

Tice, R.R., Costa, D.L., and Drew, R.T. (1980) Cytogenetic effects of inhaled benzene in murine bone marrow: induction of sister chromatid exchanges, chromosomal aberrations, and cellular proliferation inhibition in DBA/2 mice. *Proc. Natl. Acad. Sci. U. S. A.* 77(4):2148–2152.

Tough, I. and Brown, W. (1965) Chromosome aberrations and exposure to ambient benzene. *Lancet* 285(7387):684.

Tunsaringkarn, T., Soogarun, S., and Palasuwan, A. (2013) Occupational exposure to benzene and changes in hematological parameters and urinary *trans*, *trans*-muconic acid. *Int. J. Occup. Environ. Med.* 4(1):45–49.

United Sates Environmental Protection Agency (USEPA) (2009) *Integrated Risk Information System (IRIS) on Benzene*, Washington, DC: National Center for Environmental Assessment, Office of Research and Development.

Utterback, D.F. and Rinsky, R.A. (1995) Benzene exposure assessment in rubber hydrochloride workers: a critical evaluation of previous estimates. *Am. J. Ind. Med.* 27(5):661–676.

Vainio, H., Waters, M.D., and Norppa, H. (1985) Mutagenicity of selected organic solvents. *Scand. J. Work Environ. Health* 11 (Suppl. 1):75–82.

Weed, D.L. (2010) Meta-analysis and causal inference: a case study of benzene and non-Hodgkin lymphoma. *Ann. Epidemiol.* 20(5):347–355.

Whysner, J., Reddy, M.V., Ross, P.M., Mohan, M., and Lax, E.A. (2004) Genotoxicity of benzene and its metabolites. *Mutat. Res.* 566(2):99–130.

Williams, P.R., Sahmel, J., Knutsen, J., Spencer, J., and Bunge, A.L. (2011) Dermal absorption of benzene in occupational settings: estimating flux and applications for risk assessment. *Crit. Rev. Toxicol.* 41(2):111–142.

Wong, O. (1987) An industry wide mortality study of chemical workers occupationally exposed to benzene. I. General results. *Br. J. Ind. Med.* 44(6):365–381.

Wong, O., Harris, F., Rosamilia, K., and Raabe, G.K. (2001) An updated mortality study of workers at a petroleum refinery in Beaumont, Texas, 1945 to 1996. *J. Occup. Environ. Med.* 43(4):384–401.

Wong, O. (2002) Investigations of benzene exposure, benzene poisoning, and malignancies in China. *Regul. Toxicol. Pharmacol.* 35(1):126–135.

Yardley-Jones, A., Anderson, D., Lovell, D., and Jenkinson, P. (1990) Analysis of chromosomal aberrations in workers exposed to low level benzene. *Br. J. Ind. Med.* 47(1):48–51.

Yardley-Jones, A., Anderson, D., and Parke, D. (1991) The toxicity of benzene and its metabolism and molecular pathology in human risk assessment. *Br. J. Ind. Med.* 48(7):437–444.

Yin, S., Linet, M., Hayes, R., Li, G., Dosemeci, M., Wang, Y., Chow, W., Jiang, Z., Wacholder, S., and Zhang, W. (1994) Cohort study among workers exposed to benzene in China: I. General methods and resources. *Am. J. Ind. Med.* 26(3):383–400.

Zhang, L., Robertson, M.L., Kolachana, P., Davison, A.J., and Smith, M.T. (1993) Benzene metabolite, 1,2,4-benzenetriol, induces micronuclei and oxidative DNA damage in human lymphocytes and HL60 cells. *Environ. Mol. Mutag.* 21(4):339–348.

PHENOL

Target organ(s): Eyes, skin, respiratory system, liver, kidneys.

Occupational exposure limits: NIOSH REL: TWA 5 ppm (19 mg/m^3); C 15.6 ppm (60 mg/m^3); OSHA PEL: TWA 5 ppm (19 mg/m^3).

Risk/safety phrases: None.

First aid: Eye: Immediate irrigation; skin: wash with soap immediately; breath: respiratory support; swallow: seek medical attention immediately.

Conversion: 1 ppm = 3.85 mg/m^3.

BACKGROUND AND USES

Phenol (carbolic acid, hydroxybenzene, monohydroxy benzene, phenyl alcohol, phenyl hydroxide, CASRN 108-95-2) is a colorless to light pink crystalline solid with a sweet, acrid odor with a sharp burning taste (IPCS, 1999). It has a vapor pressure of 0.4 mmHg and a solubility in water by weight of 9% (25 °C [77°F]). Commonly used for the treatment of wounds in the 1800s, phenol is thought to have poisoned many patients and physicians (Merliss, 1972). Currently, phenol is found in dilute forms in some pharmaceuticals for external use and in throat lozenges. Some common products containing phenol are calamine lotion (1–2%), and Campho-Phenique (4.75% phenol, 10.86% camphor in aromatized liquid petroleum) (ATSDR, 2008).

PHYSICAL AND CHEMICAL PROPERTIES

Phenol has limited solubility in water at room temperature; however, it is completely soluble above 68 °C (IPCS, 1999). The melting point is 43 °C, while the boiling point is 181.75 °C (IPCS, 1999). The flashpoint is 80 °C. It is an A4, Group D carcinogen—it is not classifiable as a human carcinogen (HSDB, 2010). Benzene oxide undergoes a nonenzymatic rearrangement to phenol. Phenol is produced by hydrolysis to a dihydrodiol ring opening to ttMA. Additional oxidation can be catalyzed by enzyme CYP2E1 to hydroquinone and catechol.

MAMMALIAN TOXICOLOGY

Acute Effects

Acute effects of phenol intoxication include lethargy, nausea, vomiting, metabolic acidosis, diarrhea, cardiac arrhythmia and bradycardia, hypotension, tachypnea, respiratory arrest, convulsive seizures, and coma (Haddad et al., 1979). Dermal exposure to concentrated phenol may cause burns because of its defatting and caustic properties (Anderson, 2006; Philip and Marraffa, 2012). The half-life of conjugated phenol has been reported to be as low as 1 h and as high as 5 h (IPCS, 1999).

Chronic Effects

In many situations, it is often unclear whether formaldehyde is the primary causative agent of phenol marasmus or carbol marasmus can be an effect of chronic phenol exposure. This type of exposure has become rare, since the use of higher concentrations of phenol for medical purposes was discontinued in the late 1800s (Merliss, 1972).

Mechanisms of Action

Phenol is one of the major metabolites of benzene. It is also formed by the degradation of other organic wastes (Miller, 1942). It can be endogenously produced during the breakdown of certain protein metabolites, resulting in normal concentrations of phenol in urine to be <20 mg/l (Rogers et al., 2013; Wetherill et al., 2007).

In mice, more than 65% of phenol administered orally or intravenously was recovered in 48 h (Kenyon et al., 1995). Formation of various phenol metabolites was dependent on the dose and route of administration. The gastrointestinal tract is active in phenol metabolism. Renal damage is caused by unconjugated phenol, which damages the glomeruli and tubules as well as produces hypovolemic shock leading to renal ischemia (Becerro de Bengoa Vallejo et al., 2012; Kenyon et. al., 1995; Powell et al., 1974; Stasiuk and Kozubek, 2010; Zambito Marsala, 2010).

Phenol has been shown to be embryotoxic in rats when administered with a bioactivating system (Chapman et. al., 1994). Some phenol metabolites, (hydroquinone, catechol, and benzoquinone) have been demonstrated to be directly embryotoxic.

Chemical Pathology

Lethal doses can range from 1 to 15 g in adults (IPCS, 1999). Phenol can have a depressant effect on the central nervous system (Haddad et. al., 1979). Some reactive metabolites of phenol can activate leukocytes as shown in some *in vitro* studies (IPCS, 1994). No-observed-adverse-effect level (NOAEL) values in two multiple dose rat studies were 40 and 60 mg/kg/day (IPCS, 1994).

SOURCES OF EXPOSURE

Occupational

Occupational exposures occur at increased rates among laboratory workers, morticians, and house cleaners. The general population can be exposed to phenol in consumer products such as disinfectants and other household cleaners, throat lozenges, and ointments. Workers in the petroleum industry that manufacture nylon, epoxy resin, herbicides, and wood preservatives have an increased risk of phenol exposure (Bergé et al., 2012; Wasi et al., 2013). Any worker exposed to benzene is also exposed to phenol.

Environmental

Anthropogenic sources of phenol are paper mills and wood treatment plants. Phenol is rapidly degraded in air by hydroxyl radicals; degradation occurs within 19 h (Geens et al., 2012). It has an estimated half-life in air of 14.6 h. Soil can retain phenol for a half-life of <5 days (Bergé et al., 2012; Merliss, 1972). Plants can readily metabolize phenol and can be ingested by humans when plants are grown in phenol-rich soil—although exposure through this route is minimal. Phenol is also produced by first- and secondhand smoke. It is readily absorbed through the skin (Brown et al., 1975; Horch et al., 1994; Kucheki and Simi, 2010; Wasi et al., 2013) It is commonly measured in several point sources such as effluents, ambient water, drinking water, groundwater, rain, sentiment, and ambient air (Bergé et al., 2012; Hoffman et al., 2001; Ryan et al., 2001).

INDUSTRIAL HYGIENE

Phenol is still commonly used in industry for the production of other aromatic hydrocarbons, explosives, fertilizers, paints, paint removers, wood preservatives, synthetic resins, textiles, pharmaceuticals, and perfumes (Allan, 1994). In fact, total phenol production in the United States increased approximately 9% in the late 1980s.

The NIOSH, OSHA, and the ACGIH have each set the REL, PEL, and TLV for phenol at 5 ppm (19 mg/m^3) with a ceiling of 15.6 ppm (60 mg/m^3) for 15 min with a skin notation (NIOSH, 2007). Phenol levels of 250 ppm or above are considered IDLH.

The ACGIH (2009) has determined that workers exposed to the TLV for phenol will have approximately 250 mg phenol/g creatinine in their urine at the end of their shift (Heistand and Todd, 1972). Levels of phenol in urine are also monitored to estimate the exposure to benzene (Arnold et. al., 2013). However, the expected value for workers exposed to the benzene TLV would be 50 mg phenol/g creatinine. Methods for evaluating phenol in exhaled air by gas chromatography have been reported (Pendergrass, 1994). Elevated blood levels may also be used to determine the exposure for clinical purposes.

MEDICAL MANAGEMENT

Dermal exposure is most common. Contaminated areas on the body will appear red and swollen. Diagnosis may be made by examination of clinical chemistry. Normal blood concentrations of phenol are as follows: free phenol approximately 0.4 mg/100 ml; conjugated phenol approximately 0.1–2 mg/100 ml; and total phenol approximately 0.15–7.96 mg/100 ml (Horch et al., 1994; Ruedemann and Deichmann, 1953). Exposed individuals will have dramatically elevated levels of blood phenol commensurate with the intensity of exposure. Severe or prolonged exposures may result in proteinuria, oliguria, anuria, and uremia (typically dark urine) (Wasi et al., 2013; Zambito Marsala, 2010).

Treatment of any phenol exposure should include basic life-support measures as necessary. For dermal exposure, wound decontamination is recommended by sponging burns with undiluted polyethylene glycol (PEG) 300 or 400 and showering after treatment. Any wound should be treated with PEG until all traces of phenol smell disappear. Ethanolic PEG solutions are better phenol solvents but are not recommended for the treatment because of their flammability, odor, and skin absorption. A high density shower is recommended if available, low intensity without PEG treatment may increase phenol absorption. Intravenous Ringer's lactate solution may be administered. Scalds can be covered with silver sulfadiazine dressings (Horch et al., 1994). For ocular exposure, patients' eyes should be washed thoroughly with room temperature water for at least 15 min. If pain and irritation persist, an ophthalmologist must be consulted.

If ingestion occurs, dilution should be avoided as it may enhance absorption. Solutions containing >5% phenol can result in oral burns.

REFERENCES

American Conference of Governmental Industrial Hygienists (ACGIH) (2009) *Threshold Limit Values (TLVs) for Chemical Substances and Physical Agents Biological Exposure Indices (BEIs)*, Cincinnati, OH: American Conference of Governmental Industrial Hygienists.

Allan, R. (1994) *Phenols and Phenolic Compounds*, 4th ed., John Wiley & Sons.

Andersen, A. (2006) Final amended report on the safety assessment of oxyquinoline and oxyquinoline sulfate as used in cosmetics. *Int. J. Toxicol.* 25:1–9.

Arnold, S.M., Angerer, J., Boogaard, P.J., Hughes, M.F., O'Lone, R.B., Robison, S.H., and Robert Schnatter, A. (2013) The use of biomonitoring data in exposure and human health risk assessment: benzene case study. *Crit. Rev. Toxicol.* 43(2):119–153.

Agency for Toxic Substances and Disease Registry (ATSDR) (2008) *Toxicological Profile for Phenol*, Atlanta, GA: U.S. Department of Health and Human Services, Public Health Service.

Becerro de Bengoa Vallejo, R., Losa Iglesias, M., Jules, K., and Trepal, M. (2012) Renal excretion of phenol from physicians after nail matrix phenolization: an observational prospective study. *J. Eur. Acad. Dermatol. Venereol.* 26(3):344–347.

Bergé, A., Cladière, M., Gasperi, J., Coursimault, A., Tassin, B., and Moilleron, R. (2012) Meta-analysis of environmental contamination by alkylphenols. *Environ. Sci. Pollut. Res.* 19(9):3798–3819.

Brown, V.K., Box, V.L., and Simpson, B.J. (1975). Decontamination procedures for skin exposed to phenolic substances. *Arch. Environ. Health* 30(1):1–6.

Chapman, D.E., Namkung, M.J., and Juchau, M.R. (1994) Benzene and benzene metabolites as embryotoxic agents: effects on cultured rat embryos. *Toxicol. Appl. Pharmacol.* 128(1):129–137.

Geens, T., Aerts, D., Berthot, C., Bourguignon, J.-P., Goeyens, L., Lecomte, P., Maghuin-Rogister, G., Pironnet, A.-M., Pussemier, L., and Scippo, M.-L. (2012) A review of dietary and non-dietary exposure to bisphenol-A. *Food Chem. Toxicol.*. 50(10):3725–3740.

Haddad, L.M., Dimond, K.A., and Schweistris, J.E. (1979) Phenol poisoning. *J. Am. Coll. Emerg. Physicians* 8(7):267–269.

Heistand, R. and Todd, A. (1972) Automated determination of total phenol in urine. *Am. Ind. Hyg. Assoc. J.* 33(6):378–381.

Hoffman, G.M., Dunn, B.J., Morris, C.R., Butala, J.H., Dimond, S.S., Gingell, R., and Waechter, J.M. (2001) Two-week (ten-day) inhalation toxicity and two-week recovery study of phenol vapor in the rat. *Int. J. Toxicol.* 20(1):45–52.

Horch, R., Spilker, G., and Stark, G. (1994) Phenol burns and intoxications. *Burns* 20(1):45–50.

HSDB (2010) *Hazardous Substances Data Bank: Benzene. From NLM (National Library of Medicine)*, Bethesda, MD: National Institutes of Health, U.S. Department of Health and Human Services, Available at http://toxnet.nlm.nih.gov.

IPCS (1994) *Environmental Health Criteria 161: Phenol. International Programme on Chemical Safety*, Geneva, Switzerland: World Health Organization.

IPCS (1999) *Poisons Information Monograph 412: Phenol. International Programme on Chemical Safety*, Geneva, Switzerland: World Health Organization.

Kenyon, E.M., Seeley, M.E., Janszen, D., and Medinsky, M.A. (1995) Dose-, route-, and sex-dependent urinary excretion of phenol metabolites in B6C3F1 mice. *J. Toxicol. Environ. Health* 44(2):219–233.

Kucheki, M. and Simi, A. (2010) Phenol burn. *Int. J.Occup. Environ. Med.* 1(1):41–44.

Merliss, R. (1972) Phenol marasmus. *J. Occup. Environ. Med.* 14(1):55–56.

Miller, F. (1942) Poisoning by phenol. *Can. Med. Assoc. J.* 46(6):615–616.

National Institute for Occupational Safety and Health (NIOSH) (2007) *Pocket Guide to Chemical Hazards*, NIOSH 94-116, Washington, DC: US Department of Health and Human Services.

Pendergrass, S.M. (1994) An alternative method for the analysis of phenol and o-, m-, and p-cresol by capillary GC/FID. *Am. Ind. Hyg. Assoc.* 55(11):1051–1054.

Philip, A.T. and Marraffa, J.M. (2012) Death following injection sclerotherapy due to phenol toxicity. *J. Forensic Sci.* 57(5):1372–1375.

Powell, G.M., Miller, J.J., Olavesen, A.H., and Curtis, C.G. (1974) Liver as major organ of phenol detoxication? *Nature* 252(5480):234–235.

Rogers, J.A., Metz, L., and Yong, V.W. (2013) Review: endocrine disrupting chemicals and immune responses: a focus on bisphenol-A and its potential mechanisms. *Mol. Immunol.* 53(4):421–430.

Ruedemann, R. and Deichmann, W.B. (1953) Blood phenol level after topical application of phenol-containing preparations. *J. Am. Med. Assoc.* 152(6):506–509.

Ryan, B., Selby, R., Gingell, R., Waechter, J., Butala, J., Dimond, S., Dunn, B., House, R., and Morrissey, R. (2001) Two-generation reproduction study and immunotoxicity screen in rats dosed with phenol via the drinking water. *Int. J. Toxicol.* 20(3):121–142.

Stasiuk, M. and Kozubek, A. (2010) Biological activity of phenolic lipids. *Cell. Mol. Life Sci.* 67(6):841–860.

Wasi, S., Tabrez, S., and Ahmad, M. (2013) Toxicological effects of major environmental pollutants: an overview. *Environ. Monit. Assess.* 185(3):2585–2593.

Wetherill, Y.B., Akingbemi, B.T., Kanno, J., McLachlan, J.A., Nadal, A., Sonnenschein, C., Watson, C.S., Zoeller, R.T., and Belcher, S.M. (2007) *In vitro* molecular mechanisms of bisphenol A action. *Reprod. Toxicol.* 24(2):178–198.

Zambito Marsala, S. (2010) Hyperckemia after phenol intoxication. *Neurol. Sci.* 31(4):527–528.

CRESOL

Target organ(s): Eyes, skin, respiratory system, CNS, liver, kidneys, pancreas, and cardiovascular system.

Occupational exposure limits: NIOSH REL: TWA: 2.3 ppm (10 mg/m^3). OSHA PEL: TWA: 5 ppm (22 mg/m^3).

Risk/safety phrases: None.

First aid: Eye: Immediate irrigation; Skin: wash with soap immediately; breath: respiratory support; swallow: seek medical attention immediately.

Conversion: 1 ppm = 4.43 mg/m^3.

BACKGROUND AND USES

Cresol (cresylic acid, hydroxymethyl benzene, hydroxytoluene, methylphenol CASRN 1319-77-3 for all isomers combined) is a derivative of coal tar that occurs in ortho-, meta-, and para isomers (Table 60.1). In most cases, all forms exist in a mixture. All isomers have a characteristic sweet, tarry odor and a solubility in water by weight of 2%. Cresols are monomethyl derivatives of phenol, which are commonly used in household disinfectants, dyes, explosives, fragrances, pesticides, and germicides. Cresol used as a disinfectant is

TABLE 60.1 Cresol Isomers

Cresol Isomers		CAS Number	Description	Vapor Pressure at 25 °C, mmHg
ortho-	2-Cresol	95-48-7	White crystals darken with age	0.29
meta-	3-Cresol	108-39-4	Colorless to yellowish liquid	0.14
para-	4-Cresol	106-44-5	Solid	0.11

usually a mixture of the three isomers of methylphenol—2-cresol, 3-cresol, and 4-cresol—known commonly as *o*-cresol, *m*-cresol, and *p*-cresol, respectively. Isomers of cresol in air react with the hydroxyl radicals.

M-cresol is used as an antimicrobic in insulin and other pharmaceuticals. Cresol is approximately three times as effective as phenol as a bactericide. It is also found in wire enamel solvents, phenolic resins, and tricresyl phosphate.

Cresols should not be confused with creosote or cresylic acids. Creosote is a wood preservative composed of phenol derivatives, including cresols, from the distillation of coal and wood tars. Cresylic acids are mixtures of cresols, xylenols, and phenol. These compounds are part of a class of molecules known as protein-bound uremic retention toxins, which lead to nephrotoxicity. Many cresol isomers are used in cosmetics as a biocide and preservative as well as a common ingredient in fragrances (Andersen, 2006).

PHYSICAL AND CHEMICAL PROPERTIES

Cresols at room temperature are solid except *m*-cresol, which is a liquid. The melting points of *o*-cresol, *m*-cresol, and *p*-cresol are 30.9, 12.2, 34.7 °C, respectively (Andersen, 2006). The density of cresol isomers is 1.034 g/ml, except *o*-cresol which has a density of 1.047 g/ml. The concentration of cresol isomers in air may be 4.5 mg/m^3 (0.22 ppm) (ATSDR, 2008). Isomeric cresols have a vapor pressure of 0.11 ± 0.30 mmHg at 25.5 °C (ATSDR, 2008; NIOSH, 2007). Cresol isomers have a relatively high water solubility of 21.52–25.95 g/l, which allows wet deposition to remove cresols from the atmosphere; however, they do not travel long distances from the point of release. Soil absorption coefficients are 17.5–117, suggesting high mobility in soil (ATSDR, 2008). Cresols do not bio-concentrate in fish or aquatic organisms and are unlikely to concentrate in humans.

Emission studies have shown an estimated ambient mean concentration of 31.7 ng/m^3 for all cresols combined (ATSDR, 2008). Cresol has a low deposition in soil and water. Usually when cresols isomers are found in soil or water, it is due to spillage of petroleum products, although they can be degraded in water within days by microorganisms (ATSDR, 2008). Contaminated drinking water can create volatilized cresol during showering, bathing, and cooking. Cresols are degraded in air within 1–2 days and can be removed by rain. Cresols can be removed from air caused by hydroxyl radicals during the day and nitrate radicals at night.

MAMMALIAN TOXICOLOGY

Acute Effects

Acute exposures to cresol in an industry are not common. When exposures do occur they are usually dermal. Cresol is not volatile enough to pose an inhalation hazard. Skin and eye contact with cresol causes burns similar to phenol exposure only slightly more severe (Pegg and Campbell, 1985). Eye exposure may cause conjunctivitis, corneal burns, and keratitis. Cresol is absorbed more slowly than phenol, resulting in less severe systemic effects.

Cresols can be absorbed through the skin, respiratory tract, and digestive tract (Andersen, 2006). They are processed by first-pass metabolism in the liver (Dix et al., 2007). The metabolites are excreted by the kidney as glucuronide and sulfate metabolites (Andersen, 2006). Cresol poisoning can attack several systems of the body leading to corrosion of the gastrointestinal tract, respiratory distress, uremia, cardiovascular disturbances, hepatoxicity, and nephrotoxicity (Chapman et. al., 1994; Seak et al., 2010; Yan et al., 2005).

Inhalation exposure at concentrations of 6 mg/m^3 (o-cresol) can cause respiratory irritation. All isomers of cresol are neurotoxic; ingestion of household cleaning products made with mixed cresol has been shown to cause unconsciousness and coma in humans (ATSDR, 2008; Sakai et al., 1999). Exposure to p-cresol by ingestion can cause death. *P-cresol* has a high serum protein-binding affinity, which can cause uremic toxicity (Yan et al., 2005).

Chronic Effects

There have been reports of sensitization to *m*-cresol when used as a pharmaceutical preservative.

Mechanisms of Action

Cresol is metabolized by conjugation and oxidation. It is rapidly absorbed in the body where it denatures and precipitates cellular proteins causing direct toxicity to cells (Boatto et al., 2004; Monma-Ohtaki et al., 2002; Zhu et al., 2012). All modes of exposure: oral, dermal, and inhalation can be absorbed rapidly and produce cresol intoxication. Exposure to cresols may affect the CNS and respiratory system, as well as cause renal, hepatic, dermal, and ocular effects (Merliss,

1972). Rats exposed to cresol did exhibit weight gain, increased excitability, increased oxygen consumption, changes in liver and lung tissues, and decreased gamma-globulin content of serum. Cresol is metabolized primarily to 4-hydroxybenzylalcohol in the liver by oxidation of the methyl group and further oxidation forms 4-hydroxybenzaldehyde (Chang et al., 2011; Mutsaers et al., 2013; Yan et al., 2005). Oxidation of p-cresol forms a reactive electrophile quinine methide whose intermediates are primarily mediated by CYP2E1, although other P450s are also involved to a lesser extent such as 2D6, 2C19, 1A1, and 1A2 (Yan et al., 2005). Quinine methide is highly reactive, and it alkylates cellular proteins and nucleic acids (Yan et al., 2005). An alternative chemical pathway also occurs when 4-methyl-*ortho*-hydroquinone is also formed by oxidation of the benzene ring and further oxidized to a reactive intermediate as 4-methyl-*ortho*-benzoquinone (Yan et al., 2005). Cresol metabolites can form stable conjugates when they bind to proteins such as glutathione (GSH) where it can be detected (Yan et al., 2005). According to Yan et al. (2005) quinine methide is reactive with DNA forming adducts.

Cresol isomers may produce significantly different effects. In experiments using rat liver slices, p-cresol was determined to be five to ten times more potent as a cytotoxin than o-cresol and m-cresol (Thompson et al., 1994). P-cresol was also shown to rapidly deplete GSH, whereas o- and m-isomers had a less dramatic effect. This study suggests that p-cresol may have a different mechanism of action than other isomers. P-cresol is generated by protein breakdown as a metabolite of tyrosine and phenylalanine, thus it is normally found in the body (Chang et al., 2011; Cuoghi et al., 2012). It is absorbed from the intestine and excreted through urine. Healthy subjects have been found to have a mean concentration of 8.6 μmol/l (0.93 mg/l) of p-cresol in their serum (De Smet et al., 2003).

The exact mechanism of action of cresol remains unknown. In one study, a mixture of cresol isomers was negative in the Ames mutagenicity test. DeWolf et al. (1988) demonstrated that p-cresol inactivates dopamine beta-hydroxylase by covalently modifying its structure. This is significant because beta-hydroxylase catalyzes the hydroxylation of dopamine to norepinephrine. Several studies using rat liver slices and human liver microsomes confirm that rat and human metabolism of p-cresol is similar (Thompson et al., 1994; Yan et al., 2005).

Administration of o-cresol seems to lessen the carcinogenic effect of benzo-alpha-pyrene in mice (Calderón-Guzmán et al., 2005; Yanysheva et al., 1993). In an amphibian model, m-cresol was found to dramatically increase the biological activity of insulin (Gallucci and Micelli, 1992). Dietary factors, such as meat consumption, can significantly alter excretion of cresol in humans (Ling and Hänninen, 1992). Additionally, in mice there was evidence of mild atrophy of the female reproductive organs as well an extended estrous cycle (ATSDR, 2008; NTP, 1992).

Chemical Pathology

Animal studies have shown that exposures between 9 and 50 mg/m^3 produce fatty degeneration and necrosis of the liver. Cresols can impair the stratum corneum (as seen in dermal exposures) and produce coagulation necrosis. In the 1990s, Thompson et al. (1994, 1996) studied the toxicity of rat liver slices using lactate dehydrogenase that suggests a possible mechanism of cresol toxicity.

SOURCES OF EXPOSURE

Occupational

Occupational exposures to cresols can occur in coal-, wood-, or petroleum-fueled electricity-generating facilities or solid waste incinerators from all types of industry. Co-exposure can occur in conjunction with toluene. Cresols can be formed from photochemical reactions with toluene and hydroxyl radicals. Workers and residents who live near these facilities may be exposed; however, the short half-life of cresols makes them less of an exposure risk than first- and secondhand cigarette smoke.

Environmental

Cresols are emitted by automobile exhaust where low levels remain in the atmosphere. In addition cresols are emitted during the combustion of solid waste, wood, coal and in cigarette smoke. Cresols biodegrade rapidly in aerobic freshwater but degrade much slower under anaerobic conditions and in salt water. The average ambient concentration of cresol isomers is 31.7 ng/m^3 (ATSDR, 2008). Due to the short half-life and reactions during the day and at night, cresols have a short deposition in the atmosphere. At various sites around the world, cresols have been found in water at average low concentrations of 0.05 µg/l, with some areas as high as 390 µg/l (ATSDR, 2008). Cresols are bio-waste of mammals and thus can be found in sewage at high concentrations. Cresols in soil are broken down by microorganisms and are biotransformed.

INDUSTRIAL HYGIENE

The OSHA has set the PEL for all isomers of cresol at 5 ppm (22 mg/m^3). The NIOSH (2007) has set the REL for all cresol isomers significantly low at 2.3 ppm (10 mg/m^3). Cresol levels above 250 ppm are considered IDLH. The ACGIH (2009) has established a TLV for all cresol isomers at 5 ppm (22 mg/m^3) for skin exposure.

MEDICAL MANAGEMENT

Short-term exposures can cause respiratory irritation (Boatto et al., 2004). Acute exposures can cause respiratory failure,

tachycardia, and ventricular fibrillation. Additionally patients can suffer from abdominal pain, vomiting, and corrosive lesions in the gastrointestinal tract (Dix et al., 2007; Monma-Ohtaki et al., 2002). More serious complications can occur such as methemoglobinemia, leukocytosis, hemoylsis, hepatocellular injury, renal alterations, skin damage, metabolic acidosis, and even unconsciousness (Lin and Yang, 1992; Sakai et al., 1999; Seak et al., 2010). Hemodialysis has not proven to be effective for reducing blood phenol levels (Evers et al., 1994). Another method for reducing levels of cresols in the body appears to be forced diuresis (Arthurs et al., 1977). Sodium bicarbonate is indicated to reduce the patient's urinary pH, thereby decreasing precipitation of hemoglobin in the renal tubules (Lin and Yang, 1992). The use of sodium bicarbonate may be indicated anyway to reverse metabolic acidosis that results from a large burn surface area and high levels of blood phenol.

Treatment of cresol burns should include PEG in the manner described for the treatment of phenol burns. Severe poisoning can be at serum levels of 10 µg/ml and lethal at 100 µg/ml (Sakai et al., 1999). Cresol can be absorbed quickly and excreted into the urine (Boatto et al., 2004; Brown et al., 1975). Acute poisoning is usually presented clinically by black urine that can be accompanied by hemoglobinuria, melanuria, porphyriuria, alcaptonuria, myoglobinuria, and tyrosinuria (Seak et al., 2010). Treatment requires an aggressive intensive supportive care in order to manage cresol poisoning (Seak et al., 2010).

Cresol absorption can result in the depression of the CNS, hypothermia, hypotension, pulmonary edema, and intravascular hemolysis (Sakai et al., 1999). Dermal exposure to cresol can cause severe burns and should be washed thoroughly and treated with swabs of PEG sponges prior to a high density shower (Sakai et al., 1999). Dermal absorption rates are based on the surface area of the skin exposed and the duration of contact (Sakai et al., 1999).

REFERENCES

Agency for Toxic Substances and Disease Registry (ATSDR) (2008) *Toxicological Profile for Cresol*, Atlanta, GA: U.S. Department of Health and Human Services, Public Health Service.

American Conference of Governmental Industrial Hygienists (ACGIH) (2009) *Threshold Limit Values (TLVs) for Chemical Substances and Physical Agents Biological Exposure Indices (BEIs)*, Cincinnati, OH: American Conference of Governmental Industrial Hygienists.

Andersen, A. (2006) Final report on the safety assessment of sodium *p*-chloro-*m*-cresol, *p*-chloro-m-cresol, chlorothymol, mixed cresols, *m*-cresol, *o*-cresol, *p*-cresol, isopropyl cresols, thymol, *o*-cymen-5-ol, and carvacrol. *Int. J. Toxicol.* 25(Suppl. 1):29–127.

Arthurs, G., Wise, C., and Coles, G. (1977) Poisoning by cresol. *Anaesthesia* 32(7):642–643.

Boatto, G., Nieddu, M., Carta, A., Pau, A., Lorenzoni, S., Manconi, P., and Serra, D. (2004) Determination of phenol and o-cresol by GC/MS in a fatal poisoning case. *Forensic Sci. Int.* 139(2):191–194.

Brown, V.K., Box, V.L., and Simpson, B.J. (1975) Decontamination procedures for skin exposed to phenolic substances. *Arch. Environ. Health* 30(1):1–6.

Calderón-Guzmán, D., Espitia-Vázquez, I., López-Domínguez, A., Hernández-García, E., Huerta-Gertrudis, B., Coballase-Urritia, E., Juárez-Olguín, H., and García-Fernández, B. (2005) Effect of toluene and nutritional status on serotonin, lipid peroxidation levels and Na+/K+-ATPase in adult rat brain. *Neurochem. Res.* 30(5):619–624.

Chang, M.-C., Wang, T.-M., Yeung, S.-Y., Jeng, P.-Y., Liao, C.-H., Lin, T.-Y., Lin, C.-C., Lin, B.-R., and Jeng, J.-H. (2011) Antiplatelet effect by p-cresol, a uremic and environmental toxicant, is related to inhibition of reactive oxygen species, ERK/p38 signaling and thromboxane A2 production. *Atherosclerosis* 219(2):559–565.

Chapman, D.E., Namkung, M.J., and Juchau, M.R. (1994) Benzene and benzene metabolites as embryotoxic agents: effects on cultured rat embryos. *Toxicol. Appl. Pharmacol.* 128(1):129–137.

Cuoghi, A., Caiazzo, M., Bellei, E., Monari, E., Bergamini, S., Palladino, G., Ozben, T., and Tomasi, A. (2012) Quantification of p-cresol sulphate in human plasma by selected reaction monitoring. *Anal. Bioanal. Chem.* 404(6–7):2097–2104.

De Smet, R., Van Kaer, J., Van Vlem, B., De Cubber, A., Brunet, P., Lameire, N., and Vanholder, R. (2003) Toxicity of free p-cresol: a prospective and cross-sectional analysis. *Clin. Chem.* 49(3):470–478.

DeWolf, W.E. Jr., Carr, S.A., Varrichio, A., Goodhart, P.J., Mentzer, M.A., Roberts, G.D., Southan, C., Dolle, R.E., and Kruse, L.I. (1988) Inactivation of dopamine beta-hydroxylase by p-cresol: isolation and characterization of covalently modified active site peptides. *Biochemistry* 27(26), 9093–9101.

Dix, K.J., McDonald, J., Ferguson, L.-J.C., Kim, N.-C., Weber, W., Swing, S., Van Andel, R.A., and Seagrave, J. (2007) *Disposition of 5-Amino-o-cresol in Female Rats and Mice After Dermal, Oral and Intravenous Administration.*

Evers, J., Aboudan, F., Lewalter, J., and Renner, E. (1994) Hemodialysis in metacresol poisoning. *J. Mol. Med.* 72(6):472

Gallucci, E. and Micelli, S. (1992) Potentiation by m-cresol on transepithelial sodium transport across frog skin induced by insulin. *Comp. Biochem. Physiol. C* 103(3):593–595.

Lin, C.-H. and Yang, J. (1992) Chemical burn with cresol intoxication and multiple organ failure. *Burns* 18(2):162–166.

Ling, W.H. and Hänninen, O. (1992) Shifting from a conventional diet to an uncooked vegan diet reversibly alters fecal hydrolytic activities in humans. *J. Nutr.* 122(4):924–930.

Merliss, R. (1972) Phenol marasmus. *J. Occup. Environ. Med.* 14(1):55–56.

Monma-Ohtaki, J., Maeno, Y., Nagao, M., Iwasa, M., Koyama, H., Isobe, I., Seko-Nakamura, Y., Tsuchimochi, T., and Matsumoto, T. (2002) An autopsy case of poisoning by massive absorption of cresol a short time before death. *Forensic Sci. Int.* 126(1):77–81.

Mutsaers, H.A., Wilmer, M.J., Reijnders, D., Jansen, J., van den Broek, P.H., Forkink, M., Schepers, E., Glorieux, G., Vanholder, R., van den Heuvel, L.P., Hoenderop, J.G., and Masereeuw, R. (2013) Uremic toxins inhibit renal metabolic capacity through interference with glucuronidation and mitochondrial respiration. *Biochim. Biophys. Acta* 1832(1):142–150.

National Institute for Occupational Safety and Health (NIOSH) (2007) *Pocket Guide to Chemical Hazards*, NIOSH 94-116, Washington, DC: US Department of Health and Human Services.

National Toxicology Program (NTP) (1992) *Toxicity Studies of Cresols (CAS NOS. 95-48-7, 108-39-4, 106-44-5) in F344/N Rats and B6C3F1 Mice (Feed Studies)*, National Institutes of Health.. Research Triangle Park, NC: Public Health Service U.S. Department of Health and Human Services. NIH Publication No. 92-3128.

Pegg, S.P. and Campbell, D.C. (1985) Children's burns due to cresol. *Burns* 11(4):294–296.

Sakai, Y., Abo, W., Yagita, K., Tanaka, T., and Fuke, C. (1999) Chemical burn with systemic cresol intoxication. *Pediatr. Int.* 41(2):174–176.

Seak, C.-K., Lin, C.-C., Seak, C.-J., Hsu, T.-Y., and Chang, C.-C. (2010) A case of black urine and dark skin-cresol poisoning. *Clin. Toxicol.* 48(9):959–960.

Thompson, D., Perera, K., Fisher, R., and Brendel, K. (1994) Cresol isomers: comparison of toxic potency in rat liver slices. *Toxicol. Appl. Pharmacol.* 125(1):51–58.

Thompson, D.C., Perera, K., and London, R. (1996) Studies on the mechanism of hepatotoxicity of 4-methylphenol (p-cresol): effects of deuterium labeling and ring substitution. *Chem. Biol. Interact.* 101(1):1–11.

Yan, Z., Zhong, H.M., Maher, N., Torres, R., Leo, G.C., Caldwell, G.W., and Huebert, N. (2005) Bioactivation of 4-methylphenol (p-cresol) via cytochrome P450-mediated aromatic oxidation in human liver microsomes. *Drug Metab. Dispos.* 33(12):1867–1876.

Yanysheva N.Y., Balenko, N., Chernichenko, I., and Babiy, V. (1993) Peculiarities of carcinogenesis under simultaneous oral administration of benzo(a)pyrene and o-cresol in mice. *Environ. Health Perspect.* 101(Suppl. 3):341–344.

Zhu, J.Z., Zhang, J., Yang, K., Du, R., Jing, Y.J., Lu, L., and Zhang, R.Y. (2012) P-cresol, but not p-cresylsulphate, disrupts endothelial progenitor cell function *in vitro*. *Nephrol. Dial. Transplant.* 27(12):4323–4330.

TOLUENE

Target organ(s): Eyes, skin, respiratory system, CNS system, liver, and kidneys. The primary target is the CNS.

OSHA PEL: TWA 200 ppm (750 mg/m^3; ceiling 300 ppm; peak 500 ppm (10 min). Toluene levels of 500 ppm (1885 mg/m^3) or greater are considered IDLH.

The NIOSH has set the REL for toluene at 100 ppm (375 mg/m^3), with a STEL of 150 ppm (560 mg/m^3) (NIOSH, 2010).

The ACGIH has set a TLV for exposure to toluene at 20 ppm (75 mg/m^3) (ACGIH 2007). Toluene is classified as an A4 carcinogen indicating that not enough information is available to determine whether it is carcinogenic in humans. The International Agency for Research on Cancer (IARC) has classified toluene as Group 3 (*not classifiable as to its carcinogenicity in humans*) with a supporting statement that there is inadequate evidence in humans and evidence suggesting a lack of carcinogenicity of toluene in experimental animals (IARC, 1999).

Reference values for toluene: Oral reference dose (RfD) = 0.08 mg/kg-day; inhalation reference concetntration (RfC) = 5 mg/m^3. NOAEL (average): 34 ppm (128 mg/m^3); NOAEL (ADJ): 46 mg/m^3. Agency for Toxic Substances and Disease Registry (ATSDR) minimal risk level (MRL): 1 ppm (acute) .08 ppm (chronic).

Risk/safety phrases: Inhalation risk; irritant.

BACKGROUND AND USES

Millions of metric tons of toluene (toluol, phenyl methane, methylbenzene, methylbenzol; CASRN 108-88-3) are produced each year, making it one of the most abundantly produced chemicals in the United States. It is obtained primarily from the distillation of crude petroleum. It is added to gasoline, used to produce benzene, and used as a solvent. It is found in glue, paint, paint thinners, synthetic fragrances, nail polish, cigarette smoke, and transmission fluid; it has been used as a solvent in many industrial processes and as a solvent in coatings, inks, adhesives, and cleaners. It is also used in the production of nylon, plastics, plastic soda bottles, pharmaceuticals, dyes, and polyurethanes and in the manufacture of organic chemicals such as toluene diisocyanate, toluene sulfonates, nitrotoluenes, vinyl toluene, phenol, benzyl alcohol, and saccharin. As a historical note toluene was once used as a medicinal anthelmintic agent against roundworms and hookworms (ATSDR, 2000; EPA, 2012; Filley et al., 2004; Hannigan and Bowen, 2010).

PHYSICAL AND CHEMICAL PROPERTIES

At room temperature, toluene is a clear to amber colorless liquid with a pungent, benzene-like odor. It has a vapor pressure of 28.4 mmHg and solubility in water of 0.59 mg/ml at 25 °C. It has a structural formula: $C_6H_5CH_3$, molecular weight: 92.14, density: 0.867 gm/ml. Conversion factor 1 ppm = 3.77 mg/m^3, 1 mg/m^3 = 0.265 ppm (25 °C, 760 mmHg)

HEALTH EFFECTS

Acute Health Effects

Toluene is irritating to the skin, eyes, and respiratory tract. Eye contact can cause reversible conjunctivitis and keratitis. The vapor can cause eye irritation and respiratory tract irritation.

Acute inhalation exposures to toluene produce symptoms that are similar to other organic solvents and may include dizziness, headache, lethargy, inebriation, "drunkenness,"

exhilaration, drowsiness, staggering gait, tremor, euphoria, disorientation, hallucinations, and CNS depression. Other signs and symptoms may include ventricular arrhythmias, chemical pneumonitis, respiratory depression, nausea, vomiting, and electrolyte imbalances. Higher concentrations lead to collapse, loss of consciousness, coma, and death (Van Oettingen et al., 1942a, 1942b; Boor and Hurtig, 1977; Xiao and Levin, 2000; ATSDR, 2000).

At relatively low levels of ambient exposure to toluene, just above 200 ppm, fatigue, headache, paresthesias, and slowed reflexes are seen; above 600 ppm, confusion may occur; and above 800 ppm, euphoria is experienced (Greenberg, 1997). The acute encephalopathic effects of toluene are generally reversible and have not been associated with neuroimaging changes (Filley et al., 2004). A study by Deschamps et al. (2000) did not support the notion of the persistence of cognitive effects of toluene after elimination of the solvent from blood in the setting of low dose occupational exposures.

There are multiple studies of subacute and chronic occupational exposures to toluene below 200 ppm that have shown various and often inconsistent findings of subtle neurotoxic, neuropsychiatric, neurobehavioral and neuromuscular effects, color discrimination, and cognitive performance (ATSDR, 2000; USEPA, 2005b; Meyer-Baron, 2005). Limitations of the studies include lack of clear data regarding past duration and intensity of exposure, variations in subject baseline neurobehavioral and intellectual capacity, coexistent exposures and lack of adjustment for confounders, inconsistent effect sizes, diversity of study groups, the use of diverse tests of neuropsychiatric function, and the degree of test reliability (Meyer-Baron, 2005).

Chronic Health Effects

The health effects of toluene reported in the literature can be misleading unless a distinction is made between high dose exposures experienced by toluene abusers vs. the relatively low dose exposure seen in most occupational and environmental settings. A distinction also needs to be made between the acute effects of toluene exposure that are usually reversible and the long-term chronic effects of high levels of exposure (as in chronic toluene abusers) that may not be reversible.

Gericke et al. (2001) studied 1007 male volunteers in the rotogravure (printing) industry in Germany and found no clear-cut evidence of adverse health effects of long-term (more than 20 years) exposure to toluene at the lower levels experienced at work. In Germany, the initially tolerable maximal toluene concentration at the workplace (MAK-value) of 200 ppm was reduced in 1985 to 100 ppm and to 50 ppm in 1993.

Toluene leukoencephalopathy (damage to white brain matter and an ominous sequel of toluene abuse) has not been shown to be caused by toluene exposure at the permissible levels set by the OSHA.

Chronic exposure to extremely high levels of toluene may lead to the gradual development of headache, fatigue, irritability, memory impairment, depression, emotional instability, sleep disturbance, alcohol intolerance, loss of libido or potency, and loss of interest in daily activities. Most occupational settings do not generate ambient levels of toluene capable of producing these symptoms in workers.

Toluene *abusers* that have been exposed for long periods of time have been reported to exhibit a variety of neurologic manifestations, including hyperreactive tendon reflexes, ataxia, tremor, anosmia, sensorineural hearing loss, cognitive dysfunction, dementia, corticospinal tract dysfunction, bilateral sensorineural hearing loss, abnormal brainstem auditory-evoked potentials, and epileptic seizures. Neurobehavioral deficits seen include inattention, apathy, memory dysfunction, visuospacial impairment, and preserved language (Hormes et al., 1986). MRI findings in toluene abusers include generalized cerebral, cerebellar, and brainstem atrophy; atrophy of the corpus callosum; loss of gray white matter discrimination; multifocal high signal intensity in the cerebral white matter; and hypointensity of the thalami on T2-weighted images. Optic neuropathies with dyschromatopsia, blindness, changes in pattern visual-evoked potentials, pendular nystagmus, ocular flutter, opsoclonus (irregular rapid eye movement), bilateral internuclear ophthalmoplegia, and retinal impairment have been reported in those who sniffed toluene or toluene-based glue (Hormes et al., 1986; Holló and Varga, 1992; Xiao and Levin, 2000; ATSDR, 2000; USEPA, 2005b).

Risk assessment for toluene is problematic. There are ample data on human exposures because of toluene abuse; however, these are not controlled experiments, and the dose is unknown. In general, the exposure of abusers is exponentially greater than occupational exposures, and abusers often expose themselves to a variety of other recreational drugs. Extrapolation of symptoms and effects from abusers to lower level chronic exposures of workers is not appropriate.

Central Nervous System Toxicity   Damage to the brain myelin has been well documented with chronic toluene abuse; however, there have been few autopsies probably because chronic toluene abuse is under recognized by pathologists. The CNS is the primary target organ for toluene toxicity, and the acute and chronic effects are described above. Chronic encephalopathy and leukoencephalopathy are known outcomes of chronic toluene abuse. Autopsy reports include peripheral neuropathy, long tract degeneration, cerebral and cerebellar atrophy, severe thinning of the corpus callosum, diffuse demyelination, neuronal loss in the cerebral cortex, basal ganglia and cerebellum, intense reactive gliosis, and giant axonal degeneration in the long tracts of the spinal cord. The brunt of toluene toxicity is directed toward the CNS white matter. The severity of myelin damage differs from one case to another and such differences could be due to variations in the

duration and intensity of toluene exposure, coexistent abuse of other chemicals, and host factors such as genetic polymorphisms, defective metabolism, and the age of onset of abuse (Filley et al., 2004). While leukoencephalopathy has not been shown to result from occupational exposure, it has been suggested that advanced neuroimaging techniques may in the future be able to detect subtle brain changes, if they exist, caused by lower level exposures that occur in the workplace.

Reproductive Toxicity   Toluene is capable of crossing the placenta and has been detected in animal fetal-placental compartments 24 h after exposure (Ghantous and Danielsson, 1986). Lindbohm et al. (1992) found an association between occupational toluene exposure and spontaneous abortion (odds ratio 2.3); however, exposure levels were not reported in this study. Current data do not provide convincing evidence that an occupational exposure to toluene causes reproductive effects in humans. Reports that an occupational exposure to toluene may lead to an increased incidence of spontaneous abortion have not been supported by the results of animal testing. In animal studies toluene has been shown to be fetotoxic but not teratogenic. On the other hand, Hannigan and Bowen (2010) summarized the evidence to suggest that the risk for pregnancy problems, as well as developmental delays and neurobehavioral difficulties, is higher for the children of women who have been exposed to high concentrations of organic solvents (toluene abuse) during pregnancy than for those who have not. They further stated that the risks appear to be higher in cases of abuse exposure to solvents such as toluene, particularly in comparison to the risk for teratogenic outcomes with occupational solvent exposure. Children of toluene abusers exposed to toluene prenatally have been shown to have low birth weight, shorter gestation, dysmorphic features similar to fetal alcohol syndrome, microcephaly, abnormal muscle tone, and blunt fingertips with small fingernails (Hersh et al., 1985; Wilkins-Haug and Gabow, 1991; Arnold et al., 1994; Sheeres and Chudley, 2001). Children prenatally exposed to toluene may exhibit behavioral problems such as hyperactivity, head banging, and aggressive behavior. Deficiencies are also apparent in speech and cognitive and motor skills (Arnold et al., 1994).

Renal Toxicity   Studies of chronic toluene abusers, occupationally exposed workers, and laboratory animals have provided little support for serious kidney damage due to inhaled toluene. Chronic abuse of toluene can produce acidosis, but in most cases, renal dysfunction is transient and normal function returns when exposure ceases (Vyskocil et al., 1991; ATSDR, 2000).

Carcinogenicity   The IARC (IARC, 1999) and the US Environmental Protection Agency (USEPA, 2005a) consider toluene as "not classifiable" as to human carcinogenicity (Groups 3 and D, respectively).

Mechanism of Action

The mechanisms underlying the observed neurotoxicity of toluene are not fully understood but appear to be related to concentrations of the parent compound, rather than metabolites, reaching the brain (van Asperen et al., 2003; Benignus et al., 2007; Bushnell et al., 2007; Aylward et al., 2008). Toluene is distributed in lipid-rich regions of the brain; however, the pathologic action of toluene after localizing in the white matter is uncertain (Filley et al., 2004).

Unger et al. (1994) studied T2-weighted MRI scans of eight patients with histories of toluene abuse. They found marked hypointensity in thalami and moderate hypointensity in the basal ganglia. The authors suggested that toluene is partitioned in the lipid membranes of cells in the cerebral tissue, which accounts for the hypointensity of basal ganglia. Rosenberg et al. (1988) studied MRI scans of six chronic toluene abusers. They found diffuse cerebral, cerebellar, and brainstem atrophy; loss of differentiation between gray and white matter throughout the CNS; and increased periventricular white matter signal intensity on T2-weighted images. The authors suggested that these changes may be due to the increased water content of the white matter or toluene-induced metabolic changes in myelin. Further studies showed a correlation between the levels of diffuse white matter damage and impaired neuropsychological functioning (Filley et al., 1990).

In vitro work with guinea pig hippocampal slices showed that toluene had a depressive effect on postsynaptic responses (Ikeuchi and Hirai, 1994). Longer exposures produced lasting effects without visible morphological changes.

High levels of toluene have been shown to induce some cytochrome P-450 isozymes in rat livers (Nakajima and Wang, 1994). Although occupational levels are probably too low to induce P-450 isoezymes, they may be induced by an acute exposure among toluene abusers.

Bowen et al. (2006) reviewed 10 years of mechanistic studies using animal models of inhalant abuse and reported the following possible mechanisms: the action of toluene may exert effects through interaction with *N*-methyl-D-aspartate (NMDA) receptor inhibition and modulation; through inhibition of nicotinic acetylcholine receptors (nAChRs); by altering membrane permeability; by increasing the ionic currents activated by serotonin through $5HT_3$ receptors; by inhibiting voltage-gated calcium channels; by increased activity of the dopaminergic system involving gamma-aminobutyric acid (GABA) and glutamanergic systems; and through interactions with various other ion channels. Validation of these mechanisms needs further studies.

SOURCES OF EXPOSURE

Occupational

Absorption occurs via inhalation, dermal contact, and ingestions; however, inhalation of the vapor is generally the most important route. The NIOSH (2010) estimates that 4.8 million workers are exposed to toluene, the fourth largest number for an individual chemical. The primary source of occupational exposures is industrial the manufacturing of toluene diisocyanate, toluene sulfonates, nitrotoluenes, vinyl toluene, phenol, benzyl alcohol, saccharin, glue, paint, and transmission fluid. Common occupational exposures occur in printing operations and painting.

Environmental

The airborne toluene level inside and outside homes is usually <1 ppm in cities and suburbs that are not close to an industry. Automobile emissions are the principal source of toluene to the ambient air. Toluene may also be released to the ambient air during the production, use, and disposal of industrial and consumer products that contain toluene (USEPA, 2012).

Significant nonoccupational exposure to toluene occurs in hobbyists involved with model making or any craft utilizing toluene-containing glue. Drug abusers are exposed through sniffing (nasally inhaled) or huffing (orally inhaled) toluene-based products from a plastic or paper bag, can, plastic bottle, saturated rag or coat sleeve, or directly from the container. Inhalation of solvents, especially toluene, is a common and often overlooked form of chemical abuse. The inhalation of paint fumes and glue sniffing are popular forms of toluene abuse (Meadows et al., 1996; Filley et al., 2004).

INDUSTRIAL HYGIENE

Employers should use engineering controls, personal protective equipment (PPE) and clothing, and worker education programs to reduce exposures to toluene and other organic solvents at least to the concentrations specified in existing guidelines.

BIOMONITORING

Potential biomarkers of toluene exposure include blood samples for toluene; urine samples for toluene, hippuric acid, *ortho*-cresol, *S*-toluylmercapturic acid, or *S*-benzylmercapturic acid; and breathe samples for toluene (Aylward et al., 2008).

A large study of Chinese workers exposed to toluene found that the level of hippuric acid in urine per gram creatinine had a strong linear relationship with toluene exposure (Liu et al., 1992). However, an earlier study by Foo et al. (1991) noted that urine hippuric acid did not correlate well with low level exposure to airborne toluene and that measuring toluene in the expired air of workers was a more sensitive biomonitoring technique at the lower exposure levels. Additional research by Kawamoto et al.

(1994) reported that toluene metabolism and hippuric acid concentrations vary among genotypes. As would be expected, hippuric acid has been shown to be present in the urine of toluene abusers who are exposed to high concentrations of toluene (Raikhlin-Eisendraft et al. (2001). Yoshida et al. (2005) reported a semi-quantitative colorimetric method for screening of hippuric acid in toluene sniffers' urine.

Urinary o-cresol has also been investigated as a method for monitoring toluene exposure (Inoue et al., 1994, 2008). It did not prove to be as strong an indicator as hippuric acid, because the results were confounded by cigarette smoking and alcohol consumption.

Using gas chromatography, Inoue et al. (2008) evaluated blood toluene levels, against urine samples analyzed for benzyl alcohol, benzymercapturic acid, o-cresol, hippuric acid, and toluene. They concluded that urine benzyl alcohol, benzymercapturic acid, hippuric acid, and o-cresol correlated well with toluene in the air only when exposures were intense (e.g., 50 ppm or above), but did not correlate well when airborne exposures were low (e.g., 2 ppm). Blood toluene and urine toluene levels correlated well with both high and low airborne exposure levels leading to the conclusion that blood and urine toluene may be employed to estimate toluene exposure at low levels. Biomonitoring for toluene may use either urine or venous blood (ACGIH, 1996; Kawai et al., 1992). Venous blood containing 1 mg/l of toluene (end of shift) would be average for a worker exposed to the TLV. For hippuric acid in urine, a level of 2.5 g/g creatinine (end of shift—last 4 h) would be considered average for workers exposed to the TLV.

Aitio et al. (1984) suggested that the painters' habit of washing their hands with solvents after a workday may artificially elevate biomonitoring data for toluene. It is known that benzyl alcohol, benzymercapturic acid, or hippuric acid will not be formed after inhalation of benzene, hexane, or xylenes. It is also known that no metabolic interaction will take place between organic solvents when the exposure is low, that is, below occupational exposure limits (Inoue et al., 2008). Ethanol has been shown to have an effect on the metabolism of toluene (Ingren et al., 1996). Even a low blood concentration of ethanol has yielded dose-dependent decreases in toluene metabolism that could confound research and biomonitoring (Baelum et al., 1993; Imbirani and Ghittori, 1997).

MEDICAL MANAGEMENT

Adults seeking hospitalization for muscle weakness, nausea, vomiting and abdominal pain, or neurological problems should be asked about solvent inhalation. Laboratory findings of hyperchloremia, hypobicarbonatemia, hypokalemia, hypophosphatemia, and acidosis with a normal or increased anion gap associated with distal renal tubular acidosis indicate the potential involvement of toluene. Profound hypokalemia can cause muscle weakness (Streicher et al., 1981; Baskerville et al., 2001).

Management of acute toluene intoxication includes supportive care as needed, including cardiac monitoring for potential arrhythmias. The treatment should focus on correction of the resulting metabolic disorders and careful monitoring of serum electrolytes combined with aggressive electrolyte replacement with intravenous (IV) potassium, magnesium, and phosphate (Wilkins-Haug and Gabow, 1991). Bicarbonate replacement may be needed in severe acidosis with a pH < 7.2. Recovery is usually rapid with replacement of the electrolyte disorders (Baskerville et al., 2001; Camara-Lemarroy et al., 2012). Induction of vomiting should be avoided if hydrocarbon ingestion has occurred due to the risk of pulmonary aspiration. Dextrose-containing solutions and tocolysis with terbutaline should be avoided because they may further lower electrolytes. Poison control centers can provide excellent assistance for managing any overdose or intoxication. Contacting the Poison Control Center national hotline number (1-800-222-1222) connects the caller with the local poison control center in the area code where the call originates. If toluene abuse is suspected, admission into an abuse treatment program is advisable.

REFERENCES

American Conference of Governmental Industrial Hygienists (2007) *Threshold Limit Values (TLVs) for Chemical Substances and Physical Agents and Biological Exposure Indices (BEIs)*, Cincinnati, OH: American Conference of Governmental Industrial Hygienists.

Aitio, A., Pekari, K., and Jävisalo, J. (1984) Skin absorption as a source of error in biological monitoring. *Scand. J. Work Environ. Health* 10:317–320.

Arnold, G.L., et al. (1994) Toluene embryopathy: clinical delineation and developmental follow-up. *Pediatrics* 93:216–220.

Agency for Toxic Substances and Disease Registry (ATSDR) (2000) *Toxicological Profile for Toluene*, Atlanta, GA: U.S. Public Health Service, U.S. Department of Health and Human Services.

Aylward, L.L., Barton, H.A., and Hays, S.M. (2008) Biomonitoring Equivalents (BE) dossier for toluene. *Regul. Toxicol. Pharmacol.* 51 (Suppl. 3):S27–S36.

Baelum, J., et al. (1993) Hepatic metabolism of toluene after gastrointestinal uptake in humans. *Scand. J. Work Environ. Health* 19:55–62.

Baskerville, J.R., Tichenor, G.A., and Rosen, P.B. (2001) Toluene induced hypokalaemia: a case report and literature review. *Emerg. Med. J.* 18:514–516.

Benignus, V.A., et al. (2007) Quantitative comparisons of the acute neurotoxicity of toluene in rats and humans. *Toxicol. Sci.* 100:146–155.

Boor, J.W. and Hurtig, H.I. (1977) Persistent cerebellar ataxia after exposure to toluene. *Ann. Neurol.* 2:440–442.

Bowen S.E., Batis, J.C., Paez-Martinez, N., and Cruz, S.L. (2006) The last decade of solvent research in animal models of abuse: mechanistic and behavioral studies. *Neurotoxicol. Teratol.* 28(6):636–647.

Bushnell, P.J., et al. (2007) A dosimetric analysis of the acute behavioral effects of inhaled toluene in rats. *Toxicol. Sci.* 99:181–189.

Camara-Lemarroy, C.R., et al. (2012) Clinical presentation and management in acute toluene intoxication: a case series. *Inhal. Toxicol.* 24(7):434–438.

Deschamps, D., Geraud, C., and Dally, S. (2000) Cognitive functions in workers exposed to toluene: evaluation at least 48 hours after removal from exposure. *Int. Arch. Occup. Environ. Health* 74:285–288.

Filley, C.M., Heaton, R.K., and Rosenberg, N.L. (1990) White matter dementia in chronic toluene abuse. *Neurology* 40:532–534.

Filley, C.M., Halliday, W., and Kleinschmidt-DeMasters, B.K. (2004) The effects of toluene on the central nervous system. *J. Neuropathol. Exp. Neurol.* 63(1):1–12.

Foo, S.C., et al. (1991) Biological monitoring for occupational exposure to toluene. *Am. Ind. Hyg. Assoc. J.* 52:212–217.

Gericke, C. et al. (2001) Multicenter field trial on possible health effects of toluene. III. Evaluation of effects after long-term exposure. *Toxicology* 168:185–209.

Ghantous, H. and Danielsson, B.R.G. (1986) Placental transfer and distribution of toluene, xylene and benzene, and their metabolites during gestation in mice. *Biol. Res. Pregnancy Perinatol.* 7:98–105.

Greenberg M.M. (1997) The central nervous system and exposure to toluene: a risk characterization. *Environ. Res.* 72:1–7.

Hannigan, J.H. and Bowen, S.E. (2010) Reproductive toxicology and teratology of abused toluene. *Syst. Biol. Reprod. Med.* 56:184–200.

Hersh, J.H., et al. (1985) Toluene embryopathology. *J. Pediatr.* 106:922–927.

Holló, G. and Varga, M. (1992) Toluene and visual loss. *Neurology* 42:266.

Hormes, J.T., Filley, C.M., and Rosenberg, N.L. (1986) Neurologic sequelae of chronic solvent vapor abuse. *Neurology* 36:698–702.

International Agency for Research on Cancer (IARC) (1999) *Summaries and Evaluations: Toluene*, vol. 71, Lyon, France: International Agency for Research on Cancer, p. 829.

Ikeuchi, Y. and Hirai, I.H. (1994) Toluene inhibits synaptic transmission without causing gross morphological disturbances. *Brain Res.* 664:266–270.

Imbirani, M. and Ghittori, S. (1997) Effects of ethanol on toluene metabolism in man. *G. Ital. Med. Lav. Ergon.* 19(4):177–181.

Inoue, O., et al. (1994) Effects of smoking and drinking habits on urinary o-cresol excretion after occupational exposure to toluene vapor among Chinese workers. *Am. J. Ind. Med.* 25:697–708.

Inoue, O., et al. (2008) Limited value of o-cresol and benzylmercapturic acid in urine as biomarkers of occupational exposure to toluene at low levels. *Ind. Health* 46:318–325.

Iregren, A., et al. (1996) Experimental exposure to toluene in combination with ethanol intake: psychophysiological functions. *Scand. J. Work Environ. Health* 12:128–136.

Kawai, T., et al. (1992) Comparative evaluation of urinalysis and blood analysis as means of detecting exposure to organic solvents at low concentrations. *Int. Arch. Occup. Environ. Health* 64:223–234.

Kawamoto, T., et al. (1994) Distribution of urinary hippuric acid concentrations by ALDH2 genotype. *Occup. Environ. Med.* 51:817–821.

Lindbohm, M.L., et al. (1992) Effects of parental occupational exposure to solvents and lead on spontaneous abortion. *Scand. J. Work Environ. Health* 18(Suppl. 2):37–39.

Liu, S. et al. (1992) Toluene vapor exposure and urinary excretion of hippuric acid among workers in China. *Am. J. Ind. Med.* 22:313–323.

Meadows, R., et al. (1996) Medical Complications of Glue Sniffing. *South. Med. J.* 89(5):455–462.

Meyer-Baron, M. (2005) A meta-analytical approach to neurobehavioral effects of occupational toluene exposure. *Environ. Toxicol. Pharmacol.* 19(3):651–657.

Nakajima, T. and Wang, R.S. (1994) Induction of cytochrome P-450 by toluene. *Int. J. Biochem.* 26:1333–1340.

National Institute for Occupational Safety and Health (NIOSH) (2010) *Pocket Guide to Chemical Hazards*, NIOSH 94-116, Washington, DC: US Department of Health and Human Services.

Raikhlin-Eisendraft, B., et al. (2001) Determination of urinary hippuric acid in toluene abuse. *J. Toxicol. Clin. Toxicol.* 39(1):73–76.

Rosenberg, N.L., et al. (1988) Toluene abuse causes diffuse central nervous system white matter changes. *Ann. Neurol.* 23:611–614.

Sheeres, J.J. and Chudley, A.E. (2001) Solvent abuse in pregnancy: a perinatal perspective. *J. Obstet. Gynecol. Can.* 24:22–26.

Streicher, H.Z., et al. (1981) Syndromes of toluene sniffing in adults. *Ann. Intern. Med.* 94:758–762.

Unger, E., et al. (1994) Toluene abuse: physical basis for hypointensity of the basal ganglia on T2-weighted MR images. *Radiology* 193:473–476.

United States Environmental Protection Agency (USEPA) (2005a) *Guidelines for Carcinogen Risk Assessment*. March.

United States Environmental Protection Agency (USEPA) (2005b) *Toxicological Review of Toluene (CAS No. 108-88-3) In Support of Summary Information on the Integrated Risk Information System (IRIS) September.*

United States Environmental Protection Agency (2012) *Toluene Fact Sheet, (CAS No. 108-88-3) Hazard Summary-Created in April 1992; Revised in July.*

van Asperen, J., Rijcken, W.R., and Lammers, J.L. (2003) Application of physiologically based toxicokinetic modeling to study the impact of the exposure scenario on the toxicokinetics and the behavioural effects of toluene in rats. *Toxicol. Lett.* 138:51–62.

Van Oettingen, W.F., et al. (1942a) The toxicity and potential dangers of toluene: preliminary report. *JAMA* 118:579–584.

Van Oettingen, W.F., et al. (1942b) The toxicity and potential dangers of toluene with special reference to its maximal permissible concentration. *Public Health Service Bull.* 279:1–53.

Vyskocil, A., et al. (1991) Urinary excretion of proteins and enzymes in workers exposed to hydrocarbons in a shoe factory. *Int. Arch. Occup. Environ. Health* 63:359–362.

Xiao, J.Q. and Levin, S.M. (2000) The diagnosis and management of solvent-related disorders. *Am. J. Ind. Med.* 37(1):44–61.

Wilkins-Haug, L. and Gabow, P.A. (1991) Toluene abuse during pregnancy: obstetric complications and perinatal outcomes. *Obstet. Gynecol.* 77(4):504–509.

Yoshida, M., et al. (2005) Simple colorimetric semi-quantitation method of hippuric acid in urine for demonstration of toluene abuse. *Leg. Med. (Tokyo)* 7(3):198–200.

XYLENES

Target organ(s): Eyes, skin, CNS, respiratory system, liver, and kidneys. The primary target is the CNS.

OSHA PEL 100 ppm (435/m^3 as an 8-h TWA).

The NIOSH (2007) has set the REL for xylene at 100 ppm (435/m^3 TWA); with a STEL of 150 ppm (655 mg/m^3); and an IDLH concentration of 900 ppm.

The ACGIH has set a TLV for exposure to xylene at 100 ppm (435/m^3 TWA) with a STEL of 150 ppm (655 mg/m^3 STEL).

Xylene is classified as an A4 carcinogen indicating that not enough information is available to determine whether it is carcinogenic in humans.

Inhalation RfC = 0.1 mg/m^3. NOAEL 50 ppm; LOAEL 100 ppm.

Oral RfD 0.2 mg/kg-day; NOAEL 250 mg/kg-day; LOAEL 500 mg/kg-day.

ATSDR (2007) inhalation MRL: Acute: 2 ppm (≤14 days); intermediate: 0.6 ppm (15–364 days); chronic: 0.05 ppm (≥1 year). Oral MRL: Acute: 1 mg/kg/day (≤14 days); intermediate: 0.4 mg/kg/day (15–364 days); chronic: 0.2 mg/kg/day (≥1 year).

Risk/safety phrases: Inhalation risk; irritant.

BACKGROUND AND USES

Xylenes are used as industrial solvents, synthetic intermediates, and as solvents in commercial products such as paints, coatings, adhesives, adhesive removers, inks, dyes, cleaning products, paint thinners, and varnish thinners. Some paints contain more than 50% xylenes. It is used as a solvent in printing, rubber, and leather industries and is used in the production of phthalic anhydride, plasticisers, and polyesters. It is commonly found in degreasers as well as solvents for resins and gums. It is found in small quantities in airplane fuel, gasoline, and cigarette smoke. In medical technology and industry, it is widely used as a solvent. In dentistry, it is used in histological laboratories for tissue processing, staining and cover slipping, and also in endodontic retreatment (Earnstgard et al., 2002; Jacobson and McLean, 2003; Kandyala et al., 2010).

PHYSICAL AND CHEMICAL PROPERTIES

The xylene in commercial use is composed of a mixture of the three isomers ortho-xylene, meta-xylene, and para-xylene; the meta isomer predominates in these mixtures. *o*-Xylene and *m*-xylene are clear, colorless, flammable liquids that have characteristically sweet, balsam-like odors. At low temperatures (below 13 °C [56°F]), the para-isomer occurs in the form of clear, colorless plates. Common impurities include ethylbenzene, trimethylbenzene, phenol, and pyridine.

The CAS numbers, vapor pressures, and solubility for the *meta*, *ortho*, and *para* forms of xylene are listed in Table 60.2. All isomers are colorless liquids with aromatic odors at standard temperature and pressure.

Xylene has a structural formula: $C_6H_4(CH_3)_2$, molecular weight: 106, boiling point (760 torr): 135–145 °C (275–293°F), specific gravity (water = 1): 0.86 at 20 °C (68°F), vapor density (air = 1 at boiling point of xylene): 3.7, melting point: −25 °C (−13°F), vapor pressure at 20 °C (68°F): 7–9 torr, and evaporation rate (butyl acetate = 1): 0.6. It is relatively insoluble in water, but it is soluble in alcohol, ether, acetone, and benzene.

ACUTE AND CHRONIC HEALTH EFFECTS

All routes of exposure to xylene affect primarily the CNS, inhalation can affect the respiratory tract, and at high levels of oral exposure xylene can adversely affect the liver and kidney. Acute excessive inhalation exposures are rare because xylene vapor is irritating to the eyes, nose, and throat at high concentrations. Many of the observations available about the acute effects for xylene are derived

TABLE 60.2 Xylene Isomers

Xylene Isomers		CAS Number	Vapor Pressure at 68°F/mmHg	Solubility in Water at 88°F, % by Weight
ortho-	1,2 Dimethylbenzene	95-47-6	7	0.02
meta-	1,3 Dimethylbenzene	108-38-3	9	Slight
para-	1,4 Dimethylbenzene	106-42-3	9	0.02

from household accidents or suicide attempts. Commonly cited symptoms due to xylene exposure include headache, fatigue, irritability, lassitude, motor incoordination, and impairment of equilibrium. Gastrointestinal disturbances such as anorexia and nausea can occur. Flushing, redness of the face, a sensation of increased body heat, increased salivation, tremors, dizziness, confusion, and cardiac irritability have also been reported.

High levels of xylene exposure have been associated with polyuric renal failure, respiratory failure, hemorrhages, and necrosis in the brain, liver, kidneys, and heart. Other effects of high level exposures include agitation, headache, light-headedness, dysphasia, shivering, respiratory tract irritation, ataxia, slurred speech, dyspnea, hyperreflexia, vertigo, inebriation, diffuse opacity of lungs, and unconsciousness (Langman, 1994). An exposure to xylene has also been reported to cause urticaria in medical workers (Palmer and Rycroft, 1993).

Subjective symptoms of nose and throat irritation have been reported following single exposures to xylene vapor at concentrations between 50 and 700 ppm, repeated intermediate-duration exposure at 100 ppm, or chronic duration occupational exposure at 14 ppm. The NOAEL for acute CNS effects in humans has been reported to be about 70 ppm for a 4-h exposure (Earnstgard et al., 2002). Mild CNS effects (subjective symptoms of intoxication, headache, fatigue, and dizziness) have been observed following acute duration exposure of humans to *m*-xylene at 50 ppm and chronic duration exposure to mixed xylene at 14 ppm. Effects in humans exposed to 100–150 ppm *p*-xylene for 1–7.5 hs/day, 5 days per week for 4 weeks included increased subjective symptoms of irritation of the nose and throat; however, no significant alterations in objective measures of neurologic or respiratory function were found (NIOSH 2007; ATSDR, 2007). Experimental studies indicate that an acute exposure to 100 ppm mixed xylene or 200 ppm *m*-xylene causes impaired short-term memory, impaired reaction time, performance decrements in numerical ability, and alterations in equilibrium and body balance. Isolated cases of unconsciousness, amnesia, brain hemorrhage, and seizures, and even death have been associated with accidental acute inhalation exposure to unknown concentrations of xylene (estimated in one case as 10,000 ppm).

An acute exposure to xylene is typically through inhalation, although xylene vapor can be absorbed directly through the skin (McDougal et al., 1990). Skin exposures can cause drying, erythema, scaling, and blistering because of xylene's defatting action.

The signs and symptoms of chronic exposure to xylene may include conjunctivitis; dryness of the nose, throat, and skin; dermatitis; and kidney and liver damage. Shipyard workers exposed to xylene were found to have decreased peripheral nerve function (Ruijten et al., 1994). Xylene is also suspected of causing hearing loss and may result in other CNS symptoms (Gagnaire and Langlais, 2005).

A large occupational study with worker mean exposures of about 14 (geometric mean) and 21 ppm (arithmetic mean) showed an increased prevalence of subjective symptoms in the exposed workers, including local effects on the eyes, nose, and throat; however, hematology and serum biochemistry with respect to the liver and kidney functions were generally negative, indicating that xylenes are not toxic to the hematopoietic organs, liver, or kidney in low dose exposure of occupational settings (Uchida et al., 1993).

Carcinogenicity

Current data are inadequate for an assessment of the carcinogenic potential of xylenes (USEPA, 2003). Adequate human data on the carcinogenicity of xylenes are not available, and the available animal data are inconclusive as to the ability of xylenes to cause a carcinogenic response. Evaluations of the genotoxic effects of xylenes have consistently given negative results.

MECHANISM OF ACTION

Xylene is highly lipophilic and is rapidly absorbed, rapidly distributed and rapidly eliminated from the body. The lipophilic nature of xylene is thought to be responsible for the narcotic and anesthetic effects on the CNS. The mechanisms of anesthesia are not well understood but probably involve xylene interaction with the neuronal membranes resulting in alteration of nerve impulse transmission. A study by Ito et al. (2002) suggested that experimentally induced high levels of *m*-xylene in rat brains increased GABA release and/or enhanced GABA$_A$ receptor function in the cerebellum, which could cause adverse effects on motor coordination. Various studies have suggested possible adverse effects of xylene on CNS neurotransmitters and lipid compositions in areas of the brain; death of cochlear hair cells and ototoxicity; and inhibition of pulmonary microsomal enzymes. The cytotoxicity of *m*-xylene has been associated with reductions in cellular antioxidant status and studies have suggested that hydroxyl radicals may play a role in xylene toxicity (Coleman et al., 2003; Foy and Schatz, 2004).

Xylene causes an increase in cytochrome P-450 enzymes and several other liver enzymes, proliferation of the smooth endoplasmic reticulum, and enlargement of the rough endoplasmic reticulum in laboratory animals (Ungváry, 1990). Although xylene can result in these morphological changes, it is not hepatotoxic at occupational or environmental levels of exposure. A study of liver function in paint workers exposed to xylene supports this conclusion (Chen et al., 1991).

The mechanisms for xylene renal toxicity are also unknown; however, several mechanisms have been suggested, including formation of reactive metabolites and subsequent irritation or direct membrane fluidization; generation of oxidative intermediates and subsequent necrosis; and apotosis via activation of mitochondrial enzymes.

The effect of xylene exposure on pregnancy outcome has been reported. Women exposed to xylene in one case-control study had a greater risk (odds ratio 3:1) of poor pregnancy outcome (Taskinen et al., 1994). Saillenfait et al. (2003) concluded that developmental toxicity of *m*- and *p*-xylenes in standard animal bioassays is largely related to maternal toxicity. Xylene exposure was associated with a reduction in maternal weight gain and decreases in fetal body weights; however, there was no evidence of teratogenic effects. More research must be completed before xylene's effect on reproduction can be reliably determined.

Studies on rats indicate that *m*-xylene is more efficiently absorbed through the skin than toluene, benzene, or hexane (McDougal et al., 1990). Experiments with mice indicate xylene-enhanced liver damage in virus-infected mice (Selgrade et al., 1993). Ethanol may be an antagonist to xylene toxicity (Padilla et al., 1992). Xylene has been shown to cause hearing loss in rats (Rybak, 1992; Morata et al., 1994). Human experimental exposure to *m*-xylene has been reported to cause minor changes in electroencephalogram, although no adverse effects were noted (Seppäläinen et al., 1991).

SOURCES OF EXPOSURE

Possible exposure routes for xylene include inhalation, ingestion, eye contact, and to a small extent, through the skin.

Occupational

Occupational exposures to xylene occur in painters, laboratory workers, workers involved with distillation and purification of xylene, industries using xylene as a raw material, the petroleum industry, rubber and leather industries. Increased exposures have been observed for wood processing plant workers, painters, gas station employees, metal workers, and furniture refinishers (Kandyala et al., 2010). See paragraph above on background and abuses.

Environmental

Xylenes are released into the atmosphere from industrial sources (e.g., petroleum refineries, chemical plants), in automobile exhaust, and through volatilization from their use as solvents. Discharges into waterways and spills on land result primarily from use, storage and transport of petroleum products, and waste removal. Xylene concentrations range from 1 to 30 parts per billion (ppb) in outdoor air and 1–10 ppb in indoor air. Xylene has been detected in <5% of samples from groundwater surveys and the median concentrations have been less than or equal to 2 ppb. People can also be exposed to xylenes through smoking, consumption of xylene, and dermal contact with consumer products containing xylene. Inhalation is the main route of exposure to xylene. Xylene and toluene are chemicals known to be released from emission sources into the environment. A study in Japan found no difference between the indoor and outdoor concentrations of xylene and that levels of xylene in the living environment were not likely to cause adverse health effects (Sekizawa et al., 2007). No adverse health effects have been associated with background levels of xylene to which the general population is typically exposed.

INDUSTRIAL HYGIENE

Methods that can be used to control occupational exposures include process enclosure, local exhaust ventilation, general dilution ventilation, and as a last resort, PPE. The PPE can include safety goggles, a face shield, gloves, aprons, boots, and for high level ambient air concentrations a NIOSH-approved air-purifying respirator.

BIOMONITORING

The major pathway for metabolism of xylene involves mixed function oxidases in the liver, resulting mainly in the formation of isomers of methylhippuric acids (MHAs) in urine that can be measured in biomonitoring for xylene exposure. Workers exposed to the ACGIH TLV of 100 ppm should have 1.5 g MHA per gram creatinine (ACGIH, 2001) in their urine (end of shift). Other methods for biomonitoring, such as measurement of xylene in exhaled air, are also available (Glaser et al., 1990). Gas chromatography analysis of blood xylene has the lowest detection limit (Kawai et al., 1992). Jacobson and McLean (2003) reported that biological monitoring of occupational xylene exposure at levels <15 ppm using urinary MHA showed good correlation with airborne levels and is a valid complement to ambient monitoring. They added that recent exposure to xylene prior to biological monitoring can affect the results and should be considered. Their work predicted a

MHA concentration in post-shift urine of 1.3 g/g creatinine after exposure to a TWA of 100 ppm xylene, similar to the end of shift concentration previously described by the ACGIH. The highly lipophilic properties of xylene result in rapid absorption, rapid distribution, and quick elimination from the body. Measurement of blood levels of xylene is limited by the rapid metabolism of xylene. Measurement of xylene exposure by means of MHA in the urine should occur soon after exposure.

MEDICAL MANAGEMENT

Treatment of acute intoxication by xylene should include supportive care as needed, similar to other aromatic hydrocarbons. Fluid and electrolyte status as well as electrocardiogram (ECG) and vital signs must be monitored carefully. The clinician should also monitor for kidney failure and for pulmonary edema and respiratory failure, which may be delayed 24–72 h.

If xylene is ingested, induction of vomiting is generally not recommended due to the risk of hydrocarbon aspiration and chemical pneumonitis. Gastric lavage may be performed shortly after ingestion if significant amounts have been consumed (>5 ml) and the benefits outweigh the risk of aspiration. The volume of return should approximate the volume given.

In case of ocular exposure, the eyes should be irrigated for at least 15 min with warm water. If irritation persists, an ophthalmologist must be consulted.

Poison control centers can provide assistance for managing any overdose or intoxication. Contacting the Poison Control Center national hotline number (1-800-222-1222) connects the caller with the local poison control center in the area code where the call originates.

REFERENCES

American Conference of Governmental Industrial Hygienists (ACGIH) (2001) *Threshold Limit Values (TLVs) for Chemical Substances and Physical Agents and Biological Exposure Indices (BEIs)* Cincinnati, OH: 2001 American Conference of Governmental Industrial Hygienists.

ATSDR (2007) *Toxicological Profile for Xylene.*

Chen, J.D., Wang, J.D., Jang, J.P., and Chen, Y.Y. (1991) Exposure to mixtures of solvents among paint workers and biochemical alterations of liver function. *Br. Jour. Ind. Med.* 48:696–701.

Coleman, C.A., Hull, B.E., McDougal, J.N., et al. (2003) The effect of *m*-xylene on cytotoxicity and cellular antioxidant status in rat dermal equivalents. *Toxicol. Lett.* 142:133–142.

Earnstgard, L., Gullstrand, A.L., and Johanson, G. (2002) Are women more sensitive than men to 2-propanol and *m*-xylene vapours? *Occup. Environ. Med.* 59:759–767.

Foy, J.W.D. and Schatz, R.A. (2004) Inhibition of rat respiratory-tract cytochrome P-450 activity after acute low-level *m*-xylene inhalation: role in 1-nitronaphthalene toxicity. *Inhal. Toxicol.* 16:125–132.

Gagnaire, F. and Langlais, C. (2005) Relative ototoxicity of 21 aromatic solvents. *Arch. Toxicol.* 79(6):346–354.

Glaser, R.A., Arnold, J.E., and Shulman, S.A. (1990) Comparison of three sampling and analytical methods for measuring *m*-xylene in expired air of exposed humans. *Am. Ind. Hyg. Assoc. J.* 51:139–150.

Ito, T., Yoshitome, K., Horike, T., et al. (2002) Distribution of inhaled *m*-xylene in rat brain and its effect on GABA$_A$ receptor binding. *J. Occup. Health* 44(2):69–75.

Jacobson, G.A. and McLean, S. (2003) Biological monitoring of low level occupational xylene exposure and the role of recent exposure. *Ann. Occup. Hyg.* 47(4):331–336.

Kandyala, R., Raghavendra, S.C., and Rajasekharan, S.T. (2010) Xylene: an overview of its health and preventive measures. *J. Oral Maxillofac. Pathol.* 14:1–5.

Kawai, T., et al. (1992) Comparative evaluation of urinalysis and blood analysis as means of detecting exposure to organic solvents at low concentrations. *Int. Arch. Occup. Environ. Health* 64:223–234.

Langman, J.M. (1994) Xylene: its toxicity, measurement of exposure levels, absorption, metabolism and clearance. *Pathology* 26:301–309.

McDougal, J.N., et al. (1990) Dermal absorption of organic chemical vapors in rats and humans. *Fundam. Appl. Toxicol.* 14:299–308.

Morata, T.C., Dunn, D.E., and Sieber, W.K. (1994) Occupational exposure to noise and ototoxic organic solvents. *Arch. Environ. Health* 49:359–365.

National Institute for Occupational Safety and Health (NIOSH) (2007) *Pocket Guide to Chemical Hazards*, NIOSH 94-116, Washington, DC: US Department of Health and Human Services.

Padilla, S., Lyerly, D.L., and Pope, C.N. (1992) Subacute ethanol consumption reverses *p*-xylene-induced decreases in axonal transport. *Toxicology* 75:159–167.

Palmer, K.T. and Rycroft, R.J.G. (1993) Occupational airborne contact urticaria due to xylene. *Contact Dermatitis* 28:44.

Ruijten, M.W., et al. (1994) Neurobehavioral effects of long-term exposure to xylene and mixed organic solvents in shipyard spray painters. *Neurotoxicology* 15:613–620.

Rybak, L.P., (1992) Hearing: the effects of chemicals. *Otolaryngol. Head Neck Surg.* 106:677–686.

Saillenfait, A.M., Gallissot, F., Morel, G., et al. (2003) Developmental toxicities of ethylbenzene, ortho-, meta-, para-xylene and technical xylenes in rats following inhalation exposure. *Food Chem. Toxicol.* 41:415–429.

Sekizawa, J., et al. (2007) Evaluation of human health risks from exposures to four air pollutants in the indoor and the outdoor environments in Tokushima, and communication of the outcomes to the local people. *J. Risk Res.* 10(6):841–851.

Selgrade, M.D., et al. (1993) Enhanced mortality and liver damage in virus-infected mice exposed to *p*-xylene. *J. Toxicol. Environ. Health* 40:129–144.

Seppäläinen, A.M., et al. (1991) Electroencephalographic findings during experimental human exposure to *m*-xylene. *Arch. Environ. Health* 46:16–24.

Taskinen, H., et al. (1994) *Laboratory work and pregnancy outcome.* J. Occup. Environ. Med. 36:311–319.

Uchida, Y., et al. (1993) Symptoms and signs in workers exposed primarily to xylenes. *Int. Arch. Occup. Environ. Health* 64:597–605.

Ungváry, G. (1990) The effect of xylene exposure on the liver. *Acta Morphol. Hung.* 38:245–258.

United States Environmental Protection Agency (USEPA) (2003) Integrated Risk Information System (IRIS) Toxicological Reviews of Xylenes. Washington, DC: National Center for Environmental Assessment, Office of Research and Development.

POLYAROMATIC HYDROCARBONS

Target organ(s): Kidneys, liver, adipose tissue.

OSHA PEL: The OSHA PEL for PAHs in the workplace is $0.2\,mg/m^3$. OSHA has not established a substance-specific standard for occupational exposure to PAHs.

NIOSH (REL): $0.1\,mg/m^3$ for coal tar pitch volatile agents.

The ACGIH (2005) has set a TLV for exposure to $0.2\,mg/m^3$ for benzene-soluble coal tar pitch fraction.

Reference values: Vary by compound.

Risk/safety Phrases: Vary by compound.

BACKGROUND AND USES

Polycyclic aromatic hydrocarbons (PAHs), also known as polynuclear aromatics or polycyclic organic matter, are a large class of organic compounds consisting of three or more fused aromatic rings containing only carbon and hydrogen. The most common PAHs include benzo(a)anthracene, benzo(a)pyrene (BaP), benzo(e)pyrene, benzo(g,h,i)perylene, benzo(k)fluoranthene, chrysene, coronene, dibenz(a,h)acridine, dibenz(a, h)anthracene, and pyrene. They are found naturally in petroleum products such as asphalt and tar. They are produced by incomplete combustion, pyrolysis, or pyrosynthesis of organic matter. They are produced naturally by forest fires and volcanoes; in ambient air from burning of coal, wood, oil or petroleum products; and during coke production and refuse burning (Cherng et al., 1996). They are also found in motor vehicle exhaust fumes. They are found in industries that produce or use coal tar, coke, or bitumen; they are also produced in coal gasification plants, smokehouses, incinerators, and aluminum production facilities.

They can be found in a number of common domestic items. Cooking foods at high temperatures, especially grilling and smoking meats, increases the PAH content of the food. Combustion of fuels in home heating and cooking appliances contributes to domestic PAH levels. Humans expose themselves directly to PAHs in cigarette smoke, mineral oils, skin care products and shampoos, medicines, dyes, and plastics. Polycyclic aromatic hydrocarbons are found in nearly all soils, where concentrations will vary depending on whether the local surroundings are rural, urban, or agricultural.

PHYSICAL AND CHEMICAL PROPERTIES

Polycyclic aromatic hydrocarbons are typically solids with low volatility at room temperature. Solubility varies by substance; however, they are soluble in most organic solvents and are relatively insoluble in water. Most of them can be photooxidized and degraded to simpler substances. Table 60.3 provides an identification of some carcinogenic and some noncarcinogenic PAHs, including their CASRNs and the IARC classifications of the carcinogenic compounds.

ENVIRONMENTAL FATE AND BIOACCUMULATION

According to the ATSDR (1995), background PAHs concentrations in ambient air vary from 0.02–$1.2\,ng/m^3$ in rural areas to 0.15–$19.3\,ng/m^3$ in urban areas (ATSDR, 1995).

TABLE 60.3 Polycyclic Aromatic Hydrocarbons

Carcinogenic	Noncarcinogenic
Benzo(a)pyrene (CASRN 50-32-8) **IARC 1**	Acenapthene (CASRN 83-32-9)
Benzo(a]anthracene (CASRN 56-55-3) **IARC 2**	Anthracene (CASRN 120-12-7)
Benzo(b)fluoranthene (CASRN 205-99-2) **IARC 2B**	Benzo(g,h,i)perylene (CASRN 191-24-2)
Benzo(k)fluoranthene (CASRN 207-08-9) **IARC 2B**	Dibenzofuran (CASRN 132-64-9)
Benzo(g,h,i)perylene (CASRN 191-24-2) **IARC 3**	Fluoranthene (CASRN 206-44-0)
Chrysene (CASRN 218-01-9) **IARC 2B**	Fluorene (CASRN 86-73-7)
Dibenzo(a,h)anthracene (CASRN 53-70-3) **IARC 2B**	2-Methylnaphthalene (CASRN 91-57-6)
Indeno(1,2,3-c,d)pyrene (CASRN 193-39-5) **IARC 2B**	Naphthalene (CASRN 91-20-3)
	Phenanthrene (CASRN 85-01-8)
	Pyrene (CASRN 129-00-0)

Airborne deposition accounts for most of the PAHs found in soil. As previously stated, PAHs are found in nearly all soils with concentrations varying depending on the specific PAH and the environment. They can mobilize out of the soil and into ground- and surface waters. Spills may also lead to contamination of the soil and water. They do bioaccumulate.

ECOTOXICOLOGY

Quantitative information on PAHs varies from compound to compound. Based on solubility, PAHs are not expected to accumulate in water but may accumulate in sediments and soil. Soils near oil refineries and other industrial complexes may be as high as 200,000 µg/kg of dried soil.

MAMMALIAN TOXICOLOGY

Acute Effects

Polycyclic aromatic hydrocarbons have a low degree of acute toxicity in humans. There are minor hazards associated with acute occupational or environmental exposure to PAHs as a class. The NIOSH has set supplementary exposure limits for several PAHs, namely, asphalt fumes (BaP), carbon black and coal tar pitch volatiles. Supportive treatment is recommended for any acute skin, eye, or respiratory irritation.

Chronic Effects

Chronic exposure to PAHs may be associated with increased risk of lung and skin cancers and possibly urological, gastrointestinal, laryngeal, and pharyngeal cancers based on several epidemiological studies, some of which are only suggestive and not conclusive (Clavel et al., 1994; Eisen et al., 1994; Partanen and Boffetta, 1994; Murata et al., 1997; Armstrong et al., 2004; Kim et al., 2013). These studies provide only qualitative evidence of the carcinogenic potential of PAHs in humans because of the presence of multiple PAHs and other exposures to putative carcinogens and the lack of quantitative monitoring data.

Chronic PAH exposure is associated with multiple noncarcinogenic toxicities (Brucker et al., 2013). Long-term reduction in antibody responsiveness can follow administration of 7,12-dimethylbenz-(a)anthracene and BaP to mice (Stjernsward, 1966; Ward et al., 1986). Mice exposed to BaP showed depressed responses to B- and T-cell mitogens (Dean et al., 1993). Exposure of pregnant female mice to BaP resulted in severe depression of antibody responses in pups immediately after birth (Urso and Gengozian, 1980). This suppression persisted for well over 1 year and was accompanied by an increase in the incidence of tumors in adults. 3-Methylcholanthrene, another carcinogenic PAH,

may suppress formation of cytotoxic T-cells (Wodjani and Alfred, 1984). 7,12-Dimethylbenzo(a)anthracene produced long-term immunosuppression of antibody production, natural killing, and cytotoxic T-lymphocyte cytotoxicity (Ward et al., 1986). Thus, 7,12-dimethylbenzo(a)anthracene appeared to affect all phases of immunologic protection in mice, including mechanisms of tumor resistance. Immunological suppression of tumor-resistance mechanisms by PAHs tends to correlate with the carcinogenic potency of the chemical. Immunosuppressive PAHs tend to be carcinogenic, but nonimmunosuppressive congeners are not (Ward et al., 1985).

A study of roofers exposed to PAHs suggested an excess of lung, stomach, and bladder cancer (Partanen and Boffetta, 1994). A 10-year longitudinal study reported by Hansen (1992) revealed that Danish stokers exposed to PAH and benzene had an excess mortality, especially from cancer and circulatory diseases. Hansen et al., 2004, determined that bus drivers in Denmark excreted more 1-hydroxypyrene (1-OHP) in their urine than did mail carriers. This mirrored their PAH exposures.

MECHANISM OF ACTION

Cells most likely affected by PAH exposure appear to be those with rapid replicative turnover such as those in pulmonary, dermal, and reproductive tissues. Tissues with lower turnover rates, such as the liver, are less susceptible. The liver is the main site of metabolism of PAHs, although metabolism can also occur in extrahepatic tissues. Metabolic bioactivation appears to be necessary for production of the proximate carcinogenic form of PAHs.

After absorption, PAHs enter the lymph, circulate in the blood, are metabolized primarily in the liver and kidney, and are excreted in both bile and urine. Additionally, metabolism of PAHs occurs in the adrenal glands, testes, thyroid, lungs, skin, sebaceous glands, and small intestine. Metabolic activation of PAHs is primarily catalyzed by the cytochrome P-450-dependent microsomal enzyme system, aryl hydrocarbon hydroxylase (AHH) (Andrysík et al., 2007). Polycyclic aromatic hydrocarbons are metabolized by AHH initially to epoxides, which are converted by epoxide hydrolases to dihydrodiol derivatives and diol-epoxides (Levin et al., 1982). These metabolites are the reactive carcinogenic intermediates resulting from metabolism and may be detoxified by further conjugation with GSH or other substrates. Glucuronide and sulfate conjugates of these metabolites are excreted in the bile. Glutathione conjugates are further metabolized to mercapturic acid metabolites in the kidney and are excreted in the urine. Polycyclic aromatic hydrocarbons are generally nonmutagenic or very slightly mutagenic *in vitro*; however, their metabolites or derivatives may be potent mutagens. They may induce some of the isozymes of the cytochrome P-450 enzyme system leading to increased metabolism, thereby generating greater amounts of reactive intermediates (electrophiles) capable of forming PAH–DNA adducts at a site critical to cell differentiation and growth.

More than 1500 PAHs have been identified and approximately 16 are commonly monitored. A number of specific toxicities are associated with exposure to PAHs. They have been shown to suppress the immune responses in animals and induce microsomal enzymes; many are carcinogenic in animal models. Benzo(a)pyrene is one of the most potent animal carcinogens of this class. Human exposure to this compound is widespread. It is estimated that annual airborne emission of BaP exceeds 900 tons in the United States alone. Of the PAHs tested, only benzo(a,h)anthracene has been shown to have greater carcinogenic potency than BaP (Krewski et al., 1989), although others, including perylene, cyclopenta(cd) pyrene, and fluoranthene have been shown to have greater mutagenic potency than BaP (Slaga et al., 1981). With the exception of benzo(a,h)anthracene, all other PAHs are much less potent carcinogens than BaP; some lack carcinogenic potency altogether.

SOURCES OF EXPOSURE

Human exposure to PAHs can occur through inhalation, ingestion, and eye or skin contact. The National Academy of Science has estimated that a member of the general population has an average daily intake of 0.21 µg PAHs from air, 0.03 µg from water, and 1.6–16 µg from food (National Academy of Science, 1986).

Occupational

As early as 1775, Sir Percival Pott reported cancer of the scrotum in chimney sweeps as a result of exposure to soot containing PAHs. In 1992, Southam and Wilson reported that 48% of the men diagnosed with scrotal cancer were engaged in mulespinning in the cotton industry, where their thighs and scrotum came in frequent contact with the oil. A causal association between scrotal squamous cell carcinoma in the northwest of England and exposure to PAHs was suggested by Lee, Alderson, and Downes in 1972. These early reports followed by several epidemiological studies have suggested an increased risk of cancer of the skin, lung, bladder, gastrointestinal tract, and scrotum associated with PAH exposure (Southam and Wilson, 1992). Among these, the lungs and skin are the most frequently reported target organs (Lee et. al., 1972).

The atmospheric concentrations of PAHs found during occupational exposures may range from <10 to >10,000 ng/m^3, depending on the source of exposure and on the workplace (Linstedt and Sollenberg, 1982). In cities with coke ovens, the concentration of airborne PAHs may reach 150 ng/m^3 of air (the permissible workplace exposure limit for coke oven emissions is 150,000 ng/m^3). Coal tar pitch and creosote produced in coke ovens contain significant amounts of BaP and other PAHs. Water contamination may occur as a result of industrial effluents and accidental spills from oil shipments at sea. The background level of PAHs in drinking water ranges from 4 to 24 ng/l. Polycyclic aromatic hydrocarbon levels as high as 200 ng/kg of dried soil were documented in soil near oil refineries. The levels can go up as much as 10-fold in soil samples obtained near cities and areas with heavy vehicular traffic. They can leach from the soil into water. The Environmental Protection Agency's (EPA) proposed maximum level of contamination for BaP in drinking water is 0.2 ppb. The concentrations of PAHs in foodstuffs vary. Charring meat or barbecuing other food over a charcoal fire greatly increases the concentration. Smoked fish may contain up to 2.0 µg/kg of BaP. Roasted peanuts and coffee, refined vegetable oil, and many other food items contain PAHs. Cigarette smoke contains PAHs. One cigarette yields 10–50 ng of BaP, 18 ng chrysene, 40 ng dibenzo(a)anthracene, and 12–140 ng benzo(a)anthracene. Asphalt and tar, as well as medicated shampoo and ointments, may contain significant quantities of PAHs.

Industries with workers at greatest risk of PAH exposure are aluminum, asphalt, carbon black, chimney cleaning, coal gasification, coke oven operation, fisheries (coal tar nets), graphite electrodes, industries that employ machinists and mechanics (automobile and diesel engines), printing, road paving, roofing, steel foundries, and tire and rubber manufacturing. Workers exposed to creosote may be carpenters, farmers, railroad workers, tunnel construction workers, and utility workers.

The routes of exposure may be inhalation of PAH aerosols and gas or dermal absorption, and PAHs can also be ingested via the food and water, if contaminated. The predominant route of exposure will be determined by the industry involved and safety precautions taken. Alternate exposures to coal tar may be while being used to treat patients with atopic dermatitis, psoriasis, dermatitis, and chronic eczema. Studies by Hansen et al. (1993a, 1993b) of patients treated with coal tar revealed a dramatic increase in urinary PAH metabolites. Concentrations of 1-hydropyrene reached levels about 100 times that of pretreatment levels during the first week of treatment. Levels of alpha-naphthol increased 5–10 times that of 1-hydropyrene because of the higher levels of naphthalene in coal tar. Two weeks after the treatment was begun, levels of both metabolites had dropped to a third or less of their peak. Polycyclic aromatic hydrocarbon exposure may have induced liver enzymes to effectively metabolize PAH or the skin may become less permeable to PAHs.

A protective airstream helmet was shown to significantly reduce exposure and lower the dose of PAHs received (Jongeneelen et al., 1990). The TLV for PAHs, based on the benzene soluble fraction of sampled dust, is 0.2 mg/m^3 (ACGIH, 2005).

Environmental

The general population may be exposed to various PAHs through contact with contaminated soils or sediments, ingestion of drinking water and foods, and inhalation of air (HSDB, 2003). Polycyclic aromatic hydrocarbons are found in vehicle exhaust, coal, asphalt fumes, coal tar fumes, hazardous waste sites, wildfires, and cigarette smoke.

BIOMONITORING

Determination of PAH exposure can be made with HPLC analysis of 1-hydropyrene in urine (Jongeneelen et al., 1990; Buchet et al., 1992; Hansen et al., 1993a; Hansen et al., 1993b; Boogaard and van Sittert, 1994; Øvrebø et al., 1994). This measurement is usually expressed in milligram of 1-hydropyrene per milligram creatinine or nanogram per millimole of creatinine. To date, this is the most effective method for determining the level of PAH exposure in workers. However, this method is time consuming, inconvenient, and expensive if used for large-scale screening. It is also useful only for exposures that have occurred in the previous 18 h. Further research has attempted to find a more convenient method by which both long- and short-term PAH exposure could be determined. Moreover, quantitative evaluation of 1-hydropyrene in urine does not provide any information useful for evaluating the amount of PAH interaction with DNA.

Because binding of PAH metabolites to DNA is the key to the initiation of cancer, attempts have been made to quantitate PAH–DNA adducts in nucleated white blood cells and also to measure blood protein adducts (albumin and hemoglobin) and use these as surrogates for the target tissue to assess exposure and estimate risk.

Analytical techniques for monitoring DNA adducts in the peripheral blood cells of exposed subjects include immunological methods such as ultrasensitive enzymatic radioimmunoassay (USE-RIA), enzyme-linked immunosorbent assay (ELISA), and 32P-postlabeling assay, and a chemicophysical assay-synchronous fluorescence spectrometry (SFS) (dell'Omo and Lauwerys, 1993). All these methods are sensitive enough to detect PAH–DNA adducts resulting from environmental and occupational exposure. However, a direct comparison of adduct levels measured by different techniques may be misleading because they measure different end points. Reports from different laboratories cannot be compared because of the lack of interlaboratory standardization of the methodology.

Immunoassays use polyclonal antibodies (antisera) or monoclonal antibodies to detect specific DNA adducts. As known universally, ELISA uses a substrate for the enzyme-linked second antibody, which generates either colored or fluorescent products. The procedure for USE-RIA involves a radiolabeled substrate instead of a nonradiolabeled substrate

as used in ELISA. Results from human samples are calculated with reference to benzo(a)pyrene diol epoxide (BaPDE)–DNA adduct curves, and antisera recognize multiple PAH–DNA adducts. Data are usually expressed as levels of BaPDE–DNA antigenicity (in femtomole BaPDE per microgram DNA), which give the equivalent percentage of inhibition. These assays with antisera detect one BaPDE–DNA adduct per 107–108 bases in DNA. A higher sensitivity has been reached with a fiber-optic antibody-based fluoroimmunosensor used to detect BaP tetrols released from placental DNA by mild acid hydrolysis. Rabbit antisera raised against BaPDE-modified DNA have been used in biomonitoring groups occupationally exposed to PAH (i.e., coke-oven workers, foundry workers, roofers, and firefighters) (dell'Omo and Lauwerys, 1993).

Herikstad et al. (1993) have developed a relatively sensitive radioimmunoassay to determine levels of PAH in urine. Their method measures the level of eight PAH metabolites per nanomole creatinine. Although their results reflected seasonal variation (lower levels in winter, higher in summer), the assay does appear to be sensitive enough to rank levels of exposure. This method still has the shortcoming of only reflecting short-term exposure. Immunoassays couple high sensitivity with a relatively simple technique and are particularly suitable for a large number of samples. Principal drawbacks include production and characterization of specific antisera and monoclonal antibodies, as well as rigorous standardization of the methodology.

Several epidemiological studies (Phillips et al., 1988; Hemminki et al., 1990) have involved the 32P-postlabeling assay for the analysis of PAH–DNA adducts. It is a nonspecific procedure detecting several aromatic/hydrophobic adducts, hence providing total adduct levels resulting from interaction with DNA with a host of chemicals in an occupational exposure situation. This method detects the presence of 1 DNA adduct in 107–108 nucleotides (Gupta et al., 1982). A higher sensitivity of up to 1 adduct in $10'$-degree nucleotides can be achieved using the nuclease P1 procedure for the enrichment of DNA digests. However, results may be underestimated because it is not usually known how resistant unknown aromatic adducts are to nuclease P1 digestion. It is claimed to be particularly useful for the measurement of adducts derived from exposure to complex mixtures from unknown carcinogens. It is technically complex and hence can be unsuitable for screening in an occupational health practice. Moreover, different enrichment procedures should be investigated for better adduct recovery, and the final step of adduct detection should be improved to increase specificity.

Synchronous fluorescence spectrometry has been used to analyze BaP–DNA adducts in human lymphocytes because of the intrinsic fluorescence associated with BaP and its metabolites. This method can identify 1 adduct per 107 nucleotides. The specificity is sometimes not enough to detect the chemical of interest in complex mixtures. The specificity can be increased using HPLC as a separating method before quantitation (HPLC/SFS) (Weston et al., 1989). Synchronous fluorescence spectrometry has been used in several epidemiological studies for the detection of BaPDE–DNA adducts (Harris et al., 1985; Vahakangas et al., 1985; Haugen et al., 1986). After isolation using immunoaffinity chromatography, BaP tetrols from placental DNA extracts also have been quantitated by second-derivative SFS (Weston et al., 1989).

Numerous studies have attempted to find a relationship between PAH exposure and levels of PAH–DNA adducts. Results have been conflicting. In some studies, subjects exposed in the workplace to PAH had higher levels of PAH–DNA adducts, whereas in others there was no appreciable difference (Reddy et al., 1991; Szyfter and Hemminki, 1992; Øvrebø et al., 1994). Even in studies that reported differences between those occupationally exposed and controls, there was no strong association between levels of DNA adducts and actual exposure level. Studies of exposure to PAHs resulting from environmental pollution and levels of PAH–DNA adducts also failed to find a correlation (Hemminki et al., 1990). An overall evaluation indicates that these methods of detection of PAH–DNA adducts can presently be applied for assessing PAH exposure on a group basis, but they are not suitable for assessing exposure or predicting adverse health effects on an individual basis because neither normal nor toxic levels have been determined.

Studies, although few, have been conducted to determine whether a correlation between PAH exposure and PAH–hemoglobin adducts exists. Results suggest that, although those exposed to PAHs may have higher concentrations of adducts, there is not a direct mathematical correlation to exposure level (Ferreira et al., 1994; Perera et al., 1994).

Somatic gene mutation at the hypoxanthine guanine phosphoribosyl transferase (HPRT) locus in lymphocytes has been tested as an indicator for exposure to environmental mutagenic agents like PAHs. According to Perera et al. (1994), HPRT mutation frequency shows a direct correlation with PAH-DNA binding, which is consistent with experimental studies involving BaP (Arce et al., 1987; MacLeod et al., 1988). Further research is required to analyze spectra of HPRT mutation in the workers exposed to PAH, which may provide a signature mutation(s) for this exposure.

The ACGIH (2005) has recommended 1-OHP, the metabolite of the common PAH pyrene, for use as an end of shift/end of workweek indicator of PAH exposure. This assessment is complicated because 1-OHP is an accepted marker of exposure to PAHs for traffic-related air pollution. In general, biological monitoring can be an effective indicator of possible PAH exposure, but it is still difficult to assess the extent of exposure and the resultant dose, especially with low level environmental or occupational exposure. This difficulty results from several factors, including interindividual variability, confounding effects of drugs or cigarettes, and nonspecificity of monitoring techniques. Twenty-two hydroxylated PAH metabolites were measured in urine specimens as part of the National Health and Nutrition Examination Survey (NHANES) among participants aged 6 years and older during 2007–2008. The NHANES survey found the geometric mean for urinary 1-OHP was 119 ng/mg creatinine (CDC, 2013).

MEDICAL MANAGEMENT

Polycyclic aromatic hydrocarbons are associated with a low order of acute toxicity in humans. Acute exposures should be addressed based on the symptoms. Contaminated clothing should be removed from exposed persons as soon as possible. The skin is decontaminated by gently scrubbing it with soap and water. Ocular contamination can be treated with irrigation of the eye with saline or water followed by a complete eye examination. For an acute inhalation exposure, ventilatory support is tailored to the patient's clinical condition.

REFERENCES

American Conference of Governmental Industrial Hygienists (ACGIH) (2005) *Threshold Limit Values (TLVs) for Chemical Substances and Physical Agents and Biological Exposure Indices (BEIs)*, Cincinnati, OH: American Conference of Governmental Industrial Hygienists.

Andrysík, Z., Vondrácek, J., Machala, M., Krcmár, P., Svihálková-Sindlerová, L., Kranz, A., Weiss, C., Faust, D., Kozubík, A., and Dietrich, C. (2007) The aryl hydrocarbon receptor-dependent deregulation of cell cycle control induced by polycyclic aromatic hydrocarbons in rat liver epithelial cells. *Mutat. Res.* 615 (1–2):87–97.

Arce GT., et al. (1987) Relationships between benzo(a)pyrene-DNA adduct levels and genotoxic effects in mammalian cells. *Cancer Res.* 47:3388–3395.

Armstrong, B., Hutchinson, E., Unwin, J., and Fletcher, T. (2004) Lung cancer risk after exposure to polycyclic aromatic hydrocarbons: a review and meta-analysis. *Environ. Health Perspect.* 112(9):970–978.

Agency for Toxic Substances and Disease Registry (1995) *Toxicological Profile for Polycyclic Aromatic Hydrocarbons (PAHs) (Update)*, Atlanta, GA: US Department of Health and Human Services.

Boogaard, P.J. and van Sittert, N.J. (1994) Exposure to polycyclic aromatic hydrocarbons in petrochemical industries by measurement of urinary 1-hydroxypyrene. *Occup. Environ. Med.* 51:250–258.

Brucker, N., Moro, A.M., Charão, M.F., Durgante, J., Freitas, F., Baierle, M., Nascimento, S., Gauer, B., Bulcão, R.P., Bubols, G.B., Ferrari, P.D., Thiesen, F.V., Gioda, A., Duarte, M.M., de Castro, I., Saldiva, P.H., and Garcia, S.C. (2013) Biomarkers of occupational exposure to air pollution, inflammation and oxidative damage in taxi drivers. *Sci. Total Environ.* 463-464:884–893.

Buchet, J.P., et al. (1992) Evaluation of exposure to polycyclic aromatic hydrocarbons in a coke production and a graphite electrode manufacturing plant: assessment of urinary excretion of 1-hydropyrene as a biological indicator of exposure. *Occup. Environ. Med.* 49:761–768.

Centers for Disease Control and Prevention (CDC) (2013) *Fourth National Report on Human Exposure to Environmental Chemicals*, Atlanta, GA: Available at http://www.cdc.gov/exposurereport/pdf/FourthReport_UpdatedTables_Sep2013.pdf.

Cherng, S.H., et al. (1996) Modulatory effects of polycyclic aromatic hydrocarbons on the mutagenicity of 1-nitropyrene: a structure-activity relationship study. *Mutat. Res.* 367(4):177–185.

Clavel, J., et al. (1994) Occupational exposure to polycyclic aromatic hydrocarbons and the risk of bladder cancer: a French case-control study. *Int. J. Epidemiol.* 23:1145–1153.

Dean, J.H., et al. (1993) Immune suppression following exposure of mice to the carcinogen benzo(a)pyrene but not the non-carcinogenic benzo(e)pyrene. *Clin. Exp. Immunol.* 52:199–206.

dell'Omo, M. and Lauwerys, R.R. (1993) Adducts to macromolecules in the biological monitoring of workers exposed to polycyclic aromatic hydrocarbons. *Crit. Rev. Toxicol.* 23:111–126.

Eisen, E.A., et al. (1994) Mortality studies of machining fluid exposure in the automobile industry: a case-control study of larynx cancer. *Am. J. Ind. Med.* 26:185–202.

Ferreira, M., Jr., et al. (1994) Determinants of benzo-a-pyrene epoxide adducts to haemoglobin in workers exposed to polycyclic aromatic hydrocarbons. *Occup. Environ. Med.* 51:451–455.

Gupta, R.C., Reddy, M.V., and Randerath, K. (1982) 32P-post-labelling analysis of non-radioactive aromatic carcinogen-DNA adducts. *Carcinogenesis* 3:1081–1092.

Hansen, E.S. (1992) A mortality study of Danish stokers. *Br. J. Ind. Med.* 49:48–52.

Hansen, A.M., Poulsen, O.M., and Menne, T. (1993a) Longitudinal study of excretion of metabolites of polycyclic aromatic hydrocarbons in urine from two psoriatic patients. *Acta Derm. Venereol.* 73:188–190.

Hansen, A.M., Poulsen, O.M., Christensen, J.M., and Hansen, S.H. (1993b) Determination of 1-hydropyrene in human urine by high-performance liquid chromatography. *J. Anal. Toxicol.* 17:38–41.

Hansen, A.M., Wallin, H., Binderup, M.L., Dybdahl, M., Autrup, H., Loft, S., and Knudsen, L.E. (2004) Urinary 1-hydroxypyrene and mutagenicity in bus drivers and mail carriers exposed to urban air pollution in Denmark. *Mutat. Res.* 557(1):7–17.

Harris, C.C., et al. (1985) Detection of benzo(a)pyrene diol epoxide-DNA adducts in peripheral blood lymphocytes and antibodies to the adducts in serum from coke oven workers. *Proc. Natl. Acad. Sci. U. S. A.* 82:6672–6676.

Haugen, A., et al. (1986) Determination of polyaromatic hydrocarbons in urine, benzo(a)pyrene diol epoxide-DNA adducts in lymphocyte DNA, and antibodies to the adducts in sera from coke workers exposed to measured amounts of polycyclic aromatic hydrocarbons in the work atmosphere. *Cancer. Res.* 46:4178–4183.

Hemminki, K., et al. (1990) DNA adducts in humans environmentally exposed to aromatic compounds in an industrial area of Poland. *Carcinogenesis* 11:1229–1231.

Herikstad, B.V., et al. (1993) Determination of polycyclic aromatic hydrocarbons in urine from coke-oven workers with a radio-immunoassay. *Carcinogenesis* 14:307–309.

HSDB (2003) *Hazardous Substances Data Bank: Polycyclic Aromatic Hydrocarbons. From NLM (National Library of Medicine)*, Bethesda, MD: National Institutes of Health, U.S. Department of Health and Human Services. Available at http://toxnet.nlm.nih.gov.

Jongeneelen, F.J., et al. (1990) Ambient and biological monitoring of cokeoven workers: determinants of the internal dose of polycyclic aromatic hydrocarbons. *Br. J. Ind. Med.* 47:454–461.

Kim, K.H., Jahan, S.A., Kabir, E., and Brown, R.J. (2013) A review of airborne polycyclic aromatic hydrocarbons (PAHs) and their human health effects. *Environ. Int.* Oct; 60:71–80.

Krewski, D., Thorslund, T., and Withey, J. (1989) Carcinogenic risk assessment of complex mixtures. *Toxicol. Ind. Health* 5:851–867.

Lee, W.R., Alderson, M.R., and Downes, J.E. (1972) Scrotal cancer in the north-west of England. *Br. J. Ind. Med.* 29:188–195.

Levin, W., et al. (1982) Oxidative metabolism of polycyclic aromatic hydrocarbons to ultimate carcinogens. *Drug Metab. Rev.* 13:555–580.

Linstedt, G. and Sollenberg, J. (1982) Polycyclic aromatic hydrocarbons in the occupational environment with special reference to benzo(a)pyrene measurements in Swedish industry. *Scand. J. Work Environ. Health* 8:1–19.

MacLeod, M.C., Adair, G., and Humphrey, R.M. (1988) Differential efficiency of mutagenesis at three genetic loci in CHO cells by benzo(a)pyrene diol epoxide. *Mutat. Res.* 199:243–254.

Murata, Y., Denda, A., Maruyama, H., et al. (1997) Short communication. Chronic toxicity and carcinogenicity studies of 2-methylnaphthalene in B6C3F1 mice. *Fundam. Appl. Toxicol.* 36:90–93.

National Academy of Science (1986) *Drinking Water and Health*, Washington, DC: National Academy Press.

Øvrebø, S., et al. (1994) Biological monitoring of exposure to polycyclic aromatic hydrocarbon in an electrode plant. *J. Occup. Med.* 36:303–310.

Partanen, T. and Boffetta, P. (1994) Cancer risk in asphalt workers and roofers: review and metaanalysis of epidemiologic studies. *Am. J. Ind. Med.* 26:721–740.

Perera, F.P., et al. (1994) Carcinogen-DNA adducts and gene mutation in foundry workers with low-level exposure to polycyclic aromatic hydrocarbons. *Carcinogenesis* 15:2905–2910.

Phillips, D.H., et al. (1988) Monitoring occupational exposure to carcinogens: detection by 32P-postlabelling of aromatic DNA adducts in white blood cells from iron foundry workers. *Mutat. Res.* 204:531–541.

Reddy, M.V., Hemminki, K., and Randerath, K. (1991) Postlabeling analysis of polycyclic aromatic hydrocarbon-DNA adducts in white blood cells of foundry workers. *J. Toxicol. Environ. Health* 34:177–185.

Slaga, T., et al. (1981) Comparison of the skin tumor-initiating activities of dihydrodiols and diolepoxides of various polycyclic aromatic hydrocarbons. *Cancer. Res.* 40:1981–1984.

Southam, A.H. and Wilson, S.R. (1992) Cancer of the scrotum. *BMJ* 2:188–195.

Stjernsward, J. (1966) Effect of noncarcinogenic and carcinogenic hydrocarbons on antibody-forming cells measured at the cellular level *in vitro. J. Natl. Cancer Inst.* 36:1189–1195.

Szyfter, K. and Hemminki, K. (1992) Analysis of deoxyribonucleic acid adducts in workers. *Scand. Work Environ. Health* 18:22–26.

Urso, P. and Gengozian, N. (1980) Depressed humoral immunity and increased tumor incidence in mice following *in utero* exposure to benzo [alpha] pyrene. *J. Toxicol. Environ. Health* 6:569–576.

Vahakangas, K., Haugen, A., and Hams, C.C. (1985) An applied synchronous fluorescence spectrophotometry assay to study benzo (a)pyrene-diolepoxide-DNA adducts. *Carcinogenesis* 6:1109 1115.

Ward, E.C., Murray, M.J., and Dean, J.H. (1985) Immunotoxicity of nonhalogenated polycyclic aromatic hydrocarbons. In: Dean, J.H., et al. editors. *Immunotoxicology and Immunepharmacology*, New York, NY: Raven Press.

Ward, E.C., et al. (1986) Persistent suppression of humoral and cell-mediated immunity in mice following exposure to the polycyclic aromatic hydrocarbon, 7, 12-dimethyl-benz[a]anthracene. *Int. J. Immunopharmacol.* 8:13–22.

Weston, A., et al. (1989) Fluorescence and mass spectral evidence for the formation of benzo(a)pyrene antidiol-epoxide-DNA and hemoglobin adducts in humans. *Carcinogenesis* 10:251–257.

Wodjani, A. and Alfred, L. (1984) Alterations in cell-mediated immune functions induced in mouse splenic lymphocytes by polycyclic aromatic hydrocarbons. *Cancer Res.* 44:942–945.

HYDROQUINONE

Occupational exposure limits: $2 \, \text{mg/m}^3$ (OSHA PEL); $2 \, \text{mg/m}^3$ (NIOSH REL, 15-min ceiling); $2 \, \text{mg/m}^3$ (ACGIH TLV); derived no effect level (DNEL); DNEL inhalation, systemic effects, long term: $7 \, \text{mg/m}^3$; DNEL inhalation, local effects, long term: $1 \, \text{mg/m}^3$; DNEL dermal, systemic effects, long term: $128 \, \text{mg/kg-bw/day}$ (NIOSH, 1994).

Reference values: DNEL inhalation, systemic effects, long term: $1.74 \, \text{mg/m}^3$; DNEL inhalation, local effects, long term: $0.5 \, \text{mg/m}^3$; DNEL dermal, systemic effects, long term: $64 \, \text{mg/kg-bw/day}$.

Hazard statements: H302—harmful if swallowed; H317—may cause an allergic skin reaction; H318—causes serious eye damage; H341—suspected of causing genetic defects; H351—suspected of causing cancer; H410—very toxic to aquatic life with long-lasting effects (ACGIH, 1995).

Precautionary statements: P201—obtain special instructions before use; P261—avoid breathing dust/fume/gas/mist/vapours/spray; P273—avoid release to the environment; P280—wear protective gloves/protective clothing/eye protection/face protection; P305 + P351 + P338—if in eyes, rinse cautiously with water for several minutes; remove contact lenses, if present and easy to do; continue rinsing; P310—immediately call a poison center or doctor/physician.

Risk phrases: R22—harmful if swallowed; R40—limited evidence of a carcinogenic effect; R41—risk of serious damage to the eyes; R43—may cause sensitization by skin contact; R50—very toxic to the aquatic organisms; R68—possible risk of irreversible effects.

Safety phrases: S2—keep out of the reach of children; S26—in case of contact with eyes, rinse immediately with plenty of water and seek medical advice; S36/37/39—wear suitable protective clothing, gloves, and eye/face protection; S61—avoid release to the environment. Refer to special instructions/safety data sheets.

BACKGROUND AND USES

Hydroquinone (*p*-benzenediol, 1,4 benzene diol, dihydroxy benzene, 1,4 dihydroxy benzene, quinol, CASRN 123-31-9) exists as colorless to light-tan or light-gray crystals under standard conditions. It has a vapor pressure of 0.00001 mmHg and a solubility in water by weight of 7%.

Skin bleaching creams containing 4% hydroquinone are currently available for use according to a physician's supervision (Arndt and Fitzpatrick, 1965; Spencer, 1965; Brody, 1989). The bleaching effect produced is reversible. Hydroquinone is widely used in photography as a developer (Nowak et al., 1995). It is also produced *in vivo* as a metabolite of benzene. In nature, it is found in several arthropods and in the African plant Noogoora burr (*Xanthium pungens*). Concentrations of hydroquinone in this plant are adequate to poison pigs and cattle.

HEALTH EFFECTS

Acute Effects

Acute excessive exposures to hydroquinone in industry are unusual. When they occur, the symptoms produced resemble an acute exposure to phenol. Symptoms include hypersensitivity, hyperactive reflexes, dyspnea, and cyanosis. Exhaustion, hypothermia, convulsions, coma, and death may follow. Methemoglobin occurs quickly and can produce anoxia. If death occurs, it is usually as a result of respiratory failure. Dusts can adhere to the eyes and cause localized damage. Ingestion of small amounts can cause gastroenteritis (Hooper et al., 1978).

Chronic Effects

Hydroquinone is thought to have caused corneal damage and severe astigmatism in some hydroquinone workers (Naumann, 1966). There have been reports of hepatitis and anaplastic anemia in darkroom workers exposed to hydroquinone (Nowak et al., 1995).

TARGET ORGANS AND MECHNANISM OF ACTION

Hydroquinone did not produce covalent adducts in kidneys in male and female rats (English et al., 1994). No reproductive effects were noted in rats dosed with hydroquinone (Blacker et al., 1993). It has been shown to cause chromosomal breakage *in vitro* (Dobo and Eastmond, 1994).

Hydroquinone promotes skin bleaching by inhibiting melanogenesis. In an amphibian model, hydroquinone simultaneously induced indole amine 2,3-dioxygenase and inhibited tyrosinase (Chakraborty et al., 1993).

Chronic exposure to hydroquinone can cause brown discoloration of the cornea and stromal degeneration (Naumann, 1966). This may account for the severe astigmatism reported by hydroquinone workers.

In vivo hydroquinone is metabolized to quinone, a more toxic metabolite. It is also oxidized to quinone in the presence of sunlight or air.

INDUSTRIAL HYGIENE

The OSHA has set the PEL for hydroquinone at $2\,mg/m^3$. The NIOSH REL is slightly more stringent, with $2\,mg/m^3$ being the 15-min ceiling (NIOSH, 1994). Concentrations of $50\,mg/m^3$ or greater are considered to be IDLH.

The ACGIH (1995) has set a TLV for hydroquinone at $2\,mg/m^3$. The ACGIH has also proposed designating hydroquinone as an A3 carcinogen, indicating that it is a carcinogen in experimental animals.

TREATMENT

Treatment is essentially symptomatic. Because acute poisonings from hydroquinone are rare, information on treatment is limited. Exposures should be treated in a similar manner to phenol exposure. If ingested, dilution with water or milk is suggested. Vomiting may be induced if the patient is not experiencing seizures. However, it is not indicated whether seizures are likely to occur because it could induce seizures. Gastric lavage may be performed if it is done soon after ingestion or if the patient is comatose or at risk of convulsions. Other symptomatic treatment should be undertaken as necessary.

REFERENCES

American Conference of Governmental Industrial Hygienists (ACGIH) (1995) *Threshold Limit Values (TLVs) for Chemical Substances and Physical Agents and Biological Exposure Indices (BEIs)*, Cincinnati, OH: American Conference of Governmental Industrial Hygienists.

Arndt, K.A. and Fitzpatrick, T.B. (1965) Topical use of hydroquinone as a depigmenting agent. *JAMA* 194:965–967.

Blacker, A.M., et al. (1993) A two-generation reproduction study with hydroquinone in rats. *Fundam. Appl. Toxicol.* 21:420–424.

Brody, H.J. (1989) Complications of chemical peeling. *J. Dermatol. Surg. Oncol.* 15:1010–1019.

Chakraborty, A.K., Chakraborty, A., and Chakraborty, D.P. (1993) Hydroquinone simultaneously induces indole amine 2,3 dioxygenase (IOD) and inhibits tyrosinase in *Bufo melanostictus*. *Life Sci.* 52:1695–1698.

Dobo, K.L. and Eastmond, D.A. (1994) Role of oxygen radicals in the chromosomal loss and breakage induced by the quinone-forming compounds, hydroquinone and *tert*-butylhydroquinone. *Environ. Mol. Mutagen.* 24:293–300.

English, J.C., et al. (1994) Measurement of nuclear DNA modification by 32P-postlabeling in the kidneys of male and female Fischer 344 rats after multiple gavage doses of hydroquinone. *Fundam. Appl. Toxicol.* 23:391–396.

Hooper, R.R., Husted, S.R., and Smith, E.L. (1978) Hydroquinone poisoning aboard a navy ship. *MMWR* 27:237–243.

Naumann, G. (1966) Corneal damage in hydroquinone workers: a clinicopathologic study. *Arch. Ophthamol.* 76:189–194.

National Institute for Occupational Safety and Health (NIOSH) (1994) *Pocket Guide to Chemical Hazards*, NIOSH 94-116, Washington, DC: US Department of Health and Human Services.

Nowak, A.K., Shilkin, K.B., and Jeffrey, G.P. (1995) Darkroom hepatitis after exposure to hydroquinone (letter). *Lancet* 345:1187.

Spencer, M.C. (1965) Topical use of hydroquinone for depigmentation. *JAMA* 194:962–964.

OTHER HETEROCYCLIC COMPOUNDS

TURPENTINE

Occupational exposure limits: 100 ppm (OSHA PEL, NIOSH REL, ACGIH TLV); DNEL; DNEL inhalation, systemic effects, long term: $11.3\,mg/m^3$; DNEL inhalation, local effects, long term: $0.77\,mg/m^3$; DNEL dermal, systemic effects, long term: $1.6\,mg/kg$-bw/day.

Reference values: DNEL oral, systemic effects, long term: $0.57\,mg/kg$-bw/day.

Hazard statements: H225—highly flammable liquid and vapor; H302—harmful if swallowed; H304—may be fatal if swallowed and enters the airway; H312—harmful if it comes in contact with the skin; H315—causes skin irritation; H317—may cause an allergic skin reaction; H319—causes serious eye irritation; H332—

harmful if inhaled; H411—toxic to aquatic life with long-lasting effects.

Precautionary statements: P210—keep away from heat/ sparks/open flames/ . . . /hot surfaces . . . no smoking; P233—keep container tightly closed; P240—ground/ bond container and receiving equipment; P241— use explosion-proof electrical/ventilating/lighting/ . . . / equipment; P242—use only non-sparking tools; P243— take precautionary measures against static discharge; P280—wear protective gloves/protective clothing/eye protection/face protection; P303 + P361 + P353—if on skin (or hair), remove/take off immediately all contaminated clothing; rinse the skin with water/shower; P370 + P378—in case of fire, use . . . for extinction; P403 + P235—store in a well-ventilated place; keep cool; P501—dispose of contents/container to . . . ; P264—wash . . . thoroughly after handling; P270—do not eat, drink, or smoke when using this product; P301 + P312—if swallowed, call a poison center or doctor/physician if you feel unwell; P330—rinse mouth; P280—wear protective gloves/protective clothing/eye protection/face protection; P302 + P352—if on skin, wash with plenty of soap and water; P312—call a poison center or doctor/physician if you feel unwell; P322—specific measures (see . . . on the label); P363— wash contaminated clothing before reuse; P261—avoid breathing dust/fume/gas/mist/vapors/spray; P271—use only outdoors or in a well-ventilated area; P304 + P340—if inhaled, remove victim to fresh air and instruct the victim to rest in a position comfortable for breathing; P331—do not induce vomiting; P405—store locked up; P305 + P351 + P338—if in the eyes, rinse cautiously with water for several minutes; remove contact lenses, if present and easy to do; continue rinsing; P337 + P313— if eye irritation persists, get medical advice/attention; P321—specific treatment (see . . . on the label); P332 + P313—if skin irritation occurs, get medical advice/attention; P362—take off contaminated clothing and wash before reuse; P272—contaminated work clothing should not be allowed out of the workplace; P280—wear protective gloves/protective clothing/eye protection/face protection; P333 + P313—if skin irritation or rash occurs, get medical advice/attention; P321— specific treatment (see . . . on the label); P273—avoid release to the environment; P391—collect spillage.

Risk phrases: R11—highly flammable; R20/21/22—harmful by inhalation, in contact with the skin, and if swallowed; R36/38—irritating to the eyes and skin; R43— may cause sensitization by skin contact; R51/53—toxic to aquatic organisms, may cause long-term adverse effects in the aquatic environment.

Safety phrases: S16—keep away from sources of ignition . . . no smoking; S2—keep out of the reach of children; S36/37—wear suitable protective clothing and gloves; S46—if swallowed, seek medical advice immediately and show this container or label; S61—avoid release to the environment. Refer to special instructions/safety data sheets; S62—if swallowed, do not induce vomiting: seek medical advice immediately and show this container or label.

BACKGROUND AND USES

The use of turpentine (CASRN 8006-64-2) for suicidal and homicidal purposes has become quite infrequent with the increased availability of pesticides and other toxic substances (Pande et al., 1994). Available literature contains few case reports of turpentine poisoning. Turpentine has been nearly eliminated from formulas of ready-made interior oil-based paints, but it is used in shoe polishes and as a raw material in the manufacture of synthetic camphor and terpin hydrate.

HEALTH EFFECTS

Filipsson (1996) described the toxicokinetics, pulmonary effects, and subjective rating of discomfort for turpentine vapor (a mixture of monoterpenes). The subjects experienced discomfort in the throat and airway during an exposure to turpentine, and airway resistance was increased at the end of exposure.

Distilled or wood turpentine contains methyl alcohol, formaldehyde, phenols, and pyridine. Some of these compounds are irritating to the skin, and the use of this kind of turpentine has been followed by outbreaks of dermatitis in the United States, Britain, and Italy (McCord, 1926).

Chapman (1941) reported a careful review of the effect of paint on the kidneys, with particular reference to the role of turpentine. His interest was aroused by two cases, both fatal, of extensive glomerulonephritis. The changing use of turpentine in recent years may account for the absence of current reports of kidney damage associated with its use.

HISTOPATHOLOGY

Pleural effusions were produced by intrapleural turpentine instillation in mice. The fine structure of the inflammatory cells obtained from the effusions was normal except for lipid inclusions (Nielsen et al., 1992). The same type of inclusion was previously found in neutrophils from pleural effusions in patients with tuberculous infection, rheumatoid disease, or carcinomatosis.

INDUSTRIAL HYGIENE

NIOSH (1994), OSHA, and the ACGIH (1996) have set the REL, PEL, and TLV for turpentine at 100 ppm.

REFERENCES

American Conference of Governmental Industrial Hygienists (ACGIH) (1996) *Threshold Limit Values (TLVs) for Chemical Substances and Physical Agents and Biological Exposure Indices (BEIs)*, Cincinnati, OH: American Conference of Governmental Industrial Hygienists.

Chapman, E.M. (1941) Observations on the effect of paint on the kidneys with particular reference to the role of turpentine. *J. Ind. Hyg. Toxicol.* 23:277–289.

Filipsson, A.F. (1996) Short term inhalation exposure to turpentine: toxicokinetics and acute effects in men. *Occup. Environ. Med.* 53:100–105.

McCord, C.P. (1926) Occupational dermatitis from wood turpentine. *JAMA* 86:1979.

Nielsen. M.H., Faurschou, P., and Faarup, P. (1992) Fine structure of neutrophils from turpentine-induced pleuritis in mice, simulating rheumatoid arthritis cells. *APMIS* 100:188–190.

National Institute for Occupational Safety and Health (NIOSH) (1994) *Pocket Guide to Chemical Hazards*, NIOSH 94-116, Washington, DC: US Department of Health and Human Services.

Pande, T.K., et al. (1994) Turpentine poisoning: a case report. *Forensic Sci. Int.* 65:47–49.

61

AMINO AND NITRO COMPOUNDS

David Y. Lai and Yin-tak Woo

Amino and nitro compounds are important classes of industrial chemicals that contain at least one amino or nitro group attached to an aryl or aliphatic moiety. Since the mid-1800s, they have found numerous uses in various industries that manufacture dyes and pigments, rubber, polymers, agricultural chemicals, pharmaceuticals, and photographic chemicals. Hence, occupational exposure to amino and nitro compounds has been widespread. Beyond occupational exposure, public populations may also be exposed to these chemicals. For instance, the ingredients of typical permanent and semi-permanent hair colorants are primarily aromatic amines, aminophenols, nitro-substituted aromatic amines, aminoanthraquinones, and azodyes. Some commonly used pharmaceuticals also are aromatic amines and nitro compounds.

Epidemiologic evidence has indicated that exposures to aromatic amines such as benzidine and 2-naphthylamine are linked to increased incidences of cancers in exposed workers in dyes industry. Experimental studies have shown that many other amino and nitro compounds are carcinogens, mutagens, skin sensitizers, and hematotoxicants capable of inducing methemoglobinemia. Aromatic nitro compounds share many toxicological effects with aromatic amines since their nitro groups can readily be reduced to become amines. Considerable information suggests that the aromatic amines require metabolic activation to yield reactive intermediates for their toxicity and carcinogenicity, and the mechanism of activation of aromatic amines is now known in considerable detail (Woo and Lai, 2012; Lai et al., 1996; Woo et al., 1988). The first stage of metabolic activation involves *N*-hydroxylation and/or *N*-acetylation by the mixed-function oxidase system in the liver to yield the respective *N*-hydroxylated and/or

N-acetylated derivatives. The second stage involves further activation by *N,O*-transacetylation or by *O*-acylation (where the acyl group can be sulfonyl or acetyl) to acyloxyarylamines. Many of the reactive metabolites formed appear to be responsible for their toxic activity. For instance, the acyloxyarylamines are highly reactive and generate, upon departure of the acyloxy anion, electrophilic arylamidonium/arylnitrenium ions, which may readily bind covalently to cellular nucleophiles such as DNA to initiate mutagenesis and carcinogenesis. Ortho-substitution with methyl/methoxy group to unsubstituted aromatic amine or nitrobenzene can substantially enhance carcinogenic potential. *o*-Toluidine and *o*-anisidine are known to be more carcinogenic than aniline. Likewise, *o*-nitrotoluene and *o*-nitroanisole are more carcinogenic than nitrobenzene. The higher carcinogenic potency of the *o*-isomer as compared to other isomers (*m*- and *p*-) or to the unsubstituted aromatic amine or nitrobenzene is likely due to differences in binding capability of the compounds to the metabolic activation enzymes (Woo and Lai, 2012).

After peaking, during the past few decades, United States production of many hazardous amino and nitro compounds has declined or even ceased. Recent worldwide production and sales volume, however, is still growing at steady rate, and the supply of many of these compounds is shifting from the United States to East Asian countries along with textile and dye production. In this chapter, the production, exposure, toxicology, pathology, and regulations of three amino and three nitro aromatic compounds with high production volumes will be reviewed. Related information on other amine and amino compounds can be found in another review by the authors (Woo and Lai, 2012).

Hamilton & Hardy's Industrial Toxicology, Sixth Edition. Edited by Raymond D. Harbison, Marie M. Bourgeois, and Giffe T. Johnson.
© 2015 John Wiley & Sons, Inc. Published 2015 by John Wiley & Sons, Inc.

ANILINE

Occupational exposure limits: 19 mg/m^3 (OSHA PEL); 7.6 mg/m^3 (ACGIH TLV); derived no effect level (DNEL); DNEL inhalation, systemic effects, long term: 7.7 mg/m^3; DNEL inhalation, systemic effects, short term: 15.4 mg/m^3; DNEL dermal, systemic effects, long term: 2 mg/kg/bw/day; DNEL dermal, systemic effects, short term: 4 mg/kg/bw/day

Reference values: 0.001 mg/m^3 (IRIS RfC)

Hazard statements: H400—very toxic to aquatic life; H410—very toxic to aquatic life with long lasting effects; H301—toxic if swallowed; H311—toxic if contact with skin; H331—toxic if inhaled; H318—causes serious eye damage; H317—may cause an allergic skin reaction; H341—suspected of causing genetic defects; H351—suspected of causing cancer; H372—causes damage to blood and hematopoetic system through prolonged or repeated exposure

Precautionary statements: P273—avoid release to the environment; P280—wear protective gloves/protective clothing/eye protection/face protection; P301 + P310—if swallowed, immediately call a poison center or doctor/physician; P302 + P352—if on skin, wash with plenty of soap and water; P304 + P340—if inhaled, remove victim to fresh air and keep at rest in a position comfortable for breathing; P305 + P351 + P338—if in eyes, rinse cautiously with water for several minutes; remove contact lenses, if present and easy to do; continue rinsing

Risk phrases: R23/24/25—toxic by inhalation, in contact with skin and if swallowed; R41—risk of serious damage to eyes; R43—may cause sensitization by skin contact; R48/23/24/25—toxic: danger of serious damage to health by prolonged exposure through inhalation, in contact with skin, and if swallowed; R68—possible risk of irreversible effects; R40—limited evidence of a carcinogenic effect; R50—very toxic to aquatic organisms

Safety phrases: S1/2—keep locked up and out of reach of children; S26—in case of contact with eyes, rinse immediately with plenty of water and seek medical advice; S27—take off immediately all contaminated clothing, wash with water and soap; S36/37/39—wear suitable protective clothing, gloves, and eye/face protection; S46—if swallowed, seek medical advice immediately and show this container or label; S61—avoid release to the environment; refer to special instructions/safety data sheets; S63—in case of accidental inhalation: remove casualty to fresh air and keep at rest

BACKGROUND AND USES

Aniline (CASRN 62-53-3) is the simplest of the aromatic amines with an amino group attached to a benzene. It has been produced commercially for decades in the manufacture of a wide variety of products such as dyestuffs, dyestuff intermediates, rubber accelerators, and polyurethane foams. The largest users of aniline are companies that make isocyanates, especially methyl diphenyl diisocyanate. Other companies use aniline to make pesticides, dyes, and rubber. Companies also use smaller amounts of aniline to make drugs, photographic chemicals, varnishes, and explosives.

From 1969 to 1988, aniline production in the United States grew at an average annual rate of 6%; total production of aniline by eight U.S. companies amounted to 467 million kg in 1988. From 1990 to 2002, U.S. annual production of aniline had been more than 1 billion pounds (U.S. EPA, 2004). Current production figures were not available but aniline demand in the United States was expected to grow at an average annual rate of 4–6% per year (Amini, 1992). In 1979, aniline productions in Japan and Western Europe were estimated to be between 79 and 500 million kg, respectively (IARC, 1982).

Aniline derivatives such as nitro, chloro, chloronitro, *N*-alkyl, and *N*,*N*-dialkylaniline are also widely used in industry as intermediates in chemical synthesis. Aniline and many of its derivatives have a similar pattern of toxicity. They are all hematotoxic and have been shown to cause methemoglobinemia. *m*-Nitro, *o*-chloro, *m*-chloro, and 2,4-dichloroanilines are irritants to the skin and mucous membrane. Several dichloroanilines have been demonstrated to be skin sensitizers. Aniline and its monochloro derivatives are also nephrotoxic and hepatotoxic. Although there is limited or equivocal evidence of carcinogenicity in rodent studies, they are considered as suspect human carcinogens. In general, alkyl derivatives of aniline are less toxic.

However, addition of chloro or nitro groups on the phenyl ring increases the toxicity of aniline; the ortho-substitution produces the greatest enhancement. More information can be found in several toxicological reviews on aniline (U.S. EPA, 2012a, 1994; ASTDR, 2011; IARC, 1987, 1982) and its derivatives (Woo and Lai, 2012).

PHYSICAL AND CHEMICAL PROPERTIES

Aniline (C$_6$H$_7$N), also called aminobenzene, is the prototypical aromatic amine consisting of a phenyl group attached to an amino group. It is a colorless to brown oily liquid with a characteristic aromatic or fishy odor and is flammable. It has a molecular weight of 93.13, a density of 1.002, a melting point of −6.2 °C, a boiling point of 184–186 °C, a vapor density of 3.22 (air = 1), a vapor pressure of 15 mm Hg (77 °C), and a refractive index of 1.5863 (20 °C). Aniline is slightly soluble in water and is miscible with most organic solvents (Budavari et al., 1996; Woo and Lai, 2012).

MAMMALIAN TOXICOLOGY

Acute Effects

Aniline can be absorbed from the skin, lungs, and gastrointestinal tract. Contact with aniline liquid or vapor may cause mild irritation to skin or eyes. Ingestion or inhalation of aniline can lead rapidly to severe systemic toxicity. The major acute toxic effects of aniline in experimental animals are methemoglobinemia and hemolysis. An acute oral LD_{50} for rats has been reported as 440 mg/kg bw. The dermal LD_{50} on intact skin was 1320 mg/kg bw. in guinea pigs, and the 4-h LC_{50} for rats was 250 ppm (950 mg/m^3) (Woo and Lai, 2012).

In humans, the first sign of aniline poisoning is cyanosis, caused by conversion of the blood hemoglobin to methemoglobin. Other symptoms of aniline exposure include headaches, nausea and vomiting, paresthesias, tremor, colicky pain, narcosis/coma, and cardiac arrhythmia. Volunteers tolerated 1-h exposures ranging from 100 to 160 ppm with only moderate adverse health effects (von Oettingen, 1941). It has also been reported that 100–160 ppm is the maximum concentration that can be inhaled for 1 h without serious consequence (Henderson and Haggard, 1943) and that 50–100 ppm can probably be tolerated for 60 min (AIHA, 1955). Exposure to 7–53 ppm aniline vapor after several hours causes slight symptoms, and concentrations >100–160 ppm cause serious disturbances. Acute aniline exposure can cause acute kidney failure with decreased urinary output, painful urination, and blood, hemoglobin, or methemoglobin in the urine. In addition, bladder-wall irritation, kidney ulceration, and tissue death can also occur (Woo and Lai, 1994; Kimbrough et al., 1989). As little as 1 g of ingested aniline can be fatal to humans; the mean lethal dose is 5–30 g, and the minimum lethal dose for a 150 lb human is about 10 g (HSDB, 2011).

Chronic Effects

Methemoglobin formation, splenic pathology, and hemosiderosis are the principal nonneoplastic lesions reported in inhalation subchronic and chronic toxicity studies in experimental animals on aniline.

In a subchronic inhalation study (DuPont, 1982), groups of 16 male Crl:CD rats were exposed (head-only) to 0 (air-exposed controls), 17 vppm (64.7 mg/m^3), 45 ppm (171.4 mg/m^3), or 87 ppm (331.3 mg/m^3) aniline vapors, 6 h/day, 5 days/week, for 2 weeks. Significantly elevated levels in methemoglobin were observed in rats at the 45 ppm (2.2–5.4%) and at 87 ppm (4.2–23%) groups as compared to the controls (0–2.9%), but not at the 7 ppm group. This increase was accompanied by clinical symptoms as the animals exposed to 87 ppm were cyanotic. Furthermore, the animals exposed to 87 or 45 ppm aniline vapors were anemic and accompanying increases in mean relative spleen weight. There were also exposure-dependent increases in erythropoietin foci,

reticuloendothelial cell hypertrophy, and hemosiderin deposition within the spleen of the high dosed (87 ppm) rats, and slight splenic histopathology was noted in the low dose (17 ppm) group. Based on the methemoglobin and splenic effects, a LOAEL of 17 ppm was established in rats.

Oberst et al. (1956) exposed (whole body) 9 male Wistar rats to 5 ppm (19 mg/m^3; duration-adjusted value of 3.4 mg/m^3) of aniline vapor for 6 h/day, 5 days/week for 26 weeks. Blood analysis indicated a slight increase (0.6%) in methemoglobin in rats. Based on the increase of methemoglobin, the no observed adverse effect level (NOAEL) appeared to be 5 ppm (3.4 mg/m^3).

Aniline was not teratogenic to F344 rats, even at maternally toxic doses of 10, 30, or 100 mg/kg. However, methemoglobinemia, increased relative spleen weight, and decreased red blood cell count were noted in 20 dams (Price et al., 1985; CIIT, 1981). In a reproductive/developmental toxicity study, groups of 24–27 Fischer 344 rats were exposed to 10, 30, and 100 mg/kg/day aniline HCl by gavage on gestational days 7–20. A significant increase in methemoglobin (13.7% vs. 3.5% in controls) and altered hematological parameters (e.g., decrease in RBCs) were observed in the dams of the 100 mg/kg group. In addition, a dose-dependent increase in the relative spleen weight was observed in the dams of all treatment groups as compared with controls. However, there were no treatment-related alterations on various reproductive parameters examined. The methemoglobin levels in the fetuses were not significantly different from those of the controls and there was no embryotoxicity or teratogenicity observed at levels of aniline that caused maternal toxicity. Based on the increases in methemoglobin formation, a NOAEL of 30 mg/kg/day was established for maternal effects, with a NOAEL of 100 mg/kg/day for fetal effects.

Carcinogenicity studies of aniline have shown that aniline was carcinogenic in rats (CIIT, 1982). An increased incidence of primary splenic sarcomas was observed in male rats in the high dose group when aniline hydrochloride was administered in the diet for 2 years to CD-F rats (130 rats/sex/group) at levels of 0, 200, 600, and 2000 ppm. Stromal hyperplasia and fibrosis of the splenic red pulp, which may represent a precursor lesion of sarcoma, was also observed in the high dose males and, to a lesser degree, in the female rats.

In another carcinogenesis study, dietary aniline hydrochloride was administered to groups of 50 male and 50 female Fischer 344 rats at 0, 3000, or 6000 ppm and to 50 male and 50 female B6C3F1 mice at 0, 6000, or 12,000 ppm for 103 weeks (NCI, 1978). No increased tumor incidences were observed in both sexes of mice. However, the incidence of hemangiosarcomas of the spleen, and the combined incidences of fibrosarcomas and sarcoma of the spleen were statistically significantly increased in the male rats as compared to the controls. In addition, the number of female rats having fibrosarcomas of the spleen was also significantly increased (Goodman et al., 1984; NCI, 1978).

MECHANISM(S) OF ACTION

Aniline is rapidly absorbed by the skin, lungs, and the gastro-intestinal tract of experimental animals. After intravenous injection of radiolabeled aniline to rats, radioactivity is distributed throughout the body; highest concentrations were found in blood, liver, kidney, urinary bladder, and the gastro-intestinal tract. The major urinary metabolites in various animal species tested are o-, p-aminophenol, and their conjugates. p-Aminophenyl- and p-acetylaminophenylmercapturic acids are also excreted in rats and rabbits. N-Hydroxylation of aniline by liver microsomes from several species has been observed in vitro. Many of the adverse health effects of aniline are due in part to the formation of methemoglobinemia. Aniline converts the Fe^{+2} in hemoglobin to Fe^{+3} which impair its oxygen transport capacity leading to symptoms of oxygen deficiency.

Considerable information suggests that the aromatic amino and nitro compounds require metabolic activation to yield active methemoglobin-producing reactive intermediates. The formation of phenylhydroxylamine from aniline appears to be the reactive metabolite responsible for its toxic activity (IARC, 1982). Cardiac effects of acute aniline exposure, such as irregular heart rhythm, heart block, and acute congestive heart failure, may be caused by decreased oxygen delivery to the tissues. Death can result from progressive acidosis, ischemia, and cardiovascular collapse.

Humans appear to be more sensitive than rats to aniline exposure in regard to formation of methemoglobin. It was noted that after oral administration of aniline to volunteers and rats, the dose that produced increased levels of methemoglobin was 12- to 90-folds lower for humans than for rats Jenkins et al. (1972). The reason for this increased sensitivity in humans is unknown and does not appear to be related to the half-life of methemoglobin in the serum, which is three times longer in rats than in humans.

Carcinogenicity studies of aniline have shown that aniline was carcinogenic in rats, inducing sarcomas in the spleen. The genotoxicity of aniline has been extensively studied in various short-term assay systems but produced generally negative results in reverse mutation assays in Salmonella typhimurium except in strain TA98 of S. typhimurium in the presence of norharman and S-9. Aniline showed positive responses in a mouse lymphoma assay, in SCE assays in cultured mammalian cells, and in cell transformation assay in mouse Balb/3T3 cell line. DNA adduct formation has also been reported when TA98 S. typhimurium was treated with aniline and the co-mutagen norharman. DNA damage assays in E. coli and B. subtilis, however, were negative (U.S. EPA, 2012a).

Evaluation of the pathogenesis of the splenic lesions and characterization of the disposition of radiolabeled aniline in animals suggests that the spleen tumors may be a secondary response resulting from chemically-mediated erythrocyte toxicity. It was proposed that compound-derived toxicity to erythrocytes results in scavenging of damaged red blood cells by the spleen, initiating a series of events which may contribute to the development of spleen tumors. These events potentially include (i) specific accumulation of the parent compound or toxic metabolite(s) carried to the spleen by erythrocytes; (ii) deposition of erythrocytic debris, particularly iron, which may catalyse tissue-damaging free-radical reactions; and (iii) induction of splenic hyperplasia resulting from erythrocyte overload. Linkage of the splenic tumorigenicity of aniline to hematotoxicity suggests that the carcinogenicity of aniline may be determined by a definable threshold dose, i.e., the events leading to the carcinogenicity are not initiated until the capacity of the red blood cell to cope with the toxic insult is exceeded (Goodman et al., 1984).

The molecular mechanisms of splenocarcinogenic action of aniline are not clearly understood but may involve upregulation of cytokine genes (IL-1α, IL-6, and TNF-α) that contribute to induction/initiation of fibrogenic response, activation of transcription factor NF-κB as a direct signaling molecule in oxidative stress, and oxidative DNA damage as shown by biomarker 8-hydroxy-2′-deoxyguanosine (8-OHdG) (Ma et al., 2008). Along with increased oxidative stress, however, several repair enzymes of the base excision repair pathway (BER) are upregulated; these include: 1. 8-oxoguanine glycosylase 1 (OGG1), a specific DNA glycosylase involved in the removal of 8-hydroxy-2′-deoxyguanosine (8-OHdG) adducts, and 2. Niel-1 and Niel-2 DNA glycosylases (Ma et al., 2008, 2011). It is possible that the repair capacity needs to be exceeded before significant splenocarcinogenic activities can be manifested.

CHEMICAL PATHOLOGY

Aniline is irritating to the skin, eyes, and respiratory tract. All routes of aniline exposure can induce methemoglobinemia in both adults and children, impairing the delivery of oxygen to tissues. The effect is depending on how much aniline is exposed and the length and frequency of exposure. Lack of oxygen causes effects ranging from headache and light headedness to disorientation, coma, tremor, parathesis, pain, narcosis, anemia, and cardiac arrhythmia. When methemoglobin levels are 15–30%, the skin, lips, and ears of patients will become bluish in color due to the dark color of methemoglobin. Headache, dizziness, ataxia, weakness, rapid heart rate and mild shortness of breath will occur when methemoglobin levels are 30–50%. When methemoglobin levels reach 50–70%, stupor, tachycardia, dyspnea, and severe cyanosis will follow. Death can ensue when methemoglobin levels exceed 70% if untreated.

The fetal liver can also N-oxygenate aniline to form phenylhydroxylamine which has a high potency for methemoglobin production. Because fetal hemoglobin is more easily oxidized to methemoglobin than is adult hemoglobin and is less easily reduced back to normal hemoglobin, methemoglobin is likely to be at higher levels in fetuses than in exposed mothers.

Aniline exposure may also cause the destruction of red blood cells, which manifests as acute or delayed hemolytic anemia. Persons with glucose-6-phosphate dehydrogenase (G6PD) deficiency or alcoholism are at higher risk of aniline-induced hemolysis. Spleen, kidney, liver, and heart damage may occur as secondary effects of hemolysis.

Aniline can cross the placental barrier. Limited evidence suggests that aniline may also cause adverse reproductive effects in humans. A high incidence of gynecological disorders and excess frequency of spontaneous abortions have been reported in chemically exposed women. However, aniline is not a developmental toxicant or teratogen.

Bladder cancer was once referred to as "aniline tumor" because of case reports of workers in the aniline dye industries. However, because convincing evidence could not be established for aniline primarily as a result of mixed exposures to a multiplicity of compounds and dyestuff intermediates within the same work area, the International Agency for Research on Cancer (IARC) classified aniline as a Group 3 carcinogen, that is, not classifiable as to its carcinogenicity (IARC, 1987). Based on sufficient evidence of carcinogenicity in animals, the U.S. EPA classified aniline as a B2, probably human carcinogen (U.S. EPA, 2012a). The National Institute for Occupational Safety and Health (NIOSH) has considered aniline as a potential occupational carcinogen. In a more recent epidemiology study, excess bladder cancers were reported in workers exposed to o-toluidine and aniline; the authors stated that o-toluidine is an animal carcinogen, more potent than aniline and is most likely responsible, although aniline may have played a role (Ward et al., 1996).

SOURCES OF EXPOSURE

Occupational

The greatest potential for human exposure to aniline is through inhalation and dermal contact in the workplace. A specific method of measuring hemoglobin adducts of aniline has been used to biomonitor occupational exposure of workers (Ward et al., 1996). NIOSH Analytical Method 2002 is recommended for determining workplace exposures to aniline and aniline homologs (IARC, 1982). A national occupational exposure survey conducted in the U.S. from 1981 to 1983 has estimated that 41,988 workers were potentially exposed to aniline (NIOSH, 2010). Aniline was reported in the urine of workers at a chemical manufacturing plant in Niagra Falls, NY at mean concentrations of 3.9–29.8 ug/l. These workers also had blood hemoglobin levels of aniline of 3163–17,441 pg/g (Ward et al., 1996).

Environmental

The general population may be exposed to aniline by using various products such as paints, varnishes, marking inks, and shoe polishes. Exposure to aniline can occur in the environment following releases to air, water, land, or groundwater. U.S. EPA's Toxics Release Inventory reported total on- and off-site releases of aniline in 2012 as 1,731,134 pounds (U.S. EPA, 2012). Most releases of aniline in U.S. are to underground injection sites and to air. Because it is a liquid that does not bind well to soil, aniline that makes its way into the ground can move through the ground and enter groundwater. Microorganisms that live in water and in soil can also break down aniline. In air, aniline breaks down to other chemicals. Sunlight also breaks down aniline in surface water and in soil. Plants and animals are not likely to store aniline. Toxic effects appear unlikely to occur at levels of aniline that are normally found in the environment. However, aquatic toxicity studies by the aniline industry submitted to U.S. EPA show that aniline is highly toxic to aquatic life (U.S. EPA, 1994).

INDUSTRIAL HYGIENE

The OSHA-permissible exposure limit (PEL) is 5 ppm (19 mg/m^3) TWA (NIOSH, 2013). The NIOSH Immediately Dangerous to Life or Health Concentration (IDLH) is 100 ppm (NIOSH, 1994; U.S. EPA, 1994). NIOSH has recommended OSHA to require labeling aniline as a potential occupational carcinogen, and that occupational exposure is limited to lowest feasible concentration. The ACGIH TLV is 2 ppm (7.6 mg/m^3) TWA (ACGIH, 2013; U.S. EPA, 1994). The acute exposure guideline levels (AEGLs) for aniline ranges from 1 ppm (for 8 h AEGL-1, marginal) to 40 ppm (for 30 min AEGL-3, life-threatening) (U.S. EPA, 2012c).

The U.S. EPA has developed the inhalation reference concentration (RfC) is 1×10^{-3} mg/m^3 based on the NOAEL of 19 mg/m^3 (5 ppm) in a 20–26 week inhalation study in the rat. Using tumor data in the male rats of the NCI carcinogenesis bioassay (NCI, 1978), an oral slope factor of 5.7×10^{-3}/mg/kg/day, and a drinking water unit risk of 1.6×10^{-7}/μg/l were derived (U.S. EPA, 2012a).

MEDICAL MANAGEMENT

Prompt and proper treatment can usually reverse the nonlethal acute effects of aniline. If patients can walk, lead them out of the hot zone to the decontamination zone immediately and allow access for a patent airway. Patient exertion should be minimized as exposure to aniline can produce hypoxia (due to methemoglobinemia) which can be exacerbated by physical activities. Patients, who are unable to walk and if trauma is suspected, maintain cervical immobilization manually and apply a cervical collar and a backboard when feasible. Ensure adequate respiration and pulse, administer supplemental oxygen, establish intravenous access, and place on a cardiac monitor if necessary.

Because aniline is absorbed through the skin, it is important to remove wet clothing quickly. Flush exposed skin and hair with water for 2–3 min, and then wash thoroughly with mild soap. Rinse thoroughly with water. Irrigate exposed or irritated eyes with tepid water for 15 min. In cases of ingestion, do not induce emesis. Consider gastric lavage with a small nasogastric tube if a large dose has been ingested. If the patient is alert and asymptomatic, administer a slurry of activated charcoal at 1 gm/kg, usual adult dose 60–90 g, child dose 25–50 g.

Administer methylene blue antidote to patients who have signs and symptoms of hypoxia (other than cyanosis) or for patients who have methemoglobin levels >30%. The standard intravenous dose is 1–2 mg of methylene blue per kg of body weight (0.1–0.2 ml/kg of a 1% solution) over 5–10 min. Consider hyperbaric oxygen therapy in patients who are refractory to methylene blue therapy. Consider racemic epinephrine aerosol for children who develop stridor. Dose 0.25–0.75 ml of 2.25% racemic epinephrine solution in water; repeat every 20 min as needed cautioning for myocardial variability (ASTDR, 2011).

REFERENCES

ACGIH (2013) *Documentation of the Threshold Limit Values and Biological Exposure Indices*, 7th ed., Cincinnati, Ohio: American Conference of Governmental Industrial Hygienists, Inc.

AIHA (1955) Aniline. Hygienic guide series. *Am. Ind. Hyg. Assoc. Q.* 16, 331–332.

Amini, B. (1992) Aniline and its derivatives. In: *Kirk-Othmer Encyclopedia of Chemical Technology*, 4th ed., vol. 2, New York, NY: Wiley, pp. 426–442.

ASTDR (2011) Toxic Substances Portal: Aniline. In: *Medical Management for Aniline*, Atlanta, GA: Guidelines for Agency for Toxic Substances and Disease Registry. Available at http://www.atsdr.cdc.gov/mmg/mmg.asp?id=448&tid=79.

Budavari, S., et al. editors. (1996) *The Merck Index: An encyclopedia of Chemicals, Drugs, and Biologicals*, 12th ed., Rahway, NJ: Merck & Co.

CIIT (1981) *Final report: Teratological and postnatal evaluation of aniline hydrochloride in the Fischer 344 rat.* Chemical Industry Institute of Toxicology. Document No. 40+8376093.

CIIT (1982) *104-Week chronic toxicity study in rats: Aniline hydrochloride.* Chemical Industry Institute of Toxicology. Final report, (cited in: U.S. EPA, 2012).

DuPont (1982) *Subacute inhalation toxicity study of aniline in rats. DuPont deNemours and Company, Inc. OTS No. 878220240. Fiche No. 0215025.* (cited in: U.S. EPA, 2012).

Goodman, D.G., Ward, J.M., and Reichardt, W.D. (1984) Splenic fibrosis and sarcomas in F344 rats fed diets containing aniline hydrochloride, p-chloroaniline, azobenzene, o-toluidine hydrochloride, 4,4′-sulfonyldianiline, or D and C Red No. 9. *J. Natl. Cancer Inst.* 73(1):265–273.

Henderson, Y. and Haggard, H.W. (1943) *Noxious Gases*, 2nd ed., New York, NY: Reinhold Publishing Corporation, p. 228.

HSDB (2011) *Hazardous Substances Data Bank. U.S. National Library of Medicine.* Available at http://toxnet.nlm.nih.gov/cgi-bin/sis/search/f?./temp/~i2Thfi:1.

IARC (1982) *Monographs on the Evaluation of the Carcinogenic Risk of Chemicals to Man*, vol. 27, Lyons, France: International Agency for Research on Cancer.

IARC (1987) *Monographs on the Evaluation of the Carcinogenic Risk of Chemicals to Man, supplement 7: An Updating of IARC Monographs*, vols. 1–42, Lyons, France: International Agency for Research on Cancer (IARC).

Jenkins, F.P., Robinson, J.A., Gellatly, J.B.M., and Salmond, G.W.A. (1972) The no-effect dose of aniline in human subjects and a comparison of aniline toxicity in man and the rat. *Cosmet. Toxicol.* 10:671–679.

Kimbrough, R.D., et al. editors. (1989) *Clinical Effects of Environmental Chemicals*, New York, NY: Hemisphere publishers.

Lai, D.Y., et al. (1996) Cancer risk reduction through mechanism-based molecular design of chemicals. *ACS Symp. Ser.* 640:62–73.

Ma, H., Wang, J., Abdel-Rahman, S.Z., Boor, P.J., and Khan, M.F. (2008) Oxidative DNA damage and its repair in rat spleen following subchronic exposure to aniline. *Toxicol. Appl. Pharmacol.* 233(2):247–253.

Ma, H., Wang, J., Abdel-Rahman, S.Z., Hazra, T.K., Boor, P.J., and Khan, M.F. (2011) Induction of NEIL1 and NEIL2 DNA glycosylases in aniline-induced splenic toxicity. *Toxicol. Appl. Pharmacol.* 251:1–7.

NCI (1978) *Bioassay for aniline hydrochloride for possible carcinogenicity. CAS No. 142-04-1.* ITS Carcinogenesis Technical Report Ser. No. 130. National Cancer Institute, U.S. DHEW, PHS, NIH, Bethesda, MD. DHEW Publ. No. (NIH) 78-1385.

NIOSH (1994) *Documentation for Immediately Dangerous To Life or Health Concentrations (IDLHs) Chemical Listing and Documentation of Revised IDLH Values*, Cincinnati, OH: National Institute for Occupational Safety and Health. Available at http://www.cdc.gov/niosh/idlh/intridl4.html.

NIOSH (2010) National occupational exposure survey conducted from 1981–1983. In: *Estimated Numbers of Employees Potentially Exposed to Specific Agents by 2-digit Standard Industrial Classification (SIC)*, Cincinnati, OH: National Institute for Occupational Safety and Health. Available at http://www.cdc.gov/noes/.

NIOSH (2013) *Pocket Guide to Chemical Hazards, Aniline*, Cincinnati, OH: National Institute for Occupational Safety and Health. Available at http://www.cdc.gov/niosh/npg/npgd0033.html.

Oberst, F.W., Hackley, E., and Comstock, C. (1956) Chronic toxicity of aniline vapor (5 ppm) by inhalation. *Arch. Ind. Health.* 13:379–384.

Price, C.J., et al. (1985) Teratologic and postnatal evaluation of aniline hydrochloride in the Fischer 344 rat. *Toxicol. Appl. Pharmacol.* 77:465–478.

U.S. EPA (1994) *OPPT Chemical Fact Sheets: Aniline*, Washington, DC: Office of Pollution Prevention and Toxics. U.S. Environmental Protection Agency, EPA 749-F-95-002, Available at http://www.epa.gov/chemfact/anali-fs.pdf.

U.S. EPA (2004) *Non-confidential IUR Production Volume Information*, U.S. Environmental Protection Agency, Available at http://www.epa.gov/oppt/iur/tools/data/2002-vol.html.

U.S. EPA. (2012a) Aniline. In: *Integrated Risk Information System*, Washington, DC: Office of Health and Environmental Assessment, Office of Research and Development,. EPA. Available at http://www.epa.gov/iris/subst/0350.htm.

U.S. EPA (2012b) *Toxics release Inventory*, Washington, DC: Office of Pollution Prevention and Toxics, Environmental Protection Agency, Available at http://www.epa.gov/triexplorer/.

U.S. EPA (2012c) *Acute Exposure Guideline Levels (AEGLs) Program*, Washington, DC: Office of Chemical Safety and Pollution Prevention, Environmental Protection Agency, Available at http://www.epa.gov/oppt/aegl/pubs/results5.htm.

von Oettingen, W.F. (1941) *The Aromatic Amines and Nitro Compounds, their Toxicity and Potential Dangers*, Washington, D.C.: Government Printing Office, U. S. Public Health Service, Public Health Bulletin 271, 1–15.

Ward, E.M., et al. (1996) Monitoring of aromatic amine exposures in workers at a chemical plant with a known bladder cancer excess. *J. Natl. Cancer Inst.* 88:1046–1052.

Woo, Y.T. and Lai, D.Y. (2012) Aromatic amino and nitro-amino compounds and their halogenated derivatives. In: Bingham, E., Cohrssen, B., and Powell, C.H., editors., *"Patty's Industrial Hygiene and Toxicology"*, Sixth edition, vol. 2, Chapter 36, John Wiley & Sons, Inc., pp. 609–704.

Woo, Y.T., Arcos, J.C., and Lai, D.Y. (1988) Metabolic and chemical activation of carcinogens: an overview. In P. Politzer and F. J. Martin, editors. *Chemical and Carcinogens. Activation Mechanisms, Structural and Electronic Factors, and Reactivity*, Amsterdam: Elsevier, pp. 1–31.

o-TOLUIDINE

Occupational exposure limits: 5 ppm (OSHA PEL); 2 ppm (ACGIH TLV with skin notation)

Reference values: None

Hazard statements: H301–toxic if swallowed; H315–causes skin irritation; H317–may cause an allergic skin reaction; H319–causes serious eye irritation; H331–toxic if inhaled; H341–suspected of causing genetic defects; H350–may cause cancer; H400–very toxic to aquatic life; H411–toxic to aquatic life with long lasting effects

Precautionary statements: P201–obtain special instructions before use; P280–wear protective gloves/protective clothing/eye protection/face protection; P309+P311–if exposed or if you feel unwell, call a poison center or doctor/physician; P273–avoid release to the environment

Risk phrases: R45–may cause cancer; R23/25–toxic by inhalation and if swallowed; R36–irritating to eyes; R50–very toxic to aquatic organisms

Safety phrases: S45–in case of accident or if you feel unwell, seek medical advice immediately (show the label where possible); S53–avoid exposure; obtain special instructions before use; S61–avoid release to the environment; refer to special instructions/safety data sheets

BACKGROUND AND USES

o-Toluidine, is a methemoglobin-inducing chemical and a human carcinogen. Commercial production of *o*-toluidine was first reported in the United States in 1922. *o*-Toluidine and its hydrochloride salt are used primarily as intermediates in the manufacture of dyes and pigments for printing textiles, in color photography, and as biologic stains. In addition, *o*-toluidine is used as an intermediate for rubber vulcanizing chemicals, pharmaceuticals, and pesticides. Other minor uses include intermediate for organic synthesis and clinical laboratory reagent for glucose analysis (IARC, 2000, 2010; Woo and Lai, 2012).

In the late 1970s, annual production of *o*-toluidine in the United States was estimated to be 1.1–11 million pounds (IARC, 2000). The estimated annual aggregate production/import volumes were 10–50 million pounds in 1986 and 1994, peaked to 50–100 million pounds in 1998 and declined back to 10–50 million pounds in 2002 (U.S. EPA, 2004). In 1983, 34 companies were identified as suppliers. In 1990, two suppliers of *o*-toluidine and 10 suppliers of *o*-toluidine hydrochloride were identified (NTP, 2011). In Europe, *o*-toluidine is a high production volume chemical (HPVC); the estimated annual production volume is 10000–100000 tonnes (ECHA, 2014). Other countries reported to produce or supply *o*-toluidine include Canada, India, Italy, Japan, Mexico, the Netherlands, China, South Africa, and Switzerland (IARC, 2000, 2012). In 2001, the worldwide production volume of *o*-toluidine is estimated to be 59000 tonnes by 11 producers (OECD, 2004).

PHYSICAL AND CHEMICAL PROPERTIES

o-Toluidine (or *o*-methylaniline, 2-aminotoluene), is an alky aniline with a methyl group *ortho* to the amino group. It is a light yellow to reddish-brown liquid. It has a molecular weight of 107.15, a density of 0.9984, a melting point of −14.7°C, a boiling point of 200.2 °C, a refractive index of 1.57276 (20°C), a flash point of 86 °C (closed cup), and an octanol/water partition coefficient (Log P) of 1.29–1.32. *o*-Toluidine is slightly soluble in water and soluble in alcohol and ether (IARC, 2010, 2012).

MAMMALIAN TOXICOLOGY

Acute Effects

o-Toluidine can cause skin and eye irritation and is of moderate to low acute toxicity. Acute oral, dermal, or inhalation exposure to *o*-toluidine can result in methemoglobinemia, reticulocytosis, and hematuria. In rats, the LC_{50} of *o*-toluidine is 852 ppm/4 h (*ca.* 3827 mg/m^3/4 h), and the oral LD_{50} is 750–940 mg/kg bw (OECD, 2004; Woo and Lai, 2012).

Chronic Effects

Repeated-dose studies in rats show that *o*-toluidine is hematotoxic and can cause methemoglobinemia. In a 14-day feeding study in male and female Fischer rats, statistically significant increase in methemoglobin was observed in all animals. No NOAEL could be derived, and based on increases in methemoglobin formation, the LOAEL was 500 ppm (*ca.* 23.7 mg/kg bw/day and 25.5 mg/kg bw/day for males and females, respectively). In another subchronic study, hemolytic anemia was found in Wistar rats receiving daily dose of 80 mg/kg bw *o*-toluidine by gavage; toxicity of the kidney and liver was also noted at 400 mg/kg bw (HSDB, 2012). *o*-Toluidine hydrochloride caused increased incidences of hematopoiesis, hemosiderin accumulation, and capsular fibrosis in the spleen when administered to F344/N rats in feed at daily doses of 5000 ppm (*ca.* 237 mg/kg bw) for 13 or 26 weeks (NTP, 2000).

There is sufficient evidence for the carcinogenicity of *o*-toluidine in experimental animals. Chronic exposure of *o*-toluidine hydrochloride induced multiple-sites tumors in rats and mice. In a 2-year carcinogenesis bioassay (NCI, 1979), groups of 50 F344 rats of each sex and 50 B6C3 F$_1$ mice of each sex were given diets containing *o*-toluidine hydrochloride (3000 or 6000 ppm for rats and 1000 or 3000 ppm for the mice) for 101–104 weeks. In rats, it induced several types of sarcomas of the spleen and other organs in both sexes, mesotheliomas of the abdominal cavity or scrotum, fibromas of the subcutaneous tissue in males, and transitional-cell carcinomas of the urinary bladder, fibroadenomas, or adenomas of the mammary gland in females. In mice, it caused blood-vessel tumors (hemangioma and hemangiosarcoma) in both sexes and benign and malignant liver tumors (hepatocellular adenoma and carcinoma) in females. Similar tumors were induced in other strains of rat and mouse (e.g., Wistar rats, CD-1 mice) after chronic exposures to *o*-toluidine (HSDB, 2012). Interestingly, while *o*-toluidine induced leukemia and tumors of the lung, kidney, and mammary gland in rodents, urinary bladder tumors developed in two dogs after 9 and 10 years, respectively, of exposure to *o*-toluidine (Pliss, 2004).

A number of epidemiological studies have found increased incidences of urinary bladder cancer among workers exposed to mixtures of aromatic amines including *o*-toluidine in the dyestuffs and rubber industries. Carcinogenic potential of *o*-toluidine specifically cannot be definitively determined by earlier studies because of small cohorts and confounding exposures to other chemicals. In more recent follow-up studies, bladder cancer incidence remains elevated in the cohorts and despite other concurrent chemical exposures, *o*-toluidine was considered most likely responsible for the excess bladder cancer incidence (Markowitz and Levin, 2004; Baan et al., 2008; Sorahan, 2008; Carreón et al., 2010, 2014).

Mechanism of Action(s)

The metabolism and mode of action of *o*-toluidine has not yet been fully elucidated. Like other aromatic amines, *o*-toluidine is expected to be oxidized to a *N*-hydroxy and/or a *N*-/O-acetylated derivative, which interacts with the hem group of hemoglobin resulting in the formation of methaemoglobin. Studies in rats and mice have demonstrated *o*-toluidine (or its metabolites) binds to hemoglobin and that a linear dose relationship exists for hemoglobin (DeBord et al., 1992). The extent of hemoglobin binding in rats correlated well with the concentrations of methemoglobin (Birner and Neuman, 1988). There was a correlation between urinary *o*-toluidine levels and hemoglobin adduct in exposed and unexposed workers (HSDB, 2012). Prilocaine, a widely used local anesthetic, is metabolized to *o*-toluidine. Prilocaine anesthesia has resulted in a significant increase of hemoglobin adducts of *o*-toluidine in blood samples of the patients (Gaber et al., 2007).

On the basis of the *in vitro* and *in vivo* genotoxicity and DNA damages observed in various assay systems, *o*-toluidine appears to be acting as a genotoxic carcinogen. *o*-Toluidine was tested positive in a variety of *in vitro* genotoxicity assay systems that included Ames *Salmonella*, chromosome aberration, sister chromatid exchange (SCE), mouse lymphoma cell mutation, cell transformation, and unscheduled DNA synthesis tests. It was also positive in somatic mutation in *Drosophila* and *in vivo* micronucleus test in rat peripheral blood (Woo and Lai, 2012; Maire et al., 2012; HSDB, 2012; Suzuki et al., 2005). DNA adducts of *o*-toluidine have been detected in the liver and nasal mucosa of rats given prilocaine, a local anesthetic that is metabolized to *o*-toluidine (Duan et al., 2008).

It is well documented that aromatic amines require metabolic activation for their carcinogenic activities. The mechanism of activation, which involves many competing pathways in a two-stage metabolic activation, has been reviewed (Woo and Lai, 2012). 4-Amino-3-methylphenol, a major metabolite, caused DNA damage in the presence of Cu(II). *o*-Nitrosotoluene, a minor metabolite, induced DNA damage after addition of NADH. Metal-mediated DNA damage by *o*-toluidine metabolites through H_2O_2 and free radicals formation has been suggested to be responsible for the carcinogenicity of *o*-toluidine in rats and mice (Ohkuma et al., 1999). Electrophilic arylamidonium or arylnitrenium ions generated in rodents at the second stage of metabolic activation can also bind covalently to cellular nucleophiles such as DNA to initiate carcinogenesis. Competing against the second-stage activation are two other pathways involving glutathione conjugation and glucuronidation that represent detoxification pathways in rodents. However, at the acidic pH found in the urinary bladder of dogs and humans, and in the presence of urinary β-glucuronidase, they are cleaved to give the electrophilic arylnitrenium ion, which interacts with cellular macromolecules such as DNA to induce urinary bladder tumors. The differences in metabolic activation mechanism between rodents and dogs or humans are consistent with the findings that *o*-toluidine induced leukemia and tumors of the lung, kidney, and mammary gland in rodents, but urinary bladder tumors in dogs (Pliss, 2004), and support the increased risks of urinary bladder cancer observed in workers exposed to *o*-toluidine (Markowitz and Levin, 2004; Baan et al., 2008; Sorahan, 2008; Carreón et al., 2010; Carreon et al., 2014).

Chemical Pathology

o-Toluidine is readily absorbed by inhalation or through the skin and gastrointestinal tract. It can cause skin and eye irritation. The major clinical toxicity following acute and chronic/subchronic exposure to *o*-toluidine by are methemoglobinemia and related effects on the blood and spleen. Exposure to high concentrations may result in damage to kidneys and bladder. Early symptoms typical of methemoglobinemia such as headache, nausea, vomiting, diarrhea, and cyanosis may begin to appear at 6 ppm after several hours, or at 100 ppm after an hour. Continued exposure may lead to difficult breathing, dizziness, stupor, unconsciousness, and death. Ingestion of about 2 oz by a 150 lb person may also result in death (HSDB, 2012).

o-Toluidine induced multisites tumors in rats and mice and is a human carcinogen; an excess of bladder tumors has been found in workers exposed to dyestuffs and chemical mixtures containing *o*-toluidine (Markowitz and Levin,

2004; Baan et al., 2008; Sorahan, 2008; Carreón et al., 2010; Carreon et al., 2014).

SOURCES OF EXPOSURE

Occupational

Occupational exposure to *o*-toluidine and its hydrochloride salt can occur to workers in many industries; among them the greatest potential for exposure are workers in dyes and pigment industries through inhalation and dermal contact. According to a survey conducted in the United States from 1972 to 1974, 13053 workers were potentially exposed to *o*-toluidine. In another U.S. survey conducted from 1981 to 1983, 30,000 workers, including 15,500 women, were estimated to expose to *o*-toluidine (NTP, 2011). There is also evidence of dermal absorption of *o*-toluidine in workers of rubber industry (Korinth et al., 2007; Wellner et al., 2008).

Air and urine samples from automobile workers involved in rubber vulcanization have been measured. Concentrations of *o*-toluidine ranging from 26.63 to 93.93 $\mu g/m^3$ in air and from 54.65 to 242.88 $\mu g/l$ in urine were found (Riedel et al., 2006). There was a correlation between air concentrations and urinary *o*-toluidine or hemoglobin adduct levels in blood in studies on exposed and nonexposed workers (OECD, 2004).

Environmental

The general population may be exposed to low concentrations of *o*-toluidine in ambient air or by dermal contact with commercial dyes used on textiles (IARC, 2000). Between 1988 and 2009, environmental releases of *o*-toluidine in the United States ranged from 10,800 to 55,000 pounds, whereas releases of *o*-toluidine hydrochloride ranged from 0 to 265 pounds. In 2007, 16,348 pounds of *o*-toluidine were released from 12 facilities (U.S. EPA, 2009). o-Toluidine has been detected in effluents from refineries and production facilities and in river water, process water, and groundwater (IARC, 2000). In 1979, it was measured in the Rhine River at concentrations of 0.03–1.8 $\mu g/l$ and in Japan at concentrations of up to 20 $\mu g/l$ (IPCS, 1998).

o-Toluidine is present in cigarette smoke at a concentration of up to 144 ng per cigarette (Stabbert et al., 2003). Hemoglobin adduct levels of *o*-toluidine were significantly higher in children in Germany who were exposed to environmental tobacco smoke than those who were not (Richter et al., 2001). *o*-Toluidine has also been detected in blood and breast milk (Gazarian et al., 1995) and was measured in the urine of individuals with no known exposure at a median concentration of 0.12 $\mu g/l$ (Weiss and Angerer, 2002). Patients treated with prilocaine, a local anesthetic that is metabolized to *o*-toluidine, have been shown to have substantially elevated hemoglobin adducts of *o*-toluidine (Gaber et al., 2007).

INDUSTRIAL HYGIENE

The time-weighted average TLV-TWA for *o*-toluidine adopted by American Conference of Governmental Industrial Hygienists (ACGIH) is 2 ppm with skin notation. The OSHA Permissible exposure limit (PEL) is 5 ppm. The National Institute for Occupational Safety and Health (NIOSH) immediately dangerous to life or health concentration (IDLH) is 50 ppm (NTP, 2011).

o-Toluidine and *o*-toluidine hydrochloride have been changed from "reasonably anticipated to be human carcinogens" to "known to be human carcinogens" in Report on Carcinogens (NTP, 2014). The International Agency for Research on Cancer has also re-classified *o*-toluidine as a Group 1 "known human carcinogen" from Group 2A "probable human carcinogen" on the basis of increased risk of urinary-bladder cancer reported in these follow-up cohort studies of workers exposed to *o*-toluidine (IARC, 2010). NIOSH considers this compound to be an occupational carcinogen and recommends appropriate worker protection.

Standards	Values
NIOSH REL	None established; NIOSH considers *o*-toluidine to be an occupational carcinogen
OSHA PEL	5 ppm TWA
ACGIH TLV	2 ppm TWA,
IDLH	50 ppm

MEDICAL MANAGEMENT

As *o*-toluidine can be absorbed by the skin, it is important to remove all contaminated clothing immediately if splashed with *o*-toluidine. The level of methemoglobinemia needs to be consistently monitored. Administer oxygen when methemoglobinemia is above 30%. Further administer intravenously a 5% glucose solution containing ascorbic acid when methemoglobinemia is above 50%. If methemoglobinemia is above 60% methemoglobinemia, administer intravenously 10–20 cm^3 of a 1% solution of methylene blue. If there is no response to treatment with methylene blue, then perform hemodialysis or exchange transfusion (IPCS, 1998).

REFERENCES

Baan, R., Straif, K., Grosse, Y., Secretan, B., El Ghissassi, F., Bouvard, V., Benbrahim-Tallaa, L., and Cogliano, V. (2008) Carcinogenicity of some aromatic amines, organic dyes, and related exposures. *Lancet Oncol.* 9(4):322–323.

Birner, G. and Neuman, H.G. (1988) Biomonitoring of aromatic amines II: hemoglobin binding of some monocyclic aromatic amines. *Arch Toxicol* 62:110–115.

Carreón, T., Hein, M.J., Viet, S.M., Hanley, K.W., Ruder, A.M., and Ward, E.M. (2010) Increased bladder cancer risk among workers exposed to *o*-toluidine and aniline: a reanalysis. *Occup. Environ. Med.* 67:348–350.

Carreon, T., Hein, M.J., Hanley, K.W., Viet, S.M. and Ruder, A.M. (2014) Workplace Bladder cancer incidence among workers exposed to *o*-toluidine, aniline and nitrobenzene at a rubber chemical manufacturing plant. *Occup. Environ. Med.* 71:175–182.

DeBord, D.G., et al. (1992) Binding characteristics of *ortho*-toluidine to rat hemoglobin and albumin. *Arch. Toxicol.* 66:231–236.

Duan, J.D., Jeffrey, A.M., and Williams, G.M. (2008) Assessment of the medicines lidocaine, prilocaine, and their metabolites, 2,6-dimethylaniline and 2-methylaniline, for DNA adduct formation in rat tissues. *Drug Metab. Dispos.* 36:1470–1475.

ECHA. (2014) *European Chemicals Agency. Information on Chemicals,* o-*Toluidine.* Available at http://echa.europa.eu/information-on-chemicals

Gaber, K., Harreus, U.A., Matthias, C., Kleinsasser, N.H., and Richter, E. (2007) Hemoglobin adducts of the human bladder carcinogen *o*-toluidine after treatment with the local anesthetic prilocaine. *Toxicology* 229:157–164.

Gazarian, M., Taddio, A., Klein, J., Kent, G., and Koren, G. (1995) Penile absorption of EMLA cream in piglets: implications for use of EMLA in neonatal circumcision. *Biol. Neonate* 68: 334–341.

HSDB. (2012) *Hazardous Substances Data Bank. National Library of Medicine.* Available at http://toxnet.nlm.nih.gov/cgi-bin/sis/search.

IARC. (1982) *ortho*-Toluidine and ortho-toluidine hydrochloride. In: *Some Aromatic Amines, Anthraquinones and Nitroso Compounds and Inorganic Fluorides Used in Drinking Water and Dental Preparations. IARC Monographs on the Evaluation of Carcinogenic Risk of Chemicals to Humans,* vol. 27, Lyon, France: International Agency for Research on Cancer, pp. 155–175.

IARC. (2000) *ortho*-Toluidine IARC Monographs on the Evaluation of Carcinogenic Risk to Humans, In: *Some Industrial Chemicals,* vol. 77, Lyon, France: International Agency for Research on Cancer, pp. 267–322.

IARC. (2010) ortho-Toluidine. IARC monographs on the evaluation of carcinogenic risk to humans, In: *Some Aromatic Amines, Organic Dyes, and Related Exposures,* vol. 99, Lyon, France: International Agency for Research on Cancer, pp. 407–469.

IPCS. (1998) *Concise International Chemical Assessment Document No. 7.* o-*Toluidine. International Programme on Chemical Safety.* World Health Organization, Geneva. Available at http://www.inchem.org/documents/cicads/cicads/cicad07.htm.

Korinth, G., Weiss, T., Penkert, S., Schaller, K.H., Angerer, J., and Drexler, H. (2007) Percutaneous absorption of aromatic amines in rubber industry workers: Impact of impaired skin and skin barrier creams. *Occup. Environ. Med.* 64 (6):366–372.

Maire, M.A., Pant, K., Poth, A., Schwind, K.R., Rast, C., Bruce, S.W., Sly, J.E., Kunz-Bohnenberger, S., Kunkelmann, T., Engelhardt, G., Schulz, M., and Vasseur, P. (2012) Prevalidation study of the Syrian hamster embryo (SHE) cell transformation assay at pH 7.0 for assessment of carcinogenic potential of chemicals. *Mutat. Res.* 744(1):64–75.

Markowitz, S.B. and Levin, K. (2004) Continued epidemic of bladder cancer in workers exposed to ortho-toluidine in a chemical factory. *J. Occup. Environ. Med.* 46:154–160.

NCI. (1979) *Bioassay of* o-*Toluidine Hydrochloride for Possible Carcinogenicity (CAS No. 636-21-5).* Technical Report Series no. 153. DHEW (NIH) Publication No. 79-1709. Bethesda, MD: National Institutes of Health. 145 pp.

NTP. (2000) NTP Comparative Toxicity and Carcinogenicity Studies of *o*-Nitrotoluene and *o*-Toluidine Hydrochloride (CAS Nos. 88-72-2 and 636-21-5) Administered in Feed to Male F344/N Rats. National Toxicology Program. *Toxic Rep. Ser.* 44:1-C8.

NTP. (2014) Report on Carcinogens, 13th ed., National Toxicology Program. Available at http://ntp.niehs.nih.gov/ntp/roc/content/profiles/toluidine.pdf.

OECD. (2004) *Screening Information Data Set for* o-*Toluidine (95-53-4), Organization for Economic Cooperation and Development, October.* Available at http://www.chem.unep.ch/irptc/sids/OECDSIDS/sidspub.html.

Ohkuma, Y., Hiraku, Y., Oikawa, S., Yamashita, N., Murata, M., and Kawanishi, S. (1999) Distinct mechanisms of oxidative DNA damage by two metabolites of carcinogenic *o*-toluidine. *Arch. Biochem. Biophys.* 372:97–106.

Pliss, G.B. (2004) Experimental study of ortho-toluidine carcinogenicity. *Vopr. Onkol.* 50:567–571.

Richter, E., Rosler, S., Scherer, G., Gostomzyk, J.G., Grubl, A., Kramer, U., and Behrendt, H. (2001) Haemoglobin adducts from aromatic amines in children in relation to area of residence and exposure to environmental tobacco smoke. *Int. Arch. Occup. Environ. Health* 74:421–428.

Riedel, K., Scherer, G., Engl, J., Hagedorn, H.W., and Tricker, A.R. (2006) Determination of three carcinogenic aromatic amines in urine of smokers and nonsmokers. *J. Anal. Toxicol.* 30:187–195.

Sorahan, T. (2008) Bladder cancer risks in workers manufacturing chemicals for the rubber industry. *Occup. Med.* 58:496–501.

Suzuki, H., Ikeda, N., Kobayashi, K., Terashima, Y., Shimada, Y., Suzuki, T., Hagiwara, T., Hatakeyama, S., Nagaoka, K., Yoshida, J., Saito, Y., Tanaka, J., and Hayashi, M. (2005) Evaluation of liver and peripheral blood micronucleus assays with 9 chemicals using young rats. A study by the Collaborative Study Group for the Micronucleus Test (CSGMT)/Japanese Environmental Mutagen Society (JEMS)-Mammalian Mutagenicity Study Group (MMS). *Mutat. Res.* 6(583):133–145.

Stabbert, R., Schafer, K.H., Biefel, C., and Rustemeier, K. (2003) Analysis of aromatic amines in cigarette smoke. *Rapid Commun. Mass Spectrom.* 17:2125–2132.

U.S. EPA. (2004) Inventory Update Reporting. In: *Non-confidential 2002 IUR Records by Chemical, including Manufacturing, Processing and Use Information.* Washington, DC. U.S. Environmental Protection Agency.

U.S. EPA. (2009) *TRI Explorer Chemical Report*, Washington, DC. U.S. Environmental Protection Agency. Available at http://www.epa.gov/triexplorer.

Wellner, T., Lüersen, L., Schaller, K.H., Angerer, J., Drexler, H., and Korinth G. (2008) Percutaneous absorption of aromatic amines – a contribution for human health risk assessment. *Food Chem. Toxicol.* 46:1960–1968.

Weiss, T. and Angerer, J. (2002) Simultaneous determination of various aromatic amines and metabolites of aromatic nitro compounds in urine for low level exposure using gas chromatography-mass spectrometry. *J. Chromatogr. B Analyt. Technol. Biomed. Life Sci.* 778:179–192.

Woo, Y.-T. and Lai, D. Y. (2012) Aromatic amino and nitro-amino compounds and their halogenated derivatives. In: Bingham, E., Cohrssen, B., and Powell, C.H., editors. *Patty's Industrial Hygiene and Toxicology.* 6th ed., vol. 2, Chapter 36, John Wiley & Sons, Inc., pp. 609–704.

o-ANISIDINE

Occupational exposure limits: $0.5\,mg/m^3$ (OSHA PEL, NIOSH REL, ACGIH TLV)

Reference values: None

Hazard statements: H350–may cause cancer; H341–suspected of causing genetic defects; H331–toxic if inhaled; H311–toxic if contact with skin; H301–toxic if swallowed.

Precautionary statements: P201–obtain special instructions before use; P273–avoid release to the environment; P280–wear protective gloves/protective clothing/eye protection/face protection; P361–remove/take off immediately all contaminated clothing; P309 + P311–if exposed or if you feel unwell, call a poison center or doctor/physician; P404–store in a closed container.

Risk phrases: R45–may cause cancer; R68–possible risk of irreversible effects; R23/24/25–toxic by inhalation, in contact with skin and if swallowed.

Safety phrases: S53–avoid exposure—obtain special instructions before use; S45–in case of accident or if you feel unwell, seek medical advice immediately (show the label where possible).

BACKGROUND AND USES

Similar to other aromatic amines, *o*-anisidine (CASRN 90-04-0) may cause methemglobinemia and cancer in humans. It is used mainly as an intermediate for the production of azo dyes and pigments. Other industrial uses of *o*-anisidine include synthesis of other dyes and pharmaceuticals, as a corrosion inhibitor for steel, and as an antioxidant for polymercaptan resins (IARC, 1999; HSDB, 2012).

o-Anisidine was first produced commercially in the U.S. in 1922, and in 1977, one of the three U.S. companies reported production of 4.5–45.4 thousand kg (IARC, 1982). U.S. production plus imports of *o*-anisidine totaled 500,000–1 million pounds in 1986, 1990, and 2006; 1–10 million pounds in 1998; and 10,000–500,000 pounds in 2002 (U.S. EPA, 2004, 2010). In 1995, worldwide *o*-anisidine production was estimated to be 15,000 tonnes/

year, from which China alone produced about 7000 tonnes/ year. In 1997, five producing and processing companies of *o*-anisidine were identified in Western Europe, but the total amounts were <1000 tonnes/year (EU, 2002).

PHYSICAL AND CHEMICAL PROPERTIES

o-Anisidine (or *o*-methoxyaniline) is an aromatic amine with a methoxyl group ortho to the amino group of aniline. It is a colorless to yellowish, pink, or reddish liquid. It has a molecular weight of 123.2, a density of $1.0923 \, g/cm^3$ at (20 °C), a melting point of 5 °C, a boiling point of 225 °C, a vapor pressure of 0.08 mm Hg (25 °C), and an octanol/water partition coefficient (Log P) of 1.18. *o*-Anisidine is soluble in water and mineral oils, and miscible with alcohol, benzene, acetone, and diethylether. *o*-Anisidine hydrochloride, a salt of o-anisidine, is a grayish crystalline solid or powder at room temperature and is soluble in water (NTP, 2011).

MAMMALIAN TOXICOLOGY

Acute Effects

Acute dermal irritation/corrosion study and acute eye irritation/corrosion study of *o*-anisidine in rabbits were negative (EU, 2002). More recent human data, however, showed that it was very irritating to skin, eyes, and mucous membranes (HSDB, 2012). *o*-Anisidine was reported to be a weak sensitizer in the guinea pig. However, there are uncertainties about the adequacy of the study due to insufficient documentation. The oral LD_{50} of *o*-anisidine was 1890–2020 mg/ kg bw in rats, 1410 mg/kg bw in mice, and 870 mg/kg bw in rabbits. Clinical signs and some hematological changes, anemia and nephrotoxicity were only seen in the high dosed animals. The dermal LD_{50} for Wistar rats was >2000 mg/kg bw. In an acute inhalation study, no mortality and no signs of toxicity were found in rats at the highest technically feasible aerosol concentration of 3.87 mg/l *o*-anisidine; the 4-h LC_{50} was determined to be >3.87 mg/l (EU, 2002).

Similar to other aromatic amines, *o*-anisidine induces methemoglobin formation in animals. Significantly elevated methemoglobin levels were observed in rats and mice within 3–48 h after a single oral application via gavage of 690 mg/kg bw *o*-anisidine. In cats, a species with comparable methaemoglobin forming capacity to humans, a single intravenous injection of 7.7 mg/kg bw *o*-anisidine resulted in up to 15% increase of methemoglobin level within 1–5 h (EU, 2002).

Chronic Effects

From a 28-day oral study in Wistar rats, a lowest observed adverse effect level (LOAEL) of 80 mg/kg bw was derived based on hemolytic anemia, spleen toxicity, and increased liver and spleen weights. The effects observed were considered secondary to the acute methemoglobin formation. The NOAEL was 16 mg/kg bw (EU, 2002).

Except for some complaints of headache and vertigo, increased methemoglobin and sulfhemoglobin, and frequent occurrence of erythrocytic inclusion bodies (Heinz bodies), workers exposed to an air concentration of 0.4 ppm *o*-anisidine 3.5 h/day for 6 months developed no anemia or chronic toxicity (HSDB, 2012).

o-Anisidine hydrochloride was carcinogenic in rats and mice. Dietary administration of *o*-anisidine hydrochloride (at doses of 0, 5000, or 10,000 ppm) caused increased incidences of urinary bladder tumors in rats of both sexes. In male rats, it also induced transitional cell carcinomas of the kidney and follicular cell tumors of the thyroid. A significantly increased incidence of the urinary bladder tumors occurred in the high dose (5000 ppm) mice (NCI, 1978).

There are no epidemiological studies that evaluated the relationship between human cancer and exposure specifically to *o*-anisidine or *o*-anisidine hydrochloride.

Mechanism(s) of Action

The mechanisms of toxicity/carcinogenicity of *o*-anisidine have not been fully elucidated. As with other aromatic amines, it is likely that *o*-anisidine is oxidized to a *N*-hydroxy or *N*-/*O*-acetyl derivative, which interacts with the hem group of hemoglobin to cause methemoglobin formation.

o-Anisidine has been tested for genotoxicity in a variety of *in vitro* and *in vivo* assay systems. It was negative in several reverse mutation assays in *Salmonella typhimurium* and *Escherichia coli* with or without metabolic activation. *N*-acetylation appears to play a role in the metabolic activation of *o*-anisidine since mutagenic activity was detected in *Salmonella typhimurium* strain YG1029 containing elevated levels of N-acetyltransferase activity (IARC, 1999). In mammalian systems, *o*-anisidine was positive in a chromosome aberration assay and a sister chromatids exchange assay with CHO cells, and in a mouse lymphoma assay (Thompson et al., 1992; EU, 2002).

In most of the *in vivo* test systems, however, *o*-anisidine was negative (EU, 2002; IARC, 1999). Stiborová et al. (2002, 2005, 2009) showed that the genotoxic mechanisms of *o*-anisidine involve metabolic activation by *N*-hydroxylase to the proximate carcinogenic metabolite, *N*-(2-methoxyphenyl) hydroxylalmine, which spontaneously decomposes to nitrenium and/or carbonium ions responsible for the deoxyguanosine adducts formation observed in DNA incubated with *o*-anisidine and human microsomes *in vitro*. The same adducts were detected in urinary bladder, the target organ, and to a lesser extent, in liver, kidney, and spleen of rats treated with *o*-anisidine (Stiborova et al., 2005). Furthermore, an extended *in vivo* study by Naiman et al. (2012) showed that the DNA adducts in the urinary bladder were substantially more persistent than those in all the other organs. Over 39% of the adducts remained in the bladder 36 weeks after the treatment whereas no adducts could be detected in other organs. The negative *in vivo* genotoxic responses may be due to the detoxification of reactive metabolites by cytochrome P450 enzymes (e.g., CYP3A4, 2E1, and 2C) in the liver back to *o*-anisidine (Naiman et al., 2011; Dracínska et al., 2006).

Peroxidases also play a role in the metabolic activation of aromatic amines (Stiborova et al., 2002; Woo and Lai, 2012). Results of *in vitro* studies have suggested a possible role for peroxidation enzymes in the metabolic activation of *o*-anisidine (Martinkova et al., 2012). Several reactive intermediates including electrophilic diimine and quinoneimine metabolites were formed when studied of *o*-anisidine with prostaglandin H synthase, a peroxidation enzyme (Thompson et al., 1992). Formation of these reactive intermediates from *o*-anisidine and covalently binding to nucleic acids, and protein have been demonstrated using horseradish peroxidase (Thompson and Eling, 1991). Metabolic activation of *o*-anisidine by prostaglandin H synthase may play a role in the carcinogenesis of urinary bladder observed in rats and mice as prostaglandin H synthase is broadly distributed in mammalian tissues, including the urinary bladder.

In an *in vitro* study, *o*-anisidine was shown to be a strong inhibitor of thyroid peroxidase (Freyberger, 1994). The persistent thyroid peroxidase inhibition with concomitantly decreased thyroid hormone formation is a key event in thyroid tumorigenesis; the increased incidence of thyroid tumors observed in male rats in the 2-year bioassay may be resulted from thyroid peroxidase inhibition by *o*-anisidine.

Chemical Pathology

o-Anisidine is expected to be absorbed through the skin, in the gastrointestinal and the respiratory tract. Contact with *o*-anisidine may irritate the eyes, skin, and respiratory tract. Overall acute toxicity of *o*-anisidine appears to be low by all routes of exposure.

There are limited data on effects in humans after repeated exposure. Significant increased levels of methemoglobin were not seen in workers exposed to an air concentration of 0.4 ppm *o*-anisidine 3.5 h/day for 6 months (EU, 2002). However, based on acute and subchronic toxicity studies in rodents and in cats, a species with comparable methaemoglobin forming capacity to humans, chronic exposures to higher levels of *o*-anisidine may cause headache, dizziness, vertigo, cyanosis, difficulty in breathing, anemia, organs damage, and death due to methemglobinemia (HSDB, 2012).

o-Anisidine is *reasonably anticipated to be a human carcinogen* based on sufficient evidence of carcinogenicity in experimental animals (NTP, 2011). As with other aromatic amines, urinary bladder is the main target organ of cancer development in humans. Interindividual differences in susceptibility may occur due to different activities of *N*-hydroxylases and *N*-acetyltransferases, and other metabolic activation/detoxification enzyme activities for *o*-anisidine.

SOURCES OF EXPOSURE

Occupational

Occupational exposure to *o*-anisidine may occur at workplace through inhalation and dermal contact during its production or processing (IARC, 1999). The National Occupational Exposure Survey conducted from 1981 to 1983 in the U.S. estimated that 705 workers were potentially exposed to *o*-anisidine and 1108 workers were potentially exposed to *o*-anisidine hydrochloride (NIOSH, 2010). According to the survey data in 2006, 1–99 people were likely to be exposed in industrial manufacturing, processing, and use of *o*-anisidine; the data may be greatly underestimated (U.S. EPA, 2010). During the processing of *o*-anisidine, 0.05–0.15 mg/m^3 were detected from personal air sampling of some German processing companies between 1990 and 1995 (EU, 2002).

Environmental

o-Anisidine can occur as an environmental pollutant in wastewater from oil refineries and chemical plants (IARC, 1982, 1999). According to EPA's Toxics Release Inventory, about 10,000 pounds of *o*-anisidine were released annually from 1989 to 1992, mostly to air. From 1993 to 2007, releases were much lower; in 2007, total releases were down to 638 pounds (U.S. EPA, 2009). In air, *o*-anisidine is expected to be degraded by reaction with hydroxyl radicals with a half-life of 6 h. In soil, it will be subjected to rapid biodegradation under aerobic conditions. In surface water, it is expected to bind to sediment or suspended solids with high organic matter content and to volatilize from water with an estimated half-life of 31 days from streams and 350 days from lakes. *o*-Anisidine is not expected to bioaccumulate in aquatic organisms (EU, 2002; HSDB, 2012).

The general population may come into contact with *o*-anisidine during the use of consumer products such as textiles and printed packings colored with pigments or dyes based on the compound. In addition, *o*-anisidine occurs in cigarette smoke; the mean concentrations of *o*-anisidine in smoke were reported to range from <0.2 to 5.12 ng per cigarette (Stabbert et al. (2003). *o*-Anisidine was detected at concentrations ranging from <0.05 to 4.2 µg/l (median = 0.22 µg/l) in 20 urine samples from the general population in Germany (Weiss and Angerer, 2002).

INDUSTRIAL HYGIENE

The time-weighted average TLV–TWA for *o*-anisidine adopted by American Conference of Governmental

Industrial Hygienists (ACGIH) is 0.1 ppm (0.5 mg/m^3). The OSHA PEL is 0.1 ppm (0.5 mg/m^3). The NIOSH recommended exposure limit (REL) is 0.1 ppm (0.5 mg/m^3) and the IDLH is 10 ppm (50 mg/m^3) (NTP, 2011). The occupational exposure limits for *o*-anisidine in other OECD countries are similar to those of the U.S. (EU, 2002).

o-Anisidine and *o*-Anisidine hydrochloride have been listed in Report on Carcinogens (NTP, 2011) as *"reasonably anticipated to be human carcinogens"*. The International Agency for Research (IARC) on Cancer has classified *o*-anisidine as a Group 2B "possibly carcinogenic to humans" based on sufficient evidence for carcinogenicity in experimental animals and inadequate evidence for carcinogenicity in humans (IARC, 1999). NIOSH considers this compound to be an occupational carcinogen and recommends appropriate worker protection.

MEDICAL MANAGEMENT

Decontamination, symptomatic, and supportive measures are crucial for medical management of *o*-anisidine poisoning. Victims should be removed from the area of exposure immediately and if splashed with *o*-anisidine, all contaminated clothing should be removed. Early signs and symptoms of methemoglobinaemia should be monitored, and appropriate treatments for methemoglobinaemia such as administration of oxygen, ascorbic acid, or methylene blue should be used when needed (Dutta et al., 2008).

REFERENCES

Dracínska, H., Miksanová M., Svobodová M., Smrcek, S., Frei, E., Schmeiser, H.H., and Stiborová M. (2006) Oxidative detoxication of carcinogenic 2-nitroanisole by human, rat and rabbit cytochrome P450. *Neuro. Endocrinol. Lett.* 27(Suppl 2):9–13.

Dutta, R., Dube, S.K., Mishra, L.D., and Singh, A.P. (2008) Acute methemglobinemia. *Internet J. Emerg. Intensive Care Med.* 11:1092–4051.

Freyberger, A. (1994) Irreversible inhibition of thyroid peroxidase (tpo) by thyreotoxic aromatic amines *in vitro*. *Naunyn-Schmiedeberg's Arch. Pharmacol.* 349:R110.

HSDB (2012) *Hazardous Substances Data Bank. National Library of Medicine*. Available at http://toxnet.nlm.nih.gov/cgi-bin/sis/search.

IARC (1982) *ortho*-and *para*-Anisidine and their hydrochlorides. In: *Some Aromatic Amines, Anthraquinones and Nitroso Compounds and Inorganic Fluorides Used in Drinking Water and Dental Preparations*, IARC Monographs on the Evaluation of Carcinogenic Risk of Chemicals to Humans, vol. 27, Lyon, France: International Agency for Research on Cancer, pp. 63–80.

IARC. (1999) *ortho*-Anisidine. In: *Some Chemicals That Cause Tumors of the Kidney or Urinary Bladder in Rodents and Some Other Substances*, IARC Monographs on the Evaluation of Carcinogenic Risk of Chemicals to Humans, vol. 73, Lyon, France: International Agency for Research on Cancer, pp. 49–58.

Martinkova, M., Kubickova, B., and Stiborova, M. (2012) Effects of cytochrome P450 inhibitors on peroxidase activity. *Neuro. Endocrinol. Lett.* 33(Suppl 3):33–40.

Naiman, K., Dracínský, M., Hodek, P., Martínková, M., Schmeiser, H.H., Frei, E., and Stiborová, M. (2012) Formation, persistence, and identification of DNA adducts formed by the carcinogenic environmental pollutant o-anisidine in rats. *Toxicol. Sci.* 127:348–359.

Naiman, K., Martínková, M., Schmeiser, H.H., Frei, E., and Stiborová, M. (2011) Human cytochrome-P450 enzymes metabolize N-(2-methoxyphenyl)hydroxylamine, a metabolite of the carcinogens o-anisidine and o-nitroanisole, thereby dictating its genotoxicity. *Mutat. Res.* 24(726):160–168.

NCI (1978) *Bioassay of o-Anisidine Hydrochloride for Possible Carcinogenicity*. National Cancer Institute Technical Report Series no. 89. DHEW (NIH) Publication no. 78-1339. Bethesda, MD: National Institutes of Health. 130 pp.

NIOSH; NOES (2010) *National Occupational Exposure Survey conducted from 1981–1983. Estimated numbers of employees potentially exposed to specific agents by 2-digit standard industrial classification (SIC). Available from, as of June 16*. Available at http://www.cdc.gov/noes.

NTP (2011) *Report on Carcinogens*, 12th ed., National Toxicology Program. Available at http://ntp.niehs.nih.gov/go/roc12.

Stabbert, R., Schafer, K.H., Biefel, C., and Rustemeier, K. (2003) Analysis of aromatic amines in cigarette smoke. *Rapid Commun. Mass Spectrom.* 17:2125–2132.

Stiborová, M., Mikšanova, M., Havlíček, V., Schmeiser, H.H., and Frei, E. (2002) Mechanism of peroxidase-mediated oxidation of carcinogenic o-anisidine and its binding to DNA. *Mutat. Res.* 500:49–66.

Stiborová, M., Mikšanova, M., Šulc, M., Rydlova, H., Schmeiser, H.H., and Frei, E. (2005) Identification of a genotoxic mechanism for the carcinogenicity of the environmental pollutant and suspected human carcinogen o-anisidine. *Int. J. Cancer* 116:667–678.

Thompson, D.C. and Eling, T.E. (1991) Reactive intermediates formed during the peroxidative oxidation of anisidine isomers. *Chem. Res. Toxicol.* 4:474–481.

Thompson, D.C., Josephy, P.D., Chu, J.W.K., and Eling, T.E. (1992) Enhanced mutagenic of anisidine isomers in bacterial strains containing elevated N-acetyltransferase activity. *Mutat. Res.* 279:83–89.

U.S. EPA; Inventory Update Reporting (IUR) (2010) *Non-confidential 2006 IUR Records by Chemical, Including Manufacturing, Processing and Use Information*, Washington, DC: U.S. Environmental Protection Agency, Available at http://cfpub.epa.gov/iursearch/index.cfm.

U.S. EPA (2009) *TRI Explorer Chemical Report*, Washington, DC: U.S. Environmental Protection Agency, Available at http://www.epa.gov/triexplorer.

Weiss, T. and Angerer, J. (2002) Simultaneous determination of various aromatic amines and metabolites of aromatic nitro compounds in urine for low level exposure using gas chromatography-mass spectrometry. *J. Chromatogr. B Analyt. Technol. Biomed. Life Sci.* 778:179–192.

Woo, Y.-T. and Lai, D. Y. (2012) Aromatic amino and nitro-amino compounds and their halogenated derivatives. In: Bingham, E., Cohrssen, B., and Powell, C.H., editors., *Patty's Industrial Hygiene and Toxicology*, 6th ed., vol. 2, Chapter 36, John Wiley & Sons, Inc., pp. 609–704.

o-NITROTOLUENE

Occupational exposure limits: 5 ppm (OSHA PEL); 2 ppm (NIOSH REL); 2 ppm (ACGIH TLV with skin notation)

Reference values: None

Hazard statements: H302–harmful if swallowed; H340–may cause genetic defects; H350–may cause cancer; H361–suspected of damaging fertility or the unborn child; H411–toxic to aquatic life with long-lasting effects

Precautionary statements: P201–obtain special instructions before use; P301 + P310–if swallowed, immediately call a poison center or doctor/physician; P273–avoid release to the environment

Risk phrases: R45–may cause cancer; R46–may cause heritable genetic damage; R62–possible risk of impaired fertility; R22–harmful if swallowed; R51/53–toxic to aquatic organisms, may cause long-term adverse effects in the aquatic environment

Safety phrases: S53–avoid exposure—obtain special instructions before use; S45–in case of accident or if you feel unwell, seek medical advice immediately (show the label where possible); S61–avoid release to the environment; refer to special instructions/safety data sheets

BACKGROUND AND USES

o-Nitrotoluene (CASRN 88-72-2) is a high production volume chemical used primarily in the production of *o*-toluidine, 2,4-dinitrotoluene, isocyanate, and other chemicals that are intermediates for synthesizing colorants such as magenta and various azo and sulfur dyes for cotton, wool, silk, leather, and paper. In addition, it is used as an intermediate in the synthesis of agricultural chemicals, pharmaceuticals and rubber chemicals (IARC, 1996; NTP, 2011).

U.S. production of *o*-nitrotoluene was reported to be around 29 million pounds in 1981 and 35.5 million pounds in 1993 (NTP, 2011). Reports filed in 2006 under EPA's Toxic Substances Control Act Inventory Update Rule indicated that U.S. production plus imports of *o*-nitrotoluene totaled 10–50 million pounds (U.S. EPA, 2010). Since 1996, the principal producers and/or processors of *o*-nitrotoluene in Europe are located in Germany (34,400 metric tons in 2000),

the United Kingdom (3740 metric tons in 2003, imported) and Italy (49,200 metric tons in 2002) (EU, 2008).

PHYSICAL AND CHEMICAL PROPERTIES

o-Nitrotoluene (also known as 2-nitrotoluene) is a methylated nitroaromatic compound that has a methyl group ortho to the nitro group. At room temperature, it is a pale yellowish liquid which can crystallize to two solid forms (α-, and β-) at lower temperatures. It has a molecular weight of 137.15, a density of 1.1629, a melting point of −9.5 °C (needles, α-form) to −2.5 °C (crystal, β-form), a boiling point of 221.7 °C, a vapor pressure of 0.1 mm Hg at 20 °C, and an octanol/water partition coefficient (Log P) of 2.3. It is slightly soluble in water (0.54 g/l at 20 °C) and soluble in benzene, alcohol, ether, and petroleum ether (IARC, 1996; NTP, 2011).

MAMMALIAN TOXICOLOGY

Acute Effects

Acute toxicity studies in animals showed that *o*-nitrotoluene is only slightly toxic. At vapor concentrations of 191 ppm (1.1 mg/l) for 8 h or 320 ppm (1.8 mg/l) for 4 h in rats, and at 354 ppm (2 mg/l) for 4 h in mice, *o*-nitrotoluene did not produce mortalities, toxicity, and gross lesions within 14-day observation period. In dermal exposure studies, there were no mortality and clinical signs in rats at 5000 mg/kg bw, or in rabbits at 20,000 mg/kg bw. The oral LD_{50} values ranged from 890 to 2546 mg/kg bw. in rats, 970 to 2462 mg/kg bw. in mice and 1750 mg/kg bw. in rabbit. Clinical signs of toxicity observed in animals were related with methemoglobin formation. There are limited acute toxicity data in humans but effects due to methemoglobin formation can occur in humans exposed to *o*-nitrotoluene by various routes (EU, 2008).

Chronic Effects

In a repeated-dose toxicity study, *o*-nitrotoluene was administered in the feed to groups of male and female in F344 rats and B6C3F1 mice at concentrations ranging from 388 to 20,000 ppm for 14 days. There were no effects on survival or clinical signs of toxicity, although animals at the higher doses showed decreases in body weight gains relative to controls (Dunnick, 1993).

Comparative toxicity studies of *o*-, *m*-, or *p*-nitrotoluene showed that all three chemicals caused toxicity in the liver, spleen, and kidney in rats administered the test compounds in the feed at doses ranging from 625 to 10,000 ppm for 13 weeks. The most toxic effects occurred at or >2500 ppm and the toxicity was most pronounced in rats receiving *o*-nitrotoluene (Dunnick et al., 1993, 1994).

o-Nitrotoluene was clearly shown to be a multi-target carcinogen in 2-year chronic studies of rats and mice. In rats, oral exposure of *o*-nitrotoluene in the feed (approximately 0, 25, 50, or 90 mg/kg per day in males, and approximately 0, 30, 60, or 100 mg/kg per day in females) caused 1. subcutaneous skin tumors and mammary gland tumors (fibroadenoma) in both sexes, 2. malignant mesothelioma and benign or malignant tumors of the liver (hepatocellular adenoma or carcinoma, or cholangiocarcinoma) and lung (alveolar/bronchiolar adenoma or carcinoma) in males, and 3. benign liver tumors (hepatocellular adenoma) in females. In mice given *o*-nitrotoluene in the feed at approximately 0, 150, 320, or 700 mg/kg per day developed malignant blood-vessel tumors (hemangiosarcoma) in both sexes; malignant tumors of the large intestine (cecal carcinoma) in males, and benign or malignant liver tumors (hepatocellular adenoma or carcinoma) in females (NTP, 2002). Under similar conditions, the *p*-isomer was either noncarcinogenic or equivocal (Dunnick et al., 2003).

o-Nitrotoluene has also been shown to be a potent carcinogen with short tumor latency. Mesotheliomas, skin, liver, mammary gland, and liver tumors were induced in male rats administered with *o*-nitrotoluene in their feed at 125 or 315 mg/kg per day for as short as 13 weeks (NTP, 1992; Dunnick et al., 2003). Under the similar conditions the *p*- and *m*-isomers did not display any carcinogenic potential (NTP, 1992). Bile duct cancer (cholangiocarcinoma) was observed in rats exposed to *o*-nitrotoluene for 26 weeks as well as in rats exposed for 13 weeks and then observed for 13 more weeks without exposure (NTP, 1996). A concurrent comparative study showed that *o*-nitrotoluene was a more potent carcinogen than its nitroreductase-reduced metabolite, *o*-toluidine (NTP, 1996).

There are no studies on the relationship between human cancer and specific exposure to *o*-nitrotoluene. However, *o*-nitrotoluene is used to manufacture magenta and reviews of magenta manufacturing in 1987 and 1993 by the IARC working group concluded that there is sufficient evidence in humans that the manufacture of magenta entails exposures that are carcinogenic (IARC, 1996). Their assessment was based on two cohort studies and a case-control study, all of which reported an excess risk of bladder cancer, and one study specifically mentioned that the workers were exposed to *o*-nitrotoluene as a part of the magenta manufacturing process.

o-Nitrotoluene is a reproductive and developmental toxicant in rats. In a nonstandard reproductive toxicity study in rats, *o*-nitrotoluene administered in feed at 5000 ppm for 13 weeks caused damage to the testes and the epididymis with a simultaneous reduction in the sperm count and the motility, and a prolongation of the menstrual cycle among the females. Reduced sperm motility was also observed at 10,000 ppm for the mouse. The NOAEL for impaired fertility was considered to be 2500 ppm (179 mg/kg bw) in male rats.

In a developmental toxicity study in which male and female CD rats received *o*-nitrotoluene at daily doses of 0, 50, 150, or 450 mg/kg/day over a total period of approximately 10 weeks, the only effect considered as indicative of developmental toxicity was the retardation in pup growth. The LOAEL was considered to be 50 mg/kg/day (EU, 2008; Dunnick, 1994).

Mechanism of Action

o-Nitrotoluene is absorbed rapidly into the blood after oral administration to rats and mice. In the rat liver, *o*-nitrotoluene is metabolized to *o*-nitrobenzyl alcohol which can undergo several pathways: (1) glucuronidation (2) sulfation and subsequent reaction with glutathione and acetylcysteine, or (3) nitroreduction to *o*-aminobenzyl alcohol followed by oxidation to *o*-aminobenzoic acid. Most of the metabolites are eliminated primarily in the urine (NTP, 2002). The glucuronide of *o*-nitrobenzyl alcohol, via the bile, can also be excreted into the small intestine where bacteria can deconjugate and reduce the nitro group to form aminobenzyl alcohol. The aminobenzyl alcohol reabsorbed from the intestine can be further metabolized by the liver to reactive compounds (carbonium and nitrenium ions) that can covalently bind to DNA or proteins (Chism and Rickert, 1985; NTP, 2002, 2008). Thus, microbial metabolism in the intestine appears to be an important step in the carcinogenicity of *o*-nitrotoluene. The higher carcinogenic potency of the *o*-isomer as compared to *m*- and *p*-isomers is likely due to difference in metabolic activation of the isomers.

NTP studies (NTP, 2002; Dunnick, 1993) consistently showed that *o*-nitrotoluene did not cause mutations in S.*typhimirium* with and without S9. However, a study by Salamanca-Pinzon et al. (2006) showed positive results for nitroaromatic compounds using nitroreductase proficient bacteria. Mixed results were observed from assays using cultured mammalian cells. *o*-nitrotoluene caused (1) sister chromatids exchange in Chinese hamster ovary (CHO) cells, (2) chromosomal aberrations in Chinese hamster lung (CHL) cells and human peripheral lymphocytes but not in CHO cells, (3) micronucleus formation in CHL cells but not in CHO-K1 cells, and (4) DNA damage in L5178Y mouse lymphoma cells but no induction of DNA repair in rat or human hepatocytes (NTP, 2008). In rats and mice exposed *in vivo*, *o*-nitrotoluene caused a slight increase in micronucleus formation in peripheral normochromatic erythrocytes in male mice at a high dose level but this finding was not considered conclusive. *o*-Nitrotoluene did not induce micronucleus formation in peripheral normochromatic erythrocytes in female mice or in polychromatic erythrocytes in the bone marrow of male rats or mice (NTP, 2002). Following *in vivo* exposure of rats to *o*-nitrotoluene, DNA repair was increased in liver cells isolated from males, but not from females or germ-free males. These results suggest that activation of *o*-nitrotoluene to become genotoxic may be sex-specific and depends on both mammalian metabolism and metabolism by intestinal bacteria (Doolittle et al., 1983).

There is evidence that mutations of some oncogenes and tumor suppressor genes may play a role in the molecular pathogenesis of *o*-nitrotoluene-induced. Mutations in the p53 and β-catenin genes and production of these proteins in mice were detected in *o*-nitrotoluene-induced hemangiosarcomas and colon tumors from mice (Hong et al., 2003); K-*ras* gene mutations and cyclin D1 protein production also were detected in the colon tumors (Sills et al., 2004). Mutations in p53, β-catenin, and K-*ras* genes may be a result of the genotoxic effects of *o*-nitrotoluene, and these alterations probably contributed to the pathogenesis of the hemangiosarcomas and large intestinal carcinomas in mice induced by *o*-nitrotoluene. Studies in rats have also provided evidence that cellular and molecular events involved in the induction of mesotheliomas are similar in both experimental animals (rats exposed to *o*-nitrotoluene) and humans. Microarray analysis of peritoneal mesotheliomas from F344 rats treated with *o*-nitrotoluene identified the following carcinogenic pathways: insulin-like growth factor 1 (IGF-1), p38 MAPK, Wnt/β-catenin, and integrin signaling pathways (Kim et al., 2006).

o-Nitrotoluene and its metabolite, *o*-nitrosotoluene have recently been shown to cause estrogen-disrupting effects in a number of bioassays by Watanabe et al. (2010). An E-screen assay using the human breast cancer cell line MCF-7 revealed that *o*-nitrotoluene can induce estrogen-dependent cell proliferation. On the other hand, *o*-nitrosotoluene decreased the cell number and suppressed 17β-estradiol-induced cell proliferation. These disruptive effects may play a role in the reproductive toxicity of *o*-nitrotoluene.

Chemical Pathology

o-Nitrotoluene, as an aromatic, nitrogen-containing compound, is capable of forming methemoglobin. Exposures to 1 ppm (5.7 mg/m^3) or lower is not expected to show any toxic effects, and 40 ppm (228 mg/m^3) is considered as a no tolerated concentration because exposure to that level may lead to symptoms of illness. Exposure to 200 ppm (1140 mg/m^3) or higher may cause severe toxic effects (EU, 2008).

It is a fast-acting, potent carcinogen that induces tumors in multiple targets in rodents with a latency as short as 13 weeks. Although there are no studies on the relationship between human cancer and specific exposure to *o*-nitrotoluene, an excess risk of bladder cancer has been shown in workers exposed to *o*-nitrotoluene as a part of the magenta manufacturing process (NTP, 2011). There is evidence of mutations of some tumor suppressor genes and oncogenes (e.g., p53, β-catenin, and K-*ras* genes) in tumors induced by *o*-nitrotoluene. These alterations may be a result of the

genotoxic effects of *o*-nitrotoluene and probably play a role in the molecular pathogenesis of *o*-nitrotoluene in rodents and in humans (Hong et al., 2004; Sills et al., 2004; Kim et al., 2006).

There are some evidences that *o*-nitrotoluene is a reproductive and developmental toxicant in rats; estrogen disruptive effects have been suggested to play a role in the reproductive toxicity of *o*-nitrotoluene (Watanabe et al., 2010).

SOURCES OF EXPOSURE

Occupational

In factory workers exposed to *o*-nitrotoluene, *o*-nitrotoluene–hemoglobin adducts were detected in the blood (Jones et al., 2005a), and *o*-nitrobenzoic acid and *o*-nitrobenzyl alcohol were detected in the urine (Jones et al., 2005b), providing evidence that human exposure to *o*-nitrotoluene results in production of a reactive metabolite(s). In addition, adducts between hemoglobin and 2-methylaniline (a metabolite of *o*-nitrotoluene) were identified in both exposed workers and exposed rats, and the level of 2-methylaniline–hemoglobin adducts in the blood of rats was proportional to the level of 2-methylaniline–DNA adducts in the livers of rats (Jones and Sabbioni, 2003; Jones et al., 2003).

o-Nitrotoluene was detected at a concentration of 47 ng/m^3 in ambient air at a chemical manufacturing plant in New Jersey (IARC, 1996) and at air concentrations of up to 2.0 mg/m^3 in the nitrotoluene production area of a Swedish plant producing pharmaceuticals and explosives (Ahlborg et al., 1985). The highest measured value of occupational exposure in an Italian manufacturer of *o*-nitrotoluene was reported to be 0.280 mg/m^3, whereas the values based on the estimation and assessment of substance exposure (EASE) model were 0.35–0.7 mg/m^3 (EU, 2008).

Environmental

o-Nitrotoluene may enter the environment via the air and waste water during its production and/or use in the manufacture of other products. In West Europe, the major manufacturers and/or processors of *o*-nitrotoluene settled in Germany, the United Kingdom, and Italy (EU, 2008). The U.S. has one plant in Pascagoula, Mississippi (U.S. EPA, 2010).

Environmental surveys have demonstrated the presence of *o*-nitrotolueene in river and drinking water (IARC, 1996). *o*-Nitrotoluene was detected in ground water that was contaminated by munitions arsenal; the average concentrations reported in 2001 was 42.6 mg/l (42,600 μg/l) (NTP, 2011). There were some reports of detection of *o*-nitrotoluene in the smog in China (Wu et al., 2006).

INDUSTRIAL HYGIENE

The time-weighted average TLV–TWA for *o*-nitrotoluene adopted by American Conference of Governmental Industrial Hygienists (ACGIH) is 2 ppm with skin notation. The OSHA PEL is 5 ppm. The NIOSH recommended exposure limit is 2 ppm and IDLH is 200 ppm (HSBD, 2011). *o*-Nitrotoluene is considered by ACGIH to be an inducer of methemoglobin and recommends that methemoglobin in blood be used as a biological exposure index for *o*-nitrotoluene (ACGIH, 2009). It is designated as a hazardous substance under the Federal Water Pollution Control Act and further regulated by the Clean Water Act Amendments of 1977 and 1978 (HSDB, 2011).

o-Nitrotoluene has been listed in Report on Carcinogens (NTP, 2011) as "*reasonably anticipated to be human carcinogens*" based on of sufficient evidence of carcinogenicity from studies in experimental animals and supporting data on mechanisms of carcinogenesis. The IARC (IARC, 1996) has classified *o*-nitrotoluene as a Group 2B "*possible human carcinogen*" based on sufficient animal data but inadequate human data.

MEDICAL MANAGEMENT

Appropriate treatments for methemoglobinemia such as administration of oxygen, ascorbic acid or methylene blue should be performed as described (Dutta et al., 2008) if overexposure of *o*-nitrotoluene is suspected and signs of cyanosis (at the lips, tongue, and nail beds) and symptoms of methemoglobinemia occur.

REFERENCES

ACGIH (2009) TLVs and BEIs. *Cincinnati, OH: American Conference of Governmental Industrial Hygienists*. 256 pp.

Ahlborg, G., Jr., Bergstrom, B., Hogstedt, C., Einisto, P., and Sorsa, M. (1985) Urinary screening for potentially genotoxic exposures in a chemical industry. *Br. J. Ind. Med.* 42:691–699.

Chism, J.P. and Rickert, D.E. (1985) Isomer-and sex-specific bioactivation of mononitrotoluenes. Role of enterohepatic circulation. *Drug Metab. Dispos.* 13:651–657.

Doolittle, D.J., Sherrill, J.M., and Butterworth, B.E. (1983) Influence of intestinal bacteria, sex of the animal, and position of the nitro group on the hepatic genotoxicity of nitrotoluene isomers *in vivo. Cancer Res.* 43(6):2836–2842.

Dunnick, J. (1993) *NTP technical report on the toxicity studies of ortho-, meta-, and para- Nitrotoluenes (CAS Nos. 88-72-2, 99-08-1, 99-99-0) Administered in Dosed Feed to F344/N Rats And B6C3F1 Mice. Toxic Rep Ser. 23:1-E4.* Available at http://www.ncbi.nlm.nih.gov/pubmed/12209183?dopt=Abstract.

Dunnick, J.K., Elwell, M.R., and Bucher, J.R. (1994) Comparative Toxicities of o-, m-, and p-Nitrotoluene in 13-Week Feed Studies in F344 Rats and B6C3F1 Mice. *Fund Appl. Toxicol.* 22:411–421.

Dunnick, J.K., Burka, L.T., Mahler, J., and Sills, R. (2003) Carcinogenic potential of o-nitrotoluene and p-nitrotoluene. *Toxicology* 183:221–234.

Dutta, R., Dube, S.K., Mishra, L.D., and Singh, A.P. (2008) Acute methemglobinemia. *Internet J. Emerg. Intensive Care Med.* 11:1092–4051.

EU (2008) *European Union Risk Assessment Report on 2-nitrotoluene. Human Health Part, Scientific Committee on Health and Environmental Risks, (2008).* Available at http://echa.europa.eu/documents/10162/e925a928-3cc6-4448-9289-33aec88c1ead.

Hong, H.L., Ton, T.V., Devereux, T.R., Moomaw, C., Clayton, N., Chan, P., Dunnick, J.K., and Sills, R.C. (2003) Chemical-specific alterations in ras, p53, and β-catenin genes in hemangiosarcomas from B6C3F1 mice exposed to o-nitrotoluene or riddelliine for 2 years. *Toxicol. Appl. Pharmacol.* 191:227–234.

HSDB. (2011) *Hazardous Substances Data Bank. 2-Nitrotoluene. National Library of Medicine.* Available at http://toxnet.nlm.nih.gov/cgi-bin/sis/search/f?./temp/~0xLlsq:1.

IARC (1996) Printing processes and printing inks, carbon black and some nitrocompounds. In: *IARC Monographs on the Evaluation of Carcinogenic Risk of Chemicals to Humans*, vol. 65, Lyon, France: International Agency for Research on Cancer, pp. 409–435.

Jones, C.R. and Sabbioni, G. (2003) Identification of DNA adducts using HPLC/MS/MS following *in vitro* and *in vivo* experiments with arylamines and nitroarenes. *Chem. Res. Toxicol.* 16:1251–1263.

Jones, C.R., Beyerbach, A., Seffner, W., and Sabbioni, G. (2003) Hemoglobin and DNA adducts in rats exposed to 2-nitrotoluene. *Carcinogenesis* 24:779–787.

Jones, C.R., Sepai, O., Liu, Y.Y., Yan, H., and Sabbioni, G. (2005a) Hemoglobin adducts in workers exposed to nitrotoluenes. *Carcinogenesis* 26:133–143.

Jones, C.R., Sepai, O., Liu, Y.Y., Yan, H., and Sabbioni, G. (2005b) Urinary metabolites of workers exposed to nitrotoluenes. *Biomarkers* 10:10–28.

Kim, Y., Ton, T.V., DeAngelo, A.B., et al., (2006) Major carcinogenic pathways identified by gene expression analysis of peritoneal mesotheliomas following chemical in F344 rats. *Toxicol. Appl. Pharmacol.* 214:144–51.

NTP (1992) *NTP Technical Report on Toxicity Studies of o-, m-, and p-Nitrotoluenes Administered in Dosed Feed to F344/N Rats and B6C3F1 Mice*, National Toxicology Program. Available at http://ntp.niehs.nih.gov/ntp/htdocs/ST_rpts/tox023.pdf.

NTP (1996) *NTP Technical Report on Comparative Toxicity and Carcinogenicity Studies of o-Nitrotoluene and o-Toluidine Hydrochloride Administered in Feed to Male F344/N Rats*, National Toxicology Program. Available at http://ntp.niehs.nih.gov/ntp/htdocs/ST_rpts/tox044.pdf.

NTP (2002) *Toxicology and Carcinogenesis Studies of o-Nitrotoluene (CAS No. 88-72-2) in F344/N Rats and B6C3F1 Mice (Feed Studies)*, National Toxicology Program. Available at http://ntp.niehs.nih.gov/ntp/htdocs/LT_rpts/tr504.pdf.

NTP (2008) *Report on Carcinogens Background Document for o-Nitrotoluene*, National Toxicology Program. Available at http://ntp.niehs.nih.gov/files/o-NT-FINAL_508.pdf.

NTP. (2011) *Report on Carcinogens*, 12th ed., National Toxicology Program. Available at http://ntp.niehs.nih.gov/go/roc12.

Salamanca-Pinzon, S.G., Camacho-Carranza, R., Hernandez-Ojeda, S.L., and Espinosa-Aguirre, J.J. (2006) Nitrocompounds activation by cell-free extracts of nitroreductase-proficient *Salmonella typhimurium* strains. *Mutagenesis* 21:369–374.

Sills, R.C., Hong, H.L., Flake, G., Moomaw, C., Clayton, N., Boorman, G.A., Dunnick, J., and Devereux, T.R. (2004) o-Nitrotoluene-induced large intestinal tumors in B6C3F1 mice model human colon cancer in their molecular pathogenesis. *Carcinogenesis* 25:605–612.

U.S. EPA (2010) *Non-confidential IUR Production Volume Information*, U.S. Environmental Protection Agency. Available at http://www.epa.gov/oppt/iur/tools/data/2010-vol.html.

Watanabe, C., Egami, T., Midorikawa, K., Hiraku, Y., Oikawa, S., Kawanishi, S., and Murata, M. (2010) DNA damage and estrogenic activity induced by the environmental pollutant 2-nitrotoluene and its metabolite. *Environ. Health Prev. Med.* 15:319–326.

Wu, Y.F., Li, L.R., Yang, J.F., and Shi, T.R. (2006) *Detection of aromatic nitrocompounds by GC/MS. Zhongguo.*

NITROBENZENE

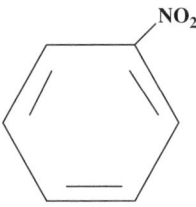

BACKGROUND AND USES

Nitrobenzene is a synthetic, volatile compound produced primarily for use to manufacture aniline. It is also used as a solvent in refining petroleum and lubricating oils, and in production of dyes, synthetic rubber, pesticides, and drugs including acetaminophen and metoclopramide. Small amounts of nitrobenzene are used as a flavoring agent for soaps and as a solvent for shoe dyes (HSDB, 2009; IARC, 1996). Dinitrobenzene isomers (1, 2-, 1, 3-, and 1, 4-) are used in organic synthesis of dyes, pesticides, and industrial solvents. 1,3-Dinitrobenzene and 1,3,5-trinitrobenzene are chemicals found in the production of explosives.

The U.S. production volume of nitrobenzene has increased from 73,000 metric tons in 1960 to 435,000 metric tons in 1986, and from 533,000 metric tons in 1990 to 740,000 metric tons in 1994 that was about one-third of the worldwide production of 2,133,800 metric tons (IARC, 1996; IPCS, 2003). Since 1990, U.S. annual production of nitrobenzene has been more than one billion pounds (U.S. EPA, 2004). In 2009, there were 5 U.S. producers and 20 U.S. suppliers of nitrobenzene (NTP, 2011).

Nitrobenzene and all isomers (1,2-, 1,3-, and 1,4-) of dinitrobenzene can be absorbed by all routes of exposures and may cause irritation in the respiratory tract and skin. 1,3-Dinitrobenzene and 1,3,5-trinitrobenzene are of severe explosive hazard. Toxicity studies on nitrobenzene, di- and trinitrobenzenes in experimental animals have revealed a similar spectrum of toxicological effects. Nitrobenzene was the subject of several toxicological reviews (ATSDR, 1990; IARC, 1996; IPCS, 2003; U.S. EPA, 2009).

PHYSICAL AND CHEMICAL PROPERTIES

Nitrobenzene has a single nitro group attached to a benzene ring. It is an oily yellow liquid with an almond-like odor. It has a molecular weight of 123.11, a density of 1.204, a melting point of 5.8 °C, a boiling point of 210.8 °C, and a vapor pressure of 0.15 mm Hg (20 °C). It is moderately soluble in water (1.9 g/l at 20 °C) and is soluble in alcohol, acetone, ether, and benzene (IARC, 1996). It has a flash point of 88 °C and an explosive limit of 1.8% by volume in air, representing a fire hazard. Its log P (octanol/water partition coefficient) is 1.85.

MAMMALIAN TOXICOLOGY

Acute Effects

Nitrobenzene can be absorbed by all routes of exposure. It is moderately irritating to the respiratory tract and skin. A 40% rate of mortality in Sprague-Dawley (CD) rats was reported after they were exposed to 125 ppm nitrobenzene by inhalation (Medinsky and Irons, 1985). After two or three dermal applications of nitrobenzene (dosage not reported) to mice resulted in 80% deaths of the animals. Most animals were in partial collapse within 15 min and were dead within the first day. Slight swelling of the glomeruli and tubular epithelium were observed upon histological examination (Shimkin, 1939). Single acute oral exposure of Fischer 344 rats to 200 mg/kg or higher nitrobenzene resulted in over 20% increase in methemoglobin (Goldstein et al., 1984). Testicular degeneration and transiently decreased sperm production were observed in rats following a single oral dose of 300 mg/kg (Levin et al., 1988). Brain damages have also been reported after a single oral administration of nitrobenzene at 550 mg/kg to rats (Morgan et al., 1985). An oral LD_{50} was estimated to be 600 mg/kg in rats (Smyth et al., 1969).

Chronic Effects

The major toxic effect of chronic and subchronic exposures to nitrobenzene is methemoglobinemia. Elevated levels of blood methemoglobin have been reported in rats exposed to nitrobenzene by inhalation at levels as low as 10 ppm for two weeks (Medinsky and Irons, 1985) or 5 ppm for 90 days or longer (CIIT, 1984, 1993; Cattley et al., 1994). In addition, dose-related lesions of the liver, kidney, spleen, brain, and lung occurred in rats and mice after exposure to 5–125 ppm nitrobenzene vapor in repeated-dose studies (Medinsky and Irons, 1985; CIIT, 1984, 1993; Cattley et al., 1994). A number of repeated dose studies by the oral route have also reported induction of methemoglobinemia with dose-related changes in the levels of hematologic parameters and histopathologic effects on various organs including spleen, liver, kidney, thymus, and the brain in rats at doses of 60–300 mg/kg bw/day nitrobenzene (Bond et al., 1981; NTP, 1983; Mitsumori et al., 1994). Differences in species, strain and sex susceptibility to the toxic effects of nitrobenzene exposure were noted. For example, despite the mice were treated with higher doses, nitrobenzene-induced pathological changes were much less pronounced in mice than in rats, suggesting that the mice were more resistant to nitrobenzene toxicity. On the basis of increases in absolute and relative organ weights and the dose-dependent increases in methemoglobin and reticulocyte count in F344 rats, a LOAEL of 9.38 mg/kg-day was determined for the subchronic oral effects of nitrobenzene (NTP, 1983).

There is strong evidence that nitrobenzene is a male reproductive toxicant. Testicular atrophy, bilateral degeneration of the seminiferous tubules, and a reduction in or absence of mature sperm in the epididymis were noted in both F-344 and Sprague-Dawley (CD) rats exposed to nitrobenzene vapor at 50 ppm for 90 days (Hamm, 1984). Dodd et al. (1987) reported a decrease in fertility indices in a two-generation inhalation study in rats exposed to 40 ppm nitrobenzene for 10 weeks resulting from atrophy of seminiferous tubules, spermatocyte degeneration and reduced testicular, and epididymal weights. Following a single oral dose of 300 mg/kg of nitrobenzene, lesions in the seminiferous tubules of the testicles, with marked necrosis of primary and secondary spermatocytes were observed in male F344 rats (Bond et al., 1981). A significant effect on testis weight in rats was seen at 75 and 150 mg/kg-day and in male mice at 300 mg/kg-day in a 90-day gavage study. In addition, sperm motility of exposed rats was adversely affected, and the incidence of abnormal sperm was increased (NTP, 1983; Morrissey et al., 1988). A dose-related increase in the incidence of sperm abnormality was seen in mice exposed to a mixture of three metabolites of nitrobenzene (Wang et al., 2011).

Several reproductive/developmental toxicity studies in Sprague-Dawley rats administered nitrobenzene by gavage have also shown significant reductions in testicular and epididymal weights, sperm count, and motility, sperm viability, atrophy of the seminiferous tubules, hyperplasia of Leydig cells, loss of intraluminal sperm in the epididymides, and fertility index in animals receiving 60 to 100 mg/kg-day of nitrobenzene (Mitsumori et al., 1994; Kawashima et al., 1995). However, studies in Sprague-Dawley rats indicate that inhalation exposure to nitrobenzene does not result in fetotoxic, embryotoxic, or teratogenic effects at concentration (40 ppm) that is maternally toxic (Tyl et al., 1987).

Nitrobenzene is carcinogenic in rats and mice, inducing multiple-sites tumors in both species. In a 2-year inhalation cancer bioassay, groups of male rats of two strains (F344/N and Sprague-Dawley), female rats of one strain (F344/N), and male and female mice (B6C3F1) were administered nitrobenzene for 6 h/day, 5 days/week for 104 weeks at concentrations of 0, 1, 5, or 25 ppm (rats) and 0, 5, 25, or 50 ppm (mice). Exposure to nitrobenzene caused liver tumors in male rats of both F344/N and Sprague-Dawley strains, kidney tumors and thyroid tumors in male F344/N rats, and endometrial polyps in female F344/N rats. In addition, positive dose–response trends in the incidences of liver tumors were observed in female F344/N rats. Exposure to nitrobenzene caused lung and thyroid tumors in male B6C3F1 mice and liver and mammary gland tumors in female B6C3F1 (CIIT, 1993; Cattley et al., 1994).

There is one case-control study describing cancer effects from exposure to nitrobenzene. Paternal exposure to nitrobenzene was associated with a statistically nonsignificant increase in the risk of childhood brain cancer, based on a small number of cancer patients whose fathers had been exposed to nitrobenzene (Wilkins and Sinks, 1990). A recent epidemiological study reported significantly increased bladder cancer incidence among workers coexposed to o-toluidine, aniline, and nitrobenzene at a rubber chemical manufacturing plant (Carreón et al., 2013).

MECHANISM(S) OF ACTION

Nitrobenzene is readily absorbed by animals and humans via inhalation, and dermal and oral routes of exposure. Metabolism studies in mammals showed that it can be reduced to aniline by intestinal microflora and hepatic microsomes. In addition, nitrobenzene can be oxidized by hepatic microsomes to o-, m-, and p-nitrophenol. Metabolism of nitrobenzene in mammals displays animal species and strain specificity, and toxicokinetics data in rats and mice after oral administration indicate that various amounts of metabolites are distributed to the blood, liver, brain, kidney, and lung. Species and/or strain differences in toxicity of nitrobenzene are likely due to the difference in oxidative and reductive metabolism of nitrobenzene to its toxic metabolites (Reddy et al., 1976; Levin and Dent, 1982; Goldstein et al., 1984). There is ample evidence that methemoglobinemia is caused by the interaction of Hb with the metabolic intermediates of nitrobenzene reduction including nitrosobenzene, phenylhydroxylamine, and aniline.

The mechanism of nitrobenzene carcinogenesis is not clearly understood. Nitrobenzene showed negative responses in most genotoxicity tests. Except in strain TA 98 in the presence of S9 and the comutagen, norharman, nitrobenzene was negative in all standard Ames assays in various strains of *Salmonella typhimurium*. It was weakly positive for the induction of chromosome aberrations in cultured human peripheral lymphocytes but negative in human spermatozoa. It failed to induce unscheduled DNA synthesis or cell transformation *in vitro*. Cytogenetic analyses of lymphocytes in the peripheral blood or in splenic blood of rats exposed to nitrobenzene did not reveal an increase in sister chromatid exchange (SCE), chromosomal aberrations or unscheduled DNA synthesis (IARC, 1996). However, a more recent study has reported DNA damages and induction of micronuclei in rat and human kidney cells by nitrobenzene (Robbiano et al., 2004). A dose-related increase in the incidence of micronucleus was seen in mice exposed to three metabolites mixture of nitrobenzene (Wang et al., 2011). It has been speculated that tumors may arise from oxidative stress resulting from oxidative and reductive metabolism of nitrobenzene and generation of free radicals (e.g., nitro anion and superoxide anion, hydronitroxide) (Hsu et al., 2007; Holder, 1999; Gutteridge, 1995). Possible events by which reactive free radicals could cause tumor formation include direct DNA oxidative damage, lipid peroxidation, damage to DNA repair enzymes, or modulation of DNA methylation (Halliwell, 2007).

CHEMICAL PATHOLOGY

Methemoglobinemia is the major adverse effect of nitrobenzene in humans. The condition arises when the rate of methemoglobin (metHb) formation exceeds the rate of reduction of oxidized heme iron from the ferrous (Fe^{2+}) state to the ferric (Fe^{3+}) state, and when a metHb concentration exceeding 2–3% of total methemoglobin (Hb) (Denshaw-Burke, 2013). Nitrobenzene and some aromatic amines can cause methemoglobinemia by accelerating the oxidation of Hb to metHb, which loses its ability to combine reversibly with and to transport oxygen (Percy et al., 2005). Following such an exposure, the skin may turn bluish in color and this may be accompanied by nausea, vomiting, and shortness of breath. Effects such as headache, irritability, dizziness, weakness, and dyspnoea may also occur. If the exposure level is extremely high, nitrobenzene can cause coma and possibly death unless prompt medical treatment is received. The lethal dose is reported to range from 1 g to 10 g (Chongtham et al., 1997).

Acute intoxication is usually asymptomatic up to the level of 10–15% of methemoglobin, showing only cyanosis. Beyond 20%, headache, dyspnea, chest pain, tachypnea, and tachycardia develop. At 40–50%, confusion, lethargy, and metabolic acidosis occur leading to coma, seizures, bradycardia, ventricular dysrhythmia, and hypertension. Fractions around 70% are fatal. Anemic or G6PD-deficient patients suffer more severe symptoms. Leukocytosis has been reported, with relative lymphopenia. Other effects include hepatosplenomegaly, altered liver functions, and Heinz body haemolytic anemia (Dutta et al., 2008). The findings of damage to the testicles and decreased levels of sperm in animals exposed to nitrobenzene is a concern to humans.

Some workers in plants producing nitrobenzene with air concentrations averaging 6 ppm (30 mg/m^3) experienced headache and vertigo with low concentrations of methemoglobin. Poisoning was reported to start at air concentrations of about 40 ppm (200 mg/m^3). A man ingested about 7 g nitrobenzene was reported to have developed severe methemoglobinemia (78% methemoglobin) (IARC, 1996). Severe methemoglobinemia (70% methemoglobin) was also observed in an 82-year-old male who had ingested 250 ml of nitrobenzene; 3.2 µg/ml of nitrobenzene was detected in the blood 48 h after ingestion (Martínez et al., 2003).

SOURCES OF EXPOSURE

Occupational

Occupational exposure is of great concern, since nitrobenzene can be absorbed readily through the skin as well as by inhalation. On the basis of a survey conducted from 1981 to 1983 by the National Institute of Occupational Safety and Health's (NIOSH's), it was estimated that 5080 employees (475 females) potentially were exposed to nitrobenzene (IARC, 1996; HSDB, 2009).

There were no recent data on occupational exposure to nitrobenzene. Direct release of nitrobenzene to air during its manufacture is minimized by passage of contaminated air through activated charcoal, and most (97–98%) of the nitrobenzene produced is retained in closed systems, thus limiting its release into air (ATSDR, 1990).

Environmental

Environmental exposure to nitrobenzene is primarily through inhalation of ambient air, ingestion of water, or dermal exposure to products and water containing nitrobenzene. Although direct release of nitrobenzene to air during its manufacture is minimized by the passage of contaminated air through activated charcoal or retained in closed manufacturing systems, as much as 8.3 million pounds/year were estimated to be released from industrial processes (ATSDR, 1990). Nitrobenzene was found in 7 of the 1177 NPL hazardous waste sites in the United States (ATSDR, 1990); a geometric mean concentration of 1000 µg/kg nitrobenzene was detected in soil/sediment samples at 4 of 862 hazardous waste sites (ATSDR, 1990). Since 1988, trends data have shown annual declines; total air emissions have varied between 25529 and 81297 pounds, with no evident upward. In 2012, EPA's Toxics Release Inventory reported total on- and off-site releases of nitrobenzene as 240,529 pounds (U.S. EPA, 2012).

Use of nitrobenzene in consumer products such as metal and shoe polishes could contribute to environmental releases via fugitive emissions, wastewater, spills, and end-product usage. However, environmental degradations including photolysis and microbial biodegradation occur within a few days after it is released, and air and water in most areas contain no detectable nitrobenzene. Nitrobenzene has been measured in fish samples in Japan (4 of 147 samples) at levels ranging from 11 to $26\,\mu g/kg$. It is poorly bioaccumulated and not biomagnified through the food chain (IPCS, 2003).

Standards	Values
NIOSH REL	1 ppm (5 mg/m^3) TWA (skin)
OSHA PEL	1 ppm (5 mg/m^3) TWA (skin)
ACGIH TLV	1 ppm (5 mg/m^3) TWA (skin)
IDLH	200 ppm
RfD	2×10^{-3} mg/kg-day
RfC	9×10^{-3} mg/m^3
Cancer unit risk (inhalation)	4×10^{-5} per $\mu g/m^3$

INDUSTRIAL HYGIENE

The American Conference of Governmental Industrial Hygienists (ACGIH, 2013) has established a threshold limit value (TLV) for nitrobenzene of 1 ppm (5 mg/m^3), with a skin notation, indicating potential absorption through skin. The Occupational Safety and Health Administration (NIOSH, 2011) has established a permissible exposure level (PEL) of 1 ppm (5 mg/m^3) as an 8-h time-weighted average (TWA) for nitrobenzene and all isomers of dinitrobenzene. According to the National Institute for Occupational Safety and Health (NIOSH) "immediately dangerous to life or health" (IDLH) concentration for nitrobenzene is 200 ppm. Nitrobenzene is grouped as 2B "Possibly carcinogenic to humans" in IARC (1996), and is listed as "Reasonably Anticipated to be a Human Carcinogen" by the National Toxicology Program (NTP, 2011).

The U.S. Environmental Protection Agency (EPA) regulates nitrobenzene under the Clean Air Act (CAA), the Clean Water Act (CWA), the Federal Insecticide, Fungicide, and Rodenticide Act (FIFRA) and the Resource Conservation and Recovery Act (RCRA). Under the Comprehensive Environmental Response, Compensation, and Liability Act (CERCLA), any release in excess of 1,000 pounds (454 kg) should be reported. Based on the toxicological data of nitrobenzene, EPA has established an oral RfD of 2×10^{-3} mg/kg-day, an inhalation RfC of 9×10^{-3} mg/m^3

and an inhalation unit cancer risk of 4×10^{-5} per $\mu g/m^3$ (U.S. EPA, 2009).

MEDICAL MANAGEMENT

A severe case of acute poisoning of a 16-year-old female with nitrobenzene was reported by Saxena and Saxena (2010). After clinical evaluation, treatment based on the principles of decontamination and symptomatic, and supportive management, a life was saved with repeated intravenous methylene blue and other treatments. Methylene blue is an exogenous cofactor, which greatly accelerates the NADPH-dependant methemoglobin reductase system, and is the antidote of choice for the acquired (toxic) methemoglobinemia. It is administered intravenously at 1–2 mg/kg as a 1% solution over 5 min. In this case, repeated low dose of methylene blue was administered which helped in tiding over the fluctuating symptoms due to the release of nitrobenzene from the body stores, without exceeding the maximum dose. Being an oxidant, methylene blue may cause methemoglobinemia in susceptible patients at levels of more than 7 mg/kg. Fresh blood transfusion was given which improved the oxygen carrying capacity and hemoglobin content, thus improving the patient symptomatically. Oral charcoal and purgation up to 5 days were administered to eliminate the body stores of nitrobenzene and prevent secondary deterioration in the patient, as reported in other cases (Dutta et al., 2008). Forced diuresis led to a rapid fall in methemoglobin levels and improved discoloration. Ascorbic acid supplements were prescribed for follow-up management of methaemoglobinaemia (Kaushik et al., 2004).

It is noteworthy that methylene blue is contraindicated in patients with G6PD deficiency, because it can lead to severe hemolysis. Ascorbic acid is an antioxidant and may be administered in such patients especially when methemoglobin levels were more than 30%. Hyperbaric oxygen is reserved only for those patients who have a methemoglobin level more than 50% or those who do not respond to standard treatment (Dutta et al., 2008). It is important to take care of nutrition, adequate urine output, and hepatoprotection to prevent possible late effects of kidney and liver failure.

REFERENCES

ACGIH. (2013) *Documentation of the threshold limit values and biological exposure indices*, 7th ed. Cincinnati, Ohio: American Conference of Governmental Industrial Hygienists, Inc.

ATSDR. (1990) *Toxicological Profile for Nitrobenzene. Agency for Toxic Substances & Disease Registry, Public Health Service, U.S. Department of Health and Human Services*, Atlanta, GA: Agency for Toxic Substances & Disease Registry Available at http://www.atsdr.cdc.gov/toxprofiles/index.asp.

Bond, J.A., Chism, J.P., Rickert, D.E., et al. (1981) Induction of hepatic and testicular lesions in Fischer 344 rats by single oral doses of nitrobenzene. *Fundam. Appl. Toxicol.* 1:389–394.

Carreón, T., Hein, M.J., Hanley, K.W., Viet, S.M., and Ruder, A.M. (2013) Bladder cancer incidence among workers exposed to *o*-toluidine, aniline and nitrobenzene at a rubber chemical manufacturing plant. *Occup. Environ. Med.* doi: 10.1136/oemed-2013-101873 [Epub ahead of print].

Cattley, R.C., Everitt, J.I., Gross, E.A., et al. (1994) Carcinogenicity and toxicity of inhaled nitrobenzene in B6C3F1 mice and F344 and CD rats. *Fundam. Appl. Toxicol.* 22:328–340.

Chongtham, D.S., Phurailatpam, Singh, M.M., and Singh, T.R. (1997) Methemoglobinemia in nitrobenzene poisoning. *J. Postgrad. Med.* 43:73–74.

CIIT (Chemical Industry Institute of Toxicology). (1984) Ninety day inhalation toxicity study of nitrobenzene in F344 rats, CD rats, and B6C3F1 mice. Chemical Industry Institute of Toxicology. Research Triangle Park, NC; Docket No. 12634, Submitted under TSCA Section 8D; EPA Document No. 878214291; NTIS No. OTS0206507.

CIIT. (1993) Initial submission: a chronic inhalation toxicity study of nitrobenzene in B6C3F1 mice, Fischer 344 rats and Sprague-Dawley (CD) rats. Chemical Industry Institute of Toxicology. Research Triangle Park, NC. EPA Document No. FYI-OTS-0794-0970; NTIS No. OTS0000970.

Denshaw-Burke, M. (2013) Methemoglobinemia over view e Medicine from Web med. Available at http://www.emedicine.medscape.com/article/956528-overview.

Dodd, D.E., Fowler, E.H., Snellings, W.M., et al. (1987) Reproduction and fertility evaluations in CD rats following nitrobenzene inhalation. *Fundam. Appl. Toxicol.* 8:493–505.

Dutta, R., Dube, S.K., Mishra, L.D., and Singh, A.P. (2008) Acute methemoglobinemia. *Internet J. Emerg. Intensive Care Med.* 11:1092–4051.

Gutteridge, J.M.C. (1995) Lipid peroxidation and antioxidants as biomarkers for tissue damage. *Clin. Chem.* 41:1819–1828.

Halliwell, B. (2007) Oxidative stress and cancer: have we moved forward? *Biochem. J.* 401:1–11.

HSDB. (2009) Hazardous substances data bank. Bethesda, Maryland, National Library of Medicine, National Toxicology Information Program. Available at http://toxnet.nlm.nih.gov/cgi-bin/sis/html.

Hsu, C.H., Stedeford, T., Okochi-Takada, E., et al. (2007) Framework analysis for the carcinogenic mode of action of nitrobenzene. *J Environ Sci Health, Part C* 25:155–184.

IARC. (1996) *IARC Monographs on the Evaluation of Carcinogenic Risks to Humans: Printing Processes and Printing Inks, Carbon Black and Some Nitro Compounds, vol 65.* International Agency for Research on Cancer. Lyon, France.

IPCS. (2003) International Programme on Chemical Safety. Nitrobenzene. Environmental health criteria 230. Geneva: WHO; (8.11). Available at http://www.inchem.org/documents/ehc/ehc/ehc230.htm#8.1.1;#7.8.

Goldstein, R.S., Chism, J.P., Sherrill, J.M., et al. (1984) Influence of dietary pectin on intestinal microfloral metabolism and toxicity of nitrobenzene. *Toxicol. Appl. Pharmacol.* 75:547–553.

Hamm, T.E. (1984) *Ninety Day Inhalation Toxicity Study of Nitrobenzene in F-344 Rats, and CD Rats and B6C3Fl Mice.* Chemical Industry Institute of Technology. Research Triangle Park, NC.

Holder, J.W. (1999) Nitrobenzene carcinogenicity in animals and human hazard evaluation. *Toxicol. Ind. Health* 15:445–457.

Kaushik, P., Zuckerman, S.J., Campo, N.J., Banda, V.R., Hayes, S.D., and Kaushik, R. (2004) Celecoxib-Induced Methemoglobinemia. *Ann. Pharmacother.* 38:1635–1638.

Kawashima, K., Usami, M., Sakemi, K., et al. (1995) Studies on the establishment of appropriate spermatogenic endpoints for male fertility disturbance in rodent induced by drugs and chemicals. I. Nitrobenzene. *J. Toxicol. Sci.* 20:15–22.

Levin, A.A. and Dent, J.G. (1082) Comparison of the metabolism of nitrobenzene by hepatic microsomes and cecal microflora from Fischer 344 rats *in vitro* and the relative importance of each *in vivo. Drug Metab. Dispos.* 10:450–454.

Levin, A.A., Bosakowski, T., Earle, L.L., et al. (1988) The reversibility of nitrobenzene-induced testicular toxicity: continuous monitoring of sperm output from vasocystotomized rats. *Toxicology* 53:219–230.

Martínez, M.A., Ballesteros, S., Almarza, E., Sánchez de la Torre, C., and Búa, S. (2003) Acute nitrobenzene poisoning with severe associated methemoglobinemia: identification in whole blood by GC-FID and GC-MS. *J. Anal. Toxicol.* 27:221–225.

Medinsky, M.A. and Irons, R.D. (1985) Sex, strain, and species differences in the response of rodents to nitrobenzene vapors. In: Rickert, D.E. editor, *Chemical Industry Institute of Toxicology Series. Toxicity of Nitroaromatic Compounds.* New York, NY: Hemisphere Publishing Corporation, pp. 35–51.

Mitsumori, K., Kodama, Y., Uchida, O., et al. (1994) Confirmation study, using nitrobenzene, of the combined repeat dose and reproductive/developmental toxicity test protocol proposed by the Organization for Economic Cooperation and Development (OECD). *J. Toxicol. Sci.* 19:141–149.

Morgan, K.T., Gross, E.A., Lyght, O., et al. (1985) Morphologic and biochemical studies of a nitrobenzene-induced encephalopathy in rats. *Neurotoxicology* 6:105–116.

Morrissey, R.E., Schwetz, B.A., Lamb, J.C., IV, et al. (1988) Evaluation of rodent sperm, vaginal cytology, and reproductive organ weight data from National Toxicology Program 13-week studies. *Fundam. Appl. Toxicol.* 11:343–358.

NIOSH. (2011) *NIOSH Pocket Guide to Chemical Hazards.* National Institute for Occupational Safety and Health, Centers for Disease Control and Prevention. Atlanta, GA: Available at http://www.cdc.gov/niosh/npg/npgd0450.html.

NTP. (1983) Report on the subchronic toxicity via gavage of nitrobenzene (C60082) in Fischer 344 rats and B6C3F1 mice [unpublished]. Prepared by the EG&G Mason Research Institute, Worcester, MA, for the National Toxicology Program, National Institute of Environmental Health Services, Public Health Service, U.S. Department of Health and Human Services, Research Triangle Park, NC; MRI-NTP 08-83-19.

NTP. (2011) Reports on Carcinogens. National Toxicology Program, 12th ed. Available at http://ntp.niehs.nih.gov/ntp/roc/twelfth/roc12.pdf.

Percy, M.J., McFerran, N.V., and Lappin, T.R. (2005) Disorders of oxidised haemoglobin. *Blood Rev.* 19:61–68.

Reddy, B.G., Pohl, T.R., and Krishna, G. (1976) The requirement of gut flora in nitrobenzene induced methemoglobinemia in rats. *Biochem. Pharmacol.* 25:1119–1122.

Robbiano, L., Baroni, D., Carrozzino, R., Mereto, E., and Brambilla, G. (2004) DNA damage and micronuclei induced in rat and human kidney cells by six chemicals carcinogenic to the rat kidney. *Toxicology* 204:187–195.

Saxena, H. and Saxena, A.P. (2010) Acute methaemoglobinaemia due to ingestion of nitrobenzene (paint solvent). *Indian J. Anaesth.* 54:160–162.

Shimkin, M.B. (1939) Acute toxicity of mononitrobenzene in mice. *Proc. Soc. Exp. Biol. Med.* 42: 844–846.

Smyth, H.F. Jr., Weil, C.S., West, J.S., et al. (1969) An exploration of joint toxic action: Twenty-seven industrial chemicals intubated in rats in all possible pairs. *Toxicol. Appl. Pharmacol.* 14:340–347.

Tyl, R.W., France, K.A., Fisher, L.C., et al. (1987) Developmental toxicity evaluation of inhaled nitrobenzene in CD rats. *Fundam. Appl. Toxicol.* 8:482–492.

U.S. EPA. (2004) Non-confidential IUR Production Volume Information. U.S. Environmental Protection Agency. Available at http://www.epa.gov/oppt/iur/tools/data/2002-vol.html.

U.S. EPA. (2009) Toxicological review of nitrobenzene (CAS No. 98-95-3) in support of summary information on the Integrated Risk Information System (IRIS). National Center for Environmental Assessment, Washington, DC. Available at http://www.epa.gov/iris/toxreviews/0079tr.pdf.

U.S. EPA. (2012) Toxics release inventory. Office of pollution prevention and toxics, environmental protection agency, Washington, DC. Available at http://www.epa.gov/triexplorer/.

Wang, G., Zhang, X., Yao, C., and Tian, M. (2011) Acute toxicity and mutagenesis of three metabolites mixture of nitrobenzene in mice. *Toxicol. Ind. Health* 27(2):167–171.

Wilkins, J.R. and Sinks, T. (1990) Parental occupation and intracranial neoplasms of childhood: results of a case-control interview study. *Am. J. Epidemiol.* 132:275–292.

2-NITROPROPANE

$$
\begin{array}{c}
CH_3 \\
| \\
CH_3 - C - NO_2 \\
| \\
H
\end{array}
$$

BACKGROUND AND USES

Several nitroalkanes, including nitromethane, nitroethane, 1-nitropropane (1-NP), and 2-nitropropane (2-NP) have been commercially available since 1940. By virtue of their unusual spectrum of industrially desirable properties, their production has rapidly increased for many applications. 2-NP is used widely as a specialty solvent for coatings, printing inks, paints, varnishes, and adhesives. It has also been used as an intermediate in the synthesis of dyes, pesticides, textile chemicals, and pharmaceutics, and an additive in explosives, propellants, and fuels. In 1977, it was estimated that about 30 million pounds of 2-NP were produced by a U.S. producer annually, and as many as 185,000 workers were exposed to 2-NP during its production and use (Bingham and Robbins, 1980). Between 1986 and 2002, U.S. production plus imports of 2-NP were reported to be about 10–50 million pounds (U.S. EPA, 2004). 2-NP is also manufactured by companies in France and other countries. The current worldwide production of 2-NP is unknown.

2-NP is a minor component of tobacco smoke and is present in smoke from other types of nitrate-rich organic matter. Residues in food products containing fatty acids separated with 2-NP and in beverage can linings and other coatings may represent additional sources of exposure to low amounts of 2-NP. Emerging evidence on carcinogenicity in experimental animals, along with various other toxic effects observed in animals and workers exposed to 2-NP, has attracted the attention of various U.S. and international regulatory agencies to the potential hazards of 2-NP and other nitroalkanes. As a result, 2-NP has been the subject of several toxicological reviews (U.S. EPA, 1991; IARC, 1999; IPCS, 1992; Woo et al., 1985).

PHYSICAL AND CHEMICAL PROPERTIES

The physicochemical properties of nitroalkanes have been extensively reviewed (Woo et al., 1985). 2-NP is an aliphatic nitro compound. It is a colorless, oily, and flammable liquid with a mild fruity odor. It has a molecular weight of 89.10, a boiling point of 120 °C, a melting point of −91.3 °C, a vapor pressure of 13 torr at 20 °C, and a flash point of 39.4 °C. It is slightly soluble in water (1.7 ml/100 ml) and miscible with many organic solvents.

As other nitroalkane, 2-NP exists in tautomeric equilibrium with its nitronic acid isomers (protonated and anionic aci forms):

$$
\underset{\text{Nitro form}}{
\begin{array}{c}
CH_3 \quad O \\
| \quad \nearrow \\
CH_3 - C - N \\
| \quad \diagdown \\
H \quad O
\end{array}}
\rightleftharpoons
\underset{\substack{\text{Aci form} \\ \text{(Protonated)}}}{
\begin{array}{c}
CH \quad O \\
| \quad \nearrow \\
CH_3 - C = N \\
| \quad \diagdown \\
H \quad OH
\end{array}}
\xrightarrow{-H^+}
\underset{\substack{\text{Aci form} \\ \text{(Anionic)}}}{
\begin{array}{c}
CH_3 \quad O \\
| \quad \nearrow \\
CH_3 - C = N \\
| \quad \diagdown \\
H \quad O^-
\end{array}}
$$

The aci form is amphoterically reactive, the protonated form being electrophilic and the anionic form being nucleophilic.

MAMMALIAN TOXICOLOGY

Acute Effects

The acute toxicity of 2-NP has mainly been studied in animals by inhalation and oral exposures (IPCS, 1992). The LC_{50} for rats following a 6-h exposure was 400 ppm ($1.5 \, g/m^3$) for males and 720 ppm ($2.6 \, g/m^3$) for females. In mice, the LC_{50} value was about 560 ppm (2038 mg) for both males and females following a 6-h exposure. Cats appeared more sensitive to acute exposure to 2-NP than other animals; the acute sensitivity of animals to inhaled 2-NP (from high to low) was cat, rat and mouse, rabbit, and guinea pig. Animals exposed to concentrations of 2,300 ppm ($8.4 \, g/m^3$) for 1 h or longer have been reported to display severe pathological changes including hepatocellular damage, pulmonary edema, and hemorrhage. The 14-day oral LD_{50} for mice was 0.40 g/kg. The minimal inhalation and oral lethal doses for the rabbit were 0.24 g/kg and 0.50–0.75 g/kg, respectively. In animals, the toxic effects of acute exposure to 2-NP are characterized by hepatotoxicity, methaemoglobin formation, and depression of the central nervous system (Zitting et al., 1981).

2-NP is Irritating to the skin, eyes, and respiratory system. Acute occupational exposures to about 20–45 ppm (73–164 mg/m³) 2-NP caused headache, nausea, vomiting, diarrhea, and anorexia in workers. Several human cases of fatalities have been reported (HSDB, 2011). One construction worker who was exposed to 2-NP while applying epoxy resin coatings died 10 days after exposure from fulminant hepatitis, and had serum concentration of 2-NP at 13 mg/l (Harrison et al., 1987). Exposures to high concentration (about 600 ppm or 2184 mg/m³) of 2-NP have also been reported to be fatal after severe liver and kidney damages and respiratory failure (Hine et al., 1978).

Chronic Effects

A number of repeated-dose studies in animals have shown that chronic exposures to 2-NP caused primarily liver toxicity and neoplasms. Increased liver weights and serum enzyme levels (e.g., SGPT), and morphological changes in the liver characterized by cytoplasmic vacuolization of hepatocytes, focal necrosis, and hypertrophic nodules with large vesiculated nuclei have been observed in a series of inhalation exposure studies in Sprague-Dawley rats or Long Evans rats (Angus Chemical Company, 1985a, 1985b; Griffin et al., 1980, 1981). These effects showed a dose–response relationship at 25–200 ppm and exposed for 2–22 months. In addition, male rats were shown to be more sensitive to 2-NP than females. Based on mild hepatic effects, a LOAEL of 25 ppm (78 mg/m³; HEC = 16.3 mg/m³) was determined (Griffin et al., 1981). In addition to hepatic effects, thickening of alveolar septa was noted in all exposed Sprague-Dawley and Long Evans rats who inhaled 200 ppm (624 mg/m³)

2-NP, 7 h/day, 5 days/week for 2–6 months, suggesting possible respiratory toxicity (Angus Chemical Company, 1985a, 1985b; Griffin et al., 1978).

In a study treating rats with oral doses of 0.002, 0.01, 0.05, and 0.25 g/kg, five times per week for 4 weeks by gavage, there was some mortality among male rats, and in both sexes there was decreased growth, anemia, increased liver and heart weights, and severe damage to the liver at 0.25 g/kg. At 0.05 g/kg, the major effect appeared to be anemia. Lower doses (0.01 and 0.002 g/kg) did not produce any toxic effects (Wester et al., 1989).

There is little information on the reproductive or developmental toxicity of 2-NP. No evidence of an increase in dominant lethality or in sperm abnormality was noted in genetic studies (McGregor, 1981). Fetal toxicity (delayed fetal heart development) in Sprague-Dawley rats following i.p. injection of 170 mg/kg 2-NP on days 1–15 of gestation was the only developmental effect reported (Hardin et al., 1981).

Several studies have demonstrated that 2-NP is carcinogenicity in the rat. In one inhalation study, groups of 50 male Sprague-Dawley rats were exposed to 0, 27, and 207 ppm 2-NP vapor 7 h/day, 5 days/week, for 24 weeks/6 months. Rats exposed at 207 ppm ($0.75 \, g/m^3$) developed various liver changes such as hepatocellular hypertrophy, hyperplasia, and necrosis after 3 months. Liver neoplasms (hepatocellular carcinoma) were found in all surviving 10 rats exposed to 207 ppm 2-NP (Lewis et al., 1979). In a similar study, groups of 10 Sprague-Dawley rats of either sex were exposed to 2-NP vapor at 0, 25, 100, or 200 ppm, 7 h/day, 5 days/week, for up to 6 months. Liver carcinomas were observed in 9 of 10 male rats exposed for 6 months at 200 ppm ($0.73 \, g/m^3$) and held unexposed for 6 more months. Liver tumors were also found in male rats exposed to 100 ppm ($0.36 \, g/m^3$) 2-NP. Only hyperplasia and vacuolar degeneration in the livers were noted in female rats exposed to 100 or 200 ppm 2-NP (Griffin & Coulston, 1983; Griffin et al., 1984; Coulston et al., 1985). Oral dosing of male rats with 89 mg/kg (1 mmol/kg) by gavage, three times a week for 16 weeks, also induced hepatocellular carcinomas in all 22 rats sacrificed 40 weeks after the last exposure; no liver carcinoma was found in 29 control rats (Fiala et al., 1987a).

MECHANISM(S) OF ACTION

The mechanisms of liver toxicity and carcinogenicity of 2-NP are not clearly understood. 2-NP has been shown to be mutagenic in a variety of test systems. It is an active genotoxic agent in rat hepatocytes both *in vitro* and *in vivo* (Andrae et al., 1988; Fiala et al., 1989; George et al., 1989; Guo et al., 1990; Conaway et al., 1991a). Studies in various strains of rats have also demonstrated that 2-NP induced nucleic acid damage in the livers but not other organs (Andrae et al., 1988; Guo et al., 1990; Conaway et al., 1991a; Robbiano et al., 1991). Damage to rat liver nucleic acids was only caused by intraperitoneal injection of secondary nitroalkanes but not with a primary or a tertiary nitroalkane. The greater genotoxicity of the secondary nitroalkanes was speculated to be due to the greater stability of their nitronate/aci forms at physiological pH values (Conaway et al., 1991b). The proportion of the nitronate/aci form present in water at 25 °C has been found to increase with the length of the side chain in the order of: $C_3H_7 > C_2H_5 > CH_3$ (Turnbsall and Maron, 1943).

In addition, 2-NP is mutagenic in bacteria both in the presence and absence of exogenous metabolic activation (Fiala et al., 1987b; Conaway et al., 1991a; Kohl et al., 1994). As the mutagenicity of 2-NP does not require metabolic activation and alkylation of DNA occurs after *in vitro* reaction with 2-NP, it has been hypothesized that 2-NP may exert its mutagenic and carcinogenic action via a direct nonenzymatic reaction between the compound and DNA (Speck et al., 1982). The nitronate tautomer of 2-NP is more mutagenic in *S. typhimurium* TA100 and TA102 than the parent compound, suggesting that the eletrophilic aci form of 2-NP may be the reactive intermediate responsible for its mutagenic and carcinogenic effects (Fiala et al., 1987b; Kohl et al., 1994).

Nitrite is the major metabolite after oxidative metabolic degradation of 2-NP. There is evidence from *in vitro* studies that 2-NP can undergo denitrification, and conversion of 2-NP into aci tautomer propane 2-nitronate, either chemically or enzymatically, was a prerequisite for rapid denitrification (Kohl et al., 1997). Denitrification has been postulated to be the molecular mechanism involved in toxicity and carcinogenicity of 2-NP (Sakurai et al., 1980; Ullrich et al., 1978). This appears to be consistent with the relative rates of "oxidative denitrification" of nitroalkanes; as the affinity toward the microsomal mixed-function oxidase system decreases with decrease of the chain length, lower nitroalkanes are noncarcinogenic or only weakly carcinogenic as compared with 2-NP. Furthermore, genotoxicity and DNA damages in liver were increased when rats were pretreated with inducers of cytochrome P-450-dependent monooxygenases, and were reduced in rats when pretreated with inhibitors of the enzymes (Robbiano et al., 1991; Roscher et al., 1990).

The formation of 8-aminodeoxyguanosine after reaction of DNA with the nitronate tautomer of 2-NP has led Sodum and coworkers (1993) to speculate that 8-aminodeoxyguanosine may be formed either by base nitrosation followed by reduction, or via an enzyme-mediated conversion of the nitronated anion to hydroxylamine-O-sulfonate or acetate, which yields the highly reactive nitrenium ion NH_2^+. There is evidence from *in vitro* and *in vivo* studies that sulfotransferases in rat liver are involved in the activation of 2-NP to a reactive species that animated guanosine at the C^8 position (Sodum et al., 1994). Recent studies have also demonstrated that human sulfotransferases play a role in the bioactivation and genotoxicity of 2-NP (Oda et al., 2012; Deng et al., 2011). Several genes involved in DNA damage, DNA repair, apoptosis, and cell cycle checkpoint control have been identified in liver of rats exposed to 2-NP for 28 days (Nakayama et al., 2006).

2-NP can induce cell proliferation in rat liver. Significant increases in the frequency of S-phase cells by measuring the incorporation of bromodeoxyuridine into newly synthesized DNA have been found in the liver of rats exposed to daily oral doses of 40 or 80 mg/kg 2-NP by gavage for 10 days. The noncarcinogenic 1-NP did not affect DNA synthesis of liver cells at these dose levels, suggesting that increased cell proliferation may contribute to the hepatocarcinogenicity of 2-NP (Cunningham and Matthews, 1991).

CHEMICAL PATHOLOGY

2-NP can be absorbed via lungs, peritoneal cavity, gastrointestinal tract, and, to a lesser extent, via skin (Nolan et al., 1982; Müller et al., 1983). High concentrations are acutely toxic and several cases of industrial fatalities have been attributed to inhalation of solvent mixtures containing 2-NP for a total of 5–16 h during the course of 1–3 days (Hine et al., 1978; Rondia, 1979; Harrison et al., 1985, 1987). The actual concentrations of 2-NP that caused deaths were not measured. In one case, the concentration of 2-NP that caused deaths was estimated to be about 2.18 g/m^3 (600 ppm), as determined by the 2-NP level in the victim's blood (0.013 g/l) (Harrison et al., 1987). Initial symptoms included headache, nausea, dizziness, anorexia, vomiting, diarrhea, and abdominal pain. Later symptoms included persistent nausea, vomiting, anorexia, jaundice, reduced urine output, diarrhea, bloody stools, and loss of reflexes. In all cases the primary cause of death was acute hepatic failure, with contributing factors such as lung edema, gastrointestinal bleeding, and respiratory and kidney failure. Liver necrosis and in some cases fatty degeneration were revealed at postmortem microscopic examination of the liver.

Exposure to nonlethal concentrations of 2-NP may exhibit symptoms such as headache, nausea, anorexia, vomiting, diarrhea, and abdominal pain (Angus Chemical Co. & Occusafe, Inc., 1986). No signs of toxic effects were noted in workers exposed to 36–109 mg/m^3 (10–30 ppm) for less than 4 h/day on not more than 3 days/week (Skinner, 1947). There is inadequate epidemiological evidence that 2-NP induces cancer or other long-term toxic effects in humans (IARC, 1982, 1999).

SOURCES OF EXPOSURE

Occupational

Occupational exposure to 2-NP may occur through inhalation and dermal contact with this compound in many industries including printing (rotogravure and flexographic inks), coatings and adhesives used in industrial construction and maintenance, highway markings, shipbuilding and mainte-nance, furniture manufacture, and food packaging. Area monitoring results in 2-NP production plants were ranged from 23 to 120 ppm. Current industrial exposure in the United States or worldwide is unknown. On the basis of a NIOSH survey between 1981 and 1983, as many as 9818 U.S. workers were potentially exposed to 2-NP (HSDB, 2011).

Although manufacture of 2-NP is an enclosed process with little worker exposure, and exposure levels in most operations are generally low, some painting and manufactur-ing operations have exposed workers to dangerously high concentrations of 2-NP vapors. Concentrations of 2-NP up to 2744 mg/m^3 (754 pm) and up to at least 265 mg/m^3 (73 ppm) were exposed by workers in a pigment production and a solvent extraction plant, respectively, and concentrations as high as 6 g/m^3 (1640 ppm) in air were recorded in a drum-filling operation (Crawford et al., 1985).

Environmental

Releases of 2-NP to the environment are mainly due to solvent evaporation from coated surfaces such as beverage can coat-ings, adhesives, and print, and from vegetable oils fractionated with 2-NP. 2-NP is a minor component of tobacco smoke and is present in smoke from other types of nitrate-rich organic matter that may represent additional sources of environment exposure to low levels of 2-NP. In 1988, environmental releases of 2-NP were estimated to be more than 655,000 lb; releases then steadily declined and in 2007, eight facilities released a total of 28,600 lb (U.S. EPA, 2009).

2-NP is not persistent in the environment. When released to air, it undergoes photodegradation, with a half-life of 9.8 days (HSDB, 2009). When released to soil or water, it will volatilize or be degraded by bacteria. It has low acute toxicity in fish and other aquatic organisms, and it is not expected to bioaccumulate in the food chain. Thus, 2-NP is not considered to pose serious environmental hazards.

INDUSTRIAL HYGIENE

A threshold limit value (TLV) for 2-NP of 10 ppm was established by the American Conference of Governmental Industrial Hygienists (ACGIH) and a permissible exposure level (PEL) of 25 ppm (90 mg/m^3) as an 8-h time-weighted average (TWA) was established by the Occupational Safety and Health Administration (OSHA). The National Institute for Occupational Safety and Health (NIOSH) "immediately dangerous to life or health" (IDLH) concentration is 100 ppm. These occupational exposure limits have been proposed to be reduced because 2-NP was found to be a potent carcinogen in rats and strong recommendations have been issued on minimizing all worker contact with 2-NP and its fumes (IRPTC, 1986; U.S. EPA, 1986; NIOSH, 1997).

NIOSH has listed 2-NP as a potential occupational carcino-gen, and has issued guidelines for handling 2-NP in the workplace as if it were a human carcinogen (Finklea, 1977). 2-NP has been listed as "reasonably anticipated to be carci-nogenic to humans" by the U.S. National Cancer Institute/National Toxicology Program (NTP, 2011) and classified as a group 2B, "possibly carcinogenic to humans" by the International Agency for Research on Cancer (IARC, 1999) based on sufficient evidence of carcinogenicity in rats.

The U.S. Environmental Protection Agency (EPA) regu-lates 2-NP under the Clean Air Act (CAA), the Emergency Planning and Community Right-To-Know Act, and the Resource Conservation and Recovery Act (RCRA). Under the Comprehensive Environmental Response, Compensa-tion, and Liability Act (CERCLA), 2-NP has a reportable quantity of 10 pounds (4.54 kg) (NTP, 2011). On the basis of the toxicological data of 2-NP, EPA has established an inhalation RfC of 2E-2 mg/m^3 (U.S. EPA, 1991) and a provisional inhalation cancer unit-risk factor of 27E-4 μg/m^3 (U.S. EPA, 1997).

Standards	Values
NIOSH REL	None established; NIOSH considers 2-NP to be a potential occupational carcinogen
OSHA PEL	25 ppm (90 mg/m^3) TWA
ACGIH TLV	10 ppm (36 mg/m^3) TWA
IDLH	100 ppm
RfC	2E-2 mg/m^3
Cancer unit risk (inhalation)	27E-4 μg/m^3

Occupational exposure limits in the air vary among other countries and range from a TWA of 3.6 mg/m^3 (1 ppm) to a short-term exposure limit (STEL) of 146 mg/m^3 (40 ppm) (IPCS, 1992). The Joint FAO/WHO Expert Committee on Food Additives (JECFA) recommended that 2-NP should not be used as a solvent in food processing or as a fractionating solvent in the production of fats and oils (FAO/WHO, 1990a, 1990b; FAO/WHO, 1984).

MEDICAL MANAGEMENT

Individuals who have been exposed to high concentrations of vapors from solvent mixtures and exhibit symptoms such as headache, nausea, dizziness, drowsiness, weakness, ano-rexia, vomiting, diarrhea, and abdominal pain may have been exposed to 2-NP and should be hospitalized immedi-ately. 2-NP exposures may produce a daily cycle in symp-toms and improvement of initial symptoms may be temporary (Skinner, 1947). Serum enzyme levels indicative of liver toxicity, lung edema, gastrointestinal bleeding, and respiratory, and kidney functions should be continuously monitored until full recovery.

REFERENCES

ACGIH. (1986) *Documentation of the Threshold Limit Values and Biological Exposure Indices*, 5th ed., Cincinnati, Ohio: American Conference of Governmental Industrial Hygienists, Inc. pp. 441–442.

Andrae, U., Homfeld, H., Vogl, L., Lichtmannegger, J., and Summer, K.H. (1988) 2-Nitropropane induces DNA repair synthesis in rat hepatocytes *in vitro* and *in vivo*. *Carcinogenesis* 9:811–815.

Angus Chemical Company. (1985) *Technical data sheet: NiPar S-20^{TM} nitropropane solvent, TDS 20B*, Northbrook, Illinois: Angus Chemical Company, 12 pp.

Angus Chemical Company. (1985a) Final report: chronic inhalation exposure of rats to vapors of 2-nitropropane at 100 ppm. By Coulston Intl. Corp. for Angus Chem. Co. OTS No. 204292. Doc No. 8EHQ-0985-0170. Fiche No. 0200504.

Angus Chemical Company. (1985b) Chronic inhalation of 200 ppm of 2-nitropropane in rats: six month report. By Coulston Intl. Corp. (May 1978) for Angus Chem. Co. OTS No. 204292. Doc No. 8EHQ-0985-0170. Fiche No. 0200504.

Angus Chemical Company & Occusafe, Incorporated. (1986) Industrial hygiene and occupational health information on 2-nitropropane, December 1982, and updates for December 1983, 1984, 1985. Northbrook, Illinois, Angus Chemical Company, 39 pp and 46 attachments (Unpublished report presented to the ACGIH TLV Committee).

Akron. (2009) The Chemical Database. The Department of Chemistry at the University of Akron. Available at http://ull.chemistry.uakron.edu/erd and search on CAS number (accessed June 4, 2009).

ChemIDplus. (2009) ChemIDplus Advanced. National Library of Medicine. Available at http://chem.sis.nlm.nih.gov/chemidplus/chemidheavy.jsp and select Registry Number and search on CAS number (accessed June 4, 2009). .

ChemSources. (2009) Chem Sources – Chemical Search. Chemical Sources International. Available at http://www.chemsources.com/chemonline.html and search on nitropropane (accessed June 4, 2009).

Crawford, G.N., Garrison, R.P., McFee, D.R. (1985) Health examination and air monitoring evaluation for workers exposed to 2-nitropropane. *Am. Ind. Hyg. Assoc. J.* 46(1):45–47.

Gaworski, C.L., Lemus-Olalde, R., and Carmines, E.L. (2008) Toxicological evaluation of potassium sorbate added to cigarette tobacco. *Food Chem. Toxicol.* 46(1):339–351.

Gargas, M.L., Burgess, R.J., Voisard, D.E., Cason, G.H., and Andersen, M.E. (1989) Partition coefficients of low-molecular-weight volatile chemicals in various liquids and tissues. *Toxicol. Appl. Pharmacol.* 98(1):87–99.

Griffin, T.B., Benitz, K.F., Coulston, F., and Rosenblum, I. (1978) Chronic inhalation toxicity of 2-nitropropane in rats. *Pharmacologist* 20(3):145.

Griffin, T.B., Coulston, F., and Stein, A.A. (1980) Chronic inhalation exposure of rats to vapors of 2-nitropropane at 25 ppm. *Ecotoxicol. Environ. Saf.* 4(3):267–281.

Griffin, T.B., Stein, A.A., and Coulston, F. (1981) Histological study of tissues and organs from rats exposed to vapors of 2-nitropropane at 25 ppm. *Ecotoxicol. Environ. Saf.* 5(2):194–201.

Hardin, B.D., Bond, G.P., Sikov, M.R., Andrew, F.D., Beliles, R.P., and Niemeier, R. (1981) Testing of selected workplace chemicals for teratogenic potential. *Scand. J. Work Environ. Health.* 7 (Suppl. 4):66–75.

Harrison, R.J., PAsternak, G., and Blanc, P. (1985) Leads from the MMWR. Acute hepatic failure after occupational exposure to 2-nitropropane. *J. Am. Med. Assoc.* 254(24):3415–3416.

Hine, C.H., Pasi, A., and Stephens, B.G. (1978) Fatalities following exposure to 2-nitropropane. *J. Occup. Med.* 20:333–337.

HSDB. (2011) Hazardous substances data bank. National Library of Medicine. Available at http://toxnet.nlm.nih.gov/cgi-bin/sis/search/f?./temp/~fLKxZm:1.

IARC. (1982) 2-Nitropropane. In: *Some Industrial Chemicals and Dyestuffs. IARC Monographs on the Evaluation of Carcinogenic Risk of Chemicals to Humans*, vol. 29, Lyon, France: International Agency for Research on Cancer, pp. 331–343.

IARC. (1999) *2-Nitropropane. In: Re-evaluation of Some Organic Chemicals, Hydrazine, and Hydrogen Peroxide. IARC Monographs on the Evaluation of Carcinogenic Risk of Chemicals to Humans*, vol. 71, Lyon, France: International Agency for Research on Cancer, pp. 1079–1094.

IPCS. (1992) Environmental Health Criteria 138. 2-Nitropropane. International Programme on Chemical Safety. Available from http://www.inchem.org/documents/ehc/ehc/ehc138.htm.

Lewis T.R., Ulrich, C.E., and Busey, W.M. (1979) Subchronic inhalation toxicity of nitromethane and 2-nitropropane. *J. Environ. Pathol. Toxicol.* 2(5):233–249.

NIOSH. (1976) *National Occupational Hazard Survey (1972-74)*. DHEW (NIOSH) Publication No. 78-114. Cincinnati, OH: National Institute for Occupational Safety and Health.

NIOSH. (1977) Current Intelligence Bulletin 17. 2-Nitropropane. National Institute for Occupational Safety and Health. Available at http://www.cdc.gov/Niosh/78127_17.htm.

NTP. (2011) Report on Carcinogens, 12th ed. Available at http://ntp.niehs.nih.gov/go/roc12.

Rondia, D. (1979) 2-Nitropropane: One more death. *Vet. Hum. Toxicol.* 21(Suppl):183–185.

U.S. EPA. (2004) Non-confidential IUR production volume information. U.S. Environmental Protection Agency. Available at http://www.epa.gov/oppt/iur/tools/data/2002-vol.html and search on CAS number (accessed April 21, 2005).

U.S. EPA. (1985) Health and environmental effects profile for 2-nitropropane. Prepared by the Office of Health and Environmental Assessment, Environmental Criteria and Assessment Office, Cincinnati, OH for the Office of Solid Waste, Washington, DC. EPA/600/X-85/112.

Woo, Y.-T., Lai, D.Y., Arcos, J.C., and Argus, M.F. (1985) Niroalkanes and nitroalkenes. In: *Chemical Induction of Cancer*, vol. IIIB, New York: Academic Press, pp. 329–341.

62

METHYL TERT-BUTYL ETHER (MTBE)

Janice S. Lee and Robyn Blain

First aid: Eye: Irrigate immediately; skin: water flush promptly; breathing: respiratory support; swallow: seek medical attention immediately

Target organ(s): Skin, eyes, CNS, liver, kidney, blood

Occupational exposure limits: ACGIH TLV: TWA 50 ppm (mg/m^3)

Reference values: EPA RfC = 3 mg/m^3

Risk phrases: 11-38

Highly flammable. Irritating to the skin

Safety phrases: 2-9-16-24-33-37

Keep out of the reach of children. Keep container in a well-ventilated place. Keep away from sources of ignition—no smoking. Avoid contact with the skin. Take precautionary measures against static discharges. Wear suitable gloves

BACKGROUND AND USES

Methyl tert-butyl ether (MTBE) (CASRN 1624-04-4) is a synthetic chemical with synonyms methyl tertiary-butyl ether, tert-butyl methyl ether, tertiary-butyl methyl ether, methyl-1,1-dimethylethyl ether, 2-methoxy-2-methylpropane, 2-methyl-2-methoxypropane, methyl t-butyl ether, and MTBE. It was primarily used as a gasoline additive in unleaded gasoline in the United States prior to 2005, in the manufacture of isobutene, and as a chromatographic eluent especially in high pressure liquid chromatography (ATSDR, 1996; HSDB, 2012). It is also a pharmaceutical agent and can be injected into the gallbladder to dissolve gallstones (ATSDR, 1996). As a result of its manufacture and use, MTBE may be released into the environment through various waste streams directly (IPCS, 1998; HSDB, 2012).

It is manufactured by the chemical reaction of methanol and isobutylene, from methanol and tert-butyl alcohol, and from tert-butyl alcohol and diazomethane (Merck, 2006). The predominant use for MTBE in the United States was as an oxygenate in unleaded gasoline promoting more complete burning of gasoline. Reformulated fuel with MTBE was widely used in the United States between 1992 and 2005 to meet the 1990 Clean Air Act Amendments (CAAA) requirements for reducing carbon monoxide (CO) and ozone (O$_3$) levels. In 1992, 9.1 billion pounds of MTBE were produced in the United States, and the capacity was estimated at 11.6 billion pounds (U.S. EPA, 1994a). The oxygenate requirement for reformulated gasoline was removed in 2005 by the Energy Policy Act. Subsequently, refiners made a wholesale switch removing MTBE and blending fuel with ethanol. It has not been used in the United States in significant quantities in reformulated gasoline or conventional gasoline since 2005.

PHYSICAL AND CHEMICAL PROPERTIES

Methyl tert-butyl ether, a colorless liquid, is an aliphatic ether and volatile organic compound (VOC). It is moderately soluble in water and very soluble in some organic solvents such as alcohol and diethylether (ATSDR, 1996). It is a flammable liquid with a characteristic odor. Selected chemical and physical properties of MTBE are presented in Table 62.1.

Hamilton & Hardy's Industrial Toxicology, Sixth Edition. Edited by Raymond D. Harbison, Marie M. Bourgeois, and Giffe T. Johnson.
© 2015 John Wiley & Sons, Inc. Published 2015 by John Wiley & Sons, Inc.

TABLE 62.1   Chemical and Physical Properties of MTBE

Characteristic	Information
Chemical name	Methyl tert-butyl ether
Synonyms/trade names	Methyl tertiary-butyl ether; tert-butyl methyl ether; tertiary-butyl methyl ether; methyl-1,1-dimethylethyl ether; 2-methoxy-2-methylpropane; 2-methyl-2-methoxypropane; methyl t-butyl ether; MTBE
Chemical formula	$C_5H_{12}O$ or $(CH_3)_3C(OCH_3)$
CASRN	1634-04-4
Molecular weight	88.15
Melting point	$-109\,°C$
Boiling point	$53.6–55.2\,°C$
Vapor pressure	245–251 mmHg at 25 °C
Density/specific gravity	0.7404–0.7578
Flashpoint	28 °C (closed cup)
Water solubility at 25 °C	51.26 g/l
Explosion limits	1.65–8.4%
Octanol/water partition coefficient (log K_{OW})	1.24
Henry's law constant	5.5×10^{-4} atm-m^3/mole at 25 °C
Odor threshold	0.32–0.47 mg/m^3
Conversion factors	1 ppm = 3.61 mg/m^3 1 mg/m^3 = 0.28 ppm

ENVIRONMENTAL FATE AND BIOACCUMULATION

It can be released during manufacturing or blending with gasoline; during the storage, distribution, and transfer of MTBE-blended gasoline; and from spills or leaks or fugitive emissions at automotive service stations (U.S. EPA, 1994a). In 1992, 3 million pounds of MTBE were released to the environment with 2.8 million pounds to the air, 100 thousand pounds to surface water, 68 thousand pounds to underground injection sites, and 288 pounds to land (U.S. EPA, 1994b). Vapor emissions from MTBE-blended gasoline may also contribute to atmospheric levels (U.S. EPA, 1988). It is not expected to persist in the atmosphere because it undergoes destruction from reactions with hydroxyl radicals. A total atmospheric lifetime for MTBE of approximately 3 and 6.1 days has been reported in polluted urban air and in non-polluted rural air, respectively (U.S. EPA, 1993a). Based upon its vapor pressure and Henry's law constant, MTBE is highly volatile and would be expected to evaporate rapidly from soil surfaces or water. It may be fairly persistent when introduced into subsurface soils or to groundwater since volatilization to the atmosphere is reduced or eliminated. It does not readily degrade in surface waters due to hydrolysis or other abiotic processes. It is also resistant to bio-degradation (U.S. EPA, 1993a). It is usually removed from surface waters very rapidly because of its high volatility. If released as part of a gasoline mixture from leaking underground storage tanks, its relatively high water solubility combined with little tendency to sorb to soil particles encourages migration to local groundwater supplies (U.S. EPA, 1993a).

The bioconcentration potential for MTBE appears to be very minor and can be considered insignificant (ATSDR, 1996).

ECOTOXICOLOGY

Aquatic toxicity data show relatively low toxicity to MTBE in freshwater and marine organisms, with lethal concentrations generally >100 mg/l (U.S. EPA, 1994a). The toxicity of MTBE has been studied in two species of freshwater algae (green algae), with 96-h EC$_{50}$ values ranging from 184 to 492 mg/l (BenKinney et al., 1994; Mancini et al., 1999). Acute studies in invertebrates (water flea, water snail, mayfly, amphipod, and midge) indicate MTBE to be of low toxicity with EC$_{50}$ values well above 100 mg/l (BenKinney et al., 1994; Mancini et al., 1999). A chronic study in the water flea reported a no observed effect concentration (NOEC) of 51 mg/l and an IC$_{20}$ of 42 mg/l for a 21-day test (Mancini et al., 1999). The toxicity of MTBE in freshwater fish has been evaluated, and all studies indicate MTBE to be of low acute toxicity with reported LC$_{50}$ values well above 250 mg/l (BenKinney et al., 1994; Geiger et al., 1988; Veith et al., 1983). The effect of MTBE on survival and development of tadpoles was evaluated and an LC$_{50}$ of >2500 mg/l was derived (Paulov, 1987).

Similar to freshwater algae, an EC$_{50}$ of 185 mg/l in marine algae also suggests low toxicity to MTBE (Mancini et al., 1999). Acute studies in marine invertebrates (copepod, mysid shrimp, grass shrimp, blue crab, mussel, and amphipod) reported LC$_{50}$ values ranging from 136 to >1000 mg/l (BenKinney et al., 1994; Bengtsson and Tarkpea, 1983; Mancini et al., 1999). In a 28 day chronic test with mysid shrimp, an IC$_{20}$ of 36 mg/l and a NOEC of 26 mg/l were reported (Mancini et al., 1999). Toxicity studies conducted in marine fish (bleak, inland silverside, sheepshead minnow) reported LC$_{50}$ values well above 500 mg/l (Bengtsson and Tarkpea, 1983; BenKinney et al., 1994; Mancini et al., 1999).

MAMMALIAN TOXICOLOGY

Acute Effects

The National Institute of Occupational Safety and Health (NIOSH) considers MTBE to be a skin and eye irritant (IPCS, 1998). However, in acute testing studies, results indicated that MTBE was not a skin sensitizer or primary skin irritant and eye irritation to be transient (Duffy et al., 1992). It does cause sensory and respiratory irritation with

1-h inhalation exposure to mice that increased with dose (Tepper et al., 1994). The NIOSH also reports acute inhalation exposures to MTBE to potentially cause other symptoms, including drowsiness, dizziness, headache, weakness, and unconsciousness (IPCS, 1998). Acute ingestion exposures to MTBE may cause abdominal pain, nausea, and vomiting (IPCS, 1998). However, acute toxicity studies in rodents have found MTBE to have low to very low acute toxicity after oral (LD_{50} = 3–3.8 g/kg), dermal (>10 g/kg), or inhalation (23,630–33,000 parts per million (ppm)) exposures (Duffy et al., 1992).

Humans are also acutely exposed to MTBE as part of a medical treatment to dissolve cholesterol gallstones (Thistle, 1992). Injecting the gallbladder with high doses of MTBE can result in nausea, vomiting, sleepiness, and minor transient mucosal damage in the gallbladder (U.S. EPA, 1993b).

A couple of epidemiology (cross-sectional and ecological) studies have been conducted to evaluate the health effects of individuals exposed to oxygenated gasoline or reformulated fuels (Mohr et al., 1994; Joseph and Weiner, 2002). Symptoms potentially associated with MTBE exposure included coughing, burning sensations of the nose and throat, headache, allergic rhinitis, nausea, dizziness, wheezing, and other nonspecific symptoms. Since the workers were also exposed to other gasoline components and/or combustion products of MTBE, the specific role of MTBE is unknown. Individuals exposed to concentrations of MTBE up to 50 ppm for up to 2 h with or without light physical work did not report any difference in symptoms, showed no difference in neurobehavioral battery tests, indicators of upper airway inflammation, or indication of eye inflammation compared to exposure to clean air (Prah et al., 1994; Cain et al., 1996; Johanson et al., 1995; Nihlén et al., 1998).

In an effort to further explore whether MTBE may cause adverse health effects from short-term exposure, controlled exposures to MTBE have been used to study sensory, symptomatic, cellular, and eye responses in healthy humans. Potential MTBE-related responses have also been studied in subjects with reported past exposures, but most have been conducted on small numbers of healthy adults and not on people who may be especially sensitive to MTBE exposure. Humans exposed to 11% MTBE (110,000 ppm) in gasoline for 15 min did not show any increase in symptoms compared to subjects that were exposed to clean air or gasoline alone (Fiedler et al., 2000). However, subjects exposed to 15% MTBE (150,000) in gasoline for 15 min and those who were self-reported sensitive to MTBE had a greater total number of symptoms compared to all other groups (Fiedler et al., 2000). Transient (1 h after exposure only) central nervous system (CNS) depression (as measured by functional observational battery tests) was observed in rats after a 6-h inhalation exposure to 4000 and 8000 ppm, which did not cause mortality or any overt signs of toxicity (Daughtrey et al., 1997). No such effects were noted at 800 ppm. These effects were not persistent with subchronic exposures lasting up to 13 weeks (Daughtrey et al., 1997).

Chronic Effects

The National Health and Nutritional Survey (NHANES) collected data on MTBE exposure using personal monitors on more than 500 participants and measured several serum enzymes associated with liver function (i.e., albumin, alkaline phosphatase [ALP], γ-glutamyl transferase [GGT], total bilirubin, alanine aminotransferase [ALT], aspartate aminotransferase [AST], and lactic dehydrogenase [LDH]) (Liu et al., 2009). Methyl tert-butyl ether was associated with increases in serum levels of albumin, ALP, and GGT. Chronic rodent studies have evaluated that all major organ systems, including the reproductive system and other systems, are affected during development. The liver and kidney appear to be the most sensitive to short- and long-term exposures to MTBE. In experimental animals, chronic MTBE exposure has the potential to induce transient neurotoxicity (Bird et al., 1997; Clary, 1997; Duffy et al., 1992), liver toxicity (Bird et al., 1997), kidney toxicity (Bird et al., 1997; Bermudez et al., 2012; Dodd et al., 2013), endocrine (de Peyster et al., 2003) and developmental toxicity in the presence of maternal toxicity (Bevan et al., 1997a, 1997b). Two-generation reproductive studies have not observed effects on reproduction (Duffy et al., 1992), but effects on the male reproductive system have been observed in male rats, including effects on spermatogenesis (Li et al., 2008). No such effects have been observed in male mice (de Peyster et al., 2008).

Longer-term exposure to MTBE induces several forms of cancer in rodents. Significant increases in hepatocellular tumors in both male and female mice, renal tubule tumors in male rats, lymphohematopoietic cancer in female rats, and Leydig cell tumors in male rats have been observed (Bird et al., 1997; Belpoggi et al., 1995, 1997, 1998). However, a meta-analysis of the literature evaluating non-Hodgkins lymphoma in workers involved in the production, distribution, or use of gasoline did not find any increase in risk (Kane and Newton, 2010). A marginal increase in brain tumors was observed in a drinking water study in male rats, which the study authors considered within the historical control and not related to MTBE treatment (Dodd et al., 2013).

Chemical Pathology

Progressive nephropathy is the main pathology associated with chronic MTBE exposure in rats (Bird et al., 1997; Bermudez et al., 2012). Although chronic progressive nephropathy is observed in both sexes, it occurs earlier in males. The same is not observed after 13 weeks of exposure (Lington et al., 1997; Bermudez et al., 2012). Oral exposures in rats have been associated with lymphoma, leukemia, and haemolymphoreticular dysplasias in female rats and interstitial cell hyperplasia and adenomas (Leydig cell tumors) accompanied by testicular degeneration, atrophy, and mineralization in male rats (Belpoggi et al., 1995, 1997, 1998). An inhalation exposure has been associated with hepatocellular tumors in both male and female mice and renal tubule tumors in male rats (Bird et al., 1997).

Hypothesized Modes of Action

There is no definitive mode of action (MOA) for MTBE-induced toxicity, but a few MOAs have been suggested. A suggested MOA for kidney toxicity in male rats, including the kidney tumors observed, has been associated with the binding of MTBE to a α_{2u}-globulin, which then accumulates in the kidney of male rats only (U.S. EPA, 1991). Kidney effects related to α_{2u}-globulin would not be relevant in humans as this MOA is not operative in humans. The presence of α_{2u}-globulin in male rats has been observed in several studies after MTBE inhalation (Bird et al., 1997; Prescott-Mathews et al., 1997, 1999; Bermudez et al., 2012). The establishment of α_{2u}-globulin as the MOA for kidney tumors is controversial with differences in opinion (Burns and Melnick, 2012; Prescott-Mathews et al., 1997). The MOA for MTBE-induced liver toxicity, including liver tumors in mice, is unknown, but MTBE has not been found to induce hepatocyte unscheduled DNA synthesis or long-term hepatocyte proliferation (Bird et al., 1997), increase nonfocal hepatocyte DNA synthesis (Moser et al. 1996), show tumor-promoting ability (Moser et al., 1996), or cause hepatic oxidative stress (de Peyster et al., 2008). Although there is no proposed MOA for the development of testicular (Leydig cell) adenoma in rats, hormonal imbalances, including reduced serum estradiol levels, decreased prolactin secretion (Cook et al., 1999), and decreased serum testosterone levels (Williams et al., 2000; de Peyster et al., 2003; Li et al., 2008) have been suggested to be involved. Others have suggested a potential role for inhibition of 5α-reductase. The combined data suggest a disruption in hormone signaling, but the data are insufficient to draw any conclusions on the specific MOA.

Methyl tert-butyl ether is metabolized to tertiary butyl alcohol (TBA), formaldehyde, 2-methyl-1,2-propanediol (MPD), and α-hydroxy isobutyric acid (HBA) (Phillips et al., 2008). Formaldehyde and TBA are the only metabolites studied in enough detail to compare to MTBE toxicity and carcinogenicity. Tertiary butyl alcohol and MTBE exposure have some similarities in their effects on rat kidney, as well as liver toxicity. However, TBA has not been found to induce liver or Leydig cell tumors (NTP, 1995). In addition, TBA exposure also has been associated with some kidney, bladder, and thyroid effects in rodents that have not been observed with MTBE exposure (NTP, 1995). In tissues exposed to exogenous formaldehyde, there is evidence for

genotoxicity observed *in vitro* in many systems with multiple end points and carcinogenesis observed in occupationally exposed humans (IARC, 2012; NTP, 2011). Methyl tert-butyl ether induces tumors in rodents, but the evidence for genotoxicity is not as clear. Genotoxicity studies have found MTBE to be nonmutagenic in Ames assays even using a sensitive strain (TA 102) of bacteria (Kado et al., 1998; McGregor et al., 2005; Vosahlikova et al., 2006), but one study found that MTBE was weakly mutagenic in the absence of S9 and moderately mutagenic in the presence of S9 using TA102, a strain proficient in excision repair (Williams-Hill et al., 1999). It has also been negative in inducing micronuclei in mouse bone marrow (McKee et al. 1997; Kado et al., 1998), Drosophila sex-linked recessive-lethal test, and rat bone marrow cytogenics test (McKee et al., 1997). However, MTBE has been found to induce DNA double-strand breaks and DNA adducts (Du et al., 2005; Yuan et al., 2007; Chen et al., 2008; Sgambato et al., 2009), but it does not appear to affect DNA repair mechanisms (Sgambato et al., 2009).

SOURCES OF EXPOSURE

Methyl tert-butyl ether has primarily been used as a gasoline additive in unleaded gasoline, in the manufacture of isobutene, as a chromatographic eluent, and as a pharmaceutical agent. Exposure to MTBE in humans can occur through inhalation, ingestion, and eye or skin contact (IPCS, 1998).

Occupational

An occupational exposure may occur through inhalation and dermal contact at workplaces where MTBE is produced or used. This includes operations outside of the United States that manufacture or blend gasoline with MTBE, as well as operations that produce isobutene and chromatographic eluents containing MTBE. It has not been used as a fuel additive in the United States since 2005. However, the National Occupational Exposure Survey (NOES), conducted from 1981–1983 by NIOSH, statistically estimated that 4213 male workers and 1783 female workers were potentially exposed to MTBE in the United States prior to the phase out of MTBE (NIOSH, 2010). Mean concentrations ($\mu g/m^3$) of MTBE in air were 1500 for manufacturing workers; 5000 for blending workers; 14,000 for transportation workers; 2600 for distribution workers; 5200 for gasoline station workers; 660 for mechanics; and 61 for professional drivers (Brown, 1997).

Environmental

The general population in the United States may have been exposed to MTBE via inhalation of ambient air, especially during refueling operations prior to 2005, ingestion of contaminated groundwater, and dermal contact with consumer products containing MTBE (HSDB, 2012). Mean concentrations ($\mu g/m^3$) of MTBE in air were 61 for commuters; 30 for other drivers; 390 for gasoline station customers; 4 for manufacturing and blending neighbors; 66 for gasoline station neighbors; and 2.6 for the general public (Brown, 1997). As the use of MTBE in gasoline has been phased out, drinking water contamination is the most likely source of exposure for the general population. Methyl tert-butyl ether can contaminate water through leaking underground gas storage tanks and pipelines, as well as from gasoline spilled on the ground. Older boat engines may also contribute small amounts of MTBE to water sources.

INDUSTRIAL HYGIENE

There is no Occupational Safety and Health Administration (OSHA) or NIOSH standard for MTBE. The American Conference of Governmental Industrial Hygienists (ACGIH) has assigned MTBE a threshold limit value (TLV) of 50 ppm as an 8-h time-weighted average (TWA) concentration based on its carcinogenicity (A3: confirmed animal carcinogen with unknown relevance to humans) (ACGIH, 2010). The ACGIH (2010) recommends excursions in worker exposure levels may exceed three times the TLV–TWA for no more than a total of 30 min during a work day, and under no circumstances should they exceed five times the TLV–TWA, provided that the TLV–TWA is not exceeded.

RISK ASSESSMENTS

The regulatory levels as well as the recommended values determined for MTBE by various government agencies and organizations are listed below.

U.S. EPA

The reference concentration (RfC) for MTBE is $3\,mg/m^3$ (0.774 ppm), and this is based on increased absolute and relative liver and kidney weights and increased severity of spontaneous renal lesions (females), increased prostration (females), and swollen periocular tissues (males and females) in a 24-month inhalation rat study (U.S. EPA, 1993b). The Environmental Protection Agency (EPA) has identified MTBE as a possible human carcinogen, Class C (U.S. EPA, 1995).

ATSDR

The toxicological profile for MTBE was released in August 1996 (ATSDR, 1996). Cancer-effect levels of MTBE were identified for inhalation and oral routes of exposure using animal studies.

WHO

The World Health Organization (WHO) concluded that MTBE should be considered a rodent carcinogen, but the data are insufficient to reach any conclusions about its potential to cause cancer in humans (IPCS, 1998). Therefore, the available data on MTBE are inconclusive and limited in their use for human carcinogenic risk assessment (IPCS, 1998). The International Agency for Research on Cancer (IARC) (IARC, 1999) has classified MTBE in Group 3, not classifiable as to its carcinogenicity to humans, based on limited evidence in experimental animals and inadequate evidence in humans.

NTP

Methyl tert-butyl ether is not included in the National Toxicology Program's (NTP) 12th report on carcinogens (NTP, 2011).

REFERENCES

ACGIH (2010) *TLVs and BEIs. Threshold Limit Values for Chemical Substances and Physical Agents and Biological Exposure Indices*, Cincinnati, OH: American Conference of Governmental Industrial Hygienists, p. 40.

ATSDR (1996) *Toxicological Profile for Methyl tert-Butyl Ether*, Atlanta, GA: U.S. Department of Health and Human Services, Division of Toxicology.

Belpoggi, F., Soffritti, M., Filippini, F., et al. (1997) Results of long-term experimental studies on the carcinogenicity of methyl *tert*-butyl ether. *Ann. N. Y. Acad. Sci.* 837(95):77–95.

Belpoggi, F., Soffritti, M., and Maltoni, C. (1995) Methyl tertiary-butyl ether (MtBE)—a gasoline additive—causes testicular and lymphohaematopoietic cancers in rats. *Toxicol. Ind. Health.* 11(2):119–149.

Belpoggi, F., Soffritti, M., and Maltoni, C. (1998) Pathological characterization of testicular tumors and lymphoma-leukaemias and their precursors observed in Sprague-Dawley rats exposed to methyl tertiary-butyl ether (MTBE). *Eur. J. Oncol.* 3(3): 201–206.

Bengtsson, E.B. and Tarkpea, M. (1983) The acute aquatic toxicity of some substances carries by ships. *Mar. Pollut. Bull.* 14:213–214.

BenKinney, M.T., Barbieri, J.F., Gross, J.S., and Naro, P.A. (1994) Acute toxicity of methyl-tertiary butyl ether (MTBE) to aquatic organisms. *Presented at SETAC, 15th annual meeting, Denver, CO, [Abstract]*.

Bermudez, E., Willson, G., Parkinson, H., and Dodd, D. (2012) Toxicity of methyl tertiary-butyl ether (MTBE) following exposure of Wistar Rats for 13 weeks or one year via drinking water. *J. Appl. Toxicol.* 2012 Sep; 32(9):687–706.

Bevan, C., Tyl, R.W., Neeper-Bradley, T.L., et al. (1997a) Developmental toxicity evaluation of methyl tertiary-butyl ether (MTBE) by inhalation in mice and rabbits. *J. Appl. Toxicol.* 17(Suppl. 1):S21–S29.

Bevan, C., Neeper-Bradley, T.L., and Tyl, R.W. (1997b) Two-generation reproductive toxicity study of methyl tertiary-butyl ether (MTBE) in rats. *J. Appl. Toxicol.* 17(Suppl. 1):S13–S19.

Bird, M.G., Burleigh-Flayer, H.D., Chun, J.S., et al. (1997) Oncogenicity studies of inhaled methyl-tertiary-butyl ether (MTBE) in CD-1 mice and F344 rats. *J. Appl. Toxicol.* 17:S45–S55.

Brown, S.L. (1997) Atmospheric and potable water exposures to methyl tert-butyl ether (MTBE). *Regul. Toxicol. Pharmacol.* 25:256–276.

Burns, K.M. and Melnick, R.L. (2012) MTBE: recent carcinogenicity studies. *Int. J. Occup. Environ. Health* 18(1):66–69.

Cain, W.C., Leaderer, B.P., Ginsberg, G.L., et al. (1996) Acute exposure to low-level methyl tertiary-butyl ether (MTBE): human reactions and pharmacokinetic response. *Inhal. Toxicol.* 8(1):21–48.

Chen, C.S., Hseu, Y.C., Liang, S.H., et al. (2008) Assessment of genotoxicity of methyl-tert-butyl ether, benzene, toluene, ethylbenzene, and xylene to human lymphocytes using comet assay. *J. Hazard Mater.* 153(1–2):351–356.

Clary, J.J. (1997) Methyl tert butyl ether systemic toxicity. *Risk Anal.* 17(6):661–672.

Cook, J.C., Klinefelter, G.R., Hardisty, J.F., et al. (1999) Rodent Leydig cell tumorogenesis: a review of the physiology, pathology, mechanisms, and relevance to human. *Crit. Rev. Toxicol.* 29:169–261.

Daughtrey, W.C., Gill, M.W., Pritts, I.M., et al. (1997) Neurotoxicological evaluation of methyl tertiary-butyl ether in rats. *J. Appl. Toxicol.* 17(Suppl. 1):S57–S64.

de Peyster, A., MacLean, K.J., Stephens, B.A., et al. (2003) Subchronic studies in Sprague-Dawley rats to investigate mechanisms of MTBE-induced Leydig cell cancer. *Toxicol. Sci.* 72:31–42.

de Peyster, A., Rodriguez, Y., Shuto, R., et al. (2008) Effect of oral methyl-t-butyl ether (MTBE) on the male mouse reproductive tract and oxidative stress in liver. *Reprod. Toxicol.* 26(3–4): 246–253.

Dodd, D., Willson, G., Parkinson, H., et al. (2013) Two-year drinking water carcinogenicity study of methyl tertiary-butyl ether (MTBE) in Wistar rats. *J. Appl. Toxicol.* 33:593–606.

Du, H.F., Xu, L.H., Wang, H.F., et al. (2005) Formation of MTBE-DNA adducts in mice measured with accelerator mass spectrometry. *Environ. Toxicol.* 20(4):397–401.

Duffy, J.S., Del Pup, J.A., and Kneiss, J.J. (1992). Toxicological evaluation of methyl tertiary-butyl ether (MTBE); testing performed under TSCA consent agreement. *J. Soil Contam.* 1(1):29–37.

Fiedler, N., Kelly-McNeil, K., Mohr, S., et al. (2000) Controlled human exposure to methyl tertiary butyl ether in gasoline: symptoms, psychophysiologic and neurobehavioral responses of self-reported sensitive persons. *Environ. Health Perspect.* 108(8):753–763.

Geiger, D.L., Call, D.J., and Brooke, L.T. (1988) *Acute Toxicities of Organic Chemicals to Fathead Minnows (*Pimephales promelas*),* vol. V IV, Superior, WI: Center for Lake Superior Environmental Studies, University of Wisconsin.

HSDB (2012) *Hazardous Substances Data Bank: Methyl t-Butyl Ether,* From NLM (National Library of Medicine), Bethesda, MD: National Institutes of Health, U.S. Department of Health and Human Services. Available at http://toxnet.nlm.nih.gov.

IARC (1999) Some chemicals that cause tumours of the kidney or urinary bladder in rodents and some other substances. In: *IARC Monographs on the Evaluation of Carcinogenic Risks to Humans,* vol. 73, Lyon, France: International Agency for Research on Cancer, pp. 339–383.

IARC (2012) Formaldehyde. In: *A Review of Human Carcinogens. Part F: Chemical Agents and Related Occupations,* Lyon, France: International Agency for Research on Cancer, pp. 401–435.

IPCS (1998) Methyl tertiary-butyl ether. In: *Environmental Health Criteria 206,* Geneva, Switzerland: World Health Organization, International Program on Chemical Safety (IPCS).

Johanson, G., Nihlén, A., and Löf, A. (1995) Toxicokinetics and acute effects of MTBE and ETBE in male volunteers. *Toxicol. Lett. (Shannon, Ireland)* 82/83:713–718.

Joseph, P.M. and Weiner, M.G. (2002) Visits to physicians after the oxygenation of gasoline in Philadelphia. *Arch. Environ. Health* 57(2):137–154.

Kado, N.Y., Kusmicky, P.A., Loarca-Piña, G., et al. (1998) Genotoxicity testing of methyl tertiary-butyl ether (MTBE) in the Salmonella microsuspension assay and mouse bone marrow micronucleus test. *Mutat. Res.* 412(2):131–138.

Kane, E.V. and Newton, R. (2010) Occupational exposure to gasoline and the risk of non-Hodgkin lymphoma: a review and meta-analysis of the literature. *Cancer Epidemiol.* 34(5): 516–522.

Li, D., Yuan, C., Gong, Y., et al. (2008) The effects of methyl tert-butyl ether (MTBE) on the male rat reproductive system. *Food Chem. Toxicol.* 46(7):2402–2408.

Lington, A.W., Dodd, D.E., Ridlon, S.A., et al. (1997) Evaluation of 13-week inhalation toxicity study on methyl *t*-butyl ether (MTBE) in Fischer 344 rats. *J. Appl. Toxicol.* 17(Suppl. 1): S37–S44.

Liu, J., Drane, W., Liu, X., and Wu, T. (2009) Examination of the relationships between environmental exposures to volatile organic compounds and biochemical liver tests: application of canonical correlation analysis. *Environ. Res.* 109(2): 193–199.

Mancini, E.R., Steen, A., Rausina, G.A., Wong, D.C.L., Arnold, W.R., Gostomski, F.E., Davies, T., Hockett, J.R., Stubblefield, W.A., Drottar, K.R., and Springer, T.A. (1999). *Preliminary calculations of freshwater and marine water quality criteria for MTBE. Presented at 20th Annual meeting of SETAC, 14–18 November 1999, Philadelphia, PA.*

McGregor, D.B., Cruzan, G., Callander, R.D., May, K., and Banton, M. (2005) The mutagenicity testing of tertiary-butyl alcohol, tertiary-butyl acetate and methyl tertiary-butyl ether in *Salmonella typhimurium. Mutat. Res.* 565:181–189.

McKee, R.H., Vergnes, J.S., Galvin, J.B., et al. (1997) Assessment of the *in vivo* mutagenic potential of methyl tertiary-butyl ether. *J. Appl .Toxicol.* 17(Suppl. 1):S31–S36.

Merck (2006) In: O'Neil, M.J., editor. *The Merck Index: An Encyclopedia of Chemicals, Drugs, and Biologicals*, 14th ed., Rahway, NJ: Merck & Co., Inc.

Mohr, S.N., Fiedler, N., Weisel, C., et al. (1994) Health effects of MTBE among New Jersey garage workers. *Inhal. Toxicol.* 6:553–562.

Moser, G.J., Wong, B.A., Wolf, D.C., et al. (1996) Methyl tertiary butyl ether lacks tumor-promoting activity in *N*-nitrosodiethyl-amine-initiated B6C3F1 female mouse liver. *Carcinogenesis* 17(12):2753–2761.

Nihlén, A., Wålinder, R., Löf, A., et al. (1998) Experimental exposure to methyl tertiary-butyl ether. II. Acute effects in humans. *Toxicol. Appl. Pharmacol.* 148:281–287.

NIOSH (2010) *National Occupational Exposure Survey (NOES) conducted from 1981–1983. Estimated numbers of employees potentially exposed to specific agents by 2-digit standard industrial classification (SIC).* Available at http://www.cdc.gov/noes/.

NTP (National Toxicology Program) (1995) *NTP Technical Report on the Toxicology and Carcinogenesis Studies of t-Butyl Alcohol (CAS No. 75-65-0) in F344/N Rats and B6C3F1 Mice (Drinking Water Studies). (NTPTR436)*, Research Triangle Park, NC: National Toxicology Program.

NTP (2011) *Report on Carcinogens*, 12th ed., Research Triangle Park, NC: U.S. Department of Health and Human Services, Public Health Service, National Toxicology Program.

Paulov, S. (1987) Action of the anti-detonation preparation of tert-butyl methyl ether on the model species *Rana temporaria. Biologia (Bratislava)* 42:185–189.

Phillips, S., Palmer, R.B., and Broady, A. (2008) Epidemiology, toxicokinetics, and health effects of methyl *tert*-butyl ether (MTBE). *J. Med. Toxicol.* 4(2):115–126.

Prah, J.D., Goldsterin, G.M., Devlin, R., et al. (1994) Sensory, symptomatic, inflammatory, and ocular responses to and the metabolism of methyl tertiary butyl ether in a controlled human exposure experiment. *Inhal. Toxicol.* 6:521–538.

Prescott-Mathews, J.S., Poet, T.S., and Borghoff, S.J. (1999) Evaluation of the *in vivo* interaction of methyl tert-butyl ether with alpha-2u globulin in male F-344 rats. *Toxicol. Appl. Pharmacol.* 157(1):60–67.

Prescott-Mathews, J.S., Wolf, D.C., Wong, B.A., et al. (1997) Methyl tert-butyl ether causes α_{2u}-globulin nephropathy and enhanced renal cell proliferation in male F-344 rats. *Toxicol. Appl. Pharmacol.* 142(2):301–314.

Sgambato, A., Iavicoli, I., De Paola, B., Bianchino, G., Boninsegna, A., Bergamaschi, A., Pietroiusti, A., and Cittadini, A. (2009). Differential toxic effects of methyl tertiary butyl ether and tert-butanol on rat fibroblasts *in vitro. Toxicol. Ind. Health* 25:141–151.

Tepper, J.S., Jackson, M.C., and McGee, J.M. (1994) Estimation or respiratory irritancy from inhaled methyl tertiary butyl ether in mice. *Inhal. Toxicol.* 6:563–569.

Thistle, J.L. (1992) Direct contact dissolution therapy. *Baillieres Clin. Gastroenterol.* 6:715–725.

U.S. EPA (1988) Fuel and fuel additives: waiver decision. Notice. *Federal Register* 44:33846–33847.

U.S. EPA (1991) *α_{2u}-Globulin Association with Chemically Induced Renal Toxicity and Neoplasia in the Male Rat*, EPA/625/3-91/019 F. Risk Assessment Forum, Washington, DC: U.S. Environmental Protection Agency, Office of Research and Development.

U.S. EPA (1993a) *Technical Information Review. Methyl tertiary Butyl Ether*, (CAS No. 1634-04-4). Washington, DC: U.S. Environmental Protection Agency, Office of Pollution Prevention and Toxics.

U.S. EPA (1993b) *Integrated Risk Information System (IRIS) Online. Summary for Methyl tert- butyl ether. IRIS, U.S. EPA, Arlington, VA, Retrieved 6/11/13.*

U.S. EPA (1994a) *Chemical Summary for Methyl-tert-butyl Ether*, Washington, DC: U.S. Environmental Protection Agency, Office of Pollution Prevention and Toxics.

U.S. EPA (1994b) *1992 Toxics Release Inventory*, EPA 745-R-94-001. Washington, DC: U.S. Environmental Protection Agency, Office of Pollution Prevention and Toxics.

U.S. EPA (1995) *Drinking Water Regulations and Health Advisories*, Washington, DC: U.S. Environmental Protection Agency, Office of Water.

Veith, G.D., Call, D.J., and Brooke, L.T. (1983) Estimating the acute toxicity of narcotic industrial chemicals to fathead minnows. In: Bishop, W.E., Cardwell, R.D., and Heidolph, B.B., editors. *Aquatic Toxicology and Hazard Assessment: Sixth Symposium*, ASTM report STP 80/2. Philadelphia, PA: American Society for Testing and Materials, pp. 90–97.

Vosahlikova, M., Cajthaml, T., Demnerova, K., and Pazlarova, J. (2006) Effect of methyl tert-butyl ether in standard tests for mutagenicity and environmental toxicity. *Environ. Toxicol.* 21(6):599–605.

Williams, T.M., Cattley, R.C., and Borghoff, S.J. (2000) Alterations in endocrine responses in male Sprague-Dawley rats following oral administration of methyl *tert*-butyl ether. *Toxicol. Sci.* 54:168–176.

Williams-Hill, D., Spears, C.P., Prakash, S., et al. (1999) Mutagenicity studies of methyl-*tert*-butyl ether using the Ames tester strain TA102. *Mutat. Res.* 446(1):15 21.

Yuan, Y., Wang, H.F., Sun, H.F., et al. (2007) Adduction of DNA with MTBE and TBA in mice studied by accelerator mass spectrometry. *Environ. Toxicol.* 22(6):630–635.

63

ETHYL TERTIARY BUTYL ETHER (ETBE)

KEITH D. SALAZAR

First aid: Respiratory: Seek fresh air and medical attention, if necessary; eyes: flush with water; skin: rinse with water and remove clothing; ingestion: rinse mouth and seek medical attention

Target organ(s): Kidney, liver, skin, lungs, eyes

Occupational exposure limits: ACGIH TLV: 5 ppm ($20.9 \, mg/m^3$)

Reference values: RIVM TDI: 0.25 mg/kg/day; RIVM TCA: $1.9 \, mg/m^3$

Risk phrases: 11-20-36/37-38-67

Highly flammable; harmful by inhalation; irritating to the eyes and respiratory system; irritating to the skin; vapors may cause drowsiness and dizziness

Safety phrases: 16-23-29-33

Keep away from sources of ignition. No smoking. Do not breathe gas/fumes/vapor/spray. Do not empty into drains. Take precautionary measures against static discharges

BACKGROUND AND USES

Ethyl tertiary butyl ether (CASRN 637-92-3) is a colorless liquid also referred to by the synonyms ethyl tert-butyl ether, ethyl tert-butyl oxide, methyl-2-ethoxypropane, tert-butyl ethyl ether, 2-ethoxy-2-methylpropane, or 2-methyl-2-ethoxypropane; and the acronym ETBE. Ethyl tertiary butyl ether is synthesized from ethanol and isobutene and is used primarily as an oxygenate that is added to gasoline to improve the automobile exhaust quality by reducing the ozone and carbon monoxide emissions (HSDB, 2012). Ethyl tertiary butyl ether has similar utility compared to another widely used oxygenate, methyl tertiary butyl ether (MTBE), and thus is a potential replacement for MTBE. Usage of ETBE as a fuel additive has halted in the United States, falling from 2 to 4 million barrels per month in 2005 to 0 barrels in 2006 (DOE, 2007). Ethyl tertiary butyl ether continues to be used widely in Europe (EFOA, 2010). Since ETBE is used almost exclusively in fuels, contamination of groundwater as a result of spillage or leakage of the underground storage tanks is a major source of environmental release.

PHYSICAL AND CHEMICAL PROPERTIES

Ethyl tertiary butyl ether is a liquid at a temperature range of −94 to 72.6 °C. It is soluble in ethanol, ethyl ether, and water. Ethyl tertiary butyl ether has a strong, highly objectionable odor and taste at relatively low concentrations. This chemical is highly flammable and reacts with strong oxidizing agents. Ethyl tertiary butyl ether is stable when stored at room temperature in tightly closed containers. Table 63.1 includes other additional physiochemical properties.

ENVIRONMENTAL FATE AND BIOACCUMULATION

Due to the usage of ETBE as a fuel additive, ETBE may be released into air, soil, and water. Ethyl tertiary butyl ether will exist as a vapor at 25 °C due to its vapor pressure of 124 mmHg and is estimated to have a half-life of 2 days. Upon release into the soil, ETBE is anticipated to have a high mobility based on the high soil organic carbon–water partitioning coefficient of 9–160. Once in the water, ETBE is not

Hamilton & Hardy's Industrial Toxicology, Sixth Edition. Edited by Raymond D. Harbison, Marie M. Bourgeois, and Giffe T. Johnson.

TABLE 63.1 Chemical and Physical Properties of Ethyl Tertiary Butyl Ether

Characteristic	Property
Chemical name	Ethyl tertiary butyl ether
Synonyms	Ethyl tert-butyl ether, ethyl tert-butyl oxide, methyl-2-ethoxypropane, tert-butyl ethyl ether, ETBE
CASRN	637-92-3
Chemical formula	$C_6H_{14}O$
Molecular weight	102.17
Melting point	$-94\,°C$
Boiling point	$72.6\,°C$
Vapor pressure	124 mmHg at the rate $25\,°C$
Density	$0.73–0.74\,g/cm^3$ at the rate $25\,°C$
Water solubility	12,000 mg/l at the rate $20\,°C$
Octanol/water partition coefficient (log k_{ow})	1.92
Henry's law constant	2.7×10^{-3} atm-m^3/mol at the rate $25\,°C$
Odor detection threshold	0.013 ppm ($0.054\,mg/m^3$)
Taste detection threshold (in water)	0.047 ppm (47 µg/l)
Conversion factors	1 ppm $= 4.18\,mg/m^3$ 1 $mg/m^3 = 0.24$ ppm 1 $mg/m^3 = 102,180$ mmol/l

predicted to adsorb onto suspended particles and is likely to resist biodegradation (Deeb et al., 2001). Based on the Henry's law constant, ETBE is likely to be volatized from the surface of the water (HSDB, 2012). A volatilization half-life of 3 h to 4 days is anticipated from water solutions. A bioconcentration factor (BCF) of 9 was estimated for ETBE to accumulate in fish, suggesting a relatively low propensity for aquatic bioaccumulation (HSDB, 2012).

ECOTOXICOLOGY

Few data are available that examine the effects of ETBE on wildlife. One study in sheepshead minnow determined that the LC_{50} following a 96-h exposure is >2500 mg/l (Boeri et al., 1994a). Another study in the shrimp species *Mysidopsis bahia* demonstrated an EC_{50} of 37 mg/l after a 96-h exposure (Boeri et al., 1994b). As mentioned above, ETBE has a low bioaccumulation potential. Altogether, the limited available data do not suggest that ETBE has high toxicity for aquatic organisms.

MAMMALIAN TOXICOLOGY

Acute Effects

Marsh and Leake (1950) reported that the AC_{50} for anesthetizing half of the white mice within 30 min was 0.7 mmol/l,

and the median lethal concentration was 1.2 mmol/l. In CD rats, no rats died after nose-only exposure at 5.88 mg/l (IIT research, 1989). When rats were treated by single-dose gavage, no mortality or abnormal physical signs were observed at concentrations up to 5000 mg/kg (MB Research, 1988a). Dermal exposure studies in rabbits did not report any lethality at concentrations up to 2 g/kg but irritation, edema, erythema, and eschar formation were observed at the application site (MB research, 1988b; IIT research, 1989). Finally, acute ocular exposure in New Zealand rabbits induced conjunctival irritation, which subsided after 7 days (MB research, 1988c); washing the eye for 20–30 s after exposure greatly reduced the iritis and conjunctival irritation. These data suggest potential irritating effects following acute dermal and ocular exposure.

Chronic Effects

Toxicity has not been assessed in humans chronically or occupationally exposed to ETBE. Three separate studies have been performed in rats to identify potential adverse effects following chronic exposure to ETBE.

Oral Exposure A single published study examined the effects of chronic drinking water exposure to ETBE (Suzuki et al., 2012). Male and female F334/N rats were exposed to ETBE in drinking water for 2 years at daily concentrations of 0, 28, 121, and 542 mg/kg in males, and 0, 46, 171, and 506 mg/kg in females. Clinical signs were monitored daily, and upon termination of the study, blood biochemistry, histopathology, and organ weights were examined. No treatment-related mortality was observed.

Both relative and absolute kidney weights were increased in males and females after 2 years; however, weights were increased to a greater extent in females than males. Relative liver weights were dose-responsively increased in males but unaffected in females. Urinalysis and blood biochemistry did not demonstrate any dose-responsive changes associated with the altered organ weights. Body weights were dose-responsively reduced in both males and females by as much as 17% in females and 10% in males at the highest dose. Water consumption was reduced in males and females throughout the experiment. Overall, water consumption was reduced more in females with the largest reduction (22%) occurring at the highest dose.

The carcinogenicity of ETBE following oral exposure has produced differing results depending on the method of exposure. No treatment-related neoplasms were observed following a 2-year drinking water study (Suzuki et al., 2012). In contrast, a 2-year oral gavage study in male and female Sprague Dawley rats to daily ETBE concentrations of 0, 250, and 1000 mg/kg significantly increased the mouth epithelium tumors at high dose in males and uterine malignancies at only the low dose in females (Maltoni et al., 1999).

Inhalation Exposure Inhalation studies elicit similar effects as observed in the oral exposures; however, data for comparison are available from a single unpublished study (Saito et al., 2013). Male and female Sprague Dawley rats were inhalationally exposed to 2090, 6270, and 20,900 mg/m^3 for 6 h/day, 5 days/week for 2 years. Clinical signs were monitored daily and identical analyses were performed as by Suzuki et al. (2012). After 2 years, survival was significantly reduced in females from 76% in controls to 60% in the two highest dose groups (Saito et al., 2013). Male survival was similarly reduced from 88% in controls compared to 60% at the highest exposure concentration.

Organ weights were increased to a greater extent in the inhalational study compared to the oral exposure. The largest organ weight changes occurred in the kidneys and liver. Relative kidney weight was dose-dependently increased in males up to 66% at the highest dose and 51% in females. Dose-responsive increases in the relative liver weights were observed in males (49% at 20,900 mg/m^3) and in females (30% at 20,900 mg/m^3). Finally, body weights were significantly reduced at all exposure concentrations with maximal reductions occurring at the highest doses in males (-26% of control) and in females (-23% of control).

Statistically significant carcinogenic effects were only observed in males. One hepatocellular carcinoma and nine hepatocellular adenomas were observed in the highest dose; statistically significant compared to controls. No additional carcinogenic effects were observed in any other organs.

Chemical Pathology

Pathological exposure data for ETBE are restricted to animal studies, specifically rats and mice. Treatment-related non-neoplastic lesions were most prominent in the kidneys and livers of males following both oral and inhalation exposures. Ethyl tertiary butyl ether exposure induced similar, but smaller, effects in females. Inhalation exposure elicited more adverse pathology than oral exposures; however, there are no published data comparing the calculated internal dose of ETBE between oral and inhalation exposures.

Following inhalation exposure, a number of pathologies related to chronic progressive nephropathy were observed with the most severe nephropathy occurring in males. A 2-year exposure increased the severity of nephropathy in both male and female rats (Saito et al., 2013). The observed nephropathy was characterized by glomeruli sclerosis, tubular basement membrane thickening, hyaline casts, tubular regeneration, and interstitial fibrosis. Shorter 13-week exposure data are supportive of the 2-year findings. The incidence and severity of hyaline droplets was increased in males but not in females (Medinsky et al., 1999; JPEC, 2008a). Proximal tubular proliferation and the incidence of papilla mineralization were also only observed in males (Saito et al., 2013; Medinsky et al., 1999). Bone marrow congestion was increased without corresponding changes in hematopoietic cell populations in one 90-day study and was not concluded to be clinically relevant (Medinsky et al., 1999); however, this result was not observed by other researchers (JPEC, 2008a).

Oral exposure induced fewer nephropathies as compared with inhalation studies. A 2-year exposure did not affect the incidence or severity of nephropathy in either males or female rats. A shorter 26-week gavage exposure increased the incidence of hyaline droplets (JPEC, 2008b) and papillary mineralization was induced following a 2-year drinking water exposure (Suzuki et al., 2012); however, these effects were only observed in males.

Liver pathologies were confined to two primary effects: centrilobular hypertrophy and acidophilic and basophilic foci. Centrilobular hypertrophy incidence was increased in males and females after a 26-week gavage exposure (JPEC, 2008b) but was not observed in a longer 2-year exposure (Suzuki et al., 2012). Similarly, a shorter 13-week inhalation exposure also increased centrilobular hypertrophy in male and female mice (Medinsky et al., 1999; Weng et al., 2012) and rats (JPEC, 2008a), but a 2-year inhalation exposure did not induce hypertrophy (Saito et al., 2013). In addition, only the inhalation exposure increased the preneoplastic lesions in the liver by increasing the incidence of acidophilic and basophilic foci in male rats (Saito et al., 2013).

Mode of Action

Available *in vivo* and *in vitro* mutagenicity studies do not provide evidence that ETBE is genotoxic (JPEC, 2007a, 2007b, 2007c, 2007d; Zeiger et al., 1992; Vergnes, 1995; Vergnes and Kubena, 1995).

Kidney Chronic progressive nephropathy, which was observed in female rats, has an unknown etiology; however, several studies have examined the possible mode of action underlying the nephropathy in male rats. Alpha$_2$u-globulin is derived from the hepatic synthesis and can be chemically induced in males to accumulate in the proximal tubule as a result of impaired renal catabolism. This accumulation leads to chronic proliferation of the renal tubule epithelium and nephrotoxicity (Hard et al., 2009). The accumulation of α_2u-globulin in hyaline droplets following ETBE exposure was examined in several studies.

Hyaline droplets were increased following oral and inhalation exposure in males only (JPEC, 2008a, 2008b). In addition, α_2u-globulin was confirmed in the hyaline droplets from multiple studies following both oral gavage and inhalational exposure (JPEC, 2008a, 2008b; Medinsky et al., 1999). Finally, mineralization of the renal papilla and urothelial hyperplasia of the renal pelvis was noted in the male rats (Suzuki et al., 2012; Saito et al., 2013). Although these data are suggestive for α_2u-globulin mode of action for nephropathy, several additional data could strengthen this hypothesized mechanism. For instance, examining the affinity of ETBE to bind to α_2u-globulin, identifying hyperplasia foci in the proximal tubules, and establishing the evidence of necrosis in the tubules could all serve to bolster the hypothesis. It should also be noted that a primary metabolite of ETBE in animals is tertiary butyl alcohol (TBA) (Dekant et al., 2001). Although TBA is known to exert kidney effects similar to ETBE, the relative contribution of ETBE and TBA in the overall toxicity observed following ETBE exposure has not been determined.

Liver   Only a single study has examined the putative mode of action for the increase in centrilobular hypertrophy and adenomas in male rats. Oral ETBE exposure increased a number of CAR and PXR nuclear receptors that are upstream of *peroxisome proliferator-activated receptors* (PPAR) activation, which is a possible mechanism for liver tumor and hepatocyte proliferation (JPEC, 2012). However, no direct measurement of PPAR activation or specific manipulation of the pathway has been performed for a definitive correlation of ETBE-induced PPAR activation.

SOURCES OF EXPOSURE

Exposure to ETBE occurs through oral, inhalation, or dermal routes. Although ETBE production has essentially ceased in the United States, ETBE is presently used as a fuel additive in Europe and Japan (Eitaki et al., 2011).

Occupational

Exposure to ETBE can occur in occupational settings that either manufacture or use ETBE. A study monitoring the exposure in gasoline tanker truck drivers and fueling-station employees whose primary route of exposure is inhalation, measured a time-weighted average (TWA) of 0.08 parts per million (ppm) in the gas station attendees and 0.04 ppm in the tanker drivers (Eitaki et al., 2011). Airborne concentrations of ETBE around the fueling pumps were 4.12 ppm (Eitaki et al., 2011).

Environmental

Groundwater contamination represents the most likely situation for the widespread oral exposure. Analysis of an aquifer beneath a gas station in the Rhone area of France found concentrations as high as 300 mg/l of ETBE in the water (Fayolle-Guichard et al., 2012). The distribution of the ETBE plume was much larger than other volatile organic compounds found in the spill, likely due to ETBE's high solubility in water and lower biodegradation (Fayolle-Guichard et al., 2012). No data are available on vapor intrusion from the contaminated groundwater. It should be noted that ETBE has taste and odor detection thresholds in water of 47 µg/l and 49 µg/l, respectively, which would make drinking water unpalatable to humans at these concentrations (Vetrano, 1993).

INDUSTRIAL HYGIENE

The American Conference of Governmental Industrial Hygienists (ACGIH lists the occupational exposure limit for ETBE to have a threshold limit value (TLV) of 5 ppm as a TWA (ACGIH, 2001). No other exposure limits have been issued by National Institute for Occupational Safety and Health (NIOSH) or Occupational Safety and Health Administration (OSHA). The International Programme on Chemical Safety (IPCS) recommends using protective gloves, eye protection, and breathing protection when handling ETBE to minimize the eye, skin, and inhalation exposures. In the case of eye or skin exposure, the affected area should be flushed with water and the contaminated clothing should be removed. If drowsiness, coughing, or unconsciousness occurs, it is recommended to seek fresh air. Since ETBE is highly flammable, maintain caution when handling and do not expose the working area to ignition sources.

RISK ASSESSMENTS

The National Institute for Public Health and the Environment (RIVM) of the Netherlands has derived noncancer values for oral and inhalation exposures. The oral noncancer tolerable daily intake (TDI) of 0.25 mg/kg/day is based on the decreased body weight, from a 2-generation rat study (Baars, 2009). The noncancer inhalation tolerable concentration in air (TCA) is 1.9 mg/m^3 (0.45 ppm) based on the increased liver and kidney weights and bone marrow congestion. The RIVM determined that there were insufficient data to evaluate the carcinogenicity of ETBE.

In 2001, the aforementioned ACGIH recommended a TLV–TWA for ETBE of 5 ppm (21 mg/m^3) based on the no-observed adverse effect levels (NOAELs) derived from short-term exposures in rodents and a 2-h human toxicokinetic study (ACGIH, 2001). No other government or independently peer-reviewed assessments for ETBE have been conducted.

REFERENCES

ACGIH (2001) *Documentation of the Threshold Limit Values and Biological Exposure Indices for Ethyl Tert-butyl Ether*, 7th ed., pp. 5 Cincinnati, OH: American Conference of Governmental Industrial Hygienists.

Baars, A.J. and Tiesjema, B. (2009) *Re-evaluation of Human-Toxicological Maximum Permissible Risk Levels*, RIVM Report 711701092, Bilthoven, the Netherlands: National Institute for Public Health and the Environment.

Boeri, R.L., Kowalski, P.L., and Ward, T.J. (1994a) *Acute Toxicity of Ethyl Tertiary Butyl Ether to the Sheepshead Minnow*, Cyprinodon variegates, Marblehead, Mass: T.R. Wilbury Laboratoreis, Inc. Study no. 425-AR, pp. 1–17.

Boeri, R.L., Kowalski, P.L., and Ward, T.J. (1994b) Acute toxicity of ethyl tertiary butyl ether to the mysid. In: *Mysidopsis Bahia*, Marblehead, Mass.: T.R. Wilbury Laboratories, Inc., Study no.426-AR, pp. 1–21.

Deeb, R.A., Hu, H.Y., Hanson, J.R., Scow, K.M., and Alvarez-Cohen, L. (2001) Substrate interactions in BTEX and MTBE mixtures by an MTBE-degrading isolate. *Environ. Sci. Technol.* 35(2):312–317.

Dekant, W., Bernauer, U., Rosner, E., and Amberg, A. (2001a) Biotransformation of MTBE, ETBE, and TAME after inhalation or ingestion in rats and humans. *Health Eff. Inst. Res. Rep.* 102:29–71.

U.S. DOE (Department of Energy) (2007) *Weekly Imports and Exports.* Available at http://tonto.eia.doe.gov/dnav/pet/pet_move_wkly_dc_nus-z00_mbblpd_w.htm.

EFOA (European Fuel Oxygenates Association) (2010) *Markets.* Available at http://www.efoa.eu/en/markets.aspx.

Eitaki, Y., Kawai, T., and Omae, K. (2011). Exposure assessment of ETBE in gas station workers and gasoline tanker truck drivers. *J. Occup. Health* 53(6):423–431.

Fayolle-Guichard, F., Durand, J., Cheucle, M., Rosell, M., Michelland, R.J., Tracol, J.P., Le Roux, F., Grundman, G., Atteia, O., Richnow, H.H., Dumestre, A., and Benoit, Y. (2012) Study of an aquifer contaminated by ethyl tert-butyl ether (ETBE): site characterization and on-site bioremediation. *J. Hazard Mater.* 201–202:236–243.

Hard, G.C., Johnson, K.J., and Cohen, S.M. (2009). A comparison of rat chronic progressive nephropathy with human renal disease-implications for human risk assessment. *Crit. Rev. Toxicol.* 39(4):332–346.

HSDB (2012) Ethyl tert-butyl ether. In: *Hazardous Substances Data Bank*, From NLM (National Library of Medicine); National Institutes of Health, Bethesda, MD: U.S. Department of Health and Human Services. Available at http://toxnet.nlm.nih.gov.

IIT Research Institute (Illinois Institute of Technology Research Institute) (1989) *Acute dermal toxicity study of ethyl tert-butyl ether (ETBE) in rabbits. IIT Research Institute, Life Sciences Research under contract to Amoco Corporation, Chicago, IL; Study No. 1495. Unpublished report.*

JPEC (Japan Petroleum Energy Center) (2007a) *Micronucleus test of ETBE using bone marrow of rats of the "13-week toxicity study of 2-ethoxy-2-methylpropane in F344 rats (drinking water study) [preliminary carcinogenicity study]." Japan Industrial Safety and Health Association. Japan Bioassay Research Center. Study No. 7046. June 29, 2007. Unpublished report.*

JPEC (Japan Petroleum Energy Center) (2007b) *Micronucleus test of ETBE using bone marrow of rats of the "13-week toxicity study of 2-ethoxy-2-methylpropane in F344 rats (inhalation study) [preliminary carcinogenicity study]." Japan Industrial Safety and Health Association. Japan Bioassay Research Center. Study No. 7047. June 29, 2007. Unpublished report.*

JPEC (Japan Petroleum Energy Center) (2007c) *Micronucleus test of 2-ethoxy-2-methylpropane (ETBE) using bone marrow of rats administered ETBE by gavage. Japan Industrial Safety and Health Association. Japan Bioassay Research Center. Study No. 7049. June 29, 2007. Unpublished report.*

JPEC (Japan Petroleum Energy Center) (2007d) *Micronucleus test of 2-ethoxy-2-methylpropane (ETBE) using bone marrow of rats administered ETBE intraperitoneally. Japan Industrial Safety and Health Association. Japan Bioassay Research Center. Study No. 7048. June 29, 2007. Unpublished report.*

JPEC (Japan Petroleum Energy Center) (2008a) *A 90-day repeat dose toxicity study of ETBE by whole-body inhalation exposure in rats. Mitsubishi Chemical Safety Institute Ltd. March, 2008. Study No. B061829. Unpublished report.*

JPEC (Japan Petroleum Energy Center) (2008b) *A 180-day repeat dose oral toxicity study of ETBE in rats. Hita Laboratory, Chemicals Evaluation and Research Institute (CERI), Japan. March, 2008. Study No. D19-0002. Unpublished report.*

JPEC (Japan Petroleum Energy Center) 2012. *Investigation of the Mechanisms of Ethyl tertiary-butyl ether (ETBE) carcinogenicity in the liver of F344 rats- Transmission Electron Microscopic Examination. Japan Industrial Safety and Health Association, Japan Bioassay Research Center. September 7, 2012. Study No. 12138. Unpublished report.*

Maltoni, C., Belpoggi, F., Soffritti, M., and Minardi, F. (1999) Comprehensive long-term experimental project of carcinogenicity bioassays on gasoline oxygenated additives: plan and first report of results from the study on ethyl-tertiary-butyl ether (ETBE). *Eur. J. Oncol.* 4:493–508.

Marsh, D.F. and Leake, C.D. (1950) Comparative anesthetic activity of the aliphatic ethers. *Anesthesiology* 11:455–463.

MB Research Laboratories, Inc., (Millennium Bioresearch Research Laboratories) (1988a) *Single dose oral toxicity in rats/LD$_{50}$ in rats. MB Research Laboratories, Inc. under contract to ARCO Chemical Company, Spinnerstown, PA; Laboratory Project ID MB 88–9137 A. Unpublished report.*

MB Research Laboratories, Inc., (Millennium Bioresearch Research Laboratories) (1988b) *Acute dermal toxicity in rabbits/LD$_{50}$ in rabbits. MB Research Laboratories, Inc. under contract to ARCO Chemical Company, Spinnerstown, PA; Laboratory Project ID MB 88–9107 B. Unpublished report.*

MB Research Laboratories, Inc., (Millennium Bioresearch Research Laboratories) (1988c) *Eye irritation in rabbits. MB Research Laboratories, Inc. under contract to ARCO Chemical Company, Spinnerstown, PA; Laboratory Project ID MB 88–9107 D. Unpublished report.*

Medinsky, M.A., Wolf, D.C., Cattley, R.C., Wong, B., Janszen, D.B., Farris, G.M., Wright, G.A., and Bond, J.A. (1999) Effects of a thirteen-week inhalation exposure to ethyl tertiary butyl ether on Fischer-344 rats and CD-1 mice. *Toxicol. Sci.* 51(1):108–118.

Saito, A., Sasaki, T., Kasai, T., Katagiri, T., Nishizawa, T., Noguchi, T., Aiso, S., Nagano, K., and Fukushima, S. (2013)

Hepatotumorigenicity of ethyl tertiary-butyl ether with 2-year inhalation exposure in F344 rats. *Arch. Toxicol.* 87(5):905–915.

Suzuki, M., Yamazaki, K., Kano, H., Aiso, S., Nagano, K., and Fukushima, S. (2012) No carcinogenicity of ethyl tertiary-butyl ether by 2-year oral administration in rats. *J. Toxicol. Sci.* 37(6):1239–1246.

Vergnes, J.S. (1995) *Ethyl tertiary butyl ether: in vitro chromosome aberrations assay in Chinese hamster ovary cells. Bush Run Research Center, Union Carbide Corporation under contract to ARCO Chemical Company, Export, PA; Laboratory Project ID 94N1425. Unpublished report.*

Vergnes, J.S. and Kubena, M.F. (1995) *Ethyl tertiary butyl ether: mutagenic potential in the CHO/HGPRT forward mutation assay. Bush Run Research Center, Union Carbide Corporation under contract to ARCO Chemical Company, Export, PA; Laboratory Project ID 94N1424. Unpublished report.*

Vetrano, K.M. (1993) *Final report to ARCO Chemical Company on the odor and taste threshold studies performed with methyl tertiary-butyl ether (MTBE) and ethyl tertiary-butyl ether (ETBE). TRC Environmental Corporation under contract to ARCO Chemical Company, Windsor, CT; Project no. 13442-M31. Unpublished report.*

Weng, Z., Suda, M., Ohtani, K., Mei, N., Kawamoto, T., Nakajima, T., and Wang, R.S. (2012) Differential genotoxic effects of subchronic exposure to ethyl tertiary butyl ether in the livers of Aldh2 knockout and wild-type mice. *Arch. Toxicol.* 86(4):675–682.

Zeiger, E., Anderson, B., Haworth, S., Lawlor, T., and Mortelmans, K. (1992) Salmonella mutagenicity tests: V. Results from the testing of 311 chemicals. *Environ. Mol. Mutagen.* 19(Suppl. 21):2–141.

64

tert-BUTYL ALCOHOL

Janice S. Lee

First aid: Eye: Irrigate immediately; skin: flush with water promptly; breathing: respiratory support; swallow: medical attention immediately

Target organ(s): Eyes, skin, respiratory system, central nervous system

Occupational exposure limits: OSHA PEL: TWA 100 ppm (300 mg/m^3), STEL 150 ppm (450 mg/m^3); NIOSH REL: TWA 100 ppm (300 mg/m^3), STEL 150 ppm (450 mg/m^3); and ACGIH TLV: TWA 100 (303 mg/m^3), STEL 150 ppm (455 mg/m^3)

Reference values: Not available

Risk phrases: 11-20-36/37

Highly flammable

Harmful by inhalation

Irritating to eyes and respiratory system

Safety phrases: 9-16-46

Keep the container in a well-ventilated place. Keep away from sources of ignition—no smoking. If swallowed, seek medical advice immediately and show the container or label

BACKGROUND AND USES

tert-Butyl alcohol (CASRN 75-65-0; NCI-C55367) is a branched-chain alcohol having the following synonyms: *tert*-butanol, *tertiary*-butanol; *t*- or *tertiary* butyl alcohol; *t*-butyl hydroxide; 2-methyl-2-propanol; 1,1-dimethyl ethanol; trimethyl methanol, trimethyl carbinol, and TBA. *tert*-Butyl alcohol is used as a solvent (e.g., for paints, lacquers, and varnishes); as a denaturant for ethanol and several other alcohols; as an octane booster in gasoline; as a dehydrating agent; as a chemical intermediate in the manufacturing of methyl methacrylate; and in the manufacturing of flotation agents, fruit essences, and perfumes. As a result of its manufacture and use, *tert*-butyl alcohol may be released into the environment through various waste streams directly (IPCS, 1987a; NIOSH, 1992; HSDB, 2007).

Several production methods exist for *tert*-butyl alcohol (HSDB, 2007). *tert*-Butyl alcohol has been prepared from the reaction of acetyl chloride with dimethylzinc, by the catalytic hydration of isobutylene, by the reduction of *tert*-butyl hydroperoxide, by the absorption of isobutene, from cracking petroleum or natural gas, and from sulfuric acid with subsequent hydrolysis by steam (Chen, 2005). It is usually purified by distillation. *tert*-Butyl alcohol is produced as a by-product from the isobutane oxidation process for producing propylene oxide and from the oxirane process for manufacturing propylene oxide. *tert*-Butyl alcohol is also a likely degradation product of methyl *tert*-butyl ether (MTBE) and has been detected in MTBE-contaminated wells (HSDB, 2007).

PHYSICAL AND CHEMICAL PROPERTIES

tert-Butyl alcohol is a white crystalline solid or colorless liquid (above 77 °F) with a camphor-like odor (IPCS, 1987a; NIOSH, 2005). It is soluble in water and miscible with alcohol, ether, and other organic solvents (IPCS, 1987a). It is highly flammable and easily ignited by heat, sparks, or flames; vapors may form explosive mixtures with air. Fire and explosion may result from contact with oxidizing agents, strong mineral acids, or strong hydrochloric acid (NIOSH,

Hamilton & Hardy's Industrial Toxicology, Sixth Edition. Edited by Raymond D. Harbison, Marie M. Bourgeois, and Giffe T. Johnson.
© 2015 John Wiley & Sons, Inc. Published 2015 by John Wiley & Sons, Inc.

TABLE 64.1 Chemical and Physical Properties of *tert*-Butyl Alcohol

Characteristic	Information
Chemical name	*tert*-Butyl Alcohol
Synonyms/trade names	*tert*-butanol, tertiary-butanol; t-, or tertiary butyl alcohol; t-butyl hydroxide; 2-methyl-2-propanol; 1,1-dimethylethanol; trimethyl methanol; trimethyl carbinol; TBA; NCI-C55367
Chemical formula	$C_4H_{10}O$
CASRN	75-65-0
Molecular weight	74.12
Melting point	25.7 °C
Boiling point	82.41 °C
Vapor pressure	40.7 mmHg at 25 °C
Density/specific gravity	0.78581
Flashpoint	11 °C (closed cup)
Water solubility at 25 °C	1×10^6 mg/l
Explosion limits	2.4–8%
Octanol/water partition coefficient (log K_{OW})	0.35
Henry's law constant	9.05×10^{-6} atm-m^3/mole
Odor threshold	219 mg/m^3
Conversion factors	1 ppm = 3.082 mg/m^3 1 mg/m^3 = 0.324 ppm

1992). Selected chemical and physical properties of *tert*-butyl alcohol are presented in Table 64.1.

ENVIRONMENTAL FATE AND BIOACCUMULATION

Based upon its log K_{OW}, *tert*-butyl alcohol is expected to have very high mobility if released into the soil. If released to air, its vapor pressure indicates that it exists solely as a vapor in the ambient atmosphere. If released into water, it is not expected to adsorb to suspended solids and sediment in water. Volatilization from moist soil surfaces and water surfaces is expected to be an important environmental fate process based upon Henry's law constant. Biodegradation is expected to occur at a slower rate for *tert*-butyl alcohol than for primary or secondary alcohols; furthermore, it does not bioaccumulate (IPCS, 1987a).

ECOTOXICOLOGY

Quantitative information on *tert*-butyl alcohol levels in the environment is not available. As *tert*-butyl alcohol is biodegradable, its substantial concentrations are only likely to occur locally when there is a major spill. *tert*-Butyl alcohol does not bioaccumulate and is not known to be toxic to ecosystems.

MAMMALIAN TOXICOLOGY

Acute Effects

The National Institute of Occupational Safety and Health (NIOSH) considers *tert*-butyl alcohol to be a skin irritant; slight skin irritation has been reported in humans following acute dermal exposure (IPCS, 1987a; NIOSH, 1992; ACGIH, 2001). In addition, NIOSH reports acute exposure to *tert*-butyl alcohol to potentially cause other symptoms, including ocular, nasal, and throat irritation; headache; nausea; fatigue; and dizziness (NIOSH, 1992).

Chronic Effects

No published data are available regarding the health effects of *tert*-butyl alcohol on humans exposed orally or by inhalation for chronic durations. However, there are chronic animal studies for oral exposures to *tert*-butyl alcohol.

Chronic Oral Exposure Studies The National Toxicology Program (NTP) (Cirvello et al., 1995; NTP, 1995) conducted a 2-year drinking water study of chronic toxicity and carcinogenesis in F344/N rats and B6C3F1 mice (60/sex/species/group) exposed to *tert*-butyl alcohol in drinking water. Male rats were exposed to 0, 1.25, 2.5, or 5 mg/ml for approximate doses of 0, 90, 200, or 420 mg/kg/day. Female rats were exposed to 0, 2.5, 5, or 10 mg/ml for approximate doses of 0, 180, 330, or 650 mg/kg/day. Mice were exposed to 0, 5, 10, or 20 mg/ml for approximate doses of 0, 540, 1040, or 2070 mg/kg/day in males and 0, 510, 1020, or 2110 mg/kg/day in females. Animals were observed twice daily. Body weights and clinical observations were recorded weekly for the first 13 weeks and monthly thereafter, as well as at interim evaluation (15 months) and at termination. Water consumption was recorded monthly.

Rats In rats exposed for 2 years, survival was significantly reduced in males at 5 mg/ml ($p = 0.01$) and females at 10 mg/ml ($p = 0.006$) by life table analysis (NTP, 1995). Body weight gain in treated rats was similar to controls during the first year and then declined in high dose groups; terminal body weights were 24% lower than controls in males at 5 mg/ml and 21% lower in females at 10 mg/ml. In males, no adverse effects of treatment were noted for water consumption, clinical signs, urinalysis, or hematology parameters. A dose-related decrease in water consumption was observed in females; during the first year, consumption was reduced by approximately 12 and 20% in the 5 and 10 mg/ml groups, respectively, but the difference from control dropped to approximately 10% in both groups during the second year of the study. Females at 10 mg/ml exhibited an increased incidence of hyperactivity (number affected not reported).

Increases in urine specific gravity (measured at 15 months) in females exposed to 5 and 10 mg/ml were consistent with their reduced water consumption. There were no treatment-related changes in hematology parameters in either sex.

Mice In male mice exposed to *tert*-butyl alcohol in drinking water for 2 years, survival was significantly reduced ($p = 0.007$) at 20 mg/ml by life table analysis (NTP, 1995). Mean body weights were 5–10% lower than controls from week 9 to week 91 in male mice exposed to 20 mg/ml; terminal body weights were similar to controls. Body weights of females exposed to 20 mg/ml were 10–15% lower than controls from week 13 through termination, and the terminal weight was 12% lower than controls. In females at lower exposures, there were dose-related decreases in body weight, but the difference was not biologically significant (<10% change from control). Exposure to *tert*-butyl alcohol had no effect on water consumption or the incidences of clinical signs in male or female mice.

Carcinogenicity The 2-year study by NTP (NTP, 1995) concluded that there was some evidence of carcinogenic activity of *tert*-butyl alcohol in male F344/N rats based on increased incidences of renal tubule adenoma and carcinoma (combined), no evidence of carcinogenic activity in female F344/N rats, equivocal evidence of carcinogenic activity in male B6C3F1 mice based on marginally increased incidences of follicular cell adenoma or carcinoma (combined) of the thyroid gland, and some evidence of carcinogenic activity in female B6C3F1 mice based on increased incidences of follicular cell adenoma of the thyroid gland.

Chemical Pathology

No published data are available regarding the chemical pathology of *tert*-butyl alcohol on humans. Potential target organs, as observed from the results of animal studies, include kidney, thyroid gland, urinary bladder, and liver.

Rats At the end of the NTP 2-year drinking water study mentioned above, kidney effects were observed in both sexes in rats (NTP, 1995). Nephropathy was observed in nearly all control and exposed rats; the severity of the lesion was moderate in all male rats, but the severity was significantly increased over controls in all female treatment groups. The incidence and severity of mineralization was dose related in males, and the incidence was significantly elevated at 5 mg/ml. Males exposed to ≥1.25 mg/ml exhibited significant dose-related increases in the incidence of a type of mineralization (linear) that is associated with α2μ-globulin (NTP, 1995). Normal, moderate-severity mineralization was present in nearly all female rats. Dose-related increases in hyperplasia of transitional epithelium were significant in males at ≥2.5 mg/ml

and females at 10 mg/ml. Renal tubule hyperplasia was observed as a minimal-to-moderate lesion in all male groups and as a minimal lesion in one female at 10 mg/ml. The incidence of this lesion was significantly increased in males exposed to 5 mg/ml. The incidence of minimal suppurative inflammation was dose related and significantly increased in females at 5 and 10 mg/ml.

Male rats were observed to have a statistically significant increase ($p \leq 0.01$) in renal tubule adenoma (multiple) and combined renal tubule adenoma or carcinoma at 2.5 mg/ml (NTP, 1995) (The incidence was significantly increased over controls when standard histopathology results were combined with an extended step-sectioning procedure used to enhance tumor detection but was not significantly elevated when evaluated by standard results). No renal tumors were observed in female rats.

Mice In mice, nonneoplastic lesions associated with chronic *tert*-butyl alcohol ingestion involved the thyroid gland, urinary bladder, and liver. Increased incidences of hyperplasia of thyroid follicular cells were statistically significant in males at all exposure levels and in females at ≥10 mg/ml. Chronic inflammation of the urinary bladder was significantly increased in incidence and severity in male and female mice at 20 mg/ml. Males exposed to 20 mg/ml also had increased incidences of hyperplasia of the transitional epithelium of the bladder and fatty change of the liver.

The incidence of follicular cell adenoma was nonsignificantly increased in males at 10 mg/ml but showed a significant ($p = 0.011$) dose-related trend in females and a significant increase over controls in females at 20 mg/ml (NTP, 1995). The historical incidence of this tumor in controls in NTP drinking water studies is 8/238 (3.4% ± 2.2%, range 0–5%; (NTP, 1995)). There were no other increases in neoplasms in mice exposed for 2 years.

The only nonneoplastic effect observed in mice exposed to 5 or 10 mg/ml *tert*-butyl alcohol for 2 years was an increased incidence of thyroid follicular cell hyperplasia in males. The NTP (NTP, 1995) noted that thyroid follicular cell tumorigenesis follows a progression from hyperplasia to adenoma and carcinoma, suggesting that hyperplasia is a preneoplastic lesion in the thyroid. Although neither adenomas nor carcinomas were statistically significantly increased in male mice, NTP noted that the incidence of adenomas in males at 10 mg/ml (7%) exceeded the upper end of the range of historical control incidence (2%). Further, carcinoma was observed in male mice at the highest concentration despite the significant early mortality in this group. Together with the evidence for a significantly increased incidence of thyroid follicular cell adenomas in high dose females, these observations support the finding that the increased hyperplasia in male mice is a preneoplastic effect rather than an adaptive response.

Hypothesized Modes of Action

The available literature indicates that *tert*-butyl alcohol is not mutagenic or genotoxic. *In vitro* studies for gene mutation, sister-chromatid exchange, and chromosomal aberrations in mammalian cells (McGregor et al., 1988; Cirvello et al., 1995), and an *in vivo* study of micronucleus induction (NTP, 1995) did not produce any evidence of mutagenicity or genotoxicity. A comprehensive review of genetic toxicity studies by McGregor (2010) revealed that most *in vitro* and *in vivo* assays for genetic effects conducted with *tert*-butyl alcohol are negative.

Kidney Tumors Two possible modes of action for the induction of renal tubule cell tumors have been suggested: consequences of excessive accumulation of α2μ-globulin and the consequences of exacerbation of chronic progressive nephropathy (CPN).

α2μ-Globulin is a low molecular weight urinary protein that is synthesized and secreted by the liver of certain strains of male rats. Approximately 50 mg of α2μ-globulin is filtered per day with 40% of the filtered protein excreted in the urine and 60% reabsorbed and catabolized (Neuhaus, 1986). Nephropathy is due to the modification of the protein in blood by a chemical ligand so that the complex reabsorbed in the P1 segment of the proximal renal tubules is rendered more resistant to proteolysis and accumulates in the phagolysosomes of the tubule cells (McGregor, 2010). The accumulation, recognized as hyaline droplets, leads to lysosomal overload and the shedding of cells into the lumen, resulting in cell population regeneration and the formation of tumors. The α2μ-globulin mode of action is an established cancer mode of action specific to male rats and not considered relevant to human health (NTP, 1995).

A possible alternative to the α2μ-globulin-associated nephropathy mode of action is exacerbation of CPN. Chronic progressive nephropathy is becoming more widely recognized as a potential mode of action by which chemicals may influence the incidence of kidney tumors in rats (McGregor, 2010). In contrast to α2μ-globulin nephropathy, CPN has no identified biochemical events and the process is not easily investigated by experiments. It must occur in untreated rats with severity increasing with age and exacerbated by exposure to the suspect chemical in comparison with untreated rats (McGregor, 2010).

Thyroid Follicular Cell Tumors There are some data to suggest the possibility of *tert*-butyl alcohol induced-thyroid follicular cell tumors but information is lacking to draw this conclusion. In a chronic drinking water study (Cirvello et al., 1995; NTP, 1995), thyroid effects were noted, including a significant increase in follicular cell hyperplasia of the thyroid gland in both male and female mice. In male mice, the increase occurred at all dose levels, while in females, it was limited to the two highest doses. The mechanism of formation of these lesions resulting from *tert*-butyl alcohol exposure has not been specifically studied. However, it is known that the metabolism of thyroxine (T4), a major thyroid hormone, involves conjugation with glucuronic acid in the liver and excretion in the bile (NSF International, 2003). *tert*-Butyl alcohol is also eliminated, at least in parts, by glucuronides conjugation, which may enhance the elimination of T4 through enzyme induction in the liver. Through a negative feedback loop, the pituitary gland increases or decreases the output of thyroid stimulating hormone (TSH) based on circulating levels of thyroid hormone. Thus, a decrease in circulating T4 may lead to sustained increases in TSH, resulting in an increase in the size of the thyroid gland or an increase in the number of thyroid follicular cells. Sustained thyroid follicular cell proliferation may eventually result in the progression of hyperplasia to adenoma and carcinoma (USEPA, 1998). There are significant differences between humans and rodents in susceptibility to this effect; the serum half-life of T4 in humans is 5–9 days, whereas in rodents, it is only 0.5–1 day. These differences have led to an uncertainty regarding the relevance of rodent thyroid follicular cell effects to human health. There are no data regarding thyroid cell proliferation after exposure to *tert*-butyl alcohol.

SOURCES OF EXPOSURE

Exposure to *tert*-butyl alcohol in humans can occur through inhalation, ingestion, and eye or skin contact (IPCS, 1987b; HSDB, 2007).

Occupational

Occupational exposure may occur through inhalation and dermal contact at workplaces where *tert*-butyl alcohol is produced or used. This includes operations that use *tert*-butyl alcohol as a solvent for paints, lacquers, varnishes, natural and synthetic resins, gums, vegetable oils, dyes, camphor, and alkaloids and as an octane booster in unleaded gasoline (OSHA, 1996). Other routes of exposure can occur through the manufacturing of artificial leather, safety glass, rubber and plastic cement, shellac, raincoats, photographic films, flotation agents, fruit essence, perfumes, cellulose esters, lacquers, paint removers, and plastics (OSHA, 1996). Occupational exposure can also occur when *tert*-butyl alcohol is used as a denaturant for alcohol and as a chemical intermediate in the manufacturing of methyl methacrylate and pharmaceuticals (OSHA, 1996).

Environmental

The general population may be exposed to *tert*-butyl alcohol through the ingestion of drinking water and foods and inhalation of air tainted with *tert*-butyl alcohol (HSDB,

2007). Although not quantified, *tert*-butyl alcohol has been found in drinking water samples from New Orleans, LA (Dowty et al., 1975) and other unspecified locations in the United States (Lucas, 1984). *tert*-Butyl alcohol was also identified, not quantified, in beer (Bavisotto and Heinisch, 1961), chickpeas (Rembold et al., 1989), and mother's milk (Erickson et al., 1980; Pellizzari et al., 1982). Snider (1985) detected *tert*-butyl alcohol in the air of Tucson, AZ at a mean concentration of 3.7 parts per billion (ppb) but not at two rural sites 40 km away.

INDUSTRIAL HYGIENE

The current Occupational Safety and Health Administration (OSHA) permissible exposure limit (PEL) for *tert*-butyl alcohol is 100 parts per million (ppm) (300 mg/m^3) as an 8-h time-weighted average (TWA) concentration and 150 ppm (450 mg/m^3) as a 15-min short-term exposure limit (STEL). NIOSH has not issued a recommended exposure limit (REL) for *tert*-butyl alcohol but has stated that they concur with OSHA's PEL (NIOSH, 1988). The American Conference of Governmental Industrial Hygienists (ACGIH) has assigned *tert*-butyl alcohol a threshold limit value (TLV) of 100 ppm (303 mg/m^3) as an 8-h TWA concentration and 150 ppm (455 mg/m^3) as a 15-min STEL (ACGIH, 2001). Recommendations by NIOSH on respiratory protection for workers exposed to *tert*-butyl alcohol are listed in Table 64.2.

OSHA (1996) recommends the following guidelines regarding respiratory protection and personal protective equipment. Government-approved respirators must be worn if the ambient concentration of *tert*-butyl alcohol exceeds prescribed exposure limits. Protective clothing should be worn to prevent skin contact with *tert*-butyl alcohol. Protective clothing made of butyl rubber has been recommended for use against permeation by *tert*-butyl alcohol and may provide protection for periods of exposure >8 h, while polyethylene ethylene/vinyl alcohol may withstand permeation for more than 4 but < 8 h. Safety glasses, goggles, or face shields should be worn during operations where *tert*-butyl alcohol may come in contact with the eyes. Eyewash fountains and emergency showers should be readily available within the immediate work area. Contact lenses should not be worn if exposure to *tert*-butyl alcohol can occur.

RISK ASSESSMENTS

No chronic reference dose (RfD) or reference concentration (RfC) for *tert*-butyl alcohol is currently available on EPA's Integrated Risk Information System (IRIS), in the Drinking Water Standards and Health Advisories (USEPA, 2006), or in the Health Effects Assessment Summary Tables (HEAST) (USEPA, 2011). The EPA's Chemical Assessments and

TABLE 64.2 NIOSH Recommended Respiratory Protection for Workers Exposed to *tert*-Butyl Alcohol

Condition/airborne concentration of *tert*-butyl alcohol	Minimum respiratory protection
100–1000 ppm (10 × PEL)	Any supplied-air respirator equipped with a half mask and operated in a demand (negative-pressure) mode
100–2500 ppm (25 × PEL)	Any supplied-air respirator equipped with a hood or helmet and operated in a continuous-flow mode
100–5000 ppm (50 × PEL)	Any supplied-air respirator equipped with a full facepiece and operated in a demand (negative-pressure) mode or Any supplied-air respirator equipped with a tight-fitting facepiece and operated in a continuous-flow mode or Any self-contained respirator equipped with a full facepiece and operated in a demand (negative-pressure) mode or Any supplied-air respirator operated in a pressure-demand or other positive-pressure mode
Entry into unknown concentrations	Any self-contained respirator equipped with a full facepiece and operated in a pressure-demand or other positive-pressure mode or Any supplied-air respirator equipped with a full facepiece and operated in a pressure-demand or other positive-pressure mode in combination with an auxiliary self-contained breathing apparatus operated in a pressure-demand or other positive pressure mode
Firefighting	Any self-contained respirator equipped with a full facepiece and operated in a pressure-demand or other positive-pressure mode
Escape	Any air-purifying, full-facepiece respirator equipped with an organic vapor canister or Any escape-type, self-contained breathing apparatus with a suitable service life (number of minutes required to escape the environment)

Source: Adapted from OSHA Occupational Safety and Health Guideline for *tert*-butyl alcohol (OSHA, 1996)

Related Activities (CARA) list (USEPA, 1994) does not include any documents relating to *tert*-butyl alcohol. ATSDR has not prepared a toxicological profile for *tert*-butyl alcohol. The World Health Organization (WHO) considered the data on *tert*-butyl alcohol to be insufficient to support derivation of an acceptable daily intake (ADI) (IPCS, 1987b). The ACGIH set a TLV of 100 ppm to protect against narcosis and also determined that *tert*-butyl alcohol was not classifiable as a human carcinogen based on the equivocal evidence from the NTP study (ACGIH, 2001). The International Agency for Research on Cancer (IARC, 1999) has not fully summarized the toxicological data on *tert*-butyl alcohol but has briefly mentioned its toxicity and genotoxicity as a metabolite of MTBE. *tert*-Butyl alcohol is not included in the National Toxicology Program's (NTP) 11th Report on Carcinogens (NTP, 2006). CalEPA and the National Institute of Public Health and the Environment (RIVM) also do not have any health or risk assessment information on *tert*-butyl alcohol.

REFERENCES

ACGIH (2001) *Documentation of the Threshold Limit Values and Biological Exposure Indices for Tert-Butanol*, 7th ed., Cincinnati, OH: American Conference of Governmental Industrial Hygienists, pp. 6.

Bavisotto, V.S., R. L.A. and Heinisch, B. (1961) *Gas chromatography of volatiles in beer. Paper presented at the Am Soc Brewing Chemists.*

Chen, M. (2005) Amended final report of the safety assessment of t-butyl alcohol as used in cosmetics. *Int. J. Toxicol.* 24 (Suppl. 2):1–20.

Cirvello, J.D., Radovsky, A., Heath, J.E., Farnell, D.R., and Lindamood, C. 3rd. (1995) Toxicity and carcinogenicity of t-butyl alcohol in rats and mice following chronic exposure in drinking water. *Toxicol. Ind. Health* 11(2):151–165.

Dowty, B.J., Carlisle, D.R., and Laseter, J.L. (1975) New Orleans drinking water sources tested by gas chromatography-mass spectrometry. *Environ. Sci. Technol.* 9:762–765.

Erickson, M.D., H.B.S., Pellizzari, E.D., Tomer, K.B., Waddell, R.D., and Whitaker, D.A. (1980) *Acquisition and Chemical Analysis of Mother's Milk for Selected Toxic Substances*, Washington, DC: U.S. Environmental Protection Agency (EPA).

HSDB (2007) *Hazardous Substances Data Bank: t-Butyl alcohol. from NLM (National Library of Medicine); National Institutes of Health, U.S. Department of Health and Human Services, Bethesda, MD.* Available at http://toxnet.nlm.nih.gov.

IARC (1999) IARC monograph on the evaluation of carcinogenic risk to humans. In: *Some Chemicals that Cause Tumours of the Kidney or Urinary Bladder in Rodents and Some Other Substances*, vol. 73, Lyon, France: IARC.

IPCS (1987a) *Environmental Health Criteria 65: Butanols: four isomers – 1 butanol 2 butanol tert butanol isobutanol* Environ Health Criter 65 *(Vol. 65, pp. 3–141). International Programme on Chemical Safety (IPCS). Geneva: World Health Organization; Albany, NY: WHO Publications Center USA.*

IPCS (1987b) *Tert-Butanol Health and Safety Guide*, No. 7 (pp. 40): World Health Organization, International Program on Chemical Safety (IPCS), Switzerland.

Lucas, S.V. (1984) *GC/MS Analysis of Organics in Drinking Water Concentrates and Advanced Waste Treatment Concentrates.* (USEPA-600/1-84-020A). Columbus.

McGregor, D. (2010) Tertiary-butanol: a toxicological review. *Crit. Rev. Toxicol.* 40(8):697–727.

McGregor, D.B., Brown, A., Cattanach, P., Edwards, I., McBride, D., and Caspary, W.J. (1988) Responses of the L5178Y tk+/tk-mouse lymphoma cell forward mutation assay. II: 18 coded chemicals. *Environ. Mol. Mutagen.* 11(1):91–118.

Neuhaus, O.W. (1986) Renal reabsorption of low molecular weight proteins in adult male rats: alpha 2μ-globulin. *Proc. Soc. Exp. Biol. Med.* 182(4):531–539.

NIOSH (1988) *Testimony of the National Institute for Occupational Safety and Health on the Occupational Safety and Health Administration's Proposed Rule: 29 CFR 1910*, (H-020, August 2, 1988.). Cincinnati: U.S. Department of Health and Human Services, Public Health Service, Centers for Disease Control.

NIOSH (1992) *Occupational Safety and Health Guidelines for Chemical Hazards (Publication No. 81–123): Tert-Butyl alcohol (pp. 7). 4676 Columbia Parkway, Cincinnati, OH: National Institute of Occupational Safety and Health (NIOSH), Centers for Disease Control and Prevention (CDC), Department of Health and Human Services (DHHS).*

NIOSH (2005) *NIOSH Pocket Guide to Chemical Hazards (2005-149), tert-Butyl Alcohol. PL Washington, DC: National Institute for Occupational Safety and Health (NIOSH), Centers for Disease Control and Prevention, Department Of Health And Human Services.*

NSF International (2003) *T-Butanol (CAS # 75-65-0) Oral Risk Assessment Document*, MI: Ann Arbor.

NTP (1995) NTP Toxicology and Carcinogenesis Studies of t-Butyl Alcohol (CAS No. 75-65-0) in F344/N Rats and B6C3F1 Mice (Drinking Water Studies). *Natl. Toxicol. Program Tech. Rep. Ser.* 436:1–305.

NTP (2006) *Report on Carcinogens. National Toxicology Program, NIH, DHHS.*

OSHA (1996) *Occupational Safety and Health Guideline for Tert-Butyl Alcohol*, Washington, DC: Occupational Safety and Health Administration, Department of Labor.

Pellizzari, E.D., Hartwell, T.D., Harris, B.S. 3rd., Waddell, R.D., Whitaker, D.A., and Erickson, M.D. (1982) Purgeable organic compounds in mother's milk. *Bull. Environ. Contam. Toxicol.* 28(3):322–328.

Rembold, H., Wallner, P., Nitz, S., Kollmannsberger, H., and Drawert, F. (1989) Volatile components of chickpea (*Cicer arietinum* L.) seed. *J. Agr. Food Chem.* 37(3):659–662.

Snider, J.R. Dawson, G.A. (1985) Tropospheric light alcohols, carbonyls, and acetonitrile concentrations in the southwestern United States and Henry's Law data. *J. Geophys. Res.* 90:3797–3805.

USEPA (1994) *Chemical Assessments and Related Activities (CARA)*, Washington, DC.

USEPA (1998) *Assessment of Thyroid Follicular Cell Tumors*, (EPA/630/R-97/002). Washington, DC.

USEPA (2006) *Drinking Water Health Advisories 2006 Edition*, (EPA 822-R-06-013). Washington, DC.

USEPA (2011) *Health Effects Assessment Summary Tables (HEAST)*, Washington, DC.

65

NAPHTHALENE

Lisa A. Bailey, Laura E. Kerper, and Lorenz R. Rhomberg

Target organ(s): Eye, skin, blood, liver, kidneys, central nervous system, respiratory tract

Occupational exposure limits: 10 ppm (50 mg/m^3) OSHA (PEL); 10 ppm (52 mg/m^3) ACGIH (TLV); 15 ppm (79 mg/m^3) ACGIH (STEL); 10 ppm (50 mg/m^3) NIOSH (REL); 250 ppm (1250 mg/m^3) NIOSH (IDLH)

Reference values: 0.02 mg/kg/day US EPA (RfD)

Risk/safety phrases: Harmful if swallowed, limited evidence of carcinogenic effect, very toxic to aquatic organisms, may cause long-term effect in the aquatic environment.

BACKGROUND AND USES

Naphthalene occurs naturally in fossil fuels such as coal and petroleum. It is commonly produced from the distillation and fractionation of coal tar. Naphthalene is used as an intermediate in the production of phthalate plasticizers, other plastics and resins, and other products such as dyes, wood preservatives, explosives, lubricants, pharmaceuticals, deodorizers, and insect repellants. Moth balls and other moth repellants, and some solid block deodorizers used for toilets and diaper pails, are made of crystalline naphthalene (ATSDR, 2005).

PHYSICAL AND CHEMICAL PROPERTIES

Naphthalene (CASRN 91-20-3) is a white solid with the chemical formula $C_{10}H_8$. Its molecular weight is 128.09. It evaporates easily and has a strong odor of tar or mothballs. The melting point of naphthalene is 80.5 °C, its boiling point is 218 °C, and its density is 1.145 g/ml at 20 °C. Solubility in water is low (31.7 mg/l at 25 °C), and it is soluble in benzene, alcohol, ether, and acetone (ATSDR, 2005).

MAMMALIAN TOXICOLOGY

Acute Effects

Dermal contact with naphthalene may cause skin irritation and an allergic skin reaction (ATSDR, 2005; Acros Organics, 2000). Contact with the eyes may cause conjunctivitis, corneal injury, diminished visual acuity, and cataracts (ATSDR, 2005; Acros Organics, 2000). Naphthalene vapor at concentrations ≥15 parts per million (ppm) may cause eye irritation. When swallowed in large amounts, it may cause liver and kidney damage, methemoglobinemia, cyanosis, convulsions, and death. Digestive tract irritation, abdominal pain, nausea, vomiting, and diarrhea may also occur. Inhalation of naphthalene may cause respiratory tract irritation, and at high concentrations, salivation, headache, confusion, nausea, vomiting, abdominal pain, fever, labored breathing, and death may occur. In individuals who have a glucose-6-phosphate dehydrogenase genetic defect, inhalation or ingestion of or dermal exposure to large amounts of naphthalene may cause hemolytic anemia (ATSDR, 2005).

Chronic Effects

In mice and rats, the target of chronic naphthalene toxicity is the respiratory tract. The National Toxicology Program (NTP) conducted 2-year bioassays in which mice (NTP, 1992; Abdo et al., 1992) or rats (NTP, 2000; Abdo et al., 2001) were exposed by inhalation to ranges of naphthalene concentrations. In mice, females (and not males) exhibited a

Hamilton & Hardy's Industrial Toxicology, Sixth Edition. Edited by Raymond D. Harbison, Marie M. Bourgeois, and Giffe T. Johnson.
© 2015 John Wiley & Sons, Inc. Published 2015 by John Wiley & Sons, Inc.

significant increase in combined incidence of alveolar/ bronchiolar adenomas and carcinomas at the highest dose (30 ppm). Inflammation of lung tissue was observed in both sexes at 10 and 30 ppm. There were no significant increases in lung tissue hyperplasia at any exposure. There was no evidence of nasal tumors in either sex at any exposure; however, all treatment groups exhibited significant increases in nasal inflammation, metaplasia of the olfactory epithelium, and hyperplasia of the respiratory epithelium. Based on the outcome of this assay, NTP concluded, there was no evidence of carcinogenic activity in male mice, and some evidence of carcinogenic activity in female mice (NTP, 1992; Abdo et al., 1992).

In rats, there was a significant increase in neuroblastomas of the olfactory tissue in females in the highest exposure group (60 ppm), and a significant trend with dose in both sexes. Males exhibited a significant increase in adenomas of the nasal respiratory epithelium at all exposure levels. For both sexes, at all exposure levels, there were significant increases in inflammation and hyperplasia in both olfactory and nasal respiratory epithelial tissues. In lung tissue, there was some evidence of hyperplasia in females but not males, and some inflammation in males but not females. No increases in lung tumors were observed. Based on the outcome of this assay, NTP concluded there was clear evidence of carcinogenic activity in male and female rats (NTP, 2000; Abdo et al., 2001).

There is very little information about cancer risk in humans associated with naphthalene exposure. The human data on naphthalene exposure and cancer risk are limited to a few case reports and one nested case-control study of oral/ oropharyngeal cancer. Lung cancer has been addressed in several occupational studies of industries where naphthalene exposure may occur, but the interpretation of these is limited due to low or unquantified naphthalene exposures and concurrent exposures to other chemical agents.

Olsson et al. (2010) conducted a case-control study of 433 lung cancer cases and 1253 controls, nested within a cohort of 38,296 asphalt workers. Although naphthalene exposure was not studied specifically, naphthalene may account for up to 90% of occupational inhalation exposure to polycyclic aromatic hydrocarbons (PAHs) for asphalt workers (NIOSH, 2000). The authors found no association between lung cancer and any exposure to asphalt fumes.

Merletti et al. (1991) conducted a population-based, case-control study of 86 oral or oropharyngeal cancer cases and 373 controls in Italy. This group evaluated the risk associated with 40 occupations, 41 industries, and 16 specific chemicals, including naphthalene. Naphthalene was not associated with the risk of oral or oropharyngeal cancer in this study.

Lewis (2012) reviewed several studies of lung cancer risk in connection with industries and occupations in which naphthalene exposure is common, including the petroleum (Wong and Raabe, 2000; Consonni et al., 1999; Lewis et al., 2003;

Rushton, 1993), asphalt (Boffetta et al., 2003a, 2003b; Olsson et al., 2010; Fayerweather, 2007), and creosote industries (Wong and Harris, 2005), as well as jet fuel handlers (D'Mello and Yamane, 2007; Yamane, 2006). No association with lung or nasal cancer was reported in any of the studies.

Mechanisms of Action

The mechanisms of acute naphthalene toxicity are known for hemolytic anemia and cataract formation. For chronic respiratory toxicity, the mechanism of action has not been definitively determined.

Hemolytic Anemia Individuals who are deficient in glucose-6-phosphate dehydrogenase are more susceptible to oxidizing agents such as naphthalene. Naphthalene and other oxidizing agents can oxidize heme iron in red blood cells to the ferric state, and the resulting free radical oxygen acts on membrane lipids to increase fragility and lysis of the red blood cell (ATSDR, 2005).

Cataracts Naphthalene is metabolized by cytochrome P-450 (CYP) enzymes to 1,2-naphthalenediol, which can reach the lens via circulation and subsequently become metabolized to 1,2-naphthoquinone. 1,2-Naphthoquinone is highly reactive and can bind to proteins, amino acids, and glutathione in the lens, affecting its transparency (ATSDR, 2005).

Respiratory Tract Lesions The respiratory toxicity of naphthalene depends on metabolic pathways that are highly specific to species, tissue, and cell type, and differences in toxicity depend on the balance between generation of toxic metabolites and the elimination of those metabolites (Bogen, 2008; Buckpitt et al., 2002). In mouse and rat respiratory tissue, CYP enzymes are largely responsible for naphthalene metabolism to reactive species, namely, CYP2F2 and CYP2A5 in mice (Bogen, 2008; Buckpitt et al., 2002; Li et al., 2011; Hu et al., 2013) and likely CYP2F4—and possibly CYP2E1—in rats (Baldwin et al., 2004; Cruzan et al., 2009). It is generally agreed that human CYP2F1 and CYP2A13 metabolize naphthalene in respiratory tissue but at a much lower rate than either mouse or rat CYP enzymes, suggesting that the highly reactive naphthalene metabolites are generated in smaller amounts with less potential to cause toxicity to human tissues (Buckpitt et al., 2002; Cruzan et al., 2009; Lewis et al., 2009; Fukami et al., 2008; Su et al., 2000). As discussed above, there are no studies that show an association between naphthalene exposure and lung or nasal cancer in humans. In fact, there is very little evidence of any cancer in humans linked to naphthalene exposure.

In the NTP mouse (NTP, 1992) and rat (NTP, 2000) studies, tumor formation was always accompanied by cytotoxicity. Other studies have shown nasal cytotoxicity in rats at lower doses of naphthalene than those associated with tumors in the NTP studies, or at similar levels but shorter exposure times (Dodd et al., 2010, 2012). These studies also identified a no observed adverse effect level (NOAEL) for cytotoxicity in the rat nose (0.1 ppm). These results suggest that cytotoxicity is an essential component of naphthalene-induced tumorigenesis, but there is a likely threshold below which cytotoxicity does not occur. In mouse nasal tissue, cytotoxicity occurs at sufficiently high doses of naphthalene, but tumor formation does not (NTP, 1992). This suggests that, while it may be necessary, cytotoxicity may not be sufficient for development of tumors, and there may be an additional step that occurs in rat nose and in mouse lung that, together with cytotoxicity, is essential for naphthalene-induced carcinogenicity.

Naphthalene metabolites are highly reactive, and the cytotoxicity observed in the respiratory tract may be due to the formation of metabolite–protein adducts and/or reactive oxygen species (ROS). Protein adducts of naphthalene metabolites have been observed in various cultured cells (Cho et al., 1994), target respiratory tissues in rodents and nonhuman primates (Boland et al., 2004; Cho et al., 1994; Lin et al., 2006; DeStefano-Shields et al., 2010), and cell-free systems (Pham et al., 2012a, 2012b). It is suggested that generation of ROS may occur by redox cycling of naphthoquinone or as a result of cytotoxicity and inflammation at the target site. ROS generation may lead to lipid peroxidation, depletion of reducing equivalents, and DNA oxidation or strand breaks (ATSDR, 2005). Glutathione (GSH) depletion has been observed with high levels of administered naphthalene, such as those that have been associated with tumor formation in mouse lung (Phimister et al., 2004). Glutathione protects cells against reactive species, including reactive metabolites and oxygen radicals that are generated from those metabolites (Bergamini et al., 2004; Phimister et al., 2004). Thus, GSH depletion could lead to cellular injury via naphthalene-metabolite-induced protein adducts and/or oxidative stress, both of which may be precursors to cytotoxicity-induced tumor formation. Several studies (although not in target respiratory tissue) have shown that naphthalene metabolites (naphthalene epoxide, 1,2-naphthoquinone, and 1,4-naphthoquinone) react readily with cellular proteins in rats and mice in a dose-dependent manner (Zheng et al., 1997; Waidyanatha et al., 2002; Waidyanatha and Rappaport, 2008; Buckpitt et al., 2002) and therefore may mediate cytotoxicity well before any 1,2-naphthoquinone is available to undergo redox cycling to generate ROS, which could react with DNA.

Most studies of naphthalene genotoxicity indicate that it occurs secondary to cytotoxicity (Brusick, 2008). Recent *in vitro* and *in vivo* assays provide supporting evidence that cytotoxicity precedes genotoxic responses to naphthalene in human lymphoblasts (Recio et al., 2012), rat nasal tissue (Meng et al., 2011), and mouse lung tissue (Karagiannis et al., 2012). Thus, the available data suggest that any positive genotoxic results induced by naphthalene are the result of (1) saturation of detoxification mechanisms and generation of downstream genotoxic metabolites (likely to occur only at high concentrations that deplete detoxification mechanisms) and/or (2) secondary genotoxic effects due to cytotoxicity. These results also suggest that there would be a practical threshold for any involvement of genotoxicity in the carcinogenic mode of action, and genotoxicity would not be an initiating event in tumor formation.

The overall weight of evidence suggests that mouse lung and rat nasal tumors following naphthalene exposure are the result of a cytotoxic or dual cytotoxic and genotoxic mode of action, and that formation of cytotoxic amounts of reactive naphthalene metabolites (via CYP enzymes) in mouse lung and rat nasal tissue also causes chronic inflammation and regenerative hyperplasia. The data suggest that genotoxic events following naphthalene treatment are secondary to naphthalene-induced cytotoxicity, or they occur concurrently but as independent events. In either scenario, DNA damage occurs at cytotoxic (or greater than cytotoxic) exposure concentrations. This mode of action, for cancer and non-cancer effects, has little or no relevance to humans at typical environmental exposures because human metabolic activation in respiratory tissue is much lower than those in rats at comparable exposure concentrations (Campbell et al., 2013). Therefore, exposure concentrations much higher than typical exposures in humans are necessary to reach a metabolic activation equivalent to those in rats that cause respiratory cytotoxicity. For some cytotoxic end points, metabolism in humans appears to saturate, reaching a maximum (Campbell et al., 2013) that is below the level needed in rats to produce cytotoxic effects. The weight-of-evidence evaluation and proposed mode of action are summarized in Bailey et al. (2015, submitted).

SOURCES OF EXPOSURE

Occupational

The highest occupational exposure to naphthalene occurs in wood treatment (creosote), coke and coal tar production, mothball production, and jet fuel industries. Exposure in these industries may be in the range of 100–3000 $\mu g/m^3$ during the course of an 8-h work day (Price and Jayjock, 2008). Lower exposures occur in the petroleum, asphalt, aluminum production, chemical manufacture, tanning, and ink and dye industries. In these occupations, work-day exposures may be in the range of 3–100 $\mu g/m^3$ (Price and Jayjock, 2008).

Environmental

Because naphthalene occurs naturally in fossil fuels and is used in the production of plasticizers and other products, it is released to the air in industrial and auto emissions. It is also produced when organic materials, such as wood and tobacco, are burned (ATSDR, 2005). Ambient air levels of naphthalene for urban and suburban areas in the U.S. have been reported to range from 0.4 to 170 μg/m³, with a median of 0.94 μg/m³ (ATSDR, 2005). Concentrations at industrial and hazardous waste sites are typically higher. Sources of exposure to naphthalene in indoor air include cooking, tobacco smoking, and moth repellants. Indoor air levels have been reported to average 1.0 μg/m³ in non-smoking households and 2.2 μg/m³ in households with smokers (ATSDR, 2005). Indoor air levels from mothball use in the home are estimated to range from 1 to 100 μg/m³ for typical use and up to 300 μg/m³ for off-label use (e.g., fumigation) (Griego et al., 2008).

REGULATORY REPORTS ON CARCINOGENESIS

In a 2011 NTP Report on Carcinogens (NTP, 2011), the agency designated naphthalene as "reasonably anticipated to be a human carcinogen." The NTP determined that the human data on naphthalene were inadequate for an evaluation of carcinogenicity, and the designation was based on the evidence of nasal tumors in rats (NTP, 2000) and lung tumors in mice (NTP, 1992) as described above. NTP noted that naphthalene exhibited genotoxicity in some, but not all test systems and the mode of action for tumor formation was not known, but oxidative damage and DNA breakage may be contributing factors.

The International Agency for Research on Cancer (IARC) (IARC, 2002) designated naphthalene as "possibly carcinogenic to humans" (Group 2B). This designation was based on inadequate human data and sufficient animal data (the NTP rat and mouse studies). The IARC noted that there was little evidence of mutagenicity, and some positive results for genotoxicity. This agency proposed a mode of action for tumor formation that resulted from the generation of cytotoxic metabolites and increased cell turnover.

Based on the likely mode of action and very low metabolic activity of naphthalene in human respiratory tissue identified as part of the PBPK model (including saturation at much lower levels compared to rodents), these classifications may only be relevant to humans at very high exposure levels (e.g., occupational).

INDUSTRIAL HYGIENE

The Occupational Safety and Health Administration (OSHA), National Institute for Occupational Safety and Health (NIOSH), and American Conference of Governmental Industrial Hygienists (ACGIH) have set the permissible exposure limit (PEL), recommended exposure limit (REL),

and threshold limit value (TLV), respectively, for naphthalene at 10 ppm (50 mg/m³) (ACGIH, 2013).

Naphthalene can be absorbed through the skin. Workers should wear protective clothing and gloves to prevent skin contact and eye protection to prevent splashing in eyes. Respirators should be worn to prevent inhalation (CDC, 2010). Naphthalene is combustible. Dispersal of dust should be prevented, as explosive vapor/air mixtures may be formed above 80 °C (IPCS and NIOSH, 2005).

MEDICAL MANAGEMENT

In the case of accidental inhalation exposure, the subject should first be removed to an area with fresh air and good ventilation and then referred for medical attention. Skin that has come into contact with naphthalene should be rinsed with soap and plenty of water. In case of eye contact, first rinse eyes with plenty of water (remove contact lenses if easily possible) and then seek medical attention (NIOSH, 2005, 2010). If naphthalene is ingested, do not induce vomiting. Seek immediate medical attention (NIOSH, 2005, 2010; Acros Organics, 2000).

REFERENCES

Abdo, K.M., Eustis, S.L., McDonald, M., Jokinen, M.P., Adkins, B., and Haseman, J.K. (1992) Naphthalene: A respiratory tract toxicant and carcinogen for mice. *Inhal. Toxicol.* 4:393–409.

Abdo, K.M., Grumbein, S., Chou, B.J., and Herbert, R. (2001) Toxicity and carcinogenicity study in F344 rats following 2 years of whole-body exposure to naphthalene vapors. *Inhal. Toxicol.* 13:931–950.

ACGIH (2013) *2013 Guide to Occupational Exposure Values*, Cincinnati, OH: American Conference of Governmental Industrial Hygienists.

Acros Organics (2000) *Material safety data sheet for naphthalene, scintillation grade, 99+%.* http://avogadro.chem.iastate.edu/MSDS/naphthalene.htm (accessed on November 6, 2013).

ATSDR (2005) *Toxicological Profile for Naphthalene, 1-Methylnaphthalene, and 2-Methylnaphthalene*, Atlanta, GA: Agency for Toxic Substances and Disease Registry.

Bailey, L.A., Nascarella, M.A., Kerper, L.E., and Rhomberg, L.R. (2015) submitted.

Baldwin, R.M., Jewell, W.T., Fanucci, M.V., Plopper, C., and Buckpitt, A. (2004) Comparison of pulmonary/nasal CYP2 F expression levels in rodents and rhesus macaque. *J. Pharmacol. Exp. Ther.* 309:127–136.

Bergamini, C.M., Gambetti, S., Dondi, A., and Cervellati, C. (2004) Oxygen, reactive oxygen species and tissue damage. *Curr. Pharm. Des.* 10:1611–1626.

Boffetta, P., Burstyn, I., Partanen, T., Kromhout, H., Svane, O., Langard, S., Jarvholm, B., Frentzel-Beyme, R., Kauppinen, T., Stucker, I., Shaham, J., Heederik, D., Ahrens, W., Bergdahl,

I.A., Cenee, S., Ferro, G., Heikkila, P., Hooiveld, M., Johansen, C., Randem, B.G., and Schill, W. (2003a) Cancer mortality among European asphalt workers: an international epidemiological study. I. Results of the analysis based on job titles. *Am. J. Ind. Med.* 43:18–27.

Boffetta, P., Burstyn, I., Partanen, T., Kromhout, H., Svane, O., Langård, S., Järvholm, B., Frentzel-Beyme, R., Kauppinen, T., Stücker, I., Shaham, J., Heederik, D., Ahrens, W., Bergdahl, I.A., Cenée, S., Ferro, G., Heikkilä, P., Hooiveld, M., Johansen, C., Randem, B.G., and Schill, W. (2003b) Cancer mortality among European asphalt workers: an international epidemiological study. II. Exposure to bitumen fume and other agents. *Am. J. Ind. Med.* 43:28–39.

Bogen, K.T. (2008) An adjustment factor for mode-of-action uncertainty with dual-mode carcinogens: the case of naphthalene-induced nasal tumors in rats. *Risk Anal.* 28:1033–1051.

Boland, B., Lin, C.Y., Morin, D., Miller, L., Plopper, C., and Buckpitt, A. (2004) Site-specific metabolism of naphthalene and 1-nitronaphthalene in dissected airways of rhesus macaques. *J. Pharmacol. Exp. Ther.* 310:546–554.

Brusick, D. (2008) Critical assessment of the genetic toxicity of naphthalene. *Regul. Toxicol. Pharmacol.* 51:37–42.

Buckpitt, A., Boland, B., Isbell, M., Morin, D., Shultz, M., Baldwin, R., Chan, K., Karlsson, A., Lin, C., Taff, A., West, J., Fanucci, M., Van Winkle, L., and Plopper, C. (2002) Naphthalene-induced respiratory tract toxicity: metabolic mechanisms of toxicity. *Drug Metab. Rev.* 34:791–820.

Campbell, J.L. Jr., Andersen, M. E., and Clewell, H.J. III. (2013) A hybrid CFD-PBPK model for naphthalene in rat and human with IVIVE for nasal tissue metabolism and cross-species dosimetry. *Inhal. Toxicol.*

Centers for Disease Control and Prevention (CDC) (2010) *NIOSH Pocket Guide to Chemical Hazards Record for Naphthalene (CAS No. 91-20-3)*. Available at http://www.cdc.gov/niosh/npg/ (accessed on November 6, 2013).

Cho, M., Chichester, C., Morin, D., Plopper, C., and Buckpitt, A. (1994) Covalent interactions of reactive naphthalene metabolites with proteins. *J. Pharmacol. Exp. Ther.* 269:881–889.

Consonni, D., Pesatori, A.C., Tironi, A., Bernucci, I., Zocchetti, C., and Bertazzi, P.A. (1999) Mortality study in an Italian oil refinery: extension of the follow-up. *Am. J. Ind. Med.* 35:287–294.

Cruzan, G., Bus, J., Banton, M., Gingell, R., and Carlson, G. (2009) Mouse specific lung tumors from CYP2F2-mediated cytotoxic metabolism: an endpoint/toxic response where data from multiple chemicals converge to support a mode of action. *Regul. Toxicol. Pharmacol.* 55:205–218.

D'Mello, T.A. and Yamane, G.K. (2007) *Occupational Jet Fuel Exposure and Invasive Cancer Occurrence in the United States Air Force, 1989–2003*, Brooks City-Base, TX: Air Force Institute for Operational Health.

DeStefano-Shields, C., Morin, D., and Buckpitt, A. (2010) Formation of covalently bound protein adducts from the cytotoxicant naphthalene in nasal epithelium: species comparisons. *Environ. Health Perspect.* 118:647–652.

Dodd, D.E., Gross, E.A., Miller, R.A., and Wong, B.A. (2010) Nasal olfactory epithelial lesions in F344 and SD rats following

1- and 5-day inhalation exposure to naphthalene vapor. *Int. J. Toxicol.* 29:175–184.

Dodd, D.E., Wong, B.A., Gross, E.A., and Miller, R.A. (2012) Nasal epithelial lesions in F344 rats following a 90-day inhalation exposure to naphthalene. *Inhal. Toxicol.* 24:70–79.

Fayerweather, W.E. (2007) Meta-analysis of lung cancer in asphalt roofing and paving workers with external adjustment for confounding by coal tar. *J. Occup. Environ. Hyg.* 4:175–200.

Fukami, T., Katoh, M., Yamazaki, H., Yokoi, T., and Nakajima, M. (2008) Human cytochrome P450 2A13 efficiently metabolizes chemicals in air pollutants: naphthalene, styrene, and toluene. *Chem. Res. Toxicol.* 21:720–725.

Griego, F.Y., Bogen, K.T., Price, P.S., and Weed, D.L. (2008) Exposure, epidemiology and human cancer incidence of naphthalene. *Regul. Toxicol. Pharmacol.* 51:22–26.

Hu, J., Sheng, L., Li, L., Zhou, X., Xie, F., D'Agostino, J., Li, Y., and Ding, X. (2013) Essential role of the cytochrome P450 enzyme CYP2A5 in olfactory mucosal toxicity of naphthalene. *Drug Metab. Dispos.* Advance Access published October 8, 2013. doi: 10.1124/dmd.113.054429.

IARC (2002) Naphthalene. In: *IARC Monographs on the Evaluation of Carcinogenic Risks to Humans. Volume 82: Some Traditional Herbal Medicines, Some Mycotoxins, Naphthalene and Styrene*, Lyon, France: World Health Organization, pp. 389–435.

International Programme on Chemical Safety (IPCS) and NIOSH (2005) *International Chemical Safety Card (ICSC) for Naphthalene (CAS No. 91-20-3)*. Available at http://www.cdc.gov/niosh/ipcs/icstart.html (accessed on November 6, 2013).

Karagiannis, T.C., Li, X., Tang, M.M., Orlowski, C., El-Osta, A., Tang, M.L., and Royce, S.G. (2012) Molecular model of naphthalene-induced DNA damage in the murine lung. *Hum. Exp. Toxicol.* 31:42–50.

Lewis, R.J. (2012) Naphthalene animal carcinogenicity and human relevancy: overview of industries with naphthalene-containing streams. *Regul. Toxicol. Pharmacol.* 62:131–137.

Lewis, D.F., Ito, Y., and Lake, B.G. (2009) Molecular modelling of CYP2 F substrates: comparison of naphthalene metabolism by human, rat and mouse CYP2 F subfamily enzymes. *Drug Metab. Drug Interact.* 24:229–257.

Lewis, R.J., Schnatter, A.R., Drummond, I., Murray, N., Thompson, F.S., Katz, A.M., Jorgensen, G., Nicolich, M.J., Dahlman, D., and Theriault, G. (2003) Mortality and cancer morbidity in a cohort of Canadian petroleum workers. *Occup. Environ. Med.* 60:918–928.

Lin, C.Y., Boland, B.C., Lee, Y.J., Salemi, M.R., Morin, D., Miller, L.A., Plopper, C.G., and Buckpitt, A.R. (2006) Identification of proteins adducted by reactive metabolites of naphthalene and 1-nitronaphthalene in dissected airways of rhesus macaques. *Proteomics* 6:972–982.

Li, L., Wei, Y., Van Winkle, L., Zhang, Q.Y., Zhou, X., Hu, J., Xie, F., Kluetzman, K., and Ding, X. (2011) Generation and characterization of a Cyp2f2-Null mouse and studies on the role of CYP2F2 in naphthalene-induced toxicity in the lung and nasal olfactory mucosa. *J. Pharmacol. Exp. Ther.* 339:62–71.

Meng, F., Wang, Y., Myers, M.B., Wong, B.A., Gross, E.A., Clewell, H.J. III Dodd, D.E., and Parsons, B.L. (2011)

p53 codon 271 CGT to CAT mutant fraction does not increase in nasal respiratory and olfactory epithelia of rats exposed to inhaled naphthalene. *Mutat. Res.* 721:199–205.

Merletti, F., Boffetta, P., Ferro, G., Pisani, P., and Terracini, B. (1991) Occupation and cancer of the oral cavity or oropharynx in Turin, Italy. *Scand. J. Work Environ. Health* 17:248–254.

NIOSH (2000) *Hazard Review: Health Effects of Occupational Exposures to Asphalt*, Cincinnati, OH: National Institute for Occupational Safety and Health.

NTP (1992) *Toxicology and Carcinogenesis Studies of Naphthalene (CAS No. 91-20-3) in B6C3F1 Mice (Inhalation Studies)*, Research Triangle Park, NC: National Toxicology Program.

NTP (2000) *NTP Technical Report on the Toxicology and Carcinogenesis Studies of Naphthalene (CAS No. 91-20-3) in F344/N Rats (Inhalation Studies)*, Research Triangle Park, NC: National Toxicology Program.

NTP. (2011) Naphthalene (CAS No. 91-20-3). In: *Report on Carcinogens*, 12th ed., Research Triangle Park, NC: U.S. Department of Health and Human Services, Public Health Service, National Toxicology Program.

Olsson, A., Kromhout, H., Agostini, M., Hansen, J., Lassen, C.F., Johansen, C., Kjaerheim, K., Langård, S., Stücker, I., Ahrens, W., Behrens, T., Lindbohm, M.L., Heikkilä, P., Heederik D., Portengen L., Shaham J., Ferro G., de Vocht F., Burstyn, I., and Boffetta P. (2010) A case-control study of lung cancer nested in a cohort of European asphalt workers. *Environ. Health Perspect.* 118:1418–1424.

Pham, N.T., Jewell, W.T., Morin, D., and Buckpitt, A.R. (2012a) Analysis of naphthalene adduct binding sites in model proteins by tandem mass spectrometry. *Chem. Biol. Interact.* 199:120–128.

Pham, N.T., Jewell, W.T., Morin, D., Jones, A.D., and Buckpitt, A.R. (2012b). Characterization of model peptide adducts with reactive metabolites of naphthalene by mass spectrometry. *PLoS ONE* 7:e42053.

Phimister, A.J., Lee, M.G., Morin, D., Buckpitt, A.R., and Plopper, C.G. (2004). Glutathione depletion is a major determinant of inhaled naphthalene respiratory toxicity and naphthalene metabolism in mice. *Toxicol. Sci.* 82:268–278.

Price, P.S. and Jayjock, M.A. (2008) Available data on naphthalene exposures: strengths and limitations. *Regul. Toxicol. Pharmacol.* 51:15–21.

Recio, L., Shepard, K.G., Hernandez, L.G., and Kedderis, G.L. (2012). Dose-response assessment of naphthalene-induced genotoxicity and glutathione detoxication in human TK6 lymphoblasts. *Toxicol. Sci.* 126:405–412.

Rushton, L. (1993). A 39-year follow-up of the U.K. oil refinery and distribution center studies: results for kidney cancer and leukemia. *Environ. Health Perspect.* 101:77–84.

Su, T., Bao, Z., Zhang, Q.Y., Smith, T.J., Hong, J.Y., and Ding, X. (2000). Human cytochrome P450 CYP2A13: predominant expression in the respiratory tract and its high efficiency metabolic activation of a tobacco-specific carcinogen, 4-(methylnitrosamino)-1-(3-pyridyl)-1-butanone. *Cancer Res.* 60:5074–5079.

Waidyanatha, S. and Rappaport, S.M. (2008) Hemoglobin and albumin adducts of naphthalene-1,2-oxide, 1,2-naphthoquinone and 1,4-naphthoquinone in Swiss Webster mice. *Chem. Biol. Interact.* 172:105–114.

Waidyanatha, S., Troester, M.A., Lindstrom, A.B., and Rappaport, S.M. (2002) Measurement of hemoglobin and albumin adducts of naphthalene-1, 2-oxide, 1, 2-naphthoquinone and 1, 4-naphthoquinone after administration of naphthalene to F344 rats. *Chem. Biol. Interact.* 141:189–210.

Wong, O. and Harris, F. (2005) Retrospective cohort mortality study and nested case-control study of workers exposed to creosote at 11 wood-treating plants in the United States. *J. Occup. Environ. Med.* 47:683–697.

Wong, O. and Raabe, G.K. (2000) A critical review of cancer epidemiology in the petroleum industry, with a meta-analysis of a combined database of more than 350,000 workers. *Regul. Toxicol. Pharmacol.* 32:78–98.

Yamane, G.K. (2006) Cancer incidence in the U.S. Air Force: 1989–2002. *Aviat. Space Environ. Med.* 77:789–794.

Zheng, J., Cho, M., Jones, A.D., and Hammock, B.D. (1997) Evidence of quinone metabolites of naphthalene covalently bound to sulfur nucleophiles of proteins of murine Clara cells after exposure to naphthalene. *Chem. Res. Toxicol.* 10:1008–1014.

66

GASOLINE (PETROL, MOTOR SPIRITS, MOTOR FUEL, NATURAL GASOLINE, BENZIN, MOGAS)

Raymond D. Harbison and Daniel A. Newfang

Note: Gasoline has many different CAS and EINICs registry numbers assigned to it due to the different distillation fractions and/or all of the different process derivatives. To name a few typically seen identifiers: CAS: 8006-61-9 gasoline, natural (EINICS 232-349-2); CAS 86290-81-5 gasoline, refinery grade components (EINICS 289-220-8); CAS 64671-46-4 gasoline, light straight run naphtha (EINICS 265-046-8); CAS 68514-15-8 gasoline, vapor recovery (EINICS 271-025-4) . . . as mentioned there are many more identifying registry numbers for gasoline, its mixtures, grades, distillation fractions, and/or process derivatives.

Target organs: Kidneys, eyes, skin, liver, CNS, respiratory system (*inhalation and aspiration risks*)

Risk/safety phrases: May cause cancer; may cause genetic damage; defects: may cause lung damage if swallowed; may be fatal if swallowed and enters airway; highly flammable

Exposure limits:

OSHA PELs (*There is no PEL for gasoline, the composition of these materials varies greatly. The OSHA recommends that the content of benzene, other aromatics, and additives should be determined individually*)

ACGIH TLVs (adopted 1982): 300 ppm (890 mg/m^3) TWA, 500 ppm (1480 mg/m^3) STEL, appendix A3 (adopted 1996) confirmed animal carcinogen with unknown relevance to humans. (ACGIH TLV Documentation Gasoline, 2014)

NIOSH RELs: No established limits, potential occupational carcinogen (i.e., "Ca" notation)

International Agency for Research on Cancer (IARC): 2B possible carcinogenic to humans

MAK (Germany): No established value

ERPG-1 (AIHA 2014): 200 ppm (654 mg/m^3) 1-h TWA. *Note: An average molecular weight of 80 g/mole conversion was used to convert mg/m^3 to ppm . . . conversion factor of 3.27*

ERPG-2 (AIHA 2014): 1000 ppm (3270 mg/m^3) 1-h TWA. *Note: An average molecular weight of 80 g/mole conversion was used to convert mg/m^3 to ppm . . . conversion factor of 3.27*

ERPG-3 (AIHA 2014): 4000 ppm (13,080 mg/m^3) 1-h TWA. *Note: An average molecular weight of 80 g/mole conversion was used to convert mg/m^3 to ppm . . . conversion factor of 3.27*

AEGL-1 USEPA (Proposed Limit 2009) (nondisabling): 730 mg/m^3 for 10 and 30 min, 1 h, 4- and 8-h timeframes

AEGL-2 USEPA (Proposed Limit 2009) (disabling): 7500 mg/m^3 for 10 and 30 min, 1 h, 4- and 8-h timeframes. *Note: The AEGL-2 value is higher than 1/10 of the lower explosive limit (LEL) of gasoline in air (which has an LEL = 14,000 ppm), therefore, safety considerations against hazard of explosion must be taken into account.*

AEGL-3 USEPA: Not determined. *Note: Although a lethal concentration was not attained in the available acute, subchronic and chronic toxicity studies, automotive gasoline vapor may act as a simple asphyxiant in sensitive individuals at 990,000 mg/m^3.*

WEEL (AIHA 2013): None established

Hamilton & Hardy's Industrial Toxicology, Sixth Edition. Edited by Raymond D. Harbison, Marie M. Bourgeois, and Giffe T. Johnson.
© 2015 John Wiley & Sons, Inc. Published 2015 by John Wiley & Sons, Inc.

IDLH: [Ca, potential occupational carcinogen] 1400 ppm: Due to all of the differing blends a specific IDLH has not been established for gasoline. A recommended IDLH of 1400 ppm based strictly on safety considerations (i.e., being 10% of the lower explosive limit of 1.4%).

Odor thresholds and recognition: Ranges from 0.1 to 1.1 ppm. The level of odor awareness (LOA), which is the concentration above which it is predicted that more than half of the exposed population will experience at least a distinct odor intensity and about 10% of the population will experience a strong odor intensity, is 7.4 ppm ($\sim$22 mg/m^3) as a representative value for a gasoline blend comprising summer and winter blends (API, 1994).

Reference values: There is no EPA reference dose (RfD) or reference concentration (RfC) for gasoline.

Typical routes of exposure: Inhalation, skin contact, and ingestion (lesser risk with normally accepted storage and use, although siphoning or accidental ingestion from unlabeled, incorrectly stored containers) (Conservation of Clean Air and Water in Europe, 1992).

BACKGROUND AND USES

Gasoline is a complex, volatile mixture of flammable liquid hydrocarbons derived chiefly from crude petroleum with specific additives such as octane improvers, antioxidants, metal deactivators, corrosion inhibitors, detergents, and demulsifiers to enhance performance characteristics. Typically gasoline contains about 150 different chemicals, including benzene, toluene, ethylbenzene, and xylene, which also are known as the BTEX compounds. Although tetraethyl lead (TEL) (used as an antiknock agent) was removed from the formulations in most industrialized nations by the early 2000s, it is still being used in some areas of the world (Taylor, 2011 and (United Nations Environmental Programme, 2014) UNEP 10th general meeting strategy presentation). Some aviation and racing fuels still contain TEL.

Environmentally, gasoline may be present in the air, groundwater, and soil. Its uses range from ground transport vehicles, aviation, agricultural, gasoline-powered equipment (i.e., lawnmowers) and stationary (i.e., generators and pumps) spark ignited, reciprocating, and Wankel (i.e., rotary) internal combustion engines.

Gasoline is a volatile liquid with characteristic odor properties. As a liquid, gasoline contains about 60–70% alkanes, 25–30% aromatics, and 6–9% alkenes. In the vapor phase, the aromatics are depleted to about 2% aromatics (which can contain up to 0.9% benzene), and the smaller chained hydrocarbons, being more volatile, are enriched to about 90% alkanes (Page and Mehlman, 1989). In many industrialized nations, there are limits on certain component concentrations specifically for the conventional, reformulated, and oxygenated blends by regulation, such as benzene, ethers (i.e., methyl tertiary butyl ether (MTBE)), and alcohols. The United States Environmental Protection Agency (USEPA) Clean Air Act Amendments of 1990 required reformulated gasoline to contain between 2.0 and 2.7% oxygen additives (i.e., MTBE) and <1% benzene. The Energy Policy Act of 2005 removed the oxygenate requirement, and the industry responded by reducing MTBE due to groundwater impacts with replacement strategies focused on alcohols (i.e., ethanol) (Weaver et al. 2014). Various additives (e.g., lubricants, antirust agents, and anti-icing agents) are blended into gasoline and may influence the properties and toxicity of specific gasoline blends (Gasoline Blending Streams Category Assessment Document, USEPA, 2014).

Oxygenating compounds have been added to the mixture to reduce carbon monoxide and soot, which is created during the burning of the gasoline. Historically, MTBE has been added but has been declining in use in the United States due to groundwater contamination (impacts taste of the water). Ethanol in the United States and ethyl tertiary butyl ether (ETBE, referred to as a bio-ether) in the European Union are now being researched and used as oxygenates (EFOA, 2013).

Holistically, gasoline components have a large boiling point in the range of 20–260 °C (−4–500°F).

Ethanol blends or gasohol (denoted by the letter "E" preceding the percentage of the mixture made up of the ethanol): New formulations of gasoline with blends from 5% ethanol denoted as E5 to 100% ethanol denoted as E100 are being introduced throughout the world. Ethanol is favored as an oxygenate additive to replace MTBE. Methyl tertiary butyl ether was used as a TEL replacement since 1979 in the United States and has been identified as a groundwater contaminant. Ethanol was identified as a renewable fuels oxygenate replacement for MTBE (*Note*: The addition of an oxygenate allows the fuel to burn more completely and therefore produce cleaner emissions). Some common ethanol blends are as follows:

E10: Mixture containing 10% anhydrous ethanol and 90% gasoline (*Note*: The anhydrous ethanol has to contain <1% water)

E15: Mixture containing 15% anhydrous ethanol and 85% gasoline

hE15: Mixture containing 15% hydrous ethanol and 85% gasoline (NOTE: The hydrous ethanol can contain up to 4.9% water)

E85: Mixture containing 51–85% anhydrous ethanol and 49–15% gasoline, respectively

ED95: Mixture containing 95% hydrous ethanol and 5% additives (there is no gasoline fraction in the ED95 blend . . . refer to ethanol toxicology for assistance to E95 exposures)

E100: 100% anhydrous ethanol (there is no gasoline fraction in the E100 blend . . . refer to ethanol toxicology for assistance to E100 exposures)

Aviation: Aviation gas (or avgas for short) is a lower volatility gasoline that is important for aircraft engines because they regularly operate at high altitudes and have high temperature engines. It is important not to confuse jet fuel with aviation gasoline; they are in no way similar. Jet fuel is similar to kerosene. As of this publishing date, organic lead (i.e., TEL) is still used in some aviation grade gasoline blends. There are two main types of leaded avgas: 100 octane, which can contain up to 4.24 grams of lead per gallon; and 100 octane low lead (100LL), which can contain up to 2.12 grams of lead per gallon. Currently, 100LL is the most commonly available and most commonly used type of avgas. Common avgas grades (current and historical) are as follows:

80/87 (80-87 octane with up to 0.5 ml TEL/US gallon—dyed red, sometimes called Grade 80 Avgas, fuel is obsolete and phased out of production now)

100LL (103-131 octane with up to 2 ml TEL/US gallon of lead—dyed blue, sometimes called Grade 100LL Avgas, the most common aviation gasoline)

100/130 (103-130 octane with up to 4 ml TEL/US gallon—dyed green, sometimes called Grade 100 Avgas, fuel is phased out of production now)

Note 1: Future Development Note, there is a concerted industry effort to phase out the use of lead compounds, which is designated as UL or UnLeaded fuels (such as 82UL, 94UL, Hjelmco 100UL, UL102/100SF-Swift Fuels or G100UL-GAMI 100UL)

Note 2: Historical Note, there have been many different grades of aviation gasoline in general use, e.g., 80/87, 91/96, 100/130, 108/135, and 115/145. With decreasing demand, many blends have become obsolete and Avgas 100/130 (a.k.a. 100LL) is the predominant fuel choice.

Note 3: Nomenclature Note, avgas grades are defined primarily by their octane rating. Two ratings are applied to avgas (the lean mixture rating first and the rich mixture rating second). For example, Avgas 100/130LL (in this case the lean mixture performance rating is 100 octane and the rich mixture rating is 130 octane). In an effort to make the nomenclature easier, it is a common practice to designate the grade by just the lean mixture performance, i.e., Avgas 100/130LL is noted as Avgas 100LL (*otherwise pronounced as Avgas one-hundred low lead*).

Mogas (*sometimes denoted by "E0" to indicate that the blend is ethanol free*) is a generic term that identifies both a lead- and ethanol-free gasoline blend.

An example of the diversity of compounds that make up automotive gasoline is given in Table 66.1.

There are no drinking water standards established for gasoline. The EPA health advisory for gasoline (unleaded) is 5 µg/l (U.S. Environmental Protection Agency [USEPA], 1991). No other health advisories are indicated. Neither cancer risk levels nor reference doses have been established by the EPA.

PHYSICAL AND CHEMICAL PROPERTIES

Note: Since gasoline is a complex mixture, all its physical and chemical properties will vary with the composition of the mixture's components.

Average molecular weight: 80–118

Color: Colorless to pale brown in its un-dyed state; aviation blends are dyed to help field personnel visually determine the octane rating or grade.

Physical state: Liquid

Boiling point (*initial*): 26–49 °C (78–120°F, NFPA 921 Table 25.3-1)

Boiling point (*final*): 171–233 °C (340–451°F, NFPA 921 Table 25.3-1)

Density: Gasoline liquid ranges from 0.70 to 0.80 kg/l (averaging ∼719.7 kg/m^3; 0.026 lb/in^3; 6.073 lb/US gal; 7.29 lb/imp gal), the heavier or higher densities having a greater volume of aromatics. Gasoline floats on water in all densities. *Gasoline vapors* are denser than air and will sink and collect at the lowest point of the area you're in such as basements, crawl spaces, tunnels, man holes, ditches, etc.

Color: In its natural state it is usually a colorless clear, red/pinkish, or light brown liquid. It can also be purposely dyed (especially with aviation fuels) to visually identify a grade or octane rating (e.g. 82UL—purple, 80/87—red, 100LL—blue, 100/130—green).

Odor: See box at the beginning of this chapter

Odor threshold: See box at the beginning of this chapter

Flash point: −35 to −40 °C (−49 to −40 °F; NFPA 921 Table 25.3-1)

Melting point: −38 to −46 °C (−36 to −51 °F)

Lower explosive limit (LEL): 1.4% (NFPA 921 Table 25.3-1)

Upper explosive limit (UEL): 7.4% (NFPA 921 Table 25.3-1)

Autoignition temperature: Minimum of 354 °C (minimum of 670°F; NFPA 921 Table 25.3-1)

Vapor pressure: 38–750 mmHg (0.6–14.5 psi; NIOSH NPG, 2007; ATSDR, 2014)

TABLE 66.1 Composition of Various Types of Gasoline (% Weight/Weight [in grams])

Chemical Name	Leaded, %	Unleaded (%) (Regular, Mid and Super grades)
Hydrocarbons (aromatic and aliphatic)	60	48.8–56.2
Benzene	3.9	0.9–4.4 (range difference is seen in the conventional, reformulated & oxygenated blends)
Toluene	4.5	4.8–6.0
Xylenes	5.6	6.6–7.4
Ethyl benzene	1.2	1.4
Other benzenes	9	6.9–7.4
Others and unknowns	15.8	20.9–24.6
Octane enhancers such as:		
Methyl *t*-butyl ether (MTBE)		
t-Butyl alcohol		
Ethanol		
Methanol		
Antioxidants		
Butylated methyl, ethyl, and dimethyl phenols		
Various other phenols and amines		
Metal deactivators		
Disalicylidene-*N*-methyldipropylene-triamine		
N,N'-disalicylidene-1,2-ethanediamine		
Other related amines		
Ignition controllers		
Tri-*o*-cresylphosphate (TOCP)		
Icing inhibitors		
Isopropyl alcohol		
Detergents/dispersants		
Various phosphates, amines, phenols, alcohols, and carboxylic acids		
Corrosion inhibitors		
Carboxylic, phosphoric, and sulfonic acids		
Ethanol	N/A	Ranges from 0 to 100% with the respect to the "E" blend number, *e.g. E15 indicates 15% ethanol and 85% unleaded gasoline*

Table is a compilation of literature sources from ATSDR and American Petroleum Institute.

Vapor density, relative to air with air = 1 : 3–4 (NFPA 921 Table 25.3-1)

Specific gravity (water = 1): 0.72–0.80 at 60°F (NIOSH NPG, 2007; ATSDR, 2014)

Water solubility: Practically insoluble (NIOSH NPG, 2007)

MAMMALIAN TOXICOLOGY

Acute Effects

The most common symptoms resulting from excessive gasoline exposure are headache, blurred vision, dizziness, inebriation, vomiting, dizziness, confusion, fever, and nausea. A level of 500–1000 parts per million (ppm) (for 30–60 min) can result in a euphoric condition followed by ataxia, drowsiness, and dizziness. Higher levels (1000–3000 ppm) result in irritation, headache, nausea, and vomiting, as well as eye, nose, and throat irritation (Davis et al., 1960). Levels in excess of 5000 ppm can result in dizziness, deep anesthesia, pulmonary edema, and death within minutes (Ainsworth, 1960; Wang and Irons, 1961). Chemical pneumonitis resulting from aspiration of gasoline after ingestion is a serious clinical concern.

Signs of severe intoxication include mild excitation followed by loss of consciousness, occasional convulsions, cyanosis, congestion, and capillary hemorrhaging of the lungs and other internal organs, and finally death from circulatory failure. In adults, ingestion of as little as 20–50 g may produce severe poisoning. Cases of accidental poisonings in the literature have noted abnormal liver function tests and acute reversible injury to the upper portions of the kidneys (Kuehnel and Fisher, 1986).

Exposure to leaded gasoline may cause lead poisoning and all of the associated symptoms (Goldings and Stewart, 1982; Kovanen et al., 1983).

Chronic Effects

Long-term exposure to gasoline has been reported to result in anorexia, weight loss, weakness and cramps, and neurological effects, including poor coordination and tremor (Poklis and Burkett, 1977). The neurological effects appear to be reversible upon cessation of exposure. However, postmortem examinations of persons with high exposure levels of gasoline (gasoline sniffers) frequently show cerebral and pulmonary edema, as well as liver and kidney necrosis (Poklis and Burkett, 1977, Dueñas-Laita, A., 2007).

Based on reported exposures, it appears that adverse hematological effects can occur following short-term, high level exposure and longer-term, lower level exposures to gasoline. The effects are most likely the result of an exposure to benzene or lead, constituents of gasoline. The effects seen after an acute exposure are most likely reversible, whereas the effects seen after a long-term exposure may not be reversible.

Carcinogenicity   Possible carcinogen. Experimental animal data provide inadequate evidence for the carcinogenicity of gasoline. The species-specific effects result from a mechanism involving the accumulation of globulin in male rat kidneys that cannot be anthropomorphized to humans. Similarly, human epidemiological studies are not adequate to support a conclusion that gasoline is carcinogenic. With this stated, the following entities have labeled gasoline as:

International Agency for Research on Cancer (IARC): Group 2B—possibly carcinogenic to humans.

American Conference for Governmental Industrial Hygienists (ACGIH): A3—confirmed animal carcinogen.

Mechanism of Action(s) and Chemical Pathology

No mechanistic or chemical pathology studies were located for gasoline. Information on the toxicokinetics of gasoline components (i.e., benzene, toluene, xylene, lead, etc.) are available separately and should be reviewed.

While gasoline is absorbed from all exposure routes, the dermal route appears to be slower than the oral and inhalation routes. Some of the gasoline components are more rapidly absorbed than others. For example, aromatic compounds (e.g., benzene, toluene, and xylene) that have both high blood/air partition coefficients and skin penetration rates are absorbed more rapidly than other gasoline components.

An acute exposure to gasoline has been associated with skin and sensory irritation, central nervous system depression, and effects on the respiratory system. A prolonged exposure also affects these organs as well as the kidney, liver, and blood system. In general, the effects that have been identified following gasoline exposure have also been identified for one or more of the components of gasoline (e.g. benzene, toluene, xylene, etc.). For example, all substances have been shown to be neurotoxic, and studies that indicate

that gasoline is hemotoxic are supported by the abundant literature on benzene hematotoxicity (Evaluation of the Health Effects From Exposure to Gasoline and Gasoline Vapors, Final Report NESCAUM, Air Toxics Committee, 1989).

Organic TEL is converted in the liver to triethyl lead, a water-soluble metabolite, which can accumulate in the brain (Robinson, 1978), which can then be further broken down to inorganic lead.

Biomarkers potentially can be used to identify exposure to gasoline:

Increased blood and urine levels for leaded gasoline

Increased urinary thioether output has a limited use due to hydrocarbon exposure (*Note*: Be aware that increased urinary thioether output is not specific to gasoline; it is also complicated by smoking)

Increased urinary phenol levels due to benzene exposure

Although gasoline vapor acts as an anesthetic at high levels (10,000 ppm), the major site of toxicity is the kidney (Kuna and Ulrich, 1984). Rats and monkeys were exposed to gasoline vapor, 284–1552 ppm (unleaded gasoline) or 103–374 ppm (leaded gasoline), both for 6 h/day, 5 days/week, for 90 days (Kuna and Ulrich, 1984). Although vomiting was noted in some monkeys after 2 weeks' exposure, no changes in body weight, hematology, or central nervous system (CNS) responses were noted in either species. Male rats in the 1552 ppm unleaded gasoline group did show regenerated epithelium and dilation of the kidney tubules, which was not present in any other group.

Gasoline vapor (100 ppm for 6 weeks) has also been reported to cause pulmonary damage in rats secondary to a decrease in production of pulmonary surfactant (Cooper, 1980). Other studies of the toxicity of chronic gasoline vapor exposure have reported eye damage in rabbits (Grant, 1974), decreased body weight in mice and rats, and a reduction in the incidence of cystic or enlarged uteri in female mice (International Research and Development Corporation [IRDC], 1983).

Chronic inhalation of unleaded gasoline vapor has also been shown to produce hepatocyte cell proliferation in mice (Tilbury et al., 1993). However, this effect was statistically significant only in mice exposed to 2056 ppm. There were no changes in liver-specific serum enzymes or histopathology in any dosage group.

The levels of gasoline that killed people are about 10,000–20,000 ppm when breathed in and about 12 ounces when swallowed (ATSDR, 2014; Harper and Liccione, 1995).

SOURCES OF EXPOSURE

Gasoline is typically in liquid form; therefore, the most likely source of exposure will be breathing its vapors when one is filling a fuel tank. Drinking contaminated groundwater (ingestion), bathing in (dermal & inhalation), and spillage onto skin are also potential routes of exposures. In the vapor phase, the aromatics are depleted to about 2% aromatics (which can contain up to 0.9% benzene), and the smaller chained hydrocarbons, being more volatile, are enriched to about 90% alkanes (Page and Mehlman, 1989).

Intentional or accidental ingestion of gasoline often results in aspiration of the gasoline into the lungs because of its high volatility and low surface tension. Therefore, the most common effect with acute gasoline ingestion is aspiration pneumonia, which is often accompanied by respiratory distress, pulmonary edema, emphysema and focal alveolar hemorrhage. Death from asphyxia is often the result in cases of gasoline ingestion when aspiration pneumonia becomes severe (ATSDR, Toxicological Profile for Automotive Gasoline, 2014).

INDUSTRIAL HYGIENE

Occupational Safety and Health Administration (OSHA): There is a partially validated monitoring method cited by the OSHA (i.e., Method PV2028)

Real time:

Detector sorbent tubes (e.g., SKC)

Flame ionizing detector (FID)

Photo ionizing detector (PID)

Multi-gas detector (O_2/LEL/H_2S/CO and some include PIDs)

BTX or BTEX sorbent tubes (e.g., Draeger, SKC)

Note: Many times exposure monitoring can be performed by monitoring for individual components such as benzene, MTBE, ethanol (for gasohol), BTX (benzene, toluene, xylene), or BTEX (benzene, toluene, ethylbenzene, xylene) collectively

MEDICAL MANAGEMENT

First Aid/Pre-hospital Measures

Eye contact: Immediately flush with large amounts of water for at least 15 min, lifting the upper and lower lids. Remove contact lenses (if worn).

Skin contact: Quickly remove any contaminated clothing. Immediately wash contaminated skin with large amounts of soap and water. *Note:* Do not place the contaminated clothing near any potential ignition sources.

Inhalation: Remove the person from the exposure.

Ingestion: Swallowing gasoline may cause damage to the linings of the mouth, throat, esophagus, stomach, and intestines. Do not induce vomiting. Pulmonary aspiration of even small amounts of *ingested* gasoline can cause chemical pneumonitis. Do not induce emesis or use gastric lavage,

and do not administer activated charcoal. Gasoline is poorly absorbed from the stomach. Cathartics with magnesium or sodium sulfate are acceptable. If spontaneous vomiting occurs, watch for signs of pulmonary aspiration. *Note:* The harsh taste of gasoline makes it unlikely that large quantities will be swallowed. However, several cases of poisoning have occurred in persons trying to suck (siphon) gas from an automobile tank using a garden hose or other tube. This practice is extremely dangerous and not advised.

There is no antidote for gasoline poisoning. Treatment consists of support of cardiovascular and respiratory functions.

Medical Care

The health-care provider will measure and monitor the patient's vital signs, including temperature, pulse, breathing rate, and blood pressure. Symptoms will be treated as appropriate. The patient may receive:

Breathing tube

Bronchoscopy—a camera down the throat to see burns in the airways and lungs

Endoscopy—a camera through the mouth to see burns in the esophagus and stomach

Fluids through a vein (IV)

Oxygen

Surgical removal of burned skin (skin debridement)

Tube through the mouth into the stomach to wash out the stomach (gastric lavage)

Washing of the skin (irrigation)—perhaps every few hours for several days

How well a patient does depends on the amount swallowed and how quickly the treatment was received. The faster a patient gets medical help, the better the chance for recovery.

Systemic effects may also include renal failure and increased susceptibility to ventricular fibrillation (Mirkin, 2007).

Delayed Effects

Patients who have ingested gasoline should be observed for at least 6 h for signs of chemical pneumonitis. Systemic effects may develop over several hours and may include hemorrhage of the pancreas and fatty degeneration of the liver and of the proximal convoluted tubules and glomeruli of the kidneys. Acute renal toxicity may persist for several weeks following ingestion of gasoline but usually resolves with treatment.

Follow-up

Patients who have aspirated gasoline should receive follow-up pulmonary function tests.

Acute renal toxicity may persist for several weeks following ingestion of gasoline but usually resolves with treatment.

Patients who have corneal injuries should be reexamined within 24 h.

See ATSDR Medical Management Guidelines for Gasoline (2014); http://www.atsdr.cdc.gov/MMG/MMG .asp?id=465&tid=83

REFERENCES

ACGIH (2014) *Threshold Limit Values for Chemical Substances in the Work Environment*, Available at http://www.acgih.org.

AIHA (2014) *American Industrial Hygiene Association, 2014 ERPG/WEEL Handbook*, AIHA Guideline Foundation, Available at http://www.aiha.orh/get-involved/AIHAGuideline Foundation/EmergencyResponsePlanningGuidelines/ Documents/2014%20ERPG%20Values.pdf.

Ainsworth, R.W. (1960) Petrol-vapour poisoning. *Br. Med. J.* 1:1547–1548.

API (1994) Available at http://www.epa.gov/oppt/aegl/pubs/ gasoline_proposed_oct_2009.v1.pdf.

ATSDR (2014) Toxicological Profile for Automotive Gasoline, Available at http://www.atsdr.cdc.gov/MMG/MMG.asp?id= 465&tid=83.

ATSDR Medical Management Guidelines for Gasoline (2014) Available at http://www.atsdr.cdc.gov/MMG/MMG.asp? id=465&tid=83.

Conservation of Clean Air and Water in Europe (1992) *Gasolines Product Dossier No. 92/103. Brussels.*

Cooper, P. (1980) A whiff of petrol. *Food Cosmet. Toxicol.* 18:433–434.

Davis, A., Schafer, L.J., and Bell, Z.G. (1960) The effects on human volunteers of exposure to air containing gasoline vapor. *Arch. Environ. Health* 1:548–554.

Dueñas-Laita, A. (2007) Freon and other inhalants. In: Shannon, M.W., Borron, S.W., and Burns, M.J., editors., *Haddad and Winchester's Clinical Management of Poisoning and Drug Overdose*, 4th ed., Chapter 95, Philadelphia, PA: Saunders Elsevier.

EFOA (European Fuel Oxygenates Association) (2013) Available at http://www.efoa.eu/en/home-99.aspx.

Gasoline Blending Streams Category Assessment Document, USEPA (2014) Available at http://www.epa.gov/hpv/pubs/ summaries/gasnecat/c13409rr3.pdf.

Goldings, A.S. and Stewart, R.M. (1982) Organic lead encephalopathy: behavioral change and movement disorder following gasoline inhalation. *J. Clin. Psychiatry* 43:70–72.

Grant, W.M. (1974) *Toxicology of the Eye*, 2nd ed., Springfield, IL: Charles C Thomas.

Harper and Liccione (1995) *Toxicological Profile for Gasoline, US Department of Health and Human Services, Public Health Service, Agency for Toxic Substances and Disease Registry, June 1995.*

International Research and Development Corporation (IRDC) (1983) *Motor fuel chronic inhalation study, API MBSD Publication No 32-32165, Sponsored by American Petroleum Institute.*

Kovanen, J., Somer, H., and Schroeder, P. (1983) Acute myopathy associated with gasoline sniffing. *Neurology* 33:629–631.

Kuehnel, E. and Fisher, P. (1986) Acute renal injury limited to the upper poles following gasoline ingestion. *Kidney. Int.* 29:304–311.

Kuna, R.A. and Ulrich, C.E. (1984) Subchronic inhalation study of two motor fuels. *J. Am. Coll. Toxicol.* 3:217–229.

Mirkin, D.B. (2007) Benzene and related aromatic hydrocarbons. In: Shannon, M.W., Borron, S.W., and Burns, M.J., editors., *Haddad and Winchester's Clinical Management of Poisoning and Drug Overdose*, 4th ed., Chapter 94, Philadelphia, PA: Saunders Elsevier.

NIOSH NPG (2007) National Institute for Occupational Safety and Health (NIOSH) *Pocket Guide to Chemical Hazards*, NIOSH 2005-149, Washington, DC: US Department of Health and Human Sciences. Available at http://www.cdc.gov/niosh/docs/ 2005-149/pdfs/2005-149.pdf.

NESCAUM (Northeast States for Coordinated Air Use Management) (1989) *Air Toxics Committee Evaluation of the Health Effects from Exposure to Gasoline and Gasoline Vapors, Final Report & Executive Summary, August.*

Page, N.P. and Mehlman, M. (1989) Health effects of gasoline refueling vapors and measured exposures at service stations. *Toxicol. Ind. Health* 5(5):869–890.

Poklis, A. and Burkett, C.D. (1977) Gasoline sniffing: a review. *Clin. Toxicol.* 11:35–41.

Robinson, R.O. (1978) Tetraethyl lead poisoning from gasoline sniffing, *J. Am. Med. Assoc.* 240:1373–1374.

Taylor, R. (2011) *Countries where Leaded Petrol is Possibly Still Sold for Road Use as at 17th June 2011.*

Tilbury, L., et al. (1993) Hepatocyte cell proliferation in mice after inhalation exposure to unleaded gasoline. *J. Toxicol. Environ. Health* 38:293–307.

(2014) *United Nations Environment Programme, UNEP 10th general meeting strategy presentation.* Available at http://www .unep.org/transport/pcfv/PDF/10gpm_CHpresentation_strategy .pdf.

USEPA (1991) *Drinking Water Regulation and Health Advisory*, Washington, DC: US Environmental Protection Agency, Office of Water.

Wade, J.F. III and Newman, L.S. (1993) Diesel asthma: reactive airways disease following overexposure m locomotive exhaust. *J. Occup. Environ. Med.* 35:149–154.

Wang, C.C. and Irons, G.V. (1961) Acute gasoline intoxication. *Arch. Environ. Health* 2:714–716.

Weaver, J., et al. (2014) *Gasoline Composition Regulations Affecting LUST Sites, Ecosystems Research Division, National Exposure Research Laboratory, Office of Research and Development, USEPA, EPA 600/R-10/001 January 2.*

AEGL's: http://response.restoration.noaa.gov/aegls
ALOHA: http://response.restoration.noaa.gov/aloha
ALOHA Tools: http://response.restoration.noaa.gov/alohatools

Ask Dr. ALOHA Articles: http://response.restoration.noaa.gov/ADA/overview

ERPG's: http://response.restoration.noaa.gov/erpgs

IDLH's: http://response.restoration.noaa.gov/idlhs

Levels of Concern: http://response.restoration.noaa.gov/locs

Public Exposure Guidelines: http://response.restoration.noaa.gov/pubexpguides

Toxic Levels of Concern: http://response.restoration.noaa.gov/toxiclocs

TEELs: http://response.restoration.noaa.gov/teels

Workplace Exposure Limits: http://response.restoration.noaa.gov/workexplimits

67

1-BROMOPROPANE

Marek Banasik

First aid: Eyes: Holding the eyelids apart, flush eyes promptly with copious flowing water for at least 20 min—get medical attention immediately; skin: remove contaminated clothing, wash the skin thoroughly with mild soap and plenty of water for at least 15 min, wash clothing before re-use—get medical attention immediately; inhalation: in case of mist inhalation or breathing fumes released from heated material, remove person to fresh air, keep him quiet and warm—apply artificial respiration if necessary, and get medical attention immediately; ingestion: if swallowed, wash mouth thoroughly with plenty of water and give water to drink—get medical attention immediately. Notes to the physician: irritant—in case of inhalation may cause central nervous system toxicity, no specific antidote, treat symptomatically and supportively.

Occupational exposure limits: Derived no effect level (DNEL) (systemic effects, long-term inhalation): 20.12 mg/m^3; DNEL (systemic effects, short-term inhalation): 4,392 mg/m^3; DNEL (local effects, short-term inhalation): 2,416 mg/m^3 (ECHA, 2014).

Reference values: ACGIH TLV time-weighted average (TWA): 0.503 mg/m^3; DNEL (systemic effects, long-term inhalation): 10.06 mg/m^3; DNEL (systemic effects, short-term inhalation): 2,196 mg/m^3; DNEL (local effects, short-term inhalation): 1,208 mg/m^3 (ECHA, 2014).

Hazard statements: H225—highly flammable liquid and vapor; H315—causes skin irritation; H319—causes serious eye irritation; H335—may cause respiratory irritation; H336—may cause drowsiness or dizziness; H360—may damage fertility or the unborn child; H373—may cause damage to organs through prolonged or repeated exposure.

Precautionary statements: P201—obtain special instructions before use; P260—do not breathe the dust/fume/gas/mist/vapors/spray; P273—avoid release to the environment; P280—wear protective gloves/protective clothing/eye protection/face protection; P310—immediately call a poison center/doctor/. . .; P303 + 361 + 353—if on skin (or hair): take off immediately all contaminated clothing, rinse skin with water/shower; P304 + 340—if inhaled: remove person to fresh air and keep comfortable for breathing; P305 + 351 + 338—if in eyes: rinse cautiously with water for several minutes, remove contact lenses, if present and easy to do—continue rinsing.

Risk phrases: R11—highly flammable; R36/37/38—irritating to eyes, respiratory system and skin; 48/20—harmful: danger of serious damage to health by prolonged exposure through inhalation; R60—may impair fertility; R63—possible risk of harm to the unborn child; R67—vapors may cause drowsiness and dizziness.

Safety phrases: S45—in case of accident or if you feel unwell seek medical advice immediately (show label where possible); S53—avoid exposure—obtain special instructions before use.

BACKGROUND AND USES

1-Bromopropane [1-BP; Chemical Abstracts Service Registry Number (CASRN): 106-94-5] is a halogenated solvent, which was initially marketed as a potential alternative to ozone-depleting chlorofluorocarbons. It is utilized in a variety of industrial sectors, including vapor decreasing, dry

Hamilton & Hardy's Industrial Toxicology, Sixth Edition. Edited by Raymond D. Harbison, Marie M. Bourgeois, and Giffe T. Johnson.
© 2015 John Wiley & Sons, Inc. Published 2015 by John Wiley & Sons, Inc.

TABLE 67.1 Chemical and Physical Properties of 1-Bromopropane

Characteristic	Information
Chemical structure	
Chemical name	1-Bromopropane
Synonyms/trade names	1-BP, CCRIS 30, n-propyl bromide, propyl bromide
CASRN	106-94-5
Chemical formula	C_3H_7Br
Molecular weight	122.99 g/mol
Physical state	Colorless liquid at 20 °C and 1,013 hPa
Melting point	−110 °C
Boiling point	71 °C
Vapor pressure	14.772 kPa at 20 °C
Vapor density	4.25
Relative density	1.353 at 20 °C
Water solubility	2,450 mg/l at 20 °C
Octanol/water partition coefficient (log K_{OW})	2.10
Henry's law constant	7.3×10^{-3} atm-m^3/mol at 20 °C

Source: ECHA (2014); HSDB (2013).

cleaning, and as a solvent carrier for adhesives in foam cushion manufacturing.

1-Bromopropane is a high production volume chemical. In the United States (US), the national production volume in 2012 was 15,348,727 pounds (*ca.* 6,962,065 kg) (EPA, 2012). In the European Union (EU), three registrations were reported on 1-BP. Two of the registrations did not report production volumes; the data were claimed as confidential or for an intermediate use only. One registration reported an annual production volume range of *ca.* 2.2–22 million pounds (1,000–10,000 tonnes) (ECHA, 2014).

1-Bromopropane has undergone extensive toxicological testing, including 2-year carcinogenicity studies. Several regulatory agencies have initiated regulatory evaluations of 1-BP, especially in the occupational setting, due to recent reports on its carcinogenic potential and adverse effects identified in exposed workers. The sections that follow provide an overview of this information and the conclusions from these regulatory evaluations.

PHYSICAL AND CHEMICAL PROPERTIES

1-Bromopropane is a clear, colorless to pale yellow compound with a strong characteristic odor that exists as a liquid at room temperature. It is considered as a volatile organic compound, based on its moderate boiling point and high vapor pressure. Its liquid density is greater than water, and its vapor density is greater than air. 1-Bromopropane is readily soluble in water. A summary of 1-BP's physicochemical properties is presented in Table 67.1.

ENVIRONMENTAL FATE AND BIOACCUMULATION

The physicochemical properties and environmental fate testing on 1-BP supports that it will migrate in soil, based on its high water solubility and low log K_{OW}. It is hydrolytically unstable at environmentally relevant pH values (i.e., pH 7–9) with half-life >19 days. In the vapor phase, 1-BP has an estimated photodegradation half-life of approximately 12 days. Biodegradation data support that 1-BP is not readily biodegradable and is not expected to bioconcentrate in aquatic organisms. A summary of the available environmental fate and bioaccumulation data on 1-BP is presented in Table 67.2.

TABLE 67.2 Environmental Fate and Bioaccumulation Data on 1-Bromopropane

Endpoint	Protocol	Result
Hydrolysis	EU method C.7; OECD[a] method 111; EPA[b] OPPTS[c] method 830.2110	$t_{1/2}$[d] at pH 4 = 25 days, pH 7 = 23 days, and pH 9 = 19 days; conclusion: hydrolytically unstable
Photodegradation	Model estimate (AOPWIN™ v1.92)[e]	$t_{1/2}$ = 11.6 days
28-day ready biodegradability study preceded by an acclimation phase	EU method C.4-E; OECD method 301D	19.2% biodegradation after 28 days; conclusion: not readily biodegradable
Bioconcentration factor (BCF) in fish	Model estimate (BCFBAF™ v3.00)[f]	BCF$_{fish}$ = 11.29 l/kg wet weight

Source: ECHA (2014).

[a]Organisation for Economic Co-operation and Development.
[b]US Environmental Protection Agency.
[c]Office of Pollution Prevention and Toxic Substances.
[d]Half-life.
[e]Atmospheric oxidation program for Microsoft Windows.
[f]Program that estimates fish BCF and its logarithm using regression method based on log K_{OW} plus any applicable correction factors and Arnot–Gobas method. It also incorporates prediction of apparent metabolism half-life in fish, and estimates BCF and BAF for three trophic levels.

TABLE 67.3 Aquatic and Terrestrial Toxicity Data on 1-Bromopropane

Test Species	Test Method	Endpoint and Exposure Duration	Result (mg/l)[a]
Toxicity to Aquatic Vertebrates			
Rainbow trout	OECD 203; EU C.1	96-h[b] LC_{50}[c]	24.3
(*Oncorhynchus mykiss*)	(semi-static test)	96-h NOEC[d]	1.77
Toxicity to Aquatic Invertebrates			
Cladoceran (*Daphnia magna*)	OECD 202 (static)	48-h EC_{50}[e] (immobilization)	99.3
		48-h NOEC	29.6
Toxicity to Aquatic Plants			
Freshwater alga	OECD 201; EU C.3; EPA	96-h EbC_{50}[f]	52.4
(*Pseudokirchneriella subcapitata*)	OPPTS 850.5400	96-h ErC_{50}[g]	72.3
	(static)	96-h NOEC	12.4
Toxicity to Microorganisms			
Activated sludge	EC C.11; EPA OPPTS	5-min EC_{50}[h]	270
	850.6800 (static)	5-min NOEC	100

Source: ECHA (2014).

[a]Measured concentration.
[b]Hour(s).
[c]Median lethal concentration.
[d]No observed effect concentration.
[e]Median effect concentration.
[f]Effect concentration measured as 50% reduction in biomass growth in algae tests.
[g]Effect concentration measured as 50% reduction in growth rate in algae tests.
[h]Test duration reduced from 3 h to 5 min due to the volatile nature of the test substance.

ECOTOXICOLOGY

The toxicity of 1-BP has been assessed in aquatic vertebrates, invertebrates, and plants. A summary of these data are presented in Table 67.3. The data support that 1-BP is moderately toxic to aquatic organisms. The manufacturers of 1-BP calculated the following predicted no effect concentrations (PNECs) for aquatic and terrestrial organisms and secondary poisoning for predators: $PNEC_{freshwater} = 0.0243$ mg/l, $PNEC_{marine} = 0.00243$ mg/l, $PNEC_{sewage\ treatment\ plant} = 10$ mg/l, $PNEC_{sediment\ freshwater} = 0.249$ mg/kg sediment dry weight (dw) (based on partition coefficient), $PNEC_{sediment\ marine} = 0.0249$ mg/kg sediment dw (based on partition coefficient), $PNEC_{soil} = 0.0361$ mg/kg soil dw (based on partition coefficient), and $PNEC_{oral\ secondary\ poisoning} = 0.084$ mg/kg food (ECHA, 2014).

MAMMALIAN TOXICOLOGY

1-Bromopropane is a member of halogenated alkanes and is a strong alkylating agent because bromine is a good leaving group, and an electrophilic intermediate is expected to be readily formed that can bind covalently with macromolecules such as proteins and DNA. Studies in experimental animals

and humans indicate that 1-BP can be absorbed following inhalation, oral, or dermal exposure (Cheever et al., 2009; Hanley et al., 2007). Metabolism studies show that oxidation by P450 enzymes (e.g., CYP2E1) and glutathione conjugation are the primary metabolic pathways (Garner et al., 2006; Ishidao et al., 2002). Over 20 metabolites have been identified in rodent studies, including the four metabolites detected in urine samples of 1-BP-exposed workers (Hanley et al., 2009) and several reactive intermediate such as glycidol, propylene oxide, and α-bromohydrin (Ishidao et al., 2002). Experimental animal studies have been performed on this substance to evaluate potential acute and repeated exposure hazards. A summary of these data and the conclusions from the US National Toxicology Program's (NTPs) evaluations are provided below.

Acute Studies

The acute toxicity of 1-BP has been evaluated in a series of guideline studies by the inhalation, dermal, and oral routes of exposure. The reported median lethal concentrations or doses in these studies were typically above classification concern levels (e.g., inhalation 4-h $LC_{50} = 72,306.04$ mg/m^3, dermal 24-h $LD_{50} > 2,000$ mg/kg body weight, and oral $LD_{50} > 2,000$ mg/kg body weight) (ECHA, 2014). 1-Bromopropane is classified as a skin and eye irritant and is not regarded as a sensitization hazard. As discussed in the next section, the hazard concerns for 1-BP are primarily aimed at the identified toxicities that occur following repeated exposures in animals and occupationally exposed humans.

Repeated Dose Studies

National Toxicology Program extensively evaluated the subchronic and chronic health effects of 1-BP, which include studies on experimental animals and case reports from occupationally exposed humans (NTP, 2003, 2011, 2013). The available data support that 1-BP may cause adverse effects to the reproduction and development in rodents, cancer in rodents, and neurotoxicity in humans.

In 2003, NTP's former Center for the Evaluation of Risks to Human Reproduction evaluated the potential reproductive and developmental effects of this substance (NTP, 2003). Center for the Evaluation of Risks to Human Reproduction identified a developmental toxicity effect level of 1,534 mg/m^3 in rats, based on a benchmark concentration (95th percentile lower confidence level) and reproductive toxicity effect levels of 1,257 mg/m^3 and 1,006 mg/m^3 in male and female rats, respectively, based on lowest-observed-adverse-effect concentrations. The no-observed-adverse-effect concentration for reproductive toxicity in male and female rats was 503 mg/m^3. Center for the Evaluation of Risks to Human Reproduction concluded that 1-BP may cause reproductive and developmental toxicity independent of systemic toxicity.

In 2011, NTP released the results of the 2-year carcinogenicity studies in male and female rats and mice in which animals were exposed to 1-BP for 6 h per day, 5 days a week (NTP, 2011). Rats were exposed to concentrations of 1-BP at

$ca.$ 0, 628.75, 1,257.5, or 2,515 mg/m^3; mice were exposed to concentrations of $ca.$ 0, 314.375, 628.75, or 1,257.5 mg/m^3. Based on the tumor incidence data presented in Table 67.4, NTP concluded that there was ". . . some evidence of carcinogenic activity . . ." in male rats, ". . . clear evidence of carcinogenic activity . . ." in female rats and mice, and ". . . no evidence of carcinogenic activity . . ." in male mice. In 2013, NTP's Office of the Report on Carcinogens concluded that 1-BP is ". . . reasonably anticipated to be a human carcinogen . . . ." This conclusion was based on the 2-year carcinogenicity studies and on data, which support that 1-BP directly or via reactive metabolites causes alterations, such as genotoxicity, oxidative stress, and glutathione depletion that are relevant to possible mechanisms of human carcinogenicity (NTP, 2013).

In addition to the adverse effects identified from experimental animal studies, several studies have reported adverse health effects in workers exposed to 1-BP. The typical findings include various neurological symptoms. For example, a case study of six foam cushion gluers reported that the worker's serum bromide concentrations ranged from 440 to 1,700 mg/l (44–170 mg/dl) while air samples taken at the workplace 1 day after the cessation of gluing operation showed concentrations of 1-BP ranging from 457.73 to 885.28 mg/m^3 [91–176 parts per million (ppm)] (Majersik et al., 2004, 2007). These workers reported neuropathies, including lower extremity pain or paresthesias, difficulty in walking, weakness and twitching in the legs, and loss of sensation in the feet and legs. Fifteen months after the initial examination, two of the most severely affected workers had regained only minimal function, and still required assistance to walk while three workers continue to experience chronic neuropathic pain.

TABLE 67.4   Summary of Tumors in Rats and Mice Exposed to 1-Bromopropane for 2 Years

Species, Sex	Tumor Type	Concentration (mg/m^3)	Incidence
F344/N rats, male	Skin (keratoacanthoma, squamous cell carcinoma, basal cell adenoma, or carcinoma combined)	0 628.75 1,257.5 2,515	1/50 (2%) 7/50* (14%) 9/50** (18%) 10/50** (20%)
F344/N rats, female	Large intestine (colon or rectum adenoma)	0 628.75 1,257.5 2,515	0/50 (0%) 1/50 (2%) 2/50 (4%) 5/50* (10%)
B6C3F1 mice, female	Lung (alveolar/bronchiolar adenoma or carcinoma combined)	0 314.375 628.75 1,257.5	1/50 (2%) 9/50** (18%) 8/50* (16%) 14/50*** (28%)

Source: NTP (2011).

Note, statistical significance as reported by NTP at *$p \leq 0.05$, **$p \leq 0.01$, or ***$p \leq 0.001$.

SOURCES OF EXPOSURE

Occupational

1-Bromopropane is utilized in a variety of industrial processes, which include emissive and non-emissive uses. Worker exposures are the greatest with emissive uses, such as dry cleaning, vapor degreasing, and spray adhesives. Several occupational exposure limits have been proposed by the government and non-government entities. In 2003, the EPA recommended an acceptable exposure limit of 25 ppm (125.75 mg/m^3) over an 8-h TWA (EPA, 2003). The American Conference of Governmental Hygienists (ACGIH) initially issued in 2009, a threshold limit value (TLV) TWA of 10 ppm (50.3 mg/m^3); however, in 2012, it released a draft TLV TWA of 0.1 ppm (0.503 mg/m^3), based on the recent NTP 2-year carcinogenicity studies (NTP, 2013). The US Occupational Safety and Health Administration (OSHA) has not issued a permissible exposure limit for 1-BP. Under its compliance monitoring program, OSHA collected personal breathing-zone data on workers from 18 facilities. Of the 164 samples, detectable measures ranged from 0.0477 to 423 ppm (*ca.* 0.24 to *ca.* 2,127.7 mg/m^3). Sixty-two of the samples collected from nine facilities exceeded the ACGIH TLV TWA value of 10 ppm; seven of these facilities utilized 1-BP for vapor degreasing or spray adhesives (NTP, 2013). The US National Institute of Occupational Safety and Health is in the process of preparing a recommended exposure limit for 1-BP.

Environmental

No environmental monitoring data were identified for non-occupational human exposures. The primary opportunities for environmental exposures to 1-BP to occur are through waste streams from industrial facilities or chemical spill incidents. Consumers may be exposed by using spot removal products where 1-BP is incorporated as a replacement to compounds such as trichloroethylene; however, no market data were identified for this use.

INDUSTRIAL HYGIENE

The personal protection equipment and mechanical controls recommended in the work place are designed to minimize eye, skin, and respiratory exposures (ECHA, 2014). Tightly sealed goggles are recommended for eye protection against liquid and aerosol. A protective suit is recommended to avoid skin exposures. Protective gloves are recommended for hand protection, and a respirator is recommended to limit the inhalation exposures. Mechanical controls are aimed at providing adequate ventilation.

RISK ASSESSMENTS

No formal risk assessments have been issued on 1-BP by the governmental agencies. The EPA is currently assessing 1-BP under its Toxic Substances Control Act work plan chemical risk assessments (EPA, 2013). No additional information is available on the status of the Agency's assessment or anticipated release date.

REFERENCES

Cheever, K.L., Marlow, K.L., B'Hymer, C., Hanley, K.W., and Lynch, D.W. (2009) Development of an HPLC-MS procedure for the quantification of N-acetyl-S-(n-propyl)-L-cysteine, the major urinary metabolite of 1-bromopropane in human urine. *J. Chromatogr. B Analyt. Technol. Biomed. Life Sci.* 877:827–832.

ECHA (2014) *Registered Substances*. Helsinki, Finland: European Chemicals Agency. Available at http://echa.europa.eu/information-on-chemicals/registered-substances. CASRN: 106-94-5.

EPA (2003) *Protection of Stratospheric Ozone: Listing of Substitutes for Ozone-Depleting Substances – n-Propyl Bromide*. 68 FR (106):33284–33316. Washington, DC: US Environmental Protection Agency.

EPA (2012) *Chemical Data Access Tool (CDAT)*. Washington, DC: Office of Pollution Prevention and Toxics, US Environmental Protection Agency. Available at http://java.epa.gov/oppt_chemical_search. CASRN: 106-94-5.

EPA (2013) *List of Chemicals for Assessment*. Washington, DC: Office of Pollution Prevention and Toxics, US Environmental Protection Agency. Available at http://www.epa.gov/oppt/existingchemicals/pubs/assessment_chemicals_list.html.

Garner, C.E., Sumner, S.C., Davis, J.G., Burgess, J.P., Yueh, Y., Demeter, J., Zhan, Q., Valentine, J., Jeffcoat, A.R., Burka, L.T., and Mathews, J.M. (2006) Metabolism and disposition of 1-bromopropane in rats and mice following inhalation or intravenous administration. *Toxicol. Appl. Pharmacol.* 215:23–36.

Hanley, K.W., Dunn, K.L., and Johnson, B. (2007) *Workers' Exposures to n-Propyl Bromide at an Adhesives and Coatings Manufacturer*. IWS-232-16. Atlanta, GA, USA: National Institute for Occupational Safety and Health, Centers for Disease Control and Prevention.

Hanley, K.W., Petersen, M.R., Cheever, K.L., and Luo, L. (2009) N-acetyl-S-(n-propyl)-L-cysteine in urine from workers exposed to 1-bromopropane in foam cushion spray adhesives. *Ann. Occup. Hyg.* 53:759–769.

HSDB (2013) 1-Bromopropane. Bethesda, MD: Hazardous Substances Data Bank (HSDB), TOXNET®–Toxicology Data Network, Specialized Information Services, US Department of Health and Human Services. Available at http://toxnet.nlm.nih.gov/cgi-bin/sis/htmlgen? HSDB. CASRN: 106-94-5.

Ishidao, T., Kunugita, N., Fueta, Y., Arashidani, K., and Hori, H. (2002) Effects of inhaled 1-bromopropane vapor on rat metabolism. *Toxicol. Lett.* 134:237–243.

Majersik, J.J., Caravati, E.M., and Steffens, J.D. (2007) Severe neurotoxicity associated with exposure to the solvent 1-bromopropane (n-propyl bromide). *Clin. Toxicol. (Phila.)* 45: 270–276.

Majersik, J.J., Steffens, J.D., and Caravati, E.M. (2004) Chronic exposure to 1-bromopropane associated with spastic paraparesis and distal neuropathy: report of six foam-cushion gluers. *Ann. Neurol.* 56:S69–S70.

NTP (2003) *NTP-CERHR Monograph on the Potential Human Reproductive and Developmental Effects of 1-Bromopropane.* NIH Publication No. 04-4479. Research Triangle Park, NC: National Toxicology Program, US Department of Health and Human Services. Available at http://ntp.niehs.nih.gov/ntp/ohat/bromopropanes/1-bromopropane/1bp_monograph.pdf.

NTP (2011) *NTP Technical Report on the Toxicology and Carcinogenesis Studies of 1-Bromopropane (CAS No. 106-94-5) in F344/N Rats and B6C3F1 Mice (Inhalation Studies).* NTP TR 564, NIH Publication No. 11-5906. Research Triangle Park, NC: National Toxicology Program, US Department of Health and Human Services. Available at http://ntp.niehs.nih.gov/ntp/htdocs/lt_rpts/tr564.pdf.

NTP (2013) *Report on Carcinogens–Monograph on 1-Bromopropane.* NIH Publication No. 13-5982; ISSN: 2331-267X. Research Triangle Park, NC: National Toxicology Program, US Department of Health and Human Services.

68

DIACETYL

DAVID J. HEWITT

Target organ: Lung

Occupational exposure limits: 0.005 ppm (NIOSH REL, proposed); 0.025 (NIOSH STEL, proposed); 0.01 ppm (ACGIH TLV); 0.02 ppm (ACGIH TLV–STEL)

Reference values: None

Hazard statements: None

Precautionary statements: None

Risk phrases: None

Safety phrases: None

BACKGROUND AND USES

Diacetyl (2,3-butanedione) is a naturally occurring product and can be found in numerous foods such as butter, milk, cheese, smoked or roasted meats, breads, fruits, vegetables, coffee, beer, and wine. Diacetyl is synthesized to be used as a food additive to impart a buttery flavor and has been designated as a generally recognized as safe (GRAS) substance with low acute toxicity (FDA, 1980). Desirable flavor concentrations in food are approximately 0.05–5.0 ppm and above that range it imparts a disagreeable taste. The most recognized recent use has been in microwave popcorn, but it has also been used for many other products (NTP, 1994, 2007). Diacetyl may be used in additives as a liquid, paste, or powder (Boylstein et al., 2006).

PHYSICAL AND CHEMICAL PROPERTIES

The physical and chemical properties of diacetyl are presented in Table 68.1 (NTP, 2007).

MAMMALIAN TOXICOLOGY

Acute Effects

Inhalation or airborne exposure is the primary exposure route of toxicological concern. At sufficient levels, acute health effects associated with airborne exposure to diacetyl include mucous membrane irritation of the eyes and respiratory tract and potential exacerbation of preexisting asthma. Severe acute irritant effects from airborne exposure have not been a prominent finding in occupational studies. A specific irritant threshold has not been established. The irritant and damaging effects of diacetyl on the respiratory tract were demonstrated in Sprague-Dawley rats in which exposure levels ranging from 99.3 to 294.6 ppm produced inflammation and necrosis. The nose showed the most sensitivity to diacetyl (Hubbs et al., 2008). Larsen et al. (2009) noted that the no observed adverse effect level (NOAEL) for mice was above 100 ppm and estimated that the sensory irritation threshold for humans was above 20 ppm.

Chronic Effects

The primary health effect of concern associated with diacetyl has been constrictive bronchiolitis obliterans (BO) after reports of the condition in microwave popcorn workers. BO is a rare pulmonary condition in which inflammation and fibrosis occurs at the bronchiolar level (i.e., small airways just proximal to the alveoli where gas exchange occurs) and severely narrows or obliterates the lumen. Causes include chemical inhalation exposures such as chlorine, nitrogen dioxide, or sulfur dioxide; rare sequelae of infections such as influenza; and connective tissue disorders such as rheumatoid arthritis or scleroderma. It is a relatively

Hamilton & Hardy's Industrial Toxicology, Sixth Edition. Edited by Raymond D. Harbison, Marie M. Bourgeois, and Giffe T. Johnson.
© 2015 John Wiley & Sons, Inc. Published 2015 by John Wiley & Sons, Inc.

TABLE 68.1 Physical and Chemical Properties of Diacetyl

Molecular Formula	$C_4H_6O_2$
Molecular weight	86.09
Physical form	Greenish-yellow liquid at room temperature
Melting point	$-2.4\,°C$
Boiling point	$88.0\,°C$
Vapor pressure	57 mmHg at $25\,°C$
Odor	Quinone, chlorine-like odor in vapors, rancid butter
Odor threshold	<1 ppb
Conversion factors	$1\ ppm = 3.52\ mg/m^3$; $1\ mg/m^3 = 0.284\ ppm$

frequent complication of lung transplantation and may also be seen with bone marrow or stem cell transplants. Presenting symptoms include gradually increasing dyspnea on exertion, which may be accompanied by cough or wheezing and may resemble asthma. A physical examination may show evidence of hyperinflation along with wheezing or end inspiratory crackles. Pulmonary function tests show airway obstruction that unlike asthma, does not respond well to bronchodilators. The diffusing capacity is minimally affected in the early stages since the gas exchange region of the lung is not initially affected. Air trapping and fibrosis may be seen on chest computed tomography (CT) and it classically shows a mosaic pattern. Diagnosis is based on a characteristic clinical history, pulmonary function testing, and chest imaging findings. In lung transplant evaluations, BO is suspected when there is a substantial unexplained decline in the forced expiratory volume (FEV)1. A lung biopsy is considered the only way to definitively diagnose BO (Chan and Allen, 2004). Treatment may include corticosteroids, prompt treatment of respiratory infections, and lung transplant for end-stage disease. Although the course may be variable, the prognosis is generally poor with gradually decreasing respiratory function (Epler, 2007).

The association of diacetyl with BO was initially identified when a cluster of eight workers at a microwave popcorn facility in Missouri were reported to have a severe fixed obstructive lung condition (Parmet and Von Essen, 2002). A review of the index cases indicated that they became symptomatic after a median of 1.5 years of employment (range 5 months to 9 years) during 1993–2000 (Akpinar-Elci et al., 2004). Five identified cases worked in the mixing room and the others worked in a packaging area. Symptoms included a dry persistent cough, dyspnea, and wheezing. Chest CT scans showed mosaic patterns. Two of the three cases that underwent lung biopsy showed findings consistent with BO. Five of the cases were reported to be awaiting lung transplant.

A National Institute of Occupational Safety and Health (NIOSH) investigation initially concluded that the fixed obstructive lung conditions identified at the facility were associated with exposure to microwave popcorn butter flavors but was unable to identify a specific substance that could explain the condition. Diacetyl was used as a marker for airborne exposure to flavoring vapors and found to be increased in certain areas of the facility, particularly the mixing room (CDC, 2002). The prevalence of airway obstruction increased with increasing cumulative exposure to diacetyl. Diacetyl exposure levels showed a mean of 32.27 ppm in the mixing room (1.34–97.94 ppm); 1.88 ppm in the packaging lines (0.26–6.80 ppm); and 0.56 ppm in the quality control or maintenance area (0.33–0.89 ppm). In the mixing room, flavorings were added by a worker who opened the lid of a heated tank (Kreiss et al., 2002). Peak air concentrations of diacetyl in the air space above the heated flavorings tank were reported as high as 1230 ppm (NIOSH, 2006).

Diacetyl has been implicated as a cause of BO in subsequent case reports or studies, including other U.S. microwave popcorn facilities (Kanwal et al., 2006), potato crisp flavoring exposure (Hendrick, 2008), California flavor manufacturing workers (CDC, 2007), a butter flavoring manufacturing worker (Modi et al., 2008), and diacetyl manufacturing workers (Van Rooy et al., 2007). Prior to the microwave popcorn cluster, BO was reported in two Indiana flavor manufacturing workers (NIOSH, 1986). A specific cause was not identified, although diacetyl was listed as one of over 100 different flavoring agents used at the facility. Egilman et al. (2007) have described several other cases of BO in flavoring industry workers prior to the microwave popcorn cases. Egilman and Schilling (2012) reported three cases of BO among consumers who had high exposure to microwave popcorn fumes and reportedly ate 1–5 bags on a daily basis. The authors suggested that the consumer exposures were comparable to the threshold cumulative diacetyl exposure that was associated with pulmonary disease in occupationally exposed individuals.

Exposure to diacetyl and associated butter flavorings has also been associated with increased respiratory complaints and a higher rate of obstructive findings on pulmonary function tests in exposed workers. In the initial study, Kreiss et al. (2002) reported that workers had 3.3 times the expected rate of airway obstruction, which was increased to 10.8 times the expected rate in never-smokers. Kanwal et al. (2006) reported that at the index facility and five other U.S. microwave popcorn plants, there was a higher prevalence of respiratory symptoms such as dyspnea, cough, and wheezing in ever-mixers compared to never-mixers. The percent predicted FEV1, a measure of airway obstruction, was also significantly decreased in the ever-mixers compared with the never-mixers. Longitudinal studies of pulmonary function in flavoring manufacturing workers correlated abnormal declines in FEV1 with greater company use of diacetyl (Kreiss et al., 2012). Lockey et al. (2009) reported decreased pulmonary function and FEV1 in mixers prior to respirator use in a study of over 700 microwave popcorn workers at four different facilities. Mean exposure levels for the mixers ranged from 0.057 to 0.860 ppm. Cumulative exposure of ≥0.8 ppm-years was significantly associated with obstructive patterns on pulmonary function testing with an adjusted odds ratio of over 8.0 compared with those with <0.8 ppm-years. The authors noted no significant effects in nonmixing room employees. The authors concluded that exposure to diacetyl by itself or in combination with other butter flavoring ingredients increased the risk of obstructive lung disease.

The specific level of diacetyl or associated flavoring chemicals, which may cause BO, has not been established. Larsen et al. (2009) concluded that sensory irritation was not a reliable warning property to protect against levels that have been associated with BO. Lockey et al. (2009) reported increased obstructive lung findings in workers with average exposure levels that were <1 ppm.

The association of diacetyl with BO has been questioned. Galbraith and Weill (2009, 2010) noted that diacetyl is neither a strong acid nor a base; has high water solubility that would not appear consistent with causing deeper lung damage; animal studies did not show evidence of distal airway damage at the levels of the bronchioles; studies of microwave popcorn workers and California flavoring company workers did not have biopsy-proven diagnoses of BO; and findings in some case reports suggested other conditions such as extrinsic allergic alveolitis. Harber et al. (2006) noted that despite compelling data that flavoring exposures were associated with BO, it was not clear if diacetyl was the sole cause, whether it was one of the many different flavoring agents causing the illness, or if it was a marker of an unidentified agent. A difficulty in analyzing the causal association is that there can be multiple other causes for obstructive lung disease such as asthma, extrinsic alveolitis, or smoking.

BO has also been associated with microwave popcorn facilities in which diacetyl use appeared minimal. A NIOSH (2007) Health Hazard Evaluation at a Montana microwave popcorn plant reported that three workers were identified with respiratory disease suggestive of asthma that developed during their work. Two of the cases were also noted to be suggestive of BO. However, diacetyl concentrations in the facility were noted as too low to be quantified (i.e., below 0.01 ppm). The maximum identified level was 0.14 ppm, which was found over a heated container of butter-flavored oil. A number of other substances were identified in air samples. The authors concluded that many flavoring chemicals other than diacetyl may be responsible for respiratory disease. Diacetyl concentrations associated with other substances used at the facility prior to the investigation were unknown.

Additional inconsistencies regarding the association of relatively low dose inhalation levels of diacetyl and BO can also be seen when evaluating exposures from cigarette smoke. The diacetyl content in mainstream cigarette smoke from regular-sized cigarettes has been reported as 0.307–0.433 mg/cigarette (Fujioka and Shibamoto, 2006). Smoking a pack of cigarettes a day and assuming an average of 0.35 mg of diacetyl/cigarette would result in an inhalation dose of 7 mg/day. In comparison, occupational exposure to diacetyl at a concentration of 0.1 ppm (i.e., $0.352 \, mg/m^3$) and assuming an upper limit inhalation rate of $20 \, m^3/day$ would also result in a comparable dose of 7 mg/day. Although cigarette smoking is associated with multiple adverse respiratory effects, it is not known to be a cause of BO.

Animal studies have demonstrated toxicity to the respiratory tract with repeated diacetyl exposure. In a subchronic study of mice, Morgan et al. (2008) noted an evidence of airway toxicity at exposures as low as 25 ppm for 12 weeks and were unable to determine a NOAEL. In order to bypass the removal of diacetyl by the nasal cavity, oropharyngeal aspiration was used to provide a higher dose to the deeper airways and resulted in fibrohistiocytic lesions in the terminal bronchioles. It was noted that the lesions developed only 4 days after a single dose and suggested that diacetyl may be highly fibrogenic in the small airways. The authors concluded that the overall findings supported the hypothesis that workplace exposures to diacetyl contributed to the development of BO in humans. Finley et al. (2008) reported several criticisms of this study, including the use of exposure levels orders of magnitude higher than typical time-weighted average(TWA) levels in mixing rooms; the lack of a control group in which animals were exposed to other heated organic vapors; the lack of distal airway damage even at levels, which caused death in some animals; and the possibility that oropharyngeal instillation of other substances would be associated with similar bronchiolar effects.

To summarize, diacetyl has been associated with serious pulmonary toxicity in workers with inhalational exposure to flavoring agents. However, there are questions as to whether diacetyl and/or some other substance with which it may be used is the primary cause of the observed toxicity.

Mechanism of Action

The specific mechanism by which diacetyl or associated substances may cause BO is unknown. Diacetyl has been hypothesized to cause direct injury to the airway epithelium in which the repair process is disorganized and produces uncontrolled fibroblastic and myoblastic proliferation. An allergic sensitization process was considered unlikely (Harber et al., 2006).

SOURCES OF EXPOSURE

Individuals may be exposed to diacetyl on a daily basis due to its natural presence and/or additive presence in various foods, beverages, cigarette smoke, and even automobile exhaust. Diacetyl has also been identified as an indoor air pollutant and in carpeting adhesive emissions (NTP, 1994). The overall use of diacetyl as a food additive has decreased due to concerns of pulmonary toxicity. The biggest makers of microwave popcorn in the United States reportedly replaced diacetyl with other substances (Popcorn Firms, 2007). Diacetyl is still considered a GRAS food ingredient by the Food and Drug Administration (FDA, 2011). The primary source of occupational exposure has been related to the use of diacetyl and other flavoring substances in the production of various foods.

INDUSTRIAL HYGIENE

Because diacetyl and related flavoring substances have been associated with serious pulmonary toxicity, engineering measures to decrease worker exposure are recommended. These include substitution of less hazardous substances when possible, using closed production processes that limit the opportunity for inhalation exposure, isolating high exposure processes from the rest of the workplace, and using the lowest possible temperatures for heated processes. Periodic air monitoring should be performed to ensure that the control efforts are effective. Respirators are recommended for individuals working directly with diacetyl. The minimum protective respirator is a half-mask, negative-pressure respirator with organic vapor cartridges and particulate filters (NIOSH, 2006).

The Occupational Safety and Health Administration (OSHA) has not published an occupational exposure standard for diacetyl. The American Conference of Governmental Industrial Hygienists (ACGIH) recommended a threshold limit value—time-weighted average (TLV—TWA) of 0.01 ppm. ACGIH noted that peak exposure may be potentially more important among the causes of BO. The European Commission, (2010) has recommended an occupational exposure limit (OEL) for diacetyl of 0.1 ppm. In a draft criteria document, NIOSH (2011) proposed a much more conservative recommended exposure limit (REL) TWA of 0.005 ppm and noted that at this concentration the risk of decreased lung function was no more than 1 in 1000 and there was a less chance for BO. A short-term exposure limit (STEL) of 0.025 ppm was recommended for a 15-min period.

MEDICAL MANAGEMENT

A medical surveillance program is recommended for individuals working with diacetyl and associated food flavorings. Components of such a program include a baseline medical evaluation focused on respiratory health, a baseline pulmonary function test prior to starting work, and a periodic pulmonary function testing. This surveillance can be included as part of the OSHA Respiratory Protection Standard for individuals who may be required to wear respirators. NIOSH (2006) has recommended at least biannual spirometry for workers with increased exposures. A decrease in FEV1 of 300 ml or 10% from baseline is recommended as a threshold for further evaluation, including worksite assessment of exposure levels.

Acute inhalation exposure to diacetyl under most circumstances is unlikely to result in significant immediate health effects requiring medical care. Irritant symptoms or aggravation of a preexisting respiratory condition should be treated symptomatically focusing on airway maintenance. Of greater concern is the potential for irreversible lung disease manifested by the gradual onset of increasing respiratory obstruction and dyspnea. The primary treatment should be prevention of exposure. Individuals suspected of having diacetyl or other work-related BO should undergo appropriate diagnostic testing, including pulmonary function testing, chest imaging, and, if necessary, a diagnostic lung biopsy. They should be removed from areas with potential exposure until their diagnosis is clarified. If confirmed to be work related, the individual should be permanently removed from positions with potential exposure. Treatment is dependent on the severity of the underlying condition and may include corticosteroids, inhaled corticosteroids, bronchodilators, and aggressive treatment of secondary pulmonary infections.

REFERENCES

Akpinar-Elci, M., Travis, W.D., Lynch, D.A., and Kreiss, K. (2004) Bronchiolitis obliterans syndrome in popcorn production plant workers. *Eur. Respir. J.* 24:298–302.

Boylstein, R., Piacitelli, C., Grote, A., Kanwal, R., Kullman, G., and Kreiss, K. (2006) Diacetyl emissions and airborne dust from butter flavorings used in microwave popcorn production. *J. Occup. Environ. Hyg.* 3:530–535.

CDC (2002) Fixed obstructive lung disease in workers at a microwave popcorn factory–Missouri, 2000–2002. *MMWR Morb. Mortal. Wkly. Rep.* 51:345–347.

CDC (2007) Fixed obstructive lung disease among workers in the flavor-manufacturing industry—California, 2004–2007. *MMWR Morb. Mortal. Wkly. Rep.* 56:389–393.

Chan, A. and Allen, R. (2004) Bronchiolitis obliterans: an update. *Curr. Opin. Pulm. Med.* 10:133–141.

Egilman, D.S., Mailloux, C., and Valentin, C. (2007) Popcorn-worker lung caused by corporate and regulatory negligence: An avoidable tragedy. *Int. J. Occup. Environ. Health* 13:85–98.

European Commission (2010) *Recommendation from the Scientific Committee on Occupational Exposure Limits for diacetyl. SCOEL/SUM/149.*

Egilman, D.S. and Schilling, J.H. (2012) Bronchiolitis obliterans and consumer exposure to butter-flavored microwave popcorn: a case series. *Int. J. Occup. Environ. Health* 18:29–42.

Epler, G.R. (2007) Constrictive bronchiolitis obliterans: the fibrotic airway disorder. *Expert Rev. Respir. Med.* 1:139–147.

FDA (1980) *Evaluation of the Health Aspects of Starter Distillate and Diacetyl and Food Ingredients. Report No. FDA/BF-80/64.*

FDA (2011) *Select Committee on GRAS Substances (SCOGS) Opinion: Diacetyl,* Available at http://www.fda.gov/Food/FoodIngredientsPackaging/GenerallyRecognizedasSafeGRAS/GRASSubstancesSCOGSDatabase/ucm261273.htm (accessed 2013).

Finley, B.L., Galbraith, D.A., and Weil, D. (2008) Comments on respiratory toxicity of diacetyl. [Re: Morgan, D.L., Flake, G.P., Kirby, P.J., and Palmer, S.M. (2008). Respiratory toxicity of diacetyl in C57Bl/6 mice. Toxicol Sci. 103, 169-180.]. *Toxicol. Sci.* 105:429–432.

Fujioka, K. and Shibamoto, T. (2006) Determination of toxic carbonyl compounds in cigarette smoke. *Environ. Toxicol.* 21:47–54.

Galbraith, D. and Weill, D. (2009) Popcorn lung and bronchiolitis oblilterans: a critical appraisal. *Int. Arch. Occup. Environ. Health* 82:407–416.

Galbraith, D.A. and Weill, D. (2010) Authors' response to Kreiss et al. (2009). *Int. Arch. Occup. Environ. Health* 83:237–240.

Harber, P., Saechao, K., and Boomus, C. (2006) Diacetyl-induced lung disease. *Toxicol. Rev.* 25:261–272.

Hendrick, D.J. (2008) "Popcorn worker's lung" in Britain in a man making potato crisp flavouring. *Thorax* 63:267–268.

Hubbs, A.F., Goldsmith, W.T., Kashon, M.L., Frazer, D., Mercer, R.R., Battelli, L.A., Kullman, G.J., Schwegler-Berry, D., Friend, S., and Castranova, V. (2008) Respiratory toxicologic pathology of inhaled diacetyl in Sprague-Dawley rats. *Toxicol. Pathol.* 36:330–344.

Kanwal, R., Kullman, G., Piacitelli, C., Boylstein, R., Sahakian, N., Martin, S., Fedan, K., and Kreiss, K. (2006) Evaluation of flavorings-related lung disease risk at six microwave popcorn plants. *J. Occup. Environ. Med.* 48:149–157.

Kreiss, K., Gomaa, A., Kullman, G., Fedan, K., Simoes, E.J., and Enright, P.L. (2002) Clinical bronchiolitis obliterans in workers at a microwave-popcorn plant. *N. Engl. J. Med.* 347:330–338.

Kreiss, K., Fedan, K.B., Nasrullah, M., Kim, T.J., Materna, B.L., Prudhomme, J.C., and Enright, P.L. (2012) Longitudinal lung function declines among California flavoring manufacturing workers. *Am. J. Ind. Med.* 55:657–668.

Larsen, S.T., Alarie, Y., Hammer, M., and Nielsen, G.D. (2009) Acute airway effects of diacetyl in mice. *Inhal. Toxicol.* 21:1123–1128.

Lockey, J.E., Hilbert, T.J., Levin, L.P., Ryan, P.H., White, K.L., Borton, E.K., Rice, C.H., McKay, R.T., and LeMasters, G.K. (2009) Airway obstruction related to diacetyl exposure at microwave popcorn production facilities. *Eur. Respir. J.* 34:63–71.

Modi, P., Yadava, V., Sreedhar, R., Khasawaneh, F., and Balk, R.A. (2008) A case of flavor-induced lung disease. *South. Med. J.* 101:541–542.

Morgan, D.L., Flake, G.P., Kirby, P.J., and Palmer, S.M. (2008) Respiratory toxicity of diacetyl in C57Bl/6 mice. *Toxicol. Sci.* 103:169–180.

NIOSH (1986) *NIOSH Health Hazard Evaluation Report. International Bakers Services, South Bend, IN, 1986. HETA-85-171-1710.*

NIOSH (2006) *NIOSH Health Hazard Evaluation Report. Glister-Mary Lee Corporation. Jasper, Missouri. HETA #2000-0401-2991.*

NIOSH (2007) *NIOSH Health Hazard Evaluation Report. Yatsko's Popcorn, Sand Coule, Montana. HETA #2006-0195-3044.*

NIOSH (2011) *Criteria for a Recommended Standard. Occupational Exposure to Diacetyl and 2,3-Pentanedione. Draft.*

NTP (1994) *2,3-Butanedione. Prepared for NCI by Technical Resources, Inc.*

NTP (2007) *Chemical Information Review Document for Artificial Butter Flavoring and Constintuents Diacetyl [CAS No. 431-03-8] and Acetoin [CAS No. 513-86-0]. NTP, Research Triangle Park. NC.*

Parmet, A.J. and Von Essen, S. (2002) Rapidly progressive, fixed airway obstructive disease in popcorn workers: a new occupational pulmonary illness? *J. Occup. Environ. Med.* 44:216–218.

Popcorn Firms (2007). Associated Press. *Popcorn firms removing flavoring chemical. Additive linked to lung ailments in factory workers.* The Boston Globe, December 18: 2007.

Van Rooy, F., Rooyackers, J.M., Prokop, M., Houba, R., Smit, L., and Heederik, D. (2007) Bronchiolitis obliterans syndrome in chemical workers producing diacetyl for food flavorings. *Am. J. Respir. Crit. Care Med.* 176:498–504.

69

PERFLUOROALKYL COMPOUNDS

Jason S. Garcia and Raymond D. Harbison

PERFLUORO BUTYL SULFONATE (PFBS)

Target organ(s): Skin, eyes, respiratory tract, liver, kidneys

Occupational exposure limits: NIOSH/OSHA/ACGIH—not established

Reference values: RfD/reference concentration (RfC)/minimal risk level (MRL)—not established

Risk/safety phrases: Corrosive

BACKGROUND AND USES

Perfluoro butyl sulfonate (PFBS, nonafluoro-1-butanesulfonic acid, perfluorobutane sulfonic acid, 1-perfluorobutane sulfonic acid; CASRN 375-73-5) is a four-carbon compound in the perfluoroalkyl family of chemicals. It can be found in stain repellents used for carpets and furniture. A specific form of the chemical, potassium perfluorobutane sulfonate, is being used as a flame retardant in place of brominated retardants. Perfluoro butyl sulfonate is being used as a substitute for other perfluoroalkyl compounds because it is not believed to bioaccumulate in the environment.

PHYSICAL AND CHEMICAL PROPERTIES

Perfluoro butyl sulfonate exists as a colorless liquid. The chemical has a molecular weight of 300.1 g/mol, a boiling point ranging from 112 °F to 114 °F, and a density of 1.81 g/cm³. Perfluoro butyl sulfonate has the ability to react violently with water, so preventing exposure to moisture should be a priority. It may also be incompatible with strong oxidizers. Upon decomposition, PFBS can form carbon oxides, sulfur oxides, and hydrogen fluoride. Additional information related to the physical and chemical properties of PFBS are not currently available.

MAMMALIAN TOXICOLOGY

Acute Effects

It is believed that exposure to PFBS can result in acute effects, including changes in blood chemistry, hepatotoxicity, and nephrotoxicity. Decreased hemoglobin and hematocrit, increased liver weight, and histological changes in the kidney have been noted at a laboratory animal reference dose (RfD) of 0.0042 mg/kg/day. Decreased serum protein, albumin, red blood cell count, as well as increased serum chloride may also result. While not related to mammalian toxicology, a study of mallard ducks and bobwhite quails exposed to dietary PFBS concentrations of 1000, 1780, 3160, 5620, and 10,000 mg PFBS/mg feed for 5 days found no associated mortalities as a result of the acute exposure (Newsted et al., 2008). Perfluoro butyl sulfonate may also be associated with neurotoxicity. Slotkin et al. (2008) found that PFBS repressed differentiation of the dopamine and acetylcholine neurotransmitter phenotypes in PC12 cells lines, indicating possible neurotoxic action. The study also noted that when compared to other perfluoroalkyl compounds, PFBS ranked lower with regard to the potential to produce adverse effects.

Chronic Effects

Chronic exposure to PFBS can result in exaggerated effects similar to the acute effects previously described. Decreased

Hamilton & Hardy's Industrial Toxicology, Sixth Edition. Edited by Raymond D. Harbison, Marie M. Bourgeois, and Giffe T. Johnson.
© 2015 John Wiley & Sons, Inc. Published 2015 by John Wiley & Sons, Inc.

hemoglobin and hematocrit may be seen along with tissue abnormalities in the kidneys and liver. The Minnesota Department of Health (2011) reported these effects as a result of a PFBS RfD of 0.0014 mg/kg/day in a laboratory animal. While not related to mammalian toxicological effects, one study examining the chronic effects of PFBS found that some adult quail exposed to the chemical developed reproductive tissue abnormalities, however, the results were not statistically significant (Newsted et al., 2008). Lou et al. (2013) found that frogs exposed to PFBS had increased expression of estrogen receptor and androgen receptor proposing a potential link between the chemical and adverse reproductive effects. The mentioned studies point to a potential link between PFBS and reproductive abnormalities, however, it is not known whether this can be extrapolated to the mammalian systems.

Mechanism of Action(s)

Most perfluoroalkyl compounds persist in the mammalian tissue and elicit adverse health effects over time. This does not seem to be the case with PFBS. A study of dairy cows fed with PFBS-contaminated feed for 28 days found that chemical concentrations in the blood were low, the chemical was barely detectable in secreted breast milk, and there were low concentrations of the chemical in the liver and kidney (Kowalczyk et al., 2013). One study estimated that PFBS distribution in rat and monkey models is primarily extracellular with urine appearing to be the major route of elimination (Olsen et al., 2009). Olsen et al. (2009) also reported that the geometric mean half-life of PFBS in human subjects followed for 180 days was 25.8 days with the chemical being primarily excreted in the urine. Perfluoro butyl sulfonate that is not excreted can lead to hematological, hepatic, and renal abnormalities. The molecular size of this chemical prevents it from bioaccumulating in the body unlike larger, longer carbon-chained perfluoroalkyls.

Chemical Pathology

It has been noted that perfluoroalkyl compounds bioaccumulate in the mammalian tissue. The larger the chemical, the more it appears to bioaccumulate in the body tissue. Perfluoro butyl sulfonate is a smaller, shorter carbon-chained compound, so accumulation is believed to be minimal. Most of the chemical encountered by the body is excreted in the urine. Based on the available information, PFBS seems to interact with the tissue in the hematologic, hepatic, reproductive, and renal systems. Information related to the exact mechanism of pathology for the different body systems is limited. It has been noted that perfluoroalkyls, including PFBS, have the ability to activate the nuclear receptor peroxisome proliferator-activated receptor alpha (PPARα), which is involved in the transcription of the genetic material (Rosen et al., 2009). Any deviation in the production of genetic material could result in adverse health effects, including cancer. There has been speculation that

PFBS may be associated with cancer, however, there is no documented evidence to support this speculation.

SOURCES OF EXPOSURE

Occupational

Individuals handling PFBS in the work environment should take precautions as the chemical can react violently with water. Since the chemical exists as a liquid, dermal contact is the most probable route of exposure. The use of personal protective equipment, including aprons, gloves, and goggles, can prevent the liquid from contacting the skin. Inhalation of PFBS can also occur. Dust in the work environment can become contaminated with the chemical. Good housekeeping procedures can prevent the buildup of dust and PFBS. Proper ventilation and personal respirators may be necessary if PFBS concentrations are high. Perfluoro butyl sulfonate and other perfluoroalkyls are used in waxes that are applied to snow skis to increase their performance. Environmental testing of ski shop waxing areas has identified PFBS in the air, making inhalation of the chemical a possibility (Nilsson et al., 2010). Ingestion of the chemical is not likely; however, employees that handle PFBS should wash their hands thoroughly before eating, drinking, or using tobacco products.

Environmental

Perfluoro butyl sulfonate is embedded into the carpet and furniture to offer protection against stains and spills. It is also applied to various household products as it may play a role in the retarding of flames during a fire. Over time as furniture, carpet, and other household products begin to degrade, the chemical may be introduced into the indoor environment in house dust. Perfluoro butyl sulfonate, along with other perfluoroalkyl compounds, has been detected in house dust samples indicating a potential route for exposure for occupants (Strynar and Lindstrom, 2008). Frequent vacuuming and dust removal can prevent buildup of the contaminant in household dust and the indoor environment. Perfluoro butyl sulfonate can also be found in water systems as a result of human waste streams. Samples of tap water collected in China, Japan, India, the United States, and Canada, all contained detectable levels of PFBS (Mak et al., 2009). While the chemical has been detected in both the indoor environment and drinking water, the lack of toxicological data related to PFBS makes it hard to determine what levels of exposure are acceptable.

INDUSTRIAL HYGIENE

In many industries, PFBS has gained popularity because it appears to be less harmful and is not known to bioaccumulate like other perfluoroalkyl compounds. The Occupational Safety and Health Administration (OSHA), the National Institute for Occupational Safety and Health (NIOSH), and the American Conference of Governmental Industrial Hygienists' (ACGIH) have not established exposure limits for PFBS. While there are no formal guidelines in place, several precautions should be taken to reduce the potential for inhalation, ingestion, and dermal exposure to PFBS. Inhalation to PFBS can be curtailed by utilizing personal respirators. Personal respirators should have proper cartridges installed to capture harmful vapors. Ventilation may also be necessary to remove vapors and particles from the atmosphere. Proper hygienic practices can prevent the ingestion of PFBS. Individuals using the chemical should ensure that they wash their hands prior to eating, drinking, applying cosmetics, and using tobacco products. Using personal protective equipment, such as gloves, rubber boots, and goggles, can prevent dermal contact with the chemical. Contaminated clothing should be removed in a proper manner to prevent additional contamination.

MEDICAL MANAGEMENT

Blood testing can be used to assess whether exposure to PFBS has occurred; however, the dose information related to adverse health effects in humans is not well understood. The route of exposure to PFBS will determine the type of medical management to be utilized. If PFBS vapors are inhaled, the exposed individual should be moved into an area with fresh air. If strenuous breathing occurs, oxygen may need to be provided. If an exposed individuals is not breathing, emergency artificial respiration will be needed. If PFBS is ingested, vomiting should not be induced and the mouth should be rinsed with water. If PFBS penetrates the skin, it should be cleansed with soap and water to remove the chemical from the skin. If PFBS contacts the eyes, they should be rinsed with water for a minimum of 15 min. In all situations, medical attention may be necessary if the exposure was excessive or if symptoms do not resolve.

REFERENCES

Kowalczyk, J., et al. (2013) Absorption, distribution and milk secretion of the perfluoroalkyl acids PFBS, PFHxS, PFOS, and PFOA by dairy cows fed naturally contaminated feed. *J. Agri. Food Chem.* 61:2903–2912.

Lou Q.Q., et al. (2013) Effects of perfluorooctanesulfonate and perfluorobutanesulfonate on the growth and sexual development of Xenopus laevis. *Exotoxicology* 22:1133–1144.

Mak, Y.L., et al. (2009) Perfluorinated compounds in tap water from China and several other countries. *Environ. Sci. Technol.* 43:4824–4829.

Minnesota Department of Health (MDH) (2011) *Chemical Name: Perfluorobutane Sulfonate*, CAS: 375-73-5, MN: St. Paul.

Newsted, J.L., et al. (2008) Acute and chronic effects of perfluorobutane sulfonate (PFBS) on the mallard and northern bobwhite quail. *Arch. Environ. Contam. Toxicol.* 54:535–545.

Nilsson, H., et al. (2010) Inhalation exposure to fluorotelomer alcohols yield perfluorocarboxylates in human blood? *Environ. Sci. Technol.* 44:7717–7722.

Olsen, G.W., et al. (2009) A comparison of the pharmacokinetics of perfluorobutanesulfonate (PFBS) in rats, monkeys, and humans. *Toxicology* 256:65–74.

Rosen, M.B., et al. (2009) Does exposure to perfluoroalkyl acids present a risk to human health? *Toxicol. Sci.* 111:1–3.

Slotkin, T.A., et al. (2008) Developmental neurotoxicity of perfluorinated chemicals modeled *in vitro*. *Environ. Health Perspect.* 116:716–722.

Strynar, M.J. and Lindstrom, A.B. (2008) Perfluorinated compounds in house dust from Ohio and North Carolina, USA. *Environ. Sci. Technol.* 42:3751–3756.

PERFLUOROBUTYRATE (PFBA)

Target organ(s): Skin, eyes, respiratory tract, thyroid, liver

Occupational exposure limits: NIOSH/OSHA/ACGIH—not established

Reference values: RfD/RfC/MRL—not established

Risk/safety phrases: Corrosive

BACKGROUND AND USES

Perfluorobutyrate (PFBA, perfluorobutyric acid, perfluorobutanoic acid, perfluoropropane carboxylic acid; CASRN 375-22-4) is a four-carbon compound in the perfluoroalkyl family of chemicals. Perfluorobutyrate is a breakdown product of stain- and grease-proof coatings applied to various consumer products. Specifically, it can be found on furniture, carpets, upholstery, and food packaging. Perfluorobutyrate can also be detected in water systems as a result of human industrial waste streams.

PHYSICAL AND CHEMICAL PROPERTIES

Perfluorobutyrate exists as a colorless liquid. The chemical has a pH of 1.0, a molecular weight of 214.04 g/mol, a boiling point of 248 °F, a vapor pressure of 10 mmHg, and a density of 1.645 g/cm^3. Perfluorobutyrate has the ability to react with bases, reducing agents, and oxidizing agents. Upon decomposition, PFBA can form carbon oxides and hydrogen fluoride. Additional information related to the physical and chemical properties of PFBA are not currently available.

MAMMALIAN TOXICOLOGY

Acute Effects

Perfluorobutyrate can result in acute effects, including changes in blood chemistry, hepatotoxicity, and thyroid dysfunction. Decreased hemoglobin and hematocrit, increased liver weight, and morphological changes in the thyroid gland have been noted at a laboratory animal RfD of 0.0038 mg/kg/day. Increased thyroid weight, decreased total thyroxine, and free thyroxine may also result from acute exposure to PFBA. Thyroid gland dysfunction can lead to hormonal imbalance thereby disrupting an organism's metabolism. Hepatomegaly and other hepatotoxic manifestations are possible with exposure to the chemical. A study of perfluorinated chemicals found that they had the ability to disrupt the mitochondrial respiration in rat liver cells (Ray, 2010). Male rats exposed to NH_4^+PFBA showed increased liver weight, hepatocellular hypertrophy, decreased serum cholesterol, and reduced serum thyroxin (Butenhoff et al., 2012). Cessation of exposure to PFBA is believed to reverse the symptoms that were described. Inhalation or dermal contact with PFBA can result in respiratory tract and skin irritation.

Chronic Effects

Chronic exposure to PFBA may result in the morphological and weight changes with respect to the liver. The thyroid gland may also be impacted by chronic exposure. Changes in the thyroid gland can result in disruption of body metabolism. Decreased hemoglobin and hematocrit may also be seen. The Minnesota Department of Health (2011) reported these effects as a result of a PFBA RfD of 0.0029 mg/kg/day in a laboratory animal. Information related to PFBA's ability to cause cancer is not very reliable. There have been some reports that indicate an increase in cases of prostate and bladder cancer in exposed individuals but there is no certainty in the exact cause. Besides exposure to PFBA, there may be other confounding factors that play a role in disease development. Rats fed with perfluoroalkyls, not including PFBA, developed tumors. While rats developed tumors in the laboratory, species differences between rats and humans may prevent the extrapolation of this information to the human population.

Mechanism of Action(s)

Unlike other perfluoroalkyl compounds, PFBA does not distribute into serum, the kidneys, and liver. The concentration of PFBA in the liver of pregnant mice has been found to be lower than in the serum by about 35–55% (Das et al., 2008). Much like PFBS, PFBA has a short half-life and is eliminated from the body via urine. The half-life of PFBA is noted to be 3 days. A study that examined the pharmacokinetics of PFBA in rats, mice, monkeys, and humans found that among individuals exposed to the chemical via drinking water, 96% of serum PFBA concentrations were <2 ng/ml (maximum = 6 ng/ml) indicating that PFBA is efficiently eliminated from serum without accumulation (Chang et al., 2008). Perfluorobutyrate has also been noted to cause peroxisome proliferation, induction of fatty acid oxidation, and hepatomegaly indicating that PFBA activates the nuclear receptor PPARα (Foreman et al., 2009). Activation of PPARα may indicate hepatotoxic response to PFBA in mammalian models.

Chemical Pathology

The specific chemical pathology of PFBA is not well understood. Some research indicates that PFBA has the ability to activate specific nuclear receptors thereby inducing adverse health responses. A study conducted by Foreman et al. (2009) found that PFBA had the ability to modify gene expression resulting in hepatomegaly and hepatocyte hypertrophy through a mechanism that requires PPARα. The researchers continued to explain that PPARα-dependent increase in PFBA-induced hepatocyte necrosis with inflammatory cell infiltrate was facilitated by the mouse PPARα but not the human PPARα (Foreman et al., 2009). These results may indicate that PFBA is successful in activating PPARα in both mice and humans, but species differences result in different hepatotoxic pathologies. Perfluorobutyrate may also play a role in the abnormal development of offspring. A study conducted by Das et al. (2008) found that mouse pups exposed to PFBA had delays in eye opening and onset of puberty in groups exposed to 350 mg/kg. Other chemical pathologies may be discovered through additional research of PFBA.

SOURCES OF EXPOSURE

Occupational

Individuals working with PFBA are at a risk for exposure through inhalation or dermal contact. Engineering controls and personal protective equipment should be used to reduce the exposure to acceptable levels. A cohort of current and former 3M employees were studied and were found to have traces of PFBA in their serum indicating exposure (Chang et al., 2008). Individuals working in professional ski shops may be at risk for PFBA exposure. Perfluoroalkyls are applied to skis to increase their performance. The rubbing of the ski wax may introduce contaminated particles into the work environment atmosphere. The highest median perfluoroalkyl concentration was found to be 50 ng/ml, which was 25 times higher than the baseline levels (Freberg et al., 2010). In a situation such as the one described, the installation of a ventilation system can capture and remove particles from the indoor air. In addition, it may be necessary to provide respirators to prevent those applying the wax from inhaling airborne PFBA.

Environmental

Perfluorobutyrate has been known to contaminate waterways when it is improperly disposed. An example would be in Minnesota where perfluoroalkyl waste at four different sites, contaminated the groundwater supplies. The Minnesota Department of Health conducted water testing and found that PFBA contamination was widespread with some concentrations over 1.0 parts per billion (ppb) (MDH, 2012). The Minnesota Department of Health (2012) uses a health-based value (HBV) of 1.0 ppb for PFBA as an advisory guideline to protect the public's health. Exposure to PFBA and other perfluoroalkyl compounds can occur from the inhalation of contaminated house dust. A study of homes in Catalonia, Spain detected 10 different perfluoroalkyl compounds, including PFBA, present in house dust (Ericson Jogsten et al., 2012). Occupants breathing in air or contacting surfaces in these indoor environments could be at risk for exposure. Individuals may also be exposed to PFBA by eating food, such as fish, contaminated with the chemical. Reducing waste streams can prevent PFBA from entering the environment and impacting the population.

INDUSTRIAL HYGIENE

Much like PFBS, PFBA has gained popularity because it appears to be less harmful and is not known to bioaccumulate like other perfluoroalkyl compounds. The Occupational Safety and Health Administration, NIOSH, and ACGIH have not established exposure limits for PFBS. While there are no formal guidelines in place, several preventative measures can be utilized to reduce the potential for inhalation, ingestion, and dermal exposure to PFBA. Inhalation of PFBA can be prevented by utilizing personal respirators. Personal respirators should have the proper accessories installed to capture harmful vapors. An additional measure to remove particles and vapors from the atmosphere would be to utilize a ventilation system. Good hygiene practices can prevent the ingestion of PFBA. Individuals using the chemical should ensure that they wash their hands prior to eating, drinking, applying cosmetics, and using tobacco products. Using personal protective equipment, such as gloves, rubber boots, and goggles, can prevent dermal contact with the chemical. Contaminated clothing should be removed in a proper manner to prevent additional contamination.

MEDICAL MANAGEMENT

Perfluorobutyrate can be detected in serum of exposed individuals. A study of current and former 3M employees found that out of 177 individuals 96.1% had traces of PFBA in their serum with the highest concentration being 6.2 ng/ml (Chang et al., 2008). Establishing a biomonitoring program can help identify employees exposed to PFBA in the work environment. Serum testing should be completed prior to employment, during employment, and at termination to get an idea of the levels of exposure that occurred. Individuals that are exposed to PFBA will be treated differently depending on their route of exposure. If PFBA vapors are encountered, the exposed individual should be moved into an area with fresh air. If breathing is labored, oxygen may need to be necessary. If an exposed individuals is not breathing, emergency artificial respiration will be needed. If PFBA is ingested, vomiting should not be induced and the mouth should be rinsed with water. If PFBA penetrates the skin, it should be cleansed with soap and water to remove the chemical from the skin. If PFBA contacts the eyes, they should be rinsed with water for a minimum of 15 min. In all situations, medical attention may be necessary if the exposure was excessive or if symptoms do not subside.

REFERENCES

Butenhoff, J.L., et al. (2012) Toxicological evaluation of ammonium perfluorobutyrate in rats: twenty-eight-day and ninety-day oral gavage studies. *Reprod. Toxicol.* 33:513–530.

Chang, S.C., et al. (2008) Comparative pharmacokinetics of perfluorobutyrate in rats, mice, monkeys, and humans and relevance to human exposure via drinking water. *Toxicol. Sci.* 104:40–53.

Das, K.P., et al. (2008) Effects of perfluorobutyrate exposure during pregnancy in the mouse. *Toxicol. Sci.* 105:173–181.

Ericson Jogsten, I., et al. (2012) Per- and polyfluorinated compounds (PFCs) in house dust and indoor air in Catalonia, Spain: implications for human exposure. *Environ. Int.* 39:172–180.

Foreman, J.E., et al. (2009) Differential hepatic effects of perfluorobutyrate mediated by mouse and human PPAR-alpha. *Toxicol. Sci.* 110:204–211.

Freberg, B.I., et al. (2010) Occupational exposure to airborne perfluorinated compounds during professional ski waxing. *Environ. Sci. Technol.* 44:7723–7728.

Minnesota Department of Health (MDH) (2011) *Chemical Name: Perfluorobutyrate*, CAS: 375-22-4, MN: St. Paul.

Minnesota Department of Health (MDH) (2012) *PFBA in the Groundwater of the South East Metro Area*, MN: St. Paul, Available at http://www.health.state.mn.us/divs/eh/hazardous/topics/pfbasemetro.html.

Ray, J.N. (2010) *Structure-activity relationships between perfluorinated chemicals and mitochondrial respiration rates,* University of Minnesota.

PERFLUOROHEXANE SULFONATE (PFHxS)

Target organ(s): Eyes, skin, respiratory system, liver, kidney, thyroid

Occupational exposure limits: NIOSH/OSHA/ACGIH—not established

Reference values: RfD/RfC/MRL—not established

Risk/safety phrases: Irritant

BACKGROUND AND USES

Perfluorohexane sulfonate (PFHxS, perfluorohexane sulfonic acid, perfluorohexane-1-sulphonic acid; CASRN 355-46-4) is a six-carbon compound in the perfluoroalkyl family of chemicals. Perfluorohexane sulfonate was once used in fire-fighting foam and carpet treatment solutions. It was once widely used by 3M as a stain and water repellent but was phased out due to U.S. Environmental Protection Agency regulations. Much like other perfluoroalkyls, PFHxS can persist in the environment.

PHYSICAL AND CHEMICAL PROPERTIES

Perfluorohexane sulfonate is slightly soluble in water, has a molecular weight of 400.1 g/mol, a boiling point of 238.5 °F, and a density of 1.841 g/cm^3 (Ark Pharm Inc., 2013). Perfluorohexane sulfonate has the ability to react with strong oxidizing agents. Upon decomposition, PFHxS can form carbon monoxide, carbon dioxide, hydrogen fluoride, and oxides of sulfur. Additional information related to the physical and chemical properties of PFHxS are not currently available.

MAMMALIAN TOXICOLOGY

Acute Effects

Perfluorohexane sulfonate has been described as an irritant. Contact with the eyes, mucous membranes, and skin may result in irritant effects. Respiratory irritation may also occur if the chemical is inhaled into the respiratory tract. Exposure to PFHxS was found to result in decreased body weight, decreased cholesterol levels, and increased prothrombin in rats. The Minnesota Department of Health (2009) also noted that microscopic changes in thyroid tissue were reported at high doses of PFHxS. Exposure to PFHxS was also found to disrupt the expression of genes related to PPARα function, including fatty acid metabolism, inflammation, peroxisome biogenesis, and proteasome biogenesis (Rosen et al., 2011).

These findings indicate that PFHxS may act similar to other perfluoroalkyl compounds, in which they disturb the PPARα function. Renal toxicity is also possible with PFHxS exposure. A 2013 study found that children living near a chemical plant exposed to PFHxS had decreased estimated glomerular filtration rates indicating possible renal impairment (Watkins et al., 2013).

Chronic Effects

As previously discussed, exposure to PFHxS may result in acute hepatic, renal, neurological, and reproductive toxicity. Chronic exposure to the chemical may result in acute effects becoming chronic in nature. Elimination of the chemical may reduce its burden on the body. The half-life of PFHxS in female rats, male rats, and monkeys has been approximated at 2 days, 1 month, and 4 months, respectively (Sundstrom et al., 2012). Gender and species differences may play a role in the chemical's elimination. The major chronic effect of concern with respect to chemical exposure is the development of cancer. A 2014 case-control study examined the exposure to perfluorinated alkyl acids and the risk of prostate cancer and found higher risks for prostate cancer when heredity was considered as a risk factor (Hardell et al., 2014). Another case-control study found a significant association between serum perfluoroalkyl levels and the risk of breast cancer (Bonefeld-Jorgensen et al., 2011). While both of these studies show a possible association between chemical exposure and disease development, additional research is needed to understand the exact mechanism of action and to make a more definitive determination on causality.

Mechanism of Action(s)

There has been some investigation into the mechanism of action of PFHxS. The chemical appears to result in renal, hepatic, and neurological effects. A study conducted in 2013 found that dairy cows fed with contaminated feed at a concentration of 4.6 μg/kg of body had increased levels of the chemical in their plasma along with high levels of PFHxS in the liver and kidney after only 29 days of exposure (Kowalczyk et al., 2013). The kidney and the liver appear to be organs where the chemical concentrates. There is also some evidence that PFHxS is associated with neurotoxic effects as well. Lee and Viberg (2013) found that neonatal mice exposed to PFHxS experienced neurotoxic effects due to increased levels of neuroproteins, including CaMKII, GAP-43, and synaptophysin in the hippocampus, and decreased levels of GAP-43 in the brain cortex. An additional study noted that there are indications that PFHxS induces apoptosis of cerebellar granule cells in the developing brain, which may signify a mechanism of action via the ERK1/2 pathway (Lee et al., 2014).

Chemical Pathology

The literature indicates that PFHxS may be associated with adverse health effects of the liver, kidney, and thyroid. Wen et al. (2013) found that elevated levels of PFHxS in 1181 subjects was associated with elevated thyroid hormones, including total T_3, total T_4, and free T_4. Disruption in hormonal activity may result in reproductive effects. Many speculate that the chemical may play a role in abnormal development and reproductive toxicity, however, research does not currently support the speculation. One study found that rats exposed to PFHxS concentrations of 0.3, 1, 3, and 10 mg/kg/day experienced no abnormal reproductive effects (Butenhoff et al., 2009). Another study found that exposure to PFHxS did not adversely affect the sperm concentration, sperm count, and semen volume in a cohort of partners of pregnant women (Toft et al., 2012). Additional research must be conducted to help determine if the chemical is associated with the development of reproductive pathologies.

SOURCES OF EXPOSURE

Occupational

Exposure to PFHxS can occur in the work environment, and employees using the chemical should take proper precautions. A study of workers in a fluorochemical plant detected several perfluoroalkyl compounds, including PFHxS, at elevated levels (Wang et al., 2012). Perfluorohexane sulfonate was also detected in plasma samples from workers responding to the World Trade Center Disaster (NYSDOH, 2008). The use of engineering controls, administrative controls, and personal protective equipment can prevent exposure in the work environment. Inhalation of PFHxS-contaminated dust may be associated with the development of adverse health effects. Employees working in contaminated areas should wear proper personal protective equipment and utilize the correct engineering controls to reduce the potential for the exposure. Splashes of PFHxS may result in dermal contact. The use of gloves, long-sleeved work apparel, and boots can prevent the chemical from penetrating the skin. If clothes become contaminated with PFHxS, they should be discarded using proper measures. Ingestion of the chemical in the work environment is not likely but could occur. Personal hygiene measures should be enforced to prevent exposure.

Environmental

Most environmental exposure to PFHxS occurs through the consumption of contaminated drinking water. A study conducted in 2008 found significant increases of PFHxS concentrations in drinking water, with age and gender being significant predictors in regression analysis (Holzer et al., 2008). Another study conducted in 2009 found PFHxS in 31 out of 40 samples with levels in drinking water up to 5.30 ng/ml (Ericson et al., 2009). Perfluorohexane sulfonate not only contaminates water supplies, but it can also impact air quality. One study found that PFHxS levels in indoor environments exceeded those of PFOS, another perfluoroalkyl compound (Goosey and Harrad, 2012). The researchers indicated that this may be due to the increased utilization of PFHxS over more persistent chemicals. While PFHxS can be detected in water and air samples, the lack of established exposure guidelines makes interpretation of the data complicated. It may be necessary to ensure that PFHxS levels stay as low as reasonably possible until additional dose-response data becomes available.

INDUSTRIAL HYGIENE

The use of PFHxS has been discontinued and other chemicals have been used in its place. The U.S. Environmental Protection Agency, in conjunction with chemical manufacturers, began phasing out the use of the chemical in 2000 with the program being completed in 2002 (USEPA, 2011). The Occupational Safety and Health Administration, NIOSH, and ACGIH have not established exposure limits for PFHxS. While there are no formal guidelines in place, several measures can be used to reduce inhalation, ingestion, and dermal exposure to PFHxS. Inhalation to PFHxS can be prevented by utilizing personal respirators. An additional measure to remove particles and vapors from the environment would be to engineer a ventilation system. Good hygiene can avert the ingestion of PFHxS. Individuals using the chemical should ensure that they wash their hands prior to eating, drinking, and using tobacco. Using personal protective equipment, such as gloves, rubber boots, and goggles, can prevent dermal contact with the chemical.

MEDICAL MANAGEMENT

Perfluorohexane sulfonate can be detected in the human blood serum. A study of 2420 donors identified PFHxS in human blood at a mean concentration of 3.1 ng/ml with children <15 years old having the highest mean concentrations (Toms et al., 2009). The institution of a biomonitoring program can help protect employees from exposure to the chemical. Testing for the chemical should occur at a frequency that will allow for the early detection of exposure. If it is determined that an individual has been exposed to PFHxS, it may be necessary to remove them from the area where the exposure has taken place until blood levels decrease. If dermal contact with the chemical occurs, it should be rinsed from the skin with contaminated clothing disposed of properly. The use of personal respiratory protection can reduce the potential for inhalation of PFHxS but respirators must be fitted properly to be effective. Ingestion of the chemical is not likely but could occur and individuals working with PFHxS should thoroughly clean their hands before eating, drinking, or smoking to prevent exposure via this route. In cases where symptoms of exposure do not subside or when individuals are exposed to high concentrations of PFHxS, prompt medical attention is recommended.

REFERENCES

Ark Pharm, Inc. (2013) *Material Safety Data Sheet–1,1,2,2,3,3,4, 4,5,5,6,6,6-Tridecafluorohexane-1-Sulfonic Acid*, Version 2.1, Libertyville, IL. Available at http://www.arkpharminc.com/files/document/B_COA_MSDS/MSDS-103042.pdf

Bonefeld-Jorgensen, E.C., et al. (2011) Perfluorinated compounds are related to breast cancer risk in Greenlandic Inuit: a case control study. *Environ. Health* 10:88.

Butenhoff, J.L., et al. (2009) Evaluation of potential reproductive and developmental toxicity of potassium perfluorohexanesulfonate in Sprague Dawley rats. *Reprod. Toxicol.* 27:331–341.

Ericson, I., et al. (2009) Levels of perfluorinated chemicals in municipal drinking water from Catalonia, Spain: public health implications. *Arch. Environ. Contam. Toxicol.* 57:631–638.

Goosey, E. and Harrad, S. (2012) Perfluoroalkyl substances in UK indoor and outdoor air: spatial and seasonal variation, and implications for human exposure. *Environ. Int.* 45:86–90.

Hardell, E., et al. (2014) Case-control study on perfluorinated alkyl acids (PFAAs) and the risk of prostate cancer. *Environ. Int.* 63:35–39.

Holzer, J., et al. (2008) Biomonitoring of perfluorinated compounds in children and adults exposed to perfluorooctanoate-contaminated drinking water. *Environ. Health Perspect.* 116:651–657.

Kowalczyk, J., et al. (2013) Absorption, distribution and milk secretion of the perfluoroalkyl acids PFBS, PFHxS, PFOS, and PFOA by dairy cows fed naturally contaminated feed. *J. Agric. Food Chem.* 61:2903–2912.

Lee, I. and Viberg, H. (2013) A single neonatal exposure to perfluorohexane sulfonate (PFHxS) affects the levels of important neuroproteins in the developing mouse brain. *Neurotoxicology* 37:190–196.

Lee, Y.J., et al. (2014) PFHxS induces apoptosis of neuronal cells via ERK1/2-mediated pathway. *Chemosphere* 94:121–127.

Minnesota Department of Health (MDH) (2009) *Chemical Name: Perfluorohexane Sulfonate*, CAS: 355-46-4, MN: St. Paul.

New York State Department of Health (NYSDOH) (2008) *Results of Blood Plasma Samples for a Group of NYS Personnel Responding to the World Trade Center Disaster*, NYSDOH 1-1, Albany, NY.

Rosen, M.B., et al. (2011) Hepatic gene expression profiling in Perfluorohexane sulfonate-exposed wild-type and PPARα-null mice, *Society of Toxicology Annual Meeting*.

Sundstrom, M., et al. (2012) Comparative pharmacokinetics of perfluorohexanesulfonate (PFHxS) in rats, mice, and monkeys. *Reprod. Toxicol.* 33:441–451.

Toft, G., et al. (2012) Exposure to perfluorinated compounds and human semen quality in arctic and European populations. *Hum. Reprod.* 27:2532–2540.

Toms, L.M., et al. (2009) Polyfluoroalkyl chemicals in pooled blood serum from infants, children, and adults in Australia. *Environ. Sci. Technol.* 43:4194–4199.

United States Environmental Protection Agency (USEPA) (2011) *Biomonitoring: Perfluorochemicals (PFCs)*, USEPA 1–38, Washington, DC:

Wang, J., et al. (2012) Association of perfluorooctanoic acid with HDL cholesterol and circulating miR-26b and miR-199-3p in workers of a fluorochemical plant and nearby residents. *Environ. Sci. Technol.* 46:9274–9281.

Watkins, D.J., et al. (2013) Exposure to perfluoroalkyl acids and markers of kidney function among children and adolescents living near a chemical plant. *Environ. Health Perspect.* 121:625–630.

Wen, L.L., et al. (2013) Association between serum perfluorinated chemicals and thyroid function in U.S. adults: the National Health and Nutrition Examination Survey 2007–2010. *J. Clin. Endocrinol. Metab.* 98:E1456–E1464.

PERFLUOROHEXANOIC ACID (PFHxA)

Target organ(s): eyes, skin, respiratory system

Occupational exposure limits: NIOSH/OSHA/ACGIH—not established

Reference values: RfD/RfC/MRL—not established

Risk/safety phrases: Corrosive

BACKGROUND AND USES

Perfluorohexanoic acid (PFHxA, perfluorocaproic acid, undecafluorohexanoic acid, undecafluoro-1-hexanoic Acid; CASRN 307-24-4) is a six-carbon compound in the perfluoroalkyl family of chemicals. Perfluorohexanoic acid is used in stain- and grease-proof coatings on furniture, carpet, and food packaging. Perfluorohexanoic acid is the six-carbon version of the highly persistent and outlawed PFOA. Much like other perfluoroalkyls, PFHxA can persist in the environment.

PHYSICAL AND CHEMICAL PROPERTIES

Perfluorohexanoic acid exists as a liquid, it is insoluble in water, has a molecular weight of 314.05, and a density of 1.759. Perfluorohexanoic acid has the ability to react with strong oxidizing agents. Upon decomposition, PFHxA can form carbon oxides and hydrogen fluoride. Additional information related to the physical and chemical properties of PFHxA are not currently available.

MAMMALIAN TOXICOLOGY

Acute Effects

Several research studies have been conducted to determine the acute effects PFHxA may have on mammalian systems. Available safety information explains that inhalation and dermal contact may result in some acute effects. Inhalation of the chemical may cause irritation to the eyes and upper respiratory tract while dermal contact with the chemical may cause irritation to the skin. One study utilized promyelocytic leukemia rat and glioma rat cell lines to determine the chemical cytoxicity and found that PFHxA has a low acute biological activity (Mulkiewicz et al., 2007). The researchers also found that the level of toxicity was dependent on perfluoroalkyl chain length (Mulkiewicz et al., 2007). Another study that used human colon carcinoma cells *in vitro* agreed that PFHxA and other perfluoroalkyl compounds do not appear to be acutely toxic at the cellular level (Kleszczynski et al., 2007). Additional research is needed to fully understand the potential acute effects related to PFHxA exposure.

Chronic Effects

There is sparse information related to the chronic effects of PFHxA exposure. Perfluoroalkyl exposure may be associated with abnormal immune system function. A study in 2012 found that high exposures to perfluoroalkyl compounds in children aged 5– 7 was associated with decreased humoral immune response to routine childhood immunizations (Grandjean et al., 2012). According to the available literature, exposure to PFHxA does not appear to lead to chronic effects associated with the liver or thyroid gland. The relationship between cancer development and exposure to PFHxA and related compounds remains unclear. A cross-sectional study conducted in 2014 found a strong inverse association between perfluoroalkyl exposure and the likelihood of colorectal cancer diagnosis (Innes et al., 2014). Hardell et al. (2014) conducted a study that indicated an elevated risk for prostate cancer, with an odds ratio of 1.8, in cases where heredity was a risk factor. Additional research into the chronic effects of PFHxA exposure is needed.

Mechanism of Action(s)

The mechanism of action of PFHxA appears to be much like that of other perfluoroalkyl compounds. Many studies

of perfluoroalkyl compounds suggest that the chemicals concentrate and disrupt cell development in the liver. However, one study found that PFHxA did not generate reactive oxygen species or DNA damage in human hepatoma HepG2 cell lines like other perfluoroalkyl compounds (Eriksen et al., 2010). The elimination of PFHxA from the body systems has been considered during the scientific investigation. A 2009 study found that serum clearance of PFHxA was more rapid when compared to PFBS and systemic exposure to PFHxA was lower when compared to PFBS in rats and monkeys (Chengelis et al., 2009). When compared to other, longer chained perfluoroalkyls, PFHxA may be eliminated from the body faster limiting the impact it may have on mammals. Additional investigation is needed to fully describe the PFHxA mechanism of action in the mammalian systems.

Chemical Pathology

Perfluorohexanoic acid has been suggested to impact the liver and the thyroid gland. A study of chicken embryos found that PFHxA accumulated in the yolk sac, liver, and cerebral cortex but did not disrupt thyroid hormone levels (Cassone et al., 2012). It remains unclear if the liver and thyroid gland are sites of pathological manifestations as a result of PFHxA exposure. Research in mammals indicates that PFHxA may be associated with other abnormalities. Human serum albumin has been identified as the site of interaction for PFHxA in human sera (D'eon et al., 2010). Molecular interactions of this nature may explain how perfluoroalkyl compounds, like PFHxA, bioaccumulate in mammalian tissue. Perfluorohexanoic acid introduced to rats orally at concentrations of 10, 50, and 200 mg/kg/day was associated with lower body weight gains, lower red blood cell parameters, higher reticulocyte counts, and lower globulin levels (Chengelis et al., 2009). Further research will help explain how PFHxA is distributed in the body and how it leads to adverse health effects.

SOURCES OF EXPOSURE

Occupational

Workers in direct contact with PFHxA are at risk for exposure. Perfluorohexanoic acid and similar compounds may be a component of ski wax. Wax is applied to skis to improve their performance when they contact the snow. Ski wax technicians may be at risk for exposure during the wax-application process when particles are introduced into the atmosphere as a result of wax buffing. Russell et al. (2013) explains that PFHxA and its salts are quickly absorbed into the body, distributed, and eliminated. If these particles are inhaled into the respiratory tract or are deposited on the skin, irritant effects may occur. One ski wax technician suffered pulmonary injury, described as acute respiratory distress syndrome (ARDS), resulting from aerosolized perfluoroalkyl compound inhalation (Bracco and Favre, 1998). Since ski waxes contain perfluoroalkyl compounds many believe that they should no longer be considered toxic, however, this is debated. In order to reduce exposure, engineering controls, administrative controls, and personal protective equipment should be utilized.

Environmental

Perfluorohexanoic acid has been used in many applications as it is less persistent than its longer chained chemical counterparts. Perfluorohexanoic acid, like other perfluoroalkyl compounds, can contaminate water systems when it is released via human waste streams. Sediment samples from a lake in Austria identified concentrations of PFHxA up to $5.1 \, \mu g/kg^{-1}$ dry weight along with concentrations of other chemically similar compounds (Clara et al., 2009). Perfluorohexanoic acid can also contaminate environments if the chemical is emitted into the atmosphere. When carpets and furniture impregnated with PFHxA begin to degrade, PFHxA can enter the air as it attaches to dust particles. A study of perfluoroalkyl compounds in the indoor environment detected PFHxA at a highest median concentration of 28 ng/g (Haug et al., 2011). The length of time a perfluoroalkyl compound persists in the environment appears to be directly related to the size of the chemical. Substituting PFHxA with other chemicals can prevent its release into the environment lessening its impact on both terrestrial and aquatic organisms.

INDUSTRIAL HYGIENE

The Occupational Safety and Health Administration, NIOSH, and ACGIH have not established exposure limits for PFHxA. While there are no formal guidelines in place, several measures can be used to reduce inhalation, ingestion, and dermal exposure to PFHxA. Individuals using the chemical should be careful not to inhale PFHxA. Inhalation of PFHxA can be prevented by utilizing personal respiratory protection. Installing a ventilation system capable of removing the chemical may also be used to prevent unnecessary exposure. Good hygiene can avert the ingestion of PFHxA. Individuals using the chemical should ensure that they wash their hands prior to eating, drinking, and using tobacco. Using personal protective equipment, such as gloves, rubber boots, and goggles, can prevent dermal contact with the chemical. A fully encapsulating suit may be necessary in environments where high concentrations and quantities of the chemical are used. If applicable, PFHxA should be substituted with less persistent, shorter carbon-chained chemicals of similar structure.

MEDICAL MANAGEMENT

Perfluorohexanoic acid can be detected in human blood samples. A study of ski wax technicians found that the half-life of PFHxA in the body of highly exposed individuals ranged from 14 to 49 days with a geometric mean of 32 days (Russell et al., 2013). Employees with the potential for exposure to PFHxA should be entered into a medical surveillance program. Examining employees at the time of hire, during employment, and at termination or transfer to different positions can help to identify exposure. If exposure is identified early, the impact on the employee could potentially be reduced. If it is determined that an individual has been exposed to PFHxA, it may be necessary to remove them from the contaminated area. Since PFHxA exists as a liquid, the chemical could be splashed into the eyes or onto the skin. If dermal contact with the chemical occurs, the eyes and skin should be rinsed thoroughly. Contaminated clothing should be disposed of properly. The use of personal respiratory protection can protect against inhalation exposure to PFHxA. Ingestion of the chemical is not likely but could occur and individuals working with PFHxA should thoroughly clean their hands before eating, drinking, or smoking to prevent exposure via this route. In cases where symptoms of exposure do not subside or when individuals are exposed to high concentrations of PFHxA, immediate medical attention should be provided.

REFERENCES

Bracco, D. and Favre, J.B. (1998) Pulmonary injury after ski wax inhalation exposure. *Ann. Emerg. Med.* 32:616–619.

Cassone, C.G., et al. (2012) In ovo effects of perfluorohexane sulfonate and perfluorohexanoate on pipping success, development, mRNA expression, and thyroid hormone levels in chicken embryos. *Toxicol. Sci.* 127:216–224.

Chengelis, C.P., et al. (2009) A 90-day dose oral (gavage) toxicity study of perfluorohexanoic acid (PFHxA) in rats (with functional observational battery and motor activity determinations). *Reprod. Toxicol.* 27:342–351.

Chengelis, C.P., et al. (2009) Comparison of the toxicokinetic behavior of perfluorohexanoic acid (PFHxA) and nonafluoro-butane-1-sulfonic acid (PFBS) in cynomolgus monkeys and rats. *Reprod. Toxicol.* 27:400–406.

Clara, M., et al. (2009) Perfluorinated alkylated substances in the aquatic environment: an Austrian case study. *Water Res.* 43:4760–4768.

D'eon, J.C., et al. (2010) Determining the molecular interactions of perfluorinated carboxylic acids with human sera and isolated human serum albumin using nuclear magnetic resonance spectroscopy. *Environ. Toxicol. Chem.* 29:1678–1688.

Eriksen, K.T., et al. (2010) Genotoxic potential of the perfluorinated chemicals PFOA, PFOS, PFBS, PFNA, and PFHxA in human HepG2 cells. *Mutat. Res.* 700:39–43.

Grandjean, P., et al. (2012) Serum vaccine antibody concentrations in children exposed to perfluorinated compounds. *JAMA* 307:391–397.

Hardell, E., et al. (2014) Case-control study on perfluorinated alkyl acids (PFAAs) and the risk of prostate cancer. *Environ. Int.* 63:35–39.

Haug, L.S., et al. (2011) Investigation on per- and polyfluorinated compounds in paired samples of house dust and indoor air from Norwegian homes. *Environ. Sci. Technol.* 45:7991–7998.

Innes, K.E., et al. (2014) Inverse association of colorectal cancer prevalence to serum levels of perfluorooctane sulfonate (PFOS) and perfluorooctanoate (PFOA) in a large Appalachian population. *BMC Cancer* 14:45.

Kleszczynski, K., et al. (2007) Analysis of structure-cytotoxicity *in vitro* relationship (SAR) for perfluorinated carboxylic acids. *Toxicol. In Vitro* 21:1206–1211.

Mulkiewicz, E., et al. (2007) Evaluation of the acute toxicity of perfluorinated carboxylic acids using eukaryotic cell lines, bacteria, and enzymatic assays. *Environ. Toxicol. Pharmacol.* 23:279–285.

Russell, M.H., et al. (2013) Elimination kinetics of perfluorohexanoic acid in humans and comparison with mouse, rat, and monkey. *Chemosphere* 93:2419–2425.

PERFLUOROHEPTANOATE (PFHpA)

Target organ(s): Eyes, skin, respiratory system, reproductive system

Occupational exposure limits: NIOSH/OSHA/ACGIH—not established

Reference values: RfD/RfC/MRL—not established

Risk/safety phrases: Corrosive

BACKGROUND AND USES

Perfluoroheptanoate (PFHpA, perfluoroheptanoic acid, perfluoro-n-heptanoic acid, tridecafluoro-1-heptanoic acid; CASRN 375-85-9) is a seven-carbon compound in the perfluoroalkyl family of chemicals. Perfluoroheptanoate is used in stain- and grease-proof coatings on furniture, carpet, and food packaging. Perfluoroheptanoate is the seven-carbon chemical that is structurally similar to the highly persistent and outlawed PFOA. Much like other perfluoroalkyls, PFHpA can persist in the environment.

PHYSICAL AND CHEMICAL PROPERTIES

Perfluoroheptanoate exists as beige crystals, it has a molecular weight of 364.06 g/mol, density of 1.792 g/cm^3, freezing point of 30 °C, boiling point of 175 °C, and a flash point >113 °C. Perfluoroheptanoate has the ability to react with strong oxidizing agents. Upon decomposition, PFHpA can form carbon oxides and hydrogen fluoride. Additional information related to the physical and chemical properties of PFHpA are not currently available.

MAMMALIAN TOXICOLOGY

Acute Effects

Perfluoroheptanoate's acute effects on mammalian systems have been the topic of several research studies. Available safety information explains that inhalation and dermal contact may result in some acute effects. Inhalation of the chemical may cause irritation to the eyes, mucous membranes, and upper respiratory tract while dermal contact with the chemical may cause irritation to the skin. The major concern related to PFHpA and related compounds are that they have the ability to bioaccumulate in the biological tissue. Perfluoroalkyl compounds were detected in several animals from the Barents Sea food web, including sea ice amphipod, polar cod, black guillemot, and glaucous gull (Haukas et al., 2007). It has been determined that the longer the chemical carbon chain, the more persistent the chemical. Perfluorinated compounds have been detected in tigers (1.18 ng/ml) and lions (2.69 ng/ml), however, there is uncertainty with regards to the actual risk and toxicity associated with the chemical (Li et al., 2008). While PFHpA is known to bioaccumulate in the biological tissue, additional research is needed to fully understand what potential acute systemic effects can occur as a result of the chemical bioaccumulation.

Chronic Effects

Some perfluoroalkyl chemicals, including PFHpA, have been linked to chronic effects associated with the liver, thyroid, and kidneys. Concrete evidence supporting these links is lacking and therefore is deserving of additional investigation. There has been some evidence that PFHpA and related compounds may be associated with adverse effects on the reproductive system. A study of 495 women found that those exposed to perfluorinated compounds had higher odds of being diagnosed with endometriosis when compared with those who did not have exposure (Louis et al., 2012). There has also been some research conducted to investigate the relationship between PFHpA exposure and the development of cancer. Bonefeld-Jorgensen et al. (2011) conducted a case-control study that examined the exposure to perfluoroalkyl compounds and the risk of breast cancer in Greenlandic Inuit woman and found a significant association between serum perfluoroalkyl compound levels and the risk of breast cancer. Additional research into the chronic effects of PFHxA exposure is needed.

Mechanism of Action(s)

It is believed that perfluoroalkyl compounds, including PFHpA, share a similar mechanism of action. The Agency for Toxic Substances & Disease Registry (ATSDR) indicates that these chemicals bind to serum albumin and other plasma proteins such as gamma globulin, alpha globulin, alpha-2-macroglobulin, transferrin, and beta-lipoproteins (ATSDR, 2013). Once in the blood, PFHpA distributes into tissues. Wolf et al. (2012) found that when compared to other perfluorinated compounds, PFHpA induced some of the highest activities for human PPARα when compared with structurally similar compounds. The same study further determined that increased PPARα activity was directly associated with increased perfluoroalkyl compound chain length (Wolf et al., 2011). Some research indicates that the thyroid and kidney may be the target organs but up to this point there is no conclusive evidence to support these speculations. Perfluoroheptanoate has a shorter half-life in the body than longer chained perfluoroalkyl compounds and is eliminated from the body in urine, which appears to be an adequate biomarker to determine the exposure.

Chemical Pathology

Like many of the other perfluoroalkyl compounds, PFHpA is believed to target the liver. Frisbee et al. (2009) explain that exposure to these chemicals has been reported to cause hepatotoxic effects, including liver enlargement, hepatocellular adenomas, and peroxisome proliferation. Some researchers believe that exposure to PFHpA and related chemicals may have an impact on the male reproductive system as a result of endocrine disruption. The semen quality of 105 Danish men was analyzed and it revealed that high levels of perfluoroalkyl compounds were related to a decrease in the amount of normal sperm (Joensen et al., 2009). Recent research indicates that exposure to perfluoroalkyl compounds may increase cardiovascular risk. Watkins et al. (2014) examined whether exposure to this class of chemicals was associated with LINE-1 DNA methylation and found that serum levels of perfluoroalkyl compounds were associated with LINE-1 DNA methylation. In order to determine the exact mechanisms of action and additional pathological manifestations, more research is required in this area.

SOURCES OF EXPOSURE

Occupational

Exposure to PFHpA can occur in the work environment and employees using the chemical should take proper precautions. Firefighters are a special population that may be at a risk for exposure to a variety of perfluoroalkyl compounds. When consumer products that contain PFHpA are burned during a fire, the chemical may be released into the atmosphere. Firefighters should take precaution by utilizing personal respiratory protection. Levels of perfluoroalkyl compounds were found to be significantly high in a study that examined 37 firefighters (Jin et al., 2011). The use of engineering controls, administrative controls, and personal protective equipment can prevent exposure in the occupational environment. Dermal contact with PFHpA is possible so the use of gloves, long-sleeved work apparel, and boots may be necessary to prevent the chemical from penetrating the skin. If clothes become contaminated with PFHpA, they should be discarded using proper measures. Ingestion of the chemical in the occupational environment is not likely but could occur.

Environmental

Environmental exposure to PFHpA occurs by contacting contaminated water and air. Improper release and disposal of perfluorinated compounds allows the chemical to impact water systems and the surrounding environments. A study of Japanese river ways found a correlation between PFHpA concentration and population density with the chemical being measured in the sewage effluent at a total sewage flux of 2.6 t/year (Murakami et al., 2008). When human consumer products containing PFHpA and related compounds begin to deteriorate, the chemical can be released into the air where it can adhere to house dust. The particles may settle on surfaces or be inhaled into the respiratory tract of building inhabitants. Kato et al. (2009) analyzed 39 house dust samples and found that PFHpA had a frequency of detection of 61.2%. Ventilating indoor settings can help prevent the buildup of perfluoroalkyl particles. In addition, purchasing consumer products that are free of PFHpA and other perfluoroalkyl compounds will eliminate exposure to the chemical.

INDUSTRIAL HYGIENE

The Occupational Safety and Health Administration, NIOSH, and ACGIH have not established exposure limits for PFHpA. While there are no formal guidelines in place, several measures can be used to reduce inhalation, ingestion, and dermal exposure to PFHpA. Individuals using the chemical should be careful to avoid inhalation of PFHpA. Inhalation of PFHpA can be prevented by utilizing personal respiratory protection. Installing a ventilation system capable of removing the chemical may also be necessary to prevent exposure. Good hygiene practices can prevent the ingestion of PFHpA. Eating, drinking, or using tobacco products should be prohibited when individuals use the chemical. Hands should be washed thoroughly with soap and water to remove the contaminant. Dermal contact with the chemical can be prevented by using personal protective equipment, such as gloves, rubber boots, and goggles. A fully encapsulating suit may be necessary in environments where high concentrations and quantities of the chemical are utilized. If applicable, PFHpA should be substituted with less persistent, shorter carbon-chained chemicals of similar structure.

MEDICAL MANAGEMENT

Perfluoroheptanoate can be detected in the blood and it has been the topic of a variety of research studies. Perfluoroheptanoate and other perfluoroalkyl compounds were detected in 30 blood samples from five cities in China (Yeung et al., 2008). It is recommended that employees that use or have the potential to contact PFHpA should be entered into a medical surveillance program. If exposure is identified early, the burden on the employee could be decreased. If it is determined that an individual has been exposed to PFHpA, it may be recommended to remove them from the contaminated work environment. Since PFHpA exists as a crystalline solid, small particles may be inhaled into the respiratory tract where they can deposit deep in the lung. The use of personal respiratory protection can protect against inhalation exposure to PFHpA. There may be some applications where PFHpA is mixed into solution that could lead to splashing or other types of dermal contact with the chemical. If dermal contact with the chemical occurs, the skin should be rinsed thoroughly. Contaminated clothing should be disposed of properly. Ingestion of the chemical is not likely but could occur and individuals working with PFHpA should thoroughly clean their hands before eating, drinking, or smoking to prevent exposure via this route. In cases where symptoms of exposure do not subside or when individuals are exposed to high concentrations of PFHpA, individuals should seek prompt medical attention.

REFERENCES

Agency for Toxic Substances and Disease Registry (ATSDR) (2013) *Perfluoroalkyls–Health Effects*, ATDSR 1–194, Atlanta, GA: US Department of Health and Human Services.

Bonefeld-Jorgensen, E.C., et al. (2011) Perfluorinated compounds are related to breast cancer risk in Greenlandic Inuit: a case control study. *Environ. Health* 10:88.

Frisbee, S.J., et al. (2009) The C8 health project: design, methods, and participants. *Environ. Health Perspect.* 117:1873–1882.

Haukas, M., et al. (2007) Bioaccumulation of per- and polyfluorinated alkyl substances (PFAS) in selected species from the Barents Sea food web. *Environ. Pollut.* 148:360–371.

Jin, C., et al. (2011) Perfluoroalkyl acids including perfluorooctane sulfonate and perfluorohexane sulfonate in firefighters. *J. Occup. Environ. Med.* 53:324–328.

Joensen, U.N., et al. (2009) Do perfluoroalkyl compounds impair human semen quality? *Environ. Health Perspect.* 117:923–927.

Kato, K., et al. (2009) Polyfluoroalkyl chemicals in house dust. *Environ. Res.* 109:518–523.

Li, X., et al. (2008) Accumulation of perfluorinated compounds in captive Bengal tigers (*Panthera tigris tigris*) and African lions (*Panthera leo Linnaeus*) in China. *Chemosphere* 73:1649–1653.

Louis, G.M., et al. (2012) Perfluorochemicals and endometriosis: the ENDO study. *Epidemiology* 23:799–805.

Murakami et al., M., et al. (2008) Occurrence and sources of perfluorinated surfactants in rivers in Japan. *Environ. Sci. Technol.* 42:6566–6572.

Watkins, D.J., et al. (2014) Associations between serum perfluoroalkyl acids and LINE-1 DNA methylation. *Environ Int.* 63:71–6.

Wolf, C.J., et al. (2012) Activation of mouse and human peroxisome proliferator-activated receptor-alpha (PPARα) by perfluorinated acids (PFAAs): further investigation of C4-C12 compounds. *Reprod. Toxicol.* 33:546–551.

Yeung, L.W., et al. (2008) Perfluorinated compounds and total and extractable organic fluorine in human blood samples from China. *Environ. Sci. Technol.* 42:8140–8145.

PERFLUOROOCTANE SULFONATE (PFOS)

Target organ(s): Eyes, skin, respiratory tract, liver, thyroid, reproductive system

Occupational exposure limits: NIOSH/OSHA/ACGIH—not established

Reference values: RfD/RfC/MRL—not established

Risk/safety phrases: Corrosive/carcinogen/teratogen

BACKGROUND AND USES

Perfluorooctane sulfonate (PFOS, 1-octanesulfonic acid, heptadecafluorooctanesulfonic acid, perfluorooctanesulfonic acid, 1-perfluorooctanesulfonic acid; CASRN 1763-23-1) is an eight-carbon compound in the perfluoroalkyl family of chemicals. Perfluorooctane sulfonate is used in a variety of applications, including textiles, leather products, cleaning products, and pesticides. Its main use was as a stain repellent on carpet, furniture, and other consumer products. 3M was the primary manufacturer of the chemical until it voluntarily phased it out in 2002 (USEPA, 2012). Perfluorooctane sulfonate is not manufactured any longer in the United States, but it can be imported for use in specific applications.

PHYSICAL AND CHEMICAL PROPERTIES

Perfluorooctane sulfonate can exist as a liquid or a powder. It has a molecular weight of 500.13 g/mol, a vapor pressure 0.002 mmHg, a boiling point of 133 °C, and is soluble in water (NTP, 2013). Perfluorooctane sulfonate has the ability to react with strong oxidizing agents. Upon decomposition, PFOS can form carbon oxides, sulfur oxides, and hydrogen fluoride. Additional information related to the physical and chemical properties of PFOS are not currently available.

MAMMALIAN TOXICOLOGY

Acute Effects

Dermal contact with PFOS has the ability to cause skin burns and eye damage. Inhalation of the chemical can reduce lung function and cause respiratory irritation. If PFOS is ingested, it can cause serious harm. Mammalian toxicology studies related to PFOS exposure have identified other acute effects. Prenatal exposure to PFOS in rats was found to cause neonatal mortality while those that survive have abnormal lung function possibly due to thickened alveolar walls (Grasty et al., 2005). Abnormal thyroid function has also been seen with PFOS exposure. Rats exposed to 3.2 mg PFOS/kg during gestation and lactation developed hypothyroxinemia (Yu et al., 2009). Studies of other organisms exposed to PFOS also showed that the chemical disrupts normal function. Short-term exposures of PFOS in fish have been found to increase hepatic fatty acyl-CoA oxidase activity and increased oxidative damage (Oakes et al., 2005). In addition, zebrafish embryos exposed to 0.1, 0.5, 1, 3, and 5 mg/l of PFOS experienced delayed hatching, reduced hatching, and reduced survivorship (Shi et al., 2008). Hypopigmentation, tail and heart malformations, and spinal curvature were also seen at exposure above 1 mg/l (Shi et al., 2008).

Chronic Effects

Cui et al. (2009) conducted a study that exposed rats to high levels of PFOS over a 28-day period and found that they experienced abnormal behavior, drastic weight loss, as well as hepatic and renal tissue abnormalities. The authors also found that the bioaccumulation in mammalian tissue appeared to target the liver followed by the heart, kidney, blood, lung, testicle, spleen, and brain, respectively (Cui et al., 2009). Concentrations of PFOS in southern sea otters was found to range from <1 to 884 ng/g wet weight (Kannan et al., 2006). The authors also found that animals with infectious diseases were found to have significantly higher concentrations of PFOS and PFOA in their tissue when compared to non-diseased otters (Kannan et al., 2006). This may indicate that PFOS degrades the mammalian immune system but additional research is needed to make a definitive conclusion. A reference dose of 0.00008 mg/kg/day in monkeys was found to promote a variety of chronic effects, including disruption of estrus cycle, decreased sperm count and quality, increased neoplasms, and increased mortality.

Mechanism of Action(s)

The mechanism of action of PFOS has been the topic of much discussion. There has been some research, which indicates that perfluorinated chemicals target the liver, thyroid, kidneys, and/or the reproductive system. Exposure to PFOS has the ability to promote the creation of reactive oxygen species, which has been another topic of investigation. Human microvascular endothelial cells exposed to PFOS increased the production of reactive oxygen species at both low and high concentrations (Qian et al., 2010). The presence of reactive oxygen species has the potential to increase oxidative stress in the exposed organism. Perfluorooctane sulfonate can also result in the dissipation of mitochondria membrane potential and the apoptosis of splenocytes and thymocytes, possibly disrupting the immune system function (Zhang et al., 2013). Perfluorooctane sulfonate is believed to be associated with the development of cancer; however, the association is not clear. Innes et al. (2014) found an inverse relationship between colorectal cancer prevalence and serum level of PFOS. Research in this area continues so that we can gain a better understanding of the relationship between PFOS exposure and the development of adverse health effects, including cancer.

Chemical Pathology

Most research related to PFOS pathology has focused on hepatotoxic effects. Perfluorooctane sulfonate has also been implicated in the development of neurological dysfunction. Mice exposed to a single dose of 21 μmol of PFOS had significantly increased levels of CaMKII, GAP-43, and synaptophysin in the hippocampus and significantly increased levels of synaptophysin and tau in the cerebral cortex (Johansson et al., 2009). Deviations in these protein levels may affect the normal development of the brain resulting in behavioral defects. There may also be an association between thyroid abnormalities and PFOS exposure. Tadpoles exposed to 0.1, 1, 10, and 100 μg/l of PFOS in water had up regulation of the thyroid hormone-regulated genes—thyroid receptor beta A (TRβA), basic transcription element-binding protein (BTEB), and type II deiodinase (DI2) mRNA expression (Cheng et al., 2011). These results show that PFOS may cause thyroid system imbalance. While PFOS is no longer used in industry, those that worked with the chemical and were acutely or chronically exposed may develop pathologies not previously considered.

SOURCES OF EXPOSURE

Occupational

Anyone working with or close to PFOS in the occupational setting is at risk for exposure. A study of 126 fluorochemical plant employees randomly selected for blood testing had a PFOS geometric mean of 0.941 that was one order of magnitude higher when compared to 60 film plant employees (Olsen et al., 2003). Animal studies of PFOS show that the chemical is well-absorbed orally, distributes in the blood and the liver, and may lead to increased mortality (Butenhoff et al., 2006). It is unknown at this time if these results can be extrapolated from animal models to humans. However, there is some literature that identifies potential associations between PFOS exposure and the development of cancer. Alexander et al. (2003) conducted a study of 2083 perfluor-ooctanesulphonyl fluoride plant workers and identified 145 deaths with 65 occurring in high PFOS exposure jobs. The risk of death from bladder cancer was increased for the entire group of plant workers with a standard mortality ratio (SMR) of 4.81 (Alexander et al., 2003). Since PFOS manufacture has been eliminated in the United States, occupational exposure to the chemical is believed to be on the decline.

Environmental

Perfluorooctane sulfonate has been known to contaminate water, soil, and air and can be found in both indoor and outdoor environments. Perfluorooctane sulfonate has been detected in environmental water at concentrations of 0.4 ng/l in remote areas and 4.0 ng/l in urban areas (Jin et al., 2009). The chemical has also been detected in rainwater at a concentration of 0.59 ng/l (Loewen et al., 2005). Contact with contaminated water is a potential route of exposure. Research studies have also been completed that have identified PFOS accumulation in the indoor environment. Moriwaki et al. (2003) examined environmental dust from vacuum cleaners that were collected in Japanese homes and identified concentrations of PFOS ranging from 11 to 2500 ng/g. Perfluorooctane sulfonate concentration of 2.2 ng/l was also detected in indoor dust samples from home and offices (D'Hollander et al., 2010). Since PFOS production in the United States has ceased, the frequency of PFOS detection in outdoor and indoor environments should be on the decline as consumer products no longer contain the chemical.

INDUSTRIAL HYGIENE

The Occupational Safety and Health Administration, NIOSH, and ACGIH have not established exposure limits for PFOS. Even though PFOS is no longer used in industry and, although there are no formal guidelines in place, several measures can be used to reduce inhalation, ingestion, and dermal exposure if one comes into contact with PFOS. Individuals using the chemical should be careful to avoid inhalation of PFOS. Personal respiratory protection may be needed if PFOS concentrations are high. Installing a ventilation system capable of removing the chemical may also be necessary to prevent exposure. Ingestion of PFOS can be prevented by using good hygiene practices. Eating, drinking, or using tobacco products should be prohibited when individuals are in contact with the chemical. Hands should be washed thoroughly with soap and water to remove the contaminant. Dermal contact with the chemical can be prevented by using personal protective equipment, such as gloves, rubber boots, and goggles. A fully encapsulating suit may be necessary in environments where high concentrations and quantities of the chemical are utilized. Perfluorooctane sulfonate has been substituted in industry and manufacturing in favor of less persistent and less harmful compounds.

MEDICAL MANAGEMENT

Perfluorooctane sulfonate can be detected in human blood samples. A study of pregnant women identified PFOS in serum ranging from 4.9 to 17.6 ng/ml with fetal samples ranging from 1.6 to 5.3 ng/ml (Inoue et al., 2004). Another study of 263 employees from two perfluorooctyl-manufacturing plants identified a mean serum concentration of PFOS to be 1.32 parts per million (ppm) and 1.78 ppm, respectively (Olsen et al., 2003). Since PFOS is no longer used in industry, there are not likely to be any new cases of exposure. However, employees that were exposed prior to the cessation of the chemicals production may be at risk for the development of adverse health outcomes. In situations where the chemical is still being used, it is recommended that employees are entered into a biomonitoring program. The goal of the biomonitoring program should be to identify the exposure and reduce it to the lowest possible level. Since PFOS can exist as a powder, small particles may be inhaled into the respiratory tract where they can deposit deep in the lung. Personal respiratory protection can protect against inhalation exposure to PFOS. There may be some applications where PFOS is in solution that could lead to splashing or other types of dermal contact with the chemical. The skin should be cleansed thoroughly if the chemical contacts it. Ingestion of the chemical is not likely but could occur and individuals working with PFOS should thoroughly clean their hands before eating, drinking, or smoking to prevent exposure via this route. In cases where symptoms of exposure do not subside or when individuals are exposed to high concentrations of PFOS, individuals should seek prompt medical attention.

REFERENCES

Alexander, B.H., et al. (2003) Mortality of employees of a perfluorooctanesulphonyl fluoride manufacturing facility. *Occup. Environ. Med.* 60:722–729.

Butenhoff, J.L., et al. (2006) The applicability of biomonitoring data for perfluorooctanesulfonate to the environmental public health continuum. *Environ. Health Perspect.* 114:1776–1782.

Cheng, Y., et al. (2011) Thyroid disruption effects of environmental level perflurooctane sulfonates (PFOS) in *Xenopus laevis*. *Ecotoxicology* 20:2069–2078.

Cui, L., et al. (2009) Studies on the toxicological effects of PFOA and PFOS on rats using histological observation and chemical analysis. *Arch. Environ. Contam. Toxicol.* 56:338–349.

D'Hollander, W., et al. (2010) Brominated flame retardants and perfluorinated compounds indoor dust from homes and offices in Flanders, Belgium. *Chemosphere* 81:478–487.

Grasty, R.C., et al. (2005) Effects of prenatal perfluorooctane sulfonate (PFOS) exposure on lung maturation in the perinatal rats. *Birth Defects Res. B Dev. Reprod. Toxicol.* 74:405–416.

Innes, K.E., et al. (2014). Inverse association of colorectal cancer prevalence to serum levels of perfluorooctane sulfonate (PFOS) and perfluorooctanoate (PFOA) in a large Appalachian population. *BMC Cancer* 14:45.

Inoue, K., et al. (2004) Perfluorooctane sulfonate (PFOS) and related perfluorinated compounds in human material and cord blood samples: assessment of PFOS exposure in a susceptible population during pregnancy. *Environ. Health Perspect.* 112:1204–1207.

Jin, Y.H., et al. (2009) PFOS and PFOA in environmental and tap water in China. *Chemosphere* 77:605–611.

Johansson, N., et al. (2009) Neonatal exposure to PFOS and PFOA in mice results in changes in proteins which are important for neuronal growth and synaptogenesis in the developing brain. *Toxicol. Sci.* 108:412–418.

Kannan, K., et al. (2006) Association between perfluorinated compounds and pathological conditions in southern sea otters. *Environ. Sci. Technol.* 40:4943–4948.

Loewen, M., et al. (2005) Fluorotelomer carboxylic acids and PFOS in rainwater from an urban center in Canada. *Environ. Sci. Technol.* 39:2944–2951.

Moriwaki, H., et al. (2003) Concentrations of perfluorooctane sulfonate (PFOS) and perfluorooctanoic acid (PFOA) in vacuum cleaner dust collected in Japanese home. *J. Environ. Monit.* 5:753–757.

National Toxicology Program (NTP) (2013) *CAS Registry Number: 1763-23-1*, National Institute of Environmental Health Sciences, Washington, DC, Available at http://ntp.niehs.nih.gov/go/25251.

Oakes, K.D., et al. (2005) Short-term exposures of fish to perfluorooctane sulfonate: acute effects on fatty acyl-coa oxidase activity, oxidative stress, and circulating sex steroids. *Environ. Toxicol. Chem.* 24:1172–1181.

Olsen, G.W., et al. (2003) An occupational exposure assessment of a perfluorooctanesulfonyl fluoride production site: biomonitoring. *AIHA J. (Fairfax, Va)* 64:651–659.

Olsen, G.W., et al. (2003) Epidemiologic assessment of worker serum perfluorooctanesulfonate (PFOS) and perfluorooctanoate (PFOA) concentrations and medical surveillance examinations. *J. Occup. Environ. Med.* 45:260–270.

Qian, Y., et al. (2010) Perfluorooctane sulfonate (PFOS) induces reactive oxygen species (ROS) production in human microvascular endothelial cells: role in endothelial permeability. *J. Toxicol. Environ. Health A* 73:819–836.

Shi, X., et al. (2008) Developmental toxicity and alteration of gene expression in zebrafish embryos exposed to PFOS. *Toxicol. Appl. Pharmacol.* 230:23–32.

United States Environmental Protection Agency (USEPA) (2012) *Emerging Contaminants–Perfluorooctane Sulfonate (PFOS) and Perfluorooctanoic Acid (PFOA)*, USEPA 505-F-11-002, Washington, DC.

Yu, W.G., et al. (2009) Prenatal and postnatal impact of perfluorooctane sulfonate (PFOS) on rat development: a cross-factor study on chemical burden and thyroid hormone system. *Environ. Sci. Technol.* 43:8416–8422.

Zhang, Y.H., et al. (2013) Mechanism of perflurooctanesulfonate (PFOS)-induced apoptosis in the immunocyte. *J. Immunotoxicol.* 10:49–58.

PERFLUOROOCTANOIC ACID (PFOA)

Target organ(s): Skin, eyes, respiratory tract, nervous system

Occupational exposure limits: NIOSH/OSHA/ACGIH—not established

Reference values: RfD/RfC/MRL—not established

Risk/safety phrases: Corrosive

BACKGROUND AND USES

Perfluorooctanoic acid (PFOA, C8, pentadecafluorooctanoic acid, perfluoro caprylic acid; CASRN 335-67-1) is an eight-carbon compound in the perfluoroalkyl family of chemicals. Perfluorooctane sulfonate is used in a variety of applications, including nonstick cookware, waterproof clothing, leather products, cleaning products, and pesticides. Its main use was as a stain repellent on carpet, furniture, and other consumer products. In 2006, the U.S. Protection Agency along with eight major companies that utilized PFOA embarked on a program to reduce emissions and use of the chemical by 2015 (USEPA, 2012).

PHYSICAL AND CHEMICAL PROPERTIES

Perfluorooctanoic acid can exist as colorless flakes. It has a molecular weight of 414.07 g/mol, a pH of 2.6, a vapor pressure 0.052 mmHg, a melting point ranging from 55 to

56 °C, a boiling point of 189 °C, and a density of 0.900 g/cm^3 (NTP, 2013). Perfluorooctanoic acid has the ability to react with bases, oxidizing agents, and reducing agents. Upon decomposition, PFOA can form carbon oxides and hydrogen fluoride. Additional information related to physical and chemical properties of PFOA are not currently available.

MAMMALIAN TOXICOLOGY

Acute Effects

Perfluorooctanoic acid is known to cause acute effects upon inhalation, ingestion, and dermal exposure. Inhalation of PFOA can occur by breathing chemical particles or vapors into the respiratory tract. Upon inhalation, PFOA is known to result in severe irritation and destruction of the mucous membranes and upper respiratory tract. Ingestion of the chemical is not likely but can be very harmful if swallowed. The Agency for Toxic Substances and Disease Registry (2009) indicates that ingestion of the chemical is likely to occur from hand-to-mouth transfer from treated carpets especially among young children. Ingestion can also occur from eating contaminated food or drinking contaminated water (ATSDR, 2009). Chemical contact with the skin will result in skin irritation and chemical burns. Franko et al. (2012) found statistically significant, dose-response increases in serum PFOA concentrations in mice exposed to a topical application of the chemical. This indicates that dermal absorption may serve as an important route of exposure.

Chronic Effects

Research on the chronic effects of PFOA in mammalian model has identified possible neurologic, physiologic, anatomic, and hormonal abnormalities as a result of chemical exposure. Mice exposed to PFOA have developed neurologic deviations. Male mice exposed to a single-oral dose of 0.58 or 8.70 mg of PFOA exhibited deranged spontaneous behavior, reduced or lack of habituation, and hyperactivity suggesting that the chemical is a developmental neurotoxicant (Johansson et al., 2008). Perfluorooctanoic acid can also disrupt normal anatomic development. Pregnant mice exposed orally to 5 mg PFOA/mg showed signs of altered milk protein gene expression and reduced mammary epithelial growth (White et al., 2007). Many have suspected that PFOA may alter thyroid function in mammalian systems. An analysis of PFOA versus disease status in 3974 adults found that higher concentrations of serum PFOA was associated with current thyroid disease after adjusting for age, sex, race/ethnicity, education, smoking status, body mass index, and alcohol intake (Melzer et al., 2010). The association between exposure to PFOA and cancer is not conclusive; however, Nakayama et al. (2005) described

an epidemiologic study conducted by 3M that showed increases in both prostate and bladder cancer.

Mechanism of Action(s)

Perfluorooctanoic acid can be absorbed into the body through the lungs, through the gastrointestinal tract from contaminated food and water, or through dermal contact with the skin. Once in the body, PFOA does not bioaccumulate in fatty tissue rather it binds to serum proteins with large concentrations residing in the liver. While ATSDR reports that the liver is the target organ for PFOA concentration, other research indicates that the chemical may concentrate elsewhere. Cui et al. (2009) found that PFOA concentration in the mammalian tissue appear to follow the order of the kidneys, liver, lung, heart/whole blood, testicle, spleen, and brain with the highest concentrations in the kidney ranging from 191 to 265 μg/g. What is known is that the liver and kidney both seem to be effected by PFOA exposure. Additional research in this area may help to define the exact mechanism of PFOA action in the mammalian liver and kidney.

Chemical Pathology

Several different pathologies have been noted following exposure(s) to PFOA. Several types of tumors, neonatal death, and toxicity of the immune, hepatic, and endocrine systems have been described in the literature (Steenland et al., 2010). A study of male rats indicated that the $LD_{50}/30$ day for PFOA was 189 (208–175) mg/kg with all lethal dose-treated rats dying during the first 5 days (Olson and Andersen, 1983). Wang et al. (2013) explain that PFOA has been found to disrupt serum lipid and lipoprotein levels and cause degeneration of hepatocytes leading to the interference with transportation and metabolism of fatty acids. Researchers believe that the abnormal localization of lipid droplets may be associated with PFOA hepatic toxicity (Wang et al., 2013). Reproductive pathologies may also develop as a result of PFOA exposure. Low doses of PFOA, 0, 0.01, 0.1, and 1 mg PFOA/kg, resulted in histopathologic changes in the uterus, cervix, and vagina of exposed mice (Dixon et al., 2012).

SOURCES OF EXPOSURE

Occupational

Individuals involved in the processing, distribution, and handling of PFOA is at risk for exposure. The ATSDR estimates that occupational intake of PFOA by workers ranges from 3.2×10^{-4} to 2.4 ng/kg/day (ATSDR, 2009). Several research studies have speculated that PFOA has powerful hepatotoxic effects. A cross-sectional study of 115 PFOA occupationally exposed workers investigated the relationship between chemical exposure and hepatic enzymes, lipoproteins, and cholesterol, and found no significant hepatic toxicity (Gilliland and Mandel, 1996). An additional study of 506 occupationally exposed workers found no significant data regarding the relationship between PFOA exposure and hepatic and thyroid toxicity (Olsen and Zobel, 2007). Based on the available information, occupational exposure to PFOA may not be related to hepatotoxic effects; however, more research must be completed before this can be confirmed. Some researchers have speculated whether or not PFOA has adverse reproductive effects on mammalian systems. A 1998 study of 191 production workers failed to describe statistically significant associations between PFOA exposure and reproductive hormone abnormalities (Olsen et al., 1998). The specific effects resulting from occupational exposure to PFOA are not conclusive and therefore deserve additional investigation.

Environmental

Perfluorooctanoic acid is a man-made compound and therefore is not naturally found in the environment. It can be found in the soil, air, and water as a result of manufacturing waste streams, improper disposal, and from the breakdown of other chemicals. Perfluorooctanoic acid has been detected in water from remote areas at median concentrations of 0.1 ng/l and maximum concentrations of 1.3 ng/l while water from urban areas has had PFOA median concentrations of 3.9 ng/l and maximum concentrations of 30.8 ng/l (Jin et al., 2009). These results appear to indicate that human interaction with the environment is directly related to elevated PFOA concentrations. Perfluorooctanoic acid has also been detected in indoor dust samples identifying another potential site for exposure. Vacuum cleaner dust samples have measured PFOA concentrations ranging from 69 to 3700 ng/g (Moriwaki et al., 2003). Individuals who breathe in this contaminated dust may experience both acute and chronic effects as a result of exposure. The cessation of PFOA use in the near future should reduce the amount of contamination in both indoor and outdoor environments.

INDUSTRIAL HYGIENE

The Occupational Safety and Health Administration, NIOSH, and ACGIH have not established exposure limits for PFOA. Since the use of PFOA in industry is on the decline, one may expect that exposure levels should also be on the decline. Several preventive measures can be taken to reduce inhalation, ingestion, and dermal exposure of PFOA. Individuals using the chemical should be careful to avoid inhalation of the chemical. Personal respiratory protection may be needed if PFOA concentrations are high. Engineering controls, such as a ventilation system capable of removing the chemical, may also be necessary to prevent exposure.

Ingestion of PFOA can be prevented by using good hygiene practices in the workplace. Eating, drinking, or using tobacco products should be prohibited when individuals are in contact with the chemical. Hands should be washed thoroughly with soap and water to remove traces of PFOA. Dermal contact with the chemical can be prevented by using personal protective equipment, such as gloves, rubber boots, and goggles. A fully encapsulating suit may be necessary in environments where high concentrations and quantities of the chemical are utilized. Perfluorooctanoic acid has been substituted in industry and manufacturing in favor of less persistent and less harmful compounds.

MEDICAL MANAGEMENT

Perfluorooctanoic acid can be measured via human serum. Among 100 serum samples from American Red Cross blood donors that were analyzed, PFOA had a geometric mean of 4.5 ng/ml in the year 2000 and a geometric mean of 2.2 ng/ml in the year 2005 (Olsen et al., 2007). The decrease in geometric mean may be a result of the chemical substitution in industrial practice; however, this is only a speculation. Since PFOA is being phased out of the industry, the number of cases reporting adverse health effects should begin to decline. However, employees that were exposed prior to the phasing out of the chemical's production may be at risk for the development of adverse health outcomes. Biomonitoring programs should be instituted in situations where the chemical is still being used. Since PFOA exists as flakes, small pieces may be inhaled into the respiratory tract where they can deposit deep in the alveolar tissue. Personal respiratory protection can protect against the inhalation exposure to PFOA. There may be some applications where PFOA is in solution that could lead to dermal contact with the chemical. The skin should be cleansed thoroughly if the chemical penetration occurs. Ingestion of the chemical is not likely but could occur and individuals working with PFOA should thoroughly clean their hands before eating, drinking, or smoking to prevent exposure via this route. In cases where symptoms of exposure do not subside or when individuals are exposed to high concentrations of PFOA, individuals should seek prompt medical attention.

REFERENCES

Agency for Toxic Substances and Disease Registry (ATSDR) (2009) *ToxGuide for Perfluoroalkyls*, Atlanta, GA: US Department of Health and Human Services. Available at http://www.atsdr.cdc.gov/toxguides/toxguide-200.pdf.

Cui, L., et al. (2009) Studies on the toxicological effects of PFOA and PFOS on rats using histological observation and chemical analysis. *Arch. Environ. Contam. Toxicol.* 56:338–349.

Dixon, D., et al. (2012) Histopathologic changes in the uterus, cervix, and vagina of immature CD-1 mice exposed to low doses of perfluorooctanoic acid (PFOA) in a uterotrophic assay. *Reprod. Toxicol.* 33:506–512.

Franko, J., et al. (2012) Dermal penetration potential of perfluorooctanoic acid (PFOA) in human and mouse skin. *J. Toxicol. Environ. Health A* 75:50–62.

Gilliland, F.D. and Mandel, J.S. (1996) Serum perfluorooctanoic acid and hepatic enzymes, lipoproteins, and cholesterol: a study of occupationally exposed men. *Am. J. Ind. Med.* 29:560–568.

Jin, Y.H., et al. (2009) PFOS and PFOA in environmental and tap water in China. *Chemosphere* 77:605–611.

Johansson, N., et al. (2008) Neonatal exposure to perfluorooctane sulfonate (PFOS) and perfluorooctanoic acid (PFOA) causes neurobehavioral defects in adult mice. *Neurotoxicology* 29:160–169.

Melzer, D., et al. (2010) Association between serum perfluorooctanoic acid (PFOA) and thyroid disease in the U.S. National Health and Nutrition Examination Survey. *Environ. Health Perspect.* 118:686–692.

Moriwaki, H., et al. (2003) Concentrations of perfluorooctane sulfonate (PFOS) and perfluorooctanoic acid (PFOA) in vacuum cleaner dust collected in Japanese home. *J. Environ. Monit.* 5:753–757.

Nakayama, S., et al. (2005) Distributions of perfluorooctanoic acid (PFOA) and perfluorooctane sulfonate (PFOS) in Japan and their toxicities. *Environ. Sci.* 12:293–313.

Olson, C.T. and Andersen, M.E. (1983) The acute toxicity of perflurooctanoic and perfluorodecanoic acids in male rats and effects on tissue fatty acids. *Toxicol. Appl. Pharmacol.* 70:362–372.

Olsen, G.W. and Zobel, L.R. (2007) Assessment of lipid, hepatic, and thyroid parameters with serum perfluorooctanoate (PFOA) concentrations in fluorochemical production workers. *Int. Arch. Occup. Environ. Health* 81:231–246.

Olsen, G.W., et al. (1998) An epidemiologic investigation of reproductive hormones in men with occupational exposure to perfluorooctanoic acid. *J. Occup. Environ. Med.* 40:614–622.

Olsen, G.W., et al. (2007) Preliminary evidence of a decline in perfluorooctanesulfonate (PFOS) and perfluorooctanoate (PFOA) concentrations in American Red Cross blood donors. *Chemosphere* 68:105–111.

Steenland, K., et al. (2010) Epidemiologic evidence on the health effects of perfluorooctanoic acid (PFOA). *Environ. Health Perspect.* 118:1100–1108.

United States Environmental Protection Agency (USEPA) (2012) *Perfluorooctanoic acid (PFOA) and fluorinated telomers, Washington, DC.* Available at http://www.epa.gov/oppt/pfoa/pubs/pfoainfo.html.

Wang, L., et al. (2013) Specific accumulation of lipid droplets in hepatocyte nuclei of PFOA-exposed BALB/c mice. *Sci. Rep.* 3:2174.

White, S.S., et al. (2007) Gestational PFOA exposure of mice is associated with altered mammary gland development in dams and female offspring. *Toxicol. Sci.* 96:133–144.

PERFLUORONONANOIC ACID (PFNA)

Target organ(s): eyes, skin, respiratory system, liver

Occupational exposure limits: NIOSH/OSHA/ACGIH—not established

Reference values: RfD/RfC/MRL—not established

Risk/safety phrases: Irritant

BACKGROUND AND USES

Perfluorononanoic acid (PFNA, heptadecafluorononanoic acid, perfluoro-n-nonanoic acid, perfluorononan-1-oic acid; CASRN 375-95-1) is a nine-carbon compound in the perfluoroalkyl family of chemicals. Perfluorononanoic acid is used in a variety of applications, including in stain- and grease-proof coatings on food packages, furniture, upholstery, and carpet. Like many other perfluoroalkyl compounds, PFNA persists in the environment due to its long carbon chain. Research related to PFNA is sparse and therefore deserve additional investigation.

PHYSICAL AND CHEMICAL PROPERTIES

Perfluorononanoic acid can exist as beige crystals. It has a molecular weight of 464.08 g/mol, a melting point ranging from 59 to 62 °C, and a boiling point of 218 °C. Perfluorononanoic acid has the ability to react with bases, oxidizing agents, and reducing agents. Upon decomposition, PFNA can form carbon oxides and hydrogen fluoride. Additional information related to the physical and chemical properties of PFNA are not currently available.

MAMMALIAN TOXICOLOGY

Acute Effects

Research indicates that PFNA and its derivatives have the potential to cause acute effects in mammalian models. Kinney et al. (1989) investigated the acute toxicity of a PFNA derivative, ammonium perfluorononanoate, in male rats. The researchers exposed male rats to a dust atmosphere of 99% ammonium perfluorononanoate. Nose-only inhalation exposure was conducted at concentrations ranging from 67 to 4600 mg/m^3 for 4 h. Mortality rates were 0, 4, 6, and 6 out of six animals at 620, 910, 1600, and 4600 mg/m^3, respectively. The LC$_{50}$ was calculated to be 820 mg/m^3 with visible clinical signs, including hunched posture, ruffled or discolored fur, red or brown discharge from the eyes, nose, and mouth, wet or stained perineum, pallor, lung noise or labored breathing, lethargy, limpness, and hair loss. In a subsequent experiment, male rats were exposed to ammonium PFNA via nose-only inhalation at concentrations ranging from 0 to 590 mg/m^3 for 4 h. The purpose was to investigate the effects of PFNA on hepatic tissue. Mortality rates were 0 and 1 out of five animals at 67 and 590 mg/m^3, respectively. Statistically significant increases in both absolute and relative liver weight at 67 mg/m^3 and at 590 mg/m^3 were seen. The LC$_{50}$ for PFNA was determined to be 820 mg/m^3 (Kinney et al., 1989).

Chronic Effects

Human HepG2 cells were exposed to PFNA at concentrations of 0.4, 4, 40, 200, 400, 1000, or 2000 μM and the production of reactive oxygen species was examined every 15 min for 3 h (Eriksen et al., 2010). Perfluorononanoic acid did not produce statistically significant increases in reactive oxygen species production. Perfluorononanoic acid propensity to induce DNA damage was the topic of a second experiment. Human HepG2 cells treated with PFNA at concentrations of 100 or 400 μM for 24 h and the amount of DNA damage was measured using the Comet assay (Eriksen et al., 2010). Perfluorononanoic acid at both concentrations was associated with statistically significant frequency of strand breaks. No treatment-related effect related to the amount of oxidative damage to purines was seen. Perfluorononanoic acid was cytotoxic to the HepG2 cells at elevated concentrations as shown by lactate dehydrogenase (LDH) release. Perfluorononanoic acid exposure was associated with genotoxicity in this assay.

Mechanism of Action(s)

There has been some research into PFNA's mechanism of action. Wild type and PPARα knockout mice were exposed to PFNA at concentrations of 0, 0.83, 1.1, 1.5, or 2 mg/kg in drinking water on gestational days (Wolf et al., 2010). Maternal weight gain, implantation, litter size, and pup weight at birth were unaffected in both strains. The number of live pups at birth and pup survival was diminished in PFNA wild-type treated animals at 1.1 and 2 mg/kg with 36% and 31% survival, respectively. Eye opening was delayed and pup weight was statistically significantly reduced in wild-type male pups at 2 mg/kg and in wild-type female pups at 2 mg/kg. These effects were not seen in the knockout mice. Relative liver weight was increased in a dose-dependent manner in dams and pups of the wild-type strain at all dose levels with only slight increases in the highest dose group in the knockout strain. These data suggest that the PPARα may play a crucial role in the developmental endpoints with other pathways in addition to PPARα available to mediate PFNA effects on liver weight.

Chemical Pathology

Many believe that PFNA concentrates in the kidney and the liver. One study conducted a comparison of PFNA pharmacokinetics in rats and mice found both species and gender differences (Tatum-Gibbs et al., 2011). Rats were exposed orally to a dose of 1, 3, or 10 mg/kg and blood was collected at 1, 2, 3, 4, 7, 16, 21, 38, 35, 42, and 50 days after treatment. Liver and kidney were collected for PFNA analysis at the end of the study. Serum elimination of PFNA in the rat was found to be linear with exposure with the rate of elimination much lengthier in males when compared to females at 30.6 days and 1.4 days, respectively. Perfluorononanoic acid was specially stored in the liver but not in the kidneys in rats with higher levels of accumulation occurring in male rats. Mice were given a single oral dose of 1 or 10 mg/kg. In the mouse, serum elimination was observed to be nonlinear with exposure dose and slightly faster in females than in males with serum half-life determined to be 25.8–68.4 days and 34.3–68.9 days in females and males, respectively. As observed in rats, PFNA was stored in the liver but not in the kidneys of mice with higher accumulation of PFNA in males than in females.

SOURCES OF EXPOSURE

Occupational

A retrospective cohort of 630 workers exposed to PFNA at a polymer production facility was investigated to observe differences in clinical measures between exposed and nonexposed employees (Mundt et al., 2007). Work histories, which were available for all employees, provided the basis for categorizing the employees into high, low, or no exposure categories. Records were abstracted for height, weight, date of exam, and 32 clinical chemistry variables. A limited number of baseline blood samples were used to validate the exposure categories, but were insufficient for use in any analysis. Five time periods were selected for cross-sectional analysis and annual measures of the 32 clinical chemistry variables were used for the longitudinal analysis. This accounted for multiple measurements per person and was conducted independently for men and women by exposure group. After adjusting for age and BMI, some small but not clinically significant differences between groups were found. These observations were not consistent between men and women or over the five analysis windows. Seven variables, including total cholesterol, GGT, AST, ALT, alkaline phosphatase, bilirubin, and triglycerides that were examined in separate longitudinal models showed no significant increase or decrease by unit increase in exposure intensity score. No significant relationships between exposure to PFNA and liver, kidney, or thyroid parameters were witnessed.

Environmental

Perfluorononanoic acid can contaminate water, air, and soil. Examination of water samples from the Cantabrian Sea identified PFNA as the perfluoroalkyl compounds with the second highest concentration level ranging from 0.04 to 1.40 ng/l (Gomez et al., 2011). The same study also identified PFNA in port waters at concentrations ranging from 0.04 to 0.20 ng/l. The results of this study indicate that the large population living near the water system may be directly responsible for perfluoroalkyl pollution. In addition, individuals living in the region will be at risk for exposure to the chemical if they contact the water. Haug et al. (2011) conducted a study to examine the presence of perfluoroalkyl compounds in paired house dust and indoor air samples. Forty-one households were examined and identified PFNA as the perfluoroalkyl compound with the second highest median concentration at 23 ng/g. Reducing the use of PFNA will decrease the accumulation of the chemical in indoor environments. Lessening the accumulation will drop the risk of exposure for building inhabitants.

INDUSTRIAL HYGIENE

The Occupational Safety and Health Administration, NIOSH, and ACGIH have not established exposure limits for PFNA. Different preventive measures can be taken to reduce inhalation, ingestion, and dermal exposure to PFNA. Individuals using the chemical should be careful to avoid breathing in the chemical. Personal respiratory protection is recommended if PFNA concentrations are high. Engineering controls, such as a ventilation system capable of removing the chemical, may also be necessary to prevent exposure. Ingestion of PFNA can occur via hand-to-mouth and can be prevented by using good hygiene practices in the workplace. Eating, drinking, or using tobacco products should be prohibited when individuals are in contact with the chemical. Hands should be cleansed thoroughly with soap and water to remove traces of PFNA. Dermal contact with the chemical can be prevented by using personal protective equipment, such as gloves, rubber boots, and goggles. Perfluorononanoic acid has been substituted in industry and manufacturing in favor of less persistent and less harmful compounds.

MEDICAL MANAGEMENT

Two studies indicate that PFNA can be detected in human breast milk. One study analyzed 45 human breast milk samples collected in Massachusetts and identified PFNA in 64% of the samples at a median concentration of 6.97 pg/ml (Tao et al., 2008). Another study analyzed 19 human breast milk samples collected from China and identified PFNA in 100% of the samples with a median concentration of 16 pg/ml (So et al., 2006). Like other perfluoroalkyl compounds, PFNA can also be detected in human serum. Employees that have been exposed to the chemical may be at risk for the development of adverse health effects. Biomonitoring programs should be instituted in situations where the chemical is still being used. Since PFOA exists as crystals, small pieces may be inhaled into the respiratory tract where they can deposit deep in the alveolar region of the lung. Personal respiratory protection can protect against inhalation exposure to PFNA. There may be some applications where PFNA is in solution that could lead to dermal contact with the chemical. The skin should be cleansed thoroughly if the chemical contact occurs. Ingestion of the chemical is not likely but could occur and individuals working with PFNA should thoroughly clean their hands before eating, drinking, or smoking to prevent exposure via this route. In cases where symptoms of exposure do not subside or when individuals are exposed to high concentrations of PFNA, individuals should seek prompt medical attention.

REFERENCES

Eriksen, K.T., et al. (2010) Genotoxic potential of the perfluorinated chemicals PFOA, PFOS, PFBS, PFNA, and PFHxA in human HepG2 cells. *Mutat. Res.* 700:39–43.

Gomez, C., et al. (2011) Occurrence of perfluorinated compounds in water, sediment, and mussels from the Cantabrian Sea (North Spain). *Mar. Pollut. Bull.* 62:948–955.

Haug, L.S., et al. (2011) Investigation on per- and polyfluorinated compounds in paired samples of house dust and indoor air from Norwegian homes. *Environ. Sci. Tech.* 45:7991–7998.

Kinney, L.A., et al. (1989) Acute inhalation toxicity of ammonium perfluorononanoate. *Food Chem. Toxicol.* 27:465–468.

Mundt, D.J., et al. (2007) Clinical epidemiological study of employees exposed to surfactant blend containing perfluorononanoic acid. *Occup. Environ. Med.* 64:589–594.

So, M.K., et al. (2006) Health risks in infants associated with exposure to perfluorinated compounds in human breast milk from Zhoushan, China. *Environ. Sci. Technol.* 40:2924–2929.

Tao, L., et al. (2008) Perfluorinated compounds in human milk from Massachusetts, U.S.A. *Environ. Sci. Technol.* 42:3096–3101.

Tatum-Gibbs, K., et al. (2011) Comparative pharmacokinetics of perfluorononanoic acid in rat and mouse. *Toxicology* 281:48–55.

Wolf, C.J., et al. (2010) Developmental effects of perfluorononanoic acid in the mouse are dependent on peroxisome proliferator-activated receptor-alpha. *PPAR Res.* 2010:1–11.

PERFLUORODECANOIC ACID (PFDA)

Target organ(s): Skin, eyes, respiratory system, liver

Occupational exposure limits: NIOSH/OSHA/ACGIH—not established

Reference values: RfD/RfC/MRL—not established

Risk/safety phrases: Irritant

BACKGROUND AND USES

Perfluorodecanoic acid (PFDA, nonadecafluorodcanoic acid, nonadecofluorocapric acid, perfluorodecanoic acid, perfluorocapric acid; CASRN 375-95-1) is a ten-carbon compound in the perfluoroalkyl family of chemicals. Perfluorodecanoic acid is used in a variety of applications, including in stain- and grease-proof coatings on food packages, furniture, upholstery, and carpet. Like many other perfluoroalkyl compounds, PFDA persists in the environment due to its long carbon chain. Some research related to PFDA has been conducted but additional investigation is recommended for better understanding of how the chemical reacts with organisms and the environment.

PHYSICAL AND CHEMICAL PROPERTIES

Perfluorodecanoic acid exists as a white powder. It has a molecular weight of 514.08 g/mol, a melting point ranging from 77 to 81 °C, and a boiling point of 218 °C. Perfluorodecanoic acid has the ability to react with bases, oxidizing agents, and reducing agents. Upon decomposition, PFDA can form carbon oxides and hydrogen fluoride. Additional information related to the physical and chemical properties of PFDA are not currently available.

MAMMALIAN TOXICOLOGY

Acute Effects

A study of the comparative toxicity of PFDA in rats, mice, hamsters, and guinea pigs was conducted by Van Rafelghem et al. (1987). Male animals were exposed once with PFDA and were monitored for up to 28 days. The study revealed that, with some deviation, the toxic potency of PFDA was the same in the four species examined. Severe body weight reduction was apparent in the four species studied. It should be noted that rats stopped eating for 5–6 days after about 6 days post dosing while hamsters continued to consume food at a decreased level. Perfluorodecanoic acid caused marked hepatomegaly in rats, mice, and hamsters, and a moderate swelling in guinea pigs. Microscopic examination of the liver showed similar alterations in the species studied consisting of a panlobular swelling of the parenchymal cells. Perfluorodecanoic acid induced thymic atrophy in hamsters, mice, and guinea pigs. Perfluorodecanoic acid also induced seminiferous tubular degeneration in the testes from rats, but not mice. The lesion in hamsters and guinea pigs was less severe than in rats. The liver from all species showed disturbance of the rough endoplasmic reticulum, rounding and swelling of the mitochondria, and mild to extensive proliferation of peroxisomes. The latter response was greatest in mice and almost absent in guinea pigs. Accumulation of lipid droplets in liver cells was more pronounced in treated hamsters and guinea pigs then in rats and mice.

Chronic Effects

Perfluorodecanoic acid is known as a peroxisome proliferator, which may result in adverse effects on the liver. Adinehzadeh and Reo (1998) explain that PFDA and related compounds encourage the development of liver peroxisomes and may result in differing degrees of hepatotoxicity and carcinogenesis in rodents. Thyroid function may also be disrupted as a result of PFDA exposure. Rats exposed to the chemical decreased their food consumption and experienced weight loss (Langley and Pilcher, 1985). The study further explains that body temperature and heart rate were also adversely affected by chemical exposure. Serum thyroxine and serum triiodothyronine both experienced reductions suggesting that PFDA disrupts normal hormone levels in mammals. Some research indicates that PFDA may be responsible for immunotoxicity as well. Nelson et al. (1992) conducted a study in which rats were injected with PFDA concentrations of 20 mg/kg or 50 mg/kg, and immunized with keyhole limpet hemocyanin. The result was a significant decrease in keyhole limpet hemocyanin specific IgG_{2a} production when the experimental group was compared with the control group indicating an immunomodulatory effect caused by PFDA.

Mechanism of Action(s)

Much like other perfluoroalkyl compounds, PFDA-induced effects in rats and mice are mediated through PPARα, and it is generally agreed that humans and nonhuman primates are less responsive than rodents to PPARα-mediated effects (Klaunig et al., 2003; Maloney and Waxman, 1999). A study with PFDA-exposed rats by intraperitoneal injections reported decreased serum testosterone and 5α-dihydrotestosterone and reductions in the weight of accessory sex organs (Bookstaff et al., 1990). Experiments in castrated rats implanted with testosterone-containing capsules displayed that the androgenic deficiency caused by PFDA was the result of diminished secretion of testosterone from the testis rather than amplified clearance from blood. Since luteinizing levels in blood were unaffected, the decreased secretion of testosterone from the testis appeared to be due to decreased testicular responsiveness to luteinizing hormone stimulation. Bookstaff et al. (1990) suggested that PFDA upsets the normal feedback relationships that occur between plasma androgen and luteinizing hormone concentrations.

Chemical Pathology

Research indicates that PFDA may play a role in hepatic pathological manifestations. Rats exposed to a single dose of 50 mg/kg of PFDA exhibited abnormal function of the endoplasmic reticulum, mitochondrial swelling, and increases in intracellular lipid droplets in hepatocytes (Harrison et al., 1988). In another study, rats exposed to a single dose of 50 mg/kg of PFDA had decreased food intake resulting in severe weight loss with late lethality (Van Rafelghem and Andersen, 1988). Langley (1990) explains that basic metabolic processes may be disrupted as a result of PFDA exposure. The researcher suggests that PFDA induces the uncoupling of electron transport and oxidative phosphorylation. Further mitochondrial respiration is affected potentially leading to the severe weight loss associated with PFDA toxicity. Perfluorodecanoic acid has been found to alter 13 proteins in rats, many of which are involved in mitochondrial processes (Witzmann and Parker, 1991). While some research has been conducted related to the chemical pathology of PFDA, additional research is needed to get a more concrete understanding of how the chemical elicits its effects on mammalian tissue.

SOURCES OF EXPOSURE

Occupational

Anyone working with or near perfluoroalkyl compounds is at risk for exposure. Ski wax technicians are a special population that may experience exposure to PFDA. Perfluoroalkyl compounds are often integrated into ski waxes. When applied to snow skis, the wax increases the performance of the skis. The rubbing and buffing of the wax can aerosolize PFDA particles capable of being inhaled by employees. Freberg et al. (2010) conducted a study to examine occupational exposure to airborne perfluorinated compounds during professional ski waxing and found that the median inhalable fraction of PFDA particles was 8.9 mg/g of dust, the median thoracic fraction was 9.2 mg/g of dust, and the median respirable fraction was 9.3 mg/g of dust. These results show that PFDA particles vary in size and have the potential to deposit in different regions of the respiratory tract. In a time-trend study of ski wax technicians exposed to PFDA, the chemical was detected in all 57 blood samples with concentrations ranging from 0.9 to 24 ng/ml (Nilsson et al., 2010).

Environmental

Perfluorodecanoic acid is distributed into the environment via human waste streams. Once in the environment, the chemical can bioaccumulate in the tissue of many different organisms. Perfluorodecanoic acid has the capability to contaminate soil, water, and air. A study of rainbow trout found that PFDA accumulated in the fish with a half-life of 11 days, a steady state bioconcentration factor of 450 l/kg, and a 12-day accumulation ratio of 350 (Martin et al., 2003). Perfluorodecanoic acid, like other perfluoroalkyl compounds, can also be found in house dust. A study that included 102 individual house dust samples detected PFDA with a mean concentration of 15.5 ng/g of dust and a median concentration of 6.65 ng/g of dust (Strynar and Lindstrom, 2008). In another study of paired indoor air and dust samples, PFDA had a mean concentration of 4.1 ng/g of dust (Haug et al., 2011). Individuals breathing in contaminated air or dust samples are at risk for acute and chronic exposure to PFDA and other perfluoroalkyl compounds.

INDUSTRIAL HYGIENE

The Occupational Safety and Health Administration, NIOSH, and ACGIH have not established exposure limits for PFOA. Various preventive measures can be taken to reduce inhalation, ingestion, and dermal exposure of PFDA. Since PFDA exists as a white powder, individuals using the chemical should be cautious of inhalation exposure. Personal respiratory protection may be needed if PFDA concentrations are high. Engineering controls, such as a ventilation system capable of removing the chemical, may also be recommended to prevent exposure. Ingestion of PFDA can be prevented by using good hygiene practices in the occupational environment. Eating, drinking, or using tobacco products should be prohibited when individuals are in contact with the chemical. Hands should be washed thoroughly with soap and water to remove traces of PFDA from the skin. Dermal contact with the chemical can be prevented by using personal protective equipment, such as gloves, rubber boots, and goggles. A fully encapsulating suit may be necessary in environments where high concentrations and quantities of the chemical are utilized. Perfluorodecanoic acid has been substituted in industry and manufacturing in favor of less persistent and less harmful compounds.

MEDICAL MANAGEMENT

Several studies indicate that PFDA can be detected in human serum. One study analyzed 85 whole human blood samples from nine cities in China and recovered PFDA in 94.6% of the samples (Yeung et al., 2006). Another study of whole blood samples from residents in Catalonia, Spain and identified PFDA concentrations of 0.30 and 0.31 ng/ml in individuals aged 25 (+/−) 5 years and age 55 (+/−) 5 years, respectively (Ericson et al., 2007). Like other perfluoroalkyl compounds, PFDA can also be detected in human serum. Employees that have been exposed to the chemical may be at risk for the development of adverse health effects. Surveillance programs should be instituted in situations where the chemical is still used. Since PFDA exists as a powder, tiny particles may be inhaled into the respiratory tract where they can deposit deep in the alveolar region of the lung. Personal respiratory protection can protect against inhalation exposure to PFDA. There may be some applications where PFDA is in solution that could lead to dermal contact with the chemical. The skin should be cleansed thoroughly if the chemical contact occurs. Ingestion of the chemical is not likely but could occur and individuals working with PFDA should thoroughly clean their hands before eating, drinking, or smoking to prevent exposure via this route. In cases where symptoms of exposure do not subside or when individuals are exposed to high concentrations of PFDA, individuals should seek prompt medical attention.

REFERENCES

Adinehzadeh, M. and Reo, N.V. (1998) Effects of peroxisome proliferators on rat liver phospholipids: sphingomyelin degradation may be involved in hepatotoxic mechanism of perfluorodecanoic acid. *Chem. Res. Toxicol.* 11:428–440.

Bookstaff, R.C., et al. (1990) Androgenic deficiency in male rats treated with perfluorodecanoic acid. *Toxicol. Appl. Pharmacol.* 104:322–333.

Ericson, I., et al. (2007) Perfluorinated chemicals in blood of residents in Catalonia (Spain) in relation to age and gender: a pilot study. *Environ. Int.* 33:616–623.

Freberg, B.I., et al. (2010) Occupational exposure to airborne perfluorinated compounds during professional ski waxing. *Environ. Sci. Technol.* 44:7723–7728.

Harrison, E.H., et al. (1988) Perfluoro-n-decanoic acid: induction of peroxisomal beta-oxidation by a fatty acid with dioxin-like toxicity. *Lipids* 23:115–119.

Haug, L.S., et al. (2011) Investigation on per- and polyfluorinated compounds in paired samples of house dust and indoor air from Norwegian homes. *Environ. Sci. Tech.* 45:7991–7998.

Klaunig, J.E., et al. (2003) PPARalpha agonist-induced rodent tumors: modes of action and human relevance. *Crit. Rev. Toxicol.* 33:655–780.

Langley, A.E. (1990) Effects of perfluoro-n-decanoic acid on the respiratory activity of isolated rat liver mitochondria. *J. Toxicol. Environ. Health* 29:329–336.

Langley, A.E. and Pilcher, G.D. (1985) Thyroid, bradycardic, and hypothermic effects of perfluoro-n-decanoic acid in rats. *J. Toxicol. Environ. Health* 15:485–491.

Maloney, E.K. and Waxman, D.J. (1999) Trans-activation of PPARalpha and PPARgamma by structurally diverse environmental chemicals. *Toxicol. Appl. Pharmacol.* 161:209–218.

Martin, J.W., et al. (2003) Bioconcentration and tissue distribution of perfluorinated acids in rainbow trout (Onocrhynchus mykiss). *Environ. Toxicol. Chem.* 22:196–204.

Nelson, D.L., et al. (1992) The effects of perfluorodecanoic acid (PFDA) on humoral, cellular, and innate immunity in Fischer 344 rats. *Immunopharmacol. Immunotoxicol.* 14:925–938.

Nilsson, H., et al. (2010) A time trend study of significantly elevated perfluorocarboxylate levels in humans after using fluorinated ski wax. *Environ. Sci. Technol.* 44:2150–2155.

Strynar, M.J. and Lindstrom, A.B. (2008) Perfluorinated compounds in house dust from Ohio and North Carolina. *Environ. Sci. Technol.* 42:3751–3756.

Van Rafelghem, M.J. and Andersen, M.E. (1988) The effects of perfluorodecanoic acid on hepatic stearoyl-coenzymes A desaturase and mixed function oxidase activities in rats. *Fundam. Appl. Toxicol.* 11:503–510.

Van Rafelghem, M.J., et al. (1987) Pathological and hepatic ultrastructural effects of a single dose of perfluoro-n-decanoic acid in the rat, hamster, mouse, and guinea pig. *Fundam. Appl. Toxicol.* 9:522–540.

Witzmann, F.A. and Parker, D.N. (1991) Hepatic protein pattern alterations following perfluorodecanoic acid exposure in rats. *Toxicol. Lett.* 57:29–36.

Yeung, L.W., et al. (2006) Perfluorooctanesulfonate and related fluorochemicals in human blood samples from China. *Environ. Sci. Technol.* 40:715–720.

PERFLUOROUNDECANOIC ACID (PFUnA)

Target organ(s): Eyes, skin, respiratory system, liver, thyroid, reproductive system

Occupational exposure limits: NIOSH/OSHA/ACGIH—not established

Reference values: RfD/RfC/MRL—not established

Risk/safety phrases: Irritant

BACKGROUND AND USES

Perfluoroundecanoic acid (PFUnA, perfluoro-n-undecanoic acid, henicosafluoroundecanoic acid; CASRN 2058-94-8) is an eleven-carbon compound in the perfluoroalkyl family of chemicals. Perfluoroundecanoic acid is used in a variety of applications, including in stain- and grease-proof coatings on food packages, furniture, upholstery, and carpet. Like many other perfluoroalkyl compounds, PFUnA persists in the environment due to its long carbon chain. Some research related to PFUnA has been conducted but additional inquiry is recommended for better understanding of how the chemical interacts with organisms and the environment.

PHYSICAL AND CHEMICAL PROPERTIES

Perfluoroundecanoic acid exists as a solid. It has a molecular weight of 564.09 g/mol, a melting point ranging from 96 to 101 °C, a flash point of 113 °C, and a boiling point of 160 °C. Perfluoroundecanoic acid has the ability to react with oxidizing agents. Upon decomposition, PFUnA can form carbon oxides and hydrogen fluoride. Additional information related to the physical and chemical properties of PFUnA are not currently available.

MAMMALIAN TOXICOLOGY

Acute Effects

Perfluoroundecanoic acid is known to cause acute effects upon inhalation, ingestion, and dermal exposure. Inhalation of PFUnA can occur by breathing aerosolized particles or vapors into the respiratory tract. Upon inhalation, PFUnA is known to result in severe irritation of the mucous membranes and upper respiratory tract. Ingestion of the chemical is not likely but can be very harmful if swallowed. Ingestion of the chemical can occur from hand-to-mouth transfer from PFUnA-treated objects especially among young children. If an individual consumes contaminated water or food, exposure via ingestion can occur. Chemical contact with the skin will result in skin irritation and possible chemical burns. If PFUnA contacts the eyes, severe irritation may occur. In many circumstances, the use of engineering controls, administrative controls, and personal protective equipment can reduce PFUnA exposure to the lowest possible levels. Additional research is needed to determine what other acute effects are possible as a result of PFUnA exposure.

Chronic Effects

Perfluoroundecanoic acid is suspected to have chronic effects on the liver, thyroid, and reproductive system in mammals. Some evidence is available to support these thoughts but more investigation is needed. One study examined perfluoroalkyl compound's effects on the thyroid system and hinted at a weak association for free thyroxine (FT_4) with PFUnA at low concentrations (Bloom et al., 2010). If PFUnA is able to disrupt the thyroid function, mammalian development may be altered. A study of PFUnA on reproductive dysfunction in rats exposed to 0, 0.1, 0.3, and 1.0 mg/kg/day found that rats exposed to 1.0 mg/kg/day experienced body weight gain and a decrease in fibrinogen (Takahashi et al., 2014). The same study revealed that pup body weight was reduced and weight gain over 4 days was inhibited at 1.0 mg/kg/day. Researchers determined that the no-observed-adverse-effect level (NOAEL) for reproductive/developmental toxicity is measured to be 0.3 mg/kg/day. While not an exposure study on a mammalian system, rainbow trout were exposed to PFUnA

and other perfluoroalkyl compounds and it was determined that the chemical was a potent inducer of the estrogen-responsive biomarker protein vitellogen at very high dietary exposures (Benninghoff et al., 2011).

Mechanism of Action(s)

Perfluoroundecanoic acid's mechanism of action is believed to be through PPARα activation similar to other perfluoroalkyl compounds. One study transfected cells with either mouse or human PPARα-luciferase reporter plasmid and exposed them to PFUnA and other related compounds at concentrations of 0.5 μM and 100 μM with the researchers finding that PFUnA ranked fourth in the induction of human PPARα and ranked second in the induction of mouse PPARα (Wolf et al., 2012). The results of this study show that the longer chain length compounds may be related to lower PPARα activity in both human and mouse models. One study examined whether PFUnA and other perfluoroalkyl compounds had an effect on the thyroid hormone system and found that PFUnA failed to elicit an activating effect on the aryl hydrocarbon receptor (AhR) (Long et al., 2013). While exposure to PFUnA is speculated to have adverse effects on the thyroid, the present evidence is inconclusive.

Chemical Pathology

The major pathological manifestation induced by PFUnA is believed to be related to hepatic abnormalities. A study of harbor seal determined that the mean whole body burden of PFUnA was 13 (+/−2.8) μg (Ahrens et al., 2009). The study determined that a majority of perfluoroalkyl compounds were concentrated in the blood and the liver. A study of polar bears with exposure to perfluoroalkyl compounds investigated whether exposure was related to the development of mononuclear cell infiltrations, lipid granulomas, and bile duct hyperplasia (Sonne et al., 2008). The study determined that the presence of liver lesions did not appear to be influenced by perfluoroalkyl concentrations; however, the present hepatic lesions were similar to some produced in the laboratory setting. One study found that rats orally exposed to perfluoroalkyl compounds had high hepatic β-oxidation and liver weight with hepatocellular hypertrophy visible at doses of 0.125 and 0.6 mg/kg/day (Mertens et al., 2010). The bioaccumulation of PFUnA in mammalian tissue may play a role in how various pathologies develop.

SOURCES OF EXPOSURE

Occupational

Workers in direct or indirect contact may be at risk for PFUnA exposure. Inhalation and dermal contact are the most probable routes of exposure with ingestion not being likely. Ski wax technicians may be at a higher risk for PFUnA if they utilize perfluoroalkyl-containing wax. A study by Nilsson et al. (2010) identified PFUnA in blood samples from eight ski wax technicians with concentrations ranging from 0.11 to 2.8 ng/ml. Samples were collected during the course of a ski season from September 2007 to March 2008. Applying and buffing of the chemical may aerosolize it increasing the potential for it to be inhaled into the respiratory tract. Freberg et al. (2010) examined occupational exposure to airborne perfluorinated compounds during professional ski waxing and found that the PFUnA inhalable aerosol fraction ranged from 1.8 to 6.5 mg/g of dust, the thoracic fraction ranged from 2.0 to 7.0 mg/g of dust, and the respirable aerosol fraction ranged from 1.9 to 8.4 mg/g of dust. The median concentrations (mg/g of dust) for inhalable, thoracic, and respirable aerosol fractions were 3.0, 3.2, and 3.2, respectively. Engineering controls, administrative controls, and personal protective equipment are recommended to reduce PFUnA to the lowest possible levels.

Environmental

Perfluoroundecanoic acid, along with other perfluoroalkyl compounds, contaminates air, water, and soil. Once air, water, and soil are contaminated, exposure may occur by way of inhalation, ingestion, or dermal contact leading to bioaccumulation of the chemical in organism tissue. A study examining effluent water and percolate from four industrial plants and two landfill sites in Denmark detected PFUnA at concentrations ranging from <2.2 to 18.8 ng/l (Bossi et al., 2008). Since the chemical is man-made, it should not be found in water systems at these concentrations. Another study detected PFUnA in water and other biological samples. The mean chemical concentrations in water, fish, crab, gastropod, and bivalve were 0.58 ng/l, 0.043 ng/g wet weight, 0.20 ng/g wet weight, 0.045 ng/g wet weight, and 0.062 ng/g wet weight, respectively (Naile et al., 2013). The literature explains that the longer carbon chain compounds tend to persist in the environment longer when compared to their shorter chain counterparts. Substituting for other, less persistent chemicals can prevent the contamination of water systems and human exposure to PFUnA.

INDUSTRIAL HYGIENE

The Occupational Safety and Health Administration, NIOSH, and ACGIH have not established exposure limits for PFUnA. Various preventive measures can be taken to reduce inhalation, ingestion, and dermal exposure of PFDA. Individuals using the chemical should be aware of the possibilities for inhalation of the chemical. Personal respiratory protection may be needed if PFUnA concentrations are high or if the chemical becomes aerosolized. Engineering controls, including ventilation systems, may be necessary to capture airborne particles and remove them from the work atmosphere. Ingestion of PFUnA can be prevented by using good hygiene practices in the occupational environment. Eating, drinking, or using tobacco products should be forbidden when individuals are in contact with the chemical. Hands should be washed thoroughly with soap and water to remove traces of PFUnA from the skin. Dermal contact with the chemical can be prevented by using personal protective equipment, such as gloves, rubber boots, and goggles. A fully encapsulating suit may be necessary in environments where high concentrations and quantities of the chemical are utilized. Perfluoroundecanoic acid has been substituted in industry and manufacturing in favor of less persistent and less harmful compounds.

MEDICAL MANAGEMENT

Several studies have been conducted that have identified PFUnA in human serum. Lindh et al. (2012) conducted a study of Greenlandic Inuit populations and detected PFUnA in human serum from 196 individuals at a mean concentration of 1.76 ng/ml and a median concentration of 1.28 ng/ml. A study of tea workers in Sri Lanka identified PFUnA from human blood at a mean concentration of 0.042 ng/ml and a median concentration of 0.043 ng/ml (Guruge et al., 2005). Individuals that have been exposed to the chemical may be at risk for the development of both acute and chronic health effects. Surveillance programs may be recommended in situations where the chemical is still used. If PFUnA becomes aerosolized, it may be inhaled into the respiratory tract where it can deposit. Personal respiratory protection can protect against inhalation exposure to PFUnA. There may be some applications where PFUnA is mixed into solution that could lead to dermal contact with the chemical. The skin should be rinsed thoroughly if the chemical contact occurs. Ingestion of the chemical is not likely but could occur and individuals working with PFUnA should be diligent about cleaning their hands before eating, drinking, or smoking to prevent exposure. In cases where symptoms of exposure do not subside or when individuals are exposed to high concentrations of PFUnA, individuals should seek prompt medical attention.

REFERENCES

Ahrens, L., et al. (2009) Total body burden and tissue distribution of polyfluorinated compounds in harbor seals (*Phoca vitulina*) from the German Bight. *Mar. Pollut. Bull.* 58:520–525.

Benninghoff, A.D., et al. (2011) Estrogen-like activity of perfluoroalkyl acids in vivo and interaction with human and rainbow trout estrogen receptors in vitro. *Toxicol. Sci.* 120:42–58.

Bloom, M.S., et al. (2010) Exploratory assessment of perfluorinated compounds and human thyroid function. *Physiol. Behav.* 99:240–245.

Bossi, R., et al. (2008) Perfluoroalkyl compounds in Danish wastewater treatment plants and aquatic environments. *Environ. Int.* 34:443–450.

Freberg, B.I., et al. (2010) Occupational exposure to airborne perfluorinated compounds during professional ski waxing. *Environ. Sci. Technol.* 44:7723–7728.

Guruge, K.S., et al. (2005) Perfluorinated organic compounds in human blood serum and seminal plasma: a study of urban and rural tea worker populations in Sri Lanka. *J. Environ. Monit.* 7:371–377.

Lindh, C.H., et al. (2012) Blood serum concentrations of perfluorinated compounds in men from Greenlandic Inuit and European populations. *Chemosphere* 88:1269–1275.

Long, M., et al. (2013) Effects of perfluoroalkyl acids on the function of the thyroid hormone and the aryl hydrocarbon receptor. *Environ. Sci. Pollut. Res. Int.* 20:8045–8056.

Mertens, J.J., et al. (2010) Subchronic toxicity of S-111-S-WB in Sprague Dawley rats. *Int. J. Toxicol.* 29:358–371.

Naile, J.E., et al. (2013) Distributions and bioconcentration characteristics of perfluorinated compounds in environmental samples collected from the west coast of Korea. *Chemosphere* 90:387–394.

Nilsson, H., et al. (2010) A time trend study of significantly elevated perfluorocarboxylate levels in humans after using fluorinated ski wax. *Environ. Sci. Technol.* 44:2150–2155.

Sonne, C., et al. (2008) Potential correlation between perfluorinated acids and liver morphology in East Greenland polar bears (*Ursus maritimus*). *Toxicol. Environ. Chem.* 90:275–283.

Takahashi, M., et al. (2014) Repeated dose and reproductive/developmental toxicity of perfluoroundecanoic acid in rats. *J. Toxicol. Sci.* 39:97–108.

Wolf, C.J., et al. (2012) Activation of mouse and human peroxisome proliferator-activated receptor-alpha (PPARα) by perfluoroalkyl acids (PFAAs): further investigation of C4-C12 compounds. *Reprod. Toxicol.* 33:546–551.

PERFLUORODODECANOIC ACID (PFDoA)

Target organ(s): Eyes, respiratory system, skin, reproductive system, liver, reproductive system

Occupational exposure limits: NIOSH/OSHA/ACGIH—not established

Reference values: RfD/TDI/MRL—not established

Risk/safety phrases: Irritant, corrosive

BACKGROUND AND USES

Perfluorododecanoic acid (PFDoA, tricosafluorododecanoic acid, perfluoric acid, tricosafluorolauric acid; CASRN 307-55-1) is a twelve-carbon compound in the perfluoroalkyl family of chemicals. It is a byproduct of stain- and grease-proof coatings on food packages, furniture, and carpets. It is known to bioaccumulate in the environment, although research related to the toxicity of the chemical is scarce.

PHYSICAL AND CHEMICAL PROPERTIES

Perfluorododecanoic acid exists as solid white, fine crystals. The chemical has a molecular weight of 614.1 g/mol, a boiling point of 473 °F, and a melting point ranging from 221 to 226 °F. Perfluorododecanoic acid may be incompatible with strong oxidizers. Upon decomposition, PFDoA can form carbon monoxide, carbon dioxide, and hydrogen fluoride. Additional information related to the physical and chemical properties of PFDoA are not currently available.

MAMMALIAN TOXICOLOGY

Acute Effects

The acute effects of PFDoA on mammalian systems have not been extensively investigated. Since the chemical has been defined as an irritant, contact with the skin, eyes, and respiratory tract can result in irritating effects. While mammalian toxicology has not been researched, some research has been completed on aquatic organisms since PFDoA is a known environmental contaminant. Acute exposure to PFDoA has been found to induce liver damage in female zebrafish, including swollen hepatocytes, vacuolar degeneration, and nuclei pycnosis (Liu et al., 2008). The study explains that fish were provided with a single injection of PFDoA at concentrations of 0, 20, 40, and 80 µg of PFDoA/g of body weight and sacrificed after 48 h, 96 h, and 7 days after the initial administration. Hepatotoxic manifestations were seen upon examination of histopathological samples. In order to completely understand the acute effects of PFDoA on mammalian systems, additional research must be completed.

Chronic Effects

Research studies have associated PFDoA with adverse reproductive effects in mammals. A study conducted in 2007 found that male rats exposed to 1, 5, and 10 mg of PFDoA/kg/day for 14 days developed decreased testicular weight, decreased luteinizing hormone, and decreased testosterone levels (Shi et al., 2007). Shi et al. (2009) also found that male rats exposed to concentrations of 0.2 PFDoA/kg/day and 0.5 PFDoA/kg/day for 110 days exhibited decreased testosterone levels, decreased protein levels, disruption in testicular steroidogenesis, and disruption of expression of related genes. Research also indicates that PFDoA may result in hepatotoxic effects. Perfluorododecanoic acid was found to decrease liver weights in rats and increase the amount of lipid accumulation in the liver (Zhang et al., 2008). Male rats exposed to PFDoA for 110 days developed hepatic lipidosis characterized by an increase in hepatic triglycerides and a decrease in serum lipoprotein levels (Ding et al., 2009). Indications of cancer development as a result of perfluorododecanoic acid have not been identified in the literature.

Mechanism of Action(s)

The available literature related to PFDoA indicates that the mechanism of action is not well understood. A study conducted by Shi et al. (2010) investigated the mechanism of action in rodents exposed to PFDoA and found that several pathways may be associated with adverse effects of PFDoA on the testes. The authors explain that PFDoA toxicity may be associated with changes in protein expression in the mitochondrial respiratory chain, oxidative stress, sperm activity, cytoskeleton, and intracellular signaling transduction in rodent testes (Shi et al., 2010). Another study conducted by Shi et al. (2010) attempted to investigate PFDoA mechanism of action and found that PFDoA inhibited steroidogenesis in adenosine $3',5'$-cyclophosphate (cAMP)-stimulated Leydig cells by decreasing the expression of steroidogenic acute regulatory protein (StAR) through a model containing oxidative stress. Current literature has focused on PFDoA and its possible association with adverse reproductive effects. Further research may identify additional mechanisms of action not currently considered.

Chemical Pathology

Based on available information, PFDoA appears to have pathological effects on the liver and reproductive system. Dose-response information related to these effects is limited. The chemical seems to elicit irritant effects on the skin, liver, and respiratory system upon acute exposure. Hepatotoxicity and reproductive toxicity are thought to occur upon chronic exposure. Shi et al. (2013) found that the MAPK pathway and CDC2 protein phosphorylation may be involved in PFDoA testicular toxicity. There is some research, which indicates that PFDoA may be associated with renal toxicity as well. Male rodents exposed to 0, 0.05, 0.2, and 0.5 mg/kg/day of PFDoA for 110 days exhibited disorders in glucose and amino acid metabolism, which researchers believed was contributing to the pathological findings in the kidneys

(Zhang et al., 2011). Additional research is needed to gain a better understanding of how PFDoA interacts with the human body thereby resulting in pathological findings. Information related to the manifestation of different cancers as a result of PFDoA exposure is inconclusive.

SOURCES OF EXPOSURE

Occupational

Any individual that uses or handles PFDoA during processing, manufacturing, and distribution is at risk for inhalation or dermal exposure to the chemical. Specific populations that may be at risk for exposure to PFDoA are firefighters. Perfluorododecanoic acid is commonly used in the manufacture of furniture and carpet. In the event that these products are burned in a fire, PFDoA fumes may be released into the surrounding atmosphere posing a risk to those nearby. Firefighters should utilize personal respiratory protection to ensure that their inhalation risk is kept to a minimum. The FOX Project, instituted by the state of California, is a biomonitoring program that has been developed for firefighters (OEHHA, 2014). The aim of the program is to investigate occupational exposures that may be encountered by this population. In addition to monitoring for various metals and pesticides, perfluoroalkyl compounds, including PFDoA, are monitored. Exposure to PFDoA should be reduced as companies that use the chemical have been encouraged by regulatory agencies to substitute PFDoA with other chemicals.

Environmental

Perfluorododecanoic acid is not found in the natural environment. Once it enters the environment, it does not normally breakdown easily. Air, water, and soil can become contaminated by the chemical from human waste streams or when similar compounds begin to degrade. Human waste streams introduce it into water systems where it has been noted to bioaccumulate in aquatic organisms. In addition to bioaccumulating in aquatic organisms, humans may be exposed to the chemical if they consume contaminated drinking water. The ATSDR explains that using bottled water for cooking and drinking as well as using carbon water filters has the potential to reduce the exposure to PFDoA and structurally similar compounds (ATSDR, 2009). Another possible source of environmental exposure involves the use of consumer products containing PFDoA. Perfluorododecanoic acid is found in many household products, including furniture and carpet. Purchasing and using household products that do not contain the chemical can help reduce the risk from environmental exposure.

INDUSTRIAL HYGIENE

The Occupational Safety and Health Administration, NIOSH, and ACGIH have not established exposure limits for PFDoA. Several precautions should be taken to prevent inhalation, ingestion, and dermal exposure to PFDoA. Inhalation to PFDoA can be prevented by utilizing personal respirators. Personal respirators will prevent inhalable particles from settling deep in lung tissue. Ventilation may also be necessary to collect particles and remove them from the work environment. Ingestion of PFDoA can be prevented by utilizing good hygiene practices. Individuals using the chemical should ensure that they wash their hands prior to eating, drinking, applying cosmetics, and using tobacco products. Dermal contact can be prevented by covering the skin with personal protective equipment, including gloves, rubber boots, and safety glasses. Any contaminated clothing should be removed to prevent secondary contamination to others. It may also be necessary to provide workers with showers at the worksite so that they can cleanse their skin and remove the chemical before returning to their place of residence.

MEDICAL MANAGEMENT

Perfluorododecanoic acid can be detected in the blood; however, it is not considered a routine test that can be performed by physicians (ATSDR, 2009). Individuals exposed to high levels of the chemical should consult a physician to determine if testing is needed. The route of exposure to PFDoA will determine the type of medical management to be used. If PFDoA is inhaled, the exposed individual should be removed out of the contaminated area and into fresh air. If labored breathing occurs, oxygen may need to be provided. If an exposed individuals is not breathing, emergency artificial respiration will be required. If PFDoA is ingested, the mouth should be rinsed out with copious amounts of water if the individual is conscious. If PFDoA contacts the skin, it should be washed with soap and large amounts of water to remove the chemical from the skin.

The eyes should be rinsed with excessive amounts of water for a minimum of 15 min if PFDoA exposure occurs. In all situations, medical attention may be necessary if the exposure was excessive or if symptoms do not subside.

REFERENCES

Agency for Toxic Substances and Disease Registry (ATDSR) (2009) *Public Health Statement-Perfluoroalkyls*, ATDSR 1–10, Atlanta, GA: US Department of Health and Human Services.

Ding, L., et al. (2009) Systems biological responses to chronic perfluorododecanoic acid exposure by integrated metabonomic and transcriptomic studies. *J. Proteome. Res.* 8:2882–2891.

Liu, Y., et al. (2008) Induction of time-dependent oxidative stress and related transcriptional effects of perfluorododecanoic acid in zebrafish liver. *Aquat. Toxicol.* 89:242–250.

Office of Environmental Health Hazard Assessment (2014) *Firefighter Occupational Exposures (FOX) Project*, Sacramento, CA: California Environmental Protection Agency.

Shi, Z., et al. (2007) Alterations in gene expression and testosterone synthesis in the testes of male rats exposed to perfluorododecanoic acid. *Toxicol. Sci.* 98:206–215.

Shi, Z., et al. (2009) Chronic exposure to perfluorododecanoic acid disrupts testicular steroidogenesis and the expression of related genes in male rats. *Toxicol. Lett.* 188:192–200.

Shi, Z., et al. (2010) Perfluorododecanoic acid-induced steroidogenic inhibition is associated with steroidogenic acute regulatory protein and reactive oxygen species in cAMP-stimulated Leydig cells. *Toxicol. Sci.* 114:285–294.

Shi, Z., et al. (2010) Proteomic analysis for testis of rats chronically exposed to perfluorododecanoic acid. *Toxicol. Lett.* 192:179–188.

Shi, Z., et al. (2013) Testicular phophoproteome in perfluorododecanoic acid-exposed rats. *Toxicol. Lett.* 221:91–101.

Zhang, H., et al. (2008) Lipid homeostasis and oxidative stress in the liver of male rats exposed to perfluorododecanoic acid. *Toxicol. Appl. Pharmacol.* 227:16–25.

Zhang, H., et al. (2011) Biological responses to perfluorododecanoic acid exposure in rat kidneys as determined by integrated proteomic and metabonomic studies. *PLoS One* 6:e20862.

70

TETRAHYDROFURAN

Ghazi A. Dannan

Target organ(s): CNS, respiratory system, liver, kidneys

Occupational exposure limits: OSHA, NIOSH, and ACGIH 200 ppm (590 mg/m³) as an 8-h TWA or 250 ppm (735 mg/m³) as a 15-min TWA

Reference values: RfD = 0.9 mg/kg/day, RfC = 2 mg/m³ (U.S. EPA, 2012)

Risk/safety phrases: DOT label: Flammable liquid; may form explosive peroxides

BACKGROUND AND USES

Tetrahydrofuran (THF; CASRN 109-99-9) is a synthetic organic compound that is not found in the natural environment (ACGIH, 2001). Most of the 2006 worldwide consumption of 439×10^3 tons of THF, with a projected annual growth rate of about 5%, was towards manufacturing poly(tetramethylene) ether glycol (PTMEG), which is essential for producing polyurethane elastomers, elastic polymers, copolyesters, or copolyamides (Müller, 2012). It is also a versatile solvent for polyvinyl chlorides and natural and synthetic resins as well as in top coating solutions, polymer coatings, cellophane, protective coatings, adhesives, magnetic strips, and printing inks. It is also a common laboratory reagent and an intermediate in chemical syntheses of consumer and industrial products such as nutritionals, pharmaceuticals, and insecticides (HSDB, 2011).

PHYSICAL AND CHEMICAL PROPERTIES

It is a colorless, volatile liquid with an ethereal or acetone-like smell and is miscible in water and most organic solvents.

It is highly flammable and may thermally decompose to carbon monoxide and carbon dioxide. Prolonged storage in contact with air and in the absence of an antioxidant may cause THF to decompose into explosive peroxides. Synonyms of THF include diethyleneoxide; tetramethyleneoxide; 1,4-epoxy butane; furanidine; oxacyclopentane. Some of the physical and chemical properties for THF are as follows.

Empirical formula	C_4H_8O
Molecular weight	72.10
Melting point	−108.5 °C
Boiling point	65/66 °C
Vapor pressure	0.173 atm at 20 °C
Density at 20 °C relative to the density of H_2O at 4 °C:	0.89
Water solubility	Miscible
Log K_{ow}	0.46
Conversion factors	1 ppm = 2.95 mg/m³
Chemical structure	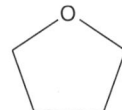

Additional properties may be found elsewhere (HSDB, 2011; Müller, 2012).

MAMMALIAN TOXICOLOGY

Acute Effects

Following single-dose gavage administration of THF, the median lethal dose (LD_{50}) values in male Sprague-Dawley rats (6–12/group) were estimated at 2038 mg/kg in 14-day-

Hamilton & Hardy's Industrial Toxicology, Sixth Edition. Edited by Raymond D. Harbison, Marie M. Bourgeois, and Giffe T. Johnson.
© 2015 John Wiley & Sons, Inc. Published 2015 by John Wiley & Sons, Inc.

old rats, 3190 mg/kg in young adult rats (80–160 g), and 2835 mg/kg in older adult rats (300–470 g) (Kimura et al., 1971). Other estimates, including a lower LD_{50} value of 1650 mg/kg in rats, have also been cited (IUCLID, 2000).

Following six daily gavage doses of 1–5 g/kg into rats (10/group, sex and strain not specified), there was no mortality in the 2 g/kg group, while 20 and 80–90% of the rats died in the groups that received 3 or 4–5 g/kg (Stasenkova and Kochetkova, 1963). Clinical signs of sedation, including immobility, drowsiness, reduced response to external stimuli, and reduced respiratory rate, were observed after 3–9 min of exposure. Mucous membranes appeared to have a cyanotic discoloration. Histopathological lesions were observed in the stomach, brain, liver, heart, spleen, and kidneys and included necrosis, edema, hemorrhage, and excess of blood or fluid in the blood vessels or tissues.

According to LaBelle and Brieger (1955), the median lethal concentration (LC_{50}) following a single 4-h exposure by inhalation in male albino rats was 18,000 ppm (53,100 mg/m^3) with narcosis (sleep) preceding death. In the same study, white mice (6/group, sex not specified) that were exposed continuously to saturated THF vapor (approximately 47,000 ppm or 138,650 mg/m^3) had a mean survival time of 41 min. Another study estimated the LC_{50} to be 21,000 ppm (61,950 mg/m^3) in male Sprague-Dawley rats following a single 3-h exposure (Horiguchi et al., 1984).

In an acute inhalation toxicity study, in which CD rats (6/sex/group) had a single 6-h exposure to THF at concentrations of 3010–20,500 ppm (8880–60,475 mg/m^3), the non-narcotic concentrations in male and female rats were identified at 5380 ppm (15,871 mg/m^3) and 5700 ppm (16,815 mg/m^3), respectively (DuPont Haskell Laboratory, 1979). During the exposure period, rats of both sexes demonstrated clinical signs of pawing and scratching and decreased or no response to sound at all concentrations. Male rats also exhibited signs of rapid respiration, and females showed signs of paralysis. Based on the clinical signs of central nervous system (CNS) toxicity, the lowest exposure concentration of 8880 mg/m^3 was the study lowest observed adverse effect level (LOAEL).

Several acute inhalation studies in animals suggest that the primary effects observed following single exposures to THF—at concentrations ranging from 300 to more than 200,000 mg/m^3 and from 30 min to several hours—are CNS toxicity and respiratory tract irritation. Symptoms of CNS toxicity, including sedation, coma, altered respiration, and decreased response to external stimuli, were observed in dogs at 150,000 mg/m^3 (the only concentration tested) (Stoughton and Robbins, 1936), mice at the lowest tested concentrations of 7000 mg/m^3 (Stasenkova and Kochetkova, 1963) or 30,000 mg/m^3 (Stoughton and Robbins, 1936), and rats at 9000–15,000 mg/m^3 (DuPont Haskell Laboratory, 1979; Horiguchi et al., 1984; Stasenkova and Kochetkova, 1963). Clinical signs of possible respiratory tract irritation,

observed only in rat studies at 9000–15,000 mg/m^3, included scratching, head shaking, face washing, tearing, salivation, and bleeding from the nose (DuPont Haskell Laboratory, 1979; Horiguchi et al., 1984). In addition, at least one acute study in rats observed structural or functional changes in respiratory tissue, including edema and hemorrhage in the lungs and bronchi (Stasenkova and Kochetkova, 1963), while studies in rabbits reported decreased ciliary beat frequency and vacuolation/degeneration of nasal mucosa at 3000 mg/m^3 (Ohashi et al., 1983) as well as decreased ciliary beat frequency and vacuolation of tracheal mucosa at 750 and 35,000 mg/m^3, respectively (Ikeoka et al., 1988).

In an acute neurotoxicity study, Crl:CD BR rats (12/sex/group) were exposed to THF vapor at concentrations of 0, 500, 2500, or 5000 ppm (0, 1475, 7375, or 14,750 mg/m^3) for a single 6-h exposure, and the animals were then observed for 2 weeks following the day of exposure (Malley et al., 2001). Motor activity assessments and functional observational battery (FOB) assessments were conducted before exposure and on test days 2, 8, and 15. Midway during the exposure at the mid and high THF concentrations, the startle response to an unexpected auditory stimulus was diminished or completely absent. Other signs of sedation in the high concentration group included a significant increase in the incidence of lethargy and abnormal gait in both male and female rats at 14,750 mg/m^3. Several parameters in the FOB were also affected in this group immediately following the exposure period only, including affected palpebral closure and increased incidence of slow or absent righting reflex in males and females. However, there were no statistically significant changes or remarkable effects on forelimb grip strength, hind limb grip strength, or foot splay. None of the FOB parameters were affected during test days 2, 8, or 15, suggesting that the sedative effects of THF were short lived. The LOAEL for this study is 7375 mg/m^3 based on the observations of the sedative effects, and the no observed adverse effect level (NOAEL) was 1475 mg/m^3.

Short-Term Effects

There are a few short-term oral toxicity studies in animals, including one in which rats were treated with a total of six daily gavage administrations of THF in distilled water resulting in increased mortality at doses >2000 mg/kg (Stasenkova and Kochetkova, 1963). Toxicity observed in this study included CNS toxicity (immobility, drowsiness, reduced response to external stimuli) and necrosis, edema, and hemorrhage of stomach, brain, liver, heart, spleen, and kidneys. In another short-term rat study, there were no changes in mortality, body weight, food or water consumption, clinical signs of CNS toxicity, hematology, or serum chemistry following administration of THF in drinking water for 4 weeks at concentrations of 1–1000 mg/l (daily doses 0.1–95.5 mg/kg/day) (Komsta et al., 1988). There were increased incidences of minimal to mild cytoplasmic homogeneity and anisokaryosis in the liver and tubular cytoplasmic inclusions in the kidneys of the high dose group males and females; however, no histopathology was performed on the lower dose groups. The study authors concluded that THF in drinking water at concentrations up to 1000 mg/l (95.5 mg/kg/day) did not produce overt toxicity and that the observed effects at the high dose were mild and adaptive and could not be related to any functional changes (Komsta et al., 1988).

The ability to irritate the respiratory tract was also evaluated following short-term inhalation exposure of male Sprague-Dawley rats (3–6/group) to 0, 100, or 5000 ppm (0, 295, or 14,750 mg/m^3) THF vapor for up to 3 weeks (Horiguchi et al., 1984). The daily exposure duration was not specified. No differences were observed in the tracheal mucosa between the treated groups and the controls following 1 day or 1 week of exposure; at week 3, however, the tracheal mucosa in the 14,750 mg/m^3 group showed a disordered cilia and epithelial cells and darkening of the cell bodies. The nasal mucosa of animals exposed to 14,750 mg/m^3 for either 1 week or 3 weeks demonstrated disruption of the epithelial architecture, congestion, and sloughing of ciliary and goblet cells, in addition to vacuolation and darkening of the cell bodies. Based on these effects, the study LOAEL was determined to be 14,750 mg/m^3 and the study NOAEL was 295 mg/m^3.

Developmental studies conducted in both rats and mice reported maternal toxicity, including decreased maternal body weights and CNS effects as well as prenatal developmental findings, including reduction in the percent live pups/litter and delayed ossification of the sternum (Mast et al., 1992). Pregnant rats and mice were exposed by inhalation to THF concentrations of 0, 600, 1800, or 5000 ppm (0, 17, 70, 5310, or 14,750 mg/m^3) for 6 h/day on gestation days 6–19 in rats or 6–17 in mice (Mast et al., 1992). Due to high toxicity and mortality, the regimen was discontinued in the 14,750 mg/m^3 female mice after 6 days of exposure. Maternal toxicity in both species included symptoms of CNS effects and significant decreases in body weight accompanied by decreases in gravid uterine weight. Decreased fetal weight was observed at the same concentration that resulted in maternal toxicity in rats. In both species, decreased fetal survival also occurred at the same concentrations that resulted in maternal toxicity (14,750 mg/m^3 in rats and 5310 mg/m^3 in mice). There were no effects on the number of implantations, the fetal sex ratio, or the incidence of abnormalities in fetuses. However, an increased incidence of incomplete or delayed sternal ossification was observed in fetuses of rats (14,750 mg/m^3) and mice (5310 mg/m^3) (Mast et al., 1992; U.S. EPA, 2012).

Chronic Effects

There are no human studies on the health effects from exposure to THF by the oral route; however, limited information is available on possible long-term effects from a two-generation reproduction toxicity study in animals (BASF, 1996; Hellwig et al., 2002). In this study, four groups of Wistar rats (25/sex/group) were continuously administered THF in drinking water through two successive generations at concentrations from 0 to 9000 ppm for a minimum of 70 days prior to and during mating, and throughout gestation, parturition, and lactation to weaning. In the high dose (approximately 700–900 mg/kg/day) parental animals, there was a slight decrease in food and water consumption as well as in body weight in addition to a slight increase in kidney weights. There were no treatment-related effects on clinical signs, necropsy, histopathology, fertility and reproductive indices, malformations, or offspring viability. However, the body weight gains of the F1 and F2 pups in the high dose group were significantly decreased during lactation. Also, the F2 pups in the high dose group had delayed opening of eyes and increased number of sloped incisors, with both findings being indicative of a slight developmental delay (BASF, 1996; Hellwig et al., 2002). The finding of decreased pup weight gain was used to set the oral reference dose (RfD) for THF (U.S. EPA, 2012).

Several case studies illustrate the potential for human health effects following an occupational exposure to THF by the inhalation and dermal routes for periods from a few weeks to a few months. In almost all of these cases, workers were exposed while performing activities where THF was present as a component of other chemicals such as solvents or adhesives. Therefore, it is not possible to conclusively attribute the observed effects to THF exposure alone. In addition, quantitative estimates of exposure were not provided in most of the published reports. Nonetheless, target organs in humans appear to be the CNS, respiratory tract, liver, and kidney. Symptoms of CNS toxicity included headache, dizziness, fatigue, and loss of sense of smell (Emmett, 1976; Garnier et al., 1989; Horiuchi et al., 1967). Symptoms of respiratory tract irritation included cough, chest pain, rhinorrhea, and dyspnea (Emmett, 1976; Garnier et al., 1989). In three cases, liver enzymes, namely alanine aminotransferase (ALT), aspartate aminotransferase (AST), and gamma-glutamyl transferase (GGT), were elevated above normal values (Edling, 1982; Garnier et al., 1989; Horiuchi et al., 1967), and in one case a liver biopsy revealed fatty changes following THF exposure (Edling, 1982). In one study, decreased white blood cell counts were reported in THF-exposed workers (Horiuchi et al., 1967). In another case study, autoimmune glomerulonephritis was observed in a man who worked with THF in adhesives for 9 years (Albrecht et al., 1987).

Some of the studies that were performed in animals reported portal-of-entry findings, including irritation of the nasal and respiratory tracts (Horiguchi et al., 1984; Kawata and Ito, 1984; Stasenkova and Kochetkova, 1963); however, there is a lack of consistency among the portal-of-entry study findings, as well as incomplete reporting of the effects and/or study design. Additionally, there is some suggestive evidence that respiratory tract responses may be transient in nature, waning with increasing exposure duration (Horiguchi et al., 1984). Several systemic effects have been observed following subchronic or chronic inhalation exposure of animals to THF. Decreased body weight has been observed in rats (Horiguchi et al., 1984; Kawata and Ito, 1984; Stasenkova and Kochetkova, 1963). Decreased blood pressure was observed in dogs (BASF, 1938) and rats (Stasenkova and Kochetkova, 1963). There were minor changes in hematology parameters in rats following subchronic exposures to relatively high THF concentrations, including decreased or increased white blood cells (Horiguchi et al., 1984; Stasenkova and Kochetkova, 1963), as well as increased hematocrit, hemoglobin, mean cell volume, mean cell hemoglobin, and neutrophils (NTP, 1998).

However, the most notable and reproducible systemic toxicity findings in subchronic and chronic inhalation rat and mouse studies (NTP, 1998) were the effects on the CNS and liver, which the U.S. EPA (2012) selected as end points for setting an inhalation reference concentration (RfC) for THF. In a chronic toxicity study, groups of F344 rats and B6C3F$_1$ mice (50/sex/group) were exposed to 0, 200, 600, or 1800 ppm (0, 590, 1,770, or 5310 mg/m^3) THF 6 h/day, 5 days/week for 105 weeks (NTP, 1998; chronic). Clinical signs of CNS toxicity (narcosis) were the only effects observed in male (but not female) mice during and up to 1 h after cessation of exposure to THF at 5310 mg/m^3. Similar effects were observed following subchronic exposure in both male and female rats, which had narcosis and ataxia at 14,750 mg/m^3 THF, and in mice, which had narcosis (stupor) at ≥5310 mg/m^3 (NTP, 1998; subchronic). As discussed above, similar CNS effects, including sedation, coma, altered respiration, and decreased response to external stimuli, were observed in a variety of acute, short-term, developmental, and subchronic animal toxicity studies. Additionally, the only findings in another subchronic study designed to assess neurotoxicity in male and female rats were sedative effects characterized as transient acute behavioral sedation, which dissipates rapidly upon termination of exposure to THF at concentrations of 4425 or 8850 mg/m^3 (Malley et al., 2001). As stated earlier, symptoms of CNS toxicity, including headache, dizziness, fatigue, and loss of sense of smell, were also reported in occupational exposure studies (Emmett, 1976; Garnier et al., 1989; Horiuchi et al., 1967).

The liver also seems to be a target organ of toxicity following chronic or subchronic inhalation exposure to THF in rats and mice. Following chronic exposure, a slight increase in liver necrosis was reported in high exposure (5310 mg/m^3) female mice (from 6% in controls to 14.6%); however, liver weights were not measured (NTP, 1998; chronic). Under the conditions of this 2-year bioassay, the NTP (1998) concluded that there was clear evidence of carcinogenic activity of THF in female B6C3F$_1$ mice due to increased incidences of hepatocellular adenomas or carcinomas. After adjusting for intercurrent mortality, the frequency of liver tumors were 61.3–93.0% in each of the three THF exposure groups compared to 46.3% in the control female mice (U.S. EPA, 2012).

In the accompanying subchronic 13-week study, female rats of the high exposure (14,750 mg/m^3) group had increased liver weight, which was accompanied by increased serum bile acids signifying a possible altered hepatic function (NTP, 1998; subchronic). Male and female mice also had increased liver weights (6–36% above control) at THF exposure concentrations of ≥1770 mg/m^3 accompanied by increased centrilobular cytomegaly (graded minimal or mild) in males and females of the high exposure group (14,750 mg/m^3) from 0/10 in the controls to 7/10 and 10/10, respectively (NTP, 1998; subchronic). The affected hepatocytes were described as having slight karyomegaly (enlarged nucleus), increased cytoplasmic volume, and granular cytoplasm with less vacuolation than that of midzonal and periportal hepatocytes (NTP, 1998; subchronic). Liver tissues from male and female mice of the three lower exposure groups were not examined, and no clinical chemistry measurements were performed in mice.

There is additional evidence that the liver is a target organ for THF exposure. In a short-term inhalation exposure study (BASF, 2001; Gamer et al., 2002), female mice had fatty liver degeneration (infiltration), which is a likely adverse finding since certain drugs may cause similar changes, which may predispose the liver to oxidative stress, lipid peroxidation, and possible mitochondrial and organ damage (Begriche et al., 2006; Letteron et al., 1996). In another subchronic inhalation toxicity study, Horiguchi et al. (1984) reported mild liver toxicity in male rats in the form of increased serum liver enzymes, bilirubin, and cholesterol at THF exposure concentrations of 2950 and 14,750 mg/m^3 in addition to increased relative liver weight at 14,750 mg/m^3, but no liver histopathology findings were reported. Other studies also reported liver effects when THF was administered in animals using exposure routes other than inhalation (Komsta et al., 1988; Stasenkova and Kochetkova, 1963). As discussed earlier, the human liver may also be a target organ, although exposure to THF might have been confounded by coexposure to other chemicals (Edling, 1982; Garnier et al., 1989; Horiuchi et al., 1967).

Kidney findings were also reported in some of the studies. A case of autoimmune glomerulonephritis was reported in a plumber who worked for over 9 years with pipe cement containing THF resulting in estimated air concentrations of 389–757 ppm (1148–2233 mg/m^3) during periods that the pipe cement was in use (Albrecht et al., 1987). In subchronic inhalation exposure rat studies, the relative kidney weights of male rats were slightly increased (by 7–8%) only in the high exposure 5000 ppm (14,750 mg/m^3) groups without changes in histopathology (Horiguchi et al., 1984; NTP, 1998). However, another subchronic rat study (Kawata and Ito,

1984) reported protein casts and hyaline droplet degeneration in the kidney tubule lumen epithelium at 3000 ppm (8850 mg/m^3). In the two-generation reproduction THF drinking water study (BASF, 1996; Hellwig et al., 2002), the relative kidney weights of the high dose F0 parental rats were also slightly increased (by 5–8%) in the absence of the histopathology findings.

Based on increased incidences of renal tubular adenoma or carcinoma of the kidney, the NTP (1998) concluded that there was *some evidence* of carcinogenic activity of THF in male F344 rats. The mortality adjusted rates for these tumors were 16.7–38.3% in each of the three THF exposure groups compared to 8.3% in the control group (U.S. EPA, 2012). Based on increased incidences of hepatocellular tumors in female mice and renal tubule tumors in male rats, the U.S. EPA (2012) concluded that there is "suggestive evidence of carcinogenic potential" by all routes of exposure to THF. According to the European Committee for Risk Assessment (RAC), THF was classified in Categories 3 (Directive 67/548/EEC) and 2 (CLP Regulation). The RAC also concluded that there were no significant grounds to limit the concern to the inhalation route of exposure (ECHA, 2010).

Mechanism of Action(s)

Studies on the absorption, distribution, metabolism, and excretion of THF were recently summarized in U.S. EPA (2012). Overall, the available data show that THF is readily absorbed by the inhalation or oral routes of exposure, is systemically distributed, and is rapidly metabolized and excreted in exhaled air and urine. Based on a number of studies, a metabolic pathway was proposed in which THF is successively metabolized to succinic acid and CO_2 (U.S. EPA, 2012). Some of the intermediate metabolites, such as γ-butyrolactone (GBL) or 5-hydroxy-THF, are expected to be unstable and may be susceptible to further metabolism or hydrolysis to γ-hydroxy-butyric acid (GHB), which undergoes successive metabolism to succinic acid and ultimately CO_2 via the citric acid cycle (TCA cycle); GHB may also be converted to the neurotransmitter γ-aminobutyric acid (GABA). Several enzymes, including CYP450, PON1, and dehydrogenases, were thought to be involved in metabolizing THF and some of the intermediate metabolites (U.S. EPA, 2012).

Several studies attempted to characterize the mode(s) of action (MOA) by which THF causes toxicity and carcinogenesis. A brief description is provided below on the scope and conclusions in some of these studies. A more comprehensive evaluation may be found in U.S. EPA (2012).

Since GBL and GHB are known to affect the nervous system, the CNS effects caused by THF were considered to be likely due to its intermediate metabolites GBL and GHB, which may be converted to GABA (U.S. EPA, 2012). For instance, the most sensitive effects in a chronic gavage

bioassay of GBL were clinical signs of CNS toxicity (lethargy) in rats and mice (NTP, 1992). In addition, GBL is considered a drug of abuse that can lead to neurotoxicity, including addiction, anxiety, depression, and tremors (Herold and Sneed, 2002). On the other hand, clinical studies of GHB reported drowsiness and unconsciousness at high doses (Metcalf et al., 1966) or transient dizziness and a sense of dullness at relatively lower doses (Ferrara et al., 1999).

A number of studies explored the ability of THF to cause mutagenicity/genotoxicity or other possible MOA that may be considered as precursors or key event(s) to rat kidney and mouse liver tumors. NTP (1998) and U.S. EPA (2012) reviewed the results of a battery of mutagenicity/genotoxicity tests, including bacterial and mammalian mutagenicity, DNA damage, chromosomal aberration, clastogenicity, and cell transformation. NTP (1998) concluded that there was little evidence of mutagenic activity with most data being conclusively negative. The U.S. EPA (2012) also concluded that THF is not likely to be mutagenic or genotoxic since nearly all the results were conclusively negative, with equivocal findings reported in a small number of assays.

Also, the ability of THF and several individual tobacco smoke compounds to cause cytotoxicity were evaluated in a number of short-term *in vitro* assays, including inhibition of cell growth of ascites sarcoma BP 8 cells, inhibition of oxidative metabolism in isolated brown fat cells, plasma membrane damage, and ciliotoxicity in cultures of trachea from unborn chickens (Curvall et al., 1984; Pettersson et al., 1982). The results of these studies suggested that THF is not cytotoxic. Similar results were found in other test systems, including human hepatoma, HepG2 cells (Dierickx, 1989), and the BALB/c-3T3 cell transformation assay (Matthews et al., 1993).

Some data suggested that male rat kidney tumors in the NTP (1998) bioassay may be due to the accumulation of α_{2u}-globulin, which is a marker of rodent kidney disease. This MOA is commonly recognized to require three criteria, namely, increased hyaline droplets in the renal proximal tubule cells, evidence that α_{2u}-globulin is in the droplets, and presence of additional pathological lesions associated with α_{2u}-globulin. While the first two criteria were met by the available data, it was concluded that an area of uncertainty is the absence of α_{2u}-globulin associated histopathological lesions characteristic of this MOA (U.S. EPA, 2012). Chronic progressive nephropathy (CPN), which is a spontaneous age-related renal disease in rats, was also suggested as a likely MOA for the rat kidney tumors (Fenner-Crisp et al., 2011a). After analyzing the available information, the U.S. EPA (2012) concluded that there is inadequate evidence to support this MOA. Tetrahydrofuran did not increase the incidence or severity of CPN, and there was no direct evidence that it may have exacerbated the development of proliferative lesions within CPN-affected tissue. Overall, the U.S. EPA (2012) concluded that the mode of carcinogenic action of THF-induced renal tumors was not established.

The existing information on the rat kidney tumors and both α_{2u}-globulin and CPN also were evaluated by the European RAC which concluded that, though the carcinogenic response was small, a positive exposure-related response was seen for adenomas and/or carcinomas and that neither α_{2u}-globulin nor CPN could account for the rat kidney tumor findings (ECHA, 2010).

An analysis was also performed on the experimental data suggesting that the development of liver tumors in the THF-exposed female mice may involve a cell proliferation-related MOA (U.S. EPA, 2012). Among the key events for this MOA are histopathological evidence of cytotoxicity/necrosis and regenerative growth as well as possible changes in cellular apoptosis that can impact the net rate of tissue growth. As commonly recognized, sustained increase in cell proliferation may also lead to the promotion of growth of preinitiated cells and subsequently to tumorigenesis. Although increased cell proliferation was noted in short-term studies, the data were not adequate to support this hypothesized MOA since there was no significant increase in cell proliferation in tissues obtained from the subchronic NTP (1998) study. Therefore, cell proliferation didn't seem to be sustained for a sufficient duration to adequately explain the late onset of tumors. Furthermore, key precursor events linked to cell proliferation, such as apoptosis, were not identified; other information, including inhibition of gap junction intercellular communication, was considered too limited to establish a MOA (U.S. EPA, 2012). Fenner-Crisp et al. (2011b) also had similar conclusions that the available data on cell proliferation, apoptosis, and genotoxicity are inadequate to support a MOA for the THF-induced rat liver tumor. The European RAC was of the opinion that the liver tumors were in the highly sensitive B6C3F1 mouse strain and that the MOA for these tumors has not been elucidated (ECHA, 2010).

SOURCES OF EXPOSURE

Potential exposures to humans are primarily from anthropogenic sources, mainly through inhalation or dermal contact. Nonoccupational exposures may occur via inhalation and oral routes from environmental (air and water) or food contamination as well as from dermal exposure through contact of THF-containing products (NTP, 1998; HSDB, 2011).

Since THF is used as a stabilizer in chlorinated solvents, such as 1,1,1-trichloroethane or trichloroethylene, THF is believed to be present at sites that are contaminated by chlorinated solvents as well as in groundwater from different states at concentrations that, in some cases, may exceed the water quality criteria and guidelines (Isaacson et al., 2006).

INDUSTRIAL HYGIENE

Exposure Limits

Occupational Safety and Health Administration (OSHA) permissible exposure limits (PEL)s: 200 ppm (590 mg/m^3) as an 8-h time-weighted average (TWA) and 250 ppm (735 mg/m^3) as a 15-min TWA short-term exposure limit (STEL)

The National Institute for Occupational Safety and Health (NIOSH) has not issued recommended exposure limits (RELs) for tetrahydrofuran; however, the NIOSH concurs with OSHA's above PELs

American Conference of Governmental Industrial Hygienists (ACGIH): Threshold limit value (TLV) 200 ppm (590 mg/m^3) as a TWA for a normal 8-h workday and a 40-h workweek; STEL 250 ppm (737 mg/m^3) for periods not to exceed 15 min

For additional information on exposure limits and acronyms, see OSHA Guideline for Tetrahydrofuran, U.S. Department of Labor; available from: http://www.osha.gov/SLTC/healthguidelines/tetrahydrofuran/recognition.html.

Respiratory and Personal Protection   Respirators must be worn if the ambient concentration of THF exceeds prescribed exposure limits. Respirators may be used (1) before engineering controls have been installed, (2) during work operations such as maintenance or repair activities that involve unknown exposures, (3) during operations that require entry into tanks or closed vessels, and (4) during emergency situations. If the use of respirators is necessary, the only respirators permitted are those that have been approved by NIOSH and the Mine Safety and Health Administration (MSHA). For additional information on protective clothing and recommended respiratory protection (based on NIOSH), see OSHA Guideline for Tetrahydrofuran, U.S. Department of Labor; available from: http://www.osha.gov/SLTC/healthguidelines/tetrahydrofuran/recognition.html.

MEDICAL MANAGEMENT

Information on medical management may be found in HSDB (2011) or in MSDS issued by manufacturers.

REFERENCES

ACGIH (2001) Tetrahydrofuran. In: *Documentation of the Threshold Limit Values and Biological Exposure Indices*, Cincinnati, OH: American Conference of Governmental Industrial Hygienists.

Albrecht, W.N., Boiano, J.M., and Smith, R.D. (1987) IgA glomerulonephritis in a plumber working with solvent-based pipe cement. *Ind. Health* 25:157–158.

BASF (Badische Anilin- und Sodafabrik) (1938) *Toxicity of tetrahydrofuran, with cover letter dated 05/10/94 (sanitized). EPA Document No. 86940000738S.* NTIS No. OTS0557148.

BASF (1996) *Tetrahydrofuran: two-generation reproduction toxicity study in Wistar rats, continuous administration in the drinking water, with cover letter dated 8/30/96. Study No. 71R0144/93038. EPA Document No. 86960000573.* NTIS No. OTS558774.

BASF (2001) *Tetrahydrofuran: subacute inhalation study in F344 rats and B6C3F1 mice 20 exposures to vapors including interim sacrifices of satellite groups after 5 exposures, with a cover letter from the Tetrahydrofuran Task Force dated 04/10/2001. Study No. 9910151/99007.*

Begriche, K., Igoudjil, A., Pessayre, D., et al. (2006) Mitochondrial dysfunction in NASH: causes, consequences and possible means to prevent it. *Mitochondrion* 6:1–28.

Curvall, M., Enzell, C.R., and Pettersson, B. (1984) An evaluation of the utility of four in vitro short term tests for predicting the cytotoxicity of individual compounds derived from tobacco smoke. *Cell Biol. Toxicol.* 1:173–193.

Dierickx, P.J. (1989) Cytotoxicity testing of 114 compounds by the determination of the protein content in HepG2 cell cultures. *Toxicol. In Vitro* 3:189–193.

DuPont Haskell Laboratory (1979) *Initial Submission: Acute Inhalation Toxicity with Tetrahydrofuran in Rats with Cover Letter Dated 061592 and Attachments*, EPA Document No. 88-920004255. Newark, DE: E.I. du Pont de Nemours and Company, HLR-848-79. NTIS No. OTS0540603.

ECHA (2010) *RAC, Annex 1: Background document to RAC opinion on tetrahydrofuran (EC Number: 203-726-8). ECHA/RAC/DOC no CLH-O-0000000954-69-03/A1. Adopted 25 May 2010. From European Chemicals Agency.* Available at http://echa.europa.eu/documents/10162/0591bacb-a600-491d-b004-ad4e1ccec91e.

Edling, C. (1982) Interaction between drugs and solvents as a cause of fatty change in the liver? *Br. J. Ind. Med.* 39:198–199.

Emmett, E.A. (1976) Parosmia and hyposmia induced by solvent exposure. *Br. J. Ind. Med.* 33:196–198.

Fenner-Crisp, P.A., Mayes, M.E., and David, R.M. (2011a) Assessing the human carcinogenic potential of tetrahydrofuran: I. Mode of action and human relevance analysis of the male rat kidney tumor. *Regul. Toxicol. Pharmacol.* 60:20–30.

Fenner-Crisp, P.A., Mayes, M.E., and David, R.M. (2011b) Assessing the human carcinogenic potential of tetrahydrofuran: II. Mode of action and human relevance analysis of the female mouse liver tumor. *Regul. Toxicol. Pharmacol.* 60:31–39.

Ferrara, S.D., Giorgetti, R., Zancaner, S., et al. (1999) Effects of single dose of gamma-hydroxybutyric acid and lorazepam on psychomotor performance and subjective feelings in healthy volunteers. *Eur. J. Clin. Pharmacol.* 54:821–827.

Gamer, A.O., Jaeckh, R., Leibold, E., et al. (2002) Investigations on cell proliferation and enzyme induction in male rat kidney and female mouse liver caused by tetrahydrofuran. *Toxicol. Sci.* 70:140–149.

Garnier, R., Rosenberg, N., Puissant, J.M., et al. (1989) Tetrahydrofuran poisoning after occupational exposure. *Br. J. Ind. Med.* 46:677–678.

Hellwig, J., Gembardt, C., and Jasti, S. (2002) Tetrahydrofuran: two-generation reproduction toxicity in Wistar rats by continuous administration in the drinking water. *Food Chem. Toxicol.* 40:1515–1523.

Herold, A.H. and Sneed, K.B. (2002) Treatment of a young adult taking gamma-butyrolactone (GBL) in a primary care clinic. *J. Am. Board Fam. Med.* 15:161–163.

Horiuchi, K., Horiguchi, S., Utsunomiya, T., et al. (1967) Toxicity of an organic solvent, tetrahydrofuran, on the basis of industrial health studies at a certain factory. *Sumitomo Bull. Ind. Health* 3:49–56.

Horiguchi, S., Teramoto, K., and Katahira, T. (1984) Acute and repeated inhalation toxicity of tetrahydrofuran in laboratory animals. *Sumitomo Sangyo Eisei* 20:141–157.

HSDB (Hazardous Substances Data Bank) (2011) *Tetrahydrofuran. National Library of Medicine, Bethesda, MD.* Available at http://toxnet.nlm.nih.gov/cgi-bin/sis/htmlgen?HSDB.

Ikeoka, H., Nakai, Y., Ohashi, Y., et al. (1988) Experimental studies on the respiratory toxicity of tetrahydrofuran in a short term exposure. *Sumitomo Sangyo Eisei* 19:113–119.

Isaacson, C., Mohr, T.K.G., and Field, J.A. (2006) Quantitative determination of 1,4-dioxane and tetrahydrofuran in groundwater by solid phase extraction GC/MS/MS. *Environ. Sci. Technol.* 40:7305–7311.

IUCLID Dataset, European Commission, ESIS (2000) *Tetrahydrofuran (109-99-9) p. 47.* Available at http://esis.jrc.ec.europa.eu/.

Kawata, F. and Ito, A. (1984) Experimental studies on the effects of organic solvents in living bodies: changes of tetrahydrofuran concentration in rats' organs and histological observations after inhalation. *Nihon Hoigaku Zasshi* 8:367–375.

Kimura, E.T., Ebert, D.M., and Dodge, P.W. (1971) Acute toxicity and limits of solvent residue for sixteen organic solvents. *Toxicol. Appl. Pharmacol.* 19:699–704.

Komsta, E., Chu, I., Secours, V.E., et al. (1988) Results of a short-term toxicity study for three organic chemicals found in Niagara River drinking water. *Bull. Environ. Contam. Toxicol.* 41:515–522.

LaBelle, C.W. and Brieger, H. (1955). The vapor toxicity of a compound solvent and its principal components. *Arch. Ind. Health* 12:623–627.

Letteron, P., Fromenty, B., Terris, B., et al. (1996) Acute and chronic hepatic steatosis lead to *in vivo* lipid peroxidation in mice. *J. Hepatol.* 24:200–208.

Malley, L.A., Christoph, G.R., Stadler, J.C., et al. (2001) Acute and subchronic neurotoxicology evaluation of tetrahydrofuran by inhalation in rats. *Drug Chem. Toxicol.* 24:201–219.

Mast, T.J., Weigel, R.J., Westerberg, R.B., et al. (1992) Evaluation of the potential for developmental toxicity in rats and mice following inhalation exposure to tetrahydrofuran. *Fundam. Appl. Toxicol.* 18:255–265.

Matthews, E.J., Spalding, J.W., and Tennant, R.W. (1993) Transformation of BALB/c-3T3 cells: V. Transformation responses of 168 chemicals compared with mutagenicity in Salmonella and carcinogenicity in rodent bioassays. *Environ. Health Perspect.* 101(Suppl. 2):347–482.

Metcalf, D.R., Emde, R.N., and Stripe, J.T. (1966) An EEG-behavioral study of sodium hydroxybutyrate in humans. *Electroencephalogr. Clin. Neurophysiol.* 20:506–512.

Müller, H. (2012) Tetrahydrofuran. In: *Ullmann's Encyclopedia of Industrial Chemistry*, Weinheim: Wiley-VCH Verlag GmbH & Co., Published Online: 15 October, 2011. doi 10.1002/14356007.a26_221.pub2.

NTP (National Toxicology Program) (1992) *Toxicology and Carcinogenesis Studies of Gamma-butyrolactone (CAS No. 96-48-0) in F344/N Rats and B6C3F1 Mice*, NTP TR-406. Available from the National Institute of Environmental Health Services, Research Triangle Park, NC: Public Health Service, U.S. Department of Health and Human Services, Available at http://ntp.niehs.nih.gov/ntp/htdocs/LT_rpts/tr406.pdf.

NTP (1998) *Toxicology and Carcinogenesis Studies of Tetrahydrofuran (CAS No. 109-99-9) in F344/N Rats and B6C3F1 Mice*, NTP TR-475. Available from the National Institute of Environmental Health Services, Research Triangle Park, NC: Public Health Service, U.S. Department of Health and Human Services, Available at http://ntp.niehs.nih.gov/ntp/htdocs/LT_rpts/tr475.pdf.

Ohashi, Y., Nakai, Y., Nakata, J., et al. (1983) Effects on the ciliary activity and morphology of rabbit's nasal epithelium exposed to tetrahydrofuran. *Osaka City Med. J.* 29:1–14.

Pettersson, B., Curvall, M., and Enzell, C.R. (1982) Effects of tobacco smoke compounds on the ciliary activity of the embryo chicken trachea *in vitro*. *Toxicology* 23:41–55.

Stasenkova, K.P. and Kochetkova, T.A. (1963) The toxicity of tetrahydrofuran. *Toksikol. Novukn. Prom. Khim.* 5:21–34.

Stoughton, R.W. and Robbins, B.H. (1936) The anesthetic properties of tetrahydrofurane. *J. Pharmacol. Exp. Ther.* 58:171–173.

U.S. EPA (2012) *Toxicological Review of Tetrahydrofuran (CAS No. 109-99-9)*, EPA/635/R-11/006F. Washington, D.C.: U.S. Environmental Protection Agency, Available at www.epa.gov/iris.

71

PERCHLOROETHYLENE (TETRACHLOROETHYLENE)

DAVID G. DODGE AND JULIE E. GOODMAN

Target organ(s): Nervous system, kidney, liver, reproductive, and developmental

Occupational exposure limits:

OSHA PEL: 100 ppm (8-h TWA); 200 ppm (ceiling)

NIOSH REL: Minimize workplace exposure concentrations (carcinogen)

ACGIH: 25 ppm (170 mg/m^3; TLV–TWA); 100 ppm (685 mg/m^3; TLV–STEL)

Reference values:

US EPA: RfD: 0.006 mg/kg/day; RfC: 0.04 mg/m^3

ATSDR: Inhalation MRL: 0.2 ppm (acute); 0.04 ppm (chronic); oral MRL: 0.05 mg/kg/day (acute)

Health Canada: TDI: 0.014 mg/kg/day

BACKGROUND AND USES

Tetrachloroethylene (PCE; CASRN 127-18-4) is also known as perchloroethylene, tetrachloroethene, and 1,1,2,2-tetrachloroethene and is also commonly abbreviated to PER or PERC. Tetrachloroethylene is a volatile, chlorinated organic hydrocarbon that is widely used as a solvent in the dry-cleaning and textile-processing industries and as an agent for degreasing metal parts. It is an environmental contaminant that has been detected in the air, groundwater, surface waters, and soil (NRC, 2010).

PHYSICAL AND CHEMICAL PROPERTIES

The molecular formula of PCE is C_2Cl_4. Tetrachloroethylene is a clear, colorless, nonflammable liquid with an ethereal odor. Physical and chemical properties include:
   Molecular weight: 165.8

Density (g/ml) at 20 °C: 1.62
Boiling point (°C): 121
Vapor pressure (mmHg) at 25 °C: 18.47
Water solubility (mg/l) at 25 °C: 150
Henry's law constant (atm-m^3/mol) at 25 °C: 1.8×10^{-2}
Log K_{ow}: 3.4

MAMMALIAN TOXICOLOGY

Acute Effects

In humans, acute accidental inhalation of high concentrations of PCE has induced central nervous system (CNS) depression, dizziness, fatigue, headache, loss of coordination, unconsciousness, narcosis, liver damage, and some deaths (WHO, 2006). Effects on vision in humans have been reported at exposures as low as 50 parts per million (ppm) (ATSDR, 1997). Human data on acute PCE ingestion at high concentrations are limited, though neurological effects have been observed at 108 mg/kg/day (ATSDR, 1997).

   In laboratory rodents, acute inhalation and oral toxicity are low. Lethal concentration 50 (LC$_{50}$) values are above 20,000 mg/m^3 and oral LD$_{50}$ values are in excess of 2 g/kg

Hamilton & Hardy's Industrial Toxicology, Sixth Edition. Edited by Raymond D. Harbison, Marie M. Bourgeois, and Giffe T. Johnson.
© 2015 John Wiley & Sons, Inc. Published 2015 by John Wiley & Sons, Inc.

body weight (WHO, 2006). Central nervous system depression is the major overt sign in treated animals. Liver and kidney toxicity also have been observed. Tetrachloroethylene is irritating to the skin in humans and rabbits. Slight, transient eye irritation was reported by volunteers exposed at about 520 mg/m^3 vapor (WHO, 2006).

Chronic Effects

Cancer   A number of epidemiology studies in human populations exposed to PCE have been conducted, principally in laundry and dry-cleaning workers and populations exposed to chlorinated solvents in drinking water. The results have been inconsistent with regard to cancer (few cancer sites were positive in more than one study) (United States Environmental Protection Agency, 2012). In experimental animals, PCE has been associated with mouse liver tumors (hepatocellular adenomas and carcinomas) in both sexes in two lifetime inhalation bioassays of different rodent strains (JISA, 1993; NTP, 1986); mouse liver tumors were also reported in both sexes in an oral bioassay (NCI, 1977). Tetrachloroethylene was associated with rat mononuclear cell leukemia (MNCL) in two studies (JISA, 1993; NTP, 1986). Tumors reported in single inhalation bioassays include kidney and testicular interstitial cell tumors in male F344 rats (NTP, 1986), brain gliomas in male and female F344 rats (NTP, 1986), and hemangiomas or hemangiosarcomas in male Crj:BDF1 mice (JISA, 1993).

Noncancer   Neurotoxicity is a key effect evaluated in studies of occupational PCE exposure. Occupational exposures have rarely been linked to clinically significant neurological illness, so researchers have instead examined more subtle endpoints (Bukowski, 2011). The most sensitive of these endpoints appear to involve visual/spatial and color perception. For example, studies have demonstrated color vision decrements at a *lowest-observed-adverse-effect level* (LOAEL) of 42 mg/m^3 (Cavalleri et al., 1994) and changes in cognitive and visuospatial function at a LOAEL of 156 mg/m^3 (Echeverria et al., 1995), both in dry-cleaning workers. Other noncancer effects of PCE identified in humans are inconclusive. Limited and/or inconsistent findings include toxicity to the kidney (i.e., chronic kidney disease), liver (i.e., some biomarker changes indicative of liver toxicity), immune and hematologic systems (i.e., some changes in cell counts), and on reproduction (e.g., spontaneous abortions) and development (e.g., oral clefts).

As with humans, neurotoxic effects of PCE have also been characterized in experimental animal studies (US EPA, 2012). Effects on vision, visuospatial function, and reaction time, as well as brain weight changes have been observed, though exposure concentrations are much higher than those experienced by humans environmentally or occupationally. For example, in rats, changes in flash-evoked potential responses were observed at subchronic PCE inhalation

exposures of 800 ppm, but not 200 ppm (Mattsson et al., 1998) and reduction in rat brain weight, DNA, and protein were observed following subchronic PCE inhalation exposures of 600 ppm, but not 300 ppm (Wang et al., 1993).

Adverse effects on the kidney have been reported in studies of rodents exposed to high concentrations of PCE by inhalation and oral exposures (US EPA, 2012). Specific nephrotoxic effects in animals include increased kidney-to-body weight ratios, hyaline droplet formation, glomerular "nephrosis," karyomegaly (enlarged nuclei), cast (cylindrical structures formed from cells and protein released from the kidney) formation, and other lesions or indicators of renal toxicity. These findings support a LOAEL of 200 ppm and a no-observed-adverse-effect level (NOAEL) of 50 ppm (US EPA, 2012).

Liver toxicity has been reported in multiple animal species by inhalation and oral exposures to PCE (US EPA, 2012). The effects are characterized by increased liver weight, fatty changes, necrosis, inflammatory cell infiltration, triglyceride increases, and proliferation. Subchronic and chronic LOAELs for liver toxicity in animals include 9 ppm for increased liver weight (Kjellstrand et al., 1984, as cited in US EPA, 2012) and 50 ppm for liver necrotic foci (JISA, 1993), respectively.

The database of experimental animal studies for PCE also includes assessments of reproductive and fertility outcomes in rats and mice following inhalation exposures; assessments of developmental toxicity in rats, mice, and rabbits following inhalation exposures during gestation; and assessments of developmental neurotoxicity in rats following pre- and/or postnatal inhalation exposures of the offspring (US EPA, 2012). According to US EPA, (2012) and supported by (NRC, 2010), the overall animal developmental and reproductive toxicity inhalation NOAEL is 100 ppm based on Tinston (1994, as cited in US EPA, 2012; NRC, 2010). In this study, increased mortality and decreased body weight of the offspring were observed at 300 ppm. Both maternal toxicity and decreases in pup viability occurred at 1000 ppm.

The available data from experimental studies assessing immunotoxic and hematological responses in animals are limited.

Agency Evaluations   United States Environmental Protection Agency (2012) classified PCE as "likely to be carcinogenic in humans by all routes of exposure," though noting that "the epidemiologic literature on cancers provides limited evidence that tetrachloroethylene is carcinogenic in humans." United States Environmental Protection Agency (2012) based its cancer risk estimation for PCE on data from animal studies because it found no available epidemiology studies of cancer that were suitable for dose-response modeling. United States Environmental Protection Agency (2012) found that mode of action (MOA) data are lacking or limited for all candidate animal cancer endpoints for PCE. Because an MOA for PCE could not be established, US EPA defaulted to using a linear, no-threshold approach to estimate the low exposure risk. Quantitative estimates of carcinogenic risk from both oral and inhalation exposures are based on the male mouse hepatocellular tumor data from the Japan Industrial Safety Association (JISA) (JISA, 1993) bioassay (US EPA, 2012). The International Agency for Research on Cancer (IARC) classified PCE as "probably carcinogenic to humans (Group 2A)" based on the limited evidence in humans and sufficient evidence in animals (Guha et al., 2012). The American Conference of Governmental Industrial Hygienists (ACGIH) (ACGIH, 2012) classifies PCE as an A3 carcinogen, meaning it is a confirmed animal carcinogen (at high doses) with unknown relevance to humans.

United States Environmental Protection Agency's (2012) chronic oral reference dose (RfD) for PCE is 0.006 mg/kg/d and its chronic inhalation reference concentration (RfC) is 0.04 mg/m^3. Both were derived from PCE exposure levels associated with measures of neurotoxicity in studies of occupationally exposed adults. In general, neurological effects were found to be associated with the lowest PCE exposures, and thus reference values based on neurotoxic effects are protective against other noncancer effects that occur at higher concentrations (US EPA, 2012; NRC, 2010). The Agency for Toxic Substances and Disease Registry's (ATSDR's) acute and chronic inhalation minimal risk levels (MRLs) are 0.2 and 0.04 ppm, respectively; both are based on neurological effects in occupationally exposed adults (ATSDR, 1997, 2013). The acute oral MRL is 0.05 mg/kg/d, based on developmental neurotoxicity in mice.

Mechanism(s) of Action

An understanding of PCE toxicokinetics is critical, for some effects, to both the qualitative and quantitative assessment of human health risks from PCE exposure because the active agent(s) for a number of endpoints are thought to be metabolites (Lash and Parker, 2001; Clewell et al., 2005; Chiu et al., 2007; NRC, 2010). Metabolism of PCE occurs through two main irreversible pathways: oxidation and glutathione (GSH) conjugation. Oxidative metabolism is thought to predominate in the liver, and the primary stable metabolite of the oxidative pathway is trichloroacetic acid (TCA) (US EPA, 2012). It is generally believed that the PCE-induced liver tumors in mice occur via the formation of metabolites in the oxidative pathway, with particular focus on TCA (US EPA, 2012). The observed species differences in liver tumor formation in response to PCE exposure could be explained by differences in the rates of oxidative metabolite formation (Lash et al., 1998; Lash and Parker, 2001). Given a similar exposure to PCE, humans are not capable of generating the same amounts of reactive oxidative metabolites as compared to rodents. A number of metabolic products can be formed in the conjugative pathway, including S-trichlorovinyl-L-cysteine (TCVC) and mercapturate *N*-acetyl trichlorovinyl cysteine (NAcTCVC). Tetrachloroethylene-induced kidney tumors

in rats are generally believed to involve metabolites from GSH conjugation (US EPA, 2012; NRC, 2010). Estimates of the extent of GSH conjugation in mice, rats, and humans are far more uncertain, especially in humans (Chiu and Ginsberg, 2011). Even less clear is the metabolite(s) contributing to the development of MNCL, brain gliomas, or testicular interstitial cell tumors in exposed rats and hemangiosarcomas in exposed mice and how metabolic differences may impact species sensitivity (Chiu and Ginsberg, 2011; US EPA, 2012). According to US EPA (2012), no mechanistic hypotheses have been advanced for the human cancers suggested to be increased with PCE exposure in epidemiology studies.

Genotoxic effects of PCE have been extensively investigated in human populations and in a variety of mammalian and nonmammalian test systems. Overall, there is no confirmed evidence that PCE has genotoxic or mutagenic activity in intact organisms (Lash and Parker, 2001; NRC, 2010). United States Environmental Protection Agency (2012) acknowledged that PCE has largely yielded negative results in standard genotoxicity assays, but it concluded nonetheless that "the hypothesis that mutagenicity contributes to the tetrachloroethylene carcinogenesis cannot be ruled out for one or more target organs, although the specific metabolic species or mechanistic effects are not known."

The potential neurotoxic mechanisms of PCE were reviewed by Bale et al. (2011). In contrast to PCE metabolites being the active agent(s) for a number of endpoints, *in vitro* studies with PCE support the theory that the parent compound interacts with discrete targets in the brain to cause neurotoxicity. It is less clear if there are key metabolites that are also involved in these changes. Based on animal toxicity and mechanistic studies, the authors suggest that PCE interacts directly with several different classes of neuronal receptors by generally inhibiting excitatory receptors/channels and potentiating the function of inhibitory receptors/channels. With regard to the visual effects of PCE, there are no associated mechanistic studies to indicate what receptor systems may be involved. The authors suggest that there may be some similarities in neurological mechanisms of PCE between animals and humans, but that it is difficult to ascertain given there is limited mechanistic information in humans.

Chemical Pathology

In experimental animals, several effects on brain pathology, including DNA- and RNA-level changes, changes in neurotransmitter levels, and changes in brain fatty acid composition have been observed in the range of 37–800 ppm (US EPA, 2012).

Pathological changes in the kidney have been observed in studies of rodents following relatively high subchronic (400–800 ppm) or chronic PCE exposures (100–200 ppm) via inhalation and oral administration. These include karyomegaly (enlarged nuclei) in proximal tubules, hyaline droplet formation, glomerular nephrosis, and cast formation (US EPA, 2012).

Pathological changes in the liver have been reported in multiple animal species by inhalation and oral exposures to PCE. The effects are characterized by increased liver weight, fatty changes, necrosis, and inflammatory cell infiltration. The LOAEL for the inhalation studies is 9 ppm for increased liver weight and morphological changes. The LOAEL for pathological changes in liver from oral administration is 100 mg/kg/d (US EPA, 2012).

SOURCES OF EXPOSURE

Atmospheric releases of PCE can occur due to evaporative losses during dry cleaning. Other atmospheric emissions may result during manufacture, from use in metal degreasing, in production of fluorocarbons and other chemicals, in the textile industry, and in miscellaneous solvent-associated applications (WHO, 2006). Tetrachloroethylene may also be disposed of to land and surface water (WHO, 2006).

Tetrachloroethylene has been detected in groundwater and surface water, air, soil, food, and breast milk (US EPA, 2012).

Occupational

Workers with potential exposure to PCE include those with occupations in the dry-cleaning industry, metal industries (cleaning machinery parts and degreasing activities), chemical-production industries that make or use PCE, and other industries with solvent-associated applications (WHO, 2006). Mean PCE concentrations inside dry-cleaning facilities have been on the decline since the 1960s, but have still been found to range from 20 to 70 mg/m^3 in the United States and Nordic countries in recent decades (US EPA, 2012).

Environmental

The most important routes of exposure to PCE for the general population include inhalation of ambient air and ingestion via drinking water. Dermal exposure is not thought to be a major route of exposure for most people (WHO, 2006; US EPA, 2012). Ambient PCE concentrations vary from source to source and with proximity to source. Median PCE concentrations in outdoor air across the United States have been estimated at about 0.3 μg/m^3 for urban areas and 0.1 μg/m^3 for rural areas. Near points of use, such as dry cleaners or industrial facilities, indoor exposure to PCE is more significant than outdoor exposure, and groundwater and surface water concentrations can be considerably higher than average (US EPA, 2012).

INDUSTRIAL HYGIENE

The Occupational Safety and Health Administration's (OSHA) permissible exposure limits (PELs) for PCE are 100 ppm as a time-weighted average (TWA), 200 ppm as a ceiling (for 5 min in any 3-h period), and 300 ppm as a maximum peak. The National Institute for Occupational Safety and Health (NIOSH) (NIOSH, 2011) does not provide a quantitative exposure limit for PCE, but instead states to minimize workplace exposure concentrations, which is consistent with their old policy for potential occupational carcinogens. Under the new policy, NIOSH will develop quantitative recommended exposure limits (RELs) that are based on risk evaluations using human and/or animal health effects data, and on an assessment of what levels can be feasibly achieved by engineering controls and measured by analytical techniques. The ACGIH (2012) threshold limit value (TLV)–TWA for PCE is 25 ppm (170 mg/m^3) to provide a margin of safety to minimize potential discomfort and subjective complaints (e.g., headache, dizziness, sleepiness, incoordination that may occur from prolonged exposure at 100–200 ppm); its TLV–short-term exposure limit (STEL) is 100 ppm (685 mg/m^3) to minimize the risk of anesthetic-like effects.

In some occupational settings, it is possible to substitute an agent less hazardous than PCE. In other settings, it may be possible to eliminate hazards by increasing ventilation (ATSDR, 2008). Personal protective equipment (PPE), including aprons, gloves, goggles, and respirators approved for use with organic chemicals, are used to help workers avoid PCE exposure (OSHA, 2005). All containers of liquid PCE should be capped; rags soaked with PCE should be stored in sealed containers (ATSDR, 2008). High levels of exposure can occur during cleanup of contaminated equipment and spills. Workers must wear respirators equipped with filters or cartridges specifically designed for organic vapors when elevated PCE exposures are anticipated (OSHA, 2005; NIOSH and IPSC, 2005).

MEDICAL MANAGEMENT

For persons with PCE toxicity, the level of exposure either must be reduced or the source eliminated. There is no antidote for PCE poisoning. Treatment consists of support of respiratory and cardiovascular functions (ATSDR, 2011). Victims exposed only to PCE vapor who have no eye or skin irritation do not need decontamination. Others may require removing contaminated clothing, washing of exposed skin and hair with mild soap and water, and/or irrigation of exposed eyes with copious amounts of tepid water or saline for at least 15 min (ATSDR, 2011; NIOSH and IPSC, 2005). In cases of ingestion, emesis should not be induced. Instead, a slurry of activated charcoal should be administered if the victim is alert, asymptomatic, and has a gag reflex. Follow-up laboratory evaluation of hepatic and renal function should be arranged for severely exposed patients. Neurologic examination for posthypoxic injury is recommended in cases of CNS or respiratory depression. Patients who have skin burns or corneal lesions should be reexamined within 24 h (ATSDR, 2011).

REFERENCES

Agency for Toxic Substances and Disease Registry (ATSDR) (1997) *Toxicological Profile for Tetrachloroethylene (Update)*, 278 pp. Available at http://www.atsdr.cdc.gov/toxprofiles/tp18.pdf (accessed February 11, 2013).

Agency for Toxic Substances and Disease Registry (ATSDR) (2008) *Case Studies in Environmental Medicine: Tetrachloroethylene Toxicity*, 46 pp. Available at http://www.atsdr.cdc.gov/csem/pce/docs/pce.pdf (accessed February 11, 2013).

Agency for Toxic Substances and Disease Registry (ATSDR) (2011) *Medical Management Guidelines for Tetrachloroethylene*, 20 pp. Available at http://www.atsdr.cdc.gov/MHMI/mmg18.pdf (accessed February 11, 2013).

Agency for Toxic Substances and Disease Registry (ATSDR) (2013) *Minimal Risk Levels (MRLs)*, 14 pp. Available at http://www.atsdr.cdc.gov/mrls/pdfs/atsdr_mrls_february_2013.pdf (accessed February 14, 2013).

American Conference of Governmental Industrial Hygienists (ACGIH) (2012) *2012 TLVs and BEIs: Threshold Limit Values for Chemical Substances and Physical Agents and Biological Exposure Indices*, ACGIH Publication No. 0112, Cincinnati, OH: ACGIH, 238 pp.

Bale, A.S., Barone, S., Scott, C.S., and Cooper, G.S. (2011) A review of potential neurotoxic mechanisms among three chlorinated organic solvents. *Toxicol. Appl. Pharmacol.* 255:113–126.

Bukowski, J.A. (2011) Review of the epidemiologic literature on residential exposure to perchloroethylene. *Crit. Rev. Toxicol.* 41:771–782.

Cavalleri, A., Gobbs, F., Paltrinieri, M., Fantuzzi, G., Righi, E., and Aggazzotti, G. (1994) Perchloroethylene exposure can induce colour vision loss. *Neurosci. Lett.* 179:162–166.

Chiu, W.A. and Ginsberg, G.L. (2011) Development and evaluation of a harmonized physiologically based pharmacokinetic (PBPK) model for perchloroethylene toxicokinetics in mice, rats, and humans. *Toxicol. Appl. Pharmacol.* 253:203–234.

Chiu, W.A., Micallef, S., Monster, A.C., and Bois, F.Y. (2007) Toxicokinetics of inhaled trichloroethylene and tetrachloroethylene in humans at 1 ppm: empirical results and comparisons with previous studies. *Toxicol. Sci.* 95:23–36.

Clewell, H.J., Gentry, P.R., Kester, J.E., and Andersen, M.E. (2005) Evaluation of physiologically based pharmacokinetic models in risk assessment: an example with perchloroethylene. *Crit. Rev. Toxicol.* 35:413–433.

Echeverria, D., White, R.F., and Sampaio, C. (1995) A behavioral evaluation of PCE exposure in patients and dry cleaners: a possible relationship between clinical and pre-clinical effects. *J. Occup. Med.* 37:667–680.

Guha, N., Loomis, D., Grosse, Y., Lauby-Secretan, B., El Ghissassi, F., Bouvard, V., Benbrahim-Tallaa, L., Baan, R., Mattock, H., Straif, K., and International Agency for Research on Cancer (IARC) Monograph Working Group (2012) Carcinogenicity of trichloroethylene, tetrachloroethylene, some other chlorinated solvents, and their metabolites. *Lancet Oncol.* 13:1192–1193.

Japan Industrial Safety Association (JISA) (1993) *Carcinogenicity Study of Tetrachloroethylene by Inhalation in Rats and Mice.* Data No. 3-1, JISA, Japan Bioassay Research Center, 53 pp.

Lash, L.H. and Parker, J.C. (2001) Hepatic and renal toxicities associated with perchloroethylene. *Pharmacol. Rev.* 53:177–208.

Lash, L.H., Qian, W., Putt, D.A., Desai, K., Elfarra, A.A., Sicuri, A.R., and Parker, J.C. (1998) Glutathione conjugation of perchloroethylene in rats and mice *in vitro*: sex-, species- and tissue-dependent differences. *Toxicol. Appl. Pharmacol.* 150:49–57.

Mattsson, J.L., Albee, R.R., Yano, B.L., Bradley, G., and Spencer, P.J. (1998) Neurotoxicologic examination of rats exposed to 1,1,2,2-tetrachloroethylene (perchloroethylene) vapor for 13 weeks. *Neurotoxicol. Teratol.* 20:83–98.

National Cancer Institute (NCI) (1977) *Bioassay of Tetrachloroethylene for Possible Carcinogenicity*, DHEW Publication No. (NIH) 77-813, Bethesda, MD: U.S. Public Health Service, National Institutes of Health.

National Institute for Occupational Safety and Health (NIOSH); International Programme on Chemical Safety (IPCS); Commission of the European Communities (2005) *International Chemical Safety Card for Tetrachloroethylene (CAS No. 127-18-4)*, 3 p., April. Available at http://www.cdc.gov/niosh/ipcsneng/neng0076.html (accessed 2013).

National Institute for Occupational Safety and Health (NIOSH) (2011) Tetrachloroethylene. In: *NIOSH Pocket Guide to Chemical Hazards*. Available at http://www.cdc.gov/niosh/npg/npgd0599.html (accessed February 14, 2013).

National Research Council (NRC) (2010) *Review of the Environmental Protection Agency's Draft IRIS Assessment of Tetrachloroethylene*. Committee to Review EPA's Toxicological Assessment of Tetrachloroethylene, Washington, DC: National Academies Press, 169 pp.

National Toxicology Program (NTP) (1986) *Toxicology and Carcinogenesis Studies of Tetrachloroethylene (Perchloroethylene) (CAS No. 127-18-4) in F344/N Rats and B6C3F(1) Mice (Inhalation Studies)*. Research Triangle Park, NC: National Toxicology ProgramNTP TR 311, 197 pp.

Occupational Safety and Health Administration (OSHA) (2005) *Reducing Worker Exposure to Perchloroethylene (PERC) in Dry Cleaning*, OSHA 3253-05N, 28 pp.

US EPA (2012) *Toxicological Review of Tetrachloroethylene (Perchloroethylene) (CAS No. 127-18-4) in Support of Summary Information on the Integrated Risk Information System (IRIS) (Final)*. EPA/635/R-08/011F, 1077 pp.

Wang, S., Karlsson, J.E., Kyrklund, T., and Haglid, K. (1993) Perchloroethylene-induced reduction in glial and neuronal cell marker proteins in rat brain. *Pharmacol. Toxicol.* 72:273–278.

World Health Organization (WHO) (2006) *Concise International Chemical Assessment Document 68: Tetrachloroethylene (Draft)*, 123 pp. Available at http://www.who.int/ipcs/publications/cicad/cicad68.pdf (accessed February 11, 2013).

72

TRICHLOROETHYLENE

DAVID G. DODGE AND JULIE E. GOODMAN

Target organ(s): Central nervous system, kidney, liver, immune system, male reproductive system, developing embryo/fetus

Occupational exposure limits:

OSHA PEL: 100 ppm (8-h TWA); 200 ppm (ceiling)

NIOSH REL: 25 ppm (as a 10-h TWA)

ACGIH: 10 ppm (54 mg/m^3; TLV–TWA); 25 ppm (135 mg/m^3; TLV–STEL)

Reference values:

US EPA: RfD: 0.0005 mg/kg/day; RfC: 0.0004 ppm (0.002 mg/m^3)

ATSDR: Oral MRL: 0.0005 mg/kg/day (chronic); Inhalation MRL: 0.0004 ppm (chronic)

Health Canada: TDI: 0.00146 mg/kg/day

BACKGROUND AND USES

Trichloroethylene (TCE; CASRN 79-01-6) is also known as trichloroethene, acetylene trichloride, 1-chloro-2,2-dichloroethylene, and ethylene trichloride, and it is also commonly abbreviated to TRI. It is a volatile, chlorinated organic hydrocarbon that is widely used for degreasing metals and as a hydrofluorocarbon (HFC-134a) intermediate (ATSDR, 2013). It is also used in adhesives, paint-stripping formulations, paints, lacquers, and varnishes. In the 1930s, TCE was introduced for use in dry cleaning, but this practice was largely discontinued in the 1950s when TCE was replaced by tetrachloroethylene (PCE). It has a number of

other past uses in cosmetics, drugs, foods, and pesticides (US EPA, 2011). It is an environmental contaminant that has been detected in air, groundwater, surface waters, and soil (US EPA, 2011; NRC, 2006).

PHYSICAL AND CHEMICAL PROPERTIES

The molecular formula of TCE is C_2HCl_3. It is a clear, colorless, nonflammable (at room temperature) liquid with chloroform-like odor (ATSDR, 2011). Physical and chemical properties include:

Molecular weight: 131.40

Density (g/ml) at 20 °C: 1.46

Boiling point (°C): 87

Vapor pressure (mmHg) at 25 °C: 69.8

Water solubility (mg/l) at 25 °C: 1280

Henry's law constant (atm-m^3/mol) at 25 °C: 9.85×10^{-3}

Log K_{ow}: 2.61

MAMMALIAN TOXICOLOGY

Acute Effects

In humans, acute inhalation of TCE has been reported to cause headache, fatigue, and drowsiness at 200 parts per million (ppm), decreased depth perception and motor skills at 1000 ppm; and unconsciousness at 3000 ppm (Stewart et al., 1970; Vernon and Ferguson, 1969; Longley and Jones, 1963). Deaths have also been reported in situations of

Hamilton & Hardy's Industrial Toxicology, Sixth Edition. Edited by Raymond D. Harbison, Marie M. Bourgeois, and Giffe T. Johnson.
© 2015 John Wiley & Sons, Inc. Published 2015 by John Wiley & Sons, Inc.

accidental exposure to unusually high levels of TCE vapors in the workplace or during TCE's use as an anesthetic (DeFalque, 1961). There have been a few reports of acute renal dysfunction in workers exposed to TCE, such as acute renal failure; these are limited by having poor or no exposure data and co-exposure to other chemicals. Human data on acute TCE ingestion at high concentrations are limited, although case studies have reported arrhythmias and heart attacks after ingestion of between 20 and 500 ml TCE (Dhuner et al. 1957; Morreale, 1976; ATSDR, 1997). Acute ingestion of smaller amounts (2 tablespoons to 16 ounces) of TCE has resulted in muscle weakness, vomiting, unconsciousness, and deliriousness, but patients recovered within 2 weeks after cessation of exposure (Morreale, 1976; ATSDR, 1997; Todd, 1954). Dermal exposure to pure TCE for several hours reportedly has caused severe exfoliative dermatitis, rashes, and other skin irritation (Gob and Ng, 1988).

In laboratory rodents, lethal concentration 50 (LC_{50}) values are reported to be between 6400 (in mice) and 12,500 ppm (in rats) after a 4-h exposure (Kylin et al., 1962; Siegel et al., 1971). Oral lethal dose 50 (LD_{50}) values have been reported as 2402 mg/kg/day in mice and 7208 mg/kg/day in rats (Tucker et al., 1982; Smyth et al., 1969). Respiratory effects in rodents after acute inhalation exposure include vacuolation and other morphological changes in Clara cells at 100–500 ppm and nasal irritation at 376 ppm (Giovanetti et al., 1998; Odum et al., 1992; ATSDR, 2013). Neurological effects have been reported in animals after acute inhalation exposure, including decreased brain ribonucleic acid (RNA) and hyperactivity at 200 ppm; decreased wakefulness at 1000 ppm; and lethargy, auditory effects, and anesthesia at concentrations of at least 3000 ppm. Transient ataxia was reported in rats after an acute oral exposure of 633 mg/kg/day (Savolainen et al., 1977; Arito et al., 1993; Adams et al., 1951; Narotsky et al., 1995). Kidney effects have been observed in rodents exposed via acute inhalation exposure of 1000 ppm or higher and oral exposures as low as 50 mg/kg/day (Chakrabarti and Tuchweber, 1988; Berman et al., 1995). Increased liver weight and other hepatic effects (e.g., hepatocellular hypertrophy) have been reported in rodents after an inhalation exposure of 1000 ppm and oral exposures of 1000–2479 mg/kg/day (Ramdhan et al., 2008; NTP, 1985). In rats, decreased fetal weight and incomplete skeletal ossification were reported after an inhalation exposure of 1800 ppm TCE from gestation days 0–20 (Dorfmueller et al., 1979). Dermal exposure to undiluted TCE caused erythema, edema, and increased epidermal thickness in guinea pigs (Anderson et al., 1986).

Chronic Effects

Cancer   A large body of epidemiology data is available on TCE and possible cancer outcomes. For example, U.S. EPA (Environmental Protection Agency) conducted a systematic review of 76 epidemiology studies on TCE and cancer (Chiu et al., 2013). Conclusions from reviews and meta-analyses of these data have varied widely, in part, due to differences in the qualitative evolution of the studies (e.g., strengths and weaknesses in exposure assessment or consideration of potential confounding factors) (see, e.g., Scott and Jinot, 2011; Kelsh et al., 2010; US EPA, 2011; Alexander et al., 2007). Modest increases in kidney cancer from high TCE exposure (mostly occupational) have received the most support generally with some, but lesser, support for non-Hodgkin lymphoma (NHL) and liver cancer because of issues of inconsistent findings, potential publication bias, weaker exposure–response results, and/or small numbers of cases. In experimental animals, it is well established that high doses of TCE in certain strains of rats and mice can cause a variety of cancers (Chiu et al., 2013; Jollow et al., 2009). The type and incidence of tumors depend on the species, strain, and sex under study (Jollow et al., 2009). Malignant liver tumor formation appears to be limited to Swiss and B6C3F1 mice and occur with greater frequency in males than females; numerous strains of rats have showed no elevated incidence of liver tumors after TCE exposure. Low incidence rates of kidney cancer occur in some strains of rats but not in mice or hamsters exposed chronically to high doses of TCE by inhalation or gavage. Lung tumors have been reported in mice but not in rats or hamsters and only after inhalation exposure. Although less well characterized and reported infrequently and/or inconsistently, TCE has been reported to induce lymphohematopoietic cancers in rats and mice and testicular tumors upon inhalation and ingestion in rats (Chiu et al., 2013).

Noncancer   Based on the available epidemiology data and experimental and mechanistic studies, TCE poses potential noncancer toxicity to the central nervous system (CNS), kidney, liver, immune system, male reproductive system, and developing fetus. The evidence for TCE toxicity to the respiratory tract and female reproductive system is more limited (US EPA, 2011).

Human and animal studies have associated TCE exposure with effects on several neurological domains. The strongest neurological evidence of human toxicological hazard, based on human and experimental studies, is for changes in trigeminal nerve function or morphology and impairment of vestibular function (e.g., with symptoms such as headaches, dizziness, and nausea). Trigeminal nerve function has been recognized as a sensitive human neurological effect of TCE, with an inhalation lowest observed adverse effect level (LOAEL) of 14 ppm (adjusted to continuous exposure from mean cumulative exposure of 704 ppm-years) (Ruijten et al., 1991; US EPA, 2011). Less, more limited evidence exists for delayed motor function; changes in auditory, visual, and cognitive function or performance; and neurodevelopmental outcomes (US EPA, 2011).

Trichloroethylene causes nephrotoxicity in the form of tubular toxicity in male and female rats and mice following oral or inhalation exposure. Toxic nephrosis, for example, was identified in mice at a LOAEL of 360 mg/kg/day (adjusted for continuous exposure) (NCI, 1976, as cited in US EPA, 2011). There are few human data pertaining to TCE-related noncancer kidney toxicity; however, several available studies reported elevated excretion of urinary proteins, considered nonspecific markers of nephrotoxicity, among TCE-exposed subjects compared to unexposed controls (US EPA, 2011).

Effects of TCE on the liver in rodents include hepatomegaly, increased serum bile acids, small transient increases in deoxyribonucleic acid (DNA) synthesis, cytomegaly in the form of swollen or enlarged hepatocytes, increased nuclear size (probably reflecting polypoidization), and proliferation of peroxisomes. Mice appear to be more sensitive to these effects than other laboratory animal species. Hepatomegaly appears to be the most sensitive indicator of toxicity for the liver; Buben and O'Flaherty (1985) reported increased liver: body weight ratios at all tested doses in mice, ranging from 100 to 3200 mg/kg/day. There are a few human studies on liver toxicity and TCE exposure; some have reported TCE exposure to be associated with significant changes in serum liver function tests—used widely in clinical settings, in part, to identify patients with liver disease—and plasma or serum bile acids (US EPA, 2011).

Human and laboratory animal studies of TCE and immune-related effects provide evidence that TCE exposure increases the risk of autoimmune disease and a specific type of generalized hypersensitivity skin disorder, with more limited evidence for immunosuppression (Chiu et al., 2013). Available human studies do not provide adequate exposure information for dose–response analysis. Among the more sensitive immune-related effects in animals are a decrease in thymus weight and an increase in early markers for autoimmune disease at a LOAEL of 0.35 mg/kg/day TCE in mice (Keil et al., 2009, as cited in US EPA, 2011 and ATSDR, 2013).

A number of human and laboratory animal studies suggest that TCE exposure has the potential for male reproductive toxicity. Human studies have reported TCE exposure to be associated with increased sperm density and decreased sperm quality, altered sexual drive or function, or altered serum endocrine levels. For example, Chia et al. (1996, as cited in US EPA, 2011 and ATSDR, 2013) identified decreased normal sperm morphology and hyperzoospermia in an occupational study in which men using TCE for electronics degreasing had a mean TCE exposure of 29.6 ppm. Several laboratory animal studies that reported effects on sperm, libido/copulatory behavior, and serum hormone levels (US EPA, 2011) provide evidence of similar effects on male reproductive toxicity.

Congenital malformations—particularly cardiac defects—have been associated with exposures to TCE and/or its metabolites, though these findings have been inconsistent in humans and rodents, and interpretation of these data have been controversial (NRC, 2006; Chiu et al., 2013). Other TCE-related developmental outcomes observed in humans and experimental animals include embryonic or fetal mortality, prenatal growth inhibition, and neurological and immunological functional deficits (Chiu et al., 2013). By far, the most sensitive developmental effect identified was heart malformations in the rat, with a LOAEL of 0.048 mg/ kg/day TCE in drinking water (NOAEL of 0.00045 mg/kg/ day) (Johnson et al., 2003). This study, however, and its use by U.S. EPA to support its noncancer reference values for TCE (see Section 3.2.3), has been criticized for having a number of limitations, such as certain study design flaws and absence of a dose–response relationship (e.g., Watson et al., 2006; Kimmel et al., 2010).

Agency Evaluations US EPA (2011) classified TCE as carcinogenic in humans by all routes of exposure based on what it characterized as convincing evidence of a causal association between TCE exposure in humans and kidney cancer. US EPA (2011) based its cancer risk estimation for TCE (both inhalation and oral) on human kidney cancer risks reported by Charbotel et al. (2006) and adjusted for potential risk for NHL and liver cancer. US EPA (2011) used linear low dose extrapolation to estimate the exposure–response relationship at low environmental exposure levels from higher occupational exposures; this was based on its conclusion that a mutagenic mode of action is operative for TCE-induced kidney tumors. The International Agency for Research on Cancer (IARC) classified TCE as carcinogenic to humans (Group 1) based on sufficient evidence in both humans and animals (Guha et al., 2012). The American Conference of Governmental Industrial Hygienists (ACGIH) (ACGIH, 2007) classifies TCE as an A2 suspected human carcinogen based on epidemiology studies of TCE.

US EPA's (2011) chronic oral reference dose (RfD) for TCE is 0.0005 mg/kg/day based on the critical effects of heart malformations (rats), adult immunological effects (mice), developmental immunotoxicity (mice), and toxic nephropathy (rats), all from oral studies. Its chronic inhalation reference concentration (RfC) is 0.0004 ppm (0.002 mg/m^3) based on route-to-route extrapolated results from oral studies for the critical effects of heart malformations (rats) and immunotoxicity (mice). In general, the critical effects were found to be associated with the lowest TCE exposures; thus, reference values based on these effects are protective against other noncancer effects that occur at higher concentrations (US EPA, 2011). The Agency for Toxic Substances and Disease Registry (ATSDR) adopted US EPA's (2011) chronic RfD and RfC as its chronic oral and inhalation minimal risk levels (MRLs), respectively (ATSDR, 2013).

Mechanism(s) of Action

An understanding of TCE toxicokinetics is critical, for some effects, to both the qualitative and quantitative assessment of human health risks from TCE exposure, because specific metabolites or metabolic pathways are associated with a number of endpoints of observed toxicity (Chiu et al., 2009). Key scientific issues for characterizing these hazards include identifying the metabolites responsible for the effects, elucidating the mode of action, and understanding the relevance of animal data for humans (NRC, 2006). Metabolism of TCE occurs through two main irreversible pathways: oxidation and glutathione (GSH) conjugation (Chiu et al., 2013). Products of the initial oxidation or conjugation step are further metabolized to a number of other metabolites (US EPA, 2011; Chiu et al., 2009). Trichloroethylene liver effects are thought to result from oxidative metabolites, whereas kidney effects generally are associated with metabolites resulting from GSH conjugation (Chiu et al., 2013; Kim et al., 2009; Evans et al., 2009). The identity of TCE metabolites involved in the induction of other effects of TCE is less clear (Chiu et al., 2013). Oxidative metabolism is thought to predominate in the liver, especially after oral exposure and, to some degree, in the respiratory tract after inhalation exposure, especially in mice (Jollow et al., 2009). The primary metabolites of the oxidative pathway that are thought to be toxicologically active are chloral hydrate, trichloroacetic acid (TCA), and dichloroacetic acid (DCA) (Chiu et al., 2013). Mice, on an average, have a greater capacity to oxidize TCE than rats or humans (Chiu et al., 2009). A number of metabolic products can be formed in the conjugative pathway, including dichlorovinyl glutathione (DCVG), S-dichlorovinyl-L-cysteine (DCVC), and mercapturate N-acetyl dichlorovinyl cysteine (NAcDCVC) (Chiu et al., 2009). Estimates of the kinetics and relative extent of GSH conjugation in mice, rats, and humans are complex and uncertain (Kim et al., 2009; Hack et al., 2006; Chiu et al., 2009).

Trichloroethylene and many of its known metabolites have been studied to varying degrees for their genotoxic potential. US EPA (2011) concluded, based on positive genotoxicity data for TCE metabolites derived from GSH conjugation (in particular DCVC), that a mutagenic mode of action is operative in TCE-induced kidney tumors. There are differing opinions, however, with regard to this conclusion (e.g., Lock and Reed, 2006). US EPA (2011) found that data are insufficient to conclude that a mutagenic mode of action is operative in TCE-induced liver tumors.

The potential neurotoxic mechanisms of TCE were reviewed by Bale et al. (2011). In contrast to TCE metabolites being the active agent(s) for a number of endpoints, *in vitro* studies with TCE support the theory that the parent compound interacts with discrete targets in the brain to cause neurotoxicity. It is less clear if there are key metabolites that are also involved in these changes. Based on animal toxicity and mechanistic studies, the authors suggest that TCE interacts directly with several different classes of neuronal receptors by generally inhibiting excitatory receptors/channels and potentiating the function of inhibitory receptors/channels. With regard to the trigeminal nerve effects of TCE, it has been hypothesized that interaction with sodium channels and the myelin sheath on peripheral neurons are the operative mechanisms. The authors suggest that there may be some similarities in neurological mechanisms of TCE between animals and humans, but they are difficult to ascertain given the limited mechanistic information in humans.

Chemical Pathology

A number of pathological changes have been observed in experimental animals after oral and inhalation exposure to TCE, including changes in the brain, kidney, liver, reproductive system, and developing organism. Some examples of these findings and the exposure levels at which they have occurred are presented below.

Several effects on brain pathology, including trigeminal nerve changes (e.g., changes in fiber diameter and myelin thickness) and decreased dopaminergic neurons in rats with subchronic exposures of 1000 mg/kg/day or more, have occurred in experimental animals after oral exposure (Barret et al., 1992).

Pathological changes in the kidney have been observed in studies of rodents following subchronic ($\geq$737 mg/kg/day oral or $\geq$75 ppm inhalation) or chronic ($\geq$500–1,100 mg/kg-day oral) TCE exposures. These effects include karyomegaly (enlarged nuclei) in renal tubule epithelial cells, increased kidney weight, nephrosis, and cytomegaly (NTP, 1985; NTP, 1988; Kjellstrand et al., 1983).

Pathological changes in the liver have been reported in rodents after inhalation and oral exposure to TCE. Effects include slight hepatocellular necrosis, centrilobular hypertrophy, enlargement and loss of vacuolization in hepatocytes, and peroxisome proliferation, with effects at lower levels in mice ($\geq$400 mg/kg/day oral or $\geq$75 ppm inhalation) (Ramdhan et al., 2008; Elcombe, 1985; Stott et al., 1982).

SOURCES OF EXPOSURE

Trichloroethylene is widely used as a degreasing agent, and it is a common environmental contaminant at Superfund sites and many industry and government facilities, including certain manufacturing operations (e.g., aircraft, spacecraft) (NRC, 2006). Most TCE used in the United States is released to the atmosphere, primarily from vapor-degreasing operations. Releases into the air also occur at treatment and disposal facilities, water treatment facilities, and landfills (US EPA, 2011). It can be found in soils and surface water as a result of direct discharges and in groundwater due to leaching from disposal operations (NRC, 2006; US EPA, 2011). Other sources of TCE include a number of consumer products, such as adhesives, paint-stripping formulations, and varnishes (US EPA, 2011). Trichloroethylene has a multitude of past uses, such as in dry cleaning (discontinued by the mid-1950s); as a food extractant, such as in the decaffeination of coffee (U.S. EPA banned the use of TCE as a food additive in 1977); as a carrier solvent for the active ingredients of insecticides and fungicides; and as an analgesic and anesthetic (Bakke et al., 2007; US EPA, 2011). Releases of TCE from non-anthropogenic activities are negligible (US EPA, 2011).

Occupational

The majority of data regarding worker exposure to TCE derives from degreasing operations, which is the primary industrial use of TCE (Bakke et al., 2007). Occupational exposure to TCE also has been identified in silk screening, taxidermy, and electronics cleaning (US EPA, 2011). In combination with the vapor source, the size and ventilation of the workroom are the main determinants of exposure intensity (US EPA, 2011). Bakke et al. (2007) reported that the arithmetic mean of industrial hygiene measurements of TCE in a variety of industries with TCE sources across several decades (mostly the 1950s, 1960s, and 1980s) was 38.2 ppm (210 mg/m^3). The highest air levels were reported in vapor degreasing (arithmetic mean of 44.6 ppm [240 mg/m^3]). Based on the development of a database of occupational air measurement data covering the years 1940–1998, Hein et al. (2010) predicted arithmetic mean exposure intensity levels for TCE ranging from 0.21 to 3700 ppm (1.1–20,000 mg/m^3) with a median of 30 ppm (160 mg/m^3). Occupational exposure to TCE has likely declined since the 1950s and 1960s due to decreased usage, better release controls, and improvements in worker protection (US EPA, 2011).

Environmental

Because of the pervasiveness of TCE in the environment, most people are likely to have some exposure via one or more pathway such as inhalation of outdoor/indoor air, ingestion of drinking water, or ingestion of food (US EPA, 2011). A substantial amount of air and groundwater data have been collected; less has been collected from other media. TCE has been detected in indoor and outdoor air throughout the United States, with the highest atmospheric levels in areas concentrated with industry and population (US EPA, 2011). U.S. EPA reported that annual mean TCE concentrations in air at approximately 300 locations across the United States between 1998 and 2008 were between 0.01 and 0.3 parts per billion (ppb) (US EPA, 2011b). Indoor air can become contaminated by vapor intrusion from contaminated ground-water, the presence of certain consumer products, and vola-tilization from contaminated water supplies (NRC, 2006). Trichloroethylene is the most frequently reported organic contaminant in groundwater; between 9 and 34% of the drinking water supply sources tested in the United States may have some TCE contamination, though generally at low levels (i.e., less than maximum contaminant levels) (NRC, 2006; US EPA, 2011). It also has been detected in a variety of foods and breast milk (NHANES III, 1997). Approximately 10% of nonoccupationally exposed persons in the United States have detectable levels of TCE in blood (US EPA, 2011).

INDUSTRIAL HYGIENE

The Occupational Safety and Health Administration (OSHA) permissible exposure limit (PEL) for TCE is 100 ppm as a time-weighted average (TWA), 200 ppm as a ceiling con-centration, and 300 ppm as a maximum peak above the ceiling concentration (for 5 min in any 2-h period). The National Institute for Occupational Safety and Health (NIOSH) considers TCE to be a potential occupational carcinogen and recommends a recommended exposure limit (REL) of 2 ppm (as a 60-min ceiling) during the usage of TCE as an anesthetic agent and 25 ppm (as a 10-h TWA) during all other exposures (CDC, 2013). The ACGIH (2012) threshold limit value–time-weighted average (TLV–TWA) for TCE is 10 ppm (54 mg/m^3) to provide a margin of safety to minimize potential discomfort and subjective complaints (e.g., headache, dizziness, sleepiness, lack of coordination that may occur from prolonged exposure) and to prevent cognitive decrements and renal toxicity; its threshold limit value–short-term exposure limit (TLV–STEL) is 25 ppm (135 mg/m^3) to minimize the risk of anesthetic-like effects and irritation.

In addition to maintaining exposure at or below permissi-ble limits, OSHA requires employers of workers occupation-ally exposed to TCE to protect employees through the use of engineering controls and proper work practices. NIOSH's (1988) occupational safety and health guideline for TCE states that workers using TCE as a cleaning solvent or degreaser should have local process enclosures (e.g., covers), local exhaust ventilation, and personal protective equipment (e.g., aprons, gloves, goggles, and respirators). NIOSH also notes that workers should be required to shower following a work shift, prior to putting on street clothes, and that clean work clothes should be provided daily. Workers not wearing protective equipment and clothing should be restricted from areas of spills or leaks until the completion of cleanup (NIOSH, 1988).

MEDICAL MANAGEMENT

For persons with TCE toxicity, the level of exposure either must be reduced (e.g., by removal from the contaminated environment) or the source must be eliminated. There is no antidote for TCE poisoning. Treatment consists of support for respiratory and cardiovascular functions (ATSDR, 2007, 2011). Victims exposed to TCE vapor only do not pose secondary contamination risks to rescuers. Victims whose clothing or skin is contaminated with liquid TCE can contaminate response personnel secondarily by direct contact or off-gassing of vapor. If TCE penetrates the clothes, they should be removed immediately and the exposed skin and hair should be flushed with water and washed with mild soap and water (ATSDR, 2011; NIOSH, 2010). If TCE gets into the eyes of workers, flush them immediately with plain water or saline for 15–20 min (ATSDR, 2011; NIOSH, 2010). In cases of ingestion, emesis should not be induced. Instead, a slurry of activated charcoal should be administered if the victim is alert, asymptomatic, and has a gag reflex. Although the efficacy of activated charcoal has not been demonstrated for TCE, it may be of assistance, particularly in cases of mixed ingestion (ATSDR, 2011). In cases of respiratory compromise, airway and respiration should be secured via endotracheal intubation. Patients who have bronchospasm may be treated with aerosolized bronchodilators. However, the use of sympathomimetic agents such as epinephrine and isoproterenol could precipitate fatal dysrhythmias and should be avoided (ATSDR, 2011).

The diagnosis of acute TCE toxicity is primarily clinical, based on symptoms of CNS disruption or respiratory distress. However, laboratory testing is useful for monitoring the patient and evaluating complications. Routine laboratory studies for all exposed patients include complete blood count (CBC) and glucose and electrolyte determinations. Additional studies for patients exposed to TCE include hepatic and renal function tests (ATSDR, 2011). Hospitalization should be considered for patients who have had significant inhalation exposure (e.g., with loss of consciousness) or ingested significant amounts of TCE. Development of cardiac dysrhythmia may be delayed for 12–24 h after exposure. Patients who have not experienced alterations in mental status or cardiac dysrhythmia or those who had mild symptoms initially and are asymptomatic 12–24 h later may be discharged. Neurologic examination for posthypoxic injury is recommended in cases of severe exposure (ATSDR, 2011).

REFERENCES

Adams, E.M., Spencer, H.C., Rowe, V.K., et al. (1951) Vapor toxicity of trichloroethylene determined by experiments on laboratory animals. *Arch. Ind. Hyg. Occup. Med.* 4:469–481.

Agency for Toxic Substances and Disease Registry (ATSDR) (1997) *Toxicological Profile for Trichloroethylene, 335 pp.* http://www.atsdr.cdc.gov/toxprofiles/tp19-p.pdf.

Agency for Toxic Substances and Disease Registry (ATSDR) (1997) *Toxicological Profile for Tetrachloroethylene (Update).* 278p., September. Available at http://www.atsdr.cdc.gov/toxprofiles/tp18.pdf. Accessed February 11, 2013.

Agency for Toxic Substances and Disease Registry (ATSDR) (2007) *ATSDR Case Study in Environmental Medicine: Trichloroethylene (TCE) Toxicity,* 59 pp.

Agency for Toxic Substances and Disease Registry (ATSDR) (2011) *Medical Management Guidelines for Trichloroethylene,* 20 pp.

Agency for Toxic Substances and Disease Registry (ATSDR) (2013) "Addendum to the Toxicological Profile for Trichloroethylene." 175p. January. Accessed on January 24, 2013 at http://www.atsdr.cdc.gov/ToxProfiles/tce_addendum.pdf

Alexander, D.D., Kelsh, M.A., Mink, P.J., Mandel, J.H., Basu, R., and Weingart, M. (2007) A meta-analysis of occupational trichloroethylene exposure and liver cancer. *Int. Arch. Occup. Environ. Health* 81:127–143.

American Conference of Governmental Industrial Hygienists (ACGIH) (2007) Trichloroethylene. In: *Documentation of the Threshold Limit Values,* 7th ed., Supplement, Cincinnati, OH: American Conference of Governmental Industrial Hygienists.

American Conference of Governmental Industrial Hygienists (ACGIH) (2012) *TLVs and BEIs: Threshold Limit Values for Chemical Substances and Physical Agents and Biological Exposure Indices,* Publication No. 0112, Cincinnati, OH: American Conference of Governmental Industrial Hygienists, 238 pp.

Anderson, C., Sundberg, K., and Groth, O. (1986) Animal model for assessment of skin irritancy. *Contact Dermatitis* 15:143–151.

Arito, H., Takahashi, M., Sotoyama, M., et al. (1993) Electroencephalographic and autonomic responses to trichloroethylene inhalation in freely moving rats. *Arch. Toxicol.* 67:193–199.

Bakke, B., Stewart, P.A., and Waters, M.A. (2007) Uses of and exposure to trichloroethylene in U.S. industry: a systematic literature review. *J. Occup. Environ. Hyg.* 4:375–390.

Bale, A.S., Barone, S., Scott, C.S., and Cooper, G.S. (2011) A review of potential neurotoxic mechanisms among three chlorinated organic solvents. *Toxicol. Appl. Pharmacol.* 255:113–126.

Barret, L., Torch, S., Leray, C.L., et al. (1992) Morphometric and biochemical studies in trigeminal nerve of rat after trichloroethylene or dichloroacetylene oral administration. *Neurotoxicology* 13:601–614.

Berman, E., Schlicht, M., Moser, V.C., et al. (1995) A multidisciplinary approach to toxicological screening: I. Systemic toxicity. *J. Toxicol. Environ. Health* 45:127–143.

Buben, J.A. and O'Flaherty, E. (1985) Delineation of the role of metabolism in the hepatotoxicity of trichloroethylene and perchloroethylene: a dose-effect study. *Toxicol. Appl. Pharmacol.* 78:105–122.

Centers for Disease Control and Prevention (CDC) (2013) Appendix C. Supplementary exposure limits. In: *NIOSH Pocket Guide to Chemical Hazards.*

Chakrabarti, S.K. and Tuchweber, B. (1988) Studies of acute nephrotoxic potential of trichloroethylene in Fischer-344 rats. *J. Toxicol. Environ. Health* 23:147–158.

Charbotel, B., Fevotte, J., Hours, M., Martin, J.L., and Bergeret, A. (2006) Case-control study on renal cell cancer and occupational exposure to trichloroethylene. Part II: Epidemiological aspects. *Ann. Occup. Hyg.* 50(8): 777–787.

Chia, S.E., Ong, C.N., Tsakok, M.F., and Ho, A. (1996) Semen parameters in workers exposed to trichloroethylene. *Reprod. Toxicol.* 10(4): 295–299.

Chiu, W.A., Jinot, J., Scott, C.S., Makris, S.L., Cooper, G.S., Dzubow, R.C., Bale, A.S., Evans, M.V., Guyton, K.Z., Keshava, N., Lipscomb, J.C., Barone, S., Jr., Fox, J.F., Gwinn, M.R., Schaum, J., and Caldwell, J.C. (2013) Human health effects of trichloroethylene: key findings and scientific issues. *Environ. Health Perspect.* 121:303–311.

Chiu, W.A., Okino, M.S., and Evans, M.V. (2009) Characterizing uncertainty and population variability in the toxicokinetics of trichloroethylene and metabolites in mice, rats, and humans using an updated database, physiologically based pharmacokinetic (PBPK) model, and Bayesian approach. *Toxicol. Appl. Pharmacol.* 241:36–60.

DeFalque, R.J. (1961) Pharmacology and toxicology of trichloroethylene: a critical review of the world literature. *Clin. Pharmacol. Ther.* 2:665–668.

Dhuner, K.G., Nordqvist, P., and Renstrom, B. (1957) Cardiac irregularities in trichloroethylene poisoning. *Acta Anaesthesiol. Scand.* 1:121–135.

Dorfmueller, M.A., Henne, S.P., York, R.G., et al. (1979) Evaluation of teratogenicity and behavioral toxicity with inhalation exposure of maternal rats to trichloroethylene. *Toxicology* 14:153–166.

Elcombe, C.R. (1985) Species differences in carcinogenicity and peroxisome proliferation due to trichloroethylene: a biochemical human hazard assessment. *Arch. Toxicol. Suppl.* 8:6–17.

Evans, M.V., Chiu, W.A., Okino, M.S., and Caldwell, J.C. (2009) Development of an updated PBPK model for trichloroethylene and metabolites in mice, and its application to discern the role of oxidative metabolism in TCE-induced hepatomegaly. *Toxicol. Appl. Pharmacol.* 236:329–340.

Giovanetti, A., Rossi, L., Mancuso, M., et al. (1998) Analysis of lung damage induced by trichloroethylene inhalation in mice fed diets with low, normal, and high copper content. *Toxicol. Pathol.* 26(5):628–635.

Gob, C.L. and Ng, S.K. (1988) A cutaneous manifestation of trichloroethylene toxicity. *Contact Dermatitis* 18:59–61.

Guha, N., Loomis, D., Grosse, Y., Lauby-Secretan, B., El Ghissassi, F., Bouvard, V., Benbrahim-Tallaa, L., Baan, R., Mattock, H., Straif, K. and International Agency for Research on Cancer (IARC) Monograph Working Group (2012) Carcinogenicity of trichloroethylene, tetrachloroethylene, some other chlorinated solvents, and their metabolites. *Lancet Oncol.* 13:1192–1193.

Hack, C.E., Chiu, W.A., Jay Zhao, Q., and Clewell, H.J. (2006) Bayesian population analysis of a harmonized physiologically based pharmacokinetic model of trichloroethylene and its metabolites. *Regul. Toxicol. Pharmacol.* 46:63–83.

Hein, M.J., Waters, M.A., Ruder, A.M., Stenzel, M.R., Blair, A., and Stewart, P.A. (2010) Statistical modeling of occupational chlorinated solvent exposures for case-control studies using a literature-based database. *Ann. Occup. Hyg.* 54:459–472.

Johnson, P.D., Goldberg, S.J., Mays, M.Z., and Dawson, B.V. (2003) Threshold of trichloroethylene contamination in maternal drinking waters affecting fetal heart development in the rat. *Environ. Health Perspect.* 111:289–292.

Jollow, D.J., Bruckner, J.V., McMillan, D.C., Fisher, J.W., Hoel, D.G., and Mohr, L.C. (2009) Trichloroethylene risk assessment: a review and commentary. *Crit. Rev. Toxicol.* 39:782–797.

Kelsh, M.A., Alexander, D.D., Mink, P.J., and Mandel, J.H. (2010) Occupational trichloroethylene exposure and kidney cancer: a meta-analysis. *Epidemiology* 21:95–102.

Kim, S., Kim, D., Pollack, G.M., Collins, L.B., and Rusyn, I. (2009) Pharmacokinetic analysis of trichloroethylene metabolism in male B6C3F1 mice: formation and disposition of trichloroacetic acid, dichloroacetic acid, S-(1,2-dichlorovinyl)glutathione and S-(1,2-dichlorovinyl)-L-cysteine. *Toxicol. Appl. Pharmacol.* 238:90–99.

Kimmel, C.A., Kimmel, G.L., and DeSesso, J.M. (2010) *Comments on the Public Review Draft of EPA's IRIS Toxicological Review for TCE: Developmental Effects.* Submitted to U.S. EPA Science Advisory Board Meeting, May 10, 2010, 11 pp.

Kjellstrand, P., Holmquist, B., Alm, P., et al. (1983) Trichloroethylene: further studies of the effects on body and organ weights and plasma butyrylcholinesterase activity in mice. *Acta Pharmacol. Toxicol. (Copenh.)* 53:375–384.

Kylin, B., Reichard, H., Sumegi, I., et al. (1962) Hepatoxic effect of tri- and tetrachloroethylene on mice. *Nature* 193:395.

Lock, E.A. and Reed, C.J. (2006) Trichloroethylene: mechanisms of renal toxicity and renal cancer and relevance to risk assessment. *Toxicol. Sci.* 91:313–331.

Longley, E.O. and Jones, R. (1963) Acute trichloroethylene narcosis. *Arch. Environ. Health* 7:249–252.

Morreale, S.A. (1976) A case of acute trichloroethylene poisoning with myocardial infarction. *Med. Lav.* 67:176–182.

National Cancer Institute (NCI) (1976) (Bethesda, MD) "Carcinogenesis bioassay of trichloroethylene (CAS No. 79-01-6)." National Technical Information Service (NTIS) NTIS PB-264122; NCI-CG-TR-2. 197p.

National Institute for Occupational Safety and Health (NIOSH) (1988) *Occupational Safety and Health Guideline for Trichloroethylene Potential Human Carcinogen,* 6 pp.

National Institute for Occupational Safety and Health (NIOSH), International Programme on Chemical Safety (IPCS), Commission of the European Communities (2010) *International Chemical Safety Card for Trichloroethylene (CAS No. 79-01-6).*

National Research Council, Committee on Human Health Risks of Trichloroethylene (2006) *Assessing the Human Health Risks of Trichloroethylene: Key Scientific Issues,* Washington, DC: National Academies Press, 444 pp.

Narotsky, M.G., Weller, E.A., Chinchilli, V.M., et al. (1995) Nonadditive developmental toxicity in mixtures of trichloroethylene, di(2-ethylhexyl) phthalate, and heptachlor in a 5×5×5 design. *Fundam. Appl. Toxicol.* 27:203–216.

NHANES III (1997) Written communication from Dr. David Ashley of CDC on levels of VOCs in blood.

NTP (1985) *Trichloroethylene: Reproduction and Fertility Assessment in CD-1 Mice When Administered in the Feed*, Final Report, Research Triangle Park, NC: National Toxicology Program, National Institute of Environmental Health Sciences.

NTP (1988) *Toxicology and Carcinogenesis Studies of Trichloroethylene (CAS No. 79-01-6) in Four Strains of Rats (ACI, August, Marshall, Osborne-Mendel) (Gavage Studies)*, National Toxicology Program—Technical Report Series No. 273, NIH Publication No. 88–2529, Research Triangle Park, NC: U.S. Department of Health and Human Services, Public Health Service, National Institutes of Health.

Odum, J., Foster, J.R., and Green, T. (1992) A mechanism for the development of Clara cell lesions in the mouse lung after exposure to trichloroethylene. *Chem. Biol. Interact.* 83:135–153.

Ramdhan, D.H., Kamijima, M., Yamada, N., et al. (2008) Molecular mechanism of trichloroethylene-induced hepatotoxicity mediated by CYP2E1. *Toxicol. Appl. Pharmacol.* 231(3):300–307.

Ruijten, M.W.W.M., Verberk, M.M., and Salle, H.J.A. (1991) Nerve function in workers with long-term exposure to trichloroethene. *Br. J. Ind. Med.* 48:87–92.

Savolainen, H., Pfaffli, P., Tengen, M., et al. (1977) Trichloroethylene and 1,1,1-trichloroethane: effects on brain and liver after five days intermittent inhalation. *Arch. Toxicol.* 38:229–237.

Scott, C.S. and Jinot, J. (2011) Trichloroethylene and cancer: systematic and quantitative review of epidemiologic evidence for identifying hazards. *Int. J. Environ. Res. Public Health* 8:4238–4272.

Siegel, J., Jones, R.A., Coon, R.A., et al. (1971) Effects on experimental animals of acute, repeated, and continuous inhalation exposures to dichloroacetylene mixtures. *Toxicol. Appl. Pharmacol.* 18:168–174.

Smyth, H.F., Carpenter, C.P., Weil, C.S., et al. (1969) Range-finding toxicity data: list VII. *Am. Ind. Hyg. Assoc. J.* 30(5): 470–476.

Stewart, R.D., Dodd, H.C., Gay, H.H., et al. (1970) Experimental human exposure to trichloroethylene. *Arch. Environ. Health* 20:64–71.

Stott, W.T., Quast, J.F., and Watanabe, P.G. (1982) Pharmacokinetics and macromolecular interactions of trichloroethylene in mice and rats. *Toxicol. Appl. Pharmacol.* 62:137–151.

Todd, J. (1954) Trichloroethylene poisoning with paranoid psychosis and Lilliputian halucination. *Br. Med. J.* 1:439–440.

Tucker, A.N., Sanders, V.M., Barnes, D.W., et al. (1982) Toxicology of trichloroethylene in the mouse. *Toxicol. Appl. Pharmacol.* 62:351–357.

US EPA (2011) *Toxicological Review of Trichloroethylene (CAS No. 79-01-6) in Support of Summary Information on the Integrated Risk Information System (IRIS) (Final)*, EPA/635/R-09/011F.

US EPA (2011b) *Monitor Values Report—Hazardous Air Pollutants.* U.S. Environmental Protection Agency. Available at http://www.epa.gov/oar/data/hapvals.html (accessed August 11, 2011).

Vernon, R.J. and Ferguson, R.K. (1969) Effects of trichloroethylene on visual-motor performance. *Arch. Environ. Health* 18: 894–900.

Watson, R.E., Jacobson, C.F., Williams, A.L., Howard, W.B., and DeSesso, J.M. (2006) Trichloroethylene-contaminated drinking water and congenital heart defects: a critical analysis of the literature. *Reprod. Toxicol.* 21:117–147.

73

ACRYLONITRILE

Jason M. Fritz and April M. Luke

Target organ(s): CNS, respiratory system, eyes, skin, mucous membranes, reproductive system, alimentary canal

Occupational exposure limits: OSHA PEL: TWA 2 ppm (4.3 mg/m^3), STEL: 10 ppm (22 mg/m^3); NIOSH REL: 1 ppm (2.2 mg/m^3), STEL: 10 ppm (22 mg/m^3); ACGIH: TLV: 2 ppm (4.3 mg/m^3)

Reference values: USEPA RfC: 0.002 mg/m^3; USEPA oral cancer slope factor: 0.54 per mg/kg/day; USEPA inhalation unit risk: 0.068 per mg/m^3 (1991); ATSDR acute oral MRL: 0.1 mg/kg/day; ATSDR chronic oral MRL: 0.04 mg/kg/day; ATSDR acute inhalation MRL: 0.2 mg/m^3 (1990); CalEPA chronic inhalation REL: 0.005 mg/m^3 (2001); CalEPA inhalation risk unit: 0.29 per mg/m^3; CalEPA oral slope factor: 1 per mg/kg/day (2009).

Risk phrases: 45, 11, 23/24/25, 37/38, 41, 43, 51/53

May cause cancer; highly flammable; toxic by inhalation, in contact with skin, and if swallowed; irritating to respiratory system and skin; risk of serious damage to eyes; may cause sensitization by skin contact; toxic to aquatic organisms, may cause long-term adverse effects in the aquatic environment

Globally Harmonized System of Classification and Labeling of Chemicals (GHS, Rev.4):

Pictograms—

Signal word—danger

Hazard statements: H225, H301, H311, H315, H317, H318, H331, H335, H2250, H411

Highly flammable liquid and vapor; toxic if swallowed; toxic in contact with skin; causes skin irritation; may cause an allergic skin reaction; causes serious eye damage; toxic if inhaled; may cause respiratory irritation; may cause cancer; toxic to aquatic life with long-lasting effects.

Precautionary statements: P201, P210, P260, P280, P284, P302 + P350, P305 + P351 + P338, P310.

Obtain special instructions before use. Keep away from heat/sparks/open flames/hot surfaces. No smoking. Do not breathe dust/fume/gas/mist/vapors/spray. Wear protective gloves/eye protection/face protection. Wear respiratory protection. *If on skin*: Gently wash with plenty of soap and water. *If in eyes*: Rinse cautiously with water for several minutes. Remove contact lenses, if present and easy to do. Continue rinsing. Immediately call a *poison center* or doctor/physician.

BACKGROUND, USES, AND GENERAL SOURCES OF EXPOSURE

Acrylonitrile (AN; CASRN 107-13-1) is a three-carbon alkene with a nitrile substituent group on carbon 1. Synonyms for acrylonitrile include: acritet; acrylnitril; acrylon; carbacryl; cyanoethene; cyanoethylene; Fumigrain; propenenitrile and 2-propenenitrile; Ventox; and vinyl cyanide. Acrylonitrile is regulated by the U.S. Environmental

Hamilton & Hardy's Industrial Toxicology, Sixth Edition. Edited by Raymond D. Harbison, Marie M. Bourgeois, and Giffe T. Johnson.
© 2015 John Wiley & Sons, Inc. Published 2015 by John Wiley & Sons, Inc.

Protection Agency (USEPA) under the Clean Air Act as a hazardous air pollutant (USEPA, 1995, 2000), and occupational exposure limits have been promulgated by various health agencies (see Section "Industrial Hygiene").

Acrylonitrile is used in the production of acrylic (>85% AN) and modacrylic fibers (35–85% AN), which are both primarily used to manufacture clothing, carpet, and upholstery (IARC, 1999). Acrylonitrile is also extensively utilized to produce thermoplastics, created by mass polymerization or batch emulsion (acrylonitrile butadiene styrene [ABS] and acrylonitrile-styrene trimer [SAN] resins), and the adiponitrile and nitrile rubbers used to create nylon. In industrial processes, acrylonitrile is an intermediate in the production of chemicals such as acrylamide, and in the synthesis of antioxidants, pharmaceuticals, dyes, and carbon fibers. Acrylonitrile is also used in surface coatings and adhesives, as well as for processes requiring the introduction of a cyanoethyl group into a molecule (HSDB, 2013). Acrylonitrile is one of the volatile organic compounds (VOC) emitted by biomass combustion (Christian et al., 2004; Warneke et al., 2011; Yokelson et al., 2007), is also present in cigarette smoke (IARC, 1999), and is found in medical equipment such as dialysis tubing (Mulvihill et al., 1992) and nitrile-butadiene rubber gloves (Wakui et al., 2001). Historically, consumer products made from acrylonitrile-based thermoplastics have included piping, refrigerator liners, carpet, automotive panels, desk-top telephones, luggage, children's building blocks and other toys, mobile homes, and housewares (ATSDR, 1990; USEPA, 1994a, 1994b). Acrylonitrile can be released by off-gassing from thermoplastics, which liberate small quantities of acrylonitrile monomer from ABS plastic in enclosed areas (WHO, 2002). Thermoplastics, including ABS, are also common feedstock used in the newly emerging field of three-dimensional printing, which is currently being employed in a variety of industrial, consumer, and biomedical research applications (Nguyen et al., 2012). Exposure could also occur from migration of residual monomer in polymeric products via contact with food or water, and was previously identified in margarine (ATSDR, 1990). However, acrylonitrile leaching from food or beverage containers is unlikely to be a significant source of exposure as modern food packaging is typically not made with acrylonitrile-based plastics, and acrylonitrile was not detected in U.S. foods tested from 1991 to 2004 (FDA, 2006). Acrylonitrile-containing fumigants were used for stored grain (fumigrain) up until the 1970s, and have been historically used as residential de-lousing agents (Ventox) (ATSDR, 1990; Grunske, 1949).

Acrylonitrile is a high production volume chemical. From 1987 to 1995 annual production volume increased from 2 to 3 billion pounds in the United States alone (ATSDR, 1990; NTP, 2011; WHO, 2002), while 2012 production levels are in the range of 3–5 billion pounds (USEPA, 2012).

TABLE 73.1 Chemical and Physical Properties of Acrylonitrile

Characteristic	Information
Chemical formula	C_3H_3N
Molecular weight	53.063
Melting point	$-83\,°C$
Boiling point	$77.4\,°C$
Density	0.806 g/ml at 20 °C
Log octanol/water partition coefficient (log K_{OW}, mean [range])	0.25 [−0.92–1.2]
Log organic carbon/water partition coefficient (log K_{OC})	−0.07
Vapor pressure	100 mmHg at 22.8 °C
Henry's law constant	8.8×10^{-5} atm-m^3/mol
Water solubility	7.9×10^4 mg/l
Flashpoint	$-1\,°C$ (closed cap)
Explosion limits	3–17%
Odor threshold	47 mg/m^3 (air)
Conversion factors	1 ppm = 2.17 mg/m^3 1 mg/m^3 = 0.46 ppm

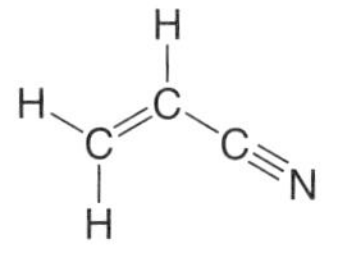

FIGURE 73.1 Chemical structure of acrylonitrile.

PHYSICAL AND CHEMICAL PROPERTIES

Acrylonitrile is a colorless liquid with a pungent onion- or garlic-like odor. It is soluble in water and miscible with most organic solvents. At room temperature, acrylonitrile is flammable and volatile, with cyanide gas formed explosively upon vapor reaction. The chemical structure of acrylonitrile is diagrammed in Figure 73.1, and select chemical and physical properties are presented in Table 73.1 (ATSDR, 1990; IPCS, 1983).

ENVIRONMENTAL FATE AND BIOACCUMULATION

Monitoring data for acrylonitrile in water and soil are limited. Acrylonitrile is soluble in water but will evaporate rapidly. In the atmosphere, acrylonitrile has a half-life of 55–96 h (range = 4–198) and is readily oxidized to formaldehyde, formic acid, formyl cyanide, and hydrogen cyanide. In soil, acrylonitrile may degrade due to microbe metabolism, although this occurs more slowly, with a soil half-life of 170 h (range = 30–552); bacterial decomposition in water occurs at a rate similar to that in soil, in the absence of evaporation. In the absence of bacterial metabolism or evaporation, acrylonitrile has a half-life of >13 years. Acrylonitrile is not anticipated to bioaccumulate in either plants or animals, given the mean log K_{OW} of 0.25 (range −0.92 to 1.2), and log bioconcentration factor (log BCF) of 0 estimated from water solubility (USEPA, 1994b; WHO, 2002).

ECOTOXICOLOGY

The calculated 96-h acute toxicity limit (96-h LC_{50}) for fish is 9.3–18.1 mg/l; the 48 h value (48 h LC_{50}) for aquatic invertebrates is 7.6–8.7 mg/l. The chronic (30 day) median tolerance limit (TLm) for fathead minnows (*Pimephales promelas*), bluegill sunfish (*Lepomis macrochirus*) and guppies (*Lebistes reticulates*) is 2.6 mg/l; the corresponding 21-day tolerance limit for reproduction in the freshwater water flea (*Daphnia magna*) is 0.71 mg/l (Hadley, 1994; USEPA, 2000). The 96 h growth inhibiting concentration (96 h EC_{50}) for duckweed (*Lemna minor*), a freshwater aquatic plant, is 1.4 mg/l (USEPA, 1994b; WHO, 2002).

MAMMALIAN TOXICOLOGY AND CHEMICAL PATHOLOGY

Acute Effects

In industrial settings, acute exposure to acrylonitrile occurs through dermal or inhalation routes, and such exposures have been associated with human health effects. Reports in the 1940s of workers in synthetic rubber manufacturing plants, who were exposed to acrylonitrile at concentrations between 16 and 100 parts per million (ppm) (34–217 mg/m^3) for 20–45 min, included complaints of dull headache, fullness in the chest, irritation of mucous membranes (e.g., eye, nose, throat), mild jaundice associated with low grade anemia, nausea, vomiting, fatigue, and subjective feelings of apprehension and nervous irritability (Wilson et al., 1948). A case report of a man accidentally sprayed with acrylonitrile in the face, eyes, and body reported symptoms of dizziness, flushing, and nausea with vomiting as well as erythema, mild conjunctivitis, and tachycardia (Vogel and Kirkendall, 1984). Chen et al. (1999) reported on 144 acute acrylonitrile poisoning cases that occurred in industrial settings in China between 1977 and 1994. Although reliable data on exposure levels were limited, 60 cases involved exposures between 18 and 258 ppm (39–560 mg/m^3) and the remaining cases may have involved exposures >461 ppm (1000 mg/m^3). Symptoms of dizziness, headache, feebleness, and chest tightness were reported in all cases. Other frequently reported symptoms included dyspnea, abdominal pain, pallor, congestion of pharynx, vomiting, and sore throat. Dermal exposure has been as associated with skin irritation, erythema, painful itching, blistering, and allergic dermatitis (Bakker et al., 1991; Wilson et al., 1948; Zeller et al., 1969).

Chronic Effects

Noncancer

Humans Most of the information on chronic acrylonitrile exposure comes from occupational studies of Chinese and Japanese workers. Increases in subjective symptoms, deficits in tests of neurobehavior, and increased adverse reproductive outcomes have been reported with chronic acrylonitrile exposure in occupational settings.

Occupational studies from chemical factory workers in China and Japan reported an increase in the prevalence of subjective symptoms in acrylonitrile-exposed workers (Chen et al., 2000; Dong et al., 2000; Kaneko and Omae, 1992; Muto et al., 1992). A broad range of subjective symptoms were reported. General information on the study populations and their exposures from representative occupational studies are listed in Table 73.2.

Neurobehavioral performance was evaluated in workers from the acrylonitrile-monomer department or the acrylic fibers department of a Chinese plant using a WHO-recommended neurobehavioral core test battery (NCTB) (Rongzhu et al., 2005). Investigators reported significant deficits in tests of negative mood (e.g., anger, confusion, depression, fatigue, and tension), attention and response speed, auditory memory, and motor steadiness with acrylonitrile exposure. Significant effects were not observed on scores of manual dexterity or perceptual motor speed. Short-term area sampling indicated a geometric mean acrylonitrile concentration of 0.11 ppm (0.2 mg/m^3) for the monomer department and 0.91 ppm (2 mg/m^3) for the acrylic fibers department. A cross-sectional health questionnaire study of Japanese acrylic fiber workers reported an increased prevalence of neurosis in acrylonitrile-exposed workers (using the Japanese version of the Cornell Medical index), although the increase was not statistically significant compared to unexposed workers (Kaneko and Omae, 1992). An average employment duration of the exposed workers was 5.6–8.6 years and exposure concentrations ranged from 1.8 ppm (4 mg/m^3) to 14.1 ppm (31 mg/m^3).

The following reproductive effects have been noted among female acrylonitrile workers: total birth defects, still births, premature delivery, pregnancy complications, and sterility (Collins et al., 2003). Increased incidences of adverse reproductive outcomes (e.g., stillbirths, spontaneous abortions, birth defects, sterility) were also reported in female spouses of male acrylonitrile workers. Reported area measurements of acrylonitrile from these studies ranged from 0.4 ppm (0.9 mg/m^3) to 42.9 ppm (93 mg/m^3); personal sample measurements were not available. Xu et al. (2003) reported decreased sperm count and significant increase in DNA damage and/or aneuploidy in sperm from male workers exposed to a mean concentration of 0.37 ppm (0.8 mg/m^3) for an average duration of 2.8 years.

TABLE 73.2 Subjective Symptoms Reported in Occupational Studies

Symptoms	Exposure	Study Design and Population
Headache, dizziness, poor memory, choking feeling in chest, loss of appetite	Area samples; average concentration reported: 0.48 ppm (1 mg/m^3)	Cross-sectional study; Chinese male and female workers from an acrylic fiber factory with an average employment duration of 13 years (Chen et al., 2000)
Headaches, dizziness, choking in the chest	Area samples; a 3-year range of concentrations reported: 0.92 ppm (2 mg/m^3)–1.3 ppm (2.8 mg/m^3)	Occupational survey; Chinese male workers from a chemical fiber factory; no information on employment duration was available (Dong et al., 2000)
Headaches, tongue trouble, choking lump in chest, fatigue, general malaise, heavy arms, heavy sweating	Area samples; factories were categorized based on concentrations, category means reported: Low—1.8 ppm (4 mg/m^3) Medium—7.4 ppm (16 mg/m^3) High—14.1 ppm (31 mg/m^3)	Cross-sectional study; Japanese male workers from seven acrylic fiber factories with employment duration ranging from 5.6 to 8.6 years (Kaneko and Omae, 1992)
Heaviness of the stomach, decreased libido, poor memory, irritability	Personal samples; factories were categorized based on concentrations, category means reported: Low—0.19 ppm (0.4 mg/m^3) High—1.13 ppm (2.5 mg/m^3)	Cross-sectional study; Japanese male workers from seven acrylic fiber factories with an average employment duration 17 years (Muto et al., 1992)

Rodents Neurological and reproductive effects have also been observed in rodent studies. Decrements in peripheral nerve function in male Sprague-Dawley rats were associated with inhalation exposure to acrylonitrile at concentrations ≥54 mg/m^3 for 24 weeks (Gagnaire et al., 1998). A concentration-dependent deficit in sensory conduction velocity and amplitude of tail nerve sensory action potential were reported. Transient deficits were also observed in tail nerve motor conduction velocity. Furthermore, these effects were consistent with results reported following 12 weeks of oral (gavage) exposure to >12.5 mg/kg/day (Gagnaire et al., 1998).

Measures of reproductive performance (e.g., mating index, fertility index, number of implantation sites per dam, number of resorptions per litter) in rats were largely unaffected by acrylonitrile exposure in two multigenerational reproductive toxicity studies (Friedman and Beliles, 2002; Nemec et al., 2008). Female B6C3F$_1$ mice exposed to ≥10 mg/kg/day acrylonitrile for 2 years were reported to have an increase in ovarian atrophy and cysts compared to controls (Ghanayem et al., 2002; NTP, 2001). Degenerative changes of the seminiferous tubules associated with a decrease in sperm count were reported in male CD-1 mice exposed to 10 mg/kg/day for 60 days (Tandon et al., 1988). These same effects were not observed in male B6C3F$_1$ mice exposed to ≥5 mg/kg/day acrylonitrile for 14 weeks or 2 years (Ghanayem et al., 2002; NTP, 2001). Increased malformations (e.g., short tail, short trunk, miss vertebrae) were observed in fetuses of female Sprague-Dawley rats exposed to maternally toxic levels of acrylonitrile by gavage or inhalation (i.e., 65 mg/kg/day and 174 mg/m^3) on GD 6–15 (Murray et al., 1978). However, similar malformations were not reported in a separate study where female Sprague-Dawley rats were exposed to acrylonitrile concentrations up to 217 mg/m^3 on GD 6–20 (Saillenfait et al., 1993).

Carcinogenicity

Humans The most informative epidemiologic literature pertaining to human cancer risk consists of cohorts from the United States, U.K., Germany, and the Netherlands (Benn and Osborne, 1998; Blair et al., 1998; Swaen et al., 2004; Symons et al., 2008; Thiess et al., 1980), as well a European case-control study (Scélo et al., 2004); together, these reports follow the mortality experience of over 35,000 workers. The results of these and other cohort studies suggest that the association between occupational acrylonitrile exposure and lung cancer mortality is not strong, and statistically significant associations are not consistently evident (IARC, 1999). Cancer of the respiratory system is the endpoint most frequently evaluated as possibly resulting from occupational exposure to acrylonitrile; more specific information regarding lung cancer histological subtype (i.e., small-cell versus non-small cell) is not available. Less information is available relating acrylonitrile exposure to mortality from other cancer types due to the small number of site-specific cancer deaths in individual reports. A meta-analysis of brain cancer incidence from 12 cohort studies found no increased risk in human workers exposed to acrylonitrile (Collins and Strother, 1999).

The U.S. National Cancer Institute (NCI) performed an epidemiology study with the largest population (Blair et al., 1998), although at time of follow-up this cohort was still young (median age 48 years, mean follow-up of 21 years), and had experienced few cancer deaths (88% overall survival). In this 8-plant cohort study that followed over 25,000 acrylonitrile workers, there was a twofold increased risk of dying from lung cancer (relative risk: 2.1, 95% confidence interval (CI) 1.2–3.8), among those with longest exposure duration (≥20 years) and highest exposure levels (≥8 ppm-years). The positive confounding effect of smoking on lung cancer incidence was ruled out using an internal control group of unexposed workers (Blair et al., 1998). While workers in two middle exposure and duration groups (11–19 years since first exposure, exposed to 0.13–0.57 or 0.57–1.5 ppm-years) also experienced a twofold increased lung cancer risk, no trend was apparent across the quintiles of exposure. Despite an association with lung cancer in the highest exposure/longest duration group, no significant association was reported with continuous exposure metrics, consistent with later re-evaluations by other investigators (Marsh et al., 2001; Starr et al., 2004). Smaller cohorts from the U.K. (~2700 workers; Benn and Osborne (1998)) and Germany (~1500 workers; Thiess et al. (1980)) also reported increased lung cancer mortality by stratified analysis, but no specific exposure monitoring was performed, and the possibility of confounding by smoking or exposure to other chemicals was not addressed. A multi-industry case-control study of 3000 lung cancer cases also found that European lung cancer patients were twice as likely to have had occupational acrylonitrile exposure versus controls (odds ratio: 2.2, 95% CI 1.1–4.4), an association that remained even after adjusting for several confounding factors such as smoking or exposure to other potential carcinogens (Scélo et al., 2004). Conversely, Dutch (~2800 workers; Swaen et al. (2004)) and other U.S. cohorts (~2500 workers; Symons et al. (2008)) with 30–40 years mean follow-up reported no significant association between lung cancer mortality and acrylonitrile exposure by any metric at exposures up to 31.5 ppm-years, an exposure level equal to or greater than that experienced by the NCI cohort (Blair et al., 1998).

Rodents The most informative experimental literature pertaining to cancer hazard consists of 2-year rodent bioassays performed in Sprague-Dawley (Johannsen and Levinskas, 2002a; Maltoni et al., 1988; Quast, 2002; Quast et al., 1980) and F344 rats (Bigner et al., 1986; Johannsen and Levinskas, 2002b), as well as B6C3F$_1$ mice (Ghanayem et al., 2002; NTP, 2001). Acrylonitrile exposure consistently induced more malignant epithelial tumors (i.e., carcinomas) at multiple tissue sites in Sprague-Dawley rats of both sexes following chronic oral or inhalation exposure, as well as in F344 rats of both sexes following chronic oral administration (IARC, 1999; NTP, 2011). Combining results across several studies, carcinogenesis in Sprague-Dawley rats has been evaluated following inhalation exposure to 43–174 mg/m^3 (6 h/day, 5 days/week for 2 years), or oral administration (primarily via drinking water) of 0.1–22 mg acrylonitrile/kg/day (7 days/week for 2 years); cancer induction in F344 rats has also been evaluated following similar drinking water concentrations.

Male and female rats of both strains developed significantly more central nervous system (CNS) tumors following chronic oral exposure to acrylonitrile at ≥3 mg/kg/day (Bigner et al., 1986; Johannsen and Levinskas, 2002a, 2002b; Quast, 2002), or inhalation exposure at ≥43 mg/m^3 (Maltoni et al., 1988; Quast et al., 1980) when compared with controls. The CNS tumors were primarily located in the cerebral cortex, and also involved the brain stem, cerebellum, and spinal cord (Bigner et al., 1986; Quast, 2002; Quast et al., 1980). While these tumors were frequently characterized as astrocytomas, recent evaluations of immune cell surface marker expression suggest that these tumors may arise from or involve heterotypic cell fusion with brain microglia and/or macrophage cells, and thus, may be more accurately classified as malignant microgliomas (Kolenda-Roberts et al., 2013; Nagatani et al., 2009). In fact, gliosis in the brain with or without perivascular cuffing was also reported in rats at these exposure levels (Quast, 2002; Quast et al., 1980). A histogenic origin similar to this may contribute to mouse CNS tumors (Huysentruyt and Seyfried, 2010) and human glioblastoma multiforme as well (Pawelek and Chakraborty, 2008; Zheng et al., 2012).

The National Toxicology Program (NTP) conducted an evaluation of chronic toxicity and carcinogenesis in B6C3F$_1$ mice of both sexes exposed to acrylonitrile via gavage at 2.5, 10, or 20 mg/kg/day (Ghanayem et al., 2002; NTP, 2001), and concluded that there was "clear evidence of carcinogenic activity" based on treatment-related increases in forestomach and harderian gland tumor incidence in both sexes. While acrylonitrile exposure induced tumorigenesis in multiple organs in both rats and mice, the only sensitive tissue common to both rodent species was the forestomach. Increased squamous cell papilloma or carcinoma incidence was observed in conjunction with increased forestomach hyperplasia following oral administration in Sprague-Dawley and F344 rats, as well as in B6C3F$_1$ mice. While the extent of forestomach tumor induction was similar between Sprague-Dawley rats and B6C3F$_1$ mice administered ≥10 mg/kg/day, the hyperplastic response was diminished in mice compared to the robust hyperplasia induced in rats (Ghanayem et al., 2002; Johannsen and Levinskas, 2002a; NTP, 2001; Quast, 2002).

Acrylonitrile exposure induces adenoma and carcinoma formation in the harderian gland of mice at ≥3 mg/kg/day and the Zymbal gland of rats at ≥1 (F344) or 5 (Sprague-Dawley) mg/kg/day. Papillomas or carcinomas of the tongue and intestinal carcinomas were reported in both male and female Sprague-Dawley rats orally exposed to ≥10 mg/kg/day and males exposed via inhalation to 174 mg/m^3. In female Sprague-Dawley rats, the incidence of benign (adenomas, fibromas, fibroadenomas, or adenofibromas) and malignant mammary gland tumors was not consistently elevated when combined; when considered separately, however, malignant carcinoma incidence was elevated following acrylonitrile exposure (Johannsen and Levinskas, 2002a, 2002b; Quast, 2002; Quast et al., 1980). In female mice, the incidence of benign or malignant granulosa cell tumors increased slightly and only at 10 mg/kg/day, an effect which may have been treatment related (Ghanayem et al., 2002; NTP, 2001).

While lung cancer has been associated with acrylonitrile exposure in some human workers, the evidence for acrylonitrile-induced lung tumors in rodents is more limited. Acrylonitrile does not induce lung tumors in rats. In female mice, acrylonitrile induced twice as many lung adenomas or carcinomas when administered at 10 mg/kg/day versus controls (Ghanayem et al., 2002; NTP, 2001). Although lung tumor incidence was not statistically significantly elevated at the highest dose evaluated, a significant dose-response trend ($p = 0.029$) was observed (Ghanayem et al., 2002; NTP, 2001). Lung tumor incidence was not significantly elevated in male mice at any dose evaluated, and the incidence of non-neoplastic lung lesions was not elevated in either sex.

Inhalation exposure has only been evaluated in rats, where carcinomas of the nasal turbinate epithelium were observed in two female Sprague-Dawley rats exposed to 174 mg/m³, and were not observed in rats administered acrylonitrile via drinking water at any dose (Quast, 2002; Quast et al., 1980). The incidence of inflammatory or degenerative lesions of the nasal epithelium (e.g., respiratory epithelial hyperplasia, squamous metaplasia, erosion of the mucous lining) was also elevated in both male and female rats exposed to ≥43 mg/m³, while similar lesions were not induced following oral exposure (Johannsen and Levinskas, 2002a, 2002b; Nemec et al., 2008). See Table 73.3 for an overview of rodent tumorigenesis as a function of species, strain, sex, and acrylonitrile exposure level.

Some evidence suggests an association between *in utero* exposure and increased susceptibility to acrylonitrile-induced tumor incidence in Sprague-Dawley rats. Lifetime inhalation exposure to 130 mg/m³ acrylonitrile beginning *in utero* at GD12 increased CNS tumor incidence in female offspring when compared to adult dams (Maltoni et al., 1988). A similar observation regarding increased CNS and Zymbal gland incidence was reported in female pups exposed to acrylonitrile trans-placentally *in utero* throughout gestation via orally exposed dams; both dams exposed as adults and *in utero*-exposed pups were orally administered 40 mg/kg/day acrylonitrile for up to 11 months (Friedman and Beliles, 2002).

Hypothesized Modes of Action

Acrylonitrile can be oxidized via cytochrome P450 2E1 (CYP2E1) to 2-cyanoethylene oxide, which can be hydrolyzed to cyanide as one of several possible downstream products of eukaryotic metabolism. Alternatively, glutathione conjugation or sulfhydryl reaction with the parent acrylonitrile compound will circumvent this metabolic route (NTP, 2001; Thier et al., 2000). Cyanide can be detected

TABLE 73.3 Rodent Tissue Tumorigenesis as a Function of Acrylonitrile Exposure

			Oral Administration (mg/kg/day)[a]						
Species	Strain	Sex[b]	0.1	0.3	1	3	5	10	22
Rat	Sprague-Dawley	M	–			cn[c]		cn, zg, fs, gi	cn, zg, fs, tg, gi
		F	–				cn, zg	cn, mg, zg, fs, gi	cn, mg, zg, fs, tg, gi
	F344	M	–	fs	fs	cn, zg, fs		cn, zg, *fs*	
		F	–	–	zg	cn, zg, fs		cn, zg, *fs*	
Mouse	B6C3F₁	M				hg		hg, fs	hg, fs
		F				–		hg, fs, lu, ov	hg, fs, *lu*

			Inhalation Exposure (mg/m³)[d]		
Species	Strain	Sex[b]	43	130[e]	174
Rat	Sprague-Dawley	M	–	cn, zg	cn, zg, fs, tg, gi
		F	cn	cn, mg	cn, mg, zg, nc

[a]Doses in drinking water converted to mg/kg/day by study authors or using study-reported mean rat weights, when presented. Daily intake doses are approximate as results are combined from several studies across multiple investigators; reported daily doses were generally ±20% of presented values. Treatment was administered 7 day/week for approximately 2 years. Empty fields represent exposure levels not evaluated; fields with "–" represent exposures evaluated that failed to elicit treatment-related tumorigenesis by pair-wise comparison as reported by study authors.

[b]Male = M, female = F.

[c]Tissues listed with statistically significantly elevated tumor incidence following acrylonitrile exposure, or otherwise described as biologically relevant and treatment-related by study authors. Abbreviated as follows: central nervous system (cn), mammary gland carcinoma (mg), zymbal gland (zg), forestomach (fs), tongue (tg), non-forestomach gastrointestinal tract (gi), harderian gland (hg), lung (lu), ovaries (ov), and nasal turbinate carcinoma (nc). In high-dose groups, tissue abbreviations in *italics* represent tumor incidences that were elevated above controls but not in a statistically significant manner, possibly due to higher mortality rate observed.

[d]Rodents were exposed for approximately 6 h/day, 5 days/week for 2 years or throughout natural lifespan.

[e]Includes data for pups exposed starting from GD12 *in utero*, and continued throughout natural lifespan; adult-only exposed males or nulliparous females were not evaluated.

in the blood and tissues of rodents experimentally administered acrylonitrile, and in the blood and urine of workers with reported dermal or inhalation acrylonitrile exposure (IARC, 1999; Kedderis et al., 1993; Major et al., 1998; Thier et al., 1999). Consistent with this observed cyanide release *in vivo*, acute exposure to acrylonitrile when used as a residential fumigant resulted in classic symptoms of cyanosis, leading to convulsions and death in exposed children (Grunske, 1949). In adults, acrylonitrile intoxication resulting from accidental overexposure can be attenuated by rapid administration of N-acetylcysteine with sodium thiosulfate, highlighting the importance of cysteine conjugation in acrylonitrile detoxification (Thier et al., 1999).

While conjugation with the free sulfhydryl groups present in therapeutic interventions can alleviate acrylonitrile toxicity by preventing cyanide generation, direct reactivity with endogenous substrates, such as reduced thiol groups in cellular proteins, contributes to the pathology observed in laboratory animals acutely exposed to lethal acrylonitrile concentrations. Seizures following initial exposure to high acrylonitrile levels were possibly mediated by cyanide release, leading to inhibition of blood oxygenation, oxidative metabolism, and mitochondrial respiration. However, the severe clonic convulsions preceding death were more likely due to acrylonitrile-protein adduction versus cyanide effects (Benz and Nerland, 2005). Brain ATP levels were not significantly decreased in rats administered an acutely toxic acrylonitrile dose, suggesting that death did not result from acute inhibition of brain metabolic activity (Campian and Benz, 2008). Furthermore, inhibiting metabolism of acrylonitrile to 2-cyanoethylene oxide by blocking CYP2E1 activity did not prevent the death of rats following acute high dose acrylonitrile administration, suggesting that acrylonitrile adduction of blood hemoglobin or other proteins may drive acute lethality (Benz and Nerland, 2005; Benz et al., 1997). Acrylonitrile exposure can be monitored by evaluating the blood levels of several such adduct species, including *N*-(2-cyanoethyl) valine-hemoglobin (Benz et al., 1997; Thier et al., 2002).

Acrylonitrile exhibits genotoxic potential and directly alkylates DNA in biochemical assays (Solomon and Segal, 1985), processes that may play a role in the mode of action for acrylonitrile carcinogenicity. However, a consistent mutagenic response in bacterial or mammalian bioassays is only observed in the presence of cytochrome P450 activity, suggesting that oxidative metabolism is a required step in acrylonitrile-induced carcinogenesis. Following bioactivation, acrylonitrile is mutagenic in *Escherichia coli* or *Salmonella* systems sensitive to point mutations, as well as in yeast and mammalian bioassays evaluating clastogenic or aneugenic endpoints *in vitro* (IARC, 1999); forward mutations are induced in human and rodent lymphocytes *in vitro*, and likewise in mouse splenocytes *in vivo* (Ghanayem and Hoffler, 2007). Induction of chromosomal aberrations (CA), sister-chromatid exchanges (SCE) and micronuclei

(MN) in rodent cells is less consistently observed *in vivo*, with changes of low magnitude reported either as negative (IARC, 1999; NTP, 2011), inconclusive (Morita et al., 1997), or positive (Fahmy, 1999; Wakata et al., 1998). Early evaluations of CAs and SCEs in acrylonitrile-exposed workers were similarly negative (Thiess and Fleig, 1978) or inconclusive (Borba et al., 1996), while more recent examinations report some MN induction in buccal mucosa and peripheral blood cells (Fan et al., 2006), and sperm aneuploidy detected by fluorescent in situ hybridization (Xu et al., 2003). In addition to genotoxicity, acrylonitrile exhibits the potential to induce oxidative stress in various cell types *in vitro* (Esmat et al., 2007; Jacob and Ahmed, 2003; Kamendulis et al., 1999), and in the CNS of rats *in vivo* (Whysner et al., 1998). Furthermore, Pu et al. (2009) reported increased levels of oxidative DNA damage in the brains of rats that correlated with the increased rat brain tumor incidence previously observed by Quast (2002) following chronic acrylonitrile exposure at similar doses.

Although acrylonitrile (or bioactive metabolites) adducts proteins, potentially suppresses both oxygen transport to and utilization in respiring tissues, displays genotoxic potential in a variety of assays and rodent tissues, and induces DNA oxidation in the rodent CNS, no comprehensive toxicity pathway has been established for any of the cancer or noncancer health effects associated with chronic acrylonitrile exposure.

POTENTIAL EXPOSURES

Historically, all environmental exposures were thought to be the result of anthropogenic release into air, water, land or groundwater, as there are no known chemical reactions that could lead to *in situ* acrylonitrile formation in the atmosphere (Grosjean, 1990). More recently, however, acrylonitrile has been detected in the smoke plume generated from biomass combustion in both controlled laboratory burns of North American biomass samples (Christian et al., 2004; Warneke et al., 2011), and South American Amazon rainforest fire emissions monitored by aircraft-mounted instrumentation (Yokelson et al., 2007), suggesting that acrylonitrile is a byproduct of natural biomass combustion. Acrylonitrile is rapidly absorbed into human systemic circulation following ingestion, inhalation, or dermal contact (IARC, 1999; WHO, 2002).

General Population

For the general public, air is likely to be the principal medium of acrylonitrile exposure, although exposure to low levels of acrylonitrile could result from direct contact with acrylic products (ATSDR, 1990; WHO, 2002). Inhalation exposure to acrylonitrile is most likely to occur among members of the general population living in the vicinity of regulated emission sources such as acrylic fiber or chemical manufacturing plants, waste sites (ATSDR, 1990), downwind of rubber production facilities (WHO, 2002), or near major populations centers (USEPA, 1994b, 2005), although the specific emission sources that contribute to urban acrylonitrile levels remain unclear. In addition to anthropogenic sources, smoke generated by combustion of plant biomass (e.g., forest, brush, or grass fires) contains numerous VOCs, including acrylonitrile (Christian et al., 2004; Warneke et al., 2011), which has a sufficiently long half-life in the atmosphere such that it could be dispersed with other combustion products as dictated by prevailing weather conditions, and could contribute to exposure in both urban and rural populations (Karl et al., 2003). Smokers are further exposed to acrylonitrile that is present in the vapor phase of sidestream smoke at 99 µg/U.S. cigarette (Miller et al., 1998) and mainstream smoke at a concentration of 3–19 µg/cigarette, depending on country of origin, brand, and nicotine yield (IARC, 1999; Laugesen and Fowles, 2005). Consequently, indoor air in the residences of smokers could be another source of potential exposure (to both smokers and nonsmokers), as acrylonitrile levels were reported to be 0.5–1.2 µg/m^3, similar to or higher than the U.S. mean urban ambient air levels of 0.7 µg/m^3, and several orders of magnitude higher than the nationwide mean estimates of 0.001 µg/m^3 (Kelly et al., 1993; Nazaroff and Singer, 2004; USEPA, 1996, 2005). Fetal exposure in any environment could occur through transplacental absorption, which has been reported in rodents and observed in pregnant human smokers (Tavares et al., 1996). Precise environmental acrylonitrile concentrations are often difficult to ascertain with a high degree of certainty, as measurements are frequently below the detection level achieved by current technology (McCarthy et al., 2009; Yokelson et al., 2007).

Occupational

Inhalation exposure to acrylonitrile may occur during its manufacture and production and in factories where it is used to create other products, where exposure levels can be the highest (ATSDR, 1990; WHO, 2002). Occupational levels of acrylonitrile exposure are likely to be several orders of magnitude greater than those experienced by the general population, as typical workplace air concentrations are reported to range from 0.1 to 4 mg/m^3 (ATSDR, 1990; IARC, 1999). Occupations with potential for exposure include acrylic resin, rubber, synthetic fiber, and textile manufacturers; synthetic organic chemists; and pesticide workers. The primary routes of acrylonitrile exposure in these and other occupational environments are inhalation and possibly dermal. From 1983 to 1996 the estimated number of U.S. workers potentially exposed to acrylonitrile increased by more than 50%, from 51,000 to 80,000, corresponding with increased acrylonitrile production and consumption during the same time frame (Boiano and Hull, 2001; NIOSH, 1990).

INDUSTRIAL HYGIENE

The American Conference of Governmental Industrial Hygienists (ACGIH) classifies acrylonitrile as a class A3 carcinogen (confirmed animal carcinogen with unknown relevance to humans) (ACGIH, 2001). The National Institute of Occupational Safety and Health (NIOSH) recommends that acrylonitrile be controlled and handled as a potential human carcinogen (NIOSH, 1988). Table 73.4 lists the permissible and recommended exposure limits for acrylonitrile.

The Occupational Safety and Health Administration (OSHA) recommends that engineering and workplace controls to be put into effect to limit employee exposure to acrylonitrile, and when these controls are insufficient to reduce exposure to or below the permissible exposure limit (PEL), employees must be supplied with respiratory protection. Recommendations on respiratory protection for workers exposed to acrylonitrile are listed in Table 73.5.

TABLE 73.4 Permissible and Recommended Occupational Exposure Limits for Acrylonitrile

Organization	Exposure Limit	Concentration	Exposure Period
Occupational Safety and Health Administration (OSHA, 2001)	PEL	2 ppm (4.3 mg/m^3)	TWA for an 8 h workday
	STEL	10 ppm (22 mg/m^3)	Up to 15 min
National Institute of Occupational Safety and Health (NIOSH, 1988)	REL	1 ppm (2.2 mg/m^3)	TWA for a 10 h workday and a 40 h workweek
	STEL	10 ppm (22 mg/m^3)	Up to 15 min
	IDLH	85 ppm (184.5 mg/m^3)	NA
American Conference of Governmental Industrial Hygienists (ACGIH, 2001)	TLV	2 ppm (4.3 mg/m^3)	TWA for an 8 h workday and a 40 h workweek

IDLH—Immediate Danger to Life and Health; PEL—permissible exposure limit; REL—recommended exposure limit; STEL—short-term exposure limit; TLV—threshold limit value, TWA—time weighted average; NA—not applicable.

TABLE 73.5   Recommended Respiratory Protection for Workers Exposed to Acrylonitrile

Condition/airborne concentration	Respiratory protection
Up to 20 ppm (10 × PEL)	A respirator with half-mask facepiece that is either: supplied-air; or equipped with organic vapor cartridges.
Up to 100 ppm (50 × PEL)	A respirator with full facepiece that is either: supplied-air; equipped with organic vapor cartridges; equipped with chin style, organic vapor gas mask; equipped with front- or back-mounted, organic vapor gas mask; or a self-contained breathing apparatus.
Up to 4000 ppm (2000 × PEL)	A supplied-air respirator with full facepiece, helmet, suit, or hood operated in positive-pressure mode.
Above 4000 ppm or unknown concentration	A respirator with full facepiece that is either: supplied-air and auxiliary self-contained breathing apparatus operated in positive-pressure mode; or self-contained breathing apparatus operated in positive-pressure mode
Firefighting	A self-contained breathing apparatus with full facepiece operated in positive-pressure mode.
Escape	Any organic vapor respirator or self-contained breathing apparatus.

Source: Adapted from OSHA guidelines for acrylonitrile (OSHA, 2012).

MEDICAL MANAGEMENT

Guidance regarding the medical management for acute acrylonitrile exposure can be found on the Hazardous Substances Data Bank and in the Medical Management Guidelines developed by Agency for Toxic Substances and Disease Registry (ATSDR) (ATSDR, 2013). General standards of care to be taken are as follows: (1) seek immediate medical help; (2) if the victim's clothing is contaminated with acrylonitrile, remove the contaminated clothing and move the victim to fresh air; (3) administer oxygen as needed, and if the victim is not breathing on their own, perform artificial respiration; (4) if the patient is unconscious, administer a cyanide antidote until the victim improves or medical response arrives; (5) do not induce vomiting following ingestion and administer a slurry of activated charcoal if the victim is not symptomatic; (6) flush exposed skin for 2–3 min and wash with mild soap; and (7) flush exposed eyes with water for at least 15 min.

RISK ASSESSMENTS

The regulatory levels and recommended values determined for acrylonitrile by various government agencies and organizations are listed below, and summarized in Table 73.6.

Agency for Toxic Substances and Disease Registry

The ATSDR developed acute values for inhalation and oral routes of exposure and a chronic value for oral exposure (ATSDR, 1990). The acute inhalation minimal risk level (MRL) for acrylonitrile was determined as 0.1 ppm (0.2 mg/m^3) based on the lack of significant neurological symptoms or other overt effects in acutely exposed human volunteers. The acute oral MRL was determined to be 0.1 mg/kg/day based on developmental effects in rats. The chronic oral MRL of 0.04 mg/kg/day was based on hematological effects in rats.

TABLE 73.6   Regulatory Standards Determined by Government Agencies and Organizations

Organization	Route	Standard	End Point	Value
ATSDR (1990)	Inhalation	Acute MRL	Lack of neurological symptoms in humans	0.2 mg/m^3
	Oral	Acute MRL	Developmental effects in rats	0.1 mg/kg/day
		Chronic MRL	Hematological effects in rats	0.04 mg/kg/day
USEPA (1991)	Inhalation	Chronic RfC	Nasal epithelium pathology in rats	0.002 mg/m^3
		Unit Risk	Respiratory cancer in humans	0.068 (mg/m^3)$^{-1}$
	Oral	Slope Factor	Tumors in rodents	0.54 (mg/kg/day)$^{-1}$
CalEPA (2001, 2009)	Inhalation	Chronic REL	Nasal epithelium pathology in rats	0.005 mg/m^3
		Unit Risk	Respiratory cancer in humans	0.29 (mg/m^3)$^{-1}$
	Oral	Slope Factor	Respiratory cancer in humans	1 (mg/kg/day)$^{-1}$
NICNAS (2000)	Inhalation	Chronic OEL	Category 2 carcinogen	4.3 mg/m^3
ECJRC (2004)	Inhalation	Chronic OEL	Category 2 carcinogen	0.5–45 mg/m^3
WHO (2002)	Inhalation	Unit Risk	CNS tumors in female rats	0.0083 (mg/m^3)$^{-1}$
Health Canada (2000)	Inhalation	TC$_{05}$LCL	CNS tumors in female rats	4.5 mg/m^3
	Oral	TD$_{05}$LCL	CNS tumors in female rats	1.4 mg/kg/day

For definition of terms and abbreviations, refer to the relevant agency in the section "Risk Assessments".

TABLE 73.7   Interim Acute Exposure Guideline Levels (AEGLs) for Acrylonitrile

AN Exposure Duration	AEGL1 (Non-disabling)	AEGL2 (Disabling)	AEGL3 (Lethal)
10 min	4.6 ppm (10 mg/m^3)	290 ppm (629 mg/m^3)	480 ppm (1042 mg/m^3)
30 min	4.6 ppm (10 mg/m^3)	110 ppm (239 mg/m^3)	180 ppm (391 mg/m^3)
1 h	4.6 ppm (10 mg/m^3)	57 ppm (124 mg/m^3)	100 ppm (217 mg/m^3)
4 h	4.6 ppm (10 mg/m^3)	16 ppm (35 mg/m^3)	35 ppm (76 mg/m^3)
8 h	4.6 ppm (10 mg/m^3)	8.6 ppm (19 mg/m^3)	19 ppm (41 mg/m^3)

U.S. Environmental Protection Agency

The reference concentration (RfC) developed by the USEPA for acrylonitrile is 0.002 mg/m^3 based on degeneration and inflammation of nasal epithelium in rats (USEPA, 1991). No reference dose has been established. The USEPA classifies acrylonitrile as a B1 carcinogen (*probable human carcinogen, based on limited evidence of carcinogenicity in humans*). USEPA derived an inhalation unit risk of 0.068 (mg/m^3)$^{-1}$ based on respiratory cancer in exposed workers and an oral slope factor of 0.54 (mg/kg/day)$^{-1}$ based on tumors in rodents (USEPA, 1991).

California Environmental Protection Agency

California EPA (CalEPA) developed a chronic inhalation value as well as cancer values for both inhalation and oral routes of exposure (CalEPA, 2001, 2009). The chronic inhalation REL was determined to be 0.0002 ppm (0.005 mg/m^3) based on degeneration and inflammation of nasal epithelium in rats. The cancer inhalation unit risk of 0.29 (mg/m^3)$^{-1}$ and cancer slope factor of 1 (mg/kg/day)$^{-1}$ are based on increased risk of lung cancer in exposed workers.

Australian National Industrial Chemical Notifications and Assessment Scheme

The Australian National Industrial Chemical Notifications and Assessment Scheme (NICNAS) set an occupational exposure limit (OEL) for acrylonitrile of 2 ppm (4.3 mg/m^3) TWA for an 8 h work day and classified acrylonitrile as a Category 2 carcinogen (*may cause cancer*) (NICNAS, 2000).

European Union Risk Assessment Report

The European Union risk assessment classifies acrylonitrile as a Category 2 carcinogen (*may cause cancer*) and reports international < OELs ranging from 0.23 ppm (0.5 mg/m^3) to 20.36 ppm (45 mg/m^3) (ECJRC, 2004).

World Health Organization

The World Health Organization (WHO) derived an inhalation cancer unit risk of 0.0083 (mg/m^3)$^{-1}$ based on increased CNS tumors in female rats (WHO, 2002).

Health Canada

Health Canada developed a tumorigenic concentration at the 5% level (95% lower CL) [TC$_{05}$(LCL)] for inhalation exposure and a tumorigenic dose at the 5% level (95% lower confidence limit) [TD$_{05}$(LCL)] oral exposure (Health Canada, 2000). The TC$_{05}$(LCL) of 4.5 mg/m^3 and the TD$_{05}$(LCL) of 1.4 mg/kg/day were based on CNS tumors in female rats.

International Agency for Research on Cancer

The International Agency for Research on Cancer (IARC) classified acrylonitrile as Group 2B (*possibly carcinogenic to humans*) based on inadequate evidence in humans and sufficient evidence of carcinogenicity in experimental animals (IARC, 1999).

National Toxicology Program

The National Toxicology Program (NTP) classified acrylonitrile as *reasonably anticipated to be a human carcinogen*, based on sufficient evidence of carcinogenicity in experimental animals (NTP, 2011).

National Advisory Committee on Acute Exposure Guideline Levels

The National Advisory Committee established interim acute exposure guideline levels (AEGLs) for acrylonitrile, which are presented in Table 73.7 (USEPA, 2009).

REFERENCES

ACGIH (2001) *Documentation of the Threshold Limit Values and Biological Exposure Indices: Acrylonitrile*, 7th ed., Cincinnati, OH: American Conference of Governmental Industrial Hygienists.

ATSDR (1990) *Toxicological Profile for Acrylonitrile*, Atlanta, GA: U.S. Department of Health and Human Services, Public Health Service. Available at http://www.atsdr.cdc.gov/toxprofiles/tp125.pdf.

ATSDR (2013) *Medical Management Guidelines for Acrylonitrile*, Atlanta, GA: Agency for Toxic Substances and Disease

Registry. Availabe at http://www.atsdr.cdc.gov/mmg/mmg.asp?id=443&tid=78.

Bakker, J.G., Jongen, S.M., Van Neer, F.C., and Neis, J.M. (1991) Occupational contact dermatitis due to acrylonitrile. *Contact Dermatitis* 24(1):50–53.

Benn, T. and Osborne, K. (1998) Mortality of United Kingdom acrylonitrile workers—an extended and updated study. *Scand. J. Work Environ. Health* 24(Suppl. 2):17–24.

Benz, F.W. and Nerland, D.E. (2005) Effect of cytochrome P450 inhibitors and anticonvulsants on the acute toxicity of acrylonitrile. *Arch. Toxicol.* 79(10):610–614.

Benz, F.W., Nerland, D.E., Li, J., and Corbett, D. (1997) Dose dependence of covalent binding of acrylonitrile to tissue protein and globin in rats. *Fundam. Appl. Toxicol.* 36(2):149–156.

Bigner, D.D., Bigner, S.H., Burger, P.C., Shelburne, J.D., and Friedman, H.S. (1986) Primary brain tumors in Fischer 344 rats chronically exposed to acrylonitrile in their drinking-water. *Food Chem. Toxicol.* 24(2):129–137.

Blair, A., Stewart, P.A., Zaebst, D.D., Pottern, L., Zey, J.N., Bloom, T.F., Miller, B., Ward, E., and Lubin, J. (1998) Mortality of industrial workers exposed to acrylonitrile. *Scand. J. Work Environ. Health* 24(Suppl. 2):25–41.

Boiano, J.M. and Hull, R.D. (2001) Development of a National Occupational Exposure Survey and Database associated with NIOSH hazard surveillance initiatives. *Appl. Occup. Environ. Hyg.* 16(2):128–34.

Borba, H., Monteiro, M., Proença, M.J., Chaveca, T., Pereira, V., Lynce, N., and Rueff, J. (1996) Evaluation of some biomonitoring markers in occupationally exposed populations to acrylonitrile. *Teratog. Carcinog. Mutagen.* 16(4):205–218.

CalEPA (2001) *Chronic Toxicity Summary: Acrylonitrile.* State of California Environmental Protection Agency. Available at http://www.oehha.ca.gov/air/chronic_rels/pdf/acrylonitrile.pdf.

CalEPA (2009) *Technical Support Document for Describing Available Cancer Potency Factors, Appendix H.* California Environmental Protection Agency, Office of Environmental Health Hazard Assessment. Available at http://www.oehha.ca.gov/air/hot_spots/2009/AppendixHexposure.pdf.

Campian, E.C. and Benz, F.W. (2008) The acute lethality of acrylonitrile is not due to brain metabolic arrest. *Toxicology* 253(1–3):104–109.

Chen, Y., Chen, C., Jin, S., and Zhou, L. (1999) The diagnosis and treatment of acute acrylonitrile poisoning: a clinical study of 144 cases. *J. Occup. Health* 41(3):172–176.

Chen, Y., Chen, C., and Zhu, P. (2000) Study on the effects of occupational exposure to acrylonitrile in workers. *China Occup. Med.* 18(3).

Christian, T.J., Kleiss, B., Yokelson, R.J., Holzinger, R., Crutzen, P.J., Hao, W.M., Shirai, T., and Blake, D.R. (2004) Comprehensive laboratory measurements of biomass-burning emissions: 2. First intercomparison of open-path FTIR, PTR-MS, and GC-MS/FID/ECD. *J. Geophys. Res. Atmos.* 109(D02311), doi: 10.1029/2003JD003874.

Collins, J.J., Cheng, R., Buck, G.M., Zhang, J., Klebanoff, M.A., Schisterman, E.F., Scheffers, T., Ohta, H., Takaya, K., Miyauchi, H., Markowitz, M., Divine, B., and Tsai, S. (2003) The feasibility of conducting a reproductive outcome study of Chinese acrylonitrile workers. *J. Occup. Environ. Med.* 1:29–32.

Collins, J.J. and Strother, D.E. (1999) CNS tumors and exposure to acrylonitrile: inconsistency between experimental and epidemiology studies. *Neuro Oncol.* 1(3):221–230.

Dong, D., Tao, D., and Yang, Y. (2000) *Study of Occupational Harmfulness to Acrylonitrile Workers.* Industrial Health Department of Safety and Technology, Daqing Petrochemical General Plant, National Technical Information Service. No. OTS0559911. Available at http://www.ntis.gov/search/product.aspx?ABBR=OTS0559911.

ECJRC (2004) *European Union risk Assessment Report: Acrylonitrile.* Brussels, Belgium: Institute for Health and Protection, European Commission and Joint Research Centre. Available at http://esis.jrc.ec.europa.eu/doc/risk_assessment/REPORT/acrylonitrilereport029.pdf.

Esmat, A., El-Demerdash, E., El-Mesallamy, H., and Abdel-Naim, A.B. (2007) Toxicity and oxidative stress of acrylonitrile in rat primary glial cells: preventive effects of *N*-acetylcysteine. *Toxicol. Lett.* 171(3):111–118.

Fahmy, M.A. (1999) Evaluation of the genotoxicity of acrylonitrile in different tissues of male mice. *Cytologia (Tokyo)* 64:1–9.

Fan, W., Wang, W.L., Ding, S., Zhou, Y.L., and Jin, F.S. (2006) Application of micronucleus test of buccal mucosal cells in assessing the genetic damage of workers exposed to acrylonitrile. *Zhonghua Lao Dong Wei Sheng Zhi Ye Bing Za Zhi* 24(2): 106–108.

FDA (2006) *Total Diet Study, Market Baskets 1991–3 and 2003–4.* Center for Food Safety and Applied Nutrition. Available at http://www.fda.gov/downloads/Food/FoodSafety/FoodContaminantsAdulteration/TotalDietStudy/UCM184304.pdf.

Friedman, M.A. and Beliles, R.P. (2002) Three-generation reproduction study of rats receiving acrylonitrile in drinking water. *Toxicol. Lett.* 132(3):249–261.

Gagnaire, F., Marignac, B., and Bonnet, P. (1998) Relative neurotoxicological properties of five unsaturated aliphatic nitriles in rats. *J. Appl. Toxicol.* 18(1):25–31.

Ghanayem, B.I. and Hoffler, U. (2007) Investigation of xenobiotics metabolism, genotoxicity, and carcinogenicity using Cyp2e1 (−/−) mice. *Curr. Drug Metab.* 8(7):728–749.

Ghanayem, B.I., Nyska, A., Haseman, J.K., and Bucher, J.R. (2002) Acrylonitrile is a multisite carcinogen in male and female B6C3F1 mice. *Toxicol. Sci.* 68(1):59–68.

Grosjean, D. (1990) Atmospheric chemistry of toxic contaminants. 3. Unsaturated aliphatics—acrolein, acrylonitrile, maleic-anhydride. *J. Air Waste Manage. Assoc.* 40(12):1664–1668.

Grunske, F. (1949) Health care and occupational medicine: ventox and ventox intoxication. *Dtsch. Med. Wochenschr.* 74 1081–1083.

Hadley, J.E., Jr. (1994) *Initial submission: Letter from Hadley and Mckenna to USEPA regarding ITC request for information on acrylonitrile with attachments, dated 06/26/87.* National Technical Information Service. No. OTS0001123. Available at http://www.ntis.gov/search/product.aspx?ABBR=OTS0001123.

Health Canada (2000) *Priority Substances List Assessment Report: Acrylonitrile.* Gatineau, Canada: Minister of Public Works and

Government Services. Available at http://www.hc-sc.gc.ca/ewh-semt/pubs/contaminants/psl2-lsp2/acrylonitrile/index-eng.php.

HSDB (2013) *Hazardous Substances Data Bank: Acrylonitrile*, Bethesda, MD: National Library of Medicine. Available at http://toxnet.nlm.nih.gov/cgi-bin/sis/htmlgen?HSDB.

Huysentruyt, L.C. and Seyfried, T.N. (2010) Perspectives on the mesenchymal origin of metastatic cancer. *Cancer Metastasis Rev.* 29(4):695–707.

IARC (1999) *Re-evaluation of Some Organic Chemicals, Hydrazine and Hydrogen Peroxide: Acrylonitrile*, Lyon, France: International Agency for Research on Cancer, pp. 43–108. Available at http://monographs.iarc.fr/ENG/Monographs/vol71/index.php.

IPCS (1983) *Environmental Health Criteria: Acrylonitrile*, Geneva, Switzerland: World Health Organization, p. 77. Available at http://www.inchem.org/documents/ehc/ehc/ehc28.htm.

Jacob, S. and Ahmed, A.E. (2003) Acrylonitrile-induced neurotoxicity in normal human astrocytes: oxidative stress and 8-hydroxy-2′-deoxyguanosine formation. *Toxicol. Mech. Methods* 13(3):169–179.

Johannsen, F.R. and Levinskas, G.J. (2002a) Chronic toxicity and oncogenic dose-response effects of lifetime oral acrylonitrile exposure to Fischer 344 rats. *Toxicol. Lett.* 132(3):221–247.

Johannsen, F.R. and Levinskas, G.J. (2002b) Comparative chronic toxicity and carcinogenicity of acrylonitrile by drinking water and oral intubation to Spartan Sprague-Dawley rats. *Toxicol. Lett.* 132(3):197–219.

Kamendulis, L.M., Jiang, J., Xu, Y., and Klaunig, J.E. (1999) Induction of oxidative stress and oxidative damage in rat glial cells by acrylonitrile. *Carcinogenesis* 20(8):1555–1560.

Kaneko, K. and Omae, K. (1992) Effect of chronic exposure to acrylonitrile on subjective symptoms. *Keio J. Med.* 41(1):25–32.

Karl, T., Jobson, T., Kuster, W.C., Williams, E., Stutz, J., Shetter, R., Hall, S.R., Goldan, P., Fehsenfeld, F., and Lindinger, W. (2003) Use of proton-transfer-reaction mass spectrometry to characterize volatile organic compound sources at the La Porte super site during the Texas Air Quality Study 2000. *J. Geophys. Res.* 108(64508), doi: 10.1029/2002JD003333.

Kedderis, G.L., Sumner, S.C., Held, S.D., Batra, R., Turner, M.J., Roberts, A.E., and Fennell, T.R. (1993) Dose-dependent urinary excretion of acrylonitrile metabolites by rats and mice. *Toxicol. Appl. Pharmacol.* 120(2):288–297.

Kelly, T.J., Callahan, P.J., Pleil, J., and Evans, G.F. (1993) Method development and field measurements for polar volatile organic compounds in ambient air. *Environ. Sci. Technol.* 27(6): 1146–1153.

Kolenda-Roberts, H.M., Harris, N., Singletary, E., and Hardisty, J.F. (2013) Immunohistochemical characterization of spontaneous and acrylonitrile-induced brain tumors in the rat. *Toxicol. Pathol.* 41(1):98–108.

Laugesen, M. and Fowles, J. (2005) Scope for regulation of cigarette smoke toxicity according to brand differences in published toxicant emissions. *N. Z. Med. J.* 118(1213):U1401.

Major, J., Hudák, A., Kiss, G., Jakab, M.G., Szaniszló, J., Náray, M., Nagy, I., and Tompa, A. (1998) Follow-up biological and genotoxicological monitoring of acrylonitrile- and

dimethylformamide-exposed viscose rayon plant workers. *Environ. Mol. Mutagen.* 31(4):301–310.

Maltoni, C., Ciliberti, A., Cotti, G., and Perino, G. (1988) Long-term carcinogenicity bioassays on acrylonitrile administered by inhalation and by ingestion to Sprague-Dawley rats. *Ann. N. Y. Acad. Sci.* 534:179–202.

Marsh, G.M., Youk, A.O., and Collins, J.J. (2001) Reevaluation of lung cancer risk in the acrylonitrile cohort study of the National Cancer Institute and the National Institute for Occupational Safety and Health. *Scand. J. Work Environ. Health* 27(1):5–13.

McCarthy, M.C., O'Brien, T.E., Charrier, J.G., and Hafner, H.R. (2009) Characterization of the chronic risk and hazard of hazardous air pollutants in the United States using ambient monitoring data. *Environ. Health Perspect.* 117(5):790–796.

Miller, S.L., Branoff, S., and Nazaroff, W.W. (1998) Exposure to toxic air contaminants in environmental tobacco smoke: an assessment for California based on personal monitoring data. *J. Expos. Anal. Environ. Epidemiol.* 8(3):287–311.

Morita, T., Asano, N., Awogi, T., Sasaki, Y.F., Sato, S., Shimada, H., Sutou, S., Suzuki, T., Wakata, A., Sofuni, T., and Hayashi, M. (1997) Evaluation of the rodent micronucleus assay in the screening of IARC carcinogens (groups 1, 2A and 2B). The summary report of the 6th collaborative study by CSGMT/JEMS MMS. Collaborative Study of the Micronucleus Group Test. Mammalian Mutagenicity Study Group [Erratum]. *Mutat. Res.* 391(3):259–267.

Mulvihill, J., Cazenave, J.P., Mazzucotelli, J.P., Crost, T., Collier, C., Renaux, J.L., and Pusineri, C. (1992) Minimodule dialyser for quantitative *ex vivo* evaluation of membrane haemocompatibility in humans: comparison of acrylonitrile copolymer, cuprophan and polysulphone hollow fibres. *Biomaterials* 13(8): 527–536.

Murray, F.J., Schwetz, B.A., Nitschke, K.D., John, J.A., Norris, J.M., and Gehring, P.J. (1978) Teratogenicity of acrylonitrile given to rats by gavage or by inhalation. *Food Cosmet. Toxicol.* 16(6):547–551.

Muto, T., Sakurai, H., Omae, K., Minaguchi, H., and Tachi, M. (1992) Health profiles of workers exposed to acrylonitrile. *Keio J. Med.* 41(3):154–160.

Nagatani, M., Ando, R., Yamakawa, S., Saito, T., and Tamura, K. (2009) Histological and immunohistochemical studies on spontaneous rat astrocytomas and malignant reticulosis. *Toxicol. Pathol.* 37(5):599–605.

Nazaroff, W.W. and Singer, B.C. (2004) Inhalation of hazardous air pollutants from environmental tobacco smoke in US residences. *J. Expos. Anal. Environ. Epidemiol.* 14 S71–S77.

Nemec, M.D., Kirkpatrick, D.T., Sherman, J., Van Miller, J.P., Pershing, M.L., and Strother, D.E. (2008) Two-generation reproductive toxicity study of inhaled acrylonitrile vapors in Crl:CD(SD) rats. *Int. J. Toxicol.* 27(1):11–29.

Nguyen, T.T., Le, H.N., Vo, M., Wang, Z., Luu, L., and Ramella-Roman, J.C. (2012) Three-dimensional phantoms for curvature correction in spatial frequency domain imaging. *Biomed. Opt. Express* 3(6):1200–14.

NICNAS (2000) *Acrylonitrile. Priority Existing Chemical Assessment Report*. National Industrial Chemicals Notification and Assessment Scheme (NICNAS). No. 10. Available at http://www.nicnas.gov.au/Publications/CAR/PEC/PEC10.asp.

NIOSH (1988) *Occupational Safety and Health Guidelines for Chemical Hazards. Acrylonitrile*, Washington, DC: National Institute for Occupational Safety and Health. Available at http://www.cdc.gov/niosh/docs/81-123/pdfs/0014.pdf.

NIOSH (1990) *National Occupational Exposure Survey (1981–1983)*, Washington, DC: National Institute for Occupational Safety and Health. Available at http://www.cdc.gov/noes/noes1/03800sic.html.

NTP (2001) *Toxicology and Carcinogenesis Studies of Acrylonitrile (CAS No. 107-13-1) in B6C3F1 Mice (Gavage Studies)*, Research Triangle Park, NC: National Toxicology Program, pp. 1–201. Available at http://www.ntis.gov/search/product.aspx?abbr=PB2002102198.

NTP (2011) Acrylonitrile. In: *Report on Carcinogens*, 12th ed., Research Triangle Park, NC: National Toxicology Program, pp. 28–30. Available at http://ntp.niehs.nih.gov/ntp/roc/twelfth/profiles/Acrylonitrile.pdf.

OSHA (2001) *Chemical Sampling and Exposure Information: Acrylonitrile*, Washington, DC: Occupational Safety and Health Administration. Available at http://www.osha.gov/dts/chemicalsampling/data/CH_217300.html.

OSHA (2012) *Occupational Safety and Health Standards, Z: Toxic and Hazardous Substances. Acrylonitrile*, Washington, DC: Occupational Safety and Health Administration. Available at http://www.osha.gov/pls/oshaweb/owadisp.show_document?p_table=STANDARDS&p_id=10065.

Pawelek, J.M. and Chakraborty, A.K. (2008) Fusion of tumour cells with bone marrow-derived cells: a unifying explanation for metastasis. *Nat. Rev. Cancer* 8(5):377–386.

Pu, X., Kamendulis, L.M., and Klaunig, J.E. (2009) Acrylonitrile-induced oxidative stress and oxidative DNA damage in male Sprague-Dawley rats. *Toxicol. Sci.* 111(1):64–71.

Quast, J.F. (2002) Two-year toxicity and oncogenicity study with acrylonitrile incorporated in the drinking water of rats. *Toxicol. Lett.* 132(3):153–196.

Quast, J.F., Schwetz, D.J., Balmer, M.F., Gushow, T.S., Park, C.N., and McKenna, M.J. (1980) *A Two-Year Toxicity and Oncogenicity Study with Acrylonitrile Following Inhalation Exposure of Rats [Final Report]*, Midland, MI: Dow Chemical Co.

Rongzhu, L., Ziqiang, C., Fusheng, J., and Collins, J. (2005) Neurobehavioral effects of occupational exposure to acrylonitrile in Chinese workers. *Environ. Toxicol. Pharmacol.* 19(3):695–700.

Saillenfait, A.M., Bonnet, P., Guenier, J.P., and de Ceaurriz, J. (1993) Relative developmental toxicities of inhaled aliphatic mononitriles in rats. *Fundam. Appl. Toxicol.* 20(3):365–375.

Scélo, G., Constantinescu, V., Csiki, I., Zaridze, D., Szeszenia-Dabrowska, N., Rudnai, P., Lissowska, J., Fabiánová, E., Cassidy, A., Slamova, A., Foretova, L., Janout, V., Fevotte, J., Fletcher, T., Mannetje, A., Brennan, P., and Boffetta, P. (2004) Occupational exposure to vinyl chloride, acrylonitrile and styrene and lung cancer risk (Europe). *Cancer Causes Control* 15(5):445–452.

Solomon, J.J. and Segal, A. (1985) Direct alkylation of calf thymus DNA by acrylonitrile. Isolation of cyanoethyl adducts of guanine and thymine and carboxyethyl adducts of adenine and cytosine. *Environ. Health Perspect.* 62:227–230.

Starr, T.B., Gause, C., Youk, A.O., Stone, R., Marsh, G.M., and Collins, J.J. (2004) A risk assessment for occupational acrylonitrile exposure using epidemiology data. *Risk Anal.* 24(3): 587–601.

Swaen, G.M., Bloemen, L.J., Twisk, J., Scheffers, T., Slangen, J.J., Collins, J.J., and ten Berge, W.F. (2004) Mortality update of workers exposed to acrylonitrile in The Netherlands. *J. Occup. Environ. Med.* 46(7):691–698.

Symons, J.M., Kreckmann, K.H., Sakr, C.J., Kaplan, A.M., and Leonard, R.C. (2008) Mortality among workers exposed to acrylonitrile in fiber production: an update. *J. Occup. Environ. Med.* 50(5):550–560.

Tandon, R., Saxena, D.K., Chandra, S.V., Seth, P.K., and Srivastava, S.P. (1988) Testicular effects of acrylonitrile in mice. *Toxicol. Lett.* 42(1):55–63.

Tavares, R., Borba, H., Monteiro, M., Proença, M.J., Lynce, N., Rueff, J., Bailey, E., Sweetman, G.M., Lawrence, R.M., and Farmer, P.B. (1996) Monitoring of exposure to acrylonitrile by determination of *N*-(2-cyanoethyl)valine at the N-terminal position of haemoglobin. *Carcinogenesis* 17(12):2655–2660.

Thier, R., Lewalter, J., and Bolt, H.M. (2000) Species differences in acrylonitrile metabolism and toxicity between experimental animals and humans based on observations in human accidental poisonings. *Arch. Toxicol.* 74(4–5):184–189.

Thier, R., Lewalter, J., Kempkes, M., Selinski, S., Brüning, T., and Bolt, H.M. (1999) Haemoglobin adducts of acrylonitrile and ethylene oxide in acrylonitrile workers, dependent on polymorphisms of the glutathione transferases GSTT1 and GSTM1. *Arch. Toxicol.* 73(4–5):197–202.

Thier, R., Lewalter, J., Selinski, S., and Bolt, H.M. (2002) Possible impact of human CYP2E1 polymorphisms on the metabolism of acrylonitrile. *Toxicol. Lett.* 128(1–3):249–255.

Thiess, A.M. and Fleig, I. (1978) Analysis of chromosomes of workers exposed to acrylonitrile. *Arch. Toxicol.* 41(2):149–152.

Thiess, A.M., Frentzel-Beyme, R., Link, R., et al. (1980) Mortality study in chemical personnel of various industries exposed to acrylonitrile. *Zentralbl. Arbeitsmed. Arbeitsschutz Prophyl. Ergon.* 30(7):259267.

USEPA (1991) *Toxicological Assessment of Acrylonitrile*. Integrated Risk Information System, National Center for Environmental Assessment. Available at http://www.epa.gov/iris/subst/0206.htm.

USEPA (1994a) *Methods for Derivation of Inhalation Reference Concentrations and Application of Inhalation Dosimetry*, Research Triangle Park, NC: U.S. Environmental Protection Agency. Available at http://cfpub.epa.gov/ncea/cfm/recordisplay.cfm?deid=71993.

USEPA (1994b) *Chemical Fact Sheet: Acrylonitrile*. Office of Pollution Prevention and Toxics. Available at http://www.epa.gov/chemfact/.

USEPA (1995) *Hazardous Air Pollutant Emissions from Process Units in the Thermoplastics Manufacturing Industry*. Basis and

Purpose Document for Proposed Standards. National Technical Information Service. No. PB95201760. Available at http://www.ntis.gov/search/product.aspx?ABBR=PB95201760.

USEPA (1996) *National-Scale Air Toxics Assessment (NATA)*, Washington, DC: United States Environmental Protection Agency, Office of Air and Radiation. Available at http://www.epa.gov/ttn/atw/nata/tablconc.html.

USEPA (2000) *Air Toxics. Hazard Summary: Acrylonitrile*. Office of Air and Radiation. Available at http://www.epa.gov/ttn/atw/hlthef/acryloni.html.

USEPA (2005) *National-Scale Air Toxics Assessment (NATA)*. United States Environmental Protection Agency, Office of Air and Radiation. Available at http://www.epa.gov/ttn/atw/natamain/index.html.

USEPA (2009) *Acute Exposure Guideline Levels (AEGLs) for Acrylonitrile (CAS Reg. No. 107-13-1) Interim*, Washington, DC: United States Environmental Protection Agency. Available at http://www.epa.gov/oppt/aegl/pubs/acrylonitrile_interim_ornl_nov2009c.pdf.

USEPA (2012) *Chemical Data Access Tool. 2-Propenenitrile*. Office of Pollution Prevention and Toxics. Available at http://java.epa.gov/oppt_chemical_search/.

Vogel, R.A. and Kirkendall, W.M. (1984) Acrylonitrile (vinyl cyanide) poisoning: a case report. *Tex. Med.* 80(5):48–51.

Wakata, A., Miyamae, Y., Sato, S., Suzuki, T., Morita, T., Asano, N., Awogi, T., Kondo, K., and Hayashi, M. (1998) Evaluation of the rat micronucleus test with bone marrow and peripheral blood: summary of the 9th collaborative study by CSGMT/JEMS. MMS. Collaborative Study Group for the Micronucleus Test. Environmental Mutagen Society of Japan. Mammalian Mutagenicity Study Group. *Environ. Mol. Mutagen.* 32(1):84–100.

Wakui, C., Kawamura, Y., and Maitani, T. (2001) Migrants from disposable gloves and residual acrylonitrile. *Shokohin Eiseigaku Zasshi* 42(5):322–328.

Warneke, C., Roberts, J.M., Veres, P., Gilman, J., Kuster, W.C., Burling, I., Yokelson, R., and de Gouw, J.A. (2011) VOC identification and inter-comparison from laboratory biomass burning using PRT-MS and PIT-MS. *Int. J. Mass Spectrom.* 303:6–14.

WHO (2002) *Acrylonitrile. Concise International Chemical Assessment Document*, Geneva, Switzerland: World Health Organisation. Available at http://www.who.int/ipcs/publications/cicad/cicad39_rev.pdf.

Whysner, J., Steward, R.E., Chen, D., Conaway, C.C., Verna, L.K., Richie, J.P., Ali, N., and Williams, G.M. (1998) Formation of 8-oxodeoxyguanosine in brain DNA of rats exposed to acrylonitrile. *Arch. Toxicol.* 72(7):429–438.

Wilson, R.H., Hough, G.V., and McCormick, W.E. (1948) Medical problems encountered in the manufacture of American-made rubber. *Ind. Med. Surg.* 17(6):199–207.

Xu, D.X., Zhu, Q.X., Zheng, L.K., Wang, Q.N., Shen, H.M., Deng, L.X., and Ong, C.N. (2003) Exposure to acrylonitrile induced DNA strand breakage and sex chromosome aneuploidy in human spermatozoa. *Mutat. Res.* 537(1):93–100.

Yokelson, R.J., Karl, T., Artaxo, P., Blake, D.R., Christian, T.J., Griffith, D.W.T., Guenther, A., and Hao, W.M. (2007) The tropical forest and fire emissions experiment: overview and airborne fire emission factor measurements. *Atmos. Chem. Phys.* 7:5175–5196.

Zeller, H., Hofmann, H.T., Thiess, A.M., and Hey, W. (1969) Toxicity of nitriles (results of animal experiments and 15 years of experience in industrial medicine). *Zentralbl. Arbeitsmed.* 19(8):225–238.

Zheng, P.P., van der Weiden, M., van der Spek, P.J., Vincent, A.J., and Kros, J.M. (2012) Isocitrate dehydrogenase 1R132H mutation in microglia/macrophages in gliomas: indication of a significant role of microglia/macrophages in glial tumorigenesis. *Cancer Biol. Ther.* 13(10):836–839.

SECTION V

ORGANIC HIGH POLYMERS, MONOMERS, AND POLYMER ADDITIVES

SECTION EDITOR: RAYMOND D. HARBISON

74

INTRODUCTION

Raymond D. Harbison

Over the past four decades the plastics and related chemical industries have experienced pronounced growth. Few aspects of daily living have not been touched in some way by the use of plastics. Superior plastic substitutes for wood, metal, and glass have found their way into medical uses, food packaging, textiles, construction, and even aircraft. The burgeoning use of plastics also has had a significant impact on waste disposal practices in areas where these materials now make up a major portion of the waste volume in landfills.

What is more, the phthalate esters, compounds used in the manufacture of plastics, have become one of the most widely found classes of laboratory and environmental contaminants. These compounds appear regularly in soil and groundwater samples taken from monitoring wells near municipal and industrial landfills. They are also found as fugitive contaminants in analytical laboratories and hospital intravenous systems, where they leach from plastic tubings. As such, they have become a source of growing concern about exposure, both in occupational and in secondary environmental settings.

Plastics and related organic high polymers, regardless of their ultimate use, are similar to the extent that they are based on the structural repetition of smaller units, called *monomers*, and they often contain added substances, which are not actually part of the polymer itself. The toxic phenomena associated with high polymers may be related to their unreacted or underreacted constituents, to various auxiliary substances, or to degradation products. It is uncommon for a fully reacted or "cured" polymer to cause illness because most of these materials have a high degree of biological inertness and insolubility.

Plastics are usually defined as high molecular weight materials that are quite soft during manufacturing and can be molded by heat or pressure, but they turn solid and permanently deform in the finished state. The term *elastomers* is applied to polymers, such as natural and synthetic rubbers, which can be stretched easily yet return to their original dimensions when the stressing force is removed.

Elasticity is thus a capacity to be reversibly deformed. Linear macromolecules with high longitudinal mechanical strength and lateral flexibility can be used to form synthetic fibers for textile uses.

HAZARDS FOLLOWING COMBUSTION

As noted previously, in their fully polymerized, or hardened, commercial form, plastics pose few if any health hazards. However, with thermal decomposition in a fire, the resultant by-products of combustion may give rise to the potential for both acute and chronic exposure to a diverse group of compounds (Table 74.1). The gases and fine-grained particulates that evolve during combustion can be extremely complex, varying greatly in their composition and concentration in direct relationship to the temperature of combustion, the composition of the source plastic, and the amount of material involved. Although these types of exposure may occur on a limited basis at the time of plastics manufacture, they are far more common for firefighters and for occupants of buildings that had previously been involved in a fire. This is a major contemporary concern for fire departments that stems from the increasing substitution of plastics for wood, metal, and natural fiber in manufactured items found in homes, commercial structures, and vehicles.

The principal toxicological concerns associated with combustion of plastics are that some of the products of thermal

Hamilton & Hardy's Industrial Toxicology, Sixth Edition. Edited by Raymond D. Harbison, Marie M. Bourgeois, and Giffe T. Johnson.
© 2015 John Wiley & Sons, Inc. Published 2015 by John Wiley & Sons, Inc.

TABLE 74.1 Possible Adverse Health Effects of Thermal Destruction of Plastics

Parent Plastic Categories	Adverse Health Effects
Thermosets	
Amino resin	MMI, RI
Epoxies	MMI, RI
Phenolics	MMI, RI, SA
Polyurethanes	MMI, RI, SA
Polyesters	MMI, RI
Silicones	MMI, RI
Thermoplastics	
Acetates and acrylics	MMI, RI
Fluoroplastics	MMI, RI
Polyvinyl chloride	MMI, RI, SA, PC
Polycarbonates	MMI, RI, SA
Polyethylene	MMI, RI
Polypropylene	MMI, RI
Polystyrene	MMI, RI, SA, PC
Polyvinylindine chloride	MMI, RI, SA, PC

MMI, mucous membrane irritant; RI, respiratory irritant; SA, systemic effects; PC, potential carcinogenic compounds.

decomposition may cause respiratory irritation and systemic effects, and some may also be carcinogenic. Although the extent and concentration of combustion by-products cannot be predicted with certainty, this can be a serious problem, particularly in large, poorly ventilated commercial structures.

The list of thermal degradation products that potentially could be liberated at the time of combustion includes compounds that can cause severe mucous membrane and respiratory effects. Other more exotic and possibly more toxic or carcinogenic compounds may be generated as a result of the special additives or coloring agents (e.g., metals and halides) added to the plastic formulation at the time of manufacture. In view of these conditions, it is often difficult to predict the toxicological effects and the presenting signs and symptoms likely to result from excessive exposure to smoke and fumes from the combustion of plastics. Furthermore, predicting the long-term health effects of exposure to the thermal degradation products of combustible plastics is problematic. Exposure to these products may cause acute, short-term symptoms that last 6–24 h. Complaints may include eye, nose, throat, and respiratory irritation, which may be accompanied by headache, intermittent cough, and production of excess mucus. In some individuals these signs and symptoms may progress to cough, chest tightness, flulike symptoms, and a marked respiratory sensitivity to the odor of smoke that lasts 2–8 weeks. The complex nature of such fire-related exposure makes it difficult to determine to what extent the health effects observed are caused by the combustion of plastics and related synthetic materials.

INDUSTRIAL HYGIENE, HEALTH, AND SAFETY STANDARDS

For the most part, no industrial hygiene, health, or safety standards have been published for plastics and related high organic polymers. In their reacted, or finished form, these materials pose few if any health hazards; the exceptions, however, are those noted for combustion by-products and materials leached from the finished polymer. The predominant concerns are the parent materials, the additives, and the related chemicals used to manufacture the product.

75

PLASTICS

Marek Banasik

The term *resin* has an imprecise meaning in the plastics industry and may be used interchangeably with the term *plastic*. In some applications, resins are short-chain uncured polymers that are subjected to further polymerization and hardening. In other applications, resins are granular, fully cured thermoplastics that can be heated for extrusion, molding, or calendering. The principal plastics may be divided into two groups. The *thermosets* cannot be reformed or melted after the initial "cure" and generally are obtained by polycondensation. The *thermoplastics* can be reheated and reshaped repeatedly and usually are produced by polymerization.

THERMOSETTING RESINS

Amino Resins

The amino resins are formed by the condensation of aldehydes with amines and are almost entirely based on the urea–formaldehyde and melamine–formaldehyde condensation products. They are typically used for electrical switch housings, knobs, molded dinnerware, adhesives, plywood glues, coatings, and industrial laminates. Crease-resistant garments are produced by pressing textiles impregnated with suitable amino resins into the desired shape. Hexamethylenetetramine, known as "hexa", is used as a stabilizer in amino resin systems to prevent premature hardening. It decomposes to form formaldehyde [Chemical Abstracts Service Registry Number (CASRN): 50-00-0] and is a possible source of irritating and sensitizing skin reactions, but it has not been a major problem in the plastics or explosives industries. Acetic acid accelerators also do not present unusual problems.

Cured, finished amino resins have not been identified as a source of occupational illness. The component of principal toxicological interest is formaldehyde, which is present during the production of the resin and in incompletely cured products. Formaldehyde is released as a gas from urea–formaldehyde foams and during heating of other polymers. Its effects are discussed in the section on aldehydes. Workers involved in the production of urea–formaldehyde resins have described some adverse health effects, most notably irritation of the respiratory tract, likely related to the strong odor and irritant properties of formaldehyde.

Urea–formaldehyde resins can be produced in three ways: as molding powders, as adhesives, or as foams. The manufacturing protocol differs depending on the particular type of resin being produced. Melamine–formaldehyde resins are easier to manufacture and are considered more water, heat, and stain resistant than polymers that contain urea. There have been many reported cases of contact dermatitis associated with the production and handling of urea– and melamine–formaldehyde resins, but these dermatitides have been found to be an outcome of exposure to unreacted formaldehyde monomer (Schwartz, 1945; Shellow and Altman, 1966). Some studies have shown cases of allergic contact dermatitis caused by formaldehyde resins that showed no response to skin patch tests with formaldehyde (Fowler et al., 1992). Respiratory effects may occur from exposure to formaldehyde, causing a sensitization reaction similar to occupational asthma. In addition, irritation of the eyes and nose may be a response during exposure to urea–formaldehyde dusts (Vale and Rycroft, 1988).

REFERENCES

Fowler, J.F. Jr., Skinner, S.M., and Belsito, D.V. (1992) Allergic contact dermatitis from formaldehyde resins in permanent press

Hamilton & Hardy's Industrial Toxicology, Sixth Edition. Edited by Raymond D. Harbison, Marie M. Bourgeois, and Giffe T. Johnson.
© 2015 John Wiley & Sons, Inc. Published 2015 by John Wiley & Sons, Inc.

clothing: an underdiagnosed cause of generalized dermatitis. *J. Am. Acad. Dermatol.* 27:962–968.

Schwartz, L. (1945) Dermatitis from synthetic resins. *J. Invest. Dermatol.* 6:239–255.

Shellow, H. and Altman, A.T. (1966) Dermatitis from formaldehyde resin textiles. *Arch. Dermatol.* 94:799–801.

Vale, P.T. and Rycroft, G. (1988) Occupational irritant contact dermatitis from fibreboard containing urea-formaldehyde resin. *Contact Dermatitis* 19:62.

Epoxy Resins

Cured epoxy resins have an outstanding resistance to heat and chemicals and find wide use in reinforced plastics, coatings, adhesives, casting, "potting," and encapsulation. The typical resin is prepared by condensation of epichlorohydrin with a diphenol in the presence of a curing agent. Epoxy resins are cured by cross-linking agents known as "hardeners," such as the polyamines (e.g., diethylenetriamine, triethylenetetramine, and piperazine) or the acid anhydrides and polybasic acids (e.g., phthalic anhydride and adipic acid). Catalysts perform a similar curing action but they do not themselves act as cross-linking agents. Common catalysts include various polyamides, monoethylamine, and tertiary amines, such as triethylamine and benzyl dimethylamine. Diluents (e.g., phenyl, allyl, and butyl glycidyl ether; styrene oxide; styrene; and other epoxides) may be used to reduce the viscosity of uncured resin systems.

Industrial exposure most commonly occurs during the production of epoxy resins. However, this is not the only opportunity for contact. Epoxy resins are often used in paints, leading to industrial as well as household exposure. They are also a component in concrete and flooring material, pipe and crack fixatives, and fiberglass. Inhalation of chemical vapors and airborne dusts related to epoxy resins is the major opportunity for contact.

Epichlorohydrin, bisphenol A (BPA), and liquid epoxy resins are skin irritants and sensitizers, as are the reactive diluent ethers and the related epoxy compounds. Epichlorohydrin is described in the section on chlorinated hydrocarbons. It has been pointed out that the epoxy group may be associated with radiomimetic effects that presumably reflect biological alkylation (Kotin and Falk, 1963). There is no evidence, however, that these substances are comparable to the nitrogen mustards in this regard or that there is an industrial hazard related to this action.

Polymerization of epoxy resins in some cases has been known to cause adverse health effects. Six workers in Japan were clinically noted to have developed a systemic sclerosis marked by skin sclerosis and muscle weakness after a short exposure (Yamakage et al., 1980). The causative agent is suspected to be an amine involved in the production of the polymer. Cases of allergic contact dermatitis involving low molecular weight monomers and oligomers of epoxy resins have been reported and confirmed with patch testing (Fregert and Thorgeirsson, 1977). These cases have been reported for almost every industrial use of epoxy resin, from its production to its use as a fiberglass coating (Holness and Nethercott, 1989; van Putten et al., 1984). These allergic reactions are the result of a sensitizing reaction capable of recurrence even with minute future exposure. Recent reports indicate that these lower molecular weight compounds are more likely to cause allergic contact dermatitis (Holmes et al., 1993).

Industrial health problems most frequently arise from the exposure to aliphatic polyamine hardeners, which can cause severe primary irritative dermatitis, as well as hypersensitivity dermatitis and asthmatic reactions. These amines are strongly alkaline (pH 13–14) and can cause chemical burns of the skin. Cutaneous amine reactions include erythema, intolerable itching, and severe facial swelling. Blistering may occur, with weeping of serous fluid, crusting, and scaling. An apparently hypersensitive state may recur as an extreme reaction upon reexposure to amines. Cases have been recorded of persons with previous amine dermatitis, who experienced a dramatic return of symptoms upon reexposure. Highly sensitive persons may also react to cured resins containing small amounts of unreacted amine hardeners.

A notable hardener used for epoxy resins is hexahydrophthalic anhydride. This reactive compound has been reported as an eye and mucous membrane irritant and is known to cause occupational asthma and hemorrhagic rhinitis (Grammer et al., 1993; Moller et al., 1985; Venables, 1989). Like many acid anhydrides with a low molecular weight, hexahydrophthalic anhydride can act as a hapten, and thus produce a variety of sensitizing effects. Studies have indicated an IgE-mediated mechanism for this chemical (Welinder et al., 1994). Once in the body, hexahydrophthalic anhydride is metabolized to hexahydrophthalic acid, which is then excreted in the urine. This acid can be measured using gas chromatography as a means of biological monitoring (Jönsson and Skarping, 1991). However, because the rate of elimination is rather fast (2–4 h), exposure can be determined only for recent contact (Jönsson and Skerfving, 1993).

Though pulmonary reactions in the modern plastics industry are most commonly associated with components of polyurethane (PUR) systems, amine hardeners for epoxy resins do cause bronchospasm and coughing episodes, which may persist for several days after cessation of exposure. Repeated attacks of acute amine-induced respiratory disease are often associated with increased responsiveness to these compounds because faint traces of amine vapors in the air have been sufficient to trigger the return of intense symptoms in persons with a history of amine-induced asthmatic conditions. Cases of asthma in workers involved in the curing process of epoxy resins have demonstrated reactions caused by phthalic acid anhydride, trimellitic acid anhydride, and triethylenetetramine (Fawcett et al., 1977).

Acid anhydrides and polyamide hardeners and catalysts are contact irritants, but industrial problems related to their use have been uncommon. Diluents such as xylene or the ketones do not sensitize but may defat the skin with repeated contact. Glass fibers, which have many uses, are mechanically irritating and cause pruritus, and they may be released when inert laminates or other forms are cut or tooled. Sawing, machining, and other manipulation of solid, cured resins that produce heat may give rise to degradation products that contain free amines or unreacted epichlorohydrin.

The carcinogenicity of epoxy resins is difficult to determine because of the array of compounds used. An updated study by Cowles et al. (1994) examined the mortality of employees at a resin-manufacturing facility. An unexpected increase in pancreatic cancer was observed, although no causative agent or process was identified. Occupational exposure at this facility included polypropylene, epoxy resins, polystyrene, and polyethylene.

Prevention of problems from epoxy exposure is based on the meticulous cleanliness of work areas. Mixing of resin and hardener and application of the liquid formulations should be done with appropriate ventilation. Material that is spilled on the skin should be removed promptly with soap and water. Organic solvents used for this purpose tend to spread contamination and irritate the skin. The use of gloves for skin protection is essential. Choosing gloves of the proper material is of great concern. An inappropriate glove can lead to exposure because different gloves are impervious to different chemicals (Holmes et al., 1993; Pegum, 1979).

REFERENCES

Cowles, S.R., Tsai, S.P., Gilstrap, E.L., and Ross, C.E. (1994) Mortality among employees at a plastics and resins research and development facility. *Occup. Environ. Med.* 51:799–803.

Fawcett, I.W., Taylor, A.J., and Pepys, J. (1977) Asthma due to inhaled chemical agents: epoxy resin systems containing phthalic acid anhydride, trimellitic acid anhydride and triethylene tetramine. *Clin. Allergy* 7, 1–14.

Fregert, S. and Thorgeirsson, A. (1977). Patch testing with low molecular oligomers of epoxy resins in humans. *Contact Dermatitis* 3:301–303.

Grammer, L.C., Shaughnessy, M.A., and Lowenthal, M. (1993) Hemorrhagic rhinitis: an immunologic disease due to hexahydrophthalic anhydride. *Chest* 104:1792–1794.

Holmes, N., Pearce, P., and Simpson, G. (1993) Prevention of epoxy resin dermatitis: failure of manufacturers to use research information. *Am. J. Ind. Med.* 24:605–617.

Holness, D.L. and Nethercott, J.R. (1989) Occupational contact dermatitis due to epoxy resin in a fiberglass binder. *J. Occup. Med.* 31:87–89.

Jönsson, B. and Skarping, G. (1991) Method for the biological monitoring of hexahydrophthalic anhydride by the determination of hexahydrophthalic acid in urine using gas chromatography and selected-ion monitoring. *J. Chromatogr.* 572:117–131.

Jönsson, B.A.G. and Skerfving, S. (1993) Toxicokinetics and biological monitoring in experimental exposure of humans to gaseous hexahydrophthalic anhydride. *Scand. J. Work Environ. Health* 19:183–190.

Kotin, P. and Falk, H.L. (1963) Organic peroxides, hydrogen peroxide, epoxides, and neoplasia. *Radiat. Res.* (Suppl. 3):193–211.

Moller, D.R. Gallagher, J.S., Bernstein, D.I., Wilcox, T.G., Burroughs, H.E., and Bernstein, I.L. (1985) Detection of IgE-mediated respiratory sensitization in workers exposed to hexahydrophthalic anhydride. *J. Allergy Clin. Immunol.* 75:663–672.

Pegum, J.S. (1979) Penetration of protective gloves by epoxy resin. *Contact Dermatitis* 5:281–283.

van Putten, P.B., Coenraads, P.J., and Nater, J.P. (1984) Hand dermatoses and contact allergic reactions in construction workers exposed to epoxy resins. *Contact Dermatitis* 10:146–150.

Venables, K.M. (1989). Low molecular weight chemicals, hypersensitivity, and direct toxicity: the acid anhydrides. *Br. J. Ind. Med.* 46:222–232.

Welinder, H.E., Jönsson, B.A.G., Nielsen, J.E., Ottosson, H.E., and Gustavsson, C.A. (1994). Exposure-response relationships in the formation of specific antibodies to hexahydrophthalic anhydride in exposed workers. *Scand. J. Work Environ. Health* 20:459–465.

Yamakage, A., Ishikawa, H., Saito, Y., and Hattori, A. (1980). Occupational scleroderma-like disorder occurring in men engaged in the polymerization of epoxy resins. *Dermatologica* 161:33–44.

Phenolics

The modern plastics industry is based on Leo Hendrik Baekeland's discovery in 1909 of phenol–formaldehyde (PF) thermosetting materials. The term *phenolic resin* now encompasses a variety of similar products made by polycondensation of phenols with aldehydes. The "phenols" used for this purpose include phenol, cresol, xylenol, p-t-butylphenol, and resorcinol. The aldehydes involved include formaldehyde, paraformaldehyde, and furfural. Alkaline catalysts include hexamethylenetetramine, ammonia, and various other amines. The uses of phenolic resins are similar to those of the amino resins, which have similar structures and properties. Large quantities of PF resins (CASRN: 9003-35-4) are used in bonded brake linings and clutch facings and as bonding agents for foundry sand models. Increases in use among certain industries, such as the aerospace and marine markets, can be attributed to the easy to produce, low viscosity liquid phenolics.

These resins are manufactured from two different processes, depending on the molecular weight desired. For lower molecular weights, an alkaline catalyst and simple heat curing are used to make a resol-type resin. Higher molecular weight resins, called novolaks, require an acid catalyst, and the addition of hexa aids in cross-linking. Resols are favored for their resistance to cracking, and novolaks are better suited for molding purposes and long-term storage. Phenolic resins in general are good electrical insulators and are resistant to chemical and biological degradation.

Dermatitides are important industrial health effects. Resins in this group have been associated with skin allergies after occupational exposure (Foussereau et al., 1976). More recent studies have shown a positive skin patch test among workers exposed to PF resin (Kieć-Świerczyńska and Szymczak, 1995). Phenol is a potent primary irritant, and resorcinol, furfural, and formaldehyde are both irritants and sensitizers. Furfural (CASRN: 98-01-1) is a photosensitizer that is found in various foods and beverages. In some applications, cashew nut shell oil has been used as a phenolic component in resin formulation. This oil is chemically related to poison ivy resins, but the oil used in industry usually is treated to render it safe for handling. Nevertheless, persons highly sensitive to Rhus-type oleoresins may be reactive. Workers who handle finely divided molding powder may be exposed to dusts that are irritating to the skin and that sometimes cause allergic dermatitis. These resins have also been shown to cause respiratory effects. Pulmonary function testing of workers exposed to PF resin fumes demonstrated airway obstruction (Schoenberg and Mitchell, 1975).

Degradation products from PFs resins are a moderate hazard. Large amounts of carbon monoxide (CO) are produced by PF resin fires, making a potentially hazardous smoke. However, these plastics are good fire retardants and do not produce large amounts of smoke.

REFERENCES

Foussereau, J., Cavelier, C., and Selig, D. (1976) Occupational eczema from para-tertiary-butylphenol formaldehyde resins: a review of the sensitizing resins. *Contact Dermatitis* 2:254–258.

Kieć-Świerczyńska, M. and Szymczak, W. (1995) The effect of the working environment on occupational skin disease development in workers processing rockwool. *Int. J. Occup. Med. Environ. Health* 8:17–22.

Schoenberg, J.B. and Mitchell, C.A. (1975) Airway disease caused by phenolic (phenol-formaldehyde) resin exposure. *Arch. Environ. Health* 30:574–577.

Polyesters and Alkyls

Materials of the polyester and alkyds group are essential to reinforced plastic technology. Reinforcement is generally provided by glass fiber textiles, and the typical structure consists of laminations of polyester and textile. Polyester resins can be saturated or unsaturated depending on the reactants used. A saturated resin will be produced from a polycondensation reaction of a carboxylic acid and an alcohol. For example, the polyester Dacron® (polyethylene terephthalate; CASRN: 25038-59-9) is formed from a reaction of methyl terephthalate and ethylene glycol. Terephthalic acid is perhaps the most commercially important acid used. It is present in most high molecular weight polyester films as polyethylene terephthalate and polybutylene terephthalate (CASRN: 24968-12-5). Saturated resins are important in the boat and automobile industries and are also used in many household furnishings, such as shower stalls.

Manufacturing an unsaturated resin requires the use of a dibasic acid, such as phthalate or maleic anhydride, and a glycol. These are low molecular weight resins that are strong and translucent. Addition of different chemicals to this resin can change its properties to best suit its intended use. Flexibility, solubility, and chemical resistance are some of the changeable characteristics of these polymers. The unsaturated resins generally are nontoxic, but contact dermatitis and skin allergy have occasionally been reported, although one of these reported exposures was not related to an occupational exposure (Lidén et al., 1984; MacFarlane et al., 1986). Poor industrial hygiene while manufacturing these resins was reported to have resulted in several cases of contact dermatitis in one plant (Lim et al., 1970). These polymers are also used in the fiberglass boat industry and are found in furniture and other products that do not require the more expensive saturated resins.

Alkyd resins are similar to polyesters but differ in that they are treated with a triglyceride oil or fatty acid to make them more useful in the coating industry. They are primarily used as a component of paints, varnishes, and lacquers. The toxicity of these compounds is not very noteworthy because industrial standards for the components and co-products of the manufacturing processes are sufficient to protect against harmful exposure to the finished product. Phthalic anhydride (CASRN: 85-44-9) is used in many alkyd resins and is a known irritant and sensitizer. Exposure to compounds such as this is more of a health concern than the actual resin itself.

In a cross-linking step, polyesters are dissolved in styrene and stabilized with an inhibitor, such as hydroquinone, to form a viscous, syrup-like liquid that does not solidify readily. When gelation is desired, a catalyst is added, usually peroxide (e.g., benzoyl peroxide or methyl ethyl ketone peroxide). Promoters or accelerators (dimethylaniline or cobalt salts) may also be added to accelerate this cross-linking solidification. These resins are insoluble in water but are soluble in some organic solvents and complex phenols. Though styrene generally is the reactive solvent and cross-linking group between the polyester macromolecules, other equivalent compounds may be used for special purposes. Methyl methacrylate (MMA) increases resistance to weathering, diallyl phthalate retards polymerization of the uncatalyzed syrup, and triallyl cyanurate increases resistance to elevated temperatures. A wide range of exposures can occur during polyester manufacturing, and proper caution should be observed in the various processes to limit exposure. Polyesters themselves do not cause skin or respiratory tract irritation or sensitization. Health concerns will most likely target styrene exposure, which is covered in the section on polystyrene.

The acid phthalates can cause burns when they come in contact with damp skin. The dusts are irritating to the eyes and upper airway, but there is no important systemic toxicity. Organic peroxides are highly irritating to epithelial tissues and are flammable oxidizing agents. There is also an evidence that benzoyl and methyl ethyl ketone peroxides are skin sensitizers, as are dimethylaniline and cobalt naphthenate. Dimethylaniline can cause most of the effects of aniline. Although the allyl group has well-known toxic properties, serious occupational hazards have not been associated with diallyl phthalate and triallyl cyanurate, both of which are irritants. Polyester dusts produced by grinding operations can result in cough, dyspnea, and eye and throat irritation in exposed workers, who may also show marked reduction in expiratory flow rates (Zuskin et al., 1979). Generally, however, industrial health problems from the polyester group are uncommon, except for the direct effects of styrene, and most of them may be attributed to the mechanical effects of glass fibers on the skin.

REFERENCES

Lidén, C., Löfström, A., and Storgårds-Hatam, K. (1984) Contact allergy to unsaturated polyester in a boatbuilder. *Contact Dermatitis* 11:262–264.

Lim, J., Balzer, J.L., Wolf, C.R., and Milby, T.H. (1970) Fiberglass reinforced plastics: associated occupational health problems. *Arch. Environ. Health* 20:540–544.

MacFarlane, A.W., Curley, R.K., and King, C.M. (1986) Contact sensitivity to unsaturated polyester resin in a limb prosthesis. *Contact Dermatitis* 15:301–303.

Zuskin, E., Saric, M., and Bouhuys, A. (1979) Airway responsiveness in workers processing polyester resins. *J. Occup. Med.* 21:825–827.

Polyurethane

Polyurethane resins and foams (CASRN: 9009-54-5) are two important industrial polymers. They can be produced as rigid, semirigid, or elastic foams or resins, which give PUR many versatile commercial uses. Polyurethane can be found in furniture, bedding material, automotive sealing material, adhesives, carpet, packaging material and coatings, and many other products. It is favored industrially because of its resistance to oil, light, and solvents, in addition to its strength and flexibility. These polymers are formed by polyaddition reactions between a diisocyanate and a polyhydroxyl compound, such as a polyol. In foams, carbon dioxide (CO_2) production is used to give rise to bubbles. Blowing and foaming agents may be used to duplicate this process. Exposure to PUR is of limited concern; the diisocyanate reactant is the main occupational health concern.

Health effects directly associated with PUR exposure are not well documented in humans. Idiomatic responses, such as thrombocytopenia, are uncommon but have been the subject of some case reports (Michelson, 1991). The main route of exposure to these dusts is inhalation, related to airborne particles and spraying of PUR coatings. Dust inhalation studies in animals have not shown a carcinogenic effect, but have demonstrated typical respiratory effects, such as pneumonitis and infiltration of macrophages (Thyssen et al., 1978). Workers exposed to dusts have reported reddening of the eyes and lacrimation (Hosein and Farkas, 1981). The hazards associated with these dusts are considered to be limited, and occupational exposure during production usually is within recommended exposure guidelines.

Industrial fires involving PUR should be managed with caution. The main degradation product at lower fire temperatures is CO, but hydrogen cyanide (HCN) is released in small amounts. At higher temperatures (600–800 °C, 1,112-1,472 °F), HCN becomes a major constituent of the smoke. Pulmonary irritation and edema can occur from exposure to lesser concentrations of these chemicals.

More hazardous than the dusts are the two most common diisocyanates used, toluene diisocyanate (TDI) and methylene diphenyl diisocyanate (MDI), which are discussed later in this section. Others of importance that may also be used are isophorone diisocyanate, hexamethylene diisocyanate, and naphthalene diisocyanate.

TDI   Toluene diisocyanate is the most important chemical in the formation of PURs. It can exist in two forms, 2,4-TDI (CASRN: 584-84-9) and 2,6-TDI (CASRN:91-08-7). Though the chemical properties vary slightly between the two isomers, the commercial mixture of 80:20 of 2,4-TDI:2,6-TDI (CASRN: 26471-62 5) is more prevalent in industry and shall be referred to unless otherwise specified. The monomer is a colorless to pale yellow liquid with a sharp, pungent odor. Soluble in organic solvents, TDI is insoluble in water but is highly reactive with water and other hydrogen-rich compounds. It has a vapor pressure of 0.04 mmHg at 20 °C (68 °F) and a relative vapor density of 6.0 (air = 1) Toluene diisocyanate has not been found to occur naturally and is used in many applications of PUR. Polyurethane resin and flexible foams account for about 90% of all TDI used.

MDI   A white to pale yellow solid, MDI (4,4′-diphenyl-methane diisocyanate; CASRN: 101-68-8) is a compound with a larger molecular weight and thus a lower vapor pressure than TDI. At 0.000005 mmHg (at 20 °C), the vapor pressure assures lower airborne concentrations of MDI than TDI. It is slightly soluble in water at 0.2% by weight and may self-polymerize at 232.2 °C (*ca.* 450 °F). Like TDI, MDI has not been found to occur in nature. The applications for MDI include coatings and lacquers, and it is used primarily to make the more rigid foams and an alternative product, polyisocyanate foam. These rigid foams provide excellent insulation properties in construction and packaging materials.

Industrial Hygiene   Toluene diisocyanate is formed in two ways: as a reaction with phosgene and toluenediamine or by carbonylation of dinitrotoluene followed by thermal conversion. To form PUR, TDI is reacted with a polyol, interacted with urea groups, and cross-linked with water as a mediator. It can also be formed with a polyol to form isocyanate end-groups, and then cross-linked. The expected route of exposure to TDI is inhalation, with dermal contact less common. Airborne concentrations of TDI have been found in the production and processing of PUR and can also be detected in emissions from processing facilities. High performance liquid chromatography and gas chromatography are capable of detecting TDI in the air. Postproduction products of PUR can contain unreacted monomer, and methods have been developed to identify these residues. The exposure guidelines set for TDI exposure are low; the United States (US) National Institute for Occupational Safety and Health (NIOSH) recommends the lowest possible exposure, and the US Occupational Safety and Health Administration (OSHA) has set a permissible exposure limit (PEL) of 0.02 parts per million (ppm) (0.14 mg/m^3) (NIOSH, 1994). The American Conference of Governmental Industrial Hygienists (ACGIH) has set a ceiling of 0.02 ppm and a threshold limit value (TLV) time-weighted average (TWA) of 0.005 ppm (0.036 mg/m^3) (ACGIH, 1995).

The production of MDI involves phosgenation of methylene dianiline that produces a mixture from which MDI can be distilled. The major routes of exposure to MDI are inhalation and dermal contact, particularly during spraying or heating. Occupational levels of MDI can be measured best with simultaneously used liquid absorbers rather than other solvent free-sampling techniques (Andersson et al., 1983). The OSHA has set its PEL for MDI at 0.02 ppm (0.2 mg/m^3), and NIOSH and ACGIH have set both the recommended exposure limit (REL) and the TLV at 0.005 ppm (0.05 mg/m^3) TWA (ACGIH, 1995; NIOSH, 1994).

Occupational protection from TDI and other diisocyanates depends on the setting in which exposure is likely to occur. When higher exposures are likely, such as when the resin temperature exceeds 70 °C (158 °F), suits that offer respiratory protection are recommended. Heating of PURs causes evaporation and release of isocyanates; therefore processing at high temperatures should be conducted in a sealed system or in exhaust-ventilated areas.

Acute and Chronic Effects   Exposure to high concentrations of inhaled TDI over a short period causes mostly irritation effects. Eye irritation and lacrimation are likely symptoms of exposure of the eyes to TDI vapor. Inflammatory changes have been noted in the mucous membranes of the respiratory tract and stomach at lower concentrations of 0.05 ppm and 0.1 ppm (Brugsch and Elkins, 1963). Systemic effects, such as headaches, nausea, and central nervous system (CNS) effects, have also been reported (Le Quesne et al., 1976).

The chronic effects of TDI occur mainly through inhalation and are manifested in the respiratory system. The unique feature of this chemical is its ability to act as a sensitizing agent in many individuals. One of the leading causes of occupational asthma, TDI is responsible for a reaction in 5–10% of exposed workers (Bernstein, 1982). Another complication involving the sensitization mechanism is extrinsic alveolitis, reported less frequently (Vandenplas et al., 1993). Both of these diseases are believed to have a latency period. A wide range of other pulmonary diseases have also been linked to TDI exposure. Loss of forced expiratory volume (FEV1) is a significant finding in some epidemiological studies (Jones et al., 1992; Wegman et al., 1982). Hypersensitivity pneumonitis and hemorrhagic pneumonia have been linked to TDI (Charles et al., 1976; Patterson et al., 1990). These obstructive airway diseases and related tissue insults are characterized by inflammation due to exposure.

Occupationally related asthma is the most common adverse outcome and is frequently reported. Unlike other industrial sensitization reactions, TDI-induced asthma can take many years to resolve. Once sensitized, workers have been known to require daily care and complete separation from TDI. Some workers react to small amounts of TDI as long as 12 years later (Banks et al., 1990; Lozewicz et al., 1987; Moller et al., 1986). This persistent sensitivity to TDI is remarkable, making transfer to another work site or complete removal necessary for the protection of the employee. Though rare, deaths from TDI-induced asthma attacks have occurred, and appropriate measures should be taken to minimize this risk (Fabbri et al., 1988).

Methylene diphenyl diisocyanate long was thought to be of low toxicity because of its low vapor pressure, and MDI-related diseases have been documented separately from TDI disorders. Diminished values for peak expiratory flow rate have been demonstrated in workers exposed to both isocyanates, together and alone (Burge et al., 1979). Asthmatic responses, extrinsic alveolitis, pneumonitis, and airway obstruction can be linked to MDI as well to TDI (Baur et al., 1984; Vandenplas et al., 1993). Some clinicians have reported findings of acute respiratory disorder, rhinoconjunctivitis, and fever associated with chronic exposure to MDI (Littorin et al., 1994).

A number of epidemiological studies have found no association between exposure to TDI, MDI, and other isocyanates and an increased incidence of cancer in the manufacturing industry (Hagmar et al., 1993; Sorahan and Pope, 1993). Animal studies have shown both positive associations and nonassociations (IARC, 1986; Loeser, 1983).

Mechanism of Action The mechanism by which TDI and MDI cause occupational asthma has been studied. An IgE-mediated mechanism has been suggested, but studies have indicated the presence of this mediator in lower numbers of affected workers than expected (Patterson et al., 1987). An IgG-mediated response has been found to be likely for TDI (Cartier et al., 1989). The most likely cause for late asthmatic responses is an inflammatory reaction (Vandenplas et al., 1993). In general, isocyanates act as haptens. A study in guinea pigs provides an *in vitro* model for evaluating the mechanism of action of isocyanates; it demonstrates direct bronchial smooth muscle-induced contractions mediated through a capsaicin-sensitive primary afferent nerve (Mapp et al., 1991).

Pulmonary injury in mice produced by exposure to isocyanates was reported to cause squamous metaplasia of the epithelium (Buckley et al., 1984). Carcinogenicity in animals is suspected with TDI exposure because of its reactive product, diaminotoluene (TDA), which is formed when TDI comes in contact with water. Diaminotoluene is a suspected human carcinogen. Similar to TDI, MDI is degraded to 4,4′-diaminodiphenylmethane, which is also a suspected human carcinogen.

Diagnosis and Treatment Occupational asthma can be defined in a variety of ways. Primarily, it is a small airway sensitization that is reversible and can be induced by either an allergic (e.g., antigen) or nonallergic response in predisposed and many atopic individuals. Isocyanate-induced asthma is diagnosed by using different methods to arrive at an acceptable clinical conclusion. The patient history has been shown to be a better method of ruling out isocyanate asthma than predicting it (Vandenplas et al., 1993). A decrease in the FEV1 and peak expiratory flow may be a good indicator of occupational asthma (Burge et al., 1979; Diem et al., 1982). Bronchial hyperresponsiveness to histamine or methacholine, although not present in all individuals, can demonstrate a sensitization reaction. These initial tests are nonspecific and do not indicate isocyanates exclusively as the primary cause of the condition. Additional pulmonary function tests, along with exposure data, would be required to identify isocyanates as the cause of occupational asthma. Workers with a diagnosis of occupational asthma should follow general treatment procedures. Removal from exposure usually results in

improvement. However, repeated exposure may cause reactions.

Primary treatment for acute effects entails removing the worker from the exposure area and flushing skin or eyes with water. Clinical evaluation should be completed immediately thereafter.

Biomonitoring Diaminotoluene is a decomposition product of TDI. Though 2,6-diaminotoluene dihydrochloride can also be formed, TDA is the main degradation product. It has been shown that measurement of plasma concentrations of TDA can be used to monitor worker exposure to TDI (Persson et al., 1993). No quantitative correlation between TDA urine levels and actual exposure has been found. For MDI, a method for monitoring dermal exposure has been developed using capillary gas chromatographic measurements of 4,4′-diaminodiphenylmethane in hydrolyzed serum and urine (Tiljander et al., 1989).

REFERENCES

ACGIH (1995) *Threshold Limit Values (TLVs) for Chemical Substances and Physical Agents and Biological Exposure Indices (BEIs)*. Cincinnati, OH: American Conference of Governmental Industrial Hygienists.

Andersson, K., Gudehn, A., Levin, J.-O., and Nilsson, C.-A. (1983) A comparative study of solvent and solvent-free sampling methods for airborne 4,4′-diphenylmethane diisocyanate (MDI) generated in polyurethane production. *Am. Ind. Hyg. Assoc. J.* 44:802–808.

Banks, D.E., Rando, R.J., and Barkman, H.W. Jr. (1990) Persistence of toluene diisocyanate-induced asthma despite negligible workplace exposures. *Chest* 97:121–125.

Baur, X., Dewair, M., and Römmelt, H. (1984) Acute airway obstruction followed by hypersensitivity pneumonitis in an isocyanate (MDI) worker. *J. Occup. Med.* 26:285–287.

Bernstein, I.L. (1982) Isocyanate-induced pulmonary disease: a current perspective. *J. Allergy Clin. Immunol.* 70:24–31.

Brugsch, H.G. and Elkins, H.B. (1963) Toluene di-isocyanate (TDI) toxicity. *N. Engl. J. Med.* 268:353–357.

Buckley, L.A., Jiang, X.Z., James, R.A., Morgan, K.T., and Barrow, C.S. (1984) Respiratory tract lesions introduced by sensory irritants at the RD$_{50}$ concentration. *Toxicol. Appl. Pharmacol.* 74:417–429.

Burge, P.S., O'Brien, I.M., and Harries, M.G. (1979) Peak flow rate records in the diagnosis of occupational asthma due to isocyanates. *Thorax* 34:317–323.

Cartier, A., Grammer, L., Malo, J.L., Lagier, F., Ghezzo, H., Harris, K., and Patterson, R. (1989) Specific serum antibodies against isocyanates: association with occupational asthma. *J. Allergy Clin. Immunol.* 84:507–514.

Charles, J., Bernstein, A., Jones, B., Jones, D.J., Edwards, J.H., Seal, R.M., and Seaton, A. (1976) Hypersensitivity pneumonitis after exposure to isocyanates. *Thorax* 31:127–136.

Diem, J.E., Jones, R.N., Hendrick, D.J., Glindmeyer, H.W., Dharmarajan, V., Butcher, B.T., Salvaggio, J.E., and Weill, H. (1982) Five-year longitudinal study of workers employed in a new toluene diisocyanate manufacturing plant. *Am. Rev. Respir. Dis.* 126:420–428.

Fabbri, L.M., Danieli, D., Crescioli, S., Bevilacqua, P., Meli, S., Saetta, M., and Mapp, C.E. (1988) Fatal asthma in a subject sensitized to toluene diisocyanate. *Am. Rev. Respir. Dis.* 137:1494–1498.

Hagmar, L., Strömberg, U., Welinder, H., and Mikoczy, Z. (1993) Incidence of cancer and exposure to toluene diisocyanate and methylene diphenyldiisocyanate: a cohort based case-referent study in the polyurethane foam manufacturing industry. *Br. J. Ind. Med.* 50:1003–1007.

Hosein, H.R. and Farkas, S. (1981) Risk associated with the spray application of polyurethane foam. *Am. Ind. Hyg. Assoc. J.* 42:663–665.

IARC (1986) Toluene diisocyanate. *IARC Monogr. Eval. Carcinog. Risk Chem. Hum.* 39, 287–323. Lyon, France: International Agency for Research on Cancer. Available at http://monographs .iarc.fr/ENG/Monographs/vol1-42/mono39.pdf.

Jones, R.N., Rando, R.J., Glindmeyer, H.W., Foster, T.A., Hughes, J.M., O'Neil, C.E., and Weill, H. (1992) Abnormal lung function in polyurethane workers: weak relationship to toluene diisocyanate exposures. *Am. Rev. Respir. Dis.* 146:871–877.

Le Quesne, P.M., Axford, A.T., McKerrow, C.B., and Jones, A.P. (1976) Neurological complications after a single severe exposure to toluene di-isocyanate. *Br. J. Ind. Med.* 33:72–78.

Littorin, M., Truedsson, L., Welinder, H., Skarping, G., Mårtensson, U., and Sjöholm, A.G. (1994) Acute respiratory disorder, rhinoconjunctivitis and fever associated with the pyrolysis of polyurethane derived from diphenylmethane diisocyanate. *Scand. J. Work Environ. Health* 20:216–222.

Loeser, E. (1983) Long-term toxicity and carcinogenicity studies with 2,4/2,6-toluene-diisocyanate (80/20) in rats and mice. *Toxicol. Lett.* 15:71–81.

Lozewicz, S., Assoufi, B.K., Hawkins, R., and Taylor, A.J. (1987) Outcome of asthma induced by isocyanates. *Br. J. Dis. Chest* 81:14–22.

Mapp, C.E., Graf, P.D., Boniotti, A., and Nadel, J.A. (1991) Toluene diisocyanate contracts guinea pig bronchial smooth muscle by activating capsaicin-sensitive sensory nerves. *J. Pharmacol. Exp. Ther.* 256:1082–1085.

Michelson, A.D. (1991) Thrombocytopenia associated with environmental exposure to polyurethane. *Am. J. Hematol.* 38:145–146.

Moller, D.R., Brooks, S.M., McKay, R.T., Cassedy, K., Kopp, S., and Bernstein, I.L. (1986) Chronic asthma due to toluene diisocyanate. *Chest* 90:494–499.

NIOSH (1994) *Pocket Guide to Chemical Hazards.* NIOSH 94-116. Washington, DC: National Institute for Occupational Safety and Health, Centers for Disease Control and Prevention, US Department of Health and Human Services.

Patterson, R., Hargreave, F.E., Grammer, L.C., Harris, K.E., and Dolovich, J. (1987) Toluene diisocyanate respiratory reactions: I. Reassessment of the problem. *Int. Arch. Allergy Appl. Immunol.* 84:93–100.

Patterson, R., Nugent, K.M., Harris, K.E., and Eberle, M.E. (1990) Immunologic hemorrhagic pneumonia caused by isocyanates. *Am. Rev. Respir. Dis.* 141:226–230.

Persson, P., Dalene, M., Skarping, G., Adamsson, M., and Hagmar, L. (1993) Biological monitoring of occupational exposure to toluene diisocyanate: measurement of toluenediamine in hydrolysed urine and plasma by gas chromatography–mass spectrometry. *Br. J. Ind. Med.* 50:1111–1118.

Sorahan, T. and Pope, D. (1993) Mortality and cancer morbidity of production workers in the United Kingdom flexible polyurethane foam industry. *Br. J. Ind. Med.* 50:528–536.

Thyssen, J., Kimmerle, G., Dickhaus, S., Emminger, E., and Mohr, U. (1978) Inhalation studies with polyurethane foam dust in relation to respiratory tract carcinogenesis. *J. Environ. Pathol. Toxicol.* 1:501–508.

Tiljander, A., Skarping, G., and Dalene, M. (1989) Chromatographic determination of amines in biological fluids with special reference to the biological monitoring of isocyanates and amines: III. Determination of 4,4′-methylenedianiline in hydrolysed human urine using derivatization and capillary gas chromatography with selected ion monitoring. *J. Chromatogr.* 479:145–152.

Vandenplas, O., Malo, J.-L., Saetta, M., Mapp, C.E., and Fabbri, L.M. (1993) Occupational asthma and extrinsic alveolitis due to isocyanates: current status and perspectives. *Br. J. Ind. Med.* 50:213–228.

Wegman, D.H., Musk, A.W., Main, D.M., and Pagnotto, L.D. (1982) Accelerated loss of FEV-1 in polyurethane production workers: a four-year prospective study. *Am J. Ind. Med.* 3:209–215.

Silicones

Silicones are composed of chains of alternate atoms of oxygen and silicon, otherwise known as a siloxane link. Various organic groups can be attached to the silicon atoms, and the amount of cross-linkage between chains by these groups determines whether the polymer will be hard, elastomeric, or fluid. The fully reacted polymers are remarkably inert and therefore are useful in medical applications such as heart valves, prostheses, heart–lung machines, and artificial kidneys. Major industrial uses are in lubricants, encapsulations, dielectric laminates, and "non-stick" release compounds.

The biological properties of the silicone intermediates have been studied. Of the silicones, the chlorosilanes are corrosive to mucous membranes, but the ethoxysilanes are less so. Various metallic soaps, acetic acid, and toluene solvents may be used in the final applications of silicones.

Polydimethylsiloxane   The polydimethylsiloxane (CASRN: 63148-62-9) materials are excellent examples of silicones because of their chemical properties and widespread industrial use. They can be manufactured as gels, resins, fluids, or elastomers, depending on the cross-linking characteristics. Toxicity testing using polydimethylsiloxane compounds on animals has found little, if any, harmful effects associated with chronic exposure. These tests include oral dosing and teratogenicity testing (Cutler et al., 1974; Kennedy et al., 1976). Human health effects associated with silicone implants have been reported. Polydimethylsiloxane is the most prevalent silicone used medically and has been incorporated into many prostheses, including breast implants. A recent epidemiological study of women with and without silicone implants revealed no association between silicone breast implants and connective tissue diseases (Sánchez-Guerrero et al., 1995). Occupational exposure to comparable amounts of silicone is highly unlikely. The manufacture of these silicone compounds has not been linked to adverse health effects except for acute minor eye irritation.

REFERENCES

Cutler, M.G., Collings, A.J., Kiss, I.S., and Sharratt, M. (1974) A lifespan study of a polydimethylsiloxane in the mouse. *Food Cosmet. Toxicol.* 12:443–450.

Kennedy, G.L. Jr., Keplinger, M.L., and Calandra, J.C. (1976) Reproductive, teratologic, and mutagenic studies with some polydimethylsiloxanes. *J. Toxicol. Environ. Health* 1:909–920.

Sánchez-Guerrero, J., Colditz, G.A., Karlson, E.W., Hunter, D.J., Speizer, F.E., and Liang, M.H. (1995) Silicone breast implants and the risk of connective-tissue diseases and symptoms. *N. Engl. J. Med.* 332:1666–1670.

THERMOPLASTICS

Acetal Resins

Acetal resins (CASRN: 9002-81-7), also known as polyoxymethylene and polyformaldehyde, are polymers of pure formaldehyde with unusually good mechanical properties. They are resistant to elongation and impact and replace light metals in many uses. The finished polymer has not been related to adverse health effects in humans or animals. They can be thermally degraded, for example, in an overheated molding machine, to produce formaldehyde gas, with its well-recognized irritant and sensitizing properties.

Acrylics

Acrylic plastics, such as Lucite® and Plexiglas®, are made from methyl and ethyl acrylate and methacrylate. Poly (methyl methacrylate) (CASRN: 9011-14-7) is the most common and best representative of this group of plastics. These substances are respiratory and cutaneous irritants, the acrylates more so than the methacrylates. Studies have shown sensitization reactions to a number of different components of photopolymerizing plates in the printing industry. Contact dermatitis from dermal exposure to uncured plates is not uncommon (Pedersen et al., 1983).

Acrylic paints are composed of a water emulsion of acrylic polymers that have no known biological hazards with occupational exposure or intended environmental use. Aside from paints and coatings, these polymers are also used in food contact items and medical implants because of their low bioavailability.

MMA Methyl methacrylate (CASRN: 80-62-6) is an important monomer and polymer with many industrial applications. As a monomer, MMA has many uses, including latex paints, lacquers, sealants, adhesives, acrylic fiber production, and surface coating resins for paper and leather. The largest application for MMA is in the acrylic sheeting industry. The polymer of MMA is widely used in dental and orthodontic devices and as cement in many orthopedic surgical procedures. Methyl methacrylate is an entirely synthetic chemical. At room temperature, it is a colorless liquid with a fruity, acrid odor. It is slightly soluble in water, 1.5% by weight, and is much more soluble in organic solvents, such as ethanol and acetone. Methyl methacrylate has a vapor pressure of 29.3 mmHg at 20 °C (68 °F) and a relative vapor density of 3.45 (air = 1). When exposed to light, heat, or ionizing radiation, MMA is likely to undergo polymerization.

Industrial Hygiene Methyl methacrylate is mainly produced industrially in two distinct processes, beginning with the chemical isobutylene. In the first process, isobutylene is oxidized to methacrylic acid with nitric acid and subsequently undergoes an esterification process with methanol. The second process involves ammoxidation of isobutylene, then hydrolyzation with sulfuric acid and a reaction with methanol to form the final product. Exposure to MMA industrially most often occurs through inhalation, usually in the production of products containing MMA rather than in the production of MMA itself. A study in one US polystyrene production facility revealed that highest concentrations of airborne MMA are at the batch reactors and the unloading dock (Samimi and Falbo, 1982). It is also liberated when acrylic plastics are cut with a CO_2 laser (IARC, 1994). A second important route of MMA exposure is through the skin. This type of exposure is most notable in hospitals and dental clinics where dermal contact with prostheses and devices containing residual amounts of the monomer occur. Gloves offer some protection but are not completely reliable, since many have been shown to be permeable to MMA (Waegemaekers et al., 1983). Respirators are an effective means of controlling inhalation exposure.

Exposure guidelines set by the ACGIH, NIOSH, and OSHA are all identical. Airborne levels of MMA are limited to a TLV, REL, and PEL of 100 ppm (410 mg/m³) (ACGIH, 1995; NIOSH, 1994).

Acute and Chronic Effects Methyl methacrylate-induced acute toxicity can occur through cutaneous and respiratory exposure routes. Methyl methacrylate has been characterized as a skin and respiratory irritant at low concentrations. At high concentrations systemic injury to other organs may occur, such as the kidneys and heart.

Chronic exposure to MMA can lead to a variety of responses, depending mostly on the exposure route and concentration. As a sensitizing agent, MMA may cause contact dermatitis after dermal exposure (Farli et al., 1990) and an asthmatic condition or other respiratory effects after inhalation (Pickering et al., 1986). There have been several cases of neurological disturbances reported following prolonged dermal exposure to MMA among dental technicians and hygienists. Symptoms described have included numbness, pain, and whitening of the fingers, as well as isolated cases of peripheral neuropathy and loss of motor function (Donaghy et al., 1991; Rajaniemi and Tola, 1985). There have been no reports of adverse reproductive effects in humans, and rodent assays have been either inconclusive or have shown no adverse developmental effects (Solomon et al., 1993).

The carcinogenicity of MMA has been evaluated in both animal studies and human epidemiological studies. Though some report increases in colorectal cancers, these results are considered insignificant, with most studies showing no significant increase, if any is seen at all (Collins et al., 1989; Walker et al., 1991). After review of the literature, the International Agency for Research on Cancer (IARC) has classified this chemical as a Group 3 compound, not classifiable as to its carcinogenicity (IARC, 1994).

Management of signs and symptoms from excessive exposure to MMA is the most likely treatment because the range of effects varies depending on the route of exposure.

Target Organs and Mechanism of Action The target organ for excessive exposure to MMA depends on the route of exposure. Inhalation of the compound is most detrimental to the lungs, the major site of tissue insult. Skin effects are most common from dermal exposure, with some systemic effects occurring in the nervous system (Bereznowski, 1994). Regardless of the exposure route, it has been shown in rats that the major metabolic path of MMA is oxidation of the chemical, leading to expiration as CO_2 (Bratt and Hathway, 1977). The study indicates that up to 65% of MMA absorbed is expired within 2 h. More recent studies have determined that MMA is quickly degraded to methacrylic acid by a serum nonspecific carboxylesterase (Bereznowski, 1995). Methacrylic acid is less toxic than MMA. This rapid hydrolysis of MMA is the reason for its low toxicity. Since the blood is the site of detoxification, this action works to limit systemic organ damage.

Biomonitoring Monitoring for MMA exposure is difficult because of its fast hydrolysis and expiration. However, analysis of blood and urine for the presence of methanol was studied and appears to be an effective procedure for monitoring occupational exposure (Mizunuma et al., 1993). The same study also estimated that only 1.5% of MMA inhaled will be excreted in the urine. Blood analysis for the presence of methacrylic acid could also be a plausible means of biomonitoring. Thus far, however, no studies have indicated this analysis as a reliable test.

REFERENCES

ACGIH (1995) *Threshold Limit Values (TLVs) for Chemical Substances and Physical Agents and Biological Exposure Indices (BEIs)*. Cincinnati, OH: American Conference of Governmental Industrial Hygienists.

Bereznowski, Z. (1994) Effect of methyl methacrylate on mitochondrial function and structure. *Int. J. Biochem.* 26:1119–1127.

Bereznowski, Z. (1995) *In vivo* assessment of methyl methacrylate metabolism and toxicity. *Int. J. Biochem. Cell Biol.* 27:1311–1316.

Bratt, H. and Hathway, D.E. (1977) Fate of methyl methacrylate in rats. *Br. J. Cancer* 36:114–119.

Collins, J.J., Page, L.C., Caporossi, J.C., Utidjian, H.M., and Saipher, J.N. (1989) Mortality patterns among men exposed to methyl methacrylate. *J. Occup. Med.* 31:41–46.

Donaghy, M., Rushworth, G., and Jacobs, J.M. (1991) Generalized peripheral neuropathy in a dental technician exposed to methyl methacrylate monomer. *Neurology* 41:1112–1116.

Farli, M., Gasperini, M., Francalanci, S., Gola, M., and Sertoli, A. (1990) Occupational contact dermatitis in 2 dental technicians. *Contact Dermatitis* 22:282–287.

IARC (1994) Methyl methacrylate. *IARC Monogr. Eval. Carcinog. Risks Hum.* 60:445–474. Lyon, France: International Agency for Research on Cancer. Available at http://monographs.iarc.fr/ENG/Monographs/vol60/mono60.pdf.

Mizunuma, K., Kawai, T., Yasugi, T., Horiguchi, S., Takeda, S., Miyashita, K., Taniuchi, T., Moon, C.S., and Ikeda, M. (1993) Biological monitoring and possible health effects in workers occupationally exposed to methyl methacrylate. *Int. Arch. Occup. Environ. Health* 65:227–232.

NIOSH (1994) *Pocket Guide to Chemical Hazards*. NIOSH 94-116. Washington, DC: National Institute for Occupational Safety and Health, Centers for Disease Control and Prevention, US Department of Health and Human Services.

Pedersen, N.B., Senning, A., and Nielsen, A.O. (1983) Different sensitising acrylic monomers in Napp printing plate. *Contact Dermatitis* 9:459–464.

Pickering, C.A., Bainbridge, D., Birtwistle, I. H., and Griffiths, D.L. (1986) Occupational asthma due to methyl methacrylate in an orthopaedic theatre sister. *Br. Med. J. (Clin. Res. Ed.)* 292:1362–1363.

Rajaniemi, R. and Tola, S. (1985) Subjective symptoms among dental technicians exposed to the monomer methyl methacrylate. *Scand. J. Work Environ. Health* 11:281–286.

Samimi, B. and Falbo, L. (1982) Monitoring of workers exposure to low levels of airborne monomers in a polystyrene production plant. *Am. Ind. Hyg. Assoc. J.* 43:858–862.

Solomon, H.M., McLaughlin, J.E., Swenson, R.E., Hagan, J.V., Wanner, F.J., O'Hara, G.P., and Krivanek, N.D. (1993) Methyl methacrylate: inhalation developmental toxicity study in rats. *Teratology* 48:115–125.

Waegemaekers, T.H., Seutter, E., den Arend, J.A., and Malten, K.E. (1983) Permeability of surgeons' gloves for methyl methacrylate. *Acta Orthop. Scand.* 54:790–795.

Walker, A.M., Cohen, A.J., Loughlin, J.E., Rothman, K.J., and DeFonso, L.R. (1991) Mortality from cancer of the colon or rectum among workers exposed to ethyl acrylate and methyl methacrylate. *Scand. J. Work Environ. Health* 17:7–19.

1,3-Butadiene   One of the many compounds used in the manufacture of plastics and elastomers is 1,3-butadiene (CASRN: 106-99-0), more commonly referred to as simply butadiene. Though butadiene is a component of many polymers, the production of acrylonitrile–butadiene–styrene and acrylonitrile–butadiene copolymers uses 72% of the butadiene produced in the US. Butadiene has also been detected in cigarette smoke, automobile exhaust, and gasoline vapor. Formed mostly from the production of thermogenic resins and rubber, butadiene has not been found to occur naturally. At room temperature, it exists as a colorless, mildly aromatic gas with a potential to explode when mixed with air, making inhalation its primary route of exposure. Butadiene is slightly soluble in water but readily dissolves in organic solvents. The compound has a vapor pressure of 2,100 mmHg at 25 °C (77 °F) and a relative vapor density of 1.87 (air = 1). Once polymerized, it may be considered biologically inert. Occupational exposure is most likely during production and polymerization of the monomer.

Industrial Hygiene   Occupational exposure to butadiene occurs mainly in the unloading and loading of crude product and during sampling (Fajen et al., 1990). A study by NIOSH has indicated that in a given year, 65,000 workers may be exposed to butadiene (Fajen et al., 1990). Exposure during processing usually is a consequence of leaks in the seals of equipment. Maintenance of valves, seals, and pumps to prevent leaks and decontamination of equipment, and monitoring of the work area, are valuable for controlling butadiene exposure. A number of studies have shown that without excessive exposure from leaks, occupational exposure is usually well below guidelines set by OSHA and ACGIH, typically below 2 ppm (Bond et al., 1995; Heseltine et al., 1993). The ACGIH has a TLV TWA of 2 ppm (4.4 mg/m^3); OSHA has a lower PEL TWA of 1 ppm (2.2 mg/m^3) (ACGIH, 1995; IARC, 2008).

Acute and Chronic Effects   Because of the relatively rare occurrence of high levels of exposure, the acute effects of butadiene exposure are seldom observed. At extremely high levels of exposure, there have been reports of respiratory tract and eye irritation. At high levels of exposure, butadiene can also produce anesthetic effects. Removal of the person from high levels of exposure will result in disappearance of symptoms.

1,3-Butadiene has been the subject of some controversy and debate concerning its potential as a human carcinogen. Two cohort studies indicate an association between industrial exposure to butadiene and increased mortality among workers from leukemia and lymphohematopoietic cancer (Divine, 1990; Matanoski et al., 1990). However, these studies have been criticized by those who do not support these findings, and conflicting results exist among similar studies (Acquavella and Cowles, 1993). A retrospective study by Cowles et al. (1994) did not demonstrate any increase in cancer among workers exposed to butadiene levels below ceiling values. A study to investigate the association between prostate cancer and butadiene exposure revealed no association between such exposure and prostate cancer deaths (Downs et al., 1987). The evidence in rodent assays is equally questionable. Most studies conclude that while butadiene is a rather potent carcinogen in mice, it is not in rats (Bond et al., 1995; Owen et al., 1987). Therefore the animal data do not support the conclusion that butadiene is a human carcinogen at occupational levels of exposure in compliance with recommended guidelines.

Target Organ or Site   The lungs are the target site of the acute effects produced by butadiene. However, chronic exposure studies in laboratory animals have revealed that this compound can affect other sites as well, some of which have been reported in humans. The studies in rodents have identified several sites for tumor induction, with the only similar location in both mice and rats being the mammary gland (Melnick and Kohn, 1995). This site has not been linked to human exposure. The suspected higher incidence of leukemia and lymphomas leads to a conclusion that butadiene may have an effect on bone marrow. It has been reported that butadiene is genotoxic in the bone marrow of mice but not rats (Melnick and Kohn, 1995).

Mechanism of Action   Studies in the past have reported that butadiene is a potential mutagen (de Meester et al., 1978). Though itself a mutagen, butadiene metabolites are of greater interest. Butadiene is converted to 1,2-epoxy-3-butene and 3-butene-l,2-diol, both of which have been shown to be carcinogenic (Bond et al., 1995; IARC, 1992, 2008). Studies indicate that mice metabolize 1,2-epoxy-3-butene to 3-butene-l,2-diol, but rats and humans are deficient in this conversion mechanism (Bond et al., 1995). These two compounds have been linked to neoplasms in laboratory animals. However, there is a lack of evidence for the metabolic production of significant amounts of these metabolites in humans at levels of exposure in compliance with recommended guidelines. Both the liver and lungs seem to be important in the detoxification of these compounds. An IARC review of current literature has led to classification of butadiene as " . . . *carcinogenic to humans (Group 1)*" (IARC, 2008).

Biomonitoring Biomonitoring exposure to butadiene is possible by measuring urinary metabolites of butadiene and not the actual compound. Two major urinary metabolites can be measured. The product of epoxybutene hydrolysis can be measured, and the product of glutathione conjugate of epoxybutene can be measured as well. The epoxybutene hydrolysis metabolite can be detected in humans who have been exposed to low levels of butadiene (Heseltine et al., 1993). Advances have occurred in recent years that have resulted in other assays. For example, occupationally exposed workers were monitored by analyzing the adducts formed by the reaction of 1,2-epoxy-3-butene with the terminal valine of hemoglobin (Neumann et al., 1995). This has also been reported to be successful in animals and is useful as a biomonitoring tool for workers.

REFERENCES

ACGIH (1995) *Threshold Limit Values (TLVs) for Chemical Substances and Physical Agents and Biological Exposure Indices (BEIs)*. Cincinnati, OH: American Conference of Governmental Industrial Hygienists.

Acquavella, J.F. and Cowles, S.R. (1993) Re: Lymphohematopoietic cancer in styrene–butadiene polymerization workers. *Am. J. Epidemiol.* 138:765–768.

Bond, J.A., Recio, L., and Andjelkovich, D. (1995) Epidemiological and mechanistic data suggest that 1,3-butadiene will not be carcinogenic to humans at exposures likely to be encountered in the environment or workplace. *Carcinogenesis* 16:165–171.

Cowles, S.R., Tsai, S.P., Snyder, P.J., and Ross, C.E. (1994) Mortality, morbidity, and haematological results from a cohort of long-term workers involved in 1,3-butadiene monomer production. *Occup. Environ. Med.* 51:323–329.

de Meester, C. Poncelet, F., Roberfroid, M., and Mercier, M. (1978) Mutagenicity of butadiene and butadiene monoxide. *Biochem. Biophys. Res. Commun.* 80:298–305.

Divine, B.J. (1990) An update on mortality among workers at a 1,3-butadiene facility: preliminary results. *Environ. Health Perspect.* 86:119–128.

Downs, T.D., Crane, M.M., and Kim, K.W. (1987) Mortality among workers at a butadiene facility. *Am. J. Ind. Med.* 12:311–329.

Fajen, J.M., Roberts, D.R., Ungers, L.J., and Krishnan, E.R. (1990) Occupational exposure of workers to 1,3-butadiene. *Environ. Health Perspect.* 86:11–18.

Heseltine, E., Peltonen, K., Sorsa, M., and Vainio, H. (1993) Assessment of the health hazards of 1,3-butadiene and styrene: meeting report. *J. Occup. Med.* 35:1089–1095.

IARC (1992) 1,3-Butadiene. *IARC Monogr. Eval. Carcinog. Risks Hum.* 54:237–285. Lyon, France: International Agency for Research on Cancer. Available at http://monographs.iarc.fr/ENG/Monographs/vol54/mono54.pdf.

IARC (2008) 1,3-Butadiene. *IARC Monogr. Eval. Carcinog. Risks Hum.* 97:45–184. Lyon, France: International Agency for Research on Cancer. Available at http://monographs.iarc.fr/ENG/Monographs/vol97/mono97.pdf.

Matanoski, G.M., Santos-Burgoa, C., and Schwartz, L. (1990) Mortality of a cohort of workers in the styrene–butadiene polymer manufacturing industry (1943–1982). *Environ. Health Perspect.* 86:107–117.

Melnick, R.L. and Kohn, M.C. (1995) Mechanistic data indicate that 1,3-butadiene is a human carcinogen. *Carcinogenesis* 16:157–163.

Neumann, H.-G., Albrecht, O., van Dorp, C., and Zwirner-Baier, I. (1995) Macromolecular adducts caused by environmental chemicals. *Clin. Chem.* 41, 1835–1840.

Owen, P.E. Glaister, J.R., Gaunt, I.F., and Pullinger, D.H. (1987) Inhalation toxicity studies with 1,3-butadiene: 3. Two year toxicity/carcinogenicity study in rats. *Am. Ind. Hyg. Assoc. J.* 48:407–413.

Polystyrene

Polystyrene (CASRN: 9003-53-6) is prepared by the polymerization of styrene under the influence of organic peroxides in solution. Addition of inhibitors such as hydroquinone or butylcatechol prevents the monomer from polymerizing in storage. The properties of the polymer vary depending on the additives involved and the fabrication method. It can be injection molded, blow molded, thermoformed, or extruded. Prepared as a resin in small spheres, polystyrene is a clear, rigid polymer that is relatively stable under most conditions. Hazards associated with the manufacture of polystyrene are associated with the monomer, styrene, and not the actual polystyrene product. Primarily a packaging material, polystyrene is used mostly as a component of a variety of food containers, such as egg cartons and hot/cold vending cups. It is also used in small appliances, medical devices, and toys. Exposure to polystyrene dusts may cause mechanical irritation of the eyes and respiratory tract. Degradation products resulting from fire release styrene at lower temperatures and mostly CO and CO_2 at higher temperatures, with a usually dense smoke being produced.

Styrene Styrene (phenylethene; CASRN: 100-42-5) is manufactured and used in large volumes. It is mainly used in plastics, latex paints and coatings, synthetic rubbers, polyesters, and styrene–alkyd coatings (Collins and Richey, 1992). Packaging is the single largest use, in the form of styrene resins, particularly foams. Unlike other organic monomers, styrene has been identified in trace amounts in the gummy exudate of tree trunks and in the spice cinnamon (Duke, 1985). At room temperature it is a colorless to yellow, oily liquid with a pungent odor. In water styrene has a low solubility of 0.03% by weight, but it is much more soluble in organic solvents, such as benzene and acetone. Styrene is a somewhat viscous liquid, with a vapor pressure of 5 mmHg at 20 °C (68 °F) and a relative vapor density of 3.6 (air = 1). Styrene is quite reactive and polymerizes rapidly, and if uncontrolled it may cause an explosion (Miller et al., 1994).

Industrial Hygiene Styrene is produced industrially by catalytic dehydrogenation of high purity phenylethene in the vapor phase. It can also be formed as a co-product from a propylene oxide process. In 1993 a study by NIOSH estimated that approximately 1.112 million workers were potentially exposed to styrene. The main route of exposure is through the air, with the highest levels of exposure occurring in the areas of polymerization, manufacturing, and purification (IARC, 1994). The most extensive exposure occurs in the fabrication of boats, tanks, wall panels, and bath and shower units from glass fiber-reinforced polyester composite plastics (Miller et al., 1994). Products with a larger surface area must be made through an open-mold process rather than a press-mold operation. Studies have shown that an open-mold process can lead to a styrene exposure of three times the magnitude of a press-mold process (Lemasters et al., 1985). The OSHA has set the PEL for styrene at 100 ppm TWA, with a 600 ppm 5-min maximum peak in any 3 h; NIOSH has set the REL at 50 ppm TWA, and the ACGIH has set the TLV TWA at 20 ppm (ACGIH, 2001; NIOSH, 1994).

Currently, occupational exposure to styrene is low, with most overexposures occurring as a result of leaks in equipment. Controlling leaks and replacing faulty machinery are the most effective controls for styrene exposure. Studies have indicated that protection of the respiratory system *via* personal respirators is beneficial. Skin protection has a negligible effect on overall exposure but is still necessary for protection against dermatitis (Brooks et al., 1980).

Acute and Chronic Effects The acute effects of excessive exposure to styrene are irritation of the skin, eyes, and airway and depression of the CNS (Härkönen, 1978; Leibman, 1975). High concentration exposures can be hepatotoxic. Disturbances detected in workers include nausea, vomiting, fatigue, and prenarcotic symptoms such as dizziness and drowsiness (Wilson, 1944). *In vitro* studies have found styrene to be acutely cytotoxic to neuronal and nonneuronal cells at high concentrations (Kohn et al., 1995), as well as reducing peroneal nerve conduction velocity (Lilis et al., 1978). The irritant properties of styrene are usually sufficient to protect workers against serious overexposure, but perception of the characteristic odor diminishes with continued exposure.

The chronic effects of styrene have been evaluated, with conflicting reports and studies emerging. A potential confounder of the epidemiological studies involves exposure to other industrial chemicals used during manufacturing, such as benzene and 1,3-butadiene. There have been reported associations between lymphatic and hematopoietic cancers and exposure to styrene (Kolstad et al., 1994; Koveginas et al., 1994). However, other cohort studies do not report an association between lymphatic and hematopoietic cancers and exposure to styrene (Ott et al., 1980; Wong et al., 1994). Studies examining prostate cancer and styrene exposure did not find any association (Coggon et al., 1987; Hodgson and Jones, 1985). An IARC review of current literature has led to classification of styrene as " . . . *possibly carcinogenic to humans (Group 2B)*", with recommendations for further study (IARC, 1994). Further, the US National Toxicology Program (NTP) classified styrene as "[r]easonably anticipated to be a human carcinogen" in its Report on Carcinogens (NTP, 2014a).

Other chronic effects of styrene exposure may include chronic respiratory effects of bronchitis, pulmonary changes, and styrene-induced asthmatic-like conditions (Hayes et al., 1991; Moscato et al., 1987). Analysis of electroencephalographic readings from some exposed workers has shown changes in cerebral activities (Matikainen et al., 1993). Changes in personality have been reported, but validation of these results has not been provided (Lindström and Martelin, 1980). Styrene can pass the placental barrier, but no adverse reproductive effects have been associated with exposure to styrene at occupational levels in compliance with guidelines (Brown, 1991).

Workers occupationally exposed to high concentrations of styrene should be removed from the area as soon as possible. Many of the styrene-induced effects are reversible once exposure is discontinued. Treatment of persistent effects should be based on management of the signs and symptoms.

Target Site and Mechanism of Action The amount of styrene absorbed *via* inhalation is about 60–70% (Stewart et al., 1968). Once styrene absorption has occurred, it is rapidly distributed throughout the body. It has been shown in animal studies to cause tumors of the forestomach and hepatocellular tumors in male mice (IARC, 1994). Styrene has also been found to be stored in adipose tissue and released slowly (IARC, 1994).

Both styrene and its metabolite, styrene-7,8-oxide (SO; CASRN: 96-09-3), are able to cross the blood–brain barrier and thus affect the nervous system (Savolainen and Vainio, 1977). This is possible because of the hydrophobic nature of both compounds.

Styrene is biotransformed to SO by a cytochrome P-450-mediated monooxygenase system (Foureman et al., 1989; Leibman, 1975). Styrene-7,8-oxide is subsequently detoxified by two different pathways: enzymatic hydration to phenylethylene glycol, or styrene glycol, *via* microsomal epoxide hydrolase (Oesch et al., 1971) and a conjugation reaction involving glutathione *S*-transferase detoxification. Further biotransformation of these metabolites yields mandelic acid and phenylglyoxylic acid, which are excreted in the urine. Styrene has also been found to be directly excreted in the urine in small amounts (Periago et al., 1996). Metabolism of styrene also occurs in the heart, kidneys, and lungs, but most styrene is metabolized in the liver (Cantoni et al., 1978).

Genetic changes have been linked to styrene exposure *in vivo* and *in vitro*. One study suggests that low, chronic exposure to styrene may be able to produce more persistent DNA adducts and mutations than acute, short-term exposures (Bastlová et al., 1995). The carcinogenicity of SO has been studied, and results suggest that the metabolite is a carcinogen and responsible for DNA adducts and sister chromatid exchanges (Bastlová et al., 1995; Uusküla et al., 1995). The NTP classified SO as "[r]easonably anticipated to be a human carcinogen" in its Report on Carcinogens (NTP, 2014b).

Biomonitoring   The most effective means of monitoring styrene exposure is through measurement of styrene and its metabolites in the urine. The measurement of mandelic acid and phenylglyoxylic acid is most common, but the determination of which metabolite to measure is subjective because both are present. These measurements have been the subject of some controversy because individual differences have been noted as the result of concurrent exposure to other solvents, alcohol consumption, and individual variation of metabolism (Guillemin and Berode, 1988). More recent studies have shown that direct measurement of styrene in urine by gas chromatography is also reliable (Gobba et al., 1993; Periago et al., 1996).

Some studies have also been reported that measure the hemoglobin adducts formed by the SO metabolite in mice and in humans. Though adducts are present in human hemoglobin, their predictive value has not been validated (Severi et al., 1994).

Risk Assessment   Many studies have been done on occupational exposure to styrene and adverse health effects in humans. Most research is in agreement that, although a long-term exposure hazard cannot be completely ruled out at this time, any carcinogenic risk is likely to be small (Coggon, 1994).

REFERENCES

ACGIH (2001) *Documentation of the Threshold Limit Values and Biological Exposure Indices*, 7th ed. Cincinnati, OH: American Conference of Governmental Industrial Hygienists.

Bastlová, T., Vodicka, P., Peterková, K., Hemminki, K., and Lambert, B. (1995) Styrene oxide-induced HPRT mutations, DNA adducts and DNA strand breaks in cultured human lymphocytes. *Carcinogenesis* 16:2357–2362.

Brooks, S.M., Anderson, L., Emmett, E., Carson, A., Tsay, J.-Y., Elia, V., Buncher, R., and Karbowsky, R. (1980) The effects of protective equipment on styrene exposure in workers in the reinforced plastics industry. *Arch. Environ. Health* 35:287–294.

Brown, N.A. (1991) Reproductive and developmental toxicity of styrene. *Reprod. Toxicol.* 5:3–29.

Cantoni, L., Salmona, M., Facchinetti, T., Pantarotto, C., and Belvedere, G. (1978) Hepatic and extrahepatic formation and hydration of styrene oxide in vitro in animals of different species and sex. *Toxicol. Lett.* 2:179–186.

Coggon, D. (1994) Epidemiological studies of styrene-exposed populations. *Crit. Rev. Toxicol.* 24:S107–S115.

Coggon, D., Osmond, C., Pannett, B., Simmonds, S., Winter, P.D., and Acheson, E.D. (1987) Mortality of workers exposed to styrene in the manufacture of glass-reinforced plastics. *Scand. J. Work Environ. Health* 13:94–99.

Collins, D.E. and Richey, F.A. Jr. (1992) Synthetic organic chemicals. In: Kent, J.A., editor. *Riegel's Handbook of Industrial Chemistry*, 9th ed., New York, NY, USA: Van Nostrand Reinhold, pp. 800–862.

Duke, J.A. (1985) *CRC Handbook of Medical Herbs*, Boca Raton, FL, USA: CRC Press, Inc.

Foureman, G.L., Harris, C., Guengerich, F.P., and Bend, J.R. (1989). Stereoselectivity of styrene oxidation in microsomes and in purified cytochrome P-450 enzymes from rat liver. *J. Pharmacol. Exp. Ther.* 248:492–497.

Gobba, F., Galassi, C., Ghittori, S., Imbriani, M., Pugliese, F., and Cavalleri, A. (1993) Urinary styrene in the biological monitoring of styrene exposure. *Scand. J. Work Environ. Health* 19:175–182.

Guillemin, M.P. and Berode, M. (1988) Biological monitoring of styrene: a review. *Am. Ind. Hyg. Assoc. J.* 49:497–505.

Härkönen, H. (1978) Styrene, its experimental and clinical toxicology: a review. *Scand. J. Work Environ. Health* 4(Suppl. 2):104–113.

Hayes, J.P., Lambourn, L., Hopkirk, J.A., Durham, S.R., and Taylor, A.J. (1991) Occupational asthma due to styrene. *Thorax* 46:396–397.

Hodgson, J.T. and Jones, R.D. (1985). Mortality of styrene production, polymerization and processing workers at a site in northwest England. *Scand. J. Work Environ. Health* 11:347–352.

IARC (1994) Styrene. *IARC Monogr. Eval. Carcinog. Risks Hum.* 60:233–320. Lyon, France: International Agency for Research on Cancer. Available at http://monographs.iarc.fr/ENG/Monographs/vol60/mono60.pdf.

Kohn, J., Minotti, S., and Durham, H. (1995) Assessment of the neurotoxicity of styrene, styrene oxide, and styrene glycol in primary cultures of motor and sensory neurons. *Toxicol. Lett.* 75:29–37.

Kolstad, H.A., Lynge, E., Olsen, J., and Breum, N. (1994) Incidence of lymphohematopoietic malignancies among styrene-exposed workers of the reinforced plastics industry. *Scand. J. Work Environ. Health* 20:272–278.

Koveginas, M., Ferro, G., Andersen, A., Bellander, T., Biocca, M., Coggon, D., Gennaro, V., Hutchings, S., Kolstad, H., Lundberg, I., Lynge, E., Partanen, T., and Saracci, R. (1994) Cancer mortality in a historical cohort study of workers exposed to styrene. *Scand. J. Work Environ. Health* 20:251–261.

Leibman, K.C. (1975) Metabolism and toxicity of styrene. *Environ. Health Perspect.* 11:115–119.

Lemasters, G.K., Carson, A., and Samuels, S.J. (1985) Occupational styrene exposure for twelve product categories in the reinforced-plastics industry. *Am. Ind. Hyg. Assoc. J.* 46:434–441.

Lilis, R., Lorimer, W.V., Diamond, S., and Selikoff, I.J. (1978) Neurotoxicity of styrene in production and polymerization workers. *Environ. Res.* 15:133–138.

Lindström, K. and Martelin, T. (1980) Personality and long term exposure to organic solvents. *Neurobehav. Toxicol.* 2:89–100.

Matikainen, E., Forsman-Grönholm, L., Pfäffli, P., and Juntunen, J. (1993) Nervous system effects of occupational exposure to styrene: a clinical and neurophysiological study. *Environ. Res.* 61:84–92.

Miller, R.R., Newhook, R., and Poole, A. (1994) Styrene production, use, and human exposure. *Crit. Rev. Toxicol.* 24:S1–S10.

Moscato, G., Biscaldi, G., Cottica, D., Pugliese, F., Candura, S., and Candura, F. (1987) Occupational asthma due to styrene: two case reports. *J. Occup. Med.* 29:957–960.

NIOSH (1994) *Pocket Guide to Chemical Hazards.* NIOSH 94-116. Washington, DC: National Institute for Occupational Safety and Health, Centers for Disease Control and Prevention, US Department of Health and Human Services.

NTP (2014a) Styrene CAS No. 100-42-5. In: *Report on Carcinogens*, 13th ed. Research Triangle Park, NC: National Toxicology Program, US Department of Health and Human Services. Available at http://ntp.niehs.nih.gov/pubhealth/roc/index.html.

NTP (2014b) Styrene-7,8-oxide CAS No. 96-09-3. In: *Report on Carcinogens*, 13th ed. Research Triangle Park, NC: National Toxicology Program, US Department of Health and Human Services. Available at http://ntp.niehs.nih.gov/pubhealth/roc/index.html.

Oesch, F., Jerina, D.M., and Daly, J. (1971). A radiometric assay for hepatic epoxide hydrase activity with [7-^{3}H] styrene oxide. *Biochim. Biophys. Acta* 227:685–691.

Ott, M.G., Kolesar, R.C., Scharnweber, H.C., Schneider, E.J., and Venable, J.R. (1980) A mortality survey of employees engaged in the development or manufacture of styrene-based products. *J. Occup. Med.* 22:445–460.

Periago, J.F., Prado, C., and Luna, A. (1996) Purge-and-trap method for the determination of styrene in urine. *J. Chromatogr. A* 719:53–58.

Savolainen, H. and Vainio, H. (1977) Organ distribution and nervous system binding of styrene and styrene oxide. *Toxicology* 8:135–141.

Severi, M., Pauwels, W., Van Hummelen, P., Roosels, D., Kirsch-Volders, M., and Veulemans, H. (1994) Urinary mandelic acid and hemoglobin adducts in fiberglass-reinforced plastics workers exposed to styrene. *Scand. J. Work Environ. Health* 20:451–458.

Stewart, R.D., Dodd, H.C., Baretta, E.D., and Schaffer, A.W. (1968) Human exposure to styrene vapor. *Arch. Environ. Health* 16:656–662.

Uusküla, M., Järventaus, H., Hirvonen, A., Sorsa, M., and Norppa, H. (1995) Influence of *GSTM1* genotype on sister chromatid exchange induction by styrene-7,8-oxide and 1,2-epoxy-3-butene in cultured human lymphocytes. *Carcinogenesis* 16:947–950.

Wilson, R.H. (1944) Health hazards encountered in the manufacture of synthetic rubber. *JAMA* 124:701–703.

Wong, O., Trent, L.S., and Whorton, M.D. (1994) An updated cohort mortality study of workers exposed to styrene in the reinforced plastics and composites industry. *Occup. Environ. Med.* 51:386–396.

Cellulose Derivatives (Cellulosics)

Cellulose acetate (CASRN: 9004-35-7), cellulose propionate, and cellulose butyrate, have almost entirely replaced industrial use of cellulose nitrate, a highly flammable source of nitrogen dioxide gas when burned. Modern cellulosics are used in films and coatings in small parts such as knobs, eyeglass frames, and pencil barrels. The properties of these compounds rely primarily on their additive and individual properties, but in general they are insoluble and are especially clear and lustrous. Direct toxicity is not associated with these polymers, but pyrolysis products and their high flammability can cause serious health hazards. An important characteristic of their burning properties is a tendency to smolder. Organic in nature, these polymers are chemically modified from wood and cotton derivatives and thermally degrade to high concentrations of CO_2 and CO. Industrial fires involving these products should be managed with caution. The manufacturing process may involve exposure to solvents and organic acid compounds that may result in skin reactions.

Fluoroplastics

Fluoroplastics vary somewhat in composition and include polytetrafluoroethylene (PTFE), polyfluorinated ethylene propylene, and polyvinylidene fluoride. They are manufactured from a fluorination of a carbon polymer backbone. These materials have high thermal stability, are biologically inert, and can be used for human organ prostheses, as well as for highly resistant insulations, chemical piping, containers, gaskets, and coatings. Their well-known high resistance to heat makes them a useful material in high temperature applications such as boiling and baking. The widely used Teflon® is a variety of fluorocarbon resin.

Polytetrafluoroethylene   One of the most notable compounds in this group of polymers is PTFE (CASRN: 9002-84-0). It is produced by a polymerization of tetrafluoroethylene. The general fluoroplastic characteristics of thermal and chemical stability are well demonstrated in this polymer. Animal studies and human evidence do not indicate that these materials are carcinogenic. Most case studies of health effects were a result of exposure to PTFE. Exposure to degradation products is of greatest health concern. Exposure to the dusts of this plastic can be harmful and are regulated as a nuisance dust by most agencies. Pyrolysis products from PTFE first become evident at temperatures above 300 °C (572 °F). These products include tetrafluoroethylene, hydrogen fluoride, silicon tetrafluoride, and an incomplete waxy substance (Okawa and Polakoff, 1974). Hexafluoropropylene and octafluoroisobutylene are released at higher temperatures, and the toxic compounds perfluoroisobutylene and carbonyl fluoride are evolved above 400 °C (752 °F). All of these compounds are potentially lethal at high concentrations, with the most toxic degradation product being perfluoroisobutylene. Workers who smoke have been the subject of many reports of "polymer fume fever" as a result of PTFE resin-contaminated cigarettes. The burning ember reaches high enough temperatures [about 875 °C (1,607 °F)] to expose the worker to sufficient toxic pyrolysis products. Illness resulting from exposure to the pyrolysis of fluorohydrocarbons is similar to influenza and was first described by the term "polymer-fume fever" by Harris (1951). Subsequent reports, by Lewis and Kerby (1965) and Wegman and Peters (1974), described the symptoms that appear 4–5 h after exposure to sublimates or thermal decomposition products of fluoroplastics. These consist of chest tightness and dyspnea, followed rapidly by general malaise, aching, weakness, headache, and cough. Chills and fever to 40 °C (104 °F) follow the onset of symptoms by about 12 h. The acute illness may be alarming and uncomfortable when it is severe, but it is of short duration, subsiding within 48 h.

Acute exposure is the main concern, but some reports indicate that repeated episodes of polymer fume fever may have serious pulmonary effects and long-term cumulative effects (Brubaker, 1977). Pulmonary edema is a rare yet serious outcome.

Harmful worker exposure is not commonly related to the production of the polymers themselves because the reactions are conducted in closed systems. The chemical intermediates include fluorohydrocarbons of the so-called Freon type, which are relatively inert or have only narcotic properties at high concentrations. More active reactants may occur in the polymerization process, and these may be highly corrosive and possibly nephrotoxic, but acute exposures are unlikely.

Worker protection is provided through warnings to the workers, prohibition of smoking in the workplace, hand washing, and attention to airborne dust and heat sources. Adequate ventilation and proper housekeeping should be sufficient to avoid severe exposures. Processing temperatures should be monitored to avoid decomposition product exposure. In the past, biomonitoring of workers involved in the production of fluoroplastics revealed fluoride concentrations in the urine below levels corresponding to a toxic exposure (Polakoff et al., 1974). These early tests led to the conclusion that occupational exposure to these polymers is not a serious threat to workers' health.

REFERENCES

Brubaker, R.E. (1977) Pulmonary problems associated with the use of polytetrafluoroethylene. *J. Occup. Med.* 19:693–695.

Harris, D.K. (1951) Polymer-fume fever. *Lancet* 2:1008–1011.

Lewis, C.E. and Kerby, G.R. (1965) An epidemic of polymer-fume fever. *JAMA* 191:375–378.

Okawa, M.T. and Polakoff, P.L. (1974) Occupational health case reports: No. 7. Teflon. *J. Occup. Med.* 16:350–355.

Polakoff, P.L., Busch, K.A., and Okawa, M.T. (1974) Urinary fluoride levels in polytetrafluoroethylene fabricators. *Am. Ind. Hyg. Assoc. J.* 35:99–106.

Wegman, D.H. and Peters, J.M. (1974) Polymer fume fever and cigarette smoking. *Ann. Intern. Med.* 81:55–57.

Polyamides

Polyamides, such as the nylons, are used in fibers, filaments, castings, and extrusions and can be classified into two families: the diamine-dibasic acid type and the condensed amino acid type. No known toxic effects are associated with the end product of either family, although nylons are not entirely inert biologically. Diamine-dibasic acid nylons are prepared by the reaction of a diamine, such as hexamethylenediamine, with an acid, such as adipic acid or sebaceous acid. The diamine has a primary irritant as well as a sensitizing effect on skin and is irritating to the eyes and upper airway. The dibasic acids are not of a significant health hazard.

Nylons are synthetics produced as a fiber, plastic, or film. They are extensively used industrially and domestically because of their stability, which renders them resistant to solvents and abrasives. Household appliances manufacturing and the automotive industry are the two largest users of nylon. Nylons 6 and 6,6 are two of the most important and widely used polyamides. The formation of nylon 6 (polycaprolactam; CASRN: 25038-54-4) involves a batch or continuous polymerization process in which caprolactam reacts with water at high temperatures. Nylon-6,6 (CASRN: 32131-17-2) is manufactured by condensing a nylon salt formed from a solution of hexamethylene diammonium and adipic acid. Special-purpose nylons produced include nylon-6,10, nylon-6,12, and nylon-11. Although nylons themselves are not completely biologically inert, the relatively few effects reported may be attributed to monomers and additives. Rare skin reactions to virgin nylon products have been described and have been attributed to residual reactants or to low molecular weight polyamides (Morris, 1960). Dyes and finishes applied to nylon for wearing apparel have also been uncommon causes of contact dermatitis.

Caprolactam   Caprolactam (hexahydro-2H-azepin-2-one; CASRN: 105-60-2) is the main component of the synthetic fiber nylon 6. It is a white, crystalline solid with an unpleasant odor. A nonvolatile compound, it has a low vapor pressure of 0.00000008 mmHg and is not found to occur in nature. Caprolactam is 53% soluble in water by weight and is also soluble in organic solvents. Production involves a conversion of cyclohexanone to caprolactam in the presence of aqueous ammonia. The standards for dust exposure to this chemical are: $5 \, mg/m^3$ TLV TWA for ACGIH and $1 \, mg/m^3$ REL TWA for NIOSH, with a 15-min short-term exposure limit of $3 \, mg/m^3$ (ACGIH, 2001; NIOSH, 1994). For vapor exposure NIOSH is recommending $1 \, mg/m^3$ REL TWA and ACGIH $5 \, mg/m^3$ TLV TWA.

The toxicity of caprolactam is relatively low. The IARC concluded that "[t]here is *evidence suggesting a lack of carcinogenicity* of caprolactam in experimental animals" and that it is "*. . . probably not carcinogenic to humans (Group4)*" (IARC, 1999). Fukushima et al. (1991) reported a wide spectrum of carcinogenicity data that demonstrated no link between exposure to caprolactam and cancer. The sites of action for caprolactam are thought to be the eyes, skin, respiratory system, CNS, cardiovascular system, liver, and kidneys. Liver damage was reported in rats as elevated serum alanine aminotransferase levels, but treatment did not cause DNA damage (Kitchin and Brown, 1989). There have been some reported cases of eye and upper airway irritation from exposure to caprolactam vapor (Ferguson and Wheeler, 1973). Uncommon contact dermatitis has been reported, but this is an infrequent outcome (Morris, 1960).

The metabolism of caprolactam in humans has not been described. In rats it has been shown that caprolactam is excreted in the urine as both unchanged caprolactam and as epsilon amino acid (Goldblatt et al., 1954).

The most prominent air contaminant in many typical polycaprolactam nylon operations is derived from eutectic mixtures of diphenyl and diphenyl oxide. This material is heated and pumped under pressure as a heat transfer fluid. In such applications the material may be heated to temperatures as high as 370 °C (698 °F); thus there is a potential for serious burns. Small leaks may give rise to vapors and mists, which reportedly induce nausea. Excessive exposure and poor hygienic conditions have resulted in liver injury and nervous system changes.

Degradation products found in smoke from nylon 6 fires have been identified. Upon combustion of nylon 6, HCN, CO, and CO_2 are produced. Nylons in general can release significant amounts of ammonia gas (Morikawa, 1978). Particular caution should be used when an industrial fire involves these polymers.

REFERENCES

ACGIH (2001) *Documentation of the Threshold Limit Values and Biological Exposure Indices*, 7th ed. Cincinnati, OH: American Conference of Governmental Industrial Hygienists.

Ferguson, W.S. and Wheeler, D.D. (1973) Caprolactam vapor exposures. *Am. Ind. Hyg. Assoc. J.* 34:384–389.

Fukushima, S., Hagiwara, A., Hirose, M., Yamaguchi, S., Tiwawech, D., and Ito, N. (1991) Modifying effects of various chemicals on preneoplastic and neoplastic lesion development in a wide-spectrum organ carcinogenesis model using F344 rats. *Jpn. J. Cancer Res.* 82:642–649.

Goldblatt, M.W., Farquharson, M.E., Bennett, G., and Askew, B.M. (1954) ε-Caprolactam. *Br. J. Ind. Med.* 11:1–10.

IARC (1999) Caprolactam. *IARC Monogr. Eval. Carcinog. Risks Hum.* 71:383–400. Lyon, France: International Agency for Research on Cancer. Available at http://monographs.iarc.fr/ENG/Monographs/vol71/mono71.pdf.

Kitchin, K.T. and Brown, J.L. (1989) Biochemical studies of promoters of carcinogenesis in rat liver. *Teratog. Carcinog. Mutagen.* 9:273–285.

Morikawa, T. (1978) Evolution of hydrogen cyanide during combustion and pyrolysis. *J. Combust. Toxicol.* 5:315–330.

Morris, G.E. (1960) Nylon dermatitis. *N. Engl. J. Med.* 263:30–32.

NIOSH (1994) *Pocket Guide to Chemical Hazards*. NIOSH 94-116. Washington, DC: National Institute for Occupational Safety and Health, Centers for Disease Control and Prevention, U.S. Department of Health and Human Services.

Polycarbonates

Polycarbonates are polyesters formed from the polymerization of bisphenol through carbonate linkages. Bisphenol A (4,4'-isopropylidenediphenol; CASRN: 80-05-7) with a phosgene carbonic acid derivative is most commonly used. There are two processes involved in producing these resins, depending on the functionality of the carbonate desired. Polycarbonate resins are amorphous and unusually stable in most solvents. The plastic is used primarily for small mechanical parts such as gears and cams and also in products such as bottles because of its strength and rigidity. Biological effects have not been associated with the finished polymer. Hazards are associated with production of the polymer, and adverse reactions may be associated with exposure to BPA, phosgene, and other co-monomers and additives used during the production of the polymer. Processing fumes of this resin may contain a variety of irritant substances. Procedures for minimizing occupational exposure include adequate ventilation and personal protection.

Bisphenol A is a halogenated, cyclic hydrocarbon that is also used as a fungicide and a component of epoxy resins. It is a white solid that is insoluble in water. Caution should be taken with this compound because it is incompatible with strong oxidizers, strong bases, acid anhydrides, and acid chlorides. The toxicity of this compound is generally low, but after dermal absorption it may affect the liver, spleen, and pancreas. A skin allergy after chronic exposure may occur, and a severe enough acute exposure may even burn the eyes and skin. Biotransformation of this compound in the rat results in metabolites primarily excreted as glucuronide conjugates in urine and as hydroxylated compounds in feces (Knaak and Sullivan, 1966). Recent *in vitro* studies have shown that oxidation of up to 70% of BPA to bisphenol *o*-quinone *via* a semiquinone intermediate is an important metabolic pathway (Atkinson and Roy, 1995a). The reactive intermediate is capable of covalently binding to DNA and producing adducts (Atkinson and Roy, 1995b). The characterization of BPA metabolism and its intermediates is not complete, and more studies are necessary. Bisphenol A, like many of the dihydroxyphenols, is structurally similar to synthetic estrogens and has demonstrated estrogenic effects in female rats. Recommendations for occupational hygiene include proper exhaust ventilation and careful washing of clothes and skin that come in contact with BPA.

REFERENCES

Atkinson, A. and Roy, D. (1995a) In vitro conversion of environmental estrogenic chemical bisphenol A to DNA binding metabolite(s). *Biochem. Biophys. Res. Commun.* 210:424–433.

Atkinson, A. and Roy, D. (1995b) In vivo DNA adduct formation by bisphenol A. *Environ. Mol. Mutagen.* 26:60–66.

Knaak, J.B. and Sullivan, L.J. (1966) Metabolism of bisphenol A in the rat. *Toxicol. Appl. Pharmacol.* 8:175–184.

Polyvinyl Chloride

Polyvinyl chloride (PVC; CASRN: 9002-86-2) is a widely used synthetic polymer made from vinyl chloride monomer (VCM). In industry it is used primarily in floor tiles, ducts, waterproof clothing and upholstery, insulation, piping, packaging, and medical devices. A pure polymer resin of PVC is rigid and somewhat unstable at high temperatures and under exposure to light. A variety of additives are used to increase the elasticity or rigidity, heat stability, and flame resistance, and other fillers and stabilizers are used as well. There are four polymerization processes, which vary in frequency of use. The molecular weight and size of the resin produced depend on the process used. As with most polymers, PVC is generally considered to be biologically inert. Exposures to PVC dusts, degradation products, and additives, and to VCM are the greatest hazards.

PVC Dusts   Exposure to PVC dust can occur during many uses of the polymer. Earlier studies have indicated that certain types of PVC dust are capable of causing cell membrane damage (Richards et al., 1975). Epidemiological and animal studies have shown significant increases in pulmonary dysfunction, most often pneumoconiosis associated with PVC dusts. This is not uncommon when chronic exposure to inert dusts occurs. The reaction occurs mostly in the lung tissues, with an inflammatory response leading eventually to fibrosis and granulomatous lesions (Agarwal, 1983). Other clinical symptoms can occur simultaneously with pneumoconiosis, such as systemic sclerosis that is thought to be caused by the presence of nondigestible particles in macrophages (Studnicka et al., 1995). Treatment of excessive exposure to dusts and the associated signs and symptoms is discussed in a separate section. Occupational exposure to PVC dusts is regulated as a respirable particulate by OSHA and a particle not otherwise specified by ACGIH.

Degradation   Products   Though   PVC   does   not depolymerize under extreme heat, it will thermally decompose to more reactive and harmful products. The most harmful product detected in the smoke from PVC fires has been hydrogen chloride, at relatively high concentrations (Baxter et al., 1995). Hydrogen chloride is a respiratory irritant that can aggravate asthma and allergic conditions, as well as irritate the mucosal membranes. Other components of PVC smoke include CO and a variety of hydrocarbons. Exposure to this smoke should be avoided.

Vinyl   Chloride   Monomer   Vinyl   chloride   monomer (chloroethylene; CASRN: 75-01-4) is the main component of polyvinyl chloride. First used as an anesthetic before the 1950s, it is today one of the most widely used monomers in the plastics industry. At room temperature it exists as a colorless, flammable gas with an ethereal odor. The explosive properties of this chemical were the major industrial hazard before the discovery of its biological effects. Slightly soluble in water, VCM is more soluble in organic solvents. It is not known to occur naturally and has a vapor pressure of 2,580 mmHg at 20 °C (68 °F) and a relative vapor density of 2.15 (air = 1). As one of the first recognized occupational carcinogens, the toxicity of VCM has been well characterized.

Industrial Hygiene   Vinyl chloride monomer is synthesized by a halogenation process involving ethylene (Budavari, 1989). Thousands of workers every year are potentially exposed to VCM occupationally. Workers involved in the production of PVC and VCM are those most likely to be exposed to the highest levels of vinyl chloride, particularly those involved in the cleaning and maintenance processes (Purchase et al., 1987). Inhalation is by far the most common route of exposure. Today, airborne levels of VCM in well-maintained settings generally are below 1 ppm because of the nature and reputation of the chemical. In the past, exposure to unreacted VCM in the postproduction areas of PVC manufacturing were of some concern, but current washing procedures have brought levels to an almost undetectable amount. Since the association of VCM with hepatic angiosarcoma (ASL), the permissible levels of exposure to vinyl chloride have been reduced. The OSHA has set its PEL TWA at 1 ppm (5 ppm for 15 min). The NIOSH has listed the chemical as an occupational carcinogen, recommending that exposure be reduced to the lowest possible levels (NIOSH, 1994). The ACGIH has set a TLV TWA value of 1 ppm (ACGIH, 2001).

Current practices in most industrial settings have maintained low levels of airborne vinyl chloride. High concentrations have a detectable odor, but concentrations likely to be encountered in the workplace are undetectable by odor. Equipment leaks and accidents are probably the cause of most cases of overexposure. Personal respirators and maintenance are recommended to prevent accidental excessive exposure.

Acute Effects   Early reports of the effects of vinyl chloride focused on the acute effects of the compound. At high concentrations VCM has a narcotic effect and depresses the CNS, beginning at levels above 8,000 ppm (Mastromatteo et al., 1960). Cases of workers becoming dizzy or unconscious, or both, have been reported, although few occupational deaths have been associated with VCM. The flammability of the gas, as well as cardiac and circulatory disturbances in patients, was a deterrent for using VCM in surgical procedures (Oster et al., 1947). Vinyl chloride is a liquid at temperatures below $-21.7\,°C$ $(-7.06\,°F)$. Dermal contact with the liquid at this temperature causes frostbite-like symptoms.

Chronic Effects and Target Organs   Chronic exposure to vinyl chloride has been well documented and may have a wide range of effects on various organs and organ systems. The primary target organ is the liver, with effects also noted in the CNS, the blood and lymphatic system, and the respiratory system. The carcinogenicity of VCM has been extensively studied in recent years.

One of the first occupational disorders associated with vinyl chloride was acroosteolysis and Raynaud's phenomenon in the hands of those who cleaned autoclaves and monomer-manufacturing vessels (Harris and Adams, 1967; Wilson et al., 1967). These reports were among the first to question the moderate toxicity associated with VCM at that time. Systemic sclerosis, affecting the skin and other organs, has also been observed in some workers (Haustein and Ziegler, 1985). Diagnosis of acroosteolysis was accomplished by use of X-rays. A combination of these signs and symptoms led to the development of a diagnosis of "vinyl chloride disease" associated with the industry. Treatment of symptoms and removal of the worker from the VCM environment were helpful in reversing some of the conditions. A study by Simonsen et al. (1994) has identified VCM as a possible neurotoxicant, with only limited evidence in humans.

The most notable and well-known chronic effect of vinyl chloride is the development of ASL. First reported by Creech and Johnson (1974), several epidemiological studies since then have confirmed this association (Smulevich et al., 1988; Wong et al., 1991; Wu et al., 1989). The extremely rare occurrence of this disease in the general population was the most prominent factor in discovering the correlation. Symptoms of this carcinoma include hepatomegaly accompanied by splenomegaly and gastrointestinal bleeding (Heath et al., 1975). Animal studies have confirmed the epidemiological

evidence and identified the mechanism for the development of the cancer (Froment et al., 1994; Viola et al., 1971).

Other cancers have also been linked to vinyl chloride exposure, although none as strongly as ASL. Associations have been reported with other liver cancers and brain and CNS cancers, and weaker and more controversial associations have been reported with lung and blood/lymphatic cancers (Beaumont and Breslow, 1981; Weinman and Chopra, 1987; Wong et al., 1991; Wu et al., 1989). These associations have not been verified at the present time. An evaluation of three teratogenicity studies does not provide any significant evidence that VCM is a cause of birth defects (Hemminki and Vineis, 1985).

Mechanism of Action   The primary site for metabolism of vinyl chloride is the liver. As with many other xenobiotics, it is not the VCM that is directly responsible for cell injury. The metabolism of VCM produces reactive intermediates, which, as a result of detoxification, lower intracellular glutathione levels after exposure (Barton et al., 1995). In an early study in rats, Hefner et al. (1975) studied the possibility that VCM is metabolized by the alcohol dehydrogenase pathway to chloroethylene oxide, which spontaneously rearranges to chloroacetaldehyde. There is a similar pathway in humans resulting in an electrophilic epoxide metabolite produced by cytochrome P-450 2E1 (Guengerich et al., 1991). Both of these metabolites produce adducts with DNA and proteins (Fedtke et al., 1990). It is these metabolites and pathways of biotransformation that are most responsible for the carcinogenic activity of vinyl chloride. Studies have suggested that these pathways and metabolites are the cause of cancers, tumors, and lymphatic tissue disorders in humans and animals (Froment et al., 1994; Fucic et al., 1995; Trivers et al., 1995). Although some of the vinyl chloride is expired in air, a larger amount has been observed to be excreted in urine in the form of subsequent polar compounds (Watanabe et al., 1976).

Complications involving Raynaud's phenomenon arise from blood vessel obstruction in the extremities. X-rays demonstrate osteolytic lesions, likened to "bullet holes," in the distal phalanges in conjunction with acroosteolysis (Haustein and Ziegler, 1985). Massive acute poisonings in rodents have identified congestion of the lungs with pulmonary edema and hemorrhages as the main cause of death (Mastromatteo et al., 1960).

Biomonitoring   Biomonitoring of radiolabeled ^{14}C has been used to examine the metabolic pathways of VCM and to measure levels exhaled as $^{14}CO_2$ in expired breath and as urine metabolites. However, these monitoring results would not be useful for assessment of the low exposures likely to occur today. Chromosomal breakage and sister chromatid exchanges have also been evaluated as possible monitoring aids. They may be useful for evaluating immediate exposure but are less likely to be accurate for long-term monitoring because of DNA repair (Du et al., 1995). Measurements of certain serum antibodies may be useful as subclinical diagnostic tools for assessing ASL and other cancer risks (Trivers et al., 1995).

Risk Assessment A risk assessment for vinyl chloride is possible with a few important assumptions made regarding animal modeling. It would not be entirely correct to assume that the dosimetry and lifetimes of an experimental animal and humans are wholly comparable. Risk assessments made for VCM have produced different results, depending on the assumptions made in the calculations. Extrapolations of past exposures confounded by the long latency period of ASL (as long as 20 years) have led some researchers to predict a relatively high number of ASL cases in the future (Purchase et al., 1987).

REFERENCES

ACGIH (2001) *Documentation of the Threshold Limit Values and Biological Exposure Indices*, 7th ed. Cincinnati, OH: American Conference of Governmental Industrial Hygienists.

Agarwal, D.K. (1983) Biochemical assessment of the bioreactivity of intratracheally administered polyvinyl chloride dust in rat lung. *Chem. Biol. Interact.* 44:195–201.

Barton, H.A., Creech, J.R., Godin, C.S., Randall, G.M., and Seckel, C.S. (1995) Chloroethylene mixtures: pharmacokinetic modeling and *in vitro* metabolism of vinyl chloride, trichloroethylene, and *trans*-1,2-dichloroethylene in rat. *Toxicol. Appl. Pharmacol.* 130:237–247.

Baxter, P.J., Heap, B.J., Rowland, M.G., and Murray, V.S. (1995) Thetford plastics fire, October 1991: the role of a preventive medical team in chemical incidents. *Occup. Environ. Med.* 52:694–698.

Beaumont, J.J. and Breslow, N.E. (1981) Power considerations in epidemiologic studies of vinyl chloride workers. *Am. J. Epidemiol.* 114:725–734.

Budavari, S. (1989) *The Merck Index: An Encyclopedia of Chemicals, Drugs and Biologicals*, 11th ed., Rahway, NJ, USA: Merck & Co., Inc.

Creech, J.L. Jr. and Johnson, M.N. (1974) Angiosarcoma of liver in the manufacture of polyvinyl chloride. *J. Occup. Med.* 16:150–151.

Du, C.-L., Kuo, M.-L., Chang, H.-L., Sheu, T.-J., and Wang, J.-D. (1995) Changes in lymphocyte single strand breakage and liver function of workers exposed to vinyl chloride monomer. *Toxicol. Lett.* 77:379–385.

Fedtke, N., Boucheron, J.A., Walker, V.E., and Swenberg, J.A. (1990) Vinyl chloride-induced DNA adducts: II. Formation and persistence of 7-(2′-oxoethyl)guanine and N^2,3-ethenoguanine in rat tissue DNA. *Carcinogenesis* 11:1287–1292.

Froment, O., Boivin, S., Barbin, A., Bancel, B., Trepo, C., and Marion, M.J. (1994) Mutagenesis of ras proto-oncogenes in rat liver tumors induced by vinyl chloride. *Cancer Res.* 54:5340–5345.

Fucic, A., Hitrec, V., Garaj-Vrhovac, V., Barkovic, D., and Kubelka, D. (1995) Relationship between locations of chromosome breaks by vinyl chloride monomer and lymphocytosis. *Am. J. Ind. Med.* 27:565–571.

Guengerich, F.P., Kim, D.-H., and Iwasaki, M. (1991) Role of human cytochrome P-450 IIE1 in the oxidation of many low molecular weight cancer suspects. *Chem. Res. Toxicol.* 4:168–179.

Harris, D.K. and Adams, W.G.F. (1967) Acro-osteolysis occurring in men engaged in the polymerization of vinyl chloride. *Br. Med. J.* 3:712–714.

Haustein, U.F. and Ziegler, V. (1985) Environmentally induced systemic sclerosis-like disorders. *Int. J. Dermatol.* 24:147–151.

Heath, C.W., Falk, H., and Creech, J.L. Jr. (1975) Characteristics of cases of angiosarcoma of the liver among vinyl chloride workers in the United States. *Ann. N. Y. Acad. Sci.* 246:231–236.

Hefner, R.E, Jr., Watanabe, P.G., and Gehring, P.J. (1975) Preliminary studies on the fate of inhaled vinyl chloride monomer (VCM) in rats. *Environ. Health Perspect.* 11:85–95.

Hemminki, K. and Vineis, P. (1985) Extrapolation of the evidence on teratogenicity of chemicals between humans and experimental animals: chemicals other than drugs. *Teratog. Carcinog. Mutagen.* 5:251–318.

Mastromatteo, E., Fisher, A.M., Christie, H., and Danziger, H. (1960) Acute inhalation toxicity of vinyl chloride to laboratory animals. *Am. Ind. Hyg. Assoc. J.* 21:394–398.

NIOSH (1994) *Pocket Guide to Chemical Hazards*. NIOSH 94-116. Washington, DC: National Institute for Occupational Safety and Health, Centers for Disease Control and Prevention, US Department of Health and Human Services.

Oster, R.H., Carr, J., Krantz, J.C. Jr., and Sauerwald, M.J. (1947) Anesthesia XXVII: narcosis with vinyl chloride. *Anesthesiology* 8:359–361.

Purchase, I.F.H., Stafford, J., and Paddle, G.M. (1987) Vinyl chloride: an assessment of the risk of occupational exposure. *Food Chem. Toxicol.* 25:187–202.

Richards, R.J., Desai, R., Hext, P.M., and Rose, F.A. (1975) Biological reactivity of PVC dust. *Nature* 256:664–665.

Simonsen, L., Johnsen, H., Lund, S.P., Matikainen, E., Midtgård, U., and Wennberg, A. (1994) Methodological approach to the evaluation of neurotoxicity data and the classification of neurotoxic chemicals. *Scand. J. Work Environ. Health* 20:1–12.

Smulevich, V.B., Fedotova, I.V., and Filatova, V.S. (1988) Increasing evidence of the rise of cancer in workers exposed to vinyl-chloride. *Br. J. Ind. Med.* 45:93–97.

Studnicka, M.J., Menzinger, G., Drlicek, M., Maruna, H., and Neumann, M.G. (1995) Pneumoconiosis and systemic sclerosis following 10 years of exposure to polyvinyl chloride dust. *Thorax* 50:583–585.

Trivers, G.E., Cawley, H.L., DeBenedetti, V.M., Hollstein, M., Marion, M.J., Bennett, W.P., Hoover, M.L., Prives, C.C., Tamburro, C.C., and Harris, C.C. (1995) Anti-p53 antibodies in sera of workers occupationally exposed to vinyl chloride. *J. Natl. Cancer Inst.* 87:1400–1407.

Viola, P.L., Bigotti, A., and Caputo, A. (1971) Oncogenic response of rat skin, lungs, and bones to vinyl chloride. *Cancer Res.* 31:516–552.

Watanabe, P.G., McGowan, G.R., and Gehring, P.J. (1976) Fate of (^{14}C)vinyl chloride after single oral administration in rats. *Toxicol. Appl. Pharmacol.* 36:339–352.

Weinman, M.D. and Chopra, S. (1987) Tumors of the liver, other than primary hepatocellular carcinoma. *Gastroenterol. Clin. North Am.* 16:627–650.

Wilson, R.H., McCormick, W.E., Tatum, C.F., and Creech, J.L. (1967) Occupational acroosteolysis: report of 31 cases. *JAMA* 201:577–581.

Wong, O., Whorton, M.D., Foliart, D.E., and Ragland, D. (1991) An industry-wide epidemiologic study of vinyl chloride workers, 1942–1982. *Am. J. Ind. Med.* 20:317–334.

Wu, W., Steenland, K., Brown, D., Wells, V., Jones, J., Schulte, P., and Halperin, W. (1989) Cohort and case-control analyses of workers exposed to vinyl chloride: an update. *J. Occup. Med.* 31:518–523.

Acrylamide

Acrylamide (2-propenamide; CASRN: 79-06-1) is a vinyl monomer produced as a white, crystalline solid. As a monomer, acrylamide is used in the manufacture of certain products such as grout material, adhesives, and biotechnical electrophoretic gels. After polymerization, polyacrylamide is primarily used as a flocculent in water treatment processes and in secondary oil recovery. One of its more favorable qualities is its ability to improve solubility, adhesion, and cross-linking of polymers. The compound is highly soluble in water and less soluble in organic solvents. As a consequence of its high solubility, acrylamide has been detected in potable water supplies. Acrylamide has a low vapor pressure of 0.007 mmHg at 25 °C (77 °F) and a vapor relative density of 2.5 (air = 1). At room temperature it is stable but may spontaneously polymerize upon exposure to ultraviolet light or upon melting. It has not been found to occur in nature.

Industrial Hygiene Acrylamide is produced from a catalytic hydration reaction of acrylonitrile with a copper-based or microbiological catalyst (IARC, 1994). Production of acrylamide may lead to exposure to either or both monomers. Dermal contact is the most probable route of exposure, whereas inhalation is of secondary importance because of its low vapor pressure. Addition of monomer and hand mixing are tasks in which possible dermal contact is high. Workers involved in the manufacturing of acrylamide, its polymer, or products containing the monomer are likely to be exposed. Surprisingly, estimations of exposure by NIOSH conclude that a larger percentage of contact results from the presence of unreacted monomer in polymer-containing products (IARC, 1994). Important factors in exposure are the lack of worker understanding of the hazards of acrylamide and

poor housekeeping and personal hygiene (Cummins et al., 1992). The chronic neurotoxicity associated with acrylamide led to the development of engineering controls, which resulted in low workplace concentrations. The NIOSH and the ACGIH have set similar values for their dermal REL TWA and TLV TWA at 0.03 mg/m3 (ACGIH, 1995; NIOSH, 1994). The OSHA has set its PEL TWA at 0.3 mg/m3.

Industrial monitoring of surface wipe samples and the presence of monomer in the polymer can be achieved with high performance liquid chromatography using water as an eluent (Skelly and Husser, 1978). Air monitoring, although not as important, may also be used with a variety of techniques. Protective clothing and gloves are a necessity for preventing overexposure to acrylamide. Personal respirators may be used for extra protection alter protective clothing has been provided.

Acute and Chronic Effects The acute and chronic effects of acrylamide exposure are similar. In some cases a large enough single exposure can result in chronic toxicity. The duration of exposure and total dose determine the severity of toxic effects. These effects, which may be delayed for several hours, include somnolence, confusion, hallucinations, tremors, and possibly seizures with cardiovascular collapse. Recovery may take a few weeks to months, and in some cases prolonged extremity weakness and loss of reflexes may be observed. Visual impairment and eye irritation may result from ocular exposure to fumes or dusts.

Chronic exposure to acrylamide is more common. Such exposure may result in effects related to the peripheral and CNS. Early effects of peripheral neuropathy may include numbness of the hands and feet, peeling skin, loss of reflexes and sensation, muscular atrophy, and loss of vibration sense (Bachmann et al., 1992; He et al., 1989). Abnormalities related to acrylamide exposure have been identified in as much as 32% of the total work force (Myers and Macun, 1991). Weight loss, anorexia, and gastrointestinal upset may be seen as well.

Epidemiological studies of worker populations have not demonstrated an increase in cancer-related mortality despite acrylamide's possible carcinogen status (Collins et al., 1989). In contrast, animal studies provide evidence of exposure-induced cancers of the thyroid gland-follicular epithelium, CNS, mammary gland, scrotal mesothelium, and uterus when acrylamide is ingested (Johnson et al., 1986). Chronic skin application of acrylamide acted as a skin tumor initiator and systemically induced lung adenomas (Bull et al., 1984). A review of present data has prompted the IARC to classify acrylamide as " . . . *probably carcinogenic to humans (Group 2A)*" (IARC, 1994). The ACGIH has changed its carcinogenic rating from A2 to A3, a change from a suspected human carcinogen to a confirmed animal carcinogen with unknown relevance to humans (ACGIH, 2001).

Diagnosis and Treatment Sites of action identified in humans are the peripheral and CNS, the eyes, and the skin, with the reproductive system being identified in animals. Diagnosis of acrylamide-induced peripheral neuropathy can be made using standard neurological procedures. Documentation of exposure to the compound is essential in concluding that acrylamide is the main cause. The clinical signs described above may be reported by an affected worker to support differential diagnosis. Decreases in vibrotactile thresholds may be quantified using a Vibratron II or similar device (Bachmann et al., 1992). Electroneuromyographic assessment may be made early by measuring decreases in sensory action amplitude, abnormalities in electromyography, and prolongation of the ankle tendon reflex (He et al., 1989).

First, an exposed worker should be removed from the area of exposure. Exposed skin should be thoroughly washed with soap and water. Rashes and other acute effects may appear and should be systematically treated as necessary. Treatment of excessive exposure to acrylamide with pyridoxine, pyruvate, and *N*-acetylcysteine is experimental, and their efficacy has not been established. Neurological manifestations will require time to gradually subside and return to normal.

Mechanism of Action The importance of the dermal route of exposure is likely a result of the ability of acrylamide to be rapidly absorbed. Chemicals that are soluble in both water and organic solvents are generally able to cross the epidermis quickly.

Metabolism of acrylamide is not well reported because further studies are required to fully identify the mechanism involved. It has been reported in some studies that acrylamide is metabolized in the body to a more reactive epoxide intermediate, glycidamide (Bergmark et al., 1991; Calleman et al., 1994). Cytochrome P-450 mixed-function oxidase is the likely enzyme for this reaction, although the results have been conflicting (Abou-Donia et al., 1993). Acrylamide-induced peripheral neuropathy has been the subject of many investigations. Fullerton and Barnes (1966) showed that acrylamide-poisoned rats exhibited degeneration of axis cylinders and myelin sheaths. This retrograde degeneration involved nerves with larger diameters and longer lengths (Spencer and Schaumburg, 1977). A model of acrylamide-induced peripheral nerve damage has been reported that involves giant axonal swelling in which the attack occurs at the nodes rather than the nerve terminal (Tilson, 1981). In more recent studies, fast axonal transport deficiencies were identified as being caused by both acrylamide and glycidamide (Harris et al., 1994). The underlying mechanisms of acrylamide-induced degenerations are likely a combination of these effects.

Biomonitoring Several methods are available for measuring occupational exposure to acrylamide. Both acrylamide and its metabolite, glycidamide, are conjugated with glutathione and excreted in the urine as mercapturic acid metabolites. Measurement of these metabolites can be used to monitor exposure. Other methods include measurement of plasma levels of acrylamide/glycidamide and the presence of hemoglobin adducts (Bergmark et al., 1993; Calleman et al., 1994). The hemoglobin adducts form by attachment to cysteine residues and the N-terminal valine. Perhaps the most complete means of biomonitoring is demonstrated in a report by Calleman et al. (1994) in which a neurotoxicity index was developed and significantly correlated with levels of mercapturic acid in the urine, hemoglobin adducts of acrylamide, accumulated *in vivo* doses of acrylamide, employment time, and vibration sensitivity.

REFERENCES

Abou-Donia, M.B., Ibrahim, S.M., Corcoran, J.J., Lack, L., Friedman, M.A., and Lapadula, D.M. (1993) Neurotoxicity of glycidamide, an acrylamide metabolite, following intraperitoneal injections in rats. *J. Toxicol. Environ. Health* 39:447–464.

ACGIH (2001) *Documentation of the Threshold Limit Values and Biological Exposure Indices,* 7th. ed. Cincinnati, OH: American Conference of Governmental Industrial Hygienists.

Bachmann, M., Myers, J.E., and Bezuidenhout, B.N. (1992) Acrylamide monomer and peripheral neuropathy in chemical workers. *Am. J. Ind. Med.* 21:217–222.

Bergmark, E., Calleman, C.J., and Costa, L.G. (1991) Formation of hemoglobin adducts of acrylamide and its epoxide metabolite glycidamide in the rat. *Toxicol. Appl. Pharmacol.* 111: 352–363.

Bergmark, E., Calleman, C.J., He, F., and Costa, L.G. (1993) Determination of hemoglobin adducts in humans occupationally exposed to acrylamide. *Toxicol. Appl. Pharmacol.* 120:45–54.

Bull, R.J., Robinson, M., Laurie, R.D., Stoner, G.D., Greisiger, E., Meier, J.R., and Stober, J. (1984) Carcinogenic effects of acrylamide in Sencar and A/J mice. *Cancer Res.* 44:107–111.

Calleman, C.J., Wu, Y., He, F., Tian, G., Bergmark, E., Zhang, S., Deng, H., Wang, Y., Crofton, K.M., Fennell, T., and Costa, L.G. (1994) Relationships between biomarkers of exposure and neurological effects in a group of workers exposed to acrylamide. *Toxicol. Appl. Pharmacol.* 126:361–371.

Collins, J.J., Swaen, G.M., Marsh, G.M., Utidjian, H.M., Caporossi, J.C., and Lucas, L.J. (1989) Mortality patterns among workers exposed to acrylamide. *J. Occup. Med.* 31:614–617.

Cummins, K., Morton, D., Lee, D., Cook, E., and Curtis, R. (1992) Exposure to acrylamide in chemical sewer grouting operations. *Appl. Occup. Environ. Hyg.* 7:385–391.

Fullerton, P.M. and Barnes, J.M. (1966) Peripheral neuropathy in rats produced by acrylamide. *Br. J. Ind. Med.* 23:210–221.

Harris, C.H., Gulati, A.K., Friedman, M.A., and Sickles, D.W. (1994) Toxic neurofilamentous axonopathies and fast axonal transport: V. Reduced bidirectional vesicle transport in cultured neurons by acrylamide and glycidamide. *J. Toxicol. Environ. Health* 42:343–356.

He, F., Zhang, S., Wang, H., Li, G., Zhang, Z., Li, F., Dong, X., and Hu, F. (1989) Neurological and electroneuromyographic assessment of the adverse effects of acrylamide on occupationally exposed workers. *Scand. J. Work Environ. Health* 15:125–129.

IARC (1994) Acrylamide. *IARC Monogr. Eval. Carcinog. Risks Hum.* 60, 389–433. Lyon, France: International Agency for Research on Cancer. Available at http://monographs.iarc.fr/ENG/Monographs/vol60/mono60.pdf.

Johnson, K.A., Gorzinski, S.J., Bodner, K.M., Campbell, R.A., Wolf, C.H., Friedman, M.A., and Mast, R.W. (1986) Chronic toxicity and oncogenicity study on acrylamide incorporated in the drinking water of Fischer 344 rats. *Toxicol. Appl. Pharmacol.* 85:154–168.

Myers, J.E. and Macun, I. (1991) Acrylamide neuropathy in a South African factory: an epidemiologic investigation. *Am. J. Ind. Med.* 19:487–493.

NIOSH (1994) *Pocket Guide to Chemical Hazards.* NIOSH 94-116. Washington, DC: National Institute for Occupational Safety and Health, Centers for Disease Control and Prevention, US Department of Health and Human Services.

Skelly, N.E. and Husser, E.R. (1978) Determination of acrylamide monomer in polyacrylamide and in environmental samples by high performance liquid chromatography. *Anal. Chem.* 50:1959–1962.

Spencer, P.S. and Schaumburg, H.H. (1977) Ultrastructural studies of the dying-back process: IV. Differential vulnerability of PNS and CNS fibers in experimental central–peripheral distal axonopathies. *J. Neuropathol. Exp. Neurol.* 36:300–320.

Tilson, H.A. (1981) The neurotoxicity of acrylamide: an overview. *Neurobehav. Toxicol. Teratol.* 3:445–461.

Polyvinylpyrrolidone

Polyvinylpyrrolidone (PVP; CASRN: 9003-39-8) is a polymer of 1-ethyl-pyrrolidone. This plastic is soluble in water and organic solvents. Typically, it is polymerized with heat or in the presence of a catalyst. The grade of PVP used, classified by its average molecular weight, is known by its K value. A K-12 value, for example, corresponds to a viscosity-average molecular weight of 3,900. Polyvinylpyrrolidone is characterized by very low toxicity and is used in an iodine complex as povidone, an antibacterial solution. It is also used in drugs, food, and cosmetics as a stabilizer or thickening agent. More industrial applications include adhesives, coatings, and cleaners. The material used in the US is PVP rather than shellac, which is used in United Kingdom. Its use in foodstuffs is an indication of the lack of toxicity of this material. Oils and propellants containing PVP have been used in hair sprays.

Some reports of toxicity upon PVP exposure have been published, although these case reports have not been validated. Bergmann et al. (1958) described pulmonary damage that followed exposure to PVP hair sprays and diagnosed the disease as thesaurosis. McLaughlin et al. (1963) reviewed the literature and reported on the 775 chest X-rays of British hairdressers, and Cares (1965) presented the US experience.

These reported cases resembled sarcoid clinically, radiographically, and by histopathological findings. Alternative diagnoses in the literature are interstitial fibrosis and foreign body granulomas. McLaughlin et al. (1963) concluded that it is possible to diagnose thesaurosis by chest X-ray, perhaps because of hair spray particles in the lung, but they could not correlate the quantity of spray inhaled and clinical findings. Their single case of a hairdresser was thought to be an example of unusual hypersensitivity. Gowdy and Wagstaff (1972) described five cases with pulmonary infiltrates on chest X-ray and four with respiratory complaints, all exposed to a variety of aerosols and all cured by discontinuation of exposure. Four biopsies and lung function tests were nonspecific. Furthermore, these investigations surveyed 227 beauty shop operators and found none ill, but 11 with increased bronchovascular markings on chest X-ray. Thesaurosis appears to be chiefly a problem of different diagnoses. When used as a plasma expander, PVP is generally thought to be safe. One reported case of a woman who received massive amounts of PVP injections clinically showed intracellular bone deposits of PVP and demonstrated pathological fractures as a result (Kepes et al., 1993). These types of massive, chronic exposures are unlikely in the workplace but are important in identifying possible long-term effects. Mutagenicity studies have not shown any significant increased frequency of sister chromatid exchanges *in vitro* (Ray et al., 1995).

REFERENCES

Bergmann, M., Flance, I.J., and Blumenthal, H.T. (1958) Thesaurosis following inhalation of hair spray: a clinical and experimental study. *N. Engl. J. Med.* 258:471–476.

Cares, R.M. (1965) Thesaurosis from inhaled hair spray? *Arch. Environ. Health* 11:80–86.

Gowdy, J.M. and Wagstaff, M.J. (1972) Pulmonary infiltration due to aerosol thesaurosis: a survey of hairdressers. *Arch. Environ. Health* 25:101–108.

Kepes, J.J., Chen, W.Y.K., and Jim, Y.F. (1993) 'Mucoid dissolution' of bones and multiple pathologic fractures in a patient with past history of intravenous administration of polyvinylpyrrolidone (PVP): a case report. *Bone Miner.* 22:33–41.

McLaughlin, A.I.G., Bidstrup, P.L., and Konstam, M. (1963) The effects of hair lacquer sprays on the lungs. *Food Cosmet. Toxicol.* 1:171–188.

Ray, B.D., Howell, R.T., McDermott, A., and Hull, M.G. (1995) Testing the mutagenic potential of polyvinylpyrrolidone and methyl cellulose by sister chromatid exchange analysis prior to use in intracytoplasmic sperm injection procedures. *Hum. Reprod.* 10:436–438.

Others

Additives The polymers previously discussed are modified through the addition of materials that impart specific, desirable properties to the finished product. These additives, not being large and inert polymeric macromolecules, may have biological characteristics that are greater than those of the basic plastic to which they are added. For the most part, the biological properties of these additives have not been well established. However, the metallic and organometallic compounds do present a known hazard. Exposure to additives tends to be small because they are introduced primarily in the early steps of formulation, and relatively few workers are exposed.

Plasticizers Plasticizers are mixed into polymers to increase flexibility and workability. Phthalate esters account for approximately 2/3rd of the total domestic plasticizer production and include di(2-ethylhexyl)phthalate (CASRN: 117-81-7), diisooctyl phthalate, and dibutyl phthalate. Di(2-ethylhexyl)phthalate is the most commonly used and is of low toxicity, although it may be a factor in pulmonary edema and in producing bronchial asthmatic conditions after excessive inhalation exposure (Roth et al., 1988). Although no such evidence exists in human data, animal studies report testicular toxicity as a result of damage to Sertoli and Leydig cells (Creasy et al., 1988; Jones et al., 1993). It is also characterized as a liver and kidney toxicant in mice after chronic exposure (Ward et al., 1988). Adipate, sebacate, and azelate esters are also commonly used, especially when low temperature flexibility is desired. These esters have low toxicity. Polyester and epoxidized plasticizers are also of low toxicity, and the epoxy materials can be sensitizers.

Phosphate esters, among the first plasticizers to be developed, impart flame resistance to polymers. Tricresyl phosphates have been the most important, and those used as plasticizers are generally free of the ortho isomer, which has prominent neurotoxic properties. Triphenyl phosphate, cresyl diphenyl phosphate, and octyl diphenyl phosphate are also used. Standard operating procedure should be used to ensure purity of materials in this group and the absence of neurotoxic components. The low vapor pressure of the aromatic phosphate esters prevents excessive exposure at ambient temperatures.

Other plasticizers include chlorinated paraffins and chlorinated biphenyls. The latter group is associated with chloracne and liver injury; the paraffins lack this toxicity.

Flame Retardants Flame-retardant properties can be inherent in certain plastics because of high chlorine content or, as in the case of the polycarbonates, because they evolve CO_2 on thermal decomposition. They are required in polymers important in electrical and high temperature applications. Additives of hygienic interest include tricresyl phosphate, chlorinated diphenyls and triphenyls, bromine compounds, halogenated phosphates, and antimony trioxide. Flame-retardant intermediates that are actually incorporated into the polymer usually contain chlorine or bromine, as discussed in the chapters on flame retardants.

Stabilizers Stabilizers retard the natural degradation of plastics caused by light and heat. Potentially hazardous stabilizers are lead salts or soaps, barium and cadmium compounds, and organotin compounds. Alternative compounds for lead and cadmium are or have been developed and put into use. Ultraviolet absorbers are added to protect polymers against those wave lengths that tend to disrupt bonds in organic material. Typical absorbers used are the benzophenones, benzotriazoles, aryl esters of salicylic acid, benzoic and terephthalic acids, and organonickel compounds. Antioxidants include alkylated phenols and bisphenols, amines, and organic phosphites.

Organic Peroxides Organic peroxides are used to cure polyester resins, to initiate polymerization of vinyl and diene monomers, and to cross-link polymer chains such as polyolefins or silicones. More than 45 organic peroxide compounds are available for industrial use. Although not all peroxides are hazardous, they are, in general, strong oxidizing agents capable of causing skin irritation or burns and severe eye damage. Peroxides on the skin or in the eyes should be removed promptly with large amounts of water. The use of face shields and protective clothing is advisable for those who handle organic peroxides. These compounds tend to be unstable and to undergo spontaneous decomposition. Fires or explosions can result from the contamination or improper storage of organic peroxides.

Inert Fillers and Fibrous Reinforcements Inert fillers and fibrous reinforcements are added to many formulations to improve their mechanical properties, to act as extenders to reduce costs, or to impart special properties to the plastic. The addition of silica, asbestos, talc, mica, glass flakes and fibers, and other materials is common. Some of these substances have inherent biological reactivity that may create an exposure hazard at the time of polymer formulation or when the finished plastic is cut or machined. Under these circumstances, potentially injurious dusts may be generated. Rapid developments within the plastics industry have created high thermal conductivity polymers (beryllia-filled epoxy), antifriction elastomers (molybdenum-disulfide filled), radiation shields (lead-filled polyolefin), and plastics reinforced with exotic fibers (boron, sapphire). Adverse biological effects may be associated with the use of some of these materials.

Foaming Agents Foaming agents may be added to increase the porosity of plastics that normally foam or to create foamed structures from plastics that have no inherent foaming properties. Foaming agents may generate gas by either physical or chemical means. Physical foaming agents are generally volatile liquids that become gaseous and create cells under the influence of exothermic reactions or externally applied heat. Aliphatic hydrocarbons of a petroleum ether, ligroin, or light spirits type are flammable but do not have prominent toxic properties. Halogenated aliphatic hydrocarbons, such as methyl chloride, methylene chloride, and trichloroethylene, are also used as foaming agents. The aliphatic fluorocarbons are essentially inert.

Chemical "blowing" agents that produce foam are compounds that decompose under the influence of heat to yield a gas, usually nitrogen. A chemical foaming agent used in industry, particularly in PVC and polyolefin resins, is azodicarboxamide (CASRN: 123-77-3). It is manufactured as a fine yellow dust that is insoluble in most organic solvents and water but dissolves readily in dimethyl sulfoxide. This compound, when inhaled as a dust, can result in a nocturnal productive cough and transient decreases in pulmonary function (Ferris et al., 1977). A study of workers exposed to azodicarboxamide during the injection-molding process reported significant associations with several respiratory symptoms, including cough, wheezing, irritation, and signs of chronic bronchitis (Whitehead et al., 1987). Chronic inhalation exposures in some workers have resulted in occupational asthma (Normand et al., 1989; Slovak, 1981). Contact dermatitis hazards are not reported with great frequency but have been reported in isolated cases (Yates and Dixon, 1988). Workplace air concentrations of azodicarboxamide can be measured *via* filter collection and subsequent chromatography (Ahrenholz and Neumeister, 1987).

Azobisisobutyronitrile (CASRN: 78-67-1), another foaming agent, releases nitrogen gas and a residue of tetramethyl succinic dinitrile that persists in the foam. The latter compound is a convulsant in animals and has been associated with headache, nausea, and other ill-defined systemic complaints in workers. The substitution of other foaming agents has eliminated many of these hazards. Azobisisobutyronitrile may also be used in small amounts for the catalytic polymerization of vinyl plastics. Dinitrosoterephthalamide, also used as a foaming agent, has not been a source of industrial illness.

REFERENCES

Ahrenholz, S.H. and Neumeister, C.E. (1987) Development and use of a sampling and analytical method for azodicarbonamide. *Am. Ind. Hyg. Assoc. J.* 48:442–446.

Creasy, D.M., Beech, L.M., and Gray, T.J.B. (1988) Effects of mono-(2-ethylhexyl) phthalate and mono-*n*-pentyl phthalate on the ultrastructural morphology of rat Sertoli cells in Sertoli/germ cell co-cultures: correlation with the *in vivo* effects of di-*n*-pentyl phthalate. *Toxicol. In Vitro* 2:83–95.

Ferris, B.G. Jr., Peters, J.M., Burgess, W.A., and Cherry, R.B. (1977) Apparent effect of an azodicarbonamide on the lungs: a preliminary report. *J. Occup. Med.* 19:424–425.

Jones, H.B., Garside, D.A., Liu, R., and Roberts, J.C. (1993) The influence of phthalate esters in Leydig cell structure and function *in vitro* and *in vivo*. *Exp. Mol. Pathol.* 58:179–193.

Normand, J.-C., Grange, F., Hernandez, C., Ganay, A., Davezies, P., Bergeret, A., and Prost, G. (1989) Occupational asthma after exposure to azodicarbonamide: report of four cases. *Br. J. Ind. Med.* 46:60–62.

Roth, B., Herkenrath, P., Lehmann, H.J., Ohles, H.D., Hömig, H.J., Benz-Bohm, G., Kreuder, J., and Younossi-Hartenstein, A. (1988) Di-(2-ethylhexyl)-phthalate as plasticizer in PVC respiratory tubing systems: indications of hazardous effects in pulmonary function in mechanically ventilated, preterm infant. *Eur. J. Pediatr.* 147:41–46.

Slovak, A.J.M. (1981) Occupational asthma caused by a plastics blowing agent, azodicarbonamide. *Thorax* 36:906–909.

Ward, J.M., Hagiwara, A., Anderson, L.M., Lindsey, K., and Diwan, B.A. (1988) The chronic hepatic or renal toxicity of di(2-ethylhexyl) phthalate, acetaminophen, sodium barbital, and phenobarbital in male B6C3F1 mice: autoradiographic, immunohistochemical, and biochemical evidence for levels of DNA synthesis not associated with carcinogenesis or tumor promotion. *Toxicol. Appl. Pharmacol.* 96:494–506.

Whitehead, L.W., Robins, T.G., Fine, L.J., and Hansen, D.J. (1987) Respiratory symptoms associated with the use of azodicarbonamide foaming agent in a plastics injection molding facility. *Am. J. Ind. Med.* 11:83–92.

Yates, V.M. and Dixon, J.E. (1988) Contact dermatitis from azodicarbonamide in earplugs. *Contact Dermatitis* 19:155–156.

76

ELASTOMERS

Marek Banasik

BACKGROUND

Elastomeric polymers are frequently constructed of the same monomers that are used in plastics. The proportions, minor constituents, catalysts, and reaction conditions are varied to yield a product with elastic properties. An important distinction between some synthetic elastomers and other macromolecular polymers is that "cured" elastomers can occasionally be a source of hazardous ingredients in the formulation stage. It has been shown that many elastomer polymers can cause occupational contact dermatitis. In the past, workers' compensation for this type of claim has been highest in the rubber industry (Kilpikari, 1982; Varigos and Dunt, 1981). However, epidemiological studies on mortality among rubber workers exposed to a variety of polymers of different concentrations have shown no increase in deaths associated with cancer or other diseases (Andjelkovich et al., 1977; Delzell et al., 1981). In fact, these studies have shown a decrease in overall mortality among these workers.

REFERENCES

Andjelkovich, D., Taulbee, J., Symons, M., and Williams, T. (1977) Mortality of rubber workers with reference to work experience. *J. Occup. Med.* 19:397–405.

Delzell, E., Louik, C., Lewis, J., and Monson, R.R. (1981) Mortality and cancer morbidity among workers in the rubber tire industry. *Am. J. Ind. Med.* 2:209–216.

Kilpikari, I. (1982) Occupational contact dermatitis among rubber workers. *Contact Dermatitis* 8:359–362.

Varigos, G.A. and Dunt, D.R. (1981) Occupational dermatitis: an epidemiological study in the rubber and cement industries. *Contact Dermatitis* 7:105–110.

POLYETHYLENES

Chlorosulfonated polyethylenes are prepared through the reaction of sulfur dioxide and chlorine with polyethylene dissolved in carbon tetrachloride, and some hazards may be associated with exposure to these ingredients. Hazards have not been identified with human use of the polymer, nor have hazards been identified by animal testing. Chlorosulfonated polyethylenes are used in white sidewall tires, hoses, and tank linings and for other products that must be resistant to heat, oxidation, and ozone. These polymers are cured with metallic oxides. Accelerators include mercaptobenzothiazole, benzothiazolyl disulfide, and dipentamethylenethiuram tetrasulfide, which as a group are known skin sensitizers and may be present in the finished rubbers.

Chlorinated polyethylenes are prepared by substituting chlorine atoms with hydrogen atoms on the ethylene polymer. This rubber may also contain residual monomer and carbon tetrachloride and should be washed thoroughly with water. Skin irritation may occur, especially during processes involving heat. Chlorinated polyethylenes are most commonly used in automobile hoses and other mechanical products.

ETHYLENE–PROPYLENE COPOLYMERS AND TERPOLYMERS

To produce ethylene–propylene copolymer [Chemical Abstracts Service Registry Number (CASRN): 9010-79-1] or terpolymer (CASRN: 25038-37-3), alkenes are reacted under the influence of a coordination catalyst such as vanadium oxychloride or diethylaluminum chloride. The terpolymer of this rubber differs in that it has small amounts of diene, which

Hamilton & Hardy's Industrial Toxicology, Sixth Edition. Edited by Raymond D. Harbison, Marie M. Bourgeois, and Giffe T. Johnson.
© 2015 John Wiley & Sons, Inc. Published 2015 by John Wiley & Sons, Inc.

aids in establishing curing sites. Peroxides may be used in curing, although incorporation of a third monomer, such as dicyclopentadiene, permits use of conventional sulfur vulcanization. Generally, these elastomers are of low toxicity, although curing compounds that contain sulfur may be released in minute concentrations, and these may cause skin or respiratory irritation. Elastomers of this group are resistant to atmospheric heat and chemical attack and are used in hoses, belts, gaskets, and footwear.

FLUORINATED ELASTOMERS

Copolymers of vinylidene fluoride (CASRN: 75-38-7) and hexafluoropropylene (CASRN: 9011-17-0), chlorotrifluoroethylene (CASRN: 79-38-9), or perfluoropropylene (CASRN: 116-15-4) are highly resistant elastomers used in gaskets, seals, and other applications in industries where high cost materials can be afforded. The elastomers are biologically inert, although extreme heating can produce harmful breakdown products such as hydrogen fluoride, monofluoroacetic acid, and other fluorocarbons. These products can cause a variety of acute and systemic effects. Skin sensitization can occur after extended contact with the polymer. Curing agents include diamines, peroxides, or ionizing radiation.

NEOPRENE

Neoprene rubber, the trade name for polychloroprene-based rubber, is obtained by the emulsion condensation of chloroprene, a derivative of butadiene. The two types of neoprene are distinguished by whether a sulfur or nonsulfur modification process was used. Neoprene is highly resistant to weathering and to oil, as well as to solvents, ozone, oxygen, and heat. It is used primarily in mechanical and automotive products, such as cable sheaths, belts, hoses, seals, and gaskets, and it also is an important additive for some adhesives. As with most rubber products, a sensitized individual can develop contact dermatitis from extended dermal exposure to neoprene (Fowler and Callen, 1988). An occupational exposure to the monomer chloroprene is the most likely contact to result in an industrial hazard.

CHLOROPRENE

Chloroprene (2-chloro-1,3-butadiene; CASRN: 126-99-8) is a flammable, colorless liquid at room temperature with a characteristic ether-like odor. It is slightly soluble in water and more soluble in organic solvents. Its vapor pressure and relative vapor density are 215 mmHg at 25°C and 3.0 (air = 1), respectively. It has not been found to occur naturally. Chloroprene is very unstable and reacts in air with oxygen and other compounds to form epoxides, peroxides, and other hazardous compounds. This instability and spontaneous formation of different compounds are thought to be the explanation for the variable findings in chloroprene studies.

Industrial Hygiene

The most probable route of chloroprene exposure in an industrial setting is inhalation by workers during manufacture of the compound or during its use. Because chloroprene may be carcinogenic, occupational exposure should be kept at a minimum. The United States (US) National Institute for Occupational Safety and Health recommends a self-contained breathing apparatus or a supplied-air respirator, both with full facepiece and operated in a positive-pressure mode (NIOSH, 1994). The US Occupational Safety and Health Administration has set a permissible exposure limit for chloroprene of 25 parts per million (ppm) (90 mg/m^3) time-weighted average (TWA); the American Conference of Governmental Industrial Hygienists has recommended a slightly lower standard threshold limit value of 10 ppm (36 mg/m^3) TWA (ACGIH, 1995; NIOSH, 1994). Dual mechanical seals on process pumps and careful attention to leaks in the machinery can keep industrial airborne levels to a minimum.

Acute and Chronic Effects

Chloroprene is a mucous membrane irritant, and an acute exposure can cause a variety of signs and symptoms, depending on the exposure route and dose (Sanotskii, 1976). Signs and symptoms caused by higher concentrations include respiratory tract and eye irritation, central nervous system depression, lacrimation, dermatitis, corneal necrosis, and hair loss from dermal contact (Gooch and Hawn, 1981). The depilatory effects of chloroprene are reversible, and hair regrows after exposure is terminated. Acute exposure studies in animals have shown similar effects, as well as adverse lung and liver effects at high concentrations (Clary et al., 1978). Low level acute exposure has had cumulative effects in rats (Plugge and Jaeger, 1979).

Chronic exposure to chloroprene yields a similar variety of systemic and local effects, including headache, irritability, dizziness, insomnia, respiratory irritation, cardiac palpitations, chest pain, gastrointestinal disorders, and dermatitis. Case reports of a long-term exposure in the rubber industry have described acute sensitizing illnesses associated with eosinophilia (Bascom et al., 1988). Reproductive effects also have been noted, including a decrease in spermatogenesis, sexual impotency, and disruption of the menstrual cycle. The diversity of both chronic and acute effects suggests that because of chloroprene's reactive nature, exposure actually involves to a variety of chemicals and compounds.

Because chloroprene has not been shown to be carcinogenic in humans, the International Agency for Research on

Cancer (IARC) has classified it in Group 3, inadequate evidence of carcinogenicity in both animal and human studies (IARC, 1987). Two epidemiological studies in Eastern Europe indicate an increase in lung and skin cancers among chloroprene workers (Lloyd et al., 1975). However, a US-based study of a similar population did not confirm these findings (Pell, 1978). Mutagenicity assays for chloroprene using bacteria have yielded positive findings, but these tests are not an indication of human carcinogenicity (Bartsch et al., 1975). Similar discrepancies have been seen in animal studies (Dong et al., 1989). These results do not demonstrate a consistent finding suggesting that chloroprene is a carcinogen.

Target Sites

The primary sites of action for chloroprene are the eyes, skin, respiratory system, and reproductive system. However, the route of exposure and reactant products of the chemical upon exposure are the two factors that determine the effects of chloroprene.

REFERENCES

ACGIH (1995) *Threshold Limit Values (TLVs) for Chemical Substances and Physical Agents and Biological Exposure Indices (BEIs)*. Cincinnati, OH: American Conference of Governmental Industrial Hygienists. ISBNs: 1882417119; 9781882417117.

Bartsch, H., Malaveille, C., Montesano, R., and Tomatis, L. (1975) Tissue-mediated mutagenicity of vinylidene chloride and 2-chlorobutadiene in *Salmonella typhimurium*. *Nature* 255:641–643.

Bascom, R., Fisher, J.F., Thomas, R.J., Yang, W.N., Baser, M.E., and Baker, J.H. (1988) Eosinophilia, respiratory symptoms and pulmonary infiltrates in rubber workers. *Chest* 93:154–158.

Clary, J.J., Feron, V.J., and Reuzel, P.G.J. (1978) Toxicity of β-chloroprene (2-chlorobutadiene-1,3): acute and subacute toxicity. *Toxicol. Appl. Pharmcol.* 46:375–384.

Dong, Q.-A., Xiao, B.-L., Hu, Y.-H., and Li, S.-Q. (1989) Short-term test for the induction of lung tumor in mouse by chloroprene. *Biomed. Environ. Sci.* 2:150–153.

Fowler, F.J. Jr. and Callen, J.P. (1988) Facial dermatitis from a neoprene rubber mask. *Contact Dermatitis* 18:310–311.

Gooch, J.J. and Hawn, W.F. (1981) Biochemical and hematological evaluation of chloroprene workers. *J. Occup. Med.* 23:268–272.

IARC (1987) Overall evaluations of carcinogenicity: an updating of *IARC Monographs* volumes 1 to 42. *IARC Monogr. Eval. Carcinog. Risks Hum.* Suppl 7. Lyon, France: International Agency for Research on Cancer. Available at http://monographs.iarc.fr/ENG/Monographs/suppl7/Suppl7.pdf.

Lloyd, J.W., Decoufle, P., and Moore, R.M. (1975) Background information on chloroprene. *J. Occup. Med.* 17:263–265.

NIOSH (1994) *Pocket Guide to Chemical Hazards*. NIOSH 94-116. Washington, DC: National Institute for Occupational Safety and Health, Centers for Disease Control and Prevention, US Department of Health and Human Services.

Pell, S. (1978) Mortality of workers exposed to chloroprene. *J. Occup. Med.* 20:21–29.

Plugge, H. and Jaeger, R.J. (1979) Acute inhalation toxicity of 2-chloro-1,3-butadiene (chloroprene): effects on liver and lung. *Toxicol. Appl. Pharmacol.* 50:565–572.

Sanotskii, I.V. (1976) Aspects of the toxicology of chloroprene: immediate and long-term effects. *Environ. Health Perspect.* 17:85–93.

POLYISOBUTYLENE

Butyl rubber (CASRN: 9010-85-9) is formed from the polymerization of isobutene (isobutylene) with small amounts of chloroprene, isoprene, or butadiene and may be halogenated. The reaction is carried out in a closed system with an aluminum chloride catalyst. Butyl rubber is highly impervious to gases, which makes the rubber best suited for inner tubes and tires, air chambers, adhesives, and dielectrics. Isobutene is an anesthetic and asphyxiant gas and can be hazardous in concentrations high enough to produce asphyxia. At these concentrations, explosion is an additional hazard.

POLYISOPRENE

Polyisoprene (CASRN: 9003-31-0) is the polymer known as natural rubber, although it can also be manufactured. The natural rubber latex is harvested from the rubber tree, *Hevea brasiliensis*. This substance has a variety of natural additives, such as proteins and sugars. The polymer from the natural latex is resistant to many solvents and also is easily processed. The synthetic form of this rubber is produced from a pure isoprene solution with a stereospecific isomer to produce the more commonly used *cis*-1,4 isomer. These rubbers are resistant to abrasion and most solvents and are commercially used in automobile tires, adhesives, and a variety of products that come in close contact with the general public. Their use in baby bottle nipples is a good indication of the extremely low toxicity associated with these elastomers.

ISOPRENE

Isoprene (2-methyl-1,3-butadiene; CASRN: 78-79-5) is the monomeric unit of polyisoprene. The structure of this molecule is very similar to that of butadiene, with a simple methyl substitution differentiating the two. Although it is mainly used in the production of polyisoprene, isoprene is also used in styrene-based polymers and in butyl rubber. At room temperature it is a clear, colorless liquid with a faint odor; it is insoluble in water but soluble in acetone and other organic solvents. The vapor pressure of isoprene is 550 mmHg, and it has a relative vapor density of 2.35 (IARC, 1994). Unlike the chemically similar butadiene, isoprene is found in abundance in nature; for example, it is a component of natural terpenes and also is expired by plants and humans.

Industrial Hygiene

In the past, different production techniques were available to make the isoprene monomer. Currently, the most commonly used process involves the formation of an isoprene by-product from petroleum cracking. Because isoprene is a naturally occurring product and its metabolism is not well characterized, most regulatory agencies have not set standards for an isoprene exposure. Such an exposure can occur in the polymerization process and in the formation of the isoprene-based synthetic rubbers and elastomers.

Acute and Chronic Effects

Inhalation is the major route of exposure to isoprene, with dermal contact and ingestion rarely occurring. Similarities between acute butadiene and isoprene exposure have been reported. At high concentrations, isoprene is a narcotic and asphyxiant, although respiratory irritation occurs at lower levels, which can warn workers to avoid higher concentrations (Dahl et al., 1987). Isoprene has also been noted to affect mucociliary flow rates (Weissbecker et al., 1971).

The chronic effects of isoprene have not been extensively reported, although animal studies have shown isoprene to be carcinogenic. Catarrhal inflammation and loss of sense of smell can occur with increasing length of exposure. A study by Melnick et al. (1994) demonstrated multiple organ neoplasia after a 6-month inhalation study in rats and mice. As a result of these studies, the IARC carcinogenicity classification for isoprene is Group 2B, sufficient animal evidence of carcinogenicity but insufficient human evidence (IARC, 1994).

Target Site and Mechanism of Action

The target organ for acute effects of isoprene is the central nervous system. Isoprene is an anesthetic at high concentrations. It can also result in asphyxia at high concentrations by displacement of breathable air. It is metabolized in rodents by hepatic cytochrome P-450, which results in two major mono-epoxide metabolites (Gervasi and Longo, 1990; Longo et al., 1985). These studies have determined that isoprene binds to the active site of cytochrome P-450 and undergoes an oxidation and hydrolysis step. The metabolites are 3,4-epoxy-3-methyl-1-butene and 3,4-epoxy-2-methyl-1-butene. The latter is the minor metabolite and the only one believed to be further converted to isoprene diepoxide, an animal carcinogen (Gervasi and Longo, 1990). The formation of a carcinogenic metabolite from isoprene exposure must be considered in the assessment of risk, although the rate of conversion to this metabolite does not appear to be high. The precursor to endogenous isoprene production is believed to be mevalonic acid, a component of cholesterol synthesis (Deneris et al., 1984). Excretion of isoprene through expiration has been measured; isoprene is the main hydrocarbon in breath (Gelmont et al., 1981).

Biomonitoring

Levels of isoprene have been detected in human breath and, more recently, in human blood samples (Cailleux et al., 1992). The correlation between these measurements and occupational exposure has not been established.

REFERENCES

Cailleux, A., Cogny, M., and Allain, P. (1992) Blood isoprene concentrations in humans and in some animal species. *Biochem. Med. Metab. Biol.* 47:157–160.

Dahl, A.R., Birnbaum, L.S., Bond, J.A., Gervasi, P.G., and Henderson, R.F. (1987) The fate of isoprene inhaled by rats: comparison to butadiene. *Toxicol. Appl. Pharmacol.* 89:237–248.

Deneris, E.S., Stein, R.A., and Mead, J.F. (1984) *In vitro* biosynthesis of isoprene from mevalonate utilizing a rat liver cytosolic fraction. *Biochem. Biophys. Res. Commun.* 123:691–696.

Gelmont, D., Stein, R.A., and Mead, J.F. (1981) Isoprene—the main hydrocarbon in human breath. *Biochem. Biophys. Res. Commun.* 99:1456–1460.

Gervasi, P.G. and Longo, V. (1990) Metabolism and mutagenicity of isoprene. *Environ. Health Perspect.* 86:85–87.

IARC (1994) Some industrial chemicals. *IARC Monogr. Eval. Carcinog. Risks Hum.* 60. Lyon, France: International Agency for Research on Cancer. Available at http://monographs.iarc.fr/ENG/Monographs/vol60/mono60.pdf.

Longo, V., Citti, L., and Gervasi, P.G. (1985) Hepatic microsomal metabolism of isoprene in various rodents. *Toxicol. Lett.* 29:33–37.

Melnick, R.L., Sills, R.C., Roycroft, J.H., Chou, B.J., Ragan, H.A., and Miller, R.A. (1994) Isoprene, an endogenous hydrocarbon and industrial chemical, induces multiple organ neoplasia in rodents after 26 weeks of inhalation exposure. *Cancer Res.* 54:5333–5339.

Weissbecker, L., Creamer, R.M., and Carpenter, R.D. (1971) Cigarette smoke and tracheal mucus transport rate: isolation of effect of components of smoke. *Ann. Rev. Respir. Dis.* 104:182–187.

POLYSULFIDE RUBBER

Polysulfide elastomers are condensation products of sodium polysulfides and dichloro compounds, such as ethylene dichloride. More complex chlorinated hydrocarbons yield products with improved properties. These elastomers are highly resistant to oils, solvents, and weathering and are found in coatings, caulking compounds, adhesives, and mechanical components. In most applications of the polysulfide rubbers, occupational health problems are related to contact dermatitis.

POLYURETHANE

The fibers commonly called spandex are polyurethane elastomers, which are biologically inert. They can be substituted without reaction for rubber by people who have become sensitized to auxiliary compounds invariably present in natural and synthetic rubbers. These polymers are noted for their resistance to oil and abrasion and for their mechanical strength. Use of these rubbers in footwear has been blamed for a limited number of cases of contact dermatitis (Grimalt and Romaguera, 1975). All spandex formulations are not identical; many are based on polytetramethylene glycol, 2,4-toluene diisocyanate, other diisocyanates, and hydrazine. The solvent for extrusion is dimethylformamide. The effects of the diisocyanates are well known and were discussed previously. The effects described for these constituents are the primary health problems associated with these elastomers as well.

REFERENCE

Grimalt, F. and Romaguera, C. (1975) New resin allergens in shoe contact dermatitis. *Contact Dermatitis* 1:169–174.

SILICONE RUBBERS

Industrial medical problems have not been prominent in the preparation or use of silicone rubbers. Benzoyl peroxide may be used as a vulcanizer. These rubbers are most appropriate for use as gaskets and seals because of their flexibility and heat resistance.

STYRENE–BUTADIENE COPOLYMER RUBBERS

Styrene–butadiene rubber (SBR; CASRN: 9003-55-8) is prepared in large quantities for tire treads and other, similar uses. The properties, molecular weight, and percentage of monomer can all be controlled in the polymerization process. Hazards of the SBR industry have been identified and related mostly to the monomers and additives. Contact dermatitis and leukemia have been associated with solvent and additive use but not with the actual polymer itself (Checkoway et al., 1984; Kilpikari, 1982). Epidemiological studies in the industry have not revealed a significant increase in disease or

causes of death (Matanoski and Schwartz, 1987; Meinhardt et al., 1982). The hazards of the butadiene and styrene monomers were discussed previously.

Polybutadiene (CASRN: 9003-17-2) is a homopolymer of butadiene (CASRN: 106-99-0). The catalysts involved in the polymerization process determine the structure and molecular weight of the rubber. It is used in tires and for other products that require abrasion resistance. At times both SBR and polybutadiene are extended with oils. There are underlying hazards associated with the individual oils, some of which are carcinogenic in animals and suspected human carcinogens.

REFERENCES

Checkoway, H., Wilcosky, T., Wolf, P., and Tyroler, H. (1984) An evaluation of the associations of leukemia and rubber industry solvent exposures. *Am. J. Ind. Med.* 5:239–249.

Kilpikari, I. (1982) Occupational contact dermatitis among rubber workers. *Contact Dermatitis* 8:359–362.

Matanoski, G.M. and Schwartz, L. (1987) Mortality of workers in styrene-butadiene polymer production. *J. Occup. Med.* 29:675–680.

Meinhardt, T.J., Lemen, R.A., Crandall, M.S., and Young, R.J. (1982) Environmental epidemiologic investigation of the styrene-butadiene rubber industry. Mortality patterns with discussion of hematopoietic and lymphatic malignancies. *Scand. J. Work Environ. Health* 8:250–259.

ADDITIVES IN ELASTOMER SYSTEMS

Various substances added to rubber formulations in the course of production may cause skin reactions in people who use the finished products, as well as in industrially exposed workers. The additives generally are similar for natural rubber, nitrile rubber, polybutadiene, neoprene, butyl rubber, and ethylene–propylene copolymers.

Vulcanizers generally contain elemental sulfur, although zinc and magnesium oxides are important in the neoprene cure, and peroxides are used for ethylene–propylene systems. The rate of vulcanization is influenced by accelerators, some of which are potent sensitizers. Common accelerators include benzothiazolyl disulfide, mercaptobenzothiazole, tetraethyl thiuramdisulfide (disulfiram), tetramethyl thiuramdisulfide (thiram), thiocarbanilide, dithiocarbamates (zinc dimethyl, zinc diethyl, zinc dibutyl, lead dimethyl compounds), diphenylguanidine phthalate, hexamethylene tetramine, and lead oxide (litharge).

Besides causing allergic contact dermatitis, the accelerators disulfiram and thiram can produce severe reactions when they are absorbed together with alcohol or paraldehyde. This reaction involves the inhibition of aldehyde dehydrogenase, presumably through the chelation of molybdenum, so that the normal metabolic degradation of aldehydes is inhibited.

Aldehydes consequently accumulate and cause flushing, palpitations, dyspnea, nausea, and hypotension. These reactions do not involve sensitization. Pharmaceutical grade disulfiram is used to produce distaste and intolerance for alcohol in the treatment of alcoholism.

Antioxidants are used to protect rubber against the destructive effects of atmospheric oxygen. The most commonly used antioxidants are monobenzyl ether of hydroquinone (AgeRite Alba), di-(β-naphthyl)-*p*-phenylenediamine, monoctyl and dioctyl diphenylamines, and phenyl alpha and beta naphthylamines.

The monobenzylether of hydroquinone, in addition to being a sensitizer, can cause loss of skin pigment in many individuals, often in the absence of any previous skin reaction. This reaction is most apparent in very dark-skinned persons in whom the response resembles vitiligo. Exposure to this compound has been significantly reduced through modifications in the rubber compounding process. Many of the amine antioxidants are also potent sensitizers.

Pigments and fillers include carbon black, whiting (calcium carbonate), clay, silica, and asbestos. The hazards of the fibrogenic dusts are well established and are discussed elsewhere. There has been some speculation that polycyclic aromatic carcinogens are adsorbed to carbon black particles handled in industry, but there is no evidence that the pigment contains these materials. However, if there is a suspicion that the carbon black contains hazardous levels of polycyclic hydrocarbons, the material should be analyzed for polycyclic hydrocarbons content.

Some additives may remain, used to arrest the polymerization process to control for elastomer characteristics and molecular weight. Residual impurities remaining after washing can cause some adverse skin reactions, which may be causally associated with the elastomer itself but are the result of the impurities and not the elastomer. Dodecyl mercapthane has been demonstrated by skin patch testing to cause contact dermatitis in association with rubber footwear (Grimalt and Romaguera, 1975).

REFERENCE

Grimalt, F. and Romaguera, C. (1975) New resin allergens in shoe contact dermatitis. *Contact Dermatitis* 1:169–174.

77

SYNTHETIC FIBERS

Marek Banasik

BACKGROUND

To a great extent, polymeric fiber textiles resemble many of the high molecular weight polymers used in plastics and elastomers. Significant chemical and physical modifications during production provide the properties necessary for fibers and filaments. In general, these polymers are not associated with occupational health hazards. However, industrial exposure to many of the chemicals and additives used with these fibers may pose health hazards. Although the mechanism of carcinogenesis has not been identified, a significant increase in the incidence of a rare colon cancer has been reported among a population of synthetic fiber workers, particularly those who worked in the extrusion departments (Vobecky et al., 1984). However, this report has not yet been verified. These fibers are manufactured for their use alone or as a component of fiber-reinforced plastics. Fiber-reinforced plastics are extremely strong and can substitute for steel in many applications. Problems may arise as the development of new materials and processes is introduced before health hazards can be adequately identified (Midtgård and Jelnes, 1991). Exposure prevention methods, such as shielding, ventilation, and personal protection, are recommended for compounds for which effects are not well characterized. Fibers usually are regulated by the United States Occupational Safety and Health Administration as nuisance dusts. The permissible exposure limits are 15 mg/m^3 for total dust and 5 mg/m^3 for the respirable fraction. The size of the fibers, including their length and diameter, and the fibers' durability and concentration are the critical determinants of their hazardous nature. Those too large to be respired may cause mechanical irritation to the eyes and skin.

ACETATES

Cellulose acetate rayon is made from the wood pulp of cotton linters treated with glacial acetic acid and acetic anhydride, along with sulfuric acid as a dehydrating agent. Both acetic acid and its anhydride have a pungent odor and are potent lacrimators. The anhydride is a severe skin and eye irritant, and it may occasionally be a sensitizer. This response is presumably the result of its reaction with amino groups of cutaneous protein.

In the production of fibers, the cellulose acetate is dissolved in a solvent for extrusion. In the case of triacetate material, the solvent may be methyl acetate, glacial acetic acid, N,N-dimethylformamide, dimethylacetamide, or dimethyl sulfoxide (DMSO). N,N-dimethylformamide [Chemical Abstracts Service Registry Number (CASRN): 68-12-2] is a powerful solvent that is irritating to the eyes, mucous membranes, and skin. It can pass through intact skin or can be absorbed from the lungs and gastrointestinal tract. It can be hepatotoxic but is rapidly metabolized by demethylation to monomethylformamide. The presence of this metabolite in urine can serve as an effective bio-monitoring tool (Maxfield et al., 1975). The unpleasant fishy odor serves as a warning and tends to limit exposure. Dimethyl sulfoxide (CASRN: 67-68-5) has been suggested as a vehicle for transporting drugs through the skin. Dermal exposure to DMSO may result in erythema, itching, and burning. For cellulose acetates other than the triacetate, the conventional solvent is acetone.

ACRYLICS

Acrylonitrile (CASRN: 107-13-1) is the chief constituent of acrylic fibers. To improve the dye and working properties of

Hamilton & Hardy's Industrial Toxicology, Sixth Edition. Edited by Raymond D. Harbison, Marie M. Bourgeois, and Giffe T. Johnson.
© 2015 John Wiley & Sons, Inc. Published 2015 by John Wiley & Sons, Inc.

the material, acrylonitrile frequently is copolymerized with small amounts of other monomers such as acrylates, amides, vinyl esters, and various hydrocarbons (e.g., styrene and isobutene). When the comonomer is vinyl chloride (CASRN: 75-01-4) or vinylidene chloride (CASRN: 75-35-4) and is present in quantities approximately equal to the acrylonitrile, the polymer is said to be a modacrylic, which is extruded in acetone. Acrylics are prepared for extrusion in solvents that may include N,N-dimethylformamide, DMSO, strong acids, and concentrated inorganic salt solutions. Although acrylics generally are inert, penetration of the skin by acrylic fibers has been reported to lead to sarcoid-like granulomas (Pimentel, 1977).

NYLON

The polyamides known as nylon are used principally as fibers and, to a lesser extent, resins in the plastics industry. In fiber applications, nylon may be modified by titanium dioxide, a delusterant, by manganese and phosphorus salts as light stabilizers, and by copper salts and acridine compounds as heat stabilizers. These substances are not released from the finished fiber. There have been only scattered reports of ill effects among workers exposed to nylon fiber. For example, sarcoid-like granulomas of the lungs were reported in a woman who was exposed to nylon dust for 20 years in a fiber bag factory (Pimentel, 1977).

ARAMID

Aramid [poly(imino-1,3-phenyleneiminocarbonyl-1,3-phenylenecarbonyl)]; CASRN: 24938-60-1) is the most common nylon fiber used, known for its strength and durability. Uses of aramid fibers include protective clothing, tires, conveyor belts, and cables. Animal studies have shown that chronic inhalation exposure to these fibers can produce pulmonary lesions (Lee et al., 1988). More recent studies have found that these lesions are not carcinomas, as originally thought; rather, they are more accurately referred to as proliferative keratin cysts (Carlton, 1994). Clearance of the fibers occurs relatively quickly, as does recovery from injury, as with other carbon fiber exposures (Kelly et al., 1993; Warheit et al., 1992). Aramid fibers are also readily cleared by rat lungs, which reduce residence time (Kelly et al., 1993). Human exposure to aramid fibers is considered to be of minimal risk.

POLYESTERS

Fibers of the polyester group are formed of polyethylene terephthalate (CASRN: 25038-59-9). This polymer is produced from dimethyl terephthalate and ethylene glycol under the influence of catalysts, which include oxides, carbonates or acetates of zinc, antimony, manganese, cobalt, calcium, or magnesium. The terephthalates have few, if any, biological properties.

POLYURETHANE

The polyurethane fiber, commonly known as spandex, can more accurately be described as an elastomeric fiber because it can be stretched 500–600% larger than its initial state. These fibers are most noted for their quick recovery after being stretched. A spandex fiber usually contains at least 85% polyurethane and is marketed under a wide variety of names. Historically, it was produced to improve the first vulcanized rubber fibers made (Hicks et al., 1965). Vulcanized rubber fibers are highly resistant to abrasion, chemicals, and sunlight. They are used in many applications but are most noticeable in their use in apparels such as brassieres and active wear. When spandex was first introduced, a few cases of allergic contact dermatitis were reported, but these were found to be caused by some trace contaminants rather than the actual fiber. Sensitization to the fiber itself may occur but is considered extremely rare. Mercaptobenzothiazole, thiram, and diphenylguanidine are some of the many accelerators and antioxidants that may have adverse effects with dermal contact (Joseph and Maibach, 1967). Production must be controlled carefully to avoid exposure to the diisocyanates used in polyurethane manufacturing.

RAYON

Rayon (CASRN: 61788-77-0) in its finished form is essentially cellulose and is not inherently toxic. The phases of production that have been hazardous are those that involve opportunities for exposure to carbon disulfide. In the production of rayon, wood pulp is treated with sodium hydroxide to produce alkali cellulose. This material is then reacted with carbon disulfide to produce cellulose xanthate. It is this exposure to carbon disulfide that is the greatest potential hazard. The resulting xanthate crumb, when dissolved in dilute sodium hydroxide to form viscose, is extrudable to form rayon fibers. In modern practice, this process takes place in closed systems. Exposure is possible, but with appropriate ventilation, a significant hazard is not likely to occur.

The biological effects of carbon disulfide are discussed elsewhere, but the health effects noted in the viscose rayon industry are described here. Epidemiological studies have described a variety of health effects associated with this industry. Neuropathy is a well-known effect produced by excessive carbon disulfide exposure and is reported in the rayon industry. The fiber cutting area of a rayon plant has been linked to higher concentrations of airborne carbon disulfide, leading to occupational exposure and outbreaks of polyneuropathies (Chu et al., 1995). In areas of lower concentrations, lesser effects, such as absentmindedness and difficulties in perception, have been noted (Cassitto et al., 1993).

The nervous system is not the only systemic effect linked to carbon disulfide exposure. One study found male workers experiencing decreased libido and potency, although sperm motility and fertility were not affected (Vanhoorne et al., 1994). Both carbon disulfide and hydrogen disulfide have been the cause of eye irritation, severe enough to cause certain workers to leave their jobs (Vanhoorne et al., 1995). Regardless of the work area involved, exposure to these chemicals must be controlled to prevent work-related illness.

REFERENCES

Carlton, W.W. (1994) "Proliferative keratin cyst," a lesion in the lungs of rats following chronic exposure to para-aramid fibrils. *Fundam. Appl. Toxicol.* 23:304–307.

Cassitto, M.G., Camerino, D., Imbriani, M., Contardi, T., Masera, L., and Gilioli, R. (1993) Carbon disulfide and the central nervous system: a 15-year neurobehavioral surveillance of an exposed population. *Environ. Res.* 63:252–263.

Chu, C.-C., Huang, C.-C., Chen, R.-S., and Shih, T.-S. (1995) Polyneuropathy induced by carbon disulphide in viscose rayon workers. *Occup. Environ. Med.* 52:404–407.

Hicks, E.M. Jr., Ultee, A.J., and Drougas, J. (1965) Spandex elastic fibers: development of a new type of elastic fiber stimulates further work in the growing field of stretch fabrics. *Science* 147:373–379.

Joseph, H.L. and Maibach, H.I. (1967) Contact dermatitis from spandex brassieres. *JAMA* 201:880–882.

Kelly, D.P., Merriman, E.A., Kennedy, G.L. Jr., and Lee, K.P. (1993) Deposition, clearance, and shortening of Kevlar para-aramid fibrils in acute, subchronic, and chronic inhalation studies in rats. *Fundam. Appl. Toxicol.* 21:345–354.

Lee, K.P., Kelly, D.P., O'Neal, F.O., Stadler, J.C., and Kennedy, G.L. Jr. (1988) Lung response to ultrafine Kevlar aramid synthetic fibrils following 2-year inhalation exposure in rats. *Fund. Appl. Toxicol.* 11:1–20.

Maxfield, M.E., Barnes, J.R., Azar, A., and Trochimowicz, H.T. (1975) Urinary excretion of metabolite following experimental human exposures to DMF or to DMAC. *J. Occup. Med.* 17:506–511.

Midtgård, U. and Jelnes, J.E. (1991) Toxicology and occupational hazards of new materials and processes in metal surface treatment, powder metallurgy, technical ceramics, and fiber-reinforced plastics. *Scand. J. Work Environ. Health* 17:369–379.

Pimentel, J.C. (1977) Sarcoid granulomas of the skin produced by acrylic and nylon fibres. *Br. J. Dermatol.* 96:673–677.

Vanhoorne, M., Comhaire, F., and de Bacquer, D. (1994) Epidemiological study of the effects of carbon disulfide on male sexuality and reproduction. *Arch. Environ. Health* 49:273–278.

Vanhoorne, M., de Rouck, A., and de Bacquer, D. (1995) Epidemiological study of eye irritation by hydrogen sulfide and/or carbon disulphide exposure in viscose rayon workers. *Ann. Occup. Hyg.* 39:307–315.

Vobecky, J., Devroede, G., and Caro, J. (1984) Risk of large-bowel cancer in synthetic fiber manufacture. *Cancer* 54:2537–2542.

Warheit, D.B., Kellar, K.A., and Hartsky, M.A. (1992) Pulmonary cellular effects in rats following aerosol exposures to ultrafine Kevlar aramid fibers: evidence for biodegradability of inhaled fibrils. *Toxicol. Appl. Pharmacol.* 116:225–239.

78

BISPHENOL A (4,4′-ISOPROPYLIDENEDIPHENOL)

Julie E. Goodman, Lorenz R. Rhomberg, and Michael K. Peterson

The following information is for the pure form of bisphenol A (BPA), generally occurring as white to light tan solid flakes, grains, or powder with a mild phenolic odor.

First aid: *Oral exposure*: Amounts that may be ingested accidentally in the workplace are considered to have low toxicity. If swallowed, wash out the mouth with water, provided the person is conscious. Call a physician. Do not induce vomiting unless directed to do so by a medical professional. *Inhalation exposure*: Exposures to vapors are minimal at room temperature due to the low vapor pressure of BPA. Vapor levels sufficient to cause irritation may be generated when BPA is in the molten state. If inhaled, move the person to fresh air. If breathing becomes difficult, call a physician. *Dermal exposure*: A short single acute exposure is not likely to cause significant skin irritation. Prolonged or repeated exposure may cause skin irritation, particularly under conditions of sweating. In case of skin contact, flush with copious amounts of water for at least 15 min. Remove the contaminated clothing and shoes. Call a physician. *Eye exposure*: BPA dust may cause moderate irritation and corneal injury. In case of contact with eyes, flush with copious amounts of water for at least 15 min. Ensure adequate flushing by separating the eyelids with fingers. Call a physician.

Target organ(s): Endocrine system, liver, kidney, bladder.

Occupational exposure limits: In the absence of specific threshold limit values (TLV) and permissible exposure limits (PEL) for BPA, OSHA has an exposure limit for particulates not otherwise regulated (15 mg/m^3 for total dust and 5 mg/m^3 for the respirable fraction). ACGIH has an exposure guideline for particulates not otherwise classified of 10 mg/m^3 for total dust and 3 mg/m^3 for the respirable fraction.

Reference values: (1) Human health: US EPA RfD, EFSA TDI, and ECHA DNEL (oral, general population): 0.05 mg/kg-bw/day; (2) Ecotoxicological: PNEC (freshwater) is 0.018 mg/l, PNEC (marine water) is 0.016 mg/l, PNEC (freshwater sediment) is 2.2 mg/kg sediment dry weight, PNEC (marine water sediment) is 0.44 mg/kg sediment dry weight, and PNEC (soil) is 3.7 mg/kg soil dry weight.

Risk/safety phrases: STOT Single Exp. 3 H335: May cause respiratory irritation. Eye damage 1 H318: Causes serious eye damage. Skin Sens. 1 H317: May cause an allergic skin reaction. Repro. 2 H361: Suspected of damaging fertility or the unborn child.

BACKGROUND AND USES

Bisphenol A (CASRN 80-05-7) is an organic compound used predominantly in the production of products made of polycarbonate plastic and epoxy resins. People are generally exposed to minute levels of BPA, mostly from the ingestion of food or beverages that have been in contact with materials manufactured with BPA. Workers may be exposed to higher levels during the manufacture of the compound. While studies have shown that BPA has weak estrogen-like activity and that it may cause effects at extremely high doses in animals, human exposures are well below the intake levels determined by most government bodies to be without harm (EFSA, 2006).

Hamilton & Hardy's Industrial Toxicology, Sixth Edition. Edited by Raymond D. Harbison, Marie M. Bourgeois, and Giffe T. Johnson.
© 2015 John Wiley & Sons, Inc. Published 2015 by John Wiley & Sons, Inc.

TABLE 78.1 Physical-Chemical Properties of Bisphenol A (ECB, 2003)

Property	Value	Comments
Form	Solid white flakes or powder	
Melting point	155–157 °C	
Boiling point	360 °C	Decomposition also likely
Density	1.1–1.2 kg/m^3 at 25 °C	
Vapor pressure	5.3 × 10^{-9} kPa	
Solubility in water	300 mg/l	
Log K$_{OW}$	3.3–3.5	
Flash point	207 °C	
Autoflammability	532 °C	
Explosive limits	0.012 g/l with O$_2$ > 5%	Not an oxidizing agent

BPA was first synthesized in 1891, but it was not used widely until applications in the plastics industry were identified in the 1950s (University of Minnesota, 2008). While the most prominent use of BPA is in the manufacture of polycarbonate plastic and epoxy resins, it is also used in the production and processing of polyvinyl chloride (PVC) and modified polyamide and in the manufacture of carbonless and thermal paper, wood filler, adhesives, printing inks, surface coatings, polyurethane, brake fluid, resin-based dental composites and sealants, flame retardants, paints, and tires (ECB, 2003; EFSA, 2006).

PHYSICAL AND CHEMICAL PROPERTIES

BPA is a white solid at room temperature with a slight phenolic odor and a very low vapor pressure (ECB, 2003). It is mildly soluble in water. It is not considered to be an explosive in the conventional sense but can pose a hazard as a finely powdered material in air (ECB, 2003). It is not considered to be a chemical oxidizer. The physical–chemical properties of BPA are provided in Table 78.1.

ENVIRONMENTAL FATE AND BIOACCUMULATION

BPA can be released into the environment during the production, processing, and use of BPA-containing materials, although levels in environmental samples are generally very low or undetectable (ECB, 2003). This is because BPA has low volatility and a short half-life in the atmosphere, is rapidly biodegraded in water, and is not expected to be stable, mobile, or bioavailable from soils (ECB, 2003; Cousins et al., 2002).

Most environmental releases of BPA are during the manufacture of BPA-containing products when residual BPA in wastewater is released from treatment plants into receiving streams (Cousins et al., 2002). BPA's half-life in

soil and water is in the order of 4.5 days while in air it is <1 day (Cousins et al., 2002). It has a low bioconcentration factor and is rapidly metabolized in fish, with a half-life of <1 day (Cousins et al., 2002).

ECOTOXICOLOGY

Numerous ecotoxicity studies have been conducted on BPA because of its prevalence in the environment (Staples et al., 2002). Under the European Union's (EU)'s registered substances database for REACH (Registration, Evaluation, Authorisation and Restriction of Chemicals), the following predicted no effect concentrations (PNECs) have been derived, namely, 0.018 mg/l (freshwater), 0.016 mg/l (marine water), 2.2 mg/kg sediment dry weight (dw) (freshwater sediment), 0.44 mg/kg sediment dw (marine sediment), and 3.7 mg/kg soil dw (ECHA, 2011). When these values are compared to the concentrations of BPA that have been reported in typical surface waters (0.000001–0.00010 mg/l), the likelihood of adverse effects on aquatic ecosystems is negligible (Staples et al., 2002). However, because the aquatic PNECs indicate that species populating this medium are the most sensitive, the sections that follow focus on the relevant acute and chronic aquatic toxicity data.

Acute Effects

Because of BPA's relatively short half-life in water, the primary aquatic toxicity concern relates to high-level acute exposures, which may occur from spills or other accidental releases. Under acute conditions, BPA has been reported to be moderately toxic to water fleas (i.e., *Daphnia magna*) (48-h EC$_{50}$s ~3.9–16 mg/l) and Mysid shrimp (i.e., *Mysidopsis bahia*) (96-h LC$_{50}$ ~1.1 mg/l; NOEC = 0.51 mg/l) (ECHA, 2011). The reported LC$_{50}$s in marine and freshwater fish also indicate a moderate acute toxicity concern with values in the following ranges, that is, approximately 7–15 mg/l (48 h) or 3–10 mg/l (96 h) (ECHA, 2011). Based on the acute aquatic hazard criteria set forth under EU's implemented version of GHS (Globally Harmonized System, the foregoing values do not meet the criteria for classification and labeling for acute aquatic hazards (ECHA, 2011).

Chronic Effects

There have been a variety of studies on BPA's effects on chronic endpoints, including survival, growth and development, and reproduction. Many of these have been discussed in detail elsewhere (e.g., Staples et al., 2002) and are summarized below.

Staples et al. (2002) evaluated a wide variety of these studies and found that eight survival studies were appropriate for assessing this endpoint. Staples et al. (2002) observed that study duration had little impact on species survival and that, for most species (including four species of fish, an amphibian, and an invertebrate), there were no effects at the highest doses tested, which ranged from 0.10 to 3.16 mg/l.

Staples et al. (2002) summarized multiple studies that evaluated BPA's effects on growth and development. Eight studies evaluating different species of fish and an amphibian species for durations of 28–431 days found that the lowest BPA concentrations that caused effects on these parameters ranged from 1.82 to 11.0 mg/l. No observed effect concentrations ranged from 0.12 to 3.64 mg/l. In aquatic invertebrate studies, growth effects were observed in a copepod, with an EC_{10} (the effect concentration for 10% of the population tested) of 0.10 mg/l (Andersen et al., 2001), but Staples et al. (2000) concluded that this study was not valid for use in hazard assessment, as few methodological details or results were reported. In a different aquatic invertebrate (*Daphnia magna*), no effects on molting were observed at concentrations up to 3.16 mg/l (the highest concentration tested). In algae, growth effects were reported for freshwater green algae (EC_{10} of 1.36–1.68 mg/l) and the marine diatom (EC_{10} of 0.40–0.69 mg/l).

Staples et al. (2002) evaluated reproductive toxicity studies in a variety of aquatic organisms, including fish, amphibians, invertebrates, and algae. In fish studies, which ranged from 14 to 100 days in duration and evaluated multiple generations as well as early life stages, lowest observed effect concentrations ranged from 0.16 to 2.28 mg/l. No adverse effect concentrations were reported in invertebrates up to the highest concentration tested of 3.16 mg/l. Adverse effects on growth were observed in green algae between 0.4 and 1.68 mg/l.

Based on the criteria for long-term aquatic hazards under the EU's GHS, the foregoing values do not support classification and labeling for chronic aquatic hazards (ECHA, 2011).

MAMMALIAN TOXICOLOGY

Standardized studies performed according to internationally validated test guidelines and conducted under good laboratory practice (GLP) standards have not shown effects from BPA at low doses (Ema et al., 2001; Tyl et al., 2002, 2008). A large number of smaller-scale non-guideline, non-GLP studies have reported an extensive and disparate set of effects in the low-dose range, i.e., <5 mg/kg-bw/day (reviewed in Gray et al., 2004; Goodman et al., 2006, 2009). Often, the effects reported in non-guideline, non-GLP studies are found only at one or two low doses but without effects at higher doses. The effects in question are often intermediary rather than apical endpoints, and their interpretation as indicators of potential frank toxicity is a matter of debate. Critics of the reliance on guideline/GLP studies cite non-guideline, non-GLP studies as evidence that guideline/GLP studies use insensitive strains of animals, fail to uncover subtle endocrine-mediated effects, and err in presuming that a lack of high-dose effects precludes lower-dose effects of concern (vom Saal and Hughes, 2005).

The counterargument is that non-guideline, non-GLP studies are inconsistent among one another (and with the GLP study results), with most examinations of validated endpoints being negative and the reported effects on nonvalidated endpoints being inconsistent and not dose-dependent (Goodman et al., 2009; Hengstler et al., 2011; Sharpe, 2010; Tyl, 2009). Under this view, the reported effects at low doses can be attributed to chance fluctuations and imprecisely defined control states, to which small studies using novel designs are prone. Moreover, many of these studies administered BPA by subcutaneous injection, which obviates the relevance of such studies to human exposures because this route of administration bypasses first-pass metabolism by the liver (Goodman et al., 2009; Hengstler et al., 2011). Though the interpretation of the body of toxicological data for BPA continues to be controversial (Myers et al., 2009a, 2009b; Becker et al., 2009), in reviews and assessments conducted by responsible government bodies around the world, the latter view—that is, the evidence does not support effects below 5 mg/kg—has prevailed, as detailed in the Risk Assessments section.

Acute Effects

In general, BPA has low acute toxicity by the oral and dermal routes, with experimental values exceeding the highest recommended dose levels from the respective test guidelines. For example, the oral LD_{50} values in laboratory animals such as mice and rats are above 2000 mg/kg-bw, and the dermal LD_{50} values in rabbits are also above 2000 mg/kg-bw (ECB, 2003). Though no deaths were reported in an acute 6-h inhalation study in rats at 170 mg/m³, minor and transient changes in the nasal epithelium were reported (ECB, 2003).

Animal studies generally report that BPA has low dermal irritation potential (ECB, 2003); however, BPA is classified under the EU's adopted version of GHS as a category 3 single exposure, specific target organ toxicity (STOT) for respiratory irritation (ECHA, 2011). In studies with rabbits, BPA has been shown to be an eye irritant (ECB, 2003) and has an EU–GHS classification of having the potential to cause serious eye damage (i.e., eye damage 1) (ECHA, 2011).

Most previous studies of BPA's sensitization potential are case reports that are confounded by co-exposure to other chemicals such as epoxy resins. Recent studies do not confirm any skin sensitization up to 30% BPA, and medical surveillance reports from five BPA manufacturing plants also did not identify any workers who had self-reported skin effects or had skin issues observed during the routine 1–3-year medical examinations (ECB, 2003). Notwithstanding the conflicting reports on this endpoint, BPA is classified as a category 1 skin sensitizer (i.e., may cause an allergic skin reaction) by the EU (ECHA, 2011).

Subchronic and Chronic Effects

The estrogenic effects of BPA at high doses are well understood and generally agreed upon by scientists, but there is controversy over BPA's effects at low doses (sometimes called the "low-dose hypothesis"). Because the lowest level at which adverse effects have been reported in guideline/GLP studies is 50 mg/kg-bw/day, studies of BPA at doses an order of magnitude lower than this (i.e., ≤5 mg/kg-bw/day) are considered relevant for addressing possible low-dose effects. Typical human exposures are several orders of magnitude lower than this dose.

When evaluating the relevance of effects reported in animal studies to humans, it is important to consider some of the pharmacokinetic similarities and differences between animals and humans exposed to BPA (reviewed in Hengstler et al., 2011). In all mammals thus far examined, BPA is readily absorbed via oral intake, with substantial first-pass metabolism by the gut wall and liver. Metabolism is generally to the glucuronide or sulfate conjugate, and these appear to be toxicologically inactive. Elimination of the metabolites is via urine in humans and the urine and feces in rodents.

Most of the BPA to which humans are exposed is ingested orally and metabolized in the intestine and liver, then eliminated in urine, so it does not reach the general circulation in appreciable amounts nor does it accumulate in the body after ingestion (Völkel et al., 2002, 2005). Though BPA in rodents is also metabolized in the intestine and liver, there is substantial enterohepatic recirculation after oral exposure (i.e., the chemical is excreted via bile from the liver into the digestive tract, where the glucuronide is cleaved and the aglycone is reabsorbed), the effect of which prolongs the presence of BPA at low levels in the systemic circulation (Kurebayashi et al., 2003). Subcutaneously injected BPA enters the systemic circulation directly, which bypasses hepatic first-pass clearance and results in substantially higher systemic levels of unconjugated BPA. Metabolized or conjugated BPA has little or no known biological or estrogenic activity (Matthews et al., 2001).

Two recent studies by the U.S. Food and Drug Administration (FDA) indicate that both adult and newborn experimental animals have the capacity to metabolize and eliminate BPA from the body; however, there are some species-specific differences. The first study examined this process in rats and found that, after ingesting BPA, adult rats could efficiently and rapidly metabolize and eliminate the chemical, but metabolism in newborn rats was not as efficient. The second study was conducted in monkeys. FDA found that adult monkeys could efficiently metabolize BPA, and that newborn monkeys had the same capacity as adult monkeys to safely metabolize and eliminate BPA (Doerge et al., 2010a, 2010b). Thus, the authors concluded that effects based on doses in rodent studies would overpredict effects in primates.

The majority of low-dose BPA studies have been conducted in rodents. Internal doses in these studies are higher than humans would experience based on the same exposure, and this is particularly true for non-oral exposures.

Reproductive and Developmental Effects   Based on hundreds of animal studies with BPA doses ≤5 mg/kg-bw/day, there is an overwhelming preponderance of lack-of-effect findings compared to findings of effect over a wide variety of reproductive and developmental toxicity endpoints (reviewed by Gray et al., 2004; Goodman et al., 2006, 2009; Hengstler et al., 2011; and multiple government agencies, discussed in the regulatory status/rationale section below). The most robust reproductive and developmental BPA evaluations are the multigeneration studies conducted with Sprague-Dawley rats (Ema et al., 2001; Tyl et al., 2002) and CD-1 Swiss mice (Tyl et al., 2008). These studies used a large number of animals with a range of oral doses and examined a wide variety of hormonally sensitive endpoints. Reproductive effects occurred only at the highest tested doses in these studies (EFSA, 2006). Effects at lower doses were observed only sporadically (e.g., in only one generation, with no dose–response pattern) and were not considered to be treatment-related.

Among all other BPA low-dose animal studies, there are some that reported responses at low doses, but no marked or consistently repeatable effects were observed. Reported effects are not consistent between rats and mice, and there are no consistent patterns among dose groups and evaluation times (Nagel et al., 1997; Gupta, 2000; Ashby et al., 1999; Cagen et al., 1999; Toyama et al., 2004; Tyl et al., 2008). The reported effects considered together lack any common pattern consistent with a hormonal mode of action. Regardless, BPA is classified under the EU's GHS as a category 2 reproductive toxicant (i.e., suspected of damaging fertility or the unborn child) (ECHA, 2011).

Neurological Effects   During fetal and childhood development, natural hormones can influence brain development and subsequent behavior, including sexual behavior and other social interactions. Thus, several investigators examined whether low doses of BPA during fetal development can affect subsequent behavior of pups (e.g., Kubo et al., 2003; Schantz and Widholm, 2001). Early studies evaluated a number of endpoints, but interpretation is hampered by a lack of understanding of whether results are applicable to humans. Even so, there is no consistent evidence among these studies that low doses of BPA (i.e., ≤5 mg/kg-bw/day) cause adverse effects on behavioral endpoints (e.g., Ema et al., 2001; Palanza et al., 2002; Aloisi et al., 2002).

Because the relevance of these studies to humans is unclear, two recent studies were conducted (one by the U.S. Environmental Protection Agency, EPA) using robust, validated methodologies to assess the effects of low doses of BPA on the developing brain. These studies found no evidence of neurological effects in animals exposed to low doses of BPA (Stump et al., 2010; Ryan et al., 2010).

Immunological Effects The natural interactions between hormones and the immune system led several investigators to study whether BPA is associated with altered immune function. The data are currently insufficient for drawing conclusions about possible immunological effects at low doses, such as those experienced by humans. In light of several guideline/ GLP multigenerational studies reporting no pathology indicative of immune system dysfunction at low doses (e.g., Tyl et al., 2002, 2008; Ema et al., 2001), the findings reported from non-guideline, non-GLP studies on immunological effects are unlikely to be indicative of potential human risks.

Carcinogenic and Mutagenic Effects Researchers have examined whether BPA could be carcinogenic via a hormonal mode of action. BPA was first tested for carcinogenicity by the U.S. National Toxicology Program (NTP), which concluded that, "under the conditions of this bioassay, there was no convincing evidence that [BPA] was carcinogenic for F344 rats or B6C3F1 mice of either sex" (NTP, 1982). In these assays, rats and mice were exposed for 103 weeks but not prenatally or during early life.

No government or international organizations classify BPA as a carcinogen. Though it is unclear whether BPA was evaluated for nomination by the NTP or the International Agency for Research on Cancer (IARC) for review of BPA's carcinogenic potential, this substance is not currently listed as a carcinogen by either agency. EPA's Integrated Risk Information System (IRIS; US EPA, 1988) developed a health assessment on BPA but did not perform a complete evaluation of its carcinogenic potential. In 2002, a panel of prominent scientists conducted a weight-of-evidence evaluation of potential BPA carcinogenicity and concluded that "BPA is not likely to be carcinogenic to humans," citing the NTP bioassay as providing "no substantive evidence to indicate that BPA is carcinogenic to rodents" (Haighton et al., 2002).

In the following year, the EU published the results of its risk assessment report on BPA, in which the EU also concluded that BPA does not have carcinogenic potential (ECB, 2003).

In the last decade, several studies have examined the association between BPA and precursor lesions to cancer. While some studies showed associations between BPA and these precursors, none actually reported cancer (although it should be noted that most of these studies did not evaluate animals for their entire lifetime; e.g., Ho et al., 2006; Prins et al., 2008). Other studies evaluated the carcinogenic potential of BPA when co-administered with other chemicals (e.g., Jenkins et al., 2009; Lamartiniere et al., 2011). While some animals developed cancer in these studies, these results are not necessarily applicable to human exposures because BPA was administered via non-oral routes and/or with chemicals known to be carcinogens. Thus, these recent findings are not indicative of BPA being a human carcinogen under relevant conditions of exposure.

Several human studies have evaluated the association between BPA and various health effects, including, for example, premature birth, miscarriages, obesity, polycystic ovarian syndrome, cancer, altered sperm and semen characteristics, and other reproductive endpoints (e.g., Cantonwine et al., 2010; Sugiura-Ogasawara et al., 2005; Takeuchi and Tsutsumi, 2002; Takeuchi et al., 2004; Hiroi et al., 2004; Yang et al., 2006). The majority of these studies have methodological limitations that make their results difficult to interpret, largely because exposures in the general population are quite low and the range of exposures among people is small. The low BPA levels measured often are close to or below laboratory detection limits. Urinary measurements are comprised largely of conjugated, biologically inactive BPA, and blood measurements in most studies are most likely due to contamination and are not reflective of actual exposure (Teeguarden et al., 2011). The majority of human studies are cross-sectional in nature, in which BPA was measured at the same time as the health effect was assessed, so it cannot be known whether past BPA exposures were causally associated with the effect. Overall, the human studies conducted thus far are insufficient to address possible effects at these environmental levels.

Mode of Action

Studies have shown that BPA has estrogen-like or endocrine activity, meaning it can mimic or alter the effects of the body's natural hormones. While toxicity tests used to develop safety standards for BPA found that the lowest dose at which adverse effects in animals were observed was 50 mg/kg-bw/day, several orders of magnitude higher than human exposures, hormonal effects were only found at much higher doses (Ashby and Odum, 2004). Still, some individual scientists support the so-called "low-dose hypothesis"—that is, at very low doses, BPA may cause adverse effects that are not observed at higher doses via an endocrine mode of action.

BPA's weak estrogenic activity has been reported since the 1930s, but it was regarded as a safe and effective plasticizer until the publication of a study by a group of investigators led by Frederick vom Saal (vom Saal et al., 1998). In this study, pregnant mice were given oral doses of BPA at 0.002 or 0.020 mg/kg-bw/day from gestation days 11 to 17. When the male offspring were 6 months old, they were reported to have permanently increased preputial glands compared with unexposed mice. Because these effects were not observed at higher doses in other studies, this led to the hypothesis that low doses of BPA could cause effects that high doses would not.

The low-dose hypothesis is inconsistent with the first tenet of toxicology—*the dose makes the poison* (Eaton and Gilbert, 2008; Faustman and Omenn, 2008). Substances generally elicit an increasing level of effect with increasing dose. That is, adverse effects occur only when the dose exceeds a threshold for a certain period of time because at low doses, biochemical or physiological mechanisms inhibit a chemical's adverse effects or the perturbations are too minor to produce a dysfunction. As the magnitude and duration of exposure to a chemical increase beyond the threshold, these protective mechanisms become overwhelmed, and an adverse response appears to a degree that corresponds to the increase in dose. Proponents of the low dose hypothesis suggest that, as the dose of BPA increases, effects occur that would not occur at higher doses. This hypothesis is based on a selective evaluation of studies, however, and not a thorough analysis of the entire toxicological database (see, e.g., reviews conducted by Goodman et al., 2006, 2009; Gray et al., 2004; Hengstler et al., 2011).

SOURCES OF EXPOSURE

Almost all nonoccupational human exposures occur via residues in food or beverages that have been in contact with polycarbonate plastic or with containers lined with epoxy resins, as these products can contain trace amounts of the original compound or additional BPA that may be generated during the breakdown of the product (ECB, 2003; FAO and WHO, 2010). Oral and dermal exposures can also occur from contact with other products made with BPA. Estimates based on urine measurements and pharmacokinetic data suggest BPA intakes generally range from <0.00003 to 0.00161 mg/kg-bw/day.

Though many industries use products that may contain BPA residues, exposures to BPA from these products are generally thought to be negligible because the BPA is incorporated into a matrix (ECB, 2003). Highest occupational exposures likely occur in the BPA manufacturing industry. BPA is generally processed under closed conditions; however,

there are several points where there may be dermal and inhalation exposures to BPA, such as during product sampling and product bagging. BPA concentrations reported in these types of processes ranged from non-detect to 23.3 mg/m^3 (8-h time-weighted average, TWA). Recent data submitted by European BPA manufacturers have reported lower concentrations, with 8-h TWAs ranging from non-detect to 3.6 mg/m^3 (ECB, 2008).

INDUSTRIAL HYGIENE

Occupational limits for BPA have not been established by the U.S. Occupational Safety and Health Administration (OSHA), the U.S. National Institute for Occupational Safety and Health, or the American Conference of Industrial Hygienists (ACGIH). Rather, BPA is covered generally under the nuisance dust standards of these organizations. For example, OSHA has a permissible exposure limit (PEL) for particulates not otherwise regulated of 15 mg/m^3 for total dust and 5 mg/m^3 for the respirable fraction, whereas ACGIH has a threshold limit value (TLV) for particulates not otherwise classified of 10 mg/m^3 for total dust and 3 mg/m^3 for the respirable fraction.

RISK ASSESSMENTS

Several government and international agencies have performed human health risk assessments on BPA. Most of these risk assessments have used a no-observed-adverse-effects level (NOAEL) of 5 mg/kg-bw/day, derived from two multigenerational rodent studies, and a composite uncertainty factor of 500 (i.e., $10 \times 10 \times 5$, for interspecies differences, interindividual variability, and uncertainties in the database on reproductive and developmental toxicity, respectively). Thus far, all of these risk assessments have concluded that exposure to BPA from its current uses poses no human health risks, i.e., either the exposure levels are far below the TDI (tolerable daily intake) or the margins of safety are far above 500 (i.e., estimated exposures are over 500 times lower than the NOAEL).

The same government and international agencies that performed risk assessments on BPA have also undertaken reviews in response to claims of low-dose BPA toxicity (i.e., ≤5 mg/kg-bw/day). Thus far, no such review has concluded that adverse effects reported to occur below the traditionally defined no-effect level are sufficiently plausible to constitute a health threat. Nonetheless, several governments have initiated bans on specific uses of BPA (e.g., in children's feeding bottles in Canada and the EU and all food contact material for children under 3 years of age in Denmark).

Risk assessments conducted by various government and international agencies and the regulatory status of BPA around the world are described below. While the agencies and countries discussed should be considered to be a comprehensive list, it may not be exhaustive.

In 2001, NTP organized a scientific panel to conduct a peer review of the evidence on low-dose reproductive and developmental effects. The panel determined that "low-dose" effects of BPA were not "conclusively established as a general or reproducible finding" (NTP, 2001). A subpanel focusing specifically on BPA concluded,

> [Th]ere is credible evidence that low doses of BPA [bisphenol A] can cause effects on specific endpoints. However, due to the inability of other credible studies in several different laboratories to observe low dose effects of BPA, and the consistency of these negative studies, the Subpanel is not persuaded that a low dose effect of BPA has been conclusively established as a general or reproducible finding. (NTP, 2001)

In other words, because the majority of credible studies found no evidence of an effect, these studies outweighed the few studies that found positive evidence.

In 2007, NTP's Center for the Evaluation of Risks to Human Reproduction (CERHR) released a report that detailed studies of reproductive and developmental effects of BPA and applied a ranking of the utility of each study for determining risks to humans using the "levels of concern" scheme (NTP, 2007). In April 2008, NTP issued a draft brief on BPA, in which it stated that it agreed with the CERHR expert panel that there was some concern for effects on neurobehavioral, prostate gland, mammary gland, and puberty endpoints in sensitive populations at current exposures levels, and that there was negligible concern for exposures in pregnant women impacting in fetal endpoints such as mortality, birth defects, or growth/weight.

> The NTP concurs with the conclusion of the CERHR expert panel on BPA that there is negligible concern that exposure to BPA causes reproductive effects in nonoccupationally exposed adults and minimal concern for workers exposed to higher levels in occupational settings. (NTP, 2008)

The highest level of concern assigned by NTP was "some" concern for certain effects and "negligible" concern for others, which indicates that NTP concluded that the data did not indicate a high level of concern for any effect. This low level of concern also indicated that NTP could not determine whether any effects observed in animal studies were applicable to humans.

In 2008, FDA issued a draft assessment (i.e., not an official FDA statement) that considered the CERHR's 2007 report on BPA. The draft assessment considered the NOAEL for systemic toxicity of 5 mg/kg-bw/day (derived from two multigenerational rodent studies) to be appropriate, and it estimated the BPA exposure from use in food contact materials in infants and adults to be 0.00242 and 0.000185 mg/kg-bw/day, respectively. Consequently, FDA concluded that there were adequate margins of safety (of approximately 2000 and 27,000 for infants and adults, respectively) at current levels of BPA exposure from uses in food contact materials. The draft assessment also concluded that the data on endpoints such as prostate effects and developmental, neural, and behavioral toxicity were insufficient to alter the NOAEL, meaning that there is no indication of low-dose effects of BPA (FDA, 2008). After further review in 2012, FDA determined that BPA "is not harmful under the intended conditions of use" (Dorsey, 2012).

In 2008, Health Canada assessed BPA exposure to newborns and infants from polycarbonate baby bottles exposed to high temperatures and from the migration of BPA from cans into infant formula. Health Canada (2008) concluded that BPA exposures to newborns and infants are below levels that may pose a risk, but the gap between exposure and potential effects concentrations is not large enough. Consequently, in April 2008, the Canadian government began action to consider banning the importation, sale, and advertising of polycarbonate baby bottles containing BPA, to develop stringent migration targets for BPA in infant formula cans, and to possibly list BPA under Schedule 1 of the Canadian Environmental Protection Act of 1999 (CEPA) (Health Canada, 2008). This action set a timeline for finalizing implementation plans, compliance strategies, and other risk management proposals by April 16, 2012 (Canada Gazette, 2010).

In 2009, the Health Canada Food Directorate issued a statement that concluded that current BPA exposure through food packaging uses was not expected to pose a health risk to the general population, including newborns and infants. This conclusion was based on the overall weight of evidence, including reaffirmation by other international regulatory agencies (in the U.S., Europe, and Japan) and was consistent with the conclusions of Health Canada's 2008 assessment. Because of the uncertainty raised by some animal studies relating to the potential effects of low levels of BPA, however, it was recommended that the general principle of *as low as reasonably achievable* (ALARA) be applied to limit BPA exposure to infants and newborns, specifically from prepackaged infant formula products as a sole source of food (INFOSAN, 2009). Consequently, on September 23, 2010, Canada listed BPA as a toxic substance under CEPA (Canada Gazette, 2010).

The use of BPA in products intended to come in contact with food was first evaluated in Europe in 1986 by the Scientific Committee for Food (SCF). SCF calculated a TDI of 0.05 mg/kg-day, based on a NOAEL of 25 mg/kg-bw/day for effects on body weight in a 90-day rat study and a composite uncertainty factor of 500 (SCF, 2002). In 2002, SCF conducted a comprehensive review of toxicity and exposure information focused on food contact applications of BPA. SCF derived a temporary TDI of 0.01 mg/kg-bw/day, based on an overall NOAEL of 5 mg/kg-bw/day in a three-generation rat study (for which liver toxicity, not a hormonal effect, was the most sensitive response) and a composite 500-fold uncertainty factor (10 for interspecies differences, 10 for interindividual variability, and 5 for uncertainties in the database on reproductive and developmental toxicity); SCF concluded that worst-case human exposures to BPA are well below the temporary TDI (SCF, 2002).

A 2003 EU Risk Assessment Report (RAR) concluded that, based on a thorough review of the evidence, no risk management actions were necessary for polycarbonate plastic or epoxy resin products. It established an overall NOAEL of 50 mg/kg-bw/day, based on the multigeneration study of BPA in rats (ECB, 2003). The EU Scientific Committee on Toxicity, Ecotoxicity, and the Environment (CSTEE), an independent scientific committee, affirmed the key conclusions of the 2003 EU RAR in its detailed opinion (CSTEE, 2002). In 2008, the EU released an addendum to the 2003 RAR (ECB, 2008), in which it stated the following:

> The worst case combined exposure would be for someone exposed via the environment near to a BPA production plant, and who is also exposed via food contact materials (oral exposure from canned food and canned beverages and from polycarbonate tableware and storage containers). Given the very large margins of safety, there are no concerns for repeated dose toxicity and reproductive toxicity . . . There is at present no need for further information and/or testing or for risk reduction measures beyond those which are being applied already.

In 2006, the European Food Safety Authority (EFSA) Scientific Panel on Food Additives, Flavorings, Processing Aids and Materials in Contact with Food (the CEF Panel) performed what is considered to be the most comprehensive risk assessment on BPA. The CEF Panel reported that the NOAEL of 5 mg/kg-bw/day was still valid and further supported by a two-generation BPA study in mice (EFSA, 2006). In addition, the panel reported that the database had been strengthened to the point that the database component of the uncertainty factor could be eliminated. EFSA established a final TDI of 0.05 mg/kg-bw/day, based on a composite 100-fold uncertainty factor. Ultimately, it stated the following:

> The Panel considered that low-dose effects of BPA in rodents have not been demonstrated in a robust and reproducible way, such that they could be used as pivotal studies for risk assessment. Moreover, the species differences in toxicokinetics, whereby BPA as parent compound is less bioavailable in humans than in rodents, raise considerable doubts about the relevance of any low-dose observations in rodents for humans. The likely high sensitivity of the mouse to oestrogens raises further doubts about the value of that particular species as a model for risk assessment of BPA in humans. (EFSA, 2006)

EFSA has reconsidered low-dose effects of BPA several times since 2006 upon the European Commission's request, and it has repeatedly concluded that evidence on potential low-dose effects of BPA remain inconclusive:

- The CEF Panel's July 2008 opinion concluded that the differences in age-dependent toxicokinetics of BPA in animals and humans would have no implication for EFSA's 2006 BPA risk assessment (EFSA, 2008).
- In response to a request to assess conclusions by Lang et al. (2008) that suggested a link between elevated levels of urinary BPA to increased occurrence of serious medical conditions (e.g., heart disease and diabetes), EFSA's October 2009 opinion concluded that the Lang et al. (2008) study did not provide sufficient proof for a causal link between BPA exposure and the serious medical conditions evaluated in the study and, therefore, the 2006 TDI need not be revised (EFSA, 2008).
- The American Chemistry Council commissioned the Stump et al. (2010) study to address safety concerns raised by the Canadian government. At the European Commission's request, EFSA reanalyzed the data from Stump et al. (2010) and showed that, statistically, the possibility of an effect of BPA on the neurotoxicity variables could neither be confirmed nor ruled out and, therefore, deemed the study to be inconclusive (EFSA, 2010b).
- In September 2010, the CEF Panel performed a detailed and comprehensive review of recent scientific literature and studies on BPA toxicity at low doses and concluded that it "could not identify any new evidence which would lead [it] to revise the current TDI for BPA of 0.05 mg/kg body weight set by EFSA in its 2006 opinion and reconfirmed in its 2008 opinion" (EFSA, 2010a). [One CEF Panel member expressed a minority opinion regarding recent studies indicating significant uncertainties in adverse health effects below the level used to determine the current TDI, but agreed with the rest of the CEF Panel members' opinion that the relevance of these recent findings for human health cannot be assessed at present (EFSA, 2010a).]

The German Federal Institute of Risk Assessment (Bundesinstitut für Risikobewertung (BfR) was involved in EFSA's reevaluation as a competent authority and confirmed the EFSA opinions on BPA. In its updated Q&A on BPA in consumer products (May 3, 2011), BfR stated that intakes of BPA pose no health risks for infants and young children during normal use (BFR, 2010).

EFSA and other agencies have repeatedly concluded that the evidence to date on low-dose effects of BPA does not indicate health risks. Nonetheless, BPA has been banned in baby bottles in France and in all materials in contact with food for children aged 0–3 years old in Denmark. Interestingly, the Danish Minister of Food, Henrik Høegh, has said:

> From DTU Food [the National Food Institute at the Technical University of Denmark], we have received an assessment based on new comprehensive studies of rats. Danish experts say there is no clear evidence that bisphenol A has harmful effects on the behaviour observed. However, the experts find that the new studies raise uncertainties about whether even small amounts of bisphenol A have an impact on the learning capacity of new-born rats. In my o[p]inion these uncertainties must benefit the consumers, so we will utilize the precautionary principle to introduce a national ban on bisphenol A in materials in contact with food for children aged 0—3 years. (Danish Ministry of Food, 2010

This opinion is based on the neurobehavioral study by Stump et al. (2010), which EFSA and other agencies have determined does not support low-dose BPA risks.

In Japan, the Ministry of Health and Welfare (MHW); the Ministry of Health, Labour and Welfare (MHLW); the Ministry of Economy, Trade and Industry (METI); the Ministry of the Environment (MOE); and others convened panels of experts in 1998, 2001, 2002, and 2004 to evaluate BPA's health risks (Nakanishi et al., 2007). None of these panels recommended prohibiting or restricting the use of BPA, but some recommended a comprehensive risk assessment (Nakanishi et al., 2007).

Based on a thorough review of safety information in 2005, the National Institute of Advanced Industrial Science and Technology (AIST), a public research organization affiliated with METI, concluded that it was unlikely that humans, including infants and children, were at unacceptable risks from possible exposures to BPA in air, food, consumer products, and water, and no adjustment to allowable exposures was deemed necessary for claimed low-dose effects (Miyamoto and Kotake, 2006). A comprehensive risk assessment conducted by this organization established a NOAEL of 50 mg/kg-bw/day for reproductive and developmental toxicity based on the results of a multigeneration study in rats (Nakanishi et al., 2007). The report further concluded, "an additional uncertainty due to the low dose effects was not incorporated because the findings in the low dose studies were not robust, while those in negative studies were consistent" (Nakanishi et al., 2007). MOE also concluded in 2005 that no clear endocrine disrupting effects of BPA were found at low doses and that no regulatory action was required to manage BPA risks (MOE, 2005). This was based on its own low-dose studies of BPA, including a comprehensive reproductive study in laboratory animals.

In 2007, AIST, together with the New Energy and Industrial Technology Development Organization (NEDO) and the Research Center for Chemical Risk Management (CRM), performed a risk assessment of BPA. This risk assessment reviewed and reanalyzed extensive data regarding BPA exposures and hazards; reduction in body weight gain, liver effects,

and reproductive toxicity were the key toxicological endpoints. Risk was characterized not only by conventional approaches (e.g., the hazard quotient or the margin of exposure method), but also by more sophisticated approaches, including Monte Carlo simulations that propagate uncertainties in exposure parameters. This 2007 risk assessment also reached the same conclusion as the 2005 AIST risk assessment—that the current BPA exposure levels "will not pose any unacceptable risk to human health" (Nakanishi et al., 2007). Regarding low-dose effects of BPA, the 2007 risk assessment mirrored the 2005 report, stating that "an additional uncertainty due to the low dose effects was not incorporated because the findings in the low dose studies were not robust while those in negative studies were consistent" (Nakanishi et al., 2007).

Food Standards Australia New Zealand (FSANZ) concluded that BPA intake levels are very low and do not pose a significant human health risk for any age group, based on a safety evaluation of BPA in food, including that consumed by infants. For example:

> [A] nine month old baby weighing 9 kg would have to eat more than 1 kg of canned baby custard containing BPA every day to reach the TDI, assuming that the custard contained the highest level of BPA found (420 parts per billion). (FSANZ, 2014)

In Australia, there is currently no mandatory standard for or ban on consumer products containing BPA. According to the Australian Competition and Consumer Commission (ACCC):

> The ACCC, in conjunction with the other relevant Australian government regulatory agencies namely; Food Standards Australia New Zealand (FSANZ), the National Industrial Chemicals Notification and Assessment Scheme (NICNAS), the Therapeutic Goods Administration (TGA) and the Australian Pesticides and Veterinary Medicines Authority (APVMA) have collectively considered the possible risks of BPA and they remain convinced that the weight of evidence, obtained from an extensive range of safety studies, indicates that BPA is safe for the whole population at the very low levels of current exposure. (ACCC, 2011).

The International Food Safety Authorities Network (INFOSAN) is a global network of 177 member states and a joint initiative between WHO and the United Nation's Food and Agriculture Organization (FAO). In a 2009 summary of the current state of knowledge on BPA, INFOSAN (2009) stated that hazard assessments by major regulatory and advisory bodies agree the overall NOAEL for BPA from robust data is 5 mg/kg-bw/day, which is minimally 500-fold above conservative estimates of human exposure, including in bottle-fed infants. The summary identified several areas of uncertainty in BPA risk assessment:

- The relevance of animal studies showing neurobehavioral effects following BPA exposure during the developmental period at doses below the overall NOAEL;

- The differences between primates, including humans, and rodents, and between exposure routes in kinetics of absorption, metabolism, and excretion of BPA and their implications for extrapolation of animal studies to humans; and
- Absence of convincing evidence from animal studies on cancer risk from BPA exposure.

In November 2010, WHO and FAO, with the support of EFSA, Health Canada, US National Institute for Environmental Health Sciences (NIEHS), and FDA, convened a panel of over 30 international experts to review toxicological and health aspects of BPA (FAO and WHO, 2010). The panel concluded that exposures to BPA occurred primarily via ingestion of food and that exposures from soil, dental sealants, and cash register receipts were of "minor relevance." It also stated:

> Studies on developmental and reproductive toxicity in which conventional end-points were evaluated have shown effects only at high doses, if at all. However, some emerging new end-points (sex-specific neurodevelopment, anxiety, preneoplastic changes in mammary glands and prostate in rats, impaired sperm parameters) in a few studies show associations at lower levels . . . There is considerable uncertainty regarding the validity and relevance of these observations. While it would be premature to conclude that these evaluations provide a realistic estimate of the human health risk, given the uncertainties, these findings should drive the direction of future research with the objective of reducing this uncertainty. (FAO and WHO, 2010)

Since the completion of the various risk and safety assessments discussed in this section, FDA has continued to study the safety of BPA in consumer products (e.g., Delclos et al., 2014). The results of these studies continue to support that BPA does not have adverse effects below 5 mg/kg-bw/day, and that, "Based on FDA's ongoing safety review of scientific evidence, the available information continues to support the safety of BPA for the currently approved uses in food containers and packaging" (FDA, 2014).

REFERENCES

Aloisi, A.M., Della Seta, D., Rendo, C., Ceccarellis, I., Scaramuzzino, A., and Farabollini, F. (2002) Exposure to the estrogenic pollutant bisphenol A affects pain behavior induced by subcutaneous formalin injection in male and female rats. *Brain Res.* 937:1–7.

Andersen, H.R., Wollenberger, L., Halling-Sørensen, B., and Kusk, K.O. (2001) Development of copepod nauplii to copepodites—a parameter for chronic toxicity including endocrine disruption. *Environ. Toxicol. Chem.* 20:2821–2829.

Ashby, J. and Odum, J. (2004) Gene expression changes in the immature rat uterus: effects of uterotrophic and sub-uterotrophic doses of bisphenol A. *Toxicol. Sci.* 82:458–467.

Ashby, J., Tinwell, H., and Haseman, J. (1999) Lack of effects for low dose levels of bisphenol A and diethylstilbestrol on the prostate gland of CF1 mice exposed *in utero. Regul. Toxicol. Pharmacol.* 30:156–166.

Australian Competition and Consumer Commission (ACCC) (2014) *Bisphenol A in consumer products. Product Safety Australia.* Available at http://www.productsafety.gov.au/content/index.phtml/itemId/971446 (accessed August 22, 2011).

Becker, R.A., Janus, E.R., White, R.D., Kruszewski, F.H., and Brackett, R.E. (2009) Good laboratory practices and safety assessments (Letter). *Environ. Health Perspect.* 11:A482–A483.

Bundesinstitut für Risikobewertung (BfR) (2010) *Ausgewählte Fragen und Antworten zu Bisphenol A in Babyfläschchen und-saugern. BfR.*

Cagen, S.Z., Waechter, J.M., Dimond, S.S., Breslin, W.J., Butala, J.H., Jekat, F.W., Joiner, R.L., Shiotsuka, R.N., Veenstra, G.E., and Harris, L.R. (1999) Normal reproductive organ development in Wistar rats exposed to bisphenol A in the drinking water. *Regul. Toxicol. Pharmacol.* 30:130–139.

Canada Gazette (2010) *Order Adding a Toxic Substance to Schedule 1 to the Canadian Environmental Protection Act.* Available at http://www.gazette.gc.ca/rp-pr/p2/2010/2010-10-13/html/sor-dors194-eng.html Published September 23, 2010 (accessed August 22, 2011).

Cantonwine, D., Meeker, J.D., Hu, H., Sánchez, B.N., Lamadrid-Figueroa, H., Mercado-García, A., Fortenberry, G.Z., Calafat, A.M., and Téllez-Rojo, M.M. (2010) Bisphenol a exposure in Mexico City and risk of prematurity: a pilot nested case control study. *Environ. Health* 18(9):62.

Cousins, I.T., Staples, C.A., Klecka, G.M., and Mackay, D. (2002). A multimedia assessment of the environmental fate of bisphenol A. *Hum. Ecol. Risk Assess.* 8:1107–1135.

Danish Ministry of Food (2010) *Danish ban on bisphenol A in materials in contact with food for children aged 0–3.*

Delclos, K.B., Camacho, L., Lewis, S.M., Vanlandingham, M.M., Latendresse, J.R., Olson, G.R., Davis, K.J., Patton, R.E., da Costa, G.G., Woodling, K.A., Bryant, M.S., Chidambaram, M., Trbojevich, R., Juliar, B.E., Felton, R.P., and Thorn, B.T. (2014) Toxicity evaluation of bisphenol A administered by gavage to Sprague Dawley rats from gestation day 6 through postnatal day 90. *Toxicol. Sci.* 139(1):174–197.

Doerge, D.R., Twaddle, N.C., Vanlandingham, M., and Fisher, J.W. (2010a) Pharmacokinetics of bisphenol A in neonatal and adult Sprague-Dawley rats. *Toxicol. Appl. Pharmacol.* 247:158–166.

Doerge, D.R., Twaddle, N.C., Woodling, K.A., and Fisher, J.W. (2010b) Pharmacokinetics of bisphenol A in neonatal and adult rhesus monkeys. *Toxicol. Appl. Pharmacol.* 248:1–11.

Dorsey, D.H. [U.S. Food and Drug Administration (FDA)]. 2012. Letter to S. Janssen and A. Colangelo (Natural Resources Defense Council) [re: Denial of petition to issue regulation prohibiting the use of bisphenol A in human food and food packaging]. FDMS Docket No. FDA-2008-P-0577-0001/CP. 15p.

Eaton, D.L. and Gilbert, D.L. (2008) Principles of toxicology. In: Klaassen,C.D., editor. *Casarett and Doull's Toxicology: The Basic Science of Poisons*, 7th ed., New York: McGraw-Hill, pp. 11–43.

Ema, M., Fujii, S., Furukawa, M., Kiguchi, M., Ikka, T., and Harazono, A. (2001) Rat two-generation reproductive toxicity study of bisphenol A. *Reprod. Toxicol.* 15:505–523.

European Chemicals Agency (ECHA) (2011) *Registered Substances*. Published March 5, 2011: Updated November 26, 2014. Available at http://echa.europa.eu/web/guest/information-on-chemicals/registered-substances (accessed December 19, 2014).

European Commission Joint Research Centre, European Chemicals Bureau (ECB) (2003) *European Union Risk Assessment Report for 4,4′-isopropylidenediphenol (Bisphenol-A) (CAS No. 80-05-7) (EINECS No. 201-245-8) (Final)*, Luxembourg: Office for Official Publications of the European Communities EUR 20843 EN; 3rd Priority List, Volume 37.

European Commission Joint Research Centre, European Chemicals Bureau (ECB) (2008) *Updated European Risk Assessment Report for 4,4′-isopropylidenediphenol (Bisphenol-A) (CAS No. 80-05-7) (EINECS No. 201-245-8) (Final)*, Luxembourg: Office for Official Publications of the European Communities, Available at http://ecb.jrc.it/documents/Existing-Chemicals/RISK_ASSESSMENT/ADDENDUM/bisphenola_add_325.pdf Published 2008 (accessed July 16, 2008).

European Commission, Health & Consumer Protection Directorate-General, Scientific Committee on Toxicity, Ecotoxicity and the Environment (CSTEE) (2002) *CSTEE Comments on: Risk Assessment Report on: Bisphenol A, Human Health (CAS No. 80-05-7) (EINECS No. 201-245-8)*. Available at http://ec.europa.eu/health/ph_risk/committees/sct/documents/out156_en.pdf Published May 22, 2002 (accessed August 22, 2011).

European Commission, Health & Consumer Protection Directorate-General, Scientific Committee on Food (SCF) (2002) *Opinion of the Scientific Committee on Food on Bisphenol A*. Available at http://ec.europa.eu/food/fs/sc/scf/out128_en.pdf p. 22. Published April 17, 2002 (accessed August 22, 2011).

European Food Safety Authority (EFSA) (2006) *Opinion of the Scientific Panel on Food Additives, Flavourings, Processing Aids and Materials in Contact with Food on a request from the Commission related to 2,2-bis(4-hydroxyphenyl)propane (bisphenol A): Question number EFSA-Q-2005-100. EFSA J.* 428, 1–75.

European Food Safety Authority (EFSA) (2008) *Toxicokinetics of Bisphenol A: Scientific Opinion of the Panel on Food Additives, Flavourings, Processing Aids and Materials in Contact with Food (AFC) (Question No EFSA-Q-2008-382). EFSA J.* 759:1–10.

European Food Safety Authority (EFSA) (2010a) Scientific opinion on bisphenol A: evaluation of a study investigating its neuro-developmental toxicity, review of recent scientific literature on its toxicity and advice on the Danish risk assessment of bisphenol A (summary). *EFSA J.* 8:1829.

European Food Safety Authority (EFSA) (2010b) Statistical re-analysis of the Biel maze data of the Stump et al. (2010) study: developmental neurotoxicity study of dietary bisphenol A in Sprague-Dawley rats. *EFSA J.* 8:1836.

Faustman, E.M. and Omenn, G.S. (2008) Risk assessment. In: Klaassen,C.D., editor. *Casarett and Doull's Toxicology: The*

Basic Science of Poisons, 7th ed., New York: McGraw-Hill, pp. 107–128.

Food and Agriculture Organization (FAO) of the United Nations, and World Health Organization (WHO) (2010) *Joint FAO/WHO Expert Meeting to Review Toxicological and Health Aspects of Bisphenol A, Summary Report including Report of Stakeholder Meeting on Bisphenol A; November 1–5, 2010; Ottawa, Canada*. Available at http://www.who.int/foodsafety/chem/chemicals/BPA_Summary2010.pdf.

Food Standards Australia New Zealand (FSANZ) (2014) *Bisphenol A (BPA) and Food Packaging*. Available at http://www.foodstandards.gov.au/consumerinformation/bisphenolabpaandfood4945.cfm Published December 2014. (accessed August 22, 2011).

Goodman, J.E., McConnell, E.E., Sipes, I.G., Witorsch, R.J., Slayton, T.M., Yu, C.J., Lewis, A.S., and Rhomberg, L.R. (2006) An updated weight of the evidence evaluation of reproductive and developmental effects of low doses of bisphenol A. *Crit. Rev. Toxicol.* 36:387–457.

Goodman, J.E., Witorsch, R.J., McConnell, E.E., Sipes, I.G., Slayton, T.M., Yu, C.J., Franz, A.M., and Rhomberg, L.R. (2009) Weight-of-evidence evaluation of reproductive and developmental effects of low doses of bisphenol A. *Crit. Rev. Toxicol.* 39:1–75.

Gray, G.M., Cohen, J.T., Cunha, G., Hughes, C., McConnell, E.E., Rhomberg, L., Sipes, I.G., and Mattison, D. (2004) Weight of the evidence evaluation of low-dose reproductive and developmental effects of bisphenol A. *Hum. Ecol. Risk Assess.* 10:875–921.

Gupta, C. (2000) Reproductive malformation of the male offspring following maternal exposure to estrogenic chemicals. *Proc. Soc. Exp. Biol. Med.* 224:61–68.

Haighton, L.A., Hlywka, J.J., Doull, J., Kroes, R., Lynch, B.S., and Munro, I.C. (2002) An evaluation of the possible carcinogenicity of bisphenol A to humans. *Regul. Toxicol. Pharmacol.* 35:238–254.

Health Canada (2008) *Government of Canada takes action on another chemical of concern: Bisphenol A*. Available by contacting info@hc-sc.gc.ca. Published April 18, 2008 (accessed May 1, 2008)..

Hengstler, J.G., Roth, H., Gebel, T., Kramer, P.J., Lilienblum, W., Schweinfurth, H., Volkel, W., Wollin, K.M., and Gundert-Remy, U. (2011) Critical evaluation of key evidence on the human health hazards of exposure to bisphenol A. *Crit. Rev. Toxicol.* 41:263–291.

Hiroi, H., Tsutsumi, O., Takeuchi, T., Momoeda, M., Ikezuki, Y., Okamura, A., Yokota, H., and Taketani, Y. (2004) Differences in serum bisphenol A concentrations in premenopausal normal women and women with endometrial hyperplasia. *Endocr. J.* 51:595–600.

Ho, S.M., Tang, W.Y., Belmonte de Frausto, J., and Prins, G.S. (2006) Developmental exposure to estradiol and bisphenol A increases susceptibility to prostate carcinogenesis and epigenetically regulates phosphodiesterase type 4 variant 4. *Cancer Res.* 66:5624–5632.

International Food Safety Authorities Network (INFOSAN) (2009) *BISPHENOL A (BPA)—Current State of Knowledge and Future*

Actions by WHO and FAO. Information Note No. 5. World Health Organization (WHO). Available at http://www.who.int/foodsafety/publications/fs_management/No_05_Bisphenol_A_Nov09_en.pdf Published November 27, 2009.

Jenkins, S., Raghuraman, N., Eltoum, I., Carpenter, M., Russo, J., and Lamartiniere, C.A. (2009) Oral exposure to bisphenol A increases dimethylbenzanthracene-induced mammary cancer in rats. *Environ. Health Perspect.* 117(6):910–915

Kubo, K., Arai, O., Omura, M., Watanabe, R., Ogata, R., and Aou, S. (2003) Low dose effects of bisphenol A on sexual differentiation of the brain and behavior in rats. *Neurosci. Res.* 45:345–356.

Kurebayashi, H., Betsui, H., and Ohno, Y. (2003) Disposition of a low dose of 14C-bisphenol A in male rats and its main biliary excretion as BPA glucuronide. *Toxicol. Sci.* 73:17–25.

Lamartiniere, C.A., Jenkins, S., Betancourt, A.M., Wang, J., and Russo, J. (2011) Exposure to the endocrine disruptor bisphenol A alters susceptibility for mammary cancer. *Horm. Mol. Biol. Clin. Investig.* 5(2):45–52.

Lang, I.A., Galloway, T.S., Scarlett, A., Henley, W.E., Depledge, M., Wallace, R.B., and Melzer, D. (2008) Association of urinary bisphenol A concentration with medical disorders and laboratory abnormalities in adults. *JAMA* 300:1303–1310.

Matthews, J.B., Twomey, K., and Zacharewski, T.R. (2001) *In vitro* and *in vivo* interactions of bisphenol A and its metabolite, bisphenol A glucuronide, with estrogen receptors a and b. *Chem. Res. Toxicol.* 14:149–157.

Ministry of the Environment (MOE) (2005) *Japan. MOE's Perspectives on Endocrine Disrupting Effects of Substances. MOE.* Available at http://www.env.go.jp/en/chemi/ed/extend2005_full.pdf Published March 2005.

Miyamoto, K. and Kotake, M. (2006) Estimation of daily bisphenol A intake of Japanese individuals with emphasis on uncertainty and variability. *Environ. Sci.* 13:15–29.

Myers, J.P., vom Saal, F.S., Taylor, J.A., Akingemi, B.T., Arizono, K., Belcher, S., Colborn, T., and Chahoud, I. (2009b) Good laboratory practices: Myers et al. respond (Letter). *Environ. Health Perspect.* 11:A483–A484.

Myers, J.P., Zoeller, R.T., and vom Saal, F.S. (2009a) A clash of old and new scientific concepts in toxicity, with important implications for public health. *Environ. Health Perspect.* 117:1652–1655.

Nagel, S.C., vom Saal, F.S., Thayer, K.A., Dhar, M.G., Boechler, M., and Welshons, W.V. (1997) Relative binding affinity-serum modified access (RBA-SMA) assay predicts the relative *in vivo* bioactivity of the xenoestrogens bisphenol A and octylphenol. *Environ. Health Perspect.* 105:70–76.

Nakanishi, J., Miyamoto, K., and Kawasaki, H. (2007) *Bisphenol A Risk Assessment Document (Summary). National Institute of Advanced Industrial Science and Technology.* Available at http://unit.aist.go.jp/riss/crm/mainmenu/BPA_Summary_English.pdf Published November 2007 (accessed August 22, 2011).

National Toxicology Program (NTP) (2007) *Center for the Evaluation of Risks to Human Reproduction. NTP-CERHR Expert Panel Report on the Reproductive and Developmental Toxicity of Bisphenol A.* Available at http://ntp.niehs.nih.gov/ntp/ohat/bisphenol/BPAFinalEPVF112607.pdf Published November 26, 2007 (accessed August 22, 2011).

National Toxicology Program (NTP) (2008) *Draft NTP Brief on Bisphenol A [CAS NO. 80-05-7].*

National Toxicology Program (NTP) (2001) *National Toxicology Program's Report of the Endocrine Disruptors Low Dose Peer Review.* Available at http://ntp-server.niehs.nih.gov/ntp/htdocs/liason/LowDosePeerFinalRpt.pdf Published August 2001 (accessed August 22, 2011).

National Toxicology Program (NTP) (1982) *NTP Technical Report on the Carcinogenesis Bioassay of Bisphenol A (CAS NO. 80-05-7) in F344 Rats and B6C3F Mice (Feed Study).* Available at http://ntp.niehs.nih.gov/ntp/htdocs/LT_rpts/tr215.pdf Published March 1982 (accessed August 22, 2011).

Palanza, P., Howdeshell, K.L., Parmigiani, S., and vom Saal, F.S. (2002) Exposure to a low dose of bisphenol A during fetal life or in adulthood alters maternal behavior of mice. *Environ. Health Perspect.* 110(Suppl. 3):415–422.

Prins, G.S., Tang, W.Y., Belmonte, J., and Ho, S.M. (2008) Perinatal exposure to oestradiol and bisphenol A alters the prostate epigenome and increases susceptibility to carcinogenesis. *Basic Clin. Pharmacol. Toxicol.* 102:134–138.

Ryan, B.C., Hotchkiss, A.K., Crofton, K.M., and Gray, L.E. Jr. (2010) *In utero* and lactational exposure to bisphenol A, in contrast to ethinyl estradiol, does not alter sexually dimorphic behavior, puberty, fertility, and anatomy of female LE rats. *Toxoicol. Sci.* 114:133–148.

Schantz, S.L. and Widholm, J.J. (2001) Cognitive effects of endocrine-disrupting chemicals in animals. *Environ. Health Perspect.* 109:1197–1206.

Sharpe, R.M. (2010) Is it time to end concerns over the estrogenic effects of bisphenol A? *Toxicol. Sci.* 114:1–4.

Staples, C.A., Dorn, P.B., Klecka, G.M., O'Block, S.T., Branson, D.R., and Harris, R.L. (2000) Bisphenol A concentrations in receiving waters near US manufacturing and processing facilities. *Chemosphere* 40:521–525.

Staples, C.A., Woodburn, K., Caspers, N., Hall, A.T., and Klecka, G.M. (2002) A weight of evidence approach to the aquatic hazard assessment of bisphenol A. *Hum. Ecol. Risk Assess.* 8:1083–1105.

Stump, D.G., Beck, M.J., Radovsky, A., Garman, R.H., Freshwater, L.L., Sheets, L.P., Marty, M.S., Waechter, J.M. Jr., Dimond, S.S., Van Miller, J.P., Shiotsuka, R.N., Beyer, D., Chappelle, A.H., and Hentges, S.G. (2010) Developmental neurotoxicity study of dietary bisphenol A in Sprague-Dawley rats. *Toxicol. Sci.* 115:167–182.

Sugiura-Ogasawara, M., Ozaki, Y., Sonta, S., Makino, T., and Suzumori, K. (2005) Exposure to bisphenol A is associated with recurrent miscarriage. *Hum. Reprod.* 20:2325–2329.

Takeuchi, T. and Tsutsumi, O. (2002) Serum bisphenol A concentrations showed gender differences, possibly linked to androgen levels. *Biochem. Biophys. Res. Commun.* 291:76–78.

Takeuchi, T., Tsutsumi, O., Ikezuki, Y., Takai, Y., and Taketani, Y. (2004) Positive relationship between androgen and the endocrine disruptor, bisphenol A, in normal women and women with ovarian dysfunction. *Endocr. J.* 51:165–169.

Teeguarden, J.G., Calafat, A.M., Ye, X., Doerge, D.R., Churchwell, M.I., Gunawan, R., Graham, M. (2011) Twenty-four hour human urine and serum profiles of bisphenol A during high dietary exposure. *Toxicol. Sci.* 123(1):48–57.

Toyama, Y., Suzuki-Toyota, F., Maekawa, M., Ito, C., and Toshimori, K. (2004) Adverse effects of bisphenol A to spermiogenesis in mice and rats. *Arch. Histol. Cytol.* 67:373–381.

Tyl, R.W., Myers, C.B., Marr, M.C., Sloan, C.S., Castillo, N.P., Veselica, M.M., Seely, J.C., Dimond, S.S., Van Miller, J.P., Shiotsuka, R. S., Stropp, G.D., Waechter, J.M., and Hentges, S.G. (2008) Two-generation reproductive toxicity evaluation of dietary 17ß-estradiol (E2; CAS No. 50-28-2) in CD-1® (Swiss) mice. *Toxicol. Sci.* 102:392–412.

Tyl, R.W., Myers, C.B., Thomas, B.F., Keimowitz, A.R., Brine, D.R., Veselica, M.M., Fail, P.A., Chang, T.Y., Seely, J.C., Joiner, R.L., Butala, J.H., Dimond, S.S., Cagen, S.Z., Shiotsuka, R.N., Stropp, G.D., and Waechter, J.M. (2002) Three-generation reproductive toxicity study of dietary bisphenol A in CD Sprague-Dawley rats. *Toxicol. Sci.* 68:121–146.

Tyl, R.W. (2009) Basic exploratory research versus guideline-compliant studies used for hazard evaluation and risk assessment: bisphenol A as a case study. *Environ. Health Perspect.* 117:1644–1651.

University of Minnesota (2008) *History of Bisphenol A. University of Minnesota.* Available at http://enhs.umn.edu/current/2008 studentwebsites/pubh6101/bpa/history.html Page last updated on December 15, 2008 (accessed August 22, 2011).

U.S. Environmental Protection Agency (EPA) (1988) *IRIS record for bisphenol A (CASRN 80-05-7).* Available at http://www.epa.gov/IRIS/subst/0356.htm (File first on-line September 26, 1988) (accessed August 22, 2011).

U.S. Food and Drug Administration (FDA) (2008) *Draft Assessment of Bisphenol A for Use in Food Contact Applications,* Available at http://www.fda.gov/ohrms/dockets/ac/08/briefing/2008-0038b1_01_02_FDA%20BPA%20Draft%20Assessment.pdf Published August 14, 2008 (accessed August 22, 2011).

U.S. Food and Drug Administration (FDA) (2014) Questions & answers on bisphenol A (BPA) use in food contact applications. Available at http://www.fda.gov/Food/IngredientsPackagingLabeling/FoodAdditivesIngredients/ucm355155.htm Updated November 5, 2014 (accessed December 19, 2014).

Völkel, W., Colnot, T., Csanady, G.A., Filser, J.G., and Dekant, W. (2002) Metabolism and kinetics of bisphenol A in humans at low doses following oral administration. *Chem. Res. Toxicol.* 15:1281–1287.

Völkel, W., Bittner, N., and Dekant, W. (2005) Quantitation of bisphenol A and bisphenol A glucuronide in biological samples by liquid chromatography-tandem mass spectrometry. *Drug Metab. Dispos.* 33:1748–1757.

vom Saal, F.S. and Hughes, C. (2005) An extensive new literature concerning low-dose effects of bisphenol A shows the need for a new risk assessment. *Environ. Health Perspect.* 113:926–933.

vom Saal, F.S., Cooke, P.S., Buchanan, D.L., Palanza, P., Thayer, K.A., Nagel, S.C., Parmigiani, S., and Welshons, W.V. (1998) A physiologically based approach to the study of bisphenol A and other estrogenic chemicals on the size of reproductive organs, daily sperm production, and behavior. *Toxicol. Ind. Health* 14:239–260.

Yang, M., Kim, S.Y., Chang, S.S., Lee, I.S., and Kawamoto, T. (2006) Urinary concentrations of bisphenol A in relation to biomarkers of sensitivity and effect and endocrine-related health effects. *Environ. Mol. Mutagen.* 47:571–578.

79

POLYSTYRENE/STYRENE

Leslie A. Beyer and Julie E. Goodman

Polystyrene (CASRN 9003-53-6) is prepared by the polymerization of styrene under the influence of organic peroxides in solution. Addition of inhibitors such as hydroquinone or butylcatechol prevents the monomer from polymerizing in storage. The properties of the polymer vary depending on the additives involved and the fabrication method. It can be injection molded, blow molded, thermoformed, or extruded. Prepared as a resin in small spheres, polystyrene is a clear, rigid polymer that is relatively stable under most conditions.

Hazards associated with the manufacture of polystyrene are associated with the monomer, styrene, and not the actual polystyrene product. Primarily a packaging material, polystyrene is used mostly as a component of a variety of food containers, such as egg cartons and hot–cold vending cups. It is also used in small appliances, medical devices, and toys.

Exposure to polystyrene dusts may cause mechanical irritation of the eyes and respiratory tract. Degradation products resulting from fire release styrene at lower temperatures and mostly CO and carbon dioxide (CO_2) at higher temperatures, with an unusually dense smoke being produced.

STYRENE

Occupational exposure limits: 100 ppm (OSHA PEL); 50 ppm (NIOSH REL); 20 ppm (ACGIH TLV).

Reference values: 0.2 mg/kg-bw/day (IRIS RfD); 1 mg/m^3 (IRIS RfC).

Hazard statements: H412—harmful to aquatic life with long lasting effects; H226—flammable liquid and vapor; H332—harmful if inhaled; H319—causes serious eye irritation; H335—may cause respiratory irritation; H315—causes skin irritation; H372—causes damage to ear through prolonged or repeated inhalation exposure; H304—may be fatal if swallowed and enters airways.

Precautionary statements: P210—keep away from heat/sparks/open flames/ . . . /hot surfaces . . . no smoking; P233—keep container tightly closed; P240—ground/bond container and receiving equipment; P241—use explosion-proof electrical/ventilating/lighting/ . . . /equipment; P242—use only non-sparking tools; P243—take precautionary measures against static discharge; P260—do not breathe dust/fume/gas/mist/vapors/spray; P261—avoid breathing dust/fume/gas/mist/vapors/spray; P264—wash . . . thoroughly after handling; additional text wash hands and open skin areas thoroughly after handling; P271—use only outdoors or in a well-ventilated area; P280—wear protective gloves/protective clothing/eye protection/face protection; P301 + P310: If swallowed—immediately call a poison center or doctor/physician; P302 + P352: If on skin—wash with plenty of soap and water; P303 + P361 + P353: If on skin (or hair)—remove/take off immediately all contaminated clothing. Rinse skin with water/shower.

P304 + P340: If inhaled—remove victim to fresh air and keep at rest in a position comfortable for breathing; P305 + P351 + P338: If in eyes—rinse cautiously with water for several minutes. Remove contact lenses, if present and easy to do. Continue rinsing; P312—call a poison center or doctor/physician if you feel unwell; P314—get medical advice/attention if you feel unwell; P321—specific treatment (see . . . on this label); P331—do not induce vomiting; P332 + P313: If skin irritation occurs—get medical advice/attention; P337 + P313: If eye irritation persists—get medical advice/attention;

Hamilton & Hardy's Industrial Toxicology, Sixth Edition. Edited by Raymond D. Harbison, Marie M. Bourgeois, and Giffe T. Johnson.
© 2015 John Wiley & Sons, Inc. Published 2015 by John Wiley & Sons, Inc.

P362—take off contaminated clothing and wash before reuse; P370 + P378: In case of fire—use . . . for extinction; additional text In case of fire: Use foam, dry chemical, carbon dioxide for extinction; P403 + P233: Store in a well-ventilated place. Keep container tightly closed; P403 + P235: Store in a well-ventilated place. Keep cool; P405—store locked up; P501—dispose of contents/container to . . . ; additional text dispose of contents/container in accordance with local/regional/national/international regulation.

Risk phrases: R10—flammable; R20—harmful by inhalation; R36/37/38—irritating to eyes, respiratory system, and skin; R48/20—harmful: danger of serious damage to health by prolonged exposure through inhalation.

Safety phrases: S23—do not breathe gas/fumes/vapor/spray (appropriate wording to be specified by the manufacturer) vapor; S62—if swallowed, do not induce vomiting: Seek medical advice immediately and show this container or label.

BACKGROUND AND USES

Styrene (monomer CASRN 100-42-5) is a viscous, highly flammable liquid that evaporates easily and polymerizes readily to polystyrene unless a stabilizer is added. Large amounts are produced in the United States with an estimated 11.4 billion pounds produced in 2006 (NTP, 2008). Estimated production capacity was 13.7 billion pounds in 2006 (NTP, 2008) and over 12 billion pounds in 2008 (ATSDR, 2010). Styrene is used in multiple industries, especially in reinforced plastics (e.g., fiberglass boats), and is widely used to make plastics and rubber, packaging materials (e.g., packing "peanuts"), insulation for buildings, plastic pipes, food containers (e.g., takeout containers), and carpet backing) (ATSDR, 2010). The industry directly employs 128,000 workers (SIRC, 2012).

EXPOSURE

People are exposed to small amounts of styrene that occur naturally in a variety of foods, such as fruits, vegetables, nuts, and meats; styrene can also migrate into food from packaging (ATSDR, 2010). Small amounts of styrene are naturally present in legumes, beef, clams, eggs, nectarines, and spices (ATSDR, 2010). Smoking and occupational exposures are the largest sources of exposure in the U.S. The highest potential exposure occurs in the reinforced plastics industry, where workers may inhale styrene and also have dermal exposure to liquid styrene or resins (ATSDR, 2010). In occupational settings, inhalation exposure is the largest source of exposure. Studies have shown dermal exposure

to be negligible from the point of view of absorption and subsequent risk of systemic effects, although dermal contact with styrene can cause irritation (ACGIH, 2001).

Air concentrations are highest in the reinforced plastic, styrene–butadiene, and styrene monomer and polymer industries, in which workers are potentially exposed to higher concentrations of styrene than the general public, who are exposed to styrene levels outdoor and indoor (including levels in most other occupational settings) that are generally below 1 part per billion (ppb) (0.001 part per million (ppm). Air levels in the reinforced plastics industry have been generally lower than 100 ppm since the 1980s, although much higher levels have frequently been measured, while levels in the styrene–butadiene industry and the styrene monomer and polymer industries have rarely been reported to exceed 20 ppm (NTP, 2008). Air concentrations of styrene in the workplace have decreased over time, but in some workplaces younger workers with little seniority are typically exposed to higher concentrations of styrene and styrene oxide than their coworkers (Serdar et al., 2006).

HEALTH EFFECTS

Noncancer

In humans, the nervous system is the most sensitive target following chronic inhalation exposure and likely the most sensitive target following shorter-term exposure (ATSDR, 2010). Styrene is a neurotoxin with acute toxicity considered to be low or moderate (NTP, 2008). The primary effects of acute exposure in both animals and humans include irritation of the skin, eyes, and respiratory tract, and central nervous system (CNS) effects. Common signs of styrene intoxication include drowsiness, listlessness, muscular weakness, and unsteadiness. A no observed adverse effect level (NOAEL) of 49 ppm has been identified by the Agency for Toxic Substances and Disease Registry (ATSDR) for acute effects based on Ska et al. (2003), who identified a NOAEL of 49 ppm in people based on a lack of significant styrene-related alterations in color discrimination, olfactory threshold, and neurobehavioral tests (ATSDR, 2010). Effects on color vision, hearing threshold, reaction time, and postural stability have also been reported in workers following long-term occupational exposure to styrene at concentrations ranging from about 20 to 30 ppm (NTP, 2008). For chronic inhalation exposure, ATSDR identified a lowest-observed adverse effect level (LOAEL) of 20 ppm based on the Benignus et al. (2005) meta-analysis of color discrimination impairment in styrene workers; the alterations were reversible and the workers were not aware of the changes (ATSDR, 2010). A multicenter study of 1620 styrene workers in the United States found that occupational exposure to styrene (at 3.5–22 ppm styrene and above) is a risk factor for hearing loss and that exposure to noise significantly increases the effect (Morata et al., 2011). A comprehensive evaluation by the National Toxicology Program (NTP) concluded that the total scientific evidence was sufficient to conclude that styrene exposures to workers (and the general population) in the U.S. were unlikely to "adversely affect human development or reproduction" (NTP, 2006).

Cancer

Excess tumors at one site—the lung—have been found following both inhalation and oral exposure in some strains of mice (NTP, 2008). In rats, however, no excess lung tumors have been observed nor has increased cancer incidence of any type has been found consistently. None of the gavage or intra-peritoneal or subcutaneous injection studies in rats reported an increased incidence in any tumor type (NTP, 2008) and inhalation studies have shown no consistent increases. The differences in effects in mice and rats are thought to be due to differences in metabolism and mode of action (see below).

Numerous epidemiology studies have been conducted in the reinforced plastics industry, manufacture of styrene monomer and polystyrene, and production of styrene–buta-diene rubber. The reinforced plastics workers have generally been exposed to higher concentrations of styrene with less potential for confounding exposures to other chemicals in the workplace; therefore, these studies are the most informative (Cohen et al., 2002; Boffetta et al., 2009). Although some epidemiology studies have reported an excess of certain lymphatic and hematopoietic cancers, the results cannot be interpreted as supporting a causal association because of the small number of observed and expected deaths from these cancers, the potential for confounding by other occupational exposures, the lack of consistent dose–response patterns, the fact that excess risks were not found in all cohorts, and the fact that the types of lymphohematopoietic cancer observed in excess varied across different cohort studies (NTP, 2011; ACGIH, 2001; Cohen et al., 2002; ATSDR, 2010).

Studies in groups with the highest exposure (e.g., workers in the reinforced plastics industry as studied by Wong et al., 1994; Kogevinas et al., 1994; Ruder et al., 2004) have not found a clear excess of hematopoietic cancers (Cohen et al., 2002; Rhomberg et al., 2013; Delzell et al., 2006). Collins et al. (2013) updated a series of epidemiology studies in which the original study examined 15,908 workers exposed between 1948 and 1977 in 30 U.S. reinforced plastic facilities in 16 states. No increased mortality was observed for all lymphatic and hematopoietic cancer combined, non-Hodg-kin lymphoma, or leukemia; and there was no trend with either cumulative exposure to styrene or number of peaks (Collins et al., 2013). However, the limited size of the populations having both high exposure and a long period of time since first exposure limits the available data (Cohen et al., 2002; NTP, 2011).

A series of epidemiology studies on North American synthetic rubber industry workers evaluated mortality from 1944 to 1998 among 17,924 men. Although the studies found an association between working in the industry and leukemia, after controlling for the effects of butadiene, no consistent exposure–response relationship was found (Delzell et al., 2006). Several other types of cancer have been associated with styrene exposure, but the overall pattern of the results demonstrates no consistent hazard and does not suggest a causal association between styrene and any form of cancer in humans (Cohen et al., 2002; Boffetta et al., 2009; Rhomberg et al., 2013; Delzell et al., 2006; ATSDR, 2010).

The differences in effects in mice, rats, and humans are thought to be due to differences in metabolism and mode of action (see below).

Absorption, Distribution, Metabolism, and Excretion

The most important route of exposure for people in occupational settings is by inhalation, which results in rapid absorption and distribution of approximately 60–70% of inhaled styrene (NTP, 2008). Styrene can also be absorbed through ingestion or skin contact.

Inhaled styrene is initially metabolized in the lung by the Clara cells, and subsequently in the liver. The initial step in styrene metabolism is catalyzed by P450 cytochromes, predominantly CYP2E1 and CYP2F2, but a number of other CYPs are involved; the degree to which they are involved and whether they predominate in the lung or liver varies between mice, rats, and humans. CYP2F2 is the primary enzyme in the mouse lung and is thought to be largely responsible for the lung cancer observed in mice but not rats or humans (NTP, 2008; Cohen et al., 2002; ATSDR, 2010). A recent study using CYP2F2 knockout mice found that styrene toxicity in the mouse lung required CYP2F2 metabolism (Cruzan et al., 2011).

Metabolic activation of styrene results in the formation of styrene-7,8-oxide, a mutagen *in vitro*, which is detoxified by glutathione conjugation or conversion to styrene glycol (NTP). Concern about the potential carcinogenicity of styrene stems largely from the ability of styrene oxide to bind covalently to DNA and to its activity in a variety of genotoxicity test systems (Henderson and Speit, 2005; Rueff et al., 2009).

Almost all absorbed styrene is excreted as urinary metabolites, primarily mandelic acid, and phenylglyoxylic acid (NTP, 2008). Generally accepted biomarkers of exposure to styrene are mandelic acid and pheylglyoxylic acid in urine and styrene in blood (IARC, 2002; ACGIH, 2003). The Biological Exposure Indices (BEIs) recommended by the American Conference of Governmental Industrial Hygienists (ACGIH) are mandelic acid plus phenylglyoxylic acid in urine (sampled at the end of the shift) of 400 mg/g creatinine and styrene in venous blood (sampled at the end of the shift) of 0.2 mg/l (ACGIH, 2012a). Up to about 150 ppm in air, extrapolation between the degree of exposure and biological levels of the metabolites is linear (ACGIH, 2003). The production of metabolites is suppressed by the presence of ethanol and is also affected by exposure to other solvents, and work load (Truchon et al., 2009; ACGIH, 2003). (For sampling and analytical methods and more information, see ACGIH, 2003). Styrene-7,8-oxide–DNA adducts and styrene-7,8-oxide–hemoglobin adducts are also generally accepted biological indices of exposure to styrene (NTP, 2008).

MODE OF ACTION, GENOTOXICITY

Styrene itself is not DNA reactive, but styrene-7,8-oxide binds covalently to macromolecules, including proteins and nucleic acids, forming stable N2 and O6 adducts of guanine in cultured mammalian cells. These adducts have also been measured in the lymphocytes of workers exposed to styrene. Other studies have shown an association between occupational exposure to styrene and DNA stand breaks, but such breaks are quickly repaired (Bastlova et al., 1995, as cited in Cohen et al., 2002), so their significance is unclear. In some animal studies (but not others), styrene has been associated with an increase in the frequency of aberrations, sister chromatid exchanges, and micronucleated cells (Cohen et al., 2002; Henderson and Speit, 2005). Human studies have shown conflicting associations between styrene exposure and the frequency of chromosomal abnormalities; there is less evidence for an association between styrene exposure and the frequency of sister chromatid exchanges; and no compelling evidence for micronuclei formation in human studies (Cohen et al., 2002; Henderson and Speit, 2005). Although some studies have shown DNA adducts and strand breaks in styrene-exposed workers, these types of damage indicate exposure, but may not result in heritable changes (Henderson and Speit, 2005). Genetic polymorphisms in the population may affect both metabolism and DNA repair (Henderson and Speit, 2005; Rueff et al., 2009).

CANCER CLASSIFICATION

Agencies have characterized/classified styrene's carcinogenicity as follows: ATSDR—may be a weak human carcinogen, (ATSDR, 2010); ACGIH—A4, not classifiable as a human carcinogen (ACGIH, 2001); IARC—possibly carcinogenic in humans (Group 2B) based on limited evidence in animals and humans (IARC, 2002); and NTP—reasonably anticipated to be a human carcinogen (NTP, 2011).

OCCUPATIONAL STANDARDS

The current Occupational Safety and Health Administration (OSHA) permissible exposure limit (PEL) is 100 ppm as a

time-weighted average (TWA) for an 8-h work day (US Office of the Federal Register, 2014, although in 1996 industries using styrene adopted a voluntary compliance program to reduce exposure to 50 ppm, the NIOSH-recommended exposure limit (NIOSH, 1983 and OSHA, 1996, both as cited in Ruder et al. 2004). ACGIH threshold limit value (TLV) is 20 ppm (ACGIH, 2012b). NIOSH (2011) and ACGIH (2012b) have recommended short-term (15-min TWA) exposure limits of 100 and 40 ppm, respectively. OSHA has a ceiling limit of 200 ppm that can be exceeded but only to an acceptable maximum peak concentration of 600 ppm for a maximum of 5 min in any 3 h (US Office of the Federal Register, 2014; OSHA 29 CFR Part 1910.1000). NIOSH has designated 700 ppm as the concentration that is Immediately Dangerous to Life or Health (IDLH) (NIOSH, 2011).

Standards are based on protection from neurological effects, including decrements in color vision and hearing.

REFERENCES

Agency for Toxic Substances and Disease Registry (ATSDR) (2010) *Toxicological Profile for Styrene*, Atlanta, GA: Agency for Toxic Substances and Disease Registry. Available at http://www.atsdr.cdc.gov/ToxProfiles/tp53.pdf (accessed March 30, 2011).

American Conference of Governmental Industrial Hygienists (ACGIH) (2001) Documentation for styrene, monomer. In: *Documentation of the Threshold Limit Values and Biological Exposure Indices*, 7th ed., Cincinnati, OH: American Conference of Governmental Industrial Hygienists (ACGIH), ACGIH Publication No. 0112.

American Conference of Governmental Industrial Hygienists (ACGIH) (2003) Documentation of the biological exposure indices for styrene. In: *Documentation of the Threshold Limit Values and Biological Exposure Indices*, 7th ed., Cincinnati, OH: American Conference of Governmental Industrial Hygienists (ACGIH), ACGIH Publication No. 0112.

American Conference of Governmental Industrial Hygienists (ACGIH) (2012a) *2012 TLVs and BEIs: Threshold Limit Values for Chemical Substances and Physical Agents and Biological Exposure Indices. American Conference of Governmental Industrial Hygienists (ACGIH)*, Cincinnati, OH: ACGIH, Publication No. 0112.

American Conference of Governmental Industrial Hygienists (ACGIH) (2012b) *2012 Guide to Occupational Exposure Values. American Conference of Governmental Industrial Hygienists (ACGIH)*, Cincinnati, OH: ACGIH, Publication No. 0390.

Bastlova, T., Vodicka, P., Peterkova, K., Hemminki, K., and Lambert, B. (1995) *Styrene oxide-induced HPRT mutations, DNA adducts and DNA strand breaks in cultured human lymphocytes*. Carcinogenesis. 16:2357–2362.

Benignus et al., (2005) *Toxicological Profile for Styrene*, Atlanta, GA: Agency for Toxic Substances and Disease Registry. Available at http://www.atsdr.cdc.gov/ToxProfiles/tp53.pdf (accessed March 30, 2011).

Boffetta, P., Adami, H.O., Cole, P., Trichopoulos, D., and Mandel, J.S. 2009. Epidemiologic studies of styrene and cancer: a review of the literature. *J. Occup. Environ. Med.* 51(11):1275–1287.

Cohen, J.T., Carlson, G., Charnley, G., Coggon, D., Delzell, E., Graham, J.D., Greim, H., Krewski, D., Medinsky, M., Monson, R., Paustenbach, D., Petersen, B., Rappaport, S., Rhomberg, L., Ryan, P.B., and Thompson, K. (2002) A comprehensive evaluation of the potential health risks associated with occupational and environmental exposure to styrene. *J. Toxicol. Environ. Health B Crit. Rev.* 5(1–2):1–263.

Collins, J.J., Bodner, K.M., and Bus, J.S. (2013) Cancer mortality of workers exposed to styrene in the U.S. reinforced plastics and composite industry. *Epidemiology* 24(2):195–203.

Cruzan, G., Bus, J., Ding, X., Hotchkiss, J., Harkema, J., and Gingell, R. (2011) *No lung toxicity from styrene in CYP2F2 knockout mice.* In: Society of Toxicology 50th Annual Meeting, Washington, DC.

Delzell, E., Sathiakumar, N., Graff, J., Macaluso, M., Maldonado, G., and Matthews, R. (2006) *An Updated Study of Mortality Among North American Synthetic Rubber Industry Workers,* HEI Research Report 132, Boston, MA: Health Effects Institute.

Henderson, L.M. and Speit, G. (2005) Review of the genotoxicity of styrene in humans. *Mutat. Res.* 589(3):158–191.

International Agency for Research on Cancer (IARC) (2002) *IARC Monographs on the Evaluation of Carcinogenic Risks to Humans. Volume 82: Some Traditional Herbal Medicines, Some Mycotoxins, Naphthalene and Styrene: Section on Styrene.* World Health Organization (WHO), pp. 437–550.

Kogevinas, M., Ferro, G., Andersen, A., Bellander, T., Biocca, M., Coggon, D., Gennaro, V., Hutchings, S., Kolstad, H., Lundberg, I., Lynge, E., Partanen, T., and Saracci, R. (1994) Cancer mortality in a historical cohort study of workers exposed to styrene. *Scand. J. Work Environ. Health* 20(4):251–261.

Morata, T.C., Sliwinska-Kowalska, M., Johnson, A.C., Starck, J., Pawlas, K., Zamyslowska-Szmytke, E., Nylen, P., Toppila, E., Krieg, E., Pawlas, N., and Prasher, D. (2011) A multicenter study on the audiometric findings of styrene-exposed workers. *Int. J. Audiol.* 50(10):652–660.

National Institute for Occupational Safety and Health (NIOSH) (1983) *Criteria for a Recommended Standard for Occupational Exposure to Styrene.* Cincinnati, OH: NIOSH.

National Institute for Occupational Safety and Health (NIOSH) (2011) *NIOSH Pocket Guide to Chemical Hazards: Styrene,* Atlanta, GA: Centers for Disease Control and Prevention. Available at http://www.cdc.gov/niosh/npg/npgd0571.html (accessed April 27, 2012).

National Toxicology Program (NTP) (2006) *NTP-CERHR Monograph on the Potential Human Reproductive and Developmental Effects of Styrene.* Center for the Evaluation of Risks to Human Production, NIH Publication No. 06-4475. Available at http://cerhr.niehs.nih.gov/evals/styrene/StyreneMono-www-s.pdf (accessed March 30, 2011).

National Toxicology Program (NTP) (2008) *Final Report on Carcinogens Background Document for Styrene*, Research Triangle Park, NC: U.S. Department of Health and Human Services, Public Health Services, National Toxicology Program.

National Toxicology Program (NTP) (2011) *Report on Carcinogens*, 12th ed., Research Triangle Park, NC: U.S. Department of Health and Human Services, Public Health Services, National Toxicology Program. Available at http://ntp.niehs.nih.gov/go/roc12 (accessed June 29, 2011).

Occupational Safety and Health Administration (OSHA) (1996) *OSHA Announces that Styrene Industry Has Adopted Voluntary Compliance Program to Improve Worker Protection.* Washington, DC: US Department of Labor, News release 96–77.

Rhomberg, L.R., Goodman, J.E., and Prueitt, R.L. (2013) The weight of evidence does not support the listing of styrene as "reasonably anticipated to be a human carcinogen" in NTP's twelfth report on carcinogens. *Hum. Ecol. Risk Assess.* 19(1):4–27.

Ruder, A.M., Ward, E.M., Dong, M., Okun, A.H., and Davis-King, K. (2004) Mortality patterns among workers exposed to styrene in the reinforced plastic boatbuilding industry: an update. *Am. J. Ind. Med.* 45(2):165–176.

Rueff, J., Teixeira, J.P., Santos, L.S., and Gaspar, J.F. (2009) Genetic effects and biotoxicity monitoring of occupational styrene exposure. *Clin. Chim. Acta* 399(1–2):8–23.

Serdar, B., Tornero-Velez, R., Echeverria, D., Nylander-French, L.A., Kupper, L.L., and Rappaport, S.M. (2006) Predictors of occupational exposure to styrene and styrene-7,8-oxide in the reinforced plastics industry. *Occup. Environ. Med.* 63(10):707–712.

Ska et al., (2003) *Toxicological Profile for Styrene*, Atlanta, GA: Agency for Toxic Substances and Disease Registry. Available at http://www.atsdr.cdc.gov/ToxProfiles/tp53.pdf (accessed March 30, 2011).

Styrene Information and Research Center (SIRC) (2012) *The Styrene Industry at a Glance,* Washington, DC: SIRC.

Truchon, G., Brochu, M., and Tardif, R. (2009) Effect of physical exertion on the biological monitoring of exposure to various solvents following exposure by inhalation in human volunteers: III. Styrene. *J. Occup. Environ. Hyg.* 6(8):460–467.

US Office of the Federal Register (2014) *Code of Federal Regulations: Labor (Title 29, Part 1910.1000),* Washington, DC: National Archives and Records Administration, Office of the Federal Register, U.S. GPO, Superintendent of Documents, July 1.

Wong, O., Trent, L.S., and Whorton, M.D. (1994) An updated cohort mortality study of workers exposed to styrene in the reinforced plastics and composites industry. *Occup. Environ. Med.* 51(6):386–396.

80

PHTHALATES

Robert W. Benson

Target organ(s): Developing male reproductive tract

Occupational exposure limits: None

Reference values: There are no occupational exposure limits for this chemical; SML of 30, 0.3, and 1.5 mg/kg food for BBP, DBP, and DEHP, respectively

BACKGROUND AND USES

The phthalate esters that are the subject of this chapter are derivatives of phthalic acid where the acid groups are in the ortho position. The two alcohol groups can be the same or different and straight or branched chain. Phthalates are manufactured by reacting phthalic anhydride with alcohols.

Phthalate esters are used to impart flexibility to plastic products. Many consumer products and food packing products contain phthalate esters. The major uses of selected phthalate esters are summarized in Table 80.1. The phthalate ester is not covalently bound in the products in which they are used and can leach into the surrounding environment.

PHYSICAL AND CHEMICAL PROPERTIES

Selected physical and chemical properties are summarized in Table 80.2.

MAMMALIAN TOXICOLOGY AND MECHANISM OF ACTION

Extensive data have been published since 2000 showing that phthalate esters with side chains of three to ten carbon units are developmental toxicants for the male reproductive tract (Foster, 2006). Johnson et al. (2012) provide a recent review. For those phthalate esters that cause malformations of the male reproductive tract, the phthalate monoester metabolite is believed to be the toxic chemical. The phthalate diester is rapidly converted to the phthalate monoester in rats and in humans. This metabolism occurs in the digestive tract prior to the absorption and after absorption by nonspecific esterases.

Matsumoto et al. (2008) summarized the studies available in human populations. A number of studies show an association between the effects (sperm and semen parameters in adults; anogenital distance in newborn males; and serum hormone levels in newborn infants) and environmental exposure to phthalates. After a thorough review of these studies, Matsumoto et al. (2008) concluded: *"some of the findings in human populations are consistent with animal data suggesting that PAEs (phthalic acid esters) and their metabolites produce toxic effects in the reproductive system. However, it is not yet possible to conclude whether phthalate exposure is harmful for human populations."*

Based on the extensive data available from laboratory animals, there is concern that an environmental exposure to phthalate esters could cause adverse effects on the developing human male reproductive system. Additional concern is raised by the demonstration that phthalate monoesters, known to cause developmental toxicity in rats, are detected in a large percentage of human amniotic fluid samples (Silva et al., 2004). The exposure window of concern for humans is the late first trimester of pregnancy, gestation weeks 8–14 (Welsh et al., 2008).

Benson (2009) conducted a literature review for the phthalate esters that have the potential to be developmental toxicants based on their structure–activity relationships. The

Hamilton & Hardy's Industrial Toxicology, Sixth Edition. Edited by Raymond D. Harbison, Marie M. Bourgeois, and Giffe T. Johnson.
© 2015 John Wiley & Sons, Inc. Published 2015 by John Wiley & Sons, Inc.

TABLE 80.1 Phthalate Diester, Its Corresponding Monoester Metabolite, and Uses

Phthalate Diester	Monoester Metabolite	Major Uses
Dibutyl phthalate (DBP) CAS # 84-74-2	Monobutyl phthalate (MBP)	Plastic products containing nitrocellulose, polyvinyl acetate, or polyvinyl chloride; pharmaceutical coatings; lubricant for aerosol valves; antifoaming agent; skin emollient; nail polishes; fingernail elongators; and hair spray
Diisobutyl phthalate (DiBP) CAS # 84-69-5	Monoisobutyl phthalate (MiBP)	Same as dibutyl phthalate but less commonly used
Butylbenzyl phthalate (BBP) CAS # 85-68-7	Monobutyl phthalate (MBP) and monobenzyl phthalate (MBzP)	Polyvinyl chloride products, including vinyl tile, food conveyer belts, carpet tile, artificial leather, tarps, automotive trim, weather strippers, traffic cones, and vinyl gloves
Diethylhexyl phthalate (DEHP) CAS # 117-81-7	Monoethylhexyl phthalate (MEHP)	Polyvinyl chloride products, including building products, car products, clothing, food packaging, children's products, and in medical devices (storage containers, bags, and flexible tubing)
Dipentyl phthalate (DPP) CAS # 131-18-0	Monopentyl phthalate (MPP)	Not found
Diisononyl phthalate (DiNP) CAS # 68515-48-0 and 28553-12-0	Monoisononyl phthalate (MiNP)	Flexible polyvinyl chloride products, including children's toys, flooring, gloves, food packaging material, drinking straws, and garden hoses

focus was on the phthalate esters with a side chain containing four to ten carbons in the ortho position. The shorter chain phthalate esters (dimethyl phthalate and diethyl phthalate) were not included because these chemicals are not developmental toxicants in laboratory animal studies. Relevant toxicological data following an *in utero* exposure were located for dibutyl phthalate (DBP), diisobutyl phthalate (DiBP), butylbenzyl phthalate (BBP), diethylhexyl phthalate (DEHP), dipentyl phthalate (DPP), and diisononyl phthalate (DiNP). Benson (2009) focused on the most sensitive effect in the most sensitive life stage, the reproductive tract of the developing male fetus, to develop the reference dose (RfD) for each of the phthalate esters.

The development of the male reproductive tract is dependent on the presence of testosterone and the androgen receptor (Sharpe, 2008; Sharpe and Skakkebaek, 2008). Any chemical that reduces the concentration of the androgen receptor–testosterone complex during the critical developmental window has the potential of causing irreversible malformations of the male reproductive tract. The mechanism of action in laboratory rats for each of the phthalate esters that are the subject of this chapter is the decrease in the concentration of fetal testosterone in Leydig cells during the critical developmental window in rats, gestation days 15.5–19.5 (Welsh et al., 2008). The decrease in testosterone concentration is triggered by a decrease in gene expression for the proteins involved in the rate limiting steps of testosterone synthesis. These steps include transport of cholesterol into the mitochondria facilitated by steroidogenic acute regulatory protein (STAR) and the conversion of cholesterol to pregnenolone by the enzyme CYP11A1 (also known as P450-SCC, side-chain cleavage enzyme).

There is experimental evidence that the decrease in fetal testosterone caused by cumulative exposure to these six

TABLE 80.2 Selected Physical and Chemical Properties

Name	Dibutyl Phthalate	Diisobutyl Phthalate	Dipentyl Phthalate	Butylbenzyl Phthalate	Diethylhexyl Phthalate	Diisononyl Phthalate
CAS #	84-74-2	84-69-5	131-18-0	85-68-7	117-81-7	28553-12-0
Molecular weight	278.34	278.344	306.397	312.360	390.56	418.61
Molecular formula	$C_{16}H_{22}O_4$	$C_{16}H_{22}O_4$	$C_{18}H_{26}O_4$	$C_{19}H_{20}O_4$	$C_{24}H_{38}O_4$	$C_{26}H_{42}O_4$
Melting point	$-35\,^{\circ}C$	$-64\,^{\circ}C$	$<-55\,^{\circ}C$	$-35\,^{\circ}C$	$-55\,^{\circ}C$	$-48\,^{\circ}C$
Boiling point	$340\,^{\circ}C$	$296.5\,^{\circ}C$	$342\,^{\circ}C$	$370\,^{\circ}C$	$384\,^{\circ}C$	$252\,^{\circ}C$
Density	1.0459 (20 °C)	1.0490 (15 °C)	1.022 (20 °C)	1.113 (25 °C)	0.986 (20 °C)	0.972 (20 °C)
Water solubility	11.2 mg/l (20 °C)	6.2 mg/l (24 °C)	0.8 mg/l (25 °C)	2.69 mg/l (25 °C)	0.27 mg/l (25 °C)	0.2 mg/l (20 °C)
Log K_{ow}	4.9	4.11	5.62	4.73	7.60	9.37
Vapor pressure	2.01×10^{-5} mmHg (25 °C)	4.76×10^{-5} mmHg (25 °C)	Not available	8.25×10^{-6} mmHg (25 °C)	1.42×10^{-7} mmHg (25 °C)	5.4×10^{-7} mmHg (25 °C)

Source: HSDB.

phthalate esters in rats follows a dose additional model (Howdeshell et al., 2008). Therefore, it is appropriate to conduct a cumulative risk assessment for simultaneous exposure to these phthalate esters using a dose addition model. This chapter provides examples of a cumulative risk assessment using exposure information from a U.S. and a German population.

SOURCES OF EXPOSURE

Occupational

Occupational exposures can occur during manufacture of the individual chemicals and their incorporation into plastic products during their manufacture. There are no quantitative data on exposures in occupational settings.

Environmental

Because the phthalate esters are used in such a wide variety of consumer products and are not covalently bound in the products in which they are used, human exposure to the phthalate esters is widespread. Some common examples are listed below.

- Inhalation of emissions from manufacturing facilities
- Dermal uptake from use in vinyl gloves and cosmetics
- Ingestion of dust from emissions from floor tile, carpet tile, and products used in automotive interiors
- Ingestion from plastic coatings on pharmaceuticals
- Ingestion from uses in children's chew toys
- Ingestion from uses in food packaging
- Intravenous exposure from uses in medical storage bags and tubing

Phthalate esters have no direct use in food. Atmospheric deposition on agricultural products from nearby manufacturing facilities is possible. However, the occurrence of a phthalate ester in food is more likely to be the result of contamination by transfer of the phthalate ester from materials in contact with the food during processing, handling, or transportation. Examples of materials that can contain a phthalate ester capable of transfer to the food include plastic bottles and containers, flexible plastic tubing, food conveyer belts, and various food packaging materials. Migration is more extensive to fatty foods because the octanol–water partition coefficient ($\log K_{ow}$) exceeds 1. It is believed that ingestion from food is the primary route of exposure for the general population. Quantitative data on total exposure to the U.S. population are available from the National Health and Nutritional Examination Survey (NHANES) reports (CDC, n.d.).

INDUSTRIAL HYGIENE

There are no occupational exposure limits for the phthalate esters.

The United States and the European Union have limited the quantity of BBP, DBP, and DEHP in children's toys that can be taken into the mouth to 0.1% by mass of the product. The European Union has also restricted the use of BBP, DBP, and DEHP to food contact surfaces only for nonfatty foods up to 0.1, 0.05, and 0.1%, respectively, in the final product. These restrictions also established specific migration limits (SML) of 30, 0.3, and 1.5 mg/kg food for BBP, DBP, and DEHP, respectively. There are no comparable restrictions in the United States on the use of BBP, DBP, and DEHP in products with contact to food.

MEDICAL MANAGEMENT

There are no generally recognized medical management strategies for the malformations of the male reproductive tract caused by an *in utero* exposure to the phthalate esters. The preventive strategy is to limit the exposure in women of childbearing age to the phthalate esters showing developmental toxicity, especially during the critical window of exposure, gestation weeks 8–14 (Welsh et al., 2008).

RISK ASSESSMENTS

The standard U.S. Environmental Protection Agency (EPA) approach was used to derive the RfD for the selected phthalate esters. The uncertainty factors included a factor of 10 for intraspecies variability, a factor of 10 for interspecies variability, and, if needed, a factor for extrapolation from a lowest observed adverse effect level (LOAEL) to a no observed adverse effect level (NOAEL).

For these phthalate esters the most vulnerable target for adverse health effects is the reproductive tract of the developing male fetus. Therefore, the RfD is based on exposure–response data from studies following an *in utero* exposure. This RfD will protect humans from other adverse effects that may occur at a higher exposure during neonatal or adult life.

Benson (2009) derived RfDs for DBP, DiBP, BBP, DEHP, DPP, and DiNP using the NOAEL or LOAEL values for each phthalate ester from the published literature. If appropriate, a benchmark dose analysis was conducted using the EPA software. These values are summarized in Table 80.3.

These RfDs are based on the assumption that the adverse health effect observed at the lowest exposure in laboratory rats (decrease in testosterone concentration in fetal Leydig cells) is relevant to humans. Some recent toxicological studies provide fairly convincing evidence that the decrease in fetal testosterone concentration is a species-specific response. Gaido et al. (2007)

TABLE 80.3   Reference Dose for Selected Phthalate Esters

Chemical	Critical Effect and Reference	POD/UF	RfD
DBP	Decreased fetal testosterone NOAEL = 30 mg/kg-day LOAEL = 50 mg/kg-day (Lehmann et al., 2004)	NOAEL/100	0.3 mg/kg-day or 0.00108 mmol/kg-day
DiBP	Decreased fetal testosterone production $BMDL_{1SD}$ = 80 mg/kg-day LOAEL = 300 mg/kg-day (Howdeshell et al., 2008)	$BMDL_{1SD}$/100	0.8 mg/kg-day or 0.00288 mmol/kg-day
BBP	Decreased fetal testosterone production $BMDL_{1SD}$ = 102 mg/kg-day LOAEL = 300 mg/kg-day (Howdeshell et al., 2008)	$BMDL_{1SD}$/100	1 mg/kg-day or 0.00327 mmol/kg-day
DEHP	Small or absent male reproductive organs $BMDL_{10}$ = 27 mg/kg-day LOAEL = 14–23 mg/kg-day (NTP-CERHR, 2006)	$BMDL_{10}$/100	0.3 mg/kg-day or 0.000692 mmol/kg-day
DPP	Decreased fetal testosterone production $BMDL_{1SD}$ = 17 mg/kg-day LOAEL = 100 mg/kg-day (Howdeshell et al., 2008)	$BMDL_{1SD}$/100	0.2 mg/kg-day or 0.000548 mmol/kg-day
DiNP	Decreased fetal testosterone LOAEL = 750 mg/kg-day (lowest dose tested) (Borch et al., 2004)	LOAEL/1000	0.8 mg/kg-day or 0.00179 mmol/kg-day

POD is the point of departure and UF is the total uncertainty factor.

$BMDL_{1SD}$ = lower 95% confidence limit of the exposure necessary to give a 1 standard deviation decrease in testosterone versus control.

$BMDL_{10}$ = lower 95% confidence limit of the exposure necessary to give a 10% increase in adverse effect versus control.

The RfD in mg/kg-day is rounded to one significant digit.

RfD (mmol/kg-day) = RfD (mg/kg-day)/molecular weight (mg/mmol).

first demonstrated that adverse developmental effects occurred in mice in the absence of a decrease in testosterone concentration and an absence in a decrease in gene expression for the proteins involved in the rate limiting steps in cholesterol transport into the mitochondria and formation of pregnenolone. A 250 or 500 mg DBP/kg-day exposure to mice from gestational day 16 to 18 significantly increased seminiferous cord diameter, the number of multinucleated gonocytes per cord, and the number of nuclei per multinucleated gonocytes. An exposure of 250 mg/kg-day was the lowest exposure tested in this study.

Two research groups have further investigated the possibility of a species-dependent response using xenografts of testicular tissue from different species implanted into rodent hosts (Mitchell et al., 2012; Heger et al., 2012). Mitchell et al. (2012) compared the effect in human fetal tissue versus rat fetal tissue xenografts in host mice. Human fetal tissue xenografts in mice showed no significant decrease in testosterone concentration following an exposure to DBP or monobutyl phthalate. In contrast rat fetal tissue xenografts showed decreases in testosterone concentration and decreased expression of rat testis CYP11A1 and STAR mRNA as found in intact rats.

Heger et al. (2012) found comparable results. These researchers implanted fetal rat, fetal mouse, or fetal human testes tissue into immunodeficient rat and mouse hosts. Host animals were treated by gavage with a range of doses of DBP. Consistent with the *in utero* response, phthalate exposure induced mononuclear gonocytes formation in rat and mouse xenografts, but only the rat xenograft exhibited suppressed expression of steroidogenesis genes. Across the range of doses tested, human fetal testis xenografts exhibited mononuclear gonocyte induction at all exposures (100, 250, and 500 mg/kg-day) but were resistant to suppression of steroidogenic genes at all exposures.

It is not known whether transport of the toxic metabolite into the testicular xenografts is the same as that in intact animals. In the studies of Heger et al. (2012,) an exposure of 100 mg/kg-day of DBP in the host animal is an adverse effect level in the human testicular xenograft. In intact rats, Lehmann et al. (2004) found a no effect level for a decrease in testosterone concentration at 30 mg/kg-day and an effect level at 50 mg/kg-day. This comparison would suggest that a hazard quotient for DBP using a decrease in testosterone concentration or using induction of mononuclear gonocytes as the end point would be in the same range. However, there are no comparable data available for the other phthalate esters.

Data on human exposure to phthalate esters is available for the U.S. population from the NHANES surveys in

TABLE 80.4   Exposure to Selected Phthalate Esters for a US Population in mg/kg-day

Chemical	Median	95th Percentile	Maximum
DBP	0.0013	0.0061	0.094
DiBP	0.0002	0.0011	0.016
BBP	0.00088	0.0040	0.029
DEHP	0.00071	0.0036	0.046
DPP	Not available	Not available	Not available
DiNP	<LOD	0.0017	0.022

Exposure to BBP, DEHP, and DiNP is directly from Kohn et al. (2000), Table 2. The original data used reported the urinary excretion of DBP and DiBP as a single summed value. Subsequent data collected by NHANES shows that urinary excretion of DiBP (creatinine corrected) is approximately 15% of the excretion of DBP (creatinine corrected). Accordingly, the DBP values have been reduced by 15%, and the remainder is attributed to DiBP.

1999–2000, 2001–2002, 2003–2004, 2005–2006, and 2007–2008 (CDC, n.d.). These studies measured the concentration of the monoester metabolites in the urine of a cross section of the U.S. population aged 6 years and above. As the monoester metabolites and additional oxidation products are rapidly excreted in urine (that is, they do not bioaccumulate), measurements of metabolites in urine can be used to estimate the total exposure to phthalate esters in the individual.

Kohn et al. (2000) used an earlier data set from 289 individuals and a linear two-compartment model to estimate the exposures (mg/kg-day) in the general population. No study has used the same method as Kohn et al. (2000) to update these exposure estimates taking into account the additional data sets available from the NHANES surveys. However, these subsequent NHANES data sets show values that are comparable to or somewhat lower than those used in the original analysis of Kohn et al. (2000). Therefore, large differences in the exposure estimates are not anticipated.

Data on human exposure to phthalate esters are also available from a German population (Wittassek and Angerer, 2008). The authors calculated the oral exposure from the urinary levels of the metabolites and urinary excretion factors determined from human metabolism studies. These data for both the United States and German populations are summarized in Table 80.4 and Table 80.5, respectively.

TABLE 80.5   Exposure to Selected Phthalate Esters for a German Population in mg/kg-day

Chemical	Median	95th Percentile	Maximum
DBP	0.0021	Not provided	0.230
DiBP	0.0015	Not provided	0.0273
BBP	0.0003	Not provided	0.0022
DEHP	0.0027	Not provided	0.0422
DPP	Not available	Not available	Not available
DiNP	0.0006	Not provided	0.0368

Source: Data are from Wittassek and Angerer (2008) for 102 individuals.

REFERENCES

Benson, R. (2009) Hazard to the developing male reproductive system from cumulative exposure to phthalate esters—dibutyl phthalate, diisobutyl phthalate, butylbenzyl phthalate, diethylhexyl phthalate, dipentyl phthalate, and diisononyl phthalate. *Regul. Toxicol. Pharmacol.* 53:90–101.

Borch, J., Ladefoged, O., and Hass, U. (2004) Steroidogenesis in fetal male rats is reduced by DEHP and DiNP, but endocrine effects of DEHP are not modulated by DEHA in fetal, pre-pubertal and adult male rats. *Reprod. Toxicol.* 18:53–61.

CDC *National Report on Human Exposure to Environmental Chemicals*, Atlanta, GA: Center for Disease Control and Prevention. Available at http://www.cdc.gov/exposurereport.

Foster, P. (2006) Disruption of reproductive development in male rat offspring following *in utero* exposure to phthalate esters. *Int. J. Androl.* 29:140–147.

Gaido, K.W., Hensley, J.B., Liu, D., Wallace, D.G., Borghoff, S., Johnson, K.J., Hall, S.J., and Boekelheide, K. (2007) Fetal mouse phthalate exposure shows that gonocyte multinucleation is not associated with decreased testicular testosterone. *Toxicol. Sci.* 97:491–503.

Heger, N., Hall, S.J., Sandrof, M.A., McDonnell, E.V., Hensley, J.B., McDowell, E.N., Martin, K.A., Gaido, K.W., Johnson, K.J., and Boekelheide, K. (2012) Human fetal testis xenografts are resistant to phthalate-induced endocrine disruption. *Environ. Health Perspect.* 120:1137–1143.

Howdeshell, K., Wilson, V., Furr, J., Lambright, C., Rider, C., Blystone, C., Hotchkiss, A., and Gray, L. (2008) A mixture of five phthalate esters inhibits fetal testicular testosterone production in the Sprague-Dawley rat in a cumulative dose additive manner. *Toxicol. Sci.* 105:153–165.

HSDB *Hazardous Substances Data Bank*. Available at http://toxnet.nlm.nih.gov (accessed January 2013).

Johnson, K., Heger, N., and Boekelheide, K. (2012) Of mice and men (and rats): phthalate-induced fetal testis endocrine disruption is species-dependent. *Toxicol. Sci.* 129:235–248.

Kohn, M., Parham, F., Masten, S., Portier, C., Shelby, M., Brock, J., and Needham, L. (2000) Human exposure estimates for phthalates. *Environ. Health Perspect.* 108:A440–A442.

Lehmann, K., Phillips, S., Sar, M., Foster, P., and Gaido, K. (2004) Dose-dependent alterations in gene expression and testosterone synthesis in the fetal testis of male rats exposed to di-n-butyl phthalate. *Toxicol. Sci.* 81:60–68.

Matsumoto, M., Hirate-Koizumi, M., and Ema, M. (2008) Potential adverse effects of phthalic acid esters on human health: a review of recent studies on reproduction. *Regul. Toxicol. Pharmacol.* 50:37–49.

Mitchell, R.T., Childs, A.J., Anderson, R.A., van den Driesche, S., Saunders, P.T.K., McKinnell, C., Wallace, W.H.B., Kelnar, C.J.H., and Sharpe, R.M. (2012) Do phthalates affect steroidogenesis by the human fetal testis? Exposure of human fetal testis xenografts to di-n-butyl phthalate. *J. Clin. Endocrinol. Metab.* 97:E341–E348.

NTP-CERHR (2006) *NTP-CERHR Monograph on the Potential Human Reproductive and Developmental Effects of Di(2-ethylhexyl)*

Phthalate, Available at http://ntp.niehs.nih.gov/ntp/ohat/phthalates/dehp/DEHP-Monograph.pdf.

Sharpe, R.M. (2008) Additional effects of phthalate mixtures on fetal testosterone production. *Toxicol. Sci.* 105:1–4.

Sharpe, R.M. and Skakkebaek, N.E. (2008) Testicular dysgenesis syndrome: mechanistic insights and potential new downstream effects. *Fertil. Steril.* 89:e33–e38.

Silva, M., Reidy, J., Herbert, A., Preau, J., Needham, L., and Calafat, A. (2004) Detection of phthalate metabolites in human amniotic fluid. *Bull. Environ. Contam. Toxicol.* 72:1226–1231.

Welsh, M., Saunders, P.T.K., Fisken, M., Scott, H.M., Hutchison, G.R., Smith, L.B., and Sharpe, R.M. (2008) Identification in rats of a programming window for reproductive tract masculinization, disruption of which leads to hypospadias and cryptorchidism. *J. Clin. Invest.* 118:1479–1490.

Wittassek, M. and Angerer, J. (2008) Phthalates: metabolism and exposure. *Int. J. Androl.* 31:131–138.

81

DECABROMODIPHENYL ETHANE

Marek Banasik

First aid: Eyes: Irrigate immediately and seek medical advice; skin: flush with water promptly; inhalation: remove to fresh air; ingestion: give 500 ml of water to drink

Occupational exposure limit: Derived no effect level (DNEL) (systemic effects, long-term inhalation): 71 mg/m³ (ECHA, 2014)

Reference values: DNEL (systemic effects, long-term inhalation): 17.4 mg/m³; DNEL (systemic effects, long-term oral): 5 mg/kg body weight (bw)/day (ECHA, 2014)

Hazard statement(s): None

Precautionary statement(s): None

Risk phrase(s): None

Safety phrase(s): None

BACKGROUND AND USES

Decabromodiphenyl ethane [DBDPEthane; Chemical Abstracts Service Registry Number (CASRN): 84852-53-9] is a brominated flame retardant used as an additive in a variety of polymer and textile applications. In polymeric applications, such as high impact polystyrene, DBDPEthane is compounded or blended into the resin. The resulting DBDPEthane-containing plastics are then converted to finished products by, for example, injection molding. For textile applications, DBDPEthane is blended with various latices and applied as a back coating to the textile.

Decabromodiphenyl ethane is a high production volume chemical. In the United States (US), the national production volume in 2012 was between 50 and 100 million pounds (*ca.* 22,680-45,359 tonnes) (EPA, 2012). In the European Union,

the production volume is between *ca.* 2.2 and *ca.* 22 million pounds (1,000–10,000 tonnes) per annum (ECHA, 2014).

Decabromodiphenyl ethane has undergone extensive toxicological testing to evaluate potential hazards to aquatic- and terrestrial-dwelling organisms and to humans through subchronic experimental animal studies. Based on these data, the United Kingdom's Environment Agency (EA) conducted an environmental and human health risk assessment on DBDPEthane. The sections that follow provide an overview of this information and the EA's risk conclusions.

PHYSICAL AND CHEMICAL PROPERTIES

Decabromodiphenyl ethane is a powder with a high molecular weight, very low water solubility, and low lipophilicity (as indicated by log K_{OW}). The particles are <15 μm in diameter, and thus, this substance is expected to be respirable after inhalation exposure. A summary of DBDPEthane's physicochemical properties is provided in Table 81.1.

ENVIRONMENTAL FATE AND BIOACCUMULATION

Decabromodiphenyl ethane's physicochemical properties suggest that it will partition predominantly in sediment and soil through binding to the organic fraction of particulate matter. In the EA environmental risk assessment, it concluded that DBDPEthane is unlikely to rapidly undergo photodegradation in the presence of hydroxyl radicals, it is not readily biodegradable in the aquatic environment

Hamilton & Hardy's Industrial Toxicology, Sixth Edition. Edited by Raymond D. Harbison, Marie M. Bourgeois, and Giffe T. Johnson.
© 2015 John Wiley & Sons, Inc. Published 2015 by John Wiley & Sons, Inc.

TABLE 81.1 Chemical and Physical Properties of Decabromodiphenyl Ethane

Characteristic	Information
Chemical structure	
Chemical name	1,1′-(Ethane-1,2-diyl)bis [pentabromobenzene]
Synonyms/trade names	DBDPE, DBDPE/RDT-3, DBDPEthane, decabromodiphenyl ethane, Ecoflame B-971, SAYTEX 8010, YCFR-03
Chemical formula	$C_{14}H_4Br_{10}$
CASRN	84852-53-9
Molecular weight	971.22 g/mol
Physical state	Odorless white powder at room temperature
Melting point	350 °C
Vapor pressure	$<1.0 \times 10^{-4}$ Pa at 20 °C
Relative density	2.67 at 20 °C
Water solubility	*ca.* 0.72 µg/l at 25 °C
Octanol/water partition coefficient (log K_{OW})	3.55 at 25 °C
Particle size distribution	Mass median diameter: 5 µm

Source: Chemical Book (2008); (ECHA, 2014).

under aerobic conditions, and it is not hydrolysable (no hydrolysable groups) (Dungey and Akintoye, 2007). No data were available to assess the biodegradation under anaerobic conditions [e.g., wastewater treatment plants (WWTPs) or in sediments]; the EA noted the possibility of reductive debromination under these conditions.

The potential for DBDPEthane to bioconcentrate was evaluated in an 8-week fish bioconcentration study (Dungey and Akintoye, 2007). When fish were exposed to DBDPEthane at concentrations of 0.5 and 0.05 mg/l, the respective fish bioconcentration factors (BCFs) were <2.5 and <25 l/kg wet weight (ww), respectively. Since the water concentrations used exceeded DBDPEthane's water solubility, the dissolved concentration was unknown. The EA appropriately considered this study invalid. No additional BCF studies are available; however, based on the physicochemical properties, low toxicity in acute and repeated dose mammalian, aquatic, and terrestrial studies, it is unlikely that DBDPEthane will bioaccumulate.

ECOTOXICOLOGY

Decabromodiphenyl ethane has been evaluated in a series of aquatic, sediment, and terrestrial studies. No toxicity

was observed when fish (*Oncorhynchus mykiss*), algae (*Pseudokirchneriella subcapitata*), and daphnia (*Daphnia magna*) were acutely exposed to water-accommodated fractions of DBDPEthane at 110 mg/l (Hardy et al., 2012). In 28-day sediment toxicity studies, no treatment-related effects were observed on midge (*Chironomus riparius*) (development times, emergence, or development rates) or oligochaeta (*Lumbriculus variegatus*) [survival, reproduction, or dry weight (dw)] at 5,000 mg/kg dw, the highest concentration was tested. Hardy et al. (2012) calculated a predicted no effect concentration (PNEC) for sediment of 100 mg/kg dw. Using these same data, the EA converted the dw measures to wws and calculated PNECs for freshwater sediment and marine sediment of ≥21 mg/kg ww and ≥6 mg/kg ww (Dungey and Akintoye, 2007).

Terrestrial toxicity evaluations have been reported on sewage sludge respiration inhibition, soil bacteria nitrification, survival, and reproduction in earthworms, seedling emergence, and growth in six plants (Hardy et al., 2011).

Decabromodiphenyl ethane caused 5.3% inhibition in a 3-h sewage sludge respiration test at 10 mg/l. 3,5-Dichlorophenol, the positive control, caused 1.9, 40.6, and 80.8% inhibition at 3, 15, and 50 mg/l, respectively. In the soil nitrification study, a 28-day exposure to DBDPEthane at 2,500 mg/kg dry soil caused 5.6% inhibition of soil nitrate (mg NO_3^-/kg dw) and 5.9% inhibition in soil nitrate formation (mg NO_3^-/kg/day). 2-Chloro-6-(trichloromethyl)pyridine, the positive control, at 500 mg/kg dw caused 83.5% inhibition of soil nitrate and 95.0% inhibition of nitrate formation rate.

In the earthworm survival (28 days) and reproduction (56 days) studies, no treatment-related effects were observed on survival or bw up to the limit dose of 3,720 mg/kg dw. Reproduction was decreased by 60% at the limit dose. A no observed effect concentration (NOEC) of 1,907 mg/kg dw was reported. Reproduction median effect concentration (EC_{50}) and EC_{10} (concentration that causes the measured effect in 10% of organisms) of 3,180 and 1,860 mg/kg dw, respectively, were calculated.

No treatment-related effects on survival or growth were observed with corn, ryegrass, or soybean up to the limit concentration of 6,250 mg/kg dw. Height and dw were decreased in onion and tomato plants at 3,125 mg/kg dw and 6,250 mg/kg dw, respectively. Survival was decreased in cucumber plants at the limit concentration. The following NOECs were reported: onion, height and dw (1,563 mg/kg dw); tomato, height and dw (3,125 mg/kg dw); and cucumber, survival (3,125 mg/kg dw).

Based on these findings, Hardy et al. (2011) calculated PNECs for sewage treatment plants and soil of 2,500 mg/kg dw and 156.3 mg/kg dw, respectively. The EA converted these dw measures to wws and calculated a PNEC for soil of 26 mg/kg ww (Dungey and Akintoye, 2007).

MAMMALIAN TOXICOLOGY

The toxicity of DBDPEthane has been evaluated by the oral and dermal routes of exposure in a number of experimental animal studies. Overall, DBDPEthane has low acute toxicity by all routes of exposure and is neither a skin/eye/respiratory irritant, nor a skin/respiratory sensitizer in animals and/or humans (ECHA, 2014). Most toxicity studies (e.g., repeated dose and developmental) conducted in rodents by the oral route of exposure indicate a low hazard concern; in the majority of these studies, no effects were found at the highest doses tested (typically ≥1,000 mg/kg bw/day). A summary of acute and repeated dose toxicology studies is provided below.

Acute Studies

The acute toxicity of DBDPEthane has been evaluated by the oral and dermal routes using standard methods. No inhalation toxicity studies have been performed due to DBDPEthane's low vapor pressure (i.e., $<1.0 \times 10^{-4}$ Pa at 20 °C) and the absence of toxicity by the inhalation route on decabromodiphenyl ether, a structurally similar compound (Hardy et al., 2009). The oral and dermal toxicity studies performed on DBDPEthane are summarized below.

Mallory (1988a) evaluated the acute oral toxicity of DBDPEthane under good laboratory practice (GLP) standards (Dungey and Akintoye, 2007; citing Mallory, 1988a). A limit test with a dose level of 5,000 mg DBDPEthane/kg bw was conducted using young adult Sprague-Dawley rats ($n = 5$/sex). The animals were fasted 18 h prior to test substance administration. No changes in bw or clinical signs of toxicity were observed during the 14-day follow-up. No animals died during the study, and no visible lesions were observed at the terminal necropsy. The estimated acute oral median lethal dose (LD_{50}) was determined to be >5,000 mg/kg bw.

The acute dermal toxicity of DBDPEthane was evaluated under GLP using New Zealand white rabbits ($n = 5$/sex) as the test species (Dungey and Akintoye, 2007; citing Mallory, 1988b). A limit dose of 2,000 mg/kg bw was applied to unabraded, shaven skin over approximately 10% of the body surface. The test substance was occluded and remained in place for 24 h. Following the 24-h period, the wrappings were removed, along with any residual test substance. No changes in bw were observed during the 14-day period. Other than diarrhea observed in one animal, no other clinical signs of toxicity were observed. No animals died during the course of the study, and no visible lesions were observed at terminal necropsy. The estimated dermal LD_{50} was determined to be >2,000 mg/kg bw.

In summary, the dose levels tested in the acute oral and dermal toxicity studies exceeded or equaled the highest dose levels (i.e., 2,000 or 5,000 mg/kg bw) recommended under current validated testing guidelines. Since no changes in bw, clinical signs, mortality, or tissues were observed at these dose levels, these studies support that DBDPEthane presents a low concern for acute exposures *via* the oral and dermal routes.

Repeated Dose Studies

Margitich (1991) performed a 28-day repeated dose toxicity study on male and female Sprague-Dawley rats ($n = 6$/sex/group), which included two 14-day recovery groups (control and high dose; $n = 6$/sex/group) (Dungey and Akintoye, 2007; citing Margitich, 1991). Decabromodiphenyl ethane was administered by gavage in a corn oil vehicle at dose levels of 0, 125, 400, and 1,250 mg/kg bw/day. At the end of 28 days, all animals were sacrificed, with the exception of the recovery groups. No treatment-related effects were reported with clinical signs of toxicity, bw, food consumption, ophthalmology, urinalysis, and histopathology of the same sex between the control and the low, middle, and high dose groups or between the recovery control and the recovery high dose groups at any point in the study. Differences were noted in some hematology and clinical chemistry measures, but were non-dose dependent and within historical control values. A dose-dependent increase in relative liver weight in female rats was reported (by 4% in the low and mid dose groups and by 10% in the high dose group). The EA concluded that there were no adverse effects up to the limit dose of 1,250 mg/kg bw/day.

Hardy et al. (2002) performed a guideline compliant 90-day repeated dose study in Sprague-Dawley rats, under GLP. Groups of male and female rats ($n = 10$/sex/group) included two 28-day recovery groups (control and high dose; $n = 10$/sex/group). Decabromodiphenyl ethane was administered daily by gavage using a corn oil vehicle at 0, 100, 320, and 1,000 mg/kg bw/day. At the end of the 90-day treatment period, all animals were sacrificed with the exception of the recovery groups. No treatment-related effects were reported with clinical signs of toxicity, body weight, food consumption, ophthalmology, or urinalysis between the controls and treated animals in the main study or the recovery groups. Differences in hematology and clinical chemistry were reported on days 30 and 90; however, they did not show a consistent pattern nor were they inconsistent with historical control ranges. In the high dose females, a statistically significant increase (by 13%) in the relative and absolute liver weight was reported at study termination. In high dose males, a statistically significant increase (by 8%) in the relative liver weight was reported at study termination. These changes were not present in recovery animals. Minimal to slight histopathological changes were observed at terminal sacrifice, which consisted of hepatocellular vacuolation (high dose males) and minimal to slight hepatocytomegaly (mid and high dose males). These changes were not observed in the recovery animals. Hardy et al. (2002) identified a no-observed-adverse-effect level (NOAEL) of

≥1,000 mg/kg bw/day. The EA concluded that the studied NOAEL was 1,000 mg/kg bw/day.

Decabromodiphenyl ethane has been evaluated in two separate, high quality prenatal developmental toxicity studies (Dungey and Akintoye, 2007; citing Mercieca, 1992a, 1992b; Hardy et al., 2010). In the first study, 25 mated female Sprague-Dawley were administered DBDPEthane by gavage with a corn oil vehicle at dose levels of 0, 125, 400, and 1,250 mg/kg bw/day from gestation day (GD) 6 through 15. The animals were sacrificed on GD 20 and subjected to cesarean section. In dams, no treatment-related effects were reported. In fetuses, no treatment-related effects were reported on weight, sex ratio, early/late resorption, or external/visceral abnormalities. Skeletal abnormalities (i.e., unossified hyoid and reduced ossification of skull) were statistically significantly increased in litters from the 400 mg/kg bw/day group. The EA considered this effect incidental to treatment because it was not seen in the high dose group, and concluded there was no evidence of developmental toxicity in rats up to 1,250 mg/kg bw/day.

In the second study, 20 artificially inseminated New Zealand white rabbits were administered DBDPEthane by gavage with a 0.5% methylcellulose vehicle at dose levels of 0, 125, 400, and 1,250 mg/kg bw/day from GD 6 through 18. The animals were sacrificed on GD 29 and subjected to cesarean section. In dams, no treatment-related effects were observed. In fetuses, no treatment-related effects were reported on weight, sex ratio, early/late resorption. Abortion occurred in one fetus from the 125 and 400 mg/kg bw/day groups and in two fetuses from the 1,250 mg/kg bw/day group. The EA considered this finding incidental given the low incidence observed in the study and the high spontaneous abortion rate in this species. A statistically significant increase in the number of litters with 27th presacral vertebra was observed in the high dose group (nine litters compared to four controls); however, the EA did not consider this effect adverse based on the historical control range (12.5–93.8%). The EA concluded that no adverse developmental effects occurred in rabbits up to the limit dose of 1,250 mg/kg bw/day.

Overall, the available database in animals supports that DBDPEthane presents a low concern for repeated subchronic exposures up to 1,000 mg/kg bw/day. The EA concurred with the foregoing study conclusions and utilized a NOAEL of 1,000 mg/kg bw/day for risk characterization. No studies have been performed to evaluate the potential reproductive or chronic/carcinogenic effects of DBDPEthane. However, DBDPEthane did not cause histological changes in the reproductive organs of male or female rats in a 90-day repeated dose study that suggests a low concern for reproductive toxicity. For chronic/carcinogenic effects, DBDPEthane was negative in standard *in vitro* mutagenicity and clastogenicity studies. The EA concluded that DBDPEthane is unlikely to be carcinogenic due to the negative genotoxicity data and the absence of proliferative lesions in the subchronic studies.

SOURCES OF EXPOSURE

Occupational

The greatest potential for worker exposure to DBDPEthane is through the inhalation of particulates in the workplace. The primary opportunities for exposure are activities that generate dust (e.g., compounding or blending of additives into the polymer) or through fumes from heated materials (e.g., conversion or injection molding). Decabromodiphenyl ethane is not combustible. It will, however, decompose at temperatures above 320 °C and generate corrosive fumes of hydrobromic acid, bromine, and oxides of carbon (ECHA, 2014).

Environmental

Environmental releases of DBDPEthane may occur during manufacture and packaging, compounding and converting, from finished articles, and from end-of-life disposal. Decabromodiphenyl ethane has been detected in virtually all environmental media, including house dust, sewage sludge, and sediment. In house dust, Stapelton et al. (2008) reported concentrations ranging from <10.0 ng/g up to 11,070 ng/g (median value = 201 ng/g). De la Torre et al. (2012) measured DBDPEthane at 47.0 ± 29.7 ng/g dw (mean ± SD) in WWTPs in Spain. Ricklund et al. (2008) reported sludge levels of DBDPEthane as high as 216 ng/g dw in Germany. Wei et al. (2012) reported sediment levels of DBDPEthane as high as 2,400 ng/g dw in water bodies surrounding US manufacturing facilities. Though present in these media, the relatively low levels are of questionable toxicological concern given the effect levels identified from the available human health and environmental toxicity studies.

INDUSTRIAL HYGIENE

Decabromodiphenyl ethane is manufactured as an odorless white powder. The exposure controls and personal protection equipment recommended in the work place are designed to minimize eye, skin, and respiratory exposures (ECHA, 2014). Local exhaust is needed to minimize dust. Personal protection equipment includes a dust respirator fitted with high efficiency (dust, fumes, and mist) filter cartridges, protective gloves resistant to chemical penetration, and chemical goggles or safety gloves.

RISK ASSESSMENTS

The EA evaluated the potential human health and environmental risks from DBDPEthane's use as a polymer additive and its use in textiles. It concluded that DBDPEthane's use as a polymer additive ". . . does not appear to pose any direct environmental risks (including risks to humans exposed via the environment) based on present knowledge." For the textile uses, the EA did not identify risks for ". . . surface water, WWTP, the atmosphere, predators or humans following environmental exposure." The EA noted the possibility of risks to sediment- and soil-dwelling organisms during the formulation and application of textile back coatings. It recommended several data needs that would aid with resolving this uncertainty, including textile site emission information, the number of user sites and geographical spread, and further sediment and soil toxicity studies to refine the PNECs, although the EA noted that ". . . [DBDPEthane] is not highly toxic."

REFERENCES

Chemical Book (2008) *1,2-Bis(pentabromophenyl) Ethane*. Beijing, China: ChemicalBook, Inc. Available at http://www.chemicalbook.com. CASRN: 84852-53-9.

De la Torre, A., Concejero, M.A., and Martinez, M.A. (2012) Concentrations and sources of an emerging pollutant, decabromodiphenylethane (DBDPE), in sewage sludge for land application. *J. Environ. Sci. (China)* 24:558–563.

Dungey, S. and Akintoye, L. (2007) *Environmental Risk Evaluation Report: 1,1'-(Ethane-1,2-diyl)bis[penta-bromobenzene]*. Science summary, Report Product Code: SCHO0507BMOR-E-P. Bristol, United Kingdom: Environment Agency, Available at http://webarchive.nationalarchives.gov.uk/2014032 8084622/http://cdn.environment-agency.gov.uk/scho0507bmos-e-e.pdf.

ECHA (2014) *Registered Substances*. Helsinki, Finland: European Chemicals Agency. Available at http://echa.europa.eu/information-on-chemicals/registered-substances. CASRN: 84852-53-9.

EPA (2012) *Chemical Data Access Tool (CDAT)*. Washington, DC: Office of Pollution Prevention and Toxics, US Environmental Protection Agency. Available at http://java.epa.gov/oppt_chemical_search. CASRN: 84852-53-9.

Hardy, M.L., Aufderheide, J., Krueger, H.O., Mathews, M.E., Porch, J.R., Schaefer, E.C., Stenzel, J.I., and Stedeford, T. (2011) Terrestrial toxicity evaluation of decabromodiphenyl ethane on organisms from three trophic levels. *Ecotoxicol. Environ. Saf.* 74:703–710.

Hardy, M.L., Banasik, M., and Stedeford, T. (2009) Toxicology and human health assessment of decabromodiphenyl ether. *Crit. Rev. Toxicol.* 39:1–44.

Hardy, M.L., Krueger, H.O., Blankinship, A.S., Thomas, S., Kendall, T.Z., and Desjardins, D. (2012) Studies and evaluation of the potential toxicity of decabromodiphenyl ethane to five aquatic and sediment organisms. *Ecotoxicol. Environ. Saf.* 75:73–79.

Hardy, M.L., Margitich, D., Ackerman, L., and Smith, R.L. (2002) The subchronic oral toxicity of ethane, 1,2-bis(pentabromophenyl) (Saytex 8010) in rats. *Int. J. Toxicol.* 21:165–170.

Hardy, M.L., Mercieca, M.D., Rodwell, D.E., and Stedeford, T. (2010) Prenatal developmental toxicity of decabromodiphenyl ethane in the rat and rabbit. *Birth Defects Res. B Dev. Reprod. Toxicol.* 89:139–146.

Mallory, V. (1988a) *Acute Exposure Oral Toxicity*, PH 402-ET-001-88. Decabromodiphenylethane, Lot #SH-6427-49. Waverly, PA, USA: Pharmakon Research International, Inc.

Mallory, V. (1988b) *Acute Exposure Dermal Toxicity*, PH 422-ET-001-88. Decabromodiphenylethane, Lot #SH-6427-49. Waverly, PA, USA: Pharmakon Research International, Inc.

Margitich, D.L. (1991) *28 Day Toxicity Study in Rats*. PH 436-ET-002-90. Saytex 402, Lot #23-014-2A. Waverly, PA, USA: Pharmakon Research International, Inc.

Mercieca, M.D. (1992a) *Developmental Toxicity (Teratology) Study in Rats with Saytex 402*. SLS study no. 3196.24. Spencerville, OH, USA: Springborn Laboratories, Inc.

Mercieca, M.D. (1992b) *Developmental Toxicity (Teratology) Study in Rabbits with Saytex 402*. SLS study no. 3196.26. Spencerville, OH, USA: Springborn Laboratories, Inc.

Ricklund, N., Kierkegaard, A., and McLachlan, M.S. (2008) An international survey of decabromodiphenyl ethane (deBDethane) and decabromodiphenyl ether (decaBDE) in sewage sludge samples. *Chemosphere* 73:1799–1804.

Stapelton, H.M., Allen, J.G., Kelly, S.M., Konstantinov, A., Klosterhaus, S., Watkins, D., McClean, M.D., and Webster, T.F. (2008) Alternate and new brominated flame retardants detected in U.S. house dust. *Environ. Sci. Technol.* 42:6910–6916.

Wei, H., Aziz-Schwanbeck, A.C., Zou, Y., Corcoran, M.B., Poghosyan, A., Li, A., Rockne, K.J., Christensen, E.R., and Sturchio, N.C. (2012) Polybrominated ethers and decabromodiphenyl ethane in aquatic sediments from southern and eastern Arkansas, United States. *Environ. Sci. Technol.* 46:8017–8024.

82

TETRABROMOBISPHENOL A

David Y. Lai

First aid: Eye: Irrigate immediately; skin: flush with water promptly; breathing: respiratory support; swallow: seek medical attention immediately

Target organ(s): Liver and uterus

Occupational exposure limits: DNEL (inhalation): 705 mg/m³; DNEL (dermal): 100 mg/kg-bw/day

Reference values: DNEL (inhalation): 174 mg/m³; DNEL (dermal): 50 mg/kg-bw/day; DNEL (oral): 5 mg/kg-bw/day

Hazard statement: H410: Very toxic to aquatic life with long-lasting effects

Precautionary statements: P273: Avoid release to the environment; P391: Collect spillage; P501: Dispose of contents/containers in accordance with appropriate local and national regulations

Risk phrases: 50/53

Very toxic to aquatic organisms, may cause long-term adverse effects in the environment

Safety phrases: 60-61

This material and its container must be disposed of as hazardous waste. Avoid release to the environment, refer to special instructions/safety data sheets.

BACKGROUND AND USES

Tetrabromobisphenol A (TBBPA; CASRN 79-94-7) is a brominated flame retardant used in a variety of reactive and additive applications. It is reacted (i.e., covalently bound) with epoxy, vinyl esters, and polycarbonate systems (e.g., high impact polystyrene (HIPS), and is used as an additive in acrylonitrile-butadiene-styrene (ABS) thermoplastic resins (Albemarle, 1999). Its primary application is in printed wire boards (PWBs) as a reactive flame retardant (BSEF, 2012).

It is a high production volume chemical. In the United States (U.S.), the national production volume, including manufacturing and importing, in 2012 was >119 million pounds (EPA, 2012). In Europe, the national production volume in 2010 was between 2.2 and 22 million pounds (ECHA, 2013). Though not manufactured in Canada, an industry survey in 2000 indicated that imports of pure TBBPA or TBBPA contained in PWBs, ABS, and HIPS were between 0.2 and 2.2 million pounds (Canada, 2013).

It has undergone extensive toxicological testing and environmental monitoring. Based on these data, several regulatory agencies from around the world have conducted risk assessments on this compound. The sections that follow provide an overview of this information and the conclusions from these regulatory evaluations.

PHYSICAL AND CHEMICAL PROPERTIES

Tetrabromobisphenol A is a crystalline particle/powder with a moderately high molecular weight, low water solubility, and moderately high lipophilicity (as indicated by log K_{OW}). Only about 4% of the particles are <15 μm in diameter, and thus, little (<4%) is expected to be respirable (<10 μm in diameter) and absorbed from the lung after inhalation exposure. A summary of TBBPA's physicochemical properties is provided in Table 82.1.

Hamilton & Hardy's Industrial Toxicology, Sixth Edition. Edited by Raymond D. Harbison, Marie M. Bourgeois, and Giffe T. Johnson.
© 2015 John Wiley & Sons, Inc. Published 2015 by John Wiley & Sons, Inc.

TABLE 82.1 Chemical and Physical Properties of tert-Butyl Alcohol (ECHA, 2013)

Characteristic	Information
Chemical name	2,2′,6,6′-Tetrabromo-4,4′-isopropylidenediphenol
Synonyms/trade names	2,2′,6,6′-Tetrabromobisphenol A; tetrabromobisphenol A; TBBPA
Chemical formula	$C_{15}H_{12}Br_4O_2$
CASRN	79-94-7
Molecular weight	543.8
Melting point	179–181 °C
Physical state	White solid crystalline powder
Particle size distribution	Mass median diameter: 52.2 μm, ca 4% with a particle size <15 μm
Vapor pressure	$<1.19 \times 10^{-5}$ Pa at 20 °C
Density/specific gravity	2.17
Water solubility at 25 °C	0.148 mg/l (pH 5); 1.26 mg/l (pH 7); 2.34 mg/l (pH 9)
Octanol/water partition coefficient (log K_{OW})	5.903 ± 0.0340 at 25 °C
Dissociation constant (pKa)	9.40

ENVIRONMENTAL FATE AND BIOACCUMULATION

Its physicochemical properties suggest that it will partition to all compartments (i.e., water, sediment, and soil), predominantly to sediment and soil through binding to the organic fraction of a particulate matter. Available environmental fate studies indicated that TBBPA is persistent in water (half-life [$t_{1/2}$] > 182 days), soil ($t_{1/2} \geq 182$ days), and sediment ($t_{1/2} \geq 365$ days) (Canada, 2013). It lacks functional groups that are expected to undergo hydrolysis (Canada, 2013). A number of laboratory studies (ECHA, 2013) showed that it can degrade to bisphenol A under aerobic conditions (Canada, 2013).

Tetrabromobisphenol A is identified as a persistent, bioaccumulative, and toxic (PBT) compound under the U.S. Environmental Protection Agency's Toxic Release Inventory (EPA, 2013). It was also placed on the State of Washington's Department of Ecology's PBT List (DOC, 2013). However, Environment Canada and Health Canada concluded that TBBPA did not meet their criteria for bioaccumulation (i.e., bioaccumulation factor >5000) (Canada, 2013). This conclusion was based on TBBPA's low bioaccumulation potential from its physicochemical properties (e.g., maximum diameter of 1.3–1.4 nm, ionization at environmentally relevant pH, and variable log K_{OW}), as well as from studies that showed TBBPA is rapidly metabolized and excreted in aquatic and terrestrial organisms (Canada, 2013).

ECOTOXICOLOGY

It has been shown to exert adverse effects on organisms in a variety of ecological toxicity test systems, with reported acute and chronic effect levels <1 mg/l in daphnia, mussels, and fish (Canada, 2013). Though the precise mode of action(s) is unknown, toxicity is expected to be higher for the non-ionized form of TBBPA than the ionized forms. In the non-ionized state, TBBPA is expected to act as a narcotic or baseline toxicant by altering membrane integrity or function, whereas the ionized forms are expected to be less bioavailable, and hence less toxic (Canada, 2013).

MAMMALIAN TOXICOLOGY

The toxicity of TBBPA has been evaluated by the oral, dermal, and inhalation routes of exposure in a number of experimental animal studies. Overall, TBBPA has low acute toxicity by all routes of exposure and is neither a skin/eye/respiratory irritant, nor a skin/respiratory sensitizer in animals and/or humans (ECHA, 2013). Most toxicity studies (e.g., repeated dose, developmental, reproductive, and neurobehavioral) conducted in rodents by all routes of exposure indicate a low hazard concern; in the majority of these studies, no effects were found at the highest doses tested. The European Union (EU) has reviewed many of the toxicological studies and has published a Risk Assessment Report (RAR) on TBBPA (EURAR, 2006), a summary of acute and repeated dose toxicology studies is provided below.

Acute Studies

Acute oral, dermal, and inhalation studies have been performed on TBBPA and submitted to regulatory agencies as part of regulatory dossiers. These studies support that TBBPA presents a low hazard concern by all routes of exposure (oral LD_{50} > 5000 mg/kg-bw in male/female Sprague-Dawley rats; dermal LD_{50} > 2000 mg/kg-bw in male/female New Zealand White rabbits; and inhalation is 8-h LC_{50} > 500 mg/m³ in male/female Wistar rats, NDMI mice, and guinea pigs) (ECHA, 2013).

Repeated Dose Studies

In general, the repeated dose toxicity studies with TBBPA have shown low hazard concerns. This is expected based on the available toxicokinetic data. Metabolism and toxicokinetic studies in rats and humans have shown that TBBPA can be absorbed following oral exposures and rapidly excreted after Phase II metabolism/conjugation to more water-soluble metabolites. Over 95% of TBBPA or metabolites are excreted in the feces within 72 h after a single administration, and there is little tissue retention or bioaccumulation.

Tetrabromobisphenol A's estimated half-life in humans is about 2 days (Sjodin et al., 2003).

The chronic/subchronic toxicity of TBBPA has been investigated in three repeated dose studies. The findings from these studies are summarized below and in Table 82.2. In a 14-day inhalation study, the EU concluded that no adverse effects were reported up to the limit dose of 18 mg/l, with the exception of signs of mechanical irritation in all treatment groups due to the high dust levels. In a 90-day oral gavage study, rats were administered TBBPA at dose levels of 0, 100, 300, or 1000 mg/kg-bw/day. No neurobehavioral effects were observed during the weekly functional observational battery (FOB) evaluations. Slight changes in hematological evaluations and clinical chemistry were reported; however, the EU concluded that these effects were not toxicologically significant. Statistically significant decreases in serum T_4 were reported in males and females, but no accompanying change in serum T_3, thyroid stimulating hormone (TSH), or histopathology of the liver, thyroid, parathyroid, and pituitary was reported. Therefore, the EU concluded that the decreases in serum T_4 were not considered to be adverse. Absolute spleen weight was decreased in males in the top two dose groups; no histopathological findings were noted. The EU concluded that these were chance findings of no toxicological significance. Finally, an increase in relative epididymis weight was reported in the middle dose group; however, no changes in relative epididymis weight or histopathology were identified in the high dose group. The EU concluded that these findings were of no toxicological significance. In a 21-day dermal study, TBBPA was administered to rabbits at dose levels up to 2500 mg/kg-bw/day for 6 h/day, 5 days/week. No toxicologically significant effects were identified.

The EU identified three principal studies and two supporting studies to assess reproductive/developmental toxicity. The principal studies are summarized below and in Table 82.3. TBBPA was evaluated in a two-generation reproductive performance and fertility study, which included a neurobehavioral and a neuropathological component. F_0 and F_1 rats were administered TBBPA by gavage at dose levels of 0, 10, 100, or 1000 mg/kg-bw/day. Animals were dosed during a 10-week premating period and during a 2-week mating period. Females were also dosed during gestation and lactation. The only treatment-related effect observed in the F_0 and F_1 generations was a statistically significant, 7% decrease in bodyweight gain during the 10-week premating period in the high dose group. Statistically significant decreases in serum T_4 were reported at various dose levels in males and females from the F_0 and F_1 generations. The EU concluded that these effects were not toxicologically significant because there were no other parameters associated with disruption of thyroid homeostasis. Detailed clinical examinations were performed on male and female F_2 pups on postnatal days (PNDs) 4, 11, 21, 35,

45, and 60; no treatment-related effects were identified. Motor activity, including horizontal and vertical activity counts and distance traveled, was assessed on male and female F_2 pups on PND 13, 17, 21, and 60. Subtle, non-dose-dependent changes were reported on PNDs 17, 21, and 60; however, the EU interpreted these as chance findings and unrelated to treatment. Learning and memory were assessed in male and female F_2 pups using a Step-through Passive Avoidance Test (SPAT) on PNDs 22 and 60. Some differences were reported for male pups on PNDs 22 and 60; no consistent pattern was observed. No differences were reported between treated and control female pups. The same animals evaluated in the SPAT testing were also evaluated in a Water M-maze beginning on PND 110 to assess short-term and long-term memory. No treatment-related differences were reported. A neuropathological evaluation on the central nervous system and peripheral nerves was performed on F_2 male and female pups on PND 60. No treatment-related differences were reported. Morphometric evaluations were also performed on F_2 males and females on PNDs 11 and 60. A statistically significant decrease in the thickness of the parietal cortex was observed in sexes on PND 11 but not on PND 60. No histologic changes were evident in the parietal cortex of F_2 male and female pups on PNDs 11 or 60. The EU interpreted the decreased thickness of the parietal cortex on PND 11 as a transient or chance finding that was unlikely to be toxicologically significant.

In a range-finder developmental study, mated female rats were administered TBBPA by gavage at 0, 30, 100, 300, 1000, 3000, or 10,000 mg/kg-bw/day from gestation days (GDs) 6 to 15. The animals were sacrificed on GD 20. No effects were reported on the number of viable and nonviable fetuses, early and late resorptions, total implantations, or corpora lutea. Three dams died at the top dose; no signs of toxicity were observed in animals receiving 3000 mg/kg-bw/day or less.

In a nonstandard developmental study, TBBPA was administered by gavage at dose levels of 0, 40, 200, or 600 mg/kg-bw/day to newborn rats from PND 4 to 21. A recovery group was maintained and sacrificed at 12 weeks of age. No changes in bodyweight, physical development, or reflex ontogeny were observed in any of the dose groups. Males and females in the 200 and 600 mg/kg-bw/day groups had sporadically occurring diarrhea. Changes were observed in hematology and clinical chemistry. Statistically significant decreases in hemoglobin and activated thromboplastin time were reported in males and females in the 600 mg/kg-bw/day groups, respectively. Total bilirubin was statistically significantly increased in both sexes in the top dose. Relative and absolute kidney weights were significantly increased in males (280%) and females (365%) compared to controls. Relative liver weight was increased by 11% in males in the 600 mg/kg-bw/day group. Histopathology of the kidney revealed moderate to severe polycystic lesions associated

TABLE 82.2 Summary of Repeated Dose Toxicity Studies on TBBPA Used by the EU

Study Type	Dose	Exposure Regimen	Species/Sex; Sample Size	Effect Levels	EURAR Conclusion
14-Day inhalation study (non-guideline/non-GLP)	2, 6, or 18 mg/l (particle size not documented)	4 h/day, 5 days/week for 2 weeks	Rat/♂ and ♀; 5 animals per sex per group	Some evidence of local irritation of eyes and upper respiratory tract at all dose levels as a consequence of mechanical irritation caused by high dust levels	No evidence of treatment-related systemic toxicity up to the highest dose of 18 mg/l
90-Day oral gavage study (guideline/GLP)	0, 100, 300, or 1000 mg/kg-bw/day (corn oil vehicle)	Daily	Rat/♂ and ♀; 10 animals per sex per group, 5 additional animals in the control and 1000 mg/kg-bw/groups for 6 week post-treatment evaluation (recovery)	<u>1000 mg/kg-bw/day:</u> 17% ↓ in platelet count, ♂; 2–3-fold ↑ in total bilirubin, ♂ and ♀; 18% ↓ in absolute spleen weight, ♂ (resolved at recovery); 1.7-fold ↑ in serum alkaline phosphatase, ♀ (resolved at recovery) <u>300 mg/kg-bw/day:</u> 2–3-fold ↑ in total bilirubin, ♀; 15% ↓ in absolute spleen weight, ♂: 13% ↑ in relative epididymis weight, ♂ <u>All doses:</u> Non-dose-dependent ↓ in serum T_4 levels on day 33 and 90, ♂ and ♀ (resolved at recovery)	No clear toxicologically significant adverse effects up to the highest dose tested, 1000 mg/kg-bw/day
21-Day dermal study (non-guideline/non-GLP)	0, 100, 500, or 2500 mg/kg-bw/day	6 h/day, 5 days/week	Rabbit/♂ and ♀; 5 animals per sex per group	No treatment-related effects reported at the highest dose tested	No toxicologically significant compound related effects were apparent

TABLE 82.3 Summary of Reproductive/Developmental Studies on TBBPA Used by the EU

Study Type	Dose	Exposure Regimen	Species/Sex; Sample Size	Effect Levels	EURAR Conclusion
2-Generation oral gavage study (guideline/GLP)	0, 10, 100, or 1000 mg/kg-bw/day (corn oil vehicle)	Daily, F_0 generation treated during a 10-week premating period and a mating period of 2 weeks; Dams also treated during gestation and lactation F_1 generation treated using same protocol	Rat/♂ and ♀; F_0 and F_1, 30 animals per sex per group; F_2, 10 animals per sex per group for neurobehavioral investigations, and 10 animals per sex per group for neuropathological examination	1000 mg/kg-bw/day: F_0: 28% ↓ and 44% ↓ in mean serum T_4 in ♂ and ♀, respectively; F_0: 19% ↓ in mean serum T_3 in ♂; F_1: 47% ↓ and 32% ↓ in mean serum T_4 in ♂ and ♀, respectively; F_1: 7% ↓ bodyweight in parental ♂ during pre-mating period F_2: PND 60, 76% ↓ in horizontal activity in the 0–5 min segment in ♂; F_2: PND 60, 68% ↓ in horizontal activity in the 5–10 min segment in ♂; F_2: PND 60, ↓ in the thickness of the parietal cortex in ♀ and ♂ 100 mg/kg-bw/day: F_0: 17% ↓ in mean serum T_4 in ♂; F_1: 38% ↓ and 32% ↓ in mean serum T_4 in ♂ and ♀, respectively; F_2: PND 17, ↓ horizontal activity in 20 min segment in ♀ F_2: PND 21, ↓ horizontal activity and distance traveled in the 5–10 min segment and over the 20 min test period in ♀ F_2: PND 60, 70% ↓ in horizontal activity in the 0–5 min segment in ♂; 10 mg/kg-bw/day: F_2: PND 17, ↓ horizontal activity in 15–20 min segment in ♀	No effects on fertility or development, including neurodevelopment, or any other effects of toxicological significance were observed in this study up to a dose of 1000 mg/kg-bw/day

TABLE 82.3 (*Continued*)

Study Type	Dose	Exposure Regimen	Species/Sex; Sample Size	Effect Levels	EURAR Conclusion
Developmental oral gavage study (range finder; non-guideline/non-GLP)	0, 15, 30, 100, 300, 1000, 3000, or 10,000 mg/kg/day	Daily, ♀, GDs 6 to 15	Rat/adult ♀ and fetuses ♂ and ♀; 5 dams assigned to each group	10,000 mg/kg-bw/day: 3 dams died; ↓ weight gain between GDs 6 and 15	There were no adverse effects on the developmental parameters assessed, up to very high dose levels, including a dose level producing severe maternal toxicity
Developmental study (non-guideline/non-GLP)	0, 40, 200, or 600 mg/kg-bw/day (0.5 w/v% carboxymethyl-cellulose vehicle)	Daily, ♂ and ♀, PNDs 4 to 21	Rat, ♂ and ♀; 6 animals per sex per group	600 mg/kg-bw/day: Sporadic diarrhea in ♂ and ♀; ↓ Hemoglobin in ♀; ↓ Activated prothrombin time in ♂; 280% and 365% ↑ in absolute/relative kidney weights in ♂ and ♀; Moderate to severe polycystic lesions associated with dilation of the tubules bilaterally from the cortico-medullary junction to the inner cortex in ♂ and ♀; 1.3 times ↑ absolute kidney weight in ♂ and ♀ after recovery; 11% ↑ in relative liver weight in ♂; Centrilobular hypertrophy in 3/6 ♂. 200 mg/kg-bw/day: Slight polycystic lesions of the kidneys in 2/6 males; Slight, multiple kidney cysts in one ♂ and ♀	It is considered that the effects on the kidneys are likely the consequence of the unconventional direct gavage administration of very high doses of TBBPA to such young animals. Therefore, the relevance to human health of this isolated finding is considered questionable

with dilation of the tubules in all animals in the 600 mg/kg-bw/day groups and slight lesions in 2/6 males in the 200 mg/kg-bw/day group. Centrilobular hypertrophy was reported in 3/6 males in the 600 mg/kg-bw/day group. In the recovery animals, absolute kidney weights were 1.3 times higher than controls. Moderate to severe cysts of the kidneys were reported in all animals in the 600 mg/kg-bw/day groups. Slight cysts of the kidneys were reported in one animal per sex in the 200 mg/kg-bw/day group. Based on these findings, the EU concluded that directly dosed young animals are more susceptible to the nephrotoxic effects of TBBPA than older animals and utilized the no-observed-adverse-effect level (NOAEL) of 40 mg/kg-bw/day for characterizing risks to infants from environmental exposures.

The U.S. National Toxicology Program (NTP) (NTP, 2013) conducted 2-year carcinogenicity studies on TBBPA in male and female rats and mice at dose levels of 0, 250, 500, or 1000 mg/kg-bw/day. The test substance was administered by gavage using a corn oil vehicle. Animals were dosed 5 days per week for the duration of the studies. In rats, no treatment-related effects were observed with survival. At a 3-month interim sacrifice, no treatment-related effects were reported, with the exception of decreased thymus weights and increased liver weights in the 1000 mg/kg-bw/day groups. At terminal sacrifice, a slight increase in the incidence of interstitial cell adenomas was observed in the 500 and 1000 mg/kg-bw/day groups. No other treatment-related effects were observed. In female rats, endometrial atypical hyperplasia was observed in female rats in all dose groups. In the 500 and 1000 mg/kg-bw/day females, the incidence of uterine adenomas, adenocarcinomas, or malignant Müllerian tumors were significantly increased. In male rats, a slight increase in the incidence of interstitial cell adenomas of the testis was observed in the 500 and 1000 mg/kg-bw/day groups.

In mice, survival at terminal sacrifice was significantly less in males and females in the 1000 mg/kg-bw/day groups compared to controls. Significant increases in the incidences of ulcer, mononuclear cell cellular infiltration, inflammation, and epithelial hyperplasia were observed in females in the 250 and 500 mg/kg-bw/day groups and in males in the 500 mg/kg-bw/day group. No other treatment-related effects were reported in female mice. In male mice, nonneoplastic lesions occurred primarily in the kidney and liver. The incidences of renal tubule cytoplasmic alteration were significantly increased in the 250 and 500 mg/kg-bw/day groups. In the liver, eosinophilic focus and clear cell focus were significantly increased in the 250 and 500 mg/kg-bw/day groups and the 500 mg/kg-bw/day groups, respectively. Several non-dose-dependent neoplastic lesions were observed. The incidences of hepatoblastoma, hepatocellular carcinoma, or hepatoplastoma and hepatocellular carcinoma (combined) were significantly increased in the 250 mg/kg-bw/day group. A significantly increased incidence of multiple hepatocellular adenoma was observed in the 500 mg/kg-bw/day group. Significant positive trends in the incidences of adenoma or carcinoma (combined) of the cecum or colon and of hemangiosarcoma in all organs were observed. Based on these findings, NTP concluded there was *clear evidence of carcinogenic activity* of TBBPA in female rats (uterine epithelial tumors), *some evidence of carcinogenic activity* in male mice (hepatoblastoma), *equivocal evidence of carcinogenic activity* in male rats (testicular adenoma), and *no evidence of carcinogenic activity* in female mice.

Modes of Action

The mechanisms/modes of action (MOA) of TBBPA toxicity/carcinogenicity are not clearly understood. Based on the toxicokinetics data, it seems likely that the thyroid hormone level changes and liver effects (i.e., enlargement of hepatocytes, increased liver weight, and slight focal necrosis of hepatocytes) observed in adult rats and/or mice after exposures to high doses of TBBPA for extended periods are due to saturated metabolic capability and diminished elimination/excretion of the compound.

Both *in vitro* and *in vivo* assays have shown that TBBPA is not genotoxic (EURAR, 2006; NTP, 2013), so a direct genotoxic MOA can be eliminated from consideration for tumorigenesis. Uterine tumors can arise in response to endogenous estrogen overstimulation in aged rats that can be exacerbated by administration of exogenous chemicals by direct and indirect pathways (Alison et al., 1994; Lax, 2004). After binding directly to estrogen receptors (ER) in a cell, estrogen/estrogen agonists can turn on hormone-responsive genes that promote DNA synthesis and cell proliferation. Therefore, estrogen/estrogen agonists can act as tumor promoters by inducing cell proliferation with preexisting mutations leading to tumor formation.

The ER-binding activity of TBBPA has been investigated in a number of *in vitro* screening assays. Review of the overall weight-of-evidence from *in vitro* assays has indicated that there is no significant estrogenic potential for TBBPA (EU, 2006; Canada, 2013). A recent study in mice has shown that TBBPA was negative for estrogenic responses by both subcutaneous injection and oral route of exposure up to 1000 mg/kg-bw/day (Ohta et al., 2012). Since *in vitro* and *in vivo* studies showed that TBBPA has no significant estrogenic potential (*reviewed in*: EU, 2006; Canada, 2013; Ohta et al., 2012), the direct ER pathway may also be eliminated from consideration as the MOA for the induction of uterine tumors by TBBPA.

Conjugations (glucuronidation and sulfation) are the major biotransformation pathways for excretion of TBBPA in rats, and these pathways are shared by estrogen, and its potentially genotoxic catechol metabolite (Raftogianis et al., 2000). Competition for glucuronosyltransferases and/or sulfotransferases by TBBPA could indirectly result in higher serum levels of estrogen and increased formation of estrogen-derived reactive radicals following exposure to high concentrations of TBBPA (NTP, 2013). Reactive oxygen radicals produced after metabolism of estrogen can cause DNA damages and if unrepaired, these mutations may lead to tumor formation.

The *Tp53* tumor suppression gene is responsible for cell cycle checkpoint maintenance and genomic stability, and loss of cell cycle checkpoint control due to inactivation of *Tp53* by mutations can result in the development of various tumor types in rodents and humans (Blagosklonny, 2000; Muller and Vousden, 2013). In the NTP bioassays (NTP, 2013), a statistically significant increase in the incidence of mutations of the *Tp53* tumor suppression gene was noted in uterine adenocarcinomas from TBBPA-treated rats (60%) compared to spontaneous tumors from control rats (20%). It is possible that increased incidence of *Tp53* mutations may be caused by the reactive oxygen radicals or metabolites produced after metabolism of the high levels of circulating estrogens due to competitive inhibition of estrogen conjugations (glucuronidation and/or sulfation) and elimination/secretion by high doses of TBBPA. However, with the exception of multiple mutations in uterine tumors of two rats treated with TBBPA, there was no difference between the mutation spectra of spontaneous tumors and those from TBBPA-treated rats. Therefore, it is more likely that increased circulating estrogen levels due to competitive inhibition of estrogen conjugations by high doses of TBBPA may cause uterine tumors by inducing cell proliferation with preexisted *Tp53* mutations.

There are two types of uterine carcinoma with respect to histology, MOA, and molecular genetic pathways. Type I carcinoma (the most common type), is associated with expression of ER, estrogen overstimulation, endometrial hyperplasia, and *Tp53* mutations in only about 10–20% of the carcinoma. Type II carcinoma is unrelated to estrogen and frequent lack of estrogen receptor activities. It is associated with atrophic endomethrium and *Tp53* mutations (90%) are the most frequent genetic alterations (Lax, 2004). In light of high incidence of *Tp53* mutations in the uterine adenocarcinomas, and lack of ER binding potential and increased levels of circulating estrogen in the TBBPA-treated rats, it is likely that TBBPA-induced type II carcinoma in the rats (rather than type I carcinoma). This notion is supported by the findings that there was no increase in endometrial hyperplasia (which is associated with type I carcinoma) in all dosed groups of female rats in the NTP bioassays when the original and residual tissues evaluations were combined; instead a new atypical hyperplasia was identified (NTP, 2013).

Uterine tumors can also be induced by dopamine receptor agonists in the rat through another indirect estrogen pathway. Dopamine serves as a key regulator of serum prolactin by activating dopamine receptors on the pituitary to inhibit the secretion of prolactin. Chronic administration of dopamine receptor agonists to rats can result in decreased serum prolactin levels after competitively binding to the dopamine receptors on the pituitary, leading to estrogen dominance due to increase of estrogen synthesis after luteolysis of the persistent corpora lutea and the formation of new follicles. This estrogen dominance then leads to estrogen activity (i.e., expression of hormone-responsive genes that promote cell proliferation) and oxidative stress, which may induce hyperplasia and tumors of the uterine. This carcinogenic effect is unique to rodents and has not been demonstrated in other species, including humans since prolactin is the luteotrophic hormone in rodents but not in primates (Neumann, 1991; Alison et al., 1994). In an *in vitro* study investigating the effects of TBBPA on the uptake of neurotransmitters into isolated rat brain synaptosomes, a concentration-dependent inhibition of dopamine uptake showing a mixed competitive and non-competitive mode of inhibition has been reported (Mariussen and Fonnum, 2003). However, it is unknown if TBBPA can also compete with dopamine in receptor binding and be a dopamine receptor agonist. Further research is warranted to investigate this MOA for TBBPA-induced uterine tumors.

SOURCES OF EXPOSURE

Occupational

Workers may be exposed to TBBPA through inhalation and dermal contact in the workplace. However, absorption of TBBPA by inhalation and dermal route of exposure is low based on its physicochemical properties. The primary opportunities for exposure are activities that generate dust or through fumes from heated materials. TBBPA is not combustable. It will, however, decompose at temperatures above 284 °C and generate corrosive fumes of hydrogen bromide.

Environmental

Tetrabromobisphenol A has been detected in virtually all environmental media. The relative contribution of TBBPA to environmental exposures depends on the end-use application. For example, when TBBPA is used as a reactive flame retardant, exposure is limited to residual quantities of the unreacted compound. Additive uses, such as those in ABS resins, represent the greatest opportunity for release due to migration or abrasion. A summary of reported levels of TBBPA in air, water, and sewage sludge is provided below.

Indoor air levels of TBBPA range from 1.6 to 8.0×10^{-7} mg/m^3 in residences in the United Kingdom and Japan, respectively (Canada, 2013). Comparable levels were reported in commercial buildings. For example, Batterman et al. (2010) measured TBBPA levels in three out of 10 buildings in Michigan, USA. The levels in two university buildings ranged from 1.2 to 2.3×10^{-8} mg/m^3. The highest level of TBBPA was measured in one computer server room/office at 8.6×10^{-8} mg/m^3.

Surface water levels of TBBPA have been measured in a variety of geographical locations, including Japan, Germany, France, and the United Kingdom (Canada, 2013). In Japan, TBBPA was detected in one sample at 5×10^{-2} µg/l out of 297 taken between 1977 and 2000. In 2000, TBBPA was detected in seven of thirty samples in German surface water downstream of a local industrial discharge at 2.04×10^{-2} µg/l. Finally, the levels of TBBPA in France and the United Kingdom ranged from <3.0 to 3.0×10^{-3} µg/l, respectively.

The levels of TBBPA in sewage sludge from 22 municipal treatment plants in Sweden ranged from non detects up to 220 ng/g wet weight(ww) (Oberg et al., 2002). The Swedish Environmental Protection Agency performed a similar study at 50 sewage treatment plants across Sweden and reported levels of TBBPA that ranged from <4 µg/kg^{-1} dry weight up to 180 µg/kg^{-1} dry weight (Law et al., 2006).

INDUSTRIAL HYGIENE

Tetrabromobisphenol A is manufactured as a white crystalline powder. Dermal absorption is a low concern for this chemical, as evidenced by its physicochemical properties and low toxicity in dermal studies. After inhalation, most of the inhaled particles will be deposited in the upper respiratory tract and some of which are expected to be swallowed into the gastrointestinal tract. The exposure controls and personal protection equipment (PPE) recommended in the work place are designed to minimize inhalation. Ventilation requirements include maintaining airborne levels below 3 mg/m^3 for respirable particles and below 10 mg/m^3 for inhalable particles. PPE include a dust respirator, protective gloves, chemical safety gloves, and body covering clothes and boots.

RISK ASSESSMENTS

The human health and environmental risks from TBBPA exposures have been evaluated by several agencies. Environment Canada and Health Canada concluded "there is currently low risk of harm to organisms or the broader integrity of the environment from TBBPA . . ." (Canada, 2013). Based on the adequacies of the margins of exposure between critical effect levels and upper bound estimates of exposure to TBBPA, Health Canada also concluded that *"TBBPA is a substance that is not entering the environment in a quantity or concentration or under conditions that constitute or may constitute a danger in Canada to human life or health"* (Canada, 2012).

A recent group of international experts has evaluated the mammary toxicology and potential risk of TBBPA (Colnot et al., 2014). Like the EU and Canada assessments, it was concluded that TBBPA has a low potential for systemic or reproductive/developmental toxicity. Review of the endocrine-mediated effects of TBBPA led to the conclusion that TBBPA should not be considered an "endocrine disruptor" in accordance with internationally accepted definitions. A health concern due to present levels of exposure of the general population with TBBPA is not supported by the available data.

In its draft environmental report, the European Union concluded "[t]here is a need for further information and/or testing" on TBBPA (EURAR, 2007). This conclusion was based on concerns over TBBPA's degradation to bisphenol A in anaerobic sewage sludge treatment processes and land application of the sludge. Though no environmental risks were identified in the EURAR on bisphenol A, additional testing on bisphenol A with an aquatic snail species formed the basis for this conclusion (EURAR, 2007). Additional environmental concerns were expressed for additive uses of TBBPA in ABS resin compounding sites. In the final human health report, the European Union concluded that "[n]o health effects of concern have been identified for [TBBPA]" (EURAR, 2006). The European Food Safety Authority concluded that the following exposures do not present a human health concern: dietary exposures in adults, breast feeding exposures in infants, and house dust exposures in young children (EFSA, 2011).

REFERENCES

Alison, R.H., Capen, C.C., and Prentice, D.E. (1994) Neoplastic lesion of questionable significance to humans. *Toxicol. Pathol.* 22:179–186.

Albemarle (1999) *"Saytex® CP-2000 Flame Retardant."* Technical data sheet: 2.

BSEF (2012) *"TBBPA Factsheet Brominated Flame Retardant October 2012; Tetrabromobisphenol A for Printed Circuit Boards and ABS plastics."* Bromine Science and Environmental Forum (BSEF): 4.

Batterman, S., Godwin, C., Chernyak, S., Jia, C., and Charles, S. (2010) Brominated flame retardants in offices in Michigan, USA. *Environ. Int.* 36(6):548–556.

Blagosklonny, M.V. (2000) p53 from complexity to simplicity: mutant p53 stabilization, gain-of-function, and dominant-negative effect. *FASEB J.* 14:1901–1907.

Canada (2013) *"Screening Assessment Report; Phenol, 4,4'-(1-methylethyllidene) bis[2,6-dibromo-, Chemical Abstracts Service Registry Number 79-94-7; Ethanol, 2,2'-[(1-methylethylidene)bis[(2,6-dibromo-4,1-phenylene)oxy]]bis, Chemical Abstracts Service Registry Number 4162-45-2; Benzene, 1,1'-(1-methylethylidene)bis[3,5-dibromo-4-(2-propenyloxy)-, Chemical Abstracts Service Registry Number 25327-89-3."* Environment Canada, Health Canada: 178.

Colnot, T., S Kacew, and W. Dekant. 2014. Mammalian toxicology and human exposures to the flame retardant 2,2', 6,6'-tetrabromo-4,4'-isopropylidenediphenol (TBBPA): implication for risk assessment. *Arch. Toxicol.* 88:553–573.

DOC (2013) *Ecology's PBT Initiative—The PBT List*, State of Washington: Department of Ecology.

ECHA (2013) *Registered Substances*, Available at http://echa.europa.eu/information-on-chemicals/registered-substances CASRN 79-94-7.

EFSA (2011) Scientific opinion on tetrabromobisphenol A (TBBPA) and its derivatives in food, EFSA Panel on Contaminants in the Food Chain (CONTAM), European Food Safety Authority (EFSA), Parma, Italy. *EFSA J.* 9(12):67.

EPA (2012) *Chemical Data Access Tool (CDAT)|US Environmental Protection Agency*. Available at http://java.epa.gov/oppt_chemical_search/ CASRN 79-94-7.

EPA (2013) *Persistent Bioaccumulative Toxic (PBT) Chemicals Covered by the TRI Program|Toxic Release Inventory (TRI) Program*, Washington, DC: U.S. Environmental Protection Agency.

EURAR (2006) *2,2',6,6'-tetrabromo-4,4'-isopropylidenediphenol (tetrabromobisphenol-A or TBBP-A) Part II—human health. European Union Risk Assessment Report* 63:170.

EURAR (2007) *"Risk Assessment Report,2',6,6'-Tetrabromo-4,4'-Isopropylidene diphenol (Tetrabromobisphenol-A), CAS Number: 79-94-7, EINECS Number: 201-236-9, Final Environmental Draft of June 2007."* 601.

Lax, S.F. (2004) Molecular genetic pathways in various types of endometrial carcinoma: from a phenotypical to a molecular-based classification. *Virchows Arch.* 444:213–223.

Law, R.J., Allchin, C.R., de Boer, J., Covanci, A., Herzke, D., Lepom, P., Morris, S., Tronczynski, J., and de Wit, C.A. (2006) Levels and trends of brominated flame retardants in the European environment. *Chemosphere* 64(2):187–208.

Mariussen, E. and Fonnum, F. (2003) The effect of brominated flame retardants on neurotransmitter uptake into rat brain synaptosomes and vesicles. *Neurochem. Int.* 43:533–542.

Muller, P.A. and Vousden, K.H. (2013) p53 mutations in cancer. *Nat. Cell Biol.* 15:2–8.

Neumann, F. (1991) Early indicators for carcinogenesis in sex-hormone-sensitive organs. *Mutat. Res.* 248:341–356.

NTP (2013) *NTP Technical Report. Toxicological studies of tetrabromobisphenol A (CAS NO. 79-94-7) in F344/NTac rats and B6C3F1/N mice and toxicology and carcinogenesis study of tetrabromobisphenol A in WISTAR HAN [Crl:WI(Han)] rats and B6C3F1/N mice (Gavage studies), NTP TR 587. National Toxicology Program, Research Triangle Park, NC.*

Oberg, K., Warman, K., and Oberg, T. (2002) Distribution and levels of brominated flame retardants in sewage sludge. *Chemosphere* 48(8):805–809.

Ohta, R., Takagi, A., Ohmukai, H., Marumo, H., Ono, A., Matsushima, Y., Inoue, T., Ono, H., and Kanno, J. (2012) Ovariectomized mouse uterotrophic assay of 36 chemicals. *J. Toxicol. Sci.* 37(5):879–889.

Raftogianis, R., Creveling, C., Weinshilboum, R., and Weisz, J. (2000) Estrogen metabolism by conjugation. *J. Natl. Cancer Inst. Monogr.* 27:113–124.

Sjodin, A., Patterson, D.G. Jr., and Bergman, A. (2003) A review on human exposure to brominated flame retardants—particularly polybrominated diphenyl ethers. *Environ. Int.* 29(6):829–839.

83

TRIS[2-CHLORO-1-(CHLOROMETHYL)ETHYL] PHOSPHATE (TDCPP)

MAREK BANASIK

First aid: Eyes: flush with copious amounts of water for at least 20 min: get medical attention immediately; skin: remove contaminated clothing, wash skin thoroughly with mild soap and plenty of water for at least 15 min—get medical attention if irritation occurs; inhalation: remove to fresh air, keep person quiet and warm, apply artificial respiration if necessary and get medical attention immediately; ingestion: if swallowed, wash mouth thoroughly with plenty of water—get medical attention immediately.

Occupational exposure limits: Derived no effect level (DNEL) (systemic effects, long-term inhalation): $0.327 \, mg/m^3$; DNEL (systemic effects, long-term dermal): 0.047 mg/kg body weight (bw)/day (ECHA, 2013).

Reference values: DNEL (systemic effects, long-term inhalation): $0.058 \, mg/m^3$; DNEL (systemic effects, long-term dermal): 0.017 mg/kg bw/day; DNEL (systemic effects, long-term oral): 0.017 mg/kg bw/day (ECHA, 2013).

Hazard statements: H351: suspected of causing cancer; H411: toxic to aquatic life with long lasting effects.

Precautionary statements: P202: do not handle until all safety precautions have been read and understood; P273: avoid release to the environment; P308 + 313: if exposed or concerned, get medical advice/attention; P391: collect spillage; P405: store locked up; P501: dispose of contents/containers to . . . (*in accordance with local/regional/national/international regulation*).

Risk phrases: R40: limited evidence of carcinogenic effect; R51/53: toxic to aquatic organisms, may cause long-term adverse effects in the aquatic environment.

Safety phrases: S2: keep out of the reach of children; S20/21: when using do not eat, drink or smoke; S36/37: wear suitable protective clothing and gloves; S61: avoid release to the environment/refer to special instructions/safety data sheet.

BACKGROUND AND USES

Tris[2-chloro-1-(chloromethyl)ethyl] phosphate [TDCPP; Chemical Abstracts Service Registry Number (CASRN): 13674-87-8] is a halogenated phosphorus flame retardant used in a variety of sectors, including manufacturing of paints/coatings, furniture and related products, building/construction materials, fabrics/textiles/leather products, and foam seating and bedding products (EPA, 2012).

It is used extensively as an additive to flexible polyurethane foams (PUFs). Its end uses include molded automotive seating foam (e.g., seat cushions and headrests), slabstock foam in furniture, automotive fabric lining, and car roofing (ECHA, 2013). It is a high production volume chemical. In the United States (US), the 2012 national production volume, including manufacturing and importing, ranged from 10 to 50 million pounds (ca. 4,536-22,680 tonnes) (EPA, 2012). In the European Union (EU), the production volume is between *ca.* 2.2 and *ca.* 22 million pounds (1,000–10,000 tonnes) per annum (ECHA, 2013).

It has undergone extensive toxicological testing to evaluate potential hazards to aquatic- and terrestrial-dwelling organisms and to humans through subchronic and chronic experimental animal studies. The EU conducted a definitive environmental and human health risk assessment on TDCPP

Hamilton & Hardy's Industrial Toxicology, Sixth Edition. Edited by Raymond D. Harbison, Marie M. Bourgeois, and Giffe T. Johnson.
© 2015 John Wiley & Sons, Inc. Published 2015 by John Wiley & Sons, Inc.

TABLE 83.1   Chemical and Physical Properties of TDCPP

Characteristic	Information
Chemical structure	
Chemical name	Tris[2-chloro-1-(chloromethyl)ethyl] phosphate
Synonyms/trade names	Amgard TDCP, FR2, Fyrol FR-2, TDCP, TDCPP, Tolgard TDCP, Tolgard TDCP MK1
Chemical formula	$C_9H_{15}Cl_6O_4P$
CASRN	13674-87-8
Molecular weight	430.91 g/mol
Physical state	Colorless liquid
Melting point	$<-20\,°C$
Vapor pressure	5.6×10^{-6} Pa at 25 °C
Relative density	1.51 at 20 °C
Water solubility	18.1 mg/l at 20 °C
Octanol/water partition coefficient (log K_{OW})	3.69 ± 0.36 at 20 °C
Henry's law constant	1.24×10^{-4} Pa m^3/mol at 25 °C

Source: ECHA (2013); EURAR (2008).

in 2008 (EURAR, 2008). The sections that follow provide an overview of this information and the EU's risk conclusions.

PHYSICAL AND CHEMICAL PROPERTIES

Tris[2-chloro-1-(chloromethyl)ethyl] phosphate is a clear colorless viscous liquid with a relatively low molecular weight, low water solubility, and low lipophilicity (as indicated by log K_{OW}). A summary of TDCPP's physicochemical properties is provided in Table 83.1.

ENVIRONMENTAL FATE AND BIOACCUMULATION

A series of studies have been performed to evaluate the environmental fate and bioaccumulation potential of TDCPP. Based on the data provided in Table 83.2, the EU concluded that TDCPP is expected to be persistent in water, sediment, sewage sludge, and soil. When emitted to the environment, the EU concluded that TDCPP will likely adsorb to particulate matter, based on its low volatility and relatively high adsorption coefficient. The Henry's law constant suggests that TDCPP will preferentially partition to water, rather than air. When assessing TDCPP's potential for persistence or bioaccumulation, the EU determined that TDCPP can potentially be persistent or very persistent and that it does not meet the criteria for bioaccumulative or very bioaccumulative compounds.

ECOTOXICOLOGY

The aquatic and terrestrial toxicity of TDCPP has been evaluated in a series of guideline studies. A summary of the studies selected by the EU as the most reliable for use in risk assessment is provided in Table 83.3. Based on these data, the EU calculated the following compartmental predicted no effect concentrations: freshwater (0.01 mg/l), freshwater sediment [0.18 mg/kg wet weight (ww)], wastewater

TABLE 83.2   Environmental Fate and Bioaccumulation Data on TDCPP

End Point	Protocol	Result
Hydrolysis	OECD[a] 111	$t_{1/2}$[b] at pH 9: >120 days at 20 °C
Photodegradation	Model estimate (AOPWIN™ program k)[c]	Reaction with hydroxyl radicals $= 18.1 \times 10^{-12}$ cm^3/molecule/s
Ready biodegradability	Modified Sturm test	Not readily biodegradable
Anaerobic biodegradation	None	<30 mg/l of chloride (limit of quantification) after 60 days
Degradation in soil	OECD 307	<6% degradation
Adsorption to three soils, sediment, and sludge	OECD 106	K_{OC}[d] $= 1{,}780$ (range 1,540–2,010), log K_{OC}[e] $= 3.25$
Adsorption to soil	Method C.19 of 2001/59/EC	Log $K_{OC} = 4.09 \pm 0.29$
Bioconcentration factor (BCF) in fish	None	BCF$_{fish}$ = 31–59, 45 (arithmetic mean) (continuous flow through system), 50–89 (static system)
BCF in earthworms	Model estimate	BCF$_{earthworm}$ = 59.61 (range 26.50–135.48)

Source: EURAR (2008).

[a]Organisation for Economic Co-operation and Development.
[b]Half-life.
[c]Atmospheric oxidation program for Microsoft Windows.
[d]Organic carbon–water partition coefficient.
[e]Logarithmic organic carbon–water partition coefficient.

TABLE 83.3   Aquatic and Terrestrial Toxicity Data on TDCPP

Toxicity to Aquatic Vertebrates

Test Species	Test Protocol	End Point and Exposure Duration	Result, mg/l
Rainbow trout (*Oncorhynchus mykiss*)	OECD 203 (semi-static test)	96-h[a] NOEC[b]	0.56 (N)[c]
		24-h LC_{50}[d]	1.8 (N)
		48-h LC_{50}	1.5 (N)
		72-h LC_{50}	1.3 (N)
		96-h LC_{50}	1.1 (N)

Toxicity to Aquatic Invertebrates

Test Species	Test Protocol	End Point and Exposure Duration	Result, mg/l
Cladoceran (*Daphnia magna*)	OECD 202; OPPTS 850.1010	48-h NOEC	1.6 (M)[e]
		24-h EC_{50}[f]	>5.1 (M)
		48-h EC_{50}	3.8 (M)
Cladoceran (*Daphnia magna*)	OECD 211 (semi-static test)	21-day LOEC[g] (reprod.[h])	1.0 (N)
		21-day LOEC (reprod.)	0.5 (N)

Toxicity to Aquatic Plants

Test Species	Test Protocol	End Point and Exposure Duration	Result, mg/l
Freshwater alga (*Pseudokirchneriella subcapitata*)	OECD 201; EEC[i] Dir 92/69/ EEC, Method C3	72-h ErC_{10}[j] (growth rate)	2.3 (M)
		72-h EbC_{10}[k] (biomass)	1.2 (M)
		72-h ErC_{50}[l] (growth rate)	4.6 (M)
		72-h EbC_{50}[m] (biomass)	2.8 (M)
		NOEC	≥1.2 (M)
Freshwater alga (*Scenedesmus subspicatus*)	OECD 201 (limit test)	72-h NOEC	≥10 (N)
		72-h ErC_{50} (growth rate)	>10 (N)
		72-h EbC_{50} (biomass)	>10 (N)

Toxicity to Microorganisms

Test Species	Test Protocol	End Point	Result, mg/l
Activated sludge	OECD 209	IC_{50}[n]	>10,000 (N)

Toxicity to Sediment-Dwelling Organisms

Test Species	Test Protocol	End Point and Exposure Duration	Result, mg/kg dw[o]
Midge (*Chironomus riparius*)	OECD 218	Day 0–3 NOEC (development)	8.8 (M)
		28-day NOEC (development)	3.9 (M)

(continued)

Toxicity to Sediment-Dwelling Organisms

Test Species	Test Protocol	End Point and Exposure Duration	Result, mg/kg dw[o]
Amphipod (*Hyallela azteca*)	ASTM[p] E 1706-00; OPPTS 850.1735	28-day LOEC (development)	8.5 (M)
		28-day EC_{50} (emergence)	16 (M)
		28-day EC_{50} (survival)	>71 (M)
		28-day NOEC (survival/reprod.)	71 (M)
		28-day LOEC (survival/reprod.)	>71 (M)
Oligochaete (*Lumbriculus variegatus*)	ASTM E 1706-00; OPPTS 850.1735	28-day EC_{50} (survival)	>60 (M)
		28-day NOEC (survival/reprod.)	60 (M)
		28-day LOEC (survival/reprod.)	>60 (M)

Toxicity to Terrestrial Invertebrates

Test Species	Test Protocol	End Point and Exposure Duration	Result, mg/kg dw
Earthworm (*Eisenia foetida*)	OECD 207	14-day NOEC	100 (N)
		7-day LC_{50}	230 (N)
		14-day LC_{50}	130 (N)
Earthworm (*Eisenia foetida*)	OECD draft guideline (January 2000)	28-day LC_{50} (adult mortality)	>100 (N)
		28-day NOEC (biomass)	100 (N)
		28-day LOEC (biomass)	>100 (N)
		57-day NOEC (reprod.)	9.6 (N)
		57-day LOEC (reprod.)	13 (N)
		57-day EC_{50} (reprod.)	67 (N)

Toxicity to Terrestrial Plants

Test Species	Test Protocol	End Point and Exposure Duration	Result, mg/kg dw
Wheat (*Triticum aestivum*)	OECD 208	NOEC (emergence)	≥202 (N)
		19-day NOEC (growth, ww)	31.5 (N)
		19-day NOEC (growth)	25.1 (N)
Red clover (*Trifolium pratense*)	OECD 208	NOEC (emergence)	≥202 (N)
		19-day NOEC (growth, ww)	28.7 (N)
		19-day NOEC (growth)	85.3
Mustard (*Sinapis alba*)	OECD 208	NOEC (emergence)	19.3
		19-day NOEC (growth, ww)	38.7
		19-day NOEC (growth)	≥202

(continued)

TABLE 83.3 *(Continued)*

Toxicity to Soil Microorganisms

Test Species	Test Protocol	End Point and Exposure Duration	Result, mg/kg
Nitrifying microorganisms in sandy loam soil	OECD 216	28-day NOEC (ww)^q 28-day NOEC (dw)	≥128 145

Source: EURAR (2008).

^{a}Hour(s).
^{b}No observed effect concentration.
^{c}Nominal concentration.
^{d}Median lethal concentration.
^{e}Measured concentration.
^{f}Median effect concentration.
^{g}Lowest observed effect concentration.
^{h}Reproduction.
^{i}European Economic Community.
^{j}Effect concentration measured as 10% reduction in growth rate in algae tests.
^{k}Effect concentration measured as 10% reduction in biomass growth in algae tests.
^{l}Effect concentration measured as 50% reduction in growth rate in algae tests.
^{m}Effect concentration measured as 50% reduction in biomass growth in algae tests.
^{n}Median inhibitory concentration.
^{o}Dry weight.
^{p}American Society for Testing and Materials.
^{q}Microorganism activity based on nitrate concentration.

treatment plant microorganisms (≥10 mg/l), marine water (0.001 mg/l, extrapolated from freshwater), marine sediment (0.036 mg/kg ww), and soil (0.29 mg/kg ww).

MAMMALIAN TOXICOLOGY

Acute and repeated dose toxicity studies on TDCPP have shown that it presents a low concern for acute adverse health effects from the inhalation, oral, and dermal routes. Repeated dose toxicity studies have only been performed by the oral route. These studies support that TDCPP presents a high concern for subchronic and chronic adverse health effects. A summary of the available acute and repeated dose toxicity studies utilized by the EU when assessing the potential risks from exposures to TDCPP is provided in the following sections.

Acute Studies

Tris[2-chloro-1-(chloromethyl)ethyl] phosphate shows low acute toxicity by the inhalation, oral, and dermal routes of exposure (Table 83.4). Skin and eye irritation studies produced minimal dermal and eye irritation, with no evidence of

TABLE 83.4 **Acute Toxicity Studies on TDCPP**

Inhalation

Test Species (n)	Test Protocol	End Point and Exposure Duration	Result, mg/l
Sprague-Dawley ♂ and ♀ rats (n = 5/sex/group)	OECD 403	4-h LC$_{50}$	>5.22 (M)

Oral

Test Species	Test Protocol	End Point	Result, mg/kg bw
Sprague-Dawley ♂ and ♀ rats (n = 5/sex/group)	OECD 401	♂ LD$_{50}$^{a} ♀ LD$_{50}$ ♂ and ♀ combined LD$_{50}$	2,236 2,489 2,359
Sprague-Dawley ♂ and ♀ rats (n = 5/sex)	OECD 401 (single dose)	♂ and ♀ combined LD$_{50}$	>2,000

Dermal

Test Species	Test Protocol	End Point	Result mg/kg bw
Sprague Dawley ♂ and ♀ rats (n = 5/sex)	OECD 402 (single dose)	♂ and ♀ combined LD$_{50}$	>2,000

Source: EURAR (2008).
^{a}Median lethal dose.

corrosivity (EURAR, 2008). No respiratory irritation studies have been performed; however, in the 4-h acute inhalation study, no signs of irritation were observed at the limit concentration of 5.22 mg/l.

Repeated Dose Studies

No repeated dose studies on TDCPP are available by the inhalation or dermal routes of exposure. Several repeated dose studies have been performed using the oral route of exposure. The EU critically evaluated these studies and utilized two developmental toxicity studies and a 2-year carcinogenicity study to evaluate the potential risks of TDCPP. A summary of these three studies is provided below.

In the first developmental toxicity study, 20 female Sprague-Dawley rats were administered TDCPP in a corn oil vehicle at dose levels of 0, 25, 100, or 400 mg/kg bw/day from gestation days (GDs) 6 through 15. Dams were observed daily, and bodyweight was measured on GDs 0, 6, 11, 15, and 19. On GD 19, all dams were sacrificed, and gross examinations were performed on dams and fetuses. Pregnancy rates, mean number of corpora lutea, implantation sites, and implantation efficiencies were unaffected by treatment. Several treatment-related effects were reported in the high dose group. The rate of resorptions was statistically significantly increased by 14.4% as compared to 6.7% in controls. Fetal viability index was statistically significantly decreased. Slightly lower mean fetal weight and crown-rump length was observed. Retarded skeletal development was considerable; however, it was unclear whether these effects were related to treatment or to maternal toxicity at the high dose. A no-observed-adverse-effect level (NOAEL) of 100 mg/kg bw/day was derived based on the increased number of resorptions and decreased fetal viability index in the high dose group.

In the second developmental toxicity study, 15–24 female Wistar rats were administered TDCPP in an olive oil vehicle at dose levels of 0, 25, 50, 100, 200, and 400 mg/kg bw/day from GDs 7 through 15. Postnatal functional tests were performed on surviving pups (date not stated). During gestation, 11 of 15 dams in the high dose group died and exhibited symptoms and signs of toxicity, including piloerection, salivation, and hematuria. Body weight gain and food consumption were significantly reduced in this group. A 17% reduction in body weight gain was observed in surviving dams. Maternal absolute kidney weights were increased by 8.7 and 35.5% in the 200 mg/kg bw/day and 400 mg/kg bw/day groups, respectively. Maternal relative kidney weights were increased by 12.2 and 65.3% in the 200 and 400 mg/kg bw/day groups, respectively. In the 400 mg/kg bw/day group, fetal death was significantly increased. Of the four surviving dams, one had total dead implants. Three had live fetuses. The number of live fetuses in all other treatment groups was comparable to controls. No treatment-related effects on skeletal formation were observed at any dose level. Postnatal functional tests were performed on surviving pups in the 0, 25, 50, 100, and 200 mg/kg bw/day groups. No changes were observed at any dose level in the following tests: open field, water maze, rota rod, inclined screen, pain reflex, and preyer's reflex examinations.

One long-term carcinogenicity study was completed on TDCPP in 1981; the complete details of the study are not available (Bio/dynamics, Inc., 1981). Freudenthal and Henrich (2000) subsequently published a summary of these data. Groups of 60 male and female Sprague-Dawley rats were fed TDCPP at dose levels of 0, 5, 20, or 80 mg/kg bw/day. An interim sacrifice was performed on 10 animals per sex per group.

A clear treatment-related effect on bodyweight was reported throughout the study. At study termination, the bodyweights of animals in the high dose group were decreased >20% of controls. At the interim and terminal sacrifice, absolute and relative liver weights in animals (both sexes) in the high dose group were increased. At terminal sacrifice, an increase in the number of foci of hepatocellular alterations and sinusoidal dilation was reported. No histological findings were reported in animals at interim sacrifice.

In the high dose groups (males and females), absolute and relative kidney weights were statistically significantly increased at the interim and terminal sacrifices. Histological findings at the interim sacrifice included discoloration and pitted surface irregularities in the mid and high dose males and in all treated females. At terminal sacrifice, an increased incidence of hyperplasia of the convoluted tubule epithelium was reported in males from all treatment groups and in females from the high dose group. Chronic nephropathy was increased in males in the mid and high dose groups and in females in the high dose group.

In high dose females, absolute thyroid weight was statistically significantly increased at terminal sacrifice. At the interim and terminal sacrifices, no changes in testes weights were observed in male rats at any dose level; however, gross findings in the mid and high dose groups at both sacrifice intervals included discoloration, masses/nodules, enlargement and flaccidity of the testes, and small seminal vesicles. Histological findings at the interim and terminal sacrifices included germinal epithelial atrophy with associated oligospermia in all dose groups. At the terminal sacrifice, the following findings were observed in all males: accumulation of amorphous eosinophilic material in the tubular lumens, sperm stasis, and periarteritis nodosa.

As shown in Table 83.5, the animals at terminal sacrifice had statistically significantly increased incidences of hepatocellular adenomas (high dose groups; male and female), hepatocellular carcinomas (high dose group; male only), hepatocellular adenomas/carcinomas (combined) (high dose groups; male and female), renal cortical adenomas (mid and high dose groups; male and female), adrenal cortical adenomas (high dose group; female only), and testicular interstitial cell tumors (mid and high dose groups; male only).

Based on the foregoing noncancer and cancer effects and the absence of positive results from *in vivo* mutagenicity studies, the EU concluded that TDCPP may be assumed to be a "threshold" carcinogen (EURAR, 2008).

The EU identified two separate NOAELs for performing its risk characterization. A NOAEL of 100 mg/kg bw/day was selected for developmental and maternal toxicity, based on the first developmental toxicity study. A lowest-observed-adverse-effect level (LOAEL) of 5 mg/kg bw/day was selected, based on preneoplastic effects observed in the kidney and testis.

SOURCES OF EXPOSURE

Occupational

The primary exposure concerns for workers include those that may occur by the inhalation and dermal routes. Tris[2-chloro-1-(chloromethyl)ethyl] phosphate is stable under normal working conditions; however, when heated to decomposition, it may release oxides of carbon and phosphorus and hydrogen

TABLE 83.5   Tumor Incidence in Sprague-Dawley Rats Fed TDCPP in a 2-Year Carcinogenicity Study

Organ	Tumor Type	Timed Evaluation, Months	Sex	Dose Group, mg/kg bw/day			
				0	5	20	80
Liver	Adenoma[a]	12	♂	0/15 (0%)	0/12 (0%)	0/13 (0%)	3/14 (21%)
			♀	0/11 (0%)	0/13 (0%)	0/9 (0%)	1/10 (10%)
		24	♂	2/45 (4%)	7/48 (15%)	1/48 (2%)	13/46[b] (28%)
			♀	1/49 (2%)	1/47 (2%)	4/46 (9%)	8/50[b] (16%)
	Carcinoma	12	♂	0/15 (0%)	0/12 (0%)	0/13 (0%)	0/14 (0%)
			♀	0/11 (0%)	0/13 (0%)	0/9 (0%)	0/10 (0%)
		24	♂	1/45 (2%)	2/48 (4%)	3/48 (6%)	7/46 (15%)
			♀	0/49 (0%)	2/47 (4%)	2/46 (4%)	4/50 (8%)
	Adenoma/carcinoma (combined)	12	♂	0/15 (0%)	0/12 (0%)	0/13 (0%)	3/14 (21%)
			♀	0/11 (0%)	0/13 (0%)	0/9 (0%)	1/10 (10%)
		24	♂	3/45 (7%)	9/48 (19%)	4/48 (8%)	20/46 (43%)
			♀	1/49 (2%)	3/47 (6%)	6/46 (13%)	12/50 (24%)
Kidney	Cortical adenoma	12	♂	0/15 (0%)	0/12 (0%)	0/13 (0%)	0/13 (0%)
			♀	0/11 (0%)	0/13 (0%)	0/9 (0%)	0/10 (0%)
		24	♂	1/45 (2%)	3/49 (6%)	9/48[b] (19%)	32/46[b] (70%)
			♀	0/49 (0%)	1/48 (2%)	8/48[b] (17%)	29/50[b] (58%)
Adrenal	Cortical adenoma	12	♂	0/15 (0%)	–[c]	–	2/13 (15%)
			♀	5/11 (45%)	–	–	1/10 (10%)
		24	♂	5/44 (11%)	3/14 (21%)	5/16 (31%)	3/44 (7%)
			♀	8/48 (17%)	5/27 (19%)	2/33 (6%)	19/49[b] (39%)
Testes	Interstitial cell tumor	12	♂	0/14 (0%)	0/12 (0%)	3/13 (23%)	3/11 (27%)
		24		7/43 (16%)	8/48 (17%)	23/47[b] (49%)	36/45[b] (80%)

[a]Note although these liver lesions are listed as adenoma, the pathologist from the Bio/dynamics, Inc. (1981) study described these lesions as "neoplastic nodules."
[b]Statistical significance ($p < 0.05$), as reported in the EURAR (2008).
[c]Animals not evaluated at 12 months.

chloride. Measured data are available on typical exposures encountered in the workplace. The EU utilized the values in Table 83.6 when assessing risks to workers and included reasonable worse case (RWC) exposures in its assessment.

Environmental

Environmental releases of TDCPP may occur during manufacture and processing, from finished articles, and from end-of-life disposal. TDCPP has been detected in virtually all environmental media, including dust, sewage sludge, groundwater, drinking water, soil, and sediment. The end uses of TDCPP,

primarily as a flame retardant foam additive, are amendable to migration from the product; therefore, its ubiquitous presence in the environment is expected.

INDUSTRIAL HYGIENE

It is manufactured as a colorless liquid. The personal protection equipment recommended in the work place is designed to minimize eye, skin, and respiratory exposures (ECHA, 2013). Chemical goggles are recommended for eye protection. Suitable protective clothing (unspecified) is recommended to

TABLE 83.6   Occupational Exposure Levels to TDCPP

Exposure Scenario	Inhalation Exposure, μg/m³		Dermal Exposure, mg/cm²/day		Dermal Exposure Area, cm²
	Typical	RWC	Typical	RWC	
Manufacture of TDCPP	2.8	5.6	5×10^{-2}	0.1	210
Manufacture of flexible PUF	0.62	5.1	2×10^{-3}	7×10^{-2}	420
Manufacture of molded foam	0.63	4.8	1.5×10^{-3}	7.5×10^{-2}	420
Cutting flexible PUF	1.9	4.1	9.8×10^{-4}	7.1×10^{-3}	420
Production of foam granules and rebonded foam	0.59	4.6	5.5×10^{-4}	1.7×10^{-3}	420
Manufacture of automotive parts	1.9	4.6	9.8×10^{-4}	7.1×10^{-3}	420

Source: EURAR (2008).

avoid skin exposures. Neoprene rubber gloves are recommended for hand protection. A respirator fitted with high efficiency (dust, fumes, and mist) filter cartridges is recommended to reduce inhalation exposures.

RISK ASSESSMENTS

The US Consumer Product Safety Commission (CPSC) and the EU completed risk assessments on TDCPP. In its screening-level assessment, CPSC assessed cancer risks for adults and children from inhalation exposures to TDCPP released from furniture foam and covered fabrics (Babich, 2006). It identified lifetime population risks in adults of 300 excess cancers per million adults. In children during the first 2 years of life, CPSC identified 20 excess cancers per million children.

The EU assessed the potential risks from exposures to TDCPP in the environment and to workers, consumers, and from exposures to the general population from environmental sources (EURAR, 2008). For the environment, the EU concluded that no further testing/information or risk reduction measures were needed, beyond those currently being applied. For workers, the EU concluded that additional information and/or testing were warranted based on possible effects on female fertility. The same concerns about female fertility were expressed for consumers and for general population exposures from environmental sources. However, it was noted that the LOAEL of 5 mg/kg bw/day from the 2-year carcinogenicity study will likely be protective for

female reproductive effects for exposures occurring under these conditions.

REFERENCES

Babich, M.A. (2006) *CPSC Staff Preliminary Risk Assessment of Flame Retardant (FR) Chemicals in Upholstered Furniture Foam*. Bethesda, MD: US Consumer Product Safety Commission. Available at http://www.cpsc.gov//PageFiles/106736/ufurn2.pdf.

Bio/dynamics, Inc. (1981) *A two-year oral toxicity/carcinogenicity study on Fyrol FR-2 in rats (final report)*. Submitted to former Stauffer Chemical Company (Westport, CT, USA) by Bio/dynamics, Inc., project no. 77–2016.

ECHA (2013) *Registered Substances*. Helsinki, Finland: European Chemicals Agency. Available at http://echa.europa.eu/information-on-chemicals/registered-substances. CASRN: 13674-87-8.

EPA (2012) *Chemical Data Access Tool (CDAT)*. Washington, DC: Office of Pollution Prevention and Toxics, US Environmental Protection Agency. Available at http://java.epa.gov/oppt_chemical_search. CASRN: 13674-87-8.

EURAR (2008) *Tris[2-chloro-1-(chloromethyl)ethyl] phosphate (TDCP), CAS No: 13674-87-8, EINECS No: 237-159-2, Risk assessment*. European Union Risk Assessment Report, prepared by Ireland and UK in the frame of Council Regulation (EEC) No. 793/93. Available at https://echa.europa.eu/documents/10162/13630/trd_rar_ireland_tdcp_en.pdf.

Freudenthal, R.I. and Henrich, R.T. (2000) Chronic toxicity and carcinogenic potential of tris-(1,3-dichloro-2-propyl) phosphate in Sprague-Dawley rat. *Int. J. Toxicol.* 19:119–125.

84

TRIS(2-CHLORO-1-METHYLETHYL) PHOSPHATE (TCPP)

Marek Banasik

First aid: Eyes: Flush eyes with large amounts of water for at least 15 min, hold eyelids apart to ensure rinsing of the entire surface of the eye and lids with water—get medical attention immediately; skin: immediately remove and discard contaminated clothing and shoes/under a safety shower, wash all affected areas with plenty of soap and water for at least 15 min/do not attempt to neutralize with chemical agents; inhalation: remove victim to fresh air/if not breathing, give artificial respiration/if breathing is difficult, give oxygen—get medical attention; ingestion: get medical attention by calling a physician or poison control center immediately/do not induce vomiting unless directed to do so by the medical personnel/if vomiting occurs, keep head below hips to reduce the risk of aspiration

Occupational exposure limits: Derived no effect level (DNEL) (systemic effects, long-term inhalation): 5.82 mg/m^3; DNEL (systemic effects, long-term dermal): 2.08 mg/kg body weight (bw)/day (ECHA, 2013)

Reference values: DNEL (systemic effects, long-term inhalation): 1.46 mg/m^3; DNEL (systemic effects, long-term dermal): 1.04 mg/kg bw/day; DNEL (systemic effects, long-term oral): 0.52 mg/kg bw/day (ECHA, 2013)

Hazard statement: H302: harmful if swallowed

Precautionary statements: P264: wash . . . thoroughly after handling; P270: do not eat, drink or smoke when using this product; P301 + 312: if swallowed: call a poison center/doctor/. . ./ if you feel unwell; P330: rinse mouth; P501: dispose of contents/container to . . . (*in accordance with local/regional/national/international regulation*)

Risk phrase: R22: harmful if swallowed

Safety phrases: S46: if swallowed, seek medical advice immediately and show this container or label; S64: if swallowed, rinse mouth with water (only if the person is conscious); S29/35: do not empty into drains; dispose of this material and its container in a safe way

BACKGROUND AND USES

Tris(2-chloro-1-methylethyl) phosphate [TCPP; Chemical Abstracts Service Registry Number (CASRN): 13674-84-5] is a halogenated phosphorus flame retardant used in a variety of industrial sectors, including construction; plastics material and resin manufacturing; furniture and related product manufacturing; textiles, apparel, leather, paint, and coating manufacturing; and in the manufacture of other chemical products and preparations. Consumer uses of TCPP include foam seating and bedding products, building/construction materials not covered elsewhere, and adhesives and sealants (EPA, 2012).

It is commonly used as a flame retardant additive to rigid polyurethane foam (PUF) in panels and laminates for insulation applications. It is typically added to pentane-blown foam (15 parts by weight), which is used in applications such as roofing laminate. It is also used in polyurethane elastomers and in flexible PUFs when combined with melamine.

It is a high production volume chemical. In the United States (US), the 2012 national production volume, including manufacturing and importing, was 54,673,933 pounds (*ca.* 24,822 tonnes) (EPA, 2012). In the European Union (EU), the annual production volume is between *ca.* 2.2 and *ca.* 220.5 million pounds (1,000–100,000 tonnes) (ECHA, 2013).

It has undergone extensive toxicological testing to evaluate potential hazards to aquatic- and terrestrial-dwelling

Hamilton & Hardy's Industrial Toxicology, Sixth Edition. Edited by Raymond D. Harbison, Marie M. Bourgeois, and Giffe T. Johnson.
© 2015 John Wiley & Sons, Inc. Published 2015 by John Wiley & Sons, Inc.

TABLE 84.1 Chemical and Physical Properties of TCPP

Characteristic	Information
Chemical structure	
Chemical name	Tris(2-chloro-1-methylethyl) phosphate
Synonyms/trade names	TCIP, TCPP, TMCP
Chemical formula	$C_9H_{18}Cl_3O_4P$
CASRN	13674-84-5
Molecular weight	327.57 g/mol
Physical state	Clear colorless liquid
Melting point	$<-20\,°C$
Boiling point	$288\,°C$
Vapor pressure	1.4×10^{-3} Pa at $25\,°C$
Relative density	1.288 at $20\,°C$
Water solubility	1,080 mg/l at $20–25\,°C$
Octanol/water partition coefficient (log K_{OW})	2.68 ± 0.36
Henry's law constant	3.96×10^{-4} Pa m^3/mol at $25\,°C$

Source: ECHA (2013); EURAR (2008).

organisms and to humans through subchronic and reproduction/developmental experimental animal studies. The EU conducted a definitive environmental and human health risk assessment on TCPP in 2008 (EURAR, 2008). The

sections that follow provide an overview of this information and the EU's risk conclusions.

PHYSICAL AND CHEMICAL PROPERTIES

Tris(2-chloro-1-methylethyl) phosphate is clear colorless liquid with high water solubility and low lipophilicity (as indicated by log K_{OW}). This substance is manufactured as a reaction mixture, which contains four isomers. It is the primary isomer in the mixture at 50–85% weight/weight (w/w), followed by bis(1-chloro-2-propyl)-2-chloropropyl phosphate [15–40% (w/w); CASRN: 76025-08-6], bis(2-chloropropyl)-1-chloro-2-propyl phosphate [<15% (w/w); CASRN: 76649-15-5], and tris(2-chloropropyl) phosphate [<1% (w/w); CASRN: 6145-73-9] (EURAR, 2008). A summary of TCPP's physicochemical properties is provided in Table 84.1.

ENVIRONMENTAL FATE AND BIOACCUMULATION

A series of studies have been performed to evaluate the environmental fate and bioaccumulation potential of TCPP. Based on the data provided in Table 84.2, the EU concluded that TCPP is expected to be persistent in water but inherently biodegradable in sewage sludge. The Henry's law constant suggests that TCPP will preferentially partition to water, rather than air. The EU determined that TCPP can potentially be

TABLE 84.2 Environmental Fate and Bioaccumulation Data on TCPP

End Point	Protocol	Result
Hydrolysis	EC[a] method C10	$t_{1/2}$[b] at pH 4, 7, and 9: $\geq$365 days at $25\,°C$
Photodegradation	Model estimate (AOPWIN™ program k)[c]	Reaction with hydroxyl radicals = 44.76×10^{-12} cm^3/molecule/s; $t_{1/2}$ = 8.6 h[d]
28-Day ready biodegradability study preceded by an acclimation phase	USEPA[e] TSCA[f] 796.3100	Not biodegradable
64-Day inherent biodegradability test	OECD[g] 302A	Inherently biodegradable
84-Day prolonged closed bottle test	EC method C6 modified	Inherently biodegradable
Adsorption–desorption test	OECD 106	K_{OC}[h] = 174 l/kg
Bioconcentration factor (BCF) in fish	OECD 305C	BCF$_{fish}$ = 0.8–4.6 l/kg, (arithmetic mean: 2.7) (flow through system)
BCF in earthworms	Model estimate	BCF$_{earthworm}$ = 6.58 l/kg (range: 3.35–14.0)

Source: EURAR (2008).

[a]European Commission.
[b]Half-life.
[c]Atmospheric oxidation program for Microsoft Windows.
[d]Hour(s).
[e]US Environmental Protection Agency.
[f]Toxic Substances Control Act.
[g]Organisation for Economic Co-operation and Development.
[h]Organic carbon–water partition coefficient.

persistent or very persistent and that it does not meet the criteria for bioaccumulative or very bioaccumulative compounds.

ECOTOXICOLOGY

The toxicity of TCPP to aquatic and terrestrial organisms has been evaluated using standard test guideline methods.

Based on the data provided in Table 84.3, the EU derived the following predicted no effect concentrations: freshwater (0.64 mg/l), freshwater sediment [2.92 mg/kg wet weight (ww), equilibrium partitioning], wastewater treatment plant microorganisms (7.84 mg/l), seawater (0.064 mg/l, extrapolated from freshwater), marine sediment (0.292 mg/kg ww, extrapolated from freshwater), and soil (1.5 mg/kg ww) (EURAR, 2008).

TABLE 84.3   Aquatic and Terrestrial Toxicity Data on TCPP

Test Species	Test protocol	End Point and Exposure Duration	Result, mg/l
Toxicity to Aquatic Vertebrates			
Zebrafish (*Brachydanio rerio*)	Not stated (static test)	96-h LC_0[a]	32 (M)[b]
		24-h LC_{50}[c]	56 (M)
		48-h LC_{50}	56 (M)
		72-h LC_{50}	56 (M)
		96-h LC_{50}	56 (M)
Bluegill sunfish (*Lepomis macrochirus*)	Not stated (static test)	96-h NOEC[d]	6.3 (M)
		96-h LC_{50}	84 (M)
Fathead minnows (*Pimephales promelas*)	Not stated (static test)	96-h NOEC	6.6 (M)
		24-h LC_{50}	>51 (M)
		48-h LC_{50}	>51 (M)
		72-h LC_{50}	51 (M)
		96-h LC_{50}	51 (M)
Toxicity to Aquatic Invertebrates			
Cladoceran (*Daphnia magna*)	Not stated	48-h NOEC	33.5 (M)
		48-h EC_{50}[e]	131 (M)
Cladoceran (*D. magna*)	OECD 202	21-day NOEC (parent mortality)	32 (N)[f]
		21-day NOEC (reproduction)	32 (N)
		14-day EC_{50} (parent immobilization)	42 (N)
		21-day EC_{50} (parent immobilization)	40 (N)
		21-day EC_{50} (reproduction)	32–56 (N)
Toxicity to Aquatic Plants			
Freshwater alga (*Pseudokirchneriella subcapitata*)	OECD 201; Commission Directive 92/69/EEC, Method C3	72-h NOEC	13 (N)
		72-h ErC_{10}[g]	42 (N)
		72-h EbC_{10}[h]	14 (N)
		72-h ErC_{50}[i]	82 (N)
		72-h EbC_{50}[j]	33 (N)
Toxicity to Microorganisms			
Activated sludge	ISO[k] 8192	EC_{50}	784 (M)

(continued)

TABLE 84.3   *(Continued)*

Toxicity to Terrestrial Invertebrates			
Test Species	Test Protocol	End Point and Exposure Duration	Result, mg/kg dw[l]
Earthworm (*Eisenia foetida*)	OECD 207	14-day NOEC	32 (N)
		7-day LC$_{50}$	131 (N)
		14-day LC$_{50}$	97 (N)
Earthworm (*E. foetida*)	OECD draft guideline (January 2000)	28-day NOEC (mortality)	≥196 (N)
		28-day NOEC (biomass)	116 (N)
		28-day LOEC[m] (biomass)	151 (N)
		EC$_{50}$ (reproduction)	71 (N)
		56-day NOEC (reproduction)	53 (N)
		56-day LOEC (reproduction)	69 (N)

Toxicity to Terrestrial Plants			
Test Species	Test Protocol	End Point and Exposure Duration	Result, mg/kg dw
Wheat (*Triticum aestivum*)	OECD 208	NOEC (emergence)	≥98 (N)
		NOEC	22 (N)
Mustard (*Sinapis alba*)	OECD 208	NOEC (emergence)	30 (N)
		21-day NOEC (ww)	28 (N)
Lettuce (*Lactuca sativa*)	OECD 208	NOEC (emergence)	17 (N)
		21-day NOEC (ww)	18 (N)

Toxicity to Soil Microorganisms			
Test Species	Test Protocol	End Point and Exposure Duration	Result, mg/kg
Nitrifying microorganisms in sandy loam soil[o]	OECD 216	28-day NOEC (ww)[n]	≥128
		28-day NOEC (dw)	145

Source: EURAR (2008).

[a]Lethal concentration 0.
[b]Measured concentration.
[c]Median lethal concentration.
[d]No observed effect concentration.
[e]Median effect concentration.
[f]Nominal concentration.
[g]Effect concentration measured as 10% reduction in growth rate in algae tests.
[h]Effect concentration measured as 10% reduction in biomass growth in algae tests.
[i]Effect concentration measured as 50% reduction in growth rate in algae tests.
[j]Effect concentration measured as 50% reduction in biomass growth in algae tests.
[k]International Organisation for Standardisation.
[l]Dry weight.
[m]Lowest observed effect concentration.
[n]Read across from tris[2-chloro-1-(chloromethyl)ethyl] phosphate (TDCPP).
[o]Microorganism activity based on nitrate concentration.

MAMMALIAN TOXICOLOGY

Tris(2-chloro-1-methylethyl) phosphate is absorbed by all routes of exposure. The EU utilized the following absorption estimates when assessing the potential risks from TCPP: inhalation (100%), oral (80%), and dermal (40%). It has been evaluated in a series of guideline studies to assess its potential hazards from acute and subchronic exposures, including reproduction and development. The studies utilized by the EU for risk characterization are summarized below.

Acute Studies

The acute toxicity of TCPP has been evaluated in rats administered the test substance by the inhalation, oral,

TABLE 84.4   Acute Toxicity Studies on TCPP

		Inhalation	
Test Species (n)	End Point and Exposure Duration	Effect	Result, mg/l
Sprague-Dawley ♂ and ♀ rats (n = 5/sex/group)	4-h LC$_{50}$ (limit test; nose only)	• No abnormalities during a 14-day observation period • No effect on bodyweight • No abnormalities post mortem	>7[a]
Sprague-Dawley ♂ and ♀ rats (n = 10/sex/group)	4-h LC$_{50}$ (limit test; whole body)	• Mild lethargy and matted fur in both sexes • No lesions in ♂ rats at necropsy • 3/10 ♀ rats had reddened lungs	>4.6 (FPA)[b]
Sprague-Dawley ♂ and ♀ rats (n = 5/sex/group)	♂ 4-h LC$_{50}$ ♀ 4-h LC$_{50}$ (limit test; whole body)	• During exposure, all rats exhibited decreased activity, partially closed eyes, wet coats, and watery salivation • During a 14-day observation period, all rats exhibited slight to severe lethargy, reddish lacrimation, and acute ↓ bodyweight, brown discharge around oral cavity, slight alopecia, convulsions, and dyspnea • 0/5 ♂ rats died • 3/5 ♀ rats died	>5.05 (FPA)[c] ~5 (FPA)[c]

		Oral	
Test Species (n)	End Point	Effect	Result, mg/kg bw
Wistar ♂ and ♀ rats (n = 5/sex/group at 200 and 500 mg/kg bw; n = 5 ♀/group at 2,000 mg/kg bw)	♂ LD$_{50}$[d] ♀ LD$_{50}$ (gavage, corn oil vehicle)	200 and 500 mg/kg bw: • No clinical signs of toxicity 2,000 mg/kg bw: • 5/5 rats died • Clinical signs included apathy, palmospasm, and blood-crusted snout • Mottled reddened lungs	>500 632

		Dermal	
Test Species (n)	End Point	Effect	Result, mg/kg bw
Sprague-Dawley rats (group size and sex not stated)	LD$_{50}$	• No clinical signs or post mortem observations.	>2,000
New Zealand albino ♂ and ♀ rabbits (n = 4/sex/group at 2,000 or 5,000 mg/kg bw)	LD$_{50}$	• No deaths occurred during the 14-day observation period in rabbits with abraded or intact skin • Slight/mild erythema observed at both dose levels • No gross abnormalities at necropsy with exception of one animal in the 2,000 mg/kg bw group (ventral surface discoloration of liver)	>5,000

Source: EURAR (2008).

[a]Measured gravimetric concentration.

[b]Fine particle aerosol (mean particle size — 2.9 mass median aerodynamic diameter).

[c]Particle size not stated.

[d]Median lethal dose.

and dermal routes of exposure. An additional study investigating TCPP's potential to cause acute delayed neurotoxicity was conducted in hens, which the EU concluded as showing ". . . no concern for this end-point." Of the available studies, the EU relied upon those summarized in Table 84.4 for determining TCPP's acute toxicity. The data support that TCPP presents a low hazard concern by the inhalation and dermal routes of exposure. The EU concluded that the oral gavage studies supported classification as ". . . R22, harmful if swallowed."

Repeated Dose Studies

In a 90-day study, male and female Sprague-Dawley rats were administered TCPP by diet at concentrations of 0, 800, 2,500, 7,500, or 20,000 parts per million. The intake concentration for males and females was 0, 52, 160, 481, or 1,349 mg/kg bw/day and 0, 62, 171, 570, or 1,745 mg/kg-bw/day, respectively. At study termination, no treatment-related effects were reported on mortality, clinical observations, clinical chemistry, hematology, urinalysis, plasma, or erythrocyte or brain cholinesterase activity. Mean body-weight was statistically significantly ($p < 0.05$) reduced from day 22 or day 35 until study termination in the respective high dose males (decreased by 7.75% compared to controls) and females (decreased by 11.8% compared to controls). Absolute and relative liver weights were statistically significantly ($p < 0.05$) increased in all treated males and in females in the 570 and 1,750 mg/kg bw/day groups. Histological findings consisted of periportal hepatocyte swelling (hypertrophy) and were limited to the high dose animals (7/20 males and 8/20 females compared to controls [0/20 males and 5/20 females]). In male rats, mild thyroid follicular cell hyperplasia was observed in all treatment groups at 2/20, 2/20, 5/20, and 8/20 compared to 0/20 controls. This effect was limited to 5/20 high dose females compared to 0/20 controls. The EU noted that the effects on the liver and thyroid could be secondary to altered hepatic metabolic activity. A lowest-observed-adverse-effect level (LOAEL) of 52 mg/kg bw/day was reported for male rats, based on the liver and thyroid effects. In female rats, a no-observed-adverse-effect level (NOAEL) of 171 mg/kg bw/day was reported, based on increased liver weights. The EU utilized the LOAEL of 52 mg/kg bw/day for risk characterization.

In a two-generation study, groups of 28 male and female Wistar rats were fed diets containing TCPP during a 10-week premating period and throughout gestation and lactation. Litters were culled on postnatal day (PND) 4 to groups of 8 animals, consisting of 4 males and 4 females. On PND 21, the next generation of 28 males and females was weaned and selected at random. F_0 and F_1 dams were sacrificed after weaning, whereas F_0 and F_1 were treated for approximately 11 weeks, prior to sacrifice. The calculated overall intake for the respective control, low, middle, or high dose in females was 0, 99, 330, or 988 mg/kg bw/day and in males was 0, 85, 293, or 925 mg/kg bw/day.

No treatment-related effects were identified in either generation for the following end points: clinical signs, pre-coital time, mating index, female fecundity index, male and female fertility index, duration of gestation and postimplantation loss, epididymal sperm motility, sperm count, sperm morphology, mean testicular sperm count. Two mortalities occurred during the pre-mating period. A male was found dead on day 41, and a female was found moribund on day 50.

At necropsy, the cause of death or moribund condition could not be determined.

As shown in Table 84.5, statistically significant changes were reported in bodyweight and organ weights of F_0 or F_1 females and males primarily in the mid and/or high dose groups. The EU considered the kidney weight changes in male rats and the uterine weight changes in female rats as the main effects on organ weight changes. The EU noted that the uterine weight change was not accompanied by histological changes, that uterine weight may fluctuate during the estrus cycle, and that the reported changes may be reflective of normal variation in cycling females. As a precautionary approach, however, it considered the uterine weight changes as a treatment-related effect. In females in the F_0 and F_1 generations, the mean length of the longest estrus cycle was statistically significantly increased in all treated F_0 females and in high dose F_1 females. The EU considered this effect as toxicologically significant in high dose animals from the F_0 and F_1 generations because the effect on the low and mid dose animals may have been due to normal variation and not a specific fertility effect.

A summary of the pup and litter data for the F_0 and F_1 generations is provided in Table 84.6. A correlation was observed between decreased maternal bodyweight during gestation and the mean number of pups delivered and the mean number of live pups per litter. Both of these end points in the offspring were decreased in the F_1 mid dose group and in the high dose groups of both generations, which may have been due to maternal toxicity. One litter from a single F_1 dam was lost on PND 4; all other pups from all treatment groups survived. The mean number of runts in the F_0 generation was statistically significantly increased on PND 1 in all treatment groups. The number of runts persisted through PND 21 in the mid and high dose groups. In the F_1 generation, the number of runts was statistically significantly increased in the high dose group on PND 14 and in all dose groups on PND 21. In the F_2 generation, there was no effect on anogenital distance in male and female pups. In the high dose group, vaginal opening was delayed, although not statistically significantly, whereas preputial separation was statistically significantly delayed.

Based on the two-generation study, the EU used the following points of departure for risk characterization. The EU identified a LOAEL of 99 mg/kg bw/day in females for parental toxicity, based on decreased bodyweight and food consumption and effects on uterus weight. A NOAEL of 85 mg/kg bw/day was identified for males for parental toxicity, based on decreased bodyweight and food consumption and organ weight changes. For developmental toxicity, the EU identified a LOAEL of 99 mg/kg bw/day, based on the number of runts in the F_0 generation. The EU also noted that there was no concern for lactational effects.

No carcinogenicity data on TCPP are currently available; however, the US National Toxicology Program is in the

TABLE 84.5 Mean Terminal Bodyweights and Organ Weights for Female and Male Rats of the F_0 and F_1 Generations

Organ	Sex	Generation	Dose Group Control	Low	Mid	High
Mean Terminal Bodyweight, g	♀	F_0	267	268	263	258
		F_1	264	265	251*	246**
	♂	F_0	416.5	400	394.9*	374.1***
		F_1	397.8	390.8	367.3**	336.1***
Mean Absolute Organ Weight, g						
Liver	♀	F_0	13.608	13.580	13.702	14.890**
		F_1	13.629	13.673	13.389	13.872
Kidney	♂	F_0	2.406	2.333	2.326	2.252**
		F_1	2.313	2.200*	2.113***	2.061***
Spleen	♀	F_0	0.508	0.490	0.466**	0.443***
		F_1	0.507	0.505	0.483	0.438***
	♂	F_0	0.742	0.730	0.703	0.629***
		F_1	0.751	0.736	0.672***	0.596***
Pituitary	♀	F_0	0.016	0.016	0.016	0.015***
		F_1	0.017	0.015**	0.016*	0.014***
	♂	F_0	0.014	0.014	0.013	0.013
		F_1	0.015	0.015	0.014	0.013***
Uterus	♀	F_0	0.46	0.375*	0.313***	0.311***
		F_1	0.455	0.369	0.367	0.295***
Ovary	♀	F_0	0.082	0.081	0.077	0.073**
		F_1	0.084	0.080	0.083	0.076
Seminal vesicles	♂	F_0	1.595	1.518	1.419*	1.388*
		F_1	1.475	1.392	1.211***	1.191***
Mean Organ Weights Relative to Terminal Bodyweight, g/kg bw						
Liver	♀	F_0	50.918	50.791	52.031	57.611***
		F_1	51.590	51.601	53.394	56.202***
Kidney	♂	F_0	5.788	5.850	5.901	6.026
		F_1	5.843	5.645	5.761	6.164*
Spleen	♀	F_0	1.9	1.833	1.770**	1.711***
		F_1	1.922	1.908	1.928	1.779*
	♂	F_0	1.781	1.823	1.782	1.683
		F_1	1.894	1.886	1.834	1.784
Pituitary	♀	F_0	0.062	0.060	0.061	0.057*
		F_1	0.065	0.057**	0.062	0.059*
	♂	F_0	0.033	0.035	0.032	0.036
		F_1	0.039	0.038	0.038	0.038
Uterus	♀	F_0	1.723	1.408*	1.192***	1.202***
		F_1	1.732	1.399	1.465	1.202**
Ovary	♀	F_0	0.309	0.304	0.293	0.285
		F_1	0.317	0.302	0.331	0.307
Seminal vesicles	♂	F_0	3.841	3.808	3.591	3.723
		F_1	3.712	3.585	3.310	3.511

Source: EURAR (2008).

Note: Statistically significantly different from the control group at: *$p < 0.05$, **$p < 0.01$, or ***$p < 0.001$.

process of completing two-year dietary studies in rats and mice (NTP, 2014). At the time the EU evaluated TCPP, it concluded that TCPP is not genotoxic *in vivo* and that there is a potential concern for carcinogenicity by non-genotoxic mechanisms, based on the 90-day repeated dose study, discussed herein, and read across from structurally similar chlorinated phosphate flame retardants (e.g., tris[2-chloro-1-(chloromethyl)ethyl] phosphate (TDCPP)).

TABLE 84.6   Delivery, Pup and Litter Data, and Clinical Observations for F_0 and F_1 Generations

	Effect		Control	Low	Mid	High
					Dose Group	
F_0	Mean number of pups delivered		10.27	10.67	9.89	9.44[*,a]
	Total number of pups delivered		267	256	277	236
	Live birth index, %		100	100	99	100
	Number of pups lost (dying, missing, and/or cannibalized) on:	Days 1–4	3	20[***]	10	14[**]
		Days 5–7	0	0	0	0
		Days 8–14	0	0	0	0
		Days 15–21	0	0	0	0
	Runts	Day 1	0	14[***] (7)[**,b]	23[***] (7)[**]	11[***] (3)
		Day 4	2 (2)	11[**] (3)	7 (5)	6 (2)
		Day 7	2 (2)	13[**] (3)	20[***] (7)	21[***] (6)
		Day 14	1	6 (2)	15[***] (7)	26[***] (9)[**]
		Day 21	1	4 (2)	30[***] (10)[**]	97[***] (19)[***]
	Mean number of live pups/litters (PND 1)		10.27	10.63	9.79	9.44[*]
	Sex ratio of PND 1 (♂/♀)		153/111	129/127	143/134	112[*]/124
	Number of pups alive on day 21		198	178	213	190
F_1	Mean number of pups delivered		10.56	10.00	9.13[*]	8.68[a,***]
	Total number of pups delivered		264	240	219	191
	Live birth index, %		100	99	100	100
	Number of pups lost (dying, missing, and/or cannibalized) on:	Days 1–4	1	0	2	12[***]
		Days 5–7	0	0	0	0
		Days 8–14	0	0	0	0
		Days 15–21	0	0	0	0
	Runts	Day 1	10 (4)	1	17 (5)	14 (4)
		Day 4	4 (3)	0	15 (3)	16 (3)
		Day 7	4 (3)	2 (2)	17 (4)	38 (8)
		Day 14	11 (6)	14 (3)	19 (5)	78[***] (13)[*]
		Day 21	5 (3)	17[**] (4)	36[***] (9)	127[***] (19)[***]
	Mean number of live pups/litters (PND 1)		10.52	9.92	9.08[**]	8.68[**]
	Sex ratio of PND 1 (♂/♀)		140/124	123/117	113/106	94/97
	Number of pups alive on day 21		198	186	181	155

Source: EURAR (2008).

Note: Statistically significantly different from the control group at: [*]$p < 0.05$, [**]$p < 0.01$, or [***]$p < 0.001$.

[a]Historical control range $= 9.40$–11.18 ($n = 19$).

[b]Numbers in parentheses represent the number of litters with pups showing the observation.

SOURCES OF EXPOSURE

Occupational

The primary exposure concerns for workers include those that may occur by the inhalation and dermal routes from vapors or aerosols produced during the manufacture of TCPP and PUF. Dust exposures may also occur during the foam conversion and cutting of PUF. Tris(2-chloro-1-methylethyl) phosphate is stable under normal working conditions. The EU utilized measured data from 10 specific exposure scenarios encountered in the workplace when assessing potential risks to workers and also derived reasonable worse case (RWC) exposure estimates (Table 84.7).

Environmental

Environmental releases of TCPP may occur during manufacture and processing, from finished articles, and from end-of-life disposal. The end uses of TCPP, primarily as a flame retardant foam additive, are amendable to migration from the product; therefore, its presence in the environment is expected. Stapleton et al. (2009) estimated the daily ingestion of flame retardants from house dust for a child and an adult at *ca.* 1.6 and

TABLE 84.7   Occupational Exposure Levels to TCPP

Exposure Scenario	Inhalation Exposure, $\mu g/m^3$		Dermal Exposure, $mg/cm^2/day$		Dermal Exposure Area, cm^2
	Typical	RWC	Typical	RWC	
1. Manufacture of TCPP	12.5	25	0.1	1	210
2. Manufacture of flexible PUF	0.62	5.1	0.002	0.07	420
3. Cutting flexible foam	1.9	4.1	9.8×10^{-4}	7.1×10^{-3}	420
4. Production of foam granules and rebonded foam	0.59	4.6	5.5×10^{-4}	1.7×10^{-3}	420
5. Formulation of systems and manufacture of spray foams	2.5	5	0.05	0.11	420
6. Use of spray foams	25	187.5	0.12	0.23	420
7. Manufacture of rigid foam	20	150	3.2×10^{-2}	6.5×10^{-2}	210
8. Use of rigid foam	1.9	4.1	6×10^{-3}	1.3×10^{-2}	210
9. Manufacture of one-component (1K) foams	6.7	12.5	1×10^{-3}	5.2×10^{-3}	210
10. Use of 1K foams	2.5×10^{-3}	5×10^{-3}	9.3×10^{-4}	1.9×10^{-3}	420

Source: EURAR (2008).

0.325 µg/day, respectively. The authors calculated these values based on the assumption that a child ingests 100 mg/day and an adult ingests 20 mg/day of dust. The EU evaluated environmental levels of TCPP and determined that the measured level in surface water from European countries, including the United Kingdom, generally ranged from 5 to 10 µg/l.

INDUSTRIAL HYGIENE

Tris(2-chloro-1-methylethyl) phosphate is manufactured as a colorless liquid. The personal protection equipment and mechanical controls recommended in the workplace are designed to minimize eye, skin, and respiratory exposures (ECHA, 2013). Chemical goggles or a face shield are recommended for eye protection against the liquid and aerosol. Suitable protective clothing (unspecified) is recommended to avoid skin exposures. Neoprene rubber gloves are recommended for hand protection. A respirator approved by the US National Institute of Occupational Safety and Health (NIOSH) for organic vapor/acid gas is recommended to reduce inhalation exposures under conditions that generate a vapor, mist, or aerosol. For conditions that warrant enhanced respiratory protection, a NIOSH-approved, positive pressure, pressure demand, air-supplied respirator is recommended. Mechanical controls are aimed at providing adequate ventilation.

RISK ASSESSMENTS

The EU assessed the potential risks from exposures to TCPP in the environment and to workers, consumers, and from exposures to the general population from environmental sources (EURAR, 2008). For the environment, the EU concluded that no further testing/information or risk reduction measures were needed, beyond those currently being applied. This conclusion applied to all environmental compartments for all local life cycle stages and at the regional scale in all compartments. For workers, the EU concluded that there is a need for limiting risks from RWC dermal exposures during the manufacture of TCPP (Table 84.7, scenario 1) for potential effects on fertility and developmental toxicity. For typical dermal and inhalation exposures, both RWC and typical, during the manufacture of TCPP, the EU concluded that no information, testing, or risk reduction measures were recommended. No information, testing, or risk reduction measures were recommended for the remaining occupational exposure scenarios listed in Table 84.7. For consumers and humans exposed *via* the environment, the EU concluded that no information, testing, or risk reduction measures were needed beyond those currently being applied.

REFERENCES

ECHA (2013) *Registered Substances.* Helsinki, Finland: European Chemicals Agency. Available at http://echa.europa.eu/information-on-chemicals/registered-substances. CASRN: 13674-84-5.

EPA (2012) *Chemical Data Access Tool (CDAT).* Washington, DC: Office of Pollution Prevention and Toxics, US Environmental Protection Agency. Available at http://java.epa.gov/oppt_chemical_search. CASRN: 13674-84-5.

EURAR (2008) *Tris(2-chloro-1-methylethyl) phosphate (TCPP), CAS No: 13674-84-5, EINECS No: 237-158-7, Risk assessment.* European Union Risk Assessment Report. Available at http://echa.europa.eu/documents/10162/13630/trd_rar_ireland_tccp_en.pdf.

NTP (2014) *Tris(chloropropyl)phosphate.* Research Triangle Park, NC: National Toxicology Program, US Department of Health and Human Services. Available at http://ntp.niehs.nih.gov/?objectid=BD724190-123F-7908-7BA185DA18C1EBB8.

Stapleton, H.M., Klosterhaus, S., Eagle, S., Fuh, J., Meeker, J.D., Blum, A., and Webster, T.F. (2009) Detection of organophosphate flame retardants in furniture foam and US house dust. *Environ. Sci. Technol.* 43:7490–7495.

SECTION VI

PESTICIDES

SECTION EDITOR: RAYMOND D. HARBISON

85

INTRODUCTION

RAYMOND D. HARBISON

The well-accepted definition of a pest is any plant, animal, fungus, or other organism that adversely affects human health or welfare. A pesticide is a chemical used to kill a species that humans have designated as pests. Most pesticides fall in one of the four categories: insecticides, herbicides, fungicides, and rodenticides. Insecticides are the most widely recognized pesticides, but the greatest volumes of pesticides used are in the herbicide category.

Early in this century, pesticide use shifted from metallic pesticides to synthetic chemical formulations. More recently, pesticide use has shifted from broad-spectrum effects to formulations with greater specificity. The list of pesticides currently registered in the United States is continually changing because of the reregistration eligibility process. Triazine herbicides are a good example of the dynamic nature of pesticide regulation and use in the United States. They are some of the most commonly used pesticides in the United States today (Environmental Protection Agency [EPA], 1994). Triazine herbicides are characterized by a low oral toxicity but are relatively persistent in most environmental conditions. The EPA has requested voluntary cancellation of the triazine herbicide cyanazine because the agency considers cyanazine as a possible human carcinogen (EPA, 1996). Other triazine herbicides are also undergoing special review. In the span of 5 years, this group of pesticides probably will go from being the most widely used herbicides as a group to near disuse.

The changing status of triazine herbicides reflects regulatory concern for chronic toxicity. In the United States regulatory attention and public concern have shifted from acute toxicity to carcinogenicity, teratogenicity, and other chronic effects. Another indication of this priority is the popularity of organophosphorus insecticides, which have largely replaced the organochlorine insecticides that was banned over two decades ago. Although the organophosphorus compounds may have severe acute effects, they are not biologically persistent and do not have a tendency to bioaccumulate.

Most modern pesticides are relatively safe when used properly; however, thousands of poisonings are attributed to pesticides each year. Many of these incidents involve intentional ingestion, but many more may be attributed to improper labeling, failure to follow label instructions, application by untrained individuals, or potent gray market pesticides, which often lack any label.

The estimated volume of pesticide used in the United States (by weight of active ingredient) has remained relatively stable for the past few years. The Environmental Protection Agency (EPA) attributes this plateau to the development of more potent pesticides, which require fewer applications, as well as to more efficient use of pesticides and lower farm commodity prices (EPA, 1994).

In spite of the long-time ban on the production and sales of organochlorine pesticides (e.g., dichlorodiphenyltrichloroethane) in the United States, they are still widely used in developing tropical and subtropical nations to control malaria through mosquito control.

Sadly, intentional and accidental ingestions remain the most common routes of overexposure to most pesticides. The most common form of occupational exposure is skin contact; for this reason, the importance of protective clothing in reducing exposure cannot be overstated.

Hamilton & Hardy's Industrial Toxicology, Sixth Edition. Edited by Raymond D. Harbison, Marie M. Bourgeois, and Giffe T. Johnson.
© 2015 John Wiley & Sons, Inc. Published 2015 by John Wiley & Sons, Inc.

The controversy over the chlorophenoxy herbicides (2,4,5-T and others) and over dioxin exposure and its health effects, particularly during the Vietnam conflict, continues to this day. Several large studies of veterans have shown no association between exposure and suspected health effects. But, a general mistrust of the U.S. government and a recent abundance of conspiracy theories have left the general public unconvinced and the scientific community somewhat divided.

REFERENCES

Environmental Protection Agency (EPA) (1994) *Pesticides Industry Sales and Usage: 1992 and 1993 Market Estimates*, Washington, DC: Environmental Protection Agency.

Environmental Protection Agency (EPA) (1996) Cyanazine: notice of preliminary determination to terminate. *Fed. Reg.* 61(42): 8185–8203.

86

INSECTICIDES

Carol S. Wood

ORGANOPHOSPHORUS AND CARBAMATE INSECTICIDES

First aid: Decontamination, symptoms

Target organ(s): Acetylcholinesterase inhibition

Occupational exposure limits: BEI® is 70% of baseline cholinesterase activity in RBC (ACGIH, 2010)

Reference values: Chemical specific, see below

Risk/safety phrases: Harmful if swallowed; avoid skin contact

BACKGROUND AND USES

Organophosphorus (OP) and carbamate insecticides have a wide variety of uses, including protection of crops, grains, gardens, and public health. Some also have application as therapeutic agents in certain disease conditions. Originally developed as chemical warfare agents, more than 30 OPs are registered for use as pesticides in the UnitedStates (Lowit, 2006). The volume of carbamates used exceeds that of the OPs because toxicity is generally less (Gupta, 2006) with 10 registered in the UnitedStates (Mortensen, 2006).

MAMMALIAN TOXICOLOGY

Acetylcholinesterase Inhibition

Acetylcholinesterase (AChE) is found primarily in the central nervous system (CNS), neuromuscular junctions, the hematopoietic system, and red blood cells (RBCs) and has a key role in cholinergic neurotransmission. Acetylcholine (ACh) is involved in neurotransmission at motor, autonomic, and central synapses. For cholinergic neurotransmission to occur properly, ACh must be inactivated rapidly (in less than a millisecond) by the enzyme AChE. This removal prevents repeated stimulation of the receptor. ACh and AChE are present at the neuromuscular junction.

OP and carbamate insecticides inhibit cholinesterase by binding to, and phosphorylating, the active site of the enzyme, preventing the inactivation of ACh. The adverse effects of these insecticides are the result of accumulation of ACh at the neuromuscular junction, which causes muscular fasciculations and ultimately paralysis. Excessive ACh also causes excessive stimulation of the autonomic nervous system; this in turn causes excessive production of mucus, which accumulates in the lungs.

Particularly with certain OPs, the phosphorylated cholinesterase is irreversibly inactivated during an "aging" process of dealkylation of the phosphyl moiety. This aged enzyme is not reactivated by antidotes. Furthermore, an individual in whom the inactivated enzyme has not been replaced by synthesis of new enzyme will be hypersusceptible to additional exposure to OP insecticides. Regeneration of new enzyme usually requires several weeks.

When organisms are exposed chronically or subchronically to OP compounds, overt signs and symptoms of exposure may be absent. This phenomenon is called "tolerance." Cholinergic muscarinic receptors (mAChR) in brain and peripheral tissues have been shown to decrease after repeated exposure to OP insecticides. This is thought to play a key role in the development of tolerance. Other mechanisms of tolerance development include the presence of

Hamilton & Hardy's Industrial Toxicology, Sixth Edition. Edited by Raymond D. Harbison, Marie M. Bourgeois, and Giffe T. Johnson.
© 2015 John Wiley & Sons, Inc. Published 2015 by John Wiley & Sons, Inc.

proteins that bind or inactivate the OP and more rapid metabolism to remove the OP (Fonnum and Sterri, 2006).

Individual variability in plasma cholinesterase activity is well documented (NRC, 2003). This variability includes age-related differences (neonates are more susceptible than are adults), gender differences (females tend to have lower plasma and RBC cholinesterase activity), and genetically determined variations in plasma cholinesterase activity. This genetic variability, which sometimes results in greatly reduced activity of plasma cholinesterase, may impart deficiencies in ability to detoxify organophosphates. Additionally, polymorphic variability in A-esterases, such as paraoxonase/arylesterase, may also contribute to individual variability in organophosphate ester detoxification processes (NRC, 2003).

Butyrylcholinesterase Inhibition

Butyrylcholinesterase (BuChE) is found in plasma, the liver, and the kidney. Its activity is also inhibited by OP and carbamate insecticides. BuChE is able to hydrolyze much larger substrates than AChE due to differences in their structure that allow BuChE to bind bulky ligands. The physiological role of BuChE is not completely understood, but it represents an important detoxification pathway for anticholinesterase compounds (Sultatos, 2006).

Acute Signs and Symptoms

Signs and symptoms of excessive exposure include headache, dizziness, and weakness. Anorexia, meiosis, impairment of vision, anxiety, and tremors of the tongue and eyelids may also be seen. Increasing exposure can result in increased salivation and lacrimation and uncontrolled urination and diarrhea. Increased sweating, abdominal cramps, nausea and vomiting, slow pulse, and muscle tremors also occur. Intoxication causes pulmonary edema, cyanosis, respiratory paralysis, coma, and heart block. With low level exposure, meiosis can be the only indication that exposure has occurred.

Occupational Monitoring

Both RBC AChE and plasma BuChE measurements are used to diagnose exposure and in occupational screening programs. Inhibition of RBC AChE to <50% of baseline, or preexposure levels, is considered diagnostic of exposure but not related to severity of poisoning. At levels of RBC AChE inhibition 30–50% of baseline, the worker should be removed from exposure while inhibition to >80% of baseline is associated with severe symptoms and mortality in the absence of treatment. Similar inhibition ranges are assigned to BuChE inhibition (Ballantyne and Salem, 2006).

Treatment

Removal from exposure and decontamination of the skin and clothing are the first treatment measures. Gastric lavage may be used in serious cases of ingestion, but induction of vomiting should be avoided especially if the victim may lose consciousness. Severe clinical intoxication requires rapid treatment with atropine and pralidoxime (2-PAM, Protopam); however, first excess airway secretions must be removed completely to maintain a patent airway. Atropine should be administered intravenously until signs of atropinization, especially reduced salivation and bronchial secretion, are observed. Atropine competitively blocks ACh and other muscarinic receptor agonists. Pralidoxime can be administered to reactivate cholinesterase. Treatment is effective when administered within 24 h of intoxication (Marrs and Vale, 2006).

While acute symptoms can be successfully treated if aggressively monitored, protection against intermediate syndrome (IMS) and organophosphate-induced delayed polyneuropathy (OPIDP) is less certain. Pralidoxime has been somewhat effective in preventing the development of IMS when given within 12 h of poisoning. No antidotal treatment for organophosphate-induced delayed neuropathy is recognized as effective (Marrs and Vale, 2006).

An individual recovering from overexposure has a hypersusceptibility to the effects of additional exposure to OP insecticides. This increased susceptibility usually lasts for several weeks, followed generally, by complete recovery.

Organophosphate-Induced Delayed Neuropathy

Organophosphate-induced delayed neurotoxicity (OPIDN) is a relatively rare toxicity of the nervous system caused by some OP compounds. It is characterized by sensory and motor axon degeneration in distal regions of the peripheral nerves and spinal cord tracts. The axonal degeneration is delayed for 1–3 weeks after exposure depending on dose. Organophosphate-induced delayed neurotoxicity is recognized as an effect independent of AChE or BuChE inhibition. Symptoms initially include cramping muscle pain in the lower limbs followed by distal numbness and paresthesia, and progressive wasting and flaccid weakness especially of the lower limbs (Moretto and Lotti, 2006). A high stepping gait associated with foot drop is the classic sign of OPIDN.

It occurs as a result of the inhibition of neuropathy target esterase (NTE). Phosphorylation and subsequent aging of at least 70% of NTE typically is sufficient to cause OPIDN (Johnson, 1990). Inhibition of NTE in the sciatic nerve and spinal cord is usually apparent within 24–48 h after exposure. This inhibition correlates with the clinical expression of OPIDN (Lotti, 1992). Neuropathy target esterase is widely distributed throughout the nervous system, except glial cells, and is also present in other tissues.

Despite evidence showing the involvement of NTE in OPIDN, the enzyme's role in axonal degeneration is unclear. Axonal degeneration has been shown to be associated with a deficit of retrograde axonal transport (Moretto et al., 1987). Other mechanistic studies have implicated protein kinase-mediated phosphorylation of cytoskeletal proteins; altered expression of neurofilament subunits; changes in protein kinases, glyceraldehyde-3-phosphate dehydrogenase, phosphorylated cAMP-response element binding protein, and α-tubulin expression; Ca^{2+} calmodulin-dependent autophosphorylation of proteins; and disruption of calcium homeostasis. These other mechanistic studies have been reviewed by Moretto and Lotti (2006).

The potency of individual OPs to inhibit AChE and NTE *in vitro* correlates with their toxicity *in vivo* and can be used as a measure of the potential of a specific compound to cause OPIDN. This fact is used in premarket testing such that commercial OPs in current use are much more acutely toxic with cholinergic effects rather than causing neuropathy (Moretto and Lotti, 2006).

Species vary greatly in their susceptibility to OPIDN with the hen being the experimental animal model because rodents are relatively resistant. In general, the young of any given species are less susceptible to OPIDN than adult animals. Case reports of OPs that cause OPIDN in humans have been reviewed by Lotti and Moretto (2005) and include poisoning by chlorpyrifos, dichlorvos, isofenphos, methamidophos, mipafox, trichlorfon, trichlornat, triaryl phosphates, and phosphamidon/mevinphos.

Intermediate Syndrome

The IMS following OP poisoning was first described in patients who developed facial, proximal limb, and respiratory muscle weakness (Senanayake and Karalliedde, 1987). The syndrome occurs after apparent recovery from the acute cholinergic crisis and before a possible OPIDN. IMS occurs in approximately 20% of patients following oral exposure to OP pesticides, with no clear association between the particular OP pesticide involved and the development of the syndrome (Karalliedde et al., 2006). It usually becomes established 1–4 days after exposure when the symptoms and signs of the acute cholinergic syndrome are no longer obvious. The characteristic features of the IMS are persistent inhibition of RBC AChE; weakness of the extraocular, bulbar, facial, respiratory, neck, and proximal limb muscles; and absent or depressed deep tendon reflexes (De Bleeker, 2006). These patients may not require ventilation but supportive care and monitoring is mandatory.

Gulf War Syndrome

After Operation Desert Storm ended in 1991, a large number of veterans who had served experienced chronic health problems, including memory loss, fatigue, cognitive problems, somatic pain, skin abnormalities, and gastrointestinal difficulties. These veterans were potentially exposed to the OP chlorpyrifos, non-OP insect repellents, pyridostigmine bromide (a nerve agent antidote enhancer) and atropine, as well an OP nerve gas from a detonated munitions bunker in Iraq. Individual chemical and mixture exposures cannot be quantified and despite years of research, a link to OP exposures and Gulf War Syndrome has not been definitively established. An historical perspective and summary of ongoing research can be found by McCauley (2006). A more recent study found association between symptoms and use of pyridostigmine bromine while being within one mile of an exploding SCUD missile for veterans near the front line; for those who remained in support areas, symptoms were significantly associated only with personal pesticide use (although specific pesticide(s) was not specified) (Steele et al., 2012).

EXAMPLES OF ORGANOPHOSPHATE AND CARBAMATE INSECTICIDES

CARBOFURAN

First aid: Decontamination, symptoms

Target organ(s): Acetylcholinesterase inhibition

Occupational exposure limits: BEI® is 70% of baseline cholinesterase activity in RBC (ACGIH, 2010)

Reference values:
1-Day and 10-Day Health Advisory (child) = 0.05 mg/l (U.S. EPA, 2012a)
DWEL Health Advisory = 0.18 mg/l (U.S. EPA, 2012a)
Life-time Health Advisory = 0.036 mg/l (U.S. EPA, 2012a)
RfD = 0.005 mg/kg/day (U.S. EPA, 1987)
REL–TWA = 0.1 mg/m³ [skin] (NIOSH, 2013)
TLV–TWA = 0.1 mg/m³ [skin] (ACGIH, 2010)

Risk/safety phrases: Harmful if swallowed; avoid skin contact

BACKGROUND AND USES

Carbofuran is a broad-spectrum N-methyl carbamate sold under the trade name Furadan. First registered in 1969, it is used to control a variety of insect pests on field, fruit, and vegetable crops. Carbofuran is not used in residential settings or food-handling establishments. The chemical is a restricted use pesticide with nearly 1 million pounds of active ingredient almost exclusively used on corn, alfalfa, and potatoes (U.S. EPA, 2006a).

PHYSICAL AND CHEMICAL PROPERTIES

Carbofuran (2,3-dihydro-2,3-dimethyl-7-benzpfuranyl-N-methylcarbamate) is a white crystal-like solid. It is slightly soluble in water, is stable under neutral and acidic conditions, but decomposes under alkaline conditions (U.S. EPA, 2006a).

ENVIRONMENTAL FATE AND BIOACCUMULATION

Carbofuran is highly mobile in soils and can leach into groundwater and enters surface water as runoff. The chemical breaks down though hydrolysis, photodegradation, and moderate bacterial degradation at rates that depend on environmental conditions. Hydrolysis is faster in water with a $pH \geq 7$, with a half-life ranging from a few hours to 28 days. Carbofuran is stable to hydrolysis in acidic water. Photodegradation is fast in a thin water layer, with a half-life of 6 days. In the top few millimeters of a sandy loam soil, carbofuran degrades in 78 days. Bioconcentration is not expected to occur (U.S. EPA, 2006a).

ECOTOXICOLOGY

Carbofuran toxicity to nontarget species has been summarized by the U.S. EPA (2006a). The chemical is characterized as highly toxic to birds, fish, and honey bees. Ducks showed an LD50 of 0.238 mg/kg and a high mortality rate when given carbofuran in the feed for several weeks. Chronic toxicity testing of freshwater fish showed larval survival as the most sensitive end point at water concentrations in the part-per-billion (ppb) range. Marine fish were more sensitive than freshwater fish.

MAMMALIAN TOXICOLOGY

Acute Effects

Acute toxicity of carbofuran is typical of other carbamates with symptoms a result of AChE inhibition. Typical of this class of pesticides inhibition is followed by rapid recovery. Young animals are more susceptible to carbofuran toxicity than older animals (U.S. EPA, 2006a).

Chronic Effects

Chronic toxicity is also due to cholinesterase inhibition. Due to the rapid recovery (within 24 h) of AChE inhibition, longer-term exposures are considered a series of acute exposures (U.S. EPA, 2006a).

SOURCES OF EXPOSURE

Occupational

Workers can be exposed to a carbofuran through mixing, loading, and/or applying the pesticide, or reentering treated sites.

Environmental

Carbofuran is released to the atmosphere and water during foliage or soil application by ground or air broadcast equipment. On soils, the chemical is highly mobile and can enter streams and surface waters.

INDUSTRIAL HYGIENE

The occupational exposure limits shown in the table above, come with a skin notation indicating dermal absorption can occur and should be avoided.

RISK ASSESSMENTS

The Reference dose(RfD) is based on RBC AChE inhibition and testicular and uterine effects in beagle dogs given carbofuran in the diet for 1 year (U.S. EPA, 1987). The U.S. Environment Protection Agency (EPA) has not evaluated the chemical for carcinogenicity.

CHLORPYRIFOS

First aid: Decontamination, symptoms

Target organ(s): Acetylcholinesterase inhibition

Occupational exposure limits: BEI® is 70% of baseline cholinesterase activity in RBC (ACGIH, 2010)

Reference values:
1-Day and 10-Day Health Advisory (child) = 0.03 mg/l (U.S. EPA, 2012a)
DWEL Health Advisory = 0.01 mg/l (U.S. EPA, 2012a)
Life-time Health Advisory = 0.002 mg/l (U.S. EPA, 2012a)

MRL (acute and intermediate) = 0.003 mg/kg/day (ATSDR, 1997)

MRL (chronic) = 0.001 mg/kg/day (ATSDR, 1997)

REL–TWA = 0.2 mg/m^3 [skin] (NIOSH, 2013)

REL-STEL = 0.6 mg/m^3 [skin] (NIOSH, 2013)

TLV-TWA = 0.1 mg/m^3 [skin] (ACGIH, 2010)

Risk/safety phrases: Harmful if swallowed; avoid skin contact

BACKGROUND AND USES

Chlorpyrifos is a broad-spectrum organophosphate insecticide sold under the trade names of Dursban, Empire 20, Equity, and Whitmire PT 270. First registered in 1965 to control foliage- and soil-borne insect pests on a variety of food and feed crops, it is one of the most widely used and one of the major insecticides used residentially (U.S. EPA, 2000). Chlorpyrifos is the active ingredient in more than 800 registered products on the market, resulting in 21–24 million pounds used annually in the United States.

PHYSICAL AND CHEMICAL PROPERTIES

Chlorpyrifos (diethyl 3,5,6-trichloro-2-pyridyl phosphorothionate) is a white crystal-like solid with a strong odor. It does not mix well with water, so it is usually mixed with oily liquids before it is applied to crops or animals (ATSDR, 1997).

ENVIRONMENTAL FATE AND BIOACCUMULATION

In the atmosphere, the chemical exists as a vapor but will partition to airborne particulates (ATSDR, 1997). In water, chlorpyrifos may slowly volatilize but the amount available for volatilization from surface water is reduced by adsorption to sediments. Volatilization does not appear to be a major route of dissipation from soil for chlorpyrifos (ATSDR, 1997; U.S. EPA, 1999). Because of its low water solubility and high soil binding capacity, there is potential for chlorpyrifos sorbed to soil to run off into surface water via erosion (U.S. EPA, 1999).

Abiotic hydrolysis and photodegradation do not contribute significantly in the removal of chlorpyrifos from the environment. Based on available data, chlorpyrifos appears to degrade slowly in soil under both the aerobic and anaerobic conditions. Under aerobic conditions, the half-life of chlorpyrifos ranged from 11 to 141 days in seven soils ranging in texture from loamy sand to clay, with the major degradate being CO_2, comprising 27–88% of the applied total after 360 days aerobic incubation (U.S. EPA, 1999).

ECOTOXICOLOGY

Birds, fish, and amphibians are susceptible to AChE inhibition similar to mammals. Studies have shown ducks exposed to chlorpyrifos in the diet lost weight and hatched significantly fewer ducklings. Similarly, bobwhite quail had decreased food consumption and weight loss (HSDB, 2008). Alterations in swimming ability and cholinesterase inhibition have been measured in tadpoles (HSDB, 2008).

Coho salmon exposed to chlorpyrifos-treated water for 96 h showed concentration-related inhibition of brain AChE activity whereas changes in swimming performance were measured when AChE inhibition reached a threshold (Tierney et al., 2007). Reductions in spontaneous swimming and feeding behaviors were also shown to correlate with AChE inhibition in juvenile coho salmon exposed to concentrations <5 µg/l (Sandahl et al., 2005).

Using the Nile tilapia, body size was shown to affect susceptibility to AChE inhibition. Following a 48-h exposure to chlorpyrifos, the concentrations resulting in 50% inhibition for fry, fingerlings, and subadults stages were 0.53, 0.75, and 3.86 µg/l, respectively (Chandrasekara and Pathiratne, 2007).

MAMMALIAN TOXICOLOGY

Acute Effects

Acute toxicity of chlorpyrifos is typical of other organophosphates with symptoms a result of AChE inhibition. Delayed neuropathy has been described in a number of human case reports following acute cholingeric symptoms from accidental or intentional poisoning (Aiuto et al., 1993; Kaplan et al., 1993; Lotti et al., 1986). Similarly, studies have demonstrated OPIDN in the hen, but only when the animals were atropinized to prevent lethality (Gaines, 1969; Capodicasa et al., 1991).

Chronic Effects

Epidemiology studies and case reports have implicated chlorpyrifos in causing birth defects in humans after the mother had been exposed to high levels during pregnancy (Whyatt et al., 2004; Perera et al., 2003). With repeat dosing to rats, generally doses of ≥1 mg/kg/day resulted in mild clinical signs such as urine staining on females; doses of ≥3 mg/kg/day caused miosis and brain cholinesterase activity inhibition; and at ≥5 mg/kg/day rats developed adrenal lesions and offspring body weight and survival were decreased (Maurissen et al., 2000a, 2000b; Yano et al., 2000).

SOURCES OF EXPOSURE

Occupational

Exposure to chlorpyrifos can occur during foliage or soil application by ground or air broadcast equipment and by reentry into treated fields. Exposure can also occur during production of the chemical and mixing prior to application.

Environmental

Chlorpyrifos is released to the atmosphere and water during foliage or soil application by ground or air broadcast equipment. Leaching and runoff from treated fields, pesticide disposal pits, or hazardous waste sites may inadvertently contaminate both groundwater and surface water with chlorpyrifos. Entry into water can also occur from accidental spills, redeposition of atmospheric chlorpyrifos, and discharge of waste water from chlorpyrifos manufacturing, formulation, and packaging facilities (ATSDR, 1997).

INDUSTRIAL HYGIENE

The occupational exposure limits shown in the table above, come with a skin notation indicating dermal absorption can occur and should be avoided.

RISK ASSESSMENTS

The acute and intermediate minimal risk levels (MRLs) were based on the dose that did not cause symptoms in human volunteers given chlorpyrifos for 21–28 days. The chronic MRL is based on a no-observed-adverse-effect level (NOAEL) for RBC AChE inhibition in rats fed diets containing chlorpyrifos for 2 years (ATSDR, 1997). The U.S. EPA has not evaluated the chemical for long-term health effects or carcinogenicity.

METHYL PARATHION

First aid: Decontamination, symptoms

Target organ(s): Acetylcholinesterase inhibition

Occupational exposure limits: BEI® is 70% of baseline cholinesterase activity in RBC (ACGIH, 2010)

Reference values:
1-Day and 10-Day Health Advisory (child) = 0.3 mg/l (U.S. EPA, 2012a)
DWEL Health Advisory = 0.007 mg/l (U.S. EPA, 2012a)
Life-time Health Advisory = 0.001 mg/l (U.S. EPA, 2012a)
RfD = 0.00025 mg/kg/day (U.S. EPA, 1991)
TWA = 0.2 mg/m^3 [skin] (NIOSH, 2013)
TLV–TWA = 0.02 mg/m^3 [skin] (ACGIH, 2010)

Risk/safety phrases: Harmful if swallowed; avoid skin contact

BACKGROUND AND USES

Methyl parathion is a restricted use organophosphate sold under a number of trade names and formulations. First registered in 1954, it is used to control a variety of insect pests on food and feed crops. Methyl parathion is not used in residential settings due to its toxicity to humans, birds, and honey bees. The chemical is a restricted use pesticide with nearly 4 million pounds of active ingredient mainly used on cotton, corn, wheat, soybeans, and rice (U.S. EPA, 2006b).

PHYSICAL AND CHEMICAL PROPERTIES

Methyl parathion (O,O-dimethyl)-*p*-nitrophenyl phosphorothioate) is a white crystalline solid. It is slightly soluble in water and has a low vapor pressure (U.S. EPA, 2006b).

ENVIRONMENTAL FATE AND BIOACCUMULATION

Methyl parathion is mobile in soils and can leach into groundwater and enter surface water as runoff. The chemical breaks down though microbial degradation, aqueous photolysis, hydrolysis, and incorporation into soil organic matter; thus, it degrades rapidly in soil and water with a half-life <5 days. Photodegradation is rapid in aquatic environments with a half-life of 49 h. Bioconcentration is not expected to occur (U.S. EPA, 2006b).

ECOTOXICOLOGY

Methyl parathion toxicity to nontarget species has been summarized by U.S. EPA (2006b). The chemical is characterized as highly toxic to birds, fish, and honey bees. Ducks showed an LD$_{50}$ of 6.6 mg/kg and reproductive toxicity occurred in bobwhite quail when given in the feed for several weeks. Chronic toxicity testing of freshwater fish showed reduced growth.

MAMMALIAN TOXICOLOGY

Acute Effects

Acute toxicity of methyl parathion is typical of other organophosphates with symptoms a result of AChE inhibition. Neuropathology may also occur (U.S. EPA, 2006b).

Chronic Effects

Chronic toxicity is also due to cholinesterase inhibition. Retinal degeneration, peripheral neuropathy of the proximal and distal sciatic nerve, and reduced RBC parameters have been seen in rats given methyl parathion for 2 years (U.S. EPA, 2006b).

SOURCES OF EXPOSURE

Occupational

Workers can be exposed to methyl parathion through mixing, loading, and/or applying the pesticide, or reentering treated sites. The use of human flaggers for aerial application is prohibited.

Environmental

Residues of methyl parathion can occur in food and drinking water.

INDUSTRIAL HYGIENE

The occupational exposure limits shown in the table above come with a skin notation indicating dermal absorption can occur and should be avoided.

RISK ASSESSMENTS

The RfD is based on RBC AChE inhibition, neuropathology, and changes in hematology parameters in rats given methyl paration in the diet for 2 years (U.S. EPA, 1991). The U.S. EPA has not evaluated the chemical for carcinogenicity.

ORGANOCHLORINE INSECTICIDES

ALDRIN

First aid: Decontamination, symptoms

Target organ(s): Nervous system, liver

Occupational exposure limits:

REL–TWA $= 0.25$ mg/m^3 [skin] (NIOSH, 2013)

TLV–TWA $= 0.05$ mg/m^3 [skin] (ACGIH, 2010)

Reference values:	1-Day and 10-Day Health Advisory (child) $= 0.0003$ mg/l (U.S. EPA, 2012a)
	DWEL Health Advisory $= 0.001$ mg/l (U.S. EPA, 2012a)
	Acute MRL $= 0.002$ mg/kg/day (ATSDR, 2002a)
	Chronic MRL $= 0.00003$ mg/kg/day (ATSDR, 2002a)
	RfD $= 0.00003$ mg/kg/day (U.S. EPA, 1988a)

Risk/safety phrases: Harmful if swallowed; avoid skin contact

DIELDRIN

First aid: Decontamination, symptoms

Target organ(s): Nervous system, liver

Occupational exposure limits:

REL–TWA $= 0.25$ mg/m^3 [skin] (NIOSH, 2013)

TLV–TWA $= 0.1$ mg/m^3 [skin] (ACGIH, 2010)

Reference values:	1-Day and 10-Day Health Advisory (child) $= 0.0005$ mg/l (U.S. EPA, 2012a)
	DWEL Health Advisory $= 0.002$ mg/l (U.S. EPA, 2012a)
	Intermediate MRL $= 0.0001$ mg/kg/day (ATSDR, 2002a)
	Chronic MRL $= 0.00005$ mg/kg/day (ATSDR, 2002a)
	RfD $= 0.00005$ mg/kg/day (U.S. EPA, 1990)

Risk/safety phrases: Harmful if swallowed; avoid skin contact

DDT

First aid: Decontamination, symptoms

Target organ(s): Nervous system, liver, reproductive system

Occupational exposure limits:

REL–TWA $= 0.5$ mg/m^3 (NIOSH, 2013)

TLV–TWA $= 1$ mg/m^3 (ACGIH, 2010)

Reference values: Acute MRL = 0.0005 mg/kg/day (ATSDR, 2002b)

Intermediate MRL = 0.0005 mg/kg/day (ATSDR, 2002b)

RfD = 0.0005 mg/kg/day (U.S. EPA, 1996)

Risk/safety phrases: Harmful if swallowed; avoid skin contact

MIREX AND CHLORDECONE

First aid: Decontamination, symptoms

Target organ(s): Nervous system, liver

Occupational exposure limits:

REL–TWA = 0.25 mg/m^3 [skin] (NIOSH, 2013)

TLV–TWA = 0.1 mg/m^3 [skin] (ACGIH, 2010)

Reference values: Acute MRL = 0.01 mg chloredecone/kg/day (ATSDR, 1995)

Chronic MRL = 0.0005 mg chlordecone/kg/day (ATSDR, 1995)

Chronic MRL = 0.0008 mg mirex/kg/day (ATSDR, 1995)

RfD = 0.0002 mg mirex/kg/day (U.S. EPA, 1992a)

RfD = 0.0003 mg chlordecone/kg/day (U.S. EPA, 2009)

Risk/safety phrases: Harmful if swallowed; avoid skin contact

BACKGROUND AND USES

Organochlorine compounds were widely used in agriculture and malaria control programs for decades, beginning in the mid-1940s. They declined in use in the 1960s and now have been nearly completely banned in the United States. This action was taken in response to evidence that these insecticides have a high potential for persistence and bio-magnification in food chains. Organochlorine insecticides are still used in many other countries for malaria control because of their low cost and high efficacy (IPCS, 1995). Aldrin, dieldrin, and chlordane were used extensively to control termites around buildings until the early 1970s when their use was canceled (ATSDR, 2002a). dichlorodiphenyltrichloroethane (DDT) was widely used during World War II to protect troops and civilians from malaria and other insect-borne diseases; it was also found to be useful against agriculture and forest pests. In some malaria endemic regions of the world, DDT is still used for mosquito control (ATSDR, 2002b; IPCS, 1995).

PHYSICAL AND CHEMICAL PROPERTIES

The organochlorine insecticides are persistent organic pollutants that are highly resistant to biological, photolytic, or chemical degradation due to the stable carbon–chlorine bond. These compounds are characterized by low water solubility, high lipid solubility, and semi-volatility. Large molecular weight compounds with cyclic, aromatic, and cyclodiene-type structures include the organochlorine insecticides such as DDT, chlordane, lindane, heptachlor, dieldrin, aldrin, toxaphene, mirex, and chlordecone (IPCS, 1995).

ENVIRONMENTAL FATE AND BIOACCUMULATION

The organochlorines strongly adsorb to soils and do not readily leach into groundwater due to low water solubility. However, the presence of organic solvents at hazardous waste sites may promote leaching. Limited biodegradation by microorganisms may occur in soils and water. Photodegradation and hydrolysis are not important pathways for degradation of these compounds. Volatilization and adsorption to atmospheric particulates results in global transport and deposition. Bioconcentration has been measured throughout the food chain from plants to domestic livestock and fish (IPCS, 1995).

Chemical properties resulting in limited environmental degradation mean these compounds are persistent pollutants. While aldrin is rapidly converted to dieldrin, the half-life of dieldrin on temperate soils is about 5 years. As much as 50% of applied DDT and related compounds can remain on soils for >10 years (IPCS, 1995; ATSDR, 2002a,b).

ECOTOXICOLOGY

In general the organochlorine insecticides have low acute toxicity to birds. For example, LD$_{50}$ values to selected avian species range from 6.6 to 520 mg/kg for aldrin, 26.6 to 381 mg/kg for dieldrin, 386 to >2200 mg/kg for DDT and its derivatives, and 1400–10,000 mg/kg for mirex. Despite low toxicity to adult birds, adverse effects on reproduction are well known, especially for DDT and dichlorodiphenyldichloroethylene (DDE). These compounds are associated with egg shell thinning, which adversely impacts reproductive success, most notably in birds of prey (IPCS, 1995). In addition, DDT has been linked to feminization and altered sex ratios in several gull populations.

The acute toxicity of organochlorines to aquatic invertebrates is highly variable, with aquatic insects the most sensitive aquatic organisms. The organochlorines are highly toxic to most species of fish with 96-h LC$_{50}$ values

0.53–42 µg/l for DDT, 1.1–41 µg/l for dieldrin, and 2.2–53 µg/l for aldrin. Developmental behavior of hatched fry has been shown to be altered following exposure of salmon eggs to DDT (IPCS, 1995).

MAMMALIAN TOXICOLOGY

Acute Effects

Acute toxicity from high doses of organochlorine insecticides involves CNS stimulation leading to convulsions. Other signs and symptoms may include headache, dizziness, nausea, general malaise, and vomiting. The mechanisms of nervous system stimulation differ and are not entirely understood; aldrin and dieldrin may block inhibitory activity within the brain while DDT interferes with ion movement through the nerve membrane (ATSDR, 2002a,b).

Oral LD_{50} values in laboratory animals range from 33 to 330 mg/kg for aldrin and dieldrin, and 113–1770 mg/kg for DDT, mirex, and chlordane (ATSDR, 2002a,b; ATSDR, 1995; IPCS, 1995).

Chronic Effects

Numerous epidemiology studies have been conducted to determine the health effects in humans from long-term exposure to organochlorine insecticides. Target organs include the nervous system and the liver for aldrin and dieldrin and the reproductive system for DDT. Long-term exposure of workers to aldrin or dieldrin resulted in headache, dizziness, hyperirritability, malaise, nausea, vomiting, and muscle twitching (ATSDR, 2002a). However, liver toxicity has not been conclusively observed in workers exposed to aldrin and dieldrin or in volunteers given up to 0.003 mg dieldrin/kg/day for up to 18 months. Dichlorodiphenyltrichloroethane is an endocrine disrupting compound that has estrogen-like properties. Studies in humans have shown associations between high body burdens of DDT and effects on lactation, maintenance of pregnancy, and fertility, as well as preterm births, small-for-gestational-age infants (ATSDR, 2002b), and decreased male reproductive parameters (ATSDR, 2008). A recent study found an increased risk of hypertension for adult daughters from mothers who had had higher levels of DDT during pregnancy (La Merrill et al., 2013).

Chronic oral studies in laboratory animals confirm the effects observed in humans. In the 2-year oral studies in rats, liver lesions were observed at 0.025 mg/kg/day of aldrin or dieldrin (ATSDR, 2002a), 7 mg DDT/kg/day (ATSDR, 2002b), and 0.7 mg mirex/kg/day (ATSDR, 1995). Higher doses of these compounds resulted in decreased offspring survival, reduced fertility, decreased growth of offspring, and

developmental neurotoxicity. While most of the organochlorines cause cancer in laboratory animals, epidemiology studies in humans are inconclusive.

SOURCES OF EXPOSURE

Past uses of the organochlorine insecticides have resulted in soil residues and uptake in a wide range of foods such that exposures are mainly through the diet. Exposure to aldrin and dieldrin can occur through ingestion of contaminated crops and also through inhalation of contaminated air around homes that have been treated in the past with either chemical for termite control (ATSDR, 2002a). Although DDT residues are declining, exposure is still possible in meat, fish, poultry, and dairy products. Consumption of fish from the Great Lakes and traditional foods such as seals and caribou result in a higher intake of DDT (ATSDR, 2002b). Organochlorines have been detected in hundreds of hazardous waste sites across the United States indicating potential for continued release to the environment.

INDUSTRIAL HYGIENE

The occupational exposure limits for several individual organochlorine insecticides are shown in the tables above. Aldrin and dieldrin come with a skin notation indicating dermal absorption can occur and should be avoided. Both particulate and vapor fractions are of concern for inhalation exposures.

RISK ASSESSMENTS

The RfD for aldrin, dieldrin, DDT, and mirex is based on liver lesions in rats given the chemical in the diet for 2 years. The RfD for chlordecone was based on kidney lesions in rats.

BOTANICALLY DERIVED INSECTICIDES

PYRETHRIN AND PYRETHRIODS

PERMETHRIN

First aid: Decontamination, symptoms
Target organ(s): Liver, skin, thyroid, nervous system

Occupational exposure limits:

TLV–TWA = 5 mg/m^3 (pyrethrum) (ACGIH, 2010)

REL–TWA = 5 mg/m^3 (pyrethrum) (NIOSH, 2013)

Reference values: MRL (acute) = 0.3 mg/kg/day (ATSDR, 2003)
MRL (intermediate) = 0.2 mg/kg/day (ATSDR, 2003)
RfD = 0.05 mg/kg/day (U.S. EPA, 1992b)

Risk/safety phrases: Avoid skin and eye contact

CYHALOTHRIN

First aid: Decontamination, symptoms

Target organ(s): Liver, skin, thyroid, nervous system

Occupational exposure limits: None found

Reference values: MRL (acute) = 0.01 mg/kg/day (ATSDR, 2003)
MRL (intermediate) = 0.01 mg/kg/day (ATSDR, 2003)
RfD = 0.005 mg/kg/day (U.S. EPA, 1988b)

Risk/safety phrases: Avoid skin and eye contact

BACKGROUND AND USES

Pyrethrins are used to kill a number of different flying and crawling insects and arthropods. First registered in the 1950s, currently over 1350 end-use products containing pyrethrins are available for agricultural, commercial, residential, and public health areas. They are used as household insecticides, as grain protectants, and to control pests on edible products just prior to harvest in a variety of locations, including residential, public, and commercial buildings, animal houses, warehouses, fields, and green houses. Pyrethrins are also extensively used in the field of veterinary medicine (U.S. EPA, 2006c; ATSDR, 2003).

Commercially available pyrethroids include allethrin, bifenthrin, bioresmethrin, cyfluthrin, cyhalothrin, cypermethrin, deltamethrin, esfenvalerate (fenvalerate), flucythrinate, flumethrin, fluvalinate, fenpropathrin, permethrin, phenothrin, resmethrin, tefluthrin, tetramethrin, and tralomethrin.

PHYSICAL AND CHEMICAL PROPERTIES

Pyrethrum (CAS #8003-34-7), derived from extracts of the *Chrysanthemum cinerariaefolinum* plant, is a combination of six pyrethrin isomers, namely, pyrethrin 1, pyrethrin 2, cinerin 1, cinerin 2, jasmolin 1, and jasmolin 2. Pyrethroids are synthetically derived commercial compounds similar to pyrethrum. Pyrethrins and pyrethriods are insoluble in water and have a low vapor pressure. Pyrethrum is subject to photodegradation and is oxidized rapidly in the presence of air (U.S. EPA, 2006c; ATSDR, 2003).

ENVIRONMENTAL FATE AND BIOACCUMULATION

If released to air, the relatively low vapor pressure indicates that the pyrethrins and pyrethroids will exist in both the vapor and particulate phases in the atmosphere. Vapor-phase compounds are rapidly degraded by direct photolysis and by reaction with photochemically produced hydroxyl radicals and ozone; the half-lives for these reactions in air are estimated to be 1.3 h and 17 min, respectively. Particulates may travel long distances and are removed from the atmosphere by wet or dry deposition (HSDB, 2013; ATSDR, 2003).

Pyrethrins and pyrethroids are strongly adsorbed to the soil surfaces so they are not expected to be mobile. The compounds also strongly adsorb to suspended solids and sediment in the water column. Thus, partitioning to solids attenuates volatilization from soil and water surfaces. Pyrethrins and pyrethroids are often used indoors in sprays or aerosol bombs, and the volatilization rates from glass or floor surfaces may be significantly faster than from soils since these compounds are not likely to adsorb as strongly to these surfaces (ATSDR, 2003). These insecticides are readily biodegraded by microorganisms.

Pyrethrins and pyrethroids bioconcentrate in aquatic organisms, including fish, oysters, and insects. The bioconcentration factor for several commercial products in three species of fish ranged from 180 to 1200 depending on the amount of dissolved organic matter in the water column (ATSDR, 2003).

ECOTOXICOLOGY

Pyrethrins and pyrethroids are extremely toxic to fish with acute LC$_{50}$ values of 5.1 µg/l to freshwater fish and 16 µg/l to estuarine/marine fish. Chronic exposure to very low levels has caused impaired growth of freshwater fish and reproduction of freshwater invertebrates (U.S. EPA, 2006c).

These compounds are practically nontoxic to birds and have low toxicity to small mammals. However, the pyrethrins are highly toxic to honey bees and other nontarget beneficial insects (U.S. EPA, 2006c).

MAMMALIAN TOXICOLOGY

Acute Effects

The pyrethrins and pythroids have low acute toxicity to mammals and overexposure is uncommon. In rats, LD_{50} is 1.4 g/kg and the LC_{50} is 3400 mg/m^3 (U.S. EPA, 2006c). However, exposure may be irritating to the eyes and nasal passages (HSDB, 2013) and dermal exposure may cause paresthesia as a result of the direct action of pyrethroids on sensory nerve endings (ATSDR, 2003).

Ingestion of large amounts may result in neurobehavioral effects, such as dizziness, headache, nausea, muscle twitching or tremors, reduced energy, and changes in awareness. The effects are due to the mechanism of action in which the compounds bind to sodium channels slowing the closing of channel gates following an initial influx of sodium during the depolarizing phase of an action potential, which results in a prolonged sodium tail current (ATSDR, 2003).

Chronic Effects

Allergic contact dermatitis is the most common health effect associated with pyrethrin and pyrethroid exposure (HSDB, 2013; ATSDR, 2003).

Chronic exposure to pyrethrins and pyrethroids may cause effects on the nervous system, thyroid, and liver. Some laboratory animal studies have shown increased liver weight and microscopic lesion in the liver at doses that also resulted in signs of neurotoxicity. Similarly, animals given a variety of commercial products by oral route had changes in thyroid hormones, including decreased T_3 and T_4 and increased thyroid stimulating hormone (TSH) (ATSDR, 2003; U.S. EPA, 2006c).

SOURCES OF EXPOSURE

Occupational

Exposure to pyrethrins and pyrethriods can occur during mixing and application for agriculture and residential use. Inhalation and dermal contact are of concern during use.

Environmental

Environmental exposure to pyrethrins and pyrethroids can occur following their application as indoor surface treatments or home gardens; exposures may be to the person applying the insecticide or others who enter the areas after application. Residues on fruits and vegetables are the main source of ingestion exposure (U.S. EPA, 2006c; ATSDR, 2003).

INDUSTRIAL HYGIENE

The occupational exposure limits for two individual pyrethrin and pyrethroid insecticides are shown in the tables above. Dermal contact should be avoided.

RISK ASSESSMENTS

The acute and intermediate MRLs for permethrin and cyhalothrin were based on neurotoxicity in rats and gastrointestinal effects in dogs, respectively, and are considered protective of chronic effects. The RfD for permethrin was based on increased liver weight in rats. For cyhalothrin, the RfD was based on decreased body weight gain observed in adult rats and their offspring when given the compound in the diet for three generations. The U.S. EPA has not evaluated the chemical for long-term health effects or carcinogenicity.

ROTENONE

First aid: Decontamination, symptoms

Target organ(s): Upper respiratory tract, eye, skin, nervous system

Occupational exposure limits: TLV–TWA = 5 mg/m^3 (ACGIH, 2010)
TWA = 5 mg/m^3 (NIOSH, 2013)

Reference values: RfD = 0.004 mg/kg/day (U.S. EPA, 1988c)

Risk/safety phrases: Harmful if swallowed, irritating

BACKGROUND AND USES

Rotenone was first registered in 1947 and is currently used exclusively to kill fish (U.S. EPA, 2007). In 2006, registrants voluntarily canceled all livestock, residential and home owner uses, domestic pet uses, and all other uses except for piscicide uses. Currently the main uses include fish management strategies to remove nonnative fish species from lakes, ponds, or streams and in catfish aquaculture prior to stocking ponds with with fry to remove undesirable fish species (U.S. EPA, 2006d).

Rotenone has been historically used by native people to paralyze fish for capture and consumption. Outside the United States, the compound is still used to control insects in fruit and vegetable cultivation and for control of fire ants and mosquito larvae in pond water (HSDB, 2012a).

PHYSICAL AND CHEMICAL PROPERTIES

Rotenone (CAS #83-79-4) is related to isoflavonoid compounds derived from the roots of *Derris* spp., *Lonchocarpus* spp., and *Tephrosia* spp., found primarily in Southeast Asia, South America, and East Africa. The isolated compound is an odorless, colorless to red crystalline solid. It is insoluble in water and has a very low vapor pressure (U.S. EPA, 2007; HSDB, 2012a).

ENVIRONMENTAL FATE AND BIOACCUMULATION

Rotenone released to the atmosphere will exist as particulates due to the extremely low vapor pressure. Particulate-phase rotenone will be removed from the atmosphere by wet and dry deposition and may be degraded by direct photolysis. It is mobile to moderately mobile in soil and sediment and volatilization from soil surfaces is not expected to occur to any extent. If released to water, rotenone generally degrades quickly through abiotic (hydrolytic and photolytic) mechanisms, with half-lives of a few days to several weeks or longer depending on water temperature (U.S. EPA, 2007; HSDB, 2012a).

Rotenone has a relatively low potential for bioconcentration in aquatic organisms (Bioconcentration Factor (BCF) < 30X) (U.S. EPA, 2007).

ECOTOXICOLOGY

Rotenone is extremely toxic to fish with an acute LC_{50} value of 1.94 ppb to freshwater fish. Chronic exposure to very low levels has caused impaired growth of freshwater fish and reproduction of freshwater invertebrates (U.S. EPA, 2007).

Rotenone has an extremely small margin between no lethality (5.0 μg/l) and 100% mortality (6.6 μg/l) for static-renewal 96-h toxicity tests with juvenile rainbow trout. Increasing the amount of dissolved organic carbon significantly increased the rotenone 96-h LC_{50} probably as a result of rotenone adsorption, which decreased its bioavailability. The threshold concentration of rotenone for impairment of critical swimming performance was 3.0 μg/l (Cheng and Farrell, 2007).

Rotenone has very low toxicity to birds, and it is unlikely that piscivorous birds could consume enough killed fish to result in a toxic dose. Similarly, it is unlikely that a mammal foraging on dead or dying fish could consume toxic levels of rotenone (U.S. EPA, 2007).

MAMMALIAN TOXICOLOGY

Acute Effects

Rotenone has relatively high toxicity by the oral and inhalation routes, but relatively nontoxic by dermal exposure. The acute oral LD_{50} in male and female rats is 102 and 39.5 mg/kg, respectively; the LC50 is 0.0212 mg/l (U.S. EPA, 2007). The mechanism of action is uncoupling oxidative phosphorylation within cell mitochondria by blocking electron transport at complex I.

Ingestion of large amounts may result in vomiting, respiratory depression and hypoglycemia. The dust of rotenone is irritation to the eyes (HSDB, 2012a). In case reports of human exposures, the most common symptom was eye irritation; other symptoms reported included dermal irritation, throat irritation, nausea, and cough/choke (U.S. EPA, 2007).

Chronic Effects

Numerous studies indicate that the mechanism of rotenone toxicity at the mitochondria can cause symptoms of Parkinson's disease (U.S. EPA, 2006d). A study of agricultural workers showed strong evidence of an association between rotenone use and Parkinson's disease. The disease developed 2.5 times as often in those who reported use of rotenone compared with nonusers (Tanner et al., 2011).

Decreased body weight was found in adult rats and their offspring were given rotenone orally at doses ≥1.5 mg/kg/day. Rotenone has not been shown to cause cancer in animals or humans (U.S. EPA, 1981, 2006d).

SOURCES OF EXPOSURE

Occupational

Workers can be exposed to rotenone during mixing, loading, and application activities.

Environmental

Environmental exposure to rotenone may occur from recreational use of treated ponds or lakes. However, due to the rapid degradation, it is unlikely that exposures would be excessive (U.S. EPA, 2006d, 2007).

INDUSTRIAL HYGIENE

The occupational exposure limits shown in the table above are generally based on contact irritation to mucus membranes.

RISK ASSESSMENTS

The RfD was based on decreased offspring body weight when rats were given the compound in the diet for two generations. The U.S. EPA has not evaluated the chemical for long-term health effects or carcinogenicity.

NEONICOTINOIDS

THIAMETHOXAM

First aid: Decontamination, flush eyes and skin with water, symptoms

Target organ(s): Liver, kidney, testes

Occupational exposure limits: None found

Reference values: Acute RfD = 0.35 mg/kg (U.S. EPA, 2010)
Chronic RfD = 0.012 mg/kg/day (U.S. EPA, 2010)

Risk/safety phrases: Harmful if swallowed; avoid skin contact

IMIDACLOPRID

First aid: Decontamination, flush eyes and skin with water, symptoms

Target organ(s): Liver, thyroid

Occupational exposure limits: None found

Reference values: ADI = 0–0.6 mg/kg (WHO, 2001)
Acute RfD = 0.4 mg/kg (WHO, 2001)

Risk/safety phrases: Harmful if swallowed; avoid skin contact

CLOTHIANIDRIN

First aid: Decontamination, flush eyes and skin with water, symptoms

Target organ(s): Liver, kidney

Occupational exposure limits: None found

Reference values: Acute RfD = 0.025 mg/kg (U.S. EPA, 2003)
Chronic RfD = 0.0098 mg/kg/day (U.S. EPA, 2003)

Risk/safety phrases: Harmful if swallowed; avoid skin contact

BACKGROUND AND USES

The neonicotinoids are broad-spectrum nitroguanidine insecticides with activity against sucking and chewing insects. They are used in foliar, soil, and seed applications on a wide variety of crops, including fruiting and root vegetables, berries, pome and stone fruits, grains, cotton, turf and sod, and ornamentals (U.S. EPA, 2010; HSDB, 2006, 2012b). When used as a seed dressing, the insecticides become systemic in the plant, migrating in the sap to all parts of the plant and providing protection against pests. Imidacloprid is the active ingredient in Advantage™ used to control fleas in dogs and cats (HSDB, 2006). Clothianidin is the major metabolite of thiamethoxam and both compounds are registered for use as insecticides.

Widespread use increased greatly in the 1990s as alternatives to organophosphorus and carbamate insecticides because of their much lower mammalian toxicity and resistance developed to other pesticides. Imidacloprid has become the most widespread insecticide used in the world.

PHYSICAL AND CHEMICAL PROPERTIES

Thiamethoxam (153719-23-4), imidacloprid (138261-41-3), and clothianidin (210880-92-5) are solid crystals or powder under ambient conditions and have a low volatility (U.S. EPA, 2010; HSDB, 2006, 2012b).

The compounds have relatively high water solubility and have been measured in runoff and surface waters (ABC, 2013).

ENVIRONMENTAL FATE AND BIOACCUMULATION

Neonicotinoids are expected to be persistent, resistant to hydrolysis, and mobile in terrestrial and aquatic environments. These fate properties indicate that the compounds have a potential to move into surface water and shallow groundwater (U.S. EPA, 2003, 2010). In general, surface water surveys suggest that concentrations of several neonicotinoids are high enough to cause impacts on aquatic food chains by affecting invertebrates (ABC, 2013).

Imidacloprid is rapidly and almost completely absorbed (>92%) from the gastrointestinal tract of rats, and is eliminated from the organism rapidly and completely, with no indication of bioaccumulation of the parent compound or its metabolites (WHO, 2001). Similarly, thiamethoxam has relatively low solubility in nonpolar organic solvents and its octanol/water partition coefficient suggests that accumulation in fatty tissues is unlikely to occur (U.S. EPA, 2010).

ECOTOXICOLOGY

Neonicotinoids are extremely toxic to pollinators such as honey bees. Because these insecticides are systemic, pollinators are potentially exposed in the nectar and pollen during the flowering period of some crops. At nonlethal levels (1.34 ng/bee), thiamethoxam has been shown to increase mortality of bee foragers due to reduced homing success, an effect that can affect the survival of the entire colony (Henry et al., 2012). Similarly, bumble bee colonies exposed to imidacloprid at levels occurring in treated crops and allowed to develop naturally in the field showed reductions in both colony growth rate and in the production of new queens (Whitehorn et al., 2012).

Seeds treated with neonicotinoids can be toxic to bees and birds. Thiamethoxam and imidicloprid are toxic to fish and aquatic invertebrates (U.S. EPA, 2012b, 2012c).

Oral LD_{50} values in various species of birds range from 15 to 152 mg imidacloprid/kg, 576 to 1552 mg thiamethoxam/kg, and 430 to >2000 mg clothianidin/kg (ABC, 2013). Exposure to treated seeds may cause adverse chronic effects in birds, such as eggshell thinning (U.S. EPA, 2003; ABC, 2013).

MAMMALIAN TOXICOLOGY

Acute Effects

Neonicotinoids bind selectively to insect nicotinic ACh receptors, although the specific binding site(s)/receptor(s) are unknown. Some evidence suggests that clothianidin operates by direct competitive inhibition, while thiamethoxam is a noncompetitive inhibitor. Structural variations between the insect and mammalian nicotinic ACh receptors produce quantitative differences in the binding affinity of the neonicotinoids towards these receptors, which confers the notably greater selective toxicity of this class towards insects, compared to mammals. These insecticides do not inhibit cholinesterase or interfere with sodium channels and, therefore, have a different mode of action than organophosphate, carbamate, and pyrethroid insecticides (U.S. EPA, 2010).

Imidacloprid was moderately toxic to rats and mice with LD_{50} values of 380–650 mg/kg and 130–170 mg/kg, respectively. Transient clinical signs of toxicity included behavioral and respiratory signs, disturbances of motility, narrowed palpebral fissures, transient trembling, and spasms in rats and mice at doses ≥200 mg/kg and ≥71 mg/kg, respectively (WHO, 2001). Clothianidrin is relatively nontoxic to rats with an oral LD_{50} >500 mg/kg and no evidence of eye or dermal irritation (U.S. EPA, 2003).

Chronic Effects

Potential chronic effects of neonicotinoids to humans are unknown. A number of laboratory animal studies suggest different target organs for the individual compounds. For example, while the insecticidal action of the neonicotinoids is neurotoxic, the most sensitive toxicological effect in mammals was testicular lesions with thiamethoxam and thyroid gland lesions with imidacloprid. Thus, there is currently no evidence to indicate that neonicotinoids share common mechanisms of toxicity (U.S. EPA, 2010).

In laboratory animal studies with imidacloprid, reduced body weight gain, often accompanied by decreased food consumption, was the most sensitive toxicological end point in mice at doses >86 mg/kg bw/day, in rats at doses >30 mg/kg/day, in rabbits at doses >24 mg/kg/day, and in dogs at doses >22 mg/kg/day. Similarly, decreased body weight gain was the main effect during lactation in rat pups of dams given doses >6.6 mg/kg/day (WHO, 2001). The commercially available flea treatment given either dermally (intended use) or orally did not cause adverse effects in dogs or cats (Cal EPA, 2012).

The liver was the other main target organ after repeated administration of imidacloprid to mice, rats, and dogs at doses >410 mg/kg/day, >17 mg/kg/day, and >31 mg/kg/day, respectively. Effects in the liver included adaptive changes such as induction of hepatic microsomal enzymes as well as disturbances of hepatic function and microscopic lesions. Changes in blood cholesterol, triglyceride, protein, and albumin concentrations as well as increased activities of alanine aminotransferase, alkaline phosphatase, and galactodehydrogenase in plasma were also observed. An increased incidence of mineralization was seen in the colloid of thyroid gland follicles in a long-term study in rats given a dietary concentration equal to 17 mg/kg/day, although the plasma concentrations of thyroid hormones remained unchanged. No evidence of a carcinogenic effect of imidacloprid was found in either mice or rats in the long-term studies of dietary administration (WHO, 2001).

Male rats given diets containing up to 1500 parts per million (ppm) thiamethoxam for 2 years had kidney lesions which are specific to rats and not likely relevant to other mammals. Higher doses to rats for two generations resulted in decreased body weight of both adults and pups and testicular lesions in adult males, but no adverse effects on reproductive parameters. Male dogs given 750 ppm in the feed for 1 year had testicular atrophy. Pregnant rats given thiamethoxam had decreased body weight gain with 200 mg/kg/day and fetuses had decreased ossification at higher doses. Developmental neurological effects on brain morphometry measurements were seen in offspring when their dams were given about 300 mg/kg/day during gestation and lactation (Cal EPA, 2008).

Liver tumors have been observed in male and female mice given ≥500 ppm thiamethoxam in the diet for 18 months. Liver tumors are not seen in rats fed diets containing up to 3000 ppm thiamethoxam. The key metabolites responsible for the events leading to liver tumors in the mouse are not formed in sufficient quantities in the rat and explain the lack of a carcinogenic response in this species. Limited data suggest that humans are more similar to the rat than the mouse in producing the active metabolites. Thus, the mouse appears to be uniquely sensitive to this mode of action and humans are unlikely to be at risk for developing tumors following exposures to thiamethoxam (U.S. EPA, 2010).

With clothianidin, decreased body weight and food consumption occurred in rats and mice given up to 98 and 200 mg/kg/day, respectively, with females of both species slightly more sensitive. At these same doses, liver and kidney lesions were observed in rats and liver hypertrophy occurred in mice. Doses of ≥180 mg/kg/day to rats for two generation resulted in decreased parental and offspring body weight but no effects on reproductive end points (Cal EPA, 2003).

SOURCES OF EXPOSURE

Occupational

The neonicotinoids are to be applied by commercial applicators only, with the exception of flea treatments for pets. Farmers handling treated seeds should wear gloves, long sleeved shirts, and long pants to avoid contact.

Environmental

Exposures in the environment may occur through spray drift from aerial application and dermal contact from reentry to treated fields. Contaminated surface waters are another source of environmental exposures. Use of neonicotinoids on turf or sod, in crack and crevice applications, and flea treatments on pets are potential routes of exposure for the homeowner/family and general public.

INDUSTRIAL HYGIENE

Occupational exposure limits for several individual neonicotinoid insecticides have not been set by National Institute for Occupational Safety and Health (NIOSH) or American Conference of Governmental Industrial Hygienists (ACGIH).

RISK ASSESSMENTS

The risk assessments shown in the tables above were developed by the U.S. EPA as part of the pesticide registration process.

REFERENCES

ABC (American Bird Conservancy) (2013) *The Impact of the Nation's Most Widely Used Insecticides on Birds*, 98 pp. Available at www.abcbirds.org/abcprograms/policy/toxins/Neonic_FINAL.pdf.

ACGIH (American Conference of Governmental Industrial Hygienists) (2010) *TLVs and BEIs Based on the Documentation of the Threshold Limit Values for Chemical Substances and Physical Agents and Biological Exposure Indices*, Cincinnati, OH: American Conference of Governmental Industrial Hygienists.

Aiuto, L.A., Paylakis, S.G., and Boxer, R.A. (1993) Life-threatening organophosphate-induced delayed polyneuropathy in a child after accidental chlorpyrifos ingestion. *J. Pediatr.* 122:658–660.

ATSDR (Agency for Toxic Substances and Disease Registry) (1995) *Toxicological Profile for Mirex and Chlordecone*, Atlanta, GA: U.S. Department of Health and Human Services, Public Health Service, 362 pp.

ATSDR (Agency for Toxic Substances and Disease Registry) (1997) *Toxicological Profile for Chlorpyrifos*. Atlanta, GA: U.S. Department of Health and Human Services, Public Health Service, 179 pp.

ATSDR (Agency for Toxic Substances and Disease Registry) (2002a) *Toxicological Profile for Aldrin/Dieldrin*, Atlanta, GA: U.S. Department of Health and Human Services, Public Health Service, 354 pp.

ATSDR (Agency for Toxic Substances and Disease Registry) (2002b) *Toxicological Profile for DDT, DDE, and DDD*, Atlanta, GA: U.S. Department of Health and Human Services, Public Health Service, 497 pp.

ATSDR (Agency for Toxic Substances and Disease Registry) (2003) *Toxicological Profile for Pyrethrins and Pyrethroids*, Atlanta, GA: U.S. Department of Health and Human Services, Public Health Service, 328 pp.

ATSDR (Agency for Toxic Substances and Disease Registry) (2008) *Addendum to the DDT/DDD/DDE Toxicological Profile*, Atlanta, GA: U.S. Department of Health and Human Services, Public Health Service, 73 pp.

Ballantyne, B. and Salem, H. (2006) Occupational toxicology and occupational hygiene aspects of organophosphate and carbamate anticholinesterases with particular reference to pesticides. In: Gupta, R.C., editor. *Toxicology of Organophosphate & Carbamate Compounds*, Burlington, MA: Elsevier Academic Press, pp. 567–598.

Cal EPA (California Environmental Protection Agency) (2003) *Toxicology Data Review Summary for Clothianidin (210880-92-5)*, Department of Pesticide Regulation. Available at http://www.cdpr.ca.gov/docs/risk/toxsums/toxsumlist.htm.

Cal EPA (California Environmental Protection Agency) (2008) *Toxicology Data Review Summary for Thiamethoxam (153719-23-4)*, Department of Pesticide Regulation. Available at http://www.cdpr.ca.gov/docs/risk/toxsums/toxsumlist.htm.

Cal EPA (California Environmental Protection Agency) (2012) *Toxicology Data Review Summary for Imidacloprid (138261-41-3)*, Department of Pesticide Regulation. Available at http://www.cdpr.ca.gov/docs/risk/toxsums/toxsumlist.htm.

Capodicasa, E., Scapellato, M.L., Moretto, A., Caroldi, S., and Lotti, M. (1991) Chlorpyrifos-induced delayed polyneuropathy. *Arch. Toxicol.* 65:150–155.

Chandrasekara, L.W. and Pathiratne, A. (2007) Body size-related differences in the inhibition of brain acetylcholinesterase activity in juvenile Nile tilapia (*Oreochromis niloticus*) by chlorpyrifos and carbosulfan. *Ecotoxicol. Environ. Saf.* 67:109–119.

Cheng, W.W. and Farrell, A.P. (2007) Acute and sublethal toxicities of rotenone in juvenile rainbow trout (*Oncorhynchus mykiss*): swimming performance and oxygen consumption. *Arch. Environ. Contam. Toxicol.* 52:388–396.

De Bleeker, J.L. (2006) Intermediate syndrome in organophosphate poisoning. In: Gupta, R.C., editor. *Toxicology of Organophosphate & Carbamate Compounds*, Burlington, MA: Elsevier Academic Press, pp. 371–380.

Fonnum, F. and Sterri, S.H. (2006) Tolerance development to toxicity of cholinesterase inhibitors. In: Gupta, R.C., editor. *Toxicology of Organophosphate & Carbamate Compounds*, Burlington, MA: Elsevier Academic Press, pp. 257–270.

Gaines, T.B. (1969) Acute toxicity of pesticides. *Toxicol. Appl. Pharmacol.* 4:515–534.

Gupta, R.C. (2006) Classification and uses of organophosphates and carbamates. In: Gupta, R.C., editor. *Toxicology of Organophosphate & Carbamate Compounds*, Burlington, MA: Elsevier Academic Press, pp. 5–24.

Henry, M., Béguin, M., Requier, F., Rollin, O., Odoux, J.F., Aupimel, P., Aptel, J., Tchamitchian, S., and Decouteye, A. (2012) A common pesticide decreases foraging success and survival in honey bees. *Science* 336:348–350.

HSDB (Hazardous Substances Data Bank) (2006) *Imidaclorpid*. HSDB, National Library of Medicine.

HSDB (Hazardous Substances Data Bank) (2008) *Chlorpyrifos*. HSDB, National Library of Medicine.

HSDB (Hazardous Substances Data Bank) (2012a) *Rotenone*. HSDB, National Library of Medicine.

HSDB (Hazardous Substances Data Bank) (2012b) *Thiamethoxam*. HSDB, National Library of Medicine.

HSDB (Hazardous Substances Data Bank) (2013) *Pyrethrum*. HSDB, National Library of Medicine.

IPCS (International Programme on Chemical Safety) (1995) *A Review of Selected Persistent Organic Pollutants*. PCS/95.39, 149 pp.

Johnson, M.K. (1990) Organophosphates and delayed neuropathy—is NTE alive and well? *Toxicol. Appl. Pharmacol.* 102:385–399.

Kaplan, J.G., Kessler, J., Rosenberg, N., Pack, D., and Schaumburg, H.H. (1993) Sensory neuropathy associated with Dursban (chlorpyrifos) exposure. *Neurology* 43:2193–2196.

Karalliedde, L., Baker, D., and Marrs, T.C. (2006) Organophosphate-induced intermediate syndrome: aetiology and relationships with myopathy. *Toxicol. Rev.* 25:1–14.

La Merrill, M., Cirillo, P.M., Terry, M.B., Krigbaum, N.Y., Flom, J.D., and Cohn, B.A. (2013) Prenatal exposure to the pesticide DDT and hypertension diagnosed in women before age 50: a longitudinal birth cohort study. *Environ. Health Perspect.* 121:594–599.

Lotti, M. (1992) The pathogenesis of organophosphate polyneuropathy. *Crit. Rev. Toxicol.* 21:465–487.

Lotti, M. and Moretto, A. (2005) Organophosphate-induced delayed polyneuropathy. *Toxicol. Rev.* 24:37–49.

Lotti, M., Moretto, A., Zoppellari, R., Dainese, R., Rizzato, N., and Barusco, G. (1986) Inhibition of lymphocytic neuropathy target esterase predicts the development of organophosphate-induced polyneuropathy. *Arch. Toxicol.* 59:176–179.

Lowit, A.B. (2006) Federal regulation and risk assessment of organophosphate and carbamate pesticides. In: Gupta, R.C., editor. *Toxicology of Organophosphate & Carbamate Compounds*, Burlington, MA: Elsevier Academic Press, pp. 617–634.

Marrs, T.C. and Vale, J.A. (2006) Management of organophosphorus pesticide poisoning. In: Gupta, R.C., editor. *Toxicology of Organophosphate & Carbamate Compounds*, Burlington, MA: Elsevier Academic Press, pp. 715–733.

Maurissen, J.P.J., Hoberman, A.M., Garman, R.H., and Hanley, T.R., Jr. (2000a) Lack of selective developmental neurotoxicity in rat pups from dams treated by gavage with chlorpyrifos. *Toxicol. Sci.* 57:250–263.

Maurissen, J.P.J., Shankar, M.R., and Mattsson, J.L. (2000b) Chlorpyrifos: lack of cognitive effects in adult Long-Evans rats. *Neurotoxicol. Teratol.* 22:237–246.

McCauley, L.A. (2006) Organophosphates and the Gulf War syndrome. In: Gupta, R.C., editor. *Toxicology of Organophosphate & Carbamate Compounds*, Burlington, MA: Elsevier Academic Press, pp. 69–78.

Moretto, A. and Lotti, M. (2006) Peripheral nervous system effects and delayed neuropathy. In: Gupta, R.C., editor. *Toxicology of Organophosphate & Carbamate Compounds*, Burlington, MA: Elsevier Academic Press, pp. 361–370.

Moretto, A., Lotti, M., Sabri, M.I., and Spencer, P.S. (1987) Progressive deficit of retrograde axonal transport is associated with the pathogenesis of di-*n*-butyl dichlorvos axonopathy. *J. Neurochem.* 49:1515–1522.

Mortensen, S.R. (2006) Toxicity of organophosphorus and carbamate insecticides using birds as sentinels for terrestrial vertebrate wildlife. In: Gupta, R.C., editor. *Toxicology of Organophosphate & Carbamate Compounds*, Burlington, MA: Elsevier Academic Press, pp. 673–680.

NIOSH (National Institute for Occupational Safety and Health) (2013) *Pocket Guide to Chemical Hazards*, Centers for Disease Control and Prevention. Available at http://www.cdc.gov/niosh/npg/.

NRC (National Research Council) (2003) *Acute Exposure Guideline Levels for Selected Airborne Chemicals*, vol. 3, Subcommittee on Acute Exposure Guideline Levels, Committee on Toxicology, National Research Council of the National Academies, Washington, DC: The National Academies Press, 497 pp.

Perera, F.P., Rauh, V., Tsai, W.Y., Kinney, P., Camann, D., Barr, D., Bernert, T., Garfinkel, R., Tu, Y.H., Diaz, D., Dietrich, J., and Whyatt, R.M. (2003) Effects of transplacental exposure to environmental pollutants on birth outcomes in a multiethnic population. *Environ. Health Perspect.* 111:201–205.

Sandahl, J.F., Baldwin, D.H., Jenkins, J.J., and Scholz, N.L. (2005) Comparative thresholds for acetylcholinesterase inhibition and behavioral impairment in coho salmon exposued to chlorpyrifos. *Environ. Toxicol. Chem.* 24:136-145.

Senanayake, N. and Karalliedde, L. (1987) Neurotoxic effects of organophosphorus insecticides. An intermediate syndrome. *N. Engl. J. Med.* 316:761–763.

Steele, L., Sastre, A., Gerkovich, M.M., and Cook, M.R. (2012) Complex factors in the etiology of Gulf War illness: wartime exposures and risk factors in veteran subgroups. *Environ. Health Perspect.* 120:112–118.

Sultatos, L.G. (2006) Interatction of organophosphorus and carbamate compounds with cholinesterases. In: Gupta, R.C., editor. *Toxicology of Organophosphate & Carbamate Compounds*, Burlington, MA: Elsevier Academic Press, pp. 209–218.

Tanner, C.M., Kamel, F., Ross, G.W., Hoppin, J.A., Goldman, S.M., Korell, M., Marras, C., Bhudhikanok, G.S., Kasten, M., Chade, A.R., Comyns, K., Richards, M.B., Meng, C., Priestley, B., Fernandez, H.H., Cambi, F., Umbach, D.M., Blair, A., Sandler, D.P., and Langston, J.W. (2011) Rotenone, paraquat, and Parkinson's disease. *Environ. Health Perspect.* 119: 866–872.

Tierney, K., Casselman, M., Takeda, S., Farrell, T., and Kennedy, C. (2007) The relationship between cholinesterase inhibition and two types of swimming performance in chlorpyrifos-exposed coho salmon (*Oncorhynchus kisutch*). *Environ. Toxicol. Chem.* 26:998–1004.

U.S. EPA (U.S. Environmental Protection Agency) (1981) *Carcinogenic Potential of Rotenone: Subchronic Oral and Peritoneal Administration to Rats and Chronic Dietary Administration to Syrian Golden Hamsters.* EPA 600/S1-81-037, 6 pp.

U.S. EPA (U.S. Environmental Protection Agency) (1987) *Carbofuran.* Integrated Risk Information System (IRIS), U.S. EPA. Available at http://www.epa.gov/iris/.

U.S. EPA (U.S. Environmental Protection Agency) (1988a) *Aldrin.* Integrated Risk Information System (IRIS), U.S. EPA. Available at http://www.epa.gov/iris/.

U.S. EPA (U.S. Environmental Protection Agency) (1988b) *Cyhalothrin.* Integrated Risk Information System (IRIS), U.S. EPA. Available at http://www.epa.gov/iris/.

U.S. EPA (U.S. Environmental Protection Agency) (1988c) *Rotenone.* Integrated Risk Information System (IRIS), U.S. EPA. Available at http://www.epa.gov/iris/.

U.S. EPA (U.S. Environmental Protection Agency) (1990) *Dieldrin.* Integrated Risk Information System (IRIS), U.S. EPA. Available at http://www.epa.gov/iris/.

U.S. EPA (U.S. Environmental Protection Agency) (1991) *Methyl Parathion.* Integrated Risk Information System (IRIS), U.S. EPA. Available at http://www.epa.gov/iris/.

U.S. EPA (U.S. Environmental Protection Agency) (1992a) *Mirex.* Integrated Risk Information System (IRIS), U.S. EPA. Available at http://www.epa.gov/iris/.

U.S. EPA (U.S. Environmental Protection Agency) (1992b) *Permethrin.* Integrated Risk Information System (IRIS), U.S. EPA. Available at http://www.epa.gov/iris/.

U.S. EPA (U.S. Environmental Protection Agency) (1996) *DDT.* Integrated Risk Information System (IRIS), U.S. EPA. Available at http://www.epa.gov/iris/.

U.S. EPA (U.S. Environmental Protection Agency) (1999) *Registration Eligibility Science Chapter for Chlorpyrifos Fate and Environmental Risk Assessment Chapter.* U.S. EPA, Office of Pesticide Programs. Special docket EPA-HQ-OPP-2007-0151.

U.S. EPA (U.S. Environmental Protection Agency) (2000) *Human Health Risk Assessment: Chlorpyrifos.* U.S. EPA, Office of Pesticide Programs. Special docket EPA-HQ-OPP-2007-0151.

U.S. EPA (U.S. Environmental Protection Agency) (2003) *Pesticide Fact Sheet: Clothianidin.* Office of Prevention, Pesticides and Toxic Substances, U.S. EPA, 19 pp.

U.S. EPA (U.S. Environmental Protection Agency) (2006a) *Interim Reregistration Eligibility Decision: Carbofuran.* EPA-738-R-06-031, 44 pp.

U.S. EPA (U.S. Environmental Protection Agency) (2006b) *Interim Reregistration Eligibility Decision for Methyl Parathion.* Memorandum dated July 31, 2006, 88 pp.

U.S. EPA (U.S. Environmental Protection Agency) (2006c) *Reregistration Eligibility Decision for Pyrethrins.* EPA 738-R-06-004, 108 pp.

U.S. EPA (U.S. Environmental Protection Agency) (2006d) *Rotenone: Final HED Chapter of the Reregistration Eligibility Decision Document (RED).* Memorandum dated June 28, 2006, 71 pp.

U.S. EPA (U.S. Environmental Protection Agency) (2007) *Reregistration Eligibility Decision for Rotenone.* EPA 738-R-07-005, 44 pp.

U.S. EPA (U.S. Environmental Protection Agency) (2009) *Chlordecone.* Integrated Risk Information System (IRIS), U.S. EPA. Available at http://www.epa.gov/iris/.

U.S. EPA (U.S. Environmental Protection Agency) (2010) *Thiamethoxam—Human Health Risk Assessment for New Seed Treatment Use on Onions, Dry Bulb, Eliminating the Current Geographic Use Restrictions for the Foliar Treatment of Barley, and Review of Other Conditional Registration Data.* Memorandum dated May 3, 2010, 51 pp.

U.S. EPA (U.S. Environmental Protection Agency) (2012a) *Fall 2012 Edition of the Drinking Water Standards and Health Advisories.* Office of Water, U.S. EPA.

U.S. EPA (U.S. Environmental Protection Agency) (2012b) *Amendment (R170) to Add New Seed Treatment Use on Small Cereal Grains (Crop Group 15 and 16). Thiamethoxam Technical, Cruiser Insecticide.* Letter dated February 22, 2012.

U.S. EPA (U.S. Environmental Protection Agency) (2012c) *Submission of Amended Labeling to Add the Following Me-Too Uses: Peanuts, Soybeans, Glove Artichoke, and Bulb Vegetables. Imidacloprid 4FL AG.* Letter dated February 24, 2012.

Whitehorn, P.R., O'Conner, S., Wackers, F.L., and Goulson, D. (2012) Neonicotinoid pesticide reduces bumble bee colony growth and queen production. *Science* 336:351–352.

WHO (World Health Organization) (2001) *WHO/FAO Joint Meeting on Pesticide Residues on Imidacloprid (138261-41-3).* Available at http://www.inchem.org/pages/jmpr.html.

Whyatt, R.M., Rauh, V., Barr, D.B., Camann, D.E., Andrews, H.F., Garfinkel, R., Hoepner, L.A., Diaz, D., Dietrich, J., Reyes, A., Tang, D., Kinney, P.L., and Perera, F.P. (2004) Prenatal insecticide exposures and birth weight and length among an urban minority cohort. *Environ. Health Perspect.* 112:1125–1132.

Yano, B.L., Young, J.T., and Mattsson, J.L. (2000) Lack of carcinogenicity of chlorpyrifos insecticide in a high-dose, 2-year dietary toxicity study in Fischer 344 rats. *Toxicol. Sci.* 53:135–144.

87

HERBICIDES

Giffe T. Johnson

INTRODUCTION

In 2007, over 15 million dollars of herbicides were sold globally, constituting 39% of total pesticide sales. More than a third of those sales were in the United States where herbicides make up to 47% of total pesticide use. The agriculture industry uses 83% of the herbicides produced, which amounted to 442 million pounds in 2007. Herbicides are among the most widely used pesticides and since 2001 in the United States glyphosate and atrazine have had the highest estimated use for pesticides in terms of total pounds applied (EPA, 2011). Exposure to these agents can occur in agricultural, commercial, and residential settings.

The purpose of this chapter is to provide a broad overview of various classes of commonly used herbicides. The most up to date label information should always be consulted before transporting, mixing, or applying pesticides.

REFERENCES

Environmental Protection Agency (EPA) (2011) *Pesticides Industry Sales and Usage: 2006 and 2007 Market Estimates*, Washington, DC: Biological and Economic Analysis Division, Office of Pesticide Programs.

TRIAZINES

ATRAZINE

BACKGROUND AND USES

Atrazine is used as a selective herbicide to control broadleaf and grassy weeds for agriculture and other land not used for crops. In agriculture, atrazine is used on corn, sugarcane, and pineapple and for orchards, sod, tree plantations, and rangeland. Atrazine is moderately persistent in the environment because of its low solubility. It can be detected in the water table and in the upper layers of the soil profile in many areas (Huang and Frink, 1989). The Environmental Protection Agency (EPA) reported that atrazine was one of the two most commonly used agricultural herbicides in 2007 (EPA, 2011). It is an active ingredient in many brands, including Actinite PK, Atranex, Atrasine, Atrataf, Atrazin, Chromozin, Cyazin, Primatol A, Primase, AAtre, Griffex, and Weedex.

PHYSICAL AND CHEMICAL PROPERTIES

Atrazine (CAS #1912-24-9) is generally found as a dibromide salt. It has a solubility of 0.003% by weight in water and a vapor pressure of <0.0000003 mmHg at 20 °C (68 °F).

MAMMALIAN TOXICOLOGY

Acute Effects

Atrazine has a low oral acute toxicity in mammals with a minimum risk level of 0.01 mg/kg/day for oral exposure. There is little clinical experience with atrazine, and adverse effects in humans have been limited to cases of intentional ingestion (Pommery et al., 1993). High doses in cattle and sheep have caused anorexia, ataxia, salivation, dyspnea, and muscle spasms. Atrazine does not pose an acute inhalation hazard and has not been reported to cause skin irritation. Exposure to concentrated amounts may result in short-term eye and skin irritation.

Chronic Effects

Most chronic effects associated with long-term atrazine exposure are related to the potential for endocrine effects, developmental toxicity, or reproductive toxicity. Outcomes related to endocrine toxicity include increases in pituitary weight and alterations in levels of reproductive hormones. However, these effects occur only after substantial exposure and differ by receptor species. Pituitary prolactin levels increased and serum prolactin and luteinizing hormone levels decreased in ovariectomized, estrogen supplemented female Long-Evans rats receiving gavage doses of 50 mg/kg/day and higher for 1–3 days (Cooper et al., 2000). Similarly treated Sprague-Dawley rats exhibited similar outcomes after receiving 300 mg/kg/day via gavage for 1–3 days (Cooper et al., 2000). Serum estradiol levels were increased in Sprague-Dawley rats after receiving gavage doses of 200 mg/kg/day on gestational days 1–8 (Cummings et al., 2000). As well, juvenile male rats exposed to 50 mg/kg/day for 3 days experienced decreases in serum and intratesticular testosterone (Friedmann, 2002). At 120 mg/kg/day and higher, in rats, reproductive outcomes such as decreased fertility and altered estrus cycle have been observed (Cooper et al., 2000; Simic et al., 1994).

Mechanism of Action(s)

Several animal experiments have been conducted to determine whether atrazine has teratogenic potential. Infurna et al. (1988) conducted an *in vivo* study of atrazine's teratogenic effects in rats and rabbits. Statistically significant teratogenic effects were observed only at a dose sufficient to cause maternal toxicity (70 mg/kg). These researchers reported that rabbits were much more sensitive to atrazine exposure than rats. In another study, rats given dietary doses up to 500 ppm did not show any adverse reproductive effects (Giknis et al., 1988). A third study showed that rats injected with atrazine at dosages of 800, 1000, and 2000 mg/kg on the third, sixth, and ninth day of gestation, respectively, did have reduced litter size and increased resorptions (Peters and Cook, 1973).

Reproductive effects do not appear to be the result of intrinsic estrogenic activity of atrazine. In the uterus of Sprague-Dawley, Long-Evans, or Donryu rats following 28 days of oral exposure up to 50 mg/kg/day atrazine there were no observed increases in BrdU-positive cells (Aso et al., 2000). A 300 mg/kg/day oral dose for 3 days in Sprague-Dawley rats demonstrated no increases in uterine weight, cytosolic progesterone receptor binding, or peroxidase activity compared to positive controls that received estradiol and demonstrated increases in all three outcomes (Connor et al., 1996).

Weisenberger et al. (1988) reported that atrazine alone was not mutagenic in the Ames assay. However, the authors did find that *N*-nitrosoatrazine, formed readily at an acidic pH when nitriles are present, was mildly mutagenic. This finding is relevant because nitrogen fertilizers are commonly found in farmlands where atrazine is also used.

Gojmerac and Kniewald (1989) examined the fate of metabolized atrazine in rats *in vivo*. The most common residues found were deethylatrazine, atrazine, and deisopropylatrazine, located in the liver, kidneys, and brain. These metabolites indicate that atrazine undergoes partial *N*-dealkylation. Total *N*-dealkylation follows with the formation of diaminoatrazine.

Atrazine apparently is not bioactivated by mammalian microsomal preparations (Dunkelberg et al., 1994).

SOURCES OF EXPOSURE

Occupational

The principal occupational exposure is for agricultural workers who may be directly applying formulations containing atrazine or workers harvesting crops in areas treated with atrazine. Atrazine is considered as a restricted use pesticide and is not available for private use. However, exposure can also occur during manufacture and transport. Other than direct exposure from manufacture, transport, mixing, and application, some exposure can occur from working in soils contaminated with atrazine after treatment. The half-life of atrazine in soils may be 2 to 14 weeks depending on environmental conditions. Workers exposed indirectly may be more susceptible to exposure due to a lack of personal protective equipment normally worn by applicators.

Environmental

In analyses conducted by the FDA and USDA atrazine residue concentrations on food products with common atrazine use was very low (0.001–0.028 µg/g) where it was detected. Occasionally, atrazine is detected in water supplies such as rural well water at low concentrations. While most people have little to no potential for exposure to atrazine, people who live near agricultural areas that regularly use atrazine have increased potential to be exposed. Airborne exposure potential results from atrazine-containing dusts or precipitation that has absorbed atrazine-containing dusts. However, these concentrations are also small; one study in Paris, France measured rainfall concentrations of atrazine as <5–400 ng/l with a median of 50 ng/l (Chevreuil et al., 1996).

INDUSTRIAL HYGIENE

The National Institute for Occupational Safety and Health (NIOSH) and the American Conference of Governmental Industrial Hygienists (ACGIH) have set the recommended exposure limit (REL) and the threshold limit value (TLV), respectively, for atrazine at 5 mg/m^3 (NIOSH, 2013; ACGIH, 2013). The allowable tolerances for atrazine in foodstuffs range from 0.02 ppm to 0.25 ppm. Allowances for fodder are significantly higher (EPA, 1988). Immuno-assays may be helpful in monitoring atrazine residues (Goh et al., 1993); however, some reagents have been shown to interfere with this method of atrazine quantitation (Stearman and Wells, 1993). Exposure to atrazine correlates with urinary diaminoatrazine and other urinary metabolites (Ikonen et al., 1988; Catenacci et al., 1990; Meli et al., 1992). Appropriate work clothing and eye protection should be used to limit exposure when mixing, transporting, and applying atrazine.

MEDICAL MANAGEMENT

Flushing skin and eyes with water to remove minor exposures is appropriate. In cases of extreme acute atrazine exposure, basic life-support measures should be taken. Emesis should not be induced, but gastric lavage may be helpful if performed shortly after ingestion. Administration of oral charcoal may enhance total body clearance of atrazine. If atrazine has been inhaled, the patient should be moved to fresh air. If the patient develops difficulty breathing, humidified oxygen should be administered.

REFERENCES

American Conference of Governmental Industrial Hygienists (ACGIH) (2013) *Threshold Limit Values (TLVs) for Chemical Substances and Physical Agents and Biological Exposure Indices (BEIs)*, Cincinnati, OH: American Conference of Governmental Industrial Hygienists (ACGIH).

Aso, S., Anai, N., Noda, S., et al. (2000) Twenty-eight-day repeated-dose toxicity studies for detection of weak endocrine disrupting effects of nonylphenol and atrazine in female rats. *J. Toxicol. Pathol.* 13:13–20.

Catenacci, G., et al. (1990) Assessment of human exposure to atrazine through determination of free atrazine in urine. *Bull. Environ. Contam. Toxicol.* 44:1–7.

Chevreuil, M., Garmouma, M., Teil, M.J., et al. (1996) Occurrence of organochlorines (PCBs, pesticides) and herbicides (triazines, phenylureas) in the atmosphere and in the fallout from urban and rural stations of the Paris area. *Sci. Total Environ.* 182:25–37.

Connor, K., Howell, J., Chen, I., et al. (1996) Failure of chloro-*s*-triazine-derived compounds to induce estrogen receptor-mediated responses *in vivo* and *in vitro*. *Fundam. Appl. Toxicol.* 30:93–101.

Cooper, R.L., Stoker, T.E., Tyrey, L., et al. (2000) Atrazine disrupts the hypothalamic control of pituitary–ovarian function. *Toxicol. Sci.* 53:297–307.

Cummings, A.M., Rhodes, B.E., and Cooper, R.L. (2000) Effect of atrazine on the implantation and early pregnancy in 4 strains of rats. *Toxicol. Sci.* 58:135–143.

Dunkelberg, H., et al. (1994) Genotoxic effects of the herbicides alachlor, atrazine, pendimethaline, and simazine in mammalian cells. *Bull. Environ. Contam. Toxicol.* 52:498–504.

Environmental Protection Agency (EPA) (1988) *Code of Federal Regulations, Title 40*.

Environmental Protection Agency (EPA) (2011) *Pesticides Industry Sales and Usage: 2006 and 2007 Market Estimates*, Washington, DC: Biological and Economic Analysis Division, Office of Pesticide Programs.

Friedmann, A.S. (2002) Atrazine inhibition of testosterone production in rat males following peripubertal exposure. *Reprod. Toxicol.* 16:275–279.

Giknis, M., et al. (1988) Two generation study on the effect of the triazine herbicide, atrazine, in rats. *Teratology* 37:460–461.

Goh, K.S., et al. (1993) ELISA regulatory application: compliance monitoring of simazine and atrazine in California soils. *Bull. Environ. Contam. Toxicol.* 51:333–340.

Gojmerac, T. and Kniewald, J. (1989) Atrazine biodegradation in rats: a model for mammalian metabolism. *Bull. Environ. Contam. Toxicol.* 43:199–206.

Huang, L.Q. and Frink, C.R. (1989) Distribution of atrazine, simazine, alachlor, and metolachlor in soil profiles in Connecticut. *Bull. Environ. Contam. Toxicol.* 43:159–164.

Ikonen, R., Kangas, J., and Savolainen, H. (1998) Urinary atrazine metabolites as indicators for rat and human exposure to atrazine. *Toxicol. Lett.* 44:109–112.

Infurna, R., et al. (1988) Teratological evaluations of atrazine technical, a triazine herbicide, in rats and rabbits. *J. Toxicol. Environ. Health.* 24:307–320.

Meli, G., et al. (1992) Metabolic profile of atrazine and *N*-nitrosoatrazine in rat urine. *Bull. Environ. Contam. Toxicol.* 48:701–708.

National Institute for Occupational Safety and Health (NIOSH) (2013) *Pocket Guide to Chemical Hazards*, NIOSH 94-116, Washington, DC: U.S. Department of Health and Human Services.

Peters, J.W. and Cook, R.M. (1973) Effects of atrazine on reproduction in rats. *Bull. Environ. Contam. Toxicol.* 9:301–304.

Pommery, J., Mathieu, M., Mathieu, D., et al. (1993) Atrazine in plasma and tissue following atrazineaminotriazole–ethylene glycol–formaldehyde poisoning. *J. Toxicol. Clin. Toxicol.* 31:323–331.

Simic, B., Kniewald, J., and Kniewald, Z. (1994) Effects of atrazine on reproductive performance in the rat. *J. Appl. Toxicol.* 14:401–404.

Stearman, G.K. and Wells, M.J.M. (1993) Enzyme immunoassay microtiter plate response to atrazine and metolachlor in potentially interfering matrices. *Bull. Environ. Contam. Toxicol.* 51:588–595.

Weisenberger, D.D., et al. (1988) Mutagenesis tests of atrazine and *N*-nitrosatrazine: compounds of special interest in the Midwest. *Carcinogenesis* 29:106.

SIMAZINE

BACKGROUND AND USES

Simazine acts by inhibiting photosynthesis. The EPA estimates that more than 5–7 million pounds of simazine were applied in 2007 (EPA, 2011). It is an active ingredient in Aquazine, Cekusan, Gesatop, Primatol/S, Princep, Simades, and Simanex. This herbicide is used primarily on fruit and maize and at industrial and aquatic sites, including near swimming pools and cooling towers. It typically is found as an 80% wettable powder or a 90% granule.

PHYSICAL AND CHEMICAL PROPERTIES

Simazine (2-chloro-4,6-bis(ethylamino)-*s*-triazine; CAS #122-34-9) is a selective, herbicide that inhibits photosynthesis. It occurs as a white crystalline solid, soluble at 5 ppm (at 20–22 °C) in water and 400 ppm in methanol. The vapor pressure of simazine is 0.0000000061 mmHg (at 20 °C [68 °F]).

MAMMALIAN TOXICOLOGY

Acute Effects

Simazine is slightly toxic. Mild to severe contact dermatitis has been reported occasionally from large dermal exposures to simazine. No systemic poisonings from simazine have been reported. Treatment of excessive exposure is the same as for atrazine. There are no published chronic effects of simazine exposure in humans.

Chronic Effects

Some 90-day feeding studies at 67–100 mg/kg/day and 150 mg/kg/day resulted in reduced body weight and kidney toxicity. A 2-year chronic oral feeding study with 5 mg/kg/day of simazine resulted in no signs of toxicity. Simazine has an oral RfD of 0.005 mg/kg/day (EPA, 1988, 2013).

Mechanism of Action(s)

An assay of *in vitro* sister chromatid exchanges in human lymphocytes did not indicate that simazine has genotoxic potential (Dunkelberg et al., 1994). Simazine did not test positive in mutagenicity assays (IARC, 1991; Trochimowicz et al., 1994). IARC has concluded that there is insufficient evidence to classify simazine as a carcinogen. There is limited evidence that simazine may be a teratogen in animals (Trochimowicz et al., 1994).

SOURCES OF EXPOSURE

Occupational

Inhalation and dermal exposure may occur during application. Exposure may occur during manufacture, transport, mixing, and application of simazine-containing pesticide formulations.

Environmental

Simazine may be encountered as a residue on crop plants or in drinking water that has been contaminated by nearby argicultural use of this herbicide. Simazine can leach rapidly through some soil profiles and penetrate the water table (Reddy et al., 1992). Acrysol adjuvants may be used to slow this leaching (Reddy and Singh, 1993). The half-life in bodies of water is approximately 30 days (EPA, 1988).

INDUSTRIAL HYGIENE

Industrial hygiene is the same for the principal triazine pesticide, atrazine. Appropriate work clothing and eye protection should be used to limit exposure when mixing, transporting, and applying atrazine. No specific TLVs or exposure limits have been established by the ACGIH, OSHA, or NIOSH.

MEDICAL MANAGEMENT

Medical management of simazine is similar to all atrazine pesticides. Flushing skin and eyes with water to remove minor exposures is appropriate. In cases of extreme acute simazine exposure, removal from exposure and basic life support measures should be taken.

REFERENCES

Dunkelberg, H., et al. (1994) Genotoxic effects of the herbicides alachlor, atrazine, pendimethaline, and simazine in mammalian cells. *Bull. Environ. Contam. Toxicol.* 52:498–504.

Environmental Protection Agency (EPA) (1988) *Code of Federal Regulations*, Title 40.

Environmental Protection Agency (EPA) (2011) *Pesticides Industry Sales and Usage: 2006 and 2007 Market Estimates*, Washington, DC: Biological and Economic Analysis Division, Office of Pesticide Programs.

Environmental Protection Agency (EPA) (2013) *Integrated Risk Information System*, Washington, DC.

International Agency for Research on Cancer (IARC) (1991) Occupational exposures in insecticide application, and some pesticides. *IARC Monogr. Eval. Carcinog. Risks Hum.* 53:495–513.

Reddy, K.N. and Singh, M. (1993) Effect of acrylic polymer adjuvants on leaching of bromacil, diuron, norflurazon and simazine in soil columns. *Bull. Environ. Contam. Toxicol.* 50:449–457.

Reddy, K.N., Singh, M., and Alva, A.K. (1992) Sorption and leaching of bromacil and simazine in Florida flatwoods soils. *Bull. Environ. Contam. Toxicol.* 48:662–670.

Trochimowicz, H.J., Kennedy, G., and Krivanek, N.D. (1994) Heterocyclic and miscellaneous nitrogen compounds. In: Clayton, G.D. and Clayton, F.E., editors. *Patty's Industrial Hygiene and Toxicology*, 4th ed., vol. 2, Part E, New York, NY: John Wiley & Sons, Inc.

QUATERNARY NITROGEN COMPOUNDS

PARAQUAT

BACKGROUND AND USES

Paraquat is a quick-acting contact herbicide, which becomes tightly bound to soil. It is largely used in the agricultural industry as an herbicide and may also be used as a desiccant for cotton. It is the active ingredient in Crisquat, Dextrone, Dexuron, Esgram, Goldquat 276, Gramoxone, Herbaxon, Osaquat Super, Dexuron, and Sweep. The EPA estimates about 2–4 million pounds of paraquat were used in 2007 (EPA, 2011).

PHYSICAL AND CHEMICAL PROPERTIES

Paraquat (paraquat dichloride; methyl viologen; CAS #1910-42-5) is a yellow solid with a faint ammonia-like odor. It has a vapor pressure of <0.0000001 mmHg and is extremely soluble in water. Preparations are a brown liquid that resemble a cola beverage.

MAMMALIAN TOXICOLOGY

Acute Effects

One of the most striking aspects of paraquat toxicity is the delayed onset of acute effects. Depending on the amount consumed, patients may survive a few hours to several weeks (Cravey, 1979). A second remarkable aspect of paraquat toxicity is its pulmonary toxicity. Pulmonary fibrosis is the most common effect of acute exposure and follows exposure by any route. Dermal exposure may cause pulmonary fibrosis in addition to severe burns (Papiris et al., 1995; Ronnen et al., 1995).

Ingestion is a common route of accidental or intentional acute exposure. Consumption of as little as 10–15 ml of a 20% solution has proved fatal (Pond, 1990). Paraquat is most often ingested with suicidal intent or sometimes mistaken for a beverage (Cravey, 1979). Once consumed, it directly attacks the gastrointestinal tract. This injury may include buccal mucosa erosion, ulceration hemorrhage, and pseudomembranous formation in the esophagus and stomach. Gastric erosion in paraquat intoxication indicates a grim prognosis (Mui, 1993). Severe acute exposure to paraquat may also lead to renal failure. Patients who reach this level of toxicity have a poor prognosis (Pond, 1990).

In one study a community exposed to paraquat drift from aerial spraying reported a higher rate of coughing, diarrhea, eye irritation, headache, nausea, rhinitis, throat irritation, trouble breathing, tiredness, and wheezing compared with other nearby communities (Ames et al., 1993).

The number of paraquat poisonings worldwide is difficult to estimate. The countries where paraquat is most readily available do not keep a registry of pesticide-related deaths (Tinoco et al., 1993). Also, many deaths may occur in rural agricultural areas, where the cause of death may not be determined and record keeping may be less reliable.

Chronic Effects

When used properly, paraquat does not appear to produce significant chronic effects. One study of Sri Lankan tea workers with long-term use of paraquat found no reduced lung, liver, or kidney function compared with controls

(Senanayake et al., 1993). The same study did find an increased incidence of nosebleeds and skin and nail damage.

Some studies suggest that chronic exposure to paraquat may result in hyperpigmented macules or hyperkeratosis, whereas others have found no such association (Wang et al., 1987; Cooper et al., 1994).

Mechanism of Action(s)

The lungs are the primary target of paraquat-induced toxicity. The lungs provide an abundant supply of electrons and oxygen for paraquat redox activity and this toxicity is further enhanced by the fact that paraquat is similar in size and chemistry to putrescine, a similarity that facilitates its uptake through alveolar receptors. Paraquat has been demonstrated to competitively inhibit uptake of putrescine in human lung tissue (Hoet et al., 1994). The lungs and muscles provide a reservoir for paraquat, which slowly redistributes to the bloodstream (Pond, 1990). Upon pathological examination, the lungs of paraquat-intoxicated patients show consolidation and bronchiectasis (Lee et al., 1995). Magnifications of lung samples reveal interstitial areas containing collagen, fibrin, platelets, mature fibroblasts, plasma cells, and numerous leukocytes (Dearden et al., 1978). Alveolar septa may be absent.

There is also evidence that paraquat is a neurotoxin. Some animal studies suggest that paraquat may have a neuromuscular blocking action, which could contribute to respiratory failure (Yn et al., 1994).

The potential teratogenicity of paraquat has also been investigated. Neonatal mice dosed with paraquat at a critical growth stage (10 days) did not exhibit weight gain or pulmonary symptoms (Fredriksson et al., 1993). However, at 60 and 120 days of age, they exhibited behavioral deficits that increased with dose and age. The authors also reported that paraquat-dosed mice had significantly lower brain levels of dopamine. There are indications that this effect may be reversible, depending on the dose (Fredriksson et al., 1993).

In vitro studies of rat gastrointestinal epithelial cells show that absorption of paraquat is gradual but linear (Grabie et al., 1993). Absorption of paraquat through porcine skin, which closely resembles human skin, is also described as gradual (Srikrishna et al., 1992). These experiments revealed increasing levels of intercellular edema with increasing dermal paraquat exposure.

In vivo studies show that lipid peroxidation by paraquat is dose dependent (Peter et al., 1992). Cell death was also dose dependent, although not necessarily as a direct result of lipid peroxidation. Exposure to paraquat induces production of metallothionein, a radical scavenger that also aids in homeostasis of trace metals (Sato, 1991; Bauman et al., 1992).

SOURCES OF EXPOSURE

Occupational

Paraquat is used in a wide variety of agricultural crops and occupational exposure may occur for those who manufacture, mix, or apply this herbicide. Secondary exposure may occur in workers who maintain or harvest crops post application.

Environmental

As paraquat is a contact herbicide, very little residue exposure is expected to occur on crops. Environmental exposures are rare as paraquat is degraded by ultraviolet light and soil bacteria. Residual paraquat in the environment may interfere with the function of various species. Experiments with amphibians show that dilute concentrations of paraquat may interfere with offspring viability (Dial and Dial, 1995). Other studies show that paraquat may significantly reduce the ability of shrimp to feed (Chu and Lau, 1994).

INDUSTRIAL HYGIENE

Individuals employed in the manufacture of paraquat may be exposed to dust, vapor, or aerosol, or to all three (Kuo et al., 1993). NIOSH has set the REL for paraquat at $0.1\,\text{mg/m}^3$ (NIOSH, 2013). The permissible exposure limit (PEL) set by OSHA is much higher at $0.5\,\text{mg/m}^3$. The ACGIH has set a TLV of $0.5\,\text{mg/m}^3$ for total particles and $0.1\,\text{mg/m}^3$ for respirable particles (ACGIH, 2013). Paraquat is considered IDLH at $1\,\text{mg/m}^3$.

Paraquat levels may be quantified directly from serum or urine. However, estimation of urinary levels should be undertaken with the knowledge that paraquat intoxication can interfere with creatinine measurement. If creatinine measurement is to be done accurately, it should be performed by an Ektachem dry chemistry slide creatinine amidohydrolase method or some other method not affected by paraquat's influence on the Jaffé reaction (Price et al., 1995).

MEDICAL MANAGEMENT

There are no specific antidotes for paraquat intoxication. Fuller's earth or other adsorbent materials may be given. For adults, 1l of a 15% aqueous solution of adsorbent is suggested (Pond, 1990). Gastric lavage may be performed if <1 h has passed since ingestion.

Plasma concentrations of paraquat should be analyzed to spare minimally exposed patients from the rigors of overtreatment (Proudfoot et al., 1979). However, emergency treatment should not be withheld while tests are performed.

Oxygen therapy must be avoided unless absolutely necessary because it increases paraquat toxicity (Rhodes et al., 1976).

Hemoperfusion, dialysis, and diuresis are not usually effective in paraquat intoxications (Mascie-Taylor et al., 1983; Pond et al., 1993). Hemoperfusion has been shown to be helpful only if performed shortly after ingestion in the case of mild intoxication. Radiotherapy is not promising as a method of controlling paraquat-induced lung damage (Saenghirunvattana et al., 1992).

REFERENCES

Ames, R.G., Howd, R.A., and Doherty, L. (1993) Community exposure to a paraquat drift. *Arch. Environ. Health* 48:47–52.

American Conference of Governmental Industrial Hygienists (ACGIH) (2013) *Threshold Limit Values (TLVs) for Chemical Substances and Physical Agents and Biological Exposure Indices (BEIs)*, Cincinnati, OH: American Conference of Governmental Industrial Hygienists.

Bauman, J.W. et al. (1992) Induction of hepatic metallothionein by paraquat. *Toxicol. Appl. Pharmacol.* 117:233–241.

Chu, K.H., and Lau, P.Y. (1994) Effects of diazinon, malathion and paraquat on the behavioral response of the shrimp *Metapenaeus ensis* to chemoattractants. *Bull. Environ. Contam. Toxicol.* 53:127–133.

Cooper, S.P., et al. (1994) A survey of actinic keratoses among paraquat production workers and a nonexposed friend reference group. *Am. J. Ind. Med.* 25:335–347.

Cravey, R.H. (1979) Poisoning by paraquat. *Clin. Toxicol.* 14:195–198.

Dearden, L.C., et al. (1978) Pulmonary ultrastructure of the late aspects of human paraquat poisoning. *Am. J. Pathol.* 93:667–680.

Dial, C.A. and Dial, N.A. (1995) Lethal effects of the consumption of field levels of paraquat-contaminated plants on frog tadpoles. *Bull. Environ. Contam. Toxicol.* 55:870–877.

Environmental Protection Agency (EPA) (2011) *Pesticides Industry Sales and Usage: 2006 and 2007 Market Estimates*, Washington, DC: Biological and Economic Analysis Division, Office of Pesticide Programs.

Fredriksson, A., Fredriksson, M., and Eriksson, P. (1993) Neonatal exposure to paraquat or MPTP induces permanent changes in striatum dopamine and behavior in adult mice. *Toxicol. Appl. Pharmacol.* 122:258–264.

Grabie, V., et al. (1993) Paraquat uptake in the cultured gastrointestinal epithelial cell line, IEC-6. *Toxicol. Appl. Pharmacol.* 122:95–100.

Hoet, P.H.M., et al. (1994) Putrescine and paraquat uptake in human lung slices and isolated type II pneumocytes. *Biochem. Pharmacol.* 48:517–524.

Kuo, H.W., Li, C.S., and Wang, J.D. (1993) Occupational exposure to 4,4′-bipyridyl vapor and aerosol during paraquat manufacturing. *Am. Ind. Hyg. Assoc. J.* 54:440–445.

Lee, S.H., et al. (1995). Paraquat poisoning of the lung: thin-section CT findings. *Radiology* 195:271–274.

Mascie-Taylor, B.H., Thompson, J., and Davison, A.M. (1983). Haemoperfusion ineffective for paraquat removal in life-threatening poisoning. *Lancet* 1:1376–1377.

Mui, P.C. (1993). Endoscopic evaluation of paraquat-induced upper gastrointestinal injury. *Gastrointest. Endosc.* 39:105–106.

National Institute for Occupational Safety and Health (NIOSH) (2013) *Pocket Guide to Chemical Hazards*, NIOSH 94-116, Washington, DC: U.S. Department of Health and Human Services.

Papiris, S.A., et al. (1995) Pulmonary damage due to paraquat poisoning through skin absorption. *Respiration* 662:101–103.

Peter, B., et al. (1992) Role of lipid peroxidation and DNA damage in paraquat toxicity and the interaction of paraquat with ionizing radiation. *Biochem. Pharmacol.* 43:705–715.

Pond, S.M. (1990) Manifestations and management of paraquat poisoning. *Med. J. Aust.* 152:256–259.

Pond, S.M., et al. (1993) Kinetics of toxic doses of paraquat and the effects of hemoperfusion in the dog. *J. Toxicol. Clin. Toxicol.* 31:229–246.

Price, L.A., et al. (1995) Paraquat and diquat interference in the analysis of creatinine by the Jaffé reaction. *Pathology* 27:154–156.

Proudfoot, A.T., et al. (1979) Paraquat poisoning: significance of plasma paraquat concentrations. *Lancet* 2:330–332.

Rhodes, M.L., Zavala, D.C., and Brown, D. (1976). Hypoxic protection in paraquat poisoning. *Lab. Invest.* 35:496–500.

Ronnen, M., Klin, B., and Suster, S. (1995). Mixed diquat/paraquat-induced burns. *Int. J. Dermatol.* 34:23–25.

Saenghirunvattana, S., et al. (1992) Effect of lung irradiation on mice following paraquat intoxication. *Chest* 101:833–835.

Sato, M. (1991) Dose-dependent increases in metallothionein synthesis in the lung and liver of paraquat-treated rats. *Toxicol. Appl. Pharmacol.* 107:98–105.

Senanayake, N., et al. (1993) An epidemiological study of the health of Sri Lankan tea plantation workers associated with long term exposure to paraquat. *Br. J. Ind. Med.* 50:257–263.

Srikrishna, V., Riviere, J.R., and Monteiro-Riviere, N.A. (1992) Cutaneous toxicity and absorption of paraquat in porcine skin. *Toxicol. Appl. Pharmacol.* 115:89–97.

Tinoco, R., et al. (1993) Paraquat poisoning in southern Mexico: a report of 25 cases. *Arch. Environ. Health* 48:78–80.

Wang, J.D., et al. (1987) Occupational risk and the development of pre-malignant skin lesions among paraquat manufacturers. *Br. J. Ind. Med.* 44:196–200.

Yn, S., Shiau, L., and Hsu, K.S. (1994). Studies on the neuromuscular blocking action of commercial paraquat in mouse phrenic nerve-diaphragm. *Neurotoxicology* 15:379–388.

DIQUAT

BACKGROUND AND USES

The brands Weedtrim D, Reglone, and Reglox contain diquat as an active ingredient. Diquat is used as both an aquatic and crop weed killer. Crop areas treated with diquat include potatoes, cotton, rapeseed, and may be used as a plant-growth regulator to suppress budding or flowering in crops such as sugarcane.

PHYSICAL AND CHEMICAL PROPERTIES

Diquat (diquat dibromide; CAS #85-00-7) is a quick-acting contact herbicide that forms white to yellow crystals. It has a vapor pressure of <0.00001 mmHg, and its solubility in water by weight is 70%. Diquat also is found as a monohydrate (CAS #6385-62-2) and an ion (CAS #2764-72-9).

MAMMALIAN TOXICOLOGY

Acute Effects

Although diquat's chemical structure is similar to that of paraquat; diquat intoxication produces different effects. Diquat ingestion may cause nausea, vomiting, abdominal discomfort and pain, and extensive diarrhea (Powell et al., 1983). Gastrointestinal symptoms may persist for 3 days. Ileus may also occur, accompanied by abdominal distention, dehydration, and hypotensive shock. Kidney function may decline during the first 3–4 days, possibly resulting in anuria. Elevation of blood urea nitrogen (BUN) and serum creatinine levels reflects the diminished kidney function. This condition typically reverses in 7–10 days, and renal function returns to normal in 2–3 weeks. Higher doses of diquat (over 3 ounces) may cause digestive tract ulcerations, pulmonary edema, loss of consciousness, and death. Some individuals acutely excessively exposed to diquat exhibit neurological symptoms, including nervousness, agitation, hyporeflexia, lethargy, stupor, and coma. These symptoms may be the result of cerebral hemorrhages. Dermal exposure to diquat has been reported to cause Parkinsonian-like symptoms (Sechi et al., 1992). Diquat does not cause the pulmonary fibrosis associated with paraquat.

Exposure and acute effects from diquat intoxication may result from ingestion. No deaths have been recorded from dermal exposure of intact skin or inhalation of diquat mist. However, in some extreme cases, dermal exposure to diquat has resulted in severe burns (Manoguerra, 1990; Ronnen et al., 1995). Inhalation may irritate the respiratory tract, causes coughing, sore throat, and breathing difficulties.

Chronic Effects

There are no consistent reports of chronic toxicity related to diquat exposure. The EPA has set a maximum contaminant level (MCL) of 0.02 milligrams per liter (mg/l) or 20 parts per billion (ppb) and maximum contaminant level goal (MCLG) = 0.02 mg/l or 20 ppb. This MCL is based on a potential association with cataracts for long-term exposure far above the MCL.

Mechanism of Action(s)

The differences between paraquat and diquat toxicity can be attributed to different target organs and a different mechanism of action. Damage to alveolar macrophages is a primary contributor to the development of pulmonary fibrosis. Wong and Stevens (1986) found that paraquat was four times more toxic to rat alveolar macrophages than diquat, even though alveolar cells exposed to both were permeated more thoroughly by diquat. Increasing the level of oxygen increased the toxicity of paraquat but not of diquat. Finally, superoxide dismutase inhibited the toxicity of paraquat but not of diquat *in vitro*. These findings indicate that diquat toxicity is not a result of the redox cycling that causes paraquat toxicity.

Rikans and Cai (1993) studied the effects of diquat dibromide on isolated hepatocytes from mature and old (27 months) rats. They found that diquat increased lipid peroxidation and toxicity in older rat hepatocytes. This finding may result from the fact that hepatocytes from older rats had more chelatable Fe^{2+} available to react with hydrogen peroxide formed by diquat.

T_2-weighted magnetic resonance imaging (MRI) scans of patients with a diquat-induced Parkinsonian-like condition reveal small, bilaterally symmetrical areas of high signal intensity (Sechi et al., 1992).

SOURCES OF EXPOSURE

Occupational

Diquat is used in a wide variety of agricultural crops and occupational exposure may occur for those who manufacture, mix, or apply this herbicide. Secondary exposure may occur in workers who maintain or harvest crops post application.

Environmental

Diquat is more resistent to microbial and photodegradation than paraquat, and the potential for exposure from soil water in nearby treated areas is greater. As well, residue on crop plants can be somewhat higher, although as diquat is a contact heribicide, contact with crop plants is indirect. The EPA has prescribed an RfD of 0.0022 mg/kg/day at which no adverse effects are expected to occur.

INDUSTRIAL HYGIENE

No PEL has been set by OSHA for any forms of diquat. The ACGIH has set a TLV for the diquat ion at 0.5 mg/m^3, and NIOSH has set the REL for diquat dibromide at 0.5 mg/m^3 (NIOSH, 2013; ACGIH, 2013). Waterproof footwear should be worn by diquat applicators to prevent skin burns

(Manoguerra, 1990). Diquat and its metabolites can be detected in urine by high-performance liquid chromatography (HPLC) (Ameno et al., 1995).

MEDICAL MANAGEMENT

Treatment is symptomatic. Hospitalization is indicated for cases of ingestion. Emesis should not be induced because this could cause further irritation or burns. Gastric lavage is indicated if it is performed soon after ingestion. Special attention must be paid to fluid and electrolyte status. Ileus contributes to dehydration and electrolyte imbalances. Diquat may be inactivated by fuller's earth, charcoal, or bentonite clay. Cathartics should be discontinued immediately if ileus develops. Charcoal hemoperfusion is not helpful for diquat intoxication unless used shortly after ingestion.

If diquat mists have been inhaled, the patient should be moved to fresh air and monitored for respiratory distress. If symptoms develop, 100% humidified oxygen with assisted ventilation should be provided. Eyes exposed to diquat should be rinsed with lukewarm water for 15 min. If irritation and pain persist, the patient must be seen in a health care facility.

REFERENCES

Ameno, K., et al. (1995) Simultaneous quantitation of diquat and its two metabolites in serum and urine by ion-paired HPLC. *J. Liquid Chromatogr.* 18:2115–2121.

American Conference of Governmental Industrial Hygienists (ACGIH) (2013) *Threshold Limit Values (TLVs) for Chemical Substances and Physical Agents and Biological Exposure Indices (BEIs)*, Cincinnati, OH: American Conference of Governmental Industrial Hygienists.

Manoguerra, A.S. (1990) Full thickness skin burns secondary to an unusual exposure to diquat dibromide. *J. Toxicol. Clin. Toxicol.* 28:107–110.

National Institute for Occupational Safety and Health (NIOSH) (2013) *Pocket Guide to Chemical Hazards*, NIOSH 94-116, Washington, DC: U.S. Department of Health and Human Services.

Powell, D., et al. (1983) Hemoperfusion in a child who ingested diquat and died from pontine infarction and hemorrhage. *J. Toxicol. Clin. Toxicol.* 20:405–420.

Rikans, L.E. and Cai, Y. (1993) Diquat-induced oxidative damage in BCNU-pretreated hepatocytes of mature and old rats. *Toxicol. Appl. Pharmacol.* 118:263–270.

Ronnen, M., Klin, B., and Suster, S. (1995) Mixed diquat/paraquat-induced burns. *Int. J. Dermatol.* 34:23–25.

Sechi, G.P., et al. (1992) Acute and persistent parkinsonism after use of diquat. *Neurology* 42:261–263.

Wong, R.C. and Stevens, J.B. (1986) Bipyridylium herbicide toxicity *in vitro*: comparative study of the cytotoxicity of paraquat and diquat toward the pulmonary alveolar macrophage. *J. Toxicol. Environ. Health* 18:393–407.

TRIAZOLE HERBICIDES

AMITROLE

BACKGROUND AND USES

Amitrole is a nonspecific herbicide that destroys chlorophyll and is not known to occur naturally. It is an active ingredient in the brands AminoTriazole Weed Kill 90, Weedazol, Cytrol, Diurol, Cytrol, Amizol, Azolan, Amitrol T, and Azole. It is frequently used in non-crop areas to prevent weed and grass growth to maintain clearings and roadways.

PHYSICAL AND CHEMICAL PROPERTIES

Amitrole (CAS #61-82-5) is a colorless to white crystalline powder. It has a solubility of 28% by weight in water (25 °C [77 °F]) and a vapor pressure of <0.000008 mmHg.

MAMMALIAN TOXICOLOGY

Acute Effects

Amitrole has low oral toxicity in mammals. In one case in which a person ingested approximately 20 mg/kg of body weight, no intoxication was observed. One case of alveolar damage after inhalation of amitrole mists has been reported (Balkisson et al., 1992). Treatment with corticosteroids following this inhalation resulted in a complete recovery. In animals amitrole has been shown to be goitrogenic. However, there is no evidence currently available to suggest that this effect occurs in humans.

Chronic Effects

There are sufficient data to determine that amitrole is carcinogenic in animals, producing tumors in the thyroid and liver under extremely high lifetime doses, but insufficient evidence exists in humans (IARC, 1987). Dietary doses of 3 or 6 kg/mg/day for 2 weeks caused enlargement of the thyroid and reduced uptake of iodine in rats. As well, a dose of 50 mg/kg/day produced enlargement of the thyroid after 3 days of feeding (NLM, 1995).

Mechanism of Action(s)

The primary target organ for amitrole in mammals is the thyroid gland. Amitrole has an antithyroid effect on the body that interferes with hormonal regulatory function and causes thyroid enlargement. At high doses, it has been shown to cause cancer in several species and strains (IARC, 1986). Rats, golden hamsters, and mice orally dosed with amitrole showed an effect on thyroid function at 100 ppm (Steinhoff et al., 1983).

Amitrole is also hepatotoxic in animals. A study by Mori et al. (1985) revealed significant differences in hepatotoxicity between various strains of mice. Amitrole does not test positive in mutagenic assays (IARC, 1986). Nor does amitrole cause deoxyribonucleic acid (DNA) fragmentation in human thyroid and liver cells *in vivo* (Mattioli et al., 1994). This suggests that amitrole may be a nongenotoxic carcinogen. In general, nongenotoxic carcinogens are significantly less potent than genotoxic carcinogens and are likely to have a threshold (Parodi et al., 1991).

There is evidence that amitrole has a short biological half-life. In one case of human ingestion, half of the amitrole consumed was excreted unchanged within a few hours (IARC, 1986). No metabolites were detected, although amitrole can be metabolized by prostaglandin H synthase (Krauss and Eling, 1987).

SOURCES OF EXPOSURE

Occupational

Amitrole is used mostly in non-crops settings and occupational exposure may occur for those who manufacture, mix, or apply this herbicide. Secondary exposure may occur in workers who work in treated clearings or roadways post application.

Environmental

There is moderate opportunity to be exposed to amitrole in the environment. The half-life of amitrole in the soil is approximately 14 days and the half-life in water is approximately 40 days. An RfD has not been established for amitrole.

INDUSTRIAL HYGIENE

The NIOSH and the ACGIH have set the REL and the TLV, respectively, for amitrole at $0.2 \, \text{mg/m}^3$ (NIOSH, 2013; ACGIH, 2013). The NIOSH designates amitrole as a possible occupational carcinogen. No allowable tolerances have been set for amitrole by the EPA. It has been prohibited from use directly on food crops since 1971 (EPA, 1984). Amitrole may still be used on fallow lands before planting potatoes, maize, kale, and wheat or on apple and pear trees. High Performance Liquid Chromatography can be used to quantitate amitrole or its metabolites (Archer, 1984).

MEDICAL MANAGEMENT

If small amounts are ingested, emesis is not required; however, if large amounts of a concentrated formula are swallowed (more than about 20 ml) the induction of vomiting should be considered. All care must be taken to prevent vomit from being inhaled. If skin contact occurs, remove contaminated clothing and wash skin thoroughly with water. If eye contact occurs, hold eyelids open and wash with copious amounts of water for at least 15 min. Treat symptoms that occur post exposure, including corticosteriods for plumonary symptoms.

REFERENCES

American Conference of Governmental Industrial Hygienists (ACGIH) (2013) *Threshold Limit Values (TLVs) for Chemical Substances and Physical Agents and Biological Exposure Indices (BEIs)*, Cincinnati, OH: American Conference of Governmental Industrial Hygienists.

Archer, A.W. (1984) Determination of 3-amino-1,2,4-trizole (amitrole) in urine by ion-pair high-performance liquid chromatography. *J. Chromatogr.* 303:267–271.

Balkisson, R., Murray, D., and Hoffstein, V. (1992) Alveolar damage due to inhalation of amitrole-containing herbicide. *Chest* 101:1174–1175.

Environmental Protection Agency (EPA) (1984) *Amitrole: Pesticide Registration Standard and Guidance Document*, Washington, DC: Environmental Protection Agency, Office of Pesticides and Toxic Substances.

International Agency for Research on Cancer (IARC) (1986) Amitrole. *IARC Monogr. Eval. Carcinog. Risks Hum.* 41:293–319.

International Agency for Research on Cancer (IARC) (1987) Overall evaluations of carcinogenicity: an updating of IARC monographs. *IARC Monogr. Eval. Carcinog. Risks Hum. Suppl.* 7:293–319.

Krauss, R.S. and Eling, T.E. (1987) Macromolecular binding of the thyroid carcinogen 3-amino-1,2,4-triazole (amitrole) catalyzed by prostaglandin H synthase, lactoperoxidase and thyroid peroxidase. *Carcinogenesis* 8:659–664.

Mattioli, F., et al. (1994) Studies on the mechanism of the carcinogenic activity of amitrole. *Fundam. Appl. Toxicol.* 23:101–106.

Mori, S., et al. (1985) Amitrole strain differences in morphological response of the liver following subchronic administration to mice. *Toxicol. Lett.* 29:145–152.

National Institute for Occupational Safety and Health (NIOSH) (2013) *Pocket Guide to Chemical Hazards*, NIOSH 94-116, Washington, DC: U.S. Department of Health and Human Services.

National Library of Medicine (1995) *Hazardous Substances Databank*, Bethesda, MD: NLM, pp. 8–17.

Parodi, S., et al. (1991) Are genotoxic carcinogens more potent than nongenotoxic carcinogens? *Environ. Health Perspect.* 95:199–204.

Steinhoff, D., et al. (1983) Evaluation of amitrole (aminotriazole) for potential carcinogenicity in orally dosed rats, mice, and golden hamsters. *Toxicol. Appl. Pharmacol.* 69:161–169.

CHLOROPHENOXY HERBICIDES

BACKGROUND AND USES

Chlorophenoxy herbicides gained notoriety after the Vietnam War (Lilienfeld and Gallo, 1989; Combs, 1989). In this conflict, more than 11 million gallons of Agent Orange, a 50:50 mixture of 2,4-dichlorophenoxyacetic acid (2,4-D) and 2,4,5-trichlorophenoxyacetic acid (2,4, 5-T), were used to defoliate battle grounds (Young and Reggiani, 1988). The debate as to the health effects of exposure to these herbicides has been as intense as it has been lengthy. To complicate matters, it is rare to find a group of persons exposed to a single chlorophenoxy herbicide to the exclusion of others and the common contaminant 2,3,7,8-tetrachlorodibenzo-p-dioxin (TCDD), found only in 2,4,5-T. Because most of the research in this area explores the relationship between mixed exposures, including varying concentrations of TCDD, these mixtures will be discussed as a class. Also, many researchers attribute chlorophenoxy herbicide toxicity to TCDD.

2,4-D is one of the most commonly used herbicides for agricultural crop production. The EPA estimates that more than 25 million pounds of active ingredient were used in 2007 (EPA, 2011). 2,4-D is an active ingredient in the brands Weedone, Weedar, Chloroxone, Esteron, and Salvo.

2,4,5-T was the active ingredient in Shell Star, Estron Brush Killer, Marks Brushwood Killer, Estron 2,4,5, Weedar 2,4,5, and Weedone 2,4,5-T. Since 1985, the use of 2,4,5-T has been prohibited in the United States.

PHYSICAL AND CHEMICAL PROPERTIES

2,4-D (dichlorophenoxyacetic acid; CAS #94-75-7) is a white powder with a slight phenolic odor. It is soluble in water at 620 ppm at 25 °C (77 °F). The isopropyl ester has a vapor pressure of 0.0105 mmHg at 25 °C (77 °F). The ester is insoluble in water, but the sodium salt is soluble up to 4.5% in water by weight. Technical grade 2,4-D is typically 90–99% pure, and the contaminant 2,3,7,8-TCDD has not been detected in 2,4-D formulations.

The herbicide 2,4,5-T (2,4,5-trichlorophenoxy-acetic acid; CAS #93-76-5) is a colorless to tan odorless crystal. Its salts are soluble at 278 ppm at 28 °C (82.4 °F). Esters are insoluble.

MAMMALIAN TOXICOLOGY

Acute Effects

Symptoms of 2,4-D ingestion include nausea and vomiting, drowsiness, hypotension, abdominal pain, slurred speech, high temperature, hyperventilation, headache, sore mouth and throat, renal dysfunction, ataxia, metabolic acidosis, agitation, pulmonary edema, hypoxia, tachycardia, bradycardia, hallucinations, and impaired coordination (Flanagan et al., 1990; Stevens and Sumner, 1991). Acute overexposure to 2,4,5-T by ingestion can cause muscle weakness and stiffness, nausea, vomiting, and diarrhea.

Chronic Effects

Soft Tissue Sarcoma An association between chlorophenoxy herbicides and soft tissue sarcomas (STS) has been suggested. Several case-control studies have found an association between reported chlorophenoxy herbicide exposure and STS (Hardell and Sandström, 1979; Hardell et al., 1981; Smith et al., 1984). These findings were based on the recall of the cases and controls as to whether they had been exposed to these herbicides. One population-based study in Northern Italy found an increased risk of soft tissue sarcomas among women rice weeders (Vineis et al., 1986). Smith and Christophers (1992) reported an increased risk of 2.7 in their case-control study; however, this figure was not statistically significant. Other cohort studies of Swedish forestry workers exposed to chlorophenoxy herbicides have failed to confirm this association (Wiklund and Holm, 1986). When Swedish and Finnish pesticide applicators were studied, no increased risk of STS was found (Riihimäki et al., 1982; Wiklund et al., 1988). Further cohort studies of workers involved in chlorophenoxy herbicide manufacture did not support an association between exposure and soft tissue sarcoma (Coggon et al., 1991; Bueno de Mesquita et al., 1993). Later studies of a large international cohort of chlorophenoxy herbicide manufacturers and sprayers revealed a twofold increase in STS (Kogevinas et al., 1992). This finding was not statistically significant. However, there was a significant increase in STS among sprayers when analyzed by time since first exposure (Saracci et al., 1991). Some of the apparent differences in these studies may be attributable to the variation in criteria used to classify STS (Lynge et al., 1987; Kauppinen et al., 1993). The scientific community remains divided over the issue of chlorophenoxy herbicide exposure and STS.

Non-Hodgkin's Lymphoma Increases in non-Hodgkin's lymphoma (NHL) have been reported among chlorophenoxy herbicide manufacturers (Coggon et al., 1991; Bueno de Mesquita et al., 1993). A population-based case-control study of Kansas farmers reported a sixfold increased risk of NHL (Hoar et al., 1986). Farmers who mixed or applied their own herbicides had an even higher ratio of NHL. These excesses were associated with use of 2,4-D and 2,4,5-T, but other exposures could not be eliminated. A case-control study found a twofold increase in malignant lymphomas in those recalling annual chlorophenoxy herbicide exposure of 30 days or longer (Smith and Christophers, 1992). A population-based study of the incidence of NHL in the Boston metropolitan area revealed an association between NHL and occupation (Scherr et al., 1992). Residents employed in agriculture, forestry, and the fishing industry had a relative risk of 3.0, whereas construction workers had a relative risk of 2.1. The authors suggest that this increased risk may be a result of chlorophenoxy pesticide exposure. However, the same study showed a considerably lower relative risk of 1.8 (95% CI 0.9–3.7) for those exposed to pesticides. Another U.S. cohort involved in chlorophenoxy pesticide production had no notable differences in mortality patterns (Ott et al., 1987). An analysis of an international cohort of workers and sprayers found no excess of malignant lymphoma (Kogevinas et al., 1992).

A review of 2,4-D's potential carcinogenicity was conducted by a panel of 13 scientists (Ibrahim et al., 1991). After reviewing all available epidemiological evidence and animal data, 11 concluded that 2,4-D is a possible human carcinogen. This conclusion was based primarily on increased rates of Hodgkin's lymphoma in case control studies of U.S. farmers (Zahm et al., 1990; Stevens and Sumner, 1991).

Neurological Effects An investigation by the National Academy of Sciences found no evidence of neurological disorders resulting from exposure to Agent Orange (Goetz et al., 1994).

Reproductive Effects Studies of Vietnam veterans exposed to Agent Orange do not show increased rates of children with birth defects (Erickson, 1984; Friedman, 1984). A retrospective study of cleft palate incidence rates by county and 2,4,5-T spraying found no significant association between the two (Nelson et al., 1979). A study of pesticide applicators in New Zealand exposed to 2,4,5-T detected no significant association with congenital defects and miscarriages (Smith et al., 1982).

Dermatological Effects Chronic chloracne was found in 52% of workers from a 2,4,5-T manufacturing plant (Moses et al., 1984). The mean duration of residual chloracne was 26 years.

Mechanism of Action(s)

Vainio et al. (1982) report that 2,4-D causes peroxisome proliferation in Chinese hamster hepatocytes. Such proliferation may cause increased DNA damage as a result of generation of oxygen radicals. This has been suggested as a mechanism for 2,4-D carcinogenesis in the absence of any direct effect on DNA.

In the case of human poisoning with a combination of 2,4-D and 2,4,5-T, 2,4-D concentrated in the kidneys and to a much lesser extent in the liver and spleen (Coutselinis et al., 1977). High blood levels were also detected (82.6 mg/dl). It

is uncertain what effect the combination of the two herbicides would have on 2,4-D metabolism and disposition.

SOURCES OF EXPOSURE

Occupational

2,4-D is used in crop and non-crop applications and workers who manufacture, transport, mix, or apply this herbicide may be exposed. Secondary exposure may occur in workers who enter a recently treated area.

Environmental

2,4-D has a half-life in soil of approximately 7 days and in water the half-life is approximately 1–2 weeks. Potential for exposure is low unless recent contamination has occurred. The EPA MCL for 2,4-D is 0.07 mg/l or 70 ppb. MCLG is 0.07 mg/l or 70 ppb, and the RfD is 0.01 mg/kg/day.

INDUSTRIAL HYGIENE

The ACGIH, NIOSH, and OSHA have set the TLV, REL, and PEL, respectively, for 2,4-D at 10 mg/m^3 (NIOSH, 2013; ACGIH, 2013). It is considered IDLH at 100 mg/m^3. The ACGIH has proposed the carcinogenicity designation of A4. This means that there is inadequate information to classify 2,4-D as to its carcinogenicity to humans or animals.

MEDICAL MANAGEMENT

Removal from exposure is prescribed. Skin and eyes should be flushed with water upon contact. Treatment of ingestion is symptomatic. Alkaline diuresis was effective for reducing the plasma half-life of chlorophenoxy herbicides (Flanagan et al., 1990).

REFERENCES

American Conference of Governmental Industrial Hygienists (ACGIH) (2013) *Threshold Limit Values (TLVs) for Chemical Substances and Physical Agents and Biological Exposure Indices (BEIs)*, Cincinnati, OH: American Conference of Governmental Industrial Hygienists (ACGIH).

Bueno de Mesquita, H.B., et al. (1993) Occupational exposure to phenoxy herbicides and chlorophenols and cancer mortality in the Netherlands. *Am. J. Ind. Med.* 23:289–300.

Coggon, D., Pannett, B., and Winter, P. (1991) Mortality and incidence of cancer at four factories making phenoxy herbicides. *Br. J. Ind. Med.* 48:173–178.

Combs, A.B. (1989) Agent Orange and its associated dioxin: assessment of a controversy. *Toxicol. Lett.* 48:321–324.

Coutselinis, A., Kentarchou, R., and Boukis, D. (1977) Concentration levels of 2,4-D and 2,4,5-T in forensic material. *Forensic Sci.* 10:203–203.

Environmental Protection Agency (EPA) (2011) *Pesticides Industry Sales and Usage: 2006 and 2007 Market Estimates*, Washington, DC: Biological and Economic Analysis Division, Office of Pesticide Programs.

Erickson, J.D. (1984) Vietnam veterans' risks for fathering babies with birth defects. *JAMA* 252:903–912.

Flanagan, R.J., et al. (1990) Alkaline diuresis for acute poisoning with chlorophenoxy herbicides and ioxynil. *Lancet* 335:454–458.

Friedman, J.M. (1984) Does Agent Orange cause birth defects? *Teratology* 29:193–221.

Goetz, C.G., Bolla, K.I., and Rogers, S.M. (1994) Neurologic health outcomes and Agent Orange: Institute of Medicine report. *Neurology* 44:801–809.

Hardell, L. and Sandström, A. (1979) Case-control study: soft-tissue sarcomas and exposure to phenoxyacetic acids or chlorophenols. *Br. J. Cancer* 39:711–717.

Hardell, L., et al. (1981) Malignant lymphoma and exposure to chemicals, especially organic solvents, chlorophenols and phenoxy acids: a case-control study. *Br. J. Cancer* 43:169–176.

Hoar, S.K., et al. (1986) Agricultural herbicide use and risk of lymphoma and soft-tissue sarcoma. *JAMA* 256:1141–1147.

Ibrahim, M.A., et al. (1991) Weight of the evidence on the human carcinogenicity of 2,4-D. *Environ. Health Perspect.* 96:213–222.

Kauppinen, T., et al. (1993) Chemical exposure in manufacture of phenoxy herbicides and chlorophenols and in spraying of phenoxy herbicides. *Am. J. Ind. Med.* 23:903–920.

Kogevinas, M., et al. (1992) Cancer mortality from soft-tissue sarcoma and malignant lymphomas in an international cohort of workers exposed to chlorophenoxy herbicides and chlorophenols. *Chemosphere* 25:1071–1076.

Lilienfeld, D.E. and Gallo, M.A. (1989) 2,4-D, 2,4,5-T and 2,3,7,8-TCDD: an overview. *Epidemiol. Rev.* 11:28–58.

Lynge, E., Storm, H.H., and Jensen, O.M. (1987) The evaluation of trends in soft tissue sarcoma according to diagnostic criteria and consumption of phenoxy herbicides. *Cancer* 60:1896–1901.

Moses, M., et al. (1984) Health status of workers with past exposure to 2,3,7,8-tetrachlorodibenzo-*p*-dioxin in the manufacture of 2,4,5-trichlorophenoxyacetic acid: comparison of findings with and without chloracne. *Am. J. Ind. Med.* 5:161–182.

National Institute for Occupational Safety and Health (NIOSH) (2013) *Pocket Guide to Chemical Hazards*, NIOSH 94-116, Washington, DC: U.S. Department of Health and Human Services.

Nelson, C.J., et al. (1979) Retrospective study of the relationship between agricultural use of 2,4,5-T and cleft palate occurrence in Arkansas. *Teratology* 19:377–384.

Ott, M.G., et al. (1987) Cohort mortality study of chemical workers with potential exposure to the higher chlorinated dioxins. *J. Occup. Med.* 29:422–429.

Riihimäki, V., Sisko, A., and Hernberg, S. (1982) Mortality of 2,4-dichlorophenoxyacetic acid and 2,4,5-trichlorophenoxyacetic acid herbicide applicators in Finland. *Scand. J. Work Environ. Health* 8:37–42.

Saracci, R., et al. (1991) Cancer mortality in workers exposed to chlorophenols. *Lancet* 338:1027–1032.

Scherr, P.A., Hutchinson, G.B., and Neiman R.S. (1992) Non-Hodgkin's lymphoma and occupational exposure. *Cancer Res.* 52:5503s–5509s.

Smith, A.H., et al. (1982) Congenital defects and miscarriages among New Zealand 2,4,5-T sprayers. *Arch. Environ. Health* 37:197–200.

Smith, A.H., et al. (1984) Soft tissue sarcoma and exposure to phenoxy herbicides and chlorophenols in New Zealand. *J. Natl. Cancer Inst.* 73:1111–1117.

Smith, J.G. and Christophers, A.J. (1992) Phenoxy herbicides and chlorophenols: a case control study on soft tissue sarcoma and malignant lymphoma. *Br. J. Cancer* 65:442–448.

Stevens, J.T. and Sumner, D.D. (1991) Herbicides. In: Hayes, W.J., Jr. and Laws, E.R., Jr., editors. *Handbook of Pesticide Toxicology*, vol. 3, New York, NY: Academic Press.

Vainio, H., Nickels, J., and Linnainmaa, K. (1982) Phenoxy acid herbicides cause peroxisome proliferation in Chinese hamsters. *Scand. J. Work Environ. Health* 8:70–73.

Vineis, P., et al. (1986) Phenoxy herbicides and soft-tissue sarcomas in female rice weeders: a population-based case referent study. *Scand. J. Work Environ. Health* 13:9–17.

Wiklund, K., Dich, J., and Holm, L.E. (1988) Soft tissue sarcoma risk in Swedish licensed pesticide applicators. *J. Occup. Med.* 30:801–804.

Wiklund, K. and Holm, L.E. (1986) Soft tissue sarcoma risk in Swedish agricultural and forestry workers. *J. Natl. Cancer Inst.* 76:229–234.

Young, A.L. and Reggiani, G.M., editors (1988) *Agent Orange and Its Associated Dioxin: Assessment of a Controversy*, Amsterdam: Elsevier.

Zahm, S.H., et al. (1990) A case-control study of non-Hodgkin's lymphoma and the herbicide 2,4-dichlorophenoxyacetic acid (2,4-D) in eastern Nebraska. *Epidemiology* 1:349–356.

AMIDES

METOLACHLOR

BACKGROUND AND USES

Metolachlor is a widely used agricultural pesticide with more than 30 million pounds of active ingredient used in 2007 (EPA, 2011). It is a preemergent herbicide used in a variety of settings, including corn, soybeans, peanuts, grain sorghum, potatoes, pod crops, cotton, safflower, stone fruits, nut trees, highway rights-of-way, and woody ornamentals.

PHYSICAL AND CHEMICAL PROPERTIES

Metolachlor (Dual; 2-chloro-2'-ethyl-6'-methyl;-N-(2-methoxy-1-methylethyl) acetanilide; CAS #51218-45-2) has a vapor pressure of 0.0000013 mmHg and a solubility of 530 ppm in water at 20 °C (68 °F). Metolachlor is also soluble in most organic solvents.

MAMMALIAN TOXICOLOGY

Acute Effects

Acute overexposure to metolachlor may cause abdominal cramps, nausea, vomiting, diarrhea, weakness, sweating, cyanosis, hypothermia, dyspnea, collapse, convulsions, dermatitis, liver damage, nephritis, and cardiovascular failure (Stevens and Sumner, 1991). However, acute effects require high doses. The reported oral LD_{50} in rats for metolachlor is from 1200 mg/kg to 2780 mg/kg. It is slightly toxic by skin exposure, with a reported dermal LD_{50} of >2000 mg/kg.

Chronic Effects

No chronic effects in humans have been reported. One case of acute ingestion of metolachlor by a pregnant woman at 36 weeks of gestation did not result in any harmful effects to the fetus (Yang et al., 1995). Metolachlor was found to induce proliferative liver changes in female rats (EPA, 1995). In rats fed metolachlor for 90 days, no effects were noted at 90 mg/kg/day. There is no evidence of carcinogenicity in humans.

Mechanism of Action(s)

Orally administered metolachlor is quickly broken down into metabolites and is almost totally eliminated in the urine and feces in animal models. There may be an immunolgically mediated mechanism involved in slight dermal or ocular irritation.

SOURCES OF EXPOSURE

Occupational

Metolachlor is used in crop and non-crop applications and workers who manufacture, transport, mix, or apply this herbicide may be exposed. Secondary exposure may occur in workers who enter a recently treated area.

Environmental

Metolachlor can be relatively persistent in aquatic systems, with a half-life of 22–205 days in temperate regions depending on the season (Kochany and Maguire, 1994). Leaching depends on the soil's cation exchange capacity, clay content, and carbon content (VanBiljon et al., 1990). Significant amounts of metolachlor can be found in runoff from agricultural areas (Albanis, 1991). Metolachlor content in water and soil can be determined by chromatography or immunoassay (Schlaeppi et al., 1991; Lawruk et al., 1993). Immunoassays are capable of measuring levels as low as 0.05 ppb (in water). The EPA prescribes an RfD of 0.1 mg/kg/day.

INDUSTRIAL HYGIENE

No exposure limits or TLVs have been set for metolachlor by NIOSH, OSHA, or the ACGIH. Appropriate work clothing and eye protection should be used to limit exposure when mixing, transporting, and applying metolachlor.

MEDICAL MANAGEMENT

There is no specific antidote for metolachlor poisoning. However, studies have shown that metolachlor has an affinity for charcoal, and administration of activated charcoal may be somewhat effective.

REFERENCES

Albanis, T.A. (1991) Runoff losses of EPTC, molinate, simazine, diuron, propanil and metolachlor in Thermaikos Gulf, northern Greece. *Chemosphere* 22:645–654.

Environmental Protection Agency (EPA) (1995) *Reregistration Eligibility Decision (RED): Metolachlor*, Washington, DC: Environmental Protection Agency.

Environmental Protection Agency (EPA) (2011) *Pesticides Industry Sales and Usage: 2006 and 2007 Market Estimates*, Washington, DC: Biological and Economic Analysis Division, Office of Pesticide Programs.

Kochany, J. and Maguire, R.J. (1994) Sunlight photodegradation of metolachlor in water. *J. Agric. Food Chem.* 42:406–412.

Lawruk, T.S., et al. (1993) Determination of metolachlor in water and soil by a rapid magnetic particle-based ELISA. *J. Agric. Food Chem.* 41:1426–1431.

Schlaeppi, J.M., Moser, H., and Ramsteiner, K. (1991) Determination of metolachlor by competitive enzyme immunoassay using a specific monoclonal antibody. *J. Agric. Food Chem.* 39:1533–1536.

Stevens, J.T. and Sumner, D.D. (1991) Herbicides. In: Hayes, W.J., Jr. and Laws, E.R., Jr., editors. *Handbook of Pesticide Toxicology*, vol. 3, New York, NY: Academic Press.

VanBiljon, J.J., Groeneveld, H.T., and Nel, P.C. (1990) Leaching depth of metolachlor in different soils. *Appl. Plant Sci.* 4:46–49.

Yang, C.-C., et al. (1995) Metobromuron/metolachlor ingestion with late onset methemoglobinemia in a pregnant woman successfully treated with methylene blue. *J. Toxicol. Clin. Toxicol.* 33:713–716.

PROPANIL

BACKGROUND AND USES

Propanil is a postemergence selective herbicide primarily used on rice. It is one of the most commonly used agricultural herbicides. Between 4 million and 5 million pounds of propanil (active ingredient) were used in 2007 (EPA, 2011; Trevisan et al., 1991). Propanil is the active ingredient in Chem Rice, Propanilo, Riselect, Strel, Sucopur, Surpur, Vertac, DPA, Erban, Herbax, Propanex, Stam Supernox, Stam F-34, Sram M-4, and FW-734. 3,3′,4,4′-Tetrachloroazobenzene (TCAB), a structural analog of TCDD, is a contaminant of commercially available propanil (Singh and Bingley, 1991).

PHYSICAL AND CHEMICAL PROPERTIES

Pure propanil (N-[3,4-dichlorophenyl] propanamide; CAS #709-98-8) is a white crystalline solid; the technical grade has a brownish color. Propanil is soluble at 225 ppm in water at room temperature and has a vapor pressure of 0.00005 mm Hg at 25 °C (77 °F).

MAMMALIAN TOXICOLOGY

Acute Effects

The most notable consequence of propanil exposure is methemoglobinemia. Symptoms may include headache, lethargy, dizziness, fatigue, syncope, dyspnea, seizures, arrhythmia, shock, and coma.

Chronic Effects

In a 2-year study, a dietary level of about 80 mg/kg/day caused a decrease in overall growth and a relative increase in the weight of the spleen and liver in female rats and of the testes in males chronic propanil exposure has been associated with chloracne (Morse et al., 1979). There is no evidence to suggest propanil is carcinogenic in humans.

Mechanism of Action(s)

Acute exposure to propanil may cause methemoglobin formation in some animal species (Singleton and Murphy, 1973; Chow and Murphy, 1975). Propanil and other arylamines do not effectively oxidize hemoglobin themselves; however, their hydroxyamine and aminophenol metabolites are potent oxidizers. *In vivo* and *in vitro* experiments with rats support the conclusion that the primary hepatic metabolite of propanil is N-hydroxy-3,4-dichloroanaline (McMillan et al., 1990b, 1991). This metabolite then enters erythrocytes and is converted to N-hydroxy-3,4-dichloronitrosobenzene, which is responsible for methemoglobin formation.

The presence of TCAB and other contaminants in propanil may induce microsomal enzyme systems and alter metabolism (McMillan et al., 1990a). Further experiments have demonstrated that propanil may contribute to cellular damage by enhancing lipid peroxidation. One study of rats showed that exposure to propanil increased serum levels of cholesterol, triglycerides, aspartate aminotransferase, and alkaline phosphatase (Santillo et al., 1995). Lipid peroxidation in propanil-treated rats was increased 95% in liver microsomes and 38% in brain microsomes compared with controls.

Propanil has been reported to have several immunomedullary effects in animals. Mice treated with propanil had reduced numbers of thymic T-cells 1 week later (Zhao et al., 1995). Treatment did not affect T-cell populations of the spleen or lymph nodes, but the size and cellularity of the spleen were increased. Macrophages of mice treated with propanil were also affected (Theus et al., 1993). Macrophages from treated mice were stimulated by interferon alone, without the usually necessary lipopolysaccharide present. Treated macrophages were also capable of producing high levels of interferon. Combined, these factors indicate that propanil may alter the mechanism of macrophage activation. There is no evidence from chronic animal studies that propanil is a teratogen or a carcinogen (Ambrose et al., 1972).

SOURCES OF EXPOSURE

Occupational

Propanil is used mostly in rice crop production and workers who manufacture, transport, mix, or apply this herbicide may be exposed. Secondary exposure may occur in workers who enter a recently treated area for the maintenance or harvesting of crops.

Environmental

Propanil breaks down quickly in the environment with a half-life in soil and water of about 1–3 days. Due to this, the potential for environmental exposure is limited. The EPA prescribes an RfD of 0.005 mg/kg/day.

INDUSTRIAL HYGIENE

Neither OSHA nor NIOSH has set exposure limits for propanil. Because of the danger of chloracne, the manufacturer recommends that applicators wear gloves, rubber shoes and overshoes, apron, goggles, and respirator and mask. Aerially applied propanil may drift and result in minimal exposure to nonapplicators and sensitive crops (Barnes et al., 1987). Propanil exposure may be monitored by measuring propanil–hemoglobin adducts (McMillan et al., 1990b).

MEDICAL MANAGEMENT

Appropriate treatment for methemoglobinemia, as well as supportive treatment, should be given. Methemoglobinemia is commonly treated with methylene blue.

REFERENCES

Ambrose, A.M., et al. (1972) Toxicological studies on 3′,4′-dichloropropionanilide. *Toxicol. Appl. Pharmacol.* 23:650–659.

Barnes, C.J., Lavy, T.L., and Mattice, J.D. (1987) Exposure of non-applicator personnel and adjacent areas to aerially applied propanil. *Bull. Environ. Contam. Toxicol.* 39:126–133.

Chow, A.Y.K. and Murphy, S.D. (1975) Propanil (3,4-dichloropropionanilide)-induced methemoglobin formation in relation to its metabolism *in vitro. Toxicol. Appl. Pharmacol.* 33:14–20.

Environmental Protection Agency (EPA) (2011) *Pesticides Industry Sales and Usage: 2006 and 2007 Market Estimates*, Washington, DC: Biological and Economic Analysis Division, Office of Pesticide Programs.

McMillan, D.C., Freeman, J.P., and Hinson, J.A. (1990a) Metabolism of the arylamide herbicide propanil. *Toxicol. Appl. Pharmacol.* 103:102–112.

McMillan, D.C., McRae, T.A., and Hinson, J.A. (1990b) Propanil-induced methemoglobinemia and hemoglobin binding in the rat. *Toxicol. Appl. Pharmacol.* 105:503–507.

McMillan, D.C., et al. (1991) Role of metabolites in propanil-induced hemolytic anemia. *Toxicol. Appl. Pharmacol.* 110:70–78.

Morse, D.L., et al. (1979) Propanil-chloracne and methomyl toxicity in workers of a pesticide manufacturing plant. *Clin. Toxicol.* 15:13–21.

Santillo, M., et al. (1995) Enhancement of tissue lipoperoxication in propanil-treated rats. *Toxicol. Lett.* 78:215–218.

Singh, J. and Bingley, R. (1991) Levels of 3,3′,4,4′-tetrachloroazobenzene in propanil herbicide. *Bull. Environ. Contam. Toxicol.* 47:822–826.

Singleton, S.D. and Murphy, S.D. (1973) Propanil (3,4-dichloropropionanilide)-induced methemoglobin formation in mice ill relation to acylamidase activity. *Toxicol. Appl. Pharmacol.* 25:20–29.

Theus, S.A., et al. (1993) Alteration of macrophage cytotoxicity through endogenous interferon and tumor necrosis factor: induction by propanil. *Toxicol. Appl. Phamacol.* 118:46–52.

Trevisan, M., et al. (1991) Evaluation of potential hazard of propanil to groundwater. *Chemosphere* 22:637–643.

Zhao, W., et al. (1995) Changes in primary and secondary lymphoid organ T-cell subpopulations resulting from acute *in vivo* exposure to propanil. *J. Toxicol. Environ. Health* 46:171–181.

NITRILE HERBICIDES

DICHLOBENIL

BACKGROUND AND USES

Dichlobenil acts as a herbicide by inhibiting germination. It can be used as a selective, total, or aquatic herbicide (Koopman, 1960). Dichlobenil is the active ingredient in Dyclomec, Prefix D, Niagra 5006, Casoron, Decabane, and Norosac. It is also used in combination with dalapon, bromacil, and simazine.

PHYSICAL AND CHEMICAL PROPERTIES

Dichlobenil (symbol DCB; 2,6-dichlorobenzonitrile; CAS #1194-65-6) is a white crystalline solid with an aromatic odor. It is soluble in water at 25 ppm and has a vapor pressure of 0.00055 mmHg at 20 °C (68 °F). Dichlobenil is slightly soluble in organic solvents. This herbicide has been used in a variety of applications, including cranberry bogs.

MAMMALIAN TOXICOLOGY

Acute Effects

There are no published reports of dichlobenil intoxication or acute health effects, though animal studies suggest the olfactory mucosa may be damaged from concentrated exposures. Dichlobenil has been reported as the cause of chloracne in workers (Deeken, 1974). The onset of symptoms may be delayed for up to 5 weeks.

Chronic Effects

No chronic effects have been reported for dichlobenil with the exception of weight loss among animal studies conducting feeding trials.

Mechanism of Action(s)

There is evidence from animal studies that the olfactory mucosa is a primary target organ for dichlobenil toxicity. In fact, dichlobenil concentrates in the olfactory mucosa whether animals are exposed by injection, intranasal instillation, or dermally (Deamer et al., 1994; Genter et al., 1995). Within 28 h after one dose of dichlobenil (12 mg/kg), signs of necrosis in the olfactory epithelium are apparent in mice (Walters et al., 1993). Four hours after injection (12 mg/kg), covalent binding in the olfactory mucosa was 26 times higher than in the liver of mice (Brittebo et al., 1992). There is evidence that olfactory epithelium injury caused by dichlobenil involves damage to the olfactory bulb (Brittebo and Eriksson, 1995). Although mice generally recovered from this effect, an atypical epithelium in the dorsomedial section of the olfactory region remains. This is the same region of the brain believed to play a key role in the etiology of Alzheimer's disease.

Inhibitors of cytochrome P-450 reduce dichlobenil olfactory toxicity in animals (Walters et al., 1993). This indicates that dichlobenil toxicity is cytochrome P-450 mediated. Studies of rats indicate that the distribution of detoxification enzymes may determine the *in situ* toxicity of systemically administered dichlobenil (Genter et al., 1995).

In one study of mice that had been injected with dichlobenil, researchers found statistically significant increases in malignant tumors (Donna et al., 1991).

SOURCES OF EXPOSURE

Occupational

Dichlobenil is used mostly in agriculture production and workers who manufacture, transport, mix, or apply this herbicide may be exposed. Secondary exposure may occur in workers who enter a recently treated area for the maintenance or harvesting of crops.

Environmental

Dichlobenil is somewhat persistent in the environment with a half-life in soils ranging up to as much as a year depending on soil conditions.

INDUSTRIAL HYGIENE

No specific exposure limits or TLVs have been set for dichlobenil by OSHA, NIOSH, or the ACGIH. Appropriate work clothing and eye protection should be used to limit exposure when mixing, transporting, and applying this herbicide.

MEDICAL MANAGEMENT

Treatment for dichlobenil intoxication is symptomatic.

REFERENCES

Britlebo, E.B., Eriksson, C., and Brant, I. (1992) Effects of glutathione-modulated agents on the covalent binding and toxicity of dichlobenil in the mouse olfactory mucosa. *Toxicol. Appl. Pharmacol.* 114:31–40.

Brittebo, E.B. and Eriksson, C. (1995) Taurine in the olfactory system: effects of the olfactory toxicant dichlobenil. *Neurotoxicology* 16:271–280.

Deamer, N.J., O'Callaghan, J.P., and Genter, M.B. (1994) Olfactory toxicity resulting from dermal application of 2,6-dichlorobenzonitrile (dichlobenil) in the C57B1 mouse. *Neurotoxicology* 15:287–294.

Deeken, J.H. (1974) Chloracne induced by 2,6-dichlorobenzonitrile. *Arch. Dermatol.* 109:245–246.

Donna, A., et al. (1991) A one-year carcinogenicity study with 2,6-dichlorobenzonitrile (dichlobenil) in male Swiss mice: preliminary note. *Cancer Detect. Prevent.* 15:41–44.

Genter, M.B., Owens, D.M., and Deamer, N.J. (1995) Distribution of microsomal epoxide hydrolase and glutathione S-transferase in the rat olfactory mucosa: relevance to distribution of lesions caused by systemically administered olfactory toxicants. *Chem. Senses* 20:385–392.

Koopman, H. (1960) 2,6-Dichlorobenzonitrile: a new herbicide. *Nature* 186:89–90.

Walters, E., Buchheit, K., and Maruniak, J.A. (1993) Olfactory cytochrome P-450 immunoreactivity in mice is altered by dichlobenil but preserved by metyrapone. *Toxicology* 81:113–122.

IOXYNIL

BACKGROUND AND USES

Ioxynil is an active ingredient in Oxytril (bromoxynil), Toxynil, Totril, Certol, Mate, Iotril, Actrilawn (a sodium salt), Bantrol, and Actfil. Ioxynil is also typically mixed with many other herbicides in various products.

PHYSICAL AND CHEMICAL PROPERTIES

Ioxynil (4-hydroxy-3,5-diiodobenzonitrile; CAS #1698-83-4) is a colorless, odorless solid. It has a solubility in water of 130 ppm at 25 °C (77 °F), but alkali salts are soluble.

MAMMALIAN TOXICOLOGY

Acute Effects

Ingestion of ioxynil may result in meiosis, eye or skin irritation, euphoria, headache, vomiting, diarrhea, pulmonary edema, fever, coma, and death. Cardiac effects, such as tachycardia and cardiac arrest, may result with exposure to an ioxynil formulation containing chlorphenoxy compounds.

Chronic Effects

There are no published reports of chronic effects in humans from exposure to ioxynil.

Mechanism of Action(s)

Although ioxynil is structurally related to dichlobenil, it does not appear to have similar toxicity to the olfactory bulb and mucosa (Eriksson et al., 1992).

SOURCES OF EXPOSURE

Occupational

Ioxynil is used mostly in agriculture production and workers who manufacture, transport, mix, or apply this herbicide may be exposed. Secondary exposure may occur in workers who enter a recently treated area for the maintenance or harvesting of crops.

Environmental

Ioxynil is somewhat persistent in the environment and environmental exposure is possible if contamination of water or soil nearby to treatment areas occurs.

INDUSTRIAL HYGIENE

No specific exposure limits or TLVs have been set for ioxynil by OSHA, NIOSH, or the ACGIH. Appropriate work clothing and eye protection should be used to limit exposure when mixing, transporting, and applying this herbicide.

MEDICAL MANAGEMENT

Treatment for ioxynil poisoning is symptomatic. Alkaline diuresis has not proven effective in increasing ioxynil clearance (Flanagan et al., 1990).

REFERENCES

Eriksson, C., Brandt, I., and Brittebo, E. (1992) Tissue binding and toxicity of compounds structurally related to the herbicide dichlobenil in the mouse olfactory mucosa. *Food Chem. Toxicol.* 30:871–877.

Flanagan, R.J., et al. (1990) Alkaline diuresis for acute poisoning with chlorophenoxy herbicides and ioxynil. *Lancet* 335:454–458.

OTHER HERBICIDES

GLYPHOSATE

BACKGROUND AND USES

Glyphosate is a broad-spectrum, nonselective herbicide. In 2007, glyphosate was the most used pesticide in the United States agriculture industry with an estimated 180–185 millions of pounds used. Glyphosate is the active ingredient in Roundup, Rondo, Rodeo, Glycel, and Glifonox. It is generally available as an aqueous solution of isopropylaniline salt and is also commonly used in commercial and residential settings (EPA, 2011).

PHYSICAL AND CHEMICAL PROPERTIES

Glyphosate (N-[phosphonomethyl] glycine; CAS # 38641-94-0) is a white, odorless solid that is 1.2% soluble in water at 25 °C (77 °F) and insoluble in most organic solvents and degrades slowly in water (Antón et al., 1993).

MAMMALIAN TOXICOLOGY

Acute Effects

In spite of low oral toxicity to other mammals in controlled experiments, human fatalities have been reported from intentional glyphosate poisoning. Pulmonary dysfunction, gastrointestinal irritation and bleeding, metabolic acidosis, hypotension, oliguria, leukocytosis, fever, renal failure, cardiac arrest, seizures, and coma have been observed after human ingestion. In one study by Tominack et al. (1991), those who died consumed an average of 108 ml of active ingredient.

Chronic Effects

Glyphosate did not have a clastogenic effect in mouse bone marrow micronucleus test (Rank et al., 1993). Roundup did have a weak mutagenic effect in the Salmonella mutagenicity assay at high concentrations. This activity did not depend on microsomal metabolism. Glyphosate isopropylamine salt did not increase chromosomal aberrations in the anaphase–telophase *Allium* test, but Roundup did increase chromosomal aberrations significantly (Rank et al., 1993). Glyphosate is widely regarded as noncarcinogenic. Glyphosate has not been demonstrated to be an endocrine disruptor (Williams et al., 2000).

Mechanism of Action(s)

The mechanism of action for glyphosate is poorly characterized. Acute toxicity from accidental or intentional exposure of very high doses of applicator formulations may be attributable to formula additives such as surfactants (Talbot et al., 1991).

SOURCES OF EXPOSURE

Occupational

The majority of glyphosate exposure occurs in the agriculture industry, with primary exposures in the application of glyphosate in crop fields and secondary exposure from maintenance and harvesting activities in those areas post-application. Several million pounds of glyphosate are also estimated to be used in commercial landscaping applications as well as personal residential use. Both primary and secondary exposures may occur from this activity (EPA, 2011).

Environmental

Some secondary exposure to glyphosate may occur through residue on food crops. The EPA has determined a 2 mg/kg/day oral reference dose for glyphosate residue as a dose that is not expected to produce adverse effects (EPA, 1993).

INDUSTRIAL HYGIENE

A study of conifer seedling nursery workers indicated that workers were not significantly exposed to glyphosate (Lavy et al., 1992). Another study of forest workers by Jauhiainen et al. (1991) determined that the workers suffered no medical effects from exposure to low levels of glyphosate. No exposure limits or TLVs have been set for glyphosate by OSHA, NIOSH, or the ACGIH. Personal protective equipment (PPE) must be used in compliance with the EPA label requirements and typically indicates coveralls, chemical-resistant gloves made of waterproof material, shoes and socks, and protective eyewear. Reentry after treatment is typically 12 h.

MEDICAL MANAGEMENT

In the absence of obvious evidence, the diagnosis of glyphosate poisoning may be made by amino acid analysis (Ekstrom and Johansson, 1975; Parrot et al., 1995). Analysis of biological fluids for glyphosate and metabolites yields rapid results and aids efforts to monitor the patient's progress. Also, direct measurement of glyphosate in blood or urine will confirm exposure. Symptoms of nausea and dermal or ocular irritation may be due to additives in the pesticide formula, including surfactants. Treatment is symptomatic and includes gastric lavage using 5% sodium bicarbonate for oral exposures. Activated charcoal may be given to reduce absorption of remaining residues. For dermal exposure, the affected area should be washed with soap and water. Ocular exposure requires washing with isotonic saline or sterile water.

REFERENCES

Antón, F.A., et al. (1993) Degradation behavior of the pesticides glyphosate and diflubenzuron in water. *Bull. Environ. Contam. Toxicol.* 51:881–888.

Ekstrom, G. and Johansson, S. (1975) Determination of glyphosate (*N*-phosphonomethyl glycine) using an amino acid analyzer. *Bull. Environ. Contam. Toxicol.* 14:295–296.

Environmental Protection Agency (EPA) (1993) *R.E.D. Facts Glyphosate. Prevention, Pesticides, and Toxic Substances.* EPA-738-F-93-011.

Environmental Protection Agency (EPA) (2011) *Pesticides Industry Sales and Usage: 2006 and 2007 Market Estimates*, Washington, DC: Biological and Economic Analysis Division, Office of Pesticide Programs.

Jauhiainen, A., et al. (1991) Occupational exposure of forest workers to glyphosate during brush saw spraying work. *Am. Ind. Hyg. Assoc. J.* 52:61–64.

Lavy, T.L., et al. (1992) Conifer seedling nursery worker exposure to glyphosate. *Arch. Environ. Contam. Toxicol.* 22:6–13.

Parrot, F., et al. (1995) Glyphosate herbicide poisoning: use of a routine amino acid analyzer appears to be a rapid method for determining glyphosate and its metabolite in biological fluids. *J. Toxicol. Clin. Toxicol.* 33:695–698.

Rank, J., et al. (1993) Genotoxicity testing of the herbicide Roundup and its active ingredient glyphosate isopropylamine using the mouse bone marrow micronucleus test, *Salmonella* mutagenicity test, and *Allium* anaphase-telophase test. *Mutat. Res.* 300:29–36.

Talbot, A.R., Shiaw, M.H., Huang, J.S., Yang, S.F., Goo, T.S., Wang, S.H., Chen, C.L., and Sanford, T.R. (1991) Acute poisoning with a glyphosate-surfactant herbicide ('Roundup'): a review of 93 cases. *Hum. Exp. Toxicol.* 10:1–8.

Tominack, R.L., et al. (1991) Taiwan National Poison Center survey of glyphosate surfactant herbicide ingestions. *J. Toxicol. Clin. Toxicol.* 29:91–109.

Williams, G.M., Kroes, R., and Munro, I.C. (2000) Safety evaluation and risk assessment of the herbicide Roundup and its active ingredient, glyphosate, for humans. *Regul. Toxicol. Pharmacol.* 31:117–165.

PICLORAM

BACKGROUND AND USES

Picloram is the active ingredient in Grazon and Tordon. Tordon 202c is a combination of 12 g/l of picloram and 200 g/l of 2,4-D. Picloram herbicides have been used for woody and broadleaf weed control in grasses, rangelands, forests, and utility rights-of-way.

PHYSICAL AND CHEMICAL PROPERTIES

Picloram (4-amino-3,5,6-trichloropiclinic acid; CAS #1918-02-1) is a colorless to white crystal with a chlorinelike odor and a vapor pressure of 0.0000006 mm Hg at 35 °C (95 °F). It is only slightly soluble in water but soluble in polar organic solvents. Commercial preparations are water-soluble potassium or amine salts.

MAMMALIAN TOXICOLOGY

Acute Effects

The acute effects of picloram in humans have not been well established. There are no published reports of human poisonings attributed to picloram.

Chronic Effects

There are currently no studies of the chronic effects of picloram exposure in humans. Animal models suggest that long-term exposure to 60 mg/kg/day may result in liver toxicity (Dow Chemical, 1986). The IARC has concluded that picloram is not classifiable as to its carcinogenicity to humans (group 3) and that there is limited evidence of carcinogenicity in animals (IARC, 1991).

Mechanism of Action(s)

Animal studies do not indicate that picloram alone has a teratogenic effect. Studies of rabbits dosed with picloram revealed no adverse developmental effects at dose levels that caused maternal toxicity (John-Greene et al., 1985; Breslin et al., 1991b). No reproductive effects were observed in rats dosed with picloram at levels causing maternal toxicity (Breslin et al., 1991a). No effect was observed when this process was repeated for two generations (Kociba et al., 1992).

Fischer rats exposed to picloram showed no increase in the rate of tumor development or the incidence of tumor types (Stott et al., 1990). Exposure to Tordon 202c, the mixture of 2,4-D and picloram, has been shown to be a reproductive toxicant in animals. Blakely et al. (1989a, 1989b) noted a dose-dependent increase in the incidence of malformed fetuses and a decrease in fertility in CD-1 mice exposed to Tordon 202c. Exposure to picloram and 2,4-D mixtures has also been shown to affect rat liver mitochondria metabolism (Pereira et al., 1994). Neither picloram nor 2,4-D alone was toxic to catfish, but the combination of the two altered catfish metabolism and increased peroxisome activity (Gallagher and DiGiulio, 1991).

SOURCES OF EXPOSURE

Occupational

This herbicide is used for clearings and roadways. Workers who manufacture, transport, mix, or apply this pesticide may be exposed. Formulations of picloram often include 2,4-D and exposure assessment should include this compound where appropriate.

Environmental

Picloram is moderately persistent in the soil with soil half-lives from 20 to 300 days. In water, sunlight exposure quickly breaks down picloram producing a half-life of 2–3 days. Picloram can be detected by HPLC analysis in water at 1 ppb and in soil at 10 ppb (Krzyszowska and Vance, 1994). Enzyme immunoassays may also be used effectively for analysis (Deschamps and Hall, 1991). The EPA prescribes an RfD of 0.07 mg/kg/day.

INDUSTRIAL HYGIENE

Applicators, such as forestry and utility workers, are at highest risk of exposure. Concerns that firefighters might be exposed to herbicides when recently treated forest lands burn have been unfounded (McMahon and Bush, 1992). Dermal exposure to picloram applicators is 4–50 times greater than inhalation, depending on the method of application (Libich et al., 1984). Forestry workers wearing backpack sprayers were the most highly exposed; whereas those using injection bar and hack squirt techniques had the lowest exposure (Lavy et al., 1987). Protective gloves and boots were shown to reduce exposure for applicators. In one study of utility workers, education combined with improved procedures and equipment was shown to reduce exposure from 1 year to the next (Libich et al., 1984).

The NIOSH has not set an REL for picloram. The PEL set by OSHA is $5 \, mg/m^3$ for respirable dust (NIOSH, 2013). The ACGIH has set the TLV at $10 \, mg/m^3$ (ACGIH, 2013). In humans orally exposed to picloram, 90% was recovered unchanged in urine (Nolan et al., 1984). The presence of picloram in urine may be determined by gas chromatography, with a lower detection limit of 10 ppb (Mattice and Lavy, 1986).

MEDICAL MANAGEMENT

The treatment for acute exposure to picloram is symptomatic.

REFERENCES

American Conference of Governmental Industrial Hygienists (ACGIH) (2013) *Threshold Limit Values (TLVs) for Chemical Substances and Physical Agents and Biological Exposure Indices (BEIs)*, Cincinnati, OH: American Conference of Governmental Industrial Hygienists (ACGIH).

Blakely, P.M., Kim, J.S., and Firneisz, G.D. (1989a) Effects of gestational exposure to Tordon 202c on fetal growth and development in CD-1 mice. *J. Toxicol. Environ. Health* 28:309–316.

Blakely, P.M., Kim, J.S., and Fimeisz, G.D. (1989b) Effects of preconceptional and gestational exposure to Tordon 202c on fetal growth and development in CD-1 mice. *Teratology* 39:547–553.

Breslin, W.J., Schroeder, R.E., and Hanley, T.R. (1991a) Developmental toxicity of picloram potassium (K) and triisopropanolamine (TIPA) salts in the rat. *Toxicologist* 11:74.

Breslin, W.J., et al. (1991b) Developmental toxicity studies with picloram triisopropanolamine salt and picloram 2-ethylhexyl ester in the rabbit. *Toxicologist* 14:162.

Deschamps, R.J.A. and Hall, J.C. (1991) Enzyme immunoassays for picloram detection: a comparison of assay formats. *Food Agric. Immunol.* 3:135–145.

Dow Chemical (1986) *MRID No. 00155940*. Washington, DC: EPA.

Gallagher, E.P. and DiGiulio, R.T. (1991) Effects of 2,4-D and picloram on biotransformation, peroxisomal and serum enzyme activities in channel catfish (*Ictalurus punctatus*). *Toxicol. Lett.* 57:65–72.

International Agency for Research on Cancer (IARC) (1991) Occupational exposures in insecticide application, and some pesticides. *IARC Monogr. Eval. Carcinog. Risks Hum.* 53:481–493.

John-Greene, J.A., et al. (1985) Teratological evaluation of picloram potassium salt in rabbits. *Food Chem. Toxicol.* 23:753–756.

Kociba, R.J., Zielke, G.J., and Breslin, W.J. (1992) Picloram herbicide: two-generation dietary reproduction study in Sprague-Dawley rats. *Toxicologist* 12:120.

Krzyszowska, A.L. and Vance, G.F. (1994) Solid-phase extraction of dicamba and picloram from water and soil samples for HPLC analysis. *J. Agric. Food Chem.* 42:1693–1696.

Lavy, I.L., et al. (1987) Exposure of forestry ground workers to 2,4-D, picloram and dichloroprop. *Environ. Toxicol. Chem.* 6:209–224.

Libich, S., et al. (1984) Occupational exposure of herbicide applicators to herbicides used along electric power transmission line right-of-way. *Am. Ind. Hyg. Assoc. J.* 45:56–62.

Mattice, J.D. and Lavy, T.L. (1986) Gas chromatographic determination of picloram in human urine. *Bull. Environ. Contam. Toxicol.* 37:938–941.

McMahon, C.K. and Bush, P.B. (1992) Forest worker exposure to airborne herbicide residues in smoke from prescribed fires in the southern United States. *Am. Ind. Hyg. Assoc. J.* 53:265–272.

National Institute for Occupational Safety and Health (NIOSH) (2013) *Pocket Guide to Chemical Hazards*, NIOSH 94-116, Washington, DC: U.S. Department of Health and Human Services.

Nolan, R.J., et al. (1984) Pharmacokinetics of picloram in male volunteers. *Toxicol. Appl. Pharmacol.* 76:264–269.

Pereira, L.F., Campello, A.P., and Silveira, O. (1994) Effect of Tordon 2,4-D 64/240 triethanolamine BR on the energy metabolism of rat liver mitochondria. *J. Appl. Toxicol.* 14:21–26.

Stott, W.T., et al. (1990) Chronic toxicity and oncogenicity of picloram in Fischer 344 rats. *J. Toxicol. Environ. Health* 30:91–104.

TRIFLURALIN

BACKGROUND AND USES

In the United States approximately 5–7 million pounds of trifluralin (active ingredient) are applied annually to agricultural lands (EPA, 2011). It is a selective preemergent herbicide used for agricultural and nonagricultural control of broadleaf weeds and annual grasses. Trifluralin at one time was applied to more than 50% of the planted acreage of tomatoes, cotton, green beans, and broccoli, though some of this usage has been replaced with glyphosate and other herbicides (EPA, 1994; EPA, 2011). Common product brands include Crisalin, Elancolan, Flurene SE, Ipersan, L-36352, M.T.F., Su Seguro Carpidor, TR-10, Trefanocide, Treficon, Treflan, Tri-4, Trifluralina 600, Triflurex Trim, and Trust.

PHYSICAL AND CHEMICAL PROPERTIES

Trifluralin (CAS #1582-09-8) is a yellow-orange crystalline solid. It has a vapor pressure of 0.000102 mmHg at 25 °C

(77 °F). Trifluralin is practically insoluble in water (<1 ppm) but soluble in organic solvents such as acetone.

MAMMALIAN TOXICOLOGY

Acute Effects

Reports of acute illness caused by trifluralin are rare, and treatment is symptomatic. In rare cases the herbicide has been reported to cause allergic contact dermatitis (Pentel et al., 1994). Because of its low oral toxicity, trifluralin has been considered as a potential antiparasitic agent (Beck, 1978; Chan and Fong, 1990).

Chronic Effects

Chronic effects in humans are poorly characterized. Trifluralin has been shown to cause liver and kidney damage in other studies of chronic oral exposure in animals (EPA, 1989). The IARC has concluded that there is inadequate evidence to classify trifluralin as a human carcinogen and limited evidence in experimental animals (IARC, 1991).

Mechanism of Action(s)

In studies of mice, trifluralin inhibited benzo[α]pyrene-induced tumorigenesis in the lung and forestomach (Triano et al., 1985). It had a protective effect whether administered before or after benzo[α]pyrene. In one study of mice prenatally exposed to trifluralin, an increased number of skeletal variants were found, such as parted frontals and 14 ribs (Beck, 1981). A study of rats dosed with trifluralin found no evidence of developmental toxicity even at levels that caused maternal toxicity (Byrd et al., 1995). In the same study, developmental toxicity in rabbits was reported only at levels that caused maternal toxicity (above 225 mg/kg). Genotoxicity evaluations of trifluralin indicate that it has a low genotoxic potential (Garriott et al., 1991). Overall, evaluations show that trifluralin has a low teratogenic and embryotoxic potential (Ebert et al., 1992).

SOURCES OF EXPOSURE

Occupational

Trifluralin is used mostly in agriculture production and workers who manufacture, transport, mix, or apply this herbicide may be exposed. Secondary exposure may occur in workers who enter a recently treated area for the maintenance or harvesting of crops.

Environmental

This herbicide is moderately persistent in the environment (Smith and Hayden, 1982). In one study, 78% of trifluralin was degraded after 4 months (Camper et al., 1980). Environmental exposure is possible from contamination of water or soils nearby treated areas.

INDUSTRIAL HYGIENE

No TLVs or exposure limits have been set for trifluralin by the ACGIH, OSHA, or NIOSH. It is considered a general use pesticide and may be used residentially.

MEDICAL MANAGEMENT

The treatment for acute exposure to trifluralin is symptomatic.

REFERENCES

Beck, S.L. (1978) Concentration and toxicity of trifluralin in CD-1 mice, presented intragastrically or intraperitoneally. *Bull. Environ. Contam. Toxicol.* 20:554–560.

Beck, S.L. (1981) Assessment of adult skeletons to detect prenatal exposure to 2,4,5-T or trifluralin in mice. *Teratology* 23:33–55.

Byrd, R.A., Markham, J.K., and Emerson, J.L. (1995) Developmental toxicity of dinitroaniline herbicides in rats and rabbits. I. Trifluralin. *Fundam. Appl. Toxicol.* 26:181–190.

Camper, N.D., Stralka, K., and Skipper, H.D. (1980) Aerobic and anaerobic degradation of profluralin and trifluralin. *J. Environ. Sci. Health B* 15:457–473.

Chan, M.M. and Fong, D. (1990) Inhibition of leishmania but not host macrophages by the antitubulin herbicide trifluralin. *Science* 249:924–926.

Ebert, E., et al. (1992) Toxicology and hazard potential of trifluralin. *Food Chem. Toxicol.* 30:1031–1044.

Environmental Protection Agency (EPA) (1989) *Health Advisory Summary: Trifluralin.* Washington, DC: Office of Drinking Water.

Environmental Protection Agency (EPA) (1994) *Pesticides Industry Sales and Usage: 1992 and 1993 Market Estimates*, Washington, DC: Environmental Protection Agency, Office of Pesticide Programs.

Environmental Protection Agency (EPA) (2011) *Pesticides Industry Sales and Usage: 2006 and 2007 Market Estimates*, Washington, DC: Biological and Economic Analysis Division, Office of Pesticide Programs.

Garriott, M.L., et al. (1991) Genotoxicity studies on the preemergence herbicide trifluralin. *Mutat. Res.* 260:187–193.

International Agency for Research on Cancer (IARC) (1991) Occupational exposures in insecticide application, and some pesticides. *IARC Monogr. Eval. Carcinog. Risks Hum.* 53:515–534.

Pentel, M.T., Andreozzi, R.J., and Marks, J.G., Jr. (1994) Allergic contact dermatitis from the herbicides trifluralin and benefin. *J. Am. Acad. Dermatol.* 31:1057–1058.

Smith, A.E. and Hayden, B.J. (1982) Carry-over of dinitramine, triallate, and trifluralin to the following spring in soils treated at different times during the fall. *Bull. Environ. Contam. Toxicol.* 29:483–486.

Triano, E.A., et al. (1985) Protective effects of trifluralin on benzo-[*a*]-pyrene-induced tumors in A/J mice. *Cancer Res.* 45:601–607.

88

FUNGICIDES

Amora Mayo-Perez and Raymond D. Harbison

DITHIOCARBAMATES

Dithiocarbamates are a structurally similar group of fungicides widely used for various fungicidal and some herbicidal and nematocidal applications. Subgroups include methyldithiocarbamates (metam sodium), ethylenebisdithiocarbamates (maneb and zineb), and dimethyldithiocarbamates (thiram). In spite of effects reported from animal experiments, dithiocarbamates have been used extensively for five decades with a good safety record. They are characterized by a low mammalian toxicity and rapid environmental degradation (McGrath 2004).

HEALTH EFFECT

Dithiocarbamates are recognized as having an antithyroid effect. In fact, when administered to animals, these compounds interfere with thyroid iodine uptake almost immediately. The goitrogenic effects of dithiocarbamates may be influenced by dietary factors.

They are virtually insoluble in water and most organic solvents. They are soluble in chloroform, benzene, and carbon disulfide. Gas chromatography may be used to detect airborne dithiocarbamates (Maini and Boni, 1986).

TARGET ORGANS AND MECHANISMS OF ACTION

Dithiocarbamate fungicides are structurally related to disulfiram (Antabuse) and have a similar effect on alcohol metabolism (Garcia de Torres et al., 1983; Römer et al., 1984). Some fungicidal dithiocarbamates may be more efficient at producing the "alcohol effect" than disulfiram, but they were rejected for this purpose because of their greater toxicity (Israeli et al., 1983). Dithiocarbamates can be used to treat acute cadmium poisonings (Hogarth, 2012). Neuropathology is common in dithiocarbamates toxicity; the result is degeneration or demyelination of the sciatic or spinal nerve tissue (Mulkey, 2001). Proteasomal dysfunction is produced in the mesencephalic neuronal cell lines (Zhou et al., 2004).

Carbon disulfide is a common metabolite of all dithiocarbamate fungicides. Ethylenebisdithiocarbamates are metabolized to ethylenethiourea (ETU), which is suspected to be a carcinogen. Free diethyldithiocarbamate formation acts as a suicide substrate inhibitor of aldehyde dehydrogenase (Tilton et al., 2006). Thus some copper complex derivatives of the molecule have been investigated as superoxide dismutase inhibitors (Hogarth, 2012). Another metabolite of dithiocarbamate is 2-thiothiazolidine-4-carboxylic acid, which is also excreted in urine (Maroni et al., 2000). Metal complexes of dithiocarbamates are being explored as antitumor agents (Nagy et al., 2012).

REFERENCES

Garcia de Torres, G., Romer, K.G., Torres Alanis, O., and Freundt, K.J. (1983) Blood acetaldehyde levels in alcohol-dosed rats after treatment with ANIT, ANTU, dithiocarbamate derivatives, or cyanamide. *Drug Chem. Toxicol.* 6:317–328.

Hogarth, G. (2012) Metal–dithiocarbamate complexes: chemistry and biological activity. *Mini Rev. Med. Chem.* 12: 1202–1215.

Hamilton & Hardy's Industrial Toxicology, Sixth Edition. Edited by Raymond D. Harbison, Marie M. Bourgeois, and Giffe T. Johnson.
© 2015 John Wiley & Sons, Inc. Published 2015 by John Wiley & Sons, Inc.

Israeli, R., Sculsky, M., and Tiberin, P. (1983) Acute intoxication due to exposure to maneb and zineb. A case with behavioral and central nervous system changes. *Scand. J. Work Environ. Health* 9:47–51.

Maini, P. and Boni, R. (1986) Gas chromatographic determination of dithiocarbamate fungicides in workroom air. *Bull. Environ. Contam. Toxicol.* 37:931–937.

Maroni, M., Colosio, C., Ferioli, A., and Fait, A. (2000) Biological monitoring of pesticide exposure: a review. Introduction. *Toxicology* 143:1–118.

McGrath, M.T. (2004) *What are fungicides?* Available at https://www.apsnet.org/edcenter/intropp/topics/Pages/Fungicides.aspx.

Mulkey, M.E. (2001) *General Subject Correspondence: The Determination of Whether Dithiocarbamate Pesticides Share a Common Mechanism of Toxicity.* Office of Pesticide Programs, Environmental Protection Agency.

Nagy, E.M., Ronconi, L., Nardon, C., and Fregona, D. (2012) Noble metal-dithiocarbamates precious allies in the fight against cancer. *Mini Rev. Med. Chem.* 12:1216–1229.

Römer, K.G., Torres Alanis, O., Garcia de Torres, G., and Freundt, K.J. (1984) Delayed ethanol elimination from rat blood after treatment with thiram, tetramethylthiuram monosulfide, ziram, or cyanamide. *Bull. Environ. Contam. Toxicol.* 32:537–542.

Tilton, F., La Du, J.K., Vue, M., Alzarban, N., and Tanguay, R.L. (2006) Dithiocarbamates have a common toxic effect on zebrafish body axis formation. *Toxicol. Appl. Pharmacol.* 216:55–68.

Zhou, Y., Shie, F.S., Picardo, P., Montine, T.J., and Zhang, J. (2004) Proteasomal inhibiton induced by manganese ethylene-bis-dithiocarbamate: relevance to Parkinson's disease. *Neuroscience* 128(2):281–291.

THIRAM

Thiram (tetramethylthiuram disulfide; CAS #137-26-8) is a colorless to yellow crystalline solid with a characteristic odor. It has a solubility in water by weight of 0.003% and a vapor pressure of 0.000008 mmHg at 20 °C (68 °F). Thiram has been produced commercially in the United States since 1925. It is primarily used as an accelerator and vulcanizing agent in rubber processing. It is used as a fungicide for seed treatment, corn, small grains, peanuts, lettuce, apples, pears, ornamentals, and sod. It also is used for slime control in paper production, as a bacteriostat in soap, and in antiseptic sprays. At higher concentrations it is used as a deer, rabbit, and bird repellent in orchards. It can be a by-product of ziram or ferbam environmental degradation (IARC, 1991). Thiram is not known to occur naturally.

INDUSTRIAL HYGIENE

The National Institute for Occupational Safety and Health (NIOSH) and the Occupational Safety and Health Administration (OSHA) have set the recommended exposure limit (REL) and the permissible exposure limit (PEL), respectively, for thiram at 5 mg/m^3 (NIOSH, 2009). The American Conference of Governmental Industrial Hygienists (ACGIH) has set the threshold limit value (TLV) at 0.05 mg/m^3 (ACGIH, 2009). Thiram is considered Immediately Dangerous to Life and Health (IDLH) at 100 mg/m^3. Gloves and respirators have been shown to reduce exposure to a minimum in seed-processing facilities (Grey et al., 1983). Agricultural personnel involved in harvesting, as well as applicators, may be exposed to thiram residues (Brouwer et al., 1992a, 1992b). Workers should be required to use dust- and splash-proof goggles whenever contact with the eyes is possible. Agricultural workers who sow treated seeds or handle other thiram-treated agricultural materials should wear gloves to prevent dermal absorption.

HEALTH EFFECTS

Thiram is readily absorbed by the dermal, oral, and inhalation routes and is distributed throughout the body. Systemic signs and symptoms of overexposure may include nausea, vomiting, headache, lethargy, dizziness, drowsiness, ataxia, anorexia, diarrhea, and coma. Alcohol has a synergistic effect on thiram toxicity. The effects of a low level exposure to thiram combined with alcohol consumption mimic the effects of a high level exposure to thiram. Thiram has been shown to be more efficient at producing alcohol intolerance than disulfiram, but because of its greater toxicity, it is not used for pharmaceutical purposes.

Allergic contact dermatitis has been reported in farmers exposed to thiram and occasionally in those exposed to soaps containing thiram as a bacteriostat (Sharma and Kaur, 1990). Dermatitis may also be aggravated by alcohol consumption (Sharma et al., 2003).

The International Agency for Research on Cancer IARC has concluded that thiram is not classifiable as to its carcinogenicity to humans (Group 3) because of inadequate evidence (OSHA, 2002; IARC, 1991). Treatment of thiram poisoning is symptomatic. Thiram causes acute systemic toxicity; it is a suspected teratogen and mutagen (OSHA, 2002).

TARGET ORGANS AND MECHANISM OF ACTION

Dietary carcinogenicity studies of male and female Fischer rats did not indicate that thiram alone has carcinogenic potential (Hasegawa et al., 1988). However, rats in a chronic dietary study dosed with thiram and sodium nitrate had high rates of nasal cavity tumors (Lijinsky, 1984). Researchers theorize that together these compounds may form carcinogenic N-nitroso compounds *in vivo*.

Thiram had a median lethal concentration (LC_{50}) of 0.00000005 M in one *in vitro* study of Chinese hamster ovary cells (Hodgson and Lee, 1977). It had greater cytotoxic potential in this study than maneb or zineb. Mosesso et al. (1994) reported that thiram treatment resulted in a statistically significant increase in clastogenic effects in Chinese hamster cell lines exposed *in vitro* in the presence of the S9 fraction.

Studies of mice and rats indicate that thiram may have teratogenic potential at high doses (Fishbein, 1976).

Male rats exposed to 25 mg/kg/day of thiram exhibited an increase in testes weight gain and other morphological changes after 90 days (Mishra et al., 1993). Male mice treated with thiram reportedly showed an increase in the incidence of abnormally shaped sperm (Hemavathi and Rahiman, 1993).

Animal studies have demonstrated that thiram may have a wide spectrum of biochemical effects *in vivo* (Shukla et al., 1996). It has been reported to inhibit the conversion of dopamine to noradrenaline in the heart and adrenal glands of rodents. Rats treated with thiram had a significant loss of cytochrome P-450 but elevated levels of other liver enzymes. Vitamin B derivatives and ascorbic acid have been reported to counteract thiram's inhibition of alcohol metabolism (Edwards et al., 1991). Thiram is a potential inhibitor of angiogenesis (Marikovsky, 2002). Chickens fed thiram display growth plate defects due to the death of chondrocytes (Rasaputra et al., 2013).

Thiram and other dithiocarbamates may form lipid-soluble complexes with metals. In one study, rats injected intravenously with lead acetate were also exposed perorally to thiram (Oskarsson, 1984). Rats treated with thiram had statistically significant increases in lead levels in the liver, lungs, and brain compared with untreated rats. Thiram-treated rats also had significantly lower concentrations of lead in the femur and kidney tissues. Of all dithiocarbamates tested, thiram most dramatically increased gastrointestinal absorption and tissue levels of nickel in mice (Borg and Tjalve, 1988). In acute studies with rats, thiram was found to be relatively toxic (single-dose oral median lethal dose [LD_{50}] of 620–640 mg/kg) compared with ferbam (Lee et al., 1978).

REFERENCES

American Conference of Governmental Industrial Hygienists (2009) *Threshold Limit Values (TLVs) for Chemical Substances and Physical Agents and Biological Exposure Indices (BEIs)*, Cincinnati, OH: American Conference of Governmental Industrial Hygienists.

Borg, K. and Tjalve, H. (1988) Effect of thiram and dithiocarbamate pesticides on the gastrointestinal absorption and distribution of nickel in mice. *Toxicol. Lett.* 42:87–98.

Brouwer, D.H., Brouwer, R., De Mik, G., Maas, C.L., and van Hemmen, J.J. (1992a) Pesticides in the cultivation of carnations in greenhouses. Part I. Exposure and concomitant health risk. *Am. Ind. Hyg. Assoc. J.* 53:575–581.

Brouwer, R., Brouwer, D.H., Tijssen, S.C., and van Hemmen, J.J. (1992b) Pesticides in the cultivation of carnations in greenhouses. Part II. Relationship between foliar residues and exposures. *Am. Ind. Hyg. Assoc. J.* 53:582–587.

Edwards, I.R., Ferry, D.G., and Temple, W.A. (1991) *Fungicides and Related Compounds*, San Diego, CA: Academic Press.

Fishbein, L. (1976) Environmental health aspects of fungicides. I. Dithiocarbamtes. *J. Toxicol. Environ. Health* 1:713–735.

Grey, W.E., Marthre, D.E., and Rogers, S.J. (1983) Potential exposure of commerical seed-treating applicators to the pesticides carboxin-thiram and lindane. *Bull. Environ. Contam. Toxicol.* 31:244–250.

Hasegawa, R., et al. (1988) Carcinogenicity study of tetramethylthiuram disulfide (thiram) in F344 rats. *Toxicology* 51:155–165.

Hemavathi, E. and Rahiman, M.A. (1993) Toxicological effects of ziram, thiram, and dithane M-45 assessed by sperm shape abnormalities in mice. *J. Toxicol. Environ. Health* 38:393–398.

Hodgson, J.R. and Lee, C.C. (1977) Cytotoxicity studies on dithiocarbamate fungicides. *Toxicol. Appl. Pharmacol.* 40:19–22.

International Agency for Research on Cancer (1991) Occupational exposures in insecticide application, and some pesticides. IARC Working Group on the Evaluation of Carcinogenic Risks to Humans, Lyon, 16–23 October 1990. *IARC Monogr. Eval. Carcinog. Risks Hum.* 53:5–586.

Lee, C.C., Russell, J.Q., and Minor, J.L. (1978) Oral toxicity of ferric dimethyl-dithiocarbamate (ferbam) and tetramethylthiuram disulfide (thiram) in rodents. *J. Toxicol. Environ. Health* 4:93–106.

Lijinsky, W. (1984) Induction of tumors of the nasal cavity in rats by concurrent feeding of thiram and sodium nitrite. *J. Toxicol. Environ. Health* 13:609–614.

Marikovsky, M. (2002). Thiram inhibits angiogenesis and slows the development of experimental tumours in mice. *Br. J. Cancer* 86(5):779–787.

Mishra, V.K., Srivastava, M.K., and Raizada, R.B. (1993) Testicular toxicity of thiram in rat: morphological and biochemical evaluations. *Ind. Health* 31:59–67.

Mosesso, P., et al. (1994) Clastogenic effects of the dithiocarbamate fungicides thiram and ziram in Chinese hamster cell lines cultured *in vitro*. *Teratog. Carcinog. Mutagen.* 14:145–155.

National Institute for Occupational Safety and Health (2009) *Pocket Guide to Chemical Hazards*, Washington, DC: U.S. Department of Health and Human Services.

Occupational Safety & Health Administration (2002) Thiram. Washington, DC: United States Department of Labor.

Oskarsson, A. (1984) Dithiocarbamate-induced redistribution and increased brain uptake of lead in rats. *Neurotoxicology* 5:283–293.

Rasaputra, K.S., Liyanage, R., Lay, J.O., Jr., Slavik, M.F., and Rath, N.C. (2013) Effect of thiram on avian growth plate chondrocytes in culture. *J. Toxicol. Sci.* 38(1):93–101.

Sharma, V.K., Aulakh, J.S., and Malik, A.K. (2003) Thiram: degradation, applications and analytical methods. *J. Environ. Monit.* 5:717–723.

Sharma, V.K. and Kaur, S. (1990) Contact sensitization by pesticides in farmers. *Contact Dermatitis* 23:77–80.

Shukla, Y., Baqar, S.M., and Mehrotra, N.K. (1996) Carcinogenic and co-carcinogenic studies of thiram on mouse skin. *Food Chem. Toxicol.* 34:283–289.

ZINEB

Zineb (zinc ethylenebisdithiocarbamate; CAS #12122-67-7) is a light-colored powder or crystal. It is considered unstable in light, moisture, and heat. Zineb has been used primarily on foliar species and as a fungicidal treatment for bulbs. All uses of zineb were recently voluntarily canceled by the registrants (EPA, 1998).

INDUSTRIAL HYGIENE

Variations in mixing and loading methods for zineb may determine the extent of worker exposure (Brouwer et al., 1992). Protective gloves are the most important tool in reducing an exposure. There are no exposure limits or TLVs from OSHA, NIOSH, or the ACGIH for zineb. Limits for thiram may be followed as a protective measure.

HEALTH EFFECTS

One study on rats suggests that zineb may have an effect on the humoral activity of the thyroid gland, although the study's results were not statistically significant (Laisi et al., 1985). The alcohol effect is much less evident with zineb than with thiram or ziram.

TARGET ORGANS AND MECHANISM OF ACTION

The primary metabolite of zineb is ETU, which may be further metabolized to other products *in vivo* (Jordan and Neal, 1979). In rats, zineb has been shown to inhibit hepatic microsomal monooxygenases, specifically cytochrome P-450 (Borin et al., 1985). In another study, rats dosed once with 250 mg/kg of zineb had reduced rates of oxidative metabolism (Siddiqui et al., 1991). This inhibition is partially attributable to zineb binding to sulfhydryl groups on cytochrome P-450 and partially to unknown mechanisms.

Zineb has been implicated as a teratogen in studies of orally dosed rats (Fishbein, 1976).

When rats were dosed orally with zineb in combination with copper sulfate (another fungicide used for seed treatment), the two compounds were reported to have a synergistic relationship (Ahmed and Shoka, 1994). When combined, these compounds demonstrated synergistic effects by elevating serum alkaline phosphatase activity, diminishing cholinesterase (ChE) activity, and increasing alanine aminotransferase activity. These effects suggest a complex relationship between these two fungicides.

Pathological changes were observed in the testes of male rats dosed with zineb at 1000 mg/kg/day for 30 days (Raizada et al., 1979). The thyroids of these animals were also reported as having marked hyperplasia and increased weight.

Nebbia et al. (1995) reported a decrease in hemoglobin concentration, hematocrit, and erythrocyte and leukocyte counts in rabbits dosed subchronically with a dietary level of 0.6% zineb. These findings suggest that zineb interferes with the synthesis of hemoglobin. However, the authors concluded that the primary effects of zineb toxicity in rabbits were in the thyroid and liver. They found a dose-related depression of circulating T3 hormones but no effect on thyroid-stimulating hormone. The thyroid gland was considerably enlarged. Effects on the thyroid gland have been observed in other species as well (Soffietti et al., 1988). Rabbits exposed subchronically to zineb also showed a dose-dependent decrease in hepatic glutathione-S-transferase activity and impaired oxidative drug metabolism (Nebbia et al., 1993). Zineb did not cause testicular damage in the rabbit, as it has in many other species, possibly because it did not accumulate in the rabbits' testes (Nebbia et al., 1995).

In one study of Chinese hamster embryo cells exposed to dithiocarbamates, zineb was a less potent cytotoxin by an order of magnitude (LD_{50} 0.0000006 M) than thiram and ferbam (Hodgson and Lee, 1977).

REFERENCES

Ahmed, H.M. and Shoka, A.A. (1994) Toxic interactions between copper sulphate and some organic agrochemicals. *Toxicol. Lett.* 70:109–119.

Borin, C., Periquet, A., and Mitjavila, S. (1985) Studies on the mechanism of nabam- and zineb-induced inhibition of the hepatic microsomal monooxygenases of the male rat. *Toxicol. Appl. Pharmacol.* 81:460–468.

Brouwer, D.H., Brouwer, E.J., and van Hemmen, J.J. (1992) Assessment of dermal and inhalation exposure to zineb/maneb in the cultivation of flower bulbs. *Ann. Occup. Hyg.* 36:373–384.

Environmental Protection Agency (1998) *Status of Pesticides in Reregistration and Special Review (Rainbow Report)*, Washington, DC: Environmental Protection Agency.

Fishbein, L. (1976) Environmental health aspects of fungicides. I. Dithiocarbamtes. *J. Toxicol. Environ. Health* 1:713–735.

Hodgson, J.R. and Lee, C.C. (1977) Cytotoxicity studies on dithiocarbamate fungicides. *Toxicol. Appl. Pharmacol.* 40:19–22.

Jordan, L.W. and Neal, R.A. (1979) Examination of the *in vivo* metabolism of maneb and zineb to ethylenethiourea (ETU) in mice. *Bull. Environ. Contam. Toxicol.* 22:271–277.

Laisi, A., Tuominen, R., Mannisto, P., Savolainen, K., and Mattila, J. (1985) The effect of maneb, zineb, and ethylenethiourea on the humoral activity of the pituitary-thyroid axis in rat. *Arch. Toxicol. Suppl.* 8:253–258.

Nebbia, C., et al. (1993) Inhibition of hepatic xenobiotic metabolism and of glutathione-dependent enzyme activities by zinc ethylene-bis-dithiocarbamate in the rabbit. *Pharmacol. Toxicol.* 73:233–239.

Nebbia, C., et al. (1995) Effects of the subchronic administration of zinc ethylene-bis-dithiocarbamate (zineb) to rabbits. *Vet. Hum. Toxicol.* 37:137–142.

Raizada, R.B., Datta, K.K., and Dikshith, T.S. (1979) Effect of zineb on male rats. *Bull. Environ. Contam. Toxicol.* 22:208–213.

Siddiqui, A., Ali, B., and Srivastava, S.P. (1991) Heterogeneous effects of ethylenebisdithiocarbamate (EBDC) pesticides on oxidative metabolism of xenobiotics. *Pharmacol. Toxicol.* 69:13–16.

Soffietti, M.G., et al. (1988) The target organ and the toxic process. In: Chambers, P.L., Chambers, C.M., and Dirheimer, G., editors. *Archives of Toxicology*, vol. 12, Chapter 16, Berlin: Springer, pp. 107–109.

MANEB

Maneb (manganese 1,2-ethanediylbiscarbamodithioate; CAS #12427-38-2) is a yellow to brown powder with a vapor pressure of <0.000000075 mmHg at 20 °C (68 °F). It is not known to occur naturally. Maneb is a broad-spectrum fungicide used extensively for the treatment of bulbs, ornamental plants, turf, field crops, and fruit and as protection against tomato and potato blights. Approximately, 2.5 million pounds of maneb are used annually in the United States. Maneb has a low mobility when applied to soil. It is degraded by abiotic hydrolytic processes and also undergoes photodegradation. The half-life of maneb in soil is approximately 1–4 weeks. Maneb is a class of dithiocarbamates known as ethylenebisdithiocarbamate (EBDC), which also includes mancozeb and metrizam (EPA, 2005).

INDUSTRIAL HYGIENE

Protective gloves are the most important tool in reducing exposure to maneb (Brouwer et al., 1992). No exposure limits or TLVs have been set by the OSHA, NIOSH, or ACGIH. Limits for thiram may be followed as a protective measure. Maneb has been voluntarily canceled by registrants for the following labels: sweet corn, grapes, apples, kadota figs, and seed treatment use for peanuts and rice (EPA, 2005).

HEALTH EFFECTS

In general, an acute overexposure to maneb is rare. One farmer exposed to a combination of zineb and maneb (Manzidan) experienced weakness, headache, nausea, and fatigue before losing consciousness and having convulsions (Israeli et al., 1983). Sensitization to maneb has been reported in flower cultivators (Crippa et al., 1990). Treatment of maneb overexposure is symptomatic. As with many EBDCs and ETUs the thyroid is the target organ of maneb, showing effects such as increased thyroid weight, thyroid hyperplasia, decreased T4 across species (EPA, 2005).

TARGET ORGANS AND MECHANISM OF ACTION

In vivo, like all EBDCs, maneb is metabolized to ETU, after which it may undergo further metabolism (Jordan and Neal, 1979). Another active ingredient of maneb, manganese ethylene-bis-dithiocarbamate, has been shown to cause neurodegeneration in the nigrostriatal region of the brain in mice and rats (Zhou et al., 2004). Maneb affects the nerve impulse transmission through inhibition of ChE enzymes (Manfo et al., 2011). Thyroid effects have been reported as a result of ETU metabolism (Zhou et al., 2004; Laisi et al., 1985). Acute renal failure and inhibition of mitochondrial function have been reported (Manfo et al., 2011). Epithelial cell necrosis is observed in rainbow trout exposed to maneb for as little as 6 h at concentrations of 1.30 mg/l (Boran et al., 2010). A decrease in thyroxine production has altered the sperm quality in rabbits administered maneb daily (Manfo et al., 2011). However, maneb's effect on testosterone production is still unclear (Manfo et al., 2011).

REFERENCES

Boran, H., Altinok, I., and Capkin, E. (2010) Histopathological changes induced by maneb and carbaryl on some tissues of rainbow trout, *Oncorhynchus mykiss. Tissue Cell* 42:158–164.

Brouwer, D.H., Brouwer, E.J., and van Hemmen, J.J. (1992) Assessment of dermal and inhalation exposure to zineb/maneb in the cultivation of flower bulbs. *Ann. Occup. Hyg.* 36:373–384.

Crippa, M., Misquith, L., Lonati, A., and Pasolini, G. (1990) Dyshidrotic eczema and sensitization to dithiocarbamates in a florist. *Contact Dermatitis* 23:203–204.

Environmental Protection Agency (2005) *Maneb Facts. Prevention, Pesticides and Toxic Substances (7508C)*, Washington, DC: United States Environmental Protection Agency.

Israeli, R., Sculsky, M., and Tiberin, P. (1983) Acute central nervous system changes due to intoxication by Manzidan (a combined dithiocarbamate of Maneb and Zineb). *Arch. Toxicol. Suppl.* 6:238–243.

Jordan, L.W. and Neal, R.A. (1979) Examination of the *in vivo* metabolism of maneb and zineb to ethylenethiourea (ETU) in mice. *Bull. Environ. Contam. Toxicol.* 22:271–277.

Laisi, A., Tuominen, R., Mannisto, P., Savolainen, K., and Mattila, J. (1985) The effect of maneb, zineb, and ethylenethiourea on the humoral activity of the pituitary-thyroid axis in rat. *Arch. Toxicol. Suppl.* 8:253–258.

Manfo, F.P., Chao, W.F., Moundipa, P.F., Pugeat, M., and Wang, P.S. (2011) Effects of maneb on testosterone release in male rats. *Drug Chem. Toxicol.* 34:120–128.

Zhou, Y., Shie, F.S., Picardo, P., Montine, T.J., and Zhang, J. (2004) Proteasomal inhibiton induced by manganese ethylene-bis-dithiocarbamate: relevance to Parkinson's disease. *Neuroscience* 128(2):281–291.

METAM SODIUM

Metam sodium (sodium methyldithiocarbamate; CAS #137-42-8) is a white crystalline solid. It is slightly soluble in acetone, xylene, and kerosene and insoluble in most organic solvents. It has a solubility in water of 722 g/l at 20 °C (68 °F). Metam sodium is typically used in a 33% aqueous solution. It includes a mixture of ammoniates of zinc-ethylene-(bis)-dithiocarbamates with ethlyene-(bis)-dithiocarbamic acid bimolecular and trimolecular cyclic anhydrides and disulfides (Mulkey, 2001).

The Environmental Protection Agency (EPA) estimates the annual use of metam sodium at more than 25 million pounds active ingredient (EPA, 2011). Metam sodium is also used as a nematocide and an insecticide. It is the third most widely used pesticide in the United States (Cox, 2006).

Methyl isothiocyanate (MITC) is formed from metam sodium through chemical breakdown by removal of the sulfa group. This process occurs within 1–5 h of application. The MITC formed is very mobile and permeates the soil, effectively killing pests. It contaminates the air that can cause symptoms as far as a mile away (Cox, 2006). Health protective concentrations of MITC have been estimated as 0.2 parts per million (ppm) for 4 hours of exposure and 0.8 ppm for a 14-min exposure (Dourson et al., 2010).

INDUSTRIAL HYGIENE

No exposure limits or TLVs have been set for metam sodium by the OSHA, NIOSH, or ACGIH. The MITC levels during production of metam sodium may be monitored by gas chromatography (Collina and Maini, 1979).

HEALTH EFFECTS

At high concentrations metam sodium vapors are irritating and may induce an asthma-like condition or reactive airways dysfunction syndrome (Cone et al., 1994). In laboratory animals, metam sodium can trigger a reduction in the hormone levels, leg strength, and immune activity (Cox, 2006).

Contact dermatitis has been reported in workers exposed to metam sodium and MITC (Schubert, 1978; Koo et al., 1995).

TARGET ORGANS AND MECHANISM OF ACTION

Like other dithiocarbamates, metam sodium interferes with alcohol metabolism. Metam sodium has not been shown to be mutagenic in *Salmonella* assays (De Lorenzo et al., 1978). Animal studies of metam sodium show some immunological hypersensitivity and developmental effects (Pruett et al., 2011).

REFERENCES

Collina, A. and Maini, P. (1979) Analysis of methylisothiocyanate derived from the soil fumigant metham-sodium in workroom air. *Bull. Environ. Contam. Toxicol.* 22:400–404.

Cone, J.E., et al. (1994) Persistent respiratory health effects after a metam sodium pesticide spill. *Chest* 106:500–508.

Cox, C. (2006) Metam sodium. *J. Pest. Ref.* 26(1):12–16.

De Lorenzo, F., Staiano, N., Silengo, L., and Cortese, R. (1978) Mutagenicity of diallate, sulfallate, and triallate and relationship between structure and mutagenic effects of carbamates used widely in agriculture. *Cancer Res.* 38:13–15.

Dourson, M.L., Kohrmann-Vincent, M.J., and Allcn, B.C. (2010) Dose response assessment for effects of acute exposure to methyl isothiocyanate (MITC). *Regul. Toxicol. Pharmacol.* 58(2): 181–188.

Environmental Protection Agency (2011) *Pesticides Industry Sales and Usage: 2006 and 2007 Market Estimates*, Washington, DC: Environmental Protection Agency.

Koo, D., Goldman, L., and Baron, R. (1995) Irritant dermatitis among workers cleaning up a pesticide spill: California 1991. *Am. J. Ind. Med.* 27:545–553.

Mulkey, M.E. (2001) *General Subject Correspondence: The Determination of Whether Dithiocarbamate Pesticides Share a Common Mechanism of Toxicity*. Office of Pesticide Programs. Environmental Protection Agency.

Pruett, S.B., Myers, L.P., and Keil, D.E. (2001) Toxicology of metam sodium. *J. Toxicol. Environ. Health B Crit. Rev.* 4(2):207–222.

Schubert, H. (1978) Contact dermatitis to sodium *N*-methyldithio-carbamate. *Contact Dermatitis* 4:370–371.

PHTHALIMIDE FUNGICIDES

CAPTAN

Captan (3A,4,7,7A-tetrahydro-2-trichloromethylthio; CAS #133-06-2)-l*H*-isoindole-l,3-(2H)-dione) is a white crystalline powder with a pungent odor. It has a solubility of 0.0001% by weight in water and a vapor pressure of 0.0 mmHg (NIOSH 2009). The TLV is 5 mg/m^3. (ACGIH, 2009) Captan is the active ingredient in the fungicides Orthocide, Orthocide 406, and Vancide 89. It is not known to occur naturally.

Captan is probably not a significant threat to aquatic flora and fauna because of its low solubility and rapid biodegradation (Antón et al., 1993). Hydrolyzed by soil microorganisms, it typically persists for a couple of days to a couple of weeks (Li and Nelson, 1985). High levels of captan may kill earthworms, but current commercial levels of application do not (Anton et al., 1990). Captan residues may persist on food and should be removed by rinsing (el-Zemaity, 1988). Catpan has also been shown to control fungal infections on certain bee species (Huntzinger et al., 2008).

INDUSTRIAL HYGIENE

Agricultural workers may be exposed to captan when spraying the fungicide or when entering a field shortly after application (McJilton et al., 1983; Ritcey et al., 1987). Dermal exposure is the most significant route of absorption. However, one study examining exposure in farm workers suggests limited absorption even through the dermal route (Berthet et al., 2012). Fruit growers wearing gloves or rubber boots had dramatically lower levels of captan exposure than those without either (de Cock et al., 1995). Application of captan from a tractor with a cab rather than an uncovered tractor reduced worker exposure by about half.

The NIOSH identifies captan as a possible occupational carcinogen. As such, exposure should be kept to a minimum. The REL and the TLV for captan have been set at 5 mg/m^3 (NIOSH, 2009). The OSHA has not set a PEL for this fungicide. The ACGIH gives captan a carcinogen ranking of A3, meaning that it causes cancer in animals (ACGIH, 2009).

HEALTH EFFECTS

There are no reports of systemic poisoning by captan (Edwards et al., 1991). Ingestion of captan may cause nausea, vomiting, and diarrhea. Dermal exposure often causes an irritant or allergic contact dermatitis (Dooms-Goossens et al., 1986). In rainbow trout exposed to captan, hypertrophy and epithelial cell necrosis leading to inflammation and necrosis

of the liver, kidney, and spleen have been reported (Boran et al., 2012). Captan is also associated with non-Hodgkin's lymphoma, spontaneous abortion, and congenital defects (Chodorowski and Anand, 2003).

As mentioned previously, the NIOSH classifies captan as a possible occupational carcinogen, and the ACGIH identifies it as an animal carcinogen. Folpet, tricholoromethylthio-related fungicide, similar to captan, has not been shown to cause cancer in humans (Cohen et al., 2010). Captan was considered not likely to cause cancer by the EPA in 2004 (Gordon, 2007). A panel concluded that a non-mutagenic mode of action is necessary through extensive irritation of the duodenal villi (Gordon, 2007). Captan has low oral, dermal, and inhalation toxicity across various animal species (NPIC, 2000). Rats and mice have decreased immune function when given large oral doses (NPIC, 2000).

TARGET ORGANS AND MECHANISM OF ACTION

The liver is one of the target organs of captan toxicity. Rats injected intraperitoneally with captan had changes in the activities of cytochrome P-450 and reduced nicotinamide adenine dinucleotide phosphate (NADPH)–cytochrome C reductase at levels of 20 mg/kg of body weight (Dalvi and Mutinga, 1990). Oxidative damage and lipid peroxidation caused by metabolites of captan can lead to cytotoxicity (Suzuki et al., 2004). Captan has tested positive in assays designed to measure genotoxicity (Snyder, 1992; Perocco et al., 1995). However, it is not generally believed to be mutagenic in higher animals.

Poor reproductive outcomes have been linked with captan exposure in animal experiments. Female rats fed captan in the postimplantation stage of pregnancy had significantly reduced litter sizes and lower survival rates (Spencer, 1984). The treated rats also had decreased uterus and spleen weights. Captan treatment did not significantly reduce liver, adrenal, or ovary weights, which indicate that, at least in rats, the uterus and spleen are the potential target organs.

The potential teratogenicity of captan has been investigated repeatedly because of its structural similarity to thalidomide. In one study of two rabbit strains, golden hamsters, and albino rats dosed with captan throughout organogenesis (Kennedy et al., 1968), no increase in teratogenic effects was noted. However, in another study of golden hamsters, one dose of captan during organogenesis did cause teratogenicity (Robens, 1970). Teratogenic studies with rhesus monkeys and stump-tailed macaques found that a dose of 25 mg/kg/day administered during organogenesis did not have any teratogenic effects (Vondruska et al., 1971).

Animal experiments suggest that captan exposure may affect the metabolism. Studies of mice indicate that captan may inhibit intestinal absorption (Iturri and Soto, 1994). Exposure to captan also increased DNA synthesis and thymidine kinase activity in rat gastric mucosa (Wahby et al., 1990). Captan has been found to suppress the immune response of rats and mice (Lafarge-Frayssinet and Decloitre, 1982).

BIOMONITORING

Urinary tetrahydrophthalimide (THPI) correlates closely with actual exposure (de Cock et al., 1995). Measurement of THPI is a key biomarker for captan; it can be detected through elimination half-life estimated to be approximately 13 h after a dermal exposure (Heredia-Ortiz and Bouchard, 2012). The limits of detection can be between 3.0 and 8.0 µg/kg using techniques such as dispersive liquid–liquid micro-extraction and gas chromatography (Zang et al., 2008).

REFERENCES

American Conference of Governmental Industrial Hygienists (2009) *Threshold Limit Values (TLVs) for Chemical Substances and Physical Agents and Biological Exposure Indices (BEIs)*, Cincinnati, OH: American Conference of Governmental Industrial Hygienists.

Anton, F., Laborda, E., and Laborda, P. (1990) Acute toxicity of the fungicide captan to the earthworm *Eisenia foetida* (Savigny). *Bull. Environ. Contam. Toxicol.* 45:82–87.

Antón, F.A., Laborda, E., and Laborda, P. (1993) Acute toxicity of technical captan to algae and fish. *Bull. Environ. Contam. Toxicol.* 50:392–399.

Berthet, A., Heredia-Ortiz, R., Vernez, D., Danuser, B., and Bouchard, M. (2012) A detailed urinary excretion time course study of captan and folpet biomarkers in workers for the estimation of dose, main route-of-entry and most appropriate sampling and analysis strategies. *Ann. Occup. Hyg.* 56:815–828.

Boran, H., Capkin, E., Altinok, I., and Terzi, E. (2012) Assessment of acute toxicity and histopathology of the fungicide captan in rainbow trout. *Exp. Toxicol. Pathol.* 64, 175–179.

Chodorowski, Z. and Anand, J.S. (2003) Acute oral suicidal intoxication with Captan—a case report. *J. Toxicol. Clin. Toxicol.* 41:603.

Cohen, S.M., Gordon, E.B., Singh, P., Arce, G.T., and Nyska, A. (2010) Carcinogenic mode of action of folpet in mice and evaluation of its relevance to humans. *Crit. Rev. Toxicol.* 40:531–545.

Dalvi, R.R. and Mutinga, M.L. (1990) Comparative studies of the effects on liver and liver microsomal drug-metabolizing enzyme system by the fungicides captan, captafol and folpet in rats. *Pharmacol. Toxicol.* 66:231–233.

de Cock, J., Heederik, D., Hoek, F., Boleij, J., and Kromhout, H. (1995) Urinary excretion of tetrahydrophtalimide in fruit growers with dermal exposure to captan. *Am. J. Ind. Med.* 28:245–256.

Dooms-Goossens, A.E., et al. (1986) Contact dermatitis caused by airborne agents. A review and case reports. *J. Am. Acad. Dermatol.* 15:1–10.

Edwards, I.R., Ferry, D.G., and Temple, W.A. (1991) *Fungicides and Related Compounds*, San Diego, CA: Academic Press.

el-Zemaity, M.S. (1988) Residues of captan and folpet on greenhouse tomatoes with emphasis on the effect of storing, washing, and cooking on their removal. *Bull. Environ. Contam. Toxicol.* 40:74–79.

Gordon, E. (2007) Captan: transition from 'B2' to 'not likely'. How pesticide registrants affected the EPA Cancer Classification Update. *J. Appl. Toxicol.* 27:519–526.

Heredia-Ortiz, R. and Bouchard, M. (2012) Toxicokinetic modeling of captan fungicide and its tetrahydrophthalimide biomarker of exposure in humans. *Toxicol. Lett.* 213:27–34.

Huntzinger, C.I., James, R.R., Bosch, J., and Kemp, W.P. (2008) Fungicide tests on adult alfalfa leafcutting bees (Hymenoptera: Megachilidae). *J. Econ. Entomol.* 101:1088–1094.

Iturri, S.J. and Soto, J. (1994) Inhibitory effect of captan in the small intestine absorption capacity of the mouse. *Bull. Environ. Contam. Toxicol.* 53:648–654.

Kennedy, G., Fancher, O.E., and Calandra, J.C. (1968) An investigation of the teratogenic potential of captan, folpet, and difolatan. *Toxicol. Appl. Pharmacol.* 13:420–430.

Lafarge-Frayssinet, C. and Decloitre, F. (1982) Modulatory effect on the pesticide captan on the immune response in rats and mice. *J. Immunopharmacol.* 4:43–52.

Li, C.Y. and Nelson, E.E. (1985) Persistence of benomyl and captan and their effects on microbial activity in field soils. *Bull. Environ. Contam. Toxicol.* 34:533–540.

McJilton, C.E., Berckman, G.E., and Deer, H.M. (1983) Captan exposure in apple orchards. *Am. Ind. Hyg. Assoc. J.* 44:209–210.

National Institute of Occupational Safety Health (2009) *Pocket Guide to Chemical Hazards*, Washington, DC: U.S. Department of Health and Human Services.

National Pesticide Information Center (2000) *Captan: General Fact Sheet*, U.S. Environmental Protection Agency.

Perocco, P., Colacci, A., Del Ciello, C., and Grilli, S. (1995) Transformation of BALB/c 3T3 cells *in vitro* by the fungicides captan, captafol and folpet. *Jpn. J. Cancer Res.* 86:941–947.

Ritcey, G., Frank, R., McEwen, F.L., and Braun, H.E. (1987) Captan residues on strawberries and estimates of exposure to pickers. *Bull. Environ. Contam. Toxicol.* 38:840–846.

Robens, J.F. (1970) Teratogenic activity of several phthalimide derivatives in the golden hamster. *Toxicol. Appl. Pharmacol.* 16:24–34.

Snyder, R.D. (1992) Effects of Captan on DNA and DNA metabolic processes in human diploid fibroblasts. *Environ. Mol. Mutagen* 20:127–133.

Spencer, F. (1984) Structural and reproductive modifications in rats following a post-implantational exposure to captan. *Bull. Environ. Contam. Toxicol.* 33:84–91.

Suzuki, T., Nojiri, H., Isono, H., and Ochi, T. (2004) Oxidative damages in isolated rat hepatocytes treated with the organochlorine fungicides captan, dichlofluanid and chlorothalonil. *Toxicology* 204:97–107.

Vondruska, J.F., Fancher, O.E., and Calandra, J.C. (1971) An investigation into the teratogenic potential of captan, folpet, and Difolatan in nonhuman primates. *Toxicol. Appl. Pharmacol.* 18:619–624.

Wahby, M., Shelef, L.A., Luk, G.D., and Majumdar, A.P. (1990) Induction of gastric mucosal cell proliferation by the fungicide captan: role of tyrosine kinases. *Toxicol. Lett.* 54:189–198.

Zang, X., et al. (2008) Analysis of captan, folpet, and captafol in apples by dispersive liquid–liquid microextraction combined with gas chromatography. *Anal. Bioanal. Chem.* 392:749–754.

CAPTAFOL

Captafol (*3a*, 4,7,7*a*-tetrahydro-2-[(1,1,2,2-tetrachloroethyl) thio]-1*H*-isoindole-1,3(2*H*)-dione; Difolatan; CAS #2425-06-1) is a white crystalline solid with a pungent odor. It has a vapor pressure of 0.000008 mmHg and a solubility of 53% in water by weight. Captafol is moderately persistent in soils (Venkatramesh and Agnihothrudu, 1988).

INDUSTRIAL HYGIENE

The ACGIH and NIOSH have set the TLV and the REL, respectively, for captafol at 0.1 mg/m^3 with a skin notation (NIOSH, 2009; ACGIH, 2009). The NIOSH considers captafol a possible occupational carcinogen. The OSHA has not set a PEL.

HEALTH EFFECTS

Respiratory sensitization and occupational asthma have also been associated with captafol (Royce et al., 1993).

Captafol has been reported to cause contact dermatitis (NTP, 2011; Brown, 1984). In one epidemiological study of pesticide poisonings in Japan, captafol was the most common cause of contact dermatitis, responsible for 29% of the cases (Matsushita et al., 1980).

The IARC has concluded that there is sufficient evidence of captafol's carcinogenicity in animals to classify it in Group 2A, probably carcinogenic in humans (IARC, 1991).

TARGET ORGANS

A study of subchronic administration of captafol in mice indicated that captafol is a hepatotoxin. Mice fed dietary levels of captafol over 0.3% had increased liver weights, histopathological changes, and reduced body weight gains (Tamano et al., 1991,1993). Rats injected intraperitoneally with captafol had changes in the activities of cytochrome P-450 and NADPH–cytochrome C reductase (Dalvi and

Mutinga, 1990). Captafol produced these effects in rats at 25% of the dose required to produce a similar effect with captan. Some captafol metabolites are known to be carcinogenic, such as dichloroacetic acid, 2-chloro-2-methyl-thio-ethylene sulfonic acid, and transient episulfonium ion are alkylating agents like captafol (NTP, 2011). Tetrahydrophthalimide is the common metabolite of multiple metabolic pathways, which is eliminated through blood, urine, and feces across animal species (NTP, 2008, 2011). Captafol with thiol groups can lead to cytotoxicity (NTP, 2008, 2011).

As with captan, captafol is structurally similar to thalidomide and therefore a suspected teratogen. In one study, captafol was not demonstrated to be teratogenic when administered to rabbits, hamsters, and rats throughout organogenesis (Kennedy et al., 1968). In another study, captafol was found to increase malformations when given once during organogenesis (Robens, 1970).

REFERENCES

American Conference of Governmental Industrial Hygienists (2009) *Threshold Limit Values (TLVs) for Chemical Substances and Physical Agents and Biological Exposure Indices (BEIs)*, Cincinnati, OH: American Conference of Governmental Industrial Hygienists.

Brown, R. (1984) Contact sensitivity to Difolatan (Captafol). *Contact Dermatitis* 10:181–182.

Dalvi, R.R. and Mutinga, M.L. (1990) Comparative studies of the effects on liver and liver microsomal drug-metabolizing enzyme system by the fungicides captan, captafol and folpet in rats. *Pharmacol. Toxicol.* 66:231–233.

International Agency for Research on Cancer (1991) Occupational exposures in insecticide application, and some pesticides. IARC Working Group on the Evaluation of Carcinogenic Risks to Humans, Lyon, 16–23 October 1990. *IARC Monogr. Eval. Carcinog. Risks Hum.* 53:353–369.

Kennedy, G., Fancher, O.E., and Calandra, J.C. (1968) An investigation of the teratogenic potential of captan, folpet, and difolatan. *Toxicol. Appl. Pharmacol.* 13:420–430.

Matsushita, T., Nomura, S., and Wakatsuki, T. (1980) Epidemiology of contact dermatitis from pesticides in Japan. *Contact Dermatitis* 6:255–259.

National Institute for Occupational Safety and Health (2009) *Pocket Guide to Chemical Hazards*, Washington, DC: U.S. Department of Health and Human Services.

National Toxicology Program (NTP) (2008) Final report on carcinogens background document for captafol. *Rep. Carcinog. Backgr. Doc.* I–xvi, 1–99.

National Toxicology Program (NTP) (2011) Captafol. *Rep. Carcinog.* 12:83–86.

Robens, J.F. (1970) Teratogenic activity of several phthalimide derivatives in the golden hamster. *Toxicol. Appl. Pharmacol.* 16:24–34.

Royce, S., Wald, P., Sheppard, D., and Balmes, J. (1993) Occupational asthma in a pesticides manufacturing worker. *Chest* 103:295–296.

Tamano, S., Kawabe, M., Sano, M., Masui, T., and Ito, N. (1993) Subchronic oral toxicity study of captafol in B6C3F1 mice. *J. Toxicol. Environ. Health* 38:69–75.

Tamano, S., et al. (1991) 13-Week oral toxicity study of captafol in F344/DuCrj rats. *Fundam. Appl. Toxicol.* 17:390–398.

Venkatramesh, M. and Agnihothrudu, V. (1988) Persistence of captafol in soils with and without amendments and its effects on soil microflora. *Bull. Environ. Contam. Toxicol.* 41:548–555.

BENOMYL

Benomyl (Benlate; methyl 1-(butylcarbamoyl)-2-benzimidazolecarbamate; CAS #17804-35-2) is a crystalline solid with a faint acrid odor. It has a solubility in water by weight of 0.0004% and a vapor pressure of <0.00001 mmHg. Benomyl is used in flora culture, on vegetables, nuts, field crops, fruit, and turf grass.

Benomyl applied to soil degrades almost completely in 10 weeks, partly because of the presence of microorganisms (Li and Nelson, 1985). Higher temperature and moisture levels speed this process. Methyl 2-benzimidazolecarbamate (MBC, or carbendazim) is the primary breakdown product of benomyl in soils and aqueous medium. It may persist in soils for 3 months to over 2 years and is responsible for benomyl's fungicidal activity.

INDUSTRIAL HYGIENE

The OSHA has set the PEL for benomyl at 15 mg/m^3 total and 5 mg/m^3 for the respirable fraction (NIOSH, 2009). The ACGIH has set the TLV at 1 mg/m^3 (ACGIH, 2009).

Benomyl does not constitute an inhalation hazard to workers involved in landscaping applications, and respirators are not necessary. As long as the label instructions for mixing and loading, such as wearing goggles and an apron, are followed, worker exposure is well below the TLV (Leonard and Nelson, 1990).

HEALTH EFFECTS

Although benomyl contains a carbamate grouping, it does not inhibit ChE activity. Research indicates that it has a low oral toxicity in mammals. No systemic effects have been reported in humans. Contact dermatitis has been attributed to benomyl exposure (Sanitt, 1972; van Ketel, 1976; van Joost

et al., 1983). Treatment of acute poisoning, if it occurred, would be symptomatic.

Benomyl has not been found to be a human teratogen, although it has been found to produce teratogenic effects in laboratory animals. Human epidemiological studies have been reported, but none have shown a strong correlation between benomyl exposure and human birth defects (Von Burg, 1993). Spagnolo et al. (1994) studied the prevalence of anophthalmia and microphthalmia in Italy and possible associations with an occupation in agriculture. They found no association between the agricultural occupation and incidence of these conditions.

TARGET ORGANS AND MECHANISM OF ACTION

The potential teratogenicity or fetotoxicity of benomyl has been evaluated in numerous studies. Pregnant Sprague-Dawley rats treated with benomyl by gavage at 62.4 mg/kg/day from day 7 of gestation to sacrifice were reported to have an increase in the craniocerebral abnormalities (Ellis et al., 1988). Fetuses were examined on gestational days 16 and 20. The severity and incidence of abnormalities were higher in the fetuses evaluated on day 20, which suggests that benomyl may be teratogenic to animals in the later stages of gestation. Craniocerebral effects, as well as ocular effects, were also reported in fetuses of pregnant rats exposed to benomyl by Hoogenboom et al. (1991).

Methyl 2-benzimidazolecarbamate, the degradation product of benomyl, has been tested for potential fetotoxicity and embryotoxicity. Wistar rats treated with MBC late in pregnancy (days 6–15) were reported to have increased dead and resorbed fetuses at dose levels of 40 and 80 mg/kg/day (Janardhan et al., 1984). Fetal malformations were observed following a gavage dose of 63.5 mg/kg/day in rats and 100 mg/kg/day in mice (Kavlock et al., 1982). Dietary exposure was fetotoxic only at a considerably higher dose and was not teratogenic.

Another study of rats treated in the early days of pregnancy (days 1–8) found a no observed adverse effect level (NOAEL) for MBC of 400 mg/kg/day (Cummings et al., 1992, 1990). Combined, these findings suggest that MBC may exert an embryotoxic effect only late in pregnancy.

Several studies indicate that benomyl causes testicular dysfunction, specifically damage to seminiferous tubules, in laboratory rodents. In one study by Hess et al. (1991), rats were given a single oral dose of 25–800 mg/kg and examined at 2 and 70 days after treatment. Two days after exposure, investigators reported premature release of germ cells (sloughing) in rats exposed to all dosage levels (Hess and Nakai, 2000). An increase in testes weight was attributed to occluded efferent ductules. Long-term effects (at 70 days) included a dose-dependent increase in atrophy of the seminiferous tubules and decrease in testes weight. Wistar rats dosed with benomyl at 45 mg/kg/day for 63 days by gavage were reported to have a diminished sperm production, higher incidence of decapitated sperm, and decrease in testes and epididymis weight (Linder et al., 1988). Treatment with benomyl did not affect the sperm motility or reproductive performance.

Benomyl reportedly reduced total sperm counts and caused testicular damage in Sprague-Dawley rats treated with 400 mg/kg/day for 10 days (Carter and Laskey, 1982). In another experiment, benomyl was administered by gavage and dietary routes to pregnant rats on gestational days 7–16 and by gavage to mice on gestational days 7–17. Further research suggests that male rats exposed before puberty were less sensitive to the effects of benomyl than those exposed during or after puberty (Carter et al., 1984).

Olfactory epithelial lesions have been reported in rats exposed to benomyl by inhalation (Warheit et al., 1989). However, further studies of rats fed benomyl indicate that the effect on the nasal mucosa depended on the route of exposure (Hurtt et al., 1993).

In chickens, also, benomyl toxicity was found to depend on the route of administration (Terse et al., 1993). Critical enzyme levels, including cytochrome P-450, cytochrome b5, and NADPH–cytochrome C reductase were significantly affected in chickens receiving one dose (100 mg/kg) of benomyl intraperitoneally. Chickens fed benomyl at 4000 ppm for 15 days did not have any significant variation in enzyme levels. Therefore benomyl is not expected to have adverse effects on wild birds.

In rats, the hepatotoxic effects of benomyl were evident at oral doses as low as 165 mg/kg (Igbedioh and Akinyele, 1992).

There is no substantial evidence to indicate that benomyl has significant mutagenic potential.

REFERENCES

American Conference of Governmental Industrial Hygienists (2009) Threshold Limit Values (TLVs) for Chemical Substances and Physical Agents and Biological Exposure Indices (BEIs), Cincinnati, OH: American Conference of Governmental Industrial Hygienists.

Carter, S.D., Hein, J.F., Rehnberg, G.L., and Laskey, J.W. (1984) Effect of benomyl on the reproductive development of male rats. *J. Toxicol. Environ. Health* 13:53–68.

Carter, S.D. and Laskey, J.W. (1982) Effect of benomyl on reproduction in the male rat. *Toxicol. Lett.* 11:87–94.

Cummings, A.M., Ebron-McCoy, M.T., Rogers, J.M., Barbee, B.D., and Harris, S.T. (1992) Developmental effects of methyl benzimidazolecarbamate following exposure during early pregnancy. *Fundam. Appl. Toxicol.* 18:288–293.

Cummings, A.M., Harris, S.T., and Rehnberg, G.L. (1990) Effects of methyl benzimidazolecarbamate during early pregnancy in the rat. *Fundam. Appl. Toxicol.* 15:528–535.

Ellis, W.G., De Roos, F., Kavlock, R.J., and Zeman, F.J. (1988) Relationship of periventricular overgrowth to hydrocephalus in brains of fetal rats exposed to benomyl. *Teratog. Carcinog. Mutagen.* 8:377–391.

Hess, R.A., Moore, B.J., Forrer, J., Linder, R.E., and Abuel-Atta, A.A. (1991) The fungicide benomyl (methyl 1-(butylcarbamoyl)-2-benzimidazolecarbamate) causes testicular dysfunction by inducing the sloughing of germ cells and occlusion of efferent ductules. *Fundam. Appl. Toxicol.* 17:733–745.

Hess, R.A. and Nakai, M. (2000) Histopathology of the male reproductive system induced by the fungicide benomyl. *Histol. Histopathol.* 15:207–224.

Hoogenboom, E.R., Ransdell, J.F., Ellis, W.G., Kavlock, R.J., and Zeman, F.J. (1991) Effects on the fetal rat eye of maternal benomyl exposure and protein malnutrition. *Curr. Eye Res.* 10:601–612.

Hurtt, M.E., Mebus, C.A., and Bogdanffy, M.S. (1993) Investigation of the effects of benomyl on rat nasal mucosa. *Fundam. Appl. Toxicol.* 21:253–255.

Igbedioh, S.O. and Akinyele, I.O. (1992) Effect of benomyl toxicity on some liver constituents of albino rats. *Arch. Environ. Health* 47:314–317.

Janardhan, A., Sattur, P.B., and Sisodia, P. (1984) Teratogenicity of methyl benzimidazole carbamate in rats and rabbits. *Bull. Environ. Contam. Toxicol.* 33:257–263.

Kavlock, R.J., Chernoff, N., Gray, L.E., Jr., Gray, J.A., and Whitehouse, D. (1982) Teratogenic effects of benomyl in the Wistar rat and CD-1 mouse, with emphasis on the route of administration. *Toxicol. Appl. Pharmacol.* 62:44–54.

Leonard, J.A. and Yeary, R.A. (1990) Exposure of workers using hand-held equipment during urban application of pesticides to trees and ornamental shrubs. *Am. Ind. Hyg. Assoc. J.* 51:605–609.

Li, C.Y. and Nelson, E.E. (1985) Persistence of benomyl and captan and their effects on microbial activity in field soils. *Bull. Environ. Contam. Toxicol.* 34:533–540.

Linder, R.E., Rehnberg, G.L., Strader, L.F., and Diggs, J.P. (1988) Evaluation of reproductive parameters in adult male Wistar rats after subchronic exposure (gavage) to benomyl. *J. Toxicol. Environ. Health* 25, 285–298.

National Institute for Occupational Safety and Health (2009) *Pocket Guide to Chemical Hazards*, Washington, DC: U.S. Department of Health and Human Services.

Sanitt, L.E. (1972) Contact dermatitis due to benomyl insecticide. *Arch. Dermatol.* 105:926–927.

Spagnolo, A. et al. (1994) Anophthalmia and benomyl in Italy: a multicenter study based on 940,615 newborns. *Reprod. Toxicol.* 8:397–403.

Terse, P.S., Sawant, S.G., and Dalvi, R.R. (1993) Benomyl toxicity in chickens. *Bull. Environ. Contam. Toxicol.* 50:817–822.

van Joost, T., Naafs, B., and van Ketel, W.G. (1983). Sensitization to benomyl and related pesticides. *Contact Dermatitis* 9:153–154.

van Ketel, W.G. (1976). Sensitivity to the pesticide benomyl. *Contact Dermatitis* 2:290–291.

Von Burg, R. (1993) Benomyl. *J. Appl. Toxicol.* 13:377–381.

Warheit, D.B., Kelly, D.P., Carakostas, M.C., and Singer, A.W. (1989) A 90-day inhalation toxicity study with benomyl in rats. *Fundam. Appl. Toxicol.* 12:333–345.

OTHER FUNGICIDES

PENTACHLOROPHENOL

Pentachlorophenol (symbol PCP; penta; CAS #87-86-5) is a colorless to white crystalline solid with a benzene-like odor. It has a vapor pressure of 0.0001 mmHg at 25 °C (77 °F) and a solubility in water by weight of 0.001%. It is used primarily as a wood preservative, mainly to prevent fungal decay but also to protect lumber from insects. It also has been used in adhesives and textiles, as a preemergence and nonselective contact herbicide, and to control the snail host of schistosomiasis.

It is persistent in the environment and may be found in surface water close to points of application and along rights-of-way (Wan, 1992). Although PCP does bioconcentrate in individual organisms, biomagnification in food chains has not been reported, most likely because PCP is rapidly metabolized.

INDUSTRIAL HYGIENE

Workers at smaller wood preservative plants are generally at a greatest risk for PCP exposure (Wood et al., 1983). Both leather and rubber gloves may become saturated with PCP (Colosio et al., 1993a). Nitrile rubber, polyvinyl chloride, or neoprene gloves are recommended (Silkowski et al., 1984).

The NIOSH, OSHA, and the ACGIH have set the REL, PEL, and TLV, respectively, for PCP at 5 mg/m^3 with a skin notation (NIOSH, 2009; ACGIH, 2009). Pentachlorophenol is considered IDLH at 2.5 mg/m^3. The ACGIH intends to add the A3 carcinogen classification to indicate that PCP is an animal carcinogen.

ACUTE EFFECTS

Acute intoxication typically occurs as a consequence of an occupational exposure during manufacture or application and by intentional ingestion (Guitart et al., 1990). Poisonings have also been reported in a neonate nursery from contaminated laundry (Armstrong et al., 1969; Robson et al., 1969). Victims of acute poisoning experience a hypermetabolic state that results from the derangement of aerobic metabolism. Symptoms of this state include weakness, changes in respiration and blood pressure, hyperpyrexia, metabolic acidosis, cardiac arrest, tachycardia, bronchitis, and tachypnea (Haley, 1977; Gray et al., 1985). Intravascular hemolysis has been reported from exposure to wood cleaners containing PCP wood cleaners (Hassan et al., 1985). Dermal

exposures to concentrations of PCP over 10% generally cause irritation and dermatitis. An exposure to PCP or its salts may also cause chloracne (Sehgal and Ghorpade, 1982; O'Malley et al., 1990). Aplastic anemia has been associated with acute poisoning (Rugman and Cosstick, 1990). Pentachlorophenol has been shown to be immunotoxic in mice (Chen et al., 2013). Survivors of intoxication may suffer from ocular damage, such as corneal opacity and inflammation of the conjunctivae. Severe exposure uncoupling of oxidative phosphorylation has been reported (Proudfoot, 2003). Acute toxic doses can range from 27 to 350 mg/kg (Black 2012).

CHRONIC EFFECTS

Pentachlorophenol has been suspected of causing cancer and abnormalities of the lymphoid system (Cooper and Jones, 2008). However, a study of workers exposed to PCP found little evidence of any significant damage to functional immune responses (Colosio et al., 1993b).

An increase in the incidence of soft tissue sarcoma and of non-Hodgkin's lymphoma in Finland has been attributed to a PCP-contaminated water supply (Lampi et al., 1992). Subjects in this study were exposed to drinking water levels of 70–140 µg/l. Clusters of Hodgkin's disease have also been associated with exposure to PCP among woodworkers (Greene et al., 1978).

Case reports have attributed aplastic anemia and leukemia to PCP (Roberts, 1983; Roberts, 1990). In one study, people exposed to PCP for over 6 months showed a decrease in lymphocyte responsiveness *in vitro* (Daniel et al., 1995). A slight increase in lymphomas was also reported among sawmill workers in Finland (Jäppinen et al., 1989). None of these reports have been confirmed. In a study of Hawaiian workers exposed to PCP and other wood preservatives for an average of 6½ years, no increase in any type of morbidity was found (Gilbert et al., 1990).

The IARC has classified PCP in Group 2B, meaning that it is considered an animal carcinogen but that there is insufficient evidence to consider it a human carcinogen (IARC, 1991). Pentachlorophenol can cause hepatocarcinogenicity (Kanno et al., 2013; Shan et al., 2013). Some pentachlorophenol metabolites have been shown to generate DNA single-strand breaks (Dahlhaus et al., 1996).

Exposure may also occur in sawmill workers and residents living in log homes treated with a wood protectant that contains PCP (Jorens and Schepen, 1993).

A study of workers in Germany reported the effects of PCP on nerve conduction velocity (Triebig et al., 1987). Workers exposed to PCP for an average of 16 years at low levels (0.3 to 180 µg/l) had no observable deficit in nerve conduction velocity.

TREATMENT

The primary treatment for PCP poisoning involves measures that promote heat loss, such as lukewarm baths, that reduce anxiety, and that replace lost fluids and electrolytes. Emesis may be induced in cases of ingestion if the patient is not at risk of coma or convulsions. Activated charcoal may also be administered. Significant symptoms are unlikely in patients with urinary levels below 36 ppm.

METABOLISM

Uhl et al. (1986) studied the pharmacokinetics of PCP in human volunteers. They determined the half-life in humans to be approximately 20 days, with about 3 months required to reach a steady state. More than 96% of PCP administered was determined to be bound to serum. Pentachlorophenol was excreted unchanged and as its glucuronide. The blood may act as a reservoir for PCP and is partly responsible for its relatively long half-life. At normal urinary pH (5–6), PCP is filtered and reabsorbed at a high rate by the kidney tubules. Oral administration of sodium bicarbonate increased the rate of excretion (Uhl et al., 1986).

The biological half-life of PCP in rodents is relatively short. It is somewhat longer in monkeys (Braun and Sauerhoff, 1976).

TARGET ORGANS

Acute systemic poisonings induced by PCP are the result of the uncoupling of oxidative phosphorylation reactions in the mitochondria. Even at low concentrations PCP has been shown to inhibit rat liver mitochondrial respiration *in vitro* (Shannon et al., 1991). This mitochondrial toxicity disrupts aerobic metabolism and precipitates the acute effects associated with PCP-induced poisoning.

It is readily absorbed through the skin, lungs, and gastrointestinal tract. It may reduce kidney function by impairing tubular function and reducing the glomerular filtration rate (Begley et al., 1977). One study on cattle suggested that the toxic effects of PCP were attributable to contaminants (McConnell et al., 1980).

Pathological examination of victims of acute poisoning has revealed centrilobular congestion in the liver and fat droplets in the myocardial fibers and proximal renal tubular epithelial cells (Gray et al., 1985). Focal swelling of myelin sheaths and cerebral edema were evident in the hemispheric white matter. Mitochondria in all organs were swollen, but this could be a result of postmortem change.

Pentachlorophenol-induced DNA damage was compared to tetrachlorohydroquinone-induced DNA damage in Chinese hamster ovary cells, and PCP was found to be less potent (Ehrlich, 1990).

Immunotoxicity in the form of depressed humoral immune response was reported in mice exposed to technical grade PCP (Kerkvliet et al., 1982). No effect was found in mice exposed to analytical grade PCP further emphasizing the role of contaminants in PCP-induced toxicity.

BIOMONITORING

The ACGIH has advised that the Biological Exposure Index (BEI) for PCP should be 2 mg/g of creatinine measured in the urine at the end of shift or, if a blood sample is used, 70% of the individual's baseline ChE activity in red cells (ACGIH, 2009).

Colosio et al. (1993a) demonstrated that urinary concentrations correlate closely with plasma concentrations after a few days of exposure. Urinary biomonitoring should be performed in the morning before the work shift begins (Barbieri et al., 1995).

During periods of no exposure, plasma levels of PCP drop more sharply in the first 20 days than the urine levels (Begley et al., 1977).

REFERENCES

American Conference of Governmental Industrial Hygienists (2009) *Threshold Limit Values (TLVs) for Chemical Substances and Physical Agents and Biological Exposure Indices (BEIs)*, Cincinnati, OH: American Conference of Governmental Industrial Hygienists.

Armstrong, R.W., et al. (1969) Pentachlorophenol poisoning in a nursery for newborn infants. II. Epidemiologic and toxicologic studies. *J. Pediatr.* 75:317–325.

Barbieri, F., Colosio, C., Schlitt, H., and Maroni, M. (1995) Urine excretion of pentachlorophenol (PCP) in occupational exposure. *J. Pestic. Sci.* 43:259–262.

Begley, J., Reichert, E.L., Rashad, M.N., and Klemmer, H.W. (1977) Association between renal function tests and pentachlorophenol exposure. *Clin. Toxicol.* 11:97–106.

Black, W.D. (2012) *Overview of Pentachlorophenol Poisoning.* Available at http://www.merckmanuals.com/vet/toxicology/pentachlorophenol_poisoning.html.

Braun, W.H. and Sauerhoff, M.W. (1976) The pharmacokinetic profile of pentachlorophenol in monkeys. *Toxicol. Appl. Pharmacol.* 38:525–533.

Chen, H.M., et al. (2013) The immunotoxic effects of dual exposure to PCP and TCDD. *Chem. Biol. Interact.* 206:166–174.

Colosio, C., Barbieri, F., Bersani, M., Schlitt, H., and Maroni, M. (1993a) Markers of occupational exposure to pentachlorophenol. *Bull. Environ. Contam. Toxicol.* 51:820–826.

Colosio, C., et al. (1993b) Toxicological and immune findings in workers exposed to pentachlorophenol (PCP). *Arch. Environ. Health* 48:81–88.

Cooper, G.S. and Jones, S. (2008) Pentachlorophenol and cancer risk: focusing the lens on specific chlorophenols and contaminants. *Environ. Health Perspect.* 116:1001–1008.

Dahlhaus, M., Almstadt, E., Henschke, P., Luttgert, S., and Appel, K.E. (1996) Oxidative DNA lesions in V79 cells mediated by pentachlorophenol metabolites. *Arch. Toxicol.* 70:457–460.

Daniel, V., Huber, W., Bauer, K., and Opelz, G. (1995) Impaired *in-vitro* lymphocyte responses in patients with elevated pentachlorophenol (PCP) blood levels. *Arch. Environ. Health* 50:287–292.

Ehrlich, W. (1990) The effect of pentachlorophenol and its metabolite tetrachlorohydroquinone on cell growth and the induction of DNA damage in Chinese hamster ovary cells. *Mutat. Res.* 244:299–302.

Gilbert, F.I., Jr., Minn, C.E., Duncan, R.C., and Wilkinson, J. (1990) Effects of pentachlorophenol and other chemical preservatives on the health of wood-treating workers in Hawaii. *Arch. Environ. Contam. Toxicol.* 19:603–609.

Gray, R.E., Gilliland, R.D., Smith, E.E., Lockard, V.G., and Hume, A.S. (1985) Pentachlorophenol intoxication: report of a fatal case, with comments on the clinical course and pathologic anatomy. *Arch. Environ. Health* 40:161–164.

Greene, M.H., Brinton, L.A., Fraumeni, J.F., and D'Amico, R. (1978) Familial and sporadic Hodgkin's disease associated with occupational wood exposure. *Lancet* 2:626–627.

Guitart, R., Abian, J., Arboix, M., and Gelpi, E. (1990) Detection and isolation of pentachlorophenol in oil samples associated with the Spanish toxic oil syndrome. *Bull. Environ. Contam. Toxicol.* 45:181–188.

Haley, T.J. (1977) Human poisoning with pentachlorophenol and its treatment. *Ecotoxicol. Environ. Saf.* 1:343–347.

Hassan, A.B., Seligmann, H., and Bassan, H.M. (1985). Intravascular haemolysis induced by pentachlorophenol. *Br. Med. J. (Clin. Res. Ed.)* 291:21–22.

International Agency for Research on Cancer (1991) Occupational exposures in insecticide application, and some pesticides. IARC Working Group on the Evaluation of Carcinogenic Risks to Humans, Lyon, 16–23 October 1990. *IARC Monogr. Eval. Carcinog. Risks Hum.* 53:5–586.

Jäppinen, P., Pukkala, E., and Tola, S. (1989) Cancer incidence of workers in a Finnish sawmill. *Scand. J. Work Environ. Health* 15:18–23.

Jorens, P.G. and Schepens, P.J. (1993) Human pentachlorophenol poisoning. *Hum. Exp. Toxicol.* 12:479–495.

Kanno, J., et al. (2013) Oral administration of pentachlorophenol induces interferon signaling mRNAs in C57BL/6 male mouse liver. *J. Toxicol. Sci.* 38:643–654.

Kerkvliet, N.I., Baecher-Steppan, L., Claycomb, A.T., Craig, A.M., and Sheggeby, G.G. (1982) Immunotoxicity of technical pentachlorophenol (PCP-T): depressed humoral immune responses to T-dependent and T-independent antigen stimulation in PCP-T exposed mice. *Fundam. Appl. Toxicol.* 2:90–99.

Lampi, P., Hakulinen, T., Luostarinen, T., Pukkala, E., and Teppo, L. (1992) Cancer incidence following chlorophenol exposure in a community in southern Finland. *Arch. Environ. Health* 47:167–175.

McConnell, E.E., et al. (1980) The chronic toxicity of technical and analytical pentachlorophenol in cattle. I. Clinicopathology. *Toxicol. Appl. Pharmacol.* 52:468–490.

National Institute for Occupational Safety and Health (2009) *Pocket Guide to Chemical Hazards*, Washington, DC: U.S. Department of Health and Human Services.

O'Malley, M.A., et al. (1990) Chloracne associated with employment in the production of pentachlorophenol. *Am. J. Ind. Med.* 17:411–421.

Proudfoot, A.T. (2003) Pentachlorophenol poisoning. *Toxicol. Rev.* 22:3–11.

Roberts, H.J. (1983) Aplastic anemia and red cell aplasia due to pentachlorophenol. *South. Med. J.* 76:45–48.

Roberts, H.J. (1990) Pentachlorophenol-associated aplastic anemia, red cell aplasia, leukemia and other blood disorders. *J. Fla. Med. Assoc.* 77:86–90.

Robson, A.M., Kissane, J.M., Elvick, N.H., and Pundavela, L. (1969) Pentachlorophenol poisoning in a nursery for newborn infants. I. Clinical features and treatment. *J. Pediatr.* 75:309–316.

Rugman, F.P. and Cosstick, R. (1990) Aplastic anaemia associated with organochlorine pesticide: case reports and review of evidence. *J. Clin. Pathol.* 43:98–101.

Sehgal, V.N. and Ghorpade, A. (1983) Fume inhalation chloracne. *Dermatologica* 167:33–36.

Shan, G., Ye, M., Zhu, B., and Zhu, L. (2013) Enhanced cytotoxicity of pentachlorophenol by perfluorooctane sulfonate or perfluorooctanoic acid in HepG2 cells. *Chemosphere* 93:2101–2107.

Shannon, R.D., Boardman, G.D., Dietrich, A.M., and Bevan, D.R. (1991) Mitochondrial response to chlorophenols as a short-term toxicity assay. *Environ. Toxicol. Chem.* 10:57–66.

Silkowski, J.B., Horstman, S.W., and Morgan, M.S. (1984) Permeation through five commercially available glove materials by two pentachlorophenol formulations. *Am. Ind. Hyg. Assoc. J.* 45:501–504.

Triebig, G., Csuzda, I., Krekeler, H.J., and Schaller, K.H. (1987) Pentachlorophenol and the peripheral nervous system: a longitudinal study in exposed workers. *Br. J. Ind. Med.* 44:638–641.

Uhl, S., Schmid, P., and Schlatter, C. (1986) Pharmacokinetics of pentachlorophenol in man. *Arch. Toxicol.* 58:182–186.

Wan, M.T. (1992) Utility and railway right-of-way contaminants in British Columbia: chlorophenols. *J. Environ. Qual.* 21:225–231.

Wood, S., Rom, W.N., White, G.L., Jr., and Logan, D.C. (1983) Pentachlorophenol poisoning. *J. Occup. Med.* 25:527–530.

NON-ANTICOAGULANTS

ZINC PHOSPHIDE

Target organ(s): Central nervous system, cardiovascular system

Occupational exposure limits (phosphine):

NIOSH REL: TWA 0.4 mg/m^3

OSHA PEL: 0.4 mg/m^3

IDLH: 50 ppm

Reference values:

Oral RfD 0.0001 mg/kg/day

Inhalation RfC for phosphine gas is 3×10^{-4} mg/m^3

Risk/safety phrases: Signal words for rodenticides containing zinc phosphide range from "Caution" to "Danger" depending upon the specific formulation

BACKGROUND AND USES

Zinc phosphide was first registered for use as a rodenticide in 1947 by the U.S. Department of Agriculture. It was subsequently registered by the U.S. Environmental Protection Agency in 1982, and underwent reregistration in 1998 (US EPA, 1998a,1998b). Formulations containing zinc phosphide can be found in agricultural and general (consumer-use) settings. These formulations include powders, dusts, granules, and bait pellets. Zinc phosphide is used to control rats and house mice, voles, ground squirrels, prairie dogs, and pocket gophers. Most end-use products contain 2% zinc phosphide.

PHYSICAL AND CHEMICAL PROPERTIES

The chemical name for zinc phosphide (CAS # 1314-84-7) is trizinc diphosphide. It has the empirical formula PzZn$_3$ and a molecular weight of 258.09. It is a gray-black crystalline powder with a faint garlic odor. The melting point is above 420 °C. The toxicity of zinc phosphide is a result of phosphine gas. Zinc phosphide is converted to phosphine gas in the presence of moisture and acidic conditions in the stomach. The chemical name for phosphine is, and it has the empirical formula. The molecular weight of phosphine is 34. It is a gas that may have a garlic or rotting fish odor. The vapor pressure of phosphine is 2.93×10^4 mmHg at 25 °C. Humans can begin to detect phosphine at concentrations of 2 ppm, but adverse effects and toxicity have been reported to occur at lower concentrations.

MAMMALIAN TOXICOLOGY

Acute Effects

As a rodenticide, the acute toxic effects of zinc phosphide require ingestion to produce phosphine gas. Incidents of fatalities to nontarget species, including birds and other wildlife, have been reported in association with zinc phosphide ingestion (Bildfell et al., 2013). The intentional ingestion of zinc phosphide rodenticides is a common cause of suicide in certain countries (Shyam et al., 2002). The phosphine gas that is liberated presents inhalation risks to bystanders, and some case reports have described transient symptoms among medical personnel providing care to patients who have ingested zinc phosphide (Stephenson, 2002).

Phosphine gas is toxic to all of the major organ systems. Early effects include gastrointestinal symptoms of nausea, vomiting, epigastric pain, and dyspnea. Cardiovascular effects are prominent and characterized by severe hypotension and impaired myocardial contractility leading to shock and central nervous system depression. Pulmonary edema is a common clinical finding, and severe metabolic acidosis has been reported in human poisonings (Proudfoot, 2009).

Zinc phosphide is highly toxic from oral and inhalation exposure. The oral LD$_{50}$ in rats ranges from 12 to 56 mg/kg (Krishnakumari et al., 1980). It is considered low in toxicity from dermal exposure, with an LD$_{50}$ >2000 mg/kg. The 4-h inhalation LC$_{50}$ of phosphine gas is 11 ppm (approximately 0.015 mg/l) in rats (US EPA, 1998b).

Chronic Effects

In a 90-day subchronic toxicity study in rats, zinc phosphide was administered by gavage at doses of 0.1, 1.0, or 3.0 mg/kg/day. At higher doses, salivation, lower body temperature, and hydronephrosis were observed. The NOAEL was established at 0.1 mg/kg/day. The inhalation reference concentration (RfC) for phosphine is 3×10^{-4} mg/m^3.

Mechanism of Action(s)

Several toxicodynamic mechanisms have been described for phosphine. Mitochondrial effects have been consistently reported, including inhibitory effects on cytochrome c oxidase (Chefurka et al., 1976). More recent experimental toxicology studies in animals have demonstrated inhibitory effects on other mitochondrial enzyme complexes, as well as the generation of oxidative stress synergistically leading to tissue injury and physiological dysfunction (Anand et al., 2012).

Chemical Pathology

Pathological findings in cases of animal intoxication from zinc phosphide include corrosive gastroenteritis with hemorrhage, pulmonary edema, and evidence of venous congestion and capillary damage in the cardiovascular system (Gray et al., 2011). In cases of severe and fatal phosphine poisoning in humans, pathological findings in humans include hepatic congestion, centrilobular necrosis, and acute pancreatitis (Saleki et al., 2007; Sarma and Narula, 1996).

SOURCES OF EXPOSURE

Zinc phosphide has indoor and outdoor uses as a rodenticide. Food uses include grapes, rangeland grasses, and regional uses on artichokes and sugar beets. There are also nonfood uses, including agricultural areas, in and around homes, on golf courses, and in orchards.

INDUSTRIAL HYGIENE

Occupational and industrial hygiene guidelines and limits for zinc phosphide are based on phosphine gas. The NIOSH REL for phosphine is $0.4\,\text{mg/m}^3$ as a time-weighted average (TWA). The OSHA PEL is $0.4\,\text{mg/m}^3$, and the IDLH is 50 ppm.

MEDICAL MANAGEMENT

The management of zinc phosphide poisoning is based on symptomatic and aggressive supportive care. Gastric lavage is discouraged, as this may result in increased production of phosphine gas. There is some evidence supporting the use of a single dose of activated charcoal to reduce the oral bioavailability. Aggressive administration of intravenous fluids and pressors may be necessary for hemodynamic instability. There are some reports of improved outcomes with the intravenous administration of magnesium; however, this

has not been validated in well-designed clinical studies (Proudfoot, 2009).

REFERENCES

Anand, R., Kumari, P., Kaushal, A., Bal, A., Wani, W.Y., Sunkaria, A., Dua, R., Singh, S., Bhalla, A., and Gill, K.D. (2012) Effect of acute aluminum phosphide exposure on rats: a biochemical and histological correlation. *Toxicol. Lett.* 215(1):62–69.

Bildfell, R.J., Rumbeiha, W.K., Schuler, K.L., Meteyer, C.U., Wolff, P.L., and Gillin, C.M. (2013) A review of episodes of zinc phosphide toxicosis in wild geese (*Branta* spp.) in Oregon (2004–2011). *J. Vet. Diagn. Invest.* 25(1):162–7.

Chefurka, W., Kashi, K.P., and Bond, E.J. (1976) The effect of phosphine on electron transport in mitochondria. *Pestic. Biochem. Physiol.* 6(1):65–84.

Gray, S.L., Lee, J.A., Hovda, L.R., and Brutlag, A.G. (2011) Potential zinc phosphide rodenticide toxicosis in dogs: 362 cases (2004–2009). *J. Am. Vet. Med. Assoc.* 239(5):646–651.

Krishnakumari, M.K., Bai, K.M., and Majumder, S.K. (1980) Toxicity and rodenticidal potency of zinc phosphide. *Bull. Environ. Contam. Toxicol.* 25(1):153–159.

Proudfoot, A.T. (2009) Aluminium and zinc phosphide poisoning. *Clin. Toxicol. (Phila.)* 47(2):89–100.

Saleki, S., Ardalan, F.A., and Javidan-Nejad, A. (2007) Liver histopathology of fatal phosphine poisoning. *Forensic Sci. Int.* 166(2–3):190–193.

Sarma, P.S. and Narula, J. (1996) Acute pancreatitis due to zinc phosphide ingestion. *Postgrad. Med. J.* 72(846):237–238.

Shyam, P.L., Kumar, B., and Bidur, O. (2002) An epidemiological study on acute zinc phosphide poisoning in Nepal. *J. Nepal Health Res. Counc.* 1(1):13–16.

Stephenson, J.B.P. (2002) Phosphine poisoning by proxy. *Lancet* 360(9338):1024.

United States Environmental Protection Agency (1998a) *Reregistration Eligibility Decision: Aluminum and Magnesium Phosphide*, Washington, DC: United States Environmental Protection Agency.

United States Environmental Protection Agency (1998b) *Reregistration Eligibility Decision (RED): Zinc Phosphide*, Washington, DC: United States Environmental Protection Agency.

89

RODENTICIDES

Daniel L. Sudakin

FIRST-GENERATION ANTICOAGULANT RODENTICIDES (WARFARIN, CHLOROPHACINONE, DIPHACINONE)

Target organ(s): Liver (coagulopathy)

Occupational exposure limits:

Warfarin: NIOSH REL: TWA $0.1\,mg/m^3$; OSHA PEL: TWA $0.1\,mg/m^3$; IDLH: $100\,mg/m^3$

Chlorophacinone: No occupational exposure limits have been established by OSHA, NIOSH, or ACGIH

Diphacinone: No occupational exposure limits have been established by OSHA, NIOSH, or ACGIH

Reference values: (list RfDs/RfCs, TDIs, TIs, MRLs, etc.)

Warfarin: Oral RFD $3 \times 10^{-4}\,mg/kg/day$

Chlorophacinone: No RfDs/RfCs, TDIs, Tis, MRLs have been established

Diphacinone: No RfDs/RfCs, TDIs, Tis, MRLs have been established

Risk/safety phrases: Technical grade warfarin, chlorophacinone, and diphacinone carry the signal word "Danger." Products containing low concentrations and read-to-use formulations carry the signal word "Caution."

BACKGROUND AND USES

Warfarin was the first anticoagulant rodenticide, and was registered for use by the US Environmental Protection Agency US EPA in 1952. The toxic effects of warfarin were originally discovered in moldy sweet clover, when cattle became sick and developed hemorrhagic effects after repeated ingestion (Link, 1959). Diphacinone was first registered for use as a rodenticide in 1960, and chlorphacinone was first registered in 1971 (United States Environmental Protection Agency, 1998).

All of the first-generation anticoagulant rodenticides share a common effect of antagonizing vitamin K, resulting in prolongation of the bleeding time and hemorrhage. The first-generation anticoagulants are sometimes referred to as "multiple-dose" because, in contrast to second-generation anticoagulants, repeated doses are usually required to result in coagulopathy.

There have been significant changes in the regulation of rodenticides as a result of recent risk assessments by the US EPA (United States Environmental Protection Agency, 1998). First-generation anticoagulants are currently registered for use in "consumer size" rodenticide products with bait stations containing less than or equal to 1 pound of bait. Target pests for these registered uses include mice and rats. They are also registered for rodent control applications outdoors, including agricultural and underground uses for moles and pocket gophers.

PHYSICAL AND CHEMICAL PROPERTIES

The chemical name for warfarin (CAS # 81-81-2) is 3(alpha acetonylbenzyl)-4-hydroxycoumarin. It has the empirical formula of $C_{19}H_{16}O_4$, and a molecular weight of 308.32.

It is tasteless, odorless, and colorless, with a melting point of 15–161 °C. Warfarin is a racemic mixture of R- and S-warfarin enantiomers. Studies in rodents and humans

Hamilton & Hardy's Industrial Toxicology, Sixth Edition. Edited by Raymond D. Harbison, Marie M. Bourgeois, and Giffe T. Johnson.
© 2015 John Wiley & Sons, Inc. Published 2015 by John Wiley & Sons, Inc.

have reported that S-warfarin is more potent in producing coagulopathy (Breckenridge et al., 1974; Breckenridge and Orme, 1972).

Chlorophacinone (CAS # 3691-35-8) is a 1,3-indandione with the chemical name 2-[4-(chlorophenyl)phenylacetyl]-1H-indene-1,3(2H)-dione. It has the empirical formula $C_{23}H_{15}ClO_3$, and a molecular weight of 364.8. It is a yellow, crystalline solid with a melting point of 140 °C.

Diphacinone (CAS # 82-66-6) is also a 1,3-indandione with the chemical name 2(Diphenylacetyl)indan-1,3-dione. It has the empirical formula $C_{23}H_{16}O_3$, and a molecular weight of 340.4. It is a yellow crystalline powder with a melting point of 145 °C. It degrades rapidly in aqueous environments from sun exposure.

MAMMALIAN TOXICOLOGY

Acute Effects

There is a characteristic delay of 24–72 h between the ingestion of a first-generation anticoagulant and the onset of bleeding. Generalized hematoma formation, pallor, and weakness are common manifestations in acute intoxication. Fatal outcomes often result from bleeding in the central nervous system. While most cases of intoxication from first-generation anticoagulants result from repeated ingestion, some animal toxicology studies have reported that significant absorption of technical grade formulations can occur from inhalation and dermal pathways.

The oral LD_{50} of chlorophacinone in rats is 6.26 mg/kg. The dermal LD_{50} in rabbits is 0.329 mg/kg, and the inhalation LC_{50} in rats range from 7 to 12 μg/l (United States Environmental Protection Agency, 1998). Chlorophacinone is not considered a dermal sensitizer or eye or skin irritant.

The oral LD_{50} of diphacinone in rats is 2.4 mg/kg. The dermal LD_{50} in male rabbits is 3.6 mg/kg. The inhalation LC_{50} in rats is less than 0.6 μg/l. Diphacinone is not considered a dermal sensitizer, but slight skin irritation and moderate eye irritation have been reported (United States Environmental Protection Agency, 1998).

As warfarin has a long history of pharmaceutical use in humans, much is known about its toxicokinetics and acute effects. Warfarin has high oral bioavailability, and peak serum concentrations are observed approximately 3 h after ingestion (Sutcliffe et al., 1987). The half-life of warfarin in humans is approximately 35 h, and on an average it takes several days of oral administration to achieve a steady-state anticoagulant effect. Genetic factors may influence individual susceptibility to the effects of warfarin, as polymorphisms have been found to exist in the CYP2C9 gene that metabolizes S-warfarin (Krynetskiy and McDonnell, 2007).

In adults, the typical starting dose to achieve anticoagulation is 5 mg/day, with subsequent doses based on nomograms, computer programs, and in some cases, pharmacogenetic data (Gage et al., 2008). Resistance to the anticoagulant effects of warfarin has been described in humans, and can occur from several factors including drug interactions and pharmacokinetic effects among individuals with ultrarapid metabolism (Osinbowale et al., 2009).

Chronic Effects

In a chronic toxicity study in Sprague-Dawley rats receiving chlorophacinone by gavage, no mortality or signs of toxicity were observed at a dose of 5 μg/kg/day for 77 days. The US EPA considers this dose as a no-observed-adverse-effect-level (NOAEL), with a lowest observable effect level (for coagulopathy) of 10 μg/kg/day (United States Environmental Protection Agency, 1998).

In a study of repeated oral exposure to diphacinone over a 14-day period, the no observed adverse effect level is 0.040 mg/kg/day, and the lowest observable adverse effect level is 0.085 mg/kg/day (United States Environmental Protection Agency, 1998).

Teratogenic effects have been described, in association with the pharmaceutical use of warfarin during pregnancy. A fetal warfarin syndrome, characterized by nasal hypoplasia, limb deformities, and respiratory distress, has been described in association with the use of warfarin between 6 and 12 weeks of gestation. Other developmental anomalies, including mental retardation, microcephaly, and blindness, have been reported in association with chronic exposure to warfarin during the second and third trimesters (Holzgreve et al., 1976).

Some investigators have suggested a possible oncogenic effect of prolonged warfarin therapy, based upon a small case series (Krauss, 1982). This hypothesis was not confirmed in an epidemiological cohort study (Annegers and Zacharski, 1980).

Mechanism of Action(s)

Warfarin, chlorophacinone, and diphacinone exert their anticoagulant activity through the inhibition of vitamin K-dependent carboxylation of clotting factors II, VII, IX, and X, as well as the anticoagulant proteins C and S. Specifically, these first-generation anticoagulants inhibit the enzyme K_1 2,-epoxide reductase complex, subunit 1 (VKORC1) (Park, 1988). Inhibition of VKORC1 prevents the conversion of vitamin K epoxide to the biologically active vitamin K hydroquinone, which carboxylates and activates these clotting factors. After exposure to warfarin, the activity of clotting factors with shorter half-lives (including factor VII) is inhibited within 12–24 h. The full anticoagulant effect is not observed until 5–7 days after initial exposure because of the longer half-lives of clotting factors X and II.

Chemical Pathology

Pathological findings at autopsy in animals poisoned by first-generation anticoagulants include pallor of the muscles and viscera, and generalized hemorrhage. Necrosis of the skin has been described in association with human exposure to warfarin as a pharmaceutical anticoagulant, in an individual with acquired protein C deficiency (Parsi et al., 2003).

SOURCES OF EXPOSURE

First-generation anticoagulants can be encountered in "consumer size" rodenticide products containing less than 1 pound of solid bait. Current EPA regulations prohibit their formulation in loose bait forms for consumer products. For agricultural applications and pesticide applicators, first-generation anticoagulants products must contain at least 4 pounds of solid bait.

INDUSTRIAL HYGIENE

Occupational exposure limits and guidelines have been established for warfarin. The National Institute for Occupational Safety and Health (NIOSH) recommended exposure limit (REL) is a time-weighted average (TWA) of $0.1\,mg/m^3$. The OSHA permissible exposure limit (PEL) is $0.1\,mg/m^3$. The Immediately Dangerous to Life and Health (IDLH) for warfarin is $100\,mg/m^3$.

Occupational exposure guidelines and limits have not been established for chlorophacinone or diphacinone.

MEDICAL MANAGEMENT

The medical management for first-generation anticoagulants depends upon the circumstances surrounding exposure. Accidental ingestion of small amounts of first-generation anticoagulants like warfarin typically present a very low risk of coagulopathy in adults and children (Katona and Wason, 1989). Intentional ingestion of these rodenticides can result in delayed onset of bleeding. For individuals presenting within several hours of intentional ingestion, a single dose of activated charcoal should be considered for gastric decontamination. The clinical laboratory assessment of individuals at risk of coagulopathy from first-generation anticoagulants should include a prothrombin time (PT-INR) at 48 h after the time of ingestion (Smolinske et al., 1989). A careful physical examination should be conducted to assess for evidence of bruising and bleeding. Life threatening hemorrhage should be managed with the infusion of fresh frozen plasma, which is rich in active vitamin K-dependent coagulation factors and will temporarily reverse coagulopathy in most patients (Cruickshank et al., 2001).

The specific antidote for coagulopathy resulting from exposure to first-generation anticoagulants is vitamin K_1. It is effective orally and via intravenous or intramuscular administration. Repeated, large doses of vitamin K (up to or greater than 60 mg/day) may be required in some cases. The health care provider should seek consultation from a Regional Poison Control Center for additional advice and treatment recommendations.

REFERENCES

Annegers, J.F. and Zacharski, L.R. (1980) Cancer morbidity and mortality in previously anticoagulated patients. *Thromb. Res.* 18(3–4):399–403.

Breckenridge, A. and Orme, M.L. (1972) The plasma half lives and the pharmacological effect of the enantiomers of warfarin in rats. *Life Sci. II* 11(7):337–345.

Breckenridge, A., Orme, M., Wesseling, H., Lewis, R.J., and Gibbons, R. (1974) Pharmacokinetics and pharmacodynamics of the enantiomers of warfarin in man. *Clin. Pharmacol. Ther.* 15(4):424–430.

Cruickshank, J., Ragg, M., and Eddey, D. (2001) Warfarin toxicity in the emergency department: recommendations for management. *Emerg. Med. (Fremantle, WA)* 13(1):91–97.

Gage, B.F., Eby, C., Johnson, J.A., Deych, E., Rieder, M.J., Ridker, P.M., Milligan, P.E., Grice, G., Lenzini, P., Rettie, A.E., Aquilante, C.L., Grosso, L., Marsh, S., Langaee, T., Farnett, L.E., Voora, D., Veenstra, D.L., Glynn, R.J., Barrett, A., and McLeod, H.L. (2008) Use of pharmacogenetic and clinical factors to predict the therapeutic dose of warfarin. *Clin. Pharmacol. Ther.* 84(3):326–331.

Holzgreve, W., Carey, J.C., and Hall, B.D. (1976) Warfarin-induced fetal abnormalities. *Lancet* 2(7991):914–915.

Katona, B., and Wason, S. (1989) Superwarfarin poisoning. *J. Emerg. Med.* 7(6):627–631.

Krauss, J.S. (1982) Warfarin, pulmonary embolism, and cancer. *Ann. Intern. Med.* 97(2):282.

Krynetskiy, E. and McDonnell, P. (2007) Building individualized medicine: prevention of adverse reactions to warfarin therapy. *J. Pharmacol. Exp. Ther.* 322(2):427–434.

Link, K.P. (1959) The discovery of dicumarol and its sequels. *Circulation* 19(1):97–107.

Osinbowale, O., Malki, M.A., Schade, A., and Bartholomew, J.R. (2009) An algorithm for managing warfarin resistance. *Cleve. Clin. J. Med.* 76(12):724–730.

Park, B.K. (1988) Warfarin: metabolism and mode of action. *Biochem. Pharmacol.* 37(1):19–27.

Parsi, K., Younger, I., and Gallo, J. (2003) Warfarin-induced skin necrosis associated with acquired protein C deficiency. *Australas. J. Dermatol.* 44(1):57–61.

Smolinske, S.C., Scherger, D.L., Kearns, P.S., Wruk, K.M., Kulig, K.W., and Rumack, B.H. (1989) Superwarfarin poisoning in children: a prospective study. *Pediatrics* 84(3):490–494.

Sutcliffe, F.A., MacNicoll, A.D., and Gibson, G.G. (1987) Aspects of anticoagulant action: a review of the pharmacology, metabolism and toxicology of warfarin and congeners. *Rev. Drug Metab. Drug Interact.* 5(4):225–272.

United States Environmental Protection Agency (1998) *Reregistration Eligibility Decision (RED): Rodenticide Cluster*, Washington, DC: United States Environmental Protection Agency, Office of Pesticide Programs.

SECOND-GENERATION ANTICOAGULANT RODENTICIDES (BRODIFACOUM, BROMADIOLONE, DIFENACOUM, DIFETHIALONE)

Target organ(s): Liver (coagulopathy)

Occupational exposure limits:

> Brodifacoum: No occupational exposure limits have been established by OSHA, NIOSH, or ACGIH

> Bromadiolone: No occupational exposure limits have been established by OSHA, NIOSH, or ACGIH

> Difenacoum: No occupational exposure limits have been established by OSHA, NIOSH, or ACGIH

> Difethialone: No occupational exposure limits have been established by OSHA, NIOSH, or ACGIH

Reference values: No reference values (RfDs/RfCs, TDIs, TIs, MRLs) have been established for second-generation anticoagulants

Risk/safety phrases: End-use pesticide formulations of second-generation anticoagulants generally carry the signal word "Caution"

BACKGROUND AND USES

The second-generation anticoagulant rodenticides were developed after the emergence of resistance to first-generation anticoagulants and were observed in certain rodent species. The second-generation anticoagulants can be distinguished from the first-generation rodenticides by their higher degree of toxicological potency, and the ability to cause lethal effects from a single ingestion. Lethal effects on nontarget species, including birds and other wildlife, have been frequently reported in association with second-generation anticoagulant rodenticides (Sánchez-Barbudo et al., 2012).

Brodifacoum was first registered for use in 1979 by the US EPA. It is a rodent control agent for use against rats and mice. It is formulated as a bait in pellet formulations, in mouse bait

ready-to-use packets, and paraffin blocks. End-use products contain 0.005% brodifacoum.

Bromadiolone was first registered for use in 1980 by the US EPA. It is a rodent control agent for mice and rats in buildings, inside transport vehicles, and in sewers. It is formulated as a meal bait, in pellet formulations, in mouse bait ready-to-use packets, and paraffin blocks. End-use products contain 0.005% bromadiolone.

Difenacoum was first registered by the US EPA in 1975. It is formulated as concentrate containing 1 g/kg, and as a ready-to-use bait containing 50 mg difenacoum per kg of bait.

Difethialone was first registered for use by the US EPA in 1995. It is formulated as ready-to-use whole grain cereals and oat grain baits containing 0.0025% difethialone.

At the current time, the U.S. EPA is in the process of banning the sale of certain consumer use products containing second-generation anticoagulant rodenticides. In addition, the EPA has recently prohibited the sale of consumer products containing brodifacoum, difethialone, and difenacoum because of their toxicity to wildlife. The US EPA website should be accessed for current information about the regulatory status of products containing second-generation rodenticides.

PHYSICAL AND CHEMICAL PROPERTIES

The chemical name for brodifacoum (CAS # 56073-10-0) is 3-[3-(4′-bromo[1,1′biphenyl]-4yl)=1,2,3,4-tetrahydro-1-napthalenyl]-4-hydroxy-2H-1-benzopyran-2-one. It has the empirical formula $C_{32}H_{23}BrO_3$, and a molecular weight of 523.4. It is an odorless powder with a melting point of 228–232 °C.

The chemical name for bromadiolone is 3-[3-(4′-bromo[1,1′-biphenyl]-4-yl)-3-hydroxy-1-phenylpropyl]-4-hydroxy-2H-1-benzopyran-2-one. It has the empirical formula $C_{30}H_{23}BrO_4$, and a molecular weight of 527.4. It is a yellowish powdered mixture of two diastereoisomers, with a melting point of 200–210 °C.

The chemical name for difenacoum is 3-[3-(1,1′-biphenyl) 4-yl-1,2,3,4-tetrahydro-1-naphthalenyl]-4-hydroxy-2H-1-benzopyran-2-one. It has the empirical formula $C_{31}H_{24}O_s$, and a molecular weight of 444.5. It is a white powder with a melting point of 215–219 °C.

The chemical name for difethialone is 3-[3-(4′-bromo[1,1′-biphenyl]-4-yl)-1,2,3,4-tetrahydro-1-napthalenyl]-4-hydroxy-2H-1-[benzothiopyran-2-one]. It has the empirical formula $C_{31}H_{23}BrO_2S$, and a molecular weight of 539.5. It is a whitish powder with a melting point of 233–236 °C.

MAMMALIAN TOXICOLOGY

Acute Effects

The acute toxicological effects of the second-generation anticoagulants are similar to those observed from the first-generation anticoagulants. The distinctions include a higher degree of potency, and a much longer duration of effects for the second-generation anticoagulants. These distinctions arise from the higher degree of lipid solubility, and ability to concentrate within the target organ tissue of the liver (Lund, 1981). Comparative studies in target species have demonstrated the ability of second-generation anticoagulants to product coagulopathy with a single ingestion of brodifacoum (0.005% active ingredient).

There is a characteristic delay of 24–72 h between the ingestion of a second-generation anticoagulant and the onset of bleeding. Generalized hematoma formation, pallor, and weakness, are common manifestations in acute intoxication. Fatal outcomes often result from bleeding in the central nervous system. Most cases of intoxication by second-generation anticoagulants result from ingestion pathways of exposure; however, a case report described significant coagulopathy resulting from smoking marijuana that had been mixed with brodifacoum (La Rosa et al., 1997). Another case report has described coagulopathy from dermal exposure to a concentrated liquid preparation of 0.106% diphacinone in a professional applicator (Spiller et al., 2003).

The oral LD_{50} of technical brodifacoum in male rats is 0.418 mg/kg. The dermal LD_{50} in female rabbits is 3.16 mg/kg, and the inhalation LC_{50} in rats ranges from 3.05 to 4.86 µg/l (United States Environmental Protection Agency, 1998). Brodifacoum is not considered as a dermal sensitizer, and some minor transient eye and skin irritation have been reported.

The oral LD_{50} of technical bromadiolone in rats is 0.56–0.84 mg/kg. The dermal LD_{50} in rabbits is 1.71 mg/kg. The inhalation LC_{50} in rats is <0.43 µg/l. Bromadiolone is not a dermal sensitizer, and minimal eye and dermal irritation have been reported (United States Environmental Protection Agency, 1998). The oral LD_{50} of difenacoum in rats is 1.8 mg/kg in male rats, and 50 mg/kg in female guinea pigs. The dermal LD_{50} is 50 mg/kg in rats and 1000 mg/kg in rabbits (Pelfrene, 2001).

The oral LD_{50} of difethialone in rats is 0.56 mg/kg in male rats, 1.29 mg/kg for mice, and 2–3 mg/kg for pigs. The dermal LD_{50} is 5.3 mg/kg in rats. The inhalation LC_{50} in rats is between 5 and 19.3 mg/m^3 in rats exposed for 4 h. Difethialone is not considered a skin sensitizer (Pelfrene, 2001).

Chronic Effects

In a 90-day study of orally administered bromadiolone in beagle dogs, the low observed effect level (LOEL) for subchronic toxicity (coagulopathy) was 15 µg/kg, and the NOEL was 10 µg/kg (United States Environmental Protection Agency, 1998).

Mechanism of Action(s)

The second-generation anticoagulant rodenticides exert their anticoagulant activity through the inhibition of vitamin K-dependent carboxylation of clotting factors II, VII, IX, and X, as well as the anticoagulant proteins C and S. Specifically, they inhibit the enzyme K_1 2,-epoxide reductase complex, subunit 1 (VKORC1) (Park, 1988 #14612). Inhibition of VKORC1 prevents the conversion of vitamin K epoxide to the biologically active vitamin K hydroquinone, which carboxylates and activates these clotting factors. The higher degree of potency of the second-generation anticoagulants may partly reflect their ability to saturate hepatic detoxification enzymes at very low concentrations, as demonstrated by zero-order elimination kinetics following overexposure (Bruno et al., 2000).

After exposure to second-generation anticoagulants, the activity of clotting factors with shorter half-lives (including factor VII) is inhibited within 12–24 h. The full anticoagulant effect is not observed until 5–7 days after initial exposure because of the longer half-lives of clotting factors X and II.

Chemical Pathology

Pathological findings at autopsy in animals and humans poisoned by first-generation anticoagulants include pallor of the muscles and viscera, and generalized hemorrhage.

SOURCES OF EXPOSURE

At the current time in the United States, second generation rodenticides may be found in bait stations for consumer-use products. The US EPA has recently announced plans to cancel these uses. Second-generation anticoagulants are also currently approved for use in bait products for agricultural and professional applicator uses in indoor and outdoor settings.

INDUSTRIAL HYGIENE

Occupational exposure guidelines and limits have not been established for second-generation anticoagulant rodenticides.

MEDICAL MANAGEMENT

As a result of the high incidence of accidental human exposures to long-acting anticoagulant rodenticides, most commonly in the pediatric population, evidence-based consensus guidelines have been established for the out-of-hospital management of ingestions (Caravati et al., 2007). Most accidental pediatric ingestions result in small doses that do not result in coagulopathy. Consultation with a Regional Poison Control Center is recommended for the triage and medical management of all exposures to second-generation anticoagulants.

Intentional ingestion of these rodenticides can result in delayed onset of bleeding. For individuals presenting within several hours of intentional ingestion, a single dose of activated charcoal should be considered for gastric decontamination. The clinical laboratory assessment of individuals at risk of coagulopathy from first-generation anticoagulants should include a prothrombin time (PT-INR) at 48 h after the time of ingestion (Smolinske et al., 1989). A careful physical examination should be conducted to assess for evidence of bruising and bleeding. Life threatening hemorrhage should be managed with the infusion of fresh frozen plasma, which is rich in active vitamin K-dependent coagulation factors and will temporarily reverse coagulopathy in most patients (Cruickshank et al., 2001).

The specific antidote for coagulopathy resulting from exposure to first-generation anticoagulants is vitamin K_1. It is effective orally and via intravenous or intramuscular administration. Repeated, large doses of vitamin K_1 (up to or >60 mg/day) may be required in some cases. In contrast to first-generation anticoagulants, the duration of coagulopathy from second-generation anticoagulants can last from weeks to months (Jones et al., 1984; Spahr et al., 2007). Careful monitoring of coagulation status and high doses of vitamin K_1 are required until the coagulopathy resolves.

REFERENCES

Bruno, G.R., Howland, M.A., McMeeking, A., and Hoffman, R.S. (2000) Long-acting anticoagulant overdose: brodifacoum kinetics and optimal vitamin K dosing. *Ann. Emerg. Med.* 36(3):262–267.

Caravati, E.M., Erdman, A.R., Scharman, E.J., Woolf, A.D., Chyka, P.A., Cobaugh, D.J., Wax, P.M., Manoguerra, A.S., Christianson, G., Nelson, L.S., Olson, K.R., Booze, L.L., and Troutman, W.G. (2007) Long-acting anticoagulant rodenticide poisoning: an evidence-based consensus guideline for out-of-hospital management. *Clin. Toxicol. (Phila.)* 45(1):1–22.

Cruickshank, J., Ragg, M., and Eddey, D. (2001) Warfarin toxicity in the emergency department: recommendations for management. *Emerg. Med. (Fremantle, WA)* 13(1):91–97.

Jones, E.C., Growe, G.H., and Naiman, S.C. (1984) Prolonged anticoagulation in rat poisoning. *JAMA* 252(21):3005–3007.

La Rosa, F.G., Clarke, S.H., and Lefkowitz, J.B. (1997) Brodifacoum intoxication with marijuana smoking. *Arch. Pathol. Lab. Med.* 121(1):67–69.

Lund, M. (1981) Comparative effect of the three rodenticides warfarin, difenacoum and brodifacoum on eight rodent species in short feeding periods. *J. Hyg. (Lond.)* 87(1):101–107.

Park, B.K. (1988) Warfarin: metabolism and mode of action. *Biochem. Pharmacol.* 37(1):19–27.

Pelfrene, A.F. (2001) Rodenticides. In: Krieger, R.I., editor. *Handbook of Pesticide Toxicology*, 2nd ed., vol. 2, Academic Press, pp. 1793–1837.

Sánchez-Barbudo, I.S., Camarero, P.R., and Mateo, R. (2012) Primary and secondary poisoning by anticoagulant rodenticides of non-target animals in Spain. *Sci. Total Environ.* 420:280–288.

Smolinske, S.C., Scherger, D.L., Kearns, P.S., Wruk, K.M., Kulig, K.W., and Rumack, B.H. (1989) Superwarfarin poisoning in children: a prospective study. *Pediatrics* 84(3):490–494.

Spahr, J.E., Maul, J.S., and Rodgers, G.M. (2007) Superwarfarin poisoning: a report of two cases and review of the literature. *Am. J. Hematol.* 82(7):656–660.

Spiller, H.A., Gallenstein, G.L., and Murphy, M.J. (2003) Dermal absorption of a liquid diphacinone rodenticide causing coagulaopathy. *Vet. Hum. Toxicol.* 45(6):313–314.

United States Environmental Protection Agency (1998) *Reregistration Eligibility Decision (RED): Rodenticide Cluster*, Washington, DC: Office of Pesticide Programs, p. 319.

NON-ANTICOAGULANTS

BROMETHALIN

Target organ(s): Central nervous system

Occupational exposure limits: No occupational exposure guidelines limits have been established by OSHA, NIOSH, or ACGIH.

Reference values: No reference values (RfDs/RfCs, TDIs, TIs, MRLs) have been established for second-generation anticoagulants

Risk/safety phrases: End-use pesticide formulations containing bromethalin generally carry the signal word "Caution"

BACKGROUND AND USES

Bromethalin is an uncoupler of oxidative phosphorylation, and available in the United States since 1985 as a rodenticide. It is registered for use in and around buildings, inside transport vehicles, and in sewers to control rats and mice. It is a single-dose poison, formulated in blocks, baits, or pellets. All registered products contain 0.01% bromethalin, with the exception of one formulation containing 0.005% active ingredient.

PHYSICAL AND CHEMICAL PROPERTIES

The chemical name for bromethalin (CAS # 63333-35-7) is N-Methyl-2,4-dinitro-N-(2,4,5-tribromophenyl)-6(trifluoromethyl)benzenamine. It has the empirical formula $Cl_3H_7Br_3F_3N_3O$. and a molecular weight of 577.93. It is white crystal with a melting point of 148–154 °C.

MAMMALIAN TOXICOLOGY

Acute Effects

The acute toxic effects target the central nervous system, and differences in toxic syndromes susceptibility have been observed in different animal species. A convulsant syndrome, characterized by tremors, seizures, CNS, and hyperthermia, has been described at higher doses within 4–18 h after ingestion. A paralytic syndrome has also been described, with a slower onset of toxicity ranging from 1 to 7 days, and consisting of ataxia, CNS depression, weakness, and progression to paralysis. Other acute effects that have been reported in animals include anorexia, positional nystagmus, muscle tremors, and abnormal pupillary light reflexes. (DeClementi and Sobczak, 2012).

The oral LD_{50} of technical bromethalin in rats is 9.1–10.7 mg/kg. The dermal LD_{50} in rabbits is 2000 mg/kg, and the inhalation LC_{50} in rats is 0.24 mg/l. In an acute neurotoxicity study conducted in rats receiving bromethalin in mineral oil via gavage, the NOEL was >3 mg/kg. Bromethalin is not considered a skin sensitizer or skin irritant, but slight eye irritation has been reported (United States Environmental Protection Agency).

Chronic Effects

In a 13 week study of subchronic bromethalin toxicity in rats, the NOEL is 25 μg/kg/day, and the LOEL is 125 μg/kg/day, baed upon findings of leukoencephalopathy in the central nervous system (United States Environmental Protection Agency, 1998).

Mechanism of Action(s)

Bromethalin is a neurotoxin that uncouples oxidative phosphorylation, leading to decreased production of ATP and failure of the sodium–potassium ATPase. Bromethalin has high oral bioavailability, and peak levels in plasma are observed within 4 h. Its immediate metabolite, desmethyl bromethalin, has higher neurotoxic potency than the parent compound (Van Lier and Cherry, 1988).

Chemical Pathology

Pathological findings at autopsy or necropsy include generalized spongy degeneration and edema of the white matter, spinal cord, and optic nerve (Van Lier and Cherry, 1988; DeClementi and Sobczak, 2012).

SOURCES OF EXPOSURE

At the current time in the United States, bromethalin can be found in consumer-use bait products, as well as rodenticide products for agricultural and professional applicator use.

INDUSTRIAL HYGIENE

Occupational exposure guidelines and limits have not been established for bromethalin.

MEDICAL MANAGEMENT

Relatively few case reports of human poisoning from bromethalin have been published, and no evidence-based guidelines exist for the medical management of intoxication. Symptomatic and supportive management are recommended, including monitoring for adverse effects in the central nervous system and aggressive treatment with benzodiazepines if seizures develop. The potential for delayed onset of serious neurological complications should be considered.

REFERENCES

DeClementi, C. and Sobczak, B.R. (2012) Common rodenticide toxicoses in small animals. *Vet. Clin. North Am. Small Anim. Pract.* 42(2):349–360.

United States Environmental Protection Agency (1998) *Reregistration Eligibility Decision (RED): Rodenticide Cluster*, Washington, DC: Office of Pesticide Programs, p. 319.

Van Lier, R.B. and Cherry, L.D. (1988) The toxicity and mechanism of action of bromethalin: a new single-feeding rodenticide. *Fundam. Appl. Toxicol.* 11(4):664–672.

NON-ANTICOAGULANTS

CHOLECALCIFEROL

Target organ(s): Kidneys, bones

Occupational exposure limits: No occupational exposure guidelines limits have been established by OSHA, NIOSH, or ACGIH

Reference values: A tolerable upper limit of 50 μg vitamin D/day for adults has been suggested by authorities in North American and the European Union

Risk/safety phrases: End-use pesticide formulations containing cholecalciferol generally carry the signal word "Caution" or "Warning"

BACKGROUND AND USES

Cholecalciferol (vitamin D-3) was first registered in the United States for use as a multiple-dose rodenticide in 1984. It is currently registered for consumer-use products, and it also has agricultural and professional applicator uses in and around buildings, and inside transport vehicles. It is typically formulated for use in bait formulations. There is typically a delay of several days between ingestion and the onset of toxic and lethal effects in target species.

PHYSICAL AND CHEMICAL PROPERTIES

The chemical name for cholecalciferol (CAS # 67-97-0) is 9,10-secocholesta-5,7,10(10)-trien-3-betaol. It has the empirical formula $C_{27}H_{44}O$ and a molecular weight of 384.62. It has a melting point of 84–86 °C.

MAMMALIAN TOXICOLOGY

Acute Effects

Rats and mice are more susceptible to the effects of cholecalciferol than humans, and to date, cases of serious human toxicity from exposure this rodenticide have not been reported. The metabolites of cholecalciferol cause acute toxicity by increasing serum calcium and phosphorus. The metabolites increase intestinal absorption of calcium, and stimulate mobilization of calcium and phosphorus from bones into the plasma. The effects of these electrolyte imbalances result in renal injury from tissue mineralization in the kidneys, and other signs and symptoms associated with severe hypercalcemia, including vomiting, polyuria, and polydipsia. The acute effects of cholecalciferol intoxication generally occur within 12–36 h after ingestion (DeClementi and Sobczak, 2012).

The acute oral LD_{50} of cholecalciferul in rats is 43.6 mg/kg. In dogs, the oral LD_{50} is 88 mg/kg. Some authors have reported fatal outcomes from ingestions of 10–20 mg/kg in rats are 9.1–10.7 mg/kg (Pelfrene, 2001).

Chronic Effects

In a 4-month study of daily doses ranging from 100 to 4000 IU cholecalciferol/kg of feed in swine, atherosclerotic lesions were observed in a dose-dependent fashion (Toda et al., 1985).

Mechanism of Action(s)

Cholecalciferol is metabolized in the liver to 25-hydroxycholecalciferol. This metabolite is subsequently metabolized by the kidney into the active metabolite 1,25-dihydroxycholecalciferol, which is responsible for the intestinal absorption of calcium and mobilization from bone tissues (Morrow, 2001). This produces hypercalcemia, osteomalacia, and calcification of tissues in the kidneys, cardiovascular system, stomach, and lungs. Death usually occurs in 2–5 days.

Chemical Pathology

Pathological findings in cases of cholecalciferol poisoning in animals include soft tissue mineralization within the kidneys, heart, GI tract, skeletal muscles, and connective tissues. Within the lungs, alveolar septal thickening has been observed, with mineralization and hemorrhage (DeClementi and Sobczak, 2012).

SOURCES OF EXPOSURE

At the current time in the United States, cholecalciferol can be found in consumer-use bait products, as well as rodenticide products for agricultural and professional applicator use. Cholecalciferol also has human uses in vitamins and dietary supplements, as the natural form of vitamin D.

INDUSTRIAL HYGIENE

Occupational exposure guidelines and limits have not been established for cholecalciferol.

MEDICAL MANAGEMENT

No reports of human poisoning from the use of cholecalciferol as a rodenticide have been published, and no evidence-based guidelines exist for the medical management of intoxication.

Veterinary treatment of acute toxicity consists of preventing and treating the hypercalcemia by aggressive hydration, the administration of loop diuretics, and glucocorticoids. In more serious intoxications, medications that inhibit bone resorption such as pamidronate have been utilized. Careful monitoring of serum calcium, phosphorus, BUN, and creatinine are warranted in the veterinary management of intoxication (Morrow, 2001).

REFERENCES

DeClementi, C. and Sobczak, B.R. (2012) Common rodenticide toxicoses in small animals. *Vet. Clin. North Am. Small Anim. Pract.* 42(2):349–360.

Morrow, C. (2001) Cholecalciferol poisoning. *Vet. Med.* 96: 905–911.

Pelfrene, A.F. (2001) Rodenticides. In: Krieger, R.I., editor. *Handbook of Pesticide Toxicology*, 2nd ed., vol. 2, Academic Press, pp. 1793–1837.

Toda, T., Ito, M., Toda, Y., Smith, T., and Kummerow, F. (1985) Angiotoxicity in swine of a moderate excess of dietary vitamin D3. *Food Chem. Toxicol.* 23(6):585–592.

SECTION VII

DUSTS AND FIBERS

SECTION EDITOR: RAYMOND D. HARBISON

90

INTRODUCTION

David R. Johnson

Dusts are dispersed solid airborne particles, capable of temporary suspension in a gaseous medium, and often produced by the breakup of larger organic or inorganic materials. Dusts are frequently generated in occupational environments and are often complex mixtures of a variety of substances and contaminants. Workers are exposed to these dusts principally by inhalation but also by direct contact with skin or conjunctiva.

Once the dust is inhaled, characteristics of particles in the dust determine where and how much of the particle is deposited in the respiratory tract. Such particulate characteristics include size, shape, diameter, aerodynamic equivalent, surface area, density, electrostatic charge and hygroscopicity (Gross, 1981; Timbrell, 1965; Brain and Valberg, 1979; Beckett, 2000; Maxim et al., 2006). Based on such characteristics, dust particles are deposited in the nose, mouth, throat, trachea, bronchi, and alveoli or promptly exhaled.

Host characteristics also affect particle deposition. Such host characteristics include physiological and anatomical factors of the respiratory system, such as breathing patterns and airway geometry (Beckett, 2000). The location and amount of deposition of various respirable particles in the respiratory tract has been subjected to considerable experimental study in humans, and statistical analysis has allowed probability calculations of the location and quantity of particle deposition (Yu, Diu, Soong, 1981; Chen, 2003; Heyder, 2004; Straum and Hofmann, 2009; Sturm, 2011).

Once deposited in the respiratory tract, factors that influence the ability of dust particles to cause disease include particle fibrogenicity, antigenicity, toxicity, propensity to combine with lung tissues, penetrability, durability, solubility, acidity, and alkalinity. Host factors influencing the particle's ability to cause disease include the individual's respiratory tract clearing mechanisms such as ciliary action and macrophage function. Host susceptibility is influenced by the individual's health status, immunologic status, nutritional status, age, and personal habits such as smoking, medications, drug use, and alcohol use.

Typical examples of dust-induced disease include dermatitis, rhinitis, conjunctivitis, urticaria, bronchitis, bronchoconstrictive disease (asthma), pneumoconiosis, hypersensitivity pneumonitis (extrinsic allergic alveolitis), interstitial lung disease, emphysema, granulomatous disease, and cancer. The greater the intensity and duration of exposure to dust, the greater is the risk of developing an acute or chronic disease.

Pneumoconioses refers to mineral dust-induced lung disease characterized by the permanent deposition of particulate matter in the lung, which may cause tissue reactions. This term is derived from the Latin pneumono (lung) and *cono* (dust) and has a terminal ending that denotes reaction to dust in the lung. A group of cases associated with the inhalation of a single or predominant etiological dust is referred to as pneumoconiosis. Coal workers pneumonconiosis, silicosis, asbestosis, baritosis, and siderosis are examples of pneumoconioses. Asbestos and silica are well-known examples of dusts that have fibrogenic properties. Asbestosis has been clearly associated with progression to cancer, including mesothelioma and bronchogenic carcinoma. If the dust elicits no inflammatory reaction in the lung, it is referred to as a benign pneumoconiosis, such as stannosis arising in tin mining. In benign pneumoconiosis, there may be striking radiologic findings with no significant pulmonary pathology. Many dusty operations involve an exposure to a mixture of dusts that leads to chronic lung illness. The term *mixed dust pneumoconioses* is best used for this category and often involves silica mixed with other types of dusts. Some authors use a diagnostic label to identify the dust in terms of the offending agents, for example, anthracosilicosis for the disease of underground coal miners inhaling both coal and silica.

Man-made mineral fibers, also called man-made vitreous fibers (MMVF), are amorphous silicates that include fibrous glass, mineral wool, refractory ceramic fibers, and alkaline earth silicate wool. They can generate irritating dusts if not handled properly; however, no consistent epidemiological findings have found them to be fibrogenic or carcinogenic to humans. They have been associated with respiratory irritation. Fiberglass dermatitis is one of the most common forms of irritant occupational dermatitis.

Organic dusts are composed of particulates of plants, animal, and/or microorganisms and often include complex mixtures of substances. Exposure occurs in a wide range of occupations and is commonly associated with bronchoconstrictive disease, asthma, and hypersensitivity pneumonitis. rhinitis, conjunctivitis, and bronchitis are also common. Mechanisms are often not clearly delineated and may involve immunological, pharmacological or genetic mechanisms, as well as airway and neurogenic inflammation (Mapp et al., 1994). A few wood dusts have been associated with nasal cancer.

Traditionally, the focus has been on occupational lung disease caused by an exposure to mineral dusts such as asbestos, silica, and coal. However, the incidence of diseases caused by mineral dusts has declined in postindustrial countries, and asthma has emerged as the principal occupational lung disease with new substances capable of inducing asthma being introduced into the workplace every year (Beckett, 2000).

Governmental regulations vary from one country to another in setting standards for acceptable dust exposure in industry. Different companies within the same industry may voluntarily implement various levels of industrial hygiene measures resulting in different levels of dust exposure and therefore different prevalence rates of disease. Environmental conditions may also vary within the same facility. The composition of dusts are often complex and will vary considerably in various settings. With all of these variables at play, and with often limited information about the exact amount of dust capable of causing disease, linking a respiratory disease to occupational exposures can be quite challenging.

Ongoing research using a variety of modern investigative tools, including bronchoalveolar lavage, gallium-67 scanning, bioassays, diagnostic imaging, and air monitoring technology have improved our understanding of the associations and mechanisms of lung disease; however, much more research is needed. Increasingly, sophisticated instrumentation, international use of recommended safe levels of exposure, improved engineering of industrial processes, medical monitoring, and personal protective equipment are rapidly contributing to the prevention of the debilitating cases of occupational lung diseases that were seen in the past.

Medical management of dust-induced respiratory disease should always involve efforts to identify the offending agent with elimination or minimization of exposure to that agent. Preventing exposure from the outset is best.

REFERENCES

Beckett, W.S. (2000) Occupational respiratory diseases. *N. Engl. J. Med.* 342(6):406–413.

Brain, J.D. and Valberg, P.A. (1979) Deposition of aerosol in the respiratory tract. *Am. Rev. Respir. Dis.* 120:1325.

Chen, Y.C. (2003) Aerosol deposition in the extrathoracic region. *Aerosol Sci. Technol.* 37(8):659–671.

Gross, P. (1981) Consideration of the aerodynamic equivalent diameter of respirable mineral fibers. *Am. Ind. Hyg. Assoc. J.* 42:449–452.

Heyder, J. (2004) Deposition of inhaled particles in the human respiratory tract and consequences for regional targeting in respiratory drug delivery. *Proc. Am. Thorac. Soc.* 1(4):315–320.

Mapp, C.E., et al. (1994) Mechanisms and pathology of occupational asthma. *Eur. Respir. J.* 7:544–554.

Maxim L.D., et al. (2006) The role of fiber durability/biopersistence of silica-based synthetic vitreous fibers and their influence on toxicology. *Regul. Toxicol. Pharmacol.* 46:42–62.

Straum, R. and Hofmann, W. (2009) A theoretical approach to the depositon and clearance of fibers with variable size in the human respiratory tract. *J. Hazard. Mater.* 170:210–218.

Sturm, R. (2011) Modeling the deposition of bioaerosols with variable size and shape in the human respiratory tract—a review. *J. Adv. Res.* 3:295–304.

Timbrell, V. (1965) Human exposure to asbestos: dust controls and standards. The inhalation of fibrous dusts. *Ann. N. Y. Acad. Sci.* 132:255–273.

Yu, C.P., Diu, C.K., and Soong, T.T. (1981) Statistical analysis of aerosol deposition in nose and mouth. *Am. Ind. Hyg. Assoc. J.* 42:726–733.

91

BENIGN DUSTS (NUISANCE DUSTS)

Humairat H. Rahman and Raymond D. Harbison

BACKGROUND

Tiny solid particles formed by disintegration processes such as crushing, grinding, polishing, sanding, or impact constitute nuisance dusts. This dust may cause non-fibrogenic pneumoconiosis. When inhaled in excessive amounts chest X-rays may present with a radio opaque appearance. Nuisance dust or inert dust contains <1% quartz. Nuisance dust may produce little adverse effects on the lungs over long periods of time because of its low silica content but excessive concentrations due to intense short-term exposure may deposit in eyes, ears, and nasal passages. This may lead to decreased visibility and can cause injury to the skin or mucus membranes by chemical or mechanical action. This may also result in damage by overwhelming the lymphatic drainage, by obstructing the anatomy of terminal bronchioles to cause infarction, necrosis, bronchiectasis, and atelectasis. Nuisance dust can be classified in three categories depending upon size. These categories are respirable dust, inhalable dust, or total dust. Nuisance dust particle size is generally measured in micrometers. Occupational Safety and Health Administration (OSHA) has categorized nuisance dust or benign dust as either inorganic or organic, with <1% of quartz, and dust that does not contain asbestos, as particulates not otherwise regulated.

BARIUM

BACKGROUND

Barium is an earth metal that is alkaline in nature. Elemental barium is a silver-white metal and constitutes nearly 0.05% of the earth's crust. Barium sulfate and barium carbonate are the two most common forms of barium. Barium compounds have a wide variety of uses in industry, including brick, ceramic, rubber, glass, and painting industries. It is used in the manufacture of paper electrodes, in fireworks, and as contrast medium in diagnostic X-ray studies. Oil and gas industries use barium as a lubricant for mud drilling.

PHYSICAL AND CHEMICAL PROPERTIES

Insoluble barium sulfate is an odorless powder compound, yellowish or white in color. It is solid and noncombustible in nature. The molecular weight, specific gravity, and boiling point of barium sulfate are 233.4, 4.25–4.5, and 2912 °F, respectively.

EXPOSURE ROUTES AND MAMMALIAN TOXICOLOGY

Workers can be exposed to barium sulfate from withering of mineral and rock and from processes such as welding (Dare et al., 1984). Exposure to barium sulfate can occur through inhalation and skin and/or eye contact.

Inhalation of barium sulfate dust may cause irritation of nose, eyes, and throat. Barium-containing dust after long and intense exposure may form bilateral lung infiltrates. This harmless condition results in a benign pneumoconiosis known as baritosis. Various case reports and studies have reported baritosis in barium-exposed workers. Baritosis is characterized by dense, circumscribed nodules profusely disseminated throughout the lung fields, though available

Hamilton & Hardy's Industrial Toxicology, Sixth Edition. Edited by Raymond D. Harbison, Marie M. Bourgeois, and Giffe T. Johnson.
© 2015 John Wiley & Sons, Inc. Published 2015 by John Wiley & Sons, Inc.

human data suggest that baritosis is without clinical significance with limited symptoms or disability. Exposure to barium sulfate may produce non-collagenous type of pneumoconiosis that generally disappears when exposure ceases. Chronic exposure to barium sulfate may lead to bronchial irritation.

INDUSTRIAL HYGIENE

The current OSHA's permissible exposure limit (PEL) is 15 mg/m^3 of air (total dust) and 5 mg/m^3 respirable fraction 8-h time-weighted average (TWA) concentration. The National Institute for Occupational Safety and Health (NIOSH) has established a recommended exposure limit (REL) of 10 mg/m^3 of air (total dust) and 5 mg/m^3 respirable fraction 8-h TWA concentration. The American Conference of Industrial Hygienists's (ACGIH) threshold limit value (TLV) for barium sulfate is 10 mg/m^3 for total dust, containing no asbestos and <1% silica.

TIN

BACKGROUND

Tin is found in powder form and is white or pale-grey in color. Tin oxide is the most common compound of tin. Tin mining has been ongoing for 3000 years (Cierny and Weisgerber, 2003). Tin oxide is brownish-black powder. Tin oxide is insoluble in water with a specific gravity and molecular weight of 6.3 and 134.7, respectively.

EXPOSURE ROUTES AND MAMMALIAN TOXICOLOGY

Exposure to tin oxide can occur due to inhalation of tin oxide dust or fumes in a work place. Exposure to tin oxide can cause irritation of eyes and skin. Inhalation of dust or fumes may lead to deposition in the lungs. This may lead to stannosis, a benign pneumoconiosis (Sluis-Cremer et al., 1989). Grinding, briquet making, smelting, and casting of tin, as well as handling of tin oxide in industry may cause stannosis (Karkhanis and Joshi, 2013). Occupational stannosis in non-mining industries may occur when molten tin is poured into heated iron hollow ware or when articles are dipped by hand into molten tin to coat them. Inhalation of tin oxide dust or fumes may cause difficulty in breathing or may result in decreased pulmonary function. The chest X-ray may show dense, bilateral infiltrates but without any tissue reaction to deposition of dust. This condition results in

non-fibrotic pneumoconiosis. Laboratory tests showed that tin oxide engulfed by macrophages is present in alveolar space, intralobular hilar nodes, and the perivascular lymphatic system. Pathologically, cut surface of lung may contain 1–3 mm of tin oxide particles that are grayish-black in color and are spongy to touch (Yilmaz et al., 2009).

INDUSTRIAL HYGIENE

The current OSHA PEL is 15 mg/m^3 of air (total dust). The NIOSH has established a REL of 2 mg/m^3 TWA. The ACGIH TLV for tin oxide is 2 mg/m^3 for total dust.

IRON

BACKGROUND

The oxide form of iron, iron oxide, is a major source of iron dust or fumes. The main use of iron oxide is in the steel industry where processes such as welding and silver finishing results in the production of iron oxide dust or fumes. Iron oxide is reddish-brown in color and solid in nature. Molecular weight, specific gravity, and melting point of iron oxide are 159.7, 5.25, and 2664 °F, respectively.

EXPOSURE ROUTES AND MAMMALIAN TOXICOLOGY

Inhalation is the major route of exposure to iron oxide fume or dust. A study on human volunteers who were exposed to iron oxide for 30 min at a standard concentration of nearly 13 mg/m^3 showed no significant reduction in lung function, permeability of alveolar epithelium, or diffusing capacity of lung (Lay et al., 2001). Human volunteers presenting with subclinical inflammation due to acute deposition of intra-pulmonary ferric oxide showed resolution of symptoms after 96 h (Lay et al., 1998). Exposure to iron oxide fume or dust may lead to siderosis in workers. Siderosis is also known as arc welder pneumoconiosis or welder siderosis (Kleinfeld et al., 1969). In 1936, siderosis was first described in welder workers exposed to iron oxide (Doherty et al., 2004; Doig and Laughlin, 1936). Siderosis is considered as a benign condition because it is not associated with fibrosis or abnormal pulmonary function tests (Chong et al., 2006). When iron is mixed with certain amounts of silica, it can cause silico-siderosis. Silico-siderosis is a pathological condition and requires medical monitoring (Billings and Howard, 1993). Workers with siderosis can reverse symptoms after they are removed from the exposure area, by elimination of iron dust from lungs in a few years (Kim et al., 2001).

In a study of 25 workers exposed to iron oxide, eight workers presented reticulo nodular patterns on chest X-ray (Harding et al., 1958). In another study the lung function tests of 16 welders were abnormal with chest X-ray demonstrating siderosis (Stanescu et al., 1967). The welders complained of dyspnea though their spirogram was normal. After removal from exposure, iron dust is gradually eliminated from lungs resulting in partial or complete disappearance of radiographic appearance of lung opacities.

INDUSTRIAL HYGIENE

The current OSHA PEL is $10 \, mg/m^3$ of air. The NIOSH has established a REL of $5 \, mg/m^3$ of air. The ACGIH TLV–TWA for iron oxide dust and fume is $5 \, mg/m^3$.

PERLITE

BACKGROUND

Expanded perlite is an amorphous material consisting of fused sodium potassium aluminum silicate. Perlite is volcanic rock and needs to be heated up to $1000 \, °C$ to form expanded perlite. Expanded perlite is used as a conditioner for soil, packing material, construction insulator for floors, walls, and as a filter aid in agriculture processes. Perlite is categorized as one of the nuisance or benign dust that does not cause pneumoconiosis (Sampatakakis et al., 2013; Du et al., 2010).

EXPOSURE ROUTES AND MAMMALIAN TOXICOLOGY

Inhalation, skin, and/or eye contact are the common routes of exposure to expanded perlite. Exposure to perlite dust may lead to irritation of eyes, skin, or upper respiratory tract. A study in Taiwan of 24 workers exposed to expanded perlite due to an accidental explosion resulted in irritation of the respiratory tract within 24–48 h. The follow-up study showed that three subjects developed reactive airway dysfunction syndrome (Du et al., 2010). However a follow-up of perlite workers for 23 years showed no decreased pulmonary function tests (Cooper and Sargent, 1986).

INDUSTRIAL HYGIENE

The current OSHA PEL for expanded perlite is $15 \, mg/m^3$ of air (total dust) and $5 \, mg/m^3$ respirable fraction as 8-h TWA

concentration. The NIOSH has established a REL of $10 \, mg/m^3$ of air (total dust) and $5 \, mg/m^3$.

CARBON BLACK

BACKGROUND

Carbon Black is a spherical form of pure carbon particles and is produced by the incomplete combustion or thermal decomposition of liquid or gaseous hydrocarbons. Carbon black is used in rubber, plastic, paper, fiber, and the electrical industry. It is also used as a reinforcing agent and as a pigment in the printing and paint industry.

PHYSICAL AND CHEMICAL PROPERTIES

Carbon black contains nearly 95% of pure carbon along with 4–5% of nitrogen, hydrogen, sulfur, and oxygen. Carbon black is a combustible, odorless solid particle and is black in color. It is a strong oxidizer, insoluble in water with molecular weight of 12 and specific gravity of 1.8–2.1 (Brockmann et al., 1998).

EXPOSURE ROUTES AND MAMMALIAN TOXICOLOGY

The common routes of exposures to carbon black are inhalation, ingestion, or dermal contact. Accidental exposure to carbon black can cause difficulty in breathing, eye, and dermal irritation. Carbon black inhalation can cause respiratory symptoms such as cough, phlegm, and chest pain. Dermal exposure can cause skin irritation. In a study of 125 Nigerian workers exposed to carbon black abnormal pulmonary symptoms, including cough with phlegm were observed though the X-ray findings were normal (Oleru et al., 1983). Chronic exposure to carbon black may result in abnormal pulmonary function and myocardial dystrophy. However, a study of 913 workers exposed to carbon black demonstrated no pulmonary function effects on spirometry (Robertson et al., 1988). In one retrospective cohort study, no significant increase in cardiac abnormality and mortality due to cardiac disease was observed in workers employed in carbon black plants. Three studies of female rats exposed to carbon black found significant increase of malignant lung tumors, though there is inadequate and inconsistent evidence to assign carcinogenicity to humans (Brockmann et al., 1998).

A study was conducted on 3027 workers that were exposed to carbon black in the United States and Europe and evaluated using pulmonary function tests, chest X-ray, and questionnaires to examine respiratory health. Among the group only six cases showed benign pneumoconiosis among those who had nearly 10 years of carbon black exposure. In addition, this study observed decreased pulmonary function in the cases that had a history of smoking (Crosbie, 1986).

INDUSTRIAL HYGIENE

The current OSHA PEL for carbon black is $3.5\,\text{mg/m}^3$ of air. The NIOSH has established a REL of $3.5\,\text{mg/m}^3$. The ACGIH TLV–TWA for carbon black is $3.5\,\text{mg/m}^3$.

MANAGEMENT OF NUISANCE DUSTS

Exposure scenarios need to be identified to evaluate and keep exposures below acceptable levels. Other measures like wet operations and appropriate ventilation can be employed. Adequate housekeeping needs to be maintained to remove nuisance dust and reduce exposure levels. Irrigation of eyes and cleaning of skin with soap and water is advisable along with respiratory support to anyone who is exposed to excessive amounts. Contact with eyes and skin should also be prevented by using appropriate personal protective equipment. Ensure fresh air by removing workers from the exposure source as well as use first aid procedures if required. For management of pneumoconiosis, further follow-up may be needed.

If excessive absorption occurs from inhalational exposure, further management is necessary to rule out other causes of complaints. Laboratory investigation of blood chemistry, respiratory function tests, and chest X-ray may be used. Workers complaining of respiratory symptoms who were exposed to benign dust for long periods should be followed up to rule out any pathological conditions. Regular monitoring and follow-up of excessively exposed individuals may be necessary for further management.

REFERENCES

Billings, C.G. and Howard, P. (1993) Occupational siderosis and welders' lung: a review. *Monaldi Arch. Chest Dis.* 48(4):304–314.

Brockmann, M., et al. (1998) Exposure to carbon black: a cancer risk? *Int. Arch. Occup. Environ. Health* 71(2):85–99.

Chong, S., et al. (2006) Pneumoconiosis: comparison of imaging and pathologic findings. *Radiographics* 26(1):59–77.

Cierny, J. and Weisgerber, G. (2003) The Bronze Age tin mines in Central Asia. In: Giumlia-Mair, A. and Lo Schiavo, F., editors. *The Problem of Early Tin*, Oxford: Archaeopress, pp. 23–31, ISBN 1-84171-564-6.

Cooper, W.C. and Sargent, E.N. (1986) Study of chest radiographs and pulmonary ventilatory function in perlite workers. *J. Occup. Med.* 28(3):199–206.

Crosbie, W.A. (1986) The respiratory health of carbon black workers. *Arch. Environ. Health* 41(6):346–353.

Dare, P.R., et al. (1984) Barium in welding fume. *Ann. Occup. Hyg.* 28(4):445–448.

Doherty, M.J., et al. (2004) Total body iron overload in welder's siderosis. *Occup. Environ. Med.* 61(1):82–85.

Doig, A.T. and Mc Laughlin A.I.G. (1936) X ray appearances of the lungs of electric arc welders. *Lancet* 1:771–775.

Du, C.L., et al. (2010) Acute expanded perlite exposure with persistent reactive airway dysfunction syndrome. *Ind. Health* 48(1):119–122.

Harding, H.E., et al. (1958) Clinical, radiographic, and pathological studies of the lungs of electric-arc and oxyacetylene welders. *Lancet* 2(7043):394–399.

Karkhanis, V.S. and Joshi, J.M. (2013) Pneumoconioses. *Indian J. Chest Dis. Allied Sci.* 55(1):25–34.

Kim, K.I., et al. (2001) Imaging of occupational lung disease. *Radiographics* 21(6):1371–1391.

Kleinfeld, M., et al. (1969) Welders' siderosis. A clinical, roentgenographic, and physiological study. *Arch. Environ. Health* 19(1):70–73.

Lay, J.C., et al. (1998) Retention and intracellular distribution of instilled iron oxide particles in human alveolar macrophages. *Am. J. Respir. Cell Mol. Biol.* 18(5):687–695.

Lay, J.C., et al. (2001) Effects of inhaled iron oxide particles on alveolar epithelial permeability in normal subjects. *Inhal. Toxicol.* 13(12):1065–1078.

Oleru, U.G., et al. (1983) Pulmonary function and symptoms of Nigerian workers exposed to carbon black in dry cell battery and tire factories. *Environ. Res.* 30(1):161–168.

Robertson, J.M., et al. (1988) A cross-sectional study of pulmonary function in carbon black workers in the United States. *Am. Ind. Hyg. Assoc. J.* 49(4):161–166.

Sampatakakis, S., et al. (2013) Respiratory disease related mortality and morbidity on an island of Greece exposed to perlite and bentonite mining dust. *Int. J. Environ. Res. Public Health* 10(10):4982–4995.

Sluis-Cremer, G.K., et al. (1989) Stannosis. A report of 2 cases. *S. Afr. Med. J.* 75(3):124–126.

Stanescu, D.C., et al. (1967) Aspects of pulmonary mechanics in arc welders' siderosis. *Br. J. Ind. Med.* 24(2):143–147.

Yilmaz A., et al. (2009) Is tin fume exposure benign or not: two case reports. *Tuberk Toraks* 57:422–426.

92

FIBROGENIC DUSTS

Charles Barton

ASBESTOS

First aid: Remove the patient from exposure. If on the body, wash the contaminated body part with copious amounts of soap and water. If the eyes are exposed, immediately irrigate the affected eye thoroughly with water or 0.9% saline. Provide treatment for any symptoms that may arise.

Target organ(s): Respiratory system, eyes

Occupational exposure limits: OSHA PEL—0.1 fiber/cm^3 TWA; NIOSH REL—0.1 fiber/cm^3 for fibers >5 μm; ACGIH TLV—0.1 fiber/cm^3 TWA (respirable fibers)

Reference values: EPA MCL—7 million fibers per liter of water (MFL)

Hazard/precautionary phrases: H350 may cause cancer; H372 causes damage to organs through prolonged or repeated exposure

BACKGROUND AND USES

The term asbestos (CAS # 1332-21-4) refers to a group of six naturally occurring fibrous minerals that have been widely used in commercial products and are divided into two groups, serpentine and amphibole. Chrysotile (white asbestos; CAS #12001-29-5) is the only member of the serpentine group; amosite (brown asbestos; CAS #12172-73-5), crocidolite (blue asbestos; CAS #12001-28-4), anthophyllite (CAS #17068-78-9), actinolite (CAS #13768-00-8), and tremolite (CAS #14567-73-8) comprise the amphibole group. Amphiboles are thin, straight fibers; whereas, serpentine

(chrysotile) fibers are curly or coiled. All forms of asbestos are composed of silicate (SiO$_4$) groups and are essentially chemically inert.

It should be noted that serpentine and amphibole minerals also occur in nonfibrous or nonasbestiform forms. These nonfibrous minerals, which are not asbestos, are much more common and widespread than the asbestiform varieties. Nonasbestiform forms are considered to have less bioreactivity and cytotoxicity than asbestiform fibers. Moreover, they are generally considered to be nonpathogenic (Mossman, 2008).

Once highly prized for its flame-retardant properties and use in countless commercial products, asbestos has become a global concern with significant economic consequences (Sly et al., 2010). Until the 1970s, asbestos was widely used in the construction, shipbuilding, and automotive industries, among others. For example, asbestos was formerly used in the following items: boilers and heating vessels; cement pipes; clutch, brake, and transmission components; conduits for electrical wiring; corrosive chemical containers; electric motor components; heat-protective pads; laboratory furniture; paper products; pipe covering; roofing products; sealants and coatings; insulation products; and textiles (including curtains). These materials remain in many buildings, ships, and automobiles built before 1975 (Seidman and Selikoff, 1990).

Asbestos was widely used commercially until the 1970s, when health concerns led to some uses being banned and some voluntary phase outs (Seidman and Selikoff, 1990). Mining and milling of the raw material and production of asbestos has declined since the early 1970s, but asbestos is still used in some construction materials. Some asbestos-containing products, such as amphibole-contaminated vermiculite insulation, remain in many homes in the United

Hamilton & Hardy's Industrial Toxicology, Sixth Edition. Edited by Raymond D. Harbison, Marie M. Bourgeois, and Giffe T. Johnson.
© 2015 John Wiley & Sons, Inc. Published 2015 by John Wiley & Sons, Inc.

States. Asbestos fibers are released into the air and dust when asbestos-containing materials are loose, crumbling, or disturbed.

Today, most asbestos used in the United States is imported. Commercial-grade asbestos is made up of fiber bundles. These bundles, in turn, are composed of extremely long and thin fibers, often with splayed ends, which can easily be separated from one another. Asbestos is still used in brake pads, automobile clutches, roofing materials, vinyl tile, imported cement pipe, and corrugated sheeting (American Thoracic Society, 2004).

In the United States some past estimates of lung cancer attributable to asbestos exposure have been reported (Churg, 1983). The estimate of 5000—10,000 cases annually has not been verified. Similar estimates have come from other countries. In Norway, asbestos has been estimated to account for about two-thirds of work-related lung cancer (Langård, 1994). In the former East Germany, asbestos has been estimated from a 1988 analysis to be responsible for the majority of occupationally induced cancer (Sturm et al., 1994). Recently, the trend in worldwide asbestos use has shifted from established industrialized nations to developing nations (Delgermaa et al., 2011; Levy and Seplow, 1992). There is concern among occupational medicine practitioners that asbestos-related diseases are being "exported" to nations that may not have sufficient industrial hygiene experience or regulatory power to prevent them.

The majority of health effects associated with asbestos are of occupational origin. In addition to being at risk of exposure from the commercial uses of asbestos, people in some areas of the world are at risk because of geological deposits of asbestos near the surface that release asbestos if disturbed. For example, there have been reports of asbestosis and lung cancer in areas such as Turkey where there are large mineral deposits of tremolite asbestos (Metintas et al., 2012).

PHYSICAL AND CHEMICAL PROPERTIES

The asbestos minerals are not classified on a mineralogical basis, but rather on a commercial basis because of their unique properties. Therefore, the asbestos variety commercially known as crocidolite is referred to in the mineralogical literature as riebeckite. The asbestos variety called amosite is known mineralogically as grunerite. All other asbestos types are referred to by their proper mineral names. However, all asbestos fibers represent a form of silica fibers. Representative chemical formulas include chrysotile $Mg_3Si_2O_5(OH)_4$ (CAS 12001-29-5) and crocidolite $[NaFe_3^{2+}Fe_2^{3+}Si_8O_{22}(OH)_2]_n$ (CAS 12001-28-4).

Asbestos fibers are basically chemically inert, or nearly so. They do not evaporate, dissolve, burn, or undergo significant reactions with most chemicals. In acid and neutral aqueous media, magnesium is lost from the outer brucite layer of chrysotile. Amphibole fibers are more resistant to acid attack and all varieties of asbestos are resistant to attack by alkalis (Chissick, 1985; WHO, 1998).

ENVIRONMENTAL FATE AND BIOACCUMULATION

Once in the environment, fibers are mainly transported and distributed via air and water. Airborne mineral fibers are stable and may travel significant distances from the site of origin. Asbestos fibers will not volatilize or degrade, although they may be resuspended to the air by vehicular traffic over unpaved soil surfaces containing asbestos or through mining and milling operations. The importance of the transport of asbestos from the surface of water is presently undetermined. Asbestos will not volatilize or degrade in water. Asbestos released to the air will eventually settle out by gravitational settling and dry deposition.

Mineral fibers are relatively stable and tend to persist under typical environmental conditions. However, asbestos fibers may undergo chemical alteration as well as changes in dimension. The comparative solubility of selected mineral fibers has been studied and a general trend determined: chrysotile > amosite > actinolite > crocidolite > anthophyllite > tremolite (NAS, 1977). Because of their high adsorptive properties, it is thought that some mineral fibers may adsorb and carry various organic agents present in the environment.

As a naturally occurring substance, asbestos can be present in surface and ground water. Because asbestos fibers in water do not evaporate into air or break down in water, small fibers and fiber-containing particles may be carried long distances by water currents before settling to the bottom; larger fibers and particles tend to settle more quickly.

Asbestos does not tend to adsorb to solids normally found in natural water systems, but some materials (trace metals and organic compounds) have an affinity for asbestos minerals. The fibers are not able to move down through soil to ground water.

Asbestos is not affected by photolytic processes and is considered to be non-biodegradable by aquatic organisms. Asbestos fibers are not broken down to other compounds in the environment and, therefore, can remain in the environment for decades or longer.

There are no data regarding the bioaccumulation of asbestos in aquatic organisms.

MAMMALIAN TOXICOLOGY

Acute Effects

There are no clinically significant acute or intermediate effects associated with exposure to asbestos.

Significant exposure to any type of asbestos will increase the risk of lung cancer, mesothelioma and nonmalignant lung and pleural disorders, including asbestosis, pleural plaques, pleural thickening, and pleural effusions. Chronic laryngitis has also been associated with asbestos exposure (Kambic et al., 1989; Parnes, 1990). This effect is likely due to the irritant properties of asbestos and the mucociliary clearance of large fibers from the upper airways.

Chronic Effects

Asbestosis   Asbestosis, also called white lung, is a lung disease that is caused by the prolonged inhalation of asbestos fibers. A type of pneumoconiosis, it is found primarily among workers whose occupations involved asbestos. Asbestos fibers that have been inhaled remain in the lungs for years and eventually cause excessive scarring and fibrosis, resulting in a stiffening of the lungs that continues long after exposure ceases.

Asbestosis is defined as diffuse pulmonary fibrosis caused by the inhalation of excessive amounts of asbestos fibers (Roggli et al., 2010). Although the finding of asbestosis may be associated with pleural abnormalities, these alone do not constitute asbestosis. The only clinical symptom commonly associated with asbestosis is dyspnea (Selikoff et al., 1965). Clubbing of the fingers may also occur, but this is a relatively late finding in the progression of asbestosis (McGavin and Hughes, 1998).

The pulmonary effect of asbestosis is primarily the reduction of lung volumes detected by reduced inspiratory capacity and vital capacity. Large airway function, represented by the forced expiratory volume (FEV1)/forced vital capacity (FVC) ratio is relatively unaffected (Murphy, 1987).

Cancer   Chrysotile asbestos is the most common form of asbestos still produced and used. In 1992 chrysotile asbestos accounted for 99% of U.S. asbestos consumption (Pigg, 1994). Chrysotile was formerly thought to be a relatively safe form of asbestos compared to the amphibole forms (Craighead and Mossman, 1982). It was believed that because chrysotile is a curled fiber and more soluble, it would be deposited in the upper airways and effectively cleared before exerting a carcinogenic effect. The mesotheliomas associated with occupational exposure to chrysotile were attributed by some researchers to relatively low concentrations of tremolite fibers also present (Huncharek, 1994). Stayner et al. (1996) found that chrysotile may be a less potent inducer of mesotheliomas than amphiboles, but they could not determine a lower lung cancer risk. Also, the amount of time a fiber must stay in the lung to contribute to lung cancer or mesothelioma has been investigated (Barrett, 1994). Because workers are generally exposed to a mixture of fibers, it is prudent to treat chrysotile with the same level of concern as the amphibole forms of asbestos despite the fact that evidence has not conclusively demonstrated that chrysotile is an independent cause of cancer in humans.

The mechanism of asbestos carcinogenesis is far from clear and is likely to be complex, depending on fiber dimensions, surface properties, and physical durability (Hei et al., 2006). The induction of reactive oxygen and nitrogen species upon phagocytosis of asbestos fibers plays an important role in fiber genotoxicity. The beta ig-H3, a secreted protein induced by the transforming growth factor-beta and essential for cell adhesion, is down regulated in asbestos-induced tumorigenic human bronchial epithelial cells. Ectopic expression of the beta ig-H3 gene abrogates the tumorigenic phenotype and suggests that the gene plays a causal role in fiber carcinogenesis.

Lung Cancer   An association between asbesto-silicosis and lung cancer was suggested as early as the mid-1930s (Lynch and Smith, 1935). Doll (1955) examined the association between asbestos and lung cancer in British workers exposed before the first asbestos dust control efforts in 1931. He found asbestos workers in dusty environments had about 10 times the risk of developing lung cancer compared with the general population. A Canadian cohort study of asbestos cement plant workers also reported a significant excess of lung cancer (Finkelstein, 1983).

Hammond et al. (1979) studied the lung cancer incidence among 12,051 asbestos insulation workers with at least 20 years of work experience. Non-smoking asbestos workers had five times the risk of dying of lung cancer as non-smoking controls. Asbestos workers who smoked had a five times greater risk of dying of lung cancer than smoking controls. Analysis of the data suggested a multiplicative rather than additive relationship between smoking and asbestos exposure. Those exposed to both had a mortality ratio of 53 compared with non-smoking, non-asbestos control workers. Similar results were reported by Meurman et al. (1974) for anthophyllite miners and millers in Finland.

A Danish study assessed the standardized incidence ratio (SIR) of different types of lung cancer in asbestos cement workers (Raffn et al., 1996). They reported an SIR of 2.6 for adenocarcinoma, 1.7 for squamous cell carcinoma, and 1.5 for anaplastic carcinoma. Also, researchers noted that the risk for different histological types of lung cancer was similar during the first 25 years after the first exposure. After 25 years, the risk of adenocarcinoma increases.

The question of whether lung cancer can be attributed to asbestos exposure in the absence of asbestosis remains controversial (Hessel et al., 2005). This area continues to generate differences of opinion. The scientific question of whether or not asbestos-related lung cancer in man arises only in the presence of pulmonary fibrosis may be unanswerable epidemiologically. Microscopic evidence of fibrosis is a great deal more sensitive in detecting asbestosis than   chest radiography (CR) or even high resolution computed tomography (HRCT). Even HRCT scans may fail to detect fibrosis evident on microscopic examination, and fibrosis may have causes other than asbestos.

Mesothelioma Mesothelioma is a relatively rare cancer of the thin membranes that line the chest and abdomen. Although rare, mesothelioma is the most common form of cancer associated with asbestos exposure. Its incidence is anticipated to increase over the next decade in both Europe and the developing nations (Remon et al., 2012). Malignant mesotheliomas were first reported by Wagner et al. (1960) in crocidolite miners of North Western Cape Province, South Africa. Epidemiological and pathological studies have repeatedly shown a strong association between mesothelioma and high levels of asbestos exposure. Because mesothelioma is a rare form of cancer, this diagnosis typically leads the physician to search for possible sources of asbestos exposure even though there does exist a background rate of non-asbestos induced mesothelioma.

This disease is characterized clinically by chest pain and pleural effusion, and CR reveals irregular pleural thickening or pleural densities. Severe pleuritic pain, breathlessness, cough, weight loss, and fatigue follow. Peritoneal mesotheliomas produce dull pain followed by swelling and weight loss. Progression is rapid, with average survival time from onset of symptoms of about 6 months for pleural tumors and 13–14 months for peritoneal tumors (Becklake, 1976). Mesothelioma may spread to all serosal tissues, including pleura, pericardium, diaphragm, peritoneum, and tunica vaginalis testis (Watanabe et al., 1994). Tumors typically have an epithelial form, but may also have sarcomatous and microcystic patterns. Immunostaining for procollagen type 1 may distinguish mesothelioma from adenocarcinoma. Hyponatremia is common among mesothelioma victims, and its treatment may improve the comfort of these patients (Perks et al., 1979).

Because of its latent period and the occupational nature of some mesotheliomas, it is typically a disease of middle age. However, there is one report of pathologically confirmed asbestos-related mesothelioma in a 17-year-old boy (Andrion et al., 1994). The asbestos exposure was presumably due to cosmetic talc contaminated with asbestos.

Mechanism of Action(s)

The presence of asbestos fibers in the lungs sets off a variety of responses leading to inflammation and cell/tissue damage, which can lead to disease. The mechanisms by which asbestos causes disease are not fully understood. Currently, there are three hypotheses to account for asbestos's pathogenicity: (1) direct interaction with cellular macromolecules, (2) generation of reactive oxygen species, and (3) other cell-mediated mechanisms (especially inflammation). Asbestos is genotoxic and carcinogenic.

Because of their surface charge, asbestos fibers can bind to cellular macromolecules, such as proteins. This binding is believed to induce changes in macromolecular conformation, thereby affecting function. Also, long asbestos fibers can interfere physically with the mitotic spindle and cause chromosomal damage, especially deletions (NAS, 2006).

The physical–chemical attributes of mineral fibers are important in determining the type of toxicity observed. The three main determinants of asbestos toxicity are fiber size, durability, and iron content. Fibrosis results from a sequence of events following lung injury, which includes inflammatory cell migration, edema, cellular proliferation, and accumulation of collagen. Fibers do migrate to the pleural space, and it has been hypothesized that a similar cascade of inflammatory events may contribute to fibrotic lesions in the visceral pleura. Thickening of the visceral pleura is more often localized to lobes of the lung with pronounced parenchymal changes, and it has also been hypothesized that the inflammatory and fibrogenic processes within the lung parenchyma in response to asbestos fibers may influence the fibrogenic process in the visceral pleura.

Asbestos fibers are elaborators of active oxygen species, whether by reactions involving iron on the surface of the fiber, or by attempted phagocytosis of fibers by cell types resident in the lung. The link between production of active oxygen species and the pathogenesis of asbestos-mediated disease has been highlighted by the use of antioxidant scavengers that inhibit the cytotoxic effects of asbestos *in vitro* and *in vivo*. The use of antioxidant enzymes ameliorates the induction of certain genes necessary for cell proliferation, such as ornithine decarboxylase, implicating oxidants as causative factors in some abnormal cell replicative events (Quinlan et al., 1994).

Cytogenetic and molecular studies of asbestos-related cancers indicate that inactivation or loss of multiple tumor suppressor genes occurs during lung cancer development. Aneuploidy and other chromosomal changes induced by asbestos fibers may also be involved in genetic alterations. Furthermore, asbestos fibers may influence the carcinogenic process by inducing cell proliferation, free radicals, or other promotional mechanisms. Therefore, asbestos fibers may act at multiple stages of the carcinogenic process by both genetic and epigenetic mechanisms. Their biopersistence also remains important in fiber carcinogenicity (Barrett, 1994).

Pathological changes resulting from the inhalation of asbestos fibers and dust vary with the quantity and kind of material. Benign pleural plaques first described in Finland by Meurman (1966) in parietal pleura are located at the sites of greatest respiratory movement. These plaques consist of collagenous connective tissue with secondary calcification. Such plaques may occur in the absence of lung fibrosis even though asbestos bodies are found in the lung.

Smoking combined with asbestos exposure has a synergistic effect on the incidence of lung cancer and asbestosis

(Hammond et al., 1979). The strength of this effect varies depending on the type, intensity, and duration of the exposure. It is generally thought to be a multiplicative effect, more than additive (Vainio and Boffetta, 1994).

SOURCES OF EXPOSURE

Occupational

In the United States, an estimated 27 million workers were exposed to aerosolized asbestos fibers between 1940 and 1979 (Nicholson et al. 1982). Secondary exposure occurred when people who did not work directly with asbestos were nevertheless exposed to fibers as a result of sharing workspace where others handled asbestos. For example, electricians who worked in shipyards were exposed because asbestos was being used to coat the ships' pipes and hulls (Pan et al., 2005).

Currently, the people most heavily exposed to asbestos in the United States are those in construction trades. This population includes an estimated 1.3 million construction workers as well as workers in building and equipment maintenance (American Thoracic Society, 2004). Because most asbestos was used in construction, and two-thirds of asbestos produced is still used in this trade, risk to these workers can be considerable if the hazard is not recognized and the Occupational Safety and Health Administration (OSHA) standards are not enforced.

The risk of exposure to asbestos is highest for people who mine and export asbestos or those who work with it in manufacturing. The risk of developing cancer is potentially highest in these groups. Today, strict government regulations and improved work practices have decreased the risk of asbestos exposure for workers.

During renovations and repairs to older buildings, construction workers, other trade workers and building maintenance workers may be exposed to high concentrations of asbestos fibers. Most tradespeople working in construction, maintenance, or the renovation of older buildings are trained in the proper handling of asbestos-containing materials and should be particularly careful when handling these materials.

Environmental

Asbestos is ubiquitous in the environment, albeit in small quantities. Small levels of asbestos fibers occur in the soil, air and water, from natural and artificial sources. Asbestos can be released into the environment from the natural erosion of asbestos-bearing rock or when asbestos-containing products are damaged or worn down.

In the past, because of a lack of proper industrial hygiene, asbestos workers went home covered in asbestos dust. This brought about a tendency for workers' families and other household contacts to be exposed via inhalation of asbestos dust from workers' skin, hair, and clothing, and during laundering of contaminated work clothes. A mortality study of 878 household contacts of asbestos workers revealed that 4 out of 115 total deaths were from pleural mesothelioma and that the rate of deaths from all types of cancer was doubled (Joubert et al., 1991). In addition, asbestos was released into the air and soil around facilities such as refineries, power plants, factories handling asbestos, shipyards, steel mills, vermiculite mines, and building demolitions. People living around these facilities were also exposed to asbestos.

Today, the risk of environmental exposure to asbestos is considered to be very low.

Asbestos was used widely in building materials such as insulation and floor and ceiling tiles. However, asbestos used in these products is very dense and does not release significant amounts of fibers unless the products are cut, damaged, or disturbed.

There is some concern about vermiculite insulation, which may contain small amounts of amphibole asbestos. However, there is currently no evidence of a health risk if the insulation is sealed behind wallboards and floorboards, isolated in an attic or kept from exposure and from being disturbed.

Some home attic insulation and many other home and building materials produced before 1975 contain asbestos. People who live in homes with these materials are at risk of exposure if the materials are loose, crumbling, or disturbed by household activities or renovations. On the other hand, asbestos contained in intact solid material poses a negligible risk of exposure.

There are many ways that people can also be exposed to asbestos through hobbies and recreational activities that entail contact with materials containing asbestos; some examples are activities such as home renovation, auto repair, and urban spelunking (exploring abandoned buildings). In places where naturally occurring asbestos is close to the earth's surface, activities such as gardening and dirt biking can cause exposures if asbestos-bearing rock is disturbed.

INDUSTRIAL HYGIENE

Historically, a number of industries have been sources of occupational exposure to asbestos, including textiles, shipyards, construction, oil refineries, railroads, insulation installation, and pipefitting.

Determinations of employee exposure should be made from breathing zone air samples that are representative of the 8-h time-weighted average (TWA) and 30 min short-term exposures of each employee. Exposure monitoring samples should be analyzed by phase contrast microscopy (PCM) for OSHA purposes. Phase contrast microscopy methods accurately assess fiber exposure levels, but PCM cannot differentiate between asbestos and non-asbestos fibers. The transmission electron microscopy (TEM) methods may be used to identify asbestos fibers specifically. Exposures should be evaluated with standard total dust sampling techniques for comparison to the OSHA permissible exposure limits (PEL). Controlling the exposure to asbestos can be done through engineering controls, administrative actions, and personal protective equipment (PPE). Engineering controls include such things as isolating the source and using ventilation systems. Administrative actions include limiting the workers exposure time and providing showers. PPE include wearing the proper respiratory protection and clothing.

The National Institute for Occupational Safety and Health (NIOSH) recommends that exposure to all asbestos fibers $<5\,\mu m$ to be reduced to the lowest feasible level. The OSHA and NIOSH have set the PEL and the recommended exposure limit (REL) for all airborne asbestos fibers $>5\,\mu m$ at 0.1 fiber/cm^3. The OSHA has set an additional 30 min ceiling at 1 fiber/cm^3. The American Conference of Governmental Industrial Hygienists (ACGIH) has set the threshold limit value (TLV) at 0.1 fiber/cm^3.

The National Toxicology Program (NTP) has classified asbestos as a known human carcinogen. The International Agency for Research on Cancer (IARC) has classified asbestos as a Group 1 carcinogen (carcinogenic to humans).

RISK ASSESSMENTS

The California Environmental Protection Agency (CalEPA) has established an oral slope factor of 0.000190 mg/kg per day. The U.S. Environmental Protection Agency (USEPA) has established an inhalation unit risk of 0.23 fibers/ml of air. An inhalation cancer risk of 1 in 10,000 is associated with a concentration of 0.0004 fibers/ml of air.

In accordance with the United Nations Globally Harmonized System of Classification and Labelling of Chemicals (GHS), asbestos is classified as a Category 1A Carcinogen (known to have carcinogenic potential for humans, based largely on human evidence) and specific target organ toxicity–repeated or prolonged exposure (STOT–RE) Category 1. Category 1 is assigned for STOT-RE because it has produced significant toxicity in humans, or that, on the basis of evidence from studies in experimental animals can be presumed to have the potential to produce significant toxicity in humans following repeated or prolonged exposure.

Studies concerning natural asbestos in the environment show that the exposure that begins at birth does not seem to affect the duration of the latency period, but the studies do not show whether early exposure increases susceptibility; they do not suggest that susceptibility differs according to sex. Solid evidence shows an increased risk of mesothelioma among people whose exposure comes from a paraoccupational or domestic source. The risk of mesothelioma associated with exposure as result of living near an industrial asbestos source (mines, mills, asbestos processing plants) is clearly confirmed. No solid epidemiological data currently justify any judgment about the health effects associated with passive exposure in buildings containing asbestos. Most of the studies on nonoccupational sources reported mainly amphibole exposure, and it is unclear if environmental exposure to chrysotile may also cause cancer.

REFERENCES

American Thoracic Society (2004) Diagnosis and initial management of nonmalignant diseases related to asbestos. *Am. J. Respir. Crit. Care Med.* 170:691–715.

Andrion, A., Bosia, S., Paoletti, L., Feyles, E., Lanfranco, C., Bellis, D., and Mollo, F. (1994) Malignant peritoneal mesothelioma in a 17-year-old boy with evidence of previous exposure to chrysotile and tremolite asbestos. *Hum. Pathol.* 25:617–622.

Barrett, J.C. (1994) Cellular and molecular mechanisms of asbestos carcinogenicity: implications for biopersistence. *Environ. Health Perspect.* 102(Suppl. 5):19–23.

Becklake, M.R. (1976) Asbestos-related diseases of the lung and other organs: their epidemiology and implications for clinical practice. *Am. Rev. Respir. Dis.* 114:187–227.

Chissick, S.S. (1985) Asbestos. In: Gerhartz, W., Yamamoto, Y.S., Campbell, F.T., et al., editors. *Ullmann's Encyclopedia of Industrial Chemistry*, Weinheim: VCH, pp. 151–167.

Churg, A. (1983) Current issues in the pathological and mineralogic diagnosis of asbestos-induced disease. *Chest* 84:275–280.

Craighead, J.E. and Mossman, B.T. (1982) The pathogenesis of asbestos-associated diseases. *N. Engl. J. Med.* 306:1446–1455.

Delgermaa, V., Takahashi, K., Park, E.K., Le, G.V., Hara, T., and Sorahan, T. (2011) Global mesothelioma deaths reported to the World Health Organization between 1994 and 2008. *Bull. World Health Organ.* 89:716–724.

Doll, R. (1955) Mortality from lung cancer in asbestos workers. *Br. J. Ind. Med.* 12:81–86.

Finkelstein, M.M. (1983) Mortality among long-term employees of an Ontario asbestos-cement factory. *Br. J. Ind. Med.* 40:138–144.

Hammond, E.C., Selikoff, I.J., and Seidman, H. (1979) Asbestos exposure, cigarette smoking and death rates. *Ann. N. Y. Acad. Sci.* 330:473–490.

Hei, T.K., Xu, A., Huang, S.X., and Zhao, Y. (2006) Mechanism of fiber carcinogenesis: from reactive radical species to silencing of the beta igH3 gene. *Inhal. Toxicol.* 18:985–990.

Hessel, P.A., Gamble, J.F., and McDonald, J.C. (2005) Asbestos, asbestosis, and lung cancer: a critical assessment of the epidemiological evidence. *Thorax* 60:433–436.

Huncharek, M. (1994) Types of asbestos fibers and disease-causing potential. *Am. Fam. Physician* 50:306–308.

Joubert, L., Seidman, H., and Selikoff, I.J. (1991) Mortality experience of family contacts of asbestos factory workers. *Ann. N. Y. Acad. Sci.* 643:416–418.

Kambic, V., Radsel, Z., and Gale, N. (1989) Alterations in the laryngeal mucosa after exposure to asbestos. *Br. J. Ind. Med.* 46:717–723.

Langård, S. (1994) Prevention of lung cancer through the use of knowledge on asbestos and other work-related causes—Norwegian experiences. *Scand. J. Work Environ. Health* 20:100–107.

Levy, B.S. and Seplow, A. (1992) Asbestos-related hazards in developing countries. *Environ. Res.* 59:167–174.

Lynch, K.M. and Smith, W.A. (1935) Pulmonary asbestosis III: carcinoma of lung in asbesto-silicosis. *Am. J. Cancer* 24:50.

McGavin, C. and Hughes, P. (1998) Finger clubbing in malignant mesothelioma and benign asbestos pleural disease. *Respir. Med.* 92:691–692.

Metintas, S., Metintas, M., Ak, G., and Kalyoncu, C. (2012) Environmental asbestos exposure in rural Turkey and risk of lung cancer. *Int. J. Environ. Health Res.* 22:468–479.

Meurman, L. (1966) Asbestos bodies and pleural plaques in a Finnish series of autopsy cases. *Acta Pathol. Microbiol. Scand.* 181:1.

Meurman, L.O., Kiviluoto, R., and Hakama, M. (1974) Mortality and morbidity among the working population of anthophyllite asbestos miners in Finland. *Br. J. Ind. Med.* 31:105–112.

Mossman, B.T. (2008) Assessment of the pathogenic potential of asbestiform vs. nonasbestiform particulates (cleavage fragments) in *in vitro* (cell or organ culture) models and bioassays. *Regul. Toxicol. Pharmacol.* 52:S200–S203.

Murphy, R.L., Jr. (1987) The diagnosis of nonmalignant diseases related to asbestos. *Am. Rev. Respir. Dis.* 136:1516–1517.

NAS (1977) *Drinking Water and Health*, Washington, DC: National Academy of Sciences, pp. 144–168.

NAS (2006) *Asbestos: Selected Cancers*, Washington, DC: National Academy of Sciences.

Nicholson, W.J., Perkel, G., and Selikoff, I.J. (1982) Occupational exposure to asbestos: populations at risk and projected mortality—1980–2030. *Am. J. Ind. Med.* 3:259–311.

Pan, X.L., Day, H.W., Wang, W., Beckett, L.A., and Schenker, M.B. (2005) Residential proximity to naturally occurring asbestos and mesothelioma risk in California. *Am. J. Respir. Crit. Care Med.* 172:1019–1025.

Parnes, S.M. (1990) Asbestos and cancer of the larynx: is there a relationship? *Laryngoscope* 100:254–261.

Perks, W.H., Stanhope, R., and Green, M. (1979) Hyponatraemia and mesothelioma. *Br. J. Dis. Chest* 73:89–91.

Pigg, B.J. (1994) The uses of chrysotile. *Ann. Occup. Hyg.* 38:453–458, 408.

Quinlan, T.R., Marsh, J.P., Janssen, Y.M., Borm, P.A., and Mossman, B.T. (1994) Oxygen radicals and asbestos-mediated disease. *Environ. Health Perspect.* 102(Suppl. 10):107–110.

Raffn, E., Villadsen, E., Engholm, G., and Lynge, E. (1996) Lung cancer in asbestos cement workers in Denmark. *Occup. Environ. Med.* 53:399–402.

Remon, J., Lianes, P., Martínez, S., Velasco, M., Querol, R., and Zanui, M. (2012) Malignant mesothelioma: new insights into a rare disease. *Cancer Treat. Rev.* Available at http://dx.doi.org/10.1016/j.ctrv.2012.12.005 [Epub ahead of print].

Roggli, V.L., Gibbs, A.R., Attanoos, R., Churg, A., Popper, H., Cagle, P., Corrin, B., Franks, T.J., Galateau-Salle, F., Galvin, J., Hasleton, P.S., Henderson, D.W., and Honma, K. (2010) Pathology of asbestosis—an update of the diagnostic criteria: Report of the asbestosis committee of the college of American pathologists and pulmonary pathology society. *Arch. Pathol. Lab. Med.* 134:462–480.

Seidman, H. and Selikoff, I.J. (1990) Decline in death rates among asbestos insulation workers 1967–1986 associated with diminution of work exposure to asbestos. *Ann. N. Y. Acad. Sci.* 609:300–318.

Selikoff, I.J., Churg, J., and Hammond, E.C. (1965) The occurrence of asbestosis among insulation workers in the United States. *Ann. N. Y. Acad. Sci.* 132:139–155.

Sly, P.D., Chase, R., Kolbe, J., Thompson, P., Gupta, L., Daube, M., Olver, I., and Vallance, D. (2010) Asbestos still poses a threat to global health: now is the time for action. *Med. J. Aust.* 193:198–199.

Stayner, L.T., Dankovic, D.A., and Lemen, R.A. (1996) Occupational exposure to chrysotile asbestos and cancer risk: a review of the amphibole hypothesis. *Am. J. Public Health* 86:179–186.

Sturm, W., Menze, B., Krause, J., and Thriene, B. (1994) Use of asbestos, health risks and induced occupational diseases in the former East Germany. *Toxicol. Lett.* 72:317–324.

Vainio, H. and Boffetta, P. (1994) Mechanisms of the combined effect of asbestos and smoking in the etiology of lung cancer. *Scand. J. Work Environ. Health* 20:235–242.

Wagner, J.C., Sleggs, C.A., and Marchand, P. (1960) Diffuse pleural mesothelioma and asbestos exposure in the North Western Cape Province. *Br. J. Ind. Med.* 17:260–271.

Watanabe, M., Kimura, N., Kato, M., Iwami, D., Takahashi, M., and Nagura, H. (1994) An autopsy case of malignant mesothelioma associated with asbestosis. *Pathol. Int.* 44:785–792.

WHO (1998) *Chrysotile Asbestos: Environmental Health Criteria*, Geneva, Switzerland: World Health Organization.

SILICA

First aid: No specific first aid is necessary since the adverse health effects associated with exposure to crystalline silica (quartz) result from chronic exposures. If there is a gross inhalation of crystalline silica (quartz), remove the person immediately to fresh air, give artificial respiration as needed, seek medical attention as needed. If on the body, wash the contaminated body part with copious amounts of soap and water. If the eyes are exposed, immediately irrigate the affected eye thoroughly with water or 0.9% saline. Provide treatment for any symptoms that may arise.

Target organ(s): Respiratory system, eyes

Occupational exposure limits: OSHA PEL—$10/(\% \; SiO_2^{+2})$ (respirable) or $30/(\% \; SiO_2^{+2})$ (total) TWA; NIOSH

REL—0.05 mg/m^3 TWA; ACGIH TLV—0.025 mg/m^3 TWA (respirable)

Hazard/precautionary phrases: H315 causes skin irritation; H319 causes serious eye irritation; H335 may cause respiratory irritation; H372 causes damage to organs through prolonged or repeated exposure

BACKGROUND AND USES

Pulmonary disease attributable to silica (silicon dioxide, quartz, SiO_2) exposure has been recorded since the mid sixteenth century. Since that time, reports of mineral dust diseases of the chest have come from all areas where mining is operational. Silica is abundant in the earth, has fibrogenic properties, and is unique in acting synergistically with acid-fast bacillary infections. Silica should not be confused with silicates, which are based on SiO_4 units. Industrially important silica-containing materials are quartzite, sandstone, flint, tripoli, and diatomaceous earth. Cristobalite and tridymite silica are considered more potent inducers of silicosis than quartz.

At least 1.7 million U.S. workers are exposed to respirable crystalline silica in a variety of industries and occupations, including construction, sandblasting, and mining. Silicosis, an irreversible but preventable disease, is the illness most closely associated with occupational exposure to the material, which is also known as silica dust. Occupational exposures to respirable crystalline silica are associated with the development of silicosis, lung cancer, pulmonary tuberculosis, and airways diseases. These exposures may also be related to the development of autoimmune disorders, chronic renal disease, and other adverse health effects.

Crystalline silica is commonly found and used in industries such as electronics; foundry; ceramics, clay, pottery, stone and glass; construction; agriculture; maritime; railroad (setting and laying track); slate and flint quarrying and flint crushing; use and manufacture of abrasives; manufacture of soaps and detergents; and mining.

Perhaps the most familiar use of quartz sand is as an abrasive blasting agent to remove surface coatings prior to repainting or treating. The NIOSH estimates that there are more than one million American workers who are at risk of developing silicosis. Of these workers, 100,000 are employed as sandblasters (NIOSH, 1992).

PHYSICAL AND CHEMICAL PROPERTIES

Silica is a group IV metal oxide, which has good abrasion resistance, electrical insulation, and high thermal stability. It is practically insoluble in water or acids. However, it will dissolve in hydrogen fluoride (HF), forming silicon tetrafluoride.

Crystalline silica has a molecular weight of 60.09. The transparent crystals are odorless and tasteless. It has a melting point of 1710 °C, and is not corrosive.

ENVIRONMENTAL FATE AND BIOACCUMULATION

Silica-bearing deposits are found to some degree in every land mass and stratum from every era and period of geological time. Silica deposits are almost uniformly quartz or derived from quartz, formed by metamorphism, sedimentation or through igneous activity. The majority of deposits mined for silica sand consist of free quartz, quartzites, and quartzose sedimentary deposits, such as sandstone.

Silicon is the second most abundant chemical element, after oxygen, in the earth's crust accounting for 28.15% of its mass. Silicate minerals (such as plagioclase, alkali feldspars, pyroxenes, amphiboles, micas and clays, excluding silica) comprise together 80% by volume of the earth's crust, while quartz, by far the most common form of silica in nature, comprises 12% by volume of the crust.

Being refractory to weathering, quartz can be carried for long distances by eolian transport (pertains to the activity of the winds). For example, one study noted that Asian dust can be transported within five days from the Gobi desert in China to the mountain ranges between British Columbia and California, a distance >8000 km (Husar et al., 1998). The mean diameter of this material was between 2 and 3 µm and was dominated by the single minerals quartz, K-feldspa,r and plagioclase. An important dust fall on northern Scandinavia originated from western North Africa, and hence, particles were transported for at least 7000 km before their final destination. Mineralogically, these particles were dominated by small round quartz grains of a mean diameter of 2.7 µm (Franzén et al., 1994). Finally, microscopic analyses of particulate matter collected during six long-range transport events led researchers to conclude that large quartz grains (>62.5 µm) can bc transported over great distances, up to 10,000 km, to oceanic and terrestrial locations (Middleton et al., 2001).

Silica does not bioaccumulate.

ECOTOXICOLOGY

No data were found with regards to the toxicity of quartz and cristobalite to aquatic organisms. The effects of respirable crystalline silica on terrestrial mammals and birds have been noted. Development of silicosis in terrestrial organisms is mainly dependent on the quantity of particulate exposure and size of the silicon dioxide crystals.

Acute Effects

Exposure to the toxic mineral dust silica produces an acute inflammatory response in the lungs of both humans and laboratory animals. The aryl hydrocarbon receptor (AhR) may regulate silica-induced inflammation, but not fibrosis (Beamer et al., 2012).

Intense crystalline silica exposure has resulted in outbreaks of acute silicosis referred to medically as silicoproteinosis or alveolar lipoproteinosis-like silicosis. Initially, crystalline silica particles produce an alveolitis that is characterized by sustained increases in the total number of alveolar cells, including macrophages, lymphocytes, and neutrophils. The alveolitis has been found to progress to the characteristic nodular fibrosis of simple silicosis. A rapid increase in the rate of synthesis and deposition of lung collagen has also been seen with the inhalation of crystalline silica particles. The collagen formed is unique to silica-induced lung disease and biochemically different from normal lung collagen (Olishifski, 1988).

Chronic Effects

Inhalation of crystalline silica-containing dusts has been associated with silicosis, chronic obstructive pulmonary disease, bronchitis, collagen vascular diseases, chronic granulomatous infections such as tuberculosis, and lung cancer. In general, aerosols of particulates can be deposited in the lungs. This can produce rapid or slow local tissue damage, eventual disease or physical plugging. Dust containing crystalline silica can cause formation of fibrosis (scar tissue) in the lungs (Markowitz and Rosner, 1995).

Animal studies have confirmed human experience that a lung dusted with silica is extremely vulnerable to infection by tubercle bacilli (Gardner and Dworski, 1922). In areas where tuberculosis is endemic, death rates among workers concurrently exposed to silica are much higher. The risk of a patient with silicosis developing tuberculosis is higher (2.8–39-fold higher, depending on the severity of silicosis) than that found in non-exposed controls (Milovanović et al., 2011).

Scleroderma (progressive systemic sclerosis) and arthritis have been reported to be associated with silica exposure. Steenland and Goldsmith (1995) suggested a link between silica exposure and autoimmune effects.

Silicosis Silicosis is a diffuse interstitial fibronodular lung disease, caused by the inhalation of crystalline silica in the forms of quartz (the most frequently found), cristobalite, trydimite, stishovite, and coesite (Santos et al., 2010). Silicosis is believed to be the oldest form of pneumoconiosis. The term was first used by Visconti in 1870, but the disease had already been seen in ancient Egyptian and Greek mummies. Hippocrates also described observing breathing difficulties in metal miners and the condition was also seen in

the sixteenth century in miners in Bohemia and in eighteenth century stonecutters.

Silicosis is the leading cause of disability among occupational respiratory diseases, classified as the sixth most frequently occurring occupational respiratory disease by the European Occupational Disease Statistics (EODS). This pathology is rarely seen before the age of 50 years old and is more commonly found in males, due to the nature of the profession exercised.

Gradually increasing dyspnea and nonproductive cough, often thought by the patient to be due to smoking habits or to age, are the usual initial complaints. Well-studied groups of workers exposed to silica have established the fact that some of them, who have no symptoms, will show densities in the lungs on routine chest X-ray.

In the usual case of silicosis and in the absence of infection, there is a slow deterioration of capacity for physical effort. If respiratory infection occurs, dyspnea and cough are often increased and become established at more severe levels after the infection has subsided. Vague chest tightness and clinically obvious pleuritis occur in some cases. Hemoptysis is unusual unless acid-fast infection complicates the silicosis. Asthmatic bronchitis characterized by wheezing and difficulty in expiring air often occurs.

Chronic bronchitis, as a result of the direct effect of finely divided silica dust, is a consequence that is hard to establish, but emphysema is a common complication of silicosis. Smoking habits, bacterial infection, and aging are all factors that can affect normal upper respiratory tract defenses or by themselves may act as complicating etiological factors. In addition, many jobs with silica exposure may involve the inhalation of irritant chemicals, such as oxides of nitrogen from blasting fumes or welding. Underground mining in hot, damp atmospheres and operations near furnaces create other modifying or aggravating influences acting on respiratory function.

Silicosis results in an increased risk of cor pulmonale (Murray et al., 1993).

Carcinogenicity Exposure of workers to respirable crystalline silica is associated with elevated rates of lung cancer (NTP, 2011). The link between human lung cancer and exposure to respirable crystalline silica has been shown to be strongest in studies of quarry and granite workers and workers involved in ceramic, pottery, refractory brick, and diatomaceous earth industries. Human cancer risks are associated with exposure to respirable quartz and cristobalite but not to amorphous silica. The overall relative risk is approximately 1.3–1.5, with higher risks found in groups with greater exposure or longer time since first exposure. Silicosis, a marker for exposure to silica dust, is associated with elevated lung cancer rates, with relative risks of 2.0–4.0. Elevated risks have been seen in studies that accounted for smoking or asbestos exposure, and confounding by co-exposure is unlikely to explain these results (IARC, 1997).

In rats, exposure to various forms of respirable crystalline silica by inhalation or intratracheal instillation consistently caused lung cancer (adenocarcinoma or squamous-cell carcinoma). Single intrapleural or intraperitoneal injections of various forms of respirable crystalline silica also caused lymphoma in rats (IARC, 1997). Silica may increase lung cancer rates through a synergistic or promoting effect rather than direct carcinogenicity (Agius, 1992).

The IARC classifies silica as a Type 1 known human carcinogen. The NTP also lists respirable crystalline silica as a known human carcinogen.

Renal Toxicity Nephrotoxicity is associated with silica exposure (Ghahramani, 2010). Exposure to silica has been associated with tubulointerstitial disease, immune-mediated multisystem disease, chronic kidney disease, and end-stage renal disease. A rare syndrome of painful, nodular skin lesions has been described in dialysis patients with excessive levels of silicon. Balkan endemic nephropathy is postulated to be due to chronic intoxication with drinking water polluted by silicates released during soil erosion. The mechanism of silica nephrotoxicity is thought to be through direct physical action, as well as silica-induced autoimmune diseases such as scleroderma and systemic lupus erythematosus.

Chemical Pathology

Inhalation of respirable silica dusts leads to deposition in distal airways. Silica can produce reactive oxygen species either directly on freshly cleaved particle surfaces or indirectly through its effect on the phagocytic cells. Scavenger receptors, especially the macrophage receptor with collagenous structure expressed in alveolar macrophages, seem to have a role in the recognition and uptake of silica (Pollard and Kono, 2013).

Phagocytosis of crystalline silica in the lung causes lysosomal damage, activating the NALP3 inflammasome and triggering the inflammatory cascade with subsequent fibrosis. Impairment of lung function increases with disease progression, even after the patient is no longer exposed.

Pathological varieties of silicosis include simple (nodular) silicosis, progressive massive fibrosis, silicoproteinosis, and diffuse interstitial fibrosis. Gross pathological examination of the lung identifies discrete hard nodules, usually with upper-lobe predominance (Leung et al., 2012). Hilar and peribronchial lymph nodes are frequently enlarged.

Mechanism of Action(s)

Respirable crystalline silica deposited in the lungs causes epithelial injury and macrophage activation, leading to inflammatory responses and proliferation of the epithelial and interstitial cells (NTP, 2011). In humans, respirable crystalline silica persists in the lungs, culminating in the development of chronic silicosis, emphysema, obstructive airway disease, and lymph-node fibrosis. Respirable crystalline silica stimulates (1) release of cytokines and growth factors from macrophages and epithelial cells, (2) release of reactive oxygen and nitrogen intermediates, and (3) oxidative stress in the lungs. All of these pathways contribute to lung disease. Marked and persistent inflammation, specifically inflammatory-cell-derived oxidants, may provide a mechanism by which respirable crystalline silica exposure can result in genetic damage in the lung parenchyma.

SOURCES OF EXPOSURE

Occupational

Respirable quartz levels exceeding $0.1\,mg/m^3$ are most frequently found in metal, nonmetal, and coal mines and mills, granite quarrying and processing, crushed-stone and related industries, foundries, the ceramics industry, construction, and sandblasting operations (IARC, 1997). Potential exposure was highest for sculptors and carvers, stencil cutters, polishers, and sandblasters; for these occupations, the silica content of respirable dust ranged from 4.8% to 12.2%. Concentrations of respirable crystalline silica ranged from 0.01 to $0.20\,mg/m^3$ in clay-pipe factories and from 0 to $0.18\,mg/m^3$ in a plant producing ceramic electronic equipment parts.

Environmental

Crystalline silica is an abundant and commonly found natural material. Human exposure to respirable crystalline silica, primarily quartz dust, occurs mainly in industrial and occupational settings. Nonoccupational exposure to respirable crystalline silica results from natural processes and anthropogenic sources; silica is a common air contaminant. Residents near quarries and sand and gravel operations potentially are exposed to respirable crystalline silica. A major source of cristobalite and tridymite in the United States is volcanic rock. Local conditions, especially in deserts and areas around recent volcanic eruptions and mine dumps, can give rise to silica-containing dust.

Consumers may be exposed to respirable crystalline silica from abrasives, sand paper, detergent, grouts, and concrete (IARC, 1997). Crystalline silica may also be an unintentional contaminant; for example, diatomaceous earth, used as a filler in reconstituted tobacco sheets, may be converted to cristobalite as it passes through the burning tip of tobacco products (IARC, 1987).

INDUSTRIAL HYGIENE

Industrial hygiene assessments should be conducted to include silica dust monitoring, hazard analysis and providing the results to exposed workers and their supervisors. Results of these assessments should include (1) recommendation of engineering and administrative controls to reduce silica dust exposure; (2) recommendation of appropriate and suitable respirators and other PPE for the job; (3) offering hands-on practical training and quantitative fit tests for respirator use; and (4) if feasible, recommending non-silica material for abrasive blasting operations.

Engineering controls may include attaching a dust control system to power tools; using a wet spray method if feasible to suppress dust during sand blasting, jack hammering or other construction activities; and when feasible, using a local exhaust ventilation system to remove dust from the work area.

A half or full-face air-purifying respirators with the appropriate filters must be used if concentrations of silica are at or above the PEL. Safety glasses or goggles should also be worn.

The NIOSH has approved four methods for determining silica in dust samples, namely, colorimetry, X-ray powder diffraction, and two infrared spectroscopy methods (NIOSH, 2010).

REFERENCES

Agius, R. (1992) Is silica carcinogenic? *Occup. Med. (Lond.)* 42:50–52.

Beamer, C.A., Seaver, B.P., and Shepherd, D.M. (2012) Aryl hydrocarbon receptor (AhR) regulates silica-induced inflammation but not fibrosis. *Toxicol. Sci.* 126:554–568.

Franzén, L.G., Hjelmroos, M., Kållberg, P., Brorström-Lundén, E., Juntto, S., and Savolainen, A. L. (1994) The 'yellow snow' episode of northern Fennoscandia, March 1991—a case study of long-distance transport of soil, pollen and stable organic compounds. *Atmos. Environ.* 28:3587–3604.

Gardner, L.U. and Dworski, M. (1922) Studies on the relation of mineral dusts to tuberculosis. *Am. Rev. Tuberculosis* 6:782–797.

Ghahramani, N. (2010) Silica nephropathy. *Int. J. Occup. Environ. Med.* 1:108–115.

Husar, R.B., Tratt, D.M., Schichtel, B.A., Falke, S.R., Li, F., Jaffe, D., Gassó, S., Gill, T., Laulainens, N.S., Lu, F., Reheis, M.C., Chun, Y., Westphal, D., Holben, B.N., Gueymard, C., McKendry, I., Kuring, N., Feldman, G.C., McClain, C., Frouin, R.J., Merrill, J., DuBois, D., Vignola, F., Murayama, T., Nickovic, S., Wilson, W.E., Sassen, K., Sugimoto, N., and Malm, W.C. (1998) Asian dust events of April 1998. *J. Geophys. Res.* 106:18,317–18,330.

IARC (1987) *Silica and Some Silicilates. IARC Monographs on the Evaluation of Carcinogenic Risk of Chemicals to Humans*, vol. 42, Lyon, France: International Agency for Research on Cancer, pp. 39–238.

IARC (1997) Silica. In: *Silica, Some Silicates, Coal Dust and Para-Armid Fibrils. IARC Monographs on the Evaluation of Carcinogenic Risk of Chemicals to Humans*, vol. 68, Lyon, France: International Agency for Research on Cancer, pp. 41–242.

Leung, C.C., Yu, I.T., and Chen, W. (2012) *Silicosis. Lancet.* 379:2008–2018.

Markowitz, G. and Rosner, D. (1995) The limits of thresholds: silica and the politics of science, 1935 to 1990. *Am. J. Public Health.* 85:253–262.

Middleton, N.J., Betzer, P.R., and Bull, P.A. (2001) Long-range transport of 'giant' aeolian quartz grains: linkage with discrete sedimentary sources and implications for protective particle transfer. *Marine Geol.* 177:411–417.

Milovanović, A., Nowak, D., Milovanović, A., Hering, K.G., Kline, J.N., Kovalevskiy, E., Kundiev, Y.I., Perunicić, B., Popević, M., Sustran, B., and Nenadović, M. (2011) Silicotuberculosis and silicosis as occupational diseases: report of two cases. *Srp. Arh. Celok. Lek.* 139:536–539.

Murray, J., Reid, G., Kielkowski, D., and de Beer, M. (1993) Cor pulmonale and silicosis: a necropsy based case-control study. *Br. J. Ind. Med.* 50:544–548.

NIOSH (1992) *NIOSH alert: Request for assistance in preventing silicosis and deaths from sandblasting (DHHS (NIOSH) Publication No. 92-102).* Cincinnati, OH: Author.

NIOSH (2010) *Pocket Guide to Chemical Hazards*, NIOSH 2010-168c, Washington, DC: US Department of Health and Human Services, 1994.

NTP (2011) *Silica, Crystalline (Respirable Size). Report on Carcinogens, Twelfth Edition.*

Olishifski, L.B. (1988) Overview of industrial hygiene (B.A. Plog, Rev.). In: Plog, B.A., editor. *Fundamentals of Industrial Hygiene*, 3rd ed., Chicago: National Safety Council, pp. 3–28.

Pollard, K.M. and Kono, D.H. (2013) Requirements for innate immune pathways in environmentally induced autoimmunity. *BMC Med.* 11:100.

Santos, C., Norte, A., Fradinho, F., Catarino, A., Ferreira, A.J., Loureiro, M., and Baganha, M.F. (2010) Silicosis—brief review and experience of a pulmonology ward. *Rev. Port. Pneumol.* 16:99–115.

Steenland, K. and Goldsmith, D.F. (1995) Silica exposure and autoimmune diseases. *Am. J. Ind. Med.* 28:603–608.

TALC

First aid: If on the body, wash the contaminated body part with copious amounts of soap and water. If the eyes are exposed, immediately irrigate the affected eye thoroughly with water or 0.9% saline. Provide treatment for any symptoms that may arise.

Target organ(s): Respiratory system, eyes

Occupational exposure limits: NIOSH REL: TWA 2 mg/m^3; ACGIH TLV: TWA 2 mg/m^3; OSHA PEL: TWA 20 mppcf (80 mg/m^3/%SiO$_2$)

Reference values:

Hazard/precautionary phrases: H319 causes serious eye irritation; H332 harmful if inhaled; H372 causes damage to organs through prolonged or repeated exposure

BACKGROUND AND USES

Talc (hydrous magnesium silicate; steatite talc; CAS #14807-96-6) is an odorless white powder. It is insoluble in water and has a vapor pressure of approximately 0 mmHg. Talc is chemically similar to anthophyllite, chrysotile, and tremolite forms of asbestos, as they are all hydrated magnesium silicates. It is commonly found in combination with asbestos, quartz, and other minerals. Two forms occur in mineral deposits, namely, nonfibrous (platy or non-asbestiform) and fibrous (tremolite). Talc is used in the manufacture of ceramics, paints, paper, roofing materials, plastics, cosmetics, and rubber.

Talc is produced by open pit or underground mining of talc rocks and processed by crushing, drying, and milling. Geological formation of talc rock results from the alteration of magnesia- and silica-rich ultramafic rocks under a range of temperatures and pressures.

Domestic talc production in 2012 was estimated to be 623,000 tons valued at $22 million. Three companies operated six talc-producing mines in three states in 2011. Montana was the leading producer state, followed by Texas and Vermont. Sales of talc were estimated to be 572,000 tons valued at $90 million. Talc produced and sold in the United States was used for ceramics, 26%; paint and paper, 20% each; plastics and roofing, 9% each; cosmetics, 4%; rubber, 3%; and other, 9% (USGS, 2013).

According to the Hazardous Substances Data Bank (HSDB) record for talc, (NLM, 2013), the compound is typically used as: (1) an additive to clay in ceramic manufacture and paper coatings, for roofing materials; (2) a carrier and diluent for insecticides; filler and pigment for paints and elastomers; additive in manufacture of refractories; (3) in ceramics; cosmetics and pharmaceuticals; filler in rubber, paints, putty, plaster, oilcloth; abherent; slate pencils and crayons; (4) in dusting powder, either alone or with starch or boric acid, for medicinal and toilet preparations; excipient and filler for pills, tablets and for dusting tablet molds; clarifying liquids by filtration; (5) as pigment in paints, varnishes, rubber; filler for paper, rubber, soap; in fireproof and cold-water paints for wood, metal, and stone; lubricating molds and machinery; glove and shoe powder; electric and heat insulator; and (6) in veterinaries as topically on vulva fold pyodermas of dogs as protectant.

PHYSICAL AND CHEMICAL PROPERTIES

Talc as a pure chemical compound is defined as hydrous magnesium silicate, $Mg_3Si_4O_{10}(OH)_2$ and consists of a brucite sheet containing magnesium ions sandwiched between two silica sheets, which are held together by relatively weak forces. Talc as a pure mineral is composed of 63.5% SiO_2, 31.7% MgO, and 4.8% H_2O. A variety of elements such as nickel and iron may be included in the talc particle lattice but are so bound within the particle that they are not free to exert any biological action. As talc dusts are obtained from different sources, the amount and specific form of talc, as well as the amount and nature of mineral contaminants, will be different for each dust.

Talc can be tabular, granular, fibrous, or platy but it is usually crystalline, flexible, and soft. It is commonly whitish to light green but it can be yellowish to reddish, the reddish color being associated with small amounts of iron bound within the talc particle.

Talc has a greasy feel and luster. It is insoluble in water, cold acids, and alkalies and has a density of 2.7–2.8 and a melting point of 900–1000 °C.

ENVIRONMENTAL FATE AND BIOACCUMULATION

Talc is commonly found in the environment. It shows no bioaccumulation potential.

Talc is not expected to undergo chemical transformation when released into the environment. Its refractory nature precludes the effects of melting/boiling point, solubility, vapor pressure, octanol/water partition coefficient, etc., on its transport. Terrestrial and fluvial transport processes affecting talc are not well characterized.

ECOTOXICOLOGY

Talc is not expected to have an ecotoxicologic effect or to be harmful to aquatic life. The 96 h LC50 in *Danio rerio* (zebrafish) is >100 g/l (semi-static).

MAMMALIAN TOXICOLOGY

Acute Effects

While ingestion or topical application of talc produces a benign effect, the substance acts as an irritant when it is inhaled. As an immediate consequence, it dries the tracheobronchial tree mucous membranes and impairs ciliary function, inhibiting the clearing of particulate matter within the airway. Talc inhalation can produce various signs and symptoms, including sneezing, coughing, dyspnea (transient or chronic), respiratory distress/failure, and death. Conditions such as diffuse pulmonary infiltrates, acute lung injury, bronchiolar obstruction, bronchiolitis, pneumonitis, or pneumonia caused by a secondary bacterial infection can develop. Any of these conditions can lead to severe respiratory compromise, such as acute respiratory distress syndrome (Lawson, 2012).

Chronic Effects

The greatest cause of pulmonary disease resulting from talc occurs in workers engaged in mining and milling so-called tremolite talc. The four distinct forms of pulmonary disease resulting from chronic exposure to talc have been defined (Marchiori et al., 2010). Three of them (talcosilicosis, talcoasbestosis, and pure talcosis) are associated with inhalation and differ in the composition of the inhaled substance. The fourth form, a result of intravenous administration of talc, is seen in drug users who inject medications intended for oral use. Presentation of patients with talc granulomatosis can range from asymptomatic to fulminant disease. Symptomatic patients typically present with nonspecific complaints, including progressive exertional dyspnea, and cough. Late complications include chronic respiratory failure, emphysema, pulmonary arterial hypertension, and cor pulmonale.

Dyspnea on effort and cough are presenting symptoms in talc pneumoconiosis. As pathological changes progress, cyanosis, clubbing, and increasing dyspnea are the usual complaints. Diminished breath sounds, diffuse rales, and limitation of chest expansion have also been noted (Kleinfeld et al., 1963). Tuberculosis, emphysema, and right heart disease are complications of pulmonary talcosis.

Postmortem studies show talc causes a nonspecific pulmonary fibrosis of the lung (Seeler et al., 1959). In addition, foreign body granulomas containing talc fibers are found in the lung, pleura, diaphragm, pericardium, and gastric wall. These findings were once attributed to the silica content of talc. However, X-ray diffraction studies of affected lungs have demonstrated the presence of <0.1% of free silica.

Wild (2006) reviewed the epidemiological evidence between lung cancer risk and talc not containing asbestos. No increased lung cancer mortality was observed among talc millers despite their high exposure experience. In populations in which talc was associated with other potential carcinogens, some lung cancer excesses were observed.

Lung damage has developed in rats exposed repeatedly, by inhalation, to talc. No fetal effects were seen in rabbits, rats, mice, or hamsters treated orally during pregnancy. In rats given oral doses, there was no evidence of activity in assays for chromosomal damage or dominant lethal mutations. Talc did not cause chromosomal damage in human cells treated in culture and it was not mutagenic to yeast or to bacteria in Ames tests (Bibra, 1991).

No conclusive evidence of pure talc-induced carcinogenicity has been demonstrated in workers (IARC, 2010). However, talc containing asbestos may be carcinogenic. The collective data from animal, cellular studies, and epidemiological studies do not support a causal relationship between industrial-grade talc and mesothelioma (Price, 2010). Data do not support the hypothesis that use of talc in the perineal area is associated with an increased risk of endometrial cancer (Neill et al., 2012).

A number of observational studies (largely case control) suggest an association between use of talc powders on the female perineum and increased risk of ovarian cancer. However, the weak statistical associations observed in a number of epidemiological studies do not support a causal association (Huncharek and Muscat, 2011).

Based on the lack of data from human studies and on limited data in lab animal studies, the IARC classifies talc not containing asbestos as "not classifiable as to carcinogenicity in humans."

Chemical Pathology

The characteristic histopathologic feature in talc pneumoconiosis is the striking appearance of birefringent, needle-shaped particles of talc seen within the giant cells and in the areas of pulmonary fibrosis with the use of polarized light. Computed tomography can play an important role in the diagnosis of pulmonary talcosis, since suggestive patterns may be observed. The presence of these patterns in drug abusers or in patients with an occupational history of exposure to talc is highly suggestive of pulmonary talcosis.

Mechanism of Action(s)

Different mechanisms are probably operative in the effects of talc on the lung and pleura, depending on the route of exposure.

Talcosilicosis and talcoasbestosis are probably due to concomitant exposure of talc with crystalline silica and asbestos, respectively. The previous sections discuss the mechanisms by which these produce disease.

Talc deposition in the lungs typically causes granulomatous inflammation characterized by formation of foreign body granulomas of varying degree within a fibrotic stroma (Scheel et al., 2012). These granulomas are typically composed of free or intracellular birefringent deposits accompanied by multinucleated giant cells. They may also appear ill defined with only few surrounding histiocytes. Distribution of these lesions is variable and they may develop intra- and perivascular as well as in the interstitium.

SOURCES OF EXPOSURE

Occupational

Exposure to talc dust occurs during its mining, crushing, separating, bagging, and loading and in various industries that use talc.

Environmental

Talc is found in small amounts in metamorphic mafic and ultramafic rocks and in carbonates. These metamorphic rocks crop out in mountain belts such as the Alps, the Appalachians, and the Himalayas and in ancient continental shields such as the Canadian shield in New York and Canada.

Talc may also be in the environment from consumer products that contain talc.

INDUSTRIAL HYGIENE

The ACGIH and NIOSH have set the TLV and REL, respectively, for respirable talc containing no asbestos and <1% quartz at 15 million particles per cubic foot (mppcf) of air (mppcf; $2\,mg/m^3$). OSHA has set the PEL at 20 mppcf ($\sim3\,mg/m^3$). The exposure limits for asbestos or silica should be used for any talc containing asbestos or silica.

According to the European Chemicals Agency (ECHA) (ECHA, 2013) Registered Substances Database, a derived no effect level (DNEL) for long-term exposure to talc via inhalation is $72\,mg/m^3$ for the general population and $245\,mg/m^3$ for workers. The DNEL for long-term dermal exposure is 83 mg/kg bw/day for the general population and 139 mg/kg bw/day for workers. The DNEL for the general population for long-term exposure via the oral route of administration is 8.3 mg/kg bw/day.

REFERENCES

Bibra Toxicology Advice and Consulting (1991) *Toxicity Profile for Talc. Profile 274.*

ECHA (European Chemicals Agency) (2013) *Registered Substances Database for Talc.* Available at http://apps.echa.europa.eu/registered/ (accessed March 26, 2013).

Huncharek, M. and Muscat, J. (2011) Perineal talc use and ovarian cancer risk: a case study of scientific standards in environmental epidemiology. *Eur. J. Cancer Prev.* 20:501–507.

IARC (2010) *Carbon Black, Titanium Dioxide, and Talc, IARC Monographs on the Evaluation of Carcinogenic Risks to Humans*, vol. 93, International Agency for Research on Cancer. pp. 277–415.

Kleinfeld, M., Giel, C.P., Majeranowski, F.F., and Messite, J. (1963) Talc pneumoconiosis: a report of six patients with postmortem findings. *Arch. Environ. Health* 7:101–115.

Lawson, R. (2012) Acute talc inhalation. *Nursing* 42:72.

Marchiori, E., Lourenço, S., Gasparetto, T.D., Zanetti, G., Mano, C.M., and Nobre, L.F. (2010) Pulmonary talcosis: imaging findings. *Lung* 188:165–171.

Neill, A.S., Nagle, C.M., Spurdle, A.B., and Webb, P.M. (2012) Use of talcum powder and endometrial cancer risk. *Cancer Causes Control* 23:513–519.

NLM (National Library of Medicine) (2013) *Hazardous Substances Data Bank Record for Talc.* Available at http://toxnet.nlm.nih.gov/cgi-bin/sis/search/f?./temp/~JcisCF:1 [accessed March 26, 2013].

Price, B. (2010) Industrial-grade talc exposure and the risk of mesothelioma. *Crit. Rev. Toxicol.* 40:513–530.

Scheel, A.H., Krause, D., Haars, H., Schmitz, I., and Junker, K. (2012) Talcum induced pneumoconiosis following inhalation of adulterated marijuana, a case report. *Diagn. Pathol.* 7:26.

Seeler, A., Gryboski, J.S., and MacMahon, H.E. (1959) Talc pneumoconiosis. *AMA Arch. Ind. Health* 19:392–402.

U.S. Geological Survey (USGS) (2013) *Mineral Commodity Summaries: Talc*, pp. 160–161.

Wild, P. (2006) Lung cancer risk and talc not containing asbestiform fibres: a review of the epidemiological evidence. *Occup. Environ. Med.* 63:4–9.

93

MIXED DUSTS

Charles Barton

COAL

First aid: Remove the patient from exposure to fresh air. If the eyes are exposed, immediately irrigate the affected eye thoroughly with water or 0.9% saline. Provide treatment for any symptoms that may arise.

Target organ(s): Respiratory system, eyes.

Occupational exposure limits: OSHA PEL—TWA 2.4 mg/m^3 [respirable, <5% SiO$_2$] or TWA (10 mg/m^3)/(%SiO$_2$ + 2) [respirable, >5% SiO$_2$]; NIOSH REL for occupations other than mining—TWA 1 mg/m^3 [measured according to MSHA method (CPSU)] or TWA 0.9 mg/m^3 [measured according to ISO/CEN/ACGIH criteria]; ACGIH TLV—TWA 2 mg/m^3 if <5% SiO$_2$ or 0.1 mg/m^3 if 5% or greater SiO$_2$.

Hazard/precautionary phrases: H$_2$0$_5$ may mass explode in fire; H$_3$7$_2$ causes damage to organs through prolonged or repeated exposure.

BACKGROUND AND USES

Coal is a generic term for a heterogeneous, carbonaceous rock of varying composition and characteristics. It is mined in over 70 different countries around the world, and utilized in many more for electricity generation, heating, steel making, and chemical processes. It varies in type from the soft and friable lignite to the hard and brittle anthracite. The term "rank," which reflects the percentage carbon content, is used for its classification. Coal typically contains variable but substantial amounts of mineral matter, of which quartz is an important component.

Throughout history, coal has been a useful resource for human exploitation. It is primarily burned as a fossil fuel to produce electricity and heat. Coal forms when dead plant matter decomposes into peat, which in turn is converted into lignite, and then anthracite. Coal is the largest source of energy for the generation of electricity worldwide, as well as one of the largest worldwide anthropogenic sources of carbon dioxide release.

Coal dust is a fine powdered form of coal, which is created by the crushing, grinding, or pulverizing of coal. Because of the brittle nature of coal, coal dust can be created during mining, transportation, or by mechanically handling coal.

A number of occupational factors influence the symptoms and signs of chest disease in addition to the individual coal miner's medical history or the health of the community where he or she lives. Because coal mining may occur near ground level in open pits, at varying depths underground, or far under the sea, a variety of risks may be encountered. Differences in the particle size distribution and the quantity of dust inhaled by the worker vary depending on the method of mining being used. Deep mines are hard to ventilate, and this problem may involve exposure not only to dust but also to irritating chemicals arising from blasting charges and to dangerous levels of carbon monoxide and methane.

Coal occurs in ore, varying in content of the fibrogenic dust silica (quartz). Silica in the form of sand may also be applied to underground tracks with compressed air to provide traction for heavily laden cars, a practice leading to a serious risk of silicosis, especially to motormen. Rock dust used to prevent explosions may create a dust hazard.

Coal dust suspended in air is explosive because coal dust has far more surface area per unit weight than lumps of coal and is more susceptible to spontaneous combustion. Some of

Hamilton & Hardy's Industrial Toxicology, Sixth Edition. Edited by Raymond D. Harbison, Marie M. Bourgeois, and Giffe T. Johnson.
© 2015 John Wiley & Sons, Inc. Published 2015 by John Wiley & Sons, Inc.

the worst mining accidents in history have been caused by coal dust explosions, such as the disaster at Senghenydd in South Wales in 1913 in which 439 miners died, the 1906 Courrières mine disaster in Northern France which killed 1099 miners, the 1962 Luisenthal Mine disaster in Germany, which claimed 299 lives, and the explosion at Benxihu Colliery, China, which killed 1549 in 1942.

PHYSICAL AND CHEMICAL PROPERTIES

Coal mine dust is a complex and heterogeneous mixture containing more than 50 different elements and their oxides. The mineral content varies with the particle size of the dust and with the coal seam.

Coal comes in four main types or ranks: lignite or brown coal, bituminous coal or black coal, anthracite, and graphite. Each type of coal has a certain set of physical parameters which are mostly controlled by moisture, volatile content, and carbon content. Coal is composed primarily of carbon along with variable quantities of other elements, chiefly hydrogen, sulfur, oxygen, and nitrogen. Relative density of the coal depends on the rank of the coal and degree of mineral impurity. The physical properties of coal dust vary depending on the specific type of coal.

ENVIRONMENTAL FATE AND BIOACCUMULATION

It is unknown whether and to what extent toxic heavy metals (arsenic, cadmium, barium, chromium, selenium, lead, mercury) from coal dust leach into water. Coal dust may negatively affect vegetation by reducing growth, reducing reproduction, and causing leaf lesions and partial defoliation (Farmer, 1993). Fluoride and sulfur compounds are also contained in coal dust, and their fate and effects on the environment are unknown.

There are no data regarding the bioaccumulation of coal dust in aquatic organisms.

ECOTOXICOLOGY

Coal and coal dust have the potential to affect marine and estuarine organisms and habitats through abrasion, smothering, alteration of sediment texture and stability, reduced availability of light, and clogging of respiratory and feeding organs (Ahrens and Morrisey, 2005). It is likely that significant coal and coal dust from transfer activity and cargo spills will become deposited in the marine/estuarine environment (Johnson and Bustin, 2006).

Ahrens and Morrisey (2005) reported that particles of coal in suspension in the water column reduced the amount and possibly the spectral quality of light reaching the sea bed. Moreover, deposition of coal dust on plants decreased photosynthesis, and killed eggs of fish and invertebrates by smothering them. Species richness and biomass were also found to decline with the change of sediment composition.

MAMMALIAN TOXICOLOGY

Acute Effects

Coal dust exposure can induce an acute alveolar and interstitial inflammation. There is a two-step inflammatory response to acute coal-dust exposure. In an acute phase, there is lymphocyte infiltration and some neutrophil recruitment followed by a chronic phase of macrophage predominance (Pinho et al., 2004).

Chronic Effects

Chronic inhalation of coal dust can cause several lung disorders, including simple coal workers pneumoconiosis (CWP), progressive massive fibrosis (PMF), chronic bronchitis, lung function loss, and emphysema.

Studies have shown some decreased pulmonary function even at permissible exposure limits (PELs) adopted after 1970. Henneberger and Attfield (1996) followed a cohort of 1915 male coal miners for 13–19 years. They noted annual declines in forced vital capacity (FVC) of 0.10 ml/year per mg/m^3/year of cumulative coal dust exposure, and a decrease in forced expiratory volume in 1 s (FEV1) of 0.07 ml/year per mg/m^3/year. Seixas et al. (1993) reported decreased FEV1 and FVC in a cohort of 977 workers. Decreases were primarily seen in workers over 25 years of age. Decreases were observed in smokers, nonsmokers, and former smokers, but the largest decreases were in former smokers. Lewis et al. (1996) reported slightly reduced pulmonary function independent of smoking in British coal miners without pneumoconiosis.

Coal Workers' Pneumoconiosis Coal workers' pneumoconiosis may be simple or complicated. Simple CWP is defined as having coal macules (discolored tissue) in the lungs <2 cm in diameter, and complicated CWP is defined as having coal macules >2 cm in diameter. Simple CWP is not associated with any clinical signs or symptoms.

There may be an increase in alveolar–arterial diffusion in simple pneumoconiosis as well as evidence of emphysema undetected by X-ray. These findings mean that X-ray pneumoconiosis categories 0 and 1 of the International Classification may be associated with disabling disease. These facts require the attention of U.S. physicians because the 1969 Federal Coal Mine Health and Safety Act awards benefits on the basis of chest X-ray abnormalities. There is a confusing and serious defect in the law inasmuch as disabling disease of coal workers can occur, as shown by lung function studies and pathological findings, without demonstrable radiographic abnormalities.

Anthracite coal is a more potent inducer of CWP than bituminous coal.

Gastrointestinal Effects   It has been suggested that coal particles concentrated in the lung are eliminated by the ciliary cleaning function and then swallowed with saliva, thus reaching the stomach (Ames, 1982). The combustion of coal releases PAHs, including benzo[a]pyrene, a known carcinogen. Hence, it is postulated that dust particles, essentially inhalable, are swallowed and thereby exert a carcinogenic effect upon the stomach.

Inhaled coal dust is swallowed and introduced in the acidic environment of the stomach, where it may interact with nitrosating agents, such as nitrite. Swaen et al. (1995) studied the association between coal dust exposure, pneumoconiosis, and mortality from gastric cancer in a cohort of 3790 miners. Workers employed underground for at least 30 years had a standard mortality ratio for gastric cancer of 154. The increased risk of gastric cancer roughly correlated with years spent underground. Risk of gastric cancer was not elevated in workers with severe pneumoconiosis. The authors proposed that the pulmonary clearance system is so impaired in this group of workers that inhaled coal dust does not reach the digestive tract.

CHEMICAL PATHOLOGY

Depending on the quantity of dust in the lungs, chest radiographs may reveal bilateral, widely disseminated shadows described as linear or nodular opacities. The accumulation of coal dust both inside and adjacent to the respiratory bronchioles may cause distortion of the normal anatomy and may lead to focal emphysema. Infection and dust further destroy the bronchioles leading into the alveoli to produce what is now called centrilobular emphysema or centriacinar emphysema. Depending on the quantity inhaled, coal dust may be distributed throughout the lungs and the lesions of centrilobular emphysema can result in disabling disease. Distortions thus produced lead to shadows on X-ray variously described as confluent, complicated, or characteristic of PMF.

Busch et al. (1981) exposed Wistar rats to bituminous coal dust. They developed lesions similar to simple CWP as described in human subjects. Comparable lesions observed were macules (focal, coal-laden macrophage accumulations associated with increased interstitial reticulin and collagen, with emphysema involving alveoli within the macule), which were of increased size, altered shape, and increased density

in animals experiencing chronic exposures. More advanced lesion types, i.e., micronodule, macronodule, silicotic nodule, Caplan's lesion, and infective granuloma, were not observed in the experimental animals. Focal bronchiolization occurred in animals receiving at least 20-months' exposure. Usually, this pathologic change is not described as a component of CWP.

MODE OF ACTION(S)

The development and progression of CWP occurs with four basic mechanisms: (a) direct cytotoxicity of coal dust or silica; (b) activation of oxidant production by pulmonary phagocytes; (c) activation of mediator release from alveolar macrophages and epithelial cells, which leads to recruitment of polymorphonuclear leukocytes (PMNs) and macrophages, resulting in the production of proinflammatory cytokines and reactive species and in further lung injury and scarring; and (d) secretion of growth factors from alveolar macrophages and epithelial cells, stimulating fibroblast proliferation and eventual scarring (Castranova and Vallyathan, 2000).

Coal particles penetrate the lower respiratory tract and are taken up by the lung macrophages if they are <5 μm in diameter. In contrast to silica, coal does not harm the macrophages that engulf the dust throughout the lungs. During a bacterial infection the coal-laden macrophage may "dump" the coal particles in the involved lung area, a process that may account for conglomerate lesions visible on X-ray. Inhaled coal containing <2% silica is held in a reticulin network without causing a fibrotic reaction.

Inhaled coal dust particles can be important sources of reactive oxygen species (ROS) in the lung, and may be significantly involved in the damage of lung target cells as well as important macromolecules, including alpha-1-antitrypsin and DNA. *In vitro* and *in vivo* studies with coal dusts showed the upregulation of important leukocyte recruiting factors, such as leukotriene B3, platelet derived growth factor, monocyte chemotactic protein-1, tumor necrosis factor-alpha, and the neutrophil adhesion factor intercellular adhesion molecule-1. Coal dust particles can stimulate the macrophage production of various factors with potential capacity to modulate lung cells and extracellular matrix.

Miners who have an abnormally high dust-stimulated release of TNF-α have an increased risk of developing CWP (Lee et al., 2010).

In vitro studies indicate strong mechanistic links between iron content in coal and ROS, which play a major role in the inflammatory response associated with CWP (McCunney et al., 2009). The ROS can inactivate alpha-1-antitrypsin. This mechanism may also play an important role in the development of coal workers' emphysema.

SOURCES OF EXPOSURE

Occupational

The major exposures to coal dust occur during mining and processing of coal. In these operations the exposure includes dusts generated not only from the coal but also from adjacent rock strata and other sources. These may increase the quartz component of the airborne dust to about 10% of the total mixed dust, or to even greater levels if significant rock cutting is being undertaken.

Dust levels in underground mines vary considerably according to location within the mine. In general, workers at the coalface receive the highest exposures, while those working progressively further away experience lower exposures. In addition, those employed in locations receiving intake (clean) air are exposed to lower dust levels than those who have to breathe returning air, which has passed the coalface. Most surface workers at underground mines experience lower dust exposures than their colleagues underground. However, some jobs, such as tipple and coal cleaning, involve dust exposures equivalent to some underground occupations.

Although dust levels in surface mines are generally lower than those at underground mines, some jobs put workers at risk for silica exposure and silicosis.

Environmental

Other than in mining, exposure to coal dust can also occur during bulk coal transfer and at sites where coal is used. These sites include power stations, steel and coke works, and plants where coal is refined to produce chemicals or liquid fuels. The domestic use of coal for heating is another potential source of exposure to coal dust. However, information on these other exposures to coal dust is limited.

INDUSTRIAL HYGIENE

Twentieth-century medical reports of lung damage first appeared in all coal mining countries at about the same time, between 1928 and 1936. After World War II, considerable effort was spent on studying and controlling diseases of coal miners, an activity continuing today. Economic interests, strengthening of unions, health legislation, and the action of citizen groups, including miners and their families, have played a part in forcing control of coal-mining diseases. Efforts culminated in the United States with the passage of the Federal Coal Mine Health and Safety Act of 1969, which created the Mine Safety and Health Administration (MSHA). From this point on, regulations regarding coal dust exposure have been promulgated by MSHA.

Although this plan has improved conditions greatly through engineering controls and routine air measurements, coal dust-related health effects have not been abolished completely.

The most common route for coal dust exposure is through inhalation. Dermal absorption and ingestion patterns are negligible. Poor ventilation and drilling procedures are the major determinants of exposure. The cost effectiveness and optimum control measures are generally site specific for mining operations. The depth and angle of the tunnel, number of workers, and drilling procedures vary among mines and will affect dust levels and engineering procedures. Individual control measures should be evaluated and implemented as the requirements indicate.

The general plan for coal dust exposure controls is a multitiered approach, including engineering and administrative controls. Ventilation is the most important step to reducing airborne concentrations of coal dust. Directional fan systems can be used to drive sources of coal dust away from machine operators. Increased automation has allowed for higher concentrations in areas where no workers are expected. MSHA requires periodic sampling of mines to monitor the concentrations. Damp conditions help keep air concentrations of coal dust low. Sprayers and water jets associated with the drilling procedure are used to keep dust settled.

Respirators are also necessary to protect worker's health. Simple paper masks are not effective in limiting exposure and should not be used. The proper respirator should be chosen to maximize protection and minimize breathing interference. A common administrative control for reducing exposure to coal dusts is to reduce the amount of time a worker spends in high coal dust areas.

The current Occupational Safety and Health Administration (OSHA) PEL for the respirable fraction of coal dust ($<5\%$silica) is $2.4\,mg/m^3$ time-weighted average (TWA) concentration. The National Institute for Occupational Safety and Health (NIOSH) recommended exposure limit (REL) for coal dust in industries other than mining is TWA $1\,mg/m^3$ [measured according to MSHA method (CPSU)] or TWA $0.9\,mg/m^3$ [measured according to ISO/CEN/ACGIH criteria]. The American Conference of Governmental Industrial Hygienists (ACGIH) has assigned the respirable fraction of coal dust containing $<5\%$ crystalline silica a threshold limit value (TLV) of $2\,mg/m^3$ as a TWA for a normal 8-h workday and a 40-h workweek, or $0.1\,mg/m^3$ if 5% or greater SiO_2. The ACGIH limit is based on the risk of pneumoconiosis.

The National Fire Protection Association (NFPA) has not assigned a flammability rating to coal dust. Other sources rate coal dust as a fire hazard and consider the airborne dust an explosion hazard when these substances are exposed to heat or open flame.

RISK ASSESSMENTS

Although current mining procedures and regulations have reduced occupational exposures and cases of CWP, chronic exposure to low levels of coal dust is still capable of causing an adverse health effect. Attfield and Morring (1992) reported a 9% theoretical risk of a worker developing CWP with a working lifetime exposure to $2 \, mg/m^3$ for coal dust. Considering the large number of mining employees in the west and central regions of the United States, there will still be many reported cases of occupational respiratory effects in the future.

The risk of CWP pneumoconiosis increases with duration of exposure, dust concentrations, and hardness of coal (Attfield and Seixas, 1995).

According to International Agency for Research on Cancer (IARC), coal dust cannot be classified as to its carcinogenicity to humans due to inadequate evidence in humans and in experimental animals.

REFERENCES

Ahrens, M.J. and Morrisey, D.J. (2005) Biological effects of unburnt coal in the marine environment. *Oceanogr. Mar. Biol.: Ann. Rev.* 43:69–122.

Attfield, M.D. and Morring, K. (1992) An investigation into the relationship between coal workers' pneumoconiosis and dust exposure in U.S. coal miners. *Am. Ind. Hyg. Assoc. J.* 53:486–492.

Attfield, M.D. and Seixas, N.S. (1995) Prevalence of pneumoconiosis and its relationship to dust exposure in a cohort of U.S. bituminous coal miners and ex-miners. *Am. J. Ind. Med.* 27:137–151.

Ames, R.G. (1982) Gastric cancer in coal miners: some hypotheses for investigation. *J. Soc. Occup. Med.* 32:73–81.

Busch, R.H., Filipy, R.E., Karagianes, M.T., and Palmer, R.F. (1981) Pathologic changes associated with experimental exposure of rats to coal dust. *Environ. Res.* 24:53–60.

Castranova, V. and Vallyathan, V. (2000) Silicosis and coal workers' pneumoconiosis. *Environ. Health Perspect.* 108(Suppl 4):675–684.

Farmer, A.M. (1993) The effects of dust on vegetation—a review. *Environ. Pollut.* 79:63–75.

Henneberger, P.K. and Attfield, M.D. (1996) Coal mine dust exposure and spirometry in experienced miners. *Am. J. Respir. Crit. Care Med.* 153:1560–1566.

Johnson, R. and Bustin, R.M. (2006) Coal dust dispersal around a marine coal terminal (1977–1999), British Columbia: The fate of coal dust in the marine environment. *Int. J. Coal Geol.* 68:57–69.

Lee, J.S., Shin, J.H., Lee, J.O., Lee, K.M., Kim, J.H., and Choi, B.S. (2010) Serum levels of interleukin-8 and tumor necrosis factor-alpha in coal workers' pneumoconiosis: one-year follow-up study. *Saf. Health Work.* 1:69–79.

Lewis, S., Bennett, J., Richards, K., and Britton, J. (1996) A cross sectional study of the independent effect of occupation on lung function in British coal miners. *Occup. Environ. Med.* 53:125–8.

McCunney, R.J., Morfeld, P., and Payne, S. (2009) What component of coal causes coal workers' pneumoconiosis? *J. Occup. Environ. Med.* 51:462–471.

Pinho, R.A., Bonatto, F., Andrades, M., Frota, M.L., Jr., Ritter, C., Klamt, F., Dal-Pizzol, F., Uldrich-Kulczynski, J.M., and Moreira, J.C. (2004) Lung oxidative response after acute coal dust exposure. *Environ. Res.* 96:290–297.

Seixas, N.S., Robins, T.G., Attfield, M.D., and Moulton, L.H. (1993) Longitudinal and cross sectional analyses of exposure to coal mine dust and pulmonary function in new miners. *Br. J. Ind. Med.* 50:929–37.

Swaen, G.M., Meijers, J.M., and Slangen, J.J. (1995) Risk of gastric cancer in pneumoconiotic coal miners and the effect of respiratory impairment. *Occup. Environ. Med.* 52:606–610.

GRAPHITE

First aid: Remove the patient from exposure to fresh air. If the eyes are exposed, immediately irrigate the affected eye thoroughly with water or 0.9% saline. Provide treatment for any symptoms that may arise.

Target organ(s): Respiratory system, cardiovascular system.

Occupational exposure limits: OSHA PEL—TWA 15 mg/m^3 (total) TWA 5 mg/m^3 (respirable); NIOSH REL—2.5 mg/m^3 TWA (respirable); ACGIH TLV—2 mg/m^3 TWA (respirable); IDLH—1250 mg/m^3.

Hazard/precautionary phrases: H319—causes serious eye irritation; H335—may cause respiratory irritation.

BACKGROUND AND USES

Graphite is a soft, crystalline form of carbon that occurs naturally and can also be produced artificially. Graphite (black lead, plumbago, mineral carbon; CAS 7782-42-5) is one of two forms of natural carbon, the other being diamond. It is an odorless solid with a steel gray to black appearance and a greasy feeling. Granite and other metamorphosed rock contain graphite and when it is mined, free silica becomes airborne. Synthetic graphite (CAS 7440-44-0) has a well-developed crystalline structure. Graphite is used in foundries; in the electrical industry; in lead pencils; in the manufacture of carbon black and electrodes; in the rubber industry; and in heat exchangers, pumps, and valves. For certain nuclear reactor operations, the graphite must be pure (i.e., free of naturally occurring traces of other metals and silica).

Natural graphite may be classified as crystalline or microcrystalline (sometimes referred to as amorphous), and contains various impurities, such as silica (quartz). The content of free silica in natural graphite varies considerably, and can be as high as 11% or more. Synthetic graphite is almost pure crystalline carbon. The quartz content of synthetic graphite is <1%. Graphite is extremely resistant to heat and chemicals, and is therefore used in metallurgy, in foundries, in the chemical industry, etc. Natural graphite is used in the production of steel and cast iron and (in powdered form) in casting sand. Natural graphite was once widely used in the production of fireproof materials for blast furnaces, crucibles, solder ladles, etc., but it has now been largely replaced by synthetic graphite. Natural graphite is also used in brushes for motors and generators. Carbon electrodes used in steel production, electrochemical processes etc., and components for atomic reactors (neutron moderators) are made with synthetic graphite. Both synthetic graphite and high purity natural graphite are used in lubricants. Pencils contain natural graphite in microcrystalline form.

PHYSICAL AND CHEMICAL PROPERTIES

Graphite exists as a black or gray crystalline mass and occurs naturally in lump, amorphous, and flake forms. It is found in most parts of the world and is usually found with impurities such as quartz, mica, iron oxide, and granite. The crystalline silica content of natural graphite can range from 2 to 25%. Synthetic graphite contains <1% silica, if any.

Graphite is one of the softest minerals known and, on rare occasions, is found in hexagonal crystal form. It is insoluble in water and has a vapor pressure of 0 mm Hg. Graphite has a molecular weight of 12.01 g/mole, a melting point of 3650 °C, a density of 2.09–2.23 g/cm^3, and a specific gravity of 2.25 (water = 1).

ENVIRONMENTAL FATE AND BIOACCUMULATION

Graphite is insoluble in water and, therefore, would not be expected to be transported in surface of ground water. Although it may be ingested by bottom feeders, it is not expected to accumulate in aquatic organisms. No volatilization is expected to occur under natural conditions. Graphite could be present in air as particulate matter which has a settling time of days.

ECOTOXICOLOGY

The LC$_{50}$ is >100 mg/l in *Daphnia magna* (water flea) and *Oncorhynchus mykiss* (rainbow trout).

MAMMALIAN TOXICOLOGY

Acute Effects

Acute health effects from graphite exposure have not been reported.

Chronic Effects

Pneumoconiosis may occur among graphite workers. The amount of graphite inhaled and the quantity of free silica determine the clinical and pathological determinants. Pneumoconiosis has been reported from graphite dust exposure in mining, milling, and grinding; in carbon electrode manufacturing; and in pure carbon production from carbon and soot dust (Hanoa, 1983). Case reports have demonstrated that high level exposure to graphite or other relatively pure forms of carbon dust produce chest X-ray and pathological effects like those caused by coal with low silica content. Because varying amounts of fibrogenic silica dust are inhaled with graphite in all but a few work exposures, the clinical picture may be more or less like silicosis. Iron dust may also be inhaled with graphite to produce a modified dust effect.

CHEMICAL PATHOLOGY

The earliest changes that appear in X-rays may be the disappearance of normal vascular markings, with the appearance later of pinpoint and macronodular densities in all lung fields. Small nodular hyperattenuating areas (ill-defined or well defined, corresponding to macular lesions along the walls of bronchioles and nodules, respectively) are most often observed in CT scans of people with graphite pneumoconiosis (Akira, 1995). Interlobular septal thickening and large hyperattenuating areas are sometimes observed.

MODE OF ACTION(S)

Pure graphite is inert (Pendergrass et al., 1968). However, graphite dust containing silica is fibrogenic. Fibrosis associated with graphite exposure is possibly due to concomitant silica exposure (Hanoa, 1983).

SOURCES OF EXPOSURE

Occupational

Occupational exposure to graphite may occur in mining and in other industries that employ the substance. A variety of occupations use graphite, due to its many uses. These uses include for "lead" pencils, refractory crucibles, stove polish, as pigment, lubricant, graphite cement, for matches and explosives, commutator brushes, anodes, arc-lamp carbons, electroplating, polishing compounds, rust and needle-paper, coating for cathode ray tubes, and as moderator in nuclear piles.

In the production of dry cells and storage cells, graphite powder is mixed with the active mass to increase the electrical conductivity. The conductivity of natural graphite powder is also used in plastics (against static charge buildup), explosives (against ignition by friction), and conductive lacquers.

Environmental

Although graphite occurs naturally, exposure to graphite is expected to be primarily occupational.

INDUSTRIAL HYGIENE

Industrial hygiene assessments should be conducted to include dust monitoring, hazard analysis, and providing the results to exposed workers and their supervisors.

Engineering controls may include: attaching a dust control system to power tools; using a wet spray method if feasible to suppress dust during sand blasting, jack hammering, or other construction activities; and when feasible, using a local exhaust ventilation system to remove dust from the work area.

A half or full-face air-purifying respirators with the appropriate filters must be used if concentrations of silica are at or above the PEL. Safety glasses or goggles should also be worn.

RISK ASSESSMENTS

The degree of risk from graphite exposure is correlated to the silica concentration. A large concentration of silica would pose more risk than dust containing a small concentration. Steenland and Brown (1995) calculated the risk of silicosis from exposure to silica based on a cohort of gold miners. They calculated that a worker exposed to the OSHA PEL [10/(% SiO_2 + 2) (respirable) or 30/(% SiO_2 + 2) (total) TWA for 45 years would have a theoretical 35–47% chance of developing silicosis.

REFERENCES

Akira, M. (1995) Uncommon pneumoconioses: CT and pathologic findings. *Radiology.* 197:403–409.

Hanoa, R. (1983) Graphite pneumoconiosis. A review of etiologic and epidemiologic aspects. *Scand. J. Work Environ. Health.* 9:303–314.

Pendergrass, E.P., Vorwald, A.J., Mishkin, M.M., Whildin, J.G., and Werley, C.W. (1968) Observations on workers in the graphite industry. II. *Med. Radiogr. Photogr.* 44:2–17.

Steenland, K., and Brown, D. (1995) Silicosis among gold miners: exposure—response analyses and risk assessment. *Am. J. Public Health.* 85:1372–1377.

MISCELLANEOUS MIXED DUSTS

CLAYS

Clays are natural fine-grained materials found in the earth, with particle sizes not exceeding 4 µm in diameter. They are formed by sedimentation, hydrothermal action, or weathering. The mineral components of clay materials are crystalline hydrous aluminum silicates. A large variety of clays are found in nature, and many have specific applications in industry.

Brick clays are used for building, and fireclay is used where resistance to high temperature is required. Kaolin (china clay; CAS 1332-58-7) is a gray powder used in the manufacture of porcelain, as a textile dressing, and as a filler in paper; it is also a well-known ingredient in certain medications. The main constituent of kaolin is aluminum silicate. Before purification, kaolin clay may contain some quartz and mica from the granite rock in which it occurs and is formed. This clay has been widely studied for its health effects. Attapulgite (CAS 12174-11-7) is a gray to yellow fibrous aluminum–silica complex with similar characteristics to kaolin. It is used as a colloidal substance in pharmaceuticals such as antidiarrheal medications, pesticides, and other products. Shales are related to clays because they are impure silicates of aluminum and are mined to extract mineral oils.

The primary route of exposure to clay dusts is inhalation. Free silica may be inhaled in some operations in which clay is the chief dust and there is little control of dust levels. Because of mechanization of some industries and especially the replacement of the older wet process for preparing kaolin, large amounts of dust may reach the lungs. Control measures for clay dusts have steadily improved since they were determined to be a health hazard years ago. The actual mining of clays is not usually considered hazardous because the clay is wet. Personal protection from inhalation exposures is important in those areas of the clay process responsible for expected high concentrations, generally the milling and bagging of fine clay particles.

NIOSH has set a REL for kaolin particles at 10 mg/m³ for total exposure and 5 mg/m³ for respiratory exposure. The OSHA has established a PEL similar to NIOSH for respiratory exposures, but a less stringent 15 mg/m³ for total exposure. The ACGIH has set a TLV–TWA for kaolin at

2 mg/m^3. Kaolin is not considered classifiable as to its carcinogenicity.

Worker exposure to clays may be hazardous because there are a few reports of pneumoconiosis and other effects after chronic exposure. A study of a large population of workers in kaolin production indicated some radiographic opacity changes associated with clay dust exposures, but no overt health effects were reported (Rundle et al., 1993). Cases of mixed dust pneumoconiosis may occur, with the fibrogenic behavior of silica dominating the clinical and pathological findings. The ensuing simple or complicated pneumoconiosis occurring similarly in coal workers may end fatally because of massive fibrosis and may be associated in some cases with pulmonary tuberculosis.

Workers exposed to attapulgite clay do not demonstrate the health effects of those exposed to kaolin. Waxweiler et al. (1988) reported no significant excesses for any cause of mortality in exposed miners. Some hemolysis was observed when this clay was incubated with human red blood cells. However, this is not a great concern because this compound will most likely not enter the bloodstream during occupational exposures (Perderiset et al., 1989). The substance is not classifiable as to its carcinogenicity.

REFERENCES

Perderiset, M., Etienne, L.S., Bignon, J., and Jaurand, M.C. (1989) Interactions of attapulgite (fibrous clay) with human red blood cells. *Toxicol. Lett.* 47:303–309.

Rundle, E.M., Sugar, E.T., and Ogle, C.J. (1993) Analyses of the 1990 Chest Health Survey of china clay workers. *Br. J. Ind. Med.* 50:913–919.

Waxweiler, R.J., Zumwalde, R.D., Ness, G.O., and Brown, D.P. (1988) A retrospective cohort mortality study of males mining and milling attapulgite clay. *Am. J. Ind. Med.* 13:305–315.

FULLER'S EARTH

Fuller's earth is a variable material that will decolorize mineral and vegetable oils. Its name is derived from an old process of cleaning raw wool by kneading it in water to which earths have been added to absorb dirt and oil, a practice known as "fulling." Fuller's earth is frequently a clay composed of an aluminum silicate (usually montmorillonite) and traces of quartz. It is also used as a filter material, in soaps and pigments, in oil-drilling muds, as an insecticide carrier, and as a paper filler. Bilateral chest X-ray changes in workers exposed over a period of years to fuller's earth demonstrate fine punctate mottling with some coalescence (McNally and Trostler, 1941). There may be scattered pneumoconiosis without the pathological changes of silicosis (Campbell and Gloyne, 1942). The pathological examination of the

lung tissues may show an interstitial collection of dust-laden macrophages associated with mild fibrosis.

REFERENCES

Campbell, A.H. and Gloyne, S.R. (1942) Case of pneumoconiosis due to inhalation of fuller's earth. *J. Pathol. Bacteriol.* 54:75–80.

McNally, W.D. and Trostler, I.S. (1941) Severe pneumoconiosis caused by inhalation of fuller's earth. *J. Ind. Hyg. Toxicol.* 23:118–126.

GRANITE

Granite is an igneous rock composed primarily of quartz and feldspar, with an admixture of mica and granules of other minerals. The chest disease of granite workers appears to be silicosis. When workers are exposed to granite dust with low silica levels, there are no differences in pulmonary function tests (Graham et al., 1994). After chronic silica exposures, the severity of disease can be measured by both lung function studies and chest radiographs. It is possible to measure diminution in lung function as long as 13.5 years before chest X-ray abnormalities appear in granite quarry workers exposed to silica.

REFERENCE

Graham, W.G., Weaver, S., Ashikaga, T., and O'Grady, R.V. (1994) Longitudinal pulmonary function losses in Vermont granite workers. A reevaluation. *Chest* 106:125–130.

MICA

Mica occurs in nature in several forms (muscovite, phlogopite, and others) and is composed of potassium-aluminum silicate. This mineral, which occurs in flat sheets, is used for stove and lantern windows, paper and paint manufacture, and in insulators and dielectrics in the electric and electronic industries. There are reports of pneumoconiosis as a result of heavy exposure to mica over a period of 5 years or more. Complaints of dyspnea and cough have been described with changes in the lungs, especially the mid-zones, visible by chest X-ray.

The NIOSH has set the REL for mica containing <1% quartz at 3 mg/m^3. The OSHA has set the PEL at 20 million particles per cubic foot of air. The ACGIH has set the TLV at 3 mg/m^3.

Landas and Schwartz (1991) reported the case of a worker who had a long history of extensive exposure to mica. He presented 30 years later with a chronic nonproductive cough and progressive shortness of breath. Pulmonary function testing revealed restrictive lung changes and a 50% reduction in diffusion capacity. A computed tomography scan showed diffuse interstitial fibrosis. Bronchoalveolar lavage revealed abundant rectangular flaking crystals. Energy-dispersive spectroscopy and electron diffraction studies confirmed the material to have the features of mica.

A controversy in the literature exists as to whether mica induces pneumoconiosis. An extensive review of the experimental and epidemiological papers, and case studies involving mica exposure up to 1985, reported only a few descriptions of pneumoconiosis due to pure mica exposure in humans and inadequate data on the potential for collagen deposition in experimental animals (Skulberg et al., 1985). Other reviews concluded that mica has a relatively low fibrogenic potential (Jones et al., 1994). However, several case reports presented with careful pathological and mineralogical analysis leave little doubt that mica with low quartz content can be associated with the development of disabling forms of pneumoconiosis (Davies and Cotton, 1983; Landas and Schwartz, 1991; Zinman et al., 2002). Furthermore, there is clear evidence that exposure to fine sericite particles is associated with the development of functional and radiological changes in workers inducing mixed dust lesions, which are distinct histologically from silicosis (Algranti et al., 2005).

REFERENCES

Algranti, E., Handar, A.M., Dumortier, P., Mendonça, E.M., Rodrigues, G.L., Santos, A.M., Mauad, T., Dolhnikoff, M., De Vuyst, P., Saldiva, P.H., and Bussacos, M.A. (2005) Pneumoconiosis after sericite inhalation. *Occup. Environ. Med.* 62:e2.

Davies, D. and Cotton, R. (1983) Mica pneumoconiosis. *Br. J. Ind. Med.* 40:22–27.

Jones, R.N., Weill, H., and Parkes, W.R. (1994) Diseases related to non-asbestos silicates. In: Parkes W.R. editor. *Occupational Lung Disorders*, 3rd ed., Oxford: Butterworth-Heinemann, pp. 536–570.

Landas, S.K. and Schwartz, D.A. (1991) Mica-associated pulmonary interstitial fibrosis. *Am. Rev. Respir. Dis.* 144:718–721.

Skulberg, K.R., Gylseth, B., Skaug, V., and Hanoa, R. (1985) Mica pneumoconiosis—a literature review. *Scand. J. Work. Environ. Health* 11;65–74.

Zinman, C., Richards, G.A., Murray, J., Phillips, J.I., Rees, D.J., and Glyn-Thomas, R. (2002) Mica dust as a cause of severe pneumoconiosis. *Am. J. Ind. Med.* 41:139–144.

SLATE

Slate is a metamorphic, fine-grained rock derived from sedimentary deposits. Its components include silica, aluminum silicates, and small amounts of chlorite, hematite, magnetite, and various carbonates. The flaky crystal structure is, in part, from mica and other constituents of slate, including quartz and other minerals.

There are a number of reports of pneumoconiosis in workers who handle slate, as well as in those who mine and quarry it. The mining and processing of slate may result in concurrent exposure to silica and silicosis. Dust control measures have been successfully implemented in many areas to reduce silica and slate dust exposure.

Oldham et al. (1986) conducted a mortality study of 725 men exposed exclusively to slate dust between 1975 and 1981. There was a high prevalence of pneumoconiosis, particularly among smokers. However, this ailment did not significantly contribute to mortality. Smokers in the cohort had a 76% greater risk of mortality than nonsmokers compared with a 50% greater mortality for smokers in the control group. Mortality associated with slate dust exposure was 0.25% for nonsmokers and 26% for smokers.

Reports vary from descriptions of slight X-ray changes to severe dust disease, inhalation of varying percentages of free silica explains this difference because the amount of such free silica is said to vary from 7% to as much as 35%.

REFERENCE

Oldham, P.D., Bevan, C., Elwood, P.C., and Hodges, N.G. (1986) Mortality of slate workers in north Wales. *Br. J. Ind. Med.* 43:550–555.

CALCIUM DUSTS

This category includes lime, limestone, gypsum, marble, and cement, which is a complex compound of lime. Many dusts of calcium are soluble in lung tissue and thus cannot cause pneumoconiosis. Most series of cases reported for workers with heavy or long exposures to calcium dusts describe ill-defined or no chest X-ray changes. Workers disabled as a result of exposure to calcium dusts are considered to suffer damage because of the silica content of such dust.

Because cement varies in composition, the literature on its effects on the lungs is confusing. Cement is described as 62% calcium oxide (CaO) and 22% silicon dioxide (SiO_2), but the amount of silica may be reduced to 6.5% in some processes and only 1% in finished cement. Although there are scattered case reports of pneumoconiosis diagnosed by X-ray or at necropsy, surveys of cement workers suggest that cement remains a relatively harmless dust. Animal studies have supported the view that cement dust is biologically harmless. Without proper engineering controls and personal protection, however, cement remains irritating and causes dermatitis, rhinitis, and pharyngitis.

Vestbo and Rasmussen (1990) and Vestbo et al. (1991); in their two studies of 546 cement factory workers, concluded that long-term exposure to cement dust did not lead to higher morbidity from severe respiratory disease and did not increase the risk of lung cancer. However, Zeleke et al. (2011) concluded that there is a higher prevalence in lung function reduction and chronic respiratory symptoms among workers in the cement industry due to high cement dust exposure.

REFERENCES

Vestbo, J., Knudsen, K.M., Raffn, E., Korsgaard, B., and Rasmussen, F.V. (1991) Exposure to cement dust at a Portland cement factory and the risk of cancer. *Br. J. Ind. Med.* 48:803–807.

Vestbo, J. and Rasmussen, F.V. (1990) Long-term exposure to cement dust and later hospitalization due to respiratory disease. *Int. Arch. Occup. Environ. Health.* 62:217–220.

Zeleke, Z.K., Moen, B.E., and Bråtveit, M. (2011) Lung function reduction and chronic respiratory symptoms among workers in the cement industry: a follow up study. *BMC Pulm. Med.* 11:50.

94

SYNTHETIC VITREOUS FIBERS

THOMAS TRUNCALE AND YEHIA Y. HAMMAD

Target organ(s): Skin, eye, respiratory

Occupational exposure limits:

OSHA:

Regulated in *General Industry* as inert or nuisance dust:

Respirable fraction: 5 mg/m^3

Total dust: 15 mg/m^3

Shipyard: Fibrous glass

Respirable fraction: 5 mg/m^3

Total dust: 15 mg/m^3

Shipyard: Mineral wool

Respirable dust: 5 mg/m^3

Total dust: 15 mg/m^3

NIOSH: Fibrous glass dust:

Total dust: 5 mg/m^3

Fibers with diameter equal or <3.5 μm, and length ≥10 μm: 3 f/cc

RCF: REP 0.5 f/cc with action level of 0.25 f/cc

ACGIH: Synthetic vitreous fibers

Continuous filament glass fibers: 1 f/cc*A4

Continuous filament glass fibers: 5/mg/m^3** A4

Glass wool fibers: 1 f/cc* A3

Rock wool fibers: 1 f/cc* A3

Slag wool fibers: 1 f/cc* A3

Special purpose glass fibers: 1 f/cc* A3

Refractory ceramic fibers: 0.2 f/cc* A2

MRLs: A minimal risk level (MRL) of 0.03 WHO fibers/cc has been derived for chronic-duration inhalation exposure to refractory ceramic fibers

Risk/safety phrases: [list information for classification and labeling]

Post warning labels and signs (in English and the predominant language of workers who do not read English, or verbal) describing the health risks associated with RCF at entrances to work areas and inside work areas where airborne concentrations of RCF may exceed the REL

Depending on work practices and the airborne concentrations of RCF's state on the signs the need to wear protective clothing and the appropriate respiratory protection for RCF exposures above the REL

BACKGROUND AND USES

Man-made mineral fibers as the name suggests do not occur in nature. It is a generic name for a group of silica-based inorganic fibers manufactured from varying concentrations of rock, slag, glass, clay, and processed inorganic oxides. Other names include man-made vitreous fibers (MMVF), vitreous fibers,

*Fibers longer than 5 μm; diameter <3 μm; aspect ratio >5 : 1 as determined by the membrane filter method at 400–450× magnification (4-mm objective) phase contrast illumination.

**Measured as inhalable particle

A2: Suspected human carcinogen

A3: Confirmed animal carcinogen

A4: Not classifiable as a human carcinogen

Hamilton & Hardy's Industrial Toxicology, Sixth Edition. Edited by Raymond D. Harbison, Marie M. Bourgeois, and Giffe T. Johnson.
© 2015 John Wiley & Sons, Inc. Published 2015 by John Wiley & Sons, Inc.

manufactured vitreous fibers, synthetic vitreous fibers, glass wool (GW), continuous glass filament, special purpose glass fiber, or mineral wool. Depending on the process, MMVF are produced as a mass of tangled, discontinuous fibers of varying length and diameters or as filaments, which are continuous fibers of intermediate length with diameters that are more uniform and typically thicker than wool (IARC, 2002). Their noncrystalline or amorphous molecular structure facilitates early removal from the lung parenchyma, which differentiates them from crystalline forms of fibers.

In an effort to reflect the changes in industry as well as consider the variability in toxicity and the potential for causing cancer among the groups of MMVF, the World Health Organization (WHO) updated and expanded the classification of man-made mineral fibers into two main categories: filaments that include continuous glass filament and wools that include glass (insulation and special purpose), rock (stone), slag, refractory ceramic fibers (RCF), and other unspecified fibers, including high temperature (HT), high alumina, low silica, or alkaline earth silicate (AES) wools. The term whisker is used for thin inorganic fibers in crystalline form, which are placed in a different category. Examples include potassium titanate, potassium octatitanate, and silicon carbide, which unlike amorphous inorganic forms of MMVF have definite fibrogenic potential in lung tissue. In addition to the inorganic form, man-made fibers also occur in the organic form. These include natural polymers, most commonly viscose made from cellulose, or synthetic polymers, which include acrylic, polyester, polyurethane, and nylon (Figure 94.1).

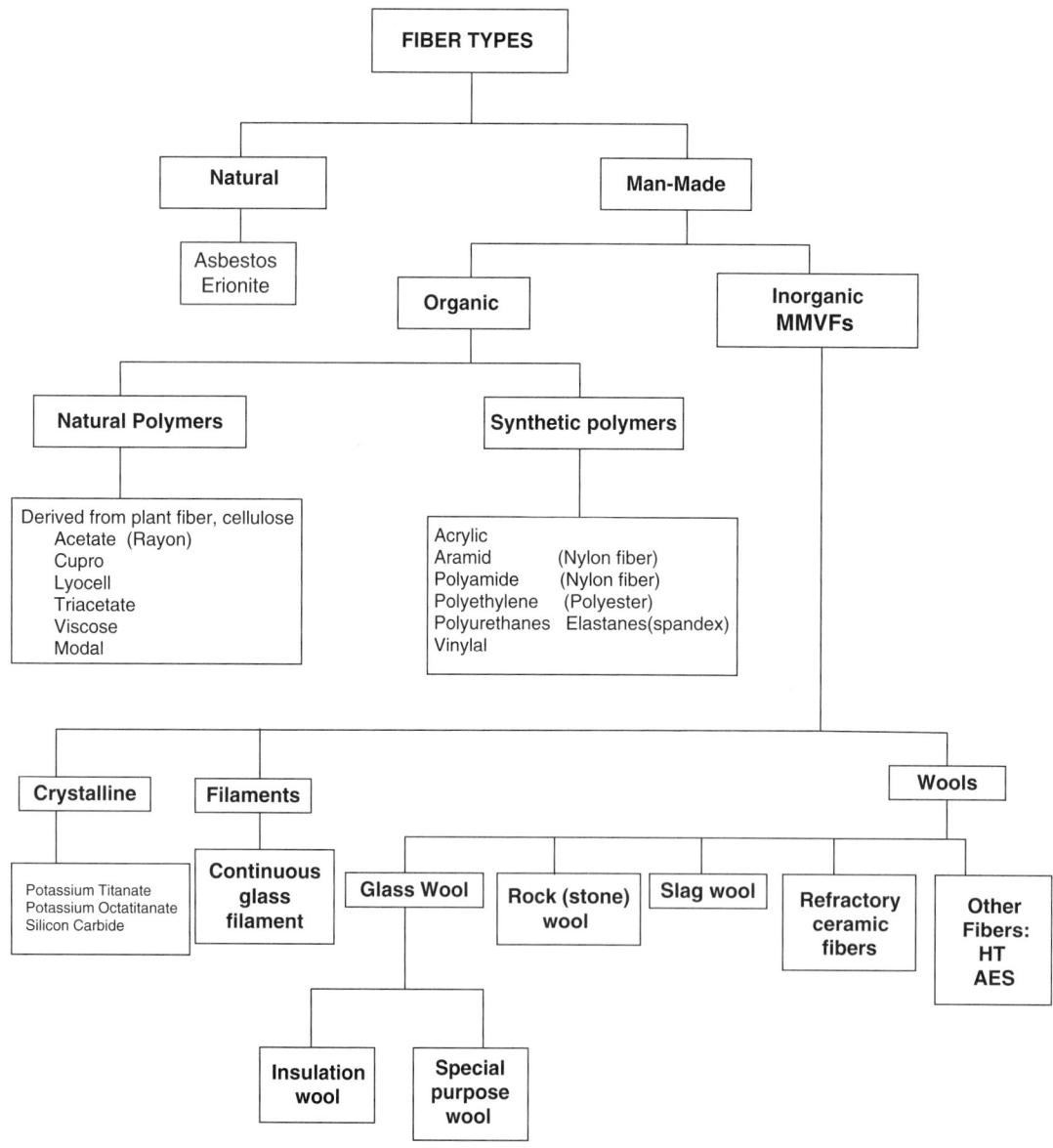

FIGURE 94.1 Classification of man-made fibers.

Historically, organic fibers have not been associated with occupational lung disorders until recently when a cluster of unexplained interstitial lung disease cases developed in the nylon flocking industry (Kern et al., 1998). In 2002, the WHO estimated that over 9 million tons of MMVF were produced annually in over 100 factories around the world. Primary applications include thermal or acoustical insulation where GW (~3 million tons, used predominately in the North America), and rock (stone) and slag wools (~3 million tons, used predominately in Europe and the rest of the world) are used in many cases as a replacement for asbestos. More recently, rock (stone) and slag wools have been replaced by high alumina, low silica wools (~1 million tons) for these applications. Special purpose glass developed for its durability is used in applications that require thermal insulation, such as the aircraft industry as well as for other purposes, including filtration media and batteries. Continuous glass fibers (~2 million tons) are used as reinforcement of plastics and textiles in the automobile, electrical, and building industries. Refractory ceramic fibers (150 thousand tons) are most often used during HT applications as insulation for furnaces and heaters (IARC, 2002). Alkaline earth silicate wools (~20,000 tons), developed in the 1980s in response over the fibrogenic and malignant potential of RCF, are a new class of fibers capable of being used as a substitute in a wide range of HT applications (Brown et al., 2012).

PHYSICAL AND CHEMICAL PROPERTIES

Man-made mineral fibers are divided based on the method of production, chemical composition, and application. Glass fibers, which include continuous filament fibers and GW, consist mainly of silicon dioxide and varying amounts of intermediate stabilizers, including aluminum, zinc, and titanium. Other oxides acting as modifiers include lithium, potassium, calcium, magnesium derived from dolomite, or boric acid from calcium borate. Varying the concentration of stabilizers and modifiers will alter the chemical and physical properties of the fiber resulting in changes in durability, heat and water resistance, and in solubility. The average diameter of continuous glass fibers ranges from 3 to 25 μm with lengths solely dependent on the industrial process. Ordinary GW diameters range from 3 to 15 μm and special purpose GW have diameters ranging from 0.1 to 10 μm. Category "c" fibers are considered special application fibers and make up <1% of the total glass fibers produced. The ability to distinguish between several types of glass fibers depends not only on the dimensions but also on their composition (Table 94.1).

Mineral wool is a term used to describe rock (stone) and slag wools in the United States, although in Europe, GW is included in the category. Rock and slag wools are composed mainly of calcium, aluminum, or magnesium silicates (- Table 94.2). Rock wool (RW) is derived by heating igneous rocks (basaltic, diabase, or olivine), which are classified based upon their alkali–silica content. Slag wool is produced by melting and fiberizing slag (wastes) from furnace iron and other raw materials, including clay, sand, and limestone. Naturally, the chemical properties will be dependent on the content of the slag that is used. Bonded wool is produced with the addition of urea–phenolic resin and is processed into ceiling tiles or used for blown insulation or other insulation material. The manufacturing process using centrifugation produces discontinuous fibers with diameters ranging from 3.5 to 7 μm.

TABLE 94.1 Composition of Various Glass Fibers

Component	Type of Glass Fiber, % by Weight				
	C	D	E	S	AR
SiO_2	55–65	72–75	52–56	55–85	60–70
AL_2O_3	1–5	0	12–16	10–35	0–5
Fe_2O_3	0.1–0.3	0	0–0.5	0–0.5	0.1
CaO	7–14	0	16–25	0	0–10
MgO	2–4	0	0–6	4–25	0
B_2O_3	6–9	0–23	7–13	0	0
ZrO_2	0	0	0	0	15–22
Na_2O	8–16	0–4	0	0	10–15
K_2O	0.4–0.7	0–4	0–0.2	0	0–2
TiO_2	0.02	0	0	0	0–5

Source: Compiled from ATSDR (2004), Lee et al. (1981), Naval Environmental Health Center (1997).

C fibers (chemical glass) typically glass wool and resistant to acids.

D fibers are sold a textile yarns.

E fibers (electrical glass) resistant to water and used for electrical applications.

S glass has the highest tensile strength and stiffness.

AR glass (alkali resistant) used for strengthening cement.

TABLE 94.2 Fiber Composition by Percent Weight of Mineral Wool and Ceramic Fibers

Component	Rock Wool	Slag Wool	Kaolin	Zirconia	Aluminum Silicate and Al Silicate	High Purity AES Wool	High Alumina Low Silica
SiO_2	37–53	32–52	49–53	5–50	95–97	50–82	33–42
AL_2O_3	6–14	5–15	43–50	5–95	95–97	<2	18–24
Fe_2O_3	0.5–11.6	0–8.2	0.06–2	<0.05	0.5–1		
CaO	10–30	10–43	<0.1	<0.05	<0.05	18–43	23–33
MgO	6–16	4–15	<0.1	0.01	0.01		
B_2O_3	0	0	0	0	0	<1	
ZrO_2	0	0	0.1	15–92	0	0–6	
Na_2O	1–3.5	0.8–3.3	0–0.5	<0.3	0.1–0.2	<1	1–10
K_2O	0.5–2	0.3–2	0.03–2	<0.01	0		
TiO_2	0.5–3.5	0.4–2.7	0.02–1	0.04	0.7–1.3		0.5–3
FeO	3–12	0–2	0	<0.05	0		3–9

Source: Compiled from ATSDR (2004), Lee et al. (1981), Naval Environmental Health Center (1997), Brown et al. (2012).

Refractory ceramic fibers sometimes called refractory fibers or ceramic fibers are characterized by their ability to withstand temperatures as high as 3000 °F and are used primarily for HT or aerospace applications. Refractory fibers consist mainly of kaolin clay, aluminum silicate with various metallic oxides (chromous, zirconia), or highly purified aluminum silicate (Table 94.2). Where HT applications are needed, fibers can contain more the 90% zirconia oxide. Ceramic fibers are unique in that they occur initially as amorphous structures but when heated the alumina–silica matrix changes to mullite, forming an alumininosilicate crystalline compound (Naval Environmental Health Center, 1997; IARC, 2002). As temperatures approach 1100 °C, excess silica is crystallized in a process known as devitrification to form crystabalite, which is considered to be carcinogenic to humans. Fibers containing higher concentrations of aluminum or zirconium oxides are able to retain their chemical structure and physical characteristics even when exposed to HTs. The average diameter of RCF ranges from 1 to 5 μm (Naval Environmental Health Center, 1997; WHO, 2000).

Alkaline earth silicate wools are produced by melting a combination of silica, calcium and magnesium oxide, alumina, titania, zirconis, and trace oxides. Commercially available in the 1990s they are primarily used in industrial equipment, fire protection, and automobile exhaust systems.

Manufacturing Process

The production of MMVF first involves liquefying raw materials, including sand, kaolin, aluminum silicate, or igneous rock, where they can be solidified and used at a later date or they can be used immediately. Continuous fibers or filaments are produced exclusively from glass, as the molten material is extruded using a set of bushings or spinerettes. Insulation wools are constructed as molten raw materials are extruded through small holes in a rotary or centrifugal method, producing fibers of variable diameters. When the molten glass emerges from the holes, they are stretched by a rotary or centrifical force and cooled by air into individual fibers. The diameters of these fibers can be adjusted with great accuracy and reproducibility to 6–25 μm. Flame attenuation refers to the special purpose glass fibers that are remelted by a jet flame blast after they are stretched into individual fibers. They are typically used for extreme temperature applications such as the skin of the space shuttle.

Once the fiber is made, various binding resins, including phenol formaldehyde, are used to give structure and rigidity to the fiber. Because glass filaments are fragile, sizings are added and used as a protective coating to increase the adhesion between the fibers. These include polyvinyl silane, epoxy silane, and polyvinyl acetate chrome chloride mixtures. Other lubricating or paraffin oils can be added to decrease dust generation during production of the end product (Naval Environmental Health Center, 1997; ATSDR, 2004).

MAMMALIAN TOXICOLOGY

Acute Effects

Skin irritation is the most common health effect associated with an exposure to MMVF. Irritation is usually related to mechanical trauma from coarse fibers measuring 4–6 μm in diameter. Irritation is reduced or resolves completely with continued exposure but returns when there is a lapse or interruption in an exposure for a few days (Lockey and Ross, 1998; Stam-Westerveld et al., 1994). Less commonly, allergic contact dermatitis resulting from sensitization to epoxy resins or hardening by-products used for finishing glass fibers or reinforced plastics have been described (Minamoto et al., 2002; Nogueira et al., 2011; Jolanki et al., 1990).

Symptoms of dry cough, eye, nose, and throat irritation have also been described (Burge et al., 1995) and appear associated with dusty working conditions >1 f/cc involving removal of fiberglass materials in closed spaces without respiratory protection (ATSDR, 2004).

Chronic Effects

Reports of airflow obstruction (Hansen et al., 1999), decrements in forced expiratory volume (FEV_1) (Trethowan et al., 1995; Clausen et al., 1993) irregular opacities, and pleural plaques (Lockey et al., 2002) have been described in workers exposed to MMVF; however, no consistent evidence for increased prevalence of any of these exists.

The National Institute for Occupational Safety and Health (NIOSH) (NIOSH, 2006) concluded that an exposure to RCF may pose a carcinogenic risk based on the results of chronic animal inhalation studies. Likewise, the National Toxicology Program (NTP) concluded that inhalable glass fibers and ceramic fibers of respirable size (3 μm in diameter and length to width (aspect) ratio of 3 : 1) were reasonably anticipated to be human carcinogens based on animal studies (National Toxicology Program). Despite this, epidemiologic studies of workers exposed to MMVF have failed to reveal an increased risk of respiratory cancer or mesotheliomas. In 2009, Lipworth conducted a systemic review and meta-analysis of risk estimates of lung and head and neck (HN) cancer in epidemiologic studies of workers exposed to MMVF, specifically RW and GW and concluded:

> despite a small elevation in RR of lung cancer among MMVF production workers, the lack of excess risk among end users, the absence of any dose-risk relation, the likelihood of detection bias, and the potential for residual confounding by smoking and asbestos exposure argue against a carcinogenic effect of MMVF, RW or GW at this time. Similar conclusions apply to HN cancer risk among workers exposed to MMVF
>
> (Lipworth et al., 2009).

Marsh performed a systemic review of the literature published following the 2001 International Agency for Cancer Research (IARC) decision to downgrade insulation GW from a Group 2B to Group 3 carcinogen. Utilizing selected Bradford Hill criteria for the evaluation of epidemiologic associations, they found no statistical significant increase in the incidence of respiratory cancer in workers exposed to MMVF (Marsh et al., 2011). Similarly, as others have mentioned (SCOEL/SUM/88 March, 2012), the NTP concludes,

> the data available from studies in humans inadequate to evaluate the relationship between human cancer and exposure to glass wool fibers. Although studies of occupational exposure found excess lung-cancer mortality or incidence, it is unclear that the excess lung cancer was due to exposure

specifically to glass wool fibers, because (1) no clear positive exposure-response relationships were observed (however, misclassification of exposure is a concern), and (2) the magnitudes of the risk estimates were small enough to potentially be explained by co-exposure to tobacco smoking.

The investigators further conclude that the data available from epidemiological studies are inadequate to evaluate the relationship between human cancer and exposure specifically to respirable ceramic fibers (National Toxicology Program).

Mechanism of Action(s)

Fiber toxicity is related to three essential factors that are as follows: An exposure to a sufficient *dose* of respirable fibers with the correct aerodynamic *dimensions* to allow deposition in the terminal portions of the lung (alveoli) and finally, after inhalation, *durability* of a fiber that will allow it to persist in the lung for long periods of time. In addition to the 3 Ds (dose, dimension, durability), alveolar macrophages, surface properties, the chemical composition and biopersistence of the fibers play a role in the elimination and half-life of the fiber.

Respirable fibers or dusts are defined by their ability to bypass the protection mechanisms of the upper airway and lodge on the alveolar surfaces of the lung. Lung deposition depends on the aerodynamic diameter of the fiber. Inhalable fibers, considered fibers with diameters of 3 μm or less with an aspect ratio (length/diameter ratio) >3, will be deposited in the lower airway. Fibers present particular problems because they can be of lengths >15–20 μm yet align parallel to the air current and be inhaled into the lower airway. Fiber dose is dependent on the length since long fibers are unable to be completely engulfed by the alveolar machrophages and thus cleared from the lung. Unlike asbestos fibers that tend to arrange in bundles and are split longitudinally, MMVF tend to occur as individual fibers and break transversely into shorter segments, which allows easier and earlier removal.

Once a fiber enters the lower respiratory tract, clearance is enhanced through removal by mucociliary transport or by the alveolar macrophage. Only those fibers <10 μm can be effectively phagocytized by alveolar machrophages (Oberdörster, 1997; Tran et al., 1996; SCOEL/SUM/88 March, 2012). Incomplete phagocytosis occurs with fibers >20 μm, which in turn leads to an inflammatory and fibrotic response that is associated with activation of various cell lines, including neutrophils, lymphocytes, mast cells, and fibroblasts. This stimulates the release of inflammatory mediators and chemokines (tumor necrosis factor α (TNFα), interleukin-1α, interleukin 6 and 8, basic fibroblast growth factor, MIP-1α, growth regulated peptide) and generates reactive oxygen and nitrogen species leading to epithelial and mesothelial cell proliferation (Ishihara, 2001; Churg et al., 2000; Driscoll, 1996; ATSDR, 2004; Morimoto et al., 1999).

Reactive oxygen species (ROS) have also been shown to induce cell injury and death by lipid peroxidation through interaction with fatty acids in the cell membrane. Alveolar macrophages isolated by bronchoalveolar lavage (BAL) and exposed to mineral and chrysotile fibers increase the production of superoxide anion and hydrogen peroxide, depleting the level of glutathione and increasing serum Ca in alveolar macrophages (Wang et al., 1999). Under more acidic conditions, the release of iron from MMVF, particularly refractory fibers, results in the production of hydroxyl free radicals in a similar way as amphibole asbestos (Brown et al., 1998). Further, the release of ROS through contact with highly durable fibers may activate signaling pathways and trigger the secretion of additional growth factors, proteases, and other proinflammatory cytokines (Nguea et al., 2008). Still other mechanisms of MMVF-induced cytotoxicity may involve the depletion of adenosine triphosphate (ATP) in the alveolar macrophage resulting in cell death (Kim et al., 2001). And finally, translocation of fibers from the lung parenchyma to the pleura may play a role in fiber-induced pleural disease (Gelzleichter et al., 1996).

The mechanism of fiber-induced genotxocity has been investigated using various types of MMVF that have been shown to produce 8-OHdG (possibly resulting in miscoding during DNA replication) (Jaurand, 1997), induce the occurrence of micronuclei and/or polynuclei abnormalities (Hart et al., 1992), stimulate morphologic transformation (Gao et al., 1995) and squamous metaplasia (preneoplastic characteristics) (Woodworth et al., 1983). Long fibers appear to induce the production of bi/trinucleated cells (an early event in asbestos-induced cancers) through incomplete or failed cytokinesis (Jenson and Watson, 1999) and result in oxidative DNA damage, all of which are considered precursors to the initiation of the cancerogenic process (Cavallo et al., 2004).

Chemical Pathology

As discussed, the dose and dimension play a key role in a fiber's ability to produce toxic effects in the lung. Fiber durability, the ability of a fiber to remain in the lung, is dependent on the dissolution rate (solubility) of the fiber. *In vitro* estimates of dissolution rate constants at varying pH for numerous fibers have been published and summarized by Maxim et al., 2006. The chemical components of the fiber influence the rate at which they dissolve. Man-made vitreous fibers composed of calcium, magnesium, and sodium tend to have faster dissolution rates, while alumina and silica tend to decrease the rate of dissolution making them less bio-soluble, thus allowing them to persist in the lung for longer periods of time. *In vitro* evidence suggests that slag wool and RCF dissolve more rapidly at a pH of 4 (simulating the environment following phagocytosis by the macrophage) as opposed to GW, which appears to dissolve more rapidly at a pH of 7.6

(simulating the extracellular environment). Thus the dissolution of fibers may involve both the intra- and extracellular environment and is accompanied by varying degrees of compositional and physical changes to the fiber. Changes in fiber morphology or dimension may in turn alter both the physical and surface chemical properties, ultimately affecting their overall biological reactivity (Bauer et al., 1994). Much of the current research regarding newer formulations of man-made mineral fibers has centered on the development of products that break down easier and faster yet retain their industrial functions. Newer fibers tend to have higher solubility, in essence, limiting the biopersistence and health effects. Biopersistence is therefore the duration a fiber remains in the lung and is determined by the physiologic/mechanical clearance as well as the chemical properties and dissolution rate of the fiber.

SOURCES OF EXPOSURE

Occupational

The IARC concluded that the worldwide average exposure levels during production, processing, and use of these fibers is thought to be generally <0.5 respirable fiber/cm^3 as an 8-h time-weighted average (TWA) (IARC, 2002). Those involved in the manufacture, installation, service, removal/repair, or use of these products are expected to be exposed to higher concentrations of the fibers.

Recently in the United States, the Health and Safety Partnership Program (HSPP), a workplace safety program for workers involved in the application of glass and mineral wool products and developed to create an occupational exposure database in the synthetic fiber industry, reported mean concentrations of 0.23–0.28 f/cc for the GW industrial sector and a mean of 0.38 f/cc for those involved in GW installation. Analysis of the mineral wool sector revealed a mean of 0.19 f/cc, with the highest levels in the manufacturing sector with a mean of 0.20 f/cc. In general, airborne concentrations of fibers measured during the production of glass and mineral wool (slag, and/or rock (stone)) are below 1 f/cc, except in areas where small diameter fibers (<1 μm) are produced (aircraft insulation, separation and filtration media) or specific products are used (blowing wool with binder) (Marchant et al., 2002).

Recently, an update to ongoing monitoring studies of an occupational exposure to RCF at plants that produce RCF and customer facilities in the United States was reported by Maxim et al. (2008), who found workers at RCF manufacturing plants to be exposed to TWA concentrations of ≤0.5 f/cc 95.8% of the time and when corrected for respirator use, 97.8% were at or below that level. The average TWA concentration of ceramic fibers adjusted for respiratory use from 2002 to 2006 was 0.28 f/cm^3 (NTP, 2011).

Environmental

Exposures to airborne MMVF in both occupational and environmental settings primarily occur as the result of construction, installation, maintenance/repair, physical damage, by degradation, or as MMVF are released to the environment over time. Much of the literature reporting on airborne levels of MMVF in commercial, residential, or ambient air fails to distinguish between organic (cellulose, cotton, nylon), inorganic mineral fibers (silicates and sulfates), or MMVF. Despite these limitations, it appears that an exposure to MMVF in indoor or environmental air is 2–3 orders of magnitude less than the exposure that occurs in an occupational environment (IARC, 2002). Mean air concentrations in 79 buildings containing MMVF products (Gaudichet et al., 1989) ranged from none detectable to $0.006\,f/cm^3$, in 16 schools none detectable to $0.08\,f/cm^3$ (Schneider, 1986), and more than 130 measurements of indoor air in various locations (office buildings, schools, laboratories, private homes) was none detectable to $0.038\,f/cm^3$ (IARC). In the 1990s, Carter et al. (1999) sampled residential and commercial buildings while taking 21 simultaneous samples from 19 locations using scanning electron microscopy (SEM) to distinguish among the fiber types. The indoor respirable fiber levels averaged 0.008 f/cm with a maximum of 0.029 f/cc using phase contrast optical microscopy (PCOM). Ninety seven percent of the respirable fibers identified by SEM were determined to be organic. Inorganic fiber levels measured indoors using SEM averaged <0.0001 f/cc. The ambient air samples revealed a mean of 0.002 f/cc by PCOM.

INDUSTRIAL HYGIENE

The Occupational Safety and Health Administration (OSHA) estimates that there are more than 225,000 workers in the United States who are exposed to MMVF in manufacturing and end-use applications (OSHA, 2002). Mineral fibers are currently regulated only as a nuisance dust in general industry as OSHA has no specific permissible exposure limit (PEL) for an occupational exposure to fibrous glass. Inert or nuisance dust is regulated to $15\,mg/m^3$ for total dust and $5\,mg/m^3$ for respirable fractions. For employees involved in ship repairing, shipbuilding, shipbreaking or related employments, fibrous glass, and mineral wool (including rock or slag), the PEL is $15\,mg/m^3$ for total dust and $5\,mg/m^3$ for respirable fractions. In 1999, the OSHA partnered with various trade organizations and major manufacturers of fiberglass, slag and rock wool insulation products to institute a voluntary HSPP for fiberglass and mineral wool. The program established a voluntary PEL of 1 f/cc 8-h TWA for respirable synthetic mineral fibers (glass/rock/slag wool). The HSPP also established a database, monitored by the North American Insulation Manufacturers Association) (NAIMA), of representative exposure limits for workers involved in manufacturing and end-use applications. It was also recommended that workers wear NIOSH-certified dust respirators when the PEL is exceeded or when performing certain tasks such as blowing SVF insulation into attics and other places and when demolishing buildings.

The current guidelines from the American Conference of Governmental Industrial Hygienists (ACGIH) adopted a threshold limit value (TLV) for an 8-h TWA of 1 f/cc for continuous glass filaments with a A4 (not classifiable as a human carcinogen) rating and 1 f/cc for GW, rock and slag wools, and special purpose fibers with an A3 (confirmed animal carcinogen with unknown relevance to humans) rating. A TLV–TWA of 0.2 f/cc was adopted for RCF with an A2 (suspected human carcinogen) rating (ACGIH, 2001).

In 1977, the NIOSH established a recommended exposure limit of (REL) of $3\,f/cc^3$ as a TWA for glass fibers with diameters $\leq 3.5\,\mu m$ and lengths $\geq 10\,\mu m$ for up to 10 h/day during a 40-h workweek. The NIOSH also recommended that airborne concentrations determined as total fibrous glass be limited to a $5\,mg/m^3$ of air as a TWA. In an effort to minimize the risk of lung cancer and irritation of the eyes and upper respiratory tract, the NIOSH proposed a REL for RCF of $0.5\,f/cm^3$ of air as a TWA and an action level of $0.25\,f/cm^3$ (used for determining when additional actions should be taken to reduce RCF exposures) (NIOSH, 2006). They further recommended that all reasonable efforts be made to reduce exposures to $<0.2\,f/cm^3$ where cancer risk estimates between 0.03/1000 and 0.47/1000 have been extrapolated from risk models (Moolgavkar et al., 1999; Maxim et al., 2003).

Exposure control and prevention is aimed at limiting exposure through a combination of engineering controls and workplace practices. Airborne concentrations can be reduced by adequate ventilation and local exhaust systems, tool selection, modifying workplace practices, and isolation of the manufacturing process to enclose airborne fibers and separate them from the workers. Personal protective equipment is geared at reducing skin, eye, and respiratory irritation and should include long sleeves, gloves, and eye protection when performing dusty activities involving MMVF. When airborne concentrations of RCF exceed the REL, the NIOSH recommends the following respiratory protection: At a minimum, use a half-mask, air-purifying respirator equipped with a 100 series particulate filter (this respirator has an assigned protection factor (APF) of 10. For a higher level of protection and for prevention of facial or eye irritation, use a full face-piece, air-purifying respirator (equipped with a 100 series filter) or any powered, air-purifying respirator equipped with a tight-fitting full facepiece. For greater respiratory protection when the work involves potentially high airborne fiber concentrations (such as removal of after-service RCF insulation such as furnace insulation), use a supplied-air respirator equipped with a full facepiece, since airborne exposure to RCFs can be high and unpredictable (NIOSH, 2006).

MEDICAL MANAGEMENT

For workers exposed to MMVF, a medical monitoring program should be established by providing a baseline medical examination within 3 months of employment. The exam should include a detailed occupational history that gathers information about past jobs, dust or fiber exposures and should include a physical exam with emphasis placed on the skin and respiratory tract, spirometry, chest X-ray, and other tests deemed necessary by the health-care provider. Suggested guidelines for the frequency of periodic examinations have been published by the NIOSH who recommend examinations every 5 years for workers exposed to RCF with fewer than 10 years since first exposure and for those with 10 years or more since first exposure, every 2 years. More frequent evaluations may be required depending on the dose, duration, and intensity of exposures, new or worsening respiratory symptoms, recurrent or chronic dermatitis, exposure to other respiratory hazards (i.e., asbestos), or changes in the job process. Programs should be monitored periodically to identify patterns where workplace practices are linked to worker health and well-being. In addition, medical monitoring programs may be useful in identifying changes needed in job processes or specific exposure conditions.

REFERENCES

ACGIH (2001) *Synthetic Vitreous Fibers. Supplement to Documentation of the Threshold Limit Values and Biological Exposure Indices*, Cincinnati, OH: American Conference of Governmental Industrial Hygienists.

ATSDR (2004) *Toxicological Profile for Synthetic Vitreous Fibers*, Atlanta, GA: U.S. Department of Health and Human Services, Public Health Service, Agency for Toxic Substances and Disease Registry.

Bauer, J.F., et al. (1994) Dual pH durability studies of man-made vitreous fiber (MMVF). *Environ. Health Perspect.* 102(Suppl. 5):61–65.

Brown, D.M., et al. (1998) Free radical activity of synthetic vitreous fibers: iron chelation inhibits hydroxyl radical generation by refractory ceramic fiber. *J. Toxicol. Environ. Health* 53:545–561.

Brown, R.C., et al. (2012) Alkaline earth silicate wools—a new generation of high temperature insulation. *Regul. Toxicol. Pharmacol.* 64:296–304.

Burge, P.S., et al. (1995) Are the respiratory health effects found in manufacturers of ceramic fibres due to the dust rather than the exposure of fibres? *Occup. Environ. Med.* 52:105–109.

Carter, C.M., et al. (1999) Indoor airborne fiber levels of MMVF in residential and commercial buildings. *Am. Ind. Hyg. Assoc. J.* 60:794–800.

Cavallo, D., et al. (2004) Cytotoxic and oxidative effects induced by man-made vitreous fibers (MMVFs) in a human mesothelial cell line. *Toxicology* 201:219–229.

Churg, A., et al. (2000) Pathogenesis of fibrosis produced by asbestos and man-made mineral fibers: what makes a fiber fibrogenic? *Inhal. Toxicol.* 12:15–26.

Clausen, J., Netterstrom, B., and Wolff, C. (1993) Lung function in insulation workers. *Br. J. Ind. Med.* 50:252–256.

Driscoll, K.E. (1996) Effects of fibres on cell proliferation, cell activation and gene expression. *IARC Sci. Publ.* 140:73–96.

Gao, H.G., et al. (1995) Morphological transformation induced by glass fibers in BALB/c-3T3 cells. *Teratog. Carcinog. Mutagen.* 15:63–71.

Gaudichet, A., et al. (1989) Levels of atmospheric pollution by man-made mineral fibres in buildings. In: Bignon, J., Peto, J., and Saracci, R., editors. *Non-Occupational Exposure to Mineral Fibres*, IARC Scientific Pub. No. 90. Lyon, France: IARC, pp. 291–298.

Gelzleichter, T.R., et al. (1996) Pulmonary and pleural responses in Fischer rats following short-term inhalation of a synthetic fiber. II. Pathobiologic responses. *Fundam. Appl. Toxicol.* 30:39–46.

Hansen, E.F., et al. (1999) Lung function and respiratory health of long-term fiber-exposed stonewool factory workers. *Am. J. Respir. Crit. Care Med.* 160:466–472.

Hart, G.A., et al. (1992) Cytotoxicity of refractory ceramic fibres to Chinese hamster ovary cells in culture. *Toxicol. In Vitro* 6:317–326.

IARC (2002) Man-made vitreous fibers. In: *IARC Monographs on the Evaluation of Carcinogenic Risks to Humans*, vol. 81, Lyon, France: International Agency for Research on Cancer.

Ishihara, Y. (2001) *In vitro* studies on biological effects of fibrous minerals. *Ind. Health* 39:94–105.

Jaurand, M.C. (1997) Mechanisms of fiber-induced genotoxicity. *Environ. Health Perspect.* 105(S5):1073–1084.

Jenson, C.G. and Watson, M. (1999) Inhibition of cytokinesis by asbestos and synthetic fibers. *Cell Biol. Int.* 23:829–840.

Jolanki, R., et al. (1990) Occupational dematoses from epoxy resin compounds. *Contact Dermatitis* 23:172–183.

Kern, D.G., et al. (1998) Flock workers lung: chronic interstitial lung disease in the nylon flocking industry. *Ann. Intern. Med.* 129:261–272.

Kim, K.A., et al. (2001) Depletion of intracellular ATP in the AM after refractory ceramic fibers or rockwool exposure diminishes cell viability and induces cytotoxicity. Mechanism of refractory ceramic fiber- and rock wool-induced cytotoxicity in alveolar macrophages. *Int. Arch. Occup. Environ. Health* 74:9–15.

Lee, K.P., et al. (1981) Pulmonary response and transmigration of inorganic fibers by inhalation exposure. *Am. J. Pathol.* 102:314–323.

Lipworth, F., et al. (2009) Occupational exposure to rock wool and glass wool and risk of cancers of the lung and the head and neck: systemic review and meta-analysis. *J. Occup. Environ. Med.* 51:1075–1087.

Lockey, J.E., et al. (2002) A longitudinal study of chest radiographic changes in workers in the refractory ceramic fiber industry. *Chest* 121:2044–2051.

Lockey, J.E. and Ross, C.S. (1998) Health effects of man-made fibres. In: Stellman, J.M., editor. *Encyclopaedia of Occupational*

Health and Safety, 4th ed., vol. 1, Geneva, Switzerland: International Labour Office, pp. 74–78.

Marchant, G.E., et al. (2002) A synthetic vitreous fiber (svr) occupational exposure database: implementing the SVF Health and Safety Partnership Program. *Appl. Occup. Environ. Hyg.* 17:276–285.

Marsh, G.M., et al. (2011) Fiberglass exposure and human respiratory system cancer risk: lack of evidence persists since 2001 IARC re-evaluation. *Regul. Toxicol. Pharmacol.* 60:84–92.

Maxim, L.D., et al. (2008) Workplace monitoring of occupational exposure to refractory ceramic fiber—a 17-year retrospective. *Inhal. Toxicol.* 20:289–309.

Maxim, L.D., et al. (2006) The role of fiber durability/biopersistence of silica-based synthetic vitreous fibers and their influence on toxicology. *Regul. Toxicol. Pharmacol.* 46:42–62.

Maxim, D.D., et al. (2003) Quantitative risk analyses for RCF: survey and synthesis. *Regul. Toxicol. Pharmacol.* 38:400–416.

Minamoto, K., et al. (2002) Occupational dermatoses among fiberglass reinforced plastics factory workers. *Contact Dermatitis* 46:339–347.

Moolgavkar, S.H., et al. (1999) Quantitative assessment of the risk of lung cancer associated with occupational exposure to refractory ceramic fibers. *Risk Anal.* 19:599–611.

Morimoto, Y., et al. (1999) Effects of mineral fibers on the gene expression of proinflammatory cytokines and inducible nitric-oxide synthase in alveolar macrophages. *Ind. Health* 37:329–334.

National Toxicology Program, Department of Health and Human Services (2011) *Report on Carcinogens*, 12th ed., Research Triangle Park, NC: National Toxicology Program, Department of Health and Human Services.

Naval Environmental Health Center (1997) *Man-Made Vitreous Fibers. Technical Manual NEHC-TM6290.91 Rev. A.*

Nguea, H.D., et al. (2008) Macorphage culture as a suitable paradigm for evaluation of synthetic vitreous fibers. *Crit. Rev. Toxicol.* 38:675–695.

NIOSH (2006) *NIOSH (DHHS) Publication No. 2006-123; Occupational Exposure to Refractory Ceramic Fibers.*

Nogueira, A., et al. (2011) Systemic allergic contact dermatitis to fiberglass in a factory worker of wind turbine blades. *Cutan. Ocul. Toxicol.* 30:228–230.

Oberdörster, G. (1997) Pulmonary carcinogenicity of inhaled particles and the maximum tolerated dose. *Environ. Health Perspect.* 105(Suppl. 5):1347–1356.

Occupational Safety & Health Administration (OSHA) (2002) *Synthetic Mineral Fibers*. Available at https://www.osha.gov/SLTC/syntheticmineralfibers/.

Schneider, T. (1986) Manmade mineral fibers and other fibers in the air and in settled dust. *Environ. Int.* 12:61–65.

SCOEL/SUM/88 March 2012: Recommendation from the Scientific Committee on Occupational Exposure Limits for man made-mineral fibers (MMMF) with no indication for carcinogencicity and not specified elsewhere.

Stam-Westerveld, E.B., et al. (1994) Rubbing test responses of the skin to man-made mineral fibres of different diameters. *Contact Dermatitis* 31:1–4.

Tran, C.L., et al. (1996) Evidence of overload, dissolution and breakage of MMVF10 fibres in the RCC chronic inhalation study. *Exp. Toxicol. Pathol.* 48:500–504.

Trethowan, M., et al. (1995) Study of the respiratory health of workers in seven European plants that manufacture ceramic fibers. *Occup. Environ. Med.* 52:97–104.

Wang, Q., et al. (1999) Biological effects of man-made mineral fibers (i): reactive oxygen species production and calcium homeostasis in alveolar macrophages. *Ind. Health* 37:62–67.

WHO (2002) Chapter 8.2: Man-made fibers. In: *Air Quality Guidelines*, 2nd ed., Copenhagen, Denmark: WHO Regional Office for Europe. Available at http://www.euro.who.int/__data/assets/pdf_file/0004/123088/AQG2ndEd_8_2MMVF.pdf.

Woodworth, C.D., et al. (1983) Induction of squemous metaplasia in organ cultures of hamster trachea by naturally occurring and synthetic fibers. *Cancer Res.* 43:4906–4911.

95

ORGANIC DUSTS

David R. Johnson

INTRODUCTION

Organic dusts are dispersed airborne particles of animal or plant protein, bacteria, fungi, amoeba, and/or molds. The dust can often be complex mixtures of organic and inorganic substances.

SOURCES OF EXPOSURE

The respiratory system is exposed through inhalation of organic dust; then, depending on their size and shape, the component particles may deposit in the respiratory tract and interact with the mucous membranes and epithelial lining of the mouth, nose, sinuses, throat, bronchial system, and/or lungs. The skin and conjunctiva can be directly exposed to airborne particles. Organic dusts can also settle on food and be ingested, however, this is not a pathway commonly associated with organic dust-induced disease.

Exposure to organic dusts occurs in many occupations such as agriculture, veterinarian practices, laboratory workers, researchers, animal caretakers and groomers, woodworkers, food producers and processors, and other manufacturing and production operations. Exposure can also occur at home, during recreational activities, while enjoying hobbies and in other nonoccupational environments.

Thorough and skillfully performed exposure histories are needed due to the complex nature of organic dusts, variations in host characteristics and habits, and the potential for obscure exposures in various occupational and nonoccupational settings.

ASSOCIATED AGENTS AND ANTIGENS

Organic dusts can be composed of a wide variety of organic substances such as animal dander, hair, feathers, serum, urine, and feces; insect parts; microorganisms; plant fibers, leaves, grains, enzymes, and foods. These organic mixtures may contain direct irritants, toxins, endotoxins, high or low molecular weight proteins acting as antigens, and even carcinogens. Examples include: an endotoxin suspected of causing inflammation in organic dust toxic syndrome (ODTS); specific sensitizing and/or precipitating antigens being linked to cases of allergic asthma, and/or hypersensitivity pneumonitis (HP); and a few wood dusts reported to increase the risk of cancer (ACGIH, 2005, Wood Dust) (See Tables 95.1, 95.2 and 95.3).

MECHANISM OF ACTION

Due to the nature and complexity of organic dusts, current science frequently cannot firmly establish the specific putative agent, the mechanism of action, and/or the chemical pathology responsible for an associated disease. For example, a case of occupational extrinsic asthma is generally accepted to be due to IgE type I-mediated immune response; however, multiple mechanisms may be involved, including additional immunologic processes, pharmacological, or genetic mechanisms, as well as airway and neurogenic inflammation (Mapp et al., 1994). Several mechanisms may be at play at any one time. It is not clear if HP is a type III immune complex-mediated or a type IV cellular-mediated disease. The former is supported by the presence of

Hamilton & Hardy's Industrial Toxicology, Sixth Edition. Edited by Raymond D. Harbison, Marie M. Bourgeois, and Giffe T. Johnson.
© 2015 John Wiley & Sons, Inc. Published 2015 by John Wiley & Sons, Inc.

TABLE 95.1   Examples of Occupational Asthma Associated with Organic Dusts

Occupational Exposure	Agent
*Laboratory workers, researchers, veterinarians, agriculture/farmers, animal caretakers, meat processing, furriers (animal handler's disease)	Animal proteins (hair, dander, serum, urine, feces from rodents, rabbits, horses, dogs, cats, farm animals, etc.)
*Poultry workers, bird handlers, pigeon breeders, feather processing, caretakers, veterinarians (bird handler's disease)	Bird proteins (parakeets, pigeons, chickens ducks, etc.)
Growers, manufacturers, food industry work	Buckwheat
Oil producers, food processors, coffee bean handlers, farmers	Castor bean
*Coffee processors, coffee roasters	Coffee bean dust
*Textile workers and food processors (Byssinosis)	Cotton dust, oils, flax, hemp
*Detergent manufacturers, plastics, food processors, laboratories (detergent workers disease)	Enzymes (e.g., *Bacillus subtilis*, proteases, amylases, lipases, cellulases)
*Fishermen, food processors, shell grinders, shellfish preparation	Fish and mollusks (crabs, shrimp other shellfish)
*Bakers, millers, food processors (baker's asthma)	Flour: Wheat, rye, barley oats
*Farmers, growers, grain handlers (grain fever)	Grain dust: Wheat, barley, oats
Food processors/handlers	Garlic
Brewery workers	Hops
*Laboratory workers, researchers, breeders, farmers, sewage workers, entomologists, grain workers	Insect proteins (butterfly, cockroach, locusts, maggots, mites, moths, silkworms, weevils)
Humidifiers, office workers, woodworkers, cheese workers, *many potential exposures*	Molds
Mushroom growers/handlers	Mushrooms
Food processors	Pepper
Gardeners, landscapers, outdoor workers	Pollens
*Tea processing/manufacturing (tea maker's asthma)	Tea
Tobacco growers and processors	Tobacco
*Carpenters, timber millers, construction workers, sawmill workers, lumber production, furniture making, cabinet making, joinery, woodworkers	Wood dusts (*many* types of wood)

Those marked with an asterisk (*) are further described in the chapter. See text. Additional references include Avila and Lacey (1974), Bielory (1982), Butikofer et al. (1970), Cockcroft et al. (1983), Cormier et al. (1998), Duchaine et al. (1999), Flannigan (1987), Greene et al. (1981), Harper et al. (1970), Huuskonen et al. (1984), Ismaili et al. (2006), Johnson et al. (1981), Kagen et al. (1981), Millon et al. (2012), Morell (2003), Morell et al. (2008), Patel et al. (2001), Pimentel (1970), Popp et al. (1987), Reed et al. (1965), Riddle et al. (1968), Schlueter et al. (1972), Terho et al. (1980), Van den Bogart et al. (1993), Vincken and Roels (1984), Warren and Tse (1974), Winck et al. (2004), Yoshida et al. (1989), and Zenz (n.d.).

high titers of antigen-specific precipitating serum IgG, the latter by combined cell infiltration and granuloma formation (Girard and Cormier, 2010).

Organic dust toxic syndrome is thought to be a nonallergic and noninfectious condition associated with endotoxin exposure, however, the dusts involved are complex and other mechanisms are likely to be involved. Exposure to wood dust has been associated with nasal cancer; again however, the mechanisms are not clearly delineated.

CLINICAL AND DIAGNOSTIC CONSIDERATIONS

Inhalation of organic dusts has been associated with a wide spectrum of acute and chronic diseases of the respiratory system, including upper respiratory allergy, irritation and inflammation, bronchitis, asthma, HP, also called extrinsic allergic alveolitis (EAA), granulomatous disease, interstitial lung disease (fibrosis) and in a few cases, cancer. Skin or mucous membrane exposure, irritation, sensitization, and subsequent reactions are also reported. The clinical picture

varies depending on the status of the host, the organic dust involved, as well as the intensity and duration of exposure.

Linking clinical findings to specific organic dust exposure in the workplace can be challenging. Exposure must be well documented and temporally defined and be associated with established clinical signs and symptoms. Given this, and depending on the situation at hand, supportive diagnostic studies may include one or more of the following: a chest X-ray (CXR), pulmonary function testing (PFT), arterial blood gas (ABG) determinations, diffusing capacity (DLCO), a complete blood count (CBC), IgE antibodies, radioallergosorbent testing (RAST), enzyme-linked immunosorbent assays (ELISA), skin allergy testing (prick test, patch testing, end point titration), magnetic resonance imaging (MRI), and/or high resolution CAT scanning (HRCT). Additional testing that may be needed includes pulmonary biopsy, provocation inhalation challenge, and bronchoalveolar lavage (BAL).

As you review the discussions of specific disease syndromes in this chapter, it is helpful to have a general understanding of occupational asthma (OA), HP, and the difference between the two. The reader is also expected to

TABLE 95.2   Examples Hypersensitivity Pneumonitis Associated with Organic Dusts

Disease	Exposure	Suspected Antigens
*Animal handlers lung	e.g., farm, veterinary, caretaker, laboratory, and researcher animals	Animal hair, dander, serum and urine proteins
*Bagassosis	Moldy pressed sugar cane (bagasse)	*Thermoactinomyces sacchari, T vulgaris*
*Bird handlers disease (bird fanciers lung, bird breeder's lung, pigeon breeder's lung, poultry workers lung, etc.)	Various birds: pigeons, parrots, parakeets, chickens, turkey, ducks, lovebirds, etc.	Avian proteins, serum, feathers, droppings
*Cheese workers lung	Moldy cheese	*Penicillium casei* or *P. roqueforti*
*Coffee workers lung	Coffee bean dust	coffee bean dust, *Thermoactinomyces spp.*
Compost lung	Compost	*Aspergillus fumigatus*
Dry rot disease	Old houses (Europe)	*Merulius lacrymans*
*Farmer's lung	Moldy grain, hay, oats, barley, millet	*Thermophilic actinomycetes, Aspergillus species, Saccharopolyspora rectivirgula, (Micropolyspora faeni)*
Fertilizer workers lung	Dirt; contaminated fertilizer	*Streptomyces albus*
*Furrier's lung	Animal fur	Animal hair protein
Horseback riders lung	Horses	*Sporobolomyces spp.*
Malt workers lung	Moldy barley	*Aspergillus clavatus, Aspergillus fumigatus*
Maple bark strippers lung	Moldy maple bark	*Cryptostroma corticale*
Mushroom workers lung	Mushroom compost	*Thermophilic actinomycetes Thermoactinomycetes vulgaris, Micropolyspora faeni, Aspergillus spp*
Papermill workers lung	Moldy wood pulp/dust	*Alternaria species*
Paprika slicer's lung	Paprika	*Mucor stolanifer*
Peat moss workers lung	Peat moss	*Monocillium species* and *Penicillium citreonigrum*
Pituitary snuff-takers lung	Pituitary snuff	pituitary protein (porcine and bovine)
Sequoiosis	Redwood bark, moldy sawdust	*Aureobasidium pullulans, Graphium species*
Suberosis	Moldy cork dust	*Penicillium glabrum, Penicillium frequentans*
Summer-type hypersensitivity pneumonitis	Japanese wood houses, damp wood and mats	Trichosporon cutaneum
Thatched roof disease	Dried grass	*Saccharomonospora viridis*
Tobacco workers lung	Tobacco dust; fungal spores	*Aspergillus species*
Wheat weevils disease (Miller's lung)	Infested grain/wheat flour	Sitophilus granaries
Wine-growers lung	Moldy grapes	*Botrytis cinerea mold*
Wood pulp workers lung	Wood pulp, dusts	*Alternaria, Penicillium*

Those marked with an asterisk (*) are further described in the chapter. See text. Additional references include ACGIH (2005) , Anonymous (1986), Johnson et al. (1981), Mapp et al. (1994), Sigsgaard and Schlunssen (2004), and Zenz (n.d.).

have a prior understanding of occupational rhinitis, conjunctivitis, bronchitis, and dermatitis.

> *Occupational asthma:* Organic dusts have been associated with occupational and nonoccupational asthma; which can be irritant or allergic. Symptoms include dyspnea, wheezing, cough, and mucous production, and usually occur immediately after exposure to an inciting agent but can be delayed for hours or days. Pulmonary

function testing typically shows an obstructive pattern but can be normal between episodes. Pulmonary function testing done pre and post bronchodilator administration can help to measure airway hyperresponsiveness. Chest X-rays are usually normal; evidence of air trapping might be seen during an acute episode. Serial peek expiratory flow rates (PEFR) performed during a work-shift may be helpful in determining the work relatedness of asthma. Various allergy tests

TABLE 95.3   Examples of Cancer Associated with Organic (Wood) Dusts

Wood Dust Exposure*	Occupation	Cancer Classification
Oak, beech	Woodworkers, carpenters, processors, lumber mills	A1—confirmed human carcinogen
Birch, mahogany, teak, walnut	Woodworkers, carpenters, processors, lumber mills	A2—suspected human carcinogen
All other wood dusts	Woodworkers, carpenters, processors, lumber mills	A4—not classifiable as a human carcinogen

Those marked with an asterisk (*) are further described in the chapter. See text. Additional references include ACGIH (2005).

may help to establish allergic asthma and an associated antigen. Sigsgaard et al. described the following steps in the clinical evaluation of OA: (1) confirm a diagnosis of asthma, (2) determine a temporospatial relationship of exposure to symptoms and lung function, (3) determine if the disease at hand is an IgE- or a non-IgE-mediated disease, and (4) perform a specific or nonspecific challenge test. Challenge tests are important in order to distinguish a causal relationship from nonspecific hyper-responsiveness in persons with pre-existing asthma. In these situations, removing the patient from exposure with subsequent challenge tests might be the only way to find the answer (Sigsgaard and Schlunssen, 2004).

Hypersensitivity pneumonitis: Also called EAA, involves the inhalation of allergen and usually develops after years of exposure to organic dusts. A wide spectrum of antigens has been associated with HP and in some cases these antigens can be identified by laboratory testing for specific precipitating antibodies. It is important to remember that while the presence of precipitating antibodies supports the diagnosis of HP, their presence does not confirm disease; likewise, the failure to identify specific precipitating antibodies does not absolutely rule out HP. Hypersensitivity pneumonitis has been reported as acute, subacute, or chronic disease. There are current debates as to whether subacute HP is a clear separate entity. There is an overlap of symptomatology and some think that the classification should only include acute and chronic disease (Lacasse et al., 2009).

Signs and symptoms of HP include fever, malaise, cough, shortness of breath, weight loss, occasional fine basilar rales, and rarely wheezing. Other clinical findings include interstitial markings on chest radiographs, a lymphocytic alveolitis on BAL, and/or a granulomatous reaction on lung biopsies (Lacasse et al., 2012). Radiologic findings have included diffuse ground-glass opacification, centrilobular ground-glass opacities, air trapping, fibrosis, lung cysts, and emphysema. Pulmonary function testing studies typically show diminished lung volumes with a restrictive pattern, however, airway obstruction may also be seen in chronic cases. Impairment of gas diffusion and hypoxemia may be present. Bronchoalveolar lavage in HP shows high total cell and lymphocyte counts, moderate neutrophilia, mild eosinophilia, and mastocytosis (Caillaud et al., 2012; Ryu et al., 2007). While there are some reports of inversion of the CD4/CD8 cell count ratio in BAL fluid, various other studies have not shown consistent findings of the CD4/CD8 inversion (Ohshimo et al. (2012). The histopathologic process of HP consists of chronic inflammation of the bronchi and peribronchiolar tissue, often with poorly defined granulomas and giant cells in the interstitium or alveoli (Hirschmann et al., 2009).

The most widely used diagnostic criteria for HP are those from Richerdson et al. (1989) and these criteria include the

following: (1) the history and physical findings and pulmonary function tests indicate an interstitial lung disease, (2) the X-ray film (CAT scan) is consistent, (3) there is an exposure to a recognized cause, (4) there is an antibody to that antigen (Richardson et al., 1989; Girard et al., 2009). In a study by Lacasse et al. (2003) six significant predictors of HP were identified: (1) exposure to a known offending antigen, (2) positive precipitating antibodies to the offending antigen, (3) recurrent episodes of symptoms, (4) inspiratory crackles on physical examination, (5) symptoms occurring 4–8 h after exposure, and (6) weight loss.

Hypersensitivity pneumonitis vs. allergic asthma: Simply put, allergic (extrinsic) asthma and allergic rhinitis are both considered an IgE-mediated disease whereas the mechanism of HP is *not* considered to be an IgE-mediated disease. Hypersensitivity pneumonitis is thought to be either a type III immune complex and/or a type IV cell-mediated reaction demonstrating some pathologic findings to support both (Blatman and Grammer 2012). It is possible for both asthma and HP to be operant simultaneously in a given patient and as stated elsewhere in this chapter, multiple underlying mechanisms can share responsibility for a clinical picture.

Cancer: In the past, associations have been made between exposure to wood dusts and the development of nasal adenocarcinoma and carcinoma (ACGIH, 2005, Wood Dusts). Due to improvements in work practices and lower levels of dust exposure, it is thought that today's conditions in wood-working occupations do not pose an excess risk of nasal cancer for the worker.

INDUSTRIAL HYGIENE

Most organic dust components do not have specific exposure limits determined by the National Institute for Occupational Safety and Health (NIOSH), the Occupational Safety and Health Administration (OSHA), or the American Conference of Governmental Industrial Hygienists (ACGIH). For dust exposure in general, OSHA establishes an 8-h time-weighted average (TWA) permissible exposure limit (PEL) of 15 mg/m^3, measured as total particulate, and 5-mg/m^3 limit for respirable particulates for all particulates not otherwise regulated. For those physical irritants for which specific toxicologic data are available, OSHA has separately identified the substance in Table Z-1-A and has also promulgated a 10-mg/m^3 8-h TWA (measured as total particulate) and a 5-mg/m^3 8-h TWA PEL (measured as the respirable fraction). The ACGIH has a threshold limit value (TLV)–TWA of 10 mg/m^3 (as total dust) for particulates having a quartz content of <1% (1988 OSHA PEL Project Documentation).

Specific dust standards exist for cotton, flour, grain, and wood dusts (see sections in this chapter). Diseases associated with organic dusts often involve an allergic component; therefore, sensitization and subsequent reactions might occur with relatively low dose exposure.

Because the most important aspect of medical management of dust-induced respiratory disease is removing exposure to the inciting agent, industrial hygiene goes hand in hand with medical intervention; prevention of exposure from the outset being the best approach.

Prevention through industrial hygiene methods involves minimizing dust exposure by a variety of means, including ongoing worker education, engineering processes that minimize the production of dust, effective ventilation, wetting techniques, encapsulation of offending agents, administrative intervention and when indicated, the use of personal protective equipment (PPE). Administrative controls (e.g., changing work hours or locations) can sometimes help to reduce exposure for the symptomatic worker. Personal protective equipment such as dust masks or respirators can be used as a last resort to prevent inhalation of the dust when it is not possible to control the level of dust in the environment.

MEDICAL MANAGEMENT

First and foremost is that treatment of diseases due to dust exposure should always consider removal from the exposure or at least minimizing dust exposure to levels that will not cause or aggravate the disease symptomatology or progression. Medical treatment should not be used to mask symptoms while significant exposure to the suspected organic dust is allowed to continue. This concept is repeated throughout this chapter due to its importance in treating disease due to dust exposure.

Anti-inflammatory agents, including corticosteroids, non-steroidal anti-inflammatory medication, bronchodilators, and/or anti-histamines may be helpful in treating organic dust respiratory disease depending on the specific clinical situation.

In severe cases of HP, systemic corticosteroids represent the only reliable pharmacologic treatment, however, it has been reported that they do not alter the long-term outcome of the disease (Lacasse and Cormier, 2006). An empiric treatment schedule put forth by Ohshimo et al. (2012) may consist of 40–50 mg per day of prednisone for 1 month, with subsequent gradual tapering during the next 2–3 months and a maintenance dose between 7.5 and 15 mg per day. In chronic progressive HP, immunosuppressants may be added as corticosteroid sparing agents, as is done in other fibrotic interstitial lung diseases. Routine follow-up should be more frequent immediately after diagnosis and during treatment (1–3 months is appropriate); later the interval can be extended to every 6–12 months. If the course is favorable, with complete remission after avoidance of further

exposure and/or corticosteroid treatment, routine follow-up can be stopped after 2–3 years. Inhaled steroids or pentoxifylline may be other options of treatment; however, their efficacy has not been evaluated. In chronic progressive HP not responding to corticosteroid and/or immunosuppressant therapy, lung transplantation should be recommended (Ohshimo et al., 2012).

With removal of exposure the long-term prognosis of HP is usually good, but some patients develop severe respiratory insufficiency, and a few die of the disease (Hirschmann et al., 2009).

Occupational rhinitis, conjunctivitis, bronchitis, and asthma can be treated using standard medical approaches if symptoms continue after eliminating exposure.

Early recognition of disease with appropriate intervention is an important aspect of prevention and treatment. Screening and monitoring employees for early signs of respiratory disease in certain occupational environments may be indicated. Removal from exposure can often stabilize or even reverse the associated disease process. It is always critical for patients with respiratory disease to stop smoking. Smoking cessation programs may be helpful along with general patient education.

REFERENCES

ACGIH (2005) Wood Dusts.

Anonymous (1986) Occupational allergic diseases. *Clin. Allergy* 16(6):35–45.

Avila, R. and Lacey, J. (1974) The role of Penicillium frequentans in suberosis (respiratory disease in workers in the cork industry). *Clin. Allergy* 4(2):109–117.

Bielory, L. (1982) Hypersensitivity pneumonitis: occupational exposure to *Sitophilus granarius* and *Thermoactinomyces vulgaris. Md. State Med. J.* 31(12):25–26.

Blatman, K.H. and Grammer, L.C. (2012) Hypersensitivity pneumonitis. Chapter 19, *Allergy Asthma Proc.* 33:S64–S66.

Butikofer, E., et al. (1970) Pituitary snuff taker's lung. *Schweiz. Med. Wochenschr.* 100(3):97–101.

Caillaud, D.M., et al. (2012) Bronchoalveolar lavage in hypersensitivity pneumonitis: a series of 139 patients. *Inflamm. Allergy Drug Targets* 11(1):15–19.

Cockcroft, D.W., et al. (1983) *Sporobolomyces:* a possible cause of extrinsic allergic alveolitis. *J. Allergy Clin. Immunol.* 72(3): 305–309.

Cormier, Y., et al. (1998) Hypersensitivity pneumonitis in peat moss processing plant workers. *Am. J. Respir. Crit. Care Med.* 158(2): 412–417.

Duchaine, C., et al. (1999) *Saccharopolyspora rectivirgula* from Quebec dairy barns: application of simplified criteria for the identification of an agent responsible for farmer's lung disease. *J. Med. Microbiol.* 48(2):173–180.

Flannigan, B. (1987) *The microfloria of barley and malt, brewing microbiology.* pp. 83–120.

Girard, M. and Cormier, Y. (2010) Hypersensitivity pneumonitis. *Curr. Opin. Allergy Clin. Immunol.* 10(2):99–103.

Girard, M., Lacasse, Y., and Cormie, Y. (2009) Hypersensitivity pneumonitis. *Allergy* 64:322–334.

Greene, J.G., et al. (1981) Hypersensitivity pneumonitis due to *Saccharomonospora viridis* diagnosed by inhalation challenge. *Ann. Allergy* 47(6):449–452.

Harper, L.O., et al. (1970) Allergic alveolitis due to pituitary snuff. *Ann. Intern. Med.* 73(4):581–584.

Hirschmann, J.V., Pipavath, S.N., and Godwin, J.D. (2009) Hypersensitivity pneumonitis: a historical, clinical, and radiologic review. *Radiographics* 29(7):1921–1938.

Huuskonen, M.S., et al. (1984) Extrinsic allergic alveolitis in the tobacco industry. *Br. J. Ind. Med.* 41:77–83.

Ismaili, T., et al. (2006) Extrinsic allergic alveolitis. *Respirology* 11(3):262–268.

Johnson, W.M., et al. (1981) Respiratory disease in a mushroom worker. *J. Occup. Med.* 23: (1) 49–51.

Kagen, S.L., et al. (1981) Streptomyces albus: a new cause of hypersensitivity pneumonitis. *J. Allergy Clin. Immunol.* 68(4): 295–299.

King, H.C. (1992) Skin endpoint titration. Still the standard? *Otolaryngol. Clin. North Am.* 25(1):13–25.

Lacasse, Y. and Cormier, Y. (2006) Hypersensitivity pneumonitis. *Orphanet. J. Rare Dis.* 1:25.

Lacasse, Y., et al. (2003) Clinical diagnosis of hypersensitivity pneumonitis. *Am. J. Respir. Crit. Care Med.* 168(8): 952–958.

Lacasse, Y., et al. (2009) Classification of hypersensitivity pneumonitis: a hypothesis. *Intern. Arch. Allergy Immunol.* 149(2): 161–166.

Lacasse, Y., Girard, M., and Cormier, Y. (2012) Recent advances in hypersensitivity pneumonitis. *Chest* 142(1):208–217.

Mapp, C.E., et al. (1994) Mechanisms and pathology of occupational asthma. *Eur. Respir. J.* 7:544–554.

Millon, L., et al. (2012) *Aspergillus* species recombinant antigens for serodiagnosis of farmer's lung disease. *J.Allergy Clin. Immunol.* 130(3):803–805.

Morell, F. (2003) Suberosis: clinical study and new etiologic agents in a series of eight patients. *Chest* 124(3):1145–1152.

Morell, F., Roger, A., Reyes, L., Cruz, M.J., Murio, C., and Muñoz, X. (2008) Bird fancier's lung: a series of 86 patients. *Medicine (Baltimore)* 87(2):110–130.

Ohshimo, S., et al. (2012) Hypersensitivity pneumonitis. *Immunol. Allergy Clin. North Am.* 32(4).

OSHA: PEL Project Documentation (1988) Available at http://www.cdc.gov/niosh/pel88/npelname.html.

Patel, A.M., et al. (2001) Hypersensitivity pneumonitis: current concepts and future questions. *J. Allergy Clin. Immunol.* 108(5): 661–670.

Pimentel, J.C. (1970) *Furrier's Lung Thorax* 25:387–398.

Popp, W., et al. (1987) "Berry sorter's lung" or wine grower's lung—an exogenous allergic alveolitis caused by Botrytis cinerea spores. *Prax. Klin. Pneumol.* 41(5):165–169.

Reed, C.E., Sosman, A., and Barbee, R.A. (1965) Pigeon breeder's lung. *JAMA* 193:261–265.

Richardson, H.B., et al. (1989) Guidelines for the clinical evaluation of hypersensitivity pneumonitis. *J. Allergy Clin. Immunol.* 84:839–844.

Riddle, H.F.V., et al. (1968) Allergic alveolitis in a malt worker. *Thorax* 23:271–280.

Ryu, J.H., et al. (2007) Diagnosis of interstitial lung diseases. *Mayo Clin. Proc.* 82(8):976–986.

Schlueter, D.P., et al. (1972) Wood-pulp workers' disease: a hypersensitivity pneumonitis caused by *Alternaria. Ann. Intern. Med.* 77:907–914.

Sigsgaard, T. and Schlunssen, V. (2004) Occupational asthma diagnosis in workers exposed to organic dust. *Ann. Agric. Environ. Med.* 11(1):1–7.

Terho, E.O., et al. (1980) Extrinsic allergic alveolitis in a sawmill worker: a case report. *Scand. J. Work Environ. Health* 6(2): 153–157.

Van den Bogart, H.G.G., et al. (1993) Mushroom worker's lung: serologic reactions to thermophilic actinomycetes present in the air of compost tunnels. *Mycopathologia* 122(1):21–28.

Vincken, W. and Roels, P. (1984) Hypersensitivity pneumonitis due to *Aspergillus fumigatus* in compost. *Thorax* 39:74–75.

Warren, C.P.W. and Tse, K.S. (1974) Extrinsic allergic alveolitis owing to hypersensitivity to chickens-significance of serum precipitins. *Am. Rev. Respir. Dis.* 109:672–677.

Winck, J.C., et al. (2004) Antigen characterization of major cork molds in suberosis (cork worker's pneumonitis) by immunoblotting. *Allergy* 59(7):739–745.

Yoshida, K., et al. (1989) Prevention of Summer-type hypersensitivity pneumonitis: effect of elimination of *Trichosporon cutaneum* from the patients' homes. *Arch. Environ. Health* 44(5):317–322.

Zenz, C. (n.d.) In: DeYoung, L., editor. *Occupational Medicine*, 3rd ed., Mosby, table pp. 208–217.

ANIMAL HANDLER DISEASES

Animal handler diseases include a variety of respiratory diseases usually involving immunologic mechanisms and allergic symptomatology. Skin irritation and sensitization can also occur in those exposed to animals.

SOURCES OF EXPOSURE

Occupations that involve exposure to animal dusts include researchers, laboratory workers and other investigative scientists, veterinarians, animal caretakers and groomers, hunters, farmers, food processors, and many others. Pet owners also experience risks of exposure to animal dusts and resultant disease.

Laboratory and research workers: Animal dander allergy has long been a significant problem in research institutions (Lutsky and Neuman, 1975; Sjostedt et al., 1993; Lincoln et al., 1974). It has been estimated that between 40,000 and 125,000 individuals are exposed to laboratory animals in the United States and that between 11 and 44% of individuals working with laboratory animals report work-related allergic symptoms (Bush and Stave, 2003). Such workers have the potential to develop skin rashes, rhinitis, conjunctivitis, asthma, and HP. Rhinitis and asthma are reported most commonly in workers exposed to animal antigens. Allergies to mice and rats are the most common clinical problems; however, allergic reactions can occur upon regular exposure to virtually all furred animals (Bush et al., 1998).

Typical cases of allergy occur in young adults who have a family history of atopy and who develop hypersensitivity within 3 years of close contact with laboratory animals. Prescreening and monitoring workers for atopy and skin reactivity to laboratory animal antigens enable those at risk to take precautions and avoid exposure.

Arthropods: Allergic rhinitis, conjunctivitis, and asthma have been reported in workers who are constantly in close contact with arthropods during the course of breeding and research. Correlations have been shown between a positive skin prick test (SPT), previous RAST positive reactions, clinical signs and symptoms, and exposure, all supporting the possibility of an occupationally associated illness. In such situations careful surveillance and environmental monitoring within the workplace may be beneficial.

Furrier's lung: A furrier's occupation involves selling, making, dressing, cleaning, or repairing fur garments. Excess complaints of rhinitis and eye symptoms have been reported in workers exposed to animal fur. Immunoglobulin E-mediated allergy, especially asthma but also rhinitis, occurs in workers exposed to fur and IgE-mediated allergy may force persons with symptoms to change jobs (Uitti et al., 1997). Zuskin et al. (1988) reported symptoms in furriers to include nasal discharge, sinusitis, chronic cough, and dyspnea, as well as decrements in forced vital capacity (FVC), forced expiratory volume (FEV1), and forced expiratory flow (FEF) 50% that were somewhat ameliorated with pre-shift administration of disodium cromoglycate. Pimental (1970) reported a case of hypersensitivity-like reaction of the lungs in a furrier exposed to animal hair for 18 years, with findings of a granulomatous interstitial pneumonia similar to previously reported cases of HP.

Food processing: There is an increasing global demand for seafood resulting in more fishing, aquaculture, and seafood-processing facilities. Workers involved with manual or automated processing of shellfish, prawns, mussels, fish, and fishmeal are commonly exposed to allergens via inhalation of aerosolized product or by conjunctiva or skin contact. Other occupations associated with exposure to seafood include oyster shuckers, laboratory technicians and researchers, restaurant chefs, fishmongers (one who sells fish), and fishermen. Due to growth of the industry, there is an increased reporting of occupational disease, including allergic rhinitis, conjunctivitis and asthma, irritant-induced asthma or reactive airways dysfunction syndrome (RADS) associated with food preservatives and ammonia used as refrigerant. Reported skin manifestations include contact urticaria, eczematous contact dermatitis, and chronic protein dermatitis with diagnosis made by SPTs; workers with pre-existing atopy are at greater risk (Jeebhay and Lopata, 2012). Environmental monitoring in seafood processing has demonstrated clam and shrimp on air sampling filters. Cross reactivity between various species within a major seafood grouping occurs (Hjorth and Roed-Petersen, 1976; Jeebhay et al., 2001). Cases of OA associated with exposure to clams have been reported and supported by specific inhalation challenge. Subjects had skin reactivity and increased IgE antibodies to clam, shrimp, or both. Bronchial asthma has been recorded in oyster workers as far back as 1969 (Nakashima, 1969) and in workers exposed to prawn aerosols in a processing plant where compressed air was used to blow the edible meat out of the tails (Gaddie et al., 1980). Occupational exposure to moldy cheese has been associated with HP. Niinimäki and Saari, (1978) reported primary irritant hand dermatitis and asthma in workers with exposure to rennet and the spores of *Penicillium casei*.

Farmer's lung, ODTS, and bird handler diseases are discussed in their own section of this chapter.

ASSOCIATED AGENTS AND ANTIGENS

Animal dander, feathers, hair, serum and other proteins, urine and droppings.

MECHANISM OF ACTION AND CHEMICAL PATHOLOGY

Disease may result from sensitization to animal antigens causing allergic rhinitis, conjunctivitis, and asthma (IgE mediated), or HP (type III immune complex and/or a type IV cell mediated). Inflammatory reactions secondary to antigen sensitization are the most commonly described mechanisms; however, irritant reactions also occur.

CLINICAL AND DIAGNOSTIC CONSIDERATIONS

Key to diagnosing disease due to animal exposure is to have a keen awareness of its possibility, to include it in the differential diagnosis when evaluating patients, and to maintain suspicion of it when completing a thorough exposure history.

There are many potential scenarios involving exposure to animals and animal products that can be occupational and nonoccupational; indoor or outdoor. It is common for people to have pets and hobbies that expose them to animal allergens and these exposures may be missed without skillfully elicited histories. Good examples involve respiratory illness related to pet birds.

Symptoms of allergic rhinitis, conjunctivitis, contact dermatitis, bronchitis, and asthma are commonly seen in workers with exposure to animal allergens (Cohen, 1974; Gorannson, 1981; Hjorth, 1978; Kirkhorn and Garry, 2000). Although not as common, angioedema and HP have also been reported. Allergy screening and testing may be helpful in identification of disease, confirming diagnosis, and implementing preventive measures. See Tables I and II and the discussion on clinical and diagnostic considerations for OA and HP found in the introduction to this chapter as well as other specific sections of this chapter.

INDUSTRIAL HYGIENE

Good ventilation, cleanliness, worker education, implementation of general dust standards, and the use of PPE will reduce the exposure and respiratory disease risks for those exposed to animals. Monitoring the level of environment contaminants when feasible is important. Due to the wide range of occupational and nonoccupational settings involving animal exposure, specific industrial hygiene measures are not included here. See specific discussions in pertinent sections of this chapter.

MEDICAL MANAGEMENT

Standard medical treatment for rhinitis, dermatitis, conjunctivitis, asthma, and HP can be implemented if needed once the worker is removed from the exposure. Pre-placement screening of workers for animal allergies and atopy followed by comprehensive medical surveillance programs may reduce the risk of disease and prevent disease progression in certain occupational settings.

REFERENCES

Bush, R.K. and Stave, G.M. (2003) Laboratory animal allergy: an update. *ILAR J.* 44(1):28–51.

Bush, R.K., et al. (1998) Laboratory animal allergy. *J. Allergy Clin. Immunol.* 102(1):99–112.

Cohen, S.R. (1974) Dermatologic hazards in the poultry industry. *J. Occup. Med.* 16:94–97.

Gaddie, J., et al. (1980) Pulmonary hypersensitivity in prawn workers. *Lancet* 2:1350–1353.

Goransson, K. (1981) Occupational contact urticaria to fresh cow and pig blood in slaughtermen. *Contact Dermatitis* 7:281–282.

Hjorth, N. (1978) Gut eczema in slaughterhouse workers. *Contact Dermatitis* 4:49–52.

Hjorth, N. and Roed-Petersen, J. (1976) Occupational protein contact dermatitis in food handlers. *Contact Dermatitis* 2:28–42.

Jeebhay, M.F. and Lopata, A.L. (2012) Occupational allergies in seafood processing workers. *Adv. Food Nutri. Res.* 66:47–73.

Jeebhay, M.F., et al. (2001) Occupational seafood allergy: a review. *Occup. Environ. Med.* 58:553–562.

Kirkhorn, S.R. and Garry, V.F. (2000) Agricultural lung diseases. *Environ. Health Perspect.* 108(Suppl. 4):705–712.

Lincoln, T.A., Bolton, N.E., and Garrett, A.S. Jr. (1974) Occupational allergy to animal dander and sera. *J. Occup. Med.* 16:465–469.

Lutsky, I.L. and Neuman, I. (1975) Laboratory animal dander allergy: an occupational disease. *Ann. Allergy* 35:201–205.

Nakashima, T. (1969) Studies of bronchial asthma observed in cultured oyster workers. *Hiroshima J. Med. Sci.* 18:141–184.

Niinimäki, A. and Saari, S. (1978) Dermatological and allergic hazards of cheese makers. *Scand. J. Work Environ. Health* 4:262–263.

Pimental, J.C. (1970) Furrier's Lung. *Thorax* 25:387–398.

Sjostedt, L., Willers, S., and Orbaek, P. (1993) A follow-up study of lab animal exposed workers: the influence of atopy for the development of occupational asthma. *Am. J. Ind. Med.* 24:459–469.

Uitti, J., et al. (1997) Respiratory symptoms, pulmonary function and allergy to fur animals among fur farmers and fur garment workers. *Scand. J. Work Environ. Health* 23(6):428–434.

Zuskin, E., et al. (1988) Respiratory symptoms and lung function in furriers. *Am. J. Ind. Med.* 14(2):187–196.

BAGASSOSIS

Bagassosis is considered as an HP associated with exposure to bagasse.

SOURCES OF EXPOSURE

Bagasse is the residual fiber remaining after sugar has been extracted from sugarcane. It has found many uses due to its low cost and wide availability. Bagasse is used in the production of paper products; the manufacture of acoustic and thermal insulation, fertilizers, building material, refractory brick, as cattle feed, as bedding in stables and for fowl, and as biofuel for electricity generation (Zhai et al., 2004;

Botha and von Blottnitz, 2006). Exposures have been studied in a variety of occupational settings, including sugarcane factories and paper mills. Bagassosis was first described by Jamison and Hopkins (1941) in a worker at a hard-board factory in New Orleans.

ASSOCIATED AGENTS AND ANTIGENS

Thermoactinomyces sacchari and *Thermoactinomyces vulgaris* have been identified as precipitating antigen associated with HP in the setting of exposure to moldy bagasse (Boiron et al., 1987; Rodriguez et al., 1990; Romeo et al., 2009).

MECHANISM OF ACTION AND CHEMICAL PATHOLOGY

Bagassosis was described decades ago as an EAA (HP) based on the findings of precipitating antibodies to bagasse in the sera of patients with typical clinical findings (Salvaggio et al., 1966; Hearn and Holford-Strevens, 1968; Lacey, 1971). Still, little is known about the crucial mediators involved in the inflammation and fibrogenesis in bagassosis (Zhai et al., 2004). Pathological material has been scarce since there have been few deaths, and most knowledge has been derived from the lung biopsy specimens (Sodeman and Pullen, 1944; Bradford et al., 1961; Boonpucknavig et al., 1973). Reported pathology in bagassosis includes bronchopneumonia, bronchiectasis, and, in a few reports, the appearance of giant cells and inflammatory changes closely resembling those of farmer's lung (Nicholson, 1968; Zaidi et al., 1983). Zhai et al. (2004) reports that endotoxin, mold cell wall constituents, lipopolysaccharide, glucan, neutrophil and lymphocyte recruitment, and the bagasse fiber itself may play a role in airway inflammation and granulomatous reactions in bagassosis and suggests that neutrophils, TNF-alpha, IL-1beta, IL-8, and IL-6 are involved in the pathogenesis in bagassosis. Still, little is known about the mediators that are involved in the processes of inflammation and fibrogenesis and further research is needed.

CLINICAL AND DIAGNOSTIC CONSIDERATIONS

Bagassosis produces influenza-like signs and symptoms, including fever, severe dyspnea, and a dry, irritating cough. Other signs and symptoms include chronic bronchitis, chest pain, sputum production, in some cases blood-tinged sputum, and moist rales are heard especially over the upper lung fields (Hargreave et al., 1968; Rodriguez et al., 1990; Zhai et al., 2004).

Exposed employees have shown positive serum precipitin to stored bagasse (Ueda et al., 1992). Chest radiographs show bilateral irregular areas of infiltration easily confused with a number of diseases, such as tuberculosis, partly because of the severity of the clinical illness. Reported CXR findings include reticulonodular infiltrates, reticular infiltrates, and miliary nodular infiltrates located diffusely with a predilection for the lower lung fields. Workers exposed to bagasse dust but who had not developed bagassosis have shown significantly reduced ventilatory capacity. Forced expiratory volume and FVC are reduced in affected workers and pulmonary function has been shown to improve when workers are moved away from worksite exposure (Hearn, 1968; Rodriguez et al., 1990). Miller et al. (1971) reported small but significant reductions in lung volumes, carbon monoxide transfer factor, and diffusion capacity of the alveolar–capillary membrane 7–10 years after acute exposure in workers with bagassosis.

Arterial blood gas may show hypoxemia (Hur et al., 1994). Bronchoalveolar lavage reveals a predominantly lymphocytic population typical of HP. In a study by Zhai et al. (2004), the BAL fluid was characterized by hypercellularity and neutrophilia; however, controls also showed an increase in lavage neutrophils, which was thought to be due to local air pollution. Biopsy specimens have shown chronic interstitial infiltrates of lymphocytes and macrophages in the alveolar walls with eventual interstitial alveolar fibrosis.

Some patients develop disabling pulmonary fibrosis and emphysema and with repeated exposure end-stage lung disease (Nicholson, 1968; Zhai et al., 2004).

INDUSTRIAL HYGIENE

Historically, studies have shown that bagassosis has been largely eliminated in many areas through better industrial hygiene practices, improved storage methods of bagasse, and increased awareness of management resulting in the implementation of greater safety measures (Anonymous, 1970; Lehrer et al., 1978). Currently there is no specific airborne dust concentration standard for bagasse; however, OSHA's general dust standard applies. In a study of sugarcane workers in Nicaragua, clinical findings by Romeo et al. (2009) led them to reemphasize the importance of implementing a surveillance and exposure prevention program for workers employed in sugarcane production and processing. Dust levels should always be brought to the lowest level possible.

MEDICAL MANAGEMENT

Satisfactory results have been reported with corticosteroid therapy in the treatment of bagassosis. As is always the case for organic dusts, prevention of exposure to bagasse dust through application of industrial hygiene principles is helpful. Early diagnosis and avoidance is a cornerstone of management.

REFERENCES

Anonymous (1970) Bagasse in the bag. *Nature* 226:489.

Boiron, P., Drouhet, E., and Dupont, B. (1987) Enzyme-linked immunosorbent-assay (ELISA) for IgG in bagasse workers' sera: comparison with counter-immunoelectrophoresis. *Clin. Allergy* 17(4):355–363.

Boonpucknavig, V., et al. (1973) Bagassosis: a histopathologic study of pulmonary biopsies from six cases. *Am. J. Clin. Pathol.* 59:461–472.

Botha, T. and von Blottnitz, H. (2006) A comparison of the environmental benefits of bagasse-derived electricity and fuel ethanol on a life-cycle basis. *Energ. Policy* 34(17):2654–2661.

Bradford, J.K., Blalock, J.B., and Wascomb, C.M. (1961) Bagasse disease of the lungs: early histopathologic changes demonstrated by lung biopsy. *Am. Rev. Respir. Dis.* 84:582–585.

Hargreave, F.E., Pepys, J., and Holford-Strevens, V. (1968) Bagassosis. *Lancet* 1:619–620.

Hearn, C.E.D. (1968) Bagassosis: an epidemiological, environmental, and clinical survey. *Br. J. Ind. Med.*. 25:267–282.

Hearn, C.E.D. and Holford-Strevens, V. (1968) Immunological aspects of bagassosis. *Br. J. Ind. Med.* 25:283–292.

Hur, T., Cheng, K.C., and Yang, G.Y. (1994) Hypersensitivity pneumonitis: bagassosis. *Gaoxiong Yi Xue Ke Xue Za Zhi* 10(10):558–564.

Jamison, S.C. and Hopkins, J. (1941) Bagassosis: a fungus disease of the lungs. *New Orleans Med. Surg. J.* 93:580–582.

Lacey, J. (1971) *Thermoactinomyces sacchari sp nov*: a thermophilic actinomycete causing Bagassosis. *J. Gen. Microbiol.* 66:327–338.

Lehrer, S.B., et al. (1978) Elimination of bagassosis in Louisiana paper manufacturing plant workers. *Clin. Allergy* 8:15–20.

Miller, G.J., Hearn, C.E.D., and Edwards, R.H.T. (1971) Pulmonary function at rest and during exercise following bagassosis. *Br. J. Ind. Med.* 28:152–158.

Nicholson, D.P. (1968) Bagasse worker's lung. *Am. Rev. Respir. Dis.* 97:546–560.

Rodriguez, L.J., et al. (1990) Clinical, epidemiological, and laboratory criteria for the diagnosis of bagassosis. *Am. J. Ind. Med.* 17(1):81–83.

Romeo, L., et al. (2009) Respiratory health effects and immunological response to thermoactinomyces among sugar cane workers in Nicaragua. *Int. J. Occup. Environ. Health* 15(3):249–254.

Salvaggio, J.E., et al. (1966) Bagassosis I: precipitins against extracts of crude bagasse in the serum of patients. *Ann. Intern. Med.* 64:748–758.

Sodeman, W.A. and Pullen, R.L. (1944) Bagasse disease of the lungs. *Arch. Intern. Med.* 73:365–374.

Ueda, A., et al. (1992) Recent trends in bagassosis in Japan. *Br. J. Ind. Med.* 49:499–506.

Zaidi, S.H., et al. (1983) Experimental bagassosis: role of infection. *Environ. Res.* 31(2):279–286.

Zhai, R.H., et al. (2004) Differences in cellular and inflammatory cytokine profiles in the bronchoalveolar lavage fluid in bagassosis and silicosis. *Am. J. Ind. Med.* 46(4):338–344.

BAKER'S ASTHMA

Ramazzini described Baker's asthma in 1713. Baker's asthma and rhinitis are among the most common and well-known types of occupational respiratory disease (Baur et al., 1998). Armentia et al. (1990) reported a prevalence of 15–30% of exposed workers; Brant (2007) reported an annual incidence of baker's asthma between 1 and 10 cases per 1000 bakery workers. The risk of developing occupational respiratory disease in any particular job environment varies considerably depending on specific working conditions, and the degree of implementation of good industrial hygiene principles.

SOURCES OF EXPOSURE

Exposure can occur via flour dust inhalation or skin contact during the work of millers, dough makers, bread formers, bakers, confectioners, and food processors. Documentation of exposure levels, the type of sensitizing allergen, and the duration of exposure are all important to link asthma symptoms to the workplace.

ASSOCIATED AGENTS AND ANTIGENS

Goehte et al. (1983) reported that wheat flour is particularly sensitizing and that comparatively low levels of exposure ($1–2$ mg/m^3) pose a significant risk for developing an allergy. Brisman (2002) reported that the most common allergens associated with baker's asthma involve IgE antibodies against cereal flours such as wheat, rye, or barley. In a study of bakery workers by Patouchas et al. (2009) a high prevalence was found for sensitization to wheat flour (4–47%), and for fungal α-amylase (4.68–24%). Less common allergens that can be considered in the absence of antibodies to common allergens include: cereal flours: hops, rice, maize; non-cereal flours: buckwheat, soybean flour; additives: cellulase, xylanase, papain, glucose oxidase; color: carmine red; spices; egg powder; milk powder; insects: flour beetle (*Tribolium confusum*), flour moth (*Ephestia kuehniella*), cockroach (*Blattella* spp.), granary weevil (*Sitophilus granarius*); molds: Alternaria, Aspergillus; and sesame seeds (Brisman, 2002).

MECHANISM OF ACTION AND CHEMICAL PATHOLOGY

Baker's asthma is a form of allergic asthma to cereal flours mediated by specific IgE antibodies (Block et al., 1983; Sander et al., 2011; Dykewicz, 2009; Mapp et al., 1994). Both the level of serum IgE antibodies and the degree of nonspecific bronchial reactivity are important factors, which may influence a baker's bronchial response upon inhalation of cereal flours.

CLINICAL AND DIAGNOSTIC CONSIDERATIONS

Symptoms usually develop after a period of months to years. Clinical findings include rhinitis, conjunctivitis, wheezing, chest tightness, and shortness of breath. The affected worker is often atopic as determined by skin or IgE testing. Important diagnostic findings for baker's asthma include a documented history of exposure to bakery allergens with a clear temporal relationship to work, characteristic PFT findings for reactive airways, and clinical tests demonstrating sensitization to specific IgE allergens by SPTs and RAST testing. Specific challenge testing may be needed in difficult diagnostic situations.

INDUSTRIAL HYGIENE

Flour dust: TLV–TWA, 0.5 mg/m^3, inhalable particulate mass (ACGIH, 2001).

Reducing worker's exposure to flour dust can be accomplished in multiple ways, including: optimal ventilation, avoiding procedures that stir up dusts, wetting techniques, use of pastes, liquids, and dust-suppressed powders, use of protective clothing and equipment including respirators, implementation of health surveillance for early disease identification, preemployment screening questionnaires, allergy screening at start of the employment and annually, and education of workers regarding risks and prevention measures (Patouchas et al., 2009).

MEDICAL MANAGEMENT

Removal from exposure is the best treatment. After removal of exposure, standard treatment modalities for allergic rhinitis and asthma can be utilized. Some workers refuse to leave employment for various reasons. Armentia et al. (1990) applied immunotherapy with a standardized wheat flour extract and showed significant decrease in hyperresponsiveness to methacholine, skin sensitivity, and specific IgE, and

an improvement in subjective symptomatology after 20 months of therapy.

REFERENCES

ACGIH (2001) *Flour dust.*

Armentia, A., et al. (1990) Baker's asthma: prevalence and evaluation of immunotherapy with a wheat flour extract. *Ann. Allergy* 65(4):265–272.

Baur, X., Degens, P.O., and Sander, I. (1998) Baker's asthma: still among the most frequent occupational respiratory disorders. *J. Allergy Clin. Immunol.* 102(6 Pt 1):984–997.

Block, G., et al. (1983) Baker's asthma. Clinical and immunological studies. *Clin. Allergy* 13(4):359–370.

Brant, A. (2007) Baker's asthma. *Curr. Opin. Allergy Clin. Immunol.* 7(2):152–155.

Brisman, J. (2002) Baker's asthma. *Occup. Environ. Med.* 59:498–502.

Dykewicz, M.S. (2009) Occupational asthma: current concepts in pathogenesis, diagnosis, and management. *J. Allergy Clin. Immunol.* 123:519–528.

Goehte, C.J., et al. (1983) Buckwheat allergy: health food, an inhalation health risk. *Allergy* 38:155–159.

Mapp, C., Saetta, M., Di Stefano, A., Chitano, P., Boschetto, P., Ciaccia, A., and Fabbri, L. (1994) Mechanisms and pathology of occupational asthma. *Eur. Respir. J.* 7(3):544–554.

Patouchas, D., et al. (2009) Determinants of specific sensitization in flour allergens in workers in bakeries with use of skin prick tests. *Eur. Rev. Med. Pharmacol. Sci.* 13:407–411.

Sander, I., et al. (2011) Multiple wheat flour allergens and cross reactive carbohydrate determinants bind IgE in baker's asthma. *Allergy* 66(9):1208–1215.

BIRD-HANDLERS OR BIRD-BREEDERS DISEASE

Bird handler's disease is also called bird fancier's disease, bird breeder's lung, pigeon breeder's disease, pigeon breeder's lung, poultry worker's lung, duck fever, and has other similar names.

SOURCES OF EXPOSURE

Bird handler diseases have been reported in connection with exposure to a wide variety of avian species, including pigeons, turkeys, chickens, canaries, cockatiels, lovebirds, parrots, parakeets, owls, and other raptors (hawks, eagles) (Kokkarinen et al., 1994; Tauer-Reich et al., 1994; Choy et al., 1995; Tanaka et al., 1995; McCluskey et al., 2002; Funke and Fellrath, 2008).

Thorough exposure histories are needed to discover and differentiate occupational and domestic exposure to birds.

ASSOCIATED AGENTS AND ANTIGENS

Sources of antigens associated with HP with birds are multiple and include bird proteins, bird droppings, and feathers. There is a large body of literature that has been developed to deal with various immunological aspects of the disease. Excreted intestinal mucin and immunoglobulins A (IgA) and G (IgG) from bird droppings and bloom (a waxy coating of feathers) are highly antigenic and are the most likely major sources for inhalant bird antigen. Additionally, bacteria, fungi, viruses, mites, and other parasites that are associated with birds may play a role (Chan et al., 2012). Inhalable feather dust contains several allergens that cross react with bird serum allergens (Tauer-Reich et al., 1994). Cessation of exposure to the provoking bird antigen results in recovery in some but not all patients (Allen et al., 1976).

MECHANISM OF ACTION(S) AND CHEMICAL PATHOLOGY

The exact pathogenesis of bird fancier's disease is unknown. A number of immunological and nonimmunological mechanisms have been described, but none is clearly or exclusively associated with the disease (Mendoza et al., 1996). Lengthy exposure to avian proteins can trigger a type III immune complex-mediated hypersensitivity reaction and type I and type IV reactions with predominant activation of alveolar macrophages and T-lymphocytes responsible for the inflammatory response leading to the disease (Morell et al., 2008). The most frequent histopathology observed in HP consists of a granulomatous interstitial bronchiolocentric pneumonitis characterized by a prominent interstitial mononuclear infiltration, with the presence of non-necrotizing poorly formed granulomas. Chronic cases also show variable degrees of fibrosis (Myers and Tazelaar, 2008; Morell et al., 2008; Miguel et al., 2011).

Interestingly, cigarette smokers show a lower incidence of HP than nonsmokers, with smokers showing lower titers of IgA and IgG and lower levels of expression of immunostimulatory molecules such as peripheral membrane protein B7 on alveolar macrophages. Additionally, various genetic factors such as polymorphisms within the major histocompatibility complex and the presence of certain HLA haplotypes may modulate lung inflammatory responses and influence disease susceptibility to avian allergens (Morell et al., 2008; Chan et al., 2012).

Studies have shown that mucociliary clearance can be significantly impaired in patients tested positive for bird antigens (Hasani et al., 1992).

CLINICAL AND DIAGNOSTIC CONSIDERATIONS

No single historical feature or laboratory test is found in the diagnosis of HP from birds. Making the diagnosis often requires a combination of findings, including clinical and exposure history, laboratory findings, CXR and chest CT, BAL, and in some cases lung biopsies. Although often misdiagnosed as influenza or viral or bacterial pneumonia, if a thorough exposure history leads the clinician to consider HP, diagnosis of the acute form is not so difficult. Four to eight hours after exposure the patient develops influenza-like symptoms, which may include cough, dyspnea, fever, chills, and nausea. Leukocytosis, lymphopenia, eosinophilia, and elevated erythrocyte sedimentation rates are common. Crackles may be heard over the lungs, and respiratory function tests typically show a restrictive abnormality with a reduction in lung volumes and alterations in the CO diffusion capacity. Chest X-ray abnormalities consist of diffuse, finely nodular shadows, and reticular linear densities, and chest CT may show ground glass areas, air trapping, and mosaic or nodular patterns (Zylak et al., 1975; Vincent et al., 1992; Remy-Jardin et al., 1993; Yoshizawa, 1995; Morais et al., 2004; Morell et al., 2008). The subacute presentation includes cough and dyspnea over days to weeks and can progress to chronic and permanent lung damage if unrecognized and untreated. The chronic form of the disease includes breathlessness, cough, obstructive as well as restrictive patterns on PFT's, and hypoxemia; symptoms may persist after removal from antigen exposure (Chan et al., 2012). As would be expected, once progression to the chronic form of the disease occurs, prognosis is reported to be worse than for patients with acute and subacute disease. Case fatalities have been reported (Edwards and Luntz, 1974). Hypersensitivity pneumonitis resulting from handling birds constitutes a spectrum of granulomatous, interstitial, and alveolar-filling lung diseases (Lacasse et al., 2009).

The finding of specific precipitating antibodies in the serum of a patient with suspected HP supports exposure. Morell et al. (2008) found positive IgG antibodies to bird antigens in 93% of patients with bird fancier's lung disease. Studies done decades ago found that agglutination reactions were mediated by antibodies directed against pigeon IgG and IgM. T-lymphocyte reactivity to pigeon antigens has also been demonstrated (Moore et al., 1974; Schatz et al., 1976; Diment and Pepys, 1977).

Examination of BAL usually shows lymphocytosis and an inversion of the CD4/CD8 T-lymphocyte ratio, however, the inversion in CD4/CD8 is not always seen.

Lung biopsy may be indicated in patients without sufficient clinical criteria for definitive diagnoses or to rule out other diseases requiring different treatment. Lung biopsy characteristically shows lymphocytic–histiocytic infiltrate, poorly formed granulomas, and bronchiolitis obliterans (Morell et al., 2008).

Chronic bronchitis is often associated with bird breeder's disease (Tanaka et al., 1995). Chan et al. (2012) reported an incidence of chronic bronchitis in 26.2% of nonsmoking pigeon breeders. In another study, Carrillo et al. (1991) reported an incidence of chronic bronchitis in 20% of a group who tested positive to bird antigens. This fraction was composed of nonsmokers with no risk factor for chronic bronchitis other than exposure to birds.

Rodriguez de Castro et al. (1993) studied 343 pigeon breeders and found that rhinitis was the most frequently reported clinical syndrome in 31%, followed by immediate bronchial symptoms in 18%.

Sansores et al. (1990) noted an association between clubbing of the fingers in bird breeder's disease patients and poor outcome. The authors suggested that digital clubbing may be used as a prognostic tool to predict clinical deterioration. Morell et al. (2008) reported that the prevalence of clubbing was low in those with bird fancier's disease overall, however, it was seen in 1/3rd of patients with chronic disease.

Bird breeder's disease should be considered whenever a patient presents with interstitial lung disease or intermittent cough and dyspnea (Mangion and Delaney, 1993). Unrecognized exposure and disease may be common. For example, a female automobile repair garage worker had respiratory symptoms thought to be due to workplace exposures. After a careful review of her history, physical examination, and laboratory testing, she was diagnosed with HP related to exposure to pet cockatiels in her home (McCluskey et al., 2002). In another case, a patient initially diagnosed as having amiodarone pulmonary toxicity (APT) was subsequently diagnosed instead with HP upon further testing. A thorough exposure history discovered that he was exposed to his daughter's lovebirds (Funke and Fellrath, 2008).

It has been estimated that the prevalence of bird fanciers' lung among bird owners in Britain is between 0.5 and 7.5%. With such a large population at risk, Judson and Sahn (2004) suggested that clinicians should ask about "bird-years" as well as cigarette pack-years.

A 98% maximum probability of diagnosing bird fancier's disease has been reported if the following six diagnostic criteria are met: (1) exposure to a known antigen, (2) recurrent symptomatic episodes, (3) symptoms occurring between 4 and 8 h after exposure, (4) weight loss, (5) positive IgG precipitating antibodies to the antigen, and (6) inspiratory crackles on chest auscultation. With these criteria being met, BAL and lung biopsies will probably not be necessary for confirmation, especially if a HRCT chest is also consistent with the diagnosis of HP (Chan et al., 2012).

INDUSTRIAL HYGIENE

It is crucial to recommend antigen avoidance when managing bird handler's disease. While complete avoidance is best, certain environmental measures may be helpful such as thorough cleaning measures, however, the antigen may stay in the environment after cleaning (Chan et al., 2012). Craig et al. (1992) demonstrated that high levels of bird antigens can be detected in the home environment for prolonged periods after removing the bird.

Indoor and outdoor environments must be considered when determining potential exposure to bird antigens. When complete removal of the patient from exposure is not acceptable, PPE such as masks may help. Changing clothes (such as a lab coat or hat) after exposure may minimize antigen contamination and exposure. Improved air circulation in working or hobby areas may also help. A study by Edwards et al. (1991) demonstrated that switching from absorbent loft litter to regular cleaning for bird enclosures decreased the amount of airborne antigens and respirable particles.

MEDICAL MANAGEMENT

When bird breeder's disease is suspected, the clinician must perform appropriate historical, environmental, and clinical evaluations to prevent both progressive and irreversible lung damage.

After removal from exposure, standard medical treatment can be utilized for rhinitis, bronchitis, asthma, and HP, see introduction to this chapter.

Overall prognosis for recovery is good. Patients removed from exposure typically improve but retain elevated levels of avian-specific antibodies (Yoshizawa, 1995; Grammer et al., 1990).

REFERENCES

Allen, D.H., Williams, G.V., and Woolcock, A.J. (1976) Bird breeder's hypersensitivity pneumonitis: progress studies of lung function after cessation of exposure to the provoking antigen. *Am. Rev. Respir. Dis.* 114:555–566.

Carrillo, T., et al. (1991) Effect of cigarette smoking on the humoral immune response in pigeon fanciers. *Allergy* 46:241–244.

Chan, A., et al. (2012) Bird fancier's lung: a state-of-the-art review. *Clin. Rev. Allergy Immunol.* 43:69–83.

Choy, A.C., et al. (1995) Hypersensitivity pneumonitis in a raptor handler and a wild bird fancier. *Ann. Allergy Asthma Immunol.* 74:437–441.

Craig, T.J., et al. (1992) Bird antigen persistence in the home environment after removal of the bird. *Ann. Allergy* 69:510–512.

Diment, J.A. and Pepys, J. (1977) Avian erythrocyte agglutination tests with sera of bird fanciers. *J. Clin. Pathol.* 30:29–34.

Edwards, C. and Luntz, G. (1974) Budgerigar-fancier's lung: report of a fatal case. *Br. J. Dis. Chest* 68:57–64.

Edwards, J.H., et al. (1991) Pigeon breeder's lung: the effect of loft litter materials on airborne particles and antigens. *Clin. Exp. Allergy* 21:49–54.

Funke, M. and Fellrath, J.M. (2008) Hypersensitivity pneumonitis secondary to lovebirds: a new cause of bird fancier's disease. *Eur. Respir. J.* 32:517–521.

Grammer, L.C., et al. (1990) Clinical and serological follow-up of four children and five adults with bird-fancier's lung. *J. Allergy Clin. Immunol.* 85:655–660.

Hasani, A., et al. (1992) Impairment of lung mucociliary clearance in pigeon fanciers. *Chest* 102:887–891.

Judson, M.A. and Sahn, S.A. (2004) Bird-years as well as pack-years. *Chest* 125(1):353–354.

Kokkarinen, J., et al. (1994) Hypersensitivity pneumonitis due to native birds in a bird ringer. *Chest* 106:1269–1271.

Lacasse, Y., et al. (2009) Classification of hypersensitivity pneumonitis; a hypothesis. *Intern. Arch. Allergy Immunol.* 149(2): 161–166.

Mangion, J.R. and Delaney, M. (1993) Pigeon-breeder's lung: extensive workup clarifies the clinical picture. *Postgrad. Med.* 93:215–217, 221–222.

McCluskey, J.D., Haight, R.H., and Brooks, S.M. (2002) Cockatiel-induced hypersensitivity pneumonitis. *Environ. Health Perspect.* 110(7):735–738.

Mendoza, F., et al. (1996) Cellular immune response to fractionated avian antigens by peripheral blood mononuclear cells from patients with pigeon breeder's disease. *J. Lab. Clin. Med.* 127:23–28.

Miguel, G., et al. (2011) Morphologic diversity of chronic pigeon breeder's disease. Clinical features and survival. *Respir. Med.* 105(4):608–614.

Moore, V.L., et al. (1974) Immunologic events in pigeon breeder's disease. *J. Allergy Clin. Immunol.* 53:319–328.

Morais, A., et al. (2004) Suberosis and bird fancier's disease: a comparative study of radiological, functional and bronchoalveolar lavage profiles. *J. Invest. Allergol. Clin. Immunol.* 14(1): 26–33.

Morell, F., et al. (2008) Bird fancier's lung: a series of 86 patients. *Medicine (Baltimore)* 87(2):110–130.

Myers, J.L. and Tazelaar, H.D. (2008) Challenges in pulmonary fibrosis: 6–problematic granulomatous lung disease. *Thorax* 63:78–84.

Remy-Jardin, M., et al. (1993) Subacute and chronic bird breeder hypersensitivity pneumonitis: sequential evaluation with CT and correlation with lung function tests and bronchoalveolar lavage. *Radiology* 189:111–118.

Rodriguez de Castro, F., et al. (1993) Relationships between characteristics of exposure to pigeon antigens. *Chest* 103: 1059–1063.

Sansores, R., et al. (1990) Clubbing in hypersensitivity pneumonitis: its prevalence and possible prognostic role. *Arch. Intern. Med.* 150:1849–1851.

Schatz, M., et al. (1976) Pigeon breeder's disease. II. Pigeon antigen induced proliferation of lymphocytes from symptomatic and asymptomatic subjects. *Clin. Allergy* 6:7–17.

Tanaka, H., et al. (1995) Budgerigar breeders' hypersensitivity pneumonitis presenting as chronic bronchitis with purulent sputum. *Intern. Med.* 34:676–678.

Tauer-Reich, I., et al. (1994) Allergens causing bird fancier's asthma. *Allergy* 49:448–453.

Vincent, J.M., et al. (1992) Extrinsic allergic alveolitis: problems in diagnosis and a potential use for computed tomography. *Respir. Med.* 86:135–141.

Yoshizawa, Y. (1995) Follow-up study of pulmonary function tests. CAL cells, and humoral and cellular immunity in birds fancier's lung. *J. Allergy Clin. Immunol.* 96:122–129.

Zylak, C.J., et al. (1975) Hypersensitive lung disease due to avian antigens. *Radiology* 114:45–49.

BYSSINOSIS

Byssinosis has also been called strippers asthma or brown lung and is a bronchoconstrictive disease associated with cotton dust exposure.

SOURCES OF EXPOSURE

The unusual and somewhat misnamed diagnostic title "byssinosis," derived from the Latin *byssinum*, linen garment, refers to a group of occupational respiratory diseases in workers exposed to the dusts of dirty cotton, flax, and hemp. The causative dusts represent a complex mixture of organic and inorganic compounds. Occupational exposures include operating cotton gins and baling cotton, opening bales, removing dust and impurities, carding, cleaning, stripping, and grinding the teeth of machines in cotton mills and textile plants.

In the United Kingdom, Schilling (1955) reported a prevalence rate of around 50% for workers in the dustiest parts of the cotton-spinning process. Later studies showed a marked improvement of the prevalence rates in the United Kingdom; 10% among the high risk workers, and an overall prevalence of 3%. However, similar prevalence rates of byssinosis as experienced in the United Kingdom in the 1950s have recently been experienced in developing countries involved with cotton production. Reported rates in textile workers vary from 30 to 50% in countries such as Indonesia, Sudan, and India (Parikh, 1992; Altin et al., 2002). As one would expect, byssinosis seems to be more of a problem in parts of the world where cheap labor can be exploited without the implementation of adequate industrial hygiene measures.

ASSOCIATED AGENTS AND ANTIGENS

The storage and handling conditions of cotton products can promote the growth of Gram-negative bacteria. Strong, yet indirect, evidence suggests that contaminating bacterial endotoxin may play a major role in the adverse pulmonary effects of cotton dust (Castellan et al., 1987; Jacobs, 1989). Many other biologically active agents are known to contaminate cotton dusts and hemp and natural constituents include such pharmacologically active agents as tannins, gossypol, and quercetin.

MECHANISM OF ACTION AND CHEMICAL PATHOLOGY

The underlying mechanisms responsible for byssinosis are not certain; however, it is thought that inflammatory reactions occur secondary to endotoxin- or immunologic-induced processes (Rooke, 1981; Niven and Pickering, 1996; Christiani et al., 1993; Kirkhorn and Garry, 2000; Thorn, 2001).

Endotoxin-contaminated organic dusts have been associated with a variety of adverse pulmonary effects, including: chronic bronchitis and increased sputum production, an increase in the volume density of mucosubstances, an increase in the secretion of mucosubstances by mucous goblet cells, and the induction of goblet cell metaplasia and hyperplasia. An increase in stored mucosubstances is accompanied by the presence of excess luminal mucus, secretory cell hyperplasia, and an influx of neutrophils (Harkema, 1991; Gordon and Harkema, 1995).

Overproduction and hypersecretion of mucin glycoproteins can contribute directly to airway obstruction, airway plugging, and harboring of bacterial infections associated with a variety of chronic pulmonary diseases, including asthma, chronic bronchitis, bronchiectasis, and cystic fibrosis (Kilburn, 1981). Several studies have shown a decrease in FEV_1 following inhalation to endotoxin (Thorn, 2001). Inhalation of lipopolysaccharide has been shown to result in an increased amount of fibronectin in BAL fluid. Recent studies of textile workers suggest an exposure-dependent effect of endotoxin on lung function impairments (Wang et al., 2003; Oldenburg et al., 2007).

Some researchers suspect airborne fungi contaminants such as Cladosporium, Alternaria, Penicillium, Fusarium, and Aspergillus as a cause of byssinosis while other researchers have suspected chemical constituents in plants such as terpenoid, quercetins, gossypol, scopoletin, and laciniliene (McPartland, 2003).

O'Neil et al. (1983) suggested that an IgE-mediated hypersensitivity against cotton or a fungal component of the dust may be important in byssinosis pathogenesis based on studies of precipitating antibodies against aqueous cotton dust extracts.

Jacobs et al. (1993) reported that atopic workers showed a greater decrease in FEV1 after cotton dust exposure than did the non-atopic workers, suggesting a possible immunologic component to the disease. Wang et al. (2003) found that the IgE level in cotton workers was not related to lung function, suggesting that IgE level might not be the basis for the airway obstruction observed in cotton workers.

CLINICAL AND DIAGNOSTIC CONSIDERATIONS

The most commonly reported health effect from cotton dust exposure is eye and nasal irritation (Fishwick and Pickering, 1994). Reports of irritation did not correlate with incidence of byssinosis or cotton dust concentration.

Byssinosis can be divided into acute and chronic forms. Short-term exposure to cotton dust for some workers is associated with acute respiratory symptoms, and long-term exposure may be associated with the classic form of byssinosis.

Acute byssinosis refers to the acute response of airways that occurs in approximately 1/3rd of volunteers exposed to cotton dust for the first time. Substantial falls in FEV_1 exceeding 30% have been reported. A study of Finnish cotton spinning mills reported that 1 in 10 employees left work within 2 weeks and one in four left within 3 months of employment (Koskela et al., 1990).

The classic form of chronic byssinosis is characterized by a feeling of chest tightness, coughing, wheezing, and dyspnea, which the worker experiences as being most severe on the first day of the workweek after a period of absence from work (Niven and Pickering, 1996; McPartland, 2003). Symptoms continue after the individual has finished work and may even progress during the evening. However, they are perceived as being less troublesome on subsequent days.

Progression of byssinosis has been described in three stages: In stage I symptoms occur when textile work is resumed after a period of several days off work ("Monday disease"). In stage II symptoms may increase and persists for 2 or more days. In stage III employees develop chronic obstructive lung disease (Baur et al., 1993).

Textile workers have a considerably higher prevalence of chronic respiratory symptoms, including cough, dyspnea, bronchitis, chest tightness, and phlegm, as well as reductions of FVC and FEV_1 (Zuskin et al., 1992; Alemu et al., 2010). Studies have shown both atopic and non-atopic persons to have a significant decrease in FEV_1 after cotton dust exposure, but the average decrease in FEV_1 was larger for the atopic group than for the non-atopic group (Sepulveda et al., 1984; Beijer et al., 1995).

Although most studies have concentrated on decreases in FEVs, some studies have shown larger changes in the caliber of the small airways, which has led to the suggestion that the disease process starts in the smaller peripheral airways. In addition to changes in lung function, cotton workers can show increases in bronchial reactivity. The changes in both reactivity and small airway caliber have given rise to the suggestion that byssinosis is an asthma-like condition of the small airways (Fishwick and Picketing, 1992; Niven and Pickering, 1996).

Bronchoalveolar lavage studies have been performed on animals, volunteers, and patients after exposure to cotton dust or dust extracts. Variable changes have been identified, but neutrophil recruitment appears to be the most consistent finding. Byssinosis bodies up to $50\,\mu m \times 200\,\mu m$ and consisting of hyalinized fibrous nodules containing a dark staining core presumably of cellulose have been described in some lung specimens of workers exposed to cotton dust (Spencer, 1977; Baur et al., 1993). There is no specific CXR picture. Baur et al. (1993) reported CXR findings of linear bronchial shadowing compatible with peribronchitis in two case reports.

The affected worker may not experience the classic symptoms of chronic byssinosis disease until he or she has worked for many years in the industry; indeed, it may be seen for the first time 25 years after starting to work in the industry and is rare in individuals continuously exposed for <10 years.

As the disease of byssinosis becomes established, its clinical course is that of chronic bronchitis and emphysema. As far back as 1975, Edwards et al. described centrilobular or paracinar emphysema in 23% and 14%, respectively, of 43 patients with byssinosis, whereas the others were free of this complication (Edwards et al., 1975). Advanced byssinosis results in severe dyspnea and disability with exacerbation of symptoms at the time of any respiratory infection. Death is usually caused by right heart failure. As might be expected, cigarette smokers have a higher prevalence and greater severity of symptoms.

INDUSTRIAL HYGIENE

Cotton dust, raw: TLV–TWA, $0.2\,mg/m^3$ (fibers <15 mm in length) (ACGIH, 2001).

In the United States the current cotton dust standard promulgated by the OSHA became effective in 1980. As found in OSHA Table Z-1 (29 CFR 1910.1000), the PEL for cotton dust (raw) is $1\,mg/m^3$ for the cotton waste processing operations of waste recycling (sorting, blending, cleaning, and willowing) and garnetting.

Permissible exposure limits for other sectors (as found in 29 CFR 1910.1043) are $0.200\,mg/m^3$ for yarn-manufacturing and cotton-washing operations, $0.500\,mg/m^3$ for textile mill waste house operations or for dust from "lower grade washed cotton" used during yarn-manufacturing, and $0.750\,mg/m^3$ for textile-slashing and -weaving operations.

The OSHA standard in 29 CFR 1910.1043 does not apply to cotton harvesting, ginning, or the handling and processing of woven or knitted materials and washed cotton.

All PELs for cotton dust are mean concentrations of lint-free, respirable cotton dust collected by the vertical elutriator or an equivalent method and averaged over an 8-h period (NIOSH, 2007).

The NIOSH (2007) recommends reducing cotton dust to the lowest level feasible to reduce the prevalence and severity of byssinosis.

The ACGIH (1996) has set the TLV for raw cotton dust (lint free) at $200\,\mu g/m^3$.

Lung function tests performed at the time of preemployment can be used for routine medical surveillance to screen for early changes associated with exposure to cotton dust. Workers with reductions in serial FEV_1 measurements should be reassigned to work areas where there is less risk of exposure to dust. Improvement in ventilatory capacity as a result of reduced dust levels may be accomplished by steaming of the cotton (Imbus and Suh, 1974), which may also remove, at least temporarily, some of the bronchoconstricting properties of the dust. Textile mills should be properly ventilated and filtered; using exhaust hoods, improving ventilation, and employing wetting procedures may help to control cotton dust. A study by Altin et al. (2002) showed that closed cotton processing and air-conditioning are very effective in preventing byssinosis. Proper retting and storage of hemp will produce a higher quality hemp fiber with less contamination that reduces the risk of byssinosis in workers (McPartland, 2003). As has been demonstrated by the cotton industry in the United States, technological investment to reduce exposure can bring both improved productivity and reduced incidence of disease (Niven and Pickering, 1996).

MEDICAL MANAGEMENT

Medical treatment for affected individuals is similar to patients with occupational asthma.

Both bronchodilators and anti-inflammatory agents can be used to ameliorate symptoms, and inhaled steroids are probably more effective than sodium cromoglycate in the chronic form of the disease. Breathing treatments, including nebulizers, may be prescribed and home oxygen therapy, physical exercise programs, breathing exercises, and patient education may be helpful for people with chronic disease.

Workers with a personal history of respiratory allergy or a history of continued cigarette smoking should be considered at increased risk of developing byssinosis (McPartland, 2003). As with all respiratory disease, it is critical for patients with byssinosis to stop smoking. A study by Yih-Ming et al. (2003) supported that smoking potentiates the effects of cotton dust exposure on respiratory symptoms and byssinosis. Smoking cessation programs may be helpful along with general patient education.

Cessation of exposure following diagnosis of byssinosis should be recommended, particularly because studies have shown increased mortality in workers with byssinosis (Niven and Pickering, 1996).

REFERENCES

ACGIH (2001) *Cotton dust.*

American Conference of Governmental Industrial Hygienists (ACGIH) (1996) *Threshold Limit Values (TLVs) for Chemical Substances and Physical Agents and Biological Exposure Indices (BEIs)*, Cincinnati, Ohio: American Conference of Governmental Industrial Hygienists.

Alemu, K., Kumie, A., and Davey, G. (2010) Byssinosis and other respiratory symptoms among factory workers in Akaki textile factory, Ethiopia. *Ethiopian J. Health Dev.* 24(2):133–139.

Altin, R., et al. (2002) Prevalence of byssinosis and respiratory symptoms among cotton mill workers. *Respiration* 69(1):52–56.

Baur, X., et al. (1993) Occupational-type exposure tests and bronchoalveolar lavage analyses in two patients with byssinosis and two asymptomatic cotton workers. *Int. Arch. Occup. Environ. Health* 65:141–146.

Beijer, L., et al. (1995) Monocyte responsiveness and a T-cell subtype predict the effects induced by cotton dust exposure. *Am. J. Respir. Crit. Care Med.* 152:1215–1220.

Castellan, R.M., et al. (1987) Inhaled endotoxin and decreased spirometric values: an exposure-response relation for cotton dust. *N. Engl. J. Med.* 317:605–610.

Christiani, D.C., et al. (1993) Cotton dust and gram-negative bacterial endotoxin correlation in two cotton textile mills. *Am. J. Ind. Med.* 23(2):333–342.

Edwards, C., et al. (1975) The pathology of the lung in byssinotics. *Thorax* 30:612–623.

Fishwick, D. and Picketing, C.A.C. (1992) Byssinosis-a form of occupational asthma. *Thorax* 47:401–403.

Gordon, T. and Harkema, J.R. (1995) Cotton dust produced an increase in intraepithelial mucosubstances in rat airways. *Am. J. Respir. Crit. Care Med.* 151:1981–1988.

Harkema, J.R. (1991) Comparative aspects of nasal airway anatomy: relevance to inhalation toxicology. *Toxicol. Pathol.* 19:321–336.

Imbus, H.R. and Suh, M.W. (1974) Steaming of cotton to prevent byssinosis: a plant study. *Br. J. Ind. Med.* 31:209–219.

Jacobs, R.R. (1989) Airborne endotoxins: an association with occupational lung disease. *Appl. Ind. Hyg.* 4:50–56.

Jacobs, R.R., et al. (1993) Bronchial reactivity, atopy and airway response to cotton dust. *Am. Rev. Respir. Dis.* 148:19–24.

Kilburn, K.H. (1981) Byssinosis. *Am. J. Ind. Med.* 2:81–88.

Kirkhorn, S.R. and Garry, V.F. (2000) Agricultural lung diseases. *Environ. Health Perspect.* 108(Suppl. 4):705–712.

Koskela, R.S., Klockars, M., and Jarven E. (1990) Mortality and disability among cotton mill workers. *Br. J. Ind. Med.* 47:384–391.

McPartland, J.M. (2003) Byssinosis in hemp mill workers. *J. Ind. Hemp* 8(1):33–44.

National Institute for Occupational Safety and Health (NIOSH) (2007) *Pocket Guide to Chemical Hazards*, NIOSH 94–116. Washington, DC: US Department of Health and Human Services.

Niven, R.McL. and Pickering, C.A.C. (1996) Byssinosis: a review. *Thorax* 51:632–637.

O'Neil, C.E., et al. (1983) Studies on the antigenic composition of aqueous cotton dust extracts. *Int. Arch. Allergy Immunol.* 72:294–298.

Oldenburg, M., Latza, U., and Baur, X. (2007) Exposure-response relationship between endotoxin exposure and lung function impairment in cotton textile workers. *Int. Arch. Occup. Environ. Health* 80(5):388–395.

Parikh, J.R. (1992) Byssinosis in developing countries. *Br. J. Ind. Med.* 49:217–219.

Fishwick, D., Fletcher, A.M., Pickering, C.A., Niven, R.M., Faragher, E.B. (1994) Respiratory symptoms and dust exposure in Lancashire cotton and man-made fiber mill operatives. *Am. J. Resp. Care Med.* 150(2):441–447.

Rooke, G.B. (1981) The pathology of byssinosis. *Chest* 79:67S–71S.

Sepulveda, M.J., et al. (1984) Acute lung function response to cotton dust in atopic and nonatopic individuals. *Br. J. Ind. Med.* 41:487–491.

Schilling, R.S.F., Hughes, J.P.W., Dingwall-Fordyce, I., Gilsoon, J.C. (1955) An Epidemiological study of byssinosis among Lancashire cotton workers. *Br. J. Ind. Med.* 12(3):219–227.

Spencer, H. (1977) *Pathology of the Lung*, 3rd ed., Philadelphia, WB: Saunders.

Thorn, J. (2001) The inflammatory response in humans after inhalation of bacterial endotoxin: a review. *Inflamm. Res.* 50(5):254–261.

Wang, X.R., et al. (2003) A longitudinal observation of early pulmonary responses to cotton dust. *Occup. Environ. Med.* 60:115–121.

Yih-Ming, S.U., et al. (2003) Additive effect of smoking and cotton dust exposure on respiratory symptoms and pulmonary function of cotton textile workers. *Ind. Health* 41:109–115.

Zuskin, E., et al. (1992) Immunological findings and respiratory function in cotton textile workers. *Int. Arch. Occup. Environ. Health* 64:31–37.

COFFEE WORKERS DISEASE

Reported diseases associated with coffee exposure include coffee workers' lung disease and coffee worker's asthma.

SOURCES OF EXPOSURE

Coffee workers are exposed to coffee bean dust during shipment, processing, and manufacturing.

ASSOCIATED AGENTS AND ANTIGENS

Occupational allergic respiratory symptoms, including asthma in coffee industry workers have been frequently reported, but the ultimate cause of sensitization is still debated; the castor bean and green coffee bean have been implicated. Work involving the immunological responses of mice (Lehrer et al., 1978) indicated that the green coffee bean is the major source of the allergen in this disease. Coffee workers lung disease is a HP that has been associated with antigens of Thermoactinomyces spp.

MECHANISM OF ACTION AND CHEMICAL PATHOLOGY

Immunoglobulin E antibodies specific for green coffee bean and castor bean have been detected in coffee workers and are capable of producing allergic asthma (Karr et al., 1978; Oldenburg et al., 2009).

Hypersensitivity pneumonitis has been reported to be associated with antigens of Thermoactinomyces spp. (Van Torn, 1970). Atopy and cigarette smoking have been suggested as promoting factors of sensitization. Larese et al. (1998) found a significant association between sensitization to green coffee beans and work-related symptoms, common allergic symptoms and atopy diagnosed by prick test.

CLINICAL AND DIAGNOSTIC CONSIDERATIONS

As far back as 1973, illnesses reported in coffee workers included allergic rhinitis, asthma, conjunctivitis, pruritus, and dermatitis as a result of exposure to dust in manufacturing establishments (Bernton, 1973). Radioallergosorbent testing and/or SP testing for green coffee bean and castor bean may be helpful in establishing a diagnosis of coffee dust-associated disease.

A total of 211 coffee workers were examined by Romano et al. (1995). Of the workers, 10% complained of conjunctivitis and rhinitis alone and 16% complained of asthma. The overall prevalence of skin sensitization was 15% for green coffee beans, 22% for castor beans, and 22% for common allergens.

Ventilatory functional changes have been demonstrated and the decreases in flow rates at low lung volumes have been interpreted decades ago to indicate that the bronchoconstrictor action of the coffee bean dust mostly affects the smaller airways (Zuskin et al., 1979). These investigators have warned that continued exposure to dust in both green coffee handling and roasted coffee processing may lead to a persistent loss of pulmonary function. Zuskin et al. (1981) reported a higher prevalence of all chronic respiratory symptoms in coffee workers compared to controls, and those workers with positive skin test to coffee allergen also had a higher prevalence of chronic cough and chronic phlegm.

A study by Jones et al. (1982) found that men with lengthy employment and exposure to green coffee bean dust had lower mean residual FEV1, likewise, those workers with serum IgE antibodies to green coffee beans had lower mean residual FEV1.

A cross-sectional study by Oldenburg et al. (2009) concluded that workers involved with the transshipment and unloading of green coffee experienced a high exposure to irritative and sensitizing dust that was clinically relevant based on their evaluation of IgE antibodies, bronchial hyper-responsiveness, chest tightness, wheezing, erythematous symptoms, conjunctivitis, and sneezing.

INDUSTRIAL HYGIENE

There have been no specific standards promulgated for coffee beans; however, reducing coffee bean dust exposure to the lowest possible levels should be implemented based on the studies associating exposure with sensitization and adverse respiratory effects.

MEDICAL MANAGEMENT

Removal from exposure is the mainstay of intervention. Standard medical treatment for rhinitis and allergic asthma can be utilized, however, treatment should not mask symptoms while continued exposure and disease progression is allowed. In the case of HP, corticosteroid treatment may be helpful as in other types of HP.

REFERENCES

Bernton, H.S. (1973) On occupational sensitization: a hazard to the coffee industry. *JAMA* 233:1146–1147.

Jones, R.N., et al. (1982) Lung function consequences of exposure and hypersensitivity in workers who process green coffee beans. *Am. Rev. Respir.* 125(2):199–202.

Karr, R.M., et al. (1978) Coffee worker's asthma: a clinical appraisal using the radioallergosorbent test. *J. Allergy Clin. Immunol.* 62(3):143–148.

Larese, F., et al. (1998) Sensitization to green coffee beans and work-related allergic symptoms in coffee workers. *Am. J. Ind. Med.* 34:623–627.

Lehrer, S.B., Karr, R.M., and Salvaggio, J.E. (1978) Extraction and analysis of coffee bean allergens. *Clin. Allergy* 8:217–226.

Oldenburg, M., Bittner, C., and Baur, X., (2009) Health risks due to coffee dust. *Chest* 136(2):536–544.

Romano, C., et al. (1995) Factors related to the development of sensitization to green coffee and castor bean allergens among coffee workers. *Clin. Exp. Allergy* 25:643–650.

Van Torn, D.W. (1970) Coffee worker's lung, a new example of extrinsic allergic alveolitis. *Thorax* 25:399–405.

Zuskin, E., Valic, F., and Skuric, Z. (1979) Respiratory function in coffee workers. *Br. J. Ind. Med.* 36:117–122.

Zuskin, E., Valic, F., and Kanceljak, B. (1981) Immunological and respiratory changes in coffee workers. *Thorax* 36:9–13.

DETERGENT WORKERS DISEASE

Enzymes are added to detergents to enhance their ability to remove certain stains. Occupational exposure to these enzymes have been associated with allergic sensitization and OA.

SOURCES OF EXPOSURE

The inhalation of the dust of microbiological products containing enzymes has led to the sensitization of workers exposed to high concentrations of enzymes. Most attention has been given to the exposures that occur in detergent manufacturing and is the focus of this discussion. Other industries with risk of exposure to enzymes include pharmaceutical manufacturing, bakers, and rubber workers. Historically, there were rare instances described where consumers experienced symptoms associated with detergent use. This problem has essentially been eliminated through improved products (encapsulation, granulation, and new formulations) that minimize consumer exposure.

ASSOCIATED AGENTS AND ANTIGENS

Enzymes used in the detergent industry are proteins derived from bacteria and fungi, and include proteases, amylases, lipases, and cellulases. The first enzyme to be associated with adverse respiratory health effects and one of the best known is the protease prepared from *Bacillus subtilis*, a nonpathogenic organism.

MECHANISM OF ACTION AND CHEMICAL PATHOLOGY

Adverse health effects associated with enzyme exposure include irritation of the skin, eye, and other mucosal sites, and antibody-mediated hypersensitivity.

Irritation has been reported after prolonged occupational exposure to high concentrations of enzymes. The irritation is mild, reversible, and resolves quickly with the removal of exposure. Irritation of consumers is not expected due to the low concentration of their exposure. Occupational controls have essentially eliminated the problem of irritation in the workplace.

Occupational allergy and asthma affecting the respiratory tract is more common and is thought to be an enzyme-specific IgE antibody-mediated response. The majority of reported allergic and asthma symptoms are immediate onset, although late onset asthma has been described. Enzymes are not contact allergens and do not induce delayed type hypersensitivity (Schweigert et al., 2000). Sensitization of consumers is not expected to occur due to low levels of exposure (Pepys et al., 1985).

Pepys (1992) noted the appearance of asthma some hours after exposure (i.e., a "late" asthmatic reaction) and in some cases, subsequent exposures caused immediate reactions. Late asthmatic reactions can result in problems in diagnosis and in identification of the cause.

CLINICAL AND DIAGNOSTIC CONSIDERATIONS

Historically, symptoms associated with heavy exposure to enzymes were striking and were progressively more severe with continued exposure. Lacrimation, rhinorrhea, cough with minimal sputum, and wheezing were the usual symptoms. The patient became increasingly breathless and had to sit upright at night to breathe. Occasionally, a worker reported mild hemoptysis and episodes of mild fever.

Before proper controls were installed, re-exposure after an interval, such as a vacation or holiday, resulted in return of symptoms. If, by accident or job change, a worker with mild symptoms received a heavy dose in a short period, he or she suffered a violent reaction with dyspnea, choking, and cyanosis severe enough to require oxygen treatment. At the time of such an episode, the CXR is said to have shown transient infiltrates.

Dermatitis resulting from sensitization was not seen in detergent factories, although palmar eczema did occur in workers producing the concentrated enzyme (Zachariae et al., 1973).

Shore et al. (1971) reported lung function abnormalities persisting in 8 of 17 workers after removal from exposure for periods up to 9 months. Weill et al. (1973) found reductions in lung volumes and in expiratory flow rates to be associated with past exposures, but they also noted in exposed workers a reversal in the downward trend in pulmonary function, presumably because of improved working conditions. Musk and Gandevia (1976) found a significant loss of pulmonary elastic recoil in workers formerly exposed to detergent proteolytic enzymes, with only partial recovery in some cases.

Pepys et al. (1969, 1973) considered the enzyme illness to be an allergic reaction on the basis of the results of skin testing and precipitins in sera. Atopic patients were more likely than non-atopic patients to have a positive SPT, and a good correlation was found between SPTs and the presence of enzyme-specific IgE in sera (Juniper et al., 1977).

J van Rooy et al. (2009) reported that workers exposed to liquid detergent enzymes reported more work-related symptoms of itching nose, sneezing, and wheezing. Additionally, 14.2% of exposed workers were sensitized to enzymes, indicating that workers exposed to liquid detergent enzymes are at risk of developing sensitization and respiratory allergy.

Diagnosis of enzyme-related OA depends on the usual methods for diagnosing asthma, including temporality of history, work-related wheezing, shortness of breath, obstructive disease on PFT, serial peak flow measurements, specific bronchial challenge, RAST, and SP testing or ELISA.

Significant correlations were found by Belin and Norman (1977) in enzyme exposed affected subjects between the SPT, clinical histories, and the RAST for specific IgE antibodies. They concluded that "a single skin prick test with a representative test reagent at appropriate test concentration can disclose specific IgE sensitivity as precisely but more cheaply and more easily than an intradermal test or a RAST." Schweigert et al. (2000) reported that, although the sensitivity and specificity of SPT is considered higher than for RAST, there is evidence demonstrating the use of serological tests as an effective alternative. For occupational health hazards, skin prick tests should be particularly useful for screening purposes.

INDUSTRIAL HYGIENE

The ACGIH established a TLV for the enzyme subtilisin (protease) of 60 ng of pure crystalline protein/m^3. Individual manufacturing operations may use even lower occupational exposure guidelines (Schweigert et al., 2000).

Minute amounts of enzyme are capable of eliciting clinical signs making it important to implement occupational controls to minimize dust exposure. Significant progress has been made through methods that include encapsulation of the enzymes as well as improved engineering of the manufacturing process. The prilled material minimizes exposure; however, enzyme dust can be liberated from the granules.

Minimizing dust exposure, applying engineering controls, and implementing medical surveillance programs are effective in creating and maintaining a safe working environment and have eliminated much of the allergy and asthma symptoms associated with enzymes.

Those manufacturing facilities that follow all of the guidelines, including industrial hygiene standards, medical prescreening (including SP testing), prospective surveillance, and continuing interventions enjoy very low or no cases of asthma and allergy among workers exposed to enzymes. With the present control of exposure to the enzymes, a working environment safe for atopic subjects can be achieved. Lessons learned in the detergent industry can be used to avoid OA due to enzyme exposure in other industries (Schweigert et al., 2000; Sarlo and Kirchner, 2002; Nicholson et al., 2001).

MEDICAL MANAGEMENT

After eliminating or minimizing enzyme exposure, OA and allergic symptoms can be treated using standard medical practice.

REFERENCES

Belin, L.G.A. and Norman, P.S. (1977) Diagnostic tests in the skin and serum of workers sensitized to *Bacillus subtilis* enzymes. *Clin. Allergy* 7:55–68.

J van Rooy, F.G.B., et al. (2009) A cross sectional survey among detergent workers exposed to liquid detergent enzymes. *Occup. Environ. Med.* 66:759–765.

Juniper, C.P., et al. (1977) *Bacillus subtilis* enzymes: a 7-year clinical, epidemiological, and immunological study of an industrial allergen. *J. Soc. Occup. Med.* 27:3–12.

Musk, A.W. and Gandevia, B. (1976) Loss of pulmonary elastic recoil in workers formerly exposed to proteolytic enzyme (alcalase) in the detergent industry. *Br. J. Ind. Med.* 33:158–165.

Nicholson, P.J., et al. (2001) Current best practice for the health surveillance of enzyme workers in the soap and detergent industry. *Occup. Med. (Lond.)* 51(2):81–92.

Pepys, J. (1992) Allergic asthma to *Bacillus subtilis* enzyme: a model for the effects of inhalable proteins. *Am. J. Ind. Med.* 21:587–593.

Pepys, J., et al. (1969) Allergic reactions of the lungs to enzymes of *Bacillus subtilis*. *Lancet* 1:1181–1184.

Pepys, J., et al. (1973) Clinical and immunological responses to enzymes of *Bacillus subtilis* in factory workers and consumers. *Clin. Allergy* 3:143–160.

Pepys, J., et al. (1985) A longitudinal study of possible allergy to enzyme detergents. *Clin. Exp. Allergy* 15(2):101–115.

Sarlo, K. and Kirchner, D.B. (2002) Occupational asthma and allergy in the detergent industry: new developments. *Curr. Opin. Allergy Clin. Immunol.* 2(2):97–101.

Schweigert, M.K., Mackenzie, D.P., and Sarlo, K. (2000) Occupational asthma and allergy associated with the use of enzymes in the detergent industry–a review of the epidemiology, toxicology and methods of prevention. *Clin. Exp. Allergy* 30(11):1511–1518.

Shore, N.S., Greene, R., and Kazemi, H. (1971) Lung dysfunction in workers exposed to *Bacillus subtilis* enzyme. *Environ. Res.* 4:512–519.

Weill, H., et al. (1973) Respiratory reactions to *B. subtilis* enzymes in detergents. *J. Occup. Med.* 15:267–271.

Zachariae, H., Thomsen, K., and Rasmussen, O.G. (1973) Occupational enzyme dermatitis: results of patch testing with alcalase. *Acta Derm. Venereol.* 53:145–148.

FARMER'S LUNG

Farmer's lung disease is one of the most common forms of HP.

SOURCES OF EXPOSURE

Farmer's lung disease has been associated with exposure to the dust of moldy hay, oats, barley, and millet grain in a variety of agricultural occupations.

FARMER'S LUNG VERSUS ORGANIC TOXIC DUST SYNDROME (ODTS)

A similar but distinct and more common respiratory illness affecting farmers has been labeled as ODTS. Making the distinction between farmer's lung and ODTS is important for both clinicians and researchers. Whereas farmer's lung disease is associated with exposure to specific antigens and is thought to be an allergic process affecting the alveoli and resulting in HP, ODTS is not considered an allergic disease; it is thought to be due to inhalation of toxins, and disease is more likely when the worker is exposed to higher concentrations of a variety of organic dusts. Organic dust toxic syndrome will be discussed in a separate section of this chapter.

ASSOCIATED AGENTS AND ANTIGENS

Farmer's lung disease has classically been associated with exposure to the antigens of thermophilic actinomycetes, *Aspergillus* species, *Saccharopolyspora rectivirgula*, (*Micropolyspora faeni*). Reboux et al. (2001) studied farmer's lung disease in France and found *Absidia corymbifera* to be the only microorganism able to discriminate farmers with disease from control farmers both in terms of exposure and of sensitization, whereas classic antigens did not seem to be involved. The authors concluded that agriculture techniques continually evolve and that there needs to be a periodic study of species and antigens involved with farmer's lung disease.

MECHANISM OF ACTION AND CHEMICAL PATHOLOGY

Farmer's lung disease is currently thought to be an EAA or HP, supported by the finding of precipitins in the sera of patients with farmer's lung and thought to be mediated by IgG antibodies (Reboux et al., 2001).

CLINICAL AND DIAGNOSTIC CONSIDERATIONS

The onset of farmer's lung disease may occur insidiously or may take place several hours after the exposure. Most commonly, severe dyspnea, cough, cyanosis, and fever are reported after exposure to the spore-laden dust of moldy hay or grain (Emanuel et al., 1964; Ames et al., 1990). These symptoms occur after the worker has been sensitized by previous exposure to the causative agents present in the dust; usually repeated exposure to high concentrations or prolonged exposure to low concentrations (Reboux et al., 2001). Wheezing, expectoration, malaise, and occasionally hemoptysis also occur (Donham, 1990).

The acute stage of farmer's lung disease may be accompanied by rales audible throughout the chest and by bilateral densities on the CXR (Barbee et al., 1968; Atwood et al., 1987). High resolution CAT scanning features are characterized by centrilobular nodules, ground-glass opacities, a mosaic pattern, air trapping on expiratory images, or some combination of these features (Ryu et al., 2007). Biopsy at this stage shows pulmonary edema as well as inflammatory changes in the alveolar walls, typical of a type III Arthus reaction (Iversen et al., 1991). Freedom from exposure will bring about relief of symptoms and reversal of the CXR changes to normal unless the disease is advanced and fibrosis has occurred. Spores of thermophilic actinomycetes can be detected by electron microscopy in pulmonary alveolar macrophages obtained by simple lavage (Romet-Lemonne et al., 1980). Both the absolute numbers and the percentages of T lymphocytes are reduced in the peripheral blood of patients with farmer's lung disease or even of those exposed to moldy hay, whereas B lymphocytes are unaffected (Flaherty et al., 1976).

Terho (1986) proposed the following *main* criteria for diagnosing farmer's lung disease: (1) exposure to antigens documented by environmental sampling or by identification of precipitating antibodies, (2) typical symptoms that worsen with exposure, and (3) lung infiltrates visible on CXR that are consistent with EAA.

Additional criteria were added to include (1) basal crepitant rales, (2) impairment of diffusing capacity, (3) decreased O_2 tension or saturation at rest or normal at rest and decreased with exercise, (4) PFT showing restrictive pattern, (5) lung biopsy histology compatible with the diagnosis, and (6) positive provocation tests by work exposure or controlled inhalation challenge. To confirm the diagnosis of farmer's lung disease using Terho's criteria, all three of the main criteria and at least two of the additional criteria must be met while other similar diseases are ruled out. If CXR findings are normal, the diagnosis can be confirmed with lung biopsy. It should be noted that radiologic (e.g., HRCT) and immunologic diagnostic tools have advanced since 1986 and can be used to enhance this diagnostic criteria.

INDUSTRIAL HYGIENE

Grain dust (oat, wheat, and barley): TLV–TWA, 4 mg/m^3 (ACGIH, 2001).

One of the main preventive measures of farmer's lung disease is to prevent the inhalation exposure to moldy hay and grain. Possible interventions include adequate ventilation in working areas, forced air ventilatory systems, mechanization of the feeding process, controlling the growth of mold on stored hay and grain, and the use of PPE. The use of hay conditioners and hay-dryer systems decreases the quantity of microorganism in hay, but their installation and usage costs are often prohibitive. Alternatively, the use of commercial additives for hay, such as propionic acid could be used to limit the growth of the mold (Reboux et al., 2002). Kusaka et al. (1993) suggested that the practical use of dust masks can effectively protect against farmer's lung disease. Dust masks with filters have been recommended. It has been debated whether farmers who have developed farmer's lung disease should leave the occupation of farming. One approach is to recommend continued farming, take all possible preventive measures, have regular evaluation of pulmonary functions, and consult physicians at the slightest suspicion of recurrence (Reboux et al., 2002). For highly sensitive patients, changing occupations may be advisable.

MEDICAL MANAGEMENT

As with other types of HP, the acute disease can be treated with corticosteroid therapy, however, its long-term efficacy has not been evaluated in prospective clinical trials (Ohshimo et al., 2012). Kokkarinen et al. (1992) found corticosteroid therapy for farmer's lung disease to help improve pulmonary function more rapidly; however, such treatment did not influence the long-term result. Early diagnosis and avoidance of exposure is the cornerstone of management.

REFERENCES

ACGIH (2001) *Grain dust.*

Ames, A.L., et al. (1990) Exposure to grain dust and changes in lung function. *Br. J. Ind. Med.* 47:466–472.

Atwood, P., Brouwer, R., and Tetal, R. (1987) A study of the relationship between airborne contaminants and environmental factors in Dutch swine confinement buildings. *Am. Ind. Hyg. Assoc. J.* 48:745–751.

Barbee, R.A., et al. (1968) The long-term prognosis in farmer's lung. *Am. Rev. Respir. Dis.* 97:223–231.

Donham, K.J. (1990) Health effects from work in swine confinement buildings. *Am. J. Ind. Med.* 17:17–25.

Emanuel, D.A., et al. (1964) Farmer's lung: clinical, pathologic, and immunologic study of twenty-four patients. *Am. J. Med.* 37:392–401.

Flaherty, D.K., et al. (1976) Lymphocyte subpopulations in the peripheral blood of patients with farmer's lung. *Am. Rev. Respir. Dis.* 114:1093–1098.

Iversen, M., et al. (1991) Mite allergy and exposure to storage mites and house dust mites in farmers. *Clin. Exp. Allergy* 20:211–219.

Kokkarinen, J.I., et al. (1992) Effect of corticosteroid treatment on the recovery of pulmonary function in farmer's lung. *Am. J. Respir. Crit. Care Med.* 145(1):3–5.

Kusaka, H., et al. (1993) Two year follow up on the protective value of dust masks against farmer's lung disease. *Intern. Med.* 32(2):106–111.

Ohshimoto, S., Bonella, F., Guzman, J., and Costabel, U. (2012) Hypersensitivity Pneumonitis. *Imm. Allergy Clinics N. Amer.* 32(4):537–558.

Reboux, G., et al. (2001) Role of molds in farmer's lung disease in Eastern France. *Am. J. Respir. Crit. Care Med.* 163(7): 1534–1539.

Reboux, G., et al. (2002) Influence of buffered propionic acid on the development of micro-organisms in hay. *Mycoses* 45:184–187.

Romet-Lemonne, J.L., Lemarie, E., and Choutet, P. (1980) Ultra-structural study of bronchopulmonary lavage liquid in farmer's lung disease. *Lancet* 1:777.

Ryu, J.H., et al. (2007) Diagnosis of interstitial lung diseases. *Mayo Clin. Proc.* 82(8):976–986.

Terho, E.O. (1986) Diagnostic criteria for farmer's lung disease. *Am. J. Ind. Med.* 10(3):329.

ORGANIC DUST TOXIC SYNDROME

Organic dust toxic syndrome is also known as pulmonary mycotoxicosis, grain fever, toxic alveolitis, and inhalation fever.

SOURCES OF EXPOSURE

Agriculture and farming involves production technology and work practices that generate substantial amounts of dusts. The usual exposures associated with ODTS include silage, moldy grain, hay, and straw. These exposures may be generated by silos, by chopping straw for bedding, unloading grain silos, shoveling feed, opening bales of hay for feed, and cleaning old animal housing structures. The physical and chemical components of organic dust vary and are affected by the dust source, geographic region, temperature, humidity, and other factors (Seifert et al., 2003). Confinement facilities generate dusts that include animal feces, endotoxins, and pollens. The occurrence of ODTS is especially common among swine workers, poultry producers, and grain workers, affecting up to 30% of exposed subjects. The higher the dust level, the more likely an exposed worker will develop signs and symptoms. Dust levels can occur up to about $15\,mg/m^3$ in swine operations, $12\,mg/m^3$ in poultry environments, and $72\,mg/m^3$ in grain operations (Kirkhorn and Garry, 2000).

The degree of respiratory dysfunction depends on the biological potency and concentration of exposure, as well as on individual susceptibility. Airborne contaminants frequently occur in concentrations and compositions that challenge the defense mechanisms of the lung.

Airborne concentrations of endotoxin in the farming environment may range between 0.01 and 50 μg/m^3 during feeding and cleaning operations on dairy farms. In swine confinement buildings, endotoxin is found in mean concentrations between 0.02 and 0.08 μg/m^3, and a similar range has been reported in poultry confinement environments. The variability of airborne endotoxin concentrations depends on the geographical location of the farm, type of ventilation, and livestock production techniques, including feeding methods (Donham, 1990).

Factors determining the intensity of exposure include meteorological conditions of the region, construction of the building, and type of farming activity.

ASSOCIATED AGENTS AND ANTIGENS

The most important airborne substances include grain dust and its constituents, bacteria and endotoxin, mycotoxins, fungal spores, fungal metabolites, glucan, and storage mites. Approximately 70 species of mites have been identified in various agricultural environments (Van Hage-Hamsten et al., 1991; Seifert et al., 2003). The most common species are *Acarus, Tyrophagus, Lepidoglyphus, Glycophagus, and Tydeus, Tarsonemus*.

The most common bacteria occurring in the farm environment are Gram-positive cocci (*Staphylococcus, Streptococcus*) and bacilli rods (*Corynebacterium*). Gram-negative species (*Proteus, Pseudomonas, Escherichia*) are, in general, less common but are potentially more pathogenic.

MECHANISM OF ACTION AND CHEMICAL PATHOLOGY

The pathogenesis of ODTS is unclear, but clinical findings and the results of BAL examination suggest a nonspecific acute inflammatory process. Direct toxic response to inhaled mycotoxin, proteinase enzymes, or endotoxin may be etiological factors (Barber, 1989; Crook et al., 1991; Seifert et al., 2003; Boehmer et al., 2009). Organic dust toxic syndrome is not considered as an infectious or allergic process. Mold and grain dust extract can contribute to inflammatory pulmonary reactions. Endotoxins are heat stable lipopolysaccharides (LPSs) found in dusts that can cause flu-like symptoms and promote bronchoconstriction. Other microbial products may be inflammatory or immunomodulatory, including: (1,3) beta-D-glucans, exotoxins from Gram-positive and Gram-negative bacteria, phytotoxins, heat shock proteins, and T-cell-activating superantigens (Kirkhorn and Garry, 2000).

CLINICAL AND DIAGNOSTIC CONSIDERATIONS

A thorough occupational history should elicit heavy exposure to organic dust. Symptoms resemble a flu-like illness and may include one or more of the following, irritation of the upper respiratory tract, eyes and skin, fever, chills, malaise, myalgia, fatigue, a dry cough, dyspnea, chest tightness, headache and nausea; the severity of which depends on the intensity and duration of the exposure (Seifert et al., 2003). It is difficult to distinguish ODTS from farmer's lung disease based on clinical symptoms alone; both occurring 4–8 h after exposure. An adequate exposure history along with diagnostic studies is needed to make the distinction.

Organic dust toxic syndrome is thought to be much more common than HP. Unlike HP, ODTS is not associated with sensitization, and neither skin tests nor serological tests identify the apparent immunological involvement (Pepys and Jenkins, 1965; Seifert et al., 2003). The lungs may be clear or rales may be heard. The CBC may show leukocytosis without infection that may be misleading.

In contrast to HP, the CXR in ODTS is usually normal without infiltrates and lung function is normal, although minor and transient changes have been described. Similar to HP, ODTS can show an increase of neutrophils on BAL (Lecours et al., 1986). Severe hypoxemia does not occur (Von Essen et al., 1990, 2005).

Respiratory symptoms of ODTS resemble those of the acute stage of HP, but the condition is considered benign, without persistent respiratory sequelae of physiologic significance. It should always be remembered that multiple disease processes may be at play when dealing with complex mixtures of organic dust particulates, and farmer's lung disease and ODTS can occur together.

INDUSTRIAL HYGIENE

Grain dust (oat, wheat, and barley): ACGIH TLV–TWA, 4 mg/m^3; NIOSH REL–TWA, 4 mg/m^3; OSHA PEL–TWA, 10 mg/mg^3.

Nuisance dust: TLV, 10 mg/m^3 inhalable, 3 mg/m^3 respirable (ACGIH, 2000).

Dust exposure should be minimized through workplace engineering and or use of respirators. General standards for dust exposure should be implemented (see introduction to this chapter). More specifically, grain dust has been assigned a TLV intended to minimize the potential for acute irritation of the upper respiratory tract, eyes, and skin, bronchitis symptoms, and chronic decrements in pulmonary function. The recommended TLV applies to the whole grain of oat, wheat, and barley before the milling operation. Data are scant or not available on which to base skin or sensitization (SEN) notations or a carcinogenicity classification. Kirkhorn and Garry (2000) reports are based on studies of post shift decrements in FEV1 where environmental dust is below current standards, recommendations have been made for the development of more stringent threshold-limit standards to adequately protect workers in the confinement industry.

MEDICAL MANAGEMENT

Most cases of ODTS are self- limiting and medical treatment is not sought. Treatment for myalgia and fever with non-steroidal anti-inflammatory agents and acetaminophen has been recommended. Corticosteroids have not been shown to be helpful. Severe cases with respiratory failure should be managed with supportive care (Von Essen et al., 2005).

REFERENCES

ACGIH (2000) *TLV's and BEI's*, Cincinnati, Ohio: American Council of Governmental Industrial Hygienists, 2000.

Barber, E.M. (1989) Air quality in swine barns. *Proceedings of the Banff Pork Seminar*, Banff Alberta.

Boehmer T.K., et al. (2009) Cluster of presumed organic dust toxic syndrome cases among urban landscape workers-Colorado, 2007. *Am. J. Ind. Med.* 52:534–538.

Crook, B., et al. (1991) Airborne dust, ammonia, microorganisms, and antigens in big confinement houses and the respiratory health of exposed farm workers. *Am. Ind. Hyg. Assoc. J.* 52:271–279.

Donham, K.J. (1990) Health effects from work in swine confinement buildings. *Am. J. Ind. Med.* 17:17–25.

Kirkhorn, S.R. and Garry, V.F. (2000) Agricultural lung diseases. *Environ. Health Perspect.* 108(Suppl. 4):705–712.

Lecours, R., Laviolette, M., and Cormier, Y. (1986) Bronchoalveolar lavage in pulmonary mycotoxicosis (organic dust toxic syndrome). *Thorax* 41:924–926.

Pepys, J. and Jenkins, P.A. (1965) Precipitin (F.L.H.) test in farmer's lung. *Thorax* 20:21–35.

Rask-Andersen, A. (1989) Organic dust toxic syndrome among farmers. *Occup. Environ. Med.* 46:233–238.

Seifert, S.A. et al. (2003) Organic dust toxic syndrome: a review. *J. Toxicol. Clin. Toxicol.* 41(2):185–193.

Von Essen S., et al. (1990) Organic dust toxic syndrome: an acute febrile reaction to organic dust exposure distinct from hypersensitivity pneumonitis. *J. Toxicol. Clin. Toxicol.* 28(4): 389–420.

Von Essen, S., et al. (2005) Organic dust toxic syndrome: a noninfectious febrile illness after exposure to the hog barn environment. *J. Swine Health Prod.* 13(5):273–276.

Van Hage-Hamsten M., et. al. (1991) Storage mites dominate the fauna in Swedish barn dust. *Allergy* 46:142–146.

TEA MAKERS ASTHMA

Exposure to tea dust has been associated with illness called tea taster's disease, tea factory cough, and tea maker's asthma.

SOURCES OF EXPOSURE

Manufacturing of tea is a dusty process that exposes workers to a fine tea fluff; occupational illness has been reported as far back as 1919 among some workers who process or taste the tea (Castellani and Chalmers, 1919).

ASSOCIATED AGENTS OR ANTIGENS

The agent associated with illness is tea dust or a fine dust called tea fluff.

MECHANISM OF ACTION AND CHEMICAL PATHOLOGY

Tea taster's disease has been thought to result from the inhalation of microorganisms while sniffing tea leaves. Both cough and asthma have been associated with inhaling the fine dust known as tea fluff and studies have suggested sensitization to tea dust with resulting respiratory symptoms (Uragoda, 1970). Currently, the underlying mechanism for tea-associated disease is unclear, although tea-induced asthma probably results from allergic sensitization similar to many other organic dust syndromes (Abramson et al., 2001).

CLINICAL AND DIAGNOSTIC CONSIDERATIONS

Rhinitis, cough, chest tightness, reductions in FEF indices, bronchitis, asthma, and chronic respiratory symptoms have been reported in tea workers (Uragoda, 1970; Zuskin and Skuric, 1984; Rask-Andersen, 1989; Mirbod et al., 1995; Hill and Waldron, 1996).

Historically, Castellani and Chalmers (1919) described two related occupational conditions, which they called tea factory cough and tea taster's disease. Workers with tea factory cough suffered from weight loss, tiredness, and cough with muco-purulent expectoration. Occasional coarse rales could be heard on auscultation. When removed from exposure, the symptoms slowly resolved. Tea tasters had similar reactions. Tea tasters not only tasted the tea but also buried their noses in the leaves inhaling not only the tea dust, but also contaminants such as the fungi Monilia, Aspergillus, and Penicillium as well as a "peculiar Streptococcus" microorganism.

Uragoda (1970) reported a case of occupational asthma in a tea maker who when exposed to tea fluff experienced sneezing, watery nasal discharge, coughing, chest tightness, shortness of breath and production of copious amounts of mucoid sputum, accompanied by wheezing, bilateral rhonchi, and crepitations. Provocative inhalation of the tea fluff produced an attack within a few minutes. A positive reaction was obtained on skin testing with tea fluff antigen.

Zuskin and Skuric (1984) found reductions in maximum expiratory flow rates at low volumes during the work shift of tea workers suggesting a bronchoconstrictor effect of the dust acting mostly on small airways. Pre-shift administration of disodium cromoglycate diminished the acute reduction in flow rates. They also concluded that exposure to tea dust in some workers may lead to chronic respiratory effects. In 1985, Zuskin E. et al. studied 26 nonsmoking female tea workers and found that those who had positive skin tests had a higher prevalence of chronic respiratory symptoms, including cough, phlegm, bronchitis, asthma, and dyspnea, however, only coughing and dyspnea showed a significant difference (Zuskin et al., 1985). Pre-shift maximum expiratory flow rates (MEFR) were found to be lower than normal in tea workers. Maximum expiratory flow rate showed significant reductions over the work shift and the acute reductions were larger in those with positive skin tests. Provocative inhalation of tea allergens caused bronchoconstriction in four out of six subjects that were tested.

A systematic study of tea workers in Sri Lanka who were exposed to tea fluff during blending operations showed a prevalence of chronic bronchitis and asthma greater than the general population (Uragoda, 1980). Roberts and Thomson (1988) reported asthma in a 55-year-old female worker in a tea-packing production line. Although SP testing was negative, diagnosis was established by serial expiratory flow rates and bronchial provocation challenge with tea dust that elicited a late asthmatic reaction. Three cases of occupational asthma in tea workers were reported by Cartier and Malo (1990).

Abramson et al. (2001) studied 192 workers in a tea-packing plant and found the prevalence of asthma, wheezing, hay fever, and atopy to be higher than has been previously reported in tea workers, but similar to the general population in the area. Tea workers in certain operations did report an excess of respiratory and nasal symptoms. The authors concluded in this study that the work-related respiratory and nasal symptoms probably represented nonspecific irritation.

Shieh et al. (2012) found tea workers in certain areas of production to have a higher prevalence of rhinorrhea, sneezing, throat irritation, cough, chronic phlegm, chronic bronchitis, dyspnea, and reductions in FEV_1/FVC and maximum mid expiratory flow rates.

Previous studies have reported an association between tea drinking as well as other dietary factors and lung cancer. This was investigated in a case-control study in the west of

Sweden by Axelsson et al. (1996). This study could not demonstrate an increased risk of cancer due to tea consumption, contrary to the findings of some previous studies.

Furthermore, a study was conducted in Okinawa, Japan, from 1988 to 1991 to evaluate the relationship between tea consumption and lung cancer risk (Ohino et al., 1995). Paradoxically, the greater the intake of Okinawan tea (a partially fermented tea), the smaller the risk, particularly in women. The risk reduction by Okinawan tea consumption was detected mainly in squamous cell carcinoma. Daily tea consumption significantly decreased the risk of squamous cell carcinoma in males and females, suggesting a protective effect of tea consumption against lung cancer in humans.

INDUSTRIAL HYGIENE

There are no specific standards for airborne tea dust. Tea dust exposure should be minimized, and for workers with disease, eliminated.

MEDICAL MANAGEMENT

Management includes removal from exposure and symptomatic treatment of rhinitis, coughing, and asthma.

REFERENCES

Abramson, M.J., et al. (2001) Respiratory disorders and allergies in tea packers. *Occup. Med. (Lond.)* 51(4):259–265.

Axelsson, G., et al. (1996) Dietary factors and lung cancer among men in West Sweden. *Int. J. Epidemiol.* 25:32–39.

Cartier, A. and Malo, J.L. (1990) Occupational asthma due to tea dust. *Thorax* 45(3):203–206.

Castellani, A. and Chalmers, A.J. (1919) *Manual of Tropical Medicine*, 3rd ed., London: Bailliere, Tindall and Cox, pp. 1890.

Hill, B. and Waldron, H. A. (1996) Respiratory symptoms and respiratory function in workers exposed to tea fluff. *Ann. Occup. Hyg.* 40:491–497.

Mirbod, S.M., et al. (1995) Some aspects of occupational safety and health in green tea workers. *Ind. Health* 33:101–117.

Ohino, Y., et al. (1995) Tea consumption and lung cancer risk: a case-control study in Okinawa. *Jpn. J. Cancer Res.* 86: 1027–1034.

Roberts, J.A. and Thomson, N.C. (1988) Tea-dust induced asthma. *Eur. Respir. J.* 1(8):769–770.

Shieh T.S., et al. (2012) Pulmonary function, respiratory symptoms, and dust exposures among workers engaged in early manufacturing processes of tea: a cohort study. *BMC Public Health* 12: 121.

Uragoda, C.G. (1970) Tea maker's asthma. *Br. J. Ind. Med.* 27:181–182.

Uragoda, C.G. (1980) Respiratory disease in tea workers in Sri Lanka. *Thorax* 35:114–117.

Zuskin, E. and Skuric, Z. (1984) Respiratory function in tea workers. *Br. J. Ind. Med.* 41:88–93.

Zuskin E., et al. (1985) Immunological and respiratory changes in tea workers. *Int. Arch. Occup. Environ. Health* 56(1):57–65.

WOOD DUSTS

SOURCES OF EXPOSURE

Exposure to wood dust occurs in many occupations such as furniture and cabinet making, carpentry, logging, lumber mills, sawmills, pulp mills, and construction. The intensity (concentration) of exposure varies depending on the types of processes (such as sawing vs. grinding), ventilation, and use of PPE; these are important considerations when implementing preventive measures or determining causation. Most studies showing associations with sinonasal cancer involved hard wood exposure in furniture workers (ACGIH, 2005, Wood Dusts; Kauppinen et al., 2006).

ASSOCIATED AGENTS AND ANTIGENS

Endotoxins of Gram-negative bacteria and the allergenic fungi that grow on wood are reported to be the main hazardous agents found in wood-processing workplaces (Dutkiewicz et al., 1988; Alwis et al., 1999). Wood contains hundreds of high and low molecular weight compounds known as "wood extractives" that protect trees from attack by bacteria and fungi, and also give color to the wood (ACGIH, 2005, Wood Dusts). Plicatic acid has been determined to be the cause of asthma associated with exposure to western red cedar. Wood dust contaminated with bacteria and fungi may contribute to irritation and sensitization. In wood pulp workers, HP has been associated with exposure to antigens of Alternaria and Penicillium. The specific causative agents of sinonasal cancer associated with wood dust exposure have not been determined (Nylander and Dement, 1993).

MECHANISM OF ACTION AND CHEMICAL PATHOLOGY

Being that wood dusts are complex mixtures of high and low molecular weight organic compounds that contain various irritants, endotoxins, and allergens, we expect the mechanisms of disease to include irritant, toxicologic, pharmacologic, and immunologic mechanisms; however, as already stated, and as is the case with most organic dusts, the mechanisms of disease for wood dust exposure have not been clearly delineated. Endotoxins have been linked to chest tightness, shortness of breath, fever, and wheezing. (1–3)-β-D-glucan, a fungal constituent and an inflammatory agent, is found in organic dusts and thought to react synergistically with endotoxin and other inflammatory agents (Fogelmark et al., 1992). (1–3)-β-D-glucan is also thought to play an important role in the development of HP (Fogelmark and Rylander, 1993).

One of the best known associations is OA due to exposure to Western Red Cedar. In this case plicatic acid, a unique component of Western Red Cedar, has been identified as a causative agent responsible for asthma, however, the pathologic mechanism is not clear. Apart from plicatic acid in Western Red Cedar wood, no causative agent for wood-related asthma has consistently been reported. Type 1 allergy is not suspected to be a major cause of wood dust-induced asthma (Jacobsen et al., 2010).

Similarly, the specific agent in hard wood dust responsible for cases of nasal cancer in furniture workers thought to be associated with wood dust exposure has not been determined.

CLINICAL AND DIAGNOSTIC CONSIDERATIONS

Health problems that have been reported to be associated with various wood dusts include dermatitis, rhinitis, conjunctivitis, sinusitis, bronchitis, OA, HP, ODTS, non-asthmatic chronic airflow obstruction, and nasal cancer (Enarson and Chan-Yeung, 1990; Chan-Yeung and Malo, 1994; ACGIH, 2005, Wood Dusts). Symptoms of cough, phlegm, chronic bronchitis, frequent headaches, eye irritation, throat irritation, and nasal symptoms have been found to be more prevalent in woodworkers (Alwis et al., 1999). While standard methods are used to diagnose these conditions, causation is not always clear. Radioallergosorbent testing, SP testing, and ELISA may be helpful in establishing immunologic vs. irritant causes.

Hildesheim et al. (2001), reported the risk for nasal cancer in those exposed to wood dust to increase with exposure for >10 years, those first exposed before the age of 25 years, and those seropositive to Ebstein–Barr virus.

Jacobsen et al. (2010) reviewed 37 articles and reported support for an association between dry wood dust exposure and asthma, rhino-conjunctivitis, coughing, bronchitis, and acute and chronic impairment of the lung function.

Associations between wood dust exposure and nasal adenocarcinomas have been consistently stronger than that between wood dust and nasopharyngeal squamous carcinoma. Additionally, it has also been reported that while nasal adenocarcinomas have been mostly associated with occupational exposure to hardwood, nasopharyngeal cancer (a squamous cell tumor) may be more closely linked to softwood exposure (Hildesheim et al., 2001).

Signs of nasal cancer may include nosebleeds, nasal obstruction, or discharge, changes in the sense of smell, headache, dysphagia, tinnitus, and cranial nerve involvement with weakness of the muscles of the face, tongue, or throat, and cervical lymphadenopathy. Prevention or early diagnosis is of course optimal when managing cancer risks.

While previous studies have been inconclusive regarding associations between wood dust exposure and cancers other than sinonasal cancer, Barcenas et al. (2005) concluded that wood dust is a potential risk for lung cancer based on their case control study of 1368 lung cancer patients.

INDUSTRIAL HYGIENE

Western Red Cedar: TLV–TWA, 0.5 mg/m^3, inhalable particulate mass. Sensitizer (SEN): A4—not classifiable as a human carcinogen.

All other species: TLV–TWA, 1 mg/m^3, inhalable particulate mass. Carcinogenicity: A1—confirmed human carcinogen: Oak, beech. A2—suspected human carcinogen: Birch, mahogany, teak, and walnut. A4—not classifiable as a human carcinogen: All other wood dusts. (ACGIH, 2005).

All TLVs for wood dust are measured as *inhalable* particulate mass because of the evidence of an increased risk of upper and lower respiratory symptoms and sinonasal cancer.

MEDICAL MANAGEMENT

Successful treatment of symptoms depends on an accurate diagnosis of irritation vs. allergic vs. carcinogenic responses. Application of standard medical care for dermatitis, rhinitis, bronchitis, asthma, and HP can be implemented. Identification and documentation of involved wood exposure is challenging yet crucial in accurately determining the cause and for effective exposure mitigation. Once the offending agent is identified and exposure eliminated some but not all cases show reversal and improvement of disease.

The treatment of diagnosed cancer is beyond the scope of this chapter. Prevention of cancer through appropriate exposure control is the best approach. Monitoring the health status of the occupationally exposed worker may help to achieve early diagnosis, treatment, and exposure intervention, which may improve outcomes.

REFERENCES

ACGIH (2005) *Wood Dusts.*

Alwis, U.A., Mandryk, J., and Hocking, A.D. (1999) Exposure to biohazards in wood dust: bacteria, fungi, endotoxins, and (1 3)-beta-D-glucans. *Appl. Occup. Environ. Hygiene* 14: (9): 598–608.

Barcenas, C.H., et al. (2005) Wood dust exposure and the association with lung cancer risk. *Am. J. Ind. Med.* 47(4):349–357.

Chan-Yeung, M. and Malo, J.L. (1994) Aetiological agents in occupational asthma. *Eur. Respir. J.* 7:346–371.

Dutkiewicz, J., Jabionski, L., and Olenchock, S. A. (1988) Occupational biohazards: a review. *Am. J. Ind. Med.* 14:605–623.

Enarson, D.A. and Chan-Yeung, M. (1990) Characterization of health effects of wood dust exposures. *Am. J. Ind. Med.* 17(1): 33–38.

Fogelmark, B. and Rylander, R. (1993) Lung inflammatory cells after exposure to mouldy hay. *Agents Action* 39:25–30.

Fogelmark, B., Goto, H., Yuasa, K., et al. (1992) Acute pulmonary toxicity of inhaled (1–3)-β-D-glucan and endotoxin. *Agent Action* 35:50–56.

Hildesheim, A., et al. (2001) Occupational exposure to wood, formaldehyde, and solvents and risk of nasopharyngeal carcinoma. *Cancer Epidemiol. Biomarkers Prev.* 10:1145–1153.

Jacobsen, G., et al. (2010) Non-malignant respiratory diseases and occupational exposure to wood dust. Part H. Dry wood industry. *Ann. Agri. Environ. Med.* 17(1):29–44.

Kauppinen, T., et al. (2006) Occupational exposure to inhalable wood dust in the member states of the European Union. *Ann. Occup. Hyg.* 50(6):549–561.

Nylander, L.A. and Dement, J.M. (1993) Carcinogenic effects of wood dust: review and discussion. *Am. J. Ind. Med.* 24(5): 619–647.

96

NATURALLY OCCURRING MINERAL FIBERS

ALI K. HAMADE, CHRISTOPHER M. LONG, AND PETER A. VALBERG

ASBESTOS

First aid: Remove to fresh air free of airborne asbestos fibers. Decontaminate clothing and skin surfaces

Target organ(s): Lung

Occupational exposure limits: OSHA 8-h TWA 0.1 f/cc; ACGIH 8-h TWA 0.1 f/cc; OSHA PEL (excursion limit) averaged over a 30 min sampling period 1.0 f/ml

Reference values: No RfC derived; inhalation unit risk = 0.23 per (f/cc); MCL = MCLG = 7 million fibers per liter (MFL); no MRLs derived

Risk/safety phrases: IARC Group 1 carcinogen; ACGIH, A1; NIOSH "considers asbestos a potential occupational carcinogen and recommends that exposures be reduced to the lowest possible concentration."

ASBESTOS MINERALS: BACKGROUND AND USES

Historical Overview

Asbestos refers to a group of naturally occurring, fibrous, hydrated mineral silicates; i.e., asbestiform minerals are made up of thin, fiber-like crystals. These minerals have been historically used in a wide variety of products where resistance to heat has been of crucial importance, for example, thermal insulation of high pressure steam pipes, fireproofing of building materials, and refractory bricks in high temperature furnaces. The fact that asbestiform minerals could be woven into textiles also allowed for their use in fire blankets, insulating gloves, and fire-resistant garments. Asbestos is also very durable, insoluble, and chemically inert, and hence, found in

siding and roofing shingles, vinyl asbestos floor tiles, asbestos boards, and friction products.

Asbestos minerals have been known for their useful properties for millennia, and were widely used in the early twentieth century because of characteristics such as superb performance as a flame retardant, low thermal conductivity, good sound absorption, tensile strength, and resistance to thermal, electrical, and chemical damage (Alleman and Mossman, 1997). However, because of serious health consequences resulting from elevated exposures to airborne asbestos, particularly in the occupational environment, the use of asbestos minerals is currently federally regulated, and is banned in many countries.

Straif (2008) estimated that 85–90% of pleural mesotheliomas and about 4.5% of lung cancers arise in men due to exposure to asbestos.. In women, the asbestos-attributable risks for pleural and peritoneal mesotheliomas combined are estimated at 23% (Spirtas et al., 1994). The role of asbestos in these diseases has changed over time due to changes in asbestos usage and control of exposure, and it appears to be decreasing at present (Craighead, 2011; Pukkala et al., 2009; Moolgavkar et al., 2009).

Asbestos use in the United States and around the world has been steadily decreasing, as described by the United States Environmental Protection Agency (US EPA), United States Geological Survey (USGS), National Institute for Occupational Safety and Health (NIOSH), and World Health Organization (WHO) on the following websites (and illustrated in Figure 96.1).

http://www.epa.gov/asbestos/pubs/ban.html

http://minerals.usgs.gov/minerals/pubs/commodity/asbestos/mcs-2011-asbes.pdf

Hamilton & Hardy's Industrial Toxicology, Sixth Edition. Edited by Raymond D. Harbison, Marie M. Bourgeois, and Giffe T. Johnson.
© 2015 John Wiley & Sons, Inc. Published 2015 by John Wiley & Sons, Inc.

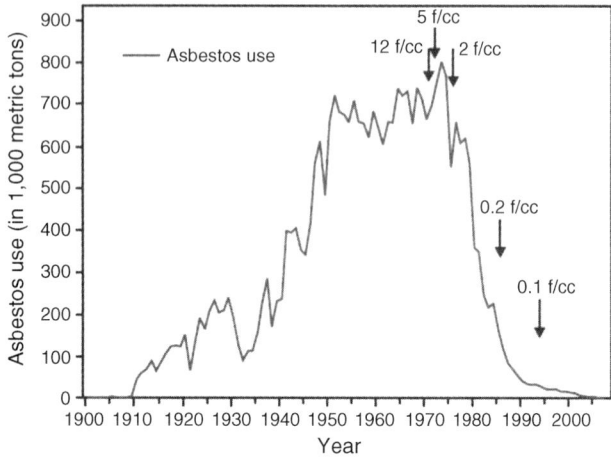

FIGURE 96.1 Annual asbestos use in the United States has changed over time (1900–2007). Permissible asbestos exposure limits (numbers next to arrows) have evolved since about 1970. (*Source:* NIOSH, March 2009).

http://minerals.usgs.gov/minerals/pubs/commodity/
asbestos/mis-2007-asbes.pdf

http://www.cdc.gov/mmwr/preview/mmwrhtml/
mm5815a3.htm

http://whqlibdoc.who.int/hq/2006/WHO_SDE_
OEH_06.03_eng.pdf

The regulatory definition of a "fiber" is a structure with a length >5 μm, diameter <3 μm, and an aspect ratio (length-to-diameter ratio) >3:1. However, many asbestos fibers found in insulation products are longer and thinner than this prescription, typically with an aspect ratio of 20:1. The number of fibers detected in air samples or in body tissues depends on the method used [e.g., phase contrast microscopy (PCM) *versus* scanning or transmission electron microscopy]. Guidelines for limiting asbestos exposure in the occupational environment are generally given in terms of fiber counts per volume of air, e.g., fibers per cubic centimeter (f/cc) of air. The Occupational Safety and Health Administration (OSHA) 8-h time-weighted average (TWA) concentration allowable in the workplace is 0.1 f/cc. However, there are numerous other guidelines from OSHA on appropriate asbestos work practices and exposure restrictions (http://www.osha.gov/SLTC/asbestos/standards.html). The American Conference of Governmental Industrial Hygienists (ACGIH) has the same guideline level for asbestos, namely an 8-h TWA of 0.1 f/cc. These airborne concentrations are specified as being measured by PCM.

Types of Asbestos Mineral Fibers

"Asbestos" includes the following six naturally occurring, hydrated-silicate minerals with fibrous habits, i.e., made up of thin, fiber-like crystals (HEI-AR, 1991). The mineralogy

of asbestos has been recently reviewed (Sporn, 2011; Case et al., 2011).

Serpentine asbestos, which forms polymeric sheets that tend to wrap into curved and flexible tubular fibers (i.e., hollow tubules):

chrysotile—"white asbestos"—CAS # 12001-29-5, e.g., $Mg_3 (OH)_4 (Si_2O_5)$

Amphibole asbestos, which is made up of thin, straight, needlelike fibers, consists of:

actinolite—CAS # 13768-00-8, e.g., $Ca_2(Mg, Fe)_5 (OH)_2 (Si_8O_{22})$

amosite[1]—"brown asbestos"—CAS # 12172-73-5, e.g., $Fe_7 (OH)_2 (Si_8O_{22})$

anthophylite—CAS # 17068-78-9 e.g., $(Mg, Fe)_7 (OH)_2 (Si_8O_{22})$

crocidolite—"blue asbestos"—CAS # 12001-28-4 e.g., $Na_2Fe_5 (OH)_2 (Si_8O_{22})$

tremolite—CAS # 14567-73-8 e.g., $Ca_2Mg_5 (OH)_2 (Si_8O_{22})$

Chrysotile asbestos is more flexible than amphibole types of asbestos, and it has often been spun and woven into fabric and embedded into various matrices (for example, joint compound, corrugated asbestos-cement roofing sheets, and friction products) so as to contribute to strength, heat resistance, and chemical inertness. Amphibole fibers were more frequently used in thermal insulation (lagging). Crocidolite and amosite are the only amphiboles that have had significant industrial uses over the years. For example, amosite was used in insulation Batts, asbestos sheets, pipe insulation, block insulation, and ceiling tiles. The other three amphiboles, actinolite, anthophyllite, and tremolite, have had only infrequent industrial application.

In the post-WWII time period, some asbestos could be found in products as diverse as pipe insulation, boiler insulation, fire retardant coatings, brake linings, bricks, cement pipe, and fireplace cement, heat-, fire-, and acid-resistant gaskets, ceiling tiles and insulation, fireproof drywall, flooring, roofing, lawn furniture, and drywall joint compound. Chrysotile has been used more than any other type of asbestos, and accounts for about 95% of the asbestos found in buildings in the United States (HEI-AR, 1991).

General Epidemiology and Impact on Occupational Health

Asbestos exposure and the subsequent development of disease have been frequently evaluated over the years, and many

---

[1] Amosite is a trade name for the amphiboles in the cummingtonite-grunerite series, i.e., asbestos that came from the Asbestos Mines of South Africa (AMOSA) mine.

reviews are available (Doll and Peto, 1985; US EPA, 1986; HEI-AR, 1991; ATSDR, 2001; Roggli et al., 2004; Craighead and Gibbs, 2008; Dodson and Hammar, 2011).

Airborne asbestos fibers can be a health hazard, if inhaled at sufficient doses, because of the physical and chemical properties of the asbestos fibers. The primary route of exposure is *via* inhalation of fibers released during asbestos mining, manufacture of asbestos-based products, and breakdown of asbestos-based materials. Depending on fiber size (length and diameter), inhaled fibers deposit onto lung airway surfaces, and subsequently are removed from the lung (cleared), retained in lung tissues, or translocated to nearby sites. Because the fibers are insoluble and resistant to chemical attack, retained fibers are not readily broken down. Of the various asbestos species, chrysotile is the most soluble and most rapidly cleared from the lungs, which contrasts to the amphibole varieties of asbestos, which are considerably more resistant to dissolution and clearance.

Inhalation of sufficient quantities of asbestos fibers over protracted periods of time can lead to serious illnesses, including asbestosis (a type of pneumoconiosis), pleural plaques, pleural effusions, lung cancer, and mesothelioma (a rare cancer of the lung or peritoneal pleura, associated with exposure to amphibole asbestos). Asbestos is classified as a known human carcinogen by International Agency for Research on Cancer (IARC) [(1977, 2012; Straif et al., 2009)]. A low concentration of asbestos fibers exists in the ambient air, due to both natural sources and society's long use of asbestos in construction and commerce (HEI-AR, 1991). Exposure at the low levels found in nonoccupational settings is less likely to cause health problems (Pierce et al., 2008).

PHYSICAL AND CHEMICAL PROPERTIES

Overall Structure

Because asbestos fibers are all silicates, they exhibit properties common to silicates, such as incombustibility, thermal stability, resistance to biodegradation, non-reactivity toward most chemicals, and low electrical conductivity. In fact, the word "asbestos" comes from the Greek, meaning "unquenchable" or "inextinguishable" (Alleman and Mossman, 1997). Asbestos minerals are found in bundles of fibers, which can be easily separated from the host matrix or cleaved into thinner fibers. Chrysotile may form very fine fibrils that range between 0.02 and 0.08 µm in diameter; amosite diameters range between 0.06 and 0.35 µm, and crocidolite between 0.04 and 0.15 µm (HEI-AR, 1991).

The basic building blocks of asbestos fibers are the silicate tetrahedra, which may occur as double chains (Si_4O_{11}), as in the amphiboles, or in sheets (Si_2O_5), as in chrysotile. In the case of chrysotile, an octahedral brucite layer having the formula ($Mg_6O_4 (OH)_8$) is intercalated between each silicate tetrahedra sheet. Asbestos fibers used in most industrial applications consist of aggregates of smaller units (fibrils). This is most evident with chrysotile, which exhibits an inherent, well-defined unit fiber.

Chemical Constituents

Asbestos fibers are virtually inert; they do not evaporate, dissolve, or burn, and react only minimally with other substances. Although asbestos minerals share the silicate tetrahedral group, the main differences among them are the metal cations. In contrast to chrysotile, where the major metal is magnesium, amphiboles possess crystal structures accommodating cations of differing size and valence, and hence, can have significant calcium and iron content. It has often been suggested that the iron cations are significant contributors to amphibole asbestos toxicity.

Shapes of Fibers

Chrysotile fibers are composed of aggregates of long, thin, and curly fibrils. In contrast, amphibole fibers are straight, needlelike, and splintery. Crocidolite fibrils are shorter and thinner than other amphibole fibrils, but not as thin as chrysotile fibrils. Amosite fibril diameter is thicker than either chrysotile or crocidolite.

Fibers *Versus* Cleavage Fragments

Some asbestos minerals can exist in forms called "cleavage fragments," which, because they are resistant to disaggregation into free fibers, are considered far less potent in their ability to trigger diseases characteristic of fibrous asbestos (Gamble and Gibbs, 2008). For example, besides the asbestiform varieties of tremolite and actinolite (which release long, thin fibers composed of fibrils), non-asbestiform varieties also occur, which release only cleavage fragments.

ENVIRONMENTAL FATE AND BIOACCUMULATION

Asbestos fibers can be released into the air when asbestos-containing materials are mined, manufactured, used, or demolished. The vast majority of asbestos exposure has taken place in asbestos-worker occupational settings (Williams et al., 2007). A small amount of exposure is known to be present in some public and commercial buildings (HEI-AR, 1991; Lee and Van Orden, 2008). Unless one is in proximity to asbestos being manipulated in construction and/or demolition of asbestos-containing structures, one's outdoor, ambient-air exposures are relevant only in a few geographical locations where asbestos materials are present in natural rocks and soils (Lee et al., 2008). Also, "environmental" exposures can sometimes be related to neighborhoods where historical manufacturing sites were located and societal uses of asbestos containing materials occurred (Case and Abraham, 2009; Craighead and Gibbs, 2008; Case et al., 2011; NIOSH, 2008; Pan et al., 2005).

Inhaled asbestos fibers deposit on lung surfaces, but because the fibers are insoluble, they do not dissolve readily into lung fluids and must be removed by physical or cellular processes. Fibers that deposit onto surfaces of major airways are removed from the lung by the upward movement of the mucociliary "escalator" to the pharynx, where they are swallowed. Fibers that deposit in the air-exchanges regions of the lung (the alveolar regions) attract alveolar macrophages, which attempt to "eat" (phagocytize) the fibers and remove them from alveolar surfaces. However, if the fibers are too long to be ingested by macrophages ($>12–17\,\mu m$), then removal from the lung is incomplete, and in fact, the "frustrated phagocytosis" can release damaging lysosomal enzymes on lung surfaces. Macrophages coat uningested fibers with a proteinaceous material and form "asbestos bodies" or "ferruginous bodies." Clumps of fibers may fragment or split and form smaller fibrils, which may then be removed or slowly dissolve. Fibers may also penetrate cells that line alveolar surfaces (epithelial cells), and reach lung interstitial tissues and the lung pleura. Thus, asbestos fibers, particularly amphiboles, may accumulate during exposure, and amphibole asbestos fiber retention in the lungs persists long after exposure ceases.

Clearance of asbestos fibers from the lungs varies by fiber type, and chrysotile short fibers clear from the lungs more rapidly than amosite long fibers (ATSDR, 2001; Bernstein et al., 2008). In fact, for nonoccupationally exposed individuals, the greater biopersistence in the lungs of amosite fibers *versus* chrysotile fibers is illustrated by the higher fiber counts for the amphibole asbestos varieties in autopsy sample of lungs (Dodson et al., 1999). The differences in fiber size and chemistry between chrysotile and amphiboles result in chrysotile clearing very rapidly from the lung ($T\frac{1}{2} = 0.3–11$ days) while amphiboles are among the slowest clearing fibers known ($T\frac{1}{2} = 500$ to $\propto$ days) (Bernstein and Hoskins, 2006; Bernstein et al., 2011).

MAMMALIAN TOXICOLOGY

Acute Effects

Due to the insoluble, chemically inert nature of asbestos, there are no clinically significant acute or intermediate effects associated with exposure to asbestos fibers. In fact, the clinically significant effects of asbestos (i.e., asbestosis and cancer) require long-duration exposure, and generally, long-duration (20–40 year) latency periods.

Chronic Effects

Introduction and Historical Overview Alice Hamilton (1869–1970) was a physician and pioneer in industrial diseases and hygiene, and she became Harvard University's first woman professor in 1919. Dr. Hamilton and Dr. Harriet Louise Hardy (1905–1993) published one of the first books on Industrial Toxicology in 1934, a 500+ page text, which was republished in the second edition in 1949 as a 574 page text, when both women were on the faculty of the Harvard School of Public Health. Dr. Hardy was also a physician and established an occupational medicine clinic at Massachusetts General Hospital in 1947 and was a clinical professor at Harvard from 1979 to 1993. Interestingly, the 1949 edition of Hamilton & Hardy's Industrial Hygiene (Hamilton and Hardy, 1949) makes no mention of asbestos and its health hazards, but Dr. Hardy was one of the very first scientists to recognize the link between exposures to elevated levels of airborne asbestos and increased lung cancer risk in her 1965 article "Asbestos related disease" in the American Journal of Medical Science (Hardy, 1965). W.C. Hueper, in his 1942 compendium Occupational Tumors and Allied Diseases, does mention asbestos in the context of pneumoconiosis and lung cancer (Hueper, 1942). Although Dr. Hueper notes that, since 1935, cases of lung cancer in workers have appeared in conjunction with asbestosis, he goes on to say that at that point in time (1942) "[a]sbestosis carcinoma of the lung is not included in any group of occupational tumors recognized by any country."

Asbestos fiber inhalation at elevated levels and over sufficiently long durations can cause lung cancer and fibrotic lung disease in exposed humans and animals. Likewise, excessive-exposure circumstances can lead to malignant mesothelioma of the thin (mesothelial) membrane lining the chest (pleura) and abdominal (peritoneum) cavities. Some studies of occupationally exposed workers also suggest an increased risk of gastrointestinal cancer. Asbestos is classified as a Group 1 carcinogen (human evidence is sufficient) by the IARC (IARC, 1977) and a Group A carcinogen (again, the human evidence is sufficient) by US EPA (US EPA, 1999). In addition to carcinogenesis, asbestos exposure can cause fibrotic lung and pleural disease, termed asbestosis. Experimental models have been used to elucidate the mechanisms by which asbestos causes cancer and lung disease and to determine what properties of the fiber are important for disease development (Aust et al., 2011; Ghio et al., 2004; Mossman et al., 2011).

Noncancer/Asbestosis/Pleural Plaques Asbestosis is a lung pneumoconiosis wherein a diffuse interstitial fibrosis develops over time. The effect of asbestosis on pulmonary function is primarily the reduction of lung volumes detected by reduced inspiratory capacity and vital capacity (Churg, 1993; Kishimoto et al., 2011). The role of asbestos in lung function decrements is revealed by either occupational history of exposure to asbestos or the presence of asbestos fibers and asbestos bodies in lung tissues. Pleural plaques are generally a benign asbestos-related disease, which may occur in the absence of lung fibrosis, and the presence of which, per se, do not relate to lung cancer or mesothelioma risk (Ameille et al., 2009). Benign pleural plaques occur on the parietal pleura in locations of greatest respiratory movement and consist of collagenous connective tissue with secondary calcification.

Lung Cancer Subsequent to the classic 1954 publication of Doll and Hill on lung cancer risks from tobacco smoking, Doll (1955) examined the association between asbestos and lung cancer in British workers exposed before asbestos dust control efforts. He found that average lung cancer risk among men employed for more than 20 years under the old, dusty conditions was on the order of 10 times that experienced by the general population. In the US, Selikoff et al. (1964) identified neoplastic asbestos diseases in asbestos-insulated workers, and in 1968 Selikoff and colleagues reported on deaths in 370 asbestos workers over a 1-year period (Selikoff et al., 1968). In this group, there were 27 observed deaths from cancer of the lungs, with only 2.3 expected. In his 1977 review covering 9 years of observation of a cohort of 17,800 asbestos workers, Selikoff noted: "Asbestos workers who have no history of cigarette smoking are not likely to die of lung cancer" (Selikoff, 1977). As a further observation, Hammond et al. (1979) reported on lung cancer incidence in about 12,000 workers with at least 20 years of asbestos-insulation work experience. The authors reported that, compared to non-smoking, non-asbestos-workers, asbestos workers in various smoking categories showed mortality ratios of 87 (for >1 pack per day), 51 (for <1 pack per day), 37 (for ex-cigarette smokers), seven (for pipe and cigar smokers), and five (for workers who never smoked regularly).

Thus, the Hammond et al. (1979) data suggested that lung cancer risk showed a multiplicative relationship between smoking and asbestos exposure, such that the "combination of cigarette smoking and asbestos exposure among amosite asbestos factory workers was calculated to have increased their risk of lung cancer death about 80 times as compared with like men of the same age who neither smoked cigarettes nor worked with asbestos." Other analyses have suggested that the interaction is "less-than-multiplicative" (Liddell and Armstrong, 2002). In Great Britain, a study of nearly 100,000 asbestos workers reported that while there is a synergistic interaction between smoking and asbestos, the combined effect of the factors is more than additive, but less than multiplicative (Frost et al., 2011).

Jones et al. (1996) examined the association of lung cancer with asbestosis in a case-control study of 271 patients with a confirmed diagnosis of primary lung cancer. After adjusting for age, sex, smoking history, and area of referral, the authors found that cases classified as having asbestosis had an odds ratio (OR) of 2.00 for lung cancer, whereas, those without asbestosis had an OR of 1.56. The authors concluded asbestosis is not a necessary step in the development of asbestos-related lung cancer. However, more recent studies have concluded that, even though asbestosis is not a prerequisite for lung cancer development, asbestosis is seemingly a better predictor of lung cancer risk than past history of asbestos exposure *per se* (Weiss, 1999; Zhong et al., 2008).

Mesothelioma Mesothelioma is a cancer of epithelial-lining cells (mesothelial cells) of the lung or peritoneal cavities. Mesotheliomas were first reported by Wagner et al. (1960) in crocidolite miners of North Western Cape Province, South Africa. The authors described 33 cases of mesothelioma with exposure to one type of asbestos, the so-called blue asbestos (crocidolite). In 1962, Wagner was able to produce mesothelial tumors of the pleura by direct implantation of asbestos dust or silica dust into laboratory rats (Wagner, 1962).

Mesothelioma has been shown to have occurred over the years in populations of asbestos workers (McDonald et al., 1997), and it has a latency period in the range of 20–40 years. Mesothelioma is strongly associated with exposure to amphibole (amosite and crocidolite) asbestos (Churg, 1982, 1998; Wagner et al., 1988; Kishimoto et al., 2004). Mesothelioma from chrysotile asbestos cannot be ruled out, but the potency appears to be much less (Pierce et al., 2008; Berman and Crump, 2008a), as discussed below.

In the U.S. studies mentioned above, Selikoff et al. (1968) reported finding 10 (pleural + peritoneal) mesothelioma deaths in a cohort of 370 asbestos workers, stating, "All three of the men who died of pleural mesothelioma had a history of regular cigarette smoking." Of the seven who died of peritoneal mesothelioma, five had a history of regular cigarette smoking, one never smoked regularly, and one smoked only pipes and cigars. In his 1977 review, Selikoff tabulated deaths among 17,800 asbestos insulation workers in the United States and Canada between 1967 and 1975, reporting 0.38 pleural mesothelioma deaths per 1000 man-years of observation in cigarette smokers compared to 0.16 such deaths per 1000 man-years of observation in "never smokers"; because of the results for a category listed as "no history of cigarette smoking," Selikoff stated that mesothelioma had not been associated with "cigarette smoking among asbestos workers" (Selikoff, 1977). However, Hammond et al. (1979) reported that, for pleural mesothelioma, among men who never smoked, "there were three observed deaths *versus*. 4.8 expected deaths" (ratio 0.63), and "five observed deaths *versus*. 3.2 expected deaths for pipe/cigar smokers (ratio 1.59)." Thus, the interaction of smoking and mesothelioma in the Selikoff studies is somewhat suggestive of a smoking effect. However, in a 1980 paper based on seven pleural mesotheliomas and seven peritoneal mesotheliomas (Selikoff et al., 1980), the authors concluded that smoking history did not affect mesothelioma risk (even though ¾ of the mesothelioma deaths were in smokers).

Later studies on the role of smoking in asbestos-associated mesothelioma risk reported a lack of interaction (Muscat and Wynder, 1991), although the fact that tobacco smoke contains radioisotopes and ionizing radiation, which themselves are risk factors for mesothelioma (Goodman et al., 2009; Nakamura et al., 2009; Prueitt et al., 2009), suggests that a role for smoking in mesothelioma risk is biologically plausible. Pleural mesotheliomas have also been produced in animals through lung exposure to radionuclides (Sanders and Jackson, 1972; Sanders, 1992).

As noted earlier, mesothelioma risk is more closely associated with amphibole than chrysotile fibers (Neuberger and Kundi, 1990; Dodson et al., 1997; Rödelsperger et al., 1999; Yarborough, 2007; Roggli and Vollmer, 2008; Sichletidis et al., 2009). A comprehensive analysis commissioned by US EPA (Berman and Crump, 2003, 2008a,b) concluded that the epidemiologic data are consistent with chrysotile exposure *per se* not contributing to risk of mesothelioma. Moreover, as discussed and summarized by Hodgson and Darnton (2000) and Berman and Crump (2003, 2008a,b), per unit fiber exposure mesothelioma risk from amphibole exposure is far more greater than that from chrysotile exposure. Hodgson and Darnton (2000) estimated that amosite is about 100 times, and crocidolite about 500 times, more potent than chrysotile in causing mesothelioma. Berman and Crump (2003, 2008a,b) concluded that amphiboles are about 750 times as potent as chrysotile in causing mesothelioma. As discussed by Pierce et al. (2008), cohorts exposed predominantly to chrysotile show that, if there is a mesothelioma risk for chrysotile at all, the dose threshold falls above 15–500 f/cc-year level; whereas, given that amphibole fibers are estimated to be about 500 times more potent, depending on fiber dimensions, the dose threshold level for amphiboles causing mesothelioma would be in the range 0.3–1.0 f/cc-year. For example, chrysotile used in brake pads does not appear to increase the risk of mesothelioma (Goodman et al., 2004; Laden et al., 2004); in those brake mechanics who develop mesothelioma, lung tissues show a presence of amphibole fibers (Butnor et al., 2003). Also, for patients believed to have developed mesothelioma from "bystander exposure" to asbestos, the predominant fiber type identified by electron microscopy is amphibole asbestos (crocidolite or amosite) (Neumann et al., 2011).

Given that the percentage of mesotheliomas attributable to asbestos exposure is about 85%, this disease has non-asbestos causes such as idiopathic origin, SV-40 virus infection, chronic inflammation, and ionizing radiation (Pelnar, 1988; Peterson et al., 1984; Toyooka et al., 2001; Huncharek, 2002; Goodman et al., 2009; Weiner and Neragi-Miandoab, 2009; de Bruin et al., 2009; Craighead, 2011). In fact, a large case-control study of mesothelioma in Great Britain (Rake et al., 2009) concluded that 14% of male and 62% of female mesothelioma cases were not attributable to occupational or domestic asbestos exposure. For individuals not exposed to asbestos, the authors estimated a lifetime risk of mesothelioma (pleural and peritoneal combined) from other causes of about 1 in 1000.

Modes of Action

The diseases of greatest concern with regard to asbestos exposure are lung cancer and mesothelioma. As is the case with all cancers, DNA errors in key areas of the cell genome, namely, in proto-oncogenes and tumor-suppressor genes, are thought to be the key events that initiate uncontrolled cell division (i.e., carcinogenesis). DNA damage is a normal consequence of metabolic processes (active oxygen species), background ionizing radiation, and errors in cell division. The consequences of DNA errors can be innocuous, lead to cell death, initiate DNA repair, or, if at critical gene loci and unrepaired, trigger changes that lead to neoplasia. Tomlinson et al. (2002) estimated that by the time we reach age 15 we have accumulated thousands of mutations in our bodies, and this number increases with age. Because of the accumulation of DNA damage, the risk of all cancers increases dramatically with age.

The mechanism by which asbestos fibers in the body add to this mutation load is unclear, but the persistence of the fibers, their resistance to phagocytosis, and their surface chemistry likely lead to cancer through epigenetic pathways such as chronic inflammation, frustrated phagocytosis, and free-radical generation. Carcinogens that are not directly genotoxic have a threshold level for carcinogenesis, requiring a certain level of exposure before promotion of cell proliferation can occur, and asbestos-related cancers likely fall in this category (Butterworth, 1990; Schins, 2002; Mossman et al., 2011).

In terms of mechanisms of action related to fiber properties, it appears that carcinogenic potency of asbestos fibers depends on fiber type (with amphibole potency about 100-fold greater than serpentine potency) and on fiber length (with potency of fibers longer than 5 μm being greater than for short fibers) (Berman et al., 1995; ATSDR, 2001; Berman and Crump, 2008a,b; Stayner et al., 2008). Likewise, thin fibers are more potent than thick fibers (ATSDR, 2001; Hesterberg and Barrett, 1984), and cleavage fragments of the minerals are least potent (Gamble and Gibbs, 2008).

Some investigators place significant emphasis on the role of iron (Fe) in asbestos toxicity, oxygen-radical generation, and carcinogenicity (Aust et al., 2011; Ghio et al., 2004; Nakamura et al., 2009; Quinlan et al., 1994), whereas, others find the role of biopersistence to be predominant (Bernstein et al., 2003; Hesterberg et al., 1998).

SOURCES OF EXPOSURE

Occupational

Historically, thermal insulation work typically led to elevated airborne concentrations of free fibers (2–30 f/cc) (Balzer and Cooper, 1968; Peto, 1980; Williams et al., 2007). Working with asbestos in friable insulation releases far more fibers than work with products having encapsulated fibers (0.02–0.10 f/cc) (Faulring et al., 1975; Mlynarek et al., 1996; Mowat et al. 2005). Moreover, compared to work with other asbestos-containing materials, thermal insulation work can result in elevated exposures to amosite/amphibole asbestos fibers *versus* serpentine (chrysotile) asbestos fibers. In occupational settings, exposures to workers who are bystanders to asbestos work has been evaluated (Donovan et al., 2011), and air dispersion models suitable to estimating such exposures have been developed (Drivas et al., 1996).

Figure 96.1, prepared by NIOSH, shows how the use of asbestos (vertical axis) and permissible asbestos exposure limits (numbers next to arrows) have changed in the United States from 1900 to 2007, with use peaking in the early 1970s (http://www.cdc.gov/mmwr/preview/mmwrhtml/mm5815a3.htm). In concordance with the latency period of asbestos-related cancers, time trends in these diseases over the past two decades have also shown a downward slope. For example, age-adjusted rates of pleural mesothelioma among men rose from about 7.5 per million person-years in 1973 to about 20 per million person-years in the early 1990s; rates appear to be stable or declining thereafter (Moolgavkar et al., 2009).

Interestingly, the geographical distribution of mesothelioma mortality in the United States (2000–2004) suggests a pattern of elevated mesothelioma rates for coastal, shipbuilding regions, where many workers outfitted military destroyers and battleships with asbestos insulation, as can be appreciated on a map: http://www2.cdc.gov/drds/WorldReportData/FigureTableDetails.asp?FigureTableID=900&GroupRefNumber=F07-03.

Environmental

Geographical Distribution Asbestos exposure occurs primarily in the "built environment," although natural sources of asbestos exposure, e.g., soils and rocks in Montana and California, are well known (Gunter, 2010; Cooper et al., 1979).

As shown by the figure above, asbestos usage in the United States manufacturing and "built environment" has declined dramatically, but occasionally nonoccupational sources of asbestos exposure raise concerns. An example is the 9/11 destruction of the World Trade Center towers, which released asbestos-related building materials, both during the initial collapse and also during the subsequent cleanup (Lowers et al., 2009; Nolan et al., 2005).

Environmental Exposure Concentrations An overview of sources and types of asbestos exposure has been published recently (Case et al., 2011). Background concentrations inside residences and public buildings have been measured in the range of 0.002–0.027 f/cc, with personal exposure measurements averaging 0.016 f/cc (Lee and Van Orden, 2008 [Table 96.1). Outdoor exposure concentrations in the ambient environment depend on the levels of human activity. Some assessments have been made in Libby, Montana, and San Jose, California (Adgate et al., 2011; Moolgavkar et al., 2010; Steiner et al., 1990). Outdoor concentrations in urban environments have been reported in the range 0.0008–0.0001 f/cc (HEI-AR, 1991; ATSDR, 2001).

INDUSTRIAL HYGIENE

Measurement Methods

Several measurement methods have been used over the years. The basic measurement principle involves optical visualization of the fibers and fiber counting to derive a concentration of airborne fibers, typically in units of f/cc of air (f/cc or f/cm^3 or f/ml). As noted earlier, the focus is on fibers longer than 5 μm.

MI: Midget impinger

PCM: Phase contrast microscopy

SEM: Scanning electron microscopy

TEM: Transmission electron microscopy

EDS: Energy dispersive X-ray detection (for determination of fiber type)

PLM: Polarized light microscopy

Sampling for asbestos in the workplace was initially motivated by findings of health effects in workers. Initially, those effects were identified in industries with heavy exposures (i.e., the textile industry). Early exposures in the textile industry were extremely high, sometimes in the 100s of f/cc. When workers with exposure to asbestos in finished products, rather than to raw asbestos, showed indications of asbestos-related disease, additional monitoring was conducted in these populations (e.g., insulation workers). Among end-users of asbestos, insulators (workers installing and removing asbestos-based thermal insulation) were found to have particularly elevated exposures to free asbestos fibers compared to other craftsmen, especially during ship building, refitting, and repair. Insulation-related asbestos concentrations in shipyards during 1940s to 1960s/early 1970s were in the 10s and 100s of f/cc (Harries, 1968; Mangold et al., 2006; Marr, 1964). Later timeframes (i.e., 1970s and 80s) were characterized by the initiation of asbestos-exposure studies in worker populations expected to have relatively lower exposures to asbestos—in terms of concentrations, frequency, and duration—than those in the textile, mining, and shipyard industries.

Quantification of exposure is crucial to assessment of asbestos-related disease risk. As already noted by Selikoff et al. (1965) more than four decades ago, exposures in "different occupations vary widely in important respects; in intimacy, intensity and duration of exposure, in variety and grade of asbestos used, in working conditions, in concomitant exposure to other dusts or inhalants." The unit of asbestos dose most used is the "fibers per cc-years" (f/cc-year), which is the average 8-h workday-free asbestos fiber concentration in f/cc multiplied by the number of work-years over which that exposure occurred.

Numerical Values of the Standards/Exposure Limits

An exposure limit for asbestos in the United States was first established in 1938, when a U.S. Public Health Service study of asbestos textile workers indicated that asbestosis was not likely to occur at fiber concentrations (type not specified) <5 million particles per cubic foot (mppcf). The value of 5 mppcf was subsequently adopted as a threshold limit value (TLV) by ACGIH as well as the United States and several state governments. The 5 mppcf exposure value remained in use in the United States until 1971, when OSHA established a regulatory permissible exposure limit (PEL) of 12 f/cc, which was officially revised to 5 f/cc in June 1972. In 1980, the ACGIH adopted fiber-type specific values of 2 f/cc for chrysotile, 0.2 f/cc for crocidolite, and 0.5 f/cc for amosite and tremolite. In 1991, ACGIH proposed a value of 0.2/cc for all fiber types, which was subsequently lowered to 0.1 f/cc (for fibers with length >5 μm) before it was adopted in 1998. OSHA did not adopt fiber-type specific exposure limits, and the 1972 OSHA value of 5 f/cc was subsequently lowered over time to the current value (set in 1994) of 0.1 f/cc for fibers >5 μm in length. Finally, the Asbestos Hazard Emergency Response Act (AHERA) "clearance level" applicable to public exposure is 0.01 f/cc (measured by PCM). The AHERA transmission electron microscopy (TEM) method requires collecting five air-filter samples (≥1199 l of air for a 25 mm filter or ≥2799 l of air for a 37 mm filter) under "aggressive" air disturbance conditions, with clearance achieved if the average result of the five filter samples is <70 s/mm^2 of filter surface (Federal Register, 1987).

RISK ASSESSMENTS

Assumptions Used in Human Health Risk Assessment

Human health risk assessment (HHRA) is a method whereby one can estimate quantitative bounds on the possible risks associated with specific exposure scenarios. A central problem in risk assessment is that, often, potential risks at low levels of exposure, i.e., levels at which no direct observations (typically occupational data) are available, need to be estimated. In the absence of relevant data, a number of upper bound assumptions are used in the process of risk assessment, i.e., assumptions more likely to overstate, rather than understate, the risks associated with exposure. For all carcinogens, the US EPA and other regulatory agencies, as a matter of policy, assume a "linear, no threshold" model of dose response. For asbestos, quantitative dose–response relations are based on occupational cohorts where average exposures were greater than about 15 f/cc-year. The quantitative risk assessments for asbestos include those of US EPA-IRIS (2001) [unit risk (UR) = 0.23 per f/cc];[2] Lash et al. (1997); Hodgson and Darnton (2000); Berman and Crump (2008a,b); Pierce et al. (2008); and Moolgavkar et al. (2010).

Threshold Considerations

Thresholds of effect, specifically for asbestos-related diseases, have been examined by numerous investigators (Doll and Peto, 1985; Ilgren and Browne, 1991; Hodgson and Darnton, 2000, 2010), with the general conclusion that such "no-adverse-effect" thresholds do exist. Pierce et al. (2008) evaluated over 350 asbestos studies to determine thresholds for mesothelioma and lung cancer from exposure to predominantly chrysotile asbestos (i.e., ≥90%) in mining and manufacturing settings. For chrysotile, they reported the "no-effects" exposure levels for lung cancer and mesothelioma to fall in a range of approximately 25–1000 f/cc-year and 15–500 f/cc-year, respectively. For mesothelioma, two studies reported no increase in mesothelioma risk at the highest exposure levels >400 and ≥112 f/cc-year, respectively, while two studies reported chrysotile thresholds between 800 f/cc-year and <15 f/cc-year, respectively. The authors acknowledge that many studies likely lacked sufficient power (e.g., due to small cohort size) to assess whether there could have been a significant increase in risk at the reported no-observed-adverse-effects level (NOAEL).

REFERENCES

Adgate, J.L., Cho, S.J., Alexander, B.H., Ramachandran, G., Raleigh, K.K., Johnson, J., Messing, R.B., Williams, A.L., Kelly, J., and Pratt, G.C. (2011) Modeling community asbestos exposure near a vermiculite processing facility: Impact of human activities on cumulative exposure. *J. Expo. Sci. Environ. Epidemiol.* doi 10.1038/jes.2011.8.

Agency for Toxic Substances and Disease Registry (ATSDR) *Toxicological Profile for Asbestos*, Available at http://www.atsdr.cdc.gov/toxprofiles/tp.asp?id=30&tid=4. Published 2001.

---

[2] S EPA Integrated Risk Information System for Asbestos: http://www.epa.gov/iris/subst/0371.htm.

Alleman, J.E. and Mossman, B.T. (1997) Asbestos revisited. *Sci. Am.* 54–57.

Ameille, J., Brochard, P., Letourneux, M., Paris, C., and Pairon, J.C. (2009) [Asbestos-related cancer risk in the presence of asbestosis or pleural plaques]. *Rev. Mal. Respir.* 26:413–421.

Aust, A.E., Cook, P.M., and Dodson, R.F. (2011) Morphological and chemical mechanisms of elongated mineral particle toxicities. *J. Toxicol. Environ. Health B Crit. Rev.* 14:40–75.

Balzer, J.L. and Cooper, W.C. (1968) The work environment of insulating workers. *Am. Ind. Hyg. Assoc. J.* 29:222–227.

Berman, D.W. and Crump, K.S. (2008a) Update of potency factors for asbestos-related lung cancer and mesothelioma. *Crit. Rev. Toxicol.* 38(Suppl. 1):1–47.

Berman, D.W. and Crump, K.S. (2008b) A meta-analysis of asbestos-related cancer risk that addresses fiber size and mineral type. *Crit. Rev. Toxicol.* 38(Suppl. 1):49–73.

Berman, D.W. and Crump, K.S.[US EPA] (2003) *Final Draft: Technical Support Document for a Protocol to Assess Asbestos-Related Risk*, EPA 9345.4-06. Washington, DC: Office of Solid Waste and Emergency Response.

Berman, D.W., Crump, K.S., Chatfield, E.J., Davis, J.M., and Jones, A.D. (1995) The sizes, shapes, and mineralogy of asbestos structures that induce lung tumors or mesothelioma in AF/HAN rats following inhalation. *Risk Anal.* 15:181–195.

Bernstein, D.M. and Hoskins, J.A. (2006) The health effects of chrysotile: current perspective based upon recent data. *Regul. Toxicol. Pharmacol.* 45:252–264.

Bernstein, D.M., Chevalier, J., and Smith, P. (2003) Comparison of Calidria chrysotile asbestos to pure tremolite: inhalation biopersistence and histopathology following short-term exposure. *Inhal. Toxicol.* 15:1387–1419.

Bernstein, D.M., Donaldson, K., Decker, U., Gaering, S., Kunzendorf, P., Chevalier, J., and Holm, S.E. (2008) A biopersistence study following exposure to chrysotile asbestos alone or in combination with fine particles. *Inhal. Toxicol.* 20:1009–1028.

Bernstein, D.M., Rogers, R.A., Sepulveda, R., Donaldson, K., Schuler, D., Gaering, S., Kunzendorf, P., Chevalier, J., and Holm, S.E. (2011) Quantification of the pathological response and fate in the lung and pleura of chrysotile in combination with fine particles compared to amosite-asbestos following short-term inhalation exposure. *Inhal. Toxicol.* 23:372–391.

Butnor, K.J., Sporn, T.A., and Roggli, V.L. (2003) Exposure to brake dust and malignant mesothelioma: a study of 10 cases with mineral fiber analyses. *Ann. Occup. Hyg.* 47:325–330.

Butterworth, B.E. (1990) Consideration of both genotoxic and nongenotoxic mechanisms in predicting carcinogenic potential. *Mutat. Res.* 239:117–132.

Case, B.W. and Abraham, J.L. (2009) Heterogeneity of exposure and attribution of mesothelioma: Trends and strategies in two American counties. *J. Phys. Conf. Ser.* 151: doi 10.1088/1742-6596/151/1/012008

Case, B.W., Abraham, J.L., Meeker, G., Pooley, F.D., and Pinkerton, K.E. (2011) Applying definitions of "asbestos" to environmental and "low-dose" exposure levels and health effects, particularly malignant mesothelioma. *J. Toxicol. Environ. Health B Crit. Rev.* 14:3–39.

Churg, A. (1982) Asbestos fibers and pleural plaques in a general autopsy population. *Am. J. Pathol.* 109:88–96.

Churg, A. (1993) Asbestos-related disease in the workplace and the environment: controversial issues. *Monogr. Pathol.* 36:54–77.

Churg, A. (1998) Nonneoplastic disease caused by asbestos. In: Churg, A. and Green, F.H., editors. *Pathology of Occupational Lung Disease*, 2nd ed., Baltimore, MD: Williams & Wilkins, pp. 277–338.

Cooper, W.C., Murchio, J., Popendorf, W., and Wenk, H.R. (1979) Chrysotile asbestos in a California recreational area. *Science* 206:685–688.

Craighead, J.E. (2011) Epidemiology of mesothelioma and historical background. *Recent Results Cancer Res.* 189:13–25.

Craighead, J.E. and Gibbs, A.R., editors. (2008) *Asbestos and Its Diseases*, New York, NY: Oxford University Press.

De Bruin, M.L., Burgers, J.A., Baas, P., van t' Veer, M.B., Noordijk, E.M., Louwman, M.W., Zijlstra, J.M., van den Berg, H., Aleman, B.M., and van Leeuwen, F.E. (2009) Malignant mesothelioma following radiation treatment for Hodgkin's lymphoma. *Blood* 113:3679–3681.

Dodson, R.F. and Hammar, S.P., editors. (2011) *Asbestos: Risk Assessment, Epidemiology, and Health Effects*, 2nd ed., Boca Raton, FL: CRC Press.

Dodson, R.F., O'Sullivan, M., Corn, C.J., McLarty, J.W., and Hammar, S.P. (1997) Analysis of asbestos fiber burden in lung tissue from mesothelioma patients. *Ultrastruct. Pathol.* 21:321–336.

Dodson, R.F., Williams, M.G., Huang, J., and Bruce, J.R. (1999) Tissue burden of asbestos in nonoccupationally exposed individuals from east Texas. *Am. J. Ind. Med.* 35:281–286.

Doll, R. (1955) Mortality from lung cancer in asbestos workers. *Br. J. Ind. Med.* 12:81–86.

Doll, R. and Peto, J. *Asbestos: Effects on Health of Exposure to Asbestos. Health and Safety Executive (UK) Web site.* Available at http://www.hse.gov.uk/asbestos/exposure.pdf. Published 1985.

Donovan, E.P., Donovan, B.L., Sahmel, J., Scott, P.K., and Paustenbach, D.J. (2011) Evaluation of bystander exposures to asbestos in occupational settings: a review of the literature and application of a simple eddy diffusion model. *Crit. Rev. Toxicol.* 41:50–72.

Drivas, P.J., Valberg, P.A., Murphy, B.L., and Wilson, R. (1996) Modeling indoor contaminant exposure from short-term point source releases. *Indoor Air* 6:271–277.

Faulring, G.M., Forgeng, W.D., Kleber, E.J., and Rhodes, H.B. (1975) Detection of chrysotile asbestos in airborne dust from thermosetting resin grinding. *J. Test. Eval.* 3:482–490.

Federal Register. (1987) Code of Federal regulations, Asbestos-Containing Materials in Schools; Final Rule and Notice, 40 CFR part 763.90(i)(4).

Frost, G., Darnton, A., and Harding, A.H. (2011) The effect of smoking on the risk of lung cancer mortality for asbestos workers in Great Britain (1971–2005). *Ann. Occup. Hyg.* 55:239–247.

Gamble, J.F. and Gibbs, G.W. (2008) An evaluation of the risks of lung cancer and mesothelioma from exposure to amphibole cleavage fragments. *Regul. Toxicol. Pharmacol.* 52(Suppl. 1): S154–S186.

Ghio, A.J., Churg, A., and Roggli, V.L. (2004) Ferruginous bodies: implications in the mechanism of fiber and particle toxicity. *Toxicol. Pathol.* 32:643–649.

Goodman, J.E., Nascarella, M.A., and Valberg, P.A. (2009) Ionizing radiation: a risk factor for mesothelioma. *Cancer Causes Control* 20:1237–1254.

Goodman, M., Teta, M.J., Hessel, P.A., Garabrant, D.H., Craven, V.A., Scrafford, C.G., and Kelsh, M.A. (2004) Mesothelioma and lung cancer among motor vehicle mechanics: a meta-analysis. *Ann. Occup. Hyg.* 48:309–326.

Gunter, M.E. (2010) Defining asbestos: differences between the built and natural environments. *Chimia (Aarau)* 64:747–752.

Hamilton, A. and Hardy, H.L., editors. (1949) *Industrial Toxicology*, 2nd ed., New York, NY: Harper & Brothers.

Hammond, E.C., Selikoff, I.J., and Seidman, H. (1979) Asbestos exposure, cigarette smoking and death rates. *Ann. NY Acad. Sci.* 330:473–490.

Hardy, H.L. (1965) Asbestos related disease. *Am. J. Med. Sci.* 250:381–389.

Harries, P.G. (1968) Asbestos hazards in naval dockyards. *Ann. Occup. Hyg.* 11:135–145.

Health Effects Institute-Asbestos Research (HEI-AR) *Asbestos in public and commercial buildings: A literature review and synthesis of current knowledge (Executive Summary). Health Effects Institute Web site.* Available at http://www.asbestos-institute.ca/reviews/hei-ar/hei-ar.html. Published 1991.

Hesterberg, T.W. and Barrett, J.C. (1984) Dependence of asbestos- and mineral dust-induced transformation of mammalian cells in culture on fiber dimension. *Cancer Res.* 44:2170–2180.

Hesterberg, T.W., Chase, G., Axten, C., Miller, W.C., Musselman, R.P., Kamstrup, O., Hadley, J., Morscheidt, C., Bernstein, D.M., and Thevenaz, P. (1998) Biopersistence of synthetic vitreous fibers and amosite asbestos in the rat lung following inhalation. *Toxicol. Appl. Pharmacol.* 151:262–275.

Hodgson, J.T. and Darnton, A. (2000) The quantitative risks of mesothelioma and lung cancer in relation to asbestos exposure. *Ann. Occup. Hyg.* 44:565–601.

Hodgson, J.T. and Darnton. A. (2010) Mesothelioma risk from chrysotile. *Occup. Environ. Med.* 67:432.

Hueper, W.C. (1942) Occupational Tumors and Allied Diseases, Springfield, Illinois: Charles Thomas.

Huncharek, M. (2002) Non-asbestos related diffuse malignant mesothelioma. *Tumori* 88:1–9.

Ilgren, E.B. and Browne, K. (1991) Asbestos-related mesothelioma: evidence for a threshold in animals and humans. *Regul. Toxicol. Pharmacol.* 13:116–132.

International Agency for Research on Cancer (IARC) (2012) Monograph Volume 100: A Review of Human Carcinogens - Part C: Arsenic, Metals, Fibres, and Dusts. Lyon, France: World Health Organization.

International Agency for Research on Cancer (IARC) (1977) *Monographs on the Evaluation of Carcinogenic Risk of Chemicals to Man: Asbestos*, vol. 14, Lyon, France: World Health Organization.

Jones, R.N., Hughes, J.M., and Weill, H. (1996) Asbestos exposure, asbestosis, and asbestos-attributable lung cancer. *Thorax* 51 (Suppl. 2):S9–S15.

Kishimoto, T., Kato, K., Arakawa, H., Ashizawa, K., Inai, K., and Takeshima, Y. (2011) Clinical, radiological, and pathological investigation of asbestosis. *Int. J. Environ. Res. Publ. Health* 8:899–912.

Kishimoto, T., Ozaki, S., Kato, K., Nishi, H., and Genba, K. (2004) Malignant pleural mesothelioma in parts of Japan in relationship to asbestos exposure. *Ind. Health* 42:435–439.

Laden, F., Stampfer, M.J., and Walker, A.M. (2004) Lung cancer and mesothelioma among male automobile mechanics: a review. *Rev. Environ. Health* 19:39–61.

Lash, T.L., Crouch, E.A., and Green, L.C. (1997) A meta-analysis of the relation between cumulative exposure to asbestos and relative risk of lung cancer. *Occup. Environ. Med.* 54:254–263.

Lee, R.J. and Van Orden, D.R. (2008) Airborne asbestos in buildings. *Regul. Toxicol. Pharmacol.* 50:218–225.

Lee, R.J., Strohmeier, B.R., Bunker, K.L., and Van Orden, D.R. (2008) Naturally occurring asbestos: a recurring public policy challenge. *J. Hazard. Mater.* 153:1–21.

Liddell, F.D. and Armstrong, B.G. (2002) The combination of effects on lung cancer of cigarette smoking and exposure in Quebec chrysotile miners and millers. *Ann. Occup. Hyg.* 46:5–13.

Lowers, H.A., Meeker, G.P., Lioy, P.J., and Lippmann, M. (2009) Summary of the development of a signature for detection of residual dust from collapse of the World Trade Center buildings. *J. Expo. Sci. Environ. Epidemiol.* 19:325–335.

Mangold, C., Clark, K., Madl, A., and Paustenbach, D. (2006) An exposure study of bystanders and workers during the installation and removal of asbestos gaskets and packing. *J. Occup. Environ. Hyg.* 3:87–98.

Marr, W.T. (1964) Asbestos exposure during naval vessel overhaul. *Am. Ind. Hyg. Assoc. J.* 25:264–268.

McDonald, A.D., Case, B.W., Churg, A., Dufresne, A., Gibbs, G.W., Sébastien, P., and McDonald, J.C. (1997) Mesothelioma in Quebec chrysotile miners and millers: epidemiology and aetiology. *Ann. Occup. Hyg.* 41:707–719.

Mlynarek, S., Corn, M., and Blake, C. (1996) Asbestos exposure of building maintenance personnel. *Regul. Toxicol. Pharmacol.* 23:213–224.

Moolgavkar, S.H., Meza, R., and Turim, J. (2009) Pleural and peritoneal mesotheliomas in SEER: age effects and temporal trends, 1973–2005. *Cancer Causes Control* 20:935–944.

Moolgavkar, S.H., Turim, J., Alexander, D.D., Lau, E.C., and Cushing, C.A. (2010) Potency factors for risk assessment at Libby, Montana. *Risk Anal.* 30:1240–1248.

Mossman, B.T., Lippmann, M., Hesterberg, T.W., Kelsey, K.T., Barchowsky, A., and Bonner, J.C. (2011) Pulmonary endpoints (lung carcinomas and asbestosis) following inhalation exposure to asbestos. *J. Toxicol. Environ. Health B Crit. Rev.* 14:76–121.

Mowat, F., Bono, M., Lee, R.J., Tamburello, S., and Paustenbach, D. (2005) Occupational exposure to airborne asbestos from phenolic molding material (Bakelite) during sanding, drilling, and related activities. *J. Occup. Environ. Hyg.* 2:497–507.

Muscat, J.E. and Wynder, E.L. (1991) Cigarette smoking, asbestos exposure, and malignant mesothelioma. *Cancer Res.* 51:2263–2267.

Nakamura, E., Makishima, A., Hagino, K., and Okabe, K. (2009) Accumulation of radium in ferruginous protein bodies formed in lung tissue: association of resulting radiation hotspots with malignant mesothelioma and other malignancies. *Proc. Jpn. Acad. Ser. B Phys. Biol. Sci.* 85:229–239.

National Institute of Occupational Safety and Health (NIOSH) (2008) *Work-related lung disease (WoRLD) surveillance system. Malignant mesothelioma: Counties with highest age-adjusted death rates (per million population), U.S. residents age 15 and over, 2000–2004.* Available at http://www2.cdc.gov/drds/WorldReportData/FigureTableDetails.asp?FigureTableID=901&GroupRefNumber=T07-10.

Neuberger, M. and Kundi, M. (1990) Individual asbestos exposure: smoking and mortality—a cohort study in the asbestos cement industry. *Br. J. Ind. Med.* 47:615–620.

Neumann, V., Löseke, S., and Tannapfel, A. (2011) Mesothelioma and analysis of tissue fiber content. *Recent Results Cancer Res.* 189:79–95.

Nolan, R.P., Ross, M., Nord, G.L., Axten, C.W., Osleeb J.P., Domnin, S.G., Price, B., and Wilson, R. (2005) Risk assessment for asbestos-related cancer from the 9/11 attack on the World Trade Center. *J. Occup. Environ. Med.* 47:817–825.

Pan, X.L., Day, H.W., Wang, W., Beckett, L.A., Schenker, M.B. (2005) Residential proximity to naturally occurring asbestos and mesothelioma risk in California. *Am. J. Respir. Crit. Care Med.* 172(8):1019–1025.

Pelnar, P.V. (1988) Further evidence of non-asbestos-related mesothelioma: a review of the literature. *Scand. J. Work Environ. Health* 14:141–144.

Peterson, J.T., Greenberg, S.D., and Buffler, P.A. (1984) Non-asbestos-related malignant mesothelioma: a review. *Cancer* 54:951–960.

Peto, J. (1980) Lung cancer mortality in relation to measured dust levels in an asbestos textile factory. In: Wagner,J. C., editor. *Biological Effects of Mineral Fibers,* Lyon, France: International Agency for Research on Cancer, pp. 829–836.

Pierce, J.S., McKinley, M.A., Paustenbach, D.J., and Finley, B.L. (2008) An evaluation of reported no-effect chrysotile asbestos exposures for lung cancer and mesothelioma. *Crit. Rev. Toxicol.* 38:191–214.

Prueitt, R.L., Goodman, J.E., and Valberg, P.A. (2009) Radionuclides in cigarettes may lead to carcinogenesis *via* $p16^{INK4a}$ inactivation. *J. Environ. Radioact.* 100:157–161.

Pukkala, E., Martinsen, J.I., Lynge, E., Gunnarsdottir, H.K., Sparen, P., Tryggvadottir, L., Weiderpass, E., and Kjaerheim, K. (2009) Occupation and cancer—follow-up of 15 million people in five Nordic countries. *Acta Oncol.* 48:646–790.

Quinlan, T.R., Marsh, J.P., Janssen, Y.M., Borm, P.A., and Mossman, B.T. (1994) Oxygen radicals and asbestos-mediated disease. *Environ. Health Perspect.* 102(Suppl. 10):107–110.

Rake, C., Gilham, C., Hatch, J., Darnton, A., Hodgson, J., and Peto, J. (2009) Occupational, domestic and environmental mesothelioma risks in the British population: a case-control study. *Br. J. Cancer* 100:1175–1183.

Rödelsperger, K., Woitowitz, H.J., Brückel, B., Arhelger, R., Pohlabeln, H., and Jöckel, K.H. (1999) Dose-response relationship between amphibole fiber lung burden and mesothelioma. *Cancer Detect Prev.* 23:183–193.

Roggli, V.L., Oury, T.D., and Sporn, T.A., editors. (2004) *Pathology of Asbestos-Associated Diseases,* 2nd ed. New York, NY: Springer.

Roggli, V.L. and Vollmer, R.T. (2008) Twenty-five years of fiber analysis: what have we learned? *Hum. Pathol.* 39(3):307–315.

Sanders, C.L. (1992) Pleural mesothelioma in the rat following exposure to 239PuO2. *Health Phys.* 63:695–697.

Sanders, C.L. and Jackson T.A. (1972) Induction of mesotheliomas and sarcomas from "hot spots" of 239PuO2 activity. *Health Phys.* 22:755–759.

Schins, R.P. (2002) Mechanisms of genotoxicity of particles and fibers. *Inhal. Toxicol.* 14:57–78.

Selikoff, I.J. (1977) Cancer risk of asbestos exposure. In: Hiatt, H.H., Watson, J.D., and Winsten, J.A., editors. *Origins of Human Cancer, Book C: Human Risk Assessment,* Cold Spring Harbor Laboratory, pp. 1765–1784.

Selikoff, I.J., Churg, J., and Hammond, E.C. (1964) Asbestos exposure and neoplasia. *JAMA* 188:22–26.

Selikoff, I.J., Churg, J., and Hammond, E.C. (1965) The occurrence of asbestosis among insulation workers in the United States. *Ann. NY Acad. Sci.* 132:139–155.

Selikoff, I.J., Hammond, E.C., and Churg, J. (1968) Asbestos exposure, smoking, and neoplasia. *JAMA* 204:106–112.

Selikoff, I.J., Seidman, H., and Hammond, E.C. (1980) Mortality effects of cigarette smoking among amosite asbestos factory workers. *J. Natl. Cancer Inst.* 65:507–513.

Sichletidis, L., Chloros, D., Spyratos, D., Haidich, A.B., Fourkiotou, I., Kakoura, M., and Patakas, D. (2009) Mortality from occupational exposure to relatively pure chrysotile: a 39-year study. *Respiration* 78:63–68.

Spirtas, R., Heineman, E.F., Bernstein, L., Beebe, G.W., Keehn, R.J., Stark, A., Harlow, B.L., and Benichou J. (1994) Malignant mesothelioma: attributable risk of asbestos exposure. *Occup. Environ. Med.* 51:804–811.

Sporn, T.A. (2011) Mineralogy of asbestos. *Recent Results Cancer Res.* 189:1–11.

Stayner, L., Kuempel, E., Gilbert, S., Hein, M., and Dement, J. (2008) An epidemiological study of the role of chrysotile asbestos fibre dimensions in determining respiratory disease risk in exposed workers. *Occup. Environ. Med.* 65:613–619.

Steiner, W.E., Koehler, J.L., and Popenuck, W.W. (1990) Guadalupe corridor transportation project asbestos health risk assessment, San Jose, California. *Sci. Total Environ.* 93:115–124.

Straif, K. (2008) The burden of occupational cancer. *Occup. Environ. Med.* 65:787–788.

Straif, K., Benbrahim-Tallaa, L., Baan, R., Grosse, Y., Secretan, B., El Ghissassi, F., et al. (2009) A review of human carcinogens—part C: metals, arsenic, dusts, and fibres. *Lancet Oncol.* 10:453–454.

Tomlinson, I., Sasieni, P., and Bodmer, W. (2002) How many mutations in a cancer? *Am. J. Pathol.* 160:755–758.

Toyooka, S., Pass, H.I., Shivapurkar, N., Fukuyama, Y., Maruyama, R., Toyooka, K.O., Gilcrease, M., Farinas, A., Minna, J.D., and

Gazdar, A.F. (2001) Aberrant methylation and simian virus 40 tag sequences in malignant mesothelioma. *Cancer Res.* 61:5727–5730.

US EPA (1986) *Airborne Asbestos Health Assessment Update*, EPA-600/8-84-003F. Research Triangle Park, NC: Office of Health and Environmental Assessment.

U.S. Environmental Protection Agency. (1999) Integrated Risk Information System (IRIS) on Asbestos. National Center for Environmental Assessment, Office of Research and Development, Washington, DC.

Wagner, J.C. (1962) Experimental production of mesothelial tumours of the pleura by implantation of dusts (silica) in laboratory animals. *Nature* 196:180–181.

Wagner, J.C., Newhouse, M.L., Corrin, B., Rossiter, C.E., and Griffiths, D.M. (1988) Correlation between fibre content of the lung and disease in east London asbestos factory workers. *Br. J. Ind. Med.* 45:305–308.

Wagner, J.C., Sleggs, C.A., and Marchand, P. (1960) Diffuse pleural mesothelioma and asbestos exposure in the North Western Cape Province. *Br. J. Ind. Med.* 17:260–271.

Weiner, S.J. and Neragi-Miandoab, S. (2009) Pathogenesis of malignant pleural mesothelioma and the role of environmental and genetic factors. *J. Cancer Res. Clin. Oncol.* 135:15–27.

Weiss, W. (1999) Asbestosis: a marker for the increased risk of lung cancer among workers exposed to asbestos. *Chest* 115:536–549.

Williams, P.R., Phelka, A.D., and Paustenbach, D.J. (2007) A review of historical exposures to asbestos among skilled craftsmen (1940–2006). *J. Toxicol. Environ. Health B Crit. Rev.* 10:319–377.

Yarborough, C.M. (2007) The risk of mesothelioma from exposure to chrysotile asbestos. *Curr. Opin. Pulm. Med.* 13:334–338.

Zhong, F., Yano, E., Wang, Z.M., Wang, M.Z., and Lan, Y.J. (2008) Cancer mortality and asbestosis among workers in an asbestos plant in Chongqing, China. *Biomed. Environ. Sci.* 21:205–211.

ERIONITE AND OTHER ZEOLITES

First aid: Inhalation: Move person to an area with fresh air. Eyes: irrigate eyes immediately with water.

Target organ(s): Lung

Occupational exposure limits: Erionite: None; other zeolites, e.g., mordenite, OSHA PEL, 15 mg/m^3 total dust and 5 mg/m^3 respirable. ACGIH TLV, 10 mg/m^3 total dust and 3 mg/m^3 respirable.

Reference values: None determined.

Risk/safety phrases: None specified.

BACKGROUND AND USES

Types of Fibers and Their Uses

Zeolites are a group of hydrated aluminosilicates of the alkaline and alkaline-earth metals, which contain exchangeable cations of groups IA and IIA, such as sodium, potassium, magnesium, and calcium. Approximately 40 natural types of zeolites have been identified, with erionite (CAS: 66733-21-9; CAS: 12510-42-8) being one of the common types. Other common natural zeolites are analcime, chabazite, mordenite, ferrierite, heulandite, faujasite, clinoptilolite, phillipsite, and laumontite. Out of the large number of natural zeolites, only some are fibrous, including erionite, mesolite, mordenite, natrolite, scolecite, and thomsonite. Although not the subject of this chapter, more than 150 types of *synthetic* zeolites exist (IARC, 1997).

Erionite has no commercial uses and is not currently mined. It is reported that erionite was one of several commercially important zeolites in the 1960s and 1970s when commercial mining of zeolites began, with some prior use as a noble metal-impregnated catalyst in a hydrocarbon-cracking process (IARC, 1987; NTP, 2011). Other natural zeolites continue to have commercial uses, owing to their thermal stability and ability to selectively adsorb molecules from both air and liquids. Virta (2010) lists a number of domestic uses for natural zeolites in the United States, including animal feed, odor control, pet litter, water purification, fungicide or pesticide carrier, wastewater cleanup, gas absorbent, horticultural applications (soil conditioners and growth media), oil absorbent, desiccant, and aquaculture. More than 75% of these uses are reported to be for animal feed, odor control, and pet litter (Virta, 2010).

General Epidemiology and Impact on Occupational Health

Analogous to asbestos, airborne free erionite fibers can be hazardous to health because of their physical and chemical properties. The primary route of exposure is by inhalation of fibers released during production and mining of other zeolites, during commercial use of other zeolites in various processes and products (e.g., pet litter, animal feed, horticultural applications), and from production and use of erionite-containing materials such as gravel. Depending on fiber size (length and diameter), inhaled fibers deposit onto lung surfaces, and subsequently are removed from the lung, are retained in lung tissues, or penetrate to nearby sites. Because the fibers are resistant to chemical attack, retained fibers are not readily broken down.

Erionite is classified as a known human carcinogen (IARC; National Toxicology Program (NTP) based on epidemiological studies reporting an excess of mortality from mesothelioma in three Turkish villages with elevated long-term exposures to airborne erionite fibers due to the use of erionite-containing building materials. Chronic inhalation exposures to airborne erionite fibers, such as those that have occurred in the Cappadocian region of Turkey, are thus believed to increase the risk of serious illnesses, including mesothelioma, lung cancer, and fibrogenic lung disease.

Fiber concentration data are cross sectional in nature and are therefore not sufficient to assess quantitative exposure–response relationship for erionite. With the exception of a recent study suggesting that occupational exposure to gravel containing erionite may be associated with chest radiographic abnormalities, including mild pleural and possible interstitial changes in the lung (Ryan et al., 2011), there remains a general absence of epidemiologic data for occupational exposures to erionite. Thus, there remains uncertainty regarding the impact of erionite on occupational health in workplaces with potential low to moderate exposure to erionite fibers, such as during mining, processing, and use of other natural zeolites and materials, such as gravel, which may contain erionite.

PHYSICAL AND CHEMICAL PROPERTIES

Consisting of brittle, wool-like fibrous masses, erionite can be found naturally in volcanic ash that has been compressed and altered by weathering (volcanic "tuff"). It is rarely found in pure form, instead occurring with other natural zeolites. Like other zeolites, it has a basic hexagonal structure that is composed of a framework of linked aluminosilicate tetrahedra (SiO_4 and AlO_4). Erionite has been classified into three main series, Na-erionite, Ca-erionite, and K-erionite, based on the respective cation prevalence. Its color varies from white to clear, and it resembles transparent, glass-like fibers. It has been reported to have a molecular weight of 715.68 (Lewis, 2004) and a density of 2.02–2.08 (IARC, 1987).

While erionite is a fibrous mineral that has a morphological resemblance to amphibole asbestos (IARC, 1987), it has different physical and chemical properties. It is silica rich, having varying amounts of sodium, calcium, and potassium cations. It consists of "bundles" that are composed of many fibers and fibrils, with the number of fibrils and the Si/Al ratio correlating with the surface area of this mineral. Its surface area has been approximated to be $200 \, m^2/g$ (Baris and Baris, 1993), a surface-area-per-gram that is 10-fold higher than that of crocidolite amphibole asbestos (Bish and Chipera, 1991).

Accurate detection and differentiation of trace amounts of erionite is difficult, due to its similarity to other minerals such as smectite and clinoptilolite (Bish and Chipera, 1991). Elemental composition analyses [e.g., energy-dispersive X-ray analysis (EDXA)] alone cannot reliably differentiate erionite from chemically similar minerals. The complexity of identifying erionite is highlighted in a recent article (Dogan and Dogan, 2008), in which the authors identified two tests, the E%-balance test (i.e., a test that compares proportions of the various cations) and the Mg-content test [which quantifies the number of Mg (magnesium) atoms per cell], which can be used to identify erionite. The authors analyzed dozens of erionite samples from Turkey and the United States, finding that some of the "erionite" samples from various parts of Turkey (e.g., Cappadocia, Karain, and Tuzkoy) failed these tests. In addition, they reported that some U.S. erionite samples (e.g., from Rome, Oregon) were consistent with true erionite while others (e.g., from Agatate, Oregon) were not.

Comparisons of erionite samples from Turkey and the United States have shown that samples are comparable in terms of physical properties and mineral and chemical content [based on various miscroscopic analyses, including X-ray diffraction (XRD), scanning electron microscope-energy dispersive spectroscopy (SEM-EDS), and TEM] (Dogan et al., 2006, 2008; Carbone et al., 2011; Lowers and Meeker, 2007). For example, Carbone et al. (2011) concluded that the physical and chemical properties of erionite from North Dakota and Cappadocia, Turkey were "very similar," including an identical average fiber width of $0.31 \, \mu m$ and a small difference in the atomic Si/(Si+Al) ratio indicating Turkish erionite to be slightly richer in Si. Lowers and Meeker (2007) concluded that there was some overlap in the chemical composition of "erionite" observed in dust samples[3] from North Dakota with Turkish erionite and/or offretite (Figure 96.2). Other studies have reported some differences in fiber size distributions, with data indicating a higher proportion of longer fibers ($>4 \, \mu m$) in erionite from Oregon than from Karain, Turkey (11 *versus* 32%; as reported in Ilgren et al., 2008a).

As mentioned above, the other 40 or so known natural zeolites share the same framework of SiO_4 and AlO_4 tetrahedra as erionite. However, in contrast to erionite that is known to only occur in fibrous form, other natural zeolites are more typically found as plates, prisms, or laths. For example, of the commonly found natural zeolites, clinoptilolite can be found as euhedral (idiomorphic) plates and laths that are several micrometers in length and $1–2 \, \mu m$ thick, and phillipsite as prisms and laths that are $3–30 \, \mu m$ in length and $0.3–3 \, \mu m$ thick, typically with pseudo-orthorhombic symmetry (IARC, 1997). Given that different natural zeolites have similar chemical compositions, crystalline structure is generally used for identification of the various zeolites.

ENVIRONMENTAL FATE AND BIOACCUMULATION

Fibrous zeolites are not expected to bioaccumulate. However, some zeolite fibers are relatively biopersistent and can stay in the lung or elsewhere for prolonged periods of time.

---

[3] Only identification given by the authors: "20 soil and roadbed samples collected from North Dakota."

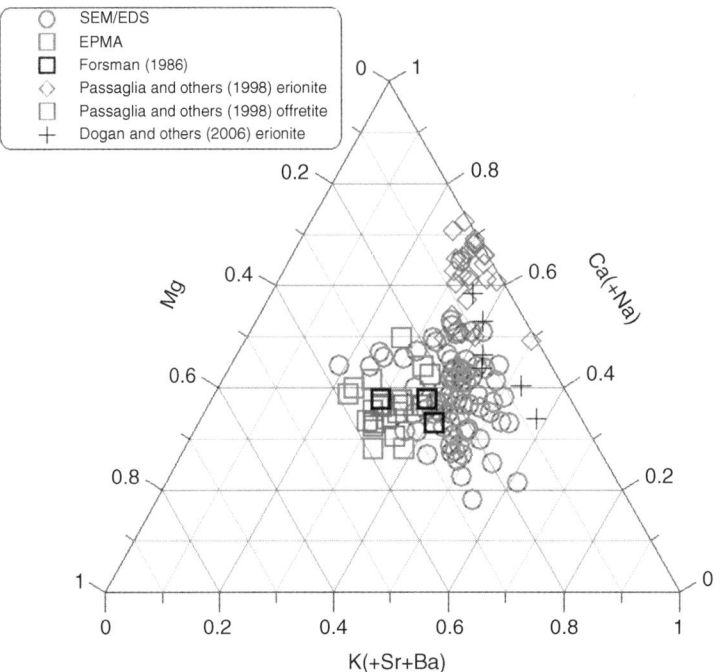

FIGURE 96.2 SEM/EDS analysis of samples from North Dakota zeolite (in red) falls between the Ca(+Na)-Mg-K erionite and offretite fields described by Passaglia and others (1998). The electron microprobe data (EPMA) fall in the offretite field. However, the best structural match based on X-ray diffraction data is to erionite. While falling overall between the erionite and offretite fields, some of the SEM/EDS data overlap compositionally with zeolite compositions that were associated with malignant diseases (Dogan et al., 2006). Note that the Forsman (1986) samples are from the Killdeer Mountains in North Dakota, while the Dogan et al. (2006) samples are from the Cappadocian villages in Turkey. Adapted from Lowers and Meeker (2007)

ECOTOXICOLOGY

Natural fibrous zeolites can be released to the environment from mining or in refined form. However, there are no published studies on the ecotoxicity of these zeolites.

MAMMALIAN TOXICOLOGY

Acute Effects

Animal Zeolites generally have not shown adverse acute effects in short-term studies, except at massive inhaled or ingested doses (Tátrai et al., 1991, 1992). There are no studies available that investigated the acute health effects of erionite or other naturally occurring fibrous zeolites in humans or in animals.

Human No studies were identified that have examined the short-term health effects of zeolite fibers in humans.

Chronic Effects

Animal Only a few animal studies examining the carcinogenicity of zeolites are available. For nonfibrous zeolites, available studies generally do not show evidence of carcinogenicity. For example, in BALB/c mice given a single intraperitoneal injection of 10 mg per animal mordenite (particle length, 94% <3 μm and 4% >3.8 μm), no tumors were observed for up to 23 months. However, the authors noted mild peritoneal fibrosis for the mordenite- *vs.* saline-treated mice (Suzuki and Kohyama, 1984).

For rats exposed by inhalation or by intrapleural instillation, albeit to very high concentrations or doses of erionite or asbestos (10 mg/m^3 and 20 mg, respectively), Wagner et al. (1985) reported findings indicating a considerably greater potency for erionite than for asbestos. For example, 27 of 28 erionite-exposed rats (erionite source: Oregon) developed mesotheliomas, compared to 11 of 648 rats for various types of asbestos and 1 of 28 for a nonfibrous zeolite having a similar chemical makeup as the Oregon erionite (Wagner et al., 1974, 1985). Moreover, Wagner et al. (1985) observed reduced survival times for erionite-associated mesotheliomas than for asbestos-associated mesotheliomas (for inhalation, mean survival times of 580 days and 917 days for Oregon erionite and crocidolite, respectively; for intrapleural inoculation, mean survival times of 390 days and 678 days for Oregon erionite and chrysotile, respectively). This reduction in survival time for erionite-induced mesotheliomas agrees with human clinical findings from Selçuk et al. (1992); who reported a significantly shorter survival time for mesothelioma patients from the erionite villages of Karain, Tuzkoy, and Sarihidir, as compared to patients with asbestos-associated mesothelioma (13.52 months for erionite-associated malignant pleural mesothelioma versus 21.56 months for erionite-associated malignant pleural mesothelioma) in their study of 135 Turkish patients with malignant pleural mesothelioma.

In the Wagner et al. (1985) experiments, erionite samples (20 mg dose) from Karain, Turkey, and Rome, Oregon, were found to be similar in carcinogenic potency when delivered intrapleurally to rats (38/40 and 40/40 exposed rats developed mesothelioma, respectively). Maltoni et al. (1982) exposed rats to 25 mg erionite intrapleurally (erionite source: Prof. Gottardi, Italy), observing lower mesothelioma rates than Wagner et al. (9/40 for erionite vs. 0/40 for crocidolite). One possible explanation for these variable findings between studies may involve the shorter monitoring period in this study (53 weeks) as compared to that of Wagner et al. (1985) (>2 years). Yet another animal study (Fraire et al., 1997) injected erionite from Pine Valley, Nevada directly into the pleural cavity of rats (20 mg per rat). For time periods of up to 480 days postexposure, the authors observed evidence of pulmonary fibrosis and pleural aplasia, as well as hyperplasia, in most erionite-exposed rats, but a lower yield of mesotheliomas (1/21) compared to previous studies. The exact determinants of the interstudy differences in biological responses are uncertain and may be related to the type and source of erionite used.

A meta-analysis of studies that exposed rats by inhalation or intrapleural administration found a much greater cancer potency for erionite by both mass and fiber size classification than for any type of asbestos (Carthew et al., 1992). For example, with intrapleural inoculation in rats, the authors estimated that erionite was 300- to 800-fold more potent than chrysotile and 100- to 500-fold more potent than crocidolite.

It is uncertain whether the specific erionite samples used in the animal carcinogenicity studies were included in the analysis by Dogan and Dogan (2008); since the study authors were not sufficiently specific about the source of the erionite used. It is possible that, even if some of these samples were not to meet the Dogan and Dogan (2008) definition of erionite, they may still possess chemical and physical attributes that can increase the risk of mesothelioma or lung cancer given sufficient exposure.

Human Epidemiologic studies from several Turkish villages in the Cappadocia region show elevated mesothelioma mortality (up to 50% of all deaths in some villages), which has generally been attributed to exposure to erionite. A study that followed immigrants from Turkey to Sweden found that risk of mesothelioma decreased with decreasing age at immigration (Metintas et al., 1999). The latest update of mortality rates from mesothelioma in the Turkish villages shows that some villages continue to have much higher rates than others (Baris et al., 1987; Baris and Grandjean, 2006). The elevated rates of mesothelioma deaths exceed those observed for *occupational* asbestos exposures (Metintas et al., 1999). A genetic component to mesothelioma risk has been suggested based on a pedigree analysis of some families from several Turkish villages (Dogan et al., 2006) and the results of human leukocyte antigen typing of biological samples from the villagers (Karakoca et al., 1998). However, the researchers who are most involved in these studies have suggested that the etiology of the mesotheliomas in not fully understood (Carbone et al., 2007), and some even suggest a larger role for fiber exposures than genetic influence (Metintas et al., 2010).

A nongenetic cause for elevated mesothelioma rates in these Turkish villages may be related to fiber exposures and properties. However, assessment of exposure to material reported to be erionite has been limited to a cross-sectional collection of samples and may not be representative of the true exposure that these villagers have experienced over the past decades (Baris et al., 1987). This cross-sectional assessment showed differences in erionite fiber airborne concentrations that were not consistent with the variation in mesothelioma rates among the villages examined in the study. However, it may be that the potency of each dust is modulated by how it is sequestered or by type and quantity of the minerals associated with it. For example, Baris and Grandjean (2006) reported age-adjusted annual mortality rates from pleural mesothelioma for the 25-year period 1979–2003 of 697, 197, and 11/100,000 for the Turkish villages of Karain, Sarihidir, and Karlik, respectively. The cross-sectional sampling scheme showed that the zeolite content of airborne street dust for these villages was 80, 60, and 20%, respectively (Baris et al., 1987), suggesting that zeolite content may be associated with mesothelioma risk.

In Mexico, two recent case studies (Kliment et al., 2009; Ilgren et al., 2008b) each claim to have reported the first case of a patient with erionite-associated pleural mesothelioma in North America. Kliment et al. (2009) described a 47-year-old male diagnosed with pleural mesothelioma and showing evidence of classic pathological changes commonly associated with asbestos-related disease, including ferruginous body formation, interstitial fibrosis, and pleural plaque development. The study authors reported an absence of any evidence indicating that this patient had significant asbestos exposure. Microscopic analyses identified 124,000 uncoated fibers and 2480 ferruginous bodies per gram of wet lung tissue, with 24 of the 30 uncoated fibers analyzed by EDXA containing various combinations of Al, Si, K, Na, and Ca in ratios consistent with erionite.

Ilgren et al. (2008b) also reported a single confirmed case of erionite-associated pleural mesothelioma in a Mexican male. Over one million fibers per gram lung dry weight of fibrous erionite were identified using EDXA and selected area electron diffraction (SAED). Ilgren et al. (2008b) stated that this confirmed mesothelioma case is one of two mesothelioma cases they identified among residents of a small village, reportedly located within a zeolite-rich plain in the northern part of the Mexican state of Jalisco. Both individuals were involved in agricultural tilling and may also have been exposed to zeolite-containing materials used as preservatives for vegetables, additives to fertilizers, and nutritional supplements for animals.

Further, Ilgren et al. (2008b) claimed to have identified an additional 11 mesothelioma cases among Mexican villagers in the state of Zacatecas, including a father and son now living in California. The Ilgren et al. (2008a) publication investigated the mesothelioma cases of the father and son (44 and 27 years old, respectively, at diagnosis), also identifying an additional nine cases "clustered" around the small ancestral village where the father was born and the son spent some time as a child. Given the absence of any detections of erionite in this area, Ilgren et al. (2008a) acknowledged that the attribution of a causal role of erionite in these cases requires additional confirmation, especially given the presence of asbestos in both roofing and slate materials in this village (Los Sabinos).

Earlier, Casey et al. (1985) reported on a patient who lived and worked in an area rich in zeolite deposits, showing evidence of pulmonary deposition of fibrous and nonfibrous aluminum silicate particles with an EDXA pattern characteristic of zeolites. This patient's histologic picture was similar to the characteristic of asbestosis, with evidence of pleuropulmonary fibrosis, pleural thickening, and infiltrate; however, no asbestos fibers could be identified. Energy-dispersive X-ray Analysis analysis of over 100 particles and fibers analyzed by EDXA revealed the presence of aluminum, the absence of magnesium, and very low iron content, findings that were all consistent with those of a sample of northern Nevada fibrous erionite. In addition, there was no history of exposure to asbestos. The authors claimed this as the first case of zeolite-associated pulmonary disease reported in the United States.

Recently, a team of investigators, including scientists from the US EPA, the University of Cincinnati, and the North Dakota Department of Health sampled road dust and gravel for fibers in response to concerns of potential erionite exposures in several western North Dakota counties known to have used gravel containing erionite to surface roads, parking lots, and other areas (Ryan et al., 2011). In addition, investigators conducted a health study of the possible health effects of erionite exposure in these western North Dakota counties, recruiting 34 residents and workers considered to represent those with the greatest potential exposure to erionite-containing gravel and using a questionnaire, chest radiograph, and high resolution computed tomography (HRCT) scan to assess the rate of interstitial and pleural changes, which have been associated with fibrous mineral exposure. Study participants included individuals having worked in gravel pits, road maintenance, or both, as well as ranchers/farmers and individuals in occupations with frequent driving (e.g., delivery truck drivers, mail carriers). Chest X-ray results did not indicate a significant increase in interstitial changes or localized pleural changes above background prevalence. Consistent with the increased sensitivity and specificity of HCRT for identifying subtle pleural and interstitial changes, the HRCT scan results did indicate a statistically significant *increase* in the rate of pulmonary interstitial changes of 12.5% ($n = 4$, $p < 0.01$), compared to 1.5% for a study of male urban transportation workers with low cumulative asbestos fiber exposure. Two of these individuals, both of whom were ex-smokers (3 and 7 pack-years) with prolonged occupational exposure to road gravel *via* gravel pits, road maintenance, or both, were also observed to have bilateral localized pleural changes with calcification. These study results thus suggest that occupational exposures to erionite *via* erionite-containing gravel in western North Dakota counties, particularly for workers employed in road maintenance or gravel pits, could result in mild pleural and possible interstitial changes that have been historically associated with commercial asbestos exposure.

Overall, a number of epidemiologic studies of Turkish populations have thus reported elevated mesothelioma rates in the Cappadocia region, which have been linked with exposure to fibrous erionite. Such reports have raised concerns regarding potential erionite exposures of health concern in other areas having erionite-like formations, including the United States in particular. Although there is now anecdotal evidence of a limited number of possible erionite-related mesothelioma cases among Mexican villagers, no U.S. cases of mesothelioma have been specifically linked to erionite exposures in western U.S. states with prevalent erionite-like formations.

Mode of Action(s)

Using Chinese hamster lung V79 cells, Palekar et al. (1988) compared the cytotoxicity of mordenite; two preparations of fibrous erionite from Rome, Oregon, including erionite with a mean length of 2.2 μm, and erionite prepared by ball milling and with a mean length of 1.4 μm; Union Internationale Contre le Cancer (UICC) standard crocidolite, with a mean length of 1.3 μm; and UICC chrysotile, with a mean length of 2.4 μm. Mordenite was found to be nontoxic, while the other minerals, which have all been observed to be tumorigenic in animal studies, exhibited evidence of cytotoxicity.

Fach et al. (2002) also compared the *in vitro* toxicity of erionite and an essentially benign mordenite in a rat macrophage model. The authors reported that oxidative burst increased with decreasing particle size regardless of zeolite chemical composition, but that the Fenton reaction depended on the type of zeolite, with erionite producing more reactive oxygen species than mordenite for the same amount of iron. Related to these findings, Baris and Baris (1993) estimated a surface area-to-mass ratio of approximately $200 \, m^2/g$ for a sample of Turkish erionite—i.e., 10-fold higher than that of amphibole crocidolite asbestos—thus suggesting that some of the reported differences in toxicity may be related to surface area.

Moreover, Timblin et al. (1998) reported that while both erionites and crocidolite exposure induced genetic damage, as supported by increased unscheduled DNA synthesis, crocidolite was much faster and more effective than erionite at inducing apoptosis. This suggests that erionite may tip the balance in favor of cell proliferation *versus* apoptosis, and crocidolite may be more likely to induce apoptosis faster and at lower doses than erionite. These findings thus provide a possible explanation for the higher mesothelioma rates of erionite *versus* crocidolite asbestos, as reported by Wagner et al. (1974, 1985). This is further supported by findings from Bertino et al. (2007) that compared cell proliferation induction by Oregon erionite and asbestos samples.

In summary, elevated rates of mesothelioma in the Turkish villages and in erionite-exposed rats suggest the potential health risk significance of some forms of erionite exposure. However, the determinants of the toxicity and carcinogenicity of various forms of erionite and/or zeolite fibers similar to erionite remain inadequately characterized. There are some indications that erionite toxicity may increase with fiber surface area. The shape of fibers and fiber bundles may also be important factors, and are candidates for further investigation. Erionite has a higher surface area than asbestos, and it has been hypothesized that this may be one factor for its greater potency due to an increased potential of adverse chemical reactions (Dogan et al., 2006, 2008). Erionite may exert its adverse effects *via* oxidative stress, cytotoxicity, genetic damage, and favoring of proliferation pathways over apoptosis, as compared to some asbestos species. Additional comparative evaluation of the chemical composition and fiber shapes of erionite from sites outside Turkey with erionite from the mesothelioma-affected villages in Turkey may provide further insights on the indicators of potential toxicity and carcinogenicity of different erionite-like materials.

SOURCES OF EXPOSURE

Geographical Distribution

Erionite and other natural zeolites can be found in different parts of the world in over 40 countries (IARC, 1997). Most notably, erionite has been reported to be highly prevalent in volcanic tuffs in the Cappadocian region of central Turkey. Erionite-like formations are known to occur in 12 states in the western U.S.; referred to as the Intermountain West, these states include Arizona, California, Idaho, Montana, Nevada, New Mexico, North Dakota, Oregon, South Dakota, Texas, Utah, and Wyoming. Areas noted to have naturally-occurring deposits of fibrous erionites include Shoshone, California; Jersey Valley, Nevada; and Rome, Oregon (Ilgren *et al.*, 2008a). Many of these states also have commercial zeolite deposits.

In addition to the United States, countries that mine natural zeolites include Japan, Hungary, Bulgaria, Cuba, Italy, and South Africa, while Australia, the former Czechoslovakia, Greece, and Turkey are reported to have large unexploited reserves of natural zeolites (IARC, 1997). Countries reported to have mineable deposits of erionite include Kenya, Mexico, New Zealand, the United Republic of Tanzania, and the former Yugoslavia (IARC, 1987). Ilgren et al. (2008a) hypothesized that fibrous erionites in central Mexico are significant contributors to mesothelioma cases in the Valley of Jalpa, which is located in the southern part of the state of Zacatecas. Although there is an absence of any reported detections of erionite in this area, Ilgren et al. (2008a) point to the scarcity of mineralogical studies in this area and to the identification of other zeolite species (clinoptilolite, chabazite, and phillipsite), which have been shown to co-occur with fibrous erionite. In addition, this part of Mexico contains the Sierra Madre Occidental (SMO), a plateau rich in zeolites that is contiguous with regions in Mexico and the United States and shown to have fibrous erionite formations.

Industries and Populations at Risk

Occupational populations at risk include those involved in the mining and processing of natural zeolites and materials, such as gravel, which may contain erionite. In addition, workers involved in the manufacture and application of natural zeolite products that may contain erionite, such as animal feed, odor control products, and pet litter, are at potential risk. Although it has been reported that in the past erionite-rich blocks were quarried for use as building materials in some western U.S. states, this is not considered to be a common current practice and construction workers are not considered to be a population at risk (IARC, 1987).

Exposure Concentrations

There are few reported measurements of airborne exposure levels of erionite fibers of relevance to occupational exposure settings. Although themselves limited, the largest number of measurements of airborne exposure levels of erionite fibers is available from studies conducted in several villages in the Cappadocian region of Turkey, where erionite is a common constituent of rocks and soil and erionite-containing building materials have historically been used (e.g., Baris et al., 1981; Baris et al., 1987). For example, Baris et al. (1987) reported airborne fiber (>5 μm in length) levels as high as 0.029 fibers/cm^3 for street samples from three Cappadocian villages where dominant portions of airborne fibers have been shown to have chemical compositions similar to erionite, and as high as 1 fiber/cm^3 for samples collected during sweeping activities inside homes. Given the common occurrence of erionite in building materials, soils, and dust that is unique to this area, these measurements are not considered relevant to other geographic areas and potential exposure settings.

Perhaps of greater relevance to potential exposure concentrations of airborne erionite fibers in occupational settings are recent measurements conducted by a multi-agency team of investigators to characterize airborne levels of erionite fibers associated with erionite-containing gravel in North Dakota. Led by US EPA, this team conducted activity-based sampling during transportation scenarios (e.g., driving cars and school buses on gravel roads), measuring fiber concentrations that ranged from 0.0107 to 0.0391 fibers/cm^3 (PCM equivalent) and averaged 0.0249 fibers/cm^3 (Ryan et al., 2011). Concentrations ranging from 0 to 0.0012 fibers/cm^3 (average of 0.0008 fibers/cm^3) were measured at stationary air samplers located near the roadways. Ryan et al. (2011) suggested that these sampling results may not fully represent the range of potential exposure concentrations that road maintenance workers may experience during gravel pit and road grading activities. For indoor air samples collected in North Dakota in areas with use of erionite-containing gravel, Carbone et al. (2011) reported average concentrations of 0 and 0.06 fibers/cm^3 (PCM equivalent) for a social services office and a road maintenance garage, respectively, during sweeping and house-keeping activities.

Few measurement data are available to assess potential exposure levels of airborne erionite fibers during mining activities. IARC (1987) reported that a 1979 monitoring study conducted at an open-pit zeolite mining operation in Bowie, Arizona did not provide evidence of significant fiber exposures levels despite the known presence of erionite.

INDUSTRIAL HYGIENE

No health standards are currently available for erionite, although it is considered by both IARC and NTP to be a known human carcinogen due to its effects on the lung, and specifically the lung pleura.

Given its similarity in both morphology and toxicity to asbestos, it is recommended that standard industrial hygiene practices for asbestos (e.g., OSHA asbestos regulations at 29 CFR 1910.1001) also be applied to occupational environments with possible erionite exposures. In particular, initial personal air sampling is recommended to assess levels of airborne erionite fibers and to determine the need for personal protective equipment such as respiratory protection.

RISK ASSESSMENTS

Erionite has been classified by IARC as a Group 1 substance, i.e., known to be carcinogenic to humans (IARC, 1987). According to IARC (1997); zeolites other than erionite (e.g., clinoptilolite, phillipsite, mordenite, nonfibrous Japanese zeolite,and synthetic zeolites) cannot be evaluated as carcinogenic to humans (Group 3). NTP, which is part of the U.S. Department of Health and Human Services, lists erionite in Part A of the 12th Report on Human Carcinogens (2011), i.e., among those substances known to be human carcinogens. The German MAK Commission (Federal Republic of Germany Maximum Concentration Values in the Workplace) classifies erionite as a "MAK-1" substance, with "adequate epidemiological evidence for a positive correlation between the exposure of humans and the occurrence of cancer," but no quantitative exposure limit value was derived.

REFERENCES

Baris, Y. I., Saracci, R., Simonato, L., Skidmore, J. W., and Artvinli, M. (1981) Malignant mesothelioma and radiological chest abnormalities in two villages in Central Turkey. An epidemiological and environmental investigation. *Lancet* 1:984–987.

Baris, I., Simonato, L., Artvinli, M., Pooley, F., Saracci, R., Skidmore, J., and Wagner, C. (1987) Epidemiological and environmental evidence of the health effects of exposure to erionite fibres: a four-year study in the Cappadocian region of Turkey. *Int. J. Cancer* 39:10–17.

Baris, E. and Baris, Y. I. (1993) Environmental exposure to fibrous zeolite in Turkey: an appraisal of the epidemiological and environmental evidence. In: Peters, G.A. and Peters, W., editors. *Asbestos Risks and Medical Advances*, Salem: Butterworth Legal Publishers, pp. 53–72.

Baris, I. and Grandjean, P. (2006) Prospective study of mesothelioma mortality in Turkish villages with exposure to fibrous zeolite. *J. Natl. Cancer Inst.* 98:414–417.

Bertino, P., Marconi, A., Palumbo, L., Bruni, B. M., Barbone, D., Germano, S., Dogan, A. U., Tassi, G. F., Porta, C., Mutti, L., and Gaudino G. (2007) Erionite and asbestos differently cause transformation of human mesothelial cells. *Int. J. Cancer* 121:12–20.

Bish, D. L. and Chipera, S. J. (1991) Detection of trace amounts of erionite using X-ray powder diffraction: erionite in tuffs of Yucca Mountain, Nevada, and Central Turkey. *Clay Clay Miner.* 39:437–445.

Carbone, M., Baris, Y. I., Bertino, P., Brass, B., Comertpay, S., Dogan, A.U., Gaudino, G., Jube, S., Kanodia, S., Partridge, C.R., Pass, H.I., Rivera, Z.S., Steele, I., Tuncer, M., Way, S., Yang, H., and Miller, A. (July 25, 2011) Erionite exposure in

North Dakota and Turkish villages with mesothelioma. *Proc. Natl. Acad. Sci. U. S. A.* doi 10.1073/pnas.1105887108.

Carbone, M., Emri, S., Dogan, A. U., Steele, I., Tuncer, M., Pass, H. I., and Baris, Y. I. (2007) A mesothelioma epidemic in Cappadocia: scientific developments and unexpected social outcomes. *Nat. Rev. Cancer* 7:147–154.

Carthew, P., Hill, R. J., Edwards, R. E., and Lee, P. N. (1992) Intrapleural administration of fibres induced mesothelioma in rats in the same relative order of hazard as occurs in man after exposure. *Hum. Exp. Toxicol.* 11:530–534.

Casey, K. R., Shigeoka, J. W., Rom, W. N., and Moatamed, F. (1985) Zeolite exposure and associated pneumoconiosis. *Chest* 87:837–840.

Dogan, A. U., Baris, Y. I., Dogan, M., Emri, S., Steele, I., Elmishad, A. G., and Carbone, M. (2006) Genetic predisposition to fiber carcinogenesis causes a mesothelioma epidemic in Turkey. *Cancer Res.* 66:5063–5068.

Dogan, A. U. and Dogan, M. (2008) Re-evaluation and re-classification of erionite series minerals. *Environ. Geochem. Health* 30:355–366.

Dogan, A.U., Dogan, M., and Hoskins, J.A. (2008) Erionite series minerals: mineralogical and carcinogenic properties. *Environ. Geochem. Health* 30:367–381.

Fach, E., Waldman, W. J., Williams, M., Long, J., Meister, R. K., and Dutta, P. K. (2002) Analysis of the biological and chemical reactivity of zeolite-based aluminosilicate fibers and particulates. *Environ. Health Perspect.* 110:1087–1096.

Fraire, A. E., Greenberg, S. D., Spjut, H. J., Dodson, R. F., Williams, G., Lach-Pasko, E., and Roggli, V. L. (1997) Effect of erionite on the pleural mesothelium of the Fischer 344 rat. *Chest* 111:1375–1380.

Ilgren, E. B., Brena, M. O., Larragoitia, J. C., Navarette, G. L., Brena, A. F., Krauss, E., and Feher, G. (2008a) A reconnaissance study of a potential emerging Mexican mesothelioma epidemic due to fibrous zeolite exposure. *Indoor Built Environ.* 17:496–515.

Ilgren, E. B., Pooley, F. D., Larragoitia, J. C., Talamantes, M., Navarette, G. L., Krauss, E., and Brena, A. F. (2008b) First confirmed erionite related mesothelioma in North America. *Indoor Built Environ.* 17:567–568.

International Agency for Research on Cancer (IARC) (1987) *Monographs on the Evaluation of Carcinogenic Risks to Humans: Silica and Some Silicates*, vol. 42, Lyon, France: World Health Organization.

International Agency for Research on Cancer (IARC) (1997) *Monographs on the Evaluation of Carcinogenic Risks to Humans: Silica, Some Silicates, Coal Dust and Para-Aramid Fibrils*, vol. 68, Lyon, France: World Health Organization.

Karakoca, Y., Emri, S., Bagci, T., Demir, A., Erdem, Y., Baris, E., and Sahin, A.A. (1998) Environmentally-induced malignant pleural mesothelioma and HLA distribution in Turkey. *Int. J. Tuberc. Lung Dis.* 2:1017–1022.

Kliment, C. R., Clemens, K., and Oury, T. D. (2009) North american erionite-associated mesothelioma with pleural plaques and pulmonary fibrosis: a case report. *Int. J. Clin. Exp. Pathol.* 2:407–410.

Lewis, R. J., editors. (2004) *Sax's Dangerous Properties of Industrial Materials*, 11th ed. Hoboken, New Jersey: Wiley-Interscience.

Lowers, H. A. and Meeker, G. P. (2007) *Denver microbeam laboratory administrative report 14012007: U.S. Geological Survey Administrative Report*. Available at http://www.ndhealth.gov/EHS/erionite/General/USGS_NDerionite_report.pdf (accessed June 7, 2011). Published 2007.

Maltoni, C., Minardi, F., and Morisi L. (1982) Pleural mesotheliomas in Sprague-Dawley rats by erionite: first experimental evidence. *Environ. Res.* 29:238–244.

Metintas, M., Hillerdal, G., and Metintas, S. (1999) Malignant mesothelioma due to environmental exposure to erionite: follow-up of a Turkish emigrant cohort. *JT Eur. Respir. J.* 13:523–526.

Metintas, M., Hillerdal, G., Metintas, S., and Dumortier P. (2010) Endemic malignant mesothelioma: exposure to erionite is more important than genetic factors. *Arch. Environ. Occup. Health* 65:86–93.

National Toxicology Program (NTP) (2011) Erionite. In: *Report on Carcinogens*, 12 ed., pp. 183–184. Available at http://ntp.niehs.nih.gov/ntp/roc/twelfth/roc12.pdf. Published June 10, 2011.

Passaglia, E., Artioli, G., and Gualtieri, A. (1998) Crystal chemistry of the zeolites erionite and offretite. *Am. Mineral.* 83:577–589.

Palekar, L.D., Most, B.M., and Coffin, D.L. (1988) Significance of mass and number of fibers in the correlation of V79 cytotoxicity with tumorigenic potential of mineral fibers. *Environ Res.* 46(2):142–152.

Ryan, P. H., Dihle, M., Griffin, S., Partridge, C., Hilbert, T. J., Taylor, R., Adjei, S., and Lockey, J. E. (2011) Erionite in road gravel associated with interstitial and pleural changes—an occupational hazard in Western United States. *J. Occup. Environ. Med.* 53:892–898.

Selçuk, Z. T., Cöplü, L., Emri, S., Kalyoncu, A. F., Sahin, A. A., and Bariş, Y. I. (1992) Malignant pleural mesothelioma due to environmental mineral fiber exposure in Turkey. Analysis of 135 cases. *Chest* 102:790–796.

Suzuki, Y. and Kohyama, N. (1984) Malignant mesothelioma induced by asbestos and zeolite in the mouse peritoneal cavity. *Environ. Res.* 35:277–292.

Tátrai, E., Bácsy, E., Kárpáti, J., and Ungváry, G. (1992) On the examination of the pulmonary toxicity of mordenite in rats. *Pol. J. Occup. Med. Environ. Health* 5:237–243.

Tátrai, E., Wojnárovits, I., and Ungváry, G. (1991) Non-fibrous zeolite induced experimental pneumoconiosis in rats. *Exp. Pathol.* 43:41–46.

Timblin, C. R., Guthrie, G. D., Janssen, Y. W., Walsh, E. S., Vacek, P., and Mossman, B. T. (1998) Patterns of c-fos and c-jun proto-oncogene expression, apoptosis, and proliferation in rat pleural mesothelial cells exposed to erionite or asbestos fibers. *Toxicol. Appl. Pharmacol.* 151:88–97.

Virta, R. L. (2010) Zeolites (advance release). In: *U.S. Geological Survey Minerals Yearbook—2009*, Available at http://minerals.usgs.gov/minerals/pubs/myb.html. Published October 2010.

Wagner, J.C., Berry, G., Skidmore J.W., and Timbrell, V. (1974) The effects of the inhalation of asbestos in rats. *Br. J. Cancer* 29:252–269.

Wagner, J. C., Skidmore, J. W., Hill, R. J., and Griffiths, D. M. (1985) Erionite exposure and mesotheliomas in rats. *Br. J. Cancer* 51:727–730.

MINERAL TALC

First aid: Inhalation: Move person to an area with fresh air. Eyes: Irrigate eyes immediately with water

Target organ(s): Occupational: lung; Cosmetic: female reproductive system

Occupational exposure limits: ACGIH: TLV = 2 mg/m^3, respirable fraction, 8-h TWA; OSHA PEL = 20 mppcf air over an 8-h workshift (for talc with <1% quartz content and no asbestos); NIOSH REL = 2 mg/m^3, respirable fraction, over a 10-h workshift; IDLH = 1000 mg/m^3

Reference values: None specified.

BACKGROUND AND USES

Types of Fibers and Their Uses

Mineral talc can be found in a variety of different forms, with the most common form consisting of nonfibrous plate-like particles. Although these plate-like particles are sometimes identified as fibers in microscopic analyses (IARC, 2010), normal or platiform talc is considered to be free of asbestiform fibers. True mineral fibers can be present in other forms of talc, either due to asbestos contamination (this form of talc is referred to as talc containing asbestos) or due to the presence of asbestiform fibers of talc (this talc is referred to as asbestiform talc). Note that asbestiform talc should not be confused with talc containing asbestos, and that the section of asbestos should be consulted regarding potential occupational exposures and health risks from talc containing asbestos. In addition, talc may also contain elongated mineral fragments that resemble fibers but are not asbestiform, with an example being the tremolitic talc found in New York State, which contains mineral fragments or cleavage fragments of nonasbestiform amphiboles such as tremolite and anthophyllite. Note that there remains some scientific disagreement regarding the classification of particles from the Gouverneur talc district of Upstate New York, with some concluding that the talc contains amphibole asbestos fibers (NIOSH, 2011a).

Mineral talc has a variety of common uses, given its many beneficial physical and chemical properties, which include its softness, purity, fragrance retention, whiteness, luster, moisture content, oil and grease adsorption, chemical inertness, low electrical conductivity, high dielectric strength, and high thermal conductivity (USGS, undated). In the United States, it is estimated that 562,000 tons of talc were sold in 2010, with major applications, including ceramics (23%), paint and paper (19% each), plastics (9%), roofing (7%), cosmetics (4%), rubber (3%), and other (16%) (Virta, 2011). Talc also has uses in construction materials (e.g., caulks, flooring, joint compounds), car parts (in ceramic substrates of catalytic converters, auto body putty, gaskets, hoses, belts), and in agricultural applications (e.g., fertilizer, seeds, insecticides, animal feed). Examples of everyday consumer products that may contain talc include pharmaceuticals, chewing gum, candy, and baby and body powders (USGS, undated; IARC, 2010).

General Epidemiology and Impact on Occupational Health

Based on a number of epidemiological studies of talc-exposed mine and mill workers, talc not containing asbestos or asbestiform fibers is generally regarded as a poorly soluble nuisance dust, which presents similar health hazards to particulate materials such as carbon black and titanium dioxide. It is well known that long-term exposures of miners and millers to elevated levels of airborne talc are associated with an increased risk of pulmonary fibrosis, also referred to as talc pneumoconiosis, as well as other forms of non-malignant respiratory disease, including pleural and lung function abnormalities. Inhaled talc that lacks asbestos or asbestiform fibers is not regarded as a human carcinogen, with the IARC re-affirming its Group 3 classification (inadequate evidence in humans) for talc not containing asbestos or asbestiform fibers in 2006, based on a lack of epidemiological evidence (IARC, 2010).

There remains scientific debate regarding the impact of occupational exposures to asbestiform talc on the risk of lung cancer and mesothelioma. It has been hypothesized that the presence of long, thin amphibole fibers in talc may elicit adverse health effects similar to those associated with asbestos, including mesothelioma and lung cancer (NIOSH, 2011a). Others question whether there are any mineral talc deposits containing sufficient quantities of asbestiform talc fibers to constitute a health hazard. While the talc deposits of the Gouverneur talc district of Upstate New York are among those that have been identified as containing asbestiform minerals (NIOSH, 2011a), Gamble and Gibbs (2008) recently concluded, based on a comprehensive literature review, that the epidemiologic evidence did not support a role of talc in the risk of lung cancer or mesothelioma among New York State talc workers.

Although of uncertain relevance to occupational health, there is mixed epidemiologic evidence linking perineal use of talc-containing products (e.g., body powders) with risk of ovarian cancer. Based on this epidemiologic evidence, IARC classified perineal use of talc-based body powder as a Group 2B, or possible, human carcinogen, during its 2006 review of talc carcinogenicity (IARC, 2010).

PHYSICAL AND CHEMICAL PROPERTIES

With an ideal formula of $Mg_3Si_4O_{10}(OH)_2$, mineral talc is a hydrous silicate material that contains magnesium (Mg), silicon and oxygen (SiO_2, silica), and water. Talc-rich rocks or mineral composites include agalite, potstone, soapstone, and talcite. Although generally very pure in composition, small amounts of impurities, including aluminum, iron, manganese, and titanium, are common. Other impurities can include fluorine, nickel, chromium, calcium, potassium, and sodium. Talc can have a variety of different colors depending on its composition, including white, green, and brown. Having a Mohs hardness of 1, talc is known as the softest mineral. Talc is also well known as a low solubility material. Pure talc has a molecular weight of 379.26 and a density of 2.58–2.83 (IARC, 2010).

As mentioned earlier, mineral talc is most commonly found in the form of plates. Generally consisting of thin tabular crystals that are up to 1 μm in width, these plates are described as massive, fine-grained, and compact (IARC, 2010). Although rarer, mineral talc can also be found in the form of long thin fibers (fibrous talc) or bundles of long, thin, flexible fibrils, which can be easily separated along grain boundaries between the fibrils (asbestiform talc). Talc can also contain mineral fibers due to asbestos contamination, but it is important to emphasize that talc-containing asbestos is not the same thing as asbestiform talc. Some talc deposits, including most notably those of the Gouverneur talc district of Upstate New York, have been shown to contain mineral fragments or cleavage fragments of nonasbestiform amphiboles such as tremolite and anthophyllite.

Importantly, the chemical and mineral compositions of talc can be highly variable between different geologic deposits, as well as within a single deposit (Table 96.1). This is due in part to differences in the geological mechanisms by which it can be formed, including from alteration of mafic and ultramafic rocks and from replacement of carbonate and sandy carbonates such as dolomite and limestone. Mineral talc is often found with a number of other minerals, including calcite, dolomite, magnesite, tremolite, anthophyllite, antigorite, quartz, pyrophyllite, micas, and chlorites. While industrial talcs can be highly variable in their talc content and amounts of other minerals, it is important to note that cosmetic talc typically contains >98% talc, while pharmaceutical-grade talc is >99% talc (IARC, 2010).

ENVIRONMENTAL FATE AND BIOACCUMULATION

Talc fibers are not expected to bioaccumulate, although some may be absorbed with massive ingested or inhaled doses. However, they are relatively biopersistent and can stay in the lung or elsewhere for prolonged periods of time (as reviewed by IARC, 2010).

ECOTOXICOLOGY

There are no known adverse ecotoxicologic effects of talc. The European Chemicals Bureau found no damageable effect of talc on the environment around large open pit talc mines. The Bureau also found no data to support possible effects of talc on environment, fauna, or flora (European Chemicals Bureau, 2000).

MAMMALIAN TOXICOLOGY

Acute and Subacute Effects

Animal Talc inhalation exposures of <1 month at concentrations as high as 18 mg/m3 have only been shown to produce mild inflammation and histopathologic changes limited to macrophage increases in the pulmonary tissue. For example, Pickrell et al. (1989) exposed F344/Crl rats and

TABLE 96.1 Variability in Mineral Composition of Talc from Different Locations

Mineral	Montana	Vermont	North Carolina	New York[a]	California	France
Talc	90–95	80–92	80–92	35 60	85–90	70–90
Tremolite	–	–	–	30–55	0–12	–
Anthophyllite	–	–	0–5	3–10	–	–
Serpentine	–	–	–	2–5	–	–
Quartz	<1	<1	1–3	1–3	<1	<1
Chlorite	2–4	2–4	5–7	–	–	10–30
Dolomite	1–3	1–3	2–4	0–2	0–3	–
Calcite	–	–	–	1–2	–	–
Magnesite	0–5	0–5	–	1–3	–	–

Source: From Harben and Kuzvart (1996).

[a]Gouverneur District.

B6C3F1 mice to asbestos-free talc aerosol concentrations of up to 20 mg/m3 (mass median aerodynamic diameter [MMAD], 3 µm), observing a modest, diffuse increase of talc-containing free macrophages in the alveolar spaces of mice and rats exposed to the highest concentration, but no clinical signs within 24 h after the end of exposure. The NTP (NTP, 1993) also conducted a 4-week exposure in which B6C3Fl mice and F344/N rats were exposed to 0, 6, or 18 mg/m3 talc. Most particles were in the respirable range for the animals and 90% of particle aspect ratios were between 1 and 1.41. Samples consisted primarily of talc, free of asbestos and SiO_2, with small quantities of chlorite and dolomite. NTP (1993) reported that inhaled talc concentrations above 18 mg/m3 of talc overwhelmed lung clearance mechanisms and impaired lung function in mice and rats after 4 weeks of exposure.

Human   There is a scarcity of reports of the adverse health effects of acute occupational and environmental exposures to talc in the literature, suggesting that such exposures do not generally pose significant health risks. However, some case reports are available for adverse effects in susceptible populations. For example, acute respiratory failure was noted in an infant (Patarino et al., 2010) and in an immunocompromised patient (Bandhakavi et al., 2010), both of whom had inhaled large quantities of talc powder.

Chronic and Subchronic Effects

Animal

Nonmalignant Disease   NTP (1993) exposed B6C3F$_l$ mice and F344/N rats to 0, 6, or 18 mg/m^3 talc (MMAD, ~3 µm) for over 24 months. Animals were examined physiologically and histopathologically at 6-, 11-, 18-, and 24-month intervals from the start of the study. This study reported time- and concentration-dependent reductions in several pulmonary function parameters in rats (NTP, 1993). The study also found concentration-dependent increases in indicators of pulmonary inflammation, cell proliferation, and fibrosis in male and female rats, but only evidence of chronic pulmonary inflammation without fibrosis in *mice* (NTP, 1993). Other studies have reported evidence of talc-dependent pulmonary fibrosis in rats with chronic inhalation exposures to talc (e.g., Wagner et al., 1977; for 10 mg/m^3 Italian 00000 grade talc, free from asbestos), but an absence of pulmonary histopathological changes in hamsters (e.g., Wehner et al., 1977; for 27.4 mg/m^3 total and 8.1 mg/m^3 respirable fraction concentrations of cosmetic-grade talc).

Malignant Disease   NTP (1993) exposed B6C3F$_l$ mice and F344/N rats to 0, 6, or 18 mg/m^3 talc (MMAD, ~3 µm) for over 24 months, finding no exposure-related tumors in mice. However, they reported "some evidence" of a talc effect on

increased benign or malignant pheochromocytoma (adrenal gland tumors) in male rats (tumors/animals in study: 26/49, 32/48, 37/47 for 0, 6, 18 mg/m^3 talc, respectively). Female rats were also reported to have talc-elevated rates of pheochromocytoma (tumors/animals in study: 13/48, 14/47, 23/49 for 0, 6, 18 mg/m^3 talc, respectively) and benign or malignant alveolar or bronchiolar adenoma or carcinoma (tumors/animals in study: 1/50, 0/48, 13/50 for 0, 6, 18 mg/m^3 talc, respectively). The NTP concluded that these findings provided sufficient evidence for carcinogenicity of talc, although others have questioned the relevance to humans of these findings. For example, in reviewing the NTP (1993) 2-year bioassay results, Oberdörster (1995) observed that lung overload was reached in both rats and mice at the talc exposure concentrations of 6 and 18 mg/m^3, and proposed that the relevancy of pulmonary tumors found only in female rats at the high exposure may be questionable for comparison to occupational exposures, since lung overload conditions frequently lead to tumor formation, which is independent of the nature of the exposure. In addition, Goodman (1995) concluded that the high rate of spontaneous pheochromocytomas in the control group precluded any reliable determination that these tumors were talc related. IARC (2010) also concluded that these tumors may not be related to talc since NTP studies have observed them in this rat strain in conjunction with other particulates.

Wagner et al. (1977) exposed Wistar rats intrapleurally with a single dose of 20 mg of either SFA chrysotile asbestos or Italian 00000 grade talc, free from asbestos, from an Italian mine. Following lifetime monitoring, no mesothelioma cases and only one small pulmonary adenoma were observed in 48 talc-exposed rats, while 18/48 of the chrysotile-exposed rats developed pleural mesothelioma. Moreover, as reviewed by IARC (2010); this finding is consistent with other studies that have administered up to 40 mg of talc intrapleurally and failed to observe a significant increase in occurrence of mesothelioma.

As illustrated in the following section, talc has surfaced as a potential risk factor for ovarian cancer among women engaging in perineal use of talc-based body powders. The NTP talc oral exposure study (NTP, 1993) found no talc-related lesions or tumors in exposed mice or rats, and a histopathological examination of ovarian tissue found no evidence of talc particles in this tissue in exposed rats (Boorman and Seely, 1995). The authors noted that there was ample opportunity for perineal and oral exposure to talc in these animals from fur and cage bars. Further, Hamilton et al. (1984) found that even injection of talc particles into rat ovary bursa did not produce ovarian tumors. Overall, these animal studies do not provide support for a causal linkage between talc exposure and risk of ovarian cancer.

Human

Nonmalignant Disease   A major issue in talc toxicology and epidemiology is the uncertainty surrounding its mineral characteristics. Four forms of pulmonary disease illustrate talc's diverse mineral characteristics and occurrence.

Talcosilicosis is caused by mined talc with high silica-content mineralogy, and its pulmonary findings are similar to those of silicosis.

Talcoasbestosis is produced by talc inhaled with asbestos fibers and is characterized by a similar set of pathologic and radiographic abnormalities as asbestosis (Bezerra et al., 2003).

Talcosis is caused by inhalation of pure talc and can manifest in bronchitis, interstitial inflammation, and pulmonary interstitial reticulations, or small, irregular nodules.

Intravenous administration of talc among drug users injecting talc-containing medications intended for oral use has been reported to contribute to a specific form of talc pulmonary disease characterized by vascular granulomas and pulmonary nodules, pleural effusions and lymph node enlargement (Lockey, 1981; Feigin, 1986; Marchiori et al., 2010).

Based on a body of epidemiologic studies, it is well known that long-term exposures of miners and millers to elevated levels of airborne talc are associated with an increased risk of pulmonary fibrosis, also referred to as talc pneumoconiosis, as well as other forms of non-malignant respiratory disease, including pleural and lung function abnormalities. For example, respiratory symptoms and lung function decrements were observed in a cohort of Vermont millers and miners of talc ore free of asbestos and silica (Wegman et al., 1982; Boundy et al., 1979). The authors reported talc effects that were independent of smoking, including lung function decrements and increases in the prevalence of cough and phlegm. The authors also observed an increased occurrence of small round and irregular opacities in the lung that were associated with length of employment and length of exposure to talc. Mean dust exposure levels were $1.8\,mg/m^3$ and generally lower than $3\,mg/m^3$ (Wegman et al., 1982). In a series of studies following a cohort of talc miners in France and Austria exposed to chlorite talc reported to be free from asbestiform fibers but containing small amounts of quartz, Wild et al. (1995, 2008) observed associations between talc exposures ranging from 8 to $20\,mg/m^3$ and prevalence of small pulmonary radiological opacities or lung function parameters, but a general absence of such associations for talc exposures below $2\,mg/m^3$. Wild et al. (2002) observed elevated rates of nonmalignant respiratory disease in a cohort of talc miners in France and Austria exposed to chlorite talc reported to be free from asbestiform fibers (OR/ 100 year-mg talc/m^3 = 1.08, 95% confidence interval (CI), CI: 1.02–1.16). Pneumoconiosis was significantly elevated in the French but not Austrian workers (OR = 5.56, 95% CI: 1.12–16.2).

Rubino et al. (1976) reported that talc miners and millers exposed to high purity talc in Piedmont, Italy had statistically significantly elevated rated of respiratory disease, particularly those miners exposed for ≥20 years to ≥1700 mppcf/ years. This finding was confirmed in a follow-up study of this cohort (Coggiola et al., 2003).

Honda et al. (2002) reported an increase in nonmalignant respiratory disease mortality among a cohort of talc miners and millers at a mining facility in Upstate New York (Honda et al., 2002). The authors noted a disease association with pulmonary fibrosis in both miners and millers and that the disease increased with duration of employment (Honda et al., 2002). The authors noted that the talc ore from this mine has high nonasbestiform amphibole content.

Malignant Disease

MESOTHELIOMA AND LUNG CANCER   Epidemologic studies do not generally support the linkage between talc exposure and mesothelioma (Honda et al., 2002; Selevan et al., 1979; Wild et al., 2002).

There is inconsistent epidemiologic evidence linking talc exposure and lung cancer. For example, studies of miners and millers exposed to pure talc in an Italian mine have not observed an increased risk of lung cancer deaths (Rubino et al., 1976; Coggiola et al., 2003). Wild et al. (2002) found no association between lung cancer mortality and talc exposure in a cohort of talc miners in France and Austria exposed to chlorite talc reported to be free from asbestiform fibers (OR/100 year-mg talc/m^3 = 0.98, 95% CI: 0.88–1.10). Chiazze et al. (1993) found no association between lung cancer or nonmalignant respiratory diseases and exposure to talc up to >1000 fibers/ml in workers at a U.S. fiberglass wool manufacturing facility, although the authors provided no information concerning the chemical characteristics of talc in this study.

Honda et al. (2002) observed a statistically significant increase in lung cancer mortality in miners and millers working at a talc mine in Upstate New York; however, as the authors specified, this mortality was limited to workers who had been working at the mine for approximately 1 year, but not longer. As the authors suggested, this finding supports a possible role of previous exposures not related to this talc mine. In addition, the authors noted that this mine has high nonasbestiform amphibole content in the talc ore, pointing to several animal and human studies that showed no association between these cleavage fragments and lung cancer. Likewise, Gamble (1993) reported no association between exposure to nonasbestiform talc cleavage fragments and lung cancer mortality in a cohort of Upstate New York talc miners. Steenland and Brown (1995) observed an association between exposure to nonasbestiform talc cleavage fragments and silica and increased lung cancer mortality among gold miners in South Dakota when county data were used for reference but not when federal data were used. However, the authors found that lung cancer mortality did not show a positive exposure-response trend with estimated cumulative dust exposure.

In Vermont, where talc is relatively free of asbestos and has small amounts of quartz, increased lung cancer risk has been reported among talc miners, but not millers; however, these associations have generally been attributed to radon and quartz exposure, and not talc, in the miners (Selevan et al., 1979; Baan et al., 2006).

OVARIAN CANCER A number of epidemiologic studies have investigated the association between cosmetic use of talc and ovarian cancer. The only cohort study (Gertig et al., 2000) to examine this association observed no statistically significant risk when "all serous ovarian cancers" were considered (relative risk (RR), RR = 1.26, 95% CI: 0.94–1.69) in 78,630 nurses in the Nurses' Health Study. However, the authors reported a statistically significant association for "serous invasive cancers" (RR = 1.40, 95% CI: 1.02–1.91). The larger number of case-control studies that have examined the association between talc use and ovarian cancer have reported mixed findings, but evidence of homogeneity in the magnitude of the observed risk. In reviewing findings from 20 of these case-control studies, Langseth et al. (2008) found a striking similarity in the outcomes of ovarian cancer among 14 population-based studies, with a pooled odds ratio of 1.40 (95% CI: 1.29–1.52), although a pooled odds ratio for the six hospital-based studies that does not support a significant talc-related risk (OR = 1.12, 95% CI: 0.92–1.36). The authors noted that studies conducted both before and after 1976, when talc became relatively free of asbestos, had similar outcomes, diminishing support for potential confounding by asbestos. An uncertainty in these studies is the potential contamination of talc by quartz, which has been shown to cause cancer or inflammatory reactions contributing to cancer occurrence. Moreover, these studies generally relied on questionnaires of cosmetic talc use without proper exposure assessment and seldom a complete history of use of talc. Overall, although there are some epidemiologic associations supporting an increased risk of ovarian cancer with perineal talc use, there remains some uncertainty regarding the causal nature of such an association.

Mode of Action(s)

In vitro experiments have shown that various samples of talc can produce cytotoxicity in mouse peritoneal macrophages (Davies et al., 1983), the release of chemokines and cytokines (Nasreen et al., 1998), and basic fibroblast growth factor (Antony et al., 2004) in human pleural mesothelial cells. These findings provide mechanistic support for the known fibrotic and inflammatory effects of talc in animals and humans exposed to sufficiently high concentrations.

SOURCES OF EXPOSURE

Geographical Distribution

Talc deposits can be found throughout the world, with the world's leading producers, including China, the United States, India, Finland, and France (Virta, 2010). World talc production is dominated by talc deposits derived from magnesium carbonates (e.g., dolomite, limestone), which are found in locations such as Austria (Leogen), Australia (Mount Seabrook and Three Springs), Brazil (Brumado), Canada (Madoc), China, France (Trimouns), Germany (Wunsiedel), India, Italy (Chisone Valley), the Republic of Korea, the Russian Federation (Onot, Krasnoyarsk), Slovakia (Gemerska Poloma), Spain (Respina), and the US (Alabama-Winterboro, California-Talc City and Death Valley-Kingston Range, Georgia-Chatsworth, Montana-Yellowstone, New York-Gouverneur District, North Carolina-Murphy Marble belt, Washington-Metaline Falls, and western Texas) (IARC, 2010). In the United States, large talc deposits are particularly prevalent in the Appalachian and Piedmont regions, from Vermont to Alabama, as well as in the western U.S. (e.g., Montana, California, Nevada, Washington, Idaho, and New Mexico) and Texas.

As discussed in the Physical and Chemical Properties section, the mineral and chemical composition of talc can vary considerably among geographic regions. For example, some New York talc has been shown to contain up to 55% by weight tremolite, while talc from other geographic locations contains no detectable levels of tremolite. Also, French talc is estimated to contain up to 30% chlorite, while other talc deposits have been shown to contain little or no chlorite.

Industries and Populations at Risk

The primary route of occupational exposure to mineral talc is by inhalation of particles and/or fibers released during mining and handling (e.g., crushing, separating, bagging, loading) of talc during its production and during its industrial use in the manufacture of a variety of commercial products. User industries with potential occupational exposures to talc include those involved in the manufacture of paints, paper, plastics, ceramics, roofing and other construction materials (e.g., caulks, flooring, joint compounds), rubber, some agricultural applications (e.g., fertilizers, seeds, insecticides, animal feed), and cosmetics and pharmaceuticals (see Background and Uses section).

Exposure to talc may also occur *via* consumer use of talc-containing products. Talc-based body powders are considered to present the most significant source of consumer exposures to talc given that talc in other product classes is generally bound within a product matrix (e.g., ceramics, paint, paper, construction materials). IARC (2010) summarized prevalence data of ever use of talc for feminine hygiene purposes in various countries; these range from 1.8% in in the United States (Washington, D.C.) to 59% in the United Kingdom.

Exposure Concentrations

Gravimetric measurements of respirable dust in talc mining and milling environments are available from a number of studies (ACGIH, 2010; IARC, 2010). These studies have reported data indicating elevated historical exposure levels of respirable dust in talc operations, in particular for European talc mills and mines (IARC, 2010). For example, in an epidemiologic study of French talc quarry and mill workers, Wild et al. (2002) reported dust levels for the 1960s and 1970s ranging from below $5 \, mg/m^3$ to more than $30 \, mg/m^3$. More recent studies have generally reported reduced respirable dust levels, with Wild et al. (2008) reporting geometric mean exposure concentrations for respirable talc dust of $0.80 \, mg/m^3$ (geometric standard deviation (GSD): 4.3) and $0.30 \, mg/m^3$ (GSD: 3.3) for the years 2003 and 1996–2003 in a French mill and an Austrian mill, respectively. Wild et al. (2008) also reported a decrease from 0.67 to $0.37 \, mg/m^3$ in exposure levels of respirable talc dust at a French mine for the period 1990–2003.

Studies of U.S. talc mines and mills have generally reported respirable dust levels of less than a few mg/m^3. For example, for a survey of 362 personal samples collected over a full shift at U.S. talc mines and mills by the Mine Safety and Health Administration, the NIOSH reported a median dust exposure level of $1.20 \, mg/m^3$, with 90% of all exposure levels being $<2.78 \, mg/m^3$ (IARC, 2010). For a study of Vermont talc miners and millers, Wegman et al. (1982) reported a geometric mean respirable dust level of $1.8 \, mg/m^3$ based on 312 personal samples. For a respiratory survey of nearly 300 talc miners and millers in three states (Montana, Texas, and North Carolina), Gamble et al. (1982) reported average talc exposures of $1.2 \, mg/m^3$ (Montana), $2.6 \, mg/m^3$ (Texas), and $0.3 \, mg/m^3$ (North Carolina). Based on analysis of bulk samples by TEM, Gamble et al. (1982) reported an absence of any fibers in samples of Montana talc, but evidence of antigorite and tremolite fibers in the Texas talc. In addition, the North Carolina talc was found to contain particles with length:diameter ratios as high as 100:1, although it was described as acicular in nature, with fragmentation of platy particles being a possible source of the high aspect ratio particles (ACGIH, 2010; IARC, 2010). ACGIH (2010) reported that none of the observed fibers were identified as being asbestiform by the study authors.

Fewer exposure measurement data are available for talc user industries (IARC, 2010), with the limited data that are available being historical in nature and unlikely to be representative of current exposure levels.

INDUSTRIAL HYGIENE

For talc *not* containing asbestos, the following occupational exposure limits are in place:

- ACGIH: TLV $= 2 \, mg/m^3$, respirable fraction, 8-h TWA
- OSHA PEL $= 20 \, mppcf$ air over an 8-h workshift (for talc with $<1\%$ quartz content and no asbestos)
- NIOSH REL $= 2 \, mg/m^3$ over a 10-h workshift; IDLH $= 1,000 \, mg/m^3$

The following maximum-use concentrations and protective respirators are recommended by NIOSH (2011b):

- $10 \, mg/m^3$: Any quarter-mask respirator
- $20 \, mg/m^3$: Any particulate respirator equipped with an N95, R95, or P95 filter or facepiece, except quarter-mask respirators
- $50 \, mg/m^3$: Any powered, air-purifying respirator with a high efficiency particulate filter
- $100 \, mg/m^3$: Any air-purifying, full-facepiece respirator with an N100, R100, or P100 filter
- $1000 \, mg/m^3$: any supplied-air respirator operated in a pressure-demand or other positive-pressure mode

For talc containing asbestos, asbestos limits and industrial hygiene practices should be followed per 29 CFR 1910.1001 (also described in the asbestos section of this chapter).

RISK ASSESSMENTS

IARC

The IARC has reviewed talc on three occasions, each time making determinations for different variations of talc:

1987: Talc not containing asbestiform fibers (Group 3) and talc containing asbestiform fibers (Group 1). The Group 1 classification for talc containing asbestiform fibers is based on early studies of New York talc miners and millers that appear to have made claims of exposures to talc containing asbestiform tremolite and anthophyllite, resulting in some uncertainty as to whether these studies provide findings of relevance to talc contaminated with asbestos—i.e., asbestiform amphibole minerals—rather than "asbestiform talc."

2006: Inhaled talc not containing asbestos or asbestiform fibers (Group 3) and perineal use of talc-based body powder (Group 2B) (Baan et al., 2006).

2009: Based on the article by Straif et al. (2009); it appears that talc containing asbestos will be classified as Group 1 while there is no mention of talc containing asbestiform fibers, which appears to have been dropped from consideration in this review due to confusion as to how to define asbestiform fibers and as to whether New York State talc contained them.

NTP

Talc was nominated for the 12th Report on Carcinogens (12th RoC) by the National Institute of Environmental Health Sciences for both cosmetic talc use and occupational exposure to talc. However, the NTP deferred consideration of listing talc (asbestiform and nonasbestiform talc) in the RoC based on its review in the year 2000 that identified persisting confusion over the mineral nature and consequences of exposure to talc, both containing asbestiform fibers and not containing asbestiform fibers.

REFERENCES

American Conference of Governmental Industrial Hygienists (ACGIH) (2010) *Talc. Documentation of the Threshold Limit Values for Chemical Substances*, 7th ed. Cincinnati, Ohio: ACGIH.

Antony, V.B., Nasreen, N., Mohammed, K.A., Sriram, P.S., Frank, W., Schoenfeld, N., and Loddenkemper, R. (2004) Talc pleurodesis: basic fibroblast growth factor mediates pleural fibrosis. *Chest* 126:1522–1528.

Baan, R., Straif, K., Grosse, Y., Secretan, B., El Ghissassi, F., Cogliano, V., and WHO International Agency for Research on Cancer (IARC) Monograph Working Group. (2006) Carcinogenicity of carbon black, titanium dioxide, and talc. *Lancet Oncol.* 7:295–296.

Bandhakavi, V., Leke-Tambo, A., Moizuddun, M., Rayasam, N., Kaleekal, T., and Majumdar, T. (2010) Acute pneumonitis and respiratory failure due to inhaled talc in a patient with human immunodeficiency virus infection. *Chest* 138:11A.

Bezerra, O.M., Dias, E.C., Galvão, M.A., and Carneiro, A.P. (2003) [Talc pneumoconiosis among soapstone handicraft workers in a rural area of Ouro Preto, Minas Gerais, Brazil]. *Cad. Saude Publica* 19:1751–1759.

Boorman, G.A. and Seely, J.C. (1995) The lack of an ovarian effect of lifetime talc exposure in F344/N rats and B6C3F1 mice. *Regul. Toxicol. Pharmacol.* 21:242–243.

Boundy, M.G., Gold, K., Martin, K.P., Burgess, W.A., and Dement, J.M. (1979) Occupational exposures to non-asbestiform talc in Vermont. Dusts and disease. In: Lemen, R. and Dement, J.M., editors. *Proceedings of the Conference on Occupational Exposures to Fibrous and Particulate Dust and Their Extension into the Environment*, Park Forest South Il: Pathotox Publishers, Inc.,

Chiazze, L. Jr., Watkins, D.K., Fryar, C., and Kozono, J. (1993) A case-control study of malignant and non-malignant respiratory disease among employees of a fiberglass manufacturing facility. II. Exposure assessment. *Br. J. Ind. Med.* 50:717–725.

Coggiola, M., Bosio, D., Pira, E., Piolatto, P.G., La Vecchia, C., Negri, E., Michelazzi, M., and Bacaloni, A. (2003) An update of a mortality study of talc miners and millers in Italy. *Am. J. Ind. Med.* 44:63–69.

Davies, R., Skidmore, J.W., Griffiths, D.M., and Moncrieff, C.B. (1983) Cytotoxicity of talc for macrophages *in vitro. Food Chem. Toxicol.* 21:201–207.

European Chemicals Bureau (2000) *IUCLID Dataset, Talc (14807-96-6)*. Available at http://esis.jrc.ec.europa.eu/doc/existing-chemicals/IUCLID/data_sheets/14807966.pdf. Published February 18, 2000.

Feigin, D.S. (1986) Talc: understanding its manifestations in the chest. *AJR Am. J. Roentgenol.* 146:295–301.

Gamble, J., Greife, A., and Hancock, J. (1982) An epidemiological-industrial hygiene study of talc workers. *Ann. Occup. Hyg.* 26:841–859.

Gamble, J.F. (1993) A nested case control study of lung cancer among New York talc workers. *Int. Arch. Occup. Environ. Health* 64:449–456.

Gamble, J.F. and Gibbs, G.W. (2008) An evaluation of the risks of lung cancer and mesothelioma from exposure to amphibole cleavage fragments. *Regul. Toxicol. Pharmacol.* 52:S154–S186.

Gertig, D.M., Hunter, D.J., Cramer, D.W., Colditz, G.A., Speizer, F.E., Willett, W.C., and Hankinson, S.E. (2000) Prospective study of talc use and ovarian cancer. *J. Natl. Cancer Inst.* 92:249–252.

Goodman, J.I. (1995) An analysis of the National Toxicology Program's (NTP) Technical Report (NTP TR 421) on the toxicology and carcinogenesis studies of talc. *Regul. Toxicol. Pharmacol.* 21:244–249.

Hamilton, T.C., Fox, H., Buckley, C.H., Henderson, W.J., and Griffiths, K. (1984) Effects of talc on the rat ovary. *Br. J. Exp. Pathol.* 65:101–106.

Harben, P.W. and Kuzvart, M. (1996) Talc and soapstone. In: Harben, P.W. and Kuzvart, M., editors. *Industrial Minerals: A Global Geology*, Metal Bulletin PLC, London: Industrial Minerals Information Ltd., pp. 407–417.

Honda, Y., Beall, C., Delzell, E., Oestenstad, K., Brill, I., and Matthews, R. (2002) Mortality among workers at a talc mining and milling facility. *Ann. Occup. Hyg.* 46:575–585.

International Agency for Research on Cancer (IARC) (2010) *Monographs on the Evaluation of Carcinogenic Risks to Humans: Carbon Black, Titanium Dioxide, and Talc*, vol. 93, Lyon, France: World Health Organization.

Langseth, H., Hankinson, S.E., Siemiatycki, J., and Weiderpass E. (2008) Perineal use of talc and risk of ovarian cancer. *J. Epidemiol. Community Health* 62:358–60.

Lockey, J.E. (1981) Nonasbestos fibrous minerals. *Clin. Chest Med.* 2:203–218.

Marchiori, E., Lourenço, S., Gasparetto, T.D., Zanetti, G., Mano, C.M., and Nobre, L.F. (2010) Pulmonary talcosis: imaging findings. *Lung* 188:165–171.

Nasreen, N., Hartman, D.L., Mohammed, K.A., and Antony, V.B. (1998) Talc-induced expression of C-C and C-X-C chemokines and intercellular adhesion molecule-1 in mesothelial cells. *Am. J. Respir. Crit. Care Med.* 158:971–978.

National Toxicology Program (1993) NTP Toxicology and Carcinogenesis Studies of Talc (CAS No. 14807-96-6)(Non-Asbestiform) in F344/N Rats and B6C3F1 Mice (Inhalation Studies). *Natl. Toxicol. Program Tech. Rep. Ser.* 421:1–287.

National Institute for Occupational Safety and Health (NIOSH) (2011a) Asbestos fibers and other elongate mineral particles: state of the science and roadmap for research (revised edition). *Curr. Intell. Bull.* 62, April. Available at http://www.cdc.gov/niosh/docs/2011-159/.

National Institute for Occupational Safety and Health (NIOSH) (2011b) *NIOSH Pocket Guide to Chemical Hazards*, Centers for Disease Control and Prevention Web site. Available at http://www.cdc.gov/niosh/npg/npgd0584.html (accessed on June 9, 2011).

Oberdörster, G. (1995) The NTP talc inhalation study: a critical appraisal focused on lung particle overload. *Regul. Toxicol. Pharmacol.* 21:233–241.

Patarino, F., Norbedo, S., Barbi, E., Poli, F., Furlan, S., and Savron, F. (2010) Acute respiratory failure in a child after talc inhalation. *Respiration* 79:340.

Pickrell, J.A., Snipes, M.B., Benson, J.M., Hanson, R.L., Jones, R.K., Carpenter, R.L., Thompson, J.J., Hobbs, C.H., and Brown, S.C. (1989) Talc deposition and effects after 20 days of repeated inhalation exposure of rats and mice to talc. *Environ. Res.* 49:233–245.

Rubino, G.F., Scansetti, G., Piolatto, G., and Romano, C.A. (1976) Mortality study of talc miners and millers. *J. Occup. Med.* 18:187–193.

Selevan, S.G., Dement, J.M., Wagoner, J.K., and Froines, J.R. (1979) Mortality patterns among miners and millers of non-asbestiform talc: preliminary report. *J. Environ. Pathol. Toxicol.* 2:273–284.

Steenland, K. and Brown, D. (1995) Mortality study of gold miners exposed to silica and nonasbestiform amphibole minerals: an update with 14 more years of follow-up. *Am. J. Ind. Med.* 27:217–229.

Straif, K., Benbrahim-Tallaa, L., Baan, R., Grosse, Y., Secretan, B., El Ghissassi, F., Bouvard, V., Guha, N., Freeman, C., Galichet, L., and Cogliano, V. (2009) A review of human carcinogens—part C: metals, arsenic, dusts, and fibres. *Lancet Oncol.* 10:453–454.

USGS, Undated. *"Industrial Minerals of the United States: U.S. Talc—Baby Powder and Much More."* p. 2.

Virta, R.L. (2010) Zeolites (advance release). In: *U.S. Geological Survey Minerals Yearbook—2009*, Available at http://minerals.usgs.gov/minerals/pubs/myb.html. Published October 2010.

Virta, R.L. (2011) Talc and pyrophyllite. *In Mineral Commodity Summaries*, pp. 160–161. Available at http://minerals.usgs.gov/minerals/pubs/mcs/2011/mcs2011.pdf. Published January 2011.

Wagner, J.C., Berry, G., Cooke, T.J., Hill, R.J., Pooley, F.D., and Skidmore, J.W. (1977) Animal experiments with talc. In: Walton,W.H. and McGovern,B., editors. *Inhaled Particles*, vol. IV, Oxford, UK: Pergamon Press, Part 2, pp. 647–654.

Wegman, D.H., Peters, J.M., Boundy, M.G., and Smith, T.J. (1982) Evaluation of respiratory effects in miners and millers exposed to talc free of asbestos and silica. *Br. J. Ind. Med.* 39:233–238.

Wehner, A.P., Zwicker, G.M., and Cannon, W.C. (1977) Inhalation of talc baby powder by hamsters. *Food Cosmet. Toxicol.* 15:121–129.

Wild, P., Réfrégier, M., Auburtin, G., Carton, B., and Moulin, J.J. (1995) Survey of the respiratory health of the workers of a talc producing factory. *Occup. Environ. Med.* 52:470–477.

Wild, P., Leodolter, K., Réfrégier, M., Schmidt, H., Zidek, T., and Haidinger, G. (2002) A cohort mortality and nested case-control study of French and Austrian talc workers. *Occup. Environ. Med.* 59:98–105.

Wild, P., Leodolter, K., Réfrégier, M., Schmidt, H., and Bourgkard, E. (2008) Effects of talc dust on respiratory health: results of a longitudinal survey of 378 French and Austrian talc workers. *Occup. Environ. Med.* 65:261–267.

97

NANOPARTICLES

Daniel A. Newfang, Giffe T. Johnson, and Raymond D. Harbison

BACKGROUND AND USES

Nanoparticles are currently defined in the literature as particles that have a diameter of 100 nanometers or less (although the scientific community has not collectively agreed with this definition). Nanomaterials or Engineered Nanomaterials (ENMs) are similarly defined as a man-made material having particles *or constiuents* in nanoscale dimentions. Nanotechnology is the process and science used to build or manipulate nanomaterials. This chapter will use the term nanoparticle, nanomaterial and nanotechnology interchangeably. Regardless of their creation (i.e. naturally occurring or man-made), they have a large surface area to volume ratio that enhances the nanoparticle's ability to interact with other molecules. Nanomaterials exist naturally in the environment, for examples, viruses, bacteria, and byproducts of nature, such as ash from volcanoes. They may also be a result from industrial in the forms of byproducts of combustion, for example, soot and air pollution. Nanoparticles may be present as gases, liquids, or dust.

Nanoparticles may enter the body via respiratory, oral, dermal routes, as well as may enter the body by intravascular penetration via dermal injections (Pompa et al., 2011). Nanoparticles may even be introduced into the body by mucous membranes and ocular routes. They can stay airborne longer, which increases exposure times and penetrate the dermal layers, which may enhance their ability to be toxic (Madl and Pinkerton, 2009). Nanoparticles are currently being used in technology, such as medical imaging, electronics, and therapeutics (Madl and Pinkerton, 2009). In the medical setting, nanoparticles are being used in both *in vivo* and *in vitro* research. In the occupational setting, nanomaterials may act as chemical catalysts and some may be considered to be highly combustible and toxic.

MAMMALIAN TOXICOLOGY

Nanomaterials are considered to be free unbound particles that are capable of mobility in the environment and in the body (Elaesser and Howard, 2012). The toxicity of a nanoparticle depends on its size, shape, surface charge, chemistry composition, and stability. A toxicological concern is that the immune system may not able to detect the presence of nanoparticles (Pompa et al., 2011). As well, there are difficulties in determining the toxic effects of nanomaterials because of impurities or other components present in the materials. Nanoparticles have an increased reactivity due to higher surface to volume ratio (Elaesser and Howard, 2012). Once in the environment, nanoparticles interact with all aspects of air, land, and water.

In vitro nanotoxicology testing methods, due to the small size of the particles, uses endocytosis to evaluate the distribution of effects from cell to cell (Elaesser and Howard, 2012). Depending on reactivity, nanoparticles generate reactive oxygen species leading to the disruption of membranes, protein aggregation, and DNA damage potentially inducing apoptosis or necrosis (Elaesser and Howard, 2012).

Nanoparticles have an effect on epithelial cells. There are chemotactic properties on the skin that generate an inflammatory response. There are nanoparticles that stimulate epithelial cells such as carbon black nanoparticles that are byproducts of incomplete combustion of petroleum products. The central nervous system is susceptible to injury by nanoparticles due to their small size, they are able to penetrate the

Hamilton & Hardy's Industrial Toxicology, Sixth Edition. Edited by Raymond D. Harbison, Marie M. Bourgeois, and Giffe T. Johnson.
© 2015 John Wiley & Sons, Inc. Published 2015 by John Wiley & Sons, Inc.

tight junctions of the blood–brain barrier and enter the central nervous system.

SOURCES OF EXPOSURE

Occupational

Nanotechnology is the use of nanomaterials to create new chemical structures, materials, and devices. Quantum dots, nanotubes are currently being investigated for use in technology. In addition, other nanoparticles such as titanium dioxide, and silver are being used in commercial products. Nanoparticles are being used to deliver medications to specific organs to reduce systematic adverse medication reactions and increase the potency of medications (Yildirimer et al., 2011).

Quantum dots are used in medical applications; because of their fluorescent properties they are able to visualize cancer cells in the body. Quantum dots are metalloid crystalline cores covered with a shell (Madl and Pinklerton, 2011). Caution should be used because they may be considered toxic due to size and electronic properties as in the case of cadmium and zinc. However, if coated with a polymer of polyethylene glycol, a quantum dot creates a barrier that is not recognized by the body, thus not initiating an inflammatory response (Stone et al., 2007). Quantum dots are being used to deliver medicals such as antibiotics and receptor ligands to specific body sites (Madl and Pinkerton, 2009). Quantum dots can penetrate the skin and may induce an irritant or contact dermatitis (Crosera et al., 2009). For Occupational Safety and Health Administration (OSHA) specific permissible exposure levels (PEL), use those for specific minerals, such as cadmium levels.

Carbon nanotubes are carbon-based nanoparticles that are hollow graphite tubes and are being investigated for use in cell growth, tumor imaging, and drug delivery as well as in industries (Madl and Pinkerton, 2009). Workers are exposed to carbon nanotubes via inhalation. Carbon nanotubes and nanofibers may cause pulmonary inflammation and fibrosis. The OSHA recommends that exposure levels do not exceed $1.0\,\mu g/m^3$ in an 8-h time weighted, as recommended by National Institute for Occupational Safety and Health (NIOSH).

Titanium dioxide nanoparticles are currently being studied because of their ability to study nanoparticles' influence on biological fate and toxicity, for example, in pulmonary disease (Madl and Pinkerton, 2009). Titanium dioxide can also penetrate the dermal layers and cause an irritant and contact dermatitis (Crosera et al., 2009). Titanium dioxide is used in sunscreen to absorb UV light. Titanium dioxide is also used in paint, paper, as well as foods. The OSHA recommends exposure levels not to exceed $0.3\,\mu g/m^3$. The NIOSH recommends exposure levels do not exceed $2.4\,\mu g/m^3$. Titanium dioxide is considered a potential occupational carcinogen.

Silver nanoparticles have antimicrobial abilities and are used in wound dressings. However, silver dust is considered harmful if inhaled, as well as oral ingestion that causes argyria and may cause liver damage or dermal exposure may cause transitional damage to the stratum corneum (Yildirimer et al., 2011). The OSHA PEL for silver dust and fumes is (including the metal dust and fume) is $0.01\,mg/m^3$, as Ag. The NIOSH does not have a recommended exposure level. The ACGIH has established a $0.1\,mg/m^3$ TLV for silver metal dust and fume.

Gold may be used as a carrier molecule because of its biocompatibilities (Yildirimer et al., 2011). Gold nanoparticles may penetrate the skin, may cause an irritant dermatitis. If inhaled, gold may induce pulmonary irritation.

Silica nanoparticles may be used as biomarkers, cancer therapeutics, and drug delivery vehicles. For OSHA permissible exposure levels, follow silica protocols.

As a final paragraph in this section, it needs to be emphasized that future-use prospects, global demands for innovative products, novel and inexpensive manufacturing processes will present new challenges for assessment, exposure monitoring, fate analysis & risk evaluation *and* what their impacts on humans and the environment will be. For example, the simple act of quantifying background concentrations from a work process (e.g. near & far field sampling) can be extremely challenging with nanomaterials. The future is exciting and the authors believe the benefits will be maximized with regard to nanomaterial design, manufacturing and use. A good strategy to follow for managing potential exposure to nanomaterials is to continue classical industrial hygiene and medical practices, keep up-to-date with guidance documents, identify the potential exposure group (for future epidemiological cohort reference/studies).

INDUSTRIAL HYGIENE

Elimination of nanoparticle exposure is optimal; however, may not be possible. Engineering controls consist of controlling aerosols such as gases, dust, chemical vapors that require local exhaust ventilation with filtration. A high efficiency particular air filter is effective in removing nanoparticles in the workplace. Administrative practices consist of good personal hygiene such as hand washing and using proper procedures in the workplace. It may be appropriate to use a wet wiping technique with solvent to solubilize the nanoparticle. Safety Data Sheets (SDS) may be available for specific compounds. Currently, data are not conclusive as per the actual workplace exposure of workers to specific nanoparticles (Brouwer, 2010).

Proper selection of personal protective equipment is required and may consist of gloves, clothing, and respirators.

Personal protective equipment should be resistant to nanoparticle penetration. Respirators may consist of filters particles of 300 nanometers or less. Depending on the nanoparticles, N-95/N-100 respirators may be adequate.

When cleaning a worksite that had been contaminated by nanoparticles, place absorbents, all towels, and other materials used to clean, place in a plastic bag, seal, then place in a double bag and label in accordance with the National Institute of Health, Division of Environmental Protection (EPA, 2013).

REFERENCES

Bakand, S., Hayes, A., and Dechasakulthorn, F. (2012) Nanoparticles: a review of particle toxicology following inhalation exposure. *Inhal. Toxicol.* 24(12):125–135.

Brouwer, D. (2010) Exposure to manufactured nanoparticles in different workplaces. *Toxicology* 269:120–127.

Crosera, M., Bovenzi, M., Maina, G., Adami, G., Zanette, C., and Florio, C. (2009) Nanoparticle dermal absorption and toxicity: a review of the literature. *Int. Arch. Occup. Environ. Health* 82:1043–1055.

Elsaesser, A. and Howard, V. (2012) Toxicology of nanoparticles. *Adv. Drug Deliv. Rev.* 64(2012):129–137.

Environmental Protection Agency. (2013) *Environmental Protection, Waste Deposal.* Available at http://orf.od.nih.gov/Environmental+Protection/Waste+Disposal (accessed 2014)

Madl, A. and Pinkerton, K. (2009) Health effects of inhaled engineered and incidental nanoparticles. *Crit. Rev. Toxicol.* 39(8): 629–658.

Pompa, G., Vecchio, A., Galeone, V., Brunetti, G., Maiorano, S., Sabella, S., and Cingolani, R. (2011) Physical assessment of toxicology at nanoscale: nano dose-metrics and toxicity factor. *Nanoscale* 3:2889–2897.

Stone, V., Johnston, H., and Clift, M. (2007) Air pollution, ultrafine, and nanotoxicology: cellular and molecular interactions. *IEEE Trans. Nanobiosci.* 6(4): doi 10.1109/TNB.2007.909005.

Yildirimer, L., Thanh, N., Loizidou, M., and Seifalian, A. (2011) Toxicology and clinical potential of nanoparticles. *Nano Today* 6(6):585–607.

SECTION VIII

PHYSICAL AGENTS

SECTION EDITOR: MARIE BOURGEOIS

98

RADIO-FREQUENCY RADIATION

R. Timothy Hitchcock

BACKGROUND AND USES

Radio-frequency radiation (RFR) is a form of nonionizing, electromagnetic energy. Electromagnetic energy is the propagation of energy through space or matter by time-varying electric and magnetic fields. These energies are distributed across the electromagnetic spectrum, which is characterized by three related quantities: wavelength, frequency, and energy. The continuum is divided into various spectral regions. By convention, RFR is described by frequency, in hertz (Hz). The RFR part of the electromagnetic spectrum extends from 300 gigahertz (GHz) down to 3 kilohertz (kHz). Usually, microwave radiation is considered to be a subset of RFR between 300 GHz and 300 megahertz (MHz).

Other nomenclature is used for RFR at times. For example, in the global community, frequencies between 300 Hz and 10 MHz are sometimes called the intermediate frequency (IF) range. A professional society, the American Conference of Governmental Industrial Hygienists (ACGIH), that publishes exposure guidelines, including those for RFR, includes a portion of the RFR spectrum (3–30 kHz) in a region they call the sub-RF spectral region.

The RF spectrum is further divided into bands, which are order-of-magnitude ranges with the designations described in Table 98.1. Frequencies in the various bands are allocated for uses such as aeronautical radio, broadcasting, citizens band radio, and cellular communications. The band that includes 30–300 Hz is the extremely low frequency (ELF) band. The ELF band contains the frequencies at which electric power is generated, transmitted, distributed, and used. This is 60 Hz in North America and 50 Hz in the rest of the world.

PHYSICAL PROPERTIES

Radio-frequency electromagnetic energies are considered nonionizing because the photon energies are well below the 12 eV needed to ionize water molecules. The energy is electromagnetic because it is characterized by two fields, the electric and the magnetic.

Electromagnetic quantities and units used are shown in Table 98.2 where it is noted if the quantity is a basic restriction or a derived limit. Basic restrictions are essentially dosimetric quantities that are used to derive the exposure limits. The basic restrictions include *in situ* electric field strength, specific absorption rate (SAR), specific absorption (SA), and power density. Power density is both a basic restriction and derived limit.

The SAR, which is the RF dose rate, and power density are basic restrictions for the portion of the RFR spectrum that can produce adverse heating in tissues, 300 GHz to 100 kHz (IEEE, 2005a). Power density is used for the highest frequencies, 300 GHz to 3 GHz, and SAR applies from 3 GHz to 100 kHz. The time integral of the SAR is the SA or the RF dose.

The *in situ* electric field strength is the basic restriction from 5 MHz to 3 kHz. This basic restriction applies to electric fields within the human body. It is not to protect from adverse heating but from a family of potential effects called electrostimulation (e.g., excitation of muscle, cardiac tissue, or neurons). There is a region of overlap where both SAR and *in situ* electric field strength apply, 5 MHz to 100 kHz.

The derived limits are in terms of power density, electric field strength, and magnetic field strength. Power density is

Hamilton & Hardy's Industrial Toxicology, Sixth Edition. Edited by Raymond D. Harbison, Marie M. Bourgeois, and Giffe T. Johnson.
© 2015 John Wiley & Sons, Inc. Published 2015 by John Wiley & Sons, Inc.

TABLE 98.1 Nomenclature of Band Designations

Frequency Range	Designation	Abbreviation
<30 Hz	Sub-extremely low frequency	Sub-ELF
30–300 Hz	Extremely low frequency	ELF
300–3000 Hz	Voice frequency	VF
3–30 kHz	Very low frequency	VLF
30–300 kHz	Low frequency	LF
300–3000 kHz	Medium frequency	MF
3–30 MHz	High frequency	HF
30–300 MHz	Very high frequency	VHF
300–3000 MHz	Ultrahigh frequency	UHF
3–30 GHz	Superhigh frequency	SHF
30–300 GHz	Extremely high frequency	EHF

TABLE 98.2 Important Electromagnetic Quantities with Their Symbols and Units

Quantity	Symbol	Units	Unit Abbreviation
Frequency	f	Hertz	Hz (GHz, MHz, kHz)
Energy	Q	Joule	J
Power	P	Watt	W
Electric field strength	E	Volt/meter	V/m, V^2/m^2
Magnetic field strength	H	Ampere/meter	A/m, A^2/m^2
Magnetic flux density	B	Tesla	T (μT, mT)
Power density[a]	W, S	Watt/square meter	W/m^2, mW/cm^2
In situ electric field[b]	E_o	Volt/meter	V/m
Specific absorption rate[b]	SAR	Watts/kilogram	W/kg
Specific absorption[b]	SA	Joules/kilogram	J/kg

[a]Power density is both basic restriction and derived limit.
[b]Basic restrictions.

the power incident on a surface, divided by the area of that surface. Electric and magnetic field strengths are vector quantities that are used to relate the intensity of the two fields. The potential for biological effects is associated with the power deposition, and the squares of the electric-and magnetic field strengths are proportional to power. Hence, the squares of the field strength units (V/m, A/m) are often used in biological sciences. Radio-frequency magnetic fields may sometimes be described in terms of the magnetic flux density, B. B is related to the magnetic field strength where $B = \mu H$, where μ is the permeability of free space, 1.26×10^{-6} T(m/A). The SI unit of magnetic flux density is tesla (T), although the cgs unit, gauss (G), may be used in the United States. The conversion factor is 1 T = 10,000 G.

MAMMALIAN TOXICOLOGY

Human data are limited and present no clear trends. Therefore scientists have had to rely on animals as models to establish the biological effects. Animal studies have found effects in most major animal systems, including the nervous, neuroendocrine, reproductive, immune, and sensory systems.[1] Combined interactions of RF fields with neuroactive drugs and chemicals have been reported (Lai et al., 1987; Frey and Wesler, 1990).

Acute Effects

Behavior: The effects established in test animals have been extrapolated to human beings and used in setting human exposure guidelines for thermal effects. Currently, the exposure criteria are based on a few well-established effects observed in studies with test animals. Reversible behavioral disruption in short-term studies is an effect often cited in the exposure guidelines. This is because this end point has been found to be a sensitive measure of RF exposure and has been demonstrated in different laboratories, at various frequencies, and with more than one animal species (DeLorge, 1976, 1978; D'Andrea et al., 1977, 2003). Generally, behavioral effects are a thermal effect resulting from significant increases in body temperature caused by the absorption of RF energy (DeLorge, 1983; D'Andrea, 1991; D'Andrea et al., 2003).

East European and Russian literature discusses the occurrence of certain nonspecific symptoms associated with the nervous system, with clinical signs extending to the cardiovascular system (Sadcikova, 1974). The signs and symptoms include headache, nervousness, dizziness, memory loss, fatigue, irritability, insomnia, loss of appetite, emotional instability, depression, thyroid enlargement, sweating, tremor of extended fingers, loss of sexual drive, and impotence (Silverman, 1973). Similar symptoms were described in Western medical literature in two case reports of apparently high, acute overexposure to microwaves (Williams and Webb, 1980; Forman et al., 1982).

Reproductive and Development: Radio-frequency overexposure can have reproductive and developmental effects in test animals if the exposures are quite high. Researchers at the National Institute for Occupational Safety and Health (NIOSH) have reproducibly demonstrated teratogenic effects at 27.12 MHz with rats, when the whole body average (WBA)-SAR (RF dose rate) is around 10 W/kg (Lary et al., 1982, 1983, 1986). These effects appear to be thermal

---

[1] Elder and Cahill, 1984; NCRP, 1986; Heynick, 1987; Beers, 1989; Saunders et al., 1991a, 1991b; Dennis et al., 1992; Murray et al., 1995; Hitchcock and Patterson 1995; Scientific Committee on Emerging and Newly Identified Health Risks, 2007; International Commission on Non-ionizing Radiation Protection, 2009a; Advisory Group on Nonionising Radiation, 2012; International Agency for Research on Cancer, 2013.

in nature because a dose–response effect was seen with high rectal temperatures (Lary et al., 1986). Researchers at the Environmental Protection Agency (EPA) observed developmental abnormalities in rodents exposed to 2450 MHz. Again, the effects appear at high values of WBA-SAR (Berman et al., 1982a, 1982b).

Epidemiological studies of reproductive end points have not demonstrated any trends.[2] Two studies reported effects on semen (Lancranjan et al., 1975; Weyandt, 1992), but the small sample of people involved and lack of exposure data make interpretation difficult.

Ocular: Cataracts have been demonstrated in laboratory animals, and the most effective frequencies were 1–10 GHz. Acute thresholds were determined for rabbits receiving ocular exposure in the near field of a 2450-MHz applicator (Carpenter and Van Ummersen, 1968; Guy et al., 1975). No cataracts were observed when the animals were unrestrained and exposed in the far field (Michaelson et al., 1971; Appleton et al., 1975), even if exposures were almost lethal. A number of epidemiological investigations and clinical evaluations have been performed, but none has found an excess of cataracts in populations purported to have received RF and microwave exposure (Cleary and Pasternack, 1966; Majewska, 1968; Siekierzynski et al., 1974; Cleary et al., 1965; Bonomi and Bellucci, 1989).

Carcinogenicity: Limited *in vivo* data suggest that microwaves may be a tumor promoter in laboratory animals (Szydzinski et al., 1982; Szmigielski et al., 1982), whereas other studies demonstrated no significant differences in RF-exposed groups and controls (Santini et al., 1988; Svedenstal and Holmberg, 1993; Wu et al., 1994). In a study funded by the U.S. Air Force and performed at the University of Washington, significant differences were seen in primary malignancies in microwave-exposed animals when the tumor data for all tissue types were combined. However, evaluation of more than 100 end points supported the conclusion that the exposed animals were generally healthy and disease free (Guy et al., 1985; Chou et al., 1992). In sum, the animal data must be viewed as inconclusive.

Other researchers used C3H/HeJ mice, a strain that has a naturally high incidence of mammary tumors. There were no statistically significant findings for neoplasms of the mammary gland, uterus, liver, adrenal gland, and lungs with exposure to pulsed, 435-MHz microwaves. Bilateral ovarian epithelial stromal tumor was significantly elevated for the ovaries, whereas epithelial stromal tumors and total epithelial stromal tumors were not. There was no difference in longevity between the two groups (Toler, 1997). An evaluation with CW RFR (SAR = 0.3 W/kg) produced no significant differences in tumors of the mammary gland, ovary, liver, or uterus

in C3H/HeJ mice (Frei et al., 1998a). At an SAR of 1.0 W/kg, there were no significant differences in mammary tumors and neoplastic lesions of other tissues. There were statistically significant differences in the incidence of five types of neoplastic tumors in the control group and of two types in the RF-exposed group (Frei et al., 1998b). C3H/HeJ mice were chronically exposed at 2450 MHz and evaluated for genotoxicity. There were no differences in the micronucleus frequency in peripheral blood and bone marrow (Vijayalaxmi et al., 1997).

Chronic Effects

Historically, epidemiological studies have provided little evidence that RF energies are carcinogenic to human beings (Hitchcock and Patterson, 1995). Differences have not been observed in cancer mortality in military personnel (Robinette et al., 1980; Garland et al., 1990). A 40-year, follow-up study of Korean War naval personnel found that microwave exposure had little effect on mortality. The risk of nonlymphocytic leukemia was doubled in one of the three high exposure occupations, electronics technicians in aviation squadrons (Groves et al., 2002). An evaluation was performed of the staff at the Massachusetts Institute of Technology (MIT Rad Lab) who worked in radar development during World War II and members of the general population. A marginally significant difference in cancer of the gallbladder and bile ducts was reported when the workers were compared to a cohort of physician specialists. No other differences were reported (Hill, 1988). A number of evaluations were performed for reproductive effects in operators of cathode-ray-tube video display terminals (VDTs). Associations reported early in the evolution were not related to the actual exposures (Hitchcock and Patterson 1995). An evaluation performed by the NIOSH using measurement data did not find meaningful differences between exposed and control groups for live births, stillbirths, and spontaneous abortion during the first trimester of pregnancy (Schnorr et al., 1991). Milham (1982) determined proportionate mortality ratio (PMR) for death by acute leukemia and leukemia for various occupational classifications. Four of the occupational classes he defined (electrical engineers, radio and telegraph operators, welders, and TV and radio repairmen) have the potential for exposure to RF (and ELF) fields. Milham reported that the PMR for acute leukemia in TV and radio repairmen was significantly different from the U.S. population. Sheikh (1986) has commented on this study, suggesting that "the apparent associations between certain electrical occupations and leukemia could be chance occurrences." He observed that this study calculated 34,444 PMRs, and at the 5% level, one would expect 1700 PMRs to exceed the null value.

A hypothesis concerning testicular cancer and use of radar was formed in a study that found an excess of this cancer in a group of policemen in Washington State who used radar

---

[2] Kallen et al., 1982; Kolmodin-Hedman et al., 1988; Brandt and Nielsen, 1990; Nielsen and Brandt, 1990; Taskinen et al., 1990; Larsen et al., 1991; Schnorr et al., 1991.

guns (Davis and Mostofi, 1993). Finkelstein (1998) determined that the standardized incidence ratios (SIR) were statistically significant for melanoma for police officers in Ontario, Canada with 10–60 years from hire (SIR = 1.49, 90% confidence interval (CI) = 1.15–1.98), and for the total cohort (SIR = 1.37, 90% CI = 1.08–1.72). For testicular cancer, the SIR was non-significantly elevated for the 10-to-60 group (SIR = 1.45, 90% CI = 0.96–2.1) and of marginal statistical significance for the total group (SIR = 1.33, 90% CI = 1.0–1.74). For the 90% CI, the lower limit corresponds to the lower 5% limit for a one-tailed test for sites with *a priori* hypotheses, which was the case for melanoma and testicular cancer.

The results from a number of evaluations of potential exposures to cell phone emissions and brain cancer have been published and reviewed (Moulder et al., 1999; Mild et al., 2003; Ahlbom et al., 2009; Swerdlow et al., 2011). The list of studies is extensive but is highlighted by the findings of the large Interphone study, which involved the participation of 16 study groups in 13 countries. The general finding was that individuals who had ever used a cell phone had a decreased risk of brain cancer, glioma (odds ratio (OR) = 0.81, 95% CI = 0.70–0.94), and meningioma (OR = 0.79, 95% CI = 0.68–0.91), with no signs of an increased risk of meningioma. There were some statistically significantly elevated findings such as an increased risk of glioma for cumulative call time and ipsilateral use (OR = 1.96, 95% CI = 1.22–3.16). Call time is determined by the number and duration of calls, which may be viewed as surrogate for total RFR dose. With respect to this, it was noted that "the lack of a consistently increasing risk with dose, duration of exposure and time since first exposure weigh against cause and effect" for glioma (The INTERPHONE Study Group, 2010).

In May 2011, the International Agency for Research on Cancer (IARC) (IARC, 2011) categorized RFR as a possible human carcinogen (Category 2B) on the basis of the association between cell phone use and glioma. In 2013, the IARC published Monograph 102 that reviewed the literature and explained the rationale for the classification, which is based on limited evidence in humans and test animals with positive associations found with glioma and acoustic neuroma (International Agency for Research on Cancer, 2013).

SOURCES OF EXPOSURE

Occupational

Dielectric heaters may be used to weld, mold, seal, and emboss plastics and to cure glues, resins, and particleboard. Many of these devices operate at 27 MHz, although some may operate as low as 13.56 MHz and others as high as 70 MHz. A number of workplace evaluations have demonstrated the potential of an overexposure to individuals who work with dielectric heaters (Stuchly et al., 1980; Cox et al., 1982; Gandhi et al., 1986; Conover et al., 1992).

Induction heating (50–60 Hz up to 27 MHz) is used to heat conducting materials. High values of magnetic field strength have been reported near induction heaters (Stuchly and Lecuyer, 1985). However, the intensity of the fields diminishes rapidly with distance from the source (Conover et al., 1980, 1986; Mild, 1980). Hence it is possible that workers who do not work near the source or who spend little time in the most intense regions will receive relatively low level exposures, when time averaged.

In the electronics industry, plasma processing (0.1–27 MHz) may be classified as either dry etching or deposition operations (e.g., sputtering). An etcher is used to etch the surface of a wafer, and a sputterer is used to metallize the surface of a wafer. Typically, evaluations have demonstrated that leakage from well-designed, well-installed, and well-maintained units is low. Viewing ports may be a potential problem if they are not shielded. One evaluation found that peripherally attached equipment could act as antennas because of conducted or coupled RF energy (Ungers et al., 1985).

Communication systems may be either mobile or fixed, or a combination of these two. Fixed systems are used in the telecommunications industry and include high frequency radio, tropospheric scatter, satellite communications (SATCOMs), and microwave radio (point-to-point radio relay) systems. Mobile systems include small portable transceivers such as radios, walkie-talkies, and vehicular units. Combination systems may be used for paging or cellular radio. Maintenance personnel who work on high power systems, such as SATCOM systems, have the greatest potential for an overexposure if they do not follow proper lockout, tagout procedures. In certain cases, it may be possible for personnel who maintain cell-site antennas to be exposed to relatively high levels, for example, while working near an energized antenna on a tower or building (Petersen and Testagrossa, 1992). In general, exposures to portable communication devices have shown that the output from two-way radios may produce high, momentary exposures (Chatterjee et al., 1985; Cleveland and Athey, 1989).

Most radar units operate in the extremely high frequency (EHF), superhigh frequency (SHF), and ultrahigh frequency (UHF) bands in a pulse-modulated mode and at high peak transmitter powers (Jordan, 1985; Skolnik, 1990). Evaluations of commercial radar (e.g., airport surveillance, airport approach traffic control) have not revealed potential overexposure during normal operation. It is possible that maintenance activities (work on or at the transmit–receive duplexer, magnetron cabinet, or trapdoor of the antenna pedestal) may result in relatively high exposure (Joyner and Bangay, 1986). Exposures in the cockpit of an aircraft (Tell and Nelson, 1974; Tell et al., 1976), near marine radar (Peak et al., 1975; Moss, 1990), and near the police radar

(Bitran et al., 1992; Fisher, 1993) have not been reported as a problem. Possible overexposures to the output from military units have been reported (Williams and Webb, 1980; Forman et al., 1982).

Diathermy units use ultrasonic, microwave (915 and 2450 MHz), or shortwave (13.56 or 27.12 MHz) frequencies. The leakage field around the applicator depends on the type of applicator used. Relatively high field strengths may be found in the vicinity of the cables (Ruggera, 1980; Stuchly et al., 1981b). If the physician or physiotherapist adjusts the equipment during operation, the greatest exposure will be to the hands (Mild, 1980; Kalliomaki et al., 1982). Service personnel may be exposed during the maintenance of an energized system. Fields near the back of an energized unit that had the access cover removed for servicing were approximately 1000 V/m and 3 A/m (Mild, 1980). A review of exposures reported in the literature found that measured levels at distances greater than a meter from the source did not exceed exposure guidelines recommended by the International Commission on Nonionizing Radiation Protection (ICNIRP) (Shah and Farrow 2013). Evaluations of stray emissions from shortwave units found that levels decreased to the exposure guidelines at 1 m for inductive units and at 2 m for capacitive units (Shields et al., 2004).

Electrosurgical units (0.5–2.4 MHz) are used for cauterizing or coagulating tissues. Evaluations of solid-state and spark-gap units demonstrated that field strengths increased with increasing output power, and levels were higher for solid-state units (Ruggera, 1977). Levels near the probe and unshielded leads may exceed the exposure criteria (Ruggera, 1977; Mild, 1980). Relatively high levels have been reported near the eye/forehead region, about 20 cm from the active lead (Paz, 1987). Exposures to the hand of the user estimated by calculation and measurement exceeded the ICNIRP criteria (Liljestrand et al., 2003).

Environmental

Naturally occurring background sources of RF include terrestrial, extraterrestrial, and atmospheric electrical discharges (lightning), and even the human body (Stuchly et al., 1981a). Although the list of artificial sources is lengthy (Hitchcock and Patterson, 1995), the major sources include dielectric heaters, induction heaters, broadcast and communications, radar, plasma processors, diathermy units, and electrosurgical devices.

An exposure to the radiation from base stations has not been shown to produce exposure to members of the general public that exceeds the exposure guidelines (Thansandote et al., 1999; Mann et al., 2000; Fuller et al., 2002; Lin et al., 2000). Evaluations of potential exposures to RFR from Wi-Fi have not found overexposures (Industry Canada, 2012; Foster and Moulder, 2013; Office of Telecommunications Authority, 2007).

Broadcasting includes standard amplitude-modulation (AM) audio broadcasting, frequency-modulation (FM) broadcasting, and educational and commercial TV. The frequencies are allocated by the Federal Communications Commission (FCC) (Stern, 1989). Areas of potential concern include work on "hot" (energized) AM towers (Tell, 1991), access on tower structures near energized FM and TV antennas, and leaks from transmitter cabinets in monitor buildings (Curtis, 1980; Mild, 1980).

Exposures to the output from mobile phones have been an issue since the early 1990s, although the technology has changed as has the manner of usage, i.e., voice communications vs. texting and voice communications, as well as the option for hands-free operation. Some early evaluations demonstrated that such exposures are within the acceptable levels (Joyner et al., 1992; Balzano et al., 1995). Kuhn reported the results of an evaluation of 668 cell phones based on compliance testing data finding that the data fit a Gaussian (normal) distribution and that the maximum SAR (RFR dose rate) in human head phantom at the peak frequency was around 1 W/kg for both 1 and 10 g SARs (Kuhn et al., 2005, 2007). Other studies have demonstrated that relatively high, localized exposures to parts of the ear and skin of the head, with decreasing exposure as the RFR penetrates more deeply and into the brain tissue nearest the antenna. The FCC adopted guidelines that include mobile communications such as cell phones and PCS in 1996 (FCC Report and Order 96-326). This includes a local SAR of 1.6 W/kg for phones sold within the United States. The FCC maintains an online SAR database for mobile communications devices as well as links to manufacturer's data bases (http://www.fcc.gov/encyclopedia/specific-absorption-rate-sar-cellular-telephones).

The introduction of smart meter technology has received a lot of attention (California Council on Technology, 2011; Vermont Department of Health, 2012; Levy and Page, 2012; Michigan Public Service Commission, 2012; Rivaldo, 2012). One evaluation (Sage Associates, 2011) reported the possibility of overexposures where multiple smart meters are used and that reflected levels of microwave radiation are high. This report has been criticized for use of unrealistically high reflection factors and exposure guidelines that are not current (Electric Power Research Institute, 2011). One of the most thorough measurement reports on this topic was performed by Tell. He evaluated 37 locations in Vermont with sources operating between 902 and 928 MHz and a smaller sample that also emitted at 2.4 GHz. Measurements as close as 1 ft from the meter and measurements within structures were well beneath the limits specified by the FCC. Tell also reported that "there was no general correlation between overall higher RF fields associated with large banks of meters." and that the peak RF fields associated with multiple meters approached levels associated with a single meter. Additionally, as would be anticipated, the strength of the field decreased with distance from the smart meters (Tell and Tell, 2013).

Whole body scanners used in airports utilize millimeter-wave microwave radiation (tens of GHz) to expose the subject. In general, at these frequencies, the penetration depth into the tissue is quite small and, when absorbed, results in localized heating of tissue. According to ICNIRP, the resulting exposure is about 1/10th of the exposure guideline for members of the general public (International Commission on Nonionizing Radiation Protection, 2012).

Industrial Hygiene

Guidelines for occupational exposure to RF radiation have been recommended by several groups, including the Institute of Electrical and Electronics Engineers (IEEE) (IEEE, 2005a), the ACGIH (ACGIH, 2013), the ICNIRP (International Commission on Nonionizing Radiation Protection, 1998, 2009b), and the National Council on Radiation Protection and Measurements (NCRP) (NCRP, 1986). The ICNIRP, NCRP, and the IEEE recommend dual limits for members of the general public and occupational limits. The IEEE also calls the occupational limits the upper tier or the controlled environment, while the general public limits also serve as the action level for the occupational setting. In general, the exposure limits for the two target groups are either the same or have a fivefold difference in their magnitudes.

The ACGIH values are called threshold limit values (TLVs). Radio-frequency radiation TLVs apply from 30 kHz to 300 GHz. The ACGIH includes the band between 3 and 30 kHz in the sub-RF exposure limit.

The exposure limits are based on the basic restrictions from which the exposure limits are derived, as discussed earlier. The derived limits recommended by the IEEE are called maximum permissible exposures (MPEs, see Tables 98.3 and 98.4). Many of the limits in Tables 98.3 and 98.4 have a 6-min averaging time or a 30-min averaging time when the exposure limits are fivefold greater in magnitude. The 6-min time is based on cooling time constants derived from partial body irradiation of test animals with microwaves at 3 GHz (Ely et al., 1964). The IEEE reduces the averaging time to protect against skin burns (Petersen, 1991) at frequencies at which the penetration depth is shallow. In this case, averaging time is reduced to 10 s at 300 GHz.

The IEEE discusses the use of the action level in the exposure guideline (IEEE, 2005a), as well as the IEEE recommended practice on establishing a radio-frequency safety program (IEEE, 2005b).

The IEEE limits are based on electrostimulation from 3 kHz to 5 MHz and thermal effects from 100 kHz to 300 GHz. This is reflected in Tables 98.3 and 98.4 most prominently in terms of averaging time where 0.2 s is used for electrostimulation and times between 30 min and 10 s used for thermal effects. There is an overlap that defines a transition region between 100 kHz and 5 MHz. In the transition region, it is necessary to apply both sets of limits.

In the spectral region from 100 kHz to 3 GHz, the basic restriction is the SAR. This region includes the resonance range where the absorption of RFR electromagnetic energy is highest. Hence, the exposure limits are based upon WBA exposure. The WBA-SAR, or dose rate, is 0.4 W/kg for occupational exposure and is 1/5th that value, 0.08 W/kg, for the general public or action level (IEEE, 2005a). This fivefold difference is based upon an additional safety factor that is applied to the general public limits.

The IEEE (2005a) also allows spatial-peak excursions above the WBA levels. These excursion factors are called relaxations, and the magnitude of the relaxation is frequency dependent. Hence, to insure compliance with the guidelines, one must determine the WBA exposure and compare that to the WBA-MPE as well as compare the spatial-peak exposure levels to the relaxations.

Although the WBA-SAR is invariant, the derived limits are frequency dependent, as indicated in Tables 98.3 and 98.4. This is because the human body couples best to wavelengths that approach a certain body height-to-wavelength ratio. For a standard man, this ratio is established at a frequency around 70 MHz. Coupling and absorption would be maximized around this frequency and decrease for higher and lower frequencies up to a point. Hence, absorption is considered to be maximized in the human whole-body resonance region, which extends from about 30 to 300 MHz, while absorption is lowest at the lower frequencies, below about 3–5 MHz, which is where the limits are the highest.

Because of this frequency dependence, exposure at different frequencies has different allowable values, but exposure at these values still produces a WBA-SAR that does not exceed the basic limit. For example, the values of allowable electric field strength at 1 and 100 MHz are 1842 and 61.4 V/m, respectively. Occupational exposure at both limits would produce an SAR of 0.4 W/kg in standard man. Hence, if the measured value of field strength or power density does not exceed the applicable exposure limit, the WBA-SAR will not be exceeded.

Exposure to low frequency RFR may result in electrostimulation, burns, shock, and RFR currents that may pose contact-current and induced-current hazards. Shock and burns are important because ungrounded, conductive objects illuminated by an RFR source couple with the field and store the energy as an electrical charge. If a grounded person touches the object, excessive current could be discharged to the body.

To control for potentially hazardous effects associated with high current density, the guidelines, in general, recommend limiting field strengths and levels of current at low frequencies. Recommendations for control of RFR current are made between 110 MHz and 3 kHz. As with field exposure, the limits are based on electrostimulation with a 0.2-s

TABLE 98.3   IEEE MPEs for Controlled Environments (upper tier): Fields in Space

Frequency Range, MHz	E, V/m*	H, A/m*	Power Density (S), W/m²*	Averaging Time, min
0.003–0.1[a]	1842	–	–	0.0033 (0.2 s)
0.003–0.0035[a]	–	1640/f_M [b] 3016/f_M [c]	–	0.0033 (0.2 s)
0.00035–5.0[a]	–	490[b] 900[c]	–	0.0033 (0.2 s)
0.1–1.0	1842	16.3/f_M	–	6
1.0–30	1842/f_M	16.3/f_M	–	
30–100	61.4	16.3/f_M	–	6
100–300	61.4	0.163	10	6
300–3000	–	–	f_M/30	6
3000–30,000	–	–	100	$19.63/f_G^{1.073}$
30,000–300,000	–	–	100	$2.524/f_G^{0.476}$

	Induced and Contact Currents		
Frequency Range, MHz	Through Both Feet, mA	Maximum Current Through Each Foot, mA	Contact Current, mA
0.003–0.1[a]	2.0f_k	1.0f_k	1.0f_k [d] 0.5f_k [e]
0.1–100	200	100	100[d] 50[e]

Source: After IEEE, 2005a.

f_M—frequency in megahertz (MHz).

f_G—frequency in gigahertz (GHz).

f_k—frequency in kilohertz (kHz).

[a]Limits based on electrostimulation.

[b]Head and torso.

[c]Limbs.

[d]Grasping contact.

[e]Touching contact.

TABLE 98.4   IEEE MPEs for General Public (action level): Fields in Space

				Averaging Time, min*	
Frequency Range, MHz	E, V/m	H, A/m	Power Density (S), W/m²	E, E^2, S	H or H^2
0.003–0.1[a]	614	–	–	0.0033 (0.2 s)*	–
0.003–0.0035[a]	–	547/f_k [b] 3016/f_k [c]	–	–	0.0033 (0.2 s)*
0.00035–5.0[a]	–	163[b] 900[c]	–	–	0.0033 (0.2 s)*
0.1–1.34	614	16.3/f_M	–	6	6
1.34–3.0	823.8/f_M	16.3/f_M	–	f_M^2/0.3	6
3–30	823.8/f_M	16.3/f_M	–	30	6
30–100	27.5	158.3/$f^{1.668}$	–	30	$0.0636/f_M^{1.337}$
100–400	27.5	0.0729	2	30	30
400–2000	–	–	f_M/200	30	–
2000–5000	–	–	10	30	–
5000–30,000	–	–	10	150/f_G	–
30,000–100,000	–	–	10	$25.24/f_G^{0.476}$	–
100,000–300,000	–	–	$(90f_G - 70000)/200$	$5048[(9f_G - 700)/f_G^{0.476}]$	–

(Continued)

TABLE 98.4   *(Continued)*

Frequency Range, MHz	Induced and Contact Currents		
	Through Both Feet	Maximum Current Through Each Foot, mA	Contact Current, mA
0.003–0.1	$0.9f_k$	$0.45f_k$	$0.167f_k^d$
0.1–100	90	45	16.7^d

Source: After IEEE, 2005a.

[*]Averaging time based on *E* or *H*.

f_M—frequency in megahertz (MHz).

f_G—frequency in gigahertz (GHz).

f_k—frequency in kilohertz (kHz).

[a]Limits based on electrostimulation.

[b]Head and torso.

[c]Limbs.

[d]Touching contact.

averaging time between 3 and 100 kHz. This is related to the sensation of warmth in the ankle that may become evident 5 s after the onset of exposure at lower frequencies (Gandhi et al., 1985). The current limits are based on thermal effects with a 6-min averaging time for frequencies between 100 kHz and 110 MHz (IEEE, 2005a). Note that the guidelines for contact current are for both grasping (whole hand) contact and touching (e.g., unintended finger contact) in the controlled environment, but just for touching contact for the general public limits.

AB HINC (FROM HERE ON . . .)

High visibility issues have become a fact of life with certain RFR emitters. This includes a number of sources such as the PAVE PAWs early warning system, cathode ray tube VDTs, cellular phones, and now, possibly, smart meters. Some of these issues will diminish in importance as technological changes replace an existing technology with newer technology (e.g., CRT-VDTs replaced by flat-panel displays).

As has been demonstrated a number of times in the past, when there are questions about possible health effects with newly introduced RFR emitters, the industry is unprepared with satisfactory answers. This occurred with cell phones when the issue of brain cancer was raised during a popular television show in the early 1990s. When this occurred, the industry response was to use information based upon RFR sources with different frequencies, modulations, operational power levels, and usage patterns than cell phones, because that was all that was available to them. In other words, there were no *a priori* evaluations of potential exposures and possible health effects of the source under question.

Following on the coattails of those inadequate responses, simple studies were designed, implemented, published and analyzed, leading to further studies using ever more sophisticated methodologies. However, as studies were planned

and implemented, the technology under evaluation evolved (analogue to digital modulation) and usage changed (voice communications to texting) so quickly that the results of studies just completed often no longer directly applied to the exposures of current users. Fast forward to 2011, and a highly respected organization, IARC, categorized RFR into a relatively weak hazard category, a possible human carcinogen.

No one has ever claimed that scientific understanding occurs with rapidity, but it is approaching 20 years since questions of brain cancer and exposure to RFR from wireless devices were first raised. The scientific understanding of how the RFR power is deposited in the tissue of the head and the brain is better understood, and a number of health effects studies using cells, animals, humans, and human populations have been performed. But, only the future will tell if RFR is truly a hazard to brain tissue, or if it is simply a source that heats the tissue.

Of course, this pattern will be repeated with some future RFR emitter. In some cases the repetition (i.e., not learning from history) will be because the industry did not perform the necessary studies of potential health hazards associated with a new product that emits "radiation." Why not? The obvious answer is because of the additional cost to product development and potential delays to introducing the product to market. And, unfortunately, because of ongoing debates about topics such as nonthermal effects of RFR exposure, electrosensitivity, and the very RFR exposure guidelines themselves; it is often not adequate simply to evaluate by measurement the potential exposures users may receive and compare those measured levels to the exposure guidelines.

Hopefully, in the future, evaluations of potential exposures to RFR will occur prior to market, especially where the product has the potential to expose a large number of people and has a somewhat unique set of output characteristics. Of course, such proactive evaluations will not mean the end to the debate, especially where people's health is potentially at stake. It will mean, however, that a prudent approach is being taken, one that recognizes the necessity of examining all of the possible ramifications of a new radiation source used in the public domain.

REFERENCES

Advisory Group on Nonionising Radiation (2012) *Health Effects from Radiofrequency Electromagnetic Fields—Report of the Independent Advisory Group on Nonionising Radiation*, Health Protection Agency.

Ahlbom, A., et al. (2009) Epidemiologic evidence on mobile phones and tumor risk—a review. *Epidemiology* 20:639–652.

American Conference of Governmental Industrial Hygienists (ACGIH) (2013) *2013 TLVs Based on the documentation of the Threshold Limit Values (TLVs) for Chemical Substances and Physical Agents and Biological Exposure Indices (BEIs)*, Cincinnati: American Conference of Governmental Industrial Hygienists.

Appleton, B., Hirsch, S.E., and Brown, P.V.K. (1975) Investigation of single-exposure microwave ocular effects at 3000 MHz. *Ann. NY Acad. Sci.* 247:125–134.

Balzano, Q., Garay, O., and Manning, T.J. (1995) Electromagnetic energy exposure of simulated users of portable cellular telephones. *IEEE Trans. Veh. Technol.* 44:390–403.

Beers, G.J. (1989) Biological effects of weak electromagnetic fields from 0 Hz to 200 MHz: a survey of the literature with special emphasis on possible magnetic resonance effects. *Magn. Reson. Imaging* 7:309–331.

Berman, E., Carter, H., and House, D. (1982a) Observations of Syrian hamster fetuses after exposure to 2450-MHz microwaves. *J. Microw. Power* 17:107–112.

Berman, E., Carter, H., and House, D. (1982b) Reduced weight in mice offspring after *in utero* exposure to 2450-MHz (cw) microwaves. *Bioelectromagnetics* 3:285–291.

Bitran, M.E., Charron, D.E., and Nishio, J.M. (1992) *Microwave Emissions and Operator Exposures from Traffic Radars Used in Ontario*, Weston, Ontario: Ministry of Labour.

Bonomi, L. and Bellucci, R. (1989) Considerations of the ocular pathology in 30,000 personnel of the Italian telephone company (SIP) using VDTs. *Bollettion Di Oculistica* 68(S7): 85–98.

Brandt, L. and Nielsen, C.V. (1990) Congenital malformations among children of women working with video display terminals. *Scand. J. Work Environ. Health* 16:329–33.

California Council on Science and Technology (2011) *Health Impacts of Radio Frequency from Smart Meters*, Sacramento, California: CCST.

Carpenter, R. and Van Ummersen, C. (1968) The action of microwave radiation on the eye. *J. Microw. Power* 3:3–19.

Chatterjee, I., Gu, Y., and Gandhi, O.P. (1985) Quantification of electromagnetic absorption in humans from body-mounted communications transceivers. *IEEE Trans. Veh. Technol.* VT-34:55–62.

Chou, C.K., et al. (1992) Long-term, low-level microwave irradiation of rats. *Bioelectromagnetics* 13:469–496.

Cleary, S.F. and Pasternack, B.S. (1966) Lenticular changes in microwave workers. A statistical study. *Arch. Environ. Health* 12:23–29.

Cleary, S.F., Pasternack, B.S., and Beebe, G.W. (1965) Cataract incidence in radar workers. *Arch. Environ. Health* 11:179–182.

Cleveland, R.F. Jr., and Athey, T.W. (1989) Specific absorption rate (SAR) in models of the human head exposed to hand-held UHF portable radios. *Bioelectromagnetics* 10:173–186.

Conover, D.L., et al. (1980) Measurement of electric- and magnetic-field strengths from industrial radio-frequency (6–38 MHz) plastic sealers. *Proc. IEEE* 68:17–20.

Conover, D.L., et al. (1986) Magnetic field measurements near RF induction heaters. *Bioelectromagnetics* 7:83–90.

Conover, D.L., et al. (1992) Foot currents and ankle SARs induced by dielectric heaters. *Bioelectromagnetics* 13:103–110.

Cox, C., Murray, W.E., and Foley, E.D. (1982) Occupational exposures to radiofrequency radiation (18–31 MHz) from RF dielectric heat sealers. *Am. Ind. Hyg. Assoc. J.* 43:149–153.

Curtis, R.A. (1980) Occupational exposures to radio-frequency radiation from FM radio and TV antennas. In: *Nonionizing Radiation: Proceedings from a Topical Symposium*, Cincinnati: American Conference of Governmental Industrial Hygienists.

D'Andrea, J., Gandhi, O., and Lords, J.L. (1977) Behavioral and thermal effects of microwave radiation at resonant and non-resonant wavelengths. *Radio Sci.* 12(6S):251–256.

D'Andrea, J.A. (1991) Microwave radiation absorption: behavioral effects. *Health Phys.* 61:29–40.

D'Andrea, J.A., Adair, E., and DeLorge, J.O. (2003) Behavioral and cognitive effects of microwave exposure. *Bioelectromagnetics* 6:S39–S62.

Davis, R.L. and Mostofi, F.K. (1993) Cluster of testicular cancer in police officers exposed to hand-held radar. *Am. J. Ind. Med.* 24:231–233.

DeLorge, J.O. (1976) The effects of microwave radiation on behavior and temperature in rhesus monkeys. In: Johnson, C.C. and Shore, M.L., editors. *Biological Effects of Electromagnetic Waves*, DHHS Pub (FDA) 77-8010, Rockville, MD: Bureau of Radiological Health.

DeLorge, J.O. (1978) Disruption of behavior in mammals of three different sizes exposed to microwaves: extrapolation to larger mammals. In: *Proceedings of the 1978 Symposium on Electro-Magnetic Fields in Biological Systems*, Edmonton, Canada: International Microwave Power Institute.

DeLorge, J.O. (1983) The thermal basis for disruption of operant behavior by microwaves in three animal species. In: Adair, E.R., editor. *Microwaves and Thermoregulation*, New York, NY: Academic Press.

Dennis, J.A., Muirhead, C.R., and Ennis, J.R. (1992) *Human Health and Exposure to Electromagnetic Radiation*, NRPB-R241. Chilton, Didcot, Oxon: National Radiological Protection Board.

Elder, J.A. and Cahill, D.F., editors. (1984) *Biological Effects of Radiofrequency Radiation*, EPA-600/8-83-026F, Springfield, VA: National Technical Information Service.

Electric Power Research Institute (EPRI) (2011) *EPRI Comment: Sage Report on Radio-Frequency (rf) Exposures from Smart Meters*, Palo Alto, California: Electric Power Research Institute.

Ely, T.S., Goldman, D.E., and Hearon, J.Z. (1964) Heating characteristics of laboratory animals exposed to ten-centimeter microwaves. *IEEE Trans. Biomed. Eng.* BME-11:123–137.

FCC Report and Order 96-326, Washington, DC, Available at http://transition.fcc.gov/Bureaus/Engineering_Technology/Orders/1996/fcc96326.pdf.

Finkelstein, M. (1998) Cancer incidence among Ontario police officers. *Am. J. Ind. Med.* 34:157–162.

Fisher, P.D. (1993) Microwave exposure levels encountered by police traffic radar operators. *IEEE Trans. Electromagn. Compat.* 35:36–45.

Forman, S.A., et al. (1982) Psychological symptoms and intermittent hypertension following acute microwave exposure. *J. Occup. Med.* 24:932–934.

Foster, K.R. and Moulder, J.E. (2013) Wi-fi and health: review of current status of research. *Health Phys.* 105:561–575.

Frei, M.R., et al. (1998a) Chronic exposure of cancer-prone mice to low-level 2450 MHz radiofrequency radiation. *Bioelectromagnetics* 19:20–31.

Frei, M.R., et al. (1998b) Chronic, low level (1.0 W/kg) exposure of mice prone to mammary cancer to 2450 MHz microwaves. *Radiat. Res.* 150:568–576.

Frey, A.H. and Wesler, L.S. (1990) Interaction of psychoactive drugs with exposure to electromagnetic fields. *J. Bioelectricity* 9:187–196.

Fuller, K., et al. (2002) *Radiofrequency Electromagnetic Fields in the Cookridge Area of Leeds*, National Radiological Protection Board report W23, Oxon, Didcot, Chilton.

Gandhi, O., Chen, J.Y., and Riazi, A. (1986) Currents induced in human beings for plane-wave exposure conditions 0–50 MHz and for RF sealers. *IEEE Trans. Biomed. Eng.* BME-33:757–767.

Gandhi, O., et al. (1985) Likelihood of high rates of energy deposition in the human legs at the ANSI recommended 3–30 MHz RF safety levels. *Proc. IEEE* 73:1145–1147.

Garland, F.C., et al. (1990) Incidence of leukemia in occupations with potential electromagnetic field exposure in United States Navy personnel. *Am. J. Epidemiol.* 132:293–303.

Groves, F.D., et al. (2002) Cancer in Korean War navy technicians: mortality survey after 40 years. *Am. J. Epidemiol.* 155:810–818.

Guy, A.W., et al. (1975) Effect of 2450-MHz radiation on the rabbit eye. *IEEE Trans. Microw. Theory Tech.* MTT-23:492–498.

Guy, A.W., et al. (1985) *Effects of Long-Term Low-Level Radiofrequency Radiation Exposure on Rats*, vol 9, summary, School of Aerospace Medicine Report USAFSAM-TR-85-64, Brooks Air Force Base, Tex: US Air Force.

Heynick, L. (1987) *Critique of the Literature on Bioeffects of Radiofrequency Radiation: A Comprehensive Review Pertinent to Air Force Operations*, USAFSAM-TR-87-3, Brooks Air Force Base, Tex: USAF School of Aerospace Medicine.

Hill, D.G. (1988) *A longitudinal study of a cohort with past exposure to radar: the MIT radiation laboratory follow-up study*, Dissertation, Ann Arbor, Michigan: Johns Hopkins University, University Microfilms International.

Hitchcock, R.T. and Patterson, R.M. (1995) *Radio-Frequency and ELF Electromagnetic Energies: A Handbook for Health Professionals*, New York, NY: Van Nostrand Reinhold.

Industry Canada (2012) *Case study: measurements of radio frequency exposure from Wi-fi devices, Industry Canada, Spectrum Management and Telecommunication Report*.

Institute of Electrical and Electronics Engineers (IEEE) (2005a) *IEEE Standard for Safety Levels with Respect to Human Exposure to Radio Frequency Electromagnetic Fields, 3 kHz to 300 GHz*, IEEE Standard C95.1-2005, New York, NY: Institute of Electrical and Electronics Engineers.

Institute of Electrical and Electronics Engineers (2005b) *IEEE Recommended Practice for Radio Frequency Safety Programs, 3 kHz to 300 GHz*, IEEE Standard C95.7-2005, New York, NY: Institute of Electrical and Electronics Engineers.

International Agency for Research on Cancer (2011) *IARC Classifies Radiofrequency electromagnetic fields as possibly carcinogenic to humans (Press Release No. 208), May 31, 2011*. Available at http://www.iarc.fr/en/media-centre/pr/2011/pdfs/pr208_E.pdf.

International Agency for Research on Cancer (2013) *Nonionizing Radiation, Part 2: Radiofrequency Electromagnetic Fields*, Monograph 102, Lyon, France: IARC.

International Commission on Nonionizing Radiation Protection (1998) Guidelines for limiting exposure to time-varying electric, magnetic, and electromagnetic fields (up to 300 GHz). *Health Phys.* 74:494–522.

International Commission on Nonionizing Radiation Protection (2009a) *Exposure to high frequency electromagnetic fields, biological effects and health consequences (100 kHz–300 GHz), ICNIRP 16*.

International Commission on Nonionizing Radiation Protection (2009b) ICNIRP statement on the "Guidelines for limiting exposure to time-varying electric, magnetic, and electromagnetic fields (up to 300 GHz)". *Health Phys.* 97:257–258.

International Commission on Nonionizing Radiation Protection (2012) ICNIRP statement—health issue associated with millimeter wave whole body imaging technology. *Health Phys.* 102:81–81.

Jordan, E.C. (1985) *Reference Data for Engineers: Radio, Electronics, Computer, and Communications*, 2nd ed., Indianapolis: Howard K Sams.

Joyner, K.H. and Bangay, M.J. (1986) Exposure survey of civilian airport radar workers in Australia. *J. Microw. Power Electromagn. Energy* 21:209–219.

Joyner, K.H., et al. (1992) Radio frequency radiation (rfr) exposures from mobile phones. In: *Worldwide Achievement in Public and Occupational Health Protection Against Radiation (IRPA8)*, New York, NY: International Radiation Protection Association.

Kallen, B., Malmquist, G., and Moritz, U. (1982) Delivery outcome among physiotherapists in Sweden: is nonionizing radiation a fetal hazard? *Arch. Environ. Health* 37:81–84.

Kalliomaki, P.L., et al. (1982) Measurements of electric and magnetic stray fields produced by various electrodes of 27-MHz diathermy equipment. *Radio Sci.* 17(5S):29S–34S.

Kolmodin-Hedman, B., et al. (1988) Health problems among operators of plastic welding machines and exposure to radio-frequency electromagnetic fields. *Int. Arch. Occup. Environ. Health* 60:243–247.

Kuhn, S., et al. (2005) *Assessment methods for demonstrating compliance with safety limits of wireless devices used in home and office environments, Proceedings of the URSI General Assembly, New Delhi, 2005.* Available at http://www.ursi.org/proceedings/procga05/pdf/KAE.7(01491).pdf.

Kuhn, S., et al. (2007) Assessment methods for demonstrating compliance with safety limits of wireless devices used in home and office environments. *IEEE Trans. Electromagn. Compat.* 49:519–525.

Lai, H., et al. (1987) A review of microwave irradiation and actions of psychoactive drugs. *IEEE Eng. Med. Biol. Mag.* 6(1):31–36.

Lancranjan, I., et al. (1975) Gonadic function in workmen with long-term exposure to microwaves. *Health Phys.* 29:381–383.

Larsen, A.I., Olsen, J., and Svane, O. (1991) Gender-specific reproductive outcome and exposure to high-frequency electromagnetic radiation among physiotherapists. *Scand. J. Work Environ. Health* 17:324–329.

Lary, J.M., et al. (1982) Teratogenic effects of 27.12 MHz radio-frequency radiation in rats. *Teratology* 26:299–309.

Lary, J.M., et al. (1983) Teratogenicity of 27.12-MHz radiation in rats is related to duration of hyperthermic exposure. *Bioelectromagnetics* 4:249–255.

Lary, J.M., et al. (1986) Dose-response relationship and birth defects in radiofrequency-irradiated rats. *Bioelectromagnetics* 7:141–149.

Levy, R. and Page, J. (2012) *Review of the January 13, 2012 County of Santa Cruz Health Services Agency memorandum: Health Risks Associated with Smart Meters*, Berkeley, CA: Lawrence Berkeley National Laboratory, Available at http://smartresponse.lbl.gov/reports/schd041312.pdf.

Liljestrand, B., Sandström, M., and Mild, K.H. (2003) RF exposure during use of electrosurgical units. *Electromagn. Biol. Med.* 22:127–132.

Lin, P., et al. (2000) *Levels of radiofrequency radiation from GSM mobile telephone base stations, Australian Radiation Protection and Nuclear Safety Agency technical report 129, Yallambie, Victoria.*

Majewska, K. (1968) Investigations on the effect of microwaves on the eye. *Pol. Med. J.* 38:989–994.

Mann, S.M., et al. *Exposure to radio waves near mobile phone base stations, National Radiological Protection Board report NRPB-R321, Chilton, Didcot, Oxon, 2000.* Available at http://www.hpa.org.uk/wcbc/HPAwcb&HPAwcbStandard/HPAweb_C/1195733833994.

Michaelson, S.M., Howland, J.W., and Deichmann, W.B. Response of the dog to 24,000 and 1285 MHz microwave exposure. *IMS Ind. Med. Surg.* 40:18–23.

Michigan Public Service Commission (2012) *U-17000 Report to the Commission,* Available at http://efile.mpsc.state.mi.us/efile/docs/17000/0455.pdf.

Mild, K.H. (1980) Occupational exposure to radio-frequency electromagnetic fields. *Proc. IEEE* 68:12–17.

Mild, K.H., et al. (2003) Mobile telephones and cancer: is there really no evidence of an association? (review). *Int. J. Mol. Med.* 12:67–72.

Milham, S. Jr. (1982) Mortality from leukemia in workers exposed to electrical and magnetic fields. *N. Engl. J. Med.* 307:249.

Moulder, J.E., et al. (1999) Cell phones and cancer: what is the evidence for a connection. *Radiat. Res.* 151:513–531.

Moss, C.E. (1990) *Hazard Evaluation and Technical Assistance Report HETA 89-284-L2029: Technical Assistance to the Federal Employees Occupational Health, Seattle, Washington,* NTIS #PB91-107920, Springfield, VA: National Technical Information Service.

Murray, W.E., et al. (1995) Nonionizing electromagnetic energies. In: Cralley, L., Cralley, L., and Bus, J., editors. *Patty's Industrial Hygiene and Toxicology,* vol. 3, pt B, 3rd ed., New York, NY: John Wiley & Sons.

National Council on Radiation Protection (NCRP) (1986) *Biological Effects and Exposure Criteria for Radiofrequency Electromagnetic Fields,* NCRP Report No 86, Bethesda, MD: National Council on Radiation Protection and Measurements.

Nielsen, C.V. and Brandt, L. (1990) Spontaneous abortion among women using video display terminals. *Scand. J. Work Environ. Health* 16:323–328.

Office of Telecommunications Authority (2007) *Radiofrequency Radiation Measurements—Public Wi-Fi Installations in Hong Kong,* Government of Hong Kong: Office of Telecommunications Authority, Available at http://www.gov.hk/en/theme/wifi/health/.

Paz, J.D. (1987) Potential ocular damage from microwave exposure during electrosurgery: dosimetric survey. *J. Occup. Med.* 29:580–583.

Peak, D.W., et al. (1975) *Measurement of Power Density from Marine Radar,* DHHS Pub (FDA) 76-8004, Washington, DC: US Government Printing Office.

Petersen, R.C. (1991) Radiofrequency/microwave protection guides. *Health Phys.* 61:59–67.

Petersen, R.C. and Testagrossa, P.A. (1992) Radio-frequency electromagnetic fields associated with cellular radio cell-site antennas. *Bioelectromagnetics* 13:527–542.

Rivaldo, A. *Health and rf emf from advanced meters—an overview of recent investigations and analyses, Project No. 40190, Public Utility Commission of Texas, 2012.* Available at http://www.puc.texas.gov/industry/electric/reports/smartmeter/smartmeter_rf_emf_health_12-14-2012.pdf.

Robinette, D., Silverman, C., and Jablon, S. (1980) Effects upon health of occupational exposure to microwave radiation (radar). *Am. J. Epidemiol.* 112:39–53.

Ruggera, P.S. (1977) Near-field measurements of RF fields. In: Hazzard, D.G., editor. *Symposium on Biological Effects and Measurement of Radio Frequency/Microwaves,* HHS Pub (FDA) 77-8026, Washington, DC: US Government Printing Office.

Ruggera, P.S. (1980) *Measurements of Emission Levels During Microwave and Shortwave Diathermy Treatments,* HHS Pub

(FDA) 80-8119, Washington, DC: US Government Printing Office.

Sadcikova, M.N. (1974) Clinical manifestations of reactions to microwave irradiation in various occupational groups. In: *Biological Effects and Health Hazards of Microwave Radiation*, Warsaw: Polish Medical Publishers.

Sage Associates (2011) *Assessment of Radiofrequency Microwave Radiation Emissions from Smart Meters*, Santa Barbara, CA: Sage Associates, Available at http://sagereports.com/smart-meter-rf/.

Santini, R., et al. (1988) B16 melanoma development in black mice exposed to low-level microwave radiation. *Bioelectromagnetics* 9:105–107.

Saunders, R.D., Sienkiewicz, Z.J., and Kowalczuk, C.I. (1991a) Biological effects of electromagnetic fields and radiation. *J. Radiol. Prot.* 11:27–42.

Saunders, R.D., Sienkiewicz, Z.L., and Kowalczuk, C.I. (1991b) *Biological Effects of Exposure to Nonionising Electromagnetic Fields and Radiation*, NRPB-R240, Chilton, Oxon, UK: National Radiological Protection Board.

Schnorr, T.M., et al. (1991) Video display terminals and the risk of spontaneous abortion. *N. Engl. J. Med.* 324:727–733.

Scientific Committee on Emerging and Newly Identified Health Risks (2007) *Possible Effects of Electromagnetic Fields (EMF) on Human Health*, Brussels: European Commission.

Shah, S.G. and Farrow, A. (2013) Assessment of physiotherapists' occupational exposure to radiofrequency electromagnetic fields from shortwave and microwave diathermy devices: a literature review. *J. Occup. Environ. Hyg.* 10:312–327.

Sheikh, K. (1986) Exposure to electromagnetic fields and the risk of leukemia. *Arch. Environ. Health* 41:56–63.

Shields, N., O'Hare, N., and Gormley, J. (2004) An evaluation of safety guidelines to restrict exposure to stray radiofrequency radiation from short-wave diathermy units. *Phys. Med. Biol.* 49:2999–3015.

Siekierzynski, M., et al. (1974) Health surveillance of personnel occupationally exposed to microwaves. III. Lens translucency. *Aerosp. Med.* 45:1146–1148.

Sienkiewicz, Z.J., et al. (1994) Effects of prenatal exposure to 50 Hz magnetic fields on development in mice. II. Postnatal development and behavior. *Bioelectromagnetics* 15:363–375.

Silverman, C. (1973) Nervous and behavioral effects of microwave radiation in humans. *Am. J. Epidemiol.* 97:219–224.

Skolnik, M.I., editor (1990) *Radar Handbook*, 2nd ed., New York, NY: McGraw-Hill.

Stern, J.L. (1989) Broadcast transmission practice. In: Fink, D.G. and Christiansen, D., editors. *Electronics Engineers' Handbook*, 3rd ed., New York, NY: McGraw-Hill.

Stuchly, M.A. and Lecuyer, D.W. (1985) Induction heating and operator exposure to electromagnetic fields. *Health Phys.* 49:693–700.

Stuchly, M.A., et al. (1980) Radiation survey of dielectric (RF) heaters in Canada. *J. Microw. Power* 15:113–121.

Stuchly, M.A., et al. (1981a) Sources and applications of radio-frequency (RF) and microwave energy. In: Grandolfo, M.,

Michaelson, S., and Rindi, A., editors. *Biological Effects and Dosimetry of Nonionizing Radiation Radiofrequency and Microwave Energies*, New York, NY: Plenum Press.

Stuchly, M.A., et al. (1981b) Exposure to the operator and patient during short wave diathermy treatments. *Health Phys.* 42:341–366.

Svedenstal, B.M. and Holmberg, B. (1993) Lymphoma development among mice exposed to X-rays and pulsed magnetic fields. *Int. J. Radiat. Biol.* 64:119–125.

Swerdlow, A.J., et al. (2011) Mobile phones, brain tumours and the Interphone study: where are we now? *Environ. Health. Perspect* 119:1534–1538.

Szmigielski, S.A., et al. (1982) Accelerated development of spontaneous and benzopyrene-induced skin cancer in mice exposed to 2450 MHz microwave radiation. *Bioelectromagnetics* 3:171–191.

Szydzinski, A., et al. (1982) Acceleration of the development of benzopyrene-induced skin cancer in mice by microwave radiation. *Arch. Dermatol. Res.* 274:303–312.

Taskinen, H., Kyyronen, P., and Hemminki, K. (1990) Effects of ultrasound, shortwaves, and physical exertion on pregnancy outcome in physiotherapists. *J. Epidemiol. Community Health* 44:196–201.

Tell, R.A. and Tell, C.A. (2013) *An Evaluation of Radio Frequency Fields Produced by Smart Meters Deployed in Vermont*, Colville, WA: Richard Tell Associates, Inc., Available at http://www.fpl.com/energysmart/pdf/vermont.pdf.

Tell, R.A. (1991) *Induced Body Currents and Hot AM Tower Climbing: Assessing Human Exposure in Relation to the ANSI Radiofrequency Protection Guide*, PB92-125186, Springfield, VA: National Technical Information Service.

Tell, R.A., Hankin, N.N., and Janes, D.E. (1976) Aircraft radar measurements in the near field. In: *Operational Health Physics: Proceedings of Tile Ninth Midyear Topical Symposium of the Health Physics Society*, McLean, VA: Health Physics Society.

Tell, R.A. and Nelson, I.C. (1974) Microwave hazard measurements near various aircraft radars. *Radiat. Data Rep.* 15:161–179.

Thansandote, A., Gadja, G.B., and Lecuyer, D.W. (1999) Radiofrequency in five Vancouver schools: exposure standards not exceeded. *CMAJ* 160:1311–1312. Available at http://www.hc-sc.gc.ca/ewh-semt/pubs/radiation/schools-vancouver-ecoles-eng.php.

The INTERPHONE Study Group (2010) Brain tumour risk in relation to mobile telephone use: results of the INTERPHONE international case-control study. *Int. J. Epidemiol.* 39:675–694.

Toler, J.C. (1997) Long-term, low-level exposure of mice prone to mammary tumors to 435 MHz radiofrequency radiation. *Radiat. Res.* 148:227–234.

Ungers, L.J., Jones, J.H., and Mihlan, G.J. (1985) *Emission of radio-frequency radiation from plasma-etching operations, Paper presented at the 1985 American Industrial Hygiene Conference, Las Vegas, May 19–24, 1985*.

Vermont Department of Health (2012) *Radio Frequency Radiation and Health: Smart Meters*, Available at http://healthvermont.gov/enviro/rad/documents/smart_meters_facts.pdf.

Vijayalaxmi, R.L., et al. (1997) Frequency of micronuclei in the peripheral blood and bone marrow of cancer-prone mice chronically exposed to 2450 MHz radiofrequency radiation. *Radiat. Res.* 147:495–500.

Weyandt, T.B. (1992) *Evaluation of Biological and Male Reproductive Function Responses to Potential Lead Exposures in 155 mm Howitzer Crewmen (AD-A247 384)*, Springfield, VA: National Technical Information Service.

Williams, R.A. and Webb, T.S. (1980) Exposure to radio-frequency radiation from an aircraft radar unit. *Aviat. Space Environ. Med.* 51:243–244.

Wu, R.Y., et al. (1994) Effects of 2.45-GHz microwave radiation and phorbol ester 12-O-tetradecanoylphorbol-13-acetate on dimethylhydrazine colon cancer in mice. *Bioelectromagnetics* 15:531–538.

99

OPTICAL RADIATION: ULTRAVIOLET, VISIBLE LIGHT, INFRARED, AND LASERS

JOSEPH M. GRECO

Target organs: Eyes, skin

Occupational exposure limits:

Ultraviolet: Wavelength dependent. For ultraviolet radiation in the UV-A region, the ACGIH TLV suggests limiting long exposures (over 1000 s, about 16 min) to an irradiance of 1 mW/cm². For shorter durations (up to 1000 s), the total energy should not exceed 1 J/cm².

For the UV-B TLV, the effective irradiance (E_{eff}) (weighted as a function of wavelength and based on the action spectra for eyes and skin) should be limited to 3 mJ/cm² per 8-h day. Ultraviolet radiometers with a built-in spectral response, which mimics the relative spectral effectiveness, can provide results directly in E_{eff}.

Visible light/infrared: ACGIH TLV is wavelength dependent and complex to determine. Blue-light hazard and infrared hazards should be assessed separately, as appropriate.

Laser: Maximum permissible exposure for eye and skin, as defined by the ANSI Z136.1 standard. Dependent on wavelength, exposure duration, mode, and other factors. Recommend that a trained laser safety professional perform these calculations due to the many parameters involved. Non-beam hazards should also be assessed.

BACKGROUND AND USES

Optical radiation is comprised of the ultraviolet (UV), visible, and infrared (IR) portions of the electromagnetic spectrum. The International Commission on Illumination (the CIE, i.e., Commission International d'Eclairage) Committee on Photobiology has designated the optical radiation spectral bands summarized in Table 99.1 (CIE, 1970). Most commonly used lasers will fall into the UV, visible, and IR wavelength regions of the electromagnetic spectrum.

While ionizing radiation is described in terms of photon energy (e.g., electron volts), and radiofrequency radiation in terms of frequency (Hz), optical radiation is normally described in terms of wavelength (i.e., nm, μm, or mm). Exposure to optical radiation is normally expressed in units of Watts/cm² (for continuous sources) or Joules/cm² (for pulsed or short-term exposure).

The intensity of radiant energy from conventional light sources usually falls off markedly as distance from the source increases, and generally follows the inverse square rule, assuming a point source. This allows a rapid falloff of intensity in a relatively short distance. In contrast, the laser process (sometimes referred to as *lasing*) can produce a highly coherent beam of optical radiation that concentrates its energy in a narrow beam over considerable distances. A typical laser emits a beam with little divergence (on the order of a milliradian), and the irradiance (energy per area) does not decrease appreciably over distances. As an example, there have been many reports of pilots of planes and helicopters being targeted by terrestrial-based low power laser pointers, creating an optical hazard (distraction or temporary flash blindness), some even resulting in arrests (Barat, 2009).

Ultraviolet Radiation

The UV radiation spectrum resides between the ionizing (gamma and X) radiation and the visible light region in the

Hamilton & Hardy's Industrial Toxicology, Sixth Edition. Edited by Raymond D. Harbison, Marie M. Bourgeois, and Giffe T. Johnson.
© 2015 John Wiley & Sons, Inc. Published 2015 by John Wiley & Sons, Inc.

TABLE 99.1 Divisions of the Optical Spectrum According to the International Commission on Illumination (CIE)

Region	Wavelength Range	Other Terminology
UV	100–400 nm	
UV-C	100–280 nm	Far UV
UV-B	280 to 315–320 nm	Middle UV
UV-A	315 to 380–400 nm	Near UV
Visible light	380–400 to 760–780 nm	Visible
IR	760–780 nm to 1 mm	
IR-A	760–780 to 1400 nm	Near IR
IR-B	1400–3000 nm	Middle IR
IR-C	3000 nm to 1 mm	Far IR

electromagnetic spectrum. The CIE defines UV wavelengths as 100–400 nanometers (nm), and divides it into three regions (UV-A, UV-B, and UV-C). The UV-C and UV-B wavelengths are collectively called the *actinic* region. These three regions vary widely with respect to the penetration and potential for causing biological effects. For example, the shorter UV wavelengths (below 200 nm) are highly attenuated by most materials (including the atmosphere), and are not considered an exposure problem because sources are uncommon. Prolonged exposure to other UV wavelengths (especially 270–320 nm) can cause deleterious bioeffects, as well as cause the degradation of many materials such as plastic and rubber (Sliney and Wolbarsht, 1980).

Visible/IR

It should be noted that, although the visible region of the spectrum is defined as the wavelengths between 400 and 700 nm, the actual range visible to the eye can range from 380 to 780 nm. Incoherent sources of visible light (400 to 760–780 nm) and IR (i.e., "heat") radiation (760–780 nm to 1 mm) are common, and most do not present a hazard. This chapter addresses the relatively few sources of intense visible/IR radiation that should be handled with care, but does not cover conditions such as heat stress or conductive heat. Visible and IR sources are discussed together, because most (but not all) IR sources will also have a visible component.

Laser Radiation

A laser is a device that produces coherent electromagnetic radiant energy within the optical spectrum, from the far UV to the far IR region of the spectrum. The term laser is an acronym for *light amplification by stimulated emission of radiation*. Based on the theoretical work by Albert Einstein in 1916, and developed by Townes, Schawlow, and others in the late 1950s, the first laser was built by T. Maiman in 1960 (Henderson, 1997). At the time, it was considered a solution waiting for problems. And since then, lasers have been developed into an essential tool, used in a multitude of diverse applications—medicine, technology, consumer items, industrial, defense, entertainment, communication, law enforcement, and a host of others.

Besides having the characteristic of coherence, lasers emit discrete wavelengths within the optical spectrum, and although most lasers are monochromatic (emitting one wavelength, or single color), many lasers emit several discrete wavelengths. For example, the argon laser emits radiation at several different wavelengths, but it is generally designed to emit only one green "line" (wavelength) at 514.5 nm or a blue line at 488 nm. When considering potential health hazards, it is crucial to establish the output wavelength or wavelengths. A list of common laser wavelengths is summarized in Table 99.2.

SOURCES OF EXPOSURE

Ultraviolet Radiation

Although the earth's ozone (O_3) layer absorbs most of the UV produced by the sun, the remaining unattenuated solar UV reaching the earth is by far the largest source of exposure for the general population (Sliney and Wolbarsht, 1980). Exposure to UV varies and is dependent on the time of day, latitude, elevation, cloud cover, and presence of reflective surfaces. As an example, measurements of the daily variation in UV energy (erythemal effective irradiance) were made in Denver, CO, on a clear summer day. The maximum value (11.3 μW/cm^2) was reached at approximately noon while low dense cloud cover somewhat reduced this intensity (Machta, 1975).

The UV-emitting equipment in the work/laboratory setting is common, and is used for a wide variety of purposes. The radiation emitted from UV sources is either deliberate (e.g., killing bacteria, ink curing), or incidental (unwanted UV from arc welding). Table 99.3 lists some processes and devices found in the workplace and other locations that emit non-coherent UV radiation (Cremonese et al., 1982; NRPB, 1977).

Visible/IR

Like UV radiation, the single largest source of visible/IR radiation is the sun. Occupational sources include lamps and lighting systems (flashbulbs, photocopiers, etc.), space heaters, welding, and industrial settings such as glass works and iron foundries. Some devices, such as remote control units and IR loops for the hearing impaired use IR signals, but the output power level is insignificant.

Lasers

As previously stated, lasers are used in a plethora of applications, and are on their way to becoming a ubiquitous

TABLE 99.2   Common Laser Wavelengths

Type	Wavelength (nm)	Optical Region
Krypton–fluoride excimer	249	UV
Xenon chloride excimer	308	UV
Helium–cadmium ("He–Cad")	325, 442	UV, visible (blue)
Krypton	351, 356, 530, 647	UV, visible, IR
Nitrogen	337	UV
Argon	457, 476, 488, 514	Visible (blue-green)
Copper vapor	511, 578	Visible (green)
Helium neon ("HeNe")	633	Visible (red)
Gold vapor	628	Visible (red)
Titanium sapphire ("Ti–Saph")	660–1060 (tunable)	Visible, IR
Semiconductor (GaInP family)	670–680	Visible (red)
Ruby	694	Visible (red)
Alexandrite	700–830	IR
Semiconductor (GaAlAs family)	750–900	IR
Neodymium: YAG	266, 355, 532, 1064	UV, Visible (green), IR
Fiber—erbium, ytterbium, neodymium, and others	1060–1610	IR
Semiconductor (InGaAsP family)	1300–1600	IR
Hydrogen fluoride	2600–3000	IR
Carbon dioxide	10,600	IR

presence and technology as they become smaller, more powerful, and cheaper to manufacture. Essentially all segments of the laser market, including communication, materials processing, optical storage, medical and aesthetic, military, R&D, instrumentation, entertainment, and others are expected to continue to grow at a steady pace in coming years (Overton et al., 2014).

BIOEFFECTS

Ultraviolet Radiation

Effects from the exposure to UV radiation are very dependent on the wavelength. The organs that are most vulnerable are the eyes (especially the cornea and lens) and skin. In addition, photosensitization to UV can occur from various substances. It should be noted that a small amount

TABLE 99.3   Processes and Devices That Emit UV Radiation

Germicidal lamps	Tanning beds
Curing equipment	Welding equipment
Fluorescent lamps and equipment	Halogen lamps
Mercury lamps	Dermatology lamps ("Wood's lamps")
Carbon and xenon arcs	Scatter from UV lasers
Metal smelting	Solar lamps
Glass processing	Projection systems
"Black lights"	

of UV exposure is beneficial, in that it promotes vitamin D production.

Effects on the skin can be divided into acute and chronic. The most common acute effect from overexposure to UV is erythema (sunburn), which is a photochemical response of the skin resulting from overexposure to the UV-C and UV-B regions (especially 200–315 nm). The severity of the effect depends on the duration and intensity of the exposure. Exposure to UV-A alone requires far greater levels to induce erythema; the UV-A dose required to produce erythema is 800–1000 times that of the UV-B dose. Maximum sensitivity of the skin occurs at 295 nm (ACGIH, 2010).

Chronic exposure to sunlight (especially wavelengths shorter than 315 nm) accelerates skin aging and increases the risk of developing skin cancer. These effects may not manifest themselves for years or even decades. Although the causal relationship of UV-B exposure and squamous cell and basal cell carcinoma seems to be well established (van der Leun, 1987; Sterenborg and van der Leun, 1987), the causal relationship for malignant melanoma is weaker (Koh et al., 1990).

The primary effect from UV-B and UV-C overexposure on the eye is keratoconjunctivitis (also known as welders flash, snow blindness, and arc eye), which involves the cornea. Symptoms include a "ground-glass" sensation in the eye, aversion to bright light, and pain/discomfort within hours of exposure. The cornea is considered as the "skin" of the eye, and the corneal cells have a very short lifespan— only about 48 h. Damage to this outermost layer, then, is usually repaired within 1–2 days by normal cell turnover (Sliney and Wolbarsht, 1980).

Excessive UV exposure has been associated with cataracto-genesis, especially in the case of brunescent (brown colored) cataract, a form of senile cataract (Hiller et al., 1977).

Photosensitivity, described as hypersensitivity to certain wavelengths in the optical spectrum, may be genetic, associated with various immune states, or induced. Certain creams, lotions, shampoos, and drugs (such as tetracycline) may induce hypersensitivity to UV radiation in many individuals. Fitzpatrick et al. lists many common photosensitizers (Fitzpatrick et al., 1974).

Visible/IR

Skin injury is usually not a concern from most conventional sources of IR radiation (such as industrial furnaces and open arcs) because the natural aversion response—moving away from the source of heat—will normally occur before injury can take place (Sliney and Wolbarsht, 1980). The eye does not have adequate thermal warning properties and is the organ most susceptible to visible/IR overexposure. Far-IR radiation (3000–1 mm) is absorbed in the cornea, and middle IR radiation (1400–3000 nm) is absorbed by the lens. Radiation between 400 and 1400 nm is focused onto the retina, and this wavelength range is referred to as the retinal hazard region (Youssef et al., 2011). It should be noted that glare can overwhelm adaptive capacity of the eye, causing visual fatigue and reduced visual perception, which can make a worker more prone to accidents.

Injury to the retina can be caused by two mechanisms—thermal injury (focusing thermal energy onto the retina), and blue-light hazard [B(λ)] photochemical reaction to chronic exposures to visible light (especially blue wavelengths between 400 and 500 nm). Figure 99.2 shows relative retinal irradiances for staring directly at various light sources, and includes exposure risks from retinal thermal injury and photochemical injury.

Although photokeratitis from welding arcs and other arc sources can be painful, it is transient. The principal risk of permanent injury from viewing bright light sources is photo-retinitis, for example, solar retinitis with an accompanying scotoma, which results from staring at the sun. Blue light retinal injury (photoretinitis) can result from either viewing an extremely bright light for a short time or viewing a less bright light for longer exposure periods; that is, reciprocity. Reciprocity helps to distinguish these effects from thermal burns, in which heat conduction requires a very intense exposure within seconds to cause a retinal coagulation; otherwise, surrounding tissue conducts the heat away from the retinal image. Injury thresholds for an acute injury in experimental animals for both corneal and retinal effects have been corroborated for the human eye from the accidental data. Occupational safety limits for exposure to UV radiation and bright light are based on this knowledge (ACGIH, 2010).

Lasers

Although skin injury may occur from higher powered lasers, the eye is the organ of greatest concern. For lower power lasers (<1 milliwatt), the natural aversion response or blink reflex, assumed to be up to 0.25 s, is sufficient to protect the eye. However, higher power lasers may produce damage to various structures of the eye in a much shorter time. Lasers in pulse mode can produce extremely high peak powers, and can have pulse durations in the femtosecond regime, which could result in numerous pulses to be absorbed into an unprotected eye before being aware of the exposure. Compared to a continuous wave laser with the same average power, or single pulse lasers with the same energy, more injury potential exists with repetitively pulsed lasers (Marshall, 2001).

The laser wavelength will determine the ocular absorption site, as shown in the Figure 99.1. Injury mechanisms are mostly either thermal, which are immediately apparent, or photochemical, which are manifested in approximately 24 h.

EXPOSURE GUIDELINES

The American Conference of Governmental Industrial Hygienists (ACGIH) annually publishes threshold limit values (TLVs©) for chemical and physical agents. Threshold

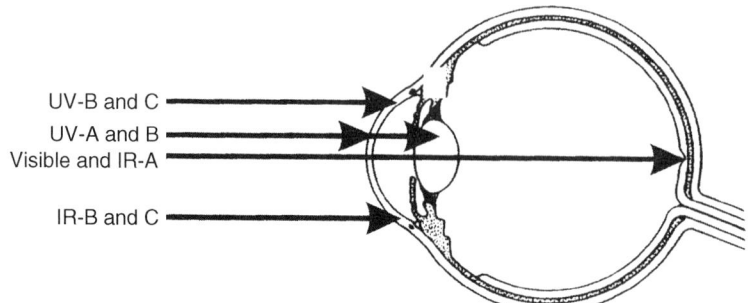

UV-B and C
UV-A and B
Visible and IR-A

IR-B and C

FIGURE 99.1   Locations in the eye where different bands of optical radiation deposit energy that may cause injury.

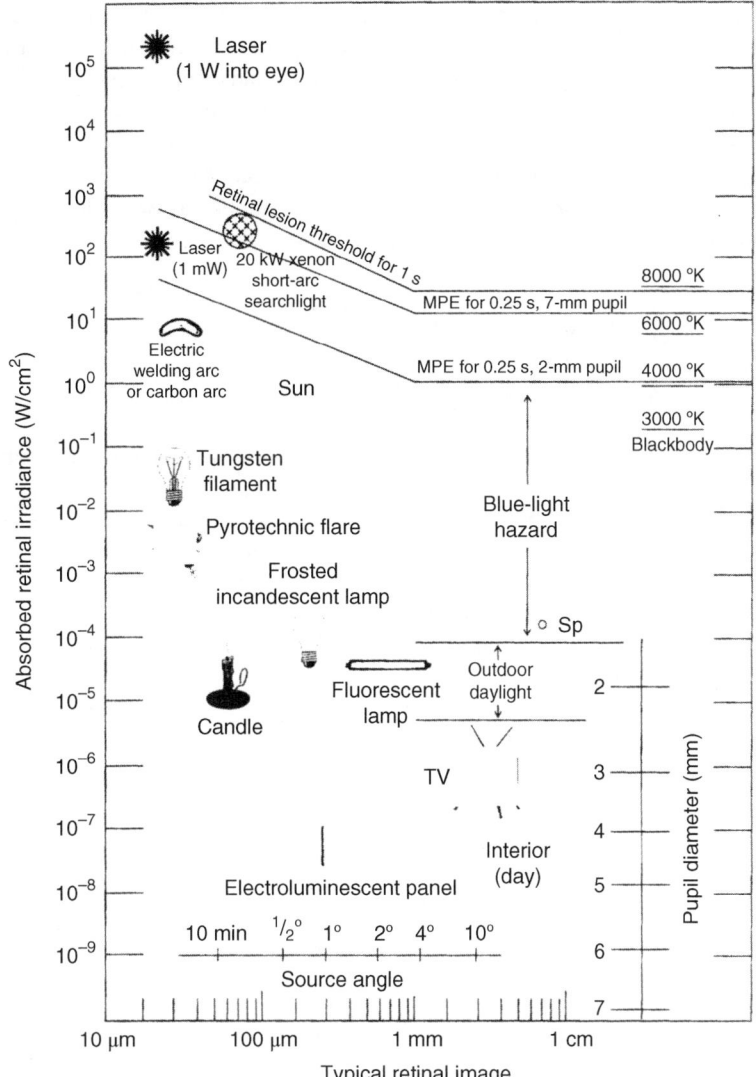

FIGURE 99.2 Relative retinal irradiances for staring directly at various light sources. Two retinal exposure risks are depicted: retinal thermal injury, which depends on the retinal image size, and photochemical injury, which depends on the degree of blue light in the source's spectrum. The horizontal scale indicates typical image sizes. Most intense sources are so small that eye movements and heat flow spread the incident energy over a larger retinal area. The range of photoretinitis ("blue-light hazard") is shown to extend from the normal outdoor light levels and above. The xenon arc lamp is clearly the most dangerous non-laser hazard. (*Source*: Sliney and Wolbarsht, 1980).

limit values are widely used for UV, visible, and IR radiation. For laser exposures, the American National Standards Institute (ANSI) Z136.1 standard is the most widely used for determining maximum permissible exposures (MPEs) for both skin and eyes (ANSI, 2014).

Ultraviolet Radiation

Threshold limit values "refer to UV radiation in the spectral region between 180 and 400 nm, and represent conditions under which it is believed that nearly all workers may be repeatedly exposed without adverse health effects." The UV TLV is highly dependent upon the wavelength, and separate TLVs are provided for UV-A and UV-B regions.

For UV radiation in the *UV-A region*, the ACGIH suggests limiting long exposures (over 1000 s, about 16 min) to an irradiance of 1 mW/cm^2. For shorter durations (up to 1000 s), the total energy should not exceed 1 J/cm^2.

In the *UV-B region*, the total irradiance is determined by multiplying the irradiance in small wavelength bands ($\Delta\lambda$)

by the relative spectral effectiveness (S_λ), a weighting factor based on the action spectra for eyes and skin:

$$E_{\text{eff}} = \Sigma E_\lambda \cdot S_\lambda \cdot \Delta\lambda$$

The effective irradiance (E_{eff}), the weighted total in the UV-B range is limited to 3 mJ/cm^2 over an 8-h day. Some UV radiometer manufacturers manufacture detectors that respond to UV radiation according to the relative spectral effectiveness. Since the measurement result using this detector is in units of effective W/cm^2, comparison of the reading to the permissible UV exposure table will readily provide a determination of the allowable exposure duration per day.

Visible/IR

The ACGIH provides TLVs for visible and near IR radiation. Biological weighting factors for both the B(λ) and retinal thermal hazard [R(λ)] are assigned to wavelengths between 305 and 1400 nm. Blue-light hazard has peak values at 435–440 nm while R(λ) values rise sharply to the 435–700 nm region, then tail off to 1400.

Calculating TLVs for visible and IR wavelengths is multi-faceted. The ACGIH includes four conditions that may apply, and each requires one or more equations for calculating the appropriate TLV:

- Retinal thermal injury from a visible light source
- Retinal photochemical injury from chronic blue light exposure
- Thermal injury to cornea and lens from IR radiation
- Retinal thermal injury from near IR radiation

Due to the complexity and number of equations for calculating TLVs (over a dozen), the reader is referred directly to the ACGIH TLV booklet.

On a more empirical note, Whillock et al. measured the radiance levels at four industrial sites (a glass works, iron foundry, large-scale float glass facility, and a small food-preparation facility). They found that the radiance did not indicate potential photochemical or burn hazards to the retina from these operations. However, the authors did note that a potential for lenticular injury (cataracts) exists for long-term exposures (10–15 years) to IR radiation (Whillock, 1990).

Lasers

Although the ACGIH publishes TLVs for lasers, the ANSI Z136.1 standard, Safe Use of Lasers, is much more widely used. The ANSI standard employs the use of the MPE, which is defined as the level of laser radiation to which an unprotected person may be exposed without adverse biological changes to the eye or skin. The MPE is dependent upon a number of factors, including wavelength, duration of exposure, whether the laser is continuous wave or pulsed, and others. The Z136.1 standard includes several pages of tables, charts, correction factors, rules, and other information to properly calculate the MPE. It is highly recommended that the MPE is calculated by a qualified individual, such as a Laser Safety Officer, due to the many possible pitfalls and caveats involved. It should be noted that several versions of software have been developed for these calculations. However, like any computations, if improper assumptions are entered, incorrect MPE values will result.

Since the early years of the ANSI Z136.1 standard, a classification scheme has been used to separate lasers into categories according to their potential for hazard. In 2007, the standard updated this classification system to harmonize with the international standards. Thus, there are a few classes that have changed only in name, but not intent. Table 99.4 shows the current and past terminology (from Rockwell Laser Industries, 2014).

INDUSTRIAL HYGIENE

For emitters in the optical spectrum, a comprehensive hazard evaluation requires precise instrumentation and complex measurements of spectral irradiance and radiance of the broad-band sources. This can often be done by an experienced industrial hygienist or health physicist. And for lasers, measurements of beam irradiance or other characteristics are rarely done for a safety evaluation. Instead, the ANSI classification will determine the types of controls needed for the protection. Class 4 lasers generally require the most stringent control measures. A qualified laser safety officer or professional is often consulted for a safety evaluation for high powered systems.

Engineering controls such as welding curtains, baffles, controlled areas, and doorway interlocks are used to control optical radiation hazards. Administrative controls, such as limiting the time of exposure, can be useful with regard to UV sources and welding arcs. Protective eye wear and face shields constitute a primary control measure in many operations such as welding or laser research.

Ultraviolet and IR Protection

Although many sources of UV radiation also emit some visible light, some do not. A germicidal lamp, for example, may only emit a faint visible glow while emitting intense UV radiation. Therefore, the amount of visible light should not be used as an indicator of the UV intensity. A combination of engineering/administrative controls and personal protection should be used to limit UV exposures. Engineering controls are preferred due to their broader effectiveness.

Shielding. UV leakage should be contained and shielded whenever possible. If viewing the process is required,

TABLE 99.4   ANSI Z136.1 Laser Classification System, Pre- and Post-2007

Class	Relative Hazard/Examples	ANSI Z136.1 (Post-2007)	ANSI Z136.1 (Pre-2007)
1	Minimal hazard. DVD player, laser printer are Class 1 laser systems (actual internal laser may be 3b)	Any laser or laser system containing a laser that cannot emit laser radiation at levels that are known to cause eye or skin injury during normal operation. Also, a *Class 1 system* may contain a Class 3 or 4 laser where the beam is fully enclosed	
1M	Safe for unprotected eye, potentially hazardous when collecting optics are used	Not known to cause eye or skin damage unless collecting optics are used	NA
2	Low hazard. Safe for unintended exposure, prolonged staring should be avoided Up to 1 mW power for visible, continuous wave (cw) lasers	Visible lasers are considered incapable of emitting laser radiation at levels that are known to cause eye injury within the time period of the human *eye aversion response* (0.25 s)	
2M	Visible low power lasers; either collimated with large beam diameter or highly divergent	Not known to cause eye injury within the aversion response time unless collecting optics are used	NA
3a	Can be hazardous. Power range for cw lasers: 1–5 mW	NA	Lasers similar to Class 2 with the exception that collecting optics cannot be used to directly view the beam
3R	Can be hazardous. Only small hazard potential for accidental exposure	Replaces Class 3a and has different limits. Up to five times the Class 2 limit for visible and five times the Class 1 limits for some invisible	NA
3b	Medium power lasers. Power range for cw lasers: 5–500 mW. Do not present a diffuse (scatter) hazard or significant skin hazard except for higher powered 3b lasers operating at certain wavelength regions	Medium powered lasers (visible or invisible) that present a potential eye hazard for intrabeam (direct) or specular (mirror-like) conditions. Class 3b lasers	
4	High power lasers. Greater than 500 mW for cw lasers. Industrial laser cutters, most R&D and surgical lasers. May be a fire hazard, and result in hazardous fumes	High powered lasers (visible or invisible) considered to present potential acute hazard to the eye and skin for both direct (intrabeam) and scatter (diffused) conditions	

Source: Rockwell Laser Industries (2014b).

observations should take place through materials such as polycarbonate plastics that attenuate UV-C and UV-B wavelengths while remaining transparent to visible light. Almost all glass and plastic lens materials block UV radiation below 300 nm and IR radiation at wavelengths above 3000 nm.

Interlocks. If access to equipment that houses UV radiation is required, the housing should be interlocked such that the UV source will be turned off when the housing or cover is removed. Interlocks should be tested regularly to ensure proper operation.

Non-reflective materials. If possible, reflective materials such as shiny metals (e.g., aluminum), glossy light-colored paint, and paint containing titanium should be avoided. It should also be noted that materials that appear dark and non-reflective in the visible portion of the spectrum may not be the same in the UV portion.

If UV radiation exposure cannot be avoided, the following will help to minimize the exposure.

Personal protection. All skin with potential for exposure (especially hands and face) should be protected. Gies et al. measured the effectiveness of various types of clothing and fabrics, and determined that clothing can provide significant protection against UV exposures (Gies et al., 1994). In general, dense, heavy fabrics with no direct transmission paths are recommended for UV protection during welding operations. Most gloves found in a laboratory environment will provide at least some protection for hands from UV exposure.

Safety eyewear with a UV-protective coating are available that will protect against UV wavelengths and still afford adequate visibility (they sometimes are light yellow in color). Moseley measured the UV and visible light transmission properties of several types of protective eyewear used in medical, industrial, and welding applications (Moseley, 1985). Of paramount importance in these types of eye wear (and for industrial sunglasses) is the peripheral protection. Side shields or wraparound designs are important to protect against the Coroneo effect, which is the focus of temporal, oblique rays into the nasal equatorial area of the lens, where cortical cataracts frequently originate (Sliney, 1996).

Spectacles and goggles designed for ocular protection against IR radiation in foundries and glass manufacturing generally have a light greenish tint, although the tint may be darker if some comfort against visible radiation is desired. Such eye protectors should not be confused with the blue lenses used in the steel and foundry operations, for which the objective is to visually check the temperature of the melt; these blue spectacles do not provide protection and should be worn only briefly (Vinger and Sliney, 1995).

Ozone production. Ozone is a colorless, toxic gas, irritating to the lungs and mucous membranes, having a characteristic pungent odor. It is formed by the photochemical reaction between short wavelength UV radiation and oxygen molecules present in the air. In particular, wavelengths in the range of 185–300 nm (the UV-C and UV-B ranges), and especially with the 254 nm common in germicidal lamps, there is a distinct hazard of O_3 production. The threshold for odor detection ranges from 0.005 to 2 ppm. The TLV–time weighted average (TWA) (2001) for O_3 ranges from 0.05 to 0.1 ppm, depending on the workload (and resultant inhalation rate). Odor detection, therefore, may not provide adequate protection from occupational exposures. Diffey and Hughes provided a way to estimate the concentration of O_3 that may exist in the vicinity of UV-emitting lamps (Diffey and Hughes, 1980).

Welding Filters

Infrared and UV radiation filtration is readily achieved in glass filters with additives such as iron oxide, but the visible attenuation determines the shade number, which is a logarithmic expression of the attenuation (the shade number is equal to $7/3$ the optical density (OD) plus 1). Normally a shade of 3–4 is used for gas welding (goggles), and 10–14 for arc welding and plasma arc operations (helmet protection required). The thumb rule is that if the welder finds the arc comfortable to view, adequate attenuation has been provided against ocular hazards. Supervisors, welders' helpers, and other persons in the work area may require filters with a relatively low shade number (e.g., 3–4) to

protect against welder's photokeratitis ("arc eye") (ANSI, 1994).

The autodarkening filter/helmet has become a popular and effective solution. Regardless of the type of filter, it should meet the ANSI Z87.1 and Z-49.1 standards for fixed welding filters specified for the dark shade (ANSI, 1989, 1994). Considering the total welding process, the welders formerly had to lower and raise the helmet or filter each time an arc was started and quenched. The welder had to work "blind" just before striking the arc. Furthermore, the helmet frequently is lowered and raised with a sharp snap of the neck and head, which can lead to neck strain or more serious injuries. Faced with this uncomfortable and cumbersome procedure, some welders often initiate the arc with a conventional helmet in the raised position—leading to photokeratitis. Under normal ambient lighting conditions, a welder wearing a helmet fitted with an autodarkening filter can see well enough with the eye protection in place to perform tasks such as aligning and fixturing the parts to be welded, precisely positioning the welding equipment and striking the arc (Eriksen, 1985). Then, in the most typical helmet designs, light sensors detect the arc flash and direct an electronic drive unit to switch a liquid crystal filter from a light shade to a preselected dark shade. The aforementioned drawbacks of fixed shade filters are largely eliminated by autodarkening filters, which explain the widespread popularity of these systems.

Lasers

The ANSI Z136.1 standard includes a comprehensive list of requirements for addressing laser safety concerns, including engineering and administrative controls, eyewear requirements, signs, entryway controls, and others. Non-beam hazards should not be ignored—in fact, numerous accidents, some fatal, have occurred in the past from lack of proper controls for these hazards. Non-beam hazards include high voltage and capacitors, laser-generated air contaminants from laser cutting and welding operations, fires, injuries from robotic arms, and others. Rockwell Laser Industries keeps a listing of laser accidents (reported voluntarily) on their website, although it is estimated that only a small percentage of laser injuries are reported (Rockwell Laser Industries, 2014a).

Laser Eye Protectors

The ANSI Z136.1 standard calls for laser eye protection when working with Class 3b or 4 lasers. The objective of laser eye-protective filters is to transmit as much visible light for observing while blocking the laser wavelength or wavelengths of concern. The protective factor used for laser eye wear is called OD. The OD is a logarithmic expression for attenuation factor, for example, a protection factor of 1000 (10^3) is equivalent to an OD of 3; a protection factor of 1 million (10^6) is equivalent to an OD of 6. The equation for determining OD is as follows:

$$OD = Log_{10}[Irradiance/MPE],$$

where irradiance = maximum power/area(W/cm^2)
MPE = maximum permissible exposure(W/cm^2).

Because most lasers emit powers or energies on the order of thousands to millions of times the MPE limit, optical densities of 3–6 are most typical. The ANSI standard Z136.1 provides detailed methods for determining the appropriate eye wear (ANSI, 2014). Polycarbonate has superior burn-through characteristics (Sliney et al., 1993) compared with other plastics and is also superior to all other lens materials for industrial impact protection.

MEDICAL MANAGEMENT

Photokeratitis

It is also called welder's flash, flash burn, arc eye, and snow-blindness. Ultraviolet photochemical injury of the cornea, occurring principally from 200 to 315 nm. This is a transient condition that lasts for 1–2 days and usually heals without leaving a scar. Symptoms include mild to severe pain, light sensitivity, blurry vision, eye irritation, bloodshot/watery eyes. Treatment may include dilation, padded dressing (cool) to immobilize the eyelids, and rest. Infection is a possibility and antibiotics may be considered.

Erythema

Ultraviolet photochemical injury of the skin occurring principally from 200 to 315 nm. This is also a transient condition that lasts for 1–2 days. Treatment may include cool wet cloths, lotions using aloe vera, topical steroids (e.g., 1% hydrocortisone cream) for more severe cases.

Laser Eye Injury

Thermal injury to the retina (occurring at 400–1400 nm); the local burn of the retina results in a blind spot (scotoma). Thermal injury of the cornea and conjunctiva (occurring at approximately 1400 nm to 1 mm). This type of injury is almost exclusively limited to laser radiation exposure. A visit to an ophthalmologist should be made immediately after a suspected laser injury.

REFERENCES

ACGIH (American Conference of Governmental Industrial Hygienists) (2010) *Documentation of the Threshold Limit Values and Biological Exposure Indices*, Cincinnati, OH.

ANSI (American National Standards Institute) (1989) *Practice for Occupational and Educational Eye and Face Protection*, ANSI Standard Z87.1, New York, NY.

ANSI (American National Standards Institute) (1994) *Safety with Welding and Cutting and Allied Processes*, ANSI Z49.1, Miami: American Welding Society.

ANSI (American National Standards Institute) (2014) *Safe Use of Lasers*, Standard Z136.1, Orlando: Laser Institute of America.

Barat, K. (2009) *Laser Safety, Tools and Training*, Boca Raton, FL: CRC Press.

CIE (1970) *Commission International de l'Eclairage (International Commission on Illumination), International Lighting Vocabulary*, Publication CIE No. 17 (E1.1), Paris.

Cremonese, M., et al. (1982) Irradiance measurements of some UV sources. *Health Phys.* 42(2):179–185.

Diffey, B.L. and Hughes, D. (1980) Estimates of ozone concentrations in the vicinity of ultraviolet emitting lamps. *Phys. Med. Biol.* 25(3):559–561.

Eriksen, P. (1985) Time resolved optical spectra from M1G welding arc ignition. *Am. Ind. Hyg. Assoc. J.* 46:101–104.

Fitzpatrick, T.B., et al. (1974) *Sunlight in Man, Normal and Abnormal Photobiologic Responses*, Tokyo: University of Tokyo Press.

Gies, H.P., et al. (1994) Ultraviolet radiation protection factors for clothing. *Health Phys.* 67(2):131–139.

Henderson, A.R. (1997) *A Guide to Laser Safety*, Chapman & Hall.

Hiller, R., et al. (1977) Sunlight and cataract: an epidemiologic investigation. *Am. J. Epidemiol.* 105:450–459.

Koh, H.K., et al. (1990) Sunlight and cutaneous malignant melanoma: evidence for and against causation. *Photochem. Photbiol.* 51(6):765–779.

Machta, L. (1975) CIAP In: Cotton, L.G., Hass, W., and Komhyr, W., editor. *Measurements of Solar Ultraviolet Radiation*, Interagency Agreement DOT-A5-20082, Washington, DC: U.S. Department of Transportation Final Report.

Marshall, W. (2001) Lasers. In: Alaimo, R.J., editor. *Handbook of Chemical Health and Safety*, Chapter 66, American Chemical Society, Oxford University Press.

Moseley, H. (1985) Ultraviolet and visible radiation transmission properties of some types of protective eyewear. *Phys. Med. Biol.* 30(2):177–181.

NRPB (1977) *National Radiation Protection Board, Protection Against Ultraviolet Radiation in the Workplace*, Harwell, UK.

Overton, G., et al. (2014) *Laser Marketplace 2014: Lasers forge 21st Century Innovations, Laser Focus World*.

Rockwell Laser Industries (2014a) Laser Accident Database. Available at https://www.rli.com/resources/accident.aspx.

Rockwell Laser Industries (2014b) Laser Standards and Classifications. Available at http://www.rli.com/resources/articles/classification.aspx.

Sliney, D. and Wolbarsht, M. (1980) *Safety with Lasers and Other Optical Sources*, New York, NY: Plenum Press.

Sliney, D.H., Sparks, S.D., and Wood, R.L. (1993) The protective characteristics of polycarbonate lenses against CO_2 laser radiation. *J. Laser Appl.* 5(1):49–52.

Sliney, D.H. (1996) Laser effects on vision and ocular exposure limits. *Appl. Occup. Environ. Hyg.* 11:313–319.

Sterenborg, H.J.C.M. and van der Leun, J.C. (1987) Action spectra for tumorigenesis by ultraviolet radiation. In: Passchier, W.F. and Bosnjacovic, B.F.M., editors. *Proceedings of Seminar on Human Exposure to Ultraviolet Radiation: Risks and Regulations*, ICS 744, Exerpta Medica. Amsterdam: Elsevier Science Publishers.

van der Leun, J.C. (1987) Interactions of different wavelengths in effects of UV radiation on skin. *Photodermatology* 4(5): 257–264.

Vinger, P.F. and Sliney, D.H. (1995) The prevention of sports and work-related injury. *Ophthalmol. Clin. NA* 8(4):709–721.

Whillock, M.J. (1990) Measurements and hazard assessment of the optical emissions from various industrial infrared sources. *J. Radiol. Prot.* 10(1):43–46.

Youssef, P.N., et al. (2011) *Retinal Light Toxicity, Eye, (London), The Royal College of Ophthalmologists, Jan 2011;* 25 (1): 1–14.

100

IONIZING RADIATION

Thomas A. Lewandowski, Juhi K. Chandalia, and Peter A. Valberg

IONIZING RADIATION

Ionizing radiation refers to subatomic particles or electromagnetic waves that have sufficient energy to break apart chemical bonds. Life on earth has always existed in a constant, low level stream of ionizing radiation from the radionuclides (elements emitting ionizing radiation) naturally present in the earth's crust, as well as cosmic rays from the sun and interstellar sources (UNSCEAR, 2000). Life has adapted to these conditions and, in fact, the low level of mutation attributable to these natural sources may have been instrumental in the evolution of life on earth. Modern technology allows for the production and exploitation of differing forms of ionizing radiation and with it, different types and intensities of exposure. This has in turn led to a greater study of exposures and the potential health effects of ionizing radiation.

Types of Ionizing Radiation

The two main forms of ionizing radiation are high energy subatomic particles (e.g., alpha and beta radiation) or high energy electromagnetic waves (gamma rays or X-rays). The energies of both types of radiation are measured in "electron-volts," or eV, where 1 eV corresponds to the energy of an electron accelerated through a potential of 1 V.

Subatomic particles are high energy fragments of atoms such as protons, neutrons, alpha particles, and beta particles (discussed below). High energy subatomic particles may originate from two sources: the decay of radioactive elements or generation by extraterrestrial sources (i.e., cosmic rays[1] and solar flare events). From an occupational standpoint, the

[1] Cosmic "rays" are actually high energy subatomic particles emitted by the sun and other stars. Their description as rays is an historical artifact.

former is of more importance. Radioactive elements are composed of inherently unstable nuclei that undergo decay, shedding energy and/or mass in the form of subatomic particles, in order to reach a stable state. Radioactive elements transform into other elements as they decay, some of which may also be unstable and emit ionizing radiation as part of a decay chain. The chain continues until a stable (i.e., non-radioactive) isotope is reached. Each radioactive isotope decays on a specific timescale called the half-life ($t_{1/2}$)— i.e., the time it takes for half of the existing material to be converted to a daughter product.

An alpha particle (α) is a highly energetic nucleus of a helium atom, ejected from a much larger nucleus during decay. It consists of two positively charged protons and two electrically neutral neutrons and is designated by the symbols $^2He_4{}^{++}$ or α. Alpha particles are emitted from the nuclei of certain radioactive isotopes (see Table 100.1). When an alpha particle is emitted, the alpha-emitting parent atom is transformed into a daughter isotope with an atomic number decreased by two, and an atomic mass decreased by four. For example, uranium-238 (^{238}U, atomic number 92) is transformed by alpha emission to thorium-234 (^{234}Th, atomic number 90).

A beta particle (β) is either an electron or a positron that is ejected from the nucleus in the process of nuclear decay. The β-particle is typically an electron formed by the decay of a neutron (yielding both an electron and proton). A beta particle can also be a positron (a positively charged electron), but the lifetime of the positron (which is essentially an "antielectron") is very short. Beta particles have very low mass (i.e., 1836 times less than the mass of a proton). Emission of a beta particle results in the transformation of the beta-emitting parent atom into a daughter, the atomic number of which is one greater (due to the gaining of one

Hamilton & Hardy's Industrial Toxicology, Sixth Edition. Edited by Raymond D. Harbison, Marie M. Bourgeois, and Giffe T. Johnson.
© 2015 John Wiley & Sons, Inc. Published 2015 by John Wiley & Sons, Inc.

TABLE 100.1   Examples of Isotopes That Undergo Alpha Decay

Isotope	Daughter	Half-life
^{238}U	^{234}Th	4.47 billion years
^{232}Th	^{228}Ra	14 billion years
^{241}Am	^{237}Np	432.7 years
^{210}Po	^{206}Pb	138 days
^{222}Rn	^{218}Po	3.8 days

proton); the atomic mass remains unchanged (because a neutron and proton have nearly identical mass). For example, ^{90}Sr decays *via* beta particle emission to ^{90}Y, which has an atomic number one greater but the same atomic mass (see Table 100.2). Beta particles emitted by radioisotopes exhibit a range of energies that extends from near zero to a maximum, which is characteristic of that radioisotope. These maximum energies extend from several kiloelectron volts (keV) to several megaelectron volts (MeV).

A more unusual form of particulate radiation involves free neutrons. Having no net charge, these particles do not interact electromagnetically with atoms but, upon collision with an atom, may cause the atom to become ionized. Gamma radiation (see below) is typically produced by this interactive process. Other types of particle decay (e.g., electron capture, neutrino emission) are also possible but less common.

The other main class of ionizing radiation consists of high energy electromagnetic waves. Although the electromagnetic spectrum includes radio waves, microwaves, visible light, and ultraviolet (UV) rays, only the most energetic portion of the electromagnetic spectrum, X-rays, and gamma rays (γ), have the ability to ionize matter (i.e., adversely affect atomic structure). Some ambiguity exists concerning the distinction between gamma rays and X-rays. Historically, they were differentiated by the way in which they were produced (i.e., from decay of radioactive elements [γ-rays] *versus* the bombarding of targets with accelerated electrons [X-rays]), but this distinction has fallen out of use under current technologies and

TABLE 100.2   Examples of Isotopes That Undergo Beta Decay

Isotope	Daughter	Half-life
^{40}K	^{40}Ca	1.25 billion years
^{14}C	^{14}N	5730 years
^{137}Cs	^{137}Ba	30.2 years
^{90}Sr	^{90}Y	29.1 years
^{3}H	^{3}He	12.3 years
^{228}Ra	^{228}Ac	5.76 years
^{60}Co	^{60}Ni	5.27 years
^{32}P	^{32}S	14.3 days
^{131}I	^{131}Xe	8 days

the terms are used interchangeably (UNSCEAR, 2000). As with subatomic particles, the energy of the X-rays and gamma rays is quantified by eV. The "soft X-ray" spectrum ranges from electromagnetic waves with energies of 120 eV (0.12 keV) to 12 keV; the "hard X-ray" spectrum ranges from 12 to 120 keV.

Historical Background

In 1895, Wilhelm Röntgen discovered the phenomenon of penetrating radiation when high voltages were applied across electrodes contained in evacuated glass containers (Crookes tubes). Even when a Crookes tube was enclosed in a black box, the rays produced could penetrate the box and cause fluorescence of certain materials external to the box. Röntgen also found that photographic plates located outside the box became exposed and displayed images influenced by intervening objects (e.g., his wife's hand) (Röntgen, 1896). He named the radiation associated with these effects "X-rays." Röntgen was awarded the 1901 Nobel Prize in physics for his discovery. In the following year, Henri Becquerel reported that a similar phenomenon was associated with uranium ore. During the same period, Pierre and Marie Curie were investigating the properties of two new highly radioactive elements, polonium and radium, which they had painstakingly separated from a uranium ore (pitchblende). Becquerel and the Curies shared the 1903 Nobel Prize in physics for their work. In 1899 and 1900, scientists Ernest Rutherford and Paul Villard developed the nomenclature of alpha, beta, and gamma radiation based on the degree to which each penetrated various barriers (although they lacked a full understanding of the exact nature of each type of radiation).

In 1905, Einstein postulated the photon model of electromagnetic radiation. According to this model, electromagnetic radiation behaves as if it consists of particles of energy called photons, which travel at the speed of light. Each particle contains a discrete quantity, or "quantum," of electromagnetic energy. The energy content of the photon (E) is proportional to the frequency of the radiation (ν), as described by $E = h\nu$, where h is called Planck's constant ($6.62606957 \times 10^{-34}$ m^2 kg/s). Concordant with the theory of electromagnetism, a general atomic theory was developed initially by Neils Bohr and completed by Erwin Schrödinger. Together, these two fundamental theories explain much of the phenomenon of ionizing radiation.

Concurrent with scientific discoveries of the properties of ionizing radiation and radioactive elements was the technologic exploitation of ionizing radiation. In medicine, the first cure of a cancer (a basal cell carcinoma) with X-rays was accomplished in 1899 (Williams, 1903). X-rays were widely used to treat various diseases (psoriasis, tuberculosis) and in imaging, including the imaging of developing fetuses (Williams, 1903). In addition, the first self-sustaining, nuclear-fission reactor was developed in 1942 by Enrico Fermi and Leo Szilard. This advance led to the Manhattan

Project and development of the atomic bomb, with the first detonation in 1945 during World War II. After the war, both military and peaceful uses of ionizing radiation continued.

After World War II, with the recognition of radiation-induced mortality and morbidity among bomb survivors, guidelines for exposure to radiation were put in place (although the levels of exposure tolerated by these guidelines were substantially higher than those considered acceptable today). Over succeeding decades, observations were made of occupationally exposed people, including scientists, medical personnel, radium dial painters, uranium miners, industrial radiographers, and nuclear energy workers. Coincident with the expansion of knowledge about the biological effects of radiation was the development of sensitive instruments that enabled accurate measurement of radiation levels as well as levels of exposures to individuals. Due to the extensive data collected on exposure and health effects of radiation, the risks of radiation exposure are better understood than those associated with most other environmental agents.

ASSESSING EXPOSURE TO IONIZING RADIATION

Dosimetry and Units

There are several ways to assess exposure to ionizing radiation. One important distinction is the difference between radioactive decay rate (or flux of decay products) *versus* tissue dose. Measures of radioactive decay can be measured as "counts per minute" (cpm) or number of radioactive events (alpha particle emissions, X-ray photons, etc.) per unit time. A small amount of a highly radioactive material could have the same radioactive decay rate as a large amount of a slightly radioactive material. Measurement units of radioactivity decay events per unit of time include curies (Ci) and becquerels (Bq). One Bq is equal to one disintegration per second, and one Ci is equal to 3.7×10^{10} disintegrations per second.[2]

Measures of radioactive decay rates are of limited relevance to evaluating potential effects on human health because they pertain only to the quantity—and not the form—of the radiation involved. To characterize the effect of ionizing radiation on tissues, a number of different measures are used: the absorbed dose, equivalent dose, effective dose, and committed dose. Each of these terms is described below.

The *absorbed dose* has units of absorbed energy per unit mass and can be calculated from the rate of radioactive emission, the specific properties of the material through which the radiation moves, and various geometric and conversion factors. Absorbed dose can therefore account for the different penetrating properties of alpha-, beta-, and X-ray radiation in air, water, or biological tissue. Because it does not take into account potential tissue damage, the absorbed dose is not usually used as an indicator of long-term health risk in humans; however, it is often used to gauge the risk of acute radiation syndrome (ARS) (See below), which involves high level and acute exposures. The two main metrics for absorbed dose are the gray (Gy) and the rad. The Gy is the Système Internationale d'unités (SI) unit for absorbed dose and thus used increasingly relative to the more historical rad; 100 rads are equal to 1 Gy (NCRP, 1993).

The *equivalent dose* is based on absorbed dose, but is more closely related to biological and health effects of radiation than absorbed dose. It is important to realize that both absorbed and equivalent dose nominally measure the same quantity of energy deposited per mass (joules per kilogram). The equivalent dose differs from the absorbed dose in that it accounts for the relative biological effectiveness of the type of radiation to cause tissue damage (e.g., α-particles *versus* X-rays). Thus, the equivalent dose uses different units of measurement: the Sievert (Sv) and the rem (roentgen equivalent man). The Sv is the SI unit and is increasingly displacing the rem in most applications. To convert an absorbed dose to an equivalent dose, one needs to know the type of radiation involved and apply the appropriate radiation weighting factor (W_R, previously called the quality factor), which is a measure of linear energy transfer (LET) of different types of radiation. The LET is an indication of the potential of a given form of radiation to interact with matter as it passes through it. For example, for alpha particles, $W_R = 20$; for X-rays, $W_R = 1$. A summary of radiation weighting factors is provided in Table 100.3.

The *effective dose* is yet another dose metric that may be encountered and is used to account for the relative susceptibility of specific tissues to radiation-induced damage. To obtain an effective dose from the equivalent dose, tissue weighting factors are applied. These factors have been

TABLE 100.3 Radiation Weighting Factors

Radiation Weighting Factor	W_R
Photons, all energies (e.g., X- and gamma rays)	1
Beta particles, muons, all energies	1
Protons and charged pions	2
Alpha particles, fission fragments, heavy ions	20
Neutrons[a]	
Energy (E_n) < 1 MeV	$2.5 + 18.2\, e^{-[\ln(E_n)]^2/6}$
Energy (E_n) 1 MeV to 50 MeV	$5 + 17.0\, e^{-[\ln(2E_n)]^2/6}$
Energy (E_n) > 50 MeV	$2.5 + 3.25\, e^{-[\ln(0.04E_n)]^2/6}$

[a]Calculated as a continuous function based on the neutron energy (E_n).
Source: ICRP, 2007.

---

[2] Another unit occasionally seen, the roentgen (R) can only be used to quantify the amount energy deposition from gamma and X-rays, and only in air. One R is equal to depositing in a kilogram of dry air enough energy to cause 0.000258 coulombs of electric charge. It is a measure of the number of ionized molecules in a mass of air.

TABLE 100.4   Tissue Weighting Factors

Organ or Tissue Weighting Factor, wT	ICRP, 2007	NRC, 2012
Red bone marrow	0.12	0.12
Colon	0.12	–
Lung	0.12	0.12
Stomach	0.12	–
Breast	0.12	0.15
Gonads	0.08	0.25
Bladder	0.04	–
Liver	0.04	–
Esophagus	0.04	–
Thyroid	0.04	0.03
Skin	0.01	–
Bone surface	0.01	0.03
Brain	0.01	–
Salivary glands	0.01	–
Remainder	0.12	0.30
Body as a whole	1.0	1.0

officially tabulated by organizations such as the National Council on Radiation Protection (NCRP), International Commission on Radiological Protection (ICRP), and U.S. Nuclear Regulatory Commission (NRC) and are given in Table 100.4. Like the equivalent dose, the effective dose is expressed in either Sv or rem.

Finally, the term *committed dose* is used for cases where radionuclides are internalized by ingestion or inhalation and reflects the cumulative or chronic dose that will be received as the radionuclides decay over a period of 50 years. Thus, the committed dose encompasses the amount and activity of the internalized material as well as time the material remains within the body before being eliminated. The concept of committed dose is particularly important for bone-seeking radionuclides (e.g., calcium, strontium, radium, and plutonium) that undergo very slow turnover in the body.

Types of Exposure to Ionizing Radiation

People are exposed to both naturally occurring and anthropogenic ionizing radiation. Naturally occurring sources include cosmic rays and radioactive isotopes found everywhere (e.g., tritium, carbon-14, potassium-40, radium-226, radon-222). Examples of anthropogenic sources of radiation include medical devices such as X-ray machines, nuclear medicine, atomic-bomb fallout, and activities that generate radioactive waste.

Natural Sources   Cosmic radiation, which consists of highly energetic subatomic particles, is a large ubiquitous source of naturally occurring ionizing radiation. Radiation from the sun, or solar radiation, is one source of cosmic radiation (ATSDR, 1999). Cosmic radiation creates additional ionizing radiation

(cosmogenic nuclides) when highly energetic cosmic particles collide with stable elements in the earth's atmosphere and on its surface. This is the mechanism of formation of ^{14}C and tritium (^{3}H), which can be incorporated into living matter as they cycle through the earth's ecosystem.

Primordial radionuclides are those that originate independent of cosmic rays and are primarily found in rocks and soil (NCRP, 1987a). Terrestrially derived radionuclides are also present in all organic matter, including food (Harley, 2008; NCRP, 1987a; Voelz, 1994) and the human body (NCRP, 1987a). For example, radioactive potassium (^{40}K) comprises about 0.012% of all terrestrial potassium, is found naturally in food, comprises the largest source of effective dose from the diet, and delivers a bone dose of approximately 0.15 mSv per year (Harley, 2008). When accounting for all radionuclides incorporated into the body tissues, as well as radionuclides in the diet that transit through the gastrointestinal (GI) tract without being absorbed/incorporated, the estimated effective dose is approximately 0.40 mSv per year (Harley, 2008).

One of the most significant sources of exposure to naturally occurring radionuclides involves inhalation of radon, a naturally occurring radioactive gas produced by the decay of uranium and thorium in soil and rock (NCRP, 1987a). The effective dose of radionuclide-containing atmospheric aerosols and gas-phase radionuclides (primarily due to radon) is estimated to be 2 mSv per year (Harley, 2000; Harley, 2008). Radon exposure varies depending on geography (e.g., radon exposures in California tend to be lower than those in the Midwest due to different soil composition; ATSDR, 1999) and housing type (e.g., radon exposure is typically lower in a house or apartment with no basement than in a house with a below-ground basement; NCRP, 1987a).

Anthropogenic Sources   In 1987, the NCRP estimated that approximately 19% of all ionizing radiation exposures were due to medical devices, consumer products, and other man-made sources (NCRP, 1987b), but as noted above, present day radiation exposures to the population are about half from natural sources and half from medical procedures.

X-rays, in particular dental X-rays, are a common source of exposure to ionizing radiation. A patient's effective dose from dental X-rays can range from 0.005 to 0.39 mSv, depending on the type of X-ray and the film speed (Ludlow et al., 2008). Airline travel is another source of exposure to ionizing radiation. On average, airline passengers receive an effective dose of 0.005 mSv per year from air travel (Wallace and Sondhaus 1978). Although the use of backscatter X-ray security scanners at airports has led to some public concerns, the typical dose associated with a normal 8-s scan is 0.03 microSv (0.00003 mSv); thus, even for frequent travelers, the cumulative dose will be well below typical background exposure levels (NCRP, 2003). People who smoke tobacco or regularly encounter secondhand smoke can be exposed to increased levels of ionizing radiation from naturally occurring radionuclides present in tobacco; the average annual effective dose has been estimated at 0.25 mSv/year (Papastefanou, 2009). In addition, there are minor exposures related to nuclear applications and consumer products containing radioisotopes (e.g., televisions, tobacco products, smoke detectors, electric thermostats, and building materials) (CIWMB, 2005). Ionizing radiation from nuclear sources, such as residual atmospheric radiation from nuclear weapons testing, was a more significant source of exposure in the past than it is today (Ropeik and Gray, 2002). Current exposure to ionizing radiation from nuclear sources is very small, comprising an effective dose of less than 0.01 mSv per year (ATSDR, 1999; Ropeik and Gray, 2002). The 1986 nuclear accident at Chernobyl in the Ukraine resulted in a very slight increase in worldwide radiation exposures (estimated at an additional 21 days of world exposure to natural background radiation levels), although, on a regional level, exposures were substantially more (i.e., the estimated dose commitment was 1.2 mSv or 1/30th of a typical average lifetime dose) (Gonzales, 1996). The nuclear incident in Fukushima, Japan, is not expected to contribute appreciably to global radiation exposure levels (being less than 0.01 mSv) (WHO, 2012).

Overall, the typical cumulative dose for the United States population from all sources of ionizing radiation is approximately 6.2 mSv per year, with about half of this coming from natural sources and half from anthropogenic sources.[3]

Notable Sources of Occupational Exposure Exposure to ionizing radiation can occur in many different occupational settings and involve both natural and anthropogenic sources. Fields with widely recognized potential for ionizing radiation exposure include dentistry, medicine, nuclear power production, and the defense industry (in terms of nuclear weapons production). Other, less well-recognized fields include commercial aviation, academic research, veterinary care, construction, and mining. Table 100.5 summarizes the annual average exposure levels associated with various industries. As shown in this table, the typical annual exposure for nearly all occupations is less than the annual background radiation exposure due to nonoccupational sources (6.2 mSv).

HEALTH EFFECTS OF IONIZING RADIATION

Interactions Between Ionizing Radiation and Matter

Subatomic particles generally interact with atoms by transferring kinetic energy to the atom. The physical mechanisms by which charged, high speed particles such as alpha and beta

---

[3] http://www.nrc.gov/images/about-nrc/radiation/factoid2-lrg.gif.

TABLE 100.5 Examples of Occupations Involving Exposures to Ionizing Radiation

Occupation	Average Annual Effective Dose, mSv
Uranium mining	4.5
Nuclear fuel production	1.0
Nuclear power generation (reactor personnel)	1.4
Medical radiology (e.g., X-ray technicians)	0.5
Dentistry (X-rays)	0.06
Heart catheterization laboratories (X-rays)[a]	4.8
Well logging (radioisotopes)	0.4
Coal mining (radon)	0.7
Metal mining (radioisotopes, including radon)	2.7
Space travel to the international space station (cosmic rays)[b]	4.1
Commercial aviation (cosmic rays)	3.0
Multiple indoor occupations (radon)	4.8
Backscatter X-ray security scanner operator[c]	"Indistinguishable from background"

[a]Venneri et al., 2009.
[b]Friedberg and Copeland, 2011. Cumulative value for a 9.8-day space shuttle mission.
[c]NCRP, 2003.
All other data from UNSCEAR, 2000.

radiation transfer their kinetic energy to matter are relatively well understood. The most likely occurrence is a collision between the charged particle and one of the orbital electrons. When the orbital electron is struck with sufficient force, it is ejected from the atom, and the atom is ionized. In air and soft tissue, alpha and beta particles lose about 35 eV of energy per ion pair produced.[4]

Because an alpha particle has a large mass and a double charge, it has a high rate of ionizing collisions and, consequently, a high rate of energy loss. The high rate of collision and energy loss explains the very low penetration of alpha radiation into matter (ANSI, 1983). As a result, it is fairly easy to protect tissue from external alpha radiation; a sheet of paper may be sufficient. However, alpha particles emitted adjacent to cells within the human body (e.g., after being ingested or injected) may be highly damaging; in this case, they deposit most of their energy in the cells and tissues lying within a millimeter or so of the emitting radionuclide.

Beta particles travel near the speed of light and have only a single charge. The depth of penetration increases as the energy of the radiation increases. In air, very low energy beta particles have a range of several centimeters, whereas high energy beta particles may travel up to several meters in

---

[4] An ion pair refers to the positively and negatively charged particles (ions) that are produced when an electrically neutral atom is ionized by a source of external energy.

air. Being smaller in size and less massive than an alpha particle, they transfer relatively less energy per target encountered. This difference in rate of energy loss by beta radiation results in a lower relative biological effectiveness of the energy transferred to viable tissue, and it accounts for the higher penetrating power of beta radiation over that of alpha radiation. Beta particles can generally be stopped by thin barriers of low density materials such as aluminum, glass, or plastics such as polymethyl methacrylate (Plexiglas™).

For very high energy beta particles, the production of X-rays when a charged particle undergoes a sudden change in velocity is an important source of ionizing radiation. This phenomenon, called *bremsstrahlung* (from the German meaning "braking radiation"), occurs when a beta particle collides with an atomic nucleus. There is an abrupt change in the particle's velocity and some of the particle's kinetic energy is converted into X-rays. This X-ray production is extremely small for low energy beta particles and low atomic-numbered target atoms, but it increases with increasing beta energy and atomic number for the target. For this reason, X-rays are associated with high energy electron beams hitting targets, as in electron microscopes, and some high energy beta-ray emitters, such as ^{60}Co.

X-rays are very penetrating compared to subatomic particles; they pass through matter fairly easily and can travel long distances in air. X-rays interact with absorbing media by several different competing mechanisms, depending on the photon energy and atomic number of the target. Two of these mechanisms, photoionizing electric absorption and Compton scattering, involve an interaction between an orbital electron and the photon that results in the ejection of the electron. For photons in which energy exceeds 1.02 MeV, a third mechanism, involving interaction with the target nucleus, results in the transformation of the photon's energy into two particles: a negatively charged electron and a positively charged electron (positron). In each of these types of interactions, high speed electrons are produced. These energetic electrons, which are called primary ionizing particles, then transfer their energy by ionizing collisions with the orbital electrons of nearby atoms.

Effects at the Cellular Level

There are two primary ways that ionizing radiation may interact with biological structures, produce damage, and lead to adverse health outcomes. Direct interactions involve the transfer of energy from the radiation directly to the target tissue, producing ionization or breakage of chemical bonds, which in turn can result in a loss of molecular structure and function. For example, radiation may ionize lipids in the cell membrane or the membranes of organelles, leading to a loss of membrane function (Giusti et al., 1998). Radiation may also damage cells *via* indirect interactions. This involves ionization of an intermediate chemical (e.g., a water or oxygen molecule),

yielding a highly reactive by-product that can then interact with critical cellular components and produce damage. Because water comprises the largest component of cells, most of the indirect effects of radiation involve the generation of hydrogen and hydroxyl radicals (H· and OH·). Hydroxyl radicals may also combine to form hydrogen peroxide, H_2O_2, a more long-lived oxidant that can also produce cellular damage (ATSDR, 1999). Finally, the degree of oxygenation of a tissue is correlated with susceptibility to radiation damage due to the generation of oxygen radicals (e.g., superoxide O_2^-). Thus, one of the treatments for individuals with high levels of radiation exposure is to administer chemicals that depress tissue levels of oxygen (ATSDR, 1999).

Both types of damage can be prevented or repaired at the cellular level. Because humans as a species have always had to contend with the threats of oxidative damage, various compensatory mechanisms exist. There are scavenger molecules (e.g., glutathione, tocopherol) that bind to free radicals and prevent them from damaging critical cell components, as well as various enzymes designed to break down oxidizing chemicals or repair the damage they cause (e.g., the enzyme catalase inactivates hydrogen peroxide, various glutathione reductases regenerate the antioxidant glutathione, and various enzymes are involved in repair of damaged DNA). Thus, it is only when damage occurs and is not repaired (either not repaired in time for critical cellular action or not repaired at all because the compensatory mechanisms are overwhelmed) that radiation exposure can lead to toxicity.

As noted previously, the degree of cell damage caused by radiation exposure depends on a number of factors: the energy of the radiation, form of the radiation (in terms of the LET), and sensitivity of the tissue involved. Radiation with a higher LET will produce more intense damage over a limited area than radiation with a lower LET, where the damage pattern may be more diffused (ICRP, 2007). The former case may pose a greater health risk if, on the local level, compensatory mechanisms become overwhelmed. At the lowest levels of exposure, single molecules may become ionized, leading to negligible effects on cell viability. As the level of damage increases, however, cell function will become increasingly compromised. In the case of the cell nucleus, sufficient damage to cellular DNA may trigger apoptosis or cell death.

Effects at the Tissue and Organism Level

Cells are most susceptible to the effects of radiation while undergoing division (Mettler and Moseley, 1985). Cell division involves a tightly coordinated series of events, many of which are susceptible to disruption by radiation or chemical insults. As a result, tissues composed of many rapidly dividing cells (e.g., skin, intestinal epithelium, bone marrow, and gonads) are more susceptible to radiation damage than tissues where cells divide very slowly or not at all. These tissues will show the greatest and earliest damage from substantial radiation exposures (ATSDR, 1999). Damage to tissues with nondividing cells (e.g., the central nervous system, CNS) may also be significant, as there is no ability for the damage cells to be replaced.

Cancer and Noncancer Effects   Radiation toxicology has generally made a distinction between the noncancer and cancer health effects of radiation exposure. The reason is twofold. First, most noncancer effects of radiation exposure tend to occur over the short term, whereas cancer is likely to manifest itself only after a latency period of at least several years and generally several decades. In addition, the way that scientists conceive of the damage done to the cell is different in each case. Noncancer effects occur only when the damage from radiation exposure is large enough to overwhelm the cell's compensatory capacity, producing cell death and/or loss of cell function at such a scale that effects become apparent at the tissue level. Thus, a substantial number of "hits" are required to produce readily apparent symptoms. This obviously requires substantial exposures. In contrast, radiation-induced cancer (or heritable mutations) may only require a single "hit" if that particular lesion occurs in a critical part of the cell's DNA and the damage is not repaired before the cell divides. The initiation of the carcinogenic/mutagenic process, therefore, involves an aspect of probability (the likelihood that damage occurs at the necessary target and is not repaired in time), and it is feasible—although highly unlikely—that a single radiation "hit" may lead to tumor formation. Because there is an element of randomness in their occurrence, these types of effects are said to be stochastic events, whereas noncancer, non-mutagenic effects are said to be deterministic (the same level of exposure will always produce the same effect). Because they are associated with a threshold, the latter effects are only seen at higher levels of exposure.

General Noncancer Effects   Acute, high level exposures to external radiation (several Gy or more) are associated with ARS, a constellation of serious symptoms that at higher exposures can include death. The time and severity of symptom onset after exposure are determined by the dose. Symptoms primarily involve cells within rapidly dividing tissues such as the skin, hair, GI tract, bone marrow, and immune system (ATSDR, 1999). Very high acute exposures (around 10 Gy) may also result in CNS symptoms (Baum et al., 1984; ATSDR, 1999). Destruction of stem cells in the bone marrow leads to a sequential depletion of different blood cell types, with white cells (critical for immune response) depleted first and platelets (associated with clotting), appearing depleted several days later. Decreased production of red cells (erythrocytes) is generally not observed as a sign of acute radiation exposure due to the long life span of the existing red cell population (100–120 days). Monitoring of peripheral blood for differential cell counts is, therefore, one of the best indicators of radiation-induced damage. At somewhat higher doses, the rapidly dividing stem cells in the intestinal villi (responsible for producing the lining cells of the intestines) are also affected, leading to sloughing off of the intestinal epithelium and the loss of its absorptive and barrier functions. This in turn produces nausea, vomiting, diarrhea, hemorrhage, septicemia (due to gut bacteria entering the intestinal blood supply) and dehydration. The GI-related symptoms typically appear several days after the initial signs seen in the blood, although the latent period may be decreased at high levels of exposure (ATSDR, 1999). In terms of cutaneous effects, an acute dermal dose of about 2 Gy leads to an initial redness (erythema) of the skin; acute doses of 3 Gy or more may produce epilation (hair loss) (IAEA, 2013; CDC, 2005). Complete recovery of skin function is possible but may take several months and will confer a long-term increased risk of skin cancer (CDC, 2005). Dermal exposures in excess of 15 Gy will pose a greater challenge for recovery and may produce persistent or recurrent ulceration and tissue necrosis (CDC, 2005). The severity of symptoms and progress of deterioration is dose dependent (Table 100.6). If damage is not too severe and some stem cells are spared, survival is possible. Treatment is generally supportive and directed

TABLE 100.6   Acute Health Effects of Radiation Exposure

Dose Range, Gy	Effects	Survival Potential
0.15–0.5	Isolated chromosome breaks, possible slight changes in blood cell profiles in sensitive individuals	Certain
0.5–1 Gy	Mild changes in blood cell profiles	Virtually certain
5 Gy	Severe bone marrow damage, substantial intestinal damage, risk of infection	Without treatment, a substantial risk of dying within several weeks. With treatment, a fair chance of recovery
8–30 Gy	Severe GI damage, loss of GI stem cells, incapacitation, damage to major organs beyond GI tract and bone marrow	Minimal chances of survival, limited treatment other than palliative care
30 Gy and above	Massive and acute damage to CNS and all other tissues, rapid collapse and loss of autonomic functions	Death within several hours, if not sooner

Source: Modified from Baum et al., 1984, and updated based on ATSDR, 1999, and UNSCEAR, 2006.

toward the immune and GI symptoms. Individuals who die from ARS typically do so as a result of infection (due to the damaged immune system), dehydration, and blood loss (both due to the damage to the GI tract). At very high doses (10–30 Gy and above), the CNS may also be specifically affected, leading to rapid unconsciousness, loss of autonomic function, and rapid death (UNSCEAR, 1988; NCRP, 1980). Acute radiation syndrome is very rarely encountered and is seen in situations such as nuclear power plant disasters (e.g., Chernobyl) or accidental and unknowing exposures to high energy radioactive sources (e.g., in cases where X-ray sources were improperly disposed of and unknowingly used by the public) (Champlin et al., 1988).

Lower levels of acute radiation exposure (approximately 1 Gy) may lead to transient immune system depression (changes in blood cell counts) (UNSCEAR, 2006). Below 0.5 Gy, the primary concern involves chromosomal damage and its potential to produce cancer or birth defects. For example, Mettler and Moseley (1985) reported that the threshold for single-strand DNA breakage under acute exposure was 0.1 Gy. Chronic, low level exposure to ionizing radiation may also produce skin lesions and cataracts (a clouding of the lens of the eye). A cumulative ionizing radiation dose of approximately 0.5 Gy is considered to be the threshold for cataract formation (ICRP, 2011).

The foregoing discussion is relevant for external radiation doses. For exposures to radionuclides *via* ingestion, inhalation, or injection, the effects will depend in part on the ability of the material to move within the body (based on its form and chemical properties), its propensity to accumulate within particular tissues (e.g., radon decay products in the lungs, iodine accumulation in the thyroid, strontium, and radium accumulation in bone) and its LET. A significant issue with internally deposited radionuclides is that exposure continues even when the individual moves away from the actual source of the material (the concept behind the notion of committed dose). Individuals with substantial internal exposures to radionuclides require long-term monitoring to assess changes in body burden and cumulative radiological dose.

Reproductive and Developmental Effects   Radiation exposure poses particular risks for both reproduction and development in utero. For example, as with other tissues with rapidly dividing cells, the testes are sensitive to the effects of ionizing radiation. Reduced sperm counts or sperm abnormalities have been observed with external radiation doses as low as 1 Gy (Birioukov et al., 1993), and "temporary sterility" has been associated with doses as low as 0.15 Sv (Upton, 2005). At these lower exposure levels, sperm production can rebound to normal levels after cessation of exposure; however, permanent sterility (due to the death of spermatogonial stem cells) can result from doses to the testes in excess of about 6 Gy (Adams and Wilson, 1993). The ovaries do not undergo the same degree of rapid cell division as is

observed in the testes. However, because half of the DNA in all of the future child's cells is attributable to the maternal contribution, damage to the DNA contained in the egg still represents a substantially more significant event than damage to a single cell elsewhere in the body. Temporary impaired fertility in women may be seen after a dose to the ovaries of 1.5–2.0 Sv (Upton, 2005). Permanent sterility may occur with acute ovarian doses in the range of 2.5–6 Sv (Upton, 1991).

Because it is undergoing rapid growth, cell division, and cell differentiation, the developing embryo and fetus is also susceptible to radiation damage (CDC, 2011). Timing of exposure is critical; early exposures (when the population of cells that will go on to form the adult is limited) may have more significant consequences than those later in pregnancy (ATSDR, 1999). The form of radiation is also an important consideration; external alpha and beta radiation would be unable to reach the developing child, but X-ray radiation or exposure to internalized radionuclides could be of concern. Small doses of external radiation (0.05 Gy, equivalent to several full-body positron emission tomography scans) at any time during pregnancy are not believed to result in an increased risk of birth defects (McCollough et al., 2007). Much larger doses, if experienced between weeks 2 and 18 of pregnancy (the most sensitive period) can result in birth defects. For example, children exposed *in utero* during the atomic bombs dropped on Hiroshima and Nagasaki were found to have a high rate of brain damage, which resulted in lower IQs and even severe mental retardation. They also suffered an increased risk of other birth defects (NRC, 1990). The no observable effect level for these individuals was reported to be 0.2–0.4 Gy (NRC, 1990). After about week 18 of pregnancy, the fetus is much more resistant to the effects of ionizing radiation and only large doses (3 Gy or more, a dose producing acute radiation sickness in the mother) will result in birth defects seen at 0.3 Gy earlier on (CDC, 2011). In addition to birth defects, exposure to radiation *in utero* may result in a higher chance of developing cancer later in life. Thus, the nuclear accident at Chernobyl (which released large amounts of radioactive iodine) is now believed to have produced an increased incidence of thyroid cancers in persons exposed *in utero* (Bromet et al., 2009; Hatch et al., 2009). However, the increased risk of cancer from *in utero* exposure is believed to be small, except at exceptionally high doses. For example, it is estimated that the probability of developing childhood cancer from *in utero* external radiation exposure is less than 1% at doses of 0.1 Gy and less than 0.1% at 0.025 Gy (McCollough et al., 2007). Note that the exposure to the fetus from a single mammography X-ray is approximately 0.001 Gy (Behrman et al., 2007).

Radiation and Cancer   For most cases, the primary concern regarding exposure to radiation is its potential to cause cancer. The carcinogenic potential of radiation exposure is in fact likely to be the greatest concern of workers who are involved with technologies with a potential for radiation exposure.

There are three major dose–response models that are used to predict biological damage that may result from exposure to ionizing radiation. The "linear no-threshold" (LNT) theory postulates a linear relationship between a response, such as cellular damage, and any concomitant increase in the dose of ionizing radiation (NRC, 2006). Under this LNT theory, any exposure to radiation is presumed to cause biological harm. An alternative model, the "threshold" model, is that disease (e.g., cancer) occurs only above a defined dose or "threshold" of ionizing radiation exposure (Cohen, 2008; Calabrese, 2012). A third dose-response model is the hormetic response (Calabrese and Baldwin, 2000). This "radiation hormesis" model is characterized as a "U-shaped" dose response (or inverted U for some response measures) that shows an opposite response at low *versus* high doses. In terms of radiation risk, this usually means that while high doses are damaging (similar to the previous two models), low level exposures may actually be more beneficial than zero exposure. Currently, both the NRC and U.S. Environmental Protection Agency (EPA) have, as a matter of policy, adopted the linear dose-response model for assessing radiation dose (NRC) and radiation risk (U.S. EPA). The NRC, however, recognizes "practical" thresholds for a safe dose that are incorporated in their exposure guidelines.

The carcinogenic properties of radiation are somewhat unique in that estimates regarding the potency of radiation to produce cancer are largely derived from studies conducted in human populations. This is quite different from cancer assessments of most chemical agents, where estimates of the carcinogenic potency have to be inferred from studies conducted in laboratory animals. The key studies that have allowed for a greater understanding of radiation-induced cancer include those of the following populations (Harley, 2008):

- *Miners with radon exposure.* Several cohorts of miners (e.g., uranium miners in Bohemia) have experienced chronic exposures to radon gas released as a by-product of the decay of naturally occurring radium found in uranium-containing ores. Comprehensive analyses and modeling of the data collected in these cohorts (e.g., NIH, 1994; NRC, 1998) have shown a statistically increased risk of lung cancer relative to control populations.
- *Radium dial painters.* In the early part of the twentieth century, female workers at the United States Radium factory in New Jersey applied radium-containing paints to watch dials and other electronic equipment to create fluorescent indicators. In order to obtain fine brush marks, the workers would use their tongue to create a fine brush point and consequently ingested considerable quantities of radium over the course of their careers. Some individuals ingested as much as several thousand Ci of radium. Consistent with the bone-seeking nature of radium, these workers developed osteogenic sarcoma, as well as symptoms of acute radiation poisoning (Rowland et al., 1978).
- *Survivors of the atomic bombs dropped on Hiroshima and Nagasaki, Japan.* Large cohorts of atomic bomb survivors were enrolled in monitoring programs beginning in the 1950s that collected cancer and other health statistics over time. Considerable modeling has been done to estimate the exposures of these individuals based on their location during the bombing. Whole-body doses varied with distance from the bomb epicenter, but a substantial number of individuals who survived the acute effects of the exposure and were enrolled in the study had experienced doses as high as 0.24 Gy. Increased risks for various types of cancer have been observed, including those of the lung, breast, thyroid, and lymphatic system (RERF, 1987; Pierce and Preston, 1993).
- *Ankylosing spondylitis patients treated with X-rays.* In the middle part of the last century, individuals with ankylosing spondylitis (a painful inflammation of the spinal joints) were treated with ^{224}Ra, which was able to relieve the debilitating pain. The average dose received during a single treatment has been estimated to be 2.6 Gy (NRC, 2006). Studies of this cohort have indicated increased risks of leukemia and lung and other solid tumors (Darby et al., 1987).
- *Children treated with radiation for ringworm.* Between 1905 and 1960, approximately 200,000 children worldwide were treated with radiation to induce hair loss and facilitate treatment for ringworm (*Tinea capitis*). The use of standard treatment protocols has allowed for derivation of reasonably good estimates of radiation exposure. Studies have included a group of 11,000 children treated in New York (Shore, 1990). Exposures to the scalp in this cohort were in the range of 0.2–0.5 Gy; doses to the thyroid were in the range of 0.06 Gy. Increased incidences of skin and thyroid cancers have been reported in these studies.

Because radiation-induced cancer cannot be distinguished from cancer arising from other causes, it is impossible to say with certainty that cancer observed in a specific individual enrolled in these studies was due to radiation exposure. One can only compare the incidence of cancer in the exposed group to that in a suitable control population to examine the possible excess risk. For example, analysis of cancer data by the Radiation Effects Research Foundation (RERF) on the survivors of the atomic bomb found no increased probability of dying from leukemia for individuals whose radiation dose did not exceed 0.1 Gy (ATSDR, 1999). Among the various types of cancer, some leukemia subtypes are most commonly associated with elevated exposures to ionizing radiation (IARC, 2000). Other cancers commonly associated with elevated doses of ionizing radiation include cancers of the thyroid, breast, lung, and pleural tissues (IARC, 2000; Ron, 1998).

In addition to the large cohorts involving historical high level exposures, a number of studies have been conducted with occupational cohorts with lower level exposures to ionizing radiation. For example, some studies of radiology technicians have suggested an increased risk of developing leukemia for technicians with cumulative whole-body doses of ionizing radiation in the range of 470 mSv (Yoshinaga et al., 1999). Individuals with lower levels of cumulative exposure, in the range of 132 mSv (primarily those who entered the workforce after the 1950s), did not evidence an increased risk of leukemia (Yoshinaga et al., 1999). Similar conclusions were drawn from an in-depth review of the epidemiology literature concerning cancer risks in medical radiation workers by Linet et al. (2010). A study of Canadian dental workers (dentists, dental assistants, and hygienists) indicated no increased risks for any cancer except skin melanoma, where the increase was modest [standardized incidence ratio (SIR) = 1.46, 90% confidence interval (CI): 1.14–1.85] (Zielinski et al., 2005). The cohort in this study included nearly 200,000 workers whose employment spanned 1951–1987. The authors noted that average annual radiation exposures in this group peaked in 1963 (at approximately 1.1 mSv per year) and declined dramatically to less than 0.1 mSv per year in 1970–1987 (Zielinski et al., 2005).

Concerns have been raised about exposures of interventional cardiologists exposed to radiation scatter while conducting cardiac catheterizations, specifically increased risks for cataracts and brain cancer (Kuon et al., 2003). The available information, however, is limited to isolated case reports, so no definitive conclusions can be drawn regarding these possible risks; cardiologists are urged to employ appropriate protection to address these concerns (Kuon et al., 2003). Studies have also investigated the effects of ionizing radiation on cohorts of pilots and/or cockpit crew members, who experience increased levels of naturally occurring radiation while flying at high altitudes. For example, Band et al. (1996) reported an increase in one type of leukemia (acute myelogenous leukemia) in a cohort of 2680 Canadian pilots (SIR = 4.72; 90% CI: 2.05–9.31) who had been employed as pilots more than 20 years and experienced estimated mean annual radiation exposures of 6 mSv. However, a review of multiple studies has concluded that the data supporting an increased risk of cancer among commercial airline personnel are equivocal and suggestive of, at most, a very low level of increased risk (Bagshaw, 2008).

RADIATION EXPOSURE LIMITS AND GUIDELINES

Some of the agencies that have established and continue to monitor acceptable limits for ionizing radiation include the ICRP, NCRP, National Academy of Sciences Committee on the Biological Effects of Ionizing Radiation (the BEIR Committee), U.S. EPA, United Nations Scientific Committee on the Effects of Atomic Radiation (UNSCEAR), and U.S. NRC.

Exposure Limits and Guidelines for the General Population

According to both the ICRP and NRC, general population exposures to ionizing radiation should be kept as low as reasonably achievable (ALARA) to protect against the stochastic effects of radiation. More generally, the ICRP and U.S. NRC have recommended dose limits for exposure to specific, nonnatural sources of radiation of 1 mSv per year, or roughly one sixth of the normal background exposure due to anthropogenic and naturally occurring radiation. In the United States, the NRC imposes more stringent exposure limits for the public from nuclear power plants (0.25 mSv/year) and waste disposal facilities (0.15 mSv/year) (CPEP, 2003). The U.S. EPA has established an "action level" for indoor radon exposure of 4 pCi/l of air (approximately 8 mSv/year) at which it recommends activities to reduce exposure be taken (CPEP, 2003). In addition to whole-body dose limits, the ICRP recommendations also indicate tissue-specific doses of 50 mSv for the skin and 15 mSv per year for the lens of the eye (and an average equivalent dose of 20 mSv averaged over 5 years) (ICRP, 2007).

Occupational Exposure Limits

Consistent with the concept of a zero threshold for stochastic effects, the ALARA principle also applies to occupational exposures to ionizing radiation. Thus, use of appropriate personal protective equipment and good hygiene practices are required in workplaces using radiological materials even when measured exposures are below established workplace limits. In terms of specific limits, the U.S. NRC has set a total effective dose limit of 50 mSv per year for whole-body occupational exposures. Specific dose limits have also been established for the lens of the eye (150 mSv per year) and the skin and extremities (500 mSv per year). The sum of the deep dose (non-skin), non-eye equivalent and committed dose equivalents cannot exceed 500 mSv for any given organ (U.S. Code of Federal Regulations, Title 10, Part 20). The dose that can be experienced by a pregnant woman is limited to 5 mSv (as the dose to the fetus) summed over the gestation period. The analogous ICRP limits allow for averaging over a 5-year time period (i.e., a maximum

5-year average total-body dose of 20 mSv, with exposure in any individual year not to exceed 50 mSv). The allowable doses to the skin and extremities are identical to those adopted by the NRC (Boice, 2012). Regarding the lens of the eye, in 2011, the IRCP recommended reducing the exposure limit to 20 mSv averaged over 5 years, with exposure to the lens in any individual year not to exceed 50 mSv (Boice, 2012).

REFERENCES

Adams, G.E. and Wilson, A. (1993) Radiation toxicology. In: Ballantyne, B., Marrs, T., and Turner, P., editors. *General and Applied Toxicology*, vol. 2, New York, NY: Grove's Dictionaries, Inc., pp. 1397–1415.

American National Standards Institute (ANSI) (1983) *Internal Dosimetry for Tritium Exposure: Minimum Requirements*, New York, NY: ANSI, ANSI NI3.14-1983.

Agency for Toxic Substances and Disease Registry (ATSDR) (1999) *Toxicological Profile for Ionizing Radiation*, Atlanta, GA: US Department of Health and Human Services, Public Health Service.

Bagshaw, M. (2008) Cosmic radiation in commercial aviation. *Travel Med. Infect. Dis.* 6:125–127.

Band, P.R., Le, N.D., Fang, R., Deschamps, M., Coldman, A.J., Gallagher, R.P., and Moody, J. (1996) Cohort study of Air Canada pilots: mortality, cancer incidence, and leukemia risk. *Am. J. Epidemiol.* 143:137–143.

Baum, S.J., Anno, G.H., Young, R.W., and Withers, H.R. (1984) *Symptomatology of Acute Radiation Effects in Humans, After Exposures to Doses of 75 to 4500 rads (cGy) Free-In-Air*, Washington, DC: Defense Nuclear Energy.

Behrman, R.H., Homer, M.J., Yang, W.T., and Whitman, G.J. (2007) Mammography and fetal dose. *Radiology* 243:605.

Birioukov, A., Meurer, M., Peter, R.U., Braun-Falco, O., and Plewig, G. (1993) Male reproductive system in patients exposed to ionizing irradiation in the Chernobyl accident. *Arch. Androl.* 30:99–104.

Boice, J.D. (2012) *NCRP and international consistency in radiation protection standards. Presented at Health Physics Society 57th Annual Meeting, July 23; Sacramento, CA: National Council on Radiation Protection & Measurements*.

Bromet, E.J., Taormina, D.P., Guey, L.T., Bijlsma, J.A., Gluzman, S.F., Havenaar, J.M., Carlson, H., and Carlson, G.A. (2009) Subjective health legacy of the Chornobyl accident: a comparative study of 19-year olds in Kyiv. *BMC Public Health* 9:417.

Calabrese, E.J. (2012) Muller's Nobel Prize Lecture: when ideology prevailed over science. *Toxicol. Sci.* 126:1–4.

Calabrese, E.J. and Baldwin, L.A. (2000) Radiation hormesis: its historical foundations as a biological hypothesis. *Hum. Exp. Toxicol.* 19:41–75.

Centers for Disease Control (CDC) (2005) *Cutaneous Radiation Injury (CRI) Factsheet*, Available at http://www.bt.cdc.gov/radiation/criphysicianfactsheet.asp (accessed December 5, 2014).

Centers for Disease Control (CDC) (2011) *Radiation and Pregnancy: A Fact Sheet for the Public*, Available at http://www.bt.cdc.gov/radiation/prenatal.asp. (accessed December 5, 2014).

Champlin, R.E., Kastenberg, W.E., and Gale, R.P. (1988) Radiation accidents and nuclear energy: medical consequences and therapy. *Ann. Intern. Med.* 109:730–744.

California Integrated Waste Management Board (CIWMB) (2005) Ionizing radiation exposure-monitoring and dosimetry. *Health and Safety Manual*, Chapter 6, Available at http://www.calrecycle.ca.gov/Safety/Manual/Chapter6/default.htm (accessed December 5, 2014).

Cohen, B.L. (2008) The linear no-threshold theory of radiation carcinogenesis should be rejected. *J. Am. Phys. Surg.* 13:70–76.

Contemporary Physics Education Project (CPEP) (2003) Radiation in the environment. In: Matis, H., editor. *Nuclear Science—A Guide to the Nuclear Science Wall Chart*, 3rd ed., Chapter 15. Berkeley, CA: Contemporary Physics Education Project. Available at http://www.lbl.gov/abc/wallchart/teachersguide/pdf/Chap15.pdf (accessed December 5, 2014).

Darby, S.C., Doll, R., Gill, S.K., and Smith, P.G. (1987) Long term mortality after a single treatment course with X-rays in patients treated for ankylosing spondylitis. *Br. J. Cancer* 55:179–190.

Friedberg, W. and Copeland, K. (2011) *Ionizing Radiation in Earth's Atmosphere and in Space near Earth*, Washington, DC: Office of Aerospace Medicine.

Giusti, A.M., Raimondi, M., Ravagnan, G., Sapora, O., and Parasassi, T. (1998) Human cell membrane oxidative damage induced by single and fractionated doses of ionizing radiation: a fluorescence spectroscopy study. *Int. J. Radiat. Biol.* 74:595–605.

Gonzales, A.J. (1996) Chernobyl—ten years after: global experts clarify the facts about the 1986 accident and its effects. *IAEA Bull.* 3:2–13.

Harley, N.H. (2000) The 1999 Lauriston S. Taylor lecture—back to background: natural radiation and radioactivity exposed. *Health Phys.* 79:121–128.

Harley, N.H. (2008) Health effects of radiation and radioactive materials. In: Klaassen, C.D., editor. *Casarett and Doull's Toxicology: The Basic Science of Poisons*, 7th ed., New York, NY: McGraw-Hill Companies, Inc., pp. 1053–1082.

Hatch, M., Brenner, A., Bogdanova, T., Derevyanko, A., Kuptsova, N., Likhtarev, I., Bouville, A., Tereshchenko, V., Kovgan, L., Shpak, V., Ostroumova, E., Greenebaum, E., Zablotska, L., Ron, E., and Tronko, M. (2009) A screening study of thyroid cancer and other thyroid diseases among individuals exposed *in utero* to iodine-131 from Chernobyl fallout. *J. Clin. Endocrinol. Metab.* 94:899–906.

International Agency for Research on Cancer (IARC) (2000) *IARC Monographs on the Evaluation of Carcinogenic Risks to Humans: Volume 75 Ionizing Radiation, Part 1: X- and Gamma- Radiation, and Neutrons*, Lyon, France: World Health Organization.

International Atomic Energy Agency (IAEA) (2013) *Skin injuries: Occurrence and Evolution*. Available at https://rpop.iaea.org/RPOP/RPoP/Content/InformationFor/HealthProfessionals/5_InterventionalCardiology/skin-injuries.htm (accessed December 5, 2014).

International Commission on Radiation Protection (ICRP) (2007) Chapter 3 (Biological aspects of radiological protection) and Chapter 4 (Quantities used in radiological protection). *Ann. ICRP* 37:49–79.

International Commission on Radiation Protection (ICRP) (2011) *Statement on tissue reactions. Approved by the Commission on April 21, 2011 [online].* Available at http://www.icrp.org/docs/ICRP%20Statement%20on%20Tissue%20Reactions.pdf (accessed December 5, 2014).

Kuon, E., Birkel, J., Schmitt, M., and Dahm, J.B. (2003) Radiation exposure benefit of a lead cap in invasive cardiology. *Heart* 89:1205–1210.

Linet, M.S., Kim, K.P., Miller, D.L., Kleinerman, R.A., Simon, S.L., and Berrington de Gonzalez, A. (2010) Historical review of occupational exposures and cancer risks in medical radiation workers. *Radiat. Res.* 174:793–808.

Ludlow, J.B., Davies-Ludlow, L.E., and White, S.C. (2008) Patient risk related to common dental radiographic examinations: the impact of 2007 International Commission on Radiological Protection recommendations regarding dose calculation. *J. Am. Dent. Assoc.* 139:1237–1243.

McCollough, C.H., Schueler, B.A., Atwell, T.D., Braun, N.N., Regner, D.M., Brown, D.L., and LeRoy, A.J. (2007) Radiation exposure and pregnancy: when should we be concerned? *Radiographics* 27:909–917.

Mettler, F.A. and Moseley, R.D. (1985) *Medical Effects of Ionizing Radiation*, Orlando, FL: Grune & Stratton, Inc.

National Council on Radiation Protection and Measurements (NCRP) (1980) *Influence of Dose and its Distribution in Time of Dose-Response Relationships of Low LET Radiations*, Washington, DC: NRCP.

National Council on Radiation Protection and Measurements (NCRP) (1987a) *Exposure of the Population in the United States and Canada from Natural Background Radiation*, Washington, DC: NCRP.

National Council on Radiation Protection and Measurements (NCRP) (1987b) *Ionizing Radiation Exposure of the Population of the United States*, Washington, DC: NCRP.

National Council on Radiation Protection and Measurements (NCRP) (1993) *Limitation of Exposure to Ionizing Radiation*, Washington, DC: NCRP.

National Council on Radiation Protection and Measurements (NCRP) (2003) *Presidential Report on Radiation Protection Advice: Screening of Humans for Security purposes Using Ionizing Radiation Scanning Systems*, Bethesda, MD: NCRP, Available at www.fda.gov/ohrms/dockets/ac/03/briefing/3987b1_pres-report.pdf (accessed December 5, 2014).

National Institutes of Health (NIH) (1994) *Radon and Lung Cancer Risk: A Joint Analysis of 11 Underground Miner Studies*, Washington, DC: US Dept. of Health and Human Services, Public Health Service, National Institutes of Health.

National Research Council (NRC), Committee on the Biological Effects of Ionizing Radiations (BEIR) (1990) *Health Effects of Exposure to Low Levels of Ionizing Radiation: BEIR V*, Washington, DC: National Academies Press.

National Research Council (NRC), Committee on Health Effects of Exposure to Radon (1998) *Health Effects of Exposure to Radon (BEIR VI)*, Washington, DC: National Academies Press.

National Research Council (NRC) (2006) *BEIR VII: Health Risks from Exposure to Low Levels of Ionizing Radiation*, Washington, DC: National Academies Press.

Papastefanou, C. (2009) Radioactivity of tobacco leaves and radiation dose induced from smoking. *Int. J. Environ. Res. Public Health* 6:558–567.

Pierce, D.A. and Preston, D.L. (1993) Joint analysis of site-specific cancer risks for the atomic bomb survivors. *Radiat. Res.* 134:134–142.

Radiation Effects Research Foundation (RERF). (1987) US Japan joint reassessment of atomic bomb radiation dosimetry in Hiroshima and Nagasaki. RERF final report DS86. Available at http://www.rerf.jp/shared/ds86/ds86a.html (accessed December 5, 2014).

Ron, E. (1998) Ionizing radiation and cancer risk: evidence from epidemiology. *Radiat. Res.* 150:S30–S41.

Röntgen, W.C. (1896) On a new kind of rays. *Science* 3:227–231.

Ropeik, D. and Gray, G. (2002) *Risk: A Practical Guide for Deciding What's Really Safe and What's Really Dangerous in the World Around You*, Boston, MA: Houghton Mifflin Co.

Rowland, R.E., Stehney, A.F., and Lucas, H.F. Jr., (1978) Dose-response relationships for female radium dial workers. *Radiat. Res.* 76:368–383.

Shore, R.E. (1990) Overview of radiation-induced skin cancer in humans. *Int. J. Radiat. Biol.* 57:809–827.

United Nations Scientific Committee on the Effects of Atomic Radiation (UNSCEAR) (1988) *Sources, Effects and Risks of Ionizing Radiation*, Available at http://www.unscear.org/docs/reports/1988/1988a_unscear.pdf (accessed December 5, 2014).

United Nations Scientific Committee on the Effects of Atomic Radiatio (UNSCEAR) (2000) *Sources and effects of ionizing radiation: Report of the United Nations Scientific Committee on the Effects of Atomic Radiation to the General Assembly, with scientific annexes. Volume I: Sources. Submitted to United Nations General Assembly.* Available at http://www.unscear.org/unscear/en/publications/2000_1.html (accessed December 5, 2014).

United Nations Scientific Committee on the Effects of Atomic Radiation (UNSCEAR) (2006) *Annex D: Effects of ionizing radiation on the immune system. In Effects of Ionizing Radiation: UNSCEAR 2006 Report to the General Assembly: Volume I.* Available at http://www.unscear.org/docs/reports/2006/07-82087_Report_Annex_B_2006_Web.pdf (accessed December 5, 2014).

United States Nuclear Regulatory Commission (NRC) (2012) *Information for Radiation Workers*. Available at http://www.nrc.gov/about-nrc/radiation/health-effects/info.html (accessed December 5, 2014).

Upton, A.C. (1991) Radiation, diagnosis, and management. *Cancer Detect. Prev.* 15(3):241–247.

Upton, A.C. (2005) Radiation. In: Frumkin, H., editor. *Environmental Health from Global to Local*, San Francisco, CA: John Wiley and Sons, Inc., pp. 683–714.

Venneri, L., Rossi, F., Botto, N., Andreassi, M.G., Salcone, N., Emad, A., Lazzeri, M., Gori, C., Vano, E., and Picano, E. (2009). Cancer risk from professional exposure in staff working in cardiac catheterization laboratory: Insights from the National Research Council's Biological Effects of Ionizing Radiation VII Report. *Am. Heart J.* 157:118–124.

Voelz, G.L. (1994) Ionizing radiation. In: Zens, C., editor. *Occupational Medicine*, 3rd ed., St. Louis, MO. Mosby: pp. 393–427.

Wallace, R.W. and Sondhaus, C.A. (1978) Cosmic radiation exposure in subsonic air transport. *Aviat. Space Environ. Med.* 49:610–623.

World Health Organization (WHO) (2012) *Preliminary dose estimation from the nuclear accident after the 2011 Great East Japan Earthquake and Tsunami.* Available at http://whqlibdoc .who.int/publications/2012/9789241503662_eng.pdf (accessed December 5, 2014).

Williams, F.H. (1903) *The Roentgen Rays in Medicine and Surgery as an Aid in Diagnosis and as a Therapeutic Agent*, 3rd ed., London: Macmillan and Co.

Yoshinaga, S., Aoyama, T., Yoshimoto, Y., and Sugahara, T. (1999) Cancer mortality among radiological technologists in Japan: updated analysis of follow-up data from 1969 to 1993. *J. Epidemiol.* 9:61–72.

Zielinski, J.M., Garner, M.J., Krewski, D., Ashmore, J.P., Band, P.R., Fair, M.E., Jiang, H., Letourneau, E.G., Semenciw, R., and Sont, W.N. (2005) Decreases in occupational exposure to ionizing radiation among Canadian dental workers. *J. Can. Dent. Assoc.* 71:29–33.

101

ELECTROMAGNETIC WAVES (EMF AND RF) AND HEALTH EFFECTS

Peter A. Valberg

First aid: Remove from RF or ELF-EMF exposure by increasing distance from the emissions source or by de-energizing the source (e.g., transmitting antenna).

Target organ(s): For high level RF: skin, cornea, inner ear. For high level power-line EMF, none known other than non-adverse "magneto-phosphenes" stimulated in retinal receptor cells. Other, putative organs: lymphocyte stem cells in the bone marrow, neurons and neural support cells in the brain, peripheral nervous system.

Occupational exposure limits: ACGIH TLV: For power-line (60 Hz) fields: 1 mT (10,000 mG); for workers with implanted cardiac pacemakers: 0.1 mT (1000 mG). For RF fields, see values tabulated in TLV handbook, which range from 10 to 100 W/m². For FCC and ICNIRP values, see Tables 101.4 and 101.5 in this chapter.

Reference values: No RfC derived for either RF or ELF-EMF exposure. Radio frequencies specific absorption limit (SAR) is 0.08 W/kg for the general public, and 0.40 W/kg for RF workers.

Risk/Safety Phrases: IARC Group 2B Carcinogen; ACGIH, A4; ACGIH has Threshold Limit Values (TLVs) for EMF exposure that are protective of worker health and are listed by frequency. There are currently no OSHA standards for worker exposure to extremely low frequency (ELF) fields. Likewise, there are no OSHA-specific standards for radiofrequency and microwave radiation exposure. The Federal Communications Commission (FCC) is the U.S. federal regulatory agency as to radiofrequency exposure limits.

BACKGROUND: THE ELECTROMAGNETIC WAVE SPECTRUM, NONIONIZING *VERSUS* IONIZING RADIATION

As illustrated in Table 101.1 below, the electromagnetic spectrum encompasses wave energy with a vast range in frequency from very low (e.g., power lines at 60 Hz), through the kilo- and megahertz radio frequencies (RF) (e.g., radio and television signals), to microwaves (gigahertz), and on up into waves of infrared, light, ultraviolet, X-rays, and gamma rays. The last three categories (from UV and higher in frequency) are considered "ionizing radiation," and this chapter focuses on electromagnetic waves with wavelengths longer, and frequencies below those of ultraviolet, namely, waves that fall into the "nonionizing" portion of the electromagnetic spectrum. Ionizing radiation is discussed in Chapter 100 of this book, and ionizing radiation is distinguished by the fact that those electromagnetic waves have sufficient energy to break apart chemical bonds in biological molecules, whereas nonionizing waves do not.

The International Agency on Research in Cancer (IARC) has classified ultraviolet light and ionizing radiation as "Group 1" or "known" carcinogens, meaning epidemiology of exposed populations is sufficiently strong to establish that elevated exposure increases cancer risk in humans. For nonionizing radiation, IARC has classified power-line (extremely low frequency) electric and magnetic fields (ELF-EMF) and RF as "Group 2B" or "possibly carcinogenic to humans," which, in the RF and EMF circumstances, refers to limited-to-inadequate evidence of cancer risk in humans

Hamilton & Hardy's Industrial Toxicology, Sixth Edition. Edited by Raymond D. Harbison, Marie M. Bourgeois, and Giffe T. Johnson.
© 2015 John Wiley & Sons, Inc. Published 2015 by John Wiley & Sons, Inc.

TABLE 101.1   The Electromagnetic Spectrum: The Columns Illustrate How Wave Properties Change as You Go Up in the Spectrum (The Second Row Gives Median Values for Wavelength, Frequency, and Photon Energy

Power Lines	Navigation, AM Radio, Ham Radio	FM Radio, UHF TV, Cell Phones	Microwave Beacons and Radar	Radiant Heating, Infrared	Sunlight, e.g., yellow light	Medical and Dental x-Rays	α-, β-, γ-Rays
5000 km	300 m	30 cm	3 mm	6 μm	600 nm	0.3 nm	0.0003 nm
50–60 Hz	0.001 GHz	1 GHz	100 GHz	50 THz	500 THz	10^{18} Hz	10^{21} Hz
0.24 peV	4 neV	4 μeV	0.0004 eV	0.2 eV	2 eV	4,000 eV	4 MeV
		◆		◆	◆	◆	◆
		Cell phones, ~1–2 GHz		Body heat	Vision		cosmic rays
−	−	−	−	←←Nonionizing←←		→Ionizing→→→→	
		(*RF heating currents*)		(*Photochemistry*)		(*molecular damage*)	

nm = nanometers = 10^{-9} meter = one-billionth of a meter.

GHz = gigahertz = 10^9 Hz = one thousand million cycles per second.

THz = terahertz = 10^{12} Hz = 10^{12} cycles per second = one million cycles per second.

eV = electron volt = energy gained by electron accelerated through 1 volt potential difference.

and limited-to-inadequate evidence of carcinogenicity in experimental animals.

The Physical Properties of Electromagnetic Waves are Frequency and Intensity

All matter contains electrically charged particles. Most objects are electrically neutral because positive and negative charges are present in equal numbers. When the balance of electric charges is altered, we experience electrical effects caused by the force between electric charges, such as the static electricity attraction between a comb and our hair, or the force between current carrying wires in an electric motor. Electric charges that accelerate back and forth ("oscillate") can lose their energy into electromagnetic radiation that propagates away at the speed of light.

Electric and magnetic fields (EMF) are essentially constructs created by scientists to help understand how electrically charged particles interact with each other. Scientists explain the forces exerted by charges by saying that each electric charge generates an electric field that exerts force on other nearby charges. That is, an electric field is a measure of force per unit charge (newtons per coulomb), but is usually expressed in units of volts per meter (V/m) or kilovolts per meter (kV/m). When electric charges move, an electric current exists, and a current generates a magnetic field. Units of electric current are amperes (A), and current measures the flow of electricity, somewhat like the flow of water in a plumbing system. The current of moving electric charges produces a magnetic field that exerts force on other moving charges. That is, a magnetic field expresses the force per unit length of current-carrying wire (newtons per ampere-meter), but is usually expressed in units of gauss (G) or milligauss (mG). Another magnetic field unit is the tesla (T), where 1 T = 10,000 G, and thus, 1 μT = 10 mG.

Oscillation Frequencies for Electromagnetic Waves

Oscillating, electrically charged particles create "waves" in the EMF lines associated with them, and these waves move outward at the speed of light. That is, electromagnetic waves have a time period or frequency equal to the rate at which the electric charges creating them are being shaken back and forth. The overall result is called an "electromagnetic wave," with the frequency given in "Hertz" (Hz), which is the same as "cycles per second." Table 101.1 (above) shows the vast range in frequencies of electromagnetic waves. The energy contained in the electromagnetic waves increases in proportion to their frequency.

The science of electromagnetic waves has been studied and tested over a very long period of time. James Clerk Maxwell described the basic interactions between electromagnetic waves and matter in the 1860s, and he showed that electromagnetic waves travel at the speed of light. In 1887, Heinrich Hertz experimentally demonstrated the existence of electromagnetic waves, and, in 1909, Guglielmo Marconi was awarded the Nobel Prize in physics for inventing the radio, i.e., showing how electromagnetic waves could be used to transmit information without wires. Notably, Maxwell's 1867 equations (with the addition of quantum mechanics) have been verified time and time again as valid predictors of how electromagnetic waves and matter interact. No exceptions to Maxwell's equations have been found, and no unexplained electromagnetic phenomena have been encountered.

Radio-wave frequencies cover the range from about 300,000 Hz (i.e., 0.3 megahertz, or 0.3 MHz) to 30,000,000,000 Hz (i.e., 30 gigahertz, or 30 GHz), and beyond. Communications signals rely on a "carrier frequency," which is different for each communication signal, and the difference in frequencies allows many RF signals to be present simultaneously, because the information carried at each frequency can be extracted by frequency-selective electronic tuners. For example, for cellular telephone technology, the carrier frequencies range from about 900 MHz up to about 2200 MHz.

By itself, an RF carrier wave is an unchanging continuous electromagnetic wave, and it carries no information. Information is imposed on the carrier wave by a modulation process that alters it by changing its amplitude, frequency, or phase in step, with the voice frequency (or other information) being imposed (amplitude modulation, AM; or frequency modulation, FM). Alternatively, information can be coded into computer bits, and the carrier wave can be modulated by changing its amplitude or frequency in discrete steps (digital modulation). The interaction of RF waves with cells and molecules depends on the frequency of the carrier wave, but not on the type information being transmitted, e.g., "voice," or "music," or "computer bits." This is because the physical energy of the RF waves depends only on the power of the carrier wave, and studies have provided no evidence that the biological impact (or lack of impact) of RF depends on the information content carried by the radio waves.

Visible light is the major source of electromagnetic energy in our daylight environment. Also, the human body, by virtue of being alive and warm, generates heat energy (electromagnetic energy in the infrared portion [IR] of the spectrum), which can be seen by a "night vision" camera, in the absence of visible light.

Absorption of Energy from Electromagnetic Waves

In considering potential health effects of EMF or RF, it is important to recognize that electromagnetic radiation, although "wave-like" in nature, can also act like "particles" when being absorbed or emitted by matter. That is, absorption and emission of radiation occurs in discrete energy units, photons or quanta, with energy content $E = h\upsilon$, where h is Plank's constant and υ is the frequency. Table 101.1 lists photon energies, and these energies can be compared to chemical bond energies, which are typically 3–12 electron-volts (eV) per bond. Thus, high frequency radiation, e.g., X-ray photons have high enough energy content to ionize (disrupt) biological molecules held together by covalent bonds. Photons in the visible and UV range can excite molecules and initiate molecular shape changes or chemical reactions. Photon energies of electromagnetic radiation in the microwave region and above can excite vibrational energy levels of molecules. Lower frequencies, including microwaves and down to 50/60-Hz EMF have small photon energies, and in fact, 50/60-Hz EMF are not considered to "radiate."

For EMF or RF exposures to cause or exacerbate disease in humans, such exposures would have to trigger a series of sequential steps that lead to a disease outcome. The causal chain would begin with human exposure to some particular frequency/intensity/duration of EMF or RF. To complete the first step, the fields would interact with biological molecules (or structures) in such a way as to alter their size, shape, charge, chemical state, function, or energy (by a mechanism currently unknown). In this energy "transduction" step, some absorption of electromagnetic energy must occur or there can be no effect. For observable biological (and possibly health adverse) effects to follow transduction, a cascade of sequential events at the molecular, cellular, and tissue level would be required, leading without interruption to the final outcome. Identifying a plausible, mechanistic or transduction step, in this multistep pathway has been one of the most challenging and elusive puzzles, despite considerable effort by biologists, chemists, and physicists.

Identifying how electromagnetic waves alter biological systems is crucial to determine the correct exposure metric. Since the health and viability of the human body depends in a fundamental way on the normal structure and function of large molecules (e.g., proteins, nucleic acids, carbohydrates, and lipids), a theory on how EMF or RF mechanisms act must predict how weak electromagnetic energy could interfere with or modify the normal synthesis, function, or degradation of these molecules. For example, a viable mechanism would predict thresholds of exposure effectiveness in terms of electromagnetic wave amplitude, frequency, time of onset, intermittency, coherence, exposure duration, polarization, etc.

ELECTRIC POWER: 60 Hz ELECTRIC AND MAGNETIC FIELDS (EMF)

Electrical effects can occur through the generation and use of electric power. The power grid creates EMF varying in time at 50 or 60 Hz (cycles per second), which are considered "extremely low frequency" (ELF) fields.

The electrical tension on utility power lines is expressed in V or kilovolts (kV; 1 kV = 1000 V). Voltage can be thought of as the pressure driving the flow of electricity. The existence of a voltage difference between power lines and ground results in an electric field, which is usually expressed in units of kV/m. The size of the electric field depends on the voltage, the separation between lines and ground, and other factors.

Power lines also carry an electric current that creates a magnetic field. The units for electric current are A and are a measure of the flow of electricity. Electric current can be envisioned as analogous to the flow of water in a plumbing system. The magnetic field produced by an electric current is usually expressed in units of G or mG, where 1 G = 1000 mG. As noted earlier, another unit for magnetic field levels is the microtesla (μT), where 1 μT = 10 mG. The size of the magnetic field depends on the electric current, the distance to the current-carrying conductor, and other factors. The units of measure are basically the same as those for static fields.

For example the steady magnetic field from the earth is 570 mG (57 μT).

Properties of Power-Line EMF

When EMF derives from different sources (e.g., adjacent wires), the size of the net EMF produced will be somewhere in the range between the sum of EMF from the individual sources and the difference of the EMF from the individual sources. Thus, EMF may partially add, or partially cancel, but generally, because adjacent wires are often carrying current in opposite directions, the EMF produced tends to be cancelled. Inside residences, typical baseline 60-Hz magnetic fields (far away from appliances) range from 0.5 to 5.0 mG. Electric and magnetic fields in the home arise from electric appliances, indoor wiring, grounding currents on pipes and ground wires, and outdoor distribution or transmission circuits.

Larger 60-Hz magnetic field levels are found near operating appliances. For example, can openers, mixers, blenders, refrigerators, fluorescent lamps, electric ranges, clothes washers, toasters, portable heaters, vacuum cleaners, electric tools, and many other appliances generate magnetic fields of size 40–300 mG at distances of 1 ft (NIEHS, 2002). Magnetic fields from personal care appliances held within ½ ft (e.g., shavers, hair dryers, massagers) can produce 600–700 mG. At school and in the workplace, lights, motors, copy machines, vending machines, video-display terminals, pencil sharpeners, electric tools, and electric heaters are all sources of 60-Hz magnetic fields.

Although the steady geomagnetic field does not have the 60-Hz time variation characteristic of power line EMF, people's movements in its presence can cause it to be experienced as a changing magnetic field. Also, moving magnets generate time-varying magnetic fields. For example, a magnet spinning at 60 times a second will produce a 60-Hz magnetic field indistinguishable from that found near electric power lines carrying the appropriate level of electric current. Even the rotating steel-belted radial tires on a car produce time-varying magnetic fields. Magnetic resonance imaging (MRI) is a diagnostic procedure that puts humans in large steady and changing magnetic fields (e.g., static fields of size 20,000,000 mG). In contrast to medical X-rays, MRIs have no known health risks (other than the large forces exerted on nearby steel objects).

Review of Power-Line EMF Bioeffects

Power-line EMF has been the focus of considerable research for more than three decades. Over this period of time, the focus has been primarily on the magnetic field component. The three major lines of investigation have involved epidemiology, laboratory animal studies, and biological mechanism studies. The scientific evidence currently accumulated does not support a clear and coherent picture whereby environmental levels of power-line EMF constitute a hazard to human health, primarily because animal studies and mechanistic investigations have not shown a consistent, deleterious effect of typical ambient power-line magnetic fields on biology.

EMF epidemiology studies focused on childhood leukemia have received considerable attention. An observational epidemiologic study published by Wertheimer and Leeper (1979) suggested that living near electric power distribution lines was linked to an increased risk of childhood cancer. In this and subsequent epidemiology studies, the actual EMF levels that children had been exposed to were unknown, so researchers developed surrogates for past EMF exposures based, for example, on the proximity, number, and size of electric-utility distribution (or transmission) lines near the homes. In the initial 1979 study, the electric utility distribution line configuration near a home was called its "wire code," and homes with high wire codes (and presumably higher EMF levels) were found to be represented in a greater proportion of the leukemia cases as compared to the control children.

During the 35 years since this first study, a large number of epidemiological studies have examined associations between disease and various proxies of power line field strength (e.g., the "wire code" classification of homes, the distance to power-line corridors, present-day EMF measurements, the field strength calculated from power-line loading). If a correlation was detected, it was generally interpreted as linking power-line EMF to increased risk for the disease being studied, but consistency of the findings was poor. Often, the associations became weaker or disappeared when actual personal-monitor measured magnetic fields were substituted in place of other surrogate measures. It was found that some surrogates used for ranking EMF exposure also correlated with non-EMF factors such as traffic density, age of the home, rental vs. ownership, and assessed value of the home. Such potential confounders made it problematic to interpret the associations as an effect of EMF exposure per se. That is, the statistical correlations did not establish that power-line EMF exposure was the "causal" factor.

Hundreds of EMF epidemiology and laboratory research studies have been published in the 35 years since the initial 1979 study reported a statistical correlation between residential "wire codes" and childhood leukemia. Generally, each study focused on a particular hypothesis, and the range of possible investigations has been immense. Some of the most important work was done under the auspices of the National Institute of Environmental Health Sciences (NIEHS). The NIEHS had a program called "EMF RAPID,"[1] which funded laboratory research to determine what, if any, aspects of power-line magnetic fields (ELF-EMF) interaction with

---

[1] RAPID = "research and public information dissemination."

biological systems had the potential to trigger adverse disease outcomes. The conclusion of this extensive laboratory research program was summarized by NIEHS (1999).

> The scientific evidence suggesting that ELF-EMF exposures pose any health risk is weak. . . . No indication of increased leukemias in experimental animals has been observed. . . . Virtually all of the laboratory evidence in animals and humans, and most of the mechanistic studies in cells fail to support a causal relationship between exposure to ELF-EMF at environmental levels and changes in biological function or disease status.

For the proposition that power-line EMF exposure leads to health effects, there continues to be a lack of supporting laboratory-animal evidence, or support as to a plausible biological mechanism (Wood, 1993; Valberg et al., 1997; Boorman et al., 1999, 2000; McCormick et al., 1999; Swanson and Kheifets, 2006; Brain et al., 2003; Foster, 2003; WHO, 2007; SCENIHR, 2009).

Epidemiologic analyses have continued over the years, and some associations continue to be reported. The following list provides examples of prominent analyses, reviews, and/or summaries of the more recent power-line EMF literature. Notably, the epidemiological associations have not become stronger over the years, i.e., following the advent of larger, more in-depth studies. There still remains considerable inconsistency among the epidemiology results, the levels of incremental risk are low, and often do not reach statistical significance. Although the listing below (15 articles, 2000–2014) is not intended as a comprehensive review, it provides a sampling of some of the more recent and more significant epidemiological results. The reader is encouraged to read some of the individual studies in more detail.

- Ahlbom et al. (2000): "When [we] pooled nine epidemiology studies, . . . [we] found a relative risk of 2.0 (1.27–3.13) for childhood leukemia in the children with average exposures of 4 mG or greater. For children with lower average exposures, no significant elevation of childhood leukemia was found in the pooled studies. . . . The explanation for the elevated risk is unknown, but selection bias may have accounted for some of the increase."

- Greenland et al. (2000): "Summary estimates from 12 studies that supplied magnetic field measures exhibited little or no association of magnetic fields with leukemia when comparing 0.1–0.2 and 0.2–0.3 microtesla (μT) categories with the 0–0.1 μT category, but the Mantel-Haenszel summary odds ratio comparing >0.3 μT versus 0–0.1 μT was 1.7." "Based on a survey of household magnetic fields, an estimate of the U.S. population attributable fraction of childhood leukemia associated with residential exposure is 3%."

- Hatch et al. (2000): "Our recent large case-control study [638 cases, 620 controls] found little association between childhood acute lymphoblastic leukemia (ALL) and electric-power-line wire codes."

- Kleinerman et al. (2000): "Neither distance nor exposure index was related to risk of childhood acute lymphoblastic leukemia, although both were associated with in-home magnetic field measurements. Residence near high-voltage lines did not increase risk."

- UKCC (2000): "Our results provide no evidence that proximity to electricity supply equipment or exposure to magnetic fields associated with such equipment is associated with an increased risk for the development of childhood leukemia nor any other childhood cancer."

- Rubin et al. (2005): "The symptoms described by "electromagnetic hypersensitivity" sufferers can be severe and are sometimes disabling. However, it has proved difficult to show under blind conditions that exposure to EMF can trigger these symptoms. This suggests that "electromagnetic hypersensitivity" is unrelated to the presence of EMF, although more research into this phenomenon is required."

- Kabuto et al. (2006): "We analyzed 312 children newly diagnosed with ALL or AML in 1999–2001. [. . .] Weekly mean MF level was determined for the child's bedroom. [. . .] The odds ratios for children whose bedrooms had MF levels $\geq$0.4 μT compared with the reference category (MF < 0.1 μT) was 2.6 (n.s., 95% CI = 0.76–8.6) for AML + ALL and 4.7 (1.15–19.0) for ALL only."

- Mezei and Kheifets (2006): "The International Agency for Research on Cancer [has] classified ELF-MF as a possible human carcinogen. Since clear supportive laboratory evidence is lacking and biophysical plausibility of carcinogenicity of MFs is questioned, a causal relationship between childhood leukaemia and magnetic field exposure is not established. Among the alternative explanations, selection bias in epidemiological studies of MFs seems to be the most plausible hypothesis. In reviewing the epidemiological literature on ELF-MF exposure and childhood leukaemia, we found evidence both for and against the existence of selection bias."

- Kavet et al. (2008): "Limits on exposures to extremely low-frequency electric fields, magnetic fields and contact currents, designated as voluntary guidelines or standards by several organizations worldwide, are specified so as to minimize the possibility of neural stimulation." "[We describe] neurostimulation thresholds and the relevance of magnetophosphenes to setting guideline levels."

- Kheifets et al. (2010): "10,865 cases and 12,853 controls were pooled from 7 studies; 24-hr meas. or calculated MF; >3 mG, compared to MF <1 mG: OR = 1.44 (n.s.,

95% CI 0.88–2.36), "the results are compatible with no effect [of EMF]. Overall, the association is weaker in the most recently conducted studies, but these studies are small and lack the methodological improvements needed to resolve the apparent association."

- Kroll et al. (2010): "For children born in England and Wales during 1962–1995; there were 28,968 complete matched case–control pairs [calculated fields for 58,162 total]." "We found no statistically significant associations between childhood-cancer risks and estimated magnetic fields from high-voltage power lines near the child's home address at birth."

- Keegan et al. (2012): In a case-control study of paternal occupation and childhood leukaemia, "results showed some support for a positive association between childhood leukaemia risk and paternal occupation involving social contact." Of the 16,764 cases of childhood leukemia, 93% were either acute myeloid leukemia or lymphoid leukemia, and neither showed an association with parents' occupational EMF exposure. Of the 7% "other leukemias," EMF exposure showed an increased odds ratio (OR = 1.6), but it was based on small numbers.

- Elliott et al. (2013): "[Our] study included 7,823 leukemia, 6,781 brain/central nervous system cancers, 9,153 malignant melanoma, 29,202 female breast cancer cases, and 79,507 controls [. . .] 15–74 years of age living within 1000 m of a high-voltage overhead power line." "We observed no meaningful excess risks and no trends of risk with magnetic field strength for the four cancers examined." "Our results do not support an epidemiologic association of adult cancers with residential magnetic fields in proximity to high-voltage overhead power lines."

- Pedersen et al. (2014): "1,698 childhood leukemia cases were compared to 3,396 controls; exposure assessment used the distance between residence at birth and the nearest 132–400 kV overhead power line; children who lived 0–199 m from the nearest power line had OR = 0.76 [0.40–1.45] when compared to children >600 m away. Overall distance to the nearest power line was not associated with a higher risk of childhood leukemia. We did not observe any association with close distance or further away."

- Bunch et al. (2014): "16,630 leukemia cases 1962–2008 compared to 20,429 matched controls; calculated distances of mother's address at child's birth to power lines used as exposure metric. Odds ratio for leukemia, 0-200 m compared with >1,000 m over the whole period OR = 1.12 (0.90–1.38) – not statistically significant. Over the whole period, there is no evidence of a distance effect for any of the three cancer groups."

As can be seen, the power line magnetic-field epidemiology studies have yielded some statistical associations, and

scientists have struggled with whether such associations can really be interpreted as having a causal basis. Over the years, EMF epidemiology studies have stimulated numerous laboratory experiments where scientists examined the adverse health effect hypothesis, i.e., can environmental power-line EMF affect biology, alter processes in living cells, or change molecules in such a way as to increase the risk of cancer or other diseases?

To date, there is neither an accepted mechanism by which power line EMF can cause disease, nor is there any animal model in which lifetime exposure to even considerably elevated 60-Hz magnetic fields has reliably produced a disease or a pre-disease condition (Valberg et al., 1997). That is, the research work has not been able to identify what aspect of EMF is the one we should potentially avoid or regulate. If adverse health effects are to be expected, would they be due specifically to the frequency of oscillation, the electric fields, the magnetic fields, continuous exposure, intermittent exposure, peak fields, transients? Despite considerable effort and many years of work, no firm evidence of adverse EMF effects has been found in the laboratory for any of the measures of EMF exposure that have been experimentally examined. Because the laboratory evidence and mechanistic analyses have not supported a causal link for the increments in risk suggested by the epidemiology studies, most scientists give less weight to the statistical correlations.

Public Health Agency Views on EMF Causing Health Effects

In 2002, the IARC classified power-line-frequency EMF as "possibly carcinogenic to humans" which refers to the circumstances where there is limited-to-inadequate evidence of carcinogenicity in humans and limited-to-inadequate evidence in experimental animals. As noted, a biological mechanism to support this carcinogenic effect has not been found, because 60-Hz EMF interact weakly with the human body, because a 60-Hz wavelength is much larger than body size, and EMF exposure results in extremely low levels of energy deposition in the body. One must also consider the many years of human experience with EMF, i.e., use of electricity at an increasing rate for more than 100 years with no indication of increasing disease at the national, population level (Jackson, 1992).

The scientific data on EMF and health have been assembled and reviewed by many independent consensus groups of research and health scientists. These groups and agencies include (among many others) the European Union (EU), International Commission on Nonionizing Radiation Protection (ICNIRP), World Health Organization (WHO), the National Academy of Sciences, (NAS) the American Cancer Society (ACS), and the Scientific Committee on Emerging and Newly Identified Health Risks (SCENIHR). As illustrated by the examples listed below, these "blue-ribbon" panels do not conclude that ambient levels of EMF are unsafe. The reports of these groups are voluminous, thorough, and evenhanded. Some of their conclusions are illustrated below, but many of the documents extend to many hundreds of pages, so a more complete view of their analyses and opinions requires going to the reports themselves.

- American Cancer Society (ACS) (2014a): "The possible link between electromagnetic fields and cancer has been a subject of controversy for several decades. It's not clear exactly how electromagnetic fields, a form of low-energy, non-ionizing radiation, could increase cancer risk. Plus, because we are all exposed to different amounts of these fields at different times, the issue has been difficult to study."

- European Union (EU) (2009): "Animal studies do not provide evidence that ELF magnetic field exposure alone causes tumours or enhances the growth of implanted tumours. Some inconsistent evidence has suggested that ELF magnetic fields might be co-carcinogenic (enhance the effects of known carcinogens) and that they may cause cancer-relevant biological changes in short-term animal studies. However, it was concluded that the data were not sufficient to challenge IARC's evaluation that the experimental evidence for carcinogenicity of ELF magnetic fields is inadequate."

- Institute of Electrical & Electronics Engineers (IEEE) (2002): "Protection is to be afforded to individuals in the general population by limiting maximum permissible exposure to magnetic field levels of 9,040 mG at 60-Hz power-line frequencies."

- International Agency for Research on Cancer (IARC) (2002): "The association between childhood leukemia and high levels of magnetic fields is unlikely to be due to chance, but it may be affected by bias. In particular, selection bias may account for part of the association." (p. 332) [Thus] there is limited evidence in humans for the carcinogenicity of extremely low-frequency magnetic fields in relation to childhood leukemia. There is inadequate evidence in humans for the carcinogenicity of extremely low-frequency magnetic fields in relation to all other cancers." (p. 338)[2]

- International Commission on Non-Ionizing Radiation Protection (ICNIRP) (2010): "[Two pooled epidemiological analyses] indicated that long-term exposure to 50–60 Hz magnetic fields might be associated with an increased risk of leukemia. . . . However, a combination of selection bias, some degree of confounding, and chance could possibly explain the results. In addition, no biophysical mechanism has been identified and the experimental results from animal and cellular laboratory studies do not support the notion that exposure to 50–60 Hz magnetic fields is a cause of childhood leukemia."

- National Academy of Sciences (NAS) (1999): "Results of the EMF-RAPID program do not support the contention that the use of electricity poses a major unrecognized public-health danger."

- National Cancer Institute (NCI) ((2005): "Currently, researchers conclude that there is limited evidence that magnetic fields from power lines cause childhood leukemia, and that there is inadequate evidence that these magnetic fields cause other cancers in children. Researchers have not found a consistent relationship between magnetic fields from power lines or appliances and childhood brain tumors."

- Scientific Committee on Emerging and Newly Identified Health Risks (SCENIHR) (2013): "Some epidemiological studies are consistent with earlier findings of an increased risk of childhood leukemia with long-term average exposure to magnetic fields above 0.3 to 0.4 µT [3 to 4 mG]. However, as stated in [SCENIHR's] previous opinions, no mechanisms have been identified that could explain these findings. The lack of experimental support and shortcomings identified for the epidemiological studies prevent a causal interpretation."

- World Health Organization (WHO) (2007): "Uncertainties in the hazard assessment [of epidemiological studies] include the role that control selection bias and exposure misclassification might have on the observed relationship between magnetic fields and childhood leukemia. In addition, virtually all of the laboratory evidence and the mechanistic evidence fail to support a relationship between low-level ELF magnetic fields and changes in biological function or disease status. Thus, on balance, the evidence is not strong enough to be considered causal, but sufficiently strong to remain a concern."

Regulatory Guidelines for Electric and Magnetic Fields

The US has no federal standards limiting occupational or residential exposure to 60-Hz EMF. Table 101.2 shows guidelines for power-line EMF suggested by national and

---

[2] In 2002, the IARC classified ELF magnetic fields as Group 2B (possibly carcinogenic) on the IARC scale of carcinogenic risk to humans. IARC uses the "possibly carcinogenic" category when talking about both cell phone RF fields and power-line magnetic fields ("EMF"), and the IARC category 2B includes many ordinary exposures as "possible carcinogens," e.g., coconut oil, gasoline, diesel fuel, fuel oil, mobile phones, "carpentry and joinery," coffee, carbon black (car tires), car-engine exhaust, surgical implants, talc-based body powder, iron supplement pills, mothballs, nickels, pickled vegetables, safrole tea, titanium dioxide, chloroform, for a total of 285 substances.

TABLE 101.2 60-Hz EMF Guidelines Established by Health and Safety Organizations

Organization	Magnetic Field	Electric Field
American Conference of Governmental and Industrial Hygienists (ACGIH) (occupational)	10,000 mG[a] 1000 mG[b]	25 kV/m[a] 1 kV/m[b]
International Commission on Non-Ionizing Radiation Protection (ICNIRP) (general public, continuous exposure)	2000 mG	4.2 kV/m
Non-Ionizing Radiation (NIR) Committee of the American Industrial Hygiene Assoc. (AIHA) endorsed (in 2003) ICNIRP's occupational EMF levels for workers	4170 mG	8.3 kV/m
Institute of Electrical and Electronics Engineers (IEEE) Standard C95.6 (general public, continuous exposure)	9040 mG	5.0 kV/m
UK, National Radiological Protection Board (NRPB) (now Health Protection Agency [HPA])	2000 mG	4.2 kV/m
Australian Radiation Protection and Nuclear Safety Agency (ARPANSA), Draft Standard, Dec. 2006[c]	3000 mG	4.2 kV/m
Comparison to steady (see text) *(DC) EMF, encountered as EMF outside the 60-Hz frequency range:*		
Earth's magnetic field and atmospheric electric fields, steady levels, typical of environmental exposure[d]	520 mG[e]	0.2 kV/m up to >12 kV/m
Magnetic resonance imaging scan, static magnetic field intensity[d]	20,000,000 mG	–

[a]The ACGIH (2014) guidelines for the general worker.
[b]The ACGIH (2014) guidelines for workers with cardiac pacemakers.
[c]ARPANSA (2006, 2008).
[d]These EMF are steady fields and do not vary in time at the characteristic 60 cycles per second that power-line fields do. However, if a person moves in the presence of these fields, the body experiences a time-varying field.
[e]At 42 degrees latitude (NOAA, 2013).

WHO. The levels shown in Table 101.2 are designed to be protective against any adverse health effects. The limit values should not be viewed as demarcation lines between safe and dangerous levels of EMF, but rather, levels that assure safety with an adequate margin of safety to allow for uncertainties in the science. Table 101.3 lists guidelines that have been adopted by various states in the United States. State guidelines are not health effect based

TABLE 101.3 State EMF Standards and Guidelines for Transmission Lines

State/Line Voltage	Electric Field		Magnetic Field	
	On ROW	Edge ROW	On ROW	Edge ROW
Y69–230 kV	8.0 kV/m	2.0 kV/m[b]		150 mG
Florida[a]	10.0 kV/m			200 mG, 250 mG[c]
Y500 kV				
Massachusetts		1.8 kV/m		85 mG
Minnesota	8.0 kV/m			
Montana	7.0 kV/m[d]	1.0 kV/m[e]		
New Jersey		3.0 kV/m		
New York[a]	11.8 kV/m 11.0 kV/m[f] 7.0 kV/m[d]	1.6 kV/m		200 mG
Oregon	9.0 kV/m			

ROW = right-of-way; mG = milligauss; kV/m = kilovolts per meter.

Sources: NIEHS (2002); FDEP (2008).

[a]Magnetic fields for winter-normal, i.e., at maximum current-carrying capability of the conductors.
[b]Includes the property boundary of a substation.
[c]500 kV double-circuit lines built on existing ROWs.
[d]Maximum for highway crossings.
[e]May be waived by the landowner.
[f]Maximum for private road crossings.

and have been typically adopted to maintain the status quo for EMF on and near transmission line rights-of-way (ROWs).

RADIOFREQUENCY (RF) WAVES AND COMMUNICATIONS TECHNOLOGIES

The RF portion of the electromagnetic spectrum lies at much higher frequencies than the ELF-EMF frequency range, but at a lower frequency range than radiation in the infrared (heat) or visible (light) portion of the spectrum (see Table 101.1). In the RF range, some sources of radio-wave energy include the following:

Commercial radio (AM, FM), television (VHF, UHF, digital), amateur (ham) radio

Marine and aviation radio services, military and weather radar, satellite TV/radio, GPS

Hospital (EMS), fire, police dispatch services

Wireless paging, routers, remote-control, baby monitors, walkie-talkies, etc.

Cordless telephones, cell phones, smart phones, smart meters, base station antennas

Microwave ovens (RF leakage); microwave computer links

RF in medicine: ablation, cautery, diathermy, MRI

As can be appreciated from this list, our society has used RF communication for more than 100 years, and RF energy has been used in medical treatments for over 75 years (Hunt, 1982). The health effects of RF have been vigorously investigated from the 1950s, when military uses of RF, and radar in particular, were greatly expanded, on up to the present day (Schwan, 1954; Guy, 1975; Adair, 1983; Lin and Michaelson, 1987; Valberg, 1997; Valberg et al., 2007; Lin and Michaelson, 2010; Foster and Moulder, 2013).

In 2011, IARC classified RF as a "possible carcinogen," which IARC describes circumstances where there is limited-to-inadequate evidence of carcinogenicity in humans and limited-to-inadequate evidence in experimental animals (IARC, 2011). But, as in the case of ELF-EMF, laboratory animal and biological mechanism evidence fail to support adverse health effects from low levels of RF exposure. Notably, MRI uses radio frequency waves to generate images of all parts of the human body, and MRI scans are not considered to pose health risks, in contrast to imaging techniques that use ionizing radiation (X-rays, CAT scans, PET scans, etc.).

Properties of Radiofrequency (RF) Electromagnetic Waves

In the above list of communications technologies, the total amount of RF energy transmitted by these sources varies widely, and it's helpful to compare the RF emissions to a "100 watt light bulb." Most commercial radio and television broadcast stations are licensed to operate at power outputs of tens of kilowatts to millions of watts; cell telephone base antennas range in power from 100 to 1000 W; a cell-phone handset typically produces less than 2–4 Wof RF energy. For any antenna, the energy emitted is spread across a wide angle (in different directions), and the RF energy level decreases rapidly with distance. At the closest publicly accessible point, all transmitters must comply with the RF safety standards and guidelines for the general public, which in the United States are set by the Federal Communications Commission (FCC), and which are overall similar world-wide (ICNIRP, 2009). Below is a list of some sources of electromagnetic energy, listed according to the power they emit into the electromagnetic spectrum (mostly in the RF spectrum, but, for perspective, including some with emissions in the "heat" and "light" part of the electromagnetic spectrum).

Electric utility "smart meters"---	<1 W
Handheld cell phones, cordless phones---	<2–4 W
Remote control toys and nursery monitors---	~3 W
Typical flashlight---	~5 W (light + heat)
"Walkie-talkies"---	~10 W
Cellular telephone base stations---	~100–1000 W
Incandescent light bulb---	~100 W (light + heat)
The living human body---	~100 W (heat [IR waves])
Inside a microwave oven---	~1500 W (some RF leaks out)
Electric space heater---	~1500 W (light + heat)
Radio and television antennas---	~50,000 to 1,000,000 W

Typical measurements of the intensity of RF waves provide "energy per unit area," and the results are given in "microwatts per square centimeter" or $\mu W/cm^2$. A microwatt is a millionth of a watt. Sometimes the units are "watts per square meter" of W/m^2 ($1\,W/m^2 = 100\,\mu W/cm^2$). The Safety-standard allowable RF exposure levels vary with the frequency of the radio waves, being lowest (most restrictive) level in the frequency range 30–300 MHz (FM radio). The RF safety standard for public exposure in the AM radio frequency band is 20,000 $\mu W/cm^2$, in the FM-radio frequency band is 200 $\mu W/cm^2$, at cellular telephone frequencies of 910 MHz is 610 $\mu W/cm^2$, and at cellular telephone frequencies of 2000 MHz and above is 1000 $\mu W/cm^2$. (Refer also to Tables 101.4 and 101.5 at the end of the chapter.) By comparison, summertime sunlight at noon bathes us with about 150,000 $\mu W/cm^2$ of electromagnetic energy in the visible light portion of the spectrum.

When considering biological effects, another useful comparison to consider is the whole-body specific absorption rate (SAR) guideline used by the FCC, ICNIRP, and IEEE for the

TABLE 101.4 FCC Limits for Maximum Permissible Exposure (MPE), 300 kHz to 100 GHz

Frequency Range (MHz)	Electric Field Strength (E) (V/m)	Magnetic Field Strength (H) (A/m)	Power Density (S) (mW/cm²)	Averaging Time $|E|^2$, $|H|^2$ or S (minutes)
(A) Limits for Occupational/Controlled Exposure[a]				
0.3–3.0	614	1.63	(100)*	6
3.0–30	1842/f	4.89/f	(900/f²)*	6
30–300	61.4	0.163	1.0	6
300–1500	–	–	f/300	6
1500–100,000	–	–	5	6
(B) Limits for General Population/Uncontrolled Exposure[b]				
0.3–1.34	614	1.63	(100)*	30
1.34–30	824/f	2.19/f	(180/f²)*	30
30–300	27.5	0.073	0.2	30
300–1500	–	–	f/1500	30
1500–100,000	–	–	1.0	30

f = frequency in MHz.

*Plane-wave equivalent power density.

[a]Occupational/controlled limits apply in situations in which persons are exposed as a consequence of their employment provided those persons are fully aware of the potential for exposure and can exercise control over their exposure. Limits for occupational/controlled exposure also apply in situations when an individual is transient through a location where occupational/controlled limits apply, provided he or she is made aware of the potential for exposure.

[b]General population/uncontrolled exposures apply in situations in which the general public may be exposed, or in which persons that are exposed as a consequence of their employment may not be fully aware of the potential for exposure or cannot exercise control over their exposure.

TABLE 101.5 IEEE Basic Restrictions (BRs) for Frequencies Between 100 kHz and 3 GHz

		Action level[a] SAR[b] (W/kg)	Persons in Controlled Environments SAR[c] (W/kg)
Whole-body exposure	Whole-body Average (WBA)	0.08	0.4
Localized exposure	Localized (peak spatial-average)	2[c]	10[c]
Localized exposure	Extremities[d] and Prinnae	4[c]	20[c]

[a]These are basic restrictions (BR) for the general public when an RF safety program is unavailable.

[b]SAR is averaged over 30 min for the general public and over 6 min for controlled (worker) environments.

[c]Averaged over any 10 g of tissue (defined as a tissue volume in the shape of a cube [the volume of the cube is approximately 10 cm³]).

[d]The "extremities" are the arms and legs distal from the elbows and knees, respectively.

general public, which is 0.08 W/kg, and which is the basis of the RF "maximum permissible exposure" guidelines.[3] How much would this amount of continuous energy input (0.08 W) heat up a kilogram of water (1 l of water) over 1 h of exposure? The answer is that, absorbing 0.08 W for a whole hour would raise the water temperature by 0.07 °C, assuming all of the heat input from the RF remained with the water, and did not get conducted, convected, or radiated away.[4] By way of comparison, the human body generates energy constantly at about 100 W, in the process of "burning" ingested food and staying warm at about 37 °C. When exercising, the energy generation rate of the human body goes up many-fold. The IR radiation from the warm human body (37 °C, or 310 K) has an intensity of about 50 mW/cm², and the IR wavelength ranges from 6 to 14 μm (Rogalski, 2010). Thus, if our bodies, organs, cells, and molecules typically function well in a 100 W bath of IR electromagnetic energy (with IR photons having much more energy than RF photons), it is hard to explain why absorbing less than a watt of RF power would disrupt physiological function.

Research Studies on Health Effects of RF

The absorption of RF energy by living organisms is well understood to cause some degree of heating, in an amount dependent upon the RF intensity and RF frequency (or wavelength). This well established effect of RF exposure (thermal effects) forms the basis of guidelines protective against adverse effects via this mechanistic pathway, for both occupational and general-public RF exposure standards (IEEE, 2006; ICNIRP, 2009). Although "nonthermal" effects of RF are regularly reported in the research literature, the consistency, reproducibility, and usefulness of the "non-thermal" results have not achieved a reliability to the point where they can form the basis of RF exposure standards.

Epidemiologic analyses have continued over the years, and more recent studies have primarily focused on cellular telephone exposures, because this technology has become so ubiquitous. The 15 articles listed below (2002–2014) provide examples of prominent analyses, reviews, and/or summaries of the more recent RF literature. Although not a comprehensive review, the summary conclusions provide a sampling of some of the more recent and more significant epidemiological results. Even though very brief, quoted conclusions are presented, the reader is of course, encouraged to consult the complete article for a more complete presentation.

- Groves et al. (2002): "This study reports on over 40 years of mortality follow-up of 40,581 Navy veterans of the

---

[3] Federal Communications Commission: http://www.fcc.gov/Bureaus/Engineering_Technology/Documents/bulletins/oet56/oet56e4.pdf

[4] 0.08 W = 0.08 joules/sec, so for 1 h, energy going in = 288 joules = 69 calories, which would raise the temperature of 1000 g of water by 0.069 °C.

Korean War." For these radar technicians, "Deaths from all diseases and all cancers were significantly below expectation overall, and [in particular] for the 20,021 sailors with high radar exposure potential. There was no evidence of increased brain cancer in the entire cohort or in high-exposure occupations." "No significant excesses were seen for lymphoid malignancies."

- Johansen (2004): "At present, there is little, if any, evidence that the use of mobile phones is associated with cancer in adults, including brain tumors, acoustic neuroma, cancer of the salivary glands, leukemia, or malignant melanoma of the eye."

- Takebayashi et al. (2008): In this study of exposure to radiofrequency electromagnetic fields from mobile phone use, and brain tumor risk: "the adjusted odds ratios (ORs) for regular mobile phone users [was] 1.22 (95% confidence interval (CI): 0.63–2.37) for glioma and 0.70 (0.42–1.16) for meningioma. When the maximal SAR value inside the tumour tissue was accounted for in the exposure indices, the overall OR was again not increased and there was no significant trend towards an increasing OR in relation to SAR-derived exposure indices. A non-significant increase in OR among glioma patients in the heavily exposed group may reflect recall bias."

- Ahlbom et al. (2009): p. 642. Glioma: "The pooled analysis of Nordic and UK Interphone studies, which to date includes the largest number of glioma cases, found an OR of 1.0 (0.7–1.2) based on 143 exposed cases, among persons who started to use a mobile phone 10 or more years before diagnosis." p. 646. Meningioma: "The largest study so far—the pooled analysis of the Nordic and UK Interphone studies— found an OR of 0.9 (0.7–1.3) for long-term use. Pooling all original studies gave risk estimates close to or below unity." p. 647. Acoustic neuroma: "For long durations of exposure (10 years or more), the Nordic-UK pooled analysis included the largest number of cases, and reported an OR of 1.0 (0.7–1.5)." "Pooling all studies gave summary risk estimates of 1.2 (0.8–2.0) for long-term use, and 1.1 (0.8–1.4) for ever-use." p. 650. Salivary gland tumors: "There is no consistent evidence of an increased risk of salivary gland tumors among mobile phone users"

- Aydin et al. (2011, 2012): "There is no plausible explanation of how a notably increased risk from use of wireless phones would correspond to the relatively stable incidence time trends for brain tumours among children and adolescents observed in the Nordic countries." "Regular users of mobile phones were not statistically significantly more likely to have been diagnosed with brain tumors compared with nonusers." "Almost 90% of the [Swedish] population had been using mobile phones for at least seven years in 2009, and the

proportion that had been using them for 10 years or even 15 years must have been substantial. Hence, the absence of a trend in the incidence of brain tumours in national statistics is reassuring [as to mobile phones not increasing risk of brain cancer]."

- de Vocht et al. (2011): "Given the widespread use and nearly two decades elapsing since mobile phones were introduced, an association should have produced a noticeable increase in the incidence of brain cancer by now. Trends in rates of newly diagnosed brain cancer cases in England between 1998 and 2007 were examined. There were no time trends in overall incidence of brain cancers for either gender, or any specific age group." "The increased use of mobile phones between 1985 and 2003 has not led to a noticeable change in the incidence of brain cancer in England between 1998 and 2007."

- INTERPHONE Study Group (2011): "There was no increase in risk of acoustic neuroma with ever regular use of a mobile phone or for users who began regular use 10 years or more before the reference date,"

- Swerdlow et al. (2011): "Although there remains some uncertainty, the trend in the accumulating evidence is increasingly against the hypothesis that mobile phone use can cause brain tumors in adults."

- Larjavaara et al. (2011): "The study included 888 gliomas from 7 European countries (2000–2004), with tumor midpoints defined on a 3-dimensional grid based on radiologic images." "[Our] results do not suggest that gliomas in mobile phone users are preferentially located in the parts of the brain with the highest radio-frequency fields from mobile phones."

- Schüz et al. (2011): "In this study including 2.9 million subjects, a long-term mobile phone subscription of ≥ 11 years was not related to an increased vestibular schwannoma risk in men (RR = 0.87, 95% CI: 0.52, 1.46), and no vestibular schwannoma cases among long-term subscribers occurred in women versus 1.6 expected. Vestibular schwannomas did not occur more often on the right side of the head, although the majority of Danes reported holding their mobile phone to the right ear."

- Deltour et al. (2012) and Little et al. (2012): Time trends in brain cancer rates do not reflect increases in mobile phone use, suggesting that there is no effect of low-level RF on brain cancer risk. "Age specific incidence rates of glioma remained generally constant in 1992–2008 (−0.02% change per year, 95% CI −0.28% to 0.25%), a period coinciding with a substantial increase in mobile phone use from close to 0% to almost 100% of the US population. If phone use [were] associated with glioma risk, we expected glioma incidence rates to be higher than those observed, even with a latency period of 10 years and low relative risks (1.5)."

- Mohler et al. (2012): "The results of [our] large cross-sectional study did not indicate an impairment of subjective sleep quality due to exposure from various sources of RF EMFs in everyday life." "We did not find evidence for adverse effects on sleep quality from RF-EMF exposure in our everyday environment." "individuals who claim to be able to detect low level RF-EMF are not able to do so under double-blind conditions"

- Barchana et al. (2012): "We found a statistically significant decrease in [gliomas] over 30-years period that correlates with introducing of mobile phones technology" "[This] is in-line with other observations and does not support the assumption that mobile phone use is a causative factor for brain gliomas."

- Kwon et al. (2012): [people with self-reported electromagnetic hypersensitivity (EHS)] "In this double-blind study, two volunteer groups of 17 EHS and 20 non-EHS subjects were simultaneously investigated for physiological changes (heart rate, heart rate variability, and respiration rate), eight subjective symptoms, and perception of RF-EMFs during real and sham exposure sessions." . . . "There was no evidence that EHS subjects perceived RF-EMFs better than non-EHS subjects." "32 min of RF radiation emitted by WCDMA mobile phones demonstrated no effects in either EHS or non-EHS subjects."

- Lagorio and Röösli (2014): "A meta-analysis of [29] studies on intracranial tumors and mobile phone use published by the end of 2012 was performed." "High heterogeneity was detected across estimates of glioma and acoustic neuroma risk in long term users, with cRRs ranging between 1.19 (95% CI 0.86–1.64) and 1.40 (0.96–2.04), and from 1.14 (0.65–1.99) to 1.33 (0.65–2.73), respectively." "Overall, the results of our study detract from the hypothesis that mobile phone use affects the occurrence of intracranial tumors."

Public Health Agency Views on RF Causing Health Effects

As with guidelines and standards, generally RF exposure standards have been developed by interdisciplinary, consensus groups, based on the scientific knowledge accumulated from many years of laboratory work and of human experience with RF waves (e.g., radio, television, navigation, telemetry, cell telephones, radar). As is the case with power-line EMF, research findings on potential health effects of RF waves have been assembled and periodically reviewed by numerous independent scientific professional groups composed of research, engineering, medical, and public health scientists. The reports of these groups, written by researchers, medical doctors, biologists, engineers, and toxicologists, are voluminous, thorough, and evenhanded.

To account for uncertainties in the data and increase confidence that adverse health effects will not occur at exposure levels below the RF standards, the established threshold of actual biological effects is generally divided by a factor of 10 to provide a margin of safety for occupational environments. For general public environments, an additional factor of 5 is applied, meaning that the RF guidelines are typically 50-fold lower than the empirically observed threshold for RF effects that might be considered adverse to health. The public health groups looking at RF health effects include the ACS, ICNIRP, WHO, and SCENIHR. As illustrated by the examples listed below, a consistent finding is that, by limiting RF exposures according to the current RF guidelines, we can expect to be protective of health.

- American Cancer Society (ACS) (2014b). "Most animal and laboratory studies have found no evidence of an increased risk of cancer with exposure to RF radiation. A few studies have reported evidence of biological effects that could be linked to cancer. Studies of people who may have been exposed to RF radiation at their jobs (such as people who work around or with radar equipment, those who service communication antennae, and radio operators) have found no clear increase in cancer risk."

- Advisory Group on Non-Ionizing Radiation (AGNIR) (2012). "Exposure of the general public to low level RF fields from mobile phones, wireless networking, TV and radio broadcasting, and other communications technologies is now almost universal and continuous." "In summary, although a substantial amount of research has been conducted in this area, there is not convincing evidence that RF exposure below internationally accepted guidance levels causes health effects in adults or children."

- Australian Radiation Protection and Nuclear Safety Agency (ARPANSA) (2012). "Laboratory studies do not provide evidence to support the notion that RF fields cause cancer. Review groups evaluating the state of knowledge about possible links between RF exposure and excess risk of cancer have concluded that there is no clear evidence for any links."

- Federal Communications Commission (FCC) (2014) "Some health and safety interest groups have interpreted certain reports to suggest that wireless device use may be linked to cancer and other illnesses, posing potentially greater risks for children than adults. While these assertions have gained increased public attention, currently no scientific evidence establishes a causal link between wireless device use and cancer or other illnesses. Those evaluating the potential risks of using wireless devices agree that more and longer-term studies should explore whether there is a better basis for RF safety standards than is currently used. The FCC closely monitors all of these study results. However, at this time, there is no basis on which to establish a different safety threshold than our current requirements."

- Food and Drug Administration (FDA) (2012) "Many people are concerned that cell phone radiation will cause cancer or other serious health hazards. The weight of scientific evidence has not linked cell phones with any health problems."

- Health Canada, Royal Society of Canada (RSC) (2003) "All of the authoritative reviews completed within the last two years have concluded that there is no clear evidence of adverse health effects associated with RF fields"

- Health Council of the Netherlands (NHC) (2011) "Available data do not indicate that exposure to radiofrequency electromagnetic fields affect brain development or health in children."

- Institute for Electrical and Electronics Engineers (IEEE) (2006) "[RF standards] protect against harmful effects in human beings exposed to electromagnetic fields in the frequency range from 3 kHz to 300 GHz."

- International Agency for Research on Cancer (IARC) ((2013). In May of 2011, the IARC Working Group determined that "There is *limited* evidence in humans for the carcinogenicity of radiofrequency radiation. Positive associations have been observed between exposure to radiofrequency radiation from wireless phones and glioma, and acoustic neuroma." "Radiofrequency electromagnetic fields are *possibly* carcinogenic to humans (Group 2B)." "There was, however, a minority opinion that current evidence in humans was *inadequate*, therefore permitting no conclusion about a causal association."[5] However, IARC's classification, because it did not include a quantitative analysis, has not led to the modification of RF guidelines and standards for safe exposure levels.

- International Commission on Non-Ionizing Radiation Protection (ICNIRP) (2009) "With regard to [RF and] non-thermal interactions, it is in principle impossible to disprove their possible existence but the plausibility of the various non-thermal mechanisms that have been proposed is very low. In addition, the recent *in vitro* and animal genotoxicity and carcinogenicity studies are

rather consistent overall and indicate that such effects are unlikely at low levels of exposure."

- National Cancer Institute (NCI) (2013). "Cell Phones and Cancer Risk." "Studies thus far have not shown a consistent link between cell phone use and cancers of the brain, nerves, or other tissues of the head or neck. More research is needed because cell phone technology and how people use cell phones have been changing rapidly.

- National Council on Radiation Protection & Measurements (NCRP) (2003). "[NCRP] concludes that the scientific literature related to modulation-dependence of biological effects of RF energy is not sufficient to draw any conclusions about possible modulation-dependent health hazards of RF fields, nor is there any apparent biophysical basis from which to anticipate such hazards apart from exposure to very intense RF pulses produced by some specialized military equipment."

- National Radiological Protection Board (NRPB) (2013) "The [Interphone] study provides no clear, or even strongly suggestive, evidence of a hazard. Moreover, it indicates that if there is any hazard of brain cancer or meningioma from use of mobile phones then the risk during the initial 10–15 years of use must be small. This conclusion is consistent with the findings of most other epidemiological studies that have examined the relation of brain tumours to use of mobile phones, and also with the absence of demonstrable effects on cancer incidence when laboratory animals have been exposed to radiofrequency radiation experimentally."

- New Zealand Ministry for the Environment (NZME) (2012). "The Ministry of Health considers there are no established adverse effects from exposures to radiofrequency fields which comply with the ICNIRP guidelines and the New Zealand Standard."

- Norwegian Institute of Public Health (2012-3) "With the exception of some case-control studies, the majority of the case-control studies and cohort studies have reported no increased risk of cancer. The results of the incidence studies show no evidence of increasing incidence of these cancers over time." "A number of studies of cancer in animals have been performed, and relevant mechanisms have also been studied using micro-organisms and cells. Overall, these studies provide further evidence that exposure to weak RF fields does not lead to cancer." Electromagnetic Hypersensitivity: "Blind trials show that symptoms also occur when subjects are not exposed. This means that electromagnetic fields do not need to be present for health problems attributed to electromagnetic fields to occur."

- Scientific Committee on Emerging and Newly Identified Health Risks (SCENIHR) (2013): "Overall, there is evidence that exposure to RF fields does not cause

---

[5] IARC uses the "possibly carcinogenic" category when talking about both cell phones and power-line magnetic fields ("EMF"), and the IARC category 2B includes "possible carcinogens" such as coconut oil, gasoline, diesel fuel, fuel oil, power-line EMF, "carpentry and joinery," coffee, carbon black (car tires), car-engine exhaust, surgical implants, talc-based body powder, iron supplement pills, mothballs, nickels, pickled vegetables, safrole tea, titanium dioxide (sunscreen), chloroform, and many other substances. http://monographs.iarc.fr/ENG/Classification/ClassificationsGroupOrder.pdf

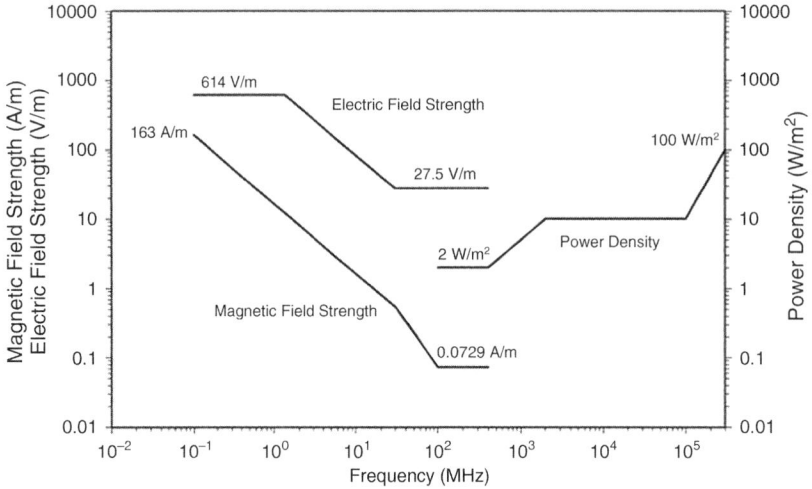

FIGURE 101.1 Graphic Representation of MPEs. Reprinted with permission from IEEE (2006), p. 27, Figure 4.

symptoms or affect cognitive function in humans. The previous SCENIHR opinion concluded that there were no adverse effects on reproduction and development from RF fields at exposure levels below existing limits. The inclusion of more recent human and animal data does not change that assessment." (p. 5) "The results [. . .] have typically not found any effect of exposure to radiofrequency fields on self-reported symptoms, are supported by a series of meta-analyses conducted by Augner, Gnambs, Winker and Barth (2012). These authors identified nine single- or double-blind provocation studies which assessed the effects of [RF] exposure on five self-reported symptoms (headache, nausea, dizziness, fatigue and skin irritation) and which were suitable for inclusion in a meta-analysis. No evidence was found in the meta-analyses that any of these endpoints were affected by exposure." (p. 109)

• World Health Organization (WHO) (2011) "A large number of studies have been performed over the last two decades to assess whether mobile phones pose a potential health risk. To date, no adverse health effects have been established as being caused by mobile phone use."

In summary, there is general agreement among a wide range of "blue ribbon" scientific review panels that the current standards used to prevent overexposure to RF levels can be expected to be health protective.

RF Electromagnetic Wave Exposure Limits and Guidelines

A number of scientific consensus groups have developed quantitative RF exposure guidelines, both for the occupational and general public environment. The groups that have produced numerical values include the following, and the standards are generally in agreement with each other.

• American Conference of Governmental Industrial Hygienists (ACGIH, 2014)
• Federal Communications Commission (FCC, 1997)
• Health Canada, Safety Code 6 (Canada, 2010)
• Institute of Electrical & Electronics Engineers/American National Standards Institute (IEEE, 2006)
• International Commission for Non-Ionizing Radiation Protection (ICNIRP, 2009)

Rather than display all of the possible standards, the FCC and IEEE standards are tabulated above to show the general manner in which these RF standards are presented.

The FCC standards were finalized in 1997 (FCC, 1997), and the agency has periodically reviewed the literature to ascertain that the standards remain current and health protective.[6]

The IEEE/ANSI standards for RF were finalized in 2005, and published in 2006, and in the table below, basic restrictions on human exposure to RF are given in terms of a SAR (energy absorbed per unit mass of tissue), over either a 6 min (occupational) or 30 min (general public) averaging time. The units of energy absorption are "watts per kilogram." The "Action Level" is the value applied for the general public, and the "Controlled Environment" figure is applied to occupational environments where workers are fully aware of the potential for exposure and can exercise control of their RF exposure levels.

Because the amount of energy absorbed from RF waves by the human body varies with frequency, the IEEE graph above (Figure 101.1) illustrates how these limitations in SAR

[6] http://www.fcc.gov/encyclopedia/faqs-wireless-phones

play out in terms of maximum permissible exposure (MPS's) limits for RF as a function of RF frequency (IEEE, 2006).

REFERENCES

Adair, E.R. (1983) *Microwaves and Thermoregulation*, New York, NY: Academic Press Inc.

Advisory Group on Non-Ionizing Radiation (AGNIR) (2012) *Health Effects of Radiofrequency Electromagnetic Fields.* Health Protection Agency (UK), RCE-20, 333 pp. Available at http://www.hpa.org.uk/webc/hpawebfile/hpaweb_c/1317133827077.

Ahlbom, A., Day, N., Feychting, M., Roman, E., Skinner, J., et al. (2000) Pooled analysis of magnetic fields and childhood leukemia. *Br. J. Cancer* 83:692–698.

Ahlbom, A., Feychting, M., Green, A., Kheifets, L., Savitz, D.A., et al. (2009) Epidemiologic evidence on mobile phones and tumor risk: a review. *Epidemiology* 20:639–652.

American Cancer Society (ACS) (2014a) *Power Lines, Electrical Devices and Extremely Low Frequency Radiation.* Available at http://www.cancer.org/cancer/cancercauses/radiationexposure andcancer/extremely-low-frequency-radiation (accessed November 12, 2014).

American Cancer Society (ACS) (2014b) *Microwaves, Radio Waves, and Other Types of Radiofrequency Radiation.* Available at http://www.cancer.org/cancer/cancercauses/radiationexposureandcan cer/radiofrequency-radiation (accessed November 12, 2014).

American Conference of Governmental Industrial Hygienists (ACGIH) (2014) *2014 TLVs and BEIs: Threshold Limit Values for Chemical Substances and Physical Agents and Biological Exposure Indices*, Cincinnati, OH: ACGIH, ACGIH Publication No. 0114. pp. 139–147. Available at http://www.acgih.org.

Augner, C., Gnambs, T., Winker, R., and Barth, A. (2012) Acute effects of electromagnetic fields emitted by GSM mobile phones on subjective well-being and physiological reactions: a meta-analysis. *Sci. Total Environ.* 424:11–15.

Australian Radiation Protection and Nuclear Safety Agency (ARPANSA) (2006) *Radiation Protection Standard; Exposure Limits for Electric & Magnetic Fields – 0 Hz to 3 kHz* 163 pp. Available at http://www.arpansa.gov.au/pubs/rps/dr_elfstd.pdf.

Australian Radiation Protection and Nuclear Safety Agency (ARPANSA) (2012) *Mobile Telephone Communication Antennas and Health Effects, Fact Sheet 4.* Available at http://www.arpansa.gov.au/pubs/factsheets/004%20is_antenna.pdf.

Australian Radiation Protection and Nuclear Safety Agency (ARPANSA) (2008) *Forum on the Development of the ELF Standard.* Available at http://www.arpansa.gov.au/News/events/elf.cfm.

Aydin, D., Feychting, M., Schüz, J., Röösli, M., et al. (2012) Childhood brain tumours and use of mobile phones: comparison of a case-control study with incidence data. *Environ. Health* 11:35.

Aydin, D., Feychting, M., Schüz, J., Tynes, T., Andersen, T.V., et al. (2011) Mobile phone use and brain tumors in children and adolescents: a multicenter case-control study. *J. Natl. Cancer Inst.* 103(16):1264–1276.

Barchana, M., Margaliot, M., and Liphshitz, I. (2012) Changes in brain glioma incidence and laterality correlates with use of mobile phones—a nationwide population based study in Israel. *Asian Pac. J. Cancer Prev.* 13:5857–5863.

Boorman, G.A., McCormick, D.L., Findlay, J.C., Hailey, J.R., Gauger, J.R., Johnson, T.R., et al. (1999) Chronic toxicity oncogenicity evaluation of 60 Hz (power frequency) magnetic fields in F344/N rats. *Toxicol. Pathol.* 27:267–278.

Boorman, G.A., McCormick, D.L., Ward, J.M., Haseman, J.K., and Sills, R.C. (2000) Magnetic fields and mammary cancer in rodents: a critical review and evaluation of published literature. *Radiat. Res.* 153(5 Pt 2):617–626.

Brain, J.D., Kavet, R., McCormick, D.L., Poole, C., Silverman, L.B., Smith, T.J., Valberg, P.A., Van Etten, R.A., and Weaver, J.C. (2003) Childhood leukemia: electric and magnetic fields (EMF) as possible risk factors. *Environ. Health Perspect.* 111:962–970.

Bunch, K.J., Keegan, T.J., Swanson, J., Vincent, T.J., and Murphy, M.F.G. (2014) Residential distance at birth from overhead high-voltage powerlines: childhood cancer risk in Britain 1962–2008. *Br. J. Cancer*, doi: 10.1038/bjc.2014.15.

Canada (Health Canada) (2010) *Radiofrequency Safety Code 6.* Available at http://www.hc-sc.gc.ca/ewh-semt/pubs/radiation/radio_guide-lignes_direct-eng.php.

Deltour, I., Auvinen, A., Feychting, M., Johansen, C., Klaeboe, L., Sankila, R., and Schüz, J. (2012) Mobile phone use and incidence of glioma in the Nordic countries 1979–2008: consistency check. *Epidemiology* 23:301–307.

de Vocht, F., Burstyn, I., and Cherrie, J.W. (2011) Time trends (1998–2007) in brain cancer incidence rates in relation to mobile phone use in England. *Bioelectromagnetics* 32:334–339.

Elliott, P., Shaddick, G., Douglass, M., de Hoogh, K., Briggs, D.J., and Toledano, M.B. (2013) Adult cancers near high-voltage overhead power lines. *Epidemiology* 24(2):184–190.

European Union (EU) (2009) *Extremely Low Frequency Fields Like Those from Power Lines and Household Appliances.* Available at http://ec.europa.eu/health/scientific_committees/opinions_layman/en/electromagnetic-fields/l-3/7-power-lines-elf.htm.

Federal Communications Commission (FCC) (2014) *Wireless Devices and Health Concerns.* Available at http://transition.fcc.gov/cgb/consumerfacts/mobilephone.pdf.

Federal Communications Commission (FCC) (1997) *Evaluating Compliance with FCC Guidelines for Human Exposure to Radiofrequency Electromagnetic Fields.* OET Bulletin 65. Available at http://transition.fcc.gov/Bureaus/Engineering_Technology/Documents/bulletins/oet65/oet65.pdf.

Florida Department of Environmental Protection (FDEP) (2008) *Electric and Magnetic Fields, Florida Rules and Statutes, Chapter 62-814.* Available at http://www.dep.state.fl.us/siting/files/rules_statutes/62_814_emf.pdf (accessed November 12, 2014).

Food and Drug Administration (US FDA) (2012) *No Evidence Linking Cell Phone Use to Risk of Brain Tumors.* Available at http://www.fda.gov/Radiation-EmittingProducts/Radiation EmittingProductsandProcedures/HomeBusinessandEntertainment/CellPhones/ucm116282.htm.

Foster, K.R. (2003) Mechanisms of interaction of extremely low frequency electric fields and biological systems. *Radiat. Prot. Dosimetry* 106(4):301–310.

Foster, K.R. and Moulder, J.E. (2013) Wi-Fi and health: review of current status of research. *Health Phys.* 105:561–575.

Greenland, S., Sheppard, A.R., Kaune, W.T., Poole, C., and Kelsh, M.A. (2000) A pooled analysis of magnetic fields, wire codes, and childhood leukemia, childhood leukemia-EMF study group. *Epidemiology* 11:624–634.

Groves, F.D., Page, W.F., Gridley, G., et al. (2002) Cancer in Korean war navy technicians: mortality survey after 40 years. *Am. J. Epidemiol.* 155:810–818.

Guy, A.W. (1975) Future research directions and needs in biologic electromagnetic radiation research. *Ann. NY Acad. Sci.* 247:539–545.

Hatch, E.E., Kleinerman, R.A., Linet, M.S., Tarone, R.E., Kaune, W.T., Auvinen, A., Baris, D., Robison, L.L., and Wacholder, S. (2000) Do confounding or selection factors of residential wiring codes and magnetic fields distort findings of electromagnetic fields studies? *Epidemiology* 11(2):189–198.

Hunt, J.W. (1982) Applications of microwave, ultrasound, and radio-frequency heating. *Natl. Cancer Inst. Monogr.* 61:447–456.

Institute of Electrical & Electronics Engineers (IEEE) (2002) *C95.6-2002 Standard for Safety Levels with Respect to Human Exposure to Electromagnetic Fields 0 to 3 kHz*, Standards Coordinating Committee 28, NY: IEEE, Inc.

Institute of Electrical & Electronics Engineers (IEEE) (2006) *C95.1-2005 RF Standard for Safety Levels with Respect to Human Exposure to Radio Frequency Electromagnetic Fields, 3 kHz to 300 GHz*. New York, NY: IEEE, Inc.

International Agency for Research on Cancer (IARC) (2011) *"IARC Classifies RF Electromagnetic Fields."* Available at http://www.iarc.fr/en/media-centre/pr/2011/pdfs/pr208_E.pdf.

International Agency for Research on Cancer (IARC) (2013) *Non-ionizing Radiation, Part 2: Radiofrequency Electromagnetic Fields*, vol. 102, Lyon, France: WHO, p. 480. Available at http://monographs.iarc.fr/ENG/Monographs/vol102/mono102.pdf.

International Agency for Research on Cancer (IARC) (2002) Non-ionizing radiation, part 1: static and extremely low-frequency (ELF) electric and magnetic fields. *IARC Monogr. Eval. Carcinog. Risks Hum.* 80:1–429.

Interphone Study Group (2011) Acoustic neuroma risk in relation to mobile telephone use: results of the INTERPHONE international case-control study. *Cancer Epidemiol.* 35(5):453–464.

International Commission on Non-Ionizing Radiation Protection (ICNIRP) (2009) ICNIRP statement on the 'Guidelines for limiting exposure to time-varying electric, magnetic, and electromagnetic fields (up to 300 GHz).' *Health Phys.* 97(3):257–258.

International Commission for Non-Ionizing Radiation Protection (ICNIRP) (2010) Guidelines for limiting exposure to time-varying electric, magnetic, and electromagnetic fields (1 Hz to 100 kHz). *Health Phys.* 99(6):818–836.

Jackson, J.D. (1992) Are the stray 60-Hz electromagnetic fields associated with the distribution and use of electric power a significant cause of cancer? *Proc. Natl. Acad. Sci.* 89:3508–3510.

Johansen, C. (2004) Electromagnetic fields and health effects—epidemiologic studies of cancer, diseases of the central nervous system and arrhythmia-related heart disease. *Scand. J. Work Environ. Health* 30(Suppl 1):1–30.

Kabuto, M., Nitta, H., Yamamoto, S., Yamaguchi, N., Akiba, S., Honda, Y., et al. (2006) Childhood leukemia and magnetic fields in Japan: a case-control study of childhood leukemia and residential power-frequency magnetic fields in Japan. *Int. J. Cancer* 119:643–650.

Kavet, R., Bailey, W.H., Bracken, T.D., and Patterson, R.M. (2008) Recent advances in research relevant to electric and magnetic field exposure guidelines. *Bioelectromagnetics* 29:499–526.

Keegan, T.J., Bunch, K.J., Vincent, T.J., King, J.C., O'Neill, K.A., et al. (2012) Case-control study of paternal occupation and childhood leukaemia in Great Britain, 1962–2006. *Br. J. Cancer* 107(9):1652–1659.

Kheifets, L., Ahlbom, A., Crespi, C.M., Draper, G., Hagihara, J., Lowenthal, R.M., et al. (2010) Pooled analysis of recent studies on magnetic fields and childhood leukaemia. *Br. J. Cancer* 103:1128–1135.

Kleinerman, R.A., Kaune, W.T., Hatch, E.E., Wacholder, S., Linet, M.S., et al. (2000) Are children living near high-voltage power lines at increased risk of acute lymphoblastic leukemia? *Am. J. Epidemiol.* 151:512–515.

Kroll, M.E., Swanson, J., Vincent, T.J., and Draper, G.J. (2010) Childhood cancer and magnetic fields from high-voltage power lines in England and Wales: a case-control study. *Br. J. Cancer* 103:1122–1127.

Kwon, M.K., Choi, J.Y., Kim, S.K., Yoo, T.K., and Kim, D.W. (2012) Effects of radiation emitted by WCDMA mobile phones on electromagnetic hypersensitive (EHS) subjects. *Environ. Health* 11:69.

Lagorio, S. and Röösli, M. (2014) Mobile phone use and risk of intracranial tumors: a consistency analysis. *Bioelectromagnetics* 35:79–90.

Larjavaara, S., Schüz, J., Swerdlow, A., Feychting, M., Johansen, C., Lagorio, S., Tynes, T., et al. (2011) Location of gliomas in relation to mobile telephone use: a case-case and case-specular analysis. *Am. J. Epidemiol.* 174(1):2–11.

Lin, J.C. and Michaelson, S.M. (1987) *Biological Effects and Health Implications of Radiofrequency Radiation*, New York, NY: Plenum Press. Available at http://getebook.org/?p=52206.

Lin, J.C. and Michaelson, S.M. (2010) *Biological Effects and Health Implications of Radiofrequency Radiation*, New York, NY: Springer.

Little, M.P., Rajaraman, P., Curtis, R.E., Devesa, S.S., Inskip, P.D., Check, D.P., and Linet, M.S. (2012) Mobile phone use and glioma risk: comparison of epidemiological study results with incidence trends in the United States. *BMJ* 344:e1147.

McCormick, D.L., Boorman, G.A., Findlay, J.C., Hailey, J.R., Johnson, T.R., Gauger, J.R., et al. (1999) Chronic toxicity/oncogenicity evaluation of 60 Hz (power frequency) magnetic fields in B6C3F1 mice. *Toxicol. Pathol.* 27:279–285.

Mezei, G. and Kheifets, L. (2006) Selection bias and its implications for case-control studies: a case study of magnetic field exposure and childhood leukaemia. *Int. J. Epidemiol.* 35:397–406.

Mohler, E., Frei, P., Fröhlich, J., Braun-Fahrländer, C., Röösli, M., et al. (2012) Exposure to radiofrequency electromagnetic fields and sleep quality: a prospective cohort study. *PLoS One* 7(5): e37455.

National Academy of Sciences (NAS) (1999) *Research on Power-Frequency Fields Completed Under the Energy Policy Act of 1992*, Washington, DC: National Academy Press, p. 107.

National Cancer Institute (NCI) (2005) *Magnetic Field Exposure and Cancer: Questions and Answers.* Available at http://www.cancer.gov/cancertopics/factsheet/Risk/magnetic-fields.

National Cancer Institute (NCI) (2013) *Cell Phones and Cancer Risk.* Available at http://www.cancer.gov/cancertopics/factsheet/Risk/cellphones.

National Council on Radiation Protection (NCRP) (2003) *Commentary No. 18, Biological Effects of Modulated Radiofrequency Fields.* Available at http://www.ncrponline.org/Publications/Commentaries/Comm18press.html.

National Institutes of Environmental Health Sciences (NIEHS) (1999) *Health Effects From Exposure to Power-Line Frequency Electric and Magnetic Fields,* NIH# 99-4493.

National Institute of Environmental Health Sciences (NIEHS) (2002) *Questions and Answers about EMF Electric and Magnetic Fields Associated with the Use of Electric Power.* 65 pp. Available at http://www.niehs.nih.gov/health/materials/electric_and_magnetic_fields_associated_with_the_use_of_electric_power_questions_and_answers_english_508.pdf.

National Oceanic and Atmospheric Administration (NOAA) (2013) *Magnetic Field Calculators: Estimated Values of [Earth's] Magnetic Field.* National Geophysical Data. Available at http://www.ngdc.noaa.gov/geomag-web/?id=igrfwmmFormId#igrfwmm (accessed November 12, 2014).

National Radiation Protection Board UK (NRPB) (2013) *UK Health Protection Agency.* Available at http://www.hpa.org.uk/web/HPAweb&HPAwebStandard/HPAweb_C/1274088317073.

Netherlands Health Council (NHC) (2011) *Radiofrequency Electromagnetic Fields and Children's Brains.* Available at http://www.gezondheidsraad.nl/en/news/infleuence-radiofrequency-telecommunication-signals-children-s-brains.

New Zealand Ministry for the Environment (NZME) (2012) *Questions and answers – National Environmental Standards for Telecommunications Facilities.* Available at http://www.mfe.govt.nz/rma/rma-legislative-tools/national-environmental-standards/nes-telecommunication-facilities-0.

Norwegian Institute of Public Health (2012-3) *Low-Lever Radiofrequency Electromagnetic Fields—An Assessment of Health Risks and Evaluation of Regulatory Practice.* Rapport 2012:3. Available at http://www.fhi.no/dokumenter/545eea7147.pdf.

Pedersen, C., Raaschou-Nielsen, O., Rod, N.H., Frei, P., Poulsen, A.H., Johansen, C., and Schuz, J. (2014) Distance from residence to power line and risk of childhood leukemia: a population-based case control study in Denmark. *Cancer Causes Control* 25:171–177.

Rogalski, A. (2010) *Infrared Detectors*, 2nd ed., Boca Raton, FL: CRC Press, p. 876.

Royal Society of Canada (RSC) (2003) *Recent Advances in Research on Radiofrequency Fields and Health, 2001–2003.* Available at https://rsc-src.ca/sites/default/files/pdf/expert_panel_radiofrequency_update2.pdf.

Rubin, G.J., Das Munshi, J., and Wessely, S. (2005) Electromagnetic hypersensitivity: a systematic review of provocation studies. *Psychosom. Med.* 67(2):224–232.

Scientific Committee on Emerging and Newly Identified Health Risks (SCENIHR) (2009) *Health Effects of Exposure to EMF.* European Commission, Directorate-General for Health & Consumers, p. 83. Available at http://ec.europa.eu/health/ph_risk/committees/04_scenihr/docs/scenihr_o_022.pdf (accessed November 12, 2014).

Scientific Committee on Emerging and Newly Identified Health Risks (SCENIHR) (2013) *Health Effects of Exposure to EMF and RF,* 219 pp. Available at http://ec.europa.eu/health/scientific_committees/emerging/docs/scenihr_o_041.pdf.

Schüz, J., Steding-Jessen, M., Hansen, S., Stangerup, S.E., Cayé-Thomasen, P., et al. (2011) Long-term mobile phone use and the risk of vestibular schwannoma: a Danish nationwide cohort study. *Am. J. Epidemiol.* 174:416–422.

Schwan, H.P. (1954) The absorption of electromagnetic energy in body tissues. *Am. J. Physiol. Med.* 33:371–404.

Swanson, J. and Kheifets, L. (2006) Biophysical mechanisms: a component in the weight of evidence for health effects of power-frequency electric and magnetic fields. *Radiat. Res.* 165(4):470–478.

Swerdlow, A.J., Feychting, M., Green, A.C., Kheifets, L.K., Savitz, D.A., et al. (2011) Mobile phones, brain tumors, and the interphone study: where are we now? *Environ. Health Perspect.* 119:1534–1538.

Takebayashi, T., Varsier, N., Kikuchi, Y., Wake, K., Taki, M., Watanabe, S., Akiba, S., and Yamaguchi, N. (2008) Mobile phone use, exposure to radiofrequency electromagnetic field, and brain tumour: a case-control study. *Br. J. Cancer* 98(3):652–659.

UKCC (2000) UK childhood cancer study investigators, childhood cancer and residential proximity to power lines. *Br. J. Cancer* 83:1573–1580.

Valberg, P.A. (1997) Radio-frequency radiation (RFR): the nature of exposure and carcinogenic potential. *Cancer Causes and Control* 8:323–332.

Valberg, P.A., Kavet, R., and Rafferty, C.N. (1997) Can low-level 50/60-Hz electric and magnetic fields (EMF) cause biological effects? *Radiat. Res.* 148:2–21.

Valberg, P.A., Van Deventer, T.E., and Repacholi, M.H. (2007) Base stations and wireless networks: radiofrequency (RF) exposures and health consequences. *Environ. Health Perspect.* 115:416–424.

Wertheimer and Leeper (1979) Electrical wiring configurations and childhood cancer. *Am J. Epidem.* 109:273–284.

Wood, A.W. (1993) Possible health effects of 50/60 Hz electric and magnetic fields: review of proposed mechanisms. *Aust. Phys. Eng. Sci. Med.* 16(1):1–21.

World Health Organization (WHO) (2007) *Extremely Low Frequency Fields. Environmental Health Criteria*, Monograph No. 238. Geneva: World Health Organization. Available at http://www.who.int/peh-emf/publications/elf_ehc/en/index.html.

World Health Organization (WHO) (2011) *Electromagnetic Fields and Public Health: Mobile Phones. Fact Sheet 193.* Available at http://www.who.int/mediacentre/factsheets/fs193/en/.

102

TRAUMATOGENS ASSOCIATED WITH CARPAL TUNNEL SYNDROME

JOAN M. WATKINS

BACKGROUND

Dorland's Illustrated Medical Dictionary defines "Traumato as a combining form denoting relativity to trauma or injury. Traumatogenic is defined as capable of causing trauma. Thus, traumatogen is an excellent description of forces or motions that can cause injury depending on their amount of exposure or dose."

The "worker disease" was first documented in 1717 by Ramazinni:

> Various and manifold is the harvest of diseases reaped by certain workers from the crafts and trades they pursue. All the profit that they get is fatal injury to their health. . . . That . . . I ascribe to certain violent and irregular motions and unnatural postures of the body, by reasons of which, the natural structure of the vital machine is so impaired that serious diseases gradually develop therefrom.

Although Ramazinni recognized the influences of forces on worker's muscular skeletal health, it was many years before the Science of Egonomics was developed and contributed to the relationship of design and disease. Ergonomics also works to determine permissible levels of work load that do not lead to injury. (P. Van Wely). Cumulative trauma disorders (CTDs) are now termed work related musculoskeletal disorders (WRMSD). The WRMSDs are physiological illnesses that may develop over a period of weeks, months, or even years as a result of prolonged mechanical stresses imposed on the musculoskeletal system, resulting in injuries recognized as physical ailments or abnormal conditions. These disorders may also be referred to as *repetitive strain injurys* (RSIs) (Kiesler and Finholt, 1988), *overuse injuries* (Green and Briggs, 1989), or *repetitive motion injuries*.

Cumulative trauma disorders (WRMSDs) are considered to be work related because they are more prevalent among the working population than the general public. Because of slow onset, both management and employees often ignore the microtrauma until the symptoms become chronic and permanent injury occurs (Putz-Anderson, 1988).

The development of such occupational illnesses in high risk industries has been recognized and monitored for some time. Although slight when compared with manual handling injuries, the RSI problem has been increasing consistently since the early 1980s (Green and Briggs, 1989) so that CTDs (WRMSDs) now represent a significant proportion of all workers' compensation claims (Brogmus and Marko, 1990; Brogmus et al., 1994). Several theories to substantiate this upward trend have been suggested, not least of which is a rise in symptomatic reporting as a result of increased public awareness.

CLASSIFICATION OF MUSCULOSKELETAL DISORDERS

Three basic types of cumulative illnesses or musculoskeletal disorders to the upper extremity musculoskeletal system are classified according to the anatomical source of irritation: tendon disorders, nerve disorders, and neurovascular disorders. Tendon disorders are those ailments associated with

Hamilton & Hardy's Industrial Toxicology, Sixth Edition. Edited by Raymond D. Harbison, Marie M. Bourgeois, and Giffe T. Johnson.
© 2015 John Wiley & Sons, Inc. Published 2015 by John Wiley & Sons, Inc.

overuse or unaccustomed use of a specific body part, examples of which include tendinitis, tenosynovitis, de Quervain's disease, and ganglionic cysts. Neurovascular disorders involve the compression of nerves and neighboring blood vessels. Thoracic outlet syndrome and vibration syndrome are commonly classified as neurovascular disorders.

Nerve disorders are attributable to repeated or sustained work activities that cause partial or complete loss of sensory, motor, or autonomic nerve function as a result of pressure against the nerve. Common examples of nerve compression disorders include neuritis, nerve entrapment syndrome, and carpal tunnel syndrome (CTS).

Carpal Tunnel Syndrome

First described by Sir James Paget in 1865, CTS is the most common example of a nerve compression disorder. Other terms used to describe this disorder include *writer's cramp*, *occupational neuritis*, *partial thenar atrophy*, and *median neuritis*.

It is caused by restriction of the median nerve as it passes through the carpal tunnel, an anatomical space in the wrist bound on the palmar side by the inelastic transverse carpal ligament and on the dorsal aspect by the carpal bones. The 10 structures that transverse the carpal tunnel include the four tendons of the flexor digitorum superficialis, the four tendons of the flexor digitorum profundus, the flexor pollicis longus, and the median nerve.

Robbins (1963) suggests three possible alternatives by which the cross-sectional area of the carpal tunnel may be compromised, thus reducing the available volume through which soft tissue structures pass: (1) increase of volume of the contents of the carpal canal, (2) an alteration of the osseous trough, and (3) thickening of the transverse carpal ligament.

The tendons that form a bridge between metacarpals in the hand and the flexor muscles of the forearm are lubricated along their path by tenosynovium, which allows the tendons to glide against each other. Personal factors or occupational conditions that cause irritation to, or inflammation of, the tendons can result in swelling and thickening of the tenosynovium. Because the median nerve is fragile compared to the surrounding structures, it compresses as the tendons swell, producing symptomatic experience of CTS.

Symptoms of CTS are usually experienced in the region of the hand served by the median nerve. This encompasses the second to fourth fingers, the base of the thumb on the palmar side, and the backs of the first four fingers on the dorsal side.

Acute CTS is often associated with nocturnal pain and tingling, episodic tingling during the workday, and gradual numbness, all of which may be encouraged by certain activities such as abnormal postures or repetitive or forceful hand motions. Symptoms usually diffuse shortly after the activity is changed.

As severity increases, the patient may experience aching, tingling, and what has been described as "painful numbness" in the fingers of the median distribution and deep in the palm. Perceptually, the patient may recall subjective feelings of uselessness related to the affected hand and wrist, mental sensation of swelling (although not apparent on inspection), clumsiness, and difficulty performing everyday tasks, such as unscrewing a bottle cap.

At the most severe level, patients with CTS may experience dull aching throughout the limb, which radiates not only distal to the site of compression, but also proximally. Patients may report pain throughout the forearm, upper arm, shoulder, and even the neck. Changes in coloration of the skin become more apparent, especially with exposure to cold. There may be excessive sweating in the palm and a possible mild degree of edema, which has been related to vasomotor imbalance. Wasting of the thenar muscles and mild weakness of the abductor pollicis brevis or opponens pollicis muscle may also be observed in severe cases.

Incidence of Work-Related CTS

The Bureau of Labor Statistics, a division of the US DOL, has defined repetitive trauma disorders for the purpose of classifying work-related incidence of these occupational illnesses:

Disorders associated with repeated trauma include conditions due to repeated motion, vibration, or pressure, such as CTS, noise-induced hearing loss, synovitis, tenosynovitis and bursitis, and Raynaud's phenomenon.

Although repetitive motion disorders represent only a small, but growing percentage of total occupational injuries and illnesses involving lost work time in the United States (1994 = 4.9%), group classification has limited the analysis of incidence for individual disorders. Because of the increasing prevalence of musculoskeletal disorders, and in particular, debilitating nerve disorders such as CTS, the Bureau of Labor Statistics redesigned their survey in 1992 to report for the first time the diagnosis of specific repetitive motion disorders that commonly affect the shoulders, arms, and other upper extremity parts.

The limited data from 1992 to 1994 suggest that incidence of CTS involving lost workdays was between 36.0% (1992) and 40.8% (1994) of the total occupational illnesses. (Note: Back injuries are excluded from these data because the US DOL considers them as accidents rather than cumulative injuries.)

From 1981 to 1990, there has been a steady increase in reporting of WRMSD, with a brief accelerated increase in 1987 and 1988. Several theories on the escalating occurrence include increased employee awareness through media attention and employer training programs; increased reliance on computer terminals in the workplace without appropriate ergonomic education and application. In addition increased female populations in the workforce (females reporting CTS is approximately twice than that of male counterparts), and aging of the U.S. worker population may have also influenced the increase. The increased incidence of reporting is likely to be a result of a combination of these reasons. The whole picture will only truly be painted as more information becomes available from the Bureau of Labor Statistics' refined survey.

More than half of all incidents of CTS involving lost work days are attributable to the manufacturing industry. Massive downsizing of the manufacturing industries in favor of semi-automated processes during the 1980s is likely to continue in the foreseeable future; therefore, only through appropriate ergonomic intervention in the workplace to minimize known occupational factors will this problem be solved.

The recent slowing of occupational reporting of repetitive motion disorders has been further supported by the Bureau of Labor Statistics' 1996 announcement, which encouragingly revealed a decline in the number of occupational CTS cases for the first time in 15 years. It is speculated that the reduced incidence is in part a result of successful ergonomic interventions within specific high risk trades, including the meat industry, where incidence rates had fallen by 15.9%.

The Bureau of Labor Statistics reissued a table in 2009 that detailed the Incidence Rate for nonfatal occupational injuries and illnesses involving days away from work per 10,000 full-time workers by selected worker occupation and selected nature of injury or illnesses. The highest incidence rate was food servers, non-restaurant at 8.9. The next highest 4.6 and 4.4 for cooks, institution and cafeteria, and welders, cutters, solderers, and refrigeration, respectively. Those occupations with incidence rates between 2.4 and 2.9 in ascending order were: Butchers and meat cutters, roofers, labor and freights stock and material movers, hand, industrial machinery mechanic and automotive service technicians and mechanics. Those occupations with incidence rates between 1 and 2 were bus and truck mechanics and diesel engine specialists, maids and housekeeping cleaners, truck drivers heavy and tractor trailer and truck drivers light and delivery services. Those occupations with incidence rates below 1 were nursing aids, orderlies and attendants, construction laborers, carpenters, janitors and cleaners, except maids, and house-keeping cleaners (Bureau of Labor Statistics, 2008). In July 1997, National Institute for Occupational Safety and Health (NIOSH) published Musculoskeletal Disorders and Work Place Factors—A Critical Review of Epidemiologic evidence for WRMSD of the neck, upper extremity, and lower back. This publication is a comprehensive review of the relationship between musculoskeletal disorders and exposures to physical factors at work. Evidence for work relatedness of carpal tunnel is reviewed in Chapter 5. The thirty epidemiologic studies reviewed involved studies of populations exposed to a combination of work factors. There was evidence of positive association between highly repetitive work alone and in combination with other factors and CTS. There was evidence of a positive association and forceful work in CTS. There was insufficient evidence between CTS and extreme postures. However, there was strong evidence of a positive association between exposure to a combination of risk factors such as force and repetitiveness or force and posture. One of the important aspects of this review is that it begins to quantify the risk factors associated with CTS.

Another important aspect of the developing research is CTS case definitions. The CTS case definition in Prevalence and Incidence of CTS in the United States working populations: Polled analysis of six prospective studies (Dale) was that both CTS hand diagrams and electrodiagnostic study results consistent with median nerve mononeuropathy of the wrist. More specifically; the carpal tunnel symptoms case definition required that subjects report symptoms of tingling, numbness, burning, or pain in one or greater than one of the first three digits (thumb, index, and long finger) on a hand symptom diagram. In addition, electrodiagnostic study results consistent with median nerve mononeuropathy at the wrist were required.

Social and Economic Costs

It is estimated that the economic burden of CTS on the U.S. industry presently exceeds $2 billion per year (Palmer and Hanrahan, 1995). A significant share of this expense is attributable to CTS surgery, which is estimated to cost between $400 and $500 million a year. A proportion of these surgeries are performed because of symptomatic recurrence in which a patient's recovery may be encumbered as a result of deficient procedures (Kern et al., 1993).

Cases of CTS also have the longest recuperation period for any reported occupational injury or illness category. More than 36% of work-related CTS cases involve more than 1 month of lost work time, with a median recovery period of 18 days, compared to the overall median for all occupational injuries and illnesses of 6 days (Box 102.1).

When determining the cost of worker injuries, it is usual to add the expenses associated with medical intervention and workers' compensation insurance premiums. However, the total costs might be as high as five times the actual direct costs (Box 102.1).

With cost-effective surveillance, workers at risk may be identified and the incidence of CTS prevented or reduced; thus the social and economic impact would be significantly lessened.

Occupational Factors Associated with CTS Prevalence

It is often difficult to identify a specific cause of CTS because many risk factors may interact simultaneously to bring about

---

BOX 102.1   DIRECT AND INDIRECT COSTS ASSOCIATED WITH CTD INCIDENCE

Direct Costs (20%)

Medical expenses
Workers' compensation premiums
Lost and light duty workdays

Indirect Costs (80%)

Loss of injured worker's production
Time lost of employee paid by employer
Time lost by uninjured employees
Temporary help
Reporting and retraining
Reporting and claims
Management time
Worker/management discussions
Litigation costs

---

the condition. It is also difficult to isolate occupational factors from leisure activities, and individual susceptibility further compounds the problem. Studies identified the following factors as being contributory to the development of musculoskeletal disorders:

1. *High task repetition.* Silverstein et al. (1986) classified a job as highly repetitive if the cycle time is <30 s or if more than 50% of the time involves performing the same kind of fundamental cycle. The more repetitive the task, the more rapid and frequent are the muscle contractions, which become less efficient and hence require increased effort, therefore, demanding greater time for recovery.

2. *Forceful exertions.* As muscle effort increases in response to high task load, circulation to the muscle decreases, causing more rapid muscle fatigue.

3. *Posture.* Awkward postures overload muscles and tendons, load joints in an asymmetrical manner, and impose a static load on the musculature (Van Wely, 1970).

4. *Mechanical pressure.* Frequent or continuous use of tools with hard or sharp edges or short handles can cause compressional ischemia, impeding nerve conduction.

5. *Vibration.* Vibrating tool operation also affects myelinated nerve fiber activity and parasympathetic activity, which depresses peripheral nerve conduction (Murata et al., 1990; Murata et al., 1991).

6. *Exposure to cold.* Working in a cold environment or handling cold tools affects circulation to the digits, reducing tactile sensitivity, thereby increasing hand force.

7. *Gloves.* Use of gloves reduces tactility, thereby increasing the amount of force exerted to hold or manipulate a given object.

8. *Insufficient recovery time.* Recovery time can exceed work time for jobs where physical demands are high. Deprived of sufficient recovery time, soft tissue injuries may occur.

9. *Lead.* Constant contact or exposure to lead has been shown to impair maximal motor and sensory nerve conduction velocities of the median nerve (Araki et al., 1986).

10. *Organic solvents.* Chronic exposure to n-hexane, xylene, and toluene correlates with suppressing activity of myelinated fibers in the peripheral nervous system (Murata et al., 1994).

11. *Underreporting.* A selective loss of symptomatic employees may occur in high risk jobs, a form of healthy worker effect (Morgenstern et al., 1991).

12. *Psychosocial factors.* Kiesler and Finholt (1988) suggest that job dissatisfaction leads to increased reporting of WRMSD.

When two or more of the preceding risk factors are present, the risk of trauma is dramatically increased. Odds ratios of developing WRMSDs appear to be multiplicative. Jobs that combine high force and high repetition may pose the greatest risk (Silverstein et al., 1986).

Personal Factors Associated with CTS Prevalence

It is widely recognized that several medical and postoperative conditions can predispose a patient to CTDs. For reasons that are unclear, hypothyroid and hyperthyroid patients, postmenopausal women, and posthysterectomy patients are more susceptible to musculoskeletal disorders. Increased incidence among these patients may be because of altered hormonal balances causing connective tissue changes similar to an inflammatory response. Other medical conditions associated with an increased incidence of WRMSDs include rheumatoid arthritis, hypertension, diabetes mellitus, kidney disorders, lipoma, amyloid disease, myxedema, acromegaly, gout, cystic fibrosis, body mass index, pregnancy, use of oral contraceptives, alcoholism, and a variety of endocrine problems.[*]

[*]Robbins, 1963; Chabon, 1985; Pascual et al., 1991; Nathan et al., 1994; Osorio et al., 1994; Werner et al., 1994; Chammas et al., 1995; O'Riordan et al., 1995.

Nonmedical personal factors have also been positively correlated with increased reporting, which might be indicative of CTS prevalence among the general population. Tanaka et al. (1995) determined, based on self-reported CTS findings, that race (odds ratio—4.2, whites higher than nonwhites), gender (odds ratio = 2.2, females higher than males), and age (odds ratio = 1.03, risk increasing per year) might predispose individuals to musculoskeletal disorders of the hand and wrist. Furthermore, Radecki (1994) discovered 75% of female and 40% of male CTS patients had a positive family history suggestive of hereditary traits.

Traumatic and Nontraumatic Causes of CTS

Median palsy can be produced as a result of traumatic compression of the median nerve resulting in CTS symptoms. Based on the in-depth anatomical studies of the carpal tunnel (Robbins, 1963), the following identifiable sources of traumatic and nontraumatic causes of CTS were developed. This has helped establish a foundation as to potential causes of hand–wrist musculoskeletal disorders.

Traumatic causes are as follows:

1. Hematoma caused by hemorrhage in the palm
2. Carpal dislocation: forward dislocation of the lunate bone
3. Recent fracture of the distal end of the radius
4. Immobilization of a Colles fracture in the position of the acute volar flexion and ulnar deviation (Cotton-Loder position)
5. Healed fracture of the distal end of the radius with excessive new bone formation or with an unreduced bone fragment, resulting in tardy median nerve palsy
6. Deformities of the carpal bones caused by recent or old fractures or gross traumatic arthritis of the wrist

Nontraumatic causes are as follows:

1. Tenosynovitis
2. Calcium deposit
3. Ganglion
4. Leri's pleonosteosis
5. Neuromass of the median nerve at the wrist
6. Pseudoneurosis of attrition of the median nerve in the carpal tunnel
7. Aberrant artery closely accompanying the median nerve in the carpal tunnel, beating against the nerve
8. Abnormally low muscle belly of the sublimis of the long finger contracting against the nerve
9. Accessory bundle of the flexor brevis minimi digiti originating proximally from the antebrachial fascia and passing as a bridge over the nerve

MEDICAL DIAGNOSIS OF UPPER EXTREMITY MUSCULOSKELETAL DISORDERS

Conservative surveillance techniques report that the occupational incidence of disorders associated with repeated trauma increased by 1500% between 1983 and 1996 (US DOL). This increase equates to an average annual increase of 24%, based on previous year's data. Such dramatic growth prompted attention by the research and medical communities. As a result, a repertoire of clinical and quantitative diagnostic utilities for upper extremity musculoskeletal disorders was developed.

Because of the nature of nerve disorders, such as CTS, two patients for whom symptomatic experience is similar may be suffering from two quite different disorders or from neurological irritation at two different sites along a nerve path. Therefore, the clinician must be confident in findings of the diagnostic tests before prescribing an appropriate course of treatment, surgical, or otherwise. A full diagnosis contains several parts.

History

The first and most important information is the patient's experience: how the problem started, how it progressed, and what activities encourage symptomatic trauma. The patient can offer a level of insight not found in any quantitative measures, the benefits of which should be fully utilized. Their perspective may be solicited through structured questionnaire or informal interview techniques. This history should include symptoms of numbness, tingling, burning, or pain in the distribution of the median nerve occurring more than 3 times or lasting 7 days or longer in the previous 12 months (Burt et al., 2011). Having the patient fill in a hand diagram is most likely the best way to illustrate or define these symptoms.

Physical Examination

A correct diagnosis requires specific examination of the affected body part. Many rapid assessment tools have been developed to aid the clinician in this process, including Phalen's test, Tinel's test, Flick test, reverse Phalen's maneuver, range of motion, strength, sensory, reflex, and circulation/pulse testing, the details of which are offered in more extensive texts. Carpal tunnel compression test adds to the clinical examination methods that can be used to determine the likelihood that CTS is the diagnosis or part of the diagnosis. In a prospective study on 200 hands diagnosed as having CTS a control group of 100 volunteers with no symptoms of CTS were also assessed. This study was done to evaluate the effectiveness of the compression test in the physical exam to determine if CTS was present. Essentially, the carpal tunnel compression test reproduces the disease itself because it increases the pressure within the canal by external manipulation. It can also be used on wrists with limited range of motion (Gonzalez Del Pina et al., 2003).

Electrodiagnostic Studies

Compression of a nerve has been shown to produce slow conduction, entirely block nerve conduction, or cause axonal destruction. Electrodiagnostic studies are used as a tool to measure the time required for an impulse to travel from a proximal stimulus to a more distal site at which a response is evoked. This measure is called nerve latency.

Both medial and ulnar nerve latencies may be recorded and then compared. If median latency is significantly greater than ulnar latency, this is suggestive of CTS. If median and ulnar nerve latencies are both slow compared with normal values, low hand temperature, or peripheral neuropathy may be suggested (Bleecker and Agnew, 1987) (Figure 72.4).

Electrodiagnositic utilities have become widely instituted and are the primary means of detecting mild nerve disorders such as CTS. Early diagnosis can facilitate dramatic reductions in pain and suffering, associated costs, and lost time.

Kirschberg et al. (1994) demonstrated the effectiveness of combining clinical and electrodiagnostic diagnostic utilities to improve the reliability of findings. In a reevaluation of 112 patients who performed similar repetitive motion tasks and had reported pain, tingling, numbness, or all three in the hand, wrist, and forearm, only 35% presented positive clinical symptoms or positive electrodiagnostic findings. Only 17% overall actually had positive diagnostic results for both tests. This is also illustrated when a patient presents with consistent symptoms of numbness and/or pain but has negative electrodiagnostic findings. The hand surgeon makes the clinical decision to release the carpal tunnel and finds that the patient's symptoms are relieved following the release.

As noted in Burte et al's cross-sectional study the nerve conduction study must show electrodiagnostic criteria for median mononeuropathy. That is criteria A and B or C.

Criteria A. Slowed latency in median nerve

- Wrist to index finger sensory latency >3.7 ms or
- Mid-palm to wrist sensory latency >2.2 ms or
- Motor latency >4.4 ms

Criteria B. Normal distal ulnar nerve latency

- Wrist to little finger sensory latency ≤3.7 ms

Criteria C. Distal median nerve latency>distal ulnar latency

- Median wrist to index finger—ulnar wrist to little finger latency difference >1.0 ms or

- Median mid-palm to wrist—ulnar mid-palm to wrist latency difference >0.5 ms

Treatment of CTS

If CTS is detected during the early stages of development, conservative treatment is recommended. This can be significantly less disruptive to the patient and less costly than surgical approaches. The patient may also be able to return to normal duties within a relatively short time and with minimal pain and suffering.

Conservative treatment of CTS combines four types of therapies: restricting motion and splinting to immobilize the affected area; applying heat or cold to facilitate the repair process; medications and injections to reduce inflammation and pain; and special exercises that promote circulation to speed recovery and increase the range of motion. A simple wrist brace will sometimes lessen symptoms of mild CTS, especially if the patient is instructed to refrain from activities that require working against the brace, which could accelerate the onset of more severe symptomatology. The brace might be most effectively worn at night during sleeping to prevent awkward wrist postures, which are thought to foster nocturnal exacerbation. Anti-inflammatory medications are used as the second line of treatment to help control swelling of soft tissue structures. An injection of cortisone into the carpal tunnel may be prescribed to decrease swelling within the carpal tunnel, thereby permitting time for recovery. Giannini et al. (1991) showed use of local steroid injection facilitated improvement in median nerve function long after the pharmacological effects of the agents ceased.

Only if the symptoms are extreme and the preceding medical treatments are ineffective should surgical means of rehabilitation be considered. Boniface et al. (1994) reported that symptomatology for 86% of CTS nerve disorder patients may be resolved without surgical intervention.

Traditional release surgery involves the making of a small incision, usually <2 inches, along the palm of the hand. This incision is continued through the palmar fascia to reveal the transverse carpal ligament, the constricting element. The transverse carpal ligament is then cut to relieve pressure from the median nerve. Only the skin incision is sutured, leaving the gap in the transverse carpal ligament open to slowly heal as scar tissue.

Improvements in surgical techniques now permit the operating physician to use endoscopic tools for this procedure. Either one or two ports may be cut, depending on the preferred variation, through which both monitoring and incision are accomplished. Although endoscopic surgeries tend to be more expensive and less successful than the traditional approach, the reported benefits include decreased surgical time, decreased postoperative attention, early return to duty, decreased pain, and increased thenar strength (Bernstein, 1994).

Several studies support aggressive postoperative rehabilitation therapy and early return to work. One cost-effective approach might be to teach self-stretching techniques to patients undergoing myofascial release surgery, which have been shown to reduce symptoms and improve electromyographic results (Sucher, 1993). In a comparative evaluation, Goodman (1992) demonstrated that 14% of a traditional treatment group, whereas only 2% of an aggressive return-to-work group failed to resume their normal duties following carpal tunnel release surgery. Furthermore, tangible costs computed were found to be 50% lower in the aggressively treated group.

If a complete and permanent recovery is expected, it is important that the worker not be returned to the same job or task that precipitated the musculoskeletal disorder unless that job or task has been appropriately modified.

Recurrence of CTS

Recurrences of CTS following surgery are predominantly the result of inadequacies of the first procedure, such as incomplete splitting of the transverse carpal ligament or compression of the median nerve caused by excessive and improperly formed scar tissue. Additionally, up to 50% of recurrent CTS patients have medical conditions such as insulin-dependent diabetes mellitus, terminal renal insufficiency, or acromegaly that increase their susceptibility to nerve disorders (Kern et al., 1993). These reported statistics suggest that the number of operations for recurrent CTS can be dramatically reduced if the first operation is performed with greater care and attention. Patients with CTS secondary to a systemic disease are particularly at risk.

ERGONOMIC EVALUATION OF WORKPLACE HAZARDS

Health and risk factor surveillance provides a means of systematic identification of patterns or trends of work-related musculoskeletal symptoms and disorders, and their risk factors. Through ongoing methodical collection, analysis, and interpretation of health and exposure data, surveillance systems often identify the effects of workplace hazards before individual employees would normally feel compelled to report them. When performing an ergonomic workplace evaluation, the following three steps should be followed:

Phase I: Passive surveillance of available records. Surveillance tools are characterized by their practicality and speed with which data collection may be conducted, rather than their accuracy. Passive surveillance relies on the analysis of information from existing databases, such as those listed in Table 102.1, and is generally accepted as a means of rapid assessment.

By calculating incidence rates and comparing them with company or industry norms, it is relatively easy to identify specific areas within a facility that warrant's further investigation.

Phase II: Active surveillance of workers. Questionnaire and informal interviewing techniques are useful for identifying new or incipient problems as well as for assessing the effectiveness of medical interventions and ergonomic controls. Specifically designed active surveillance tools can also be used to determine employee perceptions about aggravating factors and job improvement ideas.

Once the problem areas have been identified through passive and active surveillance, the proactive administration of physical examinations may prove to be a cost-effective approach for the early detection of WRMSD among operators in that area.

Phase III: Workplace risk factor analysis. Worksite assessments are usually performed by experienced ergonomists or other health specialists who have received specialized training in the identification of ergonomic hazards. Workstations, posture, tools and materials, as well as organizational and environmental factors should all be considered.

Circumstances that may raise concern include awkward postures, relatively short cycle time or high task repetition, high force demands, awkward or improperly designed tools, and employee-initiated task or tool redesigns. The ergonomist will likely use a variety of techniques to record and document these circumstances for further examination.

Once an ergonomic job analysis has established the presence of potential contributors to cumulative trauma, a plan for control and prevention should be devised. The principal technique for preventing musculoskeletal disorders is to reduce exposure to associated risk factors, although the specific amount of acceptable exposure has not been determined. To this end, a combination of administrative and engineering approaches may be considered. Hand activity level (HAL) and wrist strain index (SI), parameters may also be considered.

Engineering Controls

Engineering controls focus on the job or work environment. The aim is to redesign the job or tool to achieve control over those job factors associated with the onset of CTDs. Examples of engineering controls include the following:

1. Design jobs to reduce hand force and frequency of repetition.
2. Position the work and worker to eliminate awkward postures.

TABLE 102.1   Tools for Work-Related Musculoskeletal Disorder and Associated Risk Factor Surveillance

Surveillance Level	Method of Surveillance	
	Passive	Active
Level I (low)	Company dispensary logs	Checklists
	Insurance records	Questionnaires
	Workers' compensation records	
	Accident reports	
	Transfer requests	
	Absentee records	
	Grievances and complaints	
Level II (high)		Interviews
		Physical examinations
Workplace risk factor		Checklists
		Questionnaires
		Job analysis

Source: From *Traumatogens Associated with Carpal Tunnel Syndrome in Hamilton and Hardy's Industrial Toxicology 5th Edition*, Mosby, 1988.

3. Provide adjustable workstations and seating that permit postural changes.

4. Design workstations to minimize biomechanical loading.

5. Eliminate wrist deviations by angling the work toward the operator.

6. Use fixtures and jigs to support and position work pieces.

7. Provide parts bins below elbow height to avoid bent wrists.

8. Locate tools and components within easy reach.

9. Design or select hand tools using ergonomic principles to eliminate risk factors.

10. Use power tools to reduce exertion and repetition, but be sure to follow ergonomic tool selection guidelines.

Administrative Controls

Administrative controls refer to actions taken by management or medical staff to limit the potentially harmful effects of a stressful job, which is usually achieved by modifying existing personnel functions. Appropriate administrative interventions for a given situation might include several of the following control measures:

1. Job rotation to less hand-intensive tasks.

2. Integrate work breaks to permit physiological recovery from biomechanically demanding tasks.

3. Alternate hands so that the task is not performed primarily by one hand.

4. Reduce repetitive body motion through task automation.

5. Avoid machine pacing.

6. Avoid incentive pay scales (i.e., piece work).

7. Provide and enforce the use of adequate protective equipment.

8. Control environment to prevent extreme conditions.

9. Gradually introduce new and returning employees.

10. Train workers in correct methods.

11. Educate management and supervisory personnel about the problem.

12. Employee selection; seek medical advice about job placement.

STANDARDS AND LEGISLATION

The late 1990s was a time when American industries were grappling with significant incidence of musculoskeletal disorders and, in particular, CTS in many workplaces. Unfortunately, market forces seem to have provided insufficient incentive for top management to grasp the economic potential of workplace ergonomic investments.

During the mid-1980s, in response to the beginnings of increased CTD incidence, the NIOSH investigated appropriate strategies for the prevention and control of musculoskeletal injuries and illnesses. The following is an excerpt from their report that presently serves as the basis for the Occupational Safety and Health Administration's (OSHA's) legislative intervention endeavors.

When job demands repeatedly exceed the biomechanical capacity of the worker, the activities become trauma inducing. Hence, traumatogens are workplace sources of biomechanical strain that contribute to the onset of injuries affecting the musculoskeletal system.

In the absence of a general industry standard, OSHA continues to rely on the general safety clause 5 (a)(i) of their Health and Safety Act of 1970 to pursue legislative correction for ergonomic infringements. This approach has been successful in several well-publicized cases, including the John Morrell meatpacking facility in South Dakota (NIOSH, 1988). However, OSHA's goals lie not in seeking monetary compensation through citation, but in the prevention of occupational injuries and illnesses that can be debilitating to the employee and employer alike.

The original OSHA enactment of 1970, currently standing, is outdated as a legislative means of preventing occupational illnesses of ergonomic source. A more contemporary mandate may be required to reflect industry changes during the past two decades.

In light of the Morrell case, a guideline was set forth for the meatpacking industry (OSHA, 1990), which proved to be the most effective testing ground (see the case study). OSHA has since pursued a general industry enactment of their workplace ergonomics protection standard (EPS), the first draft of which was released in March 1995 (OSHA, 1995). Neither the profession of ergonomics nor manufacturing industries supported the first draft of the EPS because of potential financial ramifications. Efforts to refine the standard into a suitable format continue.

The OSHA again published an ergonomics program standard on 11/14/2000 that went into effect on 01/16/01. Congress overturned the rule in 2001. A major factor in defeating this standard was again the economic impact on industry. In addition, it was felt that there was a multiple factorial etiology for CTS that is the worker's outside of work activities also significantly contributed to the CTS and the industry will not have to pay for the cost of care.

Progress has been made in research and assessment tools for the effect of exertion on upper extremity muscles and structures. The American Conference of Governmental Industrial Hygienists (ACGIH) threshold limit value (TLV) for hand activity is a method for evaluation of job risk factors due to musculoskeletal disorders of the hand and wrist. The TLV for HAL is aimed for monotask jobs with 4 or more hours of repetitive handwork (www.ttl.fi/workloadexposuremethods June2009).

The level of hand activity is identified by using a scale of 0–10, where zero is virtually no activity to a level of 10 (highest imaginable hand activity) The normalized peak flow (NPF) is the relative level of effort on a scale of 0–10 that a person of average strength would exert in the same posture required by the task (v1.52/17/02 2002 Thomas E Bernard and ACGIH).

Hand Activity Level (HAL) TLV Scoring Sheet

HAL/TLV Scoring Graft Tiffany Cash described in her Thesis and Dissertations "Using the Strain Index and TLV for HAL to Predict Incidence of Aggregate Distal Upper Extremity Disorders in a Prospective Cohort" 2012 that if the job falls above the TLV line on the graph, it is said to be hazardous to most workers. If the plotted line falls below the Al (action level) line it is said to be safe for most workers.

The wrist SI includes six putative risk factors (force, repetition, percent, duration of exertion, postural, speed of work, and shift duration and was derived from epidemiology of (Moore and Garg, 1995). Garg et al. (2012) studied a cohort of 536 workers monthly for a period of 6 years to determine if the TLV for HAL and the SI were useful metrics for estimating exposure to biomechanical stressors. "Both the TLV for HAL and the SI were found to predict risk of CTS when adjusted for relevant covariates."

Workplace and industrial risk factors for CTS was a cross-sectional study published in 2011. This study concluded that the quantitative and rating space job exposure measures were each associated with CTS and that obesity increased the association between frequency and exertion and CTS (Burt et al., 2011).

The value of these entities is that they allow not only the ability to design research tools to investigate the relationship of these factors, but also the ability to assess the safety of workplace tasks. Thus, jobs can be modified to prevent the development of CTS. These research and measurement methods allow industry to design systems that are less likely to contribute to the development of CTS. The concept of action level in TLV certainly placed traumatogens in the realm of industrial toxicology.

CASE STUDY

In 1990 OSHA published its Ergonomics Guidelines for the Meatpacking Industry (OSHA, 1990). This directive was developed in cooperation with the meat industry and the American Meat Institute (AMI) trade association in an attempt to combat the extraordinarily high and ever increasing incidence of CTDs among industry employees.

It is speculated that the meat industry has the highest rates of CTS because of the high percentage of the workforce that performs hand-intensive tasks. Further, the meat industry faced severe engineering constraints on equipment development as a result of the random shapes of their products and strict sanitation requirements. Although the financial benefits of an ergonomic risk management program are understood among health and safety specialists, it was believed that direct legislation would provide the necessary impetus to initiate proactive intervention programs and keep them going. Thus, OSHA and AMI developed the first industry-specific ergonomic guidelines in an attempt to control CTD incidence.

The meatpacking ergonomics guidelines encourage management commitment as evidenced by a written company program, employee involvement, and regular program review and evaluation. The guidelines, while being flexible enough to be adapted to the needs and resources of each employer, provided policy on the following four program elements: (1) worksite analysis, (2) hazard prevention and control, (3) medical management, and (4) training.

The results of 4 years' postguideline experience are encouraging. Although still 30 times greater than the general industry norm, CTD incidence in the meat industry has demonstrated a promising trend since 1991, during which time rates have declined by almost 16%. These data support the notion that employer ergonomics programs in the meat industry are working and that OSHA's Ergonomics Guidelines for the Meatpacking Industry are having a positive impact.

CONCLUSIONS

Carpal tunnel syndrome, a compression neuropathy affecting the median nerve, may develop whenever task demands habitually exceed a worker's capacity to respond to those demands. Personal or medical factors might predispose certain individuals to the development of CTS such that, in the absence of sufficient physiological recovery, relatively inconsiderable occupational demands may impose mechanical stresses on the musculoskeletal system.

Hand activity level and wrist SI through ongoing passive and active surveillance of health and exposure data, the effects of workplace hazards may be identified before employees would normally feel compelled to report them. The solution is to balance task demands with worker capacity through the appropriate implementation of engineering and administrative controls.

REFERENCES

Araki, S., et al. (1986) Psychological performance in relation to central and peripheral nerve conduction in workers exposed to lead, zinc and copper. *Am. J. Ind. Med.* 9:535–542.

Bernstein, R.A. (1994) Endoscopic carpal tunnel release. *Conn. Med.* 58:387–394.

Bleecker, M.L. and Agnew, J. (1987) New techniques for the diagnosis of carpal tunnel syndrome. *Scand. J. Work Environ. Health* 13:385–388.

Boniface, S.J., Morris, I., and Macleod, A. (1994) How does neurophysiological assessment influence the management and outcome of patients with carpal tunnel syndrome? *Br. J. Rheumatol.* 33:1169–1170.

Brogmus, G.E. and Marko, R. (1990) Cumulative trauma disorders of the upper extremities: the magnitude of the problem in U.S. industry, *Proceedings of the IEA Conference on Human Factors in Design for Manufacturability and Process Planning, Honolulu.*

Brogmus, G.E., Webster, B., and Sorock, G. (1994) Recent trends in cumulative trauma disorders of the upper extremities in the United States: an evaluation of possible reasons. *J. Occup. Environ. Med.* 38(4):401–411.

Bureau of Labor Statistics (2008) *Summary of Occupational Illnesses and Injuries*, Washington, DC: US Department of Labor, Tables 20, 21 and 23.

Burt, S., Crombie, K., Jin, Y., et al. (2011) Work place and individual risk factors for carpal tunnel syndrome. *Occup. Environ. Med.* 68:928–933, originally published online 05/25/2011.

Chabon, S.J. (1985) *Carpal tunnel syndrome: a medical and occupational disease, PA Outlook September: 4–16.*

Chammas, M., et al. (1995) Dupuytren's disease, carpal tunnel syndrome, trigger digit and diabetes mellitus. *J. Hand Surg. Am.* 20:109–114.

Garg, A., Kapellusch, J., Hegmann, K., Wertsch, J., Merryweather, A., Deckow-Shaefer, G., Malloi, E.J., and the WISTAH Hand Study Research Team (2012) The Strain Index (SI) and Threshold Limit Value TLV for Hand Activity Level (HAL: risk of Carpal Tunnel Syndrome (CTS) and a prospective cohort. *Ergonomics* 55(4):396–414.

Giannini, F., et al. (1991) Electrophysiologic evaluation of local steroid injection in carpal tunnel syndrome. *Arch. Phys. Med. Rehabil.* 72:738–742.

Gonzalez Del Pina, J., Delgato-Martinez, A.D., Gonzalez, I., Gonzalez, and Lovic, A. (1997) Value of the carpal compression test in the diagnosis of carpal tunnel syndrome. *J. Hand Surg. Br.* 22B(1):38–41.

Goodman, R.C. (1992) An aggressive return-to-work program in surgical treatment of carpal tunnel syndrome: a comparison of costs. *Plast. Reconstr. Surg.* 89:715–717.

Green, R.A. and Briggs, C.A. (1989) Effect of overuse injury and the importance of training on the use of adjustable workstations by keyboard operators. *J. Occup. Med.* 31:557–562.

Kern, B.C., et al. (1993) The recurrent carpal tunnel syndrome. *Zentralbl. Neurochir.* 54:80–83.

Kiesler, S. and Finholt, T. (1988) The mystery of RSI. *Am. Psychol.* 43:1004–1015.

Kirschberg, G.J., et al. (1994) Carpal tunnel syndrome: classic clinical symptoms and electrodiagnostic studies in poultry workers with hand, wrist, and forearm pain. *South. Med. J.* 87:328–331.

Moore, J.S. and Garg, A. (1995) The strain index: A proposed method to analyze jobs for risk of distal upper extremity disorders. *Am. Ind. Hyg. Assoc. J.* 56:443–458 (page 447).

Morgenstern, H., et al. (1991) A cross-sectional study of hand/wrist symptoms in female grocery checkers. *Am. J. Ind. Med.* 20:209–218.

Murata, K., et al. (1994) Changes in autonomic function as determined by ECG R-R interval variability in sandal, shoe and leather workers exposed to n-hexane, xylene and toluene. *Neurotoxicology* 15:867–875.

Musculoskeletal Disorders and Workplace Factors, A Critical Review of Epidemiological Evidence for Work Related Musculoskeletal Disorders of the Neck, Upper Extremity and Low Back, 2nd Printing, US Department of Health and Human Services, Public Health Service, CDC, NIOSH.

Murata, K., Araki, S., and Aono, K. (1990) Central and peripheral nervous system effects of hand-arm vibrating tool operation: a study of brainstem auditory-evoked potential and peripheral nerve conduction velocity. *Int. Arch. Occup. Environ. Health* 62:183–187.

Murata, K., Araki, S., and Maeda, K. (1991) Autonomic and peripheral nervous system dysfunction in workers exposed to hand-arm vibration: a study of R-R interval variability. *Int. Arch. Occup. Environ. Health* 63:205–211.

Murata, K., et al. (1994) Changes in autonomic function as determined by ECG R-R interval variability in sandal, shoe and leather workers exposed to n-hexane, xylene and toluene. *Neurotoxicology* 15:867–875.

Nathan, P.A., et al. (1994) Slowing of sensory conduction of the median nerve and carpal tunnel syndrome in Japanese and American industrial workers. *J. Hand. Surg. Br.* 19B:30–34.

National Institute for Occupational Safety and Health (NIOSH) (1988) *Health Hazard Evaluation Report: John Morell*, HETA 88-180-1958, Sioux Falls, SD: National Institute for Occupational Safety and Health.

O'Riordan, J.I., et al. (1995) Peripheral nerve dysfunction in adult patients with cystic fibrosis. *Ir. J. Med. Sci.* 164:207–208.

Occupational Safety and Health Administration (OSHA) (1990) *Ergonomics Guidelines for the Meatpacking Industry*, Washington, DC: Occupational Safety and Health Administration.

Occupational Safety and Health Administration (OSHA) (1995) *Ergonomics Protection Standard (Draft)*, Washington, DC: Occupational Safety and Health Administration.

Osorio, A.M., et al. (1994) Carpal tunnel syndrome among grocery store workers. *Am. J. Ind. Med.* 25:229–245.

Palmer, D.H., and Hanrahan, L.P. (1995) Social and economic costs of carpal tunnel surgery. *Instr. Course Lect.* 44:167–172.

Pascual, E., et al. (1991) Higher incidence of carpal tunnel syndrome in oophorectomized women. *Br. J. Rheumatol.* 30:60–62.

Putz-Anderson, V. editor. (1988) *Cumulative Trauma Disorders: A Manual for Musculoskeletal Diseases of the Upper Limbs*, London: Taylor & Francis.

Radecki, P. (1994) The familial occurrence of carpal tunnel syndrome. *Muscle Nerve* 17:325–330.

Robbins, H. (1963) Anatomical study of the median nerve in the carpal tunnel and the etiologies of the carpal tunnel syndrome. *J. Bone Joint Surg. Am.* 45A:953–966.

Silverstein, B.A., Fine, L.J., and Armstrong, T.J. (1986) Hand wrist cumulative trauma disorders in industry. *Br. J. Ind. Med.* 43:779–784.

Sucher, B.M. (1993) Myofascial release of carpal tunnel syndrome. *J. Am. Osteopath. Assoc.* 93:92–94.

Tanaka, S., et al. (1995) Prevalence and work-relatedness of self-reported carpal tunnel syndrome among U.S. workers: analysis of the occupational health supplement data of 1988 national health interview survey. *Am. J. Ind. Med.* 27:451–470.

Van Wely, P. (1970) Design and disease. *Appl. Ergon.* 1:262–269.

Werner, R.A., et al. (1994) The relationship between body mass index and the diagnosis of carpal tunnel syndrome. *Muscle Nerve* 17:632–636.

103

VIBRATION

Donald E. Wasserman

DESCRIPTION OF THE HAZARD

Vibration is mechanical motion that is defined as a vector, which simply means that it describes motion as consisting of both a magnitude-intensity and a direction; vibration at any point contains six direction vectors, three linear directions, and three rotational directions. (For a complete description of vibration see: Tse et al., 1963; Wasserman, 1987) In the linear direction [i.e., moving along a line, or axis], the total linear motion is described as movement along three mutually perpendicular (orthogonal) "triaxial" directional axes: front-to-rear motion, side-to-side motion, up-down motion (designated as: x-, y-, and z-axes, respectively). Similarly, total rotational motion is described as motion around these same orthogonal axes called: pitch, roll, and yaw. However, as a practical matter, the vast majority of human and/or occupational medical and epidemiology studies and their respective data bases, and its effects on humans are virtually supported only by human vibration linear axes triaxial vibration measurements, predominantly performed as part of these biological studies. Thus, only linear vibration is discussed herein, neither rotational motion nor its measurement, because of this lack of supporting medical and epidemiology effects data to date.

How is the vibration magnitude described and expressed for each of these three axes? To answer that question, a few more vibration basics are needed: When we speak of vibratory motion we usually refer to motion of an object starting at some reference [or neutral] point and then moving off to another position along the same axis. The linear distance the object has traversed is called *displacement*. The units of displacement are given as distances in meters, millimeters, centimeters, inches, feet, miles, etc. It obviously took some

time for that object to move from its initial position to its new position. If we in effect divide the distance traversed by its travel time, we obtain the object's "*speed*" or *velocity* (more precisely mathematically: velocity is the time rate of change of displacement). The units of velocity are a ratio of distance to time, for example, ft/s, inch/s, m/s, cm/s, mm/s, miles/h, etc.

In many instances the speed of a moving object changes over time. The time rate of change of speed or velocity is called *acceleration*. The units of acceleration are distance/time/time, or m/s/s, or ft./sec/sec. It is convenient to compare vibration acceleration to gravitational or "g" units, where $1\,g = 9.81$ m/s/s. Thus displacement, velocity, and acceleration are three different descriptors of motion magnitude or "intensity" which are all mathematically linked; if you know either one of these three descriptors, the remaining two can be calculated. To answer the original question above: *acceleration* is the measurement of choice to quantify the vibration magnitude/intensity for human exposure to vibration; expressed as either "g" or "m/s^2." When performing triaxial vibration measurements we simultaneously measure all three linear axes and obtain three independent measurements for each of these axes. What is eventually sought from these triaxial acceleration measurements is a "vector sum" of the acceleration values for all three x, y, z directions; representing the overall total linear vibration exposure for the moving point.

In addition to direction and acceleration, the *frequency* of vibration must also be known. Vibration frequency is expressed in hertz (Hz). The elemental or simplest form of vibratory motion is sinusoidal (like a pure single frequency sound emanating from a vibrating tuning fork). The frequency is defined as the number of times in 1 s that a cycle is completed; 10 complete cycles/s = 10 Hz; 1000 complete

Hamilton & Hardy's Industrial Toxicology, Sixth Edition. Edited by Raymond D. Harbison, Marie M. Bourgeois, and Giffe T. Johnson.
© 2015 John Wiley & Sons, Inc. Published 2015 by John Wiley & Sons, Inc.

cycles/s = 1 kHz; 1,000,000 cycles/s = 1 mHz, etc. However, in reality most power tool or vehicle vibration is complex, that is, it is composed of a mixture of multiple varying frequency sinusoids appearing together forming a complex mixture. This vibration mixture can be decomposed into a display called a "vibration spectrum." This method is called Fourier analysis and is capable of analyzing and decomposing a complex vibration time mixture into all of its elemental (sinusoidal) frequencies and corresponding acceleration/intensity amounts which composed the original complex mixture. This analysis technique can be likened to a familiar technique used in chemistry for many years called "spectroscopy" wherein an unknown compound is similarly analyzed (mathematically decomposed) into all its elements, including the calculation of their corresponding concentrations for each element composing the original compound and then visually displaying all as the total composition spectrum of the original compound. Know that sinusoidal vibration is usually periodic and thus repeats its motion over and over again, but all vibration need not be periodic, it could be impulsive (bursts) and/or random depending on the characteristics of the vibrating source.

Finally, there is a unique vibration characteristic and parameter that particularly concerns both health professionals and engineers; it is called "resonant" frequency or natural frequency vibration. Resonant frequency vibration is an aberrant and undesirable condition whereby a small amount of input vibration impinging on a system or structure produces an uncontrolled, exaggerated, and undesirable amplified response. This occurs when the input vibration mechanically excites the natural frequency within a structure. For instance, soldiers marching in cadence across a bridge induce in-phase vibration into the bridge structure, which in turn mechanically excites the bridge structure by naturally tuning to this pounding vibration and then the structure actually makes things worse by internally amplifying this vibration causing the bridge to sway and collapse. To avoid bridge damage, soldiers *never* march across a bridge. Instead, they will break the unified rhythm of marching cadence and simply walk normally when crossing. At resonance the vibration energy transfer from the source to the structure is very efficient and is easily tuned and amplified and therefore most likely to cause damage. Unfortunately, like mechanical structures, the whole human body and individual organs have resonant vibration frequencies too which respond similarly to the structure just described and must be avoided when subjected to vibration; for example, whole-body vibration (WBV) resonances occur: 4–8 Hz vertical up-down "z"-axis; 1–2 Hz for either, front-back "x"- or side-to-side "y"-axes. It becomes increasingly obvious that human resonances usually represent the Achilles heel of human vulnerability to impinging exposure to vibration, mainly because it takes only a small amount of vibration energy at any of those resonant frequencies to trigger an exacerbated

human response to this input vibration, as compared to nonresonant vibration frequencies impinging on the human body, at the same: point, exposure time, and axis, which requires much more vibration exposure to elicit the same response.

WHOLE-BODY AND HAND–ARM VIBRATION

Human vibration exposure is customarily subdivided into ubiquitous WBV and (segmental or hand transmitted or) hand-arm vibration (HAV). Whole-body vibration affects the entire person; typical sources include cars, highway and off-highway trucks, forklift trucks, buses, trains, heavy equipment and farm vehicles, subways, aircraft, hovercraft, and fixed machinery in buildings, etc. Hand-arm vibration exposures come primarily from vibrating hand operated power tools such as gasoline-powered chain saws and brush cutters; pneumatic, hydraulic, and electrically powered tools such as chippers, grinders, drills, scalers, rivet guns, nailers, sanders, impact hammers, jackhammers, jackleg type mining tools, etc. Occasionally there are crossover situations between WBV and HAV, for example, a worker using a jackhammer ripping up a road: if the hammer is operated with the hands extended away from the workers body, it is a HAV problem; but if the worker operates the hammer where both the hands and tool are drawn in touching the body at the abdomen, chest, pelvis then it becomes a WBV problem; since vibration now can enter the body via multiple pathways. In order to cause damage to the body, vibration *must* come into physical contact with the body; where physical contact occurs defines its potential damage entrance and route.

All human triaxial vibration measurements are obtained using three, light-weight [i.e., <15 g each] accelerometers, each mounted perpendicular to each other on a small metal cube. This method allows all three perpendicular [triaxial] vibration axes to be measured simultaneously. The location and orientation of the cube are specified by internationally agreed upon methods and directions: *biodynamic coordinate systems* (Figures 103.1 and 103.2).

The biodynamic coordinate system shown in Figure 103.1 is used for WBV measurements, where the z-axis is vertical vibration defined as up-down motion from head-to-toe, the y-axis is motion across the shoulders, and the x-axis is motion from from-to-back. In practice, for example, a driving situation, all three axes (triaxial) measurements are obtained simultaneously using a special pie-plate size hard rubber disk placed between the top of the driver's seat cushion and the buttocks. Buried in the center of this disk are three tiny accelerometers and their preamplifiers each mounted orthogonal (perpendicular) to each other on a common small metal cube. For triaxial measurements each of the accelerometer output signals is simultaneously conditioned, amplified, digitized, and stored for later computer analysis.

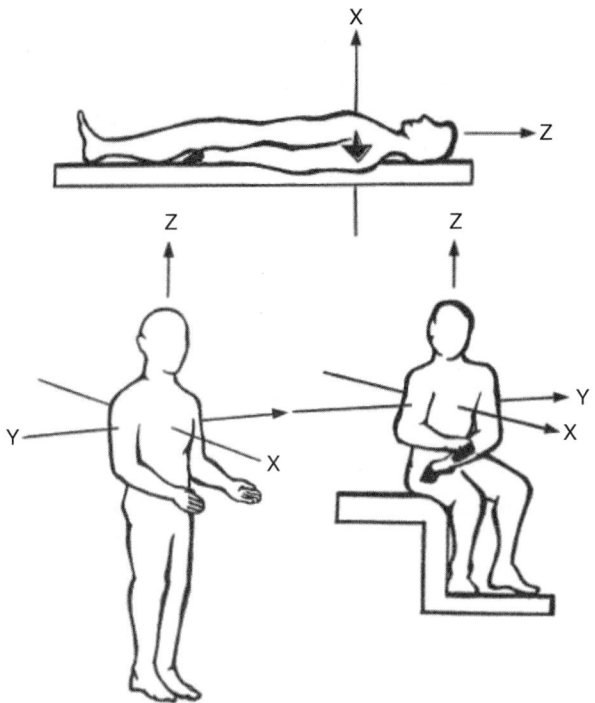

FIGURE 103.1 Biodynamic coordinate system used to assess whole-body vibration.

The biodynamic/basicentric coordinate system shown in Figure 103.2 is used for HAV measurements, where the z-axis defines motion along the long bones of the HAV, the y-axis defines motion across the knuckles, and the x-axis defines motion through the palm of the hand. Although the point for HAV measurements is defined at the third knuckle or metacarpal, as a practical matter HAV measurements are usually obtained on the actual tool handle(s) near where the hand(s) grasps the tool using only the basicentric coordinate system. In practice, three small accelerometers each with built in preamplifiers are perpendicularly orientated to each other and all are mounted to a single small metal cube, which in turn is mounted to a hose clamp. This entire clamp assembly is then oriented and clamped snugly around the vibrating tool handle, and [as before] each of the accelerometers output signals is electronically conditioned, amplified, digitized, and stored for later computer analysis. Dual handle tools usually require a second independent triaxial accelerometer clamp assembly.

In both WBV and HAV triaxial measurements, each axis of acceleration is recorded over time; as early as the 1970s it took a very sophisticated and expensive computer system to perform a Fourier spectrum analysis, which identified and separated the vibration signal into its component sinusoidal frequencies and performed other needed calculations. Today with the advent of modern solid state electronics and special software a Fourier analysis is no longer needed when performing an initial vibration survey; because this initial survey simply asks the question: is this device being vibration tested in compliance with or has it exceeded the recommended vibration exposure standard[s] criteria, programmed into the measurement computer? If the answer is that the tested device HAS met all exposure criteria then the initial survey has been successful. If the answer is that the device has exceeded the exposure criteria, then if the problem is to be solved, usually much more data analysis is required where a Fourier spectrum computer analysis, etc. is then needed as well as other engineering techniques beyond the scope of this presentation. Today several hand-held and portable basic vibration survey instrumentation, some with portable printer and/or PC interfaces, have been developed which collect data and perform these basic calculations; all are commercially available.

Finally, there are two more important critical vibration measurement concepts, briefly presented next, which the reader needs to know before a meaningful discussion of HAV and WBV standards can be given—here is our *problem*—in order for vibration workplace exposure measurements to be meaningful, these measurements must

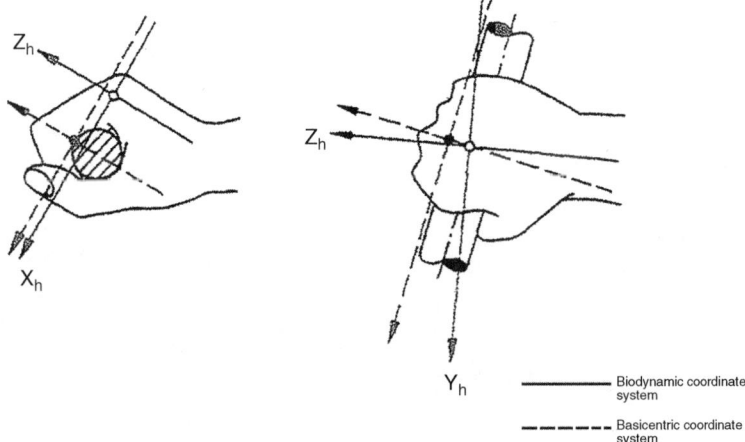

FIGURE 103.2 Biodynamic and basicentric coordinate systems used to assess hand-arm vibration.

attempt to exemplify, characterize, and quantify the [overall] typical daily vibration exposure dose impinging on a tool operators hands for HAV, or in the case of WBV the entire human body, for say a truck driver. *Answer*: Since it is neither practical, nor economically feasible to return multiple times to obtain more measurements over time, since *vibration conditions commonly change daily*! How can this practical dilemma be solved? *Answer*:

First, *we must average the triaxial acceleration data collected*, specifically, when triaxial acceleration data is originally collected, digitized and stored, for each of the three axes, separately we must mathematically calculate the so-called *root-mean-squared* (rms) acceleration values of the data, because the resulting rms value of acceleration is proportional to the energy impinging on the human body and in our case these rms values are a form of averaging the vibration exposure over time, namely, over a daily work shift, and small variations of daily exposure data are also averaged; the process gathers all the "raw" vibration, including peak vibration data and then calculates its [average] rms values; a rms example: in the case of a sinusoidal vibration its rms value is 0.707 of the peak or largest amplitude value.

Second, *a weighting function* is needed, because of resonance and that human response for either HAV or WBV is "*not* linear," rather it is frequency, axis, acceleration, and exposure time dependent. Thus, for example, one unit of vibration acceleration in a given direction/axis, time, and frequency impinging on a human, does not necessarily produce the same biological damage as compared to the same unit of acceleration, same axis, same exposure time, but at a different frequency. So what has been developed over the years, each for HAV and WBV, are mathematical curves and expressions called "weighting functions," whose purpose is to attempt to mimic and characterize human HAV response by axis, exposure time, for frequency and acceleration; and similarly human WBV response by axis, exposure time, for frequency and acceleration. All HAV standards try to use the same HAV weighting function; similarly so do all WBV standards try to use the same WBV weighting function too.

Third, know that both human HAV and WBV exposure standards use the following nomenclature to express acceleration values as:

> "Weighted rms acceleration, m/s/s" or "weighted rms acceleration, g." For each linear triaxial direction or axis; or as an overall expression of total "X, Y, Z overall weighted rms acceleration vector sum." In either m/s/s or "g" [where 1 g = 9.81 m/s/s].

To permit vibration measurements to be compared worldwide

All WBV standards use the same triaxial directions or axes [Figure 103.1].

All HAV standards use the same [basicentric] triaxial directions or axes [Figure 103.2].

OCCUPATIONAL VIBRATION STANDARDS

Standards of good practice for the evaluation and control of occupational exposures to vibration used in the United States and elsewhere are customarily divided into separate WBV standards and separate HAV standards:

> Whole-body vibration *standards* used in the United States as follows:

> International Standards Organization (Geneva), ISO 2631-1, "Guide to the Evaluation of Human Exposure to WBV" [1974 to present; use latest release version].

> American Conference of Government Industrial Hygienists (ACGIH, Cincinnati), standard for: WBV [1997 to present].

> EU Vibration Directive. European Communities (Brussels), 2002/44/EC, [2005].

> HAV *standards* used in the United States as follows:

> ACGIH (Cincinnati), standard for HAV [1984 to present]

> International Standards Organization (Geneva) ISO 5349, Parts 1 and 2 "Guide to the Measurement & Evaluation of Human Exposure to Vibration transmitted to the Hand," [1984 to present; use latest release version].

> National Institute for Occupational Safety and Health (NIOSH, Cincinnati), Document #89-106, "criteria for a recommended standard for HAV," [1989].

> EU Vibration Directive, European Communities (Brussels), 2002/44/EC, [2005].

> American National Standards Institute (New York), ANSI S2.70-[2006], "Guide for the Measurement & Evaluation of Human Exposure to Vibration Transmitted to the Hand."

Standards Notes

1. Before attempting vibration data measurements, it is imperative that the user obtain a complete, up-to-date latest version of the standard to be used and thoroughly understand it before collecting and analyzing vibration data. To do less is folly!

2. From 1971 to the present this author has actively worked and significantly contributed to this area of occupational vibration for more than four decades. Much of this time has been spent with national and international colleagues on the development and promulgation of virtually all HAV and WBV: ISO, ANSI, ACGIH, and NIOSH vibration exposure standards used in the United States and elsewhere. Please be aware herein. I will briefly discuss each of the above listed standards and try to provide insight into these standards, but the final choice of their usage is simply yours—the reader or user to make—thank you.

3. Currently in the United States there are *no* mandatory occupational protective standard; as in many other fields related to worker health and safety, all HAV and WBV standards in the United States are voluntary "consensus standards." The U.S. Occupational Safety & Health Administration [OSHA] have issued *no* standards for exposure to vibration to date. It is known that OSHA is well aware of occupational vibration issues and has been known to cite and fine employers regarding vibration using the legal provisions known as the General Duty Clause of the OSHA Law passed in 1971. The OSHA's sister organization NIOSH created under the same law, has been actively involved and working on occupational vibration problems since 1971 [when I was the NIOSH vibration group's first Chief] and continues through the present day. By contrast, the 28 nation European Union in 2005 issued and enforces a law [called a "EU Directive"] regarding [separate] exposures and rules regarding HAV and WBV.

4. ISO and ANSI each regularly review their standards every 5 years and determine if and what revisions are necessary; if needed their standards are so modified. Thus it is imperative that you use the latest version of the standard for best results.

5. For HAV: the total weighted triaxial rms acceleration vector sum is:

$$A_{TH(rms)} = \sqrt{A_{x(rms)}^2 + A_{y(rms)}^2 + A_{z(rms)}^2}.$$

6. For WBV: the total weighted triaxial acceleration vector sum for each of the three orthogonal directions (as seen in Figure 103.1) in either a supine, seated or standing operator posture is: A total = the square root of the sum of the individual squares [i.e., A_x, A_y, A_z] as modified in the following:

$$A_{T(rms)} = \sqrt{1.4A_{x(rms)}^2 + 1.4A_{y(rms)}^2 + A_{z(rms)}^2}.$$

Whole-Body Vibration Standards Used in the United States

ISO 2631 [1974-to present]: Is the first, oldest, and well-known occupational vibration exposure standard in the world. Original development by ISO/technical committee 108 experts began in the late 1960s and it was publically issued and promulgated worldwide in 1974. At that time, much of the initial research data which contributed to the formulation of ISO 2631's development arose first primarily from many diverse military situations where WBV was a serious factor determining the outcome of missions involving: high speed fighter aircraft; the effects of buffeting on bombers and air tanker crews; helicopter piloting operations

in military situations, crews operating ground vehicles such as military tanks and troop carriers; and maintaining the ability of naval gunners to accurately firing warship guns in rough seas. The civilian sector also contributed to the WBV knowledge base: data came from the auto industry describing WBV resonance research; data also was contributed by some heavy equipment and farm tractor manufacturers too. Because of the military aspects of the data, initially the main focus of this document was on human "mission critical performance proficiency" under vibratory environments, later called "fatigue decreased proficiency" or FDP. Unfortunately in the early 1970s, supporting WBV exposure chronic occupational health epidemiology and clinical data was sparse and was needed; the medical effects of WBV on the lumbar spine was discovered and elucidated after the first standard was published.

The heart of the early ISO 2631 standard [1974–1996] was the development of "weighting function curves" shown in Figure 103.3 [vertical *z*-axis] and Figure 103.4 [both *x*- and *y*-axes]; the frequency range of 1–80 Hz is designated as the WBV bandwidth as shown in both Figures. The "flat trough" of each weighting function is the resonance range for each respective axis; because humans are of different physical weights, there is a narrow resonant frequency spread of 4–8 Hz in Figure 103.3 looking at the "U"-shaped exposure time-dependent weighting curve for the vertical "*z*"-axis; similarly in Figure 103.4 from 1 to 2 Hz resonance band looking at the "hockey stick"-shaped exposure time-dependent weighting curve for either or both "x- and y"-axes. These "flat resonant frequency troughs" in both Figures 103.3 and 103.4 represents the smallest permissible allowable vibration intensity [rms acceleration] for any daily WBV exposure time; all because of the internal amplification of impinging vibration at resonance and the body's extreme sensitivity to WBV. Notice, in Figure 103.3 for example, frequencies on either side of the resonance band [i.e., <4 Hz or >8 Hz] which shows that there is increasing tolerance and thus more allowable exposure time for nonresonant vibration impinging on the body. This same description and logic applies to Figure 103.4 where resonance is 1–2 Hz, when looking at nonresonant frequencies >2 Hz.

The major downsides of this original standard are essentially as follows: When using these graphical curves, unfortunately, a Fourier spectrum analysis must be used and it is highly technical and laborious; three spectra must be generated, each separately for the *x*-, *y*-, *z*-axes. Once these spectra have been determined, each must be separately compared to their respective curves by overlaying each spectra over the appropriate curves with a knowledge of daily WBV exposure for that particular work situation. The results are usually profound and dramatic; one can immediately determine if this standard has been exceeded, if so, at what vibration frequency(s); what directional axis or axes; over what daily exposure time, etc. If the problem is to be solved said

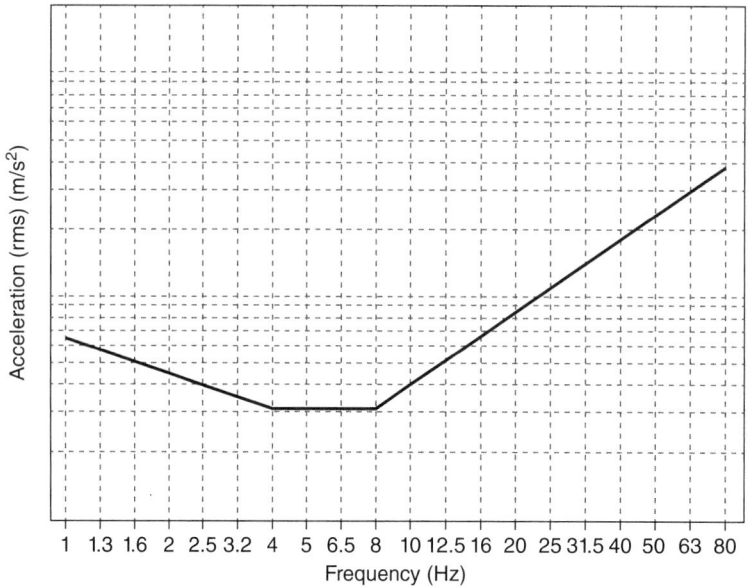

FIGURE 103.3 Early initial whole-body vibration std. weighting curve used to evaluate vertical *z*-axis root-mean-squared acceleration.

information can be used directly help to solve the problem. Unfortunately, highly trained personnel must do the work using expensive equipment. The standard also has another downside with regard to large peak or spiked WBV when measuring vehicles traversing off road conditions and travel over large road bumps. The term is called crest factor = the ratio of peak acceleration to rms acceleration. This standard does not determine human risk well; in fact it underestimates risk if the CF is >6, or a modest road bump. Why? This was simply because much of the data which became the basis of this

standard were mostly based upon sinusoidal vibration and not bumps on the road impact peaks.

Thus in 1997, for other technical reasons ISO 2631 went through major revisions and the WBV weighting function graphs described above would no longer be used as given, the weighting functions per se have been retained now extending from 0.5 to 80 Hz and the WBV calculations have increased mathematically and newer [4th.power] concepts have been introduced into later versions of ISO 2631; now more robust and employing multiple sections, including much new

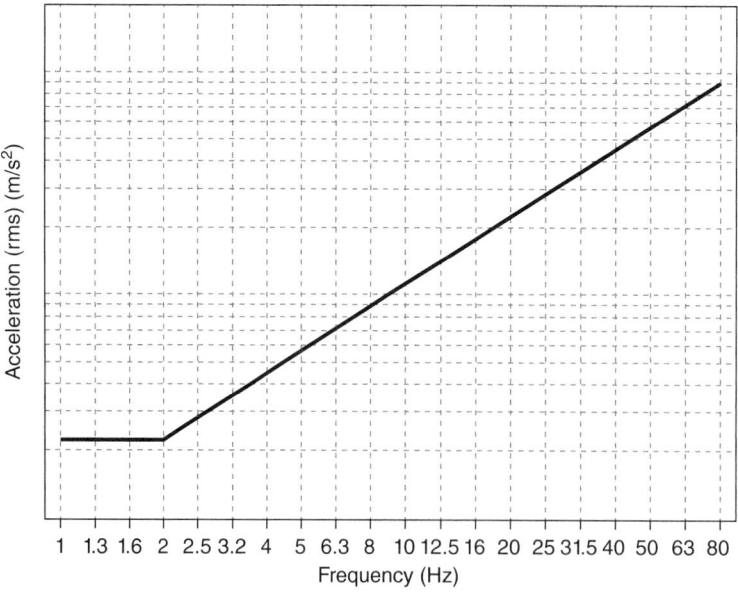

FIGURE 103.4 Early initial whole-body vibration std. weighting curve used to separately evaluate transverse *x*-axis and *y*-axis root-mean-squared acceleration.

information on sorely needed WBV impact vibration; the new focus is toward WBV and operator "posture" as an important new knowledge contribution of spinal issues resulting from WBV exposure. Thus current versions of ISO 2631 are technically much more robust, broader in scope and resulting in a more complex document.

The ACGIH [1997 to present]: WBV standard: In 1997 just as ISO chose to and replaced its original version of 2631, the ACGIH stepped in, without a WBV standard of their own, and announced they would adopt the original 2631 WBV standard, about to be replaced. ACGIH did so with full knowledge of its rich history and limitations; to this day this early version of 2631 remains the ACGIH WBV standard.

EU WBV Directive [2005]: All previous standards discussed are voluntary "consensus" standards. However, in 2005 the 28 nation European Union has taken both their WBV and HAV standards to a new level; *both* vibration standards became EU *law* ["called a Directive"] and is so enforced. The EU WBV standard uses the ISO weighting function and is briefly discussed here; the EU HAV standard also uses the ISO HAV weighting function and is discussed later with other HAV standards.

First, the EU Directive clearly states that the responsibility for the health and safety of employees exposed to either WBV or HAV rests with the employer, *not* the product manufacturer whose products are the sources of vibration. Further, the EU requires each employer to due diligence and seriously and carefully investigate [vibrating] products before purchasing said products for use by their employees, and to properly maintain these products since the employers will ultimately be held accountable! With regard to the actual WBV emanating from a product or process and daily impinging on workers, the EU requires the said products to be triaxially acceleration tested and the EU requires the results be expressed as a "weighted triaxial acceleration vector sum"; the EU WBV standard states for an 8 h/day WBV exposures: daily exposure action value [DEAV] = 0.5 m/s/s, [weighted vector sum] *and* daily exposure limit value [DELV] = 1.15 m/s/s [weighted vector sum]. If for an 8 h exposure, said weighted triaxial acceleration vector sum is <0.5 m/s/s then the vibrating device can be used for a full 8 h daily shift; if this weighted value is = 0.5 m/s/s or is >0.5 m/s/s, corrective action must begin to effectively reduce WBV exposure; if the DELV weighted vector sum value = 1.15 m/s/s, this represents the upper limit value which shall *not* be exceeded.

Hand-Arm Vibration Standards Used in the United States

For HAV, all standards listed and discussed herein [ISO 5349, Parts 1 and 2, ACGIH: HAV Std., NIOSH 89-106, EU DIRECTIVE: HAV, ANSI S2.70-2006] use the same frequency-weighting function shown in Figure 103.5 when making and analyzing HAV measurements data, but there are very significant differences among these various HAV standards as to how the results are used and interpreted; especially so with the N1OSH criteria document (#89-106) which mentions this weighting function, but does not explicitly endorse it. All other HAV standards discussed here endorse the Figure 103.5 depicted weighting function. Why there are differences among these standards? Mostly because each standard is rooted and based on the medical/

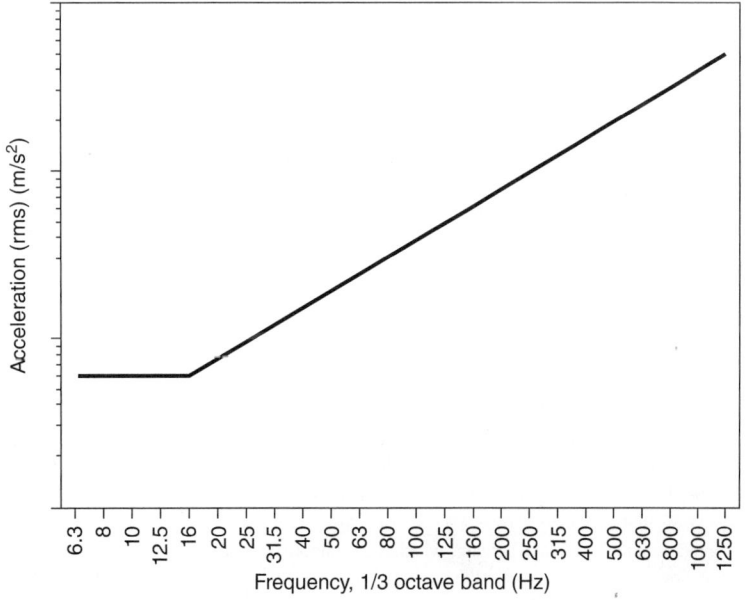

FIGURE 103.5 Per ISO 5349: Hand-arm vibration weighting curve used to evaluate *x*-, *y*-, *z*-axes root-mean-squared acceleration.

epidemiological data base available at the time of inception of each standard; over time and new studies said data bases become more robust with hopefully a better understanding of both the disease process and its etiology, but usually not without scientific probing and some controversy.

ISO 5349, Parts 1 and 2 [1984 to present]

The HAV weighting function shown in Figure 103.5, was first introduced in a very early 1980s draft version of ISO 5349; this standard was later adopted and promulgated circa 1984–1986. Here is a brief description of this weighting function: Please refer again to Figure 103.5, the graphs vertical axis depicts increasing vibration intensity in rms acceleration with the units of m/s^2. The horizontal graph axis represents vibration frequency in one-third octave bands extending from 6.3 to 1250 Hz. Within the graph is the elbow or hockey stick-shaped HAV weighting function curve. Note that there is a flat portion of the curve extending from 6.3 to 16 Hz, then the curve angles upward beginning at 16 Hz and steadily climbs upward at the same angular slope until it reaches 1250 Hz. The shape of this HAV weighting curve implies that human response to HAV is most sensitive at the lower vibration frequencies [6.3–16 Hz], and that this sensitivity steadily diminishes as the vibration frequencies increase to a maximum frequency of 1250 Hz. In actual practice, this weighting function is separately applied to each axis, thus it is applied to vibration data from the x-axis vibration measurement; then data from the y-axis vibration measurement; then again to data from the z-axis vibration measurement. Each of these measurements thus are mathematically "weighted" for each axis. Next, depending on the HAV standard being used, said weighted axial data can either be converted into a weighted vector sum [i.e., the square root of the sum of the squares of all three weighted axes] or each weighted axis can each be individually examined and evaluated.

Figure 103.5 only defines the HAV frequency bandwidth and the weighting curve shape of the mathematical weighting function for any vibrating power tool being measured; Figure 103.5 as shown does *not* define the acceptable time-dependent *daily* HAV *exposure limits*; that critical job is left to each HAV standard to determine. Why this obtuse way of determining acceptable exposure limits? ISO 5349 in 1984 originally defined this weighting curve which everyone now uses as shown in Figure 103.5. However, ISO could not reach consensus on the inclusion of acceptable regular HAV *daily* exposure limits, so they simply decided to let each nation [in 1984–1986] determine for themselves what was and was not acceptable for daily exposures, but know 1984 consensus was reached to include "yearly HAV exposures" which did became part of this standard; but a standard without daily exposure limits was an unacceptable alternative for most users who needed it. So other HAV standards later arose, which contained daily exposure limits and in doing so filled the void of this otherwise comprehensive standard. Finally, know that ISO 5349, Parts's 1&2, was a hallmark achievement for not only introducing the weighting function, but also for providing of a comprehensive description on how triaxial HAV measurements are made as well as excellent guidelines as to how to correctly analyze the HAV triaxial data.

ACGIH STD. for HAV [1984 to present]: After the aforementioned ISO 5349 decision not to include acceptable daily HAV exposure levels in their standard, the United States had a definite need to have a HAV document with daily exposure levels using the same ISO 5349 weighting function and other related criteria, thus a group of us approached and eventually collaborated with ACGIH to produce the very 1st U.S. HAV standard ever, which DID include acceptable daily HAV exposure values! The standard is shown in Table 103.1 and is briefly described as follows:

TABLE 103.1 ACGIH Threshold Limit Values for Exposure of the Hand to Vibration in either x, y, z directions.

Total Daily Exposure Duration[*]	Values of the Dominant,[†] Frequency-Weighted, rms, Component Acceleration Which Shall Not Be Exceeded	
	m/s^2	g
4 hours and <8	4	0.40
2 hours and <4	6	0.61
1 hour and <2	8	0.81
<1 hour	12	1.22

Source: From American Conference of Governmental Industrial Hygienists (ACGIH): *Threshold limit values (TLVs) for chemical substances and physical agents and Biological Exposure Indices (BEIs)*, Cincinnati, 1984-present, American Conference of Governmental Industrial Hygienists.

[*]The total time vibration enters the hand per day, whether continuously or intermittently.

[†]Usually one axis of vibration is dominant over the remaining two axes. If one or more vibration axes exceed the total daily exposure then the threshold limit value has been exceeded.

Note: 1 g = 9.81 m/s^2.

rms, Root-mean-squared.

All HAV triaxial measurements are measured, weighted, and analyzed as per ISO 5349. As before the weighted values of the rms acceleration are determined separately for each of the three vibration axes; but *no* weighted vector sum is calculated, rather each of the three axes are separately compared to Table 103.1. Refer to Table 103.1, the HAV daily exposure durations and their corresponding maximum weighted acceleration values are given as:

hours and <8; [4 m/s/s]
2 h and <4; [6 m/s/s]
1 h and <2; [8 m/s/s]
<1 h [12 m/s/s]

Usually in HAV measurements one axis of vibration is dominant over the remaining two axes. Also one must know the daily exposure time for the tool being tested. The ACGIH criteria are straight forward: If one or more vibration axes exceed the daily exposure then the [standard] threshold limit value has been exceeded. Thus the axis with the largest weighted acceleration value determines if this standard has been exceeded and for how long per 8 h shift a worker can operate the tool. Why? Because using these axis by axis criteria allows one to immediately determine which axis is most troublesome and with follow up measurements and a spectrum the vibration can be attenuated. Note that 12 m/s/s is the absolute maximum permitted [for <1 h/day use]. If you want to operate a power tool for about 8 h/day, no weighted vibration axis can exceed 4 m/s/s.

What is the downside of this ACGIH standard? Simply, it has never been updated since its inception in 1984 despite prodding by many of us to do so. Although it was the 1st U.S. standard and served us very well, it is terribly out-of- date and is in serious need of revision.

NIOSH HAV CRITERIA DOCUMENT STANDARD [89-106] [1989]: This author was NIOSH's first Chief, Occupational Vibration Group and as such I can assure the reader that since 1971 to the present NIOSH was and continues to be committed to research and the elimination of all diseases related to occupational exposure to HAV and/or WBV. In the late 1970s to early 1980s NIOSH [we] conducted multiple comprehensive pneumatic power tool HAV studies in the United States which showed very high prevalences of irreversible hand-arm vibration syndrome [HAVS] [to be discussed later]. These studies not only prompted ACGIH to develop their above cited HAV standard, but also triggered internal NIOSH activity to begin developing their own sage comprehensive technical document called "criteria document for a recommended standard" in this case for HAV. After much work, in 1989 said document was approved and initially released as NIOSH # 89-106 and has never been revised. It was the premier HAV review document of its day, but it fell short as a practical matter is unusable primarily because it did not recommend specific safe HAV exposure levels for workers. Its true value lies in fully recognizing the true causal relationship between HAV exposure and irreversible HAVS; it examined all HAV work and studies; it reviewed all existing HAV standards in use at the time and NIOSH became very concerned with the HAV "weighting function" which it acknowledged was needed but did not endorse, rather the document did endorse analyzing and comparing both unweighted and ISO weighted data to themselves and each to disease outcomes; they did endorse both administrative and engineering controls to reduce HAV workplace exposures; they did develop an excellent good work practices guide for HAV/HAVS, re. NIOSH: CIB#38, Pub.#83-110.

EU HAV Directive [2005]: All HAV standards discussed herein are voluntary "consensus" standards. However, in 2005 the 28 nation European Union has taken both their HAV and WBV standards to a new level: *both* vibration standards became EU *law* ["called a Directive"] and is so enforced. The EU HAV standard is briefly discussed here; the EU WBV standard has been previously discussed.

First, the EU directive clearly states that the responsibility for health and safety of employees exposed to either HAV or WBV rests with the employer, *not* the product manufacturer whose products are the sources of vibration. Further, the EU requires each employer to due diligence; seriously and carefully investigate [vibrating] products before purchasing said products for use by their employees and to properly maintain these products since the employers will ultimately be held accountable! With regard to the actual HAV emanating from all power tools and daily impinging on workers, the EU requires said power tools are to be triaxially acceleration tested and the EU requires that the results be expressed as a "weighted triaxial acceleration vector sum"; the EU HAV standard states for 8 h/day HAV exposures: DEAV = 2.5 m/s/s, [weighted vector sum] *and* DELV = 5.0 m/s/s [weighted vector sum]. If for example an 8 h/day exposure, said weighted triaxial acceleration vector sum is <2.5 m/s/s then the tested power tool can be used for a full 8 h daily shift; if this weighted value = 2.5 m/s/s, or is >2.5 m/s/s, corrective action must begin to effectively reduce HAV exposure; if the DELV weighted vector sum value = 5.0 m/s/s, this represents the upper limit value which shall *not* be exceeded.

ANSI: S2.70-2006 [2006]: Recall that in 1984 when ISO told member countries that it was up to each to determine acceptable daily HAV exposure criteria, while ISO 5349 chose only a yearly exposure criteria; in the United States, ACGIH responded first in 1984 with their own HAV standard which included daily HAV exposure criteria. Two years later in 1986 there followed a second equally powerful response by ANSI, HAV standard S3.34-1986. This document accepted all the measurements criteria established by ISO 5349 and then added separate sections and graphics needed to determine exposure time-dependent daily HAV exposure limits; this standard was successfully used in the United States for two decades; it had one downside; a sophisticated and expensive Fourier spectrum analysis was required to evaluate each vibration axis, the resulting spectra were each compared to the now familiar weighting function [Figure 103.5] and the vibration axis with the highest weighted vibration value peak(s) determined the maximum daily use by operators of the tested tool.

In the year 2005, this HAV standard S3.34-1986 needed to be revised or replaced and ANSI thoughtfully made a "sea changing decision," namely, the U.S. and EU needed *one* uniform HAV standard thus establishing "a flat playing field worldwide" for HAV. The ANSI did *not* want to continue on as before using differing national and international standards needed when the HAV field was young and our collective knowledge of the medical consequences of HAV exposure was an early work in progress! This decision was made in 2005 the same year the EU Directive became law in 28 EU member nations; the ANSI group were very technically and otherwise knowledgeable about the EU requirements; their excellent up-to-date data bases whence their technical criteria were derived; it became increasingly apparent that this new ANSI standard could be done and transitioned seamlessly for use in the United States because power tool manufacturers worldwide had already began complying with the EU law by producing new reduced vibration tools [called "Antivibration or A/V" power tools] to gain entry into the EU markets. The U.S. workers would also receive the same A/V tool benefits if both the EU Directive and the new ANSI HAV standard since *both* required the *same technical requirements and HAV limits,* which is what happened. There is single differing legal proviso and distinction between the EU and ANSI standards: The new ANSI HAV standard was unanimously approved and promulgated became and remains simply a "voluntary consensus standard" in the Unites States; and in the EU, their 2005 Directive remains as their law as originally intended. Thus in 2006 the new ANSI S2.70-2006 HAV standard emerged and was approved, promulgated, and remains to the present day; it is arguably the best overall HAV standard currently in use in the United States. As predicted, better A/V power tool products are now becoming increasingly more available to the U.S. work force because of the above actions.

Since the technical requirements of both the EU and ANSI S2.70 standards are the same [for an 8 h/day; DEAV = 2.5 m/s/s, weighted vector sum, and DELV = 5.0 m/s/s, weighted vector sum], it would be beneficial if the reader would now please reread the above EU Directive basic technical information and then return here to obtain additional information extracted from my review paper (Wasserman, 2008) of ANSI S2.70-2006 as follows:

Definitions: "Daily Exposure Action Value [DEAV] represents the health risk threshold to hand-transmitted vibration. For the purpose of this standard, health risk threshold is defined as the dose of hand-transmitted vibration exposure sufficient to produce abnormal signs, symptoms, & laboratory findings in the vascular, bone or joint, neurological, or muscular systems of the hands & arm in some exposed individuals."

Daily exposure limit value, "Workers who are exposed to hand-transmitted vibration at or above this level are expected to have a high health risk. For the purpose of this standard, high health risk is defined as the dose of hand-transmitted vibration exposure sufficient to produce abnormal signs, symptoms, and laboratory findings in the vascular, bone or joint, neurological, or muscular systems of the hands and arms in a high proportion of exposed individuals."

When the DEAV is exceeded, a program to reduce worker exposure to hand-transmitted vibration needs to be initiated. Further it is recommended that workers *not* be exposed to vibration above the DELV.

A *final word* about vibration standards: The standards information provided here are simply highlights of important aspects of each standard discussed; intentionally it is informative and not comprehensive. If you plan to use these standards to collect and analyze either WBV or HAV vibration data *you must* obtain a copy of the latest version of the standard you plan to use; and thoroughly understand it before attempting to use it. Next is the needed contact information you should wish to obtain any of these occupational standard(s):

[ISO and/or ANSI vibration standards]: Acoustical Society of America, Attn. Standards Secretariat, 35 Pinelawn Road, Suite 114-E, Melville, NY 11747-3177, [ph. 631-390-0215]

[NIOSH: 89-106, & 83-110]: National Institute for Occupational Safety & Health, 4676 Columbia Parkway, Cincinnati, Ohio 45226, [ph. 513-533-8287; 1-800-35-NIOSH]

[ACGIH-TLVs]: American Conference of Gov't. Industrial Hygienists, 1330 Kemper Meadow Drive, Cincinnati, Ohio 45240-4148, [ph. 513-742-2020]

[EU VIBRATION DIRECTIVE for WBV and HAV]: European Union, Brussels, Belgium

HEALTH EFFECTS OF OCCUPATIONAL EXPOSURE TO VIBRATION

As an indication of the magnitude of the occupational vibration problem in the United States, about 8 million workers are regularly exposed daily to occupational vibration (Wasserman et al., 1974). Of these, about 6.8 million workers are exposed to WBV and the remaining 1.2 million workers are exposed to HAV. These population exposure numbers represent a diversity of industries and do reflect the very large numbers of exposed workers here in the United States and many more workers worldwide as industries have spread far beyond the United States to other countries. These large numbers of exposed workers have justified: the large R&D research in this area; the enormous worldwide effort for exposure standards; as well as the worldwide ever increasing pipeline of markets for new AntiVibration power tools; as well development of special personal protective products called AntiVibration protective gloves; and finally for the enormous development of WBV protective seating called "air-ride seating" used worldwide and the United States in trucks, buses, heavy equipment, farm equipment, fork-lifts, hovercraft, trains and subways, and fixed plant installations. Why all of this activity? Simply because there are many heavily documented studies, over the past century, showing severe health effects some, including irreversible damage to workers resulting from regular daily occupational vibration exposures. It is important to know that the health effects of occupational exposure to WBV are very different from the health effects of HAV exposure and thus are addressed separately. The one common thing WBV and HAV do share is the physics of vibration.

Health and Safety Effects of Whole-Body Vibration Exposure

The major populations at risk from WBV are operators of moving vehicles such as trucks, buses, heavy equipment and farm vehicles, and forklift trucks and locomotive drivers. Other populations are helicopter and fixed wing aircraft pilots, hovercraft operators, and persons working near vibrating processes in plants. Early epidemiological studies of truck drivers (Gruber, 1976; Kelsey, 1975), bus drivers (Gruber and Zimmerman, 1974), and heavy equipment operators (Milby and Spear, 1974) often pointed to various musculoskeletal disorders and spinal disk disorders such as degenerative disk diseases and back pain (Hulshof and Veldhuijzen, 1987). Although musculoskeletal and spinal disorders were associated with and were linked to WBV exposure in the work situation, the results were sometimes confounded by other job duties such as lifting cargo among truck and long-distance bus drivers. These early studies led to laboratory studies that could help clarify some of the epidemiological results. For instance, Pope and Wilder in the United States and Dupuis in Germany started to look carefully at the spinal effects of WBV exposure under laboratory conditions. They discovered that, for a seated person subjected to vertical z-axis WBV exposure, the vibration significantly increased lumbar disk pressure and distortion, reduced disk moisture, and increased the risk of disk buckling and instability as well as slipped and herniated disks. Furthermore, in a seated person, the lumbar vertebrae facets tend to disengage, thus limiting vertebrae rotation and prestressing the disks and then adding vibration, exacerbating the stress on the lumbar disks (Wilder et al., 1994). In addition, most of the vertical z-axis spinal resonance occurs in the L3 segment down to the buttocks where it touches the seat and the resonant frequency is between 4.5 and 5.5 Hz (Wilder, 1993). This means that the lumbar disk resonance is in the same frequency band as the overall human trunk resonance of 4–8 Hz, previously discussed. This is not good news to those workers exposed to WBV who happen to be lumbar spine suffers they now have two closely allied resonances to deal with. Today the focus of much WBV research is on the spine and operator posture while working in a WBV environments. This is important because the impinging vibration must come in contact or mechanically "couple" to the person; worker posture helps determine this coupling and thus how much WBV actually reaches the spine and other parts of the body.

When the early WBV epidemiological studies were performed, few women worked on traditionally male-dominated jobs such as heavy equipment operations and truck driving. A significant number of women were driving school buses; however, it was extremely difficult to gain access to their medical records in an organized and scientific manner. Thus few epidemiological studies of women exposed to WBV have been performed. During the last few years, female truck drivers and heavy equipment operators have become more prevalent. With that influx, there are scattered reports of female drivers experiencing spontaneous abortions and other gynecological disorders (Seidel and Heide, 1986; Abrams and Wasserman, 1991) and this has prompted animal research in the United States on the effects of WBV on pregnancy.

Whole-body vibration also has a mission component that continues to be of great concern to the military worldwide (Schoenberger and Harris, 1971). Of special concern are the "performance decrements" of military personnel subjected to WBV. In particular, they are interested in ability of air crews, soldiers, and sailors to adequately perform their missions while operating, among others, high performance fighter aircraft, helicopter gunships, high performance tanks, and warships under severe storm conditions. The civilian sector is likewise concerned about the ability of a truck or bus driver to safely drive vehicles when 4–8 Hz resonant vibration appears at the driver's steering wheel. Under these conditions, their hands can momentarily decouple from the steering wheel and there is an increased risk of a serious accident (Wasserman, 1987). Recent studies have focused on locomotive drivers and spinal issues.

In summary, WBV exposure was first considered as a generalized stressor; however, epidemiological studies have targeted the spine as being at greater risk from this exposure, but not without controversy because of confounding spinal variables. To help clarify and understand WBV's etiological effects on the spine current WBV lab studies are focusing on worker spinal posture while working in a WBV environment. Now there is added concern about the effects of WBV on female workers and the effects on reproduction and the fetus.

Finally, what began as performance affects studies of WBV in military situations has pointed to the possibility of safety effects of WBV in the civilian sector. The extensive use of air-ride seats in trucks, buses, fork lifts, heavy equipment, and farm tractors has been shown to be effective in helping reduce regular daily WBV exposure.

Health Effects of Hand-Arm Vibration Exposure

In 1918 Dr. Alice Hamilton, the famous occupational health pioneer for whom this text is named, wrote in part the following after carefully investigating numerous medical complaints of finger blanching and related hand problems in a group of Oolitic limestone quarry cutters and carvers who used vibrating pneumatic power tools on the job in Bedford, Indiana:

"The [fingers & hands] trouble seems to be caused by three factors-long continued muscular contraction of the fingers in holding the tool, the vibrations of the tool, and cold. It is increased by too continuous use of the air hammer, by grasping the tool too tightly, by using a worn, loose air hammer, and by cold in the working place. If these features can be eliminated the trouble can probably be decidedly lessened." Alice Hamilton, MD, 1918

Dr. Hamilton's report became the first and earliest comprehensive occupational study causally linking workplace HAV exposure to disease. However, this story really begins much earlier in 1862 with a French physician named Maurice Raynaud who had a clinic in Paris. During the winter months some of his housewife patients complained of repeated cold-triggered attacks characterized by tingling or numbness in their fingers followed by finger "blanching" or whitening and the excruciating finger and hand pain they experienced until these attacks subsided (Raynaud, 1862). Although Raynaud was unable to explain the etiology of this condition, it is now called idiopathic and primary Raynaud's disease. In 1911 Loriga in Italy was the first to briefly and later reported Raynaud's disease-like symptoms in workers using vibrating pneumatic mining hand tools (Loriga, 1934). It was not until 1918 that the very famous Dr. Alice Hamilton provided a very complete, detailed, and the most extensive study to date describing the clinical effects of pneumatic hand-tool vibration and cold on the fingers and hands, stating that there was an 80% prevalence of a condition called Raynaud's phenomenon (Hamilton, 1918). Over time this condition was renamed vibration white finger and is now called HAVS. In 1978 my own NIOSH vibration group repeated the original Hamilton study, at the same location (Bedford, Indiana). We found the prevalence of HAVS to be nearly identical: 82% (Taylor et al., 1984). Sadly, nothing had changed in the 60 years between the original Hamilton study and our own follow-up study except for the new HAV-exposed workers who were afflicted with irreversible HAVS. Thus from this simple 1918 beginning in Bedford, Indiana and for decades to the present, this one study has triggered a virtual worldwide flood of medical, epidemiological, physiological, and engineering studies all addressing various aspects of regular daily vibrating powered hand-tool use and the subsequent disease process it produces.

Classic HAVS is precipitated by exposure to HAV triggered by cold temperatures, and exacerbated by smoking (Ekenvall and Lindblad, 1989; Pelmear and Wasserman, 1998). Symptoms begin with tingling or numbness in the fingers. As the tool vibration exposure continues, a single white fingertip usually appears during a 10–15-min attack; this is the beginning onset of the so-called blanching process. The time from beginning vibrating tool use to the appearance of this first white fingertip is the blanching "latent period," which can vary from a few months to several years depending on the types of tools used on the job. As the vibration exposure continues, the number, intensity, and severity of the blanching attacks increase, now with progressive finger involvement, but usually does not include the thumbs because of their separate and copious blood supply. In rare instances, if the patient is not promptly treated and the vibration exposure removed, finger gangrene (tissue necrosis) can occur, requiring amputation of the affected digits. For the most part HAVS is considered *irreversible*, more so as the blanching process progresses. Current treatment for HAVS is palliative and *not* a cure; *it* includes the use of calcium-channel blocker medications, which seem to be better tolerated in older rather than in younger workers. We emphasize that other conflicting medical conditions can make the HAVS differential diagnosis more difficult. Hand-arm vibration syndrome is not carpal tunnel syndrome [CTS]; there have been cases where the afflicted worker had both HAVS and CTS. In the mid-1960s Drs. William Taylor and Peter Pelmear, working in the United Kingdom, developed an HAVS classification system, which is used in the United States and many countries of the world and bears their names (Table 103.2).

In 1986 it became necessary to modify the original Taylor–Pelmear classification system because some patients afflicted with HAVS had no blanching of the fingers; only the early neurological stages of tingling or numbness persisted. The modified classification is the Stockholm system (Tables 103.3 and 103.4) which permits separate grading of the neurological and peripheral vascular components of HAVS for each hand separately (Gemne et al., 1987).

In a comprehensive study of typical metal foundry workers using vibrating pneumatic chipping and grinding hand tools, there were two distinct findings when an incentive system was used. When payment was based on piecework, there was about a 40–50% prevalence of HAVS, with a 1- to 2.4-year latent period to blanching. In the absence of piecework, hourly paid workers in a U.S. Navy shipyard, using the same tools, do contract HAVS but

TABLE 103.2   Stages of Vibration White Fingers/HAVS (Taylor-Pelmear Original System)

Stage	Condition of Fingers	Work and Social Interference
00	No tingling, numbness, or blanching of fingers	No complaints
OT	Intermittent tingling	No interference with activities
ON	Intermittent numbness	No interference with activities
TN	Intermittent tingling and numbness	No interference with activities
1	Blanching of a fingertip with our without tingling or numbness	No interference with activities
2	Blanching of one or more fingers beyond tips, usually during winter	Possible interference with activities outside work; no interference at work
3	Extensive blanching of fingers; frequent episodes in both summer and winter	Definite interference at work, at home, and with social activities; restriction of hobbies
4	Extensive blanching of most fingers; frequent episodes in both summer and winter	Occupation usually changed because of severity of signs and symptoms

at a lower prevalence of 20% with a latent period of about 19 years (Wasserman et al., 1982). Since 1918, there are numerous other HAVS studies using a variety of different power tools, most of which show a direct causal relationship between regular daily power tool use and irreversible HAVS (Pelmear and Wasserman, 1998).

Finally, crossover situations do infrequently occur between HAV and WBV; one example is the use of a jackhammer-type power tool. When working, if this tool is held and used with hands away from the body, it is an HAV problem; but if the tool is placed against the abdomen to damp vibration, it becomes a WBV problem, where the latter can result in an inflamed omentum (Shields and Chase, 1988; Wasserman, 1989) and the former as HAVS (Pelmear and Wasserman, 1998).

TABLE 103.3   The Stockholm "Peripheral Vascular" Scale for HAVS (Modified Taylor-Pelmear System)

Stage	Grade	Description
0		No attacks
1	Mild	Occasional attacks affecting only the tips of one or more fingers
2	Moderate	Occasional attacks affecting the distal and middle (rarely also proximal) phalanges of one or more fingers
3	Severe	Frequent attacks affecting all phalanges of most fingers
4	Very severe	As in stage 3, with trophic skin changes in the fingertips

HAVS: Hand-Arm Vibration Syndrome

TABLE 103.4   The Stockholm "Sensorineural" Scale for HAVS (Modified Taylor-Pelmear System)

Stages	Symptoms
OSN	Exposed to vibration but no symptoms
ISN	Intermittent numbness, with our without tingling
2SN	Intermittent or persistent numbness, reduced sensory perception
3SN	Intermittent or persistent numbness, reduced tactile discrimination, or manipulative dexterity

HAVS: Hand-Arm Vibration Syndrome

VIBRATION CONTROL

Whole-body vibration control usually includes the following: (1) using air-ride seats in vehicles when and where possible; these AR seats were developed in the early 1970s and have found wide use in vehicles such as trucks, buses, heavy construction equipment, farm vehicles, trains, hovercraft, and fixed equipment control operations. Air ride seats mechanically isolate the exposed operator by allowing this person to ride on a closed cushion of compressed air. (2) Whole-body vibration exposures can be minimized by relocating process controls away from vibrating machinery [e.g., using inexpensive closed circuit TV to remotely view and control the work process]. (3) When driving a vehicle for long periods, upon exiting the vehicle use simple motions, with minimum twisting or rotation, and then walk around for a few minutes before unloading or lifting heavy objects [Wilder et al., 1988; Wilder, 1993].

Hand-arm vibration control is multifaceted and includes the following: 1. Since 1980 various tool manufacturers have introduced newer designed power tools with reduced vibration characteristics [called "AntiVibration or A/V tools"] as alternative choices to conventional power tools. Currently not all conventional power tools have been replaced but the pipeline of new A/V tools continues to steadily grow. Try to use A/V tools when and where possible and when purchasing these tools request vibration test information showing which HAV standards they were tested and compared against.

2. In 1996, international standard ISO 10819, measurement and certification for reduced vibration gloves [called AntiVibration or A/V gloves] was introduced. A few years later ANSI in the United States adopted this same standard as ANSI S3.40/ISO10819. Both organizations require that tested gloves shall be only full finger protected and tested at an independent lab qualified to perform vibration transmissibility tests per the procedures outlined in said standard; those full finger gloves which meet or exceed this standard are called "certified A/V gloves" and are recommended for use together with A/V power tools, because not only do the A/V gloves provide vibration protection but they also keep the fingers and hands warm and dry to help avoid a HAVS attack triggered by cold temperatures. Only full-finger protected gloves are tested since HAVS virtually always begins at the fingertips and moves toward the palm; thus finger-exposed gloves are never recommended. Finally, using both A/V power tools and A/V gloves needs to be supplemented with the following good work practices guide to workers:

a] Keep fingers, hands, and body warm and dry

b] Do not smoke [vibration, cold, and nicotine all are vasoconstrictors]

c] Let the tool do the work, grasping it as lightly as possible consistent with safe work practices

d] Do not use the tool unnecessarily and keep it well maintained

e] For pneumatic tools, keep the cold exhaust air away from the fingers and hands [to avoid a possible HAVS attack]

f] If signs and symptoms of HAVS appear, seek medical help

ACKNOWLEDGEMENT

This author would like to sincerely thank Dr. Marie Bourgeois, USF for her assistance in the completion of this vibration chapter.

REFERENCES

Abrams, R. and Wasserman, D. (1991) Occupational vibration during pregnancy. *Am. J. Obstet. Gynecol.* 164:1152–1154.

Ekenvall, L. and Lindblad, L. (1989) Effect of tobacco use on vibration white finger disease. *J. Occup. Med.* 31:13–16.

Gemne, G., et al. (1987) The Stockholm workshop scale for the classification of cold-induced Raynaud's phenomenon in the hand-arm vibration syndrome (revision of the Taylor-Pelmear scale). *Scand. J. Work Environ. Health* 13:275–278.

Gruber, G. (1976) *Relationship Between Whole-body Vibration & Morbidity Patterns Among Interstate Truck Drivers*, NIOSH Report 77–167, Cincinnati: National Institute for Occupational Safety and Health.

Gruber, G. and Zimmerman, H. (1974) *Relationship Between Whole-body Vibration and Morbidity Patterns among Motor Coach Operators*, NIOSH Report 75–104, Cincinnati: National Institute for Occupational Safety and Health.

Hamilton, A. (1918) Effect of the air-hammer on the hands of stonecutters. In: *Industrial Accidents & Hygiene Series*, Report 236, #19, Washington, DC: US Bureau of Labor Statistics, Department of Labor.

Hulshof, C. and Veldhuijzen, B. (1987) Whole-body vibration and low back pain: a review of epidemiological studies. *Int. Arch. Occup. Environ. Health* 59:205–220.

Kelsey, J. (1975) An epidemiological study of the relationship between occupations and acute herniated lumbar intervertebral discs. *Int. J. Epidemiol.* 4:197–205.

Loriga, G. (1934) Pneumatic tools: occupation & health. In: *Inter Labor Office Report EHPSW*, vol. 2, Geneva: EHPSW.

Milby, T. and Spear, R. (1974) *Relationship Between Whole-body Vibration & Morbidity Patterns Among Heavy Equipment Operators*, NIOSH Report 74–131, Cincinnati: National Institute for Occupational Safety and Health.

Pelmear, P. and Wasserman, D. (1998) *Hand-arm Vibration: A Comprehensive Guide for Occupational Health Professionals*, 2nd ed., Boston: OEM Medical Press.

Raynaud, M. (1862) *Local asphyxia and symmetric gangrene of the extremities, Paris, (Monograph, translated into English by the New Sydenham Society, London, 1888).*

Schoenberger, R. and Harris, C. (1971) Psychophysical assessment of whole-body vibration. *Hum. Factors* 13:41–50.

Seidel, H. and Heide, R. (1986) Long term effects of whole-body vibration: a critical review of the literature. *Int. Arch. Occup. Health* 58:1–26.

Shields, P. and Chase, K. (1988) Primary torsion of the omentum in a jackhammer operator: another vibration related injury. *J. Occup. Med.* 31:892–894.

Taylor, W., Wasserman D. et al. (1984) Effects of the air hammer on the hands of stonecutters: the limestone quarries of Bedford, Indiana revisited. *Br. J. Ind. Med.* 41:289–295.

Tse, F., Morse, I., and Hinkle, R. (1963) *Mechanical vibrations*, Boston: Allyn & Bacon.

Wasserman, D. (1987) *Human Aspects of Occupational Vibration*, Amsterdam: Elsevier.

Wasserman, D. (1989) Jackhammer usage and the omentum. *J. Occup. Environ. Med.* 31:563–566.

Wasserman, D. (2008) Manufacturing & the new ANSI S2.70-2006 Hand-Arm Vibration exposure standard. *J. Hum. Factors Ergonomics Manuf.* 18(6):658–665.

Wasserman, D., et al. (1974) Industrial vibration: an overview. *J. Am. Soc. Saf. Eng.* 19:38–43.

Wasserman, D., et al. (1982) *Vibration White Finger Disease in U.S. Workers Using Chipping & Grinding Hand Tools*, vol. I and II, NIOSH Study Reports 82–118 & 82-10l, Cincinnati: National Institute for Occupational Safety and Health.

Wilder, D. (1993) The biomechanics of vibration & low back pain. *Am. J. Ind. Med.* 23:577–588.

Wilder, D., Pope, M., and Frymoyer, J. (1988) The biomechanics of lumbar disc herniation & the effect of overload & instability. *J. Spinal Disord.* 1:16–32.

Wilder, D., et al. (1994) Vibration. In: Wald, P.H., and Stave, G.M., editors. *Physical & Biological Hazards of the Workplace*, New York, NY: Van Nostrand Reinhold.

104

OCCUPATIONAL NOISE

David L. Dahlstrom

BACKGROUND

We perceive and interpret the series of vibrations being generated and conducted through the air or other media, which impacts our eardrums as "sound." These sounds may elicit pleasure or pain, depending upon the intensity, frequency, wavelength, and duration of the vibrations. If the vibrations are offensive or unwanted, we refer to them as "noise." When they occur at high levels as either intermittent (Impact noise) or continuous noise, they may produce harm to those exposed. Offensive noise is often associated with specific work tasks, processes, equipment, locations, and individual work practices; all can be harmful and all can be managed by administrative or engineering controls to minimize or prevent worker injury.

As we age, many people experience a gradual decrease in their hearing acuity known as *presbycusis*. Causes of presbycusis include physiological changes such as the gradual degeneration of the inner ear, changes in inner ear bone structure (otosclerosis), changes in inner ear nerve pathways to the brain, heredity, genetic factors, and more commonly these days, a history of repeated exposure to loud noise, music, or equipment (noise trauma), which damages/desensitizes the hair cells of the inner ear.

- The extent of adverse effects that result from our exposure (frequency, duration, noise type, source and level) to noisy environments in our daily lives, both occupationally and socially, are the result of our evolving modern technological advances. Short-term exposure to loud noise, such as that seen following attendance of concerts and standing too close to the speakers, can cause a temporary change in hearing (your ears may feel stopped up) or a ringing in your ears (tinnitus). These short-term problems typically resolve within a few minutes or hours after leaving the noisy area. However, repeated exposures to loud noise can lead to permanent tinnitus and/or hearing loss.

From a health and safety standpoint, occupational noise typically refers to hazardous levels of noise generated by various industrial and construction-related processes and operations. Sustained exposure to such hazardous levels can result in a permanent hearing deficit called "noise induced hearing loss" (NIHL). Exposure to occupational noise over extended periods can promote elevations in systolic blood pressure. Occupational noise is also a recognized causal factor in industrial accidents, as elevated and chronic workplace noise tends to impede concentrate, impair communication in emergency situations, and interfere with the ability to hear warning alarms. NIHL remains one of the most common work-related health injuries, largely because noise is a pervasive occupational hazard found in a wide range of industries (ACOEM, 1989). Noise induced hearing loss can be defined as a progressive loss of hearing acuity that develops from chronic exposure to continuous and/or intermittent noise above a threshold of about 85–90 dBA (decibels measured using the A-level decibel scale) (ACOEM, 2002); (ACOEM, 2003). NIOSH recommends that all worker exposures to noise should be controlled below a level equivalent to 85 dBA for eight hours to minimize occupational noise induced hearing loss (NIOSH, 1998). NIOSH also recommends a 3 dBA exchange rate so that every increase by 3 dBA doubles the amount of the noise and halves the recommended amount of exposure time. (NIOSH, 2010).

Hamilton & Hardy's Industrial Toxicology, Sixth Edition. Edited by Raymond D. Harbison, Marie M. Bourgeois, and Giffe T. Johnson.
© 2015 John Wiley & Sons, Inc. Published 2015 by John Wiley & Sons, Inc.

FACTS AND STATISTICS (NIOSH, 2010)

- Four million workers go to work each day in damaging noise. Ten million people in the United States have a noise-related hearing loss. Twenty-two million workers are exposed to potentially damaging noise each year.

- In 2008, approximately 2 million U.S. workers were exposed to noise levels at work that put them at risk of hearing loss.

- In 2007, approximately 23,000 cases were reported of occupational hearing loss that was great enough to cause hearing impairment. Reported cases of hearing loss accounted for 14% of the occupational illness in 2007.

- In 2007, approximately 82% of the cases involving occupational hearing loss were reported among workers in the manufacturing sector.

- There are an estimated 16 million people working in the manufacturing sector, which accounts for approximately 13% of the U.S. workforce. According to the Bureau of Labor Statistics, occupational hearing loss is the most commonly recorded occupational illness in manufacturing sector (17,700 cases out of 59,100 cases), accounting for 1 in 9 recordable illnesses. More than 72% of these occur among workers in manufacturing.

- Most occupational hearing losses occur so gradually that workers are unaware they are losing their hearing.

- The rate of hearing loss growth is greatest during the first 10 years of exposure. This means hearing loss prevention is especially important for new workers. However, with continued exposure, the hearing loss spreads into those frequencies most needed to understand speech. This means that preventing occupational hearing loss is also important for workers in their mid and late careers.

OCCUPATIONAL HEARING LOSS

It can be difficult to distinguish between occupational and nonoccupational hearing loss. There have been statutory changes at the state level that require agencies to create programs to target early hearing loss. Hearing loss may be temporary, permanent, or a combination of the two. Temporary hearing loss, sometimes referred to as "auditory fatigue," may become permanent with chronic exposure. The current permissible exposure limit (PEL) established by Occupational Safety and Health Administration (OSHA) is 90 dBA and the specific standard is provided in Title 29 of the Code of Federal Regulations, Part 1910.95. OSHA's Noise Exposure Standard mandates that employers establish workplace "hearing conservation program" with audiometric testing (OSHA, 2014a, c).

Audiometric testing measures the ability of each ear to hear sound at specific frequencies. This testing is used to detect the extent of hearing loss in each ear.

Hearing loss, which is classified by the part of the auditory pathway that is damaged, may be conductive, sensorineural, central, functional, or a combination of several types. *Conductive hearing loss* results from any condition that interferes with the transfer of sound pressure waves (vibration) through the external or middle ear to the inner ear. Possible causes include ossicle injury, fluid in the middle ear, foreign body, collapsed ear canal, carcinoma, cysts, congenital abnormalities, and infection of the Eustachian tube. This condition is usually reversible through surgery and the prognosis is excellent. *Sensorineural hearing loss*, the type most commonly associated with the occupational NIHL, is the result of damage to the hair cells in the inner ear. The damage is medial to the footplate of the stapes in the oval window, the auditory nerve, or both. In most cases this condition is irreversible, occurring at high frequencies around 4 kHz (kilohertz) over months or years of exposure. Millions of industrial workers and older citizens have this form of impairment (Ward et al., 2003; Sataloff and Sataloff, 1987).

Central hearing loss (central dysacusis) is often described as perceptive deafness. The damage is situated in the central nervous system at some point between the cortex and the auditory nuclei in the medulla. The function of the auditory cortex is to interpret and integrate electrical impulses and provide the listener with meaningful information or to permit the listener to react with the actual implication of the sound. It is thought that interference with the neural pattern pathways to the cortex would manifest as the reduced ability to interpret information rather than lowering of the hearing threshold for pure tones (Ward et al., 2003; Sataloff and Sataloff, 1987).

Exposure to excessive noise is not the only cause of occupational hearing loss. There is no discernable organic damage to the auditory pathways in *functional hearing loss*; rather, an underlying psychological or emotional problem is at fault. A variety of organic solvents have also been implicated in ototoxicity and hearing loss. These include toluene, styrene, and trichloroethylene. Occupational exposure to lead, mercury, carbon monoxide, and carbon disulfide can also affect the ear (Nies, 2012). In a 20-year longitudinal study of hearing sensitivity in 319 employees from different sectors of the industry, a significant proportion of the workers in the chemical sector showed pronounced hearing loss (23%), compared to groups from nonchemical environments (5–8%) (Morata et al., 1994). Currently, there are no regulations in the United States that require occupational monitoring for the employees exposed to potentially ototoxic chemicals. Additionally, they may not receive OSHA specified hearing tests if the measured noise does not exceed the regulatory standards

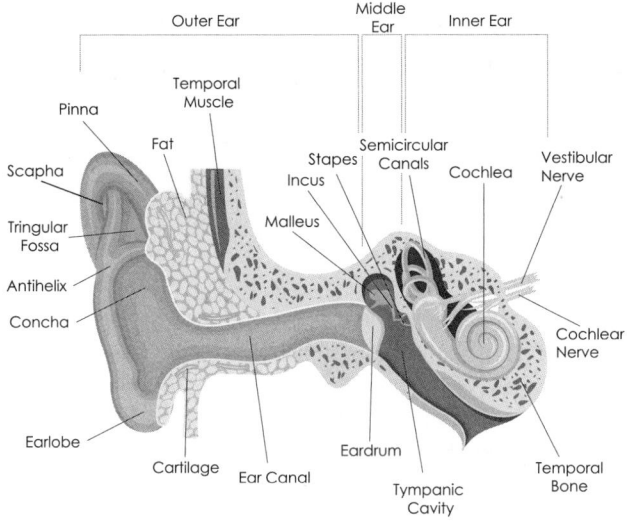

EAR

FIGURE 104.1 The Anatomy of the Human Ear.

ANATOMY OF THE EAR

Sound waves enter the pinna (external ear), pass through the auditory canal, and impinge upon the tympanic membrane (eardrum) (Figure 104.1). The eardrum is connected to the cochlea by the ossicles of the middle ear, the mallus (hammer), incus (anvil), and stapes (stirrup). As sound waves strike the eardrum, they are converted to mechanical energy (vibration) in the middle ear. This mechanical energy is transferred via the ossicles to the fenestra vestibule where it creates waves in the inner ear fluid. These waves stimulate the roughly 10,000 microscopic hair cells located on the basilar membrane along the cochlea. Different groups of hair cell sensors respond to different frequencies based on the stiffness of the hair cell. Excessive noise can damage these hair cell sensors resulting in permanent hearing loss (Silverman, 1996).

PRINCIPLES AND PHYSICS OF SOUND

Noise is a quickly modulating elastic pressure wave moving through a material medium such as a solid, liquid, or gas. When we speak of noise, we usually mean audible sound, which is a sensation detected by the ear of very small rapid changes in the air pressure above and below a static value. This static value is atmospheric pressure. When rapid variations in pressure occur between 20 and 20,000 Hz (20–20,000 times per second), sound is substantially audible even though the sound pressure variation is low. Louder sounds are caused by greater variation in pressure, provided that most of the acoustic energy is in the mid-frequencies (1000–4000 Hz) where the ear is most sensitive.

The rate at which the pressure waves are produced is the frequency measured in cycles per second in units of hertz (Hz). One cycle per second represents the number of times that an air molecule at the sound source is displaced from its equilibrium position, rebounds through the equilibrium position to its maximum displacement opposite in direction to the initial displacement and then returns to its equilibrium position (Standard, 1996). There is a period associated with the frequency of the wave, measured in seconds, which is the time required for each cycle and is merely the reciprocal of the frequency. The wavelength is measured in meters, centimeters, or feet. It is the distance measured between two equivalent points on two successive parts of the wave and is the distance the sound travels in one cycle. The amplitude of the pressure disturbance is directly related to the displacement amplitude of the sound and is expressed in units of force per unit area. Common units of pressure are newtons per meter squared (N/m^2), micropascals (μPa), dynes per centimeter squared (d/cm^2), and microbars (μbar).

The density and the compressibility of the medium through which the sound wave migrates primarily determine the speed with which sound travels. In air, the speed of sound is approximately 344 m/s at normal temperature and pressure (NTP) (25 °C and 760 mmHg) or 331.6 m/s at standard temperature and pressure (STP) (0 °C and 760 mmHg). The speed of sound in water is approximately 1500 m/s and in steel about 6100 m/s. Velocity of sound is the speed of sound equal to the wavelength times the frequency.

If we were to calculate the mean value of the sound pressure disturbances, we note the mean value to be zero because there are as many negative rarefactions as positive compressions. A practical approach to measuring sound pressure is the root-mean-square (rms). The rms measurement is obtained by squaring the value of the sound pressure at each instance in time, adding the squared values together, and then taking the mean over the given time. Because squaring values converts negative numbers to a positive value, the rms method produces nonzero measurements of the sound pressure magnitude.

The sound pressure range frequently encountered is remarkably wide. The weakest sound pressure that can be heard by a young, healthy, human is approximately 20 μPa at a reference tone of 1000 Hz and is known as the threshold of hearing. The threshold of pain, which is the greatest amount of energy that can be heard without pain, is approximately 200 million μPa. A relative scale is more convenient and easily interpreted and for this purpose a logarithmic dB scale was selected (Table X). The dB is 0.1 of a Bel and is a dimensionless unit related to the logarithm of the ratio of a measured quantity to a reference quantity. The threshold of hearing of 20 μPa becomes 0 dB, and the threshold of pain of 200,000,000 μPa becomes 140 dB. The dBA scale is used to weigh the various frequency components of the noise to

TABLE 104.1 Sound Pressure and Sound Pressure Level Values

Sound pressure in μPa	Sound pressure in dB, re: 20 μPa	Example
20	0	Threshold of hearing
63	10	
200	20	Inside bedroom
630	30	Soft whisper at 5 ft
2000	40	Quiet office
6300	50	Large office
20,000	60	Conversation at 3 ft.
63,000	70	Vacuum cleaner
200,000	80	Garbage disposal
630,000	90	Printing press plant
2,000,000	100	Electric furnace
6,300,000	110	Roll 'n' roll band
20,000,000,	120	Hydraulic press
200,000,000	140	Threshold of pain, jet aircraft, artillery fire
20,000,000,000	180	Rocket launch pad

TABLE 104.2 Reference Sound Values in Air

Sound pressure	0.00002 N/m² or 0.00002 Pa (0.0002 μbar or 0.0002 d/cm²)
Sound intensity	1×10^{-12} W/m²
Sound power	1×10^{-12} W

N/m², Newtons per meter squared; Pa, Pascal; μbar, microbars; d/cm², dyne per centimeter squared; W/m², Watts per meter squared; W, watts. 1.01 N/cm² = 0.1 Pa = 1 μbar = 1 d/cm².

approximate the response of the human ear. A change in sound pressure by a factor of 10 corresponds to a change in sound pressure level of 20 dB (Standard, 1996).

Usually the reference value is the smallest likely value of the quantity (Table 104.1). The dB scale approximates the human ear, which interprets loudness on a scale much closer to a logarithmic scale than a linear scale and the quantities of interest often exhibit such huge ranges of variation that a dB scale is more convenient than a linear.

Sound power describes the sound source in conditions of the amount of acoustic energy that is produced per unit time and is expressed in watts (W). The sound power level (Lw) is expressed in decibels, relative to a reference quantity of 10^{-12} W. It is the total sound power radiated by a sound source, regardless of the space into which the source is placed (Standard, 1996). The relationship is shown below:

$$Lw = 10 \log \frac{W}{W_0}$$

where W = sound power in watts, W_0 = reference sound power (10^{-12} W), and log = logarithm to the base 10.

As sound power in free space is radiated, sound power is scattered over a sphere surface. The rate of sound power transmitted in a specified direction per unit area normal to the direction is expressed as the power intensity (W/m²). The intensity decreases with distance from the source, but the power being the product of the intensity and area remains constant. It is not possible to measure sound power directly.

Under most conditions, sound intensity is proportional to the square of the sound pressure (Table 104.2). Sound pressure can be more easily measured than sound intensity. A sound level meter is the principal instrument for broad noise measurement. Most sound-measuring instruments are engineered to provide a reading of rms related to a sound pressure reference quantity. For sound pressure measurements in air, the common reference quantity is 20 μPa. This reference quantity is an arbitrary pressure chosen because it approximates the normal threshold of hearing for a young, healthy human at 1000 Hz. Doubling the sound pressure represents a 6 dB increase in sound pressure level, whereas doubling the sound power results in an increase of 3 dB in the sound power level. The sound pressure level (L_p) measure from an instrument is shown below:

$$L_p = 10 \log \frac{P^2}{P_0^2}$$

where P = measured rms sound pressure, P_0 = reference rms sound pressure (20 μPa), and log = logarithm to the base 10.

When using a sound level meter, weighting scales are used to provide varying sensitivities to sound of different frequencies. The most common weighting today is dBA and is the scale required by regulatory standards to measure sound levels. The dBA scale is thought to provide a rating of industrial broadband noise that indicates the injurious effects such noise has on the human ear. Another scale used is the C weighting scale and is used to measure total energy in an environment. In addition to frequency weighting, sound pressure can be weighted in time with fast, slow, or impulse response, which is seen in the reading response time. Regulatory standards require sound levels to be measured at slow response.

While conducting industrial noise surveys, it often becomes necessary to add levels of decibels. An industrial hygienist should be consulted to conduct the survey. One of the most basic equations for adding decibels from separate random noise sources applies the following equation:

$$L_p T = 10 \log(10^{\frac{L_{P_1}}{10}} + 10^{\frac{L_{P_2}}{10}} \ldots + 10^{\frac{L_{P_n}}{10}})$$

where L_pT = total sound pressure level in decibels, L_{P1}, L_{P2}, L_{Pn} = individual sound pressure levels to be added, and log = logarithm to the base 10.

NOISE STANDARDS

The OSHA has set a standard of good practice for noise exposure (OSHA, 2014b, OSHA 2014c-Construction Industry Standard). The standard states that no employee may be exposed to the equivalent of a steady sound pressure level of 90 dBA for more than 8 h, or to the equivalent of a steady sound pressure level of 95 dBA for more than 4 h, and so on. It also establishes that employees are limited to a noise exposure with a 5 dB exchange rate so that 5 dB increase in sound pressure level constitutes a halving in duration of exposure time.

The OSHA requires that employers make audiometric tests available to employees when noise exposures exceed an 85 dBA criterion. When an employee's audiogram reveals a standard threshold shift (STS), the employee may not be exposed in excess of the 85 dBA criterion. In this case it is acceptable to reduce the exposure from an amount equaling the 90 dBA criterion to the permissible amount by using personal hearing protection devices. Employers are required to use feasible engineering and administrative controls to reduce employee exposures below the 90 dBA criterion. If the feasible engineering and administrative controls are not fully effective, the remainder of the required protections must be obtained by providing and requiring employees to wear personal hearing protection devices.

The American Conference of Governmental Industrial Hygienists (ACGIH, 2013) recommends the limit for employee noise exposure to a 3 dB exchange rate. This may be a better predictor of noise hazard for most practical conditions. A 3 dB increase in sound pressure level constitutes a halving in duration of exposure time. This exchange rate is based on the equal energy hypothesis, which maintains that equal amounts of sound energy will produce equal amounts of hearing impairment, regardless of how the sound energy is distributed in time. On an energy basis, the 3 dB exchange rate provides for the calculation of a mathematically valid time-weighted average (TWA) exposure to noise.

The National Institute for Occupational Safety and Health (NIOSH) (NIOSH, 1996a) is moving to an 85 dB criterion for an 8-h exposure since scientific data reflect that the excess risk of developing occupational NIHL for a 40-year lifetime exposure at the 85 dBA is 8%, which is considerably lower than the 25% excess risk at the 90 dBA PEL currently enforced by the OSHA and the Mine Safety and Health Administration (MSHA) (NIOSH, 1996a). Included in the evaluation is a recommendation for a 3 dB exchange rate, which is more firmly supported by scientific evidence and consistent with the ACGIH recommendation.

NOISE-RELATED STANDARDS AND GUIDANCE DOCUMENTS

OSHA

- Recording and reporting occupational injuries and illness (*29 CFR 1904*)
- *1904.10*, Recording criteria for cases involving occupational hearing loss
- General Industry (*29 CFR 1910*)
- *1910.95*, Occupational noise exposure (OSHA, 2014b)
  - *Appendix A*, Noise exposure computation
  - *Appendix B*, Methods for estimating the adequacy of hearing protector attenuation
  - *Appendix C*, Audiometric measuring instruments
  - *Appendix D*, Audiometric test rooms
  - *Appendix E*, Acoustic calibration of audiometers
  - *Appendix F*, Calculations and application of age corrections to audiograms
  - *Appendix G*, Monitoring noise levels non-mandatory informational appendix
  - *Appendix H*, Availability of referenced documents
  - *Appendix I*, Definitions
- *Section 5(a)(1)* of the OSH Act, often referred to as the General Duty Clause, requires employers to "furnish to each of his employees employment and a place of employment, which are free from recognized hazards that are causing or are likely to cause death or serious physical harm to his employees." This section may be used to address hazards for which there are no specific standards, eg., noise in agricultural operations.
- *OSHA Federal Registers*
- *Preventing Occupational Hearing Loss: Stakeholder Meeting—Notice of public meeting*. OSHA, (October 2011).
- *Occupational Injury and Illness Recording and Reporting Requirements*. Final Rules 67:44037-44048, (July 2002). Revises the criteria for recording hearing loss cases in several ways, including requiring the recording of Standard Threshold Shifts (10 dB shifts in hearing acuity) that have resulted in a total 25 dB level of hearing above audiometric zero, averaged over the frequencies at 2000, 3000, and 4000 Hz, beginning in the year 2003.
- *OSHA Directives; Instructions to OSHA staff*
- *Field Operations Manual (FOM)* CPL 02-00-150, (April 2011).
- *Hearing Conservation Program*. PER 04-00-0004, (June 2008).

- *OSHA Technical Manual (OTM).* TED 01-00-015 [TED 1-0.15A], (January 1999)
  - *Technical Equipment.* 08-05 (TED 01), (June 2008). Contains information on *noise monitors and meters*, including descriptions and techniques for the use of sound level meters and dosimeters.
  - *Noise and Hearing Conservation.* eTool. Comprehensive information on evaluating and controlling noise.
- *Guidelines for Noise Enforcement; Appendix A.* CPL 02-02-035 [CPL 2-2.35A], (December 1983). Identifies factors to consider and document in the case file with regard to comparing the relative degree of attenuation of personal protectors and engineering and/or administrative controls.
- *Occupational Noise Exposure; Hearing Conservation Amendment, 29 CFR 1910.95.* CSP 01-01-016 [STP 2.21], (1981, December 12). Describes the requirements of each regional administrator under a federal program change that affects state programs.
- OSHA Enforcement Standard Interpretations

OTHER FEDERAL AGENCY'S STANDARDS AND GUIDANCE

- **Attention:** These are *not* USDOL-OSHA regulations. However, they do provide guidance from their originating organizations related to worker protection.
- *Environmental Protection Agency (EPA)*
- *40 CFR 211,* Product Noise Labeling
- *Mine Safety and Health Administration (MSHA)*
- *Health Standards for Occupational Noise Exposure.* (September 2000). Documents and resources related to MSHA's 2000 Health Standards for Occupational Noise Exposure rule.
- *National Institute for Occupational Safety and Health (NIOSH)*
- *Occupational Noise Exposure.* Publication No. 98-126, (June 1998). With revisions to previous 1972 NIOSH recommendations by adding focus to prevention of NI H L through adminstration of hearing conservation programs.
- *U.S. Department of Defense (DOD)*
- *DoD Hearing Conservation Program.* Instruction No. 6055.12, (December 2010). Provides direction and information regarding exposure limits, requirements for monitoring, administrative and engineering control methods, hearing conservation programs, and components.
- *U.S. Department of Transportation (DOT)*

- Federal Motor Carrier Safety Administration (FMCSA)
  - *49 CFR 393.94,* Interior noise levels in power units.
- Coast Guard
  - *Recommendations on Control of Excessive Noise,* Navigation and Vessel Inspection Circular No. 12-82, (June 1982).
- Federal Railroad Adminstration (FRA)
  - 49 CFR 229.121, Locomotive Cab Noise
  - *49 CFR 227, Occupational Noise Exposure for Railroad Operating Employees.* Railroads are required to conduct noise monitoring and implement a hearing conservation program for employees whose exposure to cab noise equals or exceeds an 8-h TWA of 85 dBA. Effective February 26, 2007.
  - *49 CFR 229,* Railroad Locomotive Safety Standards.

STATE STANDARDS

- There are 25 states plus Puerto Rico and the Virgin Islands granted *OSHA-approved State Plans,* required to be at least as effective as Federal OSHA, but may have added their own standards and enforcement policies.
- Noise protection and hearing conservation program requirements are addressed in specific OSHA Title 29 standards for recordkeeping, general industry, maritime, and construction employment.

CONSENSUS STANDARDS AND RECOMMENDATIONS FROM OTHER PROFESISONAL ORGANIZATIONS

- **Attention:** These are guidelines only and are not OSHA regulations. These are "guidance" documents from respective originating organizations related to worker protection.
- *American National Standards Institute (ANSI)*
- A10.46-2007, *Hearing Loss Prevention in Construction and Demolition Workers.* For all construction and demolition workers with potential noise exposures (continuous, intermittent, and impulse) of 85 dBA and above.
- S3.1-1999 (R2008), *Maximum Permissible Ambient Noise Levels for Audiometric Test Rooms.* Identifies the maximum permissible ambient noise levels (MPANLs) permitted in audiometric test rooms.
- S3.44-1996 (R2006), *Determination of Occupational Noise Exposure and Estimation of Noise-Induced Hearing Impairment.*
- S3.6-2010, *American National Standard Specification for Audiometers.* Provides the specifications and

tolerance levels for audiometers, including standard reference threshold levels for audiometric transducers.

- S12.6-2008, *Methods for Measuring the Real-Ear Attenuation of Hearing Protectors.* Provides specifications and laboratory-based procedures for measuring, analyzing, and reporting passive noise-reducing capabilities of hearing protection devices.
- S1.4-1983 (R2006), *American National Standard Specification for Sound Level Meters.* Provides performance and accuracy requirements for sound level meters.
- S1.25-1991 (R2007*), American National Standard Specification for Personal Noise Dosimeters.* Provides specifications for performance-based characteristics of personal noise dosimeters.
- *American Conference of Governmental Industrial Hygienists (ACGIH)* (ACGIH, 2013)
- ACGIH provides established exposure guidelines for occupational exposure to noise in their annually published *Threshold Limit Values for Chemical Substances and Physical Agents and Biological Exposure Indicies* (85 dBA PEL with a 3 dBA exchange rate).
- *World Health Organization* (*WHO*)
- *Occupational Noise—Assessing the Burden of Diseases from Work-Related Hearing Impairment at National and Local levels.* World Health Organization-Protection of the Human Enviropnment, Geneva 2004. Provides an in depth look at all aspects of noise.

REFERENCES

American Conference of Governmental Industrial Hygienists (2013) *American Conference of Governmental Industrial Hygienists (ACGIH):* Noise threshold limit value: documentation of the threshold limit values and biological exposure indices, *Cincinnati, 2013 American Conference of Governmental Industrial Hygienists.*

ACOEM Noise and Hearing Conservation Committee (1989) Occupational noise-induced hearing loss. *J. Occup. Med.* 31:996.

ACOEM statement developed by the ACOEM Noise and Hearing Conservation Committee under the auspices of the Council on Scientific Affairs (2002) *It was peer-reviewed by the Committee and Council and approved by the ACOEM Board of Directors on October 27.*

American Conference of Occupational and Environmental Physicians (2003) Journal of Occupational and Environmental Medicine *American Conference of Occupational and Environmental Physicians (ACOEM) Evidence-based Statement June 2003—Volume 45—Issue 6* pp. 579–581.

Morata, T., Dunn, D., and Sieber, K. (1994) Occupational exposure to noise and ototoxic organic solvents. *Arch. Environ. Health* 49:359–365.

National Institute for Occupational Safety and Health (NIOSH) (1996a) *Criteria for a Recommended Standard Occupational Noise Exposure Revised Criteria 1996,* Cincinnati, OH: National Institute for Occupational Safety and Health.

National Institute of Occupational Safety and Health (NIOSH) (1998) *Occupational Noise Exposure, DHHS (NIOSH) Publication Number 98–126, June.*

Nies, E. (2012) Ototoxic substances at the workplace: a brief update. *Arh. Hig. Rada. Toksikol.* 63(2):147–52.

NIOSH (2010) *Occupationally-Induced Hearing Loss-DHHS (NIOSH) Publication No. 2010-136, March.*

OSHA (2014) *Title 29 Code of Federal Regulation Part 1910. 95—Occupational Noise Exposure, U. S. Departmetn of Labor—Occupational Safety and Health Administration (OSHA), General Industry Standards, 2014.*

OSHA (2014) *Title 29 Code of Federal Regulation Part 1910. 95(c)—Occupational Noise Exposure- Hearing Conservation Program, U. S. Departmetn of Labor—Occupational Safety and Health Administration (OSHA), General Industry Standards, 2014.*

OSHA (2014) *Title 29 Code of Federal Regulation Part 1926.52—Occupational Noise Exposure, U. S. Departmetn of Labor—Occupational Safety and Health Administration (OSHA), Construction Industry Standards, 2014.*

Sataloff, R. and Sataloff, J. (1987) *Occupational Hearing Loss,* NY: Marcel Dekker, 13.13.

Standard, J. (1996) Industrial noise. In: Plog, B.A., Niland, J., Quinlan, P.J., editors, *Fundamentals of Industrial Hygiene,* Itasca, IL: National Safety Council.

Silverman, A. (1996) *Acoustics FAQ,* United Kingdom: EnviroMeasure.

Ward, W.D. Royster, J.D., and Royster, L.H. (2003) Auditory and nonauditory effects of noise. In: Berger, E.H., Royster, L.H., Royster, J.D., Driscoll, D.P., and Layne, M., editors. *The Noise Manual,* 5th ed., Chapter 5, AIHA Press.

105

HEAT STRESS ILLNESS

Marie M. Bourgeois and David R. Johnson

BACKGROUND

According to Occupational Safety and Health Administration (OSHA), operations involving elevated ambient temperatures, excessive humidity, strenuous activity, and exposure to radiant heat or direct contact with hot objects are capable of inducing heat stress illness (HSI) (OSHA, 1999). Occupational Safety and Health Administration does not have a permissible exposure limit (PEL) covering work in hot environments. The General Duty Clause requires employers to provide workers "a place of employment which is free from recognized hazards that are causing or are likely to cause death or serious physical harm to his employees." It also requires employers to "comply with occupational safety and health standards promulgated under this Act" (OSHA, 1970).

Heat stress discussions involve a lot of acronyms so a few definitions from OSHA will help facilitate the discussion:

- Heat is a measure of energy in terms of quantity.
- Metabolic heat is a by-product of the body's activity.
- Environmental heat is the contribution of environmental conditions to body temperature.
- Conduction is the transfer of heat between materials that contact each other.
- Convection is the transfer of heat in a moving fluid.
- Radiation is the transfer of heat energy through space.
- Evaporative cooling takes place when sweat evaporates from the skin. High humidity reduces the rate of evaporation and thus reduces the effectiveness of the body's primary cooling mechanism.
- Thermal equilibrium is achieved when both metabolic and environmental heat production are in balance with evaporative cooling (i.e., all heat loads are dissipated by sweating).
- Acclimatization is the physiological adaptation that occurs when a person is exposed to heat over an extended period of time. A well-designed acclimatization program exposes employees to work in a hot environment for progressively longer periods to decrease the risk of heat-related illnesses.
- Globe (G) temperature is the temperature inside a blackened, hollow, thin copper globe.
- Natural wet bulb (NWB) temperature is measured by exposing a wet sensor, such as a wet cotton wick fitted over the bulb of a thermometer, to the effects of evaporation and convection. The term natural refers to the movement of air around the sensor.
- Dry bulb (DB) temperature is measured by a thermal sensor, such as an ordinary mercury-in-glass thermometer that is shielded from direct radiant energy sources.

Source: OSHA and other sources.

There are numerous commercial and industrial occupations associated with elevated risk of heat stress illness (HSI), including foundry work, construction, asbestos removal, commercial kitchens, mining, agriculture, utilities, and firefighting. Additionally, any activity performed in protective clothing can increase the likelihood of heat stress illness. Heat-related illness can be defined as a range of health disorders stemming primarily from exposure to excessive heat. These conditions include but are not limited to heat stroke, heat exhaustion, heat cramps, heat syncope, and heat rashes. National Institute for Occupational Safety and Health

Hamilton & Hardy's Industrial Toxicology, Sixth Edition. Edited by Raymond D. Harbison, Marie M. Bourgeois, and Giffe T. Johnson.
© 2015 John Wiley & Sons, Inc. Published 2015 by John Wiley & Sons, Inc.

(NIOSH) defines heat stress exposure as the sum of metabolic heat (generated within the body) and environmental heat (heat gained from the environment) minus any heat lost to the environment through processes such as evaporation (NIOSH, 1986).

Attempts to reduce heat stress illness have led to the development of various indices. Methods are continually refined to improve preventive measures. The effective temperature scale (ET) was developed in the 1920s. It measures humidity, wind speed, and temperature, and remains useful in places with high humidity and a low solar load. Effective temperature scale was modified approximately 30 years later to incorporate the effects of radiant heat and it was called effective temperature, including the radiation component (ETR). The wet-bulb globe temperature index (WBGT) replaced the ETR, and it remains the most widely used index of heat stress (Rowlinson et al., 2014). Wet-bulb globe temperature index was developed in response to the episodic heat illness in armed forces training camps in the buildup to World War II (Budd, 2008). The basic principle is simple; the readings from a NWB thermometer, a G thermometer, and a DB thermometer when indicated are weighed and averaged to calculate a WBGT appropriate to the conditions (Budd, 2008). Natural wet bulb assesses evaporation while GT responds to the environmental heat load. Dry bulb is essentially a modifying factor that incorporates solar load.

- In sunny conditions: $WBGT = 0.7NWB + 0.2GT + 0.1DB$
- In all other conditions: $WBGT = 0.7NWB + 0.3GT$,

where:

NWB = Natural wet-bulb temperature
DB = Dry-bulb temperature
GT = Globe temperature
WBGT = Wet-bulb globe temperature index (OSHA, 1999)

Widespread adoption does not mean that the WBGT is without limitations. Metabolic rate, or work, is a key factor in worker heat stress. The International Organization for Standardization (ISO) estimation method calculates the total metabolic rate (TMR) as the sum of basal metabolism (B), posture (P), work type (W), walking (D), and climbing (C) (ISO, 2002). The WBGT has been criticized for inadequate sensitivity to humidity and air movement. In 2006, the American Conference of Governmental Industrial Hygienists (ACGIH) attempted to account for the contribution of clothing and personal protective equipment (PPE) to heat stress by replacing the unacclimatized threshold limit value (TLV) with an action limit (AL). The AL is designed to be used in conjunction with detailed simplified decision trees for screening. They also developed a formula for an effective

WBGT by adding a clothing adjustment factor (CAF) to the measured WBGT (Soule et al., 2006; ACGIH 2008).

$$WBGT_{eff} = WBGT_{measured} + CAF$$

Some have suggested predicted heat strain (PHS) to cover situations with indeterminate exposures (Bernard, 2014). The PHS was derived by the ISO and permits the calculation of individual contributors to heat exchange. In times when workers exceed the TLV or when the decision trees cannot be employed, ACGIH recommends heat strain monitoring (HST). Heat strain monitoring utilizes core body temperature and oral temperature measurements along with heart rate and perceived strain to assess worker risk (Soule et al., 2006; ACGIH, 2008). The Humidex is another method for combining the effects of humidity and elevated temperatures into a scale to assess the heat stress potential (Smoyer-Tomic and Rainham, 2001). The Humidex cannot be directly compared to the WBGT; however, it is considered a particularly useful parameter in unacclimatized worker discomfort.

SOURCES OF EXPOSURE

Occupational

Although heat-related illness morbidity and mortality is not well documented for multiple reasons, excessive heat exposure has been implicated in more than 9000 deaths since 1994 (Rogers et al., 2007; Jay and Kenny 2010; DeGroot et al., 2013). A preliminary report from the Bureau of Labor on fatal occupational injuries in 2012 noted 31 worker deaths due to exposure to the environmental heat (Bureau of Labor Statistics, 2013). Members of the military are historically considered to be at the greatest risk for heat stress illness; however, some commercial and industrial occupations are just as vulnerable (Larsen et al., 2007; Bernard and Barrow, 2013; DeGroot et al., 2013). Workers may be at greater risk for developing heat stress illness when they:

- are exposed to radiant heat sources,
- come into contact with hot objects,
- are exposed to direct sunlight for prolonged periods of times or,
- are required to use bulky or nonbreathable protective clothing and equipment (ACGIH, 2008).

The majority of early work on heat-related illnesses was done on members of the military. Although the total number of hospitalizations has dropped dramatically in the last few decades, the incidence of heatstrokes in armed forces members has increased eightfold (Carter et al., 2005). It is possible that deployment of poorly acclimatized troops to theaters in the Middle East accounts for much of the increase; however, it may also be that prevention allowed the personnel to return to the field prematurely. The incidence of heat exhaustion and other heat-related illnesses is quite high in mining where the heat, humidity, and metabolic demands can be physically taxing (Xiang et al., 2013). Agricultural industries have one of the highest risks of heat-related morbidity and mortality. Statistics show that farm workers are up to four times likely to experience heat-related illness than workers in nonagricultural settings (Xiang et al., 2013). Firefighters may be the most obvious cohort threatened by heat-related illness. The protective clothing of firefighters lacks ventilation and they wear a self-contained breathing apparatus (SCBA) or compressed air breathing apparatus (CABA). The majority of firefighters wear a duty uniform under their gear consisting of long pants. The rationale has been that this provided an extra layer of protection. Recent trials in Toronto and New York have shown that replacing long pants and long-sleeved shirt with shorts and a t-shirt reduced the heat strain without an increase in injuries (Prezant et al., 2000; Selkirk and McLellan 2004; McLellan and Selkirk 2006). Manufacturing workers often labor in poorly ventilated facilities that can increase their risk for heat stress illnesses. This includes industries such as automobile manufacture and assembly, foundries, steelworkers, and industrial food-processing facilities.

Environmental

Occupational and environmental exposures may overlap depending on the industry (agriculture in particular). Days with high temperature and high humidity pose the greatest risk.

INDUSTRIAL HYGIENE

There is no specific OSHA standard for exposure to high temperatures because factors such as humidity, solar load, wind and air movement, metabolic load and protective clothing also contribute to the development of HSI. The United States uses WBGT-based occupational exposure limits (OELs) that is based on the ACGIH and the NIOSH recommendations. For more information, consult ACGIH *Threshold Limit Values for Chemical Substances and Physical Agents and Biological Exposure Indices* (ACGIH, 2008). Employers can utilize a combination of good work practices/training programs, worker monitoring programs, PPE, and administrative and engineering controls to keep employees healthy. Early recognition of the heat stress illness symptoms can prevent the progression to heatstroke, so employee education is a key to an effective training program. Workers in extreme conditions (excessive humidity or temperatures, semipermeable, or impermeable clothing, high metabolic load, etc.) should also be personally monitored.

MEDICAL MANAGEMENT

Heat stress illness is best managed through prevention measures. Being aware of the environmental conditions and utilizing weather forecasts and temperature instrumentation such as the WBGT device are key components for prevention. It has been estimated that 90% of the cases of heatstroke occur when the WBGT is 85 °F (30 °C) or more (Jenkins and Braen, 2000). Knowing personal risk factors for developing heat stress illness is useful in avoiding illness in susceptible individuals. The very young (<4 years old) and the elderly (>65 years old) are at increased risk, as are obese individuals, those with chronic medical problems such as diabetes, heart, lung, kidney, or dermatologic disease, and those taking certain medications such as diuretics, sedatives, tranquilizers, antiemetics, anticholinergics, stimulants, some heart and blood pressure medications, medications for psychiatric conditions, lithium, and others. Additional factors that increase the risk of developing heat stress illness include fatigue, sleep deprivation, low fitness, poor nutritional status, acute illness, strenuous exertion, alcoholism, intoxication, and a prior history of heat illness (Druyan et al., 2011).

Educating employees on how to recognize early symptoms of heat stress illness in themselves and fellow workers is of key importance. Symptoms of impending heat illness include fatigue, headache, nausea, muscle cramps, pale skin, profuse sweating, rapid heartbeat, and dizziness. Ominous signs of heatstroke, the most serious heat stress illness, include the absence of perspiration, hot dry red or mottled skin, throbbing headache, slow deep respirations, chills, high body temperatures (≥104 °F or ≥40 °C), behavior changes such as disorientation, mental confusion, delirium, irrational behavior, feelings of euphoria, giddiness, or impending doom, slurred speech, hallucinations, and diminished level of consciousness, or loss of consciousness.

Methods to prevent heat illness include wearing loose-fitting lightweight light-colored clothing, maintaining adequate hydration by providing cool water and liquids to workers, scheduling rest periods for workers in the shade or in cool air-conditioned environments, scheduling hot jobs for the cooler part of the day, reducing physical demands of workers in hot environments, using relief workers or assigning extra workers for physically demanding jobs, avoiding sunburn, allowing for employee acclimatization to hot environments when possible and providing employee education regarding symptoms and signs, prevention, risks, treatment, and PPE.

Once heat stress illness has occurred, treatment in general includes moving the victim to a cool environment, rapid cooling if necessary, replenishment of the fluid and electrolytes, and general supportive care.

Treatment varies depending on the type and severity of illness (CDC, NIOSH, Workplace Safety and Health Topics):

Heatstroke: Symptoms and signs of heatstroke include, mental changes, confusion, hallucinations, chills, throbbing headache, and high body temperature (>104 °F or 40 °C). Classically a victim of heatstroke has hot dry skin; however, there may be profuse sweating if exertional heatstroke develops rapidly in a matter of hours. Seizures and coma are the most common severe complications of heatstroke. The body's thermoregulatory mechanisms fail with resulting high temperatures and potential damage to critical organs. Laboratory testing may reveal multiple abnormalities, including acidosis, leukocytosis, markedly elevated liver enzymes, elevated creatinine kinase, hyperkalemia, hypocalcemia, hypoglycemia, and hypophosphatemia. Coagulopathy may occur and disseminated intravascular coagulopathy is seen in severe cases. Electrocardiogram (EKG) abnormalities may include tachycardia, nonspecific ST-segment and T-wave abnormalities, as well as reversible arrhythmias and heart blocks (Jenkins and Braen, 2000).

Heatstroke is a serious medical emergency; call 911, cool the worker while awaiting transport with such methods as placing ice packs in the groin or axillae, soaking their clothing in water, showering, spraying, or sponging the victim with water in conjunction with fanning the body. Emergency room treatment protocols are well established and involve rapid cooling of the victim, IV fluids, electrolytes and dextrose (carefully administered), oxygen supplementation, intubation if necessary, pharmacologic control of agitation and shivering, and intense supportive care and management of complications such as convulsions, rhabdomyolysis, and hepatic, pulmonary, and renal complications. Despite intense emergency medical care, severe cases can be fatal or leave the victim with permanent mental or physical damage. Early recognition and treatment is critical. Prompt body temperature reduction, early rehydration, and rapid transport to the hospital emergency room can result in >95% survival rate (Druyan et al., 2011). Mazerolle et al. (2011) promoted rectal thermometry and cold water immersion as the recommended methods for recognizing and treating exertional heatstroke. After reviewing 17 papers, Smith J. E. (2005) concluded that the main predictor of outcome in exertional heatstroke is the duration and degree of hyperthermia. Where possible, patients should be cooled using iced water immersion, but, if this is not possible, a combination of other techniques may be used to facilitate rapid cooling. Furthermore, he states that there is no evidence to support the use of dantrolene in these patients.

Heat Exhaustion: Symptoms and signs include heavy sweating, clammy moist skin, excessive thirst, extreme weakness or fatigue, muscle cramps, dizziness, confusion, nausea, slightly elevated body temperature (<102 °F or 38–39 °C) and fast shallow breathing. Lab testing may indicate dehydration, elevated creatinine kinase enzymes, and occasionally hypoglycemia.

Have the worker to rest in a cool, shaded or an air-conditioned place, drink plenty of water or other cool nonalcoholic beverages, and take a cool shower, bath or sponge bath. Call 911 if the worker develops more serious signs or symptoms indicating heatstroke. If treated in the emergency room the worker may receive IV fluid and electrolyte replacement. Workers may be more susceptible to heat stress illness during the following week and precautions are indicated.

Heat Syncope: Symptoms include dizziness and fainting.

Have the worker lie down flat in a cool place if symptoms of lightheadedness or dizziness begin. When feeling a little better, sit up and slowly drink water, clear juice or a sports beverage.

Heat Cramps: Symptoms of painful cramps usually occur after long periods of heavy sweating with strenuous activity and inadequate fluid replacement.

Have the worker stop activity, rest in a cool place and don't return to work for a few hours, drink clear juice or a sports beverage, and seek medical attention if cramps do not subside or if the worker has heart problems or is on a low sodium diet. Emergency room treatment may include IV fluid and electrolyte replacement.

Heat Rash (Miliaria or Prickly Heat): Symptoms include a cluster of pimples or small blisters occurring on the face, neck and upper chest, in the groin, under the scrotum, under the breasts, or in the elbow creases. The rash may be asymptomatic or cause varying intensity of itching.

The worker with heat rash should try to work in a cooler less humid environment, avoid sweat provoking activities and occlusive clothing, keep the affected area dry, take frequent cool showers, and use talc powder for comfort. Heat rash is usually mild and self-limiting. In some cases short-term use of OTC hydrocortisone cream for small areas can help with symptoms. Avoid oil-based creams. For more severe rashes with inflammation and blockage of the sweat glands, antibiotics can be prescribed.

REFERENCES

ACGIH (2008) *Threshold Limit Values and Biological Exposure Indices for Chemical Substances and Physical Agents: Heat Stress and Strain*, Cincinnati: American Conference of Industrial Hygienists.

Bernard, T.E. (2014) Occupational heat stress in USA: whither we go? *Ind. Health* 52(1):1–4.

Bernard, T.E. and Barrow, C.A. (2013) Empirical approach to outdoor WBGT from meteorological data and performance of two different instrument designs. *Ind. Health* 51(1):79–85.

Bureau of Labor Statistics (2013) *Injuries, Illnesses, and Fatalities: 2013 Census of Fatal Occupational Injuries* (preliminary data). U.S. Department of Labor. Available at http://www.bls.gov/iif/oshwc/cfoi/cftb0276.pdf (accessed April 2014).

Budd, G.M. (2008) Wet-bulb globe temperature (WBGT)—its history and its limitations. *J. Sci. Med. Sport* 11(1):20–32.

Carter, R. 3rd, Cheuvront, S.N., Williams, J.O., Kolka, M.A., Stephenson, L.A., Sawka M.N., and Amoroso, P.J. (2005) Epidemiology of hospitalizations and deaths from heat illness in soldiers. *Med. Sci. Sports Exerc.* 37(8):1338–1344.

DeGroot, D.W., Gallimore, R.P., Thompson, S.M., and Kenefick, R.W. (2013) Extremity cooling for heat stress mitigation in military and occupational settings. *J. Thermal Biol.* 38(6): 305–310.

Druyan, A., Janovich, R., and Heled, Y. (2011) Misdiagnosis of exertional heat stroke and improper medical treatment. *Mil. Med.* 176(11):1278–1280.

ISO (2002) *International Organization for Standardization (ISO): Hot Environments—Analytical Determination and Interpretation of Thermal Stress Using Calculation of Physiological Heat Strain (PHS) 7933.*

Jay, O. and Kenny, G.P. (2010) Heat exposure in the Canadian workplace. *Am. J. Ind. Med.* 53(8):842–853.

Jenkins, J.L. and Braen, R.G. (2000) *Heat Illness and Cold Exposure. Manual of Emergency Medicine, Lippincott Williams & Wilkins.*

Larsen, T., Kumar, S., Grimmer, K., Potter, A., Farquharson, T. and Sharpe, P. (2007) A systematic review of guidelines for the prevention of heat illness in community-based sports participants and officials. *J. Sci. Med. Sport* 10(1):11–26.

Mazerolle, S. M. et al (2011) Evidence-Based Medicine and the Recognition and Treatment of Exertional Heat Stroke, Part II: A Perspective From the Clinical Athletic Trainer. *Journal of Athletic Training* 46(5):533–542.

McLellan, T.M. and Selkirk, G.A. (2006) The management of heat stress for the firefighter: a review of work conducted on behalf of the Toronto Fire Service. *Ind. Health* 44(3): 414–426.

NIOSH (National Institute for Occupational Safety and Health) (1986) *Criteria for a Recommended Standard—Occupational Exposure to Hot Environments, Revised Criteria. 86.*

OSHA (1970) *OSH Act of 1970.*

OSHA (1999) *OSHA Technical Manual (OTM): Heat Stress.*

Prezant, D.J., Freeman, K., Kelly, K.J., Malley, K.S., Karwa, M.L., McLaughlin, M.T., Hirschhorn, R., and Brown, A. (2000) Impact of a design modification in modern firefighting uniforms on burn prevention outcomes in New York city firefighters. *J. Occup. Environ. Med.* 42(8):827–834.

Rogers, B., Stiehl, K., Borst, J., Hess, A., and Hutchins, S. (2007) Heat-related illnesses: the role of the occupational and environmental health nurse. *AAOHN J.* 55(7):279–287; quiz 288-279.

Rowlinson, S., YunyanJia, A., Li, B., and ChuanjingJu, C. (2014) Management of climatic heat stress risk in construction: a review of practices, methodologies, and future research. *Accid. Anal. Prev.* 66:187–198.

Selkirk, G.A. and McLellan, T.M. (2004) Physical work limits for Toronto firefighters in warm environments. *J. Occup. Environ. Hyg.* 1(4):199–212.

Smith, J. E. (2005) Cooling Methods used in the treatment of exertional heat illness. *Br. J. Sports Med.* 39:503–507.

Smoyer-Tomic, K.E. and Rainham, D.G. (2001) Beating the heat: development and evaluation of a Canadian hot weather health-response plan. *Environ. Health Perspect.* 109(12):1241–1248.

Soule, B., Casserly, D., Bernard, T.E. Lowry, L., and Price, J. (2006) *ACGIH®: Dedicated to Development of Exposure Guidelines for the Professional.* Chicago, IL: AIHce.

Xiang, J., Bi, P., Pisaniello, D., and Hansen, A. (2013) Health impacts of workplace heat exposure: an epidemiological review. *Ind. Health* 52:91–101.

SECTION IX

SPECIAL TOPICS

SECTION EDITOR: RAYMOND D. HARBISON

106

MODE OF ACTION/HUMAN RELEVANCE FRAMEWORK

M.E. (BETTE) MEEK

The mode of action/human relevance (MOA/HR) framework is an analytical tool designed to increase transparency in the systematic consideration of the weight of evidence (WOE) of hypothesized MOA(s) for critical effects and their relevance to humans. It was developed in initiatives of the International Life Sciences Institute Risk Sciences Institute (ILSI RSI) and the International Programme on Chemical Safety (IPCS) and derives from earlier work on MOA in animals of the U.S. Environmental Protection Agency (U.S. EPA) and IPCS (Sonich-Mullin et al., 2001).

Mode of action, as previously defined, is a biologically plausible series of key events leading to an effect (Sonich-Mullin et al., 2001). Originally, MOA was considered principally in the context of late-stage key cellular, biochemical, and tissue events. A key event is an empirically observable step or its marker that is a necessary element of the MOA critical to the outcome (i.e., necessary, but not necessarily sufficient in its own right); key events are measurable and reproducible.

The MOA framework is based, then, on the premise that any human health effect caused by exposure to an exogenous substance can be described by a series of causally linked biochemical or biological key events that result in a pathological or other disease outcome. While originally and often simply conceptualized and illustrated as a linear series of key events, in reality, MOA involves interdependent networks of events with feedback loops. Disease outcomes are initiated or modified within these networks. Differences in networks between and within human and animal populations account, in part, for interspecies differences and human variability.

Early key events in hypothesized MOAs are most often related to chemical characteristics—i.e., those characteristics of structure and/or physicochemical properties that promote interaction of the substance with biological targets. Later key events are less chemical specific and more often an expected consequence of progression of earlier key events (e.g., regenerative proliferation resulting from cytotoxicity).

An adverse outcome pathway is conceptually similar to a MOA. It was initially described by the computational ecotoxicology community (Ankley et al., 2010) and has been adopted within an international initiative to document, develop, and assess the completeness of potentially predictive tools for adverse ecological and human health effects (OECD, 2012). A focus of adverse outcome pathways is on the initial associated chemically mediated "molecular initiating event," equivalent to an early key event in a MOA.

The terms MOA and adverse outcome pathway are conceptually interchangeable, representing essentially the subdivision of the pathway between exposure and effect in either individuals or populations into a series of hypothesized key events at different levels of biological organization (e.g., molecular, subcellular, cellular, tissue).

The development and evolution of the IPCS ILSI RSI MOA/HR framework, which has involved large numbers of scientists internationally, is described in several publications. It is supported by a series of templates, which promote systematic consideration of weight of evidence for MOA of adverse effects identified in traditional toxicity studies (Boobis et al., 2006; Boobis et al., 2008; Meek, 2008a; Meek et al., 2003; Seed et al., 2005; Meek and Klaunig, 2010). Potential application in a broader range of more predictive contexts (i.e., those where adverse effects might be predicted on the basis of MOA) has been considered more recently (Carmichael et al., 2011; Meek et al., 2014).

The framework has been illustrated by an increasing number of case studies ($n = 30$, currently) (Table 106.1)

Hamilton & Hardy's Industrial Toxicology, Sixth Edition. Edited by Raymond D. Harbison, Marie M. Bourgeois, and Giffe T. Johnson.
© 2015 John Wiley & Sons, Inc. Published 2015 by John Wiley & Sons, Inc.

TABLE 106.1 Case Studies Illustrating Various Modes of Action

Mode of Action	Case Study	Reference
Tumors of various organs associated with mutagenic modes of action	Ethylene oxide	Meek et al. (2003)
	4-Aminobiphenyl	Cohen et al. (2006)
Mammary tumors associated with suppression of luteinizing hormone	Atrazine	Meek et al. (2003)
Thyroid tumors associated with increased clearance of thyroxine	Phenobarbital	Meek et al. (2003)
	Thiazopyr	Dellarco et al. (2006)
Bladder tumors associated with the formation of urinary tract calculi	Melamine	Meek et al. (2003)
Liver/kidney tumors associated with sustained cytotoxicity and regenerative proliferation	Chloroform	Meek et al. (2003)
Acute renal toxicity associated with precipitation of oxalate	Ethylene glycol	Seed et al. (2005)
Androgen receptor antagonism and developmental effects	Vinclozolin	Seed et al. (2005)
Nasal tumors associated with DNA reactivity and cytotoxicity	Formaldehyde	McGregor et al. (2006)

and is widely adopted in international and national guidance and assessments (Meek et al., 2008b).

Building on this collective experience, the framework has been updated recently to more explicitly address uncertainty and to extend its utility to emerging areas in toxicity testing and non-testing methods. This update includes incorporation within a roadmap, encouraging continuous refinement of fit-for-purpose testing strategies and risk assessment, based on MOA (Meek et al., 2014).

In the framework, the WOE for hypothesized MOAs in animals is assessed based on considerations modified from those proposed originally by Bradford Hill (Hill, 1965) for assessment of causality in epidemiological studies. Human relevance or species concordance is then systematically considered, taking into account more generic information such as anatomical, physiological, and biochemical variations. If the WOE for the hypothesized MOA is sufficient and relevant to humans, implications for dose response in humans are then considered in the context of kinetic and dynamic data. Delineation of the degree of confidence in the WOE for hypothesized MOAs and their relevance to humans at each step is critical as is the delineation of critical research needs.

Establishing support for or rejection of a hypothesized MOA provides the foundation for subsequent considerations of dose response, human relevance, and estimates of risk. It involves (1) delineation of key events leading to the relevant effect in a hypothesized MOA and (2) evaluation of the available data to consider the extent of the supporting weight of evidence. Importantly, if alternative MOA(s) are supported, these are evaluated with equal rigor in separate framework analyses.

Essentially, then the framework assists in the transparent and focused incorporation of mechanistic data in risk assessment. This requires continuing interface between the risk assessment and research communities as a basis to increase common understanding and to focus additional research on relevant priorities for risk assessment.

The framework has been recently updated to more explicitly address uncertainty and take into account advances in toxicity testing (Meek et al., 2014). It has evolved to be used not only as originally envisaged, where the outcome of chemical exposure is known, but additionally in hypothesizing potential effects resulting from exposure, based on information on putative key events in established MOAs from appropriate *in vitro* or *in silico* systems and other evidence. It draws upon the considerable experience acquired in the application of the framework in addressing documented (adverse) effects to inform the more limited knowledge base in more predictive applications. This evolution addresses the need to more efficiently assess the hazards and risks associated with the wide array of chemicals to which humans are exposed, as required by more aggressive regulatory mandates in various jurisdictions, worldwide. (See, for example, Meek and Armstrong, 2007). This necessitates the development and integration of information on key events within (hypothesized) MOAs very early in the evaluation process that will enable the effective use of data collected from lower levels of biological organization and non-test methods, such as (quantitative) structure–activity relationships ((Q)SAR) and read-across *in vitro* assays.

REFERENCES

Ankley, G.T., Bennett, R.S., Erickson, R.J., Hoff, D.J., Hornung, M.W., Johnson, R.D., Mount, D.R., Nichols, J.W., Russom, C.L., Schmieder, P.K., Serrrano, J.A., Tietge, J.E., and Villeneuve, D.L. (2010) Adverse outcome pathways: a conceptual framework to support ecotoxicology research and risk assessment. *Environ. Toxicol. Chem.* 29:730–741.

Boobis, A.R., Cohen, S.M., Dellarco, V., McGregor, D., Meek, M.E., Vickers, C., Willcocks, D., and Farland, W. (2006) IPCS framework for analyzing the relevance of a cancer mode of action for humans. *Crit. Rev. Toxicol.* 36:781–792.

Boobis, A.R., Doe, J.E., Heinrich-Hirsch, B., Meek, M.E., Munn, S., Ruchirawat, M., Schlatter, J., Seed, J., and Vickers, C. (2008) IPCS framework for analyzing the relevance of a noncancer mode of action for humans. *Crit. Rev. Toxicol.* 38:87–96.

Carmichael, N., Bausen, M., Boobis, A.R., Cohen, S.M., Embry, M., Fruijtier-Pölloth, C., Greim, H., Lewis, R., Meek, M.E., Mellor, H., Vickers, C., and Doe, J. (2011) Using mode of action information to improve regulatory decision-making: an ECE-TOC/ILSI RF/HESI workshop overview. *Crit. Rev. Toxicol.* 41:175–186.

Cohen, S.M., Boobis, A.R., Meek, M.E., Preston, R.J., and McGregor, D. (2006) 4-Aminobiphenyl and DNA reactivity: case study within the context of the 2006 IPCS human relevance framework for analysis of a cancer mode of action for humans. *Crit. Rev. Toxicol.* 36:803–819.

Dellarco, V.L., McGregor, D., Berry, S.C., Cohen, S.M., and Boobis, A.R. (2006) Thiazopyr and thyroid disruption: case study within the context of the 2006 IPCS human relevance framework for analysis of a cancer mode of action. *Crit. Rev. Toxicol.* 36:793–801.

Hill, A.B. (1965) The environment and disease: association or causation? *Proc. R. Soc. Med.* 58:295–300.

McGregor, D., Bolt, H., Cogliano, V., and Richter-Reichhelm, H.B. (2006) Formaldehyde and glutaraldehyde and nasal cytotoxicity: case study within the context of the 2006 IPCS human framework for the analysis of a cancer mode of action for humans. *Crit. Rev. Toxicol.* 36:821–835.

Meek, M.E. (2008a) Recent developments in frameworks to consider human relevance of hypothesized modes of action for tumours in animals. *Environ. Mol. Mutagen.* 49:110–116.

Meek, M.E. and Armstrong, V.C. (2007) The assessment and management of industrial chemicals in Canada. In: Van Leeuwen, K. and Vermeire, T., editors. *Risk Assessment of Chemicals: An Introduction*, 2nd ed., Dordrecht, the Netherlands: Kluwer Academic Publishers, pp. 591–621.

Meek, M.E. and Klaunig, J. (2010) Proposed mode of action of benzene-induced leukemia: Interpreting available data and identifying critical data gaps for risk assessment. *Chem. Biol. Interact.* 184:279–285.

Meek, M.E., Bucher, J.R., Cohen, S.M., Dellarco, V., Hill, R.N., Lehman-McKeeman, L.D., Longfellow, D.G., Pastoor, T., Seed, J., and Patton, D.D. (2003) A framework for human relevance analysis of information on carcinogenic modes of action. *Crit. Rev. Toxicol.* 2003:591–653.

Meek, M.E., Berry, C., Boobis, A.R., Cohen, S.M., Hartley, M., Munn, S., Olin, S.S., Schlatter, J., and Vickers, C. (2008b) Mode of action frameworks: a critical analysis. *J. Toxicol. Environ. Health B Crit. Rev.* 11:16–31.

Meek, M.E., Boobis, A.R., Cote, I., Dellarco, V., Fotakis, G., Munn, S., Seed, J., and Vickers, C. (2014) New developments in the evolution and application of the WHO/IPCS framework on mode of action/species concordance analysis. *J. Appl. Toxicol.* 34:1–18.

OECD, Organisation for Economic Co-operation and Development (2012) *Proposal for a Template, and Guidance on Developing and Assessing the Completeness of Adverse Outcome Pathways. OECD, Paris.* Available at http://www.oecd.org/chemicalsafety/testingofchemicals/49963554.pdf (accessed 28 September 2012).

Seed, J., Carney, E., Corley, E.R., Crofton, K., DeSesso, J., Foster, P., Kavlock, R., Kimmel, G., Klaunig, J., Meek, E., Preston, J., Slikker, W., Tabacova, S., and Williams, G. (2005) Overview: Using mode of action and life stage information to evaluate the human relevance of animal toxicity data. *Crit. Rev. Toxicol.* 35:664–672.

Sonich-Mullin, C., Fielder, R., Wiltse, J., Baetcke, K., Dempsey, J., Fenner-Crisp, P., Grant, D., Hartley, M., Knaap, A., Kroese, D., Mangelsdorf, I., Meek, M.E., Rice, J., and Younes, M. (2001) IPCS conceptual framework for evaluating a mode of action for chemical carcinogenesis. *Regul. Toxicol. Pharmacol.* 34:146–152.

107

CARCINOGENESIS

Carlos A. Muro-Cacho

BACKGROUND

Cancer is a group of disease that affects the genetic machinery of the cell. Since cancer can occur in any nucleated cell in the body, approximately 200 different cancer types and more than 1000 subtypes are recognized by pathologists. Despite significant scientific advances and therapeutic achievements, cancer remains responsible for one in eight deaths worldwide (WHO, 2009; Howlader et al., 2012). In the United States, cancer accounts for one in four deaths, only exceeded by heart disease. In 2012, according to the most recent data from the National Cancer Institute's Surveillance Epidemiology and End Results, 13.7 million Americans had a diagnosis of cancer and 1,638,910 Americans (848,170 men and 790,740 women) were diagnosed with cancer of all sites, resulting in 577,190 deaths (SEER, 2010; Siegel et al., 2012). The most current estimates indicate that, in 2013, 1,660,290 Americans will be diagnosed with cancer and, of these, 580,350 will die of the disease (almost 1600 people per day) (ACS, 2013).

In men, three sites (prostate, lung and bronchus, and colorectal) account for 51% of all new diagnoses, with prostate cancer accounting for 28% of the cases. In women, the three most common sites (52% of the total) are breast (29%), lung and bronchus, and colorectal (Jemal et al., 2010). Of interest is that, in the United States, from 1999 to 2008, cancer deaths have declined by more than 1% per year in men and women, except in American Indians/Alaska Natives, among whom deaths have remained stable. Death rates continue to decline for all four major cancer sites (lung, colorectal, breast, and prostate), with lung cancer accounting for almost 40% of the total decline in men; and breast cancer accounting for 34% of the total decline in women. The 5-year relative survival rate, for all cancers diagnosed between 2002

and 2008, was 68%, up from 49% in the period 1975–1977. This reduction in overall cancer deaths means that more than a million deaths have been avoided since 1990. The largest increases in death rates have been for liver cancer in men and women, esophageal cancer and melanoma in men, and lung and pancreatic cancer in women (Jemal et al., 2010; Edwards et al., 2010).

About 77% of all cancers are diagnosed in persons 55 years of age and older and, in 5% of the cancers, there is a strong hereditary component. Men have slightly less than a one in two lifetime risk of developing cancer, while for women the risk is a little more than one in three. Incidence rates in children have been increasing, especially in infants (Gapstur and Thun, 2010). Children are particularly vulnerable to environmental risk factors, including numerous toxins and detrimental exposures from air, food, water, medicines, pesticides, and ionizing radiation (NCI, 2010; Bode and Dong, 2009; Collins et al., 2008). Environmental pollution prevention acts (EPA's Resource Conservation and Recovery Act, Clean Water Act, and Clean Air Act) and controls on occupational carcinogens have prevented environmentally and occupationally related cancers (EPA, 2010; Montesano and Hall, 2001; Raaschou-Nielsen et al., 2010; Fontham et al., 2009; IOM, 2001). In addition to preventing cancer through the avoidance of risk factors, screening tests allow early detection and removal of precancerous growths (cervical and colorectal cancer) while mammary and prostatic cancer can be treated at very early stages.

Besides the obvious consequences for individuals and their families, the economic impact of cancer in society is significant. According to the National Institutes of Health (NIH), the cost of cancer, in 2008, was $201.5 billion ($77.4 billion for direct medical costs and $124.0 billion

Hamilton & Hardy's Industrial Toxicology, Sixth Edition. Edited by Raymond D. Harbison, Marie M. Bourgeois, and Giffe T. Johnson.
© 2015 John Wiley & Sons, Inc. Published 2015 by John Wiley & Sons, Inc.

for indirect mortality costs (cost of lost productivity due to premature death) (NCI, 2010). The overall effects of cancer are, however, underestimated since non-invasive carcinomas and cutaneous basal and squamous cell carcinomas account for a significant percentage of the oncological care and are not reported to cancer registries. Since certain cancers are related to infectious agents and since, according to the World Cancer Research, one quarter to 1/3rd of new cancers are related to obesity, physical inactivity, and poor nutrition, between 30 and 40% of cancer deaths are actually preventable (WHO, 2009; Tomatis et al., 1997).

HISTORY

Cancer is not a modern disease. A possible lymphoma was observed in a hominid discovered by Louis Leakey, in 1932. Nasopharyngeal carcinomas and osteosarcomas have been reported in 3000 BC Egyptian mummies. A possible melanoma was found in mummified remains of a Peruvian Inca 2400 years ago. A cancer of the head and neck was observed in the skull of a female from the Bronze Age (1900–1600 BC). Early accounts of cancer were also recorded in the Babylonian Code of Hammurabi (1750 BC), ancient Egyptian papyri (1600 BC), the Chinese Rites of the Zhou Dynasty (1100–400 BC), and the Indian Ramayana manuscript (500 BC) (Hajdu, 2011). The Egyptian Edwin Smith Papyrus describes eight ulcerated, incurable breast tumors for which palliative cauterization was recommended. Hieroglyphic inscriptions and papyri manuscripts suggest that ancient physicians had already distinguished between benign and malignant tumors and suggested that superficial tumors could be surgically excised.

The father of medicine, Hippocrates (460–370 BC) also recognized the differences between benign and malignant tumors (Lloyd, 1984). To him, the blood vessels around a malignant tumor looked like the claws of a crab and named the disease "karkinos." Karkinos was a giant crab that helped the nine-headed serpent Hydra in its battle with Herakles. The Roman physician, Celsus (28–50 BC), later translated the Greek term into *cancer*, the Latin word for crab. Galen (130–200 AD) used the word *oncos* (Greek word for swelling) to describe tumors and introduced the word "neoplasia" to define "a growth of a body area adverse to nature." In 1531, Paracelsus described the term "mala metallorum," among miners of silver and other metals, later recognized as radiation-induced lung cancer. In 1775, Port attributed scrotal skin cancers to prolonged exposure to soot in chimney sweeps, and in 1823 Earle reported the requirement of a "constitutional predisposition," which would render the individual susceptible to the action of the soot. This constituted the first proposal for a "genetic–environmental" paradigm.

Over the decades, several hypotheses were proposed to explain the etiology of cancer (Diamandopoulos, 1996;

Kardinal and Yarbro, 1979). Hippocrates believed that the body had four fluids or *humors* (blood, phlegm, yellow bile, and black bile). Health was associated with a balance among these humors while too much or too little of any of them would cause the disease. Thus, an excess of black bile was thought to cause cancer. Galen embraced this theory, which remained unchallenged for over 1300 years. Among the theories that replaced the "humoral" theory was the formation of cancer by another body fluid, *lymph*. Stahl and Hoffman theorized that cancer was composed of fermenting and degenerating lymph, varying in density, acidity, and alkalinity. The "lymph" theory gained rapid support. John Hunter, in the 1700s, believed that tumors grow from lymph constantly thrown out by the blood. Once dyes and microscopic techniques were available, however, the structural detail of tissues and cells began to be understood and it became clear that tumors, as tissues, were composed of cells, and that mitoses, nucleomegaly, necrosis, increased the vasculature and loss of differentiation were characteristics of neoplasms.

In 1838, the German pathologist Johannes Muller showed that cancer is made up of cells and not lymph, but he believed that cancer cells did not come from normal cells but from the "*blastema*," a sort of primordial stroma between normal tissues. His disciple, Rudolph Virchow, the father of cellular pathology, showed that all cells are derived from other cells and proposed a cellular origin of cancer, in 1863 (Virchow, 1863). He believed that chronic irritation was the cause of cancer and that cancers spread like a liquid. In the 1860s, Thiersch showed that cancers metastasize through the spread of malignant cells and not through some unidentified fluid. The role of lymphatic vessels in metastasis was established (Hajdu, 2011) and, in 1889, the "seed and soil" theory was proposed by Paget (Paget, 1889). From the late 1800s until the 1920s, "trauma" was thought to be the cause of cancer despite the failure to induce tumors in experimental animals by injury. This view was, however, partially supported by experimental studies in mouse skin where wounding induced tumors. Cohnheim postulated that tumor cells could be "embryonic" cells not eliminated during ontogenesis and Hanseman believed that they could be "dedifferentiated" normal cells that grew abnormally (anaplasia). Kelling proposed that cancer cells originated in a different organism and penetrated the body of the diseased individual. In 1901, deVries proposed the concept of mutation and, in 1910, Johannsen proposed the term "gene" instead of "hereditary factor." In 1914, Boveri showed that chromosomal abnormalities were often present in cancer cells (Boveri, 1914).

The role of exogenous agents as potential factors in the induction of cancer became evident, in late nineteenth and early twentieth centuries, with the development of chemical industry and the use of coal as industrial fuel and of oils as lubricants. Cancers related to industrial activity were reported in several European locations. Examples were the "paraffin cancer," the "mule-spinner's cancer related to direct contact with shale-oil," and the "aniline bladder cancer." Before the industrial revolution, a variety of occupational causes of cancer had already been documented and the relationship of cancer to specific agents began to be established, boosting general interest in carcinogenesis. In 1915, Yamagiwa induced skin cancer in rabbits by painting their ears with benzene solutions of tar (Oliveira et al., 2007). By 1922, at least 100 radiologists had died of malignant tumors arising from their occupation, and, in 1926, Muller showed the mutagenic effects of X-rays. In 1947, Beremblum and Shubik observed successive stages of carcinogenesis using polycyclic aromatic hydrocarbons (PAH) and croton oil in the skin of mice (Berenblum and Shubik, 1947). Carcinogenesis was considered to have initiation and promotion phases. In 1954, the term "progression" was introduced by Foulds (1954).

Cancer was a contagious disease was proposed by Lusitani (1649) and Tulp (1652), based on observations of breast cancer in members of the same household. Cancer patients began to be isolated, outside of cities and towns, to prevent the spread of the disease. The importance of microorganisms in oncogenesis was not fully realized until Pasteur and Koch established the relationship between bacterial infection and disease leading to a search for cancer "germs" in the late nineteenth century. It was believed that tumors were caused by organisms that released toxins and emaciated the host, and that metastasis were the result of the parasites attacking different organs (ACS, 2012). In 1907, Ciuffo demonstrated a viral etiology for human warts when cell-free filtrates from these lesions transmitted the disease (Ciuffo, 1907; Ronald and Janet, 2008). In 1911, Peyton Rous showed that a sarcoma, from a Plymouth Rock chicken, could be transmitted to healthy chickens using filtered cell-free tumor extracts (Rous, 1910; Levine, 1991; Keogh, 1938). In the 1950s, Burkitt described a new childhood tumor, now known as Burkitt's lymphoma (Burkitt, 1962), and Barr visualized, by electron microscopy, herpes virus-like particles, later recognized as the Epstein–Barr virus (Epstein et al., 1965). In 1974, the link between human papilloma virus (HPV) and cervical cancer was proposed (zur Hausen et al., 1974, 1976) and, in 1975, the link between Hepatitis B virus (HBV) and hepatocellular carcinoma was established (Blumberg et al., 1975; Beasley et al., 1981). Other tumor viruses are the hepatitis C virus (HCV) (Major et al., 2001; Tan et al., 2008), Kaposi's sarcoma virus (human herpes virus type 8) (Ganem, 2006; Chang, 1994) and the first human retrovirus, HTLV-1, recognized as the cause of adult T-cell leukemia (Butel, 2000; McLaughlin-Drubin and Munger, 2008). Sequences of polyoma virus have been identified in mesothelioma, osteosarcoma, non-Hodgkin lymphoma, brain tumors, prostate cancer and Merkel cell carcinoma, and retrovirus-like particles in seminomas, breast cancer, myeloproliferative disease, ovarian cancer,

melanoma, and prostate cancer (White and Khalili, 2004; Feng et al., 2008). Particles of human mammary tumor virus, a retrovirus similar to mouse mammary tumor virus, have been detected in human breast cancer cells. Finally, proteins from a xenotropic murine leukemia virus-related virus have been found in the stroma of human prostate cancer. However, an etiological role for these viruses remains to be proven.

Recently, a more complex view of cancer has evolved, based not on the cancer cell but on the population of malignant cells that constitute cancer. Thus, cancer can be understood as a long evolutionary process driven by sequential somatic-cell mutations, within a multicellular ecosystem, that lead to subclonal selection, in a manner that mimics a Darwinian adaptive system (Navin et al., 2011; Esteller, 2008). Traditional models of clonal evolution propose that a series of clonal expansions produce a uniform cell population that dominates the neoplasm (Cairns, 1975). It appears, however, that simultaneous competitor clones, with different genetic abnormalities, engage in clonal interference (Weinberg, 2007; Greaves and Maley, 2012) and result in tumor heterogeneity, a phenomenon than can be easily observed under the microscope (Marusyk and Polyak, 2010). These clones evolve to adapt to a particular tissue microenvironement (see below) and eventually colonize distant habitats during the metastatic process (Nowell, 1976). These metastatic clones, although genetically unique, appear to carry a signature that can be traced to the original, primary tumor (Mackay et al., 2006). For a review of the significant advances in the history of cancer and research see DeVita and Rosenberg (2012).

HALLMARKS OF CANCER

The growth of normal cells is kept under control by contact with neighboring cells and the extracellular matrix. In normal cells, growth factors interact with specific receptors, mainly with intracellular tyrosine kinase activity, and activate or deactivate interdependent signaling pathways. Cancer cells become independent of these regulatory feedbacks by producing relevant growth factors themselves, increasing the density of receptors, making these receptors constitutionally active and independent of growth factor binding, and engaging in autocrine and paracrine stimulation (Lemmon and Schlessinger, 2010; Witsch et al., 2010). The result is growth advantage and sustained proliferating ability.

Hanahan and Weinberg proposed a framework to understand neoplasia based on what they named "the six hallmarks of cancer" (Hanahan and Weinberg, 2000). The proposal was based on the concept, described above, that normal cells undergo a series of successive changes to become malignant and that tumors are complex populations where cells interact with each other and their tissue environment, which is an active participant in oncogenesis (Table 107.1).

TABLE 107.1   Hallmarks of Cancer

Enabling replicative immortality
Tumor promoting inflammation
Activating invasion and metastasis
Inducing angiogenesis
Genome instability and mutation
Resisting cell death
Deregulating cellular energetics
Sustaining proliferative signaling
Evading growth suppressors
Avoiding immune destruction

Source: Hanahan and Weinberg, 2000, 2011.

The initial hallmarks were (1) self-sufficiency in growth signals, (2) insensitivity to antigrowth signals, (3) evasion of apoptosis, (4) sustained angiogenesis, (5) tissue invasion and metastasis, and (6) unlimited replicative potential. In a recent revision (Hanahan and Weinberg, 2011), the authors suggested additional hallmarks: (7) evasion of immune surveillance, and five additional hallmarks relating to stress: (8) DNA damage and DNA replication stress, (9) oxidative stress, (10) mitotic stress, (11) proteotoxic stress, and (12) metabolic stress (Wellen and Thompson, 2010). These hallmarks do not have to be acquired in any specific order and some can overlap. At the center of the neoplastic process is, therefore, a balance between proliferation and death and interactions with normal and neoplastic cells and tissues.

Cancer Cell Proliferation   Cancer cells use various mechanisms to enhance growth and circumvent death, and many tumors take advantage of the crosstalk among multiple pathways.

One example relates to "tumor suppressor genes" (TSPs) (see below). The two most important TSPs encode the RB (retinoblastoma-associated) and the TP53 proteins. The RB protein integrates signals from intra- and extracellular sources and determines if the cell should proceed through the cell cycle (Burkhar and Sage, 2008) while TP53 receives inputs from intracellular sensors to assess the degree of cellular stress and DNA damage (Sykes et al., 2006). If abnormalities are detected, the cell cycle is arrested so that the damage can be repaired or, alternatively, apoptotic programs are triggered. Another example is the avoidance of "cell-to-cell contact inhibition." The growth of normal cells is restricted by interaction with neighboring cells to maintain tissue architecture and homeostasis. Cancer cells by-pass this inhibitory control by mechanisms still to be understood. Candidates, however, are the cytoplasmic product of the NF2 gene, which couples cell-surface adhesion molecules to transmembrane receptor tyrosine kinases (Curto, 2007), and the tumor suppressor gene LKB1 epithelial polarity protein, which can overrule the mitogenic effects of the oncogene Myc (Partanen et al., 2009).

Other mechanisms to enhance proliferation are mutations that compromise Ras GTPase activity, responsible for a negative-feedback that normally limits signal transmission, loss-of-function of PTEN phosphatase, which degrades the product of PI3-kinase, phosphatidylinositol (Jiang et al., 2009) trisphosphate (PIP3) amplifying PI3K signaling and promoting oncogenesis, and activation of the mTOR kinase, which can inhibit PI3K signaling and proliferation (Sudarsanam and Johnson, 2010). Yet another mechanism is the corruption of the TGF-β pathway. TGF-β is best known for its antiproliferative effects, which are avoided by cancer cells in ways far more elaborate than the simple shutdown of its signaling circuitry (Ikushima and Miyazono, 2010; Massagué, 2008; Bierie and Moses, 2006). Cancer cells can down-regulate TGF-β receptors (Muñoz-Antonia et al., 2009; Báez et al., 2005; West et al., 2000; Muro-Cacho et al., 1999, 2001; Anderson et al., 1999) or activate what is called the "epithelial-to-mesenchymal transition" (EMT) (see below) that confers on cancer cells traits associated with high grade malignancy. In cancer cells, however, there seems to be an optimal range of pathway activation beyond which sustained proliferation is incompatible with other essential cellular functions, leading to cell senescence or death (Collado and Serrano, 2010).

Cell Death   When cells suffer an exogenous attack, they typically undergo "necrosis." In necrosis, the cytoplasmic membrane of the dying cell ruptures and intracytoplasmic, pro-inflammatory products are released to the environment. This results in the recruitment of histiocytes and inflammatory cells that, in effort to rid the tissues of the cellular debris, can damage adjacent normal tissues in a process that we recognize as inflammation. Given the extraordinary high number of cells in the body and the immense daily turnover, organisms have developed sophisticated mechanisms for the silent disposal of cellular material. An efficient mechanism is "apoptosis" or "programmed cell death," a form of cell suicide triggered by internal sensors. In contrast to necrosis, in apoptosis the DNA becomes fragmented in a controlled manner and the cytoplasmic membrane condenses, preventing the release of inflammatory products to the extracellular milieu. The cells break in what are called "apoptotic bodies" whose surface is recognized by histiocytes, which engulf them without damage to adjacent tissues. Apoptosis is therefore a natural counterpart to mitosis and cell growth (Adams and Cory, 2007) and is mediated by two major circuits, the "extrinsic" Fas ligand/Fas receptor, which receives and processes extracellular death-inducing signals, and the "intrinsic," which senses and integrates intracellular signals. Key roles are played by members of the BCL-2 family (Bcl-xL, Bcl-w, Mcl-1) and the pro-apoptotic proteins Bax and Bak. If apoptosis is to proceed, Bax and Bak will disrupt the integrity of the outer mitochondrial membrane, causing the release of pro-apoptotic signaling proteins such as cytochrome C, which activates an otherwise latent proteolytic cascade (caspase 8 in the extrinsic pathway and caspase 9 on the intrinsic) leading to DNA fragmentation. Bax and Bak share BH3 motifs with the anti-apoptotic Bcl-2-like proteins and other proteins, such as Bim, which also contribute to the regulation of apoptosis (Willis and Adams, 2005). Cancer cells circumvent apoptosis via several mechanisms. One is the loss of TP53 function that eliminates the critical damage sensor from the apoptosis-inducing circuitry.

Another mechanism is the modulation of the length of the telomeres that protect the ends of chromosomes (Park et al., 2009; Kang et al., 2004; Masutomi et al., 2005). Telomeres are composed of multiple tandem hexanucleotide repeats that are progressively shortened in cultured, non-immortalized, cells. In these cells, the ends of chromosomal DNAs form end-to-end fusions that generate unstable dicentric chromosomes incompatible with cell survival. Telomerase, the specialized DNA polymerase that adds telomere repeat segments to the ends of telomeric DNA, is almost absent in non-immortalized cells but expressed at significant levels in spontaneously immortalized cells and cancer cells. By extending telomeric DNA, telomerase eliminates the progressive telomere erosion that would otherwise occur in its absence. This correlates with a resistance to senescence and apoptosis (Blasco, 2005).

Angiogenesis Cancer cells, as normal cells, require nutrients and oxygen for proliferation. During tumor progression, angiogenesis is activated by an "angiogenic switch" that generates new vessels to sustain neoplastic growth (Hanahan and Folkman, 1996). Well-known angiogenesis inducers and inhibitors are vascular endothelial growth factor-A (VEGF-A) and thrombospondin-1 (TSP-1), respectively. Vascular endothelial growth factor signaling can be activated, via three receptor tyrosine kinases (VEGFR-1–3), by hypoxia and oncogene activation (Ferrara, 2008, 2010). Other pro-angiogenic proteins are members of the fibroblast growth factor (FGF) family (Folkman, 2006; Baeriswyl and Christofori, 2009; Ribatti, 2009).

Invasion The invasion–metastasis cascade (Fidler, 2003) is a succession of changes in cells and tissues that allows cancer cells to invade locally gaining access to adjacent blood and lymphatic vessels invading the lymphatic system and peripheral blood. Cancer cells then leave the lumina of these vessels (extravasation) to "colonize" the distant tissues where they form micrometastases that grow into macroscopic tumors. Epithelial cells take advantage of a program called EMT that allows them to acquire the abilities to invade, resist apoptosis, and metastasize (Klymkowsky and Savagner, 2009; Polyak and Weinberg, 2009; Yilmaz et al., 2007; Barrallo-Gimeno and Nieto, 2005). The expression of pleiotropically acting transcription factors (Snail, Slug, Twist, and Zeb1/2), known to regulate migratory processes during embryogenesis (Micalizzi et al., 2010; Schmalhofer et al., 2009), contribute to significant changes such as loss of adherens junction, conversion from a polygonal/epithelial to a spindly/fibroblastic morphology, expression of matrix-degrading enzymes, increased motility, and resistance to apoptosis.

Kinases and growth factors such as epidermal growth factor receptor (EGFR), fibroblast growth factor receptor (FGFR), insulin-like growth factor receptor (IGFR), and MET are likely responsible for reduced intercellular adhesion, in part via loss of E-cadherin (Perl et al., 1998; Thiery, 2002). Loss of E-cadherin function is necessary but not sufficient for EMT. The extracellular matrix provides a scaffold where cells interact with integrins, fibronectin, collagen, and laminin. Also, calcium-dependent guanosine triphosphatases (GTPases), extracellular-matrix signals induce cytoskeletal changes that facilitate cell migration via cytoplasmic extensions called filopodia that coalesce into larger lamellipodia. Various members of the matrix metalloproteinase (MMP) family (MMP1, MMP-2, and MMP-9) are also implicated in invasion (Egeblad and Werb, 2002; Lopez-Otin and Matrisian, 2007; Martin and Matrisian, 2007). Interestingly, tumor-associated macrophages activate myofibroblasts and recruit endothelial progenitor cells via the release of SDF-1 (Orimo et al., 2005). Cancer cells may also regulate the expression of other molecules [ezrin, serine-threonine kinase 11, EREG (EGFR ligand), COX-2] to target colonization in other organs (Khanna et al., 2004; Ji et al., 2007; Minn et al., 2005).

Metastasis Metastases are responsible for 90% of deaths in cancer patients (Fidler, 2003). Metastases often appear in critical anatomical locations where they compromise vital structures. More often, however, metastases are numerous and in multiple organs leading to organ and system failures (Cairns, 1975). It is known, however, that only a small fraction of the cells in the primary tumor have the capacity to metastasize and that, in general, excision of tumors that measure 2 cm or less prevents metastatic disease (Chambers et al., 2002; Koscielny et al., 1984; Kinouchi et al., 1999). Cancer cells appear in the bone marrow in small numbers very early in the disease and can be found in peripheral blood using sensitive techniques (Luzzi et al., 1998). Several hypotheses for the generation of metastases have been proposed. A "linear" model of tumor progression (Nguyen et al., 2009; Fidler, 2003), proposes that cells in the primary tumor undergo successive rounds of mutation and selection giving rise to a biologically heterogeneous cellular population in which a subset of malignant clones have accumulated the necessary genetic alterations for metastasis (Gupta and Massague, 2006; Vogelstein et al., 1988; Foulds, 1954). In contrast, a "parallel" model argues that cancer cells disseminate very early, colonize multiple distant sites, and accumulate genetic changes independently from those incurred by the primary tumor (Hynes, 2003; Klein, 2009). The metastatic competence of these cells is under investigation. Gene expression studies (Sahai, 2007; Perou et al., 2000; Sorlie et al., 2001; van Veer et al., 2002; van de Vijver et al., 2002) seem to favor the linear model. Genomic studies have revealed "metastasis signatures" in the primary tumor that allow the prediction of a poor prognosis, suggesting that metastatic capacity is an inherent property of the primary tumor and that, within the primary tumor, are clones of cells with various metastatic proclivities and expression profiles (Ding et al., 2010). Furthermore, metastasis genes are found in gene-expression signatures of primary tumors (Weil-baecher et al., 2011; Kang et al., 2003; Bos et al., 2009). Thus, breast cancers carrying a lung-metastatic signature tend to metastasize to the lung (Minn et al., 2005).

Tumor Stroma   Cancer cells grow in a mixture of cells and extracellular matrices known as "stroma." Many resident and circulating cells are observed under the microscope in intimate association with the cancer cells. The roles played by these cells in oncogenesis are now better understood. Normal fibroblasts are mesenchymal cells that provide structural support in normal tissues. The term "cancer-associated fibroblast" is used to describe cells with similarities to these fibroblasts and also the "myofibroblast," whose biological roles and properties differ markedly from those of tissue-derived fibroblasts and are identified by the expression of smooth muscle actin (SMA). Myofibroblasts increase in wounds and at sites of chronic inflammation. Although beneficial to tissue repair, myofibroblasts contribute to the pathological fibrosis observed in tissues such as lung, kidney, and liver. Recruited myofibroblasts and reprogrammed variants of normal tissue-derived fibroblastic cells have been shown to enhance cancer cell proliferation, angiogenesis, invasion, and metastasis (Pietras and Ostman, 2010; Kalluri and Zeisberg, 2006; Bhowmick et al., 2004). Because they secrete a variety of extracellular matrix components, cancer-associated fibroblasts are implicated in the formation of the desmoplastic stroma that characterizes many invasive carcinomas. This desmoplasia is one of the factors that allow pathologists to diagnose invasion even at a very early stage. The old concept of desmoplasia being a mechanical attempt of the host to protect against the local expansion of the tumor is obviously simplistic.

Also prominent, among the stromal constituents, are "endothelial" cells that form a tumor-associated vasculature. Several interconnected signaling pathways have been implicated in endothelial cell recruitment and development (Notch, Neuropilin, Robo, Eph-A/B, VEGF, angiopoietin, and FGF) (Ahmed, 2009; Dejana et al., 2009; Carmeliet and Jain, 2000). Closely related to endothelial cells are the cells lining lymphatic vessels (Tammela and Alitalo, 2010) that actively grow at the periphery of tumors in a lymphangio-genic process that provides new channels for metastatic seeding. Another important cell is the "pericyte," a

specialized mesenchymal cell type (related to smooth muscle cells) with finger-like projections that wrap around the endothelial tubing of blood vessels. In normal tissues, pericytes are known to provide paracrine support signals to the normally quiescent endothelium (Gaengel et al., 2009) but they can also collaborate with endothelial cells to synthesize the basement membrane in support of tumor endothelium (Bergers and Song, 2005).

Tumor Microenvironment   Cancers are not uniform masses of identical cells. Tumor heterogeneity can be recognized microscopically. Differences in cellular morphology that allows pathologists not only to provide a diagnosis of cancer but also to estimate the potential clinical behavior of the tumor by recognizing features that can be grouped in histological grades. Sequencing the genomes of cancer cells, microdissected from different sectors of a tumor, has revealed prominent intratumoral genetic heterogeneity. At the cellular and molecular levels, therefore, oncogenic events cannot be fully understood by studying cancer cells in isolation. The habitat where they exist, the "tumor microenvironment," is rich in cellular products, extracellular matrix signals, and complex interactions among cancer cells and non-neoplastic cell types that, together, make tumor growth possible. Incipient neoplasias may begin by recruiting and activating stromal cells that assemble into an initial pre-neoplastic stroma. Cancer cells interact with this stroma reprogramming normal stromal cells to serve the needs of the neoplasm. Ultimately, signals originating in the tumor stroma enable cancer cells to invade normal adjacent tissues and metastasize.

A complete, graphic depiction of the network of microenvironmental signaling interactions is still far beyond our reach, as the great majority of signaling molecules and pathways remain to be identified. Certain transcription factors (SNAIL, TWIST), however, are activated and initiate a stem cell transcriptional program. This results in reduced expression of epithelial genes such as E-cadherin, and increased expression of mesenchymal (vimentin, metalloproteases, N-cadherin) and "stemness" genes (Oct-4, Nanog) (Kessenbrock et al., 2010). The cancer cell microenvironment may also induce abnormal signaling pathway activation. Tumor-invading macrophages and other immune cells as well as activated fibroblasts produce cytokines and growth factor ligands that can activate EMT-inducing signaling pathways (Hynes, 2003). Circulating cancer cells, released from the primary tumor, leave this microenvironment and, upon homing to a distant organ, encounter another naive, normal, tissue microenvironment. The succession of reciprocal cancer cell to stromal cell interactions must now be repeated in this tissue. Some tissue microenvironments, however, appear to be natural "niches" that are already supportive of seeded cancer cells (Coghlin and Murray, 2010). Implicit in this term is the notion that cancer cells seeded in such sites may not need to begin by inducing a supportive stroma because it already preexists, at least in part. Such permissive state may be intrinsic to the tissue site (Talmadge and Fidler, 2010) or pre-induced by circulating factors released by the primary tumor. The best documented components of induced pre-metastatic niches are tumor-promoting inflammatory cells, although other cell types and the extracellular matrix may well prove to play important roles in different metastatic contexts.

Cancer Stem Cells   Some authors have proposed that the tumor microenvironment also serves to maintain a pool of "cancer stem cells" (CSC) in a relatively quiescent but active state (Reya et al., 2001; Gilbertson and Rich, 2007) that may self-renew as well as spawn more differentiated derivatives (Singh and Settleman, 2010). An increasing number of human tumors are reported to contain subpopulations with the properties of CSCs that appear to be more resistant to chemotherapy. Also, certain tumors may transform their own cancer cells into stromal cell types rather than relying on recruited host cells to provide this function. An important stromal cell type is the "cancer-associated fibroblast" that can enhance tumor proliferation, angiogenesis, and invasion and metastasis (Kalluri and Zeisberg, 2006; Bhowmick et al., 2004). Because they secrete a variety of extracellular matrix components, cancer-associated fibroblasts are implicated in the formation of the desmoplastic stroma that characterizes many invasive carcinomas (van de Stolpe, 2013).

Cancer stem cells can be identified in the blood in small numbers. This "circulating tumor cells" (CTCs) can originate in both primary tumors and metastases (van de Stolpe, 2013). Furthermore, contrary to previous beliefs, differentiated cells could revert to a pluripotent state from which, in theory, they could generate other cell types (Takahashi et al., 2007). Also, in epithelial tumors, cancer stem cells could generate non-epithelial cells, depending on the type of signals present in the microenvironment or niche. It is generally accepted that during tumor progression accumulation of genetic defects in cancer cells results in multiple genetically different clones. In fact, pathologists often observe metastases with different morphology than their primary tumors and, occasionally, they come across tumors that defy any attempt of phenotypic classification even with the most sophisticated techniques. This morphological diversity, in many of these cases, could be the result of stem cells undergoing differentiation across different phenotypes. In fact, Ewing sarcoma is now considered a tumor derived from cancer stem cells (Suvà et al., 2009; Meltzer, 2007).

Cancer and the Immune System   Tumors are often surrounded and infiltrated by various densities of immune system cells. Pathologists have suspected that these inflammatory infiltrates sometimes offer protection against the tumor but other times can be facilitators of tumor growth.

It is now clear that infiltrating immune cells can be both tumor antagonizing and tumor promoting (Dvorak, 1986; Grivennikov et al., 2010). Tumor-promoting inflammatory cells are macrophages, mast cells, neutrophils, and T- and B-lymphocytes. A variety of signaling molecules have been implicated in tumor promotion: EGF, VEGF, FGF2, inflammatory chemokines and cytokines, proangiogenic and/or proinvasive matrix-degrading enzymes, including MMP-9 and other matrix metalloproteinases, cysteine cathepsin proteases, and heparanase (Qian and Pollard, 2010). Tumor-infiltrating inflammatory cells have been shown to induce and help sustain tumor angiogenesis, to stimulate cancer cell proliferation, to facilitate, via their presence at the margins of tumors, tissue invasion, and to support the metastatic dissemination and seeding of cancer cells (Coffelt et al., 2010; Egeblad et al., 2010; Qian and Pollard, 2010; Joyce and Pollard, 2009). Similarly, subclasses of B- and T-lymphocytes may facilitate the recruitment, activation, and persistence of such wound-healing and tumor-promoting macrophages and neutrophils (Biswas and Mantovani, 2010) while other subclasses of B- and T-lymphocytes and innate immune cell types can mount demonstrable tumor-killing responses.

HEREDITY AND SUSCEPTIBILITY

Rare families accumulate cases of early onset cancer affecting several generations strongly indicating that inherited factors are important causes of cancer. In hereditary cancer syndromes, one abnormal copy of the gene is inherited in the germ line from either parent, whereas the other copy is inactivated in a somatic cell, typically because of random processes whereby genes, chromosomes, or both are rearranged, deleted, or replaced. As a result, "loss of heterozygosity" is frequent at the position of the TSP gene and in a cancer cell there is biallelic inactivation of the gene (Foulkes, 2008). Inherited biallelic mutations in TSP genes are very rare and often result in a phenotype that differs from the phenotype of a monoallelic mutation. In some genes, such as *BRCA1* and *APC*, biallelic truncating mutations are incompatible with intrauterine development and are therefore lethal. Recently, large-scale consortia examining thousands of cases and controls have identified 30 or more susceptibility loci for lung, prostate, breast, and colorectal cancer.

GENES AND CANCER

The unravelling of the DNA structure, by Watson and Crick, in 1953, changed the way we understood the neoplastic process shifting the emphasis from the cellular (disease of cells) to the molecular level (disease of genes). Cancer growth was then understood as initiated by a mutation and maintained by proliferation factors that generated similar cells or "clones" (Nowell, 2002). In the late 1960s, it was discovered that carcinogens have the ability to bind and damage DNA by forming adducts or causing DNA strand breaks (Poirier, 2004). In the 1970s and 1980s, studies of the role of RNA and DNA viruses in tumor induction led to the discovery of oncogenes and the birth of contemporary cancer biology (zur Hausen, 2001). To date, more than 70 cellular proto-oncogenes have been identified through studies of oncogenic retroviruses and almost all of these genes code for key cell signaling proteins involved in the control of cellular proliferation and apoptosis. In 1971, studies in inherited and sporadic forms of retinoblastoma, showed that infants with the inherited disease had a 30,000-fold higher chance of developing Rb than did children with the sporadic disease, providing the first evidence for the existence of cellular genes, now known as TSP, which in contrast to proto-oncogenes function to prevent rather than to promote cancer. In 1976, Bishop and Varmus proved that oncogenes have a cellular origin and that carcinogenic events activate cellular genes to promote cancer (Stehelin et al., 1976). The *src* oncogene and the process of reverse transcription were discovered and *src* was proven to be a cellular gene acquired by RSV from the chicken genome. In 1979, Lane and Levine showed that products of DNA tumor viruses function through physical interactions with cellular proteins, specifically p53. In 1982 and 1983, it was shown that the *ras* proto-oncogene was present in normal cells, and that *ras* oncogenes, in human bladder carcinoma cell lines, mouse sarcoma viruses, and rat mammary carcinomas, contained a mutation that was crucial for inducing cellular transformation. In 1989, Vogelstein reported a common loss-of-heterozygosity at the p53 locus in human colorectal cancers (Vogelstein and Kinzler, 2004), suggesting that p53 was actually a TSP gene rather than an oncogene.

Oncogenes Oncogenes encode proteins that control cell proliferation and apoptosis (Table 107.2). These proteins can be transcription factors, chromatin remodelers, growth factors, growth factor receptors, signal transducers, or apoptosis regulators (Santarosa and Ashworth, 2004). They can be activated by mutation or gene fusion, juxtaposition to enhancer elements, or amplification. Translocations and mutations are typically initiating events, in the early stages of neoplasia, whereas amplification usually occurs during tumor progression (Croce, 2008). Examples of translocations are Burkitt's lymphoma, where the *MYC* oncogene (chromosome 8q24) becomes activated as a result of its juxtaposition to one of the loci for immunoglobulin genes (chromosomes 14q, 22q, and 2p), and chronic myelogenous leukemia (CML), where a reciprocal t(9;22) chromosomal translocation that fuses the *ABL* proto-oncogene to the *BCR* gene creating a chimeric protein with enhanced tyrosine kinase activity (Dalla Favera et al., 1982; Groffen et al., 1984; Shtivelman et al., 1985). Since every malignant cell carries these translocations, deregulation of these genes are likely early events in the formation of the tumor. When CML converts to acute leukemia, it acquires an additional t(9;22) translocation, an isochromosome 17, or a trisomy of chromosome 8, and when follicular lymphoma progresses to a more aggressive clinical type, it often carries a t(8;14) translocation in addition to the original t(14;18) translocation (Croce, 2008; Konopka et al., 1985). In general, mesenchymal neoplasms (hematopoietic tumors and soft-tissue sarcomas) are initiated by the activation of an oncogene, followed

TABLE 107.2 List of Common and Important Genes Involved in Oncogenesis

Gene Function	Gene
Oncogene	BAX, BCL2L1, CASP8, CDK4, ELK1, ETS1, HGF, JAK2, JUNB, JUND, KIT, KITLG, MCL1, MET, MOS, MYB, NFKBIA, NRAS, PIK3CA, PML, PRKCA, RAF1, RARA, REL, ROS1, RUNX1, SRC, STAT3, ZHX2
Tumor suppressor gene (TSP)	ATM, BRCA1, BRCA2, CDH1, CDKN2B, CDKN3, E2F1, FHIT, FOXD3, HIC1, IGF2R, MEN1, MGMT, MLH1, NF1, NF2, RASSF1, RUNX3, S100A4, SERPINB5, SMAD4, STK11, TP73, TSC1, VHL, WT1, WWOX, XRCC1
Oncogene and TSP	BCR, EGF, ERBB2, ESR1, FOS, HRAS, JUN, KRAS, MDM2, MYC, MYCN, NFKB1, PIK3C2A, RB1, RET, SH3PXD2A, TGFB1, TNF, TP53
Transcription factor	ABL1, BRCA1, BRCA2, CDKN2A, CTNNB1, E2F1, ELK1, ESR1, ETS1, FOS, FOXD3, HIC1, JUN, JUNB, JUND, MDM2, MEN1, MYB, MYC, MYCN, NF1, NFKB1, PML, RARA, RB1, REL, RUNX1, RUNX3, SMAD4, STAT3, TGFB1, TNF, TP53, TP73, TSC1, VHL, WT1, ZHX2
Cell cycle	ATM, BRCA1, BRCA2, CCND1, CDK4, CDKN1A, CDKN2A, CDKN2B, CDKN3, E2F1, HGF, MEN1, STK11, TP53
Angiogenesis	AKT1, CTNNB1, EGF, ERBB2, NF1, PML, RUNX1, TGFB1
Apoptosis	BAX, BCL2, BCL2L1, BRCA1, CASP8, E2F1, MCL1, MGMT, TNF, VHL
Cell adhesion	APC, CDH1, CDKN2A, CTNNB1, KITLG, NF1, NF2, TGFB1
Epithelial–mesenchymal transition	BRCA2, CDKN2B, CTNNB1, ERBB2, HGF, JAK2, KIT, MCL1, NF1, RUNX3, S100A4, SMAD4, TGFB1, VHL

by alterations in tumor-suppressor genes and other onco-genes. In contrast, most carcinomas are initiated by the loss of function of a tumor-suppressor gene, followed by alter-ations in oncogenes and additional tumor-suppressor genes (Huebner and Croce, 2001; Croce, 2008, 8). Mutations in oncogenes (chromosomal translocations, gene amplifica-tions, or intragenic mutations) result in their activation and the mutation needs to occur in only one allele. One example of activating mutation is *BRAF*, where valine changes to glutamate at codon 599, a residue within the activation loop of the kinase domain (Croce, 2008), normally regulated by phosphorylation at adjacent residues (Thr598 and Ser601). The change to glutamate constitutively activates BRAF kinase that phosphorylates downstream targets such as extracellular signal-regulated kinase (ERK), leading to aber-rant growth (Davies et al., 2002; Wan, 2004).

Tumor Suppressor Genes (TSPs) Tumor suppressor genes can be divided into classes based on the primary function of the proteins they encode. Anti-oncogenes, such as cyclin-dependent kinase inhibitor 2A (*CDKN2A*) and retinoblastoma 1 (*RB1*), which encode p16INK4A and RB1, respectively, antagonize the growth-promoting activities of oncogenes, such as *CDK4* and *CCND1*, which encode CDK4 and cyclin D1, respectively. DNA damage checkpoint genes, such as ataxia telangiectasia mutated (*ATM*) and *TP53*, which encode ATM and p53, respectively, induce cell death or senescence in response to DNA damage or DNA replication stress. Caretaker genes (DNA repair genes and mitotic checkpoint genes, such as *MLH1*, breast cancer susceptibility 1 (*BRCA1*), *MYH* (*MUTYH*), and xeroderma pigmentosum group A (*XPA*)) encode proteins that help to maintain genomic stability. Muta-tions in these genes impair DNA repair, mitotic recombination, or chromosomal segregation.

MicroRNA GENES MicroRNA genes regulate gene expression but do not encode proteins. They are single RNA strands of 21–23 nucleotides that can bind a com-plementary sequence on a messenger RNA (mRNA) caus-ing its degradation or blocking the translation of a protein (Croce, 2008). They are often located in chromosomal regions that undergo rearrangements, deletions, and ampli-fications. MicroRNA genes can be up-regulated or down-regulated in cancer cells (Calin et al., 2005; Yanaihara et al., 2006), and their function is tissue specific. The up-regulated genes function as oncogenes by down-regulating TSP genes, whereas the down-regulated genes function as TSP genes by down-regulating oncogenes (Croce, 2008). Up-regulation of microRNA genes can be due to amplifi-cation, deregulation of a transcription factor, or demethylation of CpG islands in the promoter regions of the gene. Down regulation can be accomplished by dele-tions, epigenetic silencing, or loss of the expression of one or more transcription factors.

GENETIC ABNORMALITIES

Like all cells in the human body, a cancer cell is a direct descendant, through a lineage of mitotic cell divisions, of the fertilized egg from which the patient developed and, there-fore, carries a copy of its diploid genome. To undergo malignant trasnformation, a normal cell has to endure heri-table changes involving multiple, independent genes, a con-cept that explains the long latency period for cancer development (Barrett and Wiseman, 1987). The cancer cell genome, therefore, acquires a set of somatic mutations (substitutions of one base by another, insertions or deletions of small or large segments of DNA), rearrangements, gene amplification with increases in copy number beyond the two gene copies in the normal diploid genome, or decreases in copy number that may result in complete absence of a DNA sequence. The cancer cell may have also acquired, from exogenous sources, completely new DNA sequences, nota-bly those of viruses such as HPV, Epstein–Barr virus, hepatitis B virus, human T lymphotropic virus 1, and human herpes virus 8, each of which is known to contribute to the genesis of one or more type of cancer. In addition, the cancer genome acquires epigenetic changes that alter chromatin structure and gene expression (see Epigenetics below).

Mutations

The size of the full repertoire of human cancer genes is a matter of speculation. Some cancer genomes carry 100,000 point mutations whereas others have fewer than 1000. Studies in mice, however, suggest that more than 2000 genes, when mutated, may have the potential to contribute to carcinogenesis (Futreal et al., 2004). Since cancer is the result of multiple mutations or genetic abnormalities, mutated genes contribute, rather than cause, cancer. Mutations can occur in the germ line (hereditary predisposition to cancer) or in somatic cells (sporadic tumors) (Knudson, 2002). So far, 291 cancer genes have been reported, more than 1% of the 22,000 protein-coding genes in the human genome. Ninety percent of cancer genes show somatic mutations in cancer, 20% show germline mutations, and 10% show both (Futreal et al., 2004). Approximately 90% of the known somatically mutated cancer genes are "dominant," that is, a mutation of just one allele is sufficient to contribute to cancer develop-ment, typically by activation of the encoded protein. The remaining 10% act in a "recessive" manner, requiring muta-tions in both alleles that, typically, result in abrogation of protein function. Recessive cancer genes are characterized by diverse mutation types, ranging from single-base substitu-tions to whole gene deletions. In each dominantly acting cancer gene, however, the repertoire of cancer-causing somatic mutations is usually more constrained, both with respect to the type of mutation and its location in the gene. The most common mutation class among the known cancer genes is a chromosomal translocation that creates a chimeric gene or apposes a gene to the regulatory elements. Several genes with small insertions and deletions of coding microsatellites/mononucleotide repeats that lead to protein truncation in MMR-deficient tumors have been proposed as cancer genes (Duval and Hamelin, 2002). More than 70% of cancer genes with somatic mutations in the census are associated with leukemias, lymphomas, and sarcomas, even though these cancers account for <10% of human cancer cases. Tyrosine kinases are more common than serine/threonine kinases (27 cancer genes encode protein-kinase domains, compared with the 6.3 expected in a random selection of the human genome), accounting for approximately one quarter of all known protein kinases and 2/3rd of the protein kinases encoded by cancer genes. Interestingly, phosphatases are not prominent in the cancer gene. A substantial proportion of germ line-mutated genes that cause cancer predisposition are involved in DNA maintenance and repair.

Some mutations are causally implicated in oncogenesis and have been named "driver mutations." Driver mutations confer growth advantage to the cancer cell and, although, may not be required for maintenance of the cancer phenotype, they have been positively selected. Some studies indicate that there may be as many as 20 driver mutations in individual cancers, considerably more than the five to seven previously predicted. "Passenger mutations," on the other hand, have not been selected, do not confer clonal growth advantage, and have not contributed to the development of the cancer. Passenger mutations are found within cancer genomes because somatic mutations without functional consequences often occur during cell division. Thus, when a cell acquires a driver mutation it already has biologically inert somatic mutations within its genome that will be perpetuated via clonal expansion. Driver mutations cluster in cancer genes, whereas passenger mutations are more or less randomly distributed. Building on the success of previous multinational, collaborative initiatives such as the Human Genome Project and the HapMap consortium, the International Cancer Genome Consortium (ICGC) (http://www .icgc.org/home) will attempt to characterize somatically acquired genetic events in at least 50 classes of cancer, including those with the highest global incidence and mortality.

Loeb proposed the existence of a "mutator phenotype" to explain the fact that the mutation rate in nonmalignant cells is not sufficient to generate the large numbers of mutations observed in cancer (Loeb et al., 1974; Loeb and Loeb, 2000; Loeb and Monnat, 2008; Beckman and Loeb, 2006). It has been estimated that a cell undergoes >20,000 DNA damaging events and >10,000 replication errors per cell per day (Preston, 2005). When a solid tumor is clinically detected, it has 10^8–10^9 cells and a volume of 1 cm^3, and it has accumulated tens of thousands of mutations (Greenman et al., 2007; Fox et al., 2009). More than 1000 different genes have been

reported to be mutated in human tumors, although no set of mutated genes is diagnostic of a specific tumor. Many genes contain multiple mutations and many tumors contain as many as 100 mutations. Some genes, such as *TP53* and *KRAS*, are frequently mutated in diverse types of cancer whereas others are rare and/or restricted to one cancer type. Recent studies have shown that four genes are altered in more than 20% of the tumors analyzed. The *TP53* tumor-suppressor and DNA damage checkpoint gene was among the most frequently mutated genes. The remaining genes encode either classical oncoproteins, such as the EGFR and the small GTPase RAS, or TSP proteins, such as the cyclin-dependent kinase 4 inhibitor p16INK4A (encoded by cyclin-dependent kinase inhibitor 2A (*CDKN2A*)), the phosphatase and tensin homologue deleted on chromosome 10 (PTeN) and neurofibromatosis type 1 (NF1). Certain gene families such as protein kinases are frequent cancer genes. Furthermore, cancer genes cluster on certain signaling pathways. For example, in the classical MAPK/ERK pathway upstream mutations are found in cell membrane-bound receptor tyrosine kinases such as EGFR, ERBB2, FGFR1, FGFR2, FGFR3, PDGFRA, and PDGFRB and also in the downstream cytoplasmic components NF1, PTPN11, HRAS, KRAS, NRAS, and BRAF. Furthermore, the length of "microsatellites," repetitive nucleotide sequences that are hot spots for mutagenesis, have been found highly mutated in some tumors and have been used to establish cellular heterogeneity and as prognostic indicators of progression from a premalignant condition (Thibodeau et al., 1993; Boland et al., 1998; Risinger et al., 1993).

Chromosomal Abnormalities

Clonal chromosome aberrations (chromosomal rearrangements, chromosomal imbalances, and genomic losses) have been found in all major tumor types (http://cgap.nci .nih.gov/) and can be caused by environmental and occupational exposures and also by cytotoxic therapy (Fröhling and Döhner 2008).

Chromosomal Rearrangements Recurrent chromosomal rearrangements can form a "chimeric fusion gene" by fusion of parts of two genes (typically genes that encode tyrosine kinases or transcription factors), or alter the expression of a normal gene. The best example of a tyrosine kinase gene is the Philadelphia chromosome (Nowell and Hungerford, 1960), present in virtually all patients with CML, in 20% of patients with acute lymphoblastic leukemia (ALL), and in rare cases of acute myeloid leukemia (AML). The Philadelphia chromosome is the result of the reciprocal translocation t (9;22)(q34.1;q11.23) (Rowley, 1973) in which sequences of the *BCR* gene (22q11.23) are fused to fragments of the gene that encodes the ABL1 tyrosine kinase (9q34.1). The fusion results in the production of the chimeric protein,

BCR–ABL1, that has the catalytic domain of ABL1 fused to a BCR domain that promotes aberrant tyrosine kinase activity independent of normal regulatory controls (Fröhling and Döhner, 2008). Another example is the inversion [inv(2) (p22-p21p23)], present in 7% of Japanese patients with non-small cell carcinoma, which results in a fusion gene with fragments of *EML4* and the gene encoding the ALK receptor tyrosine kinase. One important example of a transcription factor is Ewing sarcoma (EWS), where translocations t (11;22)(q24.1-q24.3;q12.2) and t(21;22)(q22.3;q12.2) fuse the *EWSR1* gene (22q12.2) to a gene that encodes a member of the ETS family of transcription factors. In 85% of the cases, the partner is *FLI1* (11q24.1-q24.3) and in 10% of the cases *ERG* (21q22.3) (Delattre et al., 1992; Sorensen et al., 1994). In the EWSR1 portion of the fusion protein resides a potent transactivation domain that induces the transcription of various genes (Owen and Lessnick, 2006; Riggi and Stamenkovic, 2007). An example of deregulation of a normal gene is Burkitt lymphoma. When regulatory sequences, such as promoters or enhancers, are joined to the coding region of a proto-oncogene, its expression is altered. In Burkitt lymphoma, the enhancer of an immunoglobulin gene [*IGHG1* (14q32.33), *IGKC* (2p12), *IGLC1* (22q11.2)] drives the constitutive expression of the gene encoding the MYC transcription factor (8q24.21) (Küppers, 2005).

Chromosomal Imbalances Gains or losses of genetic material can involve large chromosomal segments or small intragenic duplications or deletions. *Large-scale genomic gains* typically result from non-disjunctions, unbalanced translocations, or amplifications. Analysis of the RNAs corresponding to these regions has uncovered new breast cancer genes and potential new oncogenes in melanoma [*MITF* (3p14.2-p14.1) and *NEDD9* (6p25-p24)] (Garraway et al., 2005; Kim et al., 2006) and hepatocellular carcinoma [*YAP1* (11q13) and *BIRC2* (11q22)] (60-62). Identification of *focal genomic gains* involving small regions or single genes often requires sensitive techniques (Dutt and Beroukhim, 2007). In breast cancer, array-based CGH and SNP genotyping have revealed the amplification of a small segment of 6q25.1, the location of the gene that encodes estrogen receptor 1 (*ESR1*) (Holst et al., 2007; Albertson, 2008). Overexpression of this gene correlates with increased sensitivity to tamoxifen. Also, in a subgroup of patients with non-small-cell lung carcinoma, a 480-kb segment on 14q13 contains the *NKX2-1* oncogene coding for a transcription factor (Weir, 2007), and point mutations have been reported in the catalytic domain of the EGFR receptor tyrosine kinase, implicated the response to the kinase inhibitors (Sharma et al., 2007). Genomic losses range from complete or partial chromosomal monosomies, to single gene or intragenic deletions detectable only by high resolution techniques. *Large-scale genomic losses* involving several genes are frequent in cancer (Knudson, 2001). Examples are *RB1*

(13q14.2), *TP53* (17p13.1), *APC* (5q21-q22), *NF1* (17q11.2), *PTEN* (10q23.3), and *ATM* (11q22-q23). For many recurrent genomic losses, such as 1p deletions in neuroblastoma (Okawa et al., 2008), 3p deletions in lung cancer (Zabarovsky et al., 2002), and 7q deletions in myeloid cancers (Curtiss et al., 2005; Döhner et al., 1998), the critical genes are unknown. RNA interference screening in combination with high resolution DNA copy-number analysis identified the frequent deletion in colon cancer of the *REST* gene (4q12a) (Westbrook et al., 2005). *Genomic losses that result in allelic insufficiency* are more difficult to study (Fodde and Smits, 2002) but RNA interference has identified *RPS14* as a causal gene of the 5q minus syndrome (Boultwood et al., 2002), a myelodysplastic syndrome with a deletion in 5q (List et al., 2006). *Genomic losses that affect non-coding genes*, such as microRNA genes (90, 91), have also been identified. The loss of specific microRNAs with tumor-suppressive activity may contribute to oncogenesis. This is the case of the deletion, in 50% of patients with CML, of *MIRN15A* and *MIRN16-1* (13q14.3) that negatively regulate BCL2 expression (Cimmino et al., 2005).

Genomic Instability

Genomic instability can occur in a variety of forms. "chromosomal instability" (CIN) is the rate at which changes in chromosome number and structure occur over time when compared to normal cells. In "microsatellite instability" (MSI or MIN), there are changes in the number of the oligonucleotide repeats in microsatellite sequences. Base pair mutations are increased in several hereditary, cancer-prone, syndromes such as hereditary non-polyposis colon cancer (HNPCC or Lynch syndrome) where mutations in DNA mismatch repair genes lead to microsatellite instability (MSI) (Lynch, 2012), or hereditary MYH-associated polyposis. Germline mutations in *MYH* (*MUTYH*), a DNA base excision repair (BER) gene, leads to increased GC to TA transversion frequencies. Mutations in genes that repair DNA double-strand breaks or DNA inter-strand cross links, such as breast cancer susceptibility 1 (*BRCA1*) (Wood et al., 2007), Werner syndrome helicase (*WRN*), Bloom syndrome helicase (*BLM*), and the Fanconi anaemia genes, have also been linked to breast cancer, leukemias, and lymphomas. Finally, germline mutations in nucleotide excision DNA repair genes predispose to skin cancer.

Interdependency of Signaling Pathways

Many signaling pathways are known to play a role in oncogenesis. Many of these pathways interact with each other carrying signals from different growth factors and often becoming constitutively activated and independent of normal regulatory feedbacks. In some cases, and due to the central role played at critical crossroads in the cellular machinery, some proteins are excellent targets for therapy. One example is the "signal transducer and activator of transcription" (STAT) proteins (Darnell, 1997; Yu et al., 2009). Signal transducer and activator of transcription are latent cytoplasmic transcription factors but their constitutive activation, notably Stat3 and Stat5, is often detected in various types of tumor cells and cells transformed by oncoproteins that activate tyrosine kinase signaling pathways (Yu et al., 2009). Signal transducer and activator of transcriptions are activated by tyrosine phosphorylation following the binding of cytokines or growth factors to cognate receptors on the cell surface. Tyrosine kinases that mediate STAT activation include growth factor receptors and cytoplasmic tyrosine kinases, particularly Janus kinase (JAK) and Src kinase families (Garcia et al., 2001). Once tyrosine is phosphorylated, two STAT monomers form dimers through reciprocal phosphotyrosine–SH2 interactions, translocate to the nucleus, and bind to STAT-specific DNA-response elements of target genes to induce gene transcription. To date, there are seven STAT family members identified in mammals (Stat1, Stat2, Stat3, Stat4, Stat5a, Stat5b, and Stat6). Signal transducer and activator of transcriptions with diverse biological functions include differentiation, proliferation, development, apoptosis, and inflammation. Aberrant Stat3, for instance, promotes uncontrolled growth and survival through dysregulation of gene expression, including cyclin D1, c-Myc, Bcl-xL, Mcl-1, and survivin genes, and thereby contributes to oncogenesis (Shor et al., 2007; Diaz et al., 2006; Gritsko, 2006). Moreover, recent studies reveal that persistently active Stat3 induces tumor angiogenesis by upregulation of vascular endothelial growth factor induction and immune evasion (Mora et al., 2002).

Epigenetics

Classic genetics alone cannot explain the diversity of phenotypes within a population nor the different susceptibilities to a disease in identical twins. In 1939, Waddington introduced the field of epigenetics describing "the causal interactions between genes and their products, which bring the phenotype into being" (Waddington, 1942). Epigenetics will later be defined as "heritable changes in gene expression that are not due to any alteration in the DNA sequence" (Jones and Baylin, 2007). Epigenetic modifications regulate chromatin accessibility and compactness and determine what genetic information is made available to the cellular machinery (Sharma et al., 2010) and allow the cancer cell to adapt to changes in its microenvironment. The combinatorial patterns of these modifications are collectively called "epigenome." The cancer epigenome contains global changes in DNA methylation and histone modification patterns as well as altered expression profiles of chromatin-modifying enzymes that are heritable and can promote oncogenesis (Bernstein et al., 2007; Suzuki et al., 2008; Kouzarides, 2007; Jiang and

Lui, 2009; Bird, 2002). The most important epigenetic mechanisms are DNA methylation, covalent histone modification, incorporation of histone variants and nucleosome remodeling, and non-coding RNAs, including microRNAs (miRNAs),

1. *DNA Methylation.* The low level of DNA methylation in tumors compared with normal tissue counterparts was one of the first epigenetic alterations to be found in human cancer. The loss of methylation is mainly due to hypomethylation of repetitive DNA sequences and demethylation of coding regions and introns, the regions that allow alternative versions of messenger RNA (mRNA) (Deng et al., 2008). During the development of a neoplasm, the degree of hypomethylation of genomic DNA increases as the lesion progresses from a benign proliferation of cells to an invasive cancer. The initial finding of global hypomethylation of DNA was soon followed by the identification of hypermethylated TSP genes (Suzuki et al., 2008) and the more recent discovery of the inactivation of microRNA (miRNA) genes by DNA methylation (Bird, 2002; Prendergast et al., 1991). In normal cells, DNA methylation plays a critical role in the control of gene activity and the architecture of the nucleus. In humans, DNA methylation occurs by covalent modification of cytosine residues that precede guanines within the CpG dinucleotides, which are concentrated in "CpG islands" short CpG-rich DNA stretches at the 5′ end of genes, and large repetitive sequences (centromeric repeats, retro-transposon elements, rDNA etc.) (Takai et al., 2002; Wang, 2004). Recent studies show that 100–400 hypermethylated CpG islands occur in the promoter regions of a given tumor. These islands are usually not methylated in normal cells, although methylation of certain subgroups of promoter CpG islands can be detected in normal tissues. Methylation occurs via the cooperative activity of the de novo methyltransferases DNMT3A and DNMT3B and the maintenance DNA methyltransferase DNMT1 (Okano et al., 1999). CpG islands are found in about 60% of human gene promoters but DNA methylation is also important in the regulation of non-CpG island promoters.

    Genomic imprinting also requires DNA hypermethylation at one of the two parental alleles of a gene to ensure monoallelic and a similar gene-dosage reduction is involved during X-chromosome inactivation in females. Cells that lack the stabilizing effect of DNA methylation because they have spontaneous defects in DNA methyl transferases (DNMTs) or experimentally disrupted DNMTs have prominent nuclear abnormalities. DNA methylation is the basis for epigenetic phenomena such as imprinting, X-chromosome inactivation or formation of heterochromatin (Barski et al., 2007; Wang et al., 2008). In 70–80% of CpG dinucleotides, in mammalian cells, there is a modification at the C5 position of the cytosine base (5mC) (Feinberg et al., 1983; Riggs et al., 1983). In 60% of gene promoters CpG islands are free of methylation regardless of their activite state. In cancer cells, however, promoter CpG islands tend to become hypermethylated leading to silencing of the corresponding gene (Takai et al., 2002; Wang, 2004). Recent reports have identified genomic regions, with distinct DNA methylation patterns that regulate gene expression profiles and chromatin compartmentalization. Although DNA methylation is a very stable phenomenon, reprogramming of DNA methylation can occur via oxidation by the TET family of proteins to 5-hydroxymethylcytosine (5hmC) or via deamination to thymine by AID/APOBEC1 followed by base excision repair (BER) (Riggs et al., 1983).

2. *Histone Modifications.* The coiling of DNA around nucleosome particles is the basis for the organization of eukaryotic genomes. A multitude of different post-translational modifications of the core histone proteins (H2A, H2B, H3, and H4) allows for demarcation of specific chromatin regions and states (Jenuwein, 2006). Histone modifications can be dynamically added or removed and associated with both active and repressed regions of chromatin. More than a dozen reported histone modifications can modify more than 150 conserved residues within histone proteins (Jones and Baylin, 2007). This number of different modifications has a high combinatorial potential, which would yield a hugely complex histone code and it is under debate, whether such a code exists or whether histone modifications are a consequence and mere reflection of dynamic processes altering DNA accessibility such as transcription factor or RNA polymerase II (RNAPII) binding or chromatin remodeling (Wang et al., 2008; Schones et al., 2008). Certain histone modifications such as acetylation or phosphorylation are thought to change chromatin structure by altering the net positive charge of the histone proteins, thereby rendering the underlying DNA sequence information more accessible. Alternatively, histone modifications can be recognized by specific protein domains that in turn might enforce or stabilize the chromatin signature and provide a platform for the recruitment of additional factors. Chromatin regulators, histone modifiers, and histone modification binding proteins, are present at distinct genomic loci and frequently bring together regulators associated with opposing activities. Aside from gene regulatory functions, which occur in a relatively local and confined chromatin region, histone modifications can also span large regions, as exemplified by X-chromosome inactivation in female mammal. Large chromatin blocks associated with gene silencing have been identified on mammalian autosomes (Hassler and Egger, 2012). Further, exchange of canonical histones by variants is connected to transcriptional activity as well as chromatin structure. DNA accessibility can be affected by the structure of nucleosomes and their interaction with DNA by exchanging canonical histones with histone variants or by histone modifications, respectively.

Additionally, the position and the density of nucleosomes on the DNA string can determine the level of accessibility. Active genes have characteristic nucleosome depleted regions (NDRs) flanked by positioned nucleosomes upstream of their transcription start sites, which contain binding sequences for transcription factors. Repressed genes usually lack an NDR, but DNA sequence, binding of transcription factors and the action of chromatin remodeling complexes have been suggested to act in a multistep process to determine local nucleosome composition and density (Baylin and Ohm, 2006). DNA methylation occurs in the context of chemical modifications of histone proteins (Sharma et al., 2010). Histones are not merely DNA-packaging proteins, but molecular structures that participate in the regulation of gene expression. They store epigenetic information through post-translational modifications such as lysine acetylation, arginine and lysine methylation, and serine phosphorylation. These modifications affect gene transcription and DNA repair. It has been proposed that distinct histone modifications form a "histone code" (Jenuwein, 2006). The N-terminal tail of histones can undergo post-translational covalent modifications to help regulate transcription, replication, and repair. It has been proposed that these modifications are stored in a "histone code," a form of cellular epigenetic memory that contributes to the structure and activity of different chromatin regions. Non-histone proteins would gain access to the chromatin and would decode this message. Numerous histone modifications have been identified. They are regulated by histone acetyltransferases (HATs) and histone methyltransferases (HMTs), which add acetyl and methyl groups, respectively, and histone deacetylases (HDACs) and histone demethylases (HDMs), which remove acetyl and methyl groups, respectively (Okano et al., 1999; Feinberg et al., 2006). Furthermore, histone modifications and DNA methylation interact with each other to determine the gene expression status, chromatin organization, and cellular identity. Acetylation of histone lysines, for example, is generally associated with transcriptional activation. The functional consequences of the methylation of histones depend on the type of residue (lysine or arginine) and the specific site of methylation. Methylation of H3 at K4 is closely linked to transcriptional activation (Riggs et al., 1983; Eden et al., 2003; Wilson et al., 2007) whereas methylation of H3 at K9 or K27 and of H4 at K20 is associated with transcriptional repression (Cedar and Bergman, 2009). Hypermethylation can silence genes such as Rb, p16, and BRCA. In cancer, there is also loss of histone acetylation, which is mediated by HDACs and results in gene repression. Histone deacetylases are often found to be overexpressed in various types of cancer. Hypermethylation of the CpG-island promoter can affect genes involved in the cell cycle, DNA repair, the metabolism of carcinogens, cell-to-cell interaction, apoptosis, and angiogenesis, all of which are involved in the development of cancer. Hypermethylation occurs at different stages in the development of cancer and in different cellular networks, and it interacts with genetic lesions. Such interactions can be seen when hypermethylation inactivates the CpG island of the promoter of the DNA-repair genes *hMLH1, BRCA1, MGMT* (O6-methylguanine–DNA methyltransferase), and the gene associated with Werner's syndrome (*WRN*) (26,51-53.). Each tumor type can be assigned a specific, defining DNA "hypermethylome." Such patterns of epigenetic inactivation occur not only in sporadic tumors but also in inherited cancer syndromes in which hypermethylation can be the second lesion in Knudson's two-hit model of how cancer develops (Hassler and Egger, 2012).

3. *Deregulation of non-coding RNAs.* In recent years, it has become clear that non-coding RNAs are important modulators of chromatin regulation and gene expression. Whole genome and transcriptome sequencing has revealed that at least 90% of the genome is actively transcribed, although <2% represent protein-coding genes. Thus, the non-coding part of the transcriptome has become a new focus in the study of gene expression and regulation (Baylin and Ohm, 2006). Currently, two major groups of non-coding RNAs are recognized: small ncRNAs and long ncRNAs (lncRNAs) (Lu et al., 2005; Chan et al., 2005; Deng et al., 2008).

Small ncRNAs Small nc RNAs are processed from longer precursors and comprise, in addition to transfer RNAs (tRNAs) and ribosomal RNAs (rRNAs), the microRNAs (miRNAs), piwi interacting RNAs (piRNAs), small nuclear RNAs (snoRNAs), and other less well-characterized RNAs (He et al., 2004; Saito et al., 2006; Fabbri et al., 2007; Friedman et al., 2009). MicroRNAs consist of a single RNA strand of 21–23 nucleotides that can bind a complementary sequence on a messenger RNA (mRNA) causing its degradation or blocking the translation of a protein. They are expressed in a tissue-specific manner and participate in the control of cell proliferation, apoptosis, and differentiation. They are often located in chromosomal regions that undergo rearrangements, deletions, and amplifications. As is the case for normal genes, expression of miRNAs can be regulated by epigenetic mechanisms (Jones and Baylin, 2007) but miRNAs can also modulate epigenetic regulatory mechanisms by targeting enzymes responsible for DNA methylation and histone modification (Jones and Baylin, 2002), and they can function as either TSP genes or oncogenes depending on the target genes. Thus, oncogenic miRNAs, target growth inhibitory pathways and are often upregulated in cancer. An example is miR-21, which targets PTEN in glioblastoma. DNA hypermethylation in the miRNA 5′ regulatory region can lead to down-regulation of miRNA (Barski et al., 2007). This reveals a tumor-suppressor function for miRNAs as in the examples of downregulated *let-7* and *miR-15/miR-16*, which target the *RAS* and *BCL2* oncogenes, respectively. Methylation silencing of *miR-124a* causes activation of the cyclin D–kinase 6 oncogene (*CDK6*), a common epigenetic abnormality in cancer.

Long ncRNAs (lncRNAs) Long ncRNAs are a heterogeneous class of mRNA-like transcripts from 200 nt up to 100 kb, which do not code for proteins that, excluding ribosomal and mitochondrial RNA, appear to make up the bulk of the human transcriptome. They can be transcribed from sense or antisense strand, in a bidirectional way and fulfill important regulatory roles in gene expression and regulation by assembling protein complexes and localizing them to their genomic target DNA sequence. Another class of non-coding RNAs is the transcribed ultraconserved regions (T-UCRs) that are aberrant in several cancers, including adult chronic lymphocytic leukemia, colorectal and hepatocellular carcinomas, and neuroblastomas. Although distinct T-UCR expression signatures are associated with specific cancer types and their high sequence conservation across species argues for functional properties, their precise mode of action in the cell is still unknown.

The epigenetic code is not permanently stable. In addition to spontaneous reactions, affecting chromatin and DNA methylation, genome maintenance itself involves extensive alterations in the components of chromatin. For instance, repair of double-strand breaks involves extensive phosphorylation of histone H2AX followed by modification of histones by ubiquitination and sumoylation (Jenuwein et al., 2001; Wang et al., 2008; Haberland et al., 2009; Shi, 2007) all of which are required for efficient double-strand break repair. Also, repair of single-strand breaks involves the formation of large chains of poly(adenosine diphosphate [ADP]–ribose) on target proteins by the enzyme poly(ADP–ribose) polymerase (PARP). The chains of poly (ADP–ribose) probably serve as a platform to recruit the appropriate mechanism of repair. It is likely that the degeneration of the epigenetic code that facilitates oncogenesis originates in part from both damage and the subsequent genome maintenance. Little is known about the maintenance machinery of the epigenome and its contribution to cancer (Bernstein et al., 2007). In addition to initiating carcinogenesis, genome instability also drives the progression from benign to malignant tumors by permitting additional genetic and epigenetic changes that facilitate evolution to a more aggressive state.

Cell Cycle

The cell cycle has four sequential phases. During the S phase, DNA replication occurs, and in the M phase the cell divides into two daughter cells. Separating S and M phase are two gap phases named G1 and G2 (Williams, 2012). During G1, following mitosis, the cell is sensitive to positive and negative network signals. G2 is the gap after S phase when the cell prepares for entry into mitosis (Murray, 1992). G0 is a state when cells have reversibly withdrawn from cell division (Zetterberg and Larsson, 1985). Most cells reside in these "out-of-cycle" states while some cells are actively proliferating, mainly in the stem-transit amplifying compartments of self-renewing tissues, such as epithelia and bone marrow (Potten and Loeffler, 1990). Progression through the cell cycle is driven by the cyclin-dependent kinase (CDK) family of serine/threonine kinases and their regulatory partners the cyclins (Malumbres and Barbacid, 2006). Cyclin D-CDK4, cyclin D-CDK6, and cyclin E-CDK2 drives G1 progression through the restriction point, which commits the cell to complete the cycle (Planas-Silva and Weinberg, 1997). S phase is initiated by cyclin A-CDK2, and cyclin B-CDK1 regulates progression through G2 and entry into mitosis (Nigg, 2001). Progression through each cell cycle phase and transition from one phase to the next are monitored by sensor mechanisms, called checkpoints, which maintain the correct order of events (Hartwell and Weinert, 1989). If the sensor mechanisms detect aberrant or incomplete cell cycle events (DNA damage), checkpoint pathways carry the signal to effectors that can trigger cell cycle arrest (Bartek et al., 2004; Musacchio and Salmon, 2007). Effector proteins include CDK inhibitors (CDKIs). For example, G1 arrest can be induced by the Ink4 family [INK4A (p16), INK4B (p15), INK4C (p18), and INK4D (p19)] of CDKIs, which inhibit CDK4 and CDK6, or, alternatively, via the Cip/Kip family of inhibitors (p21, p27, p57), which suppress CDK2 activity (Malumbres and Barbacid, 2009; Hanahan and Weinberg, 2000).

Deregulation of the cell cycle engine is at the center of the uncontrolled cell proliferation that characterizes the malignant phenotype. Mitogens release the brakes of cell cycle progression by stimulating G1–S CDK activities, which trigger the phosphorylation of pRb proteins, leading to the disruption of their interaction with the E2F family of transcription factors. In cancer cells, the pRb brakes are often defective, resulting in E2F-dependent G1–S gene expression even in the absence of mitogens (Harbour and Dean, 2000). Several cancer genes directly control transitions from the resting stage (G0 or G1) to the replicating phase (S) of the cell cycle. The products of these genes include cdk4 (a kinase), cyclin D1 (activates cdk4), Rb (transcription factor), and p16 (inhibits cdk4) (Bartek et al., 2004; Musacchio and Salmon, 2007; Malumbres and Barbacid, 2009). The genes encoding Rb and p16 are TSPs, inactivated by mutation, whereas those encoding cdk4 and cyclin D1 are oncogenes, activated by mutation (Barbacid et al., 2005). Many mutations are mutually exclusive allowing the cell to acquire a growth advantage when only one mutation is present. Importantly, since there are less signaling pathways than genes, there is a plethora of interconnections that lead to the same result. One example is the TP53 pathway. The p53 protein is a transcription factor that, normally, inhibits cell growth and stimulates cell death in response to cellular stress. Typically, a point mutation prevents p53 from binding its cognate recognition sequence but the p53 pathway can also be altered by *MDM2* gene and products of DNA tumor viruses (i.e., E6 protein of HPV). Adding to this complexity is the fact that the same mutation may have different effects depending on the cell type. For instance, *KRAS2* gene mutations in normal pancreatic duct cells appear to initiate the neoplastic process while the same mutation in colonic or ovarian epithelial cells produces self-limiting, hyperplastic or borderline lesions that do not progress to carcinoma. Also, in some tumors, *RAS* genes function as oncogenes while, in others, *RAS* genes can function as TSP.

CHEMICAL CARCINOGENESIS

The important role played by chemical products in carcinogenesis is reflected in the fact that up to 8% of all human cancers are of occupational origin (Fröhling and Döhner, 2008). Thus, the risk of lung cancer for men exposed to specific carcinogens in the workplace is 9.2% while for bladder cancer, the population risk is one in five men and 1 in 10 women (Fonger, 1996). So far, approximately six million chemicals have been identified and registered with the chemical abstracts service. Of those, more than 50,000 are regularly used in commerce and industry, although <1000 have been properly scrutinized as to their carcinogenic potential (Elespuru, 1996). Today, the list of known or potential carcinogens, published in the 12th Edition of Reports Carcinogens by the National Toxicology Program, Public Health Service, U.S. Department of Health and Human Services, includes more than 200 substances. In some cases, cancer may develop only after prolonged exposure (i.e., tobacco) while in others, cancer may develop even after a brief exposure (i.e., asbestos). Section 301(b)(4) of the Public Health Service Act provides that the Secretary of the

TABLE 107.3   Classification of Carcinogens (IARC Monograhs on the evaluatin of Carcinogenic Risk in Humans. International Agency for Research on Cancer – http://monographs.iarc.fr/ENG/Classification/)

Carcinogen Group	Description	Number of Carcinogens
Group 1	Carcinogenic to humans	109
Group 2A	Probably carcinogenic to humans	65
Group 2B	Possibly carcinogenic to humans	275
Group 3	Not classifiable	505
Group 4	Probably not carcinogenic to humans	1

Department of Health and Human Services shall publish a biennial report with the list of substances known to be either "carcinogens" (sufficient evidence of carcinogenicity from human studies indicating causal relationship between exposure and cancer) or "reasonably anticipated to be carcinogens" (limited evidence of causality in human studies, sufficient evidence of carcinogenicity from animal studies with increased incidence of malignant and/or malignant and benign tumors, less than sufficient evidence of carcinogenicity in humans or animals, but the substance belongs to a structurally related class of substances known to be carcinogenic, anticipated to be a carcinogen, or convincing information that the agent acts through mechanisms known to cause cancer in humans). Also, the IARC classifies the evidence for carcinogenicity of specific exposures into four categories (Table 107.3).

Pott first recognized, in 1775, the relationship between exposure to environmental substances and neoplasia. Pott described cancerous changes in the scrotal skin of London chimney sweeps as a result of repeated contamination with soot. Daily baths were recommended in 1890, a high incidence of bladder cancer in chemical and rubber industry workers was observed across Europe. By the end of the nineteenth century it had become evident that occupational exposure to certain chemicals or mixtures of chemicals had carcinogenic effects. The first experiment on chemical carcinogenesis was carried out, in 1915, by the pathologist Yamagiwa. He rubbed rabbit ears with coal tar and observed the development of papillomas and carcinomas. In the 1930s, high molecular weight polycyclic aromatic hydrocarbons (PAHs) were proven to induce skin cancer in mice and, in fact, the active component in coal tar pitch turned out to be a PAH. Beremblum and Shubik using PAHs and croton oil showed that cutaneous cancer, in mice, developed in several stages (Berenblum and Shubik, 1947) (see below).

Prior to Watson and Crick, chemical carcinogens were believed to produce cancer by interaction with proteins. By the end of the 1960s, however, a correlation between the DNA binding capacity of a particular carcinogen and its biological potency was established. Electrophilic substances, with affinity for negative charges, could bind proteins and DNA and induce cancer directly while inert, neutrophilic chemicals, on the other hand, acted as pro-carcinogens since they required enzymatic conversion. Although this metabolic conversion was known since the 1940s, it was not until the 1960s that the cytochrome P450 was discovered (Luch, 2005). Cytochrome-P450-dependent monooxygenases (CYPs) were shown to be associated with an NADPH-dependent reductase (Luch, 2005) and, today, 57 genes encoding these enzymes, are recognized. These enzymes appear to be somewhat tissue specific and operate with high stereo-selectivity forming metabolites that differ in their biological activity. Further metabolism occurs on target cells where carcinogens or their metabolic products either directly, by genotoxic mechanisms (formation of DNA adducts, chromosome breakage, fusion, deletion, mis-segregation, and non-disjunction), or indirectly, by non-genotoxic mechanisms (induction of inflammation, immunosuppression, formation of reactive oxygen species, receptor activation, and epigenetic silencing). These changes affect gene expression, cell-cycle control, DNA repair, cell differentiation, and apoptosis, resulting in hypermutability, genomic instability, loss of proliferation control, and resistance to apoptosis, all recognized features of cancer cells.

Tests that measure DNA affinity have been developed to characterize the degree of genotoxicity of carcinogens and to identify their DNA-reactive metabolites. These studies have confirmed the relationship between DNA-binding activity and both mutagenicity and carcinogenicity, also establishing that the formation of DNA adducts is a relevant bio-indicator of cancer risk. Baseline levels of carcinogen DNA adducts in normal human tissues are estimated to be in the range of 1/107–0.2/108 nucleotides. Certain mutations are common both in human cancers and in animal models of chemical carcinogenesis. An example is the codon 12 of *KRAS* that is frequently mutated in human cancers, including lung adenocarcinoma, and is also a preferential binding site for DNA-damaging metabolites of a variety of carcinogens.

Phases of Carcinogenesis

In humans, there is a 20–50-year lag from exposure to a carcinogen to the clinical detection of a tumor. Analysis of cancers at different stages suggests a sequential order of mutations and genome rearrangements, although some cancers have only a limited of relevant genetic changes. The initial targets of carcinogenesis research were "PAHs". Chrysene was identified as a carcinogenic compound in tar, and benzopyrene was later shown to be tar's most potent carcinogenic agent. Benzopyrene then became the most intensely studied chemical carcinogen because it was believed that its chemical structure was related to that of steroidal hormones, and steroidal hormones were, at the time, the only endogenous compounds reported to induce tumors in mice and humans (Carrell et al., 1997).

This structure-function concept was supported by the synthesis of 3-methylcholanthrene, a potent carcinogen that could possibly be produced *in vivo* from cholesterol-derived bile acids. Later it was shown that compounds with different chemical structures could induce tumors, and analogies began to be seen between chemical carcinogens and radiation. This raised the possibility that carcinogens acted as radiomimetic compounds by reversible binding to purines and DNA bases or through reactive metabolites, stimulating an intense research effort intended to clarify the mechanistic aspects of cancer induction. It also became clear that age and individual differences were at the center of the susceptibility to cancer, and these differences, presumably of a genetic origin, could therefore be inherited (Melnick, 1996). Furthermore, it was observed that a long latent period could lapse from exposure to carcinogens to the development of cancer. In 1941 Rous and Kidd began to paint the skin of rabbits with tar and found that if this painting was interrupted, tumors would disappear only to reappear if the application of tar was reestablished. Therefore it seemed that a reversible process was taking place in those cells that did not attain the complete neoplastic state. These cells had undergone what Friedewald and Rous called "initiation." Since then, the process of chemical carcinogenesis has been understood as the result of three distinct successive stages: initiation, promotion, and progression.

Initiation Initiation is the induction of an irreversibly altered cell through a mutational event that may differ for different tissues or for different initiators in the same tissue (Barbacid, 1986). It is induced by the genetic changes that predispose susceptible normal cells to malignant change and immortality. Cellular division remains symmetrical by creating two initiated cells protected from apoptosis. Initiation can begin with errors in replication or with spontaneous mutations, supported by normal occurrences such as DNA depurination and deamination. Spontaneous initiation, less common than induced initiation, has been confirmed in laboratory animals (Pitot and Dragan, 1991). Initiated cells can remain latent for months or years, or can grow in an autonomous and clonal fashion. DNA damage is at the center of this process. Proliferating cells have less time to repair the damaged DNA and remove DNA adducts. DNA damage in stem cells is particularly relevant because of their long lifespan and wide tissue distribution.

Promotion Promotion is the process by which the initiated cell clonally expands into a detectable cell mass that is either benign or preneoplastic. At this stage, different chemicals may have different carcinogenic consequences, depending on the effect of cell proliferation in the initiated cell. The concept of promotion was introduced when it was discovered that chemical substances with low carcinogenic activity induced cancer under experimental conditions (Berenblum and Shubik, 1947). Promoter compounds do not interact directly with DNA and can have biological effects without being metabolically activated (Yuspa et al., 1983). They may also indirectly damage the DNA by oxidation. At first, it was believed that promoters acted via epigenetic mechanisms but, today, we know that promotion also involves genetic changes (Hanahan and Weinberg, 2000). The results are increased cellular proliferation, alterations in genetic expression, and changes in the control of cellular growth (Barrett and Anderson, 1993). In order to be effective, promoters must be present for weeks, months, or years. Promotion is reversible and can be modulated by physiological factors (Yuspa et al., 1983). Prolonged exposure or high doses, however, will induce neoplasia without initiation (Pitot and Dragan, 1991).

Progression Cells must undergo additional changes in their progression to a malignant neoplasm (Barrett and Wiseman, 1987). Progression phase, characterized by rapid, irreversible growth, genetic instability and cellular changes in metabolism and morphology, is the acquisition of the neoplastic phenotype by genetic and epigenetic processes with cell proliferation occurring without exogenous stimuli. (Oliveira et al., 2007). Some tumor promoters, however, although effective in producing multiple pre-neoplastic foci, are not particularly effective in influencing the progression of these lesions to malignant neoplasms. This suggests that promotion and progression were independent phenomena. Angiogenesis, essential in the neoplastic process, precedes the development of characteristics of malignancy, and its inhibition has been shown to delay the neoplastic process. (Oliveira, 2007).

DNA Damage

Contrary to initial assumptions that DNA, given its role in genetic information transfer, is extraordinarily stable, it has been shown that DNA is a dynamic macromolecule that is subjected to continuous changes and reacts easily with a variety of chemical and physical compounds present in the environment. DNA is made up of an array of nucleophilic centers with which DNA damaging agents, typically electrophilic, form adducts through one or more covalent bonds to the base sugar of phosphate. Ultraviolet radiation, for instance, causes 2+2 cyclo-addition reactions at the 5,6 double bonds of the adjacent pyrimidines. Ionizing radiation, on the other hand, may subtract hydrogen atoms, generating free radical species, or cause an excision of an O−H bond of water to form the highly reactive hydroxyl radical •OH. DNA is also the target of reactive oxygen species, produced intracellularly, as products of mitochondrial respiration. Tumor promoters, such as 12-O-tetradecanoylphorbol-13-acetate, are known to increase the amount of intracellular oxidative damage. Furthermore, enzymatic oxidation of exogenous and endogenous toxins by the mixed function oxygenase systems, cytochrome P-450, and flavin monooxygenase, can activate xenobiotics such as polycyclic hydrocarbons, mycotoxins, and alkylating agents (Rissle, 1997). These compounds, in turn, produce electrophiles with capabilities for DNA damage (Szeliga, 1997).

DNA damage stimulates replication-dependent phenomena such as sister chromatid exchanges, chromosomal aberrations, and gene amplification. Replication of these DNA lesions is an essential factor in the generation of genetic defects and proliferating cells are more susceptible to neoplastic transformation than non-proliferating cells, following genotoxic treatment. Mutation rates increase during the S-phase of cells pre-exposed to DNA-damaging agents (Wang et al., 1993). The cumulative damage to the DNA is significant. In a single day, each normal cell can undergo more than 10,000 DNA damaging events, 10^4 single-strand breaks and spontaneous base losses, and 10^5 UV photo-products in each sun-exposed keratinocyte (De Bont and van Larebeke, 2004). The consequences are more than 900 phosphorylation events involving more than 700 proteins. The importance of DNA damage in the development of cancer is highlighted by the observation that an elevated incidence of cancer is detected in situations where DNA repair mechanisms are deficient. Several cancer-prone syndromes seem to be related to an inherent inability to recover after DNA damage. It has also been shown that when DNA repair fails, mutations accumulate.

DNA Damage Recognition and Repair

Chemical carcinogens can attack and change the structure of DNA by alkylation, oxidation, dimerization, deamination, and the formation of aromatic adducts. The carcinogen–DNA adducts that are formed have varying mutagenic abilities. Human cells use various DNA repair mechanisms to counteract the DNA damage caused both by endogenous and environmental chemicals: (1) "BER" removes products of alkylation and oxidation, (2 "nucleotide excision repair" (NER) excises oligonucleotide segments containing larger adducts, (3) "mismatch repair" scans the DNA for

misincorporation by DNA polymerases, (4) "oxidative demethylation," (5) transcription-coupled repair (TCR), repairs lesions that block transcription, (6) "double-strand break repair and recombination" avoids errors by copying the opposite DNA strand, and (7) not yet characterized mechanisms that repair cross-links between strands. These mechanisms affect signal-transduction and effector systems connected with replication, transcription, recombination, chromatin remodeling, and differentiation, and ultimately determine survival, senescence, or death. Most DNA lesions are repaired by more than one mechanism and, as a result, only few DNA lesions escape repair. Those that do, at the time of DNA replication, can direct the incorporation of non-complementary nucleotides resulting in mutation.

Repair pathways often focus on a specific category of DNA lesion (Friedberg, 2003; Friedberg et al., 2006). Thus, BER mechanisms remove a damaged base that is refilled by DNA synthesis (Slupphaug et al., 2003; Caldecott, 2008; Barnes and Lindahl, 2004). Nucleotide-excision repair eliminates helix-distorting DNA lesions. Transcription-coupled repair, strongly linked with NER and possibly with BER, targets lesions that impair transcription. Non-homologous end joining and homologous recombination repair various types of double-strand breaks. In addition, some repair proteins are used only once. For instance, O-6-methylguanine-DNA methyltransferase repairs a single O-6-methylguanine lesion by transferring the methyl from a guanine in DNA to a cysteine in the enzyme, thereby inactivating itself (Verbeek et al., 2008).

Base Excision Repair The first step in the excision repair process is the recognition of an abnormality. This recognition process must be able to detect the relatively small amount of damaged DNA that is found within the remaining normal genome. In addition, it should not interfere with DNA replication, recombination, or transcription. Following recognition, the damaged portion of the DNA is removed by the action of a glycosylase. DNA glycosylases are monomeric enzymes that catalyze the hydrolysis of the N-glycosyl bond of damaged, modified, or mismatched bases in the DNA, resulting in their detachment from the sugar–phosphate backbone (Barnes et al., 1993). The resultant abasic sugar, now without an attached purine or pyrimidine, is subjected to endonucleolytic cleavage and excision from the DNA. A DNA polymerase then synthesizes a repair patch containing the appropriate nucleotide sequence. The patch is manufactured in the 5′ to 3′ direction using the complementary, undamaged strand as a template. DNA ligase then connects the 3′ terminus of the patch to the parental strand. This type of repair, resulting in the removal of only one or a few nucleotides, is termed *BER* (Barnes, Lindahl, Sedgwick, 1993). Base excision repair is used for the repair of small alkyl or aromatic adducts. However, larger adducts require the removal of more than just one or a few bases in the vicinity of adduct. For these adducts, NER is used (Hoeijmakers, 1993).

Nucleotide Excision Repair As many as 200 bases, including the adduct, can be removed by the NER, a process similar to BER except that the multiprotein complex that directs the endonucleolytic attack on the damaged DNA excises an entire oligonucleotide containing the lesion in question (Gillet and Scharer, 2006). The process involves lesion recognition, incision of the damaged strand on both sides of the lesion at some distance, removal of the damage-containing oligomer, and gap-filling DNA synthesis. A specialized nucleotide excision pathway exists to take care of the preferential elimination of all types of injury in the transcribed strand of active genes. This process, called *transcription coupled repair,* corrects damage in structural genes transcribed by RNA polymerase II. It prevents the blockage of transcription and replication by injuries to the template. A second type of NER, designated *global genome repair,* removes lesions in the remainder of the genome preventing the long-term effects of DNA injury (Hanawalt, 2002; Hanawalt and Mellon, 1993). The NER process is mediated in *Escherichia coli* by the products of the urvA, B, and C genes. This process is presently the only system in which the precise mode of action is entirely known. In humans, the genes whose products affect this repair are mutated in the various complementation groups of the hereditary disease, xeroderma pigmentosum (Hoeijmakers, 2001, 2009). Patients with xeroderma pigmentosum have acute photosensitivity, a 2000-fold increase in the incidence of skin cancers in sun-exposed skin and, in some cases, accelerated neurodegeneration. Patients with this disorder display an inherited inability to repair DNA damage induced by ultraviolet light. The ultraviolet light causes the formation of pyrimidine dimers, which these patients are unable to excise. Xeroderma pigmentosum is genetically heterogeneous. Cell hybridization studies have led to the identification of seven excision-deficient complementation groups, designated XP-A to XP-G. Mutated genes in these groups code for the proteins that are responsible for the NER process in humans.

Mismatch Repair DNA mismatches occur when conventional, but noncomplementary, Watson–Crick bases lie opposite each other in the DNA helix. Mismatch repair systems exist to remove replication errors and reduce the mutation rate. To correct these replication errors, the repair system must be able to distinguish between parental and newly synthesized strands. In *E. coli* this process is accomplished by adenine methylation at GATC sequences (Shrivastav et al., 2008). Although methylation of these GATC sequences is complete in nonreplicating DNA, during replication methylation lags behind the replication machinery.

Thus newly synthesized strands are transiently under-methylated. The *E. coli* mismatch repair system recognizes only unmethylated DNA strands. Consequently this methyl-directed mismatch repair system operates only on newly synthesized strands in the region immediately behind the replication fork. Mismatch repair in *E. coil* is a multistep process that involves several proteins. First the MutS protein recognizes and binds to the mismatched DNA. This MutS binding is independent of methylation. Next the MutL protein recognizes and binds to the MutS–DNA complex. The MutS–DNA–MutL complex then activates the MutH protein. MutH is an unmethylated GATC-specific endonuclease, which nicks an unmethylated GATC sequence adjacent to the mismatch (Barnes and Lindahl, 2004). Helicase then unwinds the DNA, and a specific exonuclease digests the mismatch. The gap created is then filled by DNA polymerase using the opposite strand as a template. DNA ligase then seals the repaired strand.

This repair process can recognize and repair single base pair mismatches as well as small additions or deletions. However, not all mismatches are repaired equally. Transition mispairs, guanine to thymidine (G-T) or adenine to cytosine (A-C), involve purine to pyrimidine binding. Transversion mispairs (AA, G-G, G-A, T-T, C-C, and C-T) involve purine to purine or pyrimidine to pyrimidine binding. A successful repair leads to normal stacking and intrahelical configuration. Thus transition mispairs are repaired more efficiently by the mismatch process than are transversion mispairs. Furthermore, these types of mismatches are most likely to be missed by the proofreading action of DNA polymerase. Consequently, the mismatch repair process is able to repair most efficiently those errors that are most commonly made by DNA polymerase (Huen and Chen, 2008). Components of the *E. coli* mismatch repair system appear to have been conserved throughout evolution. Multiple proteins have been found in humans that are homologous to the *E. coli* MutS and MutL proteins (Kanaar et al., 2008). Indeed, with the exception of strand discrimination, the mechanism of eukaryotic mismatch repair parallels that of *E. coli*. The only exception appears to be that no MutH homologs have been found in humans. Instead of recognizing adenine methylation of GATC sequences, strand discrimination in humans occurs as a result of discontinuities in newly synthesized strands. Furthermore, in the *E. coli* repair process, GATC sequences and MutH function are not needed if the DNA strands are nicked. Thus the mismatch repair process is virtually identical in both *E. coli* and humans because, even in *E. coli,* mismatch repair can be directed by strand discontinuities. In addition to its role in repairing damaged DNA, the mismatch repair process serves to prevent chromosomal recombination between nonidentical sequences during meiosis. Indeed inhibition of mismatch repair has been shown to allow recombination not only between nonidentical sequences but also in interspecies recombination. Consequently mismatch repair controls both the fidelity of DNA replication, as well as the maintenance of a genetic barrier between species. Because MutS and MutL homologs have been shown to be involved in the prevention of replication errors and chromosomal rearrangements in humans, mutations in the genes that code for these proteins can be expected to lead to disease states in humans. In fact, mutations in these MutL and MutS genes have been shown to be associated with a high incidence of colon cancer in a condition known as "hereditary nonpolyposis colon cancer." Chemical carcinogens that damage these mismatch repair genes, therefore, may lead to a predisposition to cancer.

As stated previously, DNA repair enzymes can modify carcinogenic damage by the removal of DNA adducts. DNA repair rates can be determined by measuring unscheduled DNA synthesis and the removal of DNA adducts using a single-cell autoradiographic technique. DNA damage in some domains of the genome is processed much more efficiently than in others. As previously discussed, patients with xeroderma pigmentosum have markedly decreased excision repair rates and have increased rates of skin cancer after exposure to ultraviolet light. A reduced capacity of mononuclear cells to repair the aromatic amine adducts have also been found in individuals who have first-degree relatives with cancer. Finally a decrease in DNA repair rates has also been seen in fibroblasts of lung cancer patients compared with normal controls. Furthermore, differences in the repair response to damage may also account for some of the significant differences in carcinogenic response seen when different organisms, such as humans and rodents, are compared.

Cell Cycle Arrest

Abrogation of cell checkpoints, with agents such as methylxanthine analogs or pentoxifylline, increases the cytotoxicity of DNA-damaging agents. Replication of a damaged template would certainly result in irreversible chromosomal aberrations and a high mutation rate. Two particular checkpoints are thought to be particularly important following DNA damage: the G1 checkpoint located between the G1 and S phases (preceding DNA replication), and the G2 checkpoint, located between the G2 and M phases (preceding chromosome segregation). A complex involved in the G2 checkpoint is CyclinB–p34(Cdk2). CyclinB levels begin to accumulate in late S-phase. At the G2–M transition, Cdk25C dephosphorylates p34(Cdk2), leading to the activation of CyclinB–p34(Cdk2) kinase activity and entry in mitosis. CyclinB is then degraded near the end of mitosis. Treatment of cells with a variety of DNA-damaging agents prevents dephosphorylation of p34(Cdk2) and, therefore, leads to the activation of CyclinB–p34(Cdk2) kinase activity

(Lock, 1992; O'Connor et al., 1992; Tsao et al., 1992). Nitrogen mustard, on the other hand, prevents activation by phosphorylation of Cdk25C, which is required for the dephosphorylation and activation of p34(Cdk2) (O'Connor et al., 1992; Kohn et al., 1994). Mutations that cause loss of the G2 checkpoint increase the cytotoxicity of DNA-damaging agents. DNA damage following such mutations generally leads to cell death instead of carcinogenesis. Loss of the G1 checkpoint, on the other hand, triggers genomic instability at the time of interaction of unrepaired DNA with the DNA replication machinery, leading to deletion-type mutations and aberrant gene amplification (Slichenmeyer et al., 1993). The inactivation of the G1 checkpoint serves as an initiation step, which makes the cell susceptible to unregulated growth (initiation), one of the hallmarks of cancer, increasing the probability of subsequent genetic alterations and establishing the fully developed neoplastic phenotype. Control at the G1 checkpoint is dependent on CyclinD1 (degraded at the G1/S transition) and CyclinE (degraded in mid-S phase). Overexpression of either CyclinD1 or CyclinE and activation of the CyclinD1–Cdk and CyclinE–Cdk complexes results in entry into the S-phase and decreased G1 time (Resnitzky et al., 1994). These complexes act to phosphorylate pRb, the product of the retinoblastoma gene, a tumor-suppressor gene (Cance et al., 1990). Unphosphorylated pRb maintains cells in G1, whereas phosphorylation inactivates pRb and allows exit from G1. Another tumor-suppressor gene, p53, is necessary for G1 phase arrest after DNA damage (Kuerbitz et al., 1992). Lack of p53 permits damage-resistant DNA synthesis and increases the incidence of selected types of mutations (Smith et al., 1996). This has been shown after a variety of DNA-damage mechanisms such as ionizing radiation (strand breaks), alkylation by methylmethane sulfonate (MMS), UV radiation (photodimers), and a variety of environmental carcinogens (Hollstein et al., 1991). After DNA damage, p53 binds a consensus binding site (Kern et al., 1991; El-Deiry et al., 1992) and activates the transcription of several "downstream" genes (Perry and Levine, 1993). One of these genes codes for the p21 protein that binds and inactivates cyclin/Cdk complexes and prevents the progression of the cell cycle beyond Gl. Another gene is the "cyclin-inhibiting protein" gene CIP1 (WAF1 or SDI1). The p53 protein also activates the BAX gene, involved in the regulation of apoptosis (Hollander et al., 1993; Zhan et al., 1993). Apoptosis is a cell suicide mechanism that leads to "programmed cell death." Apoptotic cells undergo cell shrinkage and chromosomal condensation in response to DNA damage (Goldsworthy, 1996). These changes prevent the replication of cells that have sustained a degree of genetic damage beyond repair. P53 also activates the "growth arrest and DNA damage inducible gene" GADD45 (Hollander et al., 1993; Zhan et al., 1993). The GADD family of genes is composed of five members whose transcripts are increased following various types of DNA damage (Fornace, 1992), leading to transient growth arrest often in G1 phase. Alkylating agents such as MMS, some chemotherapeutic drugs, oxidizing agents, and UV radiation induce three of the members: GADD34, GADD45, GADD153 (Fornace, 1992). GADD33 and GADD7 transcripts are expressed in a more limited range of cell types. Transcript levels of the GADD45 gene are present in very low amounts in most untreated cells. Thus in human liver, GADD45 composes <0.01% of the total cellular mRNA (Hollander et al., 1993). Following DNA damage, however, GADD45 transcript levels begin to increase only 1 h after exposure, reaching maximal production after several hours post-treatment (Fornace, 1992).

Mutations in the p53 tumor-suppressor gene occur at high frequency in a variety of human tumors (Livingstone et al., 1992). These include deletions of the entire gene with complete loss of p53 expression and missense point mutations, giving rise to mutant p53 proteins lacking wild-type p53 activity (Hollstein et al., 1991). The central portion of the p53 protein binds DNA in a sequence-specific manner, inhibiting cell proliferation at the G1 transition. The majority of mutations are located in this essential region, and tumor-derived mutant forms of p53 have reduced specific DNA-binding activity. Furthermore, the wild-type p53 protein forms stable homo-oligomers and seems to bind DNA as a tetramer. Mutant p53 interferes with the function of wild-type p53 by forming heterodimers and forcing the latter into an abnormal conformation that is unable to bind DNA. The p53 protein can also repress the transcription of a variety of promoters, including c-fos, c-jun, c-myc, IL-6, and Rb. The p53 protein regulates its own function through the activation of the MDM2 gene. The MDM2 gene codes for a protein that binds and inhibits p53 (Oliner et al., 1993). This inhibition by MDM2 acts to limit the duration of p53-mediated G1 arrest and allows the cell to progress through the remainder of the cell cycle. Many human tumor types are caused by mutation or deletion of p53 and loss of this G1 checkpoint (Livingstone et al., 1992). It has been proposed that the wild-type p53 protein functions as a sensor of DNA strand breaks and that after binding to ssDNA, it may undergo a conformational change that increases its stability and favors its accumulation, probably through poorly understood phosphorylation mechanisms, resulting in G1 arrest or apoptosis. The G2 checkpoint, on the other hand, seems to function with independence of the presence of p53. Mammalian genes and gene products known to be activated at the transcriptional or post-transcriptional level in response to DNA-damaging agents include not only those related specifically to DNA-repair processes but also transcription factors, growth factors, growth factor receptors, protective enzymes, and proteins associated with inflammation and tissue injury and repair (Fornace, 1992; Herrlich and Rahmsdorf, 1994). These findings are consistent with the current understanding of the molecular basis of carcinogenesis as a multistep process, including genetic and epigenetic phenomena. Therefore it is essential to further advance our understanding of the intricate molecular mechanisms that govern chemical carcinogenesis. This will allow us to improve strategies for assessing human cancer risk.

CHEMICAL AGENTS THAT DAMAGE DNA

Chemical damage to DNA contributory to carcinogenesis can occur through both endogenous and exogenous processes. Exogenous processes arise from nutritional habits, socioeconomic strata, and lifestyle that include exposures to physical agents (e.g., ionizing radiation), chemical agents, and biological agents (e.g., *Heliobactor pylori*, Epstein-Barr virus, human papilloma virus, *Schistosoma haemotobium*), while endogenous factors include genetic makeup, age, immune system abnormalities, endocrine imbalances, and chronic inflammatory process such as ulcerative colitis or pancreatitis. (Oliveira, 2007). Endogenous modifications in the DNA include methylation by S-adenosylmethione, modification by lipid peroxidation products, chlorination, glycosylation, oxidation, and nitrosylation (Grosse et al., 2009). Reactive oxygen and nitrogen species are particularly relevant because the activated species are generated by host cells and more than 50,000 nucleotides/cell are replaced every day.

In order to form covalent adducts with DNA and with other cellular macromolecules, many carcinogens require activation by hepatic P-450 aryl hydroxylases into reactive electrophiles, which can be mutagens and carcinogens. For instance, on DNA replication, 7,8-dihydro-8-oxoguanine can pair equally well with its normal partner or with adenine, leading to a GC→TA transversion. Other lesions, such as ionizing radiation, which induces double-strand brakes, are mainly cytotoxic or cytostatic. Finally, others can have both effects. DNA damage can also interfere with normal transcription or induce replication arrest, triggering cell death or cellular senescence.

Most human carcinogens have been found to be active in genetic toxicology tests (Bartsch and Malaveille, 1990) while others can induce chromosomal changes without gene mutations or directly damaging the DNA (Barrett and Wiseman, 1987). Other compounds may act through epigenetic mechanisms, and some may act through a combination of different mechanisms (MacLeod, 1996). Because many chemicals act by a combination of direct and indirect mechanisms, and those mechanisms may differ in different cell types, classifications according to mechanism of action cannot be comprehensive or definitive. Nevertheless, chemical carcinogens have been classified in two main groups: "genotoxic" (DNA reactive) for compounds that undergo chemical reaction with DNA and "non-genotoxic" (epigenetic).

Genotoxic Carcinogens

Nucleic acids are chemically unstable complex molecules that can be altered by numerous environmental agents and by intracellular metabolism. The consequences for the living organism of this genotoxicity include short-term genetic instability, programmed cell death or apoptosis, inheritance of mutations, and cancer. Eukaryotic chromosomes in mammalian cells are organized in multiple adjacent replicons (estimated 20,000 in the human genome) with an average size of 150 kb (Kornberg and Baker, 1992). Each replicon contains a replication origin that replicates sequentially in a timely order in S-phase and completes replication within 10 h. At any given time, approximately 2000 replicons must be replicated at an average rate of 2.5 kb/min. DNA damage can affect replicon initiation by inhibiting the number of replicons being initiated or by increasing the frequency of initiation through de novo activation of normally silent replication origins. Alternatively, DNA damage may cause a block in replicon elongation. All creatures, from viruses to humans, have developed mechanisms to either repair or tolerate DNA lesions. Several hours after treatment with genotoxic agents, mammalian cells start to recover from damage-induced inhibition of DNA replication, resuming synthesis of normal size DNA at a normal rate. Furthermore, replication of mammalian genomes seems to occur in the presence of permanent lesions such as chemical adducts. Thus, psoralen mono-adducts, induced in the gene of dihydrofolate reductase by 4′ OHmethyl-4,5′,8-trimethylpsoralen, persist at high efficiency, a phenomenon that has also been observed in other models (Wauthier et al., 1990; Vos and Hanawalt, 1999).

Replication of damaged chromosomes is accomplished through a variety of redundant mechanisms. In general, DNA lesions on the leading strand block the synthesis of both strands. DNA damage on the lagging strand, however, interrupts synthesis of the damaged fragment, although fragments downstream from the lesion can still be synthesized. The mechanisms proposed for replication past-DNA lesions are direct translesion synthesis or bypass replication (expected to be error prone), post-replication repair (expected to be error free), and homologous strand exchange between sister DNA molecules. Direct translesional synthesis requires that the DNA replication complex directly copy the damaged template by incorporating a nucleotide opposite to the lesion, perhaps by a specific DNA polymerase. This nucleotide may not necessarily be the complementary base to the altered nucleotide and, as a result, mutations may occur. Post-replication repair involves the filling of 1–2 kb gaps by homologous recombination using complete undamaged daughter DNA molecules. This avoids the direct copying of the lesion. Mammalian cells exposed to DNA-damaging agents undergo a temporary inhibition of replicon initiation, perhaps because of the preferential loss of small newly

synthesized DNA molecules through DNA strand breaks, either induced directly by the DNA-damaging agent or indirectly by excision repair-dependent strand incision. This delay in replicon initiation allows time for excision repair, decreasing the chances of genetic instability and mutagenic fixation. Specific regulatory mechanisms, such as post-translational inactivation of cofactors, operate to undo the temporary shut-down of replicon initiation (Waga et al., 1994). Paradoxically, DNA damage can also stimulate initiation of DNA replication, leading to the selective amplification of critical genes (Jussila, 1996).

Complete carcinogens can qualitatively and quantitatively change the cell's genetic information. They exhibit a direct analogy between their structure and activity, are mutagenic in *in vitro* assays, are active in high doses, and may affect several animal species, and damage different organs. In high doses, they cause toxicity and cell proliferation, increasing DNA replication, and influencing its carcinogenic activity (Cohen, 1998). Following transmembranar diffusion they are metabolized in electrophilic compounds that enter the nucleus and interact with nucleophilic sites (DNA, RNA, and proteins) changing their structural integrity and establishing covalent bonds known as adducts (Straub and Burlingame, 1981). Adduct formation is the first critical step in the process of carcinogenesis. Adduct formation may differ in number and in area of DNA damage. Lack of adduct repair before DNA replication may allow for mutations in the proto-oncogenes and tumor-suppressor genes, essential elements of the initiation stage. (Oliveira, 2007).

The formation of adducts may cause mutations through deletions, frameshift, or nucleotide transposition, or the adducts may cause DNA chain breakage by their sheer numbers. A positive association has be discerned in animal models between the quantity of adducts formed and the number of neoplasms developed. (Oliveira, 2007).

Chemicals that Cross-Link DNA

Alkylating Agents   Alkylating agents are electrophilic compounds that have an affinity for nucleophilic centers in organic macromolecules (Table 107.4). Alkylating agent–DNA adducts are believed to cause the mutations that ultimately induce a significant fraction of human tumors (many chemicals that belong to this group are proven or suspected carcinogens) and are important mediators of the cytotoxic effects of anticancer drugs. In general, bifunctional alkylating agents are potent cytotoxic agents, whereas monofunctional alkylating agents exhibit lower cytotoxicity and are more efficient mutagens. Bifunctional alkylating agents can react with two different nucleophilic centers in DNA. If the two sites are on opposite polynucleotide strands, interstrand DNA crosslinks result. If the sites are on the same chain, intrastrand crosslinks result. Interstrand DNA crosslinks prevent DNA strand

TABLE 107.4   Alkylating Agents Classified as Carcinogens by the U.S. Department of Health and Human Services.

Known to be Human Carcinogens

Aflatoxins, alcoholic beverages, 4-aminobiphenyl, phenacetin, aristolochic acids, arsenic, asbestos, azathioprine, benzene, benzidine. Beryllium, bis(chloromethyl) ether, technical-grade chloromethyl methyl ether, 1,3-butadiene, cadmium, chlorambucil, 1-(2-chloroethyl)-3-(4-methylcyclohexyl)-1-nitrosourea, chromium hexavalent compounds, coal tars, coal–tar pitches, coke-oven emissions, cyclophosphamide, cyclosporin A, diethylstilbestrol, dyes metabolized to benzidine, erionite, estrogens, ethylene oxide, formaldehyde, hepatitis B virus, hepatitis C virus, human papillomavirus, melphalan, methoxsalen with ultraviolet A therapy, mineral oils, mustard gas, 2-naphthylamine, neutrons, nickel compounds, radon, silica crystalline (respirable size), solar radiation, soots, strong inorganic acid mists containing sulfuric acid, tamoxifen, 2,3,7,8-tetrachlorodibenzo-*p*-dioxin, thiotepa, thorium dioxide, tobacco, ultraviolet radiation, vinyl chloride, wood dust, X-radiation and γ-radiation

1. Reasonably Anticipated to be Human Carcinogens

Acetaldehyde, 2-acetylaminofluorene, acrylamide, acrylonitrile, adriamycin, 2-aminoanthraquinone, *o*-aminoazotoluene, 1-amino-2,4-dibromoanthraquinone, 2-amino-3,4 dimethylimidazo [4,5-*f*]quinoline 2-amino-3,8-dimethylimidazo[4,5-*f*]quinoxaline, 1-amino-2 methylanthraquinone, 2-amino-3 methylimidazo [4,5-*f*]quinoline, 2-amino-1-methyl-6 phenylimidazo[4,5-*b*]pyridine, amitrole, *o*-anisidine and its hydrochloride azacitidine, basic red 9, monohydrochloride, benz[*a*]anthracene, benzo[*b*]fluoranthene, benzo[*j*] fluoranthene, benzo[*k*]fluoranthene, benzo[*a*]pyrene benzotrichloride, 2,2 bis(bromomethyl)-1,3-propanediol, bis(chloroethyl) nitrosourea, bromodichloromethane, 1,4-butanediol dimethanesulfonate, butylated hydroxyanisole, captafol, carbon tetrachloride, ceramic fibers, chloramphenicol, chlorendic acid, chlorinated paraffins, chloroform, 1-(2-chloroethyl)-3-cyclohexyl-1-nitrosourea, 3-chloro-2-methylpropene, 4-chloro-*o*-phenylenediamine, chloroprene, *p*-chloro-*o*-toluidine and its hydrochloride, chlorozotocin, cisplatin, cobalt sulfate, cobalt–tungsten carbide: powders and hard metals, *p*-cresidine, cupferron, dacarbazine, danthron, 2,4-diaminoanisole sulfate, 2,4-diaminotoluene, diazoaminobenzene, dibenz[*a,h*]acridine, dibenz[*a,j*]acridine, dibenz[*a,h*]anthracene, 7H-dibenzo[*c,g*]carbazole, dibenzo[*a,e*]pyrene, dibenzo[*a,h*]pyrene, dibenzo[*a,i*]pyrene, dibenzo[*a,l*]pyrene, 1,2-dibromo-3-chloropropane, 1,2-dibromoethane, 2,3-dibromo-1-propanol, 1,4-dichlorobenzene, 3,3′-dichlorobenzidine and its dihydrochloride dichlorodiphenyltrichloroethane, 1,2-dichloroethane, dichloromethane, 1,3-dichloropropene, diepoxybutane, diesel exhaust particulates, di(2-ethylhexyl) phthalate, diethyl sulfate, diglycidyl resorcinol ether, 3,3′-dimethoxybenzidine (see 3,3′-dimethoxybenzidine and dyes metabolized to 3,3′-dimethoxybenzidine), 4-dimethylaminoazobenzene, 3,3′-dimethylbenzidine (see 3,3′-dimethylbenzidine and dyes metabolized to 3,3′-dimethylbenzidine), dimethylcarbamoyl chloride, 1,1-dimethylhydrazine, dimethyl sulfate, dimethylvinyl chloride, 1,6-dinitropyrene 1,8-dinitropyrene,1,4-dioxane, disperse blue 1, dyes metabolized to 3,3′-dimethoxybenzidine, dyes metabolized to 3,3′-dimethylbenzidine, epichlorohydrin, ethylene thiourea, ethyl methanesulfonate furan, glass wool fibers (inhalable), glycidol, hexachlorobenzene, hexachloroethane, hexamethylphosphoramide, hydrazine and hydrazine sulfate, hydrazobenzene, indeno[1,2,3-*cd*] pyrene, iron dextran complex, isoprene, kepone, lead, lindane, hexachlorocyclohexane 2-methylaziridine, 5-methylchrysene, 4,4′-methylenebis (2-chloroaniline), 4,4′-methylenebis(*N,N*-dimethyl)benzenamine, 4,4′-methylenedianiline and its dihydrochloride, methyleugenol, methyl methanesulfonate, *N*-methyl-*N*′-nitro-*N*-nitrosoguanidine, metronidazole, michler's ketone, mirex, naphthalene, nickel metallic, nitrilotriacetic acid, *o*-nitroanisole, nitrobenzene, 6-nitrochrysene, nitrofen, nitrogen mustard, hydrochloride, nitromethane2-nitropropane, 1-nitropyrene, 4-nitropyrene, *N*-nitrosodi-*n*-butylamine, *N*-nitrosodiethanolamine, *N*-nitrosodiethylamine, *N*-nitrosodimethylamine, *N*-nitrosodi-*n*-propylamine, *N*-nitroso-*N*-ethylurea, 4-(*N*-nitrosomethylamino)-1-(3-pyridyl)-1-butanone, *N*-nitroso-*N*-methylurea, *N*-nitrosomethylvinylamine, *N*-nitrosomorpholine, *N*-nitrosonornicotine, *N*-nitrosopiperidine, *N*-nitrosopyrrolidine, *N*-nitrososarcosinel, *o*-nitrotoluene, norethisterone, ochratoxin A, 4,4′-oxydianiline, oxymetholone, phenacetin, phenazopyridine hydrochloride, phenolphthalein, phenoxybenzamine hydrochloride, phenytoin and phenytoin sodium, polybrominated biphenyls, polychlorinated biphenyls, procarbazine and its hydrochloride, progesterone, 1,3-propane sultone, β-propiolactone, propylene oxide, propylthiouracil, reserpine, riddelliine, safrole, selenium sulfide, streptozotocin, styrene, styrene-7,8-oxide, sulfallate, tetrachloroethylene, tetrafluoroethylene, tetranitromethane, thioacetamide, 4,4′-thiodianiline, thiourea, toluene diisocyanates, *o*-toluidine and its hydrochloride, toxaphene, trichloroethylene, 2,4,6-trichlorophenol, 1,2,3trichloropropane, tris(2,3-dibromopropyl) phosphate, ultraviolet radiation, urethane, vinyl bromide, 4-vinyl-1-cyclohexene diepoxide, vinyl fluoride

Source: Modified from the 12th Report on Carcinogens (US Department of Health and Human Services) (http://ntp.niehs.nih.gov/ntp/roc/twelfth/roc12).

separation and block replication and transcription (Borowy-Borowski et al., 1990a, 1990b). Alkylating agents can be classified as species that react with nucleophiles by either SN1 or SN2 mechanism. SN2 agents (low oxyphilic), such as dialkyl sulfates and methyl methane sulfonate, transfer the alkyl group upon nucleophilic attack by DNA. SN1 agents (high oxyphilic) such as *N*-nitroso compounds (alkyl nitrosoureas, dialkyl nitrosamines, and *N*-alkyl-*N*′ nitro-*N*-nitrosoguanidine) degrade to alkyldiazohydroxides that

secondarily react with DNA. Alkylating agents may have one or two reactive groups and are able to alkylate all four bases, although with different reactivity. In general, the ring nitrogens of the bases are more nucleophilic than the oxygens. The alkylation of base residues in DNA appears to be non-random, and the relative distribution of alkylation damage may play a role in the generation of mutational hot spots. Thus the adduct O6alk–gua has been shown to be the principal biological lesion inducing a G to A transition mutation. Alternatively, steric effects may also play a role in the exposure of specific sites to the alkylating action (Kornberg, 1991). Because SN1 agents modify the exocyclic base oxygens more frequently than SN2 compounds, they are more mutagenic and carcinogenic. Furthermore, ethylating agents are more potent mutagens than methylating agents. One of the first alkylating agents studied, mustard gas (sulfur mustard) used as a chemical warfare agent, damaged external tissues and was also shown to inhibit cell division. Studies in Drosophila demonstrated the mutagenic activity of the mustards, even before DNA was discovered to be responsible for transmission of genetic information. Nitrogen mustards were shown to react as aziridinium cation intermediates with the N7 atom of guanine, forming crosslinks that join the two antiparallel strands of DNA as 1,3 interstrand adducts in 5′-GNC-3′ sequences. Haloethyl nitrosoureas have also been used as anticancer drugs because they form dC−CH2−CH2−dG interstrand crosslinks between N1 of guanine and N3 of cytosine. Recent studies indicate that a group of chemicals termed *tobacco-specific nitrosamines,* present in tobacco and in other environmental samples, may be converted by metabolic activation to carcinogenic species such as methyldiazohydroxide. The carcinogenic mechanism may be caused by O6alk–gua formation, which has been shown to activate K-ras in the mouse lung.

Psoralens Furocoumarins with planar tricyclic configurations (psoralens) can intercalate into DNA and form covalent adducts to pyrimidines upon subsequent photoactivation by long-wavelength UV radiation. This produces significant helix distortion, kinking, and unwinding of DNA leading to the arrest of DNA replication in dividing cells.

Chemicals that are Metabolized to Electrophilic Reactants A variety of relatively nonpolar, chemically unreactive compounds undergo metabolic activation to more reactive forms that interact with nucleophilic centers in DNA. This mechanism of action is dependent on specific metabolism that protects cells from cytotoxic effects, converting toxic nonpolar chemicals to water-soluble extractable forms. Some products of these reactions are particularly reactive with DNA. Examples of these type of enzymatic systems are the cytochrome P-450 system, glutathione-S-transferases, sulfotransferases, acetyltransferases, UDP-glucuronosyltransferases, and adenosylating and methylating enzymes (Hollenberg, 1992).

Aromatic Amines and Nitrosamines Aromatic amines were implicated in urinary cancers in workers of the dye industry as early as 1895. Experimental studies in dogs confirmed a causative role in 1937. Epidemiological studies implicated 2-naphthylamine as a potent carcinogen in 1954. Because these compounds acted systemically, it was suspected that some form of metabolic activation was involved in their carcinogenic action. In rats, *N*-2-acetyl-2-aminofluorene (AAF),

used as an insecticide, results in malignant tumors (liver and bladder) with latent periods of more than 100 days. N-oxidation, by the cytochrome P-450, converted this biochemically inert aromatic amide into an *in vivo*-reactive proximate carcinogen, a mechanism found to be true for carcinogenic aromatic amides and amines in general. Thus a proximate carcinogen can become a highly reactive alkylating agent, an ultimate carcinogen following formation of electrophilic metabolites by cytosolic enzymes. Dimethylnitrosamine, implicated in the development of liver cirrhosis in two workers, has been shown to induce hepatocellular carcinoma in rats. Nitropyrenes and dinitropyrenes have been found in diesel exhaust, urban air, coal fly ash, and certain broiled foods. High temperature cooking can convert tryptophan and glutamic acids to mutagenic heterocyclic aromatic amines. Many of these compounds have been shown to be carcinogenic in experimental animals and have been implicated in human cancer. Arylamines and nitroaromatic compounds form DNA adducts at C8, N2, and O6 of Gua, C8, and N6 of Ade. Aminofluorene induces mutations by base-pair substitutions, predominantly GC-to-TA transversions. Its N-acetyl cogener, AAF, induces mutations by frame shifts in GC base pairs.

Policyclic Aromatic Hydrocarbons (PAHs) Polycyclic aromatic hydrocarbons are metabolized by liver monooxygenases. It is believed that PAHs are initially converted to a reactive epoxide intermediate that subsequently reacts with a nucleophile or is hydrolyzed by epoxide hydrolase. The most extensively studied is benzo[α]pyrene (BP). Isolated from crude coal tar (Brookes, 1990), this potent carcinogenic PAH is present in the environment as a result of cigarette smoke and automobile exhaust fumes. Unmodified BP is an unreactive non-polar compound with a planar configuration that intercalates between the hydrogen-bonded base pairs in duplex DNA. It is known, however, that components of the P-450 system known as *aryl-hydrocarbon hydroxylases* metabolize BP and other PAHs to phenols and epoxides (Brookes, 1990). Phenols are oxidized to quinones. Epoxides (4,5-epoxide, 7,8-epoxide, 9,10-epoxide) are converted to excretable dihydrodiols. The ultimate carcinogenic form is an anti-diol-epoxide that binds predominantly to the exocyclic 2-amino position of guanine (Brookes, 1990). Reactions also occur at Ade-N6 and Cyt-N4, although GC4o-TA mutations seem to be the most frequent in a variety of models.

Natural Products

Aflatoxins Among the most potent liver carcinogens known, aflatoxins (difuranocoumarins) are mycotoxins produced by the fungi *Aspergillus flavus* and *Aspergillus parasiticus* that infest peanuts, corn, and other agricultural products. The most potent is aflatoxin B$_1$ (AFB$_1$), which is oxidized by the mixed function oxygenases of the P-450 system in the microsomal fraction of liver extracts, giving rise to aflatoxin B$_1$-8,9-epoxide (Kensler et al., 2011), which forms a major adduct at the N7 position of deoxyguanosine (Simonich et al., 2007). These adducts can then undergo ring opening to generate two pyrimidine adducts. These DNA modifications may then cause malignancy by the activation of proto-oncogenes or the abolishment of tumor-suppressor genes. Aflatoxin B$_1$ adducts are strong blocks to DNA replication and highly mutagenic, predominantly GC-to-TA transversion and frameshift mutations.

Mitomycin C Mitomycins are a group of antibiotics generated by *Streptomyces caespitosus*. The major adduct is the guanine-N2-1inked mono-adduct precursor to other adducts. Interstrand crosslinks are formed at 5'-GG sequences and intrastrand crosslinks at 5'-TG sequences.

N-Methyl-N'-Nitro-N-Nitrosoguanidine Interaction of N-methyl-N'nitro-N-nitrosoguanidine with cysteine residues such as those found in glutathione gives origin to highly electrophilic methylating intermediates (O'Brien et al., 1996).

4-Nitroquinoline 1-oxide 4-Nitroquinoline 1-oxide can give rise to 8-hydroxyguanine, leading to strand breakage, and can be converted to the proximate carcinogen 4-OH-aminoquinoline 1-oxide that is acylated by a seryl-AMP enzyme complex that introduces quinoline groups into DNA.

Base Analogs Analogs of the four naturally occurring bases in DNA can be incorporated during DNA replication. Thymidine analogs include 5-bromouracyl, 5-fluorouracil, and 5-iodouracil. The most important adenine analog is 2-aminopurine.

Non-Genotoxic Carcinogens

Non-genotoxic carcinogens do not require metabolic activation. Those that have their activity mediated by a receptor are considered "cytotoxic", while those that are not are considered "mitogenic" (Cohen, 1991). As they act solely as promoters, they do not react directly with DNA and generate neither adducts nor positive responses on mutagenicity tests. (Butterworth et al., 1992) The non-genotoxic carcinogens modulate cell growth and death and may potentiate the effects of genotoxic compounds; however, they do not show a direct correlation between their structure and their activity, although their concentration does limit their behavior. (Oliveira, 2007.) Two classes of chemicals that are generally inactive in assays for mutagenicity are the mineral fibers and the hormones.

Hormones Estrogens are major factors in the risk for different cancers in women (Preston-Martin et al., 1990). Thus, the natural estrogen 17-beta-estradiol increases the incidence of a variety of tumors in experimental animals, and diethyl-stilbestrol (DES) administration to pregnant women is directly related to the development of clear-cell adenocarcinoma. Diethylstilbestrol binds to microtubules disrupting tubulin assembly and may induce aneuploidy suggesting a carcinogenic mechanism independent of estrogen receptors. It has been proposed that estrogens induce cancer primarily by stimulating cell proliferation but there is evidence that estrogens can also induce heritable alterations, even after single-dose or short-term exposures (Newbold et al., 1990).

Asbestos The name "asbestos" refers to a group of naturally occurring mineral fibers, including amphiboles (crocidolite, amosite, tremolite, anthophyllite, and actinolite) and serpentine fibers (chrysotile) (Huan et al., 2011). Exposure to asbestos, used in industry and households for decades, induces asbestosis, pleural plaques, lung cancer, mesothelioma, and other non-respiratory diseases (Currie et al., 2009; Straif et al., 2009). Asbestos has been classified as a Group I human carcinogen by the International Agency for Research on Cancer. As a result, the U.S. Environmental Protection Agency has restricted the industrial use of asbestos since the early 1970s. Asbestos is, however, widely used in many developing countries (Virta, 2003) and continues to be an important health problem due to the long latency period of asbestos-induced diseases (Huang et al., 2011). In addition, there is evidence suggesting a synergistic effect of tobacco smoke and asbestos exposure on mutagenicity and lung cancer risks (Kamp et al., 1992; Nymark et al., 2006, 2008; Selikoff et al., 1968). The mechanisms by which asbestos fibers produce malignancy and fibrosis are not entirely clear but fiber dimensions, surface and crystal properties, shape, chemical composition, physical durability, and exposure route, duration and dose are important factors (Sanchez et al., 2009). All types of asbestos are carcinogenic (Straif et al., 2009).

Oncogenesis appears to be induced via both mutagenic and non-mutagenic mechanisms (Barrett, 1992; Kane, 1996; Nymark et al., 2008). Unlike chemical mutagens that generate genetic alterations by covalent binding to/modifying the DNA structures, asbestos-induced mutagenicity can be mediated, directly, through physical interaction with the mitotic machinery of dividing cells or, indirectly, as a result of damaging of DNA and chromosome by asbestos-induced reactive oxygen (ROS) and nitrogen species (RNS) (Gulumian, 2005; Hardy and Aust, 1995; Jaurand, 1997; Kamp and Weitzman, 1999; Shukla et al., 2003a, 2003b). Persistence of asbestos fibers in the lungs triggers prolonged radical production and chronic inflammation at the sites of fiber deposition. Most cell types, except lymphocytes, are able to phagocytize fibers and physical interaction of asbestos fibers with mitotic machineries leads to mis-segregation of chromosomes during mitosis and aneuploidy (Hesterberg and Barrett, 1985; Palekar et al., 1987; Pelin et al., 1992; Yegles et al., 1993, 1995). Furthermore, asbestos-generated ROS and RNS produce dose-dependent DNA and chromosomal abnormalities, such as 8-hydroxydeoxyguanosine, DNA single-strand breaks and chromosome fragments (Jaurand et al., 1987; Kamp et al., 1995). This occurs in all relevant lung and pleural target cells, although, compared to alveolar epithelial and mesothelial cells, bronchial epithelial cells show less DNA damage, possibly due to their greater antioxidant capacity (Nygren et al., 2004; Puhakka et al., 2002).

In mesothelial cells (Unfried et al., 2002) the frequency of G→T transversions is higher than the frequency of expected spontaneous mutations. In addition, target cells and cells in the microenviroment release inflammatory mediators and growth factors that activate multiple signaling cascades responsible for fibrosis and malignancy (Kamp et al., 1998; Kane, 1992, 2006; Manning et al., 2002; Mossman and Churg, 1998; Shukla et al., 2003a, Mossman, 1990, 1993). Gene expression microarray studies (Hevel et al., 2008; Nymark et al., 2007) have shown that, in mesothelial and epithelial cells, asbestos fibers alter the p53 signaling pathway, possibly explaining the resistance to apoptosis in mesothelioma cells (Fennell and Rudd, 2004; Burmeister et al., 2004; Johnson and Jaramillo, 1997; Kopnin et al., 2004; Levresse et al., 2000; Matsuoka et al., 2003; Paakko et al., 1998; Panduri et al., 2003; Topinka et al., 2004). Also of interest is that TNF-α, which prevents cell death via a nuclear factor (NF)-κB-dependent mechanism (Yang et al., 2006), is induced by asbestos. Furthermore, the AKT/mammalian target of rapamycin (mTOR) signaling pathway, which promotes the growth of apoptosis-resistant cells (Chiang and Abraham, 2007; Guertin and Sabatini, 2007), has been implicated in the increased survival of mesothelioma cells (Wilson et al., 2008). Inhibiting mTOR with rapamycin or silencing p–S6K, a major downstream target of mTOR, leads to cell death in these apoptosis-resistant cells.

TOBACCO

According to the WHO, 5.5 trillion cigarettes were consumed in 2000, and 1 billion men and 250 million women are daily smokers (Mackay and Eriksen, 2002; WHO, 2002; WHO, 2003). Between 1965 and 2004, adult cigarette smoking declined from 42 to 21% and, between 2005 and 2011, there was a modest decline from 21 to 19% accounting for 2 million people. On the other hand, the number of smokers reporting light or intermittent smoking (<10 cigarettes/day) increased from 16% in 2005 to 22% in 2011 (22%). Tobacco causes 1.2 million deaths annually and of today's world population, 500 million will die of tobacco use, with cancer one of the main causes of death (IARC, 1985, 1986). The American Cancer Society (ACS) estimates that, in 2013, about 174,100 Americans will die of tobacco-related diseases. Cigarette

smoking causes 90% of lung cancer and is associated with cancers of the oral cavity, larynx, oropharynx, hypopharynx, sinonasal cavity, esophagus, stomach, liver, pancreas, bladder, cervix, colorectal, ureter and kidney, as well as myeloid leukemia. Furthermore, smokeless tobacco is known to cause oral cavity cancer and possibly pancreatic cancer (Hoffmann et al., 2001; Vainio and Weiderpass, 2003; Hecht, 1999). Tobacco contains nicotine, responsible for the addiction, and more than 60 carcinogens, most of them formed after combustion. The total amount of carcinogens in cigarette smoke is 1–3 mg per cigarette. Strong carcinogens, such as nitrosamines [4-(methylnitrosamino)-1-(3-pyridyl)-1-butanone (NNK) and N'-nitrosonornicotine (NNN)], PAHs (benzo[a]pyrene) and aromatic amines (4-aminobiphenyl) play important roles in carcinogenesis despite occurring in smaller amounts (1–200 ng per cigarette) than weak carcinogens such as acetaldehyde (1 mg per cigarette) (Swauger et al., 2002; Phillips, 1983; Hoffmann et al., 1987, 1995; Cooper et al., 1954; Hecht and Hoffmann, 1988). Carcinogens are metabolically activated by cytochrome P450 enzymes (P450s) (Guengerich, 2001). Most metabolites are detoxified and excreted but some are electrophilic and react with DNA forming covalently bound DNA adducts (Wiencke, 2002; Phillips, 2002; Boysen and Hecht, 2003), typically leading to G–T and G–A mutations (Loechler et al., 1984; Singer and Essigmann, 1991; Seo et al., 2000) that, if occurring in crucial regions of oncogenes such as *RAS* and *MYC* or in tumor-suppressor genes such as *TP53* and *CDKN2A* (which encodes p16), will result in loss of normal cellular growth-control mechanisms (Osada and Takahashi, 2002).

Nicotine and carcinogens can also bind to the cellular receptors directly, activating enzymes such as serine threonine kinase AKT (protein kinase B) and protein kinase A (PKA). This, in turn, can result in decreased apoptosis, increased angiogenesis, and cell transformation. Tobacco products also contain tumor promoters and co-carcinogens, which can activate protein kinase C (PKC), activator protein 1 (AP1), and other cellular factors (Hoffmann Wynder, 1971; Denissenko et al., 1996; Hoffmann et al., 1994; Spiegelhalder and Bartsch, 1996; Hoffmann et al., 1994; IARC, 1985). Six important carcinogens that can form DNA adducts are: (1) benzo[a]pyrene (BaP), which is metabolized to epoxides, which are then converted to diol epoxides by epoxide hydrolases (EHs) and P450s, (2) 4-(methylnitrosamino)-1-(3-pyridyl)-1-butanone (NNK), which is metabolized to α-hydroxy NNKs, which then spontaneously decompose to aldehydes and undergo reduction to 4-(methylnitrosamino)-1-(3-pyridyl)-1-butanol (NNAL), which is converted to NNAL-glucuronides and α-hydroxylated to diazonium ions, (3) N-nitrosodimethylamine (NDMA), which undergoes α-hydroxylation to formaldehyde and α-hydroxy NDMA, which spontaneously decomposes to methyldiazonium ions that produce methyl adducts and can be detoxified to nitrite and methylamine, (4)

N'-nitrosonornicotine (NNN), which is metabolized to the unstable α-hydroxyNNNs, which produces diazonium ions and is detoxified to norcotinine, β-hydroxyNNNs and NNN-N-oxide, (5) ethylene oxide, which reacts directly with DNA and can undergo detoxification by GST catalysis to mercapturic acids, hydration to ethylene glycol, and oxidation to CO_2, and (6) 4-aminobiphenyl (4-ABP), which undergoes N-oxidation to a hydroxylamine.

THE FUTURE OF ONCOLOGY

The Human Genome Project has opened the genomics field for biology. Building on this success, the National Cancer Institute has launched the Cancer Genome Atlas Project, whit the intent of sequencing the genomes of thousands of human tumors and compared them to the genomes of normal tissues. Diagnosis and treatment of cancer is, thus, moving to a "Personalized Medicine" approach based on these genomics data. Drugs will likely be developed and used to target a specific patient's tumor. Tumor heterogeneity, the result of active proliferation of different tumoral clones, will likely require the use of multiple drugs in a successive coordinated fashion but this targeted therapy is expected to be less toxic and more successful in killing tumor cells. The road to achieve this goal is not exempt of difficulties but we are entering an exciting era in cancer research and treatment and many fields are converging to either cure most cancers or convert them in low morbidity, survivable chronic diseases.

REFERENCES

Adams, J.M. and Cory, S. (2007) The Bcl-2 apoptotic switch in cancer developmentand therapy. *Oncogene* 26:1324–1337.

Ahmed, Z. and Bicknell, R. (2009) Angiogenic signalling pathways. *Methods Mol. Biol.* 467:3–24.

Albertson, D. (2008) Conflicting evidence on the frequency of ESR1 amplification in breast cancer. *Nat. Genet.* 40:821–822.

American Cancer Society (ACS) (2013) *Cancer Facts and Figures*, Atlanta: American Cancer Society.

American Cancer Society (ACS) (2012) *The History of Cancer*, Atlanta: American Cancer Society.

Anderson, M., Muro-Cacho, C., Cordero, J., et al. (1999) Transforming growth factor beta receptors in verrucous and squamous cell carcinoma. *Arch. Otolaryngol. Head Neck Surg.* 125(8):849–854.

Baeriswyl, V. and Christofori, G. (2009) The angiogenic switch in carcinogenesis. *Semin. Cancer Biol.* 19:329–337.

Báez, A., Cantor, A., Fonseca, S., et al. (2005) Differences in Smad4 expression in human papillomavirus type 16-positive and human papillomavirus type 16-negative head and neck squamous cell carcinoma. *Clin. Cancer Res.* 11(9):3191–3197.

Barbacid, M. (1986) Mutagens, oncogenes and cancer. *Trends Genet.* 2:188–192.

Barbacid, M., et al. (2005) *Cell Cycle and Cancer: Genetic Analysis of the Role of Cyclin-Dependent Kinases, Cold Spring Harbor Symposia on Quantitative Biology, Volume LXX, Cold Spring Harbor Laboratory Press, 0-87969-773-3.*

Barnes, D.E., Lindahl, T., and Sedgwick, B. (1993) DNA repair (review). *Curr. Opin. Cell Biol.* 5:424–433.

Barnes, D.E. and Lindahl, T. (2004) Repair and genetic consequences of endogenous DNA base damage in mammalian cells. *Annu. Rev. Genet.* 38:445–476.

Barrallo-Gimeno, A. and Nieto, M.A. (2005) The Snail genes as inducers of cell movement and survival: implications in development and cancer. *Development* 132(14):3151–3161.

Barrett, J.C. and Anderson, M. (1993) Molecular mechanisms of carcinogenesis in humans and rodents. *Mol. Carcinog.* 7(1):1–13.

Barrett, J.C. and Wiseman, R.W. (1987) Cellular and molecular mechanisms of multistep carcinogenesis: relevance to carcinogen risk assessment. *Environ. Health Perspect.* 76:65–70.

Barrett, J.C. (1992) Mechanisms of action of known human carcinogens. *IARC Sci. Publ.* 116:115–134.

Barski, A., et al. (2007) High-resolution profiling of histone methylations in the human genome. *Cell* 129:823–837.

Bartek, J., Lukas, C., and Lukas, J. (2004) Checking on DNA damage in S phase. *Nat. Rev. Mol. Cell Biol.* 5:792–804.

Bartsch, H. and Malaveille, C. (1990) Screening assays for carcinogenic agents and mixtures: an appraisal based on data in the IARC Monograph Series. In: Vainio, H., Sorsa, M., and McMichael, A.J., editors. *Complex Mixtures and Cancer Risks*, Lyon. France: IARC Scientific Pub No 104, International Agency for Research on Cancer.

Baylin, S.B. and Ohm, J.E. (2006) Epigenetic gene silencing in cancer-a mechanism for early oncogenic pathway addiction? *Nat. Rev. Cancer* 6:107–116.

Beasley, R.P., Hwang, L.Y., Lin, C.C., et al. (1981) Hepatocellular carcinoma and hepatitis B virus. A prospective study of 22,707 men in Taiwan. *Lancet* 2:1129–1133.

Beckman, R.A. and Loeb, L.A. (2006) Efficiency of carcinogenesis with and without a mutator mutation. *Proc. Natl Acad. Sci. USA* 103:14140–14145.

Berenblum, I. and Shubik, P. (1947) A new, quantitative, approach to the study of the stages of chemical cartinogenesis in the mouse's skin. *Br. J. Cancer* 1(4):383–391.

Bergers, G. and Song, S. (2005) The role of pericytes in blood-vessel formation and maintenance. *Neuro-Oncology* 7(4): 452–464.

Bernstein, B.E., et al. (2007) The mammalian epigenome. *Cell* 128:669–681.

Bhowmick, N.A., Neilson, E.G., and Moses, H.L. (2004) Stromal fibroblasts in cancer initiation and progression. *Nature* 432(7015):332–337.

Bierie, B. and Moses, H.L. (2006) Tumour microenvironment: TGF-β: the molecular Jekyll and Hyde of cancer. *Nat. Rev. Cancer* 6(7):506–520.

Bird, A. (2002) DNA methylation patterns and epigenetic memory. *Genes Dev.* 16:6–21.

Biswas, S.K. and Mantovani, A. (2010) Macrophage plasticity and interaction with lymphocyte subsets: cancer as a paradigm. *Nat. Immunol.* 11(10):889–896.

Blasco, M.A. (2005) Telomeres and human disease: ageing, cancer and beyond. *Nat. Rev. Genet.* 6:611–622.

Blumberg, B.S., Larouze, B., London, W.T., et al. (1975) The relation of infection with the hepatitis B agent to primary hepatic carcinoma. *Am. J. Pathol.* 81:669–682.

Bode, A.M. and Dong, Z. (2009) Cancer prevention research—then and now. *Nat. Rev. Cancer* 9(7):508–516.

Boland, C.R., et al. (1998) A National Cancer Institute Workshop on Microsatellite Instability for cancer detection and familial predisposition: development of international criteria for the determination of microsatellite instability in colorectal cancer. *Cancer Res.* 58:5248–5257.

Borowy-Borowski, H., Lipman, R., Chowdary, D., et al. (1990a) Duplex oligodeoxyribonucleotides cross-linked by mitomycin C at a single site: synthesis, properties, and cross-link reversibility. *Biochemistry* 29(12):2992–2999.

Borowy-Borowski, H., Lipman, R., and Tomasz, M. (1990b) Recognition between mitomycin C and specific DNA sequences for cross-link formation. *Biochemistry* 29(12):2999–3006.

Bos, P.D., et al. (2009) Genes that mediate breast cancer metastasis to the brain. *Nature* 459:1005–1009.

Boultwood, J., Fidler, C., Strickson, A.J., et al. (2002) Narrowing and genomic annotation of the commonly deleted region of the 5q syndrome. *Blood* 99:4638–4641.

Boveri, T. (1914) *Zur Frage der Entwicklung Maligner Tumoren*, Jena, Germany: Gustav Fischer-Verlag.

Boysen, G. and Hecht, S.S. (2003) Analysis of DNA and protein adducts of benzo[a]pyrene in human tissues using structure-specific methods. *Mutat. Res.* 543:17–30.

Brookes, P. (1990) The early history of the biological alkylating agents, 1918–1968. *Mutat Res.* 233(1–2):3–14.

Burkhar, D.L. and Sage, J. (2008) Cellular mechanisms of tumour suppression by the retinoblastoma gene. *Nat. Rev. Cancer* 8:671–682.

Burkitt, D. (1962) A children's cancer dependent on climatic factors. *Nature* 194:232–234.

Burmeister, B., Schwerdtle, T., Poser, I., et al. (2004) Effects of asbestos on initiation of DNA damage, induction of DNA strand breaks, P53-expression and apoptosis in primary, SV40-transformed and malignant human mesothelial cells. *Mutat. Res.* 558:81–92.

Butel, J.S. (2000) Viral carcinogenesis: revelation of molecular mechanisms and etiology of human disease. *Carcinogenesis* 21:405–426.

Butterworth, B., Popp, J.A., Conolly, R.B., and Goldsworthy, T.L. (1992) Chemically induced cell proliferation in carcinogenesis. *IARC Sci. Publ.* 116:279–305.

Cairns, J. (1975) Mutation selection and the natural history of cancer. *Nature* 255:197–200.

Caldecott, K.W. (2008) Single-strand break repair and genetic disease. *Nat. Rev. Genet.* 9:619–631.

Calin, G.A., Ferracin, M., Cimmino, A., et al. (2005) A microRNA signature associated withprognosis and progression in chronic lymphocytic leukemia. *N. Engl. J. Med.* 353:1793–1801.

Cance, W.G., et al. (1990) Altered expression of the retinoblastoma gene product in human sarcomas. *N. Engl. J. Med.* 323:1457–1462.

Carmeliet, P. and Jain, R.K. (2000) Angiogenesis in cancer and other diseases. *Nature* 407(6801):249–257.

Carrell, C.J., Carrell, T.G., Carrell, H.L., et al. (1997) Benzo[a]pyrene and its analogues: structural studies of molecular strain. *Carcinogenesis* 18(2):415–422.

Cedar, H. and Bergman, Y. (2009) Linking DNA methylation and histone modification: patterns and paradigms. *Nat. Rev. Genet.* 10:295–304.

Chambers, A.F., Groom, A.C., and MacDonald, I.C. (2002) Dissemination and growth of cancer cells in metastatic sites. *Nature Rev. Cancer* 2:563–572.

Chan, J.A., et al. (2005) MicroRNA-21 is an antiapoptotic factor in human glioblastoma cells. *Cancer Res.* 65:6029–6033.

Chang, Y., Cesarman, E., Pessin, M.S., et al. (1994) Identification of herpesvirus-like DNA sequences in AIDS-associated Kaposi's sarcoma. *Science* 366(5192):1865–1869.

Chiang, G.G. and Abraham, R. (2007) Targeting the mTOR signaling network in cancer. *Trends Mol. Med.* 13:433–442.

Cimmino, A., Calin, G.A., Fabbri, M., et al. (2005) miR-15 and miR-16 induce apoptosis by targeting BCL2. *Proc. Natl. Acad. Sci. USA* 102:13944–13949.

Ciuffo, G. (1907) Innesto positivo con filtrato di verruca volgare. *Giorn Ital Mal. Venereol.* 48:12–17.

Coffelt, S.B., Lewis, C.E., Naldini, L., et al. (2010) Elusive identities and overlapping phenotypes of proangiogenic myeloid cells in tumors. *Am. J. Pathol.* 176(4):1564–1576.

Coghlin, C. and Murray, G.I. (2010) Current and emerging concepts in tumour metastasis. *J. Pathol.* 222(1):1–15.

Cohen, S.M. (1998) Cell proliferation and carcinogenesis. *Drug Metab. Rev.* 30(2):339–357.

Collado, M. and Serrano, M. (2010) Senescence in tumors: evidence from mice and humans. *Nat. Rev. Cancer* 10:51–57.

Collins, F.S., Gray, G.M., and Bucher, J.R. (2008) Toxicology. Transforming environmental health protection. *Science* 319(5865):906–907.

Cooper, R.L., Lindsey, A.J., and Waller, R.E. (1954) The presence of 3,4–benzopyrene in cigarette smoke. *Chem. Ind.* 46:1418.

Croce, C.M. (2008) Oncogenes and Cancer. *N. Engl. J. Med.* 358:502–511.

Currie, G.P., Watt, S.J., and Maskell, N.A. (2009) An overview of how asbestos exposure affects the lung. *BMJ* 339:b3209.

Curtiss, N.P., Bonifas, J.M., Lauchle, J.O., et al. (2005) Isolation and analysis of candidate myeloid tumor suppressor genes from a commonly deleted segment of 7q22. *Genomics* 85:600–607.

Curto, M., Cole, B.K., Lallemand, D., Liu, C.H., and McLatchey, A. I. (2007) Contact-dependent inhibition of EGFR signaling by Nf2/Merlin. *J. Cell Biol.* 177(5):893–903.

Dalla Favera, R., Bregni, M., Erikson, J., et al. (1982) Human c-myc oncogene is located on the region of chromosome 8 that is translocated in Burkitt lymphoma cells. *Proc. Natl. Acad. Sci. USA* 79:7824–7827.

Darnell, J.E. Jr. (1997) STATs and gene regulation. *Science* 277 (5332):1630–1635.

Davies, H., et al. (2002) Mutations of the BRAF gene in human cancer. *Nature* 417:949–954.

De Bont, R. and van Larebeke, N. (2004) Endogenous DNA damage in humans: a review of quantitative data. *Mutagenesis* 19:169–185.

Dejana, E., Tournier-Lasserve, E., and Weinstein, B.M. (2009) The control of vascular integrity by endothelial cell junctions: molecular basis and pathological implications. *Dev. Cell* 16(2):209–221.

Delattre, O., Zucman, J., Plougastel, B., et al. (1992) Gene fusion with an ETS DNA-binding domain caused by chromosome translocation in human tumours. *Nature* 359:162–165.

Deng, S., et al. (2008) Mechanisms of microRNA deregulation in human cancer. *Cell Cycle* 7:2643–2646.

Denissenko, M.F., Pao, A., Tang, M., et al. (1996) Preferential formation of benzo[a]pyrene adducts at lung cancer mutational hot spots in p53. *Science* 274:430–432.

De Vita, V. and Rosenberg, S.A. (2012) Two hundred years of cancer research. *N. Engl. J. Med.* 366:2207–2214.

Diamandopoulos, G.T. (1996) Cancer: a historical perspective. *Anticancer Res.* 16:1595–1602.

Diaz, N., Minton, S., Cox, C., Bowman, T., et al. (2006) Activation of stat3 in primary tumors from high-risk breast cancer patients is associated with elevated levels of activated SRC and survivin expression. *Clin. Cancer Res.* 12(1):20–28.

Ding, L., et al. (2010) Genome remodelling in a basal-like breast cancer metastasis and xenograft. *Nature* 464:999–1005.

Döhner, K., Brown, J., Hehmann, U., et al. (1998) Molecular cytogenetic characterization of a critical region in bands 7q35-q36 commonly deleted in malignant myeloid disorders. *Blood* 92:4031–4035.

Dutt, A. and Beroukhim, R. (2007) Single nucleotide polymorphism array analysis of cancer. *Curr. Opin. Oncol.* 19:43–49.

Duval, A. and Hamelin, R. (2002) Mutations at coding repeat sequences in mismatch repair-deficient humancancers: toward a new concept of target genes for instability. *Cancer Res.* 62:2447–2454.

Dvorak, H.F. (1986) Tumors: wounds that do not heal. Similarities between tumor stroma generation and wound healing. *N. Engl. J. Med.* 315(26):1650–1659.

Eden, A., et al. (2003) Chromosomal instability and tumors promoted by DNA hypomethylation. *Science* 300:455.

Edwards, B.K., Ward, E., Kohler, B.A., et al. (2010) Annual report to the nation on the status of cancer, 1975–2006, featuring colorectal cancer trends and impact of interventions (risk factors, screening, and treatment) to reduce future rates. *Cancer* 116:544–573.

Egeblad, M. and Werb, Z. (2002) New functions for the matrix metalloproteinases in cancer progression. *Nat. Rev. Cancer* 2:161–174.

Egeblad, M., Nakasone, E.S., and Werb, Z. (2010) Tumors as organs: complex tissues that interface with the entire organism. *Dev. Cell* 18(6):884–901.

El-Deiry, W.S., Kern, S.E., Pietenpol, J.A. et al. (1992) Definition of a consensus binding site for p53. *Nat. Genet.* 1(1):45–9.

Elespuru, R.K. (1996) Future approaches to genetic toxicology risk assessment. *Mutat. Res.* 365(1–3):191–204.

Esteller, M. (2008) Molecular origins of cancer epigenetics in cancer. *N. Engl. J. Med.* 358:1148–1159.

EPA (2010) *Benefits and Costs of the Clean Air Act. Second Prospective Study, 1990 to 2020.* Avaiable at http://www.epa .gov/air/sect812/index.html.

Epstein, M.A., Henle, G., Achong, B.G., et al. (1965) Morphological and biological studies on a virus in cultured lymphoblasts from Burkitt's lymphoma. *J. Exp. Med.* 121:761–770.

Fabbri, M., et al. (2007) MicroRNA-29 family reverts aberrant methylation in lung cancer by targeting DNA methyltransferases 3A and 3B. *Proc. Natl. Acad. Sci. USA* 104:15805–15810.

Feinberg, A.P., et al. (1983) Hypomethylation distinguishes genes of some human cancers from their normal counterparts. *Nature* 301:89–92.

Feinberg, A.P., Ohlsson, R., and Henikoff, S. (2006) The epigenetic progenitor origin of human cancer. *Nat. Rev. Genet.* 7:21–33.

Feng, H., Shuda, M., Chang, Y., et al. (2008) Clonal integration of a polyomavirus in human Merkel cell carcinoma. *Science* 319:1096–1100.

Fennell, D.A. and Rudd, R.M. (2004) Defective core-apoptosis signalling in diffuse malignant pleural mesothelioma: Opportunities for effective drug development. *Lancet Oncol.* 5:354–362.

Ferrara, N. (2008) Vascular endothelial growth factor. *Arterioscler. Thromb.Vasc. Biol.* 29:789–791.

Ferrara, N. (2010) Pathways mediating VEGF-independent tumor angiogenesis. *Cytokine Growth Factor Rev.* 21:21–26.

Fidler, I.J. (2003) The pathogenesis of cancer metastasis: the 'seed and soil' hypothesis revisited. *Nat. Rev. Cancer* 3:453–458.

Fodde, R. and Smits, R. (2002) Cancer biology: a matter of dosage. *Science* 298:761–763.

Folkman, J. (2006) Angiogenesis. *Annu. Rev. Med.* 57:1–18.

Fonger, G.C. (1996) Toxicological and environmental health information from the National Library of Medicine. *Toxicol. Ind. Health* 12(5):639–649.

Fontham, E.T., Thun, M.J., Ward, E., et al. (2009) American Cancer Society perspectives on environmental factors and cancer. *CA Cancer J. Clin.* 59:343–351.

Fornace, A.J. Jr: (1992) Mammalian genes induced by radiation: activation of genes associated with growth control. *Annu. Rev. Genet.* 26:507–526.

Foulds, L. (1954) The experimental study of tumor progression: a review. *Cancer Res.* 14(5):327–339.

Foulkes, W.D. (2008) Molecular origins of cancer inherited susceptibility to common cancers. *N. Engl. J. Med.* 359: 2143–2153.

Fox, E.J., Salk, J.J., and Loeb, L.A. (2009) Cancer genome sequencing-an interim analysis. *Cancer Res.* 69:4948–4950.

Friedberg, E.C., Walker, G.C., Siede, W., et al., editors. (2006) *DNA Repair and Mutagenesis*, Washington, DC: ASM Press.

Friedberg, E.C. (2003) DNA damage and repair. *Nature* 421:436–440.

Friedman, J.M., et al. (2009) The putative tumor suppressor micro-RNA-101 modulates the cancer epigenome by repressing the polycomb group protein EZH2. *Cancer Res.* 69:2623–2629.

Fröhling, S. and Döhner, H. (2008) Chromosomal abnormalities in cancer. *N. Eng. J. Med.* 359:722–34

Futreal, P.A., et al. (2004) A census of human cancer genes. *Nature Rev. Cancer* 4:177–183.

Gaengel, K., Genové, G., Armulik, A., et al. (2009) Endothelial mural cell signaling in vascular development and angiogenesis. *Arterioscler Thromb. Vasc. Biol.* 29(5):630–638.

Ganem, D. (2006) KSHV infection and the pathogenesis of Kaposi's sarcoma. *Annu. Rev. Pathol.* 1:273–296.

Gapstur, S.M. and Thun, M.J. (2010) Progress in the war on cancer. *JAMA* 303(11):1084–1085.

Garcia, R., Bowman, T.L., Niu, G., et al. (2001) Constitutive activation of Stat3 by the Src and JAK tyrosine kinases participates in growth regulation of human breast carcinoma cells. *Oncogene* 20:2499–2513.

Garraway, L.A., Widlund, H.R., Rubin, M.A., et al. (2005) Integrative genomic analyses identify MITF as a lineage survival oncogene amplified in malignant melanoma. *Nature* 436:117–122.

Gilbertson, R.J. and Rich, J.N. (2007) Making a tumour's bed: glioblastoma stem cells and the vascular niche. *Nat. Rev. Cancer* 7(10):733–736.

Gillet, L.C. and Scharer, O.D. (2006) Molecular mechanisms of mammalian global genome nucleotide excision repair. *Chem. Rev.* 106:253–276.

Goldsworthy, T.L. (1996) Apoptosis and cancer risk assessment. *Mutat. Res.* 365:71–90.

Greaves, M. and Maley, C.C. (2012) Clonal evolution in cancer. *Nature* 481:306–313.

Greenman, C., et al. (2007) Patterns of somatic mutation in human cancer genomes. *Nature* 446:153–158.

Gritsko, T., Williams, A., Turkson, J., et al. (2006) Persistent activation of stat3 signaling induces survivin gene expression and confers resistance to apoptosis in human breast cancer cells. *Clin. Cancer Res.* 12(1):11–19.

Grivennikov, S.I., Greten, F.R., and Karin, M. (2010) Immunity, inflammation, and cancer. *Cell* 140(6):883–899.

Groffen, J., Stephenson, J.R., Heisterkamp, N., et al. (1984) Philadelphia chromosomal breakpoints are clustered within a limited region, bcr, on chromosome 22. *Cell* 36:93–99.

Grosse, Y., Baan, R., Straif, K., et al. (2009) A review of human carcinogens-Part A: pharmaceuticals. *Lancet Oncol.* 10(1):13–14.

Guengerich, F.P. (2001) Common and uncommon cytochrome P450 reactions related to metabolism and chemical toxicity. *Chem. Res. Toxicol.* 14:611–650.

Guertin, D.A. and Sabatini, D.M. (2007) Defining the role of mTOR in cancer. *Cancer Cell* 12:9–22.

Gulumian, M. (2005) An update on the detoxification processes for silica particles and asbestos fibers: successes and limitations. *J. Toxicol. Environ. Health B* 8:453–483.

Gupta, G.P. and Massague, J. (2006) Cancer metastasis: building a framework. *Cell* 127:679–695.

Haberland, M., et al. (2009) The many roles of histone deacetylases in development and physiology: implications for disease and therapy. *Nat. Rev. Genet.* 10:32–42.

Hajdu, S.I. (2011) A note from history: landmarks in history of cancer, Parts 1 and 2. *Cancer* 117(5).

Hanahan, D. and Folkman, J. (1996) Patterns and emerging mechanisms of the angiogenic switch during tumorigenesis. *Cell* 86:353–364.

Hanahan, D. and Weinberg, R.A. (2000) The hallmarks of cancer. *Cell* 100:57–70.

Hanahan, D. and Weinberg, R.A. (2011) Hallmarks of cancer: the next generation. *Cell* 144:646–674.

Hanawalt, P.C. and Mellon, I. (1993) Stranded in an active gene. *Curr. Biol.* 3:67–69.

Hanawalt, P.C. (2002) Sub-pathways of nucleotide excision repair and their regulation. *Oncogene* 21:8949–8956.

Harbour, J.W. and Dean, D.C. (2000) The Rb/E2F pathway: expanding roles and emerging paradigms. *Genes Dev.* 14:2393–2409.

Hardy, J.A. and Aust, A.E. (1995) Iron in asbestos chemistry and carcinogenicity. *Chem. Rev.* 95:97–118.

Hartwell, L.H. and Weinert, T.A. (1989) Checkpoints: controls that ensure the order of cell cycle events. *Science* 246:629–634.

Hassler, M.R. and Egger, G. (2012) Epigenomics of cancer emerging new concepts. *Biochimie* 94:2219–2230.

He, L., et al. (2004) MicroRNAs: small RNAs with a big role in gene regulation. *Nat. Rev. Genet.* 5:522–531.

Hecht, S.S. and Hoffmann, D. (1988) Tobacco-specific nitrosamines, an important group of carcinogens in tobacco and tobacco smoke. *Carcinogenesis* 9:875–884.

Hecht, S.S. (1999) Tobacco smoke carcinogens and lung cancer. *J. Natl. Cancer Inst.* 91:1194–1210.

Herrlich, P. and Rahmsdorf, H.J. (1994) Transcriptional and post-transcriptional responses to DNA-damaging agents. *Curr. Opin. Cell Biol.* 6:425–431.

Hesterberg, T.W. and Barrett, J.C. (1985) Induction by asbestos fibers of anaphase abnormalities: Mechanism for aneuploidy induction and possibly carcinogenesis. *Carcinogenesis* 6:473–475.

Hevel, J.L., Olson-Buelow, L., Ganesan, B., et al. (2008) Novel functional view of the crocidolite asbestos-treated A549 human lung epithelial transcriptome reveals an intricate network of pathways with opposing functions. *BMC Genomics* 9:376.

Hoeijmakers, J.H. (1993) Nucleotide excision repair. II: From yeast to mammals. *Trends Genet.* 9(6):211–7.

Hoeijmakers, J.H. (2001) Genome maintenance mechanisms for preventing cancer. *Nature* 411:366–374.

Hoeijmakers, J.H. (2009) DNA damage, aging, and cancer. *N. Engl. J. Med.* 361:1475–1485.

Hoffmann Wynder, E.L.A. (1971) A study of tobacco carcinogenesis. XI. Tumor initiators, tumor accelerators, and tumor promoting activity of condensate fractions. *Cancer* 27:848–864.

Hoffmann, D., et al. (1995) Five leading US commercial brands of moist snuff in 1994: assessment of carcinogenic N-nitrosamines. *J. Natl. Cancer Inst.* 87:1862–1869.

Hoffmann, D., Adams, J.D., Lisk, D., et al. (1987) Toxic and carcinogenic agents in dry and moist snuff. *J. Natl. Cancer Inst.* 79:1281–1286.

Hoffmann, D., Brunnemann, K.D., Prokopczyk, B., et al. (1994) Tobacco-specific N-nitrosamines and Areca-derived N-nitrosamines: chemistry, biochemistry, carcinogenicity, and relevance to humans. *J. Toxicol. Environ. Health* 41:1–52.

Hoffmann, D., Hoffmann, I., and El Bayoumy, K. (2001) The less harmful cigarette: a controversial issue. A tribute to Ernst L. Wynder. *Chem. Res. Toxicol.* 14:767–790.

Hollander, M.C., et al. (1993) Analysis of the mammalian gadd45 gene and its response to DNA damage. *J. Biol. Chem.* 268:24385–24393.

Hollenberg, P.F. (1992) Mechanisms of cytochrome P450 and peroxidase-catalyzed xenobiotic metabolism. *FASEB J.* 6(2):686–694.

Hollstein, M., et al. (1991) p53 mutations in human cancers. *Science* 253:49–53.

Holst, F., Stahl, P.R., Ruiz, C., et al. (2007) Estrogen receptor alpha (ESR1) gene amplification is frequent in breast cancer. *Nat. Genet.* 39:655–660.

Howlader, N., et al. (2012) *SEER Cancer Statistics Review, 1975–2009, National Cancer Institute. Bethesda, MD.* Available at http://seer.cancer.gov/csr/1975_2009_pops09/*.

Huan, S.X.L., Jaurand, M.C., Kamp, D.W., et al. (2011) Role of mutagenicity in asbestos fiber-induced carcinogenicity and other diseases. *J. Toxicol. Environ. Health, Part B: Crit. Rev.* 14:179–245.

Huang, S.X.L., Jaurand, M.-C., Kamp, D.W., et al. (2011) Role of mutagenicity in asbestos fiber-induced carcinogenicity and other diseases. *J. Toxicol. Environ. Health, Part B: Crit. Rev.* 14(1–4):179–245.

Huebner, K. and Croce, C.M. (2001) FRA3B and other common fragile sites: the weakest links. *Nat. Rev. Cancer* 1:214–221.

Huen, M.S. and Chen, J. (2008) The DNA damage response pathways: at the crossroad of protein modifications. *Cell Res.* 18:8–16.

Hynes, R.O. (2003) Metastatic potential: generic predisposition of the primary tumor or rare, metastatic variants-or both? *Cell* 113:821–823.

Ikushima, H. and Miyazono, K. (2010) TGF-β signalling: a complex web in cancer progression. *Nat. Rev. Cancer* 10(6):415–424.

IARC (1985) *Tobacco habits other than smoking: betel quid and Areca nut chewing and some related nitrosamines.* IARC Monographs on the Evaluation of the Carcinogenic Risk of Chemicals to Humans, 37:37–202.

IARC (1986) *Monographs on the Evaluation of the Carcinogenic Risk of Chemicals to Humans,* 38:37–375.

IOM (2001) *Rebuilding the Unity of Health and the Environment: A New Vision of Environmental Health for the 21st Century. Workshop Summary,* Institute of Medicine, Washington, DC: National Academy Press.

Jaurand, M.C. (1997) Mechanisms of fiber induced genotoxicity. *Environ. Health Perspect.* 105(Suppl. 5):1073–1084.

Jaurand, M.C., Fleury, J., Monchaux, G., et al. (1987) Pleural carcinogenic potency of mineral fibers (asbestos, attapulgite) and their cytotoxicity on cultured cells. *J. Natl. Cancer Inst.* 79:797–804.

Jemal, A., Siegel, R., Xu, J., et al. (2010) Cancer Statistics. *CA Cancer J. Clin.* 60:277–300.

Jenuwein, T. (2006) The epigenetic magic of histone lysine methylation. *FEBS J.* 273:3121–3135.

Jenuwein, T., et al. (2001) Translating the histone code. *Science* 293:1074–1080.

Ji, H., Ramsey, M.R., Hayes, D.N., et al. (2007) LKB1 modulates lung cancer differentiation and metastasis. *Nature* 448:807–810.

Jiang, B.H. and Liu, L.Z. (2009) PI3K/PTEN signaling in angiogenesis and tumorigenesis. *Ad. Cancer Res.* 102:19–65.

Jiang, C., et al. (2009) Nucleosome positioning and gene regulation: advances through genomics. *Nat. Rev. Genet.* 10:161–172.

Johnson, N.F. and Jaramillo, R.J. (1997) p53, Cip1 and Gadd153 expression following treatment of A549 cells with natural and man-made vitreous fibers. *Environ. Health Perspect.* 105(Suppl. 5):1143–1145.

Jones, P. and Baylin, S.B. (2007) The epigenomics of cancer. *Cell* 128:683–692.

Jones, P.A. and Baylin, S.B. (2002) The fundamental role of epigenetic events in cancer. *Nat. Rev. Genet.* 3:415–428.

Jones, P.A., et al. (2007) The epigenomics of cancer. *Cell* 128:683–692.

Joyce, J.A. and Pollard, J.W. (2009) Microenvironmental regulation of metastasis. *Nat. Rev. Cancer* 9(4):239–252.

Jussila, T. (1996) Oncogenes and growth factors as indicators of carcinogen exposure. *Exp. Toxicol. Pathol.* 48:145–153.

Kalluri, R. and Zeisberg, M. (2006) Fibroblasts in cancer. *Nat. Rev. Cancer* 6(5):392–401.

Kamp, D.W. and Weitzman, S.A. (1999) The molecular basis of asbestos induced lung injury. *Thorax* 54:638–652.

Kamp, D.W., Graceffa, P., Pryor, W.A., et al. (1992) The role of free radicals in asbestos-induced diseases. *Free Radical Biol. Med.* 12:293–315.

Kamp, D.W., Greenberger, M.J., Sbalchierro, J.S., et al. (1998) Cigarette smoke augments asbestos-induced alveolar epithelial cell injury: role of free radicals. *Free Radic. Biol. Med.* 25:728–739.

Kamp, D.W., Israbian, V.A., Preusen, S.E., et al. (1995) Asbestos causes DNA strand breaks in cultured pulmonary epithelial cells: role of iron-catalyzed free radicals. *Am. J. Physiol.* 268:L471–L480.

Kanaar, R., Wyman, C., and Rothstein, R. (2008) Quality control of DNA break metabolism: in the 'end', it's a good thing. *EMBO J.* 27:581–588.

Kane, A.B. (1992) Animal models of mesothelioma induced by mineral fibers: implications for human risk assessment. *Prog. Clin. Biol. Res.* 374:37–50.

Kane, A.B. (1996) Mechanisms of mineral fiber carcinogenesis. *IARC Sci. Publ.* 140:11–34.

Kane, A.B. (2006) Animal models of malignant mesothelioma. *Inhal. Toxicol.* 18:1001–1004.

Kang, H.J., Choi, Y.S., Hong, S.B., et al. (2004) Ectopic expression of the catalytic subunit of telomerase protects against brain injury resulting from ischemia and NMDA-induced neurotoxicity. *J. Neurosci.* 24:1280–1287.

Kang, Y., et al. (2003) A multigenic program mediating breast cancer metastasis to bone. *Cancer Cell* 3:537–549.

Kardinal, C.G. and Yarbro, J.W. (1979) A conceptual history of cancer. *Semin Oncol.* 6:396–408.

Kensler, T.W., Roebuck, B.D., Wogan, G.N., et al. (2011) Aflatoxin: a 50-year odyssey of mechanistic and translational toxicology. *Toxicol. Sci.* 120:S28–48.

Keogh, E.V. (1938) Ectodermal lesions produced by the virus of Rous sarcoma. *Br. J. Exp. Path.* 19:1–9.

Kern, S.E., Kinzler, K.W., Bruskin, A., et al. (1991) Identification of p53 as a sequence-specific DNA-binding protein. *Science* 252(5013):1708–1711.

Kessenbrock, K., Plaks, V., and Werb, Z. (2010) Matrix metalloproteinases: regulators of the tumor microenvironment. *Cell* 141:52–67.

Khanna, C., Wan, X., Bose, S., et al. (2004) The membrane-cytoskeleton linker ezrin is necessary for osteosarcoma metastasis. *Nat. Med.* 10:182–186.

Kim, M., Gans, J.D., Nogueira, C., et al. (2006) Comparative oncogenomics identifies NEDD9 as a melanoma metastasis gene. *Cell* 125:1269–1281.

Kinouchi, T., et al. (1999) Impact of tumor size on the clinical outcomes of patients with Robson State I renal cell carcinoma. *Cancer* 85:689–695.

Klein, C.A. (2009) Parallel progression of primary tumours and metastases. *Nature Rev. Cancer* 9:302–312.

Klymkowsky, M.W. and Savagner, P. (2009) Epithelial-mesenchymal transition: a cancer researcher's conceptual friend and foe. *Am. J. Pathol.* 174(5):1588–1593.

Knudson, A.G. (2001) Two genetic hits (more or less) to cancer. *Nat. Rev. Cancer* 1:157–162.

Knudson, A.G. (2002) Cancer genetics. *Am. J. Med. Genet.* 111:96–102.

Kohn, K.W., O'Connor, P.M., and Jackman, J. (1994) Cell cycle regulation and the chemosensitivity of cancer cells. In: Hu, V.W., editor. *The Cell Cycle: Regulators, Targets and Clinical Applications*, New York, NY: Plenum Press.

Konopka, J.B., Watanabe, S.M., Singer, J.W., et al. (1985) Cell lines and clinical isolates derived from Ph1-positive chronic myelogenous leukemia patients express c-abl proteins with a common structural alteration. *Proc. Natl. Acad. Sci. USA* 82:1810–1814.

Kopnin, P.B., Kravchenko, I.V., Furalyov, V.A., et al. (2004) Cell type-specific effects of asbestos on intracellular ROS levels, DNA oxidation and G1 cell cycle checkpoint. *Oncogene* 23:8834–8840.

Kornberg, A. and Baker, T. (1992) *DNA replication, in 1-165-194, ed 2, New York.*

Kornberg, A. (1991) Understanding life as chemistry. *Clin. Chem.* 37(11):1895–1899.

Koscielny, S., et al. (1984) Breast cancer: relationship between the size of the primary tumour and the probability of metastatic dissemination. *Br. J. Cancer* 49:709–715.

Kouzarides, T. (2007) Chromatin modifications and their function. *Cell* 128:693–705.

Kuerbitz, S.J., et al. (1992) Wild-type p53 is a cell cycle checkpoint determinant following irradiation. *Proc. Nad. Acad. Sci. USA* 89:7491–7495.

Küppers, R. (2005) Mechanisms of B-cell lymphoma pathogenesis. *Nat. Rev. Cancer* 5:251–262.

Lemmon, M.A. and Schlessinger, J. (2010) Cell signaling by receptor tyrosine kinases. *Cell* 141:1117–1134.

Levine, A.J. (1991) *Viruses*, New York, NY: Scientific American Library.

Levresse, V., Renier, A., Levy, F., et al. (2000) DNA breakage in asbestos-treated normal and transformed (TSV40) rat pleural mesothelial cells. *Mutagenesis* 15:239–244.

List, A., Dewald, G., Bennett, J., et al. (2006) Lenalidomide in the myelodysplastic syndrome with chromosome 5q deletion. *N. Engl. J. Med.* 355:1456–1465.

Livingstone, L.R., et al. (1992) Altered cell cycle arrest and gene amplification potential accompany loss of wild-type p53. *Cell* 70:923–935.

Lloyd, G.E.R. (1984) *Hippocratic Writings*, London, UK; Penguin Classics.

Lock, R.B. (1992) Inhibition of p34cdc2 kinase activation, p34 cdc2 tyrosine dephosphorylation, and mitotic progression in Chinese hamster ovary cells exposed to etoposide. *Cancer Res.* 52:1817–1822.

Loeb, K.R. and Loeb, L.A. (2000) Significance of multiple mutations in cancer. *Carcinogenesis* 21:379–385.

Loeb, L. and Monnat, R. (2008) DNA polymerases and human disease. *Nature Rev. Genet.* 9:594–604.

Loeb, L.A., Springgate, C.F., and Battula, N. (1974) Errors in DNA replication as a basis of malignant change. *Cancer Res.* 34:2311–2321.

Loechler, E.L., Green, C.L., and Essigmann, J.M. (1984) In vivo mutagenesis by O6-methylguanine built into a unique site in a viral genome. *Proc. Natl. Acad. Sci. USA* 81:6271–6275.

Lopez-Otin, C. and Matrisian, L.M. (2007) Emerging roles of proteases in tumour suppression. *Nat. Rev. Cancer* 7:800–808.

Lu, J., et al. (2005) MicroRNA expression profiles classify human cancers. *Nature* 435:834–838.

Luch, A. (2005) Nature and nurture-lessons from chemical carcinogenesis. *Nat. Rev. Cancer* 5:113–125.

Lusitani Z. Praxis Medical Admiranda. Lugduni: J. Hugvetan; 1649.

Luzzi, K.J., MacDonald, I.C., Schmidt, E.E., et al. (1998) Multistep nature of metastatic inefficiency: dormancy of solitary cells after successful extravasation and limited survival of early micrometastases. *Am. J. Pathol.* 153:865–873.

Lynch, P.M. (2012) Current approaches in hereditary nonpolyposis colorectal cancer. *J. Natl. Compr. Canc. Netw.* 10(8):961–967.

Mackay, J. and Eriksen, M. (2002) *The Tobacco Atlas*, Geneva: World Health Organization.

Mackay, J., Jemal, A., Lee, N.C., and Parkin, D.M. (2006) *The Cancer Atlas*, Atlanta, GA: American Cancer Society.

MacLeod, M.C. (1996) A possible role in chemical carcinogenesis for epigenetic, heritable changes in gene expression. *Mol. Carcinog.* 15:241–250.

Major, M.E., Rehermann, B., Feinstone, S.M., et al. (2001) In: Howley, P.M. and Knipe, D.M., editors. *Fields Virology*, vol. 1, Philadelphia: Lippincott, Williams & Wilkins, pp. 1127–1162.

Malumbres, M. and Barbacid, M. (2009) Cell cycle, CDKs and cancer: a changing paradigm. *Nat. Rev. Cancer* 9:153–166.

Malumbres, M. and Barbacid, M. (2006) Is Cyclin D1-CDK4 kinase a bona fide cancer target? *Cancer Cell* 9:2–4.

Manning, C.B., Vallyathan, V., and Mossman, B.T. (2002) Diseases caused by asbestos: mechanisms of injury and disease development. *Int. Immunopharmacol.* 2:191–200.

Martin, M.D. and Matrisian, L.M. (2007) The other side of MMPs: protective roles in tumor progression. *Cancer Metastasis Rev.* 26:717–724.

Marusyk, A. and Polyak, K. (2010) Tumor heterogeneity: causes and consequences. *Biochim. Biophys. Acta* 1805:105–117.

Massagué, J. (2008) TGF-β in Cancer. *Cell* 134(2):215–230.

Masutomi, K., Possemato, R., Wong, J.M., et al. (2005) The telomerase reverse transcriptase regulates chromatin state and DNA damage responses. *Proc. Natl. Acad. Sci. USA* 102:8222–8227.

Matsuoka, M., Igisu, H., and Morimoto, Y. (2003) Phosphorylation of p53 protein in A549 human pulmonary epithelial cells exposed to asbestos fibers. *Environ. Health Perspect.* 111:509–512.

McLaughlin-Drubin, M.E. and Munger, K. (2008) Viruses associated with human cancer. *Biochim. Biophys Acta.* 1782:127–150.

Melnick, R.L., Kohn, M.C., and Portier, C.J. (1996) Implications for risk assessment of suggested nongenotoxic mechanisms of chemical carcinogenesis. *Env. Health Perspect.* 104(Suppl 1):123.

Meltzer, P.S. (2007) Is Ewing's sarcoma a stem cell tumor? *Cell Stem Cell* 1:13–5.

Micalizzi, D.S., Farabaugh, S.M., and Ford, H.L. (2010) Epithelial-mesenchymal transition in cancer: parallels between normal development and tumor progression. *J. Mammary Gland Biol. Neoplasia* (2):117–134.

Minn, A.J., Gupta, G.P. Siegel, P.M., et al. (2005) Genes that mediate breast cancer metastasis to lung. *Nature* 436: 518–524.

Montesano, R. and Hall, J. (2001) Environmental causes of human cancers. *Eur. J. Cancer* 37:67–87.

Mora, L.B., Buettner, R., Seigne, J., et al. (2002) Constitutive activation of Stat3 in human prostate tumors and cell lines: direct inhibition of Stat3 signaling induces apoptosis of prostate cancer cells. *Cancer Res.* 62(22):6659–6666.

Mossman, B.T. (1990) In vitro studies on the biologic effects of fibers: Correlation with in vivo bioassays. *Environ. Health Perspect.* 88:319–322.

Mossman, B.T. (1993) Mechanisms of asbestos carcinogenesis and toxicity: the amphibole hypothesis revisited. *Br. J. Ind. Med.* 50:673–676.

Mossman, B.T. and Churg, A. (1998) Mechanisms in the pathogenesis of asbestosis and silicosis. *Am. J. Respir Crit. Care Med.* 157:1666–1680.

Muñoz-Antonia, T., Torrellas-Ruiz, M., Clavell, J., et al. (2009) Aberrant methylation inactivates transforming growth factor Beta receptor I in head and neck squamous cell carcinoma. *Int. J. Otolaryngol.* 2009:848695.

Muro-Cacho, C.A., Anderson, M., Cordero, J., et al. (1999) Expression of transforming growth factor beta type II receptors in head and neck squamous cell carcinoma. *Clin. Cancer Res.* 5(6):1243–1248.

Muro-Cacho, C.A., Rosario-Ortiz, K., Livingston, S., et al. (2001) Defective transforming growth factor beta signaling pathway in head and neck squamous cell carcinoma as evidenced by the lack of expression of activated Smad2. *Clin. Cancer Res.* 7(6):1618–1626.

Musacchio, A. and Salmon, E.D. (2007) The spindle-assembly checkpoint in space and time. *Nat. Rev. Mol. Cell Biol.* 8:379–393.

National Cancer Institute (NCI) (2010) *2008-2009 Annual Report.* Available at http://deainfo.nci.nih.gov/advisory/pcp/pcp08-09rpt/PCP_Report_08-09_508.

Navin, N., Kendall, J., Troge, J., Andrews, P., et al. (2011) Tumour evolution inferred by single-cell sequencing. *Nature* 472:90–94.

Newbold, R.R., Bullock, B.C., and McLachlan, J.A. (1990) Uterine adenocarcinoma in mice following developmental treatment with estrogens: a model for hormonal carcinogenesis. *Cancer Res.* 50(23):7677–7681.

Nguyen, D.X., Bos, P.D., and Massague, J. (2009) Metastasis: from dissemination to organ-specific colonization. *Nature Rev. Cancer* 9:274–284.

Nigg, E.A. (2001) Mitotic kinases as regulators of cell division and its checkpoints. *Nat. Rev. Mol. Cell Biol.* 2:21–32.

Nowell, P.C. and Hungerford, D.A. (1960) Chromosome studies on normal and leukemic human leukocytes. *J. Natl. Cancer Inst.* 25:85–109.

Nowell, P.C. (1976) The clonal evolution of tumor cell populations. *Science* 194:23–28.

Nowell, P.C. (2002) Tumor progression: a brief historical perspective. *Semin. Cancer Biol.* 12:261–266.

Nygren, J., Suhonen, S., Norppa, H., et al. (2004) DNA damage in bronchial epithelial and mesothelial cells with and without associated crocidolite asbestos fibers. *Environ. Mol. Mutagen.* 44:477–482.

Nymark, P., Lindholm, P.M., Korpela, M.V., et al. (2007) Gene expression profiles in asbestos exposed epithelial and mesothelial lung cell lines. *BMC Genomics* 8:62.

Nymark, P., Wikman, H., Hienonen-Kempas, T., and Anttila, S. (2008) Molecular and genetic changes in asbestos-related lung cancer. *Cancer Lett.* 265:1–15.

Nymark, P., Wikman, H., Ruosaari, S., et al. (2006) Identification of specific gene copy number changes in asbestos-related lung cancer. *Cancer Res.* 66:5737–5743.

O'Brien, P.J., Hales, B.F., Josephy, P.D., et al. (1996) Chemical carcinogenesis, mutagenesis, and teratogenesis. *Can. J. Physiol. Pharmacol.* 74(5):565–571.

O'Connor, P.M., et al. (1992) Relationships between cdc2 kinase, DNA cross-linking, and cell cycle perturbations induced by nitrogen mustard. *Cell Growth Differ.* 3:43–52.

Okano, M., et al. (1999) DNA methyltransferases Dnmt3a and Dnmt3b are essential for de novo methylation and mammalian development. *Cell* 99:247–257.

Okawa, E.R., Gotoh, T., Manne, J., et al. (2008) Expression and sequence analysis of candidates for the 1p36.31 tumor suppressor gene deleted in neuroblastomas. *Oncogene* 27:803–810.

Oliner, J.D., et al. (1993) Oncoprotein MDM2 conceals the activation domain of tumor suppressor p53. *Nature* 362:857–860.

Oliveira, P.A., Colaço, A., Chaves, R., et al. (2007) Chemical carcinogenesis. *An. Acad. Bras. Cienc.* 79(4):593–616.

Orimo, A., Gupta, P.B., Sgroi, D.C., et al. (2005) Stromal fibroblasts present in invasive human breast carcinomas promote tumor growth and angiogenesis through elevated SDF-1/CXCL12 secretion. *Cell* 121:335–348.

Osada, H. and Takahashi, T. (2002) Genetic alterations of multiple tumor suppressors and oncogenes in the carcinogenesis and progression of lung cancer. *Oncogene* 21:7421–7434.

Owen, L.A. and Lessnick, S.L. (2006) Identification of target genes in their native cellular context: an analysis of EWS/FLI in Ewing's sarcoma. *Cell Cycle* 5:2049–2053.

Paakko, P., Ramet, M., Vahakangas, K., et al. (1998) Crocidolite asbestos causes an induction of p53 and apoptosis in cultured A-549 lung carcinoma cells. *Apoptosis* 3:203–212.

Paget, S. (1889) The distribution of secondary growths in cancer of the breast. *Lancet* 1:571–573.

Palekar, L.D., Eyre, J.F., Most, B.M., et al. (1987) Metaphase and anaphase analysis of V79 cells exposed to erionite, UICC chrysotile and UICC crocidolite. *Carcinogenesis* 8:553–560.

Panduri, V., Weitzman, S.A., Chandel, N., et al. (2003) The mitochondria regulated death pathway mediates asbestos induced alveolar epithelial cell apoptosis. *Am. J. Respir. Cell Mol. Biol.* 28:241–248.

Park, J.I., Venteicher, A.S., Hong, J.Y., et al. (2009) Telomerase modulates Wnt signaling by association with target gene chromatin. *Nature* 460:66–72.

Partanen, J.I., Nieminen, A.I., and Klefstrom, J. (2009) 3D view to tumor suppression: Lkb1, polarity and the arrest of oncogenic c-Myc. *Cell Cycle* 8:716–724.

Pelin, K., Husgafvel-Pursiainen, K., Vallas, et al. (1992) Cytotoxicity and anaphase aberrations induced by mineral fibres in cultured human mesothelial cells. *Toxicol. In Vitro* 6:445–450.

Perl, A.K., Wilgenbus, P., Dahl, U., et al. (1998) A causal role for E-cadherin in the transition from adenoma to carcinoma. *Nature* 392:190–193.

Perou, C.M., et al. (2000) Molecular portraits of human breast tumours. *Nature* 406:747–752.

Perry, M.E. and Levine, A.J. (1993) Tumor-suppressor p53 and the cell cycle. *Curr. Opin. Genet. Dev.* 3:50–54.

Phillips, D.H. (1983) Fifty years of benzo[a]pyrene. *Nature* 303:468–472.

Phillips, D.H. (2002) Smoking-related DNA and protein adducts in human tissues. *Carcinogenesis* 23:1979–2004.

Pietras, K. and Ostman, A. (2010) Hallmarks of cancer: interactions with the tumor stroma. *Exp. Cell Res.* 316(8):1324–1331.

Pitot, H.C. and Dragan, Y.P. (1991) Facts and theories concerning the mechanisms of carcinogenesis. *FASEB J.* 5(9):2280–2286.

Planas-Silva, M.D. and Weinberg, R.A. (1997) The restriction point and control of cell proliferation. *Curr. Opin. Cell Biol.* 9:768–772.

Polyak, K. and Weinberg, R.A. (2009) Transitions between epithelial and mesenchymal states: acquisition of malignant and stem cell traits. *Nat. Rev. Cancer* 9(4):265–273.

Poirier, M.C. (2004) Chemical-induced DNA damage and human cancer risk. *Nature* 4:630–637.

Potten, C.S. and Loeffler, M. (1990) Stem cells: attributes, cycles, spirals, pitfalls and uncertainties. Lessons for and from the crypt. *Development* 110:1001–1020.

Prendergast, G.C., et al. (1991) Methylation-sensitive sequence-specific DNA binding by the c-Myc basic region. *Science* 251:186–189.

Preston, R.J. (2005) Mechanistic data and cancer risk assessment: the need for quantitative molecular endpoints. *Environ. Mol. Mutagen.* 45:214–221.

Preston-Martin, S., Pike, M.C., Ross, R.K., et al. (1990) Increased cell division as a cause of human cancer. *Cancer Res.* 50(23):7415–7421.

Puhakka, A., Ollikainen, T., Soini, Y. et al. (2002) Modulation of DNA single-strand breaks by intracellular glutathione in human lung cells exposed to asbestos fibers. *Mutat. Res.* 14:7–17.

Qian, B.Z. and Pollard, J.W. (2010) Macrophage diversity enhances tumor progression and metastasis. *Cell* 141(1):39–51.

Raaschou-Nielsen, O., Bak, H., Sørensen, M., et al. (2010) Air pollution from traffic and risk for lung cancer in three Danish cohorts. *Cancer Epidemiol. Biomarkers Prev.* 19(5):1284–1291.

Resnitzky, D., et al. (1994) Acceleration of the G1/S phase transition by expression of cyclins D1 and E with an inducible system. *Mol. Cell Biol.* 14:1669–1679.

Reya, T., Morrison, S.J., Clarke, M.F., and Weissman, I.L. (2001) Stem cells, cancer, and cancer stem cells. *Nature* 414(6859):105–111.

Ribatti, D. (2009) Endogenous inhibitors of angiogenesis: a historical review. *Leuk. Res.* 33:638–644.

Riggi, N. and Stamenkovic, I. (2007) The biology of Ewing sarcoma. *Cancer Lett.* 254:1–10.

Riggs, A.D., et al. (1983) 5-methylcytosine, gene regulation, and cancer. *Adv. Cancer Res.* 40:1–30.

Risinger, J.L., et al. (1993) Genetic instability of microsatellites in endometrial carcinoma. *Cancer Res.* 53:5100–5103.

Rissle, P. (1997) Induced drug resistance inhibits selection of initiated cells and cancer development. *Carcinogenesis* 18:649–655.

Ronald, T. Javier and Janet, S. Butel (2008) The history of tumor virology. *Cancer Res.* 68(19):7693–7706.

Rous, P. (1910) A transmissible avian neoplasm (sarcoma of the common fowl). *J. Exp. Med.* 12:696–705.

Rowley, J.D. (1973) A new consistent chromosomal abnormality in chronic myelogenous leukaemia identified by quinacrine fluorescence and Giemsa staining. *Nature* 243:290–293.

Sahai, E. (2007) Illuminating the metastatic process. *Nat. Rev. Cancer* 7:737–749.

Saito, Y., et al. (2006) Epigenetic activation of tumor suppressor micro-RNAs in human cancer cells. *Cell Cycle* 5:2220–2222.

Sanchez, V.C., Pietruska, J.R., Miselis, N.R., et al. (2009) Biopersistence and potential adverse health impacts of fibrous nanomaterials: what have we learned from asbestos? *Wiley Interdiscip. Rev. Nanomed. Nanobiotechnol.* 1:511–529.

Sander, M., Cadet, J., Casciano, D.A., et al. (2005) Proceedings of a workshop on DNA adducts: biological significance and applications to risk assessment Washington, DC, April 13–14, 2004. *Toxicol. Appl. Pharmacol.* 208:1–20.

Santarosa, M. and Ashworth, A. (2004) Haploinsufficiency for tumour suppressor genes: when you don't need to go all the way. *Biochim. Biophys. Acta* 1654:105–122.

Schmalhofer, O., Brabletz, S., and Brabletz, T. (2009) E-cadherin, beta-catenin, and ZEB1 in malignant progression of cancer. *Cancer Metastasis Rev.* 28(1–2):151–166.

Schones, D.E., et al. (2008) Dynamic regulation of nucleosome positioning in the human genome. *Cell* 132:887–898.

SEER (2010) *Cancer Statistics Review, 1975–2007. Surveillance, Epidemiology, and End Results Program, National Cancer Institute.* Available at http://seer.cancer.gov/csr/1975_2007/index/html.

Selikoff, I.J., Hammond, E.C., and Churg, J. (1968) Asbestos exposure, smoking, and neoplasia. *JAMA* 204:106–112.

Seo, K.Y., Jelinsky, S.A., and Loechler, E.L. (2000) Factors that influence the mutagenic patterns of DNA adducts from chemical carcinogens. *Mutat. Res.* 463:215–246.

Sharma, S.V., Bell, D.W., Settleman, J., et al. (2007) Epidermal growth factor receptor mutations in lung cancer. *Nat. Rev. Cancer* 7:169–181.

Sharma, S., Kelly, T., and Jones, P.A. (2010) Epigenetics in cancer. *Carcinogenesis* 31:27–36.

Shi, Y. (2007) Histone lysine demethylases: emerging roles in development, physiology and disease. *Nat. Rev. Genet.* 8:829–833.

Shor, A.C., Keschman, E.A., Lee, F.Y., Muro-Cacho, C., Letson, G. D., Trent, J.C., and Jove, R. (2007) Dasatinib inhibits migration and invasion in diverse human sarcoma cell lines and induces apoptosis in bone sarcoma cells dependent on SRC kinase for survival. *Cancer Res.* 67(6):2800–2808.

Shrivastav, M., De Haro, L.P., and Nickoloff, J.A. (2008) Regulation of DNA double-strand break repair pathway choice. *Cell Res.* 18:134–147.

Shtivelman, E., Lifshitz, B., Gale, R.P., et al. (1985) Fused transcript of abl and bcr genes in chronic myelogenous leukemia. *Nature* 315:550–554.

Shukla, A., Gulumian, M., Hei, T.K., Kamp, D., Rahman, Q., and Mossman, B.T. (2003a) Multiple roles of oxidants in the pathogenesis of asbestos-induced diseases. *Free Radic. Biol. Med.* 34:1117–1129.

Shukla, A., Jung, M., Stern, M., Fukagawa, N.K., et al. (2003b) Asbestos induces mitochondrial DNA damage and dysfunction linked to the development of apoptosis. *Am. J. Physiol. Lung Cell. Mol. Physiol.* 285:L1018–L1025.

Siegel, R., Naishadham, M.A., and Jemal, A. (2012) Cancer Statistics. *CA Cancer J. Clin.* 62:10–29.

Simonich, M.T., Egner, P.A., Roebuck, B.D., et al. (2007) Natural chlorophyll inhibits aflatoxin B1-induced multi-organ carcinogenesis in the rat. *Carcinogenesis* 6:1294–1302.

Singer, B. and Essigmann, J.M. (1991) Site-specific mutagenesis: retrospective and prospective. *Carcinogenesis* 12:949–955.

Singh, A. and Settleman, J. (2010) EMT, cancer stem cells and drug resistance: an emerging axis of evil in the war on cancer. *Oncogene* 29(34):4741–4751.

Slichenmeyer, W.J., et al. (1993) Loss of a p53-associated G1 checkpoint does not decrease cell survival following DNA damage. *Cancer Res.* 53:4164–4168.

Slupphaug, G., Kavli, B., and Krokan, H.E. (2003) The interacting pathways for prevention and repair of oxidative DNA damage. *Mutat. Res.* 531:231–251.

Smith, A.J., et al. (1996) Mutant p53 protein as a biomarker of chemical carcinogenesis in humans. *J. Occup. Environ. Med.* 38:743–749.

Sorensen, P.H., Lessnick, S.L., Lopez-Terrada, D. et al. (1994) A second Ewing's sarcoma translocation, t(21;22), fuses the EWS gene to another ETS-family transcription factor, ERG. *Nat. Genet.* 6:146–151.

Sorlie, T. et al. (2001) Gene expression patterns of breast carcinomas distinguish tumor subclasses with clinical implications. *Proc. Natl. Acad. Sci. USA* 98:10869–10874.

Spivak, G. and Hanawalt, P.C. (1992) Translesion DNA synthesis in the dihydrofolate reductase domain of UV-irradiated CHO cells. *Biochemistry* 31(29):6794–6800.

Spiegelhalder, B. and Bartsch, H. (1996) Tobacco-specific nitrosamines. *Eur. J. Cancer Prev.* 5:33–38.

Stehelin, D., Guntaka, R.V., Varmus, H.E., et al. (1976) Purification of DNA complementary to nucleotide sequences required for neoplastic transformation of fibroblasts by avian sarcoma viruses. *J. Mol. Bio.* 101(3):349–365.

Straif, K., Benbrahim-Tallaa, L., Baan, R., et al. (2009) A review of human carcinogens—Part C: metals, arsenic, dusts, and fibres. *Lancet Oncol.* 10:453–454.

Straub, K.M. and Burlingame, A.L. (1981) Carcinogen binding to DNA. *Biomed. Mass Spectrom.* 9:431–435.

Sudarsanam, S. and Johnson, D.E. (2010) Functional consequences of mTOR inhibition. *Curr. Opin. Drug Discov. Devel.* 13:31–40.

Suvà, M.L., Riggi, N., Stehle, J.C., et al. (2009) Identification of cancer stem cells in Ewing's sarcoma. *Cancer Res.* 69(5):1776–1781.

Suzuki, M.M., et al. (2008) DNA methylation landscapes: provocative insights from epigenomics. *Nat. Rev. Genet.* 9:465–476.

Swauger, J.E., Steichen, T.J., Murphy, P.A., et al. (2002) An analysis of the mainstream smoke chemistry of samples of the US cigarette market acquired between 1995 and 2000. *Regul. Toxicol. Pharmacol.* 35:142–156.

Sykes, S.M., Mellert, H.S., Holbert, M.A., et al. (2006) Acetylation of the p53 DNA-binding domain regulates apoptosis induction. *Mol. Cell* 24:841–851.

Szeliga, J. (1997) Characterization of DNA adducts formed by the four configurationally isomeric 5,6-dimethylchrysene 1,2-dihydrodiol 3,4-epoxides. *Chem. Res. Toxicol.* 105:378–385.

Takahashi, K., Okita, K., Nakagawa, M., and Yamanaka, S. (2007) Induction of pluripotent stem cells from fibroblast cultures. *Nat. Protoc.* 2:3081–3089.

Takai, D., et al. (2002) Comprehensive analysis of CpG islands in human chromosomes 21 and 22. *Proc. Natl. Acad. Sci. USA* 99:3740–3745.

Talmadge, J.E. and Fidler, I.J. (2010) AACR centennial series: the biology of cancer metastasis: historical perspective. *Cancer Res.* 70(14):5649–5669.

Tammela, T. and Alitalo, K. (2010) Lymphangiogenesis: molecular mechanisms and future promise. *Cell* 140(4):460–476.

Tan, A., Yeh, S.H., Liu, C.J., et al. (2008) Viral hepatocarcinogenesis: from infection to cancer. *Liver Int.* 28:175–188.

Thibodeau, S.N., Bren, G., and Schaid, D. (1993) Microsatellite instability in cancer of the proximal colon. *Science* 260:816–819.

Thiery, J.P. (2002) Epithelial-mesenchymal transitions in tumour progression. *Nat. Rev. Cancer* 2:442–454.

Tomatis, L., Huff, J., Hertz-Picciotto, I., et al. (1997) Avoided and avoidable risks of cancer. *Carcinogenesis* 18:97–105.

Topinka, J., Loli, P., Georgiadis, P., Dusinska, M., et al. (2004) Mutagenesis by asbestos in the lung of lambda-lacI transgenic rats. *Mutat. Res.* 553:67–78.

Tsao, Y.P., D'Arpa, P., and Liu, L.F. (1992) The involvement of active DNA synthesis in camptothecin-induced G2 arrest: altered regulation of p34cdc2/cyclin B. *Cancer Res.* 52:1823–1829.

Tulp, N. (1652) *Observationes Medicae*, Amstelredami: Elzevirium.

Unfried, K., Schurkes, C., and Abel, J. (2002) Distinct spectrum of mutations induced by crocidolite asbestos: Clue for 8-hydroxydeoxyguanosine-dependent mutagenesis in vivo. *Cancer Res.* 62:99–104.

Vainio, H. and Weiderpass, E. (2003) Smokeless tobacco: harm reduction or nicotine overload? *Eur. J. Cancer Prev.* 12:89–92.

van Veer, L.J., et al. (2002) Gene expression profiling predicts clinical outcome of breast cancer. *Nature* 415:530–536.

van de Vijver, M.J., et al. (2002) A gene-expression signature as a predictor of survival in breast cancer. *N. Engl. J. Med.* 347:1999–2009.

van de Stolpe, A. (2013) On the origin and destination of cancer stem cells: a conceptual evaluation. *Am. J. Cancer Res.* 3(1):107–116.

Verbeek, B., Southgate, T.D., Gilham, D.E., et al. (2008) O6-Methylguanine-DNA methyltransferase inactivation and chemotherapy. *Br. Med. Bull.* 85:17–33.

Vincent, M.D. (2011) Cancer: beyond speciation. *Adv. Cancer Res.* 112:283–350.

Virchow, R. (1863) *Cellular Pathology as Based Upon Physiological and Pathological Histology*, Philadelphia: J.B. Lippincott.

Virta, R.L. (2003) *Worldwide Asbestos Supply and Consumption Trends From 1900 to 2000*, Reston, VA: U.S. Department of the Interior, U.S. Geological Survey.

Vogelstein, B. and Kinzler, K.W. (2004) Cancer genes and the pathways they control. *Nat. Med.* 10(8):789–799.

Vogelstein, B., et al. (1988) Genetic alterations during colorectal-tumor development. *N. Engl. J. Med.* 319:525–532.

Vos, J.M. and Hanawalt, P.C. (1989) DNA interstrand cross-links promote chromosomal integration of a selected gene in human cells. *Mol. Cell Biol.* 9(7):2897–2905.

Waddington, C.H. (1942) The epigenotype. *Endeavour* 1:18–20.

Waga, S., Bauer, G., and Stillman, B. (1994) Reconstitution of complete SV40 DNA replication with purified replication factors. *J. Biol. Chem.* 269:10923–10934.

Wan, P.T., et al. (2004) Mechanism of activation of the RAF-ERK signaling pathway by oncogenic mutations of B-RAF. *Cell* 116:855–867.

Wang, Y.C., et al. (1993) Evidence from mutation spectra that the UV hypermutability of xeroderma pigmentosum variant cells reflects abnormal, error-prone replication on a template containing photoproducts. *Mol. Cell Biol.* 13:4276–4283.

Wang, Y., et al. (2004) An evaluation of new criteria for CpG islands in the human genome as gene markers. *Bioinformatics* 20:1170–1177.

Wang, Z., et al. (2008) Combinatorial patterns of histone acetylations and methylations in the human genome. *Nat. Genet.* 40:897–903.

Wauthier, E.L., Hanawalt, P.C., and Vos, J.-M.H. (1990) Differential repair and replication of damaged DNA in ribosomal RNA genes in different CHO cell lines. *J. Cell Biochem.* 43:173–183.

Weilbaecher, K.N., Guise, T.A., and McCauley, L.K. (2011) Cancer to bone: a fatal attraction. *Nat. Rev. Cancer* 11:411–425.

Weinberg, R.A. (2007) *The Biology of Cancer*, New York, NY: Garland Science.

Weir, B.A., Woo, M.S., Getz, G., et al. (2007) Characterizing the cancer genome in lung adenocarcinoma. *Nature* 450(7171): 893–898.

Wellen, K.E. and Thompson, C.B. (2010) Cellular metabolic stress: considering how cells respond to nutrient excess. *Mol. Cell* 40(2):323–332.

West, J., Munoz-Antonia, T., Johnson, J.G., et al. (2000) Transforming growth factor-beta type II receptors and smad proteins in follicular thyroid tumors. *Laryngoscope* 110(8):1323–1327.

Westbrook, T.F., Martin, E.S., Schlabach, M.R., et al. (2005) A genetic screen for candidate tumor suppressors identifies REST. *Cell* 121:837–848.

White, M.K. and Khalili, K. (2004) Polyomaviruses and human cancer: molecular mechanisms underlying patterns of tumorigenesis. *Virology* 324:1–16.

Williams, G.H. (2012) The cell cycle and cancer. *J. Pathol.* 226:352–364.

WHO (2009) *Cancer Fact Sheet no. 297* Available at http://www.who.int/mediacentre/factsheets/fs297/en/index.html.

Wiencke, J.K. (2002) DNA adduct burden and tobacco carcinogenesis. *Oncogene* 21:7376–7391.

Willis, S.N. and Adams, J.M. (2005) Life in the balance: how BH3-only proteins induce apoptosis. *Curr. Opin. Cell Biol.* 17:617–625.

Wilson, S.M., Barbone, D., Yang, T.M., et al. (2008) mTOR mediates survival signals in malignant mesothelioma grown as tumor fragment spheroids. *Am. J. Respir. Cell Mol. Biol.* 39:576–583.

Wilson, A.S., et al. (2007) DNA hypomethylation and human diseases. *Biochim. Biophys. Acta* 1775:138–162.

Witsch, E., Sela, M., and Yarden, Y. (2010) Roles for growth factors in cancer progression. *Physiology (Bethesda)* 25:85–101.

Wood, L.D., et al. (2007) The genomic landscapes of human breast and colorectal cancers. *Science* 318:1108–1113.

WHO (2003) Stewart, B.W. and Kleihues, P., editors. *Cancer Report*, Lyon: IARC.

WHO Health Report (2002) *Reducing Risks, Promoting Healthy Life*, Geneva: World Health Organization.

Yanaihara, N., Caplen, N., Bowman, E., et al. (2006) Unique microRNA molecular profiles in lung cancer diagnosis and prognosis. *Cancer Cell* 9:189–198.

Yang, H., Bocchetta, M., Kroczynska, B., et al. (2006) TNF-alpha inhibits asbestos-induced cytotoxicity via a NF-kappaB-dependent pathway, a possible mechanism for asbestos-induced oncogenesis. *Proc. Natl. Acad. Sci. USA* 103:10397–10402.

Yegles, M., Janson, X., Dong, H.Y., et al. (1995) Role of fibre characteristics on cytotoxicity and induction of anaphase/telophase aberrations in rat pleural mesothelial cells in vitro: Correlations with in vivo animal findings. *Carcinogenesis* 16:2751–2758.

Yegles, M., Saint-Etienne, L., Renier, A., et al. (1993) Induction of metaphase and anaphase/telophase abnormalities by asbestos fibers in rat pleural mesothelial cells in vitro. *Am. J. Respir. Cell Mol. Biol.* 9:186–191.

Yilmaz, M., Christofori, G., and Lehembre, F. (2007) Distinct mechanisms of tumor invasion and metastasis. *Trends Mol. Med.* 13:535–541.

Yu, H., Pardoll, D., and Jove, R. (2009) STATs in cancer inflammation and immunity: a leading role for STAT3. *Nat. Rev. Cancer* 9:798–809.

Yuspa, S.H., Henning, H., Lichti, U., et al. (1983) Organ specificity and tumor promotion. *Basic Life Sci.* 24:157–171.

Zabarovsky, E.R., Lerman, M.I., and Minna, J.D. (2002) Tumor suppressor genes on chromosome 3p involved in the pathogenesis of lung and other cancers. *Oncogene* 21:6915–6935.

Zetterberg, A. and Larsson, O. (1985) Kinetic analysis of regulatory events in G1 leading to proliferation or quiescence of Swiss 3T3 cells. *Proc. Natl. Acad. Sci. USA* 82:5365–5369.

Zhan, Q., Carrier, F., and Fornace, A.J. Jr (1993) Induction of cellular p53 activity by DNA-damaging agents and growth arrest. *Mol. Cell Biol.* 13:4242–4250.

zur Hausen, H., Meinhof, W., Scheiber, W., et al. (1974) Attempts to detect virus-specific DNA in human tumors. I. Nucleic acid hybridizations with complementary RNA of human wart virus. *Int. J. Cancer* 13:650–656.

zur Hausen, H. (1976) Condylomata acuminata and human genital cancer. *Cancer Res.* 36:794.

zur Hausen, H. (2001) Oncogenic DNA viruses. *Oncogene* 20:7820–7823.

108

GENOTOXICITY TESTING STRATEGIES, GUIDELINES, AND METHODS

CHING-HUNG HSU AND QINGLI WANG

Genotoxicity testing is widely used to detect the ability of substances or chemicals to induce genetic damage (mutations or chromosome aberrations (CA)s or other DNA damage). Numerous genotoxic assays have been developed, and several of them are commonly used in regulatory toxicology by regulatory authorities. This chapter describes and summarizes some *in vitro* and *in vivo* genotoxicity assays that are primarily recommended by regulatory agencies and/or organizations (e.g., International Conference on Harmonisation of Technical Requirements for Registration of Pharmaceuticals for Human Use (ICH), Organization for Economic Co-operation and Development (OECD), Food and Drug Administration (FDA), Environmental Protection Agency (EPA)) for use in regulatory registration of chemicals, drugs, biological products, and food ingredients. An abbreviated strategy is presented, highlighting the current standard test battery of genotoxicity assays and the updated guidelines for testing genetic damage developed by various regulatory authorities. In addition, nonanimal alternative methods for prediction of genotoxicity and several new genotoxicity techniques and/or methods in the field of genotoxicology are also presented in this chapter.

INTRODUCTION

Genotoxic testing has been increasingly applied to detect the ability of substances or chemicals to produce mutations or CAs or other DNA damage. Mutations are heritable alterations in the structural DNA content in somatic cells or germ cells that can cause harmful effects, such as cancer, birth defects, and inherited diseases. DNA damage is also involved in genetic toxicity, which may result in mutations and the nonheritable changes associated with mutagenesis. In early years, genetic damage mainly focused on the heritable gene and chromosomal germ cell mutations in individuals. However, with the increasing development of cancer research, accumulating evidence has demonstrated that most of the carcinogenic chemicals are mutagenic. Therefore, the genotoxicity testing shifted from heritable mutations to carcinogenesis. Numerous known human carcinogens have shown to be as non-mutagenic in human germ cells, but some germ cell mutagens are shown to be both mutagenic and carcinogenic to somatic cells *in vivo*; therefore, germ cell mutagenicity is also being considered as a factor of genetic damage by the regulatory authorities. In the 1970s, the genetic toxicity test compilations were firstly published. Given the high predictive value of cancer using the *in vitro* genetic toxicity testing, the United States and other regulatory authorities began to require pre-market genotoxicity testing for chemicals and drugs. The formal guidelines on how to conduct the tests were developed and recommended by the United States and other international organizations [such as the ICH and the International OECD]. In particular, the OECD test guidelines (TGs) are considered to be a gold standard for regulatory toxicity testing, which have been worldwide accepted. The OECD TGs undergo periodic revision to reflect scientific advances and include recently adopted OECD principles. Most existing TGs on genotoxicity were last revised in the mid-1990s. Similar to OECD TGs, the Expert Working Group (Safety) of the ICH has developed a comprehensive set of safety guidelines to reveal potential risks of carcinogenicity, genotoxicity, and reproductive toxicity in accordance with the ICH Process. The

Hamilton & Hardy's Industrial Toxicology, Sixth Edition. Edited by Raymond D. Harbison, Marie M. Bourgeois, and Giffe T. Johnson.

veterinary equivalent to the ICH is the International Co-operation of Harmonisation of Technical Requirements for Registration of Veterinary Medicinal Products and Veterinary Pharmaceuticals (VICH). The OECD and ICH/VICH guidelines comprise two sets of internationally harmonized genotoxicity guidelines in regulatory purpose. The United States, as one of tripartite regulatory bodies, also follows the ICH guidelines as technical requirements for the registration of drugs and biological products. The pharmaceutical industry in the three regions has agreed to a harmonized approach through the acceptance and implementation of the ICH. In this chapter, these genotoxicity assays for testing chemicals, drugs, and biological products or food additives for humans will be emphasized. The current generalized strategy for genotoxicity testing primarily covers three areas, namely, gene mutation, CA or breakage (clastogenicity), and chromosome loss or gain (aneuploidy). To detect all of these end points, many different assays are employed *in vitro* and *in vivo*. *In vitro* genotoxic assays include the bacterial mutagenesis assay for detecting gene mutations and mammalian cell mutagenicity assays for detection of clastogenicity. On the other hand, *in vivo* genotoxicity assays include mammalian bone marrow cell micronucleus (MN) assay or CA for testing the clastogenicity and/or aneuploidy (Kirkland et al., 2005; Kirkland et al., 2006; Kirkland and Speit, 2008). To date, the genetic toxicity test battery consists of a bacterial (Ames) assay, an *in vitro* cytogenetic assay, and *in vivo* rodent genotoxicity (Claxton et al., 2010; Mahadevan et al., 2011), that has been widely adopted for regulatory purpose. Substances or chemicals that are positive in these assays (*in vitro* or *in vivo*) are considered to be of the greatest concern that have a potential to increase cancer risk to some extent. In recent years, an *in vitro* mammalian cell MN assay was adopted by the OECD. Meanwhile, the OECD is working on three TGs (*in vitro* CA, *in vivo* MN, and CA). A recent breakthrough is that the rodent single cell gel electrophoresis (SCGE) (comet assay) has been added into the ICH guidance (ICH, 2011) because of its numerous advantages, especially *in vivo* (Burlinson, 2012).

In this chapter, the authors aim to present a brief overview of the development of genotoxicity tests for regulatory purposes and support rationales for the types of tests regulated by the United States, OECD, or other regulatory authorities and the scientific and practical issues that will need to be resolved in the future. More detailed descriptions of testing strategies, test methods, and decision processes are also addressed in this chapter.

GENOTOXICITY

Overview

Genetic toxicity testing was first developed in the early 1950s (Demerec, 1951; Demerec and Hanson, 1951) to identify potential carcinogens. Genotoxic potential can be manifested in various ways; however, the most common end points studied include gene mutations, structural/numerical CAs, DNA damage or strand breakages, and DNA repair. These end points are investigated in both *in vitro* and *in vivo* assays, but any single testing has its limitation that cannot detect all these changes. Among these genotoxicity tests, the bacterial mutagenicity assay is the most widely used because it is simple, saves time, inexpensive, and least likely to falsely detect a noncarcinogen as a carcinogen. The bacterial reverse mutation assay, usually was called the Ames test, is commonly employed as an initial screen for genotoxicity and, in particular, for point mutation. The Ames test is also required in regulatory toxicology for approval and marketing of a chemical, drug, or food additive, which has the ability to predict the mutagenicity and carcinogenicity because of its reliability, robustness, and reproducibility.

The mammalian cell cytogenetic assays are primarily used to detect a chromosome damage, and they include CA, MN, and mouse lymphoma assay (MLA) (Mahadevan et al., 2011). Generally, the chromosomal mutations have two major categories associated with changes both in chromosome structure (namely, CAs) and the number of chromosomes (also called numerical aberrations). Chromosomal mutations and related events cause many human genetic diseases. There is accumulating evidence that chromosome mutations and related events are involved in cancer occurrence in humans and experimental animals. The chromosomal mutations in somatic cells play a critical role in processes leading to cancer. If mutations occur in proto-oncogenes, tumor suppressor genes, and/or DNA damage response genes, they can result in various genetic diseases. In the CA, the most common structural CAs are observed at metaphase. In the MN, the analysis of CAs at interphase is performed via scoring the forming MN, which represent a proportion of fragments or whole chromosome lagging in anaphase and are not included in the main nucleus. The MLA detects forward mutations at the thymidine kinase (*Tk*) locus of the L5178Y (*Tk*$^{+/-}$ -3.7.2C) cell line derived from a mouse thymic lymphoma. The MLA has the capability to measure a wide range of genetic events, including point mutations, deletions (intragenic) and multilocus, chromosomal rearrangements, mitotic recombination, and nondisjunction (Moore et al., 1985a, 1985b; Applegate et al., 1990; Wang et al., 2009). In these assays, mammalian cells are used, such as Chinese hamster ovary (CHO) cells, L5179Y/ *Tk*$^{+/-}$ -3.7.2C mouse lymphoma cells, Chinese hamster lung (CHL and V79) cells, or human peripheral lymphocytes.

In vivo genotoxicity assays, as *in vitro* counterpart, primarily include *in vivo* MN and CA for detecting chromosome damage effects in mammalian cells. The OECD guideline 474 for mammalian erythrocyte MN test (OECD, 1997a) and TG 475 for mammalian bone marrow CA test (OECD, 1997b) were adopted in July 1997. In these assays,

TABLE 108.1 Some Common Assays Used for Testing Genetic Toxicity

Assay name	Endpoint	In Vitro/In Vivo
Bacterial reverse mutation assay (Ames test)	Gene mutation	In vitro
Mouse lymphoma (L5178Y/$Tk^{+/-}$ -3.7.2C) forward mutation assay	Gene mutation and chromosome mutation	In vitro
Mammalian cell Hprt mutation	Gene mutation	In vitro
Rodent lymphocyte Hprt mutation assay	Gene mutation	In vivo
Rodent erythrocyte and lymphocyte Pig-a genes mutation assay	Gene mutation	In vivo
Mouse spot test	Gene mutation	In vivo
Mouse specific locus test	Gene mutation	In vivo
Tk gene mutation assay	Gene mutation and chromosome mutation	In vivo
Transgenic rodent gene mutation assay (somatic and germ cells)	Gene mutation	In vivo
Mammalian chromosomal aberration assay	Clastogenicity	In vitro
Mammalian micronucleus assay	Clastogenicity, aneugenicity	In vitro
Rodent chromosomal aberration assay (somatic and germ cells)	Clastogenicity	In vivo
Rodent micronucleus assay (bone marrow and peripheral blood)	Clastogenicity, aneugenicity	In vivo
Micronucleus assay (human lymphocytes)	Clastogenicity, aneugenicity	In vivo
Rodent dominant lethal test	Clastogenicity (germ cell)	In vivo
Mouse heritable translocation assay	Clastogenicity (germ cell)	In vivo
Mammalian spermatogonial chromosome aberration test	Chromosome aberration(germ cell)	In vivo
Comet assay	DNA strand break	In vitro/In vivo
Alkaline elution assay	DNA strand break	In vitro/In vivo
Unscheduled DNA synthesis (UDS) assay	DNA repair	In vitro/In vivo
Sister chromatid exchange in mammalian cells	DNA damage	In vitro/In vivo
DNA adduct analysis	DNA adducts	In vitro

mice and rats are usually used. Theoretically, other appropriate mammalian species may be used. Both red blood cells from peripheral lymphocytes cultured and bone marrow from treated rodents were used to analyze CAs. The *in vivo* assays, especially related to assessing a mutagenic hazard, also allow consideration of *in vivo* metabolism, pharmacokinetics, and DNA repair processes, although they may vary among species and tissues. An *in vivo* test is also useful for further investigation of a mutagenic effect detected by these *in vitro* assays (OECD TG 473 (OECD, 1997c) and TG 487 (OECD, 2010)). Generally, when these assays alone are inadequate to evaluate the genetic toxicity, more *in vivo* assays are often required to evaluate a DNA damage end point as a surrogate. These *in vivo* assays, which have the most published protocols, include the DNA strand break assays such as the comet assay and alkaline elution assay, *in vivo* transgenic rodent (TGR) mutation assay, DNA covalent binding assay, and liver unscheduled DNA synthesis (UDS) assay (ICH, 2011). Aside from chromosomal damage, assessment of gene mutation is also an important part of genotoxicity testing employed in nonclinical safety evaluation. Recently, the rodent *Pig-a* gene mutation assay is increasingly used. If there is a loss of function of the *Pig-a* gene in red blood cells, it will result in a lack of cell surface expression of specific proteins that are targeted to the surface by glycosylphosphatidylinositol (GPI) anchors. This cell surface phenotype is readily assessed by flow cytometric analysis.

Classification (or Types of Genetic Toxicity Testing)

Genetic toxicity testing is used to evaluate the agents' ability to induce DNA alteration, including gene mutations. To date, over 200 tests have been developed for detecting genotoxic effects, investigating mechanisms of mutagenesis, and predicting the carcinogenic potential. Some of these assays are listed in Table 108.1, which can detect different end points in genotoxicity, such as gene mutation, chromosomal damage, and DNA damage. For regulatory requirements of a chemical, drug, and biological product, a comprehensive assessment of their genotoxic potential is required.

Strategy and Regulation of the Genotoxicity Assessment

Genotoxicity assessment consists of a comprehensive test battery, including different end points included because of the multiplicity of mechanisms involved in genetic damage. Genotoxicity testing schemes are becoming simplified, and short-term assays have been widely used for detecting genotoxicity, investigating genetic effects, and supporting regulatory use. A limited test battery to genotoxicity testing (ICH core battery) for detecting gene mutation and CAs is preferable given the broad spectrum of potential events and the limits of single-test analysis for detecting all genotoxic mechanisms relevant in tumorigenesis (ICH, 2011). The standard test battery has some features: (1) assessment of mutagenicity in a bacterial reverse gene mutation test, which has been shown to detect relevant genetic changes and the majority of genotoxic carcinogens in rodents and humans; (2) *in vitro* and/or *in vivo* assay in mammalian cells used for genotoxicity evaluation. Generally, several *in vitro* mammalian cell systems are widely used and can be considered sufficiently validated, such as the *in vitro* CA, *in vitro* MN, and MLA. These three assays are currently considered equally appropriate and therefore are interchangeable for measurement of chromosomal damage when used together with other genotoxicity tests in a standard battery for testing pharmaceuticals (ICH, 2011). While some alternative *in vitro* tests, such as *in vitro* cell transformation, DNA breakage and repair, UDS, and DNA adduct formation, are now primarily used in basic research or as supplemental methods to support or explain results from the core battery. The *in vitro* assays should be conducted both with and without the metabolic activation (S9) and include positive control as well as solvent/vehicle controls for each treatment. Weight of evidence (WoE) method has been set up to evaluate complex response patterns from a large test battery. Several additional *in vivo* assays, such as the comet assay, alkaline elution assay, transgenic mouse mutation assay, *in vivo* DNA covalent binding assay, and liver UDS assay also can be utilized in the ICH core battery or as follow-up tests to develop WoE in assessing results of *in vitro* or *in vivo* assays. Compared to an *in vitro* assay, the *in vivo* test plays a critical role in genotoxicity testing strategy. The value of *in vivo* results is directly related to the demonstration of adequate exposure of the target tissue to the test agent. Generally, negative results from appropriate *in vivo* assays (usually two), with adequate justification for the end points measured, and demonstration of exposure, are considered sufficient to demonstrate the absence of significant genotoxic risk. However, internationally agreed protocols are not yet in place for all *in vivo* assays, although considerable experience and protocol recommendations as well as published data exist for comet assays, alkaline elution assays, UDS assays, and DNA covalent binding measurements. Genotoxicty test batteries can detect carcinogens that are considered to act primarily via causing direct genetic damage, but they are not expected to detect non-genotoxic carcinogens. In addition, a positive result in any genotoxicity assay does not always mean that the agent poses a genotoxic/carcinogenic hazard to humans.

Genotoxicity Assays

In Vitro Assays

Bacterial Reverse Mutation Assay (Ames Test)   The bacterial reverse mutation assay, also called as Ames test, is mostly widely used for detecting gene mutation induced by agents. The Ames test was developed by Professor Bruce Ames in the early 1970s. In this assay, the most often used bacteria systems are *Salmonella typhimurium* and *Escherichia coli*. The major strains utilized in the Ames test consist of amino acid requiring strains of *S. typhimurium* TA97, 98, 100, 102, 1535, 1537, and 1538, which are used for detecting base-pair substitution or frameshift mutations in genes of the histidine (*his*) operon, and *E, coli* WP2 *uvr* A and WP2 *uvr* A pKM101, which are aimed to detect a base-pair substitution mutation in the tryptophan (*trp*) operon. These strains are mutants and are unable to grow and form colonies on a medium without histidine or tryptophan. The recommended set of bacterial strains (OECD, 1997; ICH, 2011) can detect base substitution and frameshift mutations. Test strains are treated with a range of concentrations of the test chemical or agent to determine whether the agent can cause point mutations. The revert mutations present in the test strains restore the functional capability of the bacteria to synthesize an essential amino acid. The revertant bacteria are detected by their ability to grow in the absence of the amino acid required by the parent test strain. This assay is performed both with and without the metabolic activation (S9). The most commonly used S9 fraction is a mixture of microsome and cytosol prepared from the livers of rodents treated with enzyme-inducing agents, such as Aroclor 1254 or a combination of phenobarbitone and β-naphthoflavone, which contain a variety of phase I and phase II enzymes. The choice and condition of the S9 may depend on the class of chemical being tested. There following are the two methods often used: the plate incorporation and preincubation methods. Briefly, suspensions of bacterial culture are exposed to the test chemical with and without S9. In the plate incorporation method, these suspensions are mixed with an overlay agar and plated immediately onto a minimal medium. In the preincubation method, the treatment mixture is incubated and then mixed with an overlay agar before plating onto a minimal medium. For both techniques, after 2 or 3 days of incubation, revertant colonies are counted and compared to the number of spontaneous revertant colonies on the solvent. The detailed information is described in OECD TG 471 (OECD, 1997d) and ICH S2 (R1) (ICH, 2011).

Mouse Lymphoma Forward Mutation Assay   The mouse lymphoma (L5178Y/$Tk^{+/-}$ -3.7.2C) forward mutation assay, referred to as MLA, is a mammalian gene mutation assay used within a standard test battery recommended by the OECD and ICH guideline. This assay was established by Donald Clive and colleagues in 1970s (Clive, 1973; Clive and Spector, 1975). The MLA detects a broad spectrum of genetic events, including mutations in the *Tk* gene that result from both gene mutations and chromosome damage, including chromosome loss. Therefore, the MLA is widely used in the core battery of genotoxicity tests. In addition, it is also a preferred *in vitro* mammalian gene mutation assay by the US EPA (Dearfield et al., 1991), US FDA Center for Food Safety and Nutrition (FDA, 2006), and ICH (ICH, 2011). The MLA employs a strain ($Tk^{+/-}$ -3.7.2C) of L5178Y mouse lymphoma cells heterozygous at the Tk locus. When a chemical induces DNA damage leading to a mutational event (from $Tk^{+/-}$ to $Tk^{-/-}$), induced heritable loss of TK activity can easily occur. $Tk^{-/-}$ mutants can be detected by their inherent resistance to trifluorothymidine (TFT), which is incorporated into the DNA of TK-competent ($Tk^{+/-}$) cells resulting in cytotoxicity (Moore et al., 2003). In addition, $Tk^{-/-}$ mutants result in the loss of the salvage TK enzyme so that TFT is not incorporated into the cellular DNA, thereby enabling these cells to grow in the presence of TFT. This deficiency in the TK enzyme can result from a number of genetic events affecting the Tk locus, including gene mutations (point mutations, frameshift mutations, small deletions, etc.) and chromosomal events (large deletions, chromosome rearrangements, and mitotic recombination). Among these mutants, the induction of small-colony mutants is associated with chemicals inducing gross CAs, whereas the production of large mutant colonies is generally associated with chemicals causing point mutations (Moore et al., 2006; Moore et al., 2007). Chromosomal events lead to loss of heterozygosity (LOH). There are currently two equally acceptable methods for performing the MLA, namely, the soft agar and 96-microwell versions. This assay is also conducted both with and without S9.

Recently, the ICH S2 (R1) Guideline states several differences for testing pharmaceuticals, notably for the selection of the top concentration. The maximum top concentration of test article recommended decreases to 1 mM or 0.5 mg/ml from originally proposed 10 mM or 5 mg/ml, whichever is lower, when not limited by solubility in solvent or culture medium or by cytotoxicity (ICH, 2011).

Chromosome Aberration Assay   Chromosome damage is a very important indicator of genetic damage relevant to environmental and clinical studies. Chromosomal mutation is related to changes in the chromosome structure (CAs) and changes in the number of chromosomes (numerical aberrations). Structural CAs have two types: chromosome and chromatid, which are induced by a chemical mutagenic agent through various molecular mechanisms. The structural aberrations may be a consequence of direct damage to DNA (e.g., DNA strand breaks, base damage, DNA cross-links) or indirect (e.g., inhibition of DNA topoisomerases I and II, generation of reactive oxygen species) and left unrepaired, or misrepaired to produce chromosome breaks or rearrangements. The CA test is widely used to screen for possible mammalian mutagens and carcinogens. In this assay, any cell lines, strains, or primary cell cultures, even human cells, may be used, such as Chinese hamster fibroblasts, human or other mammalian peripheral blood lymphocytes. Cell cultures are treated with the test substance both with and without S9. At predetermined intervals after the test substance exposure, cell cultures are treated with a metaphase-arresting substance

(e.g., colchicine), harvested, stained, and at least 200 metaphase cells are analyzed using microscopic analysis for the detection of CAs. The detailed requirement is described in OECD TG 473 (OECD, 1997c) and ICH S2 (R1) (ICH, 2011). For analysis of structural CAs at metaphase, there are categories described by Savage (Savage, 1976), e.g., chromosome-type aberrations (chromosome intrachanges and chromosome interchanges, including asymmetrical exchanges and symmetrical exchanges) and chromatid-type aberrations (chromatid break or terminal deletion, interstitial deletions, isochromatid deletions, inter-arm intrachanges, inter-arm interchange, triradials). Notably, the guideline ICH S2 (R1) currently decreases the top concentration for testing pharmaceuticals from the originally proposed 10 mM or 5 mg/ml to 1 mM or 0.5 mg/ml, whichever is lower, when not limited by solubility in solvent or culture medium or by cytotoxicity (ICH, 2011).

Micronucleus Assay   The *in vitro* MN (MNvit) assay is a genotoxicty test newly adopted by the OECD for detecting the activity of clastogens and aneugens in divided cells after exposure to the test substance by analysis of MN in the cytoplasm of interphase cells (OECD, 2010). It provides a comprehensive basis for studying the chromosome damaging potential *in vitro*. In the TG 487, two methods, including with or without a cytokinesis blocker, actin polymerization inhibitor cytochalasin B (cyto-B) are both recommended to be used. The addition of cyto-B makes it easy to identify and analyze the MN frequency in cells because they have completed mitosis and therefore such cells are binucleate. For the method without a cytokinesis blocker, the cell population analysis is helpful to look at whether these cells have undergone mitosis. In addition, the MNvit assay also provides some value information on the mechanisms of chromosome damage, and MN formation are also provided using a cytokinesis blocker, immunochemical labeling of kinetochores, or hybridization with centromeric/telomeric probes (fluorescence *in situ* hybridization (FISH)). Theoretically, any cell type can be used in this assay in the presence or absence of cyto-B due to its robustness and effectiveness. The common use of rodent cell lines (CHO, V79, CHL/IU, and L5178Y) and human peripheral blood lymphocytes is highly recommended because of its extensively accumulating data from international validation studies (Aardema et al., 2006; Clare et al., 2006; Lorge et al., 2006) and the reports of the International Workshop on Genotoxicity Testing (IWGT) (Kirsch-Volders et al., 2000; Kirsch-Volders et al., 2003). This assay is conducted both with and without S9, but genetically engineered cell lines, which express specific human or rodent activating enzymes, may not need an exogenous S9 and may serve as the test cells.

In addition, the European Centre for the Validation of Alternative Methods (ECVAM) of the European Commission (EC) has also reevaluated the available data in a WoE retrospective validation study, and the ECVAM Scientific Advisory Committee (ESAC) has endorsed the test method as scientifically valid. The use of the human TK6 cells (Corvi et al., 2008), HepG2 cells (Ehrlich et al., 2002), and primary Syrian hamster embryo cells (Gibson et al. 1997) has been described, although they have not been used in validation studies. Cell cultures are treated with the test substance both with and without S9 for 3–6 h, then the treatment medium is removed and replaced with fresh medium that may or may not contain cyto-B, and the cells are harvested 1.5–2.0 normal cell cycles later, stained, and the MN frequencies are analyzed in at least 2000 binucleated cells (with or without cyto-B) per concentration using a microscope, or validated flow cytometry or image analysis system.

Comet Assay   The comet assay, also known as the SCGE assay, is widely used as one of the standard methods to assess DNA damage caused by a range of DNA-damaging agents. This assay was developed in the 1980s. The unique aspect of this method is its ability to detect DNA damage in individual cells. It can detect DNA single-strand breaks (SSBs) and alkali-labile sites under alkaline conditions (pH > 13), as well as DNA double-strand breaks (DSBs) under neutral conditions (Lin et al., 2014). The comet assay protocols have been adopted and optimized by many laboratories around the world. The comet assay is a rapid, simple, sensitive, and quantitative method for measuring DNA damage. Although it has some disadvantages, such as low throughput, relatively poor reproducibility (between users and laboratories), time consuming, and potentially biased image analysis (Brodsky et al., 1999), this assay has been accepted as a valuable tool for investigating DNA damage and repair by many laboratories around the world. Over the decades, the comet assay has become a basic tool to be used for *in vitro*, *ex vivo*, and *in vivo* systems and is increasingly being employed in a broad range of scientific fields, including the genotoxicity testing (Witte et al., 2007), human biomonitoring (Dusinska and Collins, 2008; Valverde and Rojas, 2009), and genetic ecotoxicology (Cotelle and Ferard, 1999). With its development, protocols for conducting *in vitro* and *in vivo* comet assays have been published by different expert panels, such as the IWGT (Tice et al., 2000; Burlinson et al., 2007), International Comet Assay Workshop (Hartmann et al., 2003), and by different laboratories (Speit and Hartmann, 2006; Liao et al., 2009; Burlinson, 2012). The *in vitro* comet assay is useful as an alternative to cytogenetic assays in early genotoxicity or photogenotoxicity screening of drug candidates (Witte et al., 2007). Theoretically, any cells can be used in the *in vitro* comet assay, e.g., mammalian cells, including cultured cells from different cell lines or isolated cells from the organism. Generally, the cell lines commonly used in the *in vitro* standard genotoxicity test battery are most widely employed in the *in vitro* comet assay (Tice et al., 2000).

In Vivo Assay

Chromosome Aberration Assay   The *in vivo* CA test is commonly used for the detection of structural CAs induced by test chemicals in mammalian bone marrow cells, usually rodents, e.g., rats, mice, and Chinese hamsters, although any appropriate mammalian species may be used (OECD, 1997). Unlike the *in vitro* assay, this assay can assess the mutagenic hazard in humans extremely well because it considers the factors of *in vivo* metabolism, pharmacokinetics, and DNA repair processes. This assay is currently included in the stand test battery. Generally, animals are treated with the test substance by an appropriate route of exposure and are sacrificed at appropriate times after exposure. Concurrent positive and negative (solvent/vehicle) controls should be included for each sex in each test. Before sacrifice, animals are treated with a metaphase-arresting agent (e.g., colchicine), and then chromosome preparations are made from the bone marrow cells and stained. At least 100 metaphase cells for each animal are analyzed for CAs.

Micronucleus Assay   The *in vivo* MN test is routinely used for detecting agent-induced cytogenetic damage to chromosomes or mitotic apparatus of erythroblasts in mammalian bone marrow and/or peripheral blood cells, usually from rodents (OECD, 1997). This assay can identify clastogens and aneugens that results in the formation of MN containing lagging chromosome fragments or whole chromosomes. Therefore, it is currently included in the stand test battery. Briefly, animals are treated with the test substance by an appropriate route of administration. The animals are sacrificed at appropriate times after exposure, the bone marrow is extracted or peripheral blood is collected at appropriate times after treatment, and slide preparations are made, and finally stained. These preparations are analyzed for the presence of MN. Although this assay is described in OECD TG 474, no standard treatment schedule is recommended as long as a positive effect can be demonstrated, and toxicity has been demonstrated or the limit dose has been used. Recently, systems for automated analysis, particularly, flow cytometric analysis, which has appropriately justified and validated, have become more acceptable alternatives to manual evaluation. In brief, the cell nuclei and MN are stained with fluorescence dye and then analyzed after the cell membrane and cytoplasm have been removed.

Comet Assay   Recently, the *in vivo* comet assay has numerous published data and internationally acceptable protocols and even the application to various genotoxic testing strategies. However, there is no internationally agreed protocol of an *in vivo* comet assay. Scientific committees and regulatory agencies have been recommending the use of the *in vivo* comet assay to develop a WoE-based approach in genetic toxicology testing. An international collaborative trial on an *in vivo* comet assay in 14 laboratories from Europe, Japan, and United States evaluated 15 chemicals in rat liver, and the results supported the notion that the liver comet assay is a reasonable alternative to liver UDS assay to detect the chemical-induced genotoxicity in liver.

The comet assay with specific modifications can also detect oxidative stress-induced DNA damage. The modifications include lesion-specific endonucleases, such as formamidopyrimidine glycosylase (FPG) and endonuclease III (Endo III), which can introduce strand breaks at the site of oxidative damage (Collins et al., 1993; Collins et al., 1996). Human 8-oxoguanine DNA glycosylase (hOGG1) has greater lesion specificity than FPG and EndoIII (Smith et al., 2006). In addition, some new or modified techniques have been developed for the *in vitro* comet assay, such as two-dimensional gel electrophoresis or two-tailed comet (2T-comet) assay (Klaude et al., 1996), comet–FISH (Santos et al., 1997), and CometChip (microwell array) technology (Wood et al., 2010; Kiraly et al., 2012). However, these new methods still need to be adequately validated to obtain reliable and reproducible results.

For registration of pharmaceuticals, it is required to conduct a comprehensive assessment of their genotoxic potential. The current generalized strategy consists of a battery approach for detecting the genetic end point relevant in tumorigenesis, including gene mutation and chromosome damage (aberration, chromosome loss or gain) using *in vitro* and *in vivo* assays. Although *in vitro* mammalian tests increase the sensitivity for detection of rodent carcinogens and broaden the spectrum of genetic events detected, they also decrease the specificity of prediction. Some agents are mutagenic *in vivo* but not *in vitro*; therefore, some *in vivo* test(s) are also included in the test battery, e.g., CA, MN, UDS. Recently, the ICH has included a second *in vivo* assay in Option 2 of the standard test battery for genotoxicity and recommended that the comet assay in liver could be one of the second *in vivo* assays (ICH, 2011). The US FDA has also released this guidance for industry in July 2012. In principle, the comet assay can be used in repeat dose toxicology assays with appropriate dose levels and sampling times. The *in vivo* comet assay is currently in final stages of validation, and it will likely be adopted by the OECD in the near future.

Liver Unscheduled DNA Synthesis The *in vivo* UDS test with mammalian liver cells is designed to identify chemicals that induce DNA repair in liver cells of treated animals. Although tissues other than the liver may be used, rat livers are preferred for this assay. This assay is commonly used to investigate genotoxic effects of chemicals in the liver. In this assay, the end point, DNA damage and subsequent repair in liver cells, is measured by determining the uptake of labeled nucleosides in cells that are not undergoing scheduled (S-phase) DNA synthesis (OECD, 1986). In addition, the rodent *in vivo* liver UDS has been chosen as a preferred second assay in the standard test battery. This assay is often used to evaluate a DNA damage end point as a surrogate-like *in vivo* comet assay. For some test substances, the UDS assay is considered useful when they induce bulky DNA adducts or are positive in the Ames test. As with its *in vivo* counterpart, the OECD guideline (OECD, 1986) adopted in 1986 describes UDS procedures *in vitro* using primary cultures, human lymphocytes, or established cell lines, to detect DNA repair synthesis after excision and removal of a stretch of DNA containing the region of damage induced by agents. However, this *in vitro* TG will be deleted on 2 April 2014, and the assay is no more recommended.

Transgenic Rodent Mutation Assay The TGR gene mutation assay is used to detect gene mutations in any rodent tissue from which suitable DNA can be isolated. The regulatory genetic toxicology currently involves a wide range of *in vitro* mutation assays for detecting gene mutations and/or chromosomal alterations. *In vivo* assays are also the required components of many genetic toxicity testing strategies, in which current OECD TGs have been limited to investigate chromosomal damage (TG 474 and 475) and UDS in a very limited number of tissues. However, there are no OECD TGs *in vivo* to measure gene mutations, aside from the little-used mouse spot test for assessing heritable germ cell mutagenicity (OECD, 1986; Wahnschaffe et al., 2005), which will be deleted on 4 April 2014. For analysis of gene mutations *in vivo*, some non-transgenic mutation assays are described, but they are limited in use and no OECD TGs are described for these assays. The *in vivo* TGR mutation assays were originally accepted by the OECD on July 2011. The current version of the OECD guideline was updated and adopted on July 2013 (OECD, 2013) and is well characterized for *in vivo* tests for gene mutation. In the test battery, it is suggested to use an *in vivo* TGR mutation assay in place of a comet assay if an agent is positive *in vitro* in the MLA, is shown to induce predominantly large colonies and does not induce chromosome breakage in an *in vitro* metaphase assay. In this assay, the transgenic rats and mice are commonly used, which contain multiple copies of chromosomally integrated plasmid or phage shuttle vectors, and the transgenes contain reporter genes for detecting various types of mutations induced by agents. It is suggested that transgenes should respond to mutagens in a similar manner as endogenous genes, especially with regard to the detection of base-pair substitutions, frameshift mutations, and small deletions and insertions. The IWGT has recommended a protocol for implementing this assay (Heddle et al., 2000; Thybaud et al., 2003). The OECD guideline TG 488 is based on these recommendations. Mutagenesis in the TGR mutation models is generally assessed as mutant frequency (OECD, 2013). However, further molecular analysis of the mutations can provide additional information. Transgenic rodent gene mutation assays available to support their use include *lacZ* bacteriophage mouse (Muta™ Mouse), *lacZ* plasmid mouse, *gpt* delta (*gpt* and Spi-) mouse and rat, *lacI* mouse and rat (Big Blue®). Moreover, the *cII* positive selection assay can be used for evaluating mutations in the Big Blue® and Muta™ Mouse models. In brief, the TGR assay includes treatment of the rodent with an agent over a period of time, and the unrepaired DNA lesions transform into stable mutations (expression time). After the animal is sacrificed, genomic DNA is isolated from the tissue(s) of interest and purified. The mutant colonies containing specific transgene are isolated using selective conditions and quantified. The mutations scored in the *lacI*, *lacZ*, *cII*, and *gpt* point mutation assays consist primarily of base-pair substitution mutations, frameshift mutations, and small insertions/deletions. Large deletions are detected only with the Spi- selection and the *lacZ* plasmid assays (Smith et al., 2009). The *in vivo* TGR assay for gene mutation is useful for further investigating a mutagenic effect detected by an *in vitro* system and to follow up the results of tests using other *in vivo* end points. It is anticipated that it may be possible to combine a TGR gene mutation assay with a repeat dose toxicity study (TG 407) in the future.

Rodent Pig-a *Genes Mutation Assay* The *Pig-a* gene mutation assay is based on the loss of function of the *Pig-a* gene, which leads to the absence of cell surface expression of specific proteins targeted to the surface by GPI anchors. The endogenous X-linked phosphatidyl inositol glycan, class A (*Pig-a*) gene, also called *Pig-a* gene, is involved in the synthesis of GPI anchors, which can gather multiple protein markers to the exterior surface of the cytoplasmic membrane (Kinoshita et al., 2008). In *Pig-a* mutant cells, there is a lack of GPI anchors and GPI-anchored markers. The marker-deficient mutants can be identified and enumerated using immunofluorescent staining and flow cytometric analysis. In addition, GPI anchor-deficient cells can also be characterized by their ability to expand in the presence of proaerolysin (proAER), a cytolytic toxin produced by the bacterium *Aeromonas hydrophila*. Proaerolysin binds GPI anchors present on wild-type cells and eventually leads to dead wild-type cells (Brodsky et al., 1999). However, ProAER does not bind to GPI-deficient *Pig-a* mutant cells, and the mutants can be selectively expanded in a growth medium containing proAER. Generally, *Pig-a* mutations can be detected in both erythrocytes and white blood cells by flow cytometrc analysis. This assay has an opportunity to be discussed at the Working Group of the National Coordinators of the Test Guidelines Programme (WNT), which was held in November in 2013.

Germ Cell Assays *In vivo* germ cell mutagenicy tests are used to assess potential heritable damage passed onto offspring, and the tests primarily include the mammalian spermatogonial CA test (OECD, 1997), mouse heritable translocation assay (OECD, 1986), mouse specific locus test, rodent dominant lethal test (OECD, 1984), spermatid MN assay, mammalian oocyte CA/aneuploidy test, UDS test in testicular cells, and TGR germ cell gene mutation assay (OECD, 2013).

In addition, some tests for detecting genotoxicity induction in somatic cells are also used to measure whether an agent has the capability to induce germ cell genotoxicity, e.g., the mammalian erythrocyte MN assay (OECD, 1997), mammalian bone marrow CA assay (OECD, 1997), and mouse spot test (OECD, 1986) are often used in predicting potential carcinogenicity.

Moreover, for assessing germ cell mutagenicity induced by chemicals, the United Nations Globally Harmonized System (GHS) classification of germ cell mutagenicity (https://www.osha.gov/dsg/hazcom/ghs.html) provides two different categories of germ cell mutagens: Category 1 chemicals are "known to induce heritable mutations or to be regarded as if they induce heritable mutations in the germ cells of humans." Category 2 chemicals are those that "cause concern for humans owing to the possibility that they may induce heritable mutations in the germ cells of humans." Among these categories, Category 1 has two subcategories: Categories 1A and 1B. The GHS recommends use of OECD TGs for genotoxicity testing and states that "if new, well validated, tests arise these may also be used in the total weight of evidence to be considered."

Novel Assays The phosphorylation of histone H2AX at serine 139 (γ-H2AX) is considered one of the first steps of DNA damage response (Tanaka et al., 2007). It is used as a biomarker for detecting DNA DSBs induced by ionizing radiation or other genotoxic agents. The immunostained γ-H2AX can be measured using flow cytometry (Smart et al., 2011). In addition, single γ-H2AX foci can be visualized and counted under the fluorescence microscope. The expression of γ-H2AX in cells can be quantitatively analyzed using western blotting. It has been reported that γ-H2AX can be measured in different cells (somatic and germ cells) from *in vivo* animal studies, such as mouse bone marrow and testicular cells (Forand et al., 2004; Rube et al., 2008; Paris et al., 2011).

It is well known that mutagens, clastogens, and aneugens cause increased expression of the human *GADD45a* gene, which was originally identified and named by the Fornace laboratory (Fornace et al., 1992), and has been shown to be associated with the response to genome damage. The *GADD45a-GFP* GreenScreen HC genotoxicity assay has been used to monitor genotoxin-induced transcription of the *GADD45a* gene (Hastwell et al., 2006; Walmsley, 2008). In this assay, the human lymphoblastoid cell line TK6 is used, which has wild-type p53 gene (Hildesheim et al., 2002), is the host for the reporter construct, in which the gene' expression is linked to the expression of green fluorescent protein (GFP). This assay is aimed to facilitate screening, which can be conducted in a microplate format (Knight et al., 2009). Data are collected through the use of the spectrophotometrical method or flow cytometry. To date, extensive validation has demonstrated that *GADD45a* assays have high sensitivity and specificity (Billinton et al., 2008; Jagger et al., 2009).

GENOTOXIC TESTING GUIDELINES

Overview

Genotoxicity tests are used to detect the genetic damage by various mechanisms in *in vitro* and *in vivo* systems. Several regulatory guidelines have been developed to provide various assays that are conducted for testing the genotoxicity. Most regulatory authorities have set their own requirements for detecting the genotoxicity that need to supply a series of genetic toxicity testing, e.g., US FDA, US EPA, and United Kingdom Environmental Mutagen Society (UKEMS). Some guidelines, such as the OECD Guidelines for the testing of chemicals, US EPA genetic toxicity TGs, and FDA Red Book, have provided various assays for detecting the genotoxicity. To date, most regulatory agencies and international authorities recommend a test scheme consisting of *in vitro* and *in vivo* methods to detect genotoxicity/mutagenicity induced by substances. The ICH recommends a standard battery test for pharmaceuticals to detect the genotoxicity. The ICH guidance optimizes the standard battery test for genetic toxicology and also provides the guidance on the interpretation of results (ICH, 2011). This guideline helps improve risk characterization for carcinogenic effects. In the sections below, some regulatory agencies or organizations are briefly described as to their own guidelines.

Regulatory Agency and Its Test Guidelines

OECD   The OECD Guidelines for testing the genotoxicity induced by new and existing chemical substances, chemical preparations, and chemical mixtures consist of the most relevant internationally agreed testing methods used by the government, industry, and independent laboratories worldwide. They are used primarily in regulatory safety testing and chemical registration (Table 108.2).

US EPA   In the US EPA, the Office of Chemical Safety and Pollution Prevention (OCSPP) has developed a series of harmonized TGs for use in the testing of pesticides and toxic substances. The EPA recommends these methods be used to generate data that are submitted to the EPA for supporting the registration. The OCSPP harmonized TGs for detecting genotoxicity are organized in Group Do Genetic toxicity test guidelines, from Series 870—Health Effects Test Guidelines (Table 108.3).

US FDA   As mentioned previously, the FDA follows the ICH guideline for drug and biological product registration. The current guidance issued by the FDA for assessing the genotoxicity in drug registration includes S2 (R1) (Genotoxicity Testing and Data Interpretation for Pharmaceuticals Intended for Human Use). This newly released guideline applies to small-molecule, nonbiological pharmaceutical products that are intended to be used in humans.

As its responsibility, the FDA provides guidance to the industry and public for evaluating the safety of food and color additives. The current guideline, Redbook 2000 (CFSAN, 2007), known as "Toxicological Principles for the Safety Assessment of Food Ingredients" is the new name for "Toxicological Principles for the Safety Assessment of Direct Food Additives and Color Additives Used in Food" originally published in 1982 ("Redbook I") and the draft revision published in 1993 ("Redbook II"). The Center for Food Safety and Applied Nutrition (CFSAN) of the FDA is responsible for revising and updating the Redbook in light of the developments in toxicological testing methods and comments from the scientific community and public. This guidance provides toxicological information submitted to the CFSAN, Office of Food Additive Safety (OFAS) regarding food ingredients. In Redbook 2000, it includes some genotoxicity tests in Chapter IV.C.1 Short-Term Tests for Genetic Toxicity listed in Table 108.4.

ICH

Genotoxicity testing is conducted for pharmaceuticals, industrial chemicals, and consumer products. The results from these tests are used to identify chemicals for heritable germ cell mutagenicity as well as carcinogenicity. Recently, ICH guideline S2 (R1) has been finalized under Step 4 in November 2011. This revised guideline replaces and combines the ICH guidelines S2A and S2B and optimizes the standard genetic toxicology battery to provide guidance on the interpretation of results. This guidance is the internationally agreed upon standard for follow-up testing and how to interpret positive results *in vitro* and *in vivo* in the standard genetic toxicology battery, including how to assess nonrelevant findings (ICH, 2011).

TABLE 108.2 OECD Test Guidelines (TG) for Genotoxicity and Mutagenicity Testing

TG No.	Guideline Name	Original Adoption	No. of Updates	Most Recently Updated
471	Bacterial reverse mutation test	26 May 1983	1	21 July 1997
472	Genetic toxicology: *Escherichia coli* reverse assay	26 May 1983	0	Date of deletion: 21 July 1997 (method merged with TG 471)
473	*In vitro* mammalian chromosome aberration test	26 May 1983	1	21 July 1997
474	Mammalian erythrocyte micronucleus test	26 May 1983	1	21 July 1997
475	Mammalian bone marrow chromosome aberration test	4 April 1984	1	21 July 1997
476[a]	*In vitro* mammalian cell gene mutation test	4 April 1984	1	21 July 1997
477	Genetic toxicology: sex-linked recessive lethal test in *Drosophilia melanogaster*	4 April 1984	0	Date of deletion: 2 April 2014
478	Genetic toxicology: rodent dominant lethal test	4 April 1984	0	-
479	Genetic toxicology: *in vitro* sister chromatid exchange assay in mammalian cells	23 October 1986	0	Date of deletion: 2 April 2014
480	Genetic toxicology: *Saccharomyces cerevisiae*, gene mutation assay	23 October 1986	0	Date of deletion: 2 April 2014
481	Genetic toxicology: *S. cerevisiae*, mitotic recombination assay	23 October 1986	0	Date of deletion: 2 April 2014
482	Genetic toxicology: DNA damage and repair, unscheduled DNA synthesis in mammalian cells *in vitro*	23 October 1986	0	Date of deletion: 2 April 2014
483	Mammalian spermatagonial chromosome aberration test	23 October 1986	1	21 July 1997
484	Genetic toxicology: mouse spot test	23 October 1986	0	Date of deletion: 2 April 2014
485	Genetic toxicology: mouse heritable translocation assay	23 October 1986	0	–[b]
486	Unscheduled DNA synthesis (UDS) test with mammalian liver vells *in vivo*	21 July 1997	0	–
487	*In vitro* mammalian cell micronucleus test	22 July 2010	0	–
488	Transgenic rodent somatic and germ cell gene mutation assays	28 July 2011	1	26 July 2013

[a]This guideline will be split into TK and Hprt TGs (revised).

[b]This guideline will be deleted. Other deletions have been approved by the OECD National coordinators (WNT) in the April 2012 meeting.

TABLE 108.3 OCSPP Harmonized Test Guidelines Group D—Genetic Toxicity Test Guidelines

OCSP[a] No.	Guideline Name	Other Reference No. OPPT	Other Reference No. OPP	EPA Pub No.	Date Issued
870.5100	Bacterial reverse mutation test	798.5100,798.5265	84-2	712-C-98-247	August 1998
870.5140	Gene mutation in *Aspergillus nidulans*	798.5140	84-2	712-C-98-215	August 1998
870.5195	Mouse biochemical specific locus test	798.5195	84-2	712-C-98-216	August 1998
870.5200	Mouse visible specific locus test	798.5200	84-2	712-C-98-217	August 1998
870.5250	Gene mutation in *Neurospora crassa*	798.5250	84-2	712-C-98-218	August 1998
870.5275	Sex-linked recessive lethal test in Drosophila	798.5275	84-2	712-C-98-220	August 1998
870.5300	*In vitro* mammalian cell gene mutation test	798.5300	84-2	712-C-98-221	August 1998
870.5375	*In vitro* mammalian chromosome aberration test	798.5375	84-2	712-C-98-223	August 1998
870.5380	Mammalian spermatogonial chromosomal aberration	798.5380	84-2	712-C-98-224	August 1998
870.5385	Mammalian bone marrow chromosomal aberration test	798.5385	84-2	712-C-98-225	August 1998
870.5395	Mammalian erythrocyte micronucleus test	798.5395	84-2	712-C-98-226	August 1998
870.5450	Rodent dominant lethal assay	798.5450	84-2	712-C-98-227	August 1998
870.5460	Rodent heritable translocation assays	798.5460	84-2	712-C-98-228	August 1998
870.5500	Bacterial DNA damage or repair tests	798.5500	84-2	712-C-98-229	August 1998
870.5550	Unscheduled DNA synthesis in mammalian cells	798.5550	84-2	712-C-98-230	August 1998
870.5575	Mitotic gene conversion in *Saccharomyces cerevisiae*	798.5575	84-2	712-C-98-232	August 1998
870.5900	*In vitro* sister chromatid exchange assay	798.900	84-2	712-C-98-234	August 1998
870.5915	*In vivo* sister chromatid exchange assay	798.5915	84-2	712-C-98-235	August 1998

Available electronically at http://www.epa.gov/ocspp/pubs/frs/home/testmeth.htm.

[a]Guidelines issued before April 22, 2010, refer to "OPPTS" because the office name changed from "Office of Prevention, Pesticides and Toxic Substances" and "OPPTS" to "Office of Chemical Safety and Pollution Prevention" and "OCSPP." This name change does not otherwise affect the guidelines.

TABLE 108.4 Short-Term Tests for Genetic Toxicity with Redbook 2000

Chapter No.	Assay Name	Most Recently Updated
IV.C.1.a.	Bacterial reverse mutation test	July 2007
IV.C.1.b.	*In vitro* mammalian chromosomal aberration test	July 2007
IV.C.1.c.	*In vitro* mouse lymphoma thymidine kinase gene mutation assay	July 2007
IV.C.1.d.	*In vivo* mammalian erythrocyte micronucleus test	July 2007

Current Testing Assays (ICH Standard Battery)

In Vitro Genotoxicity Testing

TABLE 108.5 *In Vitro* Genetic Toxicity Assays

Test Name[a]	Number/Chapter No.		
	OECD	EPA-OCSP	FDA Redbook 2000
Bacterial reverse mutation test (Ames test)	471	870.5100	IV.C.1.a.
In vitro mammalian chromosome aberration assay	473	870.5375	IV.C.1.b.
In vitro mammalian cell micronucleus test	487	None	None
In vitro mammalian cell gene Mutation test	476	870.5300	IV.C.1.c. (only included MLA)

[a]ICH standard test battery recommended.

In Vivo Genotoxicity Testing

TABLE 108.6 *InVivo* Genetic Toxicity Assays

Test Name[a]	Number/Chapter No.		
	OECD	EPA-OPPT	FDA Redbook 2000
Mammalian micronucleus test	474	870.5395	IV.C.1.d. (only erythrocyte)
Mammalian bone marrow chromosome aberration test	475	870.5385	None
Unscheduled DNA synthesis (UDS) test with mammalian liver cells *in vivo*	486	None	None
Transgenic mouse mutation assay	488	None	None
In vivo comet assay	None	None	None
Alkaline elution assay	None	None	None
In vivo DNA covalent binding assay	None	None	None

[a]ICH standard test battery recommended.

NONANIMAL ALTERNATIVE METHOD

With the development of nonanimal alternative methods, more *in vitro* methods are increasingly used for assessing the toxicity induced by substances. In general, alternative toxicity testing methods will be accepted after they have been scientifically validated and determined to be reliable (reproducible) and relevant for their intended purpose. The current OECD genotoxicity testing includes several

in vitro genetic toxicity assays adopted, and four of which are included in a standard test battery. There four *in vitro* assays include a mutagenicity test method based on bacteria: the bacterial reverse mutation test (Ames test) (OECD TG 471). The three commonly used methods are based on mammalian cells, and they are the *in vitro* mammalian CA test (OECD TG 473), *in vitro* mammalian cell MN test (OECD TG 476), and the *in vitro* mammalian cell gene mutation test (OECD TG 476). In addition, two *in vitro* mammalian cell tests (*in vitro* sister chromatid exchange assay and DNA damage and repair, UDS) will be deleted on 2 April 2014.

To date, the criteria and processes for test method validation have been developed and implemented in Europe (under the auspices of the ECVAM), the United States (through the Interagency Coordinating Committee on the Validation of Alternative Methods, or ICCVAM), Japan (through the Japanese Centre for the Validation of Alternative Methods (JaCVAM), and internationally through the OECD. These organizations are responsible for the development and scientific regulatory acceptance of alternative testing methods. After undergoing these steps, e.g., research and development, pre-validation, and validation until regulatory acceptance, it usually takes at least 5 years. A lot of *in vitro* methods for genotoxicity testing are based primarily on bacterial and mammalian cells assays, and several of which are accepted by regulatory authorities. Aside from the standard *in vitro* mammalian cell genotoxicity tests described above, there are *in vitro* assays for detecting genotoxicity in primary germ cells; however, these are not standardized or validated, and there are no ongoing coordinated activities to address them at this time. The validation of comet assay and pre-validations of genotoxicity assays in 3D-skin model are coordinated by the JaCVAM and the European Cosmetic, Toiletry and Perfumery Association (COLIPA).

PERSPECTIVE

Some *in vitro* genotoxicity tests in cell lines for predicting carcinogenicity in rodents have shown the capability of high sensitivity, especially when combined in a standard battery. *In vitro* genotoxicity assays provide an integrated approach for assessing genetic damage. In the WoE approach, more weight is generally given to the positive results obtained by *in vivo* assays than *in vitro* assays. However, currently available genotoxicity assays are unable to detect all genotoxic carcinogens. Whichever assay is chosen, an *in vitro* assay or *in vivo* assay, it is critical to accurately identify the genotoxic carcinogens in the chosen assay. Numerous factors, such as experimental conditions or biochemical and physiological stresses, can affect the evaluation of genotoxicity in *in vitro* and *in vivo* assays, which can result in positive DNA damage response. In the future, it is important to investigate a correlation between *in vitro* and *in vivo* assays, even in human experiments. In addition, models that can be used in both *in vitro* and *in vivo* assay, as well as end points/ tissues that can be evaluated in both animal and humans may contribute to a comprehensive understanding of the results and extrapolation to humans (e.g., MN and comet assay in peripheral blood). To understand the mode of action leading to the genotoxic response, the choice of assays and the data interpretation should take into account all available information. Computational prediction of genotoxicity (*in silico* approaches) using molecular structure database, structure–activity relationship (SAR) or quantitative SAR (QSAR) models is useful and often used around the world by regulatory authorities as an adjunct to risk assessment, especially when no actual test data are available (Mahadevan et al., 2011). Generally, computer-based predictions using QSAR and *in vitro* testing are the first two steps for assessing genotoxicity/mutagenicity induced by substances. In addition, the development of toxicogenomics can contribute to screening carcinogenicity, and identifying new markers, and elucidating the mechanism. For industry perspective, several high throughput screening (HTS) assays or quantitative HTS have been established for the assessment of genotoxicity (Mahadevan et al., 2011), especially in early stage of drug development. In summary, genotoxicity testing is very useful, and a major challenge in the field of genetic toxicology is developing new or modified tests for the future that are able to efficiently detect genotoxicity with high specificity and sensitivity.

REFERENCES

Aardema, M.J., Snyder, R.D., Spicer, C., Divi, K., Morita, T., Mauthe, R.J., Gibson, D.P., Soelter, S., Curry, P.T., Thybaud, V., Lorenzon, G., Marzin, D., and Lorge, E. (2006) SFTG international collaborative study on *in vitro* micronucleus test III. Using CHO cells. *Mutat. Res.* 607(1):61–87.

Applegate, M.L., Moore, M.M., Broder, C.B., Burrell, A., Juhn, G., Kasweck, K.L., Lin, P.F. Wadhams, A. and Hozier, J.C. (1990) Molecular dissection of mutations at the heterozygous thymidine kinase locus in mouse lymphoma cells. *Proc. Natl. Acad. Sci. U. S. A.* 87(1):51–55.

Billinton, N., Hastwell, P.W., Beerens, D., Birrell, L., Ellis, P., Maskell, S., Webster, T.W., Windebank, S., Woestenborghs, F., Lynch, A.M., Scott, A.D., Tweats, D.J., van Gompel, J., Rees, R.W., and Walmsley, R.M. (2008) Interlaboratory assessment of the GreenScreen HC GADD45a-GFP genotoxicity screening assay: an enabling study for independent validation as an alternative method. *Mutat. Res.* 653(1–2):23–33.

Brodsky, R.A., Mukhina, G.L., Nelson, K.L., Lawrence, T.S., Jones, R.J., and Buckley, J.T. (1999) Resistance of paroxysmal nocturnal hemoglobinuria cells to the glycosylphosphatidylinositol-binding toxin aerolysin. *Blood* 93(5):1749–1756.

Burlinson, B. (2012) The *in vitro* and *in vivo* comet assays. *Methods Mol. Biol.* 817:143–163.

Burlinson, B., Tice, R.R., Speit, G., Agurell, E., Brendler-Schwaab, S.Y., Collins, A.R., Escobar, P., Honma, M., Kumaravel, T.S., Nakajima, M., Sasaki, Y.F., Thybaud, V., Uno, Y., Vasquez, M., and Hartmann, A. (2007) Fourth international workgroup on genotoxicity testing: results of the *in vivo* comet assay workgroup. *Mutat. Res.* 627(1):31–35.

CFSAN (2007) *Redbook 2000: Guidance for Industry and Other Stakeholders Toxicological Principles for the Safety Assessment of Food Ingredients*, Washington, DC: Centerfor Food Safety and Applied Nutrition, U.S. Federal Drug Administration, Available at http://www.fda.gov/Food/GuidanceRegulation/GuidanceDocumentsRegulatoryInformation/IngredientsAdditivesGRASPackaging/ucm2006826.htm.

Clare, M. G., Lorenzon, G., Akhurst, L.C., Marzin, D., van Delft, J., Montero, R., Botta, A., Bertens, A., Cinelli, S., Thybaud, V., and Lorge, E. (2006) SFTG international collaborative study on *in vitro* micronucleus test II. Using human lymphocytes. *Mutat. Res.* 607(1):37–60.

Claxton, L.D., Umbuzeiro Gde, A., and DeMarini, D.M. (2010) The Salmonella mutagenicity assay: the stethoscope of genetic toxicology for the 21st century. *Environ. Health Perspect.* 118(11):1515–1522.

Clive, D. (1973) Recent developments with the L5178Y TK heterozygote mutagen assay system. *Environ. Health Perspect.* 6:119–125.

Clive, D. and Spector, J.F. (1975) Laboratory procedure for assessing specific locus mutations at the TK locus in cultured L5178Y mouse lymphoma cells. *Mutat. Res.* 31(1):17–29.

Collins, A.R., Dusinska, M., Gedik, C.M., and Stetina, R. (1996) Oxidative damage to DNA: do we have a reliable biomarker? *Environ. Health Perspect.* 104(Suppl. 3):465–469.

Collins, A.R., Duthie, S.J., and Dobson, V.L. (1993) Direct enzymic detection of endogenous oxidative base damage in human lymphocyte DNA. *Carcinogenesis* 14(9):1733–1735.

Corvi, R., Albertini, S., Hartung, T., Hoffmann, S., Maurici, D., Pfuhler, S., van Benthem, J., and Vanparys, P. (2008) ECVAM retrospective validation of in vitro micronucleus test (MNT). *Mutagenesis* 23(4):271–283.

Cotelle, S. and Ferard, J.F. (1999) Comet assay in genetic ecotoxicology: a review. *Environ. Mol. Mutagen.* 34(4):246–255.

Dearfield, K.L., Auletta, A.E., Cimino, M.C., and Moore, M.M. (1991) Considerations in the U.S. Environmental Protection Agency's testing approach for mutagenicity. *Mutat. Res.* 258(3):259–283.

Demerec, M. (1951) Studies of the streptomycin-resistance system of mutations in *E. Coli. Genetics* 36(6):585–597.

Demerec, M. and Hanson, J. (1951) Mutagenic action of manganese chloride. *Cold Spring Harb. Symp. Quant. Biol.* 16:215–228.

Dusinska, M. and Collins, A.R. (2008) The comet assay in human biomonitoring: gene-environment interactions. *Mutagenesis* 23(3):191–205.

Ehrlich, V., Darroudi, F., Uhl, M., Steinkellner, H., Gann, M., Majer, B.J., Eisenbauer, M., and Knasmuller, S. (2002) Genotoxic effects of ochratoxin A in human-derived hepatoma (HepG2) cells. *Food Chem. Toxicol.* 40(8):1085–1090.

FDA (2006) *IV.C.1.c Mouse Lymphoma Thymidine Kinase Gene Mutation Assay."* Available at http://www.fda.gov/Food/GuidanceRegulation/GuidanceDocumentsRegulatoryInformation/IngredientsAdditivesGRASPackaging/ucm078336.htm.

Forand, A., Dutrillaux, B., and Bernardino-Sgherri, J. (2004) Gamma-H2AX expression pattern in non-irradiated neonatal mouse germ cells and after low-dose gamma-radiation: relationships between chromatid breaks and DNA double-strand breaks. *Biol. Reprod.* 71(2):643–649.

Fornace, A.J. Jr., Jackman, J., Hollander, M.C., Hoffman-Liebermann, B., and Liebermann, D.A. (1992) Genotoxic-stress-response genes and growth-arrest genes. gadd, MyD, and other genes induced by treatments eliciting growth arrest. *Ann. N. Y. Acad. Sci.* 663:139–153.

Gibson, D.P., Brauninger, R., Shaffi, H.S., Kerckaert, G.A., LeBoeuf, R.A., Isfort, R.J., and Aardema, M.J. (1997) Induction of micronuclei in Syrian hamster embryo cells: comparison to results in the SHE cell transformation assay for National Toxicology Program test chemicals. *Mutat. Res.* 392(1–2):61–70.

Hartmann, A., Agurell, E., Beevers, C., Brendler-Schwaab, S., Burlinson, B., Clay, P., Collins, A., Smith, A., Speit, G., Thybaud, V., and Tice, R.R. (2003) Recommendations for conducting the in vivo alkaline Comet assay. 4th International Comet Assay Workshop. *Mutagenesis* 18(1):45–51.

Hastwell, P.W., Chai, L.L., Roberts, K.J., Webster, T.W., Harvey, J.S., Rees, R.W., and Walmsley, R.M. (2006) High-specificity and high-sensitivity genotoxicity assessment in a human cell line: validation of the GreenScreen HC GADD45a-GFP genotoxicity assay. *Mutat. Res.* 607(2):160–175.

Heddle, J.A., Dean, S., Nohmi, T., Boerrigter, M., Casciano, D., Douglas, G.R., Glickman, B.W., Gorelick, N.J., Mirsalis, J.C., Martus, H.J., Skopek, T.R., Thybaud, V., Tindall, K.R., and Yajima, N. (2000) *In vivo* transgenic mutation assays. *Environ. Mol. Mutagen.* 35(3):253–259.

Hildesheim, J., Bulavin, D.V., Anver, M.R., Alvord, W.G., Hollander, M.C., Vardanian, L., and Fornace, A.J. Jr. (2002) Gadd45a protects against UV irradiation-induced skin tumors, and promotes apoptosis and stress signaling via MAPK and p53. *Cancer Res.*62(24):7305–7315.

ICH (2011) *"The international conference on harmonisation (ICH) of technical requirements for registration of pharmaceuticals for human use ICH S2 R1 (2012). Guidance on Genotoxicity Testing and Data Interpretation for Pharmaceuticals intended for Human Use."*

Jagger, C., Tate, M., Cahill, P.A., Hughes, C., Knight, A.W., Billinton, N., and Walmsley, R.M. (2009) Assessment of the genotoxicity of S9-generated metabolites using the GreenScreen HC GADD45a-GFP assay. *Mutagenesis* 24(1):35–50.

Kinoshita, T., Fujita, M., and Maeda, Y. (2008) Biosynthesis, remodelling and functions of mammalian GPI-anchored proteins: recent progress. *J. Biochem.* 144(3):287–294.

Kiraly, O., Wood, D., Weingeist, D., Ge, J., Sukup-Jackson, M., Bhatia, S., and Engelward, B. (2012) Recombomice and CometChip technology shed light on gene-exposure interactions that impact genomic stability. *Environ. Mol. Mutagen.* 53(S1);S14.

Kirkland, D., Aardema, M., Henderson, L., and Muller, L. (2005) Evaluation of the ability of a battery of three *in vitro* genotoxicity tests to discriminate rodent carcinogens and non-carcinogens I. Sensitivity, specificity and relative predictivity. *Mutat. Res.* 584 (1–2):1–256.

Kirkland, D., Aardema, M., Muller, L., and Makoto, H. (2006) Evaluation of the ability of a battery of three *in vitro* genotoxicity tests to discriminate rodent carcinogens and non-carcinogens II. Further analysis of mammalian cell results, relative predictivity and tumour profiles. *Mutat. Res.* 608(1):29–42.

Kirkland, D. and Speit, G. (2008) Evaluation of the ability of a battery of three *in vitro* genotoxicity tests to discriminate rodent carcinogens and non-carcinogens III. Appropriate follow-up testing *in vivo. Mutat. Res.* 654(2):114–132.

Kirsch-Volders, M., Sofuni, T., Aardema, M., Albertini, S., Eastmond, D., Fenech, M., Ishidate, M. Jr., Kirchner, S., Lorge, E., Morita, T., Norppa, H., Surralles, J., Vanhauwaert, A., and Wakata, A. (2003) Report from the *in vitro* micronucleus assay working group. *Mutat. Res.* 540(2):153–163.

Kirsch-Volders, M., Sofuni, T., Aardema, M., Albertini, S., Eastmond, D., Fenech, M., Ishidate, M. Jr., Lorge, E., Norppa, H., Surralles, J., von der Hude, W., and Wakata, A. (2000) Report from the *in vitro* micronucleus assay working group. *Environ. Mol. Mutagen.* 35(3):167–172.

Klaude, M., Eriksson, S., Nygren, J., and Ahnstrom, G. (1996) The comet assay: mechanisms and technical considerations. *Mutat. Res.* 363(2):89–96.

Knight, A.W., Birrell, L., and Walmsley, R.M. (2009) Development and validation of a higher throughput screening approach to genotoxicity testing using the GADD45a-GFP GreenScreen HC assay. *J. Biomol. Screen* 14(1):16–30.

Liao, W., McNutt, M.A., and Zhu, W.G. (2009) The comet assay: a sensitive method for detecting DNA damage in individual cells. *Methods* 48(1):46–53.

Lin, H., Mei, N., and Manjanatha, M.G. (2014) *In vitro* comet assay for testing genotoxicity of chemicals. In: *Optimization in Drug Discovery: In Vitro Methods*, 2nd ed.

Lorge, E., Thybaud, V., Aardema, M.J., Oliver, J., Wakata, A., Lorenzon, G., and Marzin, D. (2006) SFTG international collaborative study on *in vitro* micronucleus test I. General conditions and overall conclusions of the study. *Mutat. Res.* 607(1):13–36.

Mahadevan, B., Snyder, R.D., Waters, M.D., Benz, R.D., Kemper, R.A., Tice, R.R., and Richard, A.M. (2011) Genetic toxicology in the 21st century: reflections and future directions. *Environ. Mol. Mutagen.* 52(5):339–354.

Moore, M.M., Clive, D., Howard, B.E., Batson, A.G., and Turner, N.T. (1985a) In situ analysis of trifluorothymidine-resistant (TFTr) mutants of L5178Y/TK+/− mouse lymphoma cells. *Mutat. Res.* 151(1):147–159.

Moore, M.M., Clive, D., Hozier, J.C., Howard, B.E., Batson, A.G., Turner, N.T., and Sawyer, J. (1985b) Analysis of trifluorothymidine-resistant (TFTr) mutants of L5178Y/TK+/− mouse lymphoma cells. *Mutat. Res.* 151(1):161–174.

Moore, M.M., Honma, M., Clements, J., Bolcsfoldi, G., Burlinson, B., Cifone, M., Clarke, J., Clay, P., Doppalapudi, R., Fellows, M., Gollapudi, B., Hou, S., Jenkinson, P., Muster, W., Pant, K., Kidd, D.A., Lorge, E., Lloyd, M., Myhr, B., O'Donovan, M., Riach, C., Stankowski, L.F. Jr., Thakur, A.K., and Van Goethem, F. (2007) Mouse lymphoma thymidine kinase gene mutation assay: meeting of the International Workshop on Genotoxicity Testing, San Francisco, 2005, recommendations for 24-h treatment. *Mutat. Res.* 627(1):36–40.

Moore, M.M., Honma, M., Clements, J., Bolcsfoldi, G., Burlinson, B., Cifone, M., Clarke, J., Delongchamp, R., Durward, R., Fellows, M., Gollapudi, B., Hou, S., Jenkinson, P., Lloyd, M.,

Majeska, J., Myhr, B., O'Donovan, M., Omori, T., Riach, C., San, R., Stankowski, L.F. Jr., Thakur, A.K., Van Goethem, F., Wakuri, S., and Yoshimura, I. (2006) Mouse lymphoma thymidine kinase gene mutation assay: follow-up meeting of the International Workshop on Genotoxicity Testing—Aberdeen, Scotland, 2003—Assay acceptance criteria, positive controls, and data evaluation. *Environ. Mol. Mutagen.* 47(1):1–5.

Moore, M.M., Honma, M., Clements, J., Bolcsfoldi, G., Cifone, M., Delongchamp, R., Fellows, M., Gollapudi, B., Jenkinson, P., Kirby, P., Kirchner, S., Muster, W., Myhr, B., O'Donovan, M., Oliver, J., Omori, T., Ouldelhkim, M.C., Pant, K., Preston, R., Riach, C., San, R., Stankowski, L.F. Jr., Thakur, A., Wakuri, S., and Yoshimura, I. (2003) Mouse lymphoma thymidine kinase gene mutation assay: International Workshop on Genotoxicity Tests Workgroup report—Plymouth, UK 2002. *Mutat. Res.* 540(2):127–140.

OECD (1984) Genetic toxicology: rodent dominant lethal test. In: *Guideline for Testing of Chemicals*, 478. Organization for Economic Co-operation and Development.

OECD (1986) Genetic toxicology: mouse heritable translocation assay. In: *Guideline for Testing of Chemicals*, 485. Organization for Economic Co-operation and Development.

OECD (1997a) Mammalian erythrocyte micronucleus test. In: *Guideline for the Testing of Chemicals*, 474. Organization for Economic Co-operation and Development.

OECD (1997b) Mammalian bone marrow chromosome aberration test. In: *Guideline for the Testing of Chemicals*, 475. Organization for Economic Co-operation and Development.

OECD (1997c) *In vitro* mammalian chromosome aberration test. In: *Guideline for the Testing of Chemicals*, 473. Organization for Economic Co-operation and Development.

OECD (1997d) Bacterial reverse mutation test. In: *Guideline for the Testing of Chemicals*, 471. Organization for Economic Cooperation and Development.

OECD (2010) *In vitro* mammalian cell micronucleus test. In: *Guideline for Testing of Chemicals*, 487. Organization for Economic Co-operation and Development.

OECD (2013) Transgenic rodent somatic and germ cell gene mutation assays. In: *Guideline for Testing of Chemicals*, 488 Organization for Economic Co-operation and Development.

Paris, L., Cordelli, E., Eleuteri, P., Grollino, M.G., Pasquali, E., Ranaldi, R., Meschini, R., and Pacchierotti, F. (2011) Kinetics of gamma-H2AX induction and removal in bone marrow and testicular cells of mice after X-ray irradiation. *Mutagenesis* 26(4):563–572.

Rube, C.E., Grudzenski, S., Kuhne, M., Dong, X., Rief, N., Lobrich, M., and Rube, C. (2008) DNA double-strand break repair of blood lymphocytes and normal tissues analysed in a preclinical mouse model: implications for radiosensitivity testing. *Clin. Cancer Res.* 14(20):6546–6555.

Santos, S.J., Singh, N.P., and Natarajan, A.T. (1997) Fluorescence *in situ* hybridization with comets. *Exp. Cell Res.* 232(2):407–411.

Savage, J.R. (1976) Classification and relationships of induced chromosomal structual changes. *J. Med. Genet.* 13(2):103–122.

Smart, D.J., Ahmedi, K.P., Harvey, J.S., and Lynch, A.M. (2011) Genotoxicity screening via the gammaH2AX by flow assay. *Mutat. Res.* 715(1–2):25–31.

Smith, C.C., O'Donovan, M.R., and Martin, E.A. (2006) hOGG1 recognizes oxidative damage using the comet assay with greater specificity than FPG or ENDOIII. *Mutagenesis* 21(3):185–190.

Smith, J.D., Wolf, N.J., Handler, L., and Nash, M.R. (2009) Testing the effectiveness of family therapeutic assessment: a case study using a time-series design. *J. Pers. Assess.* 91(6):518–536.

Speit, G. and Hartmann, A. (2006) The comet assay: a sensitive genotoxicity test for the detection of DNA damage and repair. *Methods Mol. Biol.* 314:275–286.

Tanaka, T., Huang, X., Halicka, H.D., Zhao, H., Traganos, F., Albino, A.P., Dai, W., and Darzynkiewicz, Z. (2007) Cytometry of ATM activation and histone H2AX phosphorylation to estimate extent of DNA damage induced by exogenous agents. *Cytometry A* 71(9):648–661.

Thybaud, V., Dean, S., Nohmi, T., de Boer, J., Douglas, G.R., Glickman, B.W., Gorelick, N.J., Heddle, J. A., Heflich, R.H., Lambert, I., Martus, H.J., Mirsalis, J.C., Suzuki, T., and Yajima, N. (2003) *In vivo* transgenic mutation assays. *Mutat. Res.* 540(2):141–151.

Tice, R.R., Agurell, E., Anderson, D., Burlinson, B., Hartmann, A., Kobayashi, H., Miyamae, Y., Rojas, E., Ryu, J.C., and Sasaki, Y.F. (2000) Single cell gel/comet assay: guidelines for *in vitro* and *in vivo* genetic toxicology testing. *Environ. Mol. Mutagen.* 35(3):206–221.

Valverde, M. and Rojas, E. (2009) Environmental and occupational biomonitoring using the Comet assay. *Mutat. Res.* 681(1):93–109.

Wahnschaffe, U., Bitsch, A., Kielhorn, J., and Mangelsdorf, I. (2005) Mutagenicity testing with transgenic mice. Part II: comparison with the mouse spot test. *J. Carcinog.* 4(1):4.

Walmsley, R.M. (2008) GADD45a-GFP GreenScreen HC genotoxicity screening assay. *Expert Opin. Drug Metab. Toxicol.* 4(6):827–835.

Wang, J., Sawyer, J.R., Chen, L., Chen, T., Honma, M., Mei, N., and Moore, M.M. (2009) The mouse lymphoma assay detects recombination, deletion, and aneuploidy. *Toxicol. Sci.* 109(1):96–105.

Witte, I., Plappert, U., de Wall, H., and Hartmann, A. (2007) Genetic toxicity assessment: employing the best science for human safety evaluation part III: the comet assay as an alternative to *in vitro* clastogenicity tests for early drug candidate selection. *Toxicol. Sci.* 97(1):21–26.

Wood, D.K., Weingeist, D.M., Bhatia, S.N., and Engelward, B.P. (2010) Single cell trapping and DNA damage analysis using microwell arrays. *Proc. Natl. Acad. Sci. U. S. A.* 107(22):10008–10013.

109

HUMAN RELEVANCE OF RODENT LEYDIG CELL TUMORS

Thomas J. Steinbach, Robert. R. Maronpot, and Jerry F. Hardisty

INTRODUCTION

Rodents have been used extensively in virtually all fields of biomedical research and have been the primary species used in toxicologic and carcinogenic research. Over many years it has become obvious that some conditions and in particular some tumors in rodents have questionable relevance in humans. Some of these include peroxisome proliferator-activator receptor-α (PPAR-α) agonist-induced liver tumors, alpha 2u-globulin-induced renal tumors in male rats, and bladder tumors induced in rats by urinary calculi. In this chapter we review the human relevance of Leydig cell tumors (LCTs), which have been induced in rodents by a number of compounds. We will consider the similarities and differences between humans and rats in the physiology of the Leydig cell (LC) and the pathology of LCTs. Most importantly, we will examine the mechanisms of action that induce LCTs in rats and humans and present data on incidence, physiology, human endocrine disease, and comparative epidemiological studies that strongly indicate LCTs in rodents, in particular the rat, are of little relevance to human health.

DEVELOPMENT

The ontogeny of the LC can be reviewed in two basic ways: by following the stages of adult LC differentiation or by examining the two recognized generations of LCs, fetal and adult. Differentiation of the LC is typically broken down into four stages: stem LCs, progenitor LCs, immature LCs, and adult LCs. Stem LCs are spindle shaped and as a cell, which has not yet committed to a lineage of development, it does not express the LC-specific markers such as luteinizing hormone receptor or steroidogenic enzymes such as 3β-hydroxysteroid dehydrogenase. In the rat, at postnatal day 14 stem LCs begin to stain positive for 3β-hydroxysteroid dehydrogenase and are identified at this point as progenitor LCs. These cells develop from about postnatal day 14 until day 28 and begin to produce androgen. (Reviewed in Chen et al., 2009) Gene expression changes most when stem LCs develop into progenitor LCs, whereas differences in gene expression from progenitor to immature LCs and from immature to adult LCs are minimal, suggesting these cells are relatively more similar (Stanley et al., 2011).

Starting at postnatal day 28, progenitor LCs transform morphologically to become more round. The immature LC has increased amounts of smooth endoplasmic reticulum and steroidogenic enzyme levels. Testosterone is not yet the major product produced by these cells because they possess high levels of androgen-metabolizing enzymes that produce 5α-androstane-3α, 17β-diol. The immature LC population doubles once from postnatal day 28 to 56, at which point they develop into adult LCs. The androgen-metabolizing enzyme activity reduces while synthesis of testosterone increases. By day 90, testosterone production in an adult LC of a rat is 150 times than that of a progenitor, and five times than that of an immature LC (Shan et al., 1993).

Development of LCs can also be discussed in reference to the two generally recognized generations: fetal and adult. In the rat, the fetal LC begins to produce testosterone at about gestation day 16 and peaks on day 19 (Habert and Picon, 1984). In humans, testosterone peaks at the end of the 1st trimester (Reyes et al., 1974). The production of androgens is critical for masculinization of the fetus during what is termed

Hamilton & Hardy's Industrial Toxicology, Sixth Edition. Edited by Raymond D. Harbison, Marie M. Bourgeois, and Giffe T. Johnson.
© 2015 John Wiley & Sons, Inc. Published 2015 by John Wiley & Sons, Inc.

the "masculinization programming window." In the rat this window is from 15.5 to 17.5 days of gestation while in the human it extends from 8 to 14 weeks of gestation. (Welsh et al., 2008) In addition to androgens, fetal LCs produce insulin-like growth factor 3, which is responsible for testicular descent (Zimmerman et al., 1999).

A number of factors that influence the differentiation of the fetal LCs have recently been identified and include: desert hedgehog, a cell signaling molecule secreted by Sertoli cells, and transcription factors such as GATA-4. Interestingly, in rodents, development of fetal LCs is not dependent on pituitary luteinizing hormone (LH) (until the hypothalamic–pituitary–gonadal axis begins functioning near birth). Primates, including humans, have a brief period of independence from hormonal stimulation but then fetal LCs become dependent on placental chorionic gonadotropins (Reviewed in O'Shaughnessy and Fowler, 2011; Svechnikov et al., 2010).

The adult LC forms mostly during puberty and produces the testosterone responsible for spermatogenesis, along with differentiation of other secondary sex characteristics. In both the rat and human there is decreased testosterone production associated with aging. In humans the decreased testosterone is associated with increased levels of LH and appears to be due to a decline in the number of LCs through degeneration. In rats, however, decreased testosterone is associated with declining LH levels and seems to be related to the inability of the LC to respond to LH stimulation (Reviewed in Chen et al., 2009; Cook et al., 1999).

STEROIDOGENESIS

The LC is responsible for producing virtually all of the steroids of the testis; testosterone being the major steroid

(Stocco, 1996). Although a number of pathways have been elucidated recently, the primary signaling method for steroidogenesis is initiated when LH binds to a G protein-coupled receptor, which in turn activates adenylate cyclase, producing cAMP (See Figure 109.1). Protein kinase A is activated by cAMP and results in phosphorylation of a number of proteins and transcription factors (Wang and Ascoli, 1990). One of the most important of these is steroidogenic acute regulatory protein (StAR), which is estimated to mediate 85–90% of steroid synthesis (Reviewed in Stocco et al., 2005). Steroidogenic acute regulatory protein and translocator protein, formerly known as peripheral-type benzodiazepine receptor, are components of a large, multi-protein complex known as the transducesome. This complex is responsible for transporting cholesterol from the outer mitochondrial membrane to the inner mitochondrial membrane (Reviewed in Rone et al., 2009). The translocation of cholesterol from the outer to inner mitochondrial membrane is the rate limiting step in the synthesis of all steroids (Miller, 2007; Manna 2009). Intracellular cholesterol is produced by three mechanisms: de novo synthesis in the endoplasmic reticulum; mobilization from the plasma membrane and uptake from circulating cholesterol esters; and mobilization from lipid droplets (Rone et al., 2009).

Mediated by StAR, free cholesterol is transported to the inner mitochondrial membrane, where it is metabolized into pregnenolone by the P450 cholesterol side-chain cleavage enzyme (CYP11A1) (Reviewed in Payne and Hales, 2004). The remaining reactions of steroidogenesis occur within the endoplasmic reticulum with small differences between rats and humans in the pathway (Cook et al., 1999). In rat pregnenolone is transformed to progesterone by the enzyme 3β-hydroxysteroid dehydrogenase (3βHSD). Next the P450

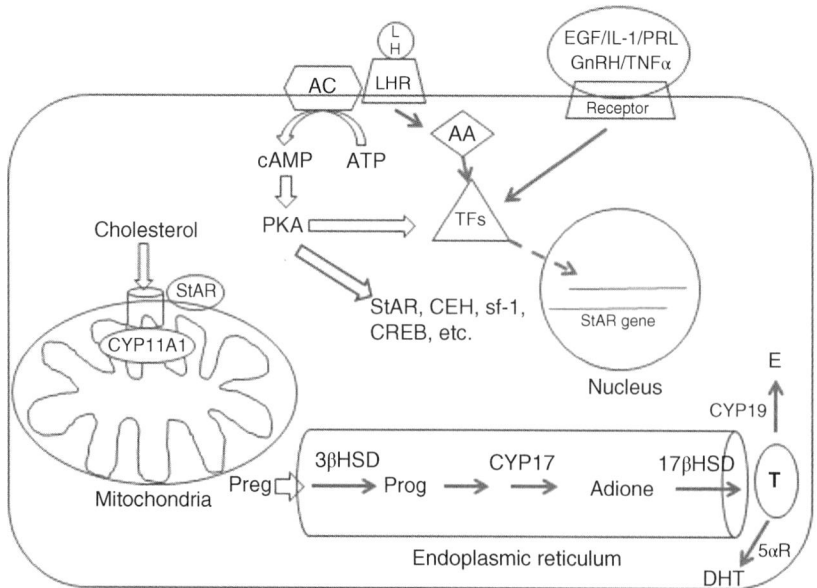

FIGURE 109.1 Mechanisms involved in the synthesis of testosterone.

enzyme, C_{17a}-hydroxylase/C_{17-20}-lysase (CYP17), catalyzes two reactions converting progesterone first to hydroxyprogesterone and then to C19 steroid, androstenedione. In humans, CYP17 prefers pregnenolone as a substrate and converts it to 17α-hydroxypregnenolone and then to C19 steroid, dehydroepiandrosterone (DHEA). Finally, 17β-hydroxysteroid dehydrogenase (17βHSD) converts the C19 steroid into testosterone (Payne and Hales, 2004). Testosterone can be further metabolized into estrogen by P450 aromatase (CYP19) (Payne and Hales, 2004) or into the more potent androgen dihydrotestosterone (DHT) (Cook et al., 1999).

As mentioned, cAMP-dependent activation of protein kinase A is essential for activation of StAR and other proteins and transcription factors involved in steroidogenesis, to include cholesterol esterase (CEH), steroidogenic factor-1 (sf-1), GATA-4, and cAMP response element-binding protein (CREB). However, there are other mechanisms by which steroidogenesis is regulated. In recent years research has focused on factors that influence and regulate StAR, since it is critical to the rate-limiting step of steroidogenesis. A number of factors can regulate StAR via cAMP-independent pathways. These include epidermal growth factor (EGF), macrophage-derived factors such as IL-1 and TNFα, hormones such as prolactin and gonadotropin-releasing hormone (GnRH), chloride ions, and calcium messenger systems. In addition, a protein kinase C pathway can be activated resulting in increased transcription and translation of StAR (Stocco et al., 2005; Manna et al., 2007). Leutinizing

hormone binding to its receptor can release arachidonic acid (AA) and metabolites of AA can regulate StAR gene expression (Wang et al., 1999).

PATHOLOGY

The mammalian testis is composed of a fibrous tunic that surrounds convoluted seminiferous tubules and supporting stroma. In a microscopic section the tubules make up the majority of the cellular structure present and contain mostly gametes in various stages of differentiation from spermatagonia to spermatozoa in addition to supportive cells, such as Sertoli cells. Surrounding the tubules is a small amount of connective tissue that contains blood vessels, nerves, and interstitial cells of Leydig (Figure 109.2). Normal LCs have a wide variation in morphology. They can be spindle shaped with little cytoplasm but are typically round and contain large amounts of lightly eosinophilic cytoplasm. In humans and the wild bush rat, the LC cytoplasm contains elongated, cigar-shaped structures known as crystals of Reinke. These crystals can be seen with light microscopy and are found only in adults. Their origin and function are unknown (Young and Heath, 2000).

In rodents, hyperplasia of LCs is characterized by aggregates of cells that focally or diffusely expand the interstitium between seminiferous tubules. Cells are typically round with centrally located nuclei and granular, eosinophilic cytoplasm. Occasionally, spindle cells with little cytoplasm and dark nuclei are also found (Boorman et al., 1990).

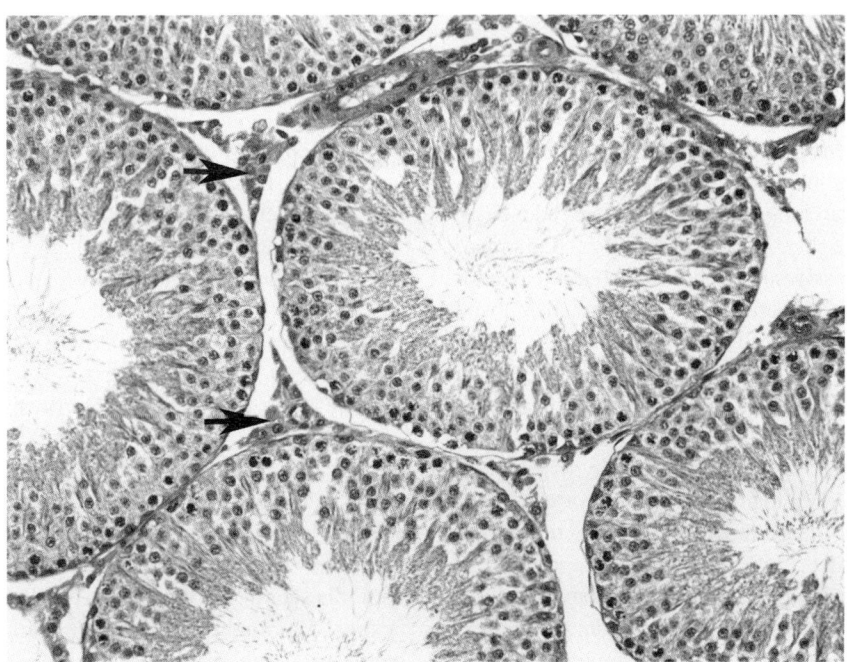

FIGURE 109.2 There are low numbers of Leydig cells (arrows) in the interstitium surrounding seminiferous tubules of this normal rat testis (H&E, 200×).

Leydig cell adenoma begins as hyperplasia and consists of one or more foci of expansile LCs that typically compress adjacent tubules. The distinction between hyperplasia and adenoma is difficult as cellular characteristics cannot distinguish between them. Most often size is used as a morphologic criteria to differentiate hyperplasia from adenoma. One recommendation from the National Toxicologic Program (NTP) is to use the size of adjacent tubules as a threshold. When the lesion grows larger than the diameter of the adjacent tubule, it is diagnosed as adenoma. In 2005 toxicologic pathology societies from around the world began an initiative to standardize nomenclature for proliferative and nonproliferative lesions in rats and mice. This initiative is termed the International Harmonization of Nomenclature and Diagnostic Criteria for Lesions in Rats and Mice (INHAND) (Mann et al., 2012). The INHAND focuses on specific organ systems. For the male reproductive system, the INHAND guidance differentiates hyperplasia from adenoma using a size of three seminiferous tubules, among other criteria (Creasy et al., 2012). This makes interpretation of incidence data in rodents difficult as published data on LC adenoma typically does not describe the methodology used in establishing the diagnosis of adenoma (Cook et al., 1999).

The incidence of LC adenoma in rodents is highly dependent on species and strain and increases with age. In mice it is typically low, <2.5%, whereas in Sprague-Dawley and Wistar rats it ranges from 4 to 7% (Cook et al., 1999). In Fisher 344 rats, however, the incidence of LC adenoma approaches 100% if the rat is allowed to live its life span (Boorman et al., 1990). LC tumors in rodents are almost always benign as malignancy is rarely reported (Boorman et al., 1990; Cook et al., 1999).

In contrast to rodents, LC tumors in humans are rare, comprising from 1 to 3% of all testicular neoplasms with an estimated incidence of 0.1–3 per million (Cook et al., 1999). Comparing the incidence of LCTs between rodents and humans is somewhat problematic. The data in rodents comes from toxicologic and carcinogenic studies, where rodents are subjected to complete histologic examination and small tumors, not evident grossly, are often discovered microscopically. In humans, testicular tumors are most often diagnosed by palpation and routine microscopic examination of testicular tissue does not occur. Advanced imaging techniques, however, have made possible the diagnosis of non-palpable testicular tumors. Recent data has shown that when ultrasound is used to diagnose testicular tumors, the incidence of LCT is significantly higher than previously reported (Leonharsberger et al., 2011). Regardless, even when considering detection bias, the incidence of LCT in rodents, especially in some strains, is markedly higher than in humans. Furthermore, LCTs in humans are found in all ages, from prepubertal boys to older men. Microscopically they resemble tumors in rodents except for the presence or Reinke crystals. While the majority of human LCTs are benign, a much higher incidence, 10%, are considered malignant in humans. (Cook et al., 1999)

MECHANISM OF TUMOR INDUCTION

In the rat there are a number of plausible mechanisms for the induction of LC tumors. These include agonists of estrogen, GnRH, dopamine receptors, and PPAR-α. In addition, androgen receptor antagonists, and inhibitors of testosterone synthesis and the enzymes aromatase and 5α-reductase can lead to LCTs. Most of these mechanisms are related to elevated levels of LH. It is well documented that prolonged elevated levels of LH induce LC hyperplasia and neoplasia in the rat (Neumann, 1991; Prentice and Meikle, 1995; Huseby, 1996). In addition, administration of testosterone to F344 rats lowers LH levels and blocks LCT formation (Chatani et al., 1990).

Because LH plays such a central role in LCTs of rats, it is useful to review the hypothalamic-pituitary-testis (HPT) axis to see how LH levels can influence the induction of LCTs. Figure 109.3, (adapted from Cook et al., 1999), highlights five mechanisms by which chemicals can disrupt LH levels and induce LCTs. These five mechanisms, and others, will be briefly reviewed in more detail and additional compounds known to induce LCTs in rats will be listed.

Flutamide is an androgen receptor antagonist. It competes with testosterone and DHT for binding at the androgen receptor (Simard et al., 1986). As a result, the androgen signal to the hypothalamus and pituitary is reduced, which stimulates an increase in LH secretion to counter the decreased androgen levels (Cook et al., 1993; Viguier-Martinez et al., 1983). Examples of other androgen receptor antagonists known to induce LCTs in rats include cimetidine (Leslie et al., 1981); procymidone (Murakami et al., 1995); and bicalutamide (Iswaran et al., 1997).

In a similar manner, 5α-reductase inhibitors, such as finasteride block the conversion of testosterone to DHT (Prahalada et al., 1994; Rittmaster et al., 1992). Dihydrotestosterone has a higher binding affinity to the androgen receptor than does testosterone (DeGroot et al., 1995). Thus a decrease in DHT reduces the net androgenic signal to the hypothalamus and pituitary resulting in a compensatory increase in LH. Interestingly, 5α-reductase inhibitors induce LCTs in mice and LC hyperplasia in rats (Cook et al., 1999).

Inhibition of testosterone synthesis also lowers the androgen signal resulting in increased LH levels. Interestingly, ketoconazole, which is the most widely known inhibitor of testosterone synthesis, did not induce LCTs but this seemingly was due to the design of the study, which exposed animals to lower levels and for shorter durations than similar bioassays (Cook et al., 1999). Examples of compounds that induce LCTs by inhibiting testosterone synthesis include lansoprazole (Fort et al., 1995); calcium

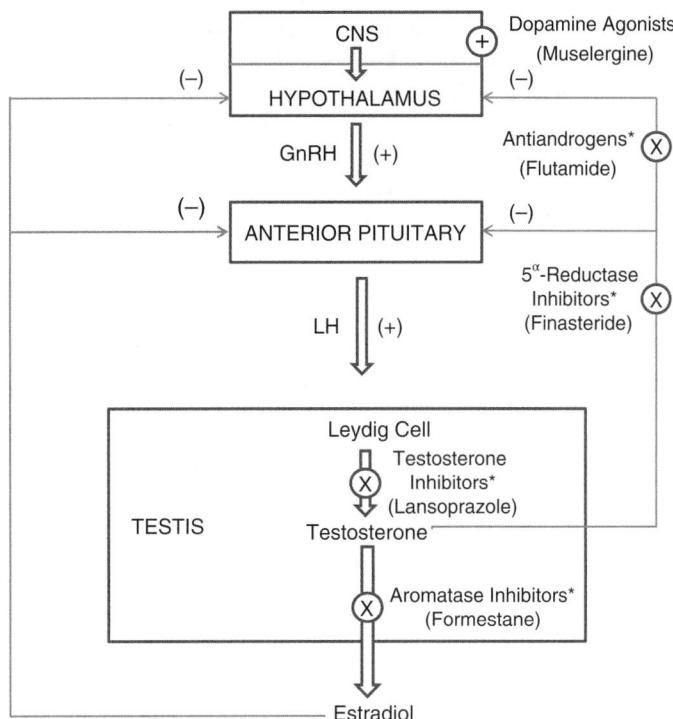

FIGURE 109.3 Regulation of the HPT axis and control points for potential disruption.

channel blockers such metronidazole (Rustia and Shubik, 1979).

Formestane is an aromatase inhibitor and as such it blocks the conversion of testosterone to estradiol. Since estradiol provides negative feedback to the hypothalamus and pituitary (see Figure 109.3), inhibition of estradiol, in theory, would increase the levels of LH and induce LCTs. However, formestane induced LC hyperplasia in beagle dogs but not rodents (Junker-Walker and Nogues, 1994) and aminoglutethimide, the most well-known aromatase inhibitor, did not induce LC hyperplasia or LCTs (Salhanick, 1982; Shaw et al., 1988). The role of aromatase inhibitors in leading to LCTs, therefore, is equivocal.

Dopamine agonists such as muselergine decrease prolactin levels, which causes downregulation of LH receptors on LCs (Prentice et al., 1992). The decrease in LH receptors lowers the overall testosterone production and LH levels increase to compensate. An alternative mechanism was proposed based on work with oxolinic acid. Yamanda et al. suggested that dopamine agonists increase levels of GnRH and thereby increase LH levels (Yamanda et al., 1995). The exact mechanism has yet to be worked out.

In rats GnRH induces LCTs at low doses but not at higher doses. This is because the LC of the rat contains GnRH receptors and can stimulate the LC directly. Low levels of GnRH can also increase LH through direct stimulation of the pituitary (Figure 109.3). At higher doses, negative feedback inhibition from GnRH would lower LH levels and thus

LCTs. (Cook et al., 1999) Mice and humans do not have GnRH receptors on their LCs and are thus not susceptible to LCTs by GnRH agonists (Hunter et al., 1982).

Estrogen agonists induce LCTs in the mouse but almost never in the rat (Cook et al., 1999). However, PPAR-α agonists are known to lead to LCTs in rats. In at least some of the cases of PPAR-α agonists there is an increase in estradiol (but not LH or testosterone) that corresponds with the potency of the compound (Biegel et al., 2001). PPAR-α agonists also can reduce testosterone levels (Cook et al., 1992; Gazouli et al., 2002). As shown previously, compounds that lower testosterone can cause a compensatory rise in LH and lead to LCTs. Most likely, both elevation of estrogen and decreased testosterone play a role in induction of LCTs by PPAR-α agonists (Klaunig et al., 2003).

HUMAN RELEVANCE

It is clear from the forgoing discussion that many compounds induce LCT in rodents, especially rats. Since the rat is one of the primary species used in toxicologic and carcinogenic risk assessment, these findings may be of importance in establishing exposure thresholds or compound safety data. However, a number of factors, when examined closely, illustrate that the occurrence of LCTs in rats is not biologically relevant to humans. These differences are thoroughly discussed and referenced in Cook et al. (1999) and the reader is referred there for a more in depth analysis. This discussion will focus on differences in comparative biology, tumor incidence, as well as cases of human genetic disease and data from epidemiology studies.

The biology of the rat and human differ in a number of important ways and some of those potentially have an impact on the relevance of LCTs in humans. These include differences in serum proteins, response to hCG, number of LH and GnRH receptors, sensitivity to prolactin, and LH-related aging changes.

The sex hormone-binding globulin is absent in the rat. It is produced in the liver in humans and binds to the majority of testosterone. This binding lowers the metabolism and clearance of testosterone in humans and dampens short-term changes in testosterone levels. In contrast, in the rat testosterone levels can be more rapidly altered and as a result, compounds that alter testosterone levels have a more dramatic effect in rats.

Rats and humans have been shown to respond differently to hCG, a hormone that has equivalent effects on LCs as does LH. *In vivo* and *in vitro* data show that hCG induces LC hyperplasia in rats while inducing hypertrophy in humans. In addition, testosterone secretion and the mitogenic response are up to 100-fold less in human LCs.

There are important differences between rats and humans in the receptors on LCs. Compared to humans, rats have over 10 times the number of LH receptors on their LCs. This spare capacity of receptors allows for a lower concentration of LH in rats that will induce a response in LCs. In addition, as mentioned, rats have GnRH receptors on LCs, while humans do not. It would appear, therefore, that induction of LCTs by GnRH agonists is not a mechanism applicable to humans. Likewise, unlike rats, humans do not express prolactin receptors on LCs. In the rat, dopamine agonists act by decreasing prolactin levels, which in turn lower the number of LH receptors on their LCs. The decrease in LH receptors results in a decrease in the testosterone levels and a compensatory increase in LH. Because humans lack prolactin receptors, dopamine agonists seemingly would not alter the number of LH receptors on human LCs and testosterone, and LH levels would remain unchanged. Therefore, it is unlikely that dopamine agonists would induce LCTs in humans.

Finally, rats and humans differ in how LH affects testosterone levels in aging. In both rats and humans testosterone levels decrease over time. However, in rats LH levels also decrease and this decrease in LH may be responsible for the decrease in testosterone. In contrast, LH levels remain unchanged in men with aging. Luteinizing hormone appears, therefore, not to be a factor in decreasing the level of testosterone in men with time.

As previously discussed, the incidence of LCT in rats and humans is markedly different. Depending on the strain, the incidence of LCT in rats ranges from nearly 100% in Fisher 344s to 5% in Wistar (Tucker, 1997) and 6.5% for Sprague-Dawley (McMartin et al., 1992) rats. In contrast, in humans

testicular cancer accounts for about 1% of all cancers in men and of these, only about 1% are of LC origin; making LCTs about 0.01% of all cancers in men (Cook et al., 1999). Consequently, the incidence of LCT in humans is in the order of 1.9 million-fold less than in F344 and 130,000-fold less than Sprague-Dawley rats (Cook et al., 1999).

There are two genetic endocrine diseases in humans that highlight the mechanistic differences in LCT induction between humans and rats. The first is androgen insensitivity syndrome, which is due to a defect in the androgen receptor. The second, familial male precocious puberty, is the result of a defect in the LH receptor. Reviewing the incidence of LCTs in individuals affected with these diseases further emphasizes that humans are clearly less sensitive to LCT formation than are rats.

In androgen insensitivity syndrome there is a genetic defect in the androgen receptor so that androgens are not recognized. The syndrome ranges from complete to partial insensitivity and phenotypically, affected individuals range from infertile males to females. Individuals with complete insensitivity and a cryptorchid testis have elevated levels of LH, but normal levels of testosterone. Testicular tumors, including LCTs, are more common in individuals with androgen insensitivity syndrome with some reports of LC adenomas as high as 2–3%. As a comparison, the incidence of LCT in rats when administered an androgen receptor antagonist, such as flutamide, is almost 100%. Administration of flutamide to rats can be thought of as a chemically induced form of androgen insensitivity syndrome. If humans were as sensitive to the formation of LCTs, then individuals with this syndrome would have a much higher incidence of LCTs, which they clearly do not. (Cook et al., 1999)

Familial male precocious puberty is the result of a genetic mutation of the LH receptor that results in constitutive activation. Affected individuals undergo puberty at about 4 years of age. They have testosterone production and LC hyperplasia but undetectable levels of LH. The continuous activation of LH receptors in individuals with familial male precocious puberty mimics the primary mechanism of LCT induction in rats, namely elevated LH levels. While LC hyperplasia is reported in cases of familial male precocious puberty, there have been no documented cases of LCTs despite a lifetime of LH-receptor activation. Once again, if humans were as sensitive to LH-induced LCT formation, individuals with familial male precocious puberty would have a much higher incidence of LCT (Cook et al., 1999).

Lastly, there have been a number of epidemiological studies conducted on compounds with human exposure for which there is comparative rodent study data available. Only three will be reviewed to illustrate the concept but for a more complete list refer to Cook et al. (1999). It must be emphasized that the epidemiological surveys relied on insensitive methods for the detection of LCT in humans but nonetheless the results are still helpful in underscoring the relative difference in LCT incidence between humans and rodents.

1,3-Butadiene is a ubiquitous product used in the plastics industry and because of its high exposure to humans, it has been extensively studied both in laboratory animals and humans. Sprague-Dawley rats exposed via inhalation had an increased incidence of LCTs (8% vs. 0% in controls) (Owen and Glaister, 1990). Using a large number of exposed workers, human epidemiological studies have failed to identify a risk of LCTs in men exposed to 1,3-butadiene (Divine and Hartman, 2001). Cadmium induces not only LCTs in rats but also adenomas in the prostate (Waalkes et al., 1988). However, a review of human exposure studies identifies increased risk of lung cancer but not prostate or LCTs (Verougstraete et al., 2003). Finally, trichloroethylene has wide human exposure as an industrial solvent and an additive in food, drugs, and consumer products. It causes marked increases in LCTs in Sprague-Dawley rats but a review of five cohort studies in humans did not identify trichloroethylene as a carcinogen (Cook et al., 1999).

CONCLUSION

In rats, LCTs are induced almost exclusively by elevated levels of LH. The exceptions seem to be LCTs induced by GnRH and dopamine agonists, which, as discussed, appear not to be applicable to humans since they lack the receptors (GnRH and prolactin) to induce LCTs. Therefore, when examining the relevance of LCTs induced in rats to human health, the question becomes what is the relative sensitivity in humans to LCT induction from elevated LH levels. As discussed, several lines of reasoning indicate that the high incidence of LCTs in rats is not relevant to humans. First, humans have a markedly lower incidence of LCTs compared to rodents, especially rats. Second, there are a number of comparative differences in the biology of the LC, especially the type and number of receptors that help explain the difference in rates of LCTs. Third, two endocrine diseases in humans do not produce LCTs at an incidence level expected, if the mechanism of LCT induction in human was similar to that in rats. And fourth, when compounds with both human epidemiological and rodent exposure data are compared, the results do not support a conclusion that rodent LCTs are relevant to human health.

REFERENCES

Biegel, L.G., et al. (2001) Mechanisms of extrahepatic tumor induction by peroxisome proliferators in male CD rats. *Toxicol. Sci.* 60(1):44–55.

Boorman, G.A., et al. (1990) Testis and epididymis. In: Boorman, G.A., et al., editors. *Pathology of the Fisher Rat*, San Diego, CA: Academic Press.

Chatani, F., et al. (1990) Stimulatory effect of a lutenizing hormone on the development and maintenance of 5α-reduced steroid-producing testicular interstitial cell tumors in Fisher 344 rats. *Anticancer Res.* 10:337–342.

Chen, H., Ge, R.S., and Zirkin, B.R. (2009) Leydig cells: from stem cells to aging. *Mol. Cell. Endocrinol.* 306(1–2):9–16.

Cook, J.C., et al. (1992) Induction of Leydig cell adenomas by ammonium perfluorooctanoate: a possible endocrine-related mechanism. *Toxicol. Appl. Pharmacol.* 113(2):209–217.

Cook, J.C., et al. (1993) Investigation of a mechanism of Leydig cell tumorigenesis by linuron in rats. *Toxicol. Appl. Pharmacol.* 119:195–204.

Cook, J.C., et al. (1999) Rodent Leydig cell tumorigenesis: a review of the physiology, pathology, mechanisms, and relevance to humans. *Crit. Rev. Toxicol.* 29(2):169–261.

Creasy, D., et al. (2012) Proliferative and nonproliferative lesions of the rat and mouse male reproductive system. *Toxicol. Pathol.* 40:40S–121S.

DeGroot, L.J., et al. (1995) *Endocrinology*, 3rd ed., W.B. Saunders, Philadelphia.

Divine, B.J. and Hartman, C.M. (2001) A cohort mortality study among workers at a 1,3 butadiene facility. *Chem. Biol. Interact.* 135–136:535–553.

Fort, F.L., et al. (1995) Mechanism for species-specific induction of Leydig cell tumors in rats by lansoprazole. *Fundam. Appl. Toxicol.* 26(2):191–202.

Gazouli, M., et al. (2002) Effect of peroxisome proliferators on Leydig cell peripheral-type benzodiazepine receptor gene expression, hormone-stimulated cholesterol transport, and steroidogenesis: role of the peroxisome proliferator-activator receptor alpha. *Endocrinology* 143(7):2571–2583.

Habert, F. and Picon, R. (1984) Testosterone, dihydrotestosterone and estradiol-17 beta levels in maternal and fetal plasma and in fetal testes in the rat. *J. Steroid Biochem.* 21(2):193–198.

Hunter, M.G., et al. (1982) Stimulation and inhibition by LHRH analogues of cultured rat Leydig cell function and lack of effect on mouse Leydig cells. *Mol. Cell Endocrinol.* 27:31–34.

Huseby, R.A. (1996) Leydig cell neoplasia. In: Payne, A.H., Hardy, M.P., and Russel, L.D., editors. *The Leydig Cell*, Vienna, IL: Cache River Press.

Iswaran, T.J., et al. (1997) An overview of animal toxicology studies with bicalutamide (ICI 176,334). *J. Toxicol. Sci.* 22(2):75–88.

Junker Walker, U. and Nogues V. (1994) Changes induced by treatment with aromatase inhibitors in testicular Leydig cells of rats and dogs. *Exp. Toxic. Pathol.* 46:211–213.

Klaunig, J.E., et al. (2003) PPAR-α agonist-induced rodent tumors: modes of action and human relevance. *Crit. Rev. Toxicol.* 33(6):655–780.

Leonharsberger, L., et al. (2011) Increased incidence of Leydig cell tumors of the testis in the era of improved imaging techniques. *BJU Int.* 108:1603–1607.

Leslie G.B., et al. (1981) A two-year study with cimetidine in the rat: assessment for chronic toxicity and carcinogenecity. *Toxicol. Appl. Pharmacol.* 61:119–137.

Mann, P.C., et al. (2012) International harmonization of toxicologic pathology nomenclature: an overview and review of basic principles. *Toxicol. Pathol.* 40:7S–13S.

Manna, P.R., Dyson, M.T., and Stocco, D.M. (2009) Regulation of the steroidogenic acute regulatory protein gene expression: present and future perspectives. *Mol. Hum. Reprod.* 15(6):321–333.

Manna, P.R., Jo, Y., and Stocco, D.M. (2007) Regulation of Leydig cell steroidogenesis by extracellular signal-regulated kinase 1/2: role of protein kinase A and protein kinase C signaling. *J. Endocrinol.* 193(1):53–63.

McMartin, D.N., et al. (1992) Neoplasms and related proliferative lesions in control Sprague-Dawley rats from carcinogenicity studies. Historical data and diagnostic considerations. *Toxicol. Pathol.* 20(2):212–225.

Miller, W.L. (2007) Steroidogenic acute regulatory protein (StAR), a novel mitochondrial cholesterol transporter. *Biochim. Biophys. Acta* 1771(6):663–676.

Murakami, M., et al. (1995) Species-specific mechanism in rat Leydig cell tumorigenesis by procymidone. *Toxicol. Appl. Pharmacol.* 131(2):244–252.

Neumann, F. (1991) Early indicators for carcinogenesis in sex hormone-sensitive organs. *Mutat. Res.* 248:341–356.

O'Shaughnessy, P.J. and Fowler, P.A. (2011) Endocrinology of the mammalian fetal testis. *Reproduction* 141:37–46.

Owen, P.E. and Glaister, J.R. (1990) Inhalation toxicity and carcinogenicity of 1,3-butadiene in Sprague-Dawley rats. *Environ. Health Perspect.* 86:19–25.

Payne, A.H. and Hales, D.B. (2004) Overview of steroidogenic enzymes in the pathway from cholesterol to active steroid hormones. *Endocr. Rev.* 25(6):947–970.

Prahalada, S., et al. (1994) Leydig cell hyperplasia and adenomas in mice treated with finastride, a 5 alpha-reductase inhibitor: a possible mechanism. *Fundam. Appl. Toxicol.* 22(2):211–219.

Prentice, D.D., et al. (1992) Mesulergine induced Leydig cell tumors, a syndrome involving the pituitary-testicular axis of the rat. *Arch. Toxicol.* 15(Suppl.):197–204.

Prentice, D.E. and Meikle, A.W. (1995) A review of drug-induced Leydig cell hyperplasia and neoplasia in the rat and some comparisons with man. *Hum. Exp. Toxicol.* 14:562–572.

Reyes, F.L., et al. (1974) Studies on human sexual development. II. Fetal and maternal serum gonadotropin and sex steroid concentrations. *J. Clin. Endocrinol. Metab.* 38(4):612–617.

Rittmaster, R.S., et al. (1992) Effect of finastride, a 5 alpha-reductase inhibitor, on serum gonadotropins in normal men. *J. Clin. Endocrinol. Metab.* 75(2):484–488.

Roberts S.A., et al. (1989) SDZ 200-100 induces Leydig cell tumors by increasing gonadotrophins in rats. *J. Am. Coll. Toxicol.* 8:487–505.

Rone, M.B., Fan, J., and Papadopoulos, V. (2009) Cholesterol transport in steroid biosynthesis: role of protein-protein interactions and implications in disease states. *Biochim. Biophys. Acta* 1791(7):646–658.

Rustia, M. and Shubik, P. (1979) Experimental induction of hepatomas, mammary tumors, and other tumors with metronidazole in noninbred Sas:MRC(WI)BR rats. *JNCI* 63:863–868.

Salhanick, H.A. (1982) Basic studies on aminoglutethimide. *Cancer Res.* 42:3315s–3321s.

Shan, L.X., et al. (1993) Differential regulation of steroidogenic enzymes during differentiation optimizes testosterone production by adult rat Leydig cells. *Endocrinology* 133(5):2277–2283.

Shaw, M.A., Nicholls, P.A., and Smith, H.J. (1988) Aminoglutethimide and ketoconazole: historical perspectives and future prospects. *J. Steroid Biochem.* 31:137–146.

Simard, J., et al. (1986) Characteristics of interaction of the antiandrogen flutamide with the androgen receptor in various target tissues. *Mol. Cell. Endocrinol.* 44(3):261–270.

Stanley, E.L., et al. (2011) Stem Leydig cell differentiation: gene expression during development of the adult rat population of Leydig cells. *Biol. Reprod.* 85:1161–1166.

Stocco, D.M. (1996) Acute regulation of Leydig cell steroidogenesis. In: Payne, A.H., Hardy, M.P., and Russel, L.D., editors. *The Leydig Cell*, Vienna, IL: Cache River Press.

Stocco, D.M., et al. (2005) Multiple signaling pathways regulating steroidogenesis and steroidogenic acute regulatory protein expression: more complicated than we thought. *Mol. Endocrinol.* 19(11):2647–2659.

Svechnikov, K., et al. (2010) Origin, development and regulation of human Leydig cells. *Horm. Res. Paediatr.* 73:93–101.

Tucker, M.J. (1997) The male genital system. In: *Diseases of the Wistar Rat*, Bristol, PA: Taylor & Francis.

Verougstraete, V., et al. (2003) Cadmium, lung and prostate cancer: a systematic review of recent epidemiological data. *J. Toxicol. Environ. Health B Crit. Rev.* 6(3):227–255.

Viguier-Martinez M.C., et al. (1983) Effect of a non-steroidal antiandrogen, flutamide, on the hypothalmo-pituitary axis, genital tract and testis in growing rats: Endocrinological and histological data. *Acta Endocrinologica* 102:299–306.

Waalkes, M.P., et al. (1988) Cadmium carcinogenesis in male Wistar [Crl:(WI)BR] rats: dose-response analysis of tumor induction in the prostate and testes and at the injection site. *Cancer Res.* 48:4656–4663.

Wang, H.Y. and Ascoli, M. (1990) Reduced gonadotropin responses in a novel clonal strain of Leydig tumor cells established by transfection of MA-10 cells with a mutant gene of a type I regulatory subunit of the cAMP-dependent protein kinase. *Mol. Endocrinol.* 4(1):80–90.

Wang, X., Walsh, L.P., and Stocco, D.M. (1999) The role of arachidonic acid on LH-stimulated steroidogenesis and steroidogenic acute regulatory protein accumulation in MA-10 mouse Leydig tumor cells. *Endocrine* 10(1):7–12.

Welsh, M., et al. (2008) Identification in rats of a programming window for reproductive tract masculinization, disruption of which leads to hypospadias and cryptorchidism. *J. Clin. Invest.* 118:1479–1490.

Yamada, T. et al. (1995) A possible mechanism for the increase of serum leutinzing hormone levels in rats by oxolinic acid. *Toxicol. Appl. Pharmacol.* 134:35–42.

Young, B. and Heath, J.W. (2000) *Wheater's Functional Histology*, London: Harcourt Publishers.

Zimmerman, S., et al. (1999) Targeted disruption of the Insl3 gene causes bilateral cryptorchidism. *Mol. Endocrinol.* 13:681–691.

110

REPRODUCTIVE TOXICOLOGY

JUDITH W. HENCK

INTRODUCTION

Adverse effects on the reproductive potential resulting from an occupational exposure to physical agents and chemicals in the workplace are important concerns for occupational medicine. Potential harm may occur to men and women, as well as to their progeny, due to an exposure prior to and during pregnancy and after birth. Adverse effects on the structural and functional components of male and female reproductive systems may lead to impaired fertility or infertility of exposed workers. According to the World Health Organization (2013), infertility is defined as: ". . . the inability to conceive a child. A couple may be considered infertile if, after two years of regular sexual intercourse, without contraception, the woman has not become pregnant (and there is no other reason, such as breastfeeding or postpartum amenorrhea). Primary infertility is infertility in a couple who have never had a child. Secondary infertility is failure to conceive following a previous pregnancy. Infertility may be caused by infection in the man or woman, but often there is no obvious underlying cause." Reduced fertility may be expressed as a reduction in the number of live births, reduced odds of conception, or increased time to pregnancy (Bellinger, 2005). One in six couples in the general population has trouble conceiving, with males and females equally responsible for infertility (Sikka and Wang, 2008). Adverse developmental effects may include death of the conceptus (embryo/fetus) or neonate, structural abnormalities, growth retardation (e.g., low birth weight), and functional deficits (e.g., in behavior, immune competence, ability to reproduce).

Reproductive hazards in the workplace have been investigated over the past several decades using workplace experience and epidemiology studies, as well as studies in laboratory animals. The complexity of the male and female reproductive systems renders this work challenging. Completion of the mammalian developmental cycle is dependent upon the following functions: gametogenesis (production of male and female germ cells); libido; transport of the germ cells to facilitate fertilization; transport, formation, and cleavage of the zygote; implantation of the embryo in the uterus; formation of a functional placenta; embryogenesis and organogenesis; fetal development and maturation; delivery, lactation, and maternal care; postnatal survival and adjustment of the neonate; growth and maturation; and sexual maturation (Johnson, 1986). The cycle then begins again with gametogenesis in the offspring. Each of these stages can potentially be disrupted by specific toxicants in humans as well as laboratory animals. Physical agents and toxicants may adversely affect various morphologic structures of the reproductive tract, as well as the hormonal balance required to maintain reproductive homeostasis. In addition, exposure of women to reproductive hazards may not only affect reproductive well-being, but also development of the conceptus if pregnancy occurs; in this circumstance exposure to toxicants may impact two individuals, rather than one.

Interpretation of data from exposure to hazardous chemicals and physical factors in the workplace may be confounded by a number of factors, including age and ethnicity, and lifestyle factors such as smoking, diet, alcohol and recreational drug use, stress, noise, and work shifts. It is often difficult to ascertain the dose/exposure. In an occupational setting, exposure levels are generally higher and more closely monitored than for environmental exposure, so individual exposure can be more reliably estimated (Schrag and Dixon, 1985). However, this is still often a difficult process, rendered more

Hamilton & Hardy's Industrial Toxicology, Sixth Edition. Edited by Raymond D. Harbison, Marie M. Bourgeois, and Giffe T. Johnson.
© 2015 John Wiley & Sons, Inc. Published 2015 by John Wiley & Sons, Inc.

complex by the fact that workers may be exposed to more than one putative toxicant, making it difficult to ascribe reproductive effects to a single physical agent or chemical. The nature and magnitude of reproductive toxicity are often dependent on the exposure levels, yet these are difficult to assess in an occupational setting. Although studies in laboratory animals have the ability to closely control and monitor exposure, they are often conducted at exposures in excess of those encountered in an occupational setting and must therefore be interpreted in that context. In female reproductive and developmental toxicity, maternal age may be a confounding factor in the evaluation of cumulative exposure to an agent such as ionizing radiation, since older women would be expected to have higher cumulative exposure than younger women (Whelan, 1997). Paternal exposure may contribute to developmental toxicity, and it may be difficult to separate maternal from paternal influences.

Often only limited information on potential human reproductive effects exists, and it may then be necessary to rely on data from laboratory animal species to inform risk assessment. Many examples exist in the scientific literature of chemicals that induce reproductive and/or developmental toxicity in humans and in one or more laboratory animal species, and the predictive value of laboratory animals has proven to be useful in the safety assessment. In most cases, more data exist in laboratory animals than in humans. Important questions related to laboratory animal studies include, how well do laboratory species predict human reproductive and developmental toxicity for a given physical or chemical agent, and, are relevant end points available in animal models? Possible sources of interspecies differences include genetic variability, metabolism, intracellular pathways of toxicity, membrane biochemistry and receptors, pharmacokinetics, and specific organ function (Working, 1988). It is important to have an understanding of these differences and recognize potential limitations in species extrapolation. In case of male reproductive toxicity, many of the same end points can be employed in humans and research animals, including measurements of sperm count and concentration, motility, and chromosome morphology (Wyrobek et al., 1983). Conversely, the number of sperm per ejaculate in humans is much less than that of laboratory species—while this means greater fertility for the laboratory species, it also renders them relatively insensitive indicators of human reproductive risk (Working, 1988). With regard to developmental toxicity, the following pre- and postnatal considerations are important in evaluating laboratory animal predictivity: (1) divergent differentiation of structure, function, and physiology across species; (2) lack of understanding of species differences in functional ontogeny; and (3) lack of common end points and milestones across species (Morford et al., 2004). It is important to keep in mind that developmental effects do not occur at the same chronological age across species; structural and functional damage demonstrates stage specificity that is dependent on developmental age.

Despite these potential limitations, human occupational and environmental experience, epidemiology studies, and often many laboratory animal studies with a given toxicant have identified physical and chemical reproductive issues, resulting in the development of health and safety regulations and guidelines designed to protect workers from reproductive hazards. The following sections describe the effects of known occupational and/or environmental toxicants on male and female reproduction, as well as on offspring development. While the agents and chemicals discussed do not comprise a comprehensive list, they are well-known examples of reproductive toxicants that may be encountered in an occupational setting. A given physical or chemical agent may appear in more than one section, highlighting the potential of some toxicants to adversely affect male and female reproduction and development. Descriptions of the occupational uses and potential environmental routes of exposure are given in the first section in which the chemical appears.

ENDOCRINE DISRUPTION

Over the past two decades, scientific and public attention has been drawn to the public health impact of chemicals known to modulate the action of hormones. These chemicals are referred to as endocrine disruptors and defined as: ". . . an exogenous agent that interferes with the production, release, transport, metabolism, binding, action, or elimination of natural hormones responsible for the maintenance of homeostasis and the regulation of developmental processes." (Kavlock et al., 1996) Reports of male reproductive tract issues have increased over the last several decades, including increased risks of testicular cancer, and congenital anomalies such as cryptorchidism and hypospadias, and declining sperm production (Oliva et al., 2001). Skakkebæk et al. (2001) have suggested that the rapid increase of reproductive disorders is due to environmental or lifestyle factors, rather than genomic structural defects. These effects have been linked to the widespread use of endocrine active chemicals. One of the lines of evidence creating this link is that the male reproductive issues of current concern in humans, with the exception of testicular cancer, can be experimentally produced in animals by pre- and perinatal exposure to chemicals known to target the endocrine system (Skakkebæk et al., 2001). Initial concerns involved chemicals that mimic estrogen or are antiestrogens; more recently chemicals that are androgens or antiandrogens, and those that can affect thyroid function, have been investigated. The Scientific Consensus Report from the International Programme on Chemical Safety (IPCS, 2002) states the following major areas of concerns surrounding endocrine disrupting chemicals:

(1) adverse effects in certain wildlife, fish, and ecosystems; (2) increased incidence of certain endocrine-related human diseases; and (3) endocrine disruption resulting from an exposure to certain environmental chemicals observed in laboratory animal experiments. Wildlife effects to which this report refers include eggshell thinning and altered gonadal development with severe population decline in birds of prey exposed to the pesticide dichlorodiphenyltrichloroethane (DDT), altered reproductive development of fish exposed to pulp and paper mill effluents, and impaired reproductive and immune function with population decline of Baltic seals exposed to the DDT metabolite DDE (IPCS, 2002). Although adverse health consequences were initially identified in environmentally exposed wildlife and domestic animal species with relatively high exposures to organochlorine compounds (Kavlock et al., 1996), recent attention has turned to reports in humans of declining sperm counts; increased incidence of hypospadias, cryptorchidism, and testicular cancer; altered sex ratios; and increased breast cancer incidence. Chemicals that target the endocrine system are capable of inducing adverse effects on adult reproductive function; however, the developing organism can be particularly vulnerable due to the importance of hormones in directing differentiation in many tissues and the enhanced potential for changes to be irreversible (Dyer, 2007). Developmental effects associated with these chemicals include congenital abnormalities such as hypospadias and cryptorchidism, testicular cancer, reduced semen quality, and neurobehavioral alterations (Barlow et al., 1999).

Organochlorine chemicals, including internationally banned polychlorinated biphenyls (PCBs) and the pesticides DDT and its metabolites, and dieldrin, were considered to act as estrogens and therefore were postulated to induce or promote breast cancer (Stockholm Convention, 2001). Although reviews of the literature have concluded that the existing data do not support a causal relationship between organochlorine exposure and breast cancer, they also do not provide sufficient information to refute the hypothesis (Barlow et al., 1999). Hoyer et al. (1998) reported results from a 17-year retrospective study that showed an increase in breast cancer with dieldrin, but not with PCBs or total DDT. Estrogenic compounds have also been shown to impact the male reproductive function. Chlordecone, a pesticide with estrogenic activity, has been reported to decrease sperm count and sperm motility in exposed workers (Sheiner et al., 2003; Jensen et al., 2006). Endocrine toxicants/disruptors that act via hormone receptors typically have a much lower binding affinity for hormone receptors compared to endogenous ligands and therefore less biologic activity; however, potency could be increased by factors such as metabolic persistence and serum protein binding affinities (Barlow et al., 1999).

Vinclozolin and its metabolites are examples of antiandrogenic endocrine toxicants. Vinclozolin is used commercially as a fungicide for fruits, vegetables, ornamental plants, and vines. It is formulated as a dry flowable and extruded granular material and is applied using various techniques such as aerial application, dip treatment, and use of thermal foggers. Experiments in laboratory animals have demonstrated that *in utero* exposure of male rats at a key stage of development alters testis development and impairs adult spermatogenic capacity and reproduction. Cowin et al. (2010) established that these adverse effects, including reductions in anogenital distance, increased testicular germ cell apoptosis, reduction in the number of elongated spermatids, and postpubertal prostatitis, are hormonally regulated via antagonism of the androgen receptor and are reversed by a high dose of testosterone administered at the time of puberty. Anway and Skinner (2006) have additionally provided evidence that vinclozolin may act transiently at the time of embryonic sex determination via an epigenetic mechanism to promote a spermatogenic cell defect and subfertility in males. According to the U.S. Environmental Protection Agency (EPA), antiandrogenic effects would be expected in humans since the androgen receptor is conserved across species (U.S. EPA, 2000a). However, human consequences of vinclozolin exposure are unknown.

Research on chemicals considered to be endocrine disruptors has led to the ban or restriction of some commercially available compounds such as the pesticides DDT and methoxychlor (MXC) and the plasticizer bisphenol A (BPA). While the subject of endocrine disruption still tends to be controversial, one idea for which there is universal agreement is that further research is required to understand its potential causes and consequences, and how the risk can be minimized for humans and animals. In many instances the functional effects of low exposure are not clear and do not provide sufficient evidence as to the possible effect on the reproductive system. Trends in the incidence of adverse effects attributed to targeting of the endocrine system, such as cryptorchidism and hypospadias, as well as potential declines in semen production and quality, do not appear to be consistent and tend to demonstrate geographical variation (Barlow et al., 1999). One area that requires better understanding is the potential consequences of low levels of exposure to synthetic endocrinologically active compounds, compared to those known to occur naturally (Safe, 2000). Also, further study is required to evaluate complex endocrine alterations induced by mixtures of chemicals. Evidence from rat studies suggests that low dose mixtures of chemicals that alter endocrine activity may have adverse effects on the male reproductive system, whereas low doses of the agents administered singly do not (Kortenkamp, 2008). The potential interaction of lifestyle factors such as diet, physical activity, and obesity with endocrine disruption, as well as disease states such as diabetes, all merit additional investigation.

MALE REPRODUCTIVE SYSTEM

The male reproductive system comprises many different cell types and is under multilevel hormonal control and autonomic nervous system input. The complexity of the structural and functional aspects provides a large number of targets for toxicants, and indeed, a substantial number of chemically diverse compounds are known to affect the male reproductive system. Specific toxicants are known to selectively target structural and/or regulatory components of the male reproductive system.

Male germ cells are produced in the testes within the seminiferous tubules. These are long, convoluted tubules lined by epithelial cells that contain differentiating populations of germ cells, from the most immature form known as spermatogonia, through increasing stages of maturation of cells known as spermatocytes and spermatids. Spermatogonia are considered to be the stem cells of the seminiferous tubules. A certain proportion of spermatogonia undergo a series of mitotic divisions to increase in number. They then move into meiotic prophase and proceed through various stages of spermatocyte maturation while becoming increasingly larger and undergoing DNA replication in preparation for meiosis. Two meiotic divisions then occur to produce haploid spermatids in preparation for fusion with the haploid female germ cell, the oocyte. Further maturation results in elongated spermatozoa, which are eventually released into the seminiferous tubule lumen (the process of spermiation) for transport to the epididymis, where final differentiation into mature sperm cells occurs (Foster and Gray, 2008). The functional morphology and actions of the sperm can also be affected by a reproductive toxicant. Motility is required for fertilization, and the energy needed is produced through specialized metabolic pathways that operate within the sperm. Agents such as mercury and lead may have a combined effect on both the dynamics of the sperm cell membrane and the epididymis.

Sperm complete their maturation in the epididymis, and the secretory and absorptive functions of the epididymal epithelium are required to maintain viable sperm (Harbison, 1998). Mature sperm must undergo a process known as capacitation following release from the epididymis and once within the oviduct of the female. This process is necessary for successful fertilization: the sperm acquires the ability to bind to the oocyte and undergo the acrosome reaction, which is a calcium-dependent process that occurs as a consequence of the interaction of capacitated spermatozoa with receptors on the zona pellucida of the oocyte. Acrosomal enzymes are thought to facilitate this process (Perreault, 1997). Interference with capacitation or the acrosomal reaction by xenobiotics can potentially have deleterious effects on fertility.

The cycle in which immature germ cells become mature sperm capable of fertilizing the female oocyte is known as the spermatogenic cycle; duration is species dependent, requiring approximately 74 days for completion in humans and 60 days in rats (Johnson et al., 1997). Cells of these various maturational stages can be selectively targeted by the reproductive toxicants. A large functional reserve of male germ cells exists in various mammalian species, and the size of this reserve pool is species dependent. Millions of cells may be produced daily due to the constant cell turnover; these large numbers may confer an advantage—there may be an increased chance of reversibility for toxic insult to the germ cells.

Additional cell types within the testis are also vulnerable to reproductive toxicants. The Sertoli cells provide nutritional and architectural support to the developing germ cells. They also serve as components of the blood–tubule barrier, which separates blood and interstitial components from the adluminal compartment, comprising Sertoli cell secreted products and germ cells. The blood–tubule barrier is considered to create a specialized environment that is believed to foster development of spermatocytes and spermatids (Richburg and Boekelheide, 1997). Modifications of the blood–tubule barrier by toxicants can lead to exposure of developing germ cells and selective destruction. Leydig cells within the testis produce testosterone, the primary hormone responsible for differentiation and development of the male reproductive tract, masculinization of the male at puberty, sperm production, secondary sex characteristics, and sexual behavior (Kelce, 1997). Chemicals that affect testosterone production can act directly on the Leydig cell or indirectly by disruption of the hypothalamic–pituitary–gonadal axis (H-P-G), which exerts hormonal control on testosterone production via regulatory feedback systems.

Testicular function is closely regulated by a series of closed-loop feedback systems involving the central nervous system (CNS), hypothalamus, pituitary, and testicular and germinal cell compartments (Sokol, 1997). Hormones important to this regulatory system, known as the H-P-G axis, include gonadotropin releasing hormone (GnRH) from the hypothalamus, which stimulates the release of the pituitary gonadotropins follicle stimulating hormone (FSH) and luteinizing hormone (LH). FSH and LH stimulate gonadal steroid secretion (testosterone and estrogen) for the maturation and maintenance of spermatogenesis.

Reproductive toxicants may also affect the transport and maturation of the sperm in the epididymis, transport of sperm in the female oviduct for the purpose of fertilization of the oocyte, and accessory glands such as the prostate and seminal vesicles, which provide secretions necessary for the survival and transport of sperm. In addition to the effects on reproductive system structure and function, exposure of males to specific mutagenic reproductive toxicants can result in embryonic death through a dominant lethal effect. This involves a genetic lesion of the germ cell that does not impair fertilization of the oocyte but is lethal to the resulting embryo, causing death before or soon after implantation

(Henck, 1997). Some chemicals require metabolic activation to produce an adverse effect. The testes have sufficient enzymatic capacity to produce harmful concentrations of some metabolites. Metabolism of *n*-hexane, acrylamide, 1,2-dibromo-3-chloropropane (DBCP), glycol ethers, and phthalates in the testes appears to contribute to their toxicity (Harbison, 1998).

Physical Agents

Radiation

The testicular tissue is highly sensitive to ionizing radiation. Radiation doses in the range of 0.11–0.15 Gy (=joule/kilogram) may temporarily decrease sperm count, while 2–5 Gy may result in long-lasting or permanent azoospermia and sterility (Sheiner et al., 2003; Jensen et al., 2006). At higher doses (>15 Gy), the Leydig cells may be affected; in addition, whole body irradiation can damage the H-P-G axis, affecting the reproductive capability (Sikka and Wang, 2008). The energy associated with X-rays and other types of radiation can cause chemical rearrangement of DNA, which may affect its ability to duplicate and direct cellular events. Some of the early stages of cell division are most susceptible to breakage of the DNA strands caused by gamma irradiation. The observable outcome of these types of breakage is malformation in the chromosomes of the germ cells. Some of the somatic cells found in the testis may be damaged in a similar manner and may be sloughed from the tubules in which spermatogenesis occurs. In this case, the result is the collection of clumps of cellular debris in the duct system that carries the sperm.

In animals, paternal exposure to gamma radiation produces an alteration in type B spermatogonia, expressed in progeny as a proliferation disadvantage in chimera assays (Warner et al., 1991). Direct X-irradiation of the testes has been shown to retard testicular growth and reduce spermatogenesis in male rats (Jensh and Brent, 1988). Breeding activity and the percentage of positive inseminations were also slightly reduced, although a significant percentage of male offspring were fertile. Head or thorax irradiation did not adversely affect testicular growth or spermatogenesis, indicating that adverse reproductive effects of X-irradiation on the testes and adverse reproductive outcome may not be mediated through the effects of X-irradiation on other parts of the body.

The sensitivity to DNA damage induced by radiation differs during different stages of spermatogenesis. Assessment of double-stranded DNA damage in various cells suggests that spermatogonia and preleptotene spermatocytes were more sensitive to gamma irradiation than other cell types. However, all did show an increase in double-stranded DNA damage. This type of damage can affect postfertilization processes as

well. X-irradiation of sperm cells produced a large number of DNA lesions in fertilized eggs at the first cleavage metaphase. The frequency of chromosome aberrations observed in eggs fertilized with sperm recovered from X-irradiated early spermatid to late spermatocyte stages was higher than in the eggs fertilized with irradiated sperm recovered from spermatozoa to late spermatid stage. Thus lesions produced in spermiogenic cells were effectively repaired in the cytoplasm of the fertilized eggs (Matsuda et al., 1989).

Recently, concerns have been raised about potential effects of radio frequency electromagnetic radiation (RFEMR)—the type of radiation employed in cell phone communication. In response to an abstract indicating that regular mobile phone use can negatively impact human semen quality, Aitken et al. (2005) exposed mice to RFEMR employed by electrical devices such as cell phones and wireless devices, such as speakers and baby monitors. Following exposure for 7 days, DNA damage to caudal epididymal spermatozoa was assessed. No adverse effects were observed on sperm number, morphology, or vitality; however, genotoxic effects on spermatozoa were observed, indicating that this is an area of study that requires further investigation.

Heat

Production of sperm requires a temperature 3–4 °C lower than normal body temperature. Decreased sperm count due to excessive heat has been reported for pathologies such as varicocele and cryptorchidism, as well as fever, in patients restricted to wheelchairs, in professional truck drivers, and in workers in the ceramics industry (Sheiner et al., 2003). Although sedentary work position is associated with increased scrotal temperature, it has not been associated with decreased sperm count (Jensen et al., 2006).

Solvents

Solvents are used to dissolve, dilute, and disperse materials that are insoluble in water; they are used as degreasers, are constituents of paints, varnishes, lacquers, inks, aerosol spray products, dyes, adhesives, fuels, and fuel additives, and are intermediates in chemical syntheses (Bruckner et al., 2008). Their small molecular size and lack of charge make inhalation a major route of exposure, and they are readily absorbed across the lungs, gastrointestinal tract, and skin. Virtually all solvents can cause adverse effects, including reproductive toxicity, as described subsequently.

Bromopropanes 2-Bromopropane (2-BP) is a solvent that was intended to be a replacement for ozone-depleting chlorofluorocarbons (CFC). Exposure of electronics factory workers in Korea via inhalation to 2-BP in a cleaning solution resulted in hematopoietic disorders in men and women, as well as azoospermia or oligospermia, and reduced sperm motility (Park et al., 1997). Reduced sperm production also occurred in rats exposed to 2-BP by inhalation for 9 weeks, as well as decreases in the weight of male reproductive organs, numbers of epididymal sperm, percent motile sperm, and the incidence of sperm with normal shape (Ichihara et al., 1996). Severe atrophy of the seminiferous tubules was observed in these rats, with reductions in all types of germ cells; Sertoli cells were unaffected. Administration of high subcutaneous doses of 2-BP to rats for 1–5 days resulted in reductions in stage 1 spermatogonia after one injection and reductions in spermatogonia at other developmental stages after multiple injections (Omura et al., 1999).

Based on observations of toxicity in Korean electronics factory workers, an alternative to 2-BP was sought to replace CFC. The selected alternative was 1-bromopropane (1-BP), which was thought to be less toxic due to its reduced mutagenic potential (Ichihara, 2005). However, 1-BP was found to be a potent neurotoxicant in laboratory animals and subsequently, in humans. Studies in rats also indicate that 1-BP is a reproductive toxicant. Rats exposed to 1-BP by inhalation for 12 weeks demonstrated reduced weights of epididymides, prostate, and seminal vesicles; reduced epididymal sperm count and motility; increased tailless sperm and sperm with immature head shape; retained, elongated spermatids; and reduced plasma testosterone (Ichihara et al., 2000). The authors concluded that the primary effect of 1-BP on reproduction in rats was to induce failure of spermiation. It is of interest that this primary effect differs from that of the structurally similar compound 2-BP, which targets spermatogonia; the reason for this difference in targets is unknown. To date there have been no reports of human reproductive toxicity with 1-BP, so the human impact remains uncharacterized.

Glycol Ethers   Glycol ethers are used as solvents for surface coatings (latex paints and varnishes), paint thinners, strippers, inks, metal cleaning products, liquid soaps, and household cleaners and are also used as jet fuel anti-icing additives and in semiconductor fabrication. Some glycol ethers and their oxidative metabolites are known to induce reproductive, developmental, hematologic, and immunologic toxicity, and a few have tested positive in rodent cancer bioassays (Bruckner et al., 2008). There appears to be an inverse relationship between the length of the alkyl chain and reproductive and developmental toxicity in monoalkyl glycol ethers and their metabolites; low molecular weight compounds such as 2-methoxyethanol (2-ME) and 2-ethoxyethanol (2-EE) are considered more toxic than ethylene glycol ethers, which are more toxic than propylene glycol ethers (U.S. Environmental Protection Agency, 1984). For this reason, industrial use has shifted away from low molecular weight compounds such as 2-ethylene glycol monomethyl ether (EGME) and 2-ethylene glycol monoethyl ether (EGEE) toward those with more favorable toxicity profiles such as ethylene, diethylene,

and triethylene glycol butyl ethers and propylene glycol monomethyl ether (Bruckner et al., 2008). Human occupational exposure can occur by inhalation, ingestion, and dermal absorption; glycol ethers are rapidly absorbed by all routes of exposure (Correa et al., 1996).

Male reproductive toxicity has been associated with glycol ethers in humans and laboratory animals. Glycol ethers target the spermatocytes, resulting in reduced sperm counts and potential infertility. Painters exposed to 2-EE and 2-ME demonstrated a reversible spermatotoxicity characterized by oligospermia and azoospermia, and reduced numbers of sperm/ejaculate were identified in foundry workers exposed to glycol ethers (Bruckner et al., 2008). Veulemans et al. (1993) found an association between impaired fertility and the detection of 2-ethoxyacetic acid (a metabolite of 2-EE and considered a biomarker of glycol ether exposure) in urine of male fertility clinic patients diagnosed as infertile or subfertile. There was no correlation with sperm quality measures, which the authors postulated might be due to a latent period between the exposure and observed effects on fertility.

In laboratory animals, glycol ethers appear to target spermatocytes, the product of the first meiotic division of spermatogenesis. Exposure of male rats to EGME for 2 or 4 weeks resulted in severe degenerative changes in the testes, including atrophy of seminiferous tubules and formation of multinucleated giant cells, abnormal sperm head morphology, degeneration of pachytene spermatocytes, and reduced numbers of germ cells (Watanabe et al., 2000). In a continuous breeding study with EGME in mice, Lamb et al. (1984) identified decreased sperm motility and count, an increased percent of abnormal sperm, and infertility. Creasy and Foster (1984) determined that primary spermatocytes undergoing pachytene development constituted the initial and major site of morphological damage in rats given oral doses of EGME or EGEE and that within this population, differential sensitivity was demonstrated depending on the precise stage of meiotic maturation. They also determined that spermatocytes in later stages of cell division, late-stage spermatids, and spermatogonia could be affected, but only if the dose was increased and exposure prolonged.

Trichloroethylene and Perchloroethylene   Trichloroethylene (TCE) and perchloroethylene (PERC) are commercially important solvents used primarily in metal degreasing, as paint and stain removers and chemical intermediates, and in the textile cleaning industry (Forkert et al., 2003; Carney et al., 2006). These solvents are volatile chemicals, and the highest human exposures occur via inhalation. From an environmental perspective, ingestion is also an important route of human exposure, as TCE and its metabolic products, including PERC, are known to be common water supply contaminants (World Health Organization, 2005).

Both TCE and PERC are well absorbed from the lungs and gastrointestinal tract and metabolized primarily by hepatic cytochrome P450 2E1 (Bruckner et al., 2008). They share a common major metabolite, trichloroacetate. Studies in laboratory animals and human experience with TCE have led to the conclusion that toxicity is attributable to metabolites, which include TCE oxide, chloral, and dichloroacetyl chloride; chloral is metabolized to chloral hydrate to form trichoroethanol (TCOH) and trichloroacetic acid (TCA). The latter two metabolites are used as biomarkers of human exposure (Forkert et al., 2002). Eight mechanics exposed to TCE occupationally and diagnosed with clinical infertility were found to have TCE and metabolites in seminal fluid but not in urine, suggesting a potential association between TCE and impaired fertility (Forkert et al., 2003). Trichloroethylene metabolites, chloral and TCOH, were also identified in the seminal fluid of mice exposed to TCE by inhalation, indicating similar metabolic disposition to that of humans. Cytochrome P450 2E1 was found to be localized in the epididymal epithelium and testicular Leydig cells in mice, with higher levels in the former. This corresponded to higher levels of chloral formed in microsomal incubations in the epididymis compared to the testis (Forkert et al., 2002). After 4 weeks of TCE exposure, structural damage, manifested as sloughing of epithelial cells, occurred in the epididymal epithelium of mice, but no morphologic changes were apparent in the testis, indicating that TCE metabolism in the epididymis resulted in adverse reproductive effects (Forkert et al., 2003).

Other Industrial Chemicals

Acrylamide   Acrylamide is a white crystalline powder used in the production of polyacrylamide polymer, gels, and binding agents; water-compatible polyacrylamides are used in water treatment, mineral processing, pulp and paper production, and in synthesis or compounding of dye materials, soil conditioners, and cosmetics. Human exposure can occur from foods (formed during frying and baking of starch-rich foods), smoking, drinking water, and from dermal contact with products that contain residual acrylamide (Centers for Disease Control, 2009). The primary adverse effect observed from a chronic occupational exposure to acrylamide is neurotoxicity. However, there are no confirmed reports in humans of acrylamide-induced testicular damage. Animal studies indicate a need for caution, in that they have identified adverse male reproductive effects, including germinal cell injury and dominant lethality. Adult and prepubertal mice demonstrated transient testicular damage after oral dosing with acrylamide (Sakamoto et al., 1988). The transient nature of the testicular effects of acute acrylamide treatment may explain the transient fertility reduction and increased pre- and postimplantation loss observed when acrylamide-treated male rats were mated with untreated females (Sublet et al.,

1989). The reproductive effects of chronic, low level exposure to acrylamide have not been assessed in animal studies. Clearly, such studies might have more relevance to occupational situations in which humans are exposed repeatedly to low levels of acrylamide (Harbison, 1998).

Binding of acrylamide to sperm DNA and to the unique sperm cell protein protamine has been observed in mice, raising questions of the potential for genotoxicity (Sega et al., 1989). Binding of acrylamide to DNA was found to be greatest in late spermatid to early spermatozoal stages and declined in later stages of spermatogenesis. *In vivo* experiments with several strains of mice suggested that acrylamide has the potential to induce gene mutations; glycidamide, the epoxide metabolite of acrylamide, is also mutagenic (Sega et al., 1989). Acrylamide is genotoxic to germ cells, affecting all stages of development with the exception of spermatogonia, and a positive dominant lethal response has been demonstrated in multiple studies conducted by the oral, parenteral, and dermal routes of administration (Dearfield et al., 1995).

Boric Acid   Boric acid is a male reproductive toxicant that specifically affects the process of spermiation, which is the interaction of Sertoli cells with elongated spermatids, the most mature testicular form of germ cells, and results in their release into the lumen of seminiferous tubules (Chapin and Ku, 1994). The element boron does not exist naturally but rather exists in combination with oxygen as, for example, boric acid or borate salts such as borax (Moore, 1997). Water-soluble forms of boron are widely distributed in surface and ground waters from natural sources and industrial contamination, and borates are known to be essential micronutrients for plants, thus introducing boron into the food supply (Murray, 1995; Moore, 1997). Industrial applications of boric acid and borax include the manufacture of glass products such as glass fiber insulation, as well as enamels, ceramics, fire retardants, insecticides, fertilizers, laundry bleaches, wood preservatives, cosmetics and personal care products, and dietary supplements. Human exposure to boron can occur by consumption of contaminated water and bottled drinking water, dietary consumption of crops and other foodstuffs, and inhalation of boron compounds during mining, manufacturing, and other industrial processing (Moore, 1997).

Few reports assessing the effects of boron on reproduction exist. One small report of workers exposed to boric acid suggested that it produced adverse effects on sperm parameters and fertility, while a larger study of borate-exposed miners revealed no adverse paternal effects from occupational exposure. Thus, data are considered insufficient to judge whether boron is a male reproductive toxicant in humans (Moore, 1997). However, studies in laboratory animals have demonstrated male reproductive toxicity. High dose boric acid has been shown to produce testicular lesions in adult rats, mice, and dogs. The specific effect has been described as inhibition of spermiation, followed by testicular atrophy. In rats, Ku et al. (1993) determined that the effect on spermiation could be separated from testicular atrophy based on the dose, with the effect on spermiation occurring first and at lower doses, and indicating that separate mechanisms may be operating. While effects on spermiation have been shown to be reversible in rats treated with boric acid, atrophy was not, and there was no return of spermatogenesis, despite the presence of a normal-sized population of dividing spermatogonia (Chapin and Ku, 1994).

Phthalates Phthalates (phthalate esters) are plasticizers added to plastics to impart flexibility and resilience or as solubilizing and stabilizing agents. They are found in a variety of products, including adhesives, detergents, lubricating oils, vinyl floors, food wraps, personal-care products, medical products, and toys. They are not chemically bound to the plastics to which they are added and can therefore be readily released into the environment (Kavlock et al., 2002). Human exposure can occur through ingestion, inhalation, dermal contact, and by intravenous or parenteral routes (medical devices and materials containing phthalates), and human milk can be a source of exposure for infants (Centers for Disease Control, 2009).

There are numerous forms of phthalates, and many of these forms have been evaluated for toxicologic potential. Male reproductive toxicity is the primary effect observed in laboratory animals. Phthalates are considered to be endocrine disruptors and have been studied extensively in laboratory animals. Although several forms were initially proposed to be estrogenic, based on *in vitro* data, they were found to not have estrogen activity in *in vivo* assays. Rather, the male reproductive toxicity attributed to the phthalates is similar to that induced by antiandrogenic compounds, although not identical to known androgen receptor antagonists such as flutamide (Mylchreest et al., 1999; Wolf et al., 1999). In a multigenerational study in which rats were exposed to diethylhexyl phthalate (DEHP) and its primary metabolite monoethylhexyl phthalate (MEHP), Parks et al. (2000) determined that exposure during the critical period of male differentiation resulted in reduced testosterone, decreased anogenital distance, and Leydig cell hyperplasia. *In vitro*, DEHP and MEHP displayed no affinity for the androgen receptor at concentrations up to 10 μM, and *in vivo*, female offspring in the same study were unaffected by prenatal exposure, indicating that perturbation of estrogenic activity was not a mechanism of action. These authors concluded that inhibition of testicular testosterone during sexual differentiation is a likely cause of DEHP- and MEHP-induced malformations in androgen-dependent tissues of males. In rats, di-*n*-butyl phthalate (DBP) is toxic to the Sertoli cells of the testis and impairs spermatogenesis by inducing widespread exfoliation of the seminiferous epithelium

(Mylchreest et al., 1999). Neonatal and pubertal rats are more sensitive to phthalate-induced testicular toxicity than are adult rats (Dostal et al., 1988).

In a review of DBP, reduced anogenital distance and increased incidence of undescended testes were reported to occur in offspring of treated rats; adult rats treated subchronically with oral doses demonstrated testicular lesions, epididymal hypospermia, and decreased sperm motility and concentration (Kavlock et al., 2002). Wolf et al. (1999) conducted a comparative developmental toxicity study in rats of several chemicals known or suspected to be antiandrogenic, which included DEHP and DBP. Reduced anogenital distance was observed, and in addition, hypospadias, cryptorchidism, ventral prostate agenesis, and testicular and/or epididymal atrophy or agenesis also occurred with each, although the effects were considerably greater with DEHP.

The relevance of these findings to human males has yet to be determined; at present human health effects from low environmental exposures of phthalates is unknown (Centers for Disease Control, 2009). A National Toxicology Program (NTP) panel expert panel reviewed human studies of phthalate exposure and found that small numbers of subjects and possible confounding factors provided insufficient evidence that DEHP exposure during pregnancy, childhood, or adulthood is causing harm (Kaiser, 2005).

Metals

Cadmium The occurrence of cadmium in the environment is widespread—it occurs naturally in the environment at low levels but can be present at higher levels due to industrial emissions from non-ferrous mining and metal refining, smelting, battery manufacture, and electroplating operations. It is found in the food chain, surface waters, soil, and ambient air and is also found in nickel/cadmium batteries, pigments (cadmium yellow), and plastics (Dyer, 2007; Massányi et al., 2007; World Health Organization, 2012). Exposure to cadmium may occur concurrently with lead and other heavy metals. Tobacco smoke is one of the most common sources of cadmium exposure because the tobacco plant concentrates cadmium (Klaassen, 2006). Cadmium absorbed through smoking has relatively high bioavailability. It has been postulated that negative effects of smoking on semen quality are the result of the observed trend for cadmium in tobacco smoke to accumulate in the genital tract of smokers (Benoff et al., 2000).

In the environment, exposure to high concentrations of airborne particles that may contain cadmium have been associated with decreased semen quality, particularly in metal workers compared to other occupations; however, dose–response relationships and/or causal mechanisms remain to be defined (Benoff et al., 2000). Cadmium is classified as a human carcinogen, with the most convincing association to lung cancer; it is also a multi-tissue animal carcinogen (Waalkes, 2003). An increased risk for human prostate cancer has also been reported; however, given the complex etiology of this disease, it is considered difficult to ascribe the cause in a small proportion of cases to a single factor (Waalkes, 2003). In addition, cohort studies in cadmium-exposed workers have not confirmed this association (Thompson and Bannigan, 2008). It should be noted, however, that the human prostatic epithelium could potentially be a target of the oncogenic effects of cadmium, based on *in vitro* evidence that human prostate epithelial cells can be malignantly transformed by cadmium (Nakamura et al., 2002). Cadmium has been shown to induce testicular tumors and preneoplastic lesions in rats at doses lower than those required to induce testicular toxicity. However, high doses of cadmium have been shown to result in substantial reduction in circulating levels of testosterone and resultant atrophy of the testis and prostate (Waalkes et al., 1997), which would likely counter a proliferative stimulus from cadmium (Waalkes, 2003).

Cadmium accumulates preferentially in reproductive organs and has been shown to adversely affect a number of cell types within the male reproductive system. Adverse effects appear to depend on the dose, route of administration, and duration of exposure. In laboratory animals, cadmium has been reported to induce irreversible necrosis of the testicular tissue, decreased spermatogenesis, testicular hemorrhage and edema, and endothelial cell damage in testicular blood vessels following oral or parenteral administration of single high doses (Thompson and Bannigan, 2008). Potential mechanisms of action include (1) indirect insult resulting from vascular disruption and (2) direct injury to spermatogenic cells (Zenick et al., 1982).

Cadmium is considered to be a Sertoli cell toxicant, primarily disrupting the blood–tubule barrier (Richburg and Boekelheide, 1997); the mechanism of action is considered to be through adverse effects on cell adhesion. Disruption of the blood–tubule barrier can result in the reduction in germ cell numbers and potential infertility. Daily subcutaneous administration of high doses of cadmium to rats for 6 weeks resulted in its accumulation in the testes, primarily in the spermatogonia and spermatocytes, with subsequent reduction of these cell types (Aoyagi et al., 2002). A study by Zenick et al. (1982) demonstrated that subchronic administration of more environmentally relevant oral doses (up to 69 parts per million (ppm) in the drinking water for 70–80 days) to rats did not result in testicular toxicity or adverse outcomes to offspring of treated males. Additional studies conducted with chronic low level cadmium administered parenterally have failed to reveal histologic damage, even with a cumulative dose higher than acutely toxic doses, possibly due to the induction of metallothionein synthesis and subsequent cadmium binding on repeated exposure (Zenick et al., 1982).

Cadmium has also been shown to induce benign Leydig cell tumors in rats. Following acute high doses of cadmium, hemorrhagic necrosis of the testis and chronic degeneration with subsequent loss of androgen production are followed by a high incidence of testicular interstitial cell tumors (Waalkes, 2003). Danielsson et al. (1984) found that cadmium accumulates in the interstitial tissues of the testis, supporting potential effects on hormone production by the Leydig cells and/or on blood supply. Loss of androgen production is thought to lead to overstimulation of remnant testicular cells by the LH (Waalkes et al., 1997). Rat testes are extremely sensitive to cadmium-induced tumorigenesis, and the mechanisms by which this occurs are likely of little relevance in humans (Waalkes, 2003).

Lead Although progress has been made in reducing environmental lead concentrations by removing or reducing sources such as leaded gasoline, lead-based paints, lead-soldered cans, and lead plumbing pipes (Goyer, 1996), lead is still considered to be a ubiquitous environmental contaminant (Dyer, 2007). Dietary intake is a common means of human exposure; however, humans are also exposed to lead aerosols occupationally and in the environment. Exposure to lead and cadmium may occur concurrently. Although the primary target of lead-induced toxicity is the CNS, lead is also known to affect the male reproductive system and preferentially accumulates in the reproductive organs. The mechanism of action of lead-induced male reproductive toxicity is not clear, but may involve both direct effects on the reproductive organs and effects on the endocrine control of reproduction. Evidence from an occupational exposure of humans to inorganic lead leading to blood lead concentrations $\geq 40\,\mu g/dl$ resulted in impaired male reproductive function characterized by reductions in sperm count and increased sperm death, reduced sperm volume, density, and motility, adverse effects on sperm morphology, and decreased secretory function of the prostate and seminal vesicles (Apostoli et al., 1998). The most prevalent effects were on sperm count and concentration. In the environment, exposure to high concentrations of airborne particles that may contain lead have been associated with decreased semen quality, particularly in metal workers compared to other occupations; however, dose–response relationships and/or causal mechanisms remain to be defined (Benoff et al., 2000). High levels of lead in humans have been associated with decreased fertility and a reduced ability to undergo the acrosome reaction (Benoff et al., 2003; Dyer, 2007).

Major risk factors for lead-induced toxicity include age, nutrition, housing and socioeconomic factors, and smoking. Tobacco smoke is considered a major source of airborne environmental lead exposure (Benoff et al., 2000), and blood lead concentrations are reported to be elevated in smokers. Reports indicate that men with marginal semen quality may become infertile by smoking. Inconsistent effects on male reproduction have been produced in laboratory animal experiments. Certain species and strains of laboratory animals are resistant to lead-induced reproductive toxicity, while others are highly sensitive (Apostoli et al., 1998). In sensitive rat strains, atrophy of the seminiferous tubules; reduced populations of spermatogonia, spermatocytes, and Leydig cells (with decreased testosterone); vacuolization of Sertoli cells; reduced sperm motility; prostate hyperplasia; and infertility have been reported to result from lead exposure. An additional effect reported in mice is reduced implantation resulting from male exposure. The reason for the difference in response among various species and strains has not been fully characterized but is thought to be due to pharmacokinetic differences in lead accumulation in the testes or to differences in function of the blood–tubule barrier (Apostoli et al., 1998).

Lead exposure may result in adverse hormonal effects. A negative effect of increased blood lead concentration on the outcome of *in vitro* fertilization was attributed to altered sperm function. This study reported an increase in spontaneous acrosome loss and a decreased ability to undergo the progesterone-stimulated acrosome reaction (Benoff et al., 2003). The question of whether lead disrupts the H-P-G axis in human males remains unanswered—some studies provide evidence while others do not. Studies in laboratory animals have shown that disruption of the H-P-G axis may actually be a primary mechanism of action for adverse effects on male reproduction. Several molecular targets have been proposed to account for the symptoms of lead poisoning; most of these targets fall into the following two primary categories: proteins that naturally bind calcium or those that bind zinc (Godwin, 2001). Interference with calcium homeostasis may play a role in lead-induced reproductive toxicity and presents a plausible association between lead and hormonal disruption in particular. Calcium has an important role in the regulation of GnRH and sperm function; as a second messenger it facilitates the transfer of extracellular hormone signals to the cell interior. Potential disruption of pulsatile GnRH release may also disrupt the release of FSH and LH (Krsmanović et al., 1992). In addition, calcium is required for sperm motility and for the steps leading to fertilization (Sokol, 1997). Blood lead levels >40 μg/dl were shown to markedly decrease serum and intratesticular testosterone in rats treated for 30 days with lead acetate, resulting in reduced spermatogenesis and impaired fertility (Sokol et al., 1985). In this study, decreased testosterone and gonadal function were associated with reduced FSH but no effect on LH, indicating that normal hormonal feedback mechanisms were impaired, and leading to the conclusion that primary targets of lead toxicity in these rats were the hypothalamus and pituitary. This effect on the H-P-G axis was reversible. McGivern et al. (1991) investigated the hypothesis that lead exposure during development would disrupt H-P-G function in adulthood. They exposed pregnant

female rats to lead acetate from gestation day 14 through parturition, considered a period of rapid differentiation of the H-P-G axis. Male offspring had reduced sperm count, less territorial scent marking and masculine behavior, azoospermia, enlarged prostate, and enlarged volume of the sexually dimorphic nucleus of the preoptic area of the hypothalamus, demonstrating disruption of hormonal feedback regulation.

Pesticides

Exposure to pesticides has been associated with male reproductive toxicity, particularly with regard to high level and/or extended duration industrial and end-use contact. Several recent epidemiology studies have evaluated exposure to a variety of pesticides and, although these studies are typically small, they have identified associations between pesticide exposure and decreased sperm concentration, motility, and morphology. Concomitant exposure to several different pesticides may also present an issue. Swan et al. (2003) measured metabolites of eight current-use, persistent pesticides to serve as potential biomarkers of pesticide exposure and found an association with semen quality in farmers from an area in the United States with high pesticide use relative to farmers from an area with low pesticide use. In contrast, a review conducted by Perry (2008) that evaluated low level occupational and environmental exposure to a variety of pesticides concluded that epidemiologic evidence is thus far equivocal due to the small number of published studies. This review evaluated sperm parameters, DNA damage, and numerical chromosome aberrations in epidemiologic studies of pyrethroids, organophosphates, phenoxyacetic acids, cabamates, organochlorines, and pesticide mixtures. Some pesticides have been definitively associated with male reproductive toxicity, however, and are described subsequently.

1,2-Dibromo-3-chloropropane The identification of infertility due to testicular failure in workers involved in the manufacture of DBCP is considered a sentinel event of major importance in occupational medicine due to a large number of cases of impaired fertility in workers exposed during the manufacturing process (Teitelbaum, 1999). 1,2-Dibromo-3-chloropropane was extensively employed as a nematocide on a global basis beginning in the 1950s. Reports of infertility in male workers engaged in the manufacture of DBCP began to emerge in 1977, leading to an eventual ban by the U.S. EPA in 1979 (ATSDR, 1995). It is still used in other countries, indicating that there is still some risk from industrial exposure. In addition, an exposure can occur from ingesting contaminated drinking water or food or from inhalation exposure to hazardous waste sites with improper disposal (ATSDR, 1995; Teitelbaum, 1999).

Workers exposed to DBCP during its manufacture were azoospermic or mildly to severely oligospermic; there was a positive correlation between duration of the occupational exposure and the extent of reduction in sperm production (Whorton et al., 1977; Potashnik et al., 1978; Whorton and Milby, 1980). The effects occurred in conjunction with elevated FSH and LH but in the absence of an effect on testosterone (Whorton et al., 1977). Follow-up of these exposed workers, as well as other cohorts exposed to DBCP during manufacturing, revealed a tendency toward reversibility of fertility effects and hormonal alterations in oligospermic men but not in azoospermic men (Whorton and Milby, 1980; Potashnik, 1983). Potashnik (1983) found that reversibility was also inversely related to previous exposure time and was most likely to occur in the presence of normal FSH values. In an evaluation of the largest cohort of DBCP-exposed workers to date, Slutsky et al. (1999) reported adverse reproductive health effects, including infertility, in men from 12 countries primarily exposed through application of DBCP, rather than through manufacturing. Azoospermia and oligospermia were still present in a high percentage of these workers many years after their direct exposure had ceased, suggesting that these conditions were permanent in these men (Slutsky et al., 1999).

1,2-Dibromo-3-chloropropane is also a male reproductive toxicant in a variety of laboratory animal species, which was actually demonstrated prior to the reports of human male infertility (Babich and Davis, 1981; Teitelbaum, 1999). Inhalation exposure of rats, rabbits, guinea pigs, and monkeys, as well as oral administration to rats, resulted in atrophy and degeneration of the testes accompanied by abnormal sperm development, reduced sperm counts, and degeneration of seminiferous tubules; at high oral doses, male rats exhibited severe reduction in both production and mobility of sperm cells (Babich and Davis, 1981). A direct effect on sperm was suggested by a study in which male rats given a single subcutaneous injection of DBCP were observed to be infertile within 2–7 days (Kluwe et al., 1983). These authors suggested that DBCP may cause nearly immediate infertility via a direct effect on post-testicular sperm, and that a possible mechanism of this infertility is inhibition of glucose metabolism in ejaculated sperm. Inhalation exposure of male rats to DBCP resulted in a dominant lethal effect, in that increased resorptions occurred in unexposed females mated to exposed males; this effect tended to reverse following termination of the exposure (Rao et al., 1983).

The metabolic pathways for DBCP following ingestion may involve epihalohydrins, other reactive epoxides, or 2-bromoacrolein as intermediates (World Health Organization, 2003). Of particular interest for reproductive toxicity is epichlorohydrin, which is also known to be a male reproductive toxicant in humans and laboratory animals and has recently been used as an experimental model to evaluate the relationship of impairment of sperm motility to fertility. Fertility in untreated female rats mated with males treated with epichlorohydrin for 19–22 days was found to be impaired due to inhibition of fertilization, rather than a dominant lethal effect (Toth et al., 1991). Reduced fertility was further correlated with impairment of sperm motility end points. Experimental evidence also suggests that DBCP may impair sperm motility and fertilizing ability, but verification is required (Perreault, 1997).

Ethylene Dibromide   Ethylene dibromide (EDB), also known as 1,2-dibromoethane, is a component of some pesticides. It is an alkylating agent and is mutagenic and carcinogenic (Ratcliffe et al., 1987). Occupational exposures have been associated with decreased sperm count, decreased production of seminal fluid, and fewer motile and morphologically normal sperm cells. One study reported reduced fertility, but this has not been reproduced (Jensen et al., 2006).

Related animal data have also shown male reproductive effects in a variety of species. Spermicidal effects, characterized by lysis of the chromatin of elongating spermatids, were evident in both rams and bulls; abnormal sperm were found in the ejaculate of bulls but not in rams (Amir, 1991). Abnormal sperm were not found in the ejaculate of rams and may have been phagocytized in the epididymis, illustrating an apparent species difference in response. Reduced litter size was observed following mating of male rats and mice treated with EDB to untreated females, indicating a potential dominant lethal effect (Ratcliffe et al., 1987). Ethylene dibromide was found to potentiate germ cell dominant lethal mutations induced by ethyl methane sulfonate (EMS), which correlated with glutathione depression. In rodent assays, EDB exposure depressed glutathione levels in the caput and cauda epididymis but not in the testes. Depression of glutathione levels in reproductive tissues correlated with potentiation of EMS-induced dominant lethal mutations and with increased binding of EMS to sperm heads, likely the cause of observed postimplantation fetal deaths (Teaf et al., 1990).

FEMALE REPRODUCTIVE SYSTEM

The female reproductive system consists of multiple anatomic and chemical components and is under strict neuroendocrine control. The large number of anatomical and chemical components provides multiple targets for physical and chemical injury. In the mammalian female reproductive system, all germ cells (ova, oocytes) are produced prior to birth. Approximately 400,000 oocytes and their surrounding primary follicles are produced by the time of birth in each human ovary (Foster and Gray, 2008). If damage to the germ cells occurs during embryo–fetal development, replacement is not possible, resulting in impaired fertility or infertility. This adverse effect is often not recognized until adulthood, when the woman tries to conceive and is unsuccessful. Many ovarian follicles undergo atresia after birth through the normal process of the menstrual cycle, so numbers are constantly decreasing, until they are exhausted after approximately three decades, and menopause ensues. If the oocyte population is diminished, these events may be accelerated, resulting in premature menopause (reproductive senescence). Alterations in the age of sexual maturation or reproductive senescence can potentially alter the reproductive capacity (Mattison et al., 1983).

The impact of occupational and industrial exposure on the female reproductive system is less well known than for males. One of the main reasons for this lack of data is that it is difficult to identify and analyze reproductive impairment in women. There is little reason to suspect impairment unless the woman has shown evidence of infertility or a prolonged inability to maintain a pregnancy. Germ cell production in a woman is difficult to monitor, and possible indicators, such as amenorrhea (lack of menstruation) or irregular menstruation, occur frequently enough and for such a variety of reasons that linking these and industrial exposure is difficult (Harbison, 1998). Pregnancy failure also occurs spontaneously, and it is therefore difficult to ascertain if the cause is due to maternal or embryonic deficiencies or abnormalities, or to an external environmental influence.

Toxicants are known to affect female reproduction primarily by disruption of the ovarian function through direct effects on the ovary or by interference with the H-P-G axis, which exerts neuroendocrine control over the female reproductive processes. The major functions of the ovary are to produce the female germ cell (oocyte) and to serve as an endocrine organ for the production of female hormones, in particular estrogen and progesterone. During fetal development, oogonia (stem cells for oocytes) proliferate and then enter meiotic cell division. At this time the supportive follicle complex is formed, which consists of the oocyte, granulosa cells, basement membrane, and thecal cells; this is the smallest functional unit of the ovary (Mattison et al., 1983). Beginning at puberty, in a cycle of specific duration depending on species (approximately 28 days for humans, 4 days for rats), ovarian follicles are recruited and selected by FSH (from the anterior pituitary) to grow and attain additional granulosa cell layers, in preparation for ovulation and, ultimately, fertilization. This cycle continues throughout the reproductive life of the female. One of the primary functions of the granulosa cells is to produce the hormone estrogen. The ovarian follicle grows through increasing stages of complexity, from primary to secondary to antral to preovulatory, and these stages can be variously affected by ovarian toxicants (Hoyer, 2005). Destruction of oocytes in small primordial follicles causes infertility since these follicles cannot be replaced, while destruction of more mature growing or antral follicles only temporarily disrupts reproductive function because they can be replaced from the existing pool of primordial follicles (Hoyer, 2005). Chemicals that selectively damage large growing or antral follicles

are known to cause transient, reversible reproductive system dysfunction (Yu et al., 1999). If fertilization occurs, granulosa cells undergo further differentiation, begin to produce the hormone progesterone, which is important in maintaining pregnancy, and ultimately become cells of the placenta. If ovulation does not occur, the ovarian follicle (including the oocyte) regresses through a process known as atresia, which is considered to occur through apoptosis (i.e., programmed cell death). Follicular atresia occurs throughout the reproductive lifetime of the female; since the number of oocytes is predetermined prior to birth the supply is eventually depleted, leading to the ovarian failure of menopause. Because menopause has been linked to a variety of disease states, premature onset induced by toxicants can potentially have a severe impact on women's health (Hoyer, 2005).

Key regulatory hormones for female reproduction are the same as those which govern the male reproductive cycle: GnRH stimulates the release of FSH and LH. Follicle stimulating hormone is responsible for the recruitment and selection of the dominant ovarian follicle for subsequent ovulation and potential fertilization; the growing follicle is then stimulated to release estrogen. Luteinizing hormone is important in ovulation and in formation of the corpus luteum, which releases progesterone and is crucial in the maintenance of pregnancy. Exposure to chemicals that act to mimic or antagonize the actions of endogenous estrogen can disrupt the balance of this feedback system and result in adverse consequences for reproduction and the maintenance of pregnancy.

Consideration of female reproductive toxicology may also involve consideration of the relationship between mother and conceptus (i.e., the developing embryo/fetus) and potential adverse outcomes. The mother may transfer chemical agents, including active metabolites, to the conceptus by a variety of mechanisms, adversely affecting development. Transport of male and female germ cells to the site of fertilization in the oviduct (fallopian tube) and the transport of the fertilized egg to the site of implantation and development in the uterus are also functions that can be disrupted by reproductive toxicants. Atrophy of the oviduct or uterus can prevent transport and exocrine support of the germ cells and embryo. Formation of the corpus luteum and the subsequent complex processes required for implantation of the embryo may also be affected by toxicants, as well as formation of the placenta to nourish and remove wastes from the conceptus. Hormonal control of the duration of pregnancy may be shortened or lengthened by toxicants, resulting in potential adverse effects on the fetus. Following birth, milk production and secretion may be affected by toxicants. Thus, a variety of targets exist for physical and chemical agents to disrupt female reproduction and development of the conceptus, pointing to the need for additional research to identify mechanisms of action to help mitigate potential risks to women in the workplace.

Physical Agents

Ionizing Radiation   Precautions are taken in the health care setting to minimize the risk of ionizing radiation exposure in women and potentially, the conceptus. Ionizing radiation is a known risk factor for ovarian toxicity and for fetal death, malformation, and growth retardation. Much of what is known about the effects of ionizing radiation on the female reproductive tissues comes from patients receiving low therapeutic doses for cancer treatment, female health workers exposed occupationally, and victims of higher environmental exposures, including survivors from the atomic bomb detonations in Hiroshima and Nagasaki, and the Chernobyl reactor accident. Therapeutic and environmental exposures have been associated with ovarian failure, premature menopause, and infertility. Depletion of primordial follicles appears to be the primary manifestation of ovarian toxicity, Adverse effects on female gonadal function occur at all ages, with the degree and persistence of damage depending on the dose, irradiation field, and age (older women are more susceptible) (Meirow and Nugent, 2001). The estimated radiation dose at which half of ovarian follicles are lost in humans is 4 Gy (Wallace et al., 1989); permanent ovarian failure is estimated to occur at 20 Gy in women <40 years of age, and at 6 Gy in older women (Lushbaugh and Casarett, 1976; Meirow and Nugent, 2001). Dose-responsive loss of primordial follicles has also been observed in mice, with a range of increasing doses producing partial to total loss (Gosden et al., 1997). With regard to occupational exposure to low levels of ionizing radiation, evidence suggests that exposure of female health-care workers to ionizing radiation within prescribed limits and prior to conception does not constitute a risk factor for reproductive health (Figà-Talamanca, 2006).

Concerns have been expressed about potential effects of RFEMR on female reproductive health, due to the prevalence of these emissions in the environment. However, no definitive conclusions have been reached. Although exposure to video display terminals is suspected to have a slight association with the risk of miscarriage, it is unclear whether the effect is due to electromagnetic frequency or to work-related conditions such as ergonomic factors, work stress, and long working hours (Figà-Talamanca, 2006).

Ergonomic Factors/Occupational Stress   A current area of interest in women's health is the impact of various occupationally related ergonomic factors and stress. In addition to exposure to various physical and chemical agents in the workplace, such factors as heavy physical work, prolonged standing or sitting, climbing stairs or ladders, cold, shift work, and irregular work schedules may also represent occupational risk factors (McDonald et al., 1988; Nurminen et al., 1989; Marbury, 1992; Figà-Talamanca, 2006). Heavy physical work is considered to be a risk factor for

spontaneous abortion, as well as low birth weight and premature delivery. Fenster et al. (1997) found that standing at work for more than 7 h was associated with an increased risk of spontaneous abortion only in women with a history of two or more spontaneous abortions. The risk of spontaneous abortion was found to increase when physical strain occurred early in pregnancy (Figà-Talamanca, 2006). Noise-induced stress has been associated with potential interference with the endocrine system and an increased risk of miscarriage in humans and laboratory animals. However, in a study conducted by Nurminen (1995), threatened abortion was not associated with noise exposure alone; when combined with shift work, the risk was enhanced twofold. Additional risks attributed in this study to noise in shift work included pregnancy-induced hypertension and shorter duration of pregnancy. Effects on gestational age have also been reported to occur from a combination of stressors, rather than a single stressor, with women in the highest exposure category having the greatest risk (Marbury, 1992).

Overall, stress is known to interfere with the endocrine system, which may be manifested by menstrual disorders and altered cycle length. However, studies of work-related stress have not provided definitive evidence that stress is an independent factor in female reproductive toxicity. It is often difficult to establish the cause-and-effect relationship, since the negative reproductive effect may be a source of stress in itself (Figà-Talamanca, 2006).

Solvents

Exposure of women to solvents in occupational settings has been associated with adverse reproductive effects. In a study of the relationship between spontaneous abortion, stillbirth without congenital defect, and working conditions, McDonald et al. (1988) found that exposure to solvents was associated with an increased risk of spontaneous abortion and stillbirth. Often exposures in the workplace occur to a mixture of solvents, so it can be difficult to attribute adverse effects to a single chemical. However, accidental contamination with specific solvents, coupled with laboratory animal data, has helped to establish the biological plausibility.

Bromopropanes   2-Bromopropane (2-BP) exerts adverse effects on both male and female reproductive systems. In addition to adverse effects on the hematopoietic (pancytopenia) and male reproductive systems, inhalation exposure of electronics factory workers in Korea to 2-BP resulted in increases in circulating FSH and LH, secondary amenorrhea, and hot flashes in women (Kim et al., 1996). Subsequent inhalation studies in rats demonstrated ovarian dysfunction, which was attributed to the destruction of the primordial follicle and its oocyte due to induction of apoptosis (Yu et al., 1999). Following the identification of toxicity associated with 2-BP, 1-BP was introduced to the workplace as a CFC replacement. As mentioned previously, 1-BP was found to be a potent neurotoxicant in laboratory animals, as well as a male reproductive toxicant (Ichihara, 2005). Female reproductive toxicity has also been identified in rats with 1-BP but not in humans. Exposure to 1-BP in a multigeneration study resulted in prolongation of the estrous cycle, decreased ovarian weight, reduced numbers of corpora lutea, and increased numbers of ovarian cysts (Yamada et al., 2003). 1-Bromopropane disrupted ovarian follicular development in these rats by targeting antral and developing follicles. Similar to the situation with males, the target is different for 1-BP and 2-BP; 2-BP targets the oocyte of the primordial follicle, whereas 1-BP disrupts follicular development (Ichihara, 2005). The reason for the difference in target between the two forms of bromopropane has not been established.

Glycol Ethers   Although glycol ethers are primarily recognized as male reproductive toxicants and developmental toxicants, limited evidence also suggests that they can adversely affect the female reproductive system. Studies of women employed in the semiconductor industry reported an increased risk for spontaneous abortion and subfertility; prolonged time to conceive has also been reported, which may be associated with evidence of prolonged menstrual cycles (Schenker et al., 1995; Correa et al., 1996; Hsieh et al., 2005).

In vitro evidence supports the potential for glycol ethers to alter ovarian function in humans. Almekinder et al. (1997) exposed cultured rat luteal cells to EGME and its proximate metabolite methoxyacetic acid (MAA) and observed cellular hypertrophy and increased progesterone production. *In vivo*, EGME has been observed to elevate serum progesterone and suppress estrous cyclicity, inhibit ovulation, and increase corpora lutea size by inhibiting luteolysis in rats (Davis et al., 1997). Exposure of cultured human luteinized granulosa cells to the same concentration of MAA to which rat luteal cells were exposed also resulted in increased progesterone production (Almekinder et al., 1997), suggesting a potential risk for human exposure. Studies in several laboratory animal species exposed to glycol ethers have reported increased embryo mortality (Hardin et al., 1986; Scott et al., 1989). Studies at lower doses in which rats have been allowed to deliver their offspring have demonstrated an association between glycol ethers and prolonged gestation (Nelson and Brightwell, 1984; Toraason and Breitenstein, 1988).

Other Industrial Chemicals

Vinylcyclohexene and Related Compounds   4-Vinylcyclohexene (VCH) is a dimer of 1,3-butadiene. Occupational exposure can potentially occur via inhalation during the production of 1,3-butadiene, VCH or 4-vinylcyclohexene diepoxide (VCD), as well as during the production of butadiene-based rubber, extrusion of electrical cable insulation, flame retardants, insecticides, and plasticizers

(IARC, 1994a). 4-Vinylcyclohexene is considered by the International Agency for Cancer Research IARC (IARC, 1994a) to be possibly carcinogenic to humans, based on laboratory animal data. 4-Vinylcyclohexene diepoxide has been extensively studied as a model for ovarian toxicity. Ovarian tumors were produced in mice in a lifetime study in which VCH was administered orally and were preceded by observations of ovarian and uterine atrophy (Hoyer and Sipes, 2007). Ovarian effects were not observed during a lifetime study in rats, and subsequent research determined that the species difference in response was due to greater conversion to mono- and di-epoxide metabolites in mice, which were also shown to induce ovarian toxicity. The ovarian toxicity of VCH has been well characterized in in vitro and in vivo experimental models and has been attributed primarily to the active metabolite VCD. The mechanism of action is a direct effect on the ovaries, with selective destruction of small pre-antral follicles. 4-Vinylcyclohexene diepoxide (VCD)-induced follicle loss is via apoptosis and results in acceleration of the natural process of ovarian follicle atresia (Hoyer and Sipes, 2007).

Vinylcyclohexene diepoxide, in addition to being an active metabolite of VCH, is also used industrially as a reactive diluent for other diepoxides and for epoxy resins derived from BPA and epichlorohydrin (IARC, 1994b). It is also considered by the IARC to be possibly carcinogenic to humans based on results from lifetime studies in which skin application resulted in benign and malignant skin tumors in mice and rats and, in one study, ovarian and lung tumors in mice (IARC, 1994b). Nonneoplastic lesions in these studies included ovarian follicular atrophy and tubular hyperplasia in mice.

Species specificity with regard to toxicant-induced ovarian atrophy is also demonstrated by 1,3-butadiene in that the mouse is the most sensitive species, likely due to pharmacokinetic differences (Christian, 1996). Mice demonstrate greater bioavailability of 1,3-butadiene, greater production of toxic intermediates, and a lower capacity for detoxification of these intermediates (U.S. EPA, 2002). Metabolic activation appears to be required for ovarian toxicity, and mice are more sensitive than rats to ovotoxic effects of the mono- and diepoxides of 1,3-butadiene (U.S. EPA, 2002). No human data are available to indicate that adverse female reproductive effects have resulted from occupational exposure to 1,3-butadiene, or to VCH or VCD.

Metals

Cadmium   Cadmium, a metal widely dispersed throughout the environment, has been linked to a range of detrimental effects on human reproduction and development. While adverse effects primarily involve male reproduction and development of children exposed in utero or after birth, evidence also points to potential disruption of the female reproductive system. Cadmium is considered to be an endocrine disruptor in that it either enhances (low doses) or inhibits (high doses) the biosynthesis of progesterone, a hormone linked to normal reproductive cyclicity in females, as well as the maintenance of pregnancy (Henson and Chedrese, 2004). A decrease in basal secretion of progesterone, as well as FSH and LH, occurred following treatment of rats with a single high dose of cadmium chloride, indicating an effect on the hypothalamic-pituitary-ovarian axis (Paksy et al., 1989). Cadmium effects on the female reproductive system appear to be species specific. A single subcutaneous injection of cadmium chloride in hamsters, four mouse strains, and two rat strains prior to and after sexual maturity revealed that hamsters had the greatest sensitivity to cadmium-induced ovarian toxicity (Rehm and Waalkes, 1988). Lesions in hamsters included ovarian hemorrhagic necrosis, with damage to small arteries of developing follicles and interstitial stroma, resulting in failure to ovulate. The most severe ovarian lesions occurred in immature hamsters, mature hamsters at high doses, and shortly before ovulation at all doses; these lesions were, however, reversible. In this study rats also demonstrated dose- and age-dependent toxicity in the ovaries, uterus, and cervix, while mice were more resistant to cadmium-induced toxicity. Massányi et al. (2007) evaluated the structures of ovary, oviduct, and uterus in rabbits after a single intraperitoneal dose or after 5 months of oral dosing with cadmium chloride. Ovarian effects included a decrease in the relative volume of growing follicles, increased stroma, and increased numbers of atretic follicles. Edema was observed in the oviduct and was considered related to the disintegration of vascular walls. Increased stroma in the uterus was considered a sign of edema caused by damage of blood vessel walls and subsequent diapedesis.

Effects on the uterus and mammary gland have also been observed in laboratory animals treated with cadmium. *In vitro*, cadmium was estrogenic in the MCF-7 human breast cancer cell line; effects on gene expression were considered to be mediated by the estrogen receptor independent of estradiol. Cadmium also induced the growth of MCF-7 cells and demonstrated signs of estrogenicity *in vivo* (Garcia-Morales et al., 1994). A single intraperitoneal injection of cadmium chloride induced an increase in uterine wet weight, promoted growth and development of mammary glands, and induced hormone-regulated genes in ovariectomized rats (Johnson et al., 2003). Female offspring of pregnant rats given intraperitoneal injections of cadmium on gestation days 12 and 17 displayed an earlier onset of puberty and increased mammary gland development (Johnson et al., 2003). Increased urinary cadmium concentrations were associated with uterine fibroids in a German cohort of infertile women (Gerhard et al., 1998).

Cadmium is known to accumulate in human ovaries. Ovarian biopsies conducted by Varga et al. (1993) revealed a linear increase in cadmium between 30 and 65 years of age, particularly in smokers versus nonsmokers. The increase in cadmium levels in ovaries of smokers is not unexpected, as tobacco smoke is a major source of cadmium contamination. Enhanced cadmium concentrations and lower progesterone levels have also been reported in follicular fluid and in placentas of smokers compared to nonsmokers (Piasek et al., 2001). Recurrent miscarriage and uterine fibroids have been directly linked with increased cadmium excretion in women.

Lead   Exposures to high levels of lead have been associated with spontaneous abortion in women. Although the evidence has in general been equivocal for low to moderate level lead exposures due to the small size of most studies, a prospective study conducted in Mexico City identified a dose–response relationship between the risk of spontaneous abortion and blood lead levels (Hertz-Picciotto, 2000). This study revealed that blood lead levels in the range of 10 ± 25 mg/dl could have adverse effects on pregnancy.

Lead is known to accumulate in the granulosa cells of the ovary and has been shown to affect follicular development and maturation. Morphological changes in the ovaries of mice treated with lead acetate were described as atresia and included a significant effect on small and medium follicles at low doses, with an effect on antral follicles at a higher dose (Junaid et al., 1997). Treatment of female rats with lead acetate for 15 days resulted in a reduction in binding of LH and FSH to granulosa cells, which altered steroidogenic enzyme activity and induced cytotoxicity (Nampoothiri and Gupta, 2006). Lead was shown to accumulate in the uterine epithelium of mice given an intraperitoneal injection of lead chloride and to disrupt morphology, potentially through interference with the activity of ovarian steroid hormones on the endometrium (Wide and Nilsson, 1979). In a subsequent study, implantation of the blastocyst was inhibited in mice given a single intravenous injection of lead chloride (Wide, 1980). This effect was counteracted by a combination of estrogen and progesterone, suggesting that implantation failure could indeed be due to an effect of lead on uterine responsiveness to ovarian steroids.

Mercury   Mercury is found in such diverse items as thermometers, traditional blood pressure meters, switches, mirrors, fluorescent lamps, and dental amalgam fillings. In addition, mercury catalysts are used in the oxidation of organic materials, and inorganic mercury is used in the pulp and paper industry as a fungicide. The most common forms of mercury to which humans have been exposed are elemental (metallic) mercury vapor, methylmercury, mercuric chloride, and phenylmercuric acetate (Seegmiller, 1997). Of these, mercury vapor has been evaluated to the greatest extent with regard to female reproductive toxicity. Reports of menstrual cycle abnormalities, including painful menstruation and changes in bleeding patterns and menstrual cycle duration, have been attributed to mercury vapor (Rowland et al., 1994). In particular, adverse effects on reproduction have been reported for female dental practitioners due to inhalation of elemental mercury vapor from dental silver amalgam fillings. A study by Rowland et al. (1994) indicated that women with exposure to 30 or more amalgams per week and with poor mercury hygiene (i.e., following practices that did not minimize exposure to mercury) showed evidence of reduced fertility, while women exposed to low levels of mercury were actually more fertile than unexposed controls.

Studies of various forms of mercury in laboratory animals, often at high doses that resulted in systemic toxicity, identified disturbances of the menstrual/estrous cycle, inhibition of ovulation, and/or infertility (Davis et al., 2001; Schuurs, 1999). In a study designed to evaluate the impact of systemic toxicity on female reproductive effects in rats, Davis et al. (2001) exposed rats to low levels of mercury vapor and found subtle changes in estrous cyclicity and corpora luteal morphology, but no adverse effects on fertility, leading these authors to suggest that reproductive effects noted in occupationally exposed women might be secondary to systemic and/or neurotoxic effects of mercury. In a review by Schuurs (1999), reproductive effects in women exposed to elemental mercury in a smelting plant, a mercury vapor lamp factory, and in dental practices were associated with mercury concentrations that likely exceeded the threshold limit value (TLV) of $25\,\mu g\ Hg^0/m^3$ for women.

Some adverse effects of mercury may be mediated through the H-P-G axis. Mercury was found to accumulate in the anterior pituitary of vervet monkeys implanted with amalgam fillings (Danscher et al., 1990) and was associated with hormonal disorders in a German cohort of women evaluated for effects of heavy metal exposures (Gerhard et al., 1998).

In animals, exposure to metallic mercury vapor prolongs the estrous cycle. Mercury also can have toxic effects on placental function. *In vitro* studies using normal, full-term human placentas reported that inorganic mercury inhibited placental amino acid transport (Iioka et al., 1987). In the rat conceptus, metallic mercury inhibited placental uptake of essential trace elements (Danielsson et al., 1984), amino acids, and nucleosides but did not directly affect the utilization of amino acids or nucleosides.

Pesticides

Organochlorine Pesticides   Pesticide exposure in women has been associated with adverse reproductive outcomes such as infertility, delay in conception, spontaneous abortions, congenital defects, and premature delivery (Figà-Talamanca, 2006). Organochlorine pesticides in particular can substitute for estradiol, resulting in disruption of female reproductive function. One example of an organochlorine pesticide that has been extensively studied in laboratory animals is MXC, which was intended as a replacement for DDT, but has now

been banned in the United States and the European Union due to its toxicity, bioaccumulation, and endocrine disruption activity (U.S. EPA, 2004). Human exposure can occur by air, soil, and water resulting from contamination with MXC. Methoxychlor is considered to be an environmental estrogen; the metabolite 2,2,bis(p-hydroxyphenyl)-1,1,1-trichloroethane (HPTE) is the active estrogenic form (Cummings, 1997). In an evaluation of a variety of short-term in vivo methods used to detect estrogenic activity in compounds known to have affinity for the estrogen receptor (although to a much lesser extent than estradiol), MXC advanced the age at vaginal opening (i.e., resulted in precocious puberty), disrupted the estrous cycle, and increased uterine weight (Laws et al., 2000). It has also been shown to induce adverse effects on fertility, early pregnancy, and early in utero development in laboratory rodents. Female mice given oral doses of MXC for 2 or 4 weeks developed persistent vaginal estrus (indicative of accelerated reproductive senescence), reduced ovarian weight, and an increase in the number of atretic ovarian follicles (Martinez and Swartz, 1991). These factors could be expected to result in reduced fertility. In rats, infertility has been demonstrated with long-term MXC exposure, while the adverse effect of short-term exposure during early pregnancy has been shown to block the decidual response of the uterus, which is necessary for successful implantation of the blastocyst (Cummings and Gray, 1989). Methoxychlor administration during days 1–3 of pregnancy resulted in a dose-dependent decrease in blastocyst implantation, while administration during days 4–8 resulted in resorption of embryos, decreased uterine weight, and a reduction in serum progesterone without affecting the number of corpora lutea or implantations, or ovarian weight (Cummings and Gray, 1989). Resorption following postimplantation dosing was considered to be a manifestation of normal decidual development. Although MXC has been thoroughly studied in laboratory animals, as per Cummings (1997), the significance of female reproductive toxicity with respect to human health remains to be determined.

One of the most well-known and well-studied organochlorine pesticides is DDT. Although DDT has been banned in many countries due to adverse effects induced in wildlife populations, it is still used in some countries and also persists in the environment. The metabolite 1,1-dichloro-2,2-*bis*(p-chlorophenyl)ethylene (DDE) is considered to be ubiquitous, is found in tissues of humans worldwide, and is consistently found in follicular fluid and serum of women (Tiemann, 2008). While DDT does not bind to the estrogen receptor, several contaminating isomers do, with the most potent being *o,p*-DDT (Das et al., 1997). Estrogenic DDT isomers induce estrogenic effects on female reproductive parameters similar to those of MXC.

γ-Hexachlorocyclohexane (lindane) is also found in air, water, and soil samples worldwide and has been shown to cause reproductive failure in laboratory animals characterized by the absence of implantation sites when given to mice during early pregnancy, total resorption of fetuses when given in mid-pregnancy, and neonatal death when given in late pregnancy (Tiemann, 2008).

DEVELOPMENTAL TOXICITY

Occupational exposure to toxicants can potentially impact the development through indirect exposure of the parents and subsequent transfer of the chemical and/or metabolites via the placenta or breast milk to the conceptus (embryo/fetus) or to the neonate. Neonates and young children who are still in the process of developing may be exposed directly to industrial chemicals in the case of contamination of air, water, foodstuffs, soil, or even contaminated clothing worn by parents. Adverse effects on development may also be induced through inheritance of genetic damage of the germ cells of one or both parents. The primary manifestations of developmental toxicity are embryo/fetal death, malformations (birth defects), growth retardation, and developmental delay. Adverse fetal outcomes may also include preterm delivery, altered sex ratio, and childhood cancer. The study of the induction of adverse structural changes during development is known as teratology; a physical or chemical agent that induces such effects is known as a teratogen. The association of blindness and deafness in children with maternal exposure to the rubella virus (German measles) and the linkage of limb abnormalities with thalidomide exposure were among the first examples of environmental factors identified as causing congenital defects. The overall incidence of birth defects in the United States is approximately 3%. Birth defects are defined as structural malformations with a significant impact on the health and development of a child and account for 20% of infant mortality in the United States (Parker et al., 2010). It is estimated that less than half of all human conceptions result in the birth of a completely normal healthy infant, and that the cause of birth defects is unknown in 65% of cases (Rogers and Kavlock, 2008), underscoring the need for continued research. For the purpose of discussing teratogenicity, human development can be divided into the following three stages: the preimplantation stage, the later embryonic stage, and the fetal stage. At any time after fertilization, the developmental process may be disrupted. Injury to the fertilized egg during the gastrula stage usually results in cell death. Damage involving minor cell death can be repaired, but major cell death cannot, and the result is abortion of the pregnancy. If the organism survives generally no structural deformity is seen. This is based on flexibility, or totipotency, of the cells, which allows them to follow various paths and replace lost cell lines. Spontaneous abortion in these early stages usually goes unrecognized and therefore rarely is associated with an exposure. The first

8-week period of organogenesis, the time of the most and the fastest growth, also is the period of greatest susceptibility to environmental influences, and most teratological effects are induced during this period. Exposure during the fetal period is most likely to result in a reduction in cell size and number. From 8 weeks to term, the fetus, generally considered less vulnerable than the embryo but more susceptible than the child or adult, can be subjected to adverse influences that lead to physical growth retardation and functional abnormalities. Some developmental processes that are not completed until after birth, such as physical growth and maturation of the nervous system, may be adversely affected by exposure during this period (Harbison, 1998).

Developmental toxicity is a subject of great concern with regard to environmental and occupational exposure, as well as pharmaceutical development, and has been well studied in laboratory animals due to the difficulties inherent in obtaining human data. Wilson (1977) proposed six scientific principles that sought to explain the roles played by critical periods of development, genotype, dose–response relationships, the various manifestations of abnormal development, access of agents to the developing tissues, and mechanisms critical to the understanding of developmental toxicity. These principles are highly valued in the risk assessment of potential developmental toxicants and can help to explain phenomena such as enhanced or diminished susceptibility during development relative to adults, responses that differ depending on developmental stage, and the role of key metabolic pathways and hormonal influences. As an example, during development, windows of vulnerability exist: these are periods during which toxic insult to specific organs and tissues can result in structural or functional damage. The age at which these windows occur and the duration of the insult required to induce damage are species specific, increasing the complexity of cross-species comparisons (Morford et al., 2004).

In addition to effects on the structure, adverse effects on function may also be induced during development and are generally manifested after birth. This is especially important for organs and tissues that have a protracted duration of development, including the postnatal period. These include the CNS and the immune, reproductive, and urinary systems. Developmental neurotoxicology, also known as behavioral teratology, is a subset of developmental toxicology, which has received a great deal of study.

An important consideration in evaluation of developmental toxicity in humans and laboratory species is maternal toxicity. Maternal factors that can affect development include genetics, disease state, nutrition, stress, and placental toxicity (Rogers and Kavlock, 2008). In addition, toxicity of a given physical or chemical agent to the maternal animal, inducing such outcomes as excessive weight loss and compromised organ-specific structure or function, can impact the development of the conceptus. Maternal effects could potentially exacerbate developmental toxicity or directly cause developmental changes.

Exposure of the mother to a carcinogen during pregnancy can result in a phenomenon known as transplacental carcinogenesis. Fetal tissues are thought to be privileged targets of neoplastic change induced by carcinogens due to the massive cell proliferation and differentiation taking place during embryogenesis (Alexandrov et al., 1990). A well-characterized example of transplacental carcinogenesis in humans and laboratory animals is the synthetic estrogen, diethylstilbestrol (DES). Diethylstilbestrol was prescribed as a drug for prevention of miscarriage and for the treatment of various hormone-dependent cancers and late complications of pregnancy; it was also used as a growth promoter in livestock (Seegmiller, 1997). Daughters of women who used DES to prevent miscarriage developed a rare form of cancer of the vagina or cervix, clear cell adenocarcinoma, which was first detected when they reached puberty. Investigative studies in mice have determined that exposure to DES during critical periods of reproductive tract differentiation permanently alters estrogen target tissues, resulting in increased susceptibility for tumor development postnatally (Newbold et al., 2006).

Physical Agents

Ionizing Radiation Exposure of the conceptus to ionizing radiation is a known risk factor for embryonic/fetal death, malformation, and growth retardation, with adverse effects dependent on the dose and developmental stage of exposure. Exposure of parents before conception may cause genetic damage to male or female germ cells and lead to abortion, stillbirth, and possibly cancer (Goldberg et al., 1998). A review by Brent (1980) described this stage-specific susceptibility. Prior to the blastocyst stage, the embryo is most sensitive to the lethal effects of ionizing radiation and insensitive to teratogencity and growth-retarding effects. During early organogenesis, the conceptus is sensitive to growth retarding, teratogenic, and lethal effects but can recover from growth retarding effects during the postpartum period. In the early fetal period, sensitivity is diminished to teratogenicity but is retained to CNS effects and growth retardation. Exposure during later fetal stages does not result in malformations, but the fetus can respond with permanent cell depletion of various organs and tissues if the exposure is sufficiently high.

The CNS is particularly vulnerable to *in utero* exposure to ionizing radiation; adverse neurologic effects such as microcephaly, mental retardation, lower intelligence scores, poor school performance, and seizures have been reported in survivors of the Nagasaki and Hiroshima nuclear bombings during World War II, who were exposed to ionizing radiation *in utero* (Yamazaki and Schull, 1990). The most critical period for these effects seemed to be 8–15 weeks after fertilization, which corresponds to the developmental stage when neuronal production is increased and immature neurons migrate to their cortical sites of function. Sensitivity to radiation is quite constant across all mammalian species during each developmental period. The animal and human data have been critically contrasted in a review article by Schull et al. (1990).

In a review on counseling of pregnant women regarding risks associated with diagnostic radiation, Brent (2009) stated that the various manifestations of radiation exposure during development have a no-observed-adverse-effect level (NOAEL), and that almost all diagnostic radiological procedures provide exposures below the NOAEL for these effects.

Solvents

Ethanol Ethanol is present as an additive in gasoline, a solvent in industry, and in many household products and pharmaceuticals. Human exposure can occur in these contexts, as well as through the consumption of intoxicating beverages. Ethanol is eliminated from the body by urinary excretion and exhalation; it is metabolized to acetaldehyde by alcohol dehydrogenase and can interact with other solvents metabolized in this way. It induces cytochrome P450 2E1, which can enhance metabolic activation and potentiate the toxicity of a considerable number of other solvents and drugs (Bruckner et al., 2008).

Fetal alcohol spectrum disorder (FASD) describes a constellation of effects that may occur in offspring of mothers who consume alcohol during pregnancy. Effects range from mild to severe and include fetal alcohol syndrome (FAS), which is at the most extreme end of the spectrum (Riley et al., 2011) and is considered the most common nonhereditary, preventable form of mental retardation (Bruckner et al., 2008). The prevalence of FAS in the United States and some Western European countries is estimated to be as high as 2–5% (May et al., 2009). Diagnostic criteria for FAS include maternal alcohol exposure during gestation, a specific pattern of craniofacial malformations (including short palpebral fissures and abnormalities in the premaxillary zone), pre- and/or postnatal growth retardation, and evidence of neurodevelopmental abnormalities, including decreased cranial size at birth, structural brain abnormalities, and/or neurological function deficits (Stratton et al., 1996). Peak maternal blood alcohol level is the most important determinant of the likelihood of FAS and severity of effects. Alcohol-related birth defects (ARBD) can also occur in which physical abnormalities of the skeleton and certain organ systems are not accompanied by the standard craniofacial malformations (May et al., 2009). These congenital anomalies can include atrial or ventricular septal defects, digit and limb anomalies, and aplastic kidneys (Stratton et al., 1996).

Developmental neurotoxicity, or alcohol-related neurodevelopmental disorder (ARND), has been shown to occur concomitantly with, or in the absence of, physical defects (including growth retardation and CNS malformations), on a continuum ranging from mild to severe. Functional consequences include a neonatal abstinence syndrome, mild to severe cognitive deficits, hyperactivity, distractibility, deficits in attention and reaction time, learning and memory disabilities, and state lability in infants. Ethanol exposure is considered to affect brain development via numerous pathways and at all stages, from neurogenesis to myelination (Riley et al., 2011).

There is a good deal of congruence between human and animal disorders resulting from maternal ethanol exposure. Developmental neurotoxicity, manifested as deficits in learning, inhibition, attention, regulatory behaviors, and motor performance, was found to be affected in both humans and laboratory animals following moderate levels of ethanol exposure during development (Driscoll et al., 1990). Although the dose required to produce an effect differed among species, blood alcohol levels were quite similar. The majority of studies conducted in humans and laboratory animals involve ingestion of relatively high doses of ethanol to mimic consumption of alcoholic beverages. While these studies provide important information, their relevance to industrial exposure is limited, since this type of exposure is typically via inhalation or dermal absorption, producing substantially lower blood alcohol levels. In a review of occupational ethanol exposure, Irvine (2003) concluded that there is no evidence that industrial exposure to ethanol is a developmental toxicity hazard.

Glycol Ethers   Some glycol ethers and their metabolites, including EGME, EGEE, 2-ME, 2-EE, and 2-MAA, are developmentally toxic in laboratory animal species and have been shown to produce malformations in a variety of organ systems (Andrew and Hardin, 1983; Hardin et al., 1986; Scott et al., 1989). Embryo mortality, as well as hydrocephaly, exencephaly, cardiovascular malformations, craniofacial anomalies, malformations of the digits and ribs, dilatation of the renal pelvis, and/or minor skeletal variations, occurred in several animal species exposed to EGME, EGEE, 2-ME, or 2-MAA, with some species specificity observed with respect to adverse effects. Functional changes have also been observed following an in utero exposure. Female rats treated with EGME or EGEE in a study in which they were allowed to deliver their offspring exhibited a slight prolongation of gestation, but no other indications of maternal toxicity (Nelson and Brightwell, 1984). Offspring from these females displayed neurobehavioral changes and regional brain alterations of several neurotransmitters. Gestation was also prolonged in rats treated with 2-ME during gestation (Toraason and Breitenstein, 1988). Electrocardiograms evaluated from offspring of these females revealed persistent, aberrant QRS waves, suggestive of an intraventricular conduction delay. These changes occurred in the absence of morphologic anomalies of the heart.

Maldonado et al. (2003) reviewed four glycol ether-related epidemiology studies. Facial malformations (e.g. oral clefts), CNS malformations, neural tube defects, and varying degrees of mental retardation in children of mothers exposed to glycol ethers occupationally were attributed to this exposure. However, these authors concluded that there were methodological issues with all four studies; in addition, the attribution of malformations to glycol ether exposure lacked biological plausibility since the glycol ethers to which the women were exposed in general have not been shown to be teratogenic in animal models. This appears to be an area requiring further research to determine potential concordance of human and animal data that may impact the risk assessment of glycol ethers.

Toluene   Toluene is present in paints, lacquers, thinners, cleaning agents, glues, and gasoline, which is the largest source of atmospheric emissions of toluene and exposure to the general population (Bruckner et al., 2008). Inhalation is the primary route of human exposure, although skin contact also occurs frequently. Toluene exposure occurs in individuals who inhale gasoline, glue, and spray paint to obtain a sense of euphoria, which occurs at an exposure of at least 500 ppm (Wilkins-Haug, 1997). The term "fetal solvent syndrome" describes the sequelae from in utero exposure, which include microcephaly, developmental delay, growth deficiency, and craniofacial defects similar to those associated with FAS (Pearson et al., 1994; Wilkins-Haug, 1997; Bruckner et al., 2008). Pearson et al. (1994) suggested that a common mechanism of action may explain the craniofacial teratogenicity induced by toluene and by ethanol.

Developmental effects were observed in a two-generation study in mice reported by Roberts et al. (2003). In this study, male or female mice were exposed by inhalation to concentrations as high as 2000 ppm toluene for 80 days prior to mating, during mating, and for females, through gestation and lactation; parental animals were evaluated for effects on fertility, and their F_1 generation offspring were evaluated for survival, growth, development, behavior, and ability to produce the F_2 generation. A separate cohort of females was exposed to toluene during gestation, and their fetuses were evaluated for developmental anomalies. F_1 offspring were subsequently directly exposed to toluene for 80 days after weaning. There were no effects on fertility or reproductive performance of males, females, or the F_1 generation, or on maternal or offspring behavior. Body weights were reduced in F_1 offspring of females exposed to toluene, as well as their F_2 offspring; body weights were unaffected in offspring of exposed males. Decreased fetal body weight and skeletal anomalies were observed in the F_1 and F_2 generation fetuses of exposed females.

Additional human studies have reported reduced fertility and increases in spontaneous abortion in women exposed to toluene in an industrial setting, but the interpretation of these studies has been confounded by unknown exposure levels, timing of exposure, and duration of exposure, as well as presence of other solvents, small sample size, and lifestyle factors such as smoking and alcohol consumption (Wilkins-Haug, 1997; Roberts et al., 2003).

Other Industrial Chemicals

Bisphenol A Bisphenol A (BPA) is used in the production of polycarbonate plastics used in food and drink containers, as well as in epoxy resins used as lacquers to coat metal products such as food cans, bottle tops, and water supply pipes. The production volume of this chemical is considered to be high. The primary source of exposure to humans is assumed to be ingestion of food or drink that has been in contact with a material containing BPA, although direct contact can also occur (Chapin et al., 2008). It is estimated that the highest intake is in infants and children. Bisphenol A has been the subject of much discussion and study due to widespread human exposure, evidence of reproductive and developmental toxicity in laboratory animals, and its status as an estrogenic compound.

Early development appears to be the greatest period of sensitivity to adverse effects in laboratory animals. Adverse developmental effects include fetal and neonatal death and decreased growth rate. Additional effects reported in rats and mice include delayed puberty in males and females, as well as altered mammary gland development in females leading to the development of preneoplastic lesions and increased susceptibility to neoplastic development, and increased susceptibility to prostate carcinogenesis (Diamanti-Kandarakis et al., 2009). Evidence for disruption of the H-P-G axis in rats exposed to BPA *in utero* has also been presented, as well as effects on brain structure and function, including behavior anomalies (Richter et al., 2007).

A NTP report on BPA, based on findings from the expert panel of the Center for the Evaluation of Risks to Human Reproduction (CERHR), found no direct evidence for health effects in humans. Although the NTP expressed minimal concern regarding developmental effects identified in laboratory animals that included an earlier age for puberty in females, and some concern regarding effects on the brain and behavior, it concluded that additional research is needed to better understand if findings from laboratory animals are of any significance to human health. New information continues to emerge based on extensive ongoing research, particularly about potential low dose effects, and recent statements of support for further clarification of BPA-associated risks have been issued by the U.S. Food and Drug Administration (FDA) and the Endocrine Society (Diamanti-Kandarakis et al., 2009; U.S. FDA, 2013).

Halogenated Hydrocarbons Polychlorinated biphenyls (PCBs) are known to be developmental neurotoxicants,

based on reports of children exposed in utero or during lactation due to maternal consumption of contaminated fish or rice oil, or to background environmental levels (Barlow et al., 1999). High levels of contamination with PCBs and polychlorinated dibenzofurans (PCDFs) reported in Yucheng, Taiwan, and Yusho, Japan, were associated with low birth weight, preterm delivery, and developmental effects in children (Whelan, 1997). Infants developed an ectodermal dysplasia syndrome, which included toxicity to skin and teeth; CNS toxicity may potentially be part of this syndrome (Peterson et al., 1993). Adverse effects occurred in children of mothers who were rendered clinically ill by exposure to PCBs; these effects included effects on motor and mental function, as well as developmental and behavioral problems. Children of mothers exposed to lower levels of PCBs in the United States exhibited smaller birth and neonatal size; weak reflexes; were less responsive to stimuli; had more jerky, unbalanced movement, and more startles; were hypotonic; had deficits in gross and fine motor coordination; and had poorer visual recognition and memory performance. These effects were presumed to occur in the absence of maternal toxicity (Tilson et al., 1990). Schantz et al. (2003) have reported that despite the decline in levels of PCBs in exposed cohorts, negative impacts on cognitive functioning are still being reported. The geometry of PCBs is such that a large number of substitutions can be made on the dual benzene rings, leading to a total of 209 theoretically possible congeners, only a few of which persist in the environment (Ulbrich and Stahlmann, 2004). Although many studies have been conducted with PCBs in laboratory animals, they have typically evaluated commercially available mixtures of congeners, different from the environmental congeners to which humans are exposed and having potentially different toxicity profiles. Schantz et al. (2003) have suggested that more complete information is needed for relevant congeners to allow for more informed scientific and risk assessment decisions.

Polychlorinated diphenyl ethers (PBDEs) are used as flame retardants, with human exposure occurring from treated surfaces and contaminated food. A prospective epidemiology study identified a positive association between congenital cryptorchidism and levels of PBDEs (Main et al., 2007). Newborn boys evaluated in this study were from an area of considerable production rates of PBDEs, resulting in increased levels of concern. It is anticipated that further investigation will occur that may also take into account other potential environmental contaminants.

Dioxins are polyhalogenated compounds that are considered to be ubiquitous environmental pollutants. The most widely studied of these is 2,3,7,8-tetrachlorodibenzo-*p*-dioxin (TCDD). The spectrum of adverse effects induced by this class varies with specific compounds and also tends to be species dependent. Developmental toxicity in mammals includes decreased growth, structural malformations, functional alterations, and prenatal mortality; at low exposure levels, structural malformations are not common (Peterson et al., 1993). Functional alterations are the most sensitive signs of developmental toxicity and include adverse effects on male reproductive structures and reproductive behavior in rats and neurobehavioral deficits in monkeys. The ectodermal dysplasia syndrome described for PCBs and PCDFs has also been observed in humans as a result of exposure to high levels of TCDD in Seveso, Italy, as well as in rats and monkeys exposed to TCDD *in utero* or via lactation (Yasuda et al., 2005). Altered sex ratio was also reported in births following the Seveso incident, with an excess of females compared to males (Yoshimura et al., 2001).

A specific pattern of structural malformations has been reported in mice exposed to TCDD *in utero*, which includes cleft palate; hydronephrosis and thymic hypoplasia have also been observed. These changes occur in the absence of maternal toxicity and embryo/fetal toxicity. Mice appear to be highly susceptible to TCDD-induced teratogenesis, as other species demonstrate maternal and embryo/fetal toxicity but are not as susceptible to malformations (Couture et al., 1990). In contrast, mice are not as sensitive to TCDD-induced developmental reproductive toxicity as compared to rats and hamsters. *In utero* exposure of male rats to low doses of TCDD resulted in reduction in numbers of sperm per cauda epididymis, daily sperm production, and sperm transit rate through the cauda epididymis when evaluated as adults (Faqi et al., 1998). These rats also had increased numbers of abnormal sperm, increased mounting and intromission latencies (measures of male sexual behavior), and, at the highest dose tested, reduced testosterone. Decreases in daily sperm production did not occur in mice exposed to TCDD during gestation and lactation; the only effect common to rats and mice was a reduction in epididymal sperm numbers (Theobald and Peterson, 1997). Reduction in anogenital distance has been reported in male rats exposed to TCDD *in utero* but not in male hamsters (Birnbaum, 1995). Reproductive alterations have also been reported in female rats and hamsters exposed to TCDD during gestation and lactation and include cleft phallus/clitoris, hypospadias, and/or delayed/incomplete vaginal opening (Birnbaum, 1995).

Trichloroethylene and Perchloroethylene Ingestion of TCE in drinking water has been associated with cardiac malformations in children exposed in utero. A study conducted in 1973 noted that congenital heart disease appeared to be increased in a small area of the Tucson Valley in Arizona; a subsequent study of a similar area found that the groundwater was contaminated with TCE, and to a lesser extent, dichloroethylene (DCE) and chromium. Follow-up of offspring from parents living in this area at the time of contamination prior to or during the first trimester of pregnancy revealed a significant association, but not a cause-and-effect relationship, between parental exposure and an increase in congenital heart disease

(Goldberg et al., 1990). Cardiac malformations were also observed in fetuses from rats administered TCE or DCE directly to the uterus during gestation days 7–22 (Dawson et al., 1990). Cardiac malformations, including atrial and ventricular septal defects and defects of aortic, pulmonary, and mitral valves, were observed with both TCE and DCE. Cardiac malformations were also observed in a study by Smith et al. (1989), in which oral intubation on gestation days 6–15 resulted in interventricular septal defect and levocardia. Subsequently, Johnson et al. (1998) studied TCE metabolites TCA, monochloroacetic acid, trichloroethanol, carboxy methylcystine, trichloroacetaldehyde, dichloroacetaldehyde, and dichlorovinyl cystine given in drinking water to rats throughout pregnancy. Cardiac malformations of the same nature as those observed with TCE and DCE were observed with TCA only. Interventricular septal defect and levocardia were observed following oral gavage administration to rats on gestation days 6–15 (Smith et al., 1989).

A number of studies in laboratory animals have evaluated effects of TCE, PERC, and metabolites on development using a variety of routes of administration and developmental periods of exposure. In addition to the cardiac malformations associated with TCE, DCE, and TCA, impaired embryo/fetal survival and growth were the most commonly observed adverse effects. Human studies have yielded equivocal effects, but there is some evidence of association between TCE exposure and intrauterine growth impairment (National Research Council, 2006).

Prenatal exposure to TCE has also been associated with postnatal functional impairment. Peden-Adams et al. (2006) evaluated the immune function in mice exposed to TCE from gestation day 0 through 3 or 8 weeks of age. Trichloroethylene caused increased hypersensitivity responses and increased numbers of thymic CD4+ and CD8+ cells at 8 weeks of age, suggesting that TCE is a developmental immunotoxicant. A cross-species comparison of PERC identified developmental neurotoxicity as an additional adverse effect and concluded that neurotoxicity was the most sensitive end point for inhalation exposure, while growth retardation was the most sensitive end point for oral exposure (Beliles, 2002). This review concluded that with regard to reproductive and developmental toxicity of PERC, in most cases there was concordance between rodent and human effects.

Hexachlorobenzene Hexachlorobenzene (HCB) was formerly manufactured as a fungicide; however, due to adverse human health effects that occurred from a contamination incident in Turkey, it is currently not in use for this purpose in most countries. At present it is found as a by-product or a contaminant from production of other chlorinated products such as industrial solvents and pesticides (Ando et al., 1985; Goldey and Taylor, 1992). The major route of human exposure to HCB is through the diet. Following absorption,

HCB accumulates in the adipose tissue and persists for many years since it is highly lipophilic and resistant to metabolism (Weisenberg, 1986). It is transferred to the human fetus and newborn through the placenta (Weisenberg, 1986) and breast milk (Watanabe et al., 1990). It crosses the blood–brain barrier and is known to be a developmental neurotoxicant. Watanabe et al. (1990) studied several neurotoxicant chemicals with pre or postnatal exposure and determined that blood–brain barrier transfer, which varies with lipid solubility, is a major factor in developmental neurotoxicity. Goldey and Taylor (1992) treated female rats with HCB for 4 days, and 2 weeks later mated them with untreated males. Offspring tested at various times during the postnatal period were hyperactive and demonstrated age-dependent changes in the acoustic startle response. The amplitude of the startle response was decreased on postnatal day 23 but increased on postnatal day 90; the reason for difference in response at these ages is unknown.

Mortality rates were high among children of lactating mothers in Turkey who ate bread contaminated with HCB. According to one report, all children born to mothers with porphyria, an effect of high dose HCB, died (ATSDR, 1990). Children nursed by mothers exposed to HCB showed pink sores on their skin and experienced weakness, convulsions, and annular erythema. Maternal blood and breast milk concentrations of HCB were unknown. Estimates of HCB consumption in bread during this period were 0.05–0.2 g/day (Ando et al., 1985).

Vinyl Chloride Vinyl chloride, also known as vinyl chloride monomer, is used primarily to produce the polymer polyvinyl chloride, which is a widely used plastic. Exposure of pregnant mice, rats, and rabbits to high concentrations of vinyl chloride resulted in no malformations; however, fetotoxicity, characterized by reduced survival, decreased body weight, and induction of anatomical variations occurred in mice and/or rats exposed during development (John et al., 1977). Vinyl chloride is mutagenic in most major short-term tests (California Environmental Protection Agency, 2000) and is known to induce a rare cancer of the liver in humans (U.S. EPA, 2000b). It is also a liver carcinogen in rats, mice, and hamsters when administered at high concentrations by inhalation to adult animals in lifetime studies. In these studies, tumors of the liver, nasal cavity, kidneys, Zymbal's gland, and blood vessels were attributed to vinyl chloride exposure. Maltoni et al. (1981) determined that in utero exposure of rats to vinyl chloride during a critical phase of development also induces tumors of the kidneys, Zymbal's gland, and blood vessels. A review by Rice (1981) concluded that vinyl chloride is a transplacental carcinogen in rats when administered at high concentrations to pregnant animals. The mechanism of action involves metabolic conversion of vinyl chloride to reactive intermediates such as chlorooxirane in the maternal and fetal tissues.

Several studies have reported significantly higher rates of developmental effects, such as fetal loss and congenital malformations, in communities containing vinyl chloride polymerization facilities. However, these studies did not show any significant correlation between developmental effects and proximity to vinyl chloride facilities or occupational exposure of either parent (Infante, 1976; Thériault et al., 1983; Rosenman et al., 1989; ATSDR, 1995). The apparent contradiction in these findings may be explained by confounding factors such as other industrial emissions.

Metals

Arsenic Arsenic is a metalloid that is ubiquitous in the environment and is also used commercially. Although both inorganic and organic forms exist in nature, the inorganic forms are considered to be more toxic (Golub et al., 1998). Exposure of humans to inorganic forms of arsenic occurs primarily in drinking water, as well as in food and from environmental contamination. Sources of environmental arsenic include glass and copper smelters, coal combustion, and uranium mining (Dyer, 2007). Inorganic forms of arsenic include trivalent (arsenite) and pentavalent (arsenate) forms, which are readily absorbed from the gastrointestinal tract (Waalkes et al., 2003; Dyer, 2007). These forms of arsenic have been most commonly used in laboratory animal experiments.

Arsenic is considered to be a human carcinogen: Chronic exposure of adults to high concentrations of at least several hundred µg/l may cause cancer of the skin, urinary bladder, lung, liver and prostate, and possibly of other tissues (Brown and Ross, 2002; Waalkes et al., 2003). Evidence also exists that arsenic is a transplacental carcinogen. Examination of a cohort of individuals from Chile who were exposed *in utero* or during early childhood to high concentrations of arsenic in contaminated drinking water revealed an increase in mortality as young adults from both malignant and nonmalignant lung disease (Smith et al., 2006). The ability of arsenic to induce transplacental carcinogenesis has been confirmed in mice, with tumors occurring at multiple sites. Offspring of female mice given sodium arsenite daily during gestation days 8–18 had dose-related increases in the incidence and multiplicity of liver tumors and adrenal tumors; the incidences of ovarian tumors and lung carcinoma were also increased (Waalkes et al., 2003). In addition, proliferative lesions were increased in the uterus and oviduct. The tumor spectrum induced by arsenic is considered to resemble that of estrogenic carcinogens and is associated with estrogen-linked genes (Liu et al., 2008). In addition, limited *in vitro* studies with MCF-7 breast cancer cells, as well as estrogen binding assays and receptor activation assays, suggest that arsenic may have an estrogenic activity (Dyer, 2007).

Arsenic has also been associated with spontaneous abortion and stillbirth in cases of human exposure to high levels, although interpretation of these case reports is complicated in some instances by potential exposure to other environmental contaminants (Golub et al., 1998). Developmental toxicity has been demonstrated in hamsters, mice, rats, and rabbits. Adverse effects included malformation, death, and growth retardation and were dependent on the dose, route of administration, developmental period of exposure, and form of arsenic tested; agents that chelate arsenic were found to protect against developmental toxicity (Golub et al., 1998).

Cadmium Cadmium is embryotoxic and teratogenic in laboratory animals. Ovarian cadmium has been associated with failure of progression of oocyte development from the primary to the secondary stage and failure to ovulate. It is also known to disrupt movement of the secondary oocyte into the oviduct, where combination with the spermatozoa normally would result in fertilization (Thompson and Bannigan, 2008). Decreased implantation of the resulting embryo has been found to result from cadmium exposure at doses that fail to disrupt early development of the embryo in vitro. Higher doses inhibit progression to the blastocyst stage and can cause degeneration and compaction in the blastocyst, with apoptosis and breakdown in cell adhesion followed by implantation failure. Cadmium has been shown to accumulate in embryos, beginning at the four-cell stage. Damage to trophoblast cells surrounding the early embryo, including inhibition of trophoblastic invasion, reduced steroidogenesis, and adjusted handling of nutritive metals such as calcium and zinc, can lead to necrosis of the placenta. While little cadmium crosses the placenta, it is known to concentrate in the placental tissue (Piasek et al., 2001; Satarug and Moore, 2004).

As mentioned previously, a predominant source of cadmium in the environment of non-polluted areas is tobacco smoke, and it has been reported that placentas from smokers have double the cadmium concentration of nonsmokers. Placental cadmium resulting from maternal exposure to industrial wastes or tobacco smoke has been associated with reduced progesterone biosynthesis by the placental trophoblast (Piasek et al., 2001). Recurrent miscarriage and uterine fibroids were linked to increased cadmium in a German cohort of infertile women (Gerhard et al., 1998). Maternal exposure to cadmium has been associated with low birth weight and an increased risk of spontaneous abortion, and cadmium has been reported to accumulate in high concentrations in the placenta (Henson and Chedrese, 2004). In perfused human placentas, the secretion of human chorionic gonadotropin (hCG), a hormone vital to early pregnancy maintenance, was reduced by exposure to cadmium (Wier et al., 1990).

Although placental transfer of cadmium is limited, a wide range of abnormalities has been reported in laboratory animals, which are dependent on the stage of developmental exposure, species, strain, dose, and route of administration (Schoeters et al., 2006). These abnormalities include craniofacial, neurological (including neural tube closure), cardiovascular, gastrointestinal, and genitourinary, as well as limb anomalies. During critical periods of development, prenatal exposure to cadmium at doses that cause no overt maternal toxicity may have postnatal, adverse behavioral sequelae that persist into adulthood (Ali et al., 1986). Limb deformities were also seen at dosages that did not cause a reduction in fetal body weight (Soukupova and Dostal, 1991). While human malformations have not been reported with cadmium, maternal exposure has been associated with low birth weight and increased preterm deliveries (Satarug and Moore, 2004). The smaller size of the infants was attributed to early delivery.

Lead Lead is considered to be a human developmental toxicant. Animal studies indicate that effects from lead absorption in developing offspring are seen at lower doses than those needed to cause overt toxicity in the adult, suggesting a more vulnerable fetus (Sikorski et al., 1989; Levallois et al., 1991; McGivern et al., 1991). Lead can potentially be transferred to a developing child via the placenta or breast milk. Occupational risks include abortion, stillbirth, postnatal death, and congenital malformations. Early fetal loss may be one of the most sensitive measures of lead reproductive toxicity (Sweeney and LaPorte, 1991). Lead is also a human developmental neurotoxicant. Lead-based paint remains the most common source of lead exposure for children <6 years of age, with a 2–4-point IQ deficit for each μg/dl increase in blood lead within the range of 5–35 μg/dl (Liu et al., 2008). Blood lead concentrations as low as 10–15 μg/dl in children have been associated with hyperactivity, poor fine motor control, decreased gestational age, low birth weight, poor academic achievement, and intelligence deficit (Suzuki and Martin, 1994). Prenatal lead exposure has also been associated with reduced intellectual development in children, with a critical period occurring around 28 weeks of gestation (Schnaas et al., 2006). Lead crosses the placenta and accumulates in fetal tissues in an amount proportional to maternal blood lead levels (Goyer, 1996). It has been measured in the human fetal brain as early as the end of the first trimester (13 weeks) (Goyer, 1990). Absorption and retention of lead may be significantly increased in later stages of pregnancy (Donald et al., 1986).

Acute lead encephalopathy, characterized by a swollen, edematous brain with proteinaceous exudates around blood vessels, petechial hemorrhage, and capillary necrosis may be produced at higher concentrations and is more severe in children than in adults; motor peripheral neuropathy may be seen as a functional consequence (Suzuki and Martin, 1994). Fetuses may be even more sensitive to cell-specific effects in the nervous system than are children. One of the proposed mechanisms of lead-induced neurotoxicity is swelling of the brain due to altered capillary permeability; immature endothelial cells may be less resistant to this effect, permitting lead to reach newly formed brain cells (Goyer, 1990). Intracellular lead can affect binding sites for calcium, which in turn can impact functions such as neurotransmitter release. Other potential mechanisms include indirect effects on cognitive development through lead-associated alterations in birth weight and gestation (Dietrich et al., 1987).

Effects on survival and development of offspring have been reported following maternal and paternal exposure to lead. Chronic paternal blood lead levels of >40 μg/dl or 25 μg/dl have been associated with increased risk of spontaneous abortion or reduced fetal growth, while maternal blood levels of approximately 10 μg/dl have been associated with increased risks of pregnancy hypertension, spontaneous abortion, and reduced offspring neurobehavioral development, with somewhat higher levels associated with reduced fetal growth (Bellinger, 2005). Delayed growth and pubertal development have been reported in girls and are likely related to disruption of the H-P-G axis (Dyer, 2007). There is currently insufficient evidence to provide a definitive association between parental lead exposure and structural malformations.

Mercury Methylmercury is a well-known human teratogen, having first been documented in 1952 following contamination of Minamata Bay, Japan, and subsequent ingestion of contaminated seafood. Methymercury poisoning was thereafter referred to as Minamata disease. Subsequent contaminations have occurred in Niigata, Japan; Sweden; the former Soviet Union; Iraq; and the United States. Methylmercury primarily targets the CNS; it is readily absorbed by the gastrointestinal tract, crosses the placenta, concentrates in the fetus, and ultimately crosses the fetal blood–brain barrier (Seegmiller, 1997). Clinical signs of toxicity resulting from *in utero* exposure included cerebral palsy, malnutrition, blindness, delayed speech development, severe hearing impairment or deafness, motor impairment, tremors, convulsions, abnormal reflexes, mental retardation, abnormal electroencephalograms (EEGs), excessive crying and irritability, and pathologic changes in the CNS (Suzuki and Martin, 1994). Most methylmercury-exposed mothers with affected infants showed no symptomatology.

Rates of spontaneous abortion among wives of 152 workers occupationally exposed to mercury vapor were reported to be higher than rates in wives of 374 control (unexposed) workers in the same plant (Cordier et al., 1991). The authors speculated that the mechanism of toxicity could include direct effects of mercury on the male reproductive system and indirect toxicity to the mother or embryo through transport of mercury from the father. Subsequent investigations of pregnant women occupationally exposed to metallic mercury in dental offices found that elevated mercury levels could be found in their placentas and fetal membranes, but these exposures were apparently not sufficient to cause developmental toxicity because there was no increase in spontaneous abortion rates and no defects found in the offspring. A questionnaire administered among men and women working in a dental environment found no differences in the rates of spontaneous abortions or congenital malformations (Brodsky et al., 1985).

Placental transfer of mercury varies according to the chemical form of the mercury compound (Messite and Bond, 1988). For example, methylmercury readily crosses the placenta, while inorganic mercury does not. In addition, mercury metabolism in fetuses appears to be much different from metabolism in mothers, possibly because of different blood circulation (Harbison, 1998). Despite these differences, laboratory animal data have demonstrated adverse effects on the reproduction and development with all chemical forms of mercury tested; these effects include impaired implantation of the embryo, spontaneous abortion, stillbirths, decreased birth weight, congenital malformations, and behavioral effects on exposed offspring (Messite and Bond, 1988; Rowland et al., 1994; Schuurs, 1999).

REFERENCES

Aitken, R.J., Bennetts, L.E., Sawyer, D., Wiklendt, A.M., and King, B.V. (2005) Impact of radio frequency electromagnetic radiation on DNA integrity in the male germline. *Int. J. Androl.* 28:171–179.

Alexandrov, V., Aiello, C., and Rossi, L. (1990) Modifying factors in prenatal carcinogenesis. *In Vivo* 4(5):327–335.

Ali, M.M., Murthy, R.C., and Chandra, S.V. (1986) Developmental and long-term neurobehavioral toxicity of low level *in-utero* cadmium exposure in rats. *Neurobehav. Toxicol. Teratol.* 8(5):463–468.

Almekinder, J.L., Lennard, D.F., and Walmer, D.K. (1997) Toxicity of methoxyacetic acid in cultured human luteal cells. *Fundam. Appl. Toxicol.* 38:191–194.

Amir, D. (1991) The spermicidal effect of ethylene dibromide in bulls and rams. *Mol. Reprod. Dev.* 28:99–109.

Ando, M., Hirano, S., and Itoh, Y. (1985) Transfer of hexachlorobenzene (HCB) from mother to newborn baby through placenta and milk. *Arch. Toxicol.* 56:195–200.

Andrew, F.D. and Hardin, B.E. (1983) Developmental effects after inhalation exposure of gravid rabbits and rats to ethylene glycol monoethyl ether. *Environ. Health Perspect.* 57:13–23.

Anway, M.D. and Skinner, M.K. (2006) Epigenetic actions of endocrine disruptors. *Endocrinology* 147(6):S43–S49.

Aoyagi, T., Ishikawa, H., Miyaji, K., Hayakawa, K., and Hata, M. (2002) Cadmium-induced testicular damage in a rat model of subchronic intoxication. *Reprod. Med. Biol.* 1(2):59–63.

Apostoli, P., Kiss, P., Porru, S., Bonde, J.P., Vanhoorne, M., and the ASCLEPIOS Study Group (1998) Male reproductive toxicity of lead in animals and humans. *Occup. Environ. Med.* 55:364–374.

ATSDR (Agency for Toxic Substances and Disease Registry) (1990) *Toxicant Profile for Hexachlorobenzene*, Atlanta, Georgia: U.S. Department of Health and Human Services, Public Health Service..

ATSDR (Agency for Toxic Substances and Disease Registry) (1995) *Tox FAQ: 1,2-Dibromo-3-chloropropane*, Atlanta, Georgia: U.S. Department of Health and Human Services, Public Health Service.

Babich, H. and Davis, D.L. (1981) Dibromochloropropane (DBCP): a review. *Sci. Total Environ.* 17:207–221.

Barlow, S., Kavlock, R.J., Moore, J.A., Schantz, S.L., Sheehan, D.M., Shuey, D.L., and Lary, J.M. (1999) Teratology society public affairs committee position paper: developmental toxicity of endocrine disruptors to humans. *Teratology* 60:365–375.

Beliles, R.P. (2002) Concordance across species in the reproductive and developmental toxicity of tetrachloroethylene. *Toxicol. Ind. Health* 18(2):91–106.

Bellinger, D.C. (2005) Teratogen update: lead and pregnancy. *Birth Defects Res. A Clin. Mol. Teratol.* 73:409–420.

Benoff, S., Jacob, A., and Hurley, I.R. (2000) Male infertility and environmental exposure to lead and cadmium. *Hum. Reprod. Update* 6(2):107–121.

Benoff, S., Centola, G.M., Millan, C., Napolitano, B., Marmar, J.L., and Hurley, I.R. (2003) Increased seminal plasma lead levels adversely affect the fertility potential of sperm in IVF. *Hum. Reprod.* 18(2):374–383.

Birnbaum, L.S. (1995) Developmental effects of dioxins. *Environ. Health Perspect.* 103(Suppl. 7):89–94.

Brent, R.L. (1980) Radiation teratogenesis. *Teratology* 21(3):281–298.

Brent, R.L. (2009) Saving lives and changing family histories: appropriate counseling of pregnant women and men and women of reproductive age, concerning the risk of diagnostic radiation exposures during and before pregnancy. *Am. J. Obstet. Gynecol.* 200(1):4–24.

Brodsky, J.B., Cohen, E.N., and Whitcher, C. (1985) Occupational exposure to mercury in dentistry and pregnancy outcome. *J. Am. Dent. Assoc.* 111:779–780.

Brown, K.G. and Ross, G.L. (2002) Arsenic, drinking water, and health: a position paper of the American Council on Science and Health. *Regul. Toxicol. Pharmacol.* 36:162–174.

Bruckner, J.V., Anand, S.S., and Warren, D.A. (2008) Toxic effects of solvents and vapors. In: Klaassen, C.D., editor. *Cassarett and Doull's Toxicology the Basic Science of Poisons*, New York, NY: McGraw-Hill. pp. 981–1051.

California Environmental Protection Agency (2000) *Public health goals for chemicals in drinking water—Vinyl chloride.*

Carney, E. W., Thorsrud, B.A., Dugard, P.H., and Zablotny, C.L. (2006) Developmental toxicity studies in Crl:CD(SD) rats following inhalation exposure to trichloroethylene and per-chloroethylene. *Birth Defects Res. B Dev. Reprod. Toxicol.* 77(5):405–412.

Centers for Disease Control (2009) *Fourth National Report on Human Exposure to Environmental Chemicals*, U.S. Department of Health and Human Services, pp. 1–529.

Chapin, R.E. and Ku, W.W. (1994) The reproductive toxicity of boric acid. *Environ. Health Perspect.* 102(Suppl. 7):87–91.

Chapin, R.E., Adams, J. Boekelheide, K., Gray, L.E., Jr., Hayward, S.W., Lees, P.S.J., McIntyre, B.S., Portier, K.M., Schnorr, T.M., Selevan, S.G., Vandenbergh, J.G., and Woskie, S.R. (2008) NTP-CERHR expert panel report on the reproductive and developmental toxicity of bisphenol A. *Birth Defects Res. B Dev. Reprod. Toxicol.* 83:157–395.

Christian, M.S. (1996) Review of reproductive and developmental toxicity of 1,3-butadiene. *Toxicology* 113:137–143.

Cordier, S., DePlan, F., Mandereau, L., and Hemon, D. (1991) Paternal exposure to mercury and spontaneous abortions. *Occup. Environ. Med.* 48:375–381.

Correa, A., Gray, R., Cohen, R.H., Cohen, R., Rothman, N., Shah, F., Seacat, H., and Com, M. (1996) Ethylene glycol ethers and risks of spontaneous abortion and subfertility. *Am. J. Epidemiol.* 143:707–717.

Couture, L.A., Abbott, B.D., and Birnbaum, L.S. (1990) A critical review of the developmental toxicity and teratogenicity of 2,3,7,8-tetrachlorodibenzo-p-dioxin: recent advances toward understanding the mechanism. *Teratology* 42:619–627.

Cowin, P.A., Gold, E., Aleksova, J., O'Bryan, M.K., Foster, P.M.D., Scott, H.S., and Risbridger, G.P. (2010) Vinclozolin exposure *in utero* induces postpubertal prostatitis and reduces sperm production via a reversible hormone-regulated mechanism. *Endocrinology* 151(2):783–792.

Creasy, D.M. and Foster, P.M.D. (1984) The morphological development of glycol ether-induced testicular atrophy in the rat. *Exp. Mol. Pathol.* 40(2):169–176.

Cummings, A.M. (1997) Methoxychlor as a model for environmental estrogens. *Crit. Rev. Toxicol.* 27(4):367–379.

Cummings, A.M. and Gray, L.E., Jr. (1989) Antifertility effect of methoxychlor in female rats: dose- and time-dependent blockade of pregnancy. *Toxicol. Appl. Pharmacol.* 97(3):454–462.

Danielsson, B.R., Dencker, L., Lindgren, A., and Tjälve, H. (1984) Accumulation of toxic metals in male reproductive organs. *Arch. Toxicol. Suppl.* 7:177–180.

Danscher, G.P., Horstedt-Bindslev, P., and Rungby, J. (1990) Traces of mercury in organs from primates with amalgam fillings. *Exp. Mol. Pathol.* 52:291–299.

Das, S.K., Paria, B.C., Johnson, D.C., and Dey, S.K. (1997) Embryo-uterine interactions during implantation: potential sites of interference by environmental toxins. In: Boekelheide, K., Chapin, R.E., Hoyer, P.B., and Harris, C., editors. *Comprehensive Toxicology*, vol. 10, New York, NY: Elsevier Science, pp. 317–328.

Davis, B.J., Price, H.C., O'Connor, R.W., Fernando, R., Rowland, A.S., and Morgan, D.L. (2001) Mercury vapor and female reproductive toxicity. *Toxicol. Sci.* 59:291–296.

Davis, B.J., Almekinder, J.L., Flagler, N., Travlos, G., Wilson, R., and Maronpot, R.R. (1997) Ovarian luteal cell toxicity of ethylene glycol monomethyl ether and methoxyacetic acid *in vivo* and *in vitro*. *Toxicol. Appl. Pharmacol.* 142(2):328–337.

Dawson, B.V., Johnson, P.D., Goldberg, S.J., and Ulreich, J.B. (1990) Cardiac teratogenesis of trichloroethylene and dichloroethylene in a mammalian model. *J. Am. Coll. Cardiol.* 16(5):1304–1309.

Dearfield, K.L. Douglas, G.R., Ehling, U.H., Moore, M.M., Sega, G.A., and Brusick, D.J. (1995) Acrylamide: a review of its genotoxicity and an assessment of heritable genetic risk. *Mutat. Res.* 330:71–99.

Diamanti-Kandarakis, E., Bourguignon, S., Soto, A.M., Zoeller, R.T., and Gore, A.C. (2009) Endocrine-disrupting chemicals: an Endocrine Society statement. *Endocr. Rev.* 30(4):293–342.

Dietrich, K.N., Krafft, K.M., Bornschein, R.L., Hammond, P.B., Berger, O., Succop, P.A., and Bier, M. (1987) Low-level fetal lead exposure effect on neurobehavioral development in early infancy. *Pediatrics* 80(5):721–730.

Donald, J.M., Cutler, M.G., and Moore, M.R. (1986) Effects of lead in the laboratory mouse. I. Influence of pregnancy upon absorption, retention, and tissue distribution of radiolabeled lead. *Environ. Res.* 41(2):420–431.

Dostal, L.A., Chapin, R.E., Stefanski, S.A., Harris, M.W., and Schwetz, B.A. (1988) Testicular toxicity and reduced Sertoli cell numbers in neonatal rats by di(2-ethylhexyl)phthalate and the recovery of fertility as adults. *Toxicol. Appl. Pharmacol.* 95:104–121.

Driscoll, C.D., Streissguth, A.P., and Riley, E.P. (1990) Prenatal alcohol exposure: comparability of effects in humans and animal models. *Neurotoxicol. Teratol.* 12(3):231–237.

Dyer, C.A. (2007) Heavy metals as endocrine-disrupting chemicals. In: Gore, A.C., editor. *Endocrine-Disrupting Chemicals: From Basic Research to Clinical Practice*, Totowa, NJ: Humana Press, Inc., pp. 111–133.

Faqi, A.S., Dalsenter, P.R., Merker, H.-J., and Chahoud, I. (1998) Reproductive toxicity and tissue concentrations of low doses of 2,3,7,8-tetrachlorodibenzo-p-dioxin in male offspring rats exposed throughout pregnancy and lactation. *Toxicol. Appl. Pharmacol.* 150:383–392.

Fenster, L., Hubbard, A.E., Windham, G.C., Waller, K.O., and Swan, S.H. (1997) A prospective study of work-related physical exertion and spontaneous abortion. *Epidemiology* 8(1):66–74.

Figà-Talamanca, I. (2006) Occupational risk factors and reproductive health of women. *Occup. Med.* 56:521–531.

Forkert, P.-G., Lash, L.H., Nadeau, V., Tardif, R., and Simmonds, A. (2002) Metabolism and toxicity of trichloroethylene in epididymis and testis. *Toxicol. Appl. Pharmacol.* 182:244–254.

Forkert, P.-G., Lash, L., Tardif, R., Tanphaichitr, N., Vandovoort, C., and Moussa, M. (2003) Identification of trichloroethylene and its metabolites in human seminal fluid of workers exposed to trichloroethylene. *Drug Metab. Dispos.* 31:306–311.

Foster, P.M.D. and Gray, L.E., Jr. (2008) Toxic responses of the reproductive system. In: Klaassen, C.D., editor. *Cassarett and Doull's Toxicology the Basic Science of Poisons*, New York, NY: McGraw-Hill, pp. 761–806.

Garcia-Morales, P., Saceda, M., Kenney, N., Kim, N., Salomon, D.S., Gottardis, M.M., Solomon, H.B., Sholler, P.F., Jordan, V.C., and Martin, M.B. (1994) Effect of cadmium on estrogen receptor levels and estrogen-induced responses in human breast cancer cells. *J. Biol. Chem.* 269(24):16896–16901.

Gerhard, I., Monga, B., Waldbrenner, A., and Runnebaum, B., (1998) Heavy metals and fertility. *J. Toxicol. Environ. Health A* 54:593–611.

Godwin, H.A. (2001) The biological chemistry of lead. *Curr. Opin. Chem. Biol.* 5(2):223–227.

Goldberg, M.S., Mayo, N.E., Levy, A.R., Scott, S.C., and Poítras, B. (1998) Adverse reproductive outcomes among women exposed to low levels of ionizing radiation from diagnostic radiography for adolescent idiopathic scoliosis. *Epidemiology* 9(3):271–278.

Goldberg, S.J., Lebowitz, M.D., Graver, E.J., and Hicks, S. (1990) An association of human congenital cardiac malformations and drinking water contaminants. *J. Am. Coll. Cardiol.* 16(1):155–164.

Goldey, E.S. and Taylor, D.H. (1992) Developmental neurotoxicity following premating maternal exposure to hexachlorobenzene in rats. *Neurotoxicol. Teratol.* 14(1):15–21.

Golub, M.S., Macintosh, M.S., and Baumrind, N. (1998) Developmental and reproductive toxicity of inorganic arsenic: animal studies and human concerns. *J. Toxicol. Environ. Health B Crit. Rev.* 1:199–241.

Gosden, R.G., Wade, J.C., Fraser, H.M., Sandow, J., and Faddy, M.J. (1997) Impact of congenital or experimental hypogonadotrophism on the radiation sensitivity of the mouse ovary. *Hum. Reprod.* 12(11):2483–2488.

Goyer, R.A. (1990) Transplacental transport of lead. *Environ. Health Perspect.* 89:101–105.

Goyer, R.A. (1996) Results of lead research: prenatal exposure and neurological consequences. *Environ. Health Perspect.* 104(10):1050–1054.

Harbison, R.D. (1998) Reproductive toxicology. In: Harbison, R.D., editor. *Hamilton and Hardy's Industrial Toxicology*, 5th ed., Maryland Heights, Missouri: Mosby, pp. 611–624.

Hardin, B.D., Goad, P.T., and Burg, J.R. (1986) Developmental toxicity of diethylene glycol monomethyl ether (diEGME). *Fundam. Appl. Toxicol.* 6:430–439.

Henck, J.W. (1997) *In vivo* animal screening systems. In: Boekelheide, K., Chapin, R.E., Hoyer, P.B., and Harris, C., editors. *Comprehensive Toxicology*, vol. 10, New York, NY: Elsevier Science Inc., pp. 621–635.

Henson, M.C. and Chedrese, P.J. (2004) Endocrine disruption by cadmium, a common environmental toxicant with paradoxical effects on reproduction. *Exp. Biol. Med.* 229:383–392.

Hertz-Picciotto, I. (2000) The evidence that lead increases the risk of spontaneous abortion. *Am. J. Ind. Med.* 38:300–309.

Hoyer, A.P., Grandjean, P., Jorgensen, T., Brock, J.W., and Hartvig, H.B. (1998) Organochlorine exposure and the risk of breast cancer. *Lancet* 352:1816–1820.

Hoyer, P.B. (2005) Damage to ovarian development and function. *Cell Tissue Res.* 322:99–106.

Hoyer, P.B. and Sipes, I.G. (2007) Development of an animal model for ovotoxicity using 4-vinylcyclohexene: a case study. *Birth Defects Res. B Dev. Reprod. Toxicol.* 80:113–125.

Hsieh, G.Y., Wang, J.D., Cheng, T.J., and Chen, P.C. (2005) Prolonged menstrual cycles in female workers exposed to ethylene glycol ethers in the semiconductor manufacturing industry. *Occup. Environ. Med.* 62:510–516.

Ichihara, G. (2005) Neuro-reproductive toxicities of 1-bromopropane and 2-bromopropane. *Int. Arch. Occup. Environ. Health* 78(2):79–96.

Ichihara, G., Asaeda, N., Kumazawa, T., Tagawa, Y., Kamijimaa, M., Yu, X., Kondo, H., Nakajima, T., Kitoh, J., Yu, J., Moon,

Y.H., Hisanaga, N., and Takeuchi, Y. (1996) Testicular toxicity of 2-bromopropane. *J. Occup. Health* 38:205–206.

Ichihara, G., Yu, X., Kitoh, J., Asaeda, N., Kumazawa, T., Iwai, H., and Shibata, E. (2000) Reproductive toxicity of 1-bromopropane, a newly introduced alternative to ozone layer depleting solvents, in male rats. *Toxicol. Sci.* 54:416–423.

Iioka, H., Moriyama, I., Oku, M., Hino, K., Itani, Y., Okamura, Y., and Ichijo, M. (1987) The effect of inorganic mercury on placental amino acid transport using microvillus membrane vesicles. *Nihon Sanka Fujinka Gakkai Zasshi* 39(2):202–206.

Infante, P.F. (1976) Oncogenic and mutagenic risks in communities with polyvinyl chloride production facilities. *Ann. N.Y. Acad. Sci.* 271:49–57.

IARC (International Agency for Research on Cancer) (1994a) 4-Vinylcyclohexene. *IARC Monogr. Eval. Carcinog. Risks Hum.* 60:347–359.

IARC (International Agency for Research on Cancer) (1994b) 4-Vinylcyclohexene diepoxide. *IARC Monogr. Eval. Carcinog. Risks Hum.* 60:361–372.

IPCS (International Programme on Chemical Safety) (2002) *Global Assessment of the State-of-the-Science of Endocrine Disruptors. International Programme on Chemical Safety*, Geneva, Switzerland: World Health Organization.

Irvine, L.F. (2003) Relevance of the developmental toxicity of ethanol in the occupational setting: a review. *J. Appl. Toxicol.* 23(5):289–299.

Jensen, T.K., Bonde, J.P., and Joffe, M. (2006) The influence of occupational exposure on male reproductive function. *Occup. Med.* 56:544–553.

Jensh, R.R. and Brent, R.L. (1988) Effects of prenatal X-irradiation on postnatal testicular development and function in the Wistar rat: development/teratology/behavior/radiation. *Teratology* 28:443–449.

John, J.A., Smith, F., Leong, B., and Schwetz, B. (1977) The effects of maternally inhaled vinyl chloride on embryonal and fetal development in mice, rats, and rabbits. *Toxicol. Appl. Pharmacol.* 39(3):497–513.

Johnson, E.M. (1986) The scientific basis for multigeneration safety evaluations. *Int. J. Toxicol.* 5:197–201.

Johnson, L., Welsh, T.H., and Wilker, C.E. (1997) Anatomy and physiology of the male reproductive system and potential targets of toxicants. In: Boelkelheide, K., Chapin, R.E., Hoyer, P.B., and Harris, C., editors. *Comprehensive Toxicology*, vol. 10, New York, NY: Elsevier Science Inc., pp. 5–61.

Johnson, M.D., Kenney, N., Stoica, A., Hilakivi-Clarke, L., Singh, B., Chepko, G., Clarke, R., Sholler, P.F., Lirio, A.A., Foss, C., Reiter, R., Trock, B., Paik, S., and Martin, M.B. (2003) Cadmium mimics the *in vivo* effect of estrogen in the uterus and mammary gland. *Nat. Med.* 9(8):1081–1084.

Johnson, P.D., Dawson, B.V., and Goldberg, S.J. (1998) Cardiac teratogenicity of trichloroethylene metabolites. *J. Am. Coll. Cardiol.* 32(2):540–545.

Junaid, M., Chowdhuri, D.K., Shanker, N.R., and Saxena, D.K. (1997) Lead-induced changes in ovarian follicular development and maturation in mice. *J. Toxicol. Environ. Health* 50(1):31–40.

Kaiser, J. (2005) Toxicology panel finds no proof that phthalates harm infant reproductive systems. *Science* 310(5747):422.

Kavlock, R.J., Daston, G.P., DeRosa, C., Fenner-Crisp, P., Gray, L.E., Kaatari, S., Lucier, G., Luster, M., Mac, M.J., Maczka, C., Miller, R., Moore, J., Rolland, R., Scott, G., Sheehan, D.M., Sinks, T., and Tilson, H.A. (1996) Research needs for the risk assessment of health and environmental effects of endocrine disruptors: a report of the U.S. EPA-sponsored workshop. *Environ. Health Perspect.* 104(Suppl. 4):715–736.

Kavlock, R., Boekelheide, K., Chapin, R., Cunningham, M., Faustman, E., Foster, P., Golub, M., Henderson, R., Hinberg, I., Little, R., Seed, J., Shea, K., Tabacova, S., Tyl, R., Williams, P., and Zacharewski, T. (2002) NTP center for the evaluation of risks to human reproduction: phthalates expert panel report on the reproductive and developmental toxicity of di-*n*-butyl phthalate. *Reprod. Toxicol.* 16:489–527.

Kelce, W.R. (1997) The Leydig cell as a target for toxicants. In: Boekelheide, K., Chapin, R.E., Hoyer, P.B., and Harris, C., editors. *Comprehensive Toxicology*, vol. 10, New York, NY: Elsevier Science, pp. 165–179.

Kim, Y., Jung, K., Hwang, T., Jung, G., Kim, H., Park, J., Kim, J., Parl, J., Park, D., Park, S., Choi, K., and Moon, Y. (1996) Hematopoietic and reproductive hazards of Korean electronic workers exposed to solvents containing 2-bromopropane. *Scand. J. Work Environ. Health* 22(5):387–391.

Klaassen, C.D. (2006) Heavy metals and heavy-metal antagonists. In: Brunton, L.L., editor. *Goodman & Gilman's The Pharmacological Basis of Therapeutics*, 11th ed., New York, NY: McGraw-Hill, pp. 1753–1775.

Kluwe, W.M., Lamb IV, J.C., Greenwell, A.E., and Harrington, F.W. (1983) 1,2-Dibromo-3-chloropropane (DBCP)-induced infertility in male rats mediated by a post-testicular effect. *Toxicol. Appl. Pharmacol.* 71(2):294–298.

Kortenkamp, A. (2008) Low dose mixture effects of endocrine disruptors: implications for risk assessment and epidemiology. *Int. J. Androl.* 31:233–240.

Krsmanović, L. Z., Stojilković, S. S., Merelli, F., DuFour, S.M., Virmani, M.A., and Catt, K.J. (1992) Calcium signaling and episodic secretion of gonadotropin-releasing hormone in hypothalamic neurons. *Proc. Natl. Acad. Sci. U. S. A.* 89(18):8462–8466.

Ku, W.W., Chapin, R.E., Wine, R.N., and Gladen, B.C. (1993) Testicular toxicity of boric acid (BA): relationship of dose to lesion development and recovery in the F344 rat. *Reprod. Toxicol.* 7(4):306–319.

Lamb, J.C., Gulato, D.K., Russell, V.S., Hommel, L., and Sabharwal, P.S. (1984) Reproductive toxicity of ethylene glycol monoethyl ether tested by continuous breeding of CD-1 mice. *Environ. Health Perspect.* 57:85–90.

Laws, S.C., Carey, S.A., Ferrell, J.M., Bodman, G.J., and Cooper, R.L. (2000) Estrogenic activity of octylphenol, nonylphenol, bisphenol A and methoxychlor in rats. *Toxicol. Sci.* 54:154–167.

Levallois, P., Lavoia, M., Goulet, L., Nantel, A.J., and Gingras, S. (1991) Blood lead levels in children and pregnant women living near a lead reclamation plant. *Can. Med. Assoc.* 44(7):877–885.

Liu, J., Goyer, R.A., and Waalkes, M.P. (2008) Metals. In: Klaassen, C.D., editor. *Cassarett and Doull's Toxicology the Basic Science of Poisons*, New York, NY: McGraw-Hill, pp. 931–979.

Lushbaugh, C.C. and Casarett, G.W. (1976) The effects of gonadal irradiation in clinical radiation therapy: a review. *Cancer* 37:1111–1125.

Main, K.M., Kiviranta, H., Virtanen, H.E., Sundqvist, E., Tuomisto, J.T., Tuomisto, J., Vartianen, T., Skakkebæk, N.E., and Toppari, J. (2007) Flame retardants in placenta and breast milk and cryptorchidism in newborn boys. *Environ. Health Perspect.* 115(10):1519–1526.

Maldonado, G., Delzell, E., Tyl, R.W., and Sever, L.E. (2003) Occupational exposure to glycol ethers and human congenital malformations. *Int. Arch. Occup. Environ. Health* 76:405–423.

Maltoni, C., Lefemine, G., Ciliberti, A., Cotti, G., and Carretti, D. (1981) Carcinogenicity bioassays of vinyl chloride monomer: a model of risk assessment on an experimental basis. *Environ. Health Perspect.* 41:3–30.

Marbury, M.C. (1992) Relationship of ergonomic stressors to birthweight and gestational age. *Scand. J. Work Environ. Health* 18:73–83.

Martinez, E.M. and Swartz, W.J. (1991) Effects of methoxychlor on the reproductive system of the adult female mouse. 1. Gross and histologic observations. *Reprod. Toxicol.* 5(2):139–147.

Massányi, P., Lukác, N., Uhrín, V., Toman, R., Pivko, J., Rafay, J., Forgács, Zs., and Somosy, Z. (2007) Female reproductive toxicology of cadmium. *Acta Biol. Hung.* 58(3):287–299.

Matsuda, Y., Seki, N., Utsugi-Takeuchi, T., and Tobari, I. (1989) X-ray- and mitomycin C (MMC)-induced chromosome aberrations in spermiogenic germ cells and the repair capacity of mouse eggs for the X-ray and MMC damage. *Mutat. Res.* 211:65–75.

Mattison, D.R., Nightingale, M.S., and Shiromizu, K. (1983) Effects of toxic substances on female reproduction. *Environ. Health Perspect.* 48:43–52.

May, P.A., Gossage, J.P., Kalberg, W.O., Robinson, L.K., Buckley, D., Manning, M., and Hoyme, H.E. (2009) Prevalence and characteristics of FASD from various research methods with an emphasis on recent in-school studies. *Dev. Disabil. Res. Rev.* 15(3):176–192.

McDonald, A.D., McDonald, J.C., Armstrong, B., Cherry, N.M., Côté, R., LaVoie, J., Nolin, A.D., and Robert, D. (1988) Fetal death and work in pregnancy. *Br. J. Ind. Med.* 45:148–157.

McGivern, R.F., Sokol, R.Z., and Berman, N.G. (1991) Prenatal lead exposure in the rat during the third week of gestation: long-term behavioral, physiological, and anatomical effects associated with reproduction. *Toxicol. Appl. Pharmacol.* 110(2):206–215.

Meirow, D. and Nugent, D. (2001) The effects of radiotherapy and chemotherapy on female reproduction. *Hum. Reprod. Update* 7(6):535–543.

Messite, J. and Bond, M.B. (1988) Reproductive toxicology and occupational exposure. In: Zenz, C., editor. *Occupational Medicine: Principles and Practical Applications*, 2nd ed., Chicago, IL: Year Book Medical Publishers, pp. 59–129.

Moore, J.A. and an Expert Scientific Committee. (1997) An assessment of boric acid and borax using the IEHR evaluative process for assessing human developmental and reproductive toxicity of agents. *Reprod. Toxicol.* 11(1):123–160.

Morford, L.L., Henck, J.W., Breslin, W.J. and DeSesso, J.M. (2004) Hazard identification and predictability of children's health risk from animal data. *Environ. Health Perspect.* 112:266–271.

Murray, F.J. (1995) A human health risk assessment of boron (boric acid and borax) in drinking water. *Regul. Toxicol. Pharmacol.* 22:221–230.

Mylchreest, E., Sar, M., Cattley, R.C., and Foster, P.M.D. (1999) Disruption of androgen-regulated male reproductive development by di(n-butyl) phthalate during late gestation in rats is different from flutamide. *Toxicol. Appl. Pharmacol.* 156:81–95.

Nakamura, K., Yasunaga, Y., Ko, D., Xu, L.L., Moul, J.W., Peehl, D.M., Srivastava, S., and Rhim, J.S. (2002) Cadmium-induced neoplastic transformation of human prostate epithelial cells. *Int. J. Oncol.* 20:543–547.

Nampoothiri, L.P. and Gupta, S. (2006) Simultaneous effect of lead and cadmium on granulosa cells: a cellular model for ovarian toxicity. *Reprod. Toxicol.* 21:179–185.

National Research Council (2006) *Assessing the Human Health Risks of Trichloroethylene: Key Scientific Issues*, Washington, DC: The National Academies Press, pp. 182–212.

Nelson, B.K. and Brightwell, W.S. (1984) Behavioral teratology of ethylene glycol monomethyl and monoethyl ethers. *Environ. Health Perspect.* 57:43–46.

Newbold, R.R., Padilla-Banks, E., and Jefferson, W.N. (2006) Adverse effects of the model environmental estrogen diethylstilbestrol are transmitted to subsequent generations. *Endocrinology* 147(Suppl. 6):S11–S17.

Nurminen, T. (1995) Female noise exposure, shift work, and reproduction. *J. Occup. Environ. Med.* 37(8):945–950.

Nurminen, T., Lusa, K., Ilmarinen, J., and Kurppa, K. (1989) Physical work load, fetal development and course of pregnancy. *Scand. J. Work Environ. Health* 15:404–414.

Oliva, A., Spira, A., and Multigner, L. (2001) Contribution of environmental factors to the risk of male infertility. *Hum. Reprod.* 8:1768–1776.

Omura, M., Romero, Y., Zhao, M., and Inoue, N. (1999) Histopathological evidence that spermatogonia are the target cells of 2-bromopropane. *Toxicol. Lett.* 104:19–26.

Paksy, K., Varga, B., Horvath, T., Tatrai, E., and Ungvary, G. (1989) Acute effects of cadmium on preovulatory serum FSH, LH, and prolactin levels and on ovulation and ovarian hormone secretion in estrous rats. *Reprod. Toxicol.* 3(4):241–247.

Park, J.-S., Kim, Y., Park, D.W., Choi, K.S., Park, S.-H., and Moon, Y.-H. (1997) An outbreak of hematopoietic and reproductive disorders due to solvents containing 2-bromopropane in an electronic factory, South Korea: epidemiological survey. *J. Occup. Health* 39:138–143.

Parker, S.E., Mai, C.T., Canfield, M.A., Rickard, R., Wang, Y., Meyer, R.E., Anderson, P., Mason, C.A., Collins, J.S., Kirby, J.S., and Correa, A. (2010) Updated national prevalence estimates for selected birth defects in the United States, 2004–2006. *Birth Defects Res. A Clin. Mol. Teratol.* 88(12):1008–1016.

Parks, L.G., Ostby, J.S., Lambright, C.R., Abbott, B.D., Klinefelter, G.R., Barlow, N.J., and Gray, L.E. (2000) The plasticizer diethyhexyl phthalate induces malformations by decreasing fetal testosterone synthesis during sexual differentiation in the male rat. *Toxicol. Sci.* 58:339–349.

Pearson, M.A., Hoyme, H.E., Seaver, L.H., and Rimsza, M.E. (1994) Toluene embryopathy: delineation of the phenotype and comparison with fetal alcohol syndrome. *Pediatrics* 93(2):211–215.

Peden-Adams, M.M., Eudaly, J.G., Heeseman, L.M., Smythe, J., Miller, J., Gilkeson, G.S., and Keil, D.E. (2006) Developmental immunotoxicity of trichloroethylene (TCE): studies in B6C3F1 mice. *J. Environ. Sci Health A Tox. Hazard. Subst. Environ. Eng.* 41:249–271.

Perreault, S.D. (1997) The mature spermatozoon as a target for reproductive toxicants. In: Boekelheide, K., Chapin, R.E., Hoyer, P.B., and Harris, C., editors. *Comprehensive Toxicology*, vol. 10, New York, NY: Elsevier Science, pp. 165–179.

Perry, M.J. (2008) Review of the effects of environmental and occupational pesticide exposure on human sperm: a systematic review. *Hum. Reprod. Update* 14(3):233–242.

Peterson, R.E., Theobald, H.M., and Kimmel, G.L. (1993) Developmental and reproductive toxicity of dioxins and related compounds: cross-species comparisons. *Crit. Rev. Toxicol.* 23(3):283–335.

Piasek, M., Blanuša, M., Kostial, K., and Laskey, J.W. (2001) Placental cadmium and progesterone concentrations in cigarette smokers. *Reprod. Toxicol.* 15:673–681.

Potashnik, G. (1983) A four-year reassessment of workers with dibromochloropropane-induced testicular dysfunction. *Andrologia* 15(2):164–170.

Potashnik, G., Ben-Aderet, N., Israeli, R., Yanai-Inbar, I., and Sober, I. (1978) Suppressive effect of 1,2-dibromo-3-chloropropane on human spermatogenesis. *Fertil. Steril.* 30:444–447.

Rao, K.S., Burek, J.D. Murray, F.J., John, J.A., Schwetz, B.A., Bell, T.J., Potts, W.J., and Parker, C.M. (1983) Toxicologic and reproductive effects of inhaled 1,2-dibromo-3-chloropropane in rats. *Fundam. Appl. Toxicol.* 3(2):104–110.

Ratcliffe, J.M., Schrader, S.M., Steenland, K., Clapp, D.E., Turner, T., and Hornung, R.W. (1987) Semen quality in papaya workers with long term exposure to ethylene dibromide. *Br. J. Ind. Med.* 44:317–326.

Rehm, S. and Waalkes, M.P. (1988) Cadmium-induced ovarian toxicity in hamsters, mice, and rats. *Fundam. Appl. Toxicol.* 10(4):635–647.

Rice, J.M. (1981) Prenatal susceptibility to carcinogenesis by xenobiotic substances including vinyl chloride. *Environ. Health Perspect.* 41:179–188.

Richburg, J.H. and Boekelheide, K. (1997) The Sertoli cell as a target for toxicants. In: Boekelheide, K., Chapin, R.E., Hoyer, P.B., and Harris, C., editors. *Comprehensive Toxicology*, vol. 10, New York, NY: Elsevier Science Inc., pp. 127–138.

Richter, C.A., Birnbaum, L.S., Farabollini, F., Newbold, R.R., Rubin, B.S., Talsness, C.E., Vandenbergh, J.G., Walser-Kuntz, D.R., and vom Saal, F.S. (2007) *In vivo* effects of bisphenol A in laboratory rodent studies. *Reprod. Toxicol.* 24(2):199–224.

Riley, E.P., Infante, M.A. and Warren, K.R. (2011) Fetal alcohol spectrum disorders: an overview. *Neuropsychol. Rev.* 21:73–80.

Roberts, L.G., Bevans, A.C., and Schreiner, C.A. (2003) Development and reproductive toxicity evaluation of toluene vapor in the rat. I. Reproductive toxicity. *Reprod. Toxicol.* 17:649–658.

Rogers, J.M. and Kavlock, R.J. (2008) Developmental toxicology. In: Klaassen, C.D., editor. *Cassarett and Doull's Toxicology the Basic Science of Poisons*, New York, NY: McGraw-Hill, pp. 415–451.

Rosenman, K.D., Rizzo, J.E., Conomos, M.G., and Halpin, G.J. (1989) Central nervous system malformations in relation to polyvinyl chloride production facilities. *Arch. Environ. Health* 44:279–282.

Rowland, A.S., Baird, D.D., Weinberg, C.R., Shore, D.L., Shy, C.M., and Wilcox, A.J. (1994) The effect of occupational exposure to mercury vapour on the fertility of female dental assistants. *Occup. Environ. Med.* 51:28–34.

Safe, S.H. (2000) Endocrine disruptors and human health—is there a problem? An update. *Environ. Health Perspect.* 108(6):487–493.

Sakamoto, J., Kurosaka, Y., and Hashimoto, K. (1988) Histological changes of acrylamide-induced testicular lesions in mice. *Exp. Mol. Pathol.* 48(3):324–334.

Satarug, S. and Moore, M.R. (2004) Adverse health effects of chronic exposure to low-level cadmium in foodstuffs and cigarette smoke. *Environ. Health Perspect.* 112(10):1099–1103.

Schantz, S.L., Widholm, J.J., and Rice, D.C. (2003) Effects of PCB exposure on neuropsychological function in children. *Environ. Health Perspect.* 111(3):357–376.

Schenker, M.B., Gold, E.B., Beaumont, J.J., Eskenazi, B., Katharine, S.H., Lasley, B.L., McCurdy, S.A., Samuels, S.J., Saiki, C.L., and Swan, S.H. (1995) Association of spontaneous abortion and other reproductive effects with work in the semiconductor industry. *Am. J. Ind. Med.* 28(6):639–659.

Schnaas, L., Rothenberg, S.J., Flores, M.-F., Martinez, S., Harnandez, C., Osorio, E., Velasco, S.R., and Perroni, E. (2006) Reduced intellectual development in children with prenatal lead exposure. *Environ. Health Perspect.* 114(5):791–797.

Schoeters, G., Den Hond, E., Zuurbier, M., Naginiene, R., Van Den Hazel, P., Stilianakis, N., Ronchetti, R., and Koppe, J.G. (2006) Cadmium and children: exposure and health effects. *Acta Paediatr. Suppl.* 95(453):50–54.

Schrag, S.D. and Dixon, R.L. (1985) Occupational exposures associated with male reproductive dysfunction. *Ann. Rev. Pharmacol. Toxicol.* 25:567–592.

Schull, W.J., Norton, S., and Jensh, R.P. (1990) Ionizing radiation and the developing brain. *Neurotoxicol. Teratol.* 12:249–260.

Schuurs, A.H.B. (1999) Reproductive toxicity of occupational mercury: a review of the literature. *J. Dent.* 27(4):249–256.

Scott, W.J., Fradkin, R., Wittfoht, W., and Nau, H. (1989) Teratologic potential of 2-methoxyethanol and transplacental distribution of its metabolite, 2-methoxyacetic acid, in non-human primates. *Teratology* 39:363–373.

Seegmiller, R.E. (1997) Selected examples of developmental toxicants. In: Boekelheide, K., Chapin, R.E., Hoyer, P.B., and Harris, C., editors. *Comprehensive Toxicology*, vol. 10, New York, NY: Elsevier Science, pp. 586–588.

Sega, G.A., Alcota, R.P.V., Tancongco, C.P., and Brimer, P.A. (1989) Acrylamide binding to the DNA and protamine of spermiogenic stages in the mouse and its relationship to genetic damage. *Mutat. Res.* 216(4):221–230.

Sheiner, E.K., Sheiner, E., Hammel, R.D., Potashnik, G., and Carel, R. (2003) Effect of occupational exposures on male fertility: literature review. *Ind. Health* 41:55–62.

Sikka, S.C. and Wang, R. (2008) Endocrine disruptors and estrogenic effects on male reproductive axis. *Asian J. Androl.* 10(1):134–145.

Sikorski, R., Paszkowski, T., Slawiński, P., Szkoda, J., Zmudzki, J. and Skawiński, S. (1989) The intrapartum content of toxic metals in maternal blood and umbilical cord blood. *Ginekol. Pol.* 60(3):151–155.

Skakkebæk, N.E., Rajpert-De Meyts, E. and Main, K.M. (2001) Testicular dysgenesis syndrome: an increasingly common developmental disorder with environmental aspects. *Hum. Reprod.* 16(5):972–978.

Slutsky, M., Levin, J.L., and Levy, B.S. (1999) Azoospermia and oligospermia among a large cohort of DBCP applicators in 12 countries. *Int. J. Occup. Environ. Health* 5:116–122.

Smith, A.H., Marshall, G., Yuan, Y., Ferreccio, C., Liaw, J., von Ehrenstein, O., Steinmaus, C., Bates, M.N., and Selvin, S. (2006) Increased mortality from lung cancer and bronchiectasis in young adults after exposure to arsenic *in utero* and in early childhood. *Environ. Health Perspect.* 114(8):1293–1296.

Smith, M.K., Randall, J.L., Read, E.J., and Stober, J.A. (1989) Teratogenic activity of trichoroacetic acid in the rat. *Teratology* 40(5):445–451.

Sokol, R.Z. (1997) The hypothalamic-pituitary-gonadal axis as a target for toxicants. In: Boekelheide, K., Chapin, R.E., Hoyer, P.B., and Harris, C., editors. *Comprehensive Toxicology*, vol. 10, New York, NY: Elsevier Science, pp. 87–98.

Sokol, R.Z., Madding, C.E., and Swerdloff, R.S. (1985) Lead toxicity in the hypothalamic-pituitary-testicular axis. *Biol. Reprod.* 33:722–728.

Soukupova, R.A. and Dostal, M. (1991) Developmental toxicity of cadmium in mice. I. Embryotoxic effects. *Funct. Dev. Morphol.* 1(2):3–9.

Stockholm Convention (2001) *Stockholm Convention on Persistent Organic Pollutants*, Geneva, Switzerland: Secretariat of the Stockholm Convention.

Stratton, K.R., Howe, C.J., and Battaglia, F.C. (1996) *Fetal Alcohol Syndrome: Diagnosis, Prevention, and Treatment*, Institute of Medicine 1996 report on FAS. Washington, DC: National Academy Press.

Sublet, V.H., Zenick, H., and Smith, M.K. (1989) Factors associated with reduced fertility and implantation rates in females mated to acrylamide-treated rats. *Toxicology* 55(1–2):53–67.

Suzuki, K. and Martin, P.M. (1994) Neurotoxicants and developing brain. In: Harry, G.J., editor. *Developmental Neurotoxicology*, New York, NY: CRC Press, pp. 9–32.

Swan, S.H., Kruse, R.L., Barr, D.B., Drobnis, E.Z., Redmon, J.B., Wang, C., Brazil, C., and Overstreet, J.W. (2003) Semen quality in relation to biomarkers of pesticide exposure. *Environ. Health Perspect.* 111(12):1478–1484.

Sweeney, A.M. and LaPorte, R.E. (1991) Advances in early fetal loss research: importance for risk assessment. *Environ. Health Perspect.* 90:165–169.

Teaf, C.M., Bishop, J.B., and Harbison, R.D. (1990) Potentiation of ethyl methane sulfonate-induced germ cell mutagenesis and depression of glutathione in male reproductive tissues by 1,2-dibromoethane. *Teratog. Carcinog. Mutagen.* 10: 427–438.

Teitelbaum, D.T. (1999) The toxicology of 1,2-dibromo-3-chloropropane (DBCP): a brief review. *Int. J. Occup. Environ. Health* 5:122–126.

Theobald, H.M. and Peterson, R.E. (1997) *In utero* and lactational exposure to 2,3,7,8-tetracholrodibenzo-p-dioxin: effects on development of the male and female reproductive system of the mouse. *Toxicol. Appl. Pharmacol.* 145:124–135.

Thériault, G., Iturra, H., and Gingras, S. (1983) Evaluation of the association between birth defects and exposure to ambient vinyl chloride. *Teratology* 27:359–370.

Thompson, J. and Bannigan, J. (2008) Cadmium: toxic effects on the reproductive system and the embryo. *Reprod. Toxicol.* 25:304–315.

Tiemann, U. (2008) *In vivo* and *in vitro* effects of the organochlorine pesticides DDT, TCPM, methoxychlor, and lindane on the female reproductive tract of mammals: a review. *Reprod. Toxicol.* 25:316–326.

Tilson, H.A., Jacobson, J.L., and Rogan, W.J. (1990) Polychlorinated biphenyls and the developing nervous system: cross-species comparisons. *Neurotoxicol. Teratol.* 12:239–248.

Toraason, M. and Breitenstein, M. (1988) Prenatal ethylene glycol ether (EGME) exposure produces electrocardiographic changes in the rat. *Toxicol. Appl. Pharmacol.* 95:321–327.

Toth, G.P., Stober, J.A., Zenick, H., Read, E.J., Christ, S.A., and Smith, M.K. (1991) Correlation of sperm motion parameters with fertility in rats treated subchronically with epichlorohydrin. *J. Androl.* 12(1):54–61.

Ulbrich, B. and Stahlmann, R. (2004) Developmental toxicity of polychlorinated biphenyls (PCBs): a systematic review of experimental data. *Arch. Toxicol.* 78:252–268.

U.S. EPA (Environmental Protection Agency) (1984) Health Effects Assessment for Glycol Ethers, Cincinnati, OH, and Washington, DC: U.S. EPA.

U.S. EPA (Environmental Protection Agency) (2000a) *EPA R.E.D. Facts: Vinclozolin*, Washington, DC: U.S. Environmental Protection Agency, EPA-738-F-00-021.

U.S. EPA (Environmental Protection Agency) (2000b) Toxicological Review of Vinyl Chloride. In: *Support of Information on the Integrated Risk Information System (IRIS)*, Washington, DC: U.S. Environmental Protection Agency.

U.S. EPA (Environmental Protection Agency) (2002) *Health Assessment of 1,3-Butadiene*, Washington, DC: U.S. Environmental Protection Agency, Office of Research and Development, National Center for Environmental Assessment, Washington Office, EPA/600/P-98/001F.

U.S. EPA (Environmental Protection Agency) (2004) *Methoxychlor Reregistration Eligibility Decision (RED) June 30, 2004 EPA Publication No. EPA 738-R-04-010.*

U.S. FDA (Food and Drug Administration) (2013) *Update on Bisphenol A (BPA) for Use in Food Contact Applications*, Silver Spring, MD: U.S. FDA.

Varga, B., Zsolnai, B., Paksy, K., Náray, M., and Ungváry, G. (1993) Age dependent accumulation of cadmium in the human ovary. *Reprod. Toxicol.* 7(3):225–228.

Veulemans, H., Steeno, O., Maschelein, R., and Groeseneken, D. (1993) Exposure to ethylene glycol ethers and spermatogenic disorders in man: a case-control study. *Br. J. Ind. Med.* 50:71–78.

Waalkes, M.P. (2003) Cadmium carcinogenesis. *Mutat. Res.* 533:107–120.

Waalkes, M.P., Rehm, S., and Devor, D.E. (1997) The effects of continuous testosterone exposure on spontaneous and cadmium-induced tumors in the male Fischer (F344/NCr) rat: loss of testicular response. *Toxicol. Appl. Pharmacol.* 142:40–46.

Waalkes, M.P., Ward, J.M., Liu, J., and Diwan, B.A. (2003) Transplacental carcinogenicity of inorganic arsenic in the drinking water: induction of hepatic, ovarian, pulmonary, and adrenal tumors in mice. *Toxicol. Appl. Pharmacol.* 186:7–17.

Wallace, W.H.B., Shalet, S.M., Hendry, J.H., Morris-Jones, P.H., and Gattananeni, H.R. (1989) Ovarian failure following ovarian irradiation in childhood: the radiosensitivity of human oocyte. *Br. J. Radiol.* 62:995–998.

Warner, P., Wiley, L.M., Oudiz, D.J., Overstreet, J.W., and Raabe, O.J. (1991) Paternally inherited effects of γ radiation on mouse preimplantation development detected by the chimera assay. *Radiat. Res.* 128(1):48–58.

Watanabe, A., Nakano, Y., Endo, T., Sato, N., Kai, K., and Shiraiwa, K. (2000) Collaborative work to evaluate toxicity on male reproductive organs by repeated dose studies in rats. 27) Repeated toxicity study on ethylene glycol monomethyl ether for 2 and 4 weeks to detect effects on male reproductive organs in rats. *J. Toxicol. Sci.* 25:259–266.

Watanabe, T, Matsuhashi, K., and Takayama, S. (1990) Placental and blood-brain barrier transfer following prenatal and postnatal exposures to neuroactive drugs: relationship with partition coefficient and behavioral teratogenesis. *Toxicol. Appl. Pharmacol.* 105(1):66–77.

Weisenberg, E. (1986) Hexachlorobenzene in human milk: a polyhalogenated risk. *IARC Sci. Pub.* 77:193–200.

Whelan, E.A. (1997) Risk assessment studies: epidemiology. In: Klaassen, C.D., editor. *Cassarett and Doull's Toxicology the Basic Science of Poisons*, New York, NY: McGraw-Hill, pp. 359–365.

Whorton, M.D., Krauss, R.M., and Marshall, S. (1977) Infertility in male pesticide workers. *Lancet* 2:1259–1261.

Whorton, M.D. and Milby, T.H. (1980) Recovery of testicular function among DBCP exposed workers. *J. Occup. Med.* 22:177–179.

Wide, M. (1980) Interference of lead with implantation in the mouse: effect of exogenous oestradiol and progesterone. *Teratology* 21(2):187–191.

Wide, M. and Nilsson, B.O. (1979) Interference of lead with implantation in the mouse: a study of the surface ultrastructure of blastocysts and endometrium. *Teratology* 20(1):101–113.

Wier, P.J., Miller, R.K., Maulik, D., and di Sant'Agnese, P.A. (1990) Toxicity of cadmium in the perfused human placenta. *Toxicol. Appl. Pharmacol.* 105:156–171.

Wilkins-Haug, L. (1997) Teratogen update: toluene. *Teratology* 55(2):145–151.

Wilson, J.G. (1977) Current status of teratology. General principles and mechanisms derived from animal studies. In: Wilson, J.G. and Fraser, F.C., editors. *Handbook of Teratology*, New York, NY: Plenum Press, pp. 47–74.

Wolf, C., Lambright, C., Mann, P., Price, M., Cooper, R.L., Ostby, J., and Gray, L.E., Jr. (1999) Administration of potentially antiandrogenic pesticides (procymidone, linuron, iprodione, chlozolinate, p,p'-DDE, and ketoconazole) and toxic substances (dibutyl- and diethylhexyl phthalate, PCB 169, and ethane dimethane sulphonate) during sexual differentiation produces diverse profiles of reproductive malformations in the male rat. *Toxicol. Ind. Health* 15(1–2):94–118.

Working, P.K. (1988) Male reproductive toxicology: comparison of the human to animal models. *Environ. Health Perspect.* 77:37–44.

World Health Organization (2003) *1,2-Dibromo-3-chloropropane in Drinking-Water. Originally Published in Guidelines for Drinking-Water Quality*, Health criteria and other supporting information. 2nd ed., vol. 2, Geneva, Switzerland: World Health Organization, p. 1996.

World Health Organization (2005) *Trichloroethylene in Drinking-Water*, Geneva, Switzerland: World Health Organization.

World Health Organization (2012) Cadmium and cadmium compounds. *IARC Monogr. Eval. Carcinog. Risks Hum.* 58:119–237.

World Health Organization (2013) *Health Topics: Infertility*, Geneva, Switzerland: Department of Reproductive Health and Research, World Health Organization.

Wyrobek, A.J., Gordon, L.A., Burkhart, J.G., Francis, M.W., Kapp, R.W., Jr., Letz, G., Malling, H.V., Topham, J.C., and Whorton, M.D. (1983) An evaluation of human sperm as indicators of chemically induced alterations of spermatogenic function. *Mutat. Res.* 115(1):73–148.

Yamada, T., Ichihara, G., Wang, H., Yu, X., Maeda, K., Tsukamura, H., Kamijima, M., Nakajima, T., and Takeuchi, Y. (2003) Exposure to 1-bromopropane causes ovarian dysfunction in rats. *Toxicol. Sci.* 71:96–103.

Yamazaki, J.N. and Schull, W.J. (1990) Perinatal loss and neurological abnormalities among children of the atomic bomb. Nagasaki and Hiroshima revisited, 1949 to 1989. *JAMA* 264:605–609.

Yasuda, I., Yasuda, M., Sumida, H., Tsusaki, H., Arima, A., Ihara, T., Kubota, S., Asaoka, K., Tsuga, K., and Akagawa, Y. (2005) *In utero* and lactational exposure to 2,3,7,8-tetrachlorodibenzo-p-dioxin (TCDD) affects tooth development in rhesus monkeys. *Reprod. Toxicol.* 20:21–30.

Yoshimura, T., Kaneko, S., and Hayabuchi, H. (2001) Sex ratio of those affected by dioxin and dioxin-like compounds: the Yusho, Seveso, and Yucheng incidents. *Occup. Environ. Med.* 58:540–541.

Yu, X., Kamijima, M., Ichihara, G., Li, W., Kitoh, J., Xie, Z., Shibata, E., Hisanaga, N., and Takeuchi, Y. (1999) 2-Bromopropane causes ovarian dysfunction by damaging primordial follicles and their oocytes in female rats. *Toxicol. Appl. Pharmacol.* 159:185–193.

Zenick, H., Hastings, L., and Goldsmith, M. (1982) Chronic cadmium exposure: relation to male reproductive toxicity and subsequent fetal outcome. *J. Toxicol. Environ. Health* 9:377–387.

111

DEVELOPMENTAL TOXICOLOGY

THOMAS A. LEWANDOWSKI

This chapter will deal with adverse effects of workplace chemical exposures on child development, starting from effects after the point of conception up through adulthood. However, one should not make too strong a distinction between the effects of exposures occurring before or after conception (i.e., reproductive versus developmental toxicology) because exposures of both female and male workers prior to conception, which do not prevent conception may nonetheless have implications for later child development (Hooiveld et al., 2006; Cordier, 2008). Thus the reader is advised to consider both reproductive and developmental toxicology when evaluating potential workplace exposures.

It is also worthwhile to define some of the terms used in this subfield of toxicology. A *teratogen* is a chemical that causes malformation (birth defects, terata) in the developing organism. Not all developmental toxicants are teratogens. Some may cause death of the embryo or fetus without producing any observable defect. Alternatively, chemicals may also produce decreases in fetal body weight, a particularly sensitive indicator of developmental toxicity. All of these effects may be attributed to *developmental toxicants*. Developmental toxicologists are particularly concerned with agents that exert a selective or greater toxicity to the developing organism relative to the adult. That is, the agent causes effects on the developing organism at doses below those associated with parental toxicity. This narrower definition is necessary because at doses that are maternally toxic, one cannot tell whether the chemical is acting specifically on the fetus or affecting maternal health to such an extent that fetal health is indirectly compromised. The focus on agents that are selectively toxic to the fetus is appropriate because general occupational exposure limits are set to protect the health of working adults and thus address the potential for indirect harm due to maternal systemic toxicity.

PRENATAL EXPOSURES

The potential impact of workplace exposures on reproduction and development is a significant concern. Approximately 75% of all women will become pregnant at least once during their working life (Alpert and Cawthorne, 2009) and 80% of all pregnant women who work will do so up until the last month of their pregnancy. It was at one time believed that the placenta effectively isolated the fetus from potentially harmful maternal chemical exposures. It is now known that the placenta is only a limited barrier and in fact may actively transport or metabolize many chemicals, potentially allowing them to accumulate in fetal tissues (Myllynen et al., 2005). Indeed for some chemicals (e.g., mercury, dioxins, chlorinated pesticides), the fetus may constitute an important sink for maternal exposures. On the other hand, maternal metabolism may significantly attenuate a chemical dose before it reaches the fetus, potentially preventing fetal exposure from reaching adverse effects levels (Figure 111.1). The nature of maternal/fetal dosimetry is unfortunately chemical specific and not easily predicted. The increasing availability of physiologically-based toxicokinetic (PBTK) models of chemical disposition during pregnancy (summarized by Corley et al., 2003) may help to resolve this particular challenge in developmental toxicology.

Hamilton & Hardy's Industrial Toxicology, Sixth Edition. Edited by Raymond D. Harbison, Marie M. Bourgeois, and Giffe T. Johnson.
© 2015 John Wiley & Sons, Inc. Published 2015 by John Wiley & Sons, Inc.

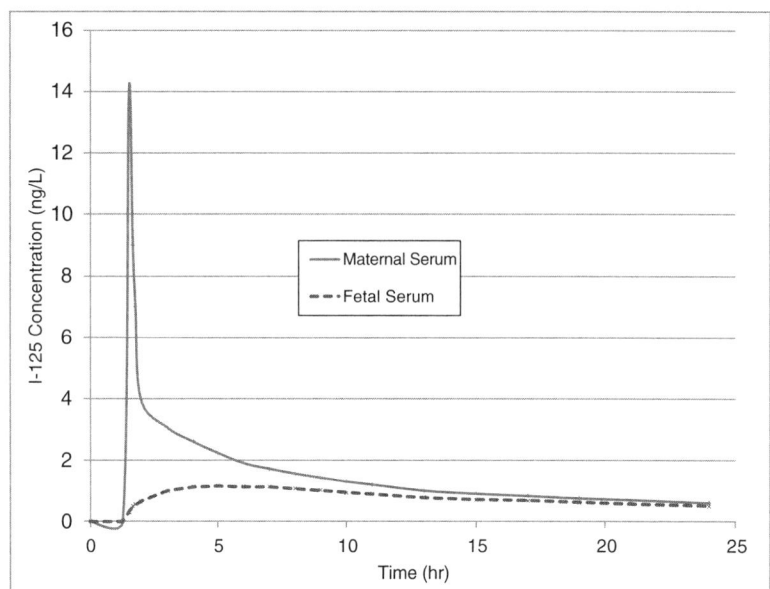

FIGURE 111.1   Effect of maternal metabolism on fetal exposure. *Source*: Data taken from the study of Clewell et al. (2003) involving radioiodine exposure.

PRENATAL DEVELOPMENT AND WINDOWS OF SUSCEPTIBILITY

The process of human development is truly an amazing process. The daughter cells of a fertilized egg, after attaching themselves to the uterine wall (thereby becoming an embryo), will undergo rapid division, rearrangement and migration, producing a fetus (at 8–9 weeks postfertilization in humans) with all of the developing tissues in their proper proportions and locations in the body. These activities are under the control of various chemical signals encoded in the embryo's DNA, which direct details such as the location and orientation of the developing limbs and the positioning of different brain regions. A large number of scientific studies have shown that if some event occurs to interfere with one of those signals, a specific birth defect related to the signal's intended target is likely to ensue. Due to the specifically timed nature of embryonic development, there exist "windows of susceptibility" when particular tissues may be affected by chemical exposure (Wilson, 1965; Rodier, 1977; Adams et al., 2000). For example, research has shown that the industrial solvent 2-methoxyethanol produces brain malformations in rats when administered on the seventh and eighth day of gestation but paw malformations when administered on gestational day 11 (Clarke et al., 1992; Terry et al., 1994). In general, chemical exposures early in development (in the first week of pregnancy) will likely result in either no effect or complete loss of the embryo, exposures during the period of main organ formation (called organogenesis, approximately weeks 3–8 in humans) may lead to visible gross deformities, and exposures during the latter part of pregnancy may result in reduced infant size, functional deficits or prematurity but not in gross malformations (Figure 111.2). As shown in Figure 111.2, for many tissues the key period of organogenesis occurs before the end of the first trimester, a key time point when pregnant women often choose to disclose their pregnancy status. It is thus important to encourage employees to inform health and safety personnel of their pregnancy (or intended pregnancy) as early as possible.

POSTNATAL EXPOSURES

The period of concern for developmental toxicity does not cease at birth. In addition to whatever chemical body burden the child receives prior to birth, the child may receive exposure via breast milk if the mother elects to breastfeed her infant. Most advisory agencies recommend that mothers breastfeed their infants for at least the first 6–24 months of life (e.g., WHO, 2003). Highly lipophilic (i.e., fat soluble) chemicals, including some pesticides and chlorinated compounds may concentrate in breast milk at levels higher than in maternal serum (Wang and Needham, 2007). Certain inorganic chemicals (e.g., iodide, thiocyanates, perchlorate) may accumulate in breast milk and may be associated with higher levels of exposure to the neonate (Azizi and Smyth, 2009). It is important to note that for the first few months of life, milk constitutes the sole source of dietary exposure for the child and the child's consumption rate of milk is high relative to his/her body weight.

Throughout the entire period of childhood, children may also experience exposures due to chemicals brought into the

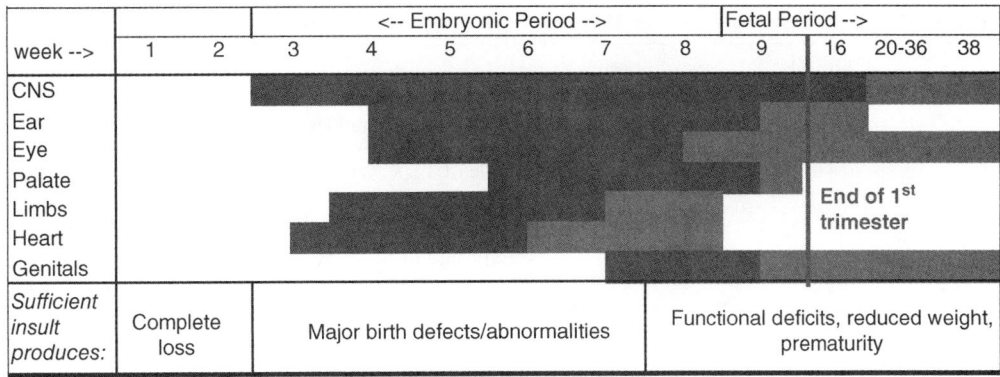

FIGURE 111.2 Windows of in utero susceptibility to chemical agents. Darker shading represents the period of greatest vulnerability. Adapted from Moore and Persaud, 1973.

home from the workplace on contaminated clothes or equipment. Because children have unique physiological and behavioral traits (e.g., proportionally higher inhalation rates, crawling and mouthing behaviors), they may be more at risk from these exposures than an adult. Although the use of dedicated work clothing left at the job site is now common in some fields, this so-called para-occupational exposure may be important in jobs where the risks of developmental toxicity are less well recognized or in cases where chemical exposure patterns are highly variable.

POSTNATAL SUSCEPTIBILITIES

From birth to the late teens the child's body grows and matures, becoming increasingly adult like. It would be incorrect, however, to assume that the progression is strictly linear. The pattern of growth and development during childhood is complex and varies among different organ systems. Of particular concern for chemical exposure, children are born with limited metabolic capacity for many chemicals compared to an adult and chemical half-life may therefore be considerably increased. The maturation of the child's metabolic capacity is substrate specific but is amenable to a few generalizations. Children are born with much less Phase I enzyme metabolic capacity (e.g., lower cytochrome P450 activity levels) than the adult, although for many enzymes children reach adult levels around age 2 (Alcorn and McNamara, 2002, 2003; de Zwart et al., 2004). Phase II (conjugating) enzyme metabolism may be slower to develop (particularly glucuronidation), although compensatory pathways (e.g., sulfation) may partially offset this metabolic limitation (Alcorn and McNamara, 2002).

TESTING FOR DEVELOPMENTAL TOXICITY

Much of the data used in assessing developmental and reproductive toxicity (DART) concern for workers comes from standard toxicity tests conducted in animals rather than studies in humans. While data from observational studies of exposed workers are highly relevant to the worker scenario, these studies are typically limited in number and are usually complicated by simultaneous exposures to multiple chemicals. Because animal testing forms the basis for many developmentally-based occupational exposure limits, it is worth describing how such tests are performed. Standard test protocols used in Western countries used to assess potential reproductive and developmental toxicity are summarized in Table 111.1.

Animal studies are used to identify no observed adverse effect levels (NOAELs) or lowest observed adverse effect levels (LOAELs), which are then divided by uncertainty factors (accounting for interspecies differences and sensitive subgroups) to arrive at an acceptable exposure level (Threshold Limit Value (TLV), WEEL, etc.). Studies in other species (fish, insects, birds) may also be available for particular compounds; however, these studies are generally not used to establish safe human exposure levels due to the greater degree of interspecies differences and the lack of standardized testing protocols. As noted above, studies that report DART effects at levels only above those causing general toxicity in the parental animals may lead one to conclude that the chemical in question does not cause selective developmental toxicity and risks are adequately controlled using an exposure limit based on preventing maternal toxicity.

SPECIFIC AGENTS

Metals

A number of metals have long been implicated as potential DART agents, most notably lead and mercury. Exposure to lead is possible in a number of industries, including metal smelting, lead mining and processing, battery reclamation, repair industries (due to soldering activities) and construction (e.g., due to exposure to existing lead-based paint) (ATSDR, 2007). Exposure to lead resulting in high blood lead levels

TABLE 111.1 Standard Animal Tests for Assessing DART Effects

Test	OECD Guideline Number	Typical Test Species	Exposure Design	Endpoints Considered
One generation reproduction toxicity study	415	Rats, mice	At least 1 spermatogenic cycle (e.g., 56 days in mice, 70 days in rats)	Litter size, litter weight, gross anomalies, microscopy of reproductive organs
Two generation reproduction toxicity study	416	Rats, mice	Exposure of first parental generation prior to mating, exposure of the first generation of offspring from birth through weaning of the second generation	Litter size, litter weight, gross anomalies, microscopy of reproductive organs, analysis of all three generations
Prenatal developmental toxicity study	414	Rat, rabbit	Exposure from implantation to just before birth (delivery by caesarean section)	Litter size, litter weight, maternal reproductive organs, skeletal and soft tissue malformations
Reproduction and developmental toxicity screening study	421	Rat	54 days (males and females) both exposed pre and postmating	Newborns examined 4 days post-birth; assesses reproductive success, fetal development and birth
Combined reproductive, developmental and repeated dose toxicity study	422	Rat	Combination of OECD 421 and OECD 408 (28-day repeated dose study for assessing general toxicity)	

Source: OECD Guidelines for the Testing of Chemicals, Section 4: Health Effects, www.oecd-ilibrary.org/books.

(>80 μg/dl) in children produces overt toxicity in the form of encephalopathy, anemia, and renal tubular damage (ATSDR, 2007), although such effects are extremely rare today. Beginning in the 1970s, studies suggested that children with blood lead concentrations well below this level (e.g., 30 μg/dl and below) had apparent deficits on various psychological tests, including IQ, attention, and behavior (notably Needleman et al., 1979). As a result of these studies, major efforts have been undertaken to reduce children's lead exposure and have met with considerable success. The most recent CDC blood lead data indicate a geometric mean of 1.9 μg/dl in children aged 1–5 (Jones et al., 2009). There is currently debate concerning the existence of a threshold for lead neurodevelopmental toxicity. Defining a threshold is difficult because at exposure levels associated with blood lead values <10 μg/dl, effects are subtle and only detectable on a population level (CDC, 2007).

Regarding occupational exposures, lead is known to cross the placenta and children born to mothers with elevated blood lead levels show elevated blood lead concentrations compared to children born to unexposed mothers (Goyer, 1990). The potential for *in utero* lead exposure is the primary basis for the Occupational Safety and Health Administration (OSHA) lead standard. It should be noted, however, that because children display a higher GI absorption capacity for lead compared to adults (O'Flaherty, 1998; Leggett, 1993) and also exhibit hand-to-mouth behaviors, significant postnatal exposure to

lead dust can occur if the material is brought into the home from the workplace (i.e., para-occupational exposure). In investigating potential occupational lead poisoning cases, the industrial hygienist should also be aware of potential worker exposures via the use of lead containing folk remedies, cosmetics, lead glazed dishware, or hobby materials, all of which can provide substantial exposures (ATSDR, 2007). For reference, the average blood lead level in pregnant women in the United States in the time period between 2003 and 2008 has been reported as 0.69 μg/dl (Jones et al., 2010).

The organic forms of mercury (notably methylmercury) are most strongly connected with neurodevelopmental problems (Aschner et al., 2010; Davidson et al., 2004; Castoldi et al., 2001), but these are unlikely to be present in the workplace outside of specific research laboratories (ATSDR, 1999). Exposures to methylmercury arise from microbial methylation of inorganic mercury in aquatic sediments, bioaccumulation in aquatic food chains and subsequent human fish consumption. Occupational exposures are therefore more likely to involve inorganic mercury salts (mercurous, mercuric) or elemental mercury. Occupations with the highest potential for non-organic mercury exposure include those involving chloralkali processes, production of mercury-containing instruments (e.g., pressure gauges, fluorescent lighting), waste disposal and recycling, and dental work with amalgam fillings (ATSDR, 1999). The data concerning the potential developmental toxicity of these forms of mercury are quite limited. In studies in which pregnant rats were exposed via inhalation to elemental mercury, it was observed to cross the placenta and accumulate in fetal tissues (Yoshida, 2002; Morgan et al., 2006). However, gross developmental effects (e.g., increased fetal loss, reduced fetal body weight) have only been observed at mercury vapor exposures which also produced maternal toxicity (maternal NOAEL = 4 mg/m^3; developmental NOAEL = 8 mg/m^3) (Morgan et al., 2002). A number of studies have indicated behavioral differences in the offspring of animals exposed to elemental mercury during pregnancy (Danielsson et al. 1993; Fredriksson et al. 1996; Newland et al., 1996). Studies among dental personnel, the best studied occupational group exposed to elemental mercury, have produced generally negative findings (Heggland et al., 2011; Lindbohm et al., 2007; Olfert, 2006). As regards the inorganic mercury salts, specific developmental toxicity data are lacking in both experimental animals and humans. Mercurous chloride (calomel) has been documented to produce skin inflammation, irritability, and other symptoms in children treated with this compound as a pharmaceutical (i.e., Pink disease) (Black, 1999). However, the doses involved (i.e., intentional treatment) are likely to be greater than those associated with potential para-occupational exposure. Unlike organic and elemental mercury, inorganic mercury does not readily cross the placenta (Yang et al., 1996), so developmental toxicity from *in utero* exposure is not likely. For reference, data collected by the CDC for the period of 2003–2008 indicate that the average blood mercury level in women aged 18–49 was 0.7 µg/l (Jones et al., 2010).

The evidence linking other industrial metals with DART effects is generally more limited and equivocal in nature. For example, animal studies of arsenic have indicated no developmental effects when exposures are via the inhalation route. Effects are observed in animals exposed to arsenic via injection or drinking water but these forms of exposure are not relevant to the occupational setting (DeSesso et al., 1998; DeSesso, 2001; Wang et al., 2006; Vahter, 2008). Arsenic is more poorly absorbed via inhalation exposure relative to absorption from the GI tract and thus the peak dose in drinking water and injection studies is likely to be much higher (DeSesso, 2001). Epidemiology studies of arsenic-exposed populations are equivocal and present contradictory findings (DeSesso et al., 1998). Thus the human relevance of arsenic's potential developmental toxicity is still a subject of debate, and risk assessments and exposure guidelines are driven by concerns regarding carcinogenicity. For other metals, the reader is directed to chemical-specific sections of this book.

Halogenated and Aromatic Hydrocarbon Solvents

Exposures to halogenated and/or aromatic hydrocarbon solvents are widespread and can be found in a variety of occupations, including those involving dry-cleaning, parts degreasing, painting and paint removal, printing, and petrochemical processing (Kumar, 2004). Studies of occupationally exposed populations have suggested associations between exposure to solvents and adverse developmental outcomes but such studies are often complicated by simultaneous exposures to multiple chemicals and a limited ability to control for important nonoccupational exposures (e.g., ethanol consumption and smoking). Among this large group of chemicals, data are most extensive and suggestive for tetrachloroethylene (PCE), trichloroethylene (TCE), and toluene.

The strongest evidence of an association between PCE and developmental effects comes from occupational studies published in the late 1980s and early 1990s, which showed an association between PCE exposure and risk of spontaneous abortion (e.g., Kyyronen et al., 1989; Olsen et al., 1990; Windham et al., 1991). These studies generally lacked quantitative estimates of dose and relied on more qualitative assessments (e.g., ever exposed, low versus high exposure). Despite methodological weaknesses in these studies, and lack of supporting evidence from animal studies (Carney et al., 2006; Schwetz et al., 1975), the UK committee of independent scientists concluded that the effects observed in the occupational studies could not necessarily be discounted. In 1994, the United Kingdom's Health and Safety Executive commissioned a large ($n = 3517$), retrospective occupational cohort study of women currently or previously employed in dry-cleaning operations. This study, by Doyle et al. (1997) observed statistically significant increased spontaneous abortion risks among workers engaged in dry-cleaning operations relative to those not engaged (OR: 1.67; 95% CI: 1.17–2.36). Risks were not increased for individuals working at dry-cleaning establishments but not running dry-cleaning equipment (OR: 1.02; 95% CI: 0.65–1.60). Although results from the larger study by Doyle et al. are consistent with results from the earlier studies in showing associations between dry-cleaning work and spontaneous abortion; a biological mechanism by which PCE could cause spontaneous abortions has not been established. Exposure levels in the study population were also not quantified but were likely higher than those encountered in workplaces today (e.g., the American Conference of Governmental Industrial Hygienists (ACGIH) TLV for PCE was 100 ppm prior to 1981, 50 ppm prior to 1992 and set to 25 ppm in 1993). An animal study by Carney et al. (2006) does not provide evidence that PCE causes spontaneous abortions, but did indicate fetotoxicity at PCE concentrations that were not maternally toxic, and thus provides evidence that PCE is a developmental toxicant in rodents. Fetal effects included reduced body weights and decreased ossification of thoracic vertebrae at 600 ppm, and minor decreases in body weight at 250 ppm; there were no developmental effects at 65 ppm.

Notable evidence for TCE-associated developmental toxicity relates to cardiac effects, which have been observed in both humans and animals. A study by Bove et al. (2002) observed a significantly increased prevalence of cardiac abnormalities in children of parents residing in areas where drinking water wells were contaminated with TCE in drinking water, compared with children of parents residing in uncontaminated areas (PR 2.58; 95% CI: 2.0–3.4). Cardiac defects were also significantly elevated in children whose parents resided in areas with high soil vapor concentrations of TCE and other VOCs (OR 1.94; 95% CI 1.21–3.12) (ATSDR, 2006). Results from some, but not all, animal studies support the epidemiological findings. For example, cardiac defects have been observed in rats exposed *in utero* to trichloroethylene, or the metabolites trichloroacetic acid (TCA) and dichloroacetic acid (DCA) in studies involving repeated exposure via the oral route (Dawson et al., 1993; Smith et al., 1989, 1992). The lowest exposure level at which effects were observed was 1.5 ppm in drinking water or approximately 0.15 mg/kg/day (Dawson et al., 1993). Cardiac effects were not observed in another study involving oral exposure (Fisher et al., 2001; TCE exposure at 500 mg/kg/day) or a study involving whole-body inhalation exposure (e.g., Carney et al., 2006). The Carney et al. study reported slight decreases in fetal and placental weight with a NOAEL of 65 ppm.

As regards the aromatic hydrocarbon solvents (benzene, ethylbenzene, toluene, and xylene) the evidence for the developmental toxicity of toluene is most clearly established. Toluene has been shown to cause congenital defects (similar to fetal alcohol syndrome) in infants born to mothers who abused toluene during pregnancy. Exposure levels in such cases would be exceptionally high (i.e., several thousand ppm; Bowen and Hannigan, 2006). Animal studies likewise have revealed malformations and reduced fetal body weight in animals exposed at concentrations in excess of 1000 ppm but no effects at concentrations in the 200–400 ppm range (Jones and Balster, 1997; Thiel and Chahoud, 1997; Jarosz et al., 2008). Thus toluene exposures at the current federal permissible exposure limit (PEL) (200 ppm, time-weighted average (TWA)) would not be expected to have developmental consequences. Data for ethylbenzene and xylene have generally been negative. For example, Saillenfait et al. (2007) reported no embryolethal or teratogenic effects in rats exposed to ethylbenzene on gestational days 6 through 20 at 250 or 1000 ppm, the higher dose producing reductions in both maternal and fetal body weights. The same research group observed a similar lack of effects for *m*- and *p*-xylene at concentrations up to 1000 ppm, a dose that produced maternal toxicity (Saillenfait et al. 2003). However, in the same study exposures to 500 ppm *o*-xylene resulted in reduced fetal body weights at concentrations below the LOAEL for maternal effects. As concerns benzene, studies of benzene-exposed pregnant animals have suggested that exposure causes retarded fetal growth, manifested mainly as decreased fetal weight and delayed ossification of the bones. These effects were observed in the presence or absence of maternal toxicity depending on the particular study (Murray et al., 1979; Mehlman et al., 1980). One group of researchers also reported that *in utero* exposure to benzene resulted in

altered hematopoietic function in neonatal animals (Keller and Snyder, 1988). The exposure levels associated with these effects are above the low exposure limits set to prevent the occurrence of benzene-induced leukemia (e.g., the current PEL is 1 ppm and the TLV is 0.5 ppm).

Glycols and Glycol Ethers

The glycols and glycol ethers are widely used industrial chemicals. Glycols (e.g., ethylene glycol, diethylene glycol, propylene glycol) may be used as chemical precursors and as heat transfer fluids ("antifreeze" in automotive usage) (NTP-CEHR, 2003a, 2003b). Propylene glycol is also used as a solvent in pharmaceutical, cosmetic and flavoring applications (NTP-CEHR, 2003b). The related glycol ethers (e.g., 2-ethoxy ethanol [EGEE], 2-methoxy ethanol [EGME], 1-methoxy-2-propanol [PGME]) and their respective acetates are primarily used as solvents for resins, lacquers, and other finishes (Nelson et al., 1984; Nagano et al., 1984). In recent years, the propyl-based ethers have displaced ethyl-based varieties due to toxicity concerns (de Ketttenis, 2005). Suitable epidemiological studies investigating the potential developmental toxicity of these compounds are lacking and thus occupational exposure guidelines are based on animal studies.

Rodent studies investigating the developmental toxicity of ethylene glycol (EG) have indicated developmental toxicity but only at exceptionally high doses, at or above 1000 mg/kg/day (Maronpot et al., 1983; Price et al., 1985; Marr et al., 1992) or 150 mg/m^3 (Tyl et al., 1995). Developmental effects were not observed in rabbits dosed by gavage up to 2000 mg/kg/day (Tyl et al., 1993). In those studies where developmental effects were noted, significant maternal toxicity was also observed at or near the dose associated with developmental toxicity, suggesting that EG is slightly more toxic to the developing organism than the adult animal. Similar findings were reported in a rodent teratology study involving diethylene glycol (DEG) (Ballantyne and Snellings, 2005). Maternal and fetal rats evidenced a similar level of sensitivity (the LOAEL in each case was 1118 mg/kg/day) whereas the fetus was found to be less sensitive than the parent in mice (fetal NOAEL—2795 mg/kg/day, maternal NOAEL—559 mg/kg/day). No developmental effects were reported at DEG doses up to 1000 mg/kg/day in rabbits (Hellwig et al., 1995). A study of triethylene glycol conducted in rats reported comparable results to the DEG and EG studies (Ballantyne and Snellings, 2007). Studies evaluating the developmental toxicity of propylene glycol (Driscoll et al., 1993; Enright et al., 2010) have not observed any developmental effects at doses ranging from 1000 to 10,000 mg/kg/day.

The glycol ethers are recognized as reproductive and developmental toxicants. For example, EGME exposure in pregnant mice produced reduction in fetal body weights and skeletal malformations at 125 and 62.5 mg/kg/day, respectively, doses that did not have systemic effects in the dams

(Nagano et al. 1984; Horton et al., 1985). Fetal viability was also affected in developmental studies involving multiple species with EGME exposures as low as 50 ppm, such effects were not observed in animals treated with concentrations as high as 3000 ppm PGME (Doe, 1984a,b; Hanley et al., 1984). The results of these studies have raised questions as to whether the OSHA PEL for EGME (25 ppm, TWA) is sufficiently protective of such effects and has led to the development of physiologically-based pharmacokinetic model of EGME metabolism (Sweeney et al., 2001). Similar to EGME, studies of EGEE have demonstrated birth defects, inhibited growth and increased embryo mortality at doses of 617 ppm in rabbits and 767 ppm in rats (Andrew and Hardin, 1984). These concentrations produced adverse effects in maternal rabbits but not in maternal rats. Comparable findings have also been reported in studies of diethylene glycol dimethyl ether (Price et al., 1987; Schwetz et al., 1992) and triethylene glycol dimethyl ether (Schwetz et al., 1992). Studies of propyl and longer chain (e.g., butyl, hexyl) glycol ethers have demonstrated either much weaker or a complete lack of developmental effects (Doe et al., 1983; Nelson et al., 1984; Tyl et al., 1984, 1989; Spencer, 2005).

Due to the developmental toxicity of EG and the EG ethers, workers may also question the safety of polymers of polyethylene glycol (PEG). Polyethylene glycols occur as a mixture of different polymeric molecules of varying length with a number appended after the name (e.g., PEG 400) to indicate the average molecular weight of the polymer. In general, such polymers are too large to be absorbed in the lungs or across the skin and have negligible vapor pressures. Although some residual monomer may exist in the polymer material, the concentration is likely to be too low to be a health concern, particularly in regards to non-ingestion exposures. One study (Vannier et al., 1989) did report developmental toxicity in mice dosed orally with 0.5 ml PEG 200 (approximately 2250 mg/kg/day) but effects were not observed in rats at the same dose. A review by Fruijtier-Pölloth (2005) concluded that PEGs did not pose a risk of developmental toxicity.

Phthalates

Much recent media concern has been focused on the potential developmental toxicity of phthalates. The term phthalates describes a class of chemicals used primarily to lend flexibility to plastics and other materials. Phthalates are used in the manufacture of a vast array of equipment, including shower curtains, shoes, IV bags and tubing, bottles, cosmetics, sex toys, adhesives and glues, detergents and surfactants, packaging, paints, printing inks and coatings, pharmaceuticals, and food wrap (NAS, 2008). Because phthalates are used so extensively, it is not surprising that they are detected in human tissues in studies of the US population (CDC, 2010). A CDC study reported that geometric mean concentrations of individual phthalate

metabolites in urine varied from non-detectable to approximately 357 µg/L, depending on age, gender and ethnicity (CDC, 2010).

The potential health effects of phthalates have been studied for a fairly long time. In the 1980s a number of animal studies indicated that high doses of phthalates could lead to testicular atrophy and other reproductive problems in rodents, particularly in male rats (Oishi, 1985, 1986; Gray et al., 1982; Gray and Gangolli, 1986; Lamb et al., 1987). While the effects were unambiguous, their meaning for humans was uncertain. Other species, such as mice and hamsters, exhibited much less sensitivity, suggesting a species-specific effect in the rat (Gray et al., 1982). The doses used in these studies were also orders of magnitude above what humans might be exposed to and showed an apparent threshold well above human exposure levels. A number of phthalates have also been shown to mimic the activity of the hormone estrogen in *in vitro* systems. However, in these tests phthalates are very weak mimics of estrogen, on the order of a million times less potent than estrogen itself (Blair et al., 2000). These results did lead to studies of the effects of phthalate exposure on reproductive development in male animals. For example, Moore et al. (2001) conducted a study in which pregnant rats were exposed to diethylhexyl phthalate and male pups were evaluated after birth for effects on male reproductive system development and sexual behavior. The authors observed a significantly shorter anogenital distance (AGD) in males dosed at 750 mg/kg/day, which was more similar to the AGD observed in female rats. They also observed increased nipple retention as well as a number of other physiological effects in the high dose group. While the effects observed were notable, it was also recognized that median diethylhexyl phthalate exposures in human children were <0.008 mg/kg/day (Wittassek et al., 2007), or more than 90,000 times lower than the level at which effects were seen in the rats studied by Moore et al. (2001). These results again suggested that the endocrine disrupting effects of phthalates occurred at doses not relevant to potential human exposures.

The current controversy surrounding phthalates arises primarily from the apparent ubiquity of phthalates in human tissues and a human study conducted by Swan et al. (2005). Swan et al. measured the concentrations of phthalate metabolites in urine samples from 85 women who recently gave birth. They also evaluated genital morphology and measured AGD in the male children. Swan et al. reported that AGD (corrected for body weight) was inversely correlated with the mother's total phthalate exposure estimate. However, only some phthalates were correlated with AGD whereas others were not (including diethylhexyl phthalate, which had been found to produce such effects in the earlier rat study). The meaning of Swan et al.'s findings has been much debated. The background level of phthalate exposure experienced by the women studied was approximately 100,000 times lower than those producing effects in rats, a rather remarkable difference in species sensitivity if true. (Marsee et al., 2006). The potential developmental effects of phthalates in humans are currently of considerable research interest and public concern.

RESOURCES FOR EVALUATING DEVELOPMENTAL HEALTH RISKS

There are a number of information resources, which can be used to evaluate potential developmental health risks. For chemicals with existing exposure limits, readers are directed to the ACGIH TLV documentation, which provides a concise summary of all of the available toxicity data (as of the date the TLV was adopted/revised), including data on reproductive and developmental toxicity. Also useful are the MAK values recommended by the Deutsche Forschungsgemeinschaft (DFG). These often contain a suffix stating whether the available toxicology data indicate the MAK value for maternal toxicity is also protective for the embryo/fetus (e.g., Group C—"no reason to fear a risk of damage to the developing embryo or fetus when MAK and BAT values are observed"). For antimicrobials (used, for example, in biological research laboratories), the U.S. Food and Drug Administration (FDA) uses a letter designation system to describe risks of exposure to the developing organism (although intake due to occupational exposure would likely be far less than that associated with medicinal use). Similar information on postnatal drug exposures can be found on the National Institutes of Health's LactMed database (http://toxnet.nlm.nih.gov/cgi-bin/sis/htmlgen?LACT). It must be recognized, however, that many of the chemicals present in the workplace lack exposure limits or appropriate toxicology data. In such cases, it might be appropriate to consult the European chemical Substances Information System (ESIS) developed by the European Commission Joint Research Center, which contains manufacturer submitted toxicity data that may not be published in scientific journals. Interpretation of such studies should be done in consultation with a scientist familiar with toxicity testing protocols and interpretation of developmental toxicity data.

ACKNOWLEDGMENT

The author would like to express his gratitude to David Dodge and Marguerite Seeley for their assistance in the preparation of this chapter.

REFERENCES

Adams, J., Barone, S., Jr., LaMantia, A., Philen, R., Rice, D.C., Spear, L., and Susser, E. (2000) Workshop to identify critical windows of exposure for children's health: neurobehavioral work group summary. *Environ. Health Perspect.* 108(Suppl. 3): 535–544.

Agency for Toxic Substances and Disease Registry (ATSDR) (1999) *Toxicological Profile for Mercury*, Atlanta, GA: U.S. Department of Health and Human Services, Public Health Service.

Agency for Toxic Substances and Disease Registry (ATSDR) (2006) *Health Statistics Review Follow-Up, Cancer and Birth Outcome Analysis: Endicott Area Investigation, Endicott Area, Town of Union, Broome County, New York*, 60 pp.

Agency for Toxic Substances and Disease Registry (ATSDR) (2007) *Toxicological Profile for Lead*, Atlanta, GA: U.S. Department of Health and Human Services, Public Health Service.

Alcorn, J. and McNamara, P.J. (2002) Ontogeny of hepatic and renal systemic clearance pathways in infants: part I. *Clin. Pharmacokinet.* 41(12):959–998.

Alcorn, J. and McNamara, P.J. (2003) Pharmacokinetics in the newborn. *Adv. Drug Deliv. Rev.* 55(5):667–686.

Alpert, M. and Cawthorne, A. (2009) *Labor Pains: Improving Employment and Income Security for Pregnant Women and New Mothers*, Washington, DC: Center for American Progress.

Andrew, F.D. and Hardin, B.D. (1984) Developmental effects after inhalation exposure of gravid rabbits and rats to ethylene glycol monoethyl ether. *Environ. Health Perspect.* 57:13–23.

Aschner, M., Onishchenko, N., and Ceccatelli, S. (2010) Toxicology of alkylmercury compounds. *Met. Ions Life Sci.* 7:403–434.

Azizi, F. and Smyth, P. (2009) Breastfeeding and maternal and infant iodine nutrition. *Clin. Endocrinol. (Oxf.)* 70(5):803–809.

Ballantyne, B. and Snellings, W.M. (2005) Developmental toxicity study with diethylene glycol dosed by gavage to CD rats and CD-1 mice. *Food Chem. Toxicol.* 43(11):1637–1646.

Ballantyne, B. and Snellings, W.M. (2005) Developmental toxicity study with triethylene glycol given by gavage to CD rats and CD-1 mice. *J. Appl. Toxicol.* 25(3):418–426.

Black, J. (1999) The puzzle of pink disease. *J. R. Soc. Med.* 92(9):478–481.

Blair, R.M., Fang, H., Branham, W.S., Hass, B.S., Dial, S.L., Moland, C.L., Tong, W., Shi, L., Perkins, R., and Sheehan, D.M. (2000) The estrogen receptor relative binding affinities of 188 natural and xenochemicals: structural diversity of ligands. *Toxicol. Sci.* 54(1):138–153.

Bove, F., Shim, Y., and Zeitz, P. (2002) Drinking water contaminants and adverse pregnancy outcomes: a review. *Environ. Health Perspect.* 110(Suppl. 1):61–74.

Bowen, S.E. and Hannigan, J.H. (2006) Developmental toxicity of prenatal exposure to toluene. *AAPS J.* 8(2):E419–E424.

Carney, E.W., Thorsrud, B.A., Dugard, P.H., and Zablotny, C.L. (2006) Developmental toxicity studies in Crl:CD (SD) rats following inhalation exposure to trichloroethylene and perchloroethylene. *Birth Defects Res. B Dev. Reprod. Toxicol.* 77(5):405–412.

Castoldi, A.F., Coccini, T., Ceccatelli, S., and Manzo, L. (2001) Neurotoxicity and molecular effects of methylmercury. *Brain Res. Bull.* 55(2):197–203.

Centers for Disease Control (CDC) (2007) Interpreting and managing blood lead levels <10 μg/dL in children and reducing childhood exposures to lead. Recommendations of CDC's Advisory Committee on Childhood Lead Poisoning Prevention. *MMWR Morb. Mortal. Wkly Rep.* 56(RR08):1–14.

Centers for Disease Control (CDC) (2010) *Fourth National Report on Human Exposure to Environmental Chemicals*, Atlanta, GA: Department of Health and Human Services.

Clarke, D.O., Duignan, J.M., and Welsch, F. (1992) 2-Methoxy-acetic acid dosimetry–teratogenicity relationships in CD-1 mice exposed to 2-methoxyethanol. *Toxicol. Appl. Pharmacol.* 114(1):77–87.

Clewell, R.A., Merrill, E.A., Yu, K.O., Mahle, D.A., Sterner, T.R., Fisher, J.W., and Gearhart, J.M. (2003) Predicting neonatal perchlorate dose and inhibition of iodide uptake in the rat during lactation using physiologically-based pharmacokinetic modeling. *Toxicol. Sci.* 74(2):416–436.

Cordier, S. (2008) Evidence for a role of paternal exposures in developmental toxicity. *Basic Clin. Pharmacol. Toxicol.* 102(2):176–181.

Corley, R.A., Mast, T.J., Carney, E.W., Rogers, J.M., and Daston, G.P. (2003) Evaluation of physiologically based models of pregnancy and lactation for their application in children's health risk assessments. *Crit. Rev. Toxicol.* 33(2):137–211.

Danielsson, B.R., Fredriksson, A., Dahlgren, L., Gårdlund, A.T., Olsson, L., Dencker, L., and Archer, T. (1993) Behavioural effects of prenatal metallic mercury inhalation exposure in rats. *Neurotoxicol. Teratol.* 15(6):391–396.

Davidson, P.W., Myers, G.J., and Weiss, B. (2004) Mercury exposure and child development outcomes. *Pediatrics* 113(4 Suppl.): 1023–1029.

Dawson, B.V., Johnson, P.D., Goldbeerg, S.J., and Ulreich, J.B. (1993) Cardiac teratogenesis of halogenated hydrocarbon-contaminated drinking water. *J. Am. Coll. Cardiol.* 21: 1466–1472.

de Ketttenis, P. (2005) The historic and current use of glycol ethers: a picture of change. *Toxicol. Lett.* 156(1):5–11.

DeSesso, J.M. (2001) Teratogen update: inorganic arsenic. *Teratology* 63:170–173.

DeSesso, J.M., Jacobson, C.F., Scialli, A.R., Farr, C.H., and Holson, J.F. (1998) An assessment of the developmental toxicity of inorganic arsenic. *Reprod. Toxicol.* 12(4):385–433.

de Zwart, L.L., Haenen, H.E.M.G., Versantvoort, C.H.M., Wolterink, G., van Engelen, J.G.M., and Sips, A.J.A.M. (2004) Role of biokinetics in risk assessment of drugs and chemicals in children. *Regul. Toxicol. Pharmacol.* 39:282–309.

Doe, J.E. (1984a) Further studies on the toxicology of the glycol ethers with emphasis on rapid screening and hazard assessment. *Environ. Health Perspect.* 57:199–206.

Doe, J.E. (1984b) Ethylene glycol monoethyl ether and ethylene glycol monoethyl ether acetate teratology studies. *Environ. Health Perspect.* 57:33–41.

Doe, J.E., Samuels, D.M., Tinston, D.J., and de Silva Wickramaratne, G.A. (1983) Comparative aspects of the reproductive toxicology by inhalation in rats of ethylene glycol monomethyl ether and propylene glycol monomethyl ether. *Toxicol. Appl. Pharmacol.* 69(1):43–47.

Doyle, P., Roman, E., Beral, V., and Brookes, M. (1997) Spontaneous abortion in dry cleaning workers potentially exposed to perchloroethylene. *Occup. Environ. Med.* 54(12):848–856.

Driscoll, C.D., Kubena, M.F., and Neeper-Bradley, T.L. (1993) *Propylene glycol: developmental toxicity gavage study III in CD-1 mice.* Danbury, CT: Industrial Chemicals Division, Union Carbide Chemicals and Plastics Company Inc..

Enright, B.P., McIntyre, B.S., Thackaberry, E.A., Treinen, K.A., and Kopytek, S.J. (2010) Assessment of hydroxypropyl methylcellulose, propylene glycol, polysorbate 80, and hydroxypropyl-β-cyclodextrin for use in developmental and reproductive toxicology studies. *Birth Defects Res. B Dev. Reprod. Toxicol.* 89(6):504–516.

Fisher, J.W., Channel, S.R., Eggers, J.S., Johnson, P.D., MacMahon, K.L., Goodyear, C.D., Sudberry, G.L., Warren, D.A., Latendresse, J.R., and Graeter, L.J. (2001) Trichloroethylene, trichloroacetic acid, and dichloroacetic acid: do they affect fetal rat heart development? *Int. J. Toxicol.* 20(5):257–267.

Fredriksson, A., Dencker, L., Archer, T., and Danielsson, B.R. (1996) Prenatal coexposure to metallic mercury vapour and methylmercury produce interactive behavioural changes in adult rats. *Neurotoxicol. Teratol.* 18(2):129–134.

Fruijtier-Pölloth, C. (2005) Safety assessment on polyethylene glycols (PEGs) and their derivatives as used in cosmetic products. *Toxicology* 2214(1–2):1–38.

Goyer, R.A. (1990) Transplacental transport of lead. *Environ. Health Perspect.* 89:101–105.

Gray, T.J. and Gangolli, S.D. (1986) Aspects of the testicular toxicity of phthalate esters. *Environ. Health Perspect.* 65:229–235.

Gray, T.J., Rowland, I.R., Foster, P.M., and Gangolli, S.D. (1982) Species differences in the testicular toxicity of phthalate esters. *Toxicol. Lett.* 11(1–2):141–147.

Hanley T.R., Jr., Young, J.T., John, J.A., and Rao, K.S. (1984) Ethylene glycol monomethyl ether (EGME) and propylene glycol monomethyl ether (PGME): inhalation fertility and teratogenicity studies in rats, mice and rabbits. *Environ. Health Perspect.* 57:7–12.

Heggland, I., Irgens, A., Tollånes, M., Romundstad, P., Syversen, T., Svendsen, K., Melø, I., and Hilt, B. (2011) Pregnancy outcomes among female dental personnel—a registry-based retrospective cohort study. *Scand. J. Work Environ. Health* 37:539–546.

Hellwig, J., Klimisch, H.J., and Jäckh, R. (1995) Investigation of the prenatal toxicity of orally administered diethylene glycol in rabbits. *Fundam. Appl. Toxicol.* 28(1):27–33.

Hooiveld, M., Haveman, W., Roskes, K., Bretveld, R., Burstyn, I., and Roeleveld, N. (2006) Adverse reproductive outcomes among male painters with occupational exposure to organic solvents. *Occup. Environ. Med.* 63(8):538–544.

Horton, V.L., Sleet, R.B., John-Greene, J.A., and Welsch F. (1985) Developmental phase-specific and dose-related teratogenic effects of ethylene glycol monomethyl ether in CD-1 mice. *Toxicol. Appl. Pharmacol.* 80(1):108–118.

Jarosz, P.A., Fata, E., Bowen, S.E., Jen, K.L., and Coscina, D.V. (2008) Effects of abuse pattern of gestational toluene exposure on metabolism, feeding and body composition. *Physiol. Behav.* 93(4–5):984–993.

Jones, H.E. and Balster, R.L. (1997) Neurobehavioral consequences of intermittent prenatal exposure to high concentrations of toluene. *Neurotoxicol. Teratol.* 19(4):305–313.

Jones, L., Parker, J.D., and Mendola, P. (2010) Blood lead and mercury levels in pregnant women in the United States, 2003–2008. *NCHS Data Brief.* 52:1–8.

Jones, R.L., Homa, D.M., Meyer, P.A., Brody, D.J., Caldwell, K.L., Pirkle, J.L., and Brown, M.J. (2009) Trends in blood lead levels and blood lead testing among US children aged 1 to 5 years, 1988–2004. *Pediatrics* 123:e376–e385.

Keller, K.A. and Snyder, C.A. (1988) Mice exposed *in utero* to 20 ppm benzene exhibit altered numbers of recognizable hematopoietic cells up to seven weeks after exposure. *Fundam. Appl. Toxicol.* 10(2):224–232.

Kumar, S. (2004) Occupational exposure associated with reproductive dysfunction. *J. Occup. Health* 46:1–19.

Kyyronen, P., Taskinen, H., Lindbohm, M.L., Hemminki, K., and Heinonen, O.P. (1989) Spontaneous abortions and congenital malformations among women exposed to tetrachloroethylene in dry cleaning. *J. Epidemiol. Community Health* 43(4):346–351.

Lamb, J.C., 4th, Chapin, R.E., Teague, J., Lawton, A.D., and Reel, J.R. (1987) Reproductive effects of four phthalic acid esters in the mouse. *Toxicol. Appl. Pharmacol.* 88(2):255–269.

Leggett, R.W. (1993) An age-specific kinetic model of lead metabolism in humans. *Environ. Health Perspect.* 101:598–616.

Lindbohm, M., Ylöstalo, P., Sallmén, M., Henriks-Eckerman, M., Nurminen, T., Forss, H., and Taskinen, H. (2007) Occupational exposure in dentistry and miscarriage. *Occup. Environ Med.* 64:127–133.

Maronpot, R.R., Zelenak, J.P., Weaver, E.V., and Smith, N.J. (1983) Teratogenicity study of ethylene glycol in rats. *Drug Chem. Toxicol.* 6(6):579–594.

Marr, M.C., Price, C.J., Myers, C.B., and Morrissey, R.E. (1992) Developmental stages of the CD (Sprague-Dawley) rat skeleton after maternal exposure to ethylene glycol. *Teratology* 46(2):169–181.

Marsee, K., Woodruff, T.J., Axelrad, D.A., Calafat, A.M., and Swan, S.H. (2006) Estimated daily phthalate exposures in a population of mothers of male infants exhibiting reduced anogenital distance. *Environ. Health Perspect.* 114(6):805–809.

Mehlman, M.A., Schreiner, C.A., and Mackerer, C.R. (1980) Current status of benzene teratology: a brief review. *J. Environ. Pathol. Toxicol.* 4(5–6):123–131.

Moore, K.L. and Persaud, T.V.N. (1973) *The Developing Human, Clinically Orientated Embryology*, Philadelphia, PA: W.B. Saunders Company.

Moore, R.W., Rudy, T.A., Lin, T.M., Ko, K., and Peterson, R.E. (2001) Abnormalities of sexual development in male rats with in utero and lactational exposure to the antiandrogenic plasticizer di(2-ethylhexyl) phthalate. *Environ. Health Perspect.* 109(3):229–237.

Morgan, D.L., Chanda, S.M., Price, H.C., Fernando, R., Liu, J., Brambila, E., O'Connor, R.W., Beliles, R.P., and Barone, S., Jr. (2002) Disposition of inhaled mercury vapor in pregnant rats: maternal toxicity and effects on developmental outcome. *Toxicol. Sci.* 66(2):261–273.

Morgan, D.L., Price, H.C., Fernando, R., Chanda, S.M., O'Connor, R.W., Barone, S.S., Jr., Herr, D.W., and Beliles, R.P. (2006) Gestational mercury vapor exposure and diet contribute to mercury accumulation in neonatal rats. *Environ. Health Perspect.* 114(5):735–739.

Murray, F.J., John, J.A., Rampy, L.W., Kuna, R.A., and Schwetz, B.A. (1979) Embryotoxicity of inhaled benzene in mice and rabbits. *Am. Ind. Hyg. Assoc. J.* 40(11):993–998.

Myllynen, P., Pasanen, M., and Pelkonen, O. (2005) Human placenta: a human organ for developmental toxicology research and biomonitoring. *Placenta* 26(5):361–371.

Nagano, K., Nakayama, E., Oobayashi, H., Nishizawa, T., Okuda, H., and Yamazaki, K. (1984) Experimental studies on toxicity of ethylene glycol alkyl ethers in Japan. *Environ. Health Perspect.* 57:75–84.

National Academies of Science (NAS) (2008) *Phthalates and Cumulative Risk Assessment: The Tasks Ahead*, Washington, DC: National Academies Press.

Needleman, H.L., Gunnoe, C., Leviton, A., Reed, R., Peresie, H., Maher, C., and Barrett, P. (1979) Deficits in psychologic and classroom performance of children with elevated dentine lead levels. *N. Engl. J. Med.* 300(13):689–695.

Nelson, B.K., Setzer, J.V., Brightwell, W.S., Mathinos, P.R., Kuczuk, M.H., Weaver, T.E., and Goad, P.T. (1984) Comparative inhalation teratogenicity of four glycol ether solvents and an amino derivative in rats. *Environ. Health Perspect.* 57:261–271.

Newland, M.C., Warfvinge, K., and Berlin, M. (1996) Behavioral consequences of in utero exposure to mercury vapor: alterations in lever-press durations and learning in squirrel monkeys. *Toxicol. Appl. Pharmacol.* 139(2):374–386.

NTP-CEHR (2003a) *NTP-CEHR Expert Panel Report on the Reproductive and Developmental Toxicity of Ethylene Glycol.* National Toxicology Program. NTP-CEHR-EG-03.

NTP-CEHR (2003b) *NTP-CEHR Expert Panel Report on the Reproductive and Developmental Toxicity of Propylene Glycol.* National Toxicology Program. NTP-CEHR-PG-03.

O'Flaherty, E.J. (1998) A physiologically based kinetic model for lead in children and adults. *Environ. Health Perspect.* 106 (Suppl. 6):1495–1503.

Oishi, S. (1985) Reversibility of testicular atrophy induced by di(2-ethylhexyl) phthalate in rats. *Environ. Res.* 36(1):160–169.

Oishi, S. (1986) Testicular atrophy induced by di(2-ethylhexyl) phthalate: changes in histology, cell specific enzyme activities and zinc concentrations in rat testis. *Arch. Toxicol.* 59(4): 290–295.

Olfert, S. (2006) Reproductive outcomes among dental personnel: a review of selected exposures. *J. Can. Dent. Assoc.* 72(9): 821–825.

Olsen, J., Hemminki, K., Ahlborg, G., Bjerkedal, T., Kyyronen, P., Taskinen, H., Lindbohm, M.L., Heinonen, O.P., Brandt, L., and Kolstad, H. (1990) Low birthweight, congenital malformations, and spontaneous abortions among dry-cleaning workers in Scandinavia. *Scand. J. Work Environ. Health* 16(3):163–168.

Price, C.J., Kimmel, C.A., George, J.D., and Marr, M.C. (1987) The developmental toxicity of diethylene glycol dimethyl ether in mice. *Fundam. Appl. Toxicol.* 8(1):115–126.

Price, C.J., Kimmel, C.A., Tyl, R.W., and Marr, M.C. (1985) The developmental toxicity of ethylene glycol in rats and mice. *Toxicol. Appl. Pharmacol.* 81(1):113–127.

Rodier, P.M. (1977) Correlations between prenatally-induced alterations in CNS cell populations and postnatal function. *Teratology* 16:235–246.

Saillenfait, A.M., Gallissot, F., Morel, G., and Bonnet, P. (2003) Developmental toxicities of ethylbenzene, *ortho-*, *meta-*, *para-*xylene and technical xylene in rats following inhalation exposure. *Food Chem. Toxicol.* 41(3):415–429.

Saillenfait, A.M., Gallissot, F., Sabaté, J.P., Bourges-Abella, N., and Muller, S. (2007) Developmental toxic effects of ethyl-benzene or toluene alone and in combination with butyl acetate in rats after inhalation exposure. *J. Appl. Toxicol.* 27(1):32–42.

Schwetz, B.A., Leong, B.K., and Gehring, P.J. (1975) The effect of maternally inhaled trichloroethylene, perchloroethylene, methyl chloroform, and methylene chloride on embryonal and fetal development in mice and rats. *Toxicol. Appl. Pharmacol.* 32:84–96.

Schwetz, B.A., Price, C.J., George, J.D., Kimmel, C.A., Morrissey, R.E., and Marr, M.C. (1992) The developmental toxicity of diethylene and triethylene glycol dimethyl ethers in rabbits. *Fundam. Appl. Toxicol.* 19(2):238–245.

Smith, M.K., Randall, J.L., Read, E.J., and Stober, J.A. (1989) Teratogenic activity of trichloroacetic acid in the rat. *Teratology* 40(5):445–451.

Smith, M.K., Randall, J.L., Read, E.J., and Stober, J.A. (1992) Developmental toxicity of dichloroacetate in the rat. *Teratology* 46(3):217–223.

Spencer, P.J. (2005) New toxicity data for the propylene glycol ethers—a commitment to public health and safety. *Toxicol. Lett.* 156(1):181–188.

Swan, S.H., Main, K.M., Liu, F., Stewart, S.L., Kruse, R.L., Calafat, A.M., Mao, C.S., Redmon, J.B., Ternand, C.L., Sullivan, S., and Teague, J.L. (2005) Decrease in anogenital distance among male infants with prenatal phthalate exposure. *Environ. Health Perspect.* 113(8):1056–1061.

Sweeney, L.M., Tyler, T.R., Kirman, C.R., Corley, R.A., Reitz, R.H., Paustenbach, D.J., Holson, J.F., Whorton, M.D., Thompson, K.M., and Gargas, M.L. (2001) Proposed occupational exposure limits for select ethylene glycol ethers using PBPK models and Monte Carlo simulations. *Toxicol. Sci.* 62(1):124–139.

Terry, K.K., Elswick, B.A., Stedman, D.B., and Welsch, F. (1994) Developmental phase alters dosimetry–teratogenicity relationship for 2-methoxyethanol in CD-1 mice. *Teratology* 49(3):218–227.

Thiel, R. and Chahoud, I. (1997) Postnatal development and behaviour of Wistar rats after prenatal toluene exposure. *Arch. Toxicol.* 71(4):258–265.

Tyl, R.W., Ballantyne, B., Fisher, L.C., Fait, D.L., Savine, T.A., Dodd, D.E., Klonne, D.R., and Pritts, I.M. (1995) Evaluation of the developmental toxicity of ethylene glycol aerosol in the CD rat and CD-1 mouse by whole-body exposure. *Fundam. Appl. Toxicol.* 24(1):57–75.

Tyl, R.W., Ballantyne, B., France, K.A., Fisher, L.C., Klonne, D.R., and Pritts, I.M. (1989) Evaluation of the developmental toxicity of ethylene glycol monohexyl ether vapor in Fischer 344 rats and New Zealand white rabbits. *Fundam. Appl. Toxicol.* 12(2):269–280.

Tyl, R.W., Millicovsky, G., Dodd, D.E., Pritts, I.M., France, K.A., and Fisher, L.C. (1984) Teratologic evaluation of ethylene glycol mono-butyl ether in Fischer 344 rats and New Zealand white rabbits following inhalation exposure. *Environ. Health Perspect.* 57:47–68.

Tyl, R.W., Price, C.J., Marr, M.C., Myers, C.B., Seely, J.C., Heindel, J.J., and Schwetz, B.A. (1993) Developmental toxicity evaluation of ethylene glycol by gavage in New Zealand white rabbits. *Fundam. Appl. Toxicol.* 20(4):402–412.

Vahter, M. (2008) Health effects of early life exposure to arsenic. *Basic Clin. Pharm. Toxicol.* 102:204–211.

Vannier, B., Bremaud, R., Benicourt, M., and Julien, P. (1989) Teratogenic effects of polyethylene glycol 200 in the mouse but not in the rat. *Teratology* 40(3):302.

Wang, A., Holladay, S.D., Wolf, D.C., Ahmed, S.A., and Robertson, J.L. (2006) Reproductive and developmental toxicity of arsenic in rodents: a review. *Int. J. Toxicol.* 25:319–331.

Wang, R.Y. and Needham, L.L. (2007) Environmental chemicals: from the environment to food, to breast milk, to the infant. *J. Toxicol. Environ. Health B Crit. Rev.* 10(8):597–609.

Wilson, J.G. (1965) Embryological considerations in teratology. In: Wilson, J.G. and Warkany, J., editors. *Teratology: Principles and Techniques*, Chicago, IL: University of Chicago, pp. 251–277.

Windham, G.C., Shusterman, D., Swan, S.H., Fenster, L., and Eskenazi, B. (1991) Exposure to organic solvents and adverse pregnancy outcomes. *Am. J. Ind. Med.* 20:241–259.

Wittassek, M., Heger, W., Koch, H.M., Becker, K., Angerer, J., and Kolossa-Gehring, M. (2007) Daily intake of di(2-ethylhexyl) phthalate (DEHP) by German children—a comparison of two estimation models based on urinary DEHP metabolite levels. *Int. J. Hyg. Environ. Health* 210(1):35–42.

World Health Organization (WHO) (2003) *Global Strategy for Infant and Young Child Feeding*, Geneva, Switzerland: World Health Organization and UNICEF.

Yang, J.M., Jiang, X.Z., Chen, Q.Y., Li, P.J., Zhou, Y.F., and Wang, Y.L. (1996) The distribution of $HgCl_2$ in rat body and its effects on fetus. *Biomed. Environ. Sci.* 9(4):437–442.

Yoshida M. (2002) Placental to fetal transfer of mercury and fetotoxicity. *Tohoku J. Exp. Med.* 196(2):79–88.

112

OTOTOXICITY

Matthew Mifsud and K. Paul Boyev

BACKGROUND

Impairment of the organs responsible for hearing and balance remains one of the most important chronic health conditions in the modern era. Given the complexity of these senses, an extensive list of underlying pathophysiologic mechanisms exists: the impact of toxic chemicals is both substantial and can be poorly understood. Ototoxicity is defined as any injury to the auditory or vestibular system caused by drugs or other chemical substances (WHO, 1994). Not surprisingly, the effects induced by these agents can be wide ranging and include tinnitus (ringing in the ear), hearing loss, dysequilibrium, or vertigo. Agents with ototoxic potential have been known to exist for over 1000 years. It was not until the nineteenth century that ototoxic substances were well described, the first two key agents being quinine and salicylate (Schacht and Hawkins, 2006). Once pure forms of these substances were synthesized and produced for evolving western medical practice, some treated with these compounds were observed to develop reversible tinnitus and hearing loss. At present the list of both known and potential ototoxins has grown expansive (see Table 112.1 for common examples). Throughout this chapter it is our hope to provide an overview of both the underlying mechanisms and context of ototoxicity.

BASIC CELLULAR BIOLOGY OF THE AUDITORY AND VESTIBULAR SYSTEM

Neural structures of both the auditory and vestibular systems are housed within the bony labyrinth of the human inner ear. It is a system of fluid-filled channels containing a series of mechanical sensors. These transduce sound pressure and linear/angular acceleration into the electrochemical nerve impulses that are then transmitted to central hearing and balance centers in the brainstem and brain (Forge and Wright, 2002). This specialized function is dependent on an intricate cellular architecture, uniquely arrayed in each component of the inner ear (Richardson et al., 2011). The key cellular structures of the auditory system are found within an area of the cochlea known as the organ of corti. This is a tetrahedral structure that runs the length of the snail-shaped cochlea. The mechanosensory function is performed by specialized "hair cells" that are rigidly organized in four separate rows. Each cell is surrounded by numerous structural cellular elements (i.e., Deiters' and pillar cells) and synapses with nerve bundles whose cell bodies are principally contained in a region known as the spiral ganglion. In the vestibular organs, hair cells perform the function of detecting acceleration and gravity (which is a type of acceleration). Three semicircular canals transduce angular acceleration, while the utricle and saccule transduce linear acceleration (Richardson et al., 2011).

Auditory and vestibular hair cells are distinguished by a specialized bundle of 20–300 stereocilia found along the apical surface (Phillips et al., 2008). They are organized by descending height into separate rows; on each cell surface a single long stereocilium (the kinocilium) is present. The arrangement is maintained by cellular cytoskeletal elements and linkage proteins (Forge and Wright, 2002). The stereocilial bundle can thus move in a coordinated manner in response to mechanical stimuli. Deflection of the hair cell bundle (i.e., in response to sound pressure or movements of fluid imparted by acceleration) will open mechanotransduction receptors along the apical surface, permitting influx of potassium, and calcium ions; altering the intracellular

Hamilton & Hardy's Industrial Toxicology, Sixth Edition. Edited by Raymond D. Harbison, Marie M. Bourgeois, and Giffe T. Johnson.
© 2015 John Wiley & Sons, Inc. Published 2015 by John Wiley & Sons, Inc.

TABLE 112.1   Agents with Commonly Reported Ototoxicity

Medicine	Antimicrobials	Aminoglycosides	Gentamicin, kanamycin, amikacin, tobramycin, neomycin, polymyxin-B
		Macrolides	Erythromycin, azithromycin, clarithromycin
		Antimalarials	Quinine, chloroquine
	Antineoplastic		Cisplatinum, carboplatinum, bleomycin
	Other drug classes	Salicylates	Aspirin, ibuprofen, naproxen
		Loop diuretics	Furosemide, ethacrynic acid
Industrial	Solvents		Toluene, styrene, ethylbenzene, xylene, trichloroethylene
	Heavy metals		Mercury, lead, trimethylin
	Asphyxiants		Carbon monoxide and hydrogen cyanide

electrical potential and generating an appropriate electrical impulse (Ciuman, 2011).

The inner ear is notable for its use of potassium as a driving force behind mechanical sensation. A relatively high potassium fluid known as endolymph fills the membranous labyrinth. Unlike most extracellular fluid, this high potassium concentration resembles that typically found within an intracellular environment, and requires significant metabolic activity to maintain. Endolymph is produced by specialized epithelial cells most prominently encountered within an area along the lateral wall of the cochlea, the stria vascularis (Forge and Wright, 2002; Raphael and Altschuler, 2003). The cell surface architecture in this region utilizes ion channels and active ion transporters (a Na-K-Cl co-transporter) to maintain this potassium concentration. This results in an electrical gradient between the apical and basal surfaces of the hair cell, which in turn creates a net positive endocochlear potential ($+80\,mV$) that further assists passive transport of potassium into these cells and enhances the sensitivity of the ear to sound vibrations (Hibino et al., 2010).

The inner ear is especially vulnerable to dysfunction due to the presence of the unique cell types, delicate architectural framework, and high metabolic demand described above (Richardson et al., 2011). Unfortunately, the human inner ear also has essentially negligible regenerative ability, and injury is often permanent. After embryonic development, hair cells are in a postmitotic state preventing duplication. This is mediated by a group of regulatory proteins known as cyclin-dependent kinase inhibitors, which cause cell cycle arrest (Kwan et al., 2009). Interestingly, supporting cells (i.e., Deiters' cells) within non-mammalian vertebrates are able to respond to injury by trans-differentiation into auditory or vestibular hair cells. However, specialization of the mammalian inner ear has sacrificed this ability (Groove, 2010). Hair cell death in humans thus leads to irreversible changes in hearing and or balance.

OXIDATIVE STRESS AND OTOTOXICITY

It had long been presumed that clinical pathology associated with ototoxic agents was caused by direct action on target tissue by these toxins (Schacht and Hawkins, 2006). In certain cases this is likely true; for example loop diuretics (i.e., furosemide) may cause a reversible hearing loss, by inhibition of ion channels within the stria vascularis (Rybak, 2010). Many known ototoxins do not have this direct impact on cellular targets within the inner ear. It is instead believed that cellular damage is instigated by generation of toxic intracellular oxygen and nitrogen free radicals within integral components of the hearing and balance organs (Evans and Halliwell, 1999).

Free radicals such as superoxide (O_2^-), hydroxyl (OH^-), and peroxynitrate ($ONOO^-$) are characterized by the presence of an unpaired electron. Reactive species are generally a by-product of cellular metabolism (particularly mitochondrial oxidative phosphorylation) and have multiple roles in normal cellular function, for example; (1) regulation of intracellular signal transduction, (2) maintenance of cell-to-cell signaling pathways, and (3) as a destructive force within phagocytic cells to assist in the human immune response (Valk et al., 2007). However, the inherent electrical instability of these structures is also linked to a propensity to cause damage to intracellular lipids, protein, and nucleic acids (Deaval et al., 2012).

Reactions involving reactive species can be depicted with the following formulas: (1) $X + HY = HX + Y$, or (2) $X + Y = [X - Y]$ (Evans and Halliwell, 1999). Consequently reactions involving free radicals will generate new reactive species, establishing a cascade of reactions. Under physiologic conditions, both enzymatic (i.e., superoxide dismutase) and non-enzymatic (i.e., glutathione) antioxidant mechanisms are utilized to neutralize this effect, producing an oxidative balance (Evans and Halliwell, 1999). However, when these defenses are overwhelmed by an abundance of reactive species, a chain of adverse reactions can propagate. Persistent damage to the key macromolecules will shut down normal cellular function, eventually dooming the cell to programmed death (apoptosis) or necrosis (Cheng et al., 2005).

Numerous studies over the past two decades have established that free radicals can be generated within the inner ear tissue in response to toxic substances and noise exposure (Kovacic and Somanathan, 2008). Some evidence has also suggested that age-related changes within the inner ear are in part related to prolonged exposure to reactive species (Li and Steyeger, 2009; Seidman et al., 1999). Oxidative stress is an attractive theory to explain acquired auditory and vestibular impairment given the end result on target tissue is cellular death. Recall that the inner ear is particularly vulnerable to this kind of injury due to the lack of regenerative ability.

DRUG INDUCED OTOTOXICITY

A number of currently used therapeutic agents are considered to have ototoxic potential. The vast majority of drug-induced hearing or vestibular impairment is associated with two drug classes; aminoglycoside antibiotics or platinum-based antineoplastic agents (Rybak, 2010). Aminoglycoside antibiotics (such as gentamycin) are bactericidal agents utilized worldwide for treatment of tuberculosis as well as for the life-threatening gram negative infections (i.e., pneumonia or endocarditis). Their notoriety for cochlear, vestibular, and renal toxicity has been a limiting factor in their use.

Ototoxicity is made possible by a tendency of these compounds to accumulate within vestibular or cochlear hair cells. Cellular entry can occur via cell surface channels found associated with stereocilia (Warchol, 2010). In animal studies, entry of aminoglycosides into hair cells is soon followed by a rapid increase in reactive oxygen species (Audo and Warchol, 2012). Free radical generation is related to their ability to chelate metal ions, including iron (Fe^{2+}). This forms highly redox-active complexes, which catalyze the generation of reactive species such as superoxide (Forge and Schacht, 2000). The importance of these reactive complexes to ototoxic effects has been well described over the last two decades. Conversely, treatment of guinea pigs exposed to gentamycin with iron chelators such as deferoxamine or dihydroxybenzoate, has a protective effect on hearing as measured by auditory brainstem responses (Song and Schact, 1996).

Aminoglycoside toxicity does not demonstrate linear dose dependency, and permanent hearing loss can be observed as soon as 4 h after a single treatment (Guthrie, 2008). It has thus been difficult to prevent the ototoxicity associated with these agents, and they are used more commonly in developing nations (due to low cost) or reserved only for the most serious infections. In contrast, platinum-based chemotherapeutic agents (i.e., cisplatin) do demonstrate dose dependency. For these agents both high acute and prolonged/accumulative dosing regimens are linked to irreversible bilateral sensorineural hearing loss (Hyppolito et al., 2006). Accordingly, close audiologic assessment during and after therapy is advised to limit the sequelae of hearing impairment (Al-Khatib et al., 2010).

Platinum-based agents also induce oxidative stress in target tissues, but through a separate pathway. The exact mechanism is still uncertain; however, it is likely that these agents primarily act by depletion of antioxidant defense mechanisms. These antineoplastic drugs are DNA-intercalating agents, and in the inner ear are thought to bind key antioxidants enzymes and/or essential cofactors, preventing their function (Rybak and Ramkumar, 2007). Prolonged exposure thus alters the cellular oxidative balance, tipping the scale in favor of free radical damage. In addition, within the embryonic kidney cells, cisplatin has been shown to upregulate synthesis of a protein known as NOX3 (Banfi et al., 2004). This is an NADPH-oxidase isoform unique to inner ear tissue. Under physiologic conditions it acts to generate a low level of superoxide anion: with enhanced function under the influence of cisplatin it likely reinforces the cascade of free radical damage (Rybak and Ramkumar, 2007).

INDUSTRIAL OTOTOXICITY—THE PROBLEM OF NOISE AND SOLVENT EXPOSURE

Since the Industrial Revolution, workers have been exposed to industrial solvents, heavy metals, and other agents with potential ototoxic effects (Schacht and Hawkins, 2006). For a long time the impact of industrial work on hearing and balance was neglected, possibly due to the subtle effect of some agents and a lack of insight on disorders of the inner ear. The National Institute of Occupational Safety and Health (NIOSH) estimates that roughly 30 million American workers are exposed to hazardous conditions that can impact hearing and balance. According to a 1996 report, roughly 25% of these workers were expected to develop permanent hearing impairment due to these exposures (Franks et al., 1996). Noise standards were put into place in the 1970s to limit the burden of occupational hearing loss, but they have yet to address the role of ototoxins.

Over the past few decades, animal models have been utilized to define the ototoxic potential of many commonly used industrial agents. The most frequently described are toluene, styrene, xylene, ethyl benzene, mercury, trichloroethylene, and carbon disulfide (Prasher, 2002 and Vyskocil et al., 2012). Toluene and styrene in particular cause necrosis or apoptosis of cochlear cell types (including hair cells) in rat models (Sliwinska-Kowalska et al., 2007). In humans, investigational studies have suggested a linear dose-response relationship with high toxin exposure and hearing loss. The best example of this is toluene exposure within rotogravure printing plants. In one study of 124 printing workers in Brazil, an increasing probability of bilateral sensorineural hearing loss was demonstrated with increasing urinary concentrations of a toluene metabolite, hippuric acid (Morata et al., 1997).

Perhaps the greatest challenge posed by industrial environments is the synergistic or additive negative impact that noise and ototoxicants generate. Noise causes both mechanical and metabolic stress on inner ear tissue, leading to a decrease in cellular integrity, neuronal excitotoxicity, a decrease in blood flow (with potential for ischemia/reperfusion injury), short- or long-term alterations of the endocochlear potential, and eventually accumulation of intracellular reactive oxygen species (Henderson et al., 2006). With data derived from animal models, ototoxic agents are thought both to modify the membranous structure of inner ear tissue (increasing susceptibility to acoustic trauma) and to enhance intracellular oxidative stress, potentially leading to greater auditory and/or vestibular impairment (Fechter et al., 2002; Sliwinska-Kowalska et al., 2007).

It is theorized that in the presence of ototoxins, lower levels of noise are required to produce auditory or vestibular impairment than previously reported. This may in part contribute to occupational hearing loss remaining the most common work-related injury in the United States, despite the introduction of permissible (noise) exposure levels by the Occupational Safety and Health Administration(OSHA) (Morata, 2002).

OTOPROTECTION

Assuming a common mechanism of ototoxicity and other forms of acquired hearing loss, it is reasonable to expect that a "cure" can be identified. Most typically, it is felt that otoprotective agents can be applied, which would act to limit cellular oxidative stress or perhaps inhibit downstream cell death pathways, in order to prevent hearing loss (Poirrier et al., 2010). Numerous animal studies demonstrate the otoprotective capability of compounds such as vitamin E, amifostine, dexamethasone, allopurinol, iron chelators (i.e., deferoxamine and dihydroxybenzoate), sodium thiosulfate, D-methionine, and N-acetyl-L-cysteine; in humans there are unfortunately few trials (Poirrier et al., 2010). The antioxidant properties of aspirin have made it one of the more promising agents thus far, despite a well-documented reversible ototoxicity at high doses. In a randomized trial of 195 patients requiring gentamycin, the use of a 14-day course of aspirin was associated with only a 3% incidence of >15 dB hearing loss from baseline levels, in comparison to 13% in the placebo group (Chen et al., 2007). Despite the clear potential of otoprotective therapeutic agents, the studies have involved smaller numbers of subjects, and the data have yet to show a benefit, which would compel more widespread clinical use (Darrat et al., 2007).

For a significant number of individuals at risk of ototoxicity, it is practical to attempt both limiting the degree of exposure to these agents and preventing the synergistic impact of ototoxins and excess noise exposure. Given the sheer number of workers exposed to ototoxins, upwards of 10 million in Europe alone, this is perhaps most essential in an occupational setting (Prasher, 2002). Organizations devoted to occupational safety such as OSHA, the European Union, or WHO have only begun to acknowledge the probable

impact of these chemical compounds. Going forward, an ideal cooperative otoprotective strategy would likely include: (1) multidisciplinary research aimed at identifying industries and toxins most likely to cause auditory or vestibular impairment, (2) rethinking acceptable thresholds for noise exposure to take into account impact of toxin exposure, (3) increased awareness on the benefits of personal protective equipment, including hearing protection, and (4) targeting workers at highest risk for periodic audiologic screening and other relevant health surveillance (Morata, 2003; Sliwinska-Kowalska et al., 2007).

CONCLUSION

Hearing loss and vestibular impairment are important problems within the developed world, currently representing one of the most common chronic health conditions. It is estimated that greater than 30 million Americans are affected in some way, with a yearly cost of $30 billion dollars when considering lost productivity, specialized needs, and medical treatment (Fausti et al., 2005). The inner ear is characterized by an elaborate cellular architecture required for optimal function. This causes it to be vulnerable to excess noise exposure as well as an expansive list of chemical compounds. Over the past few decades we have begun to understand that ototoxicity is an important factor behind the burden of modern day hearing and vestibular impairment. Agents encountered in medical and industrial applications are all unique, but likely share a common mechanism of toxic cellular damage via the generation of oxidative stress. In the future, attempts to limit the impact of these substances will likely require a multifaceted approach aimed at creating otoprotective therapeutics, increased public awareness, and decreasing exposures.

REFERENCES

Al-Khatib, T., Cohen, N., Carret, A.S., and Daniel, S. (2010) Cisplatinum ototoxity in children, long-term follow up. *Int. J. Pediatr. Otorhinolaryngol.* 74(8):913–919.

Audo, I. and Warchol, M.E. (2012) Retinal and cochlear toxicity of drugs: new insights into mechanisms and detection. *Curr. Opin. Neurol.* 25:76–85.

Banfi, B., et al. (2004) NOX3, a superoxide-generating NADPH oxidase of the inner ear. *J. Biol. Chem.* 279(44):46065–46072.

Chen, Y., et al. (2007) Aspirin attenuates gentamicin ototoxicity: from the laboratory to the clinic. *Hear. Res.* 226:178–182.

Cheng, A.G., Cunningham, L.L., and Rubel, E.W. Mechanisms of hair cell death and protection. *Curr. Opin. Otolaryngol. Head Neck Surg.* 13:343–348.

Ciuman, R.R. (2011) Auditory and vestibular hair cell stereocilia: relationship between functionality and inner ear disease. *J. Laryngol. Otol.* 125(10):991–1003.

Darrat, I., Ahmad, N., Seidman, K., and Seidman, M.D. (2007) Auditory research involving antioxidants. *Curr. Opin. Otolaryngol. Head Neck Surg.* 15:358–363.

Deaval, D.G., Martin, E.A., Horner, J.M., and Roberts, R. (2012) Drug-induced oxidative stress and toxicity. *J. Toxicol.* doi: 10.1155/2012/645460.

Evans, P. and Halliwell, B. (1999) Free radicals and hearing: cause, consequence, and criteria. *Ann. N.Y. Acad. Sci.* 884:19–40.

Fausti, S.A., et al. (2005) Hearing health and care: the need for improved hearing loss prevention and hearing conservation practices. *J. Rehabil. Res. Dev.* 42(4 Suppl. 2):45–62.

Fechter, L.D., Chen, G.D., and Johnson, D.L. (2002) Potentiation of noise-induced hearing loss by low concentrations of hydrogen cyanide in rats. *Toxicol. Sci.* 66:131–138.

Forge, A. and Schacht, J. (2000) Aminoglycoside antibiotics. *Audiol. Neurootol.* 5:3–22.

Forge, A. and Wright, T. (2002) The molecular architecture of the inner ear. *Br. Med. Bull.* 63:5–24.

Franks, J.R., Stephenson, M.R., and Merry, C.J. (1996) *Preventing Occupational Hearing Loss—A Practical Guide.* U.S. DHHS, PHS, CDC, NIOSH Publication No. 96–110.

Groove, A.K. (2010) The challenge of hair cell regeneration. *Exp. Biol. Med. (Maywood)* 235:434–446.

Guthrie, O.W. (2008) Aminoglycoside induced ototoxicity. *Toxicology* 249:91–96.

Henderson, D.H., Bielefeld, E.C., Harris, K.C., and Hu, B.H. (2006) The role of oxidative stress in noise-induced hearing loss. *Ear Hear.* 27:1–19.

Hibino, H., Nin, F., Tsuzuki, C., and Kurachi, Y. (2010) How is the highly positive endocochlear potential formed? The specific architecture of the stria vascularis and the role of the ion-transport apparatus. *Pflugers Arch.* 459:521–533.

Hyppolito, M.A., Oliveira, J.A.A., and Rossato, M. (2006) Cisplatin ototoxicity and otoprotection with sodium salicylate. *Eur. Arch. Otorhinolaryngol.* 263:798–803.

Kovacic, P. and Somanathan, R. (2008) Ototoxicity and noise trauma: electron transfer, reactive oxygen species, cell signaling, electrical effects, and protection by antioxidants: practical medical aspects. *Med. Hypotheses* 70:914–923.

Kwan, T., White, P.M., and Segil, N. (2009) Development and regeneration of the inner ear, cell cycle control and differentiation of sensory progenitors. *Ann. N.Y. Acad. Sci.* 1170: 28–33.

Li, H. and Steyeger, P. (2009) Synergistic ototoxicity due to noise exposure and aminoglycoside antibiotics. *Noise Health* 11(42): 26–39.

Morata, T.C. (2002) Interaction between noise and asphyxiants: a concern for toxicology and enviromental health. *Toxicol. Sci.* 66:1–3.

Morata, T.C. (2003) Chemical exposure as a risk factor for hearing loss. *J. Occup. Environ. Med.* 45:676–682.

Morata, T.C., et al. (1997) Toluene-induced hearing loss among rotogravure printing workers. *Scand. J. Work Environ. Health* 23(4):289–298.

Phillips, K.R., Biswas, A., and Cyr, J.L. (2008) How hair cells hear: the molecular basis of hair-cell mechanotransduction. *Curr. Opin. Otolaryngol. Head Neck Surg.* 16:445–451.

Poirrier, A.L., et al. (2010) Oxidative stress in the cochlea: an update. *Curr. Med. Chem.* 17(31):3591–3604.

Prasher, D. (2002) Toxic to your ears. *Saf. Health Practitioner* 20:36–38.

Raphael, Y. and Altschuler, R.A. (2003) Structure and innervation of the cochlea. *Brain Res. Bull.* 60:397–442.

Richardson, G.P., Monvel, J.B., and Petit, C. (2011) How the genetics of deafness illuminates auditory physiology. *Annu. Rev. Physiol.* 73:311–334.

Rybak, L.P. (2010) Vestibular and auditory ototoxicity. In: Flint, P.W., Haughey, B.H., Lund, V.J., Niparko, J.K., Richardson, M.A., Robbins, T.K., and Thomas, J.R., editors. *Cummings Otolaryngology Head & Neck Surgery*, 5th ed., Mosby, pp. 2169–2178.

Rybak, L.P. and Ramkumar, V. (2007) Ototoxicity. *Kidney Int.* 72:931–935.

Schacht, J. and Hawkins, J.E. (2006) Sketches of otohistory. Part 11. Ototoxicity: drug-induced hearing loss. *Audiol. Neurootol.* 11:1–6.

Seidman, M.D., Quirk, W.S., and Shirwany, N.A. (1999) Mechanisms of alterations in the microcirculation of the cochlea. *Ann. N.Y. Acad. Sci.* 884:226–232.

Sliwinska-Kowalska, M., et al. (2007) Ototoxicity of organic solvents-from scientific evidence to health policy. *Int. J. Occup. Med. Environ. Health* 20(2):215–222.

Song, B.B. and Schact, J. (1996) Variable efficacy of radical scavengers and iron chelators to attenuate gentamicin ototoxicity in guinea pig *in vivo. Hear. Res.* 94:87–93.

Valk, M., Leibfritz, D., Moncol, J., Cronin, M.T.D., Mazur, M., and Telser, J. (2007) Free radicals and antioxidants in normal physiological functions and human disease. *Int. J. Biochem. Cell Biol.* 39(1):44–84.

Vyskocil, A., et al. (2012) A weight of evidence approach for the assessment of ototoxic potential of industrial chemicals. *Toxicol. Ind. Health* 28(9):796–819.

Warchol, M.E. (2010) Cellular mechanisms of aminoglycoside ototoxicity. *Curr. Opin. Otolaryngol. Head Neck Surg.* 18(5):454–458.

World Health Organization (WHO) (1994) *Report of an Informal Consultation on Strategies for Prevention of Hearing Impairment from Ototoxic Drugs*, WHO/PDH/95.2, Geneva.

113

DERMAL AND OCULAR TOXICITY

Jayme P. Coyle, Alison J. Abritis, Amora Mayo-Perez, Marie M. Bourgeois, and Raymond D. Harbison

INTRODUCTION

Skin—the body's largest organ, measuring an average of around $2\,m^2$ in area—provides a substantial area for the possibility of injury and breach, and through that breach, provides a potentially large portal to internal system damage. Eyes, on the other hand, are relatively insignificant if the area dimensions are the only consideration. On face value then, it may appear counter-intuitive to combine the two organs in a discussion of the dynamics of contact damage from chemical substances. However, sensitization, irritation, and corrosive responses in both organs operate by virtually the same mechanisms.

Although the mechanisms of these three responses operate similarly within these two organs, they are distinct responses themselves. At times, they may occur simultaneously or the presence of one may mask the presence of another. However, each physiological phenomenon has its own distinct meaning.

Dermal and ocular irritation is generally accepted to be the reversible tissue damage of the organ caused by the administration of a substance. The inflammatory response from such damage is typically manifested as redness, warmth, swelling, and discomfort. Dermal or ocular corrosion is generally accepted to be the irreversible damage to tissues of the exposed organ. The inflammatory response is magnified, with loss of tissue integrity. Sensitization is a two-phase immunological response that may manifest as irritation or even corrosion in severe cases, but the tissue damage is actually a result of the body's own response to the exposure.

For the skin and eyes, risks of an exposure actually eliciting a dose response are very different. The epidermis or the skin is made up of several layers, the outer layer of which (the stratum corneum) provides a protective shield against most of the large complex molecules. In the absence of a breach, substances of exposure must be of smaller molecular structure, or be of such an irritative or corrosive nature that they can breach the layer themselves. Providing additional protection by way of garments, gloves, footwear, and headgear can generally be achieved without great difficulty, except against the more corrosive of substances.

The eyes, however, provide an optimal milieu for penetration by multiple sizes and forms of substances. The mucosal membranes allow easy absorption of soluble substances, while the fragility of the highly vascular tissues provides easy breaching by irritative substances. Of the 61% of facial injuries and injuries involving workday loss, roughly 70% involved eye injury from chemical exposure or from contact with foreign matter (Harris, 2011). Protective eye-ware is generally easily obtainable but maintaining full scope of vision without distortion or sight field restriction can often be problematic, depending upon the degree of protection needed.

Methods of testing substances to determine their capacity for tissue damage have only recently evolved into quantitative methods. Curiously, the origins of ocular and dermal testing arose not from the occupational injuries, but from consumer complaints and injuries from cosmetics, such as coal-tar-based mascara (Wilhelmus, 2001). One of the most known tests for both ocular and skin irritation/corrosion is the Draize test which exposed rabbits to irritative substances on a specific schedule of concentrations and different time spans (Draize et al., 1944). The test soon came under scrutiny as public concerns for the treatment of test animals and the conditions in which they were subjected to testing became a topic of discussion. While the Draize test is still considered "an effective, practical method of evaluating skin irritation

Hamilton & Hardy's Industrial Toxicology, Sixth Edition. Edited by Raymond D. Harbison, Marie M. Bourgeois, and Giffe T. Johnson.
© 2015 John Wiley & Sons, Inc. Published 2015 by John Wiley & Sons, Inc.

and contact sensitization reactions" (Farage et al., 2011), other testing protocols have been developed to substantially reduce the need for *in vivo* testing and to reduce the use of animals in tests for which there is no substitute for *in vivo*. Most regulatory agencies in the United States, as well as many other countries, show preferences for testing methods other than the Draize (OECD, 1992, 2010a, 2010b, 2010c).

The importance of consumer protection notwithstanding, an understanding of the nature of eye and skin responses to substance exposure and the testing protocols in use are an important facet in industrial toxicology. Exposures can come directly through industrial processes or indirectly by secondary contact. Sole reliance on a substance's material safety data sheet (MSDS) may not provide adequate protection. There is a potential for insufficient information concerning dermal and ocular damage through contact with the substance (Cashman et al, 2012). A lack of overt skin reaction upon accidental exposure may not mean that no tissue injury will happen. Certain tests may show the existence of adverse dermatological response, but not the relationship of dose to response. The most efficient way to reduce risk from a potentially harmful substance is to consider the testing protocols, workers exposure, and industrial processes.

DERMAL SENSITIZATION

OVERVIEW

Skin sensitization, also referred to as allergic dermatitis or allergic contact dermatitis, is defined as an immunological response to a substance exposure which occurs in the cutaneous tissue, manifesting itself most often as pruritus (itching), erythema (redness), edema (swelling), papules (pimples or small bumps), vesicles (fluid-filled bumps), bullae (large blisters), or some combination thereof (U.S. EPA, 2005). Skin irritation and sensitization have often been considered somewhat synonymously as an inflammatory response to an exposure with testing methods, such as patch testing, doing little to differentiate between the two (Farage et al., 2011). However, the foundations for each response are decidedly different.

The degree of skin irritation by a substance is directly related to the ability of the substance or its soluble components to damage the epidermis and/or other skin layers; is generally confined to the area of exposure; and is usually measured in seconds or minutes considering physiological response time. Skin sensitization is mediated by the degree of bodily response to the substance exposure, irrespective of the degree of substance's ability to cause direct physiological damage to the body tissue. The effects of skin sensitization, by activating a cascade of immunological responses, may display well beyond the borders of the exposure, may affect physiological systems, and may not manifest itself until minutes to several hours after exposure.

MECHANISM OF SENSITIZATION

As with immunological reactions, skin sensitization is a two-phase process. The first phase, induction, involves the passage of the substance through the stratum corneum of the epidermis and the subsequent formation of hapten–protein complexes. This complex can be formed with either the original substance or its bioactivated metabolite. The complex is then taken up by the Langerhans cells, which migrate through the lymphatic system to the draining lymph nodes where memory T-cells are created (OECD, 2012; Robinson et al., 2000).

The second phase, elicitation, is initiated upon recontact with the substance or its reactive metabolite. Memory T-cells are then triggered, inducing the formation of cytotoxic T-cells, as well as other inflammatory response cells.

The complexity of a sensitization reaction also allows for some constraints on its formation. The substance must be able to permeate the outer epidermal layer so as to make contact with the dendritic cells necessary for its transport. If the metabolite is the sensitizer, then the substance permeation must be in the physiological range for its bioactivation. Industrial organic compounds are thought to form covalent bonds with the thiols or primary amines in skin proteins (OECD, 2012) thus substrate structure and the accessibility of these residues may also play a role in the degree of induction responses.

METHODS OF SENSITIVITY TESTING

As mentioned, skin irritation and sensitization have historically been used almost interchangeably. Patch tests, sleeve tests, and other protocols of topical or intradermal administration have used visual and colorimetric measurements to define the degrees of response (Farage et al., 2011). Unfortunately, visual measures can be insufficient to determine the actual underlying mechanism of injury. In Figure 113.1, the erythema, edema, papule, and bullae formation appear quite similar even though the left image is a sensitization reaction and right image is an irritative reaction. While the general opinion is that the greater percentage of occupational dermatitis may arise from irritative properties, some studies of occupational dermatitis indicated that allergic dermatitis may be more likely to occur than irritative dermatitis depending upon the industry and worker demographics (Belsito, 2005; Cashman et al., 2012).

Many of the techniques for *in silico/in vitro/in vivo* methods are used interchangeably with studies for skin and ocular irritation/corrosion and are well-described in the following sections. Currently, there are three accepted methods for skin sensitization *in vivo* under the Federal

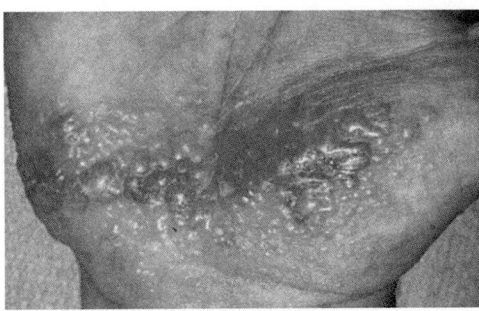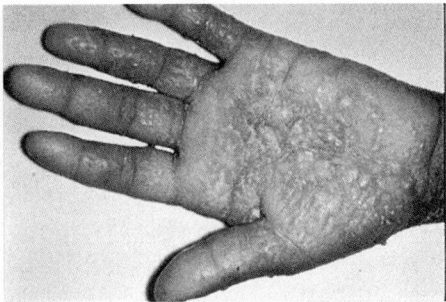

FIGURE 113.1   Visual similarities in skin reactions by different mechanisms. The left image shows skin sensitization reaction to neoprene keyboard wrist rest while the right side shows severe irritation/corrosion due to contact with alkalis (CDC, 2001).

Insecticide, Fungicide, and Rodenticide Act of 1947 (FIFRA) and the Toxic Substances Control Act of 1976 (TSCA) which are the murine local lymph node assay (LLNA), the guinea-pig maximization test (GPMT) and the Buehler Test (EPA, 2003). The LLNA protocol allows for lesser test animal distress than the GPMT and the Buehler Test, while providing greater accuracy in calculations of dose per unit area (EPA, 2003; Robinson et al., 2000). In the global community, the LLNA has been modified to compensate for testing locations with limited access to radioisotopes of thymidine or iodine or difficulties with the disposal thereof (OECD, 2010a, 2010b). Of the three, the LLNA is the preferential method for sensitization testing and thus will be discussed within this section.

Murine Local Lymph Node Assay (LLNA)

The LLNA is a sensitization test that gauges the induction phase through the measurement of lymphocytic proliferation in lymph nodes proximal to the area of substance administration. The substance is applied to the dorsum of the test mouse. Measurement of the lymphocytic proliferation is performed by β-scintillation counting of radioisotopes ^{3}H-methyl thymidine or ^{125}I-iododeoxyuridine, one of which having been injected in solution via tail vein prior to sacrifice of the animal (OECD, 2010c). Visual assessments of erythema are noted on a numerical scale of severity and post-mortem ear-punch skin samples are measured for changes in ear thickness.

The advantages of the LLNA include a lesser degree of distress to the testing animal, lower costs associated with animal acquisition and housing, and a more precise ability to calculate dose per unit area (OECD, 2010c; Robinson et al., 2000). Disadvantages, as stated, are that access to and/or disposal of the radioisotopes may be restricted in some world regions. In such cases the use of modified LLNA protocols, the LLNA-DA and the local lymph node assay-5-bromo-2-deoxyuridine-enzyme-linked immunosorbent assay (LLNA-BrdU-ELISA), are also permitted under the guidelines of the

Organisation for Economic Co-operation and Development (OECD), of which the United States is a member. Local lymph node assay-DA exchanges the use of radioisotopes with the measurement via bioluminescence to quantify the adenosine triphosphate (ATP) content of the lymphocytes (OECD, 2010a). Local lymph node assay-BrdU-ELISA similarly replaces the radioisotopes with BrdU, the uptake of which is measured by ELISA (OECD, 2010b). All three protocols are meant to assess the induction phase of skin sensitization and provide a quantitative determination of the dose–response relationship.

OCCUPATIONAL CONSIDERATIONS

The two-phase process generally requires some passage of time between phases for the complete development of the induction phase. Rarely, a dermal sensitivity response may appear to be almost immediate when the substance is readily absorbed into the skin of a highly immunoreactive individual. More commonly, the first contact with a substance elicits only the induction phase and no overt dermal response is noted. Therefore, the lack of a dermal response from workers in the occupational setting is not an indication of a lack of sensitization in the same.

Some substances are more likely than others to elicit a sensitization reaction. According to some patch test studies, nickel sulfate may be twice as likely to cause a skin sensitization as another substance such as formaldehyde (Cashman et al., 2012).

REFERENCES

Belsito, D.V. (2005) Occupational contact dermatitis: etiology, prevalence, and resultant impairment/disability. *J. Am. Acad. Dermatol.* 53:303–313.

Cashman, M.W., Reutemann, P.A., and Ehrich, A. (2012) Contact Dermatitis in the United States, Epidemiology, Economic Impact and Workplace Prevention. *Dermatol. Clin.* 30:87–98.

Centers for Disease Control and Prevention (CDC) (2001) *Occupational Dermatoses*. Slideshow. Available at http://www.cdc.gov/niosh/topics/skin/occderm-slides/ocderm.html

Draize, J.H., Woodard, G., and Calvery, H.O. (1944) Methods for the study of irritation and toxicity of substances applied to the skin and mucous membranes. *J. Pharmacol. Exp. Ther.* 82:377–390.

Environmental Protection Agency (EPA) (2003) *Health Effects Test Guideline OPPTS 870.2600—Skin Sensitization*, EPA 712-C-03-197, Washington, DC: U.S. Environmental Protection Agency.

Farage, M.A., Maibach, H.I., Andersen, K.E., Lachapelle, J.M., Kern, P., Ryan, C., Ely, J., and Kanti, A. (2011) Historical perspective on the use of visual grading scales in evaluating skin irritation and sensitization. *Contact Dermatitis* 65:65–75.

Harris, P.M. (2011) *Nonfatal Occupational Injuries Involving the Eyes, 2008*, U.S. Bureau of Labor Statistics.

Organisation for Economic Co-operation and Development (OECD) (1992) *OECD Guidelines for the Testing of Chemicals: Skin Sensitization*, OECD Testing Guideline 406, Paris, France.

Organisation for Economic Co-operation and Development (OECD) (2010a) *OECD Guidelines for the Testing of Chemicals: Skin Sensitization: Local lymph Node Assay: DA*, OECD Testing Guideline 442A, Paris, France.

Organisation for Economic Co-operation and Development (OECD) (2010b) *OECD Guidelines for the Testing of Chemicals: Skin Sensitization: Local Lymph Node Assay: BrdU-ELISA*, OECD Testing Guideline 442B, Paris, France.

Organisation for Economic Co-operation and Development (OECD) (2010c) *OECD Guidelines for the Testing of Chemicals: Skin Sensitization: Local Lymph Node Assay*, OECD Testing Guideline 429, Paris, France.

Organisation for Economic Co-operation and Development (OECD) (2012) *The Adverse Outcome Pathway for Skin Sensitisation Initiated by Covalent Binding to Proteins: Part I. Scientific Evidence*, Series on Testing and Assessment No. 168, Document ENV/JM/MONO(2012)10/PART1, Paris, France: OECD Environment, Health and Safety Publication.

Robinson, M.K., Gerberick, G.F., Ryan, C.A., McNamee, P., White, I.R., and Basketter, D.A. (2000) The importance of exposure estimation in the assessment of skin sensitization risk. *Contact Dermatitis* 42:251–259.

U.S. EPA (2005) *Specific Organ/Tissue Toxicity*, Title 40 Code of Federal Regulations, Pt. 798, 2005 ed.

Wilhelmus, K.R. (2001) The Draize eye test. *Surv. Ophthalmol.* 45:493–515.

DERMAL IRRITATION

OVERVIEW

Several compounds are known to be either dermal irritants or corrosive. Occupationally, the reported rates of new cases of skin diseases have been estimated to be between 0.5 and 1.0 case per 1000 workers in Europe, and responsible for about 33% of the suspected work-associated diseases in 2010. From 90 to 95% of the total reported skin diseases were accredited to contact dermatitis (Diepgen, 2012). In the United States, estimates from the 2010 National Health Interview Survey conducted by the National Institute for Occupational Safety and Health (NIOSH) reported a prevalence of dermatitis at 9.8% among those currently working, though the rates of allergic versus irritant were not reported (Luckhaupt et al., 2013). Both reports, however, recognize that these figures may be subjected to underreporting, possibly on the order of 50-fold or higher. Hanifin et al. (2007) reported that 10.7% of survey participants of the general population demonstrated at least one eczematous symptom consistent with contact dermatitis, though not all cases were confirmed by physician diagnosis. However, the rates of dermatitis in the general population are unknown, thus undermining the fraction credited to occupational dermatitis, especially with suspected high rates of underreporting in both the general and working populations. Therefore, objective methods for delineating agents suspected to cause dermal irritation, sensitization, and corrosion remain paramount in the regulatory as well as in the workplace setting.

Recently, a paradigmatic shift in regulatory dermal toxicity testing has been underway, moving from *in vivo* models to *in vitro* models due to animal safety and general reductions in laboratory animal use, development of sensitive *in vitro* models, and difficulty in bridging animal data with human toxicological relevance. Table 113.1 shows tests on current common use. Therefore, regulatory testing relies on *in vivo* testing typically when the weight of evidence from existing human and animal data cannot rule out the necessity of testing, especially for agents known or suspected to cause harm. Often, suspected agents, which are expected to produce substantial dermal and systemic injury resulting in pain and suffering, are tested using a standardized sequential tiered approach in order to reduce the number of test animal required to exhibit dermal irritation or corrosion.

DERMAL IRRITATION ASSAYS

In the regulatory framework, irritants are described as agents which have the potential to produce reversible dermal erythema (dermal redness), eschar (focal dermal necrosis) or edema (subcutaneous fluid accumulation). In some instances, generalized inflammatory response may be exhibited, but these responses lack humoral involvement typical of sensitization. Therefore, for an agent to be an irritant, even first-time contact can produce symptoms described above, though symptoms presented may be similar to allergic reactions.

Historically, irritants have been determined and graded based on severity, using an *in vivo* approach using albino

TABLE 113.1 Common Dermal Tests

Test Name	Dermal Endpoint Assessed	*In Vivo/In Vitro/ Ex Vivo*
Guinea pig maximization test	Sensitization	*In vivo*
Buehler test	Sensitization	*In vivo*
Local lymph node assay	Sensitization	*In vivo*
Human repeat insult patch test	Sensitization	*In vivo*
In vitro assays (*various*)	Sensitization	*In vitro*
Draize skin irritation test	Contact irritant corrosion	*In vivo*
Reconstructed human epidermis systems	Contact irritant corrosion	*In vitro*
Membrane barrier test (Corrositex®)	Corrosion	*In vitro*
Transcutaneous electrical resistance test	Corrosion	*Ex vivo*

TABLE 113.2 UN GHS Dermal Irritant Classification Criteria (Excludes Corrosion)

Category	Pretesting Considerations Warranting Classification	Draize Testing Criteria (Mean Scores)
2	- Existing data warrants classification - Positive result of valid *in vitro* test - QSAR-based classification	- Score ≥2.3 and <4.0 for erythema/eschar or edema in 2 of 3 animals - Persistent inflammation at the end of the 14-day observation period in 2 of 3 animals
3	- Existing data warrants classification	- Score ≥1.5 and <2.3 for erythema/eschar in 2 of 3 animals

rabbits called the Draize skin irritation test. Often, the Draize test combines screening for both irritant and corrosive agents, the latter which shall be discussed later. However, interest in increasing predictive power of dermal irritant detection has prompted a shift from the Draize test to several commercially available *in vitro* models individually validated by the European Centre for the Validation of Alternative Methods (ECVAM). Assessment of novel *in vitro* models depends on their comparative sensitivity and specificity against the Draize test. Paradoxically, neither the Draize test nor the *in vitro* models have been compared to direct human testing, thus confirmation of irritant class to humans is lacking for both models, but is generally supported with existing human and animal data (Basketter et al., 2012).

Dermal irritant classes have been promulgated by the United Nations (UN) Globally Harmonized System of Classification and Labelling of Chemicals (GHS), and span Category 1 (corrosive), which shall be later discussed, Category 2 (irritant), and Category 3 (mild irritant); the latter is subject to optional labeling depending upon the agency. Criteria for irritant classes are listed in Table 113.2. Any compound with a pH ≤2, pH ≥11.5, or test positive in any valid *in vitro* dermal test is immediately classified as corrosive (Category 1) (United Nations, 2013).

In Vivo Tests

Draize Skin Irritation Test   The Draize skin irritation test, referred to as the Draize test, is a standardized mammalian model developed by Draize et al. (1944) to screen for dermal irritants and corrosives. The principle of this test relies on the ability of the investigated agent to produce local dermal irritation after a single acute exposure lasting no more than 4 h. The maximum observation period is 14 days to include any latent dermal irritation. Three to six albino rabbits are of

standard sample sizes; generating preliminary results with one test animal is highly recommended, followed by confirmation with two additional animals only in instances of dubious results. The test procedure is described in the OECD Testing Guidelines 404 (2002), and has been validated for solid and liquid substances, but not other physical states. Any resulting dermal lesions are assessed for lesion type (e.g., erythema, eschar, edema, and otherwise noted) as well as reversibility and are tabulated based on a predetermined severity scale measured subjectively. Draize test scoring is provided in Table 113.3.

In Vitro Assays

Reconstructed Human Epidermis   Interest in replacement of the Draize skin irritation test has precipitated the

TABLE 113.3 Draize Skin Irritation Test Grading

	Score
Erythema and Eschar Formation	
No erythema	0
Very slight erythema (barely perceptible)	1
Well-defined erythema	2
Moderate to severe erythema	3
Severe erythema (beef redness) to eschar formation preventing grading of erythema	4
Edema Formation	
No edema	0
Very slight edema (barely perceptible)	1
Slight edema (edges of area well defined by definite raising)	2
Moderate edema (raised approximately 1 mm)	3
Severe edema (raised more than 1 mm and extending beyond area of exposure)	4

introduction of complex, physiologically relevant *in vitro* reconstructed human epidermis (RHE) tests that not only models the keratinaceous barrier of the skin, the *stratum corneum,* but also the underlying stratified layers of the human epidermis, including the strata *basale, spinosium,* and *granulosum.* Test compounds applied to the *in vitro* RHE are assessed by histological staining morphology and cellular viability via 3-(4,5-Dimethylthiazol-2-yl)-2,5-diphenyltetrazolium bromide [MTT] reduction; the latter assay is responsible for categorizing the agent according to skin irritant class. Reductions in treatment cell viability of ≤50% compared to controls warrant classification to dermal irritant Category 2 (causes skin irritation), while cytoviability counts above 50% warrant categorization into Category 3 (causes mild irritation) or simply as a nonirritant. Currently, these systems cannot distinguish between mild irritants and nonirritants, thus requiring further confirmatory testing; however, Category 2 irritants need not undergo *in vivo* confirmation. Additionally, these systems are not validated for gas or aerosolized exposures and are limited to only solid or liquid media. As the system contains an underlying nutrient medium, characterization of cytokines resultant of exposure can be performed to include inflammatory markers.

Therefore, regulatory utilization of RHE offers tangible modeling to assess dermal irritation using four validated commercially available systems (EpiSkin™ (SM); EpiDerm™ SIT (EPI-200); SkinEthic™ RHE; LabCyte EPI-MODEL24 SIT) accepted by the OECD and the United Nations Globally Harmonized system (UN GHS). Validation of the LabCyte EPI-MODEL24 SIT, for example, has determined the sensitivity and specificity of the model as 82.3% and 72.6%, respectively (OECD, 2011).

DERMAL CORROSION ASSAYS

In Vivo Assay

Draize Skin Corrosion Test The Draize skin irritation test simultaneously assesses for corrosion; therefore, the particular aspects of this test as it pertains to the regulatory framework are discussed above. However, the delineating factor between corrosion and irritation is the irreversible tissue damage (necrosis) resultant of acute exposure, e.g., between 3 min and 4 h. Subcategorization within skin irritation Category 1 (labeled superficially as corrosive) depends on the time point observed for the onset of irreversible tissue damage (Table 113.4). Given the severe nature of necrotic tissue damage, the emphasis of using one animal for preliminary assessment is strongly recommended for agents suspected to have corrosive potential based on the weight of evidence from both structural and activity relationships to known corrosives and from existing data (OECD, 2002). This test remains the gold standard for dermal corrosion categorization, but remains subject to partial replacement by in vitro methods.

TABLE 113.4 UN GHS Skin Irritation Category 1 (Corrosive)

Category 1 Subcategory	Exposure Time to Endpoint	Observation Period
1A	≤3 min	≤1 h
1B	>3 min but ≤1 h	≤14 days
1C	>1 h but ≤4 h	≤14 days

In Vitro Assays

Two *in vitro* assays have been validated by the UN GHS for determining dermal corrosion agents from non-corrosives. The first of which uses three RHE models used for skin irritation, but utilize the same endpoints (histology and MTT reduction) for analysis and subcategorization. The EpiSkin™ (SM) model provides the highest resolution to distinguish 1A verses 1B/1C versus non-corrosive. The test cannot distinguish between 1B and 1C corrosives, thus any non-1A corrosive agents are categorized as 1B and 1C. The test system, however, provides high sensitivity (100%) and specificity (80%) in delineating corrosive from non-corrosive. The EpiDerm™ SIT (EPI-200) and SkinEthic™ RHE models are also utilized for this purpose, but both tend to overpredict the severity score to produce a significant rate of false 1A results. Cytoviability thresholds for subcategorization varies depending on the system used, but are dependent upon the time point to produce the criteria for cytoviability (OECD, 2013).

Another validated method available is the membrane barrier test method, commercially known as the Corrositex® assay, and aims to delineate skin corrosives using a proteinaceous matrix supported by a dermal-like basal layer. A positive result is demonstrated by a change in the underlying solution containing an indicator dye. The sensitivity and specificity of this method to categorize known corrosive agents are 89% and 75%, respectively, based on 40 tested chemicals (OECD, 2006).

DERMAL SENSITIZATION ASSAYS

In Vivo Assays

Guinea Pig Tests The guinea pig remained the gold standard in determining sensitization agents within the regulatory framework as in vitro and QSAR methods are not sensitive enough to replace the GPMT or the Buehler test; however, the murine LLNA is now the current gold standard, except in few situations. Nevertheless, the GPMT and Buehler test are methodologically comparable, except that the GPMT uses a sensitization potentiator (Freund's complete adjuvant) to maximize sensitization, thus aiming to limit false negative results. These sensitization assays determine whether agents produce a humoral memory T-cell-mediated immune response upon topical challenge, a process which necessitates exceedance of the induction dose threshold and a 10–14 day refractory exposure period. Responses are scored according to the Magnusson and Kligman scale, while histological examination and skin fold thickness may also provide important clinical endpoints. The use of an adjuvant, irritating or inappropriate solvent vehicle, and relevant concentrations may limit translation to human sensitization potential or may over-categorize sensitizing agents (OECD, 1992).

Murine Local Lymph Node Assay The murine LLNA, created by Kimber et al. (1989), focuses resources on detecting a sensitive proxy of the humoral induction phase, rather than the elicitation phase required of the guinea pig tests. Therefore, the LLNA continues to appreciate refinement in anticipation for its continued replacement of most applications of the guinea pig sensitization model. The LLNA measures the relative lymphocyte proliferation within the auricle lymph nodes after bilateral ear exposure of the tested sensitizing agent. The test originally required lymphocyte uptake of tritiated methyl thymidine or a mixture of 125I-iododeoxyuridine and fluorodeoxyuridine as proxies of lymphocyte proliferation; however, non-radioisotope reporters, such as BrdU uptake and ATP measurement are available (OECD, 2010a, 2010b, 2010c). The index measured is referred to as the stimulation index (SI) with a positivity threshold of 3, meaning the uptake, translated to lymphocyte proliferation, is threefold above that of control animals. The LLNA demonstrates limitations as the LLNA is insensitive to some metals and high molecular weight sensitizers; has a tendency of high false positives, e.g., for surfactants and natural organic acids; requires 3 consecutive daily exposures; cannot distinguish contact dermatitis from hypersensitivity; and remains subject to variability based on the chosen vehicle (Anderson et al., 2011). Nevertheless, the LLNA does not require adjuvant application, decreases time, costs, and welfare concerns, and is a quantitative assay; all of which confer advantages in testing (OECD, 2012).

Nonvalidated Methods

Several nonvalidated and fringe methods exist for assessing sensitizing agents. The method with the highest extrapolation potential, but least favored ethically, remains the human repeat insult patch test (HRIPT). Therefore, the HRIPT is employed only when the benefits far outweigh the risks of the individual participants, especially since result variability is generally high (Basketter, 2009). In addition, several *in vitro* models have been proposed, such as the Genomic Allergen Rapid Detection Test, VITOSENS Test, Myeloid U-937 Skin Sensitization Test [MUSST], Human Cell Line Activation Test [h-CLAT], and the THP-1 cell line, but none have gained general acceptance compared to the LLNA and

guinea pig tests, despite comparable sensitivity and specificity compared to the LLNA. *In vitro* tests generally do not capture the complex nature of inter-tissue interactions, especially with regards to inflammatory responses; thus competition against the robust *in vivo* assays has not currently taken precedence. Unifying the *in vitro* studies, however, is the concept that novel biomarkers associated along the adverse outcome pathway of allergic dermatitis have been identified and may augment evidence in support of the validated methods (Aeby et al., 2010; Ashikaga et al., 2010; Lambrechts et al., 2010; OECD, 2012).

REFERENCES

Aeby, P., Ashikaga, T., Bessou-Touya, S. Schapky, A., Geberick, F., Kern, P., Marrec-Fairley, M., Maxwell, G., Ovigne, J.-M., Sakaguchi, H., Reisinger, K., Tailhardat, M., Martinozzi-Teisser, S., and Winkler, P. (2010) Identifying and characterizing chemical skin sensitizers without animal testing: Colipa's research and methods development program. *Toxicol. In Vitro* 24:1465–1473.

Anderson, S.E., Siegel, P.D., and Meade, B.J. (2011) The LLNA: a brief review of recent advances and limitations. *J. Allergy*, doi: 10.1155/2011/424203.

Ashikaga, T., Sakaguchi, H., Sono, S., Kosaka, N., Ishikawa, M., Nukada, Y., Miyazawa, M., Ito, Y., Nishiyama, N., and Itagaki, H. (2010) A comparative evaluation of *in vitro* skin sensitisation tests: the human cell-line activation test (h-CLAT) versus the local lymph node assay (LLNA). *Altern. Lab. Anim.* 38:275–284.

Basketter, D.A. (2009) The human repeated insult patch test in the 21st century: a commentary. *Cutan. Ocul. Toxicol.* 28(2): 49–53.

Basketter, D., Jírova, D., and Kandárová, H. (2012) Review of skin irritation/corrosion hazards on the basis of human data: a regulatory perspective. *Interdiscip. Toxicol.* 5:98–104.

Diepgen, T.L. (2012) Occupational skin diseases. *J. Dtsch. Dermatol. Ges.* 10(5):297–315.

Draize, J.H., Woodard, G., and Calvery, H.O. (1944) Methods for the study of irritation and toxicity of substances applied to the skin and mucous membranes. *J. Pharmacol. Exp. Ther.* 82:377–390.

Hanifin, J.M., Reed, M.L., and Impact Working Group (2007) A population-based survey of eczema prevalence in the United States. *Dermatitis* 18(2):82–91.

Kimber, I., Hilton, J., and Weisenberger, C. (1989) The murine local lymph node assay for identification of contact allergens: a preliminary evaluation of *in situ* measurement of lymphocyte proliferation. *Contact Dermatitis* 21(4):215–220.

Lambrechts, N., Vanheel, H., Nelissen, I., Witters, H., Van Den Heuval, R., Van Tendeloom, V., Schoeters, G., and Hooyberghs, J. (2010) Assessment of chemical skin sensitizing potency by an *in vitro* assay bases on human dendritic cells. *Toxicol. Sci.* 116:122–129.

Luckhaupt, S.E., Dahlhamer, J.M., Ward, B.W., Sussell, A.L., Sweeney, M.H., Sestito, J.P., and Calvert, G.M. (2013) Prevalence of dermatitis in the working population, United States, 2010 National Health Interview Survey. *Am. J. Ind. Med.* 56(6):625–634.

Organisation for Economic Co-operation and Development (OECD) (1992) *OECD Guidelines for the Testing of Chemicals: Skin Sensitization*, OECD Testing Guideline 406, Paris, France.

Organisation for Economic Co-operation and Development (OECD) (2002) *OECD Guidelines for the Testing of Chemicals: Acute Dermal Irritation/Corrosion*, OECD Testing Guideline 404, Paris, France.

Organisation for Economic Co-operation and Development (OECD) (2006) *OECD Guidelines for the Testing of Chemicals:* In Vitro *Membrane Barrier Test Method for Skin Corrosion*, OECD Testing Guideline 435, Paris, France.

Organisation for Economic Co-operation and Development (OECD) (2010a) *OECD Guidelines for the Testing of Chemicals: Skin Sensitization: Local Lymph Node Assay*, OECD Testing Guideline 429, Paris, France.

Organisation for Economic Co-operation and Development (OECD) (2010b) *OECD Guidelines for the Testing of Chemicals: Skin Sensitization: Local Lymph Node Assay: DA*, OECD Testing Guideline 442A, Paris, France.

Organisation for Economic Co-operation and Development (OECD) (2010c) *OECD Guidelines for the Testing of Chemicals: Skin Sensitization: Local Lymph Node Assay: Brd-U-ELISA*, OECD Testing Guideline 442B, Paris, France.

Organisation for Economic Co-operation and Development (OECD) (2011) *Validation Report for the Skin Irritation Test Method Using LabCyte EPI-MODEL24*. OECD Series on Testing and Assessment ENV/JM/MONO(2011)39, Paris, France: Environment, Health and Safety Publications.

Organisation for Economic Co-operation and Development (OECD) (2012) *The Adverse Outcome Pathway for Skin Sensitisation Initiated by Covalent Binding to Proteins: Part I. Scientific Evidence*, Series on Testing and Assessment No. 168, Document ENV/JM/MONO(2012)10/PART1. Paris, France: OECD Environment, Health and Safety Publication.

Organisation for Economic Co-operation and Development (OECD) (2013) *OECD Guidelines for the Testing of Chemicals:* In Vitro *Skin Corrosion: Reconstructed Human Epidermis (RHE) Test Method*, OECD Testing Guideline 431, Paris, France.

United Nations (UN) (2013) *Globally Harmonized System of Classification and Labelling of Chemicals (GHS)*, 5th revised ed., Document ST/AC.10/30/Rev.5, Geneva, Switzerland: United Nations.

OCULAR IRRITATIVE INJURIES

OVERVIEW

There are two major physiological effects that may result from test chemicals exposure on the eye—corrosion and irritation. Eye corrosion is defined as the production of irreversible tissue damage following application of a test substance to the anterior surface of the eye. Eye irritation is defined by reversible changes following the application of a test substance to the anterior surface of the eye (EPA, 1998). Strongly acidic (pH < 2) or alkaline substances (pH > 11.5) can be presumed to be an eye corrosive.

MECHANISMS OF EYE IRRITATION

For simplicity, chemicals have been classed by one of four general modes of action; cell membrane lysis, coagulation, saponification, and actions on macromolecules. Assignment to these categories is far from absolute as chemicals may possess properties from more than one. Cell membrane lysis, the breakdown of the cellular membrane, involves substances such as surfactants, organic solvents, ketones, alcohols, volatile liquids, ethers, polyethers, esters, or aromatic amines. Coagulation, the denaturation of macromolecules, involves the following chemical classes: acids, cationic surfactants, or organic solvents. Saponification, a breakdown of lipids, involves alkalis. Finally, actions on macromolecules where chemicals react with the organelles of the cell involve the chemical classes of peroxides, mustard agents, alkyl halides, epoxides, or oxidizers (Scott et al., 2010).

The U.S. Environmental Protection Agency (US EPA) and the UN GHS have determined that when a test chemical evaluated during the Tier 1 phase of the testing strategy is judged as potentially causing severe ocular damage, a single animal testing method is then considered appropriate. If anticipated or unanticipated ocular damage is actually observed then no further animal testing is required. However, additional testing may be warranted if no damage is observed in order to establish a threshold for onset of irritation. A more detailed description of this process will be discussed later.

TESTING

Current Standard: Draize Eye Test

The Draize eye test is the current ocular testing standard to evaluate eye irritation potential of test chemicals; all other tests to replace Draize are still in the validation stages. The United States and the European Union are incorporating various testing models into a tier-staging model (Adriaens et al., 2014; Barile, 2010). Draize testing was developed to identify reversible and irreversible (corrosive) effects of substances *in vivo*. Draize, a pharmacologist at the Food and Drug Administration, was tasked with developing an evaluative test when the government passed the Federal Food, Drug, and Cosmetic Act of 1939 in response to reports of ocular damage in women using eyelash dye (Wilhelmus, 2001).

Common chemicals with the potential for eye contact are evaluated using the Draize eye test. The basic protocol uses a single-dose application (0.1 ml or >100 mg) to one eye of several albino rabbits with a treatment time of 24 h before rinsing out the test substance. The treatment effects are evaluated by observation for at least 72 h and up to 21 days (Draize, 1944; EPA, 1998). The test grades ocular lesions in the cornea as well as the typical oculotoxic endpoints of corneal opacity, inflammation, and cytotoxicity (Barile, 2010; Draize, 1944; OECD, 2012b; Wilhelmus, 2001). The Draize eye test method has several limitations, including a lack of reproducibility; subjective assessment; variable interpretation; high test dosage requirements; and over-prediction of the human response to the chemical (Barile, 2010).

Alternative Testing Methods to Draize

Eye irritation testing is used for the evaluation of ocular sensitivity and corrosion test chemicals. As previously stated, no single *in vitro* method or combination of methods have been validated to fully replace Draize eye testing (Adriaens et al., 2014; Scott et al., 2010). The primary difficulty in developing an alternative testing method is the inability to predict the range of irritation across different classes of chemicals currently provided by Draize (OECD, 2013b). Put simply, most new methods fall short. The criteria for the validation of alternative testing methods include: the evaluation of toxic tissue effects; the irreversibility (corrosiveness) of eye damage; within test variability; and assessment of target values of sensitivity, specificity, and accuracy as compared to the Draize eye test values (Adriaens et al., 2014; Katoh et al., 2013).

Generally, the main reasons for method validation failure are high within test variability and low posttest sensitivity, specificity, and accuracy. All the alternative test methods discussed later demonstrated highly variable results, could not distinguish eye irritants from non-irritants or misclassified chemicals (Barile, 2010). These setbacks have made it difficult for agencies to fully replace *in vivo* testing. Using these methods as a part of a multi-tiered or "stacked" strategy has been approved by the United States and other international agencies to reduce the overall burden of *in vivo* testing and perceived animal cruelty. Many cosmetic and pharmaceutical companies have instituted these methods in-house in order to completely eliminate animal testing. Others opt to shift their product evaluations to third party laboratories to eliminate on-site animal testing.

Tier-Testing Strategy

As previously written, none of the alternative methods to Draize eye test are considered valid as a stand-alone replacement for *in vivo* rabbit eye testing. The United States and the international community have adopted a tier-testing strategy to systematically evaluate test chemicals limiting the use of *in vivo* testing until later stages (Barile, 2010; OECD, 2013a; Scott et al., 2010). A substance considered as a potential hazard at Tier 1 would advance to Tier 2 for further testing and so on. Tier 1 uses the structure activity relationship and physiochemical properties to assess the molecule under review. Tier 2 uses short-term bacterial or mammalian assays to assess the test chemicals cytotoxicity and potential inflammatory properties based on cellular release of inflammatory cytokines and other mediators. Tier 3 makes a risk assessment of human and animal eye irritation. This strategy is based on the principle of weight of evidence approach which considers all available information to evaluate eye irritation (Barile, 2010). A relatively simple decision tree permits evaluation of the test chemical given commonly available data; *in vitro* assessments of severe eye irritancy; and finally *in vitro* testing of mild to moderate irritancy (Barile, 2010).

Bottom-Up and Top-Down Approach

This approach is named for the strategy to use presumptive non-irritants in the bottom-up approach and presumptive severe eye irritants in the top-down approach. This approach is applicable in four domains of testing: categorization a test chemical's irritation severity; the test chemical's mode of action; identification of the chemical class; and the physiochemical compatibility (Barile, 2010). Scoring of test chemicals is based both on the specific tissue lesion size and persistence of lesions over time.

Several modes of action have been identified: cell membrane lysis, saponification, coagulation, and actions on macromolecules (Scott et al., 2010). Cell membrane lysis is a breakdown of membrane integrity; coagulation is the precipitation/denaturation of macromolecules such as proteins; saponification is the breakdown of lipids by alkaline action which is progressive with time; actions of macromolecules are chemicals that react with cellular constituents/organelles such as alkylation or oxidative attack on macromolecules (Scott et al., 2010). One major advantage in the bottom-up/top-down approach is that methods that include a stromal layer, as is available in isolated organ methods like isolated chicken and isolated rabbit eye testing, possess the greatest potential to identify severe eye irritants (Scott et al., 2010). The principle is based on the ability of the test chemical to create a lesion that penetrates the epithelium down to the stromal layer (Scott et al., 2010).

Cell Function and Cytotoxicity-Based Assays

Cell function and cytotoxicity-based assays rely on membrane lysis to determine the functionality of the cell based on membrane integrity (Scott et al., 2010).

Fluorescein Leakage Test Fluorescein leakage (FL) test, developed by Tchao (1988) is an in vitro test method that can classify chemicals as irritants or corrosives. It uses the principle based on the integrity of trans-epithelial permeability which is a functional characteristic of the cornea and conjunctiva of the eye (OECD, 2012a). There is a correlation between the increasing permeability of corneal epithelium and the level of inflammation and surface damage observed in the eye with the introduction of an irritant. (OECD, 2012a) The toxic effects of short-term exposure can be measured on a monolayer of Madin-Darby canine kidney cells (MDCK) by the increase of permeability of sodium fluorescein through the monolayer (OECD, 2012a). It is indicated for testing the irritancy of surfactants and surfactant-based materials. It has limited use at this time because it can only accurately classify irritation/corrosion eye risk for a limited number of chemicals. It also requires high chemical concentrations due to its short-term exposure period (OECD, 2012a). This method is currently expected to be best utilized in the top-down approach in a tier testing strategy where the FL test is used when high eye irritancy potential is suspected (OECD, 2012a; Scott et al., 2010).

Isolated Organ Method or Organotypic

Isolated Chicken Eye (ICE) or Isolated Rabbit Eye (IRE) This model was discussed in Prinsen and Koëter (1993) which uses isolated corneas from animals used for consumption. Standard application amounts are 0.3 ml or 0.3 g to the surface (OECD, 2013b). The model measures test chemical effects up to 4 h posttreatment. ICE/IRE measures increases in corneal thickness as a proxy for corneal swelling; corneal opacity to assess corneal damage; morphological changes in cornea epithelium (pitting and loosening); and permeability through corneal cell layers using fluorescein retention (Barile, 2010; OECD, 2013a). This model is limited because it only evaluates corneal effects but does not identify slow acting irritants (Barile, 2010). This model has been found to have high false positive rates for alcohols and high false negative rates of solids and surfactants, although the false negative tests are not considered as critical because of testing by other in vitro methods as part of the tier-testing strategy (OECD, 2013b).

Bovine Corneal Opacity and Permeability (BCOP) This model assumes that corneal damage is a reliable indicator of visual impairment. Using isolated eyes from slaughtered cattle as first described by Gautheron et al. (1992), the degree of a substance's irritative properties are measured by light transmission through the cornea (opacity), sodium fluorescein dye passage through the full thickness of epithelial cell layers (permeability), corneal swelling and histological evaluations of morphological alterations. (Barile, 2010; OECD, 2013a). The bovine corneal opacity and permeability

(BCOP) standard application amount for surfactants is 10% w/v in 0.9% sodium chloride solution, distilled water, or any other solvents with no adverse effect on the experiment, while liquid test chemicals remain undiluted. Non-surfactant solids are applied at 20% w/v concentration in 0.9% sodium chloride solution, distilled water, or any other solvents with no adverse effect on the experiment (OECD, 2013a). Opacity is quantitatively measured using an opacitometer, while permeability is measured by an evaluation of optical density using a UV/VIS spectrophotometer set at 490 nm (Barile, 2010). Low density, water-insoluble substances are not adequately tested using this model, as the substances may not be sufficient for the corneal contact during testing. Additionally, irritants and non-irritants are poorly differentiated by BCOP; thus it is only approved in tier-testing strategies (Barile, 2010; OECD, 2013a).

The BCOP was evaluated by the Interagency Coordinating Committee on the Validation of Alternative Methods (ICCVAM) and the ECVAM. The BCOP was recommended to be used in the bottom-up approach for the classification of chemicals that do not pose a risk of irritation or eye damage. In top-down testing strategies, BCOP is recommended as the initial step in to identify chemicals inducing serious eye damage, preventing further animal testing. (OECD, 2013a).

Porcine corneal opacity (PCOP) follows similar protocols and evaluations which can be used to evaluate eye irritation, thus will not be discussed here.

Organotypic Models

Hen's Egg Test—Chorioallantoic Membrane (HET-CAM) The Hen's *egg test*–chorioallantoic *m*embrane (HET–CAM) model tests a substance's irritative effects on the conjunctiva based on the observation that the CAM of the embryonated hen's egg is similar to the vascularized mucosal tissues of human or rabbit eye (Barile, 2010; Luepke and Kemper, 1986; Parish, 1985). A maximum of 300 µl of test substance is applied to the surface of the CAM for 20 s, after which hyperemia, hemorrhage, coagulation, and changes in blood vessels such as vessel lysis, vasoconstriction, or vasodilation are evaluated as irritation endpoints (Barile, 2010). This model has been validated to predict ocular corrosives and severe irreversible effects and is best used in a multi-tiered testing strategy to assess severe eye damage (Barile, 2010). Chorioallantoic membrane–trypan blue staining (CAM–TBS) is another chorioallantoic membrane method similar to HET–CAM. Unlike HET–CAM, CAM–TBS evaluates histological and pathological changes in the CAM with dye staining (trypan) to determine the absorbance of dye into the cell membrane, which is an indicator of cell death (ECVAM DB-ALM, 1995; Yang et al., 2010).

Cytosensor Microphysiometer (CM) Cytosensor microphysiometer is based on the observation that the measurement of small changes in cellular metabolism reflected by extracellular release of acidic byproducts of energy metabolism (Hartung et al., 2010). The process is automated to measure metabolism in living cells based on the change in the release rate of proton into the culture medium, thus it can detect small changes in pH (Hartung et al., 2010). Cytosensor microphysiometer uses seeded mouse fibroblasts (L929 cells) exposed to a 20 min dosing cycle to determine the changes in cell metabolism (IIVS, 2014). The endpoint analysis is the MRD50 which is the dose of the test material that induces a 50% decrease in metabolic rate relative to a negative control (IIVS, 2014). This testing method tends to be associated with a misclassification error in the middle range of irritancy (Curren et al., 2006).

Reconstituted Human Tissue Models

EpiOcular^{TM} (MatTek Corp.)   The EpiOcular model is an *in vitro* human corneal model in which cultured epithelium is used as a proxy for corneal epithelium. Human-derived epidermal keratinocytes are coaxed into a membrane several layers thick, creating a cultured membrane compositionally similar to corneal tissue (Barile, 2010; MatTek, 2014). The MTT colorimetric assay is used to measure cell viability. Since the matrixed cells remain mitotically and metabolically active, growth patterns and the release of cytokines continue akin to *in vivo* conditions (Barile, 2010; MatTek, 2014). This model can be used to evaluate hydrophilic and hydrophobic test chemicals in either their liquid or solid state with the additional benefit of staging irritants as mild, moderate, or severe (Barile, 2010). EpiOcular^{TM} is in the validation stage for replacement of Draize eye testing by ECVAM.

Human Corneal Epithelial (HCE^{TM}) (SkinEthic Lab)
The human corneal epithelial (HCE^{TM}) model consists of immortalized human corneal epithelial cells cultured in a fashion similar to the EpiOcular^{TM} model. Both models use an air–liquid interface and thus lack a stratum corneum, resembling the corneal mucosa. As in the EpiOcular^{TM} model, the MTT colorimetric assay is used to evaluate cell viability. *In vitro* indicators are also used with this model, such as LDH release, histology, cytokine release, and gene expression (Barile, 2010). Human corneal epithelial appears useful as an *in vitro* model for detecting corneal repair and recovery, and as a "prescreen" method to evaluate test substance ingredients (Barile, 2010). According to its developer, HCE^{TM} can be used to evaluate corneal permeability and metabolism as well as a corneal differential display, i.e., mucin production (SkinEthic, 2014). Validation testing for ocular irritative testing by EVCAM is ongoing.

Other Alternative Methods

There are some *in vitro* tests that have not been discussed because they are not used extensively in eye irritation testing;

however, they can be used for dermal, mucosal, and eye irritation studies. slug mucosal irritation assays (SMI) use the *Arion lusitanicus* as a test subject because the mucosal body surface of slugs can mimic the mucosal surfaces of humans in the ear, nose, throat, and eye. The mucosal surface of the slug contains mucus secreting cells covering a sub-epithelial connective tissue. When slugs are placed on an irritant substance they will produce mucus, and, in severe cases, tissue damage. The release of proteins and enzymes can be measured to determine the toxicity of the test chemical (Adriaens, 2006). red blood cell assay (RBC) is an *in vitro* test that evaluates hemolysis by measuring hemoglobin release in the plasma following the exposure to a test chemical to rat RBCs (ECVAM DB-ALM, 1994b). Neutral release red or neutral red uptake (NRR) is an *in vitro* assay using seeded BALB/c 3T3 cells treated with a test chemical and incubated with neutral red dye to determine the cell membrane permeability based on dye absorption in the cells (ECVAM DB-ALM, 1994a). Irritection® uses a protein matrix that mimics the biochemical components of cornea proteins. This assay measures the denaturation and degradation of matrix proteins by optical density (In Vitro International, 2007).

SUMMARY

Many *in vitro* testing methods have been developed as an alternative to *in vivo* (live animal testing) methods. The alternatives described herein have been used to study eye irritation but may also apply to dermal and mucosal irritation of the ear, nose, throat, and skin. These methods have undergone numerous independent validation studies. Although none of these alternative methods may currently be used as a stand-alone replacement for the *in vivo* Draize eye test, their use as part of a multi-tiered testing strategy can reduce the number of animals used for chemical testing. Development of these methods began in earnest over 25 years ago. Despite its documented drawbacks and even with the chemical misclassifications, no alternative to *in vivo* testing has been found to match the predictive reliability of the Draize eye test. The *in vitro* testing methods are best used for extremes on the spectrum of eye irritation. Some are best for testing severe eye irritants while others are accurate at testing non-irritants. The ICC-VAM and ECVAM validate proposed models using three major criteria: chemical classification accuracy, within test variability or interlaboratory reproducibility, specificity of eye damage and its correlation to the human eye. Unfortunately, all alternative *in vitro* methods have failed to meet the standards in one or more of these areas. Currently, all eye irritation ratings continue to use the Draize eye test outcomes when applicable to determine the eye irritant classification of a chemical.

REFERENCES

Adriaens, E. (2006) *The Slug Mucosal Irritation Assay: An Alternative Assay for Local Tolerance Testing*, London, UK: National Centre for the Replacement, Refinement and Reduction of Animals in Research.

Adriaens, E., Barroso, J., Eskes, C., Hoffman, S., McNamee, P., Alépée, N., Bessou-Touya, S., De Smedt, A., De Wever, B., Pfannenbecker, U., Tailhardat, M., and Zuang, V. (2014) Retrospective analysis of the Draize test for serious eye damage/eye irritation: importance of understanding the *in vivo* endpoints under UN GHS/EU CLP for the development and evaluation of *in vitro* test methods. *Arch. Toxicol.* 88:701–723.

Barile, F.A. (2010) Validating and troubleshooting ocular *in vitro* toxicology tests. *J. Pharmacol. Toxicol. Methods* 61:136–145.

Curren, R.D., Mun, G.C., Gibson, D.P., and Aardema, M.J. (2006) Development of a method for assessing micronucleus induction in a 3D human skin model (EpiDerm™). *Mutat. Res-Gen. Tox. Environ. Mutagen.* 607 (2):192–204.

Draize, J.H., Woodard, G., and Calvery, H.O. (1944) Methods for the study of irritation and toxicity of substances applied to the skin and mucous membranes. *J. Pharmacol. Exp. Ther.* 82:377–390.

ECVAM DB-ALM (1994a) *INVITTOX Protocol No. 100*. European Centre for the Validation of Alternative Methods, Database Service on Alternative Methods to Animal Experimentation, European Comission. Available at http://ecvam-dbalm.jrc.ec.europa.eu/ (last accessed April 10, 2014).

ECVAM DB-ALM (1994b) *INVITTOX Protocol No. 99*. European Centre for the Validation of Alternative Methods, Database Service on Alternative Methods to Animal Experimentation, European Commisson. Available at http://ecvam-dbalm.jrc.ec.europa.eu/ (last accessed April 10, 2014).

ECVAM DB-ALM (1995) *INVITTOX Protocol No. 108*. European Centre for the Validation of Alternative Methods, Database Service on Alternative Methods to Animal Experimentation, European Commisson. Available at http://ecvam-dbalm.jrc.ec.europa.eu/ (last accessed April 10, 2014).

EPA (1998) *Acute Eye Irritation. Health Effects Test Guidelines*, OPPTS 870.2400, Bethesda, MD: United States Environmental Protection Agency.

Gautheron, P., Dukic, M., Alix, D., and Sina, J.F. (1992) Bovine corneal opacity and permeability test: an *in vitro* assay of ocular irritancy. *Fundam. Appl. Toxicol.* 18:442–449.

Hartung, T., Bruner, L., Curren, R., Eskes, C., Goldberg, A., McNamee, P., Scott, L., and Zuang, V. (2010) First alternative method validated by a retrospective weight-of-evidence approach to replace the Draize eye test for the identification of non-irritant substances for a defined applicability domain. *ALTEX* 27:43–51.

IIVS (2014) *Cytosensor Microphysiometer. Ocular Irritation*. Institute for In Vitro Sciences. Available at http://www.iivs.org/home/scientific-services/laboratory-services/ocular-irritation/cytosensor/step-by-step/5/ (last accessed April 10, 2014).

In Vitro International (2007) *Irritectation. Cell Based Assays*. Available at http://www.invitrointl.com/products/irritect.php (last accessed April 10, 2014).

Katoh, M., Hamajima, F., Ogasawara, T., and Hata, K., (2013) Establishment of a new *in vitro* test method for evaluation of eye irritancy using a reconstructed human corneal epithelial model, LabCyte CORNEA-MODEL. *Toxicol. In Vitro* 27 (8):2184–2192, doi: 10.1016/j.tiv.2013.08.008. Epub 2013 Aug 30.

Luepke, N. and Kemper, F. (1986) The HET-CAM test: an alternative to the Draize eye test. *Food Chem. Toxicol.* 24:495–496.

MatTek (2014) *EpiOcular Assays*. Available at http://mattek.com/EpiOcular/data-sheet (last accessed April 10, 2014).

Organisation for Economic Co-operation and Development (OECD) (2012a) *Streamlined Summary Document Related to the Fluorescein Leakage (FL) Test Method for Identification of Ocular Corrosives and Severe Irritants*, Series on Testing and Assessment No. 180, Document ENV/JM/MONO(2012)28, Paris, France: OECD Environment, Health and Safety Publication.

Organisation for Economic Co-operation and Development (OECD) (2012b) *Acute Eye Irritation*, Testing Guidelines 405, Paris, France: OECD Environment, Health and Safety Publication.

Organisation for Economic Co-operation and Development (OECD) (2013a) *Bovine Corneal Opacity and Permeability Test Method for Identifying (i) Chemicals Inducing Serious Eye Damage and (ii) Chemicals Not Requiring Classification for Eye Irritation or Serious Eye Damage*, Testing Guidelines 437, Paris, France: OECD Environment, Health and Safety Publication.

Organisation for Economic Co-operation and Development (OECD) (2013b) *Isolated Chicken Eye Test Method for Identifying (i) Chemicals Inducing Serious Eye Damage and (ii) Chemicals Not Requiring Classification for Eye Irritation or Serious Eye Damage*, Testing Guidelines 438, Paris, France: OECD Environment, Health and Safety Publication.

Parish, W. (1985) Ability of *in vitro* (corneal injury–eye organ–and chorioallantoic membrane) tests to represent histopathological features of acute eye inflammation. *Food Chem. Toxicol.* 23:215–227.

Prinsen, M.K. and Koëter, B.W.M. (1993) Justification of the enucleated eye test with eyes of slaughterhouse animals as an alternative to the Draize eye irritation test with rabbits. *Food Chem. Toxicol.* 31:69–76.

Scott, L., ESkes, C., Hoffman, S., Adriaens, E., Alépée, N., Bufo, M., Clothier, R., Facchini, D., Faller, C., Guest, R., Harbell, J., Hartung, T., Kamp, H., Le Varlet, B., Meloni, M., McNamee, P., Osborne, R., Pape, W., Pfannenbecker, U., Prinsen, M., Seaman, C., Spielmann, H., Stokes, W., Trouba, K., Van den Berghe, C., Van Goethem, F., Vassalo, M., Vindardell, P., and Zuang, V. (2010) A proposed eye irritation testing strategy to reduce and replace *in vivo* studies using bottom-up and top-down approaches. *Toxicol. In Vitro* 24:1–9.

SkinEthic (2014) *HCE™*. Available at http://www.skinethic.com/pageLibre000101b8.asp (last accessed April 10, 2014).

Tchao, R. (1988) Trans-epithelial permeability of fluorescein in vitro as an assay to determine eye irritants. In: Goldberg, A.M., editor. *Alternative Methods in Toxicology*, vol. 6, Progress in In Vitro Toxicology, New York: Mary Ann Liebert, Inc., pp. 271–283.

Wilhelmus, K.R. (2001) The Draize eye test. *Surv. Opthalmol.* 6:493–515.

Yang, Y., Yang, X., Xue, J., Curren, R., Huang, J., Tan, X., Xie, X., and Xiong, X. (2010) A procedure for application of eye irritation alternative methods on cosmetic ingredients. *ALTEX Proceedings*, 1/12.

114

REACTIVE AIRWAYS DYSFUNCTION SYNDROME (RADS)

David J. Hewitt

BACKGROUND

Reactive Airways Dysfunction Syndrome (RADS), also termed acute irritant-induced asthma, is an asthma-like illness that develops after a single high-level exposure to a pulmonary irritant. First described by Brooks and colleagues, the criteria and clinical features with minor variations have included the following (Brooks et al., 1985; Brooks, 1992, 1998):

1. No history of asthma-like respiratory disease.
2. Onset follows a single exposure incident or accident.
3. The exposure was to an irritant gas, smoke, fume, vapor, or dust in very high concentrations.
4. The onset of symptoms was abrupt, developing within minutes or hours but always within 24 h after the exposure and persisted for at least three months.
5. Requires immediate medical assistance.
6. Symptoms are similar to asthma with cough, wheezing, and dyspnea.
7. Pulmonary function tests (PFTs) may show airflow obstruction.
8. Methacholine challenge testing was positive in the range characteristic of asthma (i.e., 8 mg/ml).
9. Other types of pulmonary diseases were ruled out.

RADS cases originally described by Brooks and subsequent reports in the medical literature have been characterized by intense high-level exposures to a pulmonary irritant, which resulted in acute respiratory symptoms of a severity generally requiring immediate medical attention and hospitalization. The exposure typically occurred in a confined space, an environment with limited ventilation, or in circumstances where the individual was not able to immediately escape the exposure. Subsequently reported symptoms include cough and bronchospasm in response to a wide variety of non-specific irritants such as cigarette smoke, cold air, traffic fumes, household cleaners, or perfumes. The diagnosis is often made in the setting of an accidental occupational exposure with resulting medicolegal ramifications.

The validity of the diagnosis of RADS has been an area of controversy. Diagnosis may be difficult due to the absence of pre-exposure baseline respiratory function data and the inability to fully exclude other potential causes of respiratory symptoms such as smoking, allergies, infections, other exposures, or other health conditions. In a review of published RADS cases, Shakeri et al. (2008) noted that the case report descriptions were remarkably inconsistent and often did not allow a diagnosis of RADS based on the information provided. The development of RADS after an irritant exposure is considered relatively rare. In a prospective study of individuals evaluated at a poison control center for accidental irritant inhalation exposures, Blanc et al. (1991) reported that 6% had symptoms for more than 2 weeks and only 2% had symptoms for more than 30 days after the exposure. Of interest was their finding that residual morbidity did not seem to be related to the degree of irritant exposure. It has been noted that most individuals diagnosed with the condition recover completely over a period of weeks to months without significant clinical or physiologic sequelae (Bardana, 1999). Further complicating the assessment of potential RADS cases is the suggestion of "not-so-sudden" irritant-induced asthma syndromes in individuals who had exposure to lower levels of irritants for an extended period (Brooks et al., 1998).

Hamilton & Hardy's Industrial Toxicology, Sixth Edition. Edited by Raymond D. Harbison, Marie M. Bourgeois, and Giffe T. Johnson.
© 2015 John Wiley & Sons, Inc. Published 2015 by John Wiley & Sons, Inc.

EXPOSURES WHICH HAVE BEEN ASSOCIATED WITH RADS

An initial case series reported by Brooks et al. (1985) included 10 individuals and illustrates characteristics of the condition. The exposures were variable and included uranium hexafluoride, floor sealant, spray paint, hydrazine, heated acid, fumigating fog, metal coat remover, and fire/smoke. It was noted that all the substances were pulmonary irritants, which was either gas or aerosol. The authors noted that in several cases, the exposure was to a mixture and the exact agent could not be determined. There was no information regarding the air concentrations of the substance. The duration of exposure ranged from a few minutes to 2–3 h. Symptoms were reported to occur within minutes to twelve hours after the exposure. Reported symptoms included cough, dyspnea, wheezing, chest pain, and chest tightness. In seven of the cases, symptoms were present for a year or more. The cases were evaluated anywhere from 4 to 140 months after the exposure for the persistent respiratory symptoms. A methacholine challenge was obtained for each of the cases and confirmed airway hyperreactivity.

Since the initial case series, there have been numerous additional case reports of RADS. Virtually any type of pulmonary irritant can be implicated as a potential cause of RADS. Incriminated substances have included various acids, ammonia, bleach, cleaning agents, dust, hydrogen sulfide, oxides of nitrogen or sulfur, paints, pesticides, smoke, and many others (Shakeri et al., 2008).

DOSE-RESPONSE RELATIONSHIPS

By definition, most RADS cases are identified retrospectively when respiratory symptoms do not seem to improve after the reported exposure. As such, there is typically no information regarding the exact concentrations of substances to which an individual may have been exposed and as noted by Brooks et al., the exact composition of the substance may not even be known. In short, there is no reliable information as to the magnitude or duration of exposure, which is necessary to produce the RADS condition.

SOURCES OF EXPOSURE

RADS cases have been reported from both non-occupational and occupational exposures. Reported cases are typically secondary to an accidental release and may involve scenarios such as residents living near an industry, road, or railway in which there was an unexpected chemical release or occupational situations in which a worker was unexpectedly exposed to high levels of an irritant chemical due to an unforeseen release, failure of personal protective equipment, or inability to immediately escape exposure.

INDUSTRIAL HYGIENE

Because RADS cases typically involve an unexpected chemical release, there are rarely any air monitoring data available to document the extent of exposure. Simulation of the reported exposure with air monitoring and/or modeling may be obtained in special situations but are rarely available. Even if such air monitoring data was immediately available, this could not reliably predict whether an individual would develop a persistent respiratory condition such as RADS. However, when assessing a situation with the potential for a RADS case, it is important to document that the individual also experienced other symptoms that would be consistent with exposure to high levels of a respiratory or mucous membrane irritant. Based on known irritant thresholds for some chemicals, this may provide some estimate of the exposures, which may have occurred. The presence of irritant symptoms and their severity in other nearby individuals may provide additional information in estimating potential exposures, which may have occurred.

MEDICAL MANAGEMENT

For a high level irritant exposure such as those associated with the development of RADS, the initial medical management is supportive, including maintenance of the airway and adequate oxygenation. Severe pulmonary irritant exposures may be associated with bronchospasm, pulmonary edema, pulmonary hemorrhage, and hypoxia. In general, there is no specific antidote or treatment, which has been identified that can decrease the risk of a subsequent persistent respiratory condition such as RADS. It is impossible to predict those who may develop a RADS condition.

There is not a specific diagnostic test, which confirms RADS. The diagnosis is one of exclusion and is based on the exposure history and the persistence of characteristic respiratory symptoms more than three months after the initial exposure. Diagnostic methods which may be used to support a diagnosis of RADS include a PFT usually showing an obstructive process and a methacholine challenge test which demonstrates increased airway reactivity. Both tests are nonspecific with multiple potential causes for abnormalities.

In particular, a positive methacholine challenge may be cited as proof of RADS. Methacholine challenge testing is not commonly obtained by primary health care providers and the significance of test results is not always understood. Methacholine is a cholinergic agonist, which causes bronchoconstriction when inhaled at sufficient concentrations. Individuals with asthma or reactive airways react to lower concentrations of methacholine. The test is performed by having an individual inhale an aerosolized mist containing methacholine at increasing concentrations through a nebulizer. The response to methacholine is measured by performing a PFT after each dose. The primary measure of interest is the forced expiratory volume in 1 second (FEV1). If the respiratory airways are narrowed, such as in an asthma attack, the FEV1 will be reduced and may be interpreted as obstruction. A positive test is usually defined as a 20% decrease in the FEV1 at a methacholine dose <16 mg/ml. Levels of 4–16 mg/ml are considered borderline (Crapo et al., 2000). A negative methacholine challenge test virtually excludes the diagnosis of asthma or RADS. However, specificity of the test is poor as there are many relatively common conditions, which may be associated with a positive test result. These include cigarette smoking, allergies, recent respiratory infection, gastroesophageal reflux, and others (Hewitt, 2008).

Caution is emphasized when making a diagnosis of RADS. Review of litigated cases has shown that some health care providers have made a diagnosis of RADS without a clear understanding of the chemical involved, the extent of exposure, and/or the known association, if any, between the chemical in question and RADS. This may result in a cascade of events including empiric respiratory treatments with significant side-effects, misinterpreted diagnostic studies, or incomplete consideration of alternative explanations for the reported symptoms. One or more of these factors may lead to an erroneous diagnosis of a chronic respiratory condition with attendant disability. The potential for respiratory medications used in the initial treatment of an irritant exposure to cause continued symptoms have been noted (Hewitt, 2011). Prior to diagnosing RADS and instituting treatment, the treating health care provider should ensure they understand the reported exposure and whether its association with RADS is plausible. This may require consultation with an occupational medicine physician, toxicologist, and/or industrial hygienist to fully assess the reported exposure.

Treatment of RADS is non-specific and is similar to any case of asthma. Medications may include bronchodilators, nebulizers, or inhaled steroids. The optimal duration of treatment has not been established. RADS cases may spontaneously resolve over time or in some cases, have symptoms, which reportedly persist for years. Because an individual with RADS is not sensitized to the causative substance, the individual can return to work in areas in which the substance is present but at levels below those, which cause aggravation of respiratory symptoms.

REFERENCES

Bardana, E.J. (1999) Reactive airways dysfunction syndrome (RADS): guidelines for diagnosis and treatment and insight into likely prognosis. *Ann. Allergy Asthma Immunol.* 83:583–586.

Blanc, P.D., Galbo, M., Hiatt, P., and Olson, K.R. (1991) Morbidity following acute irritant inhalation in a population-based study. *JAMA.* 266:664–669.

Brooks, S.M. (1998) Occupational and environmental asthma. In: Rom, W. M., editor. *Environmental and Occupational Medicine*, 3rd ed., Philadlphia, PA: Lippincott-Raven, pp. 481–524.

Brooks, S.M. (1992) Irritants, sensitisers and asthma symposium: reactive airways syndromes. *J. Occup. Health Safety-Aust. NZ.* 8:215–220.

Brooks, S.M., Hammad, Y., Richards, I., Giovinco-Barbas, J., and Jenkins, K. (1998) The spectrum of irritant-induced asthma: sudden and not-so-sudden onset and the role of allergy. *Chest.* 113:42–49.

Brooks, S.M., Weiss, M.A., and Bernstein, I.L. (1985) Reactive airways dysfunction syndrome (RADS). Persistent asthma syndrome after high level irritant exposures. *Chest.* 88:376–384.

Crapo R.O., Casaburi, R., Coates, A.L., Enright, P.L., Hankinson, J.L., Irvin C.G., Macintyre, N.R., McKay, R.T., Wanger J.S., Anderson, S.D., Cockcroft, D.W., Fish, J.E., and Sterk, P.J. (2000) Guidelines for methacholine and exercise challenge testing – 1999. *Am. J. Respir. Crit. Care Med.* 161:309–329.

Hewitt, D.J. (2008) Interpretation of the "positive" methacholine challenge. *Am. J. Indust. Med.* 51:769–781.

Hewitt, D.J. (2011) Can Reactive Airways Dysfunction Syndrome (RADS) be iatrogenic? *Resp. Care.* 56:1188–1194.

Shakeri, M.S., Dick, F.D., and Ayres, J.G. (2008) Which agents cause reactive airways dysfunction syndrome (RADS)? A systematic review. *Occup. Med. (Lond).* 58:205–211.

115

PULMONARY SENSITIZATION

David J. Hewitt

BACKGROUND

Sensitization refers to the development over time of an allergic reaction to a chemical or substance. Initial exposure to the substance does not result in adverse health effects. However, repeated exposure may cause sensitization reactions, which can become more severe with repeated exposure and may also occur at lower levels of exposure. Sensitization is generally permanent once it occurs. hypersensitivity pneumonitis (HP) and occupational asthma (OA) are the two most common conditions related to pulmonary sensitization in the occupational or industrial environment and are described in the following section.

HYPERSENSITIVITY PNEUMONITIS

Hypersensitivity pneumonitis, also called extrinsic allergic alveolitis, is a respiratory condition caused by the inhalation of a substance to which the individual has developed an immunologic sensitivity. There are numerous substances that have been identified as a cause of HP. The most common causes include bacteria, fungi, or animal proteins, although some chemicals have also been associated with the condition. Specific substances include thermophilic bacteria such as *Thermoactinomyces vulgaris* found in contaminated hay; fungi such as *Aspergillus, Penicillium, Alternaria, or Cladosporium* species; avian proteins related to bird droppings or feathers; and chemicals such as diisocyantes and trimellitic anhydride. HP may also be identified by a particular occupational syndrome or presentation such as farmer's lung, mushroom worker's lung, humidifier lung, or bird-fancier's lung (Rose, 1996). Occasional outbreaks of HP with multiple cases secondary to a common exposure may occur (Bracker et al., 2003; Robertson et al., 2007).

Prevalence of HP in the occupational environment has not been defined. Bourke et al. (2001) estimated that only 5–15% of individuals exposed to a sensitizing agent will develop the condition. Primary risk factors include the duration and magnitude of exposure. Cigarette smoking appears to have a protective effect for unknown reasons. Once a susceptible individual becomes sensitized, an acute exposure classically results in symptoms of fever, chest tightness, dyspnea, and general malaise or flu-like symptoms, which usually occur within a few hours after the exposure and resolves in 24–48 h. Alternatively, some individuals will present with progressive dyspnea from chronic exposures with no obvious temporal variations in symptoms.

In the acute form, symptoms recur with each exposure and sensitivity may increase. Repeated exposure may be associated with progressive respiratory effects, including increasing dyspnea in the absence of exposure, pulmonary function test (PFT) abnormalities (usually restrictive), and chest imaging abnormalities with patchy alveolar infiltrates, which are classically described as "ground glass" in appearance. Late stage effects may include honeycombing fibrotic patterns with irreversible respiratory effects. Lung diffusion capacity is the most severely affected physiological measurement.

OCCUPATIONAL ASTHMA

Asthma may be defined as a reversible airway obstruction in which the airways in the respiratory tract constrict and narrow in response to various stimuli. Occupational asthma can be defined as asthma that occurs secondary to

Hamilton & Hardy's Industrial Toxicology, Sixth Edition. Edited by Raymond D. Harbison, Marie M. Bourgeois, and Giffe T. Johnson.
© 2015 John Wiley & Sons, Inc. Published 2015 by John Wiley & Sons, Inc.

a work-related exposure to a pulmonary sensitizing agent. This differs from preexisting asthma that may be aggravated by exposure to respiratory irritants in the workplace that do not cause sensitization. Reactive Airways Dysfunction Syndrome (RADS), is sometimes identified as a type of OA that occurs secondary to a high level irritant exposure, does not involve an immunologic sensitization reaction, and has no latency between the exposure and the onset of the asthmatic condition. Approximately 10–15% of adult onset asthma is considered to be related to occupational exposures (Balmes et al., 2003; Blanc and Toren, 1999).

As with HP, the number of substances that have been associated with classic OA is lengthy and continues to grow. Specific substances include bacteria and molds, animal proteins, wood dusts such as Western red cedar, metals such as chromium or nickel, and chemicals such as isocyanates. Some substances have been associated with both HP and OA. It is difficult to predict who may develop OA. Potential risk factors include a history of atopy, preexisting asthma, increased duration of exposure, and exposure to increased air concentrations of the substance (Dykewicz, 2001; Mapp et al., 2005).

The condition is characterized by the onset of asthma symptoms such as dyspnea, cough, and wheezing, which occurs after repeated exposure to a pulmonary sensitizing agent. Symptoms may occur shortly after exposure, or several hours later depending on the specific substance and sensitivity. With repeated exposure, the individual may become more sensitized and react to lower airborne concentrations of the substance. Symptoms characteristically improve with removal from the exposure. However, continued exposure can result in progressively increased asthmatic symptoms, which may not fully resolve after elimination of exposure.

SOURCES OF EXPOSURE

Work environments that may be associated with pulmonary sensitization reactions are highly variable. Those that seem to be more at risk include those involving paints, plastic manufacture, electronics, welding, metal dusts, metal fluids, agriculture, hair salons, healthcare, and laboratory environments (Dykewicz, 2001).

DOSE–RESPONSE RELATIONSHIPS

The specific level of exposure necessary to result in pulmonary sensitization for HP or OA is unknown. Risks of sensitization in susceptible individuals may increase with increasing exposure; however, a threshold has not been defined for most substances.

INDUSTRIAL HYGIENE

Industrial hygiene can play an important role in identifying the potential for pulmonary sensitization conditions. This can include review of job duties and identifying potential substances in the workplace, which could cause a condition of HP or OA. In the event such exposures are identified, steps can be taken to decrease or eliminate exposures. This may include identification of alternative substances that are not associated with pulmonary sensitization, improved ventilation to decrease exposures, and identification of appropriate respiratory protection. Air testing may be considered to document that a pulmonary sensitizing agent is present and thereby assist in confirming a medical diagnosis. Repeated testing of the workplace can be performed to determine the effectiveness of control measures for decreasing exposures.

MEDICAL MANAGEMENT

Criteria for confirming suspected cases of pulmonary sensitization include identification of an agent within the workspace that has been associated with sensitization; a history that is consistent with characteristic respiratory symptoms temporally related to work (i.e., occur during or after work with the suspected substance and improve when away from work on weekends or vacations); and characteristic clinical findings. Laboratory testing may be able to support a diagnosis using immunologic tests for specific substances. Because pulmonary sensitization can have significant effects regarding employment, it is important to reliably confirm the diagnosis.

Pulmonary function testing in the form of baseline testing, serial testing, and peak flow monitoring is a primary component in attempting to verify an occupationally related condition. Ideally, a baseline PFT obtained prior to the individual starting work can be used to compare with subsequent tests obtained during or after work. Alternatively, serial PFT's can be obtained at several times during the workweek to determine if there is a change. Such testing can be performed at the beginning and end of the workweek or during several different times during a workday. Peak flow testing involves the use of a small peak flow meter that an affected individual can carry with him/her during the workday and quickly measure respiratory function throughout the workday and away from work. Individuals are generally asked to record peak flow measurements in a log or graph several times a day, which can show whether there is a consistent pattern of change with work exposures. Interpretation of peak flow measurements is confounded by reliance on self-reported data and the individual performing the test correctly with maximum effort each time (Townsend, 2000). Of these methods, serial PFT's are likely to provide the most reliable information as the test is performed according to strict criteria by a trained individual. Results can be assessed for reproducibility, validity, and effort. However, provision of on-site PFTs may be impractical.

Skin prick testing or serological testing (i.e., specific IgE) may be useful in some cases to identify evidence of sensitization and suggest a causal association (Grammer et al., 1989). However, such tests may not be readily available for the exposure of concern and may have limited sensitivity. Blood tests appear to be less sensitive than skin testing (Tan and Spector, 1999). A positive test does not automatically confirm the diagnosis and should be evaluated in the context of other information (Anonymous, 1992).

Additional diagnostic criteria specific to HP include a history of exposure to an agent known to cause HP, characteristic chest X-ray or CT findings, and precipitating antibodies to causative antigens (Rose, 1996). Significant predictors of HP have included exposure to a known offending antigen; positive precipitating antibodies to the antigen, recurrent episodes of symptoms; inspiratory crackles on exam, symptoms occurring 4–8 h after exposure; and weight loss (Lacasse et al., 2003).

For OA, a methacholine challenge test can be used to confirm increased airway reactivity in cases in which the diagnosis is unclear. A negative test virtually excludes the diagnosis of asthma. A positive test does not necessarily confirm asthma as there are numerous causes of increased airway reactivity, which may need to be considered (Hewitt, 2008). Immunological testing, if available, is generally most useful for high molecular weight substances where a negative result essentially rules out the agent as a cause of OA (Cruz and Munoz, 2012). In most cases, reliable documentation of exposure to a sensitizing agent with associated work-related decreases in pulmonary function is the most common method of verifying the diagnosis.

The definitive test for diagnosis of pulmonary sensitization is to have the individual undergo a specific challenge in which the individual is placed in a controlled environment and exposed to the substance of concern (Tan and Spector, 1999). Reproduction of symptoms or PFT decrements confirms the diagnosis. Such testing may be employed in cases in which there is uncertainty regarding the diagnosis or medicolegal issues regarding causation. However, specific airway challenge testing is rarely available to the general practitioner.

The first treatment goal for individuals with a condition related to pulmonary sensitization is to remove them from the exposure. If identified early, the chances of persistent respiratory health effects are substantially decreased. Continued exposure may result in increasing respiratory morbidity that can be permanent. Acute episodes of HP are treated symptomatically and typically resolved within 1–2 days after an exposure. Occupational asthma is treated the same as non-OA and involves medications such as bronchodilators, nebulizers, or inhaled steroids. With cessation of exposure and time, the need for respiratory medications may decrease or be eliminated. The duration of required respiratory treatment in an OA case is difficult to predict and may be affected by factors such as the chronicity of exposure prior to diagnosis.

Because an individual with HP or OA is sensitized to the causative substance, they should have a permanent restriction from working with the responsible substance. Alternative work duties or positions may need to be implemented to insure the individual does not have continued exposure.

REFERENCES

Anonymous (1992) Guidelines for the diagnosis of occupational asthma. Subcommittee on 'Occupational Allergy' of the European Academy of Allergology and Clinical Immunology. *Clin. Exp. Allergy* 22:103–108.

Balmes, J., Becklake, M., Blanc, P., Henneberger, P., Kreiss, K., Mapp, C., Milton, D., Schwartz, D., Toren, K., and Viegi, G. (2003) American Thoracic Society Statement: occupational contribution to the burden of airway disease. *Am. J. Respir. Crit. Care Med.* 167:787–797.

Blanc, P.D. and Toren, K. (1999) How much adult asthma can be attributed to occupational factors? *Am. J. Med.* 107:580–587.

Bourke, S.J., Dalphin, J.C., Boyd, G., McSharry, C., Bladwin, C.I., and Calvert, J.E. (2001) Hypersensitivity pneumonitis: current concepts. *Eur. Respir. J. Suppl.* 18(Suppl. 32):81s–92s.

Bracker, A., Storey, E., Yang, C., and Hodgson, M.J. (2003) An outbreak of hypersensitivity pneumonitis at a metalworking plant: a longitudinal assessment of intervention effectiveness. *Appl. Occup. Environ. Hyg.* 18:96–103.

Cruz, J. and Munoz, X. (2012) The current diagnostic role of the specific occupational laboratory challenge test. *Curr. Opin. Allergy Clin. Immunol.* 12:119–125.

Dykewicz, M.S. (2001) Occupational asthma: a practical approach. *Allergy Asthma Proc.* 22:225–233.

Grammer, L.E., Patterson, R., and Zeiss, C.R. (1989) Guidelines for the immunologic evaluation of occupational lung disease. Report of the Subcommittee on Immunologic Evaluation of Occupational Immunologic Lung Disease. *J. Allergy Clin. Immunol.* 5:805–814.

Hewitt, D.J. (2008) Interpretation of the "positive" methacholine challenge. *Am. J. Ind. Med.* 51:769–781.

Lacasse, Y., Selman, M., Costabel, U., Dalphin, J., Ando, M., Morrell, F., Erkinjuntti-Pekkanen, R., Muller, N., Colby, T.V., Schuyler, M., and Cormier, Y. (2003) Clinical diagnosis of hypersensitivity pneumonitis. *Am. J. Respir. Crit. Care Med.* 168:952–958.

Mapp, C.E., Boschetto, P., Maestrelli, P., and Fabbri, L.M. (2005) Occupational asthma. State of the art. *Am. J. Respir. Crit. Care Med.* 172:280–305.

Robertson, W., Robertson, A.S., Burge, C.B.S.G., Moore, V.C., Jaakkola, M.S., Dawkins, P.A., Burd, M., Rawbone, R., Gardner, I., Kinoulty, M., Crook, B., Evans, G.S., Harris-Roberts, J., Rice,

S., and Burge, P.S. (2007) Clinical investigation of an outbreak of alveolitis and asthma in a car engine manufacturing plant. *Thorax* 62:981–990.

Rose, C. (1996) Hypersensitivity pneumonitis. In: Harber, P., Schenker, M.B., and Balmes, J.R., editors. *Occupational and Environmental Respiratory Disease*, St. Louis: Mosby, pp. 201–215.

Tan, R.A. and Spector, S.L. (1999) Diagnostic testing in occupational asthma. *Ann. Allergy Asthma Immunol.* 83:587–592.

Townsend, M.C. (2000) ACOEM position statement. Spirometry in the occupational setting. *J. Occup. Environ. Med.* 42: 228–245.

116

"STREET" AND PRESCRIPTION DRUG ABUSE

Sharon S. Kelley, James Godin, and John Christie

INTRODUCTION

Drug abuse in the workplace has been a concern for occupational and environmental healthcare professionals for many reasons, including but not limited to:

Detrimental short-/long-term health effects experienced by the employee and their families

Diminished quality of job performance

Safety of the employee, fellow employees, clients, and the general public

Liability of the employer for medical and/or legal costs incurred by drug abuse

During the last century, illicit drug abuse was the focal point of scrutiny in the workplace but in the past decade, the abuse of prescription drugs and steroids has risen dramatically.

The contents of this chapter have been provided to equip occupational healthcare professionals in their understanding of the physical and psychological presentations associated with each of the listed drugs. This information, coupled with other identifying tools such as "street" names, laboratory analysis caveats, and management recommendations, specific to the drug, will hopefully serve to increase the practitioner's ability to recognize drug abuse and thereby reduce morbidity and mortality as well as corporate loss and liability.

AMPHETAMINES AND METHAMPHETAMINE

BACKGROUND AND USES

Due to concerns in the 1920s regarding the availability of naturally occurring ephedrine, specifically that sufficient supplies would not be available to meet the demands of asthmatic patients, laboratories were encouraged to design synthetic means of producing this β-adrenergic drug.

The name "amphetamine" is an acronym for α-methylphenylethylamine which belongs to the phenylethylamine family. As substitutions are made on the phenylethylamine base structure, numerous analogs are created with varying effects and potencies. The commonality among analogs is their stimulant action and high potential for abuse which has led to these drugs to be classified as Schedule II drugs per the Controlled Drugs and Substances Act (Goldfrank et al. 2007).

Medical uses include treatment for attention deficit hyperactivity disorder (ADHD), narcolepsy, depression, and weight control.

Methamphetamine originated when Japanese chemist Ogata, in his attempts to produce ephedrine, synthesized d-phenyl-isopropylmethylamine, which came to be known as methamphetamine. Methedrine™ and Benzedrine™, two forms of methamphetamine, were sold by US pharmaceutical houses as agents which would provide relief from nasal congestion with additional benefits of reduced fatigue

Hamilton & Hardy's Industrial Toxicology, Sixth Edition. Edited by Raymond D. Harbison, Marie M. Bourgeois, and Giffe T. Johnson.
© 2015 John Wiley & Sons, Inc. Published 2015 by John Wiley & Sons, Inc.

and increased alertness. Growing abuse, with drug-related deaths, resulted in the removal of methamphetamine from the market during the 1950s/1960s. A reformulated product, Desoxyn™ is a currently prescribed for ADHD and weight control (Karch 2009) (National Institute on Drug Abuse 2014).

Brand names for amphetamine include: Adderall®, Adderall XR®

Street names include: crank, uppers, kid speed, whiz

Brand names for methamphetamine include: Desoxyn® and Methampex®

Street names include: meth, crank, crystal, ice and poor man's coke

PHYSICAL AND CHEMICAL PROPERTIES

Physical Properties

While amphetamine is typically ingested in tablet form, associated analogs such as methamphetamine may appear in a variety of physical forms:

White or pink crystalline small chunk-like forms (shards) which may be utilized for smoking

White, off white, tan powder (appropriate for snorting)

Tablets

Clear liquid

Chemical Properties

Amphetamine

Chemical Name: 1-phenylpropan-2-amine

Metabolized via P450 enzymes CYP2D and CYP3A

Methamphetamine

Chemical name(s): N-α-dimethylbenzenethanamine, *d*-N-α-dimethylphen ethylamine, *d*-deoxyephedrine

Metabolized via P450 enzymes CYP2D and CYP3A

TOXICOLOGY

Acute Effects

CNS: Increased alertness, enhanced libido, anorexia, hyperactivity, mydriasis, diaphoresis, anxiety, bruxism, blurred vision, headache, hyperthermia, seizures, insomnia, dizziness, twitching, stroke, hallucinations in the presence of more psychoactive analogs such as MDMA

Gastrointestinal (GI): Diarrhea, constipation

Cardiovascular: Tachycardia, hypertension, myocardial ischemia

Chronic Effects

CNS: Psychosis, neurotoxicity due to prolonged dopamine upregulation, anxiety, post-acute withdrawal syndrome

Oropharyngeal: Dental caries ("meth mouth") (Donaldson 2006)

Cardiovascular: Hypertrophy possibly attributed to calmodulin activation, left ventricular dysfunction due to stress-induced cardiomyopathy, interstitial fibrosis from myocyte disruption, microvascular disease, multivessel coronary artery disease

Dermatological: Dermatillomania (persistent skin scratching)

Mechanism of Action

Amphetamines stimulate the central nervous system(CNS) through the neurotransmitters dopamine, serotonin, and norepinephrine specifically the facilitation of their release into the synaptic cleft, inhibition of their reuptake, and increased stimulation of postsynaptic receptors. Due to the lipophilic nature, the drugs cross the blood brain barrier more rapidly than similar stimulants. Amphetamines are metabolized by means of monoamine oxidase (MAO) degradation hence warnings are provided to patients to avoid amphetamines, and their analogs, should be taking MAO inhibitors (Horton et al. 2013).

As chronic use of methamphetamine results in damage to dopaminergic receptors, long-term neurological sequelae may be present. In addition, to loss of normal dopamine functions, serotonin, tyrosine, and tryptophan hydroxylase activity is also diminished. Methamphetamine may also exert its detrimental effects through binding to a G protein-coupled receptor $TAAR_1$. Trace amine-associated receptors are activated by amphetamine type molecules. (Hong & Amara 2013)

MODES OF ADMINISTRATION

Oral

Intravenous

Inhaled (smoked)

Insufflated (snorted)

DRUG SCREENING

Detection time in urine is $20.7\,h \pm 7.3\,h$ (amphetamine), $23.6 \pm 6.6\,h$ (methamphetamine)

Screening assays include the Marquis and Simon tests. If methamphetamine is present, the Marquis test will yield an orange/brown color while the Simon test will produce a blue color which distinguishes methamphetamine from amphetamine which manifests a red coloration.

"SAHMSA 5" drug screens include amphetamines as one of the paneled drug classifications. Tests would need to be evaluated for whether dextroamphetamine and methamphetamine would also be detected.

Care must be given during drug testing as similar stimulants may cross react and result in a "false positive" when reviewing the screen. An example would be the patient/employee using nasal inhalers. As there are two amphetamine enantiomers, the *l*- and *d*-forms, a "false positive" may occur when the active agent in the inhalers produces a "positive" result. However, inhalers possess the *l*-enantiomer which contains none of the psychoactive properties. Other drugs include commercially available cold remedies, weight reduction aids, and psychotropic medications. False positives may also occur by the metabolism of certain drugs such as selegiline as amphetamine and methamphetamine are chief metabolites (Moeller et al. 2008) (McPherson et al. 2007).

Other lab abnormalities associated with the amphetamine/analog use that may be noted are hyperglycemia, elevated CPK, elevations in liver function tests, myoglobinuria, and leukocytosis.

Mass spectroscopy (MS): Major ions are found at m/z = 42, 56, 57, 58, 59, 65, 91, and 134.

Gas chromatography (GC): Testing can be improved by N-derivatization. If using a urine sample, the limit of detection (LOD) = <10 μg/l.

MEDICAL MANAGEMENT

Supportive care is emphasized in the management of methamphetamine ingestion. Several caveats in this treatment would include safety to staff as methamphetamine can induce severe agitation and psychosis as well as the awareness of poisoning by accompanying adulterants such as heavy metals.

As in all cases of medical management, monitoring and support of airway, breathing and circulation are tantamount to successful patient outcomes. In addition, industrial healthcare specialists should also consider the following:

Specific treatments for heavy metal toxicity caused by contaminants in some methamphetamine preparations may be needed (Burton 1991).

Hyperthermia should be addressed with cooling measures, including but not limited to, the use of intravenous fluids (Santos 2005).

Electrolyte abnormalities should be considered due to anorexia and diaphoresis.

Seizure activity may benefit from the administration of benzodiazepines or related drugs (Derlet 1990).

Severe psychotic episodes may warrant the use of sedatives and appropriate antipsychotic drugs (Zorick 2008).

Antiarrhythmic medications should be utilized when indicated; however, the use of β-blocking agents should be avoided due to the risk of hypertensive crisis induced by diminishing α-adrenergic opposition (Cone et al. 2011) (Steadman & Birnbach 2003).

Rhabdomyolysis, due to long periods of immobility from taking CNS depressants to counteract the stimulatory effect of amphetamines/analogs, should be considered and CK levels utilized as part of the diagnostic process. Treatment would include IV crystalloids, sodium bicarbonate to raise pH and prevent myoglobin precipitation in the renal tubules while avoiding urinary acidity (Richards 1999).

REFERENCES

Burton, B.T. (1991) Heavy metal and organic contaminants associated with illicit methamphetamine production. *NIDA Res Monogr.* 115:47–59.

Cone, J., Jr., Harrington, M.A., Kelley, S.S., Prince, M.D., Payne, W.G., and Smith, D.J., Jr., (2011) Drug abuse in plastic surgery patients: optimizing detection and minimizing complications. *Plast. Reconstr. Surg.* 127(1):445–455.

Derlet, R.W., Albertson, T.E., and Rice, P. (1990) Protection against d-amphetamine toxicity. *Am. J. Emerg. Med.* 8(2):105–108.

Donaldson, M. and Goodchild, J.H. (2006) Oral health of the methamphetamine abuser. *Am. J. Health Syst. Pharm.* 63(21):2078–282.

Goldfrank, L., Hoffman, R., Nelson, L., Howland, M., Lewin, N., and Flomenbaum, N. (2007) *Goldfrank's Manual of Toxicologic Emergencies*, New York, NY: McGraw Hill, pp. 633–640.

Hong, W.C. and Amara, S.G. (2013) Differential targeting of the dopamine transporter to recycling or degradative pathways during amphetamine- or PKC-regulated endocytosis in dopamine neurons. *FASEB J.* 27(8):2995–3007.

Horton, D.B., Nickell, J.R., Zheng, G., Crooks, P.A., and Dwoskin, L.P. (2013) GZ-793A, a lobelane analog, interacts with the vesicular monoamine transporter-2 to inhibit the effect of methamphetamine. *J. Neurochem.* 127(2):177–186.

Karch, S. (2009) *Pathology of Drug Abuse*, 4th ed., Boca Raton, Florida: CRC Press, pp. 261–263.

McPherson, R., et al. (2007) *Drugs of Abuse in Henry's Clinical Diagnosis and Management by Laboratory Methods*, 21st ed., Philadelphia: Saunders Elsevier.

Moeller, K.E., Lee, K.C., and Kissack, J.C. (2008) Urine drug screening: practical guide for clinicians. *Mayo Clin. Proc.* 83(1):66–76.

National Institute on Drug Abuse. NIDA Info Facts: Methamphetamine. http://www.drugabuse.gov/publications/drugfacts/methamphetamine.

Richards, J.R., Johnson, E.B., Stark, R.W., and Derlet, R.W. (1999) Methamphetamine abuse and rhabdomyolysis in the ED: a 5-year study. *Am. J. Emerg. Med.* 17(7):681–685.

Santos, A.P., Wilson, A.K., Hornung, C.A., Polk, H.C., Jr., Rodriguez, J.L., and Franklin, G.A. (2005) Methamphetamine laboratory explosions: a new and emerging burn injury. *J. Burn. Care Rehabil.* 26(3):228–232.

Steadman, J.L. and Birnbach, D.J. (2003) Patients on party drugs undergoing anesthesia. *Curr. Opin. Anaesthesiol.* 16:147–152.

Zorick, T.S., Rad, D., Rim, C., and Tsuang, J. (2008) An overview of methamphetamine-induced psychotic syndromes. *Addict. Disord. Their Treat.* 7(3):143–156.

ANABOLIC STEROIDS

BACKGROUND AND USES

The use of drugs to enhance performance and increase strength dates back to as early as 776 BC when Greek Olympians were alleged to have used dried figs, mushrooms, and strychnine to enhance performance (Grivetti & Applegate 1997). Medical use of testicular extract began in the late 1800s and anabolic steroids can be traced to the 1930s when scientists synthesized a form of testosterone in order to treat men unable to produce a sufficient amount for typical development and sexual function.

In the 1940s, supraphysiological doses of anabolic androgenic steroids (AAS) were administered in eugonadal patients for anabolic benefit and in World War II artificial testosterone aided malnourished soldiers in weight gain and performance. High dose AAS have been used as therapy to promote muscle deposition post-burns, surgery, radiation therapy, and aging-related sarcopenia.

Brand names include: Androderm®, Androgel®, and Andropatch®

Street names include: Roids, gym candy, Arnolds, stackers, and juice

PHYSICAL AND CHEMICAL PROPERTIES

Physical Properties

Liquid

Gel

Chemical Properties

Chemical name of testosterone is 17-betahydroxyamndrost-4-en-3-one.

Steroids are a special class of lipids with a common feature being that of four-fused hydrocarbon rings, which offers the opportunity for a large variety of molecules.

Cholesterol, a key steroid, has within its structure a hydroxyl group that is responsible for the amphipathic character of the molecule. This is of importance as cholesterol can be incorporated into biological membranes as a lipid component.

Naturally-occurring anabolic steroids are synthesized in the testis, ovaries, and adrenal glands from cholesterol via pregnenolone.

Synthetic anabolic steroids are created through modification of testosterone in one of three ways:

Alkylation of the 17-carbon

Esterification of the 17-OH group

Modification of the steroid nucleus

TOXICOLOGY

Acute Effects

Cardiovascular: Acute myocardial infarction, arrhythmias

Cerebrovascular: Stroke

Pulmonary: Embolism

Chronic Effects

Cardiovascular: Hypertension, left ventricular hypertrophy, cardiomyopathy, peliosis hepatis (Stergiopoulos et al. 2008)

Dermatologic: Acne, male pattern baldness (Hoffman & Ratamess 2006)

Psychological: Manic symptoms associated with aggressive behavior and violence, depression

Hematologic: Polycythemia

GU: Decreased sperm count, testicular atrophy, gynecomastia, impotence (Korkia & Stimson 2007)

Musculoskeletal: Premature epiphyseal plate closure, intramuscular abscesses, possible increase in tendon tears (Casavant et al. 2007)

Apoptosis in neuronal cells (Estrada et al. 2006)

Mechanism of Action

Anabolic androgenic steroid increases lean body mass through elevated amounts of actin and myosin in skeletal muscle cells, which arises from increased protein synthesis (Patil et al. 2007). This originates from testosterone through a series of steps, which activates mRNA to increase transcription (Karch 2009). Secondary anabolic effects are achieved by an increased creatine phosphokinase activity, upregulation of insulin-like growth factor-1 receptors in conjunction with an increase of insulin-like growth factor-1 as well as reducing the catabolic effect that glucocorticoid activity has on muscle tissue.

MODES OF ADMINISTRATION

Intramuscular injection

Oral

Sublingual

Transdermal

DRUG SCREENING

Radioimmunoassay measures urinary ratios of testosterone to epitestosterone and luteinizing hormone.

Urinary T/LH ratio is a more sensitive and specific than urinary T/E ratio (Perry et al. 1997).

MEDICAL MANAGEMENT

Treatment for myocardial ischemia with oxygen therapy and nitrates dependent upon hemodynamic stability (Wysoczanski et al. 2008)

Potential phlebotomy for polycythemia

Antihypertensive agents as required

Antiarrhythmic agents as required. Note: Avoid β-blockers in the event of concomitant use of cocaine or other β-adrenergic agents (see cocaine)

Evaluation for neurological intervention in the event of cerebrovascular accident (García-Esperón et al. 2013)

Evaluation for fibrinolytic therapy in the event of pulmonary embolism (Fengler & Brady 2009)

REFERENCES

Casavant, M.J., Blake, K., et al. (2007) Consequences of use of anabolic androgenic steroids. *Pediatr. Clin. North Am.* 54(4):677–690.

Estrada, M., Varshney, A., and Ehrlich B.E. (2006) Elevated testosterone induces apoptosis in neuronal cells. *J. Biol. Chem.* 281:25492–25501.

Fengler, B.T. and Brady, W.J. (2009) Fibrinolytic therapy in pulmonary embolism: an evidence-based treatment algorithm. *Am. J. Emerg. Med.* 27(1):84–95.

García-Esperón, C., Hervás-García, J.V., Jiménez-González, M., Pérez de la Ossa-Herrero, N., Gomis-Cortina, M., Dorado-Bouix, L., López-Cancio Martinez, E., Castaño-Duque, C.H., Millán-Torné, M., and Dávalos, A. (2013) Ingestion of anabolic steroids and ischaemic stroke. A clinical case report and review of the literature. *Rev. Neurol.* 56(6):327–331.

Grivetti, L.E. and Applegate, E.A. (1997) From Olympia to Atlanta: a cultural-historical perspective on diet and athletic training. *J. Nutr.* 127(5 Suppl.):860S–868S.

Hoffman, J.R. and Ratamess, N.A. (2006) Medical issues associated with anabolic steroid use: are they exaggerated. *J. Sports Sci. Med.* 5(2):182–193.

Karch, S. (2009) *Pathology of Drug Abuse*, 4th ed., Boca Raton, Florida: CRC Press, pp. 609–610.

Korkia, P. and Stimson, G.V. (2007) Indications of prevalence, practice and effects of anabolic steroid use in Great Britain. *Int. J. Sports Med.* 18(7):557–562.

Marcel J., et al. (2007) Consequences of use of anabolic androgenic steroids, Pediatric Clinics of North America, vol.54.4, pp.677–690.

Patil, J.J., O'Donohoe, B., Loyden, C.F., and Shanahan, D. (2007) Near-fatal spontaneous hepatic rupture associated with anabolic androgenic steroid use: a case report. *Br. J. Sports Med.* 41:462–463.

Perry, P.J., MacIndoe, J.H., Yates, W.R., Scott, S.D., and Holman, T.L. (1997) Detection of anabolic steroid administration: ratio of urinary testosterone to epitestosterone vs the ratio of urinary testosterone to luteinizing hormone. *Clin. Chem.* 43(5):731–735.

Stergiopoulos, K., Brennan, J.J., Mathews, R., Setaro, J.F., and Kort, S. (2008) Anabolic steroids, acute myocardial infarction and polycythemia: a case report and review of the literature. *Vasc. Health Risk Manag.* 4(6):1475–1480.

Wysoczanski, M., Rachko, M., and Bergmann, S.R. (2008) Acute myocardial infarction in a young man using anabolic steroids. *Angiology* 59(3):376–378.

BENZODIAZEPINES

BACKGROUND AND USES

Benzodiazepines were first discovered in the 1930s by Leo Sternback, an employee of Hoffman-LaRoche but were not medically used until 1957 when Librium© was prescribed as an anxiolytic. Benzodiazepines grew to be one of the most highly prescribed drugs in the United States and reports of abuse in the 1980s led to these drugs being classified in Schedule IV as per the Controlled Substance Act (Rosenbaum et al. 2005).

Benzodiazepines are used as anxiolytics, amnestics, hypnotics, anticonvulsants, sedatives, and muscle relaxants.

Brand names include: Xanax© (alprazolam), Ativan© (lorazepam), Valium© (diazepam), and Versed© (midazolam)

Street names include: Bennies, bars, tranks, benzos, xanies, xanabars

PHYSICAL AND CHEMICAL PROPERTIES

Physical Properties

Tablet, capsule

Liquid

Chemical Properties

Benzodiazepines are composed of a benzene ring, which is joined to a seven member diazepine ring.

Unique compounds are created based upon substituents on side chains and modification of the rings.

TOXICOLOGY

Acute Effects

CNS: Bradypnea, apnea, somnolence, coma, impaired motor coordination, confusion, altered vision, dizziness/vertigo, impaired cognitive function

Cardiovascular: Bradycardia

GI: Vomiting

Psychological: Mood swings, erratic behavior, agitation, euphoria in the presence of other drugs

Neurological: Seizures, myoclonus

Chronic Effects

CNS: Memory and cognitive impairment, confusion, disorientation, slurred speech, diminished coordination

Tolerance typically for usage >6 months

Cross-tolerance with other CNS depressants, including alcohol

Mechanism of Action

Depresses CNS by enhancing $GABA_A$'s inhibitory effects

Active receptors sites are found in the periphery with the highest concentration being found in steroid-producing cells specifically the adrenal and anterior pituitary glands as well as the reproductive organs. (Goldfrank et al. 2007)

In cocaine cardiotoxicity, benzodiazepine usage may be justified due to the presence of cardiac benzodiazepine receptors

Increases the frequency and duration of opening chloride channels mediated by GABA

MODES OF ADMINISTRATION

Oral

Insufflation

DRUG SCREENING

Not part of the standard "SAMHSA-5" (UNODC 2012)

Can be detected 48–72 h after ingestion

Presumptive screens—Zimmerman test

Chromatography (e.g., TLC, GC, or HPLC)

Spectroscopy (e.g., IR, UV, MS)

MEDICAL MANAGEMENT

Respiratory support is emphasized due to CNS depression especially when other CNS depressants are present as well. Endotracheal intubation may be necessary based upon adequacy of respiratory effort.

Flumazenil can be used to reverse benzodiazepine intoxication at a starting dose of 0.2 mg intravenously over 30 min. Flumazenil cannot be given via the endotracheal tube as compared to the opioid reversal agent naloxone.

Abrupt discontinuance of benzodiazepines therapy can induce seizures, anxiety, irritability, hallucinations, and diarrhea (Tetrault & O'Connor 2008).

Withdrawal syndromes may be manifested within 2–10 days after removal of the drug.

Flumazenil can induce withdrawal

Benzodiazepines are not associated with any specific organ toxicity

Benzodiazepine overdoses are rarely lethal unless accompanied by other CNS depressants. (Goldfrank et al. 2007)

REFERENCES

Goldfrank, L., Hoffman, R., Nelson, L., Howland, M., Lewin, N., and Flomenbaum, N. (2007) *Goldfrank's Manual of Toxicologic Emergencies*, New York, NY: McGraw Hill, p. 620.

Rosenbaum, J., et al. (2005) Benzodiazepines: Revisiting Clinical Issues in Treating Anxiety Disorders, *Primary Care Companion J Clin Psychiatry*, vol. 7(1), pp. 23–30.

Tetrault, J.M. and O'Connor, P.G. (2008) Substance abuse and withdrawal in the critical care setting. *Crit. Care Clin.* 24(4):767–788.

United Nations Office on Drugs and Crime (2012) *Recommended Methods for the Identification and Analysis of Barbiturates and Benzodiazepines under International Control*, New York, NY: Manual For Use By National Drug Analysis Laboratories.

COCAINE

BACKGROUND AND USES

Cocaine is derived from the leaves of the *Erythroxylum coca* plant which is indigenous to South America, Indonesia, and the West Indies (Levine 2010). Early uses included roles in religious rites but workers in the mountainous ranges discovered benefits from cocaine's activity on the CNS that specifically increased energy, stamina, and appetite suppression. This benefit was especially helpful during long working days at high altitudes.

Cocaine was first utilized as a local anesthetic in Europe during the 1800s and the drug was legally used as a recreational drug in the United States until 1914. The vasoconstrictive properties, in addition to its anesthetic activity, made cocaine a valuable agent in both nasal and eye surgeries but its use has diminished with newer, and safer, local anesthetic agents (Karch 2009).

Cocaine is second only to marijuana, as a commonly abused illicit drug in the United States. For the occupational health practitioner, it is important to recognize "street" terms for the drug as they might be commonly heard during patient histories and reports of potential abuse from co-workers (Albertson 2014).

> Brand names for cocaine include: Pharmaceutical cocaine
>
> Street terms for cocaine include: Coke, crack, blow, snow, and nose candy

PHYSICAL AND CHEMICAL PROPERTIES

Physical Properties

Cocaine hydrochloride—the most common of the potential salts made by combining the alkaline cocaine with various acidic compounds. Typically cocaine hydrochloride will present as a white, crystalline powder but tinting with other colors may be noted due to adulterants and manufacturing methods. Polar and easily dissolved, this form is utilized primarily for insufflation and intravenous administration.

"Crack" cocaine: Appears as an irregularly shaped, amorphous "rock" or may also appear as a thin wafer depending upon the method of "cooking." Typically, "crack" or "rock" cocaine will appear as white, off-white, yellow, or potentially light brown. The methods of manufacture to convert the base and adulterants are responsible for varying tints. Easily vaporized.

Freebase cocaine: Prepared by utilizing ammonia and cocaine to produce a "base" form of the cocaine. It is typically a white powder of higher potency than the "rock" form which utilizes baking soda as a conversion agent.

Infusion: Prepared by utilizing dried coca leaves and filtering them to form a "tea." It is more common in South American countries; and used as an herbal, medicinal agent.

Chemical Properties

Chemical name: Methylbenzoylecgonine

An ester of methylecgonine and benzoic acid (Yang et al. 2001)

The ecgonine component includes four chiral carbons which can exist as four racemic forms and eight optical isomers

Subsequent to hydrolysis of the ester alkaloids from the coca plant, cocaine can be synthesized by combining the (-) ecgonine with methanol and benzoic acid (Fandiño et al. 2002)

TOXICOLOGY

Acute Effects

Cardiovascular: Myocardial ischemia, myocardial infarction, hypertension, aortic dissection, stroke, arrhythmias (i.e., sinus tachycardia, supraventricular tachycardia, ventricular tachycardia, ventricular fibrillation) (Derlet et al. 1994) (Wenzel et al. 2004)

Pulmonary: Hypoxia, pulmonary edema, pneumothorax, pneumomediastinum, pneumopericardium, tachypnea, "crack lung"[1]

ENT: Mydriasis, acute glaucoma, madarosis, oropharyngeal burns

GI: Ulcer perforation, ischemic colitis

Musculoskeletal: Rhabdomyolysis

Psychological: Restlessness, agitation, psychosis, paranoia, "Superman syndrome"

CNS: Seizures, hyperthermia, coma, focal neurologic abnormalities, cerebral hemorrhage (Larossa 2013)

Obstetrics and gynecology (OB/Gyn): abruptio placentae

Chronic Effects

Cardiac: Cardiomyopathy, myocardial spasm, ischemia, infarction

Pulmonary: Dyspnea, hemoptysis, bronchospasm, alveolar infiltrates without effusions, chest pain, lung trauma, pruritis, sore throat, asthma, hoarseness, eosinophilia

Neurological: Hemorrhagic and ischemic stroke

GI: Nausea, vomiting, abdominal pain

Dental: Gingivitis, enamel breakdown due to bruxism

ENT: Nasal septal erosion and perforation (Businco et al. 2008)

Autoimmune: Lupus, vasculitis, Goodpasture's disease (Chan et al. 2011)

Renal: Glomerulonephritis (Chan et al. 2011)

Dermatological: Burn to lips, callouses to extremities, "crack thumb" due to burns from holding the "crack" pipes

Mechanism of Action

Cocaine is a sympathomimetic that potentiates the effects of biogenic amines such as dopamine, norepinephrine, epinephrine, and serotonin by increasing release and blocking reuptake of these neurotransmitters back into the presynaptic neurons.

When dopamine accumulates in the synaptic cleft, prolonged postsynaptic effects occur at the receiving neuron, specifically psychomotor agitation. Habitual use leads to the downregulation of dopamine receptors, which can lead to depression and addictive behavior. Serotonin (5-hydroxytryptamine, 5-HT) effects include the depression and addiction listed above as serotonin is a modulator of dopamine. The 5-HT2 receptor is associated with cocaine-related hyperactivity (Baigent 2003).

Adrenally derived epinephrine is responsible for tachycardia while hypertension arises from neuronally produced norepinephrine.

Cocaine's action as a local anesthetic arises from sodium channel blockade thus interfering with action potentials. Additionally, cocaine may bind to the Kappa-opioid receptor.

Chemical Pathology

Vasospasm, increased coagulatory activity and hemorrhage from increased vascular sheering forces are all attributed with end-organ toxicity in multiple organ systems.

MODES OF ADMINISTRATION

Insufflation (snorting)

Inhalation (smoking)

Transmucosal application (e.g., nasal, gingival, vaginal, rectal)

Transdermal

Intravenous

Gastrointestinal (swallowing, "body packing")

DIAGNOSTIC METHODOLOGIES

The half-life is shortest when cocaine is smoked as "free based" or "crack" and longer when sniffed or snorted for 15 and 45 min, respectively. Urine and blood samples are the most commonly used for acute exposures while saliva, hair, and meconium may also be considered. Although the elimination half-life is 40–80 min, metabolites may be detected in urine screens for a period of 48–72 h after acute use and potentially several weeks in those chronically abusing the drug.

Commercial immunoassay testing kits may only detect the chief metabolite, benzoylecgonine (BZE), but other metabolites such as ecgonine methyl ester (EME), and ecgonine are also significant. Cobalt thiocyanate is the most common screening test for cocaine and yields a blue color with reactivity.

MS: The major ions found are m/z = 82, 182, 83, 105, 303, 77, 94, and 96

GC: In blood, the LOD in blood is 20 μg/l

Of particular interest is the metabolite methylecgonidine that would only be found in abusers who have smoked the drug. The pyrolysis induces this particular metabolite that eliminates other modes of administration (Scheidweiler et al. 2003).

Cardiac testing should include 12 lead electrocardiography (ECG), which is suggested for potential observance of acute changes, and cardiac enzymes would be considered as a substantial adjunct for the determination of cardiac injury.

Chest radiography could be helpful in determining the etiology of chest pain or related pulmonary injury.

MEDICAL MANAGEMENT

As cocaine demonstrates combined cardiac inotropic and vasoconstrictive effects, β-blockers should be avoided in the treatment of cocaine-induced tachycardia as this may result in a lack of opposition to the adrenergic, or "alpha," effects of cocaine and can exacerbate coronary vasoconstriction and systemic hypertension leading to hypertensive crisis. Preferred treatments would include nitrates, calcium channel blockers such as diltiazem, benzodiazepines, and hydralazine (Goldfrank et al. 2007) (McCord et al. 2008).

Consideration should be given to the potential adverse effects of adding epinephrine to local anesthetics as this could possibly exacerbate vasoconstriction and thereby attenuate cardiac ischemia and associated arrhythmias (Cone 2011).

As cocaine is primarily metabolized through the liver, via hepatic and plasma cholinesterase; ester-based local agents should also be avoided.

REFERENCES

Albertson, T.E. (2014) Recreational drugs of abuse. *Clin. Rev. Allergy Immunol.* 46(1):1–2.

Baigent, M. (2003) Physical complications of substance abuse: what the psychiatrist needs to know. *Curr. Opin. Psychiatry* 16(3):291–296.

Businco, L.D., Lauriello, M., Marsico, C., Corbisiero, A., Cipriani, O., and Tirelli, G.C. (2008) Psychological aspects and treatment of patients with nasal septal perforation due to cocaine inhalation. *Acta Otorhinolaryngol. Ital.* 28(5):247–251.

Chan, A.L., Louie, S., Leslie, K.O., Juarez, M.M., and Albertson, T.E. (2011) Cutting edge issues in Goodpasture's disease. *Clin. Rev. Allergy Immunol.* 41(2):151–162.

Cone, J.D., Jr., Harrington, M.A., Kelley, S.S., Prince, M.D., Payne, W.G., and Smith, D.J., Jr. (2011) Drug abuse in plastic surgery patients: optimizing detection and minimizing complications. *Plast. Reconstr. Surg.* 127(1):445–455.

Derlet, R.W., Tseng, C.C., and Albertson, T.E. (1994) Cocaine toxicity and the calcium channel blockers nifedipine and nimodipine in rats. *J. Emerg. Med.* 12(1):1–4.

Fandiño, A.S., Toennes, S.W., and Kauert, G.F. (2002) Studies on hydrolytic and oxidative metabolic pathways of anhydroecgonine methyl ester (methylecgonidine) using microsomal preparations from rat organs. *Chem. Res. Toxicol.* 15(12):1543–1548.

Goldfrank, L., Hoffman, R., Nelson, L., Howland, M., Lewin, N., and Flomenbaum, N. (2007) *Goldfrank's Manual of Toxicologic Emergencies*, New York, NY: McGraw Hill, pp. 633–640.

Karch, S. (2009) *Pathology of Drug Abuse*, 4th ed., Boca Raton, Florida: Taylor & Francis Group.

Larrosa-Campo, D., Ramon-Carbajo, C., Benavente-Fernandez, L., Alvarez-Escudero, R., et al. (2013) Diagnosis of stroke due to cocaine and its complications. *Rev. Neurol.* 57(4):167–170.

Levine, B. (2010) *Principles of Forensic Toxicology*, 3rd ed., AACT Press: Washington, DC: p. 245.

McCord, J., Jneid, H., Hollander, J.E., et al. (2008) Management of cocaine-associated chest pain and myocardial infarction: a scientific statement from the American Heart Association Acute Cardiac Care Committee of the Council on Clinical Cardiology. *Circulation* 117(14):1897–1907.

Scheidweiler, K.B., Plessinger, M.A., Shojaie, J., Wood, R.W., and Kwong, T.C. (2003) Pharmacokinetics and pharmacodynamics of methylecgonidine, a crack cocaine pyrolyzate. *J. Pharmacol. Exp. Ther.* 307(3):1179–1187.

Wenzel, V., Krismer, A.C., Arntz, H.R., Sitter, H., Stadlbauer, K.H., Lindner, K.H., and European Resuscitation Council Vasopressor during Cardiopulmonary Resuscitation Study Group. (2004) A comparison of vasopressin and epinephrine for out-of-hospital cardiopulmonary resuscitation. *N. Engl. J. Med.* 350(2):105–113.

Yang, Y., Ke, Q., Cai, J., Xiao, Y.F., and Morgan, J.P. (2001) Evidence for cocaine and methylecgonidine stimulation of M2 muscarinic receptors in cultured human embryonic lung cells. *Br. J. Pharmacol.* 132(2):451–460.

GAMMA HYDROXYBUTYRATE (GHB)

BACKGROUND AND USES

The first report of γ-hydroxybutyric acid (GHB) synthesis was in 1960 as a structural analog of the neurotransmitter gamma aminobutyric acid (GABA). This enabled research, which led to the discovery that GHB is a naturally occurring neurochemical in the human brain.

γ-hydroxybutyric acid was originally used as an anesthetic agent due to its CNS depressant effect but newer drugs with higher safety profiles have resulted in decreased use of this agent. Presently, the only medical use for GHB is in the treatment of narcolepsy under the trade name of Xyrem© (Johnson & Griffiths 2013).

Additional, nonmedical, uses include:

Weight reduction

Athletes seeking performance enhancement through increased muscle mass

Drug facilitated sexual assault: "date rape"

Brand names include: Xyrem©

Street names include: Blue thunder, blue nitro, women's Viagra, liquid ecstasy, easy lay, scoop, jib, liquid X, G

PHYSICAL AND CHEMICAL PROPERTIES

Physical Properties

Colorless, odorless liquid

Slightly salty taste

When shaken, bubbles persist on surface of fluid

Prior to dilution, the salt is found as a white/off-white powder

Chemical Properties

γ-hydroxybutyric acid is synthesized from γ-butyrolactone (GBL) by the addition of sodium hydroxide (lye) in either ETOH or H_2O (Schepp at al. 2012).

γ-hydroxybutyric acid converts to GBL dependent upon the matrix, temperature, and pH. A pH lower than 4.72, will cause GBL to predominate.

TOXICOLOGY

Acute Effects

CNS: Bradypnea, apnea, somnolence, coma, miosis (Aromatario 2012)

Cardiovascular: Bradycardia AV heart blocks

GI: Vomiting

Psychological: Hallucinations, euphoria, extreme agitation upon arousal, empathogenesis

Neurological: Seizures, myoclonus

Salivation

Amnesia

Loss of sexual inhibition

Chronic Effects

Psychosis (van Noorden et al. 2009)

Depression

Increased muscle mass

Weight reduction

Mechanism of Action

Exogenous GHB increases dopamine in the presynaptic terminal and a release of endogenous opioids. The anesthetic properties exhibited by GHB are similar to those of a structural analog, GABA (Kelley & Larison 2009).

γ-hydroxybutyric acid increases stages 3–4 and REM sleep, which is when the muscle growth hormone is predominantly released resulting in the increased muscle mass desired by athletes using the drug for performance enhancement.

MODES OF ADMINISTRATION

Oral

Transmucosal (rectal/vaginal)

DRUG SCREENING

Detectable in blood/urine up to 6 h post ingestion

Requires a special laboratory analysis typically not performed in the hospital setting (McCusker 1999)

MS: Major ions are found at: m/z 147, 233, 117, 158, 148, 149, 143, 133, 204, 234, and 235.

MEDICAL MANAGEMENT

Respiratory support due to CNS depression (Goldfrank et al. 2007)

Potential warming of hypothermia

Consideration of α/β-adrenergic agents for hypotension, and bradycardia

Naloxone (Narcan) may be helpful even though GHB doesn't bind specifically with the opioid receptor.

There is a potential need for physical/chemical restraints due to extreme agitation (Dyer 2001).

Rectal/vaginal examination for tampons

Consider sexual assault examination (Levine 2010)

Sedation, volume replacement, and cooling measures may be necessary for management of withdrawal. Note: Benzodiazepines primarily affect the GABA$_A$ whereas GHB activity is mediated by the GABA$_B$ receptor therefore may be administered.

REFERENCES

Aromatario, M., Bottoni, E., Santoni, M., and Ciallella, C. (2012) New "lethal highs": a case of a deadly cocktail of GHB and Mephedrone. *Forensic Sci. Int.* 223(1–3):e38–e41.

de Jong, C.A., Kamal, R., Van Noorden, M., and Broers, B. (2013) Treatment of GHB withdrawal syndrome: catch 22 or challenge for addiction medicine? *Addiction* 108(9):1686.

Dyer, J.E., Roth, B., and Hyma, B.A. (2001) Gamma-hydroxybutyrate withdrawal syndrome. *Ann. Emerg. Med.* 37(2):147–153.

Goldfrank, L., Flomenbaum, N., Lewin, N., Howland, M.A., Hoffman, R., and Nelson, L. (2007) *Goldfrank's Toxicologic Emergencies*, McGraw-Hill, pp. 659–664.

Johnson, M.W. and Griffiths, R.R. (2013) Comparative abuse liability of GHB and ethanol in humans. *Exp. Clin. Psychopharmacol.* 21(2):112–123.

Kelley, S., Larison, D., et al., (2009) *Pharmacology*, Lippincott's Illustrated Reviews, 4th ed., Philadelphia: Lippincott, pp. 537–538.

Levine, B. (2010) *Principles of Forensic Toxicology*, 3rd ed., Washington, DC: AACT Press, p. 207.

McCusker, R.R., Paget-Wilkes, H., Chronister, C.W., and Goldberger, B.A. (1999) Analysis of gamma-hydroxybutyrate (GHB) in urine by gas chromatography-mass spectrometry. *J. Anal. Toxicol.* 23(5):301–305.

Schep, L.J., Knudsen, K., Slaughter, R.J., Vale, J.A., and Mégarbane, B. (2012) The clinical toxicology of gamma-hydroxybutyrate, gamma-butyrolactone and 1,4-butanediol. *Clin. Toxicol. (Phila)* 50(6):458–470.

van Noorden, M.S., van Dongen, L.C., Zitman, F.G., and Vergouwen, T.A. (2009) Gamma-hydroxybutyrate withdrawal syndrome: dangerous but not well-known. *Gen. Hosp. Psychiatry* 31(4):394–396.

MARIJUANA AND SYNTHETIC CANNABINOIDS

BACKGROUND AND USES

The earliest records of marijuana usage date back to the 3000 BC and, according to the United Nations, it is the most widely abused drug in the world. Marijuana is the most commonly abused illicit drug in the United States and is derived from the hemp plant *Cannabis sativa* (Rudgley 1998) (UNODC 2006). There are many psychoactive agents found in marijuana but the primary agent is a lipid-soluble compound referred to as delta-9-tetrahydrocannabinoid (THC) (Levine 2010).

In the United States, marijuana is used for recreational purposes and, in some states, is legal for medical purposes in the treatment of glaucoma, breakthrough nausea and vomiting subsequent to chemotherapy/surgery and anorexia experienced by HIV patients. Commercially, marijuana may also be used for the production of hemp.

In 1995, a Clemson University professor conducting research on the cerebral effects from the use of cannabinoids (CB) used a synthetic compound, which was later described as part of a published research paper. Readers began reproducing the method of developing this compound, in order to get marijuana-like "high." From this one compound there are now five identified compounds (WH-018, JWH-073, JWH-200, CP-47, 497, cannabicyclohexanol—three of which list the initials of the Clemson professor) that are being sprayed on herbal blends and smoked. The effects of this "synthetic marijuana," though similar to that of THC, produce distinctive and psychoactive behavior aberrant from THC alone. And, due to these deleterious manifestations, in March 2011, the U.S. Drug Enforcement Administration (DEA) temporarily placed these five compounds into Schedule I of the Controlled Substances Act (CSA) (Macher 2012).

The main chemical observed is JWH-018 and, just like THC, it attaches to the CB receptors but with a much higher affinity and level of potency than THC (Albertson 2014).

Brand names for marijuana: Marinol®

Street names for marijuana include: pot, reefer, mary, mary j, weed, grass

Street names for "synthetic marijuana" include: K2, spice, Mr. Nice Guy

PHYSICAL AND CHEMICAL PROPERTIES

Physical Properties

Marijuana is a collective term for various parts of the *Cannabis sativa L.* plant, including the seeds, resin, and derived compounds but excluding mature stalks, fiber, and oil extractions.

Tetrahydrocannabinol is an oil, which is highly soluble in lipids but insoluble in water.

Chemical Properties

The component of marijuana that is cited as having the greatest psychological effect on users is delta-9 THC. By weight, the leaves may contain from 1to 10% THC by weight with a higher percentage, up to 15%, may be detected in hash oil/resin made from the flowering tops of the plant.

Approximately 95% of the THC exists as monocarboxylic acids, which will rapidly decarboxylate when heated. This coincides with the manner in which the drug is administered specifically through smoking and/or baking in foods.

Cannabis contains a large number of chemical compounds, including cannabinoids such as cannabinol, which is only 10% as active a psychoactive agent as THC.

TOXICOLOGY

Acute Effects

Marijuana:

CNS: Sedative effect with intense relaxation, euphoria, altered mentation, hyporeflexia, anxiety, difficulty with thinking and problem solving (El-Alfy 2010). *Note: The inclusion of adjunctive drugs such as phencyclidine or formaldehyde, known as "smoking wet," can produce hallucinations, violent behavior*

Cardiovascular: Tachycardia, bradycardia, myocardial depression, hypotension (Cone et al. 2011)

Ophthamologic: Conjunctival congestion

Pulmonary: Reactive airway exacerbation, bullae formation, laryngospasm

"Synthetic Marijuana"

CNS: Muscle twitching, hyperactivity, hallucinations, loss of consciousness, convulsions (Schneir & Baumbacher 2011)

Cardiovascular: Tachycardia, hypertension (Mir et al. 2011)

Pulmonary: Difficulty in breathing

Psychological: Paranoia, psychosis, panic, anxiety (Oluwabusi et al. 2012)

Chronic Effects

Marijuana:

Pulmonary: Increased risk of lung cancer, which is heightened with concurrent use of tobacco

Immunologic: Increased risk of infection secondary to modulation of the immune response specifically T and B lymphocytes, macrophages and immune cytokines

CNS: Increased restlessness, insomnia, irritability, sleep disturbances

OB/Gyn: Decreased fertility rates

"Synthetic" Marijuana

CNS: Increased restlessness, insomnia, irritability, sleep disturbances (Papanti et al. 2013)

Psychological: Paranoia, psychosis, panic, anxiety

Mechanism of Action

Tetrahydrocannabinoid readily crosses the blood brain barrier and binds to CB1 and CB2 receptors both of which are G-protein coupled receptors. Cannabinoid receptor type 1 is found in both the brain and periphery and acts as the presynaptic neural cannabinoid receptor. When THC binds to CB1, it is released as GABA in the cerebral cortex, hippocampus, and amygdala (Goldfrank et al. 2007).

Due to their high lipid solubility, cannabinoids may be detected in the body for long periods of time in comparison to other drugs. A single usage may be detected up to 4 weeks post ingestion. The potential accumulation in the lipid membranes of neurons has been a topic of interest with regard to manifestations by chronic users.

MODES OF ADMINISTRATION

Inhalation

Oral

DRUG SCREENING

The leaves of *Cannabis sativa* and its resin may be identified utilizing microscopy with detection of the glandular trichomes and cystolithic hairs being deemed as confirmatory (McPherson et al. 2007).

The Duquenois test is utilized for screening cannabinols and produces a blue violet color when reacting with *p*-dimethylbenzaldehyde.

MS: Major ions are m/z = 41, 43, 55, 231, 271, 295, 299, and 314.

Tetrahydrocannabinoid is stored in adipose tissue and has a half-life of approximately 56 h in occasional users and 28 h in chronic users.

Due to high lipid solubility, it may take as much as thirty (30) days to completely eliminate a single use of marijuana metabites.

Approximately 30% of THC is excreted in the urine and detection using standard urine drug screens, may occur for 1–4 weeks post ingestion.

A metabolite of THC, THC–COOH, is predominantly found in the urine. Hair and fluids such as oral samples and sweat will reflect THC predominantly. Blood samples will reflect both (Karch 2009).

The most commonly used screening test for THC is the Duquenois-Levine; however, this is a screening test and numerous examples are cited of false positives due to cross-reactivity with other agents.

MEDICAL MANAGEMENT

No specific management strategies are cited with regard to THC and there is no known withdrawal syndromes associated with cessation of drug use.

Due to the extreme agitation manifested with patients having ingested "synthetic marijuana," practitioners should be prepared to provide physical/chemical restraints in order to protect the patient, and other patients, from harm.

Practitioners should be alerted to physical manifestations associated with other drugs used in correlation with THC specifically adulterants such as "crack" cocaine or PCP and formaldehyde. The latter two agents are combined with marijuana in a "street" practice known as "smoking wet." This practice involves soaking the ends of the marijuana cigarette in these agents, which creates an increase in hallucinogenic effect but may also induce CNS stimulation, cardiovascular and pulmonary complications (EMCDDA 2012).

Obvious injuries, incurred due to altered mental status, should be prevented or treated with appropriate methodologies.

REFERENCES

Albertson, T.E. (2014) Recreational drugs of abuse. *Clin. Rev. Allergy Immunol.* 46(1):1–2.

Cone, J., Jr., Harrington, M.A., Kelley, S.S., Prince, M.D., Payne, W.G., and Smith, D.J., Jr. (2011) Drug abuse in plastic surgery patients: optimizing detection and minimizing complications. *Plast. Reconstr. Surg.* 127(1):445–455.

El-Alfy, A.T., et al. (2010) Antidepressant-like effect of delta-9-tetrahydrocannabinol and other cannabinoids isolated from *Cannabis sativa* L. *Pharmacol. Biochem. Behav.* 95(4):434–442.

European Monitoring Centre for Drugs and Drug Addiction (EMCDDA) (2012) *Drug profiles: methamphetamine*. Available at http://www.emcdda.europa.eu/publications/drug-profiles/methamphetamine#analysis (accessed: 5/21/13).

Goldfrank, L., Hoffman, R., Nelson, L., Howland, M., Lewin, N., and Flomenbaum, N. (2007) *Goldfrank's Manual of Toxicologic Emergencies*, New York, NY: McGraw Hill, pp. 633–640.

Karch, S. (2009) *Pathology of Drug Abuse*, 4th ed., Boca Raton, Florida: Taylor & Francis Group.

Levine, B. (2010) *Principles of Forensic Toxicology*, 3rd ed., Washington, DC: AACT Press, pp. 269–270.

Macher, R., Burke, T., and Owen, S. (2012) Synthetic marijuana. *FBI Law Enforc. Bull.* 81(5):17–22.

McPherson, R., et al. (2007) *Drugs of Abuse in Henry's Clinical Diagnosis and Management by Laboratory Methods*, 21st ed., Philadelphia: Saunders Elsevier.

Mir, A., Obafemi, A., Young, A., and Kane, C. (2011) Myocardial infarction associated with use of the synthetic cannabinoid K2. *Pediatrics* 128(6):e1622–e1627.

Oluwabusi, O.O., Lobach, L., Akhtar, U., Youngman, B., and Ambrosini, P.J. (2012) Synthetic cannabinoid-induced psychosis: two adolescent cases. *J. Child Adolesc. Psychopharmacol.* 22(5):393–395.

Papanti, D., Schifano, F., Botteon, G., Bertossi, F., et al. (2013) "Spiceophrenia": a systematic overview of "spice"-related psychopathological issues and a case report. *Hum. Psychopharmacol.* 28(4):379–389.

Rudgley, R. (1998) *Lost Civilisations of the Stone Age*, New York, NY: New York Free Press.

Schneir, A.B. and Baumbacher, T. (2011) Convulsions associated with the use of a synthetic cannabinoid product. *J. Med. Toxicol.* 8(1):62–64.

United Nations Office on Drugs and Crime (2006) *Cannabis: Why We Should Care*. World Drug Report 1, United Nations, p. 14.

OPIATES AND OPIOIDS

BACKGROUND AND USES

Opiates are alkaloids derived from the opium poppy *Papaver somniferum*, which was first cultivated in Mesopotamia as early as 3400 BC. The term "opiate" is often misused as it describes only those drugs made from the actual plant whereas the term "opioid" describes a much broader class of drugs, still producing an opium-like response, but also binding to the mu (μ), kappa (κ), and delta (δ) receptors. Examples of opiates include codeine and morphine where examples of semisynthetic "opioids" would include heroin, hydrocodone, hydromorphone, oxycodone, and oxymorphone. Semisynthetics are drugs derived from altering an opiate molecule as compared to synthetic opioids which are not derived from an opiate. Examples of synthetic opioids include fentanyl, meperidine, and methadone.

Whereas many opioids have established medical uses, heroin is an opioid which is illicit in its use. Heroin was first synthesized in 1874 when an English chemist added two acetyl groups to the morphine molecule thus the name diacetylmorphine. In 1895, Bayer marketed diacetylmorphine, under the trade name of "Heroin," as an over-the-counter drug which would serve as a substitute for morphine cough suppressants with their associated risk of addiction. Due to its lipophilic nature and water solubility, the effects of heroin are rapidly experienced by the user making it a very popular drug for potential addiction.

Today, opiates and opioids have a broad range of medical uses, including analgesia, anxiolysis, sedation, decreasing gastric motility, and serving as antitussives. However, prescription opioid abuse has presently risen to "epidemic" levels in the United States with a continual rise since 1999. In November 2011 the director for the Centers for Disease Control and Prevention (CDC), Dr. Thomas Frieden, was quoted as saying that "Overdoses involving prescription painkillers are at epidemic levels and now kill more Americans than heroin and cocaine combined." (CDCP 2011)

Brand names include: Oxycodone, hydrocodone, fentanyl, oxymorphone, hydromorphone

Street names for opioids/opiates include: Heroin ("H", junk, horse, tar, skag) and prescription analgesics (oxy's, roxy's, percs, vikes, blues for prescription opioids.

PHYSICAL AND CHEMICAL PROPERTIES

Physical Properties

Heroin salt (diacetylmorphine hydrochloride)—crystalline powder, typically white in color but color could vary dependent upon adulterants. This is the illicit form of the morphine and would be utilized for snorting or intravenous use due to its high boiling point.

Freebase heroin: Crystalline powder but typically a duller shade of white. This too is an illicit form of the drug and, due to its lower boiling point, is acceptable to smoke.

"Black tar" heroin: Solid, putty-like substance, which may be black or brown and is manufactured with a more rapid method utilizing morphine derivatives other than heroin.

Prescription analgesics may be found in tablet, liquid, and gel preparations.

Chemical Properties

Chemical name of diacetylmorphine (heroin) is (5α, 6α)-7,8-Didehydro-4,5-epoxy-17-methylmorphinan-3,6-diol diacetate (ester) (Karch 2009)

Pharmcodynamic effects of a specific opioid are determined by its steriospecificity when binding to certain receptors as well as the molecular structure of the drug (Levine 2010) (Katchman et al. 2002).

TOXICOLOGY

Acute Effects

CNS: Bradypnea, diminished level of consciousness, coma, miosis, hypothermia, seizures (Kleinschmidt et al. 2001) (Offiah & Hall 2008)

Pulmonary: Hypoxia, apnea, hypercarbia, decreased tidal volume, pulmonary edema, chest wall rigidity

Cardiac: Bradycardias, myocardial ischemia, hypotension, QT prolongation leading to of potentially life-threatening ventricular arrhythmias (e.g., toursades de pointes)

GI: Decreased gastric motility, ileus

Renal: Rhabdomyolysis may occur due to long periods of unconsciousness, or intra-arterial administration, resulting in potential acute renal failure

Metabolic: Hypo/hyperglycemia in diabetics secondary to vomiting

Immunological: Immunoregulatory effects of endogenous opioids may be affected by presence of exogenous opiates (Saurer et al. 2009) (Ye et al. 2001).

Chronic Effects

GI: Constipation

Metabolic: Hyponatremia secondary to increased vasopressin release

CNS: abscesses secondary intravenous-related infections

Mechanism of Action

The analgesic and euphoric effects experienced subsequent to administration of opioids is a result of agonism of the mu receptors even though other receptors have also been identified such as the kappa- and delta-opioid receptors. Hypercarbia and hypoxia risks are increased when mu agonism occurs due to decreased chemoreceptor sensitivity of these two conditions. Tidal volume is decreased and bradypnea may occur resulting in decreased oxygen saturation and increased carbon dioxide retention.

The route of administration distinguishes the metabolism of heroin. If taken intravenously, the first pass metabolism, normally occurring with an oral ingestion, will be missed. Due to the acetyl groups and subsequently the increased fat solubility compared to the morphine molecule, it will cross the blood brain barrier very quickly. As in an oral ingestion, the drug is deacetylated into the active metabolite 6-monoacetylmorphine, the inactive metabolite 3-monacetylmorphine and morphine which will bind to the μ-opioid receptors. The binding to those receptors is responsible for the analgesic and anxiolytic manifestations of the drug. An adverse effect, however, occurring in both intravenous heroin and morphine users, is pruritus which occurs subsequent to a drug-induced histamine release (Rook et al. 2006).

Repetitive use of heroin results in physiological changes, such as upregulation—an increase in μ-opioid receptors. These changes which are associated with the dependence and addiction are also responsible for withdrawal symptoms after cessation of drug use.

MODES OF ADMINISTRATION

Intravenous

Oral

Insufflation

Inhalation

DRUG SCREENING

"SAHMSA 5" tests for opiates will screen for morphine and codeine but not for opioids such as oxycontin, methadone, and fentanyl (Quest Diagnostics n.d.)

Samples may come from blood or plasma, urine, saliva, and hair

Fluorescence, microparticle and enzyme immunoassays are recommended for urine samples

ELISAs (enzyme-linked immunosorbent assays) are utilized for non-urine samples

Confirmation analysis includes MS, gas chromatography–mass spectrometry (GC–MS), gas chromatography–tandem mass spectrometry (GC–MS/MS), liquid chromatography (LC), liquid chromatography–mass spectrometry (LC–MS), and liquid chromatography–tandem mass spectrometry (LC–MS/MS)

Not all drugs detectable using saliva and hair

The half-lives of opioids depend on a number of factors, including patient tolerance, bioavailability, administration route, and chemical breakdown

Opioids taken orally or administered intravenously are typically well absorbed and peak effects are observed in 60–90 min

MEDICAL MANAGEMENT

Consideration naloxone administration −0.4–2.0 mg in adults in patients with significant respiratory compromise. Higher doses, up to 10 mg, may be necessary if the ingestion involved a synthetic opioid.

Naloxone may be given by intravenous, intramuscular, endotracheal, or subcutaneous routes. If administering by the endotracheal route the dose should be multiplied by 2–2.5 times that of the dose for intravenous administration. Though rare, a small percentage of patients may experience withdrawal symptoms, including seizures, pulmonary edema, and cardiac arrhythmias such as asystole.

Respiratory support, including endotracheal intubation in the presence of inadequate respiratory effort (Sather & Tantawy 2006)

Circulatory support with α-adrenergic agents

Observe for "track" marks especially in areas hidden from common view (e.g., between fingers, in tattoos)

Treatment of hypoglycemia

Consider associated trauma, infection, and metabolic etiologies

Utilize EKG monitoring, including 12-lead EKG's to observe for QT prolongation associated with opioids such as levoacetylmethadol and methadone. If QT prolongation occurs, the sodium channel blockade in the myocardium may be treated with lidocaine and sodium bicarbonate through competitive displacement.

In the presence of hypotension, diphenhydramine should be considered as a means of counteracting histamine release.

In clonidine withdrawal, patients may experience hypertensive crisis. Administration of vasodilators such as sodium nitroprusside, esmolol, labetalol, or nicardipine should be considered.

REFERENCES

Centers for Disease Control and Prevention (2011) *Prescription painkiller overdoses at epidemic levels.* Available at http://www. cdc.gov/media/releases/2011/p1101_flu_pain_killer_overdose. html (accessed: February 18, 2013).

Karch, S. (2009) *Pathology of Drug Abuse*, 4th ed., Boca Raton, Florida: CRC Press, p. 387.

Katchman, A.N., McGroary, K.A., Kilborn, M.J., et al. (2002) Influence of opioid agonists on cardiac human ether-a-go-go-related gene K+ currents. *J. Pharmacol. Exp. Ther.* 303:688–694.

Kleinschmidt, K.C., et al. (2001) Opioids. In: Ford, M.D., editor. *Clinical Toxicology*, 1st ed., Philadelphia: W.B. Saunders Company.

Levine, B. (2010) *Principles of Forensic Toxicology*, 3rd ed., Washington, DC: AACT Press, p. 228.

Offiah, C. and Hall, E. (2008) Heroin-induced leukoencephalopathy: characterization using MRI, diffusion-weighted imaging, and MR spectroscopy. *Clin. Radiol.* 63(2):146–152.

Rook, E.J., Van Ree, J.M., Van Den Brink, W., Hillebrand, M.J., Huitema, A.D., Hendriks, V.M., and Beijnen, J.H. (2006)

Pharmacokinetics and pharmacodynamics of high doses of pharmaceutically prepared heroin, by intravenous or by inhalation route in opioid-dependent patients. *Basic Clin. Pharmacol. Toxicol.* 98:86–96.

Sather, J. and Tantawy, H. (2006) Toxins. *Anesthesiol. Clin.* 24:647–670.

Saurer, T.B., James, S.G., and Lysle, D.T. (2009) Evidence for the nucleus accumbens as a neural substrate of heroin-induced immune alterations. *J. Pharmacol. Exp. Ther.* 3(329):1040–1047.

Quest Diagnostics n.d., *Urine Drug Testing Process and Certifications*. Available at http://www.questdiagnostics.com/home/companies/employer/drug-screening/products-services/urine-test/urine-process-certifications.

Ye, L., Wang, X., Metzger, D.S., Riedel, E., Montaner, L.J., and Ho, W. (2001) Upregulation of SOCS-3 and PIAS-3 impairs IL-12-mediated interferon-gamma response in CD56+ T cells in HCV-infected heroin users. *PLoS One* 3(5):e9602.

117

MYCOTOXINS

Clara Y. Chan and Bruce J. Kelman

Mycotoxins are secondary metabolic products produced by fungi under certain environmental conditions. The presence of fungal growth cannot be used as an indicator of mycotoxins as the production of mycotoxins by fungi depends on myriad environmental conditions, including nutrient availability, growth substrate, moisture, temperature, maturity of the fungal colony, and competition with other microorganisms (Goldblatt, 1969). A species of fungi may also produce more than one mycotoxin, and a specific mycotoxin may be produced by many different species of fungi (D'Mello, 1997). Potentially toxigenic fungi are commonly found in the outdoor environment (Sudakin and Fallah, 2008).

Potential worker exposure to mycotoxins exists through inhalation and dermal contact associated with the handling and storing of contaminated commodities. While mycotoxins are not volatile and do not evaporate from the mold spore or substrata, they may be present in agricultural environments where airborne concentrations of dusts and mold spores may be high. As of this writing, it appears that aflatoxins and ochratoxins are the primary mycotoxins for which there is potential occupational concern. There is an abundance of literature supporting exposure by the general population to mycotoxins through ingestion but only a limited number of studies on inhalation exposure in the work environment.

Considerable research has been done on the issue of inhaled mycotoxins in spores present in indoor environments. As of this writing, the weight of the scientific evidence shows no causal relationship between exposure to mycotoxins in damp indoor environments and a constellation of health effects of concern (American College of Occupational and Environmental Medicine, 2011; Hardin et al., 2009; Institute of Medicine. Committee on Damp Indoor Spaces and Health, 2004; Kelman et al., 2004).

AFLATOXINS

More than a dozen different aflatoxins are produced by several fungal species in the genus *Aspergillus*. The four major types of aflatoxin (B_1, B_2, G_1, and G_2) are regularly produced by *Aspergillus parasiticus* and *Aspergillus flavus*, and have on occasion been shown to be produced by other species of *Aspergillus* and by species of *Penicillium* and *Rhizopus* (Goldblatt, 1969). *A. flavus* growing on peanuts can produce aflatoxins at temperatures between approximately 13 °C and 41 °C at a 98% relative humidity (Goldblatt, 1969). Other types of aflatoxins (such as M_1, P_1, Q_1, B_{2a}, and G_{2a}) are not food contaminates but are mammalian metabolic products of aflatoxins B_1, B_2, G_1, or G_2 (Eaton and Groopman, 1994).

Aflatoxins exist as colorless to pale-yellow crystals and have a water solubility value of 3150 mg/l at 25 °C (Pohland et al., 1982; U.S.National Library of Medicine, 2013). Aflatoxins have melting point temperatures ranging from 237 to 320 °C and are not destroyed under normal cooking or food processing conditions (Betina, 1989).

INDUSTRIAL HYGIENE

Aflatoxins may occur in commodities such as oil seeds, grains, and nuts. Exposure to aflatoxins by the general population primarily occurs through consumption of contaminated food. *A. flavus* has been found to produce aflatoxins on numerous foods, including peanuts, corn, wheat, rice, and soybeans (Goldblatt, 1969). Consumption of milk and milk products from animals fed on aflatoxin-contaminated feed is a source of potential exposure to aflatoxin M_1. Since

Hamilton & Hardy's Industrial Toxicology, Sixth Edition. Edited by Raymond D. Harbison, Marie M. Bourgeois, and Giffe T. Johnson.
© 2015 John Wiley & Sons, Inc. Published 2015 by John Wiley & Sons, Inc.

TABLE 117.1 Maximum Tolerated Levels (µg/kg) of Aflatoxins in Foodstuffs and Dairy Products

Aflatoxin	United States	Canada	European Union	Japan
B_1, B_2, G_1, G_2	20	15	4	10 (B_1)
	All foods except milk	Nuts and nut products	Groundnuts, nuts and dried fruit, and processed products intended for direct human consumption	All foods
	15			
	Peanuts for domestic programs			
M_1	0.5		0.05	
	Milk		Milk	

Source: Food and Agriculture Organization (2004); USDA (2010).

mycotoxins are present in foods as natural contaminants and cannot be completely prevented, mycotoxins are regulated in many countries to limit the exposure by the general population (Table 117.1) (Food and Agriculture Organization, 2004).

Aflatoxin contamination is promoted by stress or damage to the crop due to drought before harvest, insect activity, poor timing of harvest, heavy rains at and after harvest, and inadequate drying of the crop before storage (Strosnider et al., 2006). Aflatoxins may be present in agricultural dusts and workers in the animal feed industry involved with grain handling and processing may experience occupational exposure to aflatoxin.

Inhalation exposure to aflatoxins occurs by exposure to airborne particles containing aflatoxins such as dust or mold fragments. Airborne aflatoxins levels in occupational environments can range between 0.002 and $3390 \, ng/m^3$ (Table 117.2). Currently there are no standards, guidelines, or regulations for workplace exposure to aflatoxins in the United States.

Analytical methods for measuring aflatoxin concentrations in agricultural commodities and food products include the use of multifunctional columns (MC), immunoaffinity columns (IC), high performance liquid chromatography (HPLC), enzyme-linked immunosorbent assay (ELISA), and thin layer chromatography (TLC). These methods are validated by the Association of Official Analytical Chemists International (AOAC) and the European Committee for Standardization (Gilbert and Anklam, 2002; Lerda, 2011; Trucksess, 2000). A complete list of approved test kits for use at field testing locations can be found through the website of USDA Federal Grain Inspection Service (USDA, 2014).

TABLE 117.2 Measured Concentrations of Aflatoxins in Air in Industrial Settings

Published Source	Aflatoxin	Concentration, ng/m^3	Environment Sampled
Brera et al. (2002)	B_1	0.002–0.045	Sample from an area in a cocoa beans warehouse
Brera et al. (2002)	B_1	0.002–0.036	Personal sampling of workers involved in cocoa bean handling
Brera et al. (2002)	B_2	0.029	Sample from an area in a nutmeg warehouse
Brera et al. (2002)	B_2	0.002–0.015	Personal sampling of workers involved in cocoa bean handling
Brera et al. (2002)	G_1	0.002	Sample from an area in a cocoa beans warehouse
Brera et al. (2002)	G_1	0.002–0.036	Personal sampling of workers involved in cocoa bean handling
Brera et al. (2002)	G_2	<0.120	Sample from an area in a black pepper warehouse
Brera et al. (2002)	G_2	<0.014–0.131	Personal sampling of workers involved in nutmeg processing
Burg et al. (1981)	$B_1 + B_2$	0.23–107	Sample from inside a storage bin during handling of aflatoxin-contaminated corn
Burg et al. (1982)	$B_1 + B_2$	55	Sample from a grain elevator during unloading 1-year-old corn
Burg and Shotwell (1984)	Total AF	12.5–1600	Personal sampling collected during harvest of contaminated corn
Selim et al. (1998)	B_1	Not detected–421	Sample from a swine building
Selim et al. (1998)	B_1	Not detected–91	Sample collected during harvest of bulk corn
Selim et al. (1998)	B_1	Not detected–4849	Sample collected during corn storage bin cleaning in swine building
Sorenson et al. (1984)	B_1	0.4–1.3	Sample collected from a mechanical peanut sheller
Sorenson et al. (1984)	B_1	7.6	Sample collected at the unloading process from drying trailers
Kussak et al. (1995)	Total AF	Not detected–1.1	Presence in dust from feed factories during unloading of copra, cottonseed, soya beans, maize gluten and sugar beets
Ghosh et al. (1997)	Total AF	0.19 and 0. 26	Work and storage area (respectively) of rice and maize processing plants in India
Nuntharatanapong et al. (2001)	Total AF	0.99, 1.55, and 6.25	Handling of animal feed in Thailand

Chromatographic techniques are commonly used to determine airborne and dust aflatoxin concentrations in occupational environments (Burg et al., 1981; Burg and Shotwell, 1984; Selim et al., 1998; Shotwell et al., 1981). Currently, there are no standardized analytical methods for mycotoxins, including aflatoxins, in environmental samples in nonagricultural environments.

Measurement of mold spores in nonagricultural environments, such as residences, schools and offices are commonly done by either culture or nonculture methods (Macher et al., 1999). These methods do not test for the presence of mycotoxins and cannot be used to assess mycotoxin exposure. Because mycotoxins production depends strongly on the environmental conditions present, the presence of toxigenic mold species measured in the indoor samples cannot be extrapolated as evidence of the presence of mycotoxins (Burge, 2001).

Measures to reduce the level of aflatoxins in specific commodities include ensuring proper post-harvest storage conditions, monitoring aflatoxin levels in susceptible commodities, and encouraging proper food preparation (Strosnider et al., 2006).

The removal of aflatoxin from contaminated foodstuffs can be achieved by several methods. Degradation of aflatoxin occurs with exposure to irradiation, autoclaving, or treatment with strong acids or bases, oxidizing agents, or bisulfate. Adsorbants, such as bentonite and activated charcoal, can physically remove aflatoxin from liquid foods. Chemical treatment with ammonia gas at high temperature and pressure can decrease aflatoxin levels in contaminated seed meals (Betina, 1989; Eaton and Groopman, 1994; Goldblatt, 1969).

ACUTE EFFECTS

Aflatoxin B_1 is acutely toxic in all species studied (Klaassen, 2013). Acute aflatoxin exposures have been associated with epidemics of human aflatoxicosis in regions of India, Southeast Asia, and Africa as a result of ingesting aflatoxin-containing foodstuffs (Krishnamachari et al., 1975a; Lewis et al., 2005; Lye et al., 1995; Ngindu et al., 1982). Clinical manifestations of aflatoxicosis include vomiting, diarrhea, abdominal pain, jaundice, low grade fever, pulmonary edema, and fatty infiltration and necrosis of the liver (Kensler et al., 2011; Krishnamachari et al., 1975b; Lye et al., 1995).

Aflatoxins are possible teratogens. A number of studies reported the presence of aflatoxin in maternal and cord blood and growth impairment in young children; however, there were confounding factors (such as poor nutritional quality) and inconsistency in the reports (Gong et al., 2004; Gong et al., 2002; Gong et al., 2003; Turner et al., 2007). Aflatoxins may also have immunomodulatory effects on humans, but this association has not been conclusively demonstrated

(Allen et al., 1992; Jiang et al., 2005; Jiang et al., 2008; Turner et al., 2000; Turner et al., 2003).

CHRONIC EFFECTS

Aflatoxins are classified as carcinogenic to humans because they are mutagenic and hepatocarcinogenic (International Agency for Research on Cancer, 2012). The liver is the principal target organ for aflatoxin B_1-induced toxicity and epidemiologic studies have shown a positive correlation between high consumption of aflatoxins and the incidence of liver cancer (National Toxicology Program, 2011). Multiple studies have found that individuals exposed to aflatoxins, as measured by urinary aflatoxin metabolites, show increased risk for hepatocellular carcinoma (Qian et al., 1994; Ross et al., 1992; Wang et al., 1996; Yu et al., 1997). Persons with hepatitis B infection have a much greater risk for hepatocellular cancer upon aflatoxin exposure compared to those without infection (Azziz-Baumgartner et al., 2005).

Although aflatoxin B_1 may play a role in the induction of human lung cancer, more research is needed to confirm this effect. Laboratory studies suggest a possible role for aflatoxin B_1 in the induction of lung tumorigenesis, but the limited number of epidemiology studies prevent conclusions about the level of risk (Baxter et al., 1981; Dickens et al., 1966; Dvorackova, 1976; Dvorackova and Pichova, 1986; Dvorackova and Polster, 1984; Dvorackova et al., 1981; Hayes et al., 1984; Massey et al., 2000; Olsen et al., 1988; Van Nieuwenhuize et al., 1973; Wieder et al., 1968).

METABOLISM

Few studies have quantified the relationship between dietary intakes of aflatoxins and the levels of aflatoxin metabolites in human body fluids following absorption and distribution. Toxicokinetic animal studies showed complete absorption after oral, intraperitoneal, and intravenous administration of aflatoxins (Eaton and Groopman, 1994). The dermal permeability rate (K_p) of aflatoxin B_1 was determined to be 2.11×10^{-4} cm/h by an *in vitro* Franz diffusion cell (FDC) infinite dose model using dermatomed split-thickness human skin (Boonen et al., 2012). Concentrations of aflatoxin M_1 in urine and human milk have been correlated to dietary aflatoxin intake. No consistent correlations have been demonstrated regarding aflatoxin concentrations in food and aflatoxin–protein or aflatoxin–DNA adducts in urine and serum (International Agency for Research on Cancer, 2002).

Aflatoxin B_1 is metabolized by cytochrome P450 CYP1A2 and CYP3A4 to aflatoxin-8,9-epoxide, which is short-lived but highly reactive. The epoxides of aflatoxin B_1 are detoxified by glutathione S-transferase (GST)-mediated conjugation to reduced glutathione (GSH) to form aflatoxin

B_1 epoxide-GSH conjugates. The epoxides can also undergo a non-enzymatic hydrolysis to form 8,9-dihydrodiol. Other metabolites formed from aflatoxin B_1 include aflatoxin Q_1 and M_1, which are produced by CYP1A2 and CYP3A4, respectively (International Agency for Research on Cancer, 2002; Wild and Turner, 2002). Aflatoxin Q_1 and M_1 are present in the urine of individuals exposed to aflatoxins (International Agency for Research on Cancer, 2002). A pilot study in three human volunteers showed rapid absorption of aflatoxin B_1 equivalents into systemic circulation with peak plasma aflatoxin concentrations reached within an hour of oral dosing (^{14}C-aflatoxin B_1, 30 ng, 5 nCi, by capsule), following a first-order process in a two-compartment model of absorption. The study reported an elimination with a rapid distribution phase followed by a slower elimination phase with >95% of the total urine aflatoxin B_1 equivalents produced within the first 24 h after exposure (Jubert et al., 2009). Rats exposed to a single dose of aflatoxins by intratracheal exposure had a slightly faster initial absorption of aflatoxin B_1 than those exposed to aflatoxin B_1 by ingestion. For both routes of exposure, excretion of aflatoxins was primarily in the feces and excreted at a similar elimination rate (Coulombe and Sharma, 1985).

Excretion of aflatoxin B_1 and its metabolites occurs primarily through the biliary pathway. A considerable amount of aflatoxin M_1 is excreted in the milk in lactating animals (Betina, 1989).

MECHANISM OF ACTION

Aflatoxin is a known genotoxin that causes genetic damage such as formation of DNA and albumin adducts, gene mutations, micronucleus formation, sister chromatic exchange, and mitotic recombination (National Toxicology Program, 2011). The hepatocarcinogencity of aflatoxin B_1 is associated with its biotransformation by cytochrome P450 to aflatoxin-8,9-epoxide, a reactive electrophilic epoxide which forms covalent adducts with DNA, RNA, and protein. The epoxide forms adducts with the DNA-N^7 position of guanine, promoting depurination and subsequently apurinic site formation. Metabolically activated aflatoxin B_1 has been shown to induce G to T transversion mutations in bacteria and G to T transversions in codon 249 of the p53 tumor-suppressor gene in human liver tumors from geographic areas with high risk of aflatoxin exposure and in experimental animals (International Agency for Research on Cancer, 2012).

BIOMONITORING

Aflatoxins and their metabolites in human tissues and body fluids may be analyzed by LC–MS/MS and ELISA, immunoaffinity purification, immunoassay, HPLC with fluorescence or ultraviolet detection, and synchronous fluorescence spectroscopy. Biomarkers that may be used to assess aflatoxin exposure include urinary metabolites such as aflatoxin–DNA adduct, aflatoxin M_1, and aflatoxin B_1-mercapturic acid, and the aflatoxin–albumin adduct in blood serum (Groopman et al., 1993; Shephard, 2009; Wild and Turner, 2002). The levels of urinary metabolites and adducts reflect intake on the previous day while the levels of aflatoxin–albumin adducts are assumed to reflect exposure to aflatoxin over the previous 2–3 months based on the half-lives of albumin (Eaton and Groopman, 1994; International Agency for Research on Cancer, 2002).

RISK ASSESSMENT

There are limited studies of occupational exposure to aflatoxins and resulting health effects. Aflatoxin exposure levels are not quantified in the available studies as of this writing. Epidemiologic studies have shown a positive correlation between high consumption of aflatoxins and the incidence of liver cancer (National Toxicology Program, 2011).

There are marked species differences in sensitivity to aflatoxin carcinogenesis, with the rat being extremely sensitive and the mouse and hamster being resistant. These variations in carcinogenicity are paralleled by differences in DNA and protein adduct formation for a given aflatoxin dose. A considerable part of the interspecies variation is a reflection of differences in expression of detoxification enzymes. Rodents express a hepatic α-class of GST that efficiently conjugates with the aflatoxin-8,9-epoxide while humans have relatively low aflatoxin-8,9-epoxide conjugation (Wild and Turner, 2002).

Differences in species susceptibility to aflatoxin make it extremely challenging to extrapolate animal toxicity data to humans.

OCHRATOXINS

The ochratoxins A (OTA) and ochratoxin B (OTB) are produced primarily by *Aspergillus ochraceus* and *Penicillium viridicatum* fungi. The minimum temperature and water activity of the substrate for ochratoxin A production by *A. ochraceus* is 12 °C and 0.83–0.87, respectively (Betina, 1989). With *P. viridicatum*, a temperature of 24 °C and a water activity of 0.95 are optimal for ochratoxin A production (Betina, 1989). Ochratoxin A is colorless and has melting points ranging from 94 to 221 °C (Betina, 1989; Pohland et al., 1982). Ochratoxin A is not destroyed by common food preparation procedures. Temperatures above 250 °C are required for several minutes to reduce ochratoxin A concentration (Boudra et al., 1995).

TABLE 117.3   Maximum Tolerated Levels (µg/kg) of Ochratoxins in Foodstuffs

	United States	Canada	European Union	Japan
Ochratoxin A	Not regulated	2000 Feed for swine and poultry	5 Raw food cereal grains 10 Dried vine fruit, e.g., currants, raisins, and sultanas	Not regulated

Source: Food and Agriculture Organization (2004).

INDUSTRIAL HYGIENE

Ochratoxins are commonly found in barley, oats, rye, wheat, coffee beans, and soy products (Council for Agricultural Science and Technology, 2003; Halsall et al., 2007). It is also found in wine made from grapes contaminated with ochratoxin A during the berry ripening phase (Codex Alimentarius Commission, 2007). There are no U.S. regulations for ochratoxin A in food commodities, but it is regulated by other countries such as those in the European Union (Table 117.3).

Occupational inhalation exposure to ochratoxins may occur by exposure to airborne particles containing ochratoxins, such as grain dust (Table 117.4). Currently there are no standards, guidelines, or regulations for workplace exposure to ochratoxins in the United States.

Analytical methods for the analysis of ochratoxins in food commodities include liquid chromatography with fluorescence detection, HPLC and TLC (Food and Drug Administration and Office of Regulatory Affairs, 2013; Lerda, 2011;

Monaci and Palmisano, 2004; Trucksess, 2000; World Health Organization, 2002).

Methods to reduce the level of ochratoxins in food commodities include adequate treatment of field crops and post-harvest storage conditions to prevent growth of ochratoxin-producing fungi. Ochratoxins may also be removed from contaminated agriculture foodstuffs by physical methods such as cleaning, mechanical sorting and separation, treatment with high temperatures, and irradiation. The use of adsorbent materials, degradation by microbes, and chemical approaches such as ammoniation can also be effective in removing ochratoxins (Amezqueta et al., 2013; Varga et al., 2010).

ACUTE EFFECTS

Most studies on ochratoxins have focused on ochratoxin A since ochratoxin B is less toxic (Betina, 1989). The kidney is the primary target of ochratoxin A toxicity. It exerts its

TABLE 117.4   Measured Concentrations of Ochratoxin A in Air in Industrial Settings

Published Source	Ochratoxin A Concentration, ng/m^3	Environment Sampled
Iavicoli et al. (2002)	<0.03–0.066	Sampling carried out in the breathing zones of trolley drivers, in selected areas of the warehouse and inside the trucks during the loading of sacks
Iavicoli et al. (2002)	<0.003–1.45	Personal and area sampling performed during nutmeg processing
Iavicoli et al. (2002)	<0.003–8.15	Personal and area sampling performed during black pepper processing
Iavicoli et al. (2002)	<0.003–0.22	Personal and area sampling in a warehouse where cocoa beans were handled performed on workers emptying and filling sacks and driving the trolley; area sampling performed in the warehouse and on the hopper-loading platform)
Halstensen et al. (2004)	0.0006–0.2	Sampling during threshing of barley, oats, and spring wheat
Halstensen et al. (2004)	0.002–0.6	Sampling during storage handling of barley, oats, and spring wheat that had been stored for 2–27 weeks
Halstensen et al. (2004)	0.003–14	Sampling during bin emptying of barley, oats, and spring wheat
Brera et al. (2002)	0.007–0.066 and 0.006–0.018	Personal and area sampling (respectively) of workers who handled and/or processed green coffee in a coffee warehouse
Brera et al. (2002)	0.004–1.414 and 0.001–0.030	Personal and area sampling (respectively) of workers engaged in the handling of nutmeg (hopper-feeding, packing, center of the workplace)
Brera et al. (2002)	5.460–8.304 and 0.001–0.431	Personal and area sampling of workers engaged in the handling of black pepper (hopper-feeding, packing, center of the workplace)
Brera et al. (2002)	0.002–0.220 and 0.002–0.044	Personal and area sampling of workers engaged in the handling of emptying and filling sacks, and driving the trolley of cocoa beans
Mayer et al. (2007)	0.00007–0.7	Airborne concentrations at 16 sites of grain elevators in Germany

primary effect on the proximal convoluted tubules and is a possible causal factor in regional endemic nephropathies, including Balkan endemic nephropathy (BEN) and chronic interstitial nephropathy (CIN), seen in Southeastern European and North Africa countries (Krogh et al., 1977; Maaroufi et al., 1995; Pavlovic et al., 1979; Smith and Moss, 1985; Zaied et al., 2011). Balkan endemic nephropathy is characterized by tubular dysfunction and atrophy, interstitial fibrosis and edema, and tubular degeneration (Plestina, 1992; Vukelic et al., 1992). Patients with BEN have a high incidence of end-stage renal failure and have a high risk of developing tumors of the urinary tract (Fuchs et al., 1991; Grosso et al., 2003; Pfohl-Leszkowicz et al., 2002). Although a direct link between BEN or CIN and ochratoxin A exposure remains to be established, epidemiology studies correlate an increase in serum and urine ochratoxin A levels with a higher incidence of nephropathy and urothelial tumors in humans (Castegnaro et al., 2006; Fuchs et al., 1991; Petkova-Bocharova and Castegnaro, 1991).

A case of acute renal failure was reported in a farmer's wife who worked 8 h sieving wheat that was stored for more than 2 years in a granary, which was closed for several months. Both the farmer and his wife reported epigastric tension, a retrosternal burning sensation, respiratory difficulty and asthenia. The farmer's condition resolved within 24 h, but his wife's condition worsened and she was admitted for hospitalization. Inhalation of ochratoxin A was postulated as the possible cause of her acute renal failure based on the isolation of ochratoxin A in the contaminated wheat. No biological samples were obtained (Di Paolo et al., 1993, 1994).

In animals, ochratoxin A is hepatotoxic, teratogenic, and immunosuppressive (Al-Anati and Petzinger, 2006; Alvarez et al., 2004; Gilani et al., 1978; Harvey et al., 1992; Mayura et al., 1983; O'Brien and Dietrich, 2005; Patil et al., 2006; Wei and Sulik, 1993; Xue et al., 2010).

CHRONIC EFFECTS

Based on sufficient evidence of carcinogenicity in animal studies, ochratoxin A is classified as "possible carcinogenic to humans" (Group 2B) by the International Agency for Research on Cancer (IARC) classification (International Agency for Research on Cancer, 1993) and as "reasonably anticipated to be a human carcinogen" in the U.S. National Toxicology Program (NTP) classification (National Toxicology Program, 2011). Ochratoxin A is thought to be the cause of urinary tract tumors in humans (National Toxicology Program, 2011; Petkova-Bocharova et al., 1991; Wafa et al., 1998). Liver and mammary-gland tumors were observed in experimental animals after oral exposure to ochratoxin A (Bendele et al., 1985; Huff, 1991). Based on a correlational study of incidence rates for testicular cancer and consumption per capita of foods commonly

contaminated with OTA and a study in which rats were given ochratoxin A by gavage, it is postulated that ochratoxin A exposure may be associated with testicular cancer (Jennings-Gee et al., 2010; Mantle, 2010; Schwartz, 2002).

METABOLISM

Studies on rodents show that ochratoxin A is readily absorbed from the upper gastrointestinal tract, in particular from the small intestine (Galtier, 1991; Li et al., 1997; Suzuki et al., 1977). Ochratoxin A was absorbed from the lungs following intratracheal administration in rats, with a bioavailability of 98% (Breitholtz-Emanuelsson et al., 1995). The dermal permeability rate (K_p) of ochratoxin A was determined to be 8.20×10^{-4} cm/h by an *in vitro* FDC infinite dose model, using dermatomed split-thickness human skin (Boonen et al., 2012).

After absorption from the gastrointestinal tract, ochratoxin A binds to serum proteins (mainly albumin) and can accumulate in body tissues and fluids. It is metabolized into 4R- and 4S-hydroxyochratoxin A (4R-OH-OTA and 4S-OH-OTA), OP-ochratoxin A, OTA–quinines, 10-hydroxy-OTA, and ochratoxin-α (Malir, 2013; Ringot et al., 2006).

Ochratoxin A is reabsorbed from the intestine, recirculated into the enterohepatic circulation, and reabsorbed in the kidney proximal and distal tubules. It is then slowly eliminated from the body by excretion into bile, feces, and urine (Galtier, 1991; Li et al., 1997; Petzinger and Ziegler, 2000; Ringot et al., 2006; Stander et al., 2001; Studer-Rohr et al., 2000; Suzuki et al., 1977; Zingerle et al., 1997). The main excretion route of ochratoxin A in rodents is the biliary route while humans excrete ochratoxin A mainly via the kidney (Pfohl-Leszkowicz and Manderville, 2007). The serum half-life of ochratoxin A in humans is about 35 days (Studer-Rohr et al., 2000).

MECHANISM OF ACTION

The carcinogenic mode of action of ochratoxin A is unclear. Based on *in vivo* and *in vitro* studies, mechanisms of ochratoxin A toxicity include oxidative stress, apoptosis, interference with the cytoskeleton, lipid peroxidation, and inhibition of mitochondrial respiration. Ochratoxin A is a competitor of phenylalanine-tRNA ligase, thereby inhibiting protein synthesis. Other enzymes are also affected by exposure to ochratoxin A (O'Brien and Dietrich, 2005).

A significantly higher kidney tumor incidence following ochratoxin A exposure has been observed in males versus females in both rats and mice. Proposed mechanisms for the gender difference in susceptibility have included differences in ochratoxin A transporters in the kidney, differences in expression of cytochrome P450 enzymes involved in the metabolism of ochratoxin A, and the presence of α-2u-globulin in the adult male rat. It is proposed that the binding of ochratoxin A to the male rat specific protein, α-2u-globulin, may result in rapid delivery of ochratoxin A to the kidney proximal tubule epithelial, concentration of ochratoxin A at the target site, and consequently nephrotoxicity (Haighton et al., 2012).

Evidence of formation of OTA–DNA adducts is inconclusive. DNA adducts from patients in the Balkans provide evidence for the involvement of ochratoxin A in ochratoxin A-induced neurotoxicity while advanced chemical analytical procedures failed to demonstrate the existence of specific OTA–DNA adducts (Mally and Dekant, 2005).

BIOMONITORING

Human exposure to ochratoxin A can be determined by plasma/serum and urine analyses for the presence of the mycotoxin or one of its subunits by HPLC (Castegnaro et al., 2006; Duarte et al., 2010; Gilbert et al., 2001). The parent compound ochratoxin A was the major compound found in blood, whereas ochratoxin α is the major compound detected in urine (Munoz et al., 2000). The long biological half-life of ochratoxin A in human blood facilitates biomonitoring studies. There are no validated methods for the analysis of ochratoxin A in human blood (World Health Organization, 2002). There has been indication of a higher correlation between urinary OTA levels and OTA consumption than between plasma levels and ochratoxin A consumption (Ministry of Agriculture, Fisheries, and Food, 1999).

Detection of the toxicological effects of ochratoxin A, such as increased urinary β2-microglobulinuria (β2M), has been suggested as one method to determine human exposure to ochratoxin A (Hassen et al., 2004). β2-microglobulin is typically filtered by the glomerulus and reabsorbed and catabolized by the proximal tubular cells. Increased urinary excretion of β2-microglobulin has been regarded as an indicator of tubular injury; however, the specificity of the detection of β2-microglobulinuria as a biomarker for ochratoxin A exposure is limited because it requires unhealthy levels of ochratoxin A contamination to be triggered and is highly non-specific since most subjects suffering from renal tubular dysfunction, regardless of cause, tend to show increased β2-microglobulin levels (Wibell, 1978). The usefulness of OTA–DNA adducts as a biomarker of ochratoxin A exposure is uncertain due to the conflicting results reported for formation of OTA–DNA adducts.

RISK ASSESSMENT

There are species differences in the toxicokinetic and toxicodynamic outcomes upon ochratoxin A exposure. The ochratoxin A elimination is slower in humans than in other species tested. The considerable variations in serum half-lives across species are apparently dependent on the affinity and extent of protein binding (Dietrich et al., 2005; Galtier, 1991).

The available epidemiology studies indicate that BEN may be associated with the consumption of ochratoxin A-contaminated foodstuffs and urinary tract tumors were observed in those with BEN; however, a direct causal role of ochratoxin A in the etiology of nephropathy and urinary tract tumors in humans has not been shown. There are limited studies of occupational exposure to ochratoxins and health effects and those that do report relevant exposure have not quantified exposure.

REFERENCES

Al-Anati, L. and Petzinger, E. (2006) Immunotoxic activity of ochratoxin A. *J. Vet. Pharmacol. Ther.* 29:79–90.

Allen, S.J., Wild, C.P., Wheeler, J.G., Riley, E.M., Montesano, R., Bennett, S., Whittle, H.C., Hall, A.J., and Greenwood, B.M. (1992) Aflatoxin exposure, malaria and hepatitis B infection in rural Gambian children. *Trans. R. Soc. Trop. Med. Hyg.* 86:426–430.

Alvarez, L., Gil, A.G., Ezpleta, O., Garcia-Jalon, J.A., and de Cerain, A.L. (2004) Immunotoxic effects of ochratoxin A in Wistar rats after oral administration. *Food Chem. Toxicol.* 42:825–834.

American College of Occupational and Environmental Medicine (2011) *Adverse Human Health Effects Associated with Molds in the Indoor Environment. Revised Position Statement.* Available at http://www.acoem.org/AdverseHumanHealthEffects_Molds.aspx.

Amezqueta, S., Gonzalez-Penas, E., Murillo-Arbizu, M., and Lopez de Cerain, A. (2013) Ochratoxin A decontamination: a review. *Food Control* 20:326–333.

Azziz-Baumgartner, E., Lindblade, K., Gieseker, K., Rogers, H.S., Kieszak, S., Njapau, H., Schleicher, R., McCoy, L.F., Misore, A., DeCock, K., Rubin, C., and Slutsker, L. (2005) Case-control study of an acute aflatoxicosis outbreak, Kenya, 2004. *Environ. Health Perspect.* 113:1779–1783.

Baxter, C.S., Wey, H.E., and Burg, W.R. (1981) A prospective analysis of the potential risk associated with inhalation of aflatoxin-contaminated grain dust. *Food Cosmet. Toxicol.* 19:765–9.

Bendele, A.M., Carlton, W.W., Krogh, P., and Lillehoj, E.B. (1985) Ochratoxin A carcinogenesis in the (C57BL/6J X C3H)F₁ mouse. *J. Natl. Cancer Inst.* 75:733–742.

Betina, V. (1989) *Mycotoxins: Chemical, Biological, and Environmental Aspects, Bioactive Molecules,* Vol. 9: Amsterdam: Elsevier.

Boonen, J., Malysheva, S.V., Taevernier, L., Di Mavungu, J.D., De Saeger, S., and De Spiegeleer, B. (2012) Human skin penetration of selected model mycotoxins. *Toxicology* 301:21–32.

Boudra, H., Le Bars, P., and Le Bars, J. (1995) Thermostability of ochratoxin A in wheat under two moisture conditions. *Appl. Environ. Microbiol.* 61:1156–1158.

Breitholtz-Emanuelsson, A., Fuchs, R., and Hult, K. (1995) Toxicokinetics of ochratoxin A in rat following intratracheal administration. *Nat. Toxins* 3:101–103.

Brera, C., Caputi, R., Miraglia, M., Iavicoli, I., Salerno, A., and Carelli, G. (2002) Exposure assessment to mycotoxins in workplaces: aflatoxins and ochratoxin A occurrence in airborne dusts and human sera. *Microchem. J.* 73:167–173.

Burg, W., Shotwell, O.L., and Saltzman, B.E. (1981) Measurements of airborne aflatoxins during the handling of contaminated corn. *Am. Ind. Hyg. Assoc. J.* 42:1–11.

Burg, W.R., and Shotwell, O.L. (1984) Aflatoxin levels in airborne dust generated from contaminated corn during harvest and at an elevator in 1980. *J. Assoc. Off. Anal. Chem.* 67:309–312.

Burg, W.R., Shotwell, O.L., and Saltzman, B.E. (1982) Measurements of airborne aflatoxins during the handling of 1979 contaminated corn. *AIHAJ* 43:580–586.

Burge, H.A. (2001) The fungi. In: Spengler, J.D., Samet, J.M., and McCarthy, J.F., editors. *Indoor Air Quality Handbook*, New York, NY: McGraw-Hill, pp. 45.1–45.33.

Castegnaro, M., Canadas, D., Vrabcheva, T., Petkova-Bocharova, T., Chernozemsky, I.N., and Pfohl-Leszkowicz, A. (2006) Balkan endemic nephropathy: role of ochratoxins A through biomarkers. *Mol. Nutr. Food Res.* 50:519–529.

Codex Alimentarius Commission (2007) *Code of practice for the prevention and reduction of ochratoxin A contamination in wine*, CAC/RCP 63–2007.

Coulombe, R.A. Jr. and Sharma, R.P. (1985) Clearance and excretion of intratracheally and orally administered aflatoxin B1 in the rat. *Food Chem. Toxicol.* 23:827–830.

Council for Agricultural Science and Technology (2003) *Mycotoxins: Risks in Plant, Animal, and Human Systems, Task Force Report No. 139*, 4420 West Lincoln Way, Ames, IA 50014-3447, Council for Agricultural Science and Technology (CAST).

D'Mello, J.P.F. (1997) *Handbook of Plant and Fungal Toxicants*, Boca Raton: CRC Press.

Di Paolo, N., Guarnieri, A., Loi, F., Sacchi, G., Mangiarotti, A.M., and Di Paolo, M. (1993) Acute renal failure from inhalation of mycotoxins. *Nephron* 64:621–625.

Di Paolo, N., Guarnieri, A., Garosi, G., Sacchi, G., Mangiarotti, A.M., and Di Paolo, M. (1994) Inhaled mycotoxins lead to acute renal failure. *Nephrol. Dial. Transplant.* 9(Suppl. 4):116–120.

Dickens, F., Jones, H.E., and Waynforth, H.B. (1966) Oral, subcutaneous and intratracheal administration of carcinogenic lactones and related substances: the intratracheal administration of cigarette tar in the rat. *Br. J. Cancer* 20:134–144.

Dietrich, D.R., Heussner, A.H., and O'Brien, E. (2005) Ochratoxin A: comparative pharmacokinetics and toxicological implications (experimental and domestic animals and humans). *Food Addit. Contam.* 22:45–52.

Duarte, S., Bento, J., Pena, A., Lino, C.M., erue-Matos, C., Oliva-Teles, T., Morais, S., Correia, M., Oliveira, M.B., Alves, M.R., and Pereira, J.A. (2010) Monitoring of ochratoxin A exposure of the Portuguese population through a nationwide urine survey—Winter 2007. *Sci. Total Environ.* 408:1195–1198.

Dvorackova, I. (1976) Aflatoxin inhalation and alveolar cell carcinoma. *Br. Med. J.* 6011:691.

Dvorackova, I. and Pichova, V. (1986) Pulmonary interstitial fibrosis with evidence of afaltoxin B1 in lung tissue. *J. Toxicol. Environ. Health* 18:153–157.

Dvorackova, I. and Polster, M. (1984) Relation between aflatoxin producing aspergilloma and lung carcinoma. *M. A. N. 2:* 187–192.

Dvorackova, I. Stora, C., and Ayraud, N. (1981) Evidence of aflatoxin B1 in two cases of lung cancer in man. *J. Cancer Res. Clin. Oncol.* 100:221–224.

Eaton, D.L. and Groopman, J.D. (1994) *The Toxicology of Aflatoxins Human Health, Veterinary, and Agricultural Significance*, xxvi, San Diego: Academic Press.

Food and Agriculture Organization (2004) *Worldwide Regulations for Mycotoxins in Food and Feed in 2003*, Rome, Italy: FAO Food and Nutrition Papers-81.

Food and Drug Administration, Office of Regulatory Affairs (2013) *Mycotoxin Analysis. In: ORA Laboratory Manual*, p. 1–23.

Fuchs, R., Radic, B., Ceovic, S., Sostaric, B., and Hult, K. (1991) Human exposure to ochratoxin A. *IARC Sci. Publ.* 131–134.

Galtier, P. (1991) Pharmacokinetics of ochratoxin A in animals. *IARC Sci. Publ.* 115:187–200.

Ghosh, S.K., Desai, M.R., Pandya, G.L., and Venkaiah, K. (1997) Airborne aflatoxin in the grain processing industries in India. *Am. Ind. Hyg. Assoc. J.* 58:583–586.

Gilani, S.H., Bancroft, J., and Reily, M. (1978) Teratogenicity of ochratoxin A in chick embryos. *Toxicol. Appl. Pharmacol.* 46:543–546.

Gilbert, J. and Anklam, E. (2002) Validation of analytical methods for determining mycotoxins in foodstuffs. *Trend Anal. Chem.* 21:468–486.

Gilbert, J., Brereton, P., and MacDonald, S. (2001) Assessment of dietary exposure to ochratoxin A in the UK using a duplicate diet approach and analysis of urine and plasma samples. *Food Addit. Contam.* 18:1088–1093.

Goldblatt, L.A. (1969) *Aflatoxin: Scientific Background, Control, and Implications*, Vol. 7: New York, NY: Academic Press. Food Science and Technology.

Gong, Y., Hounsa, A., Egal, S., Turner, P.C., Sutcliffe, A.E., Hall, A.J., Cardwell, K., and Wild, C.P. (2004) Postweaning exposure to aflatoxin results in impaired child growth: a longitudinal study in Benin, West Africa. *Environ. Health Perspect.* 112: 1334–1338.

Gong, Y.Y., Cardwell, K., Hounsa, A., Egal, S., Turner, P.C., Hall, A.J., and Wild, C.P. (2002) Dietary aflatoxin exposure and impaired growth in young children from Benin and Togo: cross sectional study. *BMJ* 325:20–21.

Gong, Y.Y., Egal, S., Hounsa, A., Turner, P.C., Hall, A.J., Cardwell, K.F., and Wild, C.P. (2003) Determinants of aflatoxin exposure in young children from Benin and Togo, West Africa: the critical role of weaning. *Int. J. Epidemiol.* 32:556–562.

Groopman, J.D., Wild, C.P., Hasler, J., Junshi, C., Wogan, G.N., and Kensler, T.W. (1993) Molecular epidemiology of aflatoxin exposures: validation of aflatoxin- N7-guanine levels in urine as a biomarker in experimental rat models and humans. *Environ. Health Perspect.* 99:107–113.

Grosso, F., Said, S., Mabrouk, I., Fremy, J.M., Castegnaro, M., Jemmali, M., and Dragacci, S. (2003) New data on the occurrence of ochratoxin A in human sera from patients affected or not by renal diseases in Tunisia. *Food Chem. Toxicol.* 41: 1133–1140.

Haighton, L.A., Lynch, B.S., Magnuson, B.A., and Nestmann, E.R. (2012) A reassessment of risk associated with dietary intake of ochratoxin A based on a lifetime exposure model. *Crit. Rev. Toxicol.* 42:147–168.

Halsall, W.J., Isham, N.C., and Ghannoum, M.A. (2007) Mycotoxins. In: Murray, P.R., Baron, E.J., Jorgensen, J.H., Landry, M.L., and Pfaller, M.A., editors. *Manual of Clinical Microbiology*, 9th edition, Washington, DC: ASM Press, pp. 1928–1935.

Halstensen, A.S., Nordby, K.C., Elen, O., and Eduard, W. (2004) Ochratoxin A in grain dust—estimated exposure and relations to agricultural practices in grain production. *Ann. Agric. Environ. Med.* 11:245–254.

Hardin, B.D., Robbins, C.A., Fallah, P., and Kelman, B.J. (2009) The concentration of no toxicologic concern (CoNTC) and airborne mycotoxins. *J. Toxicol.Environ. Health A* 72:585–598.

Harvey, R.B., Elissalde, M.H., Kubena, L.F., Weaver, E.A., Corrier, D.E., and Clement, B.A. (1992) Immunotoxicity of ochratoxin A to growing gilts. *Am. J. Vet. Res.* 53:1966–1970.

Hassen, W., Abid, S., Achour, A., Creppy, E., and Bacha, H. (2004) Ochratoxin A and beta2-microglobulinuria in healthy individuals and in chronic interstitial nephropathy patients in the centre of Tunisia: a hot spot of ochratoxin A exposure. *Toxicology* 199:185–193.

Hayes, R.B., Van Nieuwenhuize, J.P., Raatgever, J.W., and Ten Kate, F.J.W. (1984) Aflatoxin exposures in the industrial setting: an epidemiological study of mortality. *Food Chem. Toxicol.* 22:39–43.

Huff, J.E. (1991) Carcinogenicity of ochratoxin A in experimental animals. *IARC Sci. Publ.* 229–244.

Iavicoli, I., Brera, C., Carelli, G., Caputi, R., Marinaccio, A., and Miraglia, M. (2002) External and internal dose in subjects occupationally exposed to ochratoxin A. *Int. Arch. Occup. Environ. Health* 75:381–386.

Institute of Medicine, Committee on Damp Indoor Spaces and Health, Board on Health Promotion and Disease Prevention (2004) *Damp Indoor Spaces and Health*, Washington, DC: National Academy Press. National Academy of Science.

International Agency for Research on Cancer (1993) *Some Naturally Occurring Substances: Food Items and Constituents, Heterocyclic Aromatic Amines and Mycotoxins*, Vol. 56, Lyon, France: International Agency for Research on Cancer (IARC).

International Agency for Research on Cancer (2002) *Some Traditional Herbal Medicines, Some Mycotoxins, Naphthalene and Styrene*, vol. 82, Lyon, France: IARC Press, IARC Monographs on the evaluation of carcinogenic risks to humans.

International Agency for Research on Cancer (2012) *Aflatoxins*, Vol. 100F: Lyon, France: WHO.

Jennings-Gee, J.E., Tozlovanu, M., Manderville, R., Miller, M.S., Pfohl-Leszkowicz, A., and Schwartz, G.G. (2010) Ochratoxin A: in utero exposure in mice induces adducts in testicular DNA. *Toxins (Basel)* 2:1428–1444.

Jiang, Y., Jolly, P.E., Ellis, W.O., Wang, J.S., Phillips, T.D., and Williams, J.H. (2005) Aflatoxin B1 albumin adduct levels and cellular immune status in Ghanaians. *Int. Immunol.* 17:807–814.

Jiang, Y., Jolly, P.E., Preko, P., Wang, J.S., Ellis, W.O., Phillips, T.D., and Williams, J.H. (2008) Aflatoxin-related immune dysfunction in health and in human immunodeficiency virus disease. *Clin. Dev. Immunol.* doi:10.1155/2008/790309.

Jubert, C., Mata, J., Bench, G., Dashwood, R., Pereira, C., Tracewell, W., Turteltaub, K., Williams, D., and Bailey, G. (2009) Effects of chlorophyll and chlorophyllin on low-dose aflatoxin B(1) pharmacokinetics in human volunteers. *Cancer Prev. Res. (Phila.)* 2:1015–1022.

Kelman, B.J., Robbins, C.A., Swenson, L.J., and Hardin, B.D. (2004) Risk from inhaled mycotoxins in indoor office and residential environments. *Int. J. Toxicol.* 23:3–10.

Kensler, T.W., Roebuck, B.D., Wogan, G.N., and Groopman, J.D. (2011) Aflatoxin: a 50-year odyssey of mechanistic and translational toxicology. *Toxicol. Sci.* 120(Suppl. 1):S28–S48.

Klaassen, C.D. (2013) *Casarett and Doull's Toxicology: The Basic Science of Poisons*, 8th edition, New York, NY: McGraw-Hill.

Krishnamachari, K.A., Bhat, R.V., Nagarajan, V., and Tilak, T.B. (1975a) Hepatitis due to aflatoxicosis. An outbreak in Western India. *Lancet* 1061–1063.

Krishnamachari, K.A., Bhat, R.V., Nagarajan, V., and Tilak, T.B. (1975b) Investigations into an outbreak of hepatitis in parts of western India. *Indian J. Med. Res.* 63:1036–1049.

Krogh, P., Hald, B., Plestina, R., and Ceovic, S. (1977) Balkan (endemic) nephropathy and foodborn ochratoxin A: preliminary results of a survey of foodstuffs. *Acta Pathol. Microbiol. Scand. [B]* 85:238–240.

Kussak, A., Andersson, B., and Andersson, K. (1995) Determination of aflatoxins in airborne dust from feed factories by automated immunoaffinity column clean-up and liquid chromatography. *J. Chromatogr. A* 708:55–60.

Lerda, D. (2011) *Mycotoxins Factsheet*, 4th ed., C JRC 66956-2011: Belgium: European Commission.

Lewis, L., Onsongo, M., Njapau, H., Schurz-Rogers, H., Luber, G., Kieszak, S., Nyamongo, J., Backer, L., Dahiye, A.M., Misore, A., DeCock, K., and Rubin, C. (2005) Aflatoxin contamination of commercial maize products during an outbreak of acute aflatoxicosis in eastern and central Kenya. *Environ. Health Perspect.* 113:1763–1767.

Li, S., Marquardt, R.R., Frohlich, A.A., Vitti, T.G., and Crow, G. (1997) Pharmacokinetics of ochratoxin A and its metabolites in rats. *Toxicol. Appl. Pharmacol.* 145:82–90.

Lye, M.S., Ghazali, A.A., Mohan, J., Alwin, N., and Nair, R.C. (1995) An outbreak of acute hepatic encephalopathy due to severe aflatoxicosis in Malaysia. *Am. J. Trop. Med. Hyg.* 53:68–72.

Maaroufi, K., Achour, A., Betbeder, A.M., Hammami, M., Ellouz, F., Creppy, E.E., and Bacha, H. (1995) Foodstuffs and human blood contamination by the mycotoxin ochratoxin A: correlation with chronic interstitial nephropathy in Tunisia. *Arch. Toxicol.* 69:552–558.

Macher, J., Ammann, H.A., Burge, H.A., Milton, D.K., and Morey, P.R. (1999) *Bioaerosols: Assessment and Control*, Cincinnati, OH: ACGIH.

Malir, F. (2013) Toxicity of the mycotoxin ochratoxin A in the light of recent data. *Toxin Rev.* 32:19–33.

Mally, A. and Dekant, W. (2005) DNA adduct formation by ochratoxin A: review of the available evidence. *Food Addit. Contam.* 22(Suppl. 1):65–74.

Mantle, P.G. (2010) Comments on "Ochratoxin A: in utero exposure in mice induces adducts in testicular DNA. Toxins 2010, 2, 1428–1444"-Mis-citation of rat literature to justify a hypothetical role for ochratoxin A in testicular cancer. *Toxins (Basel)* 2:2333–2336.

Massey, T.E., Smith, G.B., and Tam, A.S. (2000) Mechanisms of aflatoxin B1 lung tumorigenesis. *Exp. Lung Res.* 26:673–683.

Mayer, S., Curtui, V., Usleber, E., and Gareis, M. (2007) Airborne mycotoxins in dust from grain elevators. *Mycotoxin Res.* 23:94–100.

Mayura, K., Hayes, A.W., and Berndt, W.O. (1983) Effects of dietary protein on teratogenicity of ochratoxin A in rats. *Toxicology* 27:147–157.

Ministry of Agriculture, Fisheries, and Food (MAFF) (1999) A survey of human exposure to ochratoxin A, Food Surveillance Information Sheet 172, London: Joint Food Safety and Standards Group.

Monaci, L. and Palmisano, F. (2004) Determination of ochratoxin A in foods: state-of-the-art and analytical challenges. *Anal. Bioanal. Chem.* 378:96–103.

Munoz, K., Blaszkewicz, M., and Degen, G.H. (2010) Simultaneous analysis of ochratoxin A and its major metabolite *Ochratoxin alpha* in plasma and urine for an advanced biomonitoring of the mycotoxin. *J. Chromatogr. B Analyt. Technol. Biomed. Life Sci.* 878:2623–2629.

National Toxicology Program (2011) *Report on Carcinogens*, 12th ed., Research Triangle Park, NC: U.S. Department of Health and Human Services, Public Health Service, National Toxicology Program.

Ngindu, A., Johnson, B.K., Kenya, P.R., Ngira, J.A., Ocheng, D.M., Nandwa, H., Omondi, T.N., Jansen, A.J., Ngare, W., Kaviti, J.N., Gatei, D., and Siongok, T.A. (1982) Outbreak of acute hepatitis caused by aflatoxin poisoning in Kenya. *Lancet* 1:1346–1348.

Nuntharatanapong, N., Suramana, T., Chaemthavorn, S., Zapuang, K., Ritta, E., Semathong, S., Chuamorn, S., Niyomwan, V., Dusitsin, N., Lohinavy, O., and Sinhaseni, P. (2001) Increase in tumor necrosis factor-alpha and a change in the lactate dehydrogenase isoenzyme pattern in plasma of workers exposed to aflatoxin-contaminated feeds. *Arh. Hig. Rada. Toksikol.* 52:291–298.

O'Brien, E. and Dietrich, D.R. (2005) Ochratoxin A: the continuing enigma. *Crit. Rev. Toxicol.* 35:33–60.

Olsen, J.H., Dragsted, L., and Autrup, H. (1988) Cancer risk and occupational exposure to aflatoxins in Denmark. *Br. J. Cancer* 58:392–396.

Patil, R.D., Dwivedi, P., and Sharma, A.K. (2006) Critical period and minimum single oral dose of ochratoxin A for inducing developmental toxicity in pregnant Wistar rats. *Reprod. Toxicol.* 22:679–687.

Pavlovic, M., Plestina, R., and Krogh, P. (1979) Ochratoxin A contamination of foodstuffs in an area with Balkan (endemic) nephropathy. *Acta Pathol. Microbiol. Scand. B* 87:243–246.

Petkova-Bocharova, T. and Castegnaro, M. (1991) Ochratoxin A in human blood in relation to Balkan endemic nephropathy and urinary tract tumours in Bulgaria. *IARC Sci. Publ.* 135–137.

Petkova-Bocharova, T., Castegnaro, M., Michelon, J., and Maru, V. (1991) Ochratoxin A and other mycotoxins in cereals from an area of Balkan endemic nephropathy and urinary tract tumours in Bulgaria. *IARC Sci. Publ.* 83–87.

Petzinger, E. and Ziegler, K. (2000) Ochratoxin A from a toxicological perspective. *J. Vet. Pharmacol. Ther.* 23:91–98.

Pfohl-Leszkowicz, A. and Manderville, R.A. (2007) Ochratoxin A: An overview on toxicity and carcinogenicity in animals and humans. *Mol. Nutr. Food Res.* 51:61–99.

Pfohl-Leszkowicz, A., Petkova-Bocharova, T., Chernozemsky, I.N., and Castegnaro, M. (2002) Balkan endemic nephropathy and associated urinary tract tumors: a review on aetiological causes and the potential role of mycotoxins. *Food Addit. Contam.* 19:282–302.

Plestina, R. (1992) Some features of Balkan endemic nephropathy. *Food Chem. Toxicol.* 30:177–181.

Pohland, A.E., Schuller, P.L., Steyn, P.S., and van Egmond, H.P. (1982) Physicochemical data for some selected mycotoxins. *Pure Appl. Chem.* 54:2219–2284.

Qian, G.S., Ross, R.K., Yu, M.C., Yuan, J.M., Gao, Y.T., Henderson, B.E., Wogan, G.N., and Groopman, J.D. (1994) A follow-up study of urinary markers of aflatoxin exposure and liver cancer risk in Shanghai, People's Republic of China. *Cancer Epidemiol. Biomarkers Prev.* 3:3–10.

Ringot, D., Chango, A., Schneider, Y.J., and Larondelle, Y. (2006) Toxicokinetics and toxicodynamics of ochratoxin A, an update. *Chem. Biol. Interact.* 159:18–46.

Ross, R.K., Yuan, J.M., Yu, M.C., Wogan, G.N., Qian, G.S., Tu, J.T., Groopman, J.D., Gao, Y.T., and Henderson, B.E. (1992) Urinary aflatoxin biomarkers and risk of hepatocellular carcinoma. *Lancet* 339:943–946.

Schwartz, G.G. (2002) Hypothesis: does ochratoxin A cause testicular cancer? *Cancer Causes Control* 13:91–100.

Selim, M.I., Juchems, A.M., and Popendorf, W. (1998) Assessing airborne aflatoxin B1 during on-farm grain handling activities. *Am. Ind. Hyg. Assoc. J.* 59:252–256.

Shephard, G.S. (2009) Aflatoxin analysis at the beginning of the twenty-first century. *Anal. Bioanal. Chem.* 395:1215–1224.

Shotwell, O.L., Burg, W.R., and Diller, T. (1981) Thin layer chromatographic determination of aflatoxin in corn dust. *J. Assoc. Off. Anal. Chem.* 64:1060–1063.

Smith, J.E. and Moss, M.O. (1985) *Mycotoxins: Formation, Analysis, and Significance*, Chichester: Wiley.

Sorenson, W.G., Jones, W., Simpson, J., and Davidson, J.I. (1984) Aflatoxin in respirable airborne peanut dust. *J. Toxicol. Environ. Health* 14:525–533.

Stander, M.A., Niewoudt, T.W., Steyn, P.S., Shepard, G.S., Creppy, E.E., and Sewram, V. (2001) Toxicokinetics of ochratoxin A in vervet monkeys (*Cercopithecus aethiops*). *Arch. Toxicol.* 75:262–269.

Strosnider, H., Azziz-Baumgartner, E., Banziger, M., Bhat, R.V., Breiman, R., Brune, M.N., DeCock, K., Dilley, A., Groopman, J., Hell, K., Henry, S.H., Jeffers, D., Jolly, C., Jolly, P., Kibata, G.N., Lewis, L., Liu, X., Luber, G., McCoy, L., Mensah, P., Miraglia, M., Misore, A., Njapau, H., Ong, C.N., Onsongo, M.T., Page, S.W., Park, D., Patel, M., Phillips, T., Pineiro, M., Pronczuk, J., Rogers, H.S., Rubin, C., Sabino, M., Schaafsma, A., Shephard, G., Stroka, J., Wild, C., Williams, J.T., and Wilson, D. (2006) Workgroup report: public health strategies for reducing aflatoxin exposure in developing countries. *Environ. Health Perspect.* 114:1898–1903.

Studer-Rohr, I., Schlatter, J., and Dietrich, D.R. (2000) Kinetic parameters and intraindividual fluctuations of ochratoxin A plasma levels in humans. *Arch. Toxicol.* 74:499–510.

Sudakin, D.L. and Fallah, P. (2008) Toxigenic fungi and mycotoxins in outdoor, recreational environments. *Clin. Toxicol. (Phila.)* 16:738–744.

Suzuki, S., Satoh, T., and Yamazaki, M. (1977) The pharmacokinetics of ochratoxin A in rats. *Jpn. J. Pharmacol.* 27:735–744.

Trucksess, M.W. (2000) Natural toxins. In: *Official Methods of Analysis of AOAC International*, 17th edition: Gaithersburg, MD: AOAC International.

Turner, P.C., Collinson, A.C., Cheung, Y.B., Gong, Y., Hall, A.J., Prentice, A.M., and Wild, C.P. (2007) Aflatoxin exposure *in utero* causes growth faltering in Gambian infants. *Int. J. Epidemiol.* 36:1119–1125.

Turner, P.C., Mendy, M., Whittle, H., Fortuin, M., Hall, A.J., and Wild, C.P. (2000) Hepatitis B infection and aflatoxin biomarker levels in Gambian children. *Trop. Med. Int. Health* 5:837–841.

Turner, P.C., Moore, S.E., Hall, A.J., Prentice, A.M., and Wild, C.P. (2003) Modification of immune function through exposure to dietary aflatoxin in Gambian children. *Environ. Health Perspect.* 111:217–220.

U.S. National Library of Medicine (2013) *ChemIDplus Advanced: Aflatoxins [Online]*, U.S. National Library of Medicine. Available at http://chem.sis.nlm.nih.gov/chemidplus/.

USDA *Aflatoxin Handbook*, 256,1-5-2009. Washington, DC: USDA Grain Inspection, Packers and Stockyards Administration, Federal Grain Inspection Service.

USDA *Peanut products for use in domestic programs*, PP12: 5-17-2010. USDA Commodity Requirements.

USDA Rapid Test Kit Evaluation Program (TKE) (2014) USDA *Grain Inspection, Packers and Stockyards Administration.* Available at http://www.gipsa.usda.gov/fgis/insp_weight/raptestkit.html.

Van Nieuwenhuize, J.P., Herber, R.F.M., DeBruin, A., Meyer, I.P.B., and Duba, W.C. (1973) Aflatoxins: epidemiological study on the carcinogenicity of prolonged exposure to low levels among workers of a plant. *Tijdschr. Soc. Geneeskd.* 51:754–759.

Varga, J., Kocsube, S., Peteri, Z., Vagvolgyi, C., and Toth, B. (2010) Chemical, physical and biological approaches to prevent ochratoxin induced toxicoses in humans and animals. *Toxins (Basel)* 2:1718–1750.

Vukelic, M., Sostaric, B., and Belicza, M. (1992) Pathomorphology of Balkan endemic nephropathy. *Food Chem. Toxicol.* 30:193–200.

Wafa, E.W., Yahya, R.S., Sobh, M.A., Eraky, I., el-Baz, M., el-Gayar, H.A., Betbeder, A.M., and Creppy, E.E. (1998) Human ochratoxicosis and nephropathy in Egypt: a preliminary study. *Hum. Exp. Toxicol.* 17:124–129.

Wang, L.Y., Hatch, M., Chen, C.J., Levin, B., You, S.L., Lu, S.N., Wu, M.H., Wu, W.P., Wang, L.W., Wang, Q., Huang, G.T., Yang, P.M., Lee, H.S., and Santella, R.M. (1996) Aflatoxin exposure and risk of hepatocellular carcinoma in Taiwan. *Int. J. Cancer* 67:620–625.

Wei, X. and Sulik, K.K. (1993) Pathogenesis of craniofacial and body wall malformations induced by ochratoxin A in mice. *Am. J. Med. Genet.* 47:862–871.

Wibell, L. (1978) The serum level and urinary excretion of beta2-microglobulin in health and renal disease. *Pathol. Biol. (Paris)* 26:295–301.

Wieder, R., Wogan, G.N., and Shimkin, M.B. (1968) Pulmonary tumors in strain A mice given injections of aflatoxin B1. *J. Natl. Cancer Inst.* 40:1195–1197.

Wild, C.P. and Turner, P.C. (2002) The toxicology of aflatoxins as a basis for public health decisions. *Mutagenesis* 17:471–481.

World Health Organization (2002) *Evaluation of Certain Mycotoxins in Food: Fifty-Sixth Report of the Joint FAO/WHO Expert Committee on Food Additives, 906:* Geneva: World Health Organization, WHO Technical Report Series.

Xue, C.Y., Wang, G.H., Chen, F., Zhang, X.B., Bi, Y.Z., and Cao, Y.C. (2010) Immunopathological effects of ochratoxin A and T-2 toxin combination on broilers. *Poult. Sci.* 89:1162–1166.

Yu, M.W., Lien, J.P., Chiu, Y.H., Santella, R.M., Liaw, Y.F., and Chen, C.J. (1997) Effect of aflatoxin metabolism and DNA adduct formation on hepatocellular carcinoma among chronic hepatitis B carriers in Taiwan. *J. Hepatol.* 27:320–330.

Zaied, C., Bouaziz, C., Azizi, I., Bensassi, F., Chour, A., Bacha, H., and Abid, S. (2011) Presence of ochratoxin A in Tunisian blood nephropathy patients. Exposure level to OTA. *Exp. Toxicol. Pathol.* 63:613–618.

Zingerle, M., Silbernagl, S., and Gekle, M. (1997) Reabsorption of the nephrotoxin ochratoxin A along the rat nephron *in vivo*. *J. Pharmacol. Exp. Ther.* 280:220–224.

118

PHOSPHINE

Michael W. Perkins, Benjamin Wong, Dorian Olivera, and Alfred Sciuto

First aid: Eye: Immediately wash (irrigate) the eyes with large amounts of water for at least 15 min, occasionally lifting the lower and upper lids. Get medical attention immediately. If frostbite occurs and eyes are frozen because of exposure seek immediate medical attention.

Skin: Immediately flush the contaminated skin with soap and water. Remove contaminated clothing and flush the skin with water. If frostbite occurs do not remove clothing or touch the affected area. Get medical attention promptly if irritation persists.

Inhalation: Quickly move the exposed individual to fresh air, calm and at rest in half-up right position. Monitor and maintain airway and blood pressure. If breathing has become difficult administer oxygen; if stopped, perform artificial respiration. Get medical attention as soon as possible.

Ingestion: Get medical attention as soon as possible.

Target organ(s): Lungs, heart, central nervous system (CNS), and respiratory and gastrointestinal tracts

Occupational exposure limit

OSHA PEL: 0.3 ppm ($0.4 \, \text{mg/m}^3$) TWA, 1 ppm ($1 \, \text{mg/m}^3$) STEL

NIOSH REL: 0.3 ppm ($0.4 \, \text{mg/m}^3$) TWA, 1 ppm ($1 \, \text{mg/m}^3$) STEL

ACGIH TVL: 0.3 ppm ($0.42 \, \text{mg/m}^3$) TWA, 1 ppm ($1.4 \, \text{mg/m}^3$) STEL

Risk/safety phrases:

R12: Extremely flammable

R17: Spontaneously flammable in air

R26: Very toxic by inhalation

R34: Causes burns

R50: Very toxic to aquatic organisms

S1/2: Keep locked up and out of the reach of children

S28: After contact with skin, wash immediately with plenty of . . . (to be specified by the manufacturer)

S36/37: Wear suitable protective clothing and gloves

S45: In case of accident or if you feel unwell, seek medical advice immediately (show the label where possible).

S61: Avoid release to the environment. Refer to special instructions/safety data sheets.

S63: In case of accident by inhalation, remove casualty to fresh air and keep at rest. S46: If swallowed, seek medical advice immediately and show this container or label.

BACKGROUND AND USES

Phosphine is a colorless, flammable, toxic gas and is the hydrolysis byproduct of aluminum (AlP), magnesium (MgP), and zinc phosphides (ZnP) (World Health Organization, 1988; Anand et al., 2011). When AlP, MgP, or ZnP comes into contact with water, phosphine gas is formed via the following reaction: $\text{AlP} + 3\,\text{H}_2\text{O} \rightarrow \text{PH}_3 + \text{Al(OH)}_3$. Phosphine is used to fumigate a wide variety of raw agricultural commodities, including stored grain, processed foods, and non-food agricultural products (Wilson et al., 1980; United States Environmental Protection Agency, 1998). Several fatal occupational and intentional exposures to phosphine gas originate from its use as a pesticide and rodenticide have been reported (Jones et al., 1964; Wilson et al., 1980).

Hamilton & Hardy's Industrial Toxicology, Sixth Edition. Edited by Raymond D. Harbison, Marie M. Bourgeois, and Giffe T. Johnson.
© 2015 John Wiley & Sons, Inc. Published 2015 by John Wiley & Sons, Inc.

Incidents of phosphine poisonings reported in Europe consisted of 28% by intentional ingestion and 65% by accidental inhalation (Lauterbach et al., 2005). However, the last two decades have seen a dramatic rise in the number of documented cases of phosphine use for the purpose of suicide (Chugh et al., 1991; Gupta and Ahlawat, 1995; Anand et al., 2011). National concern over phosphine's potential as a terrorist threat agent against targeted populations has increased because of the ease with which toxic phosphine fumes are produced from rodenticides in addition to its low cost, widespread use, and retail access.

The primary uses of phosphine are as a fumigant for stored food products and a reactive intermediate in the chemical industry. As a chemical intermediate, phosphine is used in the synthesis of organophosphines for oil additives and pharmaceutical applications (World Health Organization, 1988), in the preparation of organic phosphonium derivatives for the manufacture of flame retardants (Toy and Walsh, 1987), in the manufacture of solar or photovoltaic cells (Bretherick, 1990), as a dopant for silicon semiconductors, as a polymerization initiator, and as a condensation catalyst (Lewis, 2007). As a fumigant in pest control, phosphine is produced on-site by the hydrolysis of aluminum or magnesium phosphides, which are used in powder form. Phosphine is also generated as a byproduct in many other industrial processes, including the production of phosphorus (Al'zhanov et al., 1983) and ferrosilicon alloys (Anon, 1956; Lutzmann et al., 1963), machining of spheroidal graphite iron (Bowker, 1958; Mathew, 1961), steel pickling (Vdovenko et al., 1984), and other metallurgical operations (Habashi and Ismail, 1975). It is difficult to quantify the production of phosphine, since much of it is manufactured and used in relation to other processes.

PHYSICAL AND CHEMICAL PROPERTIES

Phosphine, also referred to as hydrogen phosphide, trihydrogen phosphide, phosphorus hydride, and phosphorated hydrogen, is a colorless, odorless, flammable, irritant toxic gas. It can ignite spontaneously on contact with air and is highly reactive with oxidizers, acids, halogenated hydrocarbons, and aqueous solutions. The physical and chemical properties of phosphine are listed in Table 118.1 (Thienes and Haley, 1972; Daubert and Danner, 1989; O'Neil, 2001; National Institute for Occupational Safety and Health, 2005; Lewis, 2007).

ENVIRONMENTAL FATE AND BIOACCUMULATION

Phosphine is of limited environmental contamination concern as it is extremely volatile, and quickly evaporates into

TABLE 118.1   Physical Properties of Phosphine

CAS number:	7803-51-2
Structural formula:	PH_3
Formula weight:	34.00
Boiling point:	$-87.7\,°C$
Gas density, $20\,°C$:	1.17
Specific gravity, gas, $20\,°C$:	1.185
Vapor pressure:	2.93×10^4 mmHg at $25\,°C$
Solubility:	Slightly soluble in water. Soluble in alcohol, ether, and cuprous chloride solution
Color:	Colorless
Odor:	Fish- or garlic-like odor due to impurities (pure compound is odorless)
Odor threshold:	0.03 parts per million (ppm) (200 ppm for pure)
Conversion factors:	1 ppm = 1.41 mg/m^3 1 mg/m^3 = 0.71 ppm ($20\,°C$, 760 mmHg)

the atmosphere. Natural occurrences of phosphine are extremely rare and are mostly limited to transient formation in marsh gas and other sites of decay of phosphorus-containing organic matter or waste (Ciba Foundation, 1978; O'Neil, 2001). Phosphine does occur naturally in the form of metal phosphides, but is extremely rare and only found in iron meteorites as the mineral schreibersite $(Fe,Ni)_3P$ (Van Wazer, 1961). Apart from natural sources, atmospheric phosphine results from emissions and effluents from industrial processes, such as phosphide hydrolysis and the storage of ferrous alloys (Stellman and International Labour Office, 1998), and the use of phosphides as rodenticides and fumigants (Dumas, 1980; World Health Organization, 1988). Volatilization from metal phosphides is expected to be the most important fate process, with phosphine expected to volatize rapidly from both moist and dry soil (O'Neil, 2001). However, studies have shown that sub-surface phosphine may bind with the organic matter in the soil, where it rapidly degrades (Spanggord et al., 1985). In the atmosphere, phosphine is degraded by reaction with photochemically produced hydroxyl radicals (O'Neil, 2001) and has a reported atmospheric half-life of <1 day (Spicer et al., 1993).

ECOTOXICOLOGY

Only minute amounts of phosphine and phosphides are naturally occurring, and those that have an impact on the environment are not persistent. Phosphine and phosphides are introduced into the environment for pest control, a role for which they are well suited because of their efficacy, lack of persistence, and harmless decomposition products. Careful bait positioning and low toxicity of carcasses for scavengers of poisoned animal targets minimize the toxicity on non-targeted species. Phosphine and phosphides are oxidized in the environment, and the final product, phosphate, is ubiquitous in the natural environment. However, it has been suggested that zinc phosphide may persist in aquatic sediments despite a general lack of aquatic toxicological data. A major accidental release of stored phosphine would not be expected to result in chronic environmental consequences (World Health Organization, 1988).

MAMMALIAN TOXICOLOGY

Acute Effects

Inhalation is the most common route of acute phosphine poisoning and affects the cardiovascular, respiratory, gastrointestinal, and CNS. Initially, symptoms consist of an irritated throat, tightness in the chest, dyspnea, abdominal pain, nausea, and vomiting followed by a persistent cough. These symptoms can persist for 15 min to 3 h and gradually progress to pulmonary edema and respiratory distress. However, the onset of symptoms may be delayed for days. Death primarily occurs within the first 12–24 h post-exposure because of cardiovascular failure, but can also occur after 24 h as a result of liver or kidney failure (ATSDR, 2007). Congestion of the lungs and severe pulmonary edema were the most common findings in 26 deaths following acute phosphine poisoning (Harger and Spolyar, 1958). Exposure to high concentrations of phosphine can also result in severe seizures and coma.

Phosphine poisoning can occur following accidental or intentional ingestion of metal phosphide pellets or tablets, which interact with the gastric secretions of the stomach. Acute poisoning by ingestion may induce severe gastrointestinal irritation and hemorrhage, as well as cardiovascular, respiratory, and renal failure within minutes, and hepatic damage. In one study, post-mortem findings in 10 patients who died from zinc phosphide ingestion included blood in body cavities, pulmonary congestion and edema, hemorrhage of the intestinal epithelium, and necrosis of the liver and kidneys (Stephenson, 1967).

Chronic Effects

Chronic effects of exposure to phosphine include anemia, bronchitis, and gastrointestinal, visual, speech, and motor disturbances. Consequences of long-term exposure include liver damage, jaundice, and renal failure, the onset of which may occur from 48 to 72 h after exposure. Reports of long-term occupational exposure to zinc phosphide tablets include minor symptoms such as cough, dyspnea, headache, lethargy, anorexia, and epigastric pain (Misra et al., 1988). Chronic effects of phosphine toxicity can last for several months or years and may cause weight loss, toothache, swelling of the jaw or necrosis of the jaw bone, and spontaneous fractures of bones (World Health Organization, 1988).

Furthermore, phosphine exposure may exacerbate the symptoms of any preexisting respiratory conditions, especially in patients suffering from obstructive airway diseases.

Chemical Pathology

Phosphine toxicity originates from multiple factors, but its primary symptoms are seen in the cardiovascular and respiratory systems. Following exposure to phosphine and/or metal phosphides, aerobic metabolism is inhibited, resulting in oxidative damage to tissues and cells affecting the metabolic mechanisms. Mortality is primarily the result of pulmonary edema, myocardial injury, and cardiac arrest. Post-mortem examination of brains from eight acute phosphine poisoning cases indicated multifocal small hemorrhages and marked congestion with areas of exudation (Misra et al., 1988). This study observed various histopathological changes, including pulmonary edema, desquamation of the lining epithelium of the bronchioles, vacuolar degeneration of hepatocytes, dilatation of sinusoids and hepatic veins, and nuclear fragmentation. Chromosomal abnormalities have been recorded in the blood samples of workers, who apply fumigant (Garry et al., 1996). Tests examining cardiac enzyme, liver, and kidney function, by measuring levels of blood urea, electrolytes, creatinine, bilirubin, alkaline phosphates, lactic dehydrogenases, plasma cortisol, and rennin levels, are strongly recommended following phosphine and/or metal phosphates poisoning (World Health Organization, 1988; Chugh et al., 1990). Ingestion of metal phosphides can lead to severe gastrointestinal hemorrhage and respiratory and renal failure. Although an unlikely route of exposure, direct contact with liquid or compressed phosphine can cause frostbite.

Mode of Action(s)

Phosphine is a broad-spectrum toxicant that affects a host of target organs. Cellular poisoning and oxidative stress as a result of phosphine exposure causes a variety of systemic effects, including metabolic disturbances, circulatory failure, hepatic damage, electrolyte imbalance, and cardiovascular, pulmonary, gastrointestinal, and neurological toxicity (Anand et al., 2011). The primary mode of toxicity is the reduction in the aerobic respiration by the inhibition of mitochondrial cytochrome C oxidase (Chefurka et al., 1976; Nakakita, 1987). Disruption of the mitochondrial function interferes with various cellular activities, including aerobic metabolic rates, energy utilization, and protein and enzyme synthesis. Inhibition of the endogenous antioxidants and oxidative damage caused by the formation of highly reactive hydroxyl radicals also occurs following phosphine poisoning (Chaudhry, 1997; Quistad et al., 2000). Additionally, phosphine inhibits the enzyme acetylcholinesterase (AChE), resulting in the accumulation of the neurotransmitter acetylcholine, which induces cholinergic effects such as hypersecretions, bronchoconstriction, respiratory distress, seizures, and death.

SOURCES OF EXPOSURE

Occupational

Occupational exposure to phosphine may occur in workers producing phosphine and phosphides or those in operations that can release phosphine, such as fumigators and pest-control operatives and transport workers (World Health Organization, 1988). Concentrations necessitating the use of personal respiratory protection have been reported in metal phosphide production areas (Jackson and Elias, 1984). Exposure to phosphine has been described in the operation of acetylene generators (Harger and Spolyar, 1958), in the production of phosphorus (Beloskurskaya et al., 1978), in steel pickling (Vdovenko et al., 1984), in the manufacture of magnesium powder (Cole and Bensett, 1950), and in the machining of spheroidal graphite iron (Bowker, 1958; Mathew, 1961). Although phosphine is used extensively in semi-conductor manufacture, there are no published data of occupational exposures. Since in this context, phosphine is used under precise and strictly controlled conditions, and frequently diluted with inert gases in semi-conductor manufacture, exposure of workers is unlikely (NIOSH, 1977). In the United States an estimated one million workers are at risk of inadvertent exposure; 10,000 of these workers are engaged in the grain freight trades where accidental exposure is a considerable hazard (NIOSH, 1977).

Environmental

The high volatility, limited natural sources, and negligible concentration in ambient air all ensure that any environmental exposures to phosphine are as a result of the accidental release from failed industrial processes or improper commercial usage. In most countries, strict controls exist for fumigation to prevent public exposure to phosphine, and guidelines for safe fumigation are available. These factors combine to effectively prevent nearly all environmental exposure to phosphine (World Health Organization, 1988).

INDUSTRIAL HYGIENE

Recommended personal protective equipment (PPE) for liquid phosphine includes impervious clothing, self-containing breathing apparatus, splash-proof safety goggles, rubber gloves and bands around legs, arms, and waist (United States Environmental Protection Agency, 1998). Table 118.2 shows exposure limits (NIOSH, 1977).

TABLE 118.2   Phosphine Exposure Limits

Immediately Dangerous to Life or Health (IDLH)	50 ppm
National Institute for Occupational Safety and Health (NIOSH) recommended exposure limit (REL)	Time-weighted average (TWA) 0.3 ppm (0.4 mg/m^3)
Occupational Safety and Health Administration (OSHA) permissible exposure limits (PEL)	TWA 0.3 ppm (0.4 mg/m^3)

RISK ASSESSMENTS

Phosphine is an extremely volatile, flammable gas that reacts violently with air, water, and other aqueous products. It is used extensively as a fumigant and rodenticide, and within the last decade has been increasingly used in industry for the synthesis of various chemicals (Berners-Price and Sadler, 1988; Chaudhry, 1997). The primary route of exposure to phosphine gas is via inhalation, either occupationally or accidentally, in which it is rapidly absorbed in the pulmonary system and systemically distributed. Incidents of ingestion of metal phosphides, predominantly aluminum and zinc, have also increased, as a result of homicides and suicide attempts (Goel and Aggarwal, 2007). Phosphine is a severe pulmonary irritant that has various adverse effects on the cardiovascular, respiratory, gastrointestinal, and CNS. The health effects of phosphine are primarily acute, with early indications of exposure consisting of respiratory and gastrointestinal effects that generally occur within hours, but can be delayed for days. Other adverse effects may consist of damage to the kidneys and liver, and also convulsions and pulmonary edema. The chronic effects of phosphine inhalation exposure are less defined, but can include pulmonary effects, especially in individuals with predisposing conditions. Phosphine is not classified as a carcinogenic, reproductive, or developmental toxicant because of limited and inadequate human data (U.S. Environmental Protection Agency, 2005).

Disclaimer: The views expressed in this article are those of the author(s) and do not reflect the official policy of the Department of Army, Department of Defense, or the U.S. Government.

REFERENCES

Al'zhanov, S., Moldabekov, S.M., Ershov, V., and Popov, A. (1983) *Use of Copper and Iron Sulfates in Cleaning Furnace Gas from Phosphorus Production*, Moscow: Academy of Sciences of the USSR, Institute of Scientific Information (VINITI), (Document 4802-83) (in Russian).

Anand, R., Binukumar, B.K., and Gill, K.D. (2011) Aluminum phosphide poisoning: an unsolved riddle. *J. Appl. Toxicol.* 31(6): 499–505.

Anon (1956) Ferro-silicon explosion risks. *Chem. Trade J. Chem. Eng.* (16 Nov).

ATSDR (2007) *Medical Management Guidelines for Phosphine*.

Beloskurskaya, G.I., Paraskevopulos, Y.G., and Shlygina, O.E. (1978) On the clinical and functional condition of the liver in patients with chronic phosphorus poisoning. *Tr. Inst. Kraev. Patol. Akad. Nauk Kazakhskoi SSR (in Russian)* 36:25–30.

Berners-Price, S. and Sadler, P. (1988) Phosphines and metal phosphine complexes: Relationship of chemistry to anticancer and other biological activity. In: *Bioinorganic Chemistry*, vol. 70, Berlin Heidelberg: Springer, pp. 27–102.

Bowker, J.R. (1958) The liberation of phosphine in the machining of spheroidal graphite iron. *Occup. Med.* 8(2):50–52.

Bretherick, L. (1990) *Bretherick's Handbook of Reactive Chemical Hazards*, Butterworths.

Chaudhry, M.Q. (1997) A review of the mechanisms involved in the action of phosphine as an insecticide and phosphine resistance in stored-product insects. *Pestic. Sci.* 49(3):213–228.

Chefurka, W., Kashi, K.P., and Bond, E.J. (1976) The effect of phosphine on electron transport in mitochondria. *Pestic. Biochem. Physiol.* 6(1):65–84, Available at http://dx.doi.org/10.1016/0048-3575(76)90010-9.

Chugh, S.N., Dushyant, S.R., Arora, B., and Malhotra, K.C. (1991) Incidence and outcome of aluminum phosphide poisoning in a hospital study. *Indian J. Med. Res.-B* 94:232–235.

Chugh, S.N., Singhal, H.R., Mehta, L., Chugh, K., Shankar, V., and Malhotra, K.C. (1990) Plasma renin activity in shock due to aluminium phosphide poisoning. *J. Assoc. Physicians India* 38(6):398–399.

Ciba Foundation (1978) *Phosphorus in the Environment, its Chemistry and Biochemistry*, Elsevier/Excerpta Medica/North Holland.

Cole, R. and Bensett, W. (1950) An investigation of the possibility of the occurrence of phosphine and acetylene in the atmosphere of factories for the powdering of magnesium. *J. Soc. Chem. Ind.* 69(1):29–30.

Daubert, T.E. and Danner, R.P. (1989) *Physical and Thermodynamic Properties of Pure Chemicals: Data Compilation*, Washington, DC: Taylor & Francis.

Dumas, T. (1980) Phosphine sorption and desorption by stored wheat and corn. *J. Agric. Food Chem.* 28(2):337–339.

Garry, V.F., Tarone, R.E., Long, L., Griffith, J., Kelly, J.T., and Burroughs, B. (1996) Pesticide appliers with mixed pesticide exposure: G-banded analysis and possible relationship to non-Hodgkin's lymphoma. *Cancer Epidemiol. Biomarkers Prev.* 5(1):11–16.

Goel, A. and Aggarwal, P. (2007) Pesticide poisoning. *Natl. Med. J. India* 20(4):182–191.

Gupta, S. and Ahlawat, S.K. (1995) Aluminum phosphide poisoning-a review. *J. Toxicol. Clin. Toxicol.* 33(1):19–24.

Habashi, F. and Ismail, M. (1975) Health hazards and pollution in the metallurgical industry due to phosphine and arsine. *Can. Inst. Min. (CIM) Bull.* 68:99–103.

Harger, R.N. and Spolyar, L.W. (1958) Toxicity of phosphine, with a possible fatality from this poison. *AMA Arch. Ind. Health* 18(6):497–504.

Jackson, J. and Elias, E. (1984) Hepatic function and exposure to phosphine and white phosphorus. In *Proceedings of the 12th International Congress on Occupational Health in the Chemical Industry* pp. 219–228, Dublin, Ireland.

Jones, A.T., Jones, R.C., and Longley, E.O. (1964) Environmental and clinical aspects of bulk wheat fumigation with aluminum phosphide. *Am. Ind. Hyg. Assoc. J.* 25:376–379.

Lauterbach, M., Solak, E., Kaes, J., Wiechelt, J., Von Mach, M.A., and Weilemann, L. (2005) Epidemiology of hydrogen phosphide exposures in humans reported to the poison center in Mainz, Germany, 1983–2003. *Clin. Toxicol.* 43(6):575–581.

Lewis, R.J.S. (2007) *Hawley's Condensed Chemical Dictionary*, 15th ed. New York, NY: Wiley-Interscience.

Lutzmann, L., Keinitz, M., and Klosterkotter, W. (1963) Emission of phosphine during transport of ferrosilicon. *Med. Welt (in German)* 58:114–116.

Mathew, G.G. (1961) The production of phosphine while machining spheroidal graphite iron. *Ann. Occup. Hyg.* 4(1):19–35.

Misra, U.K., Tripathi, A.K., Pandey, R., and Bhargwa, B. (1988) Acute phosphine poisoning following ingestion of aluminium phosphide. *Hum. Toxicol.* 7(4):343–345.

Nakakita, H. (1987) The mode of action of phosphine. *J. Pestic. Sci.* 12(2):299–309.

National Institute for Occupational Safety and Health (2005) *NIOSH Pocket Guide to Chemical Hazards*, U.S. Dept. of Health and Human Services, Public Health Service, Cincinnati, Ohio: Centers for Disease Control and Prevention, National Institute for Occupational Safety and Health.

NIOSH (1977) *National Occupational Hazard Survey*, vol. 3, Cincinnati, Ohio: National Institute of Occupational Safety and Health.

O'Neil, M.J. (2001) *The Merck Index*, 13th ed., Whitehouse Station, NJ: Merck and Co., Inc..

Quistad, G.B., Sparks, S.E., and Casida, J.E. (2000) Chemical model for phosphine-induced lipid peroxidation. *Pest Manag. Sci.* 56(9):779–783.

Spanggord, R.J., Rewick, R., Chou, T.W., Wilson, R., Podoll, R.T., and CA., S.I.M.P. (1985) *Environmental Fate of White Phosphorus/Felt and Red Phosphorus/Butyl Rubber Military Screening Smokes*, Defense Technical Information Center.

Spicer, C., Pollack, A., Kelly, T., and Ramamurthi, M. (1993) *A Literature Review of Atmospheric Transformation Products of Clean Air Act Title III Hazardous Air Pollutants. Final Report, Contract, 68-D80082*.

Stellman, J.M. and International Labour Office (1998) *Encyclopaedia of Occupational Health and Safety*, 4th edition, vol. 1–2: International Labor Office.

Stephenson, J.B. (1967) Zinc phosphide poisoning. *Arch. Environ. Health* 15(1):83–88.

Thienes, C.H. and Haley, T.J. (1972) *Clinical Toxicology*, Lea & Febiger.

Toy, A.D.F. and Walsh, E.N. (1987) *Phosphorus chemistry in everyday living*, American Chemical Society.

United States Environmental Protection Agency (1998) *Reregistration Eligibility Decision (RED)]: AL & MG Phosphide. U.S.*

Environmental Protection Agency, Office of Prevention, Pesticides and Toxic Substances.

U.S. Environmental Protection Agency (2005) *Phosphine (CASRN 7803-51-2).*

Van Wazer, J.R. (1961) *Phosphorus and its Compounds*, Interscience Publishers.

Vdovenko, I.D., Vakulenko, L.I., Yakovleva, L.A., Shedenka, L.I., Bereslavskaya, A.N., and Yatsenko, V.S. (1984) Effect of surfactants on phosphine formation in steel pickling. *Ukr. Khim. Zh. (in Russian)* 50(4):418–421.

Wilson, R., Lovejoy, F.H., Jaeger, R.J., and Landrigan, P.L. (1980) Acute phosphine poisoning aboard a grain freighter. Epidemiologic, clinical, and pathological findings. *J. Am. Med. Assoc.* 244(2):148–150.

World Health Organization (1988) International Program on Chemical Safety, editor, *Health and Safety Guide*, Geneva: World Health Organization, for the International Programme on Chemical Safety.

119

DIESEL EXHAUST

J. Michael Berg, Phillip T. Goad, and Thomas W. Hesterberg

First aid: Remove individual from source

Target organ(s): Eyes and respiratory system

Occupational exposure limits: OSHA PEL: Does not exist; NIOSH REL: Does not exist; ACGIH TLV: Does not exist; US MSHA PEL: 160_{TC} $\mu g/m^3$

Reference values: USEPA RfC: 5 $\mu g/m^3$ DPM

Risk/safety phrases: Not available

BACKGROUND

The diesel engine, in contrast to conventional spark ignition combustion, operates by injecting atomized fuel into compressed air under high temperature and pressure. Under these conditions, diesel fuel auto-ignites, generating power with increased efficiency and durability over conventional combustion engines (USEPA, 2002). In theory, the complete combustion of a hydrocarbon fuel in a diesel engine would yield gaseous carbon dioxide (CO_2) and water (H_2O); however, in practice, many chemical factors (e.g., fuel composition and additives) and physical processes (e.g., engine technology, type, age, and operating conditions) may influence the resultant combustion products, termed "diesel exhaust" (Hesterberg et al., 2005; Hesterberg et al., 2009).

Due to the ever evolving—and therefore more stringent—diesel exhaust regulatory environment, a review on diesel engine emissions would not be complete without a brief overview of past and present regulatory standards. During the early 1950s, experimental studies using organic extracts of diesel exhaust particulates (DEP) raised the possibility that emissions from internal combustion engines, including diesel, may elicit a tumorigenic response in laboratory mice (Kotin et al., 1954a, 1954b, 1955).

These ever-increasing concerns over human exposure to diesel exhaust prompted the enactment of the first regulatory "smoke standard[1]" on heavy-duty diesel engines in the late 1960s (33 FR 8304, 1968). Shortly thereafter (1974), the United States Environmental Protection Agency (USEPA) introduced the first regulatory limits on gaseous phase emissions, including carbon monoxide (CO), hydrocarbons (HC), and oxides of nitrogen (NO_x). During this same time, however, the particulate-component of diesel exhaust remained unregulated. The identification of solvent-extractable mutagenic compounds from DEP in 1977 prompted the EPA to issue a precautionary notice of diesel exhaust carcinogenicity, and introduce the first DEP on-road standards for light-duty and heavy-duty diesel engines in 1982 and 1988, respectively (Huisingh et al., 1978; NY Times, 1977). However, the relevancy of such organic solvent extractible mutagens is currently under debate, as a growing fraction of scientific literature suggests that these compounds have limited bioavailability in the aqueous environment of the lung (Hesterberg et al., 2012a). Since 1988, increasingly stringent emission regulations, combined with advancement in diesel engine and fuel technology, have resulted in heavy-duty diesel on-road engines, made in 2007 and later, which have 80% reductions in NO_x emissions and 98% reductions in particulate matter (PM) emissions compared to pre-1988 levels (USEPA, 2002).

As a result of significant innovations in diesel engine technology (e.g., fuel injection), modifications to fuel composition (e.g., ultralow sulfur (SO_x) diesel), and engine after-treatments; diesel exhaust emitted from 2007 and later

---

[1] Smoke emissions measured as the percent opacity of light.

Hamilton & Hardy's Industrial Toxicology, Sixth Edition. Edited by Raymond D. Harbison, Marie M. Bourgeois, and Giffe T. Johnson.
© 2015 John Wiley & Sons, Inc. Published 2015 by John Wiley & Sons, Inc.

heavy-duty on-road engines is both quantitatively and qualitatively very different from that of traditional (pre-1988) diesel exhaust (TDE). This distinct type of exhaust has been termed "New Technology Diesel Exhaust," or NTDE (Hesterberg et al., 2005). The NTDE is defined as the diesel exhaust from newly integrated systems intended to meet the stringent 2007 USEPA standards for PM and NO_x. Thus, as the majority of toxicological investigations draw conclusions based on diesel exhaust generated by the use of pre-1988 diesel engines, the applicability of such research to current and future exposures to NTDE is currently in question.

PHYSICAL AND CHEMICAL PROPERTIES

Emissions from diesel engines or diesel-engine powered vehicles are a highly variable and complex mixture of PM and gaseous-phase compounds derived from the incomplete combustion of hydrocarbon fuels, fuel additives, and lubricating components. At the rudimentary level, diesel exhaust is comprised of a few basic elements: a respirable carbon-based diesel exhaust particulate with a layer of adsorbed sulfates and solvent-extractible organics and a mixture of vapor- and gaseous-phase compounds (including, but not limited to, 1,3-butadiene, formaldehyde, acrolein, dioxins, and polycyclic aromatic hydrocarbons (PAHs)), and trace metals (Cheng et al., 1984; Kittelson, 1998; USEPA, 2002). However, the application of a strict compositional definition to the term "diesel exhaust" is not inherently possible, as a variety of chemical and physical factors, including the engine age and operating conditions, as well as the type of fuel, fuel additives, and exhaust filters, may impact the resultant emissions (Hesterberg et al., 2005; Hesterberg et al., 2009).

From a toxicological perspective, the physicochemical properties of DEP are known to largely influence the biological response (Park et al., 2011). DEP are comprised of a solid carbonaceous core with adsorbed organic compounds, sulfates, metals, and other trace elements (USEPA 2002; Wichmann, 2007). Together, these DEP components form aerosolized particulates in the fine ($<2.5\,\mu m$) and ultrafine ($<100\,nm$) size range. From a regulatory perspective, fine-sized particles dominate the particulate mass; however, by number, up to 90% of the particles are within the nanometer size range (Cheng et al., 1984; Kittelson, 1998; USEPA, 2002). The large fraction of ultrafine particulates within traditional diesel exhaust is currently attracting significant scientific attention, as particulates within this size range combine increased available surface area for organic material binding with increased potential for alveolar deposition (Geiser and Kreyling, 2010; Hesterberg et al., 2010).

TDE is comprised of a mixture of hundreds of organic and inorganic compounds (HEI, 1995; USEPA, 2002). In addition to CO_2 and CO, the organic component of the particulate phase of diesel exhaust may include, but is not inclusive of, alkane and alkene hydrocarbons, aldehydes, as well as mono- and poly-cyclic aromatic compounds and their nitro-substituted derivatives. Inorganic components, including oxides of NO and oxides of SO, are formed during the high temperature oxidation of NO and SO present in the air and fuel, respectively.

As noted in the previous section, the compositional make-up of diesel exhaust from present-day (2007 and later) diesel engines is both qualitatively and quantitatively different than that of the past TDE. These engines, able to meet the stringent 2007 USEPA NO_x and DEP emissions standards, have been documented to achieve substantial reductions in the formation and emission of both regulated (PM, NO_x, NMHC) and nonregulated species (aldehydes and aromatics) (Hesterberg et al., 2008). The representative contribution and composition of particulate matter in TDE and NTDE is provided in Figure 119.1. A more thorough review of differences between TDE and NTDE exhaust may be found elsewhere (Hesterberg et al., 2011).

ENVIRONMENTAL FATE

Diesel exhaust is released into the atmosphere where its individual constituents are subjected to physical (e.g., agglomeration and dry and wet deposition) and chemical modification (e.g., photolysis, reaction with atmospheric compounds). Chemical modification processes lead to calculated atmospheric lifetime ranging from minutes (NO_2; solar radiation) to years (aldehydes; ozone). As discussed in the Health Effects Institute Special Report on diesel exhaust (1995), these processes have the potential to define human exposure to diesel exhaust by *"(1) influencing the ambient concentrations of each specific emission products, and their reaction products, at different locations, and (2) by potentially altering the toxic, mutagenic, or carcinogenic properties of the original emission constituents as well as creating new toxic, mutagenic, or carcinogenic transformation products."*

The topics of atmospheric fate and transport of compounds present in diesel exhaust have been thoroughly discussed in previous reviews. In this regards, the reader is referred to a 1995 Health Effects Institute publication entitled "Diesel Exhaust: Critical Analysis of Emissions, Exposure, and Health Effects" (Winer and Busby, 1995).

ECOTOXICOLOGY

There exists limited information on the ecotoxicological effects of diesel exhaust as a whole. In this respect, individual

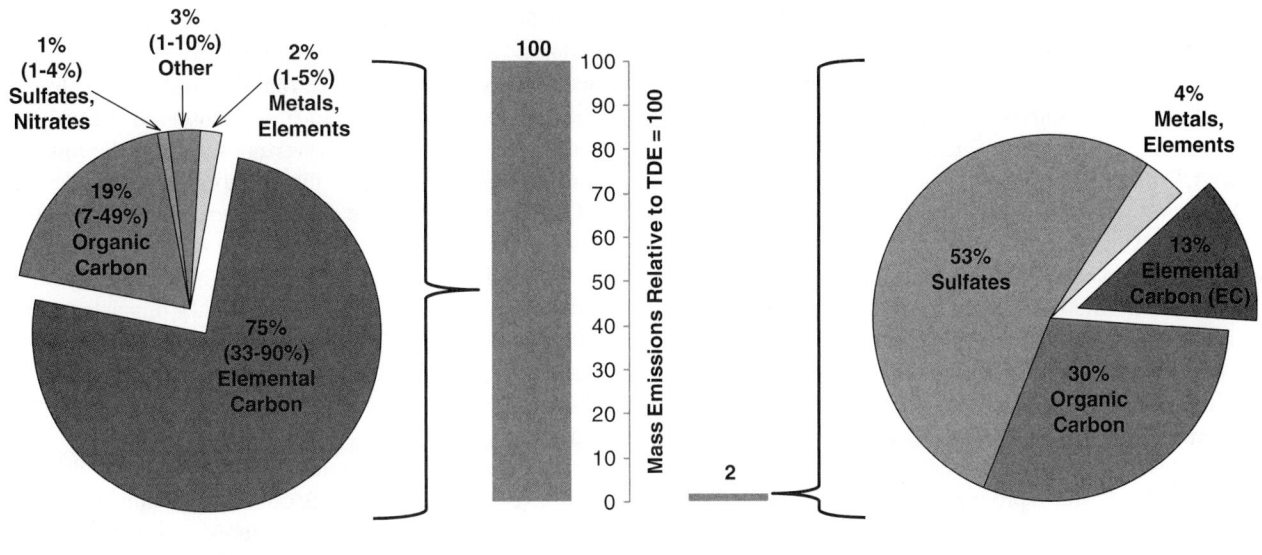

FIGURE 119.1 Quantitative and qualitative differences in the composition of diesel exhaust particulates in TDE (from pre-2000s era diesel engine technology; USEPA, 2002) versus NTDE (average data from four 2007-model-year heavy-duty diesel engines; Khalek et al., 2011) PM mass emissions derived from Hesterberg et al. (2008). Figure reproduced from Hesterberg et al. (2011).

constituents should be evaluated separately. For example, emissions of inorganic compounds, such as SO_x and NO_x, may contribute to the acidification of rain (Winer and Busby, 1995). Similarly, it has been hypothesized that the increased absorbance of solar radiation by black carbon aerosol particles, including diesel particulate matter (DPM), may influence climate change (Roberts and Jones, 2004).

MAMMALIAN TOXICOLOGY

Acute Effects

Due to the variability of diesel exhaust constituents and the lack of consistent exposure assessment methods; there exists limited information for quantifying the exposure–response relationship in acute diesel exhaust studies. In general, however, reports indicate that short-term exposure to TDE can elicit irritant (e.g., conjunctivae, nasal, and bronchial), neurophysiological (e.g., headache, dizziness) and respiratory (e.g., tightness of chest) symptoms (Battigelli, 1965; Gamble et al., 1987a; OSHA, 2013; Rudell et al., 1994; Wade and Newman, 1993). Some individuals may exhibit an unpleasant taste in the mouth (Battigelli, 1965). These symptoms are transient in nature, and can sometimes be present when levels of individual constituents in diesel exhaust are below their respective threshold limit values (TLVs) (Morgan et al., 1997).

Description of the irritant and neurophysiological symptoms, as noted above, together with reports of the characteristic odor of diesel exhaust, are often the most sensitive method of determining if an individual has been exposed to a high level of TDE. Literature indicates that the presence of general symptoms, as well as increased airway inflammatory responses, are more sensitive indicators of excessive diesel exhaust exposure than pulmonary function tests (Battigelli, 1965; Gamble et al., 1987a; Rudell et al., 1994; Salvi, 2001). A single report describes the development of asthma following exposure to high levels of diesel exhaust (Wade and Newman, 1993). It remains inconclusive as to whether or not individuals with previous allergic or respiratory problems may be more susceptible to developing acute effects (Hesterberg et al., 2009).

Chronic Effects

Noncarcinogenic Effects While limited by a number of variables (i.e., sample size, exhaust composition, etc.), several long-term human epidemiological investigations have examined the chronic noncancer respiratory effects of TDE in occupational settings. Several studies report an increased incidence of respiratory symptoms, including cough, phlegm, and/or dyspnea (Attfield et al., 1982; Gamble and Jones, 1983; Reger et al., 1982); however, others observed a notable absence in diesel exhaust-exposed groups (Ames et al., 1984; Battigelli et al., 1964). Although not consistent across all epidemiological investigations, two studies indicate statistically significant alterations in pulmonary function (FVC, FEV_1, PEF, FEV_1/FVC, FEF_{50}) (Gamble et al.,

1987b; Purdham et al., 1987). In respect to the epidemiological investigations reported, the USEPA (2002) stated that *"the absence of reported noncancerous health effects, other than infrequently occurring effects related to respiratory symptoms and pulmonary function changes, is notable"* (USEPA, 2002).

Due to the many uncertainties in diesel exhaust epidemiological investigations and the general inadequacies for characterizing potential noncancer health effects associated with long-term diesel exhaust exposure, chronic animal bioassays with well-defined and highly controlled diesel exhaust atmospheres provide health-effects evidence for chronic diesel exhaust exposure (Hesterberg et al., 2009). While a variety of toxicological endpoints have been assessed *in vivo* (e.g., cardiovascular, growth, immunological, neurological, and reproductive) the USEPA has determined that adverse respiratory effects are the most sensitive endpoint for chronic toxicity studies of diesel exhaust (USEPA, 2013); thus, further discussion on this topic will be limited as such. Using concentrations greater than those found in the ambient environment, laboratory studies undertaken in several animal species (rat, hamster, cat, and monkey) indicate that chronic exposure to traditional diesel exhaust may lead to chronic lung inflammation, pulmonary function impairment, and histological changes during the actual exposure to diesel exhaust, but these changes returned to normal levels following cessation of exposure (Hesterberg et al., 2009; USEPA, 2002). However, as will be discussed in the Modes of Action section, there is a general consensus in the scientific literature that "overload" levels of any particle, including diesel exhaust particles, can result in lung cancer in rats that may not be relevant to humans (ILSI, 2000).

Highlighted in the derivation of the USEPA reference concentration (RfC), Ishinishi et al. (1988) exposed Fisher 344 rats to exhaust from either light- or heavy-duty diesel engines for 30 months (16 h a day; 6 days a week) at DPM concentrations ranging from 110 to 3720 $\mu g/m^3$, dependent upon emission source. In addition to a dose-dependent decrease in body weights (females) at a DPM concentration of 3720 $\mu g/m^3$, significant alterations in liver and kidney function were noted. At DPM concentrations of 960 $\mu g/m^3$ or higher, significant histopathological observations were made, including shortened and absent cilia in the tracheal and bronchial epithelium, marked hyperplasia of the bronchiolar epithelium, and swelling of the Type II cellular epithelium (USEPA, 2013). Thus, the authors identified no-observable-adverse-effects-level (NOAELs) in this study of 410 and 460 $\mu g/m^3$ for light-duty and heavy-duty engines, respectively.

More recently, Ishihara and Kagawa (2003) reported on the results of rats exposed to diesel exhaust emissions for 24 months (16 h a day; 6 days per week) at doses of 210, 1180, and 3050 $\mu g/m^3$. A significant increase in mortality was observed at the highest dose examined. Indications of a localized chronic pulmonary response, including, alterations

in the bronchoalveolar lavage (BAL) fluid composition (e.g., increased protein, increased LDH activity, decreased alveolar macrophages, increased neutrophils, and elevated fructose, phospholipids, and sialic acid) were observed. As no significant physiological changes were observed in the low exposure group (210 $\mu g/m^3$) the authors determined that a NOAEL for DEP in the rat lung seems to lie between 200 and 1000 $\mu g/m^3$.

As highlighted in Hesterberg et al. (2009), current literature indicates a limited understanding of the impacts of diesel exhaust composition as it pertains to the 2007 and later regulatory environment. For example, McDonald et al. (2004) observed a significant reduction in diesel exhaust-induced lung toxicological endpoints when comparing TDE emissions to an emission-reduced exhaust, characteristic of NTDE (e.g., ultralow SO fuel, catalyzed ceramic trap). Hence, as a majority of the chronic noncancer endpoints literature is derived from TDE, it is uncertain as to the applicability of these endpoints for current and future diesel exhaust exposures.

Carcinogenicity In 2012, the International Agency for Research on Cancer (IARC) updated the carcinogenicity classification of diesel exhaust from Group 2A "probable human carcinogen" to Group 1 "known human carcinogen" (Benbrahim-Tallaa et al., 2012). One of the key studies that provided a basis for the revised IARC classification of diesel exhaust was that by Attfield et al. (2012), which suggested that the *"results of this study indicate a higher lung cancer mortality risk associated with traditional diesel exhaust exposure among [those workers with the greatest diesel exhaust exposure]."* Similarly, Silverman et al. (2012) concluded that their study *"showed a strong and consistent relation between quantitative exposure to diesel exhaust and increased risk of dying of lung cancer."* However, issues with historical exposure reconstruction (i.e., measurement of CO using colorimetric tubes and establishing the relationship between CO, horsepower and respirable elemental carbon (REC), data interpretation, and potential confounders (e.g., smoking), have led several investigators to question the validity of these conclusions (Boffetta, 2012; Borak et al., 2011; Hesterberg et al., 2012c; McClellan, 2012; Mohner et al., 2012).

The USEPA and the National Toxicology Program (NTP) have designated diesel exhaust and DEP as *"likely to be carcinogenic to humans"* and *"reasonably anticipated to be a human carcinogen,"* respectively. However, the USEPA has not developed a quantitative risk assessment for carcinogenicity for reasons discussed elsewhere (see Risk Assessment); albeit their carcinogenicity classification was based on the findings of (1) strong, but sufficient evidence for a causal association between diesel exhaust exposure and lung cancer risk in occupational studies, (2) extensive data on the mutagenicity of organic compounds adhered to the surface of the DPM, (3) evidence of carcinogenicity of DEP and adsorbed organic compounds via other routes of study (e.g., dermal) and (4) evidence of DPM carcinogenicity in animal studies (USEPA, 2013).

CHEMICAL PATHOLOGY

The use of controlled-exposure studies with human volunteers provides the most appropriate data set for evaluation of adverse pathology of clinical biochemical parameters. Of important note, however, is the potentially limited applicability of these measurements to the clinical environment, through which the American Thoracic Society has suggested that "*not all changes in biomarkers related to air pollution should be considered as indicative of injury that represents an adverse effect*" (ATS, 2000). Endpoints of specific interest with regards to diesel exhaust exposure often involve markers of pulmonary inflammation and carboxyhemoglobin measurements in addition to objective qualitative endpoints (i.e., mucomembraneous irritation).

A number of studies have evaluated airway inflammatory response in a variety of media (e.g., sputum and/or bronchoalveolar wash) following acute exposure to whole diesel exhaust or DEP. Nightingale et al. (2000) and Nordenhall et al. (2000) exposed healthy individuals to diesel exhaust particulate concentrations of 200 and 300 $\mu g/m^3$, respectively. Compositional analyses of sputum content indicate an increase in percentage neutrophils, interleukin-6 (IL-6), methylhistamine, and myeloperoxidase (MPO) over baseline (Nightingale et al., 2000; Nordenhall et al., 2000). Salvi et al. (1999, 2000) observed similar markers of pulmonary inflammation (e.g., increased IL-8, neutrophils, and MPO) in the bronchoalveolar lavage following exposure of healthy human volunteers to 300 $\mu g/m^3$ DEP. The authors noted similar observations for neutrophils and platelets in the blood. At lower diesel exhaust concentrations, however, results are conflicting; with some reports indicating that lower DEP ($\sim$100 $\mu g/m^3$) exposures are well tolerated (Behndig et al., 2006; Mudway et al., 2004) while others report extensive DE-induced airway inflammation (Stenfors et al., 2004).

Carbon monoxide poisoning from diesel exhaust is unlikely; however, in certain instances (e.g., significant acute exposure in enclosed areas) carboxyhemoglobin (COHgb) measurements may be evaluated. As a variety of factors (e.g., smoking status, administration of oxygen, and time since exposure) may influence the result, it is important to document such factors.

HYPOTHESIZED MODES OF ACTION

Diesel exhaust is a complex mixture of particulates and gases, all of which retain the potential to influence the toxicological profile through varying modes of action. While the following discussion is focused on effects of the DEP, additive or synergistic effects caused by the gaseous phase cannot be discounted. To date several mechanistic hypotheses exist (see paragraphs below). However, as most animal studies have been conducted at diesel exhaust concentrations well above those that are observed in an occupational or environmental setting, the mechanistic understanding of diesel exhaust exposure in humans remains limited.

Diesel exhaust particulates, while containing a relatively insoluble carbon core, are surrounded by a solvent-extractable organic layer containing a number of known carcinogenic and mutagenic chemicals adsorbed to their surface (USEPA, 2002). Early literature suggested that desorption of organics (including PAHs) following exposure likely played a role in diesel exhaust-induced carcinogenesis in the rat. However, more recent studies suggest that these organic compounds, which can only be extracted under laboratory conditions (e.g., strong solvents and sonication), are of limited use in predicting the bioavailability and mutagenicity of the organic compounds on DEP in the aqueous environment of the lung (Enya et al., 1997; HEI, 1995; Hesterberg et al., 2012a; NTP, 2011). This line of thinking lead Heinrich et al. (1994) to examine the carcinogenic effects of a carbonaceous particle (carbon black) alone and in combination with a PAH-laden coal tar pitch. These authors suggested that it is the carbonaceous core of diesel exhaust, and not the adsorbed PAHs, that is likely responsible for the induction of lung cancer in high level diesel exhaust-exposed rats (Heinrich et al. 1994).

Similar studies in laboratory rats repeatedly observed an increased prevalence of lung tumor formation following lifetime exposure to high level diesel exhaust ($>$2200 $\mu g/m^3$ as DEP) (Heinrich et al. 1995; Mauderly et al., 1987). However, this tumorigenic response at high particulate levels is not limited to diesel exhaust; as a variety of relatively inert particulates, including carbon black, titanium dioxide, and talc elicited a similar effect. As noted by Heinrich et al. (1995), "*lung tumor rates in the rats increased with increasing cumulative particle exposure independent of the particle employed.*" As these effects have neither been observed with consistency in other species (i.e., mice, hamster, monkeys), nor do they occur at lower DEP levels ($<$600 $\mu g/m^3$ as DEP), this specific response is commonly described as a rat "lung overload" effect (Valberg and Crouch, 1999). As proposed by Driscoll et al. (1996) and Oberdorster and Yu (1990), the formation of lung tumors through the rat "lung overload" may be driven by a progression of events, including (1) particle phagocytosis and release of inflammatory mediators, (2) pulmonary inflammation, (3) epithelial hyperplasia, and (4) neoplastic transformation (Driscoll et al., 1996; Oberdorster and Yu, 1990). Due to the aforementioned reasons, this nonspecific tumorigenic response in the rat lung should be considered of little relevance to humans exposed to diesel exhaust in either occupational or nonoccupational (i.e., ambient environment) settings (Hesterberg et al., 2012b).

A considerable body of evidence suggests that diesel exhaust exposure may induce pulmonary and systemic inflammation; albeit at concentrations significantly greater than those observed in ambient and most occupational settings. Hesterberg et al. (2009) reviewed over 100 studies on the noncancer health effects of diesel exhaust and concluded that, with few exceptions, human controlled-studies largely support the proinflammatory role of diesel exhaust (Hesterberg et al., 2009). For example, Salvi et al. (1999) noted an increase in inflammatory response endpoints in both the airway (e.g., increased neutrophils, mast cells, and lymphocytes) and the blood (e.g., increased neutrophils and platelets) following an acute exposure to high levels of diesel exhaust (Salvi et al., 1999). As proposed by Behndig et al. (2006), this inflammatory response may be due to diesel exhaust particle-induced oxidative stress resulting in the activation of redox-sensitive transcription factors (e.g., NFκb) and ultimately the release of proinflammatory cytokines; however, the authors suggest that acute human exposures of $100\,\mu g/m^3$ or less are insufficient to induce such an inflammatory response.

Current mode-of-action understanding at environmentally or occupationally relevant exposures remains limited. Additionally, the applicability of these mechanistic studies, driven largely by the presence of DEP, to current and future research is largely unknown as emissions studies of 2007 and later diesel exhaust technology (NTDE) show dramatic reductions in all components, especially the particulates, compared to TDE, which was examined in the previously discussed studies.

SOURCES OF EXPOSURE

The quantification of diesel exhaust exposure in ambient environments has been problematic, largely due to the confounding presence of airborne gases and particulates emitted from non-diesel sources. While there is no absolute marker for distinguishing diesel exhaust emissions particulates from other sources, a variety of analytes (e.g., total and respirable dust, CO, NO_x, benzene soluble fractions, specific PAHs, respirable combustible dust, elemental carbon) have been utilized as surrogates (Verma et al., 2003). Of these, however, the development and use of elemental carbon as an indicator of diesel exhaust exposure provides enhanced specificity and sensitivity over previously utilized total carbon or organic carbon methods (Verma et al., 2003).

Occupational

Occupational exposure to diesel exhaust may occur in any workplace where a diesel engine is utilized. In addition to various manufacturing operations, this includes workers in the mining, transportation, construction, agriculture, and

TABLE 119.1   Diesel Exhaust Exposure in Ambient and Occupational Settings

Exposure Environment	Elemental Carbon, $\mu g/m^{3a}$	References
Ambient levels	1.1–4	Liukonen et al. (2002)
U.S. surface miners	1–4	Coble et al. (2010)
Truckers	1.1–6	Davis et al. (2007); Garshick et al. (2002)
Locomotive crew	2.3–5.6[b]	Seshagiri (2003); Verma et al. (2003)
U.S. underground coal miners	62–202	Pronk et al. (2009)
U.S. underground metal/mineral miners	54–347	Coble et al. (2010)

[a]Geometric mean.

[b]Value represents geometric mean of combined summer and winter measurements in trailing locomotive.

maritime industries (OSHA, 2013). Historical estimates suggest that approximately 1.35 million workers (1981–1983) and 3 million workers (1990–1993) were exposed to diesel exhaust in the United States and European Union, respectively (NIOSH, 1983; Kauppinen et al., 2000). However, current practice includes the use of applicable administrative (e.g., restricted engine idling) and engineering controls (e.g., local exhaust ventilation, diesel exhaust filters) to limit occupational exposure. Table 119.1, above, compares the elemental carbon concentration (as a surrogate for diesel exhaust exposure) of various occupational settings to that of the ambient air in the United States.

Environmental

Due to the ubiquitous on- and off-road uses of diesel in technology in modern society, exposure to diesel exhaust is not limited to the occupational setting. As data suggests that diesel particulate contributes to the nationwide ambient $PM_{2.5}$ load, the general population may be exposed to diesel exhaust through the inhalation of ambient air (USEPA, 2002). In this respect, Liukonen et al. (2002) reported background elemental carbon concentrations ranging from <1 to $8\,\mu g/m^3$ (geometric mean of $3.1\,\mu g/m^3$) at numerous background locations throughout the United States. Similar ambient elemental carbon estimations have been observed by the USEPA National Air Toxics Assessment (2005). As modernization of the on-road diesel engine fleet progresses, the contribution of DPM to ambient $PM_{2.5}$ is likely to be significantly reduced.

INDUSTRIAL HYGIENE

Neither the Occupational Health and Safety Administration (OSHA) nor the American Conference of Industrial

TABLE 119.2   Occupational Standards for Major Components of Diesel Exhaust[a]

Chemical	OSHA-PELs		ACGIH TLVs	
	TWA	STEL/CEIL (c)	TWA	STEL/CEIL (c)
Carbon dioxide (ppm)	5000	–	5000	30,000
Carbon monoxide (ppm)	50	–	25	–
Nitrogen dioxide (ppm)	–	5 (c)	0.2	–
Nitric oxide (ppm)	25	–	25	–
PNOC (mg/m^3)	15[b]; 5[c]	–	–	–
Sulfur dioxide (ppm)	5	–	–	0.25

[a]As dictated by OSHA Partial List of Chemicals associated with Diesel Exhaust (www.osha.gov).
[b]Total Dust.
[c]Respirable Dust.

Hygienists (ACGIH) have established standards for diesel exhaust exposure in the general workplace. In this regard, it is emphasized that individual constituents of diesel exhaust are regulated as such (see Table 119.2).

From a historical perspective, the ACGIH has published various threshold limit values (TLV) for diesel exhaust in their Notice of Intended Changes list. The first proposed diesel exhaust specific standard, drafted in 1995, suggested a TLV of 150 µg/m^3 measured as submicron particles; with a carcinogenicity classification of A2 (suspected human carcinogen). In 1999, it was proposed that this value be decreased to 50 µg/m^3 measured as submicron particles. More recently (2000), however, the ACGIH recognized the appropriateness of elemental carbon as an "analytical surrogate" for diesel exhaust, which prompted a reanalysis of the proposed TLV. As such, the TLV of 50 µg/m^3 measured as submicron particles, was expressed as the elemental carbon equivalent of 20 µg/m^3. However, following the release of the USEPA Health Assessment Document on Diesel Exhaust in 2002, the ACGIH felt the need to reassess these values. As such, there is not ACGIH value for diesel exhaust at this time (ACGIH, 2013).

In contrast to OSHA and the ACGIH, the U.S. Mine Safety and Health Administration (MSHA) has established a diesel exhaust permissible exposure limit (PEL) of 160 µg/m^3 measured as total carbon in metal and nonmetal mines. Additionally, the mine operator must install, use, and maintain feasible engineering and administrative controls to reduce a miners exposure. In the event that these engineering and administrative controls are not effective in reducing an exposure to below the MSHA PEL, they must be supplemented with an appropriate air purifying respirator.

RISK ASSESSMENTS

Multiple agencies have reviewed the human health and diesel exhaust literature in an attempt to qualitatively assess potential noncancer and cancer risks (CALEPA, 1998a; HEI, 1995, 1999; USEPA, 2002). In the 2002 Health Assessment Document for Diesel Exhaust, the USEPA defined a RfC (a continuous exposure level for all individuals that is not likely to cause harmful effects) of 5 µg/m^3 DEP. This value has also been adopted by the California EPA (CALEPA). This RfC was set 30-fold lower than the human equivalent concentration (HEC) of the NOAEL (460 µg/m^3) observed in Ishinishi et al. (1988), and has been suggested to be quite conservative (Hesterberg et al., 2009).

At the current time, neither the USEPA nor the Health Effects Institute have developed quantitative risk assessments for carcinogenic endpoints citing, among other things, the lack of reliable exposure estimates in epidemiological studies and the invalidity of the rat model to characterize a tumorigenic response to particulates at relatively high exposure levels. However, the California Air Resources Board (CARB) felt, in relation to cancer risk assessment, that the *"epidemiological data are available upon which one can base risk estimates for humans"* (CALEPA, 1998a). In this respect, the Cal EPA has established an Inhalation Unit Risk (IUR) of 3.0×10^{-4} (3 excess cancers in 10,000 individuals if exposed daily to 1 µg/m^3 DEP). This IUR places diesel exhaust as one of the most potent carcinogens known to the State of California (CALEPA, 1998b; CALEPA, 2013). Because the CALEPA's quantitative carcinogenic risk assessment was derived from past epidemiological data that the USEPA and the HEI felt inappropriate for evaluation, combined with the fact that the diesel exhaust was of the traditional type, the applicability of such an assessment to current and future derivations of risk is of little practicability.

REFERENCES

ACGIH (2013) *Frequently Asked Questions (FAQs): Technical Questions*, Cincinnati, OH: American Conference of Government Industrial Hygienists. Available at http://www.acgih.org/faqs.htm.

Ames, R.G., Hall, D.S., and Reger, R.B. (1984) Chronic respiratory effects of exposure to diesel emissions in coal mines. *Arch. Environ. Health* 39:389–394.

ATS (2000) American Thoracic Society. What constitutes an adverse health effect of air pollution? Official statement of the American Thoracic Society. *Am. J. Respir. Crit. Care Med.* 161:665–673.

Attfield, M.D., Trabant, G.D., and Wheeler, R.W. (1982) Exposure to diesel fumes and dust at six potash mines. *Ann. Occup. Hyg.* 26:817–831.

Attfield, M.D., Schleiff, P.L., Lubin, J.H., et al. (2012) The Diesel Exhaust in Miners study: a cohort mortality study with emphasis on lung cancer. *J. Natl. Cancer Inst.* 104:869–883.

Battigelli, M.C. (1965) Effects of diesel exhaust. *Arch. Environ. Health* 10:165–167.

Battigelli, M.C., Mannella, R.J., and Hatch, T.F. (1964) Environmental and clinical investigation of workmen exposed to diesel exhaust in railroad engine houses. *Ind. Med. Surg.* 33:121–124.

Behndig, A.F., Mudway, I.S., Brown, J.L., et al. (2006) Airway antioxidant and inflammatory responses to diesel exhaust exposure in healthy humans. *Eur. Respir. J.* 27:359–365.

Benbrahim-Tallaa, L., Baan, R.A., Grosse, Y., et al. (2012) Carcinogenicity of diesel-engine and gasoline-engine exhausts and some nitroarenes. *Lancet Oncol.* 13:663–664.

Boffetta, P. (2012) Re: the diesel exhaust in miners study: a nested case-control study of lung cancer and diesel exhaust and a cohort mortality study with emphasis on lung cancer. *J. Natl. Cancer Inst.* 104:1842–1843, author reply 1848-9.

Borak, J., Bunn, W.B., Chase, G.R., et al. (2011) Comments on the Diesel Exhaust in Miners Study. *Ann. Occup. Hyg.* 55:339–432, author reply 343-6.

California Environmental Protection Agency (1998a) *Proposed Identification of Diesel Exhaust as a Toxic Air Contaminant. Part B. Health Risk Assessment for Diesel Exhaust*, Sacramento, CA: Office of Environmental Health Hazard Assessment, California Environmental Protection Agency.

California Environmental Protection Agency (1998b) *Findings of the Scientific Review Panel on the Report on Diesel Exhaust*, Sacramento, CA: California Environmental Protection Agency. Available at http://www.arb.ca.gov/toxics/dieseltac/de-fnds.htm (accessed 2/2013).

California Environmental Protection Agency (2013) *Health Effects of Diesel Exhaust Fact Sheet*, Sacramento, CA: Office of Environmental Health Hazard Assessment, California Environmental Protection Agency. Available at http://oehha.ca.gov/public_info/facts/dieselfacts.html (accessed 2/2013).

Cheng, Y.S., Yeh, H.C., Mauderly, J.L., and Mokler, B.V. (1984) Characterization of diesel exhaust in a chronic inhalation study. *Am. Ind. Hyg. Assoc. J.* 45:547–555.

Coble, J.B., Stewart, P.A., Vermeulen, R., et al. (2010) The Diesel Exhaust in Miners Study: II. Exposure monitoring surveys and development of exposure groups. *Ann. Occup. Hyg.* 54:747–761.

Davis, M.E., Smith, T.J., Laden, F., Hart, J.E., Blicharz, A.P., Reaser, P., and Garshick, E. (2007) Driver exposure to combustion particles in the U.S. Trucking industry. *J. Occup. Environ. Hyg.* 4:848–854.

Driscoll, K.E., Carter, J.M., Howard, B.W., et al. (1996) Pulmonary inflammatory, chemokine, and mutagenic responses in rats after subchronic inhalation of carbon black. *Toxicol. Appl. Pharmacol.* 136:372–380.

Enya, T., Suzuki, H., Watanabe, T., Hirayama, T., and Hisamatsu, Y. (1997) 3-Nitrobenzanthrone, a powerful bacterial mutagen and suspected human carcinogen found in diesel exhaust and airborne particulates. *Environ. Sci. Technol.* 31:2772–2776.

Gamble, J., Jones, W., and Minshall, S. (1987a) Epidemiological-environmental study of diesel bus garage workers: acute effects of NO_2 and respirable particulate on the respiratory system. *Environ. Res.* 42:201–214.

Gamble, J., Jones, W., and Minshall, S. (1987b) Epidemiological-environmental study of diesel bus garage workers: chronic effects of diesel exhaust on the respiratory system. *Environ. Res.* 44:6–17.

Gamble, J. and Jones, W.G. (1983) Chronic health effects of exposure to diesel emissions among salt miners. *Ann. Am. Conf. Ind. Hyg.* 4:73–83.

Garshick, E., Smith, T.J., and Laden, F. (2002) Quantitative assessment of lung cancer risk from diesel exhaust exposure in the US trucking industry: a feasibility study. In: HEI, editor. *Research Directions to Improve Estimates of Human Exposure and Risk from Diesel Exhaust*, Boston, MA: Health Effects Institute, pp. 113–150.

Geiser, M. and Kreyling, W.G. (2010) Deposition and biokinetics of inhaled nanoparticles. *Part Fibre Toxicol.* 7:2.

HEI (1995) *Diesel Exhaust: A Critical Analysis of Emissions, Exposure, and Health Effects. A Special Report of the Institute's Diesel Working Group*, Cambridge, MA: Health Effects Institute.

HEI (1999) *Diesel Emissions and Lung Cancer: Epidemiology and Quantitative Risk Assessment. A Special Report of the Institute's Diesel Epidemiology Expert Panel*, Cambridge, MA: Health Effects Institute.

Heinrich, U., Dungworth, D.L., Pott, F., et al. (1994) The carcinogenic effects of carbon black particles and tar-pitch condensation aerosol after inhalation exposure of rats. *Ann. Occup. Hyg.* 38:351–356.

Heinrich, U., Fuhst, R., Rittinghausen, S. et al. (1995) Chronic inhalation exposure of Wistar rats and two different strains of mice to diesel engine exhaust, carbon black, and titanium dioxide. *Inhal. Toxicol.* 7:533–556.

Hesterberg, T.W., Bunn, W.B., McClellan, R.O., Hart, G.A., and Lapin, C.A. (2005) Carcinogenicity studies of diesel engine exhausts in laboratory animals: a review of past studies and a discussion of future research needs. *Crit. Rev. Toxicol.* 35:379–411.

Hesterberg, T.W., Long, C.M., Bunn, W.B., Lapin, C.A., McClellan, R.O., and Valberg, P.A. (2012a) Health effects research and regulation of diesel exhaust: an historical overview focused on lung cancer risk. *Inhal. Toxicol.* 24:1–45.

Hesterberg, T.W., Bunn, W.B., Valberg, P.A., and Long, C.M. (2012b) Potential health effects of exposure to diesel engine exhausts. In: Friis, R.H., editor. *Praeger Handbook of Environmental Health. Volume III: Water Quality, and Solid Waste Disposal*, vol. 3, Santa Barbara, CA: ABC-CLIO, LLC, pp. 185–212.

Hesterberg, T.W., Long, C.M., and Valberg, P.A. (2012c) Re: the diesel exhaust in miners study: a nested case-control study of lung cancer and diesel exhaust and a cohort mortality study with emphasis on lung cancer. *J. Natl. Cancer Inst.* 104:1841, author reply 1848-9.

Hesterberg, T.W., Lapin, C.A., and Bunn, W.B. (2008) A comparison of emissions from vehicles fueled with diesel or compressed natural gas. *Environ. Sci. Technol.* 42:6437–6445.

Hesterberg, T.W., Long, C.M., Bunn, W.B., Sax, S.N., Lapin, C.A., and Valberg, P.A. (2009) Non-cancer health effects of diesel exhaust: a critical assessment of recent human and animal toxicological literature. *Crit. Rev. Toxicol.* 39:195–227.

Hesterberg, T.W., Long, C.M., Lapin, C.A., Hamade, A.K., and Valberg, P.A. (2010) Diesel exhaust particulate (DEP) and nanoparticle exposures: what do DEP human clinical studies tell us about potential human health hazards of nanoparticles? *Inhal. Toxicol.* 22:679–694.

Hesterberg, T.W., Long, C.M., Sax, S.N., et al. (2011) Particulate matter in new technology diesel exhaust (NTDE) is quantitatively and qualitatively very different from that found in traditional diesel exhaust (TDE). *J. Air Waste Manag. Assoc.* 61:894–913.

Huisingh, J., Bradow, R., Jungers, R., et al. (1978) Application of bioassay to the characterization of diesel particle emissions. In: Waters, M.D., Nesnow, S., Huisingh, J.L., Sandhu, S.S., and Claxton, L., editors. *Application of Short-Term Bioassays in the Fractionation and Analysis of Complex Environmental Mixtures*, Research Triangle Park, NC: U.S. Environmental Protection Agency, Health Effects Research Laboratory, pp. 381–418.

ILSI (2000) The relevance of the rat lung response to particle overload for human risk assessment: a workshop consensus report. *Inhal. Toxicol.* 12:1–17.

Ishihara, Y. and Kagawa, J. (2003) Chronic diesel exhaust exposures of rats demonstrate concentration and time-dependent effects on pulmonary inflammation. *Inhal. Toxicol.* 15:473–492.

Ishinishi, N., Kuwabara, N., Takaki, Y., et al. (1988) Long-term inhalation experiments on diesel exhaust. *Diesel Exhaust and Health Risks: Results of the HERP Studies. Japan Automobile Research Institute, Inc., Research Committee for HERP Studies*, Japan: Ibaraki, pp. 11–84.

Kauppinen, T., Toikkanen, J., Pedersen, D., et al. (2000) Occupational exposure to carcinogens in the European Union. *Occup. Environ. Med.* 57:10–18.

Khalek, I.A., Bougher, T.L., Merritt, P.M., and Zielinska, B. (2011) Regulated and unregulated emissions from highway heavy-duty diesel engines complying with U.S. Environmental Protection Agency 2007 emissions standards. *J. Air Waste Manag. Assoc.* 61:427–442.

Kittelson, D.B. (1998) Engines and nanoparticles: a review. *J. Aerosol Sci.* 29:575–588.

Kotin, P., Falk, H.L., Mader, P., and Thomas, M. (1954a) Aromatic hydrocarbons. I. Presence in the Los Angeles atmosphere and the carcinogenicity of atmospheric extracts. *AMA Arch. Ind. Hyg. Occup. Med.* 9:153–163.

Kotin, P., Falk, H.L., and Thomas, M. (1954b) Aromatic hydrocarbons. II. Presence in the particulate phase of gasoline-engine exhausts and the carcinogenicity of exhaust extracts. *AMA Arch. Ind. Hyg. Occup. Med.* 9:164–177.

Kotin, P., Falk, J.H., and Thomas, M. (1955) Aromatic hydrocarbons. III. Presence in the particulate phase of diesel engine exhausts and the carcinogenicity of exhaust extracts. *AMA Arch. Ind. Health* 11:113–120.

Liukonen, L.R., Gorgan, J.L., and Myers, W. (2002) Diesel particulate matter exposure to railroad train crews. *AIHA J. (Fairfax, Va)* 63:610–616.

Mauderly, J.L., Jones, R.K., Griffith, W.C., Henderson, R.F., and McClellan, R.O. (1987) Diesel exhaust is a pulmonary carcinogen in rats exposed chronically by inhalation. *Fundam. Appl. Toxicol.* 9:208–221.

McClellan, R.O. (2012) Re: the diesel exhaust in miners study: a nested case-control study of lung cancer and diesel exhaust, a cohort mortality study with emphasis on lung cancer, and the problem with diesel. *J. Natl. Cancer Inst.* 104:1843–1845, author reply 1847-9.

McDonald, J.D., Harrod, K.S., Seagrave, J., Seilkop, S.K., and Mauderly, J.L. (2004) Effects of low sulfur fuel and a catalyzed particle trap on the composition and toxicity of diesel emissions. *Environ. Health Perspect.* 112:1307–1312.

Mohner, M., Kersten, N., and Gellissen, J. (2012) Re: the diesel exhaust in miners study: a nested case-control study of lung cancer and diesel exhaust and a cohort mortality study with emphasis on lung cancer. *J. Natl. Cancer Inst.* 104:1846–1847, author reply 1848–9.

Morgan, W.K., Reger, R.B., and Tucker, D.M. (1997) Health effects of diesel emissions. *Ann. Occup. Hyg.* 41:643–658.

Mudway, I.S., Stenfors, N., Duggan, S.T., et al. (2004) An *in vitro* and *in vivo* investigation of the effects of diesel exhaust on human airway lining fluid antioxidants. *Arch. Biochem. Biophys.* 423:200–212.

New York Times (1977) *Diesel fumes under study as possible cancer cause.* November 13, p. 41.

Nightingale, J.A., Maggs, R., Cullinan, P., et al. (2000) Airway inflammation after controlled exposure to diesel exhaust particulates. *Am. J. Respir. Crit. Care Med* 162:161–166.

NIOSH (1983) *National Occupational Exposure Survey (1981–1983)*, Cincinnati, OH: National Institute for Occupational Safety and Health. Available at http://www.cdc.gov/noes/.

Nordenhall, C., Pourazar, J., Blomberg, A., Levin, J.O., Sandstrom, T., and Adelroth, E. (2000) Airway inflammation following exposure to diesel exhaust: a study of time kinetics using induced sputum. *Eur. Respir. J.* 15:1046–1051.

NTP (2011) *Report on Carcinogens*, 12th ed., Research Triangle Park, NC: National Toxicology Program. Available at http://ntp.niehs.nih.gov/go/roc12.

Oberdorster, G. and Yu, C.P. (1990) The carcinogenic potential of inhaled diesel exhaust: a particle effect? *J. Aerosol Sci.* 23:397–401.

OSHA/MSHA (2013) *Hazard Alert: Diesel Exhaust/Diesel Particulate Matter*, Cincinnati, OH; Arlington, VA: Occupational Health and Safety Administration and Mine Safety and Health Administration. Available at http://www.osha.gov/Publications/OSHA-3590.pdf.

Park, E.J., Roh, J., Kang, M.S., Kim, S.N., Kim, Y., and Choi, S. (2011) Biological responses to diesel exhaust particles (DEPs) depend on the physicochemical properties of the DEPs. *PLoS One* 6:e26749.

Pronk, A., Coble, J., and Stewart, P.A. (2009) Occupational exposure to diesel engine exhaust: a literature review. *J. Expo. Sci. Environ. Epidemiol.* 19:443–457.

Purdham, J.T., Holness, D.L., and Pilger, C.W. (1987) Environmental and medical assessment of stevedores employed in ferry operations. *Appl. Ind. Hyg.* 2:133–139.

Reger, R., Hancock, J., Hankinson, J., Hearl, F., and Merchant, J. (1982) Coal miners exposed to diesel exhaust emissions. *Ann. Occup. Hyg.* 26:799–815.

Roberts, D.L. and Jones, A. (2004) Climate sensitivity to black carbon aerosol from fossil fuel combustion. *J. Geophys. Res.* 109:D16202.

Rudell, B., Sandstrom, T., Hammarstrom, U., Ledin, M.L., Horstedt, P., and Stjernberg, N. (1994) Evaluation of an exposure setup for studying effects of diesel exhaust in humans. *Int. Arch. Occup. Environ. Health* 66:77–83.

Salvi, S. (2001) Pollution and allergic airways disease. *Curr. Opin. Allergy Clin. Immunol.* 1:35–41.

Salvi, S., Blomberg, A., Rudell, B., et al. (1999) Acute inflammatory responses in the airways and peripheral blood after short-term exposure to diesel exhaust in healthy human volunteers. *Am. J. Respir. Crit. Care Med.* 159:702–709.

Salvi, S.S., Nordenhall, C., Blomberg, A. et al. (2000) Acute exposure to diesel exhaust increases IL-8 and GRO-alpha production in healthy human airways. *Am. J. Respir. Crit. Care Med.* 161:550–557.

Seshagiri, B. (2003) Exposure to diesel exhaust emissions on board locomotives. *AIHA J. (Fairfax, Va)* 64:678–683.

Silverman, D.T., Samanic, C.M., Lubin, J.H., et al. (2012) The Diesel Exhaust in Miners study: a nested case-control study of lung cancer and diesel exhaust. *J. Natl. Cancer Inst.* 104:855–868.

Stenfors, N., Nordenhall, C., Salvi, S.S., et al. (2004) Different airway inflammatory responses in asthmatic and healthy humans exposed to diesel. *Eur. Respir. J.* 23:82–86.

USEPA (2002) *Health Assessment Document for Diesel Engine Exhaust*, Washington, DC: National Center for Environmental Assessment, U.S. Environmental Protection Agency.

USEPA (2013) *IRIS (Integrated Risk Information System)*, Washington, DC: U.S. Environmental Protection Agency. Available at http://www.epa.gov/iris/index.html.

USEPA (2005) *National Air Toxics Assessments (NATA)*, Washington, DC: U.S. Environmental Protection Agency. Available at http://www.epa.gov/ttn/atw/nata2005/.

Valberg, P.A. and Crouch, E.A. (1999) Meta-analysis of rat lung tumors from lifetime inhalation of diesel exhaust. *Environ. Health Perspect.* 107:693–699.

Verma, D.K., Finkelstein, M.M., Kurtz, L., Smolynec, K., and Eyre, S. (2003) Diesel exhaust exposure in the Canadian railroad work environment. *Appl. Occup. Environ. Hyg.* 18:25–34.

Wade, J.F. 3rd. and Newman, L.S. (1993) Diesel asthma. Reactive airways disease following overexposure to locomotive exhaust. *J. Occup. Med.* 35:149–154.

Wichmann, H.E. (2007) Diesel exhaust particles. *Inhal. Toxicol.* 19:241–244.

Winer, A.M. and Busby, W.F. Jr. (1995) Atmospheric transport and transformation of diesel emissions. In: HEI, editor. *Diesel exhaust: a critical analysis of emissions, exposure, and health effects. A special report of the Institute's Diesel Working Group*, Cambridge, MA: Health Effects Institute, pp. 83–105.

120

SPACEFLIGHT OPERATIONS

Michael A. Cardinale, Philip J. Scarpa, James R. Taffer, and Daniel Woodard

The space industry today is a multibillion dollar international network of government and commercial enterprises associated with designing, fabricating, processing, testing, launching, and landing of crewed and uncrewed spacecraft (Monti and D'Angelo, 1993). It includes operations common to general industry, but also many that are unique to spaceflight or even unique to single industrial sites. The industry is notable for the use of large quantities of reactive and/or toxic materials, as well as the size, complexity, and cost of major launch systems. Contingencies can be very costly and attention to detail in industrial hygiene has proven vital.

INDUSTRY OVERVIEW

The era of spaceflight began with the launch of the first satellite in 1957; competing national programs in the United States and USSR (now Russia) led to the first human landing on the moon only 12 years later. The following years have seen hundreds of launches supporting human spaceflight, numerous scientific satellites and space probes, and a myriad of satellites serve practical applications, primarily in communications and earth imaging for both commercial and military customers. An important change in the industry in recent decades has been the shift from space programs that are funded and closely managed by governmental entities to an industry in which commercial enterprise operates with increasing autonomy, launching uncrewed and ultimately crewed spacecraft.

Spaceflight is supported by an array of industrial, government, military, academic, and research organizations around the world. Major space launch facilities maintain a range of industrial and base support functions, which may be comparable to those of a small city. Physical boundaries may encompass hundreds of square miles and house dozens of government and contractor organizations. These tenant organizations are responsible for the launch of the vehicle and payload processing operations, as well as the maintenance of the infrastructure, grounds, and facilities.

The mission of any space launch facility is the preparation, test, and checkout of launch systems, spacecraft, payloads, and experiments for launch, while ensuring mission safety. For crewed spacecraft the mission also includes landing recovery operations, including postlanding evaluation, refurbishment, repair, and servicing of the space vehicle, as well as returned scientific experiments and payloads.

The diverse workforce of a space operations facility is engaged in a wide variety of unique industrial processes and operations. This workforce includes the engineers and laborers in the aerospace industries who design, manufacture, and test the spacecraft, and the technicians and specialists who process the payloads and rocket boosters and ready the vehicle for launch. Many design, logistics, construction, and maintenance activities also take place to support the infrastructure itself. These services include management of electrical power distribution, water and sewage systems, waste management, communications, buildings and structures, security and law enforcement, fire protection, grounds maintenance, and food services.

Because of the extraordinary challenges involved in reaching space, it is often necessary to choose designs and materials that achieve the highest possible performance, even if this requires accepting risks. The extremely large quantities of hazardous materials that are used contribute to the unique occupational hazards associated with this work.

Hamilton & Hardy's Industrial Toxicology, Sixth Edition. Edited by Raymond D. Harbison, Marie M. Bourgeois, and Giffe T. Johnson.
© 2015 John Wiley & Sons, Inc. Published 2015 by John Wiley & Sons, Inc.

OCCUPATIONAL MEDICINE AND ENVIRONMENTAL HEALTH PROGRAM

Workers in this industry are exposed to a variety of hazardous substances, and many of these employees are involved in potentially dangerous operations. Some hazards are inherent to spaceflight and launch activities and cannot be easily eliminated. Consequently, strong environmental health and occupational medicine programs are required to provide all employees with a safe and healthful workplace. To accomplish this task, a comprehensive, well-organized occupational medicine and environmental health plan must be in place. The management of such a far-reaching system requires a team of specialists representing diverse fields. Box 120.1 lists the components of a comprehensive occupational medicine and environmental health program.

These programs provide a basis for the management of hazardous operations and processes, the support elements required for the recognition and assessment of new hazards, and the evaluation of control measures used for the protection of personnel.

HAZARDOUS MATERIALS USED IN SPACE OPERATIONS

Space launch processing operations present some unique industrial hygiene problems. Of greatest concern is the safe use and handling of cryogenic liquid propellants, storable propellants, and solid rocket motor propellant. Table 120.1 lists some of the common propellants used in launch-processing operations.

Liquid hydrogen or nitrogen present hazards, such as frostbite from the extreme cold and potential asphyxiation from oxygen displacement. The rapid vaporization from spilled cryogenic liquids can result in oxygen deficiency in enclosed areas if the cryogen is an asphyxiant, such as nitrogen, or in excessive oxygen levels and consequent fire hazard if oxygen is spilled. Cryogenic oxygen presents special hazards from the creation of oxygen-rich atmospheres. Special ventilation and monitoring for oxygen level in air may be required, particularly in confined spaces. In addition to engineering controls, protective clothing and written procedures are used to manage cryogen hazards.

TABLE 120.1   Common Propellants

Name	TLV (ppm)[a]	OSHA PEL (ppm)[a]	Use
Aerozine 50[b]	0.01	1	Hypergolic fuel
Anhydrous hydrazine	0.01	1	Hypergolic fuel
JP-5	200	–	Hydrocarbon fuel
Liquid hydrogen	NA	NA	Cryogenic fuel
Liquid nitrogen	NA	NA	Pressurant gas
Liquid oxygen	NA	NA	Cryogenic oxidizer
Monomethylhydrazine	0.01	0.2[c]	Hypergolic fuel
Nitrogen tetroxide	0.2	5[c]	Hypergolic oxidizer
RP-1	200	–	Hydrocarbon fuel

[a]8-h time-weighted average.
[b]50/50 mixture of hydrazine and unsymmetrical dimethyl hydrazine.
[c]Ceiling limit.
American Conference of Governmental Industrial Hygienists; *TLV*, threshold limit value (ACGIH, 2014); Occupational Safety and Health Administration; *PEL*, permissible exposure limit (OSHA, 1995); *NA*, not available.

Hypergolic propellants include monomethyl or dimethyl hydrazine, which serves as fuels, and nitrogen tetroxide, used as an oxidizer, and are discussed in detail below. They are both toxic and reactive, but are used where liquid propellants must be stored at ambient temperature for extended periods.

Solid rocket propellants commonly contain a mixture of powdered aluminum fuel and ammonium perchlorate oxidizer. When ignited, these components produce large volumes of gas, which contain aluminum oxides, carbon monoxide, and hydrogen chloride gas. Hazard controls require evacuating the immediate launch area during launch in addition to the downwind corridors. Monitoring for toxic gases is performed as a part of postlaunch reentry before unprotected personnel are allowed access to the launch pad.

Although much attention is directed to the propellants associated with spacecraft, oversight is also required to manage the hazards associated with the maintenance of the facilities, grounds, infrastructure, and ground support equipment. The widespread use of asphyxiant gases, used as pressurants and for inerting propellant systems, carries a significant risk of creating oxygen-deficient atmospheres. Significant attention is also required to control exposures to asbestos, which is frequently encountered during construction and building renovations, and to silica and toxic metals that are present during the maintenance of steel structures at the launch facility when abrasive blasting is performed.

With the end of the Space Shuttle program in 2011 and the beginning of NASA's new mission to commercialize space flight, emphasis is being placed on low toxicity or "green" propellants (NASA Press Release, 2012). Green propellants, such as ammonium dinitramide (ADN) and hydroxylammonium nitrate (HAN), are being considered as low toxicity alternatives to hydrazines. With reduced handling hazards come reductions in processing costs, hazardous waste, systems complexity, and launch processing times. As reliability concerns are addressed, the use of green propellants will be expected to see more widespread use.

HYDRAZINES

Hydrazines are used in rocket engines, reaction control thrusters, the auxiliary power unit, chemical and pharmaceutical production processes, catalysts as inorganic solvents, and plastic synthesis. Hydrazine is a clear, colorless, and corrosive water-soluble liquid. Their vapors are heavier than air and have an ammonia or fishy odor.

INDUSTRIAL HYGIENE

Employees may be exposed to hydrazine via skin or eye contact, inhalation, or ingestion. During the 1990s, the American Conference of Governmental Industrial Hygienists (1996) reduced its published threshold limit value (TLV) for hydrazine fuels to a common level of 10 ppb in air, a level that required research and development of new detection methods for workplace monitoring. The NASA utilized its unique research and development resources to develop a technology enabling rapid and accurate field measurements of this air contaminant with sensitivities previously considered infeasible (Marmaro et al., 1992).

HEALTH EFFECTS

Hydrazines are caustic and produce local chemical irritation on the skin or lungs, but their primary toxic effects are the result of systemic absorption, usually after inhalation, and effects on target organs. Because spilled hydrazine frequently ignites, an acute exposure often includes combustion products as well. Systemic absorption of hydrazine can result in hypotension, irritation of the pulmonary tract mucosa, coughing, dyspnea, pulmonary edema and respiratory failure, headache, dizziness, narcosis, central nervous system depression, muscle tremors, and seizures. Hydrazine can also cause irritation of the gastrointestinal tract, nausea and vomiting, hematemesis, conjunctivitis, corneal damage, and chemical burns. Skin contact may result in irritant dermatitis and chemical burns, facial edema, and allergic sensitization after prolonged contact. Kidney and liver damage and hyperglycemia or hypoglycemia may also be observed. Red blood cell hemolysis and methemoglobinemia may result from excessive exposure to monomethylhydrazine (MMH) (NIOSH, 1994).

MEDICAL TREATMENT OF SIGNIFICANT EXPOSURES

The patient should be decontaminated by disrobing and washing, and airways, breathing, and circulation should be evaluated. If necessary, oxygen should be administered via a nonrebreather mask. An intravenous (IV) line with normal saline should be started. If the patient is hypotensive, IV fluids should be given and then the patient should be reassessed.

Hydrazines are GABA agonists and can cause seizures or central neuropathies by a mechanism similar to isoniazid toxicity. In significant exposures, or where neurologic abnormalities are seen, pyridoxine can be given at a loading dose of 25 mg/kg IV. In case of seizures, standard treatment with parenteral benzodiazepines (e.g., lorazepam or diazepam) should be administered. Based on experience with isoniazid, if seizure activity continues despite these measures, barbiturates, paralytics, and intubation should be considered.

Laboratory work, even in limited exposures, should include complete blood count, serum electrolytes, methemoglobin, liver function tests, renal function tests, chest X-ray, arterial blood gases, and pulmonary function tests. If methemoglobin is >30%, 1–2 mg/kg methylene blue can be given IV in a 1% solution (Bronstein and Currance, 1994).

NITROGEN TETROXIDE/NITROGEN DIOXIDE

Nitrogen tetroxide is used as an oxidizer for hydrazine propellants in both launch vehicles and spacecraft reaction control systems. Nitrogen tetroxide is the dimer of nitrogen dioxide (NO_2). At temperatures between $-9.3\,°C$ ($15\,°F$) and $135\,°C$ ($275\,°F$), NO_2 and nitrogen tetroxide coexist as a gaseous mixture. These vapors are clear to orange-brown with a sweetish, acrid, or bleach odor and are heavier than air. In their liquid phase, they are green to brown, highly corrosive, and water soluble.

INDUSTRIAL HYGIENE

Workers may be exposed to nitrogen tetroxide and NO_2 by inhalation, skin and eye contact. It evaporates quickly and leaves little or no residue. In 2012, American Conference of Industrial Hygienists (ACGIH) lowered its TLV for nitrogen dioxide to a concentration of 0.2 ppm in air. Detection methods and monitoring equipment are commercially available at this level.

HEALTH EFFECTS

Nitrogen dioxide absorbs water readily, becoming nitric acid. Consequently it reacts immediately on contact with human tissue and cannot be absorbed; it produces only local toxicity. It can produce significant skin and eye burns; however, life-threatening injuries are almost exclusively due to inhalation and pulmonary toxicity, which can range from mild chemical pneumonitis to frank pulmonary edema. Life-threatening exposure will generally produce immediate symptoms, but inhalation of NO_2 may produce mild effects initially, followed by a symptom-free period of approximately 5–12 h and subsequent worsening. In severe cases tachycardia, bradycardia, irritation of the pulmonary tract mucosa, coughing, dyspnea, chemical pneumonitis, pulmonary edema, respiratory failure, and severe spasm of the airways or larynx may be seen. Headache, narcosis, vertigo, somnolence, fatigue, and loss of consciousness may also be observed. Burning and irritation of the gastrointestinal tract, nausea and vomiting, as well as conjunctivitis, corneal damage, and chemical burns of the eye may be manifested. Irritant dermatitis and chemical burns of the skin, as well as methemoglobinemia, may result from excessive exposure.

MEDICAL TREATMENT OF SIGNIFICANT EXPOSURES

The patient should be decontaminated by disrobing and washing, and airways, breathing, and circulation should be evaluated. If necessary, oxygen should be administered at high concentration, i.e., by nonrebreather mask. Intravenous steroids should be administered to prevent chemical pneumonitis. Baseline laboratory studies should be performed as well as complete blood count, serum electrolytes, chest X-ray, pulmonary function tests, and arterial blood gases. Close follow-up is required during the first 24–48 h because pulmonary edema may be delayed at least 6 hours. Bronchiolitis obliterans may occur up to 6 months postexposure (Bronstein and Currance, 1994; Clayton and Clayton, 1994).

HISTORICAL TOXIC HAZARD INCIDENTS IN GROUND OPERATIONS

Operations are particularly hazardous at the launch site, where all the hazardous materials are brought together during the complex and precise procedures, which is required to prepare a rocket and spacecraft for flight. The majority of hazardous materials incidents causing injuries and fatalities at launch sites around the world have involved fires or explosions as well as chemical releases and spills. Often these contingencies have occurred while servicing rockets being prepared for launch or making repairs after a problem develops. A comprehensive program of safety, industrial hygiene, waste management, and contingency response has been essential in preventing significant health impacts.

The most serious hazards are posed by toxic propellants, particularly nitrogen tetroxide and hydrazines, generally either unsymmetrical dimethylhydrazine (UDMH) or MMH. Perhaps the first major toxic release incident in the industry occurred on 4 October, 1960 at the Baikonur launch site in the USSR, during preparations to launch an uncrewed rocket designated the R-16. A technician erroneously plugged the first stage umbilical cable into an incompatible umbilical socket on the second stage, causing a premature ignition command. The fuel tanks ignited and ruptured, releasing more than 100 tons of UDMH and nitrogen tetroxide propellants. There was a massive fire, but witnesses also described victims overcome by toxic fumes; there were about 100 fatalities (Oberg, 1988).

An even more disastrous incident occurred at the Xichang Satellite Launch Center in western China on 15 February, 1996. A Long March 3B rocket went off course almost immediately after launch and crashed a few seconds later into an area near the facility gate where a large crowd had gathered to watch. The rocket carried a nearly full load of propellants at the time of impact, including over 200 tons of UDMH and nitrogen tetroxide. While hydrazine fuels tend to burn in such a situation, the nitrogen tetroxide is largely released as nitrogen dioxide and may well have caused a significant fraction of the casualties. Although the official death toll released by the Chinese government was listed at only six, Western sources and observers at the scene estimated those killed numbered in the "hundreds" and as many as 1000 people were injured (Ferster, 1996).

In April 1994 an incident involving the spill and release of nitrogen tetroxide occurred at NASA's Johnson Space Center (JSC) at the thermochemical test area (TTA). The leak occurred while a small rocket was being readied for a test in the TTA. The total release was approximately 16 gallons of both liquid and vapor through a 25-foot-tall vent stack. The leak generated a brownish-red cloud that drifted into nearby occupied areas of the center. There were no life-threatening injuries; however, more than 80 people sought medical help for symptoms they believed were related to this exposure (Caylor, 1995).

Oxygen is essential to life, but also carries the risk of toxicity. Atmospheres containing 100% oxygen were used frequently in the early years of human flight to simplify vehicle design and avoid decompression sickness. In 1960 the USSR was conducting a 14-day test of oxygen toxicity with a human subject, a member of their cosmonaut corps, when a flash fire in the chamber left him with burns that proved fatal (Oberg, 1990). A similar hazard resulted in the most serious incident of the Apollo program, now designated Apollo 1. During a ground communications test prior to the first crewed launch, the capsule was filled with an atmosphere of 100% oxygen and pressurized to a total pressure of about 18 PSI; this was a condition that had not been anticipated in the design. A source of ignition was provided by an electrical short in the capsule, probably involving wires that were partially installed and subjected to abrasion. In the pressurized oxygen-rich atmosphere, nonhazardous materials in the cabin were ignited, including fabric netting, Velcro, polyurethane foam insulation and even ethylene glycol coolant fluids. The resulting flash fire and carbon monoxide accumulation killed all three crew members. This tragedy led to substantial modifications in materials and procedures used on Apollo and subsequent spacecraft (NASA, 1970).

Confined spaces are sometimes filled with dry nitrogen gas to reduce hazards from flammable gasses and prevent ice accumulation on cryogenic systems. This poses a significant hazard since an individual entering such an oxygen-deficient atmosphere can become hypoxic and lose consciousness in a matter of seconds. One such incident occurred at the Kennedy Space Center on 28 February, 1981 just after a 20-s engine test of the Space Shuttle Columbia on the launch pad prior to its first flight. Due to a procedural error technicians were cleared to enter the aft compartment of the Shuttle after the firing, but while the compartment was still filled with 100% nitrogen. All collapsed in the compartment before the problem was recognized; three ultimately died due to anoxic

injury (Wilcut and Harkins, 2011). On 5 May, 1995, two European Space Agency technicians died in a similar accident at the Guiana Space Center when they entered an umbilical mast which had become filled with nitrogen gas due to a leaking heat exchanger (ESA, 1995).

PREVENTION AND TREATMENT OF TOXIC HAZARDS IN GROUND OPERATIONS

Hypergolic fuels for rocket propellant systems consist primarily of hydrazine, MMH, UDMH, and Aerozine-50, which is a 50:50 mixture of hydrazine and UDMH. These hydrazine derivatives present vapor inhalation and skin contact hazards. Hypergolic oxidizer for rocket propellant systems consists of nitrogen tetroxide, which breaks down into NO_2 upon release into the atmosphere. Nitrogen tetroxide is an extremely corrosive liquid and may cause severe eye and skin burns. Nitrogen dioxide gas produces primarily alveolar pulmonary toxicity but may produce irritation to the eyes and mucous membranes as well.

When using or handling hypergolic fuels or oxidizers, the preferred method for preventing personnel exposures is to keep the toxic material fully confined in a closed system. This may not be possible when the chemicals are being transferred from a storage drum to a transport cart, during spacecraft loading procedures, or while connecting or disconnecting systems containing the liquids or vapors. In these situations personnel use fully encapsulating impermeable suits called propellant handler's ensembles (PHEs). Propellant handler's ensembles are composed primarily of butyl rubber and protect personnel by supplying clean air either from a breathing air supply located outside the hazardous area or from a tank worn on the back of the user. Propellant handler's ensembles also preclude skin contact with hypergols. Routine exposure monitoring is also performed to ensure that personnel are using adequate protection. Because of the engineering controls and protective clothing used, personnel exposures to hypergolic propellants are in most cases short term, low in concentration, and transient.

The Emergency Medical System at the Kennedy Space Center (KSC) deploys mobile medical triage and treatment forces for prelaunch and postlanding contingencies and ground mass casualty incidents. These forces are positioned upwind of hazardous exposures, assess, and monitor patient contamination, and provide onsite water-based decontamination and exposure treatment when needed.

A general strategy for exposure to a major toxic propellant release is presented in Box 120.2.

Hazardous materials can be handled safely in spaceflight and launch operations; however, the most effective risk reduction strategy remains finding less toxic alternatives. Several programs are underway in the United States to investigate less toxic storable propellants such as ammonium

BOX 120.2 GENERAL STRATEGY FOR INITIAL TREATMENT OF MAJOR EXPOSURE TO HYPERGOLIC PROPELLANTS

Respiratory protection if source still present
Move across wind or upwind away from vapor cloud to fresh air
Remove clothing
Oxygen administration
Respiratory support if required
Flush exposed areas (skin, eyes) with water for 20 min
Treat trauma if present
Pyridoxine 25 mg/kg for seizure prophylaxis in significant hydrazine exposures
Consider corticosteroids for respiratory injury
Control seizures with IV benzodiazepines
Monitor condition and transport to definitive care facility

dinitramide (Talawar et al., 2009), and to develop refrigeration methods for storing cryogenics such as liquid oxygen for long periods in space. In China, the new generation of launch vehicles will abandon hypergolic propellants and utilize kerosene, liquid hydrogen, and liquid oxygen instead.

TOXIC HAZARD INCIDENTS DURING SPACEFLIGHT

The most obvious and visible personnel in the spaceflight industry are the astronauts and cosmonauts who travel into the hostile environment of space to live and work. The time they spend in space ranges from a few minutes during suborbital flights to a year or more aboard the International Space Station. The launch into orbit and the return to the earth from space entail particularly complex and hazardous tasks.

There have been four incidents in which fatalities have occurred during spaceflight; however, all were due to mechanical failures of the spacecraft or launch vehicle rather than toxic hazards. In two of these cases, the Space Shuttles Challenger in 1986 and Columbia in 2003, there were toxicological concerns during the recovery of vehicles as recovery and reconstruction workers were exposed to friable fibrous materials and residuals of toxic chemicals (Scarpa, 2014).

On the fourth day of the Apollo 13 mission, an electrical short circuit in one of the Service Module's oxygen tanks led to an explosion and subsequent crippling of the Apollo spacecraft. The explosion terminated the mission to the moon, and the resultant loss of breathing oxygen presented an acute emergency. The crew was faced with multiple problems, including rising carbon dioxide levels, which required adapting carbon dioxide absorbent cartridges from the Apollo Command Module to a different environmental system in the Lunar Module (NASA, 1970).

The most serious toxic exposure during a spaceflight occurred in 1975 at the end of Apollo–Soyuz, the first international space mission. Shortly after the Apollo capsule began its descent to Earth, a human factors-related error left the attitude control thruster rockets firing after the parachutes were deployed. As the craft descended into the atmosphere, the cabin vent opened and thruster exhaust was drawn into the cabin. The primary toxin was nitrogen dioxide gas, although a variety of combustion products were also present. Respiratory protection was aboard (oxygen masks with eye protection); unfortunately it had been assumed it would only be needed when the crew was weightless, so the masks were stowed in an area well above the crew seats. When the incident occurred, the spacecraft was suspended by its parachutes under normal gravity, and the masks were inaccessible. The crew almost immediately displayed signs of chemical pneumonitis, including coughing and shortness of breath, as well as burning of the eyes and exposed skin. All remained conscious until the impact with the ocean, unfortunately the capsule then rolled upside down, making it impossible to open the hatch. Two of the three crewmen were incapacitated by that time, and a lethal exposure could have occurred. Fortunately the commander, who remained conscious, was finally able to reach the masks and put them on his crewmates and himself. After recovery, the crew was treated for chemical pneumonitis with oxygen and IV steroids; they were essentially free of symptoms by the following day but were hospitalized for several days as a precautionary measure (NASA, 1976; Dejournette, 1977).

There have been numerous less serious incidents involving hazardous properties of materials aboard spacecraft, including both inhalation and skin exposures. During Apollo 10, loose pieces of fiberglass insulation produced clinically significant skin, eye, and upper respiratory tract irritation. Several inhalation exposures occurred during missions of the Space Shuttle. Eye irritation from lithium hydroxide canisters has been reported on at least six separate shuttle missions. There have also been reports of nausea and headaches from outgassed ammonia and formaldehyde (Coleman and James, 1994). There have been numerous reports of hand desquamation, boils, and contact dermatitis from chemical irritants (Toback, 1991).

Exposures to toxic gases such as ethylene glycol and excessive carbon dioxide were frequent occurrences in the aging Russian space station MIR. In 1997 the life support system normally supported only three crewmembers but six were aboard the station during a crew exchange period and so supplemental breathing oxygen was needed. This was produced using a solid-fuel oxygen generator in which a cartridge containing lithium perchlorate is ignited and burned slowly, releasing oxygen. A defective cartridge ruptured, emitting a meter-long jet of flame, sparks, and particles of molten metal and rapidly filled the station with smoke. The crew was able to locate smoke masks and managed to use water from several small fire extinguishers to keep the spacecraft wall cool until the device burned itself. One crewman suffered a small third-degree burn (Morgan, 2001; Oberg 1998).

PREVENTION AND TREATMENT OF TOXIC HAZARDS DURING SPACEFLIGHT

Any toxic substance released into the sealed volume of a spacecraft can become a persistent respiratory hazard, consequently a comprehensive toxicology program has been developed to assess every material brought onto human-carrying spacecraft or used aboard the International Space Station.

The program is managed by the Space Toxicology Office at the Johnson Space Center (JSC) in Houston, Texas. Every material to be brought onboard the Space Station is assessed for toxic hazards. Spacecraft Maximum Allowable Concentrations (SMACs) are determined for each chemical constituent and used in material selection and the planning of containment procedures. In addition, air samples are collected aboard the Space Station and spacecraft during each mission, and actual concentrations of a wide range of hazardous materials are measured. Detailed documentation, including the sources for determination of maximum allowable levels, is provided in detail on the website of the JSC Space Toxicology Office, and has been published in a series of reference books which provide a useful resource for toxicology in other industrial environments (National Research Council, 1994-present).

A major concern on all spaceflights continues to be the potential presence of contaminants in the atmosphere of the cabin. Toxic substances in spaceflight can come from a number of different sources, including leaks or spills from storage containers, volatile metabolic waste products of the crew, particulate pollutants that are not easily removed from the air under weightless conditions, components of spilled food, leaks from environmental or flight control systems, thermodegradation of spacecraft materials and products produced by small electrical fires or contaminated removal systems, and outgassing of cabin materials.

The management of toxic hazards in crewed spacecraft, such as the materials selection program, is based on experiences gained during the Apollo Program. The material selection program includes a thorough screening of candidate materials for flammability, applicability, and the evaluation of the contaminant compounds that are outgassed into the crew cabin environment. All of the interior nonmetallic materials are known to outgas contaminants into the crew compartment. Some specific sources are electrical insulation, paints, lubricants, adhesives, and degradation of nonmetallic console and equipment structures. Approximately 400 compounds have been identified in the atmosphere of the shuttle cabin. Many of these compounds can be found in normal terrestrial air, but when they become part of the confined atmosphere of the spacecraft, they can concentrate to toxic levels over time (James and Coleman, 1994).

The environmental control and life support systems utilized on the Soyuz spacecraft uses an activated charcoal filter to remove most trace contaminants. This approach, although reliable, does not provide the long-term removal of volatile organics needed on the International Space Station. In the Space Station a trace contaminate control system utilizes a charcoal bed followed by an electrically heated high temperature catalytic oxidizer. Any over-oxidized chemicals are then removed by a lithium hydroxide bed (Locke, 2008). Real-time atmospheric monitoring is performed on both the US and Russian segments of the Space Station by spectroscopy which provides partial pressures of oxygen, nitrogen, carbon dioxide, hydrogen, water, and methane (Locke, 2008).

Permissible levels of a wide range of contaminates are also specified for potable water aboard the spacecraft; this becomes more important as the duration of spaceflight increases and water is to be recycled multiple times (National Research Council, 2004).

CONTINGENCY RESPONSE TO SPILLS OR SIGNIFICANT EXPOSURE

The NASA has developed equipment and procedures for the crew to follow in case a serious toxicological event occurs on orbit, and the crew receives extensive training in the effects of toxic materials, both those carried on the spacecraft and those which could result from a fire, in which toxic fumes are generated. Most of the materials selected for use in the Space Shuttle were chosen on the basis of their ability to resist combustion at elevated temperatures. This requirement resulted in the selection of many materials that contain halogenated or nitrogenated hydrocarbons. However, in many cases these materials contribute to increased combustion toxicity (Rippstein and Coleman, 1983). The potential toxic hazards associated with even small fires were demonstrated during an incident on the Shuttle in 1989, in which an electrical cable overheated, releasing a variety of chemical contaminates.

The crew has respirators that provide oxygen and eye protection for quick use in an emergency. A combustion products analyzer, now in use on the Space Station, was developed to detect and measure concentrations of some of the important contaminants released by heating or combustion of these materials, including carbon monoxide, hydrogen chloride (from polyvinyl chloride), hydrogen fluoride, and carbonyl fluoride (from Teflon), and hydrogen cyanide (from Kapton). Finally, the crew is trained in procedures for purging combustion products from the spacecraft atmosphere and cleaning up spills involving toxic materials. As with chemical exposures on the ground, the crew is provided with an extensive medical kit and training in recognition and medical treatment of crewmembers exposed to toxic materials (Hale et al., 2011).

RADIATION SAFETY CONSIDERATIONS WHEN PLANNING FOR AND SUPPORTING MAJOR RADIOLOGICAL SOURCE SPACE LAUNCHES

Radioisotope power sources (RPSs) use the heat from the radioactive decay of plutonium-238 to generate electricity the spacecraft can utilize to power science instruments, communications equipment, and general spacecraft or surface rover operations for spacecraft that cannot generate electrical power using solar panels. Typical external radiation dose rates from the gamma ray and neutron particles are approximately 100 mrem per hour 3 feet from the external surfaces of the power source with radiation areas of 2 mrem per hour dose rates extending out 16 feet in all directions from the power sources. Power source handlers must also be cognizant of the high surface temperatures of these power sources that are typically around 100 °C.

Another radiation safety issue that must be considered includes the inhalation exposure of the workforce or public to radioactive dust particles that could potentially be released as a result of a launch accident. Because plutonium-238 is an alpha particle emitter, the retention of radioactive dusts trapped in the lungs could lead to potential cancers in exposed people. In order to minimize this risk the plutonium, as plutonium-238 oxide is embedded in a ceramic matrix to help preclude formation of respirable particles. The fuel is placed into iridium capsules to help withstand crushing pressures that in turn are encased in several layers of a carbon–carbon material to absorb the heat generated from an accident environment.

Supporting major radiological source launches such as the Cassini, Pluto, and Mars Science Laboratory missions is no small task. Radiation safety management of ionizing radiation hazards must provide oversight of every step in the launch process from source receipt, ground processing, and radiological contingency launch readiness. Planning takes considerable time to outline, prepare for, and execute. Usually planning starts about 3 years prior to launch, with approximately 5–6 months involved in ground processing. Overlapping at the end of ground processing is approximately 2–3 weeks for radiological contingency preparations, training, and drilling of the contingency elements prior to launch to insure readiness to respond, as well as, have the

capability to support any protracted period postlaunch should a launch mishap occur.

Beginning 3 years prior to launch, radiological contingency planning meetings are held with appropriate federal, state, and local officials. Follow-up meetings are scheduled to status preparations, plan development, equipment procurements, and planning contingency exercises. A core planning working group meets more frequently to conduct detailed planning, submit and track procurements, develop radiological contingency plans and procedures, maintain budget tracking and forecasting, and track action items from the radiological contingency planning meetings.

Ground processing support planning typically starts 1 year out from launch. It is kicked off by a high fidelity practice dress rehearsal by the spacecraft builders. At this 1–2 day rehearsal draft procedures for ground processing source handling are walked through to validate and/or identify any issues that need resolution. During the rehearsal, health physics support is identified and time and motion studies are conducted with the source handling and support personnel in order to identify opportunities to use time, distance, or shielding to reduce radiation exposure. Data from the walkthrough are also used to define likely radiation exposures that would be received during the actual handling procedures. Prior to source arrival, all personnel involved with source handling operations are provided radiation safety training and health physics managers issue radiation work permits spelling out who is cleared to handle the sources, what source handling procedures must be used, where the activity will take place, and what radiation safety steps to follow. Radiation safety surveillance is provided for all ground processing activities involving radioactive sources. Appropriate radiation safety monitoring surveys insure compliance with issued radiation work permits, document workplace radiation exposure levels, and monitor for the presence of any radioactive contamination levels. Personnel radiation dosimeters are worn and exposures tracked usually on a monthly basis to insure radiation exposures are kept as low as reasonably achievable (ALARA). For workers who are identified as source handlers, additional direct reading electronic dosimeters are worn daily to track this potentially higher exposure work group. In order to keep exposures ALARA the RPS is loaded onto the spacecraft as late in the flow as possible. This usually means loading the power source at the launch pad.

During the radiological contingency planning period the size and scope is defined, and project millstones are established. Key exercise and rehearsal dates are established and detailed planning occurs. Specialized radiation monitoring equipment to measure air concentrations and ground deposition must be defined, procured, and tested during this period. Two to three weeks prior to launch a radiological contingency task force composed of radiation safety personnel from several Federal, state, and local agencies assemble for site-specific training and conduct appropriate level of exercises to validate readiness for launch support. A radiological contingency capability must address the following:

1. Notify appropriate organizations and agencies in the event of an accident involving potential release of radioactive material
2. Generate public information releases on the status of the launch accident that are accurate, timely, consistent, and easily understood
3. Assess whether a release of radioactive material has occurred
4. Quantify the magnitude and character, and predict the dispersion of any radioactive material released to the environment
5. Formulate appropriate and prudent protective actions to be taken onsite and provide recommendations for offsite agencies' consideration
6. Support smooth transition to Federal response model per National Response Framework
7. Address out-of-launch area accidents resulting in suborbital or orbital reentry of the spacecraft

A radiological control center is activated to coordinate radioactive material plume modeling, direct deployed ground monitoring assets, collect and analyze monitoring data, formulate and coordinate appropriate protective actions, and keep management, the media, and general public informed in a timely manner of the state of conditions, actions, and assessments that are underway.

Sufficiently in advance of any planned launch developed radiological contingency plans are briefed to various federal, state, and local officials to inform them and help allay any concerns about the specific launch. Briefings and readiness status to launch site management, and launch agency leadership is required to proceed toward launch. Finally, the White House Office of Science and Technology Policy are briefed and must grant launch approval.

In summary start planning early, develop thorough ground processing radiation safety support plans, insure development of comprehensive radiological contingency plans that are tested, and deploy an appropriately instrumented radiological contingency taskforce capable of making prompt and comprehensive radiological assessments and timely coordinated protective actions in order to protect workers and the general public.

FUTURE SPACEFLIGHT TOXIC HAZARDS

Future spaceflights will involve longer durations, the need for self-sufficiency and autonomy, and closed life support cycles. Also, effects of long exposure to the spaceflight environment, including microgravity and space radiation could generate unforeseen exposure hazards such as microbiological increased virulence and resistance already observed in spaceflight microbial payloads.

Project Apollo marked the first exposure of humans to a nonterrestrial material, lunar dust, and a potential exposure concern for future lunar explorers as well (Khan-Mayberry et al., 2011). In many respects lunar dust presents irritant and respiratory hazards similar to silicate dusts on Earth; however, in the reduced lunar gravity the deposition of inhaled dust in the lungs may follow different patterns then on earth (Darquenne and Prisk, 2008). Human travel to Mars and other natural bodies within the solar system will further expand the range of potentially toxic materials that must be considered in mission planning.

Future space missions will need to continuously monitor and respond to the long-term buildup of potentially high levels of chemical, physical, and biological contaminants. To relieve the human crews of such a burden, "smart" computer systems are being investigated to continuously monitor the spacecraft environment and automatically activate warnings and countermeasures if hazardous conditions occur.

REFERENCES

American Conference of Governmental Industrial Hygienists (ACGIH) (2014) *Threshold Limit Values (TLVs) for Chemical Substances and Physical Agents*, Cincinnati, OH: American Conference of Governmental Industrial Hygienists.

Bronstein, A. and Currance, P. (1994) *Emergency Care for Hazardous Materials Exposure*, St. Louis: Mosby Lifeline.

Caylor, G. (1995) Emergency response to nitrogen tetroxide incident at Johnson Space Center (JSC). *1995 Annual Meeting NASA Occupational Health Program*, Cape Canaveral, FL.

Clayton, G. and Clayton, F. (1994) *Patty's Industrial Hygiene and Toxicology*, New York, NY: John Wiley & Sons.

Coleman, M. and James, J. (1994) Airborne toxic hazards. In: Nicogossian, A., Huntoon, and Pool, S., editors. *Space Medicine and Physiology*, Philadelphia, PA: Lea & Febiger.

Darquenne, C. and Prisk, G. (2008) Deposition of inhaled particles in the human lung is more peripheral in lunar than in normal gravity. *Eur. J. Appl. Physiol.* 103:687–695.

Dejournette, R. (1977) Rocket propellant inhalation in the Apollo-Soyuz astronauts. *Radiology* 125:21–24.

Europen Space Agency (ESA) (1995) *European Space Agency Press Release 21–1995. Submission of enquiry board's provisional report on fatal accident at Guiana Space Center 19 May 1995.*

Ferster, W. (1996) Questions circulate about Long March death toll. *Space News* 7:28.

Hale, W., Lane, H., Chapline, G., and Lulla, K. editors. (2011) *Wings in Orbit: Scientific and Engineering Legacies of the Space Shuttle, 1971–2010*, 1st ed., NASA Johnson Space Center, NASA/SP-2010-3409.

James, J. and Coleman, M. (1994) Toxicology of airborne gaseous and particulate contaminants in space habitats. In: Nicogossian, A., Mohler, S.R., and Gazenko, O.G., editors. *Space Biology and Medicine*, Washington, DC: American Institute of Aeronautics and Astronautics.

Khan-Mayberry, N., James, J., Tyl, R., and Lam, C. (2011) Space toxicology: protecting human health during space operations. *Int. J. Toxicol.* 30:3–18.

Locke, J. (2008) Space Environments. In: Davis, J., Johnson, R., Stepanek, J., and Fogarty, J., editors. *Fundamentals of Aerospace Medicine*, 4th ed., Baltimore, MD: Lippincott Williams & Wilkins.

Marmaro, G.M., Cardinale, M.A., Summerfield, B.R., and Tipton, D.A. (1992) Kennedy Space Center environmental health program. *J. Fla. Med. Assoc.* 79:553–556.

Monti, R. and D'Angelo, L. (1993) An economic benefit deriving from space activities: integration between science institutions and industry. *Acta Astronautica* 29:743–748.

Morgan, C. (2001) *and Shuttle-Mir. NASA.* Available at http://history.nasa.gov/SP-4225/nasa4/nasa4.htm.

National Aeronautics and Space Administration (NASA) (1970) *Apollo 13 Review Commission.* Available at http://history.nasa.gov/ap13rb/ap13index.htm.

National Aeronautics and Space Administration (NASA) (1976) *Command module recorder transcript, Part 3. ASTP Gallery of Historical Documents. NASA History Program Office. NASA.* Available at http://history.nasa.gov/astp/gallery.html.

National Aeronautics and Space Administration (NASA) (2012) *Press Release 12-046, NASA Seeks Proposals for Green Propellant Technology Demonstrations,* Available at http://www.nasa.gov/home/hqnews/2012/aug/HQ_12-281_Green_Propellants.html (accessed 2014).

National Institute of Occupational Safety and Health (NIOSH) (1994) *Pocket Guide to Chemical Hazards,* Washington, DC: NIOSH: 94–116, US Department of Health and Human Services.

National Research Council (1994) *Spacecraft Maximum Allowable Concentrations for Selected Airborne Contaminants, Subcommittee on Spacecraft Maximum Allowable Concentrations,* Washington, DC: National Academies Press.

National Research Council (2004) *Spacecraft Water Exposure Guidelines for Selected Contaminants,* Washington, DC: National Academies Press.

Oberg, J. (1988) *Uncovering Soviet Disasters,* New York, N.Y.: Random House, pp. 156–176.

Oberg, J. (1990) Disaster at the Cosmodrome. *Air & Space Magazine,* December.

Oberg, J. (1998) Handling Mir misinformation. *Space News,* May 18–24.

Occupational Safety and Health Administration (OSHA) (1995) *1910.1000 Air Contaminants 29 Code of Federal Regulations,* Washington, DC: Occupational Safety and Health Administration.

Rippstein, W. and Coleman, M.E. (1983) Toxicological evaluation of the Columbia spacecraft. *Aviat. Space Environ. Med.* 54: S60–S67.

Scarpa, P. (2014) Kennedy Space Center Operations. In: Stepaniak P., editor. *Loss of Signal: Aeromedical Lessons Learned from the STS-107 Columbia Space Shuttle Mishap,* Washington, DC: US Government Printing Office, NASA/SP-2014-616, ISBN 978-0-1609-2387-6.

Talawar, M.B., Sivabalan, R., Mukundan, T., Muthurajan, H., Sikder, A.K., Gandhe, B.R., et al. (2009) Environmentally compatible next generation green energetic materials (GEMs). *J. Hazard Mater.* 161:589–607.

Toback, A. (1991) Space dermatology. *Cutis* 48:283–287.

Toxicology Office at United States National Aeronautics and Space Administration (NASA) (2013) Available at http://www.nasa.gov/centers/johnson/slsd/about/divisions/hefd/facilities/space-toxicology.html (accessed January 29, 2013).

Wilcut, T. and Harkins, W.B. (2011) *Tough Transitions: STS-1 Pre-Launch Accident.* Available at http://nsc.nasa.gov/SFCS/SystemFailureCaseStudy/Details/78 (accessed October 2, 2011).

121

REGULATORY TOXICOLOGY

David M. Polanic and Marek Banasik

BACKGROUND

Health and safety data are used for evaluating specific endpoints focused at identifying potential hazards to human health and the environment. These data are used in support of regulatory submissions and for identifying various types of potential hazards (e.g., explosivity, carcinogenicity), establishing occupational exposure limits, determining vulnerable receptors in the environment, and communicating hazards. The data requirements for regulatory submissions vary based on the type of substance being registered. For example, the requirements for industrial chemicals range from only those data in the submitter's possession to required testing based on the intended production volume.

Regardless of the jurisdiction in which a submission is made, there are two standard requirements for generating data, which include performing studies under good laboratory practice (GLP) standards and in accordance with validated testing guidelines. These requirements were established to ensure the quality, reliability, and integrity of the underlying data. Submitters and regulatory agencies may also use non-GLP, non-guideline studies for informing decision-making activities. For example, manufacturers may use this information for guiding research and development on alternative substances. On the other hand, regulatory agencies may use such information for evaluating regulatory submissions or making subsequent decisions on registered substances, which are already in the stream of commerce (e.g., bisphenol A).

The discussion that follows provides an overview of GLP standards and validated testing guidelines, followed by a discussion on various laws where these types of data are used, including chemical control laws, laws aimed at protecting workers, and hazard communication laws.

TESTING STANDARDS

GLP Standards

Good laboratory practice standards were first proposed by the United States (US) Environmental Protection Agency (EPA) in April 1980 and have since been adopted globally by inter-governmental agencies such as the Organisation for Economic Co-operation and Development (OECD) (EPA, 1983). On the OECD website, it states that the principles of GLP ". . . ensure the generation of high quality and reliable test data related to the safety of industrial chemical substances and preparations" (OECD, 2014a).

Historically, GLP standards were needed based on EPA's own experience, along with that of the US Food and Drug Administration (FDA), in dealing with the integrity of data submitted in support of the health and safety of regulated products and pesticides. Inspections by EPA and FDA of several laboratories led to the discovery of unacceptable practices that called into question the validity of submitted data. Some studies had been performed so poorly that the resulting data could not be relied upon in EPA's regulatory decision-making process (EPA, 1989a). Many studies did not adhere to specified protocols, were conducted by unqualified personnel, or were not adequately monitored by study sponsors. In some instances, results were selectively reported, underreported, or fraudulently reported. One of the most notable examples was the fraud committed by executives at

Hamilton & Hardy's Industrial Toxicology, Sixth Edition. Edited by Raymond D. Harbison, Marie M. Bourgeois, and Giffe T. Johnson.
© 2015 John Wiley & Sons, Inc. Published 2015 by John Wiley & Sons, Inc.

the Industrial Bio-Test (IBT) Laboratories (Northbrook, IL, USA) (United States v. Keplinger, 1985).

Industrial Bio-Test Laboratories was a contract research laboratory that conducted animal toxicology studies of various chemicals, drugs, and pesticides on behalf of the manufacturers in order to determine the safety and effectiveness of their products. Managers and executives of IBT Laboratories were convicted of various counts of fraud and making false statements, including suppression of relevant information and falsification of data, in connection with studies performed by IBT Laboratories. Specifically, the infractions included: (1) underreporting mortality of research animals and commingling of original and later-started research animals; (2) omission of histopathological data on research animals; (3) omission of data on post-mortem examinations; (4) failure to report the original opinion of a doctor that trichlorocarbanilide had a significant effect on research animals; and (5) failure to disclose a pathology report. Additionally, IBT Laboratories wholly fabricated blood and urine data pertaining to a study of an anti-inflammatory drug used to treat arthritis, as well as making false statements to the FDA in a written report.

The trichlorocarbanilide study performed by IBT Laboratories was plagued by an unusually high mortality rate, which resulted in the defendants commingling new animals, those that had not been fed the test material for as long as the original animals, with the existing test animals. This practice, although unacceptable because it compromises the validity of the study, was done to ensure that a sufficient number of animals survived to the end of the study so IBT Laboratories could claim a valid evaluation. At trial, the government provided evidence showing that in at least one test group the total number of animals used exceeded the number of animals that started the study by 30%. The government further proved that the defendants were aware of the commingling before the reports were submitted and, as a result, that the reports contained false statements.

After these actions, EPA needed rules to ensure the quality and integrity of data. In enacting the final rule, and keeping with its policy to reduce the burden on the regulated public, EPA found no legal or scientific reason for promulgating GLP standards that differ substantially from those of the FDA. The EPA's Toxic Substances Control Act (TSCA) GLP standards specify minimum requirements for conducting studies relating to health effects, environmental effects, and chemical fate testing to ensure the quality and integrity of data submitted under Section 4 of TSCA.

As part of the GLP standards, a testing facility is required, among other things, to permit an authorized employee of EPA or FDA to inspect the facility and all records and specimens regarding the study, or EPA will not consider the data developed by that facility as reliable. Additionally, a testing facility is required to set forth standard operating procedures, as well as specific protocol for each study being performed. Data collection, reporting, and record retention requirements are also outlined in the GLP standards to preserve the integrity of the studies performed (for discussion, see Huntsinger, 2008). Of particular note, GLPs require the retention of "raw" data, which are defined as: ". . . any laboratory worksheets, records, memoranda, notes, or exact copies thereof, that are the result of original observations and activities of a study and **are necessary for the reconstruction and evaluation of the report of that study** [emphasis added]." A failure to comply with the GLP standards may result in study rejection and possibly civil and criminal penalties (EPA, 1989a).

Compliance with GLP standards represents the first of the two major requirements, which are placed on manufacturers, who are conducting health and safety studies as part of a regulatory submission for a new chemical or as part of a testing rule issued by a regulatory agency. The second requirement consists of following the validated testing guidelines. These guidelines play a critical role in product registration testing and regulatory decision-making because they have been designed to ensure that the endpoints being evaluated (e.g., developmental neurotoxicity) are accurately assessed during the experimentation (Makris et al., 2009). In the following subsection, we discuss the available test guidelines and the process that a study design must undergo to eventually become a validated test guideline.

Validated Testing Guidelines

Validated testing guidelines provide requisite information for designing, conducting, interpreting, and reporting the results of studies. In combination with GLPs, validated testing guidelines are intended to ensure that study data are of the highest quality and reliability possible. Testing guidelines go through a rigorous validation process, which includes supporting data, commenting rounds, independent replication of results from multiple laboratories, workshop consultations, etc. Several entities issue validated testing guidelines; however, those issued by the OECD often serve as the basis for organization-specific test guidelines. For example, both the EPA's *90-Day Oral Toxicity in Rodents* test guideline and the EU's *Repeated Dose 90-Day Oral Toxicity Study in Rodents* test guideline are based on the OECD's *Repeated Dose 90-Day Oral Toxicity Study in Rodents* test guideline (EC, 2001; EPA, 1998).

Validated testing guidelines have been issued on physicochemical properties (e.g., density of liquids/solids and vapor pressure), effects on biotic systems (e.g., fish sexual development), degradation and accumulation (e.g., phototransformation of chemicals in water and bioaccumulation in fish), and health effects (e.g., skin sensitization and developmental neurotoxicity) (EPA, 2013; OECD, 2014b). Additional guidelines are available depending on the substance under evaluation (e.g., spray drift test guidelines for pesticides) (EPA, 2013).

As indicated in the foregoing discussions on GLP standards and validated test guidelines, the stringent standards placed on manufacturers and importers for testing their products are to ensure that informed decisions about the potential hazards of a chemical are realized before being placed on the market. However, manufacturers and importers are not the only sectors that perform human health or environmental effects testing on chemicals. A large portion of research on chemicals is performed in academic laboratories, which are not required to perform studies in compliance with GLP standards or in accordance with validated test guidelines. As noted earlier, GLP standards came about in part because some companies were fabricating research data, which was ultimately used in support of product registrations. However, research misconduct (e.g., data fabrication) permeates all sectors of the research community, including academic laboratories (PubMed, 2014).

Given that research on industrial chemicals is conducted by all sectors of the scientific community, it is imperative that an assessment of the quality of data from such studies be assessed. The discussion that follows provides a brief background on the differences between data generated by manufacturers and importers of chemicals versus academic and government laboratories.

Data Quality

Good laboratory practices/guideline data submissions are the primary source of experimental information from which regulatory decisions are made on a particular chemical, although basic exploratory research data from academic or government research laboratories are sometimes available to inform regulatory decision making. There are, however, several caveats with using exploratory research data, because they are typically not conducted under GLP or in accordance with a validated guideline. Rather, these data are typically reported as published summaries in the peer-reviewed literature, but the underlying raw data are difficult or sometimes impossible to obtain from the investigators (EPA, 2008a–d). Therefore, assessing the quality and reliability of basic exploratory studies is a complicated but essential requirement for the regulatory risk assessment process.

Disputes often arise between manufacturers who perform GLP/guideline studies and researchers who publish the results of exploratory studies and report hazards that were not identified in the GLP/guideline studies (Tyl, 2009). The crux of such disputes is typically focused on data quality and gaining access to the underlying "raw" data. For example, OECD's validated test guideline for a *Developmental Neurotoxicity Study* recommends using the rat as the test species with a total of 20 litters for each dose level (OECD, 2007). It states further that "[t]he statistical unit of measure should be the litter (or dam) and not the pup." However, several exploratory studies on decabromodiphenyl ether (Chemical Abstracts Service Registry Number: 1163-19-5), a brominated flame retardant, reported that this compound caused developmental neurotoxicity in mice and rats (Viberg et al., 2003, 2007). The study authors utilized from three to five litters of mice or rats at each dose level, and used the pup as the statistical unit of measure, an approach that increases the likelihood of obtaining false positives (Holson et al., 2008; Moser et al., 2005). Based on these data, the manufacturers of decabromodiphenyl ether were required to perform a GLP/guideline developmental neurotoxicity study (Biesemeier et al., 2011); no evidence of developmental neurotoxicity was identified; however, regulatory agencies initiated several measures that ultimately led to the market phase out of this substance.

When regulatory agencies utilize disputed exploratory studies for decision-making activities, administrative/legal challenges and congressional inquiries often ensue (Vitter, 2014). The standard argument proffered by impacted stakeholders is that regulatory agencies require manufacturers to follow stringent testing standards (i.e., GLP standards and validated testing guidelines) and to provide the underlying "raw" data for their studies. In contrast, the same regulatory agencies may use exploratory studies as a basis for initiating rule-making or other market deselection activities, even when the underlying "raw" data are unavailable (EPA, 2008a–d; Vitter, 2014). The US Senate's Committee on Environment and Public Works chastised EPA for this type of ". . . double standard . . ." and likened it ". . . to somebody being criminally prosecuted based on data that the defense is not allowed to see" (Vitter, 2014).

Several regulatory agencies have issued formal guidance for evaluating the quality, reliability, and integrity of data (ECHA, 2011; EPA, 2003, 2012). The European Chemicals Agency's (ECHA's) guidance is based on the approach proposed by Klimisch et al. (1997), which described the following numerical codes and descriptive categories for evaluating the quality and reliability of studies:

Reliable without Restrictions. This includes studies or data from the literature or reports which were carried out or generated according to generally valid and/or internationally accepted testing guidelines (preferably performed according to GLP) or in which the test parameters documented are based on a specific (national) testing guideline (preferably according to GLP) or in which all parameters described are closely related/comparable to a guideline method.

Reliable with Restrictions. This includes studies or data from the literature, reports (mostly not performed according to GLP), in which the test parameters documented do not totally comply with the specific testing guideline, but are sufficient to accept the data or in which investigations are described which cannot be subsumed under a testing guideline, but which are nevertheless well documented and scientifically acceptable.

Not Reliable. This includes studies or data from the literature/reports in which there are interferences between the measuring system and the test substance or in which organisms/test systems were used which are not relevant in relation to the exposure (e.g., unphysiologic pathways of application) or which were carried out or generated according to a method which is not acceptable, the documentation of which is not sufficient for an assessment and which is not convincing for an expert judgment.

Not Assignable. This includes studies or data from the literature, which do not give sufficient experimental details and which are only listed in short abstracts or secondary literature (books, reviews, etc.).

The 'Klimisch codes,' as they have since become known, are similar to the tiered approach used by the EPA for evaluating data under the High Production Volume (HPV) Challenge Program, and the OECD proposed their use in the OECD HPV program, along with the EPA's tiered approach. The ECHA recognized the importance of establishing a formal system for evaluating data quality and noted that "[t]he knowledge of how a study was carried out and consequently its relevance and reliability, is a prerequisite for the subsequent evaluation [and use] of information" (ECHA, 2011).

INDUSTRIAL CHEMICAL REGULATIONS

Regardless of the data source(s) used, regulatory agencies are tasked with administering laws that provide specific frameworks for evaluating new and existing substances. These are typically country or region specific and have different data requirements for new and existing substances. For example, in the US, manufacturers are only required to submit health and safety data in their possession at the time of filing a pre-manufacture notice (PMN) under the TSCA, whereas in European Union (EU), specific tests are mandated based on the intended production volume under the Registration, Evaluation, Authorisation and Restriction of Chemicals (REACH) law. Unlike the US, however, REACH requires manufacturers to submit regulatory submissions for new and existing substances. A general overview of TSCA and REACH is provided in the following section.

TSCA

After finding that human beings and the environment are exposed to large numbers of chemical substances and mixtures each year (15 USC § 2601 et seq.), Congress-enacted TSCA with the purpose of preventing "unreasonable risks of injury to health or the environment associated with the manufacture, processing, distribution in commerce, use, or disposal of chemical substances" (S. Rep., 1976). The TSCA defines a "chemical substance" as ". . . any organic or inorganic substance of a particular molecular identity . . . ," which includes ". . . (i) any combination of such substances occurring in whole or in part as a result of a chemical reaction or occurring in nature and (ii) any element or uncombined radical" (15 USC § 2601 et seq). However, "chemical substance" does not include, any mixture; any pesticide; tobacco or any tobacco product; any source material, special nuclear material, or byproduct material; any article, the sale of which is subject to the tax imposed by the Section 4181 of the Internal Revenue Code of 1954 or any other provision of such Code; and any food, food additive, drug, cosmetic, or device (15 USC § 2601 et seq). The regulation of chemicals under TSCA is based on the chemicals status as an existing substance or a new substance.

The TSCA defines an existing chemical substance as any chemical substance that is included in the chemical substance list compiled and published under Section 8(b) of TSCA, which includes those chemicals that were "grandfathered" in when TSCA was enacted (15 USC § 2601 et seq). Any individual who intends to manufacture or process a chemical published on this list does not need to notify EPA, so long as the use of the chemical does not constitute a significant new use. A determination by EPA that a use qualifies as a significant new use must be made by a rule promulgated after considering all relevant factors. The relevant factors include: (1) the projected volume or manufacturing and processing; (2) the extent the use changes the type or form of exposure of humans or the environment to a chemical substance; (3) the extent the use increases the magnitude and duration of exposure of humans and the environment to the chemical substance; and (4) the reasonably anticipated manner and methods of manufacture, processing, distribution, and disposal of a chemical substance. If a significant new use rule has been issued, any individual who wishes to manufacture or process for that use must submit a significant new use notice to EPA (40 CFR § 721.5).

The TSCA defines a "new chemical substance" as ". . . any chemical substance which is not included in the chemical substance list compiled and published under section 8(b)" of TSCA. Prior to manufacturing or importing, these new chemical substances are subject to specific notification requirements. However, exemptions to the notification requirements include, a showing that the substance will not present any unreasonable risk of injury to the health or the environment, or if the chemical is to be manufactured or processed only in small quantities. Any person who intends to manufacture or import a new chemical substance, which is neither exempt nor on the inventory list of existing chemicals, must submit a PMN to EPA at least 90 days prior to manufacturing.

Once EPA receives a PMN, it goes through a series of steps to make a determination of whether the substance may present an unreasonable risk to human health or the environment. The first step includes a review of the PMN to ensure that all of the required information is included. If the PMN is complete, EPA convenes a meeting known as the Chemical Review and Search Strategy (CRSS), which focuses on confirming the identity of the notified substance, evaluating the industrial synthesis process to identify by-products or impurities, evaluate or estimate physicochemical properties, and to understand the intended function and use of the substance.

After the CRSS meeting, EPA's Structure-Activity Team (SAT) performs an initial assessment of the substance's potential human health and environmental hazards. The SAT members utilize any data submitted with the PMN substance, data from previously submitted substances that may serve as analogs for informing potential hazards of the substance under review, modeling, and expert judgment to make the initial hazard calls. Thereafter, an Exposure Analysis Meeting (EXAM) is held, where the nature and magnitude of occupational exposures, environmental releases, consumer and general population exposures, and environmental fate are assessed. As with the SAT meeting, the EXAM members utilize data submitted with the PMN substance, data from previously submitted substances, modeling, and expert judgment to inform these areas.

Once the members of CRSS, SAT, and EXAM have completed their initial review, an initial risk assessment/risk management meeting is held, which is called the Focus Meeting. During the Focus Meeting, the overall potential risk of the PMN substance is discussed and a determination is made that either no further review is needed or that the substance requires a more in-depth review in a process known as Standard Review. As part of the Standard Review, EPA's scientists develop recommendations on the potential hazards and exposures that may present an unreasonable risk. These recommendations are presented at a Disposition Meeting where a decision is made to require further testing on the PMN substance (e.g., production volume based triggers or a ban pending upfront testing) or potentially to prohibit the manufacture of the PMN substance.

Once a chemical is approved under the PMN process, it is added to the TSCA inventory and is then considered an existing chemical substance. Unlike EPA's authority to regulate new chemical substances, once a chemical is designated as an existing chemical substance, EPA's options for subsequent regulatory activity are limited. For example, under TSCA Section 4, EPA must make a statutory finding of "data insufficiency" and "testing is necessary" in order to require manufacturers, by rule, to perform hazard or exposure testing. Test rules are, however, time consuming and often take years to prepare and finalize, with the subsequent testing taking additional months or years to complete before EPA obtains the required data.

Under TSCA Section 6, if EPA finds a "reasonable basis to conclude" that the manufacture or importation of a chemical substance presents or will present an unreasonable risk of injury to health or the environment, EPA shall take one or more of the following required actions: (1) prohibiting or limiting the amount of the chemical substance to be manufactured, processed, or distributed in commerce; (2) prohibiting or limiting the manufacture, processing, or distribution in commerce for a particular use or a particular use in excess of a level specified by EPA; (3) requiring that such substance be marked with or accompanied by clear warnings and instructions; (4) requires manufacturers and processors to make and retain records of the processes used to manufacture such substance; (5) prohibiting or otherwise regulating any manner or method of commercial use; (6) prohibiting or otherwise regulating any manner or method of disposal of such chemical; (7) directing manufacturers or processors to give notice of unreasonable risk of injury to distributors in commerce (15 USC § 2601 et seq). The EPA previously attempted to issue a TSCA Section 6 rule that would have effectively banned asbestos; however, on appeal, the Appellate Court overturned EPA's proposed rule for failing to consider a number of factors, including, for example, the benefits of asbestos, alternatives, and least burdensome measures (Corrosion Proof Fittings v. EPA, 1991).

The limitations on EPA's ability to regulate existing chemical substances have formed the basis for TSCA reform initiatives. Many proponents of TSCA reform favor the regulatory paradigm enforced in the EU under REACH. As discussed in the following section, REACH offers a clear advantage with regulating existing substances because it captures substances, which were grandfathered in under EU's previous chemical control law. However, it also has limitations, primarily of which, REACH does not require the submission of full study reports in support of regulatory dossiers. Rather, it requires the submission of robust data summaries, which lack the underlying "raw" data.

REACH

In December 2006, the EU enacted a new chemical control law known as the REACH (REACH, 2006). The REACH places the responsibility of assessing risks on companies that produce, import, and use chemicals, and it requires companies to take the necessary risk management measures when appropriate. In addition to regulating new chemicals (i.e., non-phase-in substances), REACH is unique in that it also requires retrospective safety evaluation of existing chemicals (i.e., phase-in substances). A phase-in system lasting up to 11 years is planned for the approximate 150,000 phase-in substances, which were pre-registered with the ECHA between June 1, 2007 and December 1, 2008 (ECHA, 2008a). Higher tonnage phase-in substances [≥1,000 tons per annum (t/a)], as well as lower tonnage phase-in substances that are very toxic to the aquatic environment (≥100 t/a) or classified as CMRs (i.e., carcinogens, mutagens, or reproductive toxicants; ≥1 t/a), must be registered before December 1, 2010. All other phase-in substances manufactured or imported in quantities $\geq 100\,t/a$ or $\geq 1\,t/a$ must be registered by June 1, 2013, or June 1, 2018, respectively (REACH, 2006). Non-phase-in substances will be evaluated on an ongoing basis and must be registered before being manufactured or imported into the EU.

The REACH employs a tiered-testing approach, meaning that the data required are dependent on the tonnage manufactured or imported into the EU, so that a higher production volume means that more information is needed. A technical dossier is required for all registered chemicals; however, only chemicals with a production volume $\geq 10\,t/a$ require a chemical safety assessment documented in a chemical safety report (REACH, 2006). A chemical safety report is a risk assessment, which must follow the general provisions of Annex I of REACH, as well as the ECHA's guidance for preparing a chemical safety report (ECHA, 2008b). No new toxicological information is required for substances between 1 and 10 t/a, unless they are considered a priority substance (e.g., CMR).

The REACH requires companies that produce or import the same substance to participate in a Substance Information Exchange Forum (REACH, 2006). Substance Information Exchange Fora are intended to facilitate data exchange, avoid duplication of studies, and reduce the number of studies on experimental animals. The basic data requirements for REACH are provided under Column 1 of the REACH Annexes, labeled "Standard Information Required," for substances produced or imported in quantities of $\geq 1\,t/a$ (Annex VII), $\geq 10\,t/a$ (Annex VIII), $\geq 100\,t/a$ (Annex IX), and $\geq 1,000\,t/a$ (Annex X). Column 2 provides specific rules for interpretation of Column 1 for when a particular study may be omitted, replaced, or performed at a later stage. For each of these data requirements, Article 10 of REACH only requires the submission of robust data summaries, which as noted previously, may present issues with evaluating the underlying quality and reliability of the submitted information.

WORKER PROTECTION LAWS

US Occupational Safety and Health Administration (OSHA)

In the US, the OSHA Act of 1970 (29 USC § 651–678) was created as a result of Congress' findings that ". . . personal injuries and illnesses arising out of work situations impose a substantial burden upon, and are a hindrance to, interstate commerce . . . ." In defining the duties of employers and employees, the OSHA Act requires employers to: (1) provide each employee with a place of employment free from recognized hazards that cause or are likely to cause death or serious physical harm; and (2) comply with occupational safety and health standards promulgated

under the OSHA Act. In turn, each employee is required to comply with the rules, regulations, and orders issued under the Act, which are applicable to the employee's own actions.

The Act also delegates broad authority to the Secretary of Labor (Secretary) to promulgate "standards to ensure safe and healthful working conditions for the Nation's workers." ". . . [O]ccupational safety and health standard . . ." is defined as a standard, which is ". . . reasonably necessary or appropriate to provide safe or healthful employment and places of employment" (Industrial Union Dept. v. American Petrol. Inst., 1980). In promulgating standards dealing with toxic materials or harmful physical agents, the Secretary shall, using the best available evidence and to the extent feasible, set standards, which most adequately assure that no employee's health or functional capacity will be materially impaired, even if employees have regular exposure to the hazard [29 USC § 655(b)(5)]. Development of standards shall be based upon research, experiments, demonstrations, and other information as appropriate. When developing standards, in addition to attaining the highest degree of health and safety protection for the employees, other considerations shall be the latest available scientific data in the field, the feasibility of the standards, and experience gained under this and other health and safety laws.

Pursuant to this granted authority, the Secretary has promulgated standards regulating the exposure limits to hazardous chemicals, known as permissible exposure limits (PELs) (29 CFR § 1910.1000). Permissible exposure limits limit the acceptable level of an employee's exposure to a particular substance. The levels for each substance are defined as an 8-h time-weighted average (29 CFR § 1910.1028), with some chemicals also having a ceiling concentration [29 CFR § 1910.1000(a)(1)] or short-term exposure limit, which are measured as an average over any 15-min period [29 CFR § 1910.1028(C)(2)].

Because PELs go through a time-consuming formal rule making process, many of the existing values may not represent the current state of the science for potential workplace hazards from chemicals. In addition, PELs do not exist for most chemicals in the workplace. Therefore, OSHA may recommend other non-binding values, such as the US National Institute for Occupational Safety and Health's (NIOSH's) recommended exposure limits or the American Conference of Governmental Industrial Hygienists' threshold limit values, as alternatives to PELs for ensuring protection of worker health (OSHA, 2014).

European Agency for Safety and Health at Work

In Europe, the European Agency for Safety and Health at Work was established by Council Regulation (EC) No 2062/94 with the aim of providing ". . . Community bodies, the Member States, the social partners and those involved in the field with the technical, scientific and economic information of use in the field of safety and health at work" (EC, 1994). European Agency for Safety and Health at Work disseminates information, including occupational exposure limits (OELs), developed by the Scientific Committee on Occupational Exposure Limits (SCOEL). The SCOEL was given formal status in 1995 and consists of 21 members from all Member States, who are tasked with evaluating risks in the workplace and developing harmonized OELs (EC, 1995). Once the SCOEL develops a recommended OEL, the Advisory Committee on Health and Safety at Work makes a feasibility determination on the recommended OEL.

For chemical substances with a clear threshold for toxicological effects, indicative OELs are established, which Member States may adopt or establish a higher or lower OEL at the national level. In contrast, binding OELs must be implemented at the established level or at a lower national level by Member States (Schenk, 2009). Under REACH, manufacturers develop worker exposure limits known as derived no effect levels for specific chemical substances; Member States may use these values at the national level in lieu of indicative OELs or binding OELs when they are lower.

HAZARD COMMUNICATION

US Hazard Communication Standards (HCS)

In addition to developing PELs, OSHA initially promulgated a HCS on November 25, 1983 to further protect employees from workplace hazards (OSHA, 1983). This standard requires chemical manufacturers and importers to "assess the hazards of chemicals, which they produce or import[.]" Originally, all employers in the manufacturing division (Standard Industrial Classification codes 20 through 39) were required to provide information to their employees concerning hazardous chemicals by way of labels, material safety data sheets, training, and access to written records. However, in 1987 OSHA revised its HCS to cover all employers with employees exposed to hazardous chemicals in their workplaces; thus expanding the scope to non-manufacturing employers as well (OSHA, 1987).

In March, 2012, OSHA modified its HCS to conform to the United Nations' Globally Harmonized System of Classification and Labelling of Chemicals (GHS) (OSHA, 2012; UNECE, 2011). This modification was made to enhance the effectiveness of the HCS in ensuring that employees are apprised of the chemical hazards to which they may be exposed, and in reducing the incidence of chemical-related occupational illnesses and injuries. This update also intends to facilitate global harmonization of standards, and includes revised criteria for classification of chemical hazards; revised labeling provisions that include

requirements for use of standardized signal words, pictograms, hazard statements, and precautionary statements; a specified format for safety data sheets; and related revisions to definitions of terms used in the standard, and requirements for employee training on labels and safety data sheets.

The EPA also has a HCS, which it adopted from OSHA (EPA, 1989b). At that time, EPA did not adopt certain sections of OSHA's HCS because they either were covered in other sections of EPA rules or were not applicable. However, as a result of changes to OSHA's HCS and changes to NIOSH respirator requirements, it is likely that EPA will initiate rule making activities to make several changes to its significant new use rule regulations under 40 CFR (Code of Federal Regulations) subparts A, B, and C. The standard designations in the sections titled "Protection in the Workplace" (40 CFR § 721.63) and "Hazard Communication Program" (40 CFR § 721.72) were modeled on OSHA regulations that were in force at the time EPA's rule was issued in 1989.

EU HCS

In 2008, the European Commission adopted parts of GHS into its classification, labeling and packaging of substances and mixtures (CLP) regulation (EC, 2008). With the exception of transport, the CLP governs hazard communication within the supply and use sectors. Mandatory compliance began on December 1, 2010 for individual substances and will begin on June 1, 2015 for mixtures. As with OSHA's HCS, the CLP sets forth specific criteria for the classification and labeling of substances based on physicochemical properties and human health hazards. However, unlike OSHA's HCS, the CLP established classification categories for aquatic toxicity, which include information on environmental fate (e.g., degradability and bioaccumulation) for classifying chronic aquatic toxicity. As part of regulatory submissions under REACH, manufacturers must classify and label their substances according to the CLP requirements at the time of submission.

REFERENCES

15 USC § 2601 et seq. *Subchapter I–Control of Toxic Substances. Chapter 53-Toxic Substances Control.* Available at http://www.gpo.gov/fdsys/pkg/USCODE-2011-title15/pdf/USCODE-2011-title15-chap53.pdf.

29 CFR § 1910.1000. *Air contaminants. Subpart Z-Toxic and Hazardous Substances.* Washington, DC: Occupational Safety and Health Administration, US Department of Labor. Available at http://www.gpo.gov/fdsys/pkg/CFR-2009-title29-vol6/pdf/CFR-2009-title29-vol6-sec1910-1000.pdf.

29 CFR § 1910.1000(a)(1). *Substances with limits preceded by "C"–Ceiling Values. Air contaminants.* Washington, DC: Occupational Safety and Health Administration, US Department of Labor.

Available at http://www.gpo.gov/fdsys/pkg/CFR-2009-title29-vol6/pdf/CFR-2009-title29-vol6-sec1910-1000.pdf.

29 CFR § 1910.1028. *Benzene. Subpart Z-Toxic and Hazardous Substances*. Washington, DC: Occupational Safety and Health Administration, US Department of Labor. Available at http://www.gpo.gov/fdsys/pkg/CFR-2013-title29-vol6/pdf/CFR-2013-title29-vol6-sec1910-1028.pdf.

29 CFR § 1910.1028(c)(2). *Short-term exposure limit (STEL). Benzene*. Washington, DC: Occupational Safety and Health Administration, US Department of Labor. Available at http://www.gpo.gov/fdsys/pkg/CFR-2013-title29-vol6/pdf/CFR-2013-title29-vol6-sec1910-1028.pdf.

29 USC § 651–678. *Chapter 15 – Occupational Safety and Health*. Available at http://www.gpo.gov/fdsys/pkg/USCODE-2011-title29/pdf/USCODE-2011-title29-chap15.pdf.

29 USC § 655(b)(5). *Procedure for Promulgation, Modification, or Revocation of Standards. Standards. Chapter 15 – Occupational Safety and Health*. Available at http://www.gpo.gov/fdsys/pkg/USCODE-2011-title29/pdf/USCODE-2011-title29-chap15.pdf.

40 CFR § 721.5. *Persons Who Must Report. Subpart A-General Provisions*. Washington, DC: US Environmental Protection Agency. Available at http://www.gpo.gov/fdsys/pkg/CFR-2013-title40-vol32/pdf/CFR-2013-title40-vol32-sec721-5.pdf.

40 CFR § 721.63. *Protection in the Workplace. Subpart B-Certain Significant New Uses*. Washington, DC: US Environmental Protection Agency. Available at http://www.gpo.gov/fdsys/pkg/CFR-2011-title40-vol31/pdf/CFR-2011-title40-vol31-sec721-63.pdf.

40 CFR § 721.72. *Hazard Communication Program. Subpart B-Certain Significant New Uses*. Washington, DC: US Environmental Protection Agency. Available at http://www.gpo.gov/fdsys/pkg/CFR-2011-title40-vol31/pdf/CFR-2011-title40-vol31-sec721-63.pdf.

Biesemeier, J.A., Beck, M.J., Silberberg, H., Myers, N.R., Ariano, J.M., Radovsky, A., Freshwater, L., Sved, D.W., Jacobi, S., Stump, D.G., Hardy, M.L., and Stedeford, T. (2011) An oral developmental neurotoxicity study of decabromodiphenyl ether (DecaBDE) in rats. *Birth Defects Res. B Dev. Reprod. Toxicol.* 92:17–35.

Corrosion Proof Fittings v. EPA (1991) 947 F.2d 1201 (5th Cir.). Available at https://law.resource.org/pub/us/case/reporter/F2/947/947.F2d.1201.89-4596.html.

EC (1994) Council Regulation (EC) No 2062/94 of 18 July 1994 Establishing a European Agency for Safety and Health at Work. OJ L 216. Available at http://www.europarl.europa.eu/document/activities/cont/201004/20100412ATT72536/20100412ATT72536EN.pdf.

EC (1995) 95/320/EC: Commission Decision of 12 July 1995 Setting Up a Scientific Committee for Occupational Exposure Limits to Chemical Agents. OJ L 188:14–15. Available at http://eur-lex.europa.eu/LexUriServ/LexUriServ.do?uri=CELEX:31995D0320:EN:HTML.

EC (2001) Annex 5D: B.26. Subchronic Oral Toxicity Test, Repeated Dose 90-day Oral Toxicity Study in Rodents. Commission Directive 2001/59/EC of 6 August 2001 Adapting to Technical Progress for the 28th time Council Directive 67/548/EEC on the Approximation of the Laws, Regulations and Administrative Provisions Relating to the Classification, Packaging and Labelling of Dangerous Substances (Text with EEA relevance). OJ L 225:150–156. Available at http://eur-lex.europa.eu/LexUriServ/LexUriServ.do?uri=OJ:L:2001:225:0001:0333:EN:PDF.

EC (2008) Regulation (EC) No 1272/2008 of the European Parliament and of the Council of 16 December 2008 on Classification, Labelling and Packaging of Substances and Mixtures, Amending and Repealing Directives 67/548/EEC and 1999/45/EC, and Amending Regulation (EC) No 1907/2006 (Text with EEA relevance). OJ L 353. Available at http://eur-lex.europa.eu/LexUriServ/LexUriServ.do?uri=OJ:L:2008:353:0001:1355:EN:PDF.

ECHA (2008a) *Press Release: List of Pre-Registered Substances Published*. ECHA/PR/08/59. Helsinki, Finland: European Chemicals Agency. Available at http://echa.europa.eu/documents/10162/13585/pr_08_59_publication_pre-registered_substances_list_20081219_en.pdf.

ECHA (2008b) *Guidance on Information Requirements and Chemical Safety Assessment, Part F: Chemical Safety Report (ver. 2). Guidance for the Implementation of REACH*. Helsinki, Finland: European Chemicals Agency. Available at http://echa.europa.eu/documents/10162/13632/information_requirements_part_f_en.pdf.

ECHA (2011) *Guidance on Information Requirements and Chemical Safety Assessment; Chapter R.4: Evaluation of Available Information (version 1.1)*. ECHA-2011-G-13-EN. Helsinki, Finland: European Chemicals Agency. Available at http://echa.europa.eu/documents/10162/13643/information_requirements_r4_en.pdf.

EPA (1983) *Good Laboratory Practice Standards. Pesticide Programs*. Washington, DC: US Environmental Protection Agency. 48 FR 53946 (codified at 40 CFR, Part 160).

EPA (1989a) *Good Laboratory Practice Standards. Toxic Substances Control Act (TSCA)*. Washington, DC: US Environmental Protection Agency. 54 FR 34043 (codified at 40 CFR, Part 792). Available at http://www.gpo.gov/fdsys/pkg/CFR-2011-title40-vol32/xml/CFR-2011-title40-vol32-part792.xml.

EPA (1989b) *Significant New Use Rules; General Provisions for New Chemical Follow-Up*. Washington, DC: US Environmental Protection Agency. 54 FR (143):31298–31317 (codified at 40 CFR, Part 721). Available at http://www.epa.gov/oppt/newchems/pubs/snur.pdf.

EPA (1998) *Health Effects Test Guidelines: OPPTS 870.3100: 90-day Oral Toxicity in Rodents*. EPA 712-C-98-199. Washington, DC: Office of Pollution Prevention and Toxics, US Environmental Protection Agency. Available at http://www.regulations.gov/#!documentDetail;D=EPA-HQ-OPPT-2009-0156-0010.

EPA (2003) *A Summary of General Assessment Factors for Evaluating the Quality of Scientific and Technical Information*. EPA 100/B-03/001. Washington, DC: Assessment Factors Workgroup, Science Policy Council, US Environmental Protection Agency. Available at http://www.epa.gov/stpc/pdfs/assess2.pdf.

EPA (2008a) *Toxicological Review of Decabromodiphenyl Ether (BDE-209) (CAS No. 1163-19-5), in Support of Summary*

Information on the Integrated Risk Information System (IRIS). EPA/635/R-07/008F, at p. 32, n. 1. ("Attempts to obtain numerical values and other information on the data from the authors were not successful"). Washington, DC: US Environmental Protection Agency. Available at http://www.epa.gov/iris/toxreviews/0035tr .pdf.

EPA (2008b) *Toxicological Review of 2,2′,4,4′,5,5′-hexabromodi-phenyl Ether (BDE-153) (CAS No. 68631-49-2), in Support of Summary Information on the Integrated Risk Information System (IRIS)*. EPA/635/R-07/007F, at p. 22, n. 1. ("Attempts to obtain numerical values and other information on the data from the neurobehavioral studies were not successful"). Washington, DC: US Environmental Protection Agency. Available at http://www .epa.gov/iris/toxreviews/1009tr.pdf.

EPA (2008c) *Toxicological Review of 2,2′,4,4′,5-pentabromodi-phenyl Ether (BDE-99) (CAS No. 60348-60-9), in Support of Summary Information on the Integrated Risk Information System (IRIS)*. EPA/635/R-07/006F, at pp. 30–31, n. 1. ("Attempts to obtain numerical values and other information on the data from the neurobehavioral studies were not successful"). Washington, DC: US Environmental Protection Agency. Available at http:// www.epa.gov/iris/toxreviews/1008tr.pdf.

EPA (2008d) *Toxicological Review of 2,2′,4,4′-tetrabromodiphenyl Ether (BDE-47) (CAS No. 5436-43-1), in Support of Summary Information on the Integrated Risk Information System (IRIS)*. EPA/635/R-07/005F, at p. 29, n. 1. ("Attempts to obtain numerical values and other information on the data were not successful"). Washington, DC: US Environmental Protection Agency. Available at http://www.epa.gov/iris/toxreviews/1010tr.pdf.

EPA (2012) *Guidance for Evaluating and Documenting the Quality of Existing Scientific and Technical Information*. Addendum to: *A Summary of General Assessment Factors for Evaluating the Quality of Scientific and Technical Information*. Washington, DC: Peer Review Advisory Group, Science and Technology Policy Council, US Environmental Protection Agency. Available at http://www.epa.gov/stpc/pdfs/assess3.pdf.

EPA (2013) *Harmonized Test Guidelines*. Washington, DC: Office of Chemical Safety and Pollution Prevention, US Environmental Protection Agency. Available at http://www.epa.gov/ocspp/ pubs/frs/home/guidelin.htm.

Holson, R.R., Freshwater, L., Maurissen, J.P., Moser, V.C., and Phang, W. (2008) Statistical issues and techniques appropriate for developmental neurotoxicity testing: a report from the ILSI Research Foundation/Risk Science Institute expert working group on neurodevelopmental endpoints. *Neurotoxicol. Teratol.* 30:326–348.

Huntsinger, D.W. (2008) OECD and USA GLP applications. *Ann. Ist. Super. Sanita* 44:403–406.

Industrial Union Dept. v. American Petrol. Inst. (1980) 448 U.S. 607. Available at http://caselaw.lp.findlaw.com/scripts/getcase.pl? court=US&vol=448&invol=607.

Klimisch, H.-J., Andreae, M., and Tillmann, U. (1997) A systematic approach for evaluating the quality of experimental toxicological and ecotoxicological data. *Regul. Toxicol. Pharmacol.* 25:1–5.

Makris, S.L., Raffaele, K., Allen, S., Bowers, W.J., Hass, U., Alleva, E., Calamandrei, G., Sheets, L., Amcoff, P., Delrue, N., and

Crofton, K.M. (2009). A retrospective performance assessment of the developmental neurotoxicity study in support of OECD test guideline 426. *Environ. Health Perspect.* 117:17–25. Available at http://www.ncbi.nlm.nih.gov/pmc/articles/PMC2627860.

Moser, V.C., Walls, I., and Zoetis, T. (2005) Direct dosing of preweaning rodents in toxicity testing and research: deliberations of an ILSI RSI Expert Working Group. *Int. J. Toxicol.* 24:87–94.

OECD (2007) Developmental neurotoxicity study. In: *OECD Guidelines for the Testing of Chemicals*. Test No. 426. Paris Cedex, France: Organisation for Economic Co-operation and Development. Available at http://www.oecd-ilibrary.org/ environment/test-no-426-developmental-neurotoxicity-study_ 9789264067394-en.

OECD (2014a) Good Laboratory Practice (GLP). In: *Testing of Chemicals*. Paris Cedex, France: Organisation for Economic Co-operation and Development. Available at http://www.oecd.org/ chemicalsafety/testing/goodlaboratorypracticeglp.htm.

OECD (2014b) *OECD Guidelines for the Testing of Chemicals*. Paris Cedex, France: Organisation for Economic Co-operation and Development. Available at http://www.oecd.org/env/ehs/ testing/oecdguidelinesforthetestingofchemicals.htm#Test_ Guidelines.

OSHA (1983) *Hazard Communication*. 48 FR 53280 (codified at 29 CFR, Part 1910).

OSHA (1987) *Hazard Communication*. 52 FR 31852 (codified at 29 CFR, Parts 1910, 1915, 1917, 1918, 1926, 1928).

OSHA (2012) *Hazard Communication*. Washington, DC: Occupational Safety and Health Administration, US Department of Labor. 77 FR (58):17574–17896. Available at https://www .osha.gov/FedReg_osha_pdf/FED20120326.pdf.

OSHA (2014) *Permissible Exposure Limits – Annotated Tables*. Washington, DC: Occupational Safety and Health Administration, US Department of Labor. Available at https://www.osha. gov/dsg/annotated-pels.

PubMed (2014) PubMed. Bethesda, MD: US National Library of Medicine, National Institutes of Health, National Center for Biotechnology Information. Available at http://www.pubmed.gov (A search using key terms "NIH Guide Grants Contracts" retrieves the outcomes of over 100 investigations undertaken by the National Institutes of Health for suspected research misconduct).

REACH (2006) Regulation (EC) No 1907/2006 of the European Parliament and of the Council, of 18 December 2006 concerning the Registration, Evaluation, Authorisation and Restriction of Chemicals (REACH), Establishing a European Chemicals Agency, Amending Directive 1999/45/EC and Repealing Council Regulation (EEC) No 793/93 and Commission Regulation (EC) No 1488/94 as well as Council Directive 76/769/EEC and Commission Directives 91/155/ EEC, 93/67/EEC, 93/105/EC and 2000/21/EC (Text with EEA relevance). OJ L 396. Available at https://osha.europa .eu/en/legislation/directives/exposure-to-chemical-agents-and-chemical-safety/osh-related-aspects/regulation-ec-no-1907-2006-of-the-european-parliament-and-of-the-council.

Schenk, L. (2009) *Managing Chemical Risk Through Occupational Exposure Limits*. Stockholm, Sweden: Royal Institute of

Technology. Available at http://www.diva-portal.org/smash/get/diva2:132945/FULLTEXT01.pdf.

S. Rep. (1976) No. 99-698, at 1.

Tyl, R.W. (2009) Basic exploratory research versus guideline-compliant studies used for hazard evaluation and risk assessment: bisphenol A as a case study. *Environ. Health Perspect.* 117:1644–1651. Available at http://www.ncbi.nlm.nih.gov/pmc/articles/PMC2801172.

UNECE (2011) *Globally Harmonized System of Classification and Labelling of Chemicals (GHS)*, 4th revised ed. Geneva, Switzerland: United Nations Economic Commission for Europe. Available at http://www.unece.org/trans/danger/publi/ghs/ghs_rev04/04files_e.html.

United States v. Keplinger (1985) 776 F.2d 678 (7th Cir.). Available at https://law.resource.org/pub/us/case/reporter/F2/776/776.F2d.678.84-1639.html.

Viberg, H., Fredriksson, A., and Eriksson, P. (2007) Changes in spontaneous behaviour and altered response to nicotine in the adult rat, after neonatal exposure to the brominated flame retardant, decabrominated diphenyl ether (PBDE 209). *Neurotoxicology* 28:136–142.

Viberg, H., Fredriksson, A., Jakobsson, E., Örn, U., and Eriksson, P. (2003) Neurobehavioral derangements in adult mice recieving decabrominated diphenyl ether (PBDE 209) during a defined period of neonatal brain development. *Toxicol. Sci.* 76:112–120.

Vitter, D. (2014) Letter dated March 17, 2014 to Francesca Grifo, Science Integrity Official, US Environmental Protection Agency, Washington, DC, from David Vitter, a ranking member on the Committee on Environment and Public Works, United States Senate. Available at http://www.epw.senate.gov/public/index.cfm?FuseAction=Files.View&FileStore_id=6de2a2b9-ad38-41bc-a0c4-c909b391a526.

INDEX

Hamilton & Hardy's Industrial Toxicology, Sixth Edition. Edited by Raymond D. Harbison, Marie M. Bourgeois, and Giffe T. Johnson.
© 2015 John Wiley & Sons, Inc. Published 2015 by John Wiley & Sons, Inc.